Mammals of South America, Volume 1

Mammals of South America, Volume 1

Marsupials, Xenarthrans, Shrews, and Bats

Edited by Alfred L. Gardner

The University of Chicago Press
Chicago and London

ALFRED L. GARDNER is a Wildlife Biologist, Research, in the U.S. Geological Survey Patuxent Wildlife Research Center's Biological Survey Unit. He is stationed at the National Museum of Natural History, Division of Mammals in Washington DC, where he is curator of North American Mammals.

The University of Chicago Press, Chicago 60637
The University of Chicago Press, Ltd., London
© 2007 by The University of Chicago
All rights reserved. Published 2007
Printed in the United States of America

16 15 14 13 12 11 10 09 08 07   1 2 3 4 5

ISBN-13: 978-0-226-28240-4 (cloth)
ISBN-10: 0-226-28240-6 (cloth)

Library of Congress Cataloging-in-Publication Data

Mammals of South America / edited by Alfred L. Gardner.
    p.   cm.
  Includes bibliographical references and index.
  ISBN-13: 978-0-226-28240-4 (cloth : alk. paper)
  ISBN-10: 0-226-28240-6 (cloth : alk. paper)
    1. Mammals—South America.   2. Mammals—South America—
Classification.   I. Gardner, Alfred L.
  QL725.A1.M36 2007
  599.098—dc22
                                    2007017496

THIS VOLUME IS DEDICATED TO A. REMINGTON KELLOGG, 1892–1969, former curator of mammals, Director of the U.S. National Museum, and Assistant Secretary of the Smithsonian Institution.

Remington Kellogg, perhaps best known for his work on marine mammals, spent many years accumulating material for a catalog of South American mammals. His work on that project stopped shortly before the appearance in 1958 of Cabrera's first volume of the *Catálogo de los mamíferos de América del Sur*. Kellogg usually spent an afternoon each week reading articles, books, and other references in the library of the U.S. National Museum, or in the nearby resources in the Library of Medicine, the Library of Congress, and library of the U.S. Department of Agriculture. Kellogg also visited many libraries outside of the Washington, DC, area to find references not available closer to home. As was the practice before xerography, Kellogg wrote down the information, which his secretary typed. He then cut the individual entries into strips and glued them on heavier paper, each in its appropriate taxonomic place, in a series of loose-leaf books. These "scrapbooks," which occupy nearly a meter of shelf space, have proven to be a treasure trove of information, the more important of which are the bibliographic citations, which have greatly aided me in locating and examining references, and in constructing synonymies.

Contents

Sections and Authors

Introduction

THE STUDY of South American mammals has a rich history, its recorded phase beginning in 1500 when Pinzón brought to Spain a number of natural history novelties from the New World that included a marsupial now known as *Didelphis marsupialis* captured on the coast of present-day Brazil. Any student of South American land mammals soon becomes familiar with the names Buffon, Cuvier, Desmarest, d'Orbigny, Fischer, Gray, Geoffroy St.-Hilaire, Lesson, Lund, Molina, Natterer, Schreber, Spix, Temminck, Tschudi, Wagner, and Wied-Neuwied, among the many names of workers who were active in the time period up to 1850 (see Hershkovitz's [1987] historical review of Neotropical mammalogy). Subsequently, the names Allen, Burmeister, Cabrera, de Miranda, Dobson, Goeldi, Gray, Hershkovitz, Lönnberg, Miller, Peters, Osgood, Sanborn, Tate, Vieira, Yepes, among many others, appear again and again. Papavero (1971) outlined the itineraries of many collectors active in South America between 1750 and 1905. Impossible to overlook, anyone familiar with South American mammals holds in awe the contributions by Oldfield Thomas. A significant portion of the approximately 2900 new names Thomas gave to genera, species, and subspecies in his nearly 1100 publications were based on mammals from South American (Hill 1990).

Husson's (1978) *The Mammals of Suriname* was the first comprehensive modern taxonomic treatment of a South American mammal fauna to appear following Cabrera's (1958–1961) *Catálogo de los mamíferos de América del Sur*. Patterned after his *The Bats of Suriname* (1962), Husson's review is of special importance because many Surinamese mammals were the basis of a number of early original descriptions of New World mammals. While certainly not the first review of a South American country's mammal fauna (see, for example, Ximenez, Langguth, and Praderi's [1972] list for Uruguay), it was the first to combine historic, taxonomic, distributional, and natural history information with keys to the orders, families, genera, and species. Husson's (1962) *The Bats of Suriname* is well illustrated with drawing of various external, cranial, and dental features useful in identification; these illustrations were incorporated into the 1978 volume, which covered all Surinamese mammals known up to that time.

Other important works on South American mammals have appeared in the last quarter century. Most have been either local, country-wide, or regional treatments on a subset of the fauna such as Hershkovitz's (1977) classic on New World platyrrhine monkeys, Albuja's work on the bats of Ecuador (1983, 1999), Simmons and Voss's (1998) review of the bats of Paracou, French Guiana; the comprehensive review of Argentine bats by Barquez, Mares, and Braun (1999), the revision of oryzomyine rodents by Musser et al. (1998), the bats of Colombia by Muñoz (2001), the review of the marsupials and insectivores of Colombia by Cuartas and Muñoz (2003), and the *Atlas of New World Marsupials* by B. Brown (2004). Other recent studies that go beyond simple lists, and are indicative of current trends in research on South American mammals, include the work by Patton, Silva, and Malcolm (2000) on the non-Primate fauna of the Amazon basin, and by Voss, Lunde, and Simmons (2001) on the fauna of French Guiana. Two recent country-wide reviews of mammal faunas cover Bolivia (S. Anderson 1997) and Venezuela (Linares 1998). The former is clearly intended for use by scientists, whereas the latter has a broader, less-technical audience.

In terms of continent-wide faunal treatments, the recent three-volume work on titled *Mammals of the Neotropics* divided the continent geographically: (Eisenberg 1989; Redford and Eisenberg 1992 [see review by Pine 1993b]; Eisenberg and Redford 1999). The field guides to rainforest mammals by Emmons and Feer (1990, 1997) include Central America and southern México as well as South America.

The present volume, one of three on the mammals of South America, divides the mammal fauna along taxonomic lines. I was charged with overseeing and editing this first volume, which covers the marsupials, shrews, armadillos, anteaters, sloths, and bats; Sydney Anderson of the American Museum of Natural History is responsible for another volume covering the carnivores, manatees, tapirs, deer, peccaries, camelids, rabbits, and marine mammals; and James Patton of the Museum of Vertebrate Zoology, University of California, Berkeley, is responsible for the volume covering the rodents. Conceived in 1977 and underway by 1980, the project has had a long gestation.

Cabrera's (1958–1961) catalog was selected as the primary point of departure because of its strong emphasis on taxonomy and nomenclature. Major differences, however, exist between the scope of our work and Cabrera's catalog. We include Trinidad and Tobago and the Netherlands Antilles in our coverage (Map 1); neither was covered by Cabrera. Along with extensive, but not exhaustive synonymies, we provide identification keys and brief descriptions of each order, family, and genus. Species accounts include synonymies, keys to identification, distributions with lists and maps of marginal localities, separate synonymies and distributions for recognized subspecies, brief summaries of natural history information, and discussions of issues related to type locality, taxonomic interpretations, or matters of nomenclatural significance. Each volume contains its own Gazetteer and Literature Cited. The latter section includes all references mentioned anywhere in the text pertinent to mammals, as well as references cited in the Gazetteer related to localities, and references cited within entries in the Literature Cited section related to dates of publication.

Authorship of contributions in the Taxonomic Accounts requires clarification. The scope of individual contributions is uneven. The authority and responsibility for an individual contribution continues until a new author line is encountered. Gardner drafted the Introduction and compiled the Gazetteer and Literature Cited sections from information provided by each author.

With some exceptions, the taxonomy of higher categories follows McKenna and Bell (1997), although, the sequence of family-, genus-, and species-group taxa follows that of the third edition of *Mammal Species of the World*, edited by Wilson and Reeder (2005). Most deviations from these two references are documented by citation of recent literature; other exceptions result from research by the contributing authors.

The synonymy for each taxon includes the citation of the original description and any unique name combination subsequently applied by any author for the first time to that taxon. In instances where the author's usage can be verified, the first name combination applied, however erroneously, is also listed. Any name combination applied to a taxon subsequent to its original description is separated by a colon from the name of the author using that name combination unless the new name combination was applied in the sense of a new name. When a name combination is a *nomen nudum*, an unavailable name, incorrect subsequent spelling, or unjustified emendation, the name is not separated by punctuation from the name of its author. Usage under any of these categories is appropriately explained. *Nomina nuda* and incorrect subsequent spellings are also unavailable names; they do not enter into homonymy and cannot be used as a substitute name (ICZN 1999:Art. 33.3). However an unjustified emendation is an available name with its own author and date, and is a junior synonym of the name in its original spelling; it enters into homonymy and is available for use as a substitute name (ICZN 1999:Art. 33.2.3).

Unless there was compelling evidence to the contrary, authors are assumed to have correctly applied each name to the taxon they reported on. Compelling evidence that the name was incorrectly applied (specimens misidentified) usually is based on the author's re-examination of the specimens themselves. Records based on obvious misidentifications are either cited under the correct name, or are ignored because the correct identity is unknown. It is obvious that the taxonomy and distribution of most South American mammals remain inadequately known and revisions of most groups are needed. The primary objective of these volumes is to facilitate that process.

Map 1 Map of the South American continent including the Netherlands Antilles, Trinidad and Tobago, and the Galapagos Islands

Acronyms and Abbreviations

The following acronyms and abbreviations are used in the text to identify collectors and museum collections.

ALG: Alfred L. Gardner field numbers

AMNH: American Museum of Natural History, New York, New York, USA

APECO: Asociación Peruana para la Conservación de la Naturaleza, Lima, Peru

BM: The Natural History Museum, London (formerly The British Museum [Natural History], London), UK

BDP: Bruce D. Patterson field numbers

CBF: Colección Boliviana de Fauna, La Paz, Bolivia

CC: Carlos Cuartos-Calle, Universidad de Antioquia, Medellín, Colombia

CM: Carnegie Museum of Natural History, Pittsburgh, Pennsylvania, USA

CNHM: Chicago Natural History Museum (see Field Museum)

CVUCV: Colección de Vertebrados, Universidad Central de Venezuela, Caracas

EBRG: Rancho Grande Biological Station, Maracay, Venezuela

EPN: Escuela Politécnica Nacional, Quito, Ecuador

FMNH: Field Museum of Natural History, Chicago, Illinois, USA

HGB: Helena Godoy Bergallo field numbers

ICN: Instituto de Ciencias Naturales, Universidad Nacional de Colombia, Bogotá

ILB: Instituto Lillo, Bogotá, Colombia

INPA: Instituto Nacional de Pesquisas da Amazônia, Manaus, Brazil

IPO: Instituto Profesional de Osorno, Osorno, Chile

IRSNB: Institut Royal des Sciences Naturelles de Belgique, Brussels, Belgium

KU: University of Kansas, Lawrence, Kansas, USA

LACM: Los Angeles County Museum of Natural History, Los Angeles, California, USA

LSUMZ: Louisiana State University Museum of Zoology, Baton Rouge, Louisiana, USA

MACN: Museo Argentino de Ciencias Naturales, Buenos Aires

MBUCV: Museum of Biology, Universidad Central de Venezuela, Caracas, Venezuela

MCNLS: Museo de Ciencias Naturales La Salle, Bogotá, Colombia

MCZ: Museum of Comparative Zoology, Harvard University, Cambridge, Massachusetts, USA

MHG: Milton H. Gallardo field numbers, to be catalogued in the Instituto de Ecología y Evolución, Universidad Austral de Chile, Valdivia.

MHNSM: Museo de Historia Natural, Universidad de San Marcos (formerly Museo de Historia Natural Xavier Prado), Lima, Peru.

MHNLS: Museo de Historia Natural La Salle, Caracas, Venezuela

MLP: Colección de Mamíferos, Museo de La Plata, La Plata, Argentina

MMNH: University of Minnesota, Bell Museum of Natural History

MNCNM: Museo Nacional de Ciencias Naturales, Madrid, Spain

MNRJ: Museu Nacional, Rio de Janeiro, Brazil.

MNURU: Museo Nacional, Montevideo, Uruguay.

MPEG: Museo Paraense Emilio Goeldi, Belém, Brazil

MPM: Milwaukee Public Museum, Milwaukee, Wisconsin, USA

MRR: Monica Romo R. field numbers

MSB: Museum of Southwestern Biology, University of New Mexico, Albuquerque, New Mexico, USA

MSU: Michigan State University, East Lansing, Michigan, USA

MUA: Museo de la Universidad de Antioquia, Medellín, Colombia

MVZ: Museum of Vertebrate Zoology, University of California, Berkeley, California, USA

MZUSP: Museu de Zoología, Universidade de São Paulo, São Paulo, Brazil

NHR: Naturhistoriska Riksmuseet, Stockholm, Sweden

NMW: Naturhistorisches Museum Wien, Vienna, Austria

OMNH: Oklahoma Museum of Natural History, Norman, Oklahoma, USA

RMNH: Rijksmuseum van Natuurlijke Historie, Leiden, The Netherlands.

ROM: Royal Ontario Museum, Toronto, Canada

TCWC: Texas A&M University, College Station, Texas, USA

TT: The Museum, Texas Tech University, Lubbock, Texas, USA

UCONN: University of Connecticut, Biological Sciences Group, Storrs, Connecticut, USA

UFMG: Universidade Federal de Minas Gerais, Minas Gerais, Brazil

UMMZ: University of Michigan Museum of Zoology, Ann Arbor, Michigan, USA

USNM: National Museum of Natural History (formerly The United States National Museum), Washington, District of Columbia, USA

UV: Universidad del Valle, Cali, Colombia.

ZMB: Zoologisches Museum der Humboldt-Universität zu Berlin ("Berlin Museum"), Berlin, Germany

ZMUC: Zoological Museum of the University, Copenhagen, Denmark

ZSM: Zoologisches Staatssammlung München, Munich, Germany

Taxonomic Accounts

Class Mammalia Linnaeus, 1758
Cohort Marsupialia Illiger, 1811

AMERICAN MARSUPIALS

Alfred L. Gardner

KEY TO THE ORDERS AND FAMILIES OF RECENT SOUTH AMERICAN MARSUPIALS:

1. Incisors 5/4; i1 similar in length to i2–i4; lower canine variable in size, but usually retaining caniniform shape 2
1'. Incisors 4/4; i1 greatly enlarged and procumbent (pseudodiprotodont); lower canine incisor-like in shape . . . . Paucituberculata, Caenolestidae
2. Upper incisors normal, one or more as simple conical pegs; centrocrista V-shaped (didelphoid dilambdodont); M4 subequal to or smaller than M3, but retaining trituberculate form; bullae incompletely ossified . . . . Didelphimorphia, Didelphidae
2'. Upper incisors with expanded, spatulate crowns; centrocrista straight; M4 much smaller than, and lacking molariform structure of, M1–M3; bullae ossified and greatly inflated . . . . Microbiotheria, Microbiotheriidae

Order Didelphimorphia Gill, 1872

The Didelphimorphia (Gill 1872b) comprises the Didelphidae, which includes the majority of the living American marsupials (Marshall 1990; McKenna and Bell 1997).

Didelphimorphia are known from the Late Cretaceous to Recent of South America; Oligocene, Pleistocene, and Recent of North America; Oligocene of western Asia; Eocene and Oligocene of Africa; and Eocene, Oligocene, and Miocene of Europe and Africa (Marshall 1990; McKenna and Bell 1997).

Family Didelphidae Gray, 1821

The family Didelphidae includes 19 Recent genera and 95 currently recognized species, of which all but four species occur in South America. Based on Reig, Kirsch, and Marshall (1985), all taxa are divided among two subfamilies (Caluromyinae and Didelphinae) and three tribes (Didelphini, Metachirini, and Monodelphini) within the Didelphinae (McKenna and Bell 1997). The fossil record is as outlined previously for the Didelphimorphia. The geographical range of the family extends from 50°S in Patagonia, northward throughout South America, Central America, Mexico, the continental shelf islands of the southern and western Caribbean, through the United States, and into southern Canada.

The Didelphidae includes the smallest (*Chacodelphys* and a few *Monodelphis*) and the largest (*Didelphis*) living New World marsupials. Head and body lengths range from less than 10 cm to over 50 cm. All members of the family are characterized by the dental formula 5/4, 1/1, 3/3, 4/4 = 50, with polyprotodont incisors; well-developed upper canines; small multicuspid, molariform, and deciduous third premolars (except in *Hyladelphys*, which has nonmolariform deciduous third premolars); and trituberculate molars that have a prominent stylar shelf. The first upper incisor is longest (crown height above alveolus) and is always separated from I2 by a gap. The auditory bullae are tripartite, consisting of unfused elements of the alisphenoid, periotic, and ectotympanic. The extent of the contribution of the ectotympanic to the ventral wall of the bulla varies greatly among genera (Reig, Kirsch, and Marshall 1987).

The manus and pes are pentadactyl with a phalangeal formula of 2–3–3–3–3 (Brown and Yalden 1973). The hallux always lacks a nail or claw, but retains a reduced eponychium. The diploid chromosome number is 22 (*Didelphis*, *Chironectes*, *Philander*, *Lutreolina*, and *Tlacuatzin*), 18 (*Monodelphis*) or 14 (all other genera for which the karyotype is known; Gardner 1973; Reig et al. 1977; Engstrom and Gardner 1988; and Voss and Jansa 2003). In general, females of the larger-bodied genera have a marsupial pouch (*Metachirus* is the exception) and the smaller taxa all lack a pouch. The number of functional mammae ranges from 4 to 27 with the higher numbers in those taxa that lack a pouch (pouchless *Hyladelphys* with four mammae is the exception).

REMARKS: Berkovitz (1967) and Archer (1978) argued that the deciduous third premolar actually is the first permanent molar, pushed out of the crowded toothrow by the erupting last premolar. Luckett (1993), however, refuted that hypothesis.

KEY TO THE SUBFAMILIES, TRIBES, AND GENERA OF RECENT DIDELPHIDAE:

1. Palatal fenestrae absent or inconspicuous; either a conspicuous medial facial stripe present (pale in *Glironia*, dark in *Caluromys*) or with dark shoulder patches extending along back as stripes; if pelage long and woolly, 1/2 or more of the tail furred . . . .
. . . . Caluromyinae, 2
1'. Palatal fenestrae large and conspicuous; no prominent medial facial stripe and no dark shoulder patch associated with dorsal stripes; if pelage long and woolly, tail not furred beyond 60 mm of base (the exception is *Lutreolina*, which has up to 1/3 of tail furred) 4
2. Tail lacking long, dense fur for distal 1/3 or more of length; medial facial stripe dark and conspicuous . . . . .
. . . . *Caluromys*
2'. Long fur, similar to body fur, extends along dorsal surface of tail for more than 90% of its length; medial facial stripe pale or absent. 3
3. Dark mask; a prominent, pale facial stripe extends from rhinarium to between ears; no shoulder patches; length of head and body less than 230 mm; length of tail less than 230 mm . . . . *Glironia*
3'. No dark mask or facial stripe; prominent dark shoulder patches extend posteriorly as parallel stripes to pelvic

region; head and body length greater than 250 mm; tail length greater than 300 mm . . . . *Caluromysiops*
4. Ectotympanic slender and "ring-like"; skull with prominent sagittal crest on frontals and parietals; marsupium developed as pouch or as prominent lateral folds in reproductive females; total length usually exceeds 500 mm
. . . . (Didelphini) 15
4'. Ectotympanic flattened and laterally expanded to form part of ventral surface of auditory bullae; sagittal crest absent or weakly developed; marsupium always absent; total length usually less than 475 mm (greater than 475 mm in *Metachirus*) 5
5. Total length greater than 475 mm; dorsal pelage pale to medium brown; face darker brown with conspicuous pale tan-colored spot above each eye; greatest length of skull more than 53 mm; cranium lacks postorbital processes; palate with a pair of maxillopalatine fenestrae
. . . . (Metachirini) *Metachirus*
5'. Total length less than 475 mm; color of dorsal pelage variable; no distinct pale spot above eyes; greatest length of skull less than 50 mm; postorbital processes present in some taxa; palatal fenestrae variable . . . .
. . . . (Monodelphini) 6
6. Total length of adults greater than 350 mm; dorsal pelage long, woolly, gray to grayish brown; tail frequently white at tip or fuscous and mottled with pale blotches for distal 1/3 or more; long, dense fur extends on to base of tail for up to 50 mm; bullae small and widely separated, ratio of distance between bullae to breadth of a bulla greater than 2.5 . . . . *Micoureus*
6'. Total length of adults less than 350 mm; dorsal pelage variable in color, but not long and woolly; tail often lacking white tip; longer dense fur extends onto base of tail for 15 mm or less; bullae variable, ratio of distance between bullae to breadth of bullae less than 2.57
7. Length of tail less than 70% length of head and body; pes with digit III longest; skull lacks postorbital processes; sagittal crest present only in larger/older individuals . . . . *Monodelphis*
7'. Length of tail greater than 75% length of head and body; pes with digit IV longest; skull with or without postorbital processes; sagittal crest absent 8
8. Tail shorter than 95% length of head and body; palmar surface densely covered with small tubercles; periotic and alisphenoid components of bulla in near or full contact 9
8'. Tail equal to or longer than head and body; palmar surface variable; alisphenoid and periotic components of bullae always separated by gap bridged by ectotympanic
. . . . 10
9. Size larger, total length more than 150 mm; fur tricolored; nasals expanded at lacrimal-frontal suture; maxillary

fenestrae absent; bullae relatively large and rounded; fenestra cochleae concealed . . . . *Lestodelphys*

9′. Size smaller, total length less than 150 mm; fur bicolored; nasals subparallel, not expanded at lacrimal-frontal suture; maxillary fenestrae present; bullae relatively small, fenestra cochleae laterally exposed . . . . . .
. . . . *Chacodelphys*

10. Ratio of tail length to head and body length between 1.0 and 1.25; tail frequently thickened (incrassate) with fat at base; palmar and plantar surfaces densely covered with small tubercles; digits of pes short, length of digit IV less than 0.45 times length of hind foot; bullae relatively large and close together, ratio of distance between bullae to breadth of a bulla 1.5 or less . . . . *Thylamys*

10′. Tail usually longer, may exceed 1.5 times length of head and body; tail not thickened at base; central palmar surfaces smooth to sparsely granulated; digit IV of pes more than 0.45 times length of hind foot; bullae relatively smaller and more widely separated, ratio of distance to breadth usually greater than 1.5 11

11. Skull with conspicuous postorbital processes, or lateral margins of frontals conspicuously ridged or beaded; bullae small, rounded, and may or may not have an anteromedial projection or strut on the alisphenoid (secondary foramen ovale present or absent) 12

11′. Skull lacks postorbital processes and lacks beaded or strongly ridged frontals (except in some old males); bullae with distinct anteromedial process of the alisphenoid (secondary foramen ovale present) 13

12. Postorbital processes present, may be greatly enlarged; lateral margin of frontals prominently ridged; postorbital region relatively broad, not conspicuously constricted behind orbits; premaxillae sometimes projecting 2.0 mm or more anterior to I1 as an acute point; mammae count more than four . . . . *Marmosa*

12′. No postorbital processes; lateral margin of frontals prominently beaded; postorbital region conspicuously constricted; premaxillae not as above; four mammae . .
. . . . *Hyladelphys*

13. Tail scales rhomboid in shape and usually spirally arranged; middle bristle of each triplet of tail hairs stout, petiolate, and darkly pigmented (in contrast to lateral hairs); manual digit III longer than digits II and IV; lower canines short and premolariform, equal to or shorter than p2 in height; bullae compressed, conical in shape, and have a stout anteromedial strut (secondary foramen ovale present) . . . . *Marmosops*

13′. Tail scales rounded to square in shape and usually annular in arrangement; tail bristles slender, lightly pigmented, and similar in thickness and length; manual digits III and IV subequal in length; lower canines normal and exceed p2 in height above the alveolus 14

14. Ratio of tail length to length of head and body 1.20 to 1.96; bullae with anteromedial process of the alisphenoid (secondary foramen ovale present); maxillary fenestrae usually present; P2 and P3 about equal in height
. . . . *Gracilinanus*

14′. Ratio of tail length to length of head and body 1.01 to 1.40, usually less than 1.20; bullae with anteromedial process of the alisphenoid usually lacking (present in some *Cryptonanus agricolai*); maxillary fenestrae usually absent (present in some *C. agricolai* and *C. unduaviensis*); P2 smaller than P3 . . . . *Cryptonanus*

15. Hind feet distinctly webbed; color pattern of dorsal pelage with alternating pale and dark gray transverse bars; males with a rudimentary pouch . . . . *Chironectes*

15′. Hind feet lack webbing; dorsal pelage lacks contrasting bands of color; males lack a pouch. 16

16. Ears short, dark, rounded, and barely projecting above surrounding fur; color of dorsum agouti to reddish brown or tawny; pelage extending on to base of tail for 1/3 or more of tail length; white on tail, if present, only at tip . . . . *Lutreolina*

16′. Ears large and conspicuous above surrounding fur; dorsal color gray to black, with or without long guard hair; long pelage on tail restricted to base; distal 1/3 to 1/4 of tail white 16

17. Dorsal pelage lacks long guard hairs, except along dorsal midline; conspicuous pale spot present above each eye . . . . *Philander*

17′. Long, prominent guard hairs present throughout dorsal pelage; extent of white on head variable; conspicuous spots above eyes absent. . . . . *Didelphis*

Subfamily Caluromyinae Kirsch, 1977

The Caluromyinae include three genera (*Caluromys*, *Caluromysiops*, and *Glironia*) that share the following features: palatal fenestrae or vacuities lacking; M2 is longer than, or about equal in length to, M3; upper molars have separated and subequal paracones and metacones; and the paracrista is not united to stylar cusps. The marsupial pouch is well developed in *Caluromysiops* and *Caluromys* (*Mallodelphys*), but rudimentary in *Caluromys* (*Caluromys*), and absent in *Glironia*. The diploid chromosome number ($2n$) is 14 in *Caluromys*, and unknown for *Glironia* and *Caluromysiops*.

Genus *Caluromys* J. A. Allen, 1900

This genus of three species occurs in southern Mexico, and Central and South America south into northern Argentina (Misiones), and on the island of Trinidad. All three species are in South America. Fossil material is known from the Pleistocene of South America (Marshall 1981; Winge 1893). The elevational range is from sea level to approximately 2,500 m. Restricted to forested habitats, these species are among the more arboreal of American marsupials, and this behavior contributes to their relative scarcity in collections.

Woolly opossums are medium-sized didelphids (length of head and body 160–300 mm, tail 250–450 mm, mass 140–410 g) and have four mammae. Other characteristics include a distinctly woolly pelage; a long, prehensile tail; large naked ears; dark lateral facial stripes and eye rings; and a dark median facial stripe, which contrasts with the gray face and extends from the rhinarium to the top of the head. Body size and the facial stripe easily distinguish members of this genus from other didelphids. The skull is short, broad, and has well-developed supraorbital processes, large orbits, a broad braincase, and a broad palate that lacks conspicuous fenestrae. The stylar shelf of the upper molars is narrow and the stylar cusps are reduced.

The species are separated into two subgenera. *Caluromys philander* is the sole member of the subgenus *Caluromys* and lacks a pouch (marsupium consists of lateral skin folds). The uniformly pale brown to buffy brown pelage is sometimes admixed with gray (especially on the face) and covers the upper and lateral surfaces of the body and limbs. Dense fur on the tail is restricted to the proximal 30 to 70 mm. *Caluromys derbianus* and *C. lanatus* make up the subgenus *Mallodelphys*; both have a well-developed marsupial pouch. The patterned dorsal and lateral pelage is reddish to yellowish brown on the shoulders, rump, and outer surfaces of the legs or pale gray on the lower legs, and sometimes includes a pale gray mid-dorsal stripe between the shoulders. The fur on the tail extends beyond 70 mm, usually covering from over 1/3 to nearly 3/4 the length of the tail.

SYNONYMS:

Didelphis Linnaeus, 1758:54; part.

Philander Beckmann, 1772:[244]; type species *Didelphis philander* Linnaeus, 1758, by absolute tautonomy (also see Kretzoi and Kretzoi 2000:309); preoccupied by *Philander* Brisson, 1762.

Didelphys Schreber, 1777:532; part; unjustified emendation of *Didelphis* Linnaeus.

Sarigua Muirhead, 1819:429; part.

Micoureus Lesson, 1842:186; part.

Philander Burmeister, 1856:74; type species *Didelphys (Philander) cayopollin* Burmeister, 1856 (= *Didelphys cayopollin* Schreber, 1778), by subsequent designation (Hershkovitz, 1949a:12); described as a subgenus of *Didelphis* Linnaeus; preoccupied by *Philander* Brisson, 1762, *Philander* Beckmann, 1772, and *Philander* Tiedemann, 1808.

Gamba Liais, 1872:330; part.

Cuica Liais, 1872:330; part.

Micoureus: Ihering, 1894:11; part.

Caluromys J. A. Allen, 1900b:189; type species *Didelphis philander* Linnaeus, 1758, by original designation.

Micoureus: Matschie, 1916:269; part; used as a subgenus of *Didelphis* Linnaeus; not *Micoureus* Lesson.

Mallodelphys O. Thomas, 1920a:195, footnote; type species *Didelphis laniger* Desmarest, 1820, by original designation; described as a subgenus of *Philander* of authors (*sensu* Burmeister 1856), not *Philander* of Brisson (1762), Beckmann (1772), or Tiedemann (1808); based on *Micoureus* as used by Matschie (1916).

Calaromys A. Miranda-Ribeiro, 1936:324; incorrect subsequent spelling of *Caluromys* J. A. Allen.

Mallodelphis Gilmore, 1941:317; incorrect subsequent spelling of *Mallodelphys* O. Thomas.

Calurosmys Ávila-Pires, 1964:11; incorrect subsequent spelling of *Caluromys* J. A. Allen.

REMARKS: Hershkovitz (1949a:12) pointed out that O. Thomas's (1888b:336) designation of *Didelphis philander* Linnaeus as the type genus of *Philander* Tiedemann is untenable because Tiedemann (1808:427–28) did not include that species or a synonym among the three he listed. With the exception of *Didelphis*, *Philander* Beckmann, 1772, is the oldest generic name applied to *Caluromys*; however, the name is preoccupied by *Philander* Brisson, 1762. Kretzoi and Kretzoi cited *Philander* Beckmann from page 244, which is the fourth page of an unnumbered index and which, if numbered, would correspond to 244. Matschie's (1916) listing of *Didelphis laniger* Desmarest, 1820, as the type species of *Micoureus* Lesson, 1842, also was incorrect because O. Thomas (1888b:340) already had designated *Didelphis cinerea* as the type species of *Micoureus* Lesson (also see Gardner and Creighton 1989). Both O. Thomas (1888b) and Matschie (1916, 1917) used *Micoureus* as a subgenus of *Didelphis* Linnaeus, but for different species.

Barbara E. Brown (2004:4) stated that the type species of Gloger's (1841) names *Asagis* and *Notagogus* is *Caluromys philander*, based on Gloger's use of the vernacular name "cayopollin" as the sole reference to kinds of American marsupials known at that time. Although it is true that Buffon's (1763:plate 55) "*Le Cayopollin*" represents *Caluromys philander* (see Remarks under that species), cayopollin also was used before 1841 by different authors as a common name for mouse opossums of the genus *Marmosa*. Therefore, reference to the "Cayopollin's" by Gloger can not be treated as restriction of the type species by monotypy. O. Thomas (1995a) equated both *Asagis and Notagogus* with *Marmosa* Gray, 1821, and earlier he (1888b:340) restricted these names to the synonymy of *Marmosa* by listing the type species as *M. murina*. Technically, if B. E. Brown's

(2004) interpretation were correct, *Asagis* would replace *Caluromys* as the oldest available name for this genus.

KEY TO THE SUBGENERA AND SPECIES OF *CALUROMYS*:

1. Tail furred on only basal 10–20%; naked portion of tail usually brown or brown mottled with paler markings; pelage short, relatively lax, and unpatterned except for gray face that contrasts with darker medial and lateral facial stripes and eye ring; marsupial pouch incomplete, consisting of two lateral abdominal folds; ears pale . . . . (subgenus *Caluromys*) *Caluromys philander*
1'. Tail furred dorsally on basal 30–70% and ventrally on basal 20–40% of length; naked portion of tail usually predominantly white; pelage long, dense, woolly, and patterned dorsally with red, orange, yellow, and (in some populations) pale gray; marsupial pouch well developed; ears pale or dark . . . . (subgenus *Mallodelphys*) 2
2. Ears dark brown or blackish; tail furred dorsally on basal 40–70% and ventrally on basal 20–35% of length, the difference between extent of dorsal and ventral fur along tail usually exceeding 65 mm . . . . . . *Caluromys lanatus*
2'. Ears whitish, usually yellowish or flesh-colored pink in preserved specimens; tail furred dorsally on basal 30–45%, and ventrally on basal 20–30%, of length, the difference between extent of dorsal and ventral fur 65 mm or less . . . . *Caluromys derbianus*

Caluromys derbianus (Waterhouse, 1841)
Derby's Woolly Opossum
SYNONYMS: The following synonyms represent either Mexican or Central American subspecies (Bucher and Hoffmann 1980; Hall 1981), or the source of the type specimen is unknown. See under Subspecies for additional synonyms.

Philander centralis Hollister, 1914:103; type locality "Talamanca, Costa Rica."

[*Didelphis (Micoureus)*] *aztecus*: Matschie, 1916:269; name combination; = *C. d. aztecus* (O. Thomas, 1913c).

[*Didelphis (Micoureus)*] *fervidus*: Matschie, 1916:269; name combination; = *C. d. fervidus* (O. Thomas, 1913c).

[*Didelphis (Micoureus)*] *centralis*: Matschie, 1916:269; name combination.

[*Didelphis (Micoureus)*] *pallidus*: Matschie, 1916:269; name combination; = *C. d. pallidus* (O. Thomas, 1899d).

[*Didelphis (Micoureus)*] *nauticus*: Matschie, 1916:269; name combination; = *C. d. nauticus* (O. Thomas, 1913c).

[*Didelphis (]Micoureus[)*] *pulcher* Matschie, 1917:281; type locality unknown, based on a zoo specimen.

[*Didelphis (]Micoureus[)*] *canus* Matschie, 1917:284; type locality "Nicaragua."

DISTRIBUTION: *Caluromys derbianus* is in western Colombia and Ecuador west of the Andes. The species also occurs in Central America and southern and eastern Mexico.

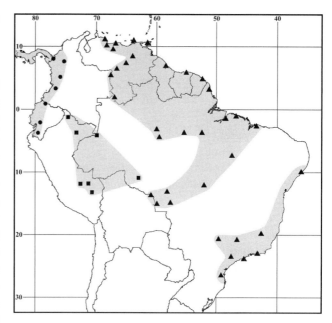

Map 2 Marginal localities for *Caluromys derbianus* ●, *Caluromys philander* ▲, and *Caluromysiops irrupta* ■

MARGINAL LOCALITIES (Map 2): COLOMBIA: Chocó, Unguía (B. E. Brown 2004); Risaralda, Apía (type locality of *Philander laniger pictus* O. Thomas); Antioquia, Cáceres (type locality of *Didelphis (Micoureus) antioquiae* Matschie, 1917); Valle del Cauca, Río Caquetá (J. A. Allen 1904a). ECUADOR: Imbabura, Paramba (B. E. Brown 2004); Pichincha, Mindo (type locality of *Philander laniger senex* O. Thomas); Guayas, Puente de Chimbo (AMNH 63525); Loja, near Punta Santa Ana (AMNH 47194).

SUBSPECIES: I am treating South American *C. derbianus* as represented by a single subspecies, pending revision. Hall (1981) recognized five subspecies in Middle America.

C. d. derbianus (Waterhouse, 1841)
SYNONYMS:

Didelphys Derbiana Waterhouse, 1841:97; type locality ["Habitat"] unknown, restricted to "Valle del Cauca, Colombia," by Cabrera (1958:2) based on similarities between the type and specimens from that region noted by J. A. Allen (1904a:57) and O. Thomas (1913c:358).

Didelphys [(Philander)] lanigera: O. Thomas, 1888b:339; part; not *Didelphis lanigera* Desmarest, 1820.

[*Didelphys (Philander) laniger*] *derbyana*: Trouessart, 1898:1238; name combination and incorrect subsequent spelling of *Didelphys derbiana* Waterhouse.

Philander laniger guayanus O. Thomas, 1899d:286–87; type locality "Balzar Mountains, Prov. Guayas, W. Ecuador."

Caluromys derbianus: J. A. Allen, 1900b:189; first use of current name combination.

Caluromys laniger guayanus: J. A. Allen, 1900b:189; name combination.

Caluromys laniger pyrrhus O. Thomas, 1901c:196; type locality "Rio Oscuro, near Cali, Cauca River, [Valle del Cauca,] Colombia."

[*Didelphys (Philander) laniger*] *derbyanus*: Trouessart, 1905: 855; name combination and incorrect subsequent spelling of *Didelphys derbiana* Waterhouse.

[*Didelphys (Philander) laniger*] *guayanus*: Trouessart, 1905: 855; name combination.

[*Didelphys (Philander) laniger*] *pyrrhus*: Trouessart, 1905: 855; name combination.

Ph[ilander]. l[aniger]. derbianus: O. Thomas, 1913c:358; name combination.

Philander laniger pictus O. Thomas, 1913c:360; type locality "Apia, Rio Apia, [Risaralda,] Cauca slopes of West Colombia."

Philander laniger senex O. Thomas, 1913d:573; type locality "Mindo, [Pichincha,] Ecuador."

[*Didelphis (Micoureus)*] *derbianus*: Matschie, 1916:269; name combination.

[*Didelphis (Micoureus)*] *pictus*: Matschie, 1916:269; name combination.

[*Didelphis (Micoureus)*] *pyrrhus*: Matschie, 1916:269; name combination.

[*Didelphis (Micoureus)*] *senex*: Matschie, 1916:269; name combination.

[*Didelphis (Micoureus)*] *guayanus*: Matschie, 1916:269; name combination.

[*Didelphis (]Micoureus[)] antioquiae* Matschie, 1917:286; type locality "Caceres am Cauca, Prov. Antioquia, Columbia."

Caluromys (Mallodelphis) derbiana: Gilmore, 1941:318; name combination.

Caluromys derbianus derbianus: Cabrera, 1958:2; first use of current name combination.

The nominate subspecies is in northern Colombia, and in the western lowlands and Pacific versant of the Andes in Colombia and Ecuador, but has not been recorded in northwestern Peru. Elsewhere, *C. d. derbianus* is in Panama. Hall (1981) recognized five additional subspecies from Mexico and Central America.

NATURAL HISTORY: *Caluromys derbianus* is arboreal, nocturnal, solitary, and consumes a variety of fruits, insects, small vertebrates, and flower parts, including nectar. The species inhabits primary and disturbed tropical humid forests from sea level to about 2,500 m. In their *Mammalian Species* account, Bucher and Hoffmann (1980) provided illustrations of the skull and a photograph of a live individual from Colombia. *Caluromys* spp. tend to live longer than other medium-sized and larger didelphids; Walker et al. (1964) gave a longevity record of 5 years and 3 months for a *C. derbianus* in the New York Zoological Park. The karyotype is $2n = 14$, FN = 24 (Reig et al. 1977), with a small acrocentric X and a minute acrocentric Y chromosome.

REMARKS: Bucher and Hoffmann's (1980) map is misleading because they show the South American distribution as separate from the population in eastern Panama. Both areas are populated by the nominate subspecies, which reaches its southern-most limit in southwestern Ecuador. Cabrera (1958) included *Didelphis (Micoureus) pulcher* Matschie, 1917 as a synonym of *C. lanatus*, but I consider the name to be a synonym of *C. derbianus*. The holotype is from a zoo and its origin is unknown.

Caluromys lanatus (Olfers, 1818)
Brown-eared Woolly Opossum

SYNONYMS: The following name cannot be assigned to a population because the source of the type specimen is unknown. See under Subspecies for additional synonyms.

Mallodelphis lanigera hemiura A. Miranda-Ribeiro, 1936: 355; type locality thought to be Brazil, but specific locality unknown.

DISTRIBUTION: *Caluromys lanatus* is in northern and central Colombia, northwestern and southern Venezuela, eastern Ecuador, eastern Peru, eastern Bolivia, eastern and southern Paraguay, northern Argentina (provincia Misiones), and western and southern Brazil.

MARGINAL LOCALITIES (Map 3): COLOMBIA: Magdalena, Minca (J. A. Allen 1904f); Guajira, Villanueva

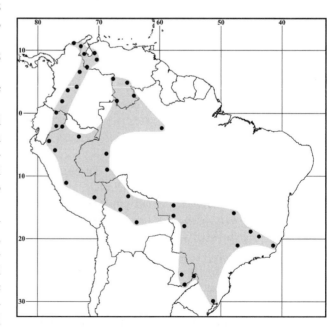

Map 3 Marginal localities for *Caluromys lanatus* ●

(B. E. Brown 2004). VENEZUELA: Zulia, Kunana (Pérez-Hernández 1990); Trujillo, 23 km NW of Valera (Handley 1976); Barinas, La Lengüeta (Pérez-Hernández 1990); Apure, Nulita (Handley 1976). COLOMBIA: Meta, Villavicencio (Bates 1944). ECUADOR: Pastaza, Montalvo (B. E. Brown 2004). PERU: Loreto, Iquitos (FMNH 87129). VENEZUELA: Amazonas, San Carlos de Río Negro (Pérez-Hernández 1990); Amazonas, 30 km S of Puerto Ayacucho (Handley, 1976); Amazonas, Cacurí (Pérez-Hernández 1990); Amazonas, mouth of Río Ocamo (Tate 1939). BRAZIL: Amazonas, ca. 80 km N of Manaus (Malcolm 1990); Amazonas, Barro Vermelho (J. L. Patton, Silva, and Malcolm 2000); Acre, Sena Madureira (B. E. Brown 2004). BOLIVIA: Beni, Piedras Blancas (S. Anderson 1997). BRAZIL: Mato Grosso, Salto da Felicidade (A. Miranda-Ribeiro 1914); Distrito Federal, Fazenda Agua Limpa (Gribel 1988); Minas Gerais, Barra do Paraopeba (type locality of *Mallodelphis laniger vitelina* A. Miranda-Ribeiro); Minas Gerais, Lagoa Santa (Winge 1893); Espírito Santo, São Pedro de Itabapoana (Ruschi 1965); São Paulo, Araraquara (C. O. C. Vieira 1955); Rio Grande do Sul, Rio Cahy (type locality of *Micoureus cahyensis* Matschie). ARGENTINA: Misiones, Campamento Yacú-poi (Massoia et al. 1987; not mapped). PARAGUAY: Itapúa, Santa María (J. A. Wagner 1847); Guairá, Villa Rica (Rengger 1830). BRAZIL: Mato Grosso do Sul, Pantanal (type locality of *Mallodelphis lanigera modesta* A. Miranda-Ribeiro); Mato Grosso, Jaurú (A. Miranda-Ribeiro 1936). BOLIVIA: Santa Cruz, Buena Vista (Crespo 1974); Beni, Las Tajibos (S. Anderson 1997). PERU: Cusco, Hacienda Cadena (Sanborn 1951b); Junín, San Ramón (B. E. Brown 2004); San Martín, Yurac Yacu (O. Thomas 1927a); Amazonas, Huampami (J. L. Patton, Berlin, and Berlin 1982). ECUADOR: Pastaza, Río Pindo Yacu (B. E. Brown 2004), Sucumbios, Santa Cecilia (R. H. Baker 1974). COLOMBIA: Huila, Pitalito (FMNH 70994); Tolima, Natagaima (B. E. Brown 2004); Santander, San Gil (B. E. Brown 2004).

SUBSPECIES: Cabrera (1958) recognized four subspecies of *Caluromys lanatus*; however, an examination of specimens, and their distributions, suggests at least six; the species needs to be revised (see Remarks).

C. l. cicur (Bangs, 1898)
SYNONYMS:
Philander cicur Bangs, 1898b:161; type locality "Pueblo Viejo, [El Cesar,] Colombia."
[*Didelphys (Philander)*] *cicur*: Trouessart, 1898:1238; name combination.
Caluromys cicur: J. A. Allen, 1900b:189; name combination.
C[aluromys]. l[aniger]. cicur: O. Thomas, 1901c:196; name combination.

Ph[ilander]. l[aniger]. cicur: O. Thomas, 1913c:358; name combination.
[*Didelphis (]Micoureus[)*] *meridensis* Matschie, 1917:285; type locality "Sierra bei Merida," Mérida, Venezuela.
[*Caluromys laniger*] *meridensis*: Tate, 1939:163; name combination.
Caluromys lanatus cicur: Cabrera, 1958:2; name combination.

This subspecies is in northeastern Colombia and northwestern Venezuela.

C. l. lanatus (Olfers, 1818)
SYNONYMS:
[*Didelphys*] *lanata* Illiger, 1815:107; *nomen nudum*.
D[idelphys]. lanata Olfers, 1818:106; type locality "Paraguay"; based on "*Micouré laineux*" of Azara (1801a: 175) from Caazapa (see Cabrera 1916).
Didelphis lanigera Desmarest, 1820:258; type localities "Caapeza [sic, = Caazapa], à cinquante lieues de la cité de l'Assomption; et l'autre, dan les champs du village de Sainte-Marie de la Foi"; type locality restricted to Caazapá, Caazapá, Paraguay by Cabrera (1916:516).
Didelphys lanigera: Waterhouse, 1841:98; name combination.
Micoureus lanigera: Lesson, 1842:186; name combination.
Didelphys [(Philander)] lanigera: O. Thomas, 1888b:339; part; name combination.
Caluromys laniger: J. A. Allen, 1900b:189; name combination.
Philander laniger: Cabrera, 1916:514; name combination.
[*Didelphis (]Micoureus[)*] *cahyensis* Matschie, 1917:288; type locality "Am Rio Cahy in Rio Grande do Sul," Brazil.
P[hilander]. lanata: Hershkovitz, 1951:552; name combination.
Philander calmensis C. O. C. Vieira, 1955:347; incorrect subsequent spelling of *cahyensis* Matschie.
Caluromys lanatus lanatus: Cabrera, 1958:2; name combination.

The nominate subspecies is in Paraguay, northern Argentina (provincia Misiones), and southern Brazil (estado do Rio Grande do Sul).

C. l. nattereri (Matschie, 1917)
SYNONYMS:
[*Didelphis (Micoureus)*] *nattereri* Matschie, 1917:291; type locality "Caissara [= Caiçara], Matto Grosso," Brazil.
Mallodelphis lanigera nattereri: A. Miranda-Ribeiro, 1936: 356; name combination.
Mallodelphis lanigera modesta A. Miranda-Ribeiro, 1936: 356; type locality "Matto-Grosso, provavelmente Pantanal," Mato Grosso do Sul, Brazil.

Caluromys lanatus lanatus: S. Anderson, 1997:140; name combination; not *Didelphys lanata* Olfers.

This subspecies is in southwestern Brazil (Mato Grosso and Mato Grosso do Sul) and adjacent Bolivia.

C. l. ochropus (J. A. Wagner, 1842)

SYNONYMS:

Didelphys ochropus J. A. Wagner, 1842:359; type locality "Barra," Amazonas, Brazil.

[*Didelphys (Philander) laniger*] *ochropus*: Trouessart, 1898: 1238; name combination; in synonymy of *Didelphis (Philander) laniger derbiana*.

[*Caluromys*] *ochropus*: O. Thomas, 1901c:196; name combination.

[*Didelphis (Micoureus)*] *ochropus*: Matschie, 1916:269; name combination.

[*Philander laniger*] *ochropus*: Cabrera, 1919:33; name combination.

Mallodelphis lanigera ochropus: A. Miranda-Ribeiro, 1936: 355; name combination.

Caluromys laniger ochropus: Tate, 1939:163; name combination.

Caluromys lanatus ochropus: Cabrera, 1958:3; name combination.

This subspecies is in southern Venezuela and western Brazil. The southern and western limits are unknown, but may include extreme southeastern Colombia and eastern Ecuador and Peru.

C. l. ornatus (Tschudi, 1845)

SYNONYMS:

D[*idelphys*]. *ornata* Tschudi, 1845:146; type locality "der mittleren und tiefen Waldregion," Peru.

P[*hilander*]. *ornatus*: Bangs, 1898b:162; name combination.

[*Didelphys (Philander) laniger*] *ornata*: Trouessart, 1898: 1238; name combination.

Caluromys derbianus ornatus: J. A. Allen, 1900b:189; name combination.

[*Caluromys*] *ornatus*: O. Thomas, 1901c:196; name combination.

[*Didelphys (Philander) laniger*] *ornatus*: Trouessart, 1905: 855; name combination and incorrect gender concordance.

Ph[*ilander*]. l[*aniger*]. *ornatus*: O. Thomas, 1913c:358; name combination.

Philander laniger jivaro O. Thomas, 1913c:360; type locality "Sarayacu on the Pastasa River," Pastaza, Ecuador.

[*Didelphis (Micoureus)*] *ornatus*: Matschie, 1916:269; name combination.

[*Didelphis (]Micoureus[)*] *juninensis* Matschie, 1917:283; "Chanchamayo in der Nähe von La Merced, Provinz Junin, Peru."

[*Didelphis (]Micoureus[)*] *bartletti* Matschie, 1917:288; type locality "Chamicaros-Fluß, südlicher Nebenfluß des Marañon zwischen Huallaga und Ucayali," Loreto, Peru.

[*Caluromys laniger*] *jivaro*: Tate, 1939:163; name combination.

Caluromys laniger ornatus: Sanborn, 1949b:277; name combination.

Caluromys lanatus ornatus: Cabrera, 1958:3; name combination.

This subspecies is in southern Colombia, and the eastern Andean valleys and adjacent lowlands of Ecuador, Peru, and Bolivia.

C. l. vitalinus (A. Miranda-Ribeiro, 1936)

SYNONYMS:

Mallodelphis lanigera vitalina A. Miranda-Ribeiro, 1936: 355; type locality "Barra do Paraopéba, Minas Gerais," Brazil.

Caluromys lanatus ochropus: Cabrera, 1958:3; part.

This subspecies is in southeastern Brazil (recorded from Distrito Federal, Minas Gerais, Espírito Santo, and São Paulo).

NATURAL HISTORY: Like other species in the genus, *C. lanatus* is arboreal and nocturnal, tends to be solitary, and eats a variety of fruits, insects, and small vertebrates. It also feeds on nectar and may consume other flower parts, including pollen (Janson, Terborgh, and Emmons 1981; Gribel 1988). The Brown-eared Woolly Opossum inhabits a variety of lowland humid forests that include primary, secondary, plantation, and gallery forest types, usually at elevations below 500 m.

Little is known of the reproductive pattern. Females with pouched young have been recorded in March, July, August, and December, and lactating females without pouched young in January, March, and July, suggesting reproductive activity throughout the year. From notations on specimen labels, the number of pouched young is usually three.

Ricardo Guerrero (1985a) reported a filarial nematode (*Skrjabinofilaria skrjabini*) from a Venezuelan specimen. The karyotype is $2n = 14$, FN $= 24$ (Reig et al., 1977), and identical to that of *C. derbianus*.

REMARKS: The extensive geographic variation in this species has resulted in a large number of names. Cabrera (1958) reduced the number of subspecies to four; however, an examination of the variation apparent between population units suggests that his treatment was too restrictive and I recognize six. Specimens are still uncommon from many areas and many more specimens are needed before the species can be adequately revised.

I do not know the basis for the identification by Marshall (1982a:260) and Reig, Kirsch, and Marshall (1987:8) of the fossil material from Lagoa Santa, Minas Gerais, Brazil, as *C. derbianus*. Winge (1893) reported this material, along with locally taken specimens, under the name *Philander laniger*, demonstrating the presence of *C. lanatus* in the modern fauna of that region. *Caluromys derbianus* is not, and probably never was, a member of the Brazilian fauna.

Caluromys philander (Linnaeus, 1758)
Bare-tailed Woolly Opossum

SYNONYMS: See under Subspecies.

DISTRIBUTION: *Caluromys philander* is in Venezuela, Trinidad, the Guianas, Brazil, and eastern Bolivia. The distribution is mapped (Map 2) as a large northern component in northern South America, including Brazil, that extends into eastern Bolivia, and a smaller eastern component restricted to Brazil.

MARGINAL LOCALITIES (Map 2): *Northern distribution*: VENEZUELA: Falcón, 14 km ENE of Mirimire (Handley 1976); Distrito Federal, Los Venados (Handley 1976); Nueva Esparta, El Valle (G. M. Allen 1902b). TRINIDAD AND TOBAGO: Trinidad, Botanic Gardens (type locality of *Philander trinitatis* O. Thomas). GUYANA: Pomeroon-Supenaam, Supenaam River (O. Thomas 1910b). SURINAM: Paramaribo, Plantation Clevia (Husson 1978). FRENCH GUIANA: Cabassou (Atramentowicz 1986). BRAZIL: Amapá, Vila Velha do Cassiporé (Carvalho 1962a); Pará, Belém (Carvalho and Toccheton 1969); Pará, Bragança (B. E. Brown 2004); Maranhão, Miritiba (C. O. C. Vieira 1955); Maranhão, Carolina (B. E. Brown 2004); Mato Grosso, Serra do Roncador (Pine, Bishop, and Jackson 1970); Mato Grosso, Tapirapoan (A. Miranda-Ribeiro 1914); Mato Grosso, Mato Grosso (type locality of *Didelphys affinis* J. A. Wagner). BOLIVIA, Santa Cruz, Lago Caimán (Emmons 1998). BRAZIL: Mato Grosso, Salto Utiarity (A. Miranda-Ribeiro 1914); Pará, 52 km SSW of Altamira (USNM 549278); Pará, Fordlândia (B. E. Brown 2004); Amazonas, Borba (J. A. Wagner 1847); Amazonas, Reserva Ducke (Ávila-Pires 1964). VENEZUELA: Amazonas, San Carlos de Río Negro (Pérez-Hernández 1990); Amazonas, 30 km S of Puerto Ayacucho (Handley, 1976); Bolívar, Serranía de los Pijiguaos (Ochoa, Sanchez, and Ibáñez 1988); Bolívar, Maripa (type locality of *Caluromys trinitatis leucurus* O. Thomas); Anzoátegui, Paso Bajito (Pérez-Hernández 1990); Guárico, Dos Caminos (Pérez-Hernández 1990); Carabobo, near Montalbán (Handley 1976). *Eastern distribution*: BRAZIL: Alagoas, Mangabeira (C. O. C. Vieira 1953); Espírito Santo (Ruschi 1965, no specific locality, not mapped); Rio de Janeiro, Rio de Janeiro (A. Miranda-Ribeiro 1936); São Paulo, São Sebastião (C. O. C. Vieira 1955); Santa Catarina, Humboldt (A. Miranda-Ribeiro 1936); São Paulo,

Ypanema (type locality of *Didelphys dichura* J. A. Wagner); São Paulo, Tanabí (C. O. C. Vieira 1955); Minas Gerais, Passos (B. E. Brown 2004); Minas Gerais, Parque Estadual da Floresta do Rio Dôce (Stallings 1989).

SUBSPECIES: I tentatively recognize four subspecies of *C. philander*; the species needs to be revised.

C. p. affinis (J. A. Wagner, 1842)
SYNONYMS:

Didelphys affinis J. A. Wagner, 1842:358; type locality "Matto grosso," Mato Grosso, Brazil.

Caluromys affinis: J. A. Allen, 1900b:189; name combination.

[*Philander philander*] *affinis*: Cabrera, 1919:34; name combination.

Caluromys philander affinis: Cabrera, 1958:3; first use of current name combination.

This subspecies is known from the state of Mato Grosso, Brazil, and adjacent Bolivia.

C. p. dichurus (J. A. Wagner, 1842)
SYNONYMS:

Didelphys dichura J. A. Wagner, 1842:358; type locality "Ypanema," São Paulo, Brazil.

Didelphys dichrura Schinz, 1844:504; incorrect subsequent spelling of *Didelphys dichura* J. A. Wagner.

[*Didelphys (]Philander[)] dichura*: Burmeister, 1856:76; name combination.

Didelphys macrura Pelzeln, 1883:111; *nomen nudum*; not *D. macrura* Olfers, 1818 (= *Thylamys macrura*).

D[idelphys]. longicaudata Pelzeln, 1883:111; *nomen nudum*.

Micoureus philander: Ihering, 1894:11; name combination.

[*Philander philander*] *dichrura* Cabrera, 1919:34; name combination and incorrect subsequent spelling of *Didelphys dichura* J. A. Wagner.

C[aluromys]. philander dichrurus C. O. C. Vieira, 1953:220; name combination and incorrect subsequent spelling of *Didelphys dichura* J. A. Wagner.

This subspecies is in eastern and southeastern Brazil and, apparently, isolated from other population units of the species.

C. p. philander (Linnaeus, 1758)
SYNONYMS:

[*Didelphis*] *philander* Linnaeus, 1758:54 type locality "America"; restricted to Surinam by O. Thomas (1911b:143).

Didelphis Cajopolin Müller, 1776:35; type locality "America"; based on Buffon (1763:plate 55) and an unidentifiable reference to Boddaert; therefore, the type locality is Surinam, the source of the specimen illustrated by Buffon (see Remarks).

Didelphys Cayopollin Schreber, 1777:plate 148; type locality "Mexico," here corrected to Surinam (see Remarks).

Didelphis Cayopollin Kerr, 1792:193; type locality "mountains of New Spain" (see Remarks).

V[*sic, = D(idelphis).*]. *Flavescens* Brongniart, 1792:115; type locality "Cayenne."

S[*arigua*] *cayopollin*: Muirhead, 1819:429; name combination.

[*Didelphis*] *dorsigera*: Desmoulins, 1824:491; not *Didelphis dorsigera* Linnaeus, 1758.

[*Didelphys (|Philander[)*] *Cayopollin*: Burmeister, 1856:76; name combination.

Gamba philander: Liais, 1872:330; name combination.

[*Didelphys (Philander)*] *philander*: Trouessart, 1898:1237; name combination.

Caluromys philander: J. A. Allen, 1900b:189; name combination.

Caluromys trinitatis leucurus O. Thomas, 1904e:36; type locality "Maripa, Caura Valley, Lower Orinoco," Bolívar, Venezuela.

[*Philander philander*] *philander*: Cabrera, 1919:34; name combination.

[*Philander trinitatis*] *leucurus*: Cabrera, 1919:34; name combination.

Caluromys philander philander: C. O. C. Vieira, 1953:220; first use of current name combination.

The nominate subspecies is in the Guianas, Venezuela south of the Orinoco, and in Brazil east of the Rio Negro.

C. p. trinitatis (O. Thomas, 1894)

SYNONYMS:

Didelphys (Philander) trinitatis O. Thomas, 1894b:438; type locality "Botanic Gardens, Trinidad," Trinidad and Tobago.

Philander trinitatis: J. A. Allen and Chapman, 1897:26; name combination.

Caluromys trinitatis: J. A. Allen, 1900b:189; name combination.

Caluromys trinitatis venezuelae O. Thomas, 1903e:493; type locality "Ypuré, Cumaná, [Sucre,] Venezuela."

[*Philander trinitatis*] *trinitatis*: Cabrera, 1919:34; name combination.

[*Philander trinitatis*] *venezuelae*: Cabrera, 1919:34; name combination.

This subspecies is on Trinidad and in Venezuela north of the Río Orinoco.

NATURAL HISTORY: This species is an arboreal, nocturnal, and solitary inhabitant of tropical lowland forests. The diet consists mainly of fruit, but includes nectar, invertebrates, and small vertebrates. Specimens of *C. philander* are relatively uncommon in collections; however, apparent rarity may be an artifact of collecting technique, as shown

by Malcolm (1990) who found the species to be the second most abundant nonflying forest mammal in his study area. He demonstrated the arboreality of this species by catching 56 (plus 23 recaptures) in 2,040 trap-station nights, in traps placed in trees at an average height of 14.7 m above ground. In the same area (ca. 80 km N of Manaus, Amazonas, Brazil), he caught five and recaptured two on the ground during 18,495 trap-station nights. Based on arboreal trapping, Malcolm (1990, 1991) estimated 0.69 *C. philander* per hectare. Atramentowicz (1986) and Julien-Laferrière and Atramentowicz (1990) found the species in French Guiana lives longer (in excess of 30 months) and has smaller litters (average 3.5), a longer gestation period (at least 21 days), and longer duration of parental care (120+ days) than do sympatric *Philander opossum* and *Didelphis marsupialis*. Although they have smaller litters, females normally have only four mammae; an average of 3.5 young per female indicates high reproductive efficiency in the species. Nowak and Paradiso (1983) gave a longevity record of 6 years, 4 months, presumably of a captive individual.

Ricardo Guerrero (1985a) reported three nematodes, two fleas, two mites, and the hemoflagellate *Trypanosoma cruzi* known to parasitize Venezuelan *C. philander*. The karyotype is $2n = 14$, FN = 24 (Reig et al., 1977), with a small submetacentric X. Reig et al. (1977) reported the Y chromosome as a minute acrocentric; however, the Y chromosome reported by Palma and Yates (1996), and confirmed by J. L. Patton, Silva, and Malcolm (2000), is a minute biarmed chromosome.

REMARKS: The Venezuelan population that Pérez-Hernández (1990:375) referred to as "*Caluromys philander* subsp. nov." is the subspecies I call *C. philander trinitatis*. This population, with that on Trinidad, were long considered a separate species. The presence of *C. philander* in Bolivia is based on an animal seen by several people during a Rapid Assessment Program survey of Parque Nacional Noël Kempff Mercado (Emmons 1998; pers. comm.).

The first use of the vernacular "cayopollin" of authors as a scientific name was by Müller (1776:35), who used the combination *Didelphis cajopolin* for Buffon's (1763:plate 55) "Le Cayopollin." Müller also cited Boddaert in his synonymy, but that reference is unidentifiable without further attribution. Schreber (1777:plate 148) used "*cayopollin*" in the name combination "*Didelphys cayopollin* B." and later (1778:544) cited several references, including "Cayopollin" attributed to Buffon (1763:350 and plate 55) in the synonymy. Schreber's (1778) synonymy clearly is composite because the species is not known to occur in Mexico or Central America. *Didelphis cayopollin* Kerr, 1792, similarly is composite based on the synonymy, but the description can refer only to the taxon known today as

C. philander, which is known from only continental South America and the islands of Trinidad and Margarita.

Didelphis cayopollin Daudin in Lacépède (1802:152), is based on "*Le Cayopollin*" described by Lacépède (1799a: 177) and associated with *Didelphis dorsigera* Linnaeus, 1758. The latter is a junior synonym of *Marmosa murina* Linnaeus, 1758; however, the former is based mainly on Hernandez's (1651) description and is not identifiable. The associated plate 19, not cited in the description, is a copy of Buffon's (1763) plate 55, which is clearly of *C. philander*.

Apparently, the first use of the vernacular "cayopollin," or any of its variant spellings, was by Hernandez (1651:10) for an animal from the mountains of southern Mexico. That animal is unidentifiable, but could have been either a *Marmosa mexicana*, *Tlacuatzin canescens*, *Philander opossum*, or *Caluromys derbianus*. A Mexican origin, as given by Schreber (1778:544) for his *Didelphys cayopollin*, was probably the basis for O. Thomas's (1888b:344) inclusion of the name under *Didelphys murina*, a name he used for most of the small, pale-bellied opossums known at that time. However, O. Thomas (1888b:337) included Burmeister's (1856:76) name combination *Didelphys (Philander) cayopollin* in the synonymy of *Didelphys philander*. Other authors (e.g., Cabrera 1919, 1958; Hershkovitz 1949a) followed O. Thomas's usage in treating Burmeister's name combination as a unique name ("*Philander cayopollin*") referring to a species different from that named by Schreber (1777). These authors must have been aware that Burmeister used *Philander* as a subgenus of *Didelphis* and, as such, if the name were applied to a species different from that named by Schreber, Burmeister's name combination *Didelphys (Philander) cayopollin* would be a junior homonym of *Didelphys cayopollin* Schreber, 1777, *Didelphis cayopollin* Kerr, 1792, and *Didelphis cayopollin* Daudin in Lacépède, 1802.

Schreber's (1777:plate 148) illustration of his "*Didelphys Cayopollin* B." is a copy of Buffon's (1763) plate 55, "*Le Cayopollin*," which was based on a specimen of *Caluromys philander*, as evidenced by the prominent medial dark stripe on the forehead and the long tail that is furred only at the base. The measurements given by Daubenton (in Buffon 1763) for this animal are too large for *Marmosa murina*, but correspond to measurements of *C. philander*, as do the illustrations of internal anatomy (plate 56), and skull and skeleton (plate 57). At least plate 57, and corresponding measurements, were based on a skeleton sent from Surinam (Daubenton in Buffon, 1763:368).

The reference to Buffon (1763:plate 55) is the only identifiable basis for *Didelphis cajopolin* Müller, and I designate the specimen illustrated in plate 55 as the type; it is clearly a *C. philander*. Schreber's (1777) plate 148 is a copy of Buffon's plate 55. The letter B following Schreber's (1777)

name *Didelphys cayopollin* on plate 148 refers to Buffon (1763:plate 55, "*Le Cayopollin*") and, because *Didelphys cayopollin* Schreber (plate 148) was published the year preceding publication of its description and synonymy, Buffon's illustration can be taken as the sole source of the name. Therefore, I also designate the specimen depicted on Buffon's plate 55 as the type of *Didelphys cayopollin* Schreber, 1777. According to Daubenton (in Buffon 1763:368), at least the skeleton of "cayopollin" came from Surinam, which can be taken as the type locality. According to É. Geoffroy St.-Hilaire (1803:143), the male in the Paris museum reported under number "CCCIII" was the basis of the figure and description in Buffon (1763). É. Geoffroy St.-Hilaire gave the locality as "La Guyane," a general reference to French Guiana.

Genus *Caluromysiops* Sanborn, 1951

Louise H. Emmons

The monotypic genus *Caluromysiops* is a large (head and body 250–330 mm, tail 310–340 mm) gray-brown opossum with furred tail (fully furred on dorsal surface to within 2 cm of the tip, but ventrally for only its proximal third) and a well-developed pouch. A characteristic large black patch covers each shoulder and extends forward down the inner surface of the forearm to the wrist or top of the manus, and dorsally along the back as a stripe on either side of the midline to the rump. The tail-tip is dirty white, and the ears are yellow. The fur is soft, long, and woolly. The skull is massively built and the greatest skull length exceeds 62 mm (see illustration in Izor and Pine 1987). The long upper canines are blade-like and unusually wide anteroposteriorly at the base. When present, the small upper premolar (P1), present in some subadults including the holotype, is crowded against the upper canine. The articulation of the third metacarpal is offset, allowing spread between the second and third digits when grasping a branch (Izor and Pine 1987).

SYNONYM:

Caluromysiops Sanborn, 1951a:473; type species *Caluromysiops irrupta* Sanborn, 1951a, by original designation.

Caluromysiops irrupta Sanborn, 1951
Black-shouldered Opossum
SYNONYM:

Caluromysiops irrupta Sanborn, 1951a:473; type locality "Quincemil, Province of Quispicanchis," Cusco, Peru.

DISTRIBUTION: *Caluromysiops irrupta* is known from southeastern Colombia and adjacent Peru, and from southeastern Peru and western Brazil. Some of the reports from Leticia, Colombia, and the vicinity of Iquitos, Peru (Izor and Pine 1987; see Remarks), may have been based on specimens in the live-animal trade.

MARGINAL LOCALITIES (Map 2): PERU: Loreto, Río Santa María (P. Soini, pers. comm.). COLOMBIA: Amazonas, Leticia (Simonetta 1979). BRAZIL: Rondônia, Upper Rio Jaru (Vivo and Gomes 1989). PERU: Cusco, Quince Mil (type locality of *Caluromysiops irrupta* Sanborn); Madre de Dios, Cocha Cashu Biological Station (Janson, Terborgh, and Emmons 1981); Madre de Dios, Itahuanía (Izor and Pine 1987, not mapped); Cusco, Pozo Cashiriari C (S. Solari et al. 1998); Loreto, lower Río Nanay (P. Soini, pers. comm.).

SUBSPECIES: *Caluromysiops irrupta* is considered to be monotypic.

NATURAL HISTORY: The species is known from tropical humid forest below 700 m. Janson, Terborgh, and Emmons (1981) recorded *C. irrupta* feeding on nectar in the forest canopy during the dry season at Cocha Cashu Biological Station, Manu National Park. At Cocha Cashu, I found *C. irrupta* to be nocturnal, solitary, arboreal, and an inhabitant of mature floodplain forest.

Females with young were received from animal dealers in July and August. No more than two young have been noted per female. Like species of *Caluromys*, individuals are long lived and longevity in captivity exceeds 6 years, 10 months (Collins 1973; Izor and Pine 1987). A wild-caught male had the following measurements (mm): head and body 279, tail 272, ear 34, mass 445 g (S. Solari and J.-J. Rodriguez, pers. comm.).

REMARKS: The genus is closely related to, and some authors (Cabrera 1958:4; Honacki, Kinman, and Koeppl 1982:18; Izor and Pine 1987) have suggested congeneric with, *Caluromys*. Recent molecular studies by Jansa and Voss (2000) confirm a close relationship with *Caluromys*. Most of the known specimens have been acquired from animal dealers and lack reliable locality information (Izor and Pine 1987; Vivo and Gomes 1989). Specimens from Leticia, Colombia (i.e., USNM 397626), and the specimen photographed by Simonetta (1979) may be of questionable origin because this port served as an export point for live animal traffic from Peru (see Izor and Pine 1987). Pekka Soini (pers. comm.) collected three specimens in northern departamento Loreto, Peru. One of these was exported alive to the United States and may have been preserved as FMNH 60698, the other two were lost. The specimen from Brazil is a skin without skull (MZUSP 11681).

Hershkovitz (1992b) placed *Caluromysiops* in a separate subfamily, Caluromysiopsinae (*nomen nudum*), to distinguish the genus from *Caluromys*. The rational for this action may have been his belief that *Caluromysiops* has a simple penis, whereas it is bifid in *Caluromys* (see

Hershkovitz (1992b:13). Apparently, he was unaware of Izor and Pine's (1987) description of the deeply-cleft, bifid penis in *Caluromysiops*. Caluromysiopsinae is an unavailable name because it lacked a description or definition purported to differentiate the taxon (ICZN 1985:Art. 13).

Genus *Glironia* O. Thomas, 1912

Linda J. Barkley

Opossums of the monotypic genus *Glironia* are small (head and body 160–225 mm, tail 195–225 mm) arboreal inhabitants of tropical humid forests of the Amazon basin from southern Colombia through Ecuador, Peru, Brazil, and Bolivia. The pelage is long, dense, woolly, and extends nearly to the tip of the tail. The color pattern of the dorsum is fawn or cinnamon brown; the venter is gray, or pale brown, and the hairs are often white tipped. A broad, dark brown, or blackish mask extends from the nares across the eyes and base of the ears to the nape. The dark mask stripes are separated by a median stripe of pale gray that extends from the rhinarium to the nape. The median facial stripe is paler (gray) than the rest of the dorsal pelage. The ears are dark, naked, and oval. The feet are whitish. The amount of white on the distal part of the tail is variable, as is the extent of the ventral bare surface (modified for a prehensile function; see illustration of dermal ridges in M. N. F. Silva and Langguth 1989) along the ventral midline. *Glironia* has five mammae (two abdominal pairs surrounding a single central teat) and lacks a pouch. Size and body proportions are similar to those of the larger members of the genus *Micoureus*, but *Glironia* can be easily recognized by the pale medial facial stripe, the bushy and nearly fully haired tail, and the posteriorly expanded nasals. In general proportions, the cranium is similar to that of mouse opossums (e.g., *Marmosa*, *Marmosops*, and *Micoureus*). However, the postorbital processes are well developed and form ledges overhanging the orbits.

Reig (1955) grouped *Glironia* with *Dromiciops* and *Caluromys* in the subfamily Microbiotheriinae. However, Kirsch (1977a) and Marshall (1978c, 1981, 1982a), having concluded that the dental similarities between *Glironia* and *Dromiciops* are convergent, placed *Glironia* with *Caluromys* and *Caluromysiops* in the subfamily Caluromyinae. Reig, Kirsch, and Marshall (1987) assessed dentition, ear structure, and other features of *Glironia* against those of other American marsupials. Hershkovitz (1992b:4) listed "Family Glironiidae (new)" for this genus; however, the name is unavailable. Hershkovitz failed to satisfy the requirements of (ICZN 1985:Art. 13; ICZN 1999:Art. 13.1.1), because he failed to accompany the name with a description or definition that differentiated the taxon from

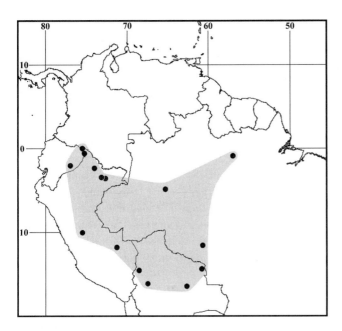

Map 4 Marginal localities for *Glironia venusta* ●

other taxa. The usage of Glironiidae by Cuartas and Muñoz (2003) does not validate the name. There is no known fossil record.

SYNONYM:

Glironia O. Thomas, 1912b:239; type species *Glironia venusta* O. Thomas, 1912b, by original designation.

Glironia venusta O. Thomas, 1912
Bushy-tailed Opossum

SYNONYMS:

Glironia venusta O. Thomas, 1912b:240; type locality "Pozuzo," Pasco, Peru.

Glironia aequatorialis H. E. Anthony, 1926:1; type locality "Boca de Lagarto Cocha, on the Rio Aguarico above its confluence with the Rio Napo," Sucumbios, Ecuador.

Glironia criniger H. E. Anthony, 1926:2; type locality "Junction of Rio Curaray with Rio Napo," Loreto, Peru.

DISTRIBUTION: *Glironia venusta* is in the Amazon lowlands of Brazil, Peru, Bolivia, Ecuador, and Colombia.

MARGINAL LOCALITIES (Map 4): COLOMBIA: Putumayo, Quebrada El Hacha (Rodríguez-Mahecha et al. 1995). ECUADOR: Sucumbios, Boca de Lagarto Cocha (type locality of *Glironia aequatorialis* H. E. Anthony). PERU: Loreto, Junction of Río Curaray with Río Napo (type locality of *Glironia criniger* H. E. Anthony); Loreto, Puerto Almendra (M. M. Díaz and Willig 2004); Loreto, Río Amazonas, ca. 10 km SSW of mouth of Río Napo on E bank of Quebrada Vainilla (B. E. Brown 2004). BRAZIL: Amazonas, 4 km inland from the upper Urucú River, ca. 180 km S of Tefé (M. R. Nogueira, Silva, and Câmara 1999); Pará, Cachoeira Porteira, mouth of Rio Mapuera

(M. N. F. Silva and Langguth 1989). BOLIVIA: La Paz, Río Machariapo (Emmons 1991). BRAZIL: Rondônia, Jaburi Farm (Bernardé and Rocha 2003); Santa Cruz, Meseta de Huanchaca (Tarifa and Anderson 1997); Santa Cruz, 12 km S of San Javier (S. Anderson 1993); La Paz, "Yungas" (O. Thomas, 1912d:47; see S. Anderson 1997). PERU: Madre de Dios, Cocha Cashu (V. Pacheco et al. 1993); Pasco, Pozuzo (type locality of *Glironia venusta* O. Thomas). ECUADOR: Pastaza, Montalvo (Marshall 1978c).

SUBSPECIES: I am treating *G. venusta* as monotypic.

NATURAL HISTORY: Almost nothing is known about this marsupial's life history. Hershkovitz (1972) likened *Glironia* to a "marsupial tree shrew" stating that *G. venusta* showed convergence with species of the squirrel-like placental genus *Tupaia*. Nowak and Paradiso (1983) speculated that the diet consists of insects, eggs, seeds, and fruit, but this is only a reflection of its anatomical similarities with some mouse opossums whose diet is better known. Emmons (1998) reported sight records of *G. venusta* in tall deciduous forest (with cactus) at El Refugio and in closed dry woodlands and dwarf evergreen forest above Lago Caimán; both localities in Parque Nacional Noël Kempff Mercado, Santa Cruz, Bolivia. Pine et al. (1985) reported fluorescence of the hair on the chin and around the mammae, the significance of which has yet to be determined. O. Thomas (1912b, 1912d), Cabrera (1919), H. E. Anthony (1926), Cabrera and Yepes (1940), and Marshall (1978c) have provided detailed morphological descriptions of *G. venusta*. M. R. Nogueira, Silva, and Câmara (1999) described the morphology of the male genital system, which is similar to that of *Caluromys lanatus*, *C. philander*, and *Gracilinanus agilis* in having two pairs of Cowper's glands, but differs from all other didelphids in its unsegmented prostate, short glans, and the near terminal openings to the urethral grooves. External measurements (mm) of five specimens are: total length 355–430, tail 190–225, hind foot 27–31, ear ($n = 2$) 23–25. Hershkovitz (1972), Marshall (1978c), and M. N. F. Silva and Langguth (1989) have provided cranial measurements.

REMARKS: The adult male from departamento Loreto, Peru, was caught in 1983 in a mist net set for bats in the primary forest. The Río Copotaza specimen (a skin and skull) was caught by the Olalla brothers on 4 April 1934; another specimen from the Río Pastaza (a skin only in the collections of the Escuela Politécnica Nacional, Quito, according to L. Albuja, *in litt.*) was taken by Spillmann on 25 September 1937. There are three sight records from Bolivia in addition to the two known specimens. One was seen along the Río Machariapo, provincia La Paz (Emmons 1991), another on the meseta above Lago Caimán, and the third at El Refugio; the latter localities are in Parque Nacional Noël Kempff Mercado, Santa Cruz (Tarifa and Anderson

1997; Emmons 1998). The report from Rondônia, Brazil (Bernardé and Rocha 2003) is based on a photographed animal that was released. These records, and the one reported by V. Pacheco et al. (1993) from Cocha Cashu, Madre de Dios, Peru, are not vouchered by specimens. The record from Colombia is based on a photograph of a captured *G. venusta* (see Rodríguez-Mahecha 1995:6, footnote). The specimen from Rio Mapuera, Brazil reported by M. N. F. Silva and Langguth (1989) was found under the roof of a house near secondary forest along a riverbank. The juvenile male from 180 km south of Tefé, Amazonas, Brazil, was shot as it clung to the trunk of a tree about 3 m above the ground (M. R. Nogueira, Silva, and Câmara 1999). The total number of known specimens of *G. venusta* is under 25, of which at least 4 were purchased from animal dealers.

Subfamily Didelphinae Gray, 1821

Tribe Didelphini Gray, 1821

Barbara R. Stein and James L. Patton

The Didelphini comprises four genera characterized by large size (total length exceeds 500 mm), moderately to well-developed pouch, and a diploid chromosome number of 22.

Genus *Chironectes* Illiger, 1811

Chironectes is a monotypic genus of large opossums (head and body 270–400 mm, tail 310–430 mm, hind foot 60–72 mm, mass 600–790 g [from Marshall 1978d]). The fur is relatively short and dense, with a few intermixed longer hairs. The ears are moderately large, naked, and rounded. Long, stout facial bristles are found in tufts above the eyes, on each cheek, at the base of the ears, and on the throat. The tail is round in cross-section, longer than head and body and, except for the furred basal portion, nearly naked and coarsely scaled. Four or five mammae are contained within a pouch, which is closed by a well-developed sphincter when the female is in water (W. L. R. Oliver 1976). Males also have a pouch-like depression into which they retract the scrotum while swimming (Enders 1937). The skull resembles that of *Philander*, with its long rostrum, rounded braincase, and strong and flaring zygomatic arches. The palate has a single pair of large maxillopalatine fenestrae opposite the molars, but lacks the smaller palatine fenestrae present in some other didelphids. The dorsal color pattern is silvery gray, overlaid with dark brown to black markings that consist of a narrow mid-dorsal stripe from the crown to the base of the tail, interconnecting four broad blackish patches located, one each, over shoulders,

center of back, hips, and lower rump. The shoulder and rump patches extend laterally over the legs. The venter is bright white, sharply contrasting with the gray color of the sides. The muzzle, crown of the head, and a band extending through the eye to below the ear, are blackish-brown. The tail is either all black or black for about 4/5 of its length and yellowish-white terminally.

Kirsch (1977b) considered *Chironectes* most closely related to *Didelphis* and *Philander*. Voss and Jansa (2003) placed *Chironectes* at the base of a clade comprising all three of the other large, $2n = 22$ opossums (*Didelphis*, *Philander*, and *Lutreolina*) in a cladistic analysis of both nonmolecular characters and nuclear DNA sequences. Marshall (1977b, 1982a) referred fossils of Montehermosan age (Pliocene) from Argentina to the living species.

SYNONYMS:

Latra Zimmermann, 1780:317; part; incorrect subsequent spelling of *Lutra* of authors; not *Lutra* Linnaeus, 1758.

Mustela: Kerr, 1792:172, 174; part; not *Mustela* Linnaeus, 1758.

Lutra: Link, 1795:84; not *Lutra* Linnaeus.

Didelphis: G. Cuvier, 1798:125; part; not *Didelphis* Linnaeus, 1758.

Lutra: G. Shaw, 1800:447; not *Lutra* Linnaeus.

Mustela: Turton, 1800:58; part; not *Mustela* Linnaeus.

Didelphis: Daudin in Lacépède, 1802:152; part; not *Didelphis* Linnaeus.

Chironectes Illiger, 1811:76; type species *Lutra minima* Zimmermann, 1780, by monotypy.

Memina G. Fischer, 1813:15; *nomen nudum*.

Memina G. Fischer, 1814:611; type species *Lutra memina* Boddaert, 1784, by monotypy (see G. Fischer 1813:579).

Sarigua Muirhead, 1819:429; part.

Cheironectes Gray, 1821:308; incorrect subsequent spelling of *Chironectes* Illiger.

Cheronectis Fleming, 1822:212; incorrect subsequent spelling of *Chironectes* Illiger.

Gamba Liais, 1872:329; type species *Gamba palmata* Liais, 1872, by subsequent designation (Hershkovitz 1949a).

Chironeytes Goeldi and Hagmann, 1904:100; incorrect subsequent spelling of *Chironectes* Illiger.

Chironectes minimus (Zimmermann, 1780)
Water Opossum

SYNONYMS: See under Subspecies.

DISTRIBUTION: *Chironectes minimus* has a disjunct distribution in South America. The northern population is known from Colombia, Ecuador, Venezuela, the Guianas, eastern Peru, and northern Brazil. The southern segment is known from Paraguay, southeastern Brazil, and northeastern Argentina. Elsewhere, the species occurs in Panama and northward into southern Mexico.

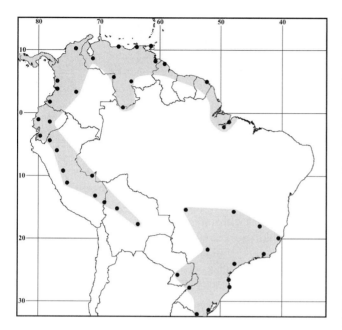

Map 5 Marginal localities for *Chironectes minimus* ●

MARGINAL LOCALITIES (Map 5): *Northern distribution.* VENEZUELA: Sucre, Marigüitar (Pérez-Hernández 1990). TRINIDAD AND TOBAGO: Trinidad (no specific locality, Goodwin and Greenhall 1961). VENEZUELA: Delta Amacuro, Río Ibaruma (Linares and Rivas 2004). GUYANA: Barima-Waini, Warapoco Mission (USNM 296645). FRENCH GUIANA: Cayenne (type locality of *Lutra minima* Zimmermann). BRAZIL: Pará, Barcarena (Pine 1973b); Pará, Cametá (B. E. Brown 2004). VENEZUELA: Bolívar, Santa María del Erebato (Pérez-Hernández 1990); Amazonas, Cerro Neblina Base Camp (Gardner 1988); Amazonas, Puerto Ayacucho (Handley 1976); Mérida, Santa Rosa (Handley 1976). COLOMBIA: Meta, Los Micos (B. E. Brown 2004). ECUADOR: Pastaza, Mera (Rageot and Albuja 1994). PERU: Ucayali, Balta (Voss and Emmons 1996). BOLIVIA: Santa Cruz, Río Pitasama (S. Anderson 1997); Beni, Campamento 6 de Agosto (Cabot 1989). PERU: Puno, Valle Grande (Sanborn 1953); Cusco, Quincemil (B. E. Brown 2004); Junín, Chanchamayo (Lyne 1959); Huánuco, Tingo María (Grimwood 1969); San Martín, Moyobamba (Osgood 1914b); Amazonas, headwaters of Río Huampami (MVZ 153307). ECUADOR: El Oro, 1 km SW of Puente de Moromoro (USNM 513428); Manabí, San José (B. E. Brown 2004). COLOMBIA: Nariño, Barbacoas (J. A. Allen 1916c); Valle del Cauca, Zabaletas (B. E. Brown 2004); Chocó, Río Bando-Río Sandó (B. E. Brown 2004); El Cesar, Colonia Agrícola de Caracolicito (USNM 280909). VENEZUELA: Distrito Federal, Caracas (Pohle 1927). *Southern distribution.* BRAZIL: Mato Grosso, Chapada (Cope 1889a);

Distrito Federal, Parque Nacional de Brasília (Alho, Pereira, and Paula 1987); Minas Gerais, Ribeirão Pedra (Kühlhorn 1953); Espírito Santo, Santa Teresa (Ruschi 1965); Rio de Janeiro, Therezopolis (type locality of *Chironectes menima bresslaui* Pohle); São Paulo, Ipanema (Pelzeln 1883); São Paulo, Rocha (B. E. Brown 2004); Santa Catarina, Rio Novo do Itapocú (A. Miranda-Ribeiro 1936); Santa Catarina, Parque Municipal da Lagoa do Peri (Graipel et al. 2004); Rio Grande do Sul, São Lourenço (C. O. C. Vieira 1949). URUGUAY: Cerro Largo, Cañada Vichadero (González and Fregueiro 1998). ARGENTINA: Misiones, San Javier (Crespo 1950). PARAGUAY: Cordillera, Salto de Pirateta (MVZ 144314); Paraguarí, Parque Nacional Ybycuí (B. E. Brown 2004). BRAZIL: Mato Grosso do Sul, Rio Pardo (Kühlhorn 1953).

SUBSPECIES: Marshall (1978d) recognized four subspecies, of which *C. m. argyrodytes* Dickey, 1928:15 (type locality Hacienda Zapotitán, departamento La Libertad, El Salvador) is confined to Central America and southern Mexico (Goodwin 1942c; Alvarez del Toro 1952; Hall 1981).

C. m. paraguensis (Kerr, 1792)
SYNONYMS:
M[ustela]. (Lutra) paraguensis Kerr, 1792:172; type locality "Rio de la Plata," Paraguay; name based, in part, on the "*saricovienne*" of Smellie's (1780) translation of Buffon (1776:159, pl. 22).
Chironectes langsdorffi Boitard, 1842:288; type locality "près Rio de Janeiro," Brazil.
Chironectes menima bresslaui Pohle 1927:242; type locality "Therezopolis," Rio de Janeiro, Brazil; incorrect subsequent spelling of *Latra minima* Zimmermann.
This subspecies represents the southern distribution in South America, where it is found in southeastern Brazil, eastern Paraguay, northern Uruguay, and northeastern Argentina.

C. m. minimus (Zimmermann, 1780)
SYNONYMS:
Latra (minima) Zimmermann, 1780:317; type locality "Gujana"; restricted by Cabrera (1958:44) to Cayenne, French Guiana.
[Lutra] memina Boddaert, 1784:168; incorrect subsequent spelling of *Latra minima* Zimmermann.
M[ustela]. (Lutra) guianensis Kerr, 1792:174; type locality "Cayenne," French Guiana; name based on "small Guiana otter" of Smellie's (1780) translation of Buffon (1776:159, pl. 22).
L[utra]. gujanensis Link, 1795:84; type locality is Cayenne, French Guiana, because the name is based on Buffon's (1776:159, pl. 22) "*petite loutre d'eau douce*

de Cayenne" and Pennant's (1781:355) "*saricovienne*," also from Cayenne.

Lutra saricovienna G. Shaw, 1800:447; type locality is Cayenne, French Guiana, because the name is based on Buffon's (1776:159, pl. 22) "*petite loutre d'eau douce de Cayenne*" and Pennant's (1781:355) "*saricovienne*," also from Cayenne.

[*Mustela*] *Cayennensis* Turton, 1800:58; type locality "*Cayenne*," French Guiana; name based on the "*petite loutre d'eau douce de Cayenne*" of Buffon (1776;159, pl. 22).

Didelphis palmata Daudin in Lacépède, 1802:152; type locality is Cayenne, French Guiana, because the name is based on Lacépède (1801:98) "*petite loutre d'eau douce de Cayenne.*"

Lutra memia Desmarest, 1803a:147; incorrect subsequent spelling of *Latra minima* Zimmermann.

Lutra memmina Desmarest, 1804a:507; incorrect subsequent spelling of *Lutra minima* Zimmermann.

Didelphis memmina Desmarest, 1804a:507, 1804b:19; name combination and incorrect subsequent spelling of *Lutra minima* Zimmermann.

[*Chironectes minimus*]: Illiger, 1811:76; name combination.

Chironectes variegatus Illiger, 1815:107; *nomen nudum*.

Didelphis lutreola Oken, 1816:1134; unavailable name (ICZN 1956)

Ch[*ironectes*]. *variegatus* Olfers, 1818:206; type locality "Südamerica."

Chironectes yapock Desmarest, 1820:261; type locality "Les bords de l'Yapock, grande rivière de la Guyane," French Guiana.

Chironectes memina F. Cuvier, 1825:252; incorrect subsequent spelling of *Lutra minima* Zimmermann, and incorrect gender concordance.

Chironectes palmata: Hamilton-Smith in Griffith, Hamilton-Smith, and Pidgeon, 1827:pl. facing p. 35; name combination.

S[*arigua*]. *memmima* Muirhead, 1819:329; name combination and incorrect subsequent spelling of *Lutra minima* Zimmermann.

Chironectes minima minima: Krumbiegel, 1940c:66; name combination and incorrect gender concordance.

Chironectes minimus minimus: Cabrera, 1958:43; name combination.

The nominate subspecies is distributed from southeastern Colombia, Venezuela south of the Río Orinoco, and the Guianas to the delta of the Rio Amazonas in the east and southeastern Peru and adjacent Bolivia in the west.

C. m. panamensis Goldman, 1914

SYNONYMS:

Chironectes panamensis Goldman, 1914a:1; type locality " Cana," Upper Río Tuyra, Darién, Panama.

Chironectes minima panamensis: Krumbiegel, 1940c:67; name combination and incorrect gender concordance.

Chironectes minimus panamensis: Cabrera, 1958:44; first use of current name combination.

The South American distribution of this subspecies is northern and western Colombia, northern Venezuela (Pérez-Hernández 1990), and western Ecuador south into northwestern Peru. Elsewhere, it occurs in Central America from southern Nicaragua through Panama (Hall 1981).

NATURAL HISTORY: *Chironectes minimus* is the only aquatic or semiaquatic marsupial and possesses webbed hind feet, a dense water-repellent pelage, a "water-proof" pouch in females (feature shared with species of *Didelphis*), and the ability to draw the scrotum tightly against the abdomen in males. Water opossums are excellent swimmers, using their webbed feet as paddles (Stein 1981) and the tail as a rudder (W. L. R. Oliver 1976); their distribution is closely tied to tropical forest streams and lakes. The den is reached through a hole in the stream bank just above water level (Nowak and Paradiso 1983). These opossums are nocturnal (Zetek 1930) and solitary (W. L. R. Oliver 1976). Augustiny (1943) described general aspects of their morphology, especially their adaptations for swimming. The tail has some prehensile capability, but is too thick for effective use in climbing (W. L. R. Oliver 1976). Water opossums capture prey in the water and feed on fish, crustaceans, insects, and frogs (Collins 1973). Their relatively high number of interramal (11–12) and submental vibrissae (7–10; Lyne 1959) may facilitate locating prey in the water. The largest litter reported is five (Mondolfi and Padilla 1957), although litters of two to three young are more usual (Enders 1966; Linares and Rivas 2004). Rosenthal (1975) described post-natal development and Thompson (1989) studied thermoregulation. W. L. R. Oliver (1976) suggested that their secretive and largely nocturnal habits, coupled with the inaccessibility of their habitat, account for the long-standing belief that these opossums are rare.

The karyotype consists of a diploid number of 22 and a totally uniarmed autosomal and sex chromosome complement (Reig et al. 1977). Marshall (1978d) illustrated the skull and dentition, and outlined available information on their natural history in his *Mammalian Species* account for *C. minimus*.

REMARKS: We do not know if the absence of records from the central Amazon is an artifact of collecting bias or reflects a distributional hiatus. In his revision of *Chironectes*, Krumbiegel (1940c) consistently used the feminine termination (-*a*) instead of the correct masculine form in his spelling of *minimus*.

Cabrera (1958) and Husson (1978) considered *C. langsdorffi* to be a synonym of *C. m. minimus*. Marshall (1978d) used *C. langsdorffi* as the name for the Brazilian,

Paraguayan, and Argentine population. However, the oldest available name for that population is *Mustela (Lutra) paraguensis* Kerr, which antedates *C. langsdorffi* by 53 years. According to Kerr (1792:172), the animal came from "Rio de la Plata," a name generally referring, in the 18th century, to the region comprising southern Paraguay, southern Brazil, western and northern Uruguay, and adjacent Argentina. Paraguay is selected as the origin of *C. m. paraguensis* on the basis of the patronymic nature of the name. The name *Didelphis lutreola*, applied by Oken (1816:1134) to water opossums from the Guiana region, is unavailable from Oken (1916) because his Lehrbuch (Vol. 3) has been rejected for nomenclatural purposes (ICZN 1956). Marshall (1978d) listed *Mustela lutris* Lacépède, 1803 (actually Daudin in Lacépède 1802:164) in the synonymy of *C. minimus*, as based on the "*saricovienne*" of Buffon. However, Buffon's *saricovienne* is a composite of river and sea otters, not the water opossum. Pennant's (1781:355) *saricovienne* is a water opossum.

The name *Didelphys alboguttata* (= *Microdelphys alboguttata* Burmeister, 1854), with type locality given as forested regions of Brazil (O. Thomas 1888b:366), sometimes has been associated with *C. minimus*. Goeldi (1894:466) found the name was based on "*Dasyurus viverrinus*, an Australian marsupial" (but see Pine's comments on this name in his "Remarks," herein under the genus *Monodelphis*).

Genus *Didelphis* Linnaeus, 1758
Rui Cerqueira and Christopher J. Tribe

South American *Didelphis* comprise two groups of species (Cerqueira 1980). The *D. marsupialis*-group (the black-eared opossums) consists of two allopatric species, *D. marsupialis* and *D. aurita*; the *D. albiventris*-group (the white-eared opossums) contains three species: *D. albiventris, D. pernigra,* and *D. imperfecta.* The numerous additional names of species and subspecies applied to South American members of this genus during the past two centuries are based mainly on minor variation insufficient to warrant formal recognition. The only other living species recognized in this genus is *D. virginiana* Kerr, 1792, which is found from Canada to Costa Rica (Gardner 1973, 1982, 1993a, 2003). *Didelphis* presently occurs over most of South America north of the 35th parallel. Fossils are known from the Middle Pleistocene (Ensenadan) in South America (Marshall 1981). The genus includes the largest living American marsupials, and some individuals (in *D. virginiana*) exceeding 1 m in total length and 7 kg in weight. Females have a well-developed marsupium usually containing from 11 to 13 teats. Long, coarse guard hairs overly soft, dense underfur.

The ears are large, oval, naked, and leathery. The tail is long and prehensile, furred at the base, but otherwise scaly, sparsely haired, and white on the terminal 1/2 or more. The skull has a well-developed sagittal crest in adults; the braincase is narrow, and the interorbital breadth is greater than the postorbital constriction. The dental formula is 5/4, 1/1, 3/3, 4/4 × 2 = 52.

Gardner (1973) showed that *D. marsupialis* (*sensu lato*) and *D. albiventris* differ karyologically from *D. virginiana* in that they have $2n = 22$, FN = 20, and a small Y chromosome, whereas *D. virginiana* has $2n = 22$, FN = 32, and an acrocentric Y. A dated review of didelphid chromosomes can be found in Reig et al. (1977). *Didelphis* spp. are important reservoirs of a number of pathogens, such as *Trypanosoma cruzi* and *Leishmania* spp. (Arias and Naiff 1981; Ribeiro et al. 1985; Jansen et al. 1997, 1999; Pinho et al. 2000; Carreira et al. 2001).

SYNONYMS:

Didelphis Linnaeus, 1758:54; part; type species *Didelphis marsupialis* Linnaeus, 1758, by subsequent selection (O. Thomas 1888b:323); placed on the Official List of Generic Names (ICZN 1926).

Didelphys Schreber, 1777:532; unjustified emendation of *Didelphis* Linnaeus.

Opossum Schmid, 1818:115; part; type species *Didelphis marsupialis* Linnaeus, 1758 by subsequent designation (B. E. Brown 2004:22).

Sarigua Muirhead, 1819:429; part; contained nine species: *Sarigua marsupialis, S. virginiana, S. opossum, S. murina, S. cayopollin, S. brachyura, S. memmima, S. crassicaudata,* and *S. pusilla; Sarigua marsupialis* (= *Didelphis marsupialis* Linnaeus, 1758) here selected as the type species.

Didelphus I. Geoffroy St.-Hilaire, 1831:139; incorrect subsequent spelling of *Didelphis* Linnaeus.

Thylacotherium Lund, 1839:233; type species *Thylacotherium ferox* Lund, 1839, by monotypy; preoccupied by *Thylacotherium* Valenciennes, 1838.

Micoureus Lesson, 1842:186; part.

Didelphus Lapham, 1853:337; incorrect subsequent spelling of *Didelphis* Linnaeus.

Gamba Liais, 1872:329; part.

Gambatherium Liais, 1872:331; replacement name for *Thylacotherium* Lund.

Dasyurotherium Liais, 1872:331; replacement name for *Thylacotherium* Lund; immediately rejected by the author in favor of *Gambatherium* Liais.

Dimerodon Ameghino, 1889:277; type species *Dimerodon mutilatus* Ameghino, 1889, by monotypy.

Leucodidelphis Ihering, 1914:347; type species *Didelphis paraguayensis* Oken, 1816 (= *Didelphis paraguayensis* A. Allen, 1902; *D. paraguayensis* is not available from

Oken [1816]), by original designation; proposed as subgenus of *Didelphis* Linnaeus.

Leucodidelphys Krumbiegel, 1941a:34; unjustified emendation of *Leucodidelphis* Ihering.

Leucodelphis Cabrera, 1958:41; incorrect subsequent spelling *Leucodidelphis* Ihering.

REMARKS: The gender of *Didelphis*, and its feminine spelling variant *Didelphys*, have been a source of confusion for taxonomists when citing specific names. No gender was given when *Didelphis* was placed on the Official List of Generic Names in Zoology (ICZN 1926), although its gender was given as masculine in the later comprehensive Official List of Generic Names in Zoology (ICZN 1958b). We believe that when the latter list was prepared, someone looked up the gender of the Greek word *delphis*, which is masculine, but did not refer to Linnaeus (1758) to verify context and content. Linnaeus clearly intended *Didelphis* to be feminine; which gave rise to the subsequent and commonly used variant spelling, *Didelphys*. Gardner (2005a) has petitioned the International Commission to correct the gender to feminine. We treat the name as feminine herein.

KEY TO SOUTH AMERICAN SPECIES OF *DIDELPHIS*:

1. Ear partially to almost all white in dried skins, white to pinkish flesh colored in live animals; black and white facial markings; jugal embracing the zygomatic process of the squamosal from below; petrosal conical; cochlear fenestra, vestibular fenestra, and the primitive foramen lying in approximately the same arc. 2
1'. Ear either entirely black in adults or black at base and white or pale pinkish at the tip in immature and subadult animals; no white facial marking; jugal embracing the squamosal zygomatic process both above and below; cochlear fenestra, vestibular fenestra, and the primitive foramen not lying in the same arc 4
2. Ear mostly black with terminal quarter completely white; face with white and black markings; third molar not peg-like; known distribution restricted to the Guiana shield and adjacent areas of northern South America . . . .*Didelphis imperfecta*
2'. Ear white in dried skins of adults, white to flesh colored in live animals; third molar variable; distribution not on the Guiana Shield or adjacent lowlands 3
3. Ear in adults with a branching dark mark in the center; third molar not peg-like . . . . *Didelphis albiventris*
3'. Ear lacking dark central markings; third molar peg-like . . . . *Didelphis pernigra*
4. Widely distributed across northern South America south to Bolivia and western Brazil. . . .*Didelphis marsupialis*
4'. Distributed in eastern and southern Brazil, northern Argentina (Misiones), and eastern Paraguay . . . . *Didelphis aurita*

Didelphis albiventris Lund, 1840
White-eared Opossum; Cassaco

SYNONYMS:

[*Didelphis*] *albiventris* Lund, 1839b:233; *nomen nudum*.

Didelphis albiventris Lund, 1840a:18 [1841b:236]; type locality "Rio das Velhas," Lagoa Santa, Minas Gerais, Brazil.

Didelphis poecilotus J. A. Wagner, 1842:358; type locality "Angaba" (= Cuiabá), Mato Grosso, Brazil; spelling emended to *Didelphis poecilotis* by J. A. Wagner, 1847:126.

D[*idelphys*]. *poecilonota* Schinz, 1844:504; type locality "Angaha in Brazilien" (= Cuiabá), Mato Grosso, Brazil.

Didelphis azarae: Tschudi, 1845:143; part; not *Didelphys azarae* Temminck, 1824b.

Didelphys leucotis J. A. Wagner, 1847:127; based on Azara (1801a:244); therefore, the type locality is Paraguay.

Gamba aurita, var. *brasiliensis* Liais, 1872:239; part; no type locality mentioned; implied type locality Brazil.

Didelphis marsupialis var. *azarae*: O. Thomas, 1888b:328; name combination; not *Didelphys azarae* Temminck, 1824b.

Didelphis marsupialis azarae: Cope, 1889a:129; name combination; not *Didelphys azarae* Temminck, 1824b.

Didelphys Azarae, m[utación]. *antiqua* Ameghino, 1889:278; type locality "Barrancas del Río Primero, en los alrededores de la ciudad de Córdoba," Córdoba, Argentina.

Didelphys lechei Ihering, 1892:98; type locality "Sul do Rio Grande," Rio Grande do Sul, Brazil; proposed as a *variedade* of *Didelphys azarae* of authors (= *Didelphis albiventris* Lund).

Didelphis marsupialis, var. *albiventris*: Winge, 1893:7; name combination.

[*Didelphys (Didelphys) marsupialis*] *Azarae*: Trouessart, 1898:1235; name combination.

Did[*elphis*]. *paraguayensis* J. A. Allen, 1902:251; based on *Didelphis paraguayensis* Oken, 1816, which in turn was based on Azara's (1801a:244) *Le Micouré premier* from Paraguay; therefore, the type locality is "Asuncion, Paraguay" (J. A. Allen 1902:268); *Didelphis paraguayensis* is not available from Oken 1816 (ICZN 1926).

[*Didelphys (Didelphys)*] *paraguayensis*: Trouessart, 1905:853; name combination.

[*Didelphis (Didelphis)*] *poecilotis*: Matschie, 1916:268; name combination.

[*Didelphis (Didelphis)*] *albiventris*: Matschie, 1916:268; name combination.

[*Didelphis (Didelphis)*] *lechei*: Matschie, 1916:268; name combination.

D[*idelphis*]. *opossum*: Larrañaga, 1923:346; name combination; not *Didelphis opossum* Linnaeus, 1758.

Didelphis paraguayensis bonariensis Marelli, 1930:2; no type locality mentioned, range given as northeastern Argentina, including Buenos Aires; subsequently restricted to Buenos Aires and Santa Fé provinces by Marelli (1932:69).

Didelphis paraguayensis dennleri Marelli, 1930:2; no type locality mentioned, range given as western Argentina, from Catamarca to southern Buenos Aires province; subsequently restricted to Buenos Aires province by Marelli (1932:69).

Didelphys azarai Ringuelet, 1954:295; incorrect subsequent spelling of, but not *Didelphys azarae* Temminck.

Didelphis lechii C. O. C. Vieira, 1955:345; incorrect subsequent spelling of *Didelphys lechei* Ihering.

Didelphis lechii Cabrera, 1958:41; incorrect subsequent spelling of *Didelphys lechei* Ihering.

Didelphis azarae azarae: Cabrera, 1958:41; name combination; not *Didelphys azarae* Temminck.

Didelphis albiventris: Hershkovitz, 1969:54; first modern use of current name combination.

DISTRIBUTION: *Didelphis albiventris* occurs from northeastern and central Brazil (Caatinga and Cerrado habitats, enclaves, and transition zones) into central and southern Paraguay, east in the state of Rio Grande do Sul, Brazil, and south into Uruguay and Argentina as far south as Buenos Aires province in the east and the Monte Desert in west. *Didelphis albiventris* also is widely distributed in eastern Bolivia (Cerqueira 1985).

MARGINAL LOCALITIES (Map 6; from Lemos and Cerqueira 2002, except as noted): BRAZIL: Ceará, Parque Nacional de Ubajara (Guedes et al. 2000); Ceará, Itapagé; Pernambuco; Caruaru; Alagoas, Quebrangulo;

Sergipe, Brejo Grande; Bahia, Feira de Santana; Minas Gerais, Lagoa Santa (type locality of *Didelphis albiventris* Lund); Minas Gerais, Pouso Alegre; São Paulo, Itapetininga. PARAGUAY: Itapuá, San Benito (Roguin 1986). BRAZIL: Santa Catarina, Concordia; Rio Grande do Sul, Taquara; Rio Grande do Sul, São Lourenço. URUGUAY: Maldonado, Maldonado. ARGENTINA: Entre Rios, Gualeguaychú; Buenos Aires, Los Ingleses de Ajo (O. Thomas 1910a); Buenos Aires, 25 de Mayo (Yepes 1944); Buenos Aires, Mar del Sur (Mares and Braun 2000); Buenos Aires, Laguna Chasicó (Contreras 1973); Córdoba, La Paz (Mares and Braun 2000); Mendoza, San Rafael (Yepes 1944); La Rioja, Villa Castelli (Yepes 1936); Catamarca, Choya (Mares et al. 1997); Salta, Cafayate (Ojeda and Mares 1989); Jujuy, Yuto. BOLIVIA: Tarija, Carlazo (S. Anderson 1997); Chuquisaca, 12 km N and 11 km E of Tarabuco; Cochabamba, Vinto; Santa Cruz, San Ignacio de Velasco (S. Anderson 1997). BRAZIL: Mato Grosso, Utiariti; Mato Grosso, São Domingos (C. O. C. Vieira 1951); Maranhão, São João dos Patos (B. E. Brown 2004).

SUBSPECIES: We treat *D. albiventris* as monotypic.

NATURAL HISTORY: *Didelphis albiventris* inhabits open and deciduous forest type from northeastern Brazil to mid-Argentina, including areas of low and irregular rainfall such as the Caatinga and Monte Desert habitats (Cerqueira 1984, 1985). It is replaced in the wetter Atlantic and *Araucaria* forests by *D. aurita*, in the Amazonian forests by *D. marsupialis*, and on the slopes of the Andes by *D. pernigra*. Sympatry with *D. aurita* or *D. marsupialis* seems rare (Cerqueira 1985), except in areas disturbed by humans (Varejão and Valle 1982).

In northeastern Brazil, reproduction in *D. albiventris* is linked to average rainfall (Cerqueira 1984). Two breeding periods have been reported in the Buenos Aires Province, Argentina (Regidor and Gorostiague 1996). Streilein (1982b, 1982c, 1982d) described diet, behavior, and the microhabitats used in western Pernambuco, Brazil. Aléssio, Pones, and Silva (2005) recorded *D. albiventris* feeding on the gummy tree-sap exudate produced when tree bark is scarified by the Common Marmoset, *Callithrix jacchus*. White-eared opossums spend the day in the tops of palm trees (Moojen 1943), in parakeet nests (Martella, Navarro, and Bucher 1985), underneath bromeliads, and in holes in the ground (Barrett 1979). These opossums commonly are infected by *Trypanosoma cruzi* (Barrett 1979). Abdala, Flores, and Giannini (2001) described the qualitative and allometric changes in the morphology of the skull of *D. albiventris* from juveniles of about 3.5 months of age to adults older than 20 months.

REMARKS: The name *D. albiventris* is only now gaining general usage. In the older literature, the species usually was referred as either *D. azarae* or *D. paraguayensis*.

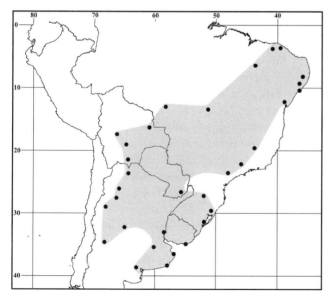

Map 6 Marginal localities for *Didelphis albiventris* ●

The first common name to be used for this species in the literature was *Le Micouré premier* (Azara, 1801a:244), on which Oken (1816) based his "*Didelphis paraguayensis*." However, Oken's "Lehrbuch" names have been ruled to be unavailable (ICZN 1956), because he did not apply the principles of binominal nomenclature (see ICZN 1999: Art. 11.4). Oken's name was not used again in the 19th century, preference having been given to *D. azarae* Temminck, 1824b, under the mistaken belief that Temminck's animals were the same species as the white-eared opossums described by Azara (1801a). J. A. Allen (1902) resurrected *D. paraguayensis*, believing it to be the senior synonym of *D. azarae* Temminck, 1824b, and its acceptance became widespread. Hershkovitz (1949b) and others rightly rejected Oken's names as not available (ICZN 1956) and *D. azarae* Temminck, 1824b, was once again adopted (e.g., Cabrera 1958). Hershkovitz (1969) later pointed out that, although Oken's names were not available, neither was *D. azarae* Temminck, 1824b, because Temminck had based his description on several black-eared opossums (i.e., members of the *D. marsupialis* group). J. A. Wagner (1847) earlier had reached the same conclusion about Temminck's material, and proposed the name *D. leucotis* for the Paraguayan white-eared form. Burmeister (1854:131) argued for retention of *D. azarae* for this species because Temminck had applied the name to Azara's (1801a) *Micouré premier*; O. Thomas (1888:328, footnote) agreed. Nevertheless, the material Temminck had before him when he described *D. azarae* is the basis for the species, not the description given by Azara (1801a). For that reason, the oldest available name is *D. albiventris* Lund, 1840a.

Cerqueira (1985) hypothesized that, together with the Andean populations, which he mistakenly called *D. albiventris azarae*, *D. albiventris* formed a rassenkreis with differentiated extremes. Further investigation (Lemos and Cerqueira 2002) has shown that these two forms are separate species. B. E. Brown (2004) did not distinguish between species in the *D. albiventris* complex. O. Thomas's (1888:329) locality "Chili" was an error and probably referred to one or more specimens from northwestern Argentina, a region that once was part of Chile.

Didelphis aurita Wied-Neuwied, 1826
Big-eared Opossum, Gambá

SYNONYMS:

Didelphis azarae Temminck, 1824b:30; type locality "Brésil"; (see Remarks).

D[*idelphys*]. *marsupialis*: Wied-Neuwied, 1826:387; part; name combination.

D[*idelphys*]. *aurita* Wied-Neuwied, 1826:395; type locality "Villa Viçosa an Flusse Peruhype," Bahia, Brazil; also see Ávila-Pires (1965) concerning the type locality.

Didelphys azarae: J. A. Wagner, 1843b:38; name combination.

Didelphys cancrivora: J. A. Wagner, 1843b:41; not *D. cancrivora* Gmelin, 1788.

Gamba aurita, var. *brasiliensis* Liais, 1872:329; part; no locality mentioned; implied type locality is Brazil.

Didelphys marsupialis aurita: Cope, 1889a:129; name combination.

Didelphis koseritzi Ihering, 1892:99; specimens credited to Sr. Th. Bischoff from "Colonia do Mundo Novo," Rio Grande do Sul, Brazil, which place may be taken as the type locality.

[*Didelphys (Marmosa)*] *koseriti* Trouessart, 1898:1240; name combination and incorrect subsequent spelling of *Didelphis koseritzi* Ihering.

[*Didelphys (Didelphys)* marsupialis] *aurita*: Trouessart, 1898:1234; name combination.

Didelphis marsupialis cancrivora: Bertoni, 1914:68; name combination.

[*Didelphis (Didelphis)*] *leucoprymnus* Matschie, 1916:268; *nomen nudum*.

Didelphis aurita longipilis A. Miranda-Ribeiro, 1935:35; type locality "Colônia Alpina," 16 km N of Teresópolis, Rio de Janeiro, Brazil.

Didelphis aurita melanoidis A. Miranda-Ribeiro, 1935,40; type locality "Therezópolis," restricted by Ávila-Pires (1968) to Colônia Alpina, 16 km N of Teresópolis, Rio de Janeiro, Brazil.

Didelphis aurita longigilis Ávila-Pires, 1968:169; incorrect subsequent spelling of *Didelphis aurita longipilis* Miranda-Ribeiro.

DISTRIBUTION: *Didelphis aurita* is found in eastern Brazil in the Tropical Atlantic and *Araucaria* Forest domains, and southward within these habitats to southeastern Paraguay and northeastern Argentina (Misiones). Populations of black-eared opossums still occur in the remnants of the Atlantic Forest of Alagoas and Pernambuco, but are rare. These populations are disjunct from populations occurring further south in Bahia.

MARGINAL LOCALITIES (Map 7; from Cerqueira 1980, except as noted): BRAZIL: Paraíba, Penha (Pohle 1927); Pernambuco, Recife; Alagoas, São Miguel dos Campos; Bahia, Itaparica; Bahia, Fazenda Beijo Grande; Bahia, Villa Viçosa (type locality of *Didelphys aurita* Wied-Neuwied); Espírito Santo, Santa Teresa; Rio de Janeiro, Floresta da Tijuca; São Paulo, Guaratuba (B. E. Brown 2004); São Paulo, Fazenda Intervales (J. L. Patton, Silva, and Malcolm 2000); Paraná, Guaratuba (J. L. Patton, Silva, and Malcolm 2000); Santa Catarina, Ilha de Santa Catarina (Graipel, Cherem, and Ximenez 2001); Rio Grande do Sul, Mundo Novo (type locality of *Didelphis koseritzi* Ihering); Rio Grande do Sul, São João do Monte Negro (Cope 1889a). ARGENTINA: Misiones, Jct. Hwy 21 and

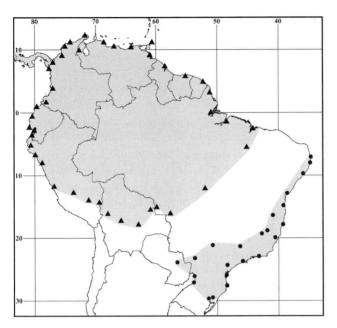

Map 7 Marginal localities for *Didelphis aurita* ● and *Didelphis marsupialis* ▲

Arroyo Oveja Negra (Mares and Braun 2000); Misiones, Reserva Natural Estricta San Antonio (Heinonen-Fortabat and Chebez 1997). PARAGUAY: San Pedro, Nueva Germania (Krumbiegel 1941a). BRAZIL: Mato Grosso do Sul, Rio Ivinheima (Krumbiegel 1941a); São Paulo, Serrinha (Krumbiegel 1941a); Minas Gerais, Monte Bello (Varejão and Valle 1982); Minas Gerais, Fazenda Esmeralda (G. A. B. Fonseca and Kierulff 1988); Minas Gerais, Governador Valadares (UFMG 950); Minas Gerais, Jequitínhonha (Varejão and Valle 1982).

SUBSPECIES: We consider *D. aurita* to be monotypic.

NATURAL HISTORY: A. Miranda-Ribeiro (1936) reviewed many anatomical and embryological features of *D. aurita*. Brain anatomy and physiology are described in Hunsaker (1977a). D. E. Davis (1945a, 1945b, 1947) studied the life history of this species in Teresópolis, Rio de Janeiro. Cerqueira et al. (1994), Bergallo (1994), Gentile, D'Andrea, and Cerqueira (1995), and Gentile et al. (2000) studied population dynamics. They found only age class III (adults) in the first half of the year and none with pouch young in April (the second quarter), indicating seasonal effects on reproduction. Gentile and Cerqueira (1995) analyzed movements. Santori, de Moraes, and Cerqueira (1996) published a study of the diet and S. R. Freitas et al. (1997) analyzed the relationship between diet and habitat. M. F. D. Motta, Carreira, and Franco (1983) and M. F. D. Motta (1988) reported on reproduction and extra-uterine development, respectively, in captives. Miles, Almeida de Souza, and Póvoa (1982) examined the ecological relationship with trypanosomiasis. Among other findings, Deane, Lenzi, and Jansen (1984, 1986), Jansen et al. (1999), Pinho et al. (2000), and Carreira et al. (2001), have shown that anal

scent glands in this species function as internal reservoirs of *Trypanosoma cruzi*. Yonenaga-Yassuda et al. (1982) reported differences in patterns of C-banded chromosomes and the nucleolus-organizer regions (NORs) that distinguish *D. aurita* (reported as *D. marsupialis*) from *D. albiventris*, *Philander frenatus* (reported as *P. opossum*), and *Lutreolina crassicaudata*. Varejão and Valle (1982) detailed the limits of distribution and the contact zone between *D. aurita* and *D. albiventris* in Minas Gerais, Brazil. Cerqueira (1985) has examined the potential limits to distribution and the ecological geography of South American *Didelphis*.

REMARKS: Cerqueira (1985) established that the black-eared opossums, then called *D. marsupialis*, have a disjunct distribution. Cerqueira (1980, 1985) considered the disjunct assemblages to represent a superspecies complex undergoing incipient allopatric speciation. However, additional work by Cerqueira and Lemos (2000) showed a clear morphometric differentiation between the Atlantic Forest and northern South America populations. Another study on rates of morphometric evolution on this clade supported a Pleistocenic cladogenesis through rapid differentiation (Lemos, Marroig, and Cerqueira 2001).

Under Remarks, in the account for *D. albiventris*, we discussed the misapplication of the name *D. azarae* Temminck to that species. Temminck's original description specifically mentioned black ears, often yellowish at the base, and refers to the gambá, or southeastern Brazilian black-eared opossum. Nevertheless, the name *D. azarae* Temminck has not been used for the gambá since 1824; on the contrary, it has been used almost universally for the white-eared opossum today known as *D. albiventris*. Although *D. azarae* Temminck, 1824b antedates *D. aurita* Wied-Neuwied, 1826, by 2 years, adoption of *D. azarae* as the valid name for the gambá would bring unnecessary instability and confusion; therefore, we continue to use the name *D. aurita* Wied-Neuwied for this species.

Yepes (1936) reported "*Didelphis marsupialis aurita*" from provincia La Rioja, Argentina based on a skin in winter pelage. B. E. Brown (2004:30) cited this record under *D. aurita*; however, the specimen was correctly cited under *D. albiventris* by Mares and Braun (2000). B. E. Brown (2004:31) listed Pohle's (1927) records of specimens, referable to *D. aurita*, from the Brazilian states of Pernambuco and Paraíba under *D. marsupialis* (also see Flores 2004).

Didelphis imperfecta Mondolfi and Pérez-Hernández, 1984

Guianan White-eared Opossum, Comadreja

SYNONYMS:

Didelphis marsupialis marsupialis: Tate, 1939:160; part; not *Didelphis marsupialis* Linnaeus, 1758.

Didelphis albiventris: Hershkovitz, 1969:54; part; not *Didelphis albiventris* Lund, 1840a.

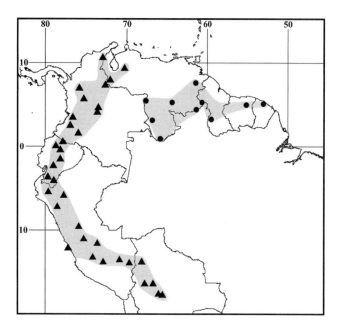

Map 8 Marginal localities for *Didelphis imperfecta* ● and *Didelphis pernigra* ▲

Didelphis albiventris imperfectus Mondolfi and Pérez-Hernández, 1984:407; type locality "km 125, Carretera El Dorado-Santa Elena, Estado Bolívar, Venezuela."

Didelphis imperfecta: Voss and Emmons, 1996:43; first use of current name combination and correct gender concordance.

DISTRIBUTION: *Didelphis imperfecta* is known from isolated populations south of the Orinoco in Venezuela and adjacent Brazil, as well as in Surinam and French Guiana. The species also undoubtedly occurs in Guyana (also see Remarks).

MARGINAL LOCALITIES (Map 8; from Pérez-Hernández 1990, except as noted): VENEZUELA: Bolívar, La Tigra. BRAZIL: Roraima, Monte Roraima (Hershkovitz 1969). SURINAM: Brokopondo, Brownsberg Nature Park (Genoways, Williams, and Groen 1981). FRENCH GUIANA: Piste de St.-Elie (Julien-Laferrière 1991, not mapped); Petit-Saut (Catzeflis et al. 1997). GUYANA: Upper Takutu-Upper Essequibo, Kanuku Mountain Region (Emmons 1993, no specific locality). VENEZUELA: Bolívar, El Paují; Bolívar, Meseta de Jaua; Amazonas, Cerro de la Neblina Camp VI (Gardner 1988); Amazonas, Campamento 2; Amazonas, 32 km S of Puerto Ayacucho (Handley 1976).

SUBSPECIES: We treat *D. imperfecta* as monotypic.

NATURAL HISTORY: Little is known about *D. imperfecta*. Specimens have been taken in elfin forest habitats on tepuis at elevations as high as 2,550 m on Cerro Marahuaca, Amazonas, Venezuela (R. Guerrero, Hoogesteijn, and Soriano 1989) to below 250 m in lowland forests of eastern Venezuela and French Guiana. Julien-Laferrière

(1991) found *D. imperfecta* (reported as *D. albiventris*) in French Guiana to be a terrestrial inhabitant of primary forests. Catzeflis et al. (1997) reported *D. imperfecta* (as *D. albiventris*) in sympatry with *D. marsupialis* in lowland primary forest along the Sinnamary River in French Guiana. Lavergne et al. (1997) used molecular techniques to confirm the ability to distinguish between these two species in the field.

REMARKS: Although the present distribution of *D. imperfecta* appears to be restricted to the Guiana Region, the species can be expected farther south in Brazil and possibly farther west in Ecuador and northeastern Peru. B. E. Brown (2004), who treated all white-eared South American *Didelphis* as *D. albiventris*, reported specimens identified as *D. albiventris* in the FMNH from Marian, Sucumbios, Ecuador, and in the AMNH from the Río Samiria region of departamento Loreto, Peru. These records might be dismissed as misidentifications were it not for another recent report of *D. albiventris* from the general area of Iquitos, Loreto, Peru by M. M. Díaz and Willig (2004). On geographic grounds, the FMNH and AMNH specimens are more likely to represent *D. pernigra* than *D. albiventris*, whose known distribution reaches its northwestern limit approximately 1,600 km SE of the rios Nanay and Samiria in Peru. Nevertheless, *D. pernigra* is unknown from the lowlands anywhere east of the Andes. *Didelphis imperfecta*, however, is known to occur approximately 1,000 km to the east-northeast, is known from both highland and lowland humid forest habitat, and has been found sympatric with *D. marsupialis* (e.g., Lavergne et al. 1997; Catzeflis et al. 1997). The identifications of these Ecuadorian and Peruvian specimens need to be verified.

Didelphis marsupialis Linnaeus, 1758
Common Opossum, Carachupa
SYNONYMS:

[*Didelphis*] *marsupialis* Linnaeus, 1758:54; type locality "America"; restricted to Surinam by O. Thomas (1991b: 143).

[*Didelphis*] *karkinophaga* Zimmermann, 1780:226; type locality "Gujana"; the name is based on "*le crabier*" of Buffon (1776); therefore, the type locality can be restricted to Cayenne, French Guiana.

Didelphis carcinophaga Boddaert, 1784:77; unjustified emendation of *Didelphis karkinophaga* Zimmerman.

[*Didelphis*] *cancrivora* Gmelin, 1788:108; based on the "*Crabier*" of Buffon (1776) from Cayenne; therefore, the type locality is Cayenne, French Guiana.

Didelphis austro-americana Oken, 1816:column 1148; unavailable name (ICZN 1956:Opinion 417).

Didelphis mes-americana Oken, 1816:column 1152; unavailable name (ICZN 1956:Opinion 417).

O[*possum*]. *marsupialis*: Schmid, 1818:116; name combination.

Gamba aurita, var. *brasiliensis* Liais, 1872:329; part; no locality mentioned; implied type locality is Brazil.

Didelphis aurita: J. A. Allen, 1897a:43; not *Didelphis aurita* Wied-Neuwied, 1826.

Didelphys marsupialis, var. *typica* O. Thomas, 1888b:323; technically a renaming of *Didelphis marsupialis marsupialis* Linnaeus, 1758.

[*Didelphys (Didelphys) marsupialis*] *cancrivora*: Trouessart, 1898:1233; name combination.

Didelphis karkinophaga caucae J. A. Allen, 1900c:192; type locality "Cali, Upper Cauca Valley," Colombia.

Didelphis karkinophaga colombica J. A. Allen, 1900c:193; type locality "Bonda," Magdalena, Colombia.

Didelphis marsupialis tabascensis J. A. Allen, 1901c:173; type locality "Teapa, Tabasco, Mexico."

Didelphis richmondi J. A. Allen, 1901c: 175; type locality "Greytown, Nicaragua."

Didelphis marsupialis battyi O. Thomas, 1902d:137; type locality "Island of Coiba, Panama."

Did[elphis]. mes-americana J. A. Allen, 1902:251; type locality "northern Mexico"; based on the large opossums of northern Mexico, and Oken (1816; *Didelphis mesamericana* is not available from Oken).

Did[elphis]. austro-americana J. A. Allen, 1902:251; type locality "South America"; based on "the large South American opossums of the *Didelphis* group" and Oken (1816; *Didelphis austroamericana* is not available from Oken).

Didelphis marsupialis insularis J. A. Allen, 1902:259; type locality "Caparo, Trinidad," Trinidad and Tobago.

Didelphis marsupialis colombica: J. A. Allen, 1902:260; name combination.

Didelphis marsupialis caucae: J. A. Allen, 1902:261; name combination.

Didelphis marsupialis etensis J. A. Allen, 1902:262; type locality "Eten, Piura, Peru."

[*Didelphys (Didelphys) marsupialis*] *insularis*: Trouessart, 1905:852; name combination.

[*Didelphys (Didelphys) marsupialis*] *colombica*: Trouessart, 1905:853; name combination.

[*Didelphys (Didelphys) marsupialis*] *caucae*: Trouessart, 1905:853; name combination.

[*Didelphys (Didelphys) marsupialis*] *etensis*: Trouessart, 1905:853; name combination.

Didelphis marsupialis particeps Goldman, 1917:107; type locality "San Miguel Island, Panama."

D[idelphis]. m[arsupialis]. richmondi: Goldman, 1920:46; name combination.

DISTRIBUTION: *Didelphis marsupialis* occurs in Trinidad and Tobago, the Guianas, and the greater Amazon basin, including the wet forest habitats of the eastern Andean slopes of Venezuela, Colombia, Ecuador, Peru, and Bolivia. The drier Caatinga, Cerrados, and Chaco habitats of Brazil and Bolivia limit its range to the east and south. The species also occurs north and west of the Andes from northwestern Venezuela, and northern and western Colombia, southward through western Ecuador to northern Peru. Elsewhere, it is found in Central America and Mexico (Gardner 1973; Hall 1981), and on the Lesser Antilles from Dominica southward, where it may be a relatively recent introduction on some islands (G. M. Allen 1911).

MARGINAL LOCALITIES (Map 7): COLOMBIA: La Guajira, Bahía Honda (B. E. Brown 2004). VENEZUELA: Zulia, Kasmera (Handley 1976); Falcón, La Pastora (B. E. Brown 2004); Distrito Federal, La Guaira (W. Robinson and Lyon 1901); Sucre, 2 km E of Cumaná (USNM 388431). TRINIDAD AND TOBAGO: Tobago, Runnemede (USNM 461885). VENEZUELA: Delta Amacuro, Güiniquina (Pérez-Hernández 1990). GUIANA: Pomeroon-Supenaam, Better Hope (O. Thomas 1888). SURINAM: Paramaribo, Paramaribo (Husson 1978). FRENCH GUIANA: Cayenne (type locality of *Didelphis karkinophaga* Zimmermann). BRAZIL: Amapá, Vila Velha do Cassiporé (C. T. Carvalho 1962a); Amapá, Macapá (C. T. Carvalho 1962a); Pará, Belém (B. E. Brown 2004); Maranhão, Anil (Cerqueira 1980); Maranhão, Barra do Corda (C. O. C. Vieira 1957); Mato Grosso, Serra do Roncador (Cerqueira 1980); Mato Grosso, Cáceres (B. E. Brown 2004); Mato Grosso, Mato Grosso (Pelzeln 1883). BOLIVIA (S. Anderson 1997): Santa Cruz, Los Palmares; Santa Cruz, Santa Cruz de la Sierra; Cochabamba, Chaparé; La Paz, El Vertigo. PERU: Puno, Sandia (Sanborn 1953); Cusco, Hacienda Erika (V. Pacheco et al. 1993); Ayacucho, Hacienda Luisiana (B. E. Brown 2004); Lima, Yangas (B. E. Brown 2004); La Libertad, Menochucho (Osgood, 1914b); Lambayeque, Lambayeque (Cerqueira 1985); Piura, Piura (Cerqueira 1985); Tumbes, Tumbes (B. E. Brown 2004). ECUADOR: Guayas, Isla Puná (J. A. Allen 1902); Guayas, Ancón (Cerqueira 1985); Manabí, Bahía de Caráquez (B. E. Brown 2004); Esmeraldas, Esmeraldas (R. H. Baker 1974). COLOMBIA: Nariño, Barbacoas (J. A. Allen 1916c); Valle del Cauca, Buenaventura (B. E. Brown 2004); Chocó, Río Curiche (Méndez 1977); Chocó, Unguía (B. E. Brown 2004); Córdoba, El Contento (Adler, Arboledo, and Travi 1997); Atlántico, La Peña (B. E. Brown 2004); Magdalena, Santa Marta (J. A. Allen 1902).

SUBSPECIES: We do not recognize subspecies in the South American population of *D. marsupialis*. Gardner (1973) and Hall (1981) recognized subspecies in Mexico and Central America.

NATURAL HISTORY: The range of this species overlaps all of Ab'Saber's (1971) morphoclimatic domains of northern South America from Amazonia through the Lesser Antilles and Middle America, but excluding higher-elevation forests, arid and semi-arid habitats (deserts, Caatinga, and Cerrado zones), and the tepuis and associated massifs of the Guiana Highlands above 1,000 m. Otherwise, *D. marsupialis* is a ubiquitous member of the fauna in tropical humid forest and gallery forest habitats. The distribution along the Pacific slope of the Andes in Peru is confined to the more humid river valleys and the cloud-mediated, humid subtropical forest, at least as far south as the Rimac Valley east of Lima. Cerqueira (1985) discussed the range of *D. marsupialis* in relation to its ecology and natural history. Arias and Naiff (1981) studied the species as a major reservoir of cutaneous leishmaniasis. Telford et al. (1981) and others have shown that *D. marsupialis* is a reservoir for *Trypanosoma cruzi*. R. Guerrero (1985a) gave references for 46 species of parasites taken from *D. marsupialis* in Venezuela. Fleming (1972, 1973), Tyndale-Biscoe and Mackenzie (1976), O'Connell (1979), and Atramentowicz (1986), reported on aspects of reproduction and population structure. Studies by Molins de la Serna and Lorenzo (1982) verified that *D. marsupialis* is an opportunistic omnivore, that feeds on vertebrates, invertebrates, leaves, and fruits in varying proportions depending on availability. Charles-Dominique (1983) also gave details of vocalizations. A general summary of New World marsupial ecology can be found in Hunsaker (1977a).

REMARKS: Cerqueira and Lemos (2000) demonstrated morphometric differentiation between *D. marsupialis* and *D. aurita*, sister groups in the *Didelphis marsupialis* species group. Cerqueira (1980) showed that variation within *D. marsupialis* is clinal and does not justify the designation of subspecies.

Didelphis austroamericana J. A. Allen, 1902:251, and *Didelphis austroamericana* O. Thomas, 1923b:604, are synonyms of *D. marsupialis* Linnaeus, 1758, and were originally based on *Didelphys austro-americana* Oken, 1816. These authors thought that *D. austroamericana* was an available name dating from Oken (1816); nevertheless, with exceptions based on decisions by the International Commission on Zoological Nomenclature, Oken's names in his 1816 "Lehrbuch," are not available (ICZN 1956).

We could not identify the source of the locality records mapped by B. E. Brown (2004:37) in the area that appears to correspond to the Brazilian states of Paraíba and Pernambuco. Nevertheless, on geographic grounds we would allocate them to *D. aurita*, along with B. E. Brown's (2004:31) listing of records from Ma[n]gabeira, Alagoas that she attributed to Veiria (= C. O. C. Vieira).

Didelphis pernigra J. A. Allen, 1900
Andean White-eared Opossum

SYNONYMS:

Didelphis azarae: Tschudi, 1845:143; not *D. azarae* Temminck, 1824b.

Didelphis pernigra J. A. Allen, 1900c:191; type locality "Juliaca," Puno, Peru; corrected to "Inca Mines" (= Santo Domingo) by J. A. Allen (1901a:41); also see O. Thomas (1901a:186).

Didelphis paraguayensis pernigra: J. A. Allen, 1902:271; name combination.

Didelphis paraguayensis andina J. A. Allen, 1902:272; type locality "Loja," Loja, Ecuador.

Didelphis paraguayensis meridensis J. A. Allen, 1902:274; type locality "Merida," Mérida, Venezuela.

[*Didelphys (Didelphys) paraguayensis*] *pernigra*: Trouessart, 1905:853; name combination.

[*Didelphys (Didelphys) paraguayensis*] *andina*: Trouessart, 1905:853; name combination.

[*Didelphys (Didelphys) paraguayensis*] *meridensis*: Trouessart, 1905:853; name combination.

[*Didelphis (Didelphis)*] *andina*: Matschie, 1916:268; name combination.

[*Didelphis (Didelphis)*] *pernigra*: Matschie, 1916:268; name combination.

Didelphis azarae pernigra: Cabrera, 1958:52; name combination.

Didelphis albiventris azarae: Cerqueira, 1985:135; name combination; not *D. azarae* Temminck, 1824b.

Didelphis albirentris S. Solari, Vivar, Rodríguez, Velazco, and Montesinos, 1999:139; incorrect subsequent spelling of *Didelphis albiventris*, but not *Didelphis albiventris* Lund, 1840.

DISTRIBUTION: *Didelphis pernigra* is found in the forested slopes of the Andes, but apparently not in the Puna zone, from northwestern Venezuela and Colombia through Ecuador and Peru into Bolivia, and possibly as far as the Andes of northern Argentina. Its distribution reaches the Pacific coast at Callao, departamento de Lima, Peru (Cerqueira 1985).

MARGINAL LOCALITIES (Map 8; from Lemos and Cerqueira 2002, except as noted): COLOMBIA: La Guajira, Villanueva. VENEZUELA: Táchira, Páramo Zumbador (Pérez-Hernández 1990); Trujillo, 15 km E of Trujillo (B. E. Brown 2004); Táchira, Buena Vista (Handley 1976, not mapped). COLOMBIA: Norte de Santander, Pamplona (B. E. Brown 2004); Cundinamarca, Guasca; Meta, Villavicencio; Huila, Aguas Claras (B. E. Brown 2004). ECUADOR: Napo, Papallacta (B. E. Brown 2004); Pastaza, Mera; Zamora-Chinchipe, Sabanilla. PERU: Amazonas, Corosha; Huánuco, Hacienda Buena Vista (B. E. Brown 2004); Junín, Perené (J. A. Allen 1902); Cusco,

RAP Camp Two (Emmons, Luna, and Romo 2001); Cusco, Limacpunco; Puno, Inca Mines. BOLIVIA (S. Anderson 1997): La Paz, Ixiamas; Cochabamba, Yungas; Cochabamba, 7.5 km SE of Rodeo; Cochabamba, 1.3 km W of Jamachuma; La Paz, Pitiguaya. PERU: Apurímac, Quebrada Matará (B. E. Brown 2004); Ayacucho, Ayacucho (B. E. Brown 2004); Lima, Callao; Cajamarca, Cajamarca; Piura, Canchaque. ECUADOR: El Oro, El Chiral; Chimborazo, San Antonio; Pichincha, Gualea (Trouessart 1910); Carchi, San Gabriel (B. E. Brown 2004). COLOMBIA: Cauca, Cerro Munchique; Valle del Cauca, 4 km NW of San Antonio (B. E. Brown 2004); Antioquia, Sonsón; Antioquia, Paramillo; Antioquia, Valdivia.

SUBSPECIES: We treat *D. pernigra* as monotypic.

NATURAL HISTORY: *Didelphis pernigra* occurs at moderate (generally above 1,500 m) to upper elevations in forested habitats in the Andes from western Venezuela and northern Colombia to southcentral Bolivia. The species also has been found at lower elevations in riparian habitats in the otherwise arid Pacific lowlands of Peru.

REMARKS: The record B. E. Brown (2004:29) cited as an AMNH *D. albiventris* caught in 1925 by Tate at Cumanacoa, Sucre, Venezuela appears to be an error. According to R. Voss (pers. comm. to A. L. Gardner), Tate's catalog does not contain an entry for a *Didelphis* from Cumanacoa, and Voss was unable to locate any white-eared opossum from that locality in the AMNH collection. Lemos and Cerqueira (2002) showed that there is a sharp discontinuity between the parapatric *D. albiventris* and *D. pernigra* in Bolivia that suggests limited or no gene flow between them. These opossums also are known in South America by the vernacular name Zarigüeya Orejiblanca.

Genus *Lutreolina* O. Thomas, 1910

Barbara R. Stein and James L. Patton

Lutreolina is a monotypic genus of medium-sized (head and body 250–400 mm, tail 210–310 mm, mass 200–540 g) terrestrial, weasel-like opossums having short, stout limbs and feet, and short, rounded ears that barely extend above the fur. Females have a weakly to moderately developed pouch and the hallux is not fully opposable. The base of the tail is as thickly haired as the body; otherwise, except for the naked ventral surface along the terminal half, the tail is covered with short hair and is either brown to black to the tip, or brown to black over the proximal 3/4 to 4/5 of its length and pale yellowish to white terminally. Like *Chironectes*, the tail is not as prehensile as in other New World opossums. The fur is soft and dense, but not water-repellent. The color pattern of the body is nearly uniform buffy yellow to dark brown above. The venter is paler, sometimes

ochraceous. Lemke et al. (1982) described a Colombian specimen as having a reddish muzzle. The pouch in females is not as well developed as it is in *Didelphis*, *Philander*, and *Chironectes*. Living or freshly killed *Lutreolina* fluoresce a bright red-orange under UV light (Pine et al. 1985).

The skull differs in general configuration from that of other didelphids in having a shorter rostrum, comparatively narrower zygomatic arches and braincase, and evenly tapered nasals (see illustrations of the skull in Marshall's [1978a] review). Reig (1958) compared dental and some cranial characters of *Lutreolina* with those of other living and extinct didelphids. Two species, based on fossil material, but possibly synonyms of *L. crassicaudata*, have been described from the Pliocene of Argentina (see Marshall 1982a).

SYNONYMS:

Didelphis: Desmarest, 1804b:19; part; not *Didelphis* Linnaeus, 1758.

Didelphys Olfers, 1818:204; part; incorrect subsequent spelling of *Didelphis* Linnaeus.

Sarigua Muirhead, 1819:429; part.

Peramys Lesson, 1842:261; part.

Micoureus: P. Gervais, 1855:287; part; not *Micoureus* Lesson, 1842.

Philander: Gerrard, 1862:139; part; not *Philander* Brisson, 1762.

Metachirus: Hensel, 1872:121; part; as a subgenus of *Didelphis* Linnaeus; not *Metachirus* Burmeister, 1854).

Lutreolina O. Thomas, 1910a:247; type species *Didelphis crassicaudata* Desmarest, 1804b, by monotypy.

Lutreolina crassicaudata (Desmarest, 1804)
Lutrine Opossum

SYNONYMS: See under Subspecies.

DISTRIBUTION: *Lutreolina crassicaudata* is endemic to South America, where it occurs as two discrete geographic units, each recognized as a subspecies. The smaller northern distribution is in western Guyana, Venezuela, and eastern Colombia; the larger southern population unit occurs in eastern Bolivia and southeastern Brazil, Paraguay, Uruguay, and northern Argentina as far south as provincia Buenos Aires. The two subspecies have greatly disjunct distributions and no specimens are known from the intervening Amazon basin.

MARGINAL LOCALITIES (Map 9): *Northern distribution.* VENEZUELA: Sucre, Caño Brea (Linares and Rivas 2004). GUYANA: Pomeroon-Supenaam, Better Hope (type locality of *Didelphys turneri* Günther). VENEZUELA: Bolívar, San Ignacio de Yuruaní (Boher-Bentti 1988); Bolívar, Hato San José (Handley 1976). COLOMBIA: Vichada, Caño Avispas (Lemke et al. 1982); Meta, Villavicencio (Lemke et al. 1982); Arauca, Puerto Gaitán (Ayala et al. 1973). VENEZUELA: Anzoátegui, Fundo la Unión

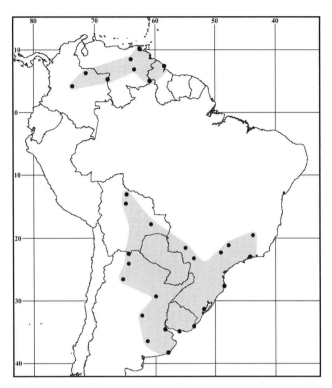

Map 9 Marginal localities for *Lutreolina crassicaudata* ●

(Pérez-Hernández 1985). *Southern distribution.* BOLIVIA: Beni, San Joaquín (Lemke et al. 1982); Santa Cruz, San José (Krumbiegel 1941b). BRAZIL: Mato Grosso do Sul, Maracajú (Ximénez 1967); Mato Grosso do Sul, Rio Ivinheima (Krumbiegel 1941b); São Paulo, Baurú (C. O. C. Vieira 1955); São Paulo, Ribeirão Prêto (Ximénez 1967); Minas Gerais, Lagoa Santa (Winge 1893); Rio de Janeiro, Angra dos Reis (C. O. C. Vieira 1955); Santa Catarina, Parque Municipal da Lagoa do Peri (Graipel et al. 2004); Rio Grande do Sul, São Lourenço do Sul (type locality of *Lutreolina crassicaudata lutrilla* O. Thomas). URUGUAY: Rocha, Camino del Indio (Ximénez 1967); Montevideo, Montevideo (Thomas 1923a). ARGENTINA: Buenos Aires, Belgrano (Waterhouse 1846); Buenos Aires, Mar del Sur (Ximénez 1967); Buenos Aires, Urdampilleta (B. E. Brown 2004); Córdoba, Noetinger (O. Thomas 1923a); Santa Fe, Malabrigo (Yepes 1944); Tucumán, Raco (Olrog 1976); Jujuy, Cerro Santa Bárbara (Olrog 1976); Salta, 24 km NW of Aguas Blancas (Ojeda and Mares 1989). BOLIVIA: Beni, 23 km W of San Xavier (S. Anderson 1997).

SUBSPECIES: There are two subspecies of *L. crassicaudata*.

L. c. crassicaudata (Desmarest, 1804)
SYNONYMS:

Didelphis crassicaudata Desmarest, 1804b:19; no locality given; based on "*Le micouré à grosse queue de l'Azara*"

(Azara 1801a:284), which came from Paraguay; Cabrera (1958:39) restricted the type locality to Asunción.

[*Didelphys*] *crassicaudis* Illiger, 1815:107; *nomen nudum*.

D[*idelphys*]. *crassicaudis* Olfers, 1818:206; type locality "Paraguay"; objective synonym of *Didelphis crassicaudata* Desmarest.

S[*arigua*]. *crassicaudata*: Muirhead, 1819:429; name combination.

Didelphis macroura Desmoulins, 1824:492; based on Azara's (1801a:284) *micouré à queue grosse*; objective synonym of *Didelphis crassicaudata* Desmarest.

Peramys crassicaudata: Lesson, 1842:187; name combination.

D[*idelphys*]. *crassicaudata*: Schinz, 1844:257; name combination.

Didelphys mustelina Waterhouse, 1846:497; in synonymy, attributed to Geoffroy; *nomen nudum*.

Micoureus crassicaudatus: P. Gervais, 1855:287; name combination.

[*Didelphys (]Metachirus[)*] *crassicaudatus*: Hensel, 1872: 121; name combination.

Lutreolina crassicaudata: O. Thomas, 1910a:247; generic description and name combination.

L[*utreolina*]. c[*rassicaudata*]. *paranalis* O. Thomas, 1923a: 584; type locality "Las Rosas, Santa Fe," Argentina.

L[*utreolina*]. c[*rassicaudata*]. *bonaria* O. Thomas, 1923a: 585; type locality "Los Yngleses, Ajo," Buenos Aires, Argentina.

L[*utreolina*]. c[*rassicaudata*]. *lutrilla* O. Thomas, 1923a: 585; type locality "San Lorenzo, R. Grande do Sul," Brazil.

Didelphis ferruginea Larrañaga, 1923:346; type locality not given, but Uruguay implied; based in part on Azara's (1802a:229) "*Coligrueso*."

Lutreolina c[*rassicaudata*]. *travassosi* A. Miranda-Ribeiro, 1936:402; type locality "Guariba, E. de S. Paulo," Brazil.

Lutreolina crassicaudus Hildebrand, 1961:244; incorrect subsequent spelling of *Didelphis crassicaudata* Desmarest.

The nominate subspecies is found in Paraguay, Uruguay, eastern Bolivia, and southeastern Brazil as far south as provincia de Buenos Aires, Argentina.

L. c. turneri (Günther, 1879)
SYNONYMS:

Didelphys turneri Günther, 1879:103; type locality "Demerara"; identified as Better Hope, Demerara [= Better Hope, Pomeroon-Supenaam, Guyana] by O. Thomas (1888b:336).

[*Didelphis (Peramys)*] *turneri*: Matschie, 1916:269; name combination.

L[*utreolina*]. c[*rassicaudata*]. *turneri*: O. Thomas, 1923a: 583; first use of current name combination.

This subspecies represents the northern South American distribution in eastern Colombia, Venezuela, and western Guyana.

NATURAL HISTORY: *Lutreolina crassicaudata* is mostly restricted to mesic savanna grassland and gallery woodland habitats. In central Argentina, Uruguay, and southernmost Brazil, *L. crassicaudata* frequents the pampas or temperate grasslands; in northern Argentina and Paraguay it occupies the more mesic savanna woodlands of the Chaco; and in Mato Grosso, Brazil, it has been taken in tropical grassland and gallery woodland environments (Marshall 1978a). Cabrera and Yepes (1940) considered the lutrine opossum to be the opossum best adapted to life on the pampas and, apparently, the species does not occur in humid tropical forests.

Lutreolina crassicaudata closely resembles certain weasels (Carnivora: *Mustela* spp.) in habit and appearance. Both kinds are terrestrial and aquatic, and frequent mesic habitats subject to periodic flooding (Lydekker 1894b). Lutrine opossums are nocturnal, agile on the ground, climb well, and have been considered as either excellent swimmers (Lydekker,1894b; Hunsaker 1977b) or clumsy ones (J. A. Davis 1966). Although a captive *L. crassicaudata* was strictly carnivorous (J. A. Davis 1966), Collins (1973) considered the natural diet to be omnivorous. Nowak and Paradiso (1983) reported two breeding periods per year, one in the spring followed by the second after the first litter is weaned and independent. Gestation is about 2 weeks, and young are raised in grass nests. A Colombian female taken 10 August carried seven attached young in the pouch (Lemke et al. 1982). The karyotype is $2n = 22$, $FN = 20$, with a metacentric X and an acrocentric Y chromosome (Reig et al. 1977; Seluja et al. 1984).

REMARKS: Although Krumbiegel (1941b) recognized all named subspecies, C. O. C. Vieira (1949) and Cabrera (1958) arranged *L. c. travassosi* Miranda-Ribeiro as a junior synonym of *L. c. crassicaudata* Desmarest. Cabrera (1958) and Ximénez (1967) treated *L. c. paranalis* O. Thomas, as a valid subspecies with *L. c. bonaria* O. Thomas, as a synonym. Cabrera (1958) also considered *L. c. lutrilla* O. Thomas, to be a valid subspecies; but Ximénez (1967) treated the name as a junior synonym of *L. c. paranalis*. Marshall (1978a) tentatively recognized only two subspecies: *L. c. crassicaudata*, a larger-bodied, grayish olivaceous to buffy brown animal found mainly in the ríos Paraguay, Paraná, and de la Plata drainage basin of south-central South America; and *L. c. turneri*, a smaller-bodied, tawny to dark brown subspecies restricted to eastern Colombia, Venezuela, and western Guyana. Ximénez (1967) suggested that *turneri* may prove to be a species separate from *L. crassicaudata*.

Cabrera's (1958:39) inclusion of *Didelphis macroura* Illiger, 1815:107 (a *nomen nudum*) in the synonymy of *L.*

crassicaudata was based on a misunderstanding of Illiger's name. Olfers (1818:205), whose names usually are the same as several of Illiger's *nomina nuda*, based Illiger's "*macroura*" on Azara's (1801a:290) "*micouré à queue longue*" and included it in the synonymy of *Didelphis murina* Linnaeus, 1758, which has not been recorded from Paraguay. However, Olfers' (1818:205) spelling *D. macrura* is not to be considered the same name as *Didelphis macroura*, which Desmoulins (1824:492) used for Azara' (1801a:284) *micouré à queue grosse*, which Desmoulins called "*Didelphe a grosse queue.*"

Olfers = (1818) *D. crassicaudis* could fit any of four categories: 1) a new name with *D. crassicaudata* Desmarest a senior synonym; 2) a replacement name for *D. crassicaudata* Desmarest; 3) an incorrect subsequent spelling of *crassicaudata* Desmarest; or 4) as claimed by Hershkovitz (1959:338), or an unjustified emendation of *D. crassicaudata* Desmarest. We reject the last two possibilities because the spelling used is consistent with that of Illiger (1815) and there is no evidence, other than the name itself, that Olfers (or Illiger 1815) intentionally modified the name *D. crassicaudata*. Because a new name may or may not have the same type, we consider *D. crassicaudis* Olfers a replacement name, thereby making it a junior objective synonym of *D. crassicaudatus*, in which case it has the same type and type locality.

Genus *Philander* Brisson, 1762

James L. Patton and Maria Nazareth F. da Silva

Members of the genus *Philander* are large (head and body length 250–350 mm, tail 250–330 mm, mass 240–600 g), gray to black opossums with a conspicuous white to cream-colored spot above each eye. These opossums are somewhat smaller and more gracile in appearance than *Didelphis*. The ears are either bicolored or black and appear naked. The tail usually slightly exceeds the combined length of head and body, is typically black over the proximal 1/2 to 2/3 of its length, and appears to be naked except for the longer fur on the proximal 5 to 6 cm. Color varies from gray above and creamy-white below, to black above and below; considerable geographic variation in color and pattern exists within the genus. Mature females have a well-developed pouch. Members of the genus superficially are most similar to the Brown Four-eyed Opossum, *Metachirus nudicaudatus*, which lacks a pouch.

The skull is similar to those of *Didelphis* spp. in general form and proportions, but the nasals are less expanded laterally at the maxillofrontal junction. The rostrum is long and slender, the zygomatic arches are broadly flared, the postorbital constriction is smoothly rounded and narrow, and the temporal ridges converge anterior to the frontal-parietal suture to form a well-defined sagittal crest.

The six living species of *Philander* we recognize are largely terrestrial, solitary inhabitants of primary and secondary lowland rainforest and gallery forest. The genus is distributed from Tamaulipas, Mexico, south into northern Argentina (Misiones). Fossils are known from the early Pliocene (Montehermosan) of Argentina and from late Quaternary cave deposits in Brazil (Marshall 1982a). The only fossil species recognized (see Reig 1957) is *P. entrerianus* (Ameghino 1899).

SYNONYMS:

Didelphis Linnaeus, 1758:54; part.

Didelphys Schreber, 1777:532; part; unjustified emendation of *Didelphis* Linnaeus.

Philander Brisson, 1762:13; type species *Didelphis opossum* Linnaeus, 1758, by plenary action (ICZN 1998).

Philander Tiedemann, 1808:426; type species *Philander virginianus* Tiedemann, 1808 (= *Didelphis opossum* Linnaeus, 1758), by subsequent designation (Hershkovitz 1949a; see Remarks).

Sarigua Muirhead, 1819:429; part.

Metachirus Burmeister, 1854:135; part; described as a subgenus of *Didelphis* Linnaeus.

Gamba Liais, 1872:329; part.

Zygolestes: Ameghino, 1899:7; part.

Metachirops Matschie, 1916:262; type species *Didelphis quica* Temminck, 1824b, by original designation (p. 268).

Holothylax Cabrera, 1919:47; type species *Didelphis opossum* Linnaeus, 1758, by original designation.

Metacherius Sanderson, 1949:787; incorrect subsequent spelling of *Metachirus* Burmeister.

Phillander Rivillas, Caro, Carvajal, and Vélez, 2004:591; incorrect subsequent spelling of *Philander* Brisson.

REMARKS: Contrary to accepted usage for more than a century, *Philander* Tiedemann, 1808 was set aside in 1998 (ICZN: Opinion 1894) as the valid generic name for the black and gray four-eyed opossums in favor of resurrecting the non-Linnaean *Philander* Brisson, 1762. This ruling by the ICZN (1998) effectively declared moot more than 50 years of often colorful contentious interpretations of the content, application, and identity of the type species of *Philander* Tiedemann, 1808 (Trouessart 1898, 1905; Hershkovitz 1949a, 1959, 1976b, 1981, 1997; Pine 1973a; Husson 1978; Hall 1981; Gardner 1981), and the context and content of *Philander* as used by Brisson (1762), Gray (1843a), Burmeister (1856); O. Thomas (1888b), J. A. Allen (1900b), Cabrera (1919, 1958), and Tate (1939). Part of even earlier controversy over application of the name *Philander* concerned its use for the woolly opossums now in

the genus *Caluromys* (see Hershkovitz 1949a, for a brief review).

Pine (1973a) and Husson (1978) considered *Metachirops* Matschie to be the correct name for the pouched four-eyed opossums, although for different reasons (also see Tate 1939). Pine (1973a) argued, on the basis of color, that *Philander* Tiedemann, referred not to the pouched, gray or black four-eyed species, but to the brown, pouchless four-eyed opossum known as *Metachirus* Burmeister, 1854. Pine also reviewed some of the history surrounding the use of *Philander* and concluded that the brown, pouchless four-eyed opossum should be called *Philander nudicaudatus* and the gray, pouched four-eyed species should be known as *Metachirops opossum*. Hall (1981:17) cited Pine (1973a) as the authority for using the generic names, *Philander* and *Metachirops*, in place of the more familiar *Metachirus* and *Philander*, respectively.

Husson (1978:27) attempted to resolve the situation generated by Pine's (1973a) conclusion that *Philander* Tiedemann was the correct name for the brown, pouchless four-eyed opossums by selecting the female from Virginia, discussed by Tyson (1698), as lectotype of *Philander virginianus* Tiedemann. Tyson (1698) is one of three secondary references included by Tiedemann under *P. virginianus*. Husson wanted to relegate *P. virginianus* Tiedemann to the synonymy of *Didelphis virginiana* Kerr, 1792, and *Philander* Tiedemann to the synonymy of *Didelphis* Linnaeus, 1758. By this action, Husson intended the pouchless four-eyed opossums to be known as *Metachirus nudicaudatus* and the pouched species as *Metachirops opossum*.

Hershkovitz (1981, 1997) pointed out that because *Philander virginianus* Tiedemann was a new name for *D. opossum* Linnaeus, the female illustrated by Seba (1734) already had been designated by Hershkovitz (1976b) as the lectotype of *D. opossum*, and that Seba's female also was the lectotype of *P. virginianus* Tiedemann. Therefore, Husson's (1978) later selection of a lectotype for *Philander* Tiedemann is invalid.

KEY TO THE SPECIES OF *PHILANDER* (SEE GARDNER AND PATTON 1972; J. L. PATTON AND SILVA 1997):

1. Dorsum black and venter dark gray to blackish; middorsal fur long or short. 2
1'. Dorsal and lateral pelage usually gray, if blackish middorsally, a well-defined dorsal stripe is lacking, the lateral pelage is distinctly paler, or the tail is entirely dark brown to black (lacks terminal pale portion); fur short 3
2. Well-defined, 3–4 cm-wide black middorsal stripe; dorsal fur short (*ca.* 10 mm); ventral coloration distinctly paler than sides of body; base of tail furred for approximately 18% of its length. . . . . *Philander andersoni*

2'. Black middorsal stripe often broadly expanded behind shoulders and lacking well-defined margins; middorsal fur mixed with longer (ca. 18 mm) coarse guard hair, often resulting in a low crest along spine; ventral coloration same color or slightly paler than sides; base of tail clothed in long fur for more than 23% of tail length . . . . *Philander mcilhennyi*

3. Color dark gray with cream-colored to white venter; fur on throat gray-based and separated in ventral midline by narrow strip of self-colored cream-colored to whitish fur; restricted to Atlantic forest region from eastern Brazil into Paraguay and northeastern Argentina . . . . *Philander frenatus*

3'. Color variable, usually pale gray; not restricted in distribution as above 4

4. Ears of adults blackish, except at base; supraorbital and postauricular spots large and strongly contrasting with color of adjacent fur; longer fur of body extending onto base of tail for more than 20% of tail length . . . . *Philander opossum*

4'. Proximal half or more of ears of adults unpigmented; postauricular spots greatly reduced in size; longer fur of body extending onto base of tail for less than 20% of tail length 5

5. Pelage short and woolly; venter broadly pale cream colored; distal half of ears pigmented; supraorbital spots large and conspicuous; postauricular spots small, but still conspicuous . . . . *Philander mondolfii*

5'. Pelage short and velvety; cream color of venter restricted laterally by gray-based fur of sides of body; pigmentation of ears restricted to distal margin; supraorbital spots relatively small; postauricular spots inconspicuous . . . . *Philander deltae*

Philander andersoni (Osgood, 1913)
Anderson's Four-eyed Opossum
SYNONYMS:

Metachirus andersoni Osgood, 1913:95; type locality "Yurimaguas, [Loreto,] Peru."

[*Didelphis* (]*Metachirops*[)] *andersoni*: Matschie, 1916: 268; name combination.

[*Holothylax*] *andersoni*: Cabrera, 1919:47; name combination.

M[*etachirus*]. *opossum andersoni*: O. Thomas, 1923b:603; name combination.

Metachirus opossum nigratus O. Thomas, 1923b: 603; type locality "Utcuyaco, Dept. Junin, 1600 m.," Peru.

Met[achirops]. opossum andersoni: Krumbiegel, 1941b: 202; name combination.

Metachirops opossum nigratus: Krumbiegel, 1941b:202; name combination.

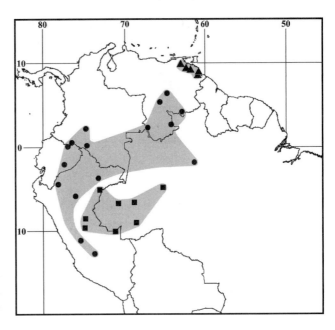

Map 10 Marginal localities for *Philander andersoni* ●, *Philander deltae* ▲, and *Philander mcilhennyi* ■

Philander opossum andersoni: Cabrera, 1958:34; name combination.

DISTRIBUTION: *Philander andersoni* occurs from the Río Tabero and Cerro Duida region of southcentral Venezuela, west to the Serra de la Macarena of Colombia, and south through eastern Ecuador and northern Peru in the eastern Andean foothill valleys at least as far south as departamento Ayacucho (Gardner and Patton 1972; J. L. Patton and Silva 1997).

MARGINAL LOCALITIES (Map 10): VENEZUELA (Lew, Pérez-Hernández, and Ventura 2006, except as noted): Bolívar, Río Tabero; Bolívar, Raudal Chalimana; Amazonas, Sierra Parima (Pérez-Hernández 1990). BRAZIL: Amazonas, Rio Jaú, above mouth (J. L. Patton, Silva, and Malcolm 2000). PERU: Loreto, Iquitos (O. Thomas 1928d); Loreto, Yurimaguas (type locality of *Metachirus andersoni* Osgood); Junín, Utcuyacu (type locality of *Metachirus opossum nigratus* O. Thomas); Ayacucho, San José (Gardner and Patton 1972; as *Philander opossum*); Amazonas, Huampami (J. L. Patton and Silva 1997). ECUADOR (Hershkovitz 1997): Pastaza, Río Copotaza; Sucumbios, Santa Cecilia. COLOMBIA: Putumayo, 17 km N of Puerto Asís (Hershkovitz 1997); Meta, Macarena (Lew, Pérez-Hernández, and Ventura 2006) ; Caquetá, Tres Troncos (Hershkovitz 1997). VENEZUELA: Amazonas, Merey (Hershkovitz 1997); Amazonas, Caño Majagua (Pérez-Hernández 1990).

SUBSPECIES: We consider *P. andersoni* to be monotypic.

NATURAL HISTORY: Other than Hershkovitz's (1997: 73–74) suggestion that *P. andersoni* may be more arboreal and aquatic than *P. opossum*, its natural history is unknown.

REMARKS: *Philander andersoni* is a dark species with a well-marked, black median dorsal stripe (see illustration in Hershkovitz 1997:9, Fig. 4) and creamy to creamy-gray venter. Gardner (1993a) included the similar *Philander mcilhennyi* Gardner and Patton, 1972 as a synonym of *P. andersoni*, anticipating Hershkovitz's (1997) review of *Philander* in which he treated *P. mcilhennyi* as a subspecies of *P. andersoni*. Based on molecular analyses of Amazon basin populations of *Philander*, J. L. Patton and Silva (1997) and J. L. Patton, Silva, and Malcolm (2000) have shown that *P. mcilhennyi* warrants recognition as a separate species.

Philander deltae Lew, Pérez-Hernández, and Ventura, 2006
Orinoco Four-eyed Opossum
SYNONYMS:

Philander opossum ssp. nov. Pérez-Hernández, 1990:373; not *Didelphis opossum* Linnaeus, 1758.

Philander opossum subspecies δ Hershkovitz, 1997:10; not *Didelphis opossum* Linnaeus.

Philander opossum subspecies nov.1 Hershkovitz, 1997:10; not *Didelphis opossum* Linnaeus.

Philander opossum ssp. nov. B Linares, 1998:29,30; not *Didelphis opossum* Linnaeus.

Philander deltae Lew, Pérez-Hernández, and Ventura, 2006:227; type locality "Sector Guanipa, Reserva Forestal de Guarapiche, 24.2 km 160° W Capure, Monagas State, Venezuela, coordinates 10°00′N, 62°49′W, 0 m elevation."

DISTRIBUTION: *Philander deltae* is known from the Río Orinoco delta of Venezuela.

MARGINAL LOCALITIES (Map 10; from Lew, Pérez-Hernández, and Ventura 2006): VENEZUELA: Monagas, Cachipo; Delta Amacuro Isla Tigre; Delta Amacuro, Los Güires; Delta Amacuro, Guayo; Delta Amacuro Curiapo.

SUBSPECIES: *Philander deltae* is monotypic.

NATURAL HISTORY: This species inhabits relatively short (15 to 25 m), evergreen permanently or seasonally flooded forest in the delta of the Río Orinoco. Addition natural history information is lacking.

REMARKS: Smaller body size; short, velvety fur; brownish dorsum and cream-colored venter (constricted laterally by gray-based hair); smaller ears pigmented only along margins; small supraorbital spots; and relatively inconspicuous post-auricular spots are features that distinguish *P. deltae* from congeners (Lew, Pérez-Hernández, and Ventura 2006). *Philander deltae* is adapted to the flooded forest habitats of the Río Orinoco delta and, apparently, is

not sympatric with any of the other four species found in Venezuela.

Philander frenatus (Olfers, 1818)
Southeastern Four-eyed Opossum
SYNONYMS:

D[*idelphys*]. *frenata* Illiger, 1815:107; *nomen nudum.*

D[*idelphys*]. *superciliaris* Illiger, 1815:107; *nomen nudum.*

D[*idelphys*]. *frenata* Olfers, 1818:204; type locality "Süd-america"; restricted to Bahia, Brazil, by J. A. Wagner (1843b: 44, footnote 25).

D[*idelphys*]. *superciliaris* Olfers, 1818:204; type locality "Südamerica."

Didelphis quica Temminck, 1824b:36; type locality "Brésil"; restricted to Sapitiba [= Sepetiba, Rio de Janeiro] by Pelzeln (1883:110; also see J. A. Allen, 1916e: 563; Hershkovitz, 1959:342).

Didelphys [*(Metachirus)*] *quica*: Burmeister, 1854:136; name combination.

Zygolestes entrerianus Ameghino, 1899:7; type locality "Argentina."

[*Didelphis (Metachirops)*] *quica*: Matschie, 1916:268; name combination.

[*Didelphis (Metachirops)*] *frenata*: Matschie, 1916:268; name combination.

[*Holothylax*] *quica*: Cabrera, 1919:48; name combination.

Metachirus opossum azaricus O. Thomas, 1923b:604; type locality "Sapucay," Paraguarí, Paraguay.

Metachirops quica: Bresslau, 1927:215; name combination.

Metachirops opossum quica: A. Miranda-Ribeiro, 1935:37; name combination.

M[*etachirops*]. *opossum quichua*: Krumbiegel, 1941b:200; name combination and incorrect subsequent spelling of *Didelphis quica* Temminck, 1824b).

Met[*achirops*]. *opossum azaricus*: Krumbiegel, 1941b:203; name combination.

Met[*achirops*]. *opossum frenatus*: Krumbiegel, 1941b:206; name combination.

Philander opossum azaricus: Cabrera, 1958:34; name combination.

Philander opossum quica: Cabrera, 1958:36; name combination.

Philander entrerrianus: Reig, 1956:220; name combination and incorrect subsequent spelling of *Zygolestes entrerianus* Ameghino, 1899.

DISTRIBUTION: *Philander frenatus* is in eastern Brazil and south to Paraguay and adjacent Argentina.

MARGINAL LOCALITIES (Map 11; from Hershkovitz 1997, except as noted): BRAZIL: Bahia, Bahia (restricted type locality of *Didelphys frenata* Olfers); Espírito Santo, Mato do Lava d'Agua (J. L. Patton and Silva 1997); Rio de Janeiro, Maricá (J. L. Patton and Silva, 1997); São

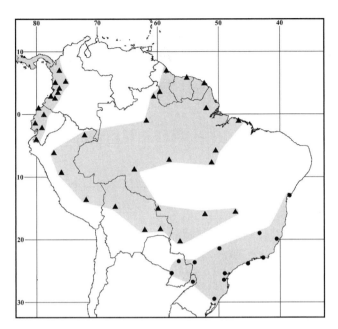

Map 11 Marginal localities for *Philander frenatus* ● and *Philander opossum* ▲

Paulo, Fazenda da Toca (J. L. Patton and Silva 1997); Paraná, Piraquara (Cáceres 2004); Santa Catarina, Hansa; Rio Grande do Sul, Taquara do Mundo Novo (A. Miranda-Ribeiro 1936). ARGENTINA: Misiones, Fracrán; Formosa, ca. 13 km S of Clorinda on Rt. 11 (AMNH 256980). PARAGUAY: San Pedro, Tacuatí. BRAZIL: Paraná, Rio Paracaí; São Paulo, Avanhandava; Minas Gerais, Conceição do Mato Dentro.

SUBSPECIES: We consider *P. frenatus* to be monotypic.

NATURAL HISTORY: These opossums are scansorial and terrestrial, and are seldom seen climbing, except on fallen logs or windfalls (D. E. Davis 1947). Like other members of the genus, they are omnivores consuming a variety of plants and animals, from fruits and flowers to invertebrates and vertebrates. Cáceres (2004) found three kinds of seeds and the remains of birds, mammals, reptiles, and invertebrates in the feces of *P. frenatus* from Atlantic forest in Paraná, Brazil. Birds and beetles were the most common prey items. D. E. Davis (1947) found *P. frenatus* to be nocturnal, solitary, and non-territorial. He also stated that average litter size varies from 3.4 to 5.5.

REMARKS: Previously treated as representing subspecies of *P. opossum* (e.g., *P. o. quica* by Cabrera, 1958; or *P. o. frenatus* and *P. o. quica* by Hershkovitz, 1997), J. L. Patton and Silva (1997) and J. L. Patton, Silva, and Malcolm (2000) concluded that the southeastern Brazilian gray *Philander* warranted recognition as the species *P. frenatus*. J. L. Patton, Silva, and Malcolm (2000) included the Paraguayan and northeastern Argentine populations Cabrera (1958:34) listed as *P. o. azaricus* in *P. frenatus*. Inexplicably, Hershkovitz (1997) treated *frenatus* and the major

population, which he included under the name *P. o. quica*, as separate taxa. Hershkovitz (1997:51) was correct in stating that the Teresópolis specimens J. L. Patton, Reis, and Silva (1995) used in their analyses of mitochondrial DNA had not been shown to represent "true" *frenatus*. However, his inclusion of southeastern Brazilian populations, under the name *P. o. quica*, with the western Brazilian populations J. L. Patton, Reis, and Silva (1995) had identified as *P. o. canus*, demonstrated his lack of understanding of the power of molecular analyses when exploring phylogenetic relationships.

Philander mcilhennyi Gardner and Patton, 1972
McIlhenny's Four-eyed Opossum

SYNONYMS:

Philander mcilhennyi Gardner and Patton, 1972:2; type locality "Balta (10°08′S, 17°13′W), Río Curanja, *ca.* 300 meters, departamento de Loreto [now departamento de Ucayali], Peru."

P[hilander]. mcilhenyi: Pérez-Hernández, 1990:373; incorrect subsequent spelling of *Philander mcilhennyi* Gardner and Patton, 1972).

Philander andersoni mcilhennyi: Hershkovitz, 1997:iii; name combination.

DISTRIBUTION: *Philander mcilhennyi* is known from the Amazon basin of central Peru and adjacent western Brazil.

MARGINAL LOCALITIES (Map 10): BRAZIL: Amazonas, alto Rio Urucú (J. L. Patton, Reis, and Silva 1995); Acre, Sena Madureira (USNM 546221). PERU: Ucayali, Balta (type locality of *Philander mcilhennyi* Gardner and Patton); Huánuco, Panguana Biological Station (Hutterer et al. 1995); Ucayali, 59 km SW of Pucallpa (Hershkovitz 1997); Loreto, Nuevo San Juan, Río Galvez (AMNH 272818). BRAZIL (J. L. Patton and Silva 1997): Amazonas, Seringal Condor; Amazonas, Altamira.

SUBSPECIES: We treat *P. mcilhennyi* as monotypic.

NATURAL HISTORY: Females with pouched young were recorded in the months of April and June at the type locality (Gardner and Patton 1972), and during July and August on the Rio Urucú in Amazonas, Brazil. The habitat at the type locality is dry tropical forest, and animals were trapped in both undisturbed and disturbed (= garden) sites. In the central Amazon of Brazil, animals were trapped in both undisturbed terra firme forest and in secondary growth. At the Peruvian localities and some of the Brazilian localities, *P. mcilhennyi* was found sympatric with the somewhat smaller subspecies of the Gray Four-eyed Opossum, *P. opossum canus*.

Hershkovitz (1997:59) referred to the dorsal pelage as "muskrat-like," and suggested that *P. mcilhennyi* had aquatic habits. However, J. L. Patton, Silva, and Malcolm

(2000), based on trap success, found *P. mcilhennyi* had a general preference for terra firme habitats; whereas *P. o. canus*, which has decidedly un-muskrat-like pelage, was more commonly taken in seasonally flooded varzea habitats. Hutterer et al. (1995) found the remains of a frog, a beetle, and 22 ants of two species in the stomach of a female from departamento Huánuco, Peru. The karyotype is $2n = 22$, FN = 20, and is identical to that of *P. opossum* (Reig et al. 1977).

REMARKS: Emmons and Feer (1990) and Gardner (1993a) followed Hershkovitz's preliminary assessments of relationships among taxa of *Philander*, finally published in 1997, in treating *P. mcilhennyi* as a synonym of *P. andersoni*. Nevertheless, J. L. Patton, Reis, and Silva (1995), J. L. Patton, Silva, and Malcolm (2000), and J. L. Patton and Silva (1997) have provided compelling evidence supporting their recognition of these taxa as separate species.

Size and color pattern characteristics have been used to assign specimens (including those described as *Metachirus opossum nigratus* O. Thomas) found along the lower Andean slopes to *P. andersoni* Osgood. If future work demonstrates that *nigratus* represents the species known today as *P. mcilhennyi*, and not *P. andersoni*, the correct name for this species would be *P. nigratus* (O. Thomas, 1923).

Philander mondolfii Lew, Pérez-Hernández, and Ventura, 2006
Mondolfi's Four-eyed Opossum
SYNONYMS:
Metachirus opossum opossum: J. A. Allen, 1911:244, 246; not *Didelphis opossum* Linnaeus, 1758.
Philander opossum: Reig, Gardner, Bianchi, and Patton, 1977:197; not *Didelphis opossum* Linnaeus.
P[*hilander*]. o[*possum*]. *grisescens*: Pérez-Hernández, 1990:373; not *Metachirus grisescens* J. A. Allen, 1901d.
P[*hilander*]. o[*possum*]. subspecies ε Hershkovitz, 1997:10; not *Didelphis opossum* Linnaeus.
Philander opossum ssp. nov. A Linares, 1998:30; not *Didelphis opossum* Linnaeus.
Philander mondolfii Lew, Pérez-Hernández, and Ventura, 2006:229: type locality "Reserva Forestal de Imataca, Unidad V, between Tumeremo and Bochinche, Bolívar State, Venezuela, coordinates 08°00′N, 61°30′W, 180 m elevation."

DISTRIBUTION: *Philander mondolfii* is known from Colombia and Venezuela where it appears to be distributed as two discrete units. The species is expected from Guyana and northern Brazil.

MARGINAL LOCALITIES (Map 12; from Lew, Pérez-Hernández, and Ventura 2006, except as noted): *Western distribution*. VENEZUELA: Aragua, Trujillo, near El Dividive; Barinas, Reserva Forestal de Ticoporo; Apure, La

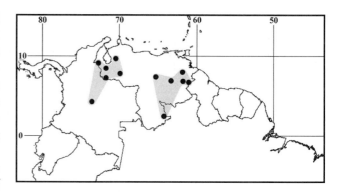

Map 12 Marginal localities for *Philander mondolfii* ●

Blanquita. COLOMBIA: Meta, Restrepo. VENEZUELA: Zulia, Boca del Rio de Oro; Mérida, Bejuquero. *Eastern distribution*. VENEZUELA: Bolívar, El Palmar; Bolívar, San Martín de Turumbán; Bolívar, Río Yuruan; Amazonas, Coyowateri; Bolívar, El Raudal; Bolívar, La Paragua.

SUBSPECIES: We treat *P. mondolfii* as monotypic.

NATURAL HISTORY: *Philander mondolfii* has been found in foothill (50 to 800 m) rain and riparian forest habitats. Other details of its natural history are lacking.

REMARKS: This species can be distinguished from congeners on the basis of average body size for the genus, but with a shorter, relatively broad skull. Other differences include short, woolly fur; pale gray dorsum; pale cream-colored venter; large ears pigmented on only distal half; and thicker fur of body extending onto base of tail for less than 20% of tail length. Although not known to be sympatric, the distribution of *P. mondolfii* may overlap that of *P. andersoni* in some parts of its range.

Hershkovitz (1997:52), under the taxon "*Philander opossum* subspecies nov.?" listed Las Bonitas, Aragua, Venezuela, as a locality, which he attributed to Pérez-Hernández (1990). This locality is not mentioned by Pérez-Hernández (1990), nor is it mentioned by Lew, Pérez-Hernández, and Ventura (2006) in their description of *P. mondolfii*, which would be the most likely identity for a *Philander* from the state of Aragua. Brown (2004:137) listed a University of Michigan specimen identified as *P. opossum*, which on basis of geographic probability also most likely represents *P. mondolfii*. This record is not mapped under either species, pending confirmation of the specimen's provenance and identity.

Philander opossum (Linnaeus, 1758)
Gray Four-eyed Opossum
SYNONYMS: The following are species synonyms and names applied to non-South American subspecies. Additional synonyms are listed under Subspecies.
Metachirus fuscogriseus pallidus J. A. Allen, 1901d:215; type locality "Orizaba, Vera Cruz, Mexico."

[*Didelphis (Metachirops)*] *pallidus*: Matschie, 1916:268; name combination.

Metachirops opossum pallidus: Miller, 1924:7; name combination.

Philander opossum pallidus: Dalquest, 1950a:2; name combination.

DISTRIBUTION: *Philander opossum* occurs from Tamaulipas in northeastern Mexico south throughout the lowland tropics of Central and South America to Bolivia and southcentral Brazil at elevations typically below 1,000 m. In South America, its range extends along the Pacific versant from Colombia to southern Ecuador, throughout the Amazon basin and the upper Paraná basin. This species is absent from the Caribbean slopes of Colombia and Venezuela, from the llanos of the Orinoco basin of Venezuela and adjacent Colombia, from the caatinga of northeastern Brazil, and from the Atlantic forest region of southeastern Brazil south to Paraguay and northern Argentina.

MARGINAL LOCALITIES (Map 11; from Hershkovitz 1997, except as noted): *Northwestern distribution*: COLOMBIA: Caldas, Samaná; Valle del Cauca, Río Frio (J. A. Allen 1916c); Valle del Cauca, Cali; Cauca, Cocal (J. A. Allen 1912). ECUADOR: Pichincha, Mindo; Guayas, Bucay; Loja, Cebollal; Manabí, Río Pescado; Esmeraldas, Esmeraldas. COLOMBIA: Cauca, Boca del Río Saijá; Chocó, Río Sandó; Antioquia, Alto Bonito. *Guayanan and Amazonian distribution*: GUYANA: Pomeroon-Supenaam, Pomeroon (B. Brown 2004). SURINAM: Paramaribo, Paramaribo (type locality of *Didelphis opossum* Linnaeus). FRENCH GUIANA: Cayenne (Brongniart 1792). BRAZIL: Amapá, Teresinha; Pará, Santa María; Pará, Carajás (J. L. Patton, Silva, and Malcolm 2000); Pará, Gradaús; Pará, Canudos; Rondônia, Pôrto Velho; Mato Grosso, Mato Grosso; Goiás, Aragarças; Goiás, Formosa; Mato Grosso do Sul, Miranda. BOLIVIA (S. Anderson 1997): Santa Cruz, Santiago (Chiquitos); Santa Cruz, 8 km SE of Tita; Beni, 45 km (by road) N of Yacuma. PERU: Cusco, San Jerónimo (O. Thomas 1928c); Huánuco, Tingo María (O. Thomas 1927b); San Martín, Rioja (O. Thomas, 1927b); Loreto, Apayacu. BRAZIL: Amazonas, Rio Xiriviny; Roraima, Boa Vista. GUYANA: Upper Takutu-Upper Essequibo, Nappi Creek (Lew, Pérez-Hernández, and Ventura 2006).

SUBSPECIES: We treat *P. opossum* as represented in South America by four subspecies, and recognize that some probably are composites of two or more species or subspecies; the species needs to be revised (see Remarks).

P. o. canus (Osgood, 1913)

SYNONYMS:

Metachirus canus Osgood, 1913:96; type locality "Moyobamba, [San Martín,] Peru."

[*Didelphis (Metachirops)*] *canus*: Matschie, 1916:268; name combination.

[*Holothylax grisescens*] *canus*: Cabrera, 1919:47; name combination.

Metachirus opossum crucialis O. Thomas, 1923b:604; type locality "Santa Cruz de la Sierra," Santa Cruz, Bolivia.

Met[achirops]. opossum canus: Krumbiegel, 1941b:203; name combination.

Met[achirops]. opossum crucialis: Krumbiegel, 1941b:203; name combination.

Philander opossum canus: Sanborn, 1949b:277; first use of current name combination.

A gray subspecies that occurs throughout the western Amazon basin of central Peru, northern Bolivia, and western Brazil.

P. o. fuscogriseus (J. A. Allen, 1900)

SYNONYMS:

Metachirus fuscogriseus J. A. Allen, 1900c:194; type locality "Central America"; restricted to Greytown, Nicaragua, by J. A. Allen (1911:247).

Metachirus grisescens J. A. Allen, 1901d:217; type locality "Rio Cauca, [Valle del Cauca,] Colombia."

[*Didelphis (Metachirops)*] *fuscogriseus*: Matschie, 1916:268; name combination.

[*Didelphis (Metachirops)*] *grisescens*: Matschie, 1916:268; name combination.

[*Holothylax*] *fuscogriseus*: Cabrera, 1919:47; name combination.

[*Holothylax grisescens*] *grisescens*: Cabrera, 1919:48; name combination.

Metachirus opossum melantho O. Thomas, 1923b:602; type locality "Condoto, 300′," Chocó, Colombia.

Metachirops opossum fuscogriseus: Miller, 1924:7; name combination.

Metachirops opossum grisescens: Krumbiegel, 1941b:202; name combination.

Metachirops opossum melantho: Krumbiegel, 1941b:202; name combination.

Philander opossum fuscogriseus: Goodwin, 1942c:113; name combination)

This is a dark gray subspecies that is known from northern and western Colombia south along the Pacific lowlands into southern Ecuador. Elsewhere, it occurs in Central America as far north as Honduras.

P. o. melanurus (O. Thomas, 1899)

SYNONYMS:

Metachirus melanurus O. Thomas, 1899d:285; type locality "Paramba, Rio Mira, N. Ecuador; alt. 1100 m."

[*Didelphis (Metachirops)*] *melanurus*: Matschie, 1916:268; name combination.

[*Holothylax opossum*] *melanurus*: Cabrera, 1919:48; name combination.

Metachirops opossum melanurus: Lönnberg, 1921:4; name combination.

Philander opossum melanurus: Cabrera, 1958:35; name combination.

This subspecies, distributed along the Pacific lowlands of Colombia and northern Ecuador, also has dark gray to blackish pelage and has a uniformly dark brown to black tail.

P. o. opossum (Linnaeus, 1758)

SYNONYMS:

[*Didelphis*] *Opossum* Linnaeus, 1758:55; type locality "America"; restricted to Surinam by J. A. Allen (1900c: 195); further restricted to Paramaribo, Surinam, by Matschie (1916:268).

Didelphis Oppossum Brongniart, 1792:115; incorrect subsequent spelling of *D. opossum* Linnaeus.

Philander virginianus Tiedemann, 1808:426; type localities "Virginien, Mexiko, Peru u. s. w."; treated as a synonym of *D. opossum* Linnaeus by Hershkovitz (1949a).

S[arigua]. opossum: Muirhead, 1819:429; name combination.

D[idelphys]. oppossum J. A. Wagner, 1843b:44; incorrect subsequent spelling of *D. opossum* Linnaeus.

[*Didelphys (]Metachirus[)*] *opossum*: Burmeister, 1856:69; name combination.

Gamba opossum: Liais, 1872:329; name combination.

[*Didelphis (Metachirops)*] *opossum*: Matschie, 1916:268; name combination.

[*Holothylax opossum*] *opossum*: Cabrera, 1919:48; name combination.

Didelphis austro-americana O. Thomas, 1923b:604; type locality "Surinam"; preoccupied by *Didelphis austro-americana* J. A. Allen, 1902:251 (= *Didelphis marsupialis* Linnaeus); both names are based on *Didelphys austro-americana* Oken, 1816 (see Remarks).

Metachirops opossum: Tate, 1939:161; name combination.

Metachirops opossum opossum: Krumbiegel, 1941b:200; name combination.

Philander opossum: Gilmore, 1941:316; first use of current name combination.

Philander opossum opossum: Reig, 1957:222; name combination.

This uniformly gray subspecies is known from eastern Venezuela and the Guianas through the northern and eastern Amazon basin of Brazil.

NATURAL HISTORY: *Philander opossum* is common in disturbed (second growth and garden plots) and undisturbed primary forest throughout the lowland Neotropics, and in gallery forests in the cerrado of southern Brazil (Fleming 1972; Charles-Dominique 1983; Redford and

Fonseca 1986). These opossums are scansorial and terrestrial, and are seldom seen climbing, except on fallen logs or windfalls. They are omnivorous, feeding on a wide variety of plant and animal products, including fruits, nectar, leaves and bark, invertebrates, and vertebrates (Seba 1734; Enders 1935, Fleming 1972; Husson 1978; Howe 1980; Tuttle, Taft, and Ryan 1981; Atramentowicz 1988). Individuals are nocturnal, solitary, and nonterritorial (Charles-Dominique 1983; Atramentowicz 1986).

Reproduction appears to peak at the height of the fruiting season (Husson, 1978; Charles-Dominique et al. 1981; Atramentowicz 1986; Julien-Laferrière and Atramentowicz 1990), and average litter size varies from 3.4 to 5.5 (Eisenberg and Wilson 1981; Julien-Laferrière and Atramentowicz 1990). Two to four litters, with a birth interval of 90 days, characterizes populations in French Guiana (Charles-Dominique et al. 1981; Julien-Laferrière and Atramentowicz 1990). Sexual maturity is achieved by 5 to 8 months of age (Charles-Dominique et al. 1981; Atramentowicz 1986). Home range size, individual movement patterns, and population turnover rates have been reported for populations in Panama (Fleming 1972) and French Guiana (Charles-Dominique 1983). Density estimates averaging nearly 150 individuals per square kilometer have been reported for French Guiana (Atramentowicz 1986), and Emmons (1984) provided relative densities based on trapping and sight censuses for Ecuador and Peru. Hershkovitz (1997) and Castro-Arellano, Zarza, and Medellín (2000) have provided detailed descriptions along with measurements and a summary of available information on natural history in their *Mammalian Species* account. The karyotype is $2n = 22$, FN $= 20$; both the X and Y chromosomes are acrocentric (Reig et al. 1977). The karyotype is similar to those of *Lutreolina*, *Chironectes*, and South American species of *Didelphis*.

REMARKS: The species concept of *P. opossum* we use here follows J. L. Patton and Silva (1997) and J. L. Patton, Silva, and Malcolm (2000). These authors suggested that *P. opossum* is composite and that future studies will likely accord species status to the Middle American, western Amazonian, and eastern Amazonian-Guayanan gray four-eyed opossums. In this preliminary review, we recognize five subspecies. Four are restricted to South America and one (*P. o. pallidus*) occurs only in eastern and southern Mexico and in northern Central America. Although not included in the South American fauna by Cabrera (1958), *P. o. fuscogriseus* is in both Central and South America. No revision has as yet examined the validity of any of these named forms and our subspecies designations should be considered provisional at best.

The species exhibits considerable geographic variation in color and color pattern, and no review of this, or other aspects of character variation has as yet been undertaken. There are anomalous distribution patterns in *P. opossum*. For example, *P. o. melanurus*, which is a distinctive melanistic form having an all black or dark-brown tail, is sandwiched between populations of *P. o. fuscogriseus* along the Pacific versant of Colombia and Ecuador (see Hershkovitz 1997:10, Fig. 5). The species is in need of revision.

Hershkovitz (1959, 1976b, 1997) consistently claimed that *Didelphis myosuros* Temminck, 1824, is a synonym of *P. opossum* and not the species known today as *Metachirus nudicaudatus* (É. Geoffroy St.-Hilaire, 1903). This same position is taken by B. Brown (2004). Hershkovitz 1997:49) acknowledged that his 1959 designation of a lectotype for *Didelphis myosuros* Temminck was invalid. In the next paragraph he stated, "It is now proposed to designate as lectotype of *myosuros* one of the pouched females used by Temminck in the original description of the taxon." Nevertheless, this statement does not constitute designation of a lectotype for *Didelphis myosuros* Temminck, 1824. Furthermore, such a designation would jeopardize the status of *Metachirus* Burmeister because *Didelphis myosuros* Temminck is its type species (see Remarks under *Metachirus nudicaudatus*). Most authors are agreed that *Didelphis myosuros* is a synonym of *Metachirus nudicaudatus* (É. Geoffroy St.-Hilaire, 1903).

Tribe Metachirini Reig, Kirsch, and Marshall, 1985

Alfred L. Gardner and Marian Dagosto

The monotypic Metachirini can be distinguished by the structure of the ear region and lateral braincase. The ossified bullae are small and incomplete with much of the middle ear open ventrally. Although the tympanic wing of the alisphenoid is not broadly inflated, it contributes to the bulla anteriorly and forms the anterior floor of tympanic cavity. The posterior bullar wall includes the tympanic process of the *pars petrosa* of the perioticum. The frontal and squamosal are in contact on the lateral aspect of the braincase (no parietal-alisphenoid contact; Voss and Jansa 2003:24).

Genus *Metachirus* Burmeister, 1854

Metachirus is a monotypic genus found from southern Mexico to northern Argentina. The only known species, *Metachirus nudicaudatus*, is the largest of the pouchless didelphids (mean and range, in mm, of total length, tail, and mass in grams, respectively, of 13 adults: 573, 476–675; 319, 251–380; and 385, 230–475). The color of the dorsum

varies geographically from reddish and yellowish brown to gray brown and is finely streaked with the paler color of the hair tips. The dorsum is darkest along the midline and the darker color extends over the flanks and outer surfaces of the legs. The color of the venter varies from buffy brown to yellowish white. A conspicuous buff to yellowish-white spot is located above each eye. A broad reddish- to dark brown stripe surrounds each eye and extends from the side of the rhinarium to the ear and merges medially on the top of the head above the pale eye spots. The upper surface of the rostrum is paler and bears a weakly defined, darker median stripe that extends from the rhinarium to the forehead. The color of the cheek is paler, like that of the venter and eye spot. Ears are dark brown and naked. The tail is longer than the head and body, sparsely haired, varies from brownish along its full length to darker above and paler below, and lacks pigment for the terminal third. The skull is slender, and older individuals have elevated supraorbital ridges that extend back onto the parietals, but do not join in the midline. The palate has a pair of large maxillopalatine fenestrae, but lacks maxillary and palatine fenestrae. The bullae are small, incompletely ossified, and widely separated. The upper canines are well developed, although less than twice the height of P3. The lower canines are shorter than p2.

SYNONYMS:

Didelphis: É. Geoffroy St.-Hilaire, 1803:142; not *Didelphis* Linnaeus, 1758.

Philander: Gray, 1843b:100; part; not *Philander* Brisson, 1762; not *Philander* Beckmann, 1772; not *Philander* Tiedemann, 1808.

Metachirus Burmeister, 1854:135; type species *Didelphys myosurus* Burmeister, 1854 (= *Didelphis myosuros* Temminck, 1824b; = *Didelphis nudicaudata* É. Geoffroy St.-Hilaire, 1803), by subsequent designation (O. Thomas 1888b:329); described as a subgenus of *Didelphis* Linnaeus.

Cuica Liais, 1872:330; part (see Remarks).

Lutreolina: Bertoni, 1939:6; not *Lutreolina* O. Thomas, 1910.

Philander: Pine, 1973a:391; not *Philander* Brisson; not *Philander* Beckmann, 1772; not *Philander* Tiedemann.

Metachirus nudicaudatus (É. Geoffroy St.-Hilaire, 1803)
Brown Four-eyed Opossum

SYNONYMS: See under Subspecies.

DISTRIBUTION: *Metachirus nudicaudatus* occurs in Colombia, Venezuela (westernmost states and south of the Río Orinoco), the Guianas, Ecuador, Brazil (except the Northeast), eastern and central Paraguay, northern Argentina (Misiones and Formosa), and eastern Bolivia and Peru. The species also is in Central America and southern Mexico.

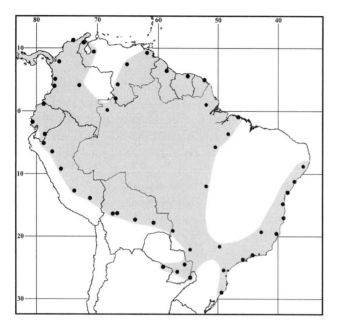

Map 13 Marginal localities for *Metachirus nudicaudatus* ●

MARGINAL LOCALITIES (Map 13; from B. E. Brown 2004, except as noted): COLOMBIA: Magdalena, Donamo (type locality of *Metachirus nudicaudatus colombianus* J. A. Allen). VENEZUELA: Zulia, Cerro Azul; Trujillo, Río Motatán, 9.8 km NNE of Motatán (Pérez-Hernández 1990). COLOMBIA: Meta, Puerto López (Ayala et al. 1973). BRAZIL: Amazonas, Comunidade Colina (J. L. Patton, Silva, and Malcolm 2000). VENEZUELA: Amazonas, San Carlos de Río Negro (Pérez-Hernández 1990); Amazonas, Salto Yureba (Pérez-Hernández 1990); Bolívar, Maripa (J. A. Allen 1904d); Delta Amacuro, Los Güires (Pérez-Hernández 1990). GUYANA: Cuyuni-Mazaruni, Kartabo (H. E. Anthony 1921b). SURINAM: Commewijne, Coropina Kreek, near Republiek (Husson 1978). FRENCH GUIANA: Cayenne (type locality of *Didelphis nudicaudata* É. Geoffroy St.-Hilaire). BRAZIL: Amapá, Serra do Navio; Pará, Jandiaí-Caratateua (USNM 393556); Pará, Ipitinga (Krumbiegel 1941b); Pará, Floresta Nacional Tapirapé-Aquiri (J. L. Patton, Silva, and Malcolm 2000); Mato Grosso, Serra do Roncador (Pine, Bishop, and Jackson 1970); São Paulo, Lins (C. O. C. Vieira 1949); Minas Gerais, Fazenda Esmeralda (Fonseca and Kierulff 1988); Pernambuco, Dois Irmãos (Pohle 1927); Sergipe, Crasto (Husband et al. 1992); Bahia, Bahia (Waterhouse 1841); Bahia, Fazenda Beijo Grande (J. L. Patton, Silva, and Malcolm 2000); Bahia, Comechatiba (Wied-Neuwied 1826); Espírito Santo, Juparanã (A. Miranda-Ribeiro 1936); Rio de Janeiro, Angra dos Reis (A. Miranda-Ribeiro 1936); São Paulo, Varjão; Paraná,

Piraquara (Krumbiegel 1941b); Santa Catarina, Araranguá (C. O. C. Vieira 1949). ARGENTINA: Misiones, Montecarlo (Massoia 1980a). PARAGUAY: Paraguarí, Sapucay (type locality of *Metachirus nudicaudatus modestus* O. Thomas). ARGENTINA: Formosa, La Urbana (Krumbiegel 1941b). PARAGUAY: Canindeyú, Curuguaty. BRAZIL: Mato Grosso do Sul, 12 km N of Dourados; Mato Grosso do Sul, Urucúm. BOLIVIA: Santa Cruz, San José (Krumbiegel 1941b); Santa Cruz, Río Surutú (S. Anderson 1997); Cochabamba, Yungas (S. Anderson 1997); La Paz, Chulumani (type locality of *Metachirus nudicaudatus bolivianus* J. A. Allen). PERU: Cusco, Hacienda Erika; Ayacucho, San José (Reig et al. 1977); Huánuco, Tingo María (O. Thomas 1927b); Amazonas, Chirimoto (O. Thomas 1882); Cajamarca, Quebrada Huarandosa. ECUADOR: Zamora-Chinchipe, 4 km ENE of Los Encuentros; Guayas, Cerro Manglaralto; Esmeraldas, St. Javier (type locality of *Metachirus nudicaudatus phaeurus* O. Thomas). COLOMBIA: Valle del Cauca, 6 km N of Buenaventura; Chocó, Río Sandó; Córdoba, Socorré.

SUBSPECIES: We tentatively recognize five subspecies of *M. nudicaudatus* in South America.

M. n. colombianus J. A. Allen, 1900
SYNONYMS:
Metachirus nudicaudatus colombianus J. A. Allen, 1900c: 196; type locality "Donamo, Santa Marta District, [Magdalena,] Colombia."
Metachirus nudicaudatus colombica J. A. Allen, 1901e:412; incorrect subsequent spelling of *Metachirus nudicaudatus colombianus* J. A. Allen.
Metachirus nudicaudatus phaeurus O. Thomas, 1901f:545; type locality "St. Javier, Lower Cachavi River," Esmeraldas, Ecuador.
[*Didelphys (Metachirus) nudicaudatus*] *columbianus*: Trouessart, 1905:854; name combination and incorrect subsequent spelling of *Metachirus nudicaudatus colombianus* J. A. Allen.
[*Didelphys (Metachirus) nudicaudatus*] *phaeurus*: Trouessart, 1905:854; name combination.
Metachirus nudicaudatus dentaneus Goldman, 1912:2; type locality "Gatun, Canal Zone, Panama."
Metachirus nudicaudatus antioquiae J. A. Allen, 1916a:83; type locality "La Frijolera (altitude 5000 feet), Antioquia, Colombia."
[*Didelphis (Metachirus)*] *columbianus*: Matschie, 1916: 268; name combination and incorrect subsequent spelling of *Metachirus nudicaudatus colombianus* J. A. Allen.
[*Didelphis (Metachirus)*] *phaeurus*: Matschie, 1916:268; name combination.

Metachirus nudicaudatus inbutus O. Thomas, 1923b:605; type locality "Mindo, [Pichincha,] Ecuador."

This subspecies is in northern and western Colombia, western Venezuela, and northwestern Ecuador. Elsewhere, *M. n. colombianus* occurs in Central America and southern Mexico.

M. n. modestus O. Thomas, 1923
SYNONYMS:
Metachirus nudicaudata: Bertoni, 1914:69; incorrect gender concordance.
Metachirus nudicaudatus modestus O. Thomas, 1923b: 606; type locality "Sapucay," Paraguarí, Paraguay.
Lutreolina nudicaudata: Bertoni, 1939:6; name combination.

This subspecies occurs in Paraguay, southern Brazil, and northern Argentina.

M. n. myosuros (Temminck, 1824)
SYNONYMS:
Didelphis myosuros Temminck, 1824b:38; type locality "Brésil"; restricted to Ipanema, São Paulo, by Pohle (1927:242).
Didelphys [(Metachirus)] myosurus Burmeister, 1854:135; name combination and incorrect subsequent spelling of *Didelphis myosuros* Temminck.
Cuica myosuros Liais, 1872:330; name combination.
M[etachirus]. n[udicaudatus]. myosurus O. Thomas; 1923: 606; first use of current name combination and incorrect subsequent spelling of *Didelphis myosuros* Temminck.
Metachirus nudicaudatus personatus A. Miranda-Ribeiro, 1936:351; type locality "Serra da Piraquara, Realengo, Distrito Federal," Rio de Janeiro, Brazil.

This subspecies is in eastern Brazil where it occurs from Pernambuco, south into Santa Catarina.

M. n. nudicaudatus (É. Geoffroy St.-Hilaire, 1803)
SYNONYMS:
Didelphis nudicaudata É. Geoffroy St.-Hilaire, 1803:142 type locality "Cayenne," French Guiana.
Didelphis nudicauda Lesson, 1827:212; incorrect subsequent spelling of *Didelphis nudicaudata* É. Geoffroy St.-Hilaire.
Philander nudicaudus Gray, 1843b:100; name combination and incorrect subsequent spelling of *Didelphis nudicaudata* É. Geoffroy St.-Hilaire.
[Didelphys (Metachirus)] nudicaudata: Trouessart, 1898: 1236; name combination.
[Philander] nudicaudatus: Pine, 1973a:391; name combination.

The nominate subspecies is in Venezuela (south and east of the Río Orinoco), the Guianas, and northern Brazil.

M. n. tschudii J. A. Allen, 1900
SYNONYMS:
Metachirus tschudii J. A. Allen, 1900c:195; type locality "Guayabamba, [Amazonas,] Peru."
Metachirus nudicaudatus bolivianus J. A. Allen, 1901e:411; type locality "Chulumani, Yungas, [La Paz,] Bolivia."
Metachirus bolivianus: J. A. Allen, 1901e:412; name combination.
[Didelphys (Metachirus) nudicaudatus] bolivianus: Trouessart, 1905:854; name combination.
Metachirus nudicaudatus tschudii: Osgood, 1914b:149; first use of current name combination.
[Didelphis (Metachirus)] tschudii: Matschie, 1916:268; name combination.
[Didelphis (Metachirus)] bolivianus: Matschie, 1916:268; name combination.
Metachirus nudicaudatus infuscus O. Thomas, 1923b:606; type locality "Rio Inambari, [Puno,] Peru."

The distribution of this subspecies encompasses the upper Amazon basin of southeastern Colombia, western Brazil, and eastern Peru and Bolivia.

NATURAL HISTORY: *Metachirus nudicaudatus* is nocturnal, solitary, mainly terrestrial, and at least partially frugivorous (Handley 1976; Hunsaker 1977b). Streilein (1982a) considered *M. nudicaudatus* to be predominantly a frugivore that also consumes insects, eggs, and small vertebrates. Cáceres (2004), based on the examination of 44 fecal samples from two sites in the Atlantic forest habitat in Paraná, Brazil, found seeds of three kinds of fruits and the remains of birds, mammals, reptiles and nine kinds of invertebrates including crayfish and snails. As compared to food items he also identified in the feces of *Didelphis aurita* and *Philander frenatus* collected in the same habitat, Cáceres (2004) stated that *M. nudicaudatus* was preponderantly insectivorous.

Metachirus nudicaudatus is found at low density in a variety of forest habitats from sea level to elevations over 2,100 m. Loretto, Ramalho, and Vieira (2005) described the nest of *M. n. myosuros* as a round ball of leaves interlaced with pieces of roots. The nest lacks a readily apparent entrance and is located on the ground within deep leaf litter, sometimes between the roots of trees. Loretto, Ramalho, and Vieira (2005) could not detect a nest from above because the surface of the leaf litter appears level. They located nests by following the thread from individual *M. nudicaudatus* that had been fitted with a spool and line device. Most individuals fled the nest upon discovery; however, two females remained in their individual nest and responded to the disturbance by clicking, hissing, and gnashing teeth while pushing on the top of the nest. Their behavior caused the leaf litter to rise and fall, and combined with the vocalizations, created the impression of a much larger animal.

Lactating females were taken on 11 April in Colombia, 30 April and 1 May in Venezuela, and on 13 September in Peru, based on specimens in the USNM. A female captured on 20 April in Paraguay had four small (eyes closed; head and body length 64 mm) young attached. Another taken 28 October in Peru had nine small (CR = 12 mm) young attached.

Husson (1978) found parts of a skull among the stomach contents of an owl in Surinam. Brennan and Reed (1975) listed two mites as ectoparasites, and R. Guerrero (1985b) listed two nematodes as endoparasites recorded from *M. nudicaudatus*. *Metachirus nudicaudatus* is known from the late Pleistocene of South America (Marshall 1981). The karyotype is $2n = 14$, FN = 20, and the X and Y chromosomes are similar small acrocentrics (Reig and Löbig 1970; Yunis, Cayon, and Ramírez 1973; Reig et al. 1977).

REMARKS: Pine (1973a), on the basis Tiedemann's (1808) description of color, argued that *Philander* Tiedemann, referred not to the pouched, gray or black four-eyed opossum, but to the brown, pouchless four-eyed genus and concluded that the correct name for this species was *Philander nudicaudatus*. This was the basis for Hall's (1981:17) use of *Philander* in place of *Metachirus* for the brown four-eyed species. Pine and Hall dated *Philander* from Tiedemann (1808), each believing that *Philander* Brisson, 1762, was non-Linnaean and, therefore, unavailable. Hershkovitz (1976b) reaffirmed his earlier (1949a) use of *Philander* Tiedemann as the correct generic name for *Didelphis opossum* Linnaeus, 1758, and reviewed the context and content of *Metachirus* Burmeister, 1854. Hershkovitz (1959:343) retracted his designation of one of Wied-Neuwied's specimens of "*Didelphis myosuros*" from "Comechatiba" (= Ponta Cumuruxatiba), Bahia, Brazil (Wied-Neuwied 1826:400), as the lectotype of *Didelphis myosuros* Temminck and concluded that *Didelphys myosurus* (incorrect subsequent spelling used by Burmeister 1854) and not *Didelphis myosuros* Temminck, is the type species (identified by O. Thomas, 1888b:329, as equivalent to *D. nudicaudata* É. Geoffroy St.-Hilaire, 1803) of *Metachirus*. Burmeister (1854) included all "four-eyed opossums" in his subgenus *Metachirus* and did not indicate (page 135) that his name combination, *Didelphys myosurus*, represented a new species. Instead he clearly equated *Didelphys myosurus* with Temminck's *Didelphis myosuros* by citing Temminck's description as his first reference in the synonymy, in which he also included *Didelphis nudicaudatus* É. Geoffroy St.-Hilaire. Hershkovitz (1959, 1976b) assumed *Didelphis myosuros* Temminck had a pouch because Temminck (1824b:40) stated it did; therefore, the name could not apply to the pouchless, Brown Four-eyed Opossum and, instead, must apply to the pouched, gray four-eyed taxon he called *Philander opossum* (currently *P.*

frenatus). Temminck, who had examined specimens in several European museums, either erroneously assumed that all opossums of that size range must have a pouch, or based his name and description on a composite, because the two syntypes of *Didelphis myosuros* Temminck in the Rijksmuseum van Natuurlijke Historie, Leiden, represent the brown, pouchless species (Husson 1978). These specimens (RMNH 26072 and 26073), both males, labeled as from Brazil, and examined by one of us (ALG), were identified by Jentink (1888:220, a; 221, b) under the name *Didelphis nudicaudata* as types of *Didelphis myosuros* Temminck. These syntypes form part of the objective basis for *Didelphis myosuros* Temminck and identify the name as a junior synonym of *Didelphis nudicaudata* É. Geoffroy St.-Hilaire. Although it is tempting to avoid further confusion by designating RMNH 26072 (Jentink, 1888:220, a) as the lectotype of *Didelphis myosurus* Temminck, Pohle (1927) had already designated one of Johann Natterer's specimens from Ypanema (= Bacaetava), São Paulo, Brazil as the type (lectotype). Pohle (1927:242) identified this specimen, a female taken on January 25, 1822, as number 166a of Natterer's collection in the Naturhistorische Museum Wein, Vienna, Austria. A. Miranda-Ribeiro's (1936:351) attempt to restrict the type locality to Bahia is invalid. Cabrera's (1958) citation of Ipanema, São Paulo, Brazil, as the type locality is based on Pohle (1927; also see Pelzeln 1887:111, 125–26).

Barbara Brown (2004:86) followed Hershkovitz (1976b) in stating that *Didelphis nudicaudatus* É. Geoffroy St.-Hilaire, not *Didelphis myosuros* Temminck, is the type species of *Metachirus* Burmeister, and cited O. Thomas's (1888b:329). O. Thomas (1888b) always cited as the type species of a genus, the valid name then in use for that species, even when that name was not available when the genus level taxon was originally described. Therefore, because *Didelphis nudicaudatus* was listed as a synonym of *Didelphys myosurus* Burmeister, which is an incorrect subsequent spelling of *Didelphis myosuros* Temminck and not a new species, and which in turn has as its lectotype a specimen that represents *Metachirus nudicaudatus* (É. Geoffroy St.-Hilaire) identified by Pohle (1927), *Didelphis myosuros* Temminck is the type species of *Metachirus*.

Husson (1978:27) attempted to resolve the situation generated by Pine's (1973a) conclusion that *Philander* Tiedemann was the correct name for the brown, pouchless four-eyed opossums by selecting the female from Virginia discussed by Tyson (1698) as lectotype of *Philander virginianus* Tiedemann. Husson wanted to relegate *Philander* Tiedemann to the synonymy of *Didelphis* Linnaeus; thereby removing the name *Philander* Tiedemann from consideration as the correct generic name for *Metachirus nudicaudatus* (also see Remarks under *Philander opossum*).

However, Hershkovitz (1976b, 1981) earlier had demonstrated that *Philander* Tiedemann referred to the gray, pouched species. The valid name of the brown, pouchless four-eyed opossums remains *Metachirus nudicaudatus* (É. Geoffroy St.-Hilaire).

Authors usually cite É. Geoffroy St.-Hilaire, 1803 as the author and date of publication of *M. nudicaudatus*. Nevertheless, Gardner (1993a) dated the name from Desmarest (1817a) because at the time, Wilson and Reeder (1993:831) considered Geoffroy St.-Hilaire's 1803 Catalog of the Museum Nacional d'Historie Naturelle to be a page proof, and not actually published (also see I. Geoffroy St.-Hilaire 1839:5[footnote], 1847b:115–18; Sherborn 1922:lviii, 1932:cxxxviii). However, É. Geoffroy St.-Hilaire's (1803) Catalog is now on the Official List of Works Approved as Available for Zoological Nomenclature (ICZN 2002).

Krumbiegel (1941b) recognized 12 subspecies, including the Central American *M. n. dentaneus* Goldman, 1912a (type locality Gatún, Canal Zone, Panama), which we are treating as a synonym of *M. n. colombianus*. Cabrera (1958) reduced this number to eight, and we have further reduced the number to five. Krumbiegel (1941b) provided a key for 11 subspecies based mainly on color (he did not include *M. n. antioquiae*). Nevertheless, the number, status, and distributional limits of populations are unresolved (see J. L. Patton, Silva, and Malcolm 2000; J. L. Patton and Costa 2003) and the species needs revision. J. L. Patton and Costa (2003) provided evidence that one or more of these named populations may represent separate species.

Kretzoi and Kretzoi (2000:91) listed *Didelphis myosuros* Liais (= *Didelphis nudicaudatus* É. Geoffroy St.-Hilaire) as the type species of *Cuica* Liais. They apparently overlooked Hershkovitz's (1949a:12) designation of *Cuica murina* Liais (= *Didelphis murina* Linnaeus, 1758) as the type species of *Cuica*.

Tribe Monodelphini Hershkovitz, 1992

Alfred L. Gardner

The Monodelphini include 11 genera characterized by small to medium size, no marsupial pouch, high number of teats in some members (especially *Thylamys* and some species of *Monodelphis*), and a diploid chromosome number of 14, 18 (*Monodelphis*), or 22 (*Tlacuatzin canescens* of Mexico). The karyotypes of *Chacodelphys* and *Hyladelphys* are unknown. Six of the genera were subsumed under the genus *Marmosa* (*sensu lato* of Tate 1933) as treated by Cabrera (1958). The relationship of the genus *Hyladelphys* to other Monodelphini or to other Didelphinae has not been satisfactorily resolved (Jansa and Voss (2005). The non-

molariform morphology of its deciduous last premolar is unique within the Didelphidae.

Genus *Chacodelphys* Voss, Gardner, and Jansa, 2004

Chacodelphys is monotypic and known from only a few specimens, mainly recovered from owl pellets, from Formosa and Chaco provinces, Argentina (Teta, Pardiñas, and D'Elía 2005, 2006). The smallest known American marsupial (total length 123 mm, tail 55 mm, condylobasal length 20.6 mm), *Chacodelphys* is rivaled in size only by the smaller species of *Monodelphis* (e.g., *M. kunsi*). In addition to small size, the following suite of characters distinguish *Chacodelphys* from other didelphids (Voss, Gardner, and Jansa 2004). The fur is bicolored; the only facial markings consist of narrow eye rings. Manual claws are short, digit II is longer that digits II and IV, the central palmar surface is densely covered with small tubercles, and the plantar surface lacks hair on the heel. The tail is shorter than head and body, densely covered in nonpetiolate hairs, and lacks a prehensile tip. Nasals are narrow and not expanded posteriorly, the interorbital region is smooth (lacks supraorbital beading and processes), and the cranium lacks a sagittal crest. The palate is perforated by maxillary, maxillopalatine, and palatine fenestrae; the posterolateral palatine foramina are small. Upper canines lack accessory cusps, P2 is taller than P3, and the anterior cingulum on M3 is incomplete.

SYNONYMS:

Marmosa: Shamel, 1930a:83; not *Marmosa* Gray, 1821.

Thylamys: Cabrera, 1958:33; not *Thylamys* Gray, 1843; used as a subgenus of *Marmosa* Gray.

Gracilinanus: Gardner and Creighton, 1989:5; part.

Chacodelphys Voss, Gardner, and Jansa, 2004:2; type species *Marmosa formosa* Shamel, 1930, by original designation.

Chacodelphys formosa (Shamel, 1930)

Pygmy Opossum

SYNONYMS:

Marmosa muscula Shamel, 1930a:83; type locality "Kilometro 182 (Central Formosa)" (= Riacho Pilagá, 10 miles northwest of Km 182), Formosa, Argentina; preoccupied by *Didelphis (Marmosa) musculus* Cabanis, 1848.

Marmosa formosa Shamel, 1930b:311; replacement name for *M. muscula* Shamel, 1930a, preoccupied.

Marmosa [(*Thylamys*)] *velutina formosa*: Cabrera, 1958:33; name combination.

[*Marmosa (Thylamys)*] *formosa*: Kirsch and Calaby, 1977:14; name combination.

Map 14 Marginal localities for *Chacodelphys formosa* ●, *Cryptonanus guahybae* ▲, and *Cryptonanus ignitus* ■

[*Thylamys*] *formosus*: Reig, Kirsch, and Marshall, 1987:7; name combination.

Gracilinanus agilis: Gardner and Creighton, 1989:5; name combination; not *Didelphys agilis* Burmeister, 1854.

Chacodelphys formosa: Voss, Gardner, and Jansa, 2004:2; first use of current name combination.

Gracilinanus formosus: Gardner, 2005b:7; name combination.

DISTRIBUTION: *Chacodelphys formosa* is known from provincias Chaco and Formosa in northern Argentina.

MARGINAL LOCALITIES (Map 14; from Teta, Pardiñas, and D'Elía 2006, except as noted): ARGENTINA: Formosa, Estancia Linda Vista near Riacho Pilagá (type locality of *Marmosa muscula* Shamel); Formosa, Reserva Ecológica El Bagual (not mapped); Formosa, Herradura; Chaco, Selvas del Río de Oro (not mapped); Chaco, 5 km N of General Vedia.

SUBSPECIES: *Chacodelphys formosa* is monotypic.

NATURAL HISTORY: Known specimens come from humid Chaco habitat. Aside from the holotype, which was likely trapped along with several rodents in a building at the type locality, specimens of *C. formosa* have been recovered from owl pellets (Teta, Pardiñas, and D'Elía 2006). The Reserva El Bagual specimen was taken in a pitfall trap in Chacoan grassland (Teta, Pardiñas, and D'Elía 2006). No additional natural history information is available.

REMARKS: Cabrera (1958:33–34) treated the holotype as a subspecies of *Marmosa velutina* (= *Thylamys*

velutinus). Based on Tate's (1933:232) statement that the specimen was a "Rather young adult male," Cabrera described it as "bastante joven," and that it had the size and cranial characteristics of juvenile *Marmosa*. Voss, Gardner, and Jansa (2004) described the unique combination of morphological features that distinguish *C. formosa* from other known small didelphids. Until recently (Teta, Pardiñas, and D'Elía 2005, 2006), this species was known by only the type specimen from Estancia Linda Vista near Riacho Pilagá, Formosa, prepared by Alexander Wetmore in 1920.

Genus *Cryptonanus* Voss, Lunde, and Jansa, 2005

The genus *Cryptonanus*, as defined by Voss, Lunde, and Jansa (2005), contains five species previously treated (e.g., Gardner 1993a, 2005b) as either species of *Gracilinanus* (*C. agricolai* and *C. ignitus*), synonyms of *G. agilis* (*C. chacoensis* and *C. unduaviensis*), or a subspecies of *G. microtarsus* (*C. guahybae*). These are small (total length less than 260 mm, condylobasal length 23.1–30.0 mm, zygomatic breadth 12.8–17.6 mm), gracile opossum most similar to species of *Gracilinanus*. Best seen in specimens of *Cryptonanus* with relatively unworn teeth, P2 is smaller (shorter in height) than P3; whereas in *Gracilinanus*, P2 and P3 are subequal. Most specimens of *Cryptonanus* lack maxillary fenestrae, and when present, these fenestrae are small. However, most *Gracilinanus* have well-marked maxillary fenestrae; exceptions have been found only in some *G. emiliae* (two of six examined *fide* Voss, Lunde, and Jansa 2005). Specimens of *Cryptonanus* lack the projection of the premaxillaries anterior to incisors (rostral process) that is present in *Gracilinanus*.

SYNONYMS:

Grymaeomys: Winge, 1893:27; part; not *Grymaeomys* Burmeister, 1854.

Marmosa: Tate, 1931:10, 11; part; not *Marmosa* Gray, 1821.

Marmosa: Moojen, 1943:2; not *Marmosa* Gray.

Marmosa (Thylamys): Cabrera, 1958:26; part.

Gracilinanus Gardner and Creighton, 1989:5, 6; part.

Gracilinanus: M. M. Díaz, Flores, and Barquez, 2002:825; not *Gracilinanus* Gardner and Creighton, 1989.

Cryptonanus Voss, Lunde, and Jansa, 2005:5; type species *Cryptonanus chacoensis* (Tate, 1931; original name *Marmosa agilis chacoensis*) by original designation.

KEY TO THE SPECIES OF *CRYPTONANUS* (BASED ON VOSS, LUNDE, AND JANSA 2005).

1. Dorsal pelage shades of brownish gray; all or part of ventral pelage self-colored, not conspicuously gray based
.... 2

1'. Dorsal pelage reddish; pelage of venter distinctly gray based . . . . *Cryptonanus guahybae*

2. Size larger; tail length more than 111 mm (112–135 mm); condylobasal length more than 25.5 mm (25.6–30.0 mm); maxillary toothrow more than 10.0 mm (10.1–11.0 mm); found only in Bolivia and northwestern Argentina 3

2'. Size generally smaller; tail length 95–117 mm; condylobasal length 23.1–26.8 mm; maxillary toothrow 9.2–10.2 mm; distribution not restricted to Bolivia and northern Argentina 4

3. Length of upper molar series (M1–M4) less than 5.5 mm; venter mainly self-colored pale orange; known from only northwestern Argentina . . . . *Cryptonanus ignitus*

3'. Length of upper molar series (M1–M4) more than 5.5 mm; venter either self-colored white, or inconspicuously gray based; currently known from only Bolivia . . . . *Cryptonanus unduaviensis*

4. Anterior cingulum of M3 complete; known from caatinga and cerrado habitats in Brazil . . . . *Cryptonanus agricolai*

4'. Anterior cingulum of M3 incomplete; known from Bolivia, Paraguay, and northern Argentina . . . . *Cryptonanus chacoensis*

Cryptonanus agricolai (Moojen, 1943)
Agricola's Opossum
SYNONYMS:

Grymaeomys pusillus: Winge, 1893:27; part; not *Didelphis pusilla* Desmarest, 1804.

Marmosa agilis agilis: Tate, 1933:195; part; not *Grymaeomys agilis* Burmeister, 1854.

Marmosa agricolai Moojen, 1943:2; type locality "Crato, Ceará," Brazil.

Marmosa [(Thylamys)] agricolai: Cabrera, 1958:28; name combination.

[*Thylamys*] *agricolai*: Reig, Kirsch, and Marshall, 1987:7; name combination.

Gracilinanus emiliae agricolai: Gardner, 1993a:17; name combination.

Gracilinanus agricolai: Gardner, 2005b:7; name combination.

Cryptonanus agricolai: Voss, Lunde, and Jansa, 2005:1; first use of current name combination.

DISTRIBUTION: *Cryptonanus agricolai* is known from the Brazilian states of Ceará, Goiás, and Minas Gerais.

MARGINAL LOCALITIES (Map 15; from Voss, Lunde, and Jansa 2005; except as noted): BRAZIL: Ceará, Crato (type locality of *Marmosa agricolai* Moojen); Goiás, 20 km NW of Colinas do Sul; Goiás, Serra Negra region (not mapped); Minas Gerais, Lagoa Santa.

SUBSPECIES: *Cryptonanus agricolai* is monotypic.

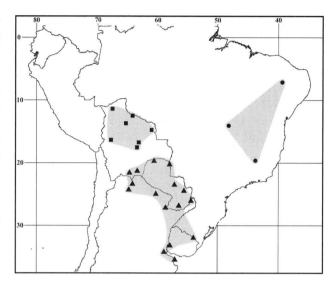

Map 15 Marginal localities for *Cryptonanus agricolai* ●, *Cryptonanus chacoensis* ▲, and *Cryptonanus unduaviensis* ■

NATURAL HISTORY: Other than its capture in xeric habitats (caatinga and cerrado), natural history information is lacking.

REMARKS: Gardner and Creighton (1989) and Gardner (1993a) included *Marmosa agricolai* Moojen, 1943, under *Gracilinanus emiliae*, but later Gardner (2005b) treated the taxon as the species *G. agricolai*. Voss, Lunde, and Simmons (2001) pointed out that the short tail, relative to length of head and body, precluded inclusion of *Marmosa agricolai* in *Gracilinanus*. Voss, Tarifa, and Yensen (2004:Table 2) stated that *agricolai* was under study and would be removed from *Gracilinanus*.

Cryptonanus chacoensis (Tate, 1931)
Chacoan Mouse Opossum
SYNONYMS:

Marmosa agilis chacoensis Tate, 1931:10; type locality "Sapucay" (= Sapucaí), Paraguarí, Paraguay.

Marmosa [(Thylamys)] agilis agilis: Cabrera, 1958:27; part; not *Grymaeomys agilis* Burmeister, 1854.

Gracilinanus agilis: Gardner and Creighton, 1989:5; part; not *Grymaeomys agilis* Burmeister.

Cryptonanus chacoensis: Voss, Lunde, and Jansa, 2005:1; first use of current name combination.

DISTRIBUTION: *Cryptonanus chacoensis* occurs in Paraguay, southern Bolivia, and northern Argentina.

MARGINAL LOCALITIES (Map 15; from Voss, Lunde, and Jansa 2005, except as noted): PARAGUAY: Alto Paraguay, Palmar de las Islas; Alto Paraguay, Estancia Doña Julia; Concepción, Concepción; Canindeyú, 13.3 km by road N of Curuguaty. ARGENTINA: Misiones, Río Urugua-í, 30 km from Puerto Libertad (Massoia and Fornes 1972). PARAGUAY: Caazapa, Estancia Dos Marías. BRAZIL: Rio Grande do Sul, Paso del Duraznero (E. M. González, Claramunt, and Saralegui 1999). URUGUAY: Río Negro, Bopicuá (J. C. González 1985). ARGENTINA: Buenos Aires, Estancia Luis Chico (Sauthier et al. 2005); Buenos Aires, Campana (Massoia and Fornes 1972); Chaco, Las Palmas (Tate 1931); Formosa, Pozo del Tigre (Massoia and Fornes 1972). BOLIVIA: Tarija, Villa Montes (Tate 1931). ARGENTINA (Flores, Días, and Barquez 2000): Salta, Ingenio San Martín del Tabacal; Jujuy, Ingenio La Esperanza. BOLIVIA: Tarija, Tablada (Hershkovitz 1992b).

SUBSPECIES: I am treating *C. chacoensis* as monotypic.

NATURAL HISTORY: Massoia and Fornes (1972) found *C. chacoensis* in wet riparian habitats where the species constructs nest in a variety of situation, such as in tree holes and among clusters of bromeliads. Seven individuals, interpreted as representing a multi-generational family unit, were found in a nest 1.6 m above ground and within 2 m of the Río Paraná. Specimens were trapped at night or caught in their nests during the day, and some records are based on remains recovered from barn owl (*Tyto alba*) pellets. Massoia and Fornes (1972) stated that females may have as many as 12 young, an observation I assume was based either on an actual count of young, or on the number of teats, either of which suggests the specimens may have been misidentified (see Remarks). Capture sites for Paraguayan specimens (Voss, Lunde, and Jansa 2005) are in a hollow log, in a wood pile, on wet ground in a marsh, on the ground in high grass at the edge of a marsh, and on the ground at the base of fruiting bromeliad stalks. E. M. González, Claramunt, and Saralegui (1999) recovered the remains of *C. chacoensis* (reported as *Gracilinanus agilis*) from barn owl pellets in Rio Grande do Sul, Brazil.

REMARKS: Many of the localities listed here under Marginal Localities represent records originally reported as for *Gracilinanus agilis*. Certain discrepancies between the diagnostic features of *Cryptonanus* and those illustrated in the line drawings of the skull provided by Massoia and Fornes (1972), and the number of young (12) they reported, suggest that at least some of the animals Massoia and Fornes called *Marmosa agilis chacoensis* were misidentified. Voss, Lunde, and Jansa (2005) used the name *C. chacoensis* for specimens of *Cryptonanus* having self-colored ventral pelage, a mammae count of 4–1–4 = 9 (not 12 as reported by Massoia and Fornes 1972), small molars, and an incomplete anterior cingulum of M3. Although I have cited Massoia and Fornes's (1972) information, Voss, Lunde, and Jansa (2005) did not, probably because of the likelihood Massoia and Fornes misidentified their specimens. Voss, Lunde, and Jansa (2005) also were unable to confirm

the identifications of three of the four Bolivian specimens S. Anderson (1997) reported as *Gracilinanus agilis chacoensis* because the specimens had been returned to Bolivia. They reidentified the specimen S. Anderson (1997) reported as *Gracilinanus agilis chacoensis* from Chuquisaca as *Marmosops* sp.

Cryptonanus guahybae (Tate, 1931)
Guahyba Mouse Opossum

SYNONYMS:

Marmosa microtarsus guahybae Tate, 1931:10; type locality "Island of Guahyba near Porto Alegre, Rio Grande do Sul, Brazil."

Marmosa [(Thylamys)] microtarsus guahybae: Cabrera, 1958:31; name combination.

Gracilinanus microtarsus: Gardner and Creighton, 1989:6; part; not *Didelphys microtarsus* J. A. Wagner, 1842.

Cryptonanus guahybae: Voss, Lunde, and Jansa, 2005:1; first use of current name combination.

[*Gracilinanus microtarsus*] *guahybae*: Gardner, 2005b:7; name combination.

DISTRIBUTION: *Cryptonanus guahybae* is verified only from the Brazilian state of Rio Grande do Sul.

MARGINAL LOCALITIES (Map 14; Voss, Lunde, and Jansa 2005, except as noted): BRAZIL: Rio Grande do Sul, Taquara; Rio Grande do Sul, Island of Guahyba, near Pôrto Alegre (type locality of *Marmosa microtarsus guahybae* Tate); Rio Grande do Sul, São Lourenço.

SUBSPECIES: *Cryptonanus guahybae* is monotypic.

NATURAL HISTORY: Hensel (1872:124) reported a specimen (the holotype of *C. guahybae*) he identified as *Grymaeomys agilis*, captured under the roots of a tree. Tate (1933:193) remarked on a specimen with 15 apparently functional mammae (7–1–7) extending "from belly to breast."

REMARKS: As Voss, Lunde, and Jansa (2005) stated, *C. guahybae* can be distinguished from congeners by its reddish dorsum and gray-based, buffy venter. The mammae count (7–1–7) is the highest for the genus and includes three pairs that are pectoral (illustrated by Tate 1933:35).

Cryptonanus ignitus (M. M. Díaz, Flores, and Barquez, 2002)
Red-bellied Mouse Opossum

SYNONYMS:

Gracilinanus ignitus M. M. Díaz, Flores, and Barquez, 2002:825; type locality "Yuto, Departamento Ledesma, Jujuy Province, Argentina (23°38′S, 64°28′W)."

Cryptonanus ignitus: Voss, Lunde, and Jansa, 2005:1; first use of current name combination.

DISTRIBUTION: *Cryptonanus ignitus* is known from only the type locality in northwestern Argentina.

MARGINAL LOCALITIES (Map 14): ARGENTINA: Jujuy, Yuto (type locality of *Gracilinanus ignitus* M. M. Díaz, Flores, and Barquez).

SUBSPECIES: *Cryptonanus ignitus* is known by only the type specimen.

NATURAL HISTORY: Unknown.

REMARKS: The holotype (AMNH 167852) was taken in 1962, and M. M. Díaz, Flores, and Barquez (2002) stated they did not know if the taxon still occurred in the region. They provided information useful for distinguishing *C. ignitus* from *Gracilinanus agilis* and *G. microtarsus*. A few of the salient features characterizing *C. ignitus* include reddish to orange, unicolored hair of underparts, except for a narrow medial patch of white on the chest; a sparsely haired, weakly bicolored tail that is a little longer than head and body. Voss, Lunde, and Jansa (2005) commented on the large size of the holotype, its long canines, broad and robust zygomatic arches, the massive development of the postorbital process of the jugal, and the small palatine fenestrae, all of which could be a function of it being an old male. Instead of treating the holotype as representing a very old male *C. chacoensis*, they retained *C. ignitus* as species on the basis of its unique ventral coloration.

Cryptonanus unduaviensis (Tate, 1931)
Unduave Mouse Opossum

SYNONYMS:

Marmosa unduaviensis Tate, 1931:11; type locality "Pitiguaya, Rio Unduavi, Yungas," La Paz, Bolivia.

Marmosa [*(Thylamys)*] *unduaviensis*: Cabrera, 1958:33; name combination.

[*Thylamys*] *unduaviensis*: Reig, Kirsch, and Marshall, 1987: 7; name combination.

Gracilinanus agilis: Gardner and Creighton, 1989:5; part; not *Didelphys agilis* Burmeister, 1854.

Marmosa undaviensis Hershkovitz, 1992b:34; incorrect subsequent spelling of *Marmosa unduaviensis* Tate; in synonymy of *Gracilinanus agilis*.

Cryptonanus unduaviensis: Voss, Lunde, and Jansa, 2005: 1; first use of current name combination.

DISTRIBUTION: *Cryptonanus unduaviensis* is known from Bolivia.

MARGINAL LOCALITIES (Map 15; from Voss, Lunde, and Jansa 2005, except as noted): BOLIVIA: Pando, Independencia; Beni, Pampa de Meio; Santa Cruz, El Refugio; Santa Cruz, Estancia Cachuela Esperanza; Santa Cruz, 2 km S of Caranda; La Paz, Pitiguaya (type locality of *Marmosa unduaviensis* Tate); Beni, Puerto Caballo.

SUBSPECIES: I treat *C. unduaviensis* as monotypic.

NATURAL HISTORY: Voss, Lunde, and Jansa (2005: Table 13) gave capture information on three specimens: one was trapped at ground level on a tree island surrounded by seasonally flooded grassland; another was trapped on the ground in grass at the edge of a marshy stream, and the third was shot in a house. No other natural history information is available.

REMARKS: Voss, Tarifa, and Yensen (2004) reidentified as *Marmosops noctivagus* three of the specimens S. Anderson (1997) listed as *Gracilinanus agilis unduaviensis*. Voss, Lunde, and Jansa (2005) found that most of the Bolivian material they referred to *C. unduaviensis* had been listed by S. Anderson (1997) under the name *Gracilinanus agilis buenavistae*.

Genus *Gracilinanus* Gardner and Creighton, 1989

G. Ken Creighton and Alfred L. Gardner

Gracilinanus contains six species (*G. aceramarcae, G. agilis, G. dryas, G. emiliae, G. marica,* and *G. microtarsus*) distributed from Colombia to northern Argentina in the tropical and subtropical zone east of the Andean cordillera below 3,000 m. Fossils referable to *Gracilinanus* are known from the late Pleistocene to Recent cave deposits in the vicinity of Lagoa Santa, Minas Gerais, Brazil (Winge 1893).

These mouse opossums are small (head and body 85–130 mm, tail 90–150 mm) and have a ratio of tail length to head-and-body length always greater than 1.3, but usually less than 1.5. Dorsal pelage ranges from bright reddish brown to a dull brownish gray. The tail has small (more than 40 per cm), rounded to square scales in annular rows. Triplet interscalar hairs are subequal in length with the middle hair two to three scale rows long. The middle hair has about twice the diameter of the lateral hairs, but is not as stout and petiolate as in *Marmosops*. The tail is weakly bicolored in some species (e.g., *G. agilis,* and *G. marica*) and unicolored fuscous in others (e.g., *G. aceramarcae, G. dryas,* and *G. microtarsus*). Claws on the manus do not extend beyond terminal digital pads. The central palmar surfaces lack the granular or tuberculate appearance of *Thylamys* and *Monodelphis*. All palmar and plantar pads (described in Creighton 1984) are present and separated by at least a double row of tubercles. Sparse tubercles are present on the central plantar surface. Tubercles on the proximal ventral surfaces of the digits are fused into transverse bars. The skull lacks postorbital processes, and the supraorbital margin of the frontals may be beaded in larger (older) individuals of some species. The lambdoidal crest is weakly developed or absent except in some older, larger individuals. The hard palate is highly fenestrated, usually with three sets of palatal fenestrae (maxillary, maxillopalatine, and palatine; Voss and Jansa 2003:Fig. 5). The posterolateral palatal foramina are moderate in size, usually about 1/3 to 1/2 the breadth of M4 in length. The nasals are moderately expanded laterally at the maxillofrontal suture. Auditory bullae are relatively large, compared to those of *Marmosa*, *Micoureus*, *Marmosops*, and *Tlacuatzin*, but proportionately smaller than in *Thylamys*. The petrosal is usually exposed between the squamosal and parietal. A slender anteromedial process of the alisphenoid portion of the auditory bulla is present, although frequently damaged during specimen preparation.

Upper incisors increase slightly in size from I2 through I5. The lower canines are relatively short compared with those of *Marmosa*, *Micoureus*, *Tlacuatzin*, and *Thylamys*, but neither as short nor as premolariform as in *Marmosops*. The second upper premolar (P2) is approximately equal in size to P3.

SYNONYMS:

Didelphis: J. A. Wagner, 1842:359; not *Didelphis* Linnaeus, 1758.

Grymaeomys Burmeister, 1854:130; part; proposed as a subgenus of *Didelphis* Linnaeus.

Marmosa: O. Thomas, 1898c:455, 456; not *Marmosa* Gray.

Marmosa: O. Thomas, 1909a:379; not *Marmosa* Gray, 1821.

Marmosa: O. Thomas, 1910c:502; not *Marmosa* Gray.

Marmosa: Cabrera, 1919:34; part; not *Marmosa* Gray.

Marmosa: Tate, 1931:10–12; not *Marmosa* Gray.

Marmosa: A. Miranda-Ribeiro, 1936:373, 382; not *Marmosa* Gray.

Thylamys: A. Miranda-Ribeiro, 1936:387; not *Thylamys* Gray, 1843b.

Marmosa: Moojen, 1943:2; not *Marmosa* Gray.

Thylamys: Cabrera, 1958:26; part; not *Thylamys* Gray.

Tylamys Ávila-Pires, 1968:167; incorrect subsequent spelling of *Thylamys*; not *Thylamys* Gray.

Gracilinanus Gardner and Creighton, 1989:4; type species *Didelphys microtarsus* J. A. Wagner, 1842, by original designation.

REMARKS: Species of *Gracilinanus* can be distinguished from *Marmosops* spp. by the shape and arrangement of tail scales and bristles, the nonpremolariform lower canine, and the presence of maxillary fenestrae. Species of *Gracilinanus* can be distinguished from *Thylamys* spp. by the absence of fat storage in the tail, the absence of densely granular central palmar and plantar surfaces, the presence of a rostral process, and the relatively longer digits and broader interdigital pads on both manus and pes. *Gracilinanus* spp. differ from members of the genera *Marmosa*, *Micoureus*, and *Tlacuatzin* by lacking postorbital processes and by the annular arrangement of minute scales on the tail. Originally described as a species of *Gracilinanus*, *Hyladelphys kalinowskii* (Hershkovitz, 1992b) lacks palatine fenestrae, has a shorter rostrum, a conspicuously constricted postorbital region, only four mammae, and (unique for the Didelphidae) smaller, nonmolariform deciduous premolars. Species of *Gracilinanus* can be distinguished from species of *Cryptonanus* by the presence of maxillary fenestrae and a rostral process and by subequal P2 and P3. Species of *Cryptonanus* lack a rostral process, lack well-developed maxillary fenestrae, and the P3 is larger than P2 (Voss, Lunde, and Jansa 2005). *Chacodelphys formosa* (Shamel, 1930b), although included in *Gracilinanus* as a synonym of *G. agilis* by Gardner and Creighton (1989) and by Gardner (1993a, 2005b), is most easily distinguished from species of *Gracilinanus* by its smaller size, much shorter tail, lack of a caudal prehensile surface, narrow nasals with parallel lateral margins, and low entoconid on the lower molars.

KEY TO SPECIES OF *GRACILINANUS*:

1. Dorsal pelage long and lax, chestnut brown in color, and with numerous, conspicuous guard hairs on rump; ventral pelage entirely gray-based with hair tips buffy brown 2
1'. Dorsal pelage reddish brown, pale brown, or grayish brown and lacking conspicuous guard hairs on rump; ventral pelage self-colored creamy white or, if gray-based, then with white or buffy-white tips 3
2. Tail faintly bicolored, paler ventrally; species known from only the Andes of central Colombia and western Venezuela . . . . *Gracilinanus dryas*
2'. Tail unicolored fuscous; species known from only southeastern Peru and the Yungas of Bolivia . . . . *Gracilinanus aceramarcae*
3. Dorsal pelage warm reddish brown; ventral pelage creamy white (hairs not gray based); tail relatively long, ratio of tail length to head-and-body length more than 1.6 . . . . *Gracilinanus emiliae*
3'. Dorsal pelage pale brown, grayish brown, or reddish brown; if reddish brown, ventral pelage entirely gray based; tail relatively shorter 4
4. Dorsal pelage reddish brown to chestnut brown; ventral pelage entirely gray based; known from only southeastern Brazil . . . . *Gracilinanus microtarsus*
4'. Dorsal pelage mottled brown or grayish brown; ventral pelage either gray based or self-colored creamy white 5
5. Dorsal pelage mottled brown to reddish brown, fur texture somewhat wavy in appearance; entire ventral pelage gray basally; distribution north of Rio Amazonas . . . . *Gracilinanus marica*
5'. Dorsal pelage short, from pale brown to grayish brown; ventral pelage self-colored creamy white or buff; or, if

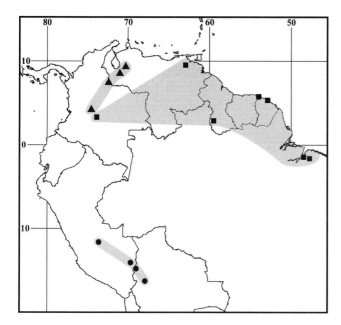

Map 16 Marginal localities for *Gracilinanus aceramarcae* ●, *Gracilinanus dryas* ▲, and *Gracilinanus emiliae* ■

ventral pelage gray based, undersides of limbs usually covered with self-colored hairs; distribution south of Rio Amazonas . . . . *Gracilinanus agilis*

Gracilinanus aceramarcae (Tate, 1931)
Aceramarca Opossum
SYNONYMS:

Marmosa aceramarcae Tate, 1931:12; type locality "Rio Aceramarca, tributary of Rio Unduavi, Yungas," La Paz, Bolivia.

Marmosa [(Thylamys)] aceramarcae: Cabrera, 1958:26; name combination.

[*Thylamys*] *aceramarcae*: Reig, Kirsch, and Marshall, 1987: 7; name combination.

Gracilinanus aceramarcae: Gardner and Creighton, 1989: 5; first use of current name combination.

DISTRIBUTION: *Gracilinanus aceramarcae* is known from southeastern Peru and departamento La Paz, Bolivia.

MARGINAL LOCALITIES (Map 16): PERU: Cusco, RAP Camp 1, Cordillera Vilcabamba (Emmons, Luna, and Romo 2001); Puno, Agualani (Salazar-Bravo et al. 2002). BOLIVIA, La Paz, Llamachaque (Salazar-Bravo et al. 2002); La Paz, Río Aceramarca, tributary of Río Unduavi (type locality of *Marmosa aceramarcae* Tate).

SUBSPECIES: We consider *G. aceramarcae* to be monotypic.

NATURAL HISTORY: According to Tate's notes (AMNH), specimens were trapped on low limbs and vines near a stream in the upper Unduavi Valley of Bolivia at tree line in disturbed cloud forest characterized by a bam-

boo (*Chusquea* sp.) understory and a mossy ground cover. Salazar-Bravo et al. (2002) mentioned trapping a specimen on a moss-covered limb about 2 m above ground near Cuticucho, at 2,800 m in the Zongo Valley of La Paz, Bolivia. The Cordillera Vilcabamba site is at 3,350 m and the habitat consists of a mixture of pajonales, sphagnum bogs, mixed-species forest, and *Polylepis* forest on a "blocky" limestone substrate.

REMARKS: Tate (1931, 1933) stated that *G. aceramarcae* strongly resembled *G. dryas* of Venezuela. The species also is similar to *G. agilis* from which it can be distinguished by its longer pelage with prominent guard hairs, smaller molars (length of upper molar series, M1–M3, less than 5 mm), relatively narrower cranium and zygomatic breadth, and relatively larger and more closely placed auditory bullae. Specimens from the type locality have been collected twice, first in 1926 and later in 1979. Despite moderate collecting efforts elsewhere in the Unduavi Valley and in the Yungas region of La Paz and Cochabamba, this Bolivian population is still known from only the type locality and may be highly endangered. The Puno, Peru record was first mentioned by V. Pacheco et al. (1995), but they did not cite a locality. S. Solari et al. (2001b) listed a *Gracilinanus* cf. *aceramarcae* from Río Abiseo (Río Abiseo National Park, San Martín, Peru); however, the identity of this specimen needs to be confirmed.

Gracilinanus agilis (Burmeister, 1854)
Agile Opossum
SYNONYMS:

Didelphys [(Grymaeomys)] agilis Burmeister, 1854:139; type locality "Lagoa Santa," Minas Gerais, Brazil.

Marmosa beatrix O. Thomas, 1910c:502; type locality "Ipu," Ceará, Brazil.

[*Didelphis (Grymaeomys)*] *agilis*: Matschie, 1916:270; name combination.

[*Didelphis (Grymaeomys)*] *beatrix*: Matschie, 1916:270; name combination.

[*Marmosa (Marmosa)*] *beatrix*: Cabrera, 1919:36; name combination.

Marmosa agilis buenavistae Tate, 1931:10; type locality "Buenavista, Department of Santa Cruz, Bolivia."

Marmosa agilis peruana Tate, 1931:11; type locality "Tingo Maria, Rio Huallaga," Huánuco, Peru.

Marmosa agils peruania Tate, 1933:legend for plate 11; incorrect subsequent spelling of *Marmosa agilis peruana* Tate.

Marmosa blaseri A. Miranda-Ribeiro, 1936:373; type locality "S. Bento, Goias," Brazil.

Thylamys rondoni A. Miranda-Ribeiro, 1936:387; type localities "Salto do Sepotuba e S. João da Serra do Norte," Mato Grosso, Brazil; type locality restricted to Salto

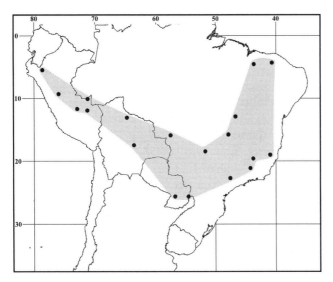

Map 17 Marginal localities for *Gracilinanus agilis* ●

do Sepotuba by selection of lectotype (Moojen in P. Miranda-Ribeiro 1955:417).

Marmosa [(Thylamys)] agilis: Cabrera, 1958:27; name combination.

Marmosa [(Thylamys)] agilis beatrix: Cabrera, 1958:27; name combination.

Marmosa [(Thylamys)] agilis buenavistae: Cabrera, 1958: 28; name combination.

Marmosa [(Thylamys)] agilis peruana: Cabrera, 1958:28; name combination.

[Thylamys] agilis: Reig, Kirsch, and Marshall, 1987:7; name combination.

Gracilinanus agilis: Gardner and Creighton, 1989:5; first use of current name combination.

DISTRIBUTION: *Gracilinanus agilis* occurs in Brazil, eastern Peru, eastern Bolivia, Paraguay, Uruguay, and adjacent Argentina.

MARGINAL LOCALITIES (Map 17): BRAZIL: Ceará, Ipu (type locality of *Marmosa beatrix* O. Thomas); Minas Gerais, Fazenda Montes Claros (Fonseca and Kierulff 1988); Minas Gerais, Lagoa Santa (type locality of *Didelphys agilis* Burmeister); Minas Gerais, Brumado (Tate 1933); São Paulo, Piracicaba (MZUSP 1542). PARAGUAY: Alto Paraná, Puerto Bertoni (MZUSP 2821); Paraguarí, Sapucay (Tate 1933). BOLIVIA: Santa Cruz, Buena Vista (type locality of *Marmosa agilis buenavistae* Tate). PERU: Madre de Dios, Pakitza (V. Pacheco et al. 1993); Cusco, Pagoreni (S. Solari et al. 2001c); Huánuco, Tingo María (type locality of *Marmosa agilis peruana* Tate); Amazonas, Bellavista (Tate 1933); Ucayali, Balta (Voss and Emmons 1996). BOLIVIA: Beni, San Joaquín (Hershkovitz 1992b [part, *fide* Voss, Lunde, and Jansa 2005]). BRAZIL: Mato Grosso, Salto do Sepotuba (type locality of *Thylamys ron-*

doni A. Miranda-Ribeiro; Goiás, São Bento (type locality of *Marmosa blaseri* Miranda-Ribeiro); Distrito Federal, Brasília (Hershkovitz 1992b); Tocantins, Cana Brava (C. O. C. Vieira 1955); Maranhão, Côcos (Hershkovitz 1992b).

SUBSPECIES: We treat *G. agilis* as monotypic. The species, even as restricted by Voss, Lunde, and Jansa (2005), likely is composite and needs revision.

NATURAL HISTORY: *Gracilinanus agilis* inhabits the lowland forest understory. The highest elevation of known occurrence is at approximately 1,800 m in Bolivia. Specimens from Paraguay have most frequently been taken on slender limbs and vines within 2 m of the ground in humid forest or gallery forest habitats. Unlike species of *Thylamys*, *G. agilis* is infrequently taken on fallen logs or on the ground. Although S. Anderson (1997) summarized the literature on fleas recorded from Bolivian specimens, the identity of these hosts needs to be verified.

REMARKS: The dorsal color pattern in *G. agilis* varies from dull or pale reddish brown to dusky gray, and the ventral pelage usually is creamy white or buff with some or all hairs inconspicuously gray based. This species is separable from the other sympatric small opossums by the annular arrangement of minute tail scales, a fully fenestrated palate (including maxillary fenestrae), relatively large and rounded auditory bullae bearing an anteromedial process of the alisphenoid, P2 approximately equal in size to P3, and the lack of supraorbital processes. Externally, the relatively longer, non-incrassate tail distinguishes *G. agilis* from *Thylamys pusillus*, and smaller size distinguishes it from *T. macrurus*.

We are not aware of any cranial features that consistently separate *G. agilis* and *G. microtarsus*. Generally, *G. agilis* inhabits the dryer woodlands of Brazil and the seasonally dry forests of Paraguay, Bolivia, and northern Argentina. *Gracilinanus microtarsus* is found in the more mesic Atlantic forest of Brazil, from Minas Gerais south into Rio Grande do Sul. The application of the name *agilis* to this species was question by J. L. Patton and Costa (2003) because they found a specimen of the quite similar *G. microtarsus* at Lagoa Santa, Minas Gerais, Brazil (type locality of *Didelphys agilis* Burmeister), where previously *G. microtarsus* was not known to occur. This complex of slender opossums needs to be revised.

We have not included the specimen Hershkovitz (1992b: 36) reported from Unguía, Chocó, Colombia, because we doubt that it represents *G. agilis*. The closest record is Bellavista, Amazonas, Peru, nearly 14° of latitude south of Unguía, and is based on three skins without skulls that were only tentatively referred to *Marmosa agilis peruana* by Tate (1933:199) and also may prove not to be true *G. agilis*. Although indicated as found in the Pacific, Amazonian, and

Orinoco zones of Colombia by Cuartas and Muñoz (2003), we are unaware of any specimens from Colombia assigned to *G. agilis*, other than the one cited by Hershkovitz (1992b) and B. E. Brown (2004). The string of unidentified Colombian localities along the border with Ecuador and Peru, as mapped by Cuartas and Muñoz (2004:Map 3.7), appears to be undocumented.

We also have included only two locality records for Bolivia (Buena Vista, departamento Santa Cruz, and San Joaquín, departamento Beni; see Voss, Lunde, and Jansa 2005). We have not confirmed the identifications of other Bolivian specimens reported by S. Anderson (1997) and Hershkovitz (1992b; e.g., Mt. Sajama, departamento Oruro); therefore, their records are not included here.

Gracilinanus dryas (O. Thomas, 1898)
Wood-sprite Opossum

SYNONYMS:

Marmosa dryas O. Thomas, 1898c:456; type locality "Culata, Merida, Venezuela."

[*Didelphys (Marmosa)*] *dryas*: Trouessart, 1898:1241; name combination.

[*Didelphis (Grymaeomys)*] *dryas*: Matschie, 1916:270; name combination.

[*Marmosa (Marmosa)*] *dryas*: Cabrera, 1919:36; name combination.

Marmosa [*(Thylamys)*] *dryas*: Cabrera, 1958:28; name combination.

[*Thylamys*] *dryas*: Reig, Kirsch, and Marshall, 1987:7; name combination.

Gracilinanus dryas: Gardner and Creighton, 1989:6; first use of current name combination.

DISTRIBUTION: *Gracilinanus dryas* is recorded only from the Andes of western Venezuela, and from one locality in the Andes of Colombia.

MARGINAL LOCALITIES (Map 16): VENEZUELA (Handley 1976): Trujillo, Hacienda Misisí, 14 to 15 km E of Trujillo; Mérida, 6 km ESE of Tabay; Táchira, Buena Vista, near Páramo de Tamá, 41 km SW San Cristobal. COLOMBIA: Cundinamarca, Boqueron de San Francisco (Handley and Gordon 1979).

SUBSPECIES: We consider *G. dryas* to be monotypic.

NATURAL HISTORY: *Gracilinanus dryas* is known from the Mérida Andes of Venezuela and one site in departamento de Cundinamarca, Colombia, where it occurs in cloud forest and montane wet forest at elevations from 2,300 to 4,000 m (Tate 1933; Handley 1976; Handley and Gordon 1979; Timm and Price 1989). Handley (1976) reported five specimens caught on the ground and five trapped in trees in Venezuela. Timm and Price (1989) described *Cummingsia micheneri* as a new species of louse recovered from Venezuelan specimens of *G. dryas* (reported as

Marmosa dryas) from Hacienda Misisí, Trujillo, and from Tabay, Mérida.

REMARKS: The dorsal pelage is long, silky, and rich chestnut brown with numerous conspicuous guard hairs on the rump that create a slightly frosted appearance. The ventral pelage is long, gray basally, and brown at the tips. The dorsal and ventral colors merge imperceptibly along the sides of the body. The species can be distinguished from *G. marica* by its longer, darker pelage, more prominent guard hairs, and by the more elongate and relatively narrower skull. The coronoid process is hooked and notably narrower and more pointed than in *G. marica* or *Marmosops parvidens*. Although similar to *G. dryas* in size and color, *M. parvidens* is easily identified by its white venter and premolariform lower canine.

The specimens from Colombia warrant reexamination to confirm their identity as *G. dryas*. Although we are unaware of additional Colombian material in collections, Cuartas and Muñoz (2003:Map 3.7) indicated several localities for the species.

Gracilinanus emiliae (O. Thomas, 1909)
Emilia's Opossum

SYNONYMS:

Marmosa emiliae O. Thomas, 1909a:379; type locality "Para, Brazil."

[*Didelphis (Grymaeomys)*] *emiliae*: Matschie, 1916:270; name combination.

[*Marmosa (Marmosa)*] *emiliae*: Cabrera, 1919:36; name combination.

Marmosa [*(Thylamys)*] *emiliae*: Cabrera, 1958:30; name combination.

Gracilinanus emiliae: Gardner and Creighton, 1989:6; first use of current name combination.

Gracilinanus longicaudus Hershkovitz, 1992b:38; type locality "Los Micos, San Juan de Arama, 03°20′N, 73°53′W, Meta, Colombia, 396 m."

DISTRIBUTION: *Gracilinanus emiliae* is known from Colombia, Venezuela, Guyana, Surinam, French Guiana, and northeastern Brazil (see Remarks).

MARGINAL LOCALITIES (Map 16; from Voss, Lunde, and Simmons 2001, except as noted): VENEZUELA: Monagas, 47 km SE of Maturín. SURINAM: Marowijne, Langamankondre. FRENCH GUIANA: Paracou. BRAZIL: Pará, Belém (type locality of *Marmosa emiliae* O. Thomas); Pará, Capim. GUYANA: Upper Takutu-Upper Essequibo, 12 km E of Dadanawa. COLOMBIA: Meta, Los Micos (type locality of *Gracilinanus longicaudus* Hershkovitz).

SUBSPECIES: We treat *G. emiliae* as monotypic.

NATURAL HISTORY: Virtually nothing is known of the natural history of *G. emiliae*. The relatively bright reddish dorsal pelage and exceptionally long tail suggest arboreality in humid forest habitat. Specimens from northeastern Brazil appear to be associated with gallery forests of major rivers. The specimen from Paracou (Voss, Lunde, and Simmons 2001) was shot from a height of about 4 m in dense secondary growth bordering a road.

REMARKS: Of the expected sympatric species, *G. emiliae* is most likely to be confused with *Marmosa lepida*, from which it can be separated by the annular pattern of tail scales, presence of an anteromedial alisphenoid process on the auditory bullae, smooth lateral margins of the frontals, and the lack of supraorbital processes. The specimens identified by B. D. Patterson (1992:8) as *G. emiliae* proved, on reexamination (Voss, Lunde, and Simmons 2001), to be *M. lepida*. B. E. Brown (2004:46) included these specimens under *G. emiliae* (the two western-most dots in Brazil in her distribution map [Fig. 20] of *G. emiliae*). B. E. Brown's eastern Brazilian locality represents Crato, Ceará, the type locality of *Marmosa agricolai* (= *Cryptonanus agricolai*). This leaves the specimens from Santarém (Tate 1933) and Km 50, Manaus-Itacoatiara railroad (Ávila-Pires 1964) along the Rio Amazonas whose identifications are as yet unconfirmed. The remaining localities mapped by B. E. Brown (2004:Fig 20) are represented by specimens identified by Voss, Lunde, and Simmons (2001) as *G. emiliae*. The identity of the male, reported by Guedes et al. (2000) from Parque Nacional de Ubajara, Ceará, Brazil, needs to be confirmed. On geographic grounds, the specimen likely represents *G. agilis*.

Gracilinanus marica (O. Thomas, 1898)
Northern Gracile Opossum
SYNONYMS:

Marmosa marica O. Thomas, 1898c:455; type locality "R. Albarregas, Merida, Venezuela."

[*Didelphys (Marmosa)*] *marica*: Trouessart, 1898:1241; name combination.

[*Didelphis (Grymaeomys)*] *marica*: Matschie, 1916:270; name combination.

[*Marmosa (Marmosa)*] *marica*: Cabrera, 1919:37; name combination.

Marmosa [*(Thylamys)*] *marica*: Cabrera, 1958:31; name combination.

[*Thylamys*] *maricus*: Reig, Kirsch, and Marshall, 1987:7; name combination.

Gracilinanus marica: Gardner and Creighton, 1989:6; first use of current name combination.

Gracilinanus perijae Hershkovitz, 1992b:41; type locality "Las Marimondas . . . , La Guajira, 1450 m," Colombia.

DISTRIBUTION: *Gracilinanus marica* occurs in Colombia and Venezuela.

MARGINAL LOCALITIES (Map 18): COLOMBIA: Guajira, Las Marimondas (type locality of *Gracilinanus*

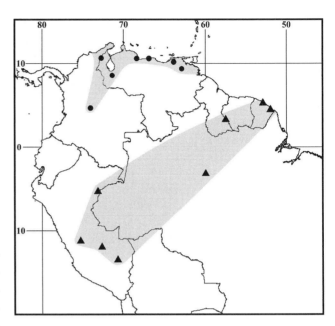

Map 18 Marginal localities for *Gracilinanus marica* ● and *Hyladelphys kalinowskii* ▲

perijae Hershkovitz). VENEZUELA: Falcón, 19 km NW of Urama (Handley 1976); Distrito Federal, 5 km N of Caracas (USNM 370045); Sucre, Mt. Turumiquire (Tate 1933); Monagas, Hato Mata de Bejuco, 55 km SSE of Maturín (Handley 1976); Mérida, Cafetal de Milla (Tate 1933). COLOMBIA: Cundinamarca, La Selva, near Bogotá (Handley and Gordon 1979).

SUBSPECIES: We consider *G. marica* to be monotypic.

NATURAL HISTORY: The seven Venezuelan specimens reported by Handley (1976) came from humid evergreen forest, deciduous forest, and savannah edge habitats at elevations from near sea level to over 2,100 m. Six of these specimens were trapped in trees. Hershkovitz (1992b) caught two at the base of trees in a coffee plantation at Las Marimondas, Colombia.

REMARKS: The dorsal pelage is brown to reddish brown and the texture, combined with the color pattern, creates a mottled and wavy appearance. The ventral pelage is conspicuously buffy, although the hairs are entirely gray-based except on the chin. Lunde and Schutt (2000) found that adult male *G. marica*, the only species of *Gracilinanus* they examined, have an enlarged ulnar tubercle resulting from hypertrophy of the pisiform bone of the carpus, a sexually dimorphic feature they also found in at least four species of *Marmosops*.

Diagnostic characters of the genus serve to separate *G. marica* from other sympatric didelphids except *G. dryas*. *Gracilinanus marica* is distinguishable from the latter by the lack of prominent guard hairs on the rump and by the broader, relatively shorter palate, and the relatively large posterolateral palatal foramina (nearly 1/2 the width of M4 in length). The skull is broader in all aspects, particularly in the widely flared zygomatic arches.

Except for slightly redder dorsal coloration and slightly paler venter, the holotype and paratype of *G. perijae* Hershkovitz are indistinguishable externally from *G. marica*. The lower, flatter cranium described for the holotype of *G. perijae* by Hershkovitz (1992b) is substantiated by additional measurements (9.1 mm versus 9.4–9.6 mm for three male *G. marica*; the skull of the paratype is fragmented). However, all other characters are shared by one or more specimens of Venezuelan *G. marica*; therefore, we consider *G. perijae* a synonym. The name is available for the northern Colombian population if it proves subspecifically distinct. Inclusion of Las Marimondas, La Guajira, Colombia, in the distributions of both *G. marica* and *G. perijae* by B. E. Brown (2004:47, 49) implies sympatry; however, both citations refer to only the specimens on which Hershkovitz (1992b) based the name *G. perijae*.

Gracilinanus microtarsus (J. A. Wagner, 1842)
Brazilian Gracile opossum

SYNONYMS:

Didelphys microtarsus J. A. Wagner, 1842:359; type locality "Ypanema," São Paulo, Brazil.

Grymaeomys microtarsus: Winge, 1893:24; name combination.

Marmosa microtarsus: O. Thomas, 1900c:546; name combination.

Marmosa microtarsus microtarsus: Tate, 1933:190; name combination.

Marmosa herhardti A. Miranda-Ribeiro, 1936:382; type locality "Humboldt," Santa Catarina, Brazil.

Marmosa [(Thylamys)] microtarsus: Cabrera, 1958:31; name combination.

[Thylamys] microtarsus: Reig, Kirsch, and Marshall, 1987: 7; name combination.

Gracilinanus microtarsus: Gardner and Creighton, 1989:6; first use of current name combination.

DISTRIBUTION: *Gracilinanus microtarsus* is endemic to southeastern Brazil, where it is known from Minas Gerais south into Santa Catarina.

MARGINAL LOCALITIES (Map 19; localities listed from north to south): BRAZIL: Minas Gerais, Lagoa Santa (J. L. Patton and Costa 2003); São Paulo, Olimpia (MZUSP 3739); Rio de Janeiro, Rio de Janeiro (Tate 1933); São Paulo, Parque Florestal do Itapetinga (Martins and Bonato 2004); São Paulo, Ypanema (type locality of *Didelphys microtarsus* J. A. Wagner); Paraná, Umbará (MZUSP 11849); Santa Catarina, Blumenau (Tate 1933).

SUBSPECIES: We treat *G. microtarsus* as monotypic, pending revision.

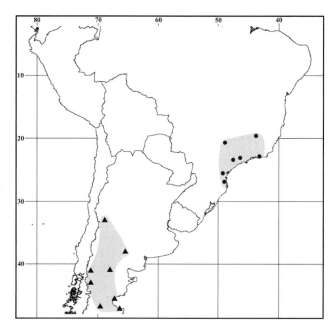

Map 19 Marginal localities for *Gracilinanus microtarsus* ● and *Lestodelphys halli* ▲

NATURAL HISTORY: The reddish brown dorsal pelage and conspicuously buffy, gray-based ventral pelage of *G. microtarsus*, along with its historical range, identify the species a member of the coastal rainforest fauna of southeastern Brazil. Martins and Bonato (2004) analyzed the stomach and gut contents of five *G. microtarsus* from secondary Atlantic rainforest in the state of São Paulo. The were able to identify the remains of a snail, three spiders, and six orders of insects, of which curculionid beetles were the most common prey item recovered. Cáceres and Pichorim (2003) found a *G. microtarsus* occupying the abandoned nest chamber of a piculet (Aves: Picidae). The opossum had carried dry leaves to the nest chamber, which was in the hollow trunk of a *Baccharis* sp. snag.

REMARKS: Tate (1933:192) noted that 19th- and early 20th-century specimens were frequently misidentified as *Marmosa pusilla* (= *Thylamys pusillus* Desmarest). *Gracilinanus microtarsus* is easily distinguished from *Thylamys velutinus* by the lack of fat deposits in the tail and by its larger, fuscous-colored feet, which lack densely tuberculate palmar and plantar surfaces. *Gracilinanus microtarsus* is distinguishable from *G. agilis* principally by the somewhat longer, more reddish dorsal pelage and by the entirely gray-based ventral pelage. We have found no trenchant cranial features to separate these species. J. L. Patton and Costa (2003) commented that finding a specimen of *G. microtarsus* at Lagoa Santa suggested the possibility that the type of *G. agilis* (Burmeister, 1854), also from Lagoa Santa, might prove to be a specimen of *G. microtarsus* as understood today. If additional research shows that both names represent *G. microtarsus* (the older name), *G. beatrix* is next available name for the species known today as *G. agilis*. This complex of small opossums needs to be revised.

Genus *Hyladelphys* Voss, Lunde, and Simmons, 2001

Alfred L. Gardner

Hyladelphys is a monotypic genus of small (total length averages 205 mm, tail 115 mm, and greatest length of skull 24.5 mm) mouse-sized opossum, uniquely characterized by a non-molariform deciduous third premolar and only four mammae. Based on the original description (Voss, Lunde, and Simmons 2001), the dorsal pelage is smooth, gray-based dull reddish brown; the ventral pelage is self-colored white to cream. The face has a broad, dark mask, and a median streak of pale orange fur from rhinarium to between eyes. The fur of the cheek and throat are white. Eyes and ears are quite large, and mammae are in two abdominal-inguinal pairs. The skull has a short rostrum, beaded supraorbital margins, but lacks supra- and postorbital processes. Premaxillae are not produced anterior to incisors (lacks rostral process), the maxillopalatine fenestrae are narrow and sometimes discontinuous, and maxillary and palatine fenestrae are lacking. This description was reiterated and extended by Jansa and Voss (2005).

SYNONYMS:

Gracilinanus: Hershkovitz, 1992b:37; part; not *Gracilinanus* Gardner and Creighton, 1989.

Hyladelphys Voss, Lunde, and Simmons, 2001:30; type species *Gracilinanus kalinowskii* Hershkovitz, 1992b, by original designation.

Hyladelphys kalinowskii (Hershkovitz, 1992)
Kalinowski's Opossum

SYNONYMS:

Gracilinanus kalinowskii Hershkovitz, 1992b:37; type locality "Hacienda Cadena, Marcapata, 13°20′S, 70° 46′W, Cuzco, Peru, 890 m."

Hyladelphys kalinowskii: Voss, Lunde, and Simmons, 2001:30; first use of current name combination.

DISTRIBUTION: *Hyladelphys kalinowskii* is known from northern Fxrench Guiana, southern Guyana, Amazonian Brazil, and eastern Peru.

MARGINAL LOCALITIES (Map 18; from Voss, Lunde, and Simmons 2001, except as noted): FRENCH GUIANA: Paracou; Route de Kaw (Moraes [Astva] 2007). BRAZIL: Amazonas, Manaus (Moraes [Astva] 2007). PERU: Cusco, Hacienda Cadena (type locality of *Gracilinanus kalinowskii* Hershkovitz); Cusco, Cashiriari-3 (S. Solari

et al. 2001c); Junín, Chanchamayo; Loreto, Nuevo San Juan. GUYANA: East Berbice-Corentyne, New River Falls.

SUBSPECIES: *Hyladelphys kalinowskii* is considered to be monotypic.

NATURAL HISTORY: Of the three specimens reported by Voss, Lunde, and Simmons (2001) from French Guiana, two were caught in pitfalls located near a small stream in well-drained primary forest. The third was shot as it sat on a palm frond about a meter above the ground in swampy primary forest.

REMARKS: The unusually small deciduous premolar brings to mind Marshall's (1982:260) comment concerning *Zygolestes paranensis* Ameghino, 1898: "It is the only known didelphoid in which the third lower premolar is greatly reduced in size relative to the other cheekteeth." Hershkovitz (1992b:38) described the permanent P3 as little more than 1/2 the bulk of P2. Jansa and Voss (2005) found a high level of molecular divergence between samples from Peru and French Guiana, suggesting that more than one species may be represented.

Genus *Lestodelphys* Tate, 1934

Oliver P. Pearson

Lestodelphys is a monotypic, small (total length less than 250 mm, mass 60–90 g) carnivorous marsupial endemic to southcentral Argentina. Similar in appearance to *Thylamys elegans*, but body, cranium, and mandibles are more robust, the tail is relatively shorter (less than 75% of head and body length), upper and lower canines are longer, and the claws on the front feet extend beyond the terminal pads of the digits. Both taxa can have fat (incrassate) tails. The fur is dense, gray dorsally and white ventrally, with pale cheeks and forehead, and a black eye ring. Cladistic analysis of nonmolecular and nuclear DNA sequence characters consistently place *Lestodelphys* as the sister to *Thylamys* (Voss and Jansa 2003).

SYNONYMS:

Notodelphys O. Thomas, 1921c:137; type species *Notodelphys halli* O. Thomas, 1921c, by original designation; preoccupied by *Notodelphys* Allman, 1847 (a copepod), and *Notodelphys* Lichtenstein and Weinland, 1854 (Amphibia).

Lestodelphys Tate, 1934:154; replacement name for *Notodelphys* O. Thomas (preoccupied).

Monodelphis [*(Lestodelphys)*]: Gilmore, 1941:315; part; not *Monodelphis* Burnett, 1830.

Lestodelphis Cabrera, 1958:11; incorrect subsequent spelling of *Lestodelphys* Tate.

Lestodelphis Crespo, 1974:5; incorrect subsequent spelling of *Lestodelphys* Tate.

Lestodelphis Olrog and Lucero, 1981: Table 1, 21; incorrect subsequent spelling of *Lestodelphys* Tate.

Lestodelphis Ojeda and Monjeau, 1995:53; incorrect subsequent spelling of *Lestodelphys* Tate.

Lestodelphys halli (O. Thomas, 1921)
Patagonian Opossum, Lestodelfo

SYNONYMS:

Notodelphys halli O. Thomas, 1921c:137; type locality "Cape Tres Puntas, S.E. Patagonia, 47°S"; subsequently emended (O. Thomas 1929:45) to "Estancia Madujada, not far from Puerto Deseado," Santa Cruz, Argentina.

Lestodelphys halli: Tate, 1934:154; first use of current name combination.

DISTRIBUTION: *Lestodelphys halli* is endemic to Argentina, where it has been found from provincia de Mendoza south into provincia de Santa Cruz.

MARGINAL LOCALITIES (Map 19): ARGENTINA: Mendoza, Chacras de Coria (MVZ 159324); La Pampa, Lihue Calel (MVZ 173727); Río Negro, 15 km SE of Los Menucos (Birney et al. 1996a); Chubut, Pico Salamanca (Birney et al. 1996a); Santa Cruz, Estancia La Madrugada (= Estancia La Madujada of O. Thomas 1929:45; type locality of *Notodelphys halli* O. Thomas); Santa Cruz, Meseta El Pedrero (Mares and Braun 2000); Chubut, Nahuel Pan (G. Martin 2003); Neuquén, Estancia Tehuel Malal (Mares and Braun 2000).

SUBSPECIES: I am treating *L. halli* as monotypic.

NATURAL HISTORY: The type was caught in midwinter in a trap set for foxes and baited with rhea meat. Budin, who collected six or more specimens at Pico Salamanca, noted on the labels that *L. halli* lives among spiny branches in quebradas. Subsequent specimens have come from traps set on the ground or in *Ctenomys* burrows in semi-arid habitats such as bushy steppe and monte. The stomach of the specimen from Lihue Calel contained small-mammal fur. Feces from two specimens trapped alive near Los Menucos contained remains of arthropods. Captives killed live mice with great ease. They always started to eat at the nose of the prey and ate everything including bones, teeth, and fur.

Tails become incrassate (swollen with fat), but hibernation may be erratic, because the southernmost specimen was trapped during the winter. Two specimens held immobile in steel traps were completely torpid when removed from the traps, but recovered without apparent harm.

The southern limit of the range of *Thylamys pallidior* (Río Collón Curá) overlaps the range of *L. halli* and the species are sympatric at Lihue Calel, provincia La Pampa. Adult *L. halli* are larger (60–90 grams), have a relatively shorter tail (tail less than 80% of head and body length), and the claws on the front and hind feet extend conspicu-ously beyond the terminal pads of the toes. The chest and abdomen of *L. halli* glows salmon-pink under longwave ultraviolet light.

Nowak and Paradiso (1983) provided photographs of a skin and skull. Marshall (1977a) illustrated the teeth, cranium, and mandible as well as gave measurements and a detailed description in his *Mammalian Species* account on *L. halli*. Most of the known material has been recovered from owl pellets. Relatively few whole specimens are known and all come from unforested habitats in Argentina that G. Martin (2003) classed as belonging to the Patagonian and Monte Phytogeographical provinces.

REMARKS: The type locality is the southernmost occurrence and was originally given as "Cape Tres Puntas, S.E. Patagonia, 47°S" (O. Thomas 1921c:138). After correspondence with the collector, O. Thomas (1929:45) emended the locality to "Estancia Madujada, not far from Puerto Deseado." This is probably Estancia La Madrugada (47°08′S, 66°27′W), which is 5 km from the coast, 47 km W of Cape Tres Puntas, and 85 km NNW of Puerto Deseado.

The specimen mentioned from provincia de Mendoza had been misidentified as a *Marmosa*, preserved in formalin, and had passed through several collections before its true identity emerged. Massoia (1982) reported on four mandibles found in a cave at Paso de los Molles, provincia Río Negro. Mandibles or fragments of crania are available from owl pellets from Pico Salamanca, provincia Chubut; from three sites in provincia Neuquén (Confluencia, Estancia Tehuel Malal, and Río Collón Curá); and from the following three localities in provincia Río Negro: Cerro Leones; 10 km WSW of Comallo; and 2 km E of Estación Perito Moreno. The numerous remains G. Martin (2003) reported from nine localities in provincia Chubut were found in owl pellets from the Barn Owl (*Tyto alba*) and the Magellan Great Horned Owl (*Bubo magellanica*). There is additional material in the British Museum (Natural History) from Comodoro Rivadavia, provincia Santa Cruz, Argentina.

Genus *Marmosa* Gray, 1821
G. Ken Creighton and Alfred L. Gardner

This genus of nine Recent species occurs in Mexico, Central America, South America (north of the Tropic of Capricorn), and on the islands of Trinidad, Tobago, and Grenada. Eight species occur in South America: *Marmosa andersoni*, *M. lepida*, *M. murina*, *M. quichua*, *M. robinsoni*, *M. rubra*, *M. tyleriana*, and *M. xerophila*. *Marmosa mexicana* Merriam, 1897a, of Mexico and Central America is the only member of the genus not found in South America.

The size of adults spans a large range (length of head and body 95–176 mm, length of tail 130–211 mm). The color, length, and density of the dorsal pelage are variable among species, but the pelage is never as long, thick, and woolly as in taxa of the genus *Micoureus*. The dorsal color varies from grayish, sandy brown (*M. xerophila*), and warm cinnamon (*M. murina* and *M. robinsoni*), to reddish brown (*M. lepida* and *M. rubra*). The ventral pelage is usually creamy white to pale buff, and in some species the ventral hairs are conspicuously gray-based, especially laterally at mid-body. Tail length always exceeds head-and-body length in a ratio greater than 1.3. Mammae range from 7 to 19 and are always confined to the abdominal-inguinal region.

The tail is unicolored, weakly bicolored, or particolored. Tail scales are of moderate size (15 to 20 per cm), arranged in spiral rows, and rhomboid in shape. Tail hairs are always three per scale with the central hair longest (from 1 to 1.5 scale lengths) and about double the thickness of each lateral hair. A ventral sulcus is usually evident on the distal 15 to 25% of the tail. Longer fur characteristic of the body extends along the base of the tail for a distance of up to 25 mm. The central palmar and plantar surfaces lack granulations, and the proximal half of the digits have transverse bars. The thenar pad and the first interdigital pad of the pes either lie close together or are fused. All palmar pads are present, separate, and distinct. The second and third interdigital pads of the manus and pes are large and triangular as in the genera *Micoureus*, *Marmosops*, and *Gracilinanus*.

Lambdoidal crests, postorbital processes, and supraorbital and temporal ridges are always present on the skull (exceptionally well developed in *M. andersoni*). The nasals are broadly expanded at the maxillofrontal suture. The palate usually has one pair of slender maxillopalatine fenestrae and a pair of small to obsolete posterolateral palatal foramina. In some taxa (notably *M. lepida* and *M. murina*) the premaxillae taper anteriorly to an acute point that projects 2 mm or more anterior to I1. The upper incisors progressively increase in breadth from I2 through I5. P2 is approximately equal to or larger than P3. The auditory bullae are essentially as in *Micoureus*, although relatively larger and closer together. Species in both genera also are similar in that the alisphenoid portion of the bulla lacks an anteromedial spine or process.

Members of the genus *Marmosa* can be distinguished from species of *Gracilinanus*, *Marmosops*, and *Thylamys* by their prominent supraorbital processes, alisphenoid portion of bulla lacking an anteromedial process, and tail scales usually arranged in spiral rows. On the tail, the middle hair of the scalar triplet is not as strongly pigmented or as petiolate as it is in *Marmosops* spp. Most species of South American *Marmosa* are distinguishable from *Micoureus* spp. by the anteriorly projecting, acutely pointed premaxillae (except in *Marmosa robinsoni*), hypothenar pad separated from interdigital pad 4 on the pes, brown or reddish-brown dorsal pelage, and the differences in relative lengths of the medial and lateral interscalar hairs on the tail.

Fossil material is known from South American Miocene deposits (Marshall 1981). Fossil material reported by Tate (1933) for *Marmosa* from cave deposits in Mexico represents *Tlacuatzin canescens* (J. A. Allen, 1893a). The known elevational range is from sea level to nearly 3,000 m. The genus contains some of the rarest (*M. lepida* and *M. andersoni*) and the most common and widespread (*M. murina* and *M. robinsoni*) species of mouse opossums. The karyotype for *M. mexicana*, *M. murina*, and *M. robinsoni* is $2n = 14$ (Reig et al. 1977).

SYNONYMS:

Didelphis Linnaeus, 1758:54; part.

Didelphis: Kerr, 1792:194; not *Didelphis* Linnaeus.

Marmosa Rafinesque, 1815:55; *nomen nudum*.

Marmosa Gray, 1821:308; type species *Didelphis marina* Gray, 1821 (= *Didelphis murina* Linnaeus), by monotypy.

Asăgis Gloger, 1841:82; no species mentioned; type species *Didelphis murina* Linnaeus, by subsequent designation (O. Thomas 1888b:340).

Notagōgus Gloger, 1841:82; no species mentioned; type species *Didelphis murina* Linnaeus, by subsequent designation (O. Thomas 1888b:340); preoccupied by *Notagogus* Agassiz, 1833 (Pisces).

Philander: Tiedemann, 1808:427; part; not *Philander* Brisson, 1762.

Opossum Schmid, 1818:115; part.

Sarigua Muirhead, 1819:429; part.

Didelphis: J. A. Wagner, 1842:359; not *Didelphis* Linnaeus.

Didelphis: Cabanis, 1848:778; not *Didelphis* Linnaeus.

Grymaeomys Burmeister, 1854:138; type species *Didelphis murina* Linnaeus, 1758, by subsequent designation (O. Thomas 1888b:340); proposed as a subgenus of *Didelphis* Linnaeus.

Didelphis: Tomes, 1860a:58; not *Didelphis* Linnaeus.

Cuica Liais, 1872:329; type species *Cuica murina*: Liais, 1872 (=*Didelphis murina* Linnaeus), by subsequent designation (Hershkovitz 1949a:12).

Didelphys O. Thomas, 1888a:258; part; incorrect subsequent spelling of *Didelphis* Linnaeus.

Micoureus: O. Thomas, 1888a:158; part; not *Micoureus* Lesson, 1842.

Quica Cabrera, 1958:12; incorrect subsequent spelling of *Cuica* Liais.

Stegomarmosa Pine, 1972a:279; type species *Marmosa andersoni* Pine, 1972a, by monotypy; described as a subgenus of *Marmosa* Gray.

Grayium Kretzoi and Kretzoi, 2000:148; replacement name for *Marmosa* Gray, erroneously presumed to be preoccupied by *Marmosa* Rafinesque.

REMARKS: Voss and Jansa (2003) erected the genus *Tlacuatzin* for the Mexican species previously known as *Marmosa canescens* J. A. Allen. *Tlacuatzin canescens* can be distinguished from *Marmosa* spp. by the lack of a prominent anterior projection of the premaxillae; presence of maxillopalatine fenestrae; annular arrangement of caudal scales; and the $2n = 22$, FN 20 karyotype, which is the highest chromosome number and only known uniarmed karyotype for any member of the Monodelphini (Engstrom and Gardner 1988; Voss and Jansa 2003).

Kretzoi and Kretzoi (2000:91) listed *Didelphis myosuros* Liais (= *Didelphis nudicaudatus* É. Geoffroy St.-Hilaire, 1803) as the type species of *Cuica* Liais. They apparently overlooked Hershkovitz's (1949a:12) earlier designation of *Cuica murina* Liais as the type species of *Cuica*.

KEY TO SOUTH AMERICAN SPECIES OF *MARMOSA*:

1. Cranium with prominent supraorbital ridges and laterally projecting postorbital processes. 2
1'. Cranium lacking prominent supra- and postorbital processes and prominent ridges; supraorbital margin of frontals finely beaded. .7
2. Postorbital processes on skull exceptionally large and flared, breadth across tips of processes greater than 10 mm and exceeding breadth of rostrum at anterior base of zygomatic arch . . . . *Marmosa andersoni*
2'. Postorbital processes not exceptional, breadth across tips of processes less than 10 mm and not wider than breadth of rostrum at anterior base of zygomatic arch 3
3. Postorbital processes broadly triangular in outline from above; postorbital ridges approximately parallel as they extend posteriorly over parietals; posterolateral palatal foramina minute (less than 0.5 mm in width); tail with sparse, short, and moderately pigmented to unpigmented bristles 4
3'. Postorbital processes distinctly beaded on margins, but less prominent and appearing to be positioned farther behind orbits; postorbital ridges converging toward midline as they extend posteriorly over parietals; posterolateral palatal foramina usually greater than 0.5 mm in width; tail with fine, relatively thick, and either unpigmented or weakly pigmented hairs. 6
4. Distributed in the western Amazon basin and eastern slopes of the Andes 5
4'. Not from the western Amazon basin and eastern slopes of the Andes. . . . . *Marmosa murina*
5. Feet brownish buff; tail long, averaging 1.5 or more than length of head and body; nasals short, usually less than 15 mm. . . . . *Marmosa quichua*
5'. Feet paler; tail shorter, usually less than 1.5 times length of head and body; nasals long, usually more than 15 mm . . . . *Marmosa murina*

6. Dorsal pelage brown to pale brown; ventral pelage yellowish buff to buffy white, usually with some hairs gray based, especially laterally; dorsal pelage long and hairs on rump longer than 5 mm . . . . . . *Marmosa robinsoni*
6'. Dorsal pelage pale sandy brown; ventral pelage white medially; dorsal and ventral pelage appearing sparse with gray basal color showing through, especially on sides; dorsal pelage on rump less than 5 mm in length . . . . *Marmosa xerophila*
7. Size small, condylobasal length less than 29 mm, total length less than 270 mm; dorsal pelage warm reddish brown. . . . . *Marmosa lepida*
7'. Size larger, condylobasal length more than 29 mm, total length more than 275 mm; dorsal pelage reddish to dark brown. 8
8. Supra- and postorbital region strongly beaded with beading extending back over parietals as prominent parallel ridges; dorsal pelage reddish brown and ventral pelage yellowish-buff with hairs gray-based, except on throat, chin, and chest; dark median line on face from between eyes to rhinarium; condylobasal length usually greater than 32 mm. . . . . *Marmosa rubra*
8'. Interorbital region faintly beaded at lateral margin; parietal ridges weakly developed; dorsal pelage dark brown and ventral pelage entirely gray-based with hairs tipped dull buff; no dark median line on face; condylobasal length less than 32 mm. . . . . *Marmosa tyleriana*

Marmosa andersoni Pine, 1972
Anderson's Mouse Opossum

SYNONYMS:

Marmosa (Stegomarmosa) andersoni Pine, 1972a:279; type locality "Hda. Villa Carmen, Cosñipata, Cuzco, Peru."
Stegomarmosa andersoni: S. Solari, Vivar, Velasco, Rodríguez, Wilson, Baker, and Mena, 2001: 179; name combination.

DISTRIBUTION: *Marmosa andersoni* is known from only departamento Cusco, Peru.

MARGINAL LOCALITIES (Map 20). PERU: Cusco, Camisea (USNM 582777; S. Solari et al. 2001c); Cusco, Hacienda Villa Carmen (type locality of *Marmosa andersoni* Pine).

SUBSPECIES: *Marmosa andersoni* is monotypic.

NATURAL HISTORY: Pine (1972a) reported three other species of small opossums (*Marmosa rubra*, *Marmosops impavidus*, and *M.* cf. *parvidens* [= Marmosops bishopi) taken at the type locality of *M. andersoni*, all in lowland humid forest. S. Solari et al. (2001c) reported two adults and four juveniles acquired at two sites in undisturbed, *terra firme* forest habitat in the lower Urubamba region, approximately 100 km north of the type locality. Both sites

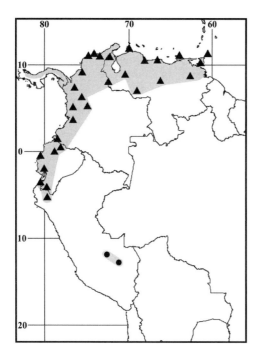

Map 20 Marginal localities for *Marmosa andersoni* ● and *Marmosa robinsoni* ▲

contained bamboo, which appeared to be correlated with higher diversity of marsupials.

REMARKS: This enigmatic species is most similar to *M. murina* in cranial and external morphology, but has brighter reddish pelage and longer, more lax fur. The principal distinguishing cranial feature appears to be the greatly enlarged supraorbital processes. Otherwise, cranial and external measurements are within the range of variation for *M. murina*. Cuartas and Muñoz (2003:97–99) described *M. andersoni* as occurring in "Colombia: en la regiones de la Amazonia y la Orinoquia, en límites con Peru." They included four dots on their distribution map for the species (p. 209), two in Colombia and two in Peru. However, the localities are not identified and no specimens are mentioned; therefore we are unable to confirm these records and have not included them in the distribution of *M. andersoni*. As far as we have been able to determine, the species is known by only the holotype and the six specimens reported by S. Solari et al. (2001c).

Marmosa lepida (O. Thomas, 1888)
Rufous Mouse Opossum

SYNONYMS:

Didelphys (Micoureus) lepida O. Thomas, 1888a:158; type locality "Peruvian Amazons"; identified as "Santa Cruz, Huallaga R.," Loreto, Peru, by O. Thomas (1888b:348).

[*Didelphys (Marmosa)*] *lepida*: Trouessart, 1898:1240; name combination.

[*Didelphis (Marmosa)*] *lepida*: Matschie, 1916:270; name combination.

[*Marmosa (Marmosa)*] *lepida*: Cabrera, 1919:37; name combination.

Marmosa lepida grandis Tate, 1931:12; type locality "Buenavista, Santa Cruz, Bolivia."

Marmosa lepida lepida: Tate, 1933:205; name combination.

[*Thylamys*] *lepida*: Reig, Kirsch, and Marshall, 1985:342; name combination.

DISTRIBUTION: *Marmosa lepida* is known from northern Surinam, French Guiana, western Brazil, and the eastern lowlands of Colombia, Ecuador, Peru, and Bolivia.

MARGINAL LOCALITIES (Map 21): SURINAM: Paramaribo, Paramaribo (MCZ 17199). FRENCH GUIANA: Les Nouragues (Voss, Lunde, and Simmons 2001). BRAZIL (B. D. Patterson 1992; as *Gracilinanus emiliae*): Amazonas, Codajás; Amazonas, Igarapé Grande. BOLIVIA: Santa Cruz, Buenavista (type locality of *Marmosa lepida grandis* Tate). PERU: Ucayali, Lagarto (Tate 1933); Huánuco, Puerto Márquez (Tate 1933); Loreto, Santa Cruz (type locality of *Didelphys (Micoureus) lepida* O. Thomas); Amazonas, Huampami (MVZ 155245). ECUADOR: Morona-Santiago, Gualaquiza (B. E. Brown 2004); Pastaza, Sarayacu (Tate 1933). COLOMBIA: Meta, Caño Entrada (Lemke et al. 1982).

SUBSPECIES: We treat *M. lepida* as monotypic.

NATURAL HISTORY: Though first described over a century ago, *M. lepida* is still known from fewer than 20 specimens, mainly from localities below 600 m around the western periphery of the Amazon basin and from the

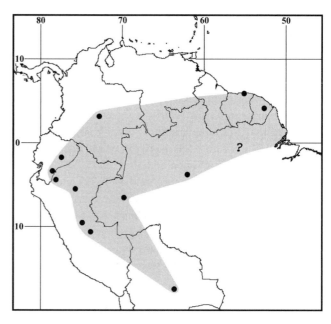

Map 21 Marginal localities for *Marmosa lepida* ●

Guianas. The conspicuously reddish-brown dorsal pelage, and the partly gray-based ventral pelage, suggest adaptations to a humid forest environment. The number of functional mammae is seven. Selected measurements in mm are: head and body 97–120, tail 140–150, hind foot 16–19, condylobasal length 23.2–26.2.

REMARKS: Tate (1933:203) consigned this species to his *microtarsus*-group (= *Gracilinanus*), though not without some misgivings. Cabrera (1958) correctly listed *M. lepida* under *Marmosa* (*sensu stricto*). *Marmosa lepida* can be distinguished from congeners by its smaller size, brighter reddish or brighter chestnut pelage, the lack of ridges on the skull, and the lack of flaring, strongly beaded supra- and postorbital processes. Absence of an anteromedial process of the alisphenoid portion of the auditory bulla, combined with the *Marmosa* pattern of scales and bristles on the tail, separate *M. lepida* from species of *Marmosops* and *Gracilinanus*. Voss, Lunde, and Simmons (2001) reidentified as *M. lepida* the specimens from Brazil that B. D. Patterson (1992) reported as *Gracilinanus emiliae*, and cautioned against uncritically accepting literature records, particularly of juveniles of both taxa, because of the ease by which they can be misidentified.

Marmosa murina (Linnaeus, 1758)
Murine Opossum
SYNONYMS:

[*Didelphis*] *murina* Linnaeus, 1758:55; type localities "Asia, America"; restricted to Surinam by O. Thomas (1911b: 144).

[*Didelphis*] *dorsigera* Linnaeus, 1758:55; type locality "America"; restricted to Surinam by O. Thomas (1911b: 144).

Didelphis guianensis Kerr, 1792:194; type locality "Guiana."

Didelphis marina É. Geoffroy St.-Hilaire, 1803:143; incorrect subsequent spelling of *Didelphis murina* Linnaeus.

Philander murinus: Tiedemann, 1808:427; name combination.

O[*possum*]. *murinus*: Schmid, 1818:117; name combination.

S[*arigua*]. *murina*: Muirhead, 1819:429; name combination.

[*Monodelphis*] *dorsigens*: Burnett, 1830:35; incorrect subsequent spelling of *Didelphis dorsigera* Linnaeus.

Micoureus murinus: Lesson, 1842:186; name combination.

Micoureus dorsigera: Lesson, 1842:186; name combination.

Didelphys macrotarsus J. A. Wagner, 1842:359; no type locality given; based on the specimen taken by Natterer on the Rio Madeira, Brazil (J. A. Wagner 1847:145); preoccupied by *Didelphys macrotarsus* Schreber, 1777:pl. 155 (Primates).

Philander dorsigera: Gray, 1843:101; name combination.

Didelphis muscula Cabanis, 1848:778; type locality "Caraiben Niederlassung Arrai am obern Pomeroon," Pomeroon-Supenaam, Guyana.

Didelphys [*(Grymaeomys)*] *murina*: Burmeister, 1854:138; name combination.

Didelphys waterhousii Tomes, 1860a:58; type locality "Gualaquiza," Morona-Santiago, Ecuador.

[*Cuica*] *murina*: Liais, 1872:330; name combination.

Didelphys [*(Micoureus)*] *murina*: O. Thomas, 1888b:343; name combination.

M[*icoureus*]. *murinus*: O. Thomas, 1894c:185; name combination.

Marmosa murina: O. Thomas, 1895b:58; first use of current name combination.

[*Didelphis (Marmosa)*] *murina*: Trouessart, 1898:1239; name combination.

Marmosa Klagesi J. A. Allen, 1900c:198; type locality "Ciudad Bolívar," Bolívar, Venezuela.

[*Didelphys (Marmosa)*] *klagesi*: Trouessart, 1905:855; name combination.

Marmosa chloe O. Thomas, 1907:167; type locality " Demerara River, 29 miles above Georgetown" (= Hyde Park according to Tate 1933:96), Demerara-Mahaica, Guyana.

Marmosa tobagi O. Thomas, 1911c:515; type locality "Island of Tobago"; restricted to Waterloo, Trinidad and Tobago by Tate (1933:120), who assumed that Waterloo was a place in the island of Tobago, not on the island of Trinidad.

Marmosa parata O. Thomas, 1911c:517; type locality "Igarapé-Assu," Pará, Brazil.

Marmosa madeirensis Cabrera, 1913:12; replacement name for *Didelphys macrotarsus* J. A. Wagner.

[*Didelphis (Caluromys)*] *waterhousei*: Matschie, 1916:269; name combination and incorrect subsequent spelling of *Didelphys waterhousii* Tomes.

[*Didelphis (Marmosa)*] *tobagi*: Matschie, 1916:270; name combination.

[*Didelphis (Marmosa)*] *musculus*: Matschie, 1916:270; name combination.

[*Didelphis (Marmosa)*] *madeirensis*: Matschie, 1916:270; name combination.

[*Didelphis (Marmosops)*] *klagesi*: Matschie, 1916:270; name combination.

[*Didelphis (Marmosops)*] *chloe*: Matschie, 1916:271; name combination.

[*Marmosa (Marmosa)*] *chloe*: Cabrera, 1919:36; name combination.

[*Marmosa (Marmosa)*] *klagesi*: Cabrera, 1919:37; name combination.

[*Marmosa (Marmosa)*] *madeirensis*: Cabrera, 1919:37; name combination.

[*Marmosa (Marmosa)*] *murina*: Cabrera, 1919:38; name combination.

[*Marmosa (Marmosa)*] *waterhousei*: Cabrera, 1919:38; name combination and incorrect subsequent spelling of *Didelphys waterhousii* Tomes.

[*Marmosa (Marmosa)*] *tobagi*: Cabrera, 1919:39; name combination.

Marmosa bombascarae H. E. Anthony, 1922:5; type locality "Zamora (junction of Rio Bombascaro with Rio Zamora)," Zamora-Chinchipe, Ecuador.

Marmosa maranii O. Thomas, 1924b:537; type locality "San Lorenzo, north side Rio Marañon, just above junction with Rio Huallaga," Loreto, Peru.

Marmosa waterhousei: O. Thomas, 1927:374; incorrect subsequent spelling of *Didelphys waterhousii* Tomes.

Marmosa murina roraimae Tate, 1931:4; type locality "Arabupu, foot of Mt. Roraima," Bolívar, Venezuela.

Marmosa murina duidae Tate, 1931:5; type locality "'Middle Camp,' foot of Mt. Duida, eight miles north of Esmeralda, upper Rio Orinoco," Amazonas, Venezuela.

Marmosa murina muscula: Tate, 1933:96; name combination.

Marmosa murina klagesi: Tate, 1933:98; name combination.

Marmosa murina madeirensis: Tate, 1933:100; name combination.

Marmosa murina maranii: Tate, 1933:102; name combination.

Marmosa murina waterhousei: Tate, 1933:103; name combination and incorrect subsequent spelling of *Didelphys waterhousii* Tomes.

Marmosa m[urina]. parata: A. Miranda-Ribeiro, 1936:370; name combination.

Marmosa murina meridionalis A. Miranda-Ribeiro, 1936: 371, 372; type locality "Sul de Mato-Grosso–(Paratudal?)," Mato Grosso do Sul, Brazil; type locality identified as "Paratubal, sul de Mato Grosso" by Langguth, Limeira, and Franco (1997:7).

Marmosa moreirae A. Miranda-Ribeiro, 1936:380; type locality "Itatiaya," Rio de Janeiro, Brazil.

Thylamys macrotarsus: A. Miranda-Ribeiro, 1936:386; name combination.

Marmosa murina tobagi: Goodwin, 1961:9; name combination.

Marmosa murina guianensis: Carvalho, 1962:285; name combination.

M[armosa]. m[urina]. waterhousei: Pérez-Hernández, 1990: 366; incorrect subsequent spelling of *Didelphys waterhousii* Tomes.

Didelphys waterhousei: B. E. Brown, 2004:63; incorrect subsequent spelling of *Didelphys waterhousii* Tomes.

Map 22 Marginal localities for *Marmosa murina* ●

DISTRIBUTION: *Marmosa murina* is known from Colombia, Venezuela, Trinidad and Tobago, the Guianas, Brazil, and eastern Ecuador, Peru, and Bolivia.

MARGINAL LOCALITIES (Map 22; from Tate 1933, except as noted): TRINIDAD AND TOBAGO: island of Tobago (type locality of *Marmosa tobagi* O. Thomas). VENEZUELA: Delta Amacuro, Winikina (Linares and Rivas 2004). GUYANA: Demerara-Mahaica, Hyde Park (type locality of *Marmosa chloe* O. Thomas). SURINAM: Brokopondo, Bergendal. FRENCH GUIANA: Cayenne. BRAZIL: Amapá, Macapá (Carvalho 1962); Pará, Belém (Ávila-Pires 1958); Maranhão, Tury-assú; Piauí, Parahyba; Ceará, Parque Nacional de Ubajara (Guedes et al. 2000); Paraíba, Penha; Pernambuco, Recife; Alagoas, Mangabeiras (C. O. C. Vieira 1953); Bahia, Reserva Biológica de Una (UFMG [RM 67]); Bahia, Helvécia (HGB 5); Espírito Santo, M7, Aracruz Florestal (UFMG [MF 31]); Rio de Janeiro, Itatiaya (type locality of *Marmosa moreirae* A. Miranda-Ribeiro); Maranhão, Côcos; Pará, Floresta Nacional Tapirapé-Aquiri (INPA [CS 44]); Mato Grosso, Serra do Roncador, 264 km N (by road) of Xavantina (Pine, Bishop, and Jackson 1970); Mato Grosso, Fazenda São Luis (MVZ 197886); Mato Grosso do Sul, Fazenda Cedro (MVZ 198038); Mato Grosso, Reserva Ecológica Cristalino (MVZ 197428). Pará, Villa Braga (O. Thomas 1920c); Amazonas, Rosarinho; Amazonas, Codajás (B. D. Patterson 1992); Amazonas, Ilha Paxiuba (J. L. Patton, Silva, and Malcolm 2000); Amazonas, Barro Vermelho (J. L. Patton, Silva, and Malcolm 2000). PERU: Ucayali, Balta (Reig et al. 1977). BOLIVIA: Beni, Puerto Salinas

(B. D. Patterson 1992); Santa Cruz, El Refugio (Emmons 1998); Santa Cruz, 2 km SW of Las Cruces (S. Anderson 1997); La Paz, Chijchipani (S. Anderson 1997). PERU: Cusco, Pagoreni (S. Solari et al. 1999); Loreto, Nuevo San Juan (AMNH 272816); Loreto, San Lorenzo (type locality of *Marmosa maranii* O. Thomas); Amazonas, Huampami (J. L. Patton, Berlin, and Berlin 1982). ECUADOR: Zamora-Chinchipe, Zamora (type locality of *Marmosa bombascarae* H. E. Anthony); Morona-Santiago, Gualaquiza (type locality of *Didelphys waterhousii* Tomes); Tungurahua, Mirador; Napo, Baeza (Tate 1933, as *Marmosa ruatanica mimetra*). COLOMBIA: Cundinamarca, Fusagasugá (see Remarks); Antioquia, Amalfi (Cuartas and Muñoz 2003). VENEZUELA (Handley 1976): Zulia, El Rosario, 48 km WNW of Encontrados; Yaracuy, 10 km NW of Urama; Monagas, San Agustín, 5 km NW of Caripe.

SUBSPECIES: For the purposes of this account, we treat *M. murina* as monotypic (see Remarks).

NATURAL HISTORY: This species occurs in humid lowland forests east of the Andes at elevations up to about 2,000 m. In Venezuela, 48% of 71 specimens were caught on the ground and 52% on logs, in trees, or in houses (Handley 1976), and the majority near streams or moist areas in evergreen forest from near sea level (25 m) up to 1,365 m. J. L. Patton, Silva, and Malcolm (2000) and Voss, Lunde, and Simmons (2001) indicated that *M. murina* was most commonly found in second growth forest. The karyotype is $2n = 14$, FN $= 24$ (Reig et al. 1977).

REMARKS: Tomes (1860a) described *Didelphys waterhousii* as having a complete pouch that, according Fraser (the collector), contained five 3-inch-long young. Tomes later (1860b:304) stated that the pouch "was about the size of a large hazel-nut." Apparently, the skin, which was sent to Tomes in "spirits," is missing and the presence or absence of a pouch can not be verified. Because of concern that, if a pouch were present, the skin and skull were mismatched, B. E. Brown (2004:63) restricted the name *Didelphys waterhousii* to the type skull.

Both dorsal and ventral pelage varies throughout the range of *M. murina* and seems, to a large degree, correlated with elevation and humidity, and how long the specimens have been in the collection. Color hues on the venter fade in time (Handley 1984; Pine, Dalby, and Matson 1985). Dorsal color varies from pale brown to chestnut brown. The ventral color varies from white to yellowish-buff with hairs self-colored (zone of self-colored hairs variably narrowed laterally by gray-based hairs) or hairs of venter entirely gray based, particularly in specimens from higher elevations. Larger specimens overlap the size range of some species of *Micoureus*, but are readily distinguished by the lack of the grayish, long, and woolly pelage, which characterizes

Micoureus. Adult *M. murina* always have broad, prominent supraorbital processes that distinguish them from sympatric *M. lepida* and *M. rubra* and species of *Marmosops* and *Gracilinanus*. The broadly flaring supraorbital processes, comparatively minute or absent posterolateral palatal foramina, relatively shorter tail bristles, and larger tail scales of *M. murina* distinguish it from *M. robinsoni*.

Barbara Brown (2004:63, 68), apparently based on Philip Hershkovitz's notes, treated Tate's (1933:103) *M. murina waterhousei* [*sic*] as composite and assigned the Colombian and Venezuelan specimens Tate listed to *M. robinsoni*. We have retained the specimen record from Fusagasugá, Cundinamarca, Colombia (BM 28.11.8.3) in *M. murina*, pending revision of the species.

Tate (1933) was able to discern 12 geographic forms of *M. murina*, which he grouped into eight subspecies. Most of these were isolated populations represented by few specimens. Although his interpretation of their distributions and degree of relatedness was highly subjective, Tate's (1933) revision remains the only study of this species based on an evaluation of morphological characteristics. Cabrera (1958) recognized seven subspecies. Although much more material is available today, the systematics of *M. murina* is still so imperfectly known that we hesitate to define subspecies at this time. As J. L. Patton and Costa (2003) commented, some of these populations will, upon revision, be recognized as species. For example the southeastern Brazilian population likely represents a species apart from *M. murina* and should bear the name *Marmosa moreirae* A. Miranda-Ribeiro. Unfortunately, the holotype of *M. moreirae* is a skin without skull. Voss and Jansa (2003:78) also commented on the composite nature of *M. murina* (*sensu lato*). D. A. Mello and Moojen (1979) were the first of several authors to report *M. murina* from the Brazil's Distrito Federal; however, we suspect that Mares, Braun, and Gettinger (1989:15) were correct in stating that the specimens may represent *M. agilis* (= *Gracilinanus agilis*).

Marmosa quichua O. Thomas, 1899
Quechuan Mouse Opossum
SYNONYMS:

Marmosa quichua O. Thomas, 1899a:43; type locality "Ocabamba" (= Valle de Ocobamba, according to Ceballos-Bendezu 1981), Cusco, Peru.

[*Didelphys (Marmosa)*] *quichua*: Trouessart, 1905:856; name combination.

Marmosa musicola Osgood, 1913:95; type locality "Moyobamba," San Martín, Peru.

[*Didelphis (Caluromys)*] *musicola*: Matschie, 1916:269; name combination.

[*Didelphis (Marmosa)*] *quichua*: Matschie, 1916:270; name combination.

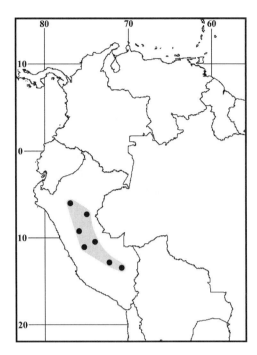

Map 23 Marginal localities for *Marmosa quichua* ●

[*Marmosa (Marmosa)*] *musicola*: Cabrera, 1919:38; name combination.

[*Marmosa (Marmosa)*] *quichua*: Cabrera, 1919:39; name combination.

DISTRIBUTION: *Marmosa quichua* occurs along the eastern slope of the Andes in Peru above 300 m.

MARGINAL LOCALITIES (Map 23; from Tate 1933): PERU: San Martín, Moyobamba (type locality of *Marmosa musicola* Osgood); Loreto, Contamana; Ucayali, Chicosa; Cusco, Marcapata; Cusco, Ocobamba (type locality of *Marmosa quichua* O. Thomas); Junín, San Ramón; Huánuco, Tingo María.

SUBSPECIES: We treat *M. quichua* as monotypic.

NATURAL HISTORY: Very little is known about the natural history of this species. Tate (1933:9) listed females as either "breeding or nursing" in July (two) and October (one); and single juveniles taken in January, February, June, and July, with two caught in May.

REMARKS: Gardner (1993a) listed *quichua* in the synonymy of *M. murina*. Voss, Lunde, and Simmons (2001:41) noted that *M. quichua* was craniodentally distinctive from *M. murina*, but did not define characters. Specimens from populations Tate (1933) identified as subspecies of *M. murina* and as representing *M. quichua* are often difficult to distinguish from one another. On average, adult *M. quichua* can be distinguished from *M. murina* by shorter nasals (5 mm), proportionally longer tail (1.5 times head and body), and darker feet. It is likely that *M. quichua* also occurs in Bolivia. B. E. Brown (2004) treated *M. quichua* as a syn-

onym of *M. murina*. The *murina* group (*sensu* Tate 1933) needs revision.

Marmosa robinsoni Bangs 1898
Robinson's Mouse Opossum

SYNONYMS: The following names represent non-South American populations; see under Subspecies for additional synonyms.

Marmosa fulviventer Bangs, 1901:632; type locality "San Miguel Island," Panamá, Panama.

[*Didelphys (Marmosa)*] *fulviventer*: Trouessart, 1905:855; name combination.

Marmosa grenadae O. Thomas, 1911c:514; type locality "Annandale," Grenada.

Marmosa ruatanica Goldman, 1911:237; type locality "Ruatan [= Roatán] Island, off the north coast of Honduras," Islas de la Bahía, Honduras.

DISTRIBUTION: *Marmosa robinsoni* occurs in northern and western Colombia, western Ecuador, northwestern Peru, northern Venezuela, and Trinidad and Tobago. Elsewhere, the species is known from the Lesser Antilles (Grenada) and Central America (Panama, El Salvador, Belize, and Isla Roatán, Honduras).

MARGINAL LOCALITIES (Map 20; from Tate 1933 [under the names *Marmosa chapmani*, *M. mitis*, *M. ruatanica*, and *M. simonsi*], except as noted): VENEZUELA: Falcón, Cerro Santa Ana, 15 km SSW of Pueblo Nuevo (Handley 1976); Carabobo, Embalse Río Morón (López-Fuster et al. 2000); Miranda, Curupao; Nueva Esparta, 2 km N, 2 km E of La Asunción. TRINIDAD AND TOBAGO: Trinidad, Caura (type locality of *Marmosa chapmani* J. A. Allen); Tobago, Speyside (type locality of *Marmosa mitis luridavolta* Goodwin). VENEZUELA: Monagas, Uverito (López-Fuster et al. 2000); Guárico, Santa Rita (Reig et al. 1977); Apure, Hato Acapulco (Reig et al. 1977); Barinas, Altamira (Handley 1976). COLOMBIA: Norte de Santander, El Guayabal (type locality of *Marmosa mitis pallidiventris* Osgood); Antioquia, Medellín; Tolima, Mariquita (Tamsitt and Valdivieso 1964, as *M. murina*). ECUADOR: Imbabura, Chota; Pichincha, Mindo; Loja, Valley of Casanga. PERU: Piura, Papayal (USNM 302980); Tumbes, near Tumbes. ECUADOR: Guayas, Cerro Baja Verde; Manabí, Río Briceño. COLOMBIA: Nariño, La Guayacana (USNM 309046); Valle del Cauca, Las Lomitas; Chocó, Condoto; Antioquia, Urabá (Cuartas and Muñoz 2004); Córdoba, El Contento (Adler, Arboledo, and Travi 1997); Atlántico, near Barranquilla (Thrasher et al. 1971); Magdalena, Bonda; La Guajira, San Miguel. VENEZUELA: Zulia, near Cerro Azul, 40 km NW of La Paz.

SUBSPECIES: We recognize five subspecies of *M. robinsoni* in South America. Hall (1981) recognized two additional subspecies in Central America.

M. r. chapmani J. A. Allen, 1900

SYNONYMS:

Marmosa chapmani J. A. Allen, 1900c:197; type locality "Caura," Trinidad, Trinidad and Tobago.

[*Didelphys (Marmosa)*] *chapmani*: Trouessart, 1905:855; name combination.

Marmosa nesaea O. Thomas, 1911c:515; type locality "Caparo," Trinidad, Trinidad and Tobago.

[*Didelphis (Marmosa)*] *chapmanni*: Matschie, 1916:270; name combination and incorrect subsequent spelling of *Marmosa chapmani* J. A. Allen.

[*Marmosa (Marmosa)*] *chapmani*: Cabrera, 1919:36; name combination.

Marmosa mitis chapmani: Hershkovitz, 1951:552; name combination.

M[*armosa*]. r[*obinsoni*]. *chapmani*: O'Connell, 1983:1; first use of current name combination.

This subspecies occurs on the island of Trinidad, Trinidad and Tobago.

M. r. isthmica Goldman, 1912

SYNONYMS:

Marmosa isthmica Goldman, 1912a:1; type locality "Rio Indio, near Gatun, Canal Zone, Panama."

[*Didelphis (Marmosa)*] *isthmica*: Matschie, 1916:270; name combination.

[*Marmosa (Marmosa) mexicana*] *isthmica*: Cabrera, 1919:37; name combination.

M[*armosa*]. *ruatanica isthmica*: Tate, 1933:123; name combination.

Marmosa robinsoni isthmica: Cabrera, 1958:24; first use of current name combination.

This subspecies is in western Colombia and, extralimitally, in Panama.

M. r. luridavolta Goodwin, 1961

SYNONYMS:

Marmosa mitis luridavolta Goodwin, 1961:5; type locality "Speyside, Tobago, the West Indies."

M[*armosa*]. r[*obinsoni*]. *luridavolta*: O'Connell, 1983:1; first use of current name combination.

This subspecies inhabits the island of Tobago, Trinidad and Tobago.

M. r. robinsoni Bangs, 1898

SYNONYMS:

Marmosa robinsoni Bangs, 1898a:95; type locality "Margarita Island," Nueva Esparta, Venezuela.

Marmosa mitis Bangs, 1898b:162; type locality "Pueblo Viejo," El Cesar, Colombia.

[*Didelphys (Marmosa)*] *mitis*: Trouessart, 1905:856; name combination.

[*Didelphis (Marmosa)*] *robinsoni*: Trouessart, 1905:856; name combination.

Marmosa mitis casta O. Thomas, 1911c:516; type locality "San Esteban, Carabobo," Venezuela.

Marmosa mitis pallidiventris Osgood, 1912:39; type locality "El Guayabal, 10 miles N. of Cucuta, Colombia."

[*Didelphis (Marmosa)*] *casta*: Matschie, 1916:270; name combination.

[*Didelphis (Marmosa)*] *mitis*: Matschie, 1916:270; name combination.

[*Marmosa (Marmosa) mitis*] *mitis*: Cabrera, 1919:38; name combination.

[*Marmosa (Marmosa) mitis*] *casta*: Cabrera, 1919:38; name combination.

[*Marmosa (Marmosa)*] *robinsoni*: Cabrera, 1919:39; name combination.

M[*armosa*]. *mitis robinsoni*: Tate, 1933:113; name combination.

Marmosa robinsoni robinsoni: Cabrera, 1958:24; first use of current name combination.

Marmosa robinsoni mitis: B. E. Brown, 2004:67; name combination.

The nominate subspecies occurs in northeastern Colombia and northern Venezuela.

M. r. simonsi O. Thomas, 1899

SYNONYMS:

Marmosa Simonsi O. Thomas, 1899d:287; type locality "Puná," Puná Island, Guayas, Ecuador.

[*Didelphys (Caluromys)*] *simonsi*: Trouessart, 1905:856; name combination.

[*Marmosa (Marmosa)*] *simonsi*: Cabrera, 1919:39; name combination.

Marmosa mimetra O. Thomas, 1921b:521; type locality "Santo Domingo, 0°13′S, 79°6′W," Pichincha, Ecuador.

M[*armosa*]. *ruatanica mimetra*: Tate, 1933:123; name combination.

Marmosa robinsoni mimetra: Cabrera, 1958:24; name combination.

Marmosa robinsoni simonsi: Cabrera, 1958:25; first use of current name combination.

M[*armosa*]. r[*obinsoni*]. *simsonsi* O'Connell, 1983:1; incorrect subsequent spelling of *Marmosa simonsi* O. Thomas.

This subspecies is in southwestern Colombia, western Ecuador (including Isla Puná), and northwestern Peru.

NATURAL HISTORY: The diet of *M. robinsoni* is mainly insects although fruit and other items are consumed (O'Connell 1979, 1983). Breeding coincides with late dry season in Panama (Fleming 1973) and Venezuela (O'Connell 1979), with a second reproductive peak during the wet season. Although occupying a variety of humid and arid habitats over an elevational range of from sea level to above 2,000 m, *M. robinsoni* prefers the dense secondary growth associated with agriculture and disturbed forests in which densities have been estimated as high as 4.25 adults per hectare (O'Connell 1979, 1983, and references cited therein). Handley (1976), reporting on 256 specimens collected in Venezuela, noted that 2/3 were caught on the ground, 65% were taken near streams or in and near other wetter sites, and 35% came from dryer sites. He also wrote that 42% of the specimens were taken in evergreen or cloud forest, 33% in openings, 25% in deciduous or thorn forest, and 62% came from below 500 m. In her *Mammalian Species* account, O'Connell (1983) summarized additional information on reproduction, parasites, physiology, ecology, and behavior, and provided illustrations and a comprehensive bibliography.

REMARKS: Ranges of selected measurements (in mm, from O'Connell 1983) are: total length 282–376, tail 152–210, hind foot 20–25, condylobasal length 30.5–41.5. These ranges undoubtedly included measurements of immature animals and, as O'Connell (1983) noted, *M. robinsoni* exhibits sexual dimorphism, with mature males attaining lengths and weights up to 25% greater than those of mature females. Generic characters serve to separate *M. robinsoni* from the other sympatric Monodelphini, with the exception of congeners (*M. murina*, *M. rubra*, *M. xerophila*, and *M. lepida*). The brighter reddish dorsal pelage, usually partially gray-based ventral pelage, and the lack of both prominent supraorbital ridges and postorbital processes distinguish *M. rubra* and *M. lepida* from *M. robinsoni*. *Marmosa robinsoni* can be distinguished from *M. murina* by its finer, weakly pigmented tail hairs and smaller scales, and the narrower supraorbital processes with associated parietal ridges that converge posteriorly (parallel in *M. murina*). The posterolateral palatal foramina of *M. robinsoni* are about 1.5 times larger in diameter than in *M. murina*. From *M. xerophila*, *M. robinsoni* is separable by its longer, browner pelage.

When first recorded in the literature, *M. robinsoni* was reported under the name *Didelphis (Micoureus) murina* by J. A. Allen and Chapman (1893) from Princestown, Trinidad. Tate (1933) erred in using *M. mitis* Bangs, 1898b, instead of *M. robinsoni* Bangs, 1898a, as the senior synonym.

Goodwin (1961) thought the population on Trinidad and the one on Tobago each represented a different subspecies of *M. robinsoni*. These, he also considered to be distinct from mainland representatives. As evidenced by the extensive synonymy we include under *M. robinsoni*, the taxon exhibits a high degree of geographic variation in addition to that resulting from sex- and age-related factors. We treat this species as monotypic, pending a revision, having concluded that a subjective sorting of the named populations into arbitrary geographic units is not advisable at this time. We also do not include the specimen from Tam-

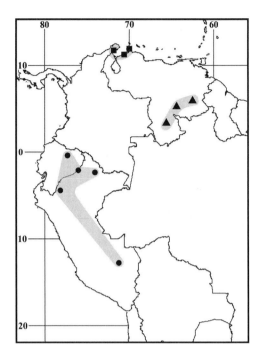

Map 24 Marginal localities for *Marmosa rubra* ●, *Marmosa tyleriana* ▲, and *Marmosa xerophila* ■

billo, Cajamarca, Peru, a juvenile (BM 81.9.7.28) listed under *Marmosa simonsi* by Tate (1933:121) with a question mark, and first reported as *Didelphys murina* by O. Thomas (1882:111). If this specimen proves to be a *M. robinsoni*, it would be the southern-most record for the species (see B. E. Brown 2004:69, Fig. 34).

Marmosa rubra Tate, 1931
Red Mouse Opossum

SYNONYM:

Marmosa rubra Tate, 1931:6; type locality "mouth of Rio Curaray," Loreto, Peru.

DISTRIBUTION: *Marmosa rubra* is known from the lowlands of eastern Ecuador, and Peru.

MARGINAL LOCALITIES (Map 24): ECUADOR: Napo, San José Abajo (Tate 1931); Pastaza, Montalvo (USNM 41451, not mapped); Pastaza, Río Pindo Yacu (USNM 274578). PERU: Loreto, mouth of Río Curaray (type locality of *Marmosa rubra* Tate); Cusco, Hacienda Villa Carmen (Pine 1972a); Amazonas, Huampami (J. L. Patton, Berlin, and Berlin 1982).

SUBSPECIES: We consider *M. rubra* to be monotypic.

NATURAL HISTORY: This species inhabits humid lowland forests where it has been taken with *M. murina* and *M. andersoni*.

REMARKS: *Marmosa rubra* can be distinguished from *M. murina* by the rich red-brown color pattern of the dorsal pelage, the dark median stripe on the forehead and rostrum, and by the less prominent postorbital processes. It differs from *M. lepida* by larger size (condylobasal length greater than 32 mm) and less projecting premaxillae. From *M. andersoni*, *M. rubra* differs by its shorter pelage, much smaller postorbital processes, and the well-defined dark medial line on the face. B. E. Brown's (2004:68) citation of V. Pacheco et al. (1993) for a Pakitza record in southeastern Peru appears to be an error. Cuartas and Muñoz (2003) included *M. rubra* as a member of the Colombian fauna, but did not cite specimen or literature records. Instead, they refer the reader to their map 3.10 where localities are indicated on the Guajira Peninsula (obviously intended for *M. xerophila*). The localities Cuartas and Muñoz (2003:209) intended for *M. rubra* (keyed in error to *M. robinsoni*) lack documentation.

Marmosa tyleriana Tate, 1931
Tyleria Mouse Opossum

SYNONYMS:

Marmosa tyleriana Tate, 1931:6; type locality "Central Camp, Mt. Duida Plateau, Upper Rio Orinoco," Amazonas, Venezuela.

Marmosa tyleriana phelpsi Tate, 1939:164; type locality "Auyan-tepui plateau, 1850 m," Bolívar, Venezuela.

DISTRIBUTION: *Marmosa tyleriana* is known only from the Guayanan Highland tepuis of Venezuela.

MARGINAL LOCALITIES (Map 24): VENEZUELA: Bolívar, Auyán-tepuí (type locality of *Marmosa tyleriana phelpsi* Tate); Bolívar, Meseta de Jaua (Ochoa 1985); Amazonas, Cerro Duida (type locality of *Marmosa tyleriana* Tate).

SUBSPECIES: We treat *Marmosa tyleriana* as monotypic.

NATURAL HISTORY: *Marmosa tyleriana* appears to be restricted to humid forests between 1,300 and 2,100 m on the northern and eastern tepuis of Venezuela.

REMARKS: The skull of *M. tyleriana* is long and narrow with slender zygomata and resembles the skull of species of *Marmosops* in outline. The supra- and postorbital region is finely beaded, but lacks lateral processes and projecting ridges. The palate is long and narrow, and has distinct palatine fenestrae. The dorsal pelage is dark brown and the ventral pelage is entirely gray based with buffy tips. The tail is weakly bicolored, dark fuscous dorsally and paler below. The tail of the holotype is unpigmented (white) for the distal half of its length, but other specimens have pigmented tails. *Marmosa tyleriana* can be distinguished from species of *Marmosops* and *Gracilinanus* by the lack of an anteromedial process on the auditory bullae. It is distinguishable from species of *Micoureus* and other *Marmosa* spp. by the narrow interorbital region and lack of laterally projecting postorbital ridges or processes.

Marmosa xerophila Handley and Gordon, 1979
Pale Mouse Opossum

SYNONYM:

Marmosa xerophila Handley and Gordon, 1979:68; type locality "La Isla, 15 m, near Cojoro, 37 km NNE Paraguaipoa, Depto. Guajira, Colombia."

DISTRIBUTION: *Marmosa xerophila* is known from the Venezuelan states of Falcón and Zulia, and adjacent departamento La Guajira, Colombia.

MARGINAL LOCALITIES (Map 24; from Handley and Gordon 1979): COLOMBIA: Guajira, near Cojoro, 34 to 37 km NNE of Paraguaipoa [Zulia, Venezuela]. VENEZUELA: Falcón, Capatárida; Falcón, 15 km SW of Pueblo Nuevo.

SUBSPECIES: *Marmosa xerophila* is monotypic.

NATURAL HISTORY: Handley (1976), reporting on 246 specimens from Venezuela and adjacent Colombia, stated that 98% were captured in xeric thorn forest, and that 81% were captured in trees and bushes, and 18% were taken on the ground. All localities are at elevations less than 100 m.

REMARKS: In their description of *M. xerophila*, Handley and Gordon (1979) provided the results of their comparisons with *M. robinsoni* and *M. murina*. As they noted, *M. xerophila* is unquestionably most similar to *M. robinsoni* in cranial and external characteristics. Smaller size and pale, thin pelage distinguishes *M. xerophila* from Venezuelan and Colombian *M. robinsoni*. Cranial features shared with *M. robinsoni*, such as less-flared, triangular-shaped supraorbital processes; relatively larger posterolateral palatal foramina; and convergent ridges on the parietals, distinguish *M. xerophila* from its other congeners.

The published type locality information, while equivalent, is not identical with that on the label of the holotype. The label bears the following locality information: "Colombia: Dpto Guajira, 119 km N. + 32 km W. of Maracaibo-Ven (la Isla) Alt 15 m."

Genus *Marmosops* Matschie, 1916
Alfred L. Gardner and G. Ken Creighton

The genus *Marmosops*, as treated here, comprises 15 species: *M. bishopi*, *M. cracens*, *M. creightoni*, *M. fuscatus*, *M. handleyi*, *M. impavidus*, *M. incanus*, *M. invictus*, *M. juninensis*; *M. neblina*; *M. noctivagus*, *M. ocellatus*, *M. parvidens*, *M. paulensis*, and *M. pinheiroi*. All except the Panamanian endemic *M. invictus* are found in tropical and subtropical South America (Pine 1981; Mustrangi and Patton 1997; Voss, Lunde, and Simmons 2001). Pleistocene and Recent fossil material are known from Lagoa Santa, Minas Gerais, Brazil (Winge 1893) and from a late

Pleistocene (Chapadmalalan) horizon in Argentina (Reig 1958). Cuartas and Muñoz (2003) included *M. invictus* as part of the Colombian fauna, cited Alberico et al. (2000) as the most recent authority, and plotted five unidentified localities in the Chocó region with the southernmost one at approximately 06°N latitude. Although we expect that *M. invictus* eventually will be recorded from Colombia, Alberico et al. (2000) listed the species only as probable. B. E. Brown (2004) did not indicate any Colombian record, and we are not aware any specimens vouchering its occurrence in Colombia at the present time.

Species span a considerable range in size (length of head and body 90–162 mm, tail 119–207 mm), in contrast to a smaller size range in species of *Micoureus*, *Cryptonanus*, and *Gracilinanus*. The tail is always longer than the head and body by a factor greater than 1.3 and may exceed 1.5 in larger individuals of some taxa. Dorsal and ventral pelage is variable among the species, but is not as long and woolly as in *Micoureus*. Mammae number 13 or fewer, and when only 9 mammae are present, they are restricted to the abdominal-inguinal region (4–1–4). When the count is 11 or 15 (5–1–5 or 7–1–7), the additional pair or pairs are located anteriorly on the body and usually referred to as pectoral mammae.

The scales on the tail are rhomboid, about equal in length and width, arranged in spiral rows, and relatively small (22–28 scales per cm). The middle hair of the interscalar hair triplets is distinctly petiolate, usually darkly pigmented, and longer and thicker than the lateral hairs. A ventral sulcus is evident along the terminal fifth of the tail.

The claws on the manus do not extend beyond the ends of the terminal digital pads. Manual digit III is longer than digits II and IV. The central palmar and plantar surfaces lack granulations. Tubercles on the proximal ventral surface of the digits are fused into transverse bars. The thenar and first interdigital pads of the pes are close together, but not fused (see Creighton 1984, for terminology of palmar and plantar features).

The skull lacks postorbital processes. The supraorbital region is either smoothly rounded, weakly ridged, or beaded with distinct temporal ridges extending posteriorly onto the parietals. In *Marmosops*, the skull is proportionally longer and narrower than skulls of *Marmosa*, *Micoureus*, *Hyladelphys*, *Cryptonanus*, and *Gracilinanus*. The rostrum is particularly long and slender. The nasals are usually widest at the maxillofrontal suture (except in *M. cracens* and *M. fuscatus*, in which the nasals have nearly straight lateral margins). The auditory bullae are relatively smaller than in *Cryptonanus*, *Gracilinanus*, or *Thylamys*. In most *Marmosops*, the alisphenoid portion of the bulla is less hemispherical than in other small Monodelphini, and appears compressed and conical. A stout anteromedial process of the alisphenoid is always present on the bulla and is per-forated by a canal leading to the foramen pseudovale. The hard palate contains a pair of maxillopalatine fenestrae and a pair of posterolateral palatal foramina, and may contain a pair of palatine fenestrae. When present, palatine fenestrae often consist of clusters of relatively small openings with irregular borders in the posterior palate. Posterolateral palatine foramina are always less than 1/2 the width of the last upper molar (M4) and are smaller than in *Cryptonanus*, *Gracilinanus*, and *Thylamys*. *Marmosops* has a distinctive dentition in which the upper canines are relatively short (height of crown) and often straight, laterally compressed, and bladelike, rather than curved, and scarcely higher than the second upper premolar (P2), except in older males, which tend to have longer upper canines. The P2 is equal to or greater than P3 in height and length. The lower canines of all *Marmosops* spp. are short and premolariform.

Marmosops spp. can be separated from all other small Monodelphini by the following combination of characteristics: strongly petiolate, darkly pigmented median interscalar tail bristles; narrow, elongated skull; conical auditory bullae that have a robust anteromedial process of the alisphenoid. *Marmosops* is most readily distinguished from *Marmosa* and *Micoureus* by the lack of laterally projecting postorbital processes, and by digit III of the manus longer than adjacent digits (digits III and IV are subequal in length in *Marmosa* and *Micoureus*). *Marmosops* can be distinguished from *Cryptonanus*, *Gracilinanus*, and *Thylamys* by the form of the auditory bullae, the shape and arrangement of tail scales and bristles, the short, premolariform lower canines, and the large P2.

SYNONYMS:

Didelphis: Lund, 1840a:19; part; not *Didelphis* Linnaeus, 1758.

Didelphys: Tschudi, 1845:148, 149; part; incorrect spelling of *Didelphis* Linnaeus; not *Didelphis* Linnaeus.

Grymaeomys Burmeister, 1854:138; part; proposed as a subgenus of *Didelphys* (= *Didelphis* Linnaeus).

Grymaeomys: Burmeister, 1856:79; part.

Marmosa: O. Thomas, 1894c:184; part; not *Marmosa* Gray, 1821.

Marmosa: O. Thomas, 1896b:313; not *Marmosa* Gray.

Thylamys: J. A. Allen and Chapman 1897:27; part; not *Thylamys* Gray, 1843.

Tylomys Trouessart, 1897:520; part; incorrect subsequent spelling of *Thylamys* Gray; not *Tylomys* W. Peters, 1866c (Rodentia).

Marmosa: O. Thomas, 1900a:221; not *Marmosa* Gray.

Thylamys: J. A. Allen, 1900c:198; not *Thylamys* Gray.

Marmosa: O. Thomas, 1911c:516; not *Marmosa* Gray.

Marmosa: Cabrera, 1913:10; not *Marmosa* Gray.

Marmosa: Osgood, 1913:94; not *Marmosa* Gray.

Marmosa: Miller, 1913a:31; not *Marmosa* Gray.

Marmosa: O. Thomas, 1913d:573; not *Marmosa* Gray.

Marmosa: Osgood, 1915:187; not *Marmosa* Gray.

Marmosops Matschie, 1916:267; type species *Didelphis incana* Lund, 1840a, by original designation (p. 271); proposed as a subgenus of *Didelphis* Linnaeus.

Marmosa: O. Thomas, 1920c:281; not *Marmosa* Gray.

Marmosa: H. E. Anthony, 1924:4; not *Marmosa* Gray.

Marmosa: O. Thomas, 1924a:236; not *Marmosa* Gray.

Marmosa: O. Thomas, 1927a:373; not *Marmosa* Gray.

Marmosa: O. Thomas, 1927b:607; not *Marmosa* Gray.

Marmosa: Tate, 1931:7–9, 13; not *Marmosa* Gray.

Marmosa: A. Miranda-Ribeiro, 1936:372; not *Marmosa* Gray.

Marmosa: Handley and Gordon, 1979:66; not *Marmosa* Gray.

Marmosa: Pine, 1981:61–63, 67; not *Marmosa* Gray.

KEY TO SOUTH AMERICAN SPECIES OF *MARMOSOPS*:

1. Nasals abruptly expanded at maxillofrontal suture. . .3
1'. Nasals either not expanded or only slightly expanded at maxillofrontal suture. 2
2. Dorsal pelage dark; ventral pelage entirely gray based . . . . *Marmosops fuscatus*
2'. Dorsal pelage pale gray brown, ventral pelage primarily self-colored white . . . . *Marmosops cracens*
3. Ventral pelage white or buffy, not entirely gray based (self-colored pelage sometimes restricted to a narrow mid-ventral strip). 7
3'. Ventral pelage gray based, hairs may be frosted with white or silvery tips; ventral pelage not buffy. 4
4. Size small, total length of adults less than 250 mm . . . . *Marmosops juninensis*
4'. Size larger, total length of adults more than 250 mm 5
5. Ventral pelage gray-based brown; white ventral pelage restricted to chin, or chin and upper throat; metacarpals and metatarsals brown, contrasting with whitish toes . . . . *Marmosops creightoni*
5'. Ventral pelage grey-based to self-colored white or cream, gray-based pelage may be restricted laterally; color of metacarpals and metatarsals not contrasting with toes 6
6. Upper canine anteriorly-posteriorly long; labial indentation on M2 and M3 shallow . . . . . *Marmosops neblina*
6'. Upper canine anteriorly-posteriorly short, labial indentation on M2 and M3 deep, sometimes cleft-like . . . . *Marmosops impavidus*
7. Size small, tail length less than 150 mm, condylobasal length less than 30 mm; upper canines short, blade-like, and crown about equal to P2 in height (canine may be longer in males). 8
7'. Size medium to large, tail length frequently exceeds 150 mm, condylobasal length more than 30 mm; upper canines curved and longer (height of crown) than P2 11

8. Dorsal pelage short and smooth, not woolly; palatine fenestrae either absent, or small and indistinct. 9
8'. Dorsal pelage woolly; palatine fenestrae distinct . . . . *Marmosops handleyi*
9. Upper canine usually with anterior and always with posterior cuspule on cingulum; anterior and posterior root of first upper premolar nearly equal in size, its tip only slightly anterior to middle of tooth. 10
9'. Upper canine lacks anterior cuspule on cingulum; anterior root of first upper premolar conspicuously smaller than posterior root, and the apex of crown well anterior to middle of tooth . . . . *Marmosops bishopi*
10. Venter broadly self-colored white, not restricted to midline by gray-based fur; lacrimal foramina contained within orbit (not opening laterally) . . . . *Marmosops parvidens*
10'. Venter with white fur restricted to midline by gray-based fur; lacrimal foramina exposed laterally (not opening within orbit) . . . . *Marmosops pinheiroi*
11. Ventral midline pelage white, may be restricted laterally by gray-based fur; tail length usually less than 170 mm; supraorbital region smooth, lacks prominent beading or strong ridges, postorbital constriction visible from above 12
11'. Ventral pelage self-colored white or creamy white, which extends onto underside of limbs; supraorbital region rounded (*M. ocellatus*), sharp edged, or with beaded ridges; tail usually longer than 160 mm 13
12. Upper canine anteriorly-posteriorly long; labial indentation (ectoflexus) on M2 and M3 shallow . . . . *Marmosops neblina*
12'. Upper canine anteriorly-posteriorly short; labial indentation (ectoflexus) on M2 and M3 deep, sometimes cleft-like. . . . . *Marmosops impavidus*
13. Dorsal pelage dark brown to reddish brown and relatively short; tail not becoming distinctly paler towards tip; supraorbital region distinctly beaded with elevated lateral ridges extending posteriorly onto parietals; postorbital constriction usually not visible from above; condylobasal length usually more than 35 mm. . . . . *Marmosops noctivagus*
13'. Dorsal pelage pale brown or grayish brown; tail fuscous, distinctly paler near the tip; supraorbital region either rounded or with a sharp edge, but lacking distinctly elevated ridges; postorbital constriction visible from above; condylobasal length variable. 14
14. Dorsal pelage pale grayish brown tinged with buff; ventral pelage self-colored cream; condylobasal length usually less than 36 mm; supraorbital region rounded; terminal 1/3 to 1/2 of tail pale . . . . *Marmosops ocellatus*

14'. Dorsal pelage darker grayish brown; condylobasal length frequently exceeds 37 mm; supraorbital margin with a sharp edge; tail variable in color 15

15. Incisive foramina short, terminating on level near posterior margin of canines; palatine fenestrae present; band of gray-based hair separates dorsal and ventral pelage . . . . *Marmosops incanus*

15'. Incisive foramina long, terminating at or behind level near middle of first premolars; palatine fenestrae absent; dorsal and ventral pelage not separated by band of gray-based hair. . . . . *Marmosops paulensis*

Marmosops bishopi (Pine, 1981)
Bishop's Slender Opossum
SYNONYMS:

Marmosa parvidens bishopi Pine, 1981:63; type locality "264 km N (by road) Xavantina (locality is at 12°51'S, 51°46'W), Serra do Roncador, Mato Grosso, Brazil."

[*Marmosops parvidens*] *bishopi*: Gardner, 1993:20; name combination.

Marmosops bishopi: Voss, Lunde, and Simmons, 2001:48; first use of current name combination.

DISTRIBUTION: *Marmosops bishopi* is known from Brazil, Peru, and Bolivia (Voss, Lunde, and Simmons 2001).

MARGINAL LOCALITIES (Map 25; from Voss, Tarifa, and Yensen 2004, except as noted): BRAZIL: Amazonas, Barro Vermelho (J. L. Patton, Silva, and Malcolm 2000); Mato Grosso, Serra do Roncador, 264 km N (by road) Xavantina (type locality of *Marmosa parvidens bishopi* Pine). BOLIVIA: Santa Cruz, El Refugio; Santa Cruz, San Rafael de Amboro; Cochabamba, Serranía Mosetenes; La

Paz, La Reserva (Mustrangi and Patton 1997; Voss, Lunde, and Simmons 2001); La Paz, Alto Río Madidi. PERU: Cusco, Hacienda Villa Carmen; Huánuco, Puerto Márquez (AMNH 67243; Voss, Lunde, and Simmons 2001).

SUBSPECIES: We consider *M. bishopi* to be monotypic.

NATURAL HISTORY: J. L. Patton, Silva, and Malcolm (2000) recorded *M. bishopi* (*M. parvidens, sensu* Pine 1981) in terra firme forest along the Rio Juruá in western Brazil. Two were caught on the ground, and four in trees at heights up to 2 m. Voss, Tarija, and Yensen (2004) documented *M. bishopi* from a variety of forest types in Bolivia, ranging from cloud forest and primary lowland rainforest to lowland dry forest.

REMARKS: The specimen from Puerto Márquez, Huánuco, Peru (AMNH 67243), identified by Tate (1933:205, as *Marmosa lepida lepida*), was referred to *M. bishopi* by Voss, Lunde, and Simmons (2001). The Rio Juruá record represents the animal from locality 12 cited by J. L. Patton, Silva, and Malcolm (2000:Table 7) under *Marmosops parvidens*. The Peruvian, Bolivian, and western Amazonian Brazil records mapped by B. E. Brown (2004:85) as *M. parvidens* represent *M. bishopi*. *Marmosops bishopi*, *M. parvidens*, and *M. pinheiroi* are similar in size and lack palatine fenestrae. They are most readily distinguishable by cingular cusps on upper canines (anterior cuspules lacking in *M. bishopi* and western populations of *M. parvidens*), shape of first upper premolar (tip conspicuously shifted anteriorly in *M. bishopi*), and ventral pelage (broadly self-colored white in *M. parvidens* and white restricted laterally by gray-based fur in *M. bishopi* and *M. pinheiroi*). Gardner (1993a) treated *M. bishopi* as a synonym of *M. parvidens*.

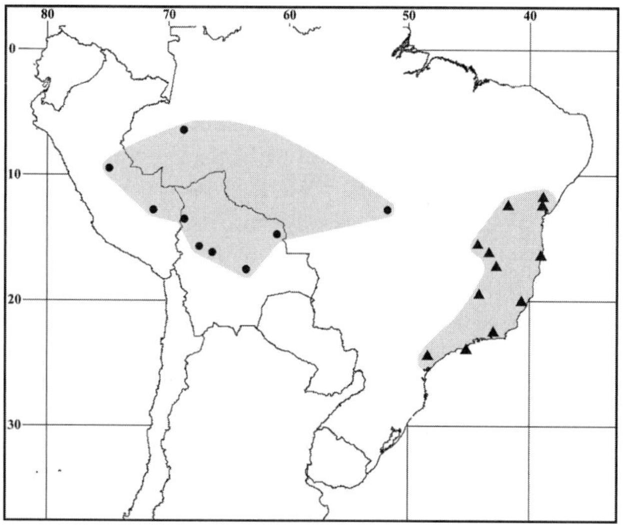

Marmosops cracens (Handley and Gordon, 1979)
Narrow-headed Slender Opossum
SYNONYMS:

Marmosa cracens Handley and Gordon, 1979:66; type locality "near Pastora (11°12'N, 68°37'W), 150 m, 14 km ENE Mirimire, Falcón, Venezuela."

[*Marmosa*] *crascens* Reig, Kirsch, and Marshall, 1985:342; incorrect subsequent spelling of *M. cracens* Handley and Gordon.

M[*armosops*]. *cracens*: Gardner and Creighton, 1989:4; first use of current name combination.

DISTRIBUTION: *Marmosops cracens* is known from only the type locality.

MARGINAL LOCALITY (Map 26): VENEZUELA: Falcón, near Pastora (type locality of *Marmosa cracens* Handley and Gordon).

SUBSPECIES: *Marmosops cracens* is monotypic.

NATURAL HISTORY: *Marmosops cracens* is known from only three specimens (Handley and Gordon 1979:66),

Map 25 Marginal localities for *Marmosops bishopi* ● and *Marmosops incanus* ▲

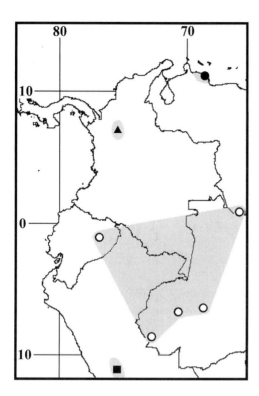

Map 26 Marginal localities for *Marmosops cracens* ●, *Marmosops handleyi* ▲, *Marmosops juninensis* ■, and *Marmosops neblina* ○

all taken on the "steep, moist, north slope of an isolated low mountain near La Pastora" at an elevation of about 150 m. Handley and Gordon (1979) described the habitat as mature evergreen forest with many vines and epiphytes, and with a closed subcanopy at 10 m and an irregular upper canopy at 25–30 m. All three specimens were caught on the ground: at the base of a tree, by a rotting stump, and in a clump of shrubs.

REMARKS: Handley's (1976:8) *Marmosa* sp. A refers to this species. Handley and Gordon (1979) compared *M. cracens* with congeners *M. fuscatus*, *M. parvidens*, *M. impavidus*, and *M. invictus*, as well as with *Gracilinanus marica* and *G. dryas*. Ranges of selected measurements (in mm) are: total length 233–237, tail 131–132, hind foot 16, ear from notch 24–26, condylobasal length 28.5–28.9. *Marmosops cracens* resembles *M. parvidens* from Colombia and Surinam (e.g., FMNH 95320). Further collecting in the coastal hills of the state of Falcón, Venezuela, should confirm whether or not local populations of *M. cracens* and *M. parvidens* maintain consistent morphological differences.

Marmosops creightoni Voss, Tarifa, and Yensen, 2004
Creighton's Slender Opossum

SYNONYMS:

"'*Marmosa*,' species not determined" S. Anderson, 1997: 151; specimens from 30 km by road N of Zongo, and from 18 km by road N of Zongo, La Paz, Bolivia.

Marmosops creightoni Voss, Tarifa, and Yensen, 2004:11; type locality "near the Saynani hydroelectric generating station (ca. 16°07′S, 68°05′W; 2500 m above sea level) in the valley of the Río Zongo, Departamento La Paz, Bolivia."

DISTRIBUTION: *Marmosops creightoni* is known from only the Valle de Zongo, between 2,000 and 3,000 m elevation, in departamento La Paz, Bolivia.

MARGINAL LOCALITIES (Map 27; from Voss, Tarifa, and Yensen 2004): BOLIVIA: La Paz, near the Saynani hydroelectric generating station (type locality of *Marmosops creightoni* Voss, Tarifa, and Yensen); La Paz, Cuticucho (not mapped).

SUBSPECIES: *Marmosops creightoni* is monotypic.

NATURAL HISTORY: According to Voss, Tarifa, and Yensen (2004), habitat in Valle de Zongo is humid montane forest. The upper collecting site, at 2,970 m, consisted of relatively undisturbed cloud forest, second growth forest dominated by bamboo, and clearings under power lines. The site is about 300 m below tree line. Two specimens were trapped here, one at the base of a tree, the other on a low branch. Forest at the second site, at 2,000 m, was mainly second growth with patches of undisturbed natural forest. The three specimens trapped at this site were taken on horizontal branches near large trees in a manner suggesting that they were accidentally caught while traveling along the branches, instead of coming to the bait. The third site is the type locality at 2,500 m on the southeastern side of the valley. The habitat consisted of disturbed forest with a ground cover of vines, bamboo, grasses, moss, and ferns. The holotype was caught on the ground in grass under a tree. Trees at all three sites were heavily festooned with mosses, lichens, ferns, and other epiphytes.

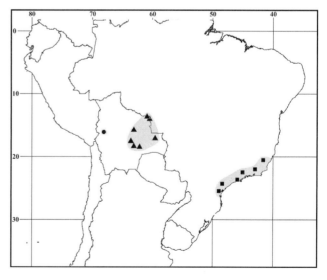

Map 27 Marginal localities for *Marmosops creightoni* ●, *Marmosops ocellatus* ▲, and *Marmosops paulensis* ■

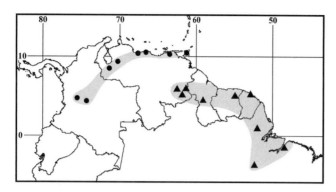

Map 28 Marginal localities for *Marmosops fuscatus* ● and *Marmosops pinheiroi* ▲

REMARKS: *Marmosops creightoni* can be distinguished from other *Marmosops* by its dark brownish coloration, lack of white or cream-colored thoracic or abdominal markings, and its dark feet with contrasting whitish toes.

Marmosops fuscatus (O. Thomas, 1896)
Dusky Slender Opossum

SYNONYMS: See under Subspecies.

DISTRIBUTION: *Marmosops fuscatus* is known from the eastern Andes of Colombia, the Mérida Andes and coastal range of Venezuela, and the island of Trinidad, Trinidad and Tobago.

MARGINAL LOCALITIES (Map 28; from Handley and Gordon 1979, except as noted; localities listed east to west): TRINIDAD AND TOBAGO: Trinidad, Caparo (type locality of *Thylamys carri* J. A. Allen and Chapman). VENEZUELA: Monagas, San Agustín, 5 km NW of Caripe; Miranda, Curupao; Aragua, Estación Biológica Rancho Grande; Trujillo, Hacienda Misisí; Mérida, Cafetal de Milla. COLOMBIA: Cundinamarca, Fusagasugá (type locality of *Marmosa perfusca* O. Thomas); Quindío, El Roble (Tate 1933).

SUBSPECIES: We recognize three subspecies of *M. fuscatus*.

M. f. carri (J. A. Allen and Chapman, 1897)
SYNONYMS:

Thylamys carri J. A. Allen and Chapman, 1897:27; type locality "Caparo," Trinidad, Trinidad and Tobago.

Tylomys carri: Trouessart, 1897:520; name combination (included under Rodentia).

[*Didelphys (Marmosa)*] *carri*: Trouessart, 1905:856; name combination.

[*Marmosa (Thylamys)*] *carri*: Cabrera, 1919:40; name combination.

This is the subspecies on the island of Trinidad.

M. f. fuscatus (O. Thomas, 1896)
SYNONYMS:

Marmosa fuscata O. Thomas, 1896b:313; type locality "Rio Abbarregas [= Río Albarregas], Merida, Venezuela."

[*Didelphys (Marmosa)*] *fuscata*: Trouessart, 1898:1240; name combination.

[*Marmosa (Marmosa)*] *fuscata*: Cabrera, 1919:36; name combination.

Marmosops fuscatus: Gardner and Creighton, 1989:4; first use of current name combination.

The nominate subspecies is known from the Mérida Andes and coastal range of Venezuela.

M. f. perfuscus (O. Thomas, 1924)
SYNONYM:

Marmosa perfusca O. Thomas, 1924a:236; type locality "Fusagasuga," Cundinamarca, Colombia.

This subspecies is known only from the eastern Andes of Colombia.

NATURAL HISTORY: *Marmosops fuscatus* inhabits montane wet forests and cloud forest from near sea level to over 2,300 m. Handley (1976) reported on specimens from the coastal mountains of northern Venezuela, where 36 were taken in humid evergreen forests, 26 in cloud forest, and 2 in gardens. In these habitats, 71% were caught on trees and vines, 29% on the ground, and all specimens were taken at sites near streams or in other moist areas. Linares (1998:46) stated that the reproductive period coincides with the rainy season and up to five young have been found with a single female. Cordero (2001) reported lactating females taken in November and December, and juveniles caught in September, November, December, and March, in lowland (40 m) forest habitat in northern Venezuela. On this basis, Cordero (2001) suggested that the breeding season probably extends from May to January or early February. He also estimated litter size as between 7 and 9, based on the number of teats secreting milk. The karyotype is $2n = 14$, $FN = 24$ (Reig and Sonnenschein 1970).

REMARKS: The dorsal pelage of *M. fuscatus* has a distinctly grayish cast, in contrast to the warm chestnut brown color of *M. noctivagus*. The hairs of the ventral pelage are entirely gray based with silvery-white tips, features that readily separates *M. fuscatus* from all other *Marmosops*. The only other *Marmosops* that has entirely gray-based ventral pelage is *M. creightoni*, which differs from *M. fuscatus* in lacking white markings behind the throat. The nasals are long and narrow, and show almost no lateral expansion at the maxillofrontal suture (in contrast with *M. creightoni* and *M. impavidus*). Males differ markedly in size from females in a large series from Venezuela at the USNM.

For example, the condylobasal lengths of skulls of males with fully-erupted dentitions are as much as 20% larger than the condylobasal lengths of adult females. Ranges in condylobasal length for 10 males and 10 females selected at random from the USNM series are: males, 34.0–39.5 mm; females, 30.3–34.9 mm. Contrary to Tate's (1933) assertion that *carri* averages larger than mainland *M. fuscata*, specimens from the mainland are larger than the island subspecies.

Cuartas and Muñoz (2003) claimed that *M. fuscatus* was known from the departments of Antioquia, Cauca, and Risaralda, but omitted Quindío. Their map (p. 210) contains several additional localities plotted; however, we are not aware of specimens of *M. fuscatus* that voucher their records.

Marmosops handleyi (Pine, 1981)
Handley's Slender Opossum

SYNONYMS:

Marmosa handleyi Pine, 1981:67; type locality "9 Km S Valdivia, Antioquia, Colombia."

Marmosops handleyi: Gardner and Creighton, 1989:4; first use of current name combination.

[*Marmosops*] *impavidus*: Hershkovitz, 1992b:14; part.

DISTRIBUTION: *Marmosops handleyi* is known from only the vicinity of the type locality in Antioquia, Colombia.

MARGINAL LOCALITIES (Map 26): COLOMBIA: Antioquia, 9 km S of Valdivia (type locality of *Marmosa handleyi* Pine).

SUBSPECIES: *Marmosops handleyi* is monotypic

NATURAL HISTORY: This species occurs in montane humid forest where it was caught at 1,400 and 1,700 m.

REMARKS: This species is most similar to *M. parvidens* with which it is sympatric; the two differ in size and coloration. *Marmosops handleyi* "is larger, with longer, slightly woolly fur with more contrastingly pale tips dorsally" (Pine 1981:68). Pine wrote that the adult *M. parvidens* he examined from the type locality of *M. handleyi* has shorter pelage, is much paler, and has an immaculate white venter. Cuartas and Muñoz Arango (2003:210) indicated four localities in their distribution map of *M. handleyi*, but did not list localities. As far as we have been able to determine, the species is represented by only the holotype and paratype.

Marmosops impavidus (Tschudi, 1845)
Tschudi's Slender Opossum

SYNONYMS:

D[*idelphys*]. *impavida* Tschudi, 1845:149; type locality "der mittleren und tiefern Waldregion" [p. 151]; interpreted by Cabrera (1958:16) as "Montaña de Vitoc, cerca de Chanchamayo," Junín, Peru.

Marmosa caucae O. Thomas, 1900a:221; type locality "R. Caquetá, a tributary of the Cauca, near Cali"; corrected by Tate (1933:178) to "Río Cauquita, near Cali, 1000 m," Valle de Cauca, Colombia.

[*Didelphys (Marmosa)*] *caucae*: Trouessart, 1905:856; name combination.

Thylamys caucae: J. A. Allen, 1912:73; name combination.

Marmosa madescens Osgood, 1913:94; type locality "Tambo Ventija, 10 miles east of Molinopampa," Amazonas, Peru.

Marmosa purui Miller, 1913a:31; type locality "Hyutanaham, upper Purus River," Amazonas, Brazil.

Marmosa sobrina O. Thomas, 1913d:573; type locality "Mindo," Pichincha, Ecuador.

[*Didelphis (Marmosa)*] *impavida*: Matschie, 1916:270; name combination.

[*Didelphis (Marmosops)*] *caucae*: Matschie, 1916:270; name combination.

[*Didelphis (Marmosops)*] *sobrina*: Matschie, 1916:270; name combination.

[*Didelphis (Marmosops)*] *madescens*: Matschie, 1916:271; name combination.

[*Didelphis (Thylamys)*] *purui*: Matschie, 1916:271; name combination.

[*Marmosa (Marmosa)*] *caucae*: Cabrera, 1919:36; name combination.

[*Marmosa (Marmosa) impavida*] *impavida*: Cabrera, 1919:37; name combination.

[*Marmosa (Marmosa)*] *madescens*: Cabrera, 1919:37; name combination.

[*Marmosa (Marmosa)*] *purui*: Cabrera, 1919:38; name combination.

[*Marmosa (Marmosa)*] *sobrina*: Cabrera, 1919:39; name combination.

Marmosa oroensis H. E. Anthony, 1922:3; type locality "Portovelo," El Oro, Ecuador.

Marmosa celicae H. E. Anthony, 1922:4; type locality "Celica," Loja, Ecuador.

Marmosa caucae ucayaliensis Tate, 1931:9; type locality "Lagarto, Rio Ucayali," Ucayali, Peru.

Marmosa caucae purui: Tate, 1933:184; name combination.

Marmosa [*(Marmosa)*] *impavida caucae*: Cabrera, 1958:16; name combination.

Marmosa [*(Marmosa)*] *impavida madescens*: Cabrera, 1958:16; name combination.

Marmosa [*(Marmosa)*] *impavida oroensis*: Cabrera, 1958:17; name combination.

M[*armosops*]. *impavidus*: Gardner and Creighton, 1989:4; first use of current name combination.

M[*armosops*]. *dorothea*: Mustrangi and Patton, 1997:21; not *Marmosa dorothea* O. Thomas,1911c.

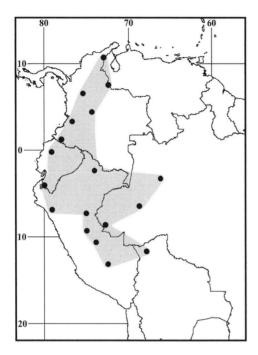

Map 29 Marginal localities for *Marmosops impavidus* •

DISTRIBUTION: *Marmosops impavidus* occurs along the western slope of the Andes from Colombia to southern Ecuador, and on the eastern slope from the Táchira Andes of Venezuela, through Colombia, Ecuador, Peru, and Bolivia, and eastward into Brazil at least as far as the Juruá basin. The species has been reported from the Darién of Panama.

MARGINAL LOCALITIES (Map 29; from Tate 1933, for *Marmosa caucae*, except as noted: COLOMBIA: Guajira, Villanueva (USNM 280889). VENEZUELA: Táchira, Buena Vista (Handley 1976). COLOMBIA: Cundinamarca, Fusagasugá. PERU: Loreto, mouth of Río Curaray. BRAZIL (J. L. Patton, Silva, and Malcolm 2000): Amazonas, Colocação Vira-Volta; Amazonas, Barro Vermelho; Acre, Igarapé Porongaba. BOLIVIA: Pando, Palmira (Voss, Tarifa, and Yensen 2004). PERU: Cusco, Torontoy (type locality of *Marmosa caucae albiventris* Tate); Ucayali, Largato (type locality of *Marmosa caucae ucayaliensis* Tate); Huánuco, Leguía; Cajamarca, Taulís; Loreto, Contamana. ECUADOR: Loja, Celica (type locality of *Marmosa celicae* H. E. Anthony); Pichincha, Santo Domingo de los Colorados. COLOMBIA: Nariño, Ricuarte; Valle del Cauca, Río Cauquita, near Cali (type locality of *Marmosa caucae* O. Thomas); Antioquia, La Bodega (USNM 293774, not mapped); Antioquia, San Francisco (Cuartas and Muñoz 2003).

SUBSPECIES: We have not delineated subspecies in this account and treat *M. impavidus* as monotypic (see Remarks).

NATURAL HISTORY: This species occurs in cloud forest, montane wet forest, and lowland rainforest at elevations from near 65 to 2,200 m. Handley (1976), reporting on eight specimens from the Táchira Andes of Venezuela, noted that five were taken in trees, vines, or shrubs and three were taken on the ground; seven came from cloud forest habitats and one from evergreen montane forest. J. L. Patton, Silva, and Malcolm (2000) took their specimens in undisturbed and second-growth terre firme forest. Three were taken at heights up to 2 m; eight were caught on the ground. Based on specimens in the AMNH that Tate (1933) reported as *Marmosa caucae*, lactating females were caught in August and October, and juveniles in all seasons. Adult females collected by J. L. Patton, Silva, and Malcolm (2000) during the rainy season were parous; a juvenile was caught during the dry season (November). The karyotype is $2n = 14$, FN = 24; the X chromosome is a small metacentric, smaller than the smallest autosome; the Y is an even smaller acrocentric (J. L. Patton, Silva, and Malcolm 2000).

REMARKS: *Marmosops impavidus* is intermediate in size between *M. noctivagus* and *M. parvidens*. The dorsal pelage is most commonly a warm chestnut brown. Most specimens have a narrow band (10–15 mm wide) of self-colored white or creamy white hair on the venter restricted laterally by gray-based fur. Some specimens from Colombia and Peru, however, have ventral pelage that is almost entirely gray based (as in *M. fuscatus*), but the tips of the fur are buff rather than bright white. Other specimens from western Brazil and southern Peru have more extensively white venters. The lack of gular glands and lack of an all-white ventral pelage distinguishes *M. impavidus* from *M. parvidens* and most *M. noctivagus*. Slightly smaller size, mid-ventral all-white pelage, and relatively larger and less widely spaced auditory bullae help to separate *M. impavidus* from *M. creightoni* and *M. fuscatus*. Selected measurements (in mm) for specimens having fully erupted dentition: length of head and body 110–125, tail 135–170, hind foot 17–20, condylobasal length 29.8–35.1. J. L. Patton, Silva, and Malcolm (2000) associated the specimen from Tita (MSB 58511) with *M. impavidus*, based on its cytochrome-b haplotype.

Cabrera (1958:16) gave the type locality of *M. impavidus* as "Montaña de Vitoc, cerca de Chanchamayo," Junín, Peru, which is the same type locality restriction given to *Didelphys noctivaga* Tschudi (1845:148), by Tate (1933:153). Because the type specimen of *M. impavidus* can not be found, O. Thomas (1927:606–07) wrote "Probably *impavida* had best be considered as a synonym of *noctivaga*, from which Tschudi gives no essential distinction, and which has not infrequently a white or whitish belly."

The Neblina, Amazonas, Venezuela record listed and mapped by B. E. Brown (2004:75, 79) refers to *M. neblina*, which she also listed on page 80 and mapped on page 83. We have not attempted to identify subspecies in this account because the kinds and extent of variation within and between populations of *M. impavidus* are insufficiently known. To list named taxa and their probable synonyms as subspecies before the species is adequately revised would only contribute more confusion.

Marmosops incanus (Lund, 1840)
Gray Slender Opossum
SYNONYMS:

[*Didelphis*] *incana* Lund, 1839b:233; *nomen nudum*.

D[*idelphis*]. *incana* Lund, 1840a:19 [1841b:237]; type locality "Rio das Velhas," Lagoa Santa, Minas Gerais, Brazil.

Didelphys [*(Metachirus)*] *incana*: Burmeister, 1854:137; name combination.

Grymaeomys scapulatus Burmeister, 1856:79; type locality "Minas gerais," Brazil; restricted to Porto Alegre, Minas Gerais, by Matschie (1916:271).

M[*icoureus*]. *incanus*: O. Thomas, 1894c:186; name combination.

M[*armosa*]. *incana*: O. Thomas, 1896b:313; name combination.

[*Didelphys (Marmosa)*] *scapanus* Trouessart, 1898:1238; in synonymy of *Didelphys (Marmosa) cinerea*; incorrect subsequent spelling of *scapulatus* Burmeister.

[*Didelphys (Marmosa)*] *incana*: Trouessart, 1898:1241; name combination.

[*Didelphis (Marmosops)*] *scapulata*: Matschie, 1916:271; name combination.

[*Marmosa (Marmosa)*] *incana*: Cabrera, 1919:37; name combination.

Marmosa incana bahiensis Tate, 1931:8; type locality "Lamarão, Bahia, Brazil."

Marmosa scapulata: Tate, 1933:167; name combination.

M[*armosops*]. *incanus*: Gardner and Creighton, 1989:4; first use of current name combination.

DISTRIBUTION: *Marmosops incanus* is endemic to Brazil, where it is known from the states of Bahia, Espírito Santo, Minas Gerais, Rio de Janeiro, and São Paulo.

MARGINAL LOCALITIES (Map 25; from Mustrangi and Patton 1997, except as noted): BRAZIL: Bahia, Lamarão (type locality of *Marmosa incana bahiensis* Tate); Bahia, São Gonçalo dos Campos; Bahia, Porto Seguro; Espírito Santo, Rio São José (MZUSP 6222); Rio de Janeiro, Theresópolis; São Paulo, Ilha de São Sebastião (MZUSP 253); São Paulo, Fazenda Intervales; Minas Gerais, Sete Lagoas (Oliveira, Lorini, and Persson 1992); Minas Gerais, Turmalina; Minas Gerais, Mocambinho (Oliveira, Lorini, and Persson 1992); Minas Gerais, Januária (Câmara, Oliveira, and Meyer 2004); Bahia, Seabra.

SUBSPECIES: We consider *M. incanus* to be monotypic.

NATURAL HISTORY: *Marmosops incanus* is a member of the forest fauna of southeastern Brazil. The species also may occur in the eastern campo cerrado and in gallery forests of the upper Rio Paraná drainage. Tate (1933) thought that *M. incanus* undergoes a seasonal molt, based on individuals taken during the austral winter (May through August) that had long, lax, and brownish-gray dorsal pelage; whereas specimens taken from late September through December showed a conspicuous molt line beginning behind the ears and progressing back to the scapular region. Fresh pelage consists of noticeably shorter and stiffer hairs, resulting in the impression of a collar. The type of *Grymaeomys scapulatus* Burmeister has this molt pattern, but the date it was taken is unknown. These pelage types were shown by Oliveira, Lorini, and Persson (1992) to be related to age. Depending on the season, the dorsal pelage of *M. incanus* is either long (about 12 mm), lax, and mousy brown (immatures), or short, stiff, and brownish gray (adults). Mustrangi and Patton (1997) suggested semelparity (individuals of both sexes live long enough to mate, bring off a litter, and then die; see Pine, Dalby, and Matson 1985) as an explanation for this phenomenon.

REMARKS: The venter is creamy white with the hairs unicolored (not gray-based). The lateral band of gray-based hair separating the dorsal from the ventral pelage, plus shorter incisive foramina, and the presence of palatine fenestrae, distinguish *M. incanus* from sympatric *M. paulensis*. Larger size, a white venter, and the pattern of tail scales and bristles (spirals of rhomboid scales; long, petiolate, and dark median bristle of each triplet) separate *M. incanus* from *Gracilinanus microtarsus* and from *G. agilis*. Short upper canines, premolariform lower canines, and the lack of postorbital processes serve to distinguish *M. incanus* from sympatric species of *Marmosa* and *Micoureus*. Specimens of *M. incanus* are most similar to those of *M. noctivagus*, which occurs in the western Amazon basin.

Cabrera's (1958:17) inclusion of *Didelphis grisea* Lydekker, and *Grymaeomys griseus* Winge, in his synonymy of *Marmosa incana*, are references to misidentifications or incorrect allocations by Lydekker (1887) and Winge (1893). The label of specimen MZUSP 253 bears the notation that it is the type of *Marmosa thomasi*. However, we can find no evidence of publication of that name and consider *Marmosa thomasi* to be a *nomen nudum*.

Marmosops juninensis (Tate, 1931)
Junín Slender Opossum
SYNONYMS:

Marmosa juninensis Tate, 1931:13; type locality "Utcuyacu, between Tarma and Chanchamayo, Province of Junín, Peru."

Marmosa parvidens juninensis: Pine, 1981:64; name combination.

[*Marmosops parvidens*] *juninensis*: Gardner, 1993:20; name combination.

Marmosops juninensis: Emmons and Feer, 1990:22; first use of current name combination.

DISTRIBUTION: *Marmosops juninensis* is known from only the Chanchamayo Valley of Peru.

MARGINAL LOCALITIES (Map 26): PERU: Junín, Utcuyacu (type locality of *Marmosa juninensis* Tate); Junín, near Tarma (Voss, Lunde, and Simmons 2001, not mapped).

SUBSPECIES: We treat *M. juninensis* as monotypic.

NATURAL HISTORY: Nothing is known concerning the natural history of *M. juninensis*.

REMARKS: According to Voss, Lunde, and Simmons (2001), *M. juninensis* can be distinguished from *M. parvidens* and *M. pinheiroi* by the presence of palatine fenestrae and the absence of accessory cuspules on upper canines. Male *M. juninensis* also have a small carpal tubercle, whereas, the carpal tubercle in males of *M. parvidens* and *M. pinheiroi* is enlarged and spoon-shaped (Lunde and Schutt 2000).

Marmosops neblina Gardner, 1990
Neblina Slender Opossum
SYNONYMS:

M[*armosa*]. *impavida neblina* Gardner, 1988:698; *nomen nudum*.

Marmosops impavidus neblina Gardner, 1990:414; type locality "Camp VII (00°50′40″N, 65°58′10″W), 1800 m, Cerro de la Neblina, Territorio Federal Amazonas, Venezuela."

M[*armosops*]. *neblina*: Mustrangi and Patton, 1997:22; first use of current name combination.

Marmosops impavidus neblinae: Linares, 1998:49; incorrect subsequent spelling of *Marmosops neblina* Gardner, 1990).

DISTRIBUTION: *Marmosops neblina* is known from Brazil, Ecuador, and southern Venezuela (J. L. Patton, Silva, and Malcolm 2000).

MARGINAL LOCALITIES (Map 26): VENEZUELA: Amazonas, Cerro de la Neblina, Camp VII (type locality of *Marmosops impavidus neblina* Gardner). BRAZIL (J. L. Patton, Silva, and Malcolm 2000): Amazonas, Jainu; Amazonas, Igarapé Nova Empressa; Acre, Igarapé Porongaba. ECUADOR, Pastaza, Tiguino (Mustrangi and Patton 1997).

SUBSPECIES: We treat *M. neblina* as monotypic.

NATURAL HISTORY: All specimens from Venezuela were caught on the ground at higher elevations on the Cerro de la Neblina tepui, primarily in elfin forest. J. L. Patton,

Silva, and Malcolm (2000) caught two 1.5 m above the ground and nine on the ground in várzea forest or in disturbed river bank habitats. The mite *Cummingsia gardneri* Price and Emerson, 1986, was described from specimens from Venezuela (host identified as *Marmosa impavida*). The karyotype is $2n = 14$, FN = 24; the X chromosome is a small metacentric and the Y an even smaller metacentric (J. L. Patton, Silva, and Malcolm 2000).

REMARKS: *Marmosops neblina* was originally described as a subspecies of *Marmosops impavidus*. Mustrangi and Patton (1997) and J. L. Patton, Silva, and Malcolm (2000) found *M. neblina* to be sympatric with *M. impavidus* at two localities along the Rio Juruá in western Brazil. One or more of the several names based on specimens from Andean foothill and eastern lowland populations in Ecuador and Peru, currently identified as *M. impavidus*, may prove to be senior synonyms of *M. neblina*.

Marmosops noctivagus (Tschudi, 1845)
White-bellied Slender Opossum
SYNONYMS:

D[*idelphys*]. *noctivaga* Tschudi, 1845:148; type locality "der mittleren und tiefen Waldregion" [p. 151]; restricted by Tate (1933:153) to Montaña de Vitoc, near Chanchamayo, Río Perené drainage, Junín, Peru.

Thylamys keaysi J. A. Allen, 1900c:198; type locality "Juliaca"; corrected by J. A. Allen (1901a:41) to Inca Mines, Río Inambari, Puno, Peru.

Marmosa Keaysi: O. Thomas, 1901a:190; name combination.

[*Didelphys (Marmosa)*] *keaysi*: Trouessart, 1905:856; name combination.

[*Didelphys (Marmosa)*] *noctivaga*: Trouessart, 1905:857; name combination.

Marmosa dorothea O. Thomas, 1911c:516; type locality "Rio Solocame, 67°W., 16°S.," La Paz, Bolivia.

Marmosa polita Cabrera, 1913:10; type locality "los alrededores de la confluencia de los rios Coca and Napo"; further restricted (p. 14) to "Río Napo, Coca," Orellana, Ecuador.

Marmosa impavida neglecta Osgood, 1915:187; type locality "Yurimaguas," Loreto, Peru.

[*Didelphis (Marmosa)*] *neglecta*: Matschie, 1916:270; name combination.

[*Didelphis (Marmosa)*] *polita*: Matschie, 1916:270; name combination.

[*Didelphis (Marmosops)*] *noctivaga*: Matschie, 1916:271; name combination.

[*Didelphis (Marmosops)*] *dorothea*: Matschie, 1916:271; name combination.

[*Didelphis (Thylamys)*] *keaysi*: Matschie, 1916:271; name combination.

[*Marmosa (Marmosa)*] *dorothea*: Cabrera, 1919:36; name combination.

[*Marmosa (Marmosa) impavida*] *neglecta*: Cabrera, 1919:37; name combination.

[*Marmosa (Marmosa)*] *noctivaga*: Cabrera, 1919:38; name combination.

[*Marmosa (Marmosa)*] *polita*: Cabrera, 1919:38; name combination.

[*Marmosa (Thylamys)*] *keaysi*: Cabrera, 1919:40; name combination.

Marmosa collega O. Thomas, 1920c:281; type locality "Villa Braga, Rio Tapajoz," Pará, Brazil.

Marmosa noctivaga lugenda O. Thomas, 1927a:373; type locality "Yurac Yacu," San Martín, Peru.

Marmosa leucastra O. Thomas, 1927b:607; type locality "Tambo carrizal, about 40 miles south of Chachapoyas," Amazonas, Peru.

Marmosa yungasensis Tate, 1931:7; type locality "Pitiguaya, Rio Unduavi, Yungas," La Paz, Bolivia.

Marmosa caucae albiventris Tate, 1931:9; type locality "Torontoy," Cusco, Peru.

Marmosa noctivaga polita: Tate, 1933:155; name combination.

Marmosa noctivaga collega: Tate, 1933:157; name combination.

Marmosa noctivaga keaysi: Tate, 1933:158; name combination.

Marmosa noctivaga neglecta: Tate, 1933:159; name combination.

Marmosa noctivaga dorothea: Tate, 1933:160; name combination.

Marmosa stollei A. Miranda-Ribeiro, 1936:372; type locality "Aripuanan," Amazonas, Brazil.

Marmosa [*(Marmosa)*] *impavida albiventris*: Cabrera, 1958:16; name combination.

Marmosa [*(Marmosa)*] *noctivaga keaysi*: Cabrera, 1958:21; name combination.

Marmosa [*(Marmosa)*] *yungasensis*: Cabrera, 1958:26; name combination.

Marmosops dorothea: Gardner and Creighton, 1989:4; name combination.

M[*armosops*]. *noctivagus*: Gardner and Creighton, 1989:4; first use of current name combination.

Marmosa (Marmosops) noctivaga keaysi: S. Anderson, Riddle, Yates, and Cook, 1993:14; name combination.

M[*armosops*] *noctivaga keaysi*: S. Anderson, 1993:6; name combination.

M[*armosops*]. *i*[*mpavidus*]. *albiventris*: S. Anderson, 1997:153; name combination.

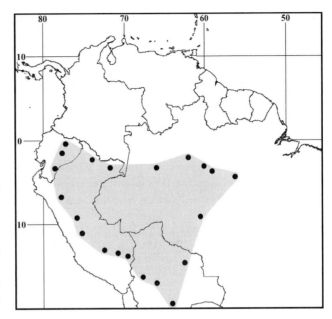

Map 30 Marginal localities for *Marmosops noctivagus* ●

DISTRIBUTION: *Marmosops noctivagus* is known from the western Amazon basin of eastern Ecuador, Peru, Bolivia, and from Brazil south of the Rio Amazonas.

MARGINAL LOCALITIES (Map 30; from Tate 1933, except as noted): ECUADOR: Napo, San José de Sumaco. PERU: Loreto, mouth of Río Curaray; Loreto, Pebas. BRAZIL: Amazonas, Lago Vai-Quem-Quer (J. L. Patton, Silva, and Malcolm 2000); Amazonas, Macaco (J. L. Patton, Silva, and Malcolm 2000); Amazonas, Igarapé Cacão Pereira; Amazonas, Rosarinho; Pará, Villa Braga; Mato Grosso, Aripuanan (type locality of *Marmosa stollei* A. Miranda-Ribeiro). BOLIVIA: Santa Cruz, Perseverencia (S. Anderson 1997); Chuquisaca, Río Limón (S. Anderson 1997); Cochabamba, 4.4 km by road N of Tablas Monte (S. Anderson 1997); La Paz, Pitiguaya (type locality of *Marmosa yungasensis* Tate). PERU: Puno, Inca Mines (type locality of *Marmosa keaysi* J. A. Allen); Cusco, Marcapata; Cusco, Machu Picchu; Junín, San Ramón; Huánuco, Tingo María; Amazonas, Tambo Carrizal. ECUADOR: Morona-Santiago, Gualaquiza; Pastaza, Canelos.

SUBSPECIES: We treat *M. noctivagus* as monotypic, pending revision (see Remarks).

NATURAL HISTORY: This species is an inhabitant of the lowland humid forests of the western Amazon basin and adjacent Andean slopes, where its elevational range is from below 300 up to about 1,500 m. Timm and Price (1988) described *Cummingsia izori* as a new species of louse, whose host is *M. noctivagus* from southeastern Peru.

REMARKS: *Marmosops noctivagus* is the larger species of *Marmosops* found in the western Amazon basin and on the lower slopes of the Andes. It can be distinguished from *M. ocellatus* by its darker dorsal pelage, longer skull (condylobasal length usually more than 36 mm), and relatively narrow, distinctly ridged supraorbital region, with the ridges occluding the postorbital constriction when viewed from above. The white or creamy-white ventral pelage and sharp lateral ridges on the margins of the frontals distinguish *M. noctivagus* from *M. fuscatus* and from most *M. impavidus*. Several authors (e.g., S. Anderson et al. 1993; S. Anderson 1997; Salazar-Bravo et al. 2003; Vargas and Simonetti 2004) recognized *M. dorothea* as a separate species endemic to Bolivia. Herein, we treat *dorothea* as a synonym of *M. noctivagus*.

Subspecies are not delineated in this account because the kinds and extent of variation within and among populations represented by the named kinds of *M. noctivagus* are insufficiently known. To try to distinguish subspecies before the species is revised would only result in further confusion.

Marmosops ocellatus (Tate, 1931)
Spectacled Slender Opossum

SYNONYMS:

Marmosa ocellata Tate, 1931:7; type locality "Buenavista, Department of Santa Cruz, Bolivia."

Marmosa [(Marmosa)] ocellata: Cabrera, 1958:23; name combination.

Marmosa (Marmosops) dorothea ocellata: S. Anderson, Riddle, Yates, and Cook, 1993:14; name combination.

[*Marmosops dorothea*] *ocellata*: Gardner, 1993:19; name combination and incorrect gender concordance.

Thylamys macrurus: S. Anderson, 1997:163; name combination, but not *Didelphys macrura* Olfers.

M[armosops]. ocellatus: Voss, Tarifa, and Yensen, 2004:2; first use of current name combination.

DISTRIBUTION: *Marmosops ocellatus* occurs in central and eastern Bolivia. The species is expected in adjacent Mato Grosso, Brazil.

MARGINAL LOCALITIES (Map 27; from Voss, Tarifa, and Yensen 2004): BOLIVIA: Santa Cruz, Lago Caimán; Santa Cruz, Mangabalito; Santa Cruz; Aserradero Pontons; Santa Cruz, Tita; Santa Cruz, Hacienda El Pelicano; Santa Cruz, Buenavista; Santa Cruz, 6 km by road W of Ascención.

SUBSPECIES: We consider *M. ocellatus* to be monotypic.

NATURAL HISTORY: Based on locality records, *M. ocellatus* inhabits relatively arid lowland deciduous forests of eastern Bolivia at elevations below 700 m. The species undoubtedly occurs in similar habitat in Brazil. The mammae count is 13 (6–1–6), with 9 teats (4–1–4) abdominal-inguinal and 2 pairs pectoral (Voss, Tarifa, and Yensen 2004). The karyotype is unknown.

REMARKS: Gardner (1993a), S. Anderson (1997), and B. E. Brown (2004) treated *M. ocellatus* as a synonym of *M. dorothea*, but later Gardner (2005b) treated *M. ocellatus* as a synonym of *M. impavidus*. Voss, Tarifa, and Yensen (2004) recognized *M. ocellatus* as a valid species, identified the Bolivian specimen J. L. Patton, Silva, and Malcolm (2000:53) called *M. impavidus* as *M. ocellatus*, and relegated *M. dorothea* to the synonymy of *M. noctivagus*. *Marmosops ocellatus*, currently known from only departamento Santa Cruz, Bolivia, is expected in adjacent Brazil. We are treating all locality records listed by S. Anderson (1997:152), under *M. dorothea* from departamento Santa Cruz, as representing *M. ocellatus*. The species can be distinguished from other *Marmosops* of similar size by its relatively smooth interorbital region; expanded nasals at the maxillary-lacrimal juncture; grayish brown dorsal pelage; gray-based ventral pelage, which is limited to lateral margins of venter at mid-body; pectoral mammae; and by the distal third or more of the tail being uniformly pale.

Marmosops parvidens (Tate, 1931)
Delicate Slender Opossum

SYNONYMS:

Marmosa parvidens Tate, 1931:13; type locality "Hyde Park, 30 miles up the Demarara River," Demerara-Mahaica, Guyana.

M[armosa]. parvidentata Tate, 1933:44; incorrect subsequent spelling of *Marmosa parvidens* Tate.

Marmosa parvidens parvidens: Pine, 1981:60; name combination.

M[armosops]. parvidens: Gardner and Creighton, 1989:4; first use of current name combination.

DISTRIBUTION: *Marmosops parvidens* is recorded from northern Venezuela, the eastern slope of the Andes in Colombia, and Guyana, French Guiana, and northern Brazil.

MARGINAL LOCALITIES (Map 31; from Pine 1981, except as noted): *Western distribution.* VENEZUELA: Falcón, 19 km NW of Urama. COLOMBIA: Boyacá,

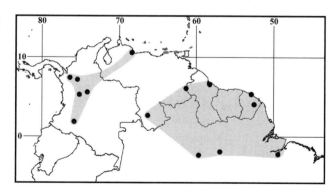

Map 31 Marginal localities for *Marmosops parvidens* ●

Muzo; Huila, Acevedo; Caldas, Samaná; Antioquia, Villa Arteaga; Antioquia, 9 km S of Valdivia. *Eastern distribution.* GUYANA: Demerara-Mahaica, Hyde Park (type locality of *Marmosa parvidens* Tate). FRENCH GUIANA (Voss, Lunde, and Simmons 2001): Paracou; Arataye. BRAZIL (Voss, Lunde, and Simmons 2001): Pará, Ilha do Taiuna; Amazonas, Boca Rio Piratucu; Amazonas, 80 km N of Manaus. VENEZUELA: Amazonas, Capibara; Bolívar, Km 125 (Handley 1976).

SUBSPECIES: We treat *M. parvidens* as monotypic, pending revision.

NATURAL HISTORY: This species is found in the Guiana Region north of the Amazon and in Colombia and northwestern Venezuela. *Marmosops parvidens* inhabits the humid tropical forests at elevations from sea level to about 1,700 m. Handley (1976) noted that three of five specimens collected in Venezuela were caught on the ground, two were trapped in trees, and all were taken near streams, or other wet sites, in evergreen forest.

Pine (1981) summarized information on the habitats of the northwestern Venezuelan material. According to L. H. Emmons (pers. comm.), *M. parvidens* frequently can be seen in the early evening in the shrub understory of tropical humid forests, where it apparently forages for arthropods and other foods.

REMARKS: Pine (1981) recognized five subspecies and gave the distribution and a morphological description for each. Here we follow Voss, Lunde, and Simmons (2001) in placing *woodalli* in the synonymy of *Marmosops pinheiroi*, and by recognizing four species (including *parvidens*) instead of the five subspecies recognized by Pine (1981). We suspect that a comprehensive revision of these slender opossums will reveal that the Colombian and western Venezuelan population unit represents an additional species. Eisenberg (1989:38) overlooked the Colombian distribution reported by Pine (1981). The western Amazonian records mapped by B. E. Brown (2004:85), along with those from Peru and Bolivia, represent records for *M. bishopi*. Voss, Tarifa, and Yensen (2004) reviewed the identity and distribution of *M. bishopi* in Bolivia.

Marmosops paulensis (Tate, 1931)
Brazilian Slender Opossum
SYNONYMS:
Marmosa incana paulensis Tate, 1931:8; type locality "Therezopolis, Rio de Janeiro, São Paulo [sic], Brazil"; corrected to "Therezopolis, Rio de Janeiro" by Tate (1933:166).
[*Marmosops incanus*] *paulensis*: Gardner, 1993:19; name combination.
Marmosops paulensis: Mustrangi and Patton, 1997:21; first use of current name combination.

DISTRIBUTION: *Marmosops paulensis* is endemic to southeastern Brazil, where it is known from the states of Minas Gerais, Rio de Janeiro, São Paulo, and Paraná.

MARGINAL LOCALITIES (Map 27; from Mustrangi and Patton 1997; localities listed from north to south): BRAZIL: Minas Gerais, Parque Nacional do Caparaó; Rio de Janeiro, Bonsucesso; Minas Gerais, Passa Quatro; São Paulo, Estação Biológica de Boracéia; São Paulo, Fazenda Intervales; Paraná, Roça Nova.

SUBSPECIES: We consider *M. paulensis* to be monotypic.

NATURAL HISTORY: Mustrangi and Patton (1997) reported *M. paulensis* from seven localities in montane and cloud forest habitats above 800 m along the coastal mountains of southeastern Brazil. They distinguished *M. paulensis* from other species of *Marmosops*, based on analyses of mtDNA sequences, and reaffirmed the identity of *M. paulensis* through morphological and color pattern comparisons. This species shows the same age differences in pelage seen in *M. incanus* with long, soft fur found in immature individuals and short fur in adults (Mustrangi and Patton 1997). This species also may be semelparous and semelseminant (see Pine, Dalby, and Matson 1985). Warneck (1937) described the louse *Cummingsia intermedia* from a host he identified as *Marmosa incana paulensis* from Itatiaya, Rio de Janeiro, Brazil. Confirmation of the identity of the type host will require reexamination of the specimen.

REMARKS: Tate (1931), mistakenly assuming that the type came from the state of São Paulo, gave the name *paulensis* to the population he described as a subspecies of *Marmosa incana*. Gardner (1993a) treated *Marmosa incana paulensis* Tate as a synonym of *Marmosops incanus* (Lund, 1940a). Mustrangi and Patton (1997) found both taxa sympatric at four of the seven localities where *M. paulensis* is known. *Marmosops paulensis* can be distinguished from *M. incanus* by the divergent posterior tips of the nasals, longer incisive foramina (terminate at or behind level of middle of first premolars, versus in line with posterior margin of canines in *M. incanus*), absence of palatine fenestrae, smaller upper canines (less than 2.5 mm), inflated auditory bullae, and the lateral band of gray-based fur that borders and restricts the white ventral pelage (Mustrangi and Patton 1997).

Marmosops pinheiroi (Pine, 1981)
Pinheiro's Slender Opossum
SYNONYMS:
Marmosa parvidens pinheiroi Pine, 1981:61; type locality "Rio Amapari, Serra do Navio (0°59′N, 52°03′W), Amapá, Brazil."

Marmosa parvidens woodalli Pine, 1981:62; type locality "'Nova Area Experimental,' Utinga (the wooded area surrounding the Belém waterworks), Belém (1°27'S, 48°29'W), Pará, Brazil."

[*Marmosops parvidens*] *pinheiroi*: Gardner, 1993:20; name combination.

Marmosops pinheiroi: Voss, Lunde, and Simmons, 2001: 49; first use of current name combination.

M[*armosops*]. *pirenhoi* Flores, 2004:133; incorrect subsequent spelling of *Marmosops pinheiroi* (Pine).

DISTRIBUTION: *Marmosops pinheiroi* is known from Venezuela, Guyana, Surinam, French Guiana, and Brazil.

MARGINAL LOCALITIES (Map 28; from Voss, Lunde, and Simmons 2001, except as noted): VENEZUELA: Bolívar, 85 km SE of El Dorado. SURINAM: Brokopondo, Brokopondo (Pine 1981). FRENCH GUIANA: Paracou. BRAZIL: Amapá, Rio Amapari, Serra do Navio (type locality of *Marmosa parvidens pinheiroi* Pine); Pará, Utinga (type locality of *Marmosa parvidens woodalli* Pine); Pará, Rio Xingu, east bank, 52 km SSW Altamira. GUYANA: Potaro-Siparuni, Iwokrama Reserve. VENEZUELA: Bolívar, Churi-tepui; Bolívar, Auyán-tepui.

SUBSPECIES: We consider *M. pinheiroi* to be monotypic.

NATURAL HISTORY: Voss, Lunde, and Simmons (2001) collected 19 *M. pinheiroi* at Paracou, French Guiana, in the following habitats: well-drained primary forest, 4; creek-side primary forest, 5; swampy primary forest, 6; and second-growth forest, 4. All were taken at night; 13 on the ground and 6 from 0.3 to 1.5 m above ground on vertical stems and lianas. The holotype of *woodalli* contained 7 embryos (ca. 3 mm CR) when caught on 14 June (Pine 1981). The karyotype is unknown.

REMARKS: Voss, Lunde, and Simmons(2001) suggest the Rio Xingu specimen (USNM 549294) may represent an undescribed species. B. E. Brown (2004) listed this record under *M. parvidens*, but did not map the locality under either *M. parvidens* or *M. pinheiroi*. She also included *Marmosa parvidens woodalli* Pine as a synonym of *Marmosops parvidens*, instead of as a synonym of *M. pinheiroi* Pine.

Genus *Micoureus* Lesson, 1842

Alfred L. Gardner and G. Ken Creighton

The genus *Micoureus* contains six species (*M. demerarae*, *M. constantiae*, *M. paraguayanus*, *M. phaeus*, *M. regina*, and *M. alstoni*) found from Belize southward through Central and South America into northern Argentina. In South America, these species occur in forested habitats from sea level to about 2,500 m. The earliest known fossils are from late Pleistocene to Recent cave deposits in the vicinity of Lagoa Santa, Minas Gerais, Brazil (Winge 1893).

Opossums of this genus are of medium size (head and body 125–220 mm, tail 175–265 mm), among living didelphids; however, some species are the largest of the Monodelphini. The dorsal pelage is relatively long, gray or grayish brown, and often appears thick and woolly. The ventral pelage is shorter, varies from pale gray to creamy yellow or buffy orange, and its texture is comparatively smooth and less woolly. The mammae are abdominal, or inguinal and abdominal, and number from 9 to 15. The tail is much longer than the combined head and body length (ratio of tail to head and body length always greater then 1.3), and the number of caudal vertebrae varies from 32 to 35. Tail scales are rhomboid, usually wider than long (width/length ratio from 1.0 to 1.5), moderately coarse (14–16 rows per cm), and arranged in a spiral pattern. Interscalar hairs of the tail are slender (never petiolate), approximately equal in length and thickness, and occur in triplets under the posterior margin of each scale. The tail is unicolored or mottled, and has a ventral sulcus on the distal quarter (distal prehensile surface). The distal third of the tail is white in *M. demerarae* and *M. constantiae*. Long pelage, like that of the body, clothes the base of the tail for a distance as far as 50 mm in some species. The manus and pes are comparatively broad, as in *Marmosa*, and the claws on the manus extend slightly beyond the terminal digital pads (see Creighton 1984, for palmar and plantar morphology). The thenar pad lies against the first interdigital pad on the manus, but these pads are fused on the pes. The fourth interdigital pad lies against the hypothenar pad on the manus, and the two are either in contact or fused on the pes. The central palmar and plantar surfaces are smooth. Digit IV of the pes is the longest and the ratio of its length to the length of the hind foot is greater than 0.45. The second and third interdigital pads of both manus and pes are triangular and about as wide as long. The ventral surface of the digits have transverse bars instead of a granulated surface.

The skull is relatively large and has broad zygomatic arches, in contrast to the narrower skulls of *Cryptonanus*, *Marmosops*, and *Gracilinanus*. Postorbital processes are well developed and beaded. The nasals are either abruptly (*M. demerarae*) or gradually and broadly (*M. regina*) expanded at the maxillofrontal suture. Temporal ridges are either subparallel or convergent posteriorly, but do not form a sagittal crest. Lambdoidal crests are prominent. The auditory bullae are comparatively small and set wide apart (ratio of distance between bullae to breadth of one bulla greater than 1.5). The alisphenoid component of the bulla is hemispherical and lacks an anteromedial process. The ectotympanic is expanded laterally to form part of the ventral wall of the bulla. A pair of maxillopalatine fenestrae and a pair of posterolateral foramina perforate the palate; the latter are moderate in length, usually about equal to 1/2 the width of the last upper molar. Palatine fenestrae also are present in some species. The crown of I1 is higher than the crown of other incisors and I1 is separated from I2 by a gap. Incisors increase in length anteroposteriorly from I2 through I5, P2 is larger than P3, and M3 is the widest of the upper molar series. Both upper and lower canines are long and curved.

Species of *Micoureus* can be distinguished from species of *Cryptonanus*, *Gracilinanus*, *Marmosops*, and *Thylamys* by the following combination of features: prominent supraorbital processes; small, hemispherical, and widely spaced auditory bullae that lack an anteromedial process on the alisphenoid; gray, woolly pelage; and particolored tails with rhomboid scales arranged in a spiral pattern, and with fine, subequal, and nonpetiolate interscalar hairs. Opossums of this genus can be separated from species of *Marmosa* by their gray, longer, and woollier pelage; short, fine, and subequal tail hairs; and rounded anterior projection of the premaxillae.

SYNONYMS:

Didelphis: Temminck, 1824b:46; part; not *Didelphis* Linnaeus, 1758.

Micoureus Lesson 1842:186; type species *Didelphis cinerea* Temminck, 1824b, by subsequent designation (O. Thomas 1888b:340).

Philander: Gray 1843b:101; part; not *Philander* Brisson, 1762; not *Philander* Beckmann, 1772; not *Philander* Tiedemann, 1808.

[*Didelphys (]Marmosa[)*]: Trouessart, 1905:855; part; incorrect subsequent spelling of *Didelphis* Linnaeus; not *Marmosa* Gray, 1821.

Caluromys: Matschie, 1916:269; part; not *Caluromys* J. A. Allen, 1900b.

Micoures Reig, Kirsch, and Marshall, 1985:337; incorrect subsequent spelling of *Micoureus* Lesson.

Micoures Massoia, 1988:6; incorrect subsequent spelling of *Micoureus* Lesson.

KEY TO SOUTH AMERICAN SPECIES OF *MICOUREUS*:

1. Ventral pelage yellow to creamy white, not gray based; tail tip white . . . . *Micoureus constantiae*
1'. Ventral pelage buffy to orange, mostly gray based; tail tip color is variable and can be white 2
2. Tail tip white or tail mottled distally; long fur covers base of tail for 30 mm or more 3
2'. Tail tip usually fuscous; long fur covers base of tail for 20 mm or less 5
3. Distribution east and south of the Amazon basin from

southern Bahia, Brazil, south into Paraguay . . . . *Micoureus paraguayanus*
3'. Distribution in Colombia, Venezuela, the Guianas, the greater Amazon basin of Brazil, and eastern Ecuador, Peru, and Bolivia 4
4. Distribution in northwestern Colombia . . . . *Micoureus alstoni*
4'. Distribution in northern and eastern Colombia and in Venezuela, the Guianas, the greater Amazon basin of Brazil, and eastern Ecuador, Peru, and Bolivia . . . . *Micoureus demerarae*
5. Pelage relatively short; palatine fenestrae present . . . . *Micoureus phaeus*
5'. Pelage longer and more woolly; palatine fenestrae lacking . . . . *Micoureus regina*

Micoureus alstoni (J. A. Allen, 1900)

Alston's Mouse Opossum

SYNONYMS:

Didelphys cinerea: Alston, 1880:199; not *Didelphis cinerea* Temminck, 1824.

Didelphys (Micoureus) cinerea: J. A. Allen, 1891:218; name combination, not *Didelphis cinerea* Temminck.

Marmosa cinerea: J. A. Allen, 1897:43; name combination, not *Didelphis cinerea* Temminck.

Caluromys alstoni J. A. Allen, 1900:189; type locality "Tres Rios," Cartago, Costa Rica.

[*Didelphys (Marmosa)*] *alstoni*: Trouessart, 1905:855; name combination.

Marmosa cinerea nicaraguae O. Thomas, 1905b:313; type locality "Bluefields," Zelaya, Nicaragua.

M[*armosa*]. c[*inerea*]. *Alstoni*: O. Thomas, 1905b:313; name combination.

[*Didelphis (Caluromys)*] *nicaraguae*: Matschie, 1916:269; name combination.

[*Didelphis (Caluromys)*] *alstoni*: Matschie, 1916:269; name combination.

M[*icoureus*]. *alstoni*: Gardner and Creighton, 1989:4; first use of current name combination.

DISTRIBUTION: *Micoureus alstoni* occurs in the northwestern lowlands of Colombia. Elsewhere, the species is found from the Bocas del Toro region of Panama north into Belize.

MARGINAL LOCALITY (Map 32): COLOMBIA: Chocó, Quibdó (Cuartas and Muñoz 2003).

SUBSPECIES: Hall (1981) recognized two subspecies; the nominate form, *M. a. alstoni*, has been recorded from Colombia.

NATURAL HISTORY: *Micoureus alstoni*, the largest mouse opossum in Central America, is an arboreal forest dweller. The diet consists of fruit, flower parts, bird eggs,

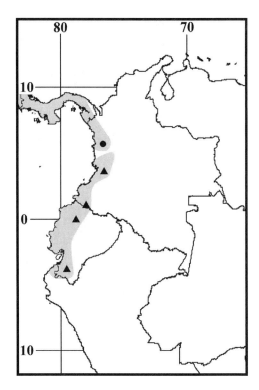

Map 32 Marginal localities for *Micoureus alstoni* ● and *Micoureus phaeus* ▲

small vertebrates, and insects. These opossums construct leaf nests in palms and in vine tangles. Tate (1933) gave the mammary formula as 5–1–5, but stated that the usual number of everted and visible teats are 4–1–4. *Micoureus alstoni* has a $2n = 14$, $FN = 24$ karyotype, and acrocentric X and Y chromosomes (Reig et al. 1977).

REMARKS: We base the occurrence of *M. alstoni* in South America on the single locality record from Quibdó, Chocó, Colombia, cited by Cuartas and Muñoz (2003:128). These authors indicated two Panamanian and two Colombian localities, the southern most of which appears to correspond to Quibdó, on their map (p. 211). However, Cuartas and Muñoz did not identify the more northern Colombian locality nor did they identify either of the Panamanian localities. The identification of their material needs to be confirmed. Tate (1933) identified specimens as *M. alstoni* from several localities in western and northcentral Colombia. These citations probably are the basis for Eisenberg's (1989) map of the species in the Magdalena Valley and along the Pacific lowlands of Colombia. Apparently based on Hershkovitz's notes on identifications, B. E. Brown (2004) divided Tate's records for Colombian *M. alstoni*, and assigned some to *M. alstoni* and the remainder to *M. demerarae meridae*.

Tate (1933:69) commented on the "Panama fault" (often referred to today as the Panama Gap), in reference to the hiatus in the distribution of animals identified as *M. alstoni*

between northern Panama and Colombia. If the identification of Colombian specimens as *M. alstoni* by Cuartas and Muñoz (2003) is correct, the hiatus exists to this day. We have assigned Tate's inter-Andean Colombian specimens to *Micoureus demerarae* O. Thomas. *Micoureus alstoni* and *M. demerarae* are similar and, while some workers may consider them to be conspecific, the evidence presented by J. L. Patton, Silva, and Malcolm (2000) supports treating them as separate species. These species are in need of revision.

Micoureus constantiae (O. Thomas, 1904)
White-bellied Woolly Mouse Opossum

SYNONYMS:

Marmosa constantiae O. Thomas, 1904b:243; type locality "Chapada," Mato Grosso, Brazil.

[*Didelphys (Marmosa)*] *constantiae*: Trouessart, 1905:856; name combination.

[*Didelphis (Caluromys)*] *constantiae*: Matschie, 1916:270; name combination.

[*Marmosa (Marmosa)*] *constantiae*: Cabrera, 1919:36; name combination.

Marmosa budini O. Thomas, 1920a:195; type locality "Altura de Yuto, Rio San Francisco," Jujuy, Argentina.

Marmosa constantiae budini: Tate, 1933:76; name combination.

M[icoureus]. constantiae: Gardner and Creighton, 1989:4; first use of current name combination.

Marmosa (Micoureus) cinerea budini: S. Anderson, Riddle, Yates, and Cook 1993:14; name combination.

Micoureus constantiae budini: S. Anderson, 1997:9; name combination.

Micoureus constantiae constantiae: S. Anderson, 1997:9; name combination.

DISTRIBUTION: *Micoureus constantiae* is known from the lowlands of eastern Bolivia and adjacent Brazil southward into northern Argentina.

MARGINAL LOCALITIES (Map 33): BOLIVIA (S. Anderson 1997): Pando, La Cruz; Beni, across river from Cascajal. BRAZIL: Rondônia, São Antônio de Guaporé (CM 1902); Mato Grosso, Chapada (type locality of *Marmosa constantiae* O. Thomas); Mato Grosso do Sul, Campo Grande (C. O. C. Vieira 1955); Mato Grosso do Sul, São Marcos Rd., 10 km NE of Urucúm (USNM 390023). ARGENTINA: Jujuy, Parque Nacional Calilegua (Heinonen-Fortabat and Chebez 1997). BOLIVIA (S. Anderson 1997, except as noted): Tarija, 5 km NNW of Entre Rios; Chuquisaca, Río Limón; Cochabamba, 4.4 km by road N of Tablas Monte; Cochabamba, Yungas (CM 5287); La Paz, Río Machariapo (Emmons 1991).

SUBSPECIES: S. Anderson (1997) tentatively recognized two subspecies; however we treat *M. constantiae* as monotypic, pending revision.

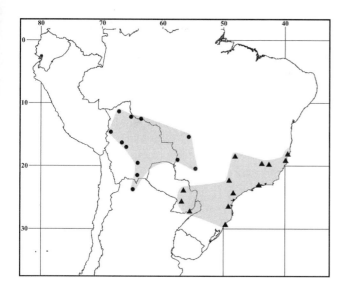

Map 33 Marginal localities for *Micoureus constantiae* ● and *Micoureus paraguayensis* ▲

NATURAL HISTORY: Based on Bolivian samples, S. Anderson (1997) reported a female with five young in August, another lactating in May, and four nonreproductive females taken in July, August. Brennan (1970) reported chiggers (*Eutrombicula alfreddugesi*) recovered from Bolivian specimens.

REMARKS: We list only four Brazilian localities for *M. constantiae*, whereas B. E. Brown (2004) included Lagoa Santa and Rio Jordão, both in Minas Gerais. These records are based on Lund (1839c) and Thomas (1901k), respectively; we assign both records to *M. paraguayanus*. *Micoureus constantiae* differs from *M. demerarae*, *M. paraguayanus*, and *M. regina*, principally in having creamy-white to yellow, self-colored ventral pelage (not gray-based). Tate (1933) reported a mammary formula of 7–1–7 in contrast to 4–1–4 or 5–1–5 in other members of the genus.

Micoureus demerarae (O. Thomas, 1905)
Woolly Mouse Opossum

SYNONYMS: See under Subspecies.

DISTRIBUTION: *Micoureus demerarae* is in Colombia, Venezuela, the Guianas, Brazil, southwestern Peru, and eastern Bolivia.

MARGINAL LOCALITIES (Map 34; most of the following records were originally cited under *Marmosa cinerea*): VENEZUELA: Carabobo, San Esteban (Tate 1933); Miranda, Curupao (Handley 1976); Monagas, Maturín (Tate 1933); Delta Amacuro, Río Acoíma (Linares and Rivas 2004). GUYANA: Cuyuni-Mazaruni, Kartabo (H. E. Anthony 1921b); Upper Demerara-Berbice, Comaccka (type locality of *Marmosa cinerea demerarae* O. Thomas). SURINAM: Paramaribo, Leonsberg (USNM 319941). FRENCH GUIANA: River Arataye, right bank, ca. 5 km downstream

from Saut Pararé (USNM 578005). BRAZIL: Amapá, Serra do Navio (USNM 392052); Amapá, Rio Araguarí, Município de Marcapá (Carvalho 1962, not mapped); Pará, Belém (Carvalho and Toccheton 1969); Ceará (type locality of *Marmosa limae* O. Thomas; precise locality unknown, not mapped); Pernambuco, Dois Irmãos (Pohle 1927); Bahia, Aritaguá (Mares et al. 1981); Tocantins, Palma (Tate 1933); Mato Grosso, Serra do Roncador (Pine, Bishop, and Jackson 1970); BOLIVIA: Santa Cruz, Flor de Oro (J. L. Patton, Silva, and Malcolm 2000); La Paz, Ixiamas (Emmons 1991). PERU: Madre de Dios, Cerros de Távara (Emmons and Romo 1994); Madre de Dios, Cocha Cashu (Terborgh, Fitzpatrick, and Emmons 1984); Cusco, Cashiriari-2 (S. Solari et al. 2001c); Cusco, 2 km SW of Tangoshiari (J. L. Patton, Silva, and Malcolm 2000). BRAZIL (J. L. Patton, Silva, and Malcolm 2000): Acre, Nova Vida; Amazonas, Seringal Condor; Amazonas, Barro Vermelho; Amazonas, Lago do Capiranga; Amazonas, Comunidade Colina. VENEZUELA: Amazonas, Cerro de la Neblina, Camp VI (Gardner 1990); Amazonas, Esmeralda, near foot of Mt. Duida (type locality of *Marmosa demerarae esmeraldae* Tate); Amazonas, 32 km S of Puerto Ayacucho (Handley 1976); Bolívar, Caicara del Orinoco (Reig et al. 1977). COLOMBIA (Tate 1933): Meta, Villavicencio; Quindío, Calarca; Antioquia, Jericó. VENEZUELA: Mérida, Hacienda San Pedro (Pérez-Hernández 1990).

SUBSPECIES: We tentatively recognize five subspecies of *M. demerarae*.

M. d. areniticola (Tate, 1931)
SYNONYMS:

Marmosa demararae areniticola Tate, 1931:2: type locality "Arabupu, foot of Mt. Roraima," Bolívar, Venezuela; incorrect subsequent spelling of *Marmosa cinerea demerarae* O. Thomas.

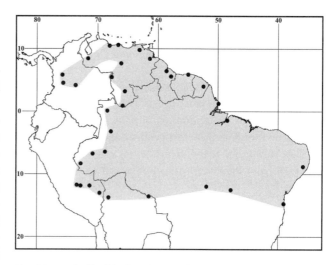

Map 34 Marginal localities for *Micoureus demerarae* ●

Marmosa cinerea areniticola: Cabrera, 1958:13; name combination.

[*Micoureus demerarae*] *areniticola*: Gardner, 1993:20; name combination.

This subspecies is known from only the Roraima region of southeastern Venezuela, and is expected in adjacent Guyana and Brazil.

M. d. demerarae (O. Thomas, 1905)
SYNONYMS:

Marmosa cinerea demerarae O. Thomas 1905b:313; type locality "Comaccka, 80 miles up Demerara River," Upper Demerara-Berbice, Guyana.

[*Didelphis (Caluromys)*] *demerarae*: Matschie, 1916:269; name combination.

[*Marmosa (Marmosa) cinerea*] *demerarae*: Cabrera, 1919:36; name combination.

Marmosa demararae demararae: Tate, 1933:62; name combination and incorrect subsequent spelling of *Marmosa cinerea demerarae* O. Thomas.

M[*icoures*]. *demararae*: Reig, Kirsch, and Marshall, 1987:39; first use of current name combination; incorrect subsequent spelling of *Micoureus* Lesson, and incorrect subsequent spelling of *Marmosa cinerea demerarae* O. Thomas.

The nominate subspecies occurs in the Guianan lowlands.

M. d. dominus (O. Thomas, 1920)
SYNONYMS:

Marmosa domina O. Thomas, 1920c:280; type locality "Villa Braga, R. Tapajoz," Pará, Brazil.

Marmosa limae O. Thomas, 1920c:282; type locality "Ceará," Brazil.

Marmosa (Micoureus) cinerea: Pohle, 1927:241; name combination, but not *Didelphis cinerea* Temminck.

Marmosa cinerea pfrimeri A. Miranda-Ribeiro, 1936:365; type locality "Palmas," Tocantins, Brazil.

Marmosa cineerea pfrimeri: Ávila-Pires, 1968:165; incorrect subsequent spelling of, but not *Didelphis cinerea* Temminck.

[*Micoureus*] *cinereus*: Reig, Kirsch, and Marshall, 1985:342; part; name combination.

[*Micoureus demerarae*] *domina*: Gardner, 1993:20; name combination and incorrect gender concordance.

This subspecies is found in the eastern Amazon basin of Brazil from Pará south into Bahia and west into Mato Grosso.

M. d. esmeraldae (Tate, 1931)
SYNONYMS:

Marmosa demararae esmeraldae Tate, 1931:2; type locality "Esmeralda, near foot of Mt. Duida, upper Rio

Orinoco," Amazonas, Venezuela; incorrect subsequent spelling of *Marmosa cinerea demerarae* O. Thomas.

Marmosa cinerea esmeraldae: Cabrera, 1958:13; name combination.

[*Micoureus demerarae*] *esmeraldae*: Gardner, 1993:20; first use of current name combination.

This subspecies is known from the state of Amazonas, Venezuela.

M. d. meridae (Tate, 1931)
SYNONYMS:

Marmosa demararae meridae Tate, 1931:3; type locality "'Cafetos de Mérida,' Mérida," Venezuela; incorrect subsequent spelling of *Marmosa cinerea demerarae* O. Thomas.

Marmosa cinerea meridae: Cabrera, 1958:13; name combination.

[*Micoureus demerarae*] *meridae*: Gardner, 1993:20; first use of current name combination.

This subspecies is known from the inter-Andean valleys and adjacent lowlands of Colombia and Venezuela.

NATURAL HISTORY: *Micoureus demerarae* lives in tropical humid forests at elevations below 1,200 m. This species is often relatively abundant in second growth and disturbed habitats, and is frequently found in plantations. Of 31 specimens reported by Handley (1976) from Venezuela, 53% were caught on trees, vines, or logs, and 47% on the ground. He stated that 22 came from evergreen forest, the remainder from openings or disturbed areas, and all but one were taken near streams or in other mesic habitats. In Venezuela, O'Connell (1979) also caught *M. demerarae* in trees 50% of the time, and when released, they escaped via trees or vines 85% of the time. She estimated densities as high as one per hectare and determined home range size to be near 0.10 hectare, approximating that of other small didelphids in the area. She found no evidence of reproduction in the winter dry season, but recorded a lactating female in April. J. L. Patton, Silva, and Malcolm (2000) found reproductively active (based on having attached young, lactating, or having orange-stained fur surrounding the teats) females along the Rio Juruá during both rainy (February to April) and dry seasons (September to November). One female had seven attached young. Eisenberg (1989) mentioned a female with a litter of nine from Surinam. The karyotype is $2n = 14$, FN $= 24$ (reported as *Marmosa cinerea* by Reig et al. 1977); or FN $= 20$ (J. L. Patton, Silva, and Malcolm 2000).

REMARKS: The extensive range of *M. demerarae* covers the lowlands along the Caribbean coast, central and eastern Amazonia, and eastern Bolivia and southeastern Peru. We suspect that the dot on B. E. Brown's (2004:99) map representing a locality east of the island of Trinidad is the result of a coordinate-coding error and not a misplaced record for Trinidad where, to our knowledge, *M. demerarae* has not been found. Cabrera (1958) recognized six subspecies under *Marmosa cinerea* (*sensu lato*). The name *Marmosa cinerea*, long used for this group of opossums, is an unavailable (preoccupied) name that corresponds, in its now restricted concept, to the taxon in Paraguay, northeastern Argentina, and southeastern Brazil, we recognize as *Micoureus paraguayanus*. Although we recognize five of the subspecies Cabrera (1958) listed under *Marmosa cinerea*, geographic variation is extensive (J. L. Patton, Silva, and Malcolm 2000; J. L. Patton and Costa 2003), and the species needs revision. Cabrera (1958) apparently overlooked the name *Marmosa limae*. We have assigned the specimens from the inter-Andean region of Colombia identified as *M. alstoni* by Tate (1933) and B. E. Brown (2004) to *M. demerarae*. The population unit of *M. demerarae* in the southwestern Amazon basin of Brazil, Peru, and Bolivia apparently lacks a name.

James L. Patton, Silva, and Malcolm (2000) found *M. demerarae* and *M. regina* to be sympatric along the Rio Juruá in western Brazil. *Micoureus demerarae* has a more extensively furred base of the tail (usually exceeds 24 mm) and can be further distinguished from *M. regina* by its extensively gray-based ventral pelage, longer ears, shorter toothrows, and narrower skull. The frontals average wider and the supraorbital shelf is positioned farther back in *M. regina* than it is in *M. demerarae*, resulting in a broader postorbital constriction (see Table 13 in J. L. Patton, Silva, and Malcolm 2000).

Characters that appear to show significant geographic variation include ventral pelage color and the extent of white on the tip of the tail. Most specimens have either a completely white tail tip or the tail is mottled fuscous and white distally. In some specimens, particularly those from central Amazonia, the tail is nearly uniform fuscous with little or no white.

Micoureus paraguayanus (Tate, 1931)
Tate's Woolly Mouse Opossum
SYNONYMS:

Didelphis cinerea Temminck, 1824b:46; type locality "Brèsil"; restricted by Tate (1933:55) to Rio Mucurí, southern Bahia near the northern boundary of Espírito Santo, Brazil, probably at Morro de Arará, to which place the type locality was further restricted by Ávila-Pires (1965:5); preoccupied by *Didelphys cinerea* Goldfuss, 1812:220.

Micoureus cinereus: Lesson, 1842:186; name combination.

Philander cinerea: Gray, 1843b:101; name combination.

Didelphys [(Metachirus)] cinerea: Burmeister, 1854:137; name combination.

Didelphys [(Micoureus)] cinerea: O. Thomas, 1888b:342; name combination.

Grymaeomys cinerea: Winge, 1893:46; name combination and incorrect gender concordance.

[Didelphys (Marmosa)] cinerea: Trouessart, 1898:1238; name combination.

Marmosa cinerea: O. Thomas, 1901k:536; name combination.

[Didelphis (Caluromys)] cinerea: Matschie 1916:269; name combination.

[Marmosa (Marmosa)] cinerea cinerea: Cabrera, 1919:36; name combination.

Marmosa (Micoureus) cinerea: Pohle, 1927:241; name combination.

Marmosa cinerea paraguayana Tate, 1931:1; type locality "Villa Rica," Guairá, Paraguay.

Marmosa cinerea travassosi A. Miranda-Ribeiro, 1936:366; type locality "Angra dos Reis," Rio de Janeiro, Brazil.

Marmosa cinerea paraguayana: Cabrera, 1958:14; name combination.

[Micoures] cinereus: Reig, Kirsch, and Marshall, 1985:342; part; name combination.

Micoures cinerea paraguayana: Massoia, 1988:6; name combination and incorrect gender concordance.

[Micoureus demerarae] paraguayana: Gardner, 1993:20; name combination and incorrect gender concordance.

Micoureus cinereus paraguayanus: J. C. González, Marques, and Pacheco, 1997:195; name combination.

[Micoureus] limae: J. L. Patton, Silva, and Malcolm, 2000:72; name combination.

[Micoureus] travassosi: J. L. Patton and Costa, 2003:75; name combination.

DISTRIBUTION: *Micoureus paraguayanus* is found in eastern Paraguay, northern Argentina (Misiones), and eastern Brazil from southern Bahia south into Rio Grande do Sul.

MARGINAL LOCALITIES (Map 33): BRAZIL: Bahia, Morro d'Arara (type locality of *Didelphis cinerea* Temminck); Espírito Santo, Fazenda Santa Terezinha (J. L. Patton, Silva, and Malcolm 2000); Rio de Janeiro, Angra dos Reis (type locality of *Marmosa cinerea travassosi* A. Miranda-Ribeiro); São Paulo, Fazenda Intervales (J. L. Patton, Silva, and Malcolm 2000); Santa Catarina, Hansa (Tate 1933); Rio Grande do Sul, Tôrres (J. C. González, Marques, and Pacheco 1997). ARGENTINA: Misiones, Arroyo Yabebyri (Massoia, Chebez, and Heinonen-Fortabat 1989b). PARAGUAY: Paraguarí, Sapucay (Tate 1933); San Pedro, Nueva Germania (Krumbiegel 1941a). BRAZIL: São Paulo, Baurú (C. O. C. Vieira 1949); Minas Gerais, Rio Jordão (O. Thomas 1901k); Minas Gerais, Lagoa Santa (Tate 1933); Minas Gerais, 13 km E of Marliéria (J. L. Patton, Silva, and Malcolm 2000).

SUBSPECIES: We treat *M. paraguayanus* as monotypic, pending revision of the *Micoureus* of eastern and southern South America.

NATURAL HISTORY: Other than being a forest dweller, very little is known concerning the natural history of *M. paraguayanus.*

REMARKS: Our understanding of the taxonomy and content of *Micoureus paraguayanus* (Tate, 1931) includes almost all of Tate's (1933:54–59) concept of *Marmosa cinerea*. However, Tate (1933:55) relied on one specimen each from the states of Pará, Pernambuco, and Rio de Janeiro for cranial characteristics; the type specimen lacked the skull. On geographic grounds, only the skull from the Rio de Janeiro animal represents *M. paraguayanus*; the first two specimens likely represent *M. demerarae dominus* (O. Thomas), as understood herein. J. L. Patton, Silva, and Malcolm (2000) used the name *Micoureus limae* and J. L. Patton and Costa (2003) used the name *Micoureus travassosi* to refer to this taxon, while acknowledging that *paraguayanus* could prove the appropriate name. Primarily on geographical grounds, we have assigned the specimens from Rio Jordão, Minas Gerais, O. Thomas (1901k) identified as *Marmosa cinerea* to *M. paraguayanus*, and not to *M. constantiae* (*contra* B. E. Brown 2004: 93).

Micoureus phaeus (O. Thomas, 1899)
Little Woolly Mouse Opossum

SYNONYMS:

Marmosa phaea O. Thomas, 1899b:44; type locality "San Pablo," Nariño, Colombia."

[*Didelphys (Marmosa)*] *phaea*: Trouessart, 1905:855; name combination.

[*Marmosa (Marmosa)*] *phaea*: Cabrera, 1919:38; name combination.

Marmosa perplexa H. E. Anthony, 1922:3; type locality "Punta Santa Ana," Loja, Ecuador."

M[*armosa*]. *pahea* Ceballos-Bendezu, 1981:3; incorrect subsequent spelling of *Marmosa phaea* O. Thomas.

[*Micoureus regina*] *phaea*: Gardner, 1993:20; name combination and incorrect gender concordance.

DISTRIBUTION: *Micoureus phaeus* occurs along the western slopes of the Andes from western Colombia to southwestern Ecuador. It probably occurs also in departamento Tumbes, Peru.

MARGINAL LOCALITIES (Map 32): COLOMBIA: Valle del Cauca, La Lomitas (J. A. Allen 1912); Nariño, San Pablo (type locality of *Marmosa phaea* O. Thomas). ECUADOR: Pichincha, Mindo (Lönnberg 1921); El Oro, Piñas (Tate 1933, not mapped); Loja, Punta Santa Ana (Tate 1933).

SUBSPECIES: We consider *M. phaeus* to be monotypic.

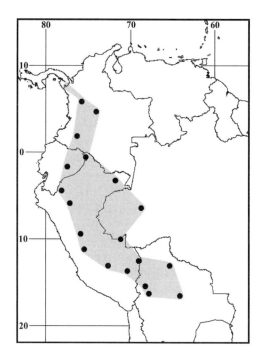

Map 35 Marginal localities for *Micoureus regina* ●

NATURAL HISTORY: A specimen (USNM 513425) captured at a place 12 km by road E of Portovelo, Ecuador, was taken from a hole in a tree limb about 3 m above and over water along the Río Puyango in otherwise xeric habitat.

REMARKS: Although similar to *M. regina* from the eastern slopes of the Colombian and Ecuadorian Andes, *M. phaeus* is smaller and has shorter fur. The palate has small palatine fenestrae, which are lacking in *M. regina*. The Panamanian specimens that Handley (1966) stated were similar to *M. phaeus* may represent *M. regina* or an undescribed taxon. Gardner (1993) and B. E. Brown (2004) treated *M. phaeus* as a synonym of *M. regina*.

Micoureus regina (O. Thomas, 1898)
Bare-tailed Woolly Mouse Opossum

SYNONYMS: See under Subspecies.

DISTRIBUTION: *Micoureus regina* is in Colombia, Ecuador, Peru, Brazil, and Bolivia. Elsewhere, the species is reported from Panama.

MARGINAL LOCALITIES (Map 35; from Tate 1933: 76–83, except as noted): COLOMBIA: Antioquia, Jericó; Cundinamarca, "Bogotá Region" (type locality of *Marmosa regina* O. Thomas). ECUADOR: Sucumbios, Largato Cocha. PERU: Loreto, Pebas (O. Thomas 1928d). BRAZIL: Amazonas, Barro Vermelho (J. L. Patton, Silva, and Malcolm 2000). PERU: Ucayali, Balta (MVZ 136377); Madre de Dios, Alberque (MVZ 165931). BOLIVIA: Beni, Villa Montes (USNM 542275); Santa Cruz, Punta Rieles

(S. Anderson 1997); La Paz, Ñequejahuira; La Paz, Ticunhuaya (type locality of *Marmosa mapiriensis* Tate). PERU: Puno, 11 km NNE of Ollachea (MVZ 172581); Cusco, Huadquiña (type locality of *Marmosa rapposa* O. Thomas; see Ceballos-Bendezu 1981); Junín, Utcuyacu; Huánuco, Chinchavita (O. Thomas 1927b); San Martín, Yurac Yacu; Amazonas, Huampami (J. L. Patton, Berlin, and Berlin 1982). ECUADOR: Pastaza, Sarayacu (type locality of *Marmosa germana* O. Thomas). COLOMBIA: Huila, La Candela (J. A. Allen 1916c).

SUBSPECIES: We recognize three subspecies of *M. regina*.

M. r. germanus (O. Thomas, 1904)
SYNONYMS:

Marmosa germana O. Thomas, 1904a:143; type locality "Sarayacu," Pastaza, Ecuador.

[*Didelphys (Marmosa)*] *germana*: Trouessart, 1905:855; name combination.

[*Didelphis (Caluromys)*] *germana*: Matschie, 1916:269; name combination.

Marmosa rutteri O. Thomas, 1924b:536; type locality "Tushemo, near Masisea," Ucayali, Peru.

Marmosa germana parda Tate, 1931:4; type locality "Huachipa, mouth of Rio Cayumba, upper Rio Huallaga," Huánuco, Peru.

Marmosa germana rutteri: Tate, 1933:81; name combination.

[*Micoureus regina*] *germana*: Gardner, 1993:20; name combination and incorrect gender concordance.

This subspecies is found from the Río Napo drainage of Ecuador, south along the lower Andean foothills and adjacent lowlands of the Ucayali, Pachitea, Huallaga, and Marañon valleys of Peru, into Bolivia and the western Amazon basin of Brazil.

M. r. rapposo (O. Thomas, 1899)
SYNONYMS:

Marmosa rapposa O. Thomas, 1899a:42; type locality "Vilcanota River, just north of Cuzco, alt. 1500 m"; restricted to Huadquiña, Distrito Santa Teresa, Valle de la Convención, Cusco, Peru, by Ceballos-Bendezu (1981:3, 12).

[*Didelphys (Marmosa)*] *rapposa*: Trouessart, 1905:855; name combination.

[*Didelphis (Caluromys)*] *rapposa*: Matschie, 1916:269; name combination.

[*Marmosa (Marmosa)*] *rapposa*: Cabrera, 1919:39; name combination.

Marmosa mapiriensis Tate, 1931:3; type locality "Ticunhuaya, road from Sorata to Guanay, Tipuani River," La Paz, Bolivia.

Marmosa mapirensis Corbet and Hill, 1991:10; incorrect subsequent spelling of *Marmosa mapiriensis* Tate.

[*Micoureus regina*] *rapposa*: Gardner, 1993:20; first use of current name combination and incorrect gender concordance.

This subspecies is found in the mid-elevation (1,000–2,500 m) forests of the eastern Andes and adjacent lowlands of Peru, Brazil, and Bolivia.

M. r. regina (O. Thomas, 1898)
SYNONYMS:

Marmosa regina O. Thomas, 1898d:274; type locality "W. Cundinamarca (Bogotá Region)," Colombia.

[*Didelphys (Marmosa)*] *regina*: Trouessart, 1898:1238; name combination.

[*Didelphis (Caluromys)*] *regina*: Matschie, 1916:269; name combination.

[*Marmosa (Marmosa)*] *regina*: Cabrera, 1919:39; name combination.

M[*icoureus*]. *regina*: Gardner and Creighton, 1989:4; first use of current name combination.

The nominate subspecies occurs along the Andean slopes and adjacent lowlands of eastern Colombia.

NATURAL HISTORY: *Micoureus regina* is found in humid tropical forest and in second growth and disturbed forests along the Andean slopes and adjacent lowlands. J. L. Patton, Silva, and Malcolm (2000) recorded this species in both várzea and terre firme forests where specimens were either shot as they perched on branches and vines above 2 m, or trapped at heights from 5 to 10 m above the ground. They caught pregnant and lactating females in February (wet season) and in September, October, and November (dry season). Attached young ranged in number from six to eight. Elsewhere, females with attached young have been taken in July and August in the upper Río Marañon drainage, Amazonas, Peru. The karyotype is $2n = 14$, $FN = 20$ (specimens from Balta, Ucayali, Peru, described under *Marmosa cinerea* by Reig et al. 1977; J. L. Patton, Silva, and Malcolm 2000)

REMARKS: The specimen J. A. Allen (1911:246) reported from San Esteban, Carabobo, Venezuela under the name *Philander regina*, and listed under *M. regina* by B. E. Brown (2004:100–01), has a tail length of 305 mm, which is too long for a *Micoureus*, but is within the range of tail lengths for *Caluromys philander*. Cabrera (1958) followed Tate (1933) in recognizing *Marmosa germana*, *M. mapiriensis*, *M. rapposa*, and *M. regina* as separate species. We consider them to be synonyms of *Micoureus regina*, with which we also include some of the specimens Tate (1933) listed under *Marmosa alstoni* from Colombia, pending a revision of the group. The distributions outlined above for the subspecies of *M. regina* are approximations; the species needs to be revised.

Although most specimens lack the distinctive white tail tip or conspicuous mottling of the distal portion of the tail characteristic of *M. demerarae*, this trait occurs sporadically in *M. regina* and *M. phaeus* from both sides of the northern Andes. Populations from higher elevations tend to have longer fur with a distinctly woolly texture. The most consistent external feature useful in separating *M. regina* from *M. demerarae* is the extent of the long pelage on the base of the tail. In *M. regina*, the longer fur usually covers less than 25 mm, and rarely more than 25 mm, of the tail. Whereas, in *M. demerarae* and *M. constantiae*, the long-furred portion of the tail usually extends beyond 25 mm.

Genus *Monodelphis* Burnett, 1830

Ronald H. Pine and Charles O. Handley, Jr.

Monodelphis, as treated here, contains 20 species. In addition, we acknowledge, but do not describe, five unnamed species as species A, B, C, D, and E. We also know of at least four other unnamed species. The need for revision of this genus has been noted by Pine, Pine, and Bruner (1981), S. Anderson (1997), Voss, Lunde, and Simmons (2001), J. L. Patton and Costa (2003), and G. A. B. Fonseca et al. (2003). The geographic range of the genus extends from extreme eastern Panama throughout South America east of the continental divide into central Argentina. These terrestrial-to-semi-fossorial opossums range in size from small to very small. The tail is always considerably shorter than the head plus body and is never bushy. Although Hershkovitz (1992b) stated, without presenting any evidence, that the tail is nonprehensile, it is prehensile, to some extent at least in some species. The ears are generally smaller than those of most didelphids. The eyes are medium-sized to small. Some species are superficially shrew-like with reduced eyes and ears, and show some modification of front feet for digging. The limbs and feet are short and stout, and, according to Hershkovitz (1992b), the ankle bones are "unspecialized." The toes are also relatively short. The claws are enlarged ("long" according to Hershkovitz [1992b]) and the pollex is reduced. As O. Thomas (1888b:354) noted, "Fifth hind toe considerably shorter than the second; third and fourth but little longer than the second, subequal." The foot pads are poorly developed (Goin and Rey 1997). J. L. Patton, Silva, and Malcolm (2000) noted that the pollex is not opposable in *M. emiliae*, and because the hind feet of all *Monodelphis* species are quite similar in appearance, this probably holds for them all. A pouch is always totally absent, as it is in the great majority of New World marsupials, notwithstanding Croft's (2003) and M. E. Jones, Dickman, and Archer's (2003) impression that all marsupials have at least a "rudimentary" or "incomplete" pouch. The dense pelage ranges

from being short and crisp to longer and densely woolly, but is never long and lax. Dark areas around the eyes are absent. There are no prominent guard hairs. Skull with "sagitally [*sic*] crested cranium" according to Hershkovitz (1992b), but this character does not hold for all species. "Palate without a second posterior pair of vacuities" (O. Thomas, 1888b:354). The nasals are widened posteriorly. The first premolars are not greatly reduced; P3 (and usually p3) are larger or much larger than P2 and p2, respectively. Usually, the premolars gradually increase in size from front to back. Tribe (1990) discussed tooth eruption sequences. According to Goin and Rey (1997), a combination of characters distinguishing *Monodelphis*, and the related extinct genus *Thylatheridium* Reig, from other marmosine didelphids is "... poorly developed orbitae and with lachrymal [*sic*] foramina opening outside the orbit limits, well developed canine fossae, alisphenoid with a lateral bony rib, well developed paroccipital [*sic*] processes, reduced palatal vacuities and posterolateral foramina, upper first incisor smaller than I2–4, M4 with vestigial metacone, lower third premolar larger than the second one, short talonids in m1–4, poorly developed hypo- and entoconids, and m4 with a well developed cingulum labial to its talonid." They also stated that, in *Monodelphis*, the median vagina is elongated anteriorly. Creighton (1984), Muizon and Argot (2003), and J. C. Nogueira and Castro (2003) provided descriptions of morphological traits, which they treated as characterizing *Monodelphis*; but because they examined specimens only of *M. domestica* and a member or members of the *M. brevicaudata* species complex, and these constitute, in our judgment, sister groups (see also J. L. Patton and Costa 2003), the characters given by them may not hold for the genus as a whole. Although Abdala, Flores, and Giannini (2001) stated that there is "strong [sexual] dimorphism in the genus," this is true of only certain species. The karyotype is $2n = 18$, FN = 20, 22, 24, 28, 30, 32, with acrocentric sex chromosomes in most species (Reig and Bianchi 1969; Reig et al. 1977; Merry, Pathak, and VandeBerg 1983; Langguth and Lima 1988; Palma and Yates 1996; B. A. Carvalho et al. 2002).

Goin and Rey (1997) wrote that no remains of *Monodelphis* have been found that can be with certainty assigned to the Pleistocene or earlier.

SYNONYMS:

Sorex: Müller, 1776:36; part; not *Sorex* Linnaeus, 1758.

Didelphis: Erxleben, 1777:80; part; not *Didelphis* Linnaeus, 1758.

Didelphys Schreber, 1777:549; part; unjustified emendation of *Didelphis* Linnaeus.

Viverra: G. Shaw, 1800:432; part; not *Viverra* Linnaeus, 1758.

Mustela: Daudin in Lacépède, 1802:163; part; not *Mustela* Linnaeus, 1758.

Philander Tiedemann, 1808:428; part.

Sarigua Muirhead, 1819, 429; part.

Monodelphis Burnett, 1830:351; type species *Monodelphis Brachyura*: Burnett, 1830:351 (= *Didelphys brachyuros* Schreber, 1777 [= *Didelphis brevicaudata* Erxleben, 1777]), by subsequent selection (Matschie 1916:271).

? *Crossopus*: Lesson, 1842:91; part; not *Crossopus* Wagler, 1832.

Micoureus Lesson, 1842:186; part.

Peramys Lesson, 1842:187; type species *Peramys brachyurus*: Lesson, 1842:187 (= *Didelphys dimidiata* J. A. Wagner, 1847, by subsequent selection [O. Thomas 1888b:354]; see Remarks); not *Didelphys brevicaudata* Erxleben, 1777, as interpreted by O. Thomas (1888b:354).

Grymaeomys Burmeister, 1854:138; part.

Hemiurus P. Gervais, 1855:287; included species *Hemiurus tristriatus*, *H. tricolor*, and *H. brachyurus*; no type species designated (see Remarks); preoccupied by *Hemiurus* Rudolphi, 1809 (Platyhelminthes).

Microdelphys Burmeister, 1856:83; type species *Didelphis (Microdelphys) tristriata*: Burmeister, 1856 (= *Didelphys tristriata* Illiger, 1815 [= *Sorex americanus* Müller, 1776]) by subsequent designation (O. Thomas 1888b:354; see Remarks); described as a subgenus of *Didelphis* Linnaeus.

Microdidelphys Trouessart, 1898:1242; incorrect subsequent spelling of *Microdelphys* Burmeister.

Monodelphiops Matschie, 1916:261; type species *Microdelphys sorex* Hensel, 1872, by original designation.

Minuania Cabrera, 1919:30; type species *Didelphys dimidiata* J. A. Wagner, 1847, by original designation.

Microdelphis Pohle, 1927:240; incorrect subsequent spelling of *Microdelphys* Burmeister.

Monodelphys Reig, 1959:57; incorrect subsequent spelling of *Monodelphis* Burnett.

Monodelhpis Tálice, Laffite de Mosera, and Machado, 1960:151; incorrect subsequent spelling of *Monodelphis* Burnett.

REMARKS: The content of *Monodelphis* has been summarized in several recent checklists: Cabrera (1958), Kirsch and Calaby (1977), Corbet and Hill (1980, 1986, 1991), Honacki, Kinman, and Koeppl (1982), Streilein (1982a), Gardner (1993a, 2005b) and Nowak (1999). See also G. A. B. Fonseca et al. (1996). Several species of *Monodelphis* have been the subject of taxonomic comment by A. Miranda-Ribeiro (1936), Pine (1975, 1976, 1977, 1980), Pine in Kirsch and Calaby (1977), Pine and Abravaya (1978), N. E. Peterson and Pine (1982), Pine in Honacki, Kinman, and Koeppl (1982), S. Anderson (1982), Pine and

Handley (1984), Pine, Dalby, and Matson (1985), Lemos, Weksler, and Bonvicino (2000), and Voss, Lunde, and Simmons (2001). The genus as a whole has not been revised since O. Thomas (1888b), who treated it as the subgenus *Peramys* Lesson, 1842, in the genus *Didelphys*. The diversity within *Monodelphis*, as shown by morphological and DNA analysis comparisons (see Kirsch, Dickerman, and Reig 1995; J. L. Patton, Reis, and Silva 1996; J. L. Patton and Costa 2003) justifies, at the least, division at the subgeneric level, but affinities between species are still too obscure to allow this at present.

As discussed by Pine (1980), Goeldi (1894) claimed that *Didelphys alboguttata* Burmeister, 1854, which, on occasion, has been placed in *Monodelphis* (or in *Peramys*), was based on an Australian marsupial of the genus *Dasyurus*. This seems odd, however, because Burmeister (1854) described *D. alboguttata* as smaller than *M. americana* (head and body length 100–130 mm); the minimum adult head plus body length is 250–350 mm in spotted Australian marsupials, which also have a noticeably hairy to bushy tail. Goeldi's (1894) hypothesis also fails to account for Wied-Neuwied's (1826) brief account of such a creature, which does not seem to have been based on a museum specimen. A. Miranda-Ribeiro (1935) discussed this matter in considerable detail.

There have been several interpretations of the content and type species of *Hemiurus* P. Gervais. In 1855, P. Gervais used three names under *Hemiurus* (*tristriatus*, *tricolor*, and *brachyurus*; see generic synonymy); whereas, in his "Mammifères" (1856b) he used five: *H. hunteri*, based on *Didelphys hunteri* Waterhouse; *H. concolor* on plate 16; and "*Didelphys tristriata*, *tricolor* et *brachyura* des auteurs." One must conclude from the content of P. Gervais's description of *H. hunteri* that he originally planned to use the name *H. concolor* for the taxon he described but, after plate 16 was completed, decided that the species was the same as that which Waterhouse had named *Didelphys hunteri*. Matschie (1916:261), Palmer (1904:319), and O. Thomas (1888b:354) thought that *Hemiurus* dated from P. Gervais's "Mammifères," published in 1855–1856, not from his "Histoire naturelle de mammifères," published in 1855. Sherborn and Woodward (1901:164) showed that the livraison including *Hemiurus* in P. Gervais's "Mammifères" was published in 1856; therefore, the "Histoire naturelle de mammifères" account has priority. O. Thomas's (1888b) designation of *Didelphys brevicaudata* as type species of *Hemiurus* is invalid because he cited (p. 357) two ("*Hemiurus brachyurus and* tricolor") of the three names P. Gervais (1855) used under *Hemiurus*. Had he cited only one, or given priority to *brachyurus* over *tricolor*, even though the two names he cited are considered synonyms, the designation would stand. Palmer's (1904) designation is likewise invalid because neither of the names he cited (*Didelphys hunteri* and *D. brevicaudata*) were mentioned by P. Gervais (1855). Kretzoi and Kretzoi's (2000:160) designation of *Didelphys hunteri* as the type species is invalid for the same reason. *Hemiurus* is unavailable, even as a subgenus name, because it is preoccupied by *Hemiurus* Rudolphi, 1809 (Platyhelminthes).

Oldfield Thomas (1888b:354) designated *Didelphys americana* as type species of *Microdelphys* Burmeister, 1856:83. In doing so, O. Thomas was indirectly designating *Microdelphys tristriatus*: Burmeister, 1856 (= *Didelphys tristriata* Illiger, 1815 [= *Sorex americanus* Müller, 1776]) as the type species of *Microdelphys*, because he included the name *Microdelphys tristriatus* referenced to Burmeister in the synonymy of *americana*.

Oldfield Thomas (1888b:354) designated *Didelphys brevicaudata* as the type species of *Peramys* Lesson, 1842. By this act, he designated, again indirectly, the type species as *Peramys brachyurus*: Lesson, 1842:187, because he (1888b:357) listed "Peramys tricolor *and* brachyurus, *Less. N. Tabl. R. A., Mamm.* pp. 186–7 (1842)" in the synonymy of *D. brevicaudata*. The type species cannot be *tricolor* because Lesson (1842:186) listed that name in the name combination *Micoureus tricolor*. When O. Thomas included *tricolor* in the name combination [*Didelphys*] *Peramys tricolor*, he was grouping *tricolor* with *brachyurus* and treating both as synonyms of *brevicaudata*. As determined by the localities given (Plata and Maldonado), *Peramys brachyurus* of Lesson (1842:187) is the same taxon reported by Waterhouse (1839:97) from "Maldonado, La Plata," Uruguay, under the name *Didelphis brachyura* and later described by J. A. Wagner (1847) as *Didelphys dimidiata*. Therefore, O. Thomas's (1888b:354) indirect designation of *Peramys brachyurus*: Lesson, 1842 (which he erroneously equated with *brevicaudata*) as type species of *Peramys* Lesson was based on a misidentification. *Peramys brachyurus*, as used by Lesson (1842), represents the species known today as *M. dimidiata*, not *M. brevicaudata* as assumed by O. Thomas (1888b)—a species unknown from Uruguay. Apparently, Thomas did not see the connection between *Didelphys brachyura*: Waterhouse (1839:97) from "Maldonado, La Plata," and Lesson's (1842) reference to "Plata; Maldonado" because he (1888b:355) listed the former as the first entry in his synonymy of *Didelphys dimidiata*. For the same reasons, Cabrera (1919:41) was in error when he gave "*Didelphis brachyuros* Schreber, = *D. brevicaudata* Erxleben" as the type species for *Peramys* Lesson.

If any reviser recognizes subgenera in *Monodelphis*, the name *Minuania* Cabrera, 1919, is unavailable because it is an objective junior synonym of *Peramys* (each has *Didelphis dimidiata* J. A. Wagner as its type species). If O. Thomas's

(1888b) designation should be set aside because he misidentified the type species, then Matschie's (1916:269) designation of *Didelphis* (*Peramys*) *crassicaudata* (Desmarest, 1804a) as the type species of *Peramys* Lesson would become valid (see O. Thomas, 1920a:195, footnote). As a result, even though Matschie (1916) used the discredited process of "fixation by elimination" for selecting type species, *Peramys* Lesson would become the senior synonym of *Lutreolina* O. Thomas, 1910a.

Eisenberg and Redford (1999) presented a table in which they attempted to contrast a classification of Brazilian species, made by N. F. Gomes (not "Gomez" as consistently given by Eisenberg and Redford) with Gardner's (1993a) classificatory scheme. The animals given certain names by Gomes, however, are not necessarily the same animals as those to which these names were given by Gardner and by the species' describers. Gomes's work (an unpublished master's dissertation) appears to have suffered from his not having non-Brazilian specimens available for him to study, and it will not be further considered here. Although J. L. Patton and Costa (2003) stated that Gomes, in contrast with other workers, had "reviewed the entire genus," Gomes had not, in fact, seen all of the named species of the genus.

KEY TO SPECIES OF *MONODELPHIS*:

1. Dorsum with longitudinal stripe or stripes 2
1′. Dorsum plain-colored, without dorsal stripes 6
2. A single, median dorsal stripe . . . . *Monodelphis unistriata*
2′. Three dorsal stripes 3
3. Dorsum reddish, with faint dark reddish stripes . . . . *Monodelphis umbristriata*
3′. Dorsum brownish to reddish, with blackish stripes 4
4. Size large (as in *M. umbristriata*); hind foot (with claw), 17–20 mm . . . . *Monodelphis americana*
4′. Size small; hind foot (with claw) 15–16 mm 5
5. Middorsal stripe beginning between ears; skull notably shallow and flattened; molars tiny; orbit small . . . . *Monodelphis iheringi*
5′. Middorsal stripe beginning posterior to ears; skull and molars of more normal proportions, neither flattened nor tiny, respectively; orbit large . . . . *Monodelphis theresa*
6. Dorsum uniformly plain-colored; neither head, rump, nor flanks contrastingly colored 7
6′. Coloration (reddish, orangish, or yellowish) of head and rump, and/or flanks differentiated from remainder of dorsum 14
7. Size small; hind foot (with claw), 13–16 mm 8
7′. Size larger; hind foot (with claw) 18–21 mm 11

8. Dorsal coloration brownish to reddish; lowlands of Bolivia and southern Brazil (and northern Argentina?) . . . . *Monodelphis kunsi*

8′. Dorsal coloration brownish to grayish; unknown from Bolivian lowlands, southern Brazil, or Argentina 9

9. Feet blackish or dark brown; cranium broad and flattened; coronoid process of mandible much broader than condylar process 10

9′. Feet whitish, buffy, or pale gray; cranium slender and not especially broad; breadth of coronoid process of mandible barely greater than that of condylar process . . . . *Monodelphis osgoodi*

10. Tail relatively short, less than 65 mm; venter conspicuously paler than dorsum, often with pale, cream-colored patches along midline; rostrum not conspicuously elongated and no conspicuous gaps between canine and premolar, or between premolars . . . . *Monodelphis adusta*

10′. Tail relatively longer, exceeds 65 mm; venter dark, similar to dorsum in color, and lacks patches of contrasting fur; rostrum elongated with conspicuous gaps between canine and first premolar, and between first and second premolars . . . . *Monodelphis reigi*

11. Dorsal coloration decidedly reddish . . . . *Monodelphis rubida*

11′. Dorsal coloration brown or gray to gray-brown. . . 12

12. Extensive orangish suffusion on flanks; tail naked and manifestly scaly; found on islands in delta of Rio Amazonas . . . . *Monodelphis maraxina*

12′. Little or no orange or yellowish on flanks; tail covered with minute bristles; found on mainland some distance south of Rio Amazonas/Solimões 13

13. Dorsal coloration uniformly brown; fur, similar to body fur, covers base of tail for less than 5 mm; dorsal profile of skull essentially flat, rostrum not conspicuously sloping downward anteriorly; zygomatic breadth more than 58% of condylobasal length . . . . . *Monodelphis ronaldi*

13′. Dorsal coloration uniformly gray to gray-brown; fur, similar to body fur, covers base of tail for more than 10 mm; dorsal profile of skull not flat, rostrum conspicuously sloping downward anteriorly; zygomatic breadth less than 58% of condylobasal length . . . . *Monodelphis domestica*

14. Flanks at mid-body colored like back, grizzled buff 15

14′. Flanks at mid-body not colored like back, buff, orange, or red 16

15. Forearm and front foot red; fur, similar to body fur, only on base of tail; snout long and narrow . . . . *Monodelphis scalops*

15′. Forearm and front foot gray-buff; fur, similar to body fur, covers tail almost to tip; snout short and broad . . . . *Monodelphis emiliae*

16. Molars tiny, relative to skull (in adult males); upper canine long and slender (in adult males); auditory bulla complete; posterolateral palatal foramina short and broad (not extending anterior to M1), interorbital-postorbital region greatly constricted 17

16′. Molars large; upper canine short and stout; auditory bulla incomplete; posterolateral palatal foramina long and narrow (extending anterior to M1); interorbital-postorbital region moderately constricted. 18

17. Feet whitish or buffy; fur long and somewhat lax (except in old males); flanks orangish, and color merging gradually with that of underparts; rump colored like back, but more or less suffused with orange . . . . *Monodelphis dimidiata*

17′. Feet reddish; fur short and smooth (lies flat); flanks reddish and contrasting with underparts; rump dark reddish . . . . *Monodelphis sorex*

18. "Body fur" covering 1/3 or more of tail dorsally; pale venter contrastingly different in color from reddish flanks; entire dorsum of head reddish or with broad reddish band above each eye . . . . *Monodelphis brevicaudata*

18′. "Body fur" covering basal 1/6 or less of tail above and below; color of venter not strikingly different from that of flanks; dorsum of head above eyes gray 19

19. Upper molar row equal to or less than 7.9 mm; found west of the Río Orinoco in northern Venezuela and northeastern Colombia . . . . *Monodelphis palliolata*

19′. Upper molar row equal to or greater than 7.9 mm; found south of the Rio Amazonas and west of the Rio Xingu . . . . *Monodelphis glirina*

Monodelphis adusta (O. Thomas, 1897)
Sepia Short—tailed Opossum

SYNONYMS: See under Subspecies.

DISTRIBUTION: *Monodelphis adusta* occurs at lower to middle elevations in the Andes from Colombia and western Venezuela into northern Bolivia. It occurs also in the eastern Darién of Panama.

MARGINAL LOCALITIES (Map 36): COLOMBIA: El Cesar, Colonia Agrícola de Caracolicito (USNM 280894). VENEZUELA: Mérida, Zea (Ramoni-Perazzi, Bianchi, and Molina 1994). COLOMBIA: Boyacá, Miraflores (ILB 622); Meta, Villavicencio (AMNH 139227). ECUADOR: Napo, San José abajo (AMNH 68136). PERU: San Martín, Moyobamba (Osgood 1913); Ucayali, Pucallpa (Ceballos-Bendezu 1959); Cusco, La Convención (S. Solari 2004); Cusco, 72 km by road NE Paucartambo (V. Pacheco et al. 1993, J. L. Patton, Reis, and Silva 1996). BOLIVIA: Pando, La Reserva (S. Anderson 1997). PERU: Huánuco, Hacienda Éxito (FMNH 23772). ECUADOR: Zamora-Chinchipe, Zamora (AMNH 47189); Pastaza, Mera (Albuja and

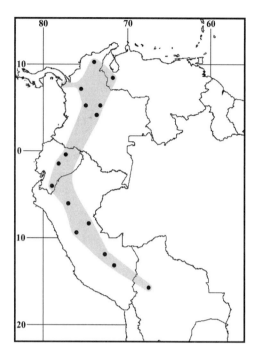

Map 36 Marginal localities for *Monodelphis adusta* ●

Rageot 1986). COLOMBIA: Tolima, Mariquita (BM 12.4.2.10); Antioquia, 9 km S of Valdivia (FMNH 70539).

ADDITIONAL LOCALITIES (not mapped): COLOMBIA: Cundinamarca, Paime (BM 23.11.13.16); Boyacá, Guaicaramo (AMNH 752332); Cundinamarca, Medina (ILB 620); Cundinamarca, Susumuco (BM 21.5.2.6); Cundinamarca, Santandercito (ROM 51866). VENEZUELA: Táchira, Potosí (Soriano 1987). ECUADOR: Pastaza, Palmera, ca. 25 km E de Baños (AMNH 67275). PERU: Loreto, Río Gálvez (Jansa and Voss 2000); Madre de Dios, Reserva Cusco Amazónico (Woodman et al. 1991); Huánuco, Hacienda Buena Vista, Río Chinchao (FMNH 23780).

SUBSPECIES: We follow S. Solari (2004) in recognizing two subspecies of *M. adusta*.

M. a. adusta (O. Thomas, 1897)
SYNONYMS:
Peramys adustus O. Thomas, 1897a:219; type locality "W. Cundinamarca, in the low-lying hot regions," Colombia.
[*Didelphys (Peramys)*] *adusta*: Trouessart, 1898:1243; name combination.
Peramys melanops Goldman, 1912b:2; type locality "Cana (altitude 2,000 feet), in the mountains of eastern Panama," Darién, Panama.
[*Didelphis (|Monodelphis[)]*] *melanops*: Matschie, 1916:271; name combination.
[*Didelphis (|Monodelphis[)]*] *adustus*: Matschie, 1916:271; name combination and incorrect gender concordance.

Monodelphis adusta: Cabrera and Yepes, 1940:33; first use of current name combination.
Monodelphis adusta adusta: Cabrera, 1958:6; name combination.

The nominate subspecies is found in Panama, Colombia, western Venezuela, Ecuador, and northern Peru.

M. a. peruviana (Osgood, 1913)
SYNONYMS:
Peramys peruvianus Osgood, 1913:93; type locality "Moyobamba," San Martín, Peru.
[*Didelphis (|Monodelphis[)]*] *peruvianus*: Matschie, 1916:271; name combination and incorrect gender concordance.
Monodelphis peruviana: Cabrera and Yepes, 1940:33; name combination.
Monodelphis adusta peruviana: Cabrera, 1958:6; first use of current name combination.
Monodelphis peruana Emmons and Feer, 1990:28; incorrect subsequent spelling of *Peramys peruvianus* Osgood.
Monodelphis kunsi: Salazar, Campbell, Anderson, Gardner, and Dunnum, 1994:127; part (specimens from La Reserva), not *Monodelphis kunsi* Pine.

This subspecies occurs in central and southern Peru, and in northern Bolivia.

NATURAL HISTORY: Albuja and Rageot (1986) provided natural history notes on two *M. adusta* from Mera, Ecuador. This region in Pastaza Province has about 5 m of rainfall annually and the forest vegetation is transitional between humid cloud and rain forest. The animals were trapped in openings, one under a pile of logs associated with fields of maize and yuca; the other on a riverbank in herbaceous ground cover. Stomach contents included adults and larvae of beetles and a small frog. One specimen contained a subcutaneous botfly larva. Albuja and Rageot (1986) thought *M. adusta* to be nocturnal and to den in tree holes. Edmund Heller's labels for specimens (USNM) he caught in central Peru reported finding the remains of insects and vegetation in the stomachs. One trapped on 24 August 1922 was suckling four young. Handley (1966c) reported trapping this species among boulders on a dry, partially open gravel bar and among rocks on a riverbank in forest in Panama. Emmons and Feer (1990) stated that this species has been trapped in a house. They also (1997:28) wrote "This is chiefly a montane species of wet forests at mid-elevations (1,400–2,200 m), but in some places it is found as low as 200 m in rainforest and wet grassland." Woodman et al. (1991, 1995) reported it at that low elevation in lowland rain forest, also called by them "floodplain forest." The two specimens from La Reserva, Bolivia, misidentified by Salazar et al. (1994:127) as *M. kunsi*, were caught in a "banana field" on a river terrace. Eisenberg (1989) reported a diet of fruits, invertebrates, and small vertebrates, but gave no evidence. Costa-Lima and Hathaway (1946) reported the flea *Adoratopsylla antiquorum discreta* from *Peramys adusta*, and Fairchild, Kohls, and Tipton (1966), the tick *Ixodes venezuelensis*. S. Solari et al. (2001c) stated that they had found *M. adusta* sympatric with *M. brevicaudata* (we presume *M. glirina*) and with *M. emiliae* in departamento Cusco, Peru. Judging from their table, however, *M. adusta* was not found truly sympatric with *M. glirina* (see also Boddicker, Rodríguez, and Amanzo 2001). Jansa and Voss (2000) reported both *M. adusta* and *M. emiliae* from "Río Gálvez," Loreto, Peru.

REMARKS: Cabrera (1958) regarded *M. osgoodi* as a subspecies of *M. adusta*. On the other hand, Handley (tentatively;1966c), Pine (in Kirsch and Calaby 1977), and Pine and Handley (1984) treated *M. osgoodi* as representing a separate species. Cabrera (1958) also listed *Peramys peruvianus* Osgood as a subspecies of *M. adusta*. Handley (1966c) cited supposed lack of geographic variation in treating *P. melanops* and *P. peruvianus* as junior synonyms of *M. adusta*. Kirsch (1977b), on the other hand, recognized *M. melanops* as a separate subspecies. Although Emmons and Feer (1990) recognized *M. peruviana* as a separate species, they did not do so in 1997. Linares (1998) used the trinomial *M. a. adusta* for Venezuelan animals, but did not specify what other forms he regarded as subspecies of *M. adusta*. The two specimens from La Reserva, La Paz, Bolivia reported by S. Anderson (1997) previously had been identified as *M. kunsi* Pine, by Salazar et al. (1994). The specimen (USNM 280894) S. Solari (2004) listed as from Santa Marta, Magdalena, Colombia, is actually from Colonia Agrícola de Caracolicito, departamento El Cesar.

Monodelphis americana (Müller, 1776)
Northern Three-striped Opossum
SYNONYMS:

Mus araneus Marcgraf, 1648:229 (see Hershkovitz 1987: 23); Marcgraf's Brazilian animals from Pernambuco (= Recife, Pernambuco, Brazil; see O. Thomas 1911b:124); not *Mus araneus* of other pre-Linnaean authors (= *Sorex araneus* Linnaeus, 1758:53).

Sorex americanus Müller, 1776:36; type locality "Brasilien"; based primarily on "*la musaraigne du Bresil*" of Buffon 1767:160, which was based in turn on *Mus araneus* Marcgraf; restricted to "Pernambuco" [= Recife], Pernambuco, Brazil, by Cabrera (1958:7).

[*Sorex*] *brasiliensis* Erxleben, 1777:127; type locality "in Brasilia"; based mostly on *Mus araneus* Marcgraf (1648); therefore, type locality restricted to Recife, Pernambuco, Brazil.

Sorex Brasiliensis Daudin in Lacépède, 1802:157; no type locality given; based on "*La musaraigne du Bresil*"

(Lacépède, 1800:339), which is based on *Mus araneus* of Marcgraf (1648); therefore, type locality is Recife, Pernambuco, Brazil; junior objective synonym and homonym of *Sorex brasiliensis* Erxleben.

[*Didelphys*] *Tristriata* Illiger, 1815:112; type locality "Brasilische"; based on *Sorex brasiliensis* (which is based on *Mus araneus* Marcgraf); therefore, the type locality is Recife, Pernambuco, Brazil.

Didelphis tristriata Kuhl, 1820:63; type locality "Brasilia"; based on a specimen in the Berlin Museum (ZMB).

[*Didelphis*] *trilineata* Lund, 1839b:233; *nomen nudum*.

D[idelphis]. trilineata Lund, 1840a:19 [1841b:237]; based on *Mus araneus* Marcgraf (1648); therefore, the type locality is Recife, Pernambuco, Brazil (see O. Thomas 1911b:124).

Peramys tristriata: Lesson, 1842:187; name combination.

Didelphys [*(Grymaeomys)*] *tristriata*: Burmeister, 1854: 140; name combination.

Hemiurus tristriatus: P. Gervais, 1855:287; name combination.

[*Didelphys (]Microdelphys[)*] *tristriata*: Burmeister, 1856: 84; name combination.

Didelphys americana: O. Thomas, 1888:363; name combination.

[*Didelphys (]Peramys[)*] *tristriatus*: Goeldi, 1894:463; part; name combination.

[*Didelphys (Peramys)*] *americana*: Trouessart, 1898:1244; name combination.

[*Didelphis (]Microdelphys[)*] *americana*: Matschie, 1916: 272; name combination.

[*Peramys*] *americanus*: Cabrera, 1919:42; part; name combination.

Monodelphis (Microdelphis) americanus: Pohle, 1927:240; name combination and incorrect gender concordance.

Peramys tristriatus: A. Miranda-Ribeiro, 1935:28; name combination.

Peramys americanus: A. Miranda-Ribeiro, 1936:414; name combination.

M[inuania?]. americanos A. Miranda-Ribeiro, 1936:423; name combination and incorrect subsequent spelling of *Sorex americanus* Müller.

Monodelphis americana: Cabrera and Yepes, 1940:30; first use of current name combination.

DISTRIBUTION: *Monodelphis americana* occurs in Brazil, east of 50° W longitude, from eastern Pará in the north to coastal Santa Catarina in the south. Most records are from eastern São Paulo, southern Minas Gerais, and adjacent states, but there are others from Bahia, Pernambuco, and near Brasilia.

MARGINAL LOCALITIES (Map 37): BRAZIL: Pará, Benevides (AMNH 37490); Paraíba, João Pessoa (Langguth and Lima 1988); Pernambuco, Dois Irmãos (ZMB 35495);

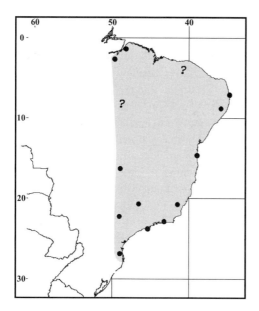

Map 37 Marginal localities for *Monodelphis americana* ●

Bahia, Urucutuca (FMNH 63768); Espírito Santo, Engenheiro Reeve (BM 3.9.4.112); Rio de Janeiro, Rio de Janeiro (ZMB 3382); São Paulo, San Sebastião (BM 2.4.6.41); Santa Catarina, Blumenau (ZMB 35492); São Paulo, Baurú (C. O. C. Vieira 1950); Minas Gerais, Passos (Pine 1976); Goiás, Annapolis (AMNH 133241); Pará, Baião (AMNH 96810).

ADDITIONAL LOCALITIES (not mapped): BRAZIL: Minas Gerais, Lagoa Santa (ZMUC 176); Minas Gerais, Mata de Prefeitura, 6 km SW of Viçosa (USNM 552402); Minas Gerais, Serra Caparão (AMNH 61836); Pará, Belém (Pine 1973b); Pará, Belém-Brasilia [Hwy] Km 96 (C. T. Carvalho and Toccheton 1969); Santa Catarina, Humboldt (BM 14.5.9.20); Rio de Janeiro, Fazenda Boa Fé (D. E. Davis 1945a); Rio de Janeiro, Angra dos Reis (L. G. Pereira et al. 2001); Rio de Janeiro, Parati (L. G. Pereira et al. 2001); São Paulo, Victoria (BM 3.7.25.13); São Paulo, Piracicaba (ZMB 14165); São Paulo, Cantareira (USNM 236676); Distrito Federal, 25 km S of Brasília (OMNH 19070).

SUBSPECIES: *Monodelphis americana* is geographically variable, but no subspecies are currently recognized.

NATURAL HISTORY: D. E. Davis (1947) noted that this species (as might be predicted on the basis of its markings) is diurnal, a conclusion supported by the findings of Nitikman and Mares (1987). Two were captured adjacent to a big log and near a windfall. One was caught in a mousetrap baited with banana, although D. E. Davis (1947) opined that this species is difficult to catch in mousetraps baited with fruit or grain. This last capture occurred in a trap from which a freshly caught *M. scalops* had just been removed, and the *M. americana* had been showing apparent

interest (predatory?) in the trapped *M. scalops*, meanwhile showing somewhat fearless behavior. D. E. Davis (1947:3) also wrote, "This species makes nests of leaves in the forks of trees or bushes about 5 meters from the ground. No young were found in the nests observed and hence they are probably only resting places." Nitikman and Mares (1987) trapped an adult male, seven subadult males, and one subadult female in central Brazil during the period January to July, and noted that one or more individuals were trapped each month and that the adult male was taken in July. They noted a more or less synchronized growth pattern in male *Gracilinanus agilis*, suggesting semelseminance (males surviving only one mating season; see Pine, Dalby, and Matson 1985), and stated that *M. americana* showed a similar pattern. Mares, Braun, and Gettinger (1989) thought that molting, near Brasilia, might be confined to the dry season. Nitikman and Mares (1987:85–86) commented that this species is terrestrial, had high trap mortality on cold nights because it lacks a torpidity response to low temperatures, and that females lack a "true pouch." Goeldi (1894) may have been discussing a composite of two or more species when stating that he found *Didelphys (Peramys) tristriatus* (= *M. americana*) to be frequently associated with water, entirely terrestrial, and by preference nocturnal. On the basis of captives, he observed (1894:465) that it "assails without hesitation birds and mammals nearly as large as itself," having noted that one killed and ate a young *Nectomys squamipes* placed in its cage. Szyszlo (1955) wrote that this species feeds on small birds, but presented no evidence to this effect. Laemmert, Ferreira, and Taylor (1946) reported animals identified as *M. americana* as trapped in both old and young forest, but not in swamp forest, near Ilhéus, Bahia, Brazil. Alho, Pereira, and Paula (1987) calculated a home range of 440 m² in gallery forest in the Parque Nacional de Brasília. Mares, Braun, and Gettinger (1989) also encountered this species only in gallery forest. It is unclear whether Alho, Pereira, and Paula (1987) found *M. americana* to occur sympatrically with *M. domestica*, or merely virtually sympatrically.

Mares, Braun, and Gettinger (1989) listed the flea *Adoratopsylla antiquorum*, and the mites/ticks *Androlaelaps* sp. and *Ixodes* sp., from this species. *Androlaelaps cuicensis*, named by Gettinger (1997), may also have been from *M. americana*, although Gettinger identified the host as *M. rubida*. The karyotype is $2n = 18$, FN $= 22$; the X chromosomes are small acrocentrics (Langguth and Lima 1988), the Y chromosome has not been described

REMARKS: Tate (1939) did not list this species for the Guiana region, Husson (1978) did not list it for Surinam, and we have not found specimens from localities outside of Brazil, but Honacki, Kinman, and Koeppl (1982) gave its range as "to French Guiana, Guyana, and

Surinam." Earlier, Cabrera (1919, 1958) had given the range as "Guayana...desde...Río [*sic*] Grande do Sul" and "Brasil...hasta las Guayanas," respectively. Shufeldt (1926a, 1926b) discussed the skeleton from Cantareira, São Paulo, listed above, under the name "*Peramys iheringi.*"

Monodelphis americana has also been listed for Misiones, Argentina (e.g., Massoia, 1980a; Olrog and Lucero 1981), but because Massoia (1980a), Chebez and Massoia (1996), and Massoia, Forasiepi, and Teta (2000) wrote of *M. americana iheringi* as occurring in Misiones, instead of just plain *M. americana*, and Heinonen-Fortabat and Chebez (1997) called the animal in question *M. iheringi*, we might assume that their records are for *M. iheringi*, except that from their known distributions, *M. americana* seems the more likely. Also, the description, dimensions, and illustrations provided by Massoia, Forasiepi, and Teta (2000) fit *M. americana* rather than *M. iheringi*. Mares and Braun (2000) treated all Argentine records as representing *M. iheringi*. Massoia (1980b) further confused the issue by writing of "*Monodelphis americana* (= *iheringi*?)." Contreras and Silvera-Avalos (1995) indicated that *M. americana* had been recorded from Misiones (and they gave specific status to both *M. americana* and *M. iheringi*) but we are unaware of any unequivocally identified specimens from definite published localities. The same goes for their statement that *M. americana* occurs in Paraguay. G. A. B. Fonseca and Kierulff (1988) interpreted small external measurements of a specimen from Minas Gerais as indicating its probable immaturity, rather than possibly representing a species other than *M. americana*. They stated that "small body size and foraging habits on the forest litter" might have explained this animal's infrequency of capture at their study sites. The "small size" of Fonseca and Kierulff's specimen was deduced in comparison with an overly restricted size range given for the genus by Nowak and Paradiso (1983).

Monodelphis brevicaudata (Erxleben, 1777)
Guianan Short-tailed Opossum
SYNONYMS:

[*Didelphis*] *brevicaudata* Erxleben, 1777:80; type locality "in Americae australis silvis"; restricted to Surinam by Matschie (1916:271); restriction corrected to Kartabo, Cuyuni-Mazaruni District, Guyana by Voss, Lunde, and Simmons (2001); see Remarks.

Didelphys brachyuros Schreber, 1777:pl. 151; type locality "Südamerika in den Wäldern" (Schreber 1778:549).

Sorex (Surinamensis) Zimmermann, 1780:386; type locality "Surinam."

Viverra Touan G. Shaw, 1800:432; type locality "Cayenne," French Guiana; based on Buffon's (1789:252, pl. 61) "*Le Touan.*"

Mustela touan Bechstein, 1800:359; type locality "Cayenne," French Guiana; based on Buffon's (1789:252, pl. 61) "*Le Touan*"; objective synonym of *Viverra touan* G. Shaw

Mustela touan Daudin in Lacépède, 1802:163; based on "*Le touan*" (Lacépède, 1801:63) acquired by de la Borde in "Cayenne," French Guiana; objective synonym of *Viverra touan* G. Shaw; objective synonym and homonym of *Mustela touan* Bechstein.

Didelphis tricolor É. Geoffroy St.-Hilaire, 1803:144; type locality "Cayenne," French Guiana.

P[*hilander*]. *brachyurus*: Tiedemann, 1808:428; name combination.

S[*arigua*]. *brachyura*: Muirhead, 1819:429; name combination.

didelphis [*sic*] *brachyura*: Desmarest, 1820:260; name combination.

D[*idelphis*]. *Sebae* Gray, 1827:190; type locality "South America"; based on "Mus Sylvestris Americana, *Seba, Mus.* I. 50" and on "Icon. *Seba, Mus.* I t. 31. f. 6."

[*Didelphys*] *Hunteri* Waterhouse, 1841:110; no locality given; description based on a spirit-preserved specimen then in the Museum of the College of Surgeons.

[*Sorex*]? *Crossopus Surinamensis*: Lesson, 1842:91; name combination.

Micoureus tricolor: Lesson, 1842:186; name combination.

Peramys brachyurus: Lesson, 1842:187; name combination.

Didelphys brachyura: Waterhouse, 1846:522; part; name combination.

Didelphys [*(Peramys)*] *brevicaudata*: O. Thomas, 1888b:356; name combination.

Peramys brevicaudatus orinoci O. Thomas, 1899c:154; type locality "Caicara, [del] Orinoco," Bolívar, Venezuela.

Peramys brevicaudatus dorsalis J. A. Allen, 1904c:327; type locality "Ciudad Bolivar," Bolívar, Venezuela.

[*Didelphys (Peramys) brevicaudata*] *orinoci*: Trouessart, 1905:857; name combination.

[*Didelphis (|Monodelphis[)*] *dorsalis*: Matschie, 1916:271; name combination.

[*Didelphis (|Monodelphis[)*] *brevicaudatus*: Matschie, 1916:271; name combination.

[*Didelphis (|Monodelphis[)*] *orinoci*: Matschie, 1916:271; name combination.

[*Didelphis (|Monodelphis[)*] *touan*: Matschie, 1916:271; name combination.

[*Peramys*] *brevicaudatus*: Cabrera, 1919:42; part; name combination.

Monodelphis brevicaudata: O. Thomas, 1920c:283; first use of current name combination.

M[*onodelphis*]. *brevicaudata dorsalis*: Pittier and Tate, 1932:254; name combination.

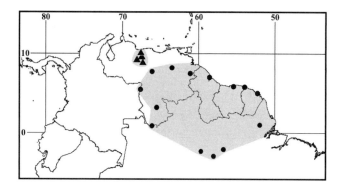

Map 38 Marginal localities for *Monodelphis brevicaudata* ● and *Monodelphis* "species A" ▲

M[*onodelphis*]. *brevicaudata orinoci*: Pittier and Tate, 1932:254; name combination.

P[*eramys*]. *tricolor tricolor*: A. Miranda-Ribeiro, 1936: 408; part; name combination.

Monodelphis touan: Cabrera and Yepes, 1940:31; name combination.

Didelphis brevicauda C. O. C. Vieira, 1950:359; incorrect subsequent spelling of *Didelphis brevicaudata* Erxleben.

Monodelphis tricolor tricolor: C. O. C. Vieira, 1955:350; name combination.

Monodelphis brevicaudata touan: Ávila-Pires, 1964:8; name combination.

DISTRIBUTION: *Monodelphis brevicaudata* is in Venezuela south of the Orinoco, in the Guianas, and in adjacent Brazil east of the Rio Negro and north of the lower Rio Amazonas.

MARGINAL LOCALITIES (Map 38; from Voss, Lunde, and Simmons 2001, except as noted): VENEZUELA: Bolívar, Ciudad Bolívar; Bolívar, Reserva Forestal Imataca. GUYANA: Essequibo Islands-West Demerara, Buck Hall. SURINAM: Saramacca, La Poule; Marowijne, Langamankondre. FRENCH GUIANA: Cayenne. BRAZIL: Amapá, Serra do Navio; Pará, Faro; Amazonas, Santo Antônio da Amatary; Amazonas, 80 km N of Manaus. VENEZUELA: Amazonas, Serra de Neblina; Amazonas, Esmeralda; Amazonas, 32 km S of Puerto Ayacucho (Handley 1976); Bolívar, Caicara.

ADDITIONAL LOCALITIES (not mapped): VENEZUELA (Handley 1976): Bolívar, Santa Lucía de Surukún; Amazonas, Boca Mavaca; Amazonas, Capibara; Bolívar, Río Supamo; Bolívar, El Manaco, 68 km SE of El Dorado. BRAZIL: Roraima, Limão (Tate 1939); Amapá, Igarapé Rio Branco, alto Rio Maracá (C. T. Carvalho 1962a); Pará, Cachoeira Porteira, near Oriximiná (USNM 546215); Pará, Serra do Tumucumaque, near Tiriós (USNM 392045). GUYANA: Potaro-Siparuni, Minnehaha Creek (AMNH 36317); East Berbice-Corentyne, Itabu Creek Head (FMNH 48418). SURINAM: Nickerie,

King Frederick Wilhelm Falls (FMNH 48416). Also see Voss, Lunde, and Simmons (2001) for additional localities.

SUBSPECIES: We are treating *M. brevicaudata* as monotypic.

NATURAL HISTORY: Eisenberg (1989) stated that *M. brevicaudata* (*sensu lato*) is more abundant in multistratal evergreen forests than in dry deciduous forest, but may be found in edge habitats around clearings. Engstrom, Lim, and Reid (1999) stated "Usually found in upland forest away from streams." According to Ventura, Pérez-Hernández, and Lopez-Fuster (1998), this species has been found in Venezuela "in lowland forests (95–400 m elevation) but occasionally in higher areas," ranging from 620 to 1,080 m. Husson (1978) wrote that *M. brevicaudata* is nocturnal; evidence for his conclusion seems sparse. Eisenberg (1989) called this animal predominantly crepuscular, while Emmons and Feer (1990), Engstrom and Lim (2002), and Engstrom, Lim, and Reid (1999) stated it is diurnal. Charles-Dominique et al. (1981) equipped a female in French Guiana with a radio transmitter and found that it was completely diurnal, spending its time searching for insects on the ground under fallen branches and lianas. Engstrom and Lim (2002) gave the litter size as seven.

Fain (1979) listed the listrophorid mite *Didelphoecius incisus* Fain as a parasite of "*M. touan*" (= *M. brevicaudata*), and *Didelphoecius callipygus* for *M. brevicaudata*. The identity of the host for *Didelphoecius monodelphis* Fain is unknown. Fain (1979) referred to it merely as "*Monodelphis touan*, Brésil." The mallophagan *Cummingsia peramydis* Ferris was reported by Emerson and Price (1975) and by Timm and Price (1985). Other ectoparasites taken from hosts identified as of this species have been reported by Brennan and Reed (1975), Costa-Lima and Hathaway (1946), Furman (1972), E. K. Jones et al. (1972), Reed and Brennan (1975), Saunders (1975), and Tipton and Machado-Allison (1972). R. Guerrero (1985a) listed parasitic nematodes and previously published records of ectoparasites. Parasites mentioned by the above authors, but not designated here by their technical names, may have been taken in whole or in part on other members of the species group.

McNab (1982) noted that *M. brevicaudata* has a low basal metabolic rate (near 65%), which, when coupled with high minimal conductance (126%) and small mass, leads to a low body temperature (33.8°C) and a propensity to enter torpor. Emmons and Feer (1990) commented that this species (*sensu lato*) possesses a glowing violaceous color on the underparts that eventually disappears from study skins. The karyotype reported for Venezuelan *M. brevicaudata* by Reig and Bianchi (1969) was actually for *M. palliolata*. Whether or not Reig et al. (1977) included any actual *M. brevicaudata* in their karyotypical observations on specimens identified by them as of this species is unknown. B. A. Carvalho et al. (2002) gave the karyotype of actual *M. brevicaudata* from Brazil as $2n = 18$, $FN = 30$, with an acrocentric X and a minute Y chromosome.

REMARKS: Some authors (see I. Geoffroy St.-Hilaire 1839:5[footnote], 1847b:115–18; Sherborn 1922:lviii, 1932:cxxxviii) considered É. Geoffroy St.-Hilaire's (1803) catalog as not published because it was a proof copy (but see ICZN 2002). Gardner (1993a) followed the guidelines for Wilson and Reeder (1993:831) and dated *Didelphis tricolor* from Desmarest (1817a). Gardner (1993a) also treated the names *Didelphys glirina* J. A. Wagner, 1842, and *Peramys palliolatus* Osgood, 1914a, as synonyms of *Monodelphis brevicaudata*. Here they are treated as separate species.

Pine in Honacki, Kinman, and Koeppl (1982), stated that *M. touan* is a probable junior synonym of *M. brevicaudata*, and Pine and Handley (1984) unequivocally included *M. touan*, *M. tricolor*, and *M. palliolata* in *M. brevicaudata*, but treated what they termed as *M. orinoci* as a separate species. Hershkovitz (1992b), however, continued to recognize *M. touan* and "*M. palliolatus*" as full species. Gardner (1993a) synonymized *M. touan*, *M. tricolor*, *M. palliolata*, *M. orinoci*, and *M. glirina*, among others, with *brevicaudata*. Linares (1998) treated *M. palliolata* as a subspecies of *brevicaudata* and treated what he referred to as *M. touan* as a separate species sympatric with *M. brevicaudata*. He gave the range of *M. touan* as including a portion of eastern Venezuela, the Guianas, and Pará, Brazil, and provided cranial characters that supposedly separated the two. *Monodelphis touan* was stated to be only slightly different from *M. emiliae*, and only in size and color, and conspecificity between the two was proposed. Voss, Lunde, and Simmons (2001) were able to distinguish two color morphs (bicolored and tricolored) of *M. brevicaudata* from *M. glirina* and *M. palliolata*. They found the bicolored morph in the region between the lower Caroni-Orinoco and the lower Mazaruni-Essequibo rivers in northeastern Venezuela and northwestern Guyana. The tricolored morph is known from elsewhere throughout the Guianas, in southern Venezuela, and in Brazil north of the Amazon and east of the Rio Negro. The type specimen of *Monodelphis brevicaudata* represents the bicolored morph; therefore, Matschie's (1916) restriction of the type locality to Surinam (where only the tricolored morph is found) was an error. Consequently, Voss, Lunde, and Simmons (2001) emended the type locality to the vicinity of Kartabo, Cuyuni-Mazaruni, Guyana.

The oldest name for the tricolored morph of *M. brevicaudata* is *Viverra touan* G. Shaw, 1800, which was based on Buffon's (1789) "*Le Touan*" from Cayenne. As the type is no longer extant, Voss, Lunde, and Simmons (2001) designated, as neotype, a male specimen (FMNH 21720) taken at Cayenne, French Guiana. Linares (1998) followed Cabr-era (1958) in recognizing *Viverra touan* Shaw, as a species, but described it as bicolored, which does not agree descriptively with a *Monodelphis* with a type locality in French Guiana. A more detailed review of the *M. brevicaudata* complex (*sensu stricto*) was provided by Voss, Lunde, and Simmons (2001).

Cabrera (1958) recognized *M. orinoci* and gave its distribution as "*Cuenca del Orinoco, en Venezuela, llegando por el este hasta la zona montañosa del extremo oeste de la Guayana Inglesa y el río Branco, en el Brasil.*" This concept of the distribution of *M. orinoci* is derived partly from Tate (1939), who synonymized *Peramys brevicaudatus dorsalis* J. A. Allen with *Peramys brevicaudatus orinoci* O. Thomas. Reig et al. (1977) also treated *M. orinoci* as a separate species. They assigned three specimens from Estación Biológica de los Llanos, Guárico, Venezuela to *M. orinoci*. Reig et al. (1977) stated that the differences between the savannah animals they regarded as *M. orinoci* and animals of the northern Coast Range of Venezuela "are of the same magnitude as those used to separate *M. dimidiata* from *M. brevicaudata*." Unfortunately, they did not explain what those differences are, nor has any subsequent author. Pérez-Hernández (1985, 1990) and Linares (1998) also applied the name *M. orinoci* to the distinctive *Monodelphis* from the llanos and Cordillera Central of Venezuela. Linares (1998:71) noted that these opossums had been previously referred to under the name *Peramys brevicaudata dorsalis*. However, *Peramys brevicaudatus orinoci* O. Thomas and *Peramys brevicaudatus dorsalis* J. A. Allen, are synonyms of *M. brevicaudata* (Erxleben)—see Voss, Lunde, and Simmons (2001). Therefore, the *Monodelphis* discussed by Pérez-Hernández (1985, 1990), Ventura, Pérez-Hernández, and Lopez-Fuster (1998), and by Linares (1998), apparently lacks a name (see *Monodelphis* [species C]). J. L. Patton and Costa (2003) treated *palliolata*, *brevicaudata*, and *glirina* as conspecific, contrary to the analysis provided by Voss, Lunde, and Simmons (2001).

Monodelphis brevicaudata can be distinguished from *M. glirina* and *M. palliolata* by its paler venter (whitish to buffy in fully adult individuals; orange in the other two), which sharply contrasts with the reddish color of the flanks; dorsal extension of fur along the proximal 1/3 to 1/2 of tail; and (if present at all) a narrower cap of grizzled-grayish fur on the head, confined laterally by the band of reddish fur above the eyes (Voss, Lunde, and Simmons 2001). A similar, but unnamed taxon (see *Monodelphis* [sp. E]) regarded by Handley as a subspecies of *M. brevicaudata* (see Voss, Lunde, and Simmons 2001), occurs south of the Amazon and east of the Xingu. The Bolivian localities mapped by B. E. Brown (2004:108) were based on S. Anderson's (1997) work and all probably correspond to collection sites for *M. glirina*; Brown's localities in west-central Venezuela represent *M. orinoci* of authors (herein = *Monodelphis* [species C]).

Monodelphis dimidiata (J. A. Wagner, 1847)
Yellow-sided Opossum

SYNONYMS:

Didelphis brachyura: Waterhouse, 1839:97; not *Didelphys brachyuros* Schreber.

Didelphys brachyura: Waterhouse, 1846:522; part; name combination.

D[*idelphys*]. *dimidiata* J. A. Wagner, 1847:151, footnote; type locality "Maldonado am la Plata," Maldonado, Uruguay; based on Darwin's specimen identified as *Didelphis brachyura* by Waterhouse (1839).

Didelphys brevicaudata: Figueira, 1894:189; not *Didelphis brevicaudata* Erxleben.

Didelphys [*(Peramys)*] *dimidiata*: O. Thomas, 1888b:355; name combination.

[*Didelphis* (]*Monodelphis*[)] *dimidiata*: Matschie, 1916: 272; name combination.

[*Minuania*] *dimidiata*: Cabrera, 1919:44; name combination.

M[*onodelphis*]. *dimidiata*: O. Thomas, 1924c:586; first use of current name combination.

Monodelphis fosteri O. Thomas, 1924c:586; type locality "Caleufú, a station on the Western Railway about 64°30′W. and 35°35′S.," La Pampa, Argentina.

Minuania fosteri: Cabrera and Yepes, 1940:33; name combination.

Monodelphis [*(Minuania)*] *dimidiata*: Cabrera, 1958:10; name combination.

Minuania dimidata Tálice, Laffite de Mosera, and Machado, 1960:150; incorrect subsequent spelling of *dimidiata* Wagner.

Monodelphis (Minuania) fosteri: Crespo, 1964:62; name combination.

M[*onodelphis*]. *d*[*imidiata*]. *fosteri*: Massoia, 1980a:16; name combination.

Monodelphis dimidiata dimidiata: E. M. González, 2001: 56; name combination.

DISTRIBUTION: *Monodelphis dimidiata* occurs in Uruguay, southern Brazil, and northern Argentina. Mares and Braun (2000) listed all, or nearly all, known Argentine localities.

MARGINAL LOCALITIES (Map 39; sequence is clockwise starting with southernmost locality): ARGENTINA: Buenos Aires, Arroyo El Pescado (Reig 1964); La Pampa, Caleufú (O. Thomas 1924c); Córdova, Río Ceballos (BM 27.3.25.6); Tucumán, Concepción (BM 27.3.25.5); Salta, Tartagal (Ojeda and Mares 1989); Buenos Aires, Pergamino (Pine, Dalby, and Matson 1985). URUGUAY: Montevideo, Colón (BM 97.1.1.9); Maldonado, Maldonado (type local-

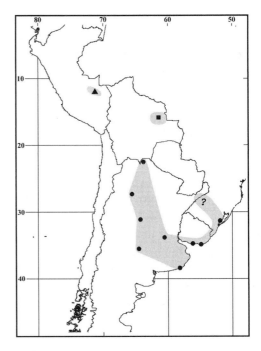

Map 39 Marginal localities for *Monodelphis dimidiata* ●, *Monodelphis ronaldi* ▲, and *Monodelphis* "species B" ■

ity of *Didelphys dimidiata* J. A. Wagner). BRAZIL: Rio Grande do Sul, São Lourenço (O. Thomas 1888b).

ADDITIONAL LOCALITIES (not mapped): ARGENTINA: Buenos Aires, 15 km NW of Balcarce (Pine, Dalby, and Matson 1985); "Misiones" (BM 26.2.11.20). URUGUAY: Montevideo, Montevideo (Sanborn 1929a).

SUBSPECIES: We treat *M. dimidiata* as monotypic. If the extent and nature of geographic variation warrant recognizing subspecies, the name *M. fosteri* O. Thomas, 1924c, is available for animals from the La Pampa area.

NATURAL HISTORY: Pine, Dalby, and Matson (1985) reported that *M. dimidiata* is diurnal in Argentina, at least during the cooler parts of the year, with an activity peak in the late afternoon. A. K. Lee and Cockburn (1985), however, stated that all didelphids are "predominantly nocturnal." Pine, Dalby, and Matson (1985) concluded on the basis of trapping records, as had Reig (1964), that these animals may live but for a year and reproduce only once in their lives, and that a burst of growth appears to take place in the spring. E. M. González (1996, 2001) also confirmed these conclusions for Uruguayan *M. dimidiata*. For unstated reasons, however, Tyndale-Biscoe (2003) is skeptical of this reproductive pattern for any species other than certain dasyurids. Sexual dimorphism is extreme (not "moderate" and comparable to that seen in *M. domestica*, as stated by Taggart et al. 2003). The adult males are larger than females and have disproportionately larger heads and much longer canines. The first observation concerning diet in the wild was provided by Darwin (in Waterhouse 1939:97), who reported the holotype as having a stomach full of insects. Busch and Kravetz (1992) analyzed the contents of 23 stomachs and found 100% frequency of insects, 22.7% of arachnids, 33.33% mammals, and 9.10% plant material. Rodent species eaten were *Bolomys obscurus*, *Oligoryzomys flavescens*, *Calomys laucha*, and *Oxymycterus rutilans*. Goin, Velázquez, and Scaglia (1992) found that, with increasing size of the animals, the "scissors-like" occlusion of the shearing crests of the molars increases, and that this is correlated with the diet becoming more carnivorous with age, as opposed to insectivorous. Pine, Dalby, and Matson (1985) reported one litter of 16 young. Mares, Braun, and Gettinger (1989a) made several unsupported claims at variance with the observations of Pine, Dalby, and Matson (1985) when they stated that *M. dimidiata* is nocturnal, semiarboreal, apparently reproduces all year, and has 8 to 14 young per litter. E. M. González and Claramunt (2000) have provided information on behavior in captivity. Fornes and Massoia (1965) listed the ectoparasitic mite *Haemolaelaps glasgowi*, and Costa-Lima and Hathaway (1946) listed the flea *Adoratopsylla antiquorum cunhai* as found on *M. dimidiata*.

Reig and Bianchi (1969) reported a $2n = 18$ karyotype having a minute telocentric X and a dot-like Y chromosome similar to the Y in the karyotype they reported for "*M. brevicaudata*" (= *M. palliolata*). Wainberg (1972), Reig et al. (1977), and B. A. Carvalho et al. (2002) also discussed the chromosomes of animals identified as *M. dimidiata*, and A. J. Solari and Bianchi (1975) discussed the synaptic behavior of the sex chromosomes.

REMARKS: B. E. Brown (2004) listed records from Paraguay and from Rio de Janeiro, Brazil, but the identity of the specimens these records represent has not been confirmed. According to Pine, Dalby, and Matson (1985), all purported records of this species from Brazil and Paraguay (e.g., Bertoni 1939) might pertain to *M. sorex*. The Brazilian specimen of *M. dimidiata* in the British Museum (BM 87.5.17.6), therefore, is of special interest. Some records for Uruguay (Sanborn 1929a; Ximénez, Langguth, and Praderi 1972) are uncertain and possibly are based on *M. sorex*.

Monodelphis fosteri O. Thomas may be worthy of recognition as a subspecies of *M. dimidiata* or even as a full species. Although Kirsch and Calaby (1977) stated that Reig (1964) had shown that "*M. fosteri* specimens were simply juveniles of *M. dimidiata*," Reig had done nothing of the sort. Certain subsequent authors have, following Kirsch and Calaby's lead, treated, without comment, *M. fosteri* as a junior synonym of *M. dimidiata*, but this has yet to be demonstrated or formally proposed (see Pine, Dalby, and Matson 1985). Pine, Dalby, and Matson (1985) tentatively treated *M. fosteri* O. Thomas as a full species, but examination of the holotype led Handley to place it in *M. dimidiata*. Pine's

subsequent examination of the holotype, however, did not rule out, in his mind, the possibility that *fosteri* could still be a separate species. The immature holotype has decidedly shorter, duller, paler, and less tawny pelage than unequivocal *M. dimidiata* of similar age. Olrog and Lucero (1981) appear to have used both the names *M. henseli* and *M. fosteri* for *M. dimidiata*. Massoia (in Massoia, Forasiepi, and Teta 2000) thought that the name *fosteri* should be applied, at the subspecific level, to all Argentine populations, with *M. d. dimidiata* being a subspecies restricted to Uruguay and Brazil. This position is completely insupportable on the basis of all specimens seen by us.

Monodelphis domestica (J. A. Wagner, 1842)
Gray Short–tailed Opossum
SYNONYMS:

Didelphys domestica J. A. Wagner, 1842:359; type locality "Cuyaba," Mato Grosso, Brazil.

Hemiurus hunteri: P. Gervais, 1856b:101; not *Didelphys hunteri* Waterhouse, 1841.

Hemiurus concolor P. Gervais, 1856b:pl. 16; type locality "province de Goyaz," Brazil.

[*Didelphys (]Microdelphys[)] domestica*: Burmeister, 1856: 87; name combination.

Didelphys [(Peramys)] domestica: O. Thomas, 1888b:358; name combination.

Hemiurus domesticus: Winge, 1893:55; name combination.

P[eramys]. domesticus: O. Thomas, 1897a:220; name combination.

[*Didelphis (]Monodelphis[)] concolor*: Matschie, 1916: 271; name combination.

[*Didelphis (]Monodelphis[)] domestica*: Matschie, 1916: 272; name combination.

Monodelphis domestica: Pohle, 1927:240; first use of current name combination.

Monodelphis domestica domestica: Cabrera, 1958:8; name combination.

Monodelphis brevicaudis brevicaudis: Wetzel and Lovett, 1974:206; not *Didelphys brevicaudis* Olfers, 1818.

Monodelphis domesticus: D. A. Mello, 1977:391; name combination and incorrect gender concordance.

DISTRIBUTION: *Monodelphis domestica* occurs in eastern Bolivia, central and northeastern Brazil, northern and central Paraguay, and northern Argentina.

MARGINAL LOCALITIES (Map 40): BRAZIL: Ceará, Ipu (O. Thomas 1910c); Ceará, Parque Nacional de Ubajara (Guedes et al. 2000); Parnaíba, Alagôa do Monteiro (Moojen 1943); Parnaíba, Penha (Pohle 1927); Pernambuco, São Lourenço (BM 3.10.1.84); Alagoas, Limoeiro da Anadia (Mares et al. 1981); Bahia, Feira (C. O. C. Vieira 1955); Bahia, Marchado Portela (Mares et al. 1981); Minas Gerais, Lagoa Santa (Winge 1893); Goiás, 7 km

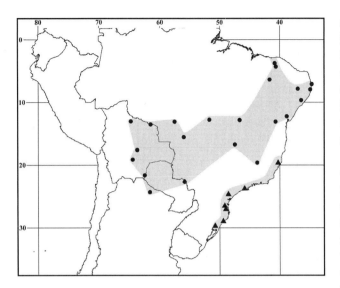

Map 40 Marginal localities for *Monodelphis domestica* ● and *Monodelphis iheringi* ▲

SE of Cristalina (Mares, Braun, and Gettinger 1989). PARAGUAY: Amambay, 4 km (by road) SW of Cerro Corá (UMMZ 125245). ARGENTINA: Formosa, Reserva Natural Formosa (Heinonen-Fortabat and Chebez 1997). BOLIVIA: Tarija, Estancia Bolívar (S. Anderson 1997); Chuquisaca, Tihumayu (Pine 1980); Santa Cruz, 4 km S and 13 km W of San Rafael de Amboró (S. Anderson 1997); Beni, San Joaquín (Pine 1980); Santa Cruz, Flor de Oro (Emmons 1998). BRAZIL: Mato Grosso, Lava Pés (Pine 1980); Mato Grosso, Cuiabá (type locality of *Didelphys domestica* J. A. Wagner); Mato Grosso, Serra do Roncador, 264 km (by road) N of Xavantina (BM 76.652); Tocantins, Cana Brava (C. O. C. Vieira 1955); Piauí, Fazenda Olho da Agua (Mares et al. 1981).

SUBSPECIES: We treat *M. domestica* as monotypic.

NATURAL HISTORY: Streilein (1982a–e) has provided information on the ecology and behavior of *M. domestica* in the Brazilian caatinga. He (1982b) found the species to be basically nocturnal, with peak activity just after dusk, and primarily terrestrial, with greatest densities in areas of granitic outcrops. Streilein (1982b) stated that the natural diet primarily consisted of invertebrates, but that a variety of live vertebrates were taken in captivity. Streilein (1982c) further stated that reproduction occurred throughout much of the year, although Bergallo and Cerqueira (1994), also working in the caatinga, claimed that it coincided with the wet season and concluded that the timing of reproduction was controlled by photoperiod. Streilein (1982b) gave litter size as from 1 to 11. He later (1982c) gave the litter size as from 6 to 11 (mean of 8.4) and reported that at least one female produced two litters in the wild. VandeBerg (1990) reported 13 as the maximum number of mammae

and the maximum number of attached young he observed in a female in his laboratory, and a litter of 14 as having been reared elsewhere. Cothran, Aivaliotis, and VandeBerg (1985) reported litter size as averaging 7.33 ± 3.20. According to O. Thomas (1910c), Snethlage found a female in the wild with 12 young attached. Bergallo and Cerqueira (1994) found a mean litter size, based on embryos, not on attached young, to be 8.2, with a maximum of 16.

Monodelphis domestica has become an important laboratory animal (and even an item in the pet trade) and its husbandry and utility have been covered by Fadem et al. (1982), VandeBerg (1983, 1990), and VandeBerg and Robinson (1997). Hayssen (1980) provided general remarks on its natural history in the laboratory, noting that the young release their hold on the nipple after an age of 14 days. Trupin and Fadem (1982) described sexual behavior in captivity, Fadem and Cole (1985) described scent marking, and Unger (1982) described nest building. J. A. Wagner (1847) and Pelzeln (1883) noted that Natterer, the collector of the type series, encountered this species in houses (hence the name) and (*fide* Pelzeln 1883) said that it ate cockroaches. O. Thomas (1910c) quoted Snethlage as having written that it "Lives in or near houses and one that we caught in our work-room had made a nest of straw and paper under the boards." Thus in size, appearance, and habits, this animal resembles the commensal shrew *Suncus murinus* of the Old World. Emerson and Price (1975) and Timm and Price (1985, 1988) listed *M. domestica* as one of the hosts of the chewing louse *Cummingsia peramydis* Ferris. Mares, Braun, and Gettinger (1989) reported mites of the genus *Androlaelaps* from this species. Macrini (2004) summarized additional information on parasites in his *Mammalian Species* account.

Merry, Pathak, and VandeBerg (1983); E. S. Robinson, Samollow, and VandeBerg (1993); Palma and Yates (1996); and B. A. Carvalho et al. (2002) reported on chromosomal studies of this species. The karyotype is $2n = 18$, FN = 30, the X is a small acrocentric and the Y is a minute acrocentric chromosome. Merry, Pathak, and VandeBerg (1983) described the Y chromosome as a metacentric.

REMARKS: *Monodelphis domestica* is most abundant in Paraguay, Bolivia, and the Mato Grosso and Northeast of Brazil, with only scattered records in central Brazil. Olrog and Lucero (1981) first listed this species for Argentina and provided a range map showing it as occurring in Formosa, but provided no localities (but see Heinonen-Fortabat and Chebez 1997). Cabrera (1958) listed *M. maraxina* O. Thomas as a subspecies of *M. domestica*. For reasons given by Pine (1980), this is not done here. Wetzel and Lovett (1974) mistakenly regarded *Didelphys brevicaudis* Olfers as a senior synonym of *M. domestica* (J. A. Wagner); see Pine (1980).

Hemiurus concolor P. Gervais first appeared as the name of the short-tailed opossum illustrated on Plate 16, Figure 2 of P. Gervais (1856b). Apparently, P. Gervais (1856b:101) changed his mind when preparing the text and misapplied the name *Hemiurus hunteri* (Waterhouse) to this animal.

Monodelphis emiliae (O. Thomas, 1912)
Emilia's Short-tailed Opossum

SYNONYMS:

Peramys emiliae O. Thomas, 1912a:89; type locality "Boim, R. Tapajoz," Pará, Brazil.

[*Didelphis (]Monodelphis[)] emiliae*: Matschie, 1916:271; name combination.

Monodelphis emiliae: O. Thomas, 1920c:283; first use of current name combination.

P[*eramys*]. *tricolor emiliae*: A. Miranda-Ribeiro, 1936:408; part; name combination.

P[*eramys*]. *tricolor emilae*: C. O. C. Vieira, 1950:360; incorrect subsequent spelling of *Peramys emiliae* O. Thomas.

Monodelphis tricolor emiliae: C. O. C. Vieira, 1955:350; name combination.

Monodelphis touan emiliae: Cabrera, 1958:9; name combination.

Monodelphis touan: Jenkins and Knutson, 1983:19; part; name combination, not *Mustela touan* Bechstein, 1800).

DISTRIBUTION: *Monodelphis emiliae* occurs in the Amazon basin from the Rio Tocantins near the mouth of the Amazon in Brazil, to the Marañón-Ucayali junction in Peru, a distance of about 2,650 km, and south to the lower Urubamba in Peru and thence into departamento Pando, Bolivia.

MARGINAL LOCALITIES (Map 41; from Pine and Handley 1984, except as noted): BRAZIL: Pará, Boim (O. Thomas 1912a); Pará, Cametá; Pará, Km 217; Pará, Villa Braga (O. Thomas 1920c); Pará, Flexal. BOLIVIA: Pando, Centro Dieciocho [*sic*] (Emmons and Smith in S. Anderson 1997). PERU: Madre de Dios, Cocha Cashu (S. Solari 2004); Cusco, San Martín-3 (S. Solari et al. 1999, 2001c). BRAZIL: Acre, Nova Vida (J. L. Patton, Silva, and Malcolm 2000). PERU: Loreto, Río Gálvez (Jansa and Voss 2000); Loreto, Quebrada Aucayo. BRAZIL: Amazonas, Tapaiúna (B. D. Patterson 1992).

ADDITIONAL LOCALITIES (not mapped): BRAZIL: Amazonas, Seringal Condor (J. L. Patton, Silva, and Malcolm 2000); Pará, Fordlândia (Pine and Handley 1984).

SUBSPECIES: *Monodelphis emiliae* is currently treated as monotypic.

NATURAL HISTORY: Fräulein Emilia Snethlage stated, concerning the first specimens collected: "no. 11 was shot by my preparator among dry palm leaves on the ground" (*in litt.*, cited by O. Thomas 1912a). J. L. Patton, Silva, and

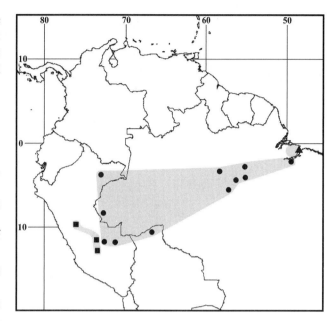

Map 41 Marginal localities for *Monodelphis emiliae* ●, *Monodelphis maraxina* ▲, and *Monodelphis* "species C" ■

Malcolm (2000) caught six adults from four localities along the central and upper Rio Juruá. Animals were trapped in *terra firme* forest, "typically" along fallen logs under a moderately dense understory. A female with three young attached was caught in February and "parous" females were taken in both February and September. In the lower Urubamba region of Peru, S. Solari et al. (2001c) trapped this species in *terra firme* forest dominated by the palm *Iriartea deltoidea* (see Comiskey et al. 2001). O. Thomas (1912a, 1920c) and Pine and Handley (1984) commented on the striking fugitive coloration of the underparts. The karyotype is $2n = 18$, FN = 30, the X chromosome is a small submetacentric; the Y, a small acrocentric (J. L. Patton, Silva, and Malcolm 2000).

REMARKS: *Monodelphis emiliae* was treated as a full species by Pine in Honacki, Kinman, and Koeppl (1982), N. E. Peterson and Pine (1982), Pine and Handley (1984), George et al. (1988), Emmons and Feer (1990, 1997), and Gardner (1993a). Linares (1998), however, seemingly fell back to the position of Cabrera (1958) that *M. emiliae* is conspecific with (a) form(s) recognized by them and to be called *M. touan* (see Voss, Lunde, and Simmons 2001). *Monodelphis emiliae* is known from only about 20 specimens, 9 of them from both banks of the lower Rio Tapajóz, and 6 from along the Rio Juruá. *Monodelphis emiliae* is sympatric with *M. glirina* on the Rio Tapajóz in Brazil (Pine and Handley 1984) and in southeastern Peru at San Martín-3 in the lower Urubamba region (S. Solari et al. 2001c), and at Cocha Cashu (S. Solari 2004). On the Rio Tocantins and eastward in Brazil (Pine and Handley 1984), *M. emiliae* is

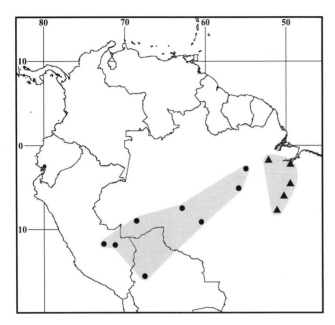

Map 42 Marginal localities for *Monodelphis glirina* ● and *Monodelphis* "species D" ▲

sympatric with an undescribed south-of-the-Amazon form (see *Monodelphis* [sp. E]) that Handley thought was a subspecies of *M. brevicaudata*. This undescribed taxon was reported by C. T. Carvalho (1960b) as *M. brevicaudata emiliae* from localities indicated by the points in southeastern Pará mapped by B. E. Brown (2004:113). For reasons that are unclear, Eisenberg and Redford (1999) regarded sympatry between *M. emiliae* and *M. brevicaudata* (*sensu lato*) as merely a possibility. At the Quebrada Aucayo in Peru, *M. emiliae* is the only known *Monodelphis*. Jansa and Voss (2000) reported sympatry between *M. emiliae* and *M. adusta*. Sympatry between these two, reported by S. Solari et al. (2001c), actually involved two localities, although not widely separated ones.

Monodelphis glirina (J. A. Wagner, 1842)
Amazonian Red-sided Opossum
SYNONYMS:
Didelphys glirina J. A. Wagner, 1842:359; type locality "Mamoré," Rondônia, Brazil; more clearly identified as "Cachoeira do Pau grande am Mamoré," by J. A. Wagner (1847:153).
[*Didelphis* (]*Monodelphis*[)] *glirina*: Matschie, 1916:271; name combination.
DISTRIBUTION: *Monodelphis glirina* is known from Brazil, south of the Rio Amazonas, into northern Bolivia.
MARGINAL LOCALITIES (Map 42): BRAZIL: Pará, Igarapé Maroi (B. D. Patterson 1992); Pará, Km 446, BR 165, Santarém-Cuiabá (USNM 546207); Mato Grosso,

Aripuanã (USNM 545576). BOLIVIA: La Paz, 4 km (by road) N of Alcoche (S. Anderson 1997). PERU: Madre de Dios, Pakitza (S. Solari 2004); Cusco, San Martín-3 (S. Solari et al. 2001c); BRAZIL: Acre, Sena Madureira (examined specimen in Museo Goeldi, Belém); Amazonas, Humaitá (USNM 545553).

ADDITIONAL LOCALITIES (not mapped): PERU: Madre de Dios, Cocha Cashu (S. Solari 2004). BOLIVIA: Pando, Santa Rosa (S. Anderson 1997); La Paz, Río Challana, 5 km (by road) SE of Guamay (UMMZ 126684).

SUBSPECIES: *Monodelphis glirina* is treated here as monotypic.

NATURAL HISTORY: *Monodelphis glirina* (under the name *M. brevicaudata*) has been reported as sympatric with *M. emiliae* on the Rio Tapajóz, Pará, Brazil (Pine and Handley 1984) and in the lower Urubamba region of Cusco, Peru (S. Solari et al. 2001c). The latter authors' report of it as sympatric with *M. adusta* does not appear to be a case of true sympatricity. The lower Urubamba habitat was *terra firme* forest dominated by the palm *Iriartea deltoidea* (Comiskey et al. 2001). Palma and Yates (1996) illustrated and described the karyotype of a Bolivian specimen (under the name *M. brevicaudata*) as $2n = 18$, FN $= 22$, with 2 metacentric, 4 submetacentric, and 10 acrocentric autosomes; the X chromosome is a small acrocentric and the Y, a minute acrocentric.

REMARKS: Voss, Lunde, and Simmons (2001) detailed morphological features useful in distinguishing *M. glirina* from its northern congener, *M. brevicaudata*. The apparent absence of *M. glirina* from the upper portions of the rivers Tapajóz and Ulgro, while being recorded in the headwaters of the Madeira and Purus, probably represents lack of collecting in the headwaters of these river systems. Although B. E. Brown (2004:111, 113) stated she that did not map the distribution of *M. glirina*, the Bolivian localities plotted in her map (p. 108) for *M. brevicaudata* appear to represent *M. glirina* as understood here.

Monodelphis iheringi (O. Thomas, 1888)
Ihering's Three-striped Opossum
SYNONYMS:
Didelphys (Peramys) Iheringi O. Thomas, 1888a:159; type locality "Rio Grande do Sul"; identified by O. Thomas (1888b:365) as Taquara, Rio Grande do Sul, Brazil.
[*Didelphys* (]P[*eramys*]. [)] *tristriatus*: Goeldi, 1894:464; part; name combination.
[*Didelphis* (]*Microdelphys*[)] *iheringi*: Matschie, 1916:272; name combination.
P[*eramys*]. [(*Microdelphis*)] *iheringi*: A. Miranda-Ribeiro, 1936:404; name combination.
Peramys iheringii A. Miranda-Ribeiro, 1936: 415; incorrect subsequent spelling of *Didelphys iheringi* O. Thomas.

Monodelphis iheringii C. O. C. Vieira, 1950:357; incorrect subsequent spelling of *Didelphys iheringi* O. Thomas.

Monodelphis americana iheringi: Cabrera, 1958:7; name combination.

Monodelphis americana: Jenkins and Knutson, 1983:14; part; name combination, not *Sorex americanus* Müller.

DISTRIBUTION: *Monodelphis iheringi* occurs in Brazil, where it was reported in Espírito Santo by A. Miranda-Ribeiro (1936); otherwise it is known from the states of São Paulo, Santa Catarina, and Rio Grande do Sul.

MARGINAL LOCALITIES (Map 40): BRAZIL: Espírito Santo, Japaraná (A. Miranda-Ribeiro 1936); São Paulo, Ribeirão da Lagoa (USNM 484015); São Paulo, Iporanga (FMNH 94736); Santa Catarina, Humboldt (BM 14.1.27.7); Santa Catarina, Blumenau (ZMB 35493); Santa Catarina, Araranguá (Pine 1977); Rio Grande do Sul, Taquara (type locality of *Didelphys iheringi* O. Thomas).

ADDITIONAL LOCALITIES (not mapped and questionable): BRAZIL: Espírito Santo, "Castelo e Alegre" (Ruschi 1965). ARGENTINA: Misiones, Parque Nacional Iguazú (Heinonen-Fortabat and Chebez 1997).

SUBSPECIES: *Monodelphis iheringi* is treated here as monotypic.

NATURAL HISTORY: Massoia, Chebez, and Heinonen-Fortabat (1989b) recorded "*Monodelphis americana iheringi*" from an owl pellet attributed to the Barn Owl, *Tyto alba*, in Misiones, Argentina," but their identification is questionable (see account of *M. americana*). Contreras and Silvera-Avalos (1995) stated that *M. iheringi* (which they regarded as a separate species from *M. americana*) occurred in Misiones, but they gave no localities.

REMARKS: *Monodelphis iheringi* was included (incorrectly) in *M. americana* by Cabrera (1958), but see Pine (1977). The second sentence of the second paragraph of Pine's article should have read, "Ihering (1894) showed uncertainty as to whether this animal deserved recognition as a species and [Goeldi (1894)] expressed strong skepticism concerning this matter." The section in brackets was, unfortunately, omitted through an error in proofreading. Pine's conclusion, on the basis of a damaged skull, that the bullae are compressed dorsoventrally, was in error.

Chebez and Massoia (1996) listed *M. iheringi* as occurring in at least four localities (see their errata section) in Misiones, Argentina, but their specimens may actually be of *M. americana* (see discussion of that species). Galliari, Pardiñas, and Goin (1996) and Mares and Braun (2000), however, unequivocally treated the Argentine records as pertaining to *M. iheringi*, but it is unclear if they themselves examined an actual specimen. The issue is further confused by the figures of skulls claimed by Chebez and Massoia (1996) to pertain to species of *Monodelphis*.

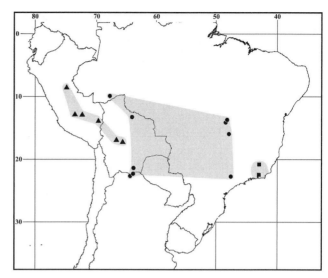

Map 43 Marginal localities for *Monodelphis kunsi* ●, *Monodelphis osgoodi* ▲, and *Monodelphis theresa* ■.

The drawing supposedly of a "*Monodelphis americana iheringi*" skull does not resemble a skull of that species, and the scale indicates that it should be about 90 mm long. In addition, the skull labeled as "*Monodelphis Henseli = M. Sorex*" appears to be that of a *Monodelphis dimidiata*, and the skull labeled as of "*Monodelphis Dimidiata* ssp" represents that of a juvenile, which still has its milk premolars and which Pine cannot identify from the drawing.

Monodelphis kunsi Pine, 1975
Pygmy Short-tailed Opossum

SYNONYM:

Monodelphis (*Monodelphis*) *kunsi* Pine, 1975:321; type locality "La Granja, W bank of Río Itonamas, 4 k N Magdalena, Provincia Itenez, Departamento Beni, Bolivia, below 200 m."

DISTRIBUTION: *Monodelphis kunsi* is known from Bolivia, Brazil, and northern Argentina.

MARGINAL LOCALITIES (Map 43): BRAZIL: Acre, Alto Acre (ZMB 35513). BOLIVIA: Beni, La Granja (type locality of *Monodelphis kunsi* Pine). BRAZIL: Goiás, 40 km SW of Minaçu (B. A. Carvalho et al. 2002); Goiás, 55 km N of Niquelândia (B. A. Carvalho et al. 2002); Distrito Federal, 20 km S of Brasilia (Mares 1986; Mares, Braun, and Gettinger 1989; given as ca. 12 km S by Gettinger 1987, 1992); São Paulo, Piracicaba (MZUSP 480, 1524). BOLIVIA: Tarija, Tapecua (Salazar et al. 1994). ARGENTINA: Salta, Finca Falcón (Jayat and Miotti 2006). BOLIVIA: Tarija, Río Lipeo (S. Anderson 1982).

Additional locality (not mapped): BRAZIL: Goiás, 20 km WW [*sic*] of Colinas do Sul (B. A. Carvalho et al. 2002).

SUBSPECIES: We regard *M. kunsi* as monotypic.

NATURAL HISTORY: The holotype was live trapped in "cut-over brush" (Pine 1975). Both of Mares, Braun, and Gettinger's (1989a) specimens were taken in "dense cerrado habitat." The specimen from Tapecua, Tarija, Bolivia (Salazar et al. 1994) was caught in a ravine under dense shrubbery on a 30° slope having 10 to 12 cm of leaves and litter covering the ground. The Argentine specimen reported by Jayat and Miotti (2006) was caught in a live trap placed on the ground in secondary forest. Palma and Yates (1996) described the karyotype as $2n = 18$, FN = 30, and submetacentric X and Y chromosomes, based on the Río Lipeo, Tarija, Bolivia specimen. B. A. Carvalho et al. (2002) reported a similar karyotype for Brazilian specimens, except they described and illustrated the Y chromosome as an acrocentric.

REMARKS: The species is known from very few specimens from widely spaced regions in Bolivia, Brazil, and Argentina. When reviewed by S. Anderson (1982), he knew only of two Bolivian specimens. Pine has examined the specimen from Río Lipeo, Tarija, Bolivia and found that the skin is very *adusta*-like. The teeth are *kunsi*-sized and what is left of the severely damaged skull also appears to be *kunsi*-like, but Pine is skeptical of this specimen's identification as *M. kunsi*. He has not seen the other specimen from Tarija, or the one from Acre. The former was the source of the chromosomes for the karyotype described and figured by Palma and Yates (1996).

Monodelphis maraxina O. Thomas, 1923
Marajó Short-tailed Opossum

SYNONYMS:

Monodelphis maraxina O. Thomas, 1923c:157; type locality "Caldeirão," Pará, Brazil.

Monodelphis domestica maraxina: Cabrera, 1958:8; name combination.

Monodelphis domestica: Jenkins and Knutson, 1983:19; part; name combination, not *Didelphys domestica* J. A. Wagner.

DISTRIBUTION: *Monodelphis maraxina* is known from Ilha de Marajó, Pará, Brazil, and possibly a second nearby island.

MARGINAL LOCALITIES (Map 41; from O. Thomas 1923c): BRAZIL: Pará, Soure, Ilha de Marajó; Pará, Caldeirão (type locality of *Monodelphis maraxina* O. Thomas, not mapped).

SUBSPECIES: We consider *M. maraxina* to be monotypic.

NATURAL HISTORY: Unknown.

REMARKS: Cabrera (1958) treated *M. maraxina* as a subspecies of *M. domestica* (J. A. Wagner) and Emmons and Feer (1990, 1997) also regarded it as conspecific with *M. domestica*. According to Pine (1980), however, it is best

regarded as a separate species, perhaps more closely related to the *M. brevicaudata* complex than to *M. domestica*. We know of only eight specimens, of which one was caught in 1911 and seven in 1923; all are in the British Museum. Pine (1980) was aware of only two of these.

Monodelphis osgoodi Doutt, 1938
Osgood's Short-tailed Opossum

SYNONYMS:

Peramys peruvianus: O. Thomas, 1920d:248; not *Peramys peruvianus* Osgood, 1913.

Monodelphis peruvianus osgoodi Doutt, 1938:100; type locality "Incachaca, Department of Cochabamba, Bolivia; altitude 2600 meters."

Monodelphis adusta osgoodi: Cabrera, 1958:6; name combination.

Monodelphis osgoodi: Handley, 1966c:755; first use of current name combination.

DISTRIBUTION: *Monodelphis osgoodi* is known from eastern Peru and central Bolivia.

MARGINAL LOCALITIES (Map 43; localities listed from north to south, west to east): PERU: Ucayali, IVITA Biological Station (Aniskin et al. 1991; see Remarks); Cusco, Wayrapata (S. Solari et al. 2001a); Cusco, Tocopoqueu (O. Thomas 1920d); Puno, Inca Mines (AMNH 16547); Puno, Oconeque (FMNH 52714, not mapped). BOLIVIA: Cochabamba, Choro (FMNH 74861); Cochabamba, Incachaca (type locality of *Monodelphis peruvianus osgoodi* Doutt).

SUBSPECIES: We treat *M. osgoodi* as monotypic.

NATURAL HISTORY: All locality records are from moderate to high elevations; otherwise the ecology and life history of *M. osgoodi* are unknown.

REMARKS: The habitat information given by S. Anderson (1997) apparently goes with *M. adusta*. A female taken in July had no embryos (S. Anderson 1997). The karyotype is $2n = 18$, FN = 32; the X chromosome is a small subtelocentric and the Y, a minute metacentric (Aniskin et al. 1991; S. Solari, *in litt.*, writes that the species in question is probably *M. adusta*). S. Solari et al. (2001a) reported *M.* cf. *osgoodi* as sympatric with the species discussed here under the name "species B" at Wayrapata, Cusco, Peru, 2,445 m.

Cabrera (1958) listed *M. osgoodi* as a subspecies of *M. adusta*, but Handley (1966c) treated it as probably representing a separate species, and Pine (in Kirsch and Calaby 1977) unequivocally treated it as such. O. Thomas (1920d) listed specimens from Peru as *Peramys peruvianus* Osgood. Our current concept of *M. osgoodi* is probably a composite of at least two, and possibly three or more species. In particular, the USNM specimen from Tocopoqueu, Peru, is doubtfully conspecific with the type series. S. Anderson (1997) placed the name *M. kunsi*, as used by Salazar et al.

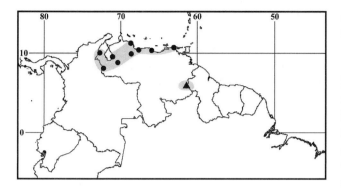

Map 44 Marginal localities for *Monodelphis palliolata* ● and *Monodelphis reigi* ▲

(1994), in the synonymy of *M. osgoodi*. The specimens Salazar et al. (1994) reported on probably represent *M. adusta*. S. Solari (*in litt.*) writes that the elevation at the IVITA Biological Station, Ucayali, Peru (see Marginal Localities above) is too low for *osgoodi* and that *M. adusta* may be the animal referred to by Aniskin et al. (1991).

Monodelphis palliolata Osgood, 1914
Hooded Red-sided Opossum
SYNONYMS:

Peramys palliolatus Osgood, 1914a:135; type locality "San Juan de Colon, State of Tachira, Venezuela."

[*Didelphis (]Monodelphis[)] palliolatus*: Matschie, 1916: 271; name combination and incorrect gender concordance.

M[onodelphis]. palliolatus: Pittier and Tate, 1932:254; first use of current name combination, but with incorrect gender concordance.

Monodelphis palliolata: Cabrera and Yepes, 1940:32; first use of name combination with correct gender concordance.

Monodelphis brevicaudata palliolata: Cabrera, 1958:8; name combination.

DISTRIBUTION: *Monodelphis palliolata* is known from northeastern Colombia and western Venezuela.

MARGINAL LOCALITIES (Map 44): VENEZUELA (Handley 1976, as *M. brevicaudata*; except as noted): Falcón, Mirimire; Aragua, Estación Biológica Rancho Grande; Miranda, 6 km SSE of Río Chico; Sucre, Manacal; Cojedes, El Candelo (Pérez-Hernández 1990); Barinas, 3 km N of Barinitas (Díaz de Pasqual 1984); Táchira, San Juan de Colón (Osgood 1914a); Zulia, Kasmera; Trujillo, La Ceiba.

ADDITIONAL LOCALITIES (not mapped): COLOMBIA (Cuervo, Hernández-Camacho, and Cadena 1986): Norte de Santander (no locality identified); Vichada (no locality identified). VENEZUELA (Handley 1976, as *M.*

brevicaudata): Falcón, Boca de Yaracuy; Yaracuy, Minas de Aroa; Miranda, Curupao.

SUBSPECIES: We treat *M. palliolata* as monotypic.

NATURAL HISTORY: Ventura, Pérez-Hernández, and Lopez-Fuster (1998) stated that this species has been taken from sea level to 1,500 m "in diverse habitats, such as tropical dry forest, premontane humid and very humid forests, and tropical humid forest." We judge, on the basis of locality, elevation, and habitat, that O'Connell (1979) was writing of this species when she reported having made all 37 of her captures and recaptures on the ground in Venezuela; in addition, her voucher specimen TTU 35994 proves to be of *M. palliolata*; also see Eisenberg, O'Connell, and August (1979). Handley (1976), reporting on Venezuelan animals (both *M. brevicaudata* and *M. palliolata*, as understood here, plus the species reported by Ventura, Pérez-Hernández, and Lopez-Fuster [1998]; Linares [1998]; and Ventura et al. [2005] as *M. orinoci*), took 96% on the ground and 4% on logs and in trees; misquoted by O'Connell (1979:77) as 4% "in trees and vines." W. Robinson and Lyon (1901) trapped a specimen "in a thicket of vines near a stream." O'Connell (1979) stated that this animal is nocturnal and more carnivorous than the six other didelphid species that she studied, but presented no supporting evidence.

O'Connell's (1979, 1989) data suggested to her that, in northern Venezuela, breeding in this species is seasonal, with lactating females taken in May, June, July, August, and November. She also reported two nursing litters, one of eight young and another of seven. Eisenberg (1989), however, wrote that the breeding season extends from May through August in northern Venezuela and that up to seven young are born in a single litter. Collins (1973) reported a Venezuelan female that contained five embryos. On the basis of the localities given, this species must be one of the hosts for the mallophagan *Cummingsia peramydis* Ferris (see Emerson and Price 1975; Timm and Price 1985). Similar considerations apply to the mites *Gigantolaelaps aitkeni* Lee and Strandtmann and *G. gilmorei* Fonseca (see Furman 1972). It is not clear that the limited natural history information purported to be for this species does not, in fact, apply to other species recently treated as representing *M. brevicaudata* (e.g., *sensu* Handley 1976) or as representing *M. orinoci* of authors. The karyotype is a $2n = 18$ with a minute telocentric X and a dot-like Y chromosome (Reig and Bianchi 1969, as *M. brevicaudata*).

REMARKS: Pine and Handley (1984) explicitly treated *M. brevicaudata* and *M. palliolata* as conspecific. Hershkovitz (1992b) continued to treat "*Monodelphis palliolatus*" as a separate species; while Linares (1998) regarded it as a subspecies of *M. brevicaudata*. Voss, Lunde, and Simmons (2001) concluded that *M. palliolata* should be recognized at the species level and gave characters by which it can be recognized. The karyotype described for *M. brevicaudata* by Reig and Bianchi (1969) actually applies to *M. palliolata*. Reig et al. (1977) stated that they had "studied several *M. brevicaudata* from distant localities and did not find any geographic variation in their karyotypes." However, they listed none of these localities, so the identities of their specimens cannot be verified. Apparently, Cuartas and Muñoz (2003) relied on the information presented by Cuervo, Hernández-Camacho, and Cadena (1986) to outline the Colombian distribution. Cuervo, Hernández-Camacho, and Cadena (1986), Alberico et al. (2000), and Cuartas and Muñoz (2003) treated *M. palliolata* as a synonym of *M. brevicaudata*. The locality records plotted by Cuartas and Muñoz (2003:211) in their map of the distribution of "*Monodelphis brevicaudata*" do not appear to be based on specimen records.

Monodelphis reigi Lew and Pérez-Hernández, 2004
Reig's Opossum
SYNONYM:

Monodelphis reigi Lew and Pérez-Hernández, 2004:9; type locality "carretera El Dorado-Santa Elena, km 134, Estado Bolívar, Venezuela, 1300m."

DISTRIBUTION: *Monodelphis reigi* is known from only the type locality in eastern Venezuela.

MARGINAL LOCALITY (Map 44): VENEZUELA: Bolívar, Km 134, carretera El Dorado-Santa Elena (type locality of *Monodelphis reigi* Lew and Pérez-Hernández).

SUBSPECIES: *Monodelphis reigi* is known from one specimen.

NATURAL HISTORY: *Monodelphis reigi*, known only by the holotype, an adult male, was taken in the Sierra de Lema, an area dominated by montane evergreen forest having a relatively open understory and a high abundance of epiphytes (Lew and Pérez-Hernández 2004). Additional natural history information is lacking.

REMARKS: Ventura et al. (2005) compared *M. reigi* with four other species of Venezuelan *Monodelphis* and retained the taxon as a species.

Monodelphis ronaldi S. Solari, 2004
Ronald's Opossum
SYNONYM:

Monodelphis ronaldi S. Solari, 2004:146; type locality "Pakitza, Manu Reserved Zone, Department of Madre de Dios, southeastern Peru."

DISTRIBUTION: *Monodelphis ronaldi* is known from only the type locality in southeastern Peru.

MARGINAL LOCALITY (Map 39): PERU: Madre de Dios, Pakitza (type locality of *Monodelphis ronaldi* S. Solari).

SUBSPECIES: *Monodelphis ronaldi* is known from one specimen.

NATURAL HISTORY: The holotype, an adult male, was taken in lowland (356 m) humid forest. There is no further natural history information.

REMARKS: A medium-sized species of *Monodelphis* (total length 214 mm), *M. ronaldi* most closely resembles *M. adusta*; both species are more or less uniformly brownish and have flattened skulls. However, *M. ronaldi* differs in larger size, paler feet and dorsum, and has a low, but conspicuous sagittal crest, which is lacking in *M. adusta*. Most external and cranial measurements of the holotype of *M. ronaldi* are from 29 to 42 % larger, on average, than corresponding measurements of *M. adusta* (compare measurements by S. Solari 2004:147, Table 1). *Monodelphis ronaldi* is smaller than, and lacks the areas of reddish pelage characteristic of *M. emiliae* and *M. glirina*. The latter two species are known from the Manu region and *M. glirina* has been captured at Pakitza.

Monodelphis rubida (O. Thomas, 1899)
Red Short-tailed Opossum
SYNONYMS:

Didelphys brevicaudata: O. Thomas, 1888b:356; part; not *Didelphis brevicaudata* Erxleben.

Peramys rubidus O. Thomas, 1899c:155; type locality "Bahia," Bahia, Brazil.

[*Didelphys (Peramys)*] *rubida*: Trouessart, 1905:857; name combination.

[*Didelphis (]Monodelphis[)*] *rubidus*: Matschie, 1916:271; name combination.

P[*eramys*]. *tricolor rubidus*: A. Miranda-Ribeiro, 1936: 408; name combination.

Monodelphis tricolor rubidus: C. O. C. Vieira, 1955:350; name combination.

Monodelphis touan rubidus: Cabrera, 1958:10; name combination.

Monodelphis touan: Jenkins and Knutson, 1983:19; part; name combination, not *Mustela touan* Bechstein, 1800.

[*Monodelphis*] *rubida*: Pine and Handley, 1984:242; part; first use of current name combination.

DISTRIBUTION: *Monodelphis rubida*, as treated here, is known from only the type locality in Bahia, Brazil.

MARGINAL LOCALITY (Map 45): BRAZIL: Bahia, Bahia (type locality of *Peramys rubidus* O. Thomas).

SUBSPECIES: *Monodelphis rubida*, as treated here, is known from only one specimen.

NATURAL HISTORY: Unknown.

REMARKS: Pine and Handley (1984) resurrected O. Thomas's *Peramys rubidus* as a full species including, in their view, *Monodelphis umbristriata*, but gave no rationale for this treatment. Nonetheless, they have been followed

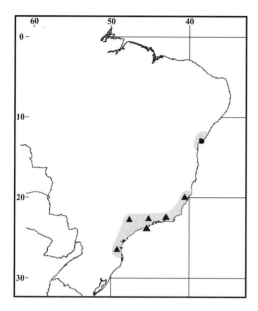

Map 45 Marginal localities for *Monodelphis rubida* ● and *Monodelphis scalops* ▲

by subsequent authors. Pine and Handley's lumping of *M. rubida* with *M. umbristriata* was based on Handley's comparison of the type of *Monodelphis rubida* with a specimen identified by Pine (1976) as of *M. umbristriata*, along with others similar to the latter in the USNM and BM. Since that time, Pine has also examined the type and compared it directly with the BM specimens. Among other things, Handley's conclusion as to conspecificity of *M. rubida* and *M. umbristriata* was based on his perceiving exceedingly faint and difficult to see *umbristriata*-like dorsal stripes in the type of *rubida*, but O. Thomas, Pine, R. S. Voss, and C. E. V. Grelle all have failed to see these stripes (see Lemos, Weksler, and Bonvicino 2000). Pine, in particular, because he knew of Handley's claim concerning these stripes, exposed the type to many angles and qualities of incident light in an effort to perceive these stripes for himself. It is possible, of course, that the stripes were actually originally present, that O. Thomas failed to see them, and that foxing has completely obscured them since Handley's visit to the BM. Stripes or no stripes, however, Pine is convinced that the holotype of *M. rubida* cannot be conspecific with other specimens in the BM and regarded by Handley, and now by Pine, as conspecific with the ones in the USNM, which Handley and, later, Pine had identified as *M. umbristriata*. Among other things, the holotype of *M. rubida* has much larger molars and an especially large and heavy upper third premolar. The BM cf. *M. umbristriata* are also much paler and less reddish than the holotype of *M. rubida*.

Lemos, Weksler, and Bonvicino (2000) identified two specimens from Brazil, Bahia, Jaguaquara, Fazenda Vaz-

ante (13°32'S, 39°58'W) as *M. rubida*. Dimensions, dentition, and pelage (as given in their table, figures, and text), however, do not correspond with those of the holotype of *M. rubida*. Therefore, Pine regards the holotype as the only known specimen of this species. In her account for *M. rubida*, B. E. Brown (2004:119) mapped two localities, which appear to represent collection sites for *M. umbristriata* (considered by her to be a synonym), but she did not include the type locality.

Gettinger (1997) described the laelapid mite *Androlaelaps cuicensis* from a host identified by him as *Monodelphis rubida* and from 12 km S Brasilia. He referred to the host as "cuica de três listas" and stated that "Voucher specimens of *Monodelphis rubida* are . . . in the Oklahoma Museum of Natural History. . . ." Pine has seen these specimens and identifies them as of *M. americana* (as had Gettinger 1987, 1992).

Monodelphis scalops (O. Thomas, 1888)
Tawny-headed Opossum

SYNONYMS:

Didelphys (Peramys) scalops O. Thomas, 1888a:158; type locality "Brazil"; restricted to Teresópolis, Rio de Janeiro, Brazil, by C. O. C. Vieira (1950:360).

[*Didelphis (]Monodelphis[)] scalops*: Matschie, 1916:271; name combination.

P[eramys]. scalops: O. Thomas, 1912a:89; name combination.

P[eramys (Peramys)]. scalops: A. Miranda-Ribeiro, 1936:404; name combination.

Monodelphis scalops: Pohle, 1927:240; first use of current name combination.

DISTRIBUTION: *Monodelphis scalops* is found in southeastern Brazil, where it occurs from Espírito Santo south into Santa Catarina. The species also has been reported from Argentina.

MARGINAL LOCALITIES (Map 45): BRAZIL: Espírito Santo, Santa Teresa (Pine and Abravaya 1978); Rio de Janeiro, Teresópolis (A. Miranda-Ribeiro 1936); São Paulo, São Sebastião (C. O. C. Vieira 1950); Santa Catarina, Humboldt (BM 14.1.27.6); São Paulo, Piracicaba (C. O. C. Vieira 1950); São Paulo, Piquete (C. O. C. Vieira 1950).

ADDITIONAL LOCALITIES (not mapped): BRAZIL: Rio de Janeiro, Fazenda Boa Fé (D. E. Davis 1945a); Rio de Janeiro, Parati, Pedra Branca (L. G. Pereira et al. 2001); ARGENTINA: Misiones, Parque Nacional Iguazú (Heinonen-Fortabat and Chebez 1997).

SUBSPECIES: We regard *M. scalops* as monotypic.

NATURAL HISTORY: Pine and Abravaya (1978) reported this animal from cloud-forest-like habitats (see Abravaya and Matson 1975) that would fall within Lower

Montane Moist Forest as defined by Holdridge (1967). D. E. Davis (1945a) captured this species in second-growth forest and later (1947:3) stated "This diurnal species frequents low brush and windfalls in second-growth forest. Individuals were captured only in September and October, suggesting a change in feeding opportunities or habits." He snap trapped at least one by using banana as bait. Massoia, Forasiepi, and Teta (2000) stated that *M. scalops* is nocturnal, but presented no evidence of this. Camardella, Abreu, and Wang (2000) reported *M. scalops* from scats of *Leopardus wiedii* and *L. tigrina* in the state of São Paulo, Brazil. The listrophorid mite *Prodidelphoecius euphallus* Fain parasitizes *M. scalops* (see Fain 1979).

REMARKS: The reddish coloration of fore and hind parts of this opossum is reminiscent of the coloration of *M. emiliae*, but the skulls of these species are so different that it is clear that they are not closely related. We have seen eight specimens of *M. scalops*. C. T. Carvalho (1960a) supposed that *M. scalops* might be conspecific with *M. brevicaudata* (*sensu lato* and including *M. emiliae*), owing to its supposed resemblance to that species. Massoia (1980a, 1980b) reported *M. scalops* from a definite locality in Misiones, Argentina. Chebez and Massoia (1996) indicated that the species was known from at least three unspecified localities in Misiones, and Heinonen-Fortabat and Chebez (1997) provided a second restricted locality for it there. Contreras and Silvera-Avalos (1995) recorded *M. scalops* from Paraguay, but the characteristics of their specimen, as described by them, leave doubts as to the correctness of their identification.

Monodelphis sorex (Hensel, 1872)
Southern Red-sided Opossum

SYNONYMS:

Didelphis tricolor: Lund, 1840a:19 [1941b:237]; not *Didelphis tricolor* É. Geoffroy St.-Hilaire, 1803.

[*Didelphys (]Microdelphys[)] sorex* Hensel, 1872:122; type locality "Provinz Rio Grande do Sul"; restricted to Taquara, Rio Grande do Sul, Brazil (Cabrera 1958:9).

Didelphys (Peramys) Henseli O. Thomas, 1888a:159; type locality "Rio Grande do Sul"; identified by O. Thomas (1888b) as "Taquara, Rio Grande do Sul," Brazil.

Peramys sorex: O. Thomas, 1909a:380; name combination.

[*Didelphis (]Monodelphis[)] henseli*: Matschie, 1916:271; name combination.

[*Didelphis (]Monodelphis[)] lundi* Matschie, 1916:271; type locality "Lagoa Santa. [Minas Gerais.] Brazilien"; based on specimens Lund (1840a:19,1841b:237) identified as *Didelphis tricolor*.

[*Didelphis (]Monodelphiops[)] sorex*: Matschie, 1916:272; name combination.

[*Peramys] henseli*: Cabrera, 1919:42; name combination.

P[eramys (Peramys)]. sorex: A. Miranda-Ribeiro, 1936:404; name combination.

Peramys henselii A. Miranda-Ribeiro, 1936:409; incorrect subsequent spelling of *Didelphys henseli* O. Thomas.

P[eramys]. d[imidiata]. itatiayae A. Miranda-Ribeiro, 1936:421; type locality "Campo Bello," Rio de Janeiro, Brazil.

Monodelphis henseli: Cabrera and Yepes, 1940:32; name combination.

Monodelphis sorex: Cabrera and Yepes, 1940:32; first use of current name combination.

Monodelphis tricolor paulensis C. O. C. Vieira, 1950:359, type locality "Pirituba, subúrbio da cidade de São Paulo," São Paulo, Brazil.

Monodelphis touan paulensis: Cabrera, 1958:9; name combination.

Monodelphis henseley Tálice, Laffite de Mosera, and Machado, 1960:151; incorrect subsequent spelling of *Didelphys henseli* O. Thomas.

Monodelphis touan: Olrog and Lucero, 1981:68; not *Mustela touan* Bechstein, 1800 (= *Didelphis brevicaudata* Erxleben).

Monodelphis henseli: Ávila-Pires, 1994:369; name combination.

Microdelphis sorex: Ávila-Pires, 1994:369; name combination.

Monodelphis brevicaudatus: Chebez and Massoia, 1996:199; not *Didelphis brevicaudata* Erxleben, 1777; specimens not seen by us, but *M. sorex* is by far their most likely identity.

DISTRIBUTION: *Monodelphis sorex* occurs in southeastern Brazil, southern Paraguay, and northeastern Argentina. Records are clustered in Minas Gerais, São Paulo, and Rio Grande do Sul (Brazil); in Paraguay; and in Misiones (Argentina).

MARGINAL LOCALITIES (Map 46): BRAZIL: Minas Gerais, Lagoa Santa (type locality of *Didelphis (Monodelphis) lundi* Matschie); Rio de Janeiro, Campo Bello (type locality of *Peramys dimidiata itatiayae* A. Miranda-Ribeiro); São Paulo, Casa Grande (Pine 1980); São Paulo, Pirituba (type locality of *M. tricolor paulensis* C. O. C. Vieira); Rio Grande do Sul, Mundo Novo (ZMB 44970); Rio Grande do Sul, San Lorenzo (BM 87.5.17.8); Rio Grande do Sul, Santa María (MCZ 35556). ARGENTINA: Misiones, Puerto Gisela (BM 24.6.6.76). PARAGUAY: Itapuá, near mouth of Río Pirapó (UCONN 16878). ARGENTINA: Misiones, Caraguatay (FMNH 44773). BRAZIL: Minas Gerais, Passos (Pine 1980).

ADDITIONAL LOCALITIES (not mapped): BRAZIL: Espírito Santo, "região de Cachoeira do Itapemirim" (Ruschi 1965); São Paulo, Mogí das Cruzes (C. O. C. Vieira 1950); São Paulo, Ribeirão da Lagoa (Pine 1980); Rio

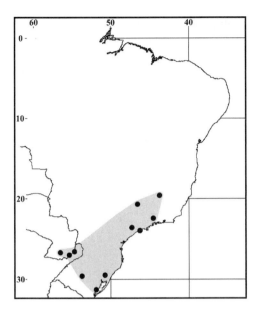

Map 46 Marginal localities for *Monodelphis sorex* ●

Grande do Sul, Taquara (type locality of *Didelphys* (*Peramys*) *Henseli* O. Thomas); São Paulo, Ibití (C. O. C. Vieira 1950); São Paulo, Piquete (C. O. C. Vieira 1950).

SUBSPECIES: We treat *M. sorex* as monotypic.

NATURAL HISTORY: In view of the extreme sexual dimorphism in *M. sorex*, along with its having the largest number of mammae known in the Mammalia (up to 27; O. Thomas 1888b:361), this species may prove to be a semelparous annual like *M. dimidiata*. This also would account for the abundance of specimens of young individuals and the scarcity of adults in museum collections. Semelparity in *M. sorex* has previously been merely hypothesized (Pine, Dalby, and Matson 1985), not reported as definitely occurring, as stated by Taggart et al. (2003). Notman (1923) named an ectoparasitic staphylinid beetle *Omaloxenus bequarti* from "a small Brazilian opossum, *Monodelphis*. Collection of the American Museum of Natural History." The type locality was given as "Alto Itatiaya, Setto [*sic*] do Itatiaya, Brazil." *Monodelphis sorex* may have been the host. The correct identification of the "*Monodelphis dimidiata*" that Fain (1979) reported as the host of the listrophorid mite *Didelphoecius paranensis* Fain from Palmeira, Paraná, Brazil, may also be *M. sorex*. Ávila-Pires and Gouvêa (1977) mentioned a specimen, probably of *M. sorex*, identified by them as of *M. dimidiata* and taken from the esophagus of a White-tailed Kite (*Elanus leucurus*).

REMARKS: The original description was based on young individuals (see figures in Ávila-Pires 1994) and, in the subsequent literature, some young *M. sorex* have been properly identified as *M. sorex*. Adult female and half-grown male *M. sorex* have been identified as *M. henseli*. Adult male *M. sorex* often have been misidentified as *M.*

dimidiata and sometimes as *M. tricolor* or *M. touan* (latter two are synonyms of *M. brevicaudata*). As late as 1998, Bonvicino et al. (1998) identified specimens presumably of *M. sorex* as *M. touan*, and stated that this designation, among those for other species, "needs no further comment." *Monodelphis sorex* occurs sympatrically with *M. americana*, *M. dimidiata*, *M. iheringi*, *M. kunsi*, *M. umbristriata*, *M. scalops*, and *M. unistriata*. At present, *M. sorex* is the only red-sided *Monodelphis* known to us that occurs in Paraguay. This makes it the most likely candidate to be Azara's (1801) "*Micouré cinquième, ou micouré à queue courte.*" If so, the oldest name (a *nomen oblitum*) for this animal would be *M. brevicaudis* Olfers, 1818 (see Hershkovitz 1959), and *Monodelphis wagneri* Matschie, 1916, would be a synonym. However, *M. dimidiata* is known from localities not far from Paraguay, and may occur there. Bertoni (1914, 1939) gave Paraguayan localities for animals he identified as *Peramys henseli*, *P. sorex*, *P. dimidiatus*, and *P. brevicaudatus*. Cabrera (1958) listed *M. sorex* for provincia Corrientes in Argentina, but we know of no records from there. The published Argentine localities have been summarized by Mares and Braun (2000). We have not listed the specimen record reported as *Monodelphis henseli* by Azevedo et al. (1982) from Corupá, Santa Catarina, because we are uncertain of its identity.

Hershkovitz (1959) treated *Monodelphis brevicaudis* (Olfers) as a valid species based on Azara's (1801) "*Micouré cinquième, ou micouré à queue courte,*" but because Hershkovitz assigned no specimens to this nominal taxon, his ideas (if any) as to its biological identity are unknown. B. E. Brown (2004:107) followed Hershkovitz in recognizing *M. brevicaudis* as a valid taxon. She remarked on the controversial nature of the name, but did not assign specimens and provided no new basis for its identity. As late as 1992b, Hershkovitz was still using both *M. henseli* and *M. sorex* as valid names for what he recognized as two species. J. A. Wagner (1855) synonymized Azara's animal with *Didelphis brachyuros* Schreber, and Matschie (1916:272) coined the name *Monodelphis wagneri* for the animals J. A. Wagner (1855:252) identified as *Didelphys brachyura* and based on Azara's *Micouré à queue courte* from Paraguay.

Alipio Miranda-Ribeiro's (1936:421) use of "*P.*" for *Peramys*, instead of "*M.*" for *Minuania*, when he coined the name *P. d. itatiayae*, appears to be a *lapsus*. Clearly he intended *dimidiata* as a species of *Minuania* as used in his key on page 419.

Monodelphis theresa O. Thomas, 1921
Southern Three-striped Opossum
SYNONYMS:
Monodelphis theresa O. Thomas, 1921g:441; type locality
   "Theresopolis, Organ Mts., [Rio de Janeiro,] Brazil."

Peramys theresae A. Miranda-Ribeiro, 1935:39; name combination and incorrect subsequent spelling of *Monodelphis theresa* O. Thomas.

P[eramys (Microdelphis)]. theresae A. Miranda-Ribeiro, 1936:404; name combination and incorrect subsequent spelling of *Monodelphis theresa* O. Thomas.

Peramys therezae A. Miranda-Ribeiro, 1936:416; name combination and incorrect subsequent spelling of *Monodelphis theresa* O. Thomas.

Monodelphis therezae Cabrera and Yepes, 1940:31; incorrect subsequent spelling of *Monodelphis theresa* O. Thomas.

Monodelphis americana theresa: Cabrera, 1958:7; name combination.

Monodelphis americana therezae Ávila-Pires, 1977:7; name combination and incorrect subsequent spelling of *Monodelphis theresa* O. Thomas.

Monodelphis americana: Jenkins and Knutson, 1983:19; part; name combination, not *Sorex americanus* Müller, 1776.

DISTRIBUTION: *Monodelphis theresa* is known to us by five specimens from the Brazilian states of Rio de Janeiro and Minas Gerais.

MARGINAL LOCALITIES (Map 43): BRAZIL: Minas Gerais, Mata da Prefeitura (USNM 552404); Rio de Janeiro, 5 miles N of Theresópolis (FMNH 25739, not mapped); Rio de Janeiro, Theresópolis, Organ Mts. (type locality of *Monodelphis theresa* O. Thomas); Rio de Janeiro, Petrópolis (MCZ 7286, not mapped).

SUBSPECIES: We treat *M. theresa* as monotypic.

NATURAL HISTORY: Camardella, Abreu, and Wang (2000) reported *M. theresa* from scats of *Leopardus pardalis*, *L. weidii*, and *L. tigrina* in the state of São Paulo, Brazil. No other natural history information is available.

REMARKS: Cabrera (1958) regarded *M. theresa* as a subspecies of *M. americana*. Pine (in Kirsch and Calaby 1977), and Pine and Handley (1984) treated *M. theresa* as a full species. Eisenberg and Redford (1999:73) reported a specimen seen by them at "the Museo, Rio de Janeiro," and which had a single "faint median dorsal stripe," but they did not give the basis of their specific identification. Their cryptic table heading, "Description (as Labeled)" may indicate that it was based only on what someone else had written on its tag. Considering the taxonomic and identification difficulties that *M. theresa* presents, even for those who have studied it and its relatives the most, there is no particular reason to think that the animal examined by Eisenberg and Redford was, in fact, assignable to *M. theresa*. Although Eisenberg and Redford hint that A. Miranda-Ribeiro (1936) may have mentioned the existence of individuals with other than three dorsal stripes, he did not do so. The species, as treated here, appears to be a composite. Handley once informed Pine that, in his view, no other specimen that he had examined was conspecific with the holotype.

Eisenberg and Redford's hint that there may be some special relationship between *M. theresa* and *M. unistriata* is without basis. Differences in size (and in other characters) between the smallest specimens that we have seen of the *americana-theresa* group and the largest ones convince us that specimens at these two extremes cannot be conspecific. Nonetheless, individuals of intermediate size exist and bridge the difference. For this and other reasons, we believe that probably at least three, and possibly more, species make up this group. On the basis of tooth alignment, Pine has concluded that the type of *M. theresa* is not a full adult. Handley concluded otherwise on the basis of tooth wear. The size of its molars, usually the best index to adult size in *Monodelphis*, is not appreciably exceeded by that of some specimens we assign to *M. americana* (e.g., ones from Engenheiro Reeve, São Paulo, in the BM). Pine has concluded that if it had reached full adulthood, the size of the holotype would have been considerably greater than that of adult *M. iheringi*. Ávila-Pires (1977) reported a specimen from Mont Serrat, Parque Nacional do Itatiaia, Minas Gerais, 815 m (not mapped) he identified as belonging to this species. Camardella, Abreu, and Wang (2000) reported *M. theresa* from Santa Virgínia Nucleous, Parque Estadual da Serra do Mar, São Paulo, 900 m (not mapped); the identity of this material needs to be verified.

Monodelphis umbristriata (A. Miranda-Ribeiro, 1936)
Faint-striped Opossum

SYNONYMS:

M[inuania]. goyana A. Miranda-Ribeiro, 1936:419; a *lapsus* and alternative name for *Minuania umbristriata* A. Miranda-Ribeiro (see Pine, Dalby, and Matson 1985).

Minuania umbristriata A. Miranda-Ribeiro, 1936:422; type locality "Goyaz"; identified as Veadeiros, Goiás, Brazil by Ávila-Pires (1968:164).

Monodelphis umbristriatus: C. O. C. Vieira, 1955:350; first use of current name combination, but with incorrect gender concordance.

DISTRIBUTION: *Monodelphis umbristriata*, as treated here, is endemic to eastern Brazil, where it is found in the states of Goiás, Minas Gerais, and São Paulo.

MARGINAL LOCALITIES (Map 47): BRAZIL: Minas Gerais, Mata da Prefeitura (USNM 552403); São Paulo, Piracicaba (ZMB 14164); São Paulo, Victoria (BM 3.7.25.12); Minas Gerais, Passos (Pine 1976); Goiás, Veadeiros (type locality of *Minuania umbristriata* A. Miranda-Ribeiro).

SUBSPECIES: We regard *M. umbristriata* as monotypic.

NATURAL HISTORY: Unknown.

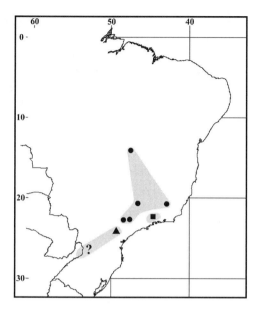

Map 47 Marginal localities for *Monodelphis umbristriata* ●,
Monodelphis unistriata ▲, and *Monodelphis* "species E" ■

REMARKS: Cabrera (1958) tentatively placed *M. umbristriata* in synonymy with *Monodelphis americana americana* (Müller), and Ávila-Pires (1968) unequivocally placed it in *M. americana*. Pine (1976) resurrected *M. umbristriata* as a full species separate from *M. americana*, but then Pine and Handley (1984), without comment, placed *M. umbristriata* in *M. rubida* (see discussion of *M. rubida* in this volume). Subsequent authors followed Pine and Handley in this arrangement until Lemos, Weksler, and Bonvicino (2000) concluded that *M. umbristriata* was a separate species from specimens identified by them as *M. rubida*. At one time, Pine and Handley had seen only six specimens that they attributed to *M. umbristriata*, all adult males. This led Pine, Dalby, and Matson (1985) to imply that this form might be no more than a color variant of adult male *M. americana*. Handley wrote in an earlier draft of this section that "...we now also have seen females and immatures with the characteristic reddish coloration and faint stripes of *M. umbristriata*." Pine, however, has not seen these and does not know what specimens Handley had in mind. Typically colored and -marked *M. americana* are known from some of the same localities as *M. umbristriata*, and the apparent absence of cranial differences leads us to suspect that *M. umbristriata* may be no more than an erythristic phase of *M. americana*. Goeldi (1894) discussed animals fitting the description of *M. umbristriata*, which he treated as representing a variant of what is now called *M. americana*, and wrote (p. 464) "Older individuals, generally more rufous, often show only pale stripes, and, not rarely, specimens will be met with apparently stripeless, the stripes only appearing against the light." Also see O. Thomas's (1888b:363) description, which is similar. Cabrera (1958) wrote, appar-

ently intending to refer to all striped forms of *Monodelphis*, "...la presencia o ausencia de rayas dorsales nos parece de poco valor taxonómico desde el momento que, cuando existen, tienden a desaparecer en los individuos muy adultos." If this sort of thing does occur, however, we have seen no indication that it happens in the forms smaller than what we are here calling *umbristriata*, which when adult, is at least as large as the largest *M. americana*. In their account of *M. americana*, Mares, Braun, and Gettinger (1989) mentioned that they kept only striped *M. americana* as voucher specimens but that "two larger males that lacked the dorsal stripes were captured and released...Pelage variation in this species needs further study." Perhaps these two males were not *M. americana*. Although Pine (1976) allied *M. umbristriata* with *M. dimidiata*, there can be no question that its affinities are with *M. americana*. A. Miranda-Ribeiro (1936) placed *M. umbristriata* in what he regarded as the separate genus *Minuania* and the first name that he (p. 419) applied to this species was *Minuania goyana*; however, Pine, Dalby, and Matson (1985:226) selected *Minuania umbristriata* as having priority over *Minuania goyana*. Cabrera (1958:11) considered *Minuania goyana* Miranda-Ribeiro to be a *nomen nudum*. Although Honacki, Kinman, and Koeppl (1982) implied that *M. umbristriata* sometimes has been treated as a subspecies of *M. americana*, we know of no instance of this.

The lectotype (MNRJ 1314) of *Minuania umbristriata* A. Miranda-Ribeiro (see Pine 1976) is here designated as the lectotype of *M. goyana* A. Miranda-Ribeiro as well. It is a specimen of an immature animal that A. Miranda-Ribeiro (1936) said had more distinct stripes than the adult paralectotype, but less so than in "*M. americanus*." Langguth, Lineira, and Franco (1997) followed Gardner (1993a) in treating A. Miranda-Ribeiro's names *Minuania goyana* and *M. umbristriata* as synonyms of *Monodelphis rubida*.

Monodelphis unistriata (J. A. Wagner, 1842)
Single-striped Opossum
SYNONYMS:

Didelphys unistriata J. A. Wagner, 1842:360; type locality "Ytarare"(= Itararé), São Paulo, Brazil.

[*Didelphys (]Microdelphys[)] unistriata*: Burmeister, 1856: 87; name combination.

Didelphys [(Peramys)] unistriata: O. Thomas, 1888b:365; name combination.

[*Peramys*] *unistriatus*: Cabrera, 1919:43; name combination.

Monodelphis unistriata: Cabrera and Yepes, 1940:31; first use of current name combination.

DISTRIBUTION: *Monodelphis unistriata* is known from estado São Paulo, Brazil, and provincia Misiones, Argentina.

MARGINAL LOCALITIES (Map 47): BRAZIL: São Paulo, Itararé (type locality of *Didelphys unistriata* J. A. Wagner). ARGENTINA: Misiones, "Misiones, alrededores" (Chebez and Massoia 1996, locality unknown).

SUBSPECIES: We consider *M. unistriata* to be monotypic.

NATURAL HISTORY: Pelzeln (1883:116), apparently getting his information from Natterer's field notes, wrote that the type specimen, a male, was found by a dog in "der Steppe bei Ytararé"; it seemed to show little fear and moved very slowly.

REMARKS: Only two specimens have been recorded with certainty. O. Thomas (1888a), who generally had a good feel for species limits in the genus, examined the male type and regarded it as distinct, as it clearly seemed to be, judging from his redescription (O. Thomas 1888b) and that of Pelzeln (1883). Pine has examined the Argentine specimen (MACN 250) and has concluded that it does, in fact, represent *Monodelphis unistriata* and that *M. unistriata* is a recognizable and quite distinctive species of *Monodelphis*. A. Miranda-Ribeiro (1936) gave measurements and a short description of a specimen, which Pine has concluded with the help of K. Bauer (*in litt.*) must be, in fact, the holotype, examined during a known visit that A. Miranda-Ribeiro made to the Vienna Museum, although A. Miranda-Ribeiro did not make this clear, and C. O. C. Vieira (1950) assumed that a second specimen had been involved. Differences between measurements of the skin, taken by A. Miranda-Ribeiro and O. Thomas, are apparently explained by the specimen having been "re-worked" by a preparator between the times that O. Thomas and when A. Miranda-Ribeiro saw it (K. Bauer, *in litt.*). Eisenberg and Redford (1999) reported having seen a specimen, or specimens, which must have been at a museum in Rio de Janeiro. Their characterization of the animal(s) does fit *M. unistriata*. They gave the range of the species as "the portions of the state of São Paulo" [*sic*]. Although reported for "Misiones, alrededores" by Chebez and Massoia (1996), Massoia, Forasiepi, and Teta (2000) did not mention *M. unistriata* in their treatment of Argentine marsupials.

Monodelphis [species A]

SYNONYMS:

Monodelphis brevicaudata: Authors; not *Didelphis brevicaudata* Erxleben, 1777.

Monodelphis orinoci: Authors; not *Peramys brevicaudatus orinoci* O. Thomas, 1899c.

DISTRIBUTION: *Monodelphis* [species A] occurs in the llanos and Cordillera Central of Venezuela.

MARGINAL LOCALITIES (Map 38; from Pérez-Hernández 1990): VENEZUELA: Aragua, Hacienda la Esperanza; Guárico, Hato La Palmita; Guárico, Estación Biológica de los Llanos; Cojedes, Hato Nuevo.

NATURAL HISTORY: Ventura, Pérez-Hernández, and Lopez-Fuster (1998) stated that this species "has been found in lowland savanna, especially in areas densely covered by grasses at elevations from 20 to 575 m." Handley (1976), reporting on Venezuelan animals (*Monodelphis* sp. A as understood here plus both *M. brevicaudata* and *M. palliolata*), took 96% on the ground and 4% on logs and in trees–misquoted by O'Connell (1979:77) as 4% "in trees and vines." Linares (1998) stated that this species is diurnal and lives in burrows or dens ("madrigueras") at the bases of trees. He also reported an instance of the American Kestrel (*Falco sparverius*) preying on this opossum.

REMARKS: This is the species reported by Pérez-Hernández (1985, 1990), Pérez-Hernández, Soriano, and Lew (1994), Ventura, Pérez-Hernández, and Lopez-Fuster (1996, 1998), Linares (1998), and Ventura et al. (2005) as *Monodelphis orinoci*. Voss, Lunde, and Simmons (2001) examined the type of *Peramys brevicaudatus orinoci* O. Thomas, 1899c, and the type of *Peramys brevicaudatus dorsalis* J. A. Allen, 1904c (another name considered by Linares [1998] to be a synonym of *M. orinoci*) and found them to represent *Monodelphis brevicaudata*. B. E. Brown (2004:107) listed *M. orinoci* O. Thomas (not *M. orinoci* of authors) in the synonymy of *M. brevicaudata*, but also included Venezuelan localities north and west of the Río Orinoco where Voss, Lunde, and Simmons (2001) showed that *M. brevicaudata* is not known to occur. Linares (1998) mapped some unspecified localities considerably to the west of those listed here. This is likely the species referred to as *M. orinoci* by Cuartas and Muñoz (2003), who gave the Colombian distribution as the departments of Meta and Vichada. Reig and Löbig (1970) and Reig et al. (1977) discussed the karyotype of this species and found it to be the same, or essentially the same, as that of specimens identified by them as *M. brevicaudata* (= *M. palliolata*).

Monodelphis [species B]

DISTRIBUTION: *Monodelphis* [species B] is known from a single locality in Santa Cruz, Bolivia.

MARGINAL LOCALITIES (Map 39): BOLIVIA: Santa Cruz, Santa Rosa de la Roca (S. Anderson 1997).

NATURAL HISTORY: Unknown.

REMARKS: This is the specimen referred to under *Monodelphis* sp. A by S. Anderson (1997:161). He commented that the specimen was "too large for *adusta*, *kunsi*, or *osgoodi* and too small for *emiliae*, *domestica*, or *brevicaudata*. In color it resembles *domestica*, but it is much smaller than other known *M. domestica*." Pine, who prepared this specimen in the field and has studied the stuffed skin and cleaned skull, as yet has been unable to discern any differences, other than size, to separate this form from *M. domestica*, but finds it difficult to believe that it could be conspecific with *M. domestica*.

Monodelphis [species C]

SYNONYMS:

Monodelphis theresa: Gardner, 1993a:22; part; not *Monodelphis theresa* O. Thomas, 1921.

M[onodelphis]. theresa: Emmons and Feer, 1997:36; part; not *Monodelphis theresa* O. Thomas (Emmons and Feer's allocation to species was tentative).

DISTRIBUTION: *Monodelphis* [species C]is known from a few localities at upper elevations on the eastern slope of the Peruvian Andes.

MARGINAL LOCALITIES (Map 41): PERU: Huánuco, E slope Cordillera Carpish, Carretera (LSUMZ 14019); Cusco, RAP Camp Two (Emmons, Luna, and Romo 2001); Cusco, Wayrapata (S. Solari et al. 2001a).

NATURAL HISTORY: The single male from the Cordillera Carpish was taken at the edge of a man-made clearing in cloud forest at 2,400 m. It was found dead in a live trap set in brush at the edge of a small clearing behind a house. The hind legs had been removed through the wire sides of the trap by a predator. A specimen reported by Emmons, Luna, and Romo (2001) was caught on the ground in a dense thicket at the edge of a bog. S. Solari et al. (2001a) reported this species as sympatric with *M.* cf. *osgoodi*.

REMARKS: This small striped species is most similar to *Monodelphis theresa* O. Thomas, 1921g, of southeastern Brazil.

Monodelphis [species D]

SYNONYMS:

Monodelphis brevicauda emiliae: C. T. Carvalho, 1960b: 122; not *Peramys emiliae* O. Thomas; incorrect subsequent spelling of *Didelphis brevicaudata* Erxleben.

Monodelphis brevicaudata emiliae: C. T. Carvalho, 1960b: 123; not *Peramys emiliae* O. Thomas.

DISTRIBUTION: *Monodelphis* [species D] is known from the state of Pará, Brazil, south of the lower Rio Amazonas and east of the Rio Xingu.

MARGINAL LOCALITIES (Map 42): BRAZIL: Pará, Pôrto de Moz (AMNH 95976); Pará, Cametá (MCZ 30385, B. D. Patterson 1992); Pará, Jatobal (USNM 519725); Pará, Serra Norte (USNM 543302); Pará, Gradaús (C. T. Carvalho (1960b).

REMARKS: This is the unnamed taxon found south of the Amazon and east of the Xingu that Voss, Lunde, and Simmons (2001:54, 56) mentioned as lacking the extension of longer fur along the proximal dorsal surface of the tail (also see Pine and Handley 1984:241). We presume that is the same *Monodelphis* referred to as *M. brevicaudata emil-*

iae by C. T. Carvalho (1960b), who noted the differences between it and typical *M. brevicaudata*.

Monodelphis [species E]

DISTRIBUTION: *Monodelphis* [species E] is known from a single locality in estado do Rio de Janeiro, Brazil.

MARGINAL LOCALITIES (Map 47): BRAZIL: estado do Rio de Janeiro, Serra do Itatiaya, Alto Itatiaya.

REMARKS: This species is represented by the second of two AMNH specimens referred to as "*Monodelphis ssp.*" by Ávila-Pires and Gouvêa (1977:9). These authors gave the elevation as 2,350 m and stated that the specimen is of a female with three small embryos. A collector's tag gives the elevation of Alto Itatiaya (= Pico das Agulhas Negras) as 7,150 ft. (2,180 m) and identifies the specimen as a male. Ávila-Pires's and Gouvêa's comments on the color of the pelage also do not fit this specimen. At minimum, there appears to have been some sort of catalog number mix up concerning this specimen, but correspondence received from the AMNH has been inadequate to resolve the confusion. There are indications that the attached field tag was, indeed, prepared for this specimen, however. Although S. Solari (*in litt.*) and Pine and Handley have all examined the specimen, and Pine had it on loan from the AMNH until he recently returned it, R. Voss (*in litt.*) has stated that that no such AMNH specimen exists. This species is similar externally to *M. osgoodi* of the Andean region.

Genus *Thylamys* Gray, 1843

G. Ken Creighton and Alfred L. Gardner

As presently understood, *Thylamys* comprises ten species (*Thylamys cinderella*, *T. elegans*, *T. karimii*, *T. macrurus*, *T. pallidior*, *T. pusillus*, *T. sponsorius*, *T. tatei*, *T. velutinus*, and *T. venustus*) of fat-tailed marsupials found only in subequatorial South America. As yet undescribed species have been indicated from Uruguay (E. M. González, Saralegui, and Fregeiro 2000) and Argentina (Flores, Díaz, and Barquez 2000). The fossil record dates from the Pliocene (Montehermosan) of Argentina (Mones 1980; Reig, Kirsch, and Marshall 1985).

Opossums of this genus are small (head plus body length 75–147 mm, tail 65–161 mm) and the ratio of tail to head and body ranges from 1.0 to 1.25, except in *T. karimii* and *T. velutinus* whose tails average shorter than the head and body. Color of the dorsum is grayish or grayish-brown agouti; the venter varies from pure white (*T. karimii*, *T. pallidior*, and *T. pusillus*) to gray-based white (*T. cinderella*, *T. elegans*, and *T. sponsorius*) or yellowish buff (*T. velutinus*, *T. venustus*, and some *T. macrurus*). Mammae are inguinal, abdominal, and pectoral, and number up to 19 (Tate 1933).

The manus and pes are white-haired and small with short digits. The claws on manual digits II–IV extend noticeably beyond the terminal digital pads. Digit IV is longest on the pes; its ratio to length of hind foot is less than 0.45. Central palmar and plantar surfaces are covered with small nodular granules to the base of the digits. The tail is usually bicolored and seasonally becomes thickened (incrassate) with fat deposits that may be evident in some species throughout the year. The exception may be *T. macrurus*, because incrassate tails have not been reported in this species. Tail scales are minute (35 or more rows per cm), rounded-square in outline, and annular in arrangement. Postorbital processes are normally lacking, but may be developed as ridges or postorbital projections in *T. karimii*, *T. pusillus*, and *T. cinderella*, especially in adults. Nasals are long, slender, and not laterally expanded at the maxillofrontal suture. Auditory bullae are relatively large, rounded, and close together (ratio of distance between bullae to breadth of a single bulla is 1.5 or less); each is buttressed by an anteromedial alisphenoid process. The palate has either two or three sets of paired fenestrae (illustrated by Voss and Jansa 2003:25). The posterolateral palatal foramina are unusually long, and the breadth of each is about equal to, or exceeds, the width of the last molar. The last upper premolar (P3) always exceeds P2 in height and anteroposterior length.

SYNONYMS:

Didelphis: Desmarest, 1804b:19; part; not *Didelphis* Linnaeus, 1758.

Didelphys Olfers, 1818:205, 206; incorrect subsequent spelling of, but not *Didelphis* Linnaeus.

Sarigua Muirhead, 1819:429; part.

Didelphis: Desmarest, 1827:398; part; not *Didelphis* Linnaeus.

Didelphys Waterhouse, 1841:106; incorrect subsequent spelling of, but not *Didelphis* Linnaeus.

Micoureus Lesson, 1842:186; part.

Didelphys J. A. Wagner, 1842:360; incorrect subsequent spelling of, but not *Didelphis* Linnaeus.

Thylamys Gray, 1843b:101; type species *Didelphis elegans* Waterhouse, 1839, by monotypy.

Didelphis: Reinhardt, 1851:v; part; not *Didelphis* Linnaeus.

Grymaeomys Burmeister, 1854:130; part.

Microdelphys Burmeister, 1856:86; part; described as a subgenus of *Didelphis* Linnaeus.

Cuica Liais, 1872:330; part.

Didelphys (Micoureus): O. Thomas, 1888b:340; part; incorrect subsequent spelling of, but not *Didelphis* Linnaeus; not *Micoureus* Lesson, 1842.

Philander: Cope, 1889a:130; part; not *Philander* Brisson, 1762.

Didelphys R. A. Philippi, 1894:36; incorrect subsequent spelling of, but not *Didelphis* Linnaeus.

Micoureus: Goeldi, 1894:462; not *Micoureus* Lesson.

Marmosa: O. Thomas 1894c:188; part; not *Marmosa* Gray, 1821.

Marmosa: O. Thomas, 1896b:313, 314, footnote; part; not *Marmosa* Gray, 1821.

[*Didelphys*] (*Peramys*): Trouessart, 1898:1244; part; not *Didelphis* Linnaeus; not *Peramys* Lesson, 1842.

Marmosa: O. Thomas, 1902f:158, 159, 161; not *Marmosa* Gray.

Marmosa: O. Thomas, 1912c:409; not *Marmosa* Gray.

Didelphis (Dromiciops): Matschie, 1916:271; not *Didelphis* Linnaeus; not *Dromiciops* O. Thomas, 1894.

Marmosa: O. Thomas, 1921a:186; not *Marmosa* Gray.

Marmosa: O. Thomas, 1921b:519, 520; not *Marmosa* Gray.

Marmosa: O. Thomas, 1926b:327; not *Marmosa* Gray.

Marmosa: Tate, 1931:14; not *Marmosa* Gray.

Thylamis A. Miranda-Ribeiro, 1936:328; incorrect subsequent spelling of *Thylamys* Gray.

Marmosa: Marelli, 1932:68; not *Marmosa* Gray.

Marmosa: Handley, 1957:402; not *Marmosa* Gray.

Didelphys (Paramys) Cabrera, 1958:29; incorrect subsequent spelling of, but not *Peramys* Lesson; in synonymy.

Dromictops Cabrera, 1958:30; incorrect subsequent spelling of, but not *Dromiciops* O. Thomas, 1894; in synonymy.

Marmosa: Petter, 1968:313; not *Marmosa* Gray.

Thulamys Reig, Kirsch, and Marshall, 1985:336; incorrect subsequent spelling of *Thylamys* Gray.

Macrodelphys B. E. Brown, 2004:145; incorrect subsequent spelling of *Microdelphys* Burmeister; in synonymy.

KEY TO SPECIES OF *THYLAMYS*:

1. Size large, tail usually longer than 145 mm, and condylobasal length more than 32 mm; no evidence of fat deposits (incrassation) in tail . . . . . *Thylamys macrurus*

1′. Size smaller, tail shorter than 145 mm, and condylobasal length less than 32 mm; tail moderately to extremely swollen (incrassate) with fat deposits 2

2. Tail shorter than 90 mm, usually less than 85 mm; hind foot less than 14 mm; ventral pelage entirely gray based . . . .*Thylamys velutinus*

2′. Tail longer than 85 mm, usually more than 90 mm; hind foot more than 14 mm; ventral pelage white, creamy white, or uniformly buff, may or may not be gray based 3

3. Ventral pelage pure white; tail always incrassate; supraorbital region smoothly rounded, lacking ridges; condylobasal length less than 27 mm; zygomatic breadth less than 14 mm; nasals less than 12 mm . . . . *Thylamys pallidior* and *Thylamys tatei*

3'. Ventral pelage white or gray-based white or buffy; tail not always incrassate; supraorbital region with distinct ridges or small processes; condylobasal length more than 27 mm; zygomatic breadth more than 14 mm; nasals more than 12 mm 4

4. Tail often lacking fat deposits (*T. pusillus*); ventral pelage all white, never gray based; zygomatic arches broadly flared, the ratio of zygomatic breadth to condylobasal length greater than 0.54 5

4'. Tail normally slightly to moderately incrassate (diameter at base greater than 5 mm); ventral pelage gray based; zygomatic arches not broadly flared, ratio of zygomatic breadth to condylobasal length less than 0.54 6

5. Posterolateral palatal foramen larger than last upper molar (M4); septum of incisive foramen extremely broad; broadly flaring lower border of orbit; evenly rounded tympanic wing of alisphenoid, upper canine usually with anterior and posterior cuspules on cingulum . . . .
. . . . *Thylamys karimii*

5'. Posterolateral palatal foramen approximately same size as last upper molar (M4); septum of incisive foramen not unusually broad; lower border of orbit not broadly flaring; tympanic wing of alisphenoid anteroventrally flattened; upper canine usually lacking anterior cuspule (and often posterior cuspule) on cingulum . . . .
. . . . *Thylamys pusillus*

6. Dorsal pelage pale brown; ventral pelage whitish; basal length more than 26 mm . . . . *Thylamys elegans*

6'. Dorsal pelage dark brown; ventral pelage buffy to yellowish; basal length less than 26 mm 7

7. Dorsal pelage cinnamon brown; ventral pelage gray based except for chin, throat, and thoracic midline; postorbital ridges may appear absent in young to weakly developed in adults . . . . *Thylamys venustus*

7'. Dorsal pelage gray brown to dark brown; hairs of chest self-colored white to yellowish, otherwise ventral pelage gray based; postorbital processes evident 8

8. Postorbital processes well developed and projecting laterally behind orbits, especially in adults . . . .
. . . . *Thylamys cinderella*

8'. Postorbital ridges weakly developed in young individuals and better developed in adults, but not as expanded lateral processes behind orbits . . . .
. . . . *Thylamys sponsorius*

Thylamys cinderella (O. Thomas, 1902)
Cinderella Thylamys

SYNONYMS:

Marmosa elegans cinderella O. Thomas, 1902f:159; type locality "Tucuman," Argentina.

[*Didelphys (Marmosa) elegans*] *cinderella*: Trouessart, 1905: 857; name combination.

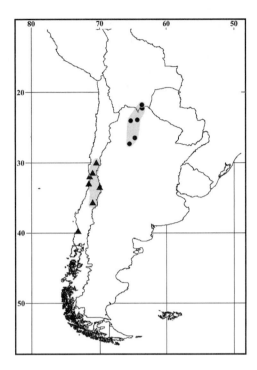

Map 48 Marginal localities for *Thylamys cinderella* ● and *Thylamys elegans* ▲

[*Didelphis (Thylamys)*] *cinderella*: Matschie, 1916:271; name combination.

[*Marmosa (Thylamys) elegans*] *cinderella*: Cabrera, 1919: 40; name combination.

Marmosa venusta cinderella: Tate, 1933:226; name combination.

[*Thylamys elegans*] *cinderella*: Gardner, 1993a:23; name combination.

Thylamys cinderella: Flores, Díaz, and Barquez, 2000:325; first use of current name combination.

DISTRIBUTION: *Thylamys cinderella* occurs in southern Bolivia and northern Argentina.

MARGINAL LOCALITIES (Map 48; from Flores, Díaz, and Barquez 2000, except as noted): BOLIVIA: Tarija, Caraparí (Tate 1933). ARGENTINA: Salta, Aguaray; Jujuy, Laguna La Brea; Tucumán, Finca El Jagüel; Tucumán, Concepción (Tate 1933); Jujuy, León (O. Thomas 1918).

SUBSPECIES: *Thylamys cinderella* is monotypic.

NATURAL HISTORY: Flores, Díaz, and Barquez (2000) reported collecting a lactating female and a young animal during February in Jujuy Province, and taking additional young individuals during April in Jujuy and Tucumán provinces, Argentina. *Thylamys cinderella* has a $2n = 14$, FN = 20 karyotype; the X is a small acrocentric, the morphology of the Y is unknown (Braun, Mares, and Stafira 2005).

REMARKS: Flores, Díaz, and Barquez (2000) emphasized the development of supraorbital ridges to distinguish

T. cinderella from similar appearing taxa; especially from *T. sponsorius*, with which it is sympatric. As also noted by Tate (1933), these ridges are weakly evident in young, but develop early to become well-developed, laterally projecting processes in older adults. B. E. Brown (2004) listed *Marmosa elegans cinderella* O. Thomas, 1902 in the synonymy of *T. venustus*.

Thylamys elegans (Waterhouse, 1839)
Elegant Thylamys
SYNONYMS:

Didelphis hortensis J. Reid, 1837:4; *nomen nudum*.

Didelphis elegans Waterhouse, 1839:95; type locality "Valparaiso," Valparaíso, Chile.

Didelphys elegans: Waterhouse, 1841:106; name combination.

Micoureus elegans: Lesson, 1842:186; name combination.

Thylamys elegans: Gray, 1843b:101; first use of current name combination.

[*Didelphys (]Grymaeomys[)] elegans*: Burmeister, 1856:83; name combination.

Didelphys [(Micoureus)] elegans: O. Thomas, 1888b:352; name combination.

M[icoureus]. elegans: O. Thomas, 1894c:186; name combination.

Didelphys soricina R. A. Philippi, 1894:36; type locality "Valdivia," Los Lagos, Chile.

[*Didelphys (Marmosa)] elegans*: Trouessart, 1898:1241; name combination.

[*Didelphys (Peramys)] soricina*: Trouessart, 1898:1244; name combination.

[*Didelphis (Dromiciops)] soricina*: Matschie, 1916:271; name combination.

[*Marmosa (Thylamys) elegans] elegans*: Cabrera, 1919:40; name combination.

Marmosa elegans coquimbensis Tate, 1931:14; type locality "Paiguano, province of Coquimbo, Chile."

Marmosa elegans soricina: Tate, 1933:216; name combination.

DISTRIBUTION: *Thylamys elegans* is endemic to Chile, where it occurs at elevations from sea level to over 3,500 m on the Pacific slope of the Andes from Región de Coquimbo south into Región de Los Lagos.

MARGINAL LOCALITIES (Map 48; from Pine, Miller, and Schamberger 1979, except as noted): Coquimbo, Paiguano (type locality of *Marmosa elegans coquimbensis* Tate); Coquimbo, Aucó (M. Lima et al. 2001); Santiago, Río Colorado, 30 km E of Guayacán; Maule, Río Maule, junction of Río Claro; Los Lagos, Valdivia (type locality of *Didelphys soricina* R. A. Philippi); Valparaíso, Valparaíso (type locality of *Didelphis elegans* Waterhouse); Coquimbo, 4 km S of Los Vilos.

SUBSPECIES: We are treating *T. elegans* as monotypic.

NATURAL HISTORY: Tate (1933) recorded wild-caught females "breeding or nursing" between September and March. Mann (1953) frequently found *T. elegans* in matorral vegetation, but not in the more mesic forests of central Chile. Mann (1953, 1956, 1978) described and illustrated the anatomy of Chilean *T. elegans*. Palma (1997) provided a rediagnosis of *T. elegans*, along with measurements and a comprehensive summary of its distribution and available life history information, in his *Mammalian Species* account. Between October 1987 and January 1992, M. Lima et al. (2001) studied demography and population dynamics of a population in the Las Chinchillas National Reserve at Aucó. They had 594 captures corresponding to 297 *T. elegans*. Numbers known to be alive fluctuated between fewer than 5/ha to over 20/ha during the 7 years of their study. Annual rainfall was positively correlated with recruitment, maturation, and reproductive rates, which they interpreted as reflecting the influence of rainfall on "environmental productivity and insect abundance." The karyotype is $2n = 14$, $FN = 24$ (Reig et al. 1977); FN corrected to 20 by Palma (1994), who found that the Y chromosome is missing from somatic cells.

REMARKS: Most authors accepted Cabrera's (1958) concept of *T. elegans* by including within it the western and northwestern Argentine populations of *Thylamys* (herein treated as *T. cinderella*, *T. sponsorius*, and *T. venustus*), as was done recently by Gardner (1993a), Mares et al. (1996), and Mares and Braun (2000). Palma (1994) distinguished *T. elegans* from other *Thylamys* and restricted the taxon to the western flank of the Andes in Chile and Peru. Recently, S. Solari (2003) concluded that the Peruvian populations represented *T. tatei* and *T. pallidior*; thereby, further restricting the known distribution of *T. elegans* to Chile. We assign the specimens Pine, Miller, and Schamberger (1979) reported as *T. elegans* from Región de Tarapacá to *T. pallidior*. The four localities where those specimens were collected are at elevations from 3,200 to 3,400 m, and are represented by the northernmost records mapped (appear as an elongate dot) for *T. elegans* by B. E. Brown (2004:Fig. 85).

Thylamys karimii (Petter, 1968)
Karimi's Thylamys
SYNONYMS:

Marmosa karimii Petter, 1968:313; type locality "région d'Exu, Pernambuco, Brasil."

Marmosa (Thylamys) karimii: Kirsch and Calaby, 1977:14; name combination.

[*Thylamys] karimii*: Reig, Kirsch, and Marshall, 1987:7; first use of current name combination.

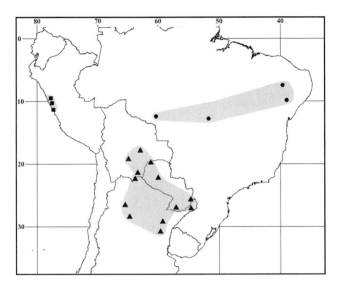

Map 49 Marginal localities for *Thylamys karimii* ●, *Thylamys pusillus* ▲, and *Thylamys tatei* ■

[*Thylamys pusillus*] *karimii*: Gardner, 1993a:23; name combination.

DISTRIBUTION: *Thylamys karimii* is found in central and northeastern Brazil.

MARGINAL LOCALITIES (Map 49): BRAZIL: Pernambuco, Exú (type locality of *Marmosa karimii* Petter); Bahia, Cocorobó (MZUSP 17456); Mato Grosso, Serra do Roncador, 264 km N by road of Xavantina (Pine, Bishop, and Jackson 1970); Rondônia, Km 55 on BR−364 (MZUSP 19899).

SUBSPECIES: We consider *Thylamys karimii* to be monotypic.

NATURAL HISTORY: Petter (1968) said the type specimen was taken on the ground near an entrance to a burrow. While in captivity, its tail enlarged (incrassation) to nearly twice its original diameter. The animal was active at night, but lethargic during the day, which was when its rectal temperature dropped to near ambient temperature (20–25°C; Petter 1968). According to Mares, Ojeda, and Kosco (1981) and Streilein (1982b), a captive was adept at catching insects, and ate insects, geckos, and hylid frogs, among other foods, which included small amounts of fruit.

REMARKS: Gardner (1993a) and B. E. Brown (2004) included *T. karimii* under *T. pusillus*. Palma (1995b) treated *T. karimii* as a synonym of *T. velutinus*, based on size. Palma (1995b:519) also concluded that the animals Streilein (1982b) reported as *T. karimii* could not be correctly identified because Streilein referred to the tail as prehensile. It is more likely, however, that Streilein either erred in his reference to a prehensile tail, or the tail has some prehensile capabilities. We recognize *T. karimii* as a species, pending revision of this genus.

Thylamys macrurus (Olfers, 1818)
Paraguayan Thylamys

SYNONYMS:

[*Didelphys*] *macroura* Illiger, 1815:107; *nomen nudum*.

[*Didelphys*] *marmota* Oken, 1816:1140; unavailable name (ICZN 1956).

D[*idelphys*]. *macrura* Olfers, 1818:205; type locality "Südamerica"; based on Azara's (1801:290) *Micouré à longue queue*; therefore, the restricted type locality is "Tapoua" (= Tapuá), Presidente Hayes, Paraguay.

Didelphis grisea Desmarest, 1827:398; type locality "Paraguay"; based on Azara's (1801:290) *Micouré à longue queue*; therefore, the restricted type locality is Tapuá, Presidente Hayes, Paraguay.

[*Didelphys (]Grymaeomys[)*] *griseus*: Burmeister, 1856:83; name combination.

T[*hylamys*]. *marmota*: J. A. Allen and Chapman, 1897:28; name combination, but not *Marmosa marmota* O. Thomas, 1896b.

[*Didelphys (Marmosa)*] *grisea*: Trouessart, 1898:1241; name combination.

Marmosa grisea: Bertoni, 1914:69; name combination.

[*Didelphis (Thylamys)*] *grisea*: Matschie, 1916:271; name combination.

Marmosa marmota marmota: Tate, 1933:218; name combination, but not *Marmosa marmota* O. Thomas.

Marmosa pusilla: Hershkovitz, 1959:338, 343; part; not *Didelphis pusilla* Desmarest

Marmosa (Thylamys) grisea: Kirsch and Calaby, 1977:14; name combination.

[*Thylamys*] *griseus*: Reig, Kirsch, and Marshall, 1987:7; name combination.

T[*hylamys*]. *macrura*: Gardner and Creighton, 1989:4; first use of current name combination and incorrect gender concordance.

Thylamys grisea: Contreras and Contreras, 1992:1; name combination and incorrect gender concordance.

DISTRIBUTION: *Thylamys macrurus* is known from Paraguay and southern Brazil.

MARGINAL LOCALITIES (Map 50): BRAZIL: Mato Grosso do Sul, Campo Grande (MZUSP 3782). PARAGUAY: Amambay, 28 km SW of Pedro Juan Caballero (UMMZ 125243); Paraguarí, Cordillera de los Altos, ca. 7 km NW of Paraguarí (Contreras and Contreras 1992); Central, Tapuá (type locality of *Didelphys macrura* Olfers; and *Didelphis grisea* Desmarest).

SUBSPECIES: We consider *T. macrurus* to be monotypic.

NATURAL HISTORY: The largest species in the genus, *T. macrurus* is known from the moist subtropical forests of eastern Paraguay and adjacent Brazil. None of the few specimens available shows evidence of fat storage in the

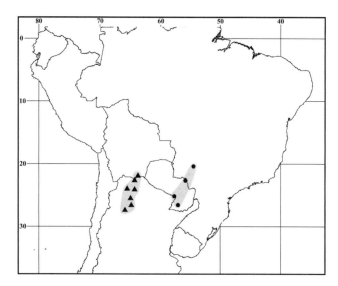

Map 50 Marginal localities for *Thylamys macrurus* ● and *Thylamys sponsorius* ▲

tail. *Thylamys macrurus* has a $2n = 14$, $FN = 20$ karyotype in which the X chromosomes are small acrocentrics (Palma 1995a). Only a single female was karyotyped by Palma (1995a); therefore, he could not determine if the Y chromosome is absent in somatic cells, a condition he found in *T. pallidior* and *T. elegans*.

REMARKS: *Thylamys macrurus* has been frequently referred to under the name *grisea*, which is a junior objective synonym, and under *marmota*, which is unavailable from Oken (1816). All three names are based on Azara's "*Micouré a longue queue*"; however, Oken's (1816) names are non-Linnaean and not available (ICZN 1956). In a footnote, while referring to a specimen from Goya, Corrientes, Argentina, O. Thomas (1896b:313–14) made the name *Marmosa marmota* available, with himself as the author. This specimen from Goya, later renamed *Marmosa citella* (see species account for *T. pusillus*), and represented by the southern-most locality mapped by B. E. Brown (2004:Fig. 86) for *T. macrurus*, represents *T. pusillus*. Although she listed the reference, B. E. Brown (2004:139–40) did not plot the specimen S. Anderson (1997) reported as *T. macrurus* from Bolivia. Voss, Tarifa, and Yensen (2004) reidentified this specimen as *Marmosops ocellatus*.

Azara's description lacked precision and could refer to this species, to *Micoureus paraguayanus*, or to *Marmosops incanus*, all of which are small, grayish, long-tailed opossums that occur in Paraguay. We are using *T. macrurus* in the same restricted sense used by Tate (1933:217) for the name *Marmosa marmota marmota*, and by Cabrera (1958:30) for the name *Marmosa grisea*. Cabrera (1958:30) included *Marmosa marmota* O. Thomas, 1902, in the synonymy of *Marmosa grisea*; however, that use of the name should have been placed in the synonymy of *Mar-*

mosa pusilla. Hershkovitz (1959:343) considered *Didelphis macrura* Olfers, and *Didelphis pusilla* Desmarest, to be conspecific (see account for *T. pusillus*).

Thylamys pallidior (O. Thomas, 1902)
White-bellied Thylamys

SYNONYMS:

Marmosa elegans pallidior O. Thomas, 1902f:161; type locality "Challapata," Oruro, Bolivia.

[*Didelphys (Marmosa) elegans*] *pallidior*: Trouessart, 1905: 856; name combination.

[*Didelphis (Thylamys)*] *pallidior*: Matschie, 1916:271; name combination.

[*Marmosa (Thylamys) elegans*] *pallidior*: Cabrera, 1919: 40; name combination.

Marmosa bruchi O. Thomas, 1921b:519; type locality "Alto Pencoso, just west of San Luis city," San Luis, Argentina.

Marmosa elegans fenestrae Marelli, 1932:68; type locality "Sierra de la Ventana (provincia de Buenos Aires)," Argentina.

Marmosa pallidior: Tate, 1933:229; name combination.

Marmosa [(Thylamys)] pusilla bruchi: Cabrera, 1958:32; name combination.

Marmosa [(Thylamys)] pusilla pallidior: Cabrera, 1958:32; name combination.

T[hylamys]. pallidior: Gardner and Creighton, 1989:4; first use of current name combination.

DISTRIBUTION: *Thylamys pallidior* is in northern Chile, western and southern Peru and Bolivia, and in Argentina as far south as Península Valdez.

MARGINAL LOCALITIES (Map 51; from Tate 1933, except as noted): PERU: Lima, 1 mile W of Canta (MVZ 119913); Huancavelica, 2 km E of Ticrapo (Reig et al. 1977, as *Marmosa elegans*); Ayacucho, 15 miles WNW of Puquio (MVZ 137896); Arequipa, 18 km E of Arequipa (Reig et al. 1977, as *Marmosa elegans*). BOLIVIA (S. Anderson 1997): La Paz, Berlin; Santa Cruz, 5 km SE of Comarapa; Tarija, Serranía Sama. ARGENTINA: Jujuy, Maimará; Salta, 30 km E of Cachi (Mares and Braun 2000); Tucumán, Tafí del Valle; Catamarca, Quirós (Mares and Braun 2000); San Luis, Potrero de los Funes (UMMZ 155898); Córdoba, Villa Valeria; Buenos Aires, Sierra de la Ventana (type locality of *Marmosa elegans fenestrae* Marelli); Buenos Aires, Laguna Chasicó (Contreras 1973); Chubut, 3 km S of Punta Norte (Mares and Braun 2000); Chubut, approximately 280 km W of Dolavón (Birney et al. 1996b); Neuquén, Collon Curá; Neuquén, Chos-Malal; Mendoza, 8 km NW of El Sosneado (Mares and Braun 2000); Mendoza, Tupungato. CHILE: Coquimbo, 7.5 km E of Illapel (Pine, Miller, and Schamberger 1979, as *Marmosa elegans elegans*); Santiago, Las Condes (Pine,

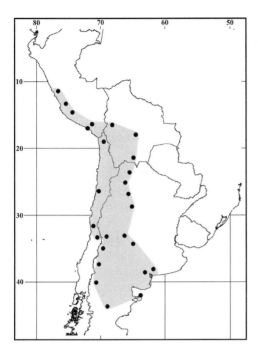

Map 51 Marginal localities for *Thylamys pallidior* ●

smaller and more delicate, and has a relatively shorter rostrum and larger auditory bullae. Relatively larger auditory bullae, a smoothly rounded supraorbital region (except is some old individuals; S. Solari, 2003), less-flared zygomata, and the greater degree of tail incrassation distinguish *T. pallidior* from *T. pusillus*.

Birney et al. (1996b) distinguished two morphs of the Argentine specimens of *Thylamys* they identified as *T. pusillus* and suggested that one might represent to *T. pallidior*. Mares and Braun (2000) treated both morphs as representatives of *T. pallidior*. Braun et al. (2005) distinguished a northern component (Peru, Bolivia, and northern Chile) and a southern component (Argentina) in their population samples of *T. pallidior*. Although museums probably contain enough material to permit a revision sufficient to detail species limits for *T. pallidior*, and to determine if the recognition of subspecies is warranted, such a revision has not been done. Therefore, for the purposes of this account, we treat the species as monotypic.

Tate (1933) identified two juveniles from the Rimac Valley of Peru as *Marmosa venusta venusta*. Although the identity of one (BM 2.1.1.120) has not been confirmed, S. Solari (2003) identified the other (FMNH 24141) as *T. pallidior*. The northern-most, elongated dot in B. E. Brown's (2004:Fig. 85) distribution map for *T. elegans* represents the Tarapacá, Chile, localities of Pine, Miller, and Schamberger (1979). Palma (1995b) reidentified these animals as *T. pallidior*. These Tarapacá specimens also are within the northern component of *T. pallidior* as outlined by Braun et al. (2005).

Miller, and Schamberger 1979, as *Marmosa elegans elegans*); Atacama, Mina Altamira (Pine, Miller, and Schamberger 1979, as *Marmosa elegans* ssp.); Tarapacá, Esquiña (USNM 391777). PERU: Arequipa, 3 miles N of Mollendo (Reig et al. 1977, as *Marmosa elegans*).

SUBSPECIES: We treat *T. pallidior* as monotypic (see Remarks).

NATURAL HISTORY: The preferred habitat of *T. pallidior*, at the northern end of its range in western and southwestern Peru, and on the altiplano of Bolivia, appears to be rocky slopes, ravines, and areas of sparse shrubs. Pearson and Ralph (1978) reported the White-bellied Thylamys (as *Marmosa elegans*) from loma, *Polylepis*, desert-scrub, and montane-scrub forest habitats in southern Peru.

Individuals from San Luis, Argentina, were caught in matorral vegetation on rocky hillsides having sparse ground cover. Specimens taken during April in Chubut, Argentina, had extremely incrassate tails. Further information on diet, habitat, and behavior can be found in Cabrera and Yepes (1940); Roig (1962); Birney et al. (1996b); Mares and Braun (2000); and Flores, Díaz, and Barquez (2000). *Thylamys pallidior* has a $2n = 14$, $FN = 20$ karyotype in which the X is a small acrocentric, and the Y is absent in somatic cells ($2n = 13$ in males; Palma 1994, 1995a; Braun, Mares, and Stafira 2005).

REMARKS: *Thylamys pallidior* has relatively long, lax, or fluffy dorsal and ventral pelage, with the latter clear white. In contrast to that of *T. elegans*, the cranium is

Thylamys pusillus (Desmarest, 1804)
Chacoan Thylamys
SYNONYMS:

Didelphis pusilla Desmarest, 1804b:19; no type locality given; based on "*le micouré nain*" of Azara (1801:290); therefore, the type locality is Saint-Ignace-Gouazou, Paraguay; identified as S. Ignacio, Misiones, Paraguay, by Tate (1933:222, footnote).

[*Didelphys*] *nana* Illiger 1815:107; *nomen nudum*.

Didelphis nana Oken, 1816:1140; unavailable name (ICZN 1956).

D[*idelphys*]. *nana* Olfers, 1818:206; no type locality given; based on *Didelphis pusilla* Desmarest, 1804b, and the "*micouré nain*" of Azara (1801); therefore, the type locality is Saint-Ignace-Gouazou, Misiones, Paraguay, the source of Azara's *micouré nain*.

Sarigua pusilla: Muirhead, 1819:429; name combination.

Didelphys [*(Grymaeomys)*] *pusilla*: Burmeister, 1854:140; name combination.

Philander pusillus: Cope, 1889a:130; name combination.

Micoureus pusillus: Ihering, 1894:11; name combination.

Micoureus griseus: O. Thomas, 1894c:184; not *Didelphis grisea* Desmarest, 1827.

M[armosa]. pusilla: O. Thomas, 1896b:314; name combination.

Marmosa marmota O. Thomas, 1896b;313–14, footnote; no type locality given; name from *Didelphys marmota* Oken, 1816 (name unavailable from Oken 1816; ICZN 1956), but indication is to O. Thomas (1894c:184–86), in which Thomas used the name *Micoureus griseus* for a specimen he described from Goya, Corrientes, Argentina, and to which he later gave the name *Marmosa citella* (see below).

[*Didelphys (Marmosa)*] *pusilla*: Trouessart, 1898:1240; name combination.

Marmosa citella O. Thomas, 1912c:409; type locality "Goya, Corrientes, Argentina."

[*Didelphis (Thylamys)*] *citella*: Matschie, 1916:271; name combination.

[*Marmosa (Marmosa)*] *pusilla*: Cabrera, 1919:38; name combination.

[*Marmosa (Thylamys)*] *citella*: Cabrera, 1919:40; name combination.

[*Marmosa (Thylamys)*] *marmota*: Cabrera, 1919:40; name combination.

Marmosa verax O. Thomas, 1921b:520; type locality "Mision, west of Concepcion," Paraguay.

Marmosa marmota verax: Tate, 1933:220; name combination.

Marmosa janetta pulchella Cabrera, 1934:126; type locality "Robles, Santiago del Estero," Argentina.

[*Marmosa (Thylamys)*] *pusilla*: Cabrera, 1958:32; name combination.

M[armosa]. pusilla verax: Wetzel and Lovett, 1974:206; name combination.

[*Thylamys*] *pusillus*: Reig, Kirsch, and Marshall, 1987:7; first use of current name combination.

Thylamys pusilla: B. E. Brown, 2004:143; incorrect gender concordance.

DISTRIBUTION: *Thylamys pusillus* occurs in Paraguay, south central Bolivia, and in Argentina south into provincia Mendoza.

MARGINAL LOCALITIES (Map 49): BOLIVIA: Santa Cruz, Santa Cruz de la Sierra (S. Anderson et al. 1993). PARAGUAY: Alto Paraguay, 28 km N and 50 km W of Mayor Pablo Lagarenza (UCONN 19181); Boquerón, 19 km N of Filadelfia (UCONN 19220); Alto Paraná, Puerto Bertoni (Bertoni 1914). ARGENTINA: Misiones, Dos de Mayo (ZSM 1966:70). PARAGUAY: Misiones, San Ignacio (type locality of *Didelphis pusilla* Desmarest). ARGENTINA: Corrientes, Goya (type locality of *Marmosa marmota* O. Thomas, and *Marmosa citella* O. Thomas); Entre Rios, La Paz (Tate 1933); Córdoba, no specific local-

ity (Morando and Polop 1997, not mapped); Santiago del Estero, Villa La Punta (Flores, Díaz, and Barquez 2000); Tucumán, Las Mesadas (Flores, Díaz, and Barquez 2000); Salta, 6 km W of Piquirenda Viejo (Mares and Braun 2000). BOLIVIA: Tarija, 8 km S and 10 km E of Villa Montes (S. Anderson 1997); Chuquisaca, 4 km N of Tarabuco (S. Anderson 1997).

SUBSPECIES: We treat *T. pusillus* as monotypic; the taxon needs to be revised.

NATURAL HISTORY: This species is relatively common in dry or seasonal thorn scrub or matorral habitats of the Chaco region of eastern Bolivia, western Paraguay, and northern Argentina. Specimens have been taken on the ground as well as in trees and shrubs. The karyotype is $2n = 14$, FN = 24 (Reig et al. 1977); FN corrected to 20 by Palma (1994), who described the X chromosome as submetacentric, and noted that the Y chromosome is missing in somatic cells. Braun, Mares, and Stafira (2005) confirmed the lack of a Y in somatic cell and described the X as a small submetacentric.

REMARKS: *Thylamys pusillus* has relatively short pelage in contrast to the longer pelage of *T. pallidior*; the ventral pelage is never gray based as in *T. cinderella*, *T. elegans*, *T. sponsorius*, *T. velutinus*, and *T. venustus*. The species is smaller than *T. macrurus* and, unlike the latter, always shows some degree of tail incrassation.

As Tate (1933:223) noted, O. Thomas (1888b) misapplied the name *pusillus* to Brazilian *Gracilinanus microtarsus* (J. A. Wagner), thereby creating much of the confusion evident in the late nineteenth and early twentieth century literature on the species. The name *Didelphys nana* Illiger, 1815:107, listed by Cabrera (1958:33), is a *nomen nudum*. The name also has no nomenclatural standing from Oken (1816), contrary to Tate (1933:27). The name *Marmosa marmota* Oken, 1816, as used by Tate (1933), is not an available name from Oken, but was made available (by indication) by O. Thomas (1896b). Names from Oken (1816) have been rejected for nomenclatural purposes because Oken did not consistently apply the principles of binominal nomenclature (ICZN 1956). Cabrera (1958:30) included *Marmosa marmota* O. Thomas, in the synonymy of *Marmosa grisea*. However, O. Thomas (1896b:footnote, 313–14) used the name *Marmosa marmota* for material from Goya, Corrientes, Argentina, that included a young mouse opossum he earlier (1894c:184) had described in detail as "*Micoureus griseus* Desmarest, 1827." Later, O. Thomas (1902f:158) reported receiving more material from Goya that he also identified as *Marmosa marmota*. These specimens became the basis for *Marmosa citella* O. Thomas. Therefore, *Marmosa citella* O. Thomas, is a junior synonym of *Marmosa marmota* O. Thomas, and both are junior synonyms of *T. pusillus*. Hershkovitz (1959:343) considered

Didelphis macrura Olfers, and *Didelphis pusilla* Desmarest, to be conspecific (see account for *T. macrurus*).

Thylamys sponsorius (O. Thomas, 1921)
Argentine Thylamys

SYNONYMS:

Marmosa elegans sponsoria O. Thomas, 1921a:186; type locality "Sunchal, 1200 m," Sierra de Santa Bárbara, Jujuy, Argentina.

Marmosa venusta sponsoria: Tate, 1933:228; name combination.

Thylamys elegans sponsoria: Gardner, 1993a:23; name combination and incorrect gender concordance.

Thylamys sponsoria: Flores, Díaz, and Barquez, 2000:330; first use of current name combination and incorrect gender concordance.

DISTRIBUTION: *Thylamys sponsorius* is known from southern Bolivia (departamento Tarija) and northwestern Argentine (provinces of Jujuy, Salta, and Tucumán; Flores, Díaz, and Barquez 2000).

MARGINAL LOCALITIES (Map 50; from Flores, Díaz, and Barquez 2000, except as noted): BOLIVIA: Tarija, Caraparí (Tate 1933). ARGENTINA: Jujuy, Palma Sola; Tucumán, Piedra Tendida, 8 km W of Dique El Caján; Tucumán, Parque provincial La Florida; Salta, Río de las Conchas, 2 km N and 6 km W of Metán; Jujuy, 9 km NW of Bárcena; Salta, Vado de Arrazayal.

SUBSPECIES: We consider *T. sponsorius* to be monotypic

NATURAL HISTORY: In northern Argentina, Flores, Díaz, and Barquez (2000) caught a female with a single attached young in December and a lactating female (no young attached) in February. They reported specimens with incrassate tails taken in May, June, and July. Young animals were caught in January, February, March, and July. *Thylamys sponsorius* has a $2n = 14$, $FN = 20$ karyotype. The X chromosome is a small acrocentric and the Y is absent in somatic cells (Braun, Mares, and Stafira 2005).

REMARKS: Flores, Díaz, and Barquez (2000) emphasized that *T. sponsorius* cannot be distinguished from *T. cinderella* based on color pattern, but can be differentiated on cranial features. The supraorbital region usually is smooth in *T. sponsorius* and only in older adults is there any indication of a supraorbital ridge. In *T. cinderella*, supraorbital ridges are evident in young individuals and develop into pronounced postorbital projections in adults. According to Flores, Díaz, and Barquez (2000), *T. sponsorius* further differs from *T. cinderella* in having a relatively narrower zygomatic breadth, a longer and narrower rostrum, and weaker lambdoidal crests. Nevertheless, Braun et al. (2005), based on their genetic assessment, concluded that *T. sponsorius* is a synonym of *T. cinderella*. B. E. Brown

(2004) treated both names as synonyms of *T. venustus*. This complex of species warrants further study.

Thylamys tatei (Handley, 1957)
Tate's Thylamys

SYNONYMS:

Marmosa tatei Handley, 1957:402; type locality "Chasquitambo (710 m, lat. 10°18′48″S., long. 77°37′20″W.)," Ancash, Peru.

[*Marmosa (Thylamys)*] *tatei*: Kirsch and Calaby, 1977:14; name combination.

[*Thylamys*] *tatei*: Reig, Kirsch, and Marshall, 1987:7; first use of current name combination.

[*Thylamys elegans*] *tatei*: Gardner, 1993a:23; name combination.

DISTRIBUTION: *Thylamys tatei* occurs in the departments of Ancash and Lima, Peru.

MARGINAL LOCALITIES (Map 49; from S. Solari 2003, except as noted): PERU: Ancash, 1 km N and 12 km E of Pariacoto; Ancash, Chasquitambo (type locality of *Marmosa tatei* Handley); Lima, Lomas de Lachay.

SUBSPECIES: *Thylamys tatei* is monotypic.

NATURAL HISTORY: This species has been found in the dry desert, lomas, and montane scrub habitats along the western slope of the central Peruvian Andes at elevations ranging from 300 m at Lomas de Lachay to 3,000 m at Huaráz, departamento Ancash (S. Solari 2003).

REMARKS: Gardner (1993a) and Palma (1997) listed *T. tatei* as a synonym of *T. elegans*.

Thylamys velutinus (J. A. Wagner, 1842)
Dwarf Thylamys

SYNONYMS:

Didelphys velutina J. A. Wagner, 1842:360; type locality "Ypanema," São Paulo, Brazil.

Didelphis pimelura Reinhardt, 1851:v; type locality "Brabreak; silien"; restricted by Tate (1933:234) to Lagoa Santa, Minas Gerais.

Microdelphys velutina: Burmeister, 1856:86; name combination.

Grymaeomys velutinus: Winge, 1893:29; name combination.

[*Didelphys (Marmosa)*] *velutina*: Trouessart, 1898:1241; name combination.

[*Didelphys (Marmosa)*] *pimelura*: Trouessart, 1898:1242; name combination; in synonymy.

[*Didelphis (Thylamys)*] *velutina*: Matschie, 1916:271; name combination.

[*Marmosa (Marmosa)*] *velutina*: Cabrera, 1919:39; name combination.

Thylamys velutinus: A. Miranda-Ribeiro, 1936:388; first use of current name combination.

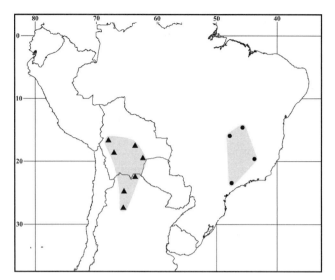

Map 52 Marginal localities for *Thylamys velutinus* ● and *Thylamys venustus* ▲

Marmosa [(Thylamys)] velutina: Cabrera, 1958:33; name combination.

DISTRIBUTION: *Thylamys velutinus* is known from only south central and southeastern Brazil.

MARGINAL LOCALITIES (Map 52): BRAZIL: Bahia, Fazenda Sertão do Formoso (Bonvicino and Bezerra 2003); Minas Gerais, Lagoa Santa (type locality of *Didelphys pimelura* Reinhardt; Tate 1933); São Paulo, Ipanema (Tate 1933); Distrito Federal, Estação Ecológica do Jardim Botânico de Brasília (E. M. Vieira and Palma 1996).

SUBSPECIES: We consider *T. velutinus* to be monotypic.

NATURAL HISTORY: *Thylamys velutinus* inhabits the moist subtropical forests of southeastern Brazil. Superficially similar to some species of *Monodelphis*, its exceptionally small feet and relatively short, seasonally incrassate tail suggest a terrestrial habitus. Apparently, in the Distrito Federal of Brazil, *T. velutinus* is a nocturnal omnivore most active immediately after sunset (E. M. Vieira and Palma 1996). Bonvicino and Bezerra (2003) recovered remains of 13 *T. velutinus* from a sample of 27 owl pellets (*Tyto alba*) at a fazenda in central Brazil where none was caught in 3,367 trap nights.

REMARKS: Its small size and gray-based ventral pelage distinguish *T. velutinus* from *T. pusillus* and *T. macrurus*. The relatively short and thickened tail, and small, white feet easily distinguish *T. velutinus* from *Gracilinanus microtarsus* and *G. agilis*.

Tate (1933:28) and Cabrera (1958:34) dated the name *Didelphis pimelura* Reinhardt as from 1849. The name appeared on page v in the prepage materials under the "minutes" of the meeting held 12 December 1849. However, these pages were not published until the release in 1851 of the complete volume for the years 1849 and 1850 of the "Videnskabeliger Middelelser." Cabrera (1958:33) listed *Marmosa formosa* Shamel, as a subspecies of *Marmosa velutina*; however, here we treat *formosa* as a species of *Chacodelphys*.

Thylamys venustus (O. Thomas, 1902)
Buff-bellied Thylamys

SYNONYMS:

Marmosa elegans venusta O. Thomas, 1902f:159; type locality "Paratani, W. of Cochabamba," Cochabamba, Bolivia.

[*Didelphys (Marmosa) elegans*] *venusta*: Trouessart, 1905: 856; name combination.

[*Didelphis (Thylamys)*] *venusta*: Matschie, 1916:271; name combination.

[*Marmosa (Thylamys) elegans*] *venusta*: Cabrera, 1919:40; name combination.

Marmosa janetta O. Thomas, 1926b:327; type locality "Carlazo, 2300 m," Tarija, Bolivia.

Marmosa venusta: Tate, 1933:225; name combination.

Thylamys elegans venusta: Gardner, 1993a; name combination and incorrect gender concordance.

Thylamys venustus: Palma, 1994:ix; first use of current name combination.

DISTRIBUTION: *Thylamys venustus* occurs along the eastern slope of the Andes, where it is found at intermediate elevations from departamento Cochabamba, Bolivia, southward into provincia Salta, Argentina.

MARGINAL LOCALITIES (Map 52): BOLIVIA (S. Anderson 1997): La Paz, Huajchilla; Santa Cruz, Buenavista; Santa Cruz, Cerro Colorado. ARGENTINA: Salta, 27 km E of Tartagal, along Tonono Rd. (Mares, Ojeda, and Kosco 1981); Tucumán, Concepción (Mares et al. 1996); Salta, Quebrada de San Lorenzo (Mares, Ojeda, and Kosco 1981). BOLIVIA: Oruro, 64 km S of Oruro (S. Anderson 1997).

SUBSPECIES: We treat *T. venustus* as monotypic (see Remarks).

NATURAL HISTORY: Most of the information on *T. venustus* has been reported under the name *T. elegans*. Roig (1971) reported on hibernation, and Housse (1953) gave information on food habits in captivity. Mares, Ojeda, and Kosco (1981) noted specimens caught in moist forests and in clumps of bamboo in northern Salta, Argentina. Mares et al. (1996) found *T. venustus* to prefer mesic agricultural areas, and transitional and humid forest below 2,000 m in provincia Tucumán. In eastern Bolivia, *T. venustus* occurs in the lower montane wet forests of the Bermejo region of Tarija and in the drier montane scrub habitats northward into Cochabamba (Tate 1933). On the arid Altiplano, *T. venustus* is replaced by *T. pallidior*, and in the lowland semiarid thorn scrub of the western Chaco, by *T. pusillus*.

REMARKS: Gardner (1993a) included *T. venustus* as a synonym of *T. elegans*. Palma (1994, 1997) restricted the distribution of *T. elegans* to the western flank of the Andes, thus separating it from *T. venustus*, which occurs along the eastern flank of the Andes. Flores, Díaz, and Barquez (2000), who separated *T. cinderella* and *T. sponsorius* from *T. venustus* (*sensu lato* as understood by S. Anderson 1997, and Mares and Braun 2000), further restricted the Argentine portion of the distribution of *T. venustus* to northern provincia Salta. Definition of the Bolivian distribution will require reexamination of Bolivian specimens previously identified as *Marmosa elegans* and *Thylamys venustus* (e.g., S. Anderson 1997). B. E. Brown (2004) listed *T. cinderella* and *T. sponsorius* as synonyms of *T. venustus*. Osgood (1943), Handley (1957), and Pine, Bishop, and Jackson (1970) have remarked on the difficulty of finding characters that consistently distinguish species of *Thylamys*. *Thylamys venustus* exhibits considerable variation in pelage length, texture, and color, which appears to covary with ambient temperature and humidity throughout its range. This species differs from *T. karimii* and *T. pusillus* in having a relatively longer rostrum and a zygomatic breadth that is less than the condylobasal length. *Thylamys venustus* can be distinguished from *T. pallidior* by its shorter pelage, larger size, larger feet, and longer tail. The skull of *T. pallidior* is smaller, has relatively larger auditory bullae, and a somewhat shorter rostrum than is typical in *T. venustus*. Distinctive morphotypes referred to these three species occur in close proximity in southern Bolivia and northern Argentina, where they appear to replace one another along an elevational gradient. We acknowledge geographic variation in this species. However, the kinds and extent of that variation are too poorly known to warrant division into subspecies at this time. The species needs revision.

Order Microbiotheria Ameghino, 1889

Bruce D. Patterson and Mary Anne Rogers

The Microbiotheria includes only the Microbiotheriidae, a family of small marsupials known from the Early Paleocene of Bolivia, the middle Paleocene of Brazil, the Eocene of Seymour Island, Antarctic Peninsula, and late Oligocene and early Miocene fossils (*Microbiotherium* spp.) from Chubut and Santa Cruz provinces, Argentina (Goin and Carlini 1995; Hershkovitz 1999). The Recent distribution is restricted to Argentina and Chile.

Family Microbiotheriidae Ameghino, 1887

The sole living representative of the family is *Dromiciops gliroides*, found in temperate rain forests of southern Chile

and adjacent Argentina between latitudes 35° and 43°S (Marshall 1982b; Hershkovitz 1999). Marshall's (1982b) diagnosis of the family reveals that it differs from other American marsupial groups in having a markedly reduced fourth or last molars, m1–m3 that have a talonid wider than the trigonid, and lower molars having a well-developed paraconid, especially on m1. Additional features (see Marshall, Case, and Woodburne 1990; Reig, Kirsch, and Marshall 1987; Segall 1969) include an elongate manubrial lever; upper incisors are spatulate, relatively large, and form a complete series (no gaps); upper molars have a straight centrocrista connecting the paracone and metacone; M4 is subtriangular and a third or less the size of M3. *Dromiciops gliroides* lacks paroccipital processes. The auditory bullae are greatly inflated and occupy a relatively large area on the basicranium. The anterior 1/3 of the bullar wall is formed by large tympanic wing of the alisphenoid; the posterior 2/3 of the wall is formed by the *pars petrosa* of the periotic and an extension of the *pars mastoidea*. The tympanic cavity is divided by two septa into three components and the ectotympanic ("tympanic") is not part of the external bulla, but is enclosed by the bulla and hidden from lateral view. In contrast to other Recent New World marsupials, sperm do not pair (conjugate). The most recent comprehensive treatment of the family is in Hershkovitz's (1999) review of *Dromiciops gliroides*.

Genus *Dromiciops* O. Thomas, 1894

This monotypic genus of mouse-sized, pouched marsupials is distributed in the southern temperate rain forests of Chile and adjacent Argentina. The single species was reviewed by Marshall (1978b, 1982b) under the name *D. australis*, and more recently by Hershkovitz (1999) under the name *D. gliroides*.

SYNONYMS:

Didelphys: Cunningham, 1871:362; incorrect subsequent spelling of, but not *Didelphis* Linnaeus, 1758.

Didelphys: F. Philippi, 1893:32; incorrect subsequent spelling of, but not *Didelphis* Linnaeus.

Didelphys: F. Philippi, 1894:33; incorrect subsequent spelling of, but not *Didelphis* Linnaeus.

Dromiciops O. Thomas, 1894c:186; type species *Dromiciops gliroides* O. Thomas, 1894c, by monotypy.

Thylamys: Matschie, 1916:271; not *Thylamys* Gray, 1843b; used as a subgenus of *Didelphis* Linnaeus.

Marmosa: O. Thomas, 1902f:160; not *Marmosa* Gray, 1821.

Dromiciops gliroides O. Thomas, 1894
Monito del Monte

SYNONYMS:

Didelphys elegans: Cunningham, 1871:362; not *Thylamys elegans* Waterhouse, 1839.

Didelphys australis F. Philippi, 1893:32; type locality "San Juan," near Unión, Región de Los Lagos, Chile; preoccupied by *Didelphys australis* Goldfuss, 1812:219.

Dromiciops gliroides O. Thomas, 1894c:187; type locality "Huite, N.E. Chiloe Island," Región de Biobío, Chile.

[*Didelphys (Peramys)*] *australis*: Trouessart, 1898:1244; name combination.

[*Marmosa*] *australis*: O. Thomas, 1902f:160; name combination.

[*Didelphis (Thylamys)*] *australis*: Matschie, 1916:271; name combination.

Dromiciops australis: O. Thomas, 1919b:212; name combination.

Dromiciops australis gliroides: Osgood, 1943:30; name combination.

DISTRIBUTION: *Dromiciops gliroides* inhabits temperate rain forests in Chile and adjacent Argentina, from about 36°S to 43°S (on Isla Chiloé) and 42°S (mainland). It is not represented in more southern collections from the Guaitecas archipelago or on the mainland south of Chaitén.

MARGINAL LOCALITIES (Map 53): CHILE: Maule, Los Queules National Reserve (Saavedra and Simonetti 2001); Biobío, Huépil (Oliver-Schneider 1946); Araucania, 15.2 km W Paso Pino Hachado (Greer 1966). ARGENTINA: Neuquén, "en la zona" del Lago Lácar (Cabrera 1958:5); Río Negro, Hotel Tunquelen (B. E. Brown 2004); Chubut, Río Azul (G. Martin 2003). CHILE: Los Lagos, 40 km S of Seno Reloncaví (Kelt and Martinez 1989);

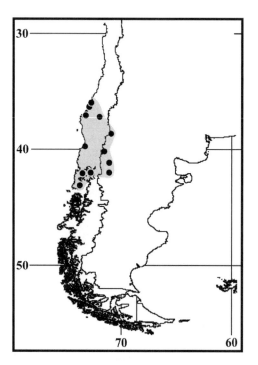

Map 53 Marginal localities for *Dromiciops gliroides* ●

Los Lagos, Huite (type locality of *Dromiciops gliroides* O. Thomas); Los Lagos, 15 km S [*sic*] Quellón near mouth of Río Yaldad (Pine, Miller, and Schamberger 1979); Los Lagos, Valdivia (Wolffsohn 1921); Biobío, Lota (Osgood 1943); Biobío, Río Itata (Oliver-Schneider 1919, *fide* Tamayo and Frassinetti 1980).

SUBSPECIES: Although two subspecies currently are recognized, we consider *D. gliroides* to be monotypic (see Remarks).

NATURAL HISTORY: *Dromiciops gliroides* typically inhabits Valdivian rainforest associations, but its range does not extend into colder, wetter northern Patagonian formations much south of 42° in continental Chile and Argentina (Meserve, Kelt, and Martínez 1991). The species appears to be more common in secondary, open habitats, that are often filled with bamboo (*Chusquea* spp.), rather than in undisturbed, primary forests. Relative to other syntopic species at La Picada, B. D. Patterson, Meserve, and Lang (1990) found *D. gliroides* in shorter-statured forest characterized by a low density of shrubs and thick herbaceous ground cover. The species occurs at sea level on Isla Chiloé in the southern part of its range, whereas in Malleco (Región de la Araucanía) to the north, it is found as high as 1,460 m (Greer 1966).

Known geographic variation in size accords with Bergmann's Rule (i.e., larger animals at higher latitudes). The male reported by Pine, Miller, and Schamberger (1979) weighed 42.0 g, whereas mass of adults examined by Greer (1966) ranged from 16.7 to 31.4 g. Adult *D. gliroides* taken in February and March, 1984, at Valle de La Picada weighed from 24.0 to 49.5 g. Heavier individuals at La Picada had accumulations of fat under the skin and at the base of the tail, which is thicker (incrassate) in winter-trapped animals than in those taken in summer (10–13 mm versus 7–9 mm; Kelt and Martinez 1989). *Dromiciops gliroides* enters torpor at about 4.5°C (Greer 1966), but readily arouses when warmed. These characteristics demonstrate it adaptation to highly seasonal environments and facilitate survival under extreme physiological stresses (Greer 1966; Kelt and Martinez 1989).

Dromiciops gliroides is poorly represented in collections, leading to the common perception that it is rare. However, at Valle de La Picada, Chile, *D. gliroides* was captured more frequently than five sympatric species of sigmodontine rodents. It was taken at all elevations along an elevational transect (425–1,135m), although in higher frequencies at higher elevations (B. D. Patterson, Meserve, and Lang 1989). Monitos del Monte climb well, using a moderately prehensile tail and opposable hallux (Pridmore 1994), and construct leafy nests 2–3 m aboveground in bamboo (Mann 1955). Predominantly scansorial habits are also indicated by their pronounced aversion to entering Sherman traps (B. D. Patterson, Meserve, and Lang 1989). Phillipi's (1894a) account is the earliest description of diet and habitat. *Dromiciops gliroides* is more animalivorous than other small mammals within its temperate forests habitat, although some seeds and vegetation are consumed. During summer at La Picada, its diet consisted primarily of arthropods (58.6%) and other invertebrates (13.2%) and lacked the annelid and hypogeous fungi components that characterize the diets of *Rhyncholestes raphanurus* and *Geoxus valdivianus* in the same forests (Meserve, Lang, and Patterson 1988). Both sexes reach sexual maturity during their second year (Collins 1973), and females have two pairs of mammae in a well-developed pouch. Young leave the pouch by early summer, attaining weights of 14.4–21.0 g by February-March at La Picada. During maturation, the sheath (*tunica vaginalis*) surrounding the testes becomes densely pigmented with melanin. Successful reproduction in captivity is unknown, although the animals are easily maintained in captivity on a variety of table foods.

REMARKS: Until the 1990s, the literature treated *D. gliroides* under the name *D. australis* with two subspecies: *D. a. australis* from all known mainland localities, and *D. a. gliroides* from Chiloé Island. The insular form is purported to be distinguished by its darker pelage, but insufficient material from Chiloé has precluded reliable definition of subspecific differences. This species was twice described by F. Philippi under the name *Didelphys australis*: first in 1893 in the Anales de la Universidad de Chile; next in 1894 in the Archiv für Naturgeschichte. Nevertheless, the name *Didelphys australis* is preoccupied by *Didelphys australis* Goldfuss, 1812, which refers to a philanger, and *Dromiciops gliroides* O. Thomas is the earliest available name for this species. Because the name *australis* is preoccupied, and *gliroides* is based on specimens from Chiloé Island, if the mainland form proves distinctive at the subspecies level, it currently lacks a name.

Affinities of *Dromiciops* to other living marsupials remain unclear, but indications suggest a close relationship to Australasian marsupials. Hershkovitz (1992b, 1999) erected the marsupial cohort Microbiotheriomorphia for the microbiotheriids and asserted that it was basal to all known marsupials. On the basis of tarsal morphology, Szalay (1982) placed *Dromiciops* in the cohort Australidelphia, a group otherwise containing exclusively Australasian forms. Aplin and Archer (1987) and Marshall, Case, and Woodburne (1990) maintained this arrangement, but Hershkovitz (1992a) refuted this, contending that the joint patterns in ankle bones are variable and intergrading, and the continuous pattern claimed to be diagnostic of the Australidelphia evolved independently more that once. *Dromiciops gliroides* is one of the few New World marsupials known to show sex-chromosome mosaicism (loss of the Y chromosome in somatic cells), an interesting trait also known in Australian peramelid and petaurid marsupials (Gallardo and Patterson 1987; Palma and Yates 1996) and suggestive of common ancestry (see sequence analyses by Palma and Spotorno 1999). Hershkovitz (1999) provided an extensive review of *D. gliroides*, including taxonomy and descriptions of its habitat, morphology, and biogeography, along with his views on its relationships with other marsupials.

Order Paucituberculata Ameghino, 1894
Bruce D. Patterson

The Paucituberculata comprise six families of which only the Caenolestidae is extant. The order is recorded from only Antarctica and South America, where the earliest fossils date from the Late Cretaceous (Marshall et al. 1990).

Family Caenolestidae Trouessart, 1898

The Caenolestidae comprise three Recent genera known from only South America. The present geographic range extends discontinuously from northern Colombia and western Venezuela southward along the Andes to south-central Chile (approximately 43° S). Living members of the family are all allocated to the generalized, prototypical tribe Caenolestini. The geological record of this tribe extends to the earliest Eocene (Casamayoran) of Patagonia, but through much of the Tertiary, the family was also represented by members of the Palaeothentinae and Abderitinae. Diagnostic characters (see Marshall 1980; Reig, Kirsch, and Marshall 1987; Marshall, Case, and Woodburne 1990) include: Unfused mandibular rami, each containing a single "gliriform" incisor; the molar series showing pronounced size reduction from Ml to M3, and the M4 s reduced but not vestigial; stylar cusps B and D are large, subequal in size, and higher than the paracone and metacone (cusp terminology follows Reig, Kirsch, and Marshall 1987:Fig. 1); and a noncuspidate hypocone is present on M1–M3. The antorbital vacuity, located at the nasal, frontal, and maxillary juncture, is usually open. Upper and lower lips have distinctive, fleshy, lateral flaps of skin (Pine, Miller, and Schamberger 1979). The digits on the pes are not syndactylous. A superficial thymus is present. The stomach is divided into three distinct compartments. The forebrain lacks a *fasciculus aberrans*. Mammae number from 4 to 7, and a marsupial pouch is lacking. The penis is bifid and shaped like a cork-screw (B. D. Patterson and Gallardo 1987). Sperm are paired in the epididymis, rectangular in shape, and have a notch on one side from which the midpiece arises. I recognize three genera, although Bublitz (1987) and

Marshall (1980) questioned the differentiation of *Lestoros* from *Caenolestes*.

KEY TO THE LIVING GENERA AND SPECIES OF CAENO-
LESTIDAE (ADAPTED FROM ALBUJA AND PATTERSON
1996):

1. Incisors 2–4 distinctly and subequally bifid along occlusal edge; upper canine separated from I4 and from
   first premolar by a gap large enough to accommodate
   M3; i1 projects beyond alveolus a distance equal to or
   greater than length of m1–m3; distance from anterior
   end of nasals to premaxilla–maxillary suture more than
   2/3 length of nasals; palatal fenestrae confluent (lack a
   median bony septum); palatal bridge separating incisive
   foramina and palatal fenestrae relatively broad, greater
   than length of M3–M4 . . . . *Rhyncholestes raphanurus*
1'. Incisors 2–4 entire, not distinctly bifid; canine nearly
    touching I4; i1 less projecting; distance from anterior
    end of nasals to premaxilla–maxillary suture about 1/2
    length of nasals; a usually complete, medial bony septum
    separates palatal fenestrae; palatal bridge separating in-
    cisive foramina and palatal fenestrae narrow, equal to
    or less than length of M3–M4 2
2. Distance of I4 from canine (and from I3) about equal to
   alveolar length of I4; canine short, double-rooted, equal
   to or shorter than Il; Pl minute, less than half the size of
   P2 . . . . *Lestoros inca*
2'. I4 tucked closely between I3 and canine; canine large,
    conical, single-rooted, and exceeds length of I1; Pl nearly
    equal in size to P2 3
3. Antorbital vacuity (at margin of expanded nasal and
   maxillary) in the shape of a parenthesis or crescent, or
   vacuity completely roofed by bone . . . .
   . . . . *Caenolestes convelatus*
3'. Antorbital vacuity comma-shaped and bounded by
    nasal, maxillary, and frontal bones 4
4. Size smaller; skull delicate; pelage silky-textured with
   faint counter-shading; canines short, cranium rests on
   bullae and incisors (except in some adult males) . . . . .
   . . . . *Caenolestes fuliginosus*
4'. Size usually larger; skull robust; pelage coarse and dis-
    tinctly counter-shaded; canines longer, cranium rests on
    bullae and canines (occasionally on labial cusps of M1
    or M2) 5
5. Head-and-body length 115 mm; tail 120 mm; condy-
   lobasal length 33 mm; coloration of ventral pelage is
   grayish with a dark and conspicuous pectoral spot; upper canines moderately long (1.9 mm); length of nasals
   less than 1/2 condylobasal length; palatal bridge curved
   . . . . *Caenolestes caniventer*
5'. Head-and-body length 135 mm; tail 125 mm; condy-
    lobasal length 39 mm; coloration of ventral pelage is

drab with an inconspicuous pectoral spot; uppers canines larger and long (3.5 mm); nasal length more
than 1/2 of condylobasal length; palatal bridge square
. . . . *Caenolestes condorensis*

Genus *Caenolestes* O. Thomas, 1895

Robert M. Timm and Bruce D. Patterson

Caenolestes contains living caenolestids that have conical,
single-rooted incisors in both sexes, I4 mostly filling the
space between I3 and canine, P1 and P2 are subequal in size,
and the infraorbital foramen opens anteriorly, not laterally
(Albuja and Patterson 1996). We recognize four species of
Caenolestes: *C. caniventer* H. E. Anthony, 1921a, known
from the western Andes in southern Ecuador and northern Peru; *C. condorensis* Albuja and Patterson, 1996, from
the Cordillera del Cóndor on the Ecuadorian and Peruvian
border; *C. convelatus* H. E. Anthony, 1924b, known from
the western Andes in Colombia and northern Ecuador; and
C. fuliginosus (Tomes, 1863a), known mainly from higher
elevations in Ecuador, Colombia, and Venezuela. Other authors (e.g., Marshall 1980; Honacki, Kinman, and Koeppl
1982; Nowak and Paradiso 1983) recognized *C. obscurus*
O. Thomas, 1895c and *C. tatei* H. E. Anthony, 1923b, as
distinct species, but herein we treat these names as synonyms of *C. fuliginosus*. In a revision of the family, Bublitz
(1987) treated *Lestoros inca* and *L. gracilis* of southern Peru
as members of the genus *Caenolestes*; however, we regard
these caenolestids as members of the distinct genus *Lestoros*
Oehser, 1934 as do Myers and Patton (this volume).

The four species of *Caenolestes* can be differentiated
into two groups that are well defined morphologically and
ecologically. Members of the first group (*C. caniventer*,
C. condorensis, and *C. convelatus*) are large and robust,
have coarse pelage that has pronounced counter-shading,
and inhabit mid-elevation subtropical forests. The single
member of the second group (*C. fuliginosus*) is smaller
and has a more slender build; has darker, glossier pelage
with reduced counter-shading; and tends to inhabit higher-
elevation temperate-zone forests and paramos. The only
localities at which two species of *Caenolestes* have been reported (i.e., Mazán, Molleturo, Las Máquinas-Pichincha,
and Caicedo; all in Ecuador) lie in subtropical-temperate
ecotones and constitute elevational replacements of members of the large-bodied group by the small-bodied *C.
fuliginosus* (*caniventer–fuliginosus* at the first two localities
and *convelatus–fuliginosus* at the last two). H. E. Anthony
(1924b) suggested that the two groups might eventually be
distinguished as subgenera.

The anatomy of *Caenolestes* has been considered at
length in attempts to interpret its peculiar combination of

polyprotodont (upper dentition) and diprotodont (lower dentition) characters. Accounts include those of O. Thomas (1896a) and subsequent redescriptions of generic characters by Dederer (1909), Broom (1911), Lönnberg (1921), Osgood (1921), Gregory (1922), and Boas (1933). Hayman et al. (1971) and Hayman and Martin (1974) described and analyzed the karyotype of *Caenolestes*, which is considered a primitive $2n = 14$ complement (see also Hayman 1990). Richardson, Bowden, and Myers (1987) and Turner-Erfort (1994) studied the cardiogastric gland and alimentary tract, and Herrick (1921) and Obenchain (1925) described the brain morphology of *C. fuliginosus* in detail. Rodger (1982) described the testes and excurrent ducts of *Caenolestes*; Temple-Smith (1987) reported on sperm morphology and pairing. Szalay (1982) offered a preliminary appraisal of tarsal morphology in the genus.

Little is known of the ecology and behavior of caenolestids. Kirsch and Waller (1979:390) found that, despite marked differences in habitat structure, microhabitats preferred by *C. fuliginosus* and *C. convelatus* were quite similar, being cool, moist, and often moss-covered (see also Osgood 1921; Bublitz 1983). The diets of *Caenolestes* have been reported on by Osgood (1921), Kirsch and Waller (1979), and Barkley and Whitaker (1984), who found them to include a wide array of arthropods as well as some fruit and smaller vertebrates. Known parasites include fleas (Smit 1953; R. E. Lewis 1974), chiggers (Brennan and Reed 1975; Brennan and Goff 1978; Goff and Timm 1985), and chewing lice (Timm and Price 1985). Hershkovitz (1972) described ecological associations and anatomical convergences between *Caenolestes* and species of sigmodontine mice (*Thomasomys* and "*Abrothrix*" [= *Akodon*]).

One of the English vernacular names, "shrew-opossums," suggests that caenolestids are shrew equivalents (ecomorphs), and they are found along the same runways as shrews. Competition between caenolestids and *Cryptotis* is undocumented, although both feed on inverterbrates. All species of *Caenolestes* are significantly larger than any species of South American *Cryptotis*; because *Caenolestes* appears to be predaceous, it is possible that they prey upon shrews. Caenolestids are sometimes called "rat opossums," in apparent reference to their rat-like appearance and ability to run quickly.

SYNONYMS:

Hyracodon Tomes, 1863a:50; type species *Hyracodon fuliginosus* Tomes, 1863a, by monotypy; preoccupied by *Hyracodon* Leidy, 1856 (Mammalia: Perissodactyla).

Caenolestes O. Thomas, 1895c:367; replacement name for *Hyracodon* Tomes.

Coenolestes O. Thomas, 1917a:3; incorrect subsequent spelling of *Caenolestes* O. Thomas.

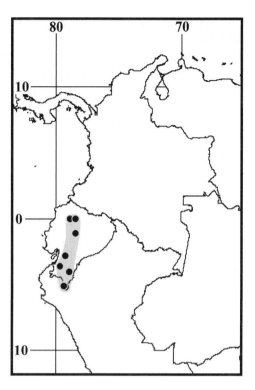

Map 54 Marginal localities for *Caenolestes caniventer* ●

Caenolestes caniventer H. E. Anthony, 1921

Gray-bellied Caenolestid

SYNONYMS:

Caenolestes caniventer H. E. Anthony, 1921a:6; type locality "El Chiral, Western Andes; altitude 5350 ft.; Prov. del Oro, Ecuador."

Caenolestes fuliginosus: Barkley and Whitaker, 1984:328; not *Caenolestes fuliginosus* Tomes.

DISTRIBUTION: *Caenolestes caniventer* is recorded from subtropical forests in western Ecuador where it is known from the provinces of Azuay, El Oro, and Pichincha, and in northwestern Peru from the department of Piura.

MARGINAL LOCALITIES (Map 54, localities listed from north to south): ECUADOR: Pichincha, San Antonio (B. E. Brown 2004); Pichincha, Río Saloya, near Mt. Cayambe (Barnett 1991); Tungurahua, San Antonio (B. E. Brown 2004); Azuay, Mazán (Bublitz 1987); El Oro, El Chiral (type locality of *C. caniventer* H. E. Anthony); Zamora-Chinchipe, Zamora (B. E. Brown 2004). PERU: Piura, Cerro Chinguela (Barkley and Whitaker 1984, as *C. fuliginosus*, not mapped); Piura, Km 30, on road from Huancabamba to San Ignacio (Albuja and Patterson 1996).

SUBSPECIES: We treat *C. caniventer* as monotypic.

NATURAL HISTORY: Barkley and Whitaker (1984) reported finding the Gray-bellied Caenolestid using small tunnels and cavities under tree roots along small streams in wet grasslands, and in patches of humid subtropical and temperate forest at 2,900 m. Barkley and Whitaker (1984) considered "*C. fuliginosus*" (= *C. caniventer*) to be an opportunistic feeder; lepidopteran larvae, centipedes, and arachnids composed over 75% of the food items in stomach contents. Other food items they found included fruit, a bird, and a wide variety of other insects and invertebrates.

Caenolestes condorensis Albuja and Patterson, 1996
Andean Caenolestid

SYNONYM:

Caenolestes condorensis Albuja and Patterson, 1996:42; type locality "'Achupallas'...[a] camp on the upper plateau of the Cordillera del Cóndor, in the Provincia de Morona-Santiago, Ecuador, coordinates 3°27′03″S, 78°29′39″W, elevation 2,080 m."

DISTRIBUTION: *Caenolestes condorensis* is known from only the type locality, which is in the Cordillera del Cóndor on the eastern versant of the Andes along the border between Ecuador and Peru.

MARGINAL LOCALITY (Map 55): ECUADOR: Morona-Santiago, "Achupallas," (type locality of *Caenolestes condorensis* Albuja and Patterson).

SUBSPECIES: *Caenolestes condorensis* is known from only the type series of three specimens.

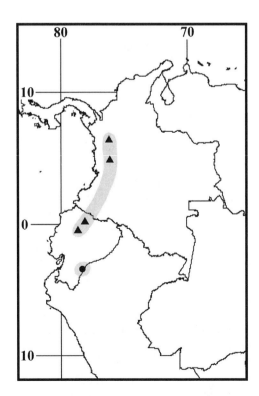

Map 55 Marginal localities for *Caenolestes condorensis* ●
and *Caenolestes convelatus* ▲

NATURAL HISTORY: The Condor Caenolestid is the largest known extant caenolestid. Adult males can attain a mass of 48 g, and a condylobasal length of at least 36 mm. The type series was taken atop a broad plateau composed of Cretaceous ash. The heath-like vegetation of the cold, humid plateau resembles that of the Venezuelan tepuis and is dominated by spiny bromeliads ("achupallas"). *Caenolestes condorensis* was captured on the ground in live traps baited with a mixture of peanut butter and oatmeal. The trapping site was at the juncture between the plateau and the surrounding forested slopes where vegetation included species of *Schefflera* (Araliaceae), *Anthurium* (Araceae), *Ugni* (Myrtaceae), *Spheradenia* (Cyclantaceae), *Clusia* (Clusiaceae), *Leandra* and *Miconia* (Melastomaceae), *Lycopodium* (Lycopodiaceae), *Bennetia* (Theaceae), *Phoradendrom* (Loranthaceae), *Gusmania* and *Tillandsia* (Bromeliaceae), *Chusquea* (Gramineae), *Piper* (Piperaceae), *Monnina* (Polygalaceae), and *Geonoma* (Palmae). Other mammals also acquired at this site include *Artibeus glauca*, *Enchisthenes hartii*, *Sturnira bidens*, *S. erythromos*, *S. oporaphilum*, *Platyrrhinus infuscus*, *P. umbratus*, *Akodon aerosus*, *Oryzomys albigularis*, and *Oryzomys* sp. (Albuja and Patterson 1994).

REMARKS: Although relationships among caenolestids remain uncertain, *C. condorensis* appears to be the only large, coarse-furred caenolestid distributed in lower-elevation forests on the eastern slope of the Andes. Both *C. caniventer* and *C. convelatus* occur on the western slopes and inter-Andean valleys.

Caenolestes convelatus H. E. Anthony, 1924
Northern Caenolestid

SYNONYMS: See under subspecies.

DISTRIBUTION: *Caenolestes convelatus* is known from the western slopes of the Andes in northwestern Ecuador and in northwestern Colombia.

MARGINAL LOCALITIES (Map 55, localities listed from north to south): COLOMBIA: Antioquia, Santa Bárbara (type locality of *Caenolestes convelatus barbarensis* Bublitz); Valle del Cauca, Alto de Galápagos (Albuja and Patterson 1996). ECUADOR: Imbabura, Hacienda La Vega (Albuja and Patterson 1996); Pichincha, 11 km W of Aloag (Kirsch and Waller 1979).

SUBSPECIES: Bublitz (1987) recognized two subspecies based on differences in overall size, especially of the skull, which is larger in *C. c. barbarensis*.

C. c. barbarensis Bublitz, 1987

SYNONYM:

Caenolestes convelatus barbarensis Bublitz, 1987:77; type locality "Santa Barbara (6° 23′0″N, 76° 7′30″W) in [Antioquia] Kolumbien, 3100 m."

This subspecies is in western Colombia.

C. c. convelatus H. E. Anthony, 1924

SYNONYMS:

Caenolestes obscurus: Lönnberg, 1921:4; not Caenolestes obscurus O. Thomas, 1895.

Caenolestes convelatus H. E. Anthony, 1924b:1; type locality "Las Maquinas, Western Andes, 7000 feet altitude, on trail from Aloag to Santo Domingo de los Colorados [Pichincha], Ecuador."

Caenolestes convelatus convelatus: Bublitz, 1987:76; name combination.

The nominate subspecies is in northwestern Ecuador.

NATURAL HISTORY: H. E. Anthony (1924b) noted that the holotype was taken in a region of heavy subtropical forest and at an elevation comparable to that inhabited by C. caniventer, but lower than elevations where C. fuliginosus has been found. R. M. Timm caught an adult male at Hacienda La Vega in thick vegetation along a small stream in pasturelands, using sardine juice as bait. The testes measured 9 × 7 mm, suggesting reproductive activity in September, when the specimen was captured. Two species of ectoparasites were named and described from this specimen: a chewing louse, Cummingsia perezi (Timm and Price, 1985) and a chigger, Peltoculus ecuadorensis (Goff and Timm, 1985). Kirsch and Waller (1979) captured an adult male (testes = 6 mm) in dense brush along a stream. They reported that a captive C. convelatus used its forepaws to direct worms into the side of the mouth where they were sheared into manageable pieces by the cheek teeth.

Caenolestes fuliginosus (Tomes, 1863)

Dusky Caenolestid

SYNONYMS: See under Subspecies.

DISTRIBUTION: Caenolestes fuliginosus is in the Andes of Colombia, Ecuador, and northwestern Venezuela.

MARGINAL LOCALITIES (Map 56; from Albuja and Patterson 1996, except as noted): VENEZUELA: Táchira, 41 km SW of San Cristóbal. COLOMBIA: Boyacá, Hacienda La Primavera; Cundinamarca, La Reserva Biológica Carpanta (López-Arevalo, Montenegro-Díaz, and Cadena 1993); Cauca, Páramo de Puracé, 1 km E of Laguna San Raphael (Kirsch and Waller 1979); Nariño, Pasto. ECUADOR: Napo, Cosanga; Azuay; Mazán (Bublitz 1987); Chimborazo, Mt. Chimborazo; Pichincha, Cerro Iliniza. COLOMBIA: Cauca, Munchique (B. E. Brown 2004); Valle del Cauca, Páramo de Barragán; Quindío, Finca Rincón Santo; Antioquia, Páramo de Sonsón, 7 km E of Sonsón; Antioquia, Páramo Frontino.

SUBSPECIES: We follow Bublitz (1987) and treat C. fuliginosus as comprising three subspecies.

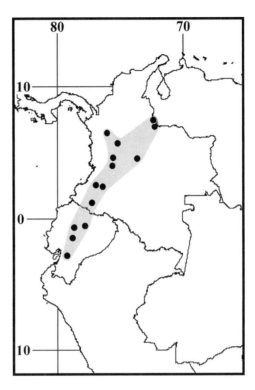

Map 56 Marginal localities for Caenolestes fuliginosus ●

C. f. centralis Bublitz, 1987

SYNONYM:

Caenolestes fuliginosus centralis Bublitz, 1987:74; type locality "Rio Termales (4° 56'N, 75° 19'W) in [Caldas] Kolumbien, 2700 m."

This subspecies occurs throughout most of Andean Colombia and in extreme western Venezuela.

C. f. fuliginosus (Tomes, 1863)

SYNONYMS:

Hyracodon fuliginosus Tomes, 1863a:51; type locality "Ecuador" (see Remarks).

Caenolestes fuliginosus: O. Thomas, 1895c:367; name combination.

Caenolestes tatei H. E. Anthony, 1923b:1; type locality "Molleturo, Provincia del Azuay, 7600 feet, Western Andes," Ecuador.

Caenolestes fuliginosus fuliginosus: Bublitz, 1987:72; first use of current name combination.

The nominate subspecies occurs in the high mountains of central Ecuador north to the Colombian border.

C. f. obscurus O. Thomas, 1895

SYNONYMS:

Caenolestes obscurus O. Thomas, 1895c:367; type locality "Bogota," Cundinamarca, Colombia.

Caenolestes fuliginosus obscurus: Bublitz, 1978:73; first use of current name combination.

This subspecies apparently is restricted to the vicinity of Bogotá, Colombia.

NATURAL HISTORY: *Caenolestes fuliginosus* has been found in tall forest with a closed canopy and little undergrowth, in dense undergrowth in scrub forest (Kirsch and Waller 1979), and in vegetated pasture within a few feet of a swiftly flowing stream (Stone 1914). At the northern end of its distribution, the species inhabits cloud forest (Handley 1976), while at its southern limits, *C. fuliginosus* has been taken from the densely forested western slope of the Ecuadorian Andes in a region that receives heavy rainfall (H. E. Anthony 1923b). The species has been documented over an elevational range from 2,150 m in Boyacá, Colombia (FMNH 92299) to 4,300 m on Volcán Pichincha, Ecuador (FMNH 53295).

What little is known concerning the ecology and behavior of *C. fuliginosus* is mainly derived from Kirsch and Waller's (1979) description of vocalizations, locomotion, mastication, and predatory behavior. Kirsch and Waller (1979) reported four lactating females (none with attached young) and one with an enlarged uterus when taken in late August in departamento Cauca, Colombia. They found *C. fuliginosus* at Finca El Soché along with *Cryptotis* and at least five species of rodents. Bublitz (1983) reported *Cryptotis equatoris*, *Reithrodontomys mexicanus*, *Thomasomys laniger*, and *T. paramorum* as associated with *C. fuliginosus* at four localities near Quito, Ecuador. Smit (1953) named and described the flea *Cleopsylla monticola* from Ecuadorian material; Brennan and Goff (1978) described a new chigger, *Crotiscus danae*, from a Venezuelan specimen; and Timm and Price (1985) described a new chewing louse, *Cummingsia albujai*, also from Ecuadorian *C. fuliginosus*.

REMARKS: Tomes's (1860b, 1863a) descriptions of *C. fuliginosus* were inadequate for differentiating this species from other small South American marsupials, and accounts for the early lack of interest that *C. fuliginosus* later excited in mammalogists (see O. Thomas 1896a). The holotype is "A young specimen preserved in spirit" (O. Thomas 1920d:246); its catalog number (BM 7.1.1.191) indicates that it was among the collection of Tomes's types found in rooms of the Zoological Society of London (O. Thomas 1898a; Gardner 1983). Tomes (1863a) gave "Ecuador" as the type locality of *H. fuliginosus*, but earlier he (1860b:211) had written "The greater portion of these [specimens] are thought to have been collected [by Fraser] at Pallatanga, on the western slope of the Cordillera; but the exact locality is not certain, from the specimens having been unfortunately mixed together." Stone (1914:18) suggested that Fraser acquired the type "on Mt. Pichincha" and Osgood (1921:17) thought that it may have come

"from the paramos of Mt. Chimborazo or Mt. Pichincha where Fraser also worked." Any of these localities is plausible, and additional information, such as date of collection or Fraser's field number, will be necessary for determining a more specific type locality.

Oldfield Thomas described *Caenolestes obscurus* based on the second known specimen of the genus, obtained near Bogotá, Colombia. The status of *C. obscurus* as a distinct species has long been in doubt. In the original description, O. Thomas (1895c) stated that *C. obscurus* was otherwise similar to, but "double the size" of, *C. fuliginosus*. O. Thomas had seen the holotype of *C. fuliginosus* and relied on Tomes's (1863a) published measurements. Stone (1914) compared external measurements of Ecuadorian and Colombian specimens of *Caenolestes* (specimens from Pichincha and Páramo de Tamá), found them to be "practically identical," and concluded "it looks very much as if *C. obscurus* O. Thomas might become a synonym of *C. fuliginosus* (Tomes)." Bublitz (1987) treated these same populations as conspecific under the names *C. f. fuliginosus* and *C. f. obscurus*. He distinguished these forms mainly through craniometric assessments. The nominate subspecies is the smallest and most gracile, and males have relatively small canines; whereas *C. f. obscurus* is the largest, and males have dagger-like canines; *C. f. centralis* is intermediate. H. E. Anthony (1924b) distinguished *C. tatei* from other *Caenolestes* by its smaller size and darker color, but Bublitz (1987) treated *C. tatei* as a junior synonym of *C. f. fuliginosus*.

Genus *Lestoros* Oehser, 1934

Philip Myers and James L. Patton

Members of the monotypic genus *Lestoros* are small, superficially shrew-like marsupials (head and body length 90–120 mm, tail 105–135 mm, mass 20–35 g), known from only southern Peru and adjacent Bolivia. Pelage color is uniformly slate gray-black or brown above and below; the fur is of medium length, thick, and lax. The pinnae are of moderate length and rounded, without an obvious tragus; the eye is very small but clearly visible. The tail is elongate, unicolored, not incrassate, about equal in length to the head and body, has visible annular scale rows arranged perpendicular to the long axis, and is thinly haired with each hair about the length of two scales. The hind feet are elongated with digits II–V long and with well-developed claws; the hallux is short and non-opposable. The forefeet have five digits, the pollex is non-opposable and about equal in length to digit V, claws are present only on digits II–IV, and nails are present on digits I and V. Vibrissae are present on the chin, sides of the nose, and on the forelimbs at the wrist.

The skull is elongate and delicate, with a rounded, globular braincase, broad interorbital region, and narrow, tapered rostrum. The palate has two pairs of large openings: the elongated incisive foramina and the longer palatal fenestrae. An expanded alisphenoid wing forms the anterior covering of the bulla, but it is not fused to the tympanic. The upper canine is usually double-rooted; P1 is either reduced in size and single-rooted, or is absent; M1–M3 are large and quadrituberculate; M4 is small, just slightly larger than P1; the first lower incisor is greatly enlarged and procumbent; the remaining lower incisors, the lower canine, and lower premolars are small and unicuspid; m1–m3 are enlarged and have a well-developed, bicuspidate talonid that is larger than the trigonid; m4 is less than 1/2 the size of the anterior molars. The dental formula is 4/3, 1/1, 2–3/3, 4/4 × 2 = 44–46.

SYNONYMS:

Orolestes O. Thomas, 1917a:3; type species *Orolestes inca* O. Thomas, 1917a, by original designation; preoccupied by *Orolestes* MacLachlan, 1895 (Insecta: Odonata).

Cryptolestes Tate, 1934:154; replacement name for *Orolestes* O. Thomas; preoccupied by *Cryptolestes* L. Ganglbauer, 1899 (Insecta: Coleoptera).

Lestoros Oehser, 1934:240; replacement name for *Cryptolestes* Tate, preoccupied.

REMARKS: Bublitz (1987) followed the suggestion by Marshall (1980) and Simpson (1970) that *Lestoros* Oehser and *Caenolestes* O. Thomas were questionably separable and synonymized *Lestoros* under *Caenolestes*. Bublitz's treatment, however, was strictly phenetic, and his decision regarding generic status was based primarily on the phenetic distance between *Lestoros* and *Caenolestes* relative to the Chilean *Rhyncholestes* Osgood, 1924. We conclude that the qualitative differences between the skulls of *Lestoros* and *Caenolestes*, while relatively slight in comparison to differences between them and the skull of *Rhyncholestes*, are still substantive. *Lestoros* differs from other caenolestids in having a double-rooted canine in both sexes that is shifted posteriorly and appears to be inserted on the maxillae; by either great reduction in the size of, or the complete loss of P1, which when present often has a single root; and by the very small M4 (also see Osgood 1924). We believe these characters justify recognizing *Lestoros* as a genus.

Lestoros inca (O. Thomas, 1917)
Incan Caenolestid

SYNONYMS:

Orolestes inca O. Thomas, 1917a:3; type locality "Torontoy, Cuzco, Peru, 4200 m."

Cryptolestes inca: Tate, 1934:154; name combination.

Lestoros inca: Oehser, 1934:240; first use of current name combination.

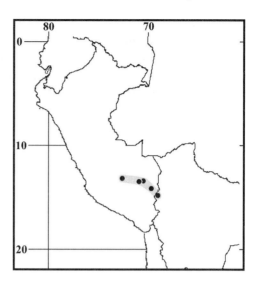

Map 57 Marginal localities for *Lestoros inca* ●

Caenolestes inca: Bublitz, 1987:77; name combination.

Caenolestes gracilis Bublitz, 1987:78; type locality "Limacpunco, in [Cusco,] Peru, 2400 m."

DISTRIBUTION: *Lestoros inca* is found in the headwater drainages of the Río Alto Urubamba, Río Alto Madre de Dios, Río Marcapata, and Río Inambari of the departments of Cusco and Puno, southeastern Peru, between elevations of approximately 3,600 and 2,000 m. The species is also in departamento La Paz, Bolivia (S. Anderson 1997).

MARGINAL LOCALITIES (Map 57; localities listed from west to east): PERU: Cusco, Machu Picchu (O. Thomas 1920d); Cusco, Km marker 112, 32 km by road NE of Paucartambo (V. Pacheco et al. 1993); Cusco, Limacpunco (type locality of *Caenolestes gracilis* Bublitz); Puno, vicinity of Limbani (MVZ 116041). BOLIVIA: La Paz, Llamachaque (S. Anderson 1997).

SUBSPECIES: We consider *L. inca* to be monotypic.

NATURAL HISTORY: Little is known of this species. According to Hunsaker (1977b), "it is terrestrial and nocturnal; stomach contents include insects and other small invertebrates." Kirsch and Waller (1979) report trapping *L. inca* in an area of trees, bushes, and grasses that seemed to them to be markedly drier than habitat preferred by *Caenolestes* spp. We have taken *Lestoros*, however, in habitats ranging from wet, mossy elfin forest to highly disturbed *Baccharis* scrub and second growth. From our experience, it is relatively common in this range of habitats, during both the wet and dry seasons. Individuals are readily caught in traps set among tree roots or in runways in moss or grass and baited with meat, sardines, or peanut butter mixed with oats. Kirsch and Waller (1979) noted that *L. inca* was "very gentle in the hand." *Lestoros inca* is known to support a number of species of ectoparasitic mites (Fain and Lukoschus 1976a, 1976b, 1976c, 1982, 1984; Goff 1981,

1982) and at least one species of chewing louse (Emerson and Price 1981; Timm and Price 1985).

Lestoros inca has a chromosomal diploid number of 14. The autosomes include three pairs of large metacentric or submetacentric chromosomes, one pair of medium-sized metacentrics, and two pairs of small metacentrics or submetacentrics. The X chromosome is a small acrocentric, and the Y appears too small in the published photograph (Hayman et al. 1971) to determine its morphology. These authors reported a small achromatic region usually near the end of the smallest submetacentric chromosome. The karyotype of *L. inca* is indistinguishable from that of *Caenolestes obscurus* and may represent the primitive morphology for all marsupials (Hayman et al. 1971).

REMARKS: *Lestoros inca* is confined to elfin forests in a narrow elevational zone at the head-waters of four drainage systems on the eastern slope of the Andes. Bublitz (1987), in the course of reviewing the systematics of the family Caenolestidae, described a second species, *Caenolestes gracilis*. The distribution of this supposed species encompassed that known for the entire genus, with the exception of the type locality of *L. inca* in the Machu Picchu region. Bublitz based his decision on a series of quantitative measurements and qualitative characters of the cranium and external body, primarily head-body to tail ratio, presence and size of the pre-orbital groove, presence of an anterior style on the upper canine, position of the canine relative to the premaxillary-maxillary suture, number of roots on P1, stoutness of the zygomatic arch, and shape of the suture between the frontal, parietal, and alisphenoid. Within the expanded series of recently collected specimens available to us, however, these characters exhibit either clinal variation or are highly variable within single population samples; in our opinion, they do not serve to differentiate taxonomic entities. Consequently, we treat *Caenolestes gracilis* Bublitz as a synonym of *L. inca* O. Thomas.

Genus *Rhyncholestes* Osgood, 1924

Bruce D. Patterson

Rhyncholestes is a monotypic genus of small shrew-like marsupials distributed in temperate rain forests of Chile and adjacent Argentina. There is no fossil record.

SYNONYM:

Rhyncholestes Osgood, 1924:169; type species *Rhyncholestes raphanurus* Osgood, 1924, by original designation.

Rhyncholestes raphanurus Osgood, 1924
Long-nosed Caenolestid

SYNONYMS: See under Subspecies.

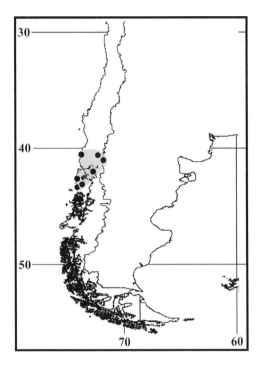

Map 58 Marginal localities for *Rhyncholestes raphanurus* ●

DISTRIBUTION: *Rhyncholestes raphanurus* occurs in the temperate Valdivian rain forests of southern Chile and adjacent Argentina (Birney et al. 1996b). Known records extend from about 40°30′ to 42°S on the mainland and to 43° S on Isla Chiloé.

MARGINAL LOCALITIES (Map 58; from B. D. Patterson and Gallardo 1987, except as noted): CHILE: Los Lagos, Maicolpue; Los Lagos, Parque Nacional Puyehue. ARGENTINA: Río Negro, Puerto Blest (Birney et al. 1996b). CHILE: Los Lagos, 40 km S of Seno Reloncaví (Kelt and Martinez 1989); Los Lagos, Puerto Carmen; Los Lagos, mouth of Río Inio (type locality of *R. raphanurus* Osgood); Los Lagos, Cucao.

SUBSPECIES: I recognize two subspecies of *R. raphanurus*.

R. r. raphanurus Osgood, 1924
SYNONYM:

Rhyncholestes raphanurus Osgood, 1924:170; type locality "mouth of Rio Inio, south end of Chiloé Island," Región de Los Lagos, Chile.

The nominate subspecies is found on Isla Chiloé.

R. r. continentalis Bublitz, 1987
SYNONYM:

Rhyncholestes continentalis Bublitz, 1987:80; type locality "Cerro La Picada (41°4′35″S, 72°24′40″W) in [Los Lagos,] Chile, 450 m."

This subspecies is known from the Chilean mainland and adjacent Argentina and is separated from the population on Isla Chiloé by the Golfo de Ancud (see Remarks).

NATURAL HISTORY: Discovered by Osgood in 1923 and represented by only three specimens until the 1970's, *R. raphanurus* remains poorly known. As summarized by B. D. Patterson and Gallardo (1987), *R. raphanurus* is semifossorial in adaptation and habits. It typically occupies cool, moist microhabitats in the southern temperate rain forests of Chile, but is also recorded from areas disturbed by logging and grazing (Pine, Miller, and Schamberger 1979; Kelt and Martinez 1989). *Rhyncholestes raphanurus* prefers tall, wet mature forest containing many tree falls and abundant bare soil (B. D. Patterson, Meserve, and Lang 1990). The species occurs from sea level on Isla Chiloé (Osgood 1924) and near Maicolpue to at least 1,135 m at Valle de La Picada (B. D. Patterson, Meserve, and Lang 1989). Pelage varies from uniform brown to a brownish gray. The largest *R. raphanurus* taken at La Picada, a female, weighed 36 g, whereas a male from Puerto Carmen, Isla Chiloé, weighed 40 g.

Females trapped by M. H. Gallardo in October 1984 and January 1985 showed swollen uteri suggesting estrous or pregnancy. Females are known to lactate from December through March (Meserve et al. 1982; B. D. Patterson and Gallardo 1987), and Kelt and Martinez (1989) also reported lactating females during the months of May, October, and November. The species exhibits seven well-developed mammae (not five as previously thought), whereas other caenolestids reportedly have four mammae (Nowak and Paradiso 1983). No young were attached to the nipples of lactating females at La Picada, suggesting that this pouchless species makes at least temporary use of a nest. Like some other austral New World marsupials, both males and females have tails variably incrassate with fat (proximal diameter up to 11 mm) that tend to be thickest during early winter. However, mid-winter captures of *R. raphanurus* on snowpack, and the species' apparent inability to arouse spontaneously from hypothermia, suggest the absence of extended hibernation or torpor (Kelt and Martinez 1989).

At La Picada, the annual diet of *R. raphanurus* consisted of invertebrates (54.7%), as well as vegetation and fungi (39.8%); annelid worms account for 8% of all foods taken (Meserve, Lang, and Patterson 1988). Among syntopic small mammals, only *Dromiciops gliroides* is more animalivorous. B. D. Patterson, Meserve, and Lang (1989) captured *R. raphanurus* with greater frequency than they did three syntopic sigmodontine rodents along the same elevational transect. B. D. Patterson, Meserve, and Lang (1990) found this species' microhabitat preferences were most similar to those of the rodents *Geoxus valdivianus* and *Abrothrix sanborni*. Rove beetles, picked from the rumps of live-trapped *R. raphanurus*, represented a distinctive genus and species, *Chilamblyopinus piceus*, which also was recorded by Ashe and Timm (1988) from sympatric *Akodon olivaceus*.

The behavioral ecology of *R. raphanurus* is unknown. In four years of live-trapping small mammals at La Picada, Meserve has recaptured only two of the 36 animals he has marked and released (Kelt and Martinez 1989). Casual behavioral observations on live-trapped animals suggest that *R. raphanurus* is slower and less agile than other caenolestids, and may be more strictly terrestrial. When caged with a newly caught *Akodon* sp., *R. raphanurus* made no attempt to prey on it, as would have *Caenolestes* spp. (personal observation; cf. Kirsch and Waller 1978). Attempts to maintain *R. raphanurus* in captivity for more than a day or two have been unsuccessful. B. D. Patterson and Gallardo (1987) provided measurements and a detailed morphological description along with a synopsis of other information on natural history in their *Mammalian Species* account.

Rhyncholestes raphanurus has a $2n = 14$ karyotype which differs in arm ratios from those of other caenolestids; moreover, the second pair of chromosomes exhibits an achromatic region, which resembles a secondary constriction (Gallardo and Patterson 1987). Differential staining procedures will be needed to determine the nature of chromosomal differences between caenolestid taxa.

REMARKS: Bublitz (1987) noted differences between the premolariform upper canines of female *R. raphanurus* from the mainland and one from Isla Chiloé. He took these differences to signify specific distinction, and named the mainland form *R. continentalis*. Having only a single specimen of each, he was unable to assess the variability of this character. The relative height of this secondary cusp in the holotype of *R. raphanurus* is 52% versus 44% (range 39–49%) in five adult female topotypes of *R. continentalis*. In addition, the entire canine of the lone female *R. raphanurus* Bublitz examined is less curved posteriorly. In the absence of detailed analysis and additional material, I only provisionally retain *R. continentalis* as a subspecies of *R. raphanurus*.

Cohort Placentalia Owen, 1837
Magnorder Xenarthra Cope, 1889
Alfred L. Gardner

The Xenarthra comprise an ancient group of predominantly Neotropical mammals endemic to the Western Hemisphere. Xenarthra Cope (1889b) was originally used as a suborder of the Edentata, which included both New and Old World taxa. The usage here follows Glass (1985) and McKenna and Bell (1997). Reports that the group is also known from the Paleocene of China (Ding 1979) and the Eocene of Europe (Storch 1981) are based on misassignment of fossils of animals similar in structure to tamanduas (see Rose and Emry 1993). Living species are divided among two orders and five families: the order Cingulata includes the family Dasypodidae (armadillos); the order Pilosa includes the families Myrmecophagidae (vermilinguas), Cyclopedidae (silky anteater), Bradypodidae (three-toed sloths), and Megalonychidae (two-toed sloths). Unifying morphological characteristics include ischial articulations with the vertebral column and intervertebral articulations on the arches of the posterior trunk vertebrae. Other features are the lack of incisors and canines; simplification of molars and premolars to homodonty or complete loss of teeth; when present, teeth are single rooted, lack enamel, and usually subcylindrical; a clavicle is present; the scapula has enlarged acromion and coracoid processes; the scaphoid and lunar are separate and the centrale is usually absent; testes are abdominal; uterus is simple; caecum is rudimentary or absent; placenta is deciduous and discoidal; and a *reta mirabilia* is present in appendages and tail (Barlow 1984; Glass 1985).

KEY TO THE ORDERS AND FAMILIES OF RECENT XENARTHRA (MODIFIED FROM WETZEL 1985A):

1. Teeth present; tongue not round, slender, or elongated; tail length variable, but never furred except when extremely short; jugals normal in size; pterygoids separate and not forming posterior margin of palate 3
1'. Teeth absent; tongue round, slender, and elongated; tail nearly as long as or longer than head and body, and partially or completely furred; jugals small; pterygoids converge at midline forming posterior margin of palate (except Cyclopedidae) . . . . (Vermilingua) 2
2. Four claw-bearing digits on manus, one claw greatly elongated; hind feet normal for terrestrial locomotion; length of skull more than 70 mm . . . . Myrmecophagidae
2'. Two claw-bearing digits on manus, claws about equal in length; fore and hind feet modified for grasping; length of skull less than 70 mm . . . . Cyclopedidae
3. Dermal armor present on head and body; forelegs much shorter than head and body; tail conspicuous and not furred; zygoma complete . . . . . Cingulata, Dasypodidae
3'. No dermal armor; forelegs as long as or longer than head and body; tail short, inconspicuous, and furred; zygoma incomplete . . . . (Phyllophaga) 4
4. Forelegs about same length as or slightly longer than hind legs; head plus body longer, total length averaging 621 mm ±48 (540–720 mm, $n = 52$); no externally visible tail; neck short with from five to seven (rarely eight) cervical vertebrae; two digits on manus, three on

pes; anterior pair of teeth caniniform, triangular in cross section, and with beveled occlusal surfaces; distinct (protruding) rostrum, prenasal bone present . . . .
. . . . Megalonychidae
4'. Forelegs 1.5 times as long as hind legs; head plus body (minus tail) smaller, total length averaging 578 mm ±59 (420–800 mm, $n = 108$); tail length averaging 58 mm ±14 (38–90 mm, $n = 101$); long, mobile neck with eight or nine cervical vertebrae; three digits on both fore and hind feet; teeth simple pegs of approximately equal size except for anterior pair, which either may be somewhat larger or smaller than the others, or missing; rostrum not protruding, no prenasal bone . . . . Bradypodidae

Order Cingulata Illiger, 1811

Ralph M. Wetzel, Alfred L. Gardner, Kent H. Redford, and John F. Eisenberg

Members of this order of xenarthrans are distinguished by having a primary body covering of bony scutes or plates. Living members are the armadillos.

Family Dasypodidae Gray, 1821

Armadillos are found from the southern United States of America into southern Patagonia. The ancient family dates from the late Paleocene of South America and spread into Central and North America after the completion of the isthmian land bridge during the Pliocene.

Living armadillos range in size from *Priodontes maximus*, with a head-and-body length averaging 900 mm, to *Chlamyphorus truncatus*, which has a head-and-body length as short as 116 mm. Dasypodids are characterized by a body armor of numerous bony dermal scutes, which are covered by horny epidermal scales and arranged in a regular pattern. The dermal scutes are organized into movable shields on the forequarters and hindquarters and arranged into movable bands in the midsection and tail (except for *Cabassous* spp. in which the tail scutes either are absent or small and widely spaced). Sparse hairs appear between the bands and on the venter, which is not armored. *Dasypus pilosus* is exceptional in having long, dense dorsal fur that hides the carapace. Teeth are peg- to plate-like and lack enamel, incisors and canines are absent, and the dental formula is highly variable.

KEY TO THE SUBFAMILIES, TRIBES, AND GENERA OF DASYPODIDAE (MODIFIED FROM WETZEL 1982, 1985A).

1. Tympanic bulla and external auditory meatus ossified; either with (1) long, white hair on sides of body below carapace, or (2) head shield broad (width:length ratios

0.69 to over 1.0); a single row of large nuchal scutes immediately behind head shield that is no wider than space between pinnae; long hair on carapace . . . . (Euphractinae) 2

1'. Bony tympanic ring instead of ossified bulla and external auditory meatus; head shield narrow; either a row of smaller scutes closer to scapular shield than to head shield, or completely lacking scutes between head and scapular shield; body lacks long, white hair . . . . (Dasypodinae and Tolypeutinae) 6

2. Eyes and pinnae vestigial; long auditory meatus with opening and pinna immediately behind orbit; body sides and venter completely covered with dense, long, and silky, white hair; large rounded rump plate independently attached to body (not attached to remainder of carapace); size small, length of head and body less than 200 mm, condylonasal length less than 50 mm; two frontal prominences on skull; teeth 7–8/8 on each side . . . . (Chlamyphorini) 3

2'. Eyes and pinnae normal; pinnae not in line with orbit; body lacks long, silky white hair; body not ending in separate rump-plate; size larger, length of head and body more than 200 mm, condylonasal length more than 50 mm; skull lacks frontal prominences; teeth 8–9/9–10 on each side . . . . (Euphractini) 4

3. Carapace attached only along spinal column; hair under carapace; tail spatulate at tip; size smaller, length of head and body averaging less than 150 mm . . . . *Chlamyphorus*

3'. Carapace attached along sides of carapace as in other armadillos, tail pointed at tip; size larger, length of head and body averaging more than 150 mm . . . . *Calyptophractus*

4. Size large, length of head and body more than 400 mm, condylonasal length more than 100 mm; carapace pale yellow or tan; dorsal hair white; complete movable band at anterior edge of scapular shield lacking; width of head shield 80% or less of length; zygomatic arch elongate and slender; jugal at symphysis never twice as high as overlying anterior projection of squamosal . . . . *Euphractus*

4'. Size small, length of head and body less than 400 mm, condylonasal length less than 100 mm; carapace and dorsal hair dark tan to dark brown; one complete movable band of scutes at anterior edge of scapular shield; width of head shield greater than 80% of length; zygomatic arch shorter, jugal at symphysis more than twice as high as overlying squamosal 5

5. Pinna short, length from notch less than 20 mm; marginal scutes with apices sharply pointed; nuchal scutes less than 5 mm in anteroposterior length; anterior rostrum slender, width 21% of length; zygomatic

arch of more or less uniform height and without distinct notch below orbit; teeth 8/9 on each side . . . . *Zaedyus*

5'. Pinna longer than 20 mm; apices of marginal scutes rounded; nuchal scutes longer than 6 mm; width of anterior rostrum 25–33% of length: zygomatic arch with distinct notch below orbit; teeth 9/10 on each side . . . . *Chaetophractus*

6. Scapular and pelvic shields flexible (in living animals), separated by six or more movable bands; greatest height of zygomatic arch more than 5 mm; no pronounced arch to toothrow and palate in lateral profile; mandible slender with tooth-bearing portion more or less uniform in height 7

6'. Scapular and pelvic shields rigid, dome-shaped, and usually separated by three movable bands (ranging from one to four); zygomatic arch slender, greatest height less than 5 mm; maxillary toothrow and palate distinctly arched dorsally in lateral profile; vertical height of mandible much greater at middle than at ends of toothrow . . . . (Tolypeutini) *Tolypeutes*

7. Relatively rigid carapace with 6 to 11 movable bands; margins of scutes of movable bands overlapped by triangular scales; rosettes of rounded scutes on scapular and pelvic shields; foreclaws moderate in size; tail long, over 55% of head-and-body length, and with slender tip; rostrum long, 55% or more of condylonasal length; condyloid process much shorter than coronoid process . . . . (Dasypodinae) *Dasypus*

7'. Flexible carapace with 11 to 14 movable bands; margins of scutes on movable bands coincide with margins of overlying scales; rectangular scutes on carapace; large, scimitar-shaped foreclaws; tail less than 1/2 of head-and-body length, and with blunt tip; rostral length less than 1/2 length of skull; condyloid process of mandible higher than coronoid process . . . . (Priodontini) 8

8. Tail armored with articulating bony scutes; body size large, rostrum long (condylonasal length more than 170 mm), and length of head and body more than 700 mm; teeth numerous (up 19 in each jaw) and transversely flattened . . . . *Priodontes*

8'. Tail lacking articulating bony scutes (may or may not have visible scales); body size smaller, rostrum shorter (condylonasal length less than 125 mm), and length of head and body less than 495 mm; comparatively fewer teeth (up to nine in each jaw) and all transversely compressed teeth confined to anterior part of toothrow . . . . *Cabassous*

Subfamily Dasypodinae Gray, 1821

These are small to large armadillos that have long pinnae and a long, slender rostrum. The smooth carapace (hidden beneath a coat of dense hair in *D. pilosus*) has 6 to 11 movable bands separating scapular and pelvic shields. Body hair is relatively sparse except in *D. pilosus*. The manus has four digits, and foreclaws are longest on digits II and III. The tympanic ring represents an otherwise unossified bulla. The subfamily is monotypic (*Dasypus*), and Wetzel and Mondolfi (1979) and Kraft (1995) recognized three subgenera (*Dasypus*, *Cryptophractus*, and *Hyperoambon*).

Genus *Dasypus* Linnaeus, 1758

Dasypus, represented by six living species, occurs from the southern United States through Mexico and Central America into northern Argentina. The genus is distinguished by a long, slender rostrum, which is 55% or more of the length of the head; long, naked pinnae; and a long tail that exceeds 1/2 the length of head and body, and tapers to a slender tip. The smooth carapace consists of the scapular and pelvic shields separated by 6 to 11 movable bands. The exception is *D. pilosus*, which has a dense coat of hair that hides the carapace. The proximal 2/3 of the tail is encased by slender rings, each formed by two or more rows of scales and scutes. Forefeet have four long claws, of which the longest are on the digits II and III. The hind feet have five claws, of which the longest is on digit III. As in *Tolypeutes*, *Priodontes*, and *Cabassous*, the bony structure of each ear consists of a tympanic ring instead of an ossified bulla and external auditory meatus. The dental formula is 7–9/7–9 × 2 = 28–36. The earliest record for the genus is the Pliocene of South America (Marshall et al. 1984; McKenna and Bell 1997).

SYNONYMS:

Dasypus Linnaeus, 1758:50; type species *Dasypus novemcinctus* Linnaeus, 1758, by Linnaean tautonomy.

Tatus Fermin, 1769:110; unavailable name (ICZN 1963).

Tatu Frisch, 1775:Table; unavailable name (ICZN 1954).

Tatu Blumenbach, 1779:74; type species *Tatu novemcinctus* (= *Dasypus novemcinctus* Linnaeus) by monotypy.

Tatus Olfers, 1818:220; part; incorrect subsequent spelling of *Tatu* Blumenbach.

Cataphractus Storr, 1780:40; no species mentioned, name included all armadillos known at that time.

Loricatus Desmarest, 1804b:28; part; type species *Loricatus niger* Desmarest, 1804 (= *Dasypus novemcinctus* Linnaeus, 1758), as designated here (see Remarks).

Tatusia Lesson, 1827:309; part; type species *Dasypus peba* Desmarest, 1822 (= *Dasypus novemcinctus* Linnaeus, 1758), as designated here (see Remarks).

Cachicamus McMurtrie, 1831:163; type species *Dasypus novemcinctus* Linnaeus, 1758, as designated here (see Remarks).

Cachicama P. Gervais in I. Geoffroy St.-Hilaire, 1835:53; invalid emendation of *Cachicamus* McMurtrie.

Zonoplites Gloger, 1841:114; no species mentioned, name proposed for armadillos with four toes on forefeet, the two middle toes being longer than the outer toes.

Praopus Burmeister, 1854:295; type species *Dasypus longicaudus* Wied-Neuwied, 1826, by monotypy.

Cryptophractus Fitzinger 1856:123; type species *Cryptophractus pilosus* Fitzinger, 1856, by monotypy.

Hyperoambon W. Peters, 1864a:180; type species *Dasypus pentadactylus* W. Peters, 1864a, by subsequent designation (Wetzel and Mondolfi, 1979:56); valid as a subgenus.

Muletia Gray, 1874:244; type species *Dasypus septemcinctus*: Gray, 1874, by monotypy (= *Loricatus hybridus* Desmarest, 1804b; not *Dasypus septemcinctus* Linnaeus, 1758).

Tatua W. Robinson and Lyon, 1901:161; incorrect subsequent spelling of *Tatu* Blumenbach.

Mulletia Yepes, 1928:506; in synonymy; incorrect subsequent spelling of *Muletia* Gray.

Mulietia Talmage and Buchanan, 1954:80; incorrect subsequent spelling of *Muletia* Gray.

REMARKS: The systematics of *Dasypus* was most recently reviewed by Wetzel and Mondolfi (1979), with additional information provided by Wetzel (1985b). Although listed in the synonymy of *Dasypus* by Cabrera (1958:222), Hamlett (1939:329), and McBee and Baker (1982:1), the name *Tatus* is unavailable from Fermin (1769), because Fermin did not apply the principles of binominal nomenclature in that work (ICZN 1963). *Tatu* Frisch, 1775, cited by Palmer (1904:664), also is unavailable because that work is nonbinominal (Sherborn 1902:xxv; O. Thomas 1905c:461; ICZN 1954). The earliest available usage of *Tatu* is by Blumenbach, 1779; and the name was cited in the emended form *Tatus* by Olfers (1818:219) and Trouessart (1905:813).

Agassiz (1842, 1847) listed the name *Armodillus* P. C. Wagner, 1763 (which he emended to *Armodillo*) under Edentata. Subsequent workers (e.g., Palmer, 1904; Schulze, Kükenthal, and Heider 1926) did not examine P. C. Wagner (1763). They assumed, based on Agassiz's (1842, 1847) citations, and the Spanish origin of the name, as well as from its usage in such nonbinominal works as Brisson (1762) and Eberhard (1768), that the name applied to the New World armadillos, and most likely to some member of the genus *Dasypus*. Through the assistance of Dr. Richard Kraft of the Zoologische Staatssammlung München (Munich), I have examined a copy of P. C. Wagner's (1763:4–5) account on *Armodillus*. P. C. Wagner used the name *Armodillus* for Old World pangolins (Pholidota, *Manis*), not for dasypodids. A copy of P. C.

Wagner (1763) resides in the library of Erlangen-Nürnberg University, Germany; additional information on the work was graciously provided by Sigrid Kohlmann of that library.

Desmarest (1804b:28) included eight species in his genus *Loricatus*, which he equated with *Dasypus* Linnaeus, and in which he included all of the then known armadillos. *Loricatus* has not been used as a valid genus-group name since shortly after Desmarest (1804) proposed it. We designate *Loricatus niger* Desmarest, 1804, as the type species of *Loricatus* Desmarest (1804), which relegates *Loricatus* to the synonymy of *Dasypus* Linnaeus, because Desmarest's *Loricatus niger* included the species "*das. septem, octo et novemcinctus* Linn." We further equate *Loricatus niger* Desmarest with *Dasypus novemcinctus* Linnaeus. Recently, Kretzoi and Kretzoi (2000:204) listed *Dasypus giganteus* É. Geoffroy St.-Hilaire, 1803 (=*Dasypus maximus* Kerr, 1792), as the type species of *Loricatus* Desmarest. The unfortunate consequence of their nomenclatural act is to relegate *Priodontes* F. Cuvier, 1825 to the synonymy of *Loricatus* Desmarest, 1804. Kretzoi and Kretzoi (2000) simply selected the first species Desmarest listed under *Loricatus* as the type species. We have not abided by Kretzoi and Kretzoi's (2000) selection, in part, because of the pervading nonstandard nomenclatural philosophy reflected in their "Index," not the least of which is that they ignored every Declaration, Direction, and Opinion issued by the International Commission on Zoological Nomenclature. We suggest that the nomenclatural acts in Kretzoi and Kretzoi (2000) be declared to be inconsistent with the intent of the International Code of Zoological Nomenclature and the goals of the International Trust for Zoological Nomenclature, and that the work be declared unavailable for nomenclatural purposes.

We also here designate *Dasypus peba* Desmarest, 1822, as the type species of *Tatusia* Lesson, 1827. Desmarest (1822) included seven species under *Tatusia*: *Dasypus apar*, *D. quadricinctus*, *D. Peba*, *D. hybridus*, *D. tatouay*, *D. villosus*, and *D. minutus*. *Dasypus peba* Desmarest was identified as the type species of *Tatusia* Lesson by Kretzoi and Kretzoi (2000). We have not found an earlier designation of type species; however, because we expect that the "Index" by Kretzoi and Kretzoi (2000) will be declared unavailable, we also designate *Dasypus peba* Desmarest, 1822, as the type species of *Tatusia* Lesson, 1827.

Cachicamus McMurtrie, 1831, was described as a subgenus of *Dasypus* Linnaeus and included *Dasypus novemcinctus* Linnaeus, and *Dasypus 7-cinctus*: Schreber, 1774b (= *Dasypus septemcinctus* Linnaeus). Kretzoi and Kretzoi (2000:57) stated that *Dasypus septemcinctus* Linnaeus was the type species. For reasons explained above, we select *Dasypus 7-cinctus*: Schreber 1774b (= *Dasypus septem-*

cinctus Linnaeus, 1758) as the type species of *Cachicamus* McMurtrie.

KEY TO THE SPECIES OF *DASYPUS*.

1. Enlarged scutes or spurs on knee; condylonasal length more than 112 mm . . . . *Dasypus kappleri*
1′. No enlarged scutes on knee; condylonasal length less than 111 mm 2
2. Carapace covered with thick coat of brown to grayish-brown fur . . . . *Dasypus pilosus*
2′. Carapace with only inconspicuous, sparse hair 3
3. Size large; condylonasal length more than 75 mm; length of head and body more than 360 mm; eight or more movable bands . . . . *Dasypus novemcinctus*
3′. Size medium to small; condylonasal length less than 75 mm; length of head and body less than 320 mm; eight or fewer movable bands 4
4. Size small; length of head and body usually less than 270 mm; ear longer, averaging 31 mm; distributed south of the Rio Amazonas into northern Argentina; number of scutes on fourth movable band average 46 . . . .
 . . . . *Dasypus septemcinctus*
4′. Size larger; length of head and body usually more than 270 mm; ear shorter, less than 30 mm; number of scutes on fourth movable band usually 50 or more 5
5. More movable bands on carapace (usually 7–9); found only in the Río Orinoco basin . . . . *Dasypus sabanicola*
5′. Fewer movable bands (usually 5–7); distributed in Uruguay, Paraguay, and Argentina south into provincia Río Negro 6
6. Size smaller; tail less than 185 mm; ear usually less than 30 mm; hind foot more than 60 mm; widely distributed in Uruguay, Paraguay, Argentina, and southern Brazil . . . . *Dasypus hybridus*
6′. Size larger; tail longer than 185 mm, usually more than 200 mm; ear more than 30 mm; hind foot 62 mm or less; known from only northwestern Argentina . . . .
 . . . . *Dasypus yepesi*

Dasypus hybridus (Desmarest, 1804)
Southern Long-nosed Armadillo
SYNONYMS:

lor[*icatus*]. *hybridus* Desmarest, 1804b:28; no locality given; based on "*Le tatou mulet de d'Azara*," which was from "Paraguay et dan celle des Missions, sans s'approche beaucoup de la rivière de la Plate. . . . vers le Sud par les Pampas de Buenos Ayres" (Azara 1801a: 186); type locality restricted to San Ignacio, Misiones, Paraguay, by Cabrera (1958:223).

[*Dasypus*] *hybridus*: G. Fischer, 1814:126; name combination.

[*Dasypus*] *auritus* Illiger, 1815:108; *nomen nudum*.

T[*atus*]. *auritus* Olfers, 1818:221; type locality "Paraguay"; based solely on "*T. mulet*" of Azara (1801a).

Tatusia hybridus: Lesson, 1827:311; name combination.

T[*atusia*] *hybrida*: Turner, 1853:213; corrected gender concordance.

Praopus hybridus: Burmeister, 1861:428; name combination.

Tatu hybridus: Lahille, 1899:203; name combination.

[*Tatusia (Muletia)*] *hybrida*: Trouessart, 1898:1140; name combination.

[*Tatus (Muletia)*] *hybridus*: Trouessart, 1905:814; name combination.

Muletia hybrida: A. Miranda-Ribeiro, 1914:46; name combination.

D[*asypus*]. *Brevi-cauda* Larrañaga, 1923:344; type locality not given, but Uruguay implied (p. 242); based on Azara's (1802b:156) "*Mulita.*"

Dasypus hibridus: Azevedo, El Achkar, Martins, and Ximénez, 1982:95; incorrect subsequent spelling of *Loricatus hybridus* Desmarest.

DISTRIBUTION: *Dasypus hybridus* is found in Uruguay, the Chaco of northern Argentina and Paraguay, and from estado Rio Grande do Sul, Brazil, south into provincia Río Negro (possibly the southernmost locality) in eastern Argentina, and provincia Mendoza in the west (Cabrera 1958; Wetzel and Mondolfi 1979).

MARGINAL LOCALITIES (Map 59; from Wetzel and Mondolfi 1979, except as noted): ARGENTINA: Formosa,

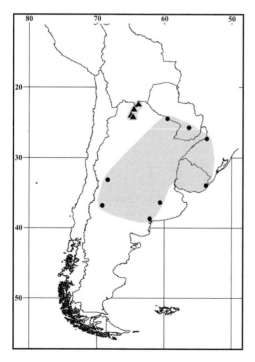

Map 59 Marginal localities for *Dasypus hybridus* ● and *Dasypus yepesi* ▲

La Victoria. PARAGUAY: Guairá, Villarrica. BRAZIL: Rio Grande do Sul, Guarita (BM 86.10.4.7). URUGUAY: Rocha, 15 miles N of San Vicente (FMNH 29331). ARGENTINA: Buenos Aires, Espigas (Yepes 1928); Buenos Aires, Bahia Blanca (BM 86.9.8.2); Río Negro (no specific locality, not mapped); Mendoza, Guayquerías (Yepes 1937); Mendoza, Rivadavia (Yepes 1937).

SUBSPECIES: We consider *D. hybridus* to be monotypic.

NATURAL HISTORY: *Dasypus hybridus* usually occurs in grassland habitats, and has been found at elevations from near sea level to as high as 2,300 m. Burrows are usually constructed in sandy soils. Feeding habits resemble those of *D. novemcinctus* (see Barlow 1965). The karyotype is $2n = 64$, FN = 76 (Saez, Drets, and Brum 1964; Benirschke, Low, and Ferm 1969; Jorge, Meritt, and Benirschke 1977; Jorge, Orsi-Souza, and Best 1985).

REMARKS: Tamayo (1968:8) wrote that the name *Dasypus undecimcinctus* G. I. Molina, 1782:305, is based on a composite of the animal known as "mulita" in the region once called antiguo Cuyo, Chile (= provincia Mendoza, Argentina), and that armadillo is known today as *Cabassous unicinctus*. Despite the inappropriate name, *D. undecimcinctus* G. I. Molina, 1782 (and obviously a *nomen oblitum*) may be the oldest available name for the species we call *D. hybridus*.

Wetzel and Mondolfi (1979) identified as *D. hybridus* the paratype Yepes (1933:Fig. 1) illustrated in his description of *D. mazzai*. However, Hamlett (1939) believed this specimen represented an undescribed species, and Vizcaíno (1995) included this specimen (MACN 13222) as a paratype of *D. yepesi*.

Wetzel and Mondolfi (1979) distinguished between *D. hybridus* and *D. septemcinctus* and determined the distributional limits of each. They identified as *D. hybridus* all specimens examined from Uruguay including the material Sanborn (1929b) included under *D. septemcinctus*. We suggest that the specimen Azevedo et al. (1982) identified as *D. hybridus* from eastern Santa Catarina, Brazil, also represents *D. septemcinctus*.

Dasypus kappleri Krauss, 1862
Greater Long-nosed Armadillo

SYNONYMS: See under Subspecies

DISTRIBUTION: *Dasypus kappleri* occurs east of the Andes in Colombia, south of the Río Orinoco in Venezuela, and in Guyana, Surinam, French Guiana, and the Amazon basin of Brazil, Ecuador, Peru, and northeastern Bolivia.

MARGINAL LOCALITIES (Map 60): VENEZUELA: Bolívar, El Palmar (Wetzel and Mondolfi 1979). GUYANA: Cuyuni-Mazaruni, Kartabo (H. E. Anthony 1921b). SURINAM: Sipaliwini, Bitagron (Husson 1978). FRENCH

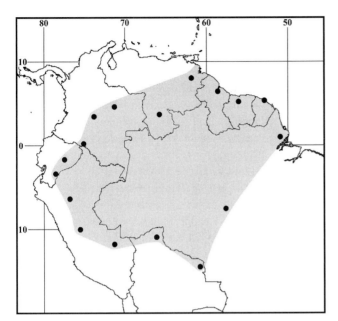

Map 60 Marginal localities for *Dasypus kappleri* ●

GUIANA: Paracou (Voss, Lunde, and Simmons 2001). BRAZIL: Amapá, Rio Tracajatuba (MNRJ 20581); Pará, Capipi (MSP 8950). BOLIVIA: Santa Cruz, Los Fierros (S. Anderson 1997); Pando, Victoria (type locality of *Dasypus kappleri beniensis* Lönnberg). PERU: Madre de Dios, Cocha Cashu (Terborgh, Fitzpatrick, and Emmons 1984); Pasco, Pozuzo (FMNH 30348); San Martín, Roque (type locality of *Dasypus kappleri peruvianus* Lönnberg). ECUADOR: Morona-Santiago, Gualaquiza (BM 14.4.25.87); Pastaza, Sarayacu (type locality of *Tatu pastasae* O. Thomas). COLOMBIA: Putumayo, Parque Nacional Natural La Paya (Polanco-Ochoa, Jaimes, and Piragua 2000); Meta, San Juan de Arama (FMNH 87914); Meta, Carimagua (Barreto, Barreto, and D'Alessandro 1985). VENEZUELA: Amazonas, Belén (Wetzel and Mondolfi 1979).

SUBSPECIES: We tentatively recognize two subspecies of *D. kappleri*.

D. k. kappleri Krauss, 1862
SYNONYMS:

Dasypus peba: Cabanis, 1848:782; not *Dasypus peba* Desmarest, 1822.

Dasypus peba: Burmeister, 1848:199; not *Dasypus peba* Desmarest, 1822.

Dasypus kappleri Krauss, 1862:20; type locality "den Urwäldern des Marowiniflusse in Surinam," which Husson (1978:261) suggested was probably in the neighborhood of Albina near the mouth of the Marowijne River.

Dasypus pentadactylus W. Peters, 1864a:179; type locality "Guiana"; name based on a specimen identified by Ca-

banis (1848:782) as *D. peba* from "am Demerara und auf der Savanne am Berbice," Guyana.

Praopus kappleri: Gray, 1873a:16; name combination.

Tatusia kappleri: O. Thomas, 1880:402; name combination.

[*Tatusia (Tatusia)*] *kappleri*: Trouessart, 1898:1139; name combination.

[*Tatusia (Tatusia)*] *pentadactylus*: Trouessart, 1898:1140; name combination.

T[atu]. kappleri: O. Thomas, 1901c:371; name combination.

[*Tatus (Tatus)*] *kappleri*: Trouessart, 1905:814; name combination.

[*Tatus (Tatus)*] *pentadactylus*: Trouessart, 1905:814; name combination.

Dasypus [(Hyperoambon)] kappleri: Wetzel and Mondolfi, 1979:56; name combination.

The nominate subspecies is in southeastern Colombia, southern Venezuela, the Guianas, and the lower Amazon basin of Brazil.

D. k. pastasae (O. Thomas, 1880)
SYNONYMS:

Tatu pastasae O. Thomas, 1901e:370; type locality "Sarayacu, Upper Pastasa River," Pastaza, Ecuador.

[*Tatus (Tatus)*] *pastasae*: Trouessart, 1905:814; name combination.

Dasypus kappleri pastasae: Lönnberg, 1928:9; name combination.

D[asypus]. k[appleri]. peruvianus Lönnberg, 1928:10; type locality (p. 1) "Roque in Eastern Peru, S.E. of Moyobamba at an altitude of about 1030 m.," San Martín, Peru.

Dasypus pastasae: Sanborn, 1929b:258; name combination.

Dasypus kappleri beniensis Lönnberg, 1942:49; type locality "near the confluence of Rio Madre de Dios with Rio Beni, Victoria, [Pando,] Bolivia."

This subspecies is in eastern Ecuador and Peru, northeastern Bolivia, and the upper Amazon basin of Brazil.

NATURAL HISTORY: Litter size in *D. kappleri* is typically two. The young are monozygotic. As is true for other species of *Dasypus*, the primary diet consists of invertebrates. Barreto, Barreto, and D'Alessandro (1985) believed the species to be a generalist feeder, without a preference for ants and termites, as has been reported for some other armadillos (Redford 1985b). Szeplaki, Ochoa, and Clavijo (1988) wrote that Venezuelan *D. kappleri* showed a marked preference for beetles. Burrows are usually constructed in well-drained forest soils (Wetzel and Mondolfi 1979; Barreto, Barreto, and D'Alessandro 1985). The karyotype is unknown.

Dasypus novemcinctus Linnaeus, 1758
Nine-banded Armadillo
SYNONYMS:

Dasypus novemcinctus var. *Mexicanus* W. Peters, 1864a:180; type locality "Mexico"; restricted to Colima, Mexico, by Bailey (1905:52, footnote), but Hollister (1925) later redefined the type locality as Matamoras [*sic*], Tamaulipas, Mexico (see Remarks).

Dasypus mexicanus: Fitzinger, 1871c:332; name combination.

Tatusia mexicana: Gray, 1873a:14; name combination.

Tatusia leptorhynchus Gray, 1873a:15; type locality "Guatemala."

Additional synonyms under Subspecies.

DISTRIBUTION: *Dasypus novemcinctus* occurs west of the Andes from Colombia into northern Peru, and east of the Andes in Colombia, Venezuela, Trinidad and Tobago, and the Guianas, southward through Ecuador, Brazil, Peru, Bolivia, Paraguay, Uruguay, and northern Argentina. The species also occurs in Grenada in the Lesser Antilles and throughout Central America and Mexico into the southern United States.

MARGINAL LOCALITIES (Map 61): TRINIDAD AND TOBAGO: Tobago (USNM 197043, no specific locality). GUYANA: Cuyuni-Mazaruni, Kartabo (H. E. Anthony 1921b). SURINAM: Para, Zanderij (Husson

Map 61 Marginal localities for *Dasypus novemcinctus* ●

1978). FRENCH GUIANA: Table du Mahury (Menegaux 1903). BRAZIL: Pará, Ilha Mexiana (Hagmann 1908); Pará, Peixe bois (Krumbiegel 1940a); Maranhão, Barra do Corda (C. O. C. Vieira 1957); Ceará, Parque Nacional de Ubajara (Guedes et al. 2000); Pernambuco, Pernambuco (type locality of *Dasypus novemcinctus* Linnaeus); Alagoas, Mangabeiras (C. O. C. Vieira 1953); Bahia, Morro d'Arara (type locality of *Dasypus longicaudus* Wied-Neuwied); Espírito Santo, Santa Lúcia Biological Station (Srbek-Araujo and Chiarello 2005); Rio de Janeiro, Rio de Janeiro (Gray 1873a); São Paulo, Ypanema (Pelzeln 1883); Santa Catarina, Ilha Santa Catarina (Graipel, Cherem, and Ximenez 2001); Rio Grande do Sul, São João do Monte Negro (Cope 1889a). URUGUAY: Canelones, Los Titanes Balneario (MNURU 959). ARGENTINA: Entre Rios, Sauce de Luna (Crespo 1974); Santa Fe, Calchaquí (Yepes 1933); Salta, Tabacal (type locality of *Dasypus mazzai* Yepes). BOLIVIA: Potosí, Uyuni (type locality of *Dasypus boliviensis* Grandidier and Neveu-Lemaire); Santa Cruz, Buena Vista (Crespo 1974); La Paz, Chulumani (S. Anderson 1997); La Paz, Bellavista (S. Anderson 1997). PERU: Madre de Dios, Cocha Cashu (Terborgh, Fitzpatrick, and Emmons 1984); Cusco, Pagoreni (Boddicker, Rodríguez, and Amanzo 1999); Huánuco, Panguana (Hutterer et al. 1995); San Martín, Yurac Yacu (O. Thomas 1927a); Amazonas, Huampami (J. L. Patton, Berlin, and Berlin 1982); Piura, provincia Huancabamba (Grimwood 1969, not mapped). ECUADOR: El Oro, Portovelo (AMNH 46600); Pichincha, Niebli (Lönnberg 1921). COLOMBIA: Cauca, Cerro Munchique (J. A. Allen 1916c); Valle del Cauca, Dagua (Barreto, Barreto, and D'Alessandro 1985); Antioquia, Concordia (type locality of *Tatusia granadiana* Gray); Magdalena, Bonda (J. A. Allen 1904f). VENEZUELA: Zulia, Maracaibo (Osgood 1912); Distrito Federal, San Julián (W. Robinson and Lyon 1901); Nueva Esparta, Isla Margarita, El Valle (Hummelinck 1940).

SUBSPECIES: We recognize four subspecies of *D. novemcinctus* in South America.

D. n. aequatorialis Lönnberg, 1913
SYNONYM:

Dasypus novemcinctus aequatorialis Lönnberg, 1913:34; type localities "Peruchu, altitude 7–9,000 feet, and . . . Niebli, altitude 7,000 feet," Pichincha, Ecuador; type locality not Perucho, Pichincha, Ecuador, as stated by Wetzel and Mondolfi's (1979:50), because of incorrect assumption of holotype. We here designate the adult male (NHR 25) from "Peruchu" (= Perucho) as the lectotype.

This subspecies occurs on the Pacific side of the Andes and adjacent lowlands from Colombia through Ecuador into northwestern Peru.

D. n. fenestratus W. Peters, 1864

SYNONYMS:

Dasypus fenestratus W. Peters, 1864a:180; type locality "Costa Rica"; restricted to San José, Costa Rica by Wetzel and Mondolfi (1979:50).

Tatusia granadiana Gray, 1873a:14; type locality "Concordia," Antioquia, Colombia.

[*Tatusia (Tatusia)*] *granadiana*: Trouessart, 1898:1140; name combination.

[*Tatus (Tatus)*] *granadianus*: Trouessart, 1905:814; name combination.

The South American distribution of this subspecies is in northern and eastern Colombia and in northern and western Venezuela. Elsewhere, the subspecies is known in Central America as far north as Honduras.

D. n. mexianae Hagmann, 1908

SYNONYM:

Tatusia novemcincta var. *mexianae* Hagmann, 1908:29; type locality "Insel Mexiana," Pará, Brazil (also see Wetzel and Mondolfi 1979:50).

This subspecies is endemic to the delta of the Rio Amazonas, Pará, Brazil.

D. n. novemcinctus Linnaeus, 1758

SYNONYMS:

[*Dasypus*] *novemcinctus* Linnaeus, 1758:51; type locality "in America Meridionali"; restricted to Pernambuco, Brazil, by Cabrera (1958:255).

Tatus minor Fermin, 1769:110; unavailable name (ICZN 1963).

Dasypus octocinctus Schreber, 1774b:pl.lxxiii; no locality given.

[*Tatu*] *Novemcincta*: Blumenbach, 1779:74; name combination.

Dasypus longicaudatus Kerr, 1792:112; type locality "America."

Dasypus novenxinctus C.W. Peale and Palisot de Beauvois, 1796:18; incorrect subsequent spelling of *Dasypus novemcinctus* Linnaeus.

Dasypus longicaudatus Daudin in Lacépède, 1802:173; no locality given; based on Buffon's "*Le tatou á longue queue*" (Lacépède 1801:168, pl. 22) of unknown provenance; preoccupied by *Dasypus longicaudatus* Kerr.

lor[*icatus*]. *niger* Desmarest, 1804b:28; no type locality given; based on "*Le tatou noir de d'Azara, le tatueté et cachicame de Buffon.*"

[*Dasypus*] *serratus* G. Fischer, 1814:128; type localities "Paraquaia, inprimis in provincia Buenos-Ayres (Boni Aëris)."

Dasypus decumanus Illiger, 1815:108; *nomen nudum*.

Dasypus decumanus Olfers, 1818:219; *nomen nudum*.

T[*atus*]. *niger* Olfers, 1818:220; type localities "Paraguay, Brasilien"; preoccupied by *Loricatus niger* Desmarest.

Dasypus niger Lichtenstein, 1818a:20; type locality not given; based on *D. novemcinctus* Linnaeus, 1758; therefore, the restricted type locality is Pernambuco, Brazil (Cabrera 1958); preoccupied by *Loricatus niger* Desmarest.

Dasypus peba Desmarest, 1822:368; type localities "Le Brésil, le Guyane, le Paraguay.... On ne le trouve pas dans la province de Buenos-Ayres."

Dasypus longicaudus Schinz, 1824:253; in synonymy, unavailable name (see Remarks).

D[*asypus*]. *longicaudus* Wied-Neuwied, 1826:531; type locality "In den Wäldern am Mucuri"; identified by Ávila-Pires (1965:12) as Morro d'Arara, Rio Mucuri, Bahia, Brazil.

Tatusia peba: Lesson, 1827:311; name combination.

Dasypus [*(Cachicamus)*] *novemcinctus*: McMurtrie, 1831:163; name combination.

Dasypus uroceras Lund, 1839c [1841a]:pl. 12, Fig. 5; type locality "Rio das Velhas Floddal" (p. 73), Lagoa Santa, Minas Gerais, Brazil.

D[*asypus*] *uroceros* Burmeister, 1848:199; incorrect subsequent spelling of *Dasypus uroceras* Lund.

Dasypus pepa Krauss, 1862:19; incorrect subsequent spelling of *Dasypus peba* Desmarest.

D[*asypus*]. *longicaudatus* W. Peters, 1864a:179; incorrect subsequent spelling of *Dasypus longicaudus* Wied-Neuwied; not *D. longicaudatus* Kerr.

Dasypus Lundii Fitzinger, 1871c:340; type locality "Brasilien."

Tatusia platycercus Hensel, 1872:105; type locality "Urwald von Rio Grande do Sul," Brazil.

Tatusia brevirostris Gray, 1873a:15; type localities "Rio de Janeiro," Brazil, and "Bolivia"; type locality not restricted to Rio de Janeiro by Wetzel and Mondolfi (1979:50), because of their false assumption of holotype (ICZN 1999:Art. 74.5). We here select the specimen from Rio de Janeiro (skin: BM 44.3.7.2; skull: BM 46.5.13.16) as the lectotype.

Tatusia boliviensis Gray, 1873a:16; type locality "Bolivia."

Tatusia leptocephala Gray, 1873a:16; type locality "Brazils."

T[*atusia*]. *leptorhinus* Gray, 1874:246; incorrect subsequent spelling of *Tatusia leptorhynchus* Gray.

Tatusia novemcincta: O. Thomas, 1880:402; name combination.

Tatusia longicaudatus: J. A. Allen, 1895a:187; name combination.

[*Tatusia (Tatusia)*] *novem-cincta*: Trouessart, 1898:1139; name combination.

[*Tatusia (Tatusia)*] *platycercus*: Trouessart, 1898:1140; name combination.

[*Tatusia (Tatusia)*] *brevirostris*: Trouessart, 1898:1140; name combination.

[*Tatusia (Tatusia)*] *leptocephala*: Trouessart, 1898:1140; name combination.

[*Tatusia (Tatusia)*] *boliviensis*: Trouessart, 1898:1140; name combination.

[*Tatus (Tatus)*] *novem-cinctus*: Trouessart, 1905:814; name combination.

[*Tatus (Tatus)*] *platycercus*: Trouessart, 1905:814; name combination.

[*Tatus (Tatus)*] *brevirostris*: Trouessart, 1905:814; name combination.

[*Tatus (Tatus)*] *leptocephalus*: Trouessart, 1905:814; name combination.

[*Tatus (Tatus)*] *boliviensis*: Trouessart, 1905:814; name combination.

Tatua novemcincta: W. Robinson and Lyon, 1901:161; name combination.

Dasypus boliviensis: Grandidier and Neveu-Lemaire, 1908: 5; type locality "environs d'Uyuni," Potosí, Bolivia; preoccupied by *Tatusia boliviensis* Gray.

D[*asypus*]. *longi-cauda* Larrañaga, 1923:343; type locality "provincia paracuarensi"; based on Azara's (1802b: 144) "*Negro*"; a junior synonym and homonym of *Dasypus longicaudus* Wied-Neuwied.

Dasypus mazzai Yepes, 1933:226; type locality "Tabacal, Departamento de Orán," Salta, Argentina.

D[*asypus*]. *brevirostris*: Yepes, 1933:230; name combination.

The nominate subspecies occurs from eastern Colombia, eastern and southern Venezuela, Trinidad and Tobago, the Guianas, Brazil, and eastern Ecuador, Peru, and Bolivia, southward into Uruguay, Paraguay, and Argentina.

NATURAL HISTORY: *Dasypus novemcinctus* occupies a wide variety of forest and savanna habitats. Individuals in the northern portions of its range exhibit delayed implantation and the interval between mating and birth may be as long as 240 days; actual gestation is approximately 70 days. Identical twins, triplets, or more commonly, quadruplets develop from a single fertilized ovum. Offspring are weaned at approximately 90 days of age and littermates may remain together in the same burrow system for an extended period. Adults forage solitarily and, depending on ambient temperature, may be active during the day or at night. As summarized by McBee and Baker (1982), *D. novemcinctus* feeds extensively on invertebrates and also consumes fruits, berries, bird eggs, small vertebrates, and carrion. Wetzel and Mondolfi (1979) stated that in tropical areas, ants and termites comprise a large portion of the diet. McBee and Baker (1982) provided illustrations of the species, and a generalized map of its distribution, along with measurements and a synopsis of available information

on natural history in their *Mammalian Species* account. The karyotype is $2n = 64$, FN = 80 (Beath, Benirschke, and Brownhill 1962; Hsu and Benirschke 1967). Barroso and Seuánez (1991) also reported a $2n = 64$ karyotype that has 10 pairs of biarmed and 21 pairs of uniarmed autosome along with a large submetacentric X and a small acroncentric Y chromosome. They gave the FN (number of autosomal arms) as 85 for males and 86 for females, which is equivalent to our calculation of the FN as 82 and confirmed from their description and illustrations.

REMARKS: Gray's (1873a:14) name combination, *Tatusia mexicana*, was based on Fitzinger's (1871c:332) name combination *Dasypus mexicanus*, which in turn was based on *Dasypus novemcinctus* var. *Mexicanus* W. Peters, 1864a (Fitzinger 1871c:364). Gray did not intend *Tatusia mexicana* as new name based on a specimen in the British Museum (see Gray 1865:372, footnote; 1873a:13); therefore, Wetzel and Mondolfi (1979:50) erred in citing BM 43.9.11.6 as the holotype of *Tatusia mexicana*. The type specimen was collected by Urde (a German entomologist ?) from an unknown locality in Mexico and sent to the Berlin Museum (ZMB). Other mammals in the ZMB collected by Urde appear to have come from western or northwestern Mexico, and that fact may have influenced Bailey (1905:52, footnote) when he restricted the type locality to Colima. Identifying Matamoros, Tamaulipas, as the type locality (Hollister's 1925) was a poor selection, as it is highly unlikely that the specimen came from eastern Mexico.

Wetzel (1980:17) and Wetzel and Mondolfi (1979) determined that *Dasypus mazzai* Yepes, 1933, is a composite species comprising *D. novemcinctus* and *D. hybridus*. But, as Hamlett (1939:335) pointed out, regardless of the identity of material Yepes (1933) identified as a paratype *D. mazzai* is not a composite because the holotype consists of only a carapace. Therefore, based on the identity of the holotype, which Wetzel and Mondolfi (1979:55) agreed is a Nine-banded Armadillo, *D. mazzai* is a junior synonym of *D. novemcinctus* Linnaeus.

Schinz (1824:253) used the name *Dasypus longicaudus*, which he attributed to Wied-Neuwied. However, as used by Schinz, the name is unavailable, because he listed it as a synonym of *Dasypus novemcinctus*. In his description of *Dasypus longicaudus*, Wied-Neuwied (1826) made no reference to the earlier usage by Schinz (1824); if he had, he would have validated Schinz's usage and the name would date from 1824 with Schinz as the author.

Dasypus pilosus (Fitzinger, 1856)
Woolly Armadillo

SYNONYMS:

Cryptophractus pilosus Fitzinger 1856:123; type locality "Peru," further restricted to montane Peru by Wetzel and Mondolfi (1979).

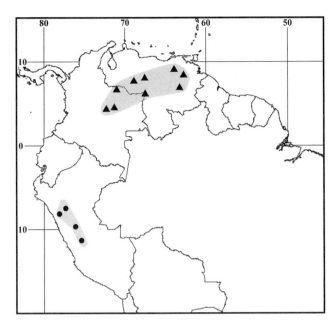

Map 62 Marginal localities for *Dasypus pilosus* ● and *Dasypus sabanicola* ▲

Praopus hirsutus Burmeister 1862:147; type locality "Guayaquil," Peru.

[*Tatusia (Cryptophractus)*] *pilosa*: Trouessart, 1898:1140; name combination.

[*Tatus (Cryptophractus)*] *pilosus*: Trouessart, 1905:814; name combination.

Tatu pilosa: O. Thomas, 1927b:605; name combination.

Dasypus pilosa: Yepes, 1928:468; first use of current name combination and incorrect gender concordance.

Dasypus pilosus: Frechkop and Yepes, 1949:27; name combination.

Dasypus (Cryptophractus) pilosus: Talmage and Buchanan, 1954:84; name combination.

DISTRIBUTION: The presence of the species has been verified only in the Andes of Peru, where it has been collected in the departments of San Martín, La Libertad, Huánuco, and Junín.

MARGINAL LOCALITIES (Map 62): PERU: San Martín, Parque Nacional Abiseo (MHNSM 7502); Huánuco, Paso Carpish (Wetzel 1985b); Junín, Maraynioc (BM 94.10.1.13); Junín, Acobamba (Grimwood 1969, not mapped); La Libertad, Santiago (Frechkop and Yepes 1949).

SUBSPECIES: We consider *D. pilosus* to be monotypic.

NATURAL HISTORY: The distinctive thick dorsal covering of long reddish tan to reddish gray hair precludes confusion with any other species of armadillo; however, nothing is known about the ecology and behavior of this species. Known collecting sites are in upper cloud forest and subparamo habitats. The karyotype is unknown.

REMARKS: Yepes (1928) gave the distribution as in Ecuador and Peru, and Cabrera (1958) also included Bo-

livia; however, the occurrence of the species has been verified only in Peru.

Dasypus sabanicola Mondolfi, 1968
Llanos Long-nosed Armadillo

SYNONYMS:

Dasypus sabanicola Mondolfi, 1968:151; type locality "Hato Macanillal, Distrito Achaguas del Estado Apure," Venezuela.

Dasypus [(Dasypus)] sabanicola: Wetzel and Mondolfi, 1979:55; name combination.

DISTRIBUTION: *Dasypus sabanicola* is known from the llanos of Colombia and Venezuela.

MARGINAL LOCALITIES (Map 62; from Wetzel and Mondolfi 1979, except as noted): VENEZUELA: Anzoátegui, Campo Petrolero Zumo (Ayarzaguena 1984); Monagas, Paso Nuevo; Bolívar, Guri (not mapped); Bolívar, Hato El Torete (Ayarzaguena 1984); Apure, 1 km W of Puerto Páez (Handley 1976). COLOMBIA: Vichada, Finca La Arepa (not mapped); Meta, Carimagua (Barreto, Barreto, and D'Alessandro 1985); Meta, Finca Cháviva; Arauca (no specific locality). VENEZUELA: Portuguesa and Barinas, between Guanarito and Arismendi (not mapped); Apure, Hato El Frío (Fergusson-Laguna and Pacheco 1981); Guárico, Km 47, NNW of San Fernando (Mondolfi 1967).

SUBSPECIES: We consider *D. sabanicola* to be monotypic.

NATURAL HISTORY: The species inhabits savannas, where its home range varies from 1.7 to 11.6 hectares. *Dasypus sabanicola* is crepuscular and feeds on invertebrates, notably ants, termites, and beetles (Mondolfi 1968; Barreto, Barreto, and D'Alessandro 1985). In Venezuela, reproduction occurs from October through March, and, as in *D. novemcinctus*, the usual litter size is four and all develop from one zygote (Mondolfi 1968; Wetzel and Mondolfi 1979). The karyotype is unknown.

REMARKS: Although most reports on *D. sabanicola* provide information on capture sites, many of these locations do not appear on maps or in gazetteers; therefore we are unable to map them. Wetzel and Mondolfi (1979) stated that the source of the specimen identified as from Tocaima, Cundinamarca, Colombia should be confirmed, because the record would extend the range to west of the Cordillera Oriental.

Dasypus septemcinctus Linnaeus, 1758
Yellow Armadillo

SYNONYMS:

[*Dasypus*] *septemcinctus* Linnaeus 1758:51; type locality "in Indiis"; corrected to "Brasilia" by Erxleben (1777:108); further restricted to Pernambuco, Brazil, by Cabrera (1958:225), as suggested by Hamlett (1939).

Tatusia (Muletia) propalatum Rhoads, 1894:111; type locality "Bahia," Brazil.

Tatusia megalolepis Cope, 1889a:134; type locality "Chapada," Mato Grosso, Brazil.

[*Tatusia (Tatusia)*] *megalolepis*: Trouessart, 1898:1140; name combination.

[*Tatusia (Muletia)*] *propalatum*: Trouessart, 1898:1141; name combination.

Tatu septemcincta: O. Thomas, 1900c:548; name combination.

Tatu megalolepis: O. Thomas, 1904b:243; name combination.

[*Tatus (Tatus)*] *megalolepis*: Trouessart, 1905:814; name combination.

[*Tatus (Muletia)*] *propalatus*: Trouessart, 1905:814; name combination.

Dasypus megalolepe: Yepes, 1928:468; name combination and incorrect subsequent spelling of *Tatusia megalolepis* Cope.

Dasypus propalatus: Yepes, 1928:468; name combination.

Dasypus [(Dasypus)] septemcinctus: Wetzel and Mondolfi, 1979:53; name combination.

DISTRIBUTION: *Dasypus septemcinctus* has been recorded from the delta of the Rio Amazonas, southward through the highlands of eastern Brazil, into the state of Rio Grande do Sul and westward into Mato Grosso and the Gran Chaco of Bolivia, Paraguay, and northern Argentina (Wetzel 1985b).

MARGINAL LOCALITIES (Map 63; from Wetzel and Mondolfi 1979, except as noted): BRAZIL: Pará, Ilha Mexiana; Maranhão, Barra do Corda; Minas Gerais, Matias Barbosa (Wetzel 1985b); São Paulo, Ypanema (Pelzeln 1883); Paraná, Palmeira (O. Thomas (1900c); Rio Grande do Sul,

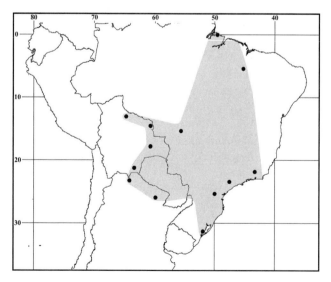

Map 63 Marginal localities for *Dasypus septemcinctus* ●

São Lourenço do Sul. ARGENTINA: Chaco, Pampa del Indio; Salta, Tabacal. BOLIVIA: Tarija, Villa Montes; Santa Cruz, San José de Chiquitos; Beni, San Joaquín (S. Anderson 1997); Santa Cruz, 45 km E of Aserradero Moira (S. Anderson 1997). BRAZIL: Mato Grosso, Chapada.

SUBSPECIES: We consider *D. septemcinctus* to be monotypic.

NATURAL HISTORY: This species inhabits savannas, grasslands, and occasionally, gallery forests. It resembles *D. novemcinctus* in ecology and diet (Wetzel and Mondolfi 1979). The diploid chromosome number is 64 (based on specimens from Botucatu, São Paulo, Brazil, misidentified as *D. hybridus* by Jorge, Orsi-Souza, and Best 1985). Barroso and Seuánez (1991) confirmed the diploid number as 64 in the majority of the specimens they analyzed. They described the karyotype as consisting of 7 pairs of biarmed and 24 pairs of uniarmed chromosomes with a large submetacentric X and a small acrocentric Y chromosome, and an FN of 79 in males and 80 in females. Their FN included the sex chromosmes; whereas, the convention we use when determining FN is to count only the autosomal arms, in which case, the FN is 76 for *D. septemcinctus*.

REMARKS: Hamlett (1939) clarified the identity of *D. septemcinctus*, which had previously been confused with *D. hybridus* and *D. novemcinctus*. Wetzel and Mondolfi (1979) provided additional comparisons with the latter two species.

Dasypus yepesi Vizcaíno, 1995
Yepes's Mulita

SYNONYMS:

Dasypus mazzai Yepes, 1933:226; part (paratype).

Dasypus hybridus: Wetzel and Mondolfi, 1979:52; not *Loricatus hybridus* Desmarest, 1804b.

Dasypus (Dasypus) yepesi Vizcaíno, 1995:7; type locality "San Andrés (1800 msnm), Dto. Orán, Salta, Argentina."

DISTRIBUTION: *Dasypus yepesi* occurs in northern Argentina.

MARGINAL LOCALITIES (Map 59): ARGENTINA: Salta, Quebrada de Acambuco (M. M. Díaz et al. 2000); Jujuy, El Palmar (Vizcaíno 1997); Jujuy, El Caulario (Vizcaíno 1995); Salta, San Andrés (type locality of *Dasypus yepesi* Vizcaíno).

SUBSPECIES: *Dasypus yepesi* is monotypic.

NATURAL HISTORY: Vizcaíno (1997) reported *D. yepesi* from xeric Chacoan habitats as low as 440 m to humid lower montane forests at elevations as high as 1,800 m. There have been no reports on other aspects of its natural history.

REMARKS: The holotype of *Dasypus mazzai* Yepes is a *D. novemcinctus*, but the paratype, as pointed out by

Hamlett (1939), represents a different species, now known as *D. yepesi*. Wetzel and Mondolfi (1979:55) identified the specimen (MACN 13222 [original number 28.225]) they assumed to be the paratype of *D. mazzai* as a *D. hybridus*. Based on correspondence from Jorge A. Crespo (MACN), Wetzel and Mondolfi (1979:Table 4) gave the locality for (MACM 13222) as Puerto Guaraní, Alto Paraguay, Paraguay. Hamlett (1939), however, stated that both specimens Yepes (1933) used in his description of *D. mazzai* came from Salta, Argentina. Yepes (1933:229) left a dash for the catalog number in his table of external measurements of the paratype; however, he gave the catalog number as 32 (p. 231) in the table of cranial measurements. Vaccaro and Piantanida (1998) gave the locality of MACN 13222 as "Tabacal, Orán Department, Salta Province," which is the type locality of *D. yepesi*. They stated that it is a mounted specimen, and that the skull is missing.

Dasypus yepesi, while considerably smaller than *D. novemcinctus*, averages larger than either *D. hybridus* or *D. septemcinctus*. Vizcaíno (1995) and M. M. Díaz (2000) identified these armadillos mainly on size.

Subfamily Euphractinae Winge, 1923

Tribe Chlamyphorini Bonaparte, 1850

These are small, fossorial armadillos with small pinnae and tiny eyes. The head shield is attached to a thin, flexible carapace, which is separate from rounded rump plate. Sides of the body, below the carapace, are covered with long white hair. The manus has five digits, and the fore claws are proportionally large. The long external auditory meatus extends forward above zygomatic arch from the ossified tympanic bulla to a position on the posterior border of the orbit. The Chlamyphorini is represented by two monotypic genera, *Calyptophractus* and *Chlamyphorus*.

Genus *Calyptophractus* Fitzinger, 1871

Calyptophractus is a monotypic genus of small armadillos, which have a head-and-body length averaging more than 150 mm. The head shield is broad and the carapace is thin, flexible, attached along sides of body, and separate from the rounded rump plate. The tail is flattened dorsoventrally and has a pointed tip. The venter, legs, and sides of the body below the carapace are covered with silky, long white hair. *Calyptophractus* closely resembles *Chlamyphorus*, and most authors treat them as congeneric (e.g., Wetzel 1985b; Corbet and Hill 1991; Gardner 1993; McKenna and Bell 1997). They are readily identifiable externally by the attachment of the carapace (along the sides of the body in *Calyptophrac-*

tus, but along the dorsal midline in *Chlamyphorus*) and by the shape of the tip of the tail (pointed in *Calyptophractus*, but spatulate in *Chlamyphorus*). The genus is known from the Pleistocene and Recent of South American (Marshall et al. 1984).

SYNONYMS:
Chlamyphorus: Burmeister, 1863:167; not *Chlamyphorus* Harlan, 1825.
Burmeisteria Gray, 1865:381; type species *Burmeisteria retusa* (= *Chlamyphorus retusus* Burmeister, 1863) by monotypy; preoccupied by *Burmeisteria* Salter, 1865, a trilobite.
Calyptophractus Fitzinger, 1871c:388; type species *Chlamyphorus retusus* Burmeister, 1863, by monotypy.

Calyptophractus retusus (Burmeister, 1863)

Greater Fairy Armadillo
SYNONYMS:
Chlamyphorus retusus Burmeister, 1863:167; type locality "Sta. Cruz de la Sierra," Santa Cruz, Bolivia.
Burmeisteria retusa: Gray, 1865:381; name combination.
Calyptophractus retusus: Fitzinger, 1871c:389; first use of current name combination.
Burmeisteria retusa clorindae Yepes, 1939:38; type locality "Tapia, en la gobernación de Formosa (Argentina)."

DISTRIBUTION: *Calyptophractus retusus* is known from the Gran Chaco of northern Argentina, western Paraguay, and southeastern Bolivia (Wetzel 1985b).

MARGINAL LOCALITIES (Map 64): BOLIVIA: Santa Cruz, Santa Cruz de la Sierra (type locality of *Chlamyphorus retusus* Burmeister). PARAGUAY: Boquerón, Teniente Enciso (UCONN 17609); Boquerón, 15 km S of Filadelfia (Wetzel 1985b). ARGENTINA: Formosa, Tapia (type locality of *Burmeisteria retusa clorindae* Yepes); Salta, Puesto Campo Grande (M. M. Díaz et al. 2000); Salta, La Angostura (Yepes 1939).

SUBSPECIES: We consider *C. retusus* to be monotypic.

NATURAL HISTORY: Presumed to be largely fossorial, *C. retusus* has not been studied in the wild. We expect its ecology to resemble that of the smaller *Chlamyphorus truncatus*. Cuéllar (2001) commented that in the Chaco Boreal biogeographic province of Bolivia, the Greater Fairy Armadillo is killed whenever encountered by the Izoceño Indians because they believe the armadillo to be an omen of bad luck. The karyotype is unknown.

REMARKS: Cabrera (1958) used the generic name *Burmeisteria* Gray, 1865, for this species; however, *Burmeisteria* Gray is not available because it is a junior homonym of *Burmeisteria* Salter, 1865, a trilobite. Burmeister (1863) used the spelling *Chlamyphorus* when describing *Calyptophractus retusus*, not the more commonly

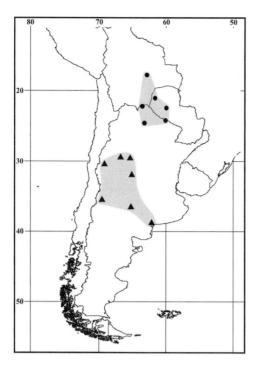

Map 64 Marginal localities for *Calyptophractus retusus* ●
and *Chlamyphorus truncatus* ▲

used incorrect spelling *Chlamydophorus* cited by Cabrera (1958:227).

Genus *Chlamyphorus* Harlan, 1825

Chlamyphorus is a monotypic genus of small armadillos having a head-and-body length averaging less than 150 mm. The head shield is broad and the carapace is thin, flexible, attached only along spinal column at mid body, and separated from the rounded rump plate. The tail is flattened dorsoventrally and has a spatulate tip. The venter, legs, sides of the body, and dorsum beneath the carapace are clothed with long, silky white hair. The single species *Chlamyphorus truncatus* closely resembles the larger *Calyptophractus retusus*, from which it can be easily distinguished by the attachment of the carapace along the spinal column and by the flattened tail with a spatulate tip (the carapace attaches along the side of the body in *Calyptophractus retusus*, as in other armadillos, and the tail tip is pointed). The genus is known from the Pleistocene and Recent of South America (Marshall et al. 1984; McKenna and Bell 1997).

SYNONYMS:

Chlamyphorus Harlan, 1825:235; type species *Chlamyphorus truncatus* Harlan, 1825, by monotypy.
Dasypus: J. B. Fischer, 1829:394; part; not *Dasypus* Linnaeus, 1758.

Chlamydophorus Wagler, 1830:35; unjustified emendation of *Chlamyphorus* Harlan.
Chlamydiphorus Bonaparte, 1831:22; incorrect subsequent spelling of *Chlamyphorus* Harlan.
Chlamydephorus Lenz, 1831:xi; incorrect subsequent spelling of *Chlamyphorus* Harlan.
Chlamiphorus Contreras, 1973:216; incorrect subsequent spelling of *Chlamyphorus* Harlan.

Chlamyphorus truncatus Harlan, 1825
Pink Fairy Armadillo

SYNONYMS:

Chlamyphorus truncatus Harlan, 1825:235; type locality "Mendoza...interior of Chili, on the east of the Cordilleras, in lat. 33°25' and long. 69°47', in the province of Cuyo"; identified by Cabrera (1958:227) as Río Tunuyán, 33°25'S, 69°45'W, Mendoza, Argentina.
Chlamydophorus truncatus: Wagler, 1830:35; name combination.
Chl[amydophorus]. pt truncatus ornatus Lahille, 1895:10; type locality "la région nord de la République Argentine."
Chl[amydophorus]. truncatus minor Lahille, 1895:10; type locality "Bahia Blanca," Buenos Aires, Argentina.
Chl[amydophorus]. truncatus typicus Lahille, 1895:10; type locality Río Tunuyán, Mendoza, Argentina, as restricted by Cabrera (1958:227); based on *Chlamyphorus truncatus* Harlan.
Chlamyphorus truncatus minor: Yepes, 1931:108; name combination.
Chlamyphorus truncatus patquiensis Yepes, 1932:16; type locality "La Rioja, Patquía, región de Guayapa," Argentina.

DISTRIBUTION: *Chlamyphorus truncatus* is endemic to Argentina, where it has been found in the provinces of Catamarca, Córdoba, Buenos Aires, La Pampa, San Luis, Mendoza, San Juan, and La Rioja.

MARGINAL LOCALITIES (Map 64): ARGENTINA: La Rioja, La Rioja (MACN 24.46); Catamarca, La Guardia (Yepes 1929); Córdoba, Villa Dolores (Yepes 1929); Buenos Aires, Bahia Blanca (Lahille 1895); La Pampa, Carro Quemado (Crespo 1974); Mendoza, Malargüe (Yepes 1937); San Juan, La Iglesia (Yepes 1929).

SUBSPECIES: We treat *C. truncatus* as monotypic.

NATURAL HISTORY: *Chlamyphorus truncatus* is almost completely fossorial and lives in burrows dug with its large foreclaws in dry sandy soils (White 1880; Minoprio 1945). Stomach contents indicate a diet primarily of invertebrates, typically beetles and their larvae. Minoprio (1945) illustrated the external, internal, and skeletal morphology of *C. truncatus* in his review of the distribution, anatomy, and life history of the species. Litter size is one (Rood 1970). The karyotype is unknown.

Tribe Euphractini Winge, 1923

Small to medium-sized armadillos that have short pinnae, a broad head, and a short snout. The head shield is broad, forms a ledge over the orbits, and is separated from scapular shield by a single nuchal (cervical) band of large scutes. Body hair is conspicuous on the carapace, sides, and venter (members of this tribe are generally known as the hairy armadillos). The anterior ventral margin of the scapular shield has two small, incomplete rows of scutes. Scutes along the ventral margins of the carapace are rounded anteriorly and pointed posteriorly. The manus has four toes; the claws on the second and third digits are approximately equal in length. The auditory bullae and the external auditory meatus are ossified. The tribe is represented by the genera *Chaetophractus*, *Euphractus*, and *Zaedyus*.

Genus *Chaetophractus* Fitzinger, 1871

Wetzel (1985b), in the most recent review of the genus, recognized three species of *Chaetophractus*. Members of the genus inhabit the Puna zone of Chile and Bolivia, and the Gran Chaco of Argentina and eastern Chile south to approximately 50°S.

Dorsal hairs are prominent and vary in color from tan to buffy, except in *C. villosus*, which has black dorsal hair. Species of *Chaetophractus* are eternally similar in appearance to species of *Euphractus*, but can be distinguished by their shorter pinnae, a proportionally broader head shield, and a separate, movable band on the anterior margin of the scapular shield. Nine teeth in the upper jaw and ten in the lower on each side, plus the conspicuous constriction (in depth) of the jugal below the orbit, serve to separate *Chaetophractus* from *Zaedyus*, which has eight teeth above and nine below on each side, and has no constriction of the jugal. The genus is known from the Pliocene to Recent of South America (McKenna and Bell 1997).

SYNONYMS:

Loricatus Desmarest, 1804b:28; part.

Dasypus: Illiger, 1811:70; part; not *Dasypus* Linnaeus, 1758.

Tatus Olfers, 1818:220; part; incorrect subsequent spelling of *Tatu* Blumenbach, 1779.

Tatusia Lesson, 1827:309; part.

Euphractus: Burmeister, 1861:427; part; not *Euphractus* Wagler.

Chaetophractus Fitzinger, 1871b:268; type species *Dasypus villosus* (= *Loricatus villosus* Desmarest, 1804b:28), by subsequent designation (Yepes 1928:494).

Dasyphractus Fitzinger, 1871b:264; type species *Dasyphractus brevirostris* Fitzinger, 1871b, by monotypy.

Choetophractus Trouessart, 1898:1146; incorrect subsequent spelling of *Chaetophractus* Fitzinger.

KEY TO THE SPECIES OF *CHAETOPHRACTUS*.

1. Size large, length of head and body averaging 330 mm (261–344 mm); darkly colored and sparsely covered with long, sparse, black hair on carapace; head shield broadest, and scutes on head shield patterned (not flat and smooth); width:length head shield ratio averaging 0.95; posterior margin of head shield not straight, but reflecting the outline of individual marginal scutes . . . . *Chaetophractus villosus*

1'. Size smaller, length of head and body averaging less than 280 mm; paler with color of hair on carapace tan or white; head shield narrower, and scutes flat and smooth; posterior margin of head shield straighter 2

2. Size small, length of head and body less than 255 mm, condylonasal length less than 80 mm; hair tan and comparatively dense; pinnae long, extending back to the first fused row of scutes on scapular shield; head shield comparatively narrow (width:length ratio averaging 0.90) and posterior margin straight . . . . *Chaetophractus vellerosus*

2'. Size intermediate, head-and-body length more than 250 mm, condylonasal length more than 80 mm; hair pale tan, sometimes white; head shield proportionally broader than in most *C. villosus* and posterior margin with indentions between individual scutes; restricted to high Andean grasslands . . . . . . *Chaetophractus nationi*

Chaetophractus nationi (O. Thomas, 1894)
Andean Hairy Armadillo

SYNONYMS:

Dasypus nationi O. Thomas, 1894a:70; type locality "Orujo [= Oruro, Depto. Oruro], Bolivia."

[*Dasypus (Choetophractus)*] *nationi*: Trouessart, 1898:1146; name combination.

Chaetophractus nationi: Yepes, 1928:500; first use of current name combination.

DISTRIBUTION: *Chaetophractus nationi* is found in Bolivia and northern Argentina.

MARGINAL LOCALITIES (Map 65; localities listed from north to south): BOLIVIA: La Paz, Viacha (Wetzel 1985b); Oruro, Oruro (type locality of *Dasypus nationi* O. Thomas); Potosí, Uyuni (Grandidier and Neveu-Lemaire 1908). ARGENTINA (Carrizo et al. 2006): Jujuy, La Quiaca; Jujuy, Río Coyaguima; Jujuy, Salinas Grandes; Tucumán, La Cienega (not mapped); Tucumán, Km 76, Ruta 307.

SUBSPECIES: We consider *C. nationi* to be monotypic.

NATURAL HISTORY: Little is known of the natural history of *C. nationi*, except that it lives in pampa and

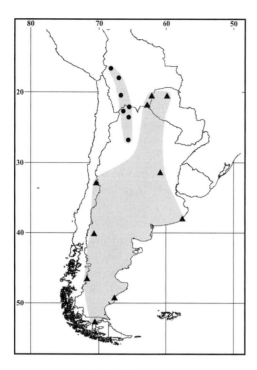

Map 65 Marginal localities for *Chaetophractus nationi* ●
and *Chaetophractus villosus* ▲

brushy habitats of the high-elevation Andean Puna zone.
The karyotype is unknown.

REMARKS: Wetzel (1985b) identified as *C. vellerosus*
specimens originally reported as *C. nationi* from Abra
Pampa, Jujuy, Argentina (O. Thomas 1919c; Yepes 1928),
even though the head shields are broad as in *C. nationi*.
Wetzel (1985b:28) also questioned the identifications of
animals reported as *C. nationi* from Chile by Mann (1945,
1978) and Tamayo (1973), writing that more material was
needed either to verify their identity as *C. nationi* or to es-
tablish their identity as a high-elevation subspecies of *C.
vellerosus*. Herein, we treat most of these Argentine and all
Chilean specimens as *C. vellerosus*.

Chaetophractus vellerosus (Gray, 1865)
Screaming Hairy Armadillo

SYNONYMS: See under Subspecies.

DISTRIBUTION: *Chaetophractus vellerosus* is known
from west-central Santa Cruz, Bolivia and the Chaco Boreal
of Bolivia and western Paraguay, south through Argentina
into the provinces of Mendoza, San Luis, La Pampa, Río
Negro, and Santa Cruz, and west to the Puna de Tarapacá,
Tarapacá, Chile.

MARGINAL LOCALITIES (Map 66): BOLIVIA: Santa
Cruz, Santa Cruz de la Sierra (type locality of *Dasypus
vellerosus* Gray); Santa Cruz, San Ramón (S. Anderson
1997). PARAGUAY: Boquerón, 24 km NW of Teniente
Ochoa (UCONN 18953); Boquerón, Fortín Nueve (MACN

43.60). ARGENTINA: Chaco, San Carlos (type locality of
Euphractus villosus desertorum Krumbiegel); Buenos Aires,
Puerto Indio (Carlini and Vizcaíno 1987); Santa Cruz (Wet-
zel 1985b, no specific locality, not mapped); Río Negro, Ar-
royo Pilcaniyeu (Wetzel 1985b); San Juan, Cañada Honda
(O. Thomas 1921e); La Rioja, Villa Unión (Yepes 1936);
Catamarca, Pastos Largos (Mares et al. 1997); Salta, San
Antonio de Los Cobres (Ojeda and Mares 1989). BOLIVIA:
Potosí, Uyuni (type locality of *Dasypus boliviensis* Gran-
didier and Neveu-Lemaire). CHILE: Tarapacá, Parinacota
(Mann 1978).

SUBSPECIES: We recognize two species of *C. vellerosus*.

C. v. pannosus (O. Thomas, 1902)
SYNONYMS:
Dasypus vellerosus pannosus O. Thomas, 1902c:244; type
locality "Cruz del Eje," Cordova, Argentina.
[*Dasypus (Chaetophractus) vellerosus*] *pannosus*: Troues-
sart, 1905:820; name combination.
Dasypus vallerosus pannosus O. Thomas, 1921e:221; in-
correct subsequent spelling of *Dasypus vellerosus* Gray.
Chaetophractus vellerosus pannosus: Yepes, 1928:500; first
use of current name combination.
Euphractus villosus desertorum Krumbiegel, 1940a:61;
type locality "San Carlos," Chaco, Argentina.

This subspecies is distributed from central Argentina
south into Mendoza Province.

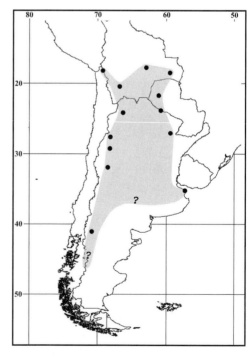

Map 66 Marginal localities for *Chaetophractus vellerosus* ●

C. v. vellerosus (Gray, 1865)

SYNONYMS:

Cryptophractus brevirostris Fitzinger, 1860:385,395; *nomen nudum.*

Dasypus vellerosus Gray, 1865:376; type locality "Santa Cruz de la Sierra," Santa Cruz, Bolivia.

Dasyphractus brevirostris Fitzinger, 1871b:264; type locality "Chili."

[*Dasypus (Choetophractus)*] *vellerosus*: Trouessart, 1898: 1146; name combination.

Dasypus boliviensis Grandidier and Neveu-Lemaire, 1908: 5; type locality "environs d'Uyuni (Bolivie), à 3,660 mètres d'altitude," Potosí, Bolivia; preoccupied by *Tatusia boliviensis* Gray, 1873a, a synonym of *Dasypus novemcinctus* Linnaeus.

Dasypus villerosus Grandidier and Neveu-Lemaire, 1908:6; incorrect subsequent spelling of *Dasypus vellerosus* Gray.

Chaetophractus vellerosus vellerosus: Yepes, 1928:500; first use of current name combination.

E[*uphractus*]. *vellerosus*: Krumbiegel, 1940a:56; name combination.

The nominate subspecies is known from eastern Bolivia, the Chaco Boreal of Paraguay, and northwestern Argentina.

NATURAL HISTORY: *Chaetophractus vellerosus* inhabits xeric habitats from low to high elevations. One animal uses several burrows located in its home range and uses dirt to plug the entrance of the burrow it is occupying. Individuals are nocturnal in the summer, diurnal in the winter, and apparently can forego drinking for long periods. The diet varies seasonally (Greegor 1985). The summer diet is predominantly insectivorous, but includes about 28% by volume of vertebrates. In winter, the diet includes fewer insects and vertebrates and consists chiefly of plant material, especially pods from *Prosopis* sp. Cassini (1993) experimentally showed that *C. vellerosus* is a systematic forager. Greegor (1985) provided ecological information on the species from his studies in the Monte desert of northern Argentina, and reviewed additional information from the literature. When handled, this species frequently emits a loud cry (Greegor 1975; Myers and Wetzel 1979), which is the basis for its common name as the Screaming Hairy Armadillo.

REMARKS: *Cryptophractus brevirostris*, as used by Fitzinger (1860:385, 395; 1871b:265), is a *nomen nudum.* The name was made available by Fitzinger (1871b:264) in the name combination *Dasyphractus brevirostris*.

We follow Wetzel (1985b) by identifying, as *C. vellerosus*, the high-elevation Puna zone armadillos from Argentina reported by O. Thomas (1919c) and Yepes (1928), and from Chile reported by Mann (1945, 1978) and Tamayo (1973) as *C. nationi*. Tamayo (1973) summarized the records of the Chilean specimens, and Mann (1978)

illustrated and gave natural history information on some of the same material.

Chaetophractus villosus (Desmarest, 1804)
Big Hairy Armadillo

SYNONYMS:

Dasypus octocinctus G. I. Molina, 1782:305: type locality "Nel Cujo," Chile (= provincia Mendoza, Argentina, see Tamayo 1968:6); preoccupied by *Dasypus octocinctus* Schreber, 1774b, a junior synonym of *Dasypus novemcinctus* Linnaeus.

lor[*icatus*]. *villosus* Desmarest, 1804b:28; based on "*Le tatou velu de d'Azara*"; therefore, the type locality is "Les Pampas" [of Buenos-Aires, Argentina], south of Río la Plata, between latitudes 35°S and 36°S (Azara 1801b: 164).

[*Dasypus*] *villosus*: G. Fischer, 1814:125; name combination.

T[*atus*]. *villosus*: Olfers, 1818:220; name combination.

Tatusia villosa: Lesson, 1827:312; name combination.

Dasypus (Tatusia) villosus: Rapp, 1852:10; name combination.

Dasypus [*(Euphractus)*] *villosus*: Burmeister, 1861:427; name combination.

Euphractus villosus: Gray, 1865:376; name combination.

Chaetophractus villosus: Fitzinger, 1871b:268; first use of current name combination.

[*Dasypus (Choetophractus)*] *villosus*: Trouessart, 1898: 1146; name combination.

D[*asypus*]. *pilosus* Larrañaga, 1923:343; type locality "campis bonaerensibus"; preoccupied by *Dasypus pilosus* (Fitzinger, 1856).

Euphractus (Chaetophractus) villosus: Moeller, 1968:514; name combination.

DISTRIBUTION: *Chaetophractus villosus* occurs from the Gran Chaco of Bolivia, Paraguay, and northern Argentina, southward into provincia Santa Cruz, Argentina, and from the regions of Valparaiso and BioBío south to Magallanes (Dazy Harbour) in Chile.

MARGINAL LOCALITIES (Map 65): PARAGUAY: Alto Paraguay, 60 km (by road) N of Madrejón (UMMZ 125579). ARGENTINA: Santa Fé, Esperanza (BM 1.2. 4.10); Buenos Aires, Mar del Plata (BM 12.2.17.10); Santa Cruz, San Julián (BM 16.8.16.1). CHILE: Magallanes, Dazy Harbour (Wetzel 1985b); Aisén, Chile Chico (Tamayo, 1973). ARGENTINA: Neuquén, Collon-Curá (BM 27.6.4.78). CHILE: Valparaiso, Río Colorado (Tamayo 1973). BOLIVIA: Tarija, 90 km SE of Villa Montes (UCONN 79574); Chuquisaca, Bolivian border near Sargento Rodriguez, Boquerón, Paraguay (S. Anderson 1997).

SUBSPECIES: We treat *C. villosus* as monotypic.

NATURAL HISTORY: Information on *C. villosus* comes primarily from the summary compiled by M. Roberts, Newman, and Peterson (1982) and is based on captives maintained in zoos. Captives ate a wide variety of foods and, seasonally, accumulated large quantities of fat. Captive *C. villosus* gave birth from February through December. Gestation is between 60 and 75 days. The female constructs a nest before giving birth to one or two young. The young open their eyes between 16 and 30 days after birth and the female lactates for approximately 55 days. Sexual maturity is attained by 9 months of age. The karyotype is $2n = 60$, FN = 88 or 90 (males have not been karyotyped; Benirschke, Low, and Ferm 1969; Jorge, Meritt, and Benirschke 1977; Jorge, Orsi-Souza, and Best 1985).

REMARKS: Tamayo (1968) pointed out that the oldest name applied to this species is *Dasypus octocinctus* G. I. Molina, 1782 (preoccupied by *Dasypus octocinctus* Schreber, 1774b). Cabrera (1958), Wetzel (1982, 1985b), and Wetzel and Mondolfi (1979) did not mention the name. We have not listed *C. villosus* from Salta and Jujuy provinces in Argentina because of the confusion in the literature between this species and *C. nationi*, although zoogeographically *C. villosus* is expected to occur in eastern provincia Salta.

Genus *Euphractus* Wagler, 1830

The monotypic genus *Euphractus* is found from southern Surinam into northern Argentina. Size is large and head-and-body lengths of adults usually exceeds 400 mm, and condylonasal lengths exceed 100 mm. This genus can be distinguished from other euphractines by the lack of a movable band of scutes at the anterior margin of the scapular shield; by the sparsely haired, pale yellow to reddish-tan carapace; the nine upper and ten lower teeth on each side; and by the presence of two to four openings for scent glands in the mid-dorsum of the pelvic shield. The genus is known from the middle Pleistocene to Recent of South America (Marshall et al. 1984; McKenna and Bell 1997).

SYNONYMS:

Dasypus Linnaeus, 1758:50; part.

Loricatus Desmarest, 1804b:28; part.

Tatus Olfers, 1818:220; part; incorrect subsequent spelling of *Tatu* Blumenbach.

Euphractus Wagler, 1830:36; type species *Dasypus sexcinctus* Linnaeus, 1758, by subsequent designation (Palmer 1904:278).

Encoubertus McMurtrie 1831:163; part; proposed as a subgenus of *Dasypus*; included *Dasypus sexcinctus* Linnaeus, 1758, and *D. octodecimcinctus* Erxleben, 1777; we select *Dasypus sexcinctus* Linnaeus, 1758, as the type species (see Remarks).

Pseudotroctes Gloger, 1841:113; type species *Dasypus setosus* Wied-Neuwied, 1826, by monotypy.

Scleropleura Milne-Edwards, 1871:178; type species *Scleropleura bruneti* Milne-Edwards, 1871, by monotypy.

Scelopleura Trouessart, 1898:1141; incorrect subsequent spelling of *Scleropleura* Milne-Edwards.

REMARKS: Moeller (1968, 1975) considered *Chaetophractus* to be a synonym of *Euphractus*; see review by Redford and Wetzel (1985). The type species of *Euphractus* was designated by Palmer (1904), not O. Thomas (1911b), as misstated by Redford and Wetzel (1985). Kretzoi and Kretzoi (2000) gave the type species of *Encoubertus* McMurtrie as *Dasypus sexcinctus* Linnaeus. We also designate *Dasypus sexcinctus* Linnaeus as the type species of *Encoubertus* McMurtrie, because of the possibility that the nomenclatural acts in Kretzoi and Kretzoi (2000) will be set aside (see Remarks under the genus *Dasypus*).

Euphractus sexcinctus (Linnaeus, 1758)
Six-banded Armadillo

SYNONYMS: See under Subspecies.

DISTRIBUTION: *Euphractus sexcinctus* has a disjunct distribution. A smaller northern population unit occurs in savanna habitats in southern Surinam and adjacent Pará, Brazil, and east into the state of Amapá. The more extensive distribution is in the drier Cerrado habitats of eastern and southern Brazil, southeastern Bolivia, Paraguay, Uruguay, and northern Argentina as far south as the provinces of Buenos Aires in the east and Catamarca in the west.

MARGINAL LOCALITIES (Map 67): *Northern distribution*: SURINAM: Sipaliwini, Sipaliwini Savanna (Husson 1973). BRAZIL: Amapá, Fazenda Itapuã (Silva-Júnior and Nunes 2001); Amapá, Ferreira Gomes (Pine 1973b); Pará, Parú Savanna (Wetzel 1985b). *Main distribution*: BRAZIL (Silva-Júnior and Nunes 2001, except as noted): Pará, Peixe bois (Krumbiegel 1940a); Maranhão, Cajual Island (Hass, Rodrigues, and Oliveira 2003); Maranhão, Caju Island (Hass, Rodrigues, and Oliveira 2003); Ceará, Parque Nacional de Ubajara (Guedes et al. 2000); Ceará, San Antônio (type locality of *Scleropleura bruneti* Milne-Edwards); Paraíba, Paraíba; Pernambuco, Poção (Moojen 1943); Bahia, Ipitinga (Krumbiegel 1940a); Bahia, Vitória da Conquista; Espírito Santo, Santa Teresa; Rio de Janeiro, Fazenda da Lapa; São Paulo, Piquete (BM 1.6.6.82); São Paulo, Itararé; Rio Grande do Sul, São Lourenço. URUGUAY: Río Negro, Bopicuá (J. C. González 1973). ARGENTINA: Buenos Aires, General Lavalle (MCZ 19502); Tucumán, Tapia (type locality of *Dasypus sexcinctus tucumanus* O. Thomas); Jujuy, Zapla (Yepes 1944). BOLIVIA: Santa Cruz, Buena Vista (AMNH 61802); Beni, Bresta (Lönnberg 1942). PERU: Cusco, Pagoreni (Boddicker, Rodríguez, and Amanzo

Map 67 Marginal localities for *Euphractus sexcinctus* ●

1999). BRAZIL: Mato Grosso, Utiarity (A. Miranda-Ribeiro 1914); Pará, Cachimbo (Silva-Júnior and Nunes 2001); Maranhão, Posto Indígena Awá (Silva-Júnior, Fernandes, and Cerqueira 2001).

SUBSPECIES: We tentatively recognize five subspecies of *E. sexcinctus*.

E. s. boliviae (O. Thomas, 1907)
SYNONYMS:

Dasypus sexcinctus boliviae O. Thomas, 1907:165); type locality "Near Santa Cruz de la Sierra," Santa Cruz, Bolivia.

Euphractus sexcinctus boliviae: Yepes, 1928:491 [p. 31 in separate]; first use of current name combination.

This subspecies is in Bolivia and northwestern Argentina.

E. s. flavimanus (Desmarest, 1804)
SYNONYMS:

lor[*icatus*]. *flavimanus* Desmarest, 1804b:28; no locality given; based on "*Le tatou poyou de d'Azara*" (1801a) and "*l'encoubert*" of Buffon (1763); type locality Paraguay, on the basis of the reference to "*le tatou poyou*" of Azara (1801a:142).

[*Dasypus*] *flavipes* G. Fischer, 1814:122; type locality "Paraguay."

Dasypus gilvipes Illiger, 1815:108; *nomen nudum*.

Das[*ypus*]. *gilvipes* Lichtenstein, 1818b:215; no locality given; based on "*Tatou-poyou*" of Azara (1801a:142); therefore, the type locality is Paraguay.

T[*atus*]. *gilvipes* Olfers, 1818:220; type localities "Paraguay, Brasilien, Guiana."

Dasypus pilosus Olfers, 1818:220; *nomen nudum*.

Dasypus encoubert Desmarest, 1822:370; type locality "Le Paraguay."

[*Dasypus*] *poyú* Larrañaga, 1923:243; type locality not given but Uruguay is implied (p. 242); based on *Dasypus sexcinctus*: Gmelin, 1788 (=*Dasypus sexcinctus* Linnaeus), and Azara's (1802b:118) "*Poyú*."

Euphractus sexcinctus gylvipes Minoprio, 1945:10; name combination and incorrect subsequent spelling of *Dasypus gilvipes* Lichtenstein.

This subspecies occurs in southern Brazil, Paraguay, Uruguay, and northeastern Argentina.

E. s. setosus (Wied-Neuwied, 1826)
SYNONYMS:

D[*asypus*]. *setosus* Wied-Neuwied, 1826:520; type localities "in den grossen Campos Geraes und den angränzenden Gegenden des Sertong" (p. 528); restricted to "Sertão da Bahia," Brazil, by Ávila-Pires (1965:13).

Euphractus setosus: Fitzinger, 1871b:251; name combination.

Scleropleura bruneti Milne-Edwards, 1871:177; type locality "du village de San-Antonio, dans la province de Ceara," Brazil.

[*Scelopleura*] *Bruneti*: Trouessart, 1898:1141; name combination.

This subspecies is in eastern and southeastern Brazil.

E. s. sexcinctus (Linnaeus, 1758)
SYNONYMS:

[*Dasypus*] *sexcinctus* Linnaeus, 1758:51; type locality "America meridionali"; restricted to Pará, Brazil, by O. Thomas (1907:165).

Tatus sexcinctus: Schinz, 1824:pl. 113, Fig. 1; name combination.

Euphractus mustelinus Fitzinger, 1871b:259; part; type localities "Süd-und Mittel-Amerika."

[*Dasypus (Dasypus)*] *sexcinctus*: Trouessart, 1898:1145; name combination.

The nominate subspecies inhabits the savannas of southern Surinam and the adjacent Brazilian states of Pará and Amapá.

E. s. tucumanus (O. Thomas, 1907)
SYNONYMS:

Dasypus sexcinctus tucumanus O. Thomas, 1907:166; type locality "Tapia, Tucuman," Argentina.

Euphractus sexcinctus tucumanus: Yepes, 1928:492 [p. 32 in separate]; first use of current name combination.

This subspecies is known only from northwestern Argentina.

NATURAL HISTORY: *Euphractus sexcinctus* usually is found in savannas and at forest edges, where it may be seen during the day. Individuals construct burrow systems, and use several. Their omnivorous diet includes carrion, small vertebrates, insects, tubers, and palm nuts. Bezerra, Rodrigues, and Carmignotto (2001) found the remains of four rodents (*Calomys* sp.) in the stomach of a specimen from the state of Goiás, Brazil. Litter size ranges from one to three, and both sexes can occur in the same litter, indicating that the species is polyovular. Gestation is 60 to 64 days and young open their eyes 22 to 25 days after birth. Sexual maturity can be attained at 9 months (Schaller 1983; Redford and Wetzel 1985). Bechara et al. (2002) reported two species of ticks found on *E. sexcinctus* in Brazil. The karyotype is $2n = 58$, FN $= 102$ (Benirschke, Low, and Ferm 1969; Hsu and Benirschke 1971a; Barroso and Seuánez 1991). Redford and Wetzel (1985) summarized information on *E. sexcinctus* in their *Mammalian Species* account.

REMARKS: Goeldi and Hagmann (1904:97) used *Dasypus setosus* as the senior synonym of *D. sexcinctus* without comment. Some authors have reported *E. sexcinctus* from Chile; however, Tamayo (1973) claimed the reports were based on misidentified *Chaetophractus villosus*. Milne-Edwards (1871) based *Scleropleura bruneti* on an aberrant specimen as confirmed by his illustration of the carapace (also see Milne-Edwards 1872a).

Husson (1978:254) interpreted O. Thomas's (1911b: 141) use of Para as the former name of the city of Belém, Pará, Brazil. O. Thomas, however, sometimes use "Para" to refer to the state of Pará and not to the city (e.g., O. Thomas 1907:165).

Genus *Zaedyus* Ameghino, 1889

This monotypic genus of small (head-and-body length averaging 277 mm), hairy armadillos can be distinguished from the similar *Chaetophractus vellerosus* by its sharply pointed marginal scutes and the extremely short pinnae (average 13.4 mm, versus 30.8 mm in *C. vellerosus*). *Zaedyus* is known from the late Pliocene to Recent of South America (Marshall et al. 1984; McKenna and Bell 1997).

SYNONYMS:

Loricatus Desmarest, 1804b:28; part.

Dasypus: G. Fischer, 1814:127; part; not *Dasypus* Linnaeus.

Tatus Olfers, 1818:220; part; incorrect subsequent spelling of *Tatu* Blumenbach.

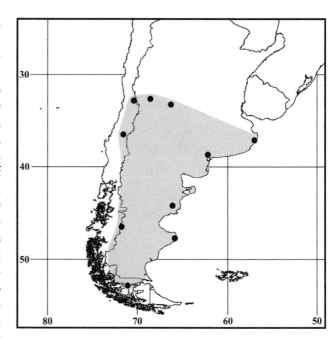

Map 68 Marginal localities for *Zaedyus pichiy* ●

Tatusia Lesson, 1827:309; part.

Euphractus Wagler, 1830:36; part.

Chaetophractus Fitzinger, 1871b:268; part.

Zaedypus Ameghino, 1889:867; type species *Dasypus minutus* Desmarest, 1822, by original designation.

Zaedypus Lydekker, 1890:50; incorrect subsequent spelling of *Zaedyus* Ameghino.

Zaedius Lydekker, 1894a:34; incorrect subsequent spelling of *Zaedyus* Ameghino.

Zaëdius Krumbiegel, 1940a:63; incorrect subsequent spelling of *Zaedyus* Ameghino.

Zaedyus pichiy (Desmarest, 1804)
Pichi

SYNONYMS: See under Subspecies.

DISTRIBUTION: *Zaedyus pichiy* has been reported from the Argentine provinces of Mendoza and Buenos Aires, south through Argentina and eastern Chile to the Straits of Magellan.

MARGINAL LOCALITIES (Map 68): ARGENTINA: Mendoza, Lavalle (Yepes 1944); San Luis, San Luis (MACN 38.253); Buenos Aires, Juancho (Frechkop and Yepes 1949); Buenos Aires, Bahía Blanca (restricted type locality of *Loricatus pichiy* Desmarest); Chubut, La Concepción (O. Thomas 1929); Santa Cruz, Port Désiré (Desmarest 1822). CHILE: Magallanes, Mina Pecket (Texera 1973); Aisén, Chile Chico (Mann 1978); Biobío, San Fabian de Alico (Wolffsohn 1921); Valparaíso, Río Colorado (Wolffsohn 1921).

SUBSPECIES: We recognize two subspecies of *Z. pichiy*.

Z. p. caurinus O. Thomas, 1928
SYNONYMS:

Dasypus minutus: Burmeister, 1879:440; part; not *Dasypus minutus* Desmarest.

Zaedyus pichiy caurinus O. Thomas, 1928b:526, type locality "Mendoza," Mendoza, Argentina.

This subspecies occurs in eastern Chile and in western Argentina, from the province of Mendoza south into Nahuel Huapi.

Z. p. pichiy (Desmarest, 1804)
SYNONYMS:

Dasypus quadricinctus G. I. Molina, 1782:302; type locality "Chili"; preoccupied by *Dasypus quadricinctus* Linnaeus, 1758, a synonym of *Tolypeutes tricinctus* Linnaeus, 1758.

lor[icatus]. *pichiy* Desmarest, 1804b:28; based on "*Le tatou pichiy de d'Azara*" (1801a:192); therefore, the type localities are "les Pampas au Sud de Buenos-Ayres, depués le parallèle de 36ᵉ. degré de latitud méridionale, jusqu'á la Terre des Patagons," Argentina (Azara 1801a:192); type locality restricted to Bahia Blanca, Buenos Aires, Argentina, by Cabrera (1958:218; not Lesson 1842:150).

[*Dasypus*] *ciliatus* G. Fischer, 1814:127; type localities "ad meridiem *Boni Aeris*, (Buenos Ayres) inde a parallela gradus 36 latitudinis meridionalis usque ad terram *Patagonum*"; junior objective synonym of *Loricatus pichiy* Desmarest.

T[atus] *fimbriatus* Olfers, 1818:220; type locality "Paraguay, en den Pampas"; based on "*T. Pichiy*" of Azara (1801a), therefore a junior objective synonym of *Loricatus pichiy* Desmarest.

Dasypus patagonicus Desmarest, 1819a:491; based on "*Le Tatou pichiy*, d'Azara" (1801a); therefore, a junior objective synonym of *Loricatus pichiy* Desmarest.

Dasypus minutus Desmarest, 1822:371; type localities "au Sud de Buenos-Ayres, depuis le parallèl de 36ᵉ. degré de latitude méridionale jusqu'a la Terre des Patagons. . . . [et] du port Désiré . . . ," Argentina; we consider Port Désiré (= Puerto Deseado, Santa Cruz) to be the restricted type locality.

Tatusia minuta: Lesson, 1827:312; name combination.

Euphractus marginatus Wagler, 1830:36; based on "*Tatou pichiy*" of Azara (1801a); therefore, the type localities are "les Pampas au Sud de Buenos-Ayres, depués le parallèle de 36ᵉ. degréde latitud Méridionale, jusqu'á la terre des Patagons," Argentina (Azara 1801a:192);

a junior objective synonym of *Loricatus pichiy* Desmarest.

Dasypus [(Euphractus)] minutus: Burmeister, 1861:427; part; name combination.

Euphractus minutus: Gray, 1865:376; name combination.

Chaetophractus minutus: Fitzinger, 1871b:272; name combination.

Zaedyus minutus: Ameghino, 1889:868; generic description and name combination.

Zaedyus cilliatus J. A. Allen, 1901b:183; name combination and incorrect subsequent spelling of *Dasypus ciliatus* G. Fischer.

[*Zaedius*] *ciliatus*: Trouessart, 1905:819; name combination.

Zaedyus pichiy: Osgood, 1919:33; first use of current name combination.

D[*asypus*]. *Australis* Larrañaga, 1923:343; type locality "australem plagam bonaerensem"; based on Azara's (1802b:158) "*Pichiy*."

Euphractus (Zaedyus) pichiy: Moeller, 1968:514; name combination.

The nominate subspecies occurs in the Pampas of eastern and southeastern Argentina.

NATURAL HISTORY: This species occurs farther south than any of the other armadillos, mostly in grassland habitats. Like other hairy armadillos, it is omnivorous and eats insects, ants, and other invertebrates, carrion, and plant material. In captivity, *Z. pichiy* becomes torpid at low ambient temperatures and is reported to hibernate in the wild. One to three young are born after a 60-day gestation period. The young weigh 95 to 115 g at birth, are weaned by 6 weeks of age, and reach sexual maturity at 9 to 12 months (Meritt 1973; Redford 1986). The karyotype is $2n = 62$, FN = 94 (Meritt, Low, and Benirschke 1973; Jorge, Meritt, and Benirschke 1977; Jorge, Orsi-Souza, and Best 1985).

REMARKS: *Dasypus quadricinctus* G. I. Molina, 1782:305, is the oldest name applied to *Loricatus pichiy* Desmarest, 1804b, with which G. I. Molina (1810:25) later equated his *D. quadricinctus*. However, Molina's name is unavailable because, as mentioned by Tamayo (1968:5), it is a junior homonym of *Dasypus quadricinctus* Linnaeus, 1758, which is a synonym of *Tolypeutes tricinctus* (Linnaeus, 1758).

Olfers (1818) based *Tatus fimbriatus* on Azara's (1801a) "*T. Pichiy*," whose distribution Azara (1801a:192) gave as the pampas south of Buenos Aires from 36 degrees latitude to as far as Patagonia. Therefore, Olfers (1818:220) erred when he gave the distribution as Paraguay. There are no confirmed records for *Z. pichiy* north of Argentina.

Squarcia and Casanave (1999) were able to discriminate between samples representing the two subspecies by analyzing 11 cranial measurements. They confirmed that the skulls of *Z. p. caurinus* are smaller and have a shorter rostrum than those of *Z. p. pichiy*.

Subfamily Tolypeutinae Gray, 1865
Tribe Priodontini Gray, 1873

The Priodontini comprise small to extremely large armadillos having a rounded, blunt snout; thick, fleshy pinnae; and relatively large eyes. The ovoid, dome-shaped carapace is separated from the head shield by three short rows of nuchal scutes. The scapular and pelvic shields are separated by 11 to 14 movable bands. The body hair is sparse. The manus has five toes; digit III has a large, long, and scimitar- or sickle-shaped claw. Frontal bones have a distinct dome-like swelling, and only the tympanic ring of each bulla is ossified. The tribe is represented by the genera *Cabassous* and *Priodontes*.

Genus *Cabassous* McMurtrie, 1831

This genus of four species occurs from southern Mexico into northern Argentina. These are medium-sized armadillos (head-and-body length 300–490 mm, tail length 90–200 mm). The carapace has dorsal plates arranged in transverse rows over its entire length. The manus bears five claws, of which the middle (digit III) is enlarged and sickle-shaped. The snout is short and broad, the eyes are proportionally smaller than in *Priodontes*, and the pinnae are moderately large and funnel shaped. The distinctive tail is slender and either lacks armor or bears small, thin, widely spaced plates. The number of teeth is variable, ranging from seven to ten above and eight to nine below on each side.

Members of the genus are sometimes misidentified as immature *Priodontes*, but can easily be distinguished by the lack of obvious and extensive covering of scutes on the tail. Available evidence suggests that species of *Cabassous* are nocturnal, usually solitary, and give birth to single young (Meritt 1985). The genus is known from the middle Pleistocene of Brazil (Marshall et al. 1984; McKenna and Bell 1997). The most recent revision of *Cabassous* was by Wetzel (1980) and most of this account is based on that report.

SYNONYMS:

Dasypus Linnaeus, 1758:50; part.

Tatus Olfers, 1818:220; part; incorrect subsequent spelling of *Tatu* Blumenbach, 1779.

Xenurus Wagler, 1830:36; type species *Dasypus gymnurus*: Wied-Neuwied, 1826 (= *Xenurus squamicaudis* Lund, 1845a) by monotypy; preoccupied by *Xenurus* Boie, 1826 (Aves).

Tatusia Lesson, 1827:309; part.

Cabassous McMurtrie, 1831:164; type species *Dasypus unicinctus* Linnaeus, 1758, by monotypy; proposed as a subgenus of *Dasypus* Linnaeus.

Arizostus Gloger, 1841:114; type species *Dasypus gymnurus* (= *Tatus gymnurus* Olfers, 1818), by monotypy.

Tatoua Gray, 1865:378; type species *Dasypus unicinctus* Linnaeus, 1758, by monotypy; proposed as a subgenus of *Dasypus* Linnaeus.

Ziphila Gray, 1873a:22; type species *Ziphila lugubris* Gray, 1873a, by monotypy.

Lysiurus Ameghino, 1891:254; replacement name for *Xenurus* Wagler, 1830, which is preoccupied by *Xenurus* Boie, 1826 (Aves).

Cabassus Trouessart, 1905:820; incorrect subsequent spelling of *Cabassous* McMurtrie.

Cabassus Neveu-Lemaire and Grandidier, 1911:103; incorrect subsequent spelling of *Cabassous* McMurtrie.

KEY TO THE SPECIES OF *CABASSOUS*:

1. Size larger, length of head and body 410–490 mm, condylonasal length greater than 100 mm; palate and rostrum actually and proportionally long, more than 48 mm and 60 mm, respectively; palate extends behind a line across anterior margins of zygomatic rami of squamosals; modal number of cephalic scutes 48 . . . . *Cabassous tatouay*

1'. Size smaller, length of head and body less than 350 mm and condylonasal length less than 90 mm (in species occurring with *C. tatouay*); palate and rostrum short, less than 45 mm and 51 mm, respectively; palate not extending behind a line connecting anterior margins of zygomatic rami of squamosals; number of cephalic scutes usually exceeds 50. 2

2. Smallest species, length of head and body averages 303 mm, condylonasal length averages 69.8 mm, and interorbital breadth less than 22 mm; pinnae shorter than 20 mm (from notch) and have fleshy anterior edges; mandibles distinctly curved on both dorsoventral and mediolateral axes; all but first and last teeth unusually broad (width:length ratio of fourth through seventh maxillary teeth from 1.2 to 1.5) . . . . *Cabassous chacoensis*

2'. Medium-sized species, length of head and body usually more than 320 mm, condylonasal length averaging more than 75 mm, and interorbital breadth greater than 23 mm; pinnae longer than 25 mm (from notch) and lack thickened anterior edge; mandibles not distinctly curved; teeth usually narrow along entire toothrow 3

3. Skull narrow, interorbital breadth averages 24.3 mm, zygomatic breadth averages 40.8 mm; body and cranial dimensions less than in adjacent population of *C. unicinctus*; range in South America north and west of

the Andes (includes inter-Andean valleys) in northern Colombia and Venezuela . . . . *Cabassous centralis*

3′. Skull broad, interorbital breadth averages 26.6 mm, zygomatic breadth usually greater than 42.0 mm; body and cranial dimensions generally larger than those dimensions in adjacent populations of *C. centralis*; range is east of the Andes . . . . *Cabassous unicinctus*

Cabassous centralis (Miller, 1899)
Northern Naked-tailed Armadillo

SYNONYMS:

Dasypus gymnurus: Frantzius, 1869:309; not *Tatus gymnurus* Olfers.

X[enurus]. hispidus: True, 1896:345; not *Dasypus hispidus* Burmeister.

Dasypus gymnurus: Alfaro, 1897:47; not *Tatus gymnurus* Olfers.

Xenurus gymnurus: J. A. Allen, 1897a:43; not *Tatus gymnurus* Olfers.

[*Lysiurus (Lysiurus)*] *unicinctus*: Trouessart, 1898:1147; part; not *Dasypus unicinctus* Linnaeus.

[*Lysiurus (Lysiurus)*] *hispidus*: Trouessart, 1898:1147; part; not *Dasypus hispidus* Burmeister.

Tatoua (Ziphila) centralis Miller, 1899a:4; type locality "Chamelecon," Cortés, Honduras.

Tatoua (Ziphila) lugubris: Miller, 1899a:6; not *Ziphila lugubris* Gray.

C[abassous]. (Ziphila) centralis: Palmer, 1899:72; name combination.

Cabassous hispidus: Bangs, 1900:89; not *Dasypus hispidus* Burmeister.

Cabassous (Ziphila) lugubris: J. A. Allen, 1904f:421; not *Ziphila lugubris* Gray.

[*Cabassous (Ziphila)*] *lugubris*: Trouessart, 1905:821; part; not *Ziphila lugubris* Gray.

Cabassous hispidus: Neveu-Lemaire and Grandidier, 1911: 106, footnote; not *Dasypus hispidus* Burmeister.

[*Cabassous*] *lugubris*: Yepes, 1928:467; part; not *Ziphila lugubris* Gray.

Cabassous lugubris: Moeller, 1968:420; part; not *Ziphila lugubris* Gray.

DISTRIBUTION: In South America, *C. centralis* is known from northern Colombia (including northern inter-Andean valleys) east of the Chocó region and north of the Andes eastward across Venezuela as far as the state of Monagas. The species also occurs in Central America and southern Mexico.

MARGINAL LOCALITIES (Map 69: from Wetzel 1980, except as noted): COLOMBIA: Magdalena, Bonda; La Guajira, Sierra Negra. VENEZUELA: Monagas, Caicara (USNM 296613); Zulia, Maracaibo; Táchira, La Fría. COLOMBIA: Antioquia, Antioquia.

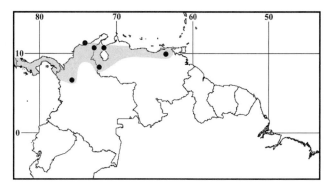

Map 69 Marginal localities for *Cabassous centralis* ●

SUBSPECIES: We consider *C. centralis* to be monotypic.

NATURAL HISTORY: This nocturnal species preferentially feeds on terrestrial ants and termites. It is highly specialized for burrowing and apparently much of its activity is below ground. Only one young is born after a gestation of unknown length. Wild-caught *C. centralis* from Panama weighed from 2.0 to 3.5 kg (Meritt 1985). When disturbed, the animals produce a variety of sounds, including a low buzz or growl and "low gurgling squeals" (Wetzel 1980:332). The karyotype is $2n = 62$, FN = 74 (Benirschke, Low, and Ferm 1969; Hsu and Benirschke 1969).

REMARKS: Most of the misidentifications of *C. centralis* listed earlier among the Synonyms are of specimens from Central America and northwestern Colombia.

Cabassous chacoensis Wetzel, 1980
Chacoan Naked-tailed Armadillo

SYNONYMS:

Xenurus gymnurus: Lahille, 1899:204; not *Tatus gymnurus* Olfers.

Cabassous loricatus: Yepes, 1935:441; part; not *Cabassous loricatus* (J. A. Wagner, 1855).

Cabassous loricatus: Cabrera, 1958:219; part; not *Cabassous loricatus* (J. A. Wagner).

Cabassous loricatus: Moeller, 1968:420; part; not *Cabassous loricatus* (J. A. Wagner).

Cabassous chacoensis Wetzel, 1980:335; type locality "Paraguay, Depto. Presidente Hayes, 5–7 km W Estancia Juan de Zalazar."

DISTRIBUTION: *Cabassous chacoensis* is endemic to the Gran Chaco of western Paraguay and northern Argentina. The species may also occur in southeastern Bolivia, and adjacent Mato Grosso do Sul, Brazil (Wetzel 1980).

MARGINAL LOCALITIES (Map 70; from Wetzel 1980): PARAGUAY: Boquerón, Filadelfia; Presidente Hayes, 5–7 km W of Juan de Zalazar (type locality of *Cabassous chacoensis* Wetzel). ARGENTINA: Santiago del Estero, Quimilí; Formosa, Ingeniero Guillermo N. Juárez.

SUBSPECIES: *Cabassous chacoensis* is monotypic.

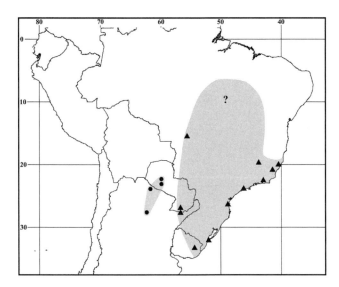

Map 70 Marginal localities for *Cabassous chacoensis* ● and *Cabassous, tatouay* ▲

NATURAL HISTORY: This species appears to be confined to the xeric habitats of the Gran Chaco. Little else is known about its natural history. When handled, males emit a loud grunt; females are not known to vocalize (Wetzel 1982).

REMARKS: A zoological park specimen (MACN 4.388), listed by Yepes (1935:444) as *Cabassous loricatus* from Mato Grosso, is the only evidence of *C. chacoensis* from Brazil. All other specimens examined by Wetzel (1980) from Mato Grosso and Mato Grosso do Sul proved to be *C. unicinctus*. We have not mapped the Brazilian record because we suspect it is an error. Moeller's (1968:420) distribution map showing *C. lugubris* in northwestern Argentina was based on specimens of *C. chacoensis*. Mares et al. (1996:107–08) suggested that *C. chacoensis* occurs in provincia Tucumán, and cited a personal observation of C. C. Olrog. Specimens vouchering the occurrence of the species in Tucumán are lacking.

Cabassous tatouay (Desmarest, 1804)
Greater Naked-tailed Armadillo

SYNONYMS:

Lor[icatus] tatouay Desmarest, 1804b:28; type locality not mentioned; considered to be Paraguay by Cabrera (1958:219), who restricted it to "a 27 de lat. sur," because he also linked the name to Azara's (1801a:155) "*Tatou tatouay*" by tautonomy.

[*Dasypus*] *dasycercus* G. Fischer, 1814:124; type locality "Paraquaia."

[*Dasypus*] *gymnurus* Illiger, 1815:108; *nomen nudum*.

T[atus] gymnurus Olfers, 1818:220; type locality "Paraguay"; based on "*T. tatouay*" (the *Tatou tatouay* of Azara 1801a:155).

Tatusia tatouay: Lesson, 1827:311; name combination.

Dasypus gymnurus: Rengger, 1830:290; name combination.

Xenurus nudicaudis Lund, 1839b:231; *nomen nudum.*

Xenurus nudicaudus Lund, 1839c:81 [1841a:143]; type locality "Rio das Velhas," Minas Gerais, Brazil (see Remarks).

Dasypus 12-cinctus: Burmeister, 1854:282; variant spelling of, but not *Dasypus duodecimcinctus* Schreber.

Xenurus tatouay: P. Gervais, 1855:254; name combination.

Xenurus unicinctus: Gray, 1865:378; part; not *Dasypus unicinctus* Linnaeus.

Xenurus gymnurus: Fitzinger, 1871b:242; name combination.

Tatoua unicinctus: Miller, 1899a:2; not *Dasypus unicinctus* Linnaeus.

Xenurus duodecimcinctus: Winge, 1915:32; not *Dasypus duodecimcinctus* Schreber.

D[asypus]. nudi-cauda Larrañaga, 1923:343; type locality "provincia paracuarensi"; based on Azara's (1802b: 131) "*Tatuaí.*"

Cabassous tatouay: Cabrera, 1958:219; first use of current name combination.

Cabassous gymnurus: Ximénez, Langguth, and Praderi, 1972: 14; name combination.

Cabassous duodecimcinctus: Paula-Couto, 1973:267; name combination.

DISTRIBUTION: *Cabassous tatouay* is found in Uruguay, the Argentine provinces of Misiones and Buenos Aires, southern Paraguay east of the Río Paraguay, and in Brazil from the states of Pará, Mato Grosso, Goiás, Minas Gerais, and Espírito Santo, south to Rio Grande do Sul.

MARGINAL LOCALITIES (Map 70: from Wetzel 1980, except as noted): BRAZIL: Pará (exact location unknown, not mapped); Minas Gerais, Lagoa Santa; Espírito Santo, Santa Lúcia Biological Station (Srbek-Araujo and Chiarello 2005); Espírito Santo, Engenheiro Reeve; Rio de Janeiro, Teresôpolis; São Paulo, Paranapiacaba; Santa Catarina, Joinvile; Rio Grande do Sul, Rio Grande. URUGUAY: Lavalleja, Río Olimar Chico. ARGENTINA: Buenos Aires (exact location unknown, not mapped; Moeller 1968); Corrientes, 15 km W of Ituzaingó (Massoia and Chebez 1985). PARAGUAY: Misiones, Curupayty. BRAZIL: Mato Grosso, Chapada.

SUBSPECIES: We treat *C. tatouay* as monotypic.

NATURAL HISTORY: This species appears to prefer savanna habitats and is highly fossorial. In southeastern Brazil, *C. tatouay* dug single-entrance burrows into active termite mounds and was never found to reuse a burrow (T. S. Carter and Encarnaçao 1983). The largest member of the genus; two specimens of *C. tatouay* from eastern Paraguay weighed over 5 kg each (Meritt 1985). Barroso and Seuánez

(1991) reported a $2n = 50$, $FN = 71$ ($FN = 68$) karyotype for a male *C. tatouay*. The autosomal complement consists of 4 pairs of large metacentrics, 6 pairs of smaller metacentrics and submetacentrics, and 14 pairs of telocentric chromosomes. The X is a small submetacentric, and the Y chromosome is minute and presumably acrocentric.

REMARKS: Desmarest (1803b:434) first described this species, under the French vernacular *"Tatou tatouay,"* in volume 21 of the "Nouveau Dictionnaire." He later (1804b:28) gave it the name *Loricatus tatouay* in his "Tableau Methodique," an independently paginated section in volume 24 of the "Nouveau Dictionnaire" published the following year.

Additional names in Cabrera's (1958) synonymy of *C. tatouay* are based on misidentifications by other authors, some of whom treated *C. tatouay* as a junior synonym of *C. unicinctus*. Wetzel (1980) gave a more extensive list of names that have been misapplied to this species.

The bones illustrated in plates 12 (Fig. 3) and 13 (Fig. 1) constitute the sample upon which the name *Xenurus nudicaudus* [= *nudicaudis*] Lund, 1839c, was based. Lund later (1845b:pl. 4) illustrated a skull he identified as from the same taxon; it may have come from the same cave deposits near Lagoa Santa. Nevertheless, Wetzel's (1980:350) selection of that skull (ZMUC L–28) as the lectotype of *Xenurus nudicaudis* is invalid because there is no evidence that the skull was part of the original sample on which Lund based his name.

Aside from being the largest of the *Cabassous*, *C. tatouay* has a higher ratio of palatal length to maxillary toothrow length (1.81; Wetzel 1980:350) than do other species. The more elongated palate is also diagnostic among living species, and extends posteriorly behind the level of the anterior margins of the squamosal root of the zygomatic arches.

Cabassous unicinctus (Linnaeus, 1758)
Southern Naked-tailed Armadillo

SYNONYMS: See under Subspecies.

DISTRIBUTION: *Cabassous unicinctus* has been recorded east of the Andes in Venezuela, the Guianas, Brazil, and in the lowlands of eastern Colombia, Ecuador, Peru, and Bolivia. The southern limits of the distribution are in southern Mato Grosso do Sul and southern Minas Gerais, Brazil.

MARGINAL LOCALITIES (Map 71: from Wetzel 1980, except as noted): VENEZUELA: Distrito Federal, San Julián; Sucre, Cumaná; Monagas, Los Araguaneyes (Linares and Rivas 2004). GUYANA: Demerara-Mahaica, Demerara. SURINAM: Para, Powakka (Husson 1978). FRENCH GUIANA: Cayenne (*Le Kabassou* of Buffon 1763). BRAZIL: Pará, Bragança; Maranhão, Barra do

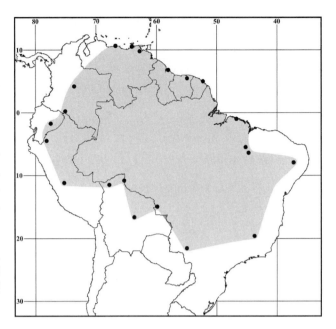

Map 71 Marginal localities for *Cabassous unicinctus* ●

Corda; Maranhão, Mirador State Park (T. G. Oliveira 1995); Pernambuco, Serra da Jabitacá (Guerra 1981c); Minas Gerais, Lagoa Santa; Mato Grosso do Sul, Maracajú; Mato Grosso, Mato Grosso. BOLIVIA: Santa Cruz, 80 km N of San Carlos (S. Anderson 1997); Beni, Guayaramerín; Pando, 15 km NW of Puerto Camacho (S. Anderson 1997). PERU: Junín, Chanchamayo; Amazonas, Huampami (J. L. Patton, Berlin, and Berlin 1982). ECUADOR: Pastaza, Sarayacu. COLOMBIA: Putumayo, Parque Nacional Natural La Paya (Polanco-Ochoa, Jaimes, and Piragua 2000); Meta, Villavicencio.

SUBSPECIES: We follow Wetzel (1980) in recognizing two subspecies of *C. unicinctus*.

C. u. squamicaudis (Lund, 1845)
SYNONYMS:

D[*asypus*]. *gymnurus*: Wied-Neuwied, 1826:529; not *Tatus gymnurus* Olfers ; name not available from Illiger (1815).

Xenurus squamicaudis Lund, 1843:lxxxiv; *nomen nudum.*

Xenurus squamicaudis Lund, 1845a:35 [1846:93]; type locality "Rio das Velhas Floddal" (Lund 1843:5), Lagoa Santa, Minas Gerais, Brazil.

Dasypus hispidus Burmeister, 1854:287; type locality "Lagoa Santa," Minas Gerais, Brazil.

Dasypus loricatus J. A. Wagner, 1855:174; cited as *"Dasypus loricatus* Natt. in mscpt." in synonymy; and equated with *"D. gymnurus* Ill[iger]." on page 175.

Xenurus hispidus: Gray, 1865:378; name combination.

Xenurus loricatus: Fitzinger, 1871b:239; name combination.

Xenurus latirostris Gray, 1873a:22; type locality "Brazils, St. Catherines.

[*Lysiurus (Lysiurus)*] *latirostris*: Trouessart, 1898:1147; name combination.

[*Lysiurus (Lysiurus)*] *loricatus*: Trouessart, 1898:1147; name combination.

[*Lysiurus (Lysiurus)*] *hispidus*: Trouessart, 1898:1147; name combination.

Tatoua (Tatoua) hispida: Miller, 1899a:5; name combination.

C[*abassous*]. *loricatus*: Palmer, 1899:72; name combination.

C[*abassous*]. *hispidus*: Palmer, 1899:72; name combination.

Lysiurus unicinctus: Goeldi and Hagmann, 1904:98; name combination.

[*Cabassus (Cabassus)*] *latirostris*: Trouessart, 1905:820; name combination.

[*Cabassus (Cabassus)*] *loricatus*: Trouessart, 1905:820; name combination.

Lysiurus hispidus: Cabrera, 1917:59; name combination.

Cabassous lugubris: Talmage and Buchanan, 1954:77; part; not *Ziphila lugubris* Gray.

Cabassous squamicaudis: Paula-Couto, 1950:537; name combination.

Cabassous unicinctus: Pine, 1973:61; name combination.

Cabassous unicinctus squamicaudis: Wetzel, 1980:323; first use of current name combination.

This subspecies is in Brazil, south of the Rio Amazonas.

C. u. unicinctus (Linnaeus, 1758)

SYNONYMS:

Dasypus unicinctus Linnaeus, 1758:50; type locality "Africa"; restricted to Surinam by O. Thomas (1911b: 141).

Dasypus duodecim cinctus Schreber, 1774b:pl. lxxiv; no locality given (described by Schreber 1774b:225); a redrawing of Buffon's (1763:pl. 40) "*Le Kabassou*" from Cayenne, French Guiana.

[*Dasypus*] *octodecimcinctus* Erxleben, 1777:113; type locality "in America australi."

Dasypus undecimcinctus Illiger, 1815:109; *nomen nudum*.

Dasypus multicinctus Thunberg, 1818:68; type locality "Brasilien."

Tatusia tatouay: Lesson, 1827:311; part; not *Loricatus tatouay* Desmarest.

Dasypus tatouay: Schomburgk, 1840:34; not *Loricatus tatouay* Desmarest.

Dasypus gymnurus var. γ J. A. Wagner, 1844:171; not *Tatus gymnurus* Olfers.

D[*asypus*]. *verrucosus* J. A. Wagner, 1844:172, footnote; type locality "den nördlichen Theil [des tropischen Südamerikas]"; name based on Buffon's (1763) *kabassu*, which came from Cayenne, French Guiana.

Xenurus [*(Tatoua)*] *unicinctus*: Gray, 1865:378; name combination.

Xenurus verrucosus: Fitzinger, 1871b:233; name combination.

Ziphila lugubris Gray, 1873a:23; type localities "Brazils, St. Catherine's . . . S. America, Demerara"; Wetzel's (1980:344) restriction of type locality to St. Catherine's is invalid because of incorrect assumption of holotype (ICZN 1999:Art. 74.5). We here select the second specimen ("*b*," BM 55.8.28.7; see Wetzel 1980:344) as the lectotype, restricting the type locality to Demerara, Guyana.

Xenurus lugubris: O. Thomas, 1880:402; name combination.

Xenurus duodecimcinctus: Jentink, 1888:213; name combination.

[*Lysiurus (Lysiurus)*] *unicinctus*: Trouessart, 1898:1146; name combination.

[*Lysiurus (Ziphila)*] *lugubris*: Trouessart, 1898:1148; name combination.

Tatoua (Ziphila) lugubris: Miller, 1899a:6; name combination.

C[*abassous*]. *(Ziphila) lugubris*: Palmer, 1899:72; name combination.

Tatoua unicincta: Miller, 1899a:2; name combination.

C[*abassous*]. *unicinctus*: Palmer, 1899:72; name combination.

[*Cabassus (Cabassus)*] *unicinctus*: Trouessart, 1905:820; name combination.

Cabassous loricatus: Yepes, 1928:467; part; not *Dasypus loricatus* J. A. Wagner.

Xenurus unicinctus: Sanderson, 1949:785; name combination.

Cabassous unicinctus unicinctus: Wetzel, 1980:343; first use of current name combination.

The nominate subspecies is found north of the Rio Amazonas-Solimões and east of the Andes in Colombia, Ecuador, and Peru.

NATURAL HISTORY: The meager information available indicates that dietary preferences and foraging strategies of *C. unicinctus* are similar to those of *C. centralis* (see Wetzel 1982). Wild-caught individuals from Guyana weighed from 2.2 to 4.8 kg (Meritt 1985), and nine from Surinam and Venezuela averaged 2.9 kg (Wetzel 1985b). The karyotype has not been reported.

REMARKS: Wetzel (1980:344) included *Ziphila lugubris* Gray, 1873a, in the synonymy of *Cabassous u. unicinctus* (Linnaeus, 1758) and, incorrectly, identified the first listed by Gray (1873a:23) under *Ziphila lugubris* as the holotype (specimen BM 51.8.25.10). Specimen BM

55.8.28.9 and BM 55.8.28.7 (not BM 55.8.28.9 as listed by Gray) are syntypes, not holotype and paratype, respectively. Wetzel (1980:346) also questioned the provenance of the first specimen, suggesting that it may have come from Sainte Catherine, French Guiana, "a nation from which other specimens were obtained by Parzudaki [the collector] for BMNH." Wetzel (1985b:35) later considered the "holotype" to be composite and associated the skin with *C. tatouay* and the skull (along with the "paratype") with *C. unicinctus*. Gray (1873:23) clearly stated on the line following specimen "1598 a," "skull of "b." We suggest that skull of *b* actually is the skull of specimen 1598 *b* (BM 55.8.28.7, which is the smaller specimen Gray referred to as having thin weak shields), and not 1598 *a* (BM 51.8.25.10). We follow Wetzel's lead in identifying the skull as representing *C. unicinctus*. However, we associate the skull with BM 55.8.28.7, and select it as the lectotype, thereby restricting the type locality to Demerara (an earlier name for Georgetown), Guyana. The type locality is not "Colômbia," as stated by C. O. C. Vieira (1955: 403).

Wetzel (1980:346) selected the skull illustrated by Lund (1845a:pl. 4, Fig. 3 [cited by Wetzel as pl. 50, Fig. 3]) as the lectotype of *C. u. squamicaudis*. Lund (1843) and Winge (1915) considered the species to be extinct, but Wetzel (1980) found Lund's fossil material to be indistinguishable from the small species of *Cabassous* still living in the same area. *Cabassous u. squamicaudis* intergrades with *C. u. unicinctus* along the Río Amazonas of Peru and Brazil (Wetzel 1980, 1985b).

Dasypus loricatus was a Natterer manuscript name that J. A. Wagner (1855:174) cited when he used it as an equivalent of his *Dasypus gymnurus* var. γ. The name became available (ICZN:1999, Art. 11.6.1) from Wagner (1855) when Fitzinger (1871b:239; not Pelzeln (1883:102) used the name combination *Xenurus loricatus* for Natterer's specimens. Pelzeln (1883:102), Trouessart (1898:1147), Palmer (1899:72), and other authors, attributed the name to Natterer. Pelzeln 1883:102) stated that Natterer's specimens came from "Matto Grosso" and "Cabeça de boi," and he (p. 103) described the specimen from Cabeça de Boi, a male, which we designate here as the lectotype. Wetzel (1980:346) erred in stating that the specimen was the holotype, and he was uncertain about the location of Cabeça de Boi; however, based on Papavero's (1971) itinerary of Natterer's travels in Brazil, it is located near Cuiabá, Mato Grosso.

Genus *Priodontes* F. Cuvier, 1825

This monotypic genus is confined to South America east of the Andes, where it occurs from northern Venezuela and the Guianas southward into Paraguay and northern Argentina. There is no fossil record (Paula-Couto 1979; McKenna and Bell 1997).

Largest of living armadillos, head-and-body lengths range from 750 to 1,000 mm, tail length approximates 500 mm, adults weigh up to 30 kg., and pinnae and eyes are large. The carapace is divided into transverse rows (bands) of small plates of which the central 11 to 13 are movable. The tail covered with closely spaced plates that are not arranged in rows. Forefeet bear well-developed claws of which the third is greatly enlarged and may measure over 200 mm along the outside curve. As in *Cabassous*, *Dasypus*, and *Tolypeutes*, the bony structure of each ear consists of a tympanic ring instead of an ossified bulla. Teeth are thin blades, their number is highly variable with up to 20 teeth in a toothrow, and many are lost as the animal gets older.

SYNONYMS:

Dasypus: Kerr, 1792:112; part; not *Dasypus* Linnaeus, 1758.

Loricatus Desmarest, 1804b:28; part (see Remarks).

Tatus Olfers, 1818:220; part; incorrect subsequent spelling of *Tatu* Blumenbach.

Priodontes F. Cuvier, 1825:257; type species *Dasypus gigas* G. Cuvier, 1817, by monotypy.

Cheloniscus Wagler, 1830:35; type species *Dasypus gigas* G. Cuvier, 1817, by monotypy.

Priodon McMurtrie, 1831:164; type species *Dasypus gigas* G. Cuvier, 1817, by monotypy; proposed as a subgenus of *Dasypus*; preoccupied by *Priodon* Quay and Gaimard, 1824 (Osteichthyes).

Polygomphius Gloger, 1841:114; type species *Dasypus gigas* G. Cuvier, 1817, by monotypy.

Prionodon Gray, 1843b:xxvii; *nomen nudum*.

Priodonta Gray, 1843b:xxvii; incorrect subsequent spelling of *Priodontes* F. Cuvier.

Prionodontes Schinz, 1845:312; incorrect subsequent spelling of *Priodontes* F. Cuvier.

Prionodos Gray, 1865:374; intended as a replacement name for *Priodontes* F. Cuvier.

REMARKS: *Prionodon* Gray, 1843b:xxvii, listed in the synonymy of *Priodontes* by many authors (e.g., Palmer 1904; Yepes 1928; Cabrera 1958) is a *nomen nudum*. Earlier, under the genus *Dasypus*, we have proposed designating *Loricatus niger* Desmarest, 1804 (= *Dasypus novemcinctus* Linnaeus) as the type species of *Loricatus* Desmarest, 1804. For reasons stated there, we ignore the recent designation by Kretzoi and Kretzoi (2000:204) of *Dasypus giganteus* É. Geoffroy St.-Hilaire, 1803, as the type species of *Loricatus* Desmarest. By setting aside the designation by Kretzoi and Kretzoi (2000), we retain *Priodontes* as the valid generic name for *P. maximus*.

Priodontes maximus (Kerr, 1792)

Giant Armadillo

SYNONYMS:

Dasypus maximus Kerr, 1792:112; type locality "Cayenne," French Guiana.

Dasypus giganteus É. Geoffroy St.-Hilaire, 1803:207; type locality "Le Paraguay" with reference to "*Le Grand Tatou* d'Azara, t. 2, p. 132"; but based on specimen "N°. CCCCXIV. *Individu qui servi de sujet pour la description précédente.*"

Dasypus gigas G. Cuvier, 1817:221; no type locality given; based on plate "xlv" [error for xli], the "*Autre Kabassou*" of volume 10 of Buffon (Daubenton in Buffon 1763). Buffon (1763) wrote that "*Le Kabassou*" was the largest tatou and came from Cayenne, which is the type locality.

D[asypus]. gigans Schmid, 1818:164; no type locality given.

T[atus] grandis Olfers, 1818:219; type locality "Paraguay."

Priodontes giganteus: Lesson, 1827:309; name combination.

D[asypus]. (P[riodontes].) Gigas: Voigt, 1831:261; name combination.

Priodonta gigas: Gray, 1843b:120; name combination.

Priodon gigas: Owen, 1845:21; name combination.

Prionodontes gigas: Schinz, 1845:316; name combination.

Prionodos gigas: Gray, 1865:374; name combination.

Prionodon gigas: Gray, 1869:380; name combination.

Cheloniscus gigas: Fitzinger, 1871b:227; name combination.

Priodontes maximus: O. Thomas, 1880:402; first use of current name combination.

Priodon maximus: J. A. Allen, 1895a:187; name combination.

D[asypus]. maximus Larrañaga, 1923:343; type locality "nemoribus septentrionalibus paraquarensibus"; based on Azara's (1802b:110) "*Maximo*"; a junior homonym and synonym of *Dasypus maximus* Kerr, 1792.

DISTRIBUTION: *Priodontes maximus* is recorded only from east of the Andes in eastern Colombia and northwestern Venezuela, the Guianas, Ecuador, Brazil, Peru, Bolivia, Paraguay, and northern Argentina. Although Wetzel (1985b:29) found no specimens or records from Brazil east of the states of Pará, Goiás, and São Paulo, Ruschi (1954b) mentioned the species as part of the fauna of Espírito Santo.

MARGINAL LOCALITIES (Map 72): VENEZUELA: Yaracuy (Wetzel 1985b, no specific locality, not mapped); Miranda, Parque Nacional Guatopo (Eisenberg, O'Connell, and August 1979). GUYANA: Cuyuni-Mazaruni, Kartabo (Beebe in H. E. Anthony 1921b). SURINAM: Sipaliwini, Kaaimanston (Husson 1978). FRENCH GUIANA: Cayenne (type locality of *Dasypus maximus* Kerr). BRAZIL (Vaz 2003, except as noted):

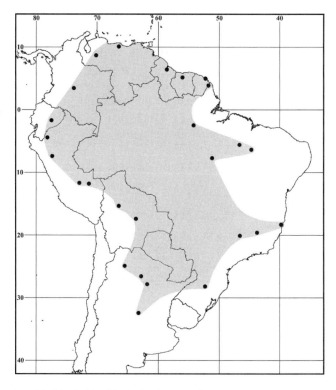

Map 72 Marginal localities for *Priodontes maximus* ●

Amapá, Oiapoque; Pará, Taperinha; Pará, Posto Indígena Kô Kraimôrô; Maranhão, Amarante do Maranhão; Maranhão, Mirador State Park (T. G. Oliveira 1995); Espírito Santo, Fazenda do Caboclo (Ruschi 1954b); Minas Gerais, Lagoa Santa, Rio das Velhas (Lund 1843); Minas Gerais, Serra da Canastra (T. S. Carter 1983); Rio Grande do Sul, Passo Fundo (AMNH 1308). ARGENTINA: Córdoba, Villa Nueva (Burmeister 1869); Chaco, Campo del Cielo (MACN 43.36); Santiago del Estero, Campo Gallo (Porini 2001); Salta, Cerillos (M. M. Díaz et al. 2000). BOLIVIA: Santa Cruz, Buena Vista (CM 20942); Beni, 75 km S of San Borja (S. Anderson 1997). PERU: Madre de Dios, Cocha Cashu (Terborgh, Fitzpatrick, and Emmons 1984); Cusco, Pagoreni (Boddicker, Rodríguez, and Amanzo 1999); San Martín, Gran Pajatén (Leo and Ortiz 1982); Amazonas, Huampami (J. L. Patton, Berlin, and Berlin 1982). ECUADOR: Pastaza, Sarayacu (O. Thomas 1880). COLOMBIA: Meta, San Juan de Arama (FMNH 87921). VENEZUELA: Barinas, Barinas (Krumbiegel 1940a).

SUBSPECIES: We consider *P. maximus* to be monotypic.

NATURAL HISTORY: Giant armadillos are nocturnal, highly fossorial, and tolerant of a range of habitats that includes tropical rainforest and open savanna. Burrows tend to be clustered and, in appropriate habitat, are often around termite mounds. The diet is almost exclusively ants and termites (Redford 1985b). One or two young may be born (Krieg 1929; T. S. Carter 1983; Redford 1985b). The karyotype is $2n = 50$, FN = 76, the X chromosome is a medium-sized metacentric; the Y a small metacentric (Benirschke and Wurster 1969; Benirschke, Low, and Ferm 1969).

REMARKS: Cabrera (1958:218) credited É. Geoffroy St.-Hilaire (1803:207) with the name *Dasypus giganteus* and restricted the type locality to Pirayú, Paraguarí, Paraguay. In addition to citing Azara's (1801b:132) "*Le Grand Tatou*," É. Geoffroy St.-Hilaire (1803:207–208) wrote that the specimen he examined (No. CCCCXIV) came from "Le Paraguay." Desmarest (1804b:28) also equated "*Le grand tatou de d'Azara ou second cabassou de Buffon*" and *Dasypus giganteus* with *Dasypus unicinctus* Linnaeus [= *Cabassous unicinctus*], which also could be considered a synonym. Gardner (1993) credited *Dasypus giganteus* to Desmarest (1804b).

Yepes (1928) cited G. Cuvier (1822 [error for 1823]) as the source for the name combination *Dasypus gigas*. Although G. Cuvier (1823:pl. 11, Figs 1, 2, 3, and 10) illustrated the skull and forelimb, nowhere in that volume did he use *Dasypus gigas* or any other scientific name for the "*tatou géant*." Therefore, we assume that Yepes meant to cite 1817, not 1822, as the source of G. Cuvier's name *Dasypus gigas*.

Tribe Tolypeutini Gray, 1865

The tribe is represented by two species in the genus *Tolypeutes*. These are the smaller of the medium-sized armadillos, and are characterized by a hard, convex carapace having scapular and pelvic shields separated by one to four (usually three) movable bands.

Genus *Tolypeutes* Illiger, 1811

Medium-sized to small, short-tailed (head-and-body length 218–277 mm, tail length 38–56 mm), armadillos with distinctive, hard, and rigid pectoral and pelvic shields separated by one to four (usually three) movable bands, and an anteriorly elongated, flat head shield that has rounded lateral and posterior margins. Body hair is conspicuous below the carapace. The manus has four or five toes; the pes has five toes, of which digits II to IV bear hoof-like claws (claws on digits I and V are normal). These are the only armadillos that can protect all unarmored areas by rolling into a ball. As in *Dasypus*, *Priodontes*, and *Cabassous*, the bony structure of the ear consists of a tympanic ring instead of an ossified bulla.

Tolypeutes comprises two species distributed in the drier regions of South America south of the Rio Amazonas. The earliest record for the genus is from the Pliocene of South America (Marshall et al. 1984); its taxonomy was reviewed most recently by Wetzel (1985b).

SYNONYMS:

Tolypeutes Illiger 1811:111; type species *Dasypus tricinctus* Linnaeus, 1758, by subsequent designation (Yepes 1928:478).

Matacus Rafinesque 1815:57; *nomen nudum.*

Tolypeutis Olfers, 1818:221; incorrect subsequent spelling of *Tolypeutes* Illiger.

Tatusia Lesson, 1827:309; part.

Apara McMurtrie 1831:163; type species *Dasypus tricinctus* Linnaeus, by monotypy; described as a subgenus of *Dasypus* Linnaeus.

Sphaerocormus Fitzinger 1871c:376; type species *Tolypeutes conurus* I. Geoffroy St.-Hilaire, 1847a, by monotypy.

Cheloniscus Gray, 1873a:23; type species *Cheloniscus tricinctus*: Gray, 1873a (= *Dasypus tricinctus* Linnaeus), by monotypy; preoccupied by *Cheloniscus* Wagler, 1830, a synonym of *Priodontes* F. Cuvier, 1825.

Tolypentes Matschie, 1894:62; incorrect subsequent spelling of *Tolypeutes* Illiger.

Tolypoïdes Grandidier and Neveu-Lemaire, 1905:370; type species *Tolypoides bicinctus* Grandidier and Neveu-Lemaire, 1905, by monotypy.

Tolypuetes Talmage and Buchanan, 1954:73; incorrect subsequent spelling of *Tolypeutes* Illiger.

KEY TO THE SPECIES OF *TOLYPEUTES*:
1. Forefeet with five digits . . . . *Tolypeutes tricinctus*
1'. Forefeet with three or four digits . . . .
. . . . *Tolypeutes matacus*

Tolypeutes matacus (Desmarest, 1804)
Southern Three-banded Armadillo
SYNONYMS:

Dasypus octodecimcinctus G. I. Molina, 1782:305; type locality "Nel Cujo," Chile (identified as provincia Mendoza, Argentina, by Tamayo, 1968); preoccupied by *Dasypus octodecimcinctus* Erxleben, 1777, a synonym of *Cabassous unicinctus* (Linnaeus, 1758).

Lor[icatus]. matacus Desmarest, 1804b:28; no locality mentioned; based on "*Le tatou mataco de d'Azara ou l'apar de Buffon*"; Sanborn (1930:62) restricted the type locality to Tucumán, Tucumán, Argentina, one of the localities Azara (1801a:197) identified for his *tatou mataco*.

[*Dasypus*] *brachyurus* G. Fischer, 1814:130; type localities "Tecumanis et circa Buenos-Ayres" (see Remarks).

Tolypeutes globules Illiger, 1815:108, 119; *nomen nudum.*

T[olypeutis]. octodecimcinctus: Olfers, 1818:221; name combination (see Remarks).

Dasypus apar Desmarest, 1822:367; type localities "Le Tecuman et les campagnes découverts dans les environs de Buenos-Ayres, à partir du 36ᵉ. degré et gagnant vers le sud" (see Remarks).

Tatusia apar: Lesson, 1827:310; name combination.

Tolypeutes conurus I. Geoffroy St.-Hilaire, 1847a:137; type localities "Tecuman et des pampas de Buenos-Ayres [Argentina], . . . et de la province de Santa-Cruz de la Sierra [Bolivia]."

Dasypus [(Tolypeutes)] conurus: Burmeister, 1861:426; name combination.

Dasypus aparoides P. Gervais, 1869:132; *nomen nudum*.

Tolypeutes muriei Garrod, 1878:223; type locality unknown "[probably] from some part of the coast of La Plata or Patagonia"; Sanborn (1930:65) restricted the type locality to Bahia Blanca, Buenos Aires, Argentina.

Tolypoides bicinctus Grandidier and Neveu-Lemaire, 1905:370; type locality "l'Amérique du Sud"; restricted to "environs de Tarija (Bolivie)" by Grandidier and Neveu-Lemaire (1908:4).

Tolypeutes matacus: Osgood, 1919:33; first use of current name combination.

D[asypus]. globosus Larrañaga, 1923:343; type locality "australem plagam bonaerensem"; based on Azara's (1802b:161) "*Mataco*."

Tolypeutes matacos Yepes, 1928:478; incorrect subsequent spelling of *Loricatus matacus* Desmarest.

Tolypeutes tricinctus matacus: Sanborn, 1930;66; name combination.

Tolypeutes tricinctus muriei: Sanborn, 1930;66; name combination.

DISTRIBUTION: *Tolypeutes matacus* is recorded from eastern Bolivia, southwestern Brazil, the Gran Chaco of Paraguay, and Argentina as far south as provincia Buenos Aires.

MARGINAL LOCALITIES (Map 73): BRAZIL: Mato Grosso, Chavantina das Mortes (Wetzel 1985b). PARAGUAY: Presidente Hayes, Puerto Pinasco (USNM 236356). ARGENTINA: Chaco, Avia Terai (BM 34.11.4.17); Buenos Aires, Bahia Blanca (type locality of *T. muriei* Garrod); Buenos Aires, Carmen de Patagones (Wetzel 1985b); La Rioja, Ávila (BM 1939.2653); Catamarca (Mares et al. 1997, no specific locality, not mapped); Tucumán, San Pedro de Colalao (Mares et al. 1996); Salta, Molinos (Ojeda and Mares 1989). BOLIVIA: Tarija, vicinity of Tarija (type locality of *Tolypoides bicinctus* Grandidier and Neveu-Lemaire); Tarija, Capirenda (Wetzel 1985b); Santa Cruz, Santa Cruz de la Sierra (Sanborn 1930); Santa Cruz, 29.5 km W of Roboré (S. Anderson 1997); Santa Cruz, Candelaria (Brooks et al. 2002). BRAZIL: Mato Grosso, Fazenda Acorizal (UCONN 20004).

SUBSPECIES: We are treating *T. matacus* as monotypic.

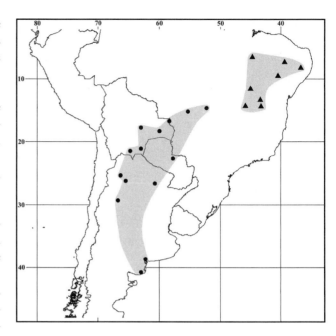

Map 73 Marginal localities for *Tolypeutes matacus* ● and *Tolypeutes tricinctus* ▲

NATURAL HISTORY: The reproductive season appears to be from October through January, and one young is born after a gestation period of 120 days. Eyes open at 22 days of age and suckling continues for approximately 2.5 months. The species is found in xeric habitats, where it uses burrows constructed by other armadillos. It feeds on invertebrates and is strongly myrmecophagous (Meritt 1973, 1976; Redford 1985b). *Tolypeutes matacus* has a $2n = 38$ karyotype (Jorge, Meritt, and Benirschke 1977; Jorge, Orsi-Souza, and Best 1985).

REMARKS: Common names include the southern domed or three-banded armadillo and tatu-bola. References including Chile in the distribution of *T. matacus* are probably based on G. I. Molina (1782), when Chile included parts of present-day northwestern Argentina. Cabrera (1958:221) credited Bravard with the name *Dasypus aparoides*; however, P. Gervais (1869:132) was simply listing names Bravard had used, some of which, including this one, are *nomina nuda*. Cabrera (1958) and Yepes (1928) overlooked the name *Dasypus brachyurus* G. Fischer, 1814, which is a replacement name for the broader meaning of *Dasypus tricinctus* Linnaeus, 1858, and when proposed, contained both Brazilian and Argentine three-banded armadillos. *Dasypus apar* Desmarest, 1822, also is a replacement name with essentially the same content. Citation of *D. brachyurus* and *D. apar* in the synonymy of *T. matacus* is based on the inclusion by their respective authors of statements paraphrasing Azara's (1801a:192) distribution for his *Tatou huitieme ou tatou mataco*, which can be considered the type basis for both names.

Tolypeutes octodecimcinctus Olfers, 1818:221, is based on *Dasypus octodecimcinctus* G. I. Molina, 1782:305, which Tamayo (1968:9) considered to be a composite based on the "*mataco*" Molina knew from antiguo Cuyo, Chile (now provincia Mendoza, Argentina), and on literature references to the "*tatou a dix huit bandes.*" The latter has been interpreted as either *Euphractus sexcinctus* or *Cabassous unicinctus.* Whatever its true identity, the name *Dasypus octodecimcinctus* G. I. Molina, 1782, is preoccupied by *Dasypus octodecimcinctus* Erxleben, 1777, which is a junior synonym of *Cabassous unicinctus* Linnaeus, 1758.

Tolypeutes tricinctus (Linnaeus, 1758)
Brazilian Three-banded Armadillo
SYNONYMS:

[*Dasypus*] *tricinctus* Linnaeus, 1758:51; type locality "in India orientali," redefined as Pernambuco, Brazil, by Sanborn (1930:62; see Remarks).

[*Dasypus*] *quadricinctus* Linnaeus, 1758:51; type locality unknown.

Tolypeutes Globulus Illiger, 1815:108, 119; *nomen nudum.*

T[*olypeutis*]. *globulus* Olfers, 1818:221; type localities "Paraguay, Brasilien, Gujana" (see Remarks).

T[*olypeutis*]. *quadricinctus* Olfers, 1818:221; type locality "Südamerica"; preoccupied by *Dasypus quadricinctus* Linnaeus, 1758.

Tatusia quadricincta: Lesson, 1827:310; name combination.

Das[*ypus (Apara)*]. *tricinctus*: McMurtrie, 1831:163; name combination.

T[*olypeutes*]. *tricinctus*: Turner, 1853:216; first use of current name combination.

Cheloniscus tricinctus: Gray, 1873a:24; name combination.

DISTRIBUTION: *Tolypeutes tricinctus* is known from the Brazilian states of Bahia, Ceará, Maranhão, and Pernambuco (Moojen 1943; Wetzel 1985b; J. M. C. Silva and Oren 1993; T. G. Oliveira 1995).

MARGINAL LOCALITIES (Map 73): BRAZIL: Maranhão, Mirador State Park (T. G. Oliveira 1995); Ceará, Crato (Moojen 1943); Pernambuco, Poção (Moojen 1943); Bahia, Juàzeiro (Vaz 2003); Bahia, Bom Jesus da Lapa (Vaz 2003); Bahia, Fazenda Boa Vista (J. M. C. Silva and Oren 1993); Bahia, Fazenda Rio Pratudão (Santos et al. 1994), Bahia, Canudos (Santos et al. 1994).

SUBSPECIES: We treat *T. tricinctus* as monotypic.

NATURAL HISTORY: Little is known about the habits of *T. tricinctus.* A behavioral trait shared with *T. matacus*, it responds to a threat by rolling into a ball. *Tolypeutes tricinctus* is endemic to the Caatinga Domain of northeastern Brazil, where it is associated with deciduous forests (J. M. C. Silva and Oren 1993). We presume its ecology is similar to that of *T. matacus*; however, this species is now

extremely rare and its existence threatened by habitat loss and subsistence hunting.

REMARKS: Although Linnaeus (1758:51, 1766:54) gave the "Habitat" as India orientali, several authors (e.g., Buffon 1763; Gmelin 1788; Kerr 1792) stated that the species occurred in Brazil. Based on Linnaeus' sole reference to Seba (1734), in his sixth edition of the *Systema Naturae* (1748), O. Thomas (1911b:141) suggested Surinam as the type locality. Sanborn (1930) designated Pernambuco, Brazil as a more suitable type locality, because the species is not known from the Guianas and Linnaeus (1758) also referred to Marcgraf (1648), who acquired material in the former Dutch colony centered in present-day eastern Pernambuco.

Several authors, including Cabrera (1958:222) have cited the name *Dasypus apar* Desmarest, 1819a:485, in the synonymy of *T. tricinctus*; however, the name Desmarest (1819a) used was *Dasypus tricinctus* Linnaeus (see Remarks under *T. matacus*). *Tolypeutes globulus* Olfers, 1818, is a replacement name for the broader meaning of *Dasypus tricinctus* Linnaeus, and in which Olfers included Azara's (1801a) *tatou mataco.* Although known from Argentina at the time, we have no evidence that the genus was known from Paraguay by 1818 and, following Hershkovitz (1959:340), we consider the name a junior synonym of *T. tricinctus* (Linnaeus).

Order Pilosa Flower, 1883
Alfred L. Gardner

Members of this order of xenarthrans are distinguished from the armadillos (Cingulata) by having a primary body covering of hair and the absence of dermal armor. Living members are the anteaters and sloths.

Suborder Folivora Desluc, Catzeflis, Stanhope, and Douzery, 2001

The Folivora consists of the two families of living sloths, the Bradypodidae and the Megalonychidae. The latter comprises the survivors of an early Neotropical radiation that include the ground sloths. Fossil evidence reveals great diversity and a former range from present day Alaska and the Greater Antilles south into Tierra del Fuego. In addition to behavioral, physiological, and anatomical features, living Folivora differ from the Vermilingua (anteaters) in having simple peg-like, ever-growing teeth that lack enamel, a well-developed coronoid process on the mandible, and the combination of a folivorous diet and arboreal habitus. However, the bradypodids and megalonychids are each so morphologically distinctive that further description of their unique anatomies is more appropriate at the family and generic level. Tardigrada and Phyllophaga are two names that have been applied to this assemblage, but these names also have been applied to other groups of organisms (e.g., see Fariña and Vizcaíno 2003). McKenna and Bell (1997) used Phyllophaga for the sloths.

Family Bradypodidae Gray, 1821

The family Bradypodidae contains the three-toed sloth genus *Bradypus*, which includes two subgenera (*Bradypus* and *Scaeopus*) and four species (*B. pygmaeus*, *B. torquatus*, *B. tridactylus*, and *B. variegatus*). Three-toed sloths inhabit lowland tropical forests across South America from the Caribbean coast into northern Argentina, except along the lower Río Orinoco drainage and in the llanos of Colombia and Venezuela. Three-toed sloths also occur in Central America north as far as Honduras.

Bradypodids have three digits on each foot; forelegs are about 1.5 times as long as hind legs. The long flexible neck has eight or nine cervical vertebrae. Teeth are simple pegs of about equal size, except for the smaller anterior maxillary pair, which sometimes are missing. The rostrum shows little development and a prenasal bone is lacking (Wetzel 1985a). Kraft (1995) provided a comprehensive description of the anatomy of bradypodid sloths.

Simpson (1945) placed the two genera of living sloths (*Bradypus* and *Choloepus*) in the Bradypodidae, a relationship supported by such features as the ball-and-socket articulation between the astragalus and fibula. This is the arrangement followed by Cabrera (1958), Hall and Kelson (1959), and Hall (1981). Although originally suggested by B. Patterson and Pascual (1968, 1972), Engelmann (1985) separated the two genera at the family level because he regarded these similarities as convergent for an arboreal existence. Wetzel and Ávila-Pires (1980) used the family name Bradypodidae for *Bradypus* and Choloepidae for *Choloepus*. More recently, however, Wetzel (1982, 1985a) and Barlow (1984) placed *Choloepus* in the Megalonychidae. Although unconfirmed because of the poor fossil record, the Bradypodidae are placed with the megatherioid sloths, based on shared derived dental and osteological characters (Stock 1925; Hoffstetter 1961, 1969; B. Patterson and Pascual 1972; Naples 1982, 1985; S. D. Webb 1985).

KEY TO THE SUBGENERA AND SOUTH AMERICAN SPECIES OF BRADYPODIDAE (MODIFIED FROM WETZEL 1985A; R. P. ANDERSON AND HANDLEY 2001):

1. Mane of long black hairs originating at nape and extending halfway down the back as two plumes; otherwise, pelage of uniform length and pale brown over head, neck, and body; males lack a speculum; pterygoids distinctly inflated; teeth anteroposteriorly compressed . . . . (subgenus *Scaeopus*) *Bradypus torquatus*
1'. No mane present; facial hair shorter, contrasting in color with dorsal body hair; pelage with blotches of pale color on darker dorsum; males have a speculum (mid-dorsal patch of short hair containing a black median stripe border by yellow- or orange-stained hair); pterygoids not inflated; teeth not obviously anteroposteriorly compressed . . . . (subgenus *Bradypus*) 2
2. Throat white or yellowish buff, continuous with pale color of forehead, no dark markings on face; skull smaller, greatest length averaging 73.8 mm (66.1–83.4 mm, $n = 46$); mandibular spout longer, averaging 2.6 mm (0.9–5.4 mm, $n = 47$); a pair of large foramina present in anterodorsal nasopharynx . . . . *Bradypus tridactylus*
2'. Throat and sides of face brown; prominent dark brown forehead and suborbital stripe outline paler color of eye patch on face; skull larger, greatest length averaging 75.8 mm (65.1–87.4 mm, $n = 108$); mandibular spout shorter, averaging 1.4 mm (0.04–3.6 mm, $n = 108$); no foramina in anterodorsal nasopharynx . . . . *Bradypus variegatus*

Genus *Bradypus* Linnaeus, 1758

Members of the genus *Bradypus* are medium-sized folivores (head-and-body length 420–800 mm) that are almost exclusively arboreal, typically cryptic, and slow moving in habit. The pelage is pale brown to gray, with long outer hair covering a short, dense black-and-white underfur. Except for *B. (Scaeopus) torquatus*, males have a dorsal speculum (a patch in which overhair is lacking) that shows the black and white underfur and, in adults, contains yellow- or orange-stained hair. The head is small, pinnae are not visible through the hair, and eyes are oriented anteriorly and usually surrounded by a mask of black hair. Facial vibrissae are sparse, and the rostrum is covered with short hair and bordered behind by the longer hair of the sides of the face and neck. The tail is short, round, mobile, covered with dense fur, and tapered distally with a fixed outward curve. The overhair has numerous transverse cracks that often contain algae, which impart a greenish tinge to the fur. Diagnostic cranial and dental characters include foramina in the anterior nasopharynx, elongated and uninflated pterygoids (partially inflated in subgenus *Scaeopus*), mandibular spout variable in length from almost lacking to short and pointed, lack of deciduous dentition, peg-like rounded teeth lacking enamel, and anterior chisel-shaped teeth 1/1 and molariform teeth 4/3. Bradypodids have eight or nine cervical vertebrae, united coracoid and acromial processes of the scapula, xenarthrous thoracic and lumbar vertebrae, and three syndactylous digits on each foot. Each foot also has a long, calloused palmar or plantar pad, and each digit bears a long, narrow, and curved claw.

The genus *Bradypus* contains four species: One is endemic to Isla Escudo de Veraguas, Panama, another occurs in Central and South America from Honduras south into southern Brazil and northern Argentina, a third is restricted to northern and northeastern South America, and the fourth is endemic to the Atlantic highland and coastal forests of eastern and southeastern Brazil from Rio Grande do Norte into Rio de Janeiro. There is no fossil record.

SYNONYMS:

Bradypus Linnaeus, 1758:34; type species *Bradypus tridactylus* Linnaeus, 1758, by subsequent designation (Miller and Rehn 1901:8).

Ignavus Blumenbach, 1779:70; type species *Ignavus tridactylus*: Blumenbach, 1779 (= *Bradypus tridactylus* Linnaeus), by monotypy.

Pradypus Ledru, 1810:257, footnote; incorrect subsequent spelling of *Bradypus* Linnaeus.

Choloepus: Desmarest, 1816:327; part; not *Choloepus* Illiger, 1811.

Acheus F. Cuvier, 1825:194; type species *Bradypus tridactylus* Linnaeus, by monotypy (see F. Cuvier 1825: 257).

Achaeus Erman, 1835:22; incorrect subsequent spelling of *Acheus* F. Cuvier, 1825; not *Achaeus* Leach, 1817 (Crustacea).

Achaeus Gray, 1843b:xxviii; incorrect subsequent spelling of *Acheus* F. Cuvier, 1825; not *Achaeus* Leach, 1817 (Crustacea).

Arctopithecus Gray, 1843b:xxviii; *nomen nudum*.

Arctopithecus Gray, 1850:65; no type species selected; preoccupied by *Arctopithecus* Virey, 1819 (Primates).

Scaeopus W. Peters, 1864c:678; type species *Bradypus torquatus* Illiger, 1811, by monotypy.

Hemibradypus R. Anthony, 1906:292; no species mentioned; type species *Hemibradypus mareyi* R. Anthony, 1907, by subsequent designation (R. Anthony 1907:220).

Eubradypus Lönnberg, 1942:5; type species *Bradypus tridactylus* Linnaeus, 1758, by original designation (see Lönnberg 1942:3).

Neobradypus Lönnberg, 1942:10; unavailable name (see Remarks).

REMARKS: The year of publication for *Scaeopus* W. Peters is 1864, as cited by Cabrera (1958), not 1865 as is sometimes cited. Tate's (1939:172) inclusion of the name "*Tardigradus* Lesson, 1762" was a lapsus for *Tardigradus* Brisson, 1762, which is unavailable (ICZN 1998). The earliest available usage of *Tardigradus* is from Boddaert (1784), and the name applies to a genus of lemurs.

Eubradypus and *Neobradypus* were proposed as subgenera of *Bradypus* by Lönnberg (1942); however, only the former is available as an objective junior synonym of *Bradypus* Linnaeus. The type species of *Eubradypus* Lönnberg, 1942, is *Bradypus tridactylus* Linnaeus, 1758, by primary designation; not *Bradypus (Eubradypus) tocantinus* Lönnberg, 1942, as listed by Hall and Kelson (1959) and Hall (1981). *Neobradypus* Lönnberg (1942) is an unavailable name because Lönnberg did not designate a type species as required by the Code of Zoological Nomenclature (ICZN 1999: Art. 13.3) for genus-group names proposed after 1930. *Neobradypus* also is not available from Cabrera (1958:207), from Hall and Kelson (1959:240), nor from Hall (1981:279), because they used the name only in synonymy.

Bradypus torquatus Illiger, 1811
Maned Sloth

SYNONYMS:

Br[adypus]. torquatus Illiger, 1811:109; type locality "Brasilia"; restricted to the Atlantic drainage of the Brazilian states of Bahia, Espírito Santo, and Rio de Janeiro, by Wetzel and Ávila-Pires (1980:834); further restricted to the vicinity of Salvador, Bahia, Brazil by Vaz (2003).

Ch[oloepus]. torquatus: Illiger, 1811:110; name combination.

Bradypus [(Choloepus)] torquatus: Desmarest, 1816:327; name combination.

[Bradypus] Cristatus Hamilton-Smith, 1827:278; in Griffith, Hamilton-Smith, and Pidgeon, 1827; no type locality given; type locality later identified as Brazil (Swainson 1835:207).

[Bradypus] melanotis Swainson, 1835:207; type locality "Brasil."

[Bradypus (]Acheus[)] torquatus: Lesson, 1840:270; name combination.

Bradypus crinitus Gray, 1850:67; type locality "British Guiana."

Bradypus affinis Gray, 1850:68; type locality "Tropical America."

Hemibradypus mareyi R. Anthony, 1907:220; no locality mentioned; type locality South America by inference.

Hemibradypus torquatus: Menegaux, 1908:702; name combination.

Bradypus (Scaeopus) torquatus: Menegaux, 1909:27; name combination.

Scaeopus torquatus: Poche, 1908:569; name combination.

DISTRIBUTION: *Bradypus torquatus* can be found in remnants of the once extensive coastal and adjacent

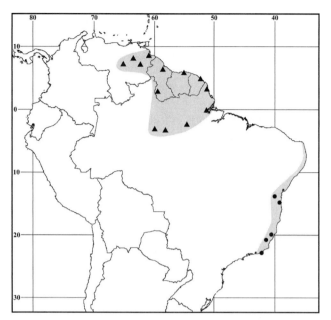

Map 74 Marginal localities for Bradypus torquatus ● and Bradypus tridactylus
▲

kg; males: total length 550–552, mass 4.05–4.90 kg). Ectoparasites included coprophagous moths (*Cryptoses* sp.), ticks (*Amblyomma varium* and *Boophilus* sp.), and beetles (*Trichilium* sp.). *Bradypus torquatus* has 50 chromosomes, the lowest diploid number known in the genus. The distribution of *B. torquatus* has been reduced to a few isolated forest fragments within its former range. Survival of the species is threatened because of habitat destruction (Coimbra-Filho 1972; Wetzel 1982).

REMARKS: The subgenus *Scaeopus* W. Peters is monotypic. *Bradypus (Scaeopus) torquatus* is distinguishable by the following features: the head and body pelage is uniformly pale brown, except for two conspicuous patches of long black hairs at base of neck that project over shoulders and extend about halfway down the back; the species lacks a speculum. The skull has a long predental mandibular spout, pterygoid flanges are elongated, pterygoid sinuses are inflated, the anterodorsal nasopharynx has more than one pair of foramina; and the teeth are anteroposteriorly compressed.

Although I consider it to be a subgenus of *Bradypus*, *Scaeopus* was recently treated as a monotypic genus (e.g., Paula-Couto 1979; Ávila-Pires in Wetzel and Ávila-Pires 1980). O. Thomas (1917b) cited Desmarest (1816) as the author of the name *Bradypus torquatus*, and stated that Illiger's (1811:109) usage was a *nomen nudum*. Wetzel (1982, 1985a) also treated Illiger's name as a *nomen nudum*; although earlier (Wetzel and Ávila-Pires 1980) he had accepted the name from Illiger (1811), as did C. O. C. Vieira (1955) and Cabrera (1958). Illiger must be credited with authorship because he (1811:109) followed the name with the letter N, which often means *novum*, but in this instance more likely referenced the "Note," beginning on the same page and ending on page 110 that contains the description of *Bradypus torquatus* under the name *Ch[oloepus]. torquatus*.

Bradypus tridactylus Linnaeus, 1758
Pale-throated Sloth

SYNONYMS:

[*Bradypus*] *tridactylus* Linnaeus, 1758:34; type locality "Americae meridionalis arboribus"; restricted to Surinam by O. Thomas (1911b:132).

Acheus aï Lesson, 1827:306; type localities "Brésil, à Cayenne, à la Nouvelle-Espagne, dans toute l'Amérique intertropicale."

Bradypus cuculliger Wagler, 1831:column 605; type localities "Surinamo, Cayenna et Guiana."

[*Bradypus (]Acheus[)*] *communis* Lesson, 1840:268; type locality "Le Brèsil."

B[*radypus*]. *tridactylus (Guianensis)* Blainville, 1840:Fig. 3; no type locality given.

highland Atlantic forests of southeastern Brazil in the states of Bahia, Espírito Santo, and Rio de Janeiro (Wetzel 1985a). The former range was from the states of Rio Grande do Norte south into Rio de Janeiro.

MARGINAL LOCALITIES (Map 74; from Wetzel and Ávila-Pires 1980, except as noted; localities listed from north to south): BRAZIL: Bahia, Jequié; Bahia, Itabuna; Espírito Santo, Santa Teresa; Espírito Santo, Engenheiro Reeve (O. Thomas 1917b); Rio de Janeiro, Lagoa de Araruama (Wied-Neuwied 1821).

SUBSPECIES: We consider *B. torquatus* to be monotypic.

NATURAL HISTORY: Except for a diet of leaves and occurring in the forests of the eastern highlands and adjacent Atlantic coast of Brazil, little else was known of the natural history of *B. torquatus* prior to Pinder's (1993) 31-month study of this species in the Poço das Antas Biological Reserve, Rio de Janeiro. Pinder (1993) determined the sex of eight females and two males by chromosomal analysis, and also identified sloths as females because they carried young. Reproduction appears to occur throughout the year. The black mane begins to appear on juveniles and is clearly evident on subadults. Pinder (1993) classed individuals into four age groups: infants ($n = 2$; total length under 250 mm, mass under 0.5 kg), juveniles ($n = 4$; total length 325–375 mm, tail 30–40 mm, mass 1.1–1.6 kg), subadults ($n = 4$, including recaptures; total length 487–552 mm, mass 2.52–3.62 kg), and adults ($n = 11$, including recaptures). Adult females are larger than adult males (females: total length 590–672 mm, mass 4.50–6.20

Bradypus gularis Rüppell, 1842a:138; type locality "die Wälder vou [sic] Guiana."

Arctopithecus gularis: Gray, 1850:70; name combination.

Arctopithecus Blainvillii Gray, 1850:71; type locality "Tropical America"; restricted to Brazil by O. Thomas (1917b:354).

Arctopithecus flaccidus Gray, 1850:72; type locality (Var. 1) "Venezuela."

[*Arctopithecus flaccidus*] var. 1. *Dysonii* Gray, 1869:365; type locality "Venezuela"; *Arctopithecus flaccidus dysonii* is an objective synonym of *Arctopithecus flaccidus* Gray.

[*Arctopithecus flaccidus*] var. 2. *Smithii* Gray, 1869:365; type locality "Pará," Pará, Brazil.

Arctopithecus cuculliger: Gray, 1871b:440; name combination.

[*Bradypus tridactylus*] Var. *Blainvillei*: Trouessart, 1898:1095; name combination.

[*Bradypus tridactylus*] Var. *flaccidus*: Trouessart, 1898:1096; name combination.

Bradypus flaccidus: Krumbiegel, 1941c:55; name combination.

Bradypus flaccions Sanderson, 1949:783; incorrect subsequent spelling of *Arctopithecus flaccidus* Gray.

Bradypus [(Bradypus)] infuscatus flaccidus: Cabrera, 1958:209; name combination.

DISTRIBUTION: *Bradypus tridactylus* is in Guyana, Surinam, French Guiana, Venezuela (south of the Río Orinoco), and Brazil south to the Rio Amazonas, where it occurs along both banks from the Rio Negro to the delta (Wetzel and Kock 1973).

MARGINAL LOCALITIES (Map 74): VENEZUELA: Delta Amacuro, Curiapo (Linares and Rivas 2004). GUYANA: Cuyuni-Mazaruni, Kalacoon (H. E. Anthony 1921b). SURINAM: Paramaribo, Paramaribo (Husson 1978). FRENCH GUIANA: Cabassou (Charles-Dominique et al. 1981). BRAZIL: Amapá, Vila Velha do Cassiporé (C. T. Carvalho 1962a); Amapá, Mazagão (Wetzel and Ávila-Pires 1980); Pará, Santarém, Mojui dos Campos (USNM 545919); Amazonas, Lago Baptista (Wetzel and Ávila-Pires 1980); Amazonas, Manaus (Jorge, Orsi-Souza, and Best 1985). GUYANA: Upper Takutu-Upper Essequibo, Dadanawa (R. P. Anderson and Handley 2001). VENEZUELA: Bolívar, Río Suapuré (J. A. Allen 1904d); Bolívar, Maripa (R. P. Anderson and Handley 2001; not mapped); Bolívar, Ciudad Bolívar (R. P. Anderson and Handley 2001); Bolívar, Los Patos (Handley 1976).

SUBSPECIES: I treat *B. tridactylus* as monotypic.

NATURAL HISTORY: *Bradypus tridactylus* is known from forested habitats and is sympatric with *B. variegatus* along both banks of the Amazon from the Rio Negro to the Atlantic (Wetzel 1985a; R. P. Anderson and Handley 2001). Details of the diet, habitat, anatomy, and physical and behavioral characteristics were described by Beebe (1926) for *B. cuculliger* (= *B. tridactylus*) from the vicinity of Kartabo, Guyana, and are the most comprehensive thus far reported. Although seldom seen, the species is a common arboreal mammal in appropriate forest habitat. Walsh and Gannon (1964) reported handling 2,104 *B. tridactylus* during Operation Gwamba in Surinam. With the exception of mother-young pairs, *B. tridactylus* usually are solitary and exhibit aggressive behavior when confined together in captivity. The species is almost exclusively arboreal, but swims well and is capable of moving distances of several miles, travel that may require moving on the ground and swimming across rivers. The habitat and behavior of all three-toed sloth taxa are similar.

Beebe (1926) gave the diet as almost exclusively the leaves of *Cecropia* spp. However, as Montgomery and Sunquist (1978) found with *B. variegatus*, further study probably will reveal, that *B. tridactylus* feeds on the leaves of a variety of tree species. Reproduction probably occurs year-round with one young born per year, following a gestation period of 4 to 6 months (Asdell 1964). Most births occur from late July to September in northern Guyana, at the beginning of the dry season (Beebe 1926). *Bradypus tridactylus* has a $2n = 52$, $FN = 56$ karyotype (Jorge, Orsi-Souza, and Best 1985).

REMARKS: The name *Bradypus cristatus*, attributed to Temminck by Fitzinger (1871a:379), was erroneously listed in the synonymy of *B. tridactylus* by Cabrera (1958:210). Hamilton-Smith (in Griffith, Hamilton-Smith, and Pidgeon 1827) was the first to use the name *cristatus*, which he also attributed to Temminck. However, Hamilton-Smith used the name for the species we know today as *B. torquatus*. The taxonomy and nomenclature of *B. tridactylus* was reviewed by Wetzel and Ávila-Pires (1980) and Wetzel (1982, 1985a). The Pale-throated Sloth can be distinguished from other *Bradypus* spp. by its paler throat (white or yellowish buff) in combination with the pale forehead and the lack of dark facial marks. The skull is distinctive in having foramina in the anterodorsal nasopharynx.

Bradypus variegatus Schinz, 1825
Brown-throated Sloth

SYNONYMS: The following synonyms are names applied to Central American representatives of the species; see under Subspecies for additional synonyms.

Arctopithecus griseus Gray, 1871a:302; type locality "Costa Rica"; corrected to Cordillera del Chucu, Veragua, Panama, by Alston (1880:183–184).

Arctopithecus castaneiceps Gray, 1871b:444; type locality "Nicaragua"; restricted to the vicinity of the Javali gold mine, Chontales, Nicaragua, by Alston (1880:184).

Map 75 Marginal localities for *Bradypus variegatus* ●

B[*radypus*]. *castaneiceps*: Alston, 1880:182; name combination.

DISTRIBUTION: *Bradypus variegatus* is known from Colombia, Venezuela, Ecuador, Brazil, eastern Peru, eastern Bolivia, Paraguay, and northern Argentina. Elsewhere, *B. variegatus* is known in Central America as far north as Honduras.

MARGINAL LOCALITIES (Map 75; from R. P. Anderson and Handley 2001, except as noted): COLOMBIA: La Guajira, Puerto Estrella (USNM 216665). VENEZUELA: Carabobo, 10 km NW Urama; Miranda, near Turgua (USNM 372831). COLOMBIA: Norte de Santander, Cúcuta; Santander, Caño Muerto; Meta, Villavicencio; Guainía, Puerto Inírida. VENEZUELA (Handley 1976): Amazonas, Río Manapiare, 163 km ESE of Puerto Ayacucho; Amazonas, Tamatama. BRAZIL: Amazonas, Iaunari; Amazonas, Codajáz (type locality of *Bradypus marmoratus codajazensis* Lönnberg); Amazonas, Itacoatiara (Wetzel and Ávila-Pires 1980); Pará, Santarém (USNM 49592); Pará, Ilha Mexiana (Hagmann 1908); Pará, Belém; Maranhão, Miritiba (type locality of *Bradypus miritibae* Lönnberg); Pernambuco, Pernambuco (type locality of *Bradypus dorsalis* Fitzinger); Alagoas, Sinimbu (Wetzel and Ávila-Pires 1980); Espírito Santo, Colatina (Wetzel and Ávila-Pires 1980); Rio de Janeiro, Sepetiba (type locality of *Bradypus pallidus* J. A. Wagner); São Paulo, Ipanema (Pelzeln 1883); Paraná, Londrina (Wetzel and Ávila-Pires 1980). ARGENTINA: Misiones, San Pedro (Bertoni 1914); Salta, Orán (Yepes 1928). BOLIVIA: Santa Cruz, Buena Vista (type locality of *Arctopithecus boliviensis* Gray); Cochabamba, Todos Santos (S. Anderson 1997); Pando, Cobija (S. Anderson 1997). PERU: Madre de Dios,

Cocha Cashu (Terborgh, Fitzpatrick, and Emmons 1984); Cusco, San Martín-3 (Boddicker 1998); Ucayali, Cumeria (O. Thomas 1928c); Loreto, Puerto Arturo; Amazonas, Huampami (J. L. Patton, Berlin, and Berlin 1982). ECUADOR: Guayas, Balzar (type locality of *Bradypus violeta* O. Thomas); Esmeraldas, Montañas de Chancameta (R. H. Baker 1974). COLOMBIA: Cauca, Isla Gorgona (type locality of *Bradypus gorgon* O. Thomas); Chocó, Condoto (O. Thomas 1917b); Chocó, Río Atrato (USNM 3352); Magdalena, Río Frio (J. A. Allen 1904f).

SUBSPECIES: I tentatively recognize seven subspecies of *B. variegatus* in South America.

B. v. boliviensis (Gray, 1871)
SYNONYMS:
Arctopithecus boliviensis Gray, 1871b:442; type locality "Bolivia"; restricted to Buena Vista, Santa Cruz, Bolivia, by Cabrera (1958:208).
[*Bradypus tridactylus*] Var. *boliviensis*: Trouessart, 1898: 1096; name combination.
Bradypus (Neobradypus) beniensis Lönnberg, 1942:22; type localities "Victoria," and "Confluencia de Rio Madre de Dios & Rio Beni," Beni, Bolivia.
Bradypus [(Bradypus)] boliviensis: Cabrera, 1958:208; name combination.
B[*radypus (Bradypus)*]. *v*[*ariegatus*]. *boliviensis*: Wetzel, 1982:353; name combination.

This subspecies is found throughout the southwestern Amazon basin of Bolivia and Brazil.

B. v. brasiliensis Blainville, 1840
SYNONYMS:
Bradypus Ai Wagler, 1831:column 610; type locality "Brasilia orientali"; restricted to Rio Matheus, Espírito Santo, Brazil, by O. Thomas (1917b:354); preoccupied by *Acheus ai* Lesson, 1827 (= *Bradypus tridactylus* Linnaeus).
Bradypus tridactylus Brasiliensis Blainville, 1840:64; type locality "Bresil"; restricted to Rio de Janeiro by O. Thomas (1917b:354).
[*Bradypus (]Acheus[)] ustus* Lesson, 1840:271; type locality "Le Brésil, la province de Rio de Janeiro."
Bradypus pallidus J. A. Wagner, 1844:143; type locality "Ostkuste von Brasilien"; restricted to "Sapitiba" [= Sepetiba], Rio de Janeiro, Brazil, by Pelzeln (1883:97); not Rio de Janeiro (O. Thomas 1917b:354) or Ipanema, São Paulo (C. O. C. Vieira 1955:401; Wetzel and Kock 1973:28).
Bradypus ustus: Liais, 1872:341; name combination.
[*Bradypus*] *brasiliensis*: O. Thomas, 1917b:354; name combination.
B[*radypus*]. *braziliensis* Sanderson, 1949:783; incorrect subsequent spelling of *Bradypus tridactylus brasiliensis* Blainville.

Bradypus [*(Bradypus)*] *infuscatus brasiliensis*: Cabrera, 1958:208; name combination.

B[*radypus (Bradypus)*]. *v*[*ariegatus*]. *brasiliensis*: Wetzel, 1982:353; name combination.

This subspecies is recorded from the highlands and adjacent Atlantic lowlands of southeastern Brazil.

B. v. ephippiger R. A. Philippi, 1870

SYNONYMS:

Bradypus ephippiger R. A. Philippi, 1870:267; type localities "das Vaterland nicht genau angeben . . . aus den Wäldern von Ostabhang der Republik Ecuador oder des nördlichsten Peru's stammt"; restricted to northwestern Colombia by O. Thomas (1917b:355); further restricted to the Río Atrato by Cabrera (1958:209).

[*Bradypus tridactylus*] Var. *ephippiger*: Trouessart, 1898:1096; name combination.

Bradypus ignavus Goldman, 1913:1; type locality "Marragantí (about 2 miles above Real de Santa Maria), near the head of tide-water on the Rio Tuyra," Darién, Panama.

Bradypus violeta O. Thomas, 1917b:357; type locality "Balzar Mountains, Guayas, W. Ecuador."

Bradypus ecuadorianus Spillmann, 1927:317; type locality "Santo Domingo de los Colorados," Pichincha, Ecuador.

Bradypus nefandus Spillmann, 1927:317; type locality "Provincia de Los Ríos," Ecuador.

Bradypus griseus ignavus: Hall and Kelson, 1952:315; name combination.

Bradypus [*(Bradypus)*] *infuscatus ephippiger*: Cabrera, 1958:209; name combination.

Bradypus variegatus ephippiger: Hall, 1981:280; first use of current name combination.

This subspecies occurs in northwestern Colombia east into northwestern Venezuela, and west of the Andes southward to the Golfo de Guayaquil, Ecuador. It also is known from eastern Panama.

B. v. gorgon O. Thomas, 1926

SYNONYMS:

Bradypus gorgon O. Thomas, 1926a:309; type locality "Gorgona Island, off the coast of Colombia."

Bradypus [*(Bradypus)*] *infuscatus gorgon*: Cabrera, 1958:209; name combination.

B[*radypus (Bradypus)*]. *v*[*ariegatus*]. *gorgon*: Wetzel, 1982:353; first use of current name combination.

This subspecies is known only from Isla Gorgona, Cauca, off the southwestern coast of Colombia.

B. v. infuscatus Wagler, 1831

SYNONYMS:

Bradypus infuscatus Wagler, 1831:column 611; type localities "Brasilia versus Peru," restricted to "la confluencia

del Solimões y el Iça," Amazonas, Brazil, by Cabrera (1958:209).

B[*radypus*]. *infumatus* Tschudi, 1845:205; incorrect subsequent spelling of *Bradypus infuscatus* Wagler.

B[*radypus*]. *rifuscatus* Cornalia, 1849:12; incorrect subsequent spelling of *Bradypus infuscatus* Wagler.

Bradypus brachydactylus J. A. Wagner, 1855:173; no type locality given; based on *Bradypus infuscatus* Var. ß., which J. A. Wagner (1844:149) described from a male specimen Natterer acquired at Borba, Rio Madeira, Amazonas, Brazil (J. A. Wagner 1844:150; Pelzeln 1883:97).

[*Bradypus infuscatus*] Var. *brachydactylus*: Trouessart, 1898:1095; name combination.

Bradypus macrodon O. Thomas, 1917b:356; type locality "Sarayacu, Upper Pastaza River, Oriente [= Pastaza] of Ecuador."

Bradypus [*(Neobradypus)*] *marmoratus codajazensis* Lönnberg, 1942:15; type locality "Codajaz, N of Rio Solimões, W of Rio Negro," Amazonas, Brazil.

[*Bradypus (Neobradypus) infuscatus*] *subjuruanus* Lönnberg, 1942:21; type locality "Lago Grande," Rio Juruá, Amazonas, Brazil.

Bradypus tridactylus infuscatus: C. O. C. Vieira, 1955:401; name combination.

Bradypus [*(Bradypus)*] *infuscatus infuscatus*: Cabrera, 1958:209; name combination.

B[*radypus (Bradypus)*]. *v*[*ariegatus*]. *infuscatus*: Wetzel, 1982:353; name combination.

This subspecies is in Colombia and Venezuela east and south of the Andes, and in the western Amazon basin of western Brazil, eastern Ecuador, and eastern Peru.

B. v. trivittatus Cornalia, 1849

SYNONYMS:

[*Bradypus*] *trivittatus* Cornalia, 1849:4; type locality "prope Gurupe ad Amazonum ripas," Pará, Brazil.

Arctopithecus marmoratus Gray, 1850:71; type locality "Brasils"; restricted to Pará by Cabrera (1958:210).

Arctopithecus problematicus Gray, 1850:73; type locality "Para," Pará, Brazil.

Bradypus problematicus: Gerrard, 1862:290; name combination.

Bradypus marmoratus: Fitzinger, 1871a:367; name combination.

Bradypus unicolor Fitzinger, 1871a:362; type locality "Süd-Amerika, Nordost-Brasilien, . . . Provinz Pará."

[*Bradypus tridactylus*] Var. *marmoratus*: Trouessart, 1898:1096; name combination.

Bradypus (Arctopithecus) marmoratus: Goeldi and Hagmann, 1904:90; name combination.

Bradypus [*(Eubradypus)*] *tocantinus* Lönnberg, 1942:5; type locality "Cametá, Rio Tocantins, [Pará,] Brazil."

Bradypus [(Eubradypus)] miritibae Lönnberg, 1942:8; type locality "Miritiba, northern coast of Brazil," Maranhão, Brazil.

Bradypus tridactylus miritibae: C. O. C. Vieira, 1955:402; name combination.

Bradypus tridactylus tocantinus: C. O. C. Vieira, 1955:402; name combination.

Bradypus [(Bradypus)] infuscatus marmoratus: Cabrera, 1958:210; name combination.

B[radypus (Bradypus)]. v[ariegatus]. marmoratus: Wetzel, 1982:353; name combination.

This subspecies occurs in the lower Amazon basin, mainly south of the Rio Amazonas.

B. v. variegatus Schinz, 1825

SYNONYMS:

Bradypus variegatus Schinz, 1824:251; *nomen nudum.*

Brad[ypus]. variegatus Schinz, 1825:510; type locality "Südamerika"; restricted to Bahia, Brazil, by Mertens (1925:23).

Bradypus dorsalis Fitzinger, 1871a:355; type localities "Süd-Amerika, Nordost-Brasilien, woselbst diese Art nicht nur zwischen dem Rio San Francisco und der Provinz Rio Grande angetroffen wird, sondern auch noch weiter nordwärts bis in die Provinz Pará hinaufreicht"; restricted to Pernambuco, Brazil, by O. Thomas (1917b: 354).

Bradypus speculiger Fitzinger, 1871a:372; *nomen nudum.*

B[radypus (Bradypus)]. v[ariegatus]. variegatus: Wetzel, 1982:353; name combination.

The nominate subspecies is in Brazil from the southeastern Amazon basin eastward to the Atlantic coast.

NATURAL HISTORY: *Bradypus variegatus* is active both day and night (Sunquist and Montgomery 1973b). Like other sloths, *B. variegatus* is almost exclusively arboreal, descending to the forest floor only to move to another tree, or to defecate, which it does at approximately 8-day intervals. Whenever possible, feces are deposited in a depression dug with the stump-like tail. Three-toed sloths move slowly on the ground, and are unable to rise to a quadrupedal posture; they use their elbows, hind feet, and hook like foreclaws in locomotion. They swim well, holding their bodies high in the water. Montgomery and Sunquist 1978) showed that three-toed sloths consumed the leaves of at least 25 species of trees and lianas. Leaves of *Cecropia* spp. are neither the most preferred, nor the most easily digested. Sloths can regulate their daytime body temperature by positional changes, and have basal metabolic rates about 42% as high as expected from their body weight (McNab 1978). Gestation lasts from 120 to 180 days, culminating in the birth of a single young. Although weaned by approximately 1 month of age, the young is carried by the mother for approximately 6 months before she leaves it within her home range. Moeller (1975) described the behavior and vocalizations of free-living "three fingered" sloths (under the name *B. tridactylus*) in Brazil. Soares and Carneiro (2002) described mother-young interactions, play, and other behaviors in *B. variegatus* in the Northeast of Brazil.

Greene (1989) described an episode of physical agonistic behavior between males that occurred in a context suggesting that social interactions, visual acuity, and activity budgets are better developed and more complex than previously suspected. Montgomery and Sunquist (1975, 1978) and Wetzel (1982) provided additional information on natural history. *Bradypus variegatus* has a $2n = 54–55$, FN $= 56–58$ karyotype, which has an unusual sex-chromosome system (Jorge, Orsi-Souza, and Best 1985).

REMARKS: The distribution in Venezuela is limited to forested habitats in the north and in the state of Amazonas. In Brazil, *B. variegatus* lives in forested habitats in the western and southern Amazon drainage. Cabrera (1958), repeated by Wetzel (1982, 1985a), gave the southern distribution in Brazil as the state of Rio Grande do Sul; however, I have not been able to verify this. The species is known from the provinces of Misiones and Salta, Argentina; but has not been recorded from Paraguay, although Bertoni (1939) mentioned unconfirmed reports (as *Bradypus tridactylus*). The species is sympatric with *B. tridactylus* along both banks of the lower Rio Amazons from the Rio Negro to the delta (Wetzel 1985a).

I am using *B. v. brasiliensis* Blainville, 1840, as the name for the southeastern Brazilian subspecies, following Wetzel's (1982) usage. Although O. Thomas (1917b) gave 1839 as the year of publication, both J. A. Wagner (1841) and Sherborn (1898) stated that the section of Blainville's "Ostéographie" on *Bradypus* (Part 4) was published in 1840; however, neither author gave a more specific date. The name *B. ustus* Lesson also was published in 1840 (November, *fide* Sherborn 1922:lxxx). If Blainville's part 4 of his Ostéographie was published after November 1840, or if the date within the year 1840 cannot be determined, the publication date must be assumed to be 31 December 1840 (ICZN 1999:Art. 21.3), and the name *B. v. ustus* Lesson would have priority and should be used for this subspecies.

The brown-throated sloth can be distinguished from *B. tridactylus* by its darker throat and facial markings, and by the absence of foramina in the anterior nasopharynx. The conspicuous facial markings, speculum (only in males), and the lack of long black hair on the neck, are easily seen features useful for distinguishing this species from *B. torquatus*.

Family Megalonychidae P. Gervais, 1855
Alfred L. Gardner and Virginia L. Naples

The family Megalonychidae includes two living species in the two-toed sloth genus *Choloepus*, and approximately 28 genera of extinct sloths (Simpson 1945; Engelmann 1985; Carroll 1988; McKenna and Bell 1997). Living megalonychids have two digits on each manus and three digits on each pes. The forelegs are only slightly longer than hind legs, and these sloths have a shorter, less flexible neck than do bradypodids, and have only five to seven (occasionally eight) cervical vertebrae. The teeth have sharp, beveled occlusal surfaces; the anterior pair is prominent and each tooth is triangular in outline. The rostrum is pronounced and includes a prenasal bone (Wetzel 1985a).

Extinct sloths typically are considered "ground sloths," although some of the smaller forms may have been arboreal (S. D. Webb 1985). Hoffstetter (1969) and B. Patterson and Pascual (1968, 1972) first suggested that all the tree sloths did not belong in the family Bradypodidae. They considered *Choloepus* to be closer to the Megalonychidae, a relationship also supported by S. D. Webb (1985). *Choloepus* is allied with the megalonychids, based on the presence of enlarged caniniform teeth, features of the zygomatic arch, and the lack of ossified auditory bullae. Engelmann (1985) listed additional features, although he did not believe them to be good derived characters. Carroll (1988) avoided the problem of claiming convergence to explain the peculiar suspensory mode of locomotion shown by tree sloths by suggesting that the two genera of living sloths were derived from primitive arboreal sloths of the late Oligocene or early Miocene, prior to the divergence of the large megatheres and megalonychoids. Carroll (1988) also considered the late Pleistocene and Recent Antillean genera *Synocnus* and *Acratocnus* to be related to *Choloepus*. McKenna and Bell (1997) included all three genera, along with *Miocnus*, in the tribe Choloepodini. Other authors (e.g., Kraft 1995) have used Choloepidae, which dates from Pocock (1924), as the name for this family.

Genus *Choloepus* Illiger, 1811

The two species of *Choloepus* are arboreal, slow-moving folivores (head-and-body length 540–720 mm). The typical color pattern is brown to pale brown, often darker on the extremities. The pinnae are not visible through the hair; the eyes are oriented forward and have small pupils. The teeth lack enamel, show differential wear of hard outer and softer inner dentine, and are designated according to shape and position, not homology. Anterior caniniform teeth are triangular in cross section and separated from the cheek-tooth row by a diastema. Sloths lack a deciduous denti-

tion and tooth homologies are unknown. Cheek teeth are simple pegs, with "facets" formed continuously through wear. Other diagnostic cranial characters include inflated pterygoid sinuses, a pair of large foramina in the anterodorsal mesopterygoid space, broad dorsal contact between the maxilla and frontal bones (not separated by the lacrimal), and a long, pointed predental mandibular spout. Cervical vertebrae number five to eight (usually seven). Forelimbs have two syndactylous digits, hind limbs have three; each digit bears a long, curved claw; and fore and hind feet have thick palmar and plantar pads, respectively. *Choloepus* lacks an external tail. The hairs have grooves that often house algae, which impart a greenish tinge to the pelage (part of this description is based on Barlow 1984). *Choloepus* spp. live in humid tropical lowland and montane forests from Central America (Nicaragua) south across northern South America to approximately 11°S latitude in Bolivia and western Brazil. There is no fossil record.

SYNONYMS:

Bradypus Linnaeus, 1758:34; part.

Choloepus Illiger, 1811:108; type species *Bradypus didactylus* Linnaeus, 1758, by subsequent designation (Gray 1827:275; misspelled *Chaelopus*).

Unaüs Rafinesque, 1815:57; no type species given, but possibly *Bradypus unau* Link, 1795, according to Palmer (1904:700); corrected spelling is *Unaues* Rafinesque, 1815 (ICZN 1999:Art. 32.5).

Choelopus Lichtenstein, 1818a:20; incorrect subsequent spelling of *Choloepus* Illiger.

Unaus Gray, 1821:305; type species *Bradypus didactylus* Linnaeus, 1758, by monotypy.

Cholaepus Schinz, 1821:328; incorrect subsequent spelling of *Choloepus* Illiger.

Cholaepus Gray, 1825b:343; incorrect subsequent spelling of *Choloepus* Illiger.

Chaelopus Gray, 1827:275; incorrect subsequent spelling of *Choloepus* Illiger.

Choelopus Tschudi, 1844:253; incorrect subsequent spelling of *Choloepus* Illiger; in synonymy.

Cholopus Agassiz, 1847:83; unjustified emendation of *Choloepus* Illiger.

Cholaepus Cornalia, 1849:4; incorrect subsequent spelling of *Choloepus* Illiger; in synonymy.

Choelopus Sclater, 1873:253; incorrect subsequent spelling of *Choloepus* Illiger.

Choelopus Sanborn, 1953:5; incorrect subsequent spelling of *Choloepus* Illiger.

KEY TO THE SPECIES OF *CHOLOEPUS* (BASED ON WETZEL AND ÁVILA-PIRES 1980, AND WETZEL 1985A):

1. Color of throat similar to that of chest; hair on cheeks neither distinctly shorter nor finer than hair of neck

and shoulders; one pair of small and one pair of large foramina in anterodorsal mesopterygoid fossa; pterygoid sinuses usually lacking posterior foramina; anterior mesopterygoid fossa broad, more than twice as wide as posterior mesopterygoid fossa; pterygoid sinuses broadly inflated (usually wider than 14 mm); maxillae in broad contact with frontals, not interrupted by lacrimals (usually evident only in immature individuals having visible sutures); cervical vertebrae usually seven (range 6−8, $n = 7$) . . . . *Choloepus didactylus*

1'. Throat paler than chest and generally clothed with finer hairs; top of the head with longer and paler hair than on neck and body; mesopterygoid fossa with one pair of small and two pairs of large foramina with the posterior pair opening into the pterygoid sinuses (anterior pair may be reduced to shallow depressions); width of anterior mesopterygoid fossa less than twice least width of posterior mesopterygoid constriction; posterior swellings of the pterygoid sinuses narrower (less than 13.5 mm in width); maxillae separated from frontals by nasal and lacrimal bones (usually visible only in skulls of immature animals having visible sutures); cervical vertebrae usually six (5−6, $n = 12$) . . . .
. . . . *Choloepus hoffmanni*

Choloepus didactylus (Linnaeus, 1758)
Linnaeus's Two-toed Sloth

SYNONYMS:

[*Bradypus*] *didactylus* Linnaeus, 1758:35; type locality "Zeylona"; corrected to Surinam by O. Thomas (1911b: 132); not British Guiana as stated by Tate (1939:174).

Bradypus unau Link, 1795:68; no locality mentioned.

Bradypus curi Link, 1795:68; no locality mentioned; based on "*Le petit Unua*" (Buffon 1782:245) from Cayenne, French Guiana.

Bradypus kouri Daudin in Lacépède, 1802:172; no locality given; based on "*Le kouri*" (Lacépède 1801:180) from Cayenne, French Guiana.

Bradypus [*(Choloepus)*] *didactylus*: Desmarest, 1816:325; name combination.

Choloepus guianensis Fitzinger, 1871a:399 type localities "Mittel-America, wo diesse Form sowohl in Guiana und Surinam, als auch—wie Gray behauptet—in West-Indien verkommt."

Choloepus brasiliensis Fitzinger, 1871a:403; type localities "Sud-Amerika, Nord-Brasilien, wo Spix und Natterer— letzterer am Rio Xié oferhalb des Äquators gegen de Grenze von Colombien—diese Form getroffen."

[*Choloepus didactylus*] Var. *columbianus* Gray, 1871b: 432; type locality "Columbia."

Choloepus didaetylus O. Thomas, 1893b:166; incorrect subsequent spelling of *Bradypus didactylus* Linnaeus.

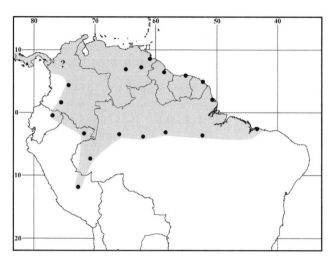

Map 76 Marginal localities for *Choloepus didactylus* ●

Choloepus florenciae J. A. Allen, 1913:469; type locality "Florencia (alt. 1000 ft.), Rio Bodoquera, Caquetá, Colombia."

Choloepus napensis Lönnberg, 1922:18; type locality "along river Napo, altitude 2000 f.," Napo, Ecuador.

Choloepus hoffmanni florenciae: Cabrera, 1958:212; first use of current name combination.

DISTRIBUTION: *Choloepus didactylus* is known from the Río Orinoco delta and south of the Orinoco in Venezuela, throughout the Guianas and eastern Brazil into Maranhão on the Atlantic Coast, and along the southern bank of the Rio Amazonas-Solimões from the Rio Xingu into the upper Amazon basin of Ecuador and Peru.

MARGINAL LOCALITIES (Map 76): VENEZUELA: Delta Amacuro, Curiapo (Linares and Rivas 2004). GUYANA: Cuyuni-Mazaruni, Kalacoon (H. E. Anthony 1921b). SURINAM: Commewijne, Plantation Clevia (Husson 1978). FRENCH GUIANA: Cabassou (Charles-Dominique et al. 1981). BRAZIL (Wetzel and Ávila-Pires 1980, except as noted): Amapá, Amapá; Maranhão, Miritiba; Pará, Cachoeira do Espelho (MZUSP 31328); Amazonas, Itacoatiara; Amazonas, Codajaz; Amazonas, Santa Cruz; Amazonas, Juruá. PERU: Cusco, Cashiriari-2 (Boddicker, Rodríguez, and Amanzo 2001); Loreto, Pebas (O. Thomas 1928d). ECUADOR: Orellana, La Coca (Cabrera 1917). COLOMBIA: Caquetá, Florencia (type locality of *Choloepus florenciae* J. A. Allen); Cundinamarca, Fusagasugá (USNM 241311). VENEZUELA: Bolívar, La Unión (J. A. Allen 1904c); Bolívar, Los Patos (Handley 1976).

SUBSPECIES: We are treating *C. didactylus* as monotypic; the species needs revision (see Remarks).

NATURAL HISTORY: *Choloepus didactylus* inhabits humid tropical lowland forests and forested second-growth habitats throughout its range, and is sympatric with *C. hoffmanni* in the western Amazon basin. This species is host to several species of amblyommid ticks and a commensal assemblage of coprophagous moths, beetles, and mites (Waage and Best 1985). These sloths also are host to a number of hemoflagellates and may be important natural reservoirs for some of these protozoan parasites (J. J. Shaw 1985). Algae inhabit grooves in the hair and are popularly assumed to aid the sloth in concealment, but the true function is unknown (Aiello 1985; Wujek and Cocuzza 1986).

Based mainly on captive individuals, Eisenberg and Malinak (1985) reviewed information on reproduction and concluded that *C. didactylus* is a "K" strategist with a long life span, reduced litter size (one), long gestation (at least 44 weeks), and long interbirth interval (average 26 months). The main predator throughout it range probably is the Harpy Eagle (*Harpia harpyja*), which selects *C. didactylus* out of proportion to its numbers in comparison to *Bradypus* (Izor 1985). *Choloepus didactylus* appears to be less common than *Bradypus tridactylus*; Walsh and Gannon (1967) reported 840 *C. didactylus* and 2,104 *B. tridactylus* relocated during Operation Gwamba in Surinam. *Choloepus didactylus* shows considerable inter- and intrapopulational karyotypic polymorphism ($2n = 52-64$) that is complicated by supernumerary chromosomes (β-chromosomes), and possibly an XO sex-chromosome system in females (Jorge, Orsi-Souza, and Best 1985; Benirschke 2006).

REMARKS: If Colombian and Ecuadorian sloths differ sufficiently from eastern populations to warrant recognition as a subspecies, of the three names available, *C. d. columbianus* Gray, 1871b has priority over both *C. d. florenciae* J. A. Allen, 1913 and *C. d. napensis* Lönnberg, 1922. Oldfield Thomas (1893b) listed the species for Trinidad based on the report by Ledru (1810). This information was repeated by Allen and Chapman (1893). Later, however, Allen and Chapman (1897) suggested that *C. didactylus* be eliminated from the known Trinidadian fauna because presence of the species had not been confirmed. The presence of *C. didactylus* in departamento Cusco, Peru (Boddicker, Rodríguez, and Amanzo 2001) needs to be substantiated by a voucher specimen.

Choloepus hoffmanni W. Peters, 1858
Hoffmann's Two-toed Sloth

SYNONYMS: See under subspecies.

DISTRIBUTION: *Choloepus hoffmanni* appears to be represented by two disjunct ranges in South America. The northern distribution includes northern and western Colombia, western Ecuador, and westernmost Venezuela. This population unit is north of the Andes and west of the continental divide, and is continuous with that of Central

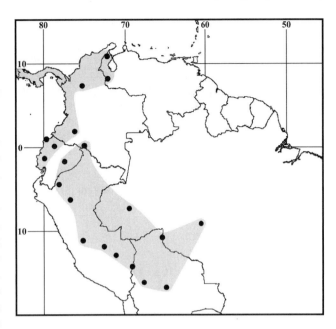

Map 77 Marginal localities for *Choloepus hoffmanni* ●

America. The southern or Amazonian distribution lies east of the Andes in eastern Ecuador, Peru, and Bolivia, and in adjacent western Brazil eastward into Mato Grosso.

MARGINAL LOCALITIES (Map 77; localities listed from north to south for northwestern portion of range and listed clockwise for the Amazonian distribution): *Northwestern distribution.* VENEZUELA: Zulia, Hacienda Platanal (USNM 443404); Táchira, Las Mesas (Handley 1976). COLOMBIA: Antioquia, Puerto Valdivia (J. A. Allen 1916c); Huila, San Agustín (type locality of *Choloepus augustinus* J. A. Allen). ECUADOR: Esmeraldas, Esmeraldas (R. H. Baker 1974); Pichincha, Gualea (Wetzel 1982); Guayas, Balzar (O. Thomas 1880). *Amazonian distribution.* COLOMBIA: Putumayo, Parque Nacional Natural La Paya (Polanco-Ochoa, Jaimes, and Piragua 2000). ECUADOR: Pastaza, Sarayacu (O. Thomas 1880). BRAZIL: Amazonas, Santo Antônio (type locality of *Choloepus juruanus* Lönnberg); Mato Grosso, Aripuanã (Wetzel and Ávila-Pires 1980). BOLIVIA: Beni, Guayaramerín (Wetzel 1985a); Cochabamba, Campamento Yuqui (Salazar, Redford, and Stearman 1990); La Paz, Mururata (Le Pont and Desjeux 1992). PERU: Puno, Valle Grande (Sanborn 1953); Madre de Dios, Hacienda Amazonía (V. Pacheco et al. 1993); Cusco, Cashiriari-3 (Boddicker, Rodríguez, and Amanzo 2001); Junín, Chanchamayo (Wetzel 1985a); San Martín, Calaveras (type locality of *Choloepus didactylus pallescens* Lönnberg); Amazonas, Huampami (J. L. Patton, Berlin, and Berlin 1982).

SUBSPECIES: We tentatively recognize five subspecies of *C. hoffmanni.*

C. h. augustinus J. A. Allen, 1913

SYNONYMS:

Choloepus augustinus J. A. Allen, 1913:470; type locality "near San Augustin (alt. 5000 ft.), Huila, Colombia."

Choloepus andinus J. A. Allen, 1913:472; type locality "Salento, West Quindio Andes (alt. 7000 ft.)," Quindío, Colombia.

Choloepus hoffmanni augustinus: Cabrera, 1958:211; name combination.

This subspecies is in the inter-Andean valleys of Colombia, northwestern Venezuela, and northern Ecuador at least as far south as Gualea, Pichincha (Wetzel 1982).

C. h. capitalis J. A. Allen, 1913

SYNONYMS:

Choloepus capitalis J. A. Allen, 1913:472; type locality "Barbacoas, [Nariño,] Colombia."

Choloepus hoffmanni capitalis: Cabrera, 1958:212; first use of current name combination.

This subspecies occurs in the Pacific lowlands west of the Andes in Colombia and northwestern Ecuador.

C. h. hoffmanni W. Peters, 1858

SYNONYMS:

Choloepus didactylus: Sclater, 1856:139; not *Bradypus didactylus* Linnaeus.

Choloepus hoffmanni W. Peters, 1858:128; type locality "Costa Rica"; restricted to Escazú, San José, by Goodwin (1946:353); corrected to Volcán Barba, Herédia, Costa Rica, by Wetzel and Ávila-Pires (1980).

The South American distribution of this subspecies is in northern and northwestern Colombia. Elsewhere, *C. h. hoffmanni* occurs in Central America as far north as Nicaragua.

C. h. juruanus Lönnberg, 1942

SYNONYMS:

Choloepus juruanus Lönnberg, 1942:29; type locality "San Antonio, E. of Rio Eiru, S. of Rio Juruá," Amazonas, Brazil.

Choloepus didactylus juruanus: C. O. C. Vieira, 1955:401; name combination.

Choloepus hoffmanni juruanus: Cabrera, 1958:212; first use of current name combination.

This subspecies is in western Brazil, southeastern Peru, and northern Bolivia.

C. h. pallescens Lönnberg, 1928

SYNONYMS:

Choloepus didactylus pallescens Lönnberg, 1928:12; type locality "Calavera, a place near Roque on the way to Moyabamba," San Martín, Peru.

Choloepus hoffmanni peruvianus Menegaux, 1906:460; type locality "Pérou."

Choloepus hoffmanni pallescens: Cabrera, 1958:212; first use of current name combination.

This subspecies is in eastern Peru and adjacent Ecuador.

NATURAL HISTORY: Hoffmann's two-toed sloth occupies a variety of lowland to higher elevation forest habitats, and has been captured at 1,800 m on Volcán de Chiriquí, Panama (Enders 1940); Mt. Sapo, Darién, Panama; and along the Cordillera Central of Costa Rica. These sloths may be better able than *Bradypus* spp. to control body temperature, perhaps because of their longer hair. *Choloepus hoffmanni* have a metabolic rate about 44% of that expected for their weight (McNab 1982). They are chiefly nocturnal, move more quickly than *Bradypus*, and descend to the ground to defecate every 6 to 8 days. Gestation is between 170 and 263 days and one young is the rule. In contrast to *Bradypus* spp., *C. hoffmanni* has a larger home range, travels farther during each activity cycle, and feeds on the leaves of approximately twice the variety of tree species.

Many accounts of the behavior and natural history of two-toed sloths have not distinguished between *C. didactylus* and *C. hoffmanni* (e.g., Moeller 1975). Therefore, specific information usually is assigned based on geographic probability. Information derived from zoo-kept animals often is impossible to allocate because identifications may not be correct and voucher specimens rarely are saved. Montgomery and Sunquist (1978) and Wetzel (1982) provide additional natural history information. *Choloepus hoffmanni* exhibits karyotypic variation ($2n = 49-54$) that includes ß-chromosomes and possibly an XX−XO sex-chromosome system (Corin-Frederic 1969; Jorge, Orsi-Souza, and Best 1985).

REMARKS: The nominate subspecies of *C. hoffmanni* (Nicaragua into northwestern South America) intergrades with *C. h. capitalis* in Pacific coastal Colombia and with *C. h. augustinus* in interior Colombia and northwestern Venezuela. *Choloepus h. pallescens* (east of the Andes in Peru) intergrades with *C. h. juruanus* in western Brazil (Wetzel 1982).

Suborder Vermilingua Illiger, 1811

Alfred L. Gardner

The Vermilingua comprise an old Neotropical radiation, reduced to two Recent Neotropical families (Myrmecophagidae and Cyclopedidae), which are distributed from eastern and southern Mexico, southward through Central and South America, including the island of Trinidad, into northern Argentina. Characteristics of the suborder include the lack of teeth, an elongate skull with weak mandibles and incomplete zygomatic arches, small premaxillae, well-developed lacrimals, and double-rolled turbinals. A coracoscapular and an entepicondylar foramen are present, and the third trochanter is undeveloped. The mouth is small, and the tongue is long, protrusible, and vermiform. The body is densely haired, and the tail is nearly as long as or longer than the head and body. The stomach and uterus are simple (Barlow 1984).

Family Cyclopedidae Pocock, 1924

Cyclopedids are small (total length less than 800 mm; greatest length of skull usually less than 60 mm) nocturnal, arboreal anteaters that have a prehensile tail, and fore and hind feet modified for grasping twigs and vines while climbing and foraging (Wetzel 1985a). The single known species has two digits on the manus and four on the pes. The family dates from the late Miocene of South America (McKenna and Bell 1997).

Kraft (1995) used Cyclothuridae, type genus *Cyclothurus* Lesson, 1842, for this family. *Cyclothurus* Lesson (the basis for Cyclothuridae) is a junior synonym of *Cyclopes* Gray. Whereas today, the family name would not be replaced on that account alone, the name was replaced before 1961 and Cyclopedidae is to be maintained as the family name (ICZN 1999:Art. 0.1).

Genus *Cyclopes* Gray, 1821

Cyclopes is a monotypic genus of small (total length of adults averaging 430 mm, mass 235 g) arboreal anteaters characterized by small clavicles, absence of jugals, and pterygoids separated by palatines on the posterior margin of the palate. Additional features include dense, woolly to silky, silvery-gray to golden-brown body pelage; two digits on the manus and four on the pes; a prehensile tail (as in *Tamandua*), which is naked along the terminal quarter of the undersurface and at the tip. The ribs are unusually broad and flattened, and the caecum is divided (Barlow 1984).

SYNONYMS:

Myrmecophaga Linnaeus, 1758:35; part.

Mirmecophaga Brongniart, 1792:115; part; incorrect subsequent spelling of *Myrmecophaga* Linnaeus

Cyclopes Gray, 1821:305; type species *Myrmecophaga didactyla* Linnaeus, 1758, by monotypy.

Cyclothurus Gray, 1825b:343; *nomen nudum*.

Didactyles F. Cuvier, 1829:501; based on "Les Didactyles," no species mentioned.

Myrmydon Wagler, 1830:36; type species *Myrmecophaga didactyla* Linnaeus, 1758, by monotypy.

Myrmecolichnus Reichenbach, 1836:51; type species *Myrmecolichnus didactylus*: Reichenbach, 1836 (= *Myrmecophaga didactyla* Linnaeus, 1758) by monotypy.

Eurypterna Gloger, 1841:112; type species *Eurypterna didactyla*: Gloger, 1841 (= *Myrmecophaga didactyla* Linnaeus, 1758), by monotypy.

Cyclothurus Lesson, 1842:152; type species *Cyclothurus didactyla*: Lesson, 1842 (= *Myrmecophaga didactyla* Linnaeus, 1758), by monotypy.

Cycloturus Sclater, 1871:546; unjustified emendation of *Cyclothurus* Lesson.

Didactyla Liais, 1872:356; type species *Myrmecophaga didactylus* Linnaeus, 1758, by monotypy.

Cycloturus Flower and Lydekker, 1891:193; incorrect subsequent spelling of *Cyclothurus* Lesson.

Mamcyclothurus A. L. Herrera, 1899:19; unavailable name because it is based on zoological formulae.

Cycloturus Goeldi and Hagmann, 1904:97; incorrect subsequent spelling of *Cyclothurus* Lesson.

Cyclopes didactylus (Linnaeus, 1758)
Silky Anteater

SYNONYMS: See under Subspecies.

DISTRIBUTION: *Cyclopes didactylus* appears to have a disjunct distribution in South America. One distribution is in the northern Andean valleys of Colombia and in the Pacific lowlands of Colombia and Ecuador. The other includes the island of Trinidad, northern Venezuela, the Guianas, and southward into the Brazilian state of Alagoas in the east and the upper Amazon basin drainage of the eastern lowlands of Colombia, Ecuador, Peru, and Bolivia in the west. Elsewhere, the species is in southern Mexico and Central America.

MARGINAL LOCALITIES (Map 78): *Northwestern distribution.* COLOMBIA: Chocó, Unguía (FMNH 69970); Antioquia, Puerto Valdivia (J. A. Allen 1916c); Nariño, 30 km E of Tumaco on Tumaco-Tuquerres Road (USNM 554227). ECUADOR: Guayas, Balzar Mts. (O. Thomas 1902g). *Greater South American distribution.* TRINIDAD AND TOBAGO: Trinidad, Princestown (J. A. Allen and Chapman 1893). VENEZUELA: Delta Amacuro, Winikina (Linares and Rivas 2004). GUYANA: Pomeroon-Supenaam, Better Hope (BM 80.11.29.10). SURINAM: Paramaribo, Paramaribo (Husson 1978). FRENCH GUIANA: Cayenne (Brongniart 1792). BRAZIL: Amapá, Taperebá (C. T. Carvalho 1962a); Pará, Ilha das Onças (Goeldi and Hagmann 1904); Maranhão, Humberto de Campos (MZUSP 3176); Paraíba, Camaratuba (MZUSP 8451); Pernambuco, São Lorenço da Mata (BM 3.10.1.80); Alagoas, Manimbu (MZUSP 7523); Pará, Villa Braga (O. Thomas 1920c); Rondônia, Pyrineus (A. Miranda-Ribeiro 1914). BOLIVIA: Santa Cruz, Parque

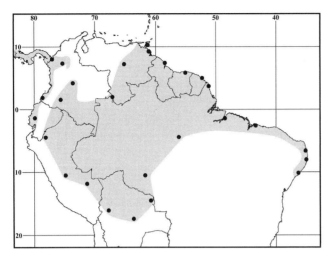

Map 78 Marginal localities for *Cyclopes didactylus* ●

Nacional Noël Kempff Mercado (Emmons 1998); Santa Cruz, Buenavista (O. Thomas 1928d); La Paz, between Suapi and Mururata (S. Anderson 1997). PERU: Madre de Dios, Cocha Cashu (Terborgh, Fitzpatrick, and Emmons 1984); Pasco, San Juan (USNM 364503); Amazonas, Huampami (J. L. Patton, Berlin, and Berlin 1982). COLOMBIA: Caquetá, Muralla (J. A. Allen 1916c); Meta, Villavicencio (BM 24.10.8.4); Guainía, Raudal Mabajate (USNM 256309). VENEZUELA: Bolívar, Suapuré (USNM 143741).

SUBSPECIES: I follow Wetzel (1982) in recognizing seven subspecies of silky anteaters; only *C. d. mexicanus* is extralimital.

C. d. didactylus (Linnaeus, 1758)
SYNONYMS:

[*Myrmecophaga*] *didactyla* Linnaeus, 1758:35; type locality "America australi"; restricted to Surinam by O. Thomas (1911b:132).

Myrmecophaga monodactyla Kerr, 1792:105; no type locality given.

Myrmecophaga unicolor Desmarest, 1822:375, footnote; no type locality given; name attributed to É. Geoffroy St.-Hilaire.

Eurypterna didactyla: Gloger, 1841:112; name combination.

Cyclothurus didactyla: Lesson, 1842:152; name combination.

C[*yclopes*]. *didactylus*: O. Thomas, 1900d:302; first use of current name combination.

The nominate subspecies occurs in Venezuela, the island of Trinidad, and the Guianas east to the limit of the known range of the species in eastern Brazil.

C. d. catellus O. Thomas, 1928
SYNONYMS:

Cyclopes didactylus catellus O. Thomas, 1928d:293; type locality "Buenavista, 500 metres," Santa Cruz, Bolivia.

Cyclopes didactylus codajazensis Lönnberg, 1942:46; type locality "Rio Solimoes, Codajaz," Amazonas, Brazil.

This subspecies is in southeastern Peru, northeastern Bolivia, and the central Amazon basin of Brazil.

C. d. dorsalis (Gray, 1865)
SYNONYMS:

Cyclothurus dorsalis Gray, 1865:385; type locality "Costa Rica."

Cyclopes dorsalis: Bangs, 1902:20; name combination.

[*Cyclopes didactylus*] *dorsalis*: Trouessart, 1905:803; first use of current name combination.

This subspecies, although mainly Central American in distribution, occurs in northern and northwestern Colombia.

C. d. eva O. Thomas, 1902
SYNONYM:

Cyclopes didactylus eva O. Thomas, 1902g:250; type locality "Rio Tapayo, N.W. Ecuador" (spelled "Tupuyo" by Cabrera 1958; probably Río Zapallo Grande, Esmeraldas, Ecuador, according to Paynter and Traylor 1977).

This subspecies inhabits the Pacific lowland forests of Colombia and Ecuador. Although expected in northwestern Peru, there is no evidence that *C. d. eva* occurs there (Grimwood 1969).

C. d. ida O. Thomas, 1900
SYNONYMS:

Cyclopes didactylus ida O. Thomas, 1900d:302; type locality "Sarayacu, Upper Pastaza River," Pastaza, Ecuador.

Cyclopes juruanus Lönnberg, 1942:47; type localities "Rio Juruá, João Pessoa . . . Rio Juruá, Rio Eirú, Santo Antonio," Amazonas, Brazil; restricted to the Rio Juruá by Cabrera (1958); the type locality can be further restricted to João Pessoa, Rio Juruá.

Cyclopes didactylus jurnanus Cabrera, 1958:207; incorrect subsequent spelling of *C. juruanus* Lönnberg, 1942, and first use of current name combination.

This subspecies occurs throughout the western Amazon drainage of Brazil and adjacent Colombia, Ecuador, and Peru, south at least to the Río Alto Purus.

C. d. melini Lönnberg, 1928
SYNONYM:

Cyclopes didactylus melini Lönnberg, 1928:15; type locality "S. Gabriel, Rio Negro," Amazonas, Brazil.

This subspecies is in the northern Amazon basin of Brazil and adjacent Colombia and Venezuela.

NATURAL HISTORY: *Cyclopes didactylus* is nocturnal and appears to be completely arboreal (Montgomery 1985b). Most nocturnal activity and daytime resting sites are on or among small stems, vines, and lianas (Sunquist and Montgomery 1973a; Montgomery 1985a). An opportunistic forager, its diet consists almost entirely of ants, which are usually extracted from hollow stems, galls, and vines (Best and Harada 1985; Montgomery 1985a, 1985b). Longevity and gestation period are unknown. One young is born and remains with the mother until attaining approximately three quarters of her size (Montgomery 1985a). The karyotype is $2n = 64$, FN $= 98-100$ (Jorge, Best, and Wetzel 1986).

REMARKS: Emmons (1998) listed *C. didactylus*, based on a sight record in semi-evergreen forest habitat in Parque Nacional Noel Kempff Mercado, Santa Cruz, Bolivia.

Family Myrmecophagidae Gray, 1825

Myrmecophagids are medium-sized to large anteaters with four claw-bearing digits on the forefeet. The hind feet have five digits and are adapted for plantigrade locomotion. Modern representatives consist of the genera *Myrmecophaga* (one species) and *Tamandua* (two species). The fossil record dates from the early Miocene of South America (McKenna and Bell 1997).

KEY TO THE GENERA AND SPECIES OF MYRMECOPHAGIDAE (BASED IN PART ON WETZEL 1985A):

1. Fourth digit on manus short and inconspicuous; rostrum usually exceeding 65% of condylonasal length; size large, total length of adults exceeding 1.5 m; tail bannerlike with long hairs. . . . . *Myrmecophaga tridactyla*
1'. Fourth digit on manus conspicuous; rostrum 50% or less than condylonasal length of skull; size intermediate, total length of adults less than 1.3 m; tail nearly naked along distal three-quarters of its length. 2
2. Four pairs of orbital foramina (*foramen rotundum* and orbital fissure separate); posterior border of infraorbital foramen symmetrical and crescent shaped; body with black vest contrasting with pale background; pinna averaging 40 to 46 mm, as measured from notch. . . . . . .
. . . . *Tamandua mexicana*
2'. Usually three pairs of orbital foramina (*foramen rotundum* and orbital fissure normally confluent); posterior border of infraorbital canal not a symmetrical crescent; color pattern variable, from uniformly golden, brown, or black to pale with partial or complete black vest; pinna longer, averaging 50 to 54 mm, as measured from notch
. . . . *Tamandua tetradactyla*

Genus *Myrmecophaga* Linnaeus, 1758

Myrmecophaga is a monotypic genus of large (total length of adults averages 2 m, mass 33 kg) terrestrial anteaters characterized by having rudimentary clavicles, a rostrum that is much longer than the braincase, and a long, nonprehensile tail that is heavily crested dorsally and ventrally with long hair. Features shared with *Tamandua* include: the posterior margin of the palate is formed by pterygoids, ribs have a double articulation with sternal elements, and the manus has four toes and the pes five. Other features of *Myrmecophaga* include pectoral mammae; long dark pelage, except on head, where hair is short and gray; color pattern of body dominated by a broad black band edged with white that extends from the throat and chest posteriorly over shoulders and back; and a short, sharp, and usually inconspicuous (may be elongate in captives) claw on manual digit I (Barlow 1984; Wetzel 1985a). The genus is known from the early Pleistocene to Recent of South America (Simpson 1945; McKenna and Bell 1997).

SYNONYMS:

Myrmecophaga Linnaeus, 1758:35; type species *Myrmecophaga tridactyla* Linnaeus, 1758, by subsequent selection (O. Thomas 1901d:143); placed on Official List of Generic Names in Zoology (ICZN 1926).

Nyrmecophaga Beckstein, 1801:1346; incorrect subsequent spelling of *Myrmecophaga* Linnaeus.

Myrmecopha G. Fischer, 1803:333; incorrect subsequent spelling of *Myrmecophaga* Linnaeus.

Myrmecophagus Gray, 1825b:343; unjustified emendation of *Myrmecophaga* Linnaeus.

Falcifer Rehn, 1900:576; type species *Myrmecophaga jubata* Linnaeus, 1766, by monotypy.

REMARKS: Rehn (1900) proposed the name *Falcifer* because he mistakenly believed that the type species of *Myrmecophaga* was *M. tetradactyla* (a *Tamandua*). Hall and Kelson (1959:237) and Hall (1981:276) credited Fleming (1822:194) with selection of *Myrmecophaga jubata* Linnaeus, 1766, as the type species of *Myrmecophaga*. However, Fleming merely gave the name *M. jubata* under *Myrmecophaga* without stating or indicating in any other manner that it was the type species; therefore, he cannot be credited with designating the type species of *Myrmecophaga*.

Myrmecophaga tridactyla Linnaeus, 1758
Giant Anteater

SYNONYMS: See under Subspecies.

DISTRIBUTION: *Myrmecophaga tridactyla* is known from lowland forest and savanna habitats in Colombia, Ecuador, Venezuela, and the Guianas southward through Peru, Brazil, Bolivia, and Paraguay into Uruguay and northern Argentina. The species is not known from the island of

Map 79 Marginal localities for *Myrmecophaga tridactyla* ●

Trinidad, but has been recorded from Central America as far north as Guatemala.

MARGINAL LOCALITIES (Map 79): COLOMBIA: La Guajira, Dibulla (Bangs 1900). VENEZUELA: Zulia, Empalado Savannas (type locality of *M. tridactyla artata* Osgood); Guárico, Masaguaral (Montgomery 1985b); Bolívar, Ciudad Bolívar (J. A. Allen 1904c). GUYANA: Cuyuni-Mazaruni, Kartabo (H. E. Anthony 1921b). SURINAM: Saramacca, Groningen (Husson 1978). FRENCH GUIANA: Kourou (Floch and Fauran 1958). BRAZIL: Amapá, Macapá (C. T. Carvalho 1962a); Pará, Lago Arary (Krumbiegel 1940b); Maranhão, Barra do Corda (Vaz 2003); Piauí, Serra da Capivara National Park (Olmos 1995); Pernambuco, Pernambuco (type locality of *M. tridactyla* Linnaeus); Espírito Santo, Lagoa de Juparanin da Praya (Wied-Neuwied 1826); São Paulo, Ypanema (Pelzeln 1883); Rio Grande do Sul, São Lourenço do Sul (Wetzel 1985a). URUGUAY: Durazno, Estancia San Jorge (Christison 1880). ARGENTINA: Buenos Aires, La Plata (Gerrard 1862). PARAGUAY: Presidente Hayes, Colonia Benjamín Aceval (Krumbiegel 1940b). ARGENTINA: Salta, Quebrachal (Ojeda and Mares 1989); Salta, Santa Bárbara (Ojeda and Mares 1989). BOLIVIA: Tarija, Samuhuate (Eisentraut 1933); Santa Cruz, Buenavista (BM 28.2.9.75); Cochabamba, Campamento Yuquí (S. Anderson 1997); Beni, El Consuelo (Lönnberg 1942). PERU: Cusco, Ha-

cienda Cadena (BM 4.12.4.17); Cusco, Pagoreni (Boddicker, Rodríguez, and Amanzo 1999); Pasco, Pozuzo (BM 8.6.17.21); Amazonas, Huampami (J. L. Patton, Berlin, and Berlin 1982). ECUADOR: Morona-Santiago, Gualaquiza (BM 14.4.26.88); Napo, Baeza (J. A. Allen 1916b); Pichincha, Río Tulipe (FMNH 95014).

SUBSPECIES: I tentatively recognize three subspecies of *M. tridactyla*.

M. t. artata Osgood, 1912
SYNONYMS:

Myrmecophaga tridactyla artatus Osgood, 1912:40; type locality "Empalado Savannas, 30 miles east of Maracaibo, [Zulia,] Venezuela."

Myrmecophaga tridactyla artata: Pittier and Tate, 1932: 255; corrected gender concordance.

Myrmecophaga trydactyla Utrera and Ramo, 1989:65; incorrect subsequent spelling of *Myrmecophaga tridactyla* Linnaeus.

This subspecies is in northeastern Colombia and northwestern Venezuela, north and west of the Mérida Andes.

M. t. centralis Lyon, 1906
SYNONYMS:

Myrmecophaga centralis Lyon, 1906:570; type locality "Pacuare," Limón, Costa Rica.

Myrmecophaga tridactyla centralis: Goldman, 1920:64; first use of current name combination.

Known primarily from Central America, this subspecies also occurs from northwestern Colombia into northern Ecuador (Hall 1981; Wetzel 1982).

M. t. tridactyla (Linnaeus, 1758)
SYNONYMS:

[*Myrmecophaga*] *tridactyla* Linnaeus, 1758:35; type locality "America meridionali"; restricted to Pernambuco, Pernambuco, Brazil (O. Thomas 1911b:132).

[*Myrmecophaga*] *jubata* Linnaeus, 1766:52; type locality "Brasilia."

M[*yrmecophaga*]. *iubata* Wied-Neuwied, 1826:537; variant spelling and incorrect subsequent spelling of *M. jubata* Linnaeus.

Tamandua tridactyla: Matschie, 1894:63; name combination.

Falcifer jubata: Rehn, 1900:576; name combination.

The nominate subspecies occurs east of the Andes from Venezuela and the Guianas into northern Argentina.

NATURAL HISTORY: Giant anteaters inhabit a variety of forest and grassland habitats, where they are terrestrial, usually active in late afternoon and evening, and feed mainly on ants (Montgomery 1985b; J. H. Shaw, Carter, and Machado-Neto 1985). Termites also are eaten, but at a much lower frequency and their importance in the diet may vary seasonally; Montgomery (1985b) found more termites were consumed during the rainy season in Venezuela. Based on an ecological study of *M. tridactyla* in western Minas Gerais, Brazil, J. H. Shaw, Carter, and Machado-Neto (1985) estimated anteater densities of from $0.17/km^2$ to $1.31/km^2$. They postulated that 432 ant nests/km^2 are raided per day based on anteater population averages of $1.5/km^2$ and a foraging rate of 36 ant nests per hour for an 8-hour activity cycle. This is comparable to Redford's (1985a) estimate of from 30 to 40 colonies visited per hour.

A single young is born following a gestation period of about 190 days (Asdell 1964). The young, often transported on its mother's back, remains with her until she again becomes pregnant. With the exception of mother-young pairs, giant anteaters appear to be solitary. J. H. Shaw, Carter, and Machado-Neto (1985) described an encounter between individuals they interpreted as agonistic, suggesting territorial behavior. A captive lived almost 26 years (Nowak and Paradiso 1983).

Myrmecophaga tridactyla is considered to be terrestrial, does not burrow, but is able to climb trees (Wetzel 1982; J. H. Shaw, Carter, and Machado-Neto 1985). J. H. Shaw, Carter, and Machado-Neto (1985) interpreted the function of the powerful fore claws as primarily defensive instead of only facilitating foraging at ant nests and termite mounds. The short, triangular inner claw on the forefeet can be an effective weapon. An immature (less than half-grown) Giant Anteater captured at Balta, Ucayali, Peru, sliced through a leather boot with one swipe of its paw.

Bechara et al. (2002) reported three species of ticks found on *T. tridactyla* in Brazil. The karyotype is $2n = 60$, FN = 104 (Hsu 1965; H. R. J. Pereira, Jorge, and Costa 2004), and shows no close resemblance to that of *Tamandua tetradactyla*. The X chromosome is a large metacentric and the Y is a small subtelocentric (Hsu 1965).

REMARKS: Hall and Kelson (1959), Hall (1981), Wetzel (1982), and Eisenberg (1989) indicated western Colombia in the distribution of *M. tridactyla* , and Alberico et al. (2000) listed the species for departamento Nariño. I have been unable to find records of specimens confirming the presence of *M. tridactyla* in western Colombia.

Genus *Tamandua* Gray, 1825

The genus *Tamandua* contains two species of medium-sized (total length of adults 770–1,300 mm, mass 3.2–7.0 kg) anteaters found from southern Tamaulipas, Mexico, south into northern Argentina. The genus is characterized by a rostrum about 1/2 (or less) as long as the greatest length of skull, ribs articulate with adjacent sternal elements, and there are four digits on the manus and five on the pes (as in *Myrmecophaga*). The tail is prehensile and the ventral portion is naked at the tip, the caecum is simple, and the body and base of the tail are covered by short, smooth hair. The color pattern varies from unicolor yellow gold, brown, or black, to pale gray with a black vest (Barlow 1984; Wetzel 1985a). The genus is known from the Pleistocene and Recent of South American (Simpson 1945; McKenna and Bell 1997).

SYNONYMS:

Myrmecophaga Linnaeus, 1758:35; part.

Tamandua Gray, 1825b:343; no species mentioned; indication is to "Gray, M. R." (= Gray 1821:305); inferred type species *Myrmecophaga tamandua* G. Cuvier, 1798, by tautonomy (Gray 1821:305).

Tamanduas F. Cuvier, 1829:501; French vernacular, unavailable name.

Uroleptes Wagler, 1830:36; type species *Myrmecophaga tetradactyla* Linnaeus, 1758, by monotypy.

Dryoryx Gloger, 1841:112; no species mentioned.

Uropeltes Alston, 1880:191; incorrect subsequent spelling of *Uroleptes* Wagler.

Tamandua mexicana (Saussure, 1860)
Northern Tamandua

SYNONYMS:

Myrmecophaga tamandua Var. *Mexicana* Saussure, 1860: 9; type locality "Tabasco," Mexico.

Tamandua tetradactyla mexicana: J. A. Allen, 1906a:200; name combination.

T[amandua]. t[etradactyla]. tetradactyla: Reeve, 1942:300; part; not *Myrmecophaga tetradactyla* Linnaeus, 1758.

Tamandua mexicana: Wetzel, 1975:104; first use of current name combination.

Additional synonyms under subspecies.

DISTRIBUTION: *Tamandua mexicana* occurs in northern and western Colombia, northwestern Venezuela (north of the Mérida Andes), western Ecuador, and northwestern Peru. Elsewhere, the species is known from Central America and Mexico.

MARGINAL LOCALITIES (Map 80): COLOMBIA: Magdalena, Bonda (type locality of *Tamandua mexicana instabilis* J. A. Allen); La Guajira, Villanueva (USNM 281335). VENEZUELA: Zulia, Empalado Savannas (Osgood 1912); Táchira, Las Mesas (USNM 443247). COLOMBIA: Antioquia, Concordia (BM 73.4.23.8); Tolima, Chicoral (J. A. Allen 1916c); Cauca, El Tambo (type locality of *Tamandua tetradactyla tambensis* Lönnberg). ECUADOR: Pichincha, Pacta (Lönnberg 1921); Guayas, Isla Puná (type locality of *Tamandua tetradactyla punensis* J. A. Allen). PERU: Lambayeque, Cabache (FMNH 80972).

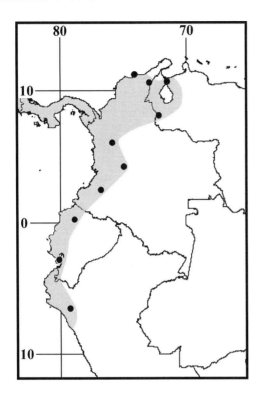

Map 80 Marginal localities for *Tamandua mexicana* •

SUBSPECIES: I recognize three subspecies in South America, and follow Wetzel (1982), except that *T. m. opistholeuca* has priority over *T. m. chiriquensis*.

T. m. instabilis J. A. Allen, 1904
SYNONYM:
Tamandua tetradactyla instabilis J. A. Allen, 1904e:392; type locality "Bonda," Magdalena, Colombia.

This subspecies is in Colombia north and east of the Cordillera Oriental, and in northwestern Venezuela.

T. m. opistholeuca Gray, 1873
SYNONYMS:
[*Tamandua bivittata*] Var. 3. *Opistholeuca* Gray, 1873a:27; several localities listed, type locality restricted to Colombia (New Grenada of Gray 1873a) by Wetzel (1975:104) based on the first (not second as stated by Wetzel) syntype in Gray's (1873) list.
Tamandua tetradactyla, var. *leucopygia* Gray, 1873b:469; *nomen nudum*.
Myrmecophaga quadridactyla True, 1884:588; incorrect subsequent spelling (*lapsus*) of *M. tetradactyla*; not *Myrmecophaga tetradactyla* Linnaeus, 1758).
Myrmecophaga sellata Cope, 1889a:133; type locality "Honduras."
[*Tamandua*] *sellata*: Trouessart, 1898:1121; name combination.

Tamandua tetradactyla chiriquensis J. A. Allen, 1904e:395; type locality "Boqueron, Chiriqui, Panama."
Tamandua tetradactyla sellata: Goldman, 1920:63, footnote; name combination.
Tamandua tetradactyla tambensis Lönnberg, 1937:25; type locality "El Tambo, Cauca, Colombia."
T[amandua]. m[exicana]. chiriquensis: Wetzel, 1982:350; name combination.

This subspecies occurs in western Ecuador, northwestern Peru, and northern and western Colombia west of the Cordillera Oriental. Elsewhere, it occurs in Central America as far north as Honduras (see Wetzel 1982, who reported this subspecies as *T. m. chiriquensis*).

T. m. punensis J. A. Allen, 1916
SYNONYMS:
Tamandua tetradactyla punensis J. A. Allen, 1916a:83; type locality "Puna Island," Guayas, Ecuador.
T[amandua]. m[exicana]. punensis: Wetzel, 1982:350; first use of current name combination.

Although Wetzel's remarks (1982:350) suggest that he considered this subspecies to be restricted to Isla Puna, I am applying this name to populations in southwestern Ecuador and northwestern Peru.

NATURAL HISTORY: *Tamandua mexicana* is relatively common from sea level to as high as 1,200 m (Cuervo-Días, Hernández-Camacho, and Cadena 1986). Active both day and night, the species forages on ants and termites on the ground and in trees (Lubin 1983; Montgomery 1985a, 1985b). Den sites include hollow trees and logs, and holes in the ground. One young is born following a gestation of from 130 to 150 days (Silveira 1968; Lubin 1983). The karyotype is $2n = 54$, FN = 104 (specimen from Mexico; Hsu and Benirschke 1969, as *T. tetradactyla*).

REMARKS: Wetzel (1975:104) selected a specimen from Colombia as the lectotype of *T. m. opistholeuca* Gray, 1873a, thereby fixing the name as the senior synonym of *T. m. chiriquensis* J. A. Allen, 1904e. Hall (1981). Nevertheless, Wetzel (1982 1985a) used the name of the junior synonym, *T. m. chiriquensis*, for the southern Central American and northwestern South American subspecies. I have included *Myrmecophaga sellata* Cope, 1889a, from Honduras, in the synonymy of *T. mexicana opistholeuca* Gray, 1873a, although Hall (1981) treated it as a synonym of the nominate subspecies occurring in Mexico and Guatemala.

Tamandua tetradactyla (Linnaeus, 1758)
Southern Tamandua
SYNONYMS: I am unable to assign the following synonyms to subspecies because the exact origins of their type specimens are unknown. I have arranged assignable synonyms (see Subspecies) according to Wetzel (1982), except

for *T. t. nigra* É. Geoffroy St.-Hilaire, which I use in place of *T. t. longicaudata* J. A. Wagner as the correct name for the northern South American subspecies.

Myrmecophaga tamandua G. Cuvier, 1798:143; no locality mentioned; identified as "L'Amérique méridionale" by É. Geoffroy St.-Hilaire (1803:217).

M[yrmecophaga] longicaudata Turner, 1853:218; no type locality given; preoccupied by *Myrmecophaga longicaudata* J. A. Wagner, 1844.

[*Tamandua bivittata*] Var. 1. *Opisthomelas* Gray, 1873a:27; based on six specimens from "Brazils" and one from "S. America"; type locality restricted to Brazil through the selection by Wetzel (1975:106) of the first in Gray's list of syntypes as the lectotype.

Tamandua longicaudata var. *nigra* Beaux, 1908:417 type locality "Brasilien (?)"; preoccupied by *Myrmecophaga nigra* É. Geoffroy St.-Hilaire, 1803:217.

[*Tamandua*] *opisthomelas*: Osgood, 1910:24; name combination.

DISTRIBUTION: *Tamandua tetradactyla* occurs on the island of Trinidad and in Venezuela, the Guianas, Brazil, Uruguay, Paraguay, and northern Argentina, as well as throughout the western Amazon basin of Colombia, Ecuador, Peru, and Bolivia.

MARGINAL LOCALITIES (Map 81): VENEZUELA: Falcón, 13 km N and 13 km E of Mirimire (USNM 443394); Miranda, Parque Nacional Guatopo (Handley

1976). TRINIDAD AND TOBAGO: Trinidad, Saint Anns (BM 92.12.15.2). GUYANA: Cuyuni-Mazaruni, Kartabo (H. E. Anthony 1921b). SURINAM: Commewijne, Plantation Clevia (Husson 1978). FRENCH GUIANA: Ouanary (Menegaux 1902). BRAZIL: Amapá, Serra do Navio (USNM 393817); Pará, Caldeirão (BM 23.8.10.10); Ceará, Parque Nacional de Ubajara (Guedes et al. 2000); Pernambuco, Pernambuco (type locality of *Myrmecophaga tetradactyla* Linnaeus); Alagoas, Mangabeiras (C. O. C. Vieira 1953); Bahia, Bahia (Pelzeln 1883); Espírito Santo, Santa Lúcia Biological Station (Srbek-Araujo and Chiarello 2005); São Paulo, Ypanema (Pelzeln 1883); Santa Catarina, Colonia Hansa (Lönnberg 1942); Santa Catarina, Araranguá (USNM 114839); Rio Grande do Sul, São João do Monte Negro (Cope 1889a). URUGUAY: Cerro Largo, Estancia La Formosa (Ximénez 1972). ARGENTINA: Corrientes, Corrientes (Krumbiegel 1940b); Santiago del Estero, Ahí Veremos (Julia, Richard, and Samaniego 1994); Catamarca, Ancasti (Julia, Richard, and Samaniego 1994); Tucumán, San Javier (Mares et al. 1996); Jujuy, Reyes (Yepes 1944); Salta, Aguaray (Yepes 1944). BOLIVIA: Santa Cruz, Comarapa (BM 34.9.2.196); La Paz, Bellavista (BM 1.2.1.35). PERU: Puno, Valle Grande (Sanborn 1953); Madre de Dios, Cocha Cashu (Terborgh, Fitzpatrick, and Emmons 1984); Cusco, Cashiriari-3 (Boddicker, Rodríguez, and Amanzo 1999); Junín, Chanchamayo (BM 7.6.15.8); San Martín, Yurac Yacu (type locality of *Tamandua tetradactyla quichua* O. Thomas); Amazonas, Huampami (J. L. Patton, Berlin, and Berlin 1982). ECUADOR: Pastaza, Andoas (BM 54.645); Sucumbios, Santa Cecilia (R. H. Baker 1974). COLOMBIA: Meta, Río Guapaya, La Macarena (Fairchild, Kohls, and Tipton 1966); Cundinamarca, Medina (USNM 544399).

SUBSPECIES: I recognize four subspecies of *T. tetradactyla*.

T. t. nigra (É. Geoffroy St.-Hilaire, 1803)

SYNONYMS:

Myrmecophaga nigra É. Geoffroy St.-Hilaire, 1803:217; type locality "La Guyane?" (= French Guiana; according to Cabrera 1958:205, "muy posiblemente de Cayena.").

[*Myrmecophaga*] *crispa* Rüppell, 1842b:179; type locality "Guiana."

M[yrmecophaga]. longicaudata J. A. Wagner, 1844:211; type locality "in dem nördlichen Theil Südamericas"; restricted "al interior de Surinam," by Cabrera (1958: 203).

Tamandua longicaudata: Gray, 1865:384; name combination.

Tamandua tamandua: Jentink, 1888:215; name combination; not *Myrmecophaga tamandua* G. Cuvier.

Map 81 Marginal localities for *Tamandua tetradactyla* ●

[*Tamandua*] *longicauda* Trouessart, 1898:1121; incorrect subsequent spelling of *Myrmecophaga longicaudata* J. A. Wagner.

Tamandua tetradactyla nigra: Menegaux, 1902:494; first use of current name combination.

Tamandua tetradactyla longicaudata: Pittier and Tate, 1932:255; name combination.

Tamandua longicauda Vesey-FitzGerald, 1936:164; incorrect subsequent spelling of *Myrmecophaga longicaudata* J. A. Wagner.

Tamandua longicauda Rode, 1937:346; incorrect subsequent spelling of *Myrmecophaga longicaudata* J. A. Wagner.

Tamandua longicaudata mexianae Cabrera, 1958:203; *nomen nudum*.

This is the subspecies found from eastern Colombia, through Venezuela, Trinidad, and the Guianas, and in the east-central and northern Amazon basin of Brazil.

T. t. quichua O. Thomas, 1927

SYNONYMS:

Tamandua tetradactyla quichua O. Thomas, 1927a:371; type locality "Yurac Yacu," San Martín, Peru.

T[*amandua*]. *quichua*: Lönnberg, 1942:43; name combination.

This subspecies inhabits the upper Amazon basin of eastern Ecuador, Peru, and adjacent Brazil.

T. t. straminea (Cope, 1889)

SYNONYMS:

Myrmecophaga bivittata straminea Cope, 1889a:132; type localities "at São João [Rio Grande do Sul] or at Chapada [Mato Grosso]"; probably Mato Grosso according to Wetzel (1975:106), which I interpret as a restriction of the type locality to Chapada, Mato Grosso, Brazil.

Tamandua tridactyla: Matschie, 1894:62; name combination; refers to *Myrmecophaga tetradactyla* Linnaeus, not *Myrmecophaga tridactyla* Linnaeus.

Tamandua tetradactyla chapadensis J. A. Allen, 1904e:392; type locality "Chapada, Matto Grosso, Brazil."

[*Tamandua tetradactyla*] *straminea*: Trouessart, 1905:803; name combination.

Tamandua tetradactyla kriegi Krumbiegel, 1940b:171; type locality "Zanja Moroti," Concepción, Paraguay.

T[*amandua*]. *kriegi*: Lönnberg, 1942:42; name combination.

[*Tamandua*] *straminea*: Osgood, 1910:24; name combination.

The distribution of this subspecies extends from the southern Amazon basin of Brazil to the southern and southwestern limits of the range of the species in Bolivia, Argentina, Paraguay, and Uruguay.

T. t. tetradactyla (Linnaeus, 1758)

SYNONYMS:

[*Myrmecophaga*] *tetradactyla* Linnaeus, 1758:35; type locality "America meridionali"; restricted to Pernambuco (= Recife), Pernambuco, Brazil, by O. Thomas (1911b: 133).

Myrmecophaga myosura Pallas, 1766:64; type locality Brazil.

Myrmecophaga bivittata Desmarest, 1817b:107; type locality "Brésil."

Uroleptes tetradactyla: Wagler, 1830:36; name combination.

Tamandua tetradactyla: Gray, 1843b:191; name combination.

Uroleptes bivittatus: Fitzinger, 1860:395; name combination.

Tamandua bivittata: Gray, 1865:384; name combination.

Tamandua brasiliensis Liais, 1872:360; type locality "Brasil"; based on Marcgraf (1648), therefore, type locality is Recife, Pernambuco, Brazil.

[*Tamandua tetradactyla*] *bivittata*: Osgood, 1910:24; name combination.

The nominate subspecies is in the Atlantic lowlands and eastern highlands of Brazil from Rio Grande do Norte south into Rio Grande do Sul.

LIFE HISTORY: *Tamandua tetradactyla* occurs in forest and woodland habitats, where it feeds on ants and termites mainly extracted from arboreal nests, although it also forages on the ground. The species may be active day or night, may visit as many as 50 to 80 colonies of ants and termites during each daily activity cycle (Montgomery 1985a), and prefers termites (Matthews in Wetzel 1982:350; Montgomery 1985a). One young is born each year (Eisenberg 1989).

REMARKS: Turner (1853) proposed the name *M. longicaudata* for the yellow, long-tailed variant lacking a black vest. Although preoccupied by *M. longicaudata* J. A. Wagner, the name is not a *nomen nudum* as stated by Wetzel (1975). Both *T. t. nigra* and *T. t. quichua* were based on melanistic individuals.

Although karyotypes have been reported for *T. tetradactyla* and "*T. longicaudata*" (2*n* = 54, FN = 104; Hsu 1965; Jorge, Meritt, and Benirschke 1977), the karyotype reported by Hsu (1965) clearly represents *T. mexicana*, and the identity of the *T. longicaudata* (= *T. tetradactyla*) from the Lincoln Park Zoological Park animal karyotyped (2*n* = 54) by Jorge, Meritt, and Benirschke (1977) needs to be confirmed. This animal was identified by Benirschke (2006:86) as "*Tamandua tetradactyla (longicaudata)*." It is likely that the 2*n* = 56, FN = 106 karyotype reported by H. R. J. Pereira, Jorge, and Costa (2004) for the male

Tamandua sp. from Brazil actually represents the karyotype for *T. tetradactyla.*

Myrmecophaga annulata Desmarest, 1817b:107, is based on a figure of a Brazilian coatimundi (*Nasua nasua*; see Turner 1853:218; Gray 1865:384). The name *Myrmecophaga striata* is based on the illustration of the "striped ant-eater" in G. Shaw (1800:pl. 51) from a composite specimen displayed in the Paris Museum of a young coatimundi (É. Geoffroy St.-Hilaire 1803:219; Desmarest 1817b:105; Gray 1865:384).

Magnorder Epitheria

Order Soricomorpha Gregory, 1910

Neal Woodman and Jaime Péfaur

The Soricomorpha originally was proposed by Gregory (1910) as a "section" containing the shrews (Soricidae) and moles (Talpidae) within the suborder Lypotyphla, order Insectivora. More recent concepts of what constitutes the Soricomorpha vary. Typically, the order comprises shrews and solenodons (Solenodontidae), and possibly tenrecs (Tenrecidae; McKenna and Bell 1997) or moles (Arnason et al. 2002) or both (MacPhee and Novacek 1993). A possible relationship with gymnures and hedgehogs (Erinacidae) also has received support (Malia, Adkins, and Allard 2002). Previously, these five families had been included with golden moles (Chrysochloridae) in the order Lypotyphla, or along with elephant shrews (Macroscelididae) in the order Insectivora. However, recent analyses of phylogenetic relationships indicate that the seven families are not as closely related as previously believed (e.g., Stanhope et al. 1998; Mouchaty et al. 2000; Arnason et al. 2002; Malia, Adkins, and Allard 2002). Although consensus has yet to be reached on their relative phyletic positions, either to each other or to other mammalian groups, recent classifications (e.g., McKenna and Bell 1997) divide the Erinacidae, Soricidae, Solenodontidae, Talpidae, and Tenrecidae among many as four orders (Macroscelidea, Chrysochloridea, Erinaceomorpha, and Soricomorpha). Only one of these families, the Soricidae, is present in South America.

Family Soricidae G. Fischer, 1814

The Soricidae comprises 23 genera and more than 320 species distributed throughout Eurasia, Africa, North and Central America, and northwestern South America. Members of the family are sometimes considered primitive or generalized, but these labels ignore a well known suite of cranial and dental specializations unique to these animals.

Shrews are small mammals (head and body length, 35–150 mm; mass, 2–106 g), typically having small pinnae often concealed by fur, minute eyes, and an elongated, pointed snout. The long, flattened skull has incomplete zygomatic arches lacking jugals, free tympanic bones (auditory bullae are lacking), and a double articulation of the articular condyle of the dentary with the cranium. The clavicle is long and slender, and the pubic symphysis is open (innominates not in contact). The deciduous dentition is shed in utero. A pincer-like foraging apparatus is formed by the large, curved, first upper incisor and the long, procumbent, first lower incisor. Behind the first upper incisor, the anterior upper dentition (incisors, canine, and anterior premolars) is comparatively simple and undifferentiated, and these teeth are often referred to as "unicuspids." Homologies of the unicuspids have been difficult to determine, and for this reason, dental formulae for individual species often disagree in the relative numbers of incisors, canines, and premolars. Only shrews of the subfamily Soricinae occur in the Americas. The red-pigmented dentition in most New World genera (except *Megasorex* and *Notiosorex*) provides an additional characteristic that aids in distinguishing these mammals as shrews.

Although most authors attribute the name Soricidae to Gray (1821:300), Palmer (1904) and McKenna and Bell (1997) dated the name from G. Fischer's ([= Fischer von Waldheim] 1817:414) "*Familia Soricinorum.*" However, G. Fischer first used "*Familia Soricinorum*" in 1814 (p. x).

Genus *Cryptotis* Pomel, 1848

The small-eared shrews of the genus *Cryptotis* include at least 28 species that are discontinuously distributed from the eastern United States and southernmost Canada to Venezuela and Peru north of the Huancabamba Depression. Individual species occupy a variety of habitats from sea-level grasslands and second-growth woodlands in northern North America to humid montane forests and paramos of northern South America. In South America, *Cryptotis* is known only above 1,200 m. Shrews have been found in the Lower Montane Moist Forest, Lower Montane Wet Forest, Montane Wet Forest, Montane Rain Forest, Subalpine Páramo, and Subalpine Rain Páramo life zones of Holdridge (1947), as well as in disturbed cloud forest, secondary forest, and pasture lands (Woodman 2002).

Cryptotis are small to medium-sized shrews (length of head and body, 50–102 mm; tail, 12–53 mm; mass, 3–19 g). In South America, head and body length ranges from 60 to 102 mm; tail length, from 20 to 46 mm; and mass, from ca. 5 to 18 g. These medium-gray to nearly black small mammals all have red-pigmented teeth, and most have four upper unicuspids. Upper unicuspids decrease in size posteriorly; the fourth unicuspid is always smaller than the third, and is occasionally absent in two species. The typical dental formula for *Cryptotis* often is given as 3/1, 1/1, 2/1, 3/3 = 30 (Hall 1981; F. A. Reid 1997). Choate (1970:208) gave the dental formula for the upper toothrow as 1 (falciform incisor), 4 (unicuspids), 1 (premolar [P4]), and 3 (molars); the lower toothrow as 1 (procumbent incisor), 1 (unicuspid), 1 (premolar [p4]), and 3 (molars). The fossil record of *Cryptotis* extends from late Miocene to Recent in North America (McKenna and Bell 1998).

Local vernacular names for shrews in South America include musarañas, ratones hocicudos, and ratones topos.

SYNONYMS:

Sorex: Say in James, 1823:163; not *Sorex* Linnaeus, 1758.

Corsira Gray, 1838a:123; part.

Brachysorex Duvernoy, 1842:37; part.

Galemys Pomel, 1848:249; part; preoccupied by *Galemys* Kaup, 1829 (Talpidae).

Musaraneus Pomel, 1848:249; part; *Musaraneus* is unavailable from Brisson (1762; see ICZN 1998), but is an available name from Pomel, 1848.

Cryptotis Pomel, 1848:249; type species "*M. cinereus* (sorex [*sic*] *cinereus* Bachm.)"; (= *Sorex parvus* Say in James, 1823); proposed as a subgenus of *Musaraneus* Pomel.

Blarina: Baird, 1857:51, 53; part; not *Blarina* Gray, 1838a.

Cryptotus Milne-Edwards, 1872b:256; incorrect subsequent spelling of *Cryptotis* Pomel.

Blarina: Coues, 1877:647; part; not *Blarina* Gray.

Soriciscus Coues, 1877:649; type species "*Sorex parvus* Say or *S. cinereus* Bachm." (= *Sorex parvus* Say in James, 1823) by original designation; proposed as a subgenus of *Blarina* Gray, 1838a.

Blarina (Soriciscus): J. A. Allen, 1895b:339, 340; not *Blarina* Gray.

Blarina: Merriam, 1897b:227; not *Blarina* Gray.

Blarina: O. Thomas, 1898c:457; not *Blarina* Gray.

Sorieiscus Elliot, 1901:382; incorrect subsequent spelling of *Soriciscus* Coues.

Cryptotis: Miller, 1911:221; first use as a genus.

Blarina: J. A. Allen, 1912:93; not *Blarina* Gray.

Blarina: O. Thomas, 1912c:409; not *Blarina* Gray.

Blarina: Stone, 1914:16; not *Blarina* Gray.

Blarina: H. E. Anthony, 1921a:5; not *Blarina* Gray.

Blarina: Hibbard, 1953:29; not *Blarina* Gray.

Cryptotys Saban, 1958:846; incorrect subsequent spelling of *Cryptotis* Pomel.

Xenosorex Schaldach, 1966:289; type species *Notiosorex phillipsii* Schaldach, 1966, by original designation; proposed as a subgenus of *Notiosorex*: Schaldach, 1966; not *Notiosorex* Coues, 1877.

Criptotis Durant and Péfaur, 1984:6; incorrect subsequent spelling of *Cryptotis* Pomel.

Cripotitis Aagaard, 1982:276; incorrect subsequent spelling of *Cryptotis* Pomel.

Criptotis Durant and Díaz, 1995:87; incorrect subsequent spelling of *Cryptotis* Pomel.

REMARKS: *Sorex surinamensis* Gmelin, 1788:114, and *Blarina pyrrhonota* Jentink, 1910b:167, are two taxa with complex taxonomic histories long associated with South American soricids. Each species was described on the basis of a single specimen believed at the time to have originated from Surinam. *Sorex surinamensis* supposedly was the first species of shrew to be described from South America or from anywhere else in the New World (Gmelin 1788:114). The type specimen has not been located, therefore, verification of its identity is not possible. O. Thomas (1888b:357), Trouessart (1898:1242), and Cabrera (1919:42; 1925:135) considered the original description of *S. surinamensis* to have been based on a marsupial, and they listed the name in their synonymies of the Guianan Short-tailed Opossum, *Monodelphis brevicaudata* (as either *Didelphys brevicaudata* or *Peramys brevicaudatus*). However, Trouessart (1897:188) also listed *S. surinamensis* as a possible synonym of *B. pyrrhonota*, another name believed to have been based on a shrew from Surinam. In contrast, Tate (1932:223) stated clearly that the descriptions of both *S. surinamensis* and *B. pyrrhonota* represented soricids and suggested that both names may have been based on the same individual. He also cast doubt on whether either species occurred in Surinam, suggesting that the specimen on which Jentink based *B. pyrrhonota* was mislabeled. Cabrera (1958:47) treated *S. surinamensis* as a species of *Cryptotis* and included *B. pyrrhonota* as a junior synonym.

Blarina pyrrhonota was first mentioned by Jentink (1888:16), but was not described until much later (Jentink 1910b:167). As noted previously, Trouessart (1897:188, 1904:138) treated *B. pyrrhonota* as a South American species and suggested that *S. surinamensis* might be a synonym. Cabrera (1925:135) listed *pyrhonota* (*sic*) as a species of *Cryptotis*, but, apparently unaware of Jentink's (1910b) subsequent validation, he said the name was a *nomen nudum* because it lacked a formal description. In accord with Tate (1932), Cabrera (1958:47) subsequently treated *B. pyrrhonota* as a synonym of *Cryptotis surinamensis*. Husson (1963) provided the first detailed redescription of the holotype of *B. pyrrhonota* and concluded that it was a *Sorex*, probably the common European shrew, *Sorex araneus*. The original label had been lost, obscuring the history of the specimen, and he noted that there was no evidence that it was also the basis for Gmelins (1788) description of *S. surinamensis*, as suggested by Tate

(1932). Modern taxonomic accounts tend to treat both *B. pyrrhonota* and *S. surinamensis* as synonyms of *S. araneus* (e.g., Hutterer 1993).

Recent taxonomic treatments of *Cryptotis* partition the species among four informal species groups based upon external, cranial, dental, and postcranial characteristics (Choate 1970; Woodman 1996, 2002; Woodman and Timm 1993, 1999, 2000; Woodman, Cuartas-Calle, and Delgado 2003): the *C. mexicana*-group in Mexico and northern Central America; the *C. nigrescens*-group distributed from Mexico to Colombia; the *C. parva*-group occurring from Canada to Costa Rica; and the *C. thomasi*-group in the Andes of northern South America. Most South American *Cryptotis* are members of the *C. thomasi*-group, which is distinguished by a number of derived characters, including modifications of the forelimbs (Woodman 1996, 2002; Woodman, Cuartas-Calle, and Delgado 2003; Woodman and Morgan, *in press*). In addition, two species (*C. colombiana*, *C. mera*) are allied with the more primitive and mostly Central American *Cryptotis nigrescens*-group (Woodman 1996; Woodman and Timm 1993; Woodman, Cuartas-Calle, and Delgado 2003).

For the most part, our current knowledge of South American shrews is limited to an incomplete understanding of their taxonomies, distributions, and associations with other small mammals, vegetation communities, and climatic zones. Few solid scientific studies of reproduction, feeding habits, and other aspects of their life histories have been undertaken. There are abundant opportunities for local researchers to undertake meaningful comparative studies of ecology, behavior, physiology, and additional aspects of the biology of these shrews.

The following key should be considered a rough guide to identifying species of shrews in South America. The genus is under active study at this time and although 11 species are included in the key, some named "species" are known to be complexes comprising two or more species. Several species are documented by only a handful of often-incomplete specimens, so the full range of morphological variation is incompletely documented. In the Cordillera Central and Cordillera Oriental of Colombia, a species from the *C. nigrescens*-group occurs sympatrically with a member of the *C. thomasi*-group, and a similar situation is expected in the poorly explored Cordillera Occidental (Woodman and Timm 1993; Woodman 1996). Through much of the highlands of northwestern South America, species of *Cryptotis* are parapatric, so location can be helpful in determining species identity. However, distributions remain poorly documented throughout the Andes, and potential contact zones between neighboring species are unsampled in nearly all cases. Specimens of shrews are relatively rare in collections, and vouchers always must be prepared and deposited

in a reputable institution where they will be available for study.

Editor's note: The gender of *Cryptotis* has been treated inconsistently (e.g., compare Miller and Kellogg 1955; Cabrera 1958). Based on his interpretation of classical usage, Woodman (1993) argued that the gender of a generic name ending in–*otis* is feminine; he has been consistent in following that interpretation (e.g., Woodman, 1996, 2002, 2003b; Woodman and Timm 1993, 1999; Woodman and Díaz de Pascual 2004). Gardner (2005a), based on his understanding of Article 30.1.4.2 (ICZN 1999) and arguments presented by Pritchard (1994) and David and Gosselin (2002), determined that the gender of *Cryptotis* is masculine. Nevertheless, pending action by the International Commission on Zoological Nomenclature, Woodman herein treats *Cryptotis* as feminine.

KEY TO SOUTH AMERICAN SPECIES OF *CRYPTOTIS*:

1. Body size small, length of head and body (HB) 60–76 mm; pelage more-or-less uniformly dark with little contrast between dorsum and ventrum; manus not enlarged; foreclaws not elongate; length of tail (LT) long relative to length of head and body (LT/HB 34–57%); typically only three unicuspids visible in lateral view of upper toothrow; unicuspids typically large in lateral view, with straight or convex postero-ventral margins; anterior element of ectoloph of M1 approximately equal in size to posterior element; anterior border of coronoid process joins horizontal ramus of mandible at a high angle . . . . (*C. nigrescens*-group, Colombia) 2

1'. Body size small to large (HB 58–102 mm); pelage variable; manus broad and foreclaws usually long; tail may be short to long (LT/HB 28–60%); three or four unicuspids visible in lateral view of upper toothrow; unicuspids slender in lateral view, with concave to deeply concave postero-ventral margins; anterior element of ectoloph of M1 smaller than posterior element; anterior border of coronoid process joins horizontal ramus of mandible at a low angle. . . . . *C. thomasi*-group) 4

2. Foramen of tympanic process of petromastoid usually absent, or if present, minute; fourth unicuspid large, averaging 50% (36–64%) of surface area of third unicuspid; tail shorter, less than 1/2 the length of HB, and LT/HB averaging 39% (34–46%); HB averaging 69 mm (67–79 mm); occurs along or near Colombia's border with Panama . . . . *Cryptotis mera*

2'. Foramen on posterior or posteromedial edge of tympanic process of petromastoid conspicuously large; fourth unicuspid smaller, averaging 42.5% (32–61%) of surface area of third unicuspid; tail longer, LT/HB averaging 47% (36–57%). 3

3. Palate broad, breadth across molars (M2–M2) 6.0–6.4 mm; posterolingual cuspules on cingulae of anterior three unicuspids obvious; lower sigmoid notch of mandible typically shallow; entoconid of m3 absent; HB averaging 65 mm (60–76 mm); occurs in Colombia . . . . *Cryptotis colombiana*

3'. Palate narrower, breadth across molars (M2–M2) 5.7–5.9 mm; posterolingual cuspules of anterior three unicuspids minute; lower sigmoid notch of mandible variable, shallow to moderately deep; entoconid of m3 present but minute; occurs in Colombia . . . . *Cryptotis brachyonyx*

4. Length of tail less than 29 mm. 5

4'. Length of tail more than 29 mm 7

5. Size large, HB averages 87 mm (74–96 mm); tail short, averaging 24 mm (20–27 mm); LT/HB averaging 29% (21–36%); dorsal pelage medium to dark brown; red pigment typically extends into hypoconal basin of M1 and M2; lower sigmoid notch shallow to very shallow; large, obvious foramen on posterior edge of tympanic process of petromastoid; fourth unicuspid large, averaging 50% of surface area of third unicuspid; known from Colombia. *Cryptotis thomasi*

5'. Medium-size to large, HB averaging 75–80 mm (range 58–89 mm); tail moderately long, both absolutely (LT averaging 21–31 mm, range 21–38 mm) and relatively (LT/HB averaging 38–39%, range 26–54%); pelage color variable; red pigment typically absent from hypoconal basin of upper molars; lower sigmoid notch shallow to deep; typically a minute foramen on posterior edge of tympanic process of petromastoid; known from Ecuador. 6

6. Head and body length averaging 80 mm (65–86 mm); dorsal pelage medium gray, appearing speckled; palate narrow relative to length-of-palate (M2–M2/PL averaging 63%, range 58–66%); fourth unicuspid large, averaging 51% of surface area of third unicuspid; coronoid process moderately high, averaging 66% (62–70%) of length of mandible; length of mandible behind m3 long, averaging 82% (77–87%) of length of mandible; lower sigmoid notch moderately deep to deep; minute entoconid occasionally present on m3; tail moderately long, both absolutely (LT averaging 31 mm, range 22–38 mm) and relatively (LT/HB averaging 38%, range 30–54%); known from southern Ecuador. . . *Cryptotis montivaga*

6'. Head and body length averaging 75 mm (58–89 mm); dorsal pelage dark chocolate brown; palate moderately broad (M2–M2/PL averaging 65%, range 58–69%); fourth unicuspid large, averaging 58% of surface area of third unicuspid; coronoid process low, averaging 63%, (range 58–69%) of length of mandible; length of mandible behind m3 moderately long, averaging 76%

(range 71–82%) of total length of mandible; lower sigmoid notch shallow to moderately deep; entoconid absent from m3; tail moderately long, both absolutely (LT averages 29 mm, range 21–34 mm) and relatively (LT/HB averages 39%, range 26–50%); known from northern and central Ecuador. . . . . *Cryptotis equatoris*

7. Fourth unicuspid typically reduced, averaging 19–29% (range 2–49%) of surface area of third unicuspid; body size large (HB averaging 84–88 mm, range 74–102 mm); medium-sized foramen normally visible on posterior edge of tympanic process of petromastoid; known from Colombia and Venezuela. 8

7'. Fourth unicuspid typically large, averaging 40–58% (25–81%) of surface area of third unicuspid; body size small to large (HB averaging 68–85 mm, range 58–98 mm); foramen on posterior edge of tympanic process of petromastoid absent, minute, or huge, but not medium-sized; known from Colombia, Ecuador, and Peru 10

8. Fourth unicuspid extremely reduced, averaging 19% (2–37%) of surface area of third unicuspid, absent on one or both sides of cranium in approximately 25% of specimens; M3 simple; length of mandible behind m3 long, averaging 81% (74–87%); HB averaging 88 mm (76–102 mm); fourth unicuspid not visible in lateral view of skull; posterolingual cuspules typically absent in anterior three unicuspids; coronoid process high, averaging 70% (63–76%) of length of mandible; known from Venezuela . . . . *C. meridensis*

8'. Fourth unicuspid reduced, averaging 29% (12–49%) of surface area of third unicuspid; M3 complex or simple; length of mandible behind m3 moderate, averaging 77–78% (72–81%) of total length of mandible 9

9. Fourth unicuspid usually visible in lateral view of skull; posterolingual cuspules usually present on anterior three unicuspids; M3 simple or complex; coronoid process moderately high, averaging 65% (62–68%) relative to length of mandible; HB averaging 86 mm (80–91 mm); known from Colombia and Venezuela . . . . *Cryptotis tamensis*

9'. Fourth unicuspid usually not visible in lateral view of skull; posterolingual cuspules usually absent from anterior three unicuspids; M3 complex; coronoid process high, averaging 71% (65–77%) of length of mandible; HB averaging 84 mm (74–92 mm); known from Colombia. . . . . *Cryptotis squamipes*

10. Size large (HB averaging 85 mm, range 77–98 mm); huge, obvious foramen on posterior edge of tympanic process of petromastoid; known from Colombia . . . . *Cryptotis medellinia*

10'. Size small to large (HB 58–89 mm); foramen on posterior edge of tympanic process small to absent. 11

11. Medium-sized to large (HB averaging 75–80 mm, range 58–89 mm); M3 typically simple; entoconid usually absent on m3 6

11′. Size small (HB averaging 68 mm, range 63–73 mm); M3 complex; entoconid present on m3; known from Peru . . . . *Cryptotis peruviensis*

Cryptotis brachyonyx Woodman, 2003
Short-clawed Colombian Shrew

SYNONYMS:

Blarina thomasi: Merriam, 1897b: 227; part.

[*Blarina (Cryptotis)*] *thomasi*: Trouessart, 1904:138; part; not *Blarina thomasi* Merriam.

Cryptotis thomasi: O. Thomas, 1921f:354; part; not *Blarina thomasi* Merriam.

Cryptotis thomasi thomasi: Cabrera, 1958:48; part; not *Blarina thomasi* Merriam.

Cryptotis colombiana: Woodman, 1996:417; part; not *Cryptotis colombiana* Woodman and Timm, 1993.

Cryptotis brachyonyx Woodman, 2003b:855; type locality "Colombia: *Department of Cundinamarca*: 'La Selva, near Bogotá.'"

DISTRIBUTION: *Cryptotis brachyonyx* is known from the Cordillera Oriental in central and eastern departamento de Cundinamarca, Colombia, between 1,300 and 2,715 m elevation.

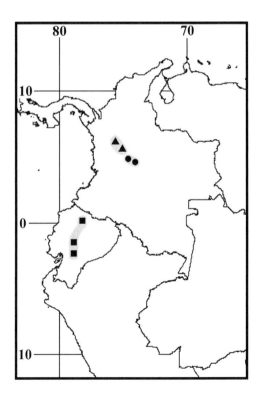

Map 82 Marginal localities for *Cryptotis brachyonyx* ●, *Cryptotis colombiana* ▲, and *Cryptotis equatoris* ■

MARGINAL LOCALITIES (Map 82; from Woodman 2003b): COLOMBIA: Cundinamarca: Plains of Bogotá; Cundinamarca, San Juan de Ríoseco.

SUBSPECIES: *Cryptotis brachyonyx* is monotypic.

NATURAL HISTORY: The habitat probably includes the Premontane Moist Forest, Premontane Wet Forest, Lower Montane Moist Forest, Montane Moist Forest, and/ or Montane Wet Forest life zones on the Cordillera Oriental in central and eastern departamento de Cundinamarca, Colombia. Its life history is unknown.

Remarks: La Selva was the name of George O. Child's estate on the Plains of Bogotá, the elevation of which he estimated as approximately 8,900 ft. Woodman (2003b) allied *Cryptotis brachyonyx* with the *Cryptotis nigrescens*-group of Central American.

Cryptotis brachyonyx is sympatric with *Cryptotis thomasi*, and they share the same general type locality. However, all four known specimens of *C. brachyonyx* were collected prior to 1925. The overall scarcity of *C. brachyonyx* and its absence in later collections led Woodman (2003b) to suggest that the species is either extinct, or restricted to specific microhabitats that have not been adequately sampled.

Cryptotis colombiana Woodman and Timm, 1993
Colombian Shrew

SYNONYMS:

Cryptotis thomasi: Hershkovitz, 1969:18; part; not *Blarina thomasi* Merriam.

Cryptotis colombiana Woodman and Timm, 1993:24; type locality "Colombia; Central Cordillera; Antioquia Dept., Sonsón; 15 km E of Río Negrito; 1750 m"; here corrected to Río Negrito, 15 km E of Sonsón, Antioquia, Colombia.

DISTRIBUTION: *Cryptotis colombiana* is known from the central portions of the Cordillera Central and the Cordillera Oriental of Colombia between 1,750 and 2,150 m elevation.

MARGINAL LOCALITIES (Map 82): COLOMBIA: Antioquia, Vereda San Antonio de Prado (MUA 060); Antioquia, Finca Los Sauces (MUA 12001).

SUBSPECIES: *Cryptotis colombiana* is monotypic.

NATURAL HISTORY: The region of the Cordillera Central where *C. colombiana* has been found corresponds to the Lower Montane Wet Forest life zone. The original vegetation is cloud forest characterized by constant fog, high humidity, and a diverse assemblage of trees covered with epiphytes. Much of the area today includes agricultural fields, pastures, secondary successional brush and woodlands, and disturbed primary forest having a thick understory. Individuals have been captured in an overgrown cattle pasture surrounded by moss-covered rock outcroppings, along the rocky bank of a small stream running through a grove of secondary-growth trees, and in minimally disturbed primary forest (C. A. Cuartas-Calle, *in litt.*). Remains of *C. colombiana* have been recovered from pellets of the Barn Owl (*Tyto alba*) and Tropical Screech-owl (*Otus choliba*) (C. A. Cuartas-Calle, *in litt.*).

REMARKS: Woodman and Timm (1993), Woodman (1996), and Woodman, Cuartas-Calle, and Delgado (2003) allied *Cryptotis colombiana* with the *Cryptotis nigrescens*-group of Central America.

Cryptotis equatoris (O. Thomas, 1912)
Ecuadorian Shrew

SYNONYMS:

Blarina equatoris O. Thomas, 1912c:409; type locality "Sinche, Guabanda [=Guaranda], 4000 m." Bolívar, Ecuador.

Blarina aequatoris: Lönnberg, 1921:4; incorrect subsequent spelling of *Blarina equatoris* O. Thomas.

Blarina osgoodi Stone, 1914:16, type locality "Hacienda Garzón, Mt. Pichincha, 10,500 ft. altitude," Pichincha, Ecuador.

[*Cryptotis*] *equatoris*: O. Thomas, 1921f:354; first use of current name combination.

C[*ryptotis*]. *osgoodi*: Tate, 1932:225; name combination.

Cryptotis thomasi equatoris: Cabrera, 1958:47; name combination.

Cryptotis e[*quatoris*]. *equatoris*: Vivar, Pacheco, and Valqui, 1997:6; name combination.

Cryptotis equatoris osgoodi: Vivar, Pacheco, and Valqui, 1997:7; name combination.

DISTRIBUTION: *Cryptotis equatoris* occurs in the Andes of central and northern Ecuador between 1,675 and 4,055 m elevation.

MARGINAL LOCALITIES (Map 82): ECUADOR: Pichincha (Pichincha-Imbabura border), Mojanda, western side (NHR A586313); Bolívar, Sinche (type locality of *Blarina equatoris* O. Thomas); Cañar, Chícal (AMNH 62923).

SUBSPECIES: *C. equatoris* is monotypic.

NATURAL HISTORY: Very little is known about the natural history of *C. equatoris*. Chapman (1926) described the vegetation of the type locality as open, with a predominance of grasses, but with several forest patches. Given the broad elevational distribution of the species, it should occur in a variety of wet montane forest and páramo habitats.

REMARKS: The holotype and one other specimen of *C. equatoris* were collected by Perry O. Simons during his stay from 12 to 25 December 1898 at a locality given simply as "Sinche, Guaranda, 4000 m" (Chubb 1919). O. Thomas (1912c) mistakenly spelled Guaranda as "Guabanda." Sinche is not listed by USBGN (1987), and the name has been interpreted as referring to Sinchig (01°32′S, 78°59′W),

a village 10 km north of the town of Guaranda (Paynter, 1993). In November 1923, G. H. H. Tate spent time collecting at a "Hacienda Sinche" (ca. 01°32′55″S, 78°56′45″W; Instituto Geográfico Militar de Ecuador, Guaranda quadrangle, ÑIV-C3; 1:50,000), at 10,400 ft (= 3,170 m) elevation. Based on hand-drawn maps in Tate's original field notes deposited in the AMNH, Hacienda Sinche is northeast of Guaranda. On specimen labels, this locality is written simply as "Sinche," and it probably is the same locality where Simons collected.

Vivar, Pacheco, and Valqui (1997) described differences in size between samples of *C. equatoris* from near the type locality and from Cerro Pichincha near the type locality of *Blarina osgoodi*, and they recognized these populations as distinct subspecies. However, they failed to investigate intervening populations, and they relied upon unnecessarily small samples to justify their separation of these taxa. *Cryptotis equatoris* may eventually prove to be a complex comprising two or more species, but understanding the interrelationships of these populations will require a comprehensive review of the taxon throughout its geographic distribution.

Cryptotis medellinia O. Thomas, 1921
Medellín Shrew

SYNONYMS:

Cryptotis medellinius O. Thomas, 1921f:354; type locality "San Pedro, 30 km. north of Medellín," Antioquia, Colombia.

Cryptotis medellinus: Tate, 1932:224; incorrect subsequent spelling of *Cryptotis medellinius* O. Thomas.

Cryptotis thomasi medellinius: Cabrera, 1958:48; name combination.

[*Cryptotis*] *medellinia*: Woodman, 1993:545; spelling changed to feminine gender.

DISTRIBUTION: *Cryptotis medellinia* is known from northern half of the Cordillera Central and the northern tip of the Cordillera Occidental of Colombia between 2,000 and 3,800 m elevation (Woodman 2002).

MARGINAL LOCALITIES (Map 83): COLOMBIA: Antioquia, Ventanas (FMNH 69812); Antioquia, Vereda Pajarito (MUA [field number CC 70]); Antioquia, Páramo Frontino (FMNH 71021). Risaralda, La Pastora (ICN, unnumbered).

SUBSPECIES: *Cryptotis medellinia* is monotypic.

NATURAL HISTORY: The distribution of *C. medellinia* may include Premontane Wet Forest, Lower Montane Wet Forest, Lower Montane Rain Forest, Montane Wet Forest (Cuartas and Muñoz 2000a, 2003b), and Montane Rain Forest life zones. Little is known of the natural history of *C. medellinia*. Specimens have been taken in a moss-lined runway in cut-over scrub vegetation, along a forest trail

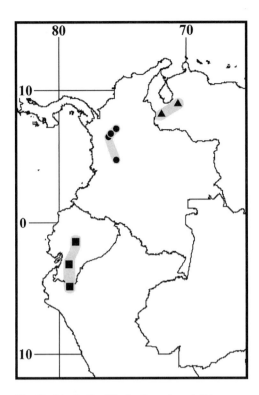

Map 83 Marginal localities for *Cryptotis medellinia* ●, *Cryptotis meridensis* ▲, and *Cryptotis montivaga* ■

(J. L. Patton field notes, MVZ), in Andean forest at 2600 m, and an area reforested with *Alnus acuminata* at 3200 m (Sánchez and Alvear 2003). In addition, an individual was found dead in an area reforested with *Pinus* (Sánchez, Sánchez-Palomino, and Cadena 2004). Data on reproduction are scarce. A lactating female was taken in February, and one pregnant with 2 fetuses was captured in October (Sánchez and Alvear 2003). Remains of *C. medellinia* have been recovered from Barn Owl (*Tyto alba*) pellets (C. A. Cuartas-Calle, *in litt.*) and from scats of the crab-eating fox (*Cerdocyon thous*; Delgado 2002).

Cryptotis mera Goldman, 1912
Goldman's Shrew
SYNONYMS:

Cryptotis merus Goldman, 1912b:17; type locality "from head of Rio Limon (altitude 4,500 feet), Mount Pirri, eastern Panama."

Cryptotis mera: Hall and Kelson, 1959:61; gender agreement.

Cryptotis nigrescens mera: Handley, 1966: 756; name combination.

Cryptotis nigrescens nigrescens: Choate, 1970:279; part; not *Blarina nigrescens* J. A. Allen, 1895b.

DISTRIBUTION: *Cryptotis mera* is known only from Cerro Pirre (Serranía Pirre), and from Cerro Tacarcuna and Cerro Malí (Serranía de Darién), between 1,370 and 1,525 m elevation (Woodman and Timm 1993). The border between Panama and Colombia passes through the Pirre and Darién highlands where this species has been taken; *C. mera* undoubtedly occurs in Colombia as well.

MARGINAL LOCALITIES: Currently known only from Panama near the border with Colombia; not mapped.

SUBSPECIES: *Cryptotis mera* is monotypic.

NATURAL HISTORY: Little is known of the natural history of *C. mera*. The two highland regions where this shrew has been found correspond to the Premontane Rain Forest and Lower Montane Rain Forest life zones (Tosi 1971). Goldman (1912b) described the Serranía de Pirre as a region of unbroken, dense cloud forest with abundant epiphytes and high annual rainfall.

REMARKS: Woodman and Timm (1993) recognized *C. mera* as a species distinct from *C. nigrescens*, but continued to ally it with the mostly Central American *Cryptotis nigrescens*-group.

Cryptotis meridensis (O. Thomas, 1898)
Mérida Shrew
SYNONYMS:

Blarina meridensis O. Thomas, 1898c:457; type locality "Merida, alt. 2165 m.," Mérida, Venezuela; corrected to "Montes del Valle Merida 2165 m" by Woodman (2002) based on label information.

C[ryptotis]. meridensis: O. Thomas, 1921f:354; first use of current name combination.

Cryptotis thomasi meridensis: Cabrera, 1958:48; part; name combination.

Cryptotis thomasi thomasi: A. Díaz, Péfaur, and Durant 1997:293; part; not *Blarina thomasi* Merriam.

Cryptotis meridensis meridensis: Linares, 1998:106; part.

DISTRIBUTION: *Cryptotis meridensis* is found in the Cordillera de los Andes of Trujillo, Mérida, and eastern Táchira, Venezuela, between 1,640 and 3,950 m elevation (Woodman and Díaz de Pascual 2004).

MARGINAL LOCALITIES (Map 83): VENEZUELA: Trujillo, Río Motatán site E-II (Durant and Díaz 1995); Mérida: Páramo de Mariño (Durant, Díaz, and Díaz de Pasqual 1994).

SUBSPECIES: *Cryptotis meridensis* is monotypic.

NATURAL HISTORY: *Cryptotis meridensis* has been studied more comprehensively than any other South American soricid, yet our knowledge of the ecology and habits of this species remains rudimentary. Woodman and Díaz de Pascual (2004) summarized the natural history of this species in their *Mammalian Species* account. The Mérida Shrew inhabits cloud forest and paramo environments and has been documented in Lower Montane Moist Forest, Lower Montane Rain Forest, Montane Wet Forest,

Montane Rain Forest, Subalpine Paramo, Subalpine Rain Páramo, and Lower Montane Wet Forest life zones (Handley 1976; Aagaard 1982; Durant, Díaz, and Díaz de Pasqual 1994; Durant and Díaz 1995). At higher elevations at which it occurs, mean annual temperatures can be as low as 2°C, with extreme daily temperature fluctuations and daily frosts in the colder seasons. In other regions that it inhabits, mean annual temperatures can reach 17°C. The species is found primarily in relatively wet environments, with mean annual precipitation from 1,023 to greater than 2,000 mm (Azócar and Monasterio 1980; Durant and Péfaur 1984; Durant and Díaz 1995; A. Díaz, Péfaur, and Durant 1997). The Mérida Shrew may require a thick and extensive moss layer in páramo in which to construct runways, tunnels, and nests. In forested habitats, it uses runways in soil under litter, fallen logs, and rocks (Durant and Péfaur 1984; Durant, Díaz, and Díaz de Pascual 1994; Durant and Díaz 1995; A. Díaz, Péfaur, and Durant 1997). The nest of *C. meridensis* comprises an inner layer of grasses and sedges and an outer layer constructed from parts of frailejón (*Espeletia schultzii*) and romerillo (*Hypericum laricifolium*) (A. Díaz, Péfaur, and Durant 1997). The Mérida Shrews consumes a diversity of invertebrates that includes centipedes, earthworms, pill bugs, snails, spiders, and the larvae, pupae, and adults of a variety of insects. Soil-dwelling invertebrates dominate the diet, and earthworms were the most frequently encountered prey, suggesting that it forages more in the subsurface than on the soil surface (Díaz de Pascual and de Ascenção 2000). Other prey may include lizards, nestlings of rodents, and eggs and chicks of ground-nesting birds (A. Díaz et al. 1995), and individuals were observed opportunistically feeding on a rice rat (*Oryzomys meridensis*) and a trap-killed conspecific (Aagaard 1982; Woodman and Díaz de Pascual 2004). *Cryptotis meridensis* is preyed upon in turn by the opossums (*Didelphis pernigra*), weasels (*Mustela frenata*), and birds of prey, such as the Barn Owl (*Tyto alba*) (A. Díaz, Péfaur, and Durant 1997; Araujo and Molinari 2000). The Mérida Shrew can be relatively abundant in mammal communities in which it occurs. It is commonly one of the three most abundant small mammals in páramo and cloud forest environments in long-term studies using snap-traps, live-traps, or pitfall-traps (Aagaard 1982; A. Díaz et al. 1995; Durant and Díaz 1995; Díaz de Pascual 1993, 1994; Woodman and Díaz de Pascual 2004). As with other small mammals, population levels fluctuate seasonally and annually. Peaks in abundance may be timed in relation to local peaks in rainfall (Aagaard 1982; Durant and Díaz 1995; Woodman 2002; Woodman and Díaz de Pascual 2004). Reproduction occurs throughout the year. Pregnant or lactating females were captured at Monte Zerpa Cloud Forest in every month of the year, with the greatest proportion of pregnant females in April and the fewest in July

and December. The typical litter consists of 3 pups, with a known range of 2–4 pups (Woodman and Díaz de Pascual 2004).

REMARKS: Woodman (2002) noted that the skin label attached to the holotype of *Cryptotis meridensis* has "Montes del Valle Merida 2165 m" as the complete locality, rather than the abbreviated type locality reported by O. Thomas (1898c). Woodman (2002) recently recognized the population of *Cryptotis* from the Tamá highlands along the Venezuela-Colombia border as *C. tamensis*; thereby, restricting the name *C. meridensis* to the population inhabiting the Cordillera de Los Andes near Mérida. These actions leave in doubt the identities of populations of *Cryptotis* from El Junquito in the Coastal Highlands west of Caracas (Ojasti and Mondolfi 1968) and from the Sierra de Perijá (Guajira), Cordillera Oriental of Colombia (Duarte and Viloria 1992), each of which is documented by a single skull (Woodman 2002). Collection and study of statistically significant samples of complete specimens of shrews from each locality (indicated by question marks in Map 83) will be required to determine the relationships of shrews comprising those two populations.

Durant and Péfaur (1984) commented on differences in pelage color and texture between *C. meridensis* from open paramo and closed cloud forest environments. Although Soriano, Utrera, and Sosa (1990) were unable to corroborate this variation, Woodman (2002) noted some differences in pelage among individuals representing different habitats, but stated that these differences were not diagnostic.

Cryptotis montivaga (H. E. Anthony, 1921)
Grizzled Ecuadorian Shrew
SYNONYMS:

Blarina montivaga H. E. Anthony, 1921a:5; type locality "Bestion, Prov. del Azuay," Ecuador.

[*Cryptotis*] *montivaga*: Cabrera, 1925:134; first use of current name combination.

Cryptotis montivagus: Cabrera, 1958:47; name combination and incorrect gender concordance.

DISTRIBUTION: *Cryptotis montivaga* occurs in the provinces of Chimborazo, Azuay, and Loja, Ecuador, at elevations between 2,500 and 3,800 m.

MARGINAL LOCALITIES (Map 83): ECUADOR: Chimborazo, Urbina (AMNH 64623); Azuay, Bestion (type locality of *Blarina montivaga* H. E. Anthony); Loja, Podocarpus National Park (Barnett 1999).

SUBSPECIES: *Cryptotis montivaga* is monotypic.

NATURAL HISTORY: Barnett (1993, 1999) reported capturing 12 *C. montivaga*: 1 at 2,700 m in montane forest dominated by *Podocarpus* and *Ocotea*, 5 in quenoa (*Polylepis*) forest at 3,700–4,000 m, and 6 in streamside scrub at 3,450–4,000 m; none were taken in grassland (paramo) habitats. Elevations recorded on skin labels of these specimens in the BMNH range from 3,300 to 3,800 m.

Reproductive data for *C. montivaga* are mostly lacking. Barnett (1999) reported a female pregnant with two embryos when captured in August. Two females, one captured in July and the other in August, were lactating. Analysis of stomach contents of five individuals revealed remains of beetles, spiders, caterpillars, and possibly arthropod larvae (Barnett 1993). Remains of *C. montivaga* have been recovered from owl pellets found in Podocarpus National Park (Barnett 1999).

Cryptotis peruviensis Vivar, Pacheco, and Valqui, 1997
Peruvian Shrew

SYNONYMS:

Cryptotis thomasi: Hutterer, 1993:109; part; not *Blarina thomasi* Merriam.

Cryptotis peruviensis Vivar, Pacheco, and Valqui, 1997:7; type locality "Peru, Department Cajamarca, Las Ashitas, 3150 m, about 42 km W of Jaén (05°42'S, 79°08'W)."

DISTRIBUTION: *Cryptotis peruviensis* is known from the departments of Cajamarca and Piura, Peru, between 2,050 and 3,150 m elevation (Vivar, Pacheco, and Valqui 1997).

MARGINAL LOCALITIES (Map 84; from Vivar,

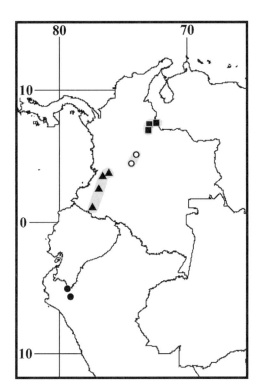

Map 84 Marginal localities for *Cryptotis peruviensis* ●, *Cryptotis squamipes* ▲, *Cryptotis tamensis* ■, and *Cryptotis thomasi* ○

Pacheco, and Valqui 1997): PERU: Piura, Machete, on Zapalache-Carmen trail; Cajamarca, Las Ashitas (type locality of *Cryptotis peruviensis* Vivar, Pacheco, and Valqui).

SUBSPECIES: *Cryptotis peruviensis* is monotypic.

NATURAL HISTORY: The habitat at the type locality, on the eastern slope of Cerro Chinguela in the Río Samaniego Valley, was described as cloud forest dominated by *Podocarpus*. The holotype was collected in elfin forest consisting of shrubby trees with abundant epiphytes and mosses, classified as representing the Tropical Montane Rain Forest life zone (Vivar, Pacheco, and Valqui 1997).

Cryptotis squamipes (J. A. Allen, 1912)
Cali Shrew

SYNONYMS:

Blarina (*Cryptotis*) *squamipes* J. A. Allen, 1912:93; type locality "crest of Western Andes (alt. 10,340 ft.), 40 miles west of Popayan, Cauca, Colombia."

C[*ryptotis*]. *squamipes*: Tate, 1932:225; first use of current name combination.

DISTRIBUTION: *Cryptotis squamipes* is known from the Cordillera del Sur of Colombia and the southern portions of the cordilleras Occidental and Central, at elevations from 1,500 to 3,375 m.

MARGINAL LOCALITIES (Map 84): COLOMBIA: Valle del Cauca, Tenerife (UV 7552); Valle del Cauca, Finca "Zingara" (UV 10143); Cauca, Cerro Munchique (FMNH 86716); Nariño, San Felipe (UV 11043)

SUBSPECIES: *Cryptotis squamipes*, as used herein, refers to a complex of at least three species that currently are under study.

NATURAL HISTORY: The holotype was collected in an area where "vegetation is scarce, scrubby and stunted" (Chapman 1917:32). The type locality is in Lower Montane Wet Forest (IGAC 1988), and the distribution of *C. squamipes* in the Cordillera Occidental probably corresponds closely to the distribution of this life zone.

REMARKS: A primary character that J. A. Allen (1912) used to distinguish *C. squamipes* was the scaliness of the feet, a feature also used by other authors to characterize this poorly known species. The feet of the holotype of *C. squamipes* have visible scales; however, the presence of obvious scales is not a diagnostic character. The upper surfaces of the feet of all species of *Cryptotis* are scaly, and the visibility of the scales depends to a large extent on the length and density of the over-laying hairs, characters that vary geographically in some species. The apparent scaliness of the feet also depends on the density and distribution of skin pigmentation, which vary among individual specimens and appear to be affected by type of preservation and subsequent exposure to light. In general, the feet of *C. thomasi* and *C. meridensis* tend to have more fur and less visible scales than do other species of Andean shrews. Among these and other Andean species, intraspecific variation in this character can equal, or exceed, interspecific variation. Moreover, the hind feet of the holotype of *C. squamipes* are aberrant and have abnormally developed claws. Malformation of the hind feet may have affected other morphological features of the feet as well, such as their hairiness, pigmentation, and apparent scaliness.

Cryptotis tamensis Woodman, 2002
Tamá Shrew
SYNONYMS:

Blarina meridensis: Osgood, 1912:62; not *Blarina meridensis* O. Thomas.

C[ryptotis]. meridensis: O. Thomas, 1921f:354; part; not *Blarina meridensis* O. Thomas.

Cryptotis thomasi meridensis: Cabrera, 1958:48; part; not *Blarina meridensis* O. Thomas.

Cryptotis meridensis meridensis: Linares, 1998:106; part; not *Blarina meridensis* O. Thomas.

Cryptotis tamensis Woodman, 2002:254; type locality: "Venezuela: State of Táchira: Buena Vista, 7°27′N, 72°26′W, 2415 m; near Páramo de Tamá; 35 km S, 22 km W of San Cristóbal."

DISTRIBUTION: *Cryptotis tamensis* occurs in the Tamá highlands in western Táchira, Venezuela, and from southeastern Norte de Santander to northeastern Santander, Colombia, at elevations from 2,385 to 3,330 m.

MARGINAL LOCALITIES (Map 84; from Woodman 2002): VENEZUELA: Táchira, Buena Vista (type locality of *Cryptotis tamensis* Woodman). COLOMBIA: Santander, above Suratá; Santander, Finca El Rasgón.

SUBSPECIES: We treat *C. tamensis* as monotypic.

NATURAL HISTORY: The distribution of *C. tamensis* includes Lower Montane Wet Forest, Montane Wet Forest, and Montane Rain Forest life zones (Woodman 2002). In describing Páramo de Tamá, where he captured most of the known specimens of this species, Osgood (1912) noted that the term "páramo" regionally denoted the entire highlands, rather than being applied more specifically to the open, high-elevation vegetational formation. Grassland paramo is limited in this region, and most specimens of *C. tamensis* were taken in dense, epiphyte-rich cloud forest, although some were captured in agricultural pasture and disturbed cloud forest (Osgood 1912; Handley 1976). A lactating female and two gravid females, one with a single fetus, the second with two fetuses, were captured in March (Woodman 2002).

REMARKS: By recognizing the population of *Cryptotis* from the Tamá highlands as a distinct species and restricting the name *C. meridensis* to the population inhabiting the

Cordillera de Los Andes near Mérida, Woodman (2002) placed in doubt the identities of populations of *Cryptotis* in the coastal highlands west of Caracas (Ojasti and Mondolfi 1968) and from the Sierra de Perijá (Duarte and Viloria 1992; also see Map 83). These populations will require additional study to determine their relationships.

Cryptotis thomasi (Merriam, 1897)
Thomas's Shrew
SYNONYMS:

Blarina thomasi Merriam, 1897b:227; type locality "Plains of Bogota, [Cundinamarca,] Colombia (on G. O. Childs estate near city of Bogota, alt. about 9000 ft.)."

[*Blarina* (*Cryptotis*)] *thomasi*: Trouessart, 1904:138; name combination.

Cryptotis thomasi: O. Thomas, 1921f:354; first use of current name combination.

Cryptotis avia G. M. Allen, 1923a:37; type locality "El Verjón, in the Andes east of Bogotá," Cundinamarca, Colombia.

Cryptotis avius Cabrera, 1958:46; incorrect gender concordance.

Cryptotis thomasi thomasi: Cabrera, 1958:48; name combination.

DISTRIBUTION: *Cryptotis thomasi* is known from the central portion of the Cordillera Oriental in Cundinamarca, Colombia, at elevations between 2,800 and 3,500 m (Woodman 2002).

MARGINAL LOCALITIES (Map 84): COLOMBIA: Cundinamarca, Represa de Neusa (ICN 9659); Cundinamarca, Páramo de Chisacá (ICN 5223).

SUBSPECIES: *Cryptotis thomasi* is monotypic.

NATURAL HISTORY: The distribution of *C. thomasi* may include Lower Montane Moist Forest, Lower Montane Wet Forest, Montane Wet Forest, Montane Rain Forest, and Lower Andean Páramo life zones. Much of the information on the natural history of *C. thomasi* comes from a short capture-recapture study of the small mammal community in Andean cloud forest and páramo at Carpanta Biological Reserve (López-Arevalo, Montenegro-Díaz, and Cadena 1993). The study site, between 3000 and 3100 m, had a unimodal rainy season with mean annual precipitation of more than 3000 mm. Mean annual temperature was 8.8°C, with daily temperatures that could fluctuate between 0° and 29°C. *Cryptotis thomasi* was the most abundant (greatest number of individuals captured) of the 11 species of small mammals taken at the site. Most captures of this species were in páramo. The largest numbers of *C. thomasi* were captured in June, suggesting an increase in abundance or activity during the period preceding the peak of the rainy season in June and July. Reproductively active females were found in April, June, and August, and two pregnant females were documented in each of the months of July and August. Pregnant females typically carried two fetuses. Examination of stomach contents revealed primarily remains of adult insects (López-Arevalo, Montenegro-Díaz, and Cadena 1993).

REMARKS: Woodman (1996) demonstrated that *Cryptotis avia* G. M. Allen, 1923a, is a synonym of *Blarina thomasi* Merriam.

Order Chiroptera Blumenbach, 1779

Alfred L. Gardner

Chiroptera, the only true flying mammals, are distinguished by wings supported primarily by the elongated radius, and elongated metacarpals and phalanges of digits II through V. The digits are connected by the wing membrane, which consists of a double layer of skin. Flight membranes include the propatagium (antebrachium) between shoulder and digit I; the dactylopatagium connecting the metacarpals and phalanges; the plagiopatagium connecting the wing to the side of the body, or in a few species, to the dorsum; and the uropatagium (interfemoral membrane), which varies between taxa from absent to extensive and extending well beyond the feet. Other adaptations for flight include lightening of the skull and post-cranial skeleton; reduction of the ulna; modifications of the shoulder girdle; thickening or fusion of thoracic vertebrae, ribs, and sternal elements; development of a keel on the sternum; and, in some taxa, fusion of anterior lumbar vertebrae and fusion of seventh cervical vertebra with the first thoracic. Modifications of the pelvic girdle appear to be related to pendant roosting posture and for maneuverability both in flight and when active on a substrate or at the roosting site. Development and specializations of the hind limbs, feet, and tail appear to be adaptations facilitating specific flight characteristics, the gathering of food, and the use of specific roosting sites (e.g., species of two families have suction disks on their thumbs and feet, permitting them to adhere to smooth surfaces). Living bats are represented by two suborders, the Megachiroptera and the Microchiroptera. Only microchiropterans occur in South American.

KEY TO THE FAMILIES OF SOUTH AMERICAN CHIROPTERA:

1. Facial appendages (noseleaf, dermal outgrowths, flaps of skin on face or lips). 2
1'. No facial appendages. 3
2. Noseleaf present (main "spear" of noseleaf missing in the Desmodontinae), may be reduced to a hood-like dermal outgrowth above the eyes in *Sphaeronycteris*, or a series of skin flaps on face in *Centurio* (these taxa lack a tail and have only a rudimentary uropatagium) . . . . Phyllostomidae
2'. No noseleaf or dermal outgrowths on face; lower lip with accessory skin folds and flaps; well-developed uropatagium and tail always present. . . . . Mormoopidae
3. Either a series of tufts of pale fur on forearms, or pocket-like glands on propatagia or center of uropatagium; postorbital processes present (fused with margins of rostral shield in *Diclidurus*); palatal branch of premaxilla reduced (incomplete anterior palate); premaxillae movable, not fused to maxillae. . . . . Emballonuridae
3'. No tufts on forearms; no pocket-like glands on propatagia or on uropatagium; no postorbital processes; palatal branch of premaxilla may or may not be reduced; premaxillae fused to maxillae. 4
4. Upper lip full, smooth, and split by a cleft or deep fold in the midline below rhinarium; pinnae long, narrow, and pointed; wing membrane attaches to leg above ankle; feet and claws may be greatly elongated; tail half or less the length of uropatagium; anterior palate complete . . . . Noctilionidae
4'. No medial cleft dividing upper lip; pinnae not notably narrow and pointed; wing membrane attaches at ankle or along foot; feet not elongated; tail usually as long as or much longer that uropatagium 5
5. Each thumb and foot bears a round or oval-shaped disc (suction cup); tragus triangular; anterior palate complete (palatal branches of maxillae fused). . . . Thyropteridae
5'. No discs on thumb or foot; tragus of various form, but not triangular; anterior palate may be complete or incomplete. 6
6. Thumb short, rudimentary, and, except for claw, entirely enclosed in propatagium. . . . . Furipteridae
6'. Thumb normal, not enclosed in propatagium. 7
7. Second digit consists only of metacarpal (phalanx lacking); tail as long as or longer than head and body, and except for tip, enclosed entirely in uropatagium . . . . Natalidae
7'. Second digit has rudimentary phalanx; tail as long as or considerably longer than uropatagium, but not longer than head and body. 8
8. Pinna of various shapes and sizes, may be very large, basal lobe present; tragus conspicuous, comparatively long, well developed, and its tip rounded or pointed . . . . Vespertilionidae
8'. Pinna wide, usually extending forward over eyes, lacking basal lobe; tragus small, truncated, and inconspicuous . . . . Molossidae

Family Emballonuridae Gervais, 1856

Craig Hood and Alfred L. Gardner

The pantropical family Emballonuridae is represented in the New World by 2 subfamilies, 8 genera, and 20 species found from northern Mexico to southern Brazil. Elsewhere, the family occurs in Africa, eastward through the Arabian Peninsula and the Indian subcontinent, to Southeast Asia and the Australian region. All 8 of the New World genera and 18 of the species are known from South America.

Characteristics of emballonurids include: wing digit II consists of only the metacarpal; digit III has two phalanges, the second longer than the first; the calcar is relatively long; the tail is slender and about 1/2 the length of the uropatagium, which it perforates dorsally; and the muzzle is smooth and lacks specialized cutaneous outgrowths. Postorbital processes are well developed (fused to supraorbital ridges in *Diclidurus*); the premaxillae are incomplete, represented only by nasal branches, and never fused to each other or to the maxillae. A baculum is present. Most species have wing sacs, usually more conspicuous in males (illustrated by Jones and Hood 1993:2), and located in the antebrachial membrane. The dental formula for all American taxa is 1/3, 1/1, 2/2, 3/3 × 2 = 32.

KEY TO GENERA OF SOUTH AMERICAN EMBALLONURIDAE:

1. Rostrum with deep, rimmed, cuplike depression occupying entire dorsal surface; combined length of distal (free) phalanges of thumb short, about 1/2 or less the length of first metacarpal; no wing sac in antebrachial membrane; a pouch-like structure in the uropatagium present or absent............................2
1'. Rostrum lacks cuplike depression; distal phalanges of thumb equal to or longer than first metacarpal; wing sac present or absent; no pouch-like structures in uropatagium..................................3
2. Postorbital processes broad, fused to supraorbital ridges; pouch-like glandular structures present on uropatagium; color white to pale brown............. *Diclidurus*
2'. Postorbital processes narrow, not fused to supraorbital ridges; uropatagium normal, lacking specialized structures; color dull smoky gray to blackish.... *Cyttarops*
3. Posterior margin of basisphenoid pit separated from basioccipital by a transverse flange; forearm naked; calcar approximately equal to, or shorter than tibia.......4
3'. No demarcation between basisphenoid pit and basioccipital; forearm sparsely haired and adorned with tufts of whitish hair; calcar much longer than tibia.......
..................................*Rhynchonycteris*

4. Upper anterior premolar tricuspidate; wing attached to side of foot near base of toes...................5
4'. Upper anterior premolar a simple spicule, not clearly tricuspidate; wing attached at or near ankle........6
5. Basisphenoid pit divided by well-developed median septum; basioccipital longer than its minimum width; wing sac absent...................... *Centronycteris*
5'. Basisphenoid pit lacks median septum; basioccipital wider than long; wing sac present......... *Cormura*
6. Rostrum inflated, appearing bulbous dorsolaterally; supraorbital region rounded, no supraorbital ridge; dorsal color uniform, without stripes...............7
6'. Rostrum relatively flat above; supraorbital ridge continuous with postorbital process; dorsum usually with two pale, more-or-less distinct, longitudinal stripes (lacking in *S. antioquensis*)....................*Saccopteryx*
7. Rostrum bulbous anteriorly with lateral inflations extending forward over roots of canines; premaxillae displaced laterally; no sagittal crest; paraoccipital processes small, not projecting ventrally........ *Balantiopteryx*
7'. Rostrum expanded laterally, approaching diamond-shape in outline when viewed from above; premaxillae in normal position with proximal portion on dorsolateral aspect of rostrum; sagittal crest well developed; paraoccipital processes conspicuous and projecting ventrally .
..................................*Peropteryx*

Subfamily Diclidurinae Gray, 1866

The subfamily Diclidurinae is confined to the New World tropics and contains two genera: *Cyttarops*, which is monotypic; and *Diclidurus*, in which four species currently are recognized. Diclidurines are characterized by a distinctive, rimmed depression on the rostrum; well-developed postorbital processes (fused to supraorbital ridges in *Diclidurus*); prominent supraorbital ridges that extend forward onto the rostrum; a strong angle between rostrum and braincase; a broad clavicle (greatest width about 1/3 its length); a deeply grooved tibia; and no wing sac. Pelage color varies from dark gray (*Cyttarops*) to pale brownish or white (*Diclidurus*).

Genus *Cyttarops* O. Thomas, 1913

The single species, *C. alecto*, is a medium-sized emballonurid (forearm 45–47 mm), with long, silky, smoky gray to almost black pelage. Unlike *Diclidurus*, the postorbital processes are long and not fused to the supraorbital ridge, the uropatagium bears no specialized sacs or glands, and there is no evident gap between the two upper premolars.

SYNONYM:

Cyttarops O. Thomas, 1913a:134; type species *Cyttarops alecto* O. Thomas, 1913a, by original designation.

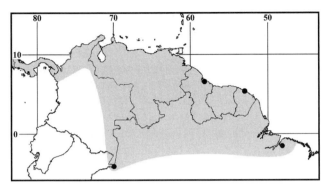

Map 85 Marginal localities for *Cyttarops alecto* ●

Cyttarops alecto O. Thomas, 1913
Short-eared Bat

SYNONYM:

Cyttarops alecto O. Thomas, 1913a:135; type locality
"Mocajatuba, near Para" (= Belém), Pará, Brazil; restricted by Starrett (1972) to Mucajatuba, Rio Guamá, 40
km E of Belém.

DISTRIBUTION: In South America, *C. alecto* has been
recorded in Colombia, Guyana, French Guiana, and Brazil.
Elsewhere, the species is known from Costa Rica (Starrett
1972) and Nicaragua (R. J. Baker and Jones 1975).

MARGINAL LOCALITIES (Map 85): GUYANA:
Mazaruni River (not mapped; O. Thomas 1913a); Demerara-Mahaica, Ceiba Biological Station (ROM 112626).
FRENCH GUIANA: Piste St.-Élie (Masson and Cosson
1992). BRAZIL: Pará, Mucajatuba (type locality of *Cyttarops alecto* O. Thomas). COLOMBIA: Amazonas, 35–
40 km SW of Leticia (Ochoa, Soriano, and Hernández-
Camacho 1994).

SUBSPECIES: We consider *C. alecto* to be monotypic.

NATURAL HISTORY: Little information is available on
the natural history of this rare species, and most of what
is known is based on Costa Rican *C. alecto* (see Starrett
1972; Reid and Langtimm 1993). Starrett (1972) found
these bats in small groups of up to four individuals roosting by day beneath fronds of coconut palms located in fairly
open areas near buildings. Starrett (1972) found both sexes
in the roosting groups and none of the specimens he examined appeared to have ectoparasites. Reid and Langtimm
(1993) observed a Costa Rican colony at irregular intervals over a 17-month period. The colony varied from 4
to 12 individuals, with most roosting under living fronds
of coconut palms. The bats hung near the midrib of the
frond with individuals spaced two to four leaflets apart.
Reid and Langtimm (1993) reported that, of two females
taken in January, one showed no signs of reproductive activity, while the other was pregnant with a small (l-mm CR)

embryo. They also cited R. K. LaVal's unpublished observations of Costa Rican female *C. alecto* suckling young
in July and August. Like other emballonurids, *C. alecto* is
insectivorous. All records of this species are from coastal
lowlands at elevations below 300 m. *Cyttarops alecto* has
a $2n = 32$, FN = 60 karyotype (R. J. Baker and Jones
1975)

REMARKS: Starrett (1972) attempted to locate the type
locality of *C. alecto* in northeastern Brazil, and found several places in the vicinity of Belém with spellings similar
to that used in the original description. He wrote (1972:2)
"The most reasonable of these is Mucajatuba, located...
some 40 km east of Belém along the south bank of the Rio
Guamá...," and this is the location to which he restricted
the type locality. D. C. Carter and Dolan's (1978:24) conclusion that the type specimen came from the Pará Zoological Garden in Belém is most likely a misinterpretation of O.
Thomas's (1913a:134) statement "caught in garden." All of
the mammals included in O. Thomas's (1913a:130) report
that came from the Pará Zoological Park were primates
kept there in captivity, and were so indicated. D. C. Carter
and Dolan (1978) gave information on and measurements
of the holotype of *Cyttarops alecto* O. Thomas.

Genus *Diclidurus* Wied-Neuwied, 1820

The genus *Diclidurus*, represented by four species (Koopman 1982), occurs from western Mexico, south into eastern
Brazil. All four species occur in South America, with three
recorded from only the northern part of that continent.
Members of this genus commonly are referred to as "ghost
bats," because of their white or very pale coloration.

These are medium-sized to large emballonurids (forearm
50–74 mm) with relatively long, soft, white or pale brownish (*D. isabella*) pelage. The first upper premolar is small
and not in contact with the second. The supraorbital ridges
are more prominent and the rostrum more distinctly dished
than in *Cyttarops*. The postorbital processes are indistinct
and fused to the supraorbital ridges (free in *Cyttarops*). The
uropatagium has a large, distinctive pouch-like glandular
structure (unique to this genus and especially prominent in
males during the breeding season) located at and behind
the exsertion of the tail from the interfemoral membrane.
The forehead has a naked area hidden under the fur above
the eyes that covers the broad rostral depression on the
skull.

SYNONYMS:

Diclidurus Wied-Neuwied, 1820a:1629; type species *Diclidurus albus* Wied-Neuwied, 1820a, by monotypy.
Depanycteris O. Thomas, 1920c:271; type species *Depanycteris isabella* O. Thomas, 1920c, by original designation.

Drepanycteris C. O. C. Vieira, 1955:341, 356; incorrect subsequent spelling of *Depanycteris* O. Thomas.

KEY TO SPECIES OF *DICLIDURUS*:

1. Color pale brownish; posterior border of palate evenly concave, mesopterygoid fossa reaching level of anterior cusps of M3 . . . . *Diclidurus isabella*
1′. Color generally white to very pale gray; posterior border of palate emarginate medially, mesopterygoid fossa not reaching level of M3 (emargination of posterior border of palate confluent with large palatine fenestra in *D. scutatus*) 2
2. Forearm more than 70 mm, greatest length of skull 20 mm or more . . . . *Diclidurus ingens*
2′. Forearm less than 70 mm; greatest length of skull less than 20 mm. 3
3. Forearm less than 60 mm; length of maxillary toothrow less than 7 mm; large palatine fenestra confluent with mesopterygoid fossa through posterior palatal emargination (evident on clean skulls) . . . .
. . . . *Diclidurus scutatus*
3′. Forearm more than 60 mm; length of maxillary toothrow more than 7 mm; no palatine fenestra . . . .
. . . . *Diclidurus albus*

Diclidurus albus Wied-Neuwied, 1820
White Ghost Bat

SYNONYMS: See under Subspecies.

DISTRIBUTION: *Diclidurus albus* is known from Colombia, Venezuela, Guyana, Surinam, and the island of Trinidad, south into Ecuador, Peru, and Brazil. Elsewhere, the species occurs in Central America and Mexico.

MARGINAL LOCALITIES (Map 86): TRINIDAD AND TOBAGO: Trinidad, Carapichaima Village (Goodwin and Greenhall 1961). GUYANA: Cuyuni-Mazaruni, Kartabo

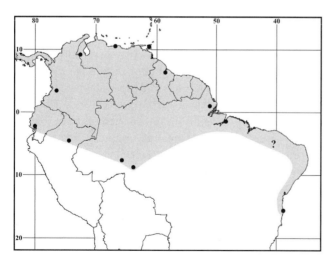

Map 86 Marginal localities for *Diclidurus albus* ●

(AMNH 142908). SURINAM: (no precise locality; Ojasti and Linares 1971). BRAZIL: Amapá, Tracajatuba (Peracchi, Raimundo, and Tannure 1984); Pará, Pará (O. Thomas 1920c); Bahia, Canavieiras (type locality of *Diclidurus albus* Wied-Neuwied); Rondônia, Pôrto Velho (C. O. C. Vieira 1955), Amazonas, Hyutanahan (Sanborn 1932b). PERU: Loreto, Parinari (Tuttle 1970). ECUADOR: Guayas, near Chongón (Albuja 1999). COLOMBIA: Valle del Cauca, Cali (Cuervo-Díaz, Hernández-Camacho, and Cadena 1986). VENEZUELA: Zulia, El Rosario, 39 km WNW of Encontrados (Handley 1976); Distrito Federal, Caracas (Ojasti and Linares 1971).

SUBSPECIES: We recognize two subspecies of *D. albus*.

D. a. albus Wied-Neuwied, 1820
SYNONYMS:

Diclidurus albus Wied-Neuwied, 1820a:column 1630; type locality "am Ausflusse des Rio Pardo"; identified as "Canavieras," Rio Pardo, Bahia, Brazil by Wied-Neuwied (1826:247; also see Ávila Pires 1965:5).

D[*iclidurus*]. *Freyreisii* Wied-Neuwied, 1820a:column 1630, footnote; type locality "am Ausflusse des Rio Pardo"; identified as "Canavieras," Rio Pardo, Bahia, Brazil by Wied-Neuwied (1826:247).

Diclid[*urus*]. *Freyreisii*: Schinz, 1821:171; selection of valid available name by first reviser (ICZN 1999:Art. 24.2.2; but see Remarks).

Diclid[*urus*]. *albus*: Schinz, 1821:171; designation of invalid available name by first reviser (ICZN 1999:Art. 24.2.2; see Remarks).

The nominate subspecies has been recorded from Guyana, Surinam, Brazil, and Peru.

D. a. virgo O. Thomas, 1903
SYNONYMS:

Diclidurus virgo O. Thomas, 1903b:377; type locality "Escazu," San José, Costa Rica.

Diclidurus albus virgo: Goodwin, 1969:48; first use of current name combination.

This subspecies is known from Colombia, Venezuela, Trinidad, and Ecuador. Elsewhere, the species occurs in Central America and Mexico.

NATURAL HISTORY: *Diclidurus albus* has been recorded roosting singly by day, except when aggregating into breeding groups, beneath the fronds of coconut palms in western Mexico and on the island of Trinidad (Goodwin and Greenhall 1961; Sánchez-Hernández and Chávez 1984). The holotype of *D. albus* also was taken from the fronds of a palm (Wied-Neuwied 1821). In Venezuela (Handley 1976), specimens were shot in flight, most often "near stream banks and other moist areas," but also in drier sites over yards, village streets, or in evergreen or cloud forest where there was a break in the canopy. This species appears to be seasonally migratory, according to evidence from Mexico (Ceballos and Medellín 1988). The known elevational range of *D. albus* is from sea level up to about 1,500 m.

At the onset of the reproductive season, small groups consisting of a male and several females have been found roosting together. In western Mexico copulation apparently takes place in January and February, and females bear a single young in May or June (Sánchez-Hernández and Chávez 1984). As in other emballonurids, *D. albus* is insectivorous. Additional natural history information was summarized by Ceballos and Medellín (1988) in their *Mammalian Species* account on *D. albus*. A specimen from Trinidad tested positive for rabies (Goodwin and Greenhall 1961). *Diclidurus albus* has a $2n = 32$, $FN = 60$ karyotype (Hood and Baker 1986).

REMARKS: Although infraspecific variation in *D. albus* has not been studied in detail, most recent authorities agree that *D. virgo*, which averages slightly larger than *D. albus*, is no more than a subspecies of the latter. Muñoz (2001:13) cited several records for Colombia; however, many contained errors of attribution. In addition to the report of this bat from unidentified localities in Surinam (Ojasti and Linares 1971), Albuja (1983, 1999) wrote of a specimen from near Chongón, Guayas, Ecuador that was sent to the National Museum of Natural History, Paris, France. The USNM has another specimen (USNM 534417) from Ecuador, but without precise locality. C. O. C. Vieira (1942, 1955) listed *D. albus* as occurring in the Brazilian state of Espírito Santo, but we know of no voucher specimens. Cabrera (1958) regarded *D. scutatus* as a synonym of *D. albus*. *Diclidurus albus* is sympatric with *D. scutatus* and, while externally similar, is easily distinguished from *D. scutatus* by its longer forearm (60 mm or longer) and by the complete posterior margin of the palate and lack of a large fenestra in the palate. D. C. Carter and Dolan (1978) gave information on and measurements of the holotypes of *Diclidurus virgo* O. Thomas, *Diclidurus freyreisii* Wied-Neuwied, and *Diclidurus albus* Wied-Neuwied.

Diclidurus freyreisii and *D. albus* were published simultaneously and are based on the same specimen. The former is the name Wied-Neuwied had intended to use for this species; however, Oken (as editor) substituted the epithet *albus* for Wied-Neuwied's choice, *freyreisii* (see Schinz 1821; Wied-Neuwied 1826; D. C. Carter and Dolan 1978:23; Gardner and Ferrell 1990). Oken (in Wied-Neuwied 1820:column 1630, footnote) did not believe in patronymics, and wrote that while his majesty thought to name the species *D. freyreisii*, we have avoided that because science does not need honors (translation from German). D. C. Carter and Dolan (1978) believed that Oken should

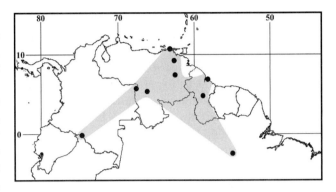

Map 87 Marginal localities for *Diclidurus ingens* ●

be considered the author of *D. albus*, an opinion in accord with Schinz's (1821) treatment of both names. Schinz (1821:171) as First Reviser (ICZN 1999: Art. 24.2.2), selected the name *Diclidurus freyreisii* over *D. albus* by listing the name *Diclidurus freyreisii* Wied-Neuwied first, followed by the name *Diclidurus albus* Oken. However, Wied-Neuwied is the author of the description of *D. albus* and, as he provided the manuscript for publication, he must be treated as the person responsible for making both names available (Gardner and Ferrell 1990; ICZN 1999:Arts. 10 and 50). While the preceding issue is of nomenclatural interest, the name *D. albus* has been in prevailing usage for this species for nearly two centuries and should not be set aside in favor of *D. freyreisii* at this late date (ICZN 1999:Art. 23.9).

Diclidurus ingens Hernández-Camacho, 1955
Greater Ghost Bat

SYNONYM:

Diclidurus ingens Hernández-Camacho, 1955:87; type locality "Puerto Leguízamo, Intendencia del Caquetá" (= Putumayo), Colombia.

DISTRIBUTION: *Diclidurus ingens* is known from Colombia, Venezuela, Guyana, and Brazil.

MARGINAL LOCALITIES (Map 87): VENEZUELA: Sucre, Carúpano (Ojasti and Linares 1971); Monagas, Laguna Guasacónica (Linares and Rivas 2004); Bolívar, El Manteco (Ochoa, Castellanos, and Ibáñez 1988). GUYANA (Lim et al. 1999): Demerara-Mahaica, Georgetown; Potaro-Siparuni, Clearwater Camp. BRAZIL, Pará, Alter do Chão (Bernard and Fenton 2002). VENEZUELA: Amazonas, San Juan (Handley 1976). COLOMBIA: Putumayo, Puerto Leguízamo (type locality of *Diclidurus ingens* Hernández-Camacho). VENEZUELA: Amazonas, Puerto Ayacucho (Handley 1976).

SUBSPECIES: *Diclidurus ingens* is monotypic.

NATURAL HISTORY: Little is known concerning this species. Handley (1976) reported *D. ingens* shot in Venezuela as these bats flew "over stream banks and in other moist areas," and also "in yards" and in evergreen forest. The holotype was taken in "*selva higrófila*" (Hernández-Camacho 1955).

REMARKS: *Diclidurus ingens* is most closely related to *D. albus*, being larger than (forearm 70 mm or longer) and differing in some proportions from the latter, but otherwise resembling it. V. Pacheco et al. (1995) listed the species for Peru. While we do not doubt the occurrence of the *D. ingens* in Peru, because the type locality is across the river from Colombia's border with Peru, we are not aware of any confirming records. Insofar as we know, fewer than a dozen specimens of this bat reside in museum collections. Hernández-Camacho (1955), Ojasti and Linares (1971), and Lim et al. (1999) provided measurements of *D. ingens*.

Diclidurus isabella (O. Thomas, 1920)
Pale-brown Ghost Bat

SYNONYMS:

Depanycteris isabella O. Thomas, 1920c:271; type locality "Manacapuru, Rio Solimoes," Amazonas, Brazil.

Drepanycteris isabellae C. O. C. Vieira, 1955:356; incorrect subsequent spelling of *Depanycteris isabella* O. Thomas.

Diclidurus isabella: Handley, 1976:12; first use of current name combination.

Diclidurus isabellus: J. K. Jones and Hood, 1993, 25; incorrect subsequent spelling of *Depanycteris isabella* O. Thomas.

Diclidurus isabellus Lim, Engstrom, Timm, Anderson, and Watson, 1999:183; incorrect subsequent spelling of *Depanycteris isabella* O. Thomas.

Diclidurus isabellus Muñoz, 2001:12, 14; incorrect subsequent spelling of *Depanycteris isabella* O. Thomas.

Diclidurus isabelus Bernard and Fenton, 2002:1132; incorrect subsequent spelling of *Depanycteris isabella* O. Thomas.

DISTRIBUTION: *Diclidurus isabella* is endemic to South America where it has been found in Guyana, southern Venezuela, and northern Brazil.

MARGINAL LOCALITIES (Map 88): GUYANA: Potaro-Siparuni, Iwokrama Reserve, S. Falls, Siparuni River (ROM 109127). BRAZIL: Amazonas, Manacapuru (type locality of *Depanycteris isabella* O. Thomas). VENEZUELA (Handley 1976): Amazonas, Boca Mavaca; Amazonas, San Juan

SUBSPECIES: We consider *D. isabella* to be monotypic.

NATURAL HISTORY: As for other members of this genus, little natural history information is available. Handley (1976) reported specimens shot in flight over and along streams in evergreen forest. Thirteen nonpregnant females and one male (testes = 1 × 2 mm) were shot as they flew low over a river in Guyana during early April (Lim et al. 1999).

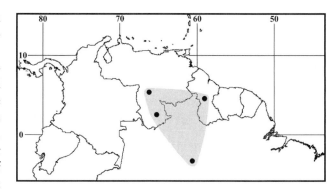

Map 88 Marginal localities for *Diclidurus isabella* ●

REMARKS: *Diclidurus isabella* originally was described as the sole member of the monotypic genus *Depanycteris*, which was regarded as distinct at the generic level by Cabrera (1958) and others, even though its close relationship to *Diclidurus* was recognized. Some authors (e.g., Handley 1976; Koopman 1982) have recognized *Depanycteris* at the subgeneric level. We believe that *D. isabella* is no more distinct than is *D. scutatus* from other members of the genus and that subgeneric separation is not warranted. The species name *isabella* is feminine and should be treated as a noun in apposition to *Diclidurus*.

Diclidurus scutatus W. Peters, 1869
Little Ghost Bat

SYNONYMS:

Diclidurus scutatus W. Peters, 1869:400; type locality "Südamerica," restricted to "Pará" (= Belém), Pará, Brazil, by Husson 1962:59.

Diclidurus albus: C. O. C. Vieira, 1942:352–54; not *Diclidurus albus* Wied-Neuwied.

Diclidurus albus: Cabrera, 1958:54; part.

D[iclidurus]. scuttatus Sodré and Uieda, 2006:897; incorrect subsequent spelling of *Diclidurus scutatus* W. Peters.

DISTRIBUTION: *Diclidurus scutatus* is known from Venezuela, Guyana, Surinam, French Guiana, Brazil, and Peru.

MARGINAL LOCALITIES (Map 89): VENEZUELA: Monagas, Laguna Guasacónica (Linares and Rivas 2004). SURINAM: Brokopondo, Brokopondo (Husson 1978). FRENCH GUIANA: Paracou (Simmons and Voss 1998). BRAZIL: Amapá, Colônia Torrão (Piccinini 1974); Pará, Para (O. Thomas 1920c); São Paulo, São Paulo (Sodré and Uieda 2006). VENEZUELA: Bolívar, Icabarú (Handley 1976); Amazonas, Cerro Neblina Base Camp (Gardner 1988). BRAZIL: Amazonas, Taracuá (C. O. C. Vieira 1942, as *D. albus*); Amazonas, Jauareté (Piccinini 1974). PERU: (Tuttle 1970; no locality information). ECUADOR: Orellana, Coca (Albuja 1999). VENEZUELA: Amazonas,

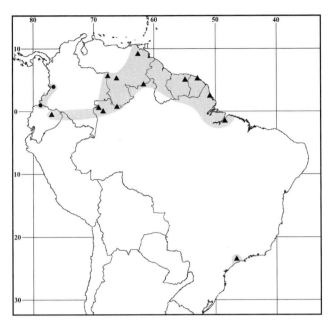

Map 89 Marginal localities for *Balantiopteryx infusca* ● and *Diclidurus scutatus* ▲

Puerto Ayacucho (Handley 1976); Amazonas, San Juan (Handley 1976).

SUBSPECIES: We regard *D. scutatus* as monotypic.

NATURAL HISTORY: As in other species of this little-known genus, little been recorded concerning *D. scutatus*. Handley (1976) reported that specimens were shot in flight "near stream banks and other moist areas" and "over yards and street in towns" in Venezuela. Simmons and Voss (1998) caught a *D. scutatus* in a net suspended 17–20 m above a narrow dirt road in well-drained primary forest. The Ecuadorian specimen had 2-cm-long moth in its mouth when found dead (Albuja and Tapia 2004). The specimen from São Paulo was found dead on a window sill of a ninth floor apartment (Sodré and Uieda 2006).According to Husson (1962:62), a Mr. Bolten observed white bats, possibly *D. scutatus*, "suspended from overhanging trees on the Marowjine River near Albina" in Surinam, and saw them flying in sunlight. This information is suspect because both *D. albus* and *D. scutatus* have been collected at the same locality.

REMARKS: The skull of *Diclidurus scutatus* is unique among diclidurine bats in having a large, median palatine fenestra confluent with the mesopterygoid fossa via a narrow posterior emargination of the palate. Also noted in the account of *D. albus*, Cabrera (1958) regarded *D. scutatus* as a synonym of the former. As pointed out by Husson (1962), the two bats recorded by C. O. C. Vieira (1942, 1955) from Taracuá, Amazonas, Brazil as *D. albus* are small (forearm 50 mm); therefore, they clearly represent *D. scutatus*. Although records are lacking, *D. scutatus* undoubtedly occurs

in eastern Colombia. The species has been found at Puerto Ayacucho, Amazonas, Venezuela, which is across the Río Orinoco from Colombia. D. C. Carter and Dolan (1978) gave information on and measurements of the holotype of *D. scutatus* W. Peters.

Subfamily Emballonurinae Gervais, 1856

This subfamily is pantropical as outlined above for the family. Of the ten Recent genera, six occur in tropical parts of the New World and four in the Old World. Emballonurines are characterized by their long, curved postorbital processes; a narrow clavicle, its greatest width about 1/6 its length; the tibia is either subterete or has the outer side flattened; and a more or less flattened rostrum except in *Balantiopteryx*, in which the anterior rostrum is inflated dorsolaterally. Wing sacs are present in four of the six American genera: *Balantiopteryx*, *Cormura*, *Peropteryx*, and *Saccopteryx* (illustrated by Jones and Hood 1993:2).

Genus *Balantiopteryx* W. Peters, 1867

Balantiopteryx comprises three species of small bats of which only *B. infusca* has been confirmed in South America. The forearm ranges from 34 to 46 mm, depending on species (37.5–40.4 mm in 16 specimens of *B. infusca*; see Hill 1987; McCarthy, Albuja, and Manzano 2000). The dorsal pelage varies from dark grayish to dark brownish with the venter slightly paler. The wing sac is located near the center of the antebrachial membrane with the opening directed toward the forearm. The rostrum is greatly inflated on each side of a median depression. The basisphenoid pit may or may not be divided by a median septum. The genus is represented in Middle America (Mexico south into Costa Rica) by *B. io* and *B. plicata*, and in northwestern Ecuador and southwestern Colombia by *B. infusca*. Cuervo-Díaz, Hernández-Camacho, and Cadena (1986:474) listed the Mexican and Middle American, *B. plicata*, as occurring in "Costa de Colombia (La Guajira, Villanueva)." Alberico et al. (2000) and Muñoz (2001) also cited Cuervo-Díaz, Hernández-Camacho, and Cadena (1986) as the information source for *B. plicata* in Colombia. Arroyo-Cabrales and Jones (1988), Jones and Hood (1993), and Lim et al. (2004b) did not indicate *B. plicata* for Colombia. We also have not included *B. plicata*, because we lack confirmation of its status in Colombia; the species is not known elsewhere on the South American continent or from Panama.

SYNONYMS:

Balantiopteryx W. Peters, 1867c:476 type species *Balantiopteryx plicata* W. Peters, 1867c, by monotypy.

Saccopteryx: Dobson, 1878:369; part; not *Saccopteryx* Illiger, 1811.

Balantiopteryx infusca (O. Thomas, 1897)
Ecuadorian Sac-winged Bat

SYNONYMS:

Saccopteryx infusca O. Thomas, 1897b:546; type locality "Cachavi," Esmeraldas, Ecuador.

B[*alantiopteryx*]. *infusca*: O. Thomas, 1904c:252; first use of current name combination.

[*Saccopteryx (Balantiopteryx)*] *infusca*: Trouessart, 1904: 98; name combination.

DISTRIBUTION: *Balantiopteryx infusca* is known from northwestern Ecuador and southwestern Colombia.

MARGINAL LOCALITIES (Map 89): COLOMBIA: Valle del Cauca, Río Chanco (Alberico et al. 2000). ECUADOR: Esmeraldas, Cachaví (O. Thomas 1897b); Imbabura, near Lita (McCarthy, Albuja, and Manzano 2000; not mapped).

SUBSPECIES: *Balantiopteryx infusca* is monotypic.

NATURAL HISTORY: Little is known of the natural history of *B. infusca*; its northern congeners roost in caves, abandoned mines, and crevices in rocks. The series of specimens that included the holotype was taken "from cave in bank of R. Cachabi" (D. C. Carter and Dolan 1978:22) on 5 January 1897. Eleven specimens were netted out of an unknown number of *B. infusca* inhabiting two railroad tunnels near Lita, Ecuador (McCarthy, Albuja, and Manzano 2000). Eight specimens collected in Colombia in September 1963 were found in a small cave among large boulders on the bank of the Río El Chanco (J. R. Tamsitt, pers. comm.). Of nine adult females taken in late December in Ecuador (McCarthy, Albuja, and Manzano 2000), four were lactating, two postlactating, and five showed no signs of reproductive activity. Arroyo-Cabrales and Jones (1988) provided a detailed description along with measurements and a summary of available information on natural history in their *Mammalian Species* account.

REMARKS: In the original description, *B. infusca* was described as similar to *B. plicata* "but rather smaller and less thickly built, much darker in colour, with decidedly narrower ears, less hairy interfemoral [membrane] and no white line along the posterior edge of the wing-membrane" (O. Thomas 1897b:546). More recently, Hill (1987:559) characterized the cranium as follows:

"As in *plicata* the braincase of *infusca* is rather elongate posteriorly and less rounded than in *io*, and there is a moderate frontal depression, less evident than in *plicata* but differing quite sharply from the very slightly depressed frontal region of *io*. The rostral swellings in *infusca* differ markedly from those of its congeners. In all the rostrum is inflated anteriorly by paired dorsal swellings, more or less separated by a longitudinal median trough.... The anterior rostral swellings of *infusca* are similar to those of *plicata* but in contrast to this species and to *io* the rostrum is more dilated posteriorly, the

posterior compartments inflated and raised on each side of the median trough that extends without interruption from the narial rim posteriorly to the frontal depression."

Hill also listed cranial measurements for four specimens. McCarthy, Albuja, and Manzano (2000) provided additional information on the color pattern, along with measurements, for two males and nine females. Lim et al. (2004b) considered *B*. io and *B. infusca* as sister species with *B. plicata* as the sister lineage to that clade. D. C. Carter and Dolan (1978:22) noted that the type locality was the Río Cachabí, 500 ft. Albuja (1983) gave the type locality as Cachabí, currently known as Urbina, a railroad station on the Río Cachabí at 150 m, Esmeraldas, Ecuador. The original spelling, Cachaví, is a variant spelling found on some maps and not to be confused with San Javier de Cachabí, also on the Río Cachabí. D. C. Carter and Dolan (1978) gave information on the holotype of *Saccopteryx infusca* O. Thomas.

Genus *Centronycteris* Gray 1838

A diagnostic feature of the genus, the dorsal pelage of the two known species of *Centronycteris* is long and soft, raw umber to tawny in color. The underparts are paler, and the interfemoral membrane and fur between the eyes are reddish. *Centronycteris* lacks wing sacs. The wing membranes (plagiopatagium) attach to the metatarsals near the base of the toes. The cranium lacks an angle or crease between the rostrum and braincase, resulting in a relatively flat dorsal profile. The lower border of the orbit is not expanded laterally and does not hide the toothrows when the cranium is viewed from above. The postorbital processes are relatively short, the sagittal crest is well developed, and the basisphenoid pit is large and subdivided by a median septum. The anterior upper premolar is tricuspidate. Simmons and Handley (1998) recognized two species in their revision of *Centronycteris*.

SYNONYMS:

Vespertilio: Schinz, 1821:180; part; not *Vespertilio* Linnaeus, 1758.

Centronycteris Gray, 1838b:499; type species *Proboscidea calcarata*: Gray, 1838b (= *Vespertilio calcaratus* Schinz, 1821), by monotypy; proposed as a subgenus of *Proboscidea* Spix, 1823.

Proboscidea: Gray, 1838b:499; part; not *Proboscidea* Spix, 1823.

Emballonura Temminck, 1838:22; part.

Saccopteryx: Dobson, 1878:376; part; not *Saccopteryx* Illiger, 1811.

KEY TO SPECIES OF *CENTRONYCTERIS*:

1. Basisphenoid pit weakly divided into anterior and posterior sections with anterior section extending forward

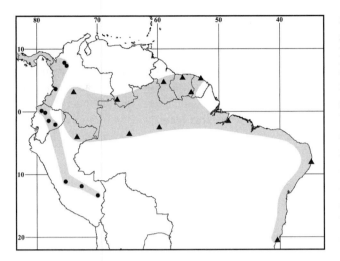

Map 90 Marginal localities for *Centronycteris centralis* ● and *Centronycteris maximiliani* ▲

between pterygoid processes; posterolateral margins of palate smoothly curved. . . . *Centronycteris maximiliani*

1'. Basisphenoid pit not divided into anterior and posterior sections and not protruding into mesopterygoid fossa; posterolateral margins of palate with an indentation extending anterior to posteromedial border of palate . . . . *Centronycteris centralis*

Centronycteris centralis O. Thomas, 1912
Thomas's Shaggy Bat

SYNONYMS:

Centronycteris centralis O. Thomas, 1912g:638; type locality "Bogava, Chiriqui, Panama."

Centronycteris maximiliani centralis: Sanborn, 1936:93; name combination.

DISTRIBUTION: *Centronycteris centralis* is known from Colombia, Ecuador, and Peru. Elsewhere, the species occurs from the Isthmus of Tehuántepec, Mexico, southeastward through Panama.

MARGINAL LOCALITIES (Map 90; from Simmons and Handley 1998, except as noted; localities listed from north to south): COLOMBIA: Córdoba, Río Ure (Lemke et al. 1982); Antioquia, La Tirana; Valle del Cauca, Río Achicaya. ECUADOR: Pichincha, Río Toachi (Albuja 1983); Manabí, Mongoya; Pastaza, Cavernas de Mera (Albuja 1998); Pastaza, Montalvo. PERU: Junín, Chanchamayo; Cusco, near Camisea (Hice and Solari 2002); Puno, Zona Reservada Tambopata-Candamo (Hice and Solari 2002).

SUBSPECIES: We consider *C. centralis* to be monotypic.

NATURAL HISTORY: This species is not commonly collected. Most specimens have come from primary or secondary forest habitats, where they have been caught in mistnets and harp traps, or shot on the wing. Females have been found in tree holes in Ecuador. In their summary of natural history information on *Centronycteris*, Simmons and Handley (1998) mentioned an individual hanging under a large *Philodendron* leaf about 5 m above ground during midday in Panama. The flight pattern has been described as slow, floppy, and highly maneuverable (Starrett and Casebeer 1968); several have been taken when flying in late afternoon at or before sundown. Hice and Solari (2002) reported a pregnant female (28-mm-CR fetus) taken on 23 September and a lactating female netted on 30 November in Peru. Based on finding the remains of plant hoppers (Homoptera) and beetles (Coleoptera) among the stomach contents of a specimen from northeastern Costa Rica, Woodman (2003a) suggested that this bat may feed by gleaning insects from the surfaces of leaves and branches, as well as taking these slow-flying insects on the wing. *Centronycteris centralis* has a $2n = 28$, FN = 46 or 48 karyotype (Greenbaum and Jones 1978).

REMARKS: Sanborn (1937) synonymized *C. centralis* with *C. maximiliani* because he could see no differences among the five specimens he examined (one *C. maximiliani*, which had a damaged skull, and four *C. centralis*). In their revision of the genus, Simmons and Handley (1998) diagnosed and described *C. centralis* and reaffirmed its validity as a species. D. C. Carter and Dolan (1978) gave information on and measurements of the holotype of *Centronycteris centralis* O. Thomas.

Centronycteris maximiliani (J. B. Fischer, 1829)
Maximilian's Shaggy Bat

SYNONYMS:

Vesp[ertilio]. calcaratus Schinz, 1821:180; type locality "Ostküste von Brasilien"; restricted to Fazenda zu Coroaba am Flüsschen Jucú, unweit des Rio Espírito Santo, by Wied-Neuwied (1826:271; also see D. C. Carter and Dolan 1978:21); preoccupied by *Vespertilio calcaratus* Rafinesque, 1818.

V[espertilio]. Maximiliani J. B. Fischer, 1829:112; replacement name for *V. calcaratus* Schinz, preoccupied.

Emballonura calcarata: Temminck, 1838:22; name combination.

Proboscidea [(Centronycteris)] calcarata: Gray, 1838b:499; name combination.

Centronycteris calcarata: Gervais, 1856a:69; name combination.

Saccopteryx [(Centronycteris)] calcarata: Dobson, 1878:376; name combination.

[Saccopteryx (Centronycteris)] calcarata: Trouessart, 1897:138; name combination.

Saccopteryx wiedi Palmer 1898:110; replacement name for *V. calcaratus* Schinz, 1821; preoccupied.

[Saccopteryx (Centronycteris)] wiedi: Trouessart, 1904:98; name combination.

Centronycteris maximiliani: Miller, 1907b:91; first use of current name combination.

DISTRIBUTION: *Centronycteris maximiliani* is known from Peru and widely scattered localities in eastern Colombia, southern Venezuela and the Guianas, and in the central Amazon basin and Atlantic coastal forests of Brazil.

MARGINAL LOCALITIES (Map 90; from Simmons and Handley 1998, except as noted): FRENCH GUIANA: Paracou. SURINAM: Marowijne, Oelemarie (Williams, Genoways, and Groen 1983). BRAZIL: Pará, Utinga (O. Thomas 1913a); Pernambuco, Recife; Espírito Santo, Fazenda Coroaba (type locality of *Vespertilio calcaratus* Schinz); Amazonas, Gavião; Amazonas, Santo Isadoro, Tefé. PERU: Loreto, Estación Biológica Allpahuayo (Hice and Solari 2002). COLOMBIA: Meta, Serranía de La Macarena (Cuervo-Díaz, Hernández-Camacho, and Cadena, 1986). VENEZUELA: Amazonas, Buena Vista (McCarthy and Ochoa 1991). GUYANA: Potaro-Siparuni, Pakatau Falls (Lim et al. 1999). SURINAM: Saramacca, Tibiti [River] (Husson 1962).

SUBSPECIES: We consider *C. maximiliani* to be monotypic.

NATURAL HISTORY: *Centronycteris maximiliani* is uncommon and information on its natural history is limited to reports of individual specimens. The species has been be found in a variety of primary and disturbed forest types. The karyotype is unknown.

REMARKS: O. Thomas (1912g) described *C. centralis* as distinct from *C. maximiliani* on the basis of slightly larger size, darker color, and shorter basisphenoid pits. Examination of four specimens of *C. centralis*, and a specimen of *C. maximiliani* (skull damaged) from Brazil, led Sanborn (1936, 1937) to conclude that the two taxa were only subspecifically distinct. Husson (1962) reviewed the characters and taxonomic history of *Centronycteris* and retained Sanborn's treatment of the genus as comprising a single species with two subspecies. In their revision of the genus, Simmons and Handley (1998) redescribed *C. maximiliani*, showed how it differs from *C. centralis*, discussed its taxonomic history, and reviewed the meager information available on its natural history. Tuttle's (1970) report of *C. maximiliani* from Peru was based on W. Peters (1872). D. C. Carter and Dolan (1978) gave information on and measurements of the type of *Vespertilio calcaratus* Schinz.

Genus *Cormura* W. Peters, 1867

Cormura is medium-sized for emballonurines (head-and-body 50–60 mm, forearm 43–50 mm). The dorsal pelage is reddish brown to blackish brown, and the underparts are paler. The wing sac is long, opens distally, and is centered in the antebrachial membrane where it extends from the margin of the antebrachium almost to the elbow. The wings are attached to the metatarsals near the base of the toes. The cranium has a relatively flat dorsal profile and lacks an angle or deep crease between the short, broad rostrum and the braincase. The orbits and zygomatic arches are broad; the postorbital processes are short; and the sagittal crest is well developed and extends forward onto the bases of the postorbital processes. The basisphenoid lacks well-marked pits and a median septum. The anterior upper premolar is tricuspidate with distinct anterior and posterior cusps. The genus is monotypic (see Remarks under *C. brevirostris*).

SYNONYMS:

Emballonura: J. A. Wagner, 1843a:367; not *Emballonura* Temminck, 1838.

Cormura W. Peters, 1867c:475; type species *Emballonura brevirostris* J. A. Wagner, 1843a, by monotypy.

Myropteryx Miller, 1906a:59; type species *Myropteryx pullus* Miller, 1906a, by original designation.

Saccopteryx: Simpson, 1945:55; part; not *Saccopteryx* Illiger, 1811.

Cormura brevirostris (J. A. Wagner, 1843)
Chestnut Sac-winged Bat

SYNONYMS:

Emballonura brevirostris J. A. Wagner, 1843a:367; type locality "Marabitanas," corrected to "Baraneiva" by Pelzeln (1883); spelling of type locality corrected to "Bananeira," on the Rio Mamoré, Rondônia, Brazil, by D. C. Carter and Dolan (1978:19).

Cormura brevirostris: W. Peters, 1867c:475; first use of current name combination.

Myropteryx pullus Miller, 1906a:59; type locality "Surinam."

DISTRIBUTION: The species is known in South America from Colombia, Venezuela, the Guianas, Ecuador, Peru, and Brazil. Elsewhere, *C. brevirostris* occurs in Central America from Nicaragua southeast into Panama.

MARGINAL LOCALITIES (Map 91): COLOMBIA: Magdalena, Don Diego (Sanborn 1932); Meta, Río Ocoa (Nicéforo-María 1947). VENEZUELA (Handley 1976, except as noted): Amazonas, 32 km SSE of Puerto Ayacucho; Amazonas, San Juan; Bolívar, 40 km NE of Icabarú; Sucre, Manacal; Delta Amacuro, Winikina (Linares and Rivas 2004). GUYANA: Cuyuni-Mazaruni, 24 miles along Potaro Rd. from Bartica (Hill 1965). SURINAM: Marowijne, Galibi (Husson 1962). FRENCH GUIANA: Cayenne (Brosset and Dubost 1968). BRAZIL: Pará, Belém (Handley 1967); Pará, Alter do Chão (Bernard 2001); Rondônia, Bananeira (type locality of *Emballonura brevirostris* J. A. Wagner). PERU: Pasco, San Juan (Tuttle 1970). ECUADOR: Napo, Cavernas de Jumandi (Albuja 1983); Esmeraldas, La Chiquita (Albuja and Mena 2004).

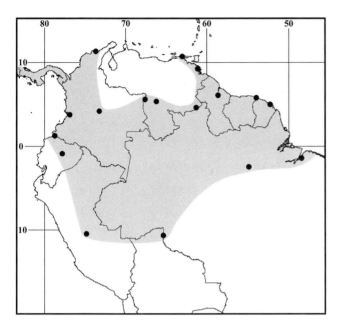

Map 91 Marginal localities for *Cormura brevirostris* ●

COLOMBIA: Valle de Cauca, Río Zabaletas (USNM 483278).

SUBSPECIES: We consider *C. brevirostris* to be monotypic.

NATURAL HISTORY: Little is known regarding the ecology and behavior of this uncommon emballonurid. Specimens have been taken from hollow trees in Surinam (Sanborn 1941) and Peru (Tuttle 1970). Simmons and Voss (1998) encountered roosting groups in the vicinity of Paracou, French Guiana, in the following situations: in dark recesses under fallen trees, clinging to a leaf of *Phenakospermum guyannensis*, in a shallow cavity in the base of a tree over a stream, and under a concrete bridge. Three roosts were over water and five were in primary or secondary forest far from water. Ten roosting groups consisted of males only; the other groups contained no more than one female each. These bats clustered tightly against one another and did not share roosts with other species. Fleming, Hooper, and Wilson (1972) examined the stomachs of nine individuals from Panama; six were empty and three contained the remains of insects. In Panama, pregnant females have been taken in April and May; whereas nonpregnant females have been taken in June, July, September, and October (Fleming, Hooper, and Wilson 1972). *Cormura brevirostris* has a $2n = 22$; FN = 40 karyotype (R. J. Baker and Jordan 1970).

REMARKS: Miller (1907b) and Sanborn (1937) treated *Myropteryx pullus* Miller as distinct from *Cormura brevirostris*, but we consider the genus to be monotypic and place *Myropteryx pullus* in synonymy with *C. brevirostris*, as was done by O. Thomas (1913a), Cabrera (1958), Hus-

son (1962), D. C. Carter and Dolan (1978), J. K. Jones and Hood (1993), and Simmons and Voss (1998). D. C. Carter and Dolan (1978:19) discussed the errors concerning the type locality, which they located in northwestern Mato Grosso (now Rondônia) on Brazil's border with Bolivia, and gave information on and measurements of the types of *Emballonura brevirostris* J. A. Wagner, and *Myropteryx pullus* Miller. There appears to be a hiatus in the distribution of *C. brevirostris* in eastern Colombia and northern Venezuela (see map in Linares 1986:36).

Genus *Peropteryx* W. Peters, 1867

These are small to medium-sized emballonurines (head and body 45–55 mm; forearm 38–54 mm); females are larger than males in most dimensions. Dorsal pelage ranges from dark buffy brown to blackish brown, with underparts paler. The wing sac is small, located near the anterior border of the antebrachial membrane, and opens distally. The wing membranes (plagiopatagium) are attached to the leg at the distal end of the tibia. The skull has an expanded rostrum, a distinct angle between the rostrum and forehead, relatively short postorbital processes, an undivided basisphenoid pit (no median septum), and a spicule-like anterior upper premolar.

The genus contains four species often treated as representing two genus-group taxa, *Peropteryx* and *Peronymus*. Some workers (Miller 1907b; Sanborn 1937; Husson 1962; Corbet and Hill 1980, 1991) recognized the latter as a separate monotypic genus represented by *Peronymus leucoptera*. Others (e.g., Cabrera 1958; J. K. Jones and Hood 1993; Koopman 1993, 1994; and McKenna and Bell 1997) have treated *Peronymus* as a subgenus of *Peropteryx*. However, we follow Simmons and Voss (1998) in not recognizing *Peronymus* even at the subgeneric level. *Peropteryx* is known from the Pleistocene of Brazil (see Mones 1986).

SYNONYMS:

Vespertilio: Wied-Neuwied, 1826:262; part, not *Vespertilio* Linnaeus, 1758.

Proboscidea: Gray, 1838b:499; part, not *Proboscidea* Spix, 1823.

Emballonura: Temminck, 1838:22; part.

Peropteryx W. Peters, 1867c:472; type species *Vespertilio caninus* Schinz, 1821, by original designation.

Peronymus W. Peters, 1868a:145; type species *Peropteryx leucoptera* W. Peters, 1867c, by monotypy; described as a subgenus of *Peropteryx* W. Peters, 1867c.

Saccopteryx: Dobson, 1878:373; part; not *Saccopteryx* Illiger, 1811.

Peropterix Trajano and Moreira, 1991:16; incorrect subsequent spelling of *Peropteryx* W. Peters.

KEY TO SOUTH AMERICAN SPECIES OF *PEROPTERYX*:

1. Wings pale, nearly transparent beyond forearms; ears connected by a low band of skin across forehead; pterygoid pits large . . . . *Peropteryx leucoptera*

1′. Wings black; ears separate, not connected by a band of skin across forehead; pterygoid pits small. 2

2. Smaller: forearm 38.3–48.2 mm (males 38.3–44.3 mm, females 43.5–48.2 mm); greatest length of skull 15 mm or less; maxillary toothrow shorter than 6.5 mm; greatest breadth across molars (M3–M3) less than 7 mm 3

2′. Larger: forearm 45.0–53.6 mm (males 45.0–51.0 mm; females 47.9–53.6 mm); greatest length of skull 16 mm or more; maxillary toothrow longer than 6.5 mm; greatest breadth across molars (M3–M3) 7 mm or more . . .
. . . . *Peropteryx kappleri*

3. Smaller: forearm shorter than 42 mm (females) or 41 mm (males); maxillary toothrow less than 5.4 mm (females) or 5.0 mm (males); first upper premolar often reduced to a simple, peg-like tooth. . . . . . . *Peropteryx trinitatis*

3′. Larger: forearm longer than 41.0 mm (females) or 39.5 mm (males); maxillary toothrow longer than 5.3 mm (females) or 5.1 mm (males); first upper premolar not peg-like, usually with a distinct posterior accessory cuspule. . . . . *Peropteryx macrotis*

Peropteryx kappleri W. Peters, 1867

Greater Dog-like Bat

SYNONYMS: See under Subspecies.

DISTRIBUTION: *Peropteryx kappleri* occurs in Colombia, Venezuela, Guyana, Surinam, French Guiana, Brazil, Ecuador, Peru, and Bolivia. Elsewhere, the species occurs north through Central America into southern Mexico.

MARGINAL LOCALITIES (Map 92): VENEZUELA (Handley 1976): Falcón, Riecito; Monagas, 3 km SW of Caripe. GUYANA: (no locality mentioned, Engstrom and Lim 2002; Lim, Engstrom, and Ochoa 2005). SURINAM: Marowijne, Albina (Husson 1978). FRENCH GUIANA: Paracou (Simmons and Voss 1998). BRAZIL: Pernambuco, Mato do Camocim (USNM 555699); Bahia, Salvador (C. O. C. Vieira 1955); Bahia, Poliraguá (Faria, Soares-Santos, and Sampaio 2006) Espírito Santo, Castelo (Ruschi 1951h); Rio de Janeiro, Terezópolis (C. O. C. Vieira 1942); São Paulo; São Sebastião (Sanborn 1937). BOLIVIA: Beni, near Espíritu (S. Anderson 1997). PERU: Cusco, Hacienda Cadena (type locality of *Peropteryx kappleri intermedius* Sanborn); Cusco, Pagoreni (S. Solari et al. 2001c); Pasco, San Juan (Tuttle 1970). ECUADOR: El Oro, Portovelo (Sanborn 1937); Carchi, Puente Piedra (Albuja 1999). COLOMBIA: Valle del Cauca, Cali (Alberico 1981); Caldas, Finca Los Naranjos (Castaño

Map 92 Marginal localities for *Peropteryx kappleri* ●

et al. 2003); Antioquia, Guapá (Morales-Alarcón et al. 1968).

SUBSPECIES: We recognize two subspecies of *P. kappleri*.

P. k. kappleri W. Peters, 1867

SYNONYMS:

Peropteryx Kappleri W. Peters, 1867c:473; type locality "Surinam."

Saccopteryx kappleri: Dobson, 1878:374; name combination.

[*Saccopteryx (Peropteryx) canina*] *kappleri*: Trouessart, 1897:138; name combination.

[*Peropteryx canina*] *kappleri*: Trouessart, 1904:98; name combination.

This subspecies is known from Ecuador and southeastern Brazil, northward through the Guianas. Elsewhere, it ranges from Veracruz, Mexico, southeastward through Central America.

P. k. intermedius Sanborn, 1951

SYNONYM:

Peropteryx kappleri intermedius Sanborn, 1951a:476:476; type locality "Hacienda Cadena, Province of Quispicanchis, Department of Cusco, Peru." This subspecies is known from Peru.

NATURAL HISTORY: Specimens have been taken at elevations from sea level to 1,500 m, from shallow caves, rock crevices, and hollow trees. *Peropteryx kappleri* roosts singly or in small groups (up to at least six individuals) of mixed sexes in Costa Rica (Bradbury and Vehrencamp 1977a). Simmons and Voss (1998) reported finding 11 roosting groups of *P. kappleri* in French Guiana. These groups consisted of one to four bats, with no more than one male found at any one roost. Roost sites were in dark spaces under fallen trees, or in large hollow logs; all were in well-drained primary forest. Simmons and Voss (1998) caught additional specimens, either in ground-level mist-nets in primary forest, or in nets placed high in a gap in the canopy 13–21 m above a narrow dirt road. Giral, Alberico, and Alvaré (1991) described social behavior and roost fidelity in several colonies inhabiting abandoned coal mines near Cali, Colombia. They found strong fidelity to particular roost sites, apparent lack of territoriality in males, and social dominance by females. Births have been recorded in January and from March to December (Sanborn 1937; Arata and Vaughn 1970; Bradbury and Vehrencamp 1977a; Rasweiler 1982; Giral, Alberico, and Alvaré 1991). Rasweiler's (1982) histological study of reproduction in a Colombian population demonstrated a highly synchronous breeding season. He reported a birth peak in the Cauca Valley during the March-May rainy season; bats conceived in May and June were born between October and December. Giral, Alberico, and Alvaré (1991) stated the main birth peak occurred in April, with a smaller peak between October and November. The karyotype appears to be unknown.

REMARKS: D. C. Carter and Dolan (1978) provided information on and measurements of the holotype of *Peropteryx kappleri* W. Peters.

Peropteryx leucoptera W. Peters, 1967
White-winged Dog-like Bat

SYNONYMS: See under Subspecies.

DISTRIBUTION: *Peropteryx leucoptera* is known from Guyana, Surinam, French Guiana, Venezuela, Colombia, Peru, and eastern Brazil.

MARGINAL LOCALITIES (Map 93): GUYANA: Upper Demerara-Berbice, Bada Creek (Lim et al. 1999). SURINAM: Para, Zanderij (Genoways, Williams, and Groen 1981). FRENCH GUIANA: Paracou (Simmons and Voss 1998). BRAZIL: Pará, Cachoeira, Ilha de Marajó (C. O. C. Vieira 1942); Pernambuco, Refúgio Ecológico Charles Darwin (L. A. M. Silva and Guerra 2000); Pernambuco, Saltinho (Guerra 1980a); Pará, Alter do Chão (Bernard 2001); Amazonas, Florestal Reserve (Sampaio et al. 2003). PERU: Ucayali, Tushemo (type locality of *Peronymus cyclops* O. Thomas). COLOMBIA: Meta, Campamento Chamusa (Lemke et al. 1982). VENEZUELA: Amazonas, Buena Vista (Sanborn 1937).

SUBSPECIES: We tentatively recognize two subspecies of *P. leucoptera*.

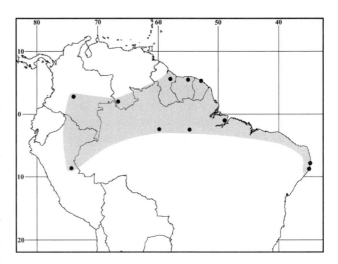

Map 93 Marginal localities for *Peropteryx leucoptera* ●

P. l. leucoptera W. Peters, 1867
SYNONYMS:

Peropteryx leucoptera W. Peters, 1867c:474; type locality "Surinam."

Peropteryx [(Peronymus)] leucoptera: W. Peters, 1868a: 145; name combination.

Saccopteryx [(Peropteryx)] leucoptera: Dobson, 1878:374; name combination.

Peronymus leucopterus: Miller, 1907b:91; name combination.

Peropteryx [(Peronymus)] leucoptera leucoptera: Cabrera, 1958:52; name combination.

Peronymus leucopterus leucopterus: Husson, 1962:54; name combination.

The nominate subspecies is known from Colombia, Venezuela, Guyana, Surinam, French Guiana, and Brazil.

P. l. cyclops O. Thomas, 1924
SYNONYMS:

Peronymus cyclops O. Thomas, 1924b:531; type locality "Tushemo, near Masisea," Río Ucayali, Ucayali, Peru.

Peronymus leucopterus cyclops: Sanborn, 1937:347; name combination.

Peropteryx [(Peronymus)] leucoptera cyclops: Cabrera, 1958:52; name combination.

This subspecies is known from only the type locality in Peru (Sanborn 1937).

NATURAL HISTORY: This bat inhabits the greater Amazon basin where it has been collected from cavities along stream banks, from dark spaces under large fallen trees, within hollow trees, and in mistnets placed in forest habitat (Sanborn 1937; Genoways, Williams, and Groen 1981; Lemke et al. 1982; Brosset and Charles-Dominique 1991; Simmons and Voss 1998). Simmons and Voss (1998) found four roosts of *P. leucoptera* in French Guiana. All were located in dark spaces between buttresses under large fallen trees. Roosting groups consisted of from two to eight individuals, each bat hanging apart from one another. Little is known regarding reproduction. Sanborn (1937) reported pregnancies in March, April, May, and June, based on specimens collected in Brazil. *Peropteryx leucoptera* has a $2n = 48$, FN = 62 karyotype (R. J. Baker, Genoways, and Seyfarth 1981).

REMARKS: O. Thomas (1924b) described *P. cyclops* based on a single individual and no additional specimens are known. Additional material from Peru could demonstrate that *P. leucopterus* is best regarded as monotypic (Sanborn 1937). D. C. Carter and Dolan (1978) gave information on and measurements of the holotype of *Peropteryx cyclops* O. Thomas.

Peropteryx macrotis (J. A. Wagner, 1843)
Lesser Dog-like Bat

SYNONYMS:

Vesp[*ertilio*]. *caninus* Schinz, 1821:179; type locality "Östküste von Brasilien"; restricted to "Timicui, on the Rio Belmonte, above Bôca d'Obu," Bahia, Brazil, by Ávila Pires (1965:8); preoccupied by *Vespertilio caninus* Blumenbach, 1797.

Proboscidea canina: Gray, 1838b:499; name combination.

Emballonura canina: Temminck, 1841:298; name combination.

Emballonura macrotis J. A. Wagner, 1843a:367; type locality "Mato grosso"; restricted to Cuiabá, Mato Grosso, Brazil, by D. C. Carter and Dolan (1978:20).

Emballonura brunnea Gervais, 1856a:66; type locality "province de Bahia," Brazil.

Peropteryx canina: W. Peters, 1867c:472; name combination.

Peropteryx macrotis: W. Peters, 1867c:472; first use of current name combination.

Saccopteryx canina: Dobson, 1878:373; name combination.

[*Saccopteryx (Peropteryx)*] *canina*: Trouessart, 1897:138; name combination.

DISTRIBUTION: Widely distributed in South America, *P. macrotis* is known from Colombia, Venezuela, Trinidad and Tobago, and the Guianas, south through Ecuador, Peru, and Brazil into Bolivia and Paraguay. Elsewhere, this species occurs on Grenada, and in Central America and southern Mexico.

MARGINAL LOCALITIES (Map 94): TRINIDAD AND TOBAGO: Tobago, Robinson Crusoe's Cave (Goodwin and Greenhall 1961). VENEZUELA: Bolívar, Piedra Virgin (Handley 1976). GUYANA: Upper Takutu-Upper Essequibo, Kanuku Mountains (BM 1.6.4.25). SURINAM:

Map 94 Marginal localities for *Peropteryx macrotis* ●

Sipaliwini, Voltzberg (Genoways, Williams, and Groen 1981). FRENCH GUIANA: Paracou (Simmons and Voss 1998). BRAZIL: Amapá, Serra do Navio (USNM 391769); Pará, Belém (USNM 393000); Rio Grande do Norte, Natal (Sanborn 1937); Pernambuco, Mata do Camocim (Mares et al. 1981); Bahia, Fazenda Lajeido (Mares et al. 1981); Bahia, Bôca d'Obu (Ávila Pires 1965); Rio de Janeiro, Angra dos Reis (Peracchi and Albuquerque 1971); São Paulo, Iguapé (C. O. C. Vieira 1942). PARAGUAY (Myers, White, and Stallings 1983): Alto Paraguay, Fuerte Olimpo; Concepción, 1 km NE of San Lazaro. BOLIVIA (S. Anderson 1997): Santa Cruz, Hacienda Cerro Colorado; Beni, 5 km NW of Río Grande. PERU: Puno, La Pampa (Sanborn 1937); Cusco, Cashiriari-2 (S. Solari et al. 1998); Pasco, San Juan (Tuttle 1970). ECUADOR: Pastaza, Alto Pastaza (Albuja 1983); Napo, Páramo de Papallacta (Jarrin 2004). COLOMBIA: Huila, La Plata (Marinkelle 1967); Valle del Cauca, Cali (Kohls, Sonenshine, and Clifford 1965); Bolívar, Cartagena (Nicéforo-María 1947); La Guajira, Villanueva (Hershkovitz 1949c). VENEZUELA: Falcón, Riecito (Handley 1976); Miranda, Cueva Ricardo Zuloaga (Handley 1976); Nueva Esparta, El Valle (Smith and Genoways 1974).

SUBSPECIES: We recognize two subspecies, of which *P. m. phaea* G. M. Allen, 1911, is endemic to the island of Grenada.

NATURAL HISTORY: *Peropteryx macrotis* is widely distributed in South America, and occurs on islands north of Venezuela. The species roosts in small groups under snags and in hollow logs, small caves, and rock crevices

(Tamsitt and Valdivieso 1963a; Mares et al. 1981; Genoways, Williams, and Groen 1981. Brosset and Charles-Dominique (1991) reported roosting groups of from one to four *P. macrotis* in French Guiana, in contrast to roosting groups numbering in excess of a hundred individuals of *P. trinitatis*. Simmons and Voss (1998) located two roosts of three bats each in well-drained primary forest in French Guiana. One group comprised two adult males and one adult female found 1.5 m above ground under the broken trunk of a tree. The second group, located in a hollow log, contained a subadult female and an adult male (the third bat escaped). Willig (1985a) noted pregnant females in September and October, and lactating females in January, in the Caatinga of northeastern Brazil. Jarrin (2004) reported on an adult male *P. macrotis* found dead, and still in relatively fresh condition, in paramo habitat near the Quito to Baeza highway, Ecuador, at an elevation between 3,300 and 3,500 m. The record is unusual because other records of *P. macrotis* are from humid tropical forest habitats at elevations usually below 1,400 m. Yee (2000) provided measurements and summarized available information in the *Mammalian Species* account on *P. macrotis*. *Peropteryx macrotis* has a $2n = 26$, $FN = 48$ karyotype (R. J. Baker, Genoways, and Seyfarth 1981).

REMARKS: Most recent authors (e.g., Sanborn 1937; Goodwin and Greenhall 1961; Jones and Hood 1993; Koopman 1993, 1994; Yee 2000) have regarded *Peropteryx trinitatis* Miller, 1898, as a subspecies of *P. macrotis*. However, we follow Handley (1976), Brosset and Charles-Dominique (1991), and Simmons and Voss (1998) in treating *P. trinitatis* as a separate species. D. C. Carter and Dolan (1978) gave information on and measurements of one of the two syntypes of *Emballonura macrotis* J. A. Wagner, 1843.

Peropteryx trinitatis Miller, 1899
Trinidadian Dog-like Bat
SYNONYMS:
Peropteryx trinitatis Miller, 1899b:178; type locality, "Port of Spain, Trinidad."
Peropteryx macrotis trinitatis: Sanborn, 1937:341; name combination.
Peropteryx canina trinitatis: Hummelinck, 1940:69; name combination.

DISTRIBUTION: *Peropteryx trinitatis* is known from Trinidad and Tobago, and Venezuela, Guyana, French Guiana, and northern Brazil.

MARGINAL LOCALITIES (Map 95): VENEZUELA (Handley 1976): Falcón, 13 km NNE of Mirimire; Miranda, 4 km SW of Birongo. TRINIDAD AND TOBAGO: Trinidad, Port of Spain (type locality of *Peropteryx trinitatis* Miller). FRENCH GUIANA: Cayenne (Brosset and Charles-Dominique 1990). BRAZIL: Pará, Utinga (BM

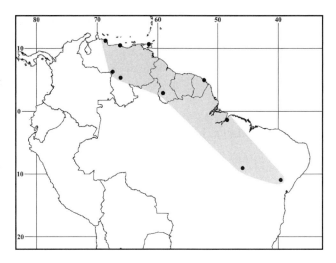

Map 95 Marginal localities for *Peropteryx trinitatis* ●

93.1.9.3); Bahia, Queimadas (Sanborn 1937, as *P. macrotis*); Maranhão, alto Parnahyba (Sanborn 1927, as *P. macrotis*). GUYANA: Upper Takutu-Upper Essequibo, 12 miles E of Dadanawa (USNM 337266). VENEZUELA: Amazonas, San Juan (Handley 1976); Apure, 8 km NW of Puerto Páez (USNM 373005).

SUBSPECIES: We treat *P. trinitatis* as monotypic.

NATURAL HISTORY: Brosset and Charles-Dominique (1990) mentioned finding a roost in the vicinity of Cayenne, French Guiana, containing over 100 individuals; whereas, roosting group size for the similar *P. macrotis* varied from one to four. Handley (1976) reported *P. trinitatis* roosting in rocks, caverns, and houses in Venezuela. Most (75%) were taken in dry areas; others were taken in savanna, pasture, yards, swamp, and evergreen forest; all at elevations under 400 m.

REMARKS: Most recent authors have treated *P. trinitatis* as a subspecies of *P. macrotis*. The two species are sympatric in many locations, and can be distinguished most readily by size (see Key). Simmons and Voss (1998:33–36) described and illustrated differences in the morphology of the first upper premolar (simple and peg-like in *P. trinitatis*) that prove useful in identifying most specimens.

Genus *Rhynchonycteris* W. Peters, 1867

Rhynchonycteris is a monotypic genus found from southern Mexico southward throughout most of the tropical regions of South America. The genus is characterized by relatively small size (forearm usually ranging from 35 to 41 mm); an elongate, pointed muzzle; tufts of whitish hair along forearms; grizzled brownish to grayish dorsum with two faint, wavy whitish stripes on the lower back and rump. The venter is paler than the dorsum, the interfemoral membrane is furred to the exsertion of tail, and wing sacs are lacking. The first upper premolar is tricuspidate, relatively large, and slightly triangular in occlusal view. The sagittal crest is not developed and there is no angle between the rostrum and braincase (dorsal profile is flat).

SYNONYMS:

Proboscidea Spix, 1823:61; type species *Proboscidea saxatilis* Spix, 1823, by subsequent designation (Miller, 1907b:88); preoccupied by *Proboscidea* Bruguière, 1791 (Nematoda).

Emballonura Temminck, 1838:22; part.

Rhynchonycteris W. Peters, 1867c:477; replacement name for *Proboscidea* Spix, 1823; preoccupied.

Rhynchiscus Miller, 1907a:65; replacement name for *Rhynchonycteris* W. Peters, on the assumption that *Rhynchonycteris* was preoccupied by *Rhinchonycteris* Tschudi, 1844b.

REMARKS: Miller (1907a) proposed the name *Rhynchiscus* to replace *Rhynchonycteris* W. Peters, which he believed was preoccupied by *Rhinchonycteris* Tschudi, 1844b, and which in turn he assumed was a synonym of *Anoura*. *Rhynchiscus* was in common usage until Simpson (1945:55) resurrected *Rhynchonycteris* W. Peters. Most recent authors have disregarded Dalquest (1957b:219, footnote) and Goodwin and Greenhall (1961:211), who claimed that *Rhynchiscus* is the valid name, in favor of the interpretation of Cabrera (1958:48) and Husson (1961:35), who correctly argued that the single letter difference in the original spellings of *Rhinchonycteris* and *Rhynchonycteris* is sufficient for the two names not to be considered homonyms. On a different basis, Griffiths and Gardner (see Remarks under *Anoura*, Glossophaginae, in this volume), agree with O. Thomas's (1928a:123) claim that *Rhinchonycteris* Tschudi, 1844b, is a *nomen nudum* and, consequently, is not an available name. They point out that *Rhynchonycteris* W. Peters is not preoccupied by *Rhinchonycteris* Tschudi, regardless of spelling, and *Rhynchiscus* Miller, is a junior objective synonym of the earliest available name, *Rhynchonycteris* W. Peters.

As is evident in W. Peters's (1867c) description of *Rhynchonycteris*, he intended the name as a replacement for *Proboscidea* Spix, 1823, which W. Peters (1866a:581, footnote) earlier noted was preoccupied. A replacement name for a genus-group taxon takes the type species of the name replaced (ICZN, 1999:Art. 67.8). Therefore, contrary to statements by Palmer (1904:610), Miller and Kellogg (1955:53), Hall (1981:78), and other authors, *Vespertilio naso* Wied-Neuwied, 1820b, is not the type species of *Rhynchonycteris* W. Peters. The type species for *Proboscidea* Spix, 1823; *Rhynchonycteris* W. Peters, 1867c; and *Rhynchiscus* Miller, 1907a, is *Proboscidea saxatilis* Spix, 1823, as designated by Miller (1907b).

Rhynchonycteris naso (Wied-Neuwied, 1820)

Proboscis Bat

SYNONYMS:

Vespertilio naso Wied-Neuwied, 1820b:251, footnote; type locality "*Mucuri*"; restricted to the vicinity of Morro d'Arara, Rio Mucuri, Bahia, Brazil, by Ávila Pires (1965:9).

Proboscidea saxatilis Spix, 1823:62; type locality "fluvium St. Francisci," Brazil.

Proboscidea rivalis Spix, 1823:62; type locality "fluvem Amazonum," Brazil.

V[espertilio]. rivalis: J. B. Fischer, 1829:116; name combination.

Emballonura saxatilis: Temminck, 1841:296; name combination.

Emballonura lineata Temminck, 1841:297; type locality "Surinam."

Emb[allonura]. naso: Schinz, 1844:201; name combination.

Proboscidea villosa Gervais, 1856a:68; type locality "la province de Goyaz," Brazil.

Peropteryx villosa: W. Peters, 1867c:473; name combination.

Rhynchonycteris naso: W. Peters, 1867c:478; first use of current name combination.

Proboscidea naso: J. A. Allen, 1904d:343; name combination.

Rhynchiscus naso: Miller, 1907b:89; name combination.

Rhynchiscus naso priscus G. M. Allen, 1914:109; type locality "Xcopen, Quintana Roo, Mexico."

Rhynchiscus nasio Dalquest, 1957b:219: incorrect subsequent spelling of *Rhynchonycteris naso* Wied-Neuwied.

DISTRIBUTION: *Rhynchonycteris naso* is found in Colombia and the island of Trinidad southward through Venezuela, the Guianas, Ecuador, eastern Peru, Brazil, and eastern Bolivia. Elsewhere, it occurs from Veracruz, Mexico, southeastward through Central America.

MARGINAL LOCALITIES (Map 96): VENEZUELA: Falcón, 20 km NNE of Mirimire (Handley 1976). TRINIDAD AND TOBAGO: Trinidad, Port-of-Spain (Goodwin and Greenhall 1961). VENEZUELA: Delta Amacuro, Caño Mariusa (Linares and Rivas 2004). GUYANA: Upper Demerara-Berbice, Kaow Island (H. E. Anthony 1921b). SURINAM: Paramaribo, Paramaribo (Husson 1962). FRENCH GUIANA: l'ilot Le Père (Menegaux 1903). BRAZIL: Pará, Belém (Handley 1967); Alagoas, Mangabeira (C. O. C. Vieira 1955); Bahia, Pôrto Seguro (Sanborn 1937); Espírito Santo, Rio Dôce (C. O. C. Vieira 1942); Rio de Janeiro, Santo Antônio de Pádua (Peracchi and Albuquerque 1971); Distrito Federal, Brasilia (Coimbra et al. 1982); Mato Grosso, Cuiabá (C. O. C. Vieira 1942). BOLIVIA: Santa Cruz, Flor de Oro

Map 96 Marginal localities for *Rhynchonycteris naso* ●

(S. Anderson 1997); La Paz, Tomonoco (Webster and Jones 1980a). PERU: Madre de Dios, Cocha Cashu (Terborgh, Fitzpatrick, and Emmons 1984); Cusco, Peruanita (S. Solari et al. 2001c); Pasco, San Juan (Tuttle 1970); Amazonas, mouth of the Cenipa [*sic*] River (Koopman 1978). ECUADOR: Morona-Santiago, Méndez (Albuja 1983); Esmeraldas, San Miguel (Albuja and Mena 2004). COLOMBIA: Chocó, Nóvita (Sanborn 1937); Magdalena, Santa Marta (Muñoz 2001); El Cesar, Caracolicito (Hershkovitz 1949c).

SUBSPECIES: We consider *R. naso* to be monotypic.

NATURAL HISTORY: This species frequents slow-moving water courses in lowland regions (up to about 900 m elevation), roosting in groups of usually no more than a dozen bats (but up to 45 have been reported) in well-lighted areas on shaded cliff facings or tree trunks; beneath overhanging logs, rocks, or exposed roots; under bridges; and occasionally beneath large leaves. The underside of large trunks overhanging rivers and streams seems to be preferred. Roosting bats usually are arranged in vertical order, each separated by a well-defined space. When disturbed, the bats take flight in the same linear "formation" and move in a rapid, weaving pattern, to another roost.

Pregnant females have been taken in most months of the year, suggesting that reproduction may be asynchronous and essentially aseasonal, and females may produce a single young twice a year (Bradbury and Vehrencamp 1977b). Dalquest (1957b) provided information on the behavior and life history based on observations of Mexican colonies. Bradbury and Vehrencamp (1977a:349, 351) reported the stomach contents of four *R. naso* from western Costa Rica as primarily "dipterans (mostly chironomids and mosquitoes), and smaller but substantial numbers of beetles and caddis flies." Albuja (1999) reported a female pregnant with an 11-mm fetus when captured on 19 March. Plumpton and Jones (1992) provided a detailed description, along with measurements, and a summary of available information on natural history in their *Mammalian Species* account. *Rhynchonycteris naso* has a $2n = 22$, FN $= 36$ karyotype (R. J. Baker and Jordan 1970).

REMARKS: D. C. Carter and Dolan (1978) provided information on and measurements of the types of *Emballonura lineata* Temminck and *Proboscidea saxatilis* Spix.

Genus *Saccopteryx* Illiger, 1811

This is a genus of five species of small to medium-sized bats (forearm 33–52 mm) in which females are larger than males in most species. The dorsal pelage is dark gray or brown to black, with two pale longitudinal lines (lacking in *S. antioquensis*). The wing sac is variable in size, located near the elbow, opens dorsally, and is especially large and seasonally active in males. Wings are attached to the tarsals in *S. bilineata*, *S. canescens*, and *S. leptura*, and attached to the metatarsals in *S. antioquensis* and *S. gymnura*. Cranial characters include large premaxillaries; a low angle between the rostrum and braincase; long, broad postorbital processes; a well-developed sagittal crest; a large basisphenoid pit, which usually is subdivided by a median septum. The anterior upper premolar is a tiny spicule. The genus occurs from Mexico southward to Bolivia and southeastern Brazil; all five species are found in South America.

SYNONYMS:

Vespertilio Schreber, 1774a:pl. lvii; part; not *Vespertilio* Linnaeus, 1758.

Saccopteryx Illiger, 1811:121; type species *Vespertilio leptura* Schreber, 1774a:pl. lvii (name), p. 173 (description), by monotypy.

Taphozous: Schinz, 1821:170; part; not *Taphozous* É. Geoffroy St.-Hilaire, 1813.

Urocryptus Temminck, 1838:31; type species *Urocryptus bilineatus* Temminck, 1838, by monotypy.

Emballonura: J. A. Wagner, 1855:694; part; not *Emballonura* Temminck, 1838.

Sacopteryx Cuervo-Díaz, Hernández-Camacho, and Cadena, 1986:475; incorrect subsequent spelling of *Saccopteryx* Illiger.

KEY TO THE SPECIES OF *SACCOPTERYX*:

1. Wings attached to metatarsals; dorsal stripes inconspicuous or lacking 2
1′. Wings attached to tibia; pair of dorsal stripes clearly evident 3

2. Size larger, forearm longer than 36 mm, zygomatic breadth greater than 8 mm; dorsal pair of stripes lacking . . . . *Saccopteryx antioquensis*

2'. Size smaller, forearm shorter than 35.5 mm, zygomatic breadth 8 mm or less; dorsal pair of stripes weak and inconspicuous. . . . . *Saccopteryx gymnura*

3. Dorsal pelage and wing membranes blackish brown; pair of longitudinal dorsal stripes conspicuous; wing sac large and prominent (especially in males); forearm more than 44 mm (males, 44.1–46.7 mm; females, 44.3–51.7 mm); maxillary toothrow usually more than 7 mm (6.4–7.4 mm); greatest width across molars (M3–M3) more than 7 mm (7.2–7.6 mm) . . . . *Saccopteryx bilineata*

3'. Dorsal pelage paler, uniformly brown or grayish brown; pair of dorsal stripes not contrasting as strongly with dorsal pelage; wing membranes paler, not black; wing sac not greatly developed (largest in males); forearm less than 44 mm; maxillary toothrow less than 6 mm; greatest width across molars (M3–M3) less than 7 mm . . . 4

4. Dorsal pelage uniformly brown; pair of dorsal stripes usually distinct; forearm 37.4–42.3 mm (males 37.4–40.0 mm, females 39.1–42.3 mm); maxillary toothrow longer than 5 mm (5.1–5.5 mm); greatest width across molars (M3–M3) usually more than 6 mm (5.9–6.3 mm) . . . . *Saccopteryx leptura*

4'. Dorsal pelage gray or brown, lightly frosted with gray; pair of dorsal stripes indistinct, but visible; forearm 35.8–40.8 mm (males 35.8–37.5 mm, females 36.5–40.8 mm); length of maxillary toothrow 5 mm or less (4.6–5.0 mm); greatest width across molars (M3–M3) less than 6 mm (5.1–5.6 mm) . . . . *Saccopteryx canescens*

Saccopteryx antioquensis Muñoz and Cuartas, 2001
Antioquian Sac-winged Bat

SYNONYM:

Saccopteryx antioquensis Muñoz and Cuartas, 2001:55; type locality "Colombia, Antioquia, municipio de Sonsón, vereda La Soledad, distante ca. 15 km, este del municipio de Sonsón; 1.200 msnm, cordillera Central, . . . coordinadas 5° 40′ N; 75° 05′ W."

DISTRIBUTION: Known only from the Cordillera Central of Antioquia, Colombia.

MARGINAL LOCALITIES (Map 97): COLOMBIA: Antioquia, vereda Las Confusas (Muñoz and Cuartas, 2001); Antioquia, vereda La Soledad (type locality of *Saccopteryx antioquensis* Muñoz and Cuartas).

SUBSPECIES: *Saccopteryx antioquensis* is monotypic.

NATURAL HISTORY: Little is known about this species. The holotype was found on the wall of a church, and the only other specimen was caught in a mistnet. Both specimens are adult males, and were taken in humid tropical forest habitat.

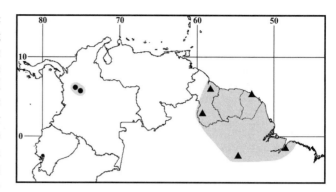

Map 97 Marginal localities for *Saccopteryx antioquensis* ● and *Saccopteryx gymnura* ▲

REMARKS: Based on the description by Muñoz and Cuartas (2001), *S. antioquensis* is similar to, but larger than *S. gymnura*, and lacks the characteristic pair of dorsal stripes found in other species of *Saccopteryx*.

Saccopteryx bilineata (Temminck, 1838)
Greater White-lined Bat

SYNONYMS:

Urocryptus bilineatus Temminck, 1838:33; type locality "Surinam."

E[mballonura]. bilineata: J. A. Wagner, 1855:694; name combination.

E[mballonura]. insignis J. A. Wagner, 1855:695; type locality "Umgebung von Rio Janeiro," Rio de Janeiro, Brazil.

Emballonura (Urocryptus) insignis: Pelzeln, 1883:39; name combination.

Saccopteryx bilineata: W. Peters, 1867c:471; first use of current name combination.

[Saccopteryx (Saccopteryx)] bilineata: Trouessart, 1897: 137; name combination.

Saccopteryx perspicillifer Miller, 1899b:176; type locality "Caura, Trinidad," Trinidad and Tobago.

[Saccopteryx (Saccopteryx)] perspicillifer: Trouessart, 1904: 98; name combination.

DISTRIBUTION: *Saccopteryx bilineata* is widespread at lower elevations in Colombia, Trinidad, the Guianas, and southward into eastern Bolivia and southcentral Brazil. Elsewhere, *S. bilineata* occurs in Mexico and Central America.

MARGINAL LOCALITIES (Map 98): COLOMBIA: Magdalena, Bonda (J. A. Allen 1900a). VENEZUELA: Falcón, Boca de Yaracuy (Handley 1976). TRINIDAD AND TOBAGO: Trinidad, Caura (type locality of *S. perspicillifer* Miller). GUYANA: Cuyuni-Mazaruni, Kalacoon (H. E. Anthony 1921b). SURINAM: Wanica, Kwatta (Husson 1978). FRENCH GUIANA: Cayenne (Brosset and Dubost 1968). BRAZIL: Amapá, Macapá (Peracchi, Raimundo, and Tannure 1984); Piauí, Deserto (Sanborn 1937); Ceará,

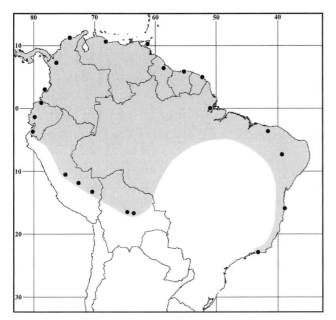

Map 98 Marginal localities for *Saccopteryx bilineata* ●

roosted under the well-shaded eaves of buildings, often on window screens where males could be observed tending their harem. Bradbury and Emmons (1974:151) described "salting," which is a behavior in which a male exposes a wing gland either toward other males in territorial defense or towards harem females in the roost. Roost sites are maintained for extended periods of time, but colonies move seasonally, presumably according to insect availability.

Reproduction is highly synchronized; females bear a single young from late May to mid-June in Trinidad, coinciding with the advent of the rainy season. Young are weaned at 10 to 12 weeks, and juvenile females disperse to new colonies at that time. *Saccopteryx bilineata* has a $2n = 26$, $FN = 36$, karyotype (R. J. Baker and Jordan 1970; Honeycutt, Baker, and Genoways 1980).

REMARKS: D. C. Carter and Dolan (1978) gave information on and measurements of the types of *Emballonura insignis* J. A. Wagner and *Urocryptus bilineatus* Temminck.

Floresta Nacional Araripe-Apodi, 9 km S of Crato (Mares et al. 1981); Bahia, Belmonte (Faria, Soares-Santos, and Sampaio 2006); Rio de Janeiro, Rio de Janeiro (Sanborn 1937). BOLIVIA: Santa Cruz, Río Palometillas (Sanborn 1933); Santa Cruz, Río Ichilo, 52 km S of mouth of Río Chaparé (S. Anderson, Koopman, and Creighton 1982). PERU: Cusco, Huajyumbe (Sanborn 1951b); Cusco, Cashiriari-2 (S. Solari et al. 2001c); Pasco, San Juan (Tuttle 1970); Tumbes, Matapalo (Koopman 1978). ECUADOR: Guayas, Pacaritambo (Brosset 1965); Esmeraldas, Playa de Oro (Albuja and Mena 2004). COLOMBIA: Cauca, Isla Gorgona (Nicéforo-María 1947); Antioquia, Santa Teresa, near Mutata (Morales-Alarcón et al. 1968).

SUBSPECIES: We recognize two subspecies. The nominate form *S. b. bilineata* (Temminck, 1838) occurs on the island of Trinidad and in mainland South America. *Saccopteryx b. centralis* O. Thomas, 1904c, is found from west-central Mexico into Nicaragua (T. Alvarez 1968; R. J. Baker and Jones 1975).

NATURAL HISTORY: Detailed information on population dynamics, social structure, and reproduction is available for *S. bilineata* from Costa Rica and the island of Trinidad (Bradbury and Emmons 1974; Bradbury and Vehrencamp 1977a, 1977c). Colonies consist of one to several well-defined social units, each including a single male with a harem of up to eight females. Males aggressively defend territories and compete for mates. These bats roost in colonies, sometimes of up to 50 or more individuals, in hollow trees and in hollows formed by the buttresses of large forest trees. On Barro Colorado Island, Panama, social units

Saccopteryx canescens O. Thomas, 1901
Gray White-lined Bat

SYNONYMS:

Saccopteryx canescens O. Thomas, 1901e:366; type locality "Obidos," Pará, Brazil.

[*Saccopteryx (Saccopteryx)*] *canescens*: Trouessart, 1904: 98; name combination.

Saccopteryx pumila O. Thomas, 1914b:410; type locality "Altagracia, Lower Orinco [sic]," Bolívar, Venezuela.

Saccopteryx (canescans) Shapley, Wilson, Warren, and Barnett, 2005:380; incorrect subsequent spelling of *Saccopteryx canescens* O. Thomas.

DISTRIBUTION: *Saccopteryx canescens* is endemic to South America, where it has been recorded from Colombia, Venezuela, Guyana, Surinam, French Guiana, northern Brazil, and eastern Peru.

MARGINAL LOCALITIES (Map 99): COLOMBIA: Magdalena, Mamatoco (Sanborn 1937). VENEZUELA (Handley 1976, except as noted): Falcón, Boca de Yaracuy; Miranda, Birongo; Sucre, Ensenada Cauranta; Delta Amacuro, Caño Araguabisi (Linares and Rivas 2004). GUYANA: Potaro-Siparuni, vicinity of Kurupukari (Smith and Kerry 1996). SURINAM: Paramaribo, Paramaribo (Husson 1962). FRENCH GUIANA: Isle le Père (Sanborn 1937). BRAZIL: Pará, Marajó Island (Sanborn 1937); Pará, Belém (USNM 460079); Pará, Rio Xingu, 52 km SSW of Altamira (Voss and Emmons 1996); Pará, Aramanay (AMNH 94366). PERU: Madre de Dios, Cocha Cashu (Tejedor 2003); Pasco, Puerto Victoria (Sanborn 1937). COLOMBIA: Tolima, Purificación (Nicéforo-María 1947); Bolívar, Norosí (Hershkovitz 1949c).

SUBSPECIES: We are treating *S. canescens* as monotypic.

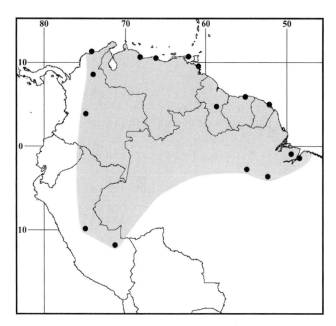

Map 99 Marginal localities for *Saccopteryx canescens* ●

NATURAL HISTORY: Little is known of the natural history of *S. canescens*. It has been captured in open areas (Handley 1976), in contrast to its congeners, which appear to prefer denser forest habitat. Brosset and Charles-Dominique (1991) located a colony of 12–15 individuals roosting under boulders along Montjoly beach, French Guiana, and commented that, when disturbed from a roost, *S. canescens* flew in tight flocks. Tejedor (2003) reported on a group of five roosting in a compact row on mosquito netting covering the ceiling in one of the Cocha Cashu Biological Station building in the Manu National Park, Peru. The group consisted of one adult male, two adult females, one subadult female, and one subadult male. The adult male had well-developed wing sacs, but had a shorter forearm (36.5 mm) and weighed less (4 g) than either of the adult females (forearms 39.0 and 39.5 mm; mass for both females, 5 g). *Saccopteryx canescens* has a $2n = 24$, FN = 38 karyotype (R. J. Baker et al. 1982).

REMARKS: O. Thomas (1914b) named *S. pumila* on the basis of the size of the molars and nature of the basisphenoid pits. Although Cabrera (1958) treated *S. pumila* as a distinct species, we follow Sanborn (1937), Husson (1962), and Koopman (1982), in treating *S. pumila* as a synonym of *S. canescens*. *Saccopteryx canescens* is similar in size to *S. gymnura*, but is readily distinguishable by its bicolored ventral pelage and its paler-brown dorsal pelage that has grayish to yellowish hair tips. In contrast, the dorsal and ventral pelage of *S. gymnura* is dark brown and lacks frosting. The record from Cocha Cashu, Manu National Park, Peru (Tejedor 2003), was not vouchered by a specimen. D. C. Carter and Dolan (1978) gave information on and measurements of the holotypes of *Saccopteryx canescens* O. Thomas and *Saccopteryx pumila* O. Thomas.

Saccopteryx gymnura O. Thomas, 1901
Little White-lined Bat

SYNONYMS:

Saccopteryx gymnura O. Thomas, 1901e:367; type locality "Santarem," Pará, Brazil.

[*Saccopteryx (Saccopteryx)*] *gymnura*: Trouessart, 1904: 98; name combination.

Saccopteryx gymura Mok, Wilson, Lacey, and Luizão, 1882:818; incorrect subsequent spelling of *Saccopteryx gymnura* O. Thomas.

DISTRIBUTION: *Saccopteryx gymnura* is known from Guyana, French Guiana, and the Brazilian state of Pará, where it has been found in the vicinity of Santarém and Belém on the lower Rio Amazonas.

MARGINAL LOCALITIES (Map 97): GUYANA: Upper Demerara-Berbice, Lucky Spot (Lim and Engstrom 2001a). FRENCH GUIANA: Paracou (Simmons and Voss 1998). BRAZIL: Pará, Belém (USNM 392995); Pará, Santarém (type locality of *Saccopteryx gymnura* O. Thomas). GUYANA: Upper Takutu-Upper Essequibo, Komawariwau river, 20 miles E of Dadanawa (Lim and Engstrom 2001).

SUBSPECIES: We consider *S. gymnura* to be monotypic.

NATURAL HISTORY: Very little is known about this species except that it inhabits humid tropical forests. Simmons and Voss (1998) netted two at ground level in French Guiana. They caught one in a small clearing bordered by well-drained primary forest; the other over a dirt road bordered by second-growth forest. Lim and Engstrom (2001) reported on six *S. gymnura* netted in Guyana, including a pregnant female (contained a 5-mm-CR embryo) taken in July at Surama Sawmill. The karyotype is unknown.

REMARKS: The attachment of the wing membranes to the proximal portion of the metatarsals is a characteristic shared with *S. antioquensis* that easily distinguishes both species from other *Saccopteryx*. D. C. Carter and Dolan (1978) gave information on and measurements of the holotype of *Saccopteryx gymnura* O. Thomas.

Saccopteryx leptura (Schreber, 1774)
Brown White-lined Bat

SYNONYMS:

Vespertilio lepturus Schreber, 1774a:pl. lvii, 173; type locality "Surinam."

Vespertilio marsupialis Müller, 1776:19; type locality "Suriname"; based solely on Schreber (1774a:pl. lvii).

S[*accopteryx*]. *lepturus*: Olfers, 1818:225; first use of current name combination, but with incorrect gender concordance.

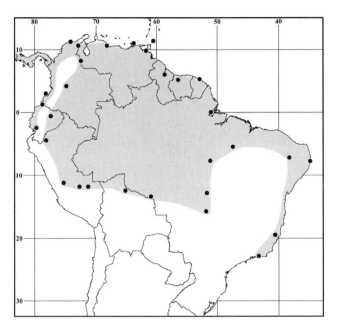

Map 100 Marginal localities for *Saccopteryx leptura* ●

Taph[ozous]. lepturus: Schinz, 1821:170; name combination.

Saccopteryx leptura: W. Peters, 1867c:471; first use of name combination with correct gender concordance.

[*Saccopteryx (Saccopteryx)*] *leptura*: Trouessart, 1897:137; name combination.

Saccopteryx Teptura N. R. Reis, 1984:247; incorrect subsequent spelling of *Vespertilio lepturus* Schreber.

DISTRIBUTION: *Saccopteryx leptura* is known from Colombia, Venezuela, Trinidad and Tobago, the Guianas, Brazil, Ecuador, eastern Peru, and northeastern Bolivia. Elsewhere, the species occurs in southern Mexico and Central America.

MARGINAL LOCALITIES (Map 100): TRINIDAD AND TOBAGO: Tobago, Pirate's Cay, 1 km N of Charlotteville (USNM 540657). VENEZUELA: Delta Amacuro, Isla Cocuina (Linares and Rivas 2004). GUYANA: Cuyuni-Mazaruni, 24 miles along Potaro Rd. from Bartica (Hill 1965). SURINAM: Sipaliwini, Stondansi Falls (Husson 1978). FRENCH GUIANA: Paracou (Simmons and Voss 1998). BRAZIL: Amapá, Horto Florestal de Macapá (Peracchi, Raimundo, and Tannure 1984); Pernambuco, Refúgio Ecológico Charles Darwin (L. A. M. Silva and Guerra 2000); Espírito Santo, Colatina (Lima 1926); Rio de Janeiro, Rio de Janeiro (M.R. Nogueira, Peracchi, and Pol 2002); Ceará, Sitio Luanda (Mares et al. 1981); Maranhão, Imperatriz (Piccinini 1974); Pará, Gradaús (Piccinini 1974); Mato Grosso, Serra do Roncador, 264 km (by road) N of Xavantina (Pine, Bishop, and Jackson 1970); Mato Grosso, Araguaiana (Piccinini 1974). BOLIVIA: Santa Cruz, Flor de Oro (Emmons 1998); Beni, Río Mamoré (S. Ander-

son, Koopman, and Creighton 1982). PERU: Madre de Dios, Cocha Cashu (Ascorra, Wilson, and Romo 1991); Cusco, Cashiriari-2 (S. Solari et al. 2001c); Junín, Chanchamayo, near Tarma (O. Thomas 1893c); Amazonas, Huampami (Patton, Berlin, and Berlin 1982). ECUADOR: Napo, Ávila (Albuja 1983). COLOMBIA: Tolima, Espinal (Nicéforo-María 1947); Norte de Santander, Río Zulia (Nicéforo-María 1947); Cauca, Isla Gorgona (Cuervo-Díaz, Hernández-Camacho, and Cadena 1986). ECUADOR (Albuja 1999): Esmeraldas, La Chiquita; Guayas, Reserva Ecológica Manglares Churute. COLOMBIA: Magdalena, Santa Marta (Bangs 1900); La Guajira, Villanueva (Hershkovitz 1949c). VENEZUELA: Falcón, Boca de Yaracuy (Handley 1976); Nueva Esparta, Isla de Margarita, El Valle (Pirlot and León 1965).

SUBSPECIES: We are treating *S. leptura* as monotypic.

NATURAL HISTORY: As revealed by the studies of Bradbury and Emmons (1974) and Bradbury and Vehrencamp (1977a, 1977c), the behavior and habitat preferences of *S. leptura* differ remarkably from those of *S. bilineata*. The species apparently is monogamous and prefers more open roost sites, such as shallow depressions on tree trunks, where they roosts singly or in small groups of up to nine individuals. Males do not appear to actively defend individual territories or to exhibit special behaviors for attracting mates. Although *S. leptura* frequently change roost location, these bats maintain group foraging territories that are actively defended. The reproductive pattern is similar to that found in *S. bilineata*, with births synchronized with the advent of the rainy season. M.R. Nogueira, Peracchi, and Pol (2002) reported the remains of Hymenoptera in feces of the specimens from Rio de Janeiro, and the remains of five flying ants of the genus *Pheidole* in the mouth of a specimen from Guapirmirim, also in the state of Rio de Janeiro. The karyotype is $2n = 28$, FN = 38 (Hood and Baker 1986).

Family Phyllostomidae Gray, 1825
Alfred L. Gardner

The Phyllostomidae are Neotropical in distribution, with a few genera reaching the southern United States (*Artibeus, Choeronycteris, Diphylla, Enchisthenes, Leptonycteris,* and *Macrotus*). Phyllostomids are characterized by a humerus having a well-developed trochiter and a double articulation with the scapula; digit II has a well-developed metacarpal and a small phalanx; digit III has three complete ossified phalanges; the seventh cervical vertebra is free from the first thoracic; the foot is normal; a fibula is present, but is cartilaginous proximally; the premaxillae are complete, fused with each other and with the maxillae; and the tragus is well developed.

The Phyllostomidae are the most diverse family of Neotropical bats. In this section, South American representatives are divided among six subfamilies comprising 44 genera and 143 species. Phyllostomids range in size from the small male of the sexually dimorphic frugivore *Ametrida centurio*, having a forearm averaging 26 mm, to the largest bat in South America, the carnivorous *Vampyrum spectrum* with a forearm averaging 106 mm.

McKenna and Bell (1997) included carolliine bats (*Carollia* and *Rhinophylla*) as a tribe within the Stenodermatinae. Herein they are maintained as a subfamily as treated by Koopman (1993) and Wetterer et al. (2000). Wetterer (2000) divided the Phyllostominae into the tribes Lonchophyllini, Micronycterini, Phyllostomini, and Vampyrini. However, because of the continued controversy over the number of natural subdivisions and their content (e.g., Baker et al. 1989, 2000; Van Den Bussche 1992; Wetterer et al. 2000), tribes herein are not recognized in the Phyllostominae. Instead, the arrangement is that of McKenna and Bell (1997), with the exception that four of the subgenera previously included in *Micronycteris* are recognized as full genera (see Simmons and Voss 1998; Wetterer et al. 2000). Tribes are recognized in the Stenodermatinae.

KEY TO THE SUBFAMILIES OF SOUTH AMERICAN PHYL-LOSTOMIDAE (MODIFIED FROM J. K. JONES AND CARTER, 1976):

1. Upper incisors and canines enlarged and bladelike; noseleaf reduced to a pair of dermal outgrowths above rhinarium . . . . Desmodontinae
1'. Upper incisors and canines not bladelike; noseleaf not reduced to a pair of outgrowths above nasal disc 2
2. Noseleaf either well developed or reduced to skin flaps on each side of nares (*Centurio*); tail absent; rostrum not narrow and elongated . . . . Stenodermatinae
2'. Noseleaf well developed; tail usually present; if tail absent, rostrum conspicuously narrow and elongate. . . . 3
3. Lower lip has a single pair of smooth tubercles divided by deep median cleft; ears short; rostrum narrow and elongate; tongue elongate with bristle-like papillae on anterodorsal surface; molars and premolars reduced; premolars usually separated from each other and from canine 4
3'. No deep median cleft in lower lip; lower lip with a pair to many tubercles; ears of various lengths, if ear short, rostrum not narrow and elongate; tongue not elongate; molars and premolars usually large and crowded in toothrow 5
4. Inner and outer upper incisors not markedly different in size; tongue with "paintbrush" tip consisting of long papillae having filamentous tips; no deep groove along each side of tongue . . . . Glossophaginae

4'. Inner upper incisors prominent with broad tips, conspicuously larger than outer incisors; tongue lacks "paintbrush" tip; a deep longitudinal groove present along each side of tongue . . . . Lonchophyllinae
5. Zygomatic arch incomplete . . . . Carolliinae
5'. Zygomatic arch complete 6
6. Upper molars dilambdodont with distinct W-shaped ectoloph . . . . Phyllostominae
6'. Upper molars lacking distinct W-shaped ectoloph . . . . Stenodermatinae, 7
7. First and second upper molars with smooth longitudinal medial groove and parallel sides; talon with hypocone of M2 not expanded lingually . . . . Sturnirini
7'. First and second upper molars lack a well-defined medial groove, sides not parallel; M2 with talon and hypocone developed lingually or posteriorly . . . . . Stenodermatini

Subfamily Carolliinae Miller, 1924

Laura J. McLellan and Karl F. Koopman

The subfamily Carolliinae contains two fruit-eating bat genera, *Carollia* and *Rhinophylla*, characterized by a skull that is generalized and intermediate compared to either the short, broad skull of most stenodermatines or the elongated skull of the glossophagines. The bony palate is unusual in that it extends posteriorly behind the maxillary toothrow as a tube-like structure enclosing the naso-pharyngeal cavity. Lateral emarginations in the outer margins of the palate separate this tube from the last molars. The *masseter medialis* originates on the maxilla and palatine bones within these emarginations. The wings are relatively long and reflect the combination of moderately long forearm and exceptionally long digit III; the fourth metacarpal is short, resulting in a relatively short digit IV. The trochanter impinges on the scapula, a shoulder characteristic also evident in the stenodermatines. The interfemoral membrane is moderately developed and completely encloses the short tail in *Carollia* (tail absent in *Rhinophylla*). Cusps of upper and lower molars are reduced in size and shifted to the lingual margin. The upper molars are so modified that the W-shaped pattern is nearly unrecognizable. The dental formula is 2/2, 1/1, 2/2, 3/3 × 2 = 32.

REMARKS: Traditionally, *Carollia* and *Rhinophylla* have been associated in their own subfamily since Miller (1907:144) grouped them in the Hemiderminae (= Carolliinae). Although R. J. Baker (1967) found *Carollia* and *Choeroniscus* (Glossophaginae) similar in standard karyotypes, Stock (1975) found no G-band autosomal homologies between these two genera and concluded that there were no data supporting a close common ancestor for *Carollia* and *Choeroniscus*. Major differences in karyotypes led R. J. Baker and Bleier (1971:221) to again suggest that *Carollia* may have its affinities with the *Choeroniscus* group and that *Rhinophylla* may be more closely related to either the "Phyllostomatine or Stenodermine group." Presumed affinity with *Choeroniscus* may be the reason J. G. Owen, Schmidly, and Davis (1984) placed *Carollia* in the Glossophaginae. In their revised classification of the Phyllostomidae, R. J. Baker, Hood, and Honeycutt (1989) did not recognize the Carolliinae and placed *Carollia* and *Rhinophylla* within the tribe Stenodermatini. Wetterer, Rockman, and Simmons (2000) maintained the traditional association of these genera, and determined that Carolliinae was monophyletic and formed a weak clade with the Stenodermatinae based on their character congruence analysis. A more recent classification of the Phyllostomidae (R. J. Baker et al. 2003) recognized the Carolliinae as monophyletic, containing only its type genus *Carollia*. They treated *Rhinophylla* as monophyletic and placed it in its own family (Rhinophyllinae, an invalid family-group name) and offered as an alternative that it be placed in its own tribe within the Stenodermatinae. Clearly, there is controversy concerning the content of the Carolliinae, which we use here in the traditional sense pending clearer resolution of the relationships of *Carollia* and *Rhinophylla* with each other and within the Phyllostomidae.

KEY TO THE GENERA AND SPECIES OF SOUTH AMERICAN CAROLLIINAE:

1. Tail short and enclosed in relatively long uropatagium; central tubercle on lower lip flanked by one or more rows of smaller tubercles; M1 and M2 with reduced protocone; metaconid present on lower molars, and lower molars do not resemble lower premolar 2
1'. Tail absent; uropatagium relatively short; central tubercle on lower lip flanked by a single large, lobate pad-like tubercle on each side; upper molars lack a protocone; lower molars lack a metaconid and resemble lower premolars. . . . 9
2. Forearm shorter than 39 mm 3
2'. Forearm longer than 39 mm 5
3. Labial outline of upper toothrow with distinct step or notch resulting from labial side of second premolar located considerably more lingual than the labial margin of first molar; second lower premolar about twice as high as first molar; first lower molar with low, smooth cuspids, tooth appears worn, even in unworn dentition; crown of first lower incisor ovoid 4
3'. Labial outline of upper toothrow evenly curved, without a distinct step or notch; second lower premolar not twice the height of first lower molar; cuspids of first lower molar as well developed as those of second; crown of first lower incisor triangular in outline . . . .
. . . . *Carollia brevicauda*

4. Anterior cingulum of last upper premolar usually well developed and projecting toward first upper premoloar; last upper premolar usually oriented in line with axis of the skull resulting in a more obvious "step" between labial margins of last premolar and first molar; anterior projection of cingulum less well developed in first upper molar, usually no in contact with last upper premolar; lateral rim of basisphenoid pits interrupted 3–4 mm anterior to posterolateral margin of basisphenoid (anterior to bulla). . . . . *Carollia benkeithi*
4'. Anterior cingulum of last upper premolar usually blunt and projecting toward lingual side of first upper premolar; last upper premolar usually directed at an angle to axis of the skull with posterolabial margin closer to anterolabial margin of first upper molar resulting in a less obvious labial "step" between last premolar and first molar; anterior projection of cingulum of first upper molar usually in contact with last upper premolar; lateral rim of basisphenoid pits continuous to posterolateral margin of basisphenoid . . . .
. . . . *Carolina castanea*
5. Pelage long, thick, fine and fluffy; forearm (and usually toes) hairy; hair on nape of neck with broad dark basal band contrasting sharply with broad dirty white band succeeding it distally 6
5'. Pelage relatively short, sparse, and coarse; forearm and toes naked or only sparsely haired; hair on nape of neck shorter, with less indistinct basal bands 7
6. Forearm 41 mm or less; tibia and toes sparsely haired; trigonid of m1 distinctly narrower than talonid . . . . . .
. . . . *Carollia brevicauda*
6'. Forearm more than 40 mm; tibia and toes conspicuously haired; trigonid of m1 approximately the same width as talonid. . . . . *Carollia manu*
7. Maxillary toothrow more than 7.4 mm 8
7'. Maxillary toothrow less than 7.4 mm . . . .
. . . . *Carollia subrufa*
8. When lower jaw viewed from directly above, 1/2 or more of outer incisors obscured by cingula of canines; lower jaw tending to be V-shaped, with straight rami . . . . . .
. . . . *Carollia perspicillata*
8'. Lower outer incisors only partially obscured by cingula of canines; lower jaw tending to be U-shaped, rami distinctly bowed laterally . . . . . *Carollia monohernandezi*
9. First upper incisor relatively broad and with 3 or 4 well-defined lobes; no space between upper incisor and canine; margin of interfemoral membrane naked
. . . . *Rhinophylla pumilio*
9'. First upper incisor relatively narrow with less than 3 well-defined lobes; space present between upper incisor and canine; conspicuous fringe of hair on margin of interfemoral membrane. 10

10. Forearm shorter than 32 mm; condylobasal length less than 16 mm; pelage brown to reddish brown . . . .
. . . . *Rhinophylla fischerae*
10'. Forearm longer than 32 mm; condylobasal length more than 16 mm; pelage blackish brown . . . .
. . . . *Rhinophylla alethina*

Genus *Carollia* Gray, 1838

The genus *Carollia*, as currently understood, contains at least seven species and is distributed from central Mexico into southern Brazil. *Carollia* differs from *Rhinophylla* in larger size (forearm 34–45 mm, versus 29–38 mm) and in having a tail, a longer muzzle, and broader molars. The tail is short, extending about 1/2 the length of the interfemoral membrane. First and second upper molars each have a protocone. Lower molars have a well-developed protoconid and hypoconid.

SYNONYMS: See Pine (1972b:14–17) for a more detailed synonymy.

Vespertilio Linnaeus, 1758:31; part.

Pteropus Erxleben, 1777:137; part.

Phyllostomus: Illiger, 1811:121; part, not *Phyllostomus* Lacépède, 1799b.

Phyllostoma: Schinz, 1821:164; part, not *Phyllostoma* G. Cuvier, 1800.

Carollia Gray, 1838b:488; type species *Carollia braziliensis* Gray, 1838b, by monotypy.

Hemiderma P. Gervais, 1856:43; type species *Phyllostoma brevicaudum* Wied-Neuwied, 1826, by monotypy.

Rhinops Gray, 1866b:115; type species *Rhinops minor* Gray, 1866b, by monotypy.

Vampyrus: Pelzeln, 1883:32; part, not *Vampyrus* Beckmann, 1772; *Vampyrus*, Ranzani, 1820; *Vampyrus* Leach, 1821b, or *Vampyrus* Spix, 1823.

Corallia Eisentraut, 1950:164; incorrect subsequent spelling of *Carollia* Gray.

Corollia Griffin, 1953:157; incorrect subsequent spelling of *Carollia* Gray.

Carrollia Bloedel, 1955:234; incorrect subsequent spelling of *Carollia* Gray.

Carrollia Cabrera, 1958:77; incorrect subsequent spelling of *Carollia* Gray.

REMARKS: The genus has presented much confusion to systematists as can be seen in its nomenclatural history (Pine 1972b). Linnaeus (1758) described a *Carollia* under the name *Vespertilio perspicillatus*. Between 1818 and 1866, additional names were proposed for various forms of *Carollia*, one of which, *brevicauda* Schinz, is currently recognized. In 1890, H. Allen named *C. castanea* from Costa Rica. Lydekker (in Flower and Lydekker 1891) urged using the junior synonym *Hemiderma* instead of *Carollia* because he believed *Carollia* to be preoccupied by *Caro-*

lia Cantraine, a fossil pelycepod. Hahn (1907) revised the genus based on 374 specimens from Mexico and recognized three species (*Hemiderma perspicillatum*, *H. subrufum*, and *H. castaneum*). Miller (1924:53) pointed out that, due to the difference in spelling between *Carolia* Cantraine (a mollusk) and *Carollia* Gray, they were not homonyms according to the International Code of Zoological Nomenclature, Article 36. Subsequently, *Carollia* was used in place of *Hemiderma*. Sanborn (1949b:281) pointed out that *Carollia* Gray, 1838b, antedated *Carolia* Cantraine, 1838, so Miller's argument, although correct, proved unnecessary.

Pine's (1972b) is the most recent revision of the genus. He recognized a fourth species, *C. brevicauda*, and clarified relationships among the other species he recognized (*C. castanea*, *C. perspicillata*, and *C. subrufa*). McLellan (1984) extended Pine's (1972b) work by using morphometric techniques to examine geographic and nongeographic variation in cranial morphology. Pine (1972b) commented on different specimens that he was unable to assign to species and he suspected that they represented undescribed species. Central American populations previously known as *C. brevicauda* have recently been redescribed by R. J. Baker, Solari, and Hoffmann (2002) as a separate species, *C. sowelli*. A specimen identified as *C. colombiana* by the authors of that name appears to be a *C. castanea*. The number of currently recognized species remains eight, seven of which are reported in South America. Of the two most recently described species, *C. manu*, is known only from southeastern Peru and adjacent Bolivia, *C. monohernandezi* is known only from Colombia, and *C. benkeithi* is know from Peru, Brazil, and Bolivia.

Members of the genus *Carollia* are primarily frugivorous, feeding on fruits such as piper, cashew, figs, and plantains (food habits summarized by Gardner 1977c). The diets of *C. perspicillata*, *C. castanea*, and *C. subrufa* also include some insects (Fleming, Hooper, and Wilson 1972; Pine 1972b). These bats frequently are accused of damaging banana and mango crops in South America. On the island of Trinidad, strychnine-laced baits have been used against *Carollia* and other frugivorous bats (Hill and Smith 1984). In reality, New World fruit-eating bats cause relatively minor damage to fruit crops because these crops are harvested before the fruit is ripe enough to attract the bats. Among the exceptions are guava and coffee, but because bat damage to these crops is rarely mentioned, their effect is probably minimal. Frugivores, including *Carollia*, attack stores of ripe bananas, but so do a variety of other vertebrates along with numerous insects.

Carollia benkeithi S. Solari and Baker, 2006
Ben Keith's Short-tailed Bat
SYNONYMS:
Carollia castanea: Tuttle, 1970:69; not *Carollia castanea* H. Allen, 1890b.

Carollia castanea: Pine, 1972:17; part.

Carollia castanea: J. L. Patton and Gardner, 1971:105; part.

Carollia castanet Ueda, 1980:937; incorrect subsequent spelling of, but not *Carollia castanea* H. Allen.

Carollia benkeithi Solari and Baker, 2006:5; type locality "2 km S of Tingo María, Province of Leoncio Prado, Department of Huánuco, Peru."

DISTRIBUTION: *Carollia benkeithi* is known from eastern Peru, northeastern Bolivia and western Brazil.

MARGINAL LOCALITIES (Map 102): BRAZIL: Pará, Ilha da Urucurituba (B. D. Patterson 1992); Rondônia, Cachoeira Nazaré (BDP 1793 [MPEG]); Mato Grosso, Fazenda São José (Ueda 1980). BOLIVIA (S. Anderson 1997, except as noted): Santa Cruz, San Miguel Rincón; Cochabamba, 12.5 km SW of Villa Tunari; La Paz, 6.6 km downstream from Caranavi; La Paz, Aserradero Moira (Emmons 1991). PERU: Madre de Dios, Cocha Salvador (Ascorra, Wilson, and Romo 1991); Cusco, Las Malvinas (S. Solari et al. 2001c); Huánuco, 19 miles S of Tingo María (Pine 1972b). BRAZIL: Acre, Rio Moa (Nogueira, Pol, and Peracchi 2000).

SUBSPECIES: We consider *C. benkeithi* to be monotypic.

NATURAL HISTORY: The natural history of *C. benkeithi* most likely closely resembles that of *C. castanea*. According to A. L. Gardner (pers. comm.), specimens from Balta in eastern Peru were most commmonly mistnetted in the brushy vegetation containing *Piper* sp. along the edges of the small air strip in the middle of the village. The karyotype is $2n = 22$, FN $= 38$ (J. L. Patton and Gardner 1971; S. Solari and Baker 2006).

REMARKS: Specimen records listed in the Marginal Localities were reported as *C. castanea*. The species *C. castanea* and *C. benkeithi* are the smallest of the *Carollia*. S. Solari and Baker (2006) mentioned another similar species from eastern Ecuador and northern Peru that they believe is undescribed. We have included this form under *C. castanea*. The features useful for distinguishing between *C. benkeithi* and *C. castanea*, as described by S. Solari and Baker (2006) are variable and difficult to apply. The teeth of *C. benkeithi*, particularly the upper and lower premolars, are narrower and less robust than those of *C. castanea*. The upper second premolar tends to be more often in alignment with the axis of the skull. The straighter alignment results in a more pronounced "notch" between the last premolar and the first molar in the labial margin of the upper toothrow than is typical for most *C. castanea*. One feature that appears to be consistent is the discontinuity in the outer rim surrounding the basisphenoid pits. In *C. benkeithi*, this ridge is briefly interrupted 3–4 mm before its terminus anterior to each auditory bulla; whereas in *C. castanea* the ridge is continuous. The $2n = 22$, FN $= 38$ karyotype lacks the X–autosomal translocation that characterizes *C. castanea* and other known species of *Carollia*.

The species appears to be uncommon in the central Amazon basin of Brazil. In addition to the marginal records listed previously for Brazil, the only other published records we have found are those of Mok et al. (1982) for the states of Acre and Rondônia.

Carollia brevicauda (Schinz, 1821)
Silky Short-tailed Bat

SYNONYMS: See Pine (1972b:29–36) for a more detailed synonymy.

Phyllost[oma]. bernicaudum Schinz, 1821:164 (printer's error for *brevicaudum*); no locality given; type locality identified by Wied-Neuwied (1826:195) as "Fazenda von Coroaba," near Rio do Espírito Santo, Brazil (also see Pine 1972b:29).

Vampyrus soricinus Spix, 1823:65; type localities "Rio de Ianiero et ad fluvium St. Francisci," Brazil; not *Vespertilio soricinus* Pallas, 1766.

Phyllostoma Grayi Waterhouse, 1838:3; type locality "Pernambuco," Pernambuco, Brazil.

Phyllostoma bicolor J. A. Wagner, 1840:400; type locality "Brasilien."

Phyllostoma lanceolatum Gray, 1843b:20; *nomen nudum*.

Phyllostoma brevicaudatum Desmarest, 1847:69; incorrect subsequent spelling of *Phyllostoma brevicaudum* Schinz.

Phyllostoma brevicaudatum Gray, 1849:58; incorrect subsequent spelling of *Phyllostoma brevicaudum* Schinz.

Hemiderma brevicaudum: P. Gervais, 1856:43; name combination.

R[hinops]. minor Gray, 1866b:115; no type locality given; type locality identified as Bahia, Brazil, by Pine (1972b:33).

Hemiderma brevicaudatum Gerrard, 1862:37; incorrect subsequent spelling of *brevicaudum* Schinz.

C[arollia]. brevicauda: W. Peters, 1865c:519; part; first use of current name combination.

DISTRIBUTION: In South America *C. brevicauda* occurs from Colombia eastward through the Guianas (Genoways and Williams, 1979b) and southward through Ecuador, Peru, and eastern Bolivia into southeastern Brazil. Elsewhere, the species is known from Panama.

MARGINAL LOCALITIES (Map 101; from Pine 1972b, except as noted): VENEZUELA (Handley 1976): Zulia, Novito; Carabobo, La Copa; Miranda, Curupao; Monagas, San Agustín, 3 km NW of Caripe; Bolívar, El Manaco, 56 km SE of El Dorado. GUYANA: Potaro-Siparuni, Kaieteur Falls (Shapley et al. 2005). SURINAM: Sipaliwini, Grassalco (Williams and Genoways 1980a). FRENCH GUIANA: Petit-Saut (Brosset and Charles-Dominique 1990). BRAZIL: Amapá, Igarapé Novo (Piccinini 1974); Pará; 14 miles N of Belém; Pernambuco,

Map 101 Marginal localities for *Carollia brevicauda* ●

Pernambuco (type locality of *Phyllostoma grayi* Waterhouse); Bahia, Bahia (type locality of *Rhinops minor* Gray); Rio de Janeiro, Rio de Janeiro; Goiás, Serra do Jagaguá (Nunes et al. 2005). BOLIVIA (S. Anderson 1997, except as noted): Santa Cruz; 27.5 km S of Campamento Los Fierros; Santa Cruz, 27 km SE of Santa Cruz; Cochabamba, Río Yanimayo, 80 km N of Monte Punco (Barquez 1984a); La Paz, 0.5 km E of Saynani. PERU: Cusco, Limacpunco; Ayacucho, San José (LSUMZ 16500); Junín, Chanchamayo; Huánuco; 3 miles N of Tingo María (TCWC 12090); Amazonas, Quebrada Kagka (J. L. Patton, Berlin, and Berlin 1982). ECUADOR: Manabí, Estero Achiote (Albuja 1999); Pichincha, Centro Científico Río Palenque (Albuja 1983). COLOMBIA: Valle del Cauca, 2 km S of Pance (M. E. Thomas 1972); Antioquia, Urrao; Magdalena, Pueblo Bello.

SUBSPECIES: *Carollia brevicauda* is considered to be monotypic.

NATURAL HISTORY: *Carollia brevicauda* feeds on a variety of fruits, which vary according to the region and season (see summary by Gardner 1977c). Wilson (1979), on finding females from Central and South America that were simultaneously pregnant and lactating, suggested that *C. brevicauda* has a bimodal polyestrous reproductive pattern. *Carollia brevicauda* is known to breed from mid-winter (Ecuador) to early summer (Peru; Pine 1972b). Brennan and Reed (1975), Reed and Brennan (1975), Wenzel (1976), R. Guerrero (1985a, 1997) listed known ectoparasites. *Carollia brevicauda* has a $2n = 20–21$, FN $= 36$ karyotype (J. L. Patton and Gardner 1971).

REMARKS: Central American and Mexican populations previously identified as *C. brevicauda* were recently described as the new species *C. sowelli* R. J. Baker, Solari, and Hoffmann, 2002. *Carollia brevicauda* has a large geographic range. Its close resemblance to *C. perspicillata* caused the two species to be confused until Pine (1972b) pointed out differences between the two species. Populations show a north-south trend in cranial and wing dimensions, the northern populations averaging larger than the southern ones (Pine 1972b; McLellan 1984). Males are larger than females, especially in rostral breadth (McLellan 1984). Pine (1972b) stated that this species is often collected in the same roost with *C. perspicillata*. Specimens reported by Husson (1962, 1978) as *C. castanea* are misidentified *C. brevicauda*. It is likely that specimens representing *C. monohernandezi* have been identified as unusually large *C. brevicauda*, because except for size and certain pelage characteristics, both species differ from *C. perspicillata* in having similar dentitions, larger outer lower incisors, and outwardly bowed rami.

Cabrera (1958) and Pine (1972b) listed *Phyllostoma lanceolatum* Gray (1843b:20) in synonymy. Apparently, *lanceolatum* is a Natterer manuscript name placed by Temminck (but not published) on the label of a specimen sent to the British Museum and subsequently listed by Gray (1843b:20). Pine (1972b:32) was correct when he cited the name as a *nomen nudum*. Pine (1972b:30) wrote that *Vampyrus soricinus* Spix was based on two specimens, one of which may be a *Glossophaga*. D. C. Carter and Dolan (1978) gave information and measurements of the types of *Phyllostoma bicolor* J. A. Wagner, *Rhinops minor* Gray, and *Carollia azteca* Saussure. Pine (1972b:32–33) considered the specimen later reported on by D. C. Carter and Dolan (1978) to be Saussure's "Variété" of *Carollia azteca* and not the holotype or a syntype of *Carollia azteca* Saussure, which he concluded represents a population of *C. perspicillata*.

Carollia castanea H. Allen, 1890
Chestnut Short-tailed Bat

SYNONYMS: See Pine (1972b:29–36) for a more detailed synonymy.

Carollia castanea H. Allen, 1890b:19; type locality "Costa Rica"; restricted to Angostura, San José, Costa Rica, by Pine (1972b:18).

[*Hemiderma*] *castaneum*: Elliot, 1904:670; name combination.

Carollia colombiana Cuartas, Muñoz, and González, 2001:65; type locality "Colombia, Antioquia, municipio de Barbosa, vereda La Cejita, 6°25′N y 75°15′W."

DISTRIBUTION: *Carollia castanea* is found from Colombia east into Guyana and south through Ecuador and

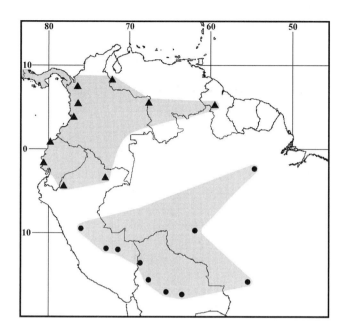

Map 102 Marginal localities for *Carollia benkeithi* ● and *Carollia castanea* ▲

northern Peru. Elsewhere, the species is in Central America as far north as Honduras.

MARGINAL LOCALITIES (Map 102): VENEZUELA (Handley 1976): Táchira, Las Mesas; Amazonas, 32 km S of Puerto Ayacucho. GUYANA: Potaro-Siparuni, Echerak River (Shapley et al. 2005). PERU: Loreto, Puerto Indiana (Pirlot 1968); Amazonas, Huampami (J. L. Patton, Berlin, and Berlin 1982). ECUADOR: Guayas, Cerro Manglar Alto (AMNH 64564); Esmeraldas, Estero Taquiama (Albuja 1999). COLOMBIA: Valle del Cauca, Río Zabaletas (M. Thomas 1972); Chocó, 2 km above Playa de Oro (AMNH 217033); Antioquia, Urabá (Pine 1972b).

SUBSPECIES: *Carollia castanea* is considered to be monotypic.

NATURAL HISTORY: *Carollia castanea* prefers tropical evergreen forests. Individuals are known to have relatively small home ranges (Handley 1966c; LaVal 1970). Fruit is the primary food (Fleming, Hooper, and Wilson 1972; Gardner 1977c). Pine (1972b) noted a polyestrous breeding pattern; but records were taken from widely separate localities and a fixed breeding pattern may characterize populations in a given area. Fleming (1973) suggested that *C. castanea* is bimodally polyestrous. R. Guerrero (1985a), Wenzel (1976), and Wenzel, Tipton, and Kiewlicz (1966) listed ectoparasites hosted by this species. The karyotype is $2n = 20–21$, FN = 36 (R. J. Baker and Bleier 1971).

REMARKS: *Carollia castanea* and *C. benkeithi* are the smallest species of *Carollia* recognized today. S. Solari and Baker (2006) suggested that the specimens from eastern Ecuador and northern Peru represent yet another unde-

scribed species. If they are correct, the distribution of *C. castanea* in South America will be restricted to the Pacific lowlands of Colombia and Ecuador, and northern Colombia eastward through northern Venezuela into Guyana.

Both *C. castanea* and *C. benkeithi* can be easily distinguished from congeners by the step or labial notch between the second upper premolar and the first upper molar. The step results from the second upper premolar being positioned toward the lingual side of the toothrow relative to the labial border of the first molar and is less pronounced in *C. castanea* than in *C. benkeithi*. The upper and lower dentition is more robust in *C. castanea*. The ridge that rims the outer margins of the basisphenoid pits is continuous, but in *C. benkeithi* the ridge is interrupted approximately 4 mm before its terminus anterior to each bulla. *Carollia castanea* shows little sexual dimorphism, but samples from different parts of the range show morphological differences corresponding with latitude. Individuals from northern parts of the range have larger cranial dimensions than do those from more southern zones (McLellan 1984).

The recently described *C. colombiana*, on the basis of a specimen in the USNM identified by the authors (Cuartas, Muñoz, and González 2001), proves to be a *C. castanea*. This raises the question, what is the taxon identified by Cuartas, Muñoz, and González (2001) as *C. castanea*. Nevertheless, based on the description and the representative specimen in the USNM from Antioquia, Colombia, *C. colombiana* is a subjective synonym of *C. castanea*.

The mapped records for *C. castanea* from Surinam and French Guiana included by Eisenberg (1989:142) represent citations by Husson (1962, 1978) and Brosset and Dubost (1968), respectively. However, specimens on which these records are based were subsequently reidentified as *C. brevicauda* by Genoways and Williams (1979) and Brosset and Charles-Dominique (1990).

Carollia manu V. Pacheco, Solari, and Velazco, 2004
Manu Short-tailed Bat
SYNONYMS:
Carollia sp.? *(3)*: Pine, 1972:76.
Carollia sp.: V. Pacheco, Macedo, Vivar, Ascorra, Arana-Cardó, and Solari, 1995:10.
Carollia manu V. Pacheco, Solari, and Velazco, 2004:3; type locality "Morro Leguía, Pucartambo-Pillcopata road, km. 134, 2250 m, Pucartambo Province, Cuzco Department, Peru."

DISTRIBUTION: *Carollia manu* is known from the departments of Cusco, Madre de Dios, and Puno, Peru, and from departamento La Paz in adjacent Bolivia.

MARGINAL LOCALITIES (Map 103; from V. Pacheco, Solari, and Velazco 2004): PERU: Madre de Dios, Cerro de Pantiacolla; Puno, Sandia. BOLIVIA: La Paz,

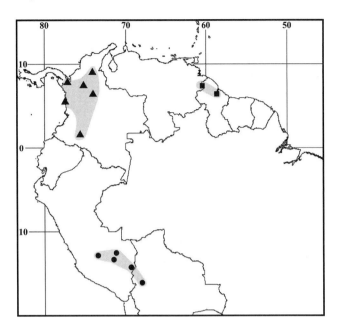

Map 103 Marginal localities for *Carollia manu* ●, *Carollia monohernandezi* ▲, and *Carollia subrufa* ■

0.5 km E of Saynani. PERU: Cusco, Bosque de las Nubes; Cusco, Llactahuamán.

SUBSPECIES: *Carollia manu* is monotypic.

NATURAL HISTORY: V. Pacheco, Solari, and Velazco (2004) reported *C. manu* from tropical montane forests (cloud forest) at elevations from 1,300 to 2,500 m. Specimens taken in the valley of the Río Cosñipata came from second growth forest dominated by thickets of climbing bamboo. At higher elevations, forest trees were heavily covered with epiphytes and the understory was dominated by bamboo.

REMARKS: Pine (1972) briefly described, but did not name this species, and wrote that it was not assignable to any of the then known taxa.

Carollia monohernandezi Muñoz, Cuartas, and González, 2004

Mono's Short-tailed Bat

SYNONYM:

Carollia monohernandezi Muñoz, Cuartas, and González, 2004, 81; type locality "República de Colombia, departamento de Caquetá, municipio de Florencia, vereda Villaraz... 01°37′N, 75°40′W.

DISTRIBUTION: *Carollia monohernandezi* is known from Colombia west of the llanos. Elsewhere, the species is found in the Darién of Panama (see Remarks).

MARGINAL LOCALITIES (Map 103; from Muñoz, Cuartas, and González 2004): COLOMBIA: Bolívar, San Martín de Loba; Antioquia, Cavernas del Nus; Caquetá, Villaraz (type locality of *Carollia monohernandezi*);

Chocó, Arusi; Chocó, Parque Natural Katíos; Antioquia, El Doce.

SUBSPECIES: We consider *C. monohernandezi* to be monotypic.

NATURAL HISTORY: *Carollia monohernandezi* has been collected in a variety of habitats ranging from tropical dry to tropical rain forest at elevations from 30 to 2,660 m (Muñoz, Cuartas, and González 2004).

REMARKS: Characters of a specimen from the Darién of Panama (USNM 309466) appear to closely match the published description of *C. monohernandezi*.

Carollia perspicillata (Linnaeus, 1758)

Seba's Short-tailed Bat

SYNONYMS: See under subspecies.

DISTRIBUTION: *Carollia perspicillata* is distributed from Colombia eastward across northern South America, including Trinidad and Tobago, and southward into Bolivia, Paraguay, southern Brazil, and northern Argentina. The species also is known from the Isthmus of Tehuantepec, Mexico, southward throughout Central America.

MARGINAL LOCALITIES (Map 104; from Pine 1972b, except as noted): COLOMBIA: Magdalena, Palomino. VENEZUELA (Handley 1976): Falcón; Río Socopito, 80 km NW of Carora; Carabobo, La Copa; Distrito Federal, Hacienda Carapiche; Nueva Esparta, 3 km NNE La Asunción. TRINIDAD AND TOBAGO: Tobago, Charlotteville. GUYANA: Demerara-Mahaica, Georgetown. SURINAM: Paramaribo, Paramaribo (Husson 1962). FRENCH GUIANA: Cayenne. BRAZIL: Amapá, Igarapé Novo (Piccinini 1974); Pará, Utinga (Handley 1967);

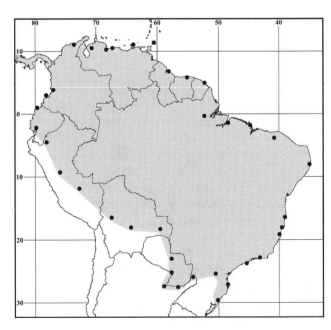

Map 104 Marginal localities for *Carollia perspicillata* ●

Maranhão, Rosario; Ceará, Parque Nacional de Ubajara (Guedes et al. 2000); Pernambuco, Estação Ecológico do Tapacurá (Mares et al. 1981); Bahia, 3 miles W of Pôrto Seguro (USNM 541467); Bahia, Morro d'Arara (type locality of *Phyllostoma brachyotos* Schinz); Espírito Santo, Linhares Forest Reserve (Peracchi and Albuquerque 1993); Rio de Janeiro, Rio de Janeiro (type locality of *Glossophaga amplexicauda* É. Geoffroy St.-Hilaire); São Paulo, São Sebastião; Paraná, Floresta Nacional do Irati (N. R. Reis et al. 2000); Santa Catarina, Porto Belo (Cherem et al. 2005); Rio Grande do Sul, Maquiné (Fabián, Rui, and Oliveira 1999). ARGENTINA: Misiones, Gobernador Lanusse (Barquez, Mares, and Braun 1999); Corrientes, Ituzaingó (Barquez, Mares, and Braun 1999); Chaco, Resistencia (Podtiaguin 1944); Formosa, Clorinda (Barquez, Mares, and Braun 1999). PARAGUAY: Presidente Hayes, Puerto Cooper (Podtiaguin 1944). BOLIVIA (S. Anderson 1997): Santa Cruz, Santiago; Santa Cruz, 5 km NE of Quiñe; La Paz, 3 km S of Irupana. PERU: Cusco, Cashiriari-2 (S. Solari et al. 2001c); Huánuco, 2 miles N of Tingo María; Amazonas, Huampami (J. L. Patton, Berlin, and Berlin 1982). ECUADOR: Guayas, Guayaquil (Brosset 1963); Esmeraldas, Esmeraldas (Albuja 1983). COLOMBIA: Cauca, Watering Bay, Gorgona Island; Valle del Cauca, Buenaventura (USNM 483414).

SUBSPECIES: Pine (1972b) did not formally recognize subspecies in *C. perspicillata*, but suggested that there were three:

C. p. azteca Saussure, 1860

SYNONYMS:

Carollia azteca Saussure, 1860:480:480; type locality "régions chaudes et tempérées du Mexique"; specimens considered to like those from near Perez, Veracruz, Mexico, by Hahn (1907:113) and type locality restricted to that place by Dalquest (1953b:29).

Hemiderma perspicillatum aztecum: Hahn, 1907:111; name combination.

Carollia perspicillata azteca: Miller, 1924:54; first use of current name combination.

In South America, *C. p. azteca* is found north and west of the Amazon Basin in Venezuela, Colombia, and Ecuador, as well as in Trinidad and Tobago. Elsewhere, the species is known throughout Central America northward at least as far as the Isthmus of Tehuantepec, Mexico.

C. p. perspicillata (Linnaeus, 1758)

SYNONYMS: See Pine (1972b:44–55) for a more detailed synonymy.

[*Vespertilio*] *perspicillatus* Linnaeus, 1758:90; type locality "America," restricted to Surinam by O. Thomas (1911b:130).

Pteropus perspicillatus: Erxleben, 1777:137; part; name combination.

Glossophaga amplexicauda É. Geoffroy St.-Hilaire, 1818a: 418; type locality "Le Brésil, aux environs de Rio-Janeiro" (see Pine, 1972b; D. C. Carter and Dolan 1978, concerning the identity of the type).

Phyllost[oma]. brachyotos Schinz, 1821:164; type locality "In den dichten Waldungen am Mucuri in Brasilien"; type locality identified by Wied-Neuwied (1826:196) as "Moro d'Arara am Mucuri," Bahia, Brazil (also see Pine 1972b:47).

Ph[yllostoma]. brachyotum Wied-Neuwied, 1826:196; unjustified emendation of *Phyllostoma brachyotos* Schinz.

Carollia Braziliensis Gray, 1838b:488; type locality "Brazils"; a new name for *Phyllostoma brachyotum* Wied-Neuwied, 1826, which is an emendation of *Phyllostoma brachyotos* Schinz, 1821; therefore, has the same type locality.

Phyllostoma calcaratum J. A. Wagner, 1843a:366; type locality "Brasil."

Carollia verrucata Gray, 1844:20; type locality "S. America," perhaps the "southern part of the species' [*C. perspicillata*] range" (Pine 1972b:48).

C[arollia]. brachyotis Gray, 1844:20; incorrect subsequent spelling of *Phyllostoma brachyotos* Schinz.

C[arollia]. brevicauda: W. Peters, 1865c:519; part; composite of *Phyllostoma brevicaudum* Schinz, 1821, and *Vespertilio perspicillatus* Linnaeus.

Vampyrus (Carollia) brevicaudum: Pelzeln, 1883:32; name combination; not *Phyllostoma brevicaudum* Schinz.

Hemiderma perspicillatum: O. Thomas, 1901h:191; name combination.

Carollia perspicillata: Miller, 1924:53; first use of current name combination.

Glossophaga amplexicaudata Rode, 1941:87; name combination and incorrect subsequent spelling of *Glossophaga amplexicauda* É. Geoffroy St.-Hilaire.

Carollia prespicillata Komeno and Linhares, 1999:152; incorrect subsequent spelling of *Vespertilio perspicillatus* Linnaeus.

The nominate subspecies occurs throughout the Amazon basin including the Guianas, Venezuela, Brazil, and eastern Colombia, Ecuador, Peru, and Bolivia.

C. p. tricolor (Miller, 1902)

SYNONYMS:

Hemiderma tricolor Miller, 1902b:408; type locality "Sapucay, Paraguay."

Carrollia tricolor: Cabrera, 1958:77; name combination.

This subspecies occurs within the Paraná drainage and probably includes Paraguay, southern Bolivia and Brazil, and northern Argentina in its range.

NATURAL HISTORY: *Carollia perspicillata* prefers tropical evergreen and tropical deciduous forests. It tolerates closer contact with humans than do the other species of *Carollia* and is often found associated with agriculture (Pine 1972b). Individuals can be found roosting alone or in colonies of as many as 1,000 individuals (Bloedel 1955). Foraging patterns of *C. perspicillata* in Costa Rica have been studied in detail (Fleming and Heithaus 1986; Fleming 1988). Fruit is usually carried off to be consumed in a temporary night roost (Pine 1972b). Individuals have spatially well-defined feeding patterns with little overlap between individuals; *Piper* spp. is the preferred food during the rainy season (Heithaus and Fleming 1978). The diet also includes other available fruits and some insects (Arata, Vaughan, and Thomas 1967; Fleming, Hooper, and Wilson 1972; Gardner 1977c). Fleming, Hooper, and Wilson (1972) noted that as much as 40% of the stomach contents of Panamanian *C. perspicillata* consisted of insect material in specimens taken in April and May. *Carollia perspicillata* shows bimodal polyestry with birth peaks occurring in February-May and June-August in Panama (Fleming, Hooper, and Wilson 1972), but earlier in Colombia (Wilson 1979), suggesting a geographically variable pattern that adjusts to local rainfall regimes. Fleming (1988) has added to and summarized information on the behavioral ecology of *C. perspicillata* in his treatise on the demography and life history of the species. See Brennan and Reed (1975), Reed and Brennan (1975), Wenzel (1976), Wenzel et al. (1966), and R. Guerrero (1985a) for lists of ectoparasites. Cloutier and Thomas (1992) provided a detailed description along with measurements and a summary of available information on natural history. *Carollia perspicillata* has a $2n = 20-21$, FN = 36 karyotype (Hsu, Baker, and Utakoji 1968; Noronha et al. 2004).

REMARKS: This is the most widely distributed and abundant species of *Carollia*. Geographic variation in size appears to be correlated with latitude; individuals from northern parts of the range are larger than those from southern populations (Pine, 1972b; McLellan, 1984). McLellan (1984) noted sexual dimorphism with males being larger than females, and having wider rostra and deeper braincases. D. C. Carter and Dolan (1978) gave information on and measurements of the types of *Carollia verrucata* Gray, *Glossophaga amplexicauda* É. Geoffroy St.-Hilaire, *Phyllostoma calcaratum* J. A. Wagner, and *Vespertilio perspicillatus* Linnaeus.

Carollia subrufa (Hahn, 1905)
Hahn's Short-tailed Bat
SYNONYMS: See Pine (1972b:22–24) for a more detailed synonymy.

Hemiderma subrufum Hahn, 1905:247; type locality "Santa Ifigenia" (= Santa Efigenia), Oaxaca, Mexico.
Carollia subrufa: Miller, 1924:54; first use of current name combination.
Carollia castanea subrufa: Felten, 1956a:199; name combination.

DISTRIBUTION: In South America, *C. subrufa* is known only by two specimens from Guyana tentatively identified as this species. Elsewhere, *C. subrufa* occurs from Costa Rica northward into western Mexico.

MARGINAL LOCALITIES (Map 103): GUYANA: Cuyuni-Mazaruni, Kartabo (AMNH 64168); Barima-Waini, Baramita (USNM 568032).

SUBSPECIES: We treat *C. subrufa* as monotypic.

NATURAL HISTORY: *Carollia subrufa* is a bat primarily of tropical deciduous forests of Mexico and Central America and may inhabit similar habitats in northern South America. Its diet includes fruit and insects (Gardner, 1977c). *Carollia subrufa* shows a bimodal breeding pattern (Wilson, 1979).

REMARKS: We are aware of only two South American specimens of *Carollia subrufa*. The Kartabo specimen was obtained by William Beebe in March 1927, and was listed as "*Carollia* sp.?" by Pine (1972b:75). McLellan (1984) identified it as *Carollia subrufa* based on the results of a morphometric analysis. The Kartabo specimen was difficult to identify because its external features resemble those of *C. brevicauda* (distinct basal banding of pelage). Overall size and tooth morphology place it with *C. subrufa*. More specimens from Guyana and neighboring areas, or examination of specimens already in collections, but unstudied by us, may reveal more about this South American population. The geographic range of *C. subrufa* previously was known to extend south from Mexico into Nicaragua, but not into South America.

Genus *Rhinophylla* W. Peters, 1865

The genus *Rhinophylla* comprises three species endemic to South America. Members of this genus are regarded as related to, but more specialized than, species of *Carollia*. The species of *Rhinophylla* differ from *Carollia* in smaller size (forearm 29.0–38.8 mm, versus 34.0–45.0 mm), no tail, a shorter muzzle, narrower molars, and a central tubercle on the lower lip flanked on each side by a large lobate tubercle. The first and second upper molars lack a protocone, lower molars lack a metaconid and closely resemble the premolars in form.

SYNONYMS:
Rhinophylla W. Peters, 1865b:355; type species *Rhinophylla pumilio* W. Peters, 1865b, by monotypy.
Rhinopylla Zortéa, 1995:2; incorrect subsequent spelling of *Rhinophylla* W. Peters.

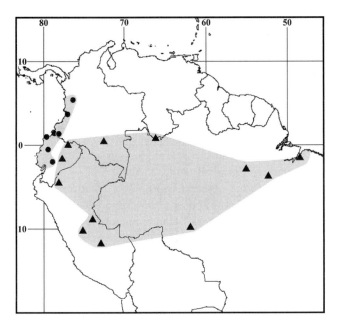

Map 105 Marginal localities for *Rhinophylla alethina* ● and *Rhinophylla fischerae* ▲

REMARKS: *Rhinophylla pumilio* was the only species known until 1966, when Handley (1966a) described *R. alethina* from the west coast of Colombia. In the same year, D. C. Carter described *R. fischerae* from eastern Peru, bringing the number of known species to three. We consider the dentition of *Rhinophylla* to be derived, because the protocone of the upper molars and the metaconid and entoconid of the lower molars have been lost, and the protoconid is positioned more centrally, thus enhancing the shearing function.

Rhinophylla alethina Handley, 1966
Hairy Little Fruit Bat

SYNONYM:

Rhinophylla alethina Handley, 1966a:86; type locality "27 km S Buenaventura, Raposo River, Valle, Colombia."

DISTRIBUTION: *Rhinophylla alethina* is known from the Pacific slope and lowlands of Colombia and Ecuador.

MARGINAL LOCALITIES (Map 105): COLOMBIA: Chocó; 2 km above Playa de Oro (AMNH 217032); Valle del Cauca, Río Raposo (type locality of *R. alethina* Handley); Nariño, 5 km E of Junín (Alberico 1987). ECUADOR (Albuja 1999, except as noted): Esmeraldas, Tabiazo; Esmeraldas, Centro Comunal Mataje; Pichincha, Centro Científico Río Palenque; Chimborazo, Pallatanga (Albuja 1983).

SUBSPECIES: We consider *R. alethina* to be monotypic.

REMARKS: *Rhinophylla alethina* has been collected from only a few localities in Colombia and Ecuador; its range does not overlap that of other members of the genus.

It is the largest member of the genus, individuals are slightly larger than corresponding *R. pumilio*; the interfemoral membrane is narrower and thickly clothed with woolly fur. The teeth, except for the first upper incisors, are smaller than those of *R. pumilio*.

NATURAL HISTORY: *Rhinophylla alethina* inhabits the foothills and lowlands on the western side of the Andes. Little is known about the ecology of this species. Limited data suggests that reproduction is either extended or asynchronous (Wilson, 1979).

REMARKS: Once considered a rare bat, Albuja (1999) reported 79 specimens from several localities in western and northwestern Ecuador, where first record for *R. alethina* outside of Colombia was collected (Baud 1982).

Rhinophylla fischerae D. C. Carter, 1966
Fischer's Little Fruit Bat

SYNONYM:

Rhinophylla fischerae D. C. Carter, 1966:235; type locality "61 mi. SE Pucallpa, about 180 m," Ucayali, Peru.

DISTRIBUTION: *Rhinophylla fischerae* is in Amazonian Colombia, Venezuela, Ecuador, Peru, and Brazil.

MARGINAL LOCALITIES (Map 105): VENEZUELA: Amazonas, Cerro de la Neblina Base Camp (Gardner 1988). BRAZIL: Pará, Belém (Piccinini 1974); Pará, Rio Xingu, 52 km SSW of Altamira (USNM 549404); Pará, Caxiricatuba (Koopman 1976); Rondônia, Cachoeira Nazaré (MZUSP 20218). PERU: Cusco, Pagoreni (S. Solari et al. 2001c); Pasco, Iscozacín (Ascorra, Gorchov, and Cornejo 1989); Ucayali, 61 miles SE of Pucallpa (type locality of *Rhinophylla fischerae* D. C. Carter); Amazonas, Huampami (J. L. Patton, Berlin, and Berlin 1982). ECUADOR: Pastaza, Canelos (Albuja 1999); Sucumbios, Santa Cecilia (Webster and Jones 1984). COLOMBIA: Caquetá, Raudal el Tubo (Montenegro and Romero-Ruiz 2000).

SUBSPECIES: We treat *R. fischerae* as monotypic.

NATURAL HISTORY: Little is known about this species. Specimens from the type locality were taken in mistnets at ground level in the open understory of mature rain forest. Specimen from southwest of Pucallpa was taken over a pond at a brick factory (D. C. Carter 1966). Available data indicate no reproductive activity between July and August (Wilson 1979). The karyotype is $2n = 34$, FN $= 56$, with a submetacentric X chromosome (R. J. Baker and Bleier 1971).

REMARKS: *Rhinophylla fischerae* is the smallest member of the genus. Its geographic range overlaps that of *R. pumilio*, from which it can be distinguished by its smaller size, a fringe of hair along the margin of interfemoral membrane, and a bilobed first incisor that lack a lateral style on the cingulum. The karyotype is $2n = 34$, FN $= 62$, with acrocentric X and Y chromosomes (R. J. Baker et al. 1982).

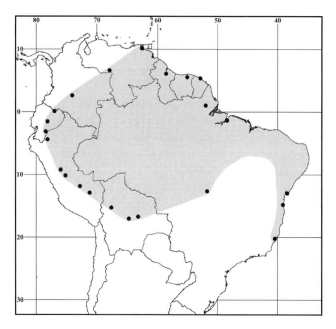

Map 106 Marginal localities for *Rhinophylla pumilio* ●

Rhinophylla pumilio W. Peters, 1865
Dwarf Little Fruit Bat

SYNONYMS:

Rhinophylla pumilio W. Peters, 1865b:355; type locality "Brasilien," restricted to Bahia, Brazil, by Dobson (1878:496) on the basis of a specimen in the British Museum.

Rhinophylla cumilis Kappler, 1881:163; incorrect subsequent spelling of *Rhinophylla pumilio* W. Peters.

DISTRIBUTION: *Rhinophylla pumilio* is known from Colombia, Venezuela, Surinam, French Guiana, Brazil, Ecuador, Peru, and Bolivia.

MARGINAL LOCALITIES (Map 106): VENEZUELA: Sucre, Caño Brea (Linares and Rivas 2004). GUYANA: Cuyuni-Mazaruni, 24 miles along Potaro Road from Bartica (Hill 1965). SURINAM: Para, Zanderij (Williams and Genoways 1980a). FRENCH GUIANA: Paracou (Simmons and Voss 1998). BRAZIL: Amapá, Serra do Navio (USNM 461513); Pará, Belém (Handley 1967), Bahia, Bahia (type locality of *R. pumilio* W. Peters); Bahia, Ilhéus (Faria, Soares-Santos, and Sampaio 2006); Espírito Santo, Duas Bocas Biological Reserve (Zortéa 1995); Mato Grosso, Suia Missu Road, 284 km N of Xavantina (Pine, Bishop, and Jackson 1970). BOLIVIA: Santa Cruz, Estancia Cachuela Esperanza (S. Anderson 1997); Cochabamba, Sajta (S. Anderson 1997); La Paz, Sararia (S. Anderson, Koopman, and Creighton 1982). PERU: Cusco, Hacienda Amazonía (V. Pacheco et al. 1993); Cusco, Cashiriari-2 (S. Solari et al. 2001c); Pasco, Iscozacín (Ascorra, Gorchov, and Cornejo 1989); Huánuco, Tingo María (Bowles, Cope, and Cope 1979); Amazonas, Huampami (J. L. Patton, Berlin, and

Berlin 1982). ECUADOR: Morona-Santiago, San Carlos de Limón (Albuja 1999); Morona-Santiago, Río Llushín (Albuja 1999); Sucumbios, Santa Cecilia (R. H. Baker 1974). COLOMBIA: Meta, Cabaña Duda (Lemke et al. 1982). VENEZUELA: Apure, Río Cinaruco, 65 km NW of Puerto Paez (Handley 1976).

SUBSPECIES: We treat *R. pumilio* as monotypic.

NATURAL HISTORY: Little is known about the natural history of this species. Most specimens have been netted; however, Marinkelle and Cadena (1972) reported two specimens from Colombia captured in a culvert. Peracchi, Raimundo, and Tannure (1984) reported a *R. pumilio* in a building, and Charles-Dominique (1993) and Zortéa (1995) commented on the use of leaf tents as day roosts and as feeding roosts by this species. One pregnant and one lactating female were taken in December, 1962, on the Río Paragua in Venezuela (Nowak and Paradiso 1983). Wenzel (1976) listed ectoparasites. *Rhinophylla pumilio* has a 2n = 34, FN = 64 (Honeycutt, Baker, and Genoways 1980) and $2n = 36$, FN = 62 karyotypes (R. J. Baker and Bleier 1971).

REMARKS: *Rhinophylla pumilio*, the first described species in the genus, has the widest geographic distribution and is sympatric with *R. fischerae*. D. C. Carter and Dolan (1978) gave information on and measurements of the holotype of *Rhinophylla pumilio* W. Peters.

Subfamily Desmodontinae J. A. Wagner, 1840
Miriam Kwon and Alfred L. Gardner

Vampire bats, the only truly sanguinivorous mammals, are found from the United States (Texas) northern Mexico, and the West Indies, south into Uruguay, northern Argentina, and central Chile. Long regarded as a separate family, the Desmodontinae were reduced to a subfamily in the Phyllostomidae by Koopman and Jones (1970). Close affinity with the Phyllostominae and Glossophaginae within the phyllostomids has been supported by karyology, sperm morphology, and immunologic studies (Forman, Baker, and Gerber 1968). More recent phylogenetic studies by Wetterer, Rockman, and Simmons (2000) and K. E. Jones et al. (2002), however, have suggested that vampires form a clade basal to all other phyllostomid subfamilies. Their results were only partially supported by R. J. Baker et al. (2003) whose assessment of mitochondrial DNA, *RAG2*, and morphological data sets placed the "Macrotinae" and "Micronycterinae" (these are unavailable family-group names) basal to the Desmodontinae and the remaining phyllostomid subfamilies.

The Desmodontinae are characterized by sharp, canine-like incisors highly specialized for slicing and the absence of crushing surfaces on the greatly reduced molars and premolars. The nostrils are surrounded by dermal growths forming a circumnasal ridge that is elevated on each side, suggesting a rudimentary noseleaf. There is no external tail, and the interfemoral membrane is reduced to a ridge-like, narrow band of skin. Long bones of the wings and legs are grooved to accommodate muscles. Wing digit III has three bony phalanges. The calcar is rudimentary or absent, except in *Diphylla ecaudata*. The stomach is sac like, elastic, and specialized for absorption (Miller 1907b; Goodwin and Greenhall 1961; Husson 1962; Nowak and Paradiso 1983).

REMARKS: Husson (1962, 1978) and Hershkovitz (1969) followed Palmer (1904) in using the name Desmodidae as the family name instead of Desmodontidae. Handley (1980) explained the derivation of the correct form Desmodontidae or Desmodontinae, and dated Desmodontinae from Gill (1884); however, the earliest use of a suprageneric name based on *Desmodus* is Desmodina (J. A. Wagner 1840:375). We follow most modern literature in treating the vampire bats as a subfamily of the Phyllostomidae, although Morgan, Linares, and Ray (1988) and Ray, Linares, and Morgan (1988) continued to recognize them as the family Desmodontidae.

KEY TO GENERA AND SPECIES OF DESMODONTINAE:
1. Thumb short (usually less than 13 mm) and lacking a basal pad; uropatagium well furred; coronoid process posterior to ventral bend in ramus; lower incisors broad with no central gap between them; two upper incisors and two lower molars on each side . . . . *Diphylla ecaudata*
1′. Thumb elongated (more than 14 mm), with one or two distinct basal tubercles; uropatagium sparsely haired; coronoid process anterior to ventral bend in ramus; narrow lower incisors separated by a median gap; one upper incisor and one lower molar on each side 2
2. Thumb greatly elongated, longer than hind foot, and with two basal pads; wing membranes normally lack white markings; inner lower incisors bilobate; one upper molar on each side . . . . *Desmodus rotundus*
2′. Thumb moderately elongated with one basal pad; tips of wings white; inner lower incisors trilobate; two upper molars on each side . . . . *Diaemus youngii*

Genus *Desmodus* Wied-Neuwied, 1826

Vampires of the genus *Desmodus* are medium-sized (head and body 69–90 mm, forearm 50–63 mm), short-haired bats, attaining adult weights of 25–40 g. The single living species is characterized by small, separate, and distinctly-pointed ears; dark grayish-brown dorsal pelage that con-

trasts with the silvery gray venter; and darkly pigmented wings, except along the forearms. Known pelage color variants range from red to blond and pale silvery gray. The forearms, legs, and extremely narrow interfemoral membrane are sparsely haired. The calcar is reduced to a wart-like excrescence. The thumb is strong, greatly elongated, and its metacarpal has a short, rounded pad at the base and a more elongate pad under its distal half. Other features include a deeply grooved lower lip, and large upper incisors having triangular tips and sharp cutting edges, and almost completely filling the space between the canines. Lower incisors are small and deeply bilobed. Canines are large, long, and acutely pointed. The upper and lower cheek teeth are small. The skull has a large, smoothly rounded braincase and a short rostrum, which is reduced to mere support for the enormous incisors and canines (Miller 1907b; Goodwin and Greenhall 1961; Greenhall et al. 1983; Koopman 1988). The dental formula is 1/2, 1/1, 1/2, 1/1 × 2 = 20.

Morgan, Linares, and Ray (1988) and Ray, Linares, and Morgan (1988) recognized four species, of which *D. rotundus* is the only one extant. Fossils of *Desmodus* have been recorded from Pleistocene and Recent deposits in the United States, Mexico, Cuba, and Venezuela (Cockerell 1930; Cushing 1945; Hatt et al. 1953; Gut 1959; Gut and Ray 1963; Hutchison 1967; Linares 1968; T. Alvarez 1972; Woloszyn and Mayo 1974; Guthrie 1980; Ray, Linares, and Morgan 1988). R. A. Martin (1972), J. D. Smith (1976), Morgan, Linares, and Ray (1988), and Ray, Linares, and Morgan (1988) have reviewed and summarized the fossil record.

SYNONYMS:
Phyllostoma É. Geoffroy St.-Hilaire, 1810b:174; part; not *Phyllostoma* G. Cuvier, 1800.
Rhino[lo]ph[us]. Schinz, 1821:168; part; not *Rhinolophus* Lacépède, 1799b.
Desmodus Wied-Neuwied, 1826:231; type species *Desmodus rufus* Wied-Neuwied, 1826, by monotypy.
Edostoma d'Orbigny, 1835:pl. 8; type species *Edostoma cinerea* d'Orbigny, 1835, by monotypy.
Desmodon Elliot, 1905b:530; unjustified emendation of *Desmodus* Wied-Neuwied.

Desmodus rotundus (É. Geoffroy St.-Hilaire, 1810)
Common Vampire Bat

SYNONYMS: See under Subspecies.

DISTRIBUTION: *Desmodus rotundus* is found on the island of Trinidad, and in every country in South America. Elsewhere, the species occurs throughout Central America and most of Mexico.

MARGINAL LOCALITIES (Map 107): VENEZUELA (Handley 1976): Falcón, 14 km ENE of Mirimire; Distrito Federal, Los Venados; Nueva Esparta, 31 km W of

Map 107 Marginal localities for *Desmodus rotundus* ●

Porlamar. TRINIDAD AND TOBAGO: Trinidad, Santa Cruz (Goodwin and Greenhall 1961). GUYANA: Cuyuni-Mazaruni, Kartabo (Beebe in H. E. Anthony 1921b). SURINAM: Paramaribo, Paramaribo (Husson 1962). FRENCH GUIANA: Saül (Webster and McGillivray 1984). BRAZIL: Amapá, Macapá (C. T. Carvalho 1962a); Pará, Fazenda Velho (USNM 361771); Ceará, Gruta de Ubajara (Uieda, Sazima, and Storti Filho 1980); Pernambuco, Estação Ecológico do Tapacurá (Mares et al. 1981); Bahia, Vila Nova (C. O. C. Vieira 1942); Espírito Santo, Estação Experimental de Linhares (Peracchi and Albuquerque 1993); Espírito Santo, Gruta do Limoeiro (Ruschi 1951b); São Paulo, São Sebastião (USNM 141381); Paraná, Palmeira (O. Thomas 1900c); Rio Grande do Sul, Pôrto Alegre (Ihering 1892); Rio Grande do Sul, São Lourenço do Sul (Fabián, Rui, and Oliveira 1999). URUGUAY: Lavalleja, Gruta de Arequita (Langguth and Achaval 1972). ARGENTINA (Barquez, Mares, and Braun 1999, except as noted): Entre Rios, Parque Nacional El Palmar; Corrientes, Goya; Chaco, 2 km NNW and 11 km NE (by road) of El Mangrullo; Santiago del Estero, Campo Alegre (Romaña and Abalos 1950); Córdoba, Corazón de María; Córdoba, Espinillo (Tiranti and Torres 1998); San Luis, 7 km E of San Francisco del Monte de Oro; La Rioja, El Barreal; La Rioja, La Rioja; San Juan, Castaño Nuevo; Catamarca, Choya; Salta, Cueva del Indio (Villa-R. and Villa-Cornejo 1971); Salta, Cueva del Murcielagallo. BOLIVIA (S. Anderson 1997): Chuquisaca,

12 km N and 11 km E of Tarabuco; Cochabamba, Laguna Alalay; La Paz, Caracato. PERU: Cusco, Puquiura (O. Thomas 1920d); Huancavelica, Lircay (Koopman 1978). CHILE: Valparaiso, Curaumilla (Wolffsohn 1921); Atacama, Domeyko (Mann 1978); Tarapacá, Guanillos (Mann 1978); Tarapacá, Caleta Cuya (Pine, Miller, and Schamberger 1979). PERU: Ica, Vieja Island (Koopman 1978); Lima, Cerro Augustino (Brack 1974); La Libertad, Otuzco (Ortiz de la Puente 1951); Piura, Mallares (Aellen 1966). ECUADOR: Guayas, Isla Silva (Albuja 1983); Esmeraldas, Malimpia (Dorst 1951). COLOMBIA: Valle del Cauca, Buenaventura (Arata et al. 1968); Antioquia, La Ceja (H. E. Anthony 1923a); Guajira, Villanueva (Hershkovitz 1949c).

SUBSPECIES: We recognize two subspecies of *D. rotundus*.

D. r. murinus J. A. Wagner, 1840
SYNONYMS:

D[esmodus]. murinus J. A. Wagner, 1840:377–79; type locality Mexico ("den mexikanischen Sendungen des Dr. Pess . . .)."

Desmodus rotundus murinus: Osgood, 1912:63; first use of current name combination.

This subspecies has been recorded from northern and western Colombia, and from the Pacific lowlands and western slopes of the Andes in Ecuador and Peru. Elsewhere, this subspecies is known from Central America north into northwestern Mexico.

D. r. rotundus (É. Geoffroy St.-Hilaire, 1810)
SYNONYMS:

Phyllostoma rotundum É. Geoffroy St.-Hilaire, 1810b:181; type locality "Paraguay"; restricted to Asunción by Cabrera (1958:93); based on Azara's (1801b) *chauve-souris troisième*.

Rhino[lo]ph[us]. ecaudatus Schinz, 1821:168; type locality "Ostküste von Brasilien."

D[esmodus]. rufus Wied-Neuwied, 1826:233; type locality "Fazenda von Muribeca am Flusse Itabapuana," Espírito Santo, Brazil.

Phyllostoma Infundibiliforme Rengger, 1830:77; type locality "Paraguay."

Edostoma cinerea d'Orbigny, 1835:pl. 8; no type locality given; type locality identified as Santa-Corazón, provincia Chiquitos, Santa Cruz, Bolivia, by d'Orbigny and Gervais (1847:11).

Desmodus D'Orbignyi Waterhouse, 1838:1; type locality "Coquimbo [Coquimbo], Chile."

Desmodus fuscus Burmeister, 1854:57; type locality "Rio das Velhas," Minas Gerais, Brazil.

Desmodus mordax Burmeister, 1879:78; intended as a replacement name for *Phyllostoma rotundum* É. Geoffroy

St.-Hilaire, and based on *El Mordedor* of Azara (1802b; = Azara's [1801b] *chauve-souris troisième*).

Desmodus rotundus: O. Thomas, 1900c:546; first use of current name combination.

The nominate subspecies is known from Venezuela, the island of Trinidad, the Guianas, and the Amazon basin of Colombia, Ecuador, Peru, Brazil, and Bolivia, south through Paraguay and Uruguay into Chile and Argentina.

NATURAL HISTORY: *Desmodus rotundus* has been found roosting in caves, wells, mine shafts, hollow trees, and abandoned buildings in practically all lowland and mid-elevation habitats in South America. Colonies up to 100 individuals are common and some roosts contain as many as 2,000 (Uieda 1987). The highest elevational records are 2,594 m at Pasto, Nariño, Colombia, in cold Andean plateau habitat (Tamsitt and Valdivieso 1962); 3,350 m in Peru (Puquiura, Cusco; O. Thomas 1920d); and 3,600 m in Bolivia (S. Anderson, Koopman, and Creighton 1982). The diet consists mainly of mammalian blood (usually from larger mammals, including humans) and larger birds. McCarthy (1989) documented increased predation on humans by *D. rotundus* following a hog cholera control program in Belize, during which pigs, the locally preferred blood source, were slaughtered. Ruschi (1953a) commented that while *D. rotundus* fed on barn owls, the owls apparently did not consume vampires. He earlier (1951b) had written that barn owls consumed young *D. rotundus*. Greenhall (1972, 1988) described prey selection and feeding behavior, including biting, bite sites, and mode of attack. The blood flows into the mouth along grooves on the under surface of the tongue. Flow results from suction created in the pharynx by lingual movements that were originally misinterpreted as lapping (Mann 1950a; Wimsatt 1959), and from capillary action in the grooves under the tongue. *Desmodus rotundus* breeds throughout the year (Wimsatt and Trapido 1952; Goodwin and Greenhall 1961; Wilson 1979; Schmidt 1988). Gestation is approximately 7 months (Schmidt 1988).

Parasites infecting *D. rotundus* include protozoans (Stiles and Nolan 1931), trypanosomes (Hoare 1972; Ubelaker, Specian, and Duszynski 1977), and nematodes (Wolfgang 1954). Wenzel, Tipton, and Kiewlicz (1966) reported 3 kinds of mites, 2 of ticks, and 12 species of batflies for *D. rotundus*. Wenzel (1976) listed an additional ten species of batflies, and Brennan and Reed (1975) found ten species of chiggers on Venezuelan specimens. Webb and Loomis (1977) added six mites, two ticks, and three batflies to the ectoparasites cited above. R. Guerrero (1985a) listed 38 species in 20 genera of ectoparasites known from *D. rotundus*, and later (1997) listed 26 species in 14 genera of streblid batflies. Méndez (1988) summarized all records of parasites and diseases known in *D. rotundus* up to the

1980s. Greenhall et al. (1983), in their *Mammalian Species* account, and Greenhall and Schmidt (1988) summarized most of the available information on the natural history of *D. rotundus*. The karyotype is $2n = 28$, FN = 52 (Forman, Baker, and Gerber 1968; R. J. Baker et al. 1982; R. J. Baker, Honeycutt, and Bass 1988).

REMARKS: The three specimens D. C. Carter and Dolan (1978:67–68) examined in the ZMB and labeled as types of *D. murinus* did not include the specimen from Mexico described by J. A. Wagner. They also were unable to locate the type specimens of *Desmodus fuscus* Burmeister and *Desmodus mordax* Burmeister. However, the type specimen of *D. mordax* is the same as that of *Phyllostoma rotundum*, because *D. mordax* was a replacement name for the latter, which was based solely on Azara's (1801b) description of his *chauve-souris troisième* (= *el mordedor* [Azara 1802b]), and no type specimen likely exists.

Numerous illustrations of *D. rotundus* have been published (e.g., Husson 1962, 1978; Greenhall et al. 1983; Koopman 1988). Morgan, Linares, and Ray (1988) and Ray, Linares, and Morgan (1988) compared the cranium of *D. rotundus* with those of other vampire taxa. D. C. Carter and Dolan (1978) presented information on and measurements of the holotype of *Desmodus dorbignyi* Waterhouse, and the presumed types of *D. murinus* J. A. Wagner and *Edostoma cinerea* d'Orbigny.

Genus *Diaemus* Miller, 1906

Bats of the monotypic genus *Diaemus* are medium-sized (head and body approximately 85 mm, forearm 50–56 mm, mass 30–45 g), pale brown to dark cinnamon brown bats. They have short ears; a narrow, sparsely haired interfemoral membrane; hairy feet and toes, and lack a visible calcar. The tips and trailing margins of the wings are creamy white. *Diaemus* can be distinguished from *Desmodus* by the larger skull and shorter thumb, which is about 1/8 as long as digit III and has a single basal pad. Inner lower incisors are trilobate and the outer lower incisors are weakly bilobate with slightly convergent tips. The thumb is slightly longer than that of *Diphylla*, and differs in having a large ventral pad on the metacarpal (Miller 1907b; Goodwin and Greenhall 1961; Nowak and Paradiso 1983; Koopman 1988). The dental formula is 1/2, 1/1, 1/2, 2/1 × 2 = 22. There is no fossil record.

SYNONYMS:

Desmodus Jentink, 1893:282; part; not *Desmodus* Wied-Neuwied, 1826.

Diæmus Miller, 1906b:84; type species *Desmodus youngii* Jentink, 1893, by original designation.

REMARKS: Handley (1976) treated *Diaemus* as a synonym of *Desmodus*. Koopman (1988), while acknowledging that *Diaemus* could be regarded a subgenus of *Desmodus*, recognized *Diaemus* as a valid genus. Koopman (1993, 1994) and Corbet and Hill (1991) treated *Diaemus* as a full genus.

Diaemus youngii (Jentink, 1893)
White-winged Vampire Bat

SYNONYMS:

Desmodus Youngii Jentink, 1893:282; type locality "Berbice," East Berbice-Corentyne, Guyana; further defined as "Upper Canje Creek," by Young (1896:46).

Diaemus youngii: Miller, 1906b:84; first use of current name combination.

Diæmus youngii cypselinus O. Thomas, 1928d:288; type locality "Pebas," Loreto, Peru.

Diaemus yougii: Piccinini, 1974:76; incorrect subsequent spelling of *Desmodus youngii* Jentink.

Desmodus youngi Honacki, Kinman, and Koeppl, 1982:155; incorrect subsequent spelling of *Desmodus youngii* Jentink.

Diaemus youngi Greenhall and Schutt, 1996:1; incorrect subsequent spelling of *Desmodus youngii* Jentink.

DISTRIBUTION: *Diaemus youngii* is known from Venezuela, the island of Trinidad, Guyana, French Guiana, Colombia, Ecuador, Peru, Brazil, Paraguay, Bolivia and northern Argentina. Elsewhere, the species is known from Central America north into Mexico.

MARGINAL LOCALITIES (Map 108): VENEZUELA: Falcón, 6 km SE of Capatárida (Handley, 1976); Nueva Esparta, El Valle (Pirlot and León 1965). TRINIDAD AND TOBAGO: Trinidad, Barataria (USNM 536938).

Map 108 Marginal localities for *Diaemus youngii* ●

VENEZUELA: Delta Amacuro, Araguaimujo (Pirlot 1965b). GUYANA: East Berbice-Corentyne, Upper Canje Creek (type locality of *Desmodus youngii* Jentink). FRENCH GUIANA: Paracou (Simmons and Voss 1998). BRAZIL: Amapá, Santa Luzia do Pacuí (Peracchi, Raimundo, and Tannure 1984); Pará, Utinga (Handley 1967); Pernambuco, Estação Ecológico do Tapacurá (Mares et al. 1981); Alagoas, Pôrto Calvo (Aguiar, Camargo, and Portella 2006); Rio de Janeiro, Rio de Janeiro (Peracchi and Albuquerque 1986); Rio de Janeiro, Bara Mansa (Peracchi and Albuquerque 1971); São Paulo, Fazenda Paraguassu (I. Sazima and Uieda 1980); Paraná, Roça Nova (Miller 1906b). ARGENTINA: Misiones, Bonpland (Massoia, Chebez, and Heinonen-Fortabat 1989b). PARAGUAY: Presidente Hayes, Estancia La Victoria (López-González et al. 1998). ARGENTINA; Jujuy, Agua Salada (Barquez 1984b). BOLIVIA (S. Anderson 1997): Santa Cruz, Buenavista; La Paz, Pasto Grande. PERU: Huánuco, Cueva de Lechuzas (Bowles, Cope, and Cope 1979); Ucayali, Yarinacocha (Sanborn 1949b); Loreto, Pebas (type locality of *Diaemus youngii cypselinus* O. Thomas). COLOMBIA: Putumayo, Río Mecaya (Lay 1962); Valle de Cauca, Río Raposo (Wenzel, Tipton, and Kiewlicz 1966).

SUBSPECIES: We treat *D. youngii* as monotypic.

NATURAL HISTORY: *Diaemus youngii* has been found in hollow trees and shallow caves in small colonies that may contain as many as 20 to 30 individuals (Uieda 1987). Aguiar, Camargo, and Portello (2006) reported mistnetting eight *D. youngii* between October 2004 and February 2005 as the bats left a cave in Distrito Federal, Brazil. Simmons and Voss (1998) reported mistnetting two specimens at ground level, one in well-drained primary forest, the other in a manmade clearing. They caught 1/3 in a mistnet strung across a gap in the forest canopy 17 to 20 m above a narrow dirt road. This species prefers avian blood to mammalian blood (Goodwin and Greenhall 1961; Gardner 1977c; I. Sazima and Uieda 1980). I. Sazima and Uieda (1980) reported *D. youngii* feeding on free-ranging chickens, turkeys, and guinea fowl in Brazil. Goodwin and Greenhall (1961) stated that this species fed on poultry, pigeons, and goats in the island of Trinidad; but in captivity, refused cattle blood and would drink only pure chicken blood. They also reported that one specimen, which had refused defibrinated bovine blood, fed readily on the blood of a live guinea pig (*Cavia porcellus*). Dickson and Green (1970), however, reported successfully keeping *D. youngii* in the laboratory on citrated bovine blood. Greenhall (1970) claimed that *D. youngii* attacks cattle and, using a precipitin test of stomach contents, determined that 13 individuals had fed solely on mammalian blood, 8 on a combination of mammalian and avian blood, and only 2 had fed solely on avian blood (see summary in Greenhall 1988).

Goodwin and Greenhall (1961) reported taking two lactating females in August, and an immature male, four pregnant females, one lactating female, and two reproductively inactive females in October in the island of Trinidad. Goodwin and Greenhall (1961) described cup-shaped glands inside the mouth, one on each side that emitted a powerful and peculiar odor when the bats were disturbed. Webb and Loomis (1977) identified two families of mites (three species) and four species of streblid batflies recovered from *D. youngii* (see summaries of known parasites in R. Guerrero [1985a] and Méndez [1988]). Greenhall and Schutt (1996) reviewed the biology and life history of *D. youngii* in their *Mammalian Species* account. The karyotype is $2n = 32$, $FN = 60$ (Forman, Baker and Gerber 1968; R. J. Baker, Honeycutt, and Bass 1988).

REMARKS: Uncommon, although widespread throughout northern South America and the greater Amazon basin, *D. youngii* has not been recorded from Uruguay, Surinam, or west of the Andes south of Colombia. Miller's (1907b:178) mention of "Dutch Guiana" as part of the geographic distribution of *D. youngii* (repeated by Lay 1962) was a *lapsus* for British Guiana, the country where the holotype was collected. O. Thomas (1928d) based his description of *D. youngii cypselinus* on the relative lengths of the digits. Husson (1962, 1978) and Koopman (1988) illustrated the skull of *D. youngii*. Morgan, Linares, and Ray (1988) and Ray, Linares, and Morgan (1988) compared and contrasted the skull morphology of *D. youngii* with those of other living and extinct species of vampires. The localities plotted by Aguiar, Camargo, and Portella (2006) on their map (Fig. 1) do not appear to correspond to the information in the figure legend. D. C. Carter and Dolan (1978) gave information on and measurements of the holotypes of *Desmodus youngii* Jentink and *Diaemus youngii cypselinus* O. Thomas. J. K. Jones and Carter (1976) regarded *D. youngii* as monotypic.

Genus *Diphylla* Spix, 1823

Vampires in the monotypic genus *Diphylla* are medium-sized bats (head and body 65–87 mm, forearm 50–56 mm, mass 24–43 g) that have long, soft fur, which is dark brown dorsally and paler ventrally; and have densely furred legs, forearms, and interfemoral membrane. These bats have short, broad ears; short, but well-formed calcars; and short thumbs, which lack metacarpal pads. The skull differs from those of *Diaemus* and *Desmodus* in having a higher and much broader interorbital region. The lower inner incisors have four lobes, and the lower outer incisors are unique among bats in being broad, fan-shaped, and seven-lobed (Miller 1907b; Nowak and Paradiso 1983; Greenhall et al.

1984; Koopman 1988). The dental formula is 2/2, 1/1, 1/2, 2/2 × 2 = 20.

SYNONYMS:
Diphylla Spix, 1823:68; type species *Diphylla ecaudata* Spix, 1823, by monotypy.
Glossophaga: J. B. Fischer, 1829:131; part; not *Glossophaga* É. Geoffroy St.-Hilaire, 1818a.
Diphydia Gray, 1829:29; incorrect subsequent spelling of *Diphylla* Spix.
Haematonycteris H. Allen, 1896a:777; type species *Diphylla ecaudata* Dobson, 1878, by monotypy.
Diphyla Coimbra, Borges, Guerra, and Mello, 1982:34; incorrect subsequent spelling of *Diphylla* Spix.

REMARKS: *Haematonycteris* was proposed by H. Allen (1896a:777) as a provisional generic name for the bat Dobson (1878) described under the name *Diphylla ecaudata*, if that bat proved to be incorrectly identified. Kretzoi and Kretzoi (2000:108) listed the name "*Diphilla* d'Orbigny and Gervais 1847" as a *lapsus* for *Diphylla* Spix, 1823. Nevertheless, the two partial illustrations (numbers 5 and 6) in d'Orbigny's plate 9 (published in 1837 according to Sherborn and Griffin [1934:130]) are of the dentition of an unidentifiable stenodermatine, not a desmodontine.

Diphylla ecaudata Spix, 1823
Hairy-legged Vampire Bat
SYNONYMS:
Diphylla ecaudata Spix, 1823:68; type locality "Brazil"; restricted to Río San Francisco, Bahia, by Cabrera (1958:94).
Gl[ossophaga]. diphylla J. B. Fischer, 1829:133 type locality "Brasilia"; based on *Diphylla ecaudata* Spix.
Diphylla centralis O. Thomas, 1903b:378; type locality "Boquete, Chiriqui," Panama.
Diphylla ecaudata centralis: Felten, 1956a:364; name combination.

DISTRIBUTION: *Diphylla ecaudata* is known from Colombia, Venezuela, Ecuador, Peru, Bolivia, and Brazil (except the central Amazon basin). Elsewhere, the species has been recorded from Central America north through Mexico into Texas, USA (Reddell 1968).

MARGINAL LOCALITIES (Map 109; arranged as eastern and western components of the South American distribution): *Western distribution.* COLOMBIA: Magdalena, Cacagualito (J. A. Allen 1900a). VENEZUELA: Mérida, Cueva de la Azulita (Ojasti and Linares 1971); Aragua, Portachuelo (Ojasti and Linares 1971); Sucre, 21 km E of Cumaná (Handley 1976). COLOMBIA: Vaupés, Cerro de las Pinturas (Wenzel, Tipton, and Kiewlicz 1966). BRAZIL: Rondônia, Pôrto Velho (USNM 562686). PERU: Cusco, San Martín-3 (S. Solari et al. 2001c); Pasco, Nevati (Tuttle 1970); Huánuco, Panguana (Hutterer et al. 1995); Ucayali,

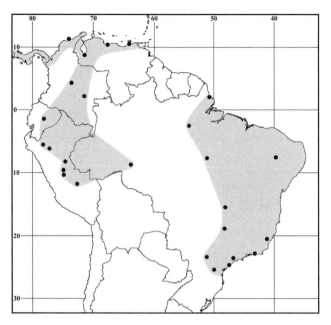

Map 109 Marginal localities for *Diphylla ecaudata* ●

confirmed. He (1953a) again reported *D. ecaudata* feeding on a hog, and illustrated the bite wounds resulting from this species feeding on a man. In the same report, Ruschi commented that vampire bats fed on barn owls, and while owls fed on a variety of bats, they apparently did not prey on vampires. Schutt and Altenbach (1997) described and illustrated the short, stout calcar of *D. ecaudata* and its use as an aid during arboreal locomotion.

Dalquest (1955) reported a well-defined breeding season during which one young is born each year, but Felten (1956a) and Wilson (1979) stated that these bats breed in both dry and wet seasons and postulated two birth periods per year. Bredt, Uieda, and Magalhães (1999) reported finding pregnant females in February, March, July, September, and October. Ectoparasites include streblid batflies (Wenzel, Tipton, and Kiewlicz 1966; Webb and Loomis 1977; R. Guerrero 1997) and mites (Brennan and Reed 1975; see reviews by R. Guerrero [1985a] and Méndez [1988]). The karyotype is $2n = 28$, FN = 52 (R. J. Baker 1973; R. J. Baker, Honeycutt, and Bass 1988).

REMARKS: O. Thomas (1903b) described *Diphylla centralis*, which he distinguished from *Diphylla ecaudata* mainly on the differences in the proportions of the lower teeth. Burt and Stirton (1961) cited a letter from O. Thomas to Dickey (*in litt.* 1927), in which O. Thomas questioned the validity of *D. centralis*, saying that he no longer considered the two groups to be distinct. Felten (1956a) considered *D. centralis* to be a subspecies of *D. ecaudata* and identified his Salvadoran specimens as such. Ojasti and Linares (1971), J. K. Jones and Carter (1976), and Greenhall et al. (1984) also recognized two subspecies, one Middle American and the other South American. Greenhall et al. (1984) and Koopman (1988) illustrated the skull and dentition of *D. ecaudata* (also see Morgan, Linares, and Ray 1988; Ray, Linares, and Morgan 1988). D. C. Carter and Dolan (1978) gave information on and measurements of the holotype of *Diphylla centralis* O. Thomas, but were unsuccessful in finding the type of *Diphylla ecaudata* Spix. The species needs to be revised.

Yarinacocha (Sanborn 1949b); San Martín, Puca Tambo (O. Thomas 1926c); Amazonas, 10 km by trail SE of La Peca (Graham and Barkley 1984). ECUADOR: Pastaza, Mera (Albuja 1983). COLOMBIA: Meta, Restrepo (Aellen 1970). *Eastern distribution.* BRAZIL: Amapá, Amapá (Peracchi, Raimundo, and Tannure 1984); Pernambuco, Fazenda Maniçoba (Mares et al. 1981); Bahia, Rio San Francisco (type locality of *Diphylla ecaudata* Spix, not mapped); Espírito Santo, Gruta do Limoeiro (Ruschi 1951c); Rio de Janeiro, Rio de Janeiro (Uieda 1986); São Paulo, Cotia (McNab 1969); São Paulo, Iguapé (Pira 1904); Paraná, Palmeira (O. Thomas 1900c); Paraná, Parque Estadual Mata dos Godoy (N. R. Reis et al. 2000); Minas Gerais, Uberlândia (Stutz et al. 2004); Distrito Federal, Gruta do Sal (Bredt, Uieda, and Magalhães 1999); Pará, Gradaús (Piccinini 1974); Pará, Taperinha (Piccinini 1974).

SUBSPECIES: Two subspecies usually are recognized. The nominate subspecies, *D. e. ecaudata* Spix, is the only subspecies reported in South America. *Diphylla e. centralis* is the name applied to US, Mexican, and Central American populations.

NATURAL HISTORY: *Diphylla ecaudata* has been found almost exclusively in caves and mines, and rarely in hollow trees (Dalquest and Hall 1947; Ruschi 1951c). Colony size usually is small, rarely numbering over 40 to 50 individuals (Uieda 1987). Avian blood is the preferred food (Koopman 1956; Villa-R. 1967; Villa-R., Silva, and Villa-Cornejo 1969), and all blood from non-avian sources was refused by the *D. ecaudata* studied by Hoyt and Altenbach (1981). Ruschi (1951c) reported that *D. ecaudata* also fed on pigs, cattle, and equines, but this remains un-

Subfamily Glossophaginae Bonaparte, 1845

Thomas A. Griffiths and Alfred L. Gardner

The Glossophaginae comprises the Neotropical nectarivorous bats, which occur in the southwestern United States southward through Mexico, Central America, and the Antilles into South America as far as Paraguay, northern Argentina, and Chile. We recognize ten genera in the subfamily of which six are found in South America.

Glossophagines are small to medium-sized bats (forearm 30–60 mm) characterized by a long, slender muzzle (greatly

elongated in some species) that bears numerous conspicuous vibrissae, and a long and highly extensible tongue having the tip clothed with hair-like papillae (paintbrush tip), but lacking a groove along the lateral margins. The ears are small and rounded; and the noseleaf is well developed, but comparatively small. The lower lip is divided by a deep vertical groove, which may be bordered laterally by small tubercles. The tail is short (absent in some *Anoura*), the tip of which usually projects free from the dorsal surface of the uropatagium. The dentition is comparatively weak, the antero-posteriorly elongated molars and premolars have reduced cusps and styles, and the upper molars lack a hypocone. The braincase is relatively large, globose, and has a large foramen magnum. The zygomatic arches are weak or incomplete, and auditory bullae are relatively small. Skeletal features include an atlas that is about twice as wide as remaining cervical vertebrae, which in turn are notably wider than postcervical vertebrae. The T-shaped manubrium (sternum) has a low ventral keel. The scapula has a reduced spinous process, a large acromion process, and a large curved coracoid process. The dental formula is 2/0–2, 1/1, 2–3/3. 2–3/2–3 × 2 = 26–34.

REMARKS: Griffiths (1982), on the basis of differences in the hyoid-lingual anatomy, removed the genera *Lonchophylla*, *Lionycteris*, and *Platalina* from the Glossophaginae and placed them in a separate subfamily, the Lonchophyllinae. Recent phylogenies of phyllostomid bats generated by R. J. Baker et al. (2000) using nuclear DNA sequences indicated that *Lonchophylla* and *Lionycteris* were, at best, only distantly related to the Glossophaginae (as suggested earlier by Griffiths 1982). The classifications by McKenna and Bell (1997) and Wetterer (2000) recognized glossophagines and lonchopyllines at the tribal level. Carstens, Lundrigan, and Myers (2002) also retained these taxa at the tribal level and divided the Glossophagini into two clades, the choeronycterines containing the genera *Anoura*, *Choeroniscus*, *Choeronycteris*, *Hylonycteris*, *Lichonycteris*, *Musonycteris*, and *Scleronycteris*; and the glossophagines comprising the genera *Glossophaga*, *Leptonycteris*, and *Monophyllus*. On the basis of substantial work supporting separation of New World pollen and nectar feeding bats into two subfamilies (see classifications by Griffiths 1982; Koopman 1993; R. J. Baker et al. 2000), R. J. Baker et al. (2003) recognized glossophagines and lonchophyllines as representing separate subfamilies. We follow their classification here.

KEY TO SOUTH AMERICAN GENERA OF GLOSSOPHAGINAE:

1. Two upper and two lower molars on each side 2
1'. Three upper and three lower molars on each side 3
2. Two lower premolars . . . . *Leptonycteris*
2'. Three lower premolars . . . . *Lichonycteris*
3. Three upper premolars . . . . *Anoura*
3'. Two upper premolars 4
4. Lower incisors well-developed in adults . . *Glossophaga*
4'. Lower incisors minute or absent 5
5. Pterygoid processes convex on inner sides; hamular processes clearly not in contact with auditory bullae . . . .
. . . . *Scleronycteris*
5'. Pterygoids deeply concave on inner sides; hamular processes in contact or nearly in contact with auditory bullae
. . . . *Choeroniscus*

Genus *Anoura* Gray, 1838

We recognize seven species of *Anoura* of which *A. candenai*, *A. caudifer*, *A. fistulata*, *A. luismanueli*, and *A. latidens* are endemic to South America (Nagorsen and Tamsitt 1981; Handley 1984; Molinari 1994; Muchala et al. 2005; Mantilla-Meluk and Baker 2006). *Anoura* spp. are rare in the lower Amazon basin, although two species have distributions that nearly encircle the basin (see Maps 110 and 113). *Anoura* are medium-sized glossophagines (forearm 33–48 mm) that lack a tail or have an inconspicuous, short tail that is completely encased in a nearly naked to well-furred, extremely narrow interfemoral membrane. The pelage is usually dark brown to blackish brown. The dorsal hairs are tricolored with a broad pale basal band and the ventral fur appears to be nearly unicolored (dark) with frosted tips. The zygomatic arches may or may not be fully ossified and thus appear incomplete (most often complete in *A. cultrata*). Upper incisors are arranged as two pairs separated by a wide gap with the outer incisors bladelike and larger than the peglike inner incisors. There is sexual dimorphism in the size of canines. The upper canines of males are conspicuously longer and more robust than those of females. The hypocone is absent, but all other major cusps are present in upper and lower molars, although reduced in size (see Phillips 1971, for a more comprehensive description of the dentition). The dental formula is 2/0, 1/1, 3/3, 3/3 × 2 = 32.

SYNONYMS:

Glossophaga É. Geoffroy St.-Hilaire, 1818a:418; part.

Anoura Gray, 1838b:490; type species *Anoura geoffroyi* Gray, 1838b, by monotypy.

Choeronycteris Tschudi, 1844b:70; part.

Rhinchonycteris Tschudi, 1844b:71; *nomen nudum*.

Anura Agassiz, 1847:27; unjustified emendation of *Anoura* Gray.

Lonchoglossa W. Peters, 1868b:364; type species *Lonchoglossa caudifera* (= *Glossophaga caudifer* É. Geoffroy St.-Hilaire, 1818a), by monotypy.

Glossonycteris W. Peters, 1868b:364; type species *Glossonycteris lasiopyga* W. Peters, 1868b, by monotypy.

Anura O. Thomas, 1893c:335; incorrect subsequent spelling of *Anoura* Gray.

Anura Trouessart, 1897:158; incorrect subsequent spelling of *Anoura* Gray.

Lonchoglosa Podtiaguin, 1944:32; incorrect subsequent spelling of *Lonchoglossa* W. Peters.

REMARKS: The name *Rhinchonycteris* Tschudi is often included in the synonymy of *Anoura* as a subjective synonym (e.g., Cabrera 1958); however, a reading of Tschudi's (1844b) text suggests that the name was a *lapsus* for *Choeronycteris* and, therefore, is a *nomen nudum* as earlier determined by O. Thomas (1928a:123). Simpson (1945:57) erred in using the junior synonym *Lonchoglossa* instead of *Anoura*. Apparently he considered *Lonchoglossa* to date from 1818 instead of 1868. Gerrard (1862:39) referred to *Anoura minor*, which appears to be a *nomen nudum*.

Koopman (1981) characterized *Anoura* as fairly primitive among New World nectar-feeding bats. However, Griffiths (1982) showed that the morphology of the tongue protrusion apparatus was comparable to that of the more derived glossophagines.

KEY TO SPECIES OF *ANOURA* (MODIFIED FROM HANDLEY 1984).

1. First lower premolar greatly enlarged and bladelike; upper canine unusually large and with a prominent, deep sulcus on its anterior face . . . . *Anoura cultrata*

1′. First lower premolar approximately the same size and shape as other premolars; upper canine not enlarged and its anterior face flat or with a shallow sulcus 2

2. Upper last premolar (P4) lacks medial internal cusp; m1 lacks anteroexternal cuspid and cristid; tail usually present 3

2′. Upper last premolar (P4) with medial internal cusp; m1 with anteroexternal cuspid and cristid; tail usually absent. 4

3. Legs and interfemoral membrane sparsely haired to nearly naked; palatal length exceeds 11.5 mm; maxillary toothrow (C-M3) more than 7.5 mm, usually exceeds 8 mm. 5

3′. Legs and interfemoral membrane hairy; palatal length less than 11.5 mm; maxillary toothrow less than 8 mm . . . . *Anoura luismanueli*

4. Medial internal cusp of P4 enclosed in the broad triangular base of tooth; upper and lower last two premolars broad . . . . *Anoura latidens*

4′. Medial internal cusp of P4 prominently protrudes from narrow base of tooth; upper and lower last two premolars narrow . . . . *Anoura geoffroyi*

5. Greatest length of skull more than 22 mm; width of braincase more than 9.1 mm; width across upper canines more than 4 mm; lower lip protrudes 3 mm or more beyond upper lip. . . . . *Anoura fistulata*

5′. Greatest length of skull less than 22 mm; width of braincase less than 9.1 mm; width across upper canines less than 4 mm; lower lip protrudes less than 3 mm beyond upper lip. 6

6. Posterior margin of uropatagium sparsely haired; keel along midline of mesopterygoid fossa flattened posteriorly; outer margin of malar portion of zygomatic arch tapering posteriorly; posterior projection of pterygoids relatively short, extending to or just beyond anterior projection of each bulla; upper canines long and robust . . . . *Anoura cadenai*

6′. Posterior margin of uropatagium usually well haired; keel along midline of mesopterygoid fossa not flattened posteriorly and usually extending onto septum between basisphenoid pits; outer margin of malar portion of zygomatic arch wide with a more pronounced shoulder; posterior projection of pterygoids relatively long, extending behind anterior projection of each bulla; upper canines moderate in size . . . . *Anoura caudifer*

Anoura cadenai Mantilla-Meluk and Baker, 2006
Cadena's Tailless Bat

SYNONYMS:

Anoura caudifera M. E. Thomas, 1972:102; incorrect subsequent spelling of, but not *Glossophaga caudifer* É. Geoffroy St.-Hilaire, 1818a.

Anoura cadenai Mantilla-Meluk and Baker, 2006:11; type locality "between the municipios of Calima and Restrepo near the Rio Bravo at 1000 m elevation at 00°00′03″S, 00°00′18″W," Valle del Cauca, Colombia.

DISTRIBUTION: *Anoura cadenai* currently is known only from the western slope of the southwestern Colombian Andes and is expected to occur in adjacent Ecuador.

MARGINAL LOCALITIES (Map 112): COLOMBIA: Valle del Cauca, Calima (Mantilla-Meluk and Baker 2006); Valle del Cauca, Pichindé (USNM 483371; not mapped); Valle del Cauca, 2 km S of Pance (USNM 483366).

SUBSPECIES: *Anoura cadenai* is monotypic.

NATURAL HISTORY: *Anoura cadenai* is known from an elevational zone between 800 and 1,600 m in the western slope of the southwestern Colombian Andes, a region containing a mosaic of primary and secondary forests and agricultural lands. Habitat at the type locality was Andean forest containing mature trees covered by epiphytes (Mantilla-Meluk and Baker 2006). The specimens from 2 km S of Pance were reported as *A. caudifera* [sic] by M. E. Thomas (1972), who recorded females pregnant in May and November. He described the site (1972:13–15) as originally "Very Humid Subtropical Forest on the eastern slope of the Western Andean Cordillera" at an elevation of 1,560 m.

REMARKS: *Anoura cadenai* is most similar to *A. caudifer*. Mantilla-Meluk and Baker (2006) reported both species sympatric with *A. cultrata* at Calima. Mantilla-Meluk and Baker (2006) commented on the presence of a sulcus on the anterior face of the upper canines. We have found that the canines of all *Anoura* have a sulcus on the anterior face of the upper canines. This sulcus is especially deep and involves the cingulum in *A. cultrata*, but is shallow, does not involve the cingulum, and is often obliterated by wear in other species of *Anoura*. A feature not mentioned by Mantilla-Meluk and Baker (2006) that is useful for distinguishing *A. cadenai* from *A. caudifer* is the larger upper canines in *A. cadenai*. Canine size is sexually dimorphic in *Anoura*; the upper canines in males are longer and somewhat more massive than in females. The upper canines are notably larger in *A. cadenai* and the shorter canines of females are as long as those of male *A. caudifer*. *Anoura cadenai* can be distinguished from *A. luismanueli* by its larger size, even more sparsely furred uropatagium, longer rostrum, and the lack of a well-marked lateral constriction of the rostrum immediately behind the canines. In addition to longer canines, *A. cadenai* can be distinguished from *A. caudifer* by the posteriorly flattened keel along the midline of the mesopterygoid fossa, the relatively shorter posterior projections of the pterygoids, and by the less robust, more tapering outline of the anterior (malar) region of the zygomatic arches. The rostrum is constricted behind the canines in both *A. cadenai* and *A. caudifer*, but less so in *A. caudifer*; in neither species is it constricted to the same extent as in *A. luismanueli*.

Anoura caudifer (É. Geoffroy St.-Hilaire, 1818)
Lesser Tailless Bat

SYNONYMS:

Glossophaga caudifer É. Geoffroy St.-Hilaire, 1818a:418, pl. 17; type locality "Rio de Janeiro," Rio de Janeiro, Brazil.

Glossophaga ecaudata É. Geoffroy St.-Hilaire, 1818a:418, pl. 18; type locality unknown.

Glossoph[aga]. caudifera: J. B. Fischer, 1829:139; incorrect subsequent spelling of *Glossophaga caudifer* É. Geoffroy St.-Hilaire.

Lonchoglossa caudifera: W. Peters, 1868b:364; name combination and incorrect subsequent spelling of *Glossophaga caudifer* É. Geoffroy St.-Hilaire.

[Lonchoglossa] ecaudata: Trouessart, 1897:158; name combination.

Lonchoglossa wiedi aequatoris Lönnberg, 1921:65; type locality "Ilambo" (= Illambo), near Gualea, Pichincha,

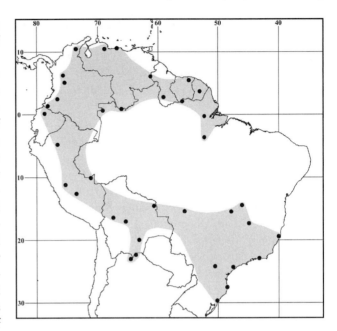

Map 110 Marginal localities for *Anoura caudifer* ●

Ecuador; incorrect subsequent spelling of *Anura wiedii* W. Peters, 1869.

Lonchoglossa caudifera aequatoris: Sanborn, 1933:27; name combination and incorrect subsequent spelling of *Glossophaga caudifer* É. Geoffroy St.-Hilaire.

Anoura caudifera: Cabrera, 1958:74; first use of current name combination and incorrect subsequent spelling of *Glossophaga caudifer* É. Geoffroy St.-Hilaire.

Lonchoglossa caudifer: Husson, 1962:136; name combination.

Anoura (Lonchoglossa) caudifer: Tamsitt and Valdivieso, 1966b:230; name combination.

A[noura]. caudira Alberico and Orejuela, 1982:34; incorrect subsequent spelling of *Glossophaga caudifer* É. Geoffroy St.-Hilaire.

Anoura caudifera: Handley, 1984:513; incorrect subsequent spelling of *Glossophaga caudifer* É. Geoffroy St.-Hilaire.

DISTRIBUTION: *Anoura caudifer* is in Colombia, Venezuela, Guyana, Surinam, French Guiana, Brazil, Ecuador, Peru, Bolivia, Paraguay, and northern Argentina. The distribution forms a large arc surrounding the central Amazon Basin.

MARGINAL LOCALITIES (Map 110): COLOMBIA: El Cesar, Pueblo Bello (Hershkovitz 1949c). VENEZUELA (Handley 1976): Yaracuy, Minas de Aroa; Distrito Federal, Los Venados; Bolívar, Km 125. SURINAM: Para, Jodensavanne (Husson 1962). FRENCH GUIANA: Saül (Brosset and Dubost 1968). BRAZIL: Amapá, Igarapé Novo (Taddei, Vizotto, and Sazima 1978); Pará, Rio Xingu, 52 km SSW of Altamira (USNM 549370). SURINAM: Sipaliwini,

Sipaliwini Airstrip (Williams and Genoways 1980a). GUYANA: Upper Takutu-Upper Essequibo, Marurawau-nawa Village (ROM 34527). VENEZUELA: Amazonas, Cerro Neblina Base Camp (Gardner 1988). COLOMBIA: Vaupés, Tahuapunta (Tamsitt and Valdivieso 1966b). PERU: Ucayali, Balta (Voss and Emmons 1996). BOLIVIA: Santa Cruz, 17 km S of Campamento Los Fierros (S. Anderson 1997). BRAZIL: Mato Grosso, Santa Anna de Chapada (O. Thomas 1904b); Distrito Federal, 40 km N of Brasília (Mares, Braun, and Gettinger 1989); Goiás, Município de Mambaí (Coimbra et al. 1982); Minas Gerais, Pirapora (C. O. C. Vieira 1955); Espírito Santo, Município de Linhares (Peracchi and Albuquerque 1993); Rio de Janeiro, Rio de Janeiro (type locality of *Glossophaga caudifer* É. Geoffroy St.-Hilaire); São Paulo, Juquiá (C. O. C. Vieira 1955); Santa Catarina, Florianópolis (Cherem et al. 2005); Rio Grande do Sul, Unidad Maquiné (Rui and Graciolli 2005); Paraná, Fazenda Monte Alegre (N. R. Reis, Peracchi, and Sekiama 2000). BOLIVIA: Santa Cruz, 72 km ESE of Monteagudo (S. Anderson 1997). ARGENTINA (Barquez, Mares, and Braun 1999): Jujuy, Arroyo Sauzalito; Salta, 6 km W of Piquirenda Viejo. BOLIVIA (S. Anderson 1997): Cochabamba, 12.5 km SW of Villa Tunari; La Paz, Irupana. PERU: Cusco, Cordillera Vilcabamba, west side (Koopman 1978); Junín, near Tarma (Sanborn 1941); Loreto, San Lorenzo (BM 24.3.1.47). ECUADOR: Pichincha, Gualea (Albuja 1983). COLOMBIA: Nariño, La Guarapería (Cadena, Anderson, and Rivas-Pava 1998); Cauca, Popayán (Dobson 1880b); Caldas, Quebrada Guayabal (Castaño et al. 2003); Antioquia, Villa Marina (Muñoz 1990).

SUBSPECIES: We are treating *A. caudifer* as monotypic (see Remarks).

NATURAL HISTORY: Handley (1976) stated that this species inhabits mountainous areas of Venezuela (most often between 500 and 1,500 m), where the majority of his specimens was taken in mistnets, though some were caught as they roosted among rocks. Most were found in humid forest, others came from drier regions. Specimens from Balta, Ucayali, Peru, were caught among tree roots in shallow erosion caves in the banks of forest streams. I. Sazima (1976) found evidence in southern Brazil that this species gleans large insects (beetles and moths) in addition to consuming nectar and pollen. Also in Brazil, *A. caudifer* pollinates *Bauhinia rufa* (see I. Sazima, Vogel, and Sazima 1989), *Vriesea morrenii* (see Vogel 1969), and *Passiflora mucronata* (see M. Sazima and Sazima 1987).

Albuja (1999) reported a female pregnant with a 15-mm–CR fetus taken on 30 August in Ecuador. Barquez and Olrog (1985) suggested that *A. caudifer* gives birth in September and October in northern Argentina. On the basis of pregnancy records from throughout the year, Wilson (1979) suggested the reproductive cycle was asynchronous.

Webb and Loomis (1977) recorded four species of mites and three species of batflies, Gettinger and Gribel (1989) recorded an additional species of spinturnicid mite. Guerrero (1985a, 1996) also recorded additional mites, and Guerrero (1997) listed 16 species in 7 genera of streblid batflies known from *A. caudifer*. Rui and Graciolli (2005) recovered two species of streblid batflies from one female from Rio Grande do Sul, Brazil. Ubelaker, Specian, and Duszynski (1977) reported two species of protozoan endoparasites. The karyotype is $2n = 30$, FN = 56 (R. J. Baker 1973).

REMARKS: Sanborn (1933) recognized *A. c. aequatoris* from western Ecuador and western Peru, and *A. c. caudifer* from elsewhere in South America. He characterized the former as being much darker in color than *A. c. caudifer*, and with a slightly shorter forearm and smaller skull. Cabrera (1958) followed Sanborn's arrangement. Tamsitt and Valdivieso (1966b) recommended that *A. caudifer* be considered monotypic because their analysis showed that the only significant difference between Andean and Amazonian populations of *A. caudifer* was in forearm length (Andean populations, 36.3–37.0 mm; Amazonian populations, 35.0–36.0 mm), a size relationship the reverse of that reported by Sanborn (1933). Albuja (1983) reported four additional specimens from western Ecuador including one from near the type locality of *A. c. aequatoris*: two females (forearms, 34.5 and 34.7 mm) and two males (forearms, 35.8 and 36.0 mm). These measurements are within the range of forearm measurements for Amazonian populations.

When C. O. C. Vieira (1955) reported *A. caudifer* from the Brazilian state of Bahia, he did not mention specific localities. N. R. Reis (1984) reported six specimens from primary forest in the vicinity of Manaus, Brazil. However, there are no other records from the central Amazon basin despite extensive collecting throughout the region. We have not included Manaus in the distribution of *A. caudifer* pending confirmation of the identification of N. R. Reis's (1984) record. Fornes (1972) recorded *A. geoffroyi* from the provincia Salta, Argentina. Barquez (1988) reidentified the specimens as *A. caudifer* and Barquez and Olrog (1985), Barquez, Mares, and Braun (1999), and M. M. Díaz et al. (2000) reported additional *A. caudifer* from other localities in northern Argentina.

Mantilla-Meluk and Baker (2006) recommended recognizing *A. caudifer aequatoris* as a full species distinguishable from *A. caudifer* on the basis of a well-furred interfemoral membrane. They stated that the type locality of *A. c. aequatoris* was in the eastern Andes of Ecuador; however, the type locality is on the western slope of the Andes. Although their determination is probably correct, we have not followed their recommendation because at this time it is difficult to adequately diagnose the taxon and the entire complex currently known as *A. caudifer* is in need of revision.

Cabrera (1958) and Tamsitt and Valdivieso (1966b) included the name *A. wiedii* (W. Peters 1869) in the synonymy of *A. caudifer*. Sanborn (1933), however, correctly associated *A. wiedii* with *A. geoffroyi* Gray, 1838b. Pine and Ruschi (1978) reviewed the allocation of these and other names used in *Anoura*. D. C. Carter and Dolan (1978) gave information and measurements for the types of *Glossophaga caudifer* É. Geoffroy St.-Hilaire, and *Lonchoglossa wiedi aequatoris* Lönnberg.

There has been confusion concerning the spelling of the name *A. caudifer*. Most authors since Husson (1962:139) have used the form *A. caudifer* because of its use by É. Geoffroy St.-Hilaire (1818a) in the original description. Handley (1984:518), however, said all authors (including É. Geoffroy St.-Hilaire) have erred when using the masculine form "*caudifer*" in combination with the generic names *Glossophaga*, *Lonchoglossa*, and *Anoura*, which are all feminine nouns. He claimed that the specific name in this example is an adjectival modifier, and as such it must agree in gender with the generic name. Nevertheless, Husson's (1982) spelling of the name is correct, and The International Code of Zoological Nomenclature is clear on this issue. The third edition (ICZN 1985) Art. 31(b)(1) and the fourth edition (ICZN 1999) require that where an author of a species group name did not indicate if the name was used as a noun or as an adjective, and where it may be regarded as either and the evidence of usage is not decisive, it is to be treated as a noun in apposition to the name of its genus and the name is not to be changed if combined with a generic name of a different gender. The examples given specifically use names ending in -fer and–ger.

Anoura cultrata Handley, 1960
Handley's Tailless Bat
SYNONYMS:

Anoura cultrata Handley, 1960:463; type locality "Tacarcuna Village, 3,200 ft., Río Pucro, Darién, Panama."

Anoura brevirostrum D. C. Carter, 1968:427; type locality "31 km S Tingo María, 850 m, Huánuco, Perú."

Anoura werckleae Starrett, 1969:l; type locality "6.8 mi. S restaurant 'La Georgina' along Interamerican Highway, 2500 m, Cerro de la Muerte massif, San José, Costa Rica."

DISTRIBUTION: *Anoura cultrata* is distributed from northern and western Venezuela through Colombia, Ecuador, and Peru into departamento La Paz, Bolivia. Elsewhere, the species occurs in Panama and Costa Rica.

MARGINAL LOCALITIES (Map 111; sequence north to south): VENEZUELA (Handley 1976): Miranda, Cueva

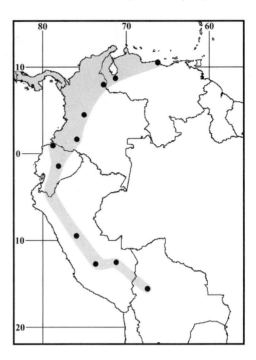

Map 111 Marginal localities for *Anoura cultrata* ●

Walter Dupouy; Mérida, La Carbonera. COLOMBIA: Norte de Santander, Gramales (D. C. Carter 1968); Tolima, El Boquerón (Nagorsen and Tamsitt 1981); Huila, Cueva del Indio (Lemke and Tamsitt 1980). ECUADOR: Esmeraldas, Luis Vargas Torres (Albuja 1989); Pastaza, Mera (Albuja 1999). PERU: Huánuco, 31 km S of Tingo María (type locality of *A. brevirostrum* D. C. Carter); Madre de Dios, Cerro de Pantiacolla (V. Pacheco et al. 1993); Ayacucho, Huanhuachayo (Nagorsen and Tamsitt 1981). BOLIVIA: La Paz, 47 km by road N of Caranavi at Serranía Bella Vista (S. Anderson, Koopman, and Creighton 1982).

SUBSPECIES: We treat *A. cultrata* as monotypic.

NATURAL HISTORY: This species has been taken at elevations ranging from about 50 m in very humid forest in Ecuador (Albuja 1989), at 200 m in Venezuela (roosting in a cave; Handley 1976), and to at least 2,600 m in the montane rain forests of Costa Rica and Peru (Nagorsen and Tamsitt 1981). Koopman (1978) commented that the range of *A. brevirostrum* (= *A. cultrata*) was peculiar in that it apparently did not extend eastward from the Andes, despite the fact that Peruvian specimens have been caught at low elevations.

Albuja (1989) found a female pregnant with a 12-mm-CR fetus in November. Webb and Loomis (1977) listed two species of mites and two species of streblid batflies, and Guerrero (1997) listed three species of streblid batflies as ectoparasites known from *A. cultrata*. Lemke and Tamsitt (1980) reported Colombian specimens taken from a railroad tunnel and a large limestone cave. Tamsitt and

Nagorsen (1982) provided a description with measurements and a summary of information on natural history in their *Mammalian Species* account. The karyotype is $2n = 30$, $FN = 54$ (R. J. Baker 1967; Nagorsen and Tamsitt 1981).

REMARKS: Nagorsen and Tamsitt (1981) found morphological variation to be continuous from Central America into Peru. Therefore, we follow them in treating *A. brevirostrum* and *A. werckleae* as synonyms of *A. cultrata* and do not recognize subspecies.

Anoura cultrata is easily distinguished from all known congeners on the basis of the extremely deep sulcus in the anterior face of the upper canines and by the unusually large first lower premolar. Whereas all species of *Anoura* have a sulcus on the anterior face of the upper canines, with the exception of *A. cultrata*, the sulcus is shallow, does not extend through the cingulum, and is often obliterated through wear. In *A. cultrata*, the anterolabial margin of the sulcus extends farther anteriorly than does the anteromedial margin of the canine.

Anoura fistulata
Long-lipped Bat

SYNONYM:

Anoura fistulata Muchhala, Mena, and Albuja, 2005:458; type locality "Condor Mirador (near the Destacamento Militar; 3°38′08″S, 78°23′22″W) of the Cordillera del Condor, 1,750 m, Zamora Chinchipe Province, Ecuador."

DISTRIBUTION: *Anoura fistulata* is known only from Ecuador.

MARGINAL LOCALITIES (Map 112; from Muchhala, Mena and Albuja 2005): ECUADOR: Pichincha, Bellavista; Pichincha, Yanayacu; Napo, El Salado; Morona-Santiago, Uuntsuants; Zamora-Chinchipe, Destacamento Militar Condor Mirador; Zamora-Chinchipe, Cuevas de Numbala.

SUBSPECIES: We consider *A. fistulata* to be monotypic.

NATURAL HISTORY: Muchhala, Mena, and Albuja (2005) commented on the unusually long tongue of *A. fistulata* that appears to be an adaptation for feeding on large flowers and flowers with deep corollas. They found pollen in the feces and on the fur representative of *Centropogon nigricans*, *Markea* sp., *Marcgravia* spp., and *Meriania* sp., all plants with large flowers. A fecal sample contained wing scales of Lepidoptera, along with other insect parts, and pollen from bromeliads and other plants. The intestinal contents of an additional bat contained insect parts as well as pollen from *Pitcairnia brogniartiana*, *Marcgravia coriaceae*, and *Markea* sp. The specimen from Numbala Cave was roosting with four other individuals. All other specimens were mistnetted in mature cloud forest at

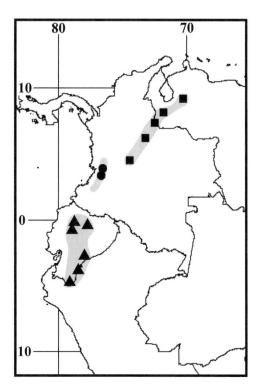

Map 112 Marginal localities for *Anoura cadenai* ●, *Anoura fistulata* ▲, and *Anoura luismanueli* ■

middle elevations on the eastern and western slopes of the Andes.

REMARKS: *Anoura fistulata* is intermediate in size between *A. caudifer* and *A. geoffroyi*. In addition to the unusually long tongue and conspicuously longer lower lip, *A. fistulata* can be most easily distinguished from *A. caudifer* by its longer skull (condylobasal length more than 22 mm) wider braincase (more than 9.1 mm), and greater breadth across canines (more than 4 mm); whereas, *A. caudifer* is smaller in all of these dimensions. *Anoura fistulata* can be distinguished from *A. geoffroyi* by its conspicuously longer lower lip, an obvious tail that protrudes slightly from the interfemoral membrane, and shorter forearm (less than 42 mm). *Anoura geoffroyi* has a comparatively shorter lower lip, may or may not have a visible tail, and has a longer forearm (exceeds 40 mm).

Anoura geoffroyi Gray, 1838
Geoffroy's Tailless Bat

SYNONYMS: See under Subspecies.

DISTRIBUTION: *Anoura geoffroyi* is in Colombia, Venezuela, the island of Trinidad, Guyana, Surinam, French Guiana, Ecuador, Peru, Bolivia, and Brazil. It also occurs from western Mexico southeastward into Panama.

MARGINAL LOCALITIES (Map 113): TRINIDAD AND TOBAGO: Trinidad, Blanchisseuse (C. H. Carter

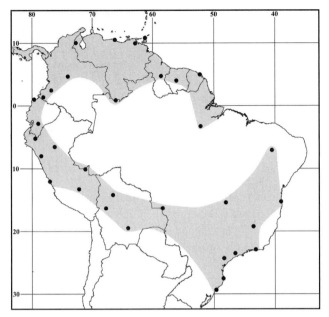

Map 113 Marginal localities for *Anoura geoffroyi* ●

et al. 1981). GUYANA: Potaro-Siparuni, Kurupukari (P. G. Smith and Kerry 1996) FRENCH GUIANA: Cayenne (Husson 1962). BRAZIL: Pará, Caverna do Tatajuba (USNM 549371). SURINAM: Saramacca, Anton van Aerde Cave (Husson 1962). VENEZUELA: Amazonas, Cerro Neblina Base Camp (Gardner 1988). COLOMBIA: Cundinamarca, Bogotá (Sanborn 1933). PERU: San Martín, Shapaja (Acha and Zapatel 1957); Ucayali, Balta (Voss and Emmons 1996). BOLIVIA: Beni, Espíritu (S. Anderson 1997); Santa Cruz, San Matías (Sanborn 1933). BRAZIL: Goiás, Gruta Morro (Bredt, Uieda, and Magalhães 1999); Ceará, km 19 on Route CE 96 (Mares et al. 1981); Bahia, Una (Faria, Soares-Santos, and Sampaio 2006); Minas Gerais, Serra do Cipó (I. Sazima 1976); Rio de Janeiro, Rio de Janeiro (type locality of *A. geoffroyi* Gray); São Paulo, São Paulo (Sanborn 1933); São Paulo, Finca Intervales (Fenton et al. 1999); Santa Catarina, Florianópolis (Cherem et al. 2005); Rio Grande do Sul, Tôrres (Fabián, Rui, and Oliveira 1999). BOLIVIA: Chuquisaca, 34 km SE of Padilla (S. Anderson 1997); La Paz, 1 km S of Chuspipata (S. Anderson, Koopman, and Creighton 1982). PERU (Tuttle 1970, except as noted): Cusco, Ollantaytambo; Lima, Lima; La Libertad, Machi (Ortiz de la Puente 1951); Piura, Huancabamba. ECUADOR: Azuay, Cuenca (Sanborn 1933); Esmeraldas, Esmeraldas (Albuja 1983). COLOMBIA: Nariño, La Guarapería (Cadena, Anderson, and Rivas-Pava 1998); Cauca, El Tambo (Tamsitt and Valdivieso 1966a). VENEZUELA: Zulia, Kasmera (Handley 1976); Miranda, Birongo (Handley 1976); Monagas, Caño Colorado (Linares and Rivas 2004).

SUBSPECIES: We tentatively recognize three subspecies of *A. geoffroyi*.

A. g. geoffroyi Gray, 1838
SYNONYMS:

Anoura geoffroyi Gray, 1838b:490; type locality "Brazil"; restricted to Rio de Janeiro by C. O. C. Vieira (1942: 324).

Anura wiedii W. Peters, 1869:398; type locality "Rio Janeiro," Brazil.

Lonchoglossa wiedii: Dobson, 1878:507; name combination.

Glossonycteris geoffroyi: Dobson, 1878:508; name combination.

Anura geoffroyi: O. Thomas, 1893c:335; name combination.

[*Anoura*] *geoffroyi geoffroyi*: H. E. Anthony, 1921a:5; first use of current name combination.

Anoura geofroyi C. O. C. Vieira, 1942:324; incorrect subsequent spelling of *Anoura geoffroyi* Gray.

Anoura geofroyii M. A. R. Mello and Schittini, 2005:208; incorrect subsequent spelling of *Anoura geoffroyi* Gray.

The nominate subspecies occurs at lower elevations in Venezuela, the island of Trinidad, the Guianas, and northeastern Brazil west and southward around the Amazon basin through Bolivia into northern Argentina, and southern and eastern Brazil (see Remarks).

A. g. lasiopyga W. Peters, 1868
SYNONYMS:

Glossonycteris lasiopyga W. Peters, 1868b:365; type locality "Mexico," restricted to the state of Veracruz, Mexico, by Arroyo-Cabrales and Gardner (2003:740).

Anoura geoffroyi lasiopyga: Sanborn, 1933:27; name combination.

This subspecies has been recorded in South America only from western Ecuador (Albuja 1983). Specimens from western Colombia probably are *A. g. lasiopyga* as well. Elsewhere, *A. geoffroyi* occurs from northwestern Mexico through Central America.

A. g. peruana (Tschudi, 1844)
SYNONYMS:

Rhinchonycteris peruana Tschudi, 1844b:71; *nomen nudum*.

[*Glossophaga* (]*Ch*[*oeronycteris*)]. *peruana* Tschudi, 1844b: 71; type locality "Hacienda der Cejaregion 5000′ ü. M. am Ostabhange der Binnencordillera," Junín, Peru (see Tschudi 1845:85).

Choeronycteris peruanus: Tschudi, 1844b:pl. 3, figs. 1 and 2; name combination.

Glossonycteris geoffroyi: Dobson, 1878:508; name combination.

Glossophaga apolinari J. A. Allen, 1916a:86; type locality "Boqueron de San Francisco," 2,730 m, near Bogotá, Cundinamarca, Colombia.

Anoura geoffroyi antricola H. E. Anthony, 1921a:5; type locality "Loja," Loja, Ecuador.

Anoura geoffroyi peruana: Sanborn, 1933:26; first use of current name combination.

This is an Andean subspecies found in Colombia, Ecuador, Peru, and Bolivia.

NATURAL HISTORY: Tuttle (1970) found *A. geoffroyi* roosting in a manmade tunnel. He also reported netting specimens near flowering shrubs that were "exceptionally attractive to hummingbirds during the day." M. Sazima and Sazima (1975) described the hovering behavior of *A. geoffroyi* at the flowers of *Lafoensia pacari* and observed that the bats remained only 0.5–1.0 second at each flower. Renner (1989) found *A. geoffroyi* visiting the flowers of *Macrocarpaea neblinae* (Gentianaceae) at 2,100 m on Cerro Neblina in southern Venezuela, and I. Sazima (1980) observed the species at flowers of *Bauhinia rufa*. M. Sazima, Buzato, and Sazima (1999) illustrated this bat hovering at flowers in their discussion of bat-pollinated flowers in rainforest habitat in southeastern Brazil, and described the foraging mode for glossophagines as traplining. Zortéa (2003) observed *A. geoffroyi* consuming arthropods, pollen and nectar, and the fruit of at least three species of plants during the rainy season in central Brazil. L. G. Herrera and Martínez del Río (1998) found that *A. geoffroyi* was more efficient than either *Artibeus jamaicensis* or *Sturnira lilium* in emptying pollen grains of *Pseudobombax ellipticum* (Bombacaceae), based on intact versus empty pollen grains found in fecal samples. However, in a comparison that also included *Leptonycteris curasoae*, *A. geoffroyi* was less efficient at emptying pollen grains of columnar cacti (Cactaceae) than was *L. curasoae*, but remained more efficient than the two frugivores, *A. jamaicensis* and *S. lilium*. Handley (1976) reported this species taken in humid forests and in caves in Venezuela, where. *A. geoffroyi* was found at elevations up to 2,550 m, although they were most commonly caught at lower elevations, usually below 1,500 m. Large highland caves are used in Peru (Graham 1988) and the species was collected in a cave south of Altamira in the lower Amazon drainage basin in Pará, Brazil. Wilson (1979) suggest partial sexual segregation and one breeding season a year, based primarily on the data of Goodwin and Greenhall (1961) from the island of Trinidad. This is corroborated by the information of Baumgarten and Vieira (1994) and Zortéa (2003) for central Brazil. I. Sazima (1976) found nectar, pollen, and the remains of large insects (beetles and moths) in the stomach contents of Brazilian specimens. Albuja (1983) recorded fruit pulp in stomachs of Ecuadorian specimens. Machado-Allison (1965) reported two species of spinturnicid mites from *A. geoffroyi* in Venezuela. In their summary of ectoparasites known from this bat, Webb and Loomis (1977) listed eight additional species of mites representing different families as well as ten species of streblid batflies, a species of tick, and a nycteribiid batfly. Guerrero (1985a) listed five additional species of mites and eight additional species of batflies. Guerrero's (1997) list of 8 genera and 18 species of streblid batflies known from *A. geoffroyi* was augmented by another species reported by Graciolli and Rui (2001). The karyotype is $2n = 30$, FN $= 56$ (R. J. Baker 1967).

REMARKS: Wied-Neuwied (1826) thought the material he described represented the same species É. Geoffroy St.-Hilaire (1818a) named *Glossophaga ecaudata*. Gray (1838b) evidently was of the same opinion because he included references to both É. Geoffroy St.-Hilaire and Wied-Neuwied in his synonymy of *Anoura geoffroyi*. W. Peters (1869:398) recognized that Wied-Neuwied's bats, which included specimens from Rio de Janeiro, were not *Glossophaga ecaudata* É. Geoffroy St.-Hilaire and proposed the name *Anura [sic] wiedii* for them. Although Gray (1838b:490) simply wrote "Inhabits Brazil" as the habitat, Sanborn (1933:24) stated that the type locality "is probably Rio de Janeiro." C. O. C. Vieira's (1942:324) statement "Rio de Janeiro," is considered to have restricted the type locality to that place. D. C. Carter and Dolan (1978) gave information and measurements on the types of *Anoura geoffroyi* Gray and *Glossonycteris lasiopyga* W. Peters.

We lack confirmed records of this species from Paraguay, and northern Argentina. Fornes (1972) reported *A. geoffroyi* from northern Argentina, but Barquez (1988) examined the specimens and reidentified them as *A. caudifer*. C. O. C. Vieira (1955) recorded *A. geoffroyi* from the state of Santa Catarina, Brazil, but did not mention specific localities. Also we doubt that the nominate subspecies represents the populations inhabiting Venezuela, Surinam, French Guiana, northern Brazil (Roraima and Pará), and probably northern Colombia. The *A. geoffroyi* from this region may represent one or more undescribed subspecies.

Cabrera (1958:75) listed *Rhinchonycteris peruana* Tschudi, in his synonymy of *A. g. peruana*; however, the context in which *Rhinchonycteris* was used by Tschudi suggests that its use was a *lapsus*. The descriptive similarity in the meaning of *Rhinchonycteris* and *Choeronycteris* is obvious. *Choeronycteris* is used elsewhere in Tschudi's accounts where *Rhinchonycteris* would seem equally appropriate; therefore, we agree with O. Thomas's (1928a:123) opinion that *Rhinchonycteris* Tschudi, 1844b:71, is a *nomen nudum*.

Rengger (1830) described six specimens from northern Paraguay as *Glossophaga villosa*. The measurements included in his description most nearly apply to either *A. geoffroyi* (if this allocation were correct, *G. villosa* is a senior synonym) or *A. caudifer* (to which *G. villosa* would be a junior synonym). The tail was described as absent, the interfemoral membrane as covered with short hairs and with a fringe of hairs, and the dental formula as 2/2, 1/1, 3/3, 3/3. If the German "Linie" is taken as 1/12 inch, as Rengger apparently intended, the total length of each specimen was about 72.0 mm and the head length about 25.4 mm. Rengger, however, also described the outer upper incisors as being shorter and thinner than the two inner incisors, a condition found normally only in *Glossophaga* and in the lonchophylline genera *Lonchophylla*, *Lionycteris*, and *Platalina*. Rengger may have been describing a sample of mixed species, or he may have had a series of large *Glossophaga* of which some had supernumerary upper premolars (see Phillips 1971). Bertoni (1939:13) tentatively used the name combination *Lonchoglossa villosa* Rengger for a bat from Concepción, Paraguay. Podtiaguin (1944:32) listed *Lonchoglosa*? [*sic*] *villosa* in the synonymy of *Lonchoglossa ecaudata* (Wied-Neuwied). Nevertheless, because Rengger's material no longer exists and because his description cannot be applied with certainty to any known species of bat, we consider *Glossophaga villosa* to be a *nomen dubium* and recommend its suppression.

Anoura latidens Handley, 1984
Broad-toothed Tailless Bat

SYNONYM:

Anoura latidens Handley, 1984:513; type locality "Pico Avila, 2150 m, 5 km NNE of Caracas (= 'Hotel Humbolt [*sic*], 9.4 km N Caracas'), Distrito Federal, Venezuela."

DISTRIBUTION: *Anoura latidens* is widely distributed in Venezuela and also is known from Guyana and Colombia on the basis of a single specimen from each country, and five from Peru.

MARGINAL LOCALITIES (Map 114; from Handley 1984, except as noted): VENEZUELA: Sucre, Manacal. GUYANA: Cuyuni-Mazaruni, Kuwaima Falls (Lim and Engstrom 2001a). VENEZUELA: Bolívar, El Manaco; Amazonas, San Juan. PERU: Junín, 2 km NW of San Ramón; Pasco, San Alberto (S. Solari, Pacheco, and Vivar 1999). COLOMBIA: Cundinamarca, San Juan de Río Seco. VENEZUELA: Zulia, Novito; Falcón, near La Pastora, 16 km ENE of Mirimire.

SUBSPECIES: *Anoura latidens* is monotypic.

NATURAL HISTORY: Handley (1976) reported this species from forested areas in Venezuela, mostly near streams or other places near water. Handley (1984) added that, although *A. latidens* was taken in a variety of lowlands to montane habitats, it was most commonly found at

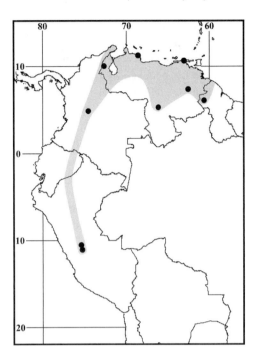

Map 114 Marginal localities for *Anoura latidens* ●

elevations between 1,000 and 1,500 m. The karyotype is unknown.

REMARKS: *Anoura latidens* was cited by Handley (1976:22) as "*Anoura* sp. A."

Anoura luismanueli Molinari, 1994
Molinari's Tailless Bat

SYNONYM:

Anoura luismanueli Molinari, 1994:76; type locality "Cueva del Salado, 4 km E Bailadores, Estado Mérida, Venezuela, elevation 2000 m."

DISTRIBUTION: *Anoura luismanueli* is known only from the Cordillera de Mérida of Venezuela in the states of Mérida, Táchira, and Trujillo. The species is expected from the Sierra de Perijá and other areas in the Andean Cordillera Oriental of Colombia.

MARGINAL LOCALITIES (Map 112; from Molinari 1994): Trujillo, Macizo de Guaramacal, 6.5 km ENE of Boconó; Mérida, Cueva del Salado (type locality of *Anoura luismanueli* Molinari); Táchira, Betania.

SUBSPECIES: *Anoura luismanueli* is monotypic.

NATURAL HISTORY: An inhabitant of cloud forest (seven of nine localities), dry evergreen forest (one locality) and deciduous forest (one locality), only 2 of the 48 known specimens were collected below 2,000 m (Molinari 1994). Known diurnal roosts are caves. Fecal samples from nine specimens contained pollen, but none contained evidence of fruit pulp, seeds, or insects. The karyotype and reproductive biology are unknown.

REMARKS: With a forearm length of 33.6–36.9 mm, greatest length of skull of 20.4–21.6 mm, and a mass (weight) of 7.5–10.0 g, *A. luismanueli* is substantially smaller than all congeners except *A. caudifer* and is smaller than *A. caudifer* in length of rostrum (palatal length 9.6–11.0 mm, versus 12.2–13.3 mm; length of maxillary toothrow 7.0–7.9 mm, versus 8.2–8.9 mm). It also is hairier than *A. caudifer*, especially along the legs, and trailing edge and ventral surface of the interfemoral membrane (these are nearly naked in *A. caudifer*). *Anoura luismanueli* has been captured together with each of the other five known species of *Anoura*. Alberico et al. (2000:52) listed the species for departamento Cundinamarca, Colombia, but did not provide the apparent authority ("Pérez ?") in the literature cited.

Genus *Choeroniscus* O. Thomas, 1928

The genus *Choeroniscus* is among the most highly derived Neotropical nectarivorous bats. In addition to an elongated rostrum and small, delicate dentition, *Choeroniscus* possess a long, extensible tongue having a number of anatomical modifications for hyperprotrusion. Relatively few specimens of this genus have been collected to date, and the systematics, distributions, and ecology of the South American representatives are poorly known.

Of the three known species in South America, only *C. godmani* has been collected elsewhere (see Hall 1981, for Central American and Mexican distribution). *Choeroniscus* spp. are small to medium-sized glossophagines (forearm 31–42 mm) with dark brown to blackish pelage. The basal color of the hair varies from pale whitish to orange brown and the overall ventral coloration is only slightly paler than that of the dorsum. The muzzle is elongated (rostrum longer than braincase in *C. periosus*, but shorter in the other species) and bears prominent short vibrissae and a small noseleaf. The ears are small and rounded. The interfemoral membrane is long, and the tail is short, about 1/3 as long as the interfemoral membrane and protrudes from its dorsal surface. The zygomatic arches are incomplete and the braincase appears conspicuously large because of the slender rostrum. The pterygoid hamulae are inflated and usually in contact with auditory bullae (nearly in contact in *C. periosus*). *Choeroniscus godmani* has an acute notch (sometimes cleftlike) in the posterior margin of the palate, whereas the mesopterygoid notch is broad and rounded in the other two species of *Choeroniscus*. The upper incisors are small, all approximately the same size, and separated as pairs on each side of the fused premaxillae. Upper molars have an enlarged parastyle, lack a hypocone and paracone, and have lost the W-shaped ectoloph pattern by a lateral shift of the paracone; thereby, opening the angle formed by the paracrista and the precentrocrista. All cusps

are present on lower molars, although reduced in size (see Phillips 1971, for a more comprehensive description of the dentition). The dental formula is 2/0, 1/1, 2/3, 3/3 × 2 = 30.

SYNONYMS:

Choeronycteris: W. Peters, 1868b:366; not *Choeronycteris* Tschudi, 1844b.

Chœronycteris: J. A. Allen and Chapman, 1893:207; not *Choeronycteris* Tschudi.

Chœronycteris: O. Thomas, 1903a:288; not *Choeronycteris* Tschudi.

Chyronycteris Festa, 1906:1; incorrect subsequent spelling of, but not *Choeronycteris* Tschudi.

Chœronycteris O. Thomas, 1912f:403; not *Choeronycteris* Tschudi.

Choeroniscus O. Thomas, 1928a:122; type species *Chœroniscus minor*: O. Thomas, 1928a (= *Choeronycteris minor* W. Peters, 1868b), by original designation.

Cheroniscus Albuja, 1989:109; incorrect subsequent spelling of *Choeroniscus* O. Thomas.

KEY TO SOUTH AMERICAN SPECIES OF *CHOERONISCUS*.

1. Rostrum longer than braincase; forearm longer than 40 mm; greatest length of skull more than 30 mm; maxillary toothrow more than 10 mm. . . . *Choeroniscus periosus*
1′. Rostrum shorter than braincase; forearm 38 mm or less; greatest length of skull less than 25 mm; maxillary toothrow less than 10 mm. 2
2. Posterolateral palate with a conspicuous notch; cranium markedly elevated from the basicranial plane; greatest length of skull less than 21 mm; maxillary toothrow 7.5 mm or less . . . . *Choeroniscus godmani*
2′. Posterolateral palate unnotched; cranium not markedly elevated from the basicranial plane; greatest length of skull usually more than 21 mm; maxillary toothrow 7.5 mm or more . . . . *Choeroniscus minor*

Choeroniscus godmani (O. Thomas, 1903)

Godman's Long-nosed Bat

SYNONYMS:

Chœronycteris godmani O. Thomas, 1903a:288; type locality "Guatemala."

Chœroniscus godmani: O. Thomas, 1928a:122; first use of current name combination.

DISTRIBUTION: *Choeroniscus godmani* is recorded from Colombia, Venezuela, Guyana, and Surinam. The species also occurs in Central America and western Mexico.

MARGINAL LOCALITIES (Map 115): VENEZUELA (Handley 1976): Falcón, Boca de Yaracuy; Distrito Federal, Caracas (Santa Monica). GUYANA: Potaro-Siparuni, Paramakatoi (Shapley et al. 2005). SURINAM: Sipaliwini, Sipaliwini (Williams and Genoways 1980a). VENEZUELA:

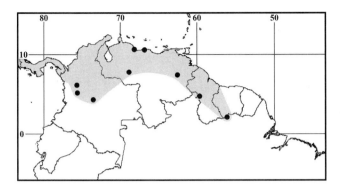

Map 115 Marginal localities for *Choeroniscus godmani* ●

Bolívar, El Manaco (Handley 1976); Apure, Hato El Frío (Ochoa and Ibáñez 1985). COLOMBIA: Meta, Restrepo (Handley 1966a); Caldas, Finca Los Naranjos (Castaño et al. 2003); Antioquia, El Porvenir (Muñoz 1990).

SUBSPECIES: We consider *C. godmani* to be monotypic.

NATURAL HISTORY: Venezuelan *C. godmani* were netted near streams and other mesic habitats, mostly in lowland forests (Handley 1976). Muñoz (1990) reported three from humid temperate-zone forest at 1,250 m in Colombia. Arends, Bonaccorso, and Genoud (1995) reported specimens taken in the shallow shafts of abandoned mines in northern Venezuela. Wilson's (1979) limited data suggested that *C. godmani* wean their young at the beginning of the rainy season. Guerrero (1985a) listed a macronyssid mite and later (1997) listed one species of streblid batfly known from *C. godmani*. The karyotype is $2n = 19/20$, FN = 32 or 36 (R. J. Baker 1967; J. L. Patton and Gardner 1971; R. J. Baker et al. 1982).

REMARKS: D. C. Carter and Dolan (1978) gave information and measurements on the holotype of *Choeronycteris godmani* O. Thomas.

Choeroniscus minor (W. Peters, 1868)
Little Long-nosed Bat
SYNONYMS:

Choeronycteris minor W. Peters, 1868b:366; type locality "Surinam."

Choeronycteris intermedia J. A. Allen and Chapman, 1893: 207; type locality "Princestown," Trinidad, Trinidad and Tobago.

Choeronycteris inca O. Thomas, 1912f:403; type locality "Yahuarmayo" (= Río Yahuarmayo), 1,200 ft., Puno, Peru.

[*Choeroniscus*] *intermedia*: O. Thomas, 1928a:122; name combination.

[*Choeroniscus*] *inca*: O. Thomas, 1928a:122; name combination.

Choeroniscus minor: O. Thomas, 1928a:123; first use of current name combination.

Choeroniscus intermedius: Cabrera, 1958:73; name combination; with agreement in gender concordance.

DISTRIBUTION: *Choeroniscus minor* is endemic to South America where it is known from the island of Trinidad, eastern Venezuela, Guyana, Surinam, and French Guiana as far south as the southern bank of the Rio Amazonas in Brazil, and west into eastern Colombia, eastern Peru, and northern Bolivia. Another population has been recoreded in western Ecuador and extreme northwestern Peru (Tumbes).

MARGINAL LOCALITIES (Map 116): TRINIDAD AND TOBAGO: Trinidad, Princes Town (type locality of *Choeronycteris intermedia* J. A. Allen and Chapman). VENEZUELA: Monagas, Isla Tigre (Linares and Rivas 2004); Bolívar, Río Supamo (Handley 1976). GUYANA: Cuyuni-Mazaruni, Kartabo (Sanborn 1943). SURINAM: Para, Zanderij (Genoways, Williams, and Groen 1981). FRENCH GUIANA: Cayenne (Brosset and Charles-Dominique 1991). BRAZIL: Pará, Belém (Handley 1966a); Bahia, Ilhéus (Faria, Soares-Santos, and Sampaio 2006); Espírito Santo, Município de Linhares (Peracchi and Albuquerque 1993); Minas Gerais, Caratinga Biological Station (Aguiar, Zortéa, and Taddei 1996); Pará, Rio Xingu, 52 km SSW of Altamira (USNM 549373); Pará, Rio Cupari (Sanborn 1954); Amazonas, Dimona Reserve (Sampaio et al. 2003). PERU: Loreto, Centro de Investigaciones Jenaro Herrera (Ascorra, Gorchov, and Cornejo 1994). BOLIVIA: Pando, Río Nareuda (S. Anderson 1997). BRAZIL: Rondônia, Ouro Preto D'Oeste (Marques 1989).

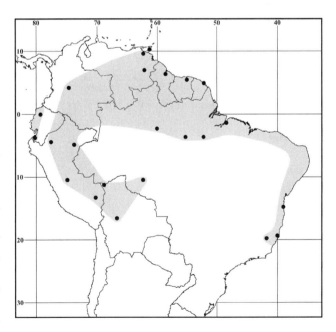

Map 116 Marginal localities for *Choeroniscus minor* ●

BOLIVIA: Cochabamba, Seque Rancho (Yensen, Tarifa, and Anderson 1994). PERU: Puno, Río Yahuarmayo (type locality of *Choeronycteris inca* O. Thomas); Pasco, San Juan (Tuttle 1970); Loreto, Puerto Meléndez (Handley 1966a). COLOMBIA: Tolima, Melgar (Valdivieso 1964). ECUADOR: Pichincha, Plan Piloto (Albuja 1983). PERU: Tumbes, Quebrada Faical (Graham and Barkley 1984).

SUBSPECIES: We are treating *C. minor* as monotypic.

NATURAL HISTORY: Tuttle (1970) netted two specimens in secondary growth around gardens at the edge of primary evergreen forest and another over a trail in mature evergreen forest in Peru. Handley (1976) reported this species in forested regions near streams and other wet areas in Venezuela. A female from the Rio Xingu, Pará, Brazil, was pregnant with a 22-mm-CR fetus when taken on 19 August. Ubelaker, Specian, and Duszynski (1977) listed five species of protozoan endoparasites known from *C. minor*. Webb and Loomis (1977) reported a tick, Guerrero (1985a) added another tick, and Guerrero (1996) reported trombiculid, labidocarpid, and myobiid mites as ectoparasites. The karyotype is $2n = 20$, FN = 36 (see summary in R. J. Baker 1979).

REMARKS: Koopman (1978) treated *C. inca* as a synonym of *C. minor* in his review of the relationships in the *C. intermedius-minor-inca* problem. He did not find consistent differences in the relative sizes of premolars and molars, characteristics that O. Thomas (1912t) used to distinguish *C. inca* from *C. minor*. Koopman (1978) differentiated *C. minor* from *C. intermedius* on the basis of rostral length (as reflected in measurements of the maxillary toothrow), but he did not give the ranges of these measurements for the two species. However, the evidence presented by Simmons and Voss (1998) supports our suspicion that the *C. inca-intermedius-minor* represents a single species for which the name *C. minor* has priority. We treat *C. minor* as a monotypic species, though unquestionably, this species and the entire genus warrants revision. D. C. Carter and Dolan (1978) gave information and measurements on the holotypes of *Choeronycteris inca* O. Thomas and *Choeronycteris minor* W. Peters.

Choeroniscus periosus Handley, 1966
Handley's Long-nosed Bat
SYNONYM:
Choeroniscus periosus Handley, 1966a:84; type locality "Río Raposo, near sea level, 27 km south of Buenaventura," Valle del Cauca, Colombia.

DISTRIBUTION: *Choeroniscus periosus* is known only from western Colombia and northwestern Ecuador.

MARGINAL LOCALITIES (Map 117): COLOMBIA: Chocó, Río San Juan, 2 km above Playa de Oro (AMNH 217038); Valle del Cauca, Río Raposo (type locality

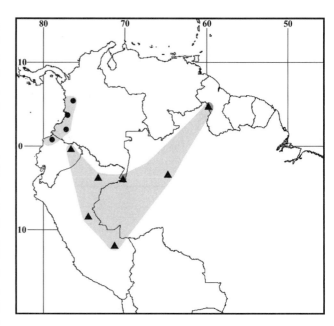

Map 117 Marginal localities for *Choeroniscus periosus* ● and *Glossophaga commissarisi* ▲

of *Choeroniscus periosus* Handley); Nariño, Finca San Marino (Alberico and Negret 1992). ECUADOR: Esmeraldas, San Miguel (Albuja 1989).

SUBSPECIES: *Choeroniscus periosus* is monotypic.

NATURAL HISTORY: The species is known from only a few specimens from two localities in western Colombia and one specimen from northwestern Ecuador. The Ecuadorian specimen was lactating when netted in September 1984 over a small stream bordered by *Heliconia* sp. in secondary forest (Albuja 1989). The karyotype is unknown.

Genus *Glossophaga* É. Geoffroy St.-Hilaire, 1818

We recognize five species of *Glossophaga*, three of which are in South America. Of these, *Glossophaga soricina*, is among the most common and widely distributed Neotropical bats (Webster 1983, 1993). The other two species have restricted South American distributions.

Species of *Glossophaga* are small (forearm 31.0–42.0 mm), brown to reddish-brown bats with a long interfemoral membrane. The tail is short, about 1/3 to 1/2 as long as the interfemoral membrane. The rostrum is shorter than the braincase, except in *G. longirostris* where they are about equal in length. The zygomatic arches are complete. Outer upper incisors are smaller and more pointed than the inner. Upper and lower molars lack a hypocone(-id); however, other cusps and lophs are relatively well developed (see Phillips 1971, for a more comprehensive description of the dentition). The dental formula is 2/2, 1/1, 2/3, 3/3 × 2 = 34.

SYNONYMS:

Vespertilio: Pallas, 1766:48; not *Vespertilio* Linnaeus, 1758.

Phyllostoma: É. Geoffroy St.-Hilaire, 1810b:179; not *Phyllostoma* G. Cuvier, 1'800.

Glossophaga É. Geoffroy St.-Hilaire, 1818a:418; type species *Vespertilio soricinus* Pallas, 1766, by monotypy.

Phyllostomus: Olfers, 1818:224; part; not *Phyllostomus* Lacépède, 1799b.

Phyllophora Gray, 1838b:489; type species *Phyllophora amplexicaudata*: Gray, 1838b (= *Glossophaga amplexicaudata* Spix, 1823), by monotypy; preoccupied by *Phyllophora* Thunberg, 1812 (Orthoptera).

Monophyllus: Gray, 1844:18; not *Monophyllus* Leach, 1821b.

Nicon Gray, 1847a:15; type species *Nicon caudifer*: Gray (= *Monophyllus leachii* Gray, 1844), by monotypy; not *Glossophaga caudifer* É. Geoffroy St.-Hilaire, 1818a.

Glassophoga Petit, 1997:221; incorrect subsequent spelling of *Glossophaga* É. Geoffroy St.-Hilaire.

Glossiphaga Shapley, Wilson, Warren, and Barnett, 2005: 382, 383; incorrect subsequent spelling of *Glossophaga* É. Geoffroy St.-Hilaire.

KEY TO SOUTH AMERICAN SPECIES OF *GLOSSOPHAGA*:

1. Tip of I1 protrudes well beyond the tip of I2; lower incisors crowded, usually in contact with each other and canines . . . . *Glossophaga soricina*
1'. Tip of I1 even with tip of I2; lower incisors not crowded 2
2. Rostrum approximately equal to braincase in length; lower incisors well developed, usually spaced evenly, and not separated medially by a conspicuous gap . . . . *Glossophaga longirostris*
2'. Rostrum shorter than braincase; lower incisors small and separated into two pairs by a conspicuous gap . . . . *Glossophaga commissarisi*

Glossophaga commissarisi Gardner, 1962

Commissaris's Long-tongued Bat

SYNONYMS:

Glossophaga commissarisi Gardner, 1962:1; type locality "10 km. S.E. Tonalá, Chiapas, Mexico."

Glossiphaga commissarissi Shapley, Wilson, Warren, and Barnett, 2005:382, 383; incorrect subsequent spelling of *Glossophaga commissarisi* Gardner.

DISTRIBUTION: *Glossophaga commissarisi* is recorded from a few localities in the upper Amazon basin of Colombia, Brazil, and Peru. The species also is known from Mexico and Central America.

MARGINAL LOCALITIES (Map 117): GUYANA: Potaro-Siparuni, Kato (USNM 565513). BRAZIL: Amazonas, Lago de Tefé (USNM 562720). PERU: Madre de Dios, Pakitza (V. Pacheco et al. 1993); Ucayali, 11 km SE of Pucallpa (Graham and Barkley 1984). ECUADOR: Sucumbios, Limoncocha (Albuja 1983). PERU: Loreto, Iquitos (Graham and Barkley 1984). COLOMBIA: Amazonas, Isla Santa Sofia (Webster and Jones 1983).

SUBSPECIES: We recognize two subspecies; only *G. c. commissarisi* is known from South America.

NATURAL HISTORY: Gardner (1962) collected this species in tropical lowland habitats in western and southern Mexico that included savanna, arid thorn forest, tropical evergreen and deciduous forest, and tropical rain forest. Specimens were caught most commonly in mist-nets set across trails, roads, open water, or in banana groves. Interestingly, South American *G. commissarisi* have not been collected in day roosts, unlike specimens of sympatric *G. soricina*. Webster and Jones (1983) reported that their South American specimens were netted in "an abandoned grove of fruit trees" in southern Colombia. Webster and Jones (1983) and Graham and Barkley (1984) both remarked that South American representatives were larger than comparable Central American material. Webster and Jones (1983) explicitly chose not to describe the South American specimens as representing a new subspecies. Graham and Barkley (1984) reported a female taken in August 1964 that contained a 12-mm-CR fetus. Two species of streblid batflies (Guerrero 1997), spinturnicid mites (Guerrero 1996), and three species of trombiculid mites (Webb and Loomis 1977) are known from *G. commissarisi*. Webster and Jones (1993) provided a description with measurements and summarized available information on natural history in their *Mammalian Species* account. The karyotype is $2n = 32$, FN = 60 (R. J. Baker 1967).

Glossophaga longirostris Miller, 1898

Greater Long-tongued Bat

SYNONYMS: See under Subspecies.

DISTRIBUTION: *Glossophaga longirostris* occurs in Colombia, Venezuela, Trinidad and Tobago, the Netherlands Antilles, Guyana, Brazil, and Ecuador. It is also in the Lesser Antilles as far north as the island of St. Vincent.

MARGINAL LOCALITIES (Map 118): NETHERLANDS ANTILLES: Aruba, Ex Cave (USNM 345005). VENEZUELA: Falcón, Boca de Yaracuy (Handley 1976); Sucre, Caño Brea (Linares and Rivas 2004). TRINIDAD AND TOBAGO: Tobago, Charlotteville (USNM 540666). VENEZUELA: Delta Amacuro, Guayo (Linares and Rivas 2004). GUYANA: Potaro-Siparuni, Paramakatoi (Shapley et al. 2005); Upper Takutu-Upper Essequibo, Dadanawa, Bush Island (AMNH 182724). BRAZIL: Roraima, Lucetania Ranch (Webster and Handley 1986; not mapped). VENEZUELA: Amazonas, San Juan (Handley 1976). COLOMBIA: Cundinamarca, Girardot (Tamsitt and Valdivieso

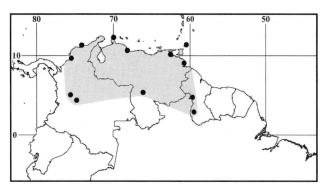

Map 118 Marginal localities for *Glossophaga longirostris* ●

1963a); Caldas, Cenicafé Reserva Natural Planalto (Castaño et al. 2003); Sucre, Tulu Viejo (USNM 431787); Magdalena, Bonda (Miller 1913c).

SUBSPECIES: Webster and Handley (1986) and Webster, Handley, and Soriano (1998) recognized six subspecies of *G. longirostris*, and Soriano, Fariñas, and Naranjo (2000) added a seventh; all are found in South America and adjacent islands.

G. l. campestris Webster and Handley, 1986
SYNONYM:
Glossophaga longirostris campestris Webster and Handley, 1986:8; type locality "Hato San José, 20 km. W. Paragua (=146 km. S, 7 km. E Ciudad Bolívar), Bolívar, Venezuela, 300 m."

This subspecies occurs from central Venezuela into southern Guyana and adjacent Brazil.

G. l. elongata Miller, 1900
SYNONYMS:
Glossophaga elongata Miller, 1900a:124; type locality "Willemstad," Curaçao.
G[lossophaga]. l[ongirostris]. elongata: Koopman, 1958:437; first use of current name combination.

This subspecies is found on all three islands that make up the Netherlands Antilles.

G. l. longirostris Miller, 1898
SYNONYM:
Glossophaga longirostris Miller, 1898:330; type locality "Santa Marta Mountains (near Santa Marta)," Magdalena, Colombia.

The nominate form is in northern Colombia and northwestern Venezuela.

G. l. major Goodwin, 1958
SYNONYMS:
Glossophaga major Goodwin, 1958b:5; type locality "Ariapita Avenue, Woodbrook, Port of Spain, Trinidad."

G[lossophaga]. [longirostris] major: Koopman, 1958:438; first use of current name combination.

This subspecies is on the island of Trinidad and from the northcentral coast of Venezuela south into the llanos of Venezuela and Colombia.

G. l. maricelae Soriano, Fariñas, and Naranjo, 2000
SYNONYM:
Glossophaga longirostris maricelae Soriano, Fariñas, and Naranjo, 2000:373; type locality "Laguna de Caparí, 3 km SE San Juan de Lagunillas, Mérida state, Venezuela, 900 m."

This subspecies is endemic to the "Lagunillas pocket" located in the Chama and Nuestra Señora river basins between Estanques and El Morro in the state of Mérida, Venezuela.

G. l. reclusa Webster and Handley, 1986
SYNONYM:
Glossophaga longirostris reclusa Webster and Handley, 1986:14:14; type locality "4 km. E Villavieja, Huila, Colombia, 1400 ft."

This subspecies is in the upper Magdalena valley of Colombia where it occurs from Cundinamarca south into Huila.

G. l. rostrata Miller, 1913
SYNONYMS:
Glossophaga rostrata Miller, 1913a:32; type locality "Westerhall Estate," Grenada.
Glossophaga longirostris rostrata: Miller, 1913c:423; first use of current name combination.

This subspecies is on Tobago and the Lesser Antilles (except Barbados) north to St. Vincent (Hall 1981; Webster and Handley 1986).

NATURAL HISTORY: Handley (1976) reported this species from the arid lowlands of northern Venezuela, usually near streams or other sources of water where they were netted or found roosting in houses, caverns, and crevices. Webster and Handley (1986) reported *G. longirostris* from savanna, arid thorn forest, and deciduous and evergreen forest habitats. They noted two periods of pregnancy (December to April and June to October) and found lactating females in every month except February. Soriano, Fariñas, and Naranjo (2000) gave an elevational range of 500–2,000 m for their subspecies *G. l. maricelae* and said it was the main pollinator and seed disperser of the local columnar cacti. Soriano, Sosa, and Rossell (1991), based on an analysis of fecal samples, found that consumption of fruits of Cactaceae and Moraceae peaked during the wet season (March and April; September and October) in semiarid, shrub habitat in the state of Mérida, Venezuela. In contrast, they found that percentages of Cactaceae pollen reached their highest values during the dry season. Petit (1997) stated that *G. longirostris* on Curaçao depend on columnar cacti flowers and fruit for survival and reproduction. She found that pregnancy (May-July) and lactation (July-September) coincided with the flowering phenology of two of the three species of columnar cacti on Curaçao. Webster, Handley, and Soriano (1998) referred to the reproductive pattern as uniparous bimodal polyestry. Ectoparasites include labidocarpid, spinturnicid, and trombiculid mites, argasid ticks, and streblid batflies (Brennan and Reed 1975; Herrin and Tipton 1975; Reed and Brennan 1975; Wenzel 1976; Webb and Loomis 1977; Guerrero 1985a). Guerrero (1997) listed 7 genera and 11 species of streblid batflies reported from *G. longirostris*. The karyotype is $2n = 32$, FN = 60 (R. J. Baker 1979).

REMARKS: Cabrera (1958) treated *G. apolinari* J. A. Allen 1916a, as a subspecies of *G. longirostris*. However, H. E. Anthony (1921a) had examined Allen's holotype of *G. apolinari* and identified it as *Anoura geoffroyi*, as affirmed by Sanborn (1933) and Webster and Handley (1986). Albuja (1983) reported *G. longirostris* from the Palenque Biological Station in western Ecuador. However, he does not mention the taxon in his 1999 revision. Webster (1993) referred these specimens to *G. soricina valens* on the basis of Albuja's (1983) measurements and photographs of the skull.

The locality record for the species in Roraima, Brazil (Webster 1983:159), is in savanna habitat at the base of the Kanaku Mountains of Guyana. *Glossophaga l. reclusa*, in the upper Magdalena valley, appears to be geographically isolated from other members of the species.

Glossophaga soricina (Pallas, 1766)
Pallas's Long-tongued Bat

SYNONYMS: See under Subspecies.

DISTRIBUTION: *Glossophaga soricina* occurs in Colombia, Venezuela, the island of Trinidad, Guyana, Surinam, French Guiana, Brazil, Ecuador, Peru, Bolivia, Paraguay, and northern Argentina. The species also occurs in Mexico and throughout Central America.

MARGINAL LOCALITIES (Map 119): VENEZUELA: Nueva Esparta, 10 km WSW of La Asunción (Handley 1976). TRINIDAD AND TOBAGO: Trinidad, Blanchisseuse (C. H. Carter et al. 1981). VENEZUELA: Delta Amacuro, Guayo (Linares and Rivas 2004). GUYANA: Upper Demerara-Berbice, Berbice (Miller 1913c). SURINAM: Paramaribo, Paramaribo (Husson 1962). FRENCH GUIANA: Cayenne (Brosset and Dubost 1968). BRAZIL: Amapá, Amapá (Piccinini 1974); Pará, between Anajás and Muaná (Marques-Aguiar et al. 2002); Pará, Belém (C.

Map 119 Marginal localities for *Glossophaga soricina* ●

O. C. Vieira 1942); Maranhão, Maranhão City (Miller 1913c); Ceará, Parque Nacional de Ubajara (Guedes et al. 2000); Ceará, Fortaleza (Piccinini 1971); Pernambuco, Estação Ecológica do Tapacurá (Mares et al. 1981); Bahia, Ilha Madre de Deus (C. O. C. Vieira 1942); Espírito Santo, Rio Dôce (C. O. C. Vieira 1942); Rio de Janeiro, Rio de Janeiro (Spix 1823); São Paulo, San Sebastião (Miller 1913c); Santa Catarina, Ilha Santa Catarina (Graipel, Cherem, and Ximenez 2001); Rio Grande do Sul, Dom Pedro de Alcântara (Fabián, Rui, and Oliveira 1999); Rio Grande do Sul, São Lourenço (Ihering 1892) Rio Grande do Sul, Caçapava do Sul (Fabián, Rui, and Oliveira 1999). ARGENTINA: Buenos Aires, La Plata (Cabrera 1930); Misiones, Teyú Cuaré (Vaccaro and Massoia 1988b); Chaco, Isla del Cerrito (Fornes and Massoia 1967). PARAGUAY: Presidente Hayes, Puerto Cooper (Podtiaguin 1944). BOLIVIA: Santa Cruz, Santiago (S. Anderson 1997). ARGENTINA Salta, Quebrada Tartagal (Barquez 1985). BOLIVIA: Chuquisaca, 70 km SE of Padilla, Río Azuero (S. Anderson, Koopman, and Creighton 1982); Cochabamba, 0.5 km NE of Villa Tunari (S. Anderson 1997); La Paz, Yanacachi (S. Anderson 1997). PERU: Arequipa, Valle del Tambo (Ortiz de la Puente 1951); Lima, Lima (Ortiz de la Puente 1951); La Libertad, Trujillo (USNM 283178); Tumbes, Zorritos (G. M. Allen 1908). ECUADOR: Manabí, Vueltas Largas (Albuja 1999); Esmeraldas, San Lorenzo (Albuja 1983). COLOMBIA: Valle del Cauca, near Buenaventura (Esslinger, Vaughn, and

Arata 1968); Cordoba, Jaraquiel (Sanborn 1932b); El Cesar, Río Guaimaral (Hershkovitz 1949c). VENEZUELA: Carabobo, Montalbán (Handley 1976).

SUBSPECIES: Three of the five subspecies of *G. soricina* recognized by Webster (1983) are known from South America.

G. s. mutica Merriam, 1898

SYNONYMS:

Glossophaga mutica Merriam, 1898:18; type locality "Maria Madre Id., Tres Marias Ids., [Nayarit,] Mexico."

Glossophaga soricina handleyi Webster and Jones 1980b:5; type locality "Colegio Peninsular, Mérida, Yucatán, México."

This subspecies is distributed in South America only in northern and western Colombia; otherwise, it is common and widespread in Central America and Mexico.

G. s. soricina Pallas, 1766

SYNONYMS:

Vespertilio soricinus Pallas, 1766:48, pls. 5 and 6; no type locality given; Pallas (1767:24) described a specimen of the opposite sex (male) under the same name and said that he had seen specimens from Surinam and the Caribbean Islands; type locality restricted to Surinam by Miller (1912c:39).

Phyllostoma soricinum: É. Geoffroy St.-Hilaire, 1810b:179; name combination.

Glossophaga soricina: É. Geoffroy St.-Hilaire, 1818a:418; first use of current name combination.

Ph[yllostomus]. soricinus: Olfers, 1818:224; name combination.

Glossophaga amplexicaudata Spix, 1823:66 (pl. 36, Fig. 4); type locality "Rio de Ianeiro [*sic*]," Rio de Janeiro, Brazil.

Phyllophora nigra Gray, 1843b:20; *nomen nudum*.

Phyllophora nigra Gray, 1844:18; type locality "tropical America"; based on the same Brazilian specimen listed by Gray (1843b) as *Phyllophora nigra* "from Brazils" (a *nomen nudum*); therefore, the restricted type locality is Brazil.

Glossophaga villosa H. Allen, 1896b:779; type locality "La Guayra, Venezuela" (see Lyon and Osgood 1909:264, concerning the type locality); preoccupied by *Glossophaga villosa* Rengger, 1830.

Glossophaga truei H. Allen, 1897:153; replacement name for *Glossophaga villosa* H. Allen.

Glossophaga soricina microtis Miller, 1913c:419; type locality "Sapucay," Paraguarí, Paraguay.

The nominate subspecies occurs from the island of Trinidad southward through South America east of the An-

des into Bolivia, Paraguay, northern Argentina, and southern Brazil.

G. s. valens Miller, 1913

SYNONYM:

Glossophaga soricina valens Miller, 1913c:420; type locality "Balsas," Amazonas, Peru.

This subspecies inhabits the Pacific slope of the Andes and adjacent lowlands from northern Ecuador into southern Peru. The Peruvian component includes populations from the arid and semiarid inter-Andean valleys of the departments of Cajamarca, Amazonas, and La Libertad.

NATURAL HISTORY: *Glossophaga soricina* uses a variety of structures as diurnal roosts that include mine tunnels, road culverts, drainage pipes, buildings, and caves that it often shares with a number of other bat species (e.g., Graham 1988). Handley (1976) reported this species throughout the lowlands of Venezuela, mostly in more humid areas, though occasionally in xeric regions. Tuttle (1970) netted *G. soricina* among blooming cashew trees in Peru. Barquez (1985) found flower parts and chitin (from thysanopterans and other insects) in stomachs of bats from northern Argentina. Throughout its range, the species consumes a variety of insects, fruits, and flower parts and is an important pollinator of a large number of plants (see summary in Gardner 1977c, and additional reports by M. Sazima and Sazima 1987; I. Sazima, Vogel, and Sazima 1989; and Zortéa 2003). Ubelaker, Specian, and Duszynski (1977) listed seven protozoan, two cestode, and four nematode endoparasites known from *G. soricina*. Machado-Allison (1965) reported two species of spinturnicid mites as ectoparasites. Yunker, Lukoschus, and Giesen (1990) recovered protonymphs of macronyssid mites from Surinamese *G. soricina*. Webb and Loomis (1977) listed 21 species of mites along with 2 species of ticks and 9 species of batflies. Guerrero (1985a) listed additional mites, ticks and batflies; he later (1997) listed 6 genera and 18 species of streblid batflies recovered from *G. soricina*. J. Alvarez et al. (1991) provided a detailed description along with measurements and a summary of available information on natural history in their *Mammalian Species* account. The karyotype is $2n = 32$, FN = 60 (R. J. Baker 1967).

REMARKS: *Glossophaga amplexicauda* É. Geoffroy St.-Hilaire, 1818a:418 (pl. 18), a name often cited as a synonym of *G. soricina*, is based on specimens of *Carollia* according to D. C. Carter and Dolan (1978:46). Cabrera (1930) recorded *G. soricina* from the cities of Buenos Aires and La Plata; however, Barquez (1985) suggested that Cabrera's records lacked sufficient documentation. Later, Barquez, Mares, and Braun (1999:33) located some of Cabrera's (1930) specimens in the collections of the Museo de Ciencias Naturales de La Plata that verified his records.

D. C. Carter and Dolan (1978) provided measurements and other information on the syntypes of *Glossophaga amplexicaudata* Spix.

Miller (1913c) gave the type locality of *G. s. valens* as Balsas, Cajamarca; whereas, Balsas (06°50'S, 78°00'W) is located in the departamento de Amazonas, Peru (Ortiz de la Puente 1951). The specimen (MNHN A 278) D. C. Carter and Dolan (1978) identified as the holotype of *Phyllostoma soricinum* É. Geoffroy St.-Hilaire, is not a type specimen and *Phyllostoma soricinum* is nothing more than a name combination used for a specimen from Surinam that É. Geoffroy St.-Hilaire (1810b) had at hand when he augmented Spix's (1823) description of *Vespertilio soricinus*.

Genus *Leptonycteris* Lydekker in Flower and Lydekker, 1891

Bats of this nectar-feeding genus are found in arid and semiarid habitats from the southwestern United States into northern South America. Of the two species known in the genus, only *Leptonycteris curasoae* occurs in South America.

Species of the genus *Leptonycteris* are large glossophagines (forearm 45–60 mm and conspicuously longer than the metacarpal of digit III) characterized by an often overlooked, minute tail, and a narrow and sparsely haired interfemoral membrane. The zygomatic arches are slender, but complete. Two lower incisors are present on each side; but third molars are lacking. The dental formula is 2/2, 1/1, 2/3, 2/2 × 2 = 30.

SYNONYMS:

Ischnoglossa Saussure, 1860:491; type species *Ischnoglossa nivalis* Saussure, 1860, by monotypy; preoccupied by *Ischnoglossa* Kraatz, 1856 (Coleoptera).

Leptonycteris Lydekker, 1891:674, in Flower and Lydekker (1891); replacement name for *Ischnoglossa* Saussure.

Leptonycteris curasoae Miller, 1900

Curaçaoan Long-nosed Bat

SYNONYMS:

Leptonycteris curasoae Miller, 1900a:126; type locality "Curaçao, West Indies"; restricted to Willemstad, Curaçao, by Lyon and Osgood (1909).

Leptonycteris nivalis yerbabuenae Martínez and Villa-R., 1940:313; type locality "Yerbabuena, Estado de Guerrero," Mexico.

Leptonycteris nivalis sanborni Hoffmeister, 1957:456; type locality "mouth of Miller Canyon, Huachuca Mountains, 10 mi. SSE Fort Huachuca, Cochise County, Arizona," U.S.A.

Leptonycteris sanborni: W. B. Davis and Carter, 1962b:196; name combination.

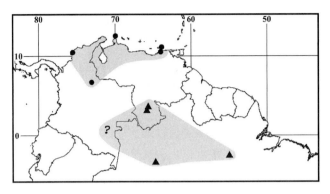

Map 120 Marginal localities for *Leptonycteris curasoae* ● and *Scleronycteris ega* ▲

Leptonycteris yerbabuenae: Villa-R., 1967:252; name combination.

Leptonycteris curasoae tarlosti Pirlot, 1965a:6; type locality "El Valle, Ile de Margarita (Estado Nueva Esparta), Venezuela."

DISTRIBUTION: *Leptonycteris curasoae* is found in northern Colombia, northern Venezuela, and the Netherlands Antilles. The species also occurs from El Salvador and Honduras northwestward into the southwestern United States.

MARGINAL LOCALITIES (Map 120): NETHERLANDS ANTILLES: Aruba, Quaridikiri Cave (W. B. Davis and Carter 1962b). VENEZUELA (Handley 1976): Nueva Esparta, 3 km NE of La Asunción; Sucre, 16 km E of Cumaná. COLOMBIA (Marinkelle and Cadena 1972): Santander, Macaregua Cave; Bolívar, old fortifications at Cartagena.

SUBSPECIES: Of the two subspecies recognized by Arita and Humphrey (1989), only the nominate form, *L. c. curasoae* Miller, is known from South America.

NATURAL HISTORY: This species is found in seasonally arid habitats along the northern coast of South America and on adjacent islands. Handley (1976) wrote that most of the Venezuelan specimens came from xeric regions (frequently found roosting in caves or houses), although a few specimens were mist-netted near open water. J. D. Smith and Genoways (1974) found a colony of approximately 4,000 females nursing nearly full-grown young in July on Isla Margarita, Venezuela. By November, the colony contained no young; adult males were in breeding condition (testes 6–8 mm in length) and 7 of 34 females examined were pregnant. Martino, Arends, and Aranguren (1998) reported finding a peak in pregnancies during May, and a peak in numbers of lactating females in June, in northern Venezuela.

In their study of the feeding habits of *L. curasoae* on the Paraguaná Peninsula, Falcón, Venezuela, Martino, Aranguren, and Arends (2002) found pollen on the fur and pollen and seeds in fecal samples, but no insects. These bats were feeding on the pollen and nectar from flowers of Cactaceae, Bombacaceae, Caricaceae, Agavaceae, Capparaceae, and Leguminosae (listed in decreasing order of importance). All seeds came from fruits of Cactaceae. While in greater quantities in the dry season, pollen was found throughout the year; whereas, seeds were present mainly during the rainy season.

Petit (1997) stated that *L. curasoae*, along with *G. longirostris* on the island of Curaçao, depend on columnar cacti flowers and fruit for survival and reproduction. She stated that *L. curasoae* also depended on flowering *Ceiba* and *Agave* during pregnancy, and that pregnancy (March-June) and lactation (June-August) coincided with the flowering phenology of two of the three species of columnar cacti on Curaçao. L. G. Herrera and Martínez del Río (1998) found that *L. curasoae*, in a comparison with *Anoura geoffroyi*, *Artibeus jamaicensis*, and *Sturnira lilium*, was the most efficient at emptying pollen grains of columnar cacti, based on the ratio between intact and empty pollen grains found in feces. Sánchez and Cadena (2000) discussed the possibility that *L. curasoae* is migratory in northern Colombia and northwestern Venezuela. Known ectoparasites include two species of ticks and two species of mites (Webb and Loomis 1977; Guerrero 1985a), and seven species in three genera of streblid batflies (Guerrero 1985a, 1997). Rincón (2001) enumerated subfossil remains recovered from guano deposits in northeastern Venezuela.

REMARKS: Pirlot (1965a) gave the name *L. c. tarlosti* to a small series of *L. curasoae* from El Valle, Isla Margarita, Nueva Esparta, Venezuela, but did not compare his material with specimens from elsewhere in the range of *L. curasoae*. J. D. Smith and Genoways (1974) concluded that an analysis of geographic variation would show island and South American mainland populations to be undifferentiated. In revising the genus *Leptonycteris*, Arita and Humphrey (1989) recognized *L. c. yerbabuenae* from the United States, Mexico, Guatemala, and El Salvador, as being disjunct from the nominate subspecies, which is known only from South America. Most authors previously treated *L. curasoae* as monotypic and restricted to South America (e.g., Cabrera 1958; J. D. Smith and Genoways 1974). J. D. Smith and Genoways (1974) gave measurements of *L. curasoae* from the population on Isla Margarita and of samples from the Netherlands Antilles.

Genus *Lichonycteris* O. Thomas, 1895

This genus of two species is characterized by small size (forearm 31.0–35.5 mm), incomplete zygomatic arches, and the lack of lower incisors and third molars. The uropatagium is relatively long and encases the tail, which is about 1/2 as long as the uropatagium. The skull resembles the skull of *Glossophaga*, except that the rostrum is longer than the braincase, and the upper incisors are small and nearly evenly spaced between the canines. Externally *Lichonycteris* is readily identifiable by the narrow dark basal band on dorsal and ventral fur, by the presence of long guard hairs scattered over the dorsum, and by the attachment of the wing membrane to the foot at about mid-length of the metatarsals. The dental formula is 2/0, 1/1, 2/3, 2/2 × 2 = 26.

SYNONYM:

Lichonycteris O. Thomas, 1895b:55; type species *Lichonycteris obscura* O. Thomas, 1895b, by original designation.

REMARKS: Little is known about the ecology, systematics, and distribution of this genus. *Lichonycteris* is easily misidentified in the field. Useful features for distinguishing species of this genus from similar taxa include the tricolored dorsal pelage, which has a dark basal band; the well-furred elbow region of the forearm; and the wing membrane attachment to the foot at about mid-length of the metatarsals. *Lichonycteris* has been regarded as a relatively derived glossophagine, except by Griffiths (1982) who found that the hyoid/lingual morphology was not developed beyond the *Glossophaga* grade of specialization. He suggested that *Lichonycteris* has its closest affinity with *Glossophaga* and *Monophyllus*. Baker et al. (2003), however, placed *Lichonycteris* with the more specialized glossophagines (under the subtribe Choeronycterina) on the basis of molecular genetics

KEY TO THE SPECIES OF *LICHONYCTERIS*.

1. Anteroposterior length of lower canine shorter than anteroposterior length of adjacent premolar; anteroposterior length of m1 more than 1.3 mm . . . . *Lichonycteris obscura*
1'. Anteroposterior length of lower canine about equal to or longer than anteroposterior length of adjacent premolar; anteroposterior length of m1 equal to or less than 1.3 mm . . . . *Lichonycteris degener*

Lichonycteris degener Miller, 1931
Pale Brown Long-nosed Bat
SYNONYMS:

Lichonycteris obscura: Miller, 1900c:156; not *Lichonycteris obscura* O. Thomas.

Lichonycteris degener Miller, 1931:411; type locality "Para," Pará, Brazil.

Lichonycteris deneger N. R. Reis, 1984:248; incorrect subsequent spelling of *Lichonycteris degener* Miller.

Lichonycteris obscura: Hill, 1986:580–81; part, not *Lichonycteris obscura* O. Thomas.

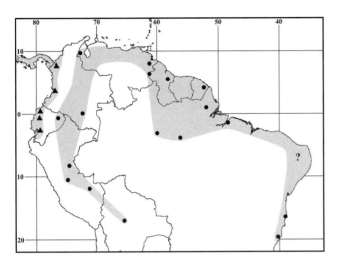

Map 121 Marginal localities for *Lichonycteris degener* ● and *Lichonycteris obscura* ▲

DISTRIBUTION: *Lichonycteris degener* is recorded south and east of the Andes from Colombia, Venezuela, Guyana, Surinam, French Guiana, Ecuador, Peru, Bolivia, and Brazil.

MARGINAL LOCALITIES (Map 121): VENEZUELA: Zulia, El Tukuco (Soriano, Ruiz, and Zambrano 2005, as *L. obscura*); Bolívar, Imataca Forest Reserve, Unit V (Ochoa et al. 1993); Bolívar, El Manaco, 59 km SE of El Dorado (Handley 1976). GUYANA: Upper Demerara-Berbice, 3 km SSW of Ituni (Lim and Engstrom 2001a). FRENCH GUIANA: Ipousin (Hill 1986). BRAZIL: Amapá, Serra do Navio (MZUSP 18925); Pará, Belém (type locality of *Lichonycteris degener* Miller); Bahia, BR–367, between Porto Seguro and BR–101 (Taddei and Pedro 1993); Espírito Santo, Lagoa Juparanã (Zortéa, Gregorin, and Ditchfield 1998); Pará, Parque Nacional da Amazônia, 54 km by road S of Itaituba (George et al. 1988); Amazonas, Igarapé Acará (N. R. Reis and Peracchi 1987). PERU: Loreto, Yarinacocha (Gardner 1976); Madre de Dios, Pakitza (S. Solari, Pacheco, and Vivar 1999). BOLIVIA: Cochabamba, 13 km SW of Villa Tunari (S. Anderson, Koopman, and Creighton 1982). PERU: Pasco, San Juan (Tuttle 1970). ECUADOR: Napo, 38 km S of Pompeya Sur (F. A. Reid, Engstrom, and Lim 2000). COLOMBIA: Caquetá, Estación Puerto Abeja (Montenegro and Romero-Ruiz 2000).

SUBSPECIES: We are treating *L. degener* as monotypic.

NATURAL HISTORY: Tuttle (1970) netted a specimen over a trail in mature evergreen forest near the edge of a village clearing in Peru. The other Peruvian specimen (Gardner 1976) was netted near flowering bushes in a yard at Yarinacocha. The specimen representing the Venezuelan locality listed above was netted in humid tropical forest at 150 m elevation (Handley 1976). F. A. Reid, Engstrom, and Lim (2000) reported capturing a nonreproductive female in June

and another that had a 20-mm–CR fetus when taken in October in eastern Ecuador. The karyotype is unknown.

REMARKS: Most of the citations listed above under Marginal Localities used the name *L. obscura* for this species. As pointed out by F. A. Reid, Engstrom, and Lim (2000), *L. degener* (reported as *L. obscura*) is easily confused with other glossophagines of similar size, especially in the field. Anteriorly-posteriorly shorter lower molars and relatively thicker and longer (anteriorly-posteriorly) lower canines distinguish *L. degener* from *L. obscura*, which is a species of the Middle American biogeographic region. The specimen record from Pakitza, Peru (S. Solari, Pacheco, and Vivar 1999; reported as *L. obscura*) had been reported previously (Ascorra, Solari, and Wilson 1996) as *Choeroniscus minor*.

Although Gardner (1976) reported that the genus was monotypic, most reviewers continued to recognize two species: *L. degener* and *L. obscura*. Hill (1986) also came to the conclusion that the genus was monotypic, and referred all known specimens of *Lichonycteris* to *L. obscura*. He noted that specimens from Costa Rica, Surinam, and French Guiana having unusually wide talonids on their molars, and having closely adpressed premolars, might represent an undescribed species. When Miller (1931) described *L. degener*, he compared the holotype with a specimen of the same taxon from "Dutch Guiana" he earlier (Miller 1900c) mistakenly identified as *L. obscura* O. Thomas, and which he assumed came from Surinam. The dental differences Miller (1931) used to distinguish *L. degener* from the specimen he identified as *L. obscura* are within the range of variation for *L. degener*. The specimen Miller (1900c:156) misidentified as *L. obscura* "was entered in the Museum register, as No. 14815 on March 6, 1885" under the name *Glossophaga soricina* (USNM 14815/37553). The specimen consists of a dry skin and partial skull and, according to Miller (1900c), the skin and partial skull was originally labeled "Surinam, Edw. Koebel" [*sic*]. Today, the skin label attributes the specimen to Surinam with the collector identified as Albert Koeble. One of us (ALG), in an attempt to learn more about the places visited by Koeble, found that Koeble was in the Brazilian states of Pernambuco and Bahia, but there is no evidence that he ever visited Surinam during any of his travels in South America. Other mammals acquired by the U.S. National Museum from Koeble were a specimen of *Kerodon rupestris* (a rodent of the dry Caatinga habitats of the Northeast of Brazil), and a specimen of *Pygoderma bilabiatum*, all accessioned as having come from Surinam. The museum also acquired a number of birds, many of which came from the vicinity of La Bonita in southeastern Pernambuco, but none from Surinam. Eastern Pernambuco was once part of a Dutch Colony; therefore, it is possible that Koeble mentioned that fact to the recipient of the mammal specimens and the information was misinterpreted as meaning the mammal specimens came from the Dutch Colony Surinam. With the exception of this specimen of *L. degener* of questionable origin, the species has not been recorded for Surinam (e.g., Husson 1962, 1978; Genoways and Williams 1979b; Williams and Genoways 1980a; Williams, Genoways, and Groen 1983).

Lichonycteris obscura O. Thomas, 1895
Dark Brown Long-nosed Bat
SYNONYMS:

Lichonycteris obscura O. Thomas, 1895b:56; type locality "Managua," Managua, Nicaragua.

Lichonycteris obscurus: Miller, 1900c:156; incorrect subsequent spelling of *Lichonycteris obscura* O. Thomas.

DISTRIBUTION: *Lichonycteris obscura* is recorded from western Colombia and Ecuador. Elsewhere, the species ranges from Panama northwestward into Belize and Mexico.

MARGINAL LOCALITIES (Map 121; localities listed from north to south): COLOMBIA: Antioquia, Chigorodó (Marinkelle and Cadena 1972); Valle del Cauca, Río Zabaletas (USNM 483374). ECUADOR (Albuja 1999): Esmeraldas, Sade; Pichincha, Estación Científico Río Palenque; Azuay, Manta Real.

SUBSPECIES: We treat *L. obscura* as monotypic.

NATURAL HISTORY: Little is known about the biology of *L. obscura* other than its having been found in disturbed forest habitats. The karyotype is $2n = 24$, FN = 44 (R. J. Baker 1979).

REMARKS: Both species of *Lichonycteris* are easily confused with other glossophagines of similar size, especially in the field. Useful features for distinguishing *Lichonycteris* spp. from similar taxa include its tricolored dorsal pelage having a dark basal band; well-furred forearm at the elbow; and wing membrane attaching to the side of the foot at about mid-length of metatarsals. The anteriorly-posteriorly longer lower molars readily distinguish *L obscura* from *L. degener*. Both species are uncommon in collections and apparently have disjunct distributions. D. C. Carter and Dolan (1978) gave information and measurements on the holotype of *Lichonycteris obscura* O. Thomas.

Although Gardner (1976) and Hill (1986) reported that the genus was monotypic, most reviewers have continued to recognize two species: *L. degener* and *L. obscura*. Miller (1900c) misidentified a specimen he assumed came from Surinam (see Remarks under *L. degener*) as *L. obscura* that he later (Miller 1931) compared with his new species, *L. degener*.

Genus *Scleronycteris* O. Thomas, 1912

This monotypic genus, known from less than a half dozen specimens, is probably the least known of all glos-

sophagines. *Scleronycteris* can be distinguished from other glossophagines by the combination of small size (forearm 35.0–35.5 mm), dark blackish-brown dorsal color pattern with a slightly paler venter, a tail about 1/3 the length of a relatively long uropatagium, and a skull with incomplete zygomatic arches. The dentition consists of four small, subequal, and evenly spaced (not in contact) upper incisors of which the inner are peg-like and the outer pointed; two upper premolars; and three upper and lower molars of which the upper molars lack the paracone, hypocone, and any trace of the anterior part of the W-shaped ectoloph. Lower incisors are absent. The dental formula is 2/0, 1/1, 2/3, 3/3 × 2 = 30.

SYNONYM:

Scleronycteris O. Thomas, 1912f:404; type species *Scleronycteris ega* O. Thomas, 1912f, by original designation.

Scleronycteris ega O. Thomas, 1912
Ega Long-tongued Bat
SYNONYM:

Scleronycteris ega O. Thomas, 1912f:405; type locality "Ega," Amazonas, Brazil.

DISTRIBUTION: *Scleronycteris ega* is known from Colombia, Brazil, and Venezuela.

MARGINAL LOCALITIES (Map 120): VENEZUELA: Amazonas, Caño Culebra (Ochoa et al. 1993); Amazonas, Tamatama (Handley 1976). BRAZIL: Pará, Alter do Chão (Bernard and Fenton 2002); Amazonas, Ega (type locality of *Scleronycteris ega* O. Thomas). COLOMBIA: Vaupés (Alberico et al. 2000; no locality given, not mapped).

SUBSPECIES: *Scleronycteris ega* is monotypic.

NATURAL HISTORY: One Venezuelan specimen was mist-netted in a yard near a stream in evergreen forest at 135 m (Handley 1976); the other in lowland primary forest (Ochoa et al. 1993). Three were taken in savanna habitat at Alter do Chão (Bernard and Fenton 2002). The karyotype is unknown.

REMARKS: The identity of a specimen Nowak and Paradiso (1982:263) reported as in the National Museum in Rio de Janeiro, Brazil, has not been confirmed. The Colombian record is based on a specimen in the collections of the Instituto de Investigaciones Alexander von Humboldt (Alberico et al. 2000).

Subfamily Lonchophyllinae Griffiths, 1982
Thomas A. Griffiths and Alfred L. Gardner

The Lonchophyllinae are small to medium-sized bats (forearm 30–50 mm) characterized by an elongated skull and a long, extensible tongue that has a deep longitudinal groove along each lateral surface, but lacks a "paintbrush" tip. The grooves are bordered with short, hair-like papillae. Members of the subfamily also can be recognized by the following combination of characters: a long muzzle (greatly elongated in *Platalina* and *Xeronycteris*) bearing short vibrissae; a wide, tear-drop-shaped noseleaf; a thick, prominent tragus; and tail about 1/3 the length of the relatively long interfemoral membrane. Other features include incomplete zygomatic arches and a reduced spinous process on the scapula. The molars and premolars are reduced in robustness and complexity, reflecting the nectarivorous feeding habits of the group. This reduction is minimal in *Lonchophylla* and *Lionycteris*, advanced in *Platalina* so that it parallels the condition of some of the most derived Glossophaginae, and so advanced in *Xeronycteris* that upper and lower molars appear as fragments of these same teeth in other lonchophyllines. All Lonchophyllinae have relatively large upper incisors, particularly the inner incisors, which usually are more than twice the size of the outer. The lower incisors are also well developed and have spatulate trifid tips (weakly trifid in *Platalina*). The dental formula is 2/2, 1/1, 2/3, 3/3 × 2 = 34.

The Lonchophyllinae can be distinguished from the Glossophaginae by the morphology of the tongue and associated hyoid musculature. The lonchophylline tongue has two conspicuous types of papillae: 1) the fleshy papillae that cover the upper surface, and 2) the short, hair-like papillae arranged in rows bordering the diagnostic longitudinal groove present on each side of the tongue. These grooves extends from just behind the tip to the base of the tongue. In contrast, the glossophagine tongue has seven types of papillae (see Griffiths 1982:28–29), the most prominent of which are the hair-like papillae distributed over the anterolateral surface of the tongue and forming the diagnostic "paintbrush" tip. Lateral grooves are not present in the glossophagine tongue. The lonchophylline tongue has a pair of lingual arteries and veins in the ventral portion of the tongue, along with dorsal lingual arteries. In contrast, the glossophagine tongue has a single, medial lingual artery, and enlarged lingual veins, one on each side of the tongue, but lacks dorsal lingual arteries. The hyoid musculature of lonchophyllines differs from the musculature of the glossophagines in that the *styloglossus* inserts on the lateral tongue surface (inserts on the lateral corner of the tongue in glossophagines), the *cricopharyngeus* has only two slips (it has several slips in glossophagines).

The Lonchophyllinae comprises four genera of Neotropical nectarivorous bats that occur from Central America (southeastern Nicaragua) southward into northern Chile, central Bolivia, and Brazil. Eight of the 11 species in the subfamily are in the genus *Lonchophylla*. The other three

represent the monotypic genera *Lionycteris*, *Platalina*, and *Xeronycteris*. There is no known fossil record.

REMARKS: Griffiths' (1982) placement of the genera *Lonchophylla*, *Lionycteris*, and *Platalina* in the Lonchophyllinae was questioned (particularly by J. D. Smith and Hood 1984) on the basis of whether proper cladistic procedure was followed. Others suggested that the tongue morphology of one subfamily was derived from the other, hence the separation was invalid. Nevertheless, there is no evidence that one type of tongue evolved directly into the other (see Griffiths 1978, 1982; Griffiths and Criley 1989). Recent phylogenies of phyllostomid bats generated by R. J. Baker et al. (2000) using nuclear DNA sequences indicated that *Lonchophylla* and *Lionycteris* were, at best, only distantly related to the Glossophaginae. In the recent assessment of relationships within the Phyllostomidae, R. J. Baker et al. (2003) recognized glossophagines and lonchophyllines as representing separate subfamilies. Their classification is followed here. *Xeronycteris*, the most derived of the Lonchophyllinae, shares the tongue structure characteristic of the subfamily.

KEY TO THE GENERA OF LONCHOPHYLLINAE.

1. Rostrum approximately as long as or markedly longer than braincase. 2
1'. Rostrum shorter than braincase. 3
2. Upper and lower molars normal for lonchophyllines, not greatly reduced; lingual cuspule often present on cingulum of last upper premolar. . . . . . . *Lonchophylla* (part)
2'. Upper and lower molars greatly reduced, lingual portion of each upper molar less than 1/2 the total length of tooth; upper premolars narrow and lack lingual cusps 4
3. Premolars elongated anteroposteriorly; central cuspid of lower premolars not deflected toward labial side of tooth and not preceded and followed by prominent cuspulids on the cingulum; cingula of lower premolars reduced or absent; base of dorsal fur paler than tips; uropatagium not conspicuously furred. . . . . *Lonchophylla* (part)
3'. Premolars not conspicuously elongated, cingula present; central cuspid of lower premolars prominent, deflected labially, and preceded and followed by prominent cuspulids on the cingulum; base of dorsal fur darker than tips; medial portions of the uropatagium conspicuously clothed with fur . . . . *Lionycteris*
4. Upper outer and inner incisors in contact; rostrum longer than braincase; forearm longer than 40 mm; canines have a cingulum; molars reduced in complexity, but retain normal cusp pattern; distributed in western Peru and northwestern Chile . . . . *Platalina*
4'. Upper outer and inner incisors separated by a conspicuous gap; rostrum approximately equal in length to

braincase; canines lack a cingulum; molars greatly reduced in complexity with lower molars lacking metacristids, entocristids and trigonids; forearm shorter than 40 mm; distributed in eastern Brazil. . . . . *Xeronycteris*

Genus *Lionycteris* O. Thomas, 1913

The monotypic genus *Lionycteris* is known from eastern Panama (Hall 1981) into central Peru and northern Brazil. *Lionycteris* shares with *Lonchophylla* the distinction of being among the least derived of the Neotropical nectar-feeding bats. The cusps of the molars are comparatively well developed and the tongue protrusion apparatus is relatively primitive in contrast to those structures in members of the Glossophaginae (Griffiths 1982). *Lionycteris* are small (forearm 33–37 mm), dark brown bats with unusual fur because a pale basal band is not present on all hairs (when present, it is narrow and inconspicuous); therefore, the pelage appears to be darker basally than at the tips. The muzzle is narrow, elongated, and bears short vibrissae and a small noseleaf. The tail is about 1/3 the length of the long, sparsely furred interfemoral membrane and the tail tip protrudes above the membrane. *Lionycteris* can be recognized by its broad, dark brown, basal band of the dorsal pelage, and conspicuously furred medial portions of the uropatagium. The rostrum is shorter than the braincase and the palatal margin of the mesopterygoid fossa is comparatively deep and U-shaped. The diagnostic lower premolars are trident with an unusually high and robust main cusp that arises from the labial margin of the tooth.

SYNONYM:

Lionycteris O. Thomas, 1913b:270; type species *Lionycteris spurrelli* O. Thomas, 1913b, by original designation.

Lionycteris spurrelli O. Thomas, 1913
Spurrell's Long-nosed Bat
SYNONYMS:

Lionycteris spurrelli O. Thomas, 1913b:271; type locality "Condoto," Chocó, Colombia.

Lionycteris spurreli Coimbra, Borges, Guerra, and Mello, 1982:37; incorrect subsequent spelling of *Lionycteris spurrelli* O. Thomas.

Lionycteris spurelli Cuervo-Díaz, Hernández-Camacho, and Cadena, 1986:478; incorrect subsequent spelling of *Lionycteris spurrelli* O. Thomas.

L[ionycteris]. spurelli Sánchez-Palomino, Rivas-Pava, and Cadena, 1993:308; incorrect subsequent spelling of *Lionycteris spurrelli* O. Thomas.

Lionycteris spurelli Carstens, Lundrigan, and Myers, 2002:24; incorrect subsequent spelling of *Lionycteris spurrelli* O. Thomas.

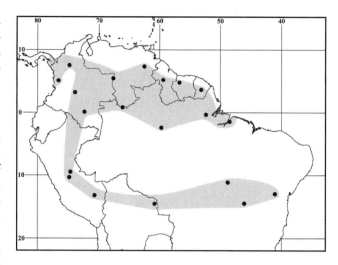

Map 122 Marginal localities for *Lionycteris spurrelli* ●

Lionycteris spurrelii Gregorin and Mendes, 1999:122; incorrect subsequent spelling of *Lionycteris spurrelli* O. Thomas.

Lionycteris sperurreli Montenegro and Romero-Ruiz, 2000:648; incorrect subsequent spelling of *Lionycteris spurrelli* O. Thomas.

Lionycteris spurelii Shapley, Wilson, Warren, and Barnett, 2005:382, 383; incorrect subsequent spelling of *Lionycteris spurrelli* O. Thomas.

DISTRIBUTION: *Lionycteris spurrelli* is known from Colombia, southern Venezuela, Guyana, Surinam, French Guiana, northeastern Brazil, and eastern Peru. Elsewhere the species occurs in eastern Panama (Handley 1966b; Hall 1981).

MARGINAL LOCALITIES (Map 122): COLOMBIA: Antioquia, Zaragoza (USNM 499771). VENEZUELA (Handley 1976): Amazonas, 32 km S of Puerto Ayacucho; Bolívar, El Manaco. GUYANA: Potaro-Siparuni, Kaieteur Falls (USNM 565516). SURINAM: Sipaliwini, Grassalco (Williams and Genoways 1980a). FRENCH GUIANA: Saül (Webster and McGillivray 1984). BRAZIL: Amapá, Igarapé Novo (Taddei, Vizotto, and Sazima 1978); Pará, Bosque Rodrigues Alves (Taddei, Vizotto, and Sazima 1978); Amazonas, Gavião (Sampaio et al. 2003). VENEZUELA: Amazonas, Serranía de la Neblina Base Camp (Gardner 1988). COLOMBIA: Caquetá, Estación Puerto Abeja (Montenegro and Romero-Ruiz 2000). PERU (Koopman 1978): Huánuco, Cerros del Sira; Pasco, Nevati; Cusco, Quince Mil. BOLIVIA: Santa Cruz, Los Fierros (L. H. Emmons, pers. comm.). BRAZIL: Tocantins, Município Aliança do Tocantins (Nunes et al. 2005); Goiás, Município de Mambaí (Coimbra et al. 1982); Bahia, Caverna Poço Encantado (Gregorin and Mendes 1999). COLOMBIA: Meta, Município de San Juan de Arama (Sánchez-Palomino, Rivas-Pava, and

Cadena 1993); Chocó, Condoto (type locality of *Lionycteris spurrelli* O. Thomas).

SUBSPECIES: We are treating *L. spurrelli* as monotypic.

NATURAL HISTORY: Tuttle (1970) collected *L. spurrelli* at the edges of Indian villages and among blooming cashew trees in eastern Peru. Handley (1976) reported this species taken in mistnets or from caves, generally in more humid regions in Venezuela, and frequently near streams. Most specimens have been captured in forests. Based on label information for USNM specimens from Antioquia, Colombia, several were caught from roosts in culverts or rock tunnels. Again, based on the USNM collection, 2 Colombian and 11 Venezuelan females showed no sign of reproduction when taken in July. One Colombian female was lactating when captured in April, one was pregnant on 29 August, and two were pregnant in September. Another Venezuelan female was pregnant in January, two in July, one in September. One Venezuelan female was lactating in February, and two were lactating in June. These data suggest reproductive activity throughout the year. Known ectoparasites include four species of streblid batflies (Guerrero 1985a, 1997) and four species of mites (Webb and Loomis 1977; Guerrero 1985a). The karyotype is $2n = 28$, FN = 50 (R. J. Baker 1979).

REMARKS: The color pattern of the dorsal fur of those *L. spurrelli* whose fur lacks a white basal color band is similar to the pattern seen in *Lichonycteris* (Glossophaginae) except, in *L. spurrelli*, the dark basal color band is the widest color band, whereas in *Lichonycteris obscura* and *L. degener*, it is narrow. D. C. Carter and Dolan (1978) gave information and measurements for the holotype of *Lionycteris spurrelli* O. Thomas.

Genus *Lonchophylla* O. Thomas, 1903

There are ten species currently recognized in *Lonchophylla* and all occur in South America. Three species (*L. concava*, *L. robusta*, and *L. thomasi*) also range into Central America (Hall 1981). Morphologically, *Lonchophylla* is among the least-derived of nectar-feeding genera. The tooth morphology is primitive in that the cusps are relatively unreduced (in contrast to the marked reduction seen in most other nectarivorous genera). The tongue protrusion apparatus is not as well developed as in *Platalina* and is primitive when compared with that of the Glossophaginae (Griffiths 1982). *Lonchophylla* are small to medium-sized (forearm 30–48 mm), pale brown to reddish-brown bats that can be distinguished on the basis of the broad, pale basal band on the body pelage, the lack of conspicuous fur on uropatagium, and the enlarged, procumbent upper, inner incisors. The rostrum is shorter to approximately the same length as the braincase; the anterior margin of the mesopterygoid fossa is relatively shallow and V-shaped; and the premolars are narrow and anteroposteriorly elongated.

SYNONYM:

Lonchophylla O. Thomas, 1903f:458; type species *Lonchophylla mordax* O. Thomas, 1903f, by original designation.

KEY TO THE SPECIES OF *LONCHOPHYLLA*:

1. Greatest length of skull more than 24.5 mm. 2
1′. Greatest length of skull less than 24.5 mm. 7
2. Size larger, greatest length of skull more than 29 mm; palatal length more than 16.5 mm; postorbital breadth more than 5.5 mm . . . . *Lonchophylla orcesi*
2′. Size smaller, greatest length of skull less than 29 mm; palatal length less than 16.5 mm; postorbital breadth usually less than 5.5 mm 3
3. Greatest length of skull 27.2–28.3 mm; M1 and M2 differ in form and lingual side of M2 is anteriorly-posteriorly shorter than lingual side of M1; outer margins of upper incisors transcribe a smooth arc; forearm usually longer than 45 mm; thumb (digit I) longer than 7.5 mm . . . . *Lonchophylla chocoana*
3′. Greatest length of skull 24.9–27.1 mm; M1 and M2 essentially alike in form and size; conspicuous gap between outer margins of I1 and I2, and outline of margins of upper incisors not transcribing a smooth arc; forearm usually shorter than 45 mm; thumb (digit I) shorter than 7.0 mm 4
4. Width across molars more than 6 mm (6.1–7.0 mm); mastoid breadth more than 10 mm (10.2–12.0); length of maxillary toothrow (C-M3) more than 9 mm (9.1–11.0 mm) 5
4′. Width across molars less than 6 mm (5.2–5.8); mastoid breadth less than 10 mm (9.4–9.7 mm); length of maxillary toothrow usually less than 9 mm (7.8–9.0 mm) 6
5. Second upper premolar with posterointernal basal cusp reduced (occasionally obsolete); palatal length more than 15 mm (15.4–17.5 mm); greatest length of skull 26.9–29.2 mm; forearm 44.4–47.9 mm . . . . *Lonchophylla handleyi*
5′. Second upper premolar with posterointernal basal cusp well developed; palatal length usually less than 15 mm (13.6–15.2 mm); greatest length of skull 24.7–27.4 mm; forearm 39.7–45.5 mm . . . . *Lonchophylla robusta*
6. Skull narrow and elongated; greatest length of skull 26–28 mm; condylobasal length 24.5–26.1 mm; length of maxillary toothrow (C-M3) 8.3–9.0 mm; forearm 36.0–40.6 mm. . . . . *Lonchophylla hesperia*
6′. Skull shorter; greatest length of skull 25.2–26.3 mm; condylobasal length 23.9–25.0 mm; length of maxillary

toothrow 7.8–8.6 mm; forearm 38.7–41.3 mm . . . . . .
. . . . *Lonchophylla bokermanni*

7. Length of maxillary toothrow (C-M3) more than 7 mm (7.2–8.1 mm); greatest length of skull 22 mm or more; forearm 32.5 mm or more 8

7′. Length of maxillary toothrow equal to or less than 7 mm (6.2–7.0 mm); greatest length of skull 19.6–21.8 mm; forearm 31.0–34.0 mm . . . *Lonchophylla thomasi*

8. Greatest length of skull usually less than 22.6 mm (22.0–22.6 mm); ratio of total skull length to forearm length less than 0.65; ratio of maxillary toothrow length (C-M3) to forearm length less than 0.21; forearm 34.7–37.7 mm . . . . *Lonchophylla dekeyseri*

8′. Greatest length of skull usually more than 22.4 mm (22.4–24.2 mm); ratio of greatest skull length to forearm length usually more than 0.65; ratio of maxillary toothrow length to forearm length usually more than 0.21; forearm 32.5–36.7 mm 9

9. High, well-defined coronoid process on mandible; distributed in Brazil and eastern Bolivia . . . .
. . . . *Lonchophylla mordax*

9′. Low coronoid process on mandible; distributed in western Colombia and northwestern Ecuador . . . .
. . . . *Lonchophylla concava*

Lonchophylla bokermanni I. Sazima, Vizotto, and Taddei, 1978

Bokermann's Nectar Bat

SYNONYM:

Lonchophylla bokermanni I. Sazima, Vizotto, and Taddei, 1978:82; type locality "Serra do Cipó," Minas Gerais, Brazil.

DISTRIBUTION: *Lonchophylla bokermanni* is endemic to Brazil where it is known from the states of Bahia, Minas Gerais, and Rio de Janeiro.

MARGINAL LOCALITIES (Map 123): BRAZIL: Bahia, Cocorobó, (USP 14170); Minas Gerais, Serra do Cipó (type locality of *Lonchophylla bokermanni* I. Sazima, Vizotto, and Taddei); Rio de Janeiro, Vila do Abraão, Ilha Grande (Taddei, Souza, and Manuzzi 1988).

SUBSPECIES: We consider *L. bokermanni* to be monotypic.

NATURAL HISTORY: I. Sazima, Vizotto, and Taddei (1978) found pollen, nectar, and fragments of ants (Formicidae, Hymenoptera) in the stomachs of three specimens. The pollen represented at least two plants, one of which was *Bauhinia rufa*. I. Sazima, Vogel, and Sazima (1989) recorded *L. bokermanni* visiting the flowers of *Encholirium glaziovii* and *Bauhinia rufa* in the Serra do Cipó, Minas Gerais. They described the behavior of bats visiting the inflorescences of *E. glaziovii* as trap line foraging. Females (six) examined by I. Sazima, Vizotto, and Taddei (1978)

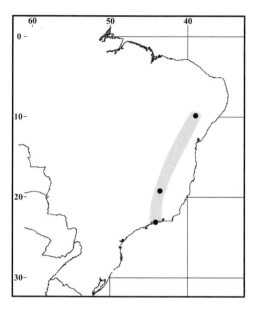

Map 123 Marginal localities for *Lonchophylla bokermanni* ●

were not visibly pregnant in September, October, or December; one male had a few sperm in the epididymides in April; whereas in September, the epididymides of another contained great quantities of sperm. The two females Taddei, Souza, and Manuzzi (1988) captured on 9 December on Ilha Grande, Rio de Janeiro, were in advanced stages of pregnancy. The karyotype is unknown.

Lonchophylla chocoana Dávalos, 2004

Chocoan Nectar Bat

SYNONYMS:

Lonchophylla sp. A., Albuja, 1999:96; an undescribed species from Ecuador.

Lonchophylla chocoana Dávalos, 2004:4; type locality "2 km south of Alto Tambo (00°54′N, 78°33′W; 700 m) in Provincia Esmeraldas, Ecuador."

DISTRIBUTION: *Lonchophylla chocoana* occurs in western Colombia and northwestern Ecuador.

MARGINAL LOCALITIES (Map 124): COLOMBIA: Valle del Cauca, Río Zabaletas, 29 km SE of Buenaventura (USNM 483361); Valle del Cauca, Río Cajambre, approximately 60 km S of Buenaventura (Alberico 1987, as *L. handleyi*); Nariño, La Guarapería (Dávalos 2004). ECUADOR: Esmeraldas, Los Pambiles (Dávalos 2004:11).

SUBSPECIES: *Lonchophylla chocoana* is monotypic.

NATURAL HISTORY: All known specimens were mistnetted at ground level in primary and secondary tropical rain forests along the western slope of the Andes and Pacific lowlands of Colombia and northwestern Ecuador. According to Dávalos (2004), the specimen from La Guarapería was captured along with three species of the glossophagine

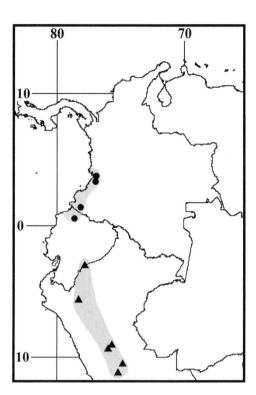

Map 124 Marginal localities for *Lonchophylla chocoana* ●
and *Lonchophylla handleyi* ▲

genus *Anoura*. She also reported netting *L. robusta* and *L. concava* (identified as *L. mordax*) in the vicinity of Alta-quer, which is southeast of Junín, Colombia. Albuja and Gardner (2005) reported the capture of *L. chocoana* at Los Pambiles, Ecuador, along with *L. robusta* and *L. orcesi*. The karyotype is unknown.

REMARKS: *Lonchophylla chocoana* is most similar to *L. robusta* from which it can be distinguished by its slightly larger body size; longer thumb (7.5−8.3 mm, ver-sus 5.2−6.9 mm); dark chestnut to chocolate brown color of dorsum (pale chestnut to orange yellow in *L. robusta*); and the even arc formed by the upper incisors (outer upper incisors of *L. robusta* are comparatively shorter than those of *L. chocoana*. *Lonchophylla chocoana* also is larger than *L. handleyi*, which occurs east of the Andes in the west-ern Amazon basin of Ecuador and Peru, and is exceeded in size only by the sympatric congener *L. orcesi*. M. E. Thomas (1972), Alberico and Orejuela (1982) and Alberico (1987) identified specimens from southwestern Colombia as *L. handleyi*.

Lonchophylla concava Goldman, 1914
Goldman's Nectar Bat

SYNONYMS:

Lonchophylla concava Goldman, 1914a:2; type locality "Cana," Darién, Panama.

L[onchophylla]. m[ordax]. concava: Handley, 1966c:763; name combination.

DISTRIBUTION: *Lonchophylla concava* occurs in western Colombia and Ecuador. Elsewhere it is found from Panama north into southern Nicaragua.

MARGINAL LOCALITIES (Map 125; localities listed from north to south): COLOMBIA (Marinkelle and Ca-dena 1972, except as noted): Bolívar, old fortifications at Cartagena; Santander, El Hoyo; Quindío, Armenia; Valle del Cauca, Anchicayá (Cuervo-Díaz, Hernández-Camacho, and Cadena 1986). ECUADOR: Esmeraldas, Mataje (Al-buja 1999); Pichincha, Centro Científico Río Palenque (Al-buja 1999); Manabí, 45 km NE of Chone (R. H. Baker 1974).

SUBSPECIES: We consider *L. concava* to be monotypic.

NATURAL HISTORY: An inhabitant of lowland tropi-cal humid forests, *L. concava* has been caught in a variety of habitat types including pasture edges and avocado plan-tations. Howell and Burch (1974) recovered nectar, pollen, and the remains of adult moths from Costa Rican speci-mens. Ubelaker, Specian, and Duszynski (1977) attributed two protozoan endoparasites, and Guerrero (1997) listed a streblid batfly as from *L. mordax*, but the hosts are most likely *L. concava*. The karyotype is unknown.

REMARKS: Ascorra, Gorchov, and Cornejo (1989) listed *L. concava* from two localities on the lower Palcazú,

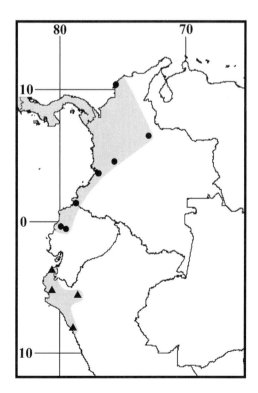

Map 125 Marginal localities for *Lonchophylla concava* ●
and *Lonchophylla hesperia* ▲

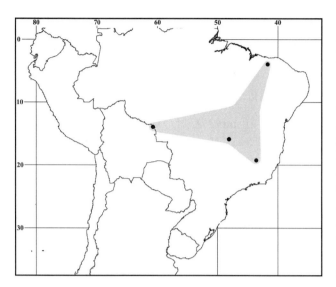

Map 126 Marginal localities for *Lonchophylla dekeyseri* ●

Pasco, Peru, and they later (1994) listed *L. mordax* along with *L. thomasi* from the Centro de Investigaciones Jenaro Herrera, Loreto, Peru. The specimens they identified as either *L. concava* or *L. mordax* were recently reidentified as *L. thomasi* by S. Solari (pers. comm.).

Lonchophylla concava can be easily distinguished from *L. mordax* by the relatively low, broadly rounded coronoid process of the mandible and the longer, relatively attenuated angular process. The septum dividing the basisphenoid pits is about equal to, or narrower than, the width of either pterygoid process(wider in *L. mordax*); the first lower premolar is relatively and absolutely larger than the first lower premolar in *L. mordax*; and the second upper premolar has a small to obsolete internal cingular cuspule that lacks any sign of root support (lingual cingular cuspule well developed and supported by a root in *L. mordax*). *Lonchophylla* is proving to be much more diverse than assumed by Handley (1966c) when he synonymized *L. concava* under *L. mordax*, and the genus needs to be revised.

Lonchophylla dekeyseri Taddei, Vizotto, and Sazima, 1983
Dekeyser's Nectar Bat
SYNONYM:
Lonchophylla dekeyseri Taddei, Vizotto, and Sazima, 1983:626; type locality "Parque Nacional de Brasília, Distrito Federal, Brasil."

DISTRIBUTION: *Lonchophylla dekeyseri* is known from Brazil and eastern Bolivia.

MARGINAL LOCALITIES (Map 126): BRAZIL: Piauí, Parque National de Sete Cidades (Taddei, Vizotto, and Sazima 1983); Distrito Federal, Toca do Falcão (Bredt, Uieda, and Magalhães 1999); Minas Gerais, Serra do Cipó (Tad-

dei, Vizotto, and Sazima 1983). BOLIVIA: Santa Cruz, Parque Nacional Noël Kempff Mercado, Huanchaca I (L. H. Emmons, pers. comm.).

SUBSPECIES: We treat *L. dekeyseri* as monotypic.

NATURAL HISTORY: *Lonchophylla dekeyseri* uses caves, tunnels, and crevasses as day roosts. I. Sazima, Vogel, and Sazima (1989) reported *L. dekeyseri* feeding on the flowers of *Bauhinia rufa* in the Serra de Cipó, Minas Gerais, Brazil. In their 12-month study of the activity pattern and diet of *L. dekeyseri*, Coelho and Marinho-Filho (2002) netted bats monthly at the entrances to caves. The found this species primarily feeds on flowers of the tree genera *Pseudobombax*, *Bauhinia*, *Lafoensia*, *Ruellia*, and *Inga*, during the dry season (April-December) and consumes more fruits (including *Piper* sp. and *Cecropia* sp.) and insects (identifiable fragments were of beetles) during the rainy season. Bredt, Uieda, and Magalhães (1999) reported pregnant females in March, April, May, and June, but commented that they never found females with young. Graciolli and Coelho (2001) and Graciolli and Aguiar (2002) reported *Trichobius lonchophyllae* (Streblidae) from *L. dekeyseri*. The karyotype is unknown.

Lonchophylla handleyi Hill, 1980
Handley's Nectar Bat
SYNONYMS:
Lonchophylla robusta: Tuttle, 1970:68; not *Lonchophylla robusta* Miller, 1912.
Lonchophylla handleyi Hill, 1980:233; type locality "Los Tayos, Morona Santiago Province, Ecuador."

DISTRIBUTION: *Lonchophylla handleyi* occurs in eastern Ecuador, and eastern Peru.

MARGINAL LOCALITIES (Map 124): ECUADOR: Morona Santiago, Yaupi (Hill 1980). PERU: Amazonas, 10 trail km SW of La Peca (Graham and Barkley 1984); Ucayali, 3 km NE of Abra Divisoria (Graham and Barkley 1984); Pasco, San Juan (Hill 1980); Junín, 3.2 km N of Vitoc (Gardner 1976); Huánuco, Cueva de las Lechuzas (Bowles, Cope, and Cope 1979).

SUBSPECIES: We treat *L. handleyi* as monotypic.

NATURAL HISTORY: Tuttle (1970) netted *L. handleyi* (identified as *L. robusta*) in primary and secondary tropical humid forest and near a banana grove in Peru (see Hill 1980). The testes of the male measured 3.0 × 2.0 mm on 5 August. The specimen Gardner (1976) reported (as *L. robusta*) was taken at the edge of an avocado orchard in second-growth humid forest. The type locality is a large cave, and Graham (1988) associated the species with large caves in Peru. The karyotype is unknown.

REMARKS: The specimens Alberico and Orejuela (1982) and Alberico (1987) reported as *L. handleyi* are now known as *L. chocoana*.

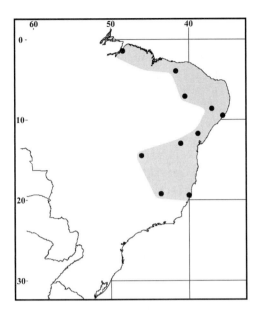

Map 127 Marginal localities for *Lonchophylla mordax* ●

Lonchophylla hesperia G. M. Allen, 1908
Western Nectar Bat

SYNONYM:

Lonchophylla hesperia G. M. Allen, 1908:35; type locality "Zorritos," Tumbes, Peru.

DISTRIBUTION: *Lonchophylla hesperia* is endemic to northwestern Peru.

MARGINAL LOCALITIES (Map 125): PERU: Tumbes, Zorritos (type locality of *Lonchophylla hesperia* G. M. Allen); Amazonas, 15 km WNW Bagua Grande (Koopman 1978); La Libertad, Trujillo (Gardner 1976); Piura, Piura (Koopman 1978).

SUBSPECIES: *Lonchophylla hesperia* is monotypic.

NATURAL HISTORY: Gardner (1976) reported a specimen from subtropical dry forest in a rain-shadow desert of the Río Marañon Valley. The karyotype is unknown.

Lonchophylla mordax O. Thomas, 1903
Brazilian Nectar Bat

SYNONYM:

Lonchophylla mordax O. Thomas, 1903f:459; type locality "Lamarão," 300 m, Bahia, Brazil.

DISTRIBUTION: *Lonchophylla mordax* is known from Brazil.

MARGINAL LOCALITIES (Map 127; Brazilian localities listed from north to south): BRAZIL: Pará, Belém (Koopman 1981); Piauí, Piracuruca (Taddei, Vizotto, and Sazima 1978); Ceará, Km 19 on Route CE 96, 4 km SE of Nova Olinda (Mares et al. 1981); Pernambuco, Buique (I. Sazima, Vizotto, and Taddei 1978); Alagoas, Canoas (C. O. C. Vieira 1953); Bahia, Lamarão (type locality of *Lonchophylla mordax* O. Thomas); Bahia, Caverna Poço

Encantado (Gregorin and Mendes 1999); Minas Gerais, Serra do Cipó, Jaboticatubas (Taddei, Vizotto, and Sazima 1978); Goiás, Mambaí (USP 13662); Espírito Santo, Município de Linhares (Pedro and Passos 1995).

SUBSPECIES: We treat *L. mordax* as monotypic.

NATURAL HISTORY: This species inhabits lowland tropical humid forests. Ruschi (1953a) stated that *L. mordax* consumes insects, nectar, pollen, and succulent fruits. Vogel, Lopes, and Machado (2005) reported *L. mordax* taking nectar from *Mimosa lewisi* in the Catimbau Valley, Pernambuco, Brazil. The hosts of the two protozoan endoparasites reported by Ubelaker, Specian, and Duszynski (1977), and the host of the streblid batfly Guerrero (1997) listed as from *L. mordax*, most likely represent *L. concava*. The karyotype is unknown.

REMARKS: Cabrera (1958:75) included Bolivia in the distribution *L. mordax*; however, S. Anderson (1997) did not include the species in the Bolivian fauna. Ascorra, Gorchov, and Cornejo (1994) listed *L. mordax* along with *L. thomasi* from the Centro de Investigaciones Jenaro Herrera, Loreto, Peru. However, S. Solari (pers. comm.) recently reidentified these specimens as *L. thomasi*.

Lonchophylla mordax differs from *L. concava* by having a broader septum separating basisphenoid pits, a higher and more angular apex on the coronoid process of the mandible, and by a well-developed and rooted lingual cuspule on the medial cingulum of the second upper premolar (see Remarks under *L. concava*). D. C. Carter and Dolan (1978) gave information on and measurements of the holotype of *L. mordax* O. Thomas.

Lonchophylla orcesi Albuja and Gardner, 2004
Orces's Nectar Bat

SYNONYMS:

Lonchophylla sp. A., Albuja, 1999:96; an undescribed species from Ecuador.

Lonchophylla orcesi Albuja and Gardner, 2005:443; type locality "Los Pambiles, 00°32′N, 78°38′W, Río Piedras, Cordillera de Toisán, Esmeraldas Province, Ecuador."

DISTRIBUTION: *Lonchophylla orcesi* is known only from the type locality in northwestern Ecuador.

MARGINAL LOCALITIES (Map 128): ECUADOR: Esmeraldas, Los Pambiles (type locality of *Lonchophylla orcesi* Albuja and Gardner).

SUBSPECIES: *Lonchophylla orcesi* is known only from the holotype.

NATURAL HISTORY: *Lonchophylla orcesi* was netted together with specimens of *L. robusta* and *L. chocoana* on a small, flat, and densely forested river terrace at Los Pambiles on the banks of the Río Piedras at an elevation of 1,200 m. The collection site is in the Reserva Ecológica Cotacachi-Cayapas. Albuja et al. (1980) described the forest as part

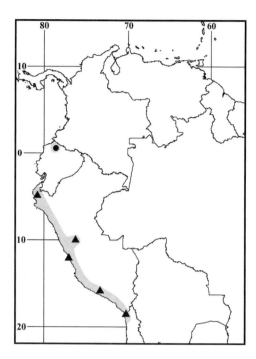

Map 128 Marginal localities for *Lonchophylla orcesi* ● and *Platalina genovensium* ▲

of the Subtropical Zoogeographic Region, where the mean annual temperature is about 20°C, and annual precipitation can exceed 4,000 mm.

REMARKS: *Lonchophylla orcesi*, with a forearm length of 47 mm and a greatest length of skull (excluding incisors) of 30.4 mm, is the largest member of the genus described to date. The rostrum is approximately as long as the braincase, and the upper inner incisors are especially long, procumbent, and have spatulate tips. The interfemoral membrane is relatively long and lacks a fringe of hair along its border. Unlike sympatric *L. chocoana* and *L. robusta*, both of which have a dorsal pelage approximating some shade of chestnut or reddish brown, the dorsal pelage is pale brown and the basal band of the hair is paler, nearly white. Although known from a single specimen, *L. orcesi* is expected elsewhere in northwestern Ecuador and adjacent Colombia.

Lonchophylla robusta Miller, 1912
Big Nectar Bat
SYNONYMS:
Lonchophylla robusta Miller, 1912a:23; type locality "cave on Chilibrillo River," Panama, Panama.
?*Choeronycteris mexicana ponsi* Pirlot, 1967:269; type locality "Kasmera," Zulia, Venezuela.
DISTRIBUTION: *Lonchophylla robusta* is known from northwestern Venezuela, western and central Colombia, Ecuador, and Peru. Elsewhere the species occurs in Central America northward at least into Nicaragua.

MARGINAL LOCALITIES (Map 129): VENEZUELA (Handley 1976): Zulia, near Cerro Azul, 33 km NW of La Paz; Barinas, near Altamira. COLOMBIA: Santander, San Gil (Sanborn 1941); Meta, Municipio de San Juan de Arama (Sánchez-Palomino, Rivas-Pava, and Cadena 1993). ECUADOR: Morona-Santiago, Yaupi (Hill 1980). PERU: Amazonas, 20 km SW Chiriaco (Graham and Barkley 1984); Pasco, Cerro Jonatán (S. Solari, Pacheco, and Vivar 1999). ECUADOR: Bolívar, Barraganete (Albuja 1983); Esmeraldas, San Lorenzo (Albuja 1983). COLOMBIA: Valle del Cauca, Zabaletas (M. E. Thomas 1972).

SUBSPECIES: We treat *L. robusta* as monotypic.

NATURAL HISTORY: Handley (1976) reported *L. robusta* from the eastern slopes and foothills of the Venezuelan Andes and the Sierra de Perijá where all specimens collected were caught in mistnets. The majority was taken in tropical humid forest. Howell and Burch (1974) found the remains of insects in the digestive tracts of Costa Rican specimens, and Fleming, Hooper, and Wilson (1972) reported 10% plant material and 90% insect remains in the stomach of a Panamanian specimen. Webb and Loomis (1977) reported four species of mites and nine species of streblid batflies as ectoparasites. Later, Guerrero (1997) listed 7 genera and 11 species of batflies known to parasitize *L. robusta*. The karyotype is $2n = 28$, FN = 50 (R. J. Baker 1973).

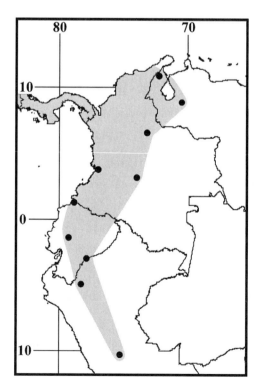

Map 129 Marginal localities for *Lonchophylla robusta* ●

REMARKS: Pirlot (1967) reported capturing two specimens (a male and female) of *Choeronycteris mexicana* in Venezuela at Kasmera, the Biological Station of the University of Maracaibo, 150 km SW of Maracaibo, Zulia. Pirlot identified the specimens as *Choeronycteris* on the basis of body size, shape of the cusps of the lower premolars, and his interpretation of the presence or absence of lower incisors in adults. Pirlot (1967:268–69) apparently interpreted Hall and Kelson's (1959:119) statement, "lower incisors absent in adults," to mean that incisors were present, but weak and subject to loss in adults because he noted that the female had four lower incisors and the male, missing one of the outer, had only three. Because of the red coloration of the fur, he named these as a new subspecies, *C. mexicana ponsi*. *Choeronycteris* had not been reported from South America previously; indeed, this record would represent approximately a 2,000-km range extension for the species. Unfortunately, Pirlot did not publish other data that would have helped to confirm his claim and the type specimens have been lost (P. Pirlot, pers. comm.). We do not believe that Pirlot's two specimens represent *Choeronycteris*. Although we considered possible identities other than *L. robusta*, we have placed *Choeronycteris mexicana ponsi* in the synonymy of this species because of the combination of large size, reddish pelage, and presence of lower incisors. Furthermore, although *L. robusta* subsequently has been reported from Kasmera (Handley 1976), Pirlot (1967) did not mention it in his report and he may have been unfamiliar with the species at that time.

Lonchophylla thomasi J. A. Allen, 1904
Thomas's Nectar Bat

SYNONYM:

Lonchophylla thomasi J. A. Allen, 1904b:230; type locality "Cuidad [*sic*] Bolivar," Bolívar, Venezuela.

DISTRIBUTION: *Lonchophylla thomasi* is known from southern and southeastern Venezuela, Guyana, French Guiana, northern Bolivia, northwestern Ecuador, eastern Peru, and Brazil as far south as the southern margin of the Rio Amazonas. Elsewhere, the species occurs in Panama (Handley 1966c).

MARGINAL LOCALITIES (Map 130): VENEZUELA: Bolívar, Ciudad Bolívar (type locality of *Lonchophylla thomasi* J. A. Allen); Bolívar, El Manaco, 59 km SE of El Dorado (Handley 1976). GUYANA: Cuyuni-Mazaruni, 27 Mile Camp (Hill 1965). SURINAM: Para, Carolina Kreek (FMNH 95390). FRENCH GUIANA: Paracou (Simmons and Voss 1998). BRAZIL: Amapá, Igarapé Novo (Taddei, Vizotto, and Sazima 1983); Pará, Belém (I. Sazima, Vizotto, and Taddei 1978); Tocantins, Município Aliança do Tocantins (Nunes et al. 2005); Rondônia, Ji-Paraná (Marques

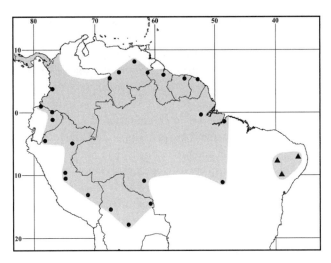

Map 130 Marginal localities for *Lonchophylla thomasi* ● and *Xeronycteris vieirai* ▲

1989). BOLIVIA: Santa Cruz, Los Fierros (L. H. Emmons, pers. comm.); Santa Cruz, El Tunal (S. Anderson 1997); La Paz, 1 mile W of Puerto Linares (S. Anderson, Koopman, and Creighton 1982). PERU: Cusco, Consuelo (V. Pacheco et al. 1993); Pasco, San Juan (Koopman 1978); Huánuco, Panguana Biological Station (Hutterer et al. 1995); Loreto, Centro de Investigaciones Jenaro Herrera (Ascorra, Gorchov, and Cornejo 1994); Amazonas, Huampami (J. L. Patton, Berlin, and Berlin 1982). ECUADOR: Pastaza, Tiguino (USNM 574510); Sucumbíos, Santa Cecilia (Webster and Jones 1984); Esmeraldas, Urbina (Albuja 1983). COLOMBIA: Valle de Cauca, Río Zabaletas (USNM 483359). VENEZUELA: Amazonas, 25 km S of Puerto Ayacucho (Handley 1976); Bolívar, 20 km S of Guaniamo (Lee, Lim, and Hanson 2000).

SUBSPECIES: We consider *L. thomasi* to be monotypic.

NATURAL HISTORY: Handley (1976) reported Venezuelan specimens collected from hollow trees and netted in forest clearings. Gardner (1977c) observed this species feeding at banana flowers. Ascorra, Solari, and Wilson (1996) found pollen and the remains of insects in the feces of *L. thomasi* from southeastern Peru. Ubelaker, Specian, and Duszynski (1977) listed a *Trypanosoma cruzi*-like hemoflagellate known from *L. thomasi*. Guerrero (1985a, 1997) listed four species of streblid batflies. The karyotype is $2n = 30$–32, FN 34–38 (see summary in R. J. Baker et al. 1982).

Genus *Platalina* O. Thomas, 1928

This monotypic genus, known only from western Peru and northern Chile, is the largest lonchophylline bat (forearm 46.0–50.0 mm, length of skull 31.1–32.7 mm). The pelage is pale and comparatively long; the basal 2/3 of each hair is whitish and the terminal 1/3 brown. *Platalina genovensium* may be identified by its greatly elongated muzzle that bears numerous short vibrissae, its long and sparsely furred interfemoral membrane, and the short tail (about 1/3 the length of the uropatagium). The rostrum is longer than the braincase and the mesopterygoid fossa is relatively long and its anterior margin U-shaped. Upper and lower incisors are large, spatulate, and procumbent. The canines are slender and lack a conspicuous cingulum. Premolars and molars are elongated and the upper molars lack a paracone and mesostyle, but retain a comparatively well-developed parastyle (see Phillips 1971 for a more comprehensive description of the dentition). The dental formula is 2/2, 1/1, 2/3, 3/3 × 2 = 34.

SYNONYM:

Platalina O. Thomas, 1928a:120; type species *Platalina genovensium* O. Thomas, 1928a, by original designation.

Platalina genovensium O. Thomas, 1928
Peruvian Long-tongued Bat
SYNONYM:

Platalina genovensium O. Thomas, 1928a:121; type locality "neighborhood of Lima," Lima, Peru.

DISTRIBUTION: *Platalina genovensium* occurs in western Peru and northern Chile.

MARGINAL LOCALITIES (Map 128): PERU: Piura, Angola (Aellen 1966); Huánuco, Huánuco (Sanborn 1936); Lima, Lima (type locality of *Platalina genovensium* O. Thomas); Arequipa, Cariveli [*sic*] (Walker et al. 1964). CHILE: Tarapacá, Valle de Azapa (Galaz, Torres-Mura, and Yáñez 1999).

SUBSPECIES: *Platalina genovensium* is monotypic.

NATURAL HISTORY: *Platalina genovensium* has been collected in mines and grottos (Sanborn 1936; Ortiz de la Puente 1951), or netted in xeric habitats (Jiménez and Péfaur 1982). With the exception of Sanborn's (1936) Huánuco record (taken in a deserted mine along with a *Desmodus rotundus*), all specimens have come from localities west of the Andes. The elevational range is from near sea level (Lima and Angolo in the departments of Lima and Piura, respectively) to near 2,300 m in the Department of Arequipa.

Platalina genovensium has acquired a highly derived feeding apparatus independent of the specialized glossophagine genera (e.g., *Choeroniscus*). Although little is known of its feeding ecology, the hyoid-lingual morphology suggests that it is well adapted for the nectarivorous niche (Griffiths 1982). Jiménez and Péfaur (1982) stated that *P. genovensium* consumed the pollen and nectar of the flowers of a large columnar cactus (*Weberbaurocereus* sp.).

Sahley and Baraybar (1996) also found a strong association between *P. genovensium* and columnar cacti. T. H. Fleming (1995:90) attributed to C. Sahley the characterization of *P. genovensium* as "a migratory cactus specialist."

Swanepoel and Genoways (1979) gave measurements of the five specimens they examined. Jiménez and Péfaur (1982) reported on additional specimens and provided measurements along with a summary of the known biology of *P. genovensium*. The karyotype is unknown.

REMARKS: The specimen from Caravalí, Arequipa, in the National Museum of Natural History (USNM 268765), first reported by Walker et al. (1964) and catalogued as a female as indicated by Swanepoel and Genoways (1979:94), is a male preserved in alcohol, with the skull removed. There is another fluid-preserved specimen (intact) at the National Museum (USNM 268766) from the same locality that apparently is the only *P. genovensium* female known from Peru. One of the three specimens Galaz, Torres-Mura, and Yáñez (1999) reported from Chile is a female. D. C. Carter and Dolan (1978) gave information and measurements on the holotype of *Platalina genovensium*, O. Thomas.

Genus *Xeronycteris* Gregorin and Ditchfield, 2005

Xeronycteris is a recently described monotypic genus known only from four specimens collected at three localities in the Northeast of Brazil. It is a medium-sized pollen and nectar-feeding bat (forearm 35.4–38.1 mm, condyloincisive length 24.2–25.2 mm, breadth of braincase 8.6–8.8 mm) characterized by dentition that lack many of the structures present in other lonchophyllines. Distinctive features of the skull include a long rostrum (approximately as long as the braincase) and a long palate that extends posteriorly to the level of the foramen for the optic nerve. The dentition is unusual in that the M3 is located well anterior to malar root of the zygoma. The premolars lack a cingulum, and the structural elements of molars (styles, cristae, and flexae in upper and stylids, cristids, and flexids in lower) are greatly reduced or absent (Gregorin and Ditchfield 2005). The dental formula is 2/2. 1/1, 2/3, 3/3 × 2 = 34.

SYNONYM:

Xeronycteris Gregorin and Ditchfield, 2005:405; type species *Xeronycteris vieirai* Gregorin and Ditchfield, 2005, by original designation.

Xeronycteris vieirai Gregorin and Ditchfield, 2005
Vieira's Flower Bat
SYNONYM:

Xeronycteris vieirai Gregorin and Ditchfield, 2005:405; type locality "Fazenda Espírito Santo, Município de Soledade, state of Paraíba, Brazil (07°05′S, 36°21′W)."

DISTRIBUTION: *Xeronycteris vieirai* is known from the Northeast of Brazil where it has been taken in the states of Bahia, Paraíba, and Pernambuco.

MARGINAL LOCALITIES (Map 130; from Gregorin and Ditchfield 2005): BRAZIL: Paraíba, Fazenda Espírito Santo (type locality of *Xeronycteris vieirai* Gregorin and Ditchfield); Pernambuco, Serra da Gritadeira, 18 km SSW of Exu; Bahia, Município de Cocorobó.

SUBSPECIES: *Xeronycteris vieirai* is monotypic.

NATURAL HISTORY: Aside from occupying semiarid caatinga habitat and having a diet of pollen, nectar, and probably some insects, the natural history of this species is unknown.

Subfamily Phyllostominae Gray, 1825
Stephen L. Williams and Hugh H. Genoways

The subfamily Phyllostominae is distributed from the southern United States (Arizona, California, and southern Nevada), southward into northern Argentina, Paraguay, and southern Brazil. South American phyllostomines are primarily restricted to the mainland, but also occur on a few major islands off the coast of South America, such as Margarita Island (Venezuela), Trinidad and Tobago, and the Netherlands Antilles, as well as the Greater and Lesser Antilles.

The number of genera and species recognized in the subfamily depends on the taxonomic interpretations of the content of the genera *Lophostoma*, *Micronycteris*, *Mimon*, *Phyllostomus*, and *Tonatia*. In near agreement with the interpretation by Simmons and Voss (1998), plus a few subsequent additions, we here recognize 16 genera and 42 species. Fifteen genera and 41 species occur in South America, where 12 species are endemic. The following nine genera are considered to be monotypic: *Chrotopterus*, *Lampronycteris*, *Macrophyllum*, *Neonycteris*, *Phylloderma*, *Trachops*, *Trinycteris*, and *Vampyrum*. In contrast, *Lophostoma* contains at least seven species and *Micronycteris* contains at least eight.

Most publications providing useful information about the subfamily Phyllostominae are restricted to specific taxa or to geographical areas. Other references, such as Goodwin and Greenhall (1961), Husson (1962, 1978), Hall (1981), and Koopman (1993, 1994) either are or contain good general references to the subfamily. Specific but dated information is available in "Biology of bats of the New World family Phyllostomatidae" (R. J. Baker, Jones, and Carter 1976, 1977, 1979) and in "Mammalian biology in South America" (Mares and Genoways 1982). A number of competing classifications attempting to arrange taxa phylogenetically within the Phyllostominae have appeared in recent years

(e.g., R. J. Baker, Hood, and Honeycutt 1989; Van Den Bussche 1992; Koopman 1993; R. J. Baker et al. 2000; Wetterer, Rockman, and Simmons 2000; K. E. Jones et al. 2002; R. J. Baker et al. 2003). These highlight the strong research interest in the phyllostomids; however, we have not segregated the genera of the Phyllostominae according to the phylogenies suggested in these publications.

In preparing this review, we examined specimens from the National Museum of Natural History; American Museum of Natural History; Field Museum; Museum of Comparative Zoology, Harvard University; Royal Ontario Museum; and Carnegie Museum of Natural History. We extend our appreciation to the individuals at these institutions who provided assistance, particularly K. F. Koopman, R. D. Fisher, R. M. Timm, J. L. Eger, R. L. Honeycutt; T. A. Hiener, M. A. Schmidt, and K. D. Williams typed the manuscript; A. L. Gardner provided information for the synonymies; K. D. Williams assisted with the final preparation of the manuscript.

The subfamily Phyllostominae is characterized by a well-defined noseleaf and a molar configuration in which the cusps and commissures maintain a W-pattern (Miller 1907b: 118, 122–23). The interfemoral membrane is usually well developed. The tail may differ among taxa, from total absence, to being long and extending to the margin of the interfemoral membrane.

KEY TO SOUTH AMERICAN GENERA OF PHYLLOSTOMINAE:

1. One pair of lower incisors 2
1'. Two pairs of lower incisors 5
2. Tail rudimentary, not readily visible; forearm more than 70 mm; three lower premolars; greatest length of skull more than 35 mm . . . . *Chrotopterus*
2'. Tail well developed; forearm less than 70 mm; greatest length of skull less than 35 mm 3
3. Two lower premolars . . . . *Mimon*
3'. Three lower premolars 4
4. Postorbital constriction less than 5 mm . . . *Lophostoma*
4'. Postorbital constriction more than 5 mm . . . . . *Tonatia*
5. Tail enclosed in, and extending to posterior margin, of interfemoral membrane 6
5'. Tail enclosed in interfemoral membrane, but not extending to posterior margin 7
6. Length of noseleaf more than three times its width; forearm more than 40 mm; rostrum elongated with a distinct concavity in interorbital region . . . . *Lonchorhina*
6'. Length of noseleaf less than three times its width; forearm less than 40 mm; rostrum not elongated and lacks a distinct concavity in interorbital region . . . .
. . . . *Macrophyllum*
7. Two lower premolars . . . . *Phyllostomus*

7'. Three lower premolars. 8
8. Forearm longer than 100 mm; tail absent; rostrum as long as braincase . . . . *Vampyrum*
8'. Forearm shorter than 75 mm; tail present; rostrum shorter than braincase. 9
9. Conspicuous, papilla-like protuberances on lips and chin; margin of noseleaf finely serrated; forearm 55–65 mm. . . . . *Trachops*
9'. Lips and chin without papilla-like protuberances; margin of noseleaf lacking fine serrations. 10
10. Forearm 65–74 mm; tips of wings whitish; greatest length of skull 30–33 mm. . . . . *Phylloderma*
10'. Forearm 33–58 mm; greatest length of skull 17–28 mm 11
11. First upper incisors similar to canines in length; first upper premolar (P3) having accessory cusps on lingual and posterior margins. . . . . *Glyphonycteris*
11'. First upper incisors distinctly shorter and narrower than canines; first upper premolar (P3) lacking accessory cusps, only the main cusp present. 12
12. Forearm longer than 35 mm; greatest length of skull more than 20 mm 13
12'. Forearm shorter than 35 mm; greatest length of skull less than 20 mm . . . . *Neonycteris*
13. Length of ear (from notch) less than 16 mm; calcar about the same length as foot; upper incisors chisel-shaped and in line with canines. . . . . . *Lampronycteris*
13'. Length of ear (from notch) more than 16 mm; calcar shorter than foot; faint gray medial stripe often present on lower back; upper incisors not chisel-shaped; upper incisors projected forward and not in line with canines . . . . *Trinycteris*

Genus *Chrotopterus* W. Peters, 1865

The monotypic genus *Chrotopterus* is represented by *Chrotopterus auritus*, one of the larger bats in South America (forearm 74–83 mm, greatest length of skull 34–37 mm). The genus is characterized by relatively long, woolly pelage; a rudimentary tail (may appear absent); a calcar that is longer than the foot; and a lower tooth row that has three premolars and one incisor. The dental formula is 2/1, 1/1, 2/3, 3/3 × 2 = 32 (also in *Tonatia* and *Lophostoma*).

SYNONYMS:

Vampyrus: W. Peters, 1856a:305; not *Vampyrus* Leach, 1821b:79.

Chrotopterus W. Peters, 1865c:505; type species *Vampyrus auritus* W. Peters, 1856a, by original designation; described as a subgenus of *Vampyrus* Leach, 1821b.

Chrotoptems Dobson, 1878:471; in synonymy; incorrect subsequent spelling of *Chrotopterus* W. Peters.

Chrotoperus R. J. Baker, R. M. Fonseca, D. A. Parish,

C. J. Phillips, and F. G. Hoffmann, 2004:4; incorrect subsequent spelling of *Chrotopterus* W. Peters.

Chrotopterus auritus (W. Peters, 1856)
Great Woolly Bat

SYNONYMS:

Vampyrus auritus W. Peters, 1856a:305, type localities "Mexico et Guiana"; restricted to Mexico by W. Peters (1856b:415).

[*Vampyrus (]Chrotopterus[)] auritus*: W. Peters, 1865c:505; name combination.

Chrotopterus auritus: Hensel, 1872:20; first use of current name combination.

[*Vampyrus (Vampyrus)] auritus*: Trouessart, 1897:153; name combination.

Chrotopterus auritus guianae O. Thomas, 1905b:308; type locality "La Vuelta, Lower Orinoco," Bolívar, Venezuela.

Chrotopterus auritus australis O. Thomas, 1905b:308; type locality "Concepcion," Concepción, Paraguay.

DISTRIBUTION: *Chrotopterus auritus* is known from Colombia, Venezuela, Guyana, Surinam, French Guiana, Brazil, Ecuador, Peru, Bolivia, Paraguay, and northern Argentina. The species also occurs in Central America and southern Mexico.

MARGINAL LOCALITIES (Map 131): VENEZUELA (Handley 1976): Falcón, 12 km ENE of Mirimire; Bolívar, El Manaco, 56 km SE of El Dorado. GUYANA: Cuyuni-Mazaruni, 24 miles from Bartica on Potaro Road (Hill 1965). SURINAM: Sipaliwini, Raleigh Falls (S. L. Williams

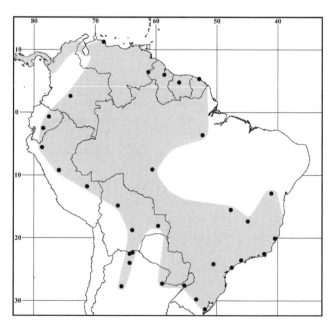

Map 131 Marginal localities for *Chrotopterus auritus* ●

and Genoways 1980a). FRENCH GUIANA: Paracou (Simmons and Voss 1998). BRAZIL: Pará, Rio Xingu, 52 km SW of Altamira (Voss and Emmons 1996); Mato Grosso, Aripuanã (Mok, Luizão, and Silva 1982); Distrito Federal, Gruta Dança dos Vampiros (Bredt, Uieda, and Magalhães 1999); Minas Gerais, Pirapora (C. O. C. Vieira 1955); Bahia, Caverna Poço Encantado (Gregorin and Mendes 1999); Espírito Santo, Santa Leopoldina (Taddei 1975); Rio de Janeiro, Poço das Antas Biological Reserve (Baptista and Mello 2001);São Paulo, Biritiba Mirim (McNab 1969); São Paulo, Iguapé (C. O. C. Vieira 1955); Paraná, Fazenda Monte Alegre (N. R. Reis, Peracchi, and Sekiama 1999); Rio Grande do Sul, São Lourenço (C. O. C. Vieira 1955); Rio Grande do Sul, Restinga Seca (Fabián, Rui, and Oliveira 1999). ARGENTINA: Misiones, Leandro N. Alem (Barquez, Mares, and Braun 1999); Chaco, Colonia Benitez (Cabrera 1938). BOLIVIA: Santa Cruz, Río Tucavaca, 24 km by road N of Santiago de Chiquitos (S. Anderson 1997). ARGENTINA: Salta, 6 km W of Piquirenda Viejo (Barquez, Mares, and Braun 1999); Jujuy, Palma Sola (Barquez and Ojeda 1992); Tucumán, Dique San Ignacio, La Cocha (Barquez, Mares, and Braun 1999); Salta, Vado de Arrazayal (Barquez 1984a). BOLIVIA: Santa Cruz, 14.5 km by road NW of Masicuri (S. Anderson 1997); Beni, Río Matos (S. Anderson 1997). PERU: Madre de Dios, Cocha Cashu (Terborgh, Fitzpatrick, and Emmons 1984); Cusco, Cashiriari-2 (S. Solari et al. 2001c); Huánuco, vicinity of Tingo María (Bowles, Cope, and Cope 1979); Cajamarca, Bellavista (MCZ 17060). ECUADOR (Albuja and Mena 1991): Azuay, San José Grande; Napo, Huamaní. COLOMBIA: Meta, Cabaña Duda (Lemke et al. 1982).

SUBSPECIES: We treat C. auritus as monotypic pending revision of the species (see Remarks).

NATURAL HISTORY: Chrotopterus auritus is associated with forested habitats (Hill 1965; Handley 1966c, 1976; S. L. Williams and Genoways 1980a). This species roosts in caves (W. B. Davis, Carter, and Pine 1964; Handley 1976; Bowles, Cope, and Cope 1979), hollow termite nests (Sanborn 1932b), and hollow trees (Medellín 1988, 1989). Although C. auritus is considered to be carnivorous, because of reports of its eating small vertebrates (Tuttle 1967; Villa-R. and Villa-Cornejo 1969; Gardner 1977c; I. Sazima 1978), insects and fruit are also included as part of the diet (Gardner 1977c). Medellín (1988) found the remains of insects (Cerambycidae, Scarabeidae, and Sphingidae), birds (3 species identified and 11 unidentified), and mammals (shrews and five species of identified rodents plus several unidentified) dropped under a tree roost of C. auritus in the state of Chiapas, Mexico. Delpietro, Contreras, and Konolaisen (1992) commented on foraging in groups and behavior they interpreted as mobbing when a group of C. auritus milled around a group member

as it vocalized after being caught in a mistnet. Webb and Loomis (1977) listed a mite, a tick, and one nycteribiid and two streblid batflies reported from C. auritus. R. Guerrero (1985a) added two streblid batflies, and later (1997) listed five species of which one had not been reported earlier. Graciolli and Carvalho (2001) listed the nycteribiid batfly Basilia ortizi as known from Brazilian C. auritus. Medellín (1989) reviewed this species in his Mammalian Species account and provided measurements, illustrations, and a synopsis of natural history information. The karyotype is $2n = 28$, FN = 52 (R. J. Baker et al. 1982; Moreille-Versute, Taddei, and Varella-Garcia 1992).

REMARKS: Handley (1966c), Myers and Wetzel (1983), Koopman (1994), and Simmons and Voss (1998) questioned the validity of the recognized subspecies. However, other recent authors (Hill 1965; Taddei 1975; J. K. Jones and Carter 1976; S. L. Williams and Genoways 1980a; Lemke et al. 1982) continued to recognize subspecies. Simmons and Voss (1998) found, represented in their small sample of five from Paracou, French Guiana, the pelage and color-pattern characters that O. Thomas (1905b) used to distinguish australis and guianae from each other, and from auritus. A thorough study of geographic variation in this species is needed.

Dilford C. Carter and Dolan (1978:37) claimed that the type locality of Vampyrus auritus was Santa Catarina, Brazil, and not Mexico as had been presumed. However, W. Peters (1856b:415) definitely stated that the specimen came from Mexico. Earlier, W. Peters (1856a:310) indicated that he had two specimens. The first, a male from Mexico collected by Deppe, and the other collected by Schomburgk in British Guiana. According to A. L. Gardner (pers. comm.), the lectotype of Vampyrus auritus W. Peters is ZMB 10058, and consists of a cleaned skull and skeleton with parts of the body still in alcohol. The specimen, originally catalogued as An. 18795/18796 in the anatomical collection, was part of Deppe's sixth shipment (inventory dated 21 Oct. 1825), which consisted of specimens collected in Oaxaca and Veracruz, Mexico. The exact source of the holotype is unknown.

Chrotopterus colombianus H. E. Anthony, 1920:84, is a junior synonym of Lophostoma silvicolum. According to K. F. Koopman (pers. comm.) the erroneous generic assignment for this taxon was the result of confusion concerning the number of lower premolars in Chrotopterus, perhaps caused by the error in Miller's (1907b:122) key to the genera, indicating that Chrotopterus had only two lower premolars. Measurements for C. auritus have been provided by Taddei (1975), Bowles, Cope, and Cope (1979), Swanepoel and Genoways (1979), S. L. Williams and Genoways (1980a), Hall (1981), Lemke et al. (1982), Myers and Wetzel (1983), and Simmons and Voss (1998).

Genus *Glyphonycteris* O. Thomas, 1896

There are three species recognized in *Glyphonycteris*, varying in size from *G. sylvestris* (forearm 38–44 mm) to *G. daviesi* (forearm 53–58 mm). The genus is characterized by the lack of a cutaneous band across the forehead connecting the ears; a domed braincase; caniniform upper inner incisors; short upper canines that are only a little longer than the incisors; a distinct lingual cingulum on the upper premolars; slightly recurved cusps on P4; and trifid lower incisors. The fourth metacarpal is the shortest and the fifth is longest. The dental formula is 1–2/3, 1/1, 2/3, 3/3 × 2 = 34–36.

SYNONYMS:

Glyphonycteris O. Thomas, 1896b:301; type species *Glyphonycteris sylvestris* O. Thomas, 1896b, by monotypy.

Micronycteris (Glyphonycteris): Sanborn, 1949a:233; name combination.

Barticonycteris Hill, 1965:556; type species *Barticonycteris daviesi* Hill, 1965, by original designation.

REMARKS: As revised by Andersen (1906a), *Glyphonycteris* also included *Schizostoma brachyote* Dobson, 1879 (= *Lampronycteris brachyotis*). Simmons and Voss (1998) compared and contrasted the several taxa traditionally treated as subgenera in *Micronycteris* and explained their reasons for assigning full generic rank to *Glyphonycteris* and for recognizing a more restricted assemblage of species in *Micronycteris*.

KEY TO THE SPECIES OF *GLYPHONYCTERIS*:

1. Forearm longer than 50 mm; dorsal hair brownish throughout; greatest length of skull more than 25 mm; one pair of upper incisors; crowns of lower incisors anteriorly-posteriorly long and transversely narrow .. . . *Glyphonycteris daviesi*
1′. Forearm shorter than 50 mm; dorsal hair tricolored; greatest length of skull less than 25 mm; two pairs of upper incisors, outer incisor hidden by cingulum of canine; lower incisors normal 2
2. Forearm shorter than 44 mm; greatest length of skull less than 22 mm. . . . . *Glyphonycteris sylvestris*
2′. Forearm longer than 44 mm; greatest length of skull more than 21 mm . . . . *Glyphonycteris behnii*

Glyphonycteris behnii (W. Peters, 1865)

Behn's Graybeard Bat

SYNONYMS:

Schizostoma Behnii W. Peters, 1865c:505; type locality "Cuyaba" (= Cuiabá), Mato Grosso, Brazil.

M[icronycteris]. behnii: Miller, 1898:330; name combination.

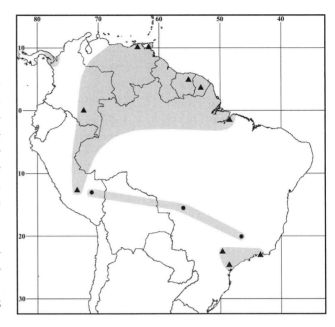

Map 132 Marginal localities for *Glyphonycteris behnii* ● and *Glyphonycteris sylvestris* ▲

Micronycteris (Glyphonycteris) behni: Sanborn, 1949a:231; name combination and incorrect subsequent spelling of *Schizostoma behnii* W. Peters.

[*Glyphonycteris*] *behnii*: Simmons and Voss, 1998:61; first use of current name combination.

DISTRIBUTION: *Glyphonycteris behnii* is known from Brazil and eastern Peru.

MARGINAL LOCALITIES (Map 132): PERU: Cusco, Río Cosñipata (Andersen 1906a). BRAZIL: Mato Grosso, Cuiabá (type locality of *Schizostoma behnii* W. Peters); Minas Gerais, Serra da Canastra (Peracchi and Albuquerque 1985).

SUBSPECIES: We regard *G. behnii* as monotypic.

NATURAL HISTORY: *Glyphonycteris behnii* is the type host for the dichrocoeliid trematode *Parametadelphis compactus* Travassos, 1955; however, we suspect that the host (from Cachimbo, Pará, Brazil) is a misidentified *G. sylvestris*. As determined by the diet of its congeners, *G. behnii* likely consumes a variety of insects and small fruits. Additional information on natural history is unavailable.

REMARKS: *Glyphonycteris behnii*, along with *G. sylvestris* and *G. daviesi*, constitutes a group previously included in *Micronycteris (Glyphonycteris)*. This rare bat is known from four specimens, two from Brazil and two from Peru. The Peruvian locality was cited as located in departamento Puno by Andersen (1906a) and Tuttle (1970), but Koopman (1978) correctly placed Río Cosñipata in departamento Cusco. Simmons and Voss (1998:61) commented on examining several specimens in museums labeled "*Micronycteris behnii*" that proved on examination to be either *G. sylvestris* or some other taxon. They suggested that *G. behnii* could prove to be a senior synonym of *G. sylvestris*. Measurements of *G. behnii* can be found in Andersen (1906a) and in Swanepoel and Genoways (1979). D. C. Carter and Dolan (1978) gave information on and measurements of the holotype of *Schizostoma Behnii* W. Peters.

Glyphonycteris daviesi (Hill, 1965)
Davies's Graybeard Bat

SYNONYMS:

Barticonycteris daviesi Hill, 1965:557; type locality "Forest reserve 24 miles from Bartica, along the Potaro Road" (= 38.4 km S of Bartica), Cuyuni-Mazaruni, Guyana.

Glyphonycteris daviesi: Handley, 1976:2; first use of current name combination.

Micronycteris daviesi: J. K. Jones and Carter, 1976:10; name combination.

Micronycteris (Barticonycteris) daviesi: Eisenberg, 1989: 109; name combination.

Micronycteris [(Glyphonycteris)] daviesi: Simmons, 1996: 4; name combination.

DISTRIBUTION: *Glyphonycteris daviesi* is known from the island of Trinidad, Venezuela, Guyana, Surinam, French Guiana, Brazil, Ecuador, Peru, and Bolivia. Elsewhere, it has been found in Panama, Costa Rica, and Honduras (Pine et al. 1996).

MARGINAL LOCALITIES (Map 133; from Pine et al. 1996, except as noted): TRINIDAD AND TOBAGO: Trinidad, Victoria-Mayaro Forest Reserve (Clarke and Racey 2003). VENEZUELA: Bolívar, Reserva Florestal de Imataca (McCarthy and Ochoa 1991). GUYANA: Cuyuni-Mazaruni, 24 miles from Bartica (type locality of *Barticonycteris daviesi* Hill). SURINAM: Sipaliwini, Raleigh

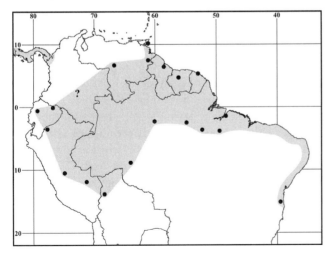

Map 133 Marginal localities for *Glyphonycteris daviesi* ●

Falls (S. L. Williams and Genoways 1980a). FRENCH GUIANA: Piste St. Élie (Brosset and Charles-Dominique 1991). BRAZIL: Pará, Belém; Bahia, Fazenda Serra do Teimoso (Gregorin and Rossi 2005); Pará, Area do Caraipe; Pará, 52 km SSW of Altamira; Pará, Alter do Chão (Bernard and Fenton 2002); Amazonas, Dimona Reserve (Sampaio et al. 2003); Rondônia, 20 km SW of Porto Velho on road to Cachoeira Teotônio. BOLIVIA: La Paz, 25 km W of Ixiamas (S. Anderson 1997). PERU: Madre de Dios, Pakitza (S. Solari, Pacheco, and Vivar 1999); Pasco, San Juan (Tuttle 1970) Amazonas, Soledad (S. Solari, Pacheco, and Vivar 1999). ECUADOR: Pichincha, Río Palenque Field Station; Sucumbios, Zafiro (Albuja 1999). VENEZUELA: Bolívar, Serranía de los Pijiguaos.

SUBSPECIES: We consider *G. daviesi* to be monotypic.

NATURAL HISTORY: Little is known about the natural history of *G. daviesi*. Pine et al. (1996:188) wrote that the species was "restricted or virtually restricted to lowland (0–500 m) primary wet tropical forest." Tuttle (1970) reported three individuals roosting in a hollow tree in mature forest. Hill (1965) and S. L. Williams and Genoways (1980a) reported netting *G. daviesi* in mature lowland rainforest. S. Solari, Pacheco, and Vivar (1999) captured an adult male, two adult females (one lactating), a subadult male and a young female in August that were roosting together in a tree hollow about 3 m above ground The tree was in mature forest on a mountain slope in departamento Amazonas, Peru. Pine et al. (1996) summarized the limited information on ecology, behavior, food habits, and reproduction. Information on reproduction is limited to finding lactating females in March (Panama) and August (Brazil), and a female pregnant with a 33-mm fetus in August (Peru). Of the two females Gregorin and Rossi (2005) reported from Bahia, Brazil, neither was pregnant or lactating when taken in July, although one showed nipple development indicating that she had reproduced in her lifetime, whereas the other had not. Pine et al. (1996) mentioned food items described as the remains of a frog, parts of a moth larva, and remains of other insects, thus suggesting gleaning habits, which are also indicated by the bat's morphology. Captured individuals held in cloth bags have escaped by chewing through the bag; R. H. Pine (pers. comm.) wrote that *G. daviesi* comes closer than any other bat to being a "semi-gnawing" animal. Webb and Loomis (1977) listed only one ectoparasite, a trombiculid mite. The karyotype is $2n = 28$, FN = 52 (Honeycutt, Baker, and Genoways 1980).

REMARKS: According to S. Solari, Pacheco, and Vivar (1999), two of the three bats Ascorra, Solari, and Wilson (1996) reported as *Phylloderma stenops* from Pakitza, Madre de Dios, Peru proved to be *G. daviesi*. This species was originally described as the only species in the genus *Barticonycteris*, and Hill (1965) based his diagnosis on its massive dentition. Koopman and Cockrum (1967) treated *Barticonycteris* as a synonym of *Micronycteris*. This name combination has been used by many recent authors (see J. K. Jones and Carter 1976), although Hall (1981) treated *Barticonycteris* as a separate genus. Koopman (1978), in explaining his reasons for treating *Barticonycteris* as a subgenus of *Micronycteris*, stated that the characteristics of *Barticonycteris* "are simply those of *M. (Glyphonycteris)*, the subgenus including *sylvestris* and *behnii*, carried one step further." Hill (1965) also recognized that the closest relatives of *Barticonycteris* were species of *Glyphonycteris*.

Glyphonycteris daviesi is the largest member of the genus (forearm 53.8–58.1 mm, greatest length of skull 25.8–27.3 mm). Hill (1965), Swanepoel and Genoways (1979), Tuttle (1970), S. L. Williams and Genoways (1980a), McCarthy and Ochoa (1991), Pine et al. (1996), and Gregorin and Rossi (2005) provided measurements of *G. daviesi*. D. C. Carter and Dolan (1978) gave additional information on and measurements of the holotype of *Barticonycteris daviesi* Hill.

Glyphonycteris sylvestris (O. Thomas, 1896)
Little Graybeard Bat
SYNONYMS:

Glyphonycteris sylvestris O. Thomas, 1896b:302; type locality "Imravalles" (= Hacienda. Miravalles), Guanacaste, Costa Rica.

Micronycteris (Glyphonycteris) sylvestris: Sanborn, 1949a: 231; name combination.

DISTRIBUTION: *Glyphonycteris sylvestris* is known east of the Andes in Colombia, Venezuela, Trinidad, Surinam, French Guiana, Peru, and from an apparently isolated population in southeastern Brazil. Elsewhere, it is recorded from Panama north to western Mexico.

MARGINAL LOCALITIES (Map 132): TRINIDAD AND TOBAGO: Trinidad, Salazar Trace (Goodwin and Greenhall 1961). SURINAM: Brokopondo, 8 km S and 2 km W of Brownsweg (S. L. Williams and Genoways 1980a). FRENCH GUIANA: Saül (Brosset and Dubost 1968). BRAZIL: Pará, Guama Ecological Research Area (Handley 1967). PERU: Cusco, W side of Cordillera Vilcabamba (Koopman 1978). COLOMBIA: Caquetá, Estación Puerto Abeja (Montenegro and Romero-Ruiz 2000). VENEZUELA: Monagas, Cueva del Guácharo, Caripe (Linares 1969). *Southeastern Brazilian records.* BRAZIL: São Paulo, Iporanga (Trajano 1982); Rio de Janeiro, Parque Estadual da Pedra Branca (Dias, Silva, and Peracchi 2003); São Paulo, Estação Ecológica dos Caetetus (Pedro, Passos, and Lim 2001).

SUBSPECIES: We regard *G. sylvestris* as monotypic.

NATURAL HISTORY: This species inhabits forest habitats (Handley 1966c 1976; S. L. Williams and Genoways 1980a), and is known to roost in caves, tunnels (Hall and Dalquest 1963; Villa-R. 1967; Linares 1969), and hollow trees (Goodwin and Greenhall 1961; Handley 1976; S. L. Williams and Genoways 1980a). *Glyphonycteris sylvestris* feeds on fruit and insects (Goodwin and Greenhall 1961). Webb and Loomis (1977) listed only one ectoparasite, a streblid batfly. We suspect that the type host of the dichrocoeliid trematode *Parametadelphis compactus* Travassos, 1955, is a *G. sylvestris* misidentified as a *G. behnii*. The karyotype is $2n = 22$, FN $= 40$ (Honeycutt, Baker, and Genoways 1980).

REMARKS: Until recently, *G. sylvestris* was regarded as a species of *Micronycteris*, subgenus *Glyphonycteris*, which had been treated as a genus prior to Sanborn's (1949a) revision. Handley (1976) was the first author in recent years to use the name combination *Glyphonycteris sylvestris*, but did not explain his reason for recognizing *Glyphonycteris* as a full genus. Goodwin and Greenhall (1961) commented briefly on geographic variation in this species. Because *G. sylvestris* is rare in collections, and capture sites are widely separated, it has been difficult to assess any geographic trends in variation. Measurements of *G. sylvestris* have been reported by Goodwin and Greenhall (1961), Linares (1969), Swanepoel and Genoways (1979), S. L. Williams and Genoways (1980a), Simmons (1996), and Simmons and Voss (1998). D. C. Carter and Dolan (1978) gave information on and measurements of the holotype of *Glyphonycteris sylvestris* O. Thomas.

Genus *Lampronycteris* Sanborn, 1949

Lampronycteris, represented by *Lampronycteris brachyotis*, is a monotypic genus formerly treated as a subgenus of *Micronycteris*. These are medium sized to smaller phyllostomines (forearm 38.3–42.5 mm, mean = 40.5 mm, $n = 23$; greatest length of skull 20.2–22.8 mm, mean = 21.6 mm, $n = 21$) most easily identified by the pointed ears, which have a concave upper outer rim, and by the yellow-orange to reddish fur on the throat and upper chest. Additional features (Sanborn 1949a; Simmons and Voss 1998) include the lack of a cutaneous band connecting the ears, lower rim of narial horseshoe defined by a ridge, and the lower lip and chin having a pair of smooth tubercles divided in the midline by a V-shaped groove. The third metacarpal is longer than the fourth, which is longer than the fifth. The calcar is shorter than the foot. Cranial and dental features include an inflated rostrum, especially in the lacrimal region; relatively shallow basisphenoid pits; upper inner incisors less than 1/2 the height of upper canines; upper outer incisors visible in the toothrow; and trifid lower incisors. The dental formula is 2/2, 1/1, 2/3, 3/3 × 2 = 34.

SYNONYMS:

Schizostoma: Dobson, 1879:880; not *Schizostoma* P. Gervais, 1856a.

Lampronycteris Sanborn, 1949a:223; type species *Micronycteris (Lampronycteris) platyceps* Sanborn, 1949a, by monotypy; described as a subgenus of *Micronycteris* Gray, 1866b.

Lampronycteris brachyotis (Dobson, 1879)
Yellow-throated Bat

SYNONYMS:

Schizostoma brachyote Dobson, 1879:880; type locality "Cayenne," French Guiana.

Micronycteris brachyotis: Miller, 1900b:154; name combination.

Glyphonycteris brachyotis: Andersen, 1906a: 60; name combination.

Micronycteris (Lampronycteris) platyceps Sanborn, 1949a: 224; type locality "Guanapo, Trinidad, British West Indies."

Micronycteris (Glyphonycteris) brachyotis: Sanborn, 1949a: 233; name combination.

Micronycteris (Lampronycteris) brachyotis: Goodwin and Greenhall, 1961:230; name combination.

Micronycteris branchyotis Brosset, Charles-Dominique, Cockle, Cosson, and Masson 1996b:1977; incorrect subsequent spelling of *Schizostoma brachyote* Dobson.

[*Lampronycteris*] *brachyotis*: Simmons and Voss, 1998:62; first use of current name combination.

Micronycteris (L[ampronycteris].) platyops Kretzoi and Kretzoi, 2000:192; incorrect subsequent spelling of *Micronycteris (Lampronycteris) platyceps* Sanborn.

DISTRIBUTION: *Lampronycteris brachyotis* occurs east of the Andes in Colombia, Venezuela, Trinidad, Guyana, Surinam, French Guiana, Peru, and northwestern Brazil. The species also is in Central America and southern Mexico.

MARGINAL LOCALITIES (Map 134): COLOMBIA: Guajira, Nasaret (Marinkelle and Cadena 1972). VENEZUELA: Falcón 19 km NW of Urama (Handley 1976); Sucre, 40 km NW of Caripito [Monagas] (Arnold, Baker, and Honeycutt 1983). TRINIDAD AND TOBAGO: Trinidad, Blanchisseuse (C. H. Carter et al. 1981). GUYANA: Potaro-Siparuni, Pakatau Falls (Lim and Engstrom 2001a). SURINAM: Brokopondo, Gros (Husson 1978). FRENCH GUIANA: Cayenne (type locality of *Schizostoma brachyote* Dobson). BRAZIL: Pará, Igarapé Brabo (Koopman 1976); Bahia, Una (Faria, Soares-Santos, and Sampaio 2006); Espírito Santo, Município de Linhares (Peracchi and Albuquerque 1985); São Paulo, Gruta São José do Ribeira (Taddei and Pedro 1996). PERU: Huánuco, Panguana Biological Station (Hutterer et al.

Map 134 Marginal localities for *Lampronycteris brachyotis* ●

1995); Loreto, Jenaro Herrera (S. Solari, Pacheco, and Vivar 1999). COLOMBIA: Vaupés, Durania (Marinkelle and Cadena 1972).

SUBSPECIES: We consider *L. brachyotis* to be monotypic.

NATURAL HISTORY: Handley (1966c, 1976) associated *L. brachyotis* with forest habitats. Weinbeer and Kalko (2004), in their study on Barro Colorado Island, Isthmus of Panama, found this species to forage primarily in the forest canopy, where it gleaned insects from vegetation. Foraging strategy included perch hunting and continuous flight; and may include "hawking" insects above the canopy. Seven of the nine bats Weinbeer and Kalko (2004) studied day-roosted in a hollow tree (*Dipteryx* sp.) and each foraged in a discrete area of forest, some as far as 2.7 km from the day roost. Husson (1978) reported individuals captured in an old gold mine in an area of savannah. *Lampronycteris brachyotis* roosts in caves, mine shafts, and hollow trees (Goodwin and Greenhall 1961; Handley 1966c, 1976; Marinkelle and Cadena 1972; Husson 1978). Goodwin and Greenhall (1961) found insects and fruit in the stomachs of *L. brachyotis*. Webb and Loomis (1977) listed three streblid batflies as known ectoparasites, to which R. Guerrero (1985a) added another streblid batfly. R. Guerrero (1997) listed five species of streblids verified as parasitizing *L. brachyotis*, two of which had not been reported by Webb and Loomis (1977) and R. Guerrero (1985a). Medellín, Wilson, and Navarro (1985) reviewed the natural history and illustrated the skull of this species in their *Mammalian Species* account. The karyotype is $2n = 32$, FN $= 60$ (Patton and Baker 1978).

REMARKS: Until recently, *Lampronycteris brachyotis* was treated as the sole member of the subgenus *Lampronycteris* in *Micronycteris* Gray, 1866b. Goodwin and Greenhall (1961) synonymized *Micronycteris (Lampronycteris) platyceps* under *Micronycteris brachyotis*. Measurements for *L. brachyotis* have been provided by Sanborn (1949a), Goodwin and Greenhall (1961), Davis, Carter, and Pine (1964), Marinkelle and Cadena (1972), Swanepoel and Genoways (1979), C. H. Carter et al. (1981), and Taddei and Pedro (1996). D. C. Carter and Dolan (1978) gave information on and measurements of the holotype of *Schizostoma brachyote* Dobson.

Genus *Lonchorhina* Tomes, 1863

The genus *Lonchorhina* contains five medium-sized species (forearm 41–60 mm, greatest length of skull 17–26 mm). These bats are easily recognized among the phyllostomines by their unusually long noseleaf, proportionally long ears, and long tail that extends to the posterior margin of the long interfemoral membrane. The calcar is longer than the foot. Cranially, these bats are readily distinguished by the presence of a concavity near the base of the rostrum, between the orbits. The dental formula is 2/2, 1/1, 2/3, 3/3 × 2 = 34.

SYNONYMS:

Lonchorhina Tomes, 1863b:81; type species *Lonchorhina aurita* Tomes, 1863b, by monotypy.

Lonchoyhina Villa-R., 1967:209; incorrect subsequent spelling of *Lonchorhina* Tomes.

Lonchohyna Villa-R., 1967:467; incorrect subsequent spelling of *Lonchorhina* Tomes.

Lonchorrina Coimbra, Borges, Guerra, and Mello, 1982: 34, 35; incorrect subsequent spelling of *Lonchorhina* Tomes.

Lonchorrhina Portfors, Fenton, Aguiar, Baumgarten, Vonhof, Bouchard, Faria, Pedro, Rautenbach, and Zortéa, 2000:535; incorrect subsequent spelling of *Lonchorhina* Tomes.

KEY TO SOUTH AMERICAN SPECIES OF *LONCHORHINA* (MAINLY BASED ON HANDLEY AND OCHOA 1997):

1. Size large: greatest length of skull more than 22 mm; rostrum relatively long, wide, and deep with C–M3 more than 7.5 mm; breadth across upper canines more than 4.9 mm; rostral breadth more than 6.0 mm, and rostral depth more than 6.8 mm 2

1′. Smaller: greatest length of skull less than 21.5 mm; rostrum shorter, narrower, and low with C–M3 less than 7.1 mm; breadth across upper canines less than 4.7 mm; rostral breadth less than 5.6 mm, and rostral depth less than 6.6 mm 3

2. Forearm more than 59 mm; greatest length of skull more than 21.3 mm; depth of cranium less than depth of rostrum; basisphenoid pits deep anteriorly . . . . *Lonchorhina marinkellei*

2′. Forearm 52–57 mm; greatest length of skull 22–23 mm; depth of cranium greater than depth of rostrum; basisphenoid pits shallow anteriorly . . . . *Lonchorhina inusitata*

3. Forearm more than 47 mm; third metacarpal more than 48 mm; greatest length of skull more than 20.5 mm; forearm and proximal portion of noseleaf hairy . . . . *Lonchorhina aurita*

3′. Forearm less than 45 mm; third metacarpal less than 43 mm; greatest length of skull less than 20 mm; forearm and proximal portion of noseleaf naked 4

4. First phalanx of digit III longer than 13 mm; dorsal pelage pale at base of hair; anterior border of ear with granulated surface; sella ovoid and short; basisphenoid pits deep posteriorly . . . . *Lonchorhina orinocensis*

4′. First phalanx of digit III shorter than 13 mm; dorsal pelage lacks pale base; anterior border of ear lacks granulation; sella relatively large and filiform; basisphenoid pits deep posteriorly . . . . *Lonchorhina fernandezi*

Lonchorhina aurita Tomes, 1863
Tomes's Sword-nosed Bat

Lonchorhina aurita Tomes, 1863b:83; type locality "West Indies"; later restricted to Trinidad by O. Thomas (1893b:162).

Lonchorhina occidentalis H. E. Anthony, 1923b:13; type locality "Puente de Chimbo, Provincia del Guayas, Ecuador, altitude 1200 feet."

L[onchorhina]. a[urita]. occidentalis: Linares and Naranjo, 1973:180; name combination.

DISTRIBUTION: *Lonchorhina aurita* occurs in Colombia, Ecuador, Trinidad, eastern and northern Venezuela, eastern Peru, eastern Bolivia, and eastern Brazil. The species also is known from Central America, southern Mexico, and the Bahamas (Hall 1981).

MARGINAL LOCALITIES (Map 135): VENEZUELA (Handley 1976): Falcón, Riecito; Miranda, Birongo. TRINIDAD AND TOBAGO: Trinidad, Saut d'Eau Cave (Goodwin and Greenhall 1961). BRAZIL: Pará, Parque Nacional da Amazonia (George et al. 1988); Pará, Caverna do Valdeci (Handley and Ochoa 1997); Maranhão, Alto Parnahyba (Sanborn 1932b); Piauí, Fazenda Olho da Agua (Mares et al. 1981); Pernambuco, Pedrados dos Pontais (Handley and Ochoa 1997); Bahia, Ilhéus (Faria, Soares-Santos, and Sampaio 2006); Espírito Santo, Gruta do Limoeiro (Hernández-Camacho and Cadena 1978); São Paulo, Iporanga (Trajano 1982). BOLIVIA: Santa Cruz, San Matías (Sanborn 1932b). PERU: Pasco, San Juan (Tuttle

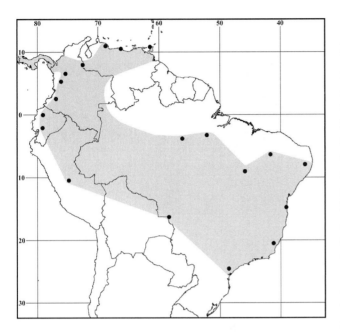

Map 135 Marginal localities for *Lonchorhina aurita* ●

1970). ECUADOR: Guayas, Puente de Chimbo (type locality of *Lonchorhina occidentalis* H. E. Anthony); Manabí, Chontillal (Solmsen 1985). COLOMBIA: Cauca, Tambito Nature Reserve (Dávalos and Guerrero 1999); Risaralda, Pueblorrico (Handley and Ochoa 1997); Antioquia, San Francisco (Muñoz 2001); Norte de Santander, Cúcuta (Sanborn 1949a).

SUBSPECIES: We are treating *L. aurita* as monotypic.

NATURAL HISTORY: This species is associated with forest (Tuttle 1970; Handley 1976) and disturbed (agricultural) habitats (Handley 1976). Common roosting sites of *L. aurita* are caves or tunnels (Sanborn 1932b, 1936, 1949a; Goodwin and Greenhall 1961; Greenhall and Paradiso 1968; Handley 1976; Solmsen 1985). Most reports indicate that this species is primarily insectivorous (Fleming, Hooper, and Wilson 1972; Howell and Burch 1974; Gardner 1977c); however, Fleming, Hooper, and Wilson (1972) found one individual with some fruit pulp in its stomach. Webb and Loomis (1977) listed three argasid ticks; one psorergatid, one spinturnicid, and one trombiculid mite; and one nycteribiid and seven streblid batflies as ectoparasites on *L. aurita* in their summary of ectoparasites of leaf-nosed bats. R. Guerrero (1985a) added a species of macronyssid mite and six species of streblid batflies. R. Guerrero (1997) listed 13 species of streblids, three of which had not been reported by Webb and Loomis (1977) and R. Guerrero (1985a). S. Solari, Pacheco, and Vivar (1999) reported *Trichobius petersoni* (Streblidae) recovered from a Peruvian specimen. The karyotype is $2n = 32$, $FN = 60$ (R. J. Baker et al. 1982).

REMARKS: *Lonchorhina aurita* has been confused with *L. inusitata*, which is known from the Guayanan lowlands of Venezuela, Surinam, and French Guiana, southward at least to the state of Rondônia, Brazil. Although several reports of *L. aurita* may yet prove to be based on misidentifications, the species nonetheless appears to be widely distributed throughout northwestern and western South America, with another segment of the species found in eastern and southeastern Brazil. Three of the other four species occur either together or in close proximity in southwestern Venezuela and eastern Colombia where *L. orinocensis* has been collected with *L. marinkellei* and *L. fernandezi*, but not with *L. aurita* or *L. inusitata* (see Ochoa and Ibáñez 1984; Ochoa and Sánchez 1988; and Handley and Ochoa 1997). Cabrera (1958) overlooked *Lonchorhina occidentalis* Anthony, 1923b, a name most authors have treated as a synonym of *L. aurita*. However, Linares and Naranjo (1973) suggested that *occidentalis* should be recognized as a subspecies of *L. aurita*, at least until additional material becomes available.

Lonchorhina aurita is medium-sized for the genus, being larger than *L. fernandezi* and *L. orinocensis*, and smaller than *L. inusitata* and *L. marinkellei*. Greatest length of skull varies from 19.8–21.3 mm, and zygomatic breadth varies from 10.0–11.1 mm (measurements combined from Ochoa and Ibáñez 1984, and Handley and Ochoa 1997). These dimensions and a forearm length of 48.8–51.9 mm (Handley and Ochoa 1997) should separate most, if not all, *L. aurita* from its morphologically closest congeners. Tuttle (1970), Linares and Ojasti (1971), Hernández-Camacho and Cadena (1978), Swanepoel and Genoways (1979), C. H. Carter et al. (1981), Ochoa and Ibáñez (1984), and Solmsen (1985) have provided measurements for *L. aurita*. D. C. Carter and Dolan (1978) gave information on and measurements of the holotype of *Lonchorhina aurita* Tomes.

Lonchorhina fernandezi Ochoa and Ibáñez, 1984
Fernandez's Sword-nosed Bat

SYNONYMS:

Lonchorhina fernandezi Ochoa and Ibáñez, 1984:147; type locality "entre Puerto Ayacucho y El Burro," 40–50 km NE of Puerto Ayacucho (along the highway), Amazonas, Venezuela.

Lonchorina fernandez: Eisenberg and Redford, 1999:146; incorrect subsequent spelling of *Lonchorhina fernandezi* Ochoa and Ibáñez.

DISTRIBUTION: *Lonchorhina fernandezi* is known from the Venezuelan states of Bolívar and Amazonas, along the Río Orinoco opposite the Río Meta, which is in the general vicinity of the type locality.

MARGINAL LOCALITIES (Map 136; from Ochoa and Sánchez 1988): VENEZUELA: Bolívar, Hacienda Sagitario; Amazonas, 40–50 km NE of Puerto Ayacucho.

SUBSPECIES: We treat *L. fernandezi* as monotypic.

NATURAL HISTORY: Little is known about the ecology of *L. fernandezi*. Ochoa and Ibáñez (1984) found that the stomachs of their specimens contained parts of noctuid moths, arachnids, and unidentifiable insects. Ochoa and Sánchez (1988) examined the stomach contents of 16 specimens in which 14 contained remains of moths, 4 contained beetles, and 3 unidentified insect remains. Ectoparasites include at least one species of streblid batflies (R. Guerrero 1997) and spinturnicid mites. Nothing is known concerning reproduction or karyotype. Based on the original description (Ochoa and Ibáñez 1984) and the extensive report by Ochoa and Sánchez (1988), known specimens number 45, which are all males. This is the smallest species in the genus, with the greatest length of skull of 17.1–17.7 mm and a zygomatic breadth of 9.2–9.6 mm (Ochoa and Sánchez 1988). Additional measurements were provided by Linares (1998).

Lonchorhina inusitata Handley and Ochoa, 1997
Hairy-faced Sword-nosed Bat

SYNONYMS:

Lonchorhina aurita: Gardner, 1988:709; not *Lonchorhina aurita* Tomes, 1863b.

Lonchorhina inusitata Handley and Ochoa, 1997:73; type locality "Boca Mavaca, 84 km SSE Esmeralda, 2°30′N–65°13′W, 138 m, Amazonas, Venezuela."

DISTRIBUTION: *Lonchorhina inusitata* has been found at low elevation on the Guayana Shield in southern Venezuela, western Surinam, and central French Guiana, south into estado de Rondônia, Brazil.

MARGINAL LOCALITIES (Map 136; from Handley and Ochoa 1997): VENEZUELA: Bolívar, 12 km S of El Manteco; Bolívar, Km 85. SURINAM: Sipaliwini, Avanavero. FRENCH GUIANA: Grotte du Bassin du Tapir. BRAZIL: Rondônia, Porto Velho ["19 km da Cidade"]. VENEZUELA: Amazonas, Río Mawarinuma, Parque Nacional Serranía de la Neblina,; Amazonas: Belén.

SUBSPECIES: We treat *L. inusitata* as monotypic.

NATURAL HISTORY: Brosset and Charles-Dominique (1991) located a colony estimated at 300 individuals in a tight cluster (mixed with approximately 50 *Phyllostomus latifolius*) over a pool in a cave in French Guiana. Elsewhere, most of the known specimens of *L. inusitata* have been mist-netted over water. Simmons, Voss, and Peckham (2000), who considered *L. inusitata* to be a cave dweller, mist-netted nine in primary forest in French Guiana. A few have been caught in mistnets placed in the canopy of both primary and secondary forests. Stomach contents (remains of spiders and scales of moths or mosquitoes; Brosset and Charles-Dominique 1991) suggest gleaning habits, as do the morphologies of all species of *Lonchorhina*. Gardner (1988) and Brosset and Charles-Dominique (1991) reported lactat-

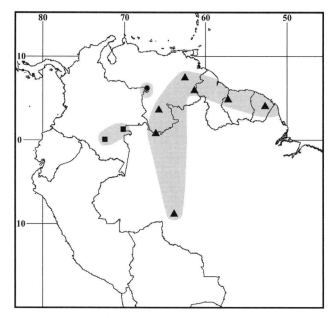

Map 136 Marginal localities for *Lonchorhina fernandezi* ●, *Lonchorhina inusitata* ▲, and *Lonchorhina marinkellei* ■

ing females taken in November. Handley and Ochoa (1997) mentioned a post-lactating female caught in October.

REMARKS: Many specimens of *L. inusitata* have been either misidentified as *L. aurita*, which is smaller, or more commonly as *L. marinkellei*, which is larger. Brosset et al. (1996b) reidentified as *L. inusitata* the specimens from French Guiana that Brosset and Charles-Dominique (1991) reported as *L. marinkellei*. Genoways, Williams, and Groen (1981) misidentified specimens from Avanavero, Surinam, as *L. aurita*.

Lonchorhina inusitata is much larger than either *L. fernandezi* or *L. orinocensis*. It is also larger than *L. aurita* (forearm 52.4–56.8 mm, *n* = 10, versus 48.8–51.9 mm, *n* = 18; greatest length of skull 22.0–22.9 mm, *n* = 10, versus 19.9–21.1 mm, *n* = 18) and has a longer tail. Although similar to *L. marinkellei* in shape of rostrum, and in the size, shape, and degree of hairiness of ears, noseleaf, and facial excrescences; the underparts are dark in *L. inusitata* and the basisphenoid pits are shallow anteriorly in contrast to *L. marinkellei* in which the venter is heavily washed with white and the basisphenoid pits are deep anteriorly. The rostrum is higher than the braincase in *L. marinkellei*, but of about equal height or slightly lower than the braincase in *L. inusitata*.

Lonchorhina marinkellei Hernández-Camacho and Cadena, 1978
Marinkelle's Sword-nosed Bat

SYNONYM:

Lonchorhina marinkellei Hernández-Camacho and Cadena, 1978:229; type locality "Durania (tambien conocida como Urania), cerca a Mitu, Comisaría del Vaupés, Colombia."

DISTRIBUTION: *Lonchorhina marinkellei* is known only from two localities in Colombia.

MARGINAL LOCALITIES (Map 136): COLOMBIA: Vaupés, Urania (type locality of *Lonchorhina marinkellei* Hernández-Camacho and Cadena); Caquetá, Estación Puerto Abeja (Montenegro and Romero-Ruiz 2000).

SUBSPECIES: *Lonchorhina marinkellei* is monotypic.

NATURAL HISTORY: The holotype of *L. marinkellei* was caught, along with two specimens of *L. orinocensis*, in a small cave located in a humid forest. Montenegro and Romero-Ruiz (2000) captured five specimens in open savanna habitat, and categorized *L. marinkellei* as an aerial insectivore. Stomach contents consisted of the remains of insects. The holotype, taken in August, contained 5.8-g fetus.

REMARKS: Originally known only by the female holotype, subsequent surveys in the same area have produced additional specimens; Handley and Ochoa (1997) had two males and two females of *L. marinkellei* at hand when they described *L. inusitata*. Apparently, *L. marinkellei* is known from only two localities, both associated with tepui formations in the eastern llanos of Colombia. This is the largest species in the genus (forearm 59.1–62.3 mm, $n = 4$; greatest length of skull 25.2–25.9 mm, $n = 4$; zygomatic breadth 13.8–14.2 mm, $n = 3$; Hernández-Camacho and Cadena 1978; Handley and Ochoa 1997).

Lonchorhina orinocensis Linares and Ojasti, 1971
Orinoco Sword-nosed Bat

SYNONYM:

Lonchorhina orinocensis Linares and Ojasti, 1971:2; type locality "Boca de Villacoa, Río Orinoco (50 km. NE de Puerto Páez)," Bolívar, Venezuela.

DISTRIBUTION: *Lonchorhina orinocensis* is recorded from eastern Colombia and southwestern Venezuela.

MARGINAL LOCALITIES (Map 137): VENEZUELA: Bolívar, Hato La Florida, 35 km ESE of Caicara (Lee, Lim, and

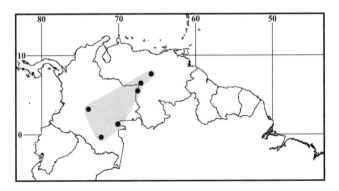

Map 137 Marginal localities for *Lonchorhina orinocensis* ●

Hanson 2000); Apure: La Villa (Ochoa and Sánchez 1988); Amazonas, 28 km SSE of Puerto Ayacucho (Handley 1976). COLOMBIA: Vaupés, Durania (Hernández-Camacho and Cadena 1978); Amazonas, Caserío Araracuara (Ochoa and Ibáñez 1984); Meta, Reserva Nacional de la Macarena, cerca de Caño Cristales (Hernández-Camacho and Cadena 1978).

SUBSPECIES: We consider *L. orinocensis* to be monotypic.

NATURAL HISTORY: Little is known about the natural history of *L. orinocensis*. Handley (1976) reported this species as roosting among large rocks in llanos habitat. Individuals also were netted in yards and forests. Others have been collected in small caves located in forest habitat (Hernández-Camacho and Cadena 1978). Gardner (1977c) suggested that the food habits of *L. orinocensis* were similar to those of *L. aurita*, but this has not been confirmed. Webb and Loomis (1977) listed three argasid ticks as known ectoparasites. R. Guerrero (1985a) added one species of spinturnicid mite and five species of streblid batflies.

REMARKS: This species has been taken sympatrically with *L. fernandezi*, *L. inusitata*, and *L. marinkellei*, but not with *L. aurita* (see Ochoa and Ibáñez 1984; Handley and Ochoa 1997). *Lonchorhina orinocensis* (greatest length of skull 18.6–19.7 mm, zygomatic breadth 9.4–10.2 mm) is intermediate in size between *L. fernandezi* and *L. aurita*. Linares and Ojasti (1971), Hernández-Camacho and Cadena (1978), Swanepoel and Genoways (1979), and Ochoa and Ibáñez (1984) have provided measurements.

Genus *Lophostoma*

As currently understood, *Lophostoma* comprises seven species of small to medium-sized bats (forearm 33–56 mm, greatest length of skull 18–31 mm). Distinguishing features include the relatively narrow postorbital constriction (less than 5 mm, and less than 90% of breadth across cingula of canines), and one incisor and three premolars in each mandible (dental formula shared with species in the genera *Tonatia* and *Chrotopterus*). As in *Tonatia*, the fur is short, the calcar is longer than the foot, and the tail extends to the middle of the interfemoral membrane. However, in contrast to *Tonatia*, the fur on the face is short, and the muzzle may appear nearly naked. Simmons and Voss (1998) noted a behavioral difference between *Tonatia saurophila* and the species now included in *Lophostoma*. All species of *Lophostoma* curled their ears when handled, but the *Tonatia saurophila* did not. The dental formula is 2/1, 1/1, 2/3, 3/3 × 2 = 32 (also in *Chrotopterus* and *Tonatia*).

SYNONYMS:

Lophostoma d'Orbigny, 1836: pl. 6; type species *Lophostoma silvicolum* d'Orbigny, 1836: pl. 6, by monotypy.

Phyllostoma: J. A. Wagner, 1843a:365; part; not *Phyllostoma* G. Cuvier, 1800.

Vampyrus: Pelzeln, 1883:32; not *Vampyrus* Leach, 1821b.

Tonatia: Palmer, 1898:110; part; not *Tonatia* Gray in Griffith, Hamilton-Smith, and Pidgeon 1827.

Chrotopterus: J. A. Allen, 1910b:147; part; not *Chrotopterus* W. Peters, 1865c.

Tonatia: O. Thomas, 1910b:184; part; not *Tonatia* Gray in Griffith, Hamilton-Smith, and Pidgeon.

Tonatia: Goodwin, 1942b:205, 209; part; not *Tonatia* Gray in Griffith, Hamilton-Smith, and Pidgeon.

Tonatia: W. B. Davis and Carter, 1978:6; part; not *Tonatia* Gray in Griffith, Hamilton-Smith, and Pidgeon.

Tonatia: Genoways and Williams, 1980:205; part; not *Tonatia* Gray in Griffith, Hamilton-Smith, and Pidgeon.

Tonakia Ascorra, Gorchov, and Cornejo, 1994:537; incorrect subsequent spelling of *Tonatia* of authors, not *Tonatia* Gray in Griffith, Hamilton-Smith, and Pidgeon.

KEY TO SOUTH AMERICAN SPECIES OF *LOPHOSTOMA*:

1. Forearm longer than 49 mm; greatest length of skull 26 mm or more 5

1'. Forearm shorter than 49 mm; greatest length of skull 26 mm or less 2

2. Forearm shorter than 40 mm; greatest length of skull less than 21 mm . . . . *Lophostoma brasiliense*

2'. Forearm longer than 40 mm; greatest length of skull more than 21 mm 3

3. Forearm longer than 45 mm; fur of underparts entirely white, except on chin and sides of abdomen; lacking small wart-like granulations on head, wings, and legs; greatest length of skull more than 24 mm 4

3'. Forearm shorter than 45 mm; hair of underparts drab to grayish olive; small wart-like granulations on dorsal surfaces of forearms, digits, and legs, and on ears and noseleaf; greatest length of skull less than 24 mm . . . . *Lophostoma schulzi*

4. Lacks post-auricular patches; margin of ear not white; basisphenoid pits shallow, almost imperceptible; sagittal crest relatively low; known only from eastern Ecuador . . . . *Lophostoma yasuni*

4'. Has pale post-auricular patches; margin of ear usually white; basisphenoid pits well developed and divided by median septum; sagittal crest usually high and well developed . . . . *Lophostoma carrikeri*

5. Lacks white postauricular patches; ventral pelage pale brownish olive, lightly frosted with white; second upper premolar not occluded by first upper premolar; known only from western Ecuador . . . . *Lophostoma aequatorialis*

5'. May or may not have white postauricular patches; ventral pelage paler, often strongly frosted white; anterior

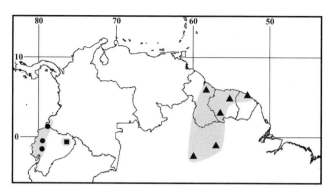

Map 138 Marginal localities for *Lophostoma aequatorialis* ●, *Lophostoma schulzi* ▲, and *Lophostoma yasuni* ■

surface of second upper premolar overlaid by first upper premolar; population in western Ecuador has conspicuous white postauricular patches and a whitish venter . . . . *Lophostoma silvicolum*

Lophostoma aequatorialis R. J. Baker, Fonseca, Parish, Phillips, and Hoffmann, 2004
Ecuadorian Round-eared Bat

SYNONYM:

Lophostoma aequatorialis R. J. Baker, Fonseca, Parish, Phillips, and Hoffmann, 2004:1; type locality "Ecuador, Province of Esmeraldas, Estación Experimental La Chiquita, near San Lorenzo town (1°16'60"N, 78°49'60"W) (UTM zone 17: 748935 E 0136902 N; 979 m)."

DISTRIBUTION: *Lophostoma aequatorialis* is known from the provinces of Esmeraldas, Los Ríos, and Pichincha, in the western lowlands of Ecuador.

MARGINAL LOCALITIES (Map 138; from R. J. Baker et al. 2004; localities listed from north to south): ECUADOR: Esmeraldas, Estación Experimental La Chiquita (type locality of *Lophostoma aequatorialis* R. J. Baker et al.); Pichincha, Estación Científica Río Palenque; Los Ríos, El Papayo.

SUBSPECIES: *Lophostoma aequatorialis* is monotypic.

NATURAL HISTORY: This species is associated with the evergreen lowland forest of the Pacific lowlands of Ecuador. Specimens from the type locality were netted at ground level in well-drained second-growth forest habitat containing shrubs and small palms (R. J. Baker et al. 2004). A lactating female was taken on 31 January at Estación Científica Río Palenque. Molecular data (R. J. Baker et al. 2004) supports a close relationship between *L. aequatorialis* and *L. schulzi*, but morphologically, *L. aequatorialis* is most similar to *L. silvicolum*. The karyotype is $2n = 34$, $FN = 62$ (R. J. Baker et al. 2004).

REMARKS: Three species of *Lophostoma* are known from western Ecuador, although as yet, none have been collected together. *Lophostoma brasiliense* is the smallest species and, with a forearm length less than 40 mm, is unlikely to be confused with the other two species whose forearm lengths usually exceed 50 mm. *Lophostoma aequatorialis* is similar in size and general morphology to *L. silvicolum occidentalis* of southwestern Ecuador and northern Peru. These two taxa appear to be separated ecologically, with *L. aequatorialis* found in the remnant and second growth stands of the tropical rain forest of western and northwestern Ecuador, and *L. s. occidentalis* occupying the drier tropical deciduous forests farther south. *Lophostoma aequatorialis* also lacks the white postauricular patches and the paler, heavily frosted venter that characterizes *L. s. occidentalis*.

Lophostoma brasiliense W. Peters, 1867
Pygmy Round-eared Bat

SYNONYMS:

Lophostoma brasiliense W. Peters, 1867a:674; type locality "Baía" (= Salvador), Bahia, Brazil.

Lophostoma venezuelæ W. Robinson and Lyon, 1901:154; type locality "Macuto, [Distrito Federal,] Venezuela."

[*Tonatia*] *brasiliense*: Trouessart, 1904:111; name combination.

T[*onatia*]. *venezuelæ*: Miller, 1907b:129; name combination.

Tonatia nicaraguae Goodwin, 1942b:205; type locality "Kanawa Creek, near Cukra, north of Bluefields, [Zelaya,] Nicaragua."

Tonatia minuta Goodwin, 1942b:209; type locality "Boca Curaray, Ecuador" (= Boca del Río Curaray, Loreto, Peru).

Tonatia brasiliensis: Handley, 1976:16; name combination and correct gender concordance.

[*Lophostoma*] *brasiliense*: Lee, Hoofer, and Van Den Bussche, 2002:55; first modern use of current name combination.

DISTRIBUTION: *Lophostoma brasiliense* occurs on the island of Trinidad and in Colombia, Venezuela, Guyana, Surinam, French Guiana, Brazil, Ecuador, Peru; Bolivia, and Paraguay. The species also is in Mexico and Central America.

MARGINAL LOCALITIES (Map 139): TRINIDAD AND TOBAGO: Trinidad, Santa María (C. H. Carter et al. 1981). GUYANA: Potaro-Siparuni, Clearwater Camp (Lim et al. 1999). SURINAM: Commewijne, Nieuwe Grond Plantation (Genoways and Williams 1984). FRENCH GUIANA: Paracou (Simmons and Voss 1998); Les Eaux Claires (Simmons, Voss, and Peckham 2000). BRAZIL: Pará, Ilha Caratateua (Piccinini 1974)); Ceará, Faculdade de Veterinária do Ceará, Fortaleza (Piccinini 1974); Pernambuco, Refúgio Ecológico Charles Darwin (Silva and Guerra 2000); Bahia, Salvador (type locality of *Lophos-*

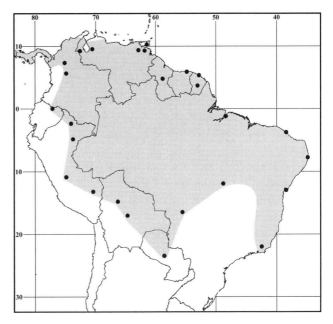

Map 139 Marginal localities for *Lophostoma brasiliense* ●

toma brasiliense W. Peters); Rio de Janeiro (no specific locality; Esbérard and Bergallo 2006); Tocantins, Município Sucupira (Nunes et al. 2005); Mato Grosso, Reserva Particular do Patrimônio Natural (Escarlate-Tavares and Pessôa 2005). PARAGUAY: Presidente Hayes, Estancia La Victoria (López-González et al. 1998). BOLIVIA: Cochabamba, Sajta (S. Anderson 1997); Beni, Estación Biológica del Beni (Wilson and Salazar 1990). PERU: Cusco, Huajyumbe (Koopman 1978); Junín, Río Perené, ca. 32 km N of Satipo (Gardner 1976); Loreto, Jenaro Herrera (Ascorra, Gorchov, and Cornejo 1994); Loreto, Boca Curaray (type locality of *Tonatia minuta* Goodwin). ECUADOR: Sucumbíos, Duvuno (Albuja 1999). COLOMBIA: Caldas, Hacienda Riomanso (Castaño et al. 2003); Antioquia, La Tirana (USNM 499293). VENEZUELA (Handley 1976, except as noted): Zulia, El Rosario, 39 km WNW of Encontrados; Trujillo, 19 km N of Valera; Monagas, Hato Mata de Bejuco; Delta Amacuro, Los Güires (Ojasti and Naranjo 1974).

SUBSPECIES: We are treating *L. brasiliense* as monotypic.

NATURAL HISTORY: Ecological information on *L. brasiliense*, while limited and obscured by unresolved questions concerning content of the species, has been published under the name *Tonatia brasiliensis* (e.g., see Handley 1976). Genoways and Williams (1984) associated *L. brasiliense* with areas having secondary vegetation or savannah forests in Surinam. Handley (1966c) associated the species (as *Tonatia minuta*) with evergreen and deciduous forests and fruit groves in Panama, where he found it roosting in hollow termite nests. Handley (1976) obtained specimens from forests and agricultural areas in Venezuela. Mares et al. (1981) collected *L. brasiliense* "either in Caatinga Alta habitats or near serrotes" in the Northeast of Brazil. Goodwin and Greenhall (1961) reported that termite nests are used as roosting sites for the species (as *Tonatia minuta*), also in Brazil. Gardner (1977c) wrote that the diet of *L. brasiliense* was probably fruit and insects. Graciolli and Bernard (2002) captured a female pregnant with a single 34-mm-CR fetus on 22 August along the lower Rio Tapajóz, Pará, Brazil. They also reported recovering three species of streblid flies on *L. brasiliense* from this area. Webb and Loomis (1977) listed one sarcoptid mite, one labidocarpid mite, and five streblid batflies as ectoparasites; they reported the host for the mites under the name *Tonatia venezuelae* and the host for four of the batflies under the name *Tonatia nicaraguae*. R. Guerrero (1985a) added one spinturnicid mite, one macronyssid mite, and one streblid batfly to the number of known ectoparasites. R. Guerrero (1997) listed 12 species of streblids, 5 of which had not been reported by Webb and Loomis (1977) and R. Guerrero (1985a). The karyotype is $2n = 30$, FN $= 56$ (Gardner 1977a; R. J. Baker 1979).

REMARKS: The common spelling in the literature is *L. brasiliense*, which is correct under the neuter generic name *Lophostoma*; but, when used under the generic name *Tonatia* should have been written as *Tonatia brasiliensis*. Apparently, W. Peters (1867:674) based the name *L. brasiliense* on the single specimen in the British Museum from Bahia that Gray had labeled *Tylostoma brasiliense*, but had not yet described. The date of publication of the name *L. brasiliense* W. Peters is usually given as 1866 (e.g., D. C. Carter and Dolan 1978); however, the pages from 657 to the end of volume 1867 (for the year 1866) of the *Monatsberichte* were published in 1867.

There has been controversy about the taxonomic status of the four nominal, small species of *Lophostoma*: *L. nicaraguae*, *L. brasiliense*, *L. minuta*, and *L. venezuelae*. We follow those authors who have treated these taxa as synonyms under the name *Tonatia brasiliensis* (see Sanborn 1932b; Ojasti and Naranjo 1974; Gardner 1976; Handley 1976; Koopman 1978, 1982; J. K. Jones and Carter 1979; C. H. Carter et al. 1981; Lemke et al. 1982; Genoways and Williams 1984). However, R. J. Baker (1979) and R. J. Baker et al. (1982) retained the names *L. brasiliense*, *L. minuta*, and *L. venezuelae*, when they reported karyotype information. Goodwin and Greenhall (1961), Swanepoel and Genoways (1979), Genoways and Williams (1980), and Simmons and Voss (1998) have provided measurements of *L. brasiliense*. D. C. Carter and Dolan (1978) gave information on and measurements of the holotype of *Lophostoma brasiliense* W. Peters.

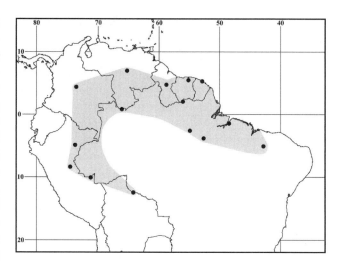

Map 140 Marginal localities for *Lophostoma carrikeri* ●

Lophostoma carrikeri (J. A. Allen, 1910)
Carriker's Round-eared Bat

SYNONYMS:

Chrotopterus carrikeri J. A. Allen, 1910b:147; type locality "Rio Mocho," Bolívar, Venezuela.

Tonatia carrikeri: Goodwin, 1942b:207; name combination.

[*Lophostoma*] *carrikeri*: Lee, Hoofer, and Van Den Bussche, 2002:55; first use of current name combination.

DISTRIBUTION: *Lophostoma carrikeri* is endemic to South America, where it is known from Colombia, Venezuela, Guyana, Surinam, French Guiana, Brazil, Peru, and Bolivia.

MARGINAL LOCALITIES (Map 140): VENEZUELA: Bolívar, Río Mocho (type locality of *Chrotopterus carrikeri* J. A. Allen). GUYANA: Potaro-Siparuni, Clearwater Camp (Lim et al. 1999). SURINAM: Para, Zanderij (Genoways and Williams 1984). FRENCH GUIANA: Paracou (Simmons and Voss 1998). SURINAM: Sipaliwini, Sipaliwini Airstrip (Genoways and Williams 1984). BRAZIL: Pará, Belém (McCarthy and Handley 1988); Piauí, Município de Teresina (Vizotto, Dumbra, and Rodrigues 1980); Pará, Rio Iriri, 85 km SW of Altamira (McCarthy and Handley 1988); Pará, Belterra (Gribel and Taddei 1989). VENEZUELA: Amazonas, Cerro Neblina Base Camp (Gardner 1988). BOLIVIA: Beni, Río Iténez, bank opposite Costa Marquez, Brazil (Koopman 1976); PERU: Ucayali, Balta (Gardner 1976); Ucayali, Cerro Tihuayo (R. M. Fonseca and Pinto 2004); Loreto, Jenaro Herrera (R. M. Fonseca and Pinto 2004). COLOMBIA: Meta, Hacienda Los Guaduales (McCarthy, Cadena, and Lemke 1983).

SUBSPECIES: We treat *L. carrikeri* as monotypic.

NATURAL HISTORY: *Lophostoma carrikeri* is associated with a variety of forested habitats, ranging from rainforests to savannah scrub forests and orchards (Handley

1976; McCarthy, Cadena, and Lemke 1983; Genoways and Williams 1984), and roosts in hollowed-out termite nests (M. A. Carriker in J. A. Allen 1911:267; McCarthy, Cadena, and Lemke 1983). Gardner (1977c) stated that the diet was unknown, and suggested that it includes a variety of arthropods and possibly fruit. Ochoa, Castellanos, and Ibáñez (1988) found unidentifiable insect remains in the stomach of the specimen from Cerro Neblina. In their review of endoparasites in leaf-nosed bats, Ubelaker, Specian, and Duszynski (1977) mentioned a nematode reported from *L. carrikeri*. R. Guerrero (1985a) listed one species of spinturnicid mite and three species of streblid batflies recovered from this species. McCarthy, Gardner, and Handley (1992) reviewed additional records of parasites in their *Mammalian Species* account on *L. carrikeri* (under the name *Tonatia carrikeri*). The karyotype is $2n = 26$, FN = 46 (Gardner 1977a; R. J. Baker et al. 1982).

REMARKS: The type locality, as explained by J. A. Allen (1910b:145), is on a tributary of "the Rio Mato, the largest western tributary of the Rio Caura," which in turn is a major southern tributary of the Orinoco. Genoways and Williams (1984) compared the measurements of specimens from Surinam with those of specimens from elsewhere and found little variation. Males averaged larger than females in all 13 measurements (significantly larger in 9) tested by McCarthy, Cadena, and Lemke (1983). Coefficients of variation in their sample (from Colombia) ranged from 1.2 to 5.3. Goodwin (1942b, 1953), Husson (1962, 1978), Gardner (1976), Swanepoel and Genoways (1979), S. L. Williams and Genoways (1980a), Genoways and Williams (1980, 1984), Genoways, Williams, and Groen (1981), McCarthy, Cadena, and Lemke (1983), and McCarthy, Gardner, and Handley (1992) have provided additional measurements for *L. carrikeri*.

Lophostoma schulzi (Genoways and Williams, 1980)
Schulz's Round-eared Bat

SYNONYMS:

Tonatia schulzi Genoways and Williams, 1980:205; type locality "3 km SW Rudi Kappelvliegveld, 320 m, Brokopondo, Suriname."

[*Lophostoma*] *schulzi*: Lee, Hoofer, and Van Den Bussche, 2002:55; first use of current name combination.

DISTRIBUTION: *Lophostoma schulzi* is endemic to South America where it is known from Guyana, Surinam, French Guiana, and Brazil.

MARGINAL LOCALITIES (Map 138): GUYANA: Upper Demerara-Berbice, 3 miles S of Linden (ROM 67468; McCarthy and Handley 1988). SURINAM: Brokopondo, Brownsberg (Genoways and Williams 1984). FRENCH GUIANA: Paracou (Simmons and Voss 1998). SURINAM:

Sipaliwini, Kayserberg Airstrip (Genoways, Williams, and Groen 1981). BRAZIL: Pará, Cachoeira de Porteira (Marques and Oren 1987); Amazonas, Fazenda Esteio (Griebel and Taddei 1989).

SUBSPECIES: We consider *L. schulzi* to be monotypic.

NATURAL HISTORY: Feeding habits and roosting sites of *L. schulzi* are unknown. Genoways and Williams (1984) wrote that this species has been collected only in dense, undisturbed, lowland rainforest. The specimen from Fazenda Esteio, Brazil, was mistnetted above a trail in secondary vegetation at a site surrounded by mature tropical rain forest (Gribel and Taddei 1989). McCarthy, Robertson, and Mitchell (1989) reported a 28-mm-CR fetus from a female taken 23 August in French Guiana. The karyotype is $2n = 26$, FN = 36 (Honeycutt, Baker, and Genoways 1980; R. J. Baker, Genoways, and Seyfarth 1981; R. J. Baker et al. 1982).

REMARKS: *Lophostoma schulzi* is a medium-sized *Lophostoma*, most closely approaching *L. carrikeri* in size. Although measurements reported by McCarthy, Cadena, and Lemke (1983) and Genoways and Williams (1984) show some overlap with *L. carrikeri* in external and cranial dimensions, *L. schulzi* averages smaller in all measurements. A unique specific character is the presence of small, wart-like granulations on the ears, noseleaf, and on the dorsal surfaces of the forearms, digits, and legs. Simmons and Voss (1998) provided measurements and repeated the advice of McCarthy, Robertson, and Mitchell (1989) about being careful to accurately sex specimens because the clitoris is elongated and can be mistaken for a penis.

Lophostoma silvicolum d'Orbigny, 1836
d'Orbigny's Round-eared Bat

SYNONYMS: See under Subspecies.

DISTRIBUTION: *Lophostoma silvicolum* is in Colombia, Venezuela, Guyana, Surinam, French Guiana, Brazil, Ecuador, Peru, Bolivia, and Paraguay. The species also occurs in Central America.

MARGINAL LOCALITIES (Map 141): VENEZUELA: Falcón 19 km NW of Urama (Handley 1976); Barinas, Reserva Forestal de Ticoporo, Unidad II (Ochoa et al. 1988); Bolívar, Serranía de lo Pijiguaos (Ochoa et al. 1988); Bolívar, Las Patos (Handley 1976). GUYANA: Cuyuni-Mazaruni, 24 miles from Bartica on Potaro Road (Hill 1965). SURINAM: Marowijne, 10 km N and 24 km W of Moengo (Genoways and Williams 1984). FRENCH GUIANA: Paracou (Simmons and Voss 1998). BRAZIL: Amapá, Macapá (Mok and Lacey 1980); Pará, Tauary (Goodwin 1942b); Pernambuco, Refúgio Ecológico Charles Darwin (L. A. M. Silva and Guerra 2000); Bahia, Bahia (C. O. C. Vieira 1955); Pernambuco, Fazenda Maniçoba (Willig 1983). PARAGUAY: Cordillera, 12 km (by road)

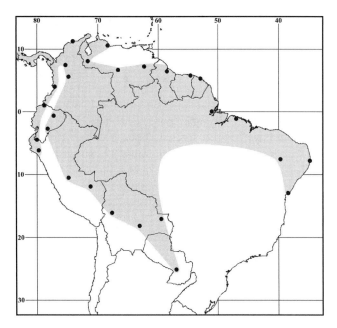

Map 141 Marginal localities for *Lophostoma silvicolum* ●

N of Tobatí (Myers and Wetzel 1979). BOLIVIA: Santa Cruz, Aserradero Pontons (Brooks et al. 2002); Santa Cruz, 10 km E of Ingeniero Mora (S. Anderson 1997); La Paz, Chijchipa (S. Anderson 1997). PERU: Madre de Dios, Pakitza (Ascorra, Wilson, and Romo 1991); Pasco, San Juan (Tuttle 1970); Lambayeque, 7 km S of Motupe (Graham and Barkley 1984); Piura, 4 miles W of Suyo (type locality of *Tonatia silvicola occidentalis* W. B. Davis and Carter). ECUADOR: Morona-Santiago, Mendéz (Albuja 1999); Napo, near Loreto (Albuja 1983); Esmeraldas, Urbina (Albuja 1999). COLOMBIA: Valle del Cauca, Cuartel B-V-83 (Alberico 1994); Caldas, Hacienda Riomanso (Castaño et al. 2003); Antioquia, Puri (J. A. Allen 1900a); Magdalena, Bonda (J. A. Allen 1900a).

SUBSPECIES: Three of the four recognized subspecies of *L. silvicolum* occur in South America. *Lophostoma s. centralis*, W. B. Davis and Carter, 1978, of Central America, is the subspecies not found in South America.

L. s. laephotis (O. Thomas, 1910)
SYNONYMS:
Tonatia læphotis O. Thomas, 1910b:184; type locality "River Supinaam, a tributary of the Lower Essequibo," Pomeroon-Supenaam, Guyana.
Tonatia loephotis Goodwin, 1942b:209; incorrect subsequent spelling of *T. laephotis* O. Thomas.
Tonatia sylvicola laephotis: Cabrera, 1958:64; name combination and incorrect subsequent spelling of *Lophostoma silvicolum* d'Orbigny.

This subspecies occurs from the Guianas to the lower Amazon basin of Brazil.

L. s. occidentalis (W. B. Davis and Carter, 1978)
SYNONYMS:
Tonatia silvicola occidentalis W. B. Davis and Carter, 1978:6; type locality "4 mi. W Suyo, 1000 ft., department of Piura, Perú."
This subspecies is known only from the Río Chira drainage basin of northwestern Peru and southwestern Ecuador.

L. s. silvicolum d'Orbigny, 1836
SYNONYMS:
Lophostoma silvicola d'Orbigny, 1836:pl. 6; type locality "le pied oriental de la Cordillere bolivienne, au pays des Juracarès" (according to Sanborn 1936:97, this is "in the Department of Yungas, between the headwaters of the rivers Secure and Isibara"); restricted by S. Anderson, Koopman, and Creighton (1982:17) to Yungas, between Río Secure and Río Isibara, 1,845 m, 15°45′S, 65°15′W, Beni, Bolivia.
Lophostoma sylvicolum d'Orbigny and Gervais, 1847:11; correction of gender agreement and incorrect subsequent spelling of *Lophostoma silvicolum* d'Orbigny.
Lophostoma sylvicola: Gray, 1838b:489; incorrect gender agreement and incorrect subsequent spelling for *Lophostoma silvicolum* d'Orbigny.
Phyllostoma amblyotis J. A. Wagner, 1843a:365; type locality "Matto Grosso, Brazil."
Ph[yllostoma]. silvicola: Schinz, 1844:238; name combination.
Lophostoma amblyotis: W. Peters, 1865c:509; name combination.
Vampyrus (Lophostoma) amblyotis: Pelzeln, 1883:32; name combination.
Phyllostoma midas Pelzeln 1883:32; in synonymy, *nomen nudum*.
Tonatia amblyotis: O. Thomas, 1902e:54; name combination.
Tonatia sylvicola: Cabrera, 1917:11; name combination and incorrect subsequent spelling of *Lophostoma silvicola* d'Orbigny.
Chrotopterus columbianus H. E. Anthony, 1920:84; type locality "Rio Quatequia, near Bogota, [Cundinamarca,] Colombia."
Tonatia silvicola: Husson, 1962:88; name combination.
Tonatia sylvicola: S. Solari, Rodriguez, Vivar, and Velasco, 2002:97; incorrect subsequent spelling of *Lophostoma silvicola* d'Orbigny.
[*Lophostoma*] *silvicolum*: Lee, Hoofer, and Van Den Bussche, 2002:55; first modern use of name combination.
L[*ophostoma*]. *silvicoulm*: R. J. Baker, R. M. Fonseca, D. A. Parish, C. J. Phillips, and F. G. Hoffmann, 2004:12; incorrect subsequent spelling of *Lophostoma silvicolum* d'Orbigny

The nominate subspecies is in Colombia, Venezuela, Paraguay, Brazil, and east of the Andes in Ecuador, Peru, and Bolivia. The subspecies also is in Panama.

NATURAL HISTORY: Hill (1965), Handley (1966c, 1976), Tuttle (1970), and Genoways and Williams (1984) have reported *L. silvicolum* (as *Tonatia silvicola*) from a variety of forest habitats. Handley (1976) also reported the species taken in agricultural clearings. This species uses termite nests, which have been hollowed out from below, for roosting sites (Sanborn 1951b; Tuttle 1970; Handley 1976; McCarthy, Cadena, and Lemke 1983; Kalko et al. 1999). The diet of *L. silvicolum* includes insects (Fleming, Hooper, and Wilson 1972; Kalko et al. 1999) and fruit (Howell and Burch 1974). Willig (1983) classified *L. silvicolum* as a foliage-gleaning insectivore. Webb and Loomis (1977), in their summary of ectoparasites, listed (under the name *Tonatia silvicola*) one argasid tick, and one nycteribiid and six streblid batflies known to infest this species. R. Guerrero (1985a) added a spinturnicid mite and three species of streblid batflies. R. Guerrero (1997) listed ten streblids, including one not previously reported by Webb and Loomis (1977) and R. Guerrero (1985a). Medellín and Arita (1989) provided a detailed description (under the name *Tonatia silvicola*), along with measurements and a summary of available information on natural history, in their *Mammalian Species* account. Kalko et al. (1999) described roosting and foraging behavior on Barro Colorado Island, Panama. The karyotype of *L. silvicolum* is $2n = 34$, FN $= 60$ (Honeycutt, Baker, and Genoways 1980; R. J. Baker et al. 1982).

W. B. Davis and Carter (1978), based on specimens from Panama, and Genoways and Williams (1984), based on material from Surinam, found males to be significantly larger than females in nearly 1/2 of the measurements tested. Males averaged equal to or larger than females in all other measurements in these samples, except for length of ear in the Panamanian sample. The coefficients of variation in the Surinam sample ranged from 2.0 to 4.5 and were consistently higher than for the same measurements of a sample of *Tonatia saurophila* (reported as *Tonatia bidens*) from Surinam.

REMARKS: W. B. Davis and Carter (1978) reviewed the geographic variation within *Tonatia silvicola* and concluded that there is only one species of large, round-eared bat in the *Lophostoma-Tonatia* complex in South America that has an extremely narrow postorbital constriction. They reduced *T. laephotis* to subspecific rank under *T. silvicola* and placed *T. amblyotis* in the synonymy of *T. s. silvicola*. W. B. Davis and Carter (1978) also discussed the correct spellings of the specific names *silvicola* and *laephotis*. B. D. Patterson (1992) presented evidence in his argument that d'Orbigny's (1836) spelling *silvicola* on Plate 6 was an error, and S. Anderson (1997) used the spelling *Tonatia sylvicola* based on Patterson's (1992) usage. Nevertheless, according to Article 32.2 (ICZN 1999:39), "the original spelling of a name is the 'correct original spelling,' unless it is demonstrably incorrect as provided in Article 32.5." Article 32.5 requires clear evidence in the original publication that there is an error. Article 32.5 further states that "incorrect transliteration or latinization,...are not to be considered inadvertent errors." Therefore, we consider the spelling *Lophostoma silvicolum* to be correct.

As noted by Cabrera (1958:65), the name *Phyllostoma midas* Pelzeln, 1883:32, was published in synonymy; therefore, the name is unavailable. *Phyllostoma midas* is a name Natterer had written on the labels of one or more specimens he collected in Brazil and the name was included in an unpublished catalogue of his collections.

Western Ecuadorian *L. silvicolum occidentalis* can be distinguished from *L. aequatorialis* by its whitish postauricular patches and heavily frosted ventral pelage. *Lophostoma aequatorialis* lacks the postauricular patches and has a pale olive brown venter. Otherwise, the two are similar in size and appearance.

All but three of the specimens reported as *T. bidens* by Mares et al. (1981) from northeastern Brazil represent *L. silvicolum*. Both species were taken at Fazenda Maniçoba (M. Willig, pers. comm.). Specimens reported by Fornes et al. (1967) and by Villa-R. and Villa-Cornejo (1973) as *T. silvicola* from Argentina are *T. bidens* according to Barquez, Mares, and Braun (1999). *Lophostoma silvicolum* has not been verified in Argentina, but is expected to be found there. Sanborn (1936, 1941), Goodwin (1942b, 1953), Husson (1962, 1978), W. B. Davis and Carter (1978), Swanepoel and Genoways (1979), Genoways and Williams (1980, 1984), Barquez (1984a), and Simmons and Voss (1998) have provided measurements. D. C. Carter and Dolan (1978) gave additional information on and measurements of the types of *Phyllostoma amblyotis* J. A. Wagner and *Tonatia laephotis* O. Thomas. Czaplewski, Rincón, and Morgan (2005) referred late Pleistocene material recovered from a tar seep in the Venezuelan state of Zulia to *L. silvicolum*.

Lophostoma yasuni Fonseca and Pinto, 2004
Yasuni Round-eared Bat

SYNONYM:

Lophostoma yasuni Fonseca and Pinto, 2004:1; type locality "vicinity of the Yasuní Research Station (00°30'S, 75°55'W, 220 m), Yasuní National Park and Biosphere Reserve, Province of Orellana, Ecuador."

DISTRIBUTION: *Lophostoma yasuni* is known only from the type locality in eastern Ecuador.

MARGINAL LOCALITY (Map 138): ECUADOR: Orellana, vicinity of Yasuní Research Station, Yasuní National

Park and Biosphere Reserve (type locality of *Lophostoma yasuni* Fonseca and Pinto).

SUBSPECIES: *Lophostoma yasuni* is monotypic.

NATURAL HISTORY: Fonseca and Pinto (2004) caught the holotype, an adult male, about 9 m above ground in a mistnet suspended between trees in terra firme evergreen lowland forest. No additional information is available.

Genus *Macrophyllum* Gray, 1838

The monotypic genus *Macrophyllum* is one of the smaller phyllostomines (forearm 34–39 mm, greatest length of skull 16–18 mm). *Macrophyllum* has a long tail, which extends to the posterior margin of the long interfemoral membrane, and elongated feet, which are proportionally like those of *Noctilio* and *Myotis* (*Pizonyx*). Cranially, these bats can be distinguished by the presence of three lower premolars and molars, and by the size and position of the incisors and premolars. The second lower premolar is tiny and easily overlooked, and the first and third are crowded together and in or near contact. The first upper premolar is about the same size as the second (outer) upper incisor, the first upper incisor is procumbent, and the crowns of the lower incisors are broad. The dental formula is 2/2, 1/1, 2/3, 3/3 × 2 = 34.

SYNONYMS:

Phyllostoma G. Cuvier, 1800: Tab. I; part; unjustified emendation of *Phyllostomus* Lacépède, 1799b; placed on Official Index of Rejected and Invalid Names (ICZN 1955:Direction 24).

Macrophyllum Gray, 1838b:489; type species *Macrophyllum nieuwiedii* Gray, 1838b, by monotypy.

Dolichophyllum Lydekker, 1891:673, in Flower and Lydekker, 1891; replacement name for *Macrophyllum* Gray, presumed by Lydekker to be preoccupied by *Macrophylla* Hope, 1837 (Coleoptera).

Dolychophyllum Trouessart, 1904:110; incorrect subsequent spelling of *Dolichophyllum* Lydekker in Flower and Lydekker.

Mesophyllum C. O. C. Vieira, 1942:311; in synonymy; incorrect subsequent spelling of *Macrophyllum* Gray.

Maerophyllum George, Marques, Vivo, Branch, Gomes, and Rodrigues, 1988:39; incorrect subsequent spelling of *Macrophyllum* Gray.

Macrophyllum macrophyllum (Schinz, 1821)
Long-legged Bat

SYNONYMS:

Phyllost[oma]. macrophyllum Schinz, 1821:163; type locality "In den Wäldern von Brasilien," further defined as "Flusse Mucuri," (= Rio Mucurí, Bahia, Brazil) by Wied-Neuwied (1826:192).

Map 142 Marginal localities for *Macrophyllum macrophyllum* ●

Ph[yllostoma]. macrophyllum Wied-Neuwied, 1826:188; objective synonym and primary homonym of *Phyllostoma macrophyllum* Schinz.

Macrophyllum nieuwiedii Gray, 1838b:489; type locality "Brazil"; junior objective synonym of *Phyllostoma macrophyllum* Schinz (see Dobson 1878:136, footnote; D. C. Carter and Dolan 1978); therefore, the type locality is the Rio Mucurí, Bahia, Brazil.

Macrophyllum neuwiedii P. Gervais, 1856a:50; incorrect subsequent spelling of *Macrophyllum nieuwiedii* Gray.

Dolichophyllum macrophyllum: J. A. Allen, 1900a:91; name combination.

Macrophyllum macrophyllum: Nelson, 1912:93; first use of current name combination.

DISTRIBUTION: *Macrophyllum macrophyllum* is in Colombia, Venezuela, the Guianas, Ecuador, Peru, Bolivia, Brazil, Paraguay, and northern Argentina. The species also occurs in Central America and southern Mexico.

MARGINAL LOCALITIES (Map 142): COLOMBIA: Magdalena, Bonda (J. A. Allen 1900a). VENEZUELA (Handley 1976, except as noted): Zulia, El Rosario, 65 km WNW of Los Encontrados; Guarico, Embalse de Guárico; Monagas, Cueva de Saffont (Linares, 1966); Delta Amacuro, Caño Araguabisi (Linares and Rivas 2004); Bolívar, El Manaco, 59 km SE de El Dorado. GUYANA: Cuyuni-Mazaruni, 24 miles from Bartica on Potaro Road (Hill 1965). SURINAM: Wanica, Santo Boma Locks (Husson 1978). FRENCH GUIANA: Sinnamary (Brosset and Dubost 1968). BRAZIL: Amapá, Km 160, Rodovia Perimetral Norte (Peracchi, Raimundo, and Tannure 1984); Pará, Ilha do Taiuna (Harrison 1975); Bahia, Bahia (P. Gervais 1856a); Bahia, Rio Mucuri (type locality of *Phyllostoma macrophyllum* Schinz); Minas Gerais, Caratinga (Taddei 1975); São Paulo, Emas (C. O. C. Vieira 1955). ARGENTINA: Misiones, San Ignacio (Fornes and Massoia 1969). PARAGUAY: Amambay, Arroyo Tacuara (Wilson and Gamarra de Fox 1991); Concepción, Belén (Baud 1989). BOLIVIA (S. Anderson 1997): Santa Cruz, Río Negrillo; Beni, Campamento El Trapiche. PERU: Madre de Dios, Pakitza (Ascorra, Wilson, and Romo 1991); Pasco, San Juan (Tuttle 1970); Amazonas, Río Cenepa (Tuttle 1970). ECUADOR: Pastaza, Putsu (Hill and Bown 1963). COLOMBIA: Valle del Cauca, Buenaventura (ROM 62588).

SUBSPECIES: We consider *M. macrophyllum* to be monotypic.

NATURAL HISTORY: *Macrophyllum macrophyllum* is commonly associated with forest habitats (Hill 1965; Tuttle 1970; Handley 1976). This species roosts in caves (Bloedel 1955; Felten 1956b; Linares 1966; Greenhall and Paradiso 1968) and in various man-made structures, including road culverts (Goldman 1920; Hill and Bown 1963; Greenhall and Paradiso 1968; Harrison and Pendleton 1975; Handley 1976; Husson 1978). Peracchi, Raimundo, and Tannure (1984) found a colony of about 50 in a 10-m-long culvert. The proximity of roosting sites to water, and the external morphological similarities this species shares with some piscivorous chiropterans, have led some authors to suggest that *M. macrophyllum* may feed on aquatic animals (W. B. Davis, Carter, and Pine 1964; Harrison 1975). Harrison and Pendleton (1975) reported *M. macrophyllum* as feeding on flying insects. Gardner (1977c) reported water striders (Hemiptera, Gerridae) as part of the diet of Panamanian *M. macrophyllum*, based on examination of stomach contents, and suggested that this species could be the only obligate insectivore in the Phyllostominae. Linares (1966) discussed the morphology and color pattern of Venezuelan specimens. Webb and Loomis (1977) listed five batflies (one nycteribiid and four streblids), one argasid tick, and one labidocarpid, one spinturnicid, and one trombiculid mite as known ectoparasites. To this, R. Guerrero (1985a) added a tick, a mite, and three streblid batflies. R. Guerrero (1997) listed ten species of streblids, four of which had not been reported by Webb and Loomis (1977) and R. Guerrero (1985a). The karyotype is $2n = 32$, FN $= 56$ (R. J. Baker et al. 1982).

REMARKS: The literature has been inconsistent in the placement of the type locality in Brazil. The Rio Mucurí flows from the state of Minas Gerais to the Atlantic across the southern tip of the state of Bahia. The literature refers to either Bahia (Ávila Pires 1965; Cabrera 1958; C. O. C. Vieira 1942 1955) or Minas Gerais (Hall 1981; Harrison 1975; Husson 1978) as the state where the type locality is located. The type locality is most likely in the vicinity of Morro da Arara, an old fazenda on the Rio Mucurí in the state of Bahia, and a place that Wied-Neuwied is known to have visited (see account for *Rhynchonycteris naso*). The type species of *Macrophyllum* Gray is *Macrophyllum nieuwiedii* Gray, which is a junior objective synonym of *Phyllostoma macrophyllum* Wied-Neuwied, which is, in turn, a junior objective synonym and primary homonym of *Phyllostoma macrophyllum* Schinz.

Davis, Carter, and Pine (1964), Hill (1965), Harrison (1975), Taddei (1975), and Swanepoel and Genoways (1979) have provided measurements of *M. macrophyllum*. Harrison (1975) reviewed this species and provided illustrations of the skull and dentition in his *Mammalian Species* account. He also described the longitudinal rows of denticles on the distal lower surface of the uropatagium that appear to be unique to this species.

Genus *Micronycteris* Gray, 1866

The genus *Micronycteris* is represented by at least nine small species (forearm 33–46 mm, greatest length of skull 17–25 mm). Although there is considerable variation between species, the genus is characterized by bicolored dorsal fur and relatively large, rounded ears that are connected by an interauricular band of skin (usually having a median cleft). The lower margin of the narial horseshoe is defined by a ridge, and the center of the lower lip and chin have a pair of smooth tubercles partly divided by a medial groove. The tail extends only to the middle of the interfemoral membrane. The length of the calcar, relative to the length of the foot, varies between species and is useful for identification. The upper canines are more than twice the height of inner upper incisors, the lower incisors are bifid, and there are three relatively large premolars in the lower toothrow. The dental formula is 2/2, 1/1, 2/3, 3/3 $\times 2 = 34$ (shared by 10 of the 15 phyllostomine genera recognized in South America).

SYNONYMS:

Phyllophora: Gray, 1842:257; not *Phyllophora* Gray, 1838b.

Phyllostoma: Gray, 1842:257; not *Phyllostoma* G. Cuvier, 1800.

Mimon Gray, 1847:14; part.

Schizostoma P. Gervais, 1856a:38; type species *Schizostoma minutum* P. Gervais, 1856a:70, by monotypy; preoccupied by *Schizostoma* Bronn, 1835 (Mollusca).

Schizastoma Gerrard, 1862:38; incorrect subsequent spelling of *Schizostoma* P. Gervais.

Micronycteris Gray, 1866b:113; type species *Micronycteris megalotis* (Gray, 1842), by monotypy.

Vampyrella Reinhardt, 1872:iii; no type species selected; intended for those species of *Schizostoma* P. Gervais that have the ears connected by a band; preoccupied by *Vampyrella* Cienkowski, 1865 (Protozoa).

Vampyrus: Pelzeln, 1883:32; not *Vampyrus* Leach, 1821b.

Mycronycteris Festa, 1906:1; incorrect subsequent spelling of *Micronycteris* Gray.

Xenoctenes Miller, 1907b:124; type species *Schizostoma hirsutum* W. Peters, 1869, by original designation.

Chronycteris Taddei, 1976:325; incorrect subsequent spelling of *Micronycteris* Gray.

Mycronicteris Coimbra, Borges, Guerra, and Mello, 1982: 34; incorrect subsequent spelling of *Micronycteris* Gray.

Mycronycteris: Albuja, 1989:107; incorrect subsequent spelling of *Micronycteris* Gray.

Ñ*Micronycteris* Cuartas and Muñoz, 1999:16; incorrect subsequent spelling of *Micronycteris* Gray.

REMARKS: *Micronycteris* was reviewed by Andersen (1906a), Sanborn (1949a), and more recently by Simmons (1996) and Simmons and Voss (1998). These authors, along with Honacki, Kinman, and Koeppl (1982) and Koopman (1993), overlooked the name *Vampyrella* Reinhardt, the type species of which has not been designated, *contra* Kretzoi and Kretzoi (2000:423). Following Sanborn's (1949a) revision, *Micronycteris* (*sensu stricto*) contained three species. His broader definition of the genus included six subgenera (*Glyphonycteris* O. Thomas, *Lampronycteris* Sanborn, *Micronycteris* Gray, *Neonycteris* Sanborn, *Trinycteris* Sanborn, and *Xenoctenes* Miller). Simmons and Voss (1998) added one species (*M. brosseti*); elevated *M. homezi* Pirlot to species level; retained *Xenoctenes* Miller, 1907b (*Micronycteris hirsuta*) as a synonym of *Micronycteris* (*sensu stricto*); and elevated the other four subgenera to full generic rank. Barquez et al. (2000) reported an as yet undescribed species, the first record for the genus in Argentina, from the provincia de Salta. Simmons, Voss, and Fleck (2002) described *M. matses* from northern Peru, increasing the number of species recognized in *Micronycteris* to nine. Czaplewski, Rincón, and Morgan (2005) identified several late Pleistocene fossils from northern Venezuela as *Micronycteris*, but were unable to allocate them to species.

KEY TO THE SPECIES OF *MICRONYCTERIS*:

1. Color of venter white, pale gray, or pale buff, conspicuously paler than dorsum 5
1'. Color of venter dark, approximates color of dorsum 2
2. Forearm longer than 37 mm; greatest length of skull more than 19 mm 3
2'. Forearm shorter than 37 mm; greatest length of skull less than 19 mm 4
3. Forearm longer than 41 mm; greatest length of skull more than 21 mm; lower incisors narrow and high crowned . . . . *Micronycteris hirsuta*

3'. Forearm shorter than 41 mm; greatest length of skull less than 21 mm; lower incisors broad and low crowned . . . . *Micronycteris matses*
4. Ear from notch usually more than 22 mm; fur on lower third of medial surface of pinna 8–10 mm in length . . . . *Micronycteris megalotis*
4'. Ear less than 22 mm; fur on lower third of medial edge of pinna 8 mm or shorter . . . . . . *Micronycteris microtis*
5. Calcar longer than foot; zygomatic breadth narrower than breadth of braincase 6
5'. Calcar equal to or shorter than foot; zygomatic breadth wider than braincase 7
6. Tibia shorter than 14.5 mm; fur on lower third of medial edge of pinna 4 mm or less . . . . . *Micronycteris brosseti*
6'. Tibia longer than 14.5 mm; fur on lower third of medial edge of pinna 5 mm or more . . . . *Micronycteris schmidtorum*
7. First phalanx of digit IV longer than second; hairless fossa or pit between ears on top of head behind interauricular band . . . . *Micronycteris homezi*
7'. First and second phalanges of digit IV approximately equal in length; no hairless fossa or pit on top of head 8
8. Color of venter pale gray or pale buff; basal 1/3 to 1/2 of fur on upper back white; calcar shorter than foot, thumb normal . . . . *Micronycteris minuta*
8'. Color of venter mainly clear white; basal 2/3 to 3/4 of fur on upper back white; calcar approximately equal to length of foot; thumb small, 7.5 mm or less . . . . *Micronycteris sanborni*

Micronycteris brosseti Simmons and Voss, 1998
Brosset's Big-eared Bat

SYNONYM:

Micronycteris brosseti Simmons and Voss, 1998:62; type locality "Paracou," French Guiana.

DISTRIBUTION: *Micronycteris brosseti* is known from Guyana, French Guiana, Brazil, and Peru.

MARGINAL LOCALITIES (Map 143; from Simmons and Voss 1998, except as noted): FRENCH GUIANA: Paracou (type locality of *Micronycteris brosseti* Simmons and Voss). PERU: Loreto, Puerto Indiana; Loreto, Jenaro Herrera; Loreto, Quebrada Esperanza. GUYANA: Potaro-Siparuni, Three-Mile Camp (Lim et al. 1999). *Isolated southeastern record*: BRAZIL: São Paulo, Barra.

SUBSPECIES: We treat *M. brosseti* as monotypic.

NATURAL HISTORY: Simmons and Voss (1998) mistnetted an individual at ground level and caught seven others from a roost in a hollow tree. Both captures were within well-drained primary forest. Lim et al. (1999) reported a batfly from Guyanan *M. brosseti*. The karyotype is unknown.

Map 143 Marginal localities for *Micronycteris brosseti* ●

REMARKS: Ascorra, Wilson, and Gardner (1991) reported the specimens from Jenaro Herrera, Loreto, Peru, as *M. schmidtorum*. *Micronycteris brosseti* can be distinguished from congeners by its pale gray to pale buff venter, which is distinctly paler than the dorsum (but not white as in other congeners with pale venters). Other features include the short (3 mm) fur on lower third of medial side of pinna; first phalanx of digit IV longer than second; a normal-sized thumb; and a calcar longer than the foot (Simmons and Voss 1998).

Micronycteris hirsuta (W. Peters, 1869)
Crested Big-eared Bat

SYNONYMS:

Schizostoma hirsutum W. Peters, 1869:396; type locality unknown; subsequently designated as Pozo Azul, San José, Costa Rica, by Goodwin (1946:302).

M[icronycteris]. hirsuta: Miller, 1898:330; first use of current name combination.

Xenoctenes hirsutus: Miller, 1907b:125; name combination.

Micronycteris (Xenoctenes) hirsuta: Sanborn, 1949a:223; name combination.

DISTRIBUTION: *Micronycteris hirsuta* is known in South America from Colombia, Venezuela, Trinidad, Guyana, Surinam, French Guiana, Ecuador, Peru, Bolivia, and from an apparently disjunct population in southeastern Brazil. The species also occurs in Central America.

MARGINAL LOCALITIES (Map 144): *Main distribution.* COLOMBIA: Magdalena, Mamatoco (Sanborn 1932b). VENEZUELA: Falcón 19 km NW of Urama (Han-

dley 1976); Monagas, Caripito (Arnold, Baker, and Honeycutt 1983). TRINIDAD AND TOBAGO: Trinidad, Las Cuevas (C. H. Carter et al. 1981). GUYANA: Cuyuni-Mazaruni, 24 miles from Bartica on Potaro Road (Hill 1965). SURINAM: Para, Zanderij (Genoways, Williams, and Groen 1981). FRENCH GUIANA: Petit-Saut (Brosset and Charles-Dominique 1991). BRAZIL: Pará, Aramanay (AMNH 94534); Pará, Area do Caraipe (Pine et al. 1996); BOLIVIA: Santa Cruz, Los Fierros (L. Emmons, pers. comm.). PERU: Madre de Dios, 12 km above mouth of Río Palotoa (V. Pacheco et al. 1993); Pasco, San Juan (Tuttle 1970). ECUADOR (Albuja 1999): Manabí, Espuela Perdida; Esmeraldas, Hacienda La Granada. COLOMBIA: Cauca, El Papayo (Simmons 1996). *Disjunct distribution in southeastern Brazil*: Bahia, Região de Conquista (Falcão, Soares-Santos, and Drummond 2005); Bahia, Una (Faria, Soares-Santos, and Sampaio 2006); Espírito Santo, Município de Linhares (Peracchi and Albuquerque 1985); Rio de Janeiro, Paraíso do Tobias (Esbérard 2004); Rio de Janeiro, Estação Ecológica Estadual de Paraíso (Esbérard 2004).

SUBSPECIES: We treat *M. hirsuta* as monotypic; the species is in need of revision.

NATURAL HISTORY: In South America, this species is most often reported from forested habitats (Hill 1965; Tuttle 1970; Handley 1976; S. L. Williams and Genoways 1980a; Pine et al. 1996), but also has been collected in cleared areas around dwellings (Handley 1976). *Micronycteris hirsuta* is known to roost in hollow trees, under bridges, and in buildings (Goodwin and Greenhall 1961; Handley 1966c; Greenhall and Paradiso 1968. Goodwin

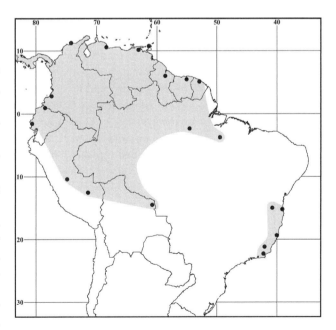

Map 144 Marginal localities for *Micronycteris hirsuta* ●

and Greenhall (1961) stated that this species feeds primarily on fruit, but may eat some insects; however, Wilson (1971b) found *M. hirsuta* to be primarily insectivorous, and suggested its diet fluctuated with the seasons and availability of fruit. In their review of endoparasites in leaf-nosed bats, Ubelaker, Specian, and Duszynski (1977) reported a trematode found in the small intestine. Webb and Loomis (1977) listed one labidocarpid and four trombiculid mites; R. Guerrero (1997) listed one species of streblid batfly known from *M. hirsuta.*

Robert J. Baker et al. (1973) described two karyotypes for *M. hirsuta.* One chromosomal form ($2n = 30$, FN $= 32$) is known from Middle America and Surinam (R. J. Baker, Genoways, and Seyfarth 1981). The other cytotype, which R. J. Baker, Genoways, and Seyfarth (1981) found only in Trinidadian *M. hirsuta*, has a $2n = 28$, FN $= 30$ karyotype.

REMARKS: Miller (1907b) described *Xenoctenes* as a genus, but Sanborn (1949a) reduced it to a monotypic subgenus of *Micronycteris. Xenoctenes* can be distinguished from other *Micronycteris* by the following features: the ears are connected across the forehead by a low unnotched band of skin, and the metacarpal of digit III is the shortest and the metacarpal of digit V is the longest. Adult males have a conspicuous tuft, or crest, of long hair on the top of the head between the ears. The shape of the skull is similar to that of members of the subgenus *Micronycteris.* The upper outer incisors are small, but the upper inner incisors are large, separated at their bases, and in contact at their tips. Lower incisors are bifid, high crowned, and wedged tightly between canines. The cingula of the canines are in or near contact. Simmons and Voss (1998) subsumed *Xenoctenes* under *Micronycteris* and did not recognize subgenera. R. J. Baker et al. (1973) demonstrated that *M. hirsuta* from Trinidad average significantly larger in length of forearm and greatest length of skull than Central American bats of the $2n = 30$ chromosomal form. Other reports of measurements for *M. hirsuta* are by Sanborn (1932b), Hershkovitz (1949c), Goodwin and Greenhall (1961), Hill (1965), Swanepoel and Genoways (1979), C. H. Carter et al. (1981), and Genoways, Williams, and Groen (1981). D. C. Carter and Dolan (1978) provided measurements and other information on the holotype of *Schizostoma hirsutum* W. Peters. The southeastern Brazilian distribution is based on one specimen from the state of Espírito Santo (Peracchi and Albuquerque 1985) and three from the state of Rio de Janeiro (Esbérard 2004).

Micronycteris homezi Pirlot, 1967
Pirlot's Big-eared Bat
SYNONYMS:

Micronycteris megalotis homezi Pirlot, 1967:265; type locality "Hato El Cedral, Northern border, Apure State, 80 m," Venezuela, based on neotype (Ochoa and Sánchez

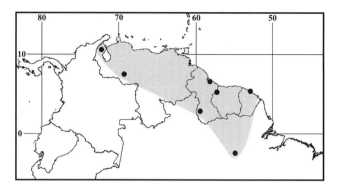

Map 145 Marginal localities for *Micronycteris homezi* ●

2005; original type localities "El Laberinto et . . . Rio Palmar," Zulia, Venezuela.

Micronycteris homezi: Simmons and Voss, 1998; first use of current name combination.

DISTRIBUTION: *Micronycteris homezi* has been recorded from Guyana, Brazil, western Venezuela, and northern French Guiana.

MARGINAL LOCALITIES (Map 145): VENEZUELA: Zulia, El Laberinto (restricted original type locality of *Micronycteris megalotis homezi* Pirlot). GUYANA: Demerara-Mahaica, Ceiba Biological Station (ROM 112573); East Berbice-Corentyne, Mango Landing (Lim and Engstrom 2001a). FRENCH GUIANA: Paracou (Simmons and Voss 1998). BRAZIL: Pará, Alter do Chão (Bernard 2001b). GUYANA: Upper Takutu-Upper Essequibo, Arakwai River, 5 km upstream from Dadanawa (Lim and Engstrom 2001a). VENEZUELA: Apure, Hato El Cedral (type locality based on neotype; Ochoa and Sánchez 2005).

SUBSPECIES: We regard *M. homezi* as monotypic (see Remarks).

NATURAL HISTORY: This was the only *Micronycteris* taken by Simmons and Voss (1998) in canopy-level mistnets, their specimen of *M. homezi* was caught at a height of 18–21 m in a gap in the canopy above a narrow dirt road in well-drained primary forest. Bernard (2001) caught an adult male in a net set at ground level within a small forest fragment surrounded by grasslands. Lim and Engstrom (2001) reported five specimens from Guyana, at least some of which were caught in ground-level nets. Apparently, although one of the paratypes described by Pirlot (1967) was identified as a female, all extant specimens are males (Ochoa and Sánchez 2005). The karyotype is unknown.

REMARKS: *Micronycteris homezi* is only tentatively recognized here. Apparently, Pirlot's (1967) original material no longer exists. The Paracou specimen Simmons and Voss (1998) identified as *M. homezi* has a bare depression located on top of the head between the ears, just behind the interauricular band. Pirlot (1967:264) illustrated this structure with a drawing. Simmons and Voss (1998) provided a vivid color photograph of their specimen while it was alive

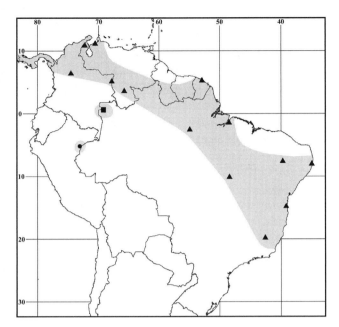

Map 146 Marginal localities for *Micronycteris matses* ●, *Micronycteris schmidtorum* ▲, and *Neonycteris pusilla* ■

that shows the fossa, which appears to be glandular. Simmons and Voss (1998) rediagnosed *M. homezi*, provided measurements, and described its morphology in their comparisons with sympatric congeners. Bernard (2001) and Lim and Engstrom (2001) provided measurements and the latter authors commented on the similarity of their *M. homezi* with *M. minuta*. Ochoa and Sánchez (2005) rejected the interauricular fossa as diagnostic for *M. homezi* because they found a similar fossa in some *M. megalotis* and *M. microtis*. They considered *M. homezi* to be a synonym of *M. minuta*, designated a neotype, and provided evidence that Pirlot's (1967) and Simmons and Voss's (1998) descriptions of *M. homezi* were based on breeding male *M. minuta* in which the interauricular fossa becomes well developed.

Micronycteris matses Simmons, Voss, and Fleck, 2002
Matses Big-eared Bat

SYNONYM:

Micronycteris matses Simmons, Voss, and Fleck, 2002:5; type locality "Nuevo San Juan," Loreto, Peru.

DISTRIBUTION: *Micronycteris matses* is known only from the type locality, a Matses Indian village on the Río Gálvez, Loreto, Peru.

MARGINAL LOCALITY (Map 146): PERU: Loreto, Nuevo San Juan (type locality of *Micronycteris matses* Simmons, Voss, and Fleck).

SUBSPECIES: *Micronycteris matses* is monotypic.

NATURAL HISTORY: Of the eight known specimens, one was mist-netted at ground level and the others were taken from diurnal roosts, which were holes in stream banks, either dug by armadillos (assumed to be *Dasypus*

kappleri) or resulting from erosion (Simmons, Voss, and Fleck 2002). Simmons, Voss, and Fleck (2002) described the habitat as lowland (150 m elevation) primary rainforest containing small tree-fall gaps and Indian gardens, which had been cleared after 1984 when the village was established. Fleck and Harder (2000) provided a more complete description of the vegetation and topography.

REMARKS: *Micronycteris matses* is larger than all congeners except *M. hirsuta*. The venter is approximately the same color as the dorsum, a feature also shared with *M. hirsuta* and the other dark-bellied *Micronycteris* (*M. megalotis* and *M. microtis*). The crowns of the lower incisors are short and broad as in other congeners with the exception of *M. hirsuta*, whose narrow, high-crowned lower incisors appear to be unique in the genus.

Micronycteris megalotis (Gray, 1842)
Brazilian Big-eared Bat

SYNONYMS:

Phyllophora megalotis Gray, 1842:257; type locality "Brazils"; restricted to Perequé, São Paulo, Brazil, by Cabrera (1958:60).

Phyllostoma elongata Gray, 1842:257; type locality "Brazils"; preoccupied by *Phyllostoma elongatum* É. Geoffroy St.-Hilaire, 1810b.

Phyllostoma elongatum: Gray, 1844:19; emendation of *Phyllostoma elongata* Gray; not *Phyllostoma elongatum* É. Geoffroy St.-Hilaire.

Mimon megalotis: Gray, 1847:14; name combination.

Phyllostoma scrobiculatum J. A. Wagner, 1855:627; replacement name for *Phyllostoma elongatum* Gray; preoccupied by *Phyllostoma elongatum* É. Geoffroy St.-Hilaire.

M[icronycteris]. megalotis: Gray, 1866b:113; first use of current name combination.

S[chizostoma]. elongatum: Gray, 1866b:115; name combination.

Schizostoma megalotis: W. Peters, 1867a:674; name combination.

Vampyrus (Schizostoma) elongatus: Pelzeln, 1883:32; name combination.

Phyllostoma Nattereri Pelzeln, 1883:32; in synonymy, *nomen nudum*.

DISTRIBUTION: *Micronycteris megalotis* is in Colombia, Venezuela, Trinidad and Tobago, the Guianas, Brazil, Ecuador, Peru, and Bolivia. The species also occurs in Central America and the Lesser Antilles.

MARGINAL LOCALITIES (Map 147): COLOMBIA: Magdalena, Santa Marta (Bangs 1900). VENEZUELA (Handley 1976): Falcón, Capatárida; Falcón, Boca de Yaracuy, 28 km WNW of Puerto Cabello; Sucre, Manacal. TRINIDAD AND TOBAGO: Tobago, Richmond (Ibáñez 1984a). VENEZUELA: Delta Amacuro, Raudales del Río

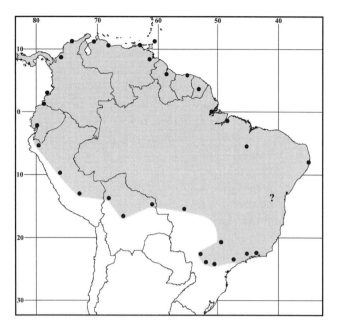

Map 147 Marginal localities for *Micronycteris megalotis* ●

Acoíma (Rivas 2000). GUYANA: Cuyuni-Mazaruni, 24 miles from Bartica on Potaro Road (Hill 1965). SURINAM: Paramaribo, Paramaribo (Husson 1978). FRENCH GUIANA: Saül (Brosset and Dubost 1968). BRAZIL: Amapá, Macapá (Carvalho 1962a); Pará, Belém (Peracchi and Albuquerque 1985); Maranhão, Barra do Corda (C. O. C. Vieira 1957); Pernambuco, São Lorenço (Andersen 1906a); Rio de Janeiro, Sacra Família do Tinguá (Peracchi and Albuquerque 1985); São Paulo, Piquete (C. O. C. Vieira 1955); São Paulo, Sorocaba (C. O. C. Vieira 1955); Paraná, Fazenda Monte Alegre (N. R. Reis, Peracchi, and Sekiama 1999); Paraná, Fazenda Cagibi (Bianconi, Mikich, and Pedro 2004); Paraná, Estação Ecológica do Caiuá (Miretzki and Margarido 1999); São Paulo, Grota de Mirassol (Taddei 1975); Mato Grosso, Chapada (Andersen 1906a). BOLIVIA: Santa Cruz, El Refugio (Emmons 1998, as *M. microtis*); Cochabamba, Puerto Patiño (S. Anderson, Koopman, and Creighton 1982); La Paz, Ixiamas (S. Anderson 1997). PERU (Koopman 1978): Cusco, Cordillera Vilcabamba; Huánuco, Carpish Pass; Piura, Palambla. ECUADOR: Guayas, Guayaquil (Brosset 1965); Esmeraldas, La Chiquita (Albuja 1999). COLOMBIA (Sanborn 1932b): Cauca, Isla Gorgona; Cordoba, Jaraquiel.

SUBSPECIES: We treat *M. megalotis* as monotypic.

NATURAL HISTORY: *Micronycteris megalotis* is locally abundant in many habitats. Roosting sites include caves, mines, buildings, bridges, wells, culverts, and hollows in trees, logs, and stumps (Goodwin and Greenhall 1961; Handley 1966c, 1976; Greenhall and Paradiso 1968). Unusual roosting sites have included spaces under exposed roots along streams (Tuttle 1970), inside large mammal burrows (Hall and Dalquest 1963), and in rock piles (Handley 1976). Label information on specimens in the USNM from the island of Tobago, Trinidad and Tobago, indicates capture from under an overhang in dirt bank along a trail, and from under tree roots in a mud bank.

Valdivieso and Tamsitt (1962) categorized *M. megalotis* as a nectar-eating species, but this is unsupported (Gardner 1977c). Lasso and Jarrín (2005) compared the remains of insects gatherer from under two roosts of *M. megalotis*, each located in different habitat, during a year-long study in northwestern Ecuador. The remains of lepidopterans predominated (monthly average of 52%) among the remains from the roost in disturbed habitat (pasture land and nearby patches of disturbed forest). Beetles (Coleoptera) proved the most common prey items (monthly average of 85%) recovered from the second roost, which was at the base of a tree in primary forest at least 2 km from areas of human disturbance.

A female from the island of Tobago was pregnant with an 11-mm–CR fetus on 30 March and four were lactating in early July. Cuartas and Muñoz (1999) reported a richulariid nematode in the small intestine of a Colombian specimen. In their review, Ubelaker, Specian, and Duszynski (1977) mentioned a nematode, a trematode, and trypanosomes as endoparasites in this species. Webb and Loomis (1977) listed 11 trombiculid mites, 1 sarcoptid mite, 1 labidocarpid mite and 2 spinturnicid mites along with 1 nycteribiid and 2 streblid batflies as known ectoparasites. R. Guerrero (1985a) added a species of spinturnicid mite and a streblid batfly, and later (1997) listed eight streblids verified as parasitizing *M. megalotis* of which six had not been reported previously by Webb and Loomis (1977) or R. Guerrero (1985a). Some of the hosts for these ectoparasites likely are misidentified *M. microtis*. Alonso-Mejía and Medellín (1991) did not distinguish between *M. megalotis* and *M. microtis* in their *Mammalian Species* account on *M. megalotis*. The karyotype is $2n = 40$, FN = 68 (R. J. Baker et al. 1982).

REMARKS: We do not know how much of the foregoing information attributed to *M. megalotis* actually applies to this species. Many author have indiscriminately applied the name *M. megalotis* to all of the smaller *Micronycteris* lacking a pale venter. Although similar to congeners that have dark venters, *M. megalotis* has a longer ear (exceeds 22 mm) that also has longer fur (8–10 mm) along its lower medial margin. Longer fur on the lower medial margin of the ear is the external character emphasized by Simmons (1996) and Simmons and Voss (1998) for distinguishing *M. megalotis* from *M. microtis*. The *M. microtis* reported from the state of Bahia, Brazil by Faria, Soares-Santos, and Sampaio (2006) may prove to represent *M. megalotis*.

Taddei's (1975) coefficients of variation for a sample of *M. megalotis* from Brazil varied from 0.66 to 3.18 for cranial measurements, and from 1.77 to 5.48 for external measurements. Sanborn (1949a), Hershkovitz (1949c), Goodwin and Greenhall (1961), Husson (1962, 1978), Tamsitt and Valdivieso (1963a), Brosset (1965), J. D. Smith and Genoways (1974), Taddei (1975), Swanepoel and Genoways (1979), and C. H. Carter et al. (1981) have all reported measurements for South American specimens of this species, but their samples most likely included specimens of *M. microtis*. D. C. Carter and Dolan (1978) gave information on and measurements of the types of *Phyllostoma elongatum* Gray, *Phyllophora megalotis* Gray, and *Phyllostoma scrobiculatum* J. A. Wagner.

Micronycteris microtis Miller, 1898
Little Big-eared Bat

SYNONYMS:

Micronycteris microtis Miller, 1898:328; type locality "Greytown," (= San Juan del Norte), San Juan del Norte, Nicaragua.

Macrotus pygmaeus Rehn, 1904:444; type locality "Izamal, Yucatan," Mexico.

Micronycteris megalotis mexicana: Andersen, 1906a:54; part.

Micronycteris (Micronycteris) megalotis microtis: Sanborn, 1949a:219; name combination.

M[icronycteris]. microtis microtis: Simmons, 1996:4; name combination.

M[icronycteris]. microtis mexicana: Simmons, 1996:4; name combination.

Micronycteris microti Simmons, Voss, and Peckham, 2000: 31; incorrect subsequent spelling of *Micronycteris microtis* Miller.

DISTRIBUTION: *Micronycteris microtis* is known from Colombia, Venezuela, Guyana, French Guiana, and southern Brazil. Elsewhere, the species occurs in Central America north into western Mexico.

MARGINAL LOCALITIES (Map 148; from Simmons 1996, except as noted): VENEZUELA (Handley 1976): Falcón, Capatárida; Distrito Federal, Los Venados; Sucre, 21 km E of Cumaná. GUYANA: Potaro-Siparuni, Clearwater Camp (Lim et al. 1999). FRENCH GUIANA: Paracou. BRAZIL: Pará, Ilha do Taiuna; Amazonas, Iucali; Amazonas, Tauá. VENEZUELA: Apure, Puerto Páez (Handley 1976). COLOMBIA: Cundinamarca, Mesitas del Colegio. VENEZUELA: Táchira, Las Mesas, 17 km NE of San Juan de Colón (Handley 1976).

SUBSPECIES: Simmons (1996) recognized two subspecies: *M. microtis mexicana* in Mexico and Central America south into western Costa Rica, and the nominate subspecies in Central America from eastern Nicaragua through Panama and into South America.

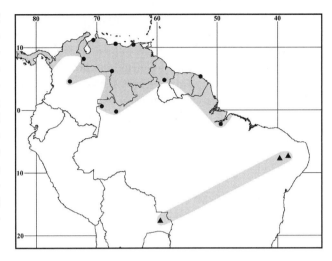

Map 148 Marginal localities for *Micronycteris microtis* ● and *Micronycteris sanborni* ▲

NATURAL HISTORY: Handley (1976) found *M. microtis* roosting in hollow trees and logs, and in rock piles in Venezuela. These bats also were mist-netted, mainly near streams and other wetter areas, in a variety of sites including yards, pastures, and deciduous thorn forest. Culverts are a common roosting site in Mexico (e.g., Pine 1972a:25). The six specimens reported by Simmons and Voss (1998) were caught in ground-level mistnets: one in well-drained primary forest, four in swampy primary forest, and one in a clearing. Information on reproduction has been reported under the name *M. megalotis*, particularly for Mexican and Central American populations (see Alonso-Mejía and Medellín 1991). Brennan and Reed (1975) reported three species of trombiculid mites found on Venezuelan *M. microtis*. The karyotype ($2n = 40$, $FN = 68$) also has been reported under *M. megalotis* (e.g., Baker 1967:408; also see Simmons 1996:16, footnote 5).

REMARKS: Andersen (1906a) assigned specimens from the Bogotá region of Colombia to *M. megalotis mexicana* Miller, but Simmons (1996) and Simmons and Voss (1998) restricted this name, as a subspecies of *M. microtis*, to Mexican and Central American populations. Hershkovitz (1949c) stated that northern Colombian specimens of *M. megalotis megalotis* (= *M. microtis microtis*) graded into the larger *M. megalotis mexicana*. Handley (1966c) used the name *M. megalotis microtis* for Panamanian specimens, an arrangement followed by Hall (1981) and J. K. Jones and Carter (1976), who described the range of this subspecies as continuing into northwestern South America. Handley (1976) used the name *Micronycteris microtis* for specimens from Venezuela, but did not explain why he elevated the name to a full species. Simmons (1996) and Simmons and Voss (1998) reassigned *Micronycteris megalotis mexicana* Miller, to *M. microtis mexicana*. Pedro, Passos, and Lim

(2001) recorded three *M. microtis* from Estação Ecológica dos Caetetus, São Paulo, Brazil; however, considering the hiatus between that record and the known distribution of this species, we suspect these specimens, plus those reported by Faria, Soares-Santos, and Sampaio (2006) from the state of Bahia, are misidentified *M. megalotis*. The same likely is true for the *M. microtis* reported by Emmons (1998) from Parque Nacional Noël Kempff Mercado, Santa Cruz, Bolivia, and commented on by Salazar-Bravo et al. (2003); therefore, we have included her record under *M. megalotis*.

Micronycteris minuta (P. Gervais, 1856)
White-bellied Big-eared Bat
SYNONYMS:

Schizostoma minutum P. Gervais, 1856a:50; type locality "Capella-Nova," Minas Gerais, Brazil.

Micronycteris hypoleuca J. A. Allen, 1900a:90; type locality "Bonda," Magdalena, Colombia.

Micronycteris minuta: O. Thomas, 1901b:191; first use of current name combination.

DISTRIBUTION: *Micronycteris minuta* is in Colombia, Venezuela, Trinidad, Guyana, Surinam, French Guiana, and Brazil, and in the western Amazon basin of eastern Ecuador, Peru, and Bolivia. Elsewhere, it occurs in Central America as far north as Nicaragua.

MARGINAL LOCALITIES (Map 149): COLOMBIA: Magdalena; Bonda (type locality of *Micronycteris hypoleuca* J. A. Allen). VENEZUELA (Handley 1976): Zulia, near Cerro Azul, 33 km NW of La Paz; Sucre, 21 km E of Cumaná. TRINIDAD AND TOBAGO: Trinidad, Las Cuevas (C. H. Carter et al. 1981). VENEZUELA: Bolívar,

Map 149 Marginal localities for *Micronycteris minuta* •

El Palmar (Ochoa 1995). GUYANA: Upper Demerara-Berbice, Dubulay Ranch (USNM 582263). SURINAM: Marowijne, 10 km N and 24 km W of Moengo (Genoways and Williams 1979b). FRENCH GUIANA: Piste St. Élie (Brosset and Charles-Dominique 1991). BRAZIL: Pará, Utinga (Handley 1967); Bahia, Bahia (C. O. C. Vieira 1955); Bahia, Itagibá (Peracchi and Albuquerque 1985); Espírito Santo, Linhares (Peracchi and Albuquerque 1985); Rio de Janeiro, Piraí (Peracchi and Albuquerque 1985); Minas Gerais, Capella-Nova (type locality of *Schizostoma minutum* P. Gervais); Distrito Federal, Gruta Toca do Falcão (Bredt, Uieda, and Magalhães 1999); Pará, Alter do Chão (Bernard 2001b); Amazonas, Florestal Reserve (Sampaio et al. 2003). BOLIVIA: Santa Cruz, 27.5 km S of Campamento Los Fierros (S. Anderson 1997); La Paz, 20 km (by road) NNE of Caranavi (S. Anderson, Koopman, and Creighton 1982). PERU: Cusco, Río Mapintunare (Koopman 1978); Pasco, San Pablo (Tuttle 1970). ECUADOR: Napo, San José abajo (Simmons 1996); Esmeraldas, Hacienda La Granada (Albuja 1999). COLOMBIA: Nariño, Reserva Natural La Planada (Ospina-Ante and Gómez 2000); Caldas, Charca Guarinocito (Castaño et al. 2003).

SUBSPECIES: We regard *M. minuta* as monotypic.

NATURAL HISTORY: This species, while often reported from forest habitats (Handley 1966c; Tuttle 1970), also is known from agricultural clearings (Handley 1976; S. L. Williams and Genoways 1980a). *Micronycteris minuta* roosts in caves, tunnels, and hollow trees (Sanborn 1949a; Goodwin and Greenhall 1961; Greenhall and Paradiso 1968; Bowles, Cope, and Cope 1979; López-González 1998). Fleming, Hooper, and Wilson (1972) described the diet as insects and plant material. Webb and Loomis (1977) listed only three ectoparasites, a streblid batfly and two spinturnicid mites. R. Guerrero (1997) increased the number of streblid batflies known to occur on *M. minuta* to four. López-González (1998) provided a detailed description, along with measurements, and a summary of available information on natural history in her *Mammalian Species* account. The karyotype is $2n = 28$, FN = 50 or 52 (R. J. Baker 1973; R. J. Baker et al. 1982).

REMARKS: Andersen (1906a) was the first to treat *Micronycteris hypoleuca* J. A. Allen as a synonym of *M. minuta* (also see Goodwin 1953). Andersen's (1906a) specimen from Santa Catherina (= Santa Catarina), Brazil represents the southernmost known distribution record for *M. minuta*. According to Simmons (1996) the *M. minuta* reported by Mares et al (1981) from the Northeast of Brazil were correctly identified as *M. megalotis* by Willig (1983). However, all of the specimens in the Carnegie Museum that Willig (1983) reported as *M. minuta* from the Brazilian states of Ceará and Pernambuco were misidentified and represent a mixed sample of *M. sanborni* and *M. schmidto-*

rum (see Simmons 1996:16, footnote 6). Sanborn (1949a), Goodwin and Greenhall (1961), Linares (1969), Bowles, Cope, and Cope (1979), Genoways and Williams (1979b), Swanepoel and Genoways (1979), S. L. Williams and Genoways (1980a), C. H. Carter et al. (1981), and Simmons and Voss (1998) provided measurements of *M. minuta*.

Micronycteris sanborni Simmons, 1996
Sanborn's Big-eared Bat

SYNONYMS:

Micronycteris sp. Mares, Willig, Streilein, and Lacher, 1981:104.

Micronycteris schmidtorum: Ascorra, Wilson, and Gardner, 1991:351; part; not *Micronycteris schmidtorum* Sanborn, 1935.

Micronycteris sanborni Simmons, 1996:6; type locality "Sitio Luanda, Itaitera, 4 km S of Crato," Ceará, Brazil.

DISTRIBUTION: *Micronycteris sanborni* is known from the Chapada do Araripe plateau of Ceará and Pernambuco, Brazil, and a single record from departamento Santa Cruz, Bolivia. .

MARGINAL LOCALITIES (Map 148; from Simmons 1996): BRAZIL: Ceará, Sitio Luanda (type locality of *Micronycteris sanborni* Simmons); Pernambuco, Serrote das Lajes, 17 km S of Exú. BOLIVIA: Santa Cruz, Estancia Patuju (Brooks et al. 2002).

SUBSPECIES: *Micronycteris sanborni* is monotypic.

NATURAL HISTORY: The holotype was caught in a ground-level mistnet set adjacent to a dam in relatively humid woodland. Willig (1983; also see description in Simmons 1996) described the Chapada and Caatinga habitats in the Northeast of Brazil. The Bolivian specimen came from Cerrado forest described as a mosaic of dry palm, savannah, and forest with rocky outcrops and rolling terrain (Brooks et al. 2002). The karyotype is $2n = 28$, FN $= 50$ (Simmons 1996).

REMARKS: Willig (1987:Table 7) gave measurements of *M. sanborni* along with those of *M. schmidtorum*, both species confused under the name *M. minuta*, which also has a white venter. *Micronycteris sanborni* is most easily distinguished from its pale-bellied congeners by its clear white venter; the conspicuously short, small thumb (7.0–7.3 mm); a calcar that is about equal to the length of the foot; and the conspicuous gap between the upper canine and adjacent incisor.

Micronycteris schmidtorum (Sanborn, 1935)
Schmidt's Big-eared Bat

SYNONYM:

Micronycteris schmidtorum Sanborn, 1935:81; type locality "Bobos, Izabal, Guatemala."

DISTRIBUTION: *Micronycteris schmidtorum* occurs in Colombia, Venezuela, French Guiana, and Brazil. Else-

where, it is known from Central America and southern Mexico.

MARGINAL LOCALITIES (Map 146): VENEZUELA: Falcón, Capatárida (Handley 1976). FRENCH GUIANA: Paracou (Simmons 1996). BRAZIL: Pará, Belém (Simmons 1996); Pernambuco, Fazenda Alto do Ferreira (Ascorra, Wilson, and Gardner 1991); Pernambuco, Estação Ecológica do Tapacurá (Ascorra, Wilson, and Gardner 1991); Bahia, Ilhéus (Faria, Soares-Santos, and Sampaio 2006); Minas Gerais, Rio Dôce State Park (Tavares 1999; Tavares and Taddei 2003); Tocantins, Paraíso do Tocantins (Nunes et al. 2005); Pará, Alter do Chão (Bernard 2001b). VENEZUELA: Amazonas, Belén (Handley 1976). COLOMBIA: Vichada, Maipures (Simmons 1996); Antioquia, Puerto Nare (Cuartas and Muñoz 1999). VENEZUELA: Zulia, near Cerro Azul, 40 km NW of La Paz (Handley 1976).

SUBSPECIES: We consider *M. schmidtorum* to be monotypic.

NATURAL HISTORY: Little is known about the ecology of South American populations. Handley (1976) reported *M. schmidtorum* roosting in tree holes and recorded captures of this species from a variety of habitats including evergreen forest, thorn forest, swamps, pastures, and orchards. Howell and Burch (1974) stated that Costa Rican animals ate insects; Gardner (1977c) suggested that the diet also included fruit. Cuartas and Muñoz (1999) found trichostrongylid nematodes in the small intestine of Colombian *M. schmidtorum*. R. Guerrero (1997) listed two species of streblid batflies as ectoparasites. The karyotype is $2n = 38$, FN $= 66$ (R. J. Baker 1973; R. J. Baker et al. 1982).

REMARKS: *Micronycteris schmidtorum* can be distinguished from congeners by the following combination of characters: the calcar is longer than the foot; the interfemoral membrane is more than twice the length of the tail; fur on venter is paler than on the dorsum, usually grayish white to pale buff; upper premolars are about equal in height; and the second lower premolar is about 3/4 the size of the first. The specimens from Sitio Luanda, Ceará, Brazil, and from Jenaro Herrera, Loreto, Peru, reported by Ascorra, Wilson, and Gardner (1991) as *M. schmidtorum* were reidentified by Simmons (1996) and Simmons and Voss (1998) as *M. sanborni* and *M. brosseti*, respectively. Swanepoel and Genoways (1979) gave measurements for South American specimens. Tavares and Taddei (2003) summarized additional measurements for a total of 35 specimens from throughout the range of the species.

Genus *Mimon* Gray, 1847

The genus *Mimon* is represented by four medium-sized species (forearm 45–61 mm, greatest length of skull 21–28 mm).

These are the only bats in the subfamily that have a lower toothrow with only one incisor and two premolars. The tail extends to the middle of the long interfemoral membrane, and the calcar is longer than the foot. The dental formula is 2/1, 1/1, 2/2, 3/3 × 2 = 30.

SYNONYMS:

Mimon Gray, 1847:14; type species *Phyllostoma bennettii* Gray, 1838b, by subsequent designation (Palmer 1904:426).

Tylostoma P. Gervais, 1856a:49; part.

Vampirus Saussure, 1860:487; part; incorrect subsequent spelling of *Vampyrus* Leach, 1821b.

Tylostomum Pelzeln, 1883:31; incorrect subsequent spelling of *Tylostoma* P. Gervais.

Anthorhina Lydekker, 1891:674, in Flower and Lydekker 1891; part.

Mimon [*(Anthorhina)*]: Cabrera, 1958:66; name combination.

Mimmon Dias, Peracchi, and Silva, 2002:113; incorrect subsequent spelling of *Mimon* Gray.

KEY TO SPECIES OF *MIMON*:

1. Margin of noseleaf crenulated and fringed with straight hairs; pale dorsal median stripe present or absent; wing membrane attached to side of foot; forearm 45–51 mm; greatest length of skull 21–23 mm 2
1'. Margin of noseleaf entire; dorsal coloration fulvous brown, no white line on back; wing membrane attached to ankle; forearm 54–61 mm; greatest length of skull 25–28 mm 3
2. Dorsal fur grayish to blackish brown; pale, usually white, dorsal stripe along midline (may be faint in some individuals); no vertical groove between protocone and hypocone on lingual cingulum of M1 and M2 . . . . . .
. . . . *Mimon crenulatum*
2'. Dorsal fur reddish to golden brown; dorsal stripe absent; shallow vertical groove between protocone and hypocone on lingual cingulum of M1 and M2 . . . .
. . . . *Mimon koepckeae*
3. Mesopterygoid fossa V-shaped; mandible longer than 16.7 mm; second lower premolar (p4) with enlarged posterior cingulum and lacking a small cuspulid on posterior cristid of main cusp next to posterior cingulum; post-auricular patches grayish white and distinct; 34 chromosomes . . . . *Mimon cozumelae*
3'. Mesopterygoid fossa U-shaped; mandible shorter than 16.7 mm; second lower premolar (p4) with posterior cingulum reduced, but with a small cuspulid on posterior cristid near posterior cingulum; post-auricular patches buff, often not distinct; 30 chromosomes . . . .
. . . . *Mimon bennettii*

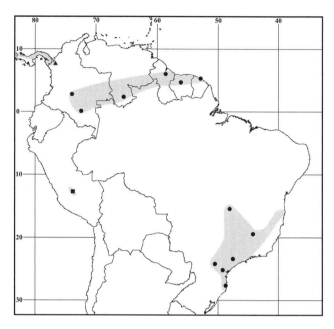

Map 150 Marginal localities for *Mimon bennettii* ●, *Mimon cozumelae* ▲, and *Mimon koepckeae* ■

Mimon bennettii (Gray, 1838)
Bennett's Spear-nosed Bat

SYNONYMS:

Phyllostoma bennettii Gray, 1838b:488; type locality "S. America"; subsequently restricted to Ypanema, São Paulo, Brazil, by Hershkovitz (1951:555).

V[*ampirus*]. *auricularis* Saussure, 1860:487; type locality "Brésil."

M[*imon*]. *bennettii*: Gray, 1847:14; first use of current name combination.

Mimon bennetti N. R. Reis, Peracchi, and Sekiama, 1999:503; incorrect subsequent spelling of *Phyllostoma bennettii* Gray.

DISTRIBUTION: *Mimon bennettii* is endemic to South America where it occurs as two separate populations: one in Colombia, Venezuela, and the Guianas, the other in southeastern Brazil.

MARGINAL LOCALITIES (Map 150): *Northern distribution*. GUYANA: Cuyuni-Mazaruni, 24 miles from Bartica on Potaro Road (Hill 1965). SURINAM: Sipaliwini, Voltzberg (Genoways, Williams, and Groen 1981). FRENCH GUIANA: Paracou (Simmons and Voss 1998). VENEZUELA: Amazonas, western border of Río Mavaca (Molina, García, and Ochoa 1995). COLOMBIA: Caquetá, Estación Puerto Abeja (Montenegro and Romero-Ruiz 2000); Meta, Serranía de La Macarena (Cuervo-Díaz, Hernández-Camacho, and Cadena 1986). *Southern distribution*. BRAZIL: Bahia, Pardo River Valley (Faria, Soares-Santos, and Sampaio 2006, not mapped); Minas Gerais,

3 miles ESE of Sete Lagoas (Molina, García, and Ochoa 1995); São Paulo, Ipanema (type locality of *Phyllostoma bennettii* Gray); Paraná, Gruta da Lancinha II (Graciolli 2004); Santa Catarina, Santo Amaro da Imperatriz (Cherem et al. 2005); Paraná, Fazenda Monte Alegre (N. R. Reis, Peracchi, and Sekiama 1999); Goiás, Gruta Morro (Bredt, Uieda, and Magalhães 1999).

SUBSPECIES: We treat *M. bennettii* as monotypic, although the unusual distribution pattern suggests that the northern component represents an unnamed subspecies.

NATURAL HISTORY: Hill (1965) and Genoways, Williams, and Groen (1981) found *M. bennettii* associated with forest habitats. Genoways, Williams, and Groen (1981) reported this species roosting in caves and in crevices in granitic extrusions in Surinam. Simmons and Voss (1998) found one *M. bennettii* roosting in a large hollow tree in French Guiana. Montenegro and Romero-Ruiz (2000) found *M. bennettii* in *terra firme* forests in southern Colombia. Ortega and Arita (1997) provided a detailed description, along with measurements and a summary of available information on natural history, in their *Mammalian Species* account. Gardner (1977c) gave the diet as insects and fruits. Molina, García, and Ochoa (1995) found the remains of carabid beetles in the stomach of their Venezuelan specimen. Ubelaker, Specian, and Duszynski (1977) listed a *Trypanosoma cruzi*-like endoparasite for Colombian specimens. Graciolli (2004) listed a nycteribiid batfly recovered from specimens from Paraná, Brazil. The karyotype is $2n = 30$, FN = 56 (R. J. Baker, Genoways, and Seyfarth 1981; R. J. Baker et al. 1982).

REMARKS: As understood herein, *M. bennettii* is confined to the lowlands of Venezuela and the Guianas, with a separate distribution in southeastern Brazil. Some recent authors, beginning with Schaldach (1965), have treated *M. cozumelae* of Middle America as a subspecies of *M. bennettii* (also see J. K. Jones 1966; Goodwin 1969; Hall 1981; Koopman 1982, 1993, 1994). Other authors have continued to treat *M. cozumelae* as a separate species (D. C. Carter, Pine, and Davis 1966; Handley 1966c; Gardner, LaVal, and Wilson 1970; Marinkelle and Cadena 1972; J. K. Jones, Smith, and Genoways 1973; J. K. Jones and Carter 1976; Genoways, Williams, and Groen 1981; Molina, Garcia, and Ochoa 1995; Simmons and Voss 1998). Those who have synonymized these two taxa, claim there is little morphological difference between them and that their relationships are best expressed by placing them in a single species. On the other hand, authors who have continued to recognize both as different species contended that no intergradation has been demonstrated between them and each has a different karyotype. Simmons and Voss (1998) illustrated the skulls and dentition, and noted that adult *M. bennet-*

tii, in contrast to *M. cozumelae*, are redder dorsally and have dark wing tips, upper inner incisors tapering to points, narrower lower incisors, a larger talonid on m3, and a U-shaped anterior border of the mesopterygoid fossa (a feature we use in the identification key). Measurements of *M. bennettii* can be found in C. O. C. Vieira (1942), Dalquest (1957a), Husson (1962, 1978), Hill (1965), Swanepoel and Genoways (1979), and Genoways, Williams, and Groen (1981).

The known distributions of *M. bennettii* and *M. cozumelae* are separated by a broad gap in northeastern Colombia and northern Venezuela. Although this cannot be taken as proof that these taxa do not occur in this region, it should be remembered that the chiropteran fauna of Venezuela has been studied extensively in recent years (Butterworth and Starrett 1964; Ojasti 1966; Linares 1969, 1998; J. D. Smith and Genoways 1974; Handley 1976). We believe it is best to treat these taxa as separate species. The J. A. Allen (1911) report of *M. bennettii* from Venezuela was based on a misidentified *Chrotopterus auritus*.

Mimon cozumelae Goldman, 1914
Cozumel Spear-nosed Bat

SYNONYMS:

Mimon cozumelae Goldman, 1914b:75; type locality "Cozumel Island," Quintana Roo, Mexico.

Mimon bennettii cozumelae: Schaldach, 1965:132; name combination.

M[imon]. cazumelae Molina, García, and Ochoa, 1995:264; incorrect subsequent spelling of *Mimon cozumelae* Goldman.

DISTRIBUTION: The only South American record for *M. cozumelae* is from Chigorodó, Antioquia, Colombia. Elsewhere, the species is known from southern Mexico and Central America.

MARGINAL LOCALITY (Map 150): COLOMBIA: Antioquia, Chigorodó (Marinkelle and Cadena 1972).

SUBSPECIES: We consider *M. cozumelae* to be monotypic.

NATURAL HISTORY: The habitats and feeding habits of *M. cozumelae* are probably similar to those of *M. bennettii*. In Mexico and Central America, *M. cozumelae* has been found roosting in caves, tunnels, culverts, and similar structures (Hall and Dalquest 1963; Handley 1966c; Villa-R. 1967; Greenhall and Paradiso 1968). Gardner (1977c) characterized the diet as "plant material and various arthropods." The species also may be carnivorous (Hall and Dalquest 1963). Webb and Loomis (1977) listed two trombiculid mites as known ectoparasites of *M. cozumelae*. The karyotype is $2n = 34$, FN = 56 (R. J. Baker 1979) or FN = 60 (Patton and Baker 1978).

REMARKS: *Mimon cozumelae*, as understood here, is known in South America only from one locality in northwestern Colombia near the Panamanian border, where three individuals were collected. One was an adult male with a forearm of 56.5 mm, another a juvenile male with a forearm of 33.0 mm (Marinkelle and Cadena 1972). No other measurements are known for South American representatives of this species, but see D. C. Carter, Pine, and Davis (1966) and Swanepoel and Genoways (1979) for measurements of individuals from Middle America.

Mimon cozumelae and *M. bennettii* are similar morphologically, but we have noted the following differences between specimens from Belize and Surinam, respectively, in the collections of the Carnegie Museum of Natural History. The lower jaw is longer in *M. cozumelae* (greater than 16.7 mm) and a small cuspulid is present on the posterior lingual cristid of the second lower premolar (p4) near the junction with the posterior cingulum, but missing in *M. bennettii*. The anterior border of the mesopterygoid fossa of *M. cozumelae* tends to be V-shaped as opposed to U-shaped in *M. bennettii*. The postauricular patches of *M. cozumelae* are grayish white and conspicuous; whereas, these patches in *M. bennettii* tend to be reddish to buff and are less distinct. Simmons and Voss (1998) mentioned and illustrated some of these features and concluded that *M. bennettii* and *M. cozumelae* were separate species. *Mimon cozumelae* has 34 chromosomes, but *M. bennettii* has 30, a difference requiring at least two chromosomal rearrangements to derive one from the other (R. J. Baker, Genoways, and Seyfarth 1981).

Mimon crenulatum (É. Geoffroy St.-Hilaire, 1803)
Striped Spear-nosed Bat

SYNONYMS: See under Subspecies.

DISTRIBUTION: *Mimon crenulatum* is in Colombia, Venezuela, Trinidad, the Guianas, Brazil, Ecuador, Peru, and Bolivia. The species also occurs in Mexico and Central America.

MARGINAL LOCALITIES (Map 151): VENEZUELA (Handley 1976): Falcón, Boca de Yaracuy; Miranda, 7 km E of Río Chico. TRINIDAD AND TOBAGO: Trinidad, Las Cuevas (C. H. Carter et al. 1981). VENEZUELA: Delta Amacuro, Río Acoíma (Ochoa 1995). GUYANA: Cuyuni-Mazaruni, 24 miles from Bartica on Potaro Road (Hill 1965). SURINAM: Para, Powaka (Genoways and Williams 1979b). FRENCH GUIANA: Paracou (Simmons and Voss 1998); Les Eaux Claires (Simmons, Voss, and Peckham 2000). BRAZIL: Amapá, Macapá (Mok and Lacey 1980); Pará, Altar do Chão (Bernard 2001b); Pará, Area do Caraipe (Pine et al. 1996); Pernambuco, Fazenda Paus Grandes (Mares et al. 1981); Bahia, Lamarão (type locality of *Anthorhina picta* O. Thomas); Bahia, Ilhéus (Faria,

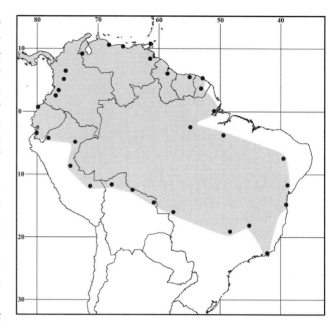

Map 151 Marginal localities for *Mimon crenulatum* ●

Soares-Santos, and Sampaio 2006); Rio de Janeiro, Poço das Antas Biological Reserve (Baptista and Mello 2001); Minas Gerais, Três Marias (Mares, Braun, and Gettinger 1989); Minas Gerais, Panga Ecological Reserve (Pedro and Taddei 1997); Mato Grosso, Villa María (type locality of *Phyllostoma longifolium* J. A. Wagner). BOLIVIA: Santa Cruz, Los Fierros (Emmons 1998); Beni, mouth of Río Baures (Koopman 1976); Pando, San Miguel (Aguirre and Urioste 1994). PERU: Madre de Dios, Pakitza (Ascorra, Wilson, and Romo 1991); Huánuco, Río Pachitea (type locality of *Anthorhina peruana* O. Thomas); Loreto, Jenaro Herrera (Ascorra, Gorchov, and Cornejo 1994); Amazonas, Quebrada Kagka (J. L. Patton, Berlin, and Berlin 1982). ECUADOR (Albuja 1999): El Oro, Cayancas; Esmeraldas, Bilsa. COLOMBIA: Cauca, Tambito Nature Preserve (Dávalos and Guerrero 1999); Valle del Cauca, Hormiguero (M. E. Thomas 1972); Caldas, Finca Los Naranjos (Castaño et al. 2003); Antioquia, Barbosa (Muñoz 2001). VENEZUELA: Zulia, El Rosario, 39 km WNW of Encontrados (Handley 1976).

SUBSPECIES: We recognize four subspecies of *M. crenulatum*.

M. c. crenulatum (É. Geoffroy St.-Hilaire, 1803)

SYNONYMS:

Phyllostoma crenulata É. Geoffroy St.-Hilaire, 1803:61; type locality unknown; given as Brazil by Schinz (1844) and further restricted to Baía (= Bahia) by Cabrera (1958:66).

Ph[yllostomus]. crenulatus: Olfers, 1818:224; name combination.

Tylostoma crenulatum: P. Gervais, 1856a:183; name combination.

[*Anthorhina*] *crenulatum*: Trouessart, 1905:112; name combination.

Mimon [*(Anthorhina)*] *crenulatum*: Cabrera, 1958:66; name combination.

The nominate subspecies occurs in eastern Venezuela, the island of Trinidad, and the Guianas south into the lower Amazon basin and along the Atlantic Coast of Brazil as far south as the state of Bahia.

M. c. keenani Handley, 1960

SYNONYM:

Mimon crenulatum keenani Handley, 1960:460; type locality "Fort Gulick, Panama Canal Zone," Panama.

This subspecies is distributed from Panama through Colombia into northwestern Venezuela and western Ecuador. Elsewhere, *M. c. keenani* is found in Central America and southern Mexico.

M. c. longifolium (J. A. Wagner, 1843)

SYNONYMS:

Phyllostoma longifolium J. A. Wagner, 1843a:365; type locality "Villa Maria," Mato Grosso, Brazil.

Phyllostoma [*(Tylostoma)*] *crenulatum*: W. Peters, 1865:514; name combination.

Tylostoma longifolium: W. Peters, 1866c:398; name combination.

Phyllostoma (Tylostoma) longifolium: Pelzeln, 1883:31; name combination.

A[*nthorhina*]. *longifolium*: O. Thomas, 1903f:458; name combination.

Anthorhina peruana O. Thomas, 1923e:693; type locality "Rio Pachitea," Huánuco, Peru.

Mimon peruanum: Cabrera, 1958:66; name combination.

Mimon crenulatum longifolium: Handley, 1966c:463; first use of current name combination.

This subspecies is in southern Colombia, eastern Ecuador, eastern Peru, western and central Brazil, and northern Bolivia.

M. c. picatum (O. Thomas, 1903)

SYNONYMS:

Anthorhina picata O. Thomas, 1903f:457; type locality "Lamarão, Bahia," Brazil.

Mimon picatum: Cabrera, 1958:66; name combination.

Mimon crenulatum picatum: Handley, 1966c:463; first use of current name combination.

This subspecies is known from the type locality in Bahia, south into the state of Rio de Janeiro.

NATURAL HISTORY: Hill (1965), Handley (1966c, 1976), Tuttle (1970), and Pine et al. (1996) associated *M.*

crenulatum with forest habitats; the species also has been reported from agricultural areas (Handley 1966c, 1976; S. L. Williams and Genoways 1980a). This species roosts in buildings and hollow trees (Goodwin and Greenhall 1961; Handley 1966c, 1976). According to Gardner (1977c), the insectivorous feeding habits of *M. crenulatum* are based on the report by Dobson (1878). Pedro, Komeno, and Taddei (1994) found pollen and insect remains in feces from specimens caught in ground-level mistnets in the vicinity of Uberlândia, Minas Gerais, Brazil. Pine et al. (1996) reported specimens taken in ground level nets placed in tall terra firme forest in southeastern Pará State in Brazil. Mares, Braun, and Gettinger (1989) reported two females, each carrying a large fetus (CR = 30 and 35 mm, respectively) when taken in October in Minas Gerais, Brazil. Also, in Minas Gerais, Pedro, Komeno, and Taddei (1994) found a pregnant female (16.6-mm-CR fetus) in September, and another in October (29.2-mm-CR fetus). They interpreted these dates as indicating births at the beginning of the rainy season. Brennan and Reed (1975) listed three species of trombiculid mites recovered from Venezuelan *M. crenulatum*. Webb and Loomis (1977) listed one nycteribiid batfly, one myobiid mite, one spinturnicid mite, one trombiculid mite, and three argasid ticks known from this species. R. Guerrero (1985a) reported another spinturnicid mite and a streblid batfly. Komeno and Linhares (1999) added a nycteribiid batfly to the diverse ectoparasitic fauna known from this species.

The karyotype is $2n = 32$, FN = 60 (R. J. Baker and Hsu 1970). R. J. Baker, Gardner, and Patton (1972) reported a chromosomal polymorphism involving the fifth largest pair of autosomes.

REMARKS: Koopman (1993) dated *M. crenulatum* from É. Geoffroy St.-Hilaire (1810b) and not from É. Geoffroy St.-Hilaire's (1803) "Catalogue des mammifères," because the latter work was considered to be an unpublished proof (I. Geoffroy St.-Hilaire 1839:5 [footnote], 1847b:115–18; Sherborn 1922:lviii, 1932:cxxxviii; Wilson and Reeder 1993:831). A recent ruling (ICZN 2002) has confirmed that É. Geoffroy St.-Hilaire's (1803) *Catalogue des mammifères du Muséum National d'Histoire Naturelle* is available for nomenclatural purposes.

When Handley (1960) reviewed the systematics of *M. crenulatum*, he recognized four subspecies, all of which occurred in South America. Subsequently, Gardner and Patton (1972) described *M. koepckeae* based upon specimens from cloud forest habitat on the eastern slope of the Andes in departamento Ayacucho, Peru. Koopman (1976, 1978) treated *M. koepckeae* as "only a Peruvian highland subspecies of *M. crenulatum*." However, *M. koepckeae* is a valid species, as confirmed by several investigators (e.g., Simmons and Voss 1998).

Some authors placed *M. crenulatum* in the genus *Anthorhina*; however, Simpson (1945) considered *Anthorhina* and *Mimon* to be congeneric. Husson (1962, 1978) argued for the continued use of *Anthorhina*, whereas Cabrera (1958), Goodwin and Greenhall (1961), Hill (1965), Gardner and Patton (1972), and Corbet and Hill (1991) all used *Anthorhina* at the subgeneric level. Handley (1960) wrote "the nominal genera *Anthorhina* and *Mimon* are not distinguishable even as subgenera." Also, as pointed out by Gardner and Ferrell (1990), *Anthorhina* is a junior synonym of *Tonatia* Gray, 1827, and is not available for taxa now included in the genus *Mimon*. *Anthorhina* Lydekker in Flower and Lydekker (1891), because it is a replacement name for *Tylostoma* P. Gervais, 1856a (preoccupied), takes the same type species as was subsequently designated by Palmer (1904) for *Tylostoma* P. Gervais.

Cabrera's (1958) secondary restriction of the type locality to Bahia, Brazil, is unfortunate. The eastern and southeastern Brazilian populations probably should bear the name *M. c. picatum*, whereas the Guianan, Venezuelan, and adjacent Brazilian populations most likely should be known as the nominate subspecies. Cabrera (1958), when dealing with a species for which the origin of its type specimen was unknown, often followed the practice of designating as the type locality the first place where a specimen was subsequently collected. Schinz (1844) restricted the type locality of *Phyllostoma crenulatum* Geoffroy St. Hilaire to Brazil. Then Cabrera (1958) secondarily restricted the type locality to Bahia based on the first specimen recorded from Brazil. Sanborn (1949b), Handley (1960), Goodwin and Greenhall (1961), Hill (1965), Gardner and Patton (1972); Husson (1978), Genoways and Williams (1979b); Swanepoel and Genoways (1979), C. H. Carter et al. (1981), and Simmons and Voss (1998) have provided measurements of South American *M. crenulatum*. D. C. Carter and Dolan (1978) gave information on and measurements of the holotypes of *Phyllostoma crenulata* É. Geoffroy St.-Hilaire, *Anthorhina peruana* O. Thomas, and *Anthorhina picata* O. Thomas.

Mimon koepckeae Gardner and Patton, 1972
Koepcke's Spear-nosed Bat
SYNONYMS:

Mimon koepckeae Gardner and Patton, 1972:7; type locality "Huanhuachayo (12°44′S, 73° 47′W, elevation *ca.* 1,660 meters, Departamento de Ayacucho, Perú."

[*Mimon crenulatum*] *koepkeae*: Koopman, 1976:46; name combination and incorrect subsequent spelling of *Mimon koepckeae* Gardner and Patton.

DISTRIBUTION: *Mimon koepckeae* is known from the type locality and vicinity.

Marginal Localities (Map 150; from Gardner and Patton 1972): PERU: Ayacucho, Huanhuachayo (type locality of

Mimon koepckeae Gardner and Patton); Ayacucho, Estera Ruana (not mapped).

SUBSPECIES: *Mimon koepckeae* is monotypic.

NATURAL HISTORY: Specimens were caught in ground-level mistnets placed in Andean slope forest adjacent to clearings (Huanhuachayo and Estera Ruana) along the trail from San José to Tambo.

REMARKS: Koopman (1976, 1978) treated *M. koepckeae* as "only a Peruvian highland subspecies of *M. crenulatum*." However, *M. koepckeae* can be distinguished by its pale color, lack of dorsal stripe, narrow auditory bullae, and by the vertical groove on the cingulum between the paracone and hypocone of the first and second upper molars. Gardner and Patton (1972, Table 3) gave measurements demonstrating the smaller size of *M. koepckeae* in comparison to Peruvian *M. crenulatum*. Although their sample of *M. koepckeae* was small (one female and two males) there was no overlap in measurements of greatest length of skull, condylobasal length, zygomatic breadth, palatal length, maxillary and mandibular toothrows, and breadth across upper molars.

Genus *Neonycteris* Sanborn, 1949

The monotypic *Neonycteris* is represented by *N. pusilla*, a small (forearm 34.3 mm, greatest length of skull 17.9 mm), dark brown phyllostomine with a brown venter. Distinguishing features include a rounded ear; the lack of an interauricular cutaneous band; lower border of narial horseshoe defined by a ridge; and lower lip with single pair of smooth tubercles separated in the midline by a shallow cleft. The metacarpal of digit IV is the shortest and the metacarpal of digit III is longest; the second phalanx of digit III is longer than the first, and the second phalanx of digit IV is about equal in length to the first phalanx. The calcar is shorter than the foot. Canines are relatively short, less than twice as high as the upper inner incisors. The occlusal surfaces of the upper outer incisors are visible in the toothrow, and not hidden under the cingula of the canines. The lower incisors are trifid. The dental formula is 2/2, 1/1, 2/3, 3/3 × 2 = 34.
SYNONYMS:

Neonycteris Sanborn 1949a:226; type species *Micronycteris (Neonycteris) pusilla* Sanborn, 1949a, by monotypy; described as a subgenus of *Micronycteris* Gray.

Neonycteris: Simmons and Voss, 1998:62; first use as a full genus.

Neonycteris pusilla (Sanborn, 1949)
Least Big-eared Bat
SYNONYMS:

Micronycteris (Neonycteris) pusilla Sanborn, 1949a:228; type locality "Tahuapunta, Rio Vaupes, at the Colombian border, Amazonas, Brazil."

[*Neonycteris*] *pusilla*: Simmons and Voss, 1998:62; first use of current name combination.

DISTRIBUTION: *Neonycteris pusilla* is known only from the type locality in northwestern Brazil.

MARGINAL LOCALITY (Map 146): BRAZIL: Amazonas, Tahuapunta, Rio Vaupés (type locality of *Micronycteris (Neonycteris) pusilla* Sanborn).

SUBSPECIES: *Neonycteris pusilla* is monotypic.

NATURAL HISTORY: Unknown.

REMARKS: *Neonycteris pusilla* is known on the basis of two males collected at the type locality. Sanborn (1949a), Swanepoel and Genoways (1979), and Simmons and Voss (1998) have provided measurements of these two specimens.

Genus *Phylloderma* W. Peters, 1865

The monotypic *Phylloderma* is represented by *P. stenops*, a medium-sized bat (forearm 65–74 mm, greatest length of skull 30–33 mm) morphologically similar to *Phyllostomus* spp. *Phylloderma* can be distinguished by the obviously expanded braincase, bifid first lower incisor, three lower premolars (third premolar small), and narrow-crowned molars. The calcar is about equal to or shorter than the foot, and the tail extends to mid-length of the interfemoral membrane. The dental formula is 2/2, 1/1, 2/3, 3/3 × 2 = 34.

SYNONYMS:

Phylloderma W. Peters, 1865c:513; type species *Phylloderma stenops* W. Peters, 1865c, by monotypy.

Guandira Gray, 1866b:114; type species *Guandira cayanensis* Gray, 1866b, by monotypy.

Phyllostomus: R. J. Baker, Dunn, and Nelson, 1988:13; part; not *Phyllostomus* Lacépède, 1899.

Guyira Muñoz, 2001, 71; incorrect subsequent spelling of *Guandira* Gray.

Phylloderma stenops W. Peters, 1865
Spear-nosed Bat

SYNONYMS: See under Subspecies.

DISTRIBUTION: *Phylloderma stenops* occurs in Colombia, Venezuela, the Guianas, Ecuador, Peru, Brazil, and Bolivia. The species also occurs in Mexico and Central America.

MARGINAL LOCALITIES (Map 152): VENEZUELA (Handley 1976): Falcón 19 km NW of Urama; Sucre, 21 km E of Cumaná. TRINIDAD AND TOBAGO: Trinidad, Blanchisseuse (C. H. Carter et al. 1981). GUYANA: Cuyuni-Mazaruni, 24 miles from Bartica on Potaro Road (Hill 1965). SURINAM: Coronie, Totness (Genoways and Williams 1979b). FRENCH GUIANA: Cayenne (type locality of *Phylloderma stenops* W. Peters, and *Guandira cayanensis* Gray); Les Eaux Claires (Simmons, Voss, and Peckham

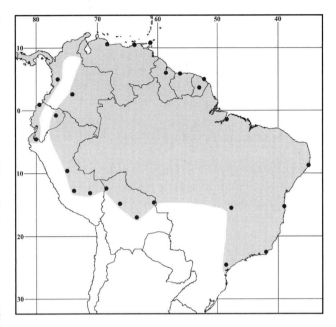

Map 152 Marginal localities for *Phylloderma stenops* ●

2000). BRAZIL: Pará, Utinga (Handley 1967); Pernambuco, Estação Florestal de Experimentação de Saltinho (Guerra 1981b); Bahia, Reserva Biológica de Una (Esbérard and Faria 2006); Rio de Janeiro, Fazenda Reunidas (Esbérard and Faria 2006); São Paulo, Iporanga (Trajano 1982); Distrito Federal, Gruta Água Rasa (Bredt, Uieda, and Magalhães 1999). BOLIVIA (S. Anderson 1997, except as noted): Santa Cruz, 23 km S of Campamento Los Fierros; Santa Cruz, 7 km N of Santa Rosa (type locality of *Phylloderma septentrionalis boliviensis* Barquez and Ojeda); Beni, Totaisal; Pando, Chivé. PERU: Cusco, Consuelo (V. Pacheco et al. 1993); Ayacucho, Yuraccyacu (Gardner 1976); Huánuco, Panguana (Hutterer et al. 1995). ECUADOR: Orellana, 30 km SW of Onkone Gare (F. A. Reid, Engstrom, and Lim 2000). COLOMBIA: Meta, Cabaña Duda (Lemke et al. 1982); Chocó, 4 km N of La Italia (Alberico 1994). ECUADOR: Esmeraldas, El Porvenir (Albuja and Mena 2004). PERU: Piura, 4 miles W of Suyo (TCWC 16412).

SUBSPECIES: We recognize three subspecies of *P. stenops*, two of which occur in South America. *Phylloderma stenops septentrionalis* Goodwin, 1940, is found in Central America.

P. s. boliviensis Barquez and Ojeda, 1979
SYNONYMS:

Phylloderma stenops boliviensis Barquez and Ojeda, 1979:84; type locality "7 km al norte de Santa Rosa, Provincia de Sara, Departamento Santa Cruz," Bolivia.

Phyllostomus stenops boliviensis: S. Anderson, 1997:202; name combination.

This subspecies is known only from southeastern Bolivia.

P. s. stenops W. Peters, 1865

SYNONYMS:

Ph[*yllostoma (Phylloderma)*]. *stenops* W. Peters, 1865c: 513; type locality "Cayenne," French Guiana.
Guandira cayanensis Gray, 1866b:114; type locality "Cayenne," French Guiana.

The nominate subspecies occurs in Colombia, Venezuela, the Guianas, Ecuador, Peru, and northern and eastern Brazil.

NATURAL HISTORY: Little is known about the ecology and behavior of *P. stenops*. It is associated with a variety of primary and disturbed forest habitats (Hill 1965; D. C. Carter, Pine, and Davis 1966; Handley 1966c, 1976; S. L. Williams and Genoways 1980a). Gardner (1977c) suggested that this species feeds on insects and plant material. However, the only reference to food habits is of one individual feeding on the larvae and pupae from the nest of a social wasp (Jeanne 1970). Brennan and Reed (1975) listed a trombiculid mite, Webb and Loomis (1977) listed a streblid batfly, and R. Guerrero (1985a) added two genera of batflies and four species of mites, representing three genera, as ectoparasites known from *P stenops*. R. Guerrero (1997) listed six streblid batflies, five of which had not been reported by either Webb and Loomis (1977) or R. Guerrero (1985a). The karyotype is $2n = 32$, FN = 58 (R. J. Baker et al. l982).

REMARKS: Handley (1966c) stated that specimens from Panama combined the characteristics of the two nominal species of *Phylloderma* (*P. stenops* and *P. septentrionalis*); therefore, he synonymized them under the name *P. stenops*. Goodwin and Greenhall (1961), Husson (1962), Hill (1965), Gardner (1976), Ojeda and Barquez (1978), Barquez and Ojeda (1979), Genoways and Williams (1979b), Swanepoel and Genoways (1979), S. L. Williams and Genoways (1980a), C. H. Carter et al. (1981), and Barquez (1984a) have provided measurements of South American specimens. D. C. Carter and Dolan (1978) gave information on and measurements of the holotypes of *Phylloderma stenops* W. Peters and *Guandira cayanensis* Gray.

Genus *Phyllostomus* Lacépède, 1799

The genus *Phyllostomus* is represented by four species of medium-sized to large bats (forearm 58–88 mm, greatest length of skull 28–39 mm). These bats can be distinguished from other phyllostomines by the presence of two incisors and two premolars in the lower toothrow. The tail extends to the middle of the interfemoral membrane. The length of the calcar, as compared to the length of the foot, varies between species. A gular gland, well developed in males but rudimentary in females, is present in all species. The dental formula is 2/2, 1/1, 2/2, 3/3 × 2 = 32.

SYNONYMS:

Vespertilio: Pallas, 1767:7; part; not *Vespertilio* Linnaeus, 1758.
Pteropus: Erxleben, 1777:136; not *Pteropus* Brisson, 1762.
Phyllostomus Lacépède, 1799b:16; type species *Vespertilio hastatus* Pallas, 1767, by monotypy; later ruled (ICZN 1955:Direction 24) to be by subsequent selection by Miller and Rehn (1901).
Phyllostoma G. Cuvier, 1800:Tab. I; unjustified emendation of *Phyllostomus* Lacépède, 1799b; placed on Official List of Rejected and Invalid Names (ICZN 1955).
Phylostoma G. Fischer, 1817:373; incorrect subsequent spelling of *Phyllostoma* G. Cuvier.
Alectops Gray, 1866b:114; type species *Alectops ater* Gray, 1866b, by monotypy.
Phyllostomum Thenius, 1969:257; incorrect subsequent spelling of *Phyllostomus* Lacépède.

KEY TO SPECIES OF *PHYLLOSTOMUS*:

1. Length of calcar equal to or longer than length of hind foot; length of ear (from notch) more than 25 mm; sagittal crest well developed; first upper and first lower incisors higher (longer) than wide. 2
1′. Length of calcar shorter than length of hind foot; length of ear (from notch) less than 25 mm; sagittal crest absent or weakly developed; first upper and lower incisors broad (wider than high) . . . . *Phyllostomus discolor*
2. Forearm shorter than 75 mm; wing membrane attached to ankles; greatest length of skull less than 35 mm . . . 3
2′. Forearm longer than 75 mm; wing membrane attached to side of foot; greatest length of skull more than 35 mm . . . . *Phyllostomus hastatus*
3. Forearm longer than 61 mm; tibia longer than 23 mm; greatest length of skull more than 29 mm . . . . *Phyllostomus elongatus*
3′. Forearm shorter than 61 mm; tibia shorter than 24 mm; greatest length of skull less than 29 mm . . . . *Phyllostomus latifolius*

Phyllostomus discolor (J. A. Wagner, 1843)

Pale Spear-nosed Bat

SYNONYMS:

Phyllostoma discolor J. A. Wagner, 1843a:366; type locality "Cuyaba," Mato Grosso, Brazil.
Ph[*yllostoma*]. *innominatum* Tschudi, 1844b:62; type locality "Maynas" [p. 68], Loreto, Peru.
Phyllostoma angusticeps P. Gervais, 1856a:47; type locality "province de Bahia," Brazil.
Phyllostoma verrucossum Elliot, 1905a:236; type locality "Niltepec, Oaxaca, Mexico."
P[*hyllostomus*]. *verrucosus*: Miller, 1907b:131; name combination and emendation of *Phyllostoma verrucossum* Elliot.

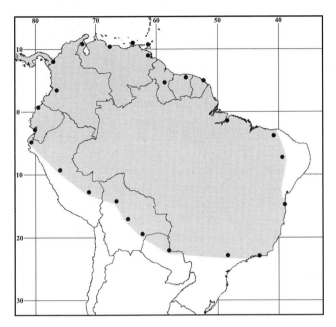

Map 153 Marginal localities for *Phyllostomus discolor* ●

Phyllostomus discolor: J. A. Allen, 1904d:344; first use of current name combination.

DISTRIBUTION: *Phyllostomus discolor* is in Colombia, Venezuela, the Guianas, Ecuador, Peru, Brazil, Bolivia, Paraguay, and possibly northern Argentina. The species also occurs in Mexico and Central America.

MARGINAL LOCALITIES (Map 153): VENEZUELA: Nueva Esparta, Margarita Island, El Valle (J. D. Smith and Genoways 1974). TRINIDAD AND TOBAGO: Trinidad, Las Cuevas (C. H. Carter et al. 1981). VENEZUELA: Delta Amacuro, Caño Araguaito (Linares and Rivas 2004). GUYANA: Potaro-Siparuni, Kurupukari (P. G. Smith and Kerry 1996). SURINAM: Para, Zanderij (Genoways, Williams, and Groen 1981). FRENCH GUIANA: Cayenne (Sanborn 1936). BRAZIL: Pará, Belém (Carvalho 1960a); Ceará, Horto (S. S. P. Silva, Guedes, and Peracchi 2001); Ceará, Aeroporto de Crato (Mares et al. 1981); Bahia Ilhéus (Faria. Soares-Santos, and Sampaio 2006); Rio de Janeiro, Reserva Florestal do Grajaú (Esbérard 2003); São Paulo, Botucatu region (Uieda and Chaves 2005). PARAGUAY: Alto Paraguay, Puerto Sastre (Myers and Wetzel 1983). BOLIVIA: Santa Cruz, Hacienda Cerro Colorado (Ibáñez and Ochoa 1989); Cochabamba, Sajta (S. Anderson 1997); Beni, Espíritu (S. Anderson 1997). PERU: Madre de Dios, Itahuana (Koopman 1978); Huánuco, Tingo María (Gardner 1976); Piura, Salitral (Koopman 1978). ECUADOR: Guayas, Isla Puná (Power and Tamsitt 1973); Esmeraldas, 3 km W of Majua (USNM 513434). COLOMBIA: Valle del Cauca, Cali (Power and Tamsitt 1973); Chocó, Unguía (Cuartas and Muñoz 1999). VENEZUELA: Zulia,

near Cerro Azul, 33 km NW of La Paz (Handley 1976); Aragua, Portachuelo (USNM 562892).

SUBSPECIES: We recognize two subspecies of *P. discolor*; the nominate form, *P. d. discolor* (J. A. Wagner), is the only South American subspecies.

NATURAL HISTORY: *Phyllostomus discolor* is associated with a variety of habitats ranging from primary forest to agricultural areas (Handley 1966c, 1976; S. L. Williams and Genoways 1980a). Goodwin and Greenhall (1961) suggested that this species prefers hollow trees for roosting sites, but it also roosts in caves (Greenhall and Paradiso 1968; Handley 1976; Gardner 1977a). Gardner (1977c) cited references documenting an omnivorous diet of insects and a variety of plant material, including fruit, pollen, nectar, and other flower parts. Goodwin and Greenhall (1961) and Valdez (1970) commented that this species has an extensible and grooved tongue that could facilitate feeding on pollen and nectar. In their summary of endoparasites, Ubelaker, Specian, and Duszynski (1977) mentioned trypanosomes and nematodes as occurring in *P. discolor*. Brennan and Reed (1975) listed six trombiculid mites, and Webb and Loomis (1977) added eight streblid batflies, one myobiid mite, one labidocarpid mite, and two spinturnicid mites as ectoparasites. R. Guerrero (1985a) added one endoparasite (a nematode) and ten ectoparasites (one trombiculid mite, three spinturnicid mites, and six streblid batflies). R. Guerrero (1997) listed 18 species of streblids known from *P. discolor*. Cuartas and Muñoz (1999) found trichostrongylid nematodes in the small intestine of Colombian specimens. The karyotype is $2n = 32$, FN $= 60$ (R. J. Baker et al. 1982).

REMARKS: Based on the results of a multivariate morphometric analysis, Power and Tamsitt (1973) did not recognize subspecies within *P. discolor*. Although they delineated morphological differences between populations living west of the Andes in South America and into Central America, from those living east of the Andes, these differences were not unambiguous. W. B. Davis and Carter (1962a) also questioned the use of subspecies in *P. discolor*, stating "that measurements hitherto considered as expressions of geographic variation are in reality expressions of individual variation." However, recent authors have not treated *P. discolor* as monotypic. A comprehensive revision, combining data from molecular genetics, has yet to be done.

Taddei (1975) found that Brazilian specimens of *P. discolor* had coefficients of variation for external measurements of from 2.38 to 6.51, and for cranial dimensions, from 0.96 to 4.45. Power and Tamsitt (1973) and Taddei (1975) documented secondary sexual variation in this species; differences that primarily involved breadth measurements of the skull. Other authors recording measurements of *P. discolor* include Sanborn (1936), Dalquest

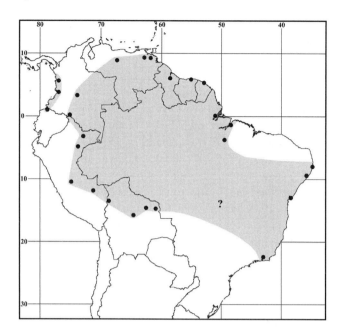

Map 154 Marginal localities for *Phyllostomus elongatus* ●

(1951), Goodwin and Greenhall (1961), W. B. Davis and Carter (1962a), Tamsitt and Valdivieso (1963a), Gardner (1976), J. D. Smith and Genoways (1974), Husson (1978), and Swanepoel and Genoways (1979). D. C. Carter and Dolan (1978) gave information on and measurements of the holotype of *Phyllostoma discolor* J. A. Wagner.

Phyllostomus elongatus (É. Geoffroy St.-Hilaire, 1810)
Lesser Spear-nosed Bat

SYNONYMS:

Phyllostoma elongatum É. Geoffroy St.-Hilaire, 1810b: 182; type locality "en Amérique" (p. 185); restricted to Brazil by Schinz (1844:325); further restricted to Rio Branco, Mato Grosso, Brazil, by Cabrera (1958), the origin of Natterer's specimen cited by Pelzeln (1883: 31).

Ph[yllostomus]. elongatus: Olfers, 1818:224; first use of current name combination.

Alectops ater Gray, 1866b:114; type locality "Surinam."

Phyllostoma lanceolatum Pelzeln, 1883:31; in synonymy, *nomen nudum*.

Phyllostomus elongatum: C. O. C. Vieira, 1942:281; incorrect subsequent spelling of *Phyllostomus elongatus*.

P[hyllostomus]. ater: Muñoz, 2001:76; name combination.

DISTRIBUTION: *Phyllostomus elongatus* occurs in Colombia, Venezuela, the Guianas, Peru, Bolivia, and Brazil, with an isolated population unit in western Colombia and northwestern Ecuador.

MARGINAL LOCALITIES (Map 154): *Primary distribution*. VENEZUELA: Monagas, Hato Mata de Bejuco (Handley 1976); Delta Amacuro, Los Güires (Ojasti and

Naranjo 1974). GUYANA: Cuyuni-Mazaruni, 24 miles from Bartica on Potaro Road (Hill 1965). SURINAM: Paramaribo, Paramaribo (Husson 1978). FRENCH GUIANA: Paracou (Simmons and Voss 1998). BRAZIL (C. O. C. Vieira 1955, except as noted): Amapá, Marcapá (Mok and Lacey 1980); Pará, Utinga (Handley 1967); Pará, Area do Caraipe (Pine et al. 1996); Pernambuco, Recife (Souza-Lopes 1978); Alagoas, Rio Largo; Bahia, Salvador; Rio de Janeiro, Terezopolis. BOLIVIA: Santa Cruz, El Refugio (Emmons 1998); Santa Cruz, Perseverencia (S. Anderson 1997); Beni, 5 km NW of mouth of Río Grande (S. Anderson, Koopman, and Creighton 1982); La Paz, Alto Río Madidi (Emmons 1991). PERU: Madre de Dios, Pakitza (Ascorra, Wilson, and Romo 1991); Pasco, San Juan (Tuttle 1970); Loreto, Jenaro Herrera (Ascorra, Gorchov, and Cornejo 1994); Loreto, Quebrada Sucusari (Ascorra and Wilson 1992). COLOMBIA: Putumayo, Parque Nacional Natural La Paya (Polanco-Ochoa, Jaimes, and Piragua 2000); Meta, Los Micos (Furman 1966). VENEZUELA: Guarico, Estación Biológica de los Llanos (Handley 1976). *Northwestern distribution*. COLOMBIA: Chocó, near Rio Patio, approximately 15 km S and 30 km W of Quibdó (Alberico 1994); Valle del Cauca, Zabaletas (M. E. Thomas 1972). ECUADOR: Esmeraldas, Urbina (Albuja 1983).

SUBSPECIES: We are treating *P. elongatus* as monotypic; the species needs to be revised.

NATURAL HISTORY: *Phyllostomus elongatus* is associated with forest habitats (Hill 1965, Tuttle 1970, Handley 1976; S. L. Williams and Genoways 1980a; Pine et al. 1996), and agricultural areas (Handley 1976). Tuttle (1970) and Handley (1976) reported this species roosting in hollow trees and culverts. Gardner (1977c), while commenting that nothing was known about its food habits, suggested that the diet includes "flower parts, fruits, insects, and small vertebrates." His suggestion was based on documentation concerning the food habits of *P. discolor* and *P. hastatus*, and on Tuttle's (1970) report of finding pollen on Peruvian *P. elongatus*. *Phyllostomus elongatus* occasionally is carnivorous; E. Fischer et al. (1997) reported watching a *P. elongatus* eating a juvenile *Carollia perspicillata* in Maruaga Cave, Presidente Figueiredo, Amazonas, Brazil. Trypanosomes were listed by Ubelaker, Specian, and Duszynski (1977) as endoparasites reported from this species. Brennan and Reed (1975) identified a trombiculid mite recovered from Venezuelan specimens. Webb and Loomis (1977) listed two streblid batflies, one spinturnicid mite, and two labidocarpid mites as ectoparasites. According to R. Guerrero (1985a), *P. elongatus* is host to an additional species of labidocarpid mite, two more spinturnicid mites, and nine streblid batflies. R. Guerrero (1997) listed 14 streblids confirmed from *P. elongatus*. The karyotype is $2n = 32$, FN = 58 (R. J. Baker et al. 1982).

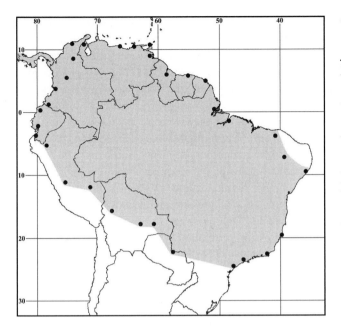

Map 155 Marginal localities for *Phyllostomus hastatus* ●

REMARKS: Husson (1962) found little variation in measurements among the samples he examined from Surinam and Guyana, and the measurements of Peruvian specimens reported by Sanborn (1951b). Sanborn (1936), Butterworth and Starrett (1964), Hill (1965), Husson (1978), Swanepoel and Genoways (1979), and Simmons and Voss (1998) have provided measurements of *P. elongatus*. D. C. Carter and Dolan (1978) gave information on and measurements of the type of *Phyllostoma elongatum* É. Geoffroy St.-Hilaire.

Phyllostomus hastatus (Pallas, 1767)
Great Spear-nosed Bat

SYNONYMS: See under Subspecies.

DISTRIBUTION: *Phyllostomus hastatus* occurs in Colombia, Venezuela, Trinidad, Guyana, Surinam, French Guiana, Ecuador, Brazil, Peru, Bolivia, and Paraguay.

MARGINAL LOCALITIES (Map 155): COLOMBIA: Magdalena, Río Frío (USNM 240093). VENEZUELA: Zulia, near Cerro Azul, 33 km NW of La Paz (Handley 1976); Miranda, Birongo (Handley 1976); Sucre, 21 km E of Cumaná (Handley 1976); Delta Amacuro, Caño Araguaito (Linares and Rivas 2004). TRINIDAD AND TOBAGO: Trinidad, Las Cuevas (C. H. Carter et al. 1981). GUYANA: Cuyuni-Mazaruni, 24 miles from Bartica on Potaro Road (Hill 1965). SURINAM: Paramaribo, Paramaribo (Husson 1978). FRENCH GUIANA: vicinity of Cayenne (Brosset and Dubost 1968). BRAZIL: Amapá, São Joaquim do Pacuí (Peracchi, Raimundo, and Tannure 1984); Pará, Belém (Carvalho 1960a); Ceará, Gruta do Tião (S. S. P. Silva, Guedes, and Peracchi 2001); Ceará, Colegio Agrícola de

Crato (Mares et al. 1981); Alagoas, Rio Largo (J. L. Lima 1926); Espírito Santo, Rio Dôce (J. L. Lima 1926); Rio de Janeiro, Poço das Antas Biological Reserve (Baptista and Mello 2001); São Paulo, Mogi das Cruzes (C. O. C. Vieira 1955); São Paulo, Registro (McNab 1969). PARAGUAY: Concepción, Entre Estancia Estrellas et Estancia Primavera (Baud 1981). BOLIVIA: Santa Cruz, San José de Chiquitos (Barquez 1984a); Santa Cruz, Santa Cruz de la Sierra (S. Anderson et al. 1993); La Paz, 4 km (by road) NW of Alcoche (S. Anderson, Koopman, and Creighton 1982). PERU: Madre de Dios, Pakitza (Ascorra, Wilson, and Romo 1991); Junín, San Ramón (Koopman 1978); Amazonas, Pomará (Koopman 1978); Tumbes, Matapalo (Koopman 1978). ECUADOR: Guayas, Guayaquil (Albuja 1983); Esmeraldas, Quinindé (Albuja 1999). COLOMBIA: Nariño, Altaquer (Cadena, Anderson, and Rivas-Pava 1998); Valle del Cauca, Zabaletas (M. E. Thomas 1972); Caldas, Charca de Guarinocito (Castaño et al. 2003); Bolívar, Norosí (Hershkovitz 1949c).

SUBSPECIES: We recognize three subspecies of *P. hastatus*.

P. h. aruma O. Thomas, 1924
SYNONYM:

Phyllostomus hastatus aruma O. Thomas, 1924a:236; type locality "Taguatinga," Tocantins, Brazil.

This subspecies is known only from the type locality (see Taddei 1975).

P. h. hastatus (Pallas, 1767)
SYNONYMS:

V[espertilio]. hastatus Pallas, 1767:7; type locality "Amérique," restricted to Surinam by J. A. Allen (1904b: 233); based on *La chauvesouris fer-de-lance* of Buffon (1765).

[Pteropus] hastatus: Erxleben, 1777:136; name combination.

Phyllostomus hastatus: Lacépède, 1799b:16; first use of current name combination.

Phyllostoma emarginata É. Geoffroy St.-Hilaire, 1803:60; type locality "La Guiane."

Phyllostoma hastatum: É. Geoffroy St.-Hilaire, 1810b:177; name combination.

Phyllostomus maximus Wied-Neuwied, 1821:242, footnote; type locality "Die Wäldern an den Ufern des Rio das Contas," Bahia, Brazil.

Phyllostomus hastatus curaca Cabrera, 1917:12; type locality "Archidona, sobre el río Napo," Napo, Ecuador.

The nominate subspecies occurs throughout most of South America east of Lake Maracaibo and south of Cordillera de Mérida (Venezuela) to Bolivia, Paraguay, and southeastern Brazil.

P. h. panamensis J. A. Allen, 1904

SYNONYMS:

Phyllostomus hastatus panamensis J. A. Allen, 1904b:233; type locality "Boqueron, Chiriqui," Panama.

Phyllostomus hastatus caurae J. A. Allen, 1904b:234; type locality "Cali, upper Cauca Valley," Valle del Cauca, Colombia; incorrect original spelling of *Phyllostomus hastatus caucae* J. A. Allen.

Phyllostomus hastatus caucae: J. A. Allen, 1916c:225; corrected spelling of *Phyllostomus hastatus caurae* J. A. Allen.

Phyllostomus hastatus paeze O. Thomas, 1924a:235; type locality "Bogota," Cundinamarca, Colombia.

This subspecies occurs in Colombia west of the Andes and north and west of Lake Maracaibo, Venezuela. Elsewhere, it is found in Central America (J. K. Jones and Carter 1976).

NATURAL HISTORY: *Phyllostomus hastatus* is a widespread species that appears to occupy the full range of habitats within its distribution. Documented roosting sites include caves, hollow trees, foliage, buildings, hollow termite nests, and thatched roofs (Goodwin and Greenhall 1961; Handley 1966c, 1976; Tuttle 1970; McCarthy, Cadena, and Lemke 1983). Gardner (1977c) documented a diet of fruits, flower parts, nectar, pollen, and a variety of insects and small vertebrates. Kalko and Condon (1998) considered *P. hastatus* the primary dispersal agent of the seeds of *Gurania spinulosa* (Cucurbitaceae), a neotropical vine with flagellichorous fruits. In their summary of endoparasites found in leaf-nosed bats, Ubelaker, Specian, and Duszynski (1977) noted that trypanosomes, cestodes, nematodes, and trematodes have been recorded from *P. hastatus*. Brennan and Reed (1975) identified four trombiculid mites from Venezuelan *P. hastatus*. Webb and Loomis (1977) listed 15 species of streblid batflies, 1 ereynetid mite, 1 demodicid mite, 1 labidocarpid mite, 2 spinturnicid mites, 5 trombiculid mites, and 2 argasid ticks as known ectoparasites found on this species. R. Guerrero (1985a) added three trypanosomes, two intestinal nematodes, a flea, two macronyssid mites, four spinturnicid mites, and eight streblid batflies to the parasites known from *P. hastatus*. R. Guerrero (1997) listed 26 species of streblids confirmed from this species. Santos et al. (2003) summarized natural history information in their *Mammalian Species* account. The karyotype is $2n = 32$, FN = 58 (R. J. Baker et al. 1982).

REMARKS: Although *P. hastatus* is a widespread and, in some areas, relatively common, little has been written about its geographic variation. Hershkovitz (1949c) made brief comments on a comparison of Venezuelan specimens of *P. h. hastatus* with Colombian and Panamanian representatives of *P. h. panamensis*, and in which *P. h. panamensis*

averaged larger. Taddei (1975) found that the type of *P. h. aruma* was smaller than the other Brazilian specimens he identified as *P. h. hastatus* (forearm 72 mm versus 78–88 mm, greatest length of skull 34.0 mm versus 36.0–38.8 mm). He also remarked that other specimens previously identified by C. O. C. Vieira (1942) as *P. h. aruma* were misidentified because they were too small to be *P. hastatus*.

Taddei (1975) gave coefficients of variation of from 1.28 to 6.04 for external measurements, and from 1.06 to 2.84 for cranial dimensions of specimens from Brazil. Using this same sample, he also found that males were larger than females in all cranial measurements and in 8 of 17 external measurements. Dalquest (1951), Goodwin and Greenhall (1961), Husson (1962, 1978), Swanepoel and Genoways (1979), Barquez (1984a), and Simmons and Voss (1998) have provided measurements of *P. hastatus*.

Phyllostomus latifolius (O. Thomas, 1901)
Guianan Spear-nosed Bat

SYNONYMS:

Phyllostoma latifolium O. Thomas, 1901g:142; type locality "Kanuku Mountains," Upper Takutu-Upper Essequibo, Guyana.

P[hyllostomus]. latifolius: Miller, 1907b:131; first use of current name combination.

DISTRIBUTION: *Phyllostomus latifolius* is endemic to South America where it is known from southeastern Colombia, southern Guyana, northern Surinam, French Guiana, and north central Brazil. There is one dubious record from the Pacific Coast of Colombia (see Remarks).

MARGINAL LOCALITIES (Map 156): SURINAM: Brokopondo, 8 km S and 2 km W of Brownsweg (S. L. Williams and Genoways 1980a). FRENCH GUIANA: Les Eaux Claires (Simmons, Voss, and Peckham 2000). BRAZIL: Pará, Tauary (FMNH 42736); Amazonas, 5 km SE of Presidente Figueiredo (Sampaio et al. 2003). COLOMBIA: Caquetá, Estación Puerto Abeja (Montenegro and Romero-Ruiz 2000); Vaupés, Mitú (Marinkelle and

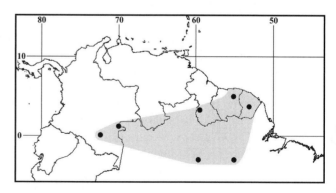

Map 156 Marginal localities for *Phyllostomus latifolius* ●

Cadena 1972). GUYANA: Upper Takutu-Upper Essequibo, Kanuku Mountains (type locality of *Phyllostoma latifolium* O. Thomas).

SUBSPECIES: We regard *P. latifolius* as monotypic.

NATURAL HISTORY: Little is known about the biology of *P. latifolius*. S. L. Williams and Genoways (1980a) collected several in mature tropical forests in Surinam. Marinkelle and Cadena (1972) reported specimens from small caves along the banks of a river in Colombia. A female from 5 km SE of Presidente Figueiredo, Amazonas, Brazil contained a 10-mm-CR embryo when taken on 20 November; the testes of a male taken on the same date measured 4 by 3 mm. Montenegro and Romero-Ruiz (2000) took one specimen in seasonally flooded forest, and eight in *terra firme* forest habitat in southern Colombia. Brosset and Charles-Dominique (1991) reported finding approximately 50 *P. latifolius* mixed in a tight cluster with an estimated 300 *Lonchorhina inusitata* over a pool in a cave in French Guiana. Based on the food habits of other species of *Phyllostomus*, Gardner (1977c) suggested that *P. latifolius* feeds on "flower parts, fruits, insects, and small vertebrates." R. Guerrero (1997) listed one species of streblid batfly known to parasitize *P. latifolius*. The karyotype is $2n = 32$, FN = 58 (Honeycutt, Baker, and Genoways 1980; R. J. Baker et al. 1982).

REMARKS: Oldfield Thomas (1901g) had eight specimens from the type locality when he described *P. latifolius*. He distinguished it from *P. elongatus* primarily on smaller size in addition to certain qualitative features, which Valdez (1970) claimed are unreliable when assessed against non-Guyanan populations. The record for *P. latifolius* from Río Zabaletas, Valle del Cauca, Colombia, listed by Muñoz (2001), is suspect. That record probably represents material from the Pacific region cited in Alberico et al. (2000) and, if correctly identified, is the only record of the species west of the Andes. Measurement of *P. latifolius* have been provided by O. Thomas (1901g), Husson (1962), Marinkelle and Cadena (1972), Swanepoel and Genoways (1979), S. L. Williams and Genoways (1980a), and Simmons and Voss (1998). D. C. Carter and Dolan (1978) gave information on and measurements of the holotype of *Phyllostoma latifolium* O. Thomas.

Genus *Tonatia* Gray, 1827

In South America, *Tonatia* is represented by two species of medium-sized, round-eared bats (forearm 51–62 mm, greatest length of skull 26–31 mm). The head, including the muzzle, is conspicuously furred. *Tonatia* is readily separated from other phyllostomines (except *Lophostoma* spp.) by the combination of a lower toothrow having only one incisor and three premolars, and by the presence of a tail that extends to the middle of the interfemoral membrane. Two features useful for distinguishing the two species of *Tonatia* from similar-sized species of *Lophostoma* with which they can be confused, are the relatively well-furred face and the comparatively broad postorbital constriction (wider than 5 mm and exceeding 90% of breadth across cingula of the canines, versus narrower than 5 mm and less than 90% of breadth across cingula of canines in *Lophostoma* spp.). As in *Chrotopterus* and *Lophostoma*, the calcar is longer than the foot. The dental formula is 2/1, 1/1, 2/3, 3/3 × 2 = 32 (also in *Chrotopterus* and *Lophostoma*).

SYNONYMS:

Tonatia Gray, 1827:71, in Griffith, Hamilton-Smith, and Pidgeon, 1827; type species *Vampyrus bidens* Spix, 1823, by monotypy.

Tylostoma P. Gervais, 1856a:49; type species *Phyllostoma bidens* (Spix, 1823), by subsequent designation (Palmer 1904:698); preoccupied by *Tylostoma* Sharpe, 1849 (Mollusca).

Anthorhina Lydekker, 1891:674, in Flower and Lydekker, 1891; replacement name for *Tylostoma* P. Gervais.

Anthorina Palmer, 1904:108; incorrect subsequent spelling of *Anthorhina* Lydekker in Flower and Lydekker.

REMARKS: Goodwin (1942b) reviewed the genus *Tonatia* (*sensu lato*) and, more recently, Genoways and Williams (1984) reviewed South American species (also *sensu lato*). In the most recent review, Lee, Hoofer, and Van Den Bussche (2002) transferred all of the species previously included in *Tonatia*, except *T. bidens* and *T. saurophila*, to the genus *Lophostoma*.

Gardner and Ferrell (1990) pointed out that several authors have placed *Mimon crenulatum* within *Anthorhina*, either at the generic level (Miller 1907b; Husson 1962, 1978) or the subgeneric level (Cabrera 1958, Goodwin and Greenhall 1961, Hill 1965, and Gardner and Patton 1972). These authors were unaware that *Anthorhina*, as a replacement name for *Tylostoma*, is a junior synonym of *Tonatia* Gray with the same type species (Palmer 1904); therefore, *Anthorhina* is not available for the taxa now included in the genus *Mimon*. As ICZN (1999) Article 67.8 makes clear, a name proposed to replace a previously established genus-group name has the same type species as the name replaced.

KEY TO SOUTH AMERICAN SPECIES OF *TONATIA*:
1. Postorbital constriction usually 5.0–5.5 mm (may approach 5.9 mm in northern Colombian populations); secondary process on mastoid partially obscuring base of mastoid bulla; distinct gap between cingula of lower canines when viewed from above; pale patch or short stripe (sometimes faint) on top of head between ears . . . . *Tonatia saurophila*

1'. Postorbital constriction 5.6–6.1 mm; no secondary process on mastoid; posterior medial margins of cingula of lower canines nearly touching; no pale patch or stripe on top of head . . . . *Tonatia bidens*

Tonatia bidens (Spix, 1823)
Spix's Round-eared Bat

SYNONYMS:

Vampyrus bidens Spix, 1823:65; type locality "fluvium St. Francisci" (= Rio São Francisco), Bahia, Brazil.

[*Tonatia*] *bidens*: Gray, 1827:71, footnote, in Griffith, Hamilton-Smith, and Pidgeon, 1827; first use of current name combination.

Phyllostoma childreni Gray, 1838b:488; type locality "S. America."

Phyllostoma bidens: Schinz, 1844:236; name combination.

Tylostoma bidens: P. Gervais, 1856a:49; name combination.

V[ampyrus]. (Tylostoma) bidens: W. Peters, 1856a:304; name combination.

Lophostoma bidens: W. Peters, 1865c:509; name combination.

Ph[yllostoma (Tylostoma)]. Childreni: W. Peters, 1865c: 514; name combination.

Tylostoma childreni: Gray, 1866b:114; name combination.

DISTRIBUTION: *Tonatia bidens* is known from eastern Brazil, Paraguay, and northern Argentina.

MARGINAL LOCALITIES (Map 157; from S. L. Williams, Willig, and Reid 1995, except as noted): BRAZIL: Ceará, Reserva Particular do Patrimônio Natural Serra das Almas (S. S. P. Silva et al. 2004); Pernambuco, Fazenda Maniçoba, 13.7 km S of Exu (Mares et al. 1981; Willig 1983); Bahia, Pardo River Valley (Faria, Soares-Santos, and

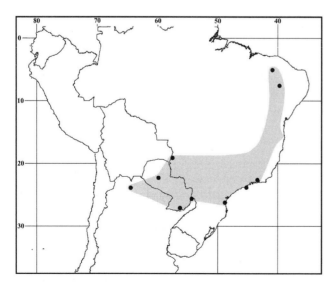

Map 157 Marginal localities for *Tonatia bidens* ●

Sampaio 2006, not mapped); Rio de Janeiro, Tinguá (Peracchi and Albuquerque 1986); São Paulo, Ilha de São Sebastião (C. O. C. Vieira 1955); Santa Catarina, Joinville. ARGENTINA: Misiones, Cataratas del Iguazú (Villa-R. and Villa-Cornejo 1973, as *T. silvicola*). PARAGUAY: Itapúa, Arroyo San Rafael. ARGENTINA: Jujuy, Laguna La Brea. PARAGUAY: Boquerón, Orloff (Myers, White, and Stallings 1983). BRAZIL, Mato Grosso do Sul, Urucúm.

SUBSPECIES: We treat *T. bidens* as monotypic.

NATURAL HISTORY: *Tonatia bidens* has been reported from dry-land forest habitats (Mares et al. 1981), but judging from its distribution, occupies a variety of forest types. Willig (1983) classed *T. bidens* as a foliage gleaning insectivore. Esbérard and Bergallo (2004) recovered remains of a variety of insects, as well as the remains of four classes of vertebrates, under feeding roosts in southeastern Brazil. They found *T. bidens* using grottos, a hole in a palm tree, and caves and hollows associated with springs as day roosts; a drain pipe, buildings (both abandoned and in use), a hollow tree, a grotto, and a spring were used as feeding roosts. Esbérard and Bergallo (2004) found females pregnant in November, lactating in January and May, and postlactating in May. Webb and Loomis (1977) reported streblid batflies found on bats identified as *T. bidens*, but based on geographic distribution, many if not most of the hosts were *T. saurophila*. This likely is true for the nine species of streblids listed by R. Guerrero (1997) for *T. bidens*. However, Esbérard and Bergallo (2004) correctly identified *T. bidens* as the host of the five species of streblids they in recovered from *Tonatia* in Rio de Janeiro, Brazil.

REMARKS: Stephen L. Williams, Willig, and Reid (1995) found little secondary sexual dimorphism in ten measurements, with males averaging slightly larger than females. Specimens reported as *T. silvicola* from Jujuy (Fornes, Massoia, and Forrest 1967) and from Misiones (Villa-R. and Villa-Cornejo 1973) are *T. bidens* according to Barquez, Giannini, and Mares (1993). References to *T. bidens* from Central America and northern South America apply to *Tonatia saurophila* Koopman and Williams. Myers and Wetzel (1983) and S. L. Williams, Willig, and Reid (1995) also have provided measurements for this species. D. C. Carter and Dolan (1978) provided information on and measurements of the apparent holotype of *Vampyrus bidens* Spix and the holotype of *Phyllostoma childreni* Gray.

Tonatia saurophila Koopman and Williams, 1951
Pale-crowned Bat

SYNONYMS:

Tonatia saurophila Koopman and Williams, 1951:11; type locality "Wallingford Roadside Cave, Balaclava, St. Elizabeth Parish, Jamaica."

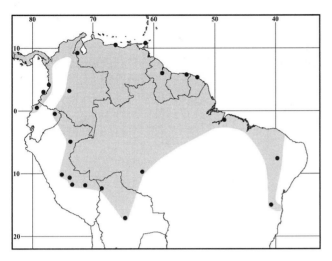

Map 158 Marginal localities for *Tonatia saurophila* ●

Additional synonyms under Subspecies.

DISTRIBUTION: *Tonatia saurophila* is known from Colombia, Venezuela, the Guianas, northern Brazil, and Amazonian Ecuador and Peru. The species also occurs in Central America.

MARGINAL LOCALITIES (Map 158): TRINIDAD AND TOBAGO: Trinidad, Las Cuevas (C. H. Carter et al. 1981). GUYANA: Cuyuni-Mazaruni, 24 miles from Bartica on Potaro Road (Hill 1965). SURINAM: Marowijne, Perica (Genoways and Williams 1984). FRENCH GUIANA: Paracou (Simmons and Voss 1998). BRAZIL: Pará, Guama Ecological Research Area (Handley 1967); Pernambuco, Fazenda Maniçoba (Mares et al. 1981; Willig 1983); Bahia, Região de Conquista (Falcão, Soares-Santos, and Drummond 2005); Bahia, Pardo River Valley (Faria, Soares-Santos, and Sampaio 2006, not mapped); Rondônia, Cachoeira Nazaré (MPEG 20918). BOLIVIA: Cochabamba, Sajta (S. Anderson 1997); Pando, Chivé (Sawada and Harada 1986). PERU: Madre de Dios, Cocha Cashu (Terborgh, Fitzpatrick, and Emmons 1984, as *Tonatia bidens*); Cusco, Pagorini (S. Solari et al. 1999); Ucayali, Largato (Tuttle 1970); Pasco, Iscozacín (Ascorra, Gorchov, and Cornejo 1989); Loreto, Jenaro Herrera (Ascorra, Gorchov, and Cornejo 1994). ECUADOR: Orellana, Laguna Garzacocha (Albuja 1999). COLOMBIA: Meta, Municipio de San Juan de Arama (Sánchez-Palomino and Rivas-Pava 1993); Valle del Cauca, North side of Bahía Málaga (Alberico 1994); Cauca, Gorgona Island (Cuervo-Díaz, Hernández-Camacho, and Cadena 1986). ECUADOR: Esmeraldas, Valle de Sade (Albuja 1999). VENEZUELA (Handley 1976): Zulia, El Rosario, 48 km WNW of Encontrados; Miranda, Birongo.

SUBSPECIES: Two of the three recognized subspecies of *T. saurophila* occur in South America. The nominate subspecies was described from material recovered from a cave deposit in Jamaica, where the taxon no longer exists.

T. s. bakeri S. L. Williams, Willig, and Reid, 1995
SYNONYM:

Tonatia saurophila bakeri S. L. Williams, Willig, and Reid, 1995:622; type locality "6 km SW Cana, ca, 1,200 m, Darién, Panama."

This subspecies is known from northern Colombia and from northern Venezuela north of the Cordillera de Mérida.

T. s. maresi S. L. Williams, Willig, and Reid, 1995
SYNONYMS:

Tonatia saurophila maresi S. L. Williams, Willig, and Reid, 1995:623; type locality "Blanchisseuse, Trinidad, Trinidad and Tobago."

Tonatia saurophilla Falcão, Soares-Santos, and Drummond, 2005:222; incorrect subsequent spelling of *Tonatia saurophila* Koopman and Williams, 1951.

This subspecies is known from Venezuela (east and south of the Cordillera de Mérida), northern Brazil, the Guianas, and along the eastern slope of the Andes and adjacent lowlands of Colombia, Ecuador, Peru, and Bolivia.

NATURAL HISTORY: All recent information on Central and South American populations, prior to the revision by S. L. Williams, Willig, and Reid (1995), has appeared under the name *T. bidens*. Hill (1965), D. C. Carter, Pine, and Davis (1966), Handley (1966c, 1976), Mares et al. (1981), Genoways and Williams (1984), reported *T. saurophila* (as *T. bidens*) from a variety of forest habitats. Handley (1976) also reported collecting individuals in swamps and in agricultural areas. Goodwin and Greenhall (1961, as *T. bidens*) found the species roosting in hollow trees. Goodwin and Greenhall (1961) gave fruit as the diet, to which Gardner (1977c) added insects. Although the name implies an affinity for lizards, we have no evidence that *T. saurophila* eats vertebrates.

Gardner (1976) reported (as *T. bidens*) two females from eastern Peru, taken on 3 April and a third on 23 July, that were neither pregnant nor lactating. He also caught two pregnant females on 23 July; one contained a very small embryo, the other contained a 35-mm-CR, near-term fetus. Redford and Eisenberg (1992), misinterpreting Gardner's (1976) report, stated that each pregnant female contained two embryos.

Webb and Loomis (1977) reported five streblid batflies recovered from *T. bidens*; however, most if not all of these host records probably represented *T. saurophila*. R. Guerrero (1985a) added two spinturnicid mites and one streblid batfly. *Tonatia saurophila* is the host for at least three of the streblids R. Guerrero (1997) listed for *T. bidens*.

The karyotype is $2n = 16$, FN $= 20$ (R. J. Baker and Hsu 1970).

REMARKS: *Tonatia saurophila* is based on subfossil material from cave deposits. The name refers to the "Lizard layers," which are the strata in Wallingford Roadside Cave where the type and paratype were found. Koopman (1976) later treated this taxon as a subspecies of *T. bidens. Tonatia saurophila* can be distinguished from its similar-sized congener, *T. bidens*, by the presence of a pale patch or stripe on the top of the head; a wider gap separating the cingulae of the lower canines; and by the greater development of the mastoid region, in which the mastoid overlies the posterior region of the bulla and hides the posterior foramen of the cochlea when viewed ventrally (S. L. Williams, Willig, and Reid 1995). Goodwin (1942b), Goodwin and Greenhall (1961), Hill (1965), D. C. Carter, Pine, and Davis (1966), Gardner (1976), Swanepoel and Genoways (1979), Genoways and Williams (1980, 1984), C. H. Carter et al. (l981), and S. L. Williams, Willig, and Reid (1995) have provided measurements for this species (all except the latter, reported under the name *T. bidens*). Genoways and Williams (1984) found no secondary sexual variation based on nine measurements of *T. saurophila* (coefficients of variation ranged from 0.8 to 3.4) in their sample from Surinam.

Genus *Trachops* Gray, 1847

The monotypic genus *Trachops* is represented by the medium-sized *T. cirrhosus* (forearm 58–65 mm, greatest length of skull 27–31 mm), which is easily recognized by the presence of papilla-like protuberances on the chin and lips and by the finely serrated margins of the noseleaf. The tail extends to the middle of the interfemoral membrane and the calcar is about the same length as the foot. The dental formula is 2/2, 1/1, 2/3, 3/3 × 2 = 34.

SYNONYMS:

Vampyrus Spix, 1823:64; type species here designated as *Vampyrus cirrhosus* Spix, 1823; preoccupied by *Vampyrus* Beckmann, 1772; *Vampyrus* Ranzani, 1820; and *Vampyrus* Leach, 1821b.

Istiophorus Gray, 1825a:242; replacement name for *Vamyprus* Spix, 1823; not *Vampyrus* Leach, 1821b; not *Vampyrus* Beckmann, 1772; preoccupied by *Istiophorus* Lacépède, 1802 (Osteichthyes).

Histiophorus Agassiz, 1847:183; unjustified emendation of *Istiophorus* Gray.

Trachops Gray, 1847:14; type species *Trachops fuliginosus* Gray, 1847, by original designation.

Trachyops W. Peters, 1865c:512; incorrect subsequent spelling of *Trachops* Gray.

Trachyops Trouessart, 1897:154; incorrect subsequent spelling of *Trachops* Gray.

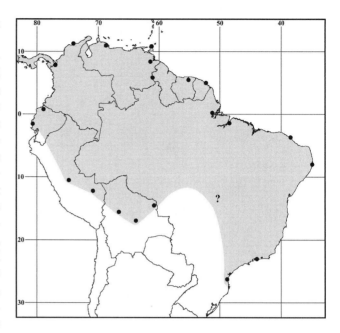

Map 159 Marginal localities for *Trachops cirrhosus* ●

Trachops cirrhosus (Spix, 1823)
Fringe-lipped Bat

SYNONYMS:

T[ylostoma]. mexicana Saussure, 1860:484; type locality "les régions chaudes du Mexique."

Additional synonyms under Subspecies.

DISTRIBUTION: *Trachops cirrhosus* is known from Colombia, Venezuela, Trinidad, Guyana, Surinam, French Guiana, Ecuador, Peru, Brazil, and Bolivia. Elsewhere, the species occurs in Mexico and Central America.

MARGINAL LOCALITIES (Map 159): COLOMBIA: Magdalena, Bonda (J. A. Allen 1900a). VENEZUELA: Falcón, Riecito (Handley 1976). TRINIDAD AND TOBAGO: Trinidad, Blanchisseuse (C. H. Carter et al. 1981). VENEZUELA: Delta Amacuro, Río Acoíma (Ochoa 1995). GUYANA: Cuyuni-Mazaruni, Namai Creek (Lim and Engstrom 2000). SURINAM: Para, Zanderij (Genoways, Williams, and Groen 1981). FRENCH GUIANA: Cayenne (Brosset and Dubost 1968). BRAZIL: Amapá, Rio Maruanum (Carvalho 1962a); Pará, Guama Ecological Research Area (Handley 1967); Ceará, Fortaleza (C. O. C. Vieira 1955); Pernambuco, Pernambuco (type locality of *Trachops fuliginosus* Gray); Rio de Janeiro, Mangaratiba (Peracchi, Albuquerque, and Raimundo 1982); Santa Catarina, Joinvile (type locality of *Trachops cirrhosus ehrhardti* Felten). BOLIVIA: Santa Cruz, Los Fierros (S. Anderson 1997); Santa Cruz, 25 km NW of Santa Rosa de Sara (S. Anderson 1997); Beni, Serranía Eva Eva (Hinojosa, S. Anderson 1997). PERU: Madre de Dios, Manu (Koopman 1978); Pasco, San Juan (Tuttle 1970). ECUADOR (Albuja

1999): Manabí, Estero Achiote; Esmeraldas, San Miguel. COLOMBIA: Chocó, Sautata (Sanborn 1932b).

SUBSPECIES: We recognize three subspecies of *T. cirrhosus*, two of which occur in South America.

T. c. cirrhosus (Spix, 1823)
SYNONYMS:

Vampyrus cirrhosus Spix, 1823:64; no type locality stated in Spix's description, but on page 53 (as noted by Husson 1962) Spix stated that the bats were collected in Brazil; type locality restricted to the state of Pará, Brazil, by Husson (1962:115); previous restrictions to Pernambuco, Brazil, based on the type locality of *Trachops fuliginosus* Gray are invalid.

Ph[yllostoma]. cirrhosum: J. B. Fischer, 1829:126; name combination.

Vampyris cirrhosum Gray, 1847:14; incorrect subsequent spelling of *Vampyrus cirrhosus* Spix.

Trachops fuliginosus Gray, 1847:14; type locality "Pernambuco," Brazil.

Trachyops cirrhosus: W. Peters, 1865c:512; first use of current name combination and incorrect subsequent spelling of *Trachops* Gray.

Phyllostoma (Trachops) fuliginosum: Pelzeln, 1883:32; name combination.

Trachops cirrhosa Cabrera 1958:69; incorrect subsequent spelling of *Trachops cirrhosus* Spix.

This subspecies is distributed in Colombia, Venezuela, the Guianas, eastern Peru, and northern and central Brazil.

T. c. ehrhardti Felten 1956
SYNONYM:

Trachops cirrhosus ehrhardti Felten 1956b:369; type locality "Joinvile," Santa Catarina, Brazil.
This subspecies is in southern Brazil.

NATURAL HISTORY: *Trachops cirrhosus* is commonly associated with forest and stream habitats (Handley 1966c, 1976; Tuttle 1970), but also has been reported from open savannah and agricultural areas. This species roosts in caves, hollow trees, houses, and culverts (Goodwin and Greenhall 1961; Handley 1966c, 1976; Kalko et al. 1999). Gardner (1977c) summarized evidence that *T. cirrhosus* feeds on insects and small vertebrates, including lizards and bats, and classed it as a carnivore. Subsequent authors have elaborated on the feeding habits of this species (Pine and Anderson 1980; Peracchi, Albuquerque, and Raimundo 1982), and its predation behavior on frogs is well documented (Barclay et al. 1981). Bonato and Facure (2000) found the remains of a thumbless bat (*Furipterus horrens*) in the stomach of an adult male collected in the Northeast of Brazil. Ubelaker, Specian, and Duszynski (1977) and Cuartas and Muñoz (1999) reported trichostrongylid nematodes

found in the small intestine of *T. cirrhosus*. Brennan and Reed (1975) identified four species of trombiculid mites recovered from Venezuelan specimens. Webb and Loomis (1977) listed a labidocarpid mite, two spinturnicid mites, one trombiculid mite, two argasid ticks, and eight streblid batflies as known ectoparasites on *T. cirrhosus*. R. Guerrero (1985a) listed an additional species of spinturnicid mite, another macronyssid mite, a flea, and six species of batflies. R. Guerrero (1997) listed 15 species of streblid batflies confirmed from *T. cirrhosus*, 5 of which had not been listed by Webb and Loomis (1977) and R. Guerrero (1985a), suggesting that some of the earlier batfly identifications had been erroneous. Kalko et al. (1999) described roosting and foraging behavior during the dry season on Barro Colorado Island, Panama. Cramer, Willig, and Jones (2001) have a detailed description, along with measurements and a summary of natural history information, in their *Mammalian Species* account. The karyotype is $2n = 30$, FN $= 56$ (R. J. Baker et al. 1982).

REMARKS: Felten (1956b) described specimens of *T. c. ehrhardti* as being smaller than specimens of *T. c. cirrhosus* from northern South America in some cranial measurements. Husson (1962) wrote that, while external measurements of a specimen from Surinam were much larger than those of the sample from Colombia reported on by Hershkovitz (1949c), the cranial measurements corresponded well. W. B. Davis and Carter (1962a), comparing published measurements of Central American specimens with those of Hershkovitz's (1949c) sample, reported considerable overlap in the ranges of variation. Goodwin (1946), Goodwin and Greenhall (1961), W. B. Davis and Carter (1962a), Husson (1962, 1978), and Swanepoel and Genoways (1979) have provided measurements of South American specimens.

The statement by Cramer, Willig, and Jones (2001), "Type *Vampyrus cirrhosus* Spix or possible *V. soricinus* Spix," does not constitute fixation of the type species of *Istiophorus* Gray, 1825. We herein select *Vampyrus cirrhosus* Spix 1923, one of the two species Spix included under the generic name *Vampyrus*, as the type species of *Vampyrus* Spix, 1823. *Istiophorus* Gray, 1825, is a replacement name for *Vampyrus* Spix, which is preoccupied by *Vampyrus* Beckmann, 1772; *Vampyrus* Ranzani, 1820; and *Vampyrus* Leach, 1821b. The type species of a genus-level replacement name is the same as the type species of the name replaced; therefore, *Vampyrus cirrhosus* Spix also is the type species of *Istiophorus* Gray.

Czaplewski, Rincón, and Morgan (2005) reported late Pleistocene fossils recovered from a tar seep in the state of Zulia, Venezuela. D. C. Carter and Dolan (1978) gave information on and measurements of the holotype of *Trachops cirrhosus ehrhardti* Felten.

Genus *Trinycteris* Sanborn, 1949

This monotypic genus is represented by *Trinycteris nicefori*, a small bat (forearm 37.1–40.2 mm, $n = 24$; greatest length of skull 20.7–22.0 mm, $n = 24$) distinguished by four-banded dorsal pelage (the pale basal band is narrow and inconspicuous) and the pale median dorsal stripe usually evident on the lower back. The narrow, pale basal band on the dorsal fur is followed by a dark band, a broad pale band, and then the dark terminal band of the hair tips. The ears are pointed and have broadly concave outer margins. Other identifying features include the lack of a band of skin across the top of the head connecting the ears; the throat has numerous long, dark, stiff hairs interspersed in the pelage; the lower margin of narial horseshoe merges smoothly with the upper lip; and the chin and lower lip bear a pair of smooth tubercles, which are partly divided by a median groove. The metacarpal of digit IV is shorter than the metacarpal of digit V, which is shorter than the third; the calcar is shorter than foot; the rostrum is not inflated; the basisphenoid pits are deep; and the lower incisors are trifid. The dental formula is 2/2, 1/1, 2/3, 3/3 × 2 = 34.

SYNONYMS:

Trinycteris Sanborn, 1949:226; type species *Micronycteris (Trinycteris) nicefori* Sanborn, 1949, by monotypy; described as a subgenus of *Micronycteris* Gray.

Trinycteris: Simmons and Voss, 1998:92; first use as a full genus.

Trinycteris nicefori Sanborn, 1949
Niceforo's Bat

SYNONYMS:

Micronycteris (Trinycteris) nicefori Sanborn, 1949a:230; type locality "Cúcuta," Norte de Santander, Colombia.
[*Trinycteris*] *nicefori*: Simmons and Voss, 1998:62; first use of current name combination.

DISTRIBUTION: *Trinycteris nicefori* occurs on the island of Trinidad and east of the Andes in Colombia, Venezuela, Guyana, Surinam, and Peru, with a disjunct population in the state of Espírito Santo, Brazil. Elsewhere, *T. nicefori* occurs in southern Mexico and Central America.

MARGINAL LOCALITIES (Map 160): TRINIDAD AND TOBAGO: Trinidad, Las Cuevas (C. H. Carter et al. 1981). VENEZUELA: Bolívar, Unidad V, Reserva Forestal Imataca (Ochoa 1995). GUYANA: Cuyuni-Mazaruni, 24 miles from Bartica on Potaro road (Hill 1965). SURINAM: Marowijne, 10 km N and 24 km W of Moengo (Genoways and Williams 1979b). FRENCH GUIANA: Paracou (Simmons and Voss 1998). BRAZIL: Pará, Utinga (Handley 1967); Pará, Area do Caraipe (Pine et al. 1996); Tocantins, Município Palmeirante (Nunes et al. 2005); Pará, 52 km SSW of Altamira (Voss and Emmons 1996); Pará, Alter do Chão (Bernard 2001b). PERU: Ucayali, Balta

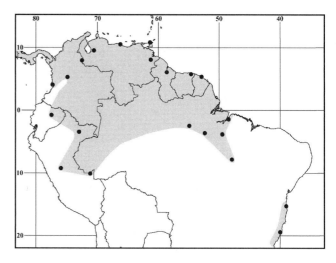

Map 160 Marginal localities for *Trinycteris nicefori* ●

(Voss and Emmons 1996); Huánuco, vicinity of Tingo María (Bowles, Cope, and Cope 1979); Loreto, Puerto Indiana (Pirlot 1968). ECUADOR: Orellana, Santa Rosa de Arapino (Albuja 1999). COLOMBIA: Tolima, Mariquita (Muñoz 2001); Valle del Cauca, Bahía Málaga Naval Base (Alberico 1994); Norte de Santander, Cúcuta (type locality of *Micronycteris (Trinycteris) nicefori* Sanborn). VENEZUELA (Handley 1976): Trujillo 19 km N of Valera; Miranda, Birongo. *Disjunct distribution in southeastern Brazil*: Bahia, Una (Faria, Soares-Santos, and Sampaio 2006); Espírito Santo, Município de Linhares (Peracchi and Albuquerque 1985).

SUBSPECIES: We treat *T. nicefori* as monotypic.

NATURAL HISTORY: *Trinycteris nicefori* is known from a variety of habitats ranging from forest to agricultural areas (Hill 1965; Handley 1966c, 1976; S. L. Williams and Genoways 1980a). Pine et al. (1996) caught this species in ground-level nets in tall terra-firme forest in southeastern Pará State in Brazil. The species roosts in hollow trees, tunnels, and buildings (Goodwin and Greenhall 1961; Handley 1966c). Sanborn (1949) wrote that the type specimen was caught in a rock tunnel along with *Lonchorhina aurita* and *Micronycteris minuta*. Goodwin and Greenhall (1961) suggested that *T. nicefori* feeds on fruit and insects. Brennan and Reed (1975) identified a species of trombiculid mite found on Venezuelan specimens. Webb and Loomis (1977) listed four species of streblid batflies as known ectoparasites, to which R. Guerrero (1985a) added a species of spinturnicid mite and two streblid batflies. R. Guerrero (1997) increased the number of streblid species known from *T. nicefori* to seven. The karyotype is $2n = 28$, FN = 52 (Honeycutt, Baker, and Genoways 1980; R. J. Baker, Genoways, and Seyfarth 1981; R. J. Baker et al. 1982).

REMARKS: Starrett (1976) claimed that *T. nicefori* from Costa Rica agreed in all essential characters and in most measurements with the specimens described by Sanborn

(1949a) from Colombia, but lacked the faint gray stripe on the lower back reported for Colombian specimens. Thirteen (18%) of 74 skins from Costa Rica, Panama, Venezuela, Guyana, and Brazil, in the USNM lack a clear indication of the dorsal stripe, demonstrating that the stripe is a variable character. Sanborn (1949a), Goodwin and Greenhall (1961), Hill (1965), Genoways and Williams (1979b), Swanepoel and Genoways (1979), and S. L. Williams and Genoways (1980a) have provided measurements of South American specimens.

Genus *Vampyrum* Rafinesque, 1815

The monotypic genus *Vampyrum* is represented by *V. spectrum*, the largest bat in the Western Hemisphere (forearm 101–110 mm, greatest length of skull 49–54 mm). In addition to its size, *Vampyrum* (along with *Chrotopterus*) is unusual among the phyllostomines in lacking a visible tail. The calcar is longer than the foot. *Vampyrum* and *Chrotopterus* also share an unusual occlusion pattern in which the first upper incisor contacts the broad cingular shelf of the posterolingual base of the lower canine when the mouth is closed (Wetterer, Rockman, and Simmons 2000:77). The dental formula is 2/2, 1/1, 2/3, 3/3 × 2 = 34.

SYNONYMS:

Vampyrum Rafinesque, 1815:54, type species *Phyllostoma spectrum* É. Geoffroy St.-Hilaire, 1810b (= *Vespertilio spectrum* Linnaeus, 1758), by subsequent selection (Andersen 1908a:433).

Vampyrus Ranzani, 1820:190; type species *Vampyrus spectrum*: Ranzani, 1820 (= *Vespertilio spectrum* Linnaeus, 1758), by monotypy; preoccupied by *Vampyrus* Beckmann, 1772.

Vampyrus Leach, 1821b:74; type species *Vampyrus spectrum*: Leach, 1821b:80 (= *Vespertilio spectrum* Linnaeus, 1758), by monotypy; primary objective homonym of *Vampyrus spectrum*: Ranzani, 1820; preoccupied by *Vampyrus* Beckmann, 1772.

Vampirus Lesson, 1827:78; incorrect subsequent spelling of *Vampyrus* of authors.

Vampyris Gray, 1847:14; incorrect subsequent spelling of *Vampyrus* of authors.

Vampyrun Ospina-Ante and Gómez, 2000; incorrect subsequent spelling of *Vampyrum* Rafinesque.

Vampirum Montenegro and Romero-Ruiz, 2000:645; incorrect subsequent spelling of *Vampyrum* Rafinesque.

Vampyrum spectrum (Linnaeus, 1758)
Great Spectral Bat
SYNONYMS:

[*Vespertilio*] *spectrum* Linnaeus, 1758:31; type locality "America *australi*," restricted to Surinam by O. Thomas (1911b:130).

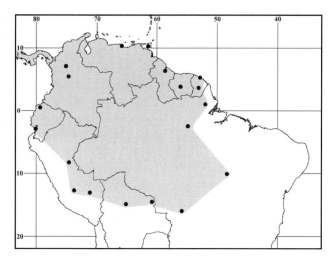

Map 161   Marginal localities for *Vampyrum spectrum* ●

Phyllostomus spectrum: Daudin in Lacépède, 1802:189; name combination.

Phyllostoma spectrum: É. Geoffroy St.-Hilaire, 1810b:174; name combination.

Vampyrus spectrum: Ranzani, 1820:190; name combination.

[*Vampyrus (Vampyrus)*] *spectrum*: Trouessart, 1897:153; name combination.

Vampyrus spectrum nelsoni Goldman, 1917:115; type locality "Coatzacoalcos, Vera Cruz, Mexico.

Vampyrum spectrum: Goodwin 1942c:128; first use of current name combination.

DISTRIBUTION: *Vampyrum spectrum* occurs in Colombia, Venezuela, Trinidad, Surinam, French Guiana, Ecuador, Brazil, Peru, and Bolivia. Elsewhere, it is known from Mexico and Central America.

MARGINAL LOCALITIES (Map 161): TRINIDAD AND TOBAGO: Trinidad, Santa María (C. H. Carter et al. 1981). GUYANA: Cuyuni-Mazaruni, Kartabo (H. E. Anthony 1921b). SURINAM: Brokopondo, Rudi Kappel-liegveld (S. L. Williams and Genoways 1980a). FRENCH GUIANA: Paracou (Simmons and Voss 1998); Les Eaux Claires (Simmons, Voss, and Peckham 2000). BRAZIL: Amapá, Serra do Navio (Peracchi, Raimundo, and Tunnare 1984); Pará, Alter do Chão (Bernard 2001b); Tocantins, Município Paraíso do Tocantins (Nunes et al. 2005); Mato Grosso, Barra do Arica (C. O. C. Vieira 1955). BOLIVIA: Santa Cruz, Los Fierros (Emmons 1998); Beni, Río Tajacuchi (S. Anderson 1991). PERU: Cusco, Cosñipata (Sanborn 1949b); Ayacucho, San José (LSUMZ 16458); Ucayali, Yarinacocha (Sanborn 1949b). ECUADOR: Guayas, San Ramón (AMNH 66815); Pichincha, Patricia Pilar (Albuja and Mena 2004). COLOMBIA (Muñoz 2001): Caldas, La Dorada; Antioquia, Anorí. VENEZUELA: Miranda, 7 km E of Río Chico (Handley 1976).

SUBSPECIES: We are treating *V. spectrum* as monotypic.

NATURAL HISTORY: Handley (1976) reported five specimens of *V. spectrum* from yards, swamps, and evergreen-forest habitats. S. L. Williams and Genoways (1980a) collected one in a clearing bordered by lowland rainforest. This species is known to roost in hollow trees (Goodwin and Greenhall 1961). Gardner (1977c) and Navarro and Wilson (1982) summarized references documenting the diet, which includes warm-blooded vertebrates such as small birds and mammals (including other bats), and possibly some insects and fruits. In their summary of endoparasites found in leaf-nosed bats, Ubelaker, Specian, and Duszynski (1977) mentioned only trypanosomes as infecting *V. spectrum*. Webb and Loomis (1977) listed one streblid batfly and two trombiculid mites as known ectoparasites. R. Guerrero (1985a) listed an additional species of spinturnicid mite, and later (1997) added another species of streblid batfly. Navarro and Wilson (1982) presented measurements, illustrations, and a synopsis of natural history information in their *Mammalian Species* account. The karyotype is $2n = 30$, FN $= 56$ (R. J. Baker et al. 1982).

REMARKS: As Hall (1981:118) pointed out, the names *Vespertilio nasutus* G. Shaw, 1800, *V. guianensis* Daudin in Lacépède, 1802:188, and *V. maximus* É. Geoffroy St.-Hilaire, 1806 were listed as synonyms of *V. spectrum* by Sanborn (1949b), Cabrera (1958), and Hall and Kelson (1959). These names are based on Buffon's *La Grande Serotine de la Guyanne*, which may be a composite of several species of large bat; but, because it is illustrated as having an obvious and well-developed tail, is not a *Vampyrum*.

Two subspecies of *V. spectrum* have been recognized: the nominate subspecies in South America and *V. s. nelsoni* in Mexico and Central America. However, Husson (1962), Handley (1966c), and R. L. W. Peterson and Kirmse (1969) have provided evidence supporting their conclusions that the species is monotypic, a determination we follow here. The locality "Boquiron, Colombia" for the skull illustrated in Hall and Kelson (1959) was corrected to Boqueron, Panama by Hall (1981).

Subfamily Stenodermatinae P. Gervais, 1856
Alfred L. Gardner

The primarily frugivorous Stenodermatinae, the most diverse phyllostomid subfamily, is represented in this account by 14 genera (including *Sturnira*) and at least 60 species in South America. Stenodermatine bats are characterized by a short, round face; round oral cavity; upper molars have reduced cusps resulting in at least partial obliteration of the W-shaped pattern of the ectoloph; and the crowns of the lower molars are usually much reduced with well-developed cusps. An external tail is lacking, a fully developed noseleaf is present (except in *Centurio*), and the trochiter impinges on the scapula.

The content of the Stenodermatinae, and relationships to other subfamilies of the Phyllostomidae, remains unclear. Slaughter (1970) noted dental similarities between stenodermatines and glossophagines (*sensu lato*). J. D. Smith (1976) favored an early differentiation of the stenodermatines from phyllostomine ancestors (perhaps the *Phyllostomus* lineage; see J. D. Smith 1972). Karyotypic and morphological characteristics suggest that the Stenodermatinae comprises three groups (J. D. Smith 1976): long-faced bats (*Chiroderma, Ectophylla, Sturnira, Uroderma, Vampyressa, Vampyriscus Vampyrodes,* and *Platyrrhinus*), short-faced bats (*Artibeus* [including *Enchisthenes*] and the four endemic Antillean genera), and if not included in the latter group, bats with a distinctly modified dental arcade (*Ametrida, Centurio, Sphaeronycteris,* and *Pygoderma*).

The phylogenetic interpretation of R. J. Baker et al. (2003) suggests three tribal units within the Stenodermatinae (Sturnirini, Stenodermatini, and Mesostenodermatini [unavailable family-group name]). Wetterer, Rockman, and Simmons (2000) recognized two tribes of stenodermatines, the Sturnirini and the Stenodermatini, which they subdivided into the Ectophyllina (short-faced bats lacking shoulder spots) and the Stenodermatina (flat-faced bats having shoulder spots). While both interpretations recognized the Sturnirini, and the generic composition of the Stenodermatina, R. J. Baker et al. (2003) separated the Ectophyllina of Wetterer, Rockman, and Simmons (2000) into the Vampyressina (the long-faced fruit bat genera *Chiroderma, Mesophylla, Platyrrhinus, Vampyressa, Vampyriscus, Vampyrodes,* and *Uroderma*) and the Mesostenodermatini (the short-faced fruit bat genera *Artibeus, Dermanura, Enchisthenes, Ectophylla,* and the Stenodermatina [corresponds to the Stenodermatina of Wetterer, Rockman, and Simmons 2000]). Herein, as much for convenience as to suggest general relationships, the Stenodermatinae are sorted into three tribes: the Sturnirini, the Ectophyllini (short- and long-faced fruit bats that lack shoulder spots), and the Stenodermatini (flat-faced fruit bats that have a white spot on each shoulder). The Ectophyllini (as interpreted here) appears to includes two general morphotypes: taxa that have a rounded dental arcade (*Artibeus* [sensu lato], *Enchisthenes,* and *Uroderma*), and those that have nearly straight toothrows, several have specialized molars, and four genera have a pale-to-white dorsal stripe (*Chiroderma, Ectophylla, Mesophylla, Platyrrhinus, Vampyressa, Vampyriscus,* and *Vampyrodes*).

KEY TO SOUTH AMERICAN GENERA OF STENODERMATI-
NAE (BASED IN PART ON J. K. JONES AND CARTER 1976):

1. Molars 2/2 2
1′. Molars 2/3 or 3/3 6
2. Upper dental arcade semicircular, rostrum less than 1/2 as long as braincase . . . . *Centurio*
2′. Upper dental arcade not semicircular, rostrum more than 1/2 as long as braincase 3
3. Rostrum inflated, nearly cuboid in form . . . *Pygoderma*
3′. Rostrum not inflated or cuboid in form 4
4. Nasals lacking, posterior margin of external nares with marked, lyre-shaped emargination on the skull where nasals would normally lie . . . . *Chiroderma*
4′. Nasals present, posterior margin of external nares lacking lyre-shaped emargination 5
5. Posterior margin of external nares more or less straight; second upper molar much smaller than first and differing in form . . . . *Artibeus* (part)
5′. Posterior margin of external nares broadly V-shaped; second upper molar resembling first in size and form . . . . *Vampyressa*
6. Molars 2/3 7
6′. Molars 3/3 11
7. First upper incisor markedly bifid, less than twice size of second incisor . . . . *Artibeus* (part)
7′. First upper incisor not bifid or only weakly so, more than twice size of second incisor 8
8. Second upper molar noticeably larger than first; upper premolars separated from each other and from adjacent teeth by evident gaps . . . . *Mesophylla*
8′. Second upper molar equal in size to, or smaller than first; no gaps between anterior upper cheekteeth 9
9. Incisors 2/1 or 2/2; height of first incisor more than height of first premolar; greatest length of skull less than 24 mm. 10
9′. Incisors 2/2; height of first incisor much less than height of first premolar; greatest length of skull more than 24 mm. . . . . *Vampyrodes*
10. Lower first premolar caniniform; basisphenoid pits deep . . . . *Vampyressa*
10′. Lower first premolar not caniniform; basisphenoid pits shallow . . . . *Vampyriscus*
11. Upper dental arcade expanded laterally to form semicircular arc 12
11′. Upper dental arcade not expanded laterally, U-shaped in occlusal view 13
12. Orbital space wider than long; interorbital constriction less than 5 mm . . . . *Ametrida*
12′. Orbital space longer than wide; interorbital constriction more than 5 mm . . . . *Sphaeronycteris*
13. Upper molars distinctly grooved longitudinally, the first two subquadrate in outline and lacking well-developed

cusps; first upper incisor approximately 1/2 as high as canine . . . . *Sturnira*
13′. Upper molars lacking longitudinal groove, the first two not subquadrate in outline and possessing well-developed cusps; first upper incisor much less than 1/2 as high as canine 14
14. First upper incisor less than twice size of second and resembling it in shape; upper incisors in contact and filling space between canines 15
14′. First upper incisor more than twice size of second and differing from it in shape; evident spaces present between upper incisors 16
15. First upper incisor clearly bifid; m3, if present, minute and peglike . . . . *Artibeus* (part)
15′. First upper incisor not bifid; m3 relatively large and well developed . . . . *Enchisthenes*
16. Crowns of first upper incisors parallel, deeply bifid; lower incisors in contact with each other . . . *Uroderma*
16′. Crowns of first upper incisors converge at their tips, not deeply bifid; lower incisors separated by distinct gaps . . . . *Platyrrhinus*

Tribe Ectophyllini

The Ectophyllini, as used here, comprise an assemblage that includes both short-faced and long-faced fruit bats. These bats have facial stripes, lack specialized growths or excrescent structures on the face, lack numerous papillae in the buccal cavity, and lack a white spot on each shoulder. Several genera have a white mid-dorsal stripe and others have reduced or absent third molars. Some genera exhibit specialized molar structure, but none have the broad longitudinal depression or groove on the molars that characterize the Sturnirini.

Genus *Artibeus* Leach, 1821

Suely A. Marques-Aguiar

The genus *Artibeus* comprises approximately 22 species occurring from Mexico to Argentina; at least 14 species are in South America (Koopman 1976, 1982, 1993; Hall 1981; Honacki, Kinman, and Koeppl 1982; Nowak and Paradiso 1983; Nowak 1991, 1999; Lim and Wilson 1993; Lim 1997). Size ranges from small to large (length of head and body 47–104 mm, forearm 34–76 mm, mass 10–87 g). Coloration varies from pale brown to dark blackish brown with venter usually paler than dorsum. Most species have pale facial stripes, and all lack dorsal stripes. The fur is soft, 4–10 mm on lower back; the ears are medium-sized and triangular in outline; and the skull is characteristically short and broad. Upper incisors are larger than the lower and both are crowded between canines. Inner upper incisors are larger than the outer and distinctly bilobed. Canines are well developed and lack secondary cusps. Upper premolars have a large triangular outer cusp and a low, broad, and somewhat concave inner basin. Lower premolars are like the upper, but broader relative to length, and lack a posteromedial basin. Molars are broad with low cusps except for the third molar, which is either reduced in size or absent. The dental formula is 2/2 1/1 2/2 2–3/2–3 × 2 = 28–32 (Dobson 1878; Miller 1907b).

Recent tradition has been to treat *Artibeus* as monophyletic, and several authors (e.g., Koopman 1993; Ascorra, Gorchov, and Cornejo 1994; Marques-Aguiar 1994) have followed Handley (1976, 1987) in including *Enchisthenes* Andersen, 1906b, as a subgenus. Other authors, however, working at the molecular level, have suggested that *Artibeus* is paraphyletic (Straney et al. 1979; Straney 1981; Koop and Baker 1983; Tandler, Nagato, and Phillips 1986). R. D. Owen (1987) expanded on that theme and placed the smaller species in the genus *Dermanura* P. Gervais, 1856a, while commenting that the relationships of *Enchisthenes hartii* and *Artibeus concolor* may not lie within *Dermanura*. Although retaining *hartii* and *concolor* under *Dermanura* in his discussion, elsewhere in the same publication (pp. 61–63) R. D. Owen listed *Enchisthenes* and *A. concolor* apart from both *Artibeus* and *Dermanura*. In 1991, R. D. Owen proposed *Koopmania* as the name of a new genus to include *A. concolor*. For practical reasons, my treatment herein of *Artibeus* is in the traditional sense as a polytypic genus. I follow R. D. Owen (1987, 1991), McKenna and Bell (1997), and Wetterer, Rockman, and Simmons (2000) to the extent that I recognize *Dermanura* and *Koopmania* as subgenera of *Artibeus*, but follow Van Den Bussche et al. (1993) and Arroyo-Cabrales and Owen (1996) in treating *Enchisthenes* as a separate genus.

Robert D. Owen's (1987) assessment of relationships based on comparative morphology, grouped *Artibeus* most closely with the genera *Uroderma* and *Ectophylla*. Gardner (1977a), based primarily on karyotypic features, placed *Artibeus* closer to *Vampyrops* (= *Platyrrhinus*) than to *Uroderma* and *Ectophylla*. *Artibeus* is known from late Pleistocene and sub-Recent fossils (J. D. Smith 1976; McKenna and Bell 1997). Some important papers concerning the systematics of this genus are Dobson (1878), Andersen (1906b, 1908b), Miller (1907b), Hershkovitz (1949c), Husson (1962), W. B. Davis (1969, 1970a, 1970c, 1984), Taddei (1969, 1979), Koopman (1978, 1982), Kraft (1982), Koepcke and Kraft (1984), Handley (1987, 1990, 1991), R. D. Owen (1987, 1988, 1991), Lim and Wilson (1993), Marques-Aguiar (1994); Lim (1993), J. A. Guerrero, Luna, and Sánchez-Hernández (2003), and Lim et al. (2004a).

SYNONYMS:

Phyllostoma: É. Geoffroy St.-Hilaire, 1810b:176; part; not *Phyllostoma* G. Cuvier, 1800.

Phyllostomus: Olfers, 1818:224; part; not *Phyllostomus* Lacépède, 1799.

Phyllostoma: Spix, 1823:66; not *Phyllostoma* G. Cuvier.

Phyllostoma: Schinz, 1821:164; part; not *Phyllostoma* G. Cuvier.

Artibeus Leach, 1821b:75; type species *Artibeus jamaicensis* Leach, 1821b, by monotypy.

Madataeus Leach, 1821b:81; type species *Madataeus lewisii* Leach, 1821b, by monotypy.

Medateus Gray, 1825:424; incorrect subsequent spelling of *Madataeus* Leach.

Madaeteus Burnett, 1829:269; incorrect subsequent spelling of *Madataeus* Leach.

Arctibeus Gray, 1838b:487; incorrect subsequent spelling of *Artibeus* Leach.

Madateus Chenu, 1853:311; incorrect subsequent spelling of *Madataeus* Leach.

Pteroderma P. Gervais, 1856a:34; type species *Pteroderma perspicillatum* P. Gervais, 1856a (= *Phyllostoma perspicillatum* É. Geoffroy St.-Hilaire, 1810b; not *Vespertilio perspicillatus* Linnaeus, 1758), by monotypy.

Artibaeus P. Gervais, 1856a:34; incorrect subsequent spelling of *Artibeus* Leach.

Dermanura P. Gervais, 1856a:36; type species *Dermanura cinereum* P. Gervais, 1856a, by monotypy; valid as a subgenus.

Artibacus Saussure, 1860:429; incorrect subsequent spelling of *Artibeus* Leach.

Artibaeus Saussure, 1860:429; incorrect subsequent spelling of *Artibeus* Leach.

Artiboeus Trouessart, 1878:214; incorrect subsequent spelling of *Artibeus* Leach.

Desmanura Kappler, 1881:163; incorrect subsequent spelling of *Dermanura* P. Gervais.

Artobius Winge, 1892:10; incorrect subsequent spelling of *Artibeus* Leach.

Artiheus Alberico and Orejuela, 1982:38; incorrect subsequent spelling of *Artibeus* Leach.

Koopmania R. D. Owen, 1991:21; type species *Artibeus concolor* W. Peters, 1865b, by original designation; valid as a subgenus.

KEY TO THE SOUTH AMERICAN SPECIES OF *ARTIBEUS*:

1. Paraoccipital process indistinct or absent; dorsal hair tricolored . . . . *Artibeus (Koopmania) concolor*
1'. Paraoccipital process small to moderately developed; dorsal hair bicolored 2
2. Forearm shorter than 46 mm . . . . *Artibeus (Dermanura)* 3

2'. Forearm longer than 46 mm . . . *Artibeus (Artibeus)* 10

3. Molars 2/3 4

3'. Molars 2/2 6

4. Dorsum dark grayish or blackish; ears dark; posterior margin of interfemoral membrane V-shaped . . . . *Artibeus glaucus* (part, *A. g. glaucus*)

4'. Dorsum pale brownish or grayish brown; ears pale; posterior margin of interfemoral membrane broadly U-shaped. 5

5. Base of noseleaf pale; posterior margin of palate deeply incised, V-shaped with nearly straight sides; mesopterygoid fossa constricted posteriorly on basicranium (internal margin of pterygoid with a ridge on each side) . . . . *Artibeus gnomus*

5'. Base of noseleaf dark, outline of posterior margin of palate shallow, U-shaped; mesopterygoid fossa not constricted posteriorly on basicranium (internal pterygoid ridge obsolete) . . . . *Artibeus watsoni* (part)

6. Interfemoral membrane appearing practically hairless . . . . *Artibeus cinereus*

6'. Interfemoral membrane conspicuously hairy 7

7. Narrow talon on M1, skull with inflated frontal and supraorbital regions and moderately to strongly arched rostrum. 8

7'. Broad talon on M1, frontal and supraorbital region of skull not inflated, rostrum flattened 9

8. Outer edge of ear slightly emarginated; maxillary toothrows convergent anteriorly, rostrum moderately arched . . . . *Artibeus glaucus* (part, *A. g. bogotensis*)

8'. Outer edge of ear deeply emarginated; maxillary toothrows nearly parallel to each other, not convergent anteriorly, rostrum strongly arched . . . . *Artibeus watsoni* (part)

9. Interfemoral membrane thickly haired, maxillary toothrow convergent anteriorly, rostrum usually not elevated (tilted up) anteriorly. . . . . *Artibeus phaeotis*

9'. Interfemoral membrane thinly haired, maxillary toothrows nearly parallel, rostrum usually elevated (tilted up) anteriorly . . . . *Artibeus anderseni*

10. Molars 3/3 11

10'. Molars 2/3 13

11. Fur long, 8–10 mm on mid-dorsum, blackish; facial stripes weak to absent; chin tubercles usually 3–4 on each side. . . . . *Artibeus obscurus* (part)

11'. Fur short, less than 8 mm on mid-dorsum, coloration paler; facial stripes usually evident; chin tubercles always more than 4 on each side. 12

12. Interfemoral membrane slightly hairy and fringed medially; horseshoe of noseleaf lacks lower rim (noseleaf fused to upper lip); wing tips dark; maxillary toothrows nearly parallel; lateral margins of rostral shield nearly parallel. . . . . *Artibeus amplus*

12'. Interfemoral membrane practically naked, and not fringed medially; horseshoe of noseleaf with lower rim and usually free from upper lip; wing tips pale; maxillary toothrows convergent anteriorly; lateral margins of rostrum convergent posteriorly 17

13. Upper parts grayish-brown or yellowish-brown (palest of the South American *Artibeus*); shape of M1 almost triangular in occlusal view; hypocone of M1 well developed . . . . *Artibeus fraterculus*

13'. Upper parts usually brown to dark brown; shape of M1 not triangular in occlusal view; hypocone of M1 not well developed. 14

14. Facial stripes prominent and well defined; underparts brown, well furred, and not conspicuously frosted with white . . . . *Artibeus lituratus*

14'. Facial stripes present but weakly defined; underparts frosted with white. 15

15. Interfemoral membrane and legs distinctly hairy, horseshoe of noseleaf bound down to upper lip with lower margin of horseshoe indicated by a transverse ridge . . . . *Artibeus fimbriatus*

15'. Interfemoral membrane and legs slightly haired or appearing naked, horseshoe of noseleaf free. 16

16. Fur long, 8–10 mm on mid-dorsum, blackish; tubercles on lower lip small, usually 3–4 on each side of chin . . . . *Artibeus obscurus* (part)

16'. Fur shorter, gray to gray brown; tubercles on lower lip larger, always more than 4 on each side of chin 17

17. Forearm more than 61 mm; length of skull more than 29.5 mm; breadth across upper molars more than 14 mm; distributed south and east of the Río Orinoco . . . . *Artibeus planirostris*

17'. Forearm less than 63 mm; length of skull less than 29.5 mm; breadth across upper molars less than 14 mm; distributed north and west of the Orinoco . . . . *Artibeus jamaicensis*

Subgenus *Artibeus* Leach, 1821

The nominate subgenus of large (forearm 50–77 mm) *Artibeus* contains seven species in South America. Coloration ranges from pale brownish to dark blackish brown with underparts usually paler than dorsum. The fur is pale basally, and facial stripes vary from prominent to indistinct or absent. The hypocone of M1 is moderately to well developed. Other characters shared by one or more other subgenera include: interfemoral membrane variably broad and sometimes hairy; plagiopatagium attached either to the ankle or to the side of the foot; the noseleaf is longer than wide; and the tragus lacks a pointed, sub-apical projection. The crowns of the inner upper incisors are bifid (bilobed), M3 (when present) and m3 are small and located lingually in

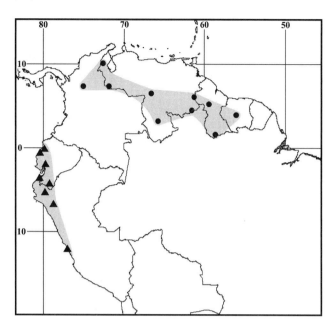

Map 162 Marginal localities for *Artibeus amplus* ● and *Artibeus fraterculus* ▲

toothrow, and the paraoccipital processes are small to moderately developed. The dental formula is 2/2 1/1 2/2 2–3/3 × 2 = 30–32.

Artibeus amplus Handley, 1987
Giant Artibeus
SYNONYMS:

Artibeus "sp. D" Handley, 1976:33.

Artibeus amplus Handley, 1987:164; type locality "Kasmera, 21 km SW Machiques, Estado Zulia, Venezuela, 270 m."

Artibeus [(Artibeus)] amplus: Marques-Aguiar, 1994:26; name combination.

DISTRIBUTION: *Artibeus amplus* is in northeastern Colombia, western and southern Venezuela (south of the rios Orinoco and Apure), Guyana, and Surinam. The species likely occurs in adjacent Brazil.

MARGINAL LOCALITIES (Map 162; from Handley 1987, except as noted): VENEZUELA: Zulia, 15 km W of Machiques; Bolívar, Mina de bauxita de la Serranía de Los Pijiguaos (Ochoa et al. 1988); Bolívar, Km 125. GUYANA: Potaro-Siparuni, Kaieteur Falls (Marques-Aguiar 1994). SURINAM, Saramacca, Tafelberg (Lim, Genoways, and Engstrom 2003). GUYANA, Upper Takutu-Upper Essequibo, Kamoa River, 50 km WSW of Gunn's Strip VENEZUELA: Bolívar, 21 km NE of Icabarú (USNM 441176); Amazonas, Tamatama; Apure, Nulita. COLOMBIA: Antioquia, La Tirana.

SUBSPECIES: I consider *A. amplus* to be monotypic.

NATURAL HISTORY: *Artibeus amplus* has been found in Venezuela mainly in evergreen forest, near streams and other wetter areas, as well as in forest openings such as yards and orchards. Elsewhere, the species has been collected in savannah (llanos), gallery forest, and lowland rainforest (Lim, Genoways, and Engstrom 2003). Handley (1987) reported the species roosting in the "twilight" zone of caves. The known elevational range is from 24 to 1,200 m. The diet of *A. amplus* probably is primarily tree fruits. Little is known about its reproduction or other aspects of its biology. The holotype is an adult female that was suckling a young when collected in April in a damp cave in a cliff along the Río Yasa, across from the Kasmera Biological Station, in the eastern foothills of the Sierra de Perijá, Venezuela (Handley 1987:164). Based on material at the National Museum of Natural History (USNM), two females were lactating when taken in Colombia in May. In Venezuela, seven were pregnant in February; one was lactating in February and another in April. The specimen from Surinam is a female that was pregnant with 16-mm–CR fetus when caught on 3 November. Handley (1987) mentioned *Strebla paramirabilis* Wenzel and *Trichobius assimilis* Wenzel, and R. Guerrero (1997) added two more species of streblid batflies known from *A. amplus*. The only other known ectoparasite is a species of argasid tick (E. K. Jones et al. 1972). No information is available on endoparasites or the karyotype.

REMARKS: *Artibeus amplus* is the "*Artibeus* sp. D" of Handley (1976).

Artibeus fimbriatus Gray, 1838
Fringed-lipped Artibeus
SYNONYMS:

Arctibeus fimbriatus Gray, 1838b:487; type locality "Brazil"; restricted to "Morretes, at the coastal foot of the Serra do Mar, state of Paraná," by Handley (1990:455).

Artibeus perspicillatus: Dobson, 1878:519; part; not *Vespertilio perspicillatus* Linnaeus, 1758.

Artibeus grandis Dobson, 1878:520; *nomen nudum*.

Artibeus lituratus: O. Thomas, 1902a:59; part; specimens from Morretes, Paraná, Brazil (Handley 1990:458); not *Artibeus lituratus* Olfers, 1818.

Artibeus jamaicensis lituratus: Andersen, 1908b:272; part; all *Artibeus grandis* in synonymy and all BM specimens of *Artibeus fimbriatus* (see Handley 1990:458); not *Artibeus lituratus* Olfers, 1818.

Artibeus fimbriatus: Myers, 1982:86; first modern use of current name combination.

Artibeus [(Artibeus)] fimbriatus: Marques-Aguiar, 1994:26; name combination.

DISTRIBUTION: *Artibeus fimbriatus* occurs in eastern and southern Brazil, Paraguay, and northeastern Argentina.

MARGINAL LOCALITIES (Map 163; Handley 1990,

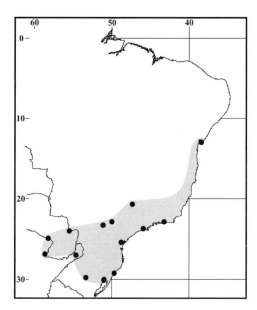

Map 163 Marginal localities for *Artibeus fimbriatus* ●

except as noted): BRAZIL: Bahia, Salvador; Rio de Janeiro, Rio de Janeiro; São Paulo, Guaratuba; Paraná, Morretes (type locality of *Artibeus fimbriatus* Gray); Santa Catarina (locality unknown; not mapped); Rio Grande do Sul, Dom Pedro de Alcântara (Graciolli and Rui 2001); Rio Grande do Sul, Viamão (Fabián, Rui, and Oliveira 1999); Rio Grande do Sul, Restinga Seca (Fabián, Rui, and Oliveira 1999). ARGENTINA (Barquez, Mares, and Braun 1999): Misiones, Dos de Mayo; Chaco, General Vedia; Formosa, Parque Nacional Río Pilcomayo. PARAGUAY: Canindeyú, Igatimí. BRAZIL: Paraná, Londrina (N. R. Reis et al. 1998); São Paulo, Salto Grande; São Paulo, Cruzeiro.

SUBSPECIES: I consider *A. fimbriatus* to be monotypic.

NATURAL HISTORY: *Artibeus fimbriatus* is a common bat in the tropical humid forest near latitude 25°S and has been found from near sea level in several localities to 530 m at Cruzeiro, São Paulo, Brazil (Handley 1990). Although few data are available, bimodal polyestry is the probable reproductive pattern. In Paraguay, a juvenile male was recorded in January and a pregnant female in July; in Brazil (Paraná), a juvenile female was taken in March and a juvenile male in April (MVZ and USNM collections). Passos et al. (2003) found the seeds of *Cecropia* spp., Cucurbitaceae, *Ficus* spp., *Piper* sp., and *Solanum sp.* and *Vassovia breviflora* in the feces of 17 of 74 *A. fimbriatus* collected in the Parque Estadual de Intervales, São Paulo, Brazil, between January 1999 and January 2000. They also recorded four *A. fimbriatus* whose feces contained unidentifiable seeds, and four whose feces consisted of fruit pulp without seeds. Autino, Claps, and Bertolini (1998) reported a species of streblid batfly from *A. fimbriatus* from Misiones, Argentina. Graciolli and Rui (2001) reported three

species of streblid batflies, including the species reported from Argentina, they found on *A. fimbriatus* taken in the state of Rio Grande do Sul, Brazil. Rui and Graciolli (2005) recovered two species of streblids from the same female, also from Rio Grande do Sul. No further information is available on parasites, diseases, or the karyotype.

REMARKS: *Artibeus fimbriatus* was overlooked by systematists for more than 150 years and Patten (1971:23) regarded the name as a *nomen nudum*. Handley (1990:455), after studying the holotype in the Natural History Museum of London (BM), resurrected and redescribed Gray's *A. fimbriatus*.

The apparent hiatus in distribution between Paraguayan and Brazilian localities may result either from confusion with other large *Artibeus* or from inadequate collecting. Handley (1990:455) wrote "the range of the species may skirt the *Araucaria* forest of the interior of southern Brazil." In 1991, Handley suggested that the low number of specimens recorded from eastern Brazil was due to the extensive loss of the coastal forest. *Artibeus grandis*, as listed by Dobson (1878), is one of Gray's unpublished manuscript names and, therefore, a *nomen nudum*.

Artibeus fraterculus Anthony, 1924
Fraternal Artibeus

SYNONYMS:

Artibeus fraterculus H. E. Anthony, 1924a:5; type locality "Portovelo, Provincia del Oro, Ecuador; altitude, 2000 ft."

A[rtibeus]. j[amaicensis]. fraterculus: Hershkovitz, 1949c: 447; name combination.

Artibeus [(Artibeus)] fraterculus: Marques-Aguiar, 1994: 26; name combination.

DISTRIBUTION: *Artibeus fraterculus* is in western Ecuador and western Peru.

MARGINAL LOCALITIES (Map 162; localities listed from north to south): ECUADOR: Manabí, San Vicente (Albuja 1983); Manabí, Puerto López (Albuja 1983); Guayas, Sanbrondon (BM 15.1.1.16). PERU: Tumbes, Corrales (Ortiz de la Puente 1951). ECUADOR: Loja, Malacatos (FMNH 53489). PERU: Piura, Hacienda Bigotes (FMNH 81137); Cajamarca, Grutas de Ninabamba (Ortiz de la Puente 1951); Lima, Lima (Ortiz de la Puente 1951).

SUBSPECIES: I consider *A. fraterculus* to be monotypic.

NATURAL HISTORY: *Artibeus fraterculus* inhabits the arid, or seasonally arid, coastal zone of southern Ecuador (H. E. Anthony 1924a) and western Peru, from sea level to 1,600 m. In northwestern Peru, R. Thomas and Thomas (1977) recovered 58 crania from an assemblage of skeletal remains presumed to represent decayed owl pellets. The diet is insects and fruits (Albuja 1983). Based on Ecuadorian material in the USNM, a female was pregnant and lactating when taken in November, others were either lactating in July and November or pregnant in October and November. These data suggest bimodal polyestry. No information is available on the karyotype or on parasites and diseases.

REMARKS: *Artibeus fraterculus* appears to be the only *Artibeus* widely distributed on the arid Pacific side of Peru (Koopman 1978).

Artibeus jamaicensis Leach, 1821
Jamaican Artibeus

SYNONYMS: The following synonyms refer to extralimital populations. See under Subspecies for additional synonyms.

Artibeus jamaicensis Leach, 1821b:75; type locality "Jamaica."

Madataeus lewisii Leach, 1821b:81; type locality "Jamaica."

Artibeus carpolegus Gosse, 1851:271; type locality "Content," Jamaica.

Artibeus macleayii Dobson, 1878:520; *nomen nudum*.

Dermanura eva Cope, 1889a:130; type locality "Island of Saint Martins, West Indies."

Artibeus coryi J. A. Allen, 1890:173; type locality "St. Andrew's Island, Caribbean Sea."

[*Artibeus (Artibeus)*] *Coryi*: Trouessart, 1897:160; name combination.

[*Artibeus (Dermanura)*] *eva*: Trouessart, 1897:160; name combination.

Artibeus parvipes Rehn, 1902b:639; type locality "Santiago de Cuba, Cuba."

Artibeus insularis J. A. Allen, 1904b:231; type locality "Island of St. Kitts, W.I."

Artibeus yucatanicus J. A. Allen, 1904b:232; type locality "Chichenitza, Yucatan," Mexico.

[*Artibeus (Artibeus)*] *jamaicensis*: Trouessart, 1904:116; name combination.

[*Artibeus (Artibeus)*] *parvipes*: Trouessart, 1904:116; name combination.

Artibeus jamaicensis praeceps Andersen, 1906b:421; type locality "Guadeloupe, W.I."

Artibeus jamaicensis parvipes: Andersen, 1908b:261; name combination.

Artibeus jamaicensis yucatanicus: Andersen, 1908b:263; name combination.

Artibeus jamaicensis richardsoni J. A. Allen, 1908b:669; type locality "Matagalpa, Nicaragua."

A[rtibeus]. l[ituratus]. praeceps: Hershkovitz, 1949c:447; name combination.

Artibeus jamaicensis paulus W. B. Davis, 1970a:119; type locality "7 ¹/₂ km WNW La Libertad, elevation *ca.* 500 feet, Department of La Libertad, El Salvador."

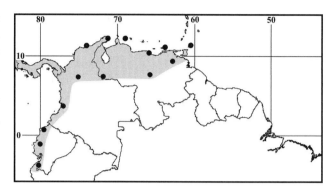

Map 164 Marginal localities for *Artibeus jamaicensis* ●

DISTRIBUTION: *Artibeus jamaicensis* is found in Mexico, south throughout Central America, into northern and western Colombia, western Ecuador, northwestern Peru, northern Venezuela, the Netherlands Antilles, and Trinidad and Tobago. It occurs also on the Bahamas, the Greater and Lesser Antilles, and has been reported by Lazell and Koopman (1985), and Nowak (1991) from the lower Florida Keys, USA (but see Humphrey and Brown 1986).

MARGINAL LOCALITIES (Map 164): COLOMBIA: La Guajira, 5 km W of Nazareth (USNM 483840). NETHERLANDS ANTILLES: Curaçao, Klein St. Martha (Husson 1960). VENEZUELA: Miranda, Río Chico (Handley 1976); Nueva Esparta, 3 km NNE of La Asunción (USNM 405230). TRINIDAD AND TOBAGO: Tobago, Charlotteville (Goodwin and Greenhall 1961). VENEZUELA: Monagas, Hato Mato de Bejuco (Handley 1976); Bolívar, Florida (Lim 1997); Apure, Nula (USNM 441083). PERU: Tumbes, Quebrada Faical (Graham and Barkley 1984). ECUADOR: Manabí, San José (FMNH 53496); Esmeraldas, Viche (Albuja 1983). COLOMBIA: Valle del Cauca, Río Raposo (USNM 334692); Antioquia, La Tirana (USNM 499488); Magdalena, Cacagualito (AMNH 23657).

SUBSPECIES: Hall (1981) recognized seven subspecies (*A. j. jamaicensis, A. j. parvipes, A. j. paulus, A. j. richardsoni, A. j. triomylus, A. j. yucatanicus,* and *A. j. schwartzi*) in tropical North America (Mexico, Central America, and the West Indies). *Artibeus j. schwartzi* is currently recognized as a subspecies of *A. planirostris* (see Pumo et al. 1996) and *A. j. triomylus* Handley, 1966b, has recently been elevated to full species status (J. A. Guerrero, Luna, and González 2004). I recognize two subspecies in South America.

A. j. aequatorialis Andersen, 1906
SYNONYMS:

Artibeus jamaicensis aequatorialis Andersen, 1906b:421; type locality "Zaruma, Loja, S. Ecuador, 1000 m."

A[rtibeus]. l[ituratus]. *aequatorialis*: Hershkovitz, 1949c: 447; name combination.

This subspecies is in northern Colombia and west of the Andes in Colombia, Ecuador, and Peru.

A. j. trinitatis (Andersen, 1906)

SYNONYMS:

Artibeus planirostris trinitatis Andersen, 1906b:420; type locality "St. Anns, Trinidad," Trinidad and Tobago.

Artibeus planirostris grenadensis Andersen, 1906b:420; type locality "Grenada, W.I."

A[rtibeus]. j[amaicensis]. *trinitatis*: Hershkovitz, 1949c: 447; name combination.

This subspecies is in northern Venezuela, Trinidad and Tobago, and on Grenada (Lesser Antilles).

NATURAL HISTORY: In South America, *A. jamaicensis* occurs on the northern and western slopes and adjacent lowlands of the Andes over an elevational range from sea level to over 2,300 m in all but the most arid habitats. Nevertheless, it is most common at lower elevations (e.g., 83% of specimens reported by Handley (1976) from Venezuela were collected below 500 m). This species roosts in tunnels, hollow logs, culverts, houses, under bridges, under tree roots, under palm leaves, in the dark foliage of trees such as mango and bread fruit, and occasionally in shallow, well-illuminated caves (Goodwin and Greenhall 1961; Tuttle 1968, 1976; Handley 1976; Kunz 1982; Morrison and Handley 1991). According to Kunz (1982), this bat makes opportunistic use of a wide variety of roost sites. Goodwin and Greenhall (1961) reported *A. jamaicensis* in Trinidad and Tobago roosting in association with *Saccopteryx leptura*, *Noctilio leporinus*, *Carollia perspicillata*, *Uroderma bilobatum*, *Artibeus lituratus*, and *A. "cinereus"* (= *A. glaucus*).

Artibeus jamaicensis eats the fruits of canopy trees and, less commonly, flowers, leaves, and insects (Tuttle 1968; Gardner 1977c; Albuja 1983). Goodwin and Greenhall (1961) recorded 51 species of fruits consumed by the Jamaican Artibeus, including a number of economically important species. Throughout most of its range, figs are probably the most important food consumed by *A. jamaicensis* in terms of quantity and because of general year-round availability. Handley, Gardner, and Wilson (1991) found the fruit of *Ficus insipida* to be favored by *A. jamaicensis* among the 17 species of fruits they recorded as eaten by this species on Barro Colorado Island, Panama. Handley and Morrison (1991) suggested group foraging at fruiting trees by *A. jamaicensis*.

Howler monkeys (*Alouatta palliata*) and kinkajous (*Potos flavus*) compete with *A. jamaicensis* for figs (Handley and Leigh 1991) along with other bats, opossums, and an array of frugivorous birds. Eisenberg (1989) suggested that

owls, snakes, and the bat falcon, *Falco rufigularis*, may be significant predators.

Artibeus jamaicensis exhibits seasonal polyestry in Panama and has alternate periods of normal and delayed embryonic development (T. H. Fleming 1971; Wilson, Handley, and Gardner 1991). Wilson (1979) categorized the reproductive cycle as bimodal polyestry. The peaks of reproduction, however, seem to vary temporally according to latitude and longitude, and may reflect rain cycles and food abundance (T. H. Fleming, Hooper, and Wilson 1972; Heithaus, Fleming, and Opler 1975; Wilson 1979; Handley and Leigh 1991; Wilson, Handley, and Gardner 1991). From west to east across South America, there is an apparent delay in breeding with pregnancy peaks in November for Ecuador, December-January for Colombia, and January-February for Venezuela. Pregnancy occurs even later (March-April) on Trinidad and Tobago. There is a second pregnancy peak in June for Ecuador, June-July for Venezuela, and July for Trinidad and Tobago (pregnancy data from Goodwin and Greenhall 1961; Wilson 1979; Albuja 1983; C. O. Handley, Jr., pers. comm.; and from information on labels of specimens in the USNM collection). Nowak (1991), Tamsitt (1966), and M. E. Thomas (1972) have suggested that *A. jamaicensis* has a continuous or acyclic breeding cycle in Colombia.

Morrison (1979) described the mating system in Panamanian and Mexican populations of *A. jamaicensis* as having harems composed of from 3 to 14 adult females, their 0–6 young, and a single adult male. Morrison (1979) and Morrison and Handley (1991) determined that the system is based on male defense of tree hole roosts used by females. In Puerto Rico, however, harem groups occupy caves, suggesting that environmental factors promoting harem social organization in this species varies between populations and is influenced by the availability of roost sites (Morrison 1979; Kunz, August, and Burnett 1983). Bachelor males and juveniles roost in foliage or leaf tents (Emmons and Feer 1990). Goodwin and Greenhall (1961) found colonies of up to 25 individuals of both sexes roosting in foliage in Trinidad and Tobago.

The karyotype is $2n = 30–31$, FN = 56, with a Y_1, Y_2 chromosome system (X-autosome fusion; R. J. Baker 1967, 1973). The X chromosome is a large subtelocentric; the Y_1 and Y_2 are small acrocentrics, one about 1/2 the size of the other.

Goodwin and Greenhall (1961) reported rabies in the species on the island of Trinidad but not on Tobago. J. P. Webb and Loomis's (1977) summary of ectoparasites listed four species of ticks (Ixodidae, Argasidae), six species of mites (Trombiculidae, Macronyssidae, Gastronyssidae, Spinturnicidae, Ereynetidae), and four species of batflies (Nycteribiidae, Streblidae) found on *A. jamaicensis*. Other reports and summaries (Furman 1972; Guimarães 1972; Tipton and Machado-Allison 1972; Herrin and Tipton 1975; Brennan and Reed 1975; Reed and Brennan 1975; Saunders 1975; Wenzel 1976; R. Guerrero 1997) indicate that *A. jamaicensis* hosts a large ectoparasite fauna that includes at least 13 species of mites (Laelapidae, Macronyssidae, Spinturnicidae, Trombiculidae), 40 of batflies (Nycteribiidae, Streblidae), and 2 of fleas (Hystrichopsyllidae, Rhopalopsyllidae); however, the actual hosts for some of these parasites probably represents misidentified *A. planirostris* and *A. obscurus*. Cuartas and Muñoz (1999) reported two onchocercid and three trichostrongylid species of nematodes from the abdominal cavity and intestinal tract, respectively, of Colombian *A. jamaicensis*. The *Mammalian Species* account by Ortega and Castro-Arellano (2001) treated *A. jamaicensis* and *A. planirostris* as conspecific.

REMARKS: Handley (1991:16) treated *A. planirostris* as a subspecies of *A. jamaicensis* "intergrading with *A. j. fallax* W. Peters, but still recognizable as *A. j. planirostris* south to Mato Grosso [Brazil] and eastern Paraguay." Handley (1990) showed that *A. obscurus* (called either *A. fuliginosus* or *A. davisi* in the recent literature) is distinct from *A. jamaicensis*; as Koopman (1993) concurred. However, in other publications Koopman (1982, 1994) considered *A. obscurus* to be a subspecies of the *A. jamaicensis* and this assumption was the basis for stating that the South American distribution of *A. jamaicensis* was south into Paraguay. Koopman (1994) treated *A. planirostris* as a species with *A. p. hercules* and *A. p. fallax* as subspecies. Here I treat *A. obscurus* as a species and follow Lim and Wilson (1993), Koopman (1994), and Lim (1997) in treating *A. planirostris* as a species separate from *A. jamaicensis*.

Patten (1971) and Handley (1990) considered *Artibeus lobatus* Gray (1838b) to be unrecognizable (a *nomen dubium*). *Artibeus macleayii*, listed by Dobson (1878), is one of Gray's unpublished manuscript names and, therefore, a *nomen nudum*. Although Hershkovitz (1949c) considered *A. j. praeceps* to be a synonym of *A. lituratus*, Koopman (1968, 1993) treated the name as a synonym of *A. jamaicensis*. D. C. Carter and Dolan (1978) provided information and measurements on types of *Artibeus carpolegus* Gosse, *Artibeus jamaicensis* Leach, *Artibeus jamaicensis aequatorialis* Andersen, *Artibeus planirostris trinitatis* Andersen, and *Madataeus lewisii* Leach. J. A. Guerrero, Luna, and González (2004) elevated *A. j. triomylus* Handley, 1966b, one of the subspecies Hall (1981) and Koopman (1994) recognized in Mexico, to full species status.

Artibeus lituratus (Olfers, 1818)
Great Artibeus
SYNONYMS: See under Subspecies.

Map 165 Marginal localities for *Artibeus lituratus* ●

DISTRIBUTION: *Artibeus lituratus* occurs from central Mexico, the Lesser Antilles, Trinidad and Tobago, south into South America as far as northcentral Argentina (Nowak and Paradiso 1983; Nowak 1991, 1999; Koopman 1993).

MARGINAL LOCALITIES (Map 165): TRINIDAD AND TOBAGO: Tobago, Charlotteville (Goodwin and Greenhall 1961). GUYANA: Essequibo Islands-West Demerara, Bonasica (BM 13.5.23.13). SURINAM: Brokopondo, Brownsberg (Brennan and Bronswijk 1975). FRENCH GUIANA: 1.5 km NE of Rémire (S. L. Williams, Phillips, and Pumo 1990). BRAZIL: Amapá, Santa Luzia do Pacuí (Peracchi, Raimundo, and Tannure 1984); Pará, Ilha Caratateua (Piccinini 1974); Ceará, São Paulo (O. Thomas 1910c); Pernambuco, Estação Ecológica do Tapacurá (Mares et al. 1981); Alagoas, Canoas (C. O. C. Vieira 1953); Bahia, Samarão (Andersen 1908b); Espírito Santo, Santa Teresa (Ruschi 1953a); Rio de Janeiro, Fazenda von Tapebuçú (type locality of *Phyllostoma superciliatum* Schinz); São Paulo, Guaratuba (USNM 542693); Santa Catarina, Joinville (FMNH 34390); Rio Grande do Sul, Dom Pedro de Alcântara (Graciolli and Rui 2001); Rio Grande do Sul, Porto Alegre (Fabián, Rui, and Oliveira 1999); Rio Grande do Sul, Santa Maria (Fabián, Rui, and Oliveira 1999); Rio Grande do Sul, Santo Cristo (Handley 1990). ARGENTINA: Corrientes, Itá Ibaté (Lord, Delpietro, and Lazaro 1973); Formosa, Bouvier (Barquez, Mares, and Braun 1999). PARAGUAY: Presidente Hayes, Rincón Charrúa (Myers and Wetzel 1983). BOLIVIA (S. Anderson, Koopman, and Creighton 1982): Santa Cruz, Caucaya; La Paz, 5 km by road SE of Guanay. PERU: Cusco, Huajyumbe

(FMNH 84402); Pasco, San Juan (Tuttle 1970); Loreto, Contamana (BM 28.5.2.135); Cajamarca, Jaén (Koopman 1978); Tumbes, Matapalo (FMNH 81065). ECUADOR: Esmeraldas, Viche (Albuja 1983). COLOMBIA: Cauca, Gorgona Island (BM 24.12.6.4); Valle del Cauca, near Buenaventura (Arata and Vaughan 1970); Magdalena, Bonda (J. A. Allen 1900a); La Guajira, La Concepción (type locality of *Artibeus femurvillosum* Bangs). VENEZUELA: Falcón, 16 km ENE of Mirimire (USNM 441187); Nueva Esparta, Las Piedras (J. D. Smith and Genoways 1974).

SUBSPECIES: There are two subspecies of *A. lituratus* in South America. A third (*A. lituratus intermedius*, considered a distinct species by W. B. Davis 1984), apparently is confined to Mexico and Central America (Hall 1981).

A. l. lituratus (Olfers, 1818)
SYNONYMS:

Phyllostoma perspicillatum: É. Geoffroy St.-Hilaire, 1810b:176; not *Vespertilio perspicillatus* Linnaeus, 1758.

[*Phyllostomus*] *lituratus* Illiger, 1815:109; *nomen nudum.*

[*Phyllostomus*] *frenatus* Illiger, 1815:109; *nomen nudum.*

Ph[*yllostomus*]. *lituratus* Olfers, 1818:224; type locality "Paraguay"; restricted to Asunción by Cabrera (1958: 90; see Remarks).

Ph[*yllostomus*]. *frenatus* Olfers, 1818:224; type locality "Brasilien."

Phyll[*ostoma*]. *superciliatum* Schinz, 1821:163; type locality "Ostküste von Brasilien"; identified by Wied-Neuwied (1826:200) as "Fazenda von Tapebuçú aufgehängt, welche etwas nördlich von Cabo Frio zwischen den Flüssen S. Joao und Macaché gelegen ist," Rio de Janeiro, Brazil.

Arctibeus perspicillatus: Gray, 1838b:487; part; not *Vespertilio perspicillatus* Linnaeus; name combination.

Stenoderma perspicillatum: d'Orbigny and Gervais, 1847:41; not *Vespertilio perspicillatus* Linnaeus; name combination.

Pteroderma perspicillatum: P. Gervais, 1856a:34; not *Vespertilio perspicillatus* Linnaeus; name combination.

[*Artibeus (Artibeus)*] *perspicillatus*: Trouessart, 1897:160; name combination.

Artibeus lituratus: O. Thomas, 1900c:547; name combination.

Artibeus rusbyi J. A. Allen, 1904b:230; type locality "Yungas, Peru (alt. 6000 ft.)"; identified as "18°S in southeastern Cochabamba," Bolivia, by S. Anderson, Koopman, and Creighton (1982:9).

[*Artibeus (Artibeus)*] *lituratus*: Trouessart, 1904:116; name combination.

[*Artibeus (Artibeus) lituratus*] *superciliatus*: Trouessart, 1904:116; name combination.

Artibeus jamaicensis lituratus: Andersen, 1908b:272; part; name combination.

Artibeus lituratus lituratus: Hershkovitz, 1949c:447; first modern use of current name combination.

The nominate subspecies is distributed south of the Orinoco basin in Venezuela, east through the Guianas and southward, east of the Andes, through Colombia, Ecuador, Brazil, Peru, and Bolivia into Paraguay and northern Argentina.

A. l. palmarum (J. A. Allen and Chapman, 1897)
SYNONYMS:

Artibeus palmarum J. A. Allen and Chapman, 1897:16; type locality "Port-of-Spain, Trinidad," Trinidad and Tobago; restricted to Royal Botanic Gardens, Port-of-Spain, by Goodwin and Greenhall (1961:261).

Artibeus femurvillosum Bangs, 1899:73; type locality "La Concepcion, [La Guajira,] Colombia, 3000 feet altitude."

[*Artibeus (Artibeus)*] *palmarum*: Trouessart, 1904:116; name combination.

A[*rtibeus*]. *j*[*amaicensis*]. *dominicanus* Andersen, 1908b: 249; *nomen nudum.*

Artibeus jamaicensis palmarum: Andersen, 1908b:278; part; name combination.

Artibeus lituratus palmarum: Hershkovitz, 1949c:445; part; first use of current name combination.

The South American distribution of this subspecies includes northern and western Colombia, Venezuela north of the Orinoco basin, and Trinidad and Tobago. Elsewhere, it occurs in Mexico, Central America (W. B. Davis 1984), and on the Lesser Antilles (St. Vincent and Grenada; Hall 1981).

NATURAL HISTORY: *Artibeus lituratus*, like *A. jamaicensis* and *A. planirostris*, occupies a variety of habitats at elevations from sea level to at least 2,620 m (Santa Elena, Colombia; Muñoz 1990); although, both species appear to be more abundant at elevations below 500 m. (Handley 1976; Mares et al. 1981; N. R. Reis and Peracchi 1987; Brosset and Charles-Dominique 1991; Bernard and Fenton 2002). These three species also are common in humid forest habitats; however, *A. lituratus* appears better able to tolerate more arid habitats (e.g., 34%, versus 16% of *A. jamaicensis*, were caught in Venezuelan dry uplands; Handley 1976). Handley (1967) found that *A. lituratus* was the most abundant *Artibeus* in the forest canopy in the vicinity of Belém, Brazil, whereas *A. planirostris* was more common at ground level. Goodwin and Greenhall (1961), Handley (1976), and Tuttle (1976) claimed that *A. lituratus* and *A. jamaicensis* use similar roost sites, and Goodwin and Greenhall (1961) reported *A. lituratus* on Trinidad and Tobago roosting in association with *Saccopteryx bilineata*,

S. leptura, Uroderma bilobatum, A. jamaicensis, and *Desmodus rotundus*. Morrison (1980) showed that *A. lituratus* commonly roosts in sheltered sites in the foliage of trees and vines, and does not roost in tree holes, at least not on Barro Colorado Island, Panama.

The diet of *A. lituratus* consists of the fruits of canopy trees and, less commonly, flowers, leaves, and insects (Gardner 1977c). Zortéa and Mendes (1993) found that *A. lituratus* chewed the leaves of at least six species of plants. Their observations, on the grounds of the Museu de Biología Mello Leitáo, Santa Teresa, Espírito Santo, Brazil, showed the bats chewing the leaves, ingesting the fluids, and dropping the fibrous material as pellets. N. R. Reis and Peracchi (1987) stated that *A. lituratus* is an important seed disperser of at least ten species of rainforest trees. Passos et al. (2003) found the seeds of *Cecropia* spp., *Ficus* spp., and *Solanum scuticum* in the feces of 7 of 29 *A. lituratus* caught in the Parque Estadual de Intervales, São Paulo, Brazil between January 1999 and January 2000. Willig (1983) found the diet of populations in Ceará, Brazil to be mainly *Vismia* sp. Figs (*Ficus* spp.) are an important food throughout the range of this species. In French Guiana, Brosset and Charles-Dominique (1991) found that *A. lituratus* has a larger home-range than either *A. planirostris* or *A. obscurus* and which, they suggested, was linked to the ability to flock together to exploit the fruit crops of large trees (also see Handley and Morrison 1991).

Wilson (1979) classified the reproductive cycle as one of bimodal polyestry; although, Tamsitt and Valdivieso (1963c, 1965) claimed that reproduction occurs throughout the year in Colombia. Peaks of pregnancy occur in October-November for Ecuador and Colombia, in February for Venezuela and Brazil (Pará), and in April-May for Trinidad and Tobago. There is a second pregnancy peak in June-July for Venezuela; July for Minas Gerais, Brazil; and August for Amazonas, Brazil (Goodwin and Greenhall 1961; Mumford and Knudson 1978; Wilson 1979; N. R. Reis and Peracchi 1987; C. O. Handley, Jr., pers. comm.; information from the USNM collection). Based on histological examination of reproductive tracts, S. F. Reis (1989) concluded that southeastern Brazilian *A. lituratus* exhibited post-partum estrus and bimodal polyestry with peaks in births in November and March, coinciding with periods of greater rainfall. Marinho-Filho (2003) recorded pregnant *A. lituratus* in October and February in southeastern Brazil. Morrison (1980), observing only one adult male in each group of Panamanian *A. lituratus*, concluded that the species has a harem-based mating system. Willig (1983) found limited secondary sexual variation in populations in the Northeast of Brazil.

The karyotype is $2n = 30$–31, FN = 56, with a Y_1, Y_2 sex chromosome system (X-autosome fusion; R. J. Baker 1967).

R. J. Baker (1979) listed the sex chromosomes as a subtelocentric or submetacentric X, a subtelocentric or acrocentric Y_1, and a submetacentric or acrocentric Y_2 chromosome. The differences he described may reflect normal variation, be an artifact of technique, or result from misidentification of taxa. L. R. R. Rodrigues et al (2003) also listed $2n = 30$ chromosomes in females and 31 in male specimens, with the chromosomal sexual system being XY_1, Y_2 type. However, they found the X chromosome to be a large submetacentric and described both Y_1 and Y_2 as acrocentric elements.

Goodwin and Greenhall (1961) found *A. lituratus* infected with rabies in Trinidad, but not in Tobago. Woodall (in Constantine 1970) reported isolating the antibodies to five arboviruses in Brazilian *A. lituratus*; three (Catu, Carapuru, and Utinga viruses) are infectious to humans. *Salmonella sandiego* (see Arata et al. 1968), *Blastomyces (Paracoccidioides) brasiliensis* (Grose and Tamsitt 1965), and *Scopulariopsis* sp. (Marinkelle and Grose 1966) have been found in Colombian *A. lituratus*. Mok, Luizão, and Silva (1982) reported 11 fungal species of the genera *Candida* and *Kluyveromyces* they isolated from Amazonian Brazilian specimens. Endoparasites listed by Ubelaker, Specian, and Duszynski (1977) included *Toxoplasma gondii*; *Trypanosoma pifanoi, T. cruzi, T. cruzi*-like, and *T. rangeli*-like flagellates; and *Vampyrolepis elongatus* (Cestoda). Cuartas and Muñoz (1999) reported finding a species of onchocercid filarial nematode in the abdominal cavity of a Colombian *A. lituratus*. J. P. Webb and Loomis's (1977) summary of ectoparasites listed two species of ticks (Ixodidae), five species of mites (Macronyssidae, Labidocarpidae, Spinturnicidae, Gastronyssidae, Psorergatidae), and four species of batflies (Streblidae). Other reports (E. K. Jones et al. 1972; Tipton and Machado-Allison 1972; Brennan and Reed 1975; Herrin and Tipton 1975; Wenzel 1976; Chavez 1998) have added a species of argasid tick, four spinturnicid and trombiculid mites, nine species of streblid batflies, and one species of rhopalopsyllid flea to the known ectoparasite fauna. R. Guerrero (1997) listed 27 species in 11 genera of streblid batflies hosted by *A. lituratus*. Graciolli and Aguiar (2002) added another species of streblid found on specimens from the Distrito Federal, Brazil.

REMARKS: Olfers based *Phyllostomus lituratus* on Azara's (1801b:269) *"chauve-souris obscure et rayée,"* which Azara did not collect. The bat came from "Pueblo mi amigo Don Pedro Blas Noseda" (Azara 1802b:291), which Azara stated was near 27°S, more closely approximating the latitude of San Ignacio, but not that of Asunción. Therefore, Cabrera's (1958:90) restriction of the type locality to Asunción may be an error.

William B. Davis (1984) considered *A. intermedius* to be a species distinct from *A. lituratus*. Other authors have considered *A. intermedius* to be a subspecies of *A. litura-tus* (e.g., Cabrera 1958; Goodwin 1969; J. K. Jones and Carter 1976). I am treating the name as a synonym of *A. lituratus palmarum* only because the problem of identity is not completely resolved. W. B. Davis (1984:10) considered individuals from the zone sympatry with *A. lituratus* to be "probable hybrids" of the two.

Artibeus obscurus (Schinz, 1821)
Black Artibeus

SYNONYMS:

Phyllost[oma]. obscurum Schinz, 1821:164; type locality "Ostküste von Brasilien"; identified by Wied-Neuwied (1826:206) as "Villa Viçosa [= Marobá; according to Bokermann, 1957:223] am Flusse Peruhype," Bahia, Brazil.

?*Arctibeus fuliginosus* Gray, 1838b:487; type locality "S. America."

Artibeus davisi Patten, 1971:12; *nomen nudum*.

A[rtibeus]. jamaicensis: Koopman, 1982:275; part; not *Artibeus jamaicensis* Leach, 1823.

A[rtibeus]. j[amaicensis]. fuliginosus: S. Anderson, Koopman, and Creighton 1982:9; name combination.

A[rtibeus]. obscurus: Handley, 1990:449; first use of current name combination.

Artibeus [(Artibeus)] obscurus: Marques-Aguiar, 1994:26; name combination.

DISTRIBUTION: *Artibeus obscurus* occurs from Venezuela (south of the Orinoco), south and east through the Guianas and Brazil, and west to the eastern lowlands of Colombia, Ecuador, Peru, and Bolivia.

MARGINAL LOCALITIES (Map 166; from Handley 1990, except as noted): VENEZUELA (Linares and Rivas 2004): Monagas, Caño Colorado; Delta Amacuro, Río Ibaruma. GUYANA: Demerara-Mahaica, Georgetown. FRENCH GUIANA: St. Elie (Brosset and Charles-Dominique 1991). BRAZIL: Amapá, Horto Florestal de Macapá (Peracchi, Raimundo, and Tannure 1984); Pará, Igarapé Assu; Piauí, Km 18 on Route BR 316, S of Teresina (Mares et al. 1981); Ceará, Parque Nacional de Ubajara (Guedes et al. 2000); Bahia, Região de Boa Nova (Falcão, Soares-Santos, and Drummond 2005); Bahia, Villa Viçosa (type locality of *Phyllostoma obscurum* Schinz); Rio de Janeiro, Parque Estadual da Pedra Branca (Dias, Peracchi, and Silva 2002); São Paulo, Guaratuba; São Paulo, Cananéia; Paraná, Fazenda Monte Alegre (N. R. Reis, Peracchi, and Sekiama 1999); Paraná, Pôrto Belo (Cherem et al. 2004); Mato Grosso, Serra do Roncador; Rondônia, Ji-Paraná (MPEG 17077). BOLIVIA (S. Anderson, Koopman, and Creighton 1982): Santa Cruz, 7 km N of Santa Rosa; La Paz, 4 km by road NW of Alcoche. PERU: Madre de Dios, Hacienda Erika (Ascorra, Wilson, and Romo

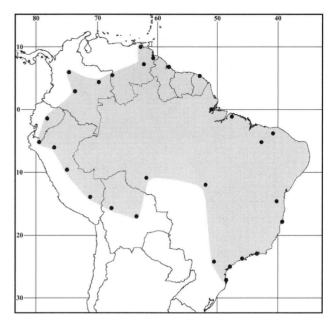

Map 166 Marginal localities for *Artibeus obscurus* ●

1991); Huánuco, Panguana (Koepcke and Kraft 1984); San Martín, Moyobamba; Piura, Huancabamba (Koopman 1978). ECUADOR: Pastaza, Mera. COLOMBIA: Meta, Río Guapaya, La Macarena; Antioquia, Vereda Río Claro (Muñoz 1986); Vichada, Santa Teresita. VENEZUELA: Amazonas, Paría; Bolívar, Los Patos (Handley 1976).

SUBSPECIES: I am treating *A. obscurus* as monotypic.

NATURAL HISTORY: *Artibeus obscurus* is found mainly in tropical humid forests and secondary habitats including palm groves, orchards, croplands, pastures, and yards in the greater Amazon basin (Handley 1990). However, Bernard and Fenton (2002) found 33% of this species also in savanna areas in central Amazonia. The Black Artibeus appears to be irregularly distributed in the arid Northeast of Brazil where Mares et al. (1981) reported the species in palm groves, but not from caatinga and cerrado habitats. The elevational range is sea level to 1,350 m (Bolivia); although in Venezuela, 97% of specimens captured came from below 500 m (Handley 1976, 1990; Koopman 1978; Mares et al. 1981).

Little is known about the diet. A. L. Gardner (pers. comm.) caught individuals in Colombia and Peru carrying figs (*Ficus* spp.) as did C. O. Handley, Jr. (pers. comm.) at Belém, Pará, Brazil. Brosset and Charles-Dominique (1991) reported finding a complete toothrow of an *A. obscurus* in the nest of the falcon *Falco deiroleucus* in French Guiana. The sparse information on reproduction suggests a pattern of bimodal polyestry. Pregnant females have been collected from February to May in Venezuela and Brazil (Pará), and from July to November in Peru, Venezuela, and Brazil (Mato Grosso); a lactating and pregnant female was taken in February in Ecuador (based on reproductive data from C. O. Handley, Jr., pers. comm., and the USNM collection). The karyotype is $2n = 30-31$, FN $= 56$, with a Y_1, Y_2 chromosome system (X-autosome fusion); the X is a subtelocentric and Y_1 and Y_2 are both acrocentric chromosomes (Gardner 1977a).

Antibodies to the Catu virus have been isolated from Brazilian *A. obscurus* (Woodall in Constantine 1970). Ectoparasites include four species of mites (Macronyssidae and Spinturnicidae) and at least ten species of streblid batflies (Herrin and Tipton 1975; Saunders 1975; Wenzel 1976; R. Guerrero 1997; Chavez 1998). Ueshima (1972:17) recorded *Hesperoctenes* sp. (Hemiptera: Polyctenidae) in Venezuelan *A. obscurus* (which they identified as *A fuliginosus*), but cautioned that the record "should be regarded as a possible contamination." Haynes and Lee (2004) provided additional natural history information in their *Mammalian Species* account.

REMARKS: Patten (1971) considered *Phyllostoma obscurum* Schinz and *Arctibeus fuliginosus* Gray as *nomina dubia* because the types were not known to exist. Cited as "in press," Patten (1971:12) redescribed the Black Artibeus in his Ph.D. dissertation under the name *Artibeus davisi*. Nevertheless, *Artibeus davisi* is a *nomen nudum* because the name was never published.

Handley (1976) began using *Artibeus fuliginosus* Gray, for the Black Artibeus and other authors followed (e.g., W. B. Davis and Dixon 1976:748; Koopman 1978:14; Taddei and Reis 1980:365; Koepcke and Kraft 1984:76). Koopman (1982:275) treated *A. fuliginosus* as a synonym of *A. jamaicensis*, but later (Koopman 1993) recognized the name as a synonym of *A. obscurus*. Handley (1990:447), like Patten (1971:24), regarded *Artibeus fuliginosus* Gray as a *nomen dubium*. Handley (1990:448) resurrected *Phyllostoma obscurum* Schinz after concluding that a specimen (AMNH 1334) of the Black Artibeus in the Wied-Neuwied collection acquired by the American Museum of Natural History "...must be the holotype of *Phyllostoma obscurum* Schinz."

Artibeus planirostris (Spix, 1823)
Spix's Artibeus
SYNONYMS: The following probable synonym (Handley 1990) can not be assigned to subspecies. See under Subspecies for additional synonyms.
? *Arctibeus lobatus* Gray 1838b:487; type locality unknown.

DISTRIBUTION: *Artibeus planirostris* occurs in southeastern Colombia; Venezuela, mainly south of the Orinoco River; the Guianas; Brazil; eastern Ecuador, Peru, and Bolivia; and Paraguay and northern Argentina. Elsewhere, it is known from the island of St. Vincent, West Indies (Pumo et al. 1996).

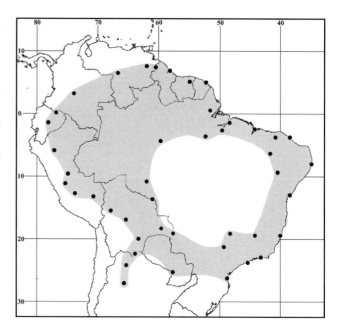

Map 167 Marginal localities for *Artibeus planirostris* ●

MARGINAL LOCALITIES (Map 167, listed as peripheral and internal marginal localities). *Peripheral localities*: VENEZUELA: Bolívar, 5 km NNW of Guasipati (Handley 1976). GUYANA: Barima-Waini, Baramita (Lim 1997); Demerara-Mahaica, Demerara (Andersen 1908b). SURINAM: Brokopondo, Brokopondo (Phillips, Nagato, and Tandler 1987). FRENCH GUIANA: Cayenne (type locality of *Uroderma validum* Elliot). BRAZIL: Amapá, Santa Luzia do Pacuí (Peracchi, Raimundo, and Tannure 1984); Pará, Pará (Andersen 1908b); Maranhão, São Luis (Handley 1991); Ceará, Parque Nacional de Ubajara (Guedes et al. 2000); Ceará, Faculdade de Veterinária do Ceará, Sítio do Itaperí, Fortaleza (Piccinini 1971); Pernambuco, Pernambuco (Handley 1991); Bahia, Salvador (type locality of *Phyllostoma planirostre* Spix); Espírito Santo, Estação Experimental de Linhares (Peracchi and Albuquerque 1993); Rio de Janeiro, Rio de Janeiro (Handley 1991); São Paulo, São Sebastião (USNM 141389); Santa Catarina, Joinville (FMNH 34389). PARAGUAY: Central, Asunción (Myers and Wetzel 1979). ARGENTINA: Salta, 6 km W of Piquirenda Viejo (Barquez, Mares, and Braun 1999); Tucumán, Playa Larga (Barquez, Mares, and Braun 1999); Jujuy, Finca La Carolina (Villa-R. and Villa-Cornejo 1971, as *A. lituratus*). BOLIVIA (S. Anderson, Koopman, and Creighton 1982): Santa Cruz, 72 km ESE of Monteagudo; Cochabamba, San Antonio; La Paz, 5 km by road SE of Guanay. PERU: Cusco, Quincemil (FMNH 84404); Ayacucho, Huahuachayo (Koopman 1978); Junín, Chanchamayo Valley (Koepcke and Kraft 1984); Huánuco, Panguana (Koepcke and Kraft 1984); San Martín, Yurac Yacu (BM 27.1.1.54). ECUADOR: Pastaza, Mera (Rageot and Albuja 1994); Sucumbíos, Lago Agrio (Albuja 1999). COLOMBIA:

Meta, Reserva National Natural La Macarena (sector norte) (Sánchez-Palomino, Rivas-Pava, and Cadena 1996). VENEZUELA, Bolívar, Mina de bauxita de la Serranía de los Pijiguaos (Ochoa et al. 1988). *Internal marginal localities*: BRAZIL: Pará, Baião (AMNH 97043); Piauí, Fazenda Olho da Agua (Mares et al. 1981); Bahia, Fazenda Barrinha (Mares et al 1981); Minas Gerais, 3 miles ESE of Sete Lagoas (USNM 391090); Minas Gerais, Panga Ecological Reserve (Pedro and Taddei 1997); São Paulo, Irapuã (Taddei 1979); Mato Grosso do Sul, Urucúm (AMNH 36992). BOLIVIA: Santa Cruz, Santiago (S. Anderson, Koopman, and Creighton 1982); Santa Cruz, Flor de Oro (Emmons 1998). BRAZIL: Rondônia, Ji-Paraná (Marques 1989); Amazonas, Igarapé Auará (AMNH 91881); Pará, Rio Xingu, 52 km SSW of Altamira (USNM 549474).

SUBSPECIES: I recognize three subspecies of *A. planirostris*.

A. p. fallax (W. Peters, 1865)

SYNONYMS:

Artibeus fallax W. Peters, 1865b:355; type localities "Guiana . . . Surinam"; subsequently restricted to Cayenne, French Guiana, by Cabrera (1958:89) whose restriction was invalidated when Husson (1962:175) designated the specimen from Surinam in the Leiden Museum as the lectotype, thereby fixing the type locality as Surinam.

Uroderma validum Elliot, 1907:537; type localities "Cayenne, French Guiana, South America."

Artibeus planirostris fallax: Andersen, 1908b:242; first use of current name combination.

A[*rtibeus*]. *l*[*ituratus*]. *fallax*: Hershkovitz, 1949c:447; name combination.

A[*rtibeus*]. *j*[*amaicensis*]. *fallax*: Handley, 1987:164; name combination.

This subspecies is distributed in Venezuela (south and east of the Orinoco), Guyana, Surinam, French Guiana, and the lower Amazon basin of Brazil.

A. p. hercules (Rehn, 1902)

SYNONYMS:

Artibeus hercules Rehn, 1902b:638; type locality "Eastern Peru."

[*Artibeus* (*Uroderma*)] *hercules*: Trouessart, 1904:116; name combination.

Artibeus lituratus hercules: Cabrera, 1958:89; name combination.

Artibeus hercules literatus Tamsitt and Valdivieso, 1963b: 263; name combination, and incorrect subsequent spelling of *lituratus* Olfers, 1818.

A[*rtibeus*]. *planirostris*: Koopman, 1982:275; not *Phyllostoma planirostre* Spix.

A[*rtibeus*]. *j*[*amaicensis*]. *hercules*: Handley, 1987:164; name combination.

This subspecies occurs in southeastern Colombia and the eastern lowlands of Ecuador, Peru, and Bolivia.

A. p. planirostris (Spix, 1823)

SYNONYMS:

Phyllostoma planirostre Spix, 1823:66; type locality "suburbiis Bahiae"; identified as Salvador, Bahia, Brazil, by Carvalho (1965:61).

Arctibeus planirostris: Gray, 1838b:487; part; name combination.

[*Artibeus* (*Uroderma*)] *planirostre*: Trouessart, 1897:159; name combination.

Uroderma planirostris: Bangs, 1900:101; name combination.

A[*rtibeus*]. *j*[*amaicensis*]. *planirostris*: Hershkovitz, 1949c: 447; name combination.

This subspecies is in southern Bolivia, northern Argentina, Paraguay, and eastern and southern Brazil.

NATURAL HISTORY: Husson (1962) reported (as *A. lituratus fallax*) a cluster of 16 roosting under a frond of a coconut palm. One female from this cluster was pregnant when taken on 4 November. Females have been found pregnant in February in southeastern Colombia; in January, February, and August in Brazil; in February, March, and April in southern Venezuela; in March in Guyana; in November in Ecuador; and in September and November in Peru (information on specimen labels, USNM). Seven *A. planirostris* were pregnant (fetuses measured 7–28 mm CR) when caught between 12 and 22 September, 1988 in the vicinity of Pakitza, southern Peru. However, in central Peru eleven females were pregnant or lactating (fetuses measured 5–25 mm), and 17 were immature (wing epiphyses open), when taken between 7 and 25 November 1972. Willig (1985a) found pregnant females in most months of the year in caatinga and cerrado habitats in the Northeast of Brazil. He often found females both pregnant and lactating, which indicates postpartum estrus and seasonal polyestry. Marinho-Filho (2003) recorded pregnant females taken in October in eastern São Paulo, and Taddei (1976) stated that *A. planirostris* did not have a well-defined annual reproductive period in northwestern São Paulo, Brazil. Willig (1983) found evidence of sexual dimorphism with males larger than females in *A. planirostris* from the Northeast of Brazil.

Food habits likely are similar to those of *A. jamaicensis*, which eats the fruits of canopy trees and, less commonly, flowers, leaves, and insects. Willig (1983) found the fruits of *Vismia* sp. to be the predominant food consumed in Ceará, Brazil. However, figs (*Ficus* spp.) of several species are probably the most important food consumed by *A. planirostris* in terms of quantity and because of their general year-round availability (e.g., N. R. Reis, Peracchi, and Onuki 1993; N. R. Reis et al. 1996). Piccinini (1971) found *A. planirostris*, captured during October in Ceará, Brazil, with their fur stained yellow from the pollen of *Anacardium occidentale*.

Antibodies to six arboviruses have been found in Brazilian *A. planirostris*; three (Catu, Eastern equine encephalitis, and Mucambo virus) can infect humans (Woodall in Constantine 1970). Other infectious agents in the Amazonian region include yeasts and yeast-like fungi (Mok, Luizão, and Silva 1982); protozoans (*Trypanosoma cruzi*-like flagellates), and nematodes (*Litomosoides chandleri* and *L. colombiensis*; see Ubelaker, Specian, and Duszynski 1977). The combined reports of R. Guerrero (1996) and Chavez (1998) listed nine species of streblid batflies found on Peruvian *A. planirostris*. Distributional records suggest that this species is absent from large areas of Cerrado habitat in southcentral Brazil (see Map 167). Hollis (2005) provided additional natural history information in her *Mammalian Species* account.

REMARKS: *Artibeus planirostris* shows high geographic variation in morphology. Many authors have considered *A. p. fallax* and *A. p. hercules* to be synonyms of *A. lituratus* because of their relatively larger size. However, these taxa differ from *A. lituratus* by having weakly defined facial stripes, a practically naked interfemoral membrane, sparsely furred forearm, 3/3 molars, weakly developed postorbital processes, and an arched rostrum. Based on these differences, Simmons and Voss (1998) likely are correct in their suspicion that Husson's (1962, 1978) sample from Surinam, reported as *A. lituratus fallax*, is a composite of *A. planirostris* and *A. lituratus*. Simmons and Voss (1998) reported 71 ground-level captures versus 2 captures in the forest canopy in French Guiana, echoing Handley's (1967) results in the vicinity of Belém, Brazil, where *A. planirostris* was most commonly caught near the forest floor.

There are a few published records for *A. planirostris* (most reported as *A. jamaicensis*) from central Brazil: Fazenda Moinho, Goiás (Coimbra et al. 1982) and the Distrito Federal (Gettinger and Gribel 1989; M. G. R. Rodrigues, Bredt, and Uieda 1994; Bredt and Uieda 1996). However, I have not mapped those records because *A. obscurus* also occurs in this area and these authors may have followed Koopman's (1982, 1994) treatment of *A. obscurus* (or *A. fuliginosus*) as a synonym of *A. jamaicensis*. Nevertheless, the gap in the distribution of Brazilian and eastern Bolivian *A. planirostris* may be more apparent than real and result from the inability to distinguish among the species of larger *Artibeus* found in that region. Other records of *A. planirostris* (usually reported as *A. jamaicensis*) from eastern Brazil are also suspect. For example, McNab and Morrison (1963:22) described the bats they identified as *A. jamaicensis* from São Salvador, Bahia, as "smaller and darker" than *A. lituratus*, a description that better fits *A. obscurus*. Hollis's (2005:3) distribution map does not indicate a hiatus in the distribution in Brazil.

Kraft (1982:315) designated Zoologische Staatssammlung, Munich (ZSM), specimen ([66] 1903/9438) as the lectotype of *Phyllostoma planirostre* Spix, because he considered the corresponding one 1903/9437 as possibly juvenile or subadult. He designated specimen no. 65 (1903/9437) as the paralectotype. However, the lectotype of *Phyllostoma planirostre* Spix had already been designated by Andersen (1908b:239) as specimen no. 65 (1903/9437). Handley (1991:14) pointed out that Kraft's designation was invalid because of the earlier selection by Andersen (1908b). In addition, ZSM specimen no. 66 (1903/9438) actually represents the species *A. fimbriatus* Gray, not *A. planirostris* (Spix) as understood today. D. C. Carter and Dolan (1978) provided information and measurements on types of *Artibeus fallax* W. Peters and *Phyllostoma planirostre* Spix.

Subgenus *Dermanura* P. Gervais, 1856

Dermanura is a subgenus of small (forearm 34.0–40.5 mm) *Artibeus* comprising at least eight species, six of which occur in South America. Coloration ranges from gray-brown to blackish brown with underparts usually paler than the dorsum. Facial stripes vary from indistinct to prominent, and the dorsal fur is either unicolored or has two color bands. The hypocone of M1 is weakly developed or absent. Characters shared by one or more species in other subgenera of *Artibeus* include a variably broad and (in some species) hairy interfemoral membrane; plagiopatagium attaching either at the ankle or on the side of the foot; a noseleaf that is longer than wide; a tragus that lacks a subterminal projection; bilobed inner upper incisors; a reduced or missing m3; and small to moderately developed paraoccipital processes. The dental formula is 2/2 1/1 2/2 2/2–3 × 2 = 28–30.

Artibeus anderseni Osgood, 1916
Andersen's Fruit-eating Bat
SYNONYMS:

Artibeus anderseni Osgood, 1916:212, type locality "Porto Velho," Rondônia, Brazil.

Artibeus cinereus anderseni: Hershkovitz, 1949c:449; name combination.

Dermanura anderseni: R. D. Owen, 1987:47; name combination.

DISTRIBUTION: *Artibeus anderseni* is in the western Amazon basin where it is recorded from eastern Colombia, western and northwestern Brazil, and the eastern lowlands of Ecuador, Peru, and Bolivia.

MARGINAL LOCALITIES (Map 168; from Handley 1987, except as noted): COLOMBIA: Córdoba, Catival; Antioquia, Aljibos. PERU: Loreto, Boca del Río Curaray. COLOMBIA: Amazonas, 7 km N of Leticia (ROM

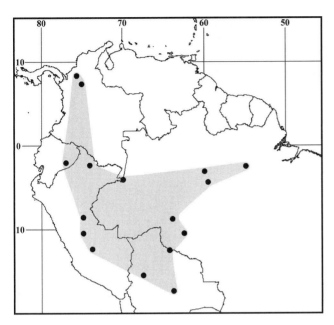

Map 168 Marginal localities for *Artibeus andersoni* ●

63061). BRAZIL: Amazonas, Lago Janauacá (USNM 531087); Pará, Alter do Chão (Bernard and Fenton, 2002); Amazonas, Borba; Rondônia, Pôrto Velho (type locality of *Artibeus anderseni* Osgood); Rondônia, Ouro Preto D'Oeste (Marques 1989). BOLIVIA: Beni, Río Iténez, 4 km above Costa Marquez, Brazil (S. Anderson, Koopman, and Creighton 1982); Santa Cruz, Buena Vista (Sanborn 1932b); La Paz, Tomonoco (Webster and Jones 1980a). PERU: Cusco, Río Mapitunari (Koopman 1978); Pasco, San Juan; Ucayali, 59 km W of Pucallpa. ECUADOR: Pastaza, Montalvo.

SUBSPECIES: I consider *A. anderseni* to be monotypic.

NATURAL HISTORY: *Artibeus anderseni* occurs from near sea level to 1,350 m (Serranía Bella Vista, Bolivia; S. Anderson, Koopman, and Creighton 1982). In central Amazonia, Brazil, Bernard and Fenton (2002) found this species in forest fragments, primary forests, and savannas. In Ecuador, *A. anderseni* constructs tent-like roosts by modifying the leaves of several species of *Heliconia* (Musaceae; Timm 1987). The diet of *A. anderseni* is unknown, but presumably includes a variety of fruits. Although few data are available, bimodal polyestry is the probable reproductive pattern. Pregnant or lactating females have been found from March to April in Colombia and Brazil (Acre), and from September to November in Bolivia and Ecuador (Timm 1987; Taddei, Rezende, and Camora 1990; USNM collection). R. Guerrero (1997) listed two species of streblid flies as ectoparasites. No further information is available on parasites and diseases, or on the karyotype.

REMARKS: The much wider distribution indicated in the literature (e.g., Honacki, Kinman, and Koeppl 1982; Koopman 1982; Nowak 1991) for *A. anderseni* resulted from confusion between this species and the recently described *Artibeus gnomus* Handley, 1987.

Artibeus cinereus (P. Gervais, 1856)
Gervais's Fruit-eating Bat

SYNONYMS: See under Subspecies.

DISTRIBUTION: *Artibeus cinereus* is recorded from southeastern Venezuela, the Guianas, Brazil, and eastern Peru.

MARGINAL LOCALITIES (Map 169); from Handley 1987, except as noted): *Eastern distribution.* VENEZUELA: Monagas, Isla Tigre (Linares and Rivas 2004). GUYANA: Barima-Waini, Cart Market (ROM 67470). SURINAM: Sipaliwini, Bitagron (CMNH 63786). FRENCH GUIANA: Kaw (Brosset and Charles-Dominique 1991). BRAZIL: Amapá, Horto Florestal de Macapá (Mok and Lacey 1980); Pará, Belém (type locality of *Dermanura cinerea* P. Gervais); Maranhão, Juryassú; Piauí, Km 18 on Route BR 316, S of Teresina (Mares et al. 1981); Ceará, Faculdade de Veterinária do Ceará (Piccinini 1971); Rio Grande do Norte, Natal; Pernambuco, Pernambuco; Alagoas, Mangabeiras (C. O. C. Vieira 1953); Bahia, Una (Faria, Soares-Santos, and Sampaio 2006); São Paulo, Guaratuba (Varjão) (USNM 542617); Minas Gerais, Panga Ecological Reserve (Komeno and Linhares 1999); Distrito Federal, 20 km S of Brasília (Mares, Braun, and Gettinger 1989); Pará, E bank Rio Iriri, 85 km SW of Altamira (USNM 549438); Pará, Fordlândia; Amazonas,

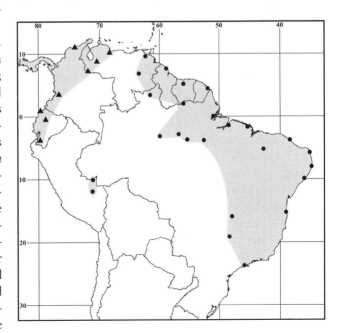

Map 169 Marginal localities for *Artibeus cinereus* ● and *Artibeus phaeotis* ▲

Santa Clara; Amazonas, Manaus (Mok and Lacey 1980). SURINAM: Sipaliwini, 1 km S and 3.5 km E of Sipaliwini Airstrip (CMNH 63785). BRAZIL: Roraima, Ilha de Maracá (USNM 531083). VENEZUELA: Bolívar, Hato San José. *Western distribution.* PERU: Ucayali, Balta (LSUMZ 12190); Madre de Dios, Pakitza (Ascorra, Wilson, and Romo 1991).

SUBSPECIES: I recognize two subspecies of *A. cinereus.*

A. c. cinereus (P. Gervais, 1856)

SYNONYMS:

Dermanura cinerea P. Gervais, 1856a:36; type locality: "Brésil"; restricted to Pará by Cabrera (1958:87); further restricted to Belém, Pará, Brazil, by Honacki, Kinman, and Koeppl (1982:151).

[*Artibeus (Dermanura)*] *cinereum*: Trouessart, 1897:160; name combination.

Artibeus cinereus: O. Thomas, 1901g:143; name combination.

Artibeus cinereus solimoesi Pirlot, 1972:73; type locality "Codajas," Amazonas, Brazil.

Artibeus cincereus L. A. M. Silva and Guerra, 2000:125; incorrect subsequent spelling of *Dermanura cinerea* P. Gervais, 1856.

The nominate subspecies is found in the lower Amazon basin and southward in eastern Brazil.

A. c. quadrivittatus (W. Peters, 1865)

SYNONYMS:

Artibeus (Dermanura) quadrivittatus W. Peters, 1865b: 358; type locality "Surinam."

Artibeus cinereus quadrivittatus: Handley, 1987:166; name combination.

This subspecies is in Venezuela (south of the Orinoco), the Guianas, and northern Brazil.

NATURAL HISTORY: Information in the literature attributed to *A. cinereus* should be treated with caution because, until Handley's (1987) review, two or more species were included under that name. In central Amazonia, Brazil, Bernard and Fenton (2002) found *A. cinereus* in forest fragments, primary forests, and savannas. The species also is present in remnant Atlantic Tropical Forest and in the palm forests of northeastern Brazil (Pernambuco and Piauí), but appears to be absent from the other habitats studied (Mares et al. 1981). The elevational range is below 350 m. The diet is predominantly fruits, although insects are consumed (Gardner 1977c). In Brazil (Pará), a pregnant female was taken in February, and two lactating females and a juvenile female were found in March (C. O. Handley, Jr., pers. comm.).

The karyotype is $2n = 30$–31, FN $= 56$, with a Y_1, Y_2 chromosome system (X-autosome fusion); the X chromosome is subtelocentric, Y_1 is submetacentric, and Y_2

metacentric (R. J. Baker 1973; R. J. Baker and Hsu 1970). Souza and Araújo (1990), and Noronha et al. (2001) reported a $2n = 30$ for males with neo-XY sex-chromosome system (translocation of autosomes on both the X and Y chromosomes).

Mok, Luizão, and Silva (1982) reported a fungus infecting *A. cinereus* in Amazonian Brazil. J. P. Webb and Loomis (1977) recorded two species of spinturnicid mites, and R. Guerrero (1997) listed five species of streblid batflies recorded from *A. cinereus*, one of which was also reported by Graciolli and Aguiar (2002).

REMARKS: The much wider distribution reported in the literature for *A. cinereus* (e.g., Andersen 1908b; Hershkovitz 1949c; Cabrera 1958; Walker et al. 1964; J. K. Jones and Phillips 1970; J. K. Jones and Carter 1976; Honacki, Kinman, and Koeppl 1982; Timm 1987) is based on a composite of *A. anderseni, A. cinereus, A. glaucus, A. gnomus,* and *A. watsoni.* Specimens from Trinidad and Tobago identified as *A. cinereus* are actually *A. glaucus.* Brosset and Charles-Dominique (1991:533) listed the species for French Guiana, but cautioned that "our specimens recorded as *cinereus* may belong to different species." Handley (1987) restricted *Artibeus cinereus* to include only the nominate subspecies and *A. c. quadrivittatus.* Future work may demonstrate that *A. cinereus,* as currently understood, is still composite. D. C. Carter and Dolan (1978) provided information and measurements on the holotype of *Artibeus (Dermanura) quadrivittatus* W. Peters.

Artibeus glaucus O. Thomas, 1893
Silvery Fruit-eating Bat

SYNONYMS: See under Subspecies.

DISTRIBUTION: *Artibeus glaucus* is in Venezuela, Surinam, Guyana, northern Brazil, Colombia, Ecuador, Bolivia, Peru, and Trinidad and Tobago.

MARGINAL LOCALITIES (Map 170; from Handley 1987, except as noted): TRINIDAD AND TOBAGO: Tobago, Speyside (AMNH 184738). VENEZUELA: Bolívar, Río Supamo. BRAZIL: Roraima, Paulo (AMNH 75537). SURINAM: Sipaliwini, Avanavero (CMNH 68419); Commewijne, Nieuwe Grond Plantation (CMNH 63779). GUYANA: Upper Takutu-Upper Essequibo, Kanuku Mountains. VENEZUELA: Amazonas, Cerro Neblina, Camp VII (Gardner 1988). COLOMBIA: Meta, Villavicencio (ROM 53609). ECUADOR: Sucumbios, Limoncocha (R. H. Baker 1974). PERU: Ucayali, Tushemo (O. Thomas 1924b); Madre de Dios, Cocha Cashu (Ascorra, Wilson, and Romo 1991); Puno, Santo Domingo (Koopman 1978). BOLIVIA: La Paz, Sararia (S. Anderson, Koopman, and Creighton 1982); Santa Cruz, Buena Vista. PERU: Cusco, Cordillera Vilcabamba (Koopman 1978); Junín, Huacapistana; Huánuco, Tingo María (Bowles, Cope, and Cope

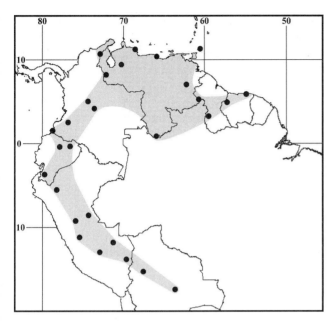

Map 170 Marginal localities for *Artibeus glaucus* •

1979); Amazonas, 12 km E by trail from La Peca (LSUMZ 20902). ECUADOR: El Oro, 1 km SW of Puente de Moromoro (USNM 513472); Napo, Baeza. COLOMBIA: Nariño, La Victoria (FMNH 114133); Cauca, Quisquio (Tamsitt and Valdivieso 1966a); Cundinamarca, Sasaima (Valdivieso 1964). VENEZUELA: Táchira, Las Mesas (USNM 440775). COLOMBIA: La Guajira, Sierra Negra (Hershkovitz 1949c). VENEZUELA: Trujillo, 13 km E of Trujillo (USNM 373841); Falcón, 5 km N and 13 km E of Mirimire (USNM 440773); Miranda, 5 km E of Río Chico (USNM 387363).

SUBSPECIES: I recognize two subspecies of *A. glaucus.*

A. g. bogotensis (Andersen, 1906)
SYNONYMS:

Artibeus quadrivittatus: Dobson, 1880b:465; not *Artibeus quadrivittatus* W. Peters.

Dermanura quadrivittata: Bangs, 1900:101; name combination, not *Artibeus quadrivittatus* W. Peters.

Artibeus cinereus bogotensis Andersen, 1906b:421; type locality "Curiche, [Cundinamarca,] near Bogota, Colombia" (see Remarks).

A[rtibeus]. glaucus bogotensis: Handley, 1987:168; name combination.

This subspecies is found in Venezuela, Surinam, Guyana, northern Brazil, Colombia, and Trinidad and Tobago.

A. g. glaucus O. Thomas, 1893
SYNONYMS:

Artibeus glaucus O. Thomas, 1893c:336; type locality "Chanchamayo," Junín, Peru.

[Artibeus (Artibeus)] glaucus: Trouessart, 1897:160; name combination.

Artibeus pumilio O. Thomas, 1924b:531; type locality "Tushemo, near Masisea, R. Ucayali [Peru.] Alt. 1000'."

Artibeus cinereus glaucus: Hershkovitz, 1949c:449; name combination.

Artibeus cinereus pumilio: Hershkovitz, 1949c:449; name combination.

Dermanura glauca: R. D. Owen, 1987:47; name combination.

A[rtibeus]. g[laucus]. glaucus: Handley, 1987:166; first use of current name combination.

The nominate subspecies is in Ecuador, Peru, and Bolivia.

NATURAL HISTORY: ALTHOUGH *A. glaucus* can be found in lowland habitats, the species is mainly found at intermediate elevations in the Andean and Guianan highlands at elevations up to 2,800 m (Monte Roraima, Brazil). Timm (1987) found this frugivore roosting under the cut leaves of *Xanthosoma* sp. (Araceae) in Ecuador. Reproductive data suggest a pattern of bimodal polyestry. Pregnant or lactating females have been found from January to August in Ecuador, Venezuela, and Trinidad and Tobago; from October to December in Ecuador and Venezuela; and a simultaneously lactating and pregnant female was taken in May in Venezuela (Albuja 1983; Wilson 1979; USNM collection).

Ectoparasites known from *A. glaucus* (as *A. cinereus* in reports by Tipton and Machado-Allison 1972; Wenzel 1976; as *A. glaucus* by R. Guerrero 1997) include two species of batflies (Streblidae) and a species of flea (Rhopalopsyllidae). The karyotype is $2n = 30–31$, FN $= 56$, with a Y_1, Y_2 chromosome system (X-autosome fusion); the X chromosome is subtelocentric and the Y_1 and Y_2 chromosomes are both acrocentrics (Gardner 1977a).

REMARKS: The type locality of *A. cinereus bogotensis* was given as "Curiche, near Bogota, Colombia" (Andersen 1906b). According to Tamsitt, Valdivieso, and Hernández-Camacho (1964:112), Curiche is a suburb of the small town of El Peñón, 75 km northwest of Bogotá and at a much lower elevation (1,311 m).

Handley (1987:166) commented that *Artibeus pumilio* O. Thomas "is an enigmatic taxon . . . perhaps the only specimen properly associated with the name is the holotype." He also suggested (pers. comm.) that *A. pumilio* may be a small representative of *A. glaucus*; therefore, I have provisionally included the taxon under *A. glaucus.*

Cabrera (1958:87) listed *Phyllostoma pusillum*, as used by Tschudi (1844b:63), in the synonymy of *A. cinereus pumilio*, apparently in accord with W. Peters's (1866c:396) determination that the specimen is a "*Dermanura quadrivittatum*," and not a *Vampyressa pusilla* J. A.

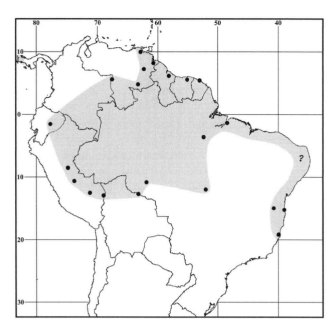

Map 171 Marginal localities for *Artibeus gnomus* ●

Wagner. Muñoz (2001:130) followed Koopman's (1993) concept of *A. glaucus* by treating *A. gnomus* and *A. watsoni* as synonyms. D. C. Carter and Dolan (1978) provided information and measurements on the types of *Artibeus cinereus bogotensis* Andersen, *Artibeus glaucus* O. Thomas, and *Artibeus pumilio* O. Thomas.

Artibeus gnomus Handley, 1987
Dwarf Fruit-eating Bat
SYNONYMS:
Artibeus gnomus Handley, 1987:167; type locality "El Manaco (=Km 74), 59 km SE El Dorado, Bolívar, Venezuela, 150m."
Dermanura gnoma: R. D. Owen, 1991:19; name combination.
A[rtibeus]. (D[ermanura].) gnomus: Marques-Aguiar, 1994:21; name combination.

DISTRIBUTION: *Artibeus gnomus* is in Venezuela (south of the Río Orinoco), the Guianas, Brazil, and east of the Andes in Ecuador, Peru, Bolivia, and probably southeastern Colombia.

MARGINAL LOCALITIES (Map 171; from Handley 1987, except as noted): VENEZUELA (Linares and Rivas 2004): Monagas, Caño Colorado; Delta Amacuro, Río Ibaruma. GUYANA: Upper Demerara-Berbice, 6 miles NE of Linden (ROM 68303). SURINAM: Para, Zanderij (CMNH 68426). FRENCH GUIANA: 3.5 km S and 10 km W of Sinnamary (S. L. Williams, Phillips, and Pumo 1990). BRAZIL: Pará, Utinga; Espírito Santo, Linhares Forest Reserve (Aguiar, Zortéa, and Taddei 1996); Bahia, Região de Conquista (Falcão, Soares-Santos, and Drummond 2005); Bahia, Una (Faria, Soares-Santos, and Sampaio 2006); Pará, Rio Xingu, 52 km SSW of Altamira (USNM 549473); Mato Grosso, Serra do Roncador; Rondônia, Ji-Paraná (MPEG 17126). BOLIVIA: Beni, Versalles (Koopman 1978). PERU: Madre de Dios, Pampas de Heath, ca. 50 km (by river) S of Puerto Pardo (LSUMZ 20901); Madre de Dios, Cerro de Pantiacolla (Timm 1987); Ucayali, Santa Rosa; Ucayali, 59 km SW of Pucallpa. ECUADOR: Pastaza, Canelos. VENEZUELA: Amazonas, 14 km SSE of Puerto Ayacucho; Bolívar, Salto Ichun; Bolívar, Los Patos.

SUBSPECIES: I am treating *A. gnomus* as monotypic.

NATURAL HISTORY: *Artibeus gnomus* is found in evergreen forest and in open areas such as savannas, yards, and orchards at elevations below 530 m (Handley 1987; Aguiar, Zortéa, and Taddei 1996; Bernard and Fenton 2002). The Dwarf Fruit Eating Bat is abundant in the forest of French Guiana (Brosset and Charles-Dominique 1991). Aguiar, Zortéa, and Taddei (1996) found the seeds of *Ficus* sp. in the stomach of the specimen from Espírito Santo, Brazil. Timm (1987) located this species roosting under a cut leaf of *Monstera lechleriana* (Araceae) in Peru.

Bimodal polyestry appears to be the reproductive pattern. Pregnant or lactating females have been found from January to April in Venezuela, French Guiana, and eastern Brazil (Pará); in June and July in Venezuela and Brazil (Mato Grosso); and in October and November in Venezuela and Peru (S. L. Williams, Phillips, and Pumo 1990; C. O. Handley, Jr., pers. comm.; USNM collection).

Ectoparasites recorded from *A. gnomus* (under the name *Artibeus pumilis*) include a species of laelapid mite and two species of streblid batflies (Furman 1972; Wenzel 1976). R. Guerrero (1997), however, listed only one species of streblid batfly for *A. gnomus*. No information is available on the karyotype or on endoparasites and diseases.

REMARKS: *Artibeus gnomus* is the "*Artibeus* sp. A" of Handley (1976). Handley (1987:170) commented on the peculiar circular distribution "completely ringing the Amazon basin but apparently not extending into its interior." However, specimens at the USNM confirm the presence of this species in the central Amazon basin at Manaus and Tefé, Amazonas, Brazil. Koopman (1993) and Muñoz (2001) treated *A. gnomus* as a synonym of *A. glaucus*.

Artibeus phaeotis (Miller, 1902)
Pygmy Fruit-eating Bat
SYNONYMS: The following synonyms refer to extralimital populations. See under Subspecies for additional synonyms.
Artibeus nanus Andersen, 1906b: 423; type locality "Tierra Colorada, Sierra Madre del Sur, Guerrero, Mexico."
Artibeus phaeotis nanus: W. B. Davis, 1970c:399; name combination.

Artibeus phaeotis palatinus W. B. Davis, 1970c:400; type locality "15 kilometers southwest of Retalhuleu, elevation 240 feet, Retalhuleu, Guatemala."

DISTRIBUTION: *Artibeus phaeotis* is recorded from northwestern South America (Handley 1987), including northwestern, and western Venezuela, Colombia, and western Ecuador. Elsewhere, it is known from Mexico and Central America. There is no evidence, *contra* Koopman (1982), Timm (1985), and Nowak (1991, 1999), that the species is in northern Peru.

MARGINAL LOCALITIES (Map 169): COLOMBIA: Magdalena, Hacienda El Recuerdo (ROM 79885). VENEZUELA: Carabobo, La Copa (USNM 440838); Barinas, La Vega del Río Santo Antonio (USNM 440845); Apure, 3 km N of Nula (Nulita) (USNM 440852). COLOMBIA: Valle del Cauca, Bellavista (ROM 64057). ECUADOR: Pichincha, Río Blanco (BM 34.9.10.56); El Oro, 1 km SW of Puente de Moromoro (USNM 513477); Esmeraldas, Esmeraldas (J. A. Allen 1916b).

SUBSPECIES: Only the nominate subspecies occurs in South America. W. B. Davis (1970c) and Hall (1981) recognized three subspecies (*A. phaeotis nanus, A. p. palatinus,* and *A. p. phaeotis*) in Mexico and Central America.

A. p. phaeotis (Miller, 1902)

SYNONYMS:

Dermanura phaeotis Miller, 1902b:405; type locality "Chichen Itza, Yucatan," Mexico.

Dermanura rava Miller, 1902b:404; type locality "San Javier, [Esmeraldas,] northern Ecuador."

[*Artibeus (Dermanura)*] *rava*: Trouessart, 1904:117; name combination.

[*Artibeus (Dermanura)*] *phaeotis*: Trouessart, 1904:117; name combination.

Artibeus turpis Andersen, 1906b:422; type locality "Teapa, Tabasco, S. Mexico."

Artibeus toltecus ravus: Andersen, 1908b:300; name combination.

Artibeus cinereus phaeotis: Hershkovitz, 1949c:449; name combination.

Artibeus phaeotis phaeotis: J. K. Jones and Lawlor, 1965:412; first modern use of name combination.

Dermanura phaeotis: R. D. Owen, 1987:47; name combination.

The South American distribution of this subspecies is north and west of the Andes in Colombia, Ecuador, and northwestern Venezuela. Elsewhere, *A. p. phaeotis* occurs along the Caribbean versant of Mexico and Central America (W. B. Davis 1970c; J. K. Jones and Carter 1976; Hall 1981; Timm 1985).

NATURAL HISTORY: *Artibeus phaeotis* is found in forested habitats at low elevations (W. B. Davis 1970c) up to at least 2,050 m in Colombia. Tuttle (1976) stated that this species roosts in caves and tunnels; Timm (1985, 1987) found *A. phaeotis* roosting under banana leaves (Musaceae) in Costa Rica. This species is frugivorous (Gardner 1977c; Albuja 1983, 1999); Timm (1985) also reported pollen and insects in its diet. Wilson (1979) suggested a reproductive cycle of bimodal polyestry. Pregnant females have been taken in July and from September to December in Colombia and Ecuador (USNM collection).

The karyotype is $2n = 30$, FN = 56, with a subtelocentric X and a submetacentric Y chromosome (R. J. Baker 1967). Koop and Baker (1983), based on electrophoretic data, suggested that *A. phaeotis* along with *A. cinereus, A. toltecus,* and *A. watsoni* shared a common ancestor and could represent recent speciation events.

Nematodes (*Cheiropteronema globocephala*) parasitize *A. phaeotis* in Mexico and Central America (Ubelaker, Specian, and Duszynski 1977). Timm's (1985) summary of ectoparasites in his *Mammalian Species* account on *A. phaeotis* listed four species of mites (Myobiidae, Spinturnicidae, Trombiculidae), and one species of streblid batfly. Timm (1985) credited R. Wenzel with the suggestion that the streblid *Neotrichobius stenopterus,* is a host-specific parasite of *A. phaeotis.*

REMARKS: Although Miller's (1902b) description of *A. ravus* appeared before the description of *A. phaeotis* and in the same publication, *A. phaeotis* is the name in current use, and should be retained as the senior synonym (see Timm 1985; Handley 1987). Muñoz's (2001:139) inclusion of *A. toltecus* (Saussure, 1860) in the Colombian bat fauna is difficult to understand. He may have included *A. toltecus* because he listed *Dermanura rava* Miller as a synonym, but in the next section under "Comentarios," Muñoz stated "no incluye a *Artibaeus ravus.*" I have seen no evidence confirming the presence of *A. toltecus* in South America, and I follow Koopman (1982, 1993) and Timm (1985) in treating *A. ravus* as a synonym of *A. phaeotis.* D. C. Carter and Dolan (1978) provided information and measurements on the types of *Artibeus turpis* Andersen and *Artibeus nanus* Andersen. Mantilla-Meluk and Baker (2006) used the name *Dermanura rava* for small *Artibeus* from the Calima River basin of western Colombia that represent the southern distribution of *A. phaeotis* as interpreted herein. The small *Artibeus* of western South America need to be revised.

Artibeus watsoni O. Thomas, 1901
Watson's Fruit-eating Bat

SYNONYMS:

Artibeus (Dermanura?) Rosenbergi O. Thomas, 1897b:545; type locality "Cachavi, [Esmeraldas,] N. Ecuador."

Artibeus Watsoni O. Thomas, 1901f:542; type locality "Bogava [=Bugaba,] Chiriqui, Panama. Altitude 250 m."

[*Artibeus (Artibeus)*] *watsoni*: Trouessart, 1904:116; name combination.

[*Artibeus (Dermanura)*] *rosenbergi*: Trouessart, 1904:117; name combination.

Dermanura jucundum Elliot, 1906:50; type locality "Achotal, State of Vera Cruz, Mexico."

Artibeus cinereus rosenbergi: Hershkovitz, 1949c:449; name combination.

Artibeus cinereus watsoni: Hershkovitz, 1949c:449; name combination.

Artibeus cinereus rosenbargi Cabrera, 1958:88; incorrect subsequent spelling of *Artibeus rosenbergi* O. Thomas.

Dermanura watsoni: R. D. Owen, 1987:47; name combination.

[*Artibeus glaucus*] *rosenbergii* Koopman, 1993:188; name combination and incorrect subsequent spelling of *Artibeus rosenbergi* O. Thomas.

A[*rtibeus*]. (D[*ermanura*].) *watsoni*: Marques-Aguiar, 1994:21; first use of current name combination.

A[*rtibeus*]. *rosembergi* Muñoz, 2001:130; incorrect subsequent spelling of *Artibeus rosenbergi* O. Thomas.

DISTRIBUTION: *Artibeus watsoni* is in western Ecuador and Colombia; elsewhere, it occurs northward through Central America into southern Mexico.

MARGINAL LOCALITIES (Map 172): COLOMBIA: Antioquia, Amalfi (BM 12.2.12.1); Chocó, Andagoya (BM 15.10.5.5); Valle del Cauca, La Bocana (ROM 75234); Nariño, Buenavista (AMNH 34243). ECUADOR: Esmeraldas, Alto Tambo (EPN 84.157); Loja, Valle (EPN 85.944).

SUBSPECIES: I treat *A. watsoni* as monotypic, pending revision of the species.

NATURAL HISTORY: *Artibeus watsoni* is a frugivore (Gardner 1977c) found in humid lowland and midelevation habitats (W. B. Davis 1970c), as high as 1,500 m (Colombia). This species commonly modifies leaves to

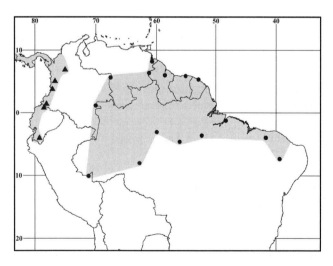

Map 172 Marginal localities for *Artibeus watsoni* ● and *Artibeus concolor* ▲

make tents for diurnal roosts. Choe and Timm (1985) and Timm (1987) found the leaves of 19 species of plants used for tents by Costa Rican *A. watsoni*, which were especially abundant in the secondary forest and coastal-strand vegetation communities at Sirenia, Costa Rica. Bimodal polyestry appears to be the reproductive pattern (J. K. Jones 1966; W. B. Davis 1970c; J. K. Jones, Smith, and Turner 1971; T. H. Fleming, Hooper, and Wilson 1972). I found only two records of pregnant females taken in South America, one in November and the other in January, both from Colombia (USNM collection).

The only ectoparasite reported from *A. watsoni* is *Paratrichobius lowei* (Diptera: Streblidae) from Panama (Wenzel, Tipton, and Kiewlicz 1966). Squirrel monkeys (*Saimiri oerstedi*) and double-toothed kites (*Harpagus bidentatus*) appear to be the major predators in Costa Rica and are a significant cause of mortality (Boinski and Timm 1985). The karyotype is $2n = 30$, FN = 56, the X chromosome is subtelocentric, and the Y is submetacentric (R. J. Baker 1973).

REMARKS: C. O. Handley, Jr. (pers. comm.) considered *A. rosenbergi* O. Thomas, 1897b, to be undeterminable. Koopman (1993) and Muñoz (2001) considered *A. rosenbergi* to be a synonym of *A. glaucus*; however, Koopman (1994) treated *A. rosenbergi* as a subspecies of *A. cinereus*. D. C. Carter and Dolan (1978) provided information and measurements on the types of *Artibeus (Dermanura) rosenbergi* O. Thomas and *Artibeus watsoni* O. Thomas.

Subgenus *Koopmania* R. D. Owen, 1991

Koopmania is a monotypic subgenus of medium-sized (forearm 43–52 mm) *Artibeus*. Dorsal coloration is gray-brown to brown with individual hairs tricolored, but imparting a uniform color because of the broad terminal brown band. Shoulders and venter are paler than the dorsum, and facial stripes are lacking. The plagiopatagium attaches to leg at the base of the toes. The rostrum is short and broad, and has a rostral shield. The paraoccipital processes are indistinct or absent. The hypocone of M1 is small to moderately developed. Characters shared with some species representing other subgenera of *Artibeus* include a relatively broad and hairy interfemoral membrane; a longer than wide noseleaf; the lack of a pointed, subapical projection on the tragus; a small M3; and bifid upper inner incisors. The dental formula is 2/2, 1/1, 2/2, 3/3 × 2 = 32.

Artibeus concolor W. Peters, 1865
Brown Fruit-eating Bat

SYNONYMS:

Artibeus concolor W. Peters, 1865b:357; type locality "Paramaribo (Surinam)."

[*Artibeus planirostris*] Var. a. Dobson, 1878:518.

[*Artibeus (Uroderma) planirostre*] *concolor*: Trouessart, 1897:159; name combination.

Dermanura concolor: R. D. Owen, 1987:47; name combination.

Koopmania concolor: R. D. Owen, 1991:21; generic description and name combination.

Artibeus [(Koopmania)] concolor: Koopman, 1993:188; first use of current name combination.

DISTRIBUTION: *Artibeus concolor* is in Peru, Colombia, Venezuela, Guyana, Surinam, French Guiana, and northern Brazil.

MARGINAL LOCALITIES (Map 172): VENEZUELA: Delta Amacuro, Río Ibaruma (Linares and Rivas 2004). GUYANA: Cuyuni-Mazaruni, 24 miles along the Potaro Road from Bartica (Hill 1965). SURINAM: Paramaribo, Paramaribo (type locality of *Artibeus concolor* W. Peters). FRENCH GUIANA: Paracou (Simmons and Voss 1998). BRAZIL: Pará, Ilha Caratateua (Piccinini 1974); Piauí, Piracuruca (I. Sazima and Ueda 1978); Ceará, Aeroporto de Crato (Mares et al. 1981); Pará, Rio Xingu, 52 km SSW of Altamira (USNM 549440); Pará, Uruá (N. R. Reis and Schubart 1979); Amazonas, Manaus (Mok and Lacey 1980); Rondônia, Calama (J. A. Allen 1916e). PERU: Ucayali, Balta (Gardner 1976). COLOMBIA: Vaupés, El Internado de María Reina, Mitú (Barriga-Bonilla 1965). VENEZUELA: Amazonas, Raudal de Atures (Linares 1969); Bolívar, El Manaco, 59 km SE of El Dorado (Handley 1976).

SUBSPECIES: I consider *A. concolor* to be monotypic.

NATURAL HISTORY: *Artibeus concolor* is a frugivore (I. Sazima and Ueda 1978; N. R. Reis and Guillaumet 1983) that also consumes pollen, nectar, and other flower parts. The species can be found in a variety of forest, cerrado, and secondary habitats such as pastures, orchards, and clearing. However, *A. concolor* appears to be absent from seasonally arid habitats such as savanna and caatinga (Handley 1976; Mares et al. 1981; N. R. Reis and Peracchi 1987; Eisenberg 1989; Bernard and Fenton 2002). The elevational range is from sea level to at least 1,030 m (Bolívar, Venezuela).

Pregnant females have been found in January, April, May, and July in Venezuela; in January in eastern Brazil (Piauí); in September and November in the vicinity of Manaus, Brazil; and in February in Colombia (M. E. Thomas 1972; I. Sazima and Ueda 1978; N. R. Reis and Peracchi 1987; USNM collection). Willig (1983) stated that he was able to distinguish sexual dimorphism in size in a series of six *A. concolor* from the state of Ceará, Brazil.

The karyotype is $2n = 30–31$, $FN = 56$, with a Y_1, Y_2 chromosome (X-autosome fusion). The X chromosome is subtelocentric and the Y_1 and Y_2 are both acrocentrics (R.

J. Baker, Genoways, and Seyfarth 1981; R. J. Baker et al. 1982).

Based on the reports of Saunders (1975), J. P. Webb and Loomis (1977), and R. Guerrero (1997), the only ectoparasites known to infest *A. concolor* are three species of spinturnicid mites and one species of streblid batfly. No data are available on endoparasites or other pathogens. Acosta and Owen (1993) provided a detailed description along with measurements and a summary of available information on natural history in their *Mammalian Species* account on *Koopmania concolor*.

REMARKS: Koopman (1976) stated that *A. concolor* appeared to be endemic to the Amazon basin. However, with the Guianan records and those from the Brazilian states of Ceará and Piauí (I. Sazima and Ueda 1978; Mares et al. 1981; Willig 1983), the known distribution extends beyond the Amazon basin. Brosset and Charles-Dominique (1991) found enough variation in size among specimens collected in French Guiana to suggest that *A. concolor*, as understood today, could consist of two or more sibling species. D. C. Carter and Dolan (1978) provided information on either the holotype or a syntype located in the ZMB.

Genus *Chiroderma* W. Peters, 1860
Alfred L. Gardner

Chiroderma comprises five species of small to medium-sized (forearm 38–54 mm; mass 12–26 g) bats superficially similar to other white-lined bats of the genera *Uroderma*, *Platyrrhinus*, and *Vampyrodes*, and the species *Vampyriscus bidens* and *V. nymphaea*. *Chiroderma* is in Mexico, Central and South America (south to Bolivia and Brazil), and the Lesser Antilles. The West Indian endemic *C. improvisum* is the only species not known from South America.

Color varies from pale reddish to olive brown above and olive brown below with the venter sometimes appearing paler because of grayish frosting on the fur, especially on the throat. Some individuals may be paler above than below. The broad, dusky eye stripe is bordered above and below by white stripes that vary from well-marked and distinct in most species to weakly defined or obsolete in *C. villosum*. Supraorbital stripes extend from behind the noseleaf, where they sometimes coalesce, to between the ears. The dorsal stripe extends from between the shoulders to the uropatagium and, although conspicuous in most species, may be faint to absent in *C. villosum*. Forearms and legs are well furred and fur extends onto the dorsal surface of interfemoral membrane, where medially, it may be especially dense. The head is short and broad with a deep muzzle, and with the basal rim of the noseleaf free from the upper lip. Nasal bones are lacking, resulting in the dorsally emarginate external nares (nasal cleft), which is a salient cranial characteristic of the genus. The upper inner incisors are long and slender; the lower incisors are smaller and subequal; and the P1 is small and in contact with the canine, but separated from P2 by a gap. The dental formula is 2/2, 1/1, 2/2, 2/2 × 2 = 28. The karyotype (2*n* = 26, FN = 48; see summary in R. J. Baker et al. 1982) is known for all species except *C. doriae*. Tucker (1986) and Tucker and Bickham (1986) showed that both the X and Y chromosomes are fused to autosomes (neo-XY type) in *C. villosum* and *C. trinitatum*.

Species of *Chiroderma* can be distinguished from other white-lined bats (*Platyrrhinus* spp., *Uroderma* spp., *Vampyrodes caraccioli*, *Vampyriscus bidens*, and *V. nymphaea*) by a combination of the following features. The dorsal stripe arises between the shoulders; the muzzle is short, broad, and deep; the legs are furred; and the interfemoral membrane is at least partially furred and lacks a conspicuous fringe of hair on its trailing edge. The upper inner incisors are long and pointed, and the external nares are dorsally emarginate (nasals are lacking).

SYNONYMS:

Chiroderma W. Peters, 1860:747; type species *Chiroderma villosum* W. Peters, 1860, by monotypy.

Mimetops Gray, 1866b:117; in synonymy, *nomen nudum*.

Chirodesma Thenius, 1989:113; incorrect subsequent spelling of *Chiroderma* W. Peters.

KEY TO SOUTH AMERICAN SPECIES OF *CHIRODERMA*:

1. Size small, forearm less than 42 mm, greatest length of skull less than 23 mm . . . . *Chiroderma trinitatum*
1'. Size moderate to large, forearm more than 43 mm, greatest length of skull more than 23 mm 2
2. Upper inner incisors slender and parallel to each other (not convergent at tips); dorsal and facial stripes relatively inconspicuous to absent . . . .
. . . . *Chiroderma villosum*
2'. Upper inner incisors comparatively thicker, either in contact along most of their length, or convergent and in contact at tips; facial stripes conspicuous; dorsal stripe present and usually conspicuous 3
3. Size intermediate, forearm usually less than 52 mm, mastoid breadth less than 13.5 mm . . . .
. . . . *Chiroderma salvini*
3'. Size large, forearm usually more than 52 mm, mastoid breadth more than 13.5 mm . . . . . . *Chiroderma doriae*

Chiroderma doriae O. Thomas, 1891
Brazilian Big-eyed Bat

SYNONYMS:

[*Phyllostoma*] *dorsale* Lund, 1842a:134; *nomen nudum*.
[*Phyllostoma*] *dorsale* Lund, 1842b:200; *nomen nudum*.

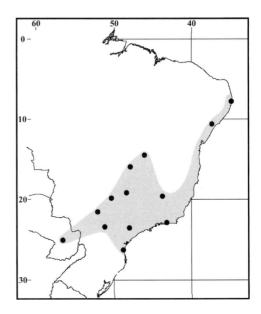

Map 173 Marginal localities for *Chiroderma doriae* ●

Chiroderma villosum: Dobson, 1878:534; not *Chiroderma villosum* W. Peters, 1860.

Ch[*iroderma*]. *doriae* O. Thomas, 1891:881; type locality "Minas Geraes," Brazil.

DISTRIBUTION: *Chiroderma doriae* is known from eastern and southern Brazil and departamento La Cordillera, Paraguay.

MARGINAL LOCALITIES (Map 173): BRAZIL: Pernambuco, Refúgio Ecológico Charles Darwin (L. A. M. Silva and Guerra 2000); Sergipe, Estação Serra de Itabaiana (Mikalauskas, Moratelli, and Peracchi 2006); Minas Gerais, Lagoa Santa (O. Thomas 1893a); Rio de Janeiro, Rio de Janeiro (Esbérard et al. 1996); São Paulo, Itapetininga (USNM 542616); Santa Catarina, Joinville (BM 9.11.19.15). PARAGUAY: La Cordillera, Estancia Sombrero (López-González et al. 1998). BRAZIL: Paraná, Parque Estadual Mata dos Godoy (N. R. Reis et al. 2000); Mato Grosso do Sul, Barma Farm (Gregorin 1998b); Minas Gerais, Salto da Água Vermelha (Pedro and Taddei 1998); Minas Gerais, Panga Ecological Reserve (Pedro and Taddei 1997); Distrito Federal, Fazenda Agua Limpa (Coimbra et al. 1982); Goiás, Município de Mambaí (Coimbra et al. 1982).

SUBSPECIES: I am treating *C. doriae* as monotypic.

NATURAL HISTORY: *Chiroderma doriae* has been found in tropical rainforest and tropical deciduous forest habitats of eastern and southeastern Brazil, and in more open marsh and grassland habitats farther south and west. Taddei (1976) found pregnant females in all months except April, May, October, and December, and stated that reproduction is not seasonally defined. He later (1980) estimated gestation at 3.5 months and suggested bimodal polyestry as the reproductive pattern. He netted *C. doriae* in cultivated areas containing small, disjunct remnants of the original broad-leafed forest. In Rio de Janeiro, Esbérard et al. (1996) found pregnant females in August, September, and October; a lactating female in January; and females with swollen mammae in March, June, August, November, and December. They also suggested bimodal polyestry as the reproductive regimen. Taddei (1980) reported a diet mainly of figs and suggested that pollen and nectar may be consumed as well. Esbérard et al. (1996) recorded the fruits of *Ficus*, *Cecropia*, *Muntingia*, and *Piper* as foods, and mentioned an individual whose head was covered with pollen. M. R. Nogueira and Peracchi (2002) stated that *C. doriae* is a fig specialist that takes some other fruits and may feed on flower parts. On the basis of fecal analysis and feeding experiments, they concluded that *C. doriae* is a seed predator, rather than a seed disperser. Taddei (1979) illustrated the skin and skull, and summarized morphometric data on the species (also see Swanepoel and Genoways 1979). The karyotype is unknown.

REMARKS: Apparently, the holotype either lacks or has poorly defined facial and dorsal stripes (Dobson 1878:534; misidentified as *C. villosum*); however, those stripes are well marked in specimens examined by Taddei (1979). *Chiroderma doriae* has the largest body size of any species in the genus. D. C. Carter and Dolan (1978) provided information on and measurements of the holotype of *Chiroderma doriae* O. Thomas.

Chiroderma salvini Dobson, 1878
Salvin's Big-eyed Bat
SYNONYMS:

Chiroderma salvini Dobson, 1878:532; type locality "Costa Rica."

Chiroderma salvini salvini: Handley, 1966b:297; name combination.

Chiroderma salvini scopaeum Handley, 1966b:297; type locality "Pueblo Juárez, Colima, México."

DISTRIBUTION: *Chiroderma salvini* is in northern and western Colombia, northern Venezuela, in Ecuador on both sides of the Andes, and there is a disjunct population east of the Andes in southern Peru and northern Bolivia. Elsewhere, the species occurs in Mexico and Central America.

MARGINAL LOCALITIES (Map 174): *Northern distribution*. VENEZUELA (Handley 1976): Distrito Federal, Pico Ávila, 5 km NNW of Caracas; Monagas, San Agustín, 5 km N of Caripe; Barinas, 2 km SW of Altamira. COLOMBIA: Antioquia, La Tebaida (Muñoz 1990); Cauca, Popayán (Arata et al. 1968). ECUADOR: Sucumbios, Limoncocha (R. H. Baker 1974); Guayas, Pacaritambo (Brosset 1965); Pichincha, Mindo (Albuja 1999). COLOMBIA: Valle del Cauca, Río Zabaletas (M. E.

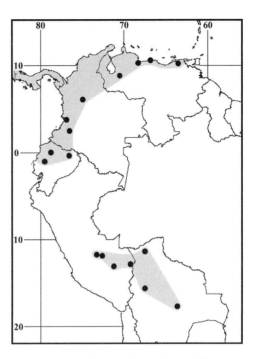

Map 174 Marginal localities for *Chiroderma salvini* ●

Thomas 1972). VENEZUELA: Carabobo, La Copa (Handley 1976). *Southern distribution*. BOLIVIA: Pando, Independencia (S. Anderson 1997); Santa Cruz, Río Pitasama, 4.5 km N and 1.5 km E of Cerro Amboró (S. Anderson 1997); La Paz, Serranía Bella Vista (S. Anderson, Koopman, and Creighton 1982). PERU: Madre de Dios, Río La Torre (Graham and Barkley 1984); Cusco, Consuelo (V. Pacheco et al. 1993); Cusco, Cashiriari-3 (S. Solari et al. 1998); Cusco, Ridge Camp (Emmons, Luna, and Romo 2001).

SUBSPECIES: Of the two recognized, only the nominate subspecies, *C. s. salvini* Dobson, occurs in South America. *Chiroderma s. salvini* also is found in eastern Mexico and Central America (Hall 1981). The second subspecies, *C. s. scopaeum* Handley, is found in the tropical deciduous forests of western Mexico and is in western Central America south into Costa Rica.

NATURAL HISTORY: Although little is known about the life history of this species, *C. salvini* appears to be a typical frugivore (Gardner 1977c), with a bimodal polyestry reproductive pattern (Wilson 1979). In Venezuela, the species is known from subtropical humid, and lower montane dry-forest habitats at elevations ranging from 600 m to 2,240 m, but most commonly above 1,000 m (Handley 1976). The Colombian specimens from Río Zabaletas came from lowland tropical rain forest (M. E. Thomas 1972). J. P. Webb and Loomis (1977) reported two mites, a tick, and a streblid batfly as ectoparasites. R. Guerrero (1985a, 1997)

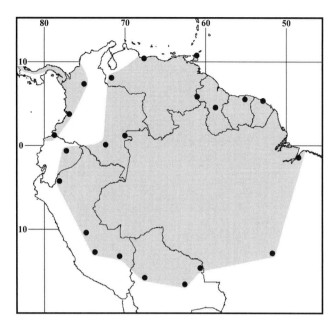

Map 175 Marginal localities for *Chiroderma trinitatum* ●

listed a tick, a mite, and two streblid batflies; no endoparasites have been reported.

REMARKS: D. C. Carter and Dolan (1978) provided information on and measurements of the holotype of *Chiroderma salvini* Dobson.

Chiroderma trinitatum Goodwin, 1958
Little Big-eyed Bat

SYNONYMS: See under Subspecies.

DISTRIBUTION: The South American population appears to be represented by two disjunct distributions, each recognized as a subspecies. The range of *C. t. trinitatum* includes Venezuela, Guyana, Surinam, French Guiana, the island of Trinidad, and the greater Amazon basin of Brazil and adjacent Bolivia, Peru, Ecuador, and Colombia. *Chiroderma t. gorgasi* is in western Colombia and northwestern Ecuador, and also occurs in Panama.

MARGINAL LOCALITIES (Map 175): *Northwestern distribution.* COLOMBIA: Antioquia, La Tirana (USNM 499479); Valle del Cauca, Río Zabaletas (USNM 483765). ECUADOR: Esmeraldas, La Chiquita (Albuja 1989). *Central distribution.* TRINIDAD AND TOBAGO: Trinidad, Cumaca (type locality of *Chiroderma trinitatum* Goodwin). GUYANA: Cuyuni-Mazaruni, Namai Creek (Lim and Engstrom 2000); Potaro-Siparuni, Iwokrama Forest (Lim and Engstrom 2001a). SURINAM: Para, Zanderij (Genoways, Williams, and Groen 1981). FRENCH GUIANA: Paracou (Simmons and Voss 1998). BRAZIL: Pará, Belém (Handley 1967); Mato Grosso, Serra do Roncador, ca. 264 km by road N of Xavantina (Pine, Bishop, and Jackson 1970). BOLIVIA: Santa Cruz, 55 km E de Aserradero Moira (S.

Anderson 1997); Santa Cruz, 10 km N of San Ramón (S. Anderson 1997); La Paz, Caranavi (Webster and Fugler 1984). PERU: Cusco, Quincemil (FMNH 93547); Ayacucho, San José (Gardner 1976); Pasco, San Pablo (Tuttle 1970); Amazonas, Río Kagka (J. L. Patton, Berlin, and Berlin 1982). ECUADOR: Napo, Río Huataracu (Albuja 1999). COLOMBIA: Caquetá, Estación Puerto Abeja (Montenegro and Romero-Ruiz 2000); Vaupés, Mitú (Barriga-Bonilla 1965). VENEZUELA: Barinas, Reserva Floresta de Ticoporo (Ochoa et al. 1988); Aragua, Portachuelo (Ojasti and Linares 1971).

SUBSPECIES: I recognize two subspecies of *C. trinitatum*.

C. t. gorgasi Handley, 1960

SYNONYMS:

Chiroderma gorgasi Handley, 1960:464; type locality "Tacarcuna Village, 3,200 ft., Río Pucro, Darién, Panama."

Chiroderma trinitatum gorgasi: Barriga-Bonilla, 1965:246; first use of current name combination.

This subspecies occurs in western Colombia and northwestern Ecuador. Elsewhere, it occurs in Panama.

C. t. trinitatum Goodwin, 1958

SYNONYMS:

Chiroderma trinitatus Goodwin, 1958b:1; type locality "Cumaca, Trinidad," Trinidad and Tobago.

Chiroderma trinitatum trinitatum: Barriga-Bonilla, 1965: 247; first use of current name combination.

This subspecies occurs on the island of Trinidad and in Venezuela, eastern Colombia, the Guianas, and the greater Amazon basin of Brazil, Peru, and Bolivia.

NATURAL HISTORY: This frugivore inhabits a variety of dry and humid tropical and subtropical forests at elevations up to 1,000 m, but is more commonly found below 500 m (Gardner 1976; Handley 1976). C. H. Carter et al. (1981) reported pregnant females in August on the island of Trinidad. R. Guerrero (1985a) reported a spinturnicid mite and a streblid batfly as the known ectoparasites from this species.

REMARKS: Barriga-Bonilla (1965) identified material from the vicinity of Mitú, Colombia, as *C. t. gorgasi*; however, I assign that material to *C. t. trinitatum* on geographic grounds.

Chiroderma villosum W. Peters, 1860
Hairy Big-eyed Bat

SYNONYMS: See under Subspecies.

DISTRIBUTION: *Chiroderma villosum* is in Colombia, Venezuela, Trinidad and Tobago, the Guianas, Brazil, Ecuador, Peru, and Bolivia. Elsewhere, the species occurs in Mexico and Central America.

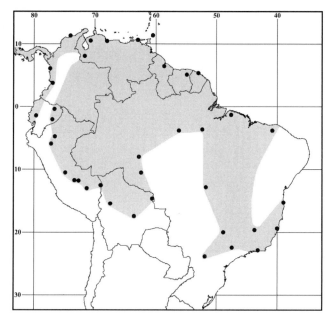

Map 176 Marginal localities for *Chiroderma villosum* ●

MARGINAL LOCALITIES (Map 176): TRINIDAD AND TOBAGO: Tobago, Charlotteville (USNM 540676). GUYANA: Cuyuni-Mazaruni, Mazaruni (Hill 1965). SURINAM: Brokopondo, Brokopondo (Husson 1978). FRENCH GUIANA: St. Elie (Brosset and Charles-Dominique 1991). BRAZIL: Pará, S. Antonio do Prata (O. Thomas 1920c); Ceará, Centro de Visitantes, Parque Nacional de Ubajara (S. S. P. Silva, Guedes, and Peracchi 2001); Minas Gerais, Lagoa Santa (Winge 1892); Bahia, Una (McNab and Morrison 1963); Espírito Santo, Estação Experimental de Linhares (Peracchi and Albuquerque 1993); Rio de Janeiro, Rio de Janeiro (Esbérard et al. 1996); São Paulo, Santa Gertrudes (Taddei 1979); Paraná, Fazenda Guajuvira (Bianconi, Mikich, and Pedro 2004); Minas Gerais, Frutal (Pedro and Taddei 1998); Mato Grosso, Serra do Roncador, ca. 264 km by road N of Xavantina (Pine, Bishop, and Jackson 1970). Pará, Rio Xingu, 52 km SSW of Altamira (MZUSP 27166); Pará, Parque Nacional da Amazônia, 54 km by road S of Itaituba (George et al. 1988); Rondônia, Calama (Handley 1960); Rondônia, Ouro Preto D'Oeste (Marques 1989). BOLIVIA: Santa Cruz, 27.5 km S of Campamento Los Fierros (S. Anderson 1997); Santa Cruz, Buenavista (S. Anderson, Koopman, and Creighton 1982); La Paz, Tomonoco (S. Anderson, Koopman, and Creighton 1982). PERU: Madre de Dios, Reserva Cuzco Amazónico (Woodman et al. 1991); Madre de Dios, Tono (V. Pacheco et al. 1993); Cusco, San Martín-3 (Wilson et al. 1997); Cusco, Ridge Camp (Emmons, Luna, and Romo 2001); Pasco, San Juan (Tuttle 1970); San Martín, Yurac Yacu (O. Thomas 1927a); Loreto, San Lorenzo (O. Thomas 1927a). ECUADOR: Pas-

taza, Montalvo (Albuja 1983); Sucumbios, Limoncocha (R. H. Baker 1974). VENEZUELA: Barinas, Reserva Floresta de Ticoporo (Ochoa et al. 1988). ECUADOR: Los Rios, Abras de Mantequilla (USNM 522438). COLOMBIA: Valle del Cauca, Río Zabaletas (M. E. Thomas 1972); Chocó, Ensenada de Utría (Alberico 1994); Magdalena, Cacagualito (type locality of *Chiroderma jesupi* J. A. Allen). VENEZUELA: Falcón, Río Socopito, 80 km NW of Carora (Handley 1976); Carabobo, San Esteban (O. Thomas 1891); Sucre, Manacal (Handley 1976).

SUBSPECIES: I recognize two subspecies of *C. villosum*.

C. v. jesupi J. A. Allen, 1900
SYNONYMS:

Chiroderma jesupi J. A. Allen, 1900a:88; type locality "Cacagualito, [Magdalena,] Colombia."

Chiroderma isthmicum Miller, 1912a:25; type locality "Cabima, [Panamá,] Panama."

Chiroderma villosum jesupi: Handley, 1960:466; first use of current name combination.

This subspecies occurs in western Ecuador and northern and western Colombia; elsewhere, it is in Central America and Mexico.

C. v. villosum W. Peters, 1860
SYNONYMS:

Chiroderma villosum W. Peters, 1860:748; type locality "*Brasilia.*"

Chiroderma villosum villosum: Handley, 1960:466; name combination.

The nominate subspecies occurs on Trinidad and Tobago, in Venezuela, the Guianas, Brazil, and the upper Amazon basin of Colombia, Ecuador, Peru, and northern Bolivia.

NATURAL HISTORY: Although common, but nowhere particularly abundant, this species is found in tropical forest habitats at elevations below 1,000 m; most commonly below 500 m (e.g., Handley 1976). *Chiroderma villosum* appears to prefer the velvety-skinned ciconia of strangler-type fig trees on Barro Colorado Island, Panama, where Handley, Gardner, and Wilson (1991) found these bats eating the fruits of at least three species of *Ficus*. M. R. Nogueira and Peracchi (2002) suggested that *C. villosum*, along with *C. doriae*, is a seed predator, not a seed disperser. Wilson (1979) summarized the reproductive pattern as polyestry and suggested that additional information may reveal the same bimodal annual pattern common to other stenodermatines. J. P. Webb and Loomis (1977) reported a tick and three genera of streblid batflies. R. Guerrero (1985a) added five mites and five streblid batflies to the number of species of ectoparasites recorded from *C. villosum*. Later, R. Guerrero (1997) increased the number of streblids known from this species to 13 species in 8 genera.

REMARKS: Cabrera's (1958:85) statement that the type locality is Venezuela was a *lapsus*, because W. Peters (1860) clearly wrote "Brasilia" as the type locality. D. C. Carter and Dolan (1978) gave information on a syntype of *Chiroderma villosum* W. Peters.

Genus *Enchisthenes* Andersen, 1906

Suely A. Marques-Aguiar

Enchisthenes is a monotypic genus of small (forearm 36–42 mm) frugivorous bat. Coloration is dark chocolate brown, darker (blackish in some individuals) on the head and paler on the venter. The facial stripes are faint to prominent and always buff colored. The dorsal fur has two color bands. The tragus usually has a pointed projection along the inner margin near the tip. The noseleaf is about as long as wide, and the lower margin of the "horseshoe" merges imperceptibly with the upper lip. The interfemoral membrane is narrow and hairy, and the plagiopatagium attaches at the ankle. The rostrum is relatively short and broad, and paraoccipital processes are moderately developed. Crowns of the inner upper incisors are simple and pointed (not bifid). The hypocone of M1 is slightly to moderately developed, and M3 is relatively large and aligned directly behind M2. The dental formula is 2/2, 1/1, 2/2, 3/3 × 2 = 32.

SYNONYMS:

Artibeus: O. Thomas,1892b:409; not *Artibeus* Leach, 1821b.

Enchisthenes Andersen, 1906b:419; type species *Artibeus hartii* O. Thomas, 1892b, by monotypy.

Dermanura: R. D. Owen, 1987:47; part; not *Dermanura* P. Gervais, 1856a.

Enchisthenes hartii (O. Thomas, 1892)

Hart's Fruit-eating Bat

SYNONYMS:

Artibeus hartii O. Thomas, 1892b:409; type locality "Trinidad," Trinidad and Tobago; restricted to Botanic Gardens, Port-of-Spain by O. Thomas (1893b:163).

[*Artibeus (Artibeus)*] *Harti*: Trouessart, 1897:160; name combination and incorrect subsequent spelling of *Artibeus hartii* O. Thomas.

Enchisthenes harti: Andersen, 1906b:419; generic description, name combination, and incorrect subsequent spelling of *Artibeus hartii* O. Thomas.

Artibeus (Enchisthenes) harti: Koopman, 1978:13; name combination and incorrect subsequent spelling of *Artibeus hartii* O. Thomas.

Dermanura hartii: R. D. Owen, 1987:47; name combination.

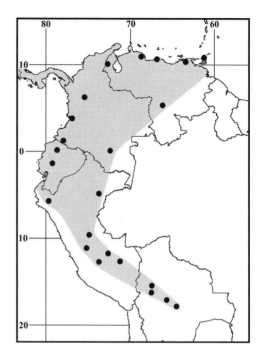

Map 177 Marginal localities for *Enchisthenes hartii* ●

DISTRIBUTION: *Enchisthenes hartii* is known from Trinidad and Tobago, northern and western Venezuela, and from eastern Colombia, Ecuador, Peru, and Bolivia. Elsewhere, the species occurs in Central America and Mexico, and there is a single record for southern Arizona, US (Irwin and R. J. Baker 1967; J. K. Jones and Carter 1976; Hall 1981; Nowak 1991).

MARGINAL LOCALITIES (Map 177): VENEZUELA: Falcón, Riecito (Handley 1976); Distrito Federal, Pico Ávila, near Hotel Humboldt (USNM 370767); Monagas, San Agustín, 5 km NW of Caripe (Handley 1976). TRINIDAD AND TOBAGO: Trinidad, Blanchisseuse (C. H. Carter et al. 1981). VENEZUELA: Amazonas, San Juan (Handley 1976). COLOMBIA: Caquetá, Estación Puerto Abeja (Montenegro and Romero-Ruiz 2000). PERU: Loreto, Jenaro Herrera (Ascorra, Gorchov, and Cornejo 1994); Huánuco, Panguana Biological Station (Hutterer et al. 1995); Cusco, San Martín-3 (Wilson et al. 1997); Madre de Dios, Aguas Calientes (V. Pacheco et al. 1993). BOLIVIA (S. Anderson 1997, except as noted): La Paz, Tomonoco (Webster and Jones 1980a); Santa Cruz, El Tunal; Cochabamba, El Sillar; La Paz, Río Solocama, 14 km on road from Chulumani to Irupana. PERU: Ayacucho, Huanhuachayo (Koopman 1978); Junín, Río Tulumayo, 3.2 km N of Vitoc (Gardner 1976); Piura, Las Juntas (Graham and Barkley 1984). ECUADOR: Bolívar, Barraganete (Albuja 1983); Pichincha, Gualea (Albuja 1983). COLOMBIA: Nariño, Ricaurte (Alberico and Orejuela 1982); Valle del Cauca, Río Zabaletas (M. E. Thomas 1972); Antioquia,

Santa Elena (Muñoz 1990). VENEZUELA: Zulia, Novito (Handley 1976).

SUBSPECIES: I am treating *E. hartii* as monotypic.

NATURAL HISTORY: *Enchisthenes hartii* has been mistnetted near streams and other wet sites, or occasionally in more xeric areas in a variety of habitats in Venezuela (Handley 1976). Graham and Barkley (1984) netted three males over the Río La Pachinga in otherwise arid habitat in the Pacific lowlands of northern Peru. The species is known from the tropical and subtropical zones on both sides of the Andes (Albuja 1983, 1999). The elevational range is from near sea level to 3,540 m at the summit of the Cordillera Vilcabamba, Peru (Koopman 1978). Most of Handley's (1976) Venezuelan specimens were taken at elevations between 1,000 and 2,250 m.

This bat is frugivorous (Gardner 1977c; Albuja 1983). Goodwin (1940) reported more than 20 skulls recovered from owl pellets in southern Ecuador, and owls may be among the most important predators. Few data are available on reproduction. Wilson (1979) mentioned reproductive activity from January to September, with a probable period of inactivity late in the year. A female from Ecuador was pregnant in December and others were either pregnant or lactating when taken in January. Pregnant or lactating females have been found from March to May and in July and August in Venezuela; and in April and May, and during July and September in Colombia (M. E. Thomas 1972; USNM collection). Arroyo-Cabrales and Owen (1997) provided a detailed description, along with measurements, and a summary of available information on natural history in their *Mammalian Species* account.

Enchisthenes hartii is host to at least five species of mites (Labidocarpidae, Spinturnicidae, and Trombiculidae) and two species of streblid batflies (B. McDaniel 1972; Brennan and Reed 1975; Herrin and Tipton 1975; J. P. Webb and Loomis 1977). R. Guerrero (1997) listed nine species of streblid batflies found on this species. Guimarães (1972) considered the record of a nycteribiid batfly found on a Venezuelan *E. hartii* to be doubtful. Nevertheless, Gracioli and Carvalho (2001) listed the nycteribiid batfly *Basilia ortizi* as recorded from *E. hartii*. No data are available on other parasites.

The karyotype is 2n = 30–31, FN = 56, with a Y_1, Y_2 chromosome system (X–autosome fusion). The X chromosome is a medium-sized subtelocentric; Y_1 is a small submetacentric, and Y_2 a minute acrocentric (R. J. Baker 1967; R. J. Baker and Hsu 1970; Gardner 1977a).

REMARKS: The "Las Juntas, Lambayeque, Peru" locality listed by Arroyo-Cabrales and Owen (1996:80) actually is in departamento Piura (L. Barkly pers. comm.). D. C. Carter and Dolan (1978) provided information on and measurements of the holotype of *Artibeus hartii* O. Thomas.

Genus *Mesophylla* O. Thomas, 1901
Joaquín Arroyo-Cabrales

Monotypic *Mesophylla* is one of the smallest frugivorous bats (forearm 29.5–33.5 mm, n = 8; greatest length of skull 16.3–17.1 mm, n = 8; Swanepoel and Genoways 1979) in South America. Only *Ectophylla alba* (a Central American species) and male *Ametrida centurio* are smaller. Color varies from pale gray on the head and shoulders to pale brownish gray posteriorly. The ears and noseleaf are yellow in life, but are paler in museum specimens. Facial stripes are obsolete, and body stripes are lacking. A characteristic, small accessory noseleaf-like structure is present behind the noseleaf.

The skull is short; the rostrum is narrow, tapering, and shorter than the braincase; and the nasals are depressed medially (in frontal view). The palate extends well behind the last molars and has a relatively shallow mesopterygoid emargination. The upper inner incisors are long, convergent, and usually have weakly bifid tips. The first lower molar is not conspicuously larger than the second lower premolar (p4), m2 lacks a posterior cuspulid, and m3 is minute. The dental formula is 2/2, 1/1, 2/2, 2/3 × 2 = 30.

SYNONYMS:

Mesophylla O. Thomas, 1901g:143; type species *Mesophylla macconnelli* O. Thomas, 1901g, by original designation.

Ectophylla: Laurie, 1955:269; not *Ectophylla* H. Allen, 1892b.

Vampyressa: R. D. Owen, 1987:46; not *Vampyressa* O. Thomas, 1900b.

Mesophylla macconnelli (O. Thomas, 1901)
MacConnell's Bat
SYNONYMS:

Mesophylla macconnelli O. Thomas, 1901g:143; type locality "the Kanuku Mountains, about 59°W. and 3°N. ... 2000 feet," Upper Takutu-Upper Essequibo, Guyana.

Ectophylla macconnelli: Laurie, 1955:269; name combination.

Ectophylla macconnelli flavescens Goodwin and Greenhall, 1962:2; type locality "Talparo, Trinidad," Trinidad and Tobago.

E[ctophylla]. m[acconnelli]. macconnelli: Goodwin and Greenhall, 1962:6; name combination.

Ectophylla (Mesophylla) macconnelli: Harrison and Horne, 1971:245; name combination

Ectophylla (Mesophylla) m[acconnelli]. macconnelli: R. L. Peterson, 1972:468; name combination.

Vampyressa macconnelli flavescens: R. D. Owen, 1987:64; name combination.

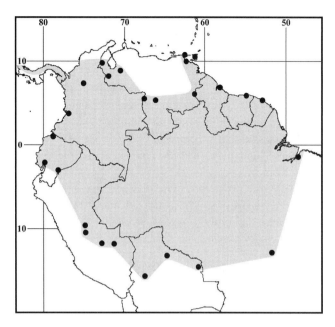

Map 178 Marginal localities for *Mesophylla macconnellii* ●

Vampyressa macconnelli macconnelli: R. D. Owen, 1987: 64; name combination.

Ectophylla maconelli Wetterer, Rockman, and Simmons, 2000:154; incorrect subsequent spelling of *Mesophylla macconnelli* O. Thomas.

DISTRIBUTION: *Mesophylla macconnelli* is known from Colombia, Venezuela, the island of Trinidad, Guyana, Brazil, Ecuador, Peru, and Bolivia. The species also occurs in Central America from Panama north into Nicaragua (Hall 1981; F. A. Reid 1997).

MARGINAL LOCALITIES (Map 178): TRINIDAD AND TOBAGO: Trinidad, Talparo (type locality of *Ectophylla macconnelli flavescens* Goodwin and Greenhall). GUYANA: Demerara-Mahaica, Georgetown (Goodwin and Greenhall 1962). SURINAM: Commewijne, Nieuwe Grond Plantation (S. L. Williams and Genoways 1980a). FRENCH GUIANA: Paracou (Simmons and Voss 1998). BRAZIL: Pará, Utinga (Handley 1967); Mato Grosso, Serra do Roncador, 264 km (by road) N of Xavantina (Pine, Bishop, and Jackson 1970). BOLIVIA: Santa Cruz, 3 km S of Campamento Los Fierros (S. Anderson 1997); Beni, Yutiole (S. Anderson 1997); La Paz, 35 km by road N of Caravani at Serranía Bella Vista (S. Anderson, Koopman, and Creighton 1982). PERU: Madre de Dios, Cocha Cashu (Terborgh, Fitzpatrick, and Emmons 1984); Cusco, Konkari-ari (S. Solari et al. 1998); Pasco, San Juan (Tuttle 1970); Huánuco, Panguana Biological Station (Hutterer et al. 1995). ECUADOR: Morona-Santiago, Los Tayos (Albuja 1983); Guayas, Guayaquil (Albuja 1983); Esmeraldas, Río Cachabí (Albuja 1983). COLOMBIA: Valle del Cauca, Río Zabaletas

(M. E. Thomas 1972); Antioquia, La Tirana, 25 km S and 22 km W of Zaragoza (USNM 499483). VENEZUELA (Handley 1976, except as noted): Zulia, El Tukuco (Soriano, Ruiz, and Zambrano 2005); Táchira, El Hatico (Soriano, Ruiz, and Zambrano 2005); Barinas, 2 km SW of Altamira; Amazonas, 25 km S of Puerto Ayacucho; Amazonas, San Juan; Bolívar, Km 125; Monagas, 2 km W of mouth of Río Guanipa (Linares and Rivas 2004). Sucre, Las Melenas (Bisbal 1998).

SUBSPECIES: I consider *M. macconnelli* to be monotypic.

NATURAL HISTORY: This species occupies humid tropical forests (see summary of habitats and roost sites in Kunz and Pena 1992). *Mesophylla macconnelli* constructs tents, often from the leaves of *Anthurium* sp., which grow in the denser vegetation found along streams in humid forests (Goodwin and Greenhall 1962; Handley 1976; Koepcke 1984). Wilson (1979) characterized *M. macconnelli* as seasonally polyestrous. Pregnant females have been captured in Peru in January (M. E. Thomas 1972), in May (Graham 1987), and in August (Tuttle 1970). Also in August, C. H. Carter et al. (1981) found seven pregnant and five lactating females on the island of Trinidad. Label information from specimens in the USNM shows pregnant females taken in January and March (Panama), January, February, and July (Venezuela), February (Brazil), and in August (Ecuador and Peru). Lactating females have been taken in February and September (Brazil), and in May (Venezuela). Additional information on reproduction, along with measurements and a synopsis of natural history information can be found in the *Mammalian Species* account by Kunz and Pena (1992). The karyotype is $2n = 21–22$, FN = 20 (R. J. Baker and Hsu 1970).

Herrin and Tipton (1975) reported two species of spinturnicid mites found on *M. macconnelli*, which also is the type host of the streblid fly, *Neotrichobius ectophyllae* Wenzel, 1976. R. Guerrero (1997) listed two additional species of streblid flies known from this species.

REMARKS: *Mesophylla macconnelli* was regarded for many years as representing the monotypic genus *Mesophylla*, until Laurie (1955) synonymized it under *Ectophylla*. Goodwin and Greenhall (1962) followed Laurie (1955) in treating *Mesophylla* as a subgenus of *Ectophylla* when they described *Ectophylla macconnelli flavescens* from the island of Trinidad. Handley (1966c) stated that nongeographic variation among Panamanian *Ectophylla macconnelli* exceeded variation interpreted by Goodwin and Greenhall (1962) as geographic; thereby implying that he considered the species to be monotypic. Starrett and Casebeer (1968) also considered the species to be monotypic, but recognized *Mesophylla* as a valid genus, and pointed out its similarities with *Vampyressa*. Honacki, Kinman, and Koeppl (1982), J. K. Jones and Carter (1976), and Koopman (1982), however, continued to treat *M. macconnelli* as a species of *Ectophylla*. The weight of morphological and karyological data (e.g., see Greenbaum, Baker, and Wilson 1975) led Gardner (1977a), C. H. Carter et al. (1981), and Hall (1981) to treat *Mesophylla* as a monotypic genus distinct from *Ectophylla*. R. D. Owen (1987) included *Mesophylla* within *Vampyressa* (*sensu lato*) and recognized *Ectophylla* as distinct. Corbet and Hill (1991), and Koopman (1993, 1994) recognized *Mesophylla* and *Ectophylla* as valid genera separate from *Vampyressa*. Recently, Simmons and Voss (1998), Wetterer, Rockman, and Simmons (2000), and Sampaio et al. (2003) have again treated *Mesophylla* as a synonym of *Ectophylla*. However, R. J. Baker et al. (2003) recognized *Mesophylla* as valid at the generic level.

Mesophylla is easily distinguished from *Ectophylla* by their contrasting cranial and dental morphologies. Nasals are depressed medially in *Mesophylla*, but elevated in *Ectophylla*. The lateral margins of the mesopterygoid fossa are broadly expanded in *Mesophylla*, but the fossa is constricted anterior to the pterygoid processes in *Ectophylla*. There are no specialized bony ridges associated with pterygoid process in *Mesophylla*, whereas *Ectophylla* has a bony ridge extending laterally on the basicranium from each pterygoid process, and this ridge partly occludes the foramen ovale (ridge forms lateral margin of a shallow concavity not seen in other stenodermatines). Bony excrescences, medial to the eustachian canal and anterior to the auditory bullae, are well developed in *Mesophylla*, but are only weakly developed and shifted laterally in *Ectophylla*. Dentally, the taxa are similar in their lack of a well-defined ectoloph on the molars, and in their spaced and relatively weak dentitions anterior to the last molars. The dentition of *Mesophylla* more closely resembles that of the generalized stenodermatine pattern. *Ectophylla* differs from *Mesophylla* in a number of other dental features that include terete, conical upper inner incisors; higher, more pointed cuspids of lower premolars; presence of a posterior cuspid on the last lower premolar; greatly expanded, nearly round, last lower molar (m2); and loss of m3. The crown of the last lower molar of *Ectophylla* has a large shallow crater with a small longitudinally oriented cuspulid in its center (see Timm 1982:Fig. 3). In *Ectophylla*, the more medial of the two anterior cuspids on the anterior margin of the last lower molar stands higher than the outer one; the lingual margin is crenulated, almost pectinate, and appears to be much higher than the outer margin because the plane of the crown is tilted laterally. The second lower molar is the largest tooth in the toothrow in *Mesophylla*, and structurally resembles that molar in some other stenodermatines. Goodwin and Greenhall (1962) were not able to appreciate the basicranial differences between *Mesophylla* and *Ectophylla* because the pterygoid region of the only skull of *Ectophylla* they examined was damaged. When taken alone, any of these differences may not appear significant at the generic level, but when taken in the aggregate, along with other features (e.g., melanized fascia covering cranial musculature in *Ectophylla*; Gardner and Wilson 1971), they support generic separation for *Mesophylla* and *Ectophylla*. D. C. Carter and Dolan (1978) provided information on and measurements of the holotype of *Mesophylla macconnelli* O. Thomas.

Genus *Platyrrhinus* Saussure, 1860

Alfred L. Gardner

Platyrrhinus comprises at least 14 species and occurs from Mexico southward into northern Argentina. Its greatest diversity is along the eastern slopes of the Andes, where as many as four species may be found at the same location.

Body size ranges from small to large (forearm 35–65 mm, greatest length of skull 20–34 mm, mass 13–65 g). Members of this genus can be identified by their comparatively narrow interfemoral membrane thickly fringed with hair, a prominent white dorsal stripe (less obvious in *P. infuscus*), comparatively large inner upper incisors that are convergent at the tips, and three upper and three lower molars. Dorsal color varies by species from buff or pale gray-brown to dark blackish brown. Supraorbital and malar facial stripes are paler than the surrounding fur. Facial stripes are white and most prominent in *P. helleri* and *P. lineatus*, and buffy brown and least conspicuous in *P. chocoensis* and *P. infuscus*, in which the malar stripe may not be evident on some specimens. The dorsal stripe is always whitish and usually conspicuous, although thin and sometimes difficult to see in *P. infuscus*. The rostrum is approximately as wide as, and almost as long as the braincase. Upper inner incisors are long and convergent at the tips, which may be weakly bifid, trifid, or entire in unworn teeth. Upper outer incisors are bifid, less than 1/2 the length of inner incisors, but fill the gap between inner incisors and canines. Lower incisors are small, bifid or weakly trifid, and fill the space between lower canines. The first two upper molars are essentially alike in form and have nearly smooth crushing surfaces. The morphology of the second lower premolar (pm4) is useful for distinguishing among species (see key and individual species accounts). The dental formula is 2/2, 1/1, 2/2, 3/3 × 2 = 32. The karyotype is $2n = 30$, $FN = 56$ (R. J. Baker 1979) in all species thus far examined.

SYNONYMS:

Phyllostoma: É. Geoffroy St.-Hilaire, 1810b:180; part; not *Phyllostoma* G. Cuvier, 1800.

Arctibeus Gray, 1838b:487; part; incorrect subsequent spelling of, but not *Artibeus* Leach, 1821b.

Artibaeus P. Gervais, 1856a:35; part; incorrect subsequent spelling of, but not *Artibeus* Leach.

Platyrrhinus Saussure, 1860:429; type species *Phyllostoma lineatum* É. Geoffroy St.-Hilaire, 1810b, by subsequent designation (O. Thomas 1900b:269).

Stenoderma: W. Peters, 1865a:257; part; not *Stenoderma* É. Geoffroy St.-Hilaire, 1818b.

Vampyrops W. Peters, 1865a:257; proposed as a subgenus of *Stenoderma* É. Geoffroy St.-Hilaire, 1818b, as a replacement name for *Platyrrhinus* Saussure, 1860, on the assumption that the name was preoccupied by *Platyrrhinus* Schellenberg, 1798; type species *Phyllostoma lineatum* É. Geoffroy St.-Hilaire, 1810b, by subsequent designation (O. Thomas 1900b:269).

Vampirops Festa, 1906:6; incorrect subsequent spelling of *Vampyrops* W. Peters.

Vampirops Piccinini, 1974:16; incorrect subsequent spelling of *Vampyrops* W. Peters.

Vamryrops Muñoz, 1993:86; incorrect subsequent spelling of *Vampyrops* W. Peters.

REMARKS: Gardner and Ferrell (1990) explained the basis for using *Platyrrhinus* instead of *Vampyrops* for this genus. O. Thomas, (1900b:269) designated *Phyllostoma lineatum* É. Geoffroy St.-Hilaire, as the type species of *Vampyrops* W. Peters. Because *Vampyrops* W. Peters (1865a:257) is replacement name for *Platyrrhinus* Saussure, 1860, *Phyllostoma lineatum* É. Geoffroy St.-Hilaire also is the type species of *Platyrrhinus* (ICZN 1999, Art. 67.8). Palmer (1904:545) identified the type species of *Platyrrhinus* Saussure as *Phyllostoma lineatum* É. Geoffroy St.-Hilaire, although he also believed the name to be preoccupied by *Platyrhinus* Schellenberg, 1798. Palmer (1904), de la Torre and Starrett (1959), and Hall (1981) erroneously attributed the authorship of *Platyrhinus* (an anthribid beetle) to Clairville, who is responsible only for the French translation from the original German. The argument by de la Torre and Starrett (1959), repeated by Goodwin and Greenhall (1961:255), that Fabricius's (1801: 408) spelling *Platyrrhinus* was an unjustifiable emendation (therefore, a senior homonym) that invalidated Saussure's (1860) name is incorrect. Fabricius's spelling is simply an incorrect subsequent spelling of *Platyrhinus* Schellenberg and has no status. Furthermore, the misspelling appeared in a synonymy (see ICZN 1999:Art. 11.6). The arguments by Alberico and Velasco (1991) on the validity of *Platyrrhinus* versus *Vampyrops* for this genus, and whether or not the latter was used by W. Peters (1865a) as a replacement name for *Platyrrhinus* reflect misunderstanding of the International Code of Zoological Nomenclature (ICZN, 1985, 1999). The most recent review of *Platyrrhinus* is by Velazco (2005).

KEY TO THE SPECIES OF *PLATYRRHINUS*:

1. Forearm longer than 54 mm; condylobasal length more than 26 mm. 2
1'. Forearm equal to or shorter than 56 mm, usually shorter than 54 mm; condylobasal length usually less than 26 mm. 5
2. Pelage dark, usually blackish brown; four facial stripes and a conspicuous dorsal stripe; pm4 with long posterior "heel" (posterolabial cuspulid). 3
2'. Dorsal pelage buffy to dark brown, not blackish; facial and dorsal stripes inconspicuous; pm4 with a short posterior "heel" (posterolabial cuspulid). . . . .
. . . . *Platyrrhinus infuscus*
3. Forearm 55 mm or longer; condylobasal length 28 mm or longer. 4
3'. Forearm 55 mm or shorter; condylobasal length less than 28 mm. 5
4. Forearm usually shorter than 62 mm; accessory cuspulid present on anterolingual cristid of pm4. . . . .
. . . . *Platyrrhinus vittatus*
4'. Forearm longer than 61 mm; anterolingual cristid of pm4 smooth, accessory cuspulids lacking on pm4. . . . .
. . . . *Platyrrhinus albericoi*
5. Pelage pale brown to dark brown; facial stripes always white and usually conspicuous; forearm 32–48 mm. . .
. . . . 6
5'. Pelage brown to blackish brown; facial stripes buff, sometimes inconspicuous; forearm 40–56 mm. 10
6. Size smaller, forearm 42 mm or less. 7
6'. Size larger, forearm 42 mm or more. 9
7. Anterolingual cristid of last lower premolar (pm4) bears one or two tubercle-like cuspulids. 8
7'. Anterolingual cristid of last lower premolar (pm4) essentially smooth, lacking accessory cuspulids. . . . .
. . . . *Platyrrhinus matapalensis*
8. Anterolingual cristid of pm4 bears two well-developed accessory cuspulids; outline of lateral margin of pterygoid process narrowly concave (small c-shaped) in posterior view because of well-developed thin ridge of bone that extends posterolaterally from near mid-height of pterygoid process to base of braincase. . . . .
. . . . *Platyrrhinus brachycephalus*
8'. Anteromedial cristid of pm4 has one or two small accessory cuspulids; posterior cristid of pm4 bears one or more erect cuspulids anterior to and in addition to a short "heel"; outline of lateral margin of pterygoid process broadly concave (large C-shaped) in posterior view, lateral bony ridge not well developed or elevated
. . . . *Platyrrhinus helleri*
9. Size larger, forearm more than 45 mm; crown of pm2 relatively narrow in labial outline and triangular in cross-sectional outline (posterolingual cristid well developed), labial surface conspicuously convex, and "heel" 20% or

more of anterior-posterior length of tooth in lateral view
. . . . *Platyrrhinus lineatus*

9′. Size smaller, forearm 42–46 mm, usually less than 45 mm; crown of pm2 "sail-shaped" in lateral outline and blade-like in cross-sectional outline (posterolingual cristid weakly developed), labial surface nearly flat, and "heel" less than 20% of anterior-posterior length of tooth in lateral view . . . . *Platyrrhinus recifinus*

10. Size larger, forearm 49–55 mm; upper inner incisors slender and long, with tips parallel for more than 1/2 their length (crown height) . . . . . *Platyrrhinus aurarius*

10′. Size small to large, forearm 40–56; upper inner incisors broad with convergent tips 11

11. Forearm 40–56 mm; stylid cuspulid present between protoconid and metaconid of m2; ventral pelage has three color bands; dorsal coloration dark 12

11′. Forearm 46–51 mm; stylid cuspulid absent between protoconid and metaconid of m2; ventral pelage with two color bands, dorsal coloration relatively pale . . . .
. . . . *Platyrrhinus chocoensis*

12. Size larger, forearm 49–56 mm; lingual cingulum not continuous from paracone to metacone (sulcus not continuous between paracone and metacone) of M1; pinnae lack lateral folds . . . . *Platyrrhinus ismaeli*

12′. Size smaller, forearm 40–51, lingual cingulum continuous from paracont to metacone (sulcus continuous between paracone and metacone) of M1; pinnae have lateral folds. 13

13. Size smaller, forearm 40–47 mm; M2 lacks lingual stylar cuspule on paracone; fossa on squamosal branch of zygomatic arch (lateral to glenoid fossa) usually deep .
. . . . *Platyrrhinus nigellus*

13′. Size larger, forearm 46–51 mm; M2 has a stylar cuspule on lingual margin of paracone; depth of fossa on squamosal branch of zygomatic arch (lateral to glenoid fossa) varies from deep to nearly imperceptible 14

14. Fold lines in pinnae relatively weak; dorsal stripe relatively narrow, fur of dorsum long (more than 8 mm); fossa in squamosal branch of zygomatic arch deep; upper outer incisors unlobed; lingual cingulum of M2 metacone continuous to paracone . . . . *Platyrrhinus dorsalis*

14′. Fold lines in pinnae well marked; dorsal stripe conspicuous and wide; fur of dorsum shorter (less than 8 mm); fossa in squamosal branch of zygomatic arch shallow to nearly imperceptible; upper outer incisors bilobed; lingual cingulum of M2 metacone does not continue to paracone . . . . *Platyrrhinus masu*

Platyrrhinus albericoi Velazco, 2005
Alberico's Broad-nosed Bat

SYNONYMS:

Vampyrops vittatus: Sanborn, 1955:405–07; part.
Platyrrhinus albericoi Velazco, 2005:21; type locality "San

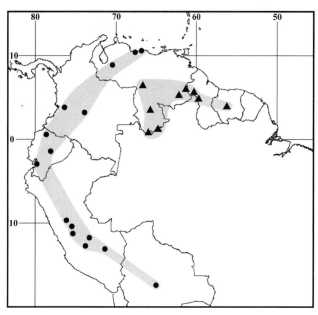

Map 179 Marginal localities for *Platyrrhinus albericoi* ● and *Platyrrhinus aurarius* ▲

Pedro, Paucartambo-Pilcopata road, 1480 m in elevation, Province of Paucartambo, Department of Cuzco, Peru, approximately 13°0′1′ [*sic*] S, 71°32′46′ [*sic*] W."

DISTRIBUTION: *Platyrrhinus albericoi* is known from northern Venezuela, western Colombia and Ecuador, eastern Peru, and northern Bolivia.

MARGINAL LOCALITIES (Map 179; localities listed from north to south): VENEZUELA: Distrito Federal, Pico Ávila, 6 km NNW of Caracas (Handley 1976); Aragua, Portachuelo (USNM 517467); Barinas, Río Santo Domingo, 2 km SW of Altamira (Handley 1976). COLOMBIA: Meta, Município de San Juan de Arama (Sánchez-Palomino and Rivas-Pava 1993); Valle del Cauca, El Tambor (USNM 483646). ECUADOR: Esmeraldas, Las Pambiles (Albuja 1999); Pastaza, Cavernas de Mera (Rageot and Albuja 1994); Guayas, Huerta Negra (USNM 522431). PERU: Huánuco, eastern slope Cordillera Carpish, Carretera Central (Gardner and Carter 1972a); Junín, Vitoc Valley (Sanborn 1955); Pasco, Palmira (Velazco 2005); Cusco, Ridge Camp (Emmons, Luna, and Romo 2001); Ayacucho, Yuraccyacu (Gardner and Carter 1972a); Cusco, Pillahuata (Velazco 2005). BOLIVIA: Cochabamba, Yungas de Totora (Barquez and Olrog 1980)

SUBSPECIES: I am treating *P. albericoi* as monotypic.

NATURAL HISTORY: Handley (1976) recorded this species in lower montane humid and subtropical wet forest habitats at elevations ranging from 600 to 2,000 m in Venezuela. Gardner and Carter (1972b) reported this frugivore from Peru in habitats and at elevations similar to those frequented in Venezuela. The bats Tuttle (1970, 1976; as *P. vittatus*) reported finding in a large, shallow cave may have been *P. infuscus* and not *P. albericoi* (Tuttle treated *infuscus* as a synonym of *vittatus*). Based on label information for specimens in the USNM and on published accounts based on Ecuadorian and Peruvian specimens, pregnant females have been found in January, February, March, and August; lactating females recorded in March and September. These data suggest postpartum estrus and bimodal polyestry, as generally characterizes the reproductive regimen in other stenodermatines. The $2n = 30$, FN $= 56$ karyotype reported by R. J. Baker (1973) for *P. vittatus* may have been based in part on specimens of *P. albericoi*. Apparantly, species of *Platyrrhinus* all have the same basic karyotype.

REMARKS: Although some of the specimens reported as *P. vittatus* by Tuttle (1970) and Albuja (1983) from Peru and Ecuador, respectively, represent *P. infuscus*, the remainder represent *P. albericoi*. Previous to Velazco's (2005) description, most authors used the name *Vampyrops vittatus* or *Platyrrhinus vittatus* for this taxon. Velazco (2005) gave the distribution of *P. albericoi* as the eastern slope of the Andes of Ecuador, Peru, and Bolivia; however, the Ecuadorian specimen he listed is from the western drainage. I have identified Venezuelan and Colombian USNM specimens, including some from the Pacific drainage, as *P. albericoi* on the basis of large size and the lack of an accessory cuspulid on the anterolingual cristid of the last lower premolar (see Velazco 2005:25, Fig. 16).

Platyrrhinus aurarius (Handley and Ferris, 1972)
Eldorado Broad-nosed Bat

SYNONYMS:

Vampyrops aurarius Handley and Ferris, 1972:522; type locality "Km 125, 1,000 m, 85 km SSE El Dorado, Bolívar, Venezuela."

Platyrrhinus aurarius: Koopman, 1993:190; first use of current name combination.

DISTRIBUTION: Known from the Guianan Highlands of Venezuela, Guyana, and Surinam; undoubtedly present in adjacent northern Brazil.

MARGINAL LOCALITIES (Map 179): VENEZUELA: Bolívar, La Serranía de Los Pijiguaos (Ochoa et al. 1988); Bolívar, Chimantá Tepuy (Ochoa et al. 1993); Bolívar, Km 125 (type locality of *Vampyrops aurarius* Handley and Ferris 1972). GUYANA: Cuyuni-Mazaruni, Karowrieng River, within 12 miles of Maipuri Falls (BM 80.744). SURINAM: Saramacca, Tafelberg (S. L. Williams, Genoways, and Groen 1983). GUYANA: Potaro-Siparuni, Mount Kowa (Shapley et al. 2005). VENEZUELA: Amazonas, Cerro de Tamacuare (Ojasti, Guerrero, and Hernández 1992); Amazonas, Camp II, 2.5 km NE of Pico Phelps, (Gardner 1990); Amazonas, Cabecera del Caño Negro, Cerro Duida (Handley 1976).

SUBSPECIES: I am treating *P. aurarius* as monotypic.

NATURAL HISTORY: This species has been most commonly found in subtropical wet forest habitats in Venezuela at elevations ranging from 700 to 2,100 m (Handley 1976; Gardner 1990). Specimens were taken from a shallow cave, the source of a stream, in a steep slope of Mt. Kowa, Guyana (Shapley et al. 2005). Male *P. aurarius* with enlarged testes were recorded in May, and pregnant females, each with a single fetus, were taken on 23 March (one), 23 May (two), 30 November (one), 1 December (one), 2 December (one), and 31 January (one). The largest fetus (41 mm CR) was from a specimen captured in May. Two females were lactating when captured in March. Other females taken in May, November and January did not show signs of reproduction. R. Guerrero (1997) listed four species of streblid batflies recovered from *P. aurarius*. The karyotype is unknown.

REMARKS: Specimens from Colombia tentatively identified as of this species (Lemke et al. 1982) represent *P. ismaeli*. The suggestion by D. C. Carter and Rouk (1973), J. K. Jones and Carter (1976), and Linares (1986), that *P. aurarius* is a synonym of *P. dorsalis*, is unsupported.

Platyrrhinus aurarius is exceeded in size only by *P. vittatus*, *P. albericoi*, and *P. infuscus*. Forearm length averages 52.1 mm (49.6–54.8 mm), $n = 36$; condylobasal length averages 25.9 mm (24.5–27.0 mm), $n = 37$; mass of males averages 34.1 g (28–39 mm), $n = 20$, and of females averages 36.3 g (24–47 g), $n = 15$. Males and females weighing 32 g or less are considered to be subadults or juveniles. Each of four females in the sample of 15 weighed 47 g, and each was pregnant. Measurements of *P. aurarius* are all exceeded by those of *P. vittatus* and *P. albericoi*, but overlap measurements of *P. infuscus*. However, *P. infuscus* is much paler than *P. aurarius*, has a weakly defined dorsal stripe, and *P. infuscus*, *P. albericoi*, and *P. vittatus* have much broader and blunt-tipped upper inner incisors that converge at the tips. *Platyrrhinus aurarius* is as dark as *P. vittatus*, and *P. albericoi*, has a well-defined dorsal stripe, and has comparatively longer, narrower upper inner incisors that are parallel for most of their length.

Platyrrhinus brachycephalus (Rouk and Carter, 1972)
Short-headed Broad-nosed Bat

SYNONYMS: See under Subspecies.

DISTRIBUTION: The species is in Colombia, Venezuela, Guyana, Surinam, French Guiana, Brazil, Ecuador, Peru, and Bolivia.

MARGINAL LOCALITIES (Map 180): *Northern distribution (P. b. saccharus)*. VENEZUELA: Sucre, Santa Rosa, 20 km (by road) SE of Casanay (Rouk and Carter 1972); Delta Amacuro, El Toro (Linares and Rivas 2004). GUYANA: Demerara-Mahaica, Demerara (Handley and Ferris 1972); Upper Demerara-Berbice, Dubulay Ranch (USNM 582324). SURINAM: Commewijne, Nieuwe Grond Plantation (S. L. Williams and Genoways 1980a);

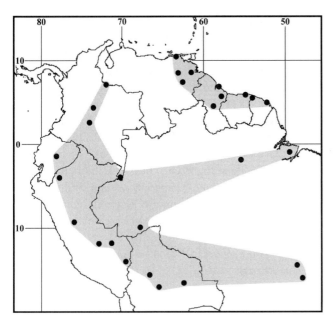

Map 180 Marginal localities for *Platyrrhinus brachycephalus* ●

Marowijne, 3 km SW of Albina (CM 63875). FRENCH GUIANA: Cayenne (Brosset and Charles-Dominique 1991). GUYANA: Potaro-Siparuni, Iwokrama Forest (Lim and Engstrom 2001b). VENEZUELA: Bolívar, El Manteco (USNM 534248); Anzoátegui, Mamo (Ochoa and Ibáñez 1985). *Central distribution (P. b. brachycephalus).* COLOMBIA: Arauca, Río Arauca (Velazco 2005); Meta, Upin Salt Mine (Rouk and Carter 1972); Meta, Cabaña Duda (Lemke et al. 1982); Amazonas, 3 miles W of Isla Santa Sofia (Rouk and Carter 1972). BRAZIL: Amazonas, Tabatinga (Velazco 2005, not mapped); Pará, Óbidos (Handley and Ferris 1972); Pará, Ilha de Marajó, between Anajá and Muaná (Marques-Aguiar et al. 2002); Acre, Rio Branco (Mok et al. 1982); Goiás, Fazenda Moinho (Coimbra et al. 1982); Distrito Federal, Fazenda Água Limpa (Coimbra et al. 1982). BOLIVIA (S. Anderson 1997): Santa Cruz, Castedo; Cochabamba, 12.5 km SW of Villa Tunari; Beni, Serranía Eva Eva. PERU: Puno, Coasa (Velazco 2005); Madre de Dios, Cocha Cashu (Ascorra, Wilson, and Romo 1991); Cusco, Las Malvinas (S. Solari et al. 1998); Pasco, Huancabamba (Velazco 2005); Huánuco, 3 miles S of Tingo María (type locality of *Vampyrops brachycephalus* Rouk and Carter); Amazonas, Santa Rosa (Velazco 2005). ECUADOR: Pastaza, Mera (Rageot and Albuja 1994).

SUBSPECIES: I recognize two subspecies of *P. brachycephalus.*

P. b. brachycephalus (Rouk and Carter, 1972)
SYNONYMS:

Vampyrops brachycephalus Rouk and Carter, 1972:1; type locality "3 mi. S Tingo Maria, 2400 ft, Huanuco, Peru."

Vampyrops latus Handley and Ferris, 1972:519; type locality "San Juan, 900 ft., Province of Oxapampa, Departamento de Pasco, Peru."

Platyrrhinus brachycephalus: Ascorra, Wilson, and Romo, 1991a:10; first use of current name combination.

This subspecies is found in Amazonian Colombia, Ecuador, Peru, Brazil, and Bolivia.

P. b. saccharus (Handley and Ferris, 1972)
SYNONYMS:

Vampyrops recifinus: Sanborn, 1955:413; part; not *Vampyrops recifinus* O. Thomas, 1901h).

Vampyrops latus saccharus Handley and Ferris, 1972:521; type locality "Manacal, 300 m, 5 km S and 25 km E Carúpano, Sucre, Venezuela."

This subspecies is known from northeastern Venezuela and northern Guyana, Surinam, and French Guiana.

NATURAL HISTORY: *Platyrrhinus brachycephalus* is found in subtropical humid forest habitats at elevations from sea level to 750 m in both Venezuela (Handley and Ferris 1972) and Peru (Gardner and Carter 1972b). Limited reproductive data (Wilson 1979; S. Anderson 1997) is consistent with bimodal polyestry. Ibáñez (1984a) found seeds of *Cecropia peltata* in the stomach of a Venezuelan specimen. R. Guerrero (1997) listed two species of streblid batflies found on *P. brachycephalus*. No data on roosting or on other aspects of its natural history are available. The karyotype is $2n = 30$, FN = 56 (R. J. Baker 1973).

REMARKS: Before its description by Rouk and Carter (1972), individuals were confused with the similar and sympatric congener *P. helleri*; therefore, natural history information gathered prior to 1972 is difficult to ascribe to *P. brachycephalus*. Although superficially similar, these species are easily identified on the basis of the morphology of pm4 and configuration of the ridge of bone extending posterolaterally from the base of the pterygoid processes (see couplet 8 in the key to the species of *Platyrrhinus*).

Although treated here as represented by two subspecies, D. C. Carter and Rouk (1973) considered *P. brachycephalus* to be monotypic. Velazco (2005) did not comment on subspecies in *P. brachycephalus*. Handley and Ferris (1972) suggested that *latus* (= *brachycephalus*) may prove to be conspecific with *P. recifinus*, but I have found no supporting evidence. Nevertheless, similarities led Sanborn (1955) to identify the Demerara, Guyana specimen of *P. brachycephalus* as *P. recifinus*.

Platyrrhinus chocoensis Alberico and Velasco, 1991
Chocoan Broad-nosed Bat
SYNONYMS:

Vampyrops dorsalis: M. E. Thomas, 1972:21; not *Vampyrops dorsalis* O. Thomas, 1900b.

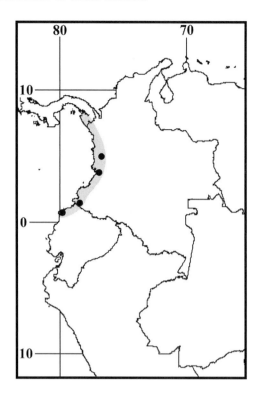

Map 181 Marginal localities for *Platyrrhinus chocoensis* ●

Vampyrops chocoensis Ziegler, 1989:319; *nomen nudum*.
Vampyrops chocoensis Alberico, 1990:348; *nomen nudum*.
Platyrrhinus chocoensis Alberico and Velasco, 1991:238: type locality "Quebrada El Platinero, 12 km W Istmina (by road), 5° 00′N, 76° 45′W, 100 m, Departamento del Chocó, Colombia."

DISTRIBUTION: In South America, *P. chocoensis* is known from the Pacific coastal plain of Colombia and northwestern Ecuador. Elsewhere, the species occurs in southern Panama (Darién, Tacarcuna Village).

MARGINAL LOCALITIES (Map 181): COLOMBIA: El Chocó, Quebrada El Platinero (type locality of *Vampyrops chocoensis* Alberico and Velasco); Valle del Cauca, Río Zabaletas (M. E. Thomas 1972, as *Vampyrops dorsalis*); Nariño, La Guayacana (USNM 309065). ECUADOR: Esmeraldas, Estero Taquiama (Albuja 1999).

SUBSPECIES: I consider *P. chocoensis* to be monotypic.

NATURAL HISTORY: *Platyrrhinus chocoensis* appears to be restricted to the Pacific coastal rain forests from southern Panama to northwestern Ecuador where all months are wet and annual average rainfall exceeds 7,000 mm. Population samples from Río Zabaletas, Colombia (M. E. Thomas 1972; identified as *Vampyrops dorsalis*) exhibited postpartum estrus and bimodal polyestry with pregnant females found in all months of the year except July, August, and September. The number of lactating females peaked in February, early in the comparatively "dry season" (January to April); followed by a lower peak in June during the win-

ter wet season. No lactating females were recorded in April or from August to January when from 75% to 100% of females caught were clearly pregnant. At least some males with enlarged testes were found in all months.

The type host of the nematode *Litomosoides colombiensis* Esslinger, 1973, identified as *Vampyrops dorsalis* from "vicinity of Buena Ventura, Valle [*sic*], Colombia" (Ubelaker, Specian, and Duszynski 1977), is a *P. chocoensis*.

REMARKS: *Platyrrhinus chocoensis* has long been confused with *P. dorsalis*, from which it is most easily distinguished by paler color, generally weaker and less contrasting dorsal stripe, and thinner (laterally compressed), sail-shaped main cuspid on last lower premolar.

Platyrrhinus dorsalis (O. Thomas, 1900)
Thomas's Broad-nosed Bat

SYNONYMS:

Vampyrops dorsalis O. Thomas, 1900b:269; type locality "Paramba, [Imbabura,] N. Ecuador. Alt. 1100 m."

Vampyrops lineatus: Bangs, 1900:100; not *Phyllostoma lineatum* É. Geoffroy St.-Hilaire, 1810b.

Vampyrops umbratus Lyon, 1902:151; type locality "San Miguel," La Guajira, Magdalena, Colombia.

Vampyrops oratus O. Thomas, 1914b:411; type locality "Galifari, Sierra del Ávila, [Distrito Federal,] N. Venezuela. Alt. 6500′."

Vampyrops aquilus Handley and Ferris, 1972:521; type locality "head of the Río Pucro, 4,100 ft., Cerro Malí, Darién, Panamá."

Platyrrhinus dorsalis: Alberico and Velasco, 1991:237; part; first use of current name combination.

Platyrrhinus umbratus: Koopman, 1993:191; name combination.

DISTRIBUTION: *Platyrrhinus dorsalis* is found at moderate to higher elevations in northern Colombia and northern Venezuela, south along the Andes into Ecuador. Elsewhere, the species is known from the Darién of Panama.

MARGINAL LOCALITIES (Map 182): COLOMBIA: Magdalena, Palamina (Lyon 1902). VENEZUELA (Handley 1976): Yaracuy, Minas de Aroa; Miranda, Curupao; Monagas, San Agustín, 3 to 5 km NW of Caripe; Mérida, La Carbonera. COLOMBIA: Meta, Restrepo (KU 158265); Huila, El Parque Nacional de la Cueva de los Guácharos (Lemke et al. 1982). ECUADOR: Napo, Alto Coca (Albuja 1999); Pastaza, Mera (Albuja 1999); Pichincha, Nanegal (Velazco 2005); El Oro, 1 km SW of Puente de Moromoro (USNM 513465); Esmeraldas, Borbón (Velazco 2005). COLOMBIA: Cauca, Munchique (Tamsitt and Valdivieso 1966a); Valle del Cauca, El Silencio (Velazco 2005); Antioquia, Betania (Muñoz 2001).

SUBSPECIES: I am treating *P. dorsalis* is as monotypic (see Remarks).

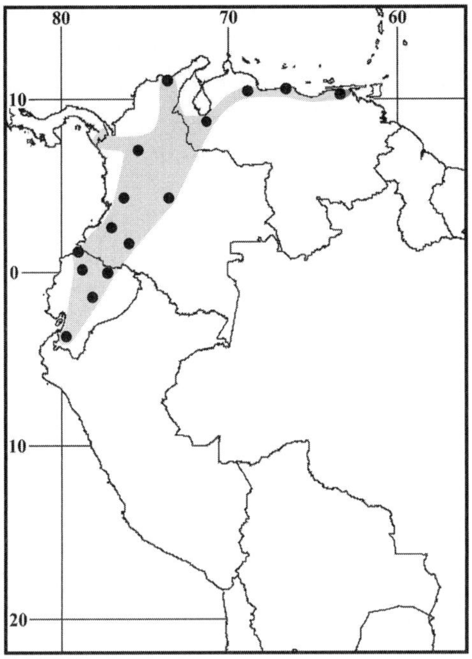

Map 182 Marginal localities for *Platyrrhinus dorsalis* ●

NATURAL HISTORY: In Venezuela this species was netted in humid to very humid forest habitats at elevations from 395 to 2,550 m, though rarely below 1,000 m (Handley 1976). Pregnant Venezuelan *P. dorsalis* have been taken in January, March, April, July, and November; lactating females in May, July, August, and October. *Platyrrhinus dorsalis* displays post-partum estrus and bimodal polyestry (Wilson 1979). This species consumes fruits and insects (Gardner 1977c), and has been found roosting in fissures in limestone cliffs and under overhanging roots at the top of a canyon wall (Sanborn 1955). Individuals were netted at the mouth of a limestone cave in Colombia (Lemke et al. 1982).

REMARKS: Velazco (2005) restricted the name *P. dorsalis* to populations in Venezuela, Colombia, and Ecuador. He described populations in Peru and Bolivia, previously identified as *P. dorsalis*, as representing his new species, *P. ismaeli* and *P. masu*. Although herein I have followed Velazco's (2005) basic concept of the species, preliminary unpublished information suggests that *P. dorsalis* has a more restricted distribution (southern Colombia and northwestern Ecuador). If that proves to be true, the correct name for the larger distribution in Venezuela, Colombia, and Ecuador is *P. umbratus*. Most authors followed Sanborn (1955) in treating *P. umbratus* as a junior synonym of *P. dorsalis*. Others (Handley 1976; Honacki, Kinman, and Koeppl 1982; Koopman 1982, 1993, 1994; Corbet and Hill 1991), treated *P. umbratus* as separate species. The type specimen is one of three first reported (Bangs 1900) from northern Colombia under the name *Vampyrops lineatus*. Unfortunately, the skulls of two of the three specimens are

missing, and the proximal ends of the forearms appear to have been removed from the holotype during preparation, rendering the correct forearm length difficult to determine.

Literature records for *P. dorsalis* are untrustworthy. Clearly, reports of *P. dorsalis* from Pacific coastal Colombia (e.g., Aellen 1970; M. Thomas 1972) and from Peru and Bolivia are based on misidentifications. The Paraguayan record mapped by Redford and Eisenberg (1992) may represent the specimen reported by Sherman (1955) as "*Vampyrops zarhinus*," which was later reidentified as a *Pygoderma bilabiatum*. Although Gardner and Carter (1972b) treated *P. umbratus* and *P. oratus* as synonyms of *P. dorsalis*, Koopman (1982) and Handley (1976 allocated *P. oratus* (along with *P. aquilus*) to *P. umbratus*, and considered *P. dorsalis* to be a separate species. J. K. Jones and Carter (1976) suggested that both *P. aurarius* and *P. umbratus* could be either subspecies or junior synonyms of *P. dorsalis*. References to *P. dorsalis* (under the name *Vampyrops dorsalis*) as the host species of parasites cited by Ubelaker, Specian, and Duszynski (1977) and by J. P. Webb and Loomis (1977) actually apply to *P. chocoensis*. R. J. Baker (1973) gave the karyotype for *P. dorsalis* as $2n = 30$; $FN = 56$ with a subtelocentric X and a submetacentric Y chromosome. Although this appears to be the uniform karyotype known for species of *Platyrrhinus*, it is not clear that Baker's specimen was correctly identified.

Platyrrhinus helleri (W. Peters, 1866)
Heller's Broad-nosed Bat

SYNONYMS: See under Subspecies.

Distribution: *Platyrrhinus helleri* occurs in Colombia, Venezuela, Guyana, Surinam, French Guiana, Ecuador, Brazil, Peru, and Bolivia. Elsewhere, *P. helleri* is found in southern Mexico and throughout Central America.

MARGINAL LOCALITIES (Map 183): COLOMBIA: Magdalena, Cacagualito (Velazco 2005). VENEZUELA: Falcón, 16 km ENE of Mirimire (Handley 1976); Distrito Federal, San Julián (W. Robinson and Lyon 1901); Sucre, Ensenada Cauranta, 9 km NE of Güiria (Handley 1976). TRINIDAD AND TOBAGO: Trinidad, Las Cuevas (C. H. Carter et al. 1981). VENEZUELA: Delta Amacuro, Los Güires (Ojasti and Naranjo 1974). GUYANA: Upper Demerara-Berbice, Dubulay Ranch (USNM 582325). SURINAM: Paramaribo, Paramaribo (Husson 1962). FRENCH GUIANA: 1 km N of Rémire (S. L. Williams, Phillips, and Pumo 1990). BRAZIL (Taddei and Vicente-Tranjan 1998, except as noted): Amapá, Horto Florestal de Macapá (Peracchi, Raimundo, and Tannure 1984); Pará, Belém (Handley 1967); Pará, Rio Xingu, 52 km SSW of Altamira (USNM 5495422); Mato Grosso, Serra do Roncador, 264 km N by road of Xavantina (Pine, Bishop, and Jackson 1970); Mato Grosso do Sul, Aporé; Minas Gerais, Uberaba; São Paulo,

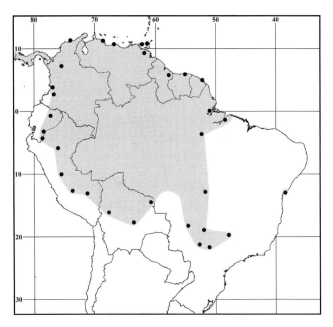

Map 183 Marginal localities for *Platyrrhinus helleri* •

Adamantina; São Paulo, Xavantina; Mato Grosso do Sul, 25 km N of Coxim. BOLIVIA: Santa Cruz, Los Fierros (L. H. Emmons, pers. comm.); Santa Cruz, Río Pitasama, 4.5 km N and 1.5 km E of Cerro Amboró (S. Anderson et al. 1993); La Paz, Chijchipa (S. Anderson 1997). PERU: Cusco, Consuelo (V. Pacheco et al. 1993); Ayacucho, Apurímac (Koopman 1978); Pasco, Pozuzo (type locality of *Vampyrops zarhinus incarum* O. Thomas); Loreto, Yurimaguas (Sanborn 1949b). ECUADOR (Albuja 1999): Zamora-Chinchipe, Miazi; Morona-Santiago, San Carlos de Limón; Napo, Río Pucuno. COLOMBIA: Cauca, Betania (Velazco 2005); Valle del Cauca, Río Zabaletas (M. E. Thomas 1972); Antioquia, Valdivia (BM 98.10.3.34). *Isolated eastern record*: BRAZIL: Bahia, Bahia (BM 49.10.15.42).

SUBSPECIES: I recognize two subspecies of *P. helleri* in South America.

P. h. helleri (W. Peters, 1866)
SYNONYMS:
Vampyrops Helleri W. Peters, 1866c:392; type locality "Mexico."
Vampyrops zarhinus H. Allen, 1891b:400; type locality "Brazil," corrected by Goldman (1920:200) to "Bas Obispo, Canal Zone," Panama.
Platyrrhinus helleri: Hall and Kelson, 1959:131; first use of current name combination.

This subspecies occurs in northern Colombia, east into Venezuela, the island of Trinidad, the Guianas, and eastern Brazil. Elsewhere, the nominate subspecies is found in Central America and southern Mexico.

P. h. incarum (O. Thomas, 1912)
SYNONYMS:
Vampyrops zarhinus incarum O. Thomas, 1912f:409, type locality "Pozuzo, [Pasco,] Peru.
[*Vampyrops helleri*] *incarum*: J. K. Jones and Carter, 1976: 23; name combination.
P[*latyrrhinus*]. *h*[*elleri*]. *incarum*: Ferrell and Wilson, 1991:1; first use of current name combination.

This subspecies is known from eastern Colombia south through eastern Ecuador, eastern Peru, and western Brazil, to south-central Bolivia.

NATURAL HISTORY: *Platyrrhinus helleri* is a frugivore known to consume insects (Gardner 1977c). The species inhabits humid tropical and subtropical forest habitats at elevations from 24 to 1,200 m (Gardner and Carter 1972b; Handley 1976). Known roosting sites include foliage, branches, hollow trees and logs, as well as caves, tunnels, buildings, bridges, and road culverts (Tuttle 1976). The species exhibits postpartum estrus and bimodal polyestry (T. H. Fleming, Hooper, and Wilson 1972; Wilson 1979; Taddei and Vincente-Tranjan 1998). R. J. Baker (1967) and R. J. Baker et al. (1982) reported a $2n = 30$, FN = 56 karyotype having a subtelocentric X and a submetacentric Y chromosome. Ferrell and Wilson (1991) provided measurements and additional natural history information in their *Mammalian Species* account. Taddei and Vincente-Tranjan (1998) supplemented this information in their review of Brazilian *P. helleri*.

REMARKS: Cabrera (1958) listed this species under the name *Vampyrops zarhinus*; while acknowledging that *zarhinus* probably was a synonym of *helleri*. Ferrell and Wilson (1991) pointed out that the supposed Paraguayan specimen (Sherman 1955) was subsequently reidentified as a *Pygoderma bilabiatum*. They erred, however, in stating (page 4) that *helleri* is "a patronym for Edmund Heller, who collected the type specimen in Mexico." The Dr. Heller referred to by W. Peters (1866c:394) was not Edmund Heller. The earlier literature referred to specimens from Mexico and Central America as "*Vampyrops lineatus*" (e.g., Alston 1879; Ferrari-Pérez 1886; J. A. Allen 1893b; Elliot 1904). The record of *P. helleri* from Bahia, Brazil, is based on a specimen in the Natural History Museum, London; its identity has been confirmed by J. Arroyo-Cabrales (pers. comm.).

Platyrrhinus infuscus (W. Peters, 1880)
Buffy Broad-nosed Bat
SYNONYMS:
Vampyrops infuscus W. Peters, 1880:259; type locality "2 mi. N Tingo María, 2000 ft., Provincia de Leoncio Prado, Departamento de Huánuco, Peru" (Gardner and Carter 1972b:72; neotype designation); original type locality "Grotte von Ninabamba," Cajamarca, Peru.

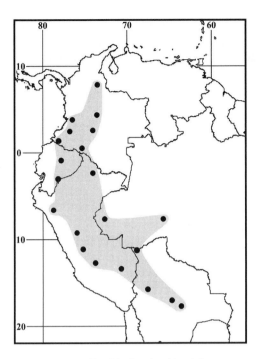

Map 184 Marginal localities for *Platyrrhinus infuscus* ●

Vampyrops fumosus Miller, 1902b:405; type locality "Purus River, Brazil"; identified as Hyutanaham, Rio Purus, Amazonas, Brazil by Lyon and Osgood (1909:266).

Vampyrops intermedius Marinkelle, 1970:49; type locality "'Mina de Upin,' near Restrepo (4° 18'N, lat., 73° 34'W, long., 550 m elev.), Dept. del Meta, Colombia."

Platyrrhinus infuscus: Ascorra, Wilson, and Romo, 1991a: 10; first use of current name combination.

DISTRIBUTION: *Platyrrhinus infuscus* is in Colombia, Ecuador, Peru, Brazil, and Bolivia.

MARGINAL LOCALITIES (Map 184): COLOMBIA: El Cesar, La Palma (Marinkelle 1970); Meta, Mina de Upin (type locality of *Vampyrops intermedius* Marinkelle); Meta, Cabaña Duda (Lemke et al. 1982); Putumayo, Río Mecaya (FMNH 72125). PERU: Loreto, Boca del Río Curaray (Sanborn 1955). BRAZIL: Acre, Seringal Lagoinha (Taddei, Rezende, and Camora 1990); Amazonas, Hyutanaham (type locality of *Vampyrops fumosus* Miller). BOLIVIA: Pando, Río Nareuda (S. Anderson 1997); Santa Cruz, Río Pitasama, 4.5 km N and 1.5 km E of Cerro Amboró (S. Anderson et al. 1993); Cochabamba, Sajta (S. Anderson 1997); La Paz, 6.6 km by road downstream from Caranavi (S. Anderson, Koopman, and Creighton 1982). PERU: Cusco, Hacienda Cadena (Sanborn 1951b); Ayacucho, San José (Gardner and Carter 1972b); Junín, Río Tulumayo, 3.2 km N of Vitoc, (USNM 507196); Huánuco, 2 miles N of Tingo María (type locality of *Vampyrops infuscus*: Gardner and Carter 1972b); Cajamarca, Gruta de Niñabamba (original type locality of *Vampyrops infuscus* W. Peters). ECUADOR

(Albuja 1983): Morona-Santiago, Los Tayos; Napo, Cavernas de Jumandi. COLOMBIA: Nariño, Junín (Alberico and Orejuela 1982); Cauca, Mina California (Marinkelle 1970); Valle del Cauca, Alto del Oso (Marinkelle 1970).

SUBSPECIES: I consider *P. infuscus* to be monotypic.

NATURAL HISTORY: This species has been captured in mist nets or obtained from shallow caves and grottoes in humid tropical forests (Marinkelle 1970; Gardner and Carter 1972b). Gardner and Carter (1972b) suggested that the weakly defined facial and dorsal stripes reflected the use of caves and grottoes as roosting sites. The species obviously uses other types of roosts because caves and grottoes are scarce in the western Amazon basin of Brazil, Colombia, Ecuador, Peru, and Bolivia. Marinkelle (1970) took one pregnant and three lactating females on 19 March 1969 at Mina de Upin, the type locality of *Vampyrops intermedius*. The karyotype is $2n = 30$, $FN = 56$, with a subtelocentric X and a submetacentric Y chromosome (Gardner 1977a).

REMARKS: The holotype of *Vampyrops infuscus* W. Peters, 1880 was deposited in the Warsaw Museum and is considered to have been destroyed by fire during WWII (see Gardner and Carter 1972b:72). The original type locality, although located on the western drainage of the Peruvian Andes, is in cloud-mediated, humid tropical forest habitat.

Platyrrhinus ismaeli Velazco, 2005
Ismael's Broad-nosed Bat
SYNONYMS:

Vampyrops dorsalis: J. A. Allen, 1916c:226; not *Vampyrops dorsalis* O. Thomas, 1900.

[*Vampyrops*] *vittatus*: Sanborn, 1955: 409; not *Artibeus vittatus* W. Peters, 1860.

cf. *Vampyrops aurarius*: Lemke, Cadena, Pine, and Hernández-Camacho, 1982:230; not *Vampyrops aurarius* Handley and Ferris, 1972.

Platyrrhinus dorsalis: Pacheco and Patterson, 1991:101; part; not *Vampyrops dorsalis* O. Thomas.

Platyrrhinus "dorsalis norte:" Solari, Vivar, Velazco, and Rodríguez, 2001:263; not *Vampyrops dorsalis* O. Thomas.

Platyrrhinus dorsalis "Norte:" Velazco and Solari, 2003: 312; not *Vampyrops dorsalis* O. Thomas.

Platyrrhinus ismaeli Velazco, 2005:27; type locality "19 km E of Balsas, 1945 m (6380 ft) in elevation, Province of Chachapoyas, Department of Amazonas, Peru."

DISTRIBUTION: *Platyrrhinus ismaeli* is known from southern Colombia, Ecuador, and north-central Peru.

MARGINAL LOCALITIES (Map 185): COLOMBIA: Cauca, 1 mile N of Moscopas (USNM 483575); Huila, El Parque Nacional Natural de la Cueva de los Guácharos (Lemke et al. 1982, as cf *Vampyrops aurarius*). ECUADOR:

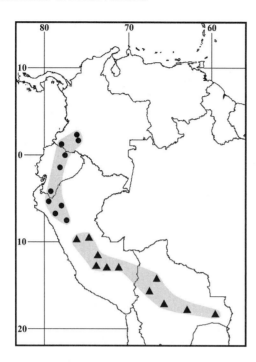

Map 185 Marginal localities for *Platyrrhinus ismaeli* ● and *Platyrrhinus masu* ▲

Napo, Cascada de San Rafael (Albuja 1999, as *P. dorsalis*); Pastaza, Mera (USNM 548220). PERU: Amazonas, 6 km SW of Laguna Pomacochas (LSUMZ 19056); San Martín, Las Palmas (Velazco 2005); Cajamarca, 35 mi WNW of Cajamarca (MVZ 137916); Piura, 15 road km E of Canchaque (LSUMZ 19053). ECUADOR: Loja, San Pedro de Vilcabamba (Velazco 2005). COLOMBIA: Nariño, Ricuarte (J. A. Allen 1916).

SUBSPECIES: I treat *P. ismaeli* as monotypic.

NATURAL HISTORY: Based on label information on USNM specimens from eastern Ecuador (Mera), a female contained a 20 mm fetus when captured on 25 January; another contained a 40 mm fetus when taken on 8 March. Additional natural history information clearly attributable to this taxon is lacking.

REMARKS: The specimen from "Ricarte" (AMNH 34232), southern Colombia reported by Sanborn 1955 as a skin of *dorsalis* and skull of *vittatus* is the same specimen J. A. Allen (1916c) listed as *Vampyrops dorsalis*. Velazco (2005:50) included this specimen under *Platyrrhinus vittatus*.

Platyrrhinus lineatus (É. Geoffroy St.-Hilaire, 1810)
White-lined Broad-nosed Bat

SYNONYMS:

Phyllostoma lineatum É. Geoffroy St.-Hilaire, 1810b:180; type locality "Paraguay"; based on Azara's (1801b:271) "*Chauve-souris brune et rayée*"; type locality restricted to Asunción, Central, Paraguay, by Cabrera (1958:80).

Phyllostomus lineatus Illiger, 1815; *nomen nudum*.

Ph[yllostomus]. lineatus Olfers, 1818:224; type locality "Paraguay"; attributed to Illiger, 1815:109; based on Azara's (1801b:271) "*Chauve-souris brune et rayée.*"

Arctibeus lineatus: Gray, 1838b:487; name combination.

Artibaeus lineatus: P. Gervais, 1856a:35; name combination.

[Vampyrops] lineatus: W. Peters, 1865b:356; name combination.

Stenoderma (Vampyrops) lineatum: Pelzeln, 1883:34; name combination.

Vampyrops lineatus sacrillus O. Thomas, 1924a:236; type locality "Rio Doce, Espiritu Santo," Brazil.

Platyrrhinus lineatus: Koopman, 1993:191; first use of current name combination.

Platyrrhinus (Vampyrops) lineatus: Zortéa, 1996:59; name combination.

DISTRIBUTION: An isolated segment of the population of *P. lineatus* is represented by a few specimens from localities in French Guiana and a single specimen from a locality in southern Surinam. The major distribution is from Bolivia across Brazil, Paraguay, and northern Argentina into northern Uruguay, and northward along the Atlantic Coast of Brazil at least as far as northern Ceará.

MARGINAL LOCALITIES (Map 186): *Northern Distribution:* FRENCH GUIANA: Les Nouragues (Brosset and Charles-Dominique 1991). SURINAM: Sipaliwini, Sipaliwini airstrip (S. L. Williams, Genoways, and Groen 1983). *Southern Distribution:* BRAZIL: Ceará, Parque Nacional de Ubajara (S. S. P. Silva, Guedes, and Peracchi 2001); Ceará, Faculdade de Veterinária do Ceará, Fortaleza

Map 186 Marginal localities for *Platyrrhinus lineatus* ●

(Piccinini 1974); Pernambuco, São Lourenço da Mata (Souza and Araújo 1990); Bahia, Bahia (P. Gervais 1856a); Espírito Santo, Município de Linhares (Peracchi and Albuquerque 1993); Rio de Janeiro, Rio de Janeiro (C. O. C. Vieira 1942); São Paulo, Ilha Victoria (J. L. Lima 1926); São Paulo, São Paulo (M. M. S. Silva, Harmani, and Gonçalves 1996); São Paulo, Fazenda Intervales (Portfors et al. 2000); Santa Catarina, Colonia Hansa (J. L. Lima 1926); Paraná, Alto Tibagi (N. R. Reis et al. 2000); Paraná, Salto Grande (USNM 141392). ARGENTINA: Misiones, Colonia Mártires (Barquez, Mares, and Braun 1999). URUGUAY: Artigas, Isla Rica (Ximénez 1969). ARGENTINA: Chaco, Resistencia (Fornes and Massoia 1966); Formosa, Parque Nacional Río Pilcomayo (Barquez, Mares, and Braun 1999). PARAGUAY: Concepción, Belén (Baud 1989). BRAZIL: Mato Grosso do Sul, Urucúm (J. A. Allen 1916e). BOLIVIA (S. Anderson 1997, except as noted): Santa Cruz, Santiago (S. Anderson, Koopman, and Creighton 1982); Santa Cruz, 27 km SE of Santa Cruz (S. Anderson et al. 1993); Beni, Puerto San Lorenzo, Río Sécure; Beni, Espíritu; Beni, 2 km from mouth of Río Yacuma (S. Anderson, Koopman, and Creighton 1982); Santa Cruz, El Encanto. BRAZIL: Mato Grosso, Chapada (Cope 1889a); Mato Grosso, Serra do Roncador, 264 km N by road of Xavantina (Pine, Bishop, and Jackson 1970); Piauí, Paranaguá (Toldt 1908); Ceará, Floresta Nacional Araripe-Apodi, 21 km SSW of Crato (Mares et al. 1981).

SUBSPECIES: I am treating *P. lineatus* as monotypic. The western populations in Bolivia and the northern isolate in Surinam and French Guiana warrant description; the taxon needs to be revised.

NATURAL HISTORY: *Platyrrhinus lineatus* has been found in a variety of habitats ranging from the humid Atlantic forests to the arid cerrado and caatinga habitats of Brazil. Willig (1985c) wrote that this species exhibited bimodal polyestry in northeastern Brazil; whereas Taddei (1976) reported nearly continuous reproductive activity throughout the year. Willig and Hollander (1986) summarized the diet, which includes fruits, flowers, and some insects. Zortéa (1996) and Aguiar (2005) documented folivory by Brazilian *P. lineatus*, which eat the leaves of *Solanum* spp. Pieces of the green leaves are chewed, the liquid component consumed, and the fibrous residue dropped as pellets (Aguiar 2005). Roosting sites include branches and foliage (Taddei 1973; Tuttle 1976) and caves (Willig 1983). Willig (1983) also suggested that males maintained small harems consisting of from 7 to 15 females; Peracchi and Albuquerque (1971) reported roosting groups of from 6 to 20 individuals. The listing of *Trypanosoma lineatus* as a parasite of *P. lineatus* by Ubelaker, Specian, and Duszynski (1977) is based on a misidentified host, possibly *P. helleri*, from Venezuela. J. P. Webb and Loomis

(1977) listed a spinturnicid mite and three streblid flies as ectoparasites. The karyotype of a male (USNM 555706) from Floresta Nacional Araripe (near Crato), Ceará, Brazil, is $2n = 30$; FN = 56, with a subtelocentric X and a subtelocentric Y chromosome (also see Souza and Araújo 1990).

REMARKS: When Cabrera (1958:81) restricted the type locality to Asunción, Paraguay, he stated that Azara sent two specimens to Madrid and these specimens were later destroyed. Rode (1941), perhaps misinterpreting the basis for É. Geoffroy St.-Hilaire's (1810b) description, mistakenly assumed the holotype was in the Paris Museum (MNHN). Apparently overlooking Cabrera's remarks, D. C. Carter and Dolan (1978:52) reported on MNHN 953, which they also assumed was the holotype. This situation is identical to that of *Phyllostoma lilium* (= *Sturnira lilium*), also described by É. Geoffroy St.-Hilaire (1810b) on the page after his description of *Phyllostoma lineatum*. Cabrera (1958:81) was correct in writing that É. Geoffroy St.-Hilaire (1810b) based *Phyllostoma lineatum* solely on Azara's (1801b:271) description of "*chauve-souris seconde, ou chauve-souris brune et rayée*." The specimen (MNHN 953) in the Paris museum, whose features É. Geoffroy St.-Hilaire (1818a:416–417) later contrasted with Azara's description, was acquired after Geoffroy described *Phyllostoma lineatum*. This is the specimen erroneously cited by Rode (1941) and D. C. Carter and Dolan (1978) as the "holotype," but it is not a type specimen.

Koopman (1978) enlarged the distribution of *P. lineatus* to include Colombia, Ecuador, and Peru when he synonymized *Vampyrops nigellus* Gardner and Carter, 1972a under *P. lineatus*. Although *P. nigellus* is easily distinguished from *P. lineatus* (see Velazco and Solari 2003; Velazco 2005), most authors have followed Koopman's (1976) treatment of *P. nigellus* as a junior synonym of *P. lineatus* (e.g., Koopman 1982, 1993, 1994; Willig and Hollander 1986; Eisenberg 1989; Eisenberg and Redford 1999; Corbet and Hill 1991; S. Anderson 1997; Albuja 1999; Simmons 2005). Some of the earlier literature (e.g., Dobson 1878; J. L. Lima 1926; Pittier and Tate 1932; Podtiaguin 1944) also erroneously included northwestern South America, Central America, and Mexico in the distribution of *P. lineatus*. As late as 1946, Goodwin (p.320) included *Vampyrops lineatus* in the fauna of Costa Rica. The type specimen of *Vampyrops umbratus* was originally identified by Bangs (1900) as *Vampyrops lineatus*. An earlier report of *Vampyrops lineatus* from Peru (J. A. Allen 1897b) was based on five misidentified *Uroderma bilobatum*. Marinkelle and Cadena (1972:55) attribute to Goodwin ("1953:259") a record of *Vampyrops lineatus* from 2,000 m in Peru. I have not found mention of any *Vampyrops lineatus* in Goodwin (1953), and suspect that

Marinkelle and Cadena meant to cite the J. A. Allen (1897b) report in which the specimens were recorded as taken at an elevation of 6,000 ft. The specimen Marinkelle and Cadena (1972:55) reported from Laguna de Pedro Palo, Cundinamarca, Colombia, most likely is a *P. dorsalis*. Contrary to the listings by Cuervo-Díaz, Hernández-Camacho, and Cadena (1986), Eisenberg (1989), and Muñoz (2001), *P. lineatus* has not been found in Colombia.

Platyrrhinus masu Velazco, 2005
Quechuan Broad-nosed Bat

SYNONYMS:

Vampyrops dorsalis: Sanborn, 1951:10; not *Vampyrops dorsalis* O. Thomas, 1900.

Platyrrhinus dorsalis: S. Anderson, 1993:8; not *Vampyrops dorsalis* O. Thomas.

Platyrrhinus "dorsalis sur:" Solari, Vivar, Velazco, and Rodríguez, 2001:263; not *Vampyrops dorsalis* O. Thomas.

Platyrrhinus dorsalis "Centro-Sur:" Velazco and Solari, 2003:312; not *Vampyrops dorsalis* O. Thomas.

Platyrrhinus masu Velazco, 2005:32; type locality "Consuelo, km 165, 17 km by road west of Pilcopata, Province of Paucarambo, Department of Cuzco, Peru."

DISTRIBUTION: *Platyrrhinus masu* is known from Peru and Bolivia.

MARGINAL LOCALITIES (Map 185): PERU: Huánuco, Cerros del Sira (AMNH 233631); Cusco, Camp Two (Emmons, Luna, and Romo 2001); Madre de Dios, Hacienda Amazonía (Velazco 2005). BOLIVIA: Beni, Espíritu (S. Anderson 1997); Santa Cruz, Santiago (S. Anderson, Koopman, and Creighton 1982); Santa Cruz, Santa Cruz de la Sierra (S. Anderson et al. 1993); Cochabamba, El Sillar (Ibáñez 1985a); La Paz, 35 km by road N of Caranavi at Serranía Bella Vista (S. Anderson, Koopman, and Creighton 1982). PERU: Cuzco, Hacienda Huyro (LSUMZ 19060); Ayacucho, Yuraccyacu (Gardner and Carter 1972b); Huánuco, eastern slope Cordillera Carpish, Carretera Central (Gardner and Carter 1972b).

SUBSPECIES: I consider *P. masu* as monotypic.

NATURAL HISTORY: *Platyrrhinus masu* has been taken along the eastern Andean slope of Peru and Bolivia in forest habitat at elevations between approximately 600 and 2,500 m.

REMARKS: This species has been reported most commonly as a large southern morph of *P. dorsalis* (e. g., S. Anderson 1997; Velazco and Solari 2003).

Platyrrhinus matapalensis Velazco, 2005
Matapalo Broad-nosed Bat

SYNONYMS:

Vampyrops helleri: Sanborn, 1955:412; not *Vampyrops helleri* W. Peters, 1866c.

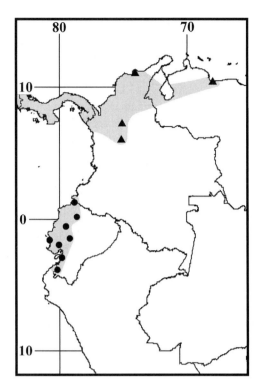

Map 187 Marginal localities for *Platyrrhinus matapalensis* ● and *Platyrrhinus vittatus* ▲

Platyrrhinus helleri: Ferrell and Wilson, 1991:1; part; not *Vampyrops helleri* W. Peters.

Platyrrhinus matapalensis Velazco, 2005:37; type locality "Matapalo, 54 m in elevation, Province of Zarumilla, Department of Tumbes, Peru, approximately 3°40′S, 80°12′W."

DISTRIBUTION: *Platyrrhinus matapalensis* is known from western Ecuador and northwestern Peru, and is expected from southwestern Colombia.

MARGINAL LOCALITIES (Map 187; from Velazco 2005, except as noted): ECUADOR: Esmeraldas, San Lorenzo (Albuja 1983); Pichincha, Playa Rica; Pichincha, Centro Científico Río Palenque (USNM 528532); Bolívar, 3 km SW of Echeandía; Guayas, San Rafael (USNM 522422). PERU: Tumbes, Campo Verde. ECUADOR: Guayas, La Unión; Manabí, Río Blanco (Albuja 1983).

SUBSPECIES: *Platyrrhinus matapalensis* is monotypic.

NATURAL HISTORY: *Platyrrhinus matapalensis* has been collected in primary forest and in a variety of secondary habitats on the Pacific lowlands of eastern Ecuador and northwestern Peru, apparently all below 800 m.

REMARKS: Until Velazco's (2005) description, *P. matapalensis* attracted little attention because the population was assumed to represent *P. helleri*, which it most closely resembles in size and other characteristics. *Platyrrhinus matapalensis* is easily distinguishable from *P. helleri*, and the also similar *P. brachycephalus*, by the smooth anterolingualr cristid of the last lower premolar. *Platyrrhinus helleri* usually has one small cuspulid on the anterolingual cristid, and *P. brachycephalus* always has two (see Velazco 2005:50, Fig. 27). The morphology of the lateral margin of the pterygoid process matches that of *P. helleri* (open C-shaped; see cuplet 8 in the key to the species of *Platyrrhinus*).

Platyrrhinus nigellus (Gardner and Carter, 1972)
Little Black Broad-nosed Bat

SYNONYMS:

Vampyrops nigellus Gardner and Carter, 1972a:1; type locality "Huanhuachayo (12°44'S, 73°47'W), about 1660 m, Departamento de Ayacucho, Peru."

Vampyrops lineatus nigellus: Koopman, 1978:11; name combination.

P[latyrrhinus]. lineatus nigellus: S. Anderson, 1997:12; name combination.

P[latyrrhinus]. nigellus: Velazco and Solari, 2003: 303; first use of current name combination.

DISTRIBUTION: *Platyrrhinus nigellus* is found from northern Colombia through Ecuador, Peru, and Bolivia.

MARGINAL LOCALITIES (Map 188; from Velazco, except as noted): COLOMBIA: El Cesar San Sebastián; Nariño, El Carmen. ECUADOR: Napo, Cascade de San Rafael; Pastaza, Mera (USNM 548208). PERU: Amazonas, Puesto Vigilancia 3, Alfonso Ugarte (USNM 581960); San Martín, Las Palmas; Huánuco, Divisoria, Cordillera Azul

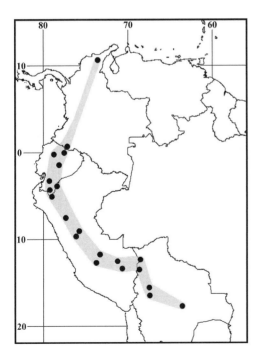

Map 188 Marginal localities for *Platyrrhinus nigellus* ●

(Gardner and Carter 1972a); Cusco, Ridge Camp (USNM 588031); Madre de Dios, Cerro de Pantiacolla (Velazco and Solari 2003). BOLIVIA: Pando, Chivé (S. Anderson 1997); La Paz, 47 km by road N of Caranavi at Serranía Bella Vista (S. Anderson, Koopman, and Creighton 1982); Santa Cruz, Río Pitasama, 4.5 km N and 1.5 km E of Cerro Amboro; La Paz, Lavi Grande (S. Anderson 1997); La Paz, Aserradero Moira (S. Anderson 1997). PERU: Cusco, Hacienda Cadena; Ayacucho, Huanhuachayo (type locality of *Vampyrops nigellus* Gardner and Carter); Huánuco, eastern slope Cordillera Carpish, Carretera Central (Gardner and Carter 1972a); Cajamarca, San Ignacio (Velazco and Solari 2003). ECUADOR: Loja, Masanamaca; Azuay, Valle Yunguilla; Pichincha, Estación Forestal La Favorita.

SUBSPECIES: I am treating *P. nigellus* as monotypic.

NATURAL HISTORY: *Platyrrhinus nigellus* has been found in tropical primary and disturbed forest habitats along the lower slopes of the Sierra de Santa Marta in northern Colombia and along both slopes of the Andes in Colombia and Ecuador. In Peru and Bolivia, the species appears to be restricted to the eastern slopes and valleys of the Andes. A pregnant female was caught in July in Peru. There is no published information on reproductive pattern, roosting behavior, and dietary habits (other than general frugivory) on this species.

REMARKS: Following Koopman's (1976, 1978) treatment of *Vampyrops nigellus* as a junior synonym of *Vampyrops lineatus*, most authors (e.g., Koopman 1982, 1993; Eisenberg 1989; Corbet and Hill 1991; Eisenberg and Redford 1999; Albuja 1999) uncritically included Colombia, Ecuador, and Peru in the distribution of *P. lineatus* (also see Remarks in the species account for *P. lineatus*). Although several authors have treated *nigellus* as a subspecies of *P. lineatus* (e.g., Koopman 1978, 1994; Willig and Hollander 1986; S. Anderson 1997), Velazco and Solari (2003) and Velazco (2005) clearly distinguished each name as applying to separate species.

Platyrrhinus nigellus can be distinguished from similar congeners by its generally smaller size and a variety of cranial and dental characteristics, of which the most readily usable when examining prepared specimens is the lack of a stylar cusp[ule] on the lingual cingulum of the M2 metacone (terminology follows Velazco 2005). Some *P. nigellus* lack both the cuspule and the cingulum on the M2 metacone.

Platyrrhinus recifinus (O. Thomas, 1901)
Recife Broad-nosed Bat

SYNONYMS:

Vampyrops recifinus O. Thomas, 1901h:192 (footnote): type locality "Pernambuco," Pernambuco, Brazil.

Platyrrhinus recifinus: Koopman, 1993:191; first use of current name combination.

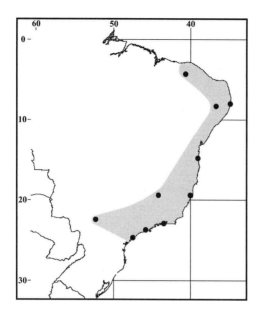

Map 189 Marginal localities for *Platyrrhinus recifinus* ●

DISTRIBUTION: *Platyrrhinus recifinus* is endemic to eastern and southeastern Brazil.

MARGINAL LOCALITIES (Map 189): BRAZIL: Ceará, Ipu (Piccinini 1974); Pernambuco, Pernambuco (type locality of *Vampyrops recifinus* O. Thomas); Bahia, Ilhéus (Faria, Soares-Santos, and Sampaio 2006); Espírito Santo, Município de Linhares (Peracchi and Albuquerque 1993); Rio de Janeiro, Parque Estadual da Pedra Branca (Dias, Peracchi, and Silva 2002); São Paulo, Guaratuba (USNM 542611); São Paulo, Iguapé (USNM 542613); São Paulo, Pontal do Paranapanema (N. R. Reis et al. 1996); Minas Gerais, Sete Lagoas (Sanborn 1955); Pernambuco, Pesqueira (C. O. C. Vieira 1942).

SUBSPECIES: I am treating *P. recifinus* as monotypic.

NATURAL HISTORY: Little is known about *P. recifinus*. A specimen from the state of Ceará, Brazil, was shot in a banana plantation.

REMARKS: Although J. K. Jones and Carter (1976) suggested that *P. recifinus* may prove to be a junior synonym of *P. lineatus*, D. C. Carter and Dolan (1978) stated that *P. recifinus* and *P. lineatus* were clearly distinguishable by the lobation of the lower incisors. Apparently, they were referring to the often trilobed upper incisors, a feature that is helpful when distinguishing *P. recifinus* from the larger *P. lineatus*, and from the smaller *P. helleri*, *P. matapalensis*, and *P. brachycephalus*.

Platyrrhinus vittatus (W. Peters, 1859)

Greater Broad-nosed Bat

SYNONYMS:

Artibeus vittatus W. Peters, 1859:225; type locality "Puerto Cabello," Carabobo, Venezuela.

V[ampyrops]. vittatus: W. Peters, 1865b:356; name combination.

Platyrrhinus vittatus: Hall and Kelson, 1959:132; first use of current name combination.

DISTRIBUTION: *Platyrrhinus vittatus* is found in Venezuela and Colombia. Elsewhere, the species is known from Panama and Costa Rica.

MARGINAL LOCALITIES (Map 187): COLOMBIA: Magdalena, Valparaiso (J. A. Allen 1900a). VENEZUELA: Carabobo, Puerto Cabello (type locality of *Vampyrops vittatus* W. Peters); COLOMBIA: Antioquia, Los Ríos (Muñoz 1990); Antioquia, Buenos Aires (USNM 499455).

SUBSPECIES: I am treating *P. vittatus* as monotypic.

NATURAL HISTORY: I have taken *P. vittatus* from under a stone railroad bridge in Costa Rica. J. P. Webb and Loomis (1977) listed three mites and three streblid flies known from *P. vittatus*. Wenzel (1976) listed two of these streblids from Venezuelan specimens. R. Guerrero (1985a) added a spinturnicid mite to the list of ectoparasites recovered from this species. *Platyrrhinus albericoi* most likely is the host for at least some of these ectoparasite records. Lactating females were taken in April in Colombia. Although it is unclear that true *P. vittatus* were represented among the specimens sampled, R. J. Baker (1973) and R. J. Baker et al. (1982) reported a $2n = 30$, FN = 56 karyotype with a subtelocentric X and acrocentric Y chromosome.

REMARKS: When Sanborn (1955) reviewed the genus *Vampyrops*, *P. vittatus* was known from Colombia and Venezuela, *P. infuscus* from northern Peru, and *fumosus* (= *P. infuscus*) from Brazil and southeastern Peru. Sanborn treated them all as synonyms of *vittatus*, believing they represented geographic variants of the same species. Reliance on Sanborn's (1955) locality records for *P. vittatus* has created confusion for some authors (e.g., Tuttle 1970; Albuja 1983) who used published locality records when they could not examine specimens while attempting to distinguish the distribution of *P. vittatus* from that of *P. infuscus*. The Ecuadorian, Peruvian, and Bolivian material previously reported as *P. vittatus* has recently been described as *P. albericoi* by Velazco (2005). He (p. 25) illustrated the accessory cuspulid on the last lower premolar of *P. vittatus*, one of the trenchant features useful for distinguishing *P. vittatus* from *P. albericoi*, which lacks the accessory cuspulid.

Genus *Uroderma* W. Peters, 1865

Alfred L. Gardner

Uroderma includes two species of medium-sized (forearm 36.0–46.6 mm, mass 14–23 g) frugivorous bats found from western Mexico south through Central and northern South America into northern Bolivia and southeastern

Brazil. These bats are characterized by having white facial stripes; a white mid-dorsal stripe; and relatively large, bifid upper inner incisors. Both species of *Uroderma* can be distinguished from other "white-lined bats" (*Chiroderma, Platyrrhinus, Vampyressa, Vampyriscus,* and *Vampyrodes*) by their evenly bifid upper inner incisors, three upper and lower molars, and the lack of a fringe of hairs along the trailing edge of a deeply notched interfemoral membrane. The dental formula is 2/2, 1/1, 2/2, 3/3 × 2 = 30.

SYNONYMS:

Phyllostoma: Rüppell, 1842b:155; not *Phyllostoma* G. Cuvier, 1800.

Uroderma W. Peters, 1866a:588, footnote; type species *Phyllostoma personatum*: W. Peters, 1866a (= *Uroderma bilobatum* W. Peters, 1866c; not *Phyllostoma personatum* J. A. Wagner, 1843a), by subsequent designation (Rehn 1901a:757).

Artibeus: Dobson, 1878:514; not *Artibeus* Leach, 1821b.

KEY TO THE SPECIES OF *URODERMA*:

1. Color dark, facial stripes prominent, ear margin yellowish in life (pale yellow to white in dry museum specimens), interfemoral membrane nearly naked, rostrum shallow, mesethmoid narrow and strap-like in frontal view . . . . *Uroderma bilobatum*

1'. Color paler, facial stripes not prominent, ears uniform in color, upper surface of interfemoral membrane hairy to beyond level of knees, rostrum deep, mesethmoid broad and shield-shaped in frontal view . . . .
. . . . *Uroderma magnirostrum*

Uroderma bilobatum W. Peters, 1866
Common Tent-making Bat

SYNONYMS: See under Subspecies.

DISTRIBUTION: *Uroderma bilobatum* is known from Colombia, Venezuela, Trinidad and Tobago, Guyana, Surinam, French Guiana, Ecuador, Brazil, Peru, and Bolivia. Elsewhere, the species is known throughout Central America and as far north as Oaxaca and Veracruz, Mexico.

MARGINAL LOCALITIES (Map 190): COLOMBIA: Magdalena, Cacagualito (J. A. Allen 1900a). VENEZUELA (Handley 1976, except as noted): Zulia, near Cerro Azul, 33 km NW of La Paz; Carabobo, El Central (USNM 373820); Miranda, Birongo; Sucre, Ensenada Cauranta, 11 km NE of Güiria. TRINIDAD AND TOBAGO: Trinidad, Blanchisseuse (R. J. Baker, Atchley, and McDaniel 1972). VENEZUELA: Delta Amacuro, Río Ibaruma (Rivas 2000). GUYANA: Upper Demerara-Berbice, Wikki River (USNM 521619). SURINAM: Paramaribo, Paramaribo (Husson 1978). FRENCH GUIANA: Pointe Combi (Brosset and Charles-Dominique 1991). BRAZIL: Amapá, Piratuba (Harrison and Horne 1971); Pará, Belém (Handley 1967);

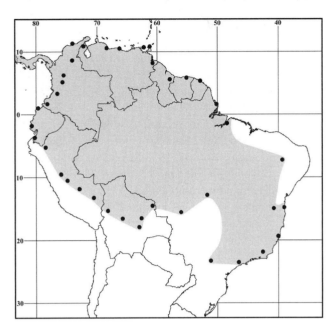

Map 190 Marginal localities for *Uroderma bilobatum* ●

Ceará, Floresta Nacional Araripe-Apodí, 8 km SSW of Crato (Mares et al. 1981); Bahia, Ilhéus (Faria, Soares-Santos, and Sampaio 2006); Espírito Santo, Município de Linhares (Peracchi and Albuquerque 1993); Rio de Janeiro, Fazenda Providência (Avilla, Rozenzstranch, and Abrantes 2001); São Paulo, São Paulo (type locality of *Uroderma bilobatum* W. Peters); Paraná, Londrina (N. R. Reis et al. 1998); Mato Grosso, Serra do Roncador, 264 km N by road of Xavantina (Pine, Bishop, and Jackson 1970); Mato Grosso, Cuiabá (C. O. C. Vieira 1945). BOLIVIA: Santa Cruz, Parque Nacional Noël Kempff Mercado (S. Anderson 1997); Santa Cruz; La Laguna, 10 km N of San Ramón (Ibáñez 1985a); Santa Cruz, 27 km SE of Santa Cruz (S. Anderson 1997); Cochabamba, Puerto Patiño (S. Anderson, Koopman, and Creighton 1982); La Paz, Bellavista (type locality of *Uroderma thomasi* Andersen). PERU: Cusco, Nusiniscato River (Sanborn 1951b); Cusco, Las Malvinas (S. Solari et al. 2001c); Pasco, San Juan (Tuttle 1970); Huánuco, San Antonio (W. B. Davis 1968); Amazonas, Pomará (W. B. Davis 1968); Tumbes, Matapalo (Tuttle 1970). ECUADOR: Guayas, near Manglar Alto (W. B. Davis 1968); Esmeraldas, Esmeraldas (J. A. Allen 1916b). COLOMBIA: Nariño, Barbacoas (J. A. Allen 1916c); Valle del Cauca, Hormiguero (M. E. Thomas 1972); Caldas, Finca los Naranjos (Castaño et al. 2003); Antioquia, Santa Elena (J. A. Allen 1916c); Bolívar, Río San Pedro (W. B. Davis 1968).

SUBSPECIES: In accord with W. B. Davis (1968), I recognize four subspecies of *U. bilobatum* in South America. Two additional subspecies are known from southern Mexico and Central America.

U. b. bilobatum W. Peters, 1866

SYNONYMS:

[*Phyllostoma*] species inedita? Rüppell, 1842b:155.

Ph[*yllostoma*]. *personatum*: W. Peters, 1866a:587, footnote; not *Phyllostoma personatum* J. A. Wagner, 1843a.

Uroderma bilobatum W. Peters, 1866c:394; type localities "St. Paulo in Brasilien . . . [and] Cayenne"; restricted to São Paulo, São Paulo, Brazil by Andersen (1908b:220).

Artibeus bilobatus: Dobson, 1878:518; name combination

Uroderma bilobatum bilobatum: Cabrera, 1958:79; first use of current name combination.

The nominate subspecies is found in Venezuela, the Guianas, northeastern Bolivia, and Brazil south into São Paulo.

U. b. convexum Lyon, 1902

SYNONYMS:

Uroderma convexum Lyon, 1902:83; type locality "Colon, Colombia [=Colón, Colón, Panama]."

Artibeus bilobatus: O. Thomas, 1903d:40; name combination.

Uroderma bilobatum: Goldman, 1920:198; name combination.

Uroderma bilobatum convexum: W. B. Davis, 1968:693; first use of current name combination.

This subspecies has been recorded from Colombia; elsewhere, the subspecies is known from Panama into western Nicaragua, and probably southwestern Honduras (R. J. Baker and McDaniel 1972).

U. b. thomasi Andersen, 1906

SYNONYMS:

Artibeus (Uroderma) bilobatus: O. Thomas, 1880:396; name combination.

Uroderma thomasi Andersen, 1906b:419; type locality "Bellavista, Bolivia, 15°S., 68°W., 1400 m."

Uroderma bilobatum thomasi: Sanborn, 1949b:281; first use of current name combination.

This subspecies is known from Ecuador, Peru, and northwestern Bolivia.

U. b. trinitatum W. B. Davis, 1968

SYNONYMS:

Artibeus bilobatus: J. A. Allen and Chapman, 1897:15; name combination.

Uroderma bilobatum bilobatum: Cabrera, 1958:79; not *Uroderma bilobatum bilobatum* W. Peters.

Uroderma bilobatum trinitatum W. B. Davis, 1968:690; type locality "Caparo 150 ft, Coroni County, Trinidad," Trinidad and Tobago.

This subspecies occurs on the island of Trinidad, Trinidad and Tobago.

NATURAL HISTORY: *Uroderma bilobatum* is a frugivore that includes insects and flower parts in its diet (T. H. Fleming, Hooper, and Wilson 1972; Gardner 1977c; M. R. Nogueira, Tavares, and Peracchi 2003). Well known as a "tent" maker, *U. bilobatum* partially severs the midveins and "ribs" of the leaves of several species of plants to create shelters (T. Barbour 1932; Foster and Timm 1976; Kunz 1982). The species is widely distributed throughout the neotropics. Females exhibit post-partum estrus and normally give birth to a single young twice a year (T. H. Fleming 1973; Wilson 1979). Known parasites include trypanosomes (see Ubelaker, Specian, and Duszynski 1977) and a variety of ticks, mites, and streblid batflies (J. P. Webb and Loomis 1977; R. Guerrero 1997). Three different karyotypes are known; $2n = 38$, FN $= 44$; $2n = 42$, FN $= 48$–50; and $2n = 43$–44, FN $= 48$ (R. J. Baker, Atchley, and McDaniel 1972). Additional information on the natural history of *Uroderma bilobatum* is summarized in the *Mammalian Species* account by R. J. Baker and Clark (1987).

REMARKS: When W. Peters (1866a:footnote, pp. 587–88) proposed *Uroderma*, he used the name for species of *Artibeus* having five upper and five lower cheek teeth (three upper and lower molars). This application suggests that W. Peters considered *Uroderma* to be a subgenus of *Artibeus*. It was in this sense that Dobson (1878:515) used *Uroderma* for "*A. planirostris*" and *A. bilobatus*.

Rehn (1901a) restricted *Uroderma* to the species represented by the bat W. Peters (1866a) first identified as *Phyllostoma personatum* J. A. Wagner. This was the fresh specimen ["ein frisches Exemplar"] from "S. Paulo in Brasilien" received from Dr. Rüppells [*sic*]. *Phyllostoma personatum* J. A. Wagner, 1843a, was based on a specimen acquired by Natterer at "Ypanema," Brazil. Later, W. Peters (1866c:394–95) wrote that Wagner's specimen of *P. personatum* could not be located and, doubting his previous identification because Wagner's measurements did not fit those of the specimens at hand (Wagner's description could also apply to *Chiroderma villosum* or to a large *Platyrrhinus lineatus*), he described the species under the name *Uroderma bilobatum*. W. Peters (1866c) mentioned four specimens: two adults from Cayenne; the young bat from São Paulo, Brazil, he had described earlier (W. Peters 1866a); and the Frankfurt Museum specimen reported by Rüppell (1842b:155) as "II. D. 3. a. [*Phyllostoma*] species inedita? (in Weingeist). Brasilien?" Andersen (1908b:220) fixed the type locality as São Paulo, São Paulo, Brazil, and identified the lectotype as the São Paulo specimen W. Peters first described in 1865. Cabrera (1958:79) repeated C. O. C. Vieira's (1942:363) error when he gave the type locality as "Ipanema, São Paulo, Brazil," a designation that Husson (1978:140) treated as a further restriction of type locality;

contrary to his earlier statement (Husson, 1962:159) that "the type locality must be restricted to São Paulo, Brazil." However, Cabrera did not state that "Ipanema" was a restriction of locality. Therefore, I believe Cabrera's mention of "Ipanema" was an error based on the origin of *Phyllostoma personatum* J. A. Wagner, and possibly influenced by C. O. C. Vieira (1942), because elsewhere in his "Catalogo," when Cabrera (1958) restricted a type locality he clearly stated his intention and gave a rationale for the restriction. Taddei and Reis (1980:364) gave the type locality as "Ipanema (atualmente Varnhagen), São Paulo"; whereas most authors cite the type locality as São Paulo, São Paulo (e.g., W. B. Davis 1968; Koopman 1993).

Pelzeln (1883) considered *Phyllostoma personatum* J. A. Wagner to be a senior synonym of *Uroderma bilobatum*, even though Wagner's type was missing. D. C. Carter and Dolan (1978:52) mentioned Wagner's name ("'var. *personatus* Natterer' [and] *Artibeus (Uroderma) personatus*") in their remarks on three of the original syntypes of *Uroderma bilobatum* in the Zoologisches Museum der Humboldt-Universität, Berlin. However, they did not list the name elsewhere in their report, including Appendix 6, which is a list of species names for which types were not found. Koopman (1993) included *personatum* as a junior synonym of *bilobatum*; however, if the type is found and proves to be a *bilobatum*, the name *personatum* has priority.

Uroderma magnirostrum W. B. Davis, 1968
Brown Tent-making Bat

SYNONYMS:

Uroderma bilobatum: Andersen, 1908b:213 (Fig. 42); not *Uroderma bilobatum* W. Peters, 1866c.

Uroderma magnirostrum W. B. Davis, 1968:679; type locality "10 km E San Lorenzo, 25 ft, Departamento de Valle, Honduras."

DISTRIBUTION: *Uroderma magnirostrum* is known from Colombia, Venezuela, Guyana, Ecuador, Peru, Bolivia, and Brazil (but not from Surinam, French Guiana, and Trinidad and Tobago). Elsewhere, the species is known from Michoacán, Mexico, southeastward through Central America.

MARGINAL LOCALITIES (Map 191): VENEZUELA (Handley 1976, except as noted): Falcón, La Pastora; Miranda, Curupao; Sucre, Manacal; Delta Amacuro, Desembocadora del Río Acoíma (Rivas 2000); Bolívar, Independencia (USNM 440457). GUYANA: Upper Takutu–Upper Essequibo, eastern Kanuku Mountain Region (Emmons 1993; not mapped). BRAZIL (M. R. Nogueira, Tavares, and Peracchi 2003, except as noted): Roraima, Ilha De Maracá (USNM 531067); Amapá, Colônia Agrícola de Matapi (Peracchi, Raimundo, and Tannure 1984); Pará, Belém (Handley 1967); Piauí, Km 18, Highway BR 316,

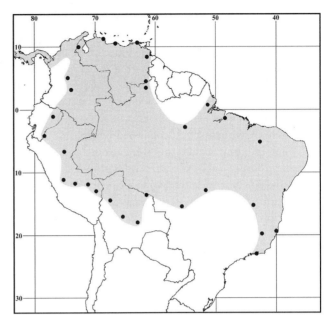

Map 191 Marginal localities for *Uroderma magnirostrum* ●

18 km S of Terezina (Mares et al. 1981); Espírito Santo, Estação Experimental de Linhares (Peracchi and Albuquerque 1993); Rio de Janeiro, Parque Arruda Câmara; Minas Gerais, Parque Estadual do Rio Dôce; Minas Gerais, Morro Solto; Mato Grosso, Serra do Roncador, 264 km N by road of Xavantina (Pine, Bishop, and Jackson 1970); Mato Grosso, Chapada. BOLIVIA (S. Anderson 1997): Santa Cruz, Flor de Oro; Santa Cruz, 27 km SE of Santa Cruz; Cochabamba, Parque Nacional Carrasco; El Beni, Rurrenabaque. PERU: Madre de Dios, Camp 8, left bank Río Molinowski (Baud 1986); Madre de Dios, Pakitza (Ascorra, Wilson, and Romo 1991); Cusco, Ridge Camp (Emmons, Luna, and Romo 2001); Junín, 3 km N of Vitoc (USNM 507186); Loreto, Sarayacu (W. B. Davis 1968); Amazonas, Falso Paquisha (Vivar and Arana-Cardó 1994). ECUADOR: Pastaza, Tiguino (USNM 574636). COLOMBIA: Meta, Municipio de San Juan de Arama (Sánchez-Palomino and Rivas-Pava 1993); Cundinamarca, Villeta (W. B. Davis 1968). VENEZUELA: Zulia, Kasmera (Handley 1976).

SUBSPECIES: I am treating *U. magnirostrum* as monotypic.

NATURAL HISTORY: *Uroderma magnirostrum* is much less commonly captured than is its congener *U. bilobatum*, and the two species have similar distributions, except that *U. magnirostrum* has yet to be recorded from the Surinam, French Guiana, and Trinidad and Tobago. Its food habits apparently also are similar in that it eats fruit, insects, and flower products (Gardner 1977c). M. R. Nogueira, Tavares, and Peracchi (2003) suggested that the fruits of *Cecropia* spp. are an important food. Females pregnant with a single fetus have been captured in January, February, July, September, and October, suggesting postpartum estrus and bimodal polyestry. M. R. Nogueira, Tavares, and Peracchi (2003) reported a lactating female taken in April, at the end of the rainy season. Graham (1987) found no pregnant females among 14 examined during the wet season in Peru, and only one that was lactating (in March). Marques (1985) confirmed post-partum estrus by finding lactating females that were also pregnant. M. R. Nogueira, Tavares, and Peracchi (2003) provided measurements and life history information in their report on the distribution of *U. magnirostrum* in Brazil. The karyotype is $2n = 35–36$, FN = 60–62 (R. J. Baker and Lopez 1970; Hsu and Benirschke 1971b).

REMARKS: Andersen (1908b:220) included three specimens of *U. magnirostrum* in his revision of *U. bilobatum*; one from each of these three localities: Valencia, Venezuela (BM 94. 9.25.9), Para (= Belém), Brazil (BM 1.7.19.4), and Ega, Brazil (BM 7.1.1.686). Specimen BM 1.7.19.4 was illustrated in Figures 42 and 44.

Genus *Vampyressa* O. Thomas, 1900

Joaquín Arroyo-Cabrales

Vampyressa comprises three species, all occurring in South America. Among the smaller of stenodermatines (forearm 29.3–39.2 mm; Swanepoel and Genoways 1979; Lim, Pedro, and Passos 2003), these bats are pale gray to reddish-brown in color, have inconspicuous facial stripes, and lack a dorsal stripe. The noseleaf is well developed and has a well-defined median rib. The interfemoral membrane is relatively narrow and sparsely haired, except for *V. melissa*, which has a hairy interfemoral membrane.

The skull is short and broad, the rostrum is shorter than the braincase, and the palate extends well behind the molars. Inner upper incisors are long, separated basally, and converge at the bifid tips. Outer upper incisors are short and fill the space between the inner incisors and canines. The lower incisors are small. The first upper premolar is much smaller than the second. The height of the inner cusp (on the heel) is reduced in M1 and almost absent on M2, resulting in an irregularly pyriform profile. The first lower premolar is distinctly caniniform and smaller than the second (p4), which has a single, high cusp and is similar in shape to the first lower molar. The dental formula is 2/2, 1/1, 2/2, 2/2–3 × 2 = 28–30.

The genus is known from southern Mexico through Middle and South America to southern Brazil. *Vampyressa* occupies five of the seven faunal provinces recognized by Koopman (1976) for phyllostomid bats. The genus was reviewed by Goodwin (1963) and R. L. Peterson (1968),

who expanded the concept of *Vampyressa* to include the three species now included in *Vampyriscus*. The relatively long and narrow mesopterygoid fossa (broader in *V. melissa*), deep basisphenoid pits, and small caniniform first lower premolar are useful for distinguishing species of *Vampyressa* from *Vampyriscus*. Gardner (1977a) reported standard karyotypes for the then recognized species and analyzed chromosomal variation in the genus (*sensu* R. L. Peterson 1968). There is no fossil record.

SYNONYMS:

Phyllostoma J. A. Wagner, 1843a:366; part; not *Phyllostoma* G. Cuvier, 1800.

Arctibeus: Tomes, 1860c:260; incorrect subsequent spelling of, but not *Artibeus* Leach, 1821b.

Vampyressa O. Thomas, 1900b:270; type species *Phyllostoma pusillum* J. A. Wagner, 1843a, by original designation; described as a subgenus of *Vampyrops* W. Peters, 1865a.

Vampiressa Festa, 1906:7; incorrect subsequent spelling of *Vampyressa* O. Thomas; first use as a valid genus.

Vamqyressa Cabrera, 1958:83; incorrect subsequent spelling of *Vampyressa* O. Thomas.

Vampiressa F. Silva, 1975:52; incorrect subsequent spelling of *Vampyressa* O. Thomas.

Vampuressa Coimbra, Borges, Guerra, and Mello, 1982: 34; incorrect subsequent spelling of *Vampyressa* O. Thomas.

Vampiressa Montenegro and Romero-Ruiz, 2000:646; incorrect subsequent spelling of *Vampyressa* O. Thomas.

Vapyressa Contreras-Vega and Cadena, 2000:287; incorrect subsequent spelling of *Vampyressa* O. Thomas.

REMARKS: Cabrera (1958) recognized *Vampyressa* and *Vampyriscus* as valid genera. However, R. L. Peterson (1968, 1972), synonymized them, and recognized *Metavampyressa* and *Vampyriscus* as subgenera of *Vampyressa*. Karyotypic analysis of the five then recognized species of the genus *Vampyressa* led Gardner (1977a) to state that *Vampyriscus* (*V. bidens*, *V. brocki*, and *V. nymphaea*) differed from *Vampyressa* (*sensu stricto*) in having an entirely biarmed autosomal chromosomes in addition to a comparatively low-crowned first lower premolar, a relatively weakly developed posterior cusp on m2, dark body coloration, a pale mid-dorsal stripe, and conspicuous facial stripes.

In his phylogenetic assessment of stenodermatines, R. D. Owen (1987:62) suggested establishing a new generic name for *Vampyressa melissa* and placing it in a separate subtribe. His "Linnaean classification" united *V. pusilla* (*sensu lato*), along with *Vampyriscus bidens* and *V. brocki*, in the subtribe Vampyrissatini (unavailable family-group name). Nevertheless in his recommended generic classification, R. D. Owen (1987:65) took the conservative approach and treated the five then recognized species along with *Meso-phylla macconnelli*, as members of the genus *Vampyressa*. Wetterer, Rockman and Simmons (2000) did not include *Mesophylla macconnelli* within *Vampyressa*, and deferred assessment of generic and subgeneric relationships within *Vampyressa*. Based on mitochondrial DNA (mtDNA) and Recombination Activating Gene-2 (*RAG2*) analyses, R. J. Baker et al. (2003) recognized *Mesophylla*, *Vampyressa*, and *Vampyriscus* as separate genera, which they grouped with *Chiroderma*, *Platyrrhinus*, *Uroderma*, and *Vampy-rodes* under the subtribe Vampyressina (unavailable family-group name). Although they lacked samples of *Vampyressa melissa* for their analysis of cytochrome-*b* gene sequences, Porter and Baker (2004) also concluded that *Vampyressa* (with *V. pusilla* and *V. thyone*) was distinct at the generic level and warranted separation from *Vampyriscus*, *Meso-phylla*, and *Ectophylla*.

KEY TO SPECIES OF *VAMPYRESSA*:

1. Size larger, forearm longer than 35 mm . . . . *Vampyressa melissa*
1′. Size smaller, forearm shorter than 36 mm 2
2. Size larger, forearm longer than 33 mm; greatest length of skull more than 19 mm; zygomatic breadth more than 11.1 mm; mandibular toothrow more than 6.3 mm . . . . *Vampyressa pusilla*
2′. Size smaller, forearm 34 mm, or less; greatest length of skull less than 19 mm; zygomatic breadth 11.1 mm, or less; mandibular toothrow less than 6.3 mm . . . . *Vampyressa thyone*

Vampyressa melissa O. Thomas, 1926
Melissa's Yellow-eared Bat

SYNONYMS:

Vampyressa melissa O. Thomas, 1926c:157, type locality "Puca Tambo, Peru, altitude 7100′," San Martín, Peru.

Vampyressa (Vampyressa) melissa: R. L. Peterson, 1968: 14; name combination.

DISTRIBUTION: This species occupies the eastern slopes of the northern Andes (Koopman 1976) in South America, where it is known only from Colombia, Ecuador, and Peru.

MARGINAL LOCALITIES (Map 192, localities listed from north to south): COLOMBIA: Huila, Parque Nacional Natural de la Cueva de los Guacharos (Lemke et al. 1982); Nariño, La Reserva Natural La Planada (Ospina-Ante and Gómez 2000). ECUADOR: Pastaza, vicinity of the Colegio Agropecuario de Mera (Rageot and Albuja 1994). PERU: Amazonas, Quebrada Kagka (J. L. Patton, Berlin, and Berlin 1982); San Martín, Puca Tambo (type locality of *Vampyressa melissa* O. Thomas); Ayacucho, Huanhuachayo (Gardner 1976); Cusco, Bosque Aputinye

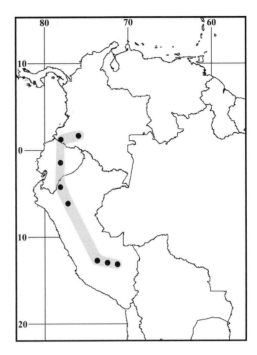

Map 192 Marginal localities for *Vampyressa melissa* ●

(LSUMZ 19103); Cusco, Puente Unión (V. Pacheco et al. 1993).

SUBSPECIES: I am treating *V. melissa* as monotypic.

NATURAL HISTORY: Peruvian specimens were netted in disturbed and undisturbed lower humid forest habitat on the eastern slope of the Andes (Gardner 1976; J. L. Patton, Berlin, and Berlin 1982). The Colombian specimen came from undisturbed cloud forest at 1,900 m (Lemke et al. 1982). The Ecuadorian specimen was caught in disturbed habitat (Albuja 1991; Rageot and Albuja 1994). Graham (1983) gave elevational limits for the species in Peru as 1,000–1,800 m. The diet is fruit (Gardner 1977c). Males had small testes, and females showed no overt signs of reproductive activity in early May in Peru (Gardner 1976). This species has $2n = 14$, $FN = 24$ karyotype with a subtelocentric X chromosome (Gardner 1977a).

REMARKS: Tuttle (1970) identified the type locality as Puca Tambo, 7,100 ft., ESE of Chachapoyas, departamento Amazonas, but Gardner (1976) placed Puca Tambo in departamento San Martín. D. C. Carter and Dolan (1978:56) listed the locality as "Puca Tambo [approximately 6°09′S, 77°16′W, 1480 m., municipality of Chachapoyas, department of Amazonas], 5100 ft., Perú." Although their coordinates and elevation are correct according to Stephens and Traylor (1983), Puca Tambo is in provincia Moyobamba, which was part of Amazonas before 1900, but now is in departamento San Martín (Stiglich 1918, 1922). Koopman's (1978) citation of 2,200 m for the type locality was based on O. Thomas's (1926c) "7,100 feet," which does not correspond to the actual elevation (1,480 m). D. C. Carter and Dolan (1978) gave measurements of and other information on the holotype of *Vampyressa melissa* O. Thomas.

Gardner (1977a) considered *V. melissa* to be the morphologically least derived member of the genus *Vampyressa* (*sensu* R. L. Peterson 1968) because it retains the maximum number of incisors and molars, and its karyotype can be derived from his hypothetical ancestral configuration (2*n* = 26, FN = 24) by the least number of chromosomal rearrangements (six Robertsonian fusions). Panamanian *V. melissa*, first reported as "*Vampyressa* species" (Handley 1966c:767), differs from South American populations in lacking the minute third lower molar. Brosset and Charles-Dominique (1991) reported *V. melissa* from French Guiana; however, their material was misidentified (P. Charles-Dominique, *in litt.*), and possibly represents *Vampyriscus brocki*.

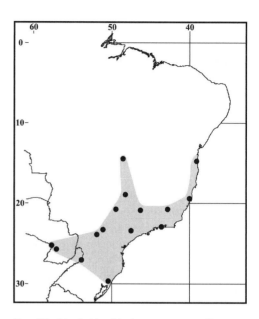

Map 193 Marginal localities for *Vampyressa pusilla* ●

Vampyressa pusilla (J. A. Wagner, 1843)
Southern Yellow-eared Bat

SYNONYMS:

Phyllostoma pusillum J. A. Wagner, 1843a:366; type locality "Sapitiva" (= Sepetiba), Rio de Janeiro, Brazil.

Chiroderma pusillum: W. Peters, 1866c:395; name combination.

Stenoderma (Chiroderma) pusillum: Pelzeln, 1883:34; name combination.

V[ampyrops]. pusillus: O. Thomas, 1889:170; name combination.

V[ampyrops (Vampyressa)]. pusillus: O. Thomas, 1900b: 270; name combination.

Vampyressa pusilla: Miller, 1907b:156; name combination.

Vampyressa pusilla pusilla: Goodwin, 1963:8; name combination.

Vampyressa nattereri Goodwin, 1963:16; type locality "probably Ipanema, district of São Paulo, Brazil."

V[ampyressa]. natteri R. L. Peterson, 1968:7; incorrect subsequent spelling of *Vampyressa nattereri* Goodwin.

Vampyressa pussilla N. R. Reis and Muller, 1995:33; incorrect subsequent spelling of *Phyllostoma pusillum* J. A. Wagner.

Vampyressa pussilla R. Guerrero, 1996, 649; incorrect subsequent spelling of *Phyllostoma pusillum* J. A. Wagner.

DISTRIBUTION: *Vampyressa pusilla* is known from southeastern Brazil, adjacent Paraguay, and provincia Misiones, Argentina.

MARGINAL LOCALITIES (Map 193): BRAZIL: Goiás, Fazenda Moinho (Coimbra et al. 1982); Distrito Federal, "rural area," (Bredt and Ueda 1996, not mapped); Minas Gerais, Alpinópolis (Taddei 1979); Minas Gerais, Viçosa (Mumford and Knudson 1978); Bahia, Ilhéus (Faria, Soares-Santos, and Sampaio 2006); Espírito Santo, Estação Experimental de Lin-

hares (Peracchi and Albuquerque 1993); Rio de Janeiro, Sapitiva (type locality of *Phyllostoma pusillum* J. A. Wagner); São Paulo, Ipanema (Goodwin 1963); Rio Grande do Sul, Rolante (F. Silva 1975). ARGENTINA: Misiones, Jct. Hwy 21 and Arroyo Oveja Negra (Barquez, Mares, and Braun 1999). PARAGUAY: Paraguarí, Parque Nacional Ybycuí (Myers, White, and Stallings 1983); Central, Asunción (Myers 1982). BRAZIL: Paraná, Parque Estadual Vila Rica (Bianconi, Mikich, and Pedro 2004); Paraná, Arthur Thomas Park (N. R. Reis and Muller 1995); São Paulo, Mirassol (Taddei 1979); Minas Gerais, Uberlândia (Stutz et al. 2004).

SUBSPECIES: I am treating *V. pusilla* as monotypic.

NATURAL HISTORY: Myers, White, and Stallings (1983) reported this frugivorous species from subtropical forests in Paraguay. Zortéa and Brito (2000) described the tents made from the leaves of plants of the genera *Simira* (Rubiaceae), *Heliconia* (Heliconiaceae), *Piper* (Piperaceae) and *Philodendron* (Araceae) and used as roosts in primary and secondary Atlantic rain forest of Espírito Santo, Brazil. They found *V. pusilla* in tents made only from *Simira* and *Heliconia* leaves. Some tents were used as feeding roost as indicated by feces and food debris found on the ground below. Zortéa and Brito (2000) reported finding a pregnant female together with a lactating female and her nonflying young in a *Simira* sp. leaf tent.

Coimbra et al. (1982) reported a *V. pusilla* caught in grassland habitat at Fazenda Moinho, Goiás, Brazil; Bredt and Ueda (1996) reported the species from "rural" areas in the Distrito Federal. These locations are in Cerrado habitat and appear to be separated from the humid forest habitat where the majority of specimens has been recorded. The karyotype is $2n = 20$, FN $= 36$ (Myers, White, and Stallings 1983).

REMARKS: Handley (1966c), whose concept of *V. pusilla* included *V. thyone* Thomas, *V. minuta* Miller, *V. nattereri* Goodwin, and *V. venilla* Thomas as synonyms, claimed that differences between "subspecies" of *V. pusilla* were mainly age-related. He considered the species mono-typic, a recommendation followed by J. K. Jones and Carter (1976) and Koopman (1982, 1993, 1994). R. L. Peterson (1968) showed that the holotype of *V. nattereri* was fully adult and that the holotype of *V. pusilla* was obviously young with open epiphyseal sutures in the wing bones. Therefore, he treated *nattereri* as a synonym because both represented the southeastern Brazilian population, but rec-ognized *V. p. pusilla* and *V. p. thyone* (with *V. venilla* and *V. minuta* as synonyms) as subspecies. In their *Mammalian Species* account, S. E. Lewis and Wilson (1987) also treated *V. thyone* as a subspecies of *V. pusilla*. Lim, Pedro, and Passos (2003) are the first modern workers to recognize *V. pusilla* in the same restricted sense I use the name here.

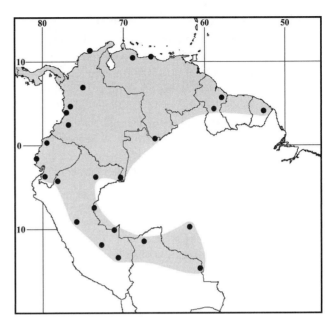

Map 194 Marginal localities for *Vampyressa thyone* ●

Vampyressa thyone O. Thomas, 1909
Little Yellow-eared Bat

SYNONYMS:

Arctibeus pusillus: Tomes, 1860c:260; name combination; not *Phyllostoma pusillum* J. A. Wagner.

Vampyressa thyone O. Thomas, 1909b:231; type local-ity "Chimbo, near Guayaquil, [Bolívar,] Ecuador. Alt. 1000'."

Vampyressa minuta Miller, 1912a:25; type locality "Ca-bima, [Panamá,] Panama."

Vampyressa venilla O. Thomas, 1924b:533; type locality "San Lorenzo, 500'," Loreto, Peru.

Vampyressa pusilla venilla: Goodwin, 1963:12; name com-bination.

Vampyressa pusilla thyone: Goodwin, 1963:14; name com-bination.

Vampyressa pussilla Ascorra, Wilson, and Romo, 1991:5; incorrect subsequent spelling of, but not *Phyllostoma pusillum* J. A. Wagner.

DISTRIBUTION: *Vampyressa thyone* occurs from western Colombia and Ecuador eastward through Venezuela into Guyana and French Guiana, and south-westward through the upper Amazon basin of Venezuela, Colombia, Ecuador, Peru, Bolivia and Brazil. The species has not been recorded from Surinam and central Brazil. Elsewhere, *V. thyone* is known from Central America and southern Mexico (Hall 1981; S. E. Lewis and Wilson 1987, as *V. pusilla thyone*).

MARGINAL LOCALITIES (Map 194): COLOMBIA: Magdalena, Santa Marta (R. J. Baker et al. 1973). VENEZUELA (Handley 1976): Yaracuy, Minas de Aroa;

Miranda, Curupao. GUYANA: Upper Demerara-Berbice, Dubulay Ranch (USNM 582327). FRENCH GUIANA: Les Nouragues (Brosset and Charles-Dominique 1991). GUYANA: Potaro-Siparuni, Cowfly Camp (Lim and En-gstrom 2001a). VENEZUELA: Amazonas, Cerro de la Neblina Base Camp (Gardner 1988). COLOMBIA: Ama-zonas, Puerto Patiño (USNM 483735). PERU: Loreto, Mishana (W. B. Davis and Dixon 1976). BRAZIL: Acre, Serra da Jaquirana (M. R. Nogueira, Pol, and Peracchi 2000). PERU: Ucayali, Balta (Voss and Emmons 1996). BOLIVIA: Pando, Independencia (S. Anderson 1997). BRAZIL: Rondônia, Cachoeira Nazaré (MPEG 21232). BOLIVIA: Santa Cruz, 27.5 km S of Campamento Los Fierros (S. Anderson 1997). PERU: Cusco, Hacienda Ca-dena (Sanborn 1953); Cusco, Cashiriari-2 (S. Solari et al. 2001c); Huánuco, Santa Elena (Gardner 1977a); Ama-zonas, Quebrada Kagka (J. L. Patton, Berlin, and Berlin 1982). ECUADOR: El Oro, 1 km SW of Puente de Moro-moro (USNM 513468); Manabí, Río Blanco (Albuja 1983); Esmeraldas, Rosa Zárate (USNM 522432). COLOMBIA: Cauca, La Costa (Tamsitt and Valdivieso 1966a); Valle del Cauca, Buenaventura (Esslinger, Vaughn, and Arata 1968); Chocó, Sipí (Goodwin 1963); Antioquia, Amalfi (BM 12.2.12.2).

SUBSPECIES: I consider *V. thyone* to be monotypic.

NATURAL HISTORY: In Peru and Venezuela, *V. thy-one* is known from a variety of tropical forest habitats from near sea level to over 1,500 m (W. B. Davis 1975; W. B. Davis and Dixon 1976; Handley 1976; Koopman 1978, J. L. Patton, Berlin, and Berlin 1982; Graham 1983;

Terborgh, Fitzpatrick, and Emmons 1984). Known roosting sites include bushes and tree branches (Hall and Jackson 1953). Based on label information on specimens in the USNM, two females from northern Antioquia, Colombia, were pregnant when captured, one with a 17-mm-CR fetus on 10 August, the other with a 22-mm-CR fetus on 7 January. Five females from the same region were lactating when taken, one on 24 August, two on 14 September, and one on 18 February. Another female caught on 28 September in the vicinity of Villavicencio, Colombia, contained a small embryo.

Wenzel, Tipton, and Kiewlicz (1966) described a streblid batfly (*Neotrichobius stenopterus*), based in part on specimens recovered from Panamanian *V. thyone*. Reed and Brennan (1975) recorded one trombiculid mite and Brennan and Reed (1975) added two more species of trombiculid mites recorded from Venezuelan *V. thyone*. J. P. Webb and Loomis (1977) listed four families, with one species each, of ectoparasitic mites, along with a streblid batfly, from *V. thyone* (also see list by R. Guerrero 1985a, reported as *V. pusilla*). R. Guerrero (1997) listed three species of streblid batflies known from this species.

Robert J. Baker et al. (1973), Goodwin (1963), and Swanepoel and Genoways (1979), gave forearm lengths (29.3–33.2 mm) and several cranial dimensions for Central and South American samples (see review in S. E. Lewis and Wilson 1987, whose measurements are of *V. thyone*). The average mass of a sample of 27 Costa Rican *V. thyone* was 8.2 g (LaVal and Fitch 1977).

This species is frugivorous (Howell and Burch 1974; Gardner 1977c; Bonaccorso 1979). T. H. Fleming, Hooper, and Wilson (1972) reported 100% plant material as the stomach contents of a Panamanian *V. thyone*; but nevertheless, they listed the species among those that also eat insects. Wilson (1979) suggested bimodal polyestry as the reproductive pattern. Panamanian and Costa Rican females (LaVal and Fitch 1977) are pregnant or lactating early in the rainy season. Colombian and Peruvian records indicate reproductive activity later in the year at the end of the dry season (e.g., almost all Peruvian females were pregnant at the beginning of November; W. B. Davis and Dixon 1976). Lim and Engstrom (2001a) took a female with a 16-mm-CR embryo on 14 October in Guyana.

Several karyotypic variants are known for *V. thyone* (see summary in R. J. Baker et al. 1982). R. J. Baker (1979) recorded two cytotypes, one from Central America with a $2n = 18$, FN $= 20$ karyotype having two pairs of submetacentric and six pairs of acrocentric autosomes, a subtelocentric X, and a small biarmed Y chromosome. The second form, found in Colombia, has a $2n = 24$ (females), $2n = 23$ (males), FN $= 22$ karyotype with a non-XX/XY sex-chromosome system and no submetacentric autosomes.

Gardner (1977a) described a third variant from eastern Peru characterized by a $2n = 22$ (males) and $2n = 23$ (females), FN $= 22$ karyotype that has one large unpaired metacentric among otherwise acrocentric autosomes, and a multiple sex-chromosome system. R. J. Baker and Bickham (1980) recognized four chromosomal variants within *V. thyone*.

REMARKS: Cabrera (1958) recognized *V. pusilla*, *V. thyone*, and *V. venilla* as species, However, Handley (1966c), whose concept of *V. pusilla* included these taxa along with *Vampyressa minuta* Miller and *V. nattereri* Goodwin, considered the supposed differences between them to be mainly age related, and he treated them as a single monotypic species. In his review of the genus, R. L. Peterson (1968) considered *V. thyone* (with *V. venilla* and *V. minuta* as synonyms) to be a Central and northwestern South American subspecies of *V. pusilla*. This taxonomic treatment was followed by subsequent workers until Lim, Pedro, and Passos (2003) recognized *V. thyone* as a valid species, separate from *V. pusilla*.

Robert J. Baker et al. (1973) found two cytotypes in their samples of "*V. pusilla*" from Central and South America in which the magnitude of variation was greater than that characteristic of most congeneric phyllostomid species (also see R. J. Baker 1979). However, because they could not find any exomorphological or cranial differences that would distinguish the cytotypes, they treated all northern specimens as representing the same taxon, which they identified as *V. p. thyone*.

The specimen (a skin only) identified by Cuervo-Díaz, Hernández-Camacho, and Cadena (1986) as *Ectophylla alba* from "Río Raposo, Valle," Colombia, is a small, pale *V. thyone* (the identity verified by A. L. Gardner). Rodríguez-Mahecha et al. (1995) and Alberico et al. (2000) included *E. alba* in their lists of Colombian mammals based on the misidentified Río Raposo specimen. Inclusion of the generic name *Ectophylla* in the South American fauna dates from Laurie (1955), who synonymized the uncommon, but widespread, *Mesophylla macconnelli* under *Ectophylla*.

Genus *Vampyriscus* O. Thomas, 1900

Joaquín Arroyo-Cabrales

Vampyriscus comprises three species, all occurring in South America. These bats are among the smaller stenodermatines (forearm 32.1–39.2 mm; Swanepoel and Genoways 1979) and reddish-brown to darker brown in color with prominent facial stripes and a thin, pale mid-dorsal stripe (often weak in *V. nymphaea* and not visible in some *V. brocki*). The noseleaf is well developed and has a well-defined median rib. The interfemoral membrane is relatively narrow and sparsely haired. The skull is short and broad, the rostrum is shorter than the braincase, and the palate extends well behind the molars. Inner upper incisors are long, separated basally, have entire or weakly bifid tips. Outer upper incisors are short and fill the space between the inner incisors and canines. The lower incisors are small; there are two on each side, except for *V. bidens*, in which normally they are reduced to a single tooth in each ramus. The first upper premolar is much smaller than the second. The height of the inner cusp (on the heel) is reduced in Ml and almost absent on M2; as a result, M2 is irregularly pyriform in outline. The first lower premolar is anteriorly-posteriorly long and smaller than the second (p4), which has a single high cusp and is similar in shape to the first lower molar. The dental formula is 2/1–2, 1/1, 2/2, 2/2–3 × 2 = 28–30.

The genus is known from Nicaragua through Middle and South America to southern Brazil. Species of *Vampyriscus* can be distinguished from those of *Vampyressa* by the pointed or weakly bifid upper inner incisors (weakly bifid in *Vampyressa pusilla*, otherwise clearly bifid in other *Vampyressa*); anteriorly-posteriorly long, noncaniniform first lower premolar (shorter and distinctly caniniform in *Vampyressa* spp.); shallow basisphenoid pits; and a relatively short and broad mesopterygoid fossa. Gardner (1977a) reported standard karyotypes for all known species and analyzed chromosomal variation (treated as a subgenus of *Vampyressa*. There is no fossil record.

SYNONYMS:

Arctibius Bonaparte, 1847:115; incorrect subsequent spelling of, but not *Artibeus* Leach 1821b.

Chiroderma: Dobson, 1878:535; part; not *Chiroderma* W. Peters, 1860.

Vampyriscus O. Thomas, 1900b:270; type species *Chiroderma bidens* Dobson, 1878, by original designation; described as a subgenus of *Vampyrops* W. Peters, 1865a.

Vampyressa: R. L. Peterson, 1968:1; part; not *Vampyressa* O. Thomas, 1900b.

Metavampyressa R. L. Peterson, 1968:13; type species *Vampyressa nymphaea* O. Thomas, 1909b, by original designation; described as a subgenus of *Vampyressa* O. Thomas.

Vampiressa Soriano, Ruiz, and Zambrano, 2005:253; incorrect subsequent spelling of, but not *Vampyressa* O. Thomas.

REMARKS: On the basis of morphological similarities between *Vampyriscus bidens* and species of *Vampyressa*, R. L. Peterson (1968, 1972), treated *Vampyriscus*, represented by *V. bidens*, as a subgenus of *Vampyressa*. He erected *Metavampyressa* as a subgenus of *Vampyressa* to contain *V. brocki* and *V. nymphaea*, with the latter as type species. Noting variation in number of incisors and molars in a large sample of *V. bidens*, W. B. Davis (1975) stated that characters other than dental formulae should be emphasized in

differentiating species of *Vampyressa* (*sensu* R. L. Peterson 1968). He concluded that the allocation of *V. bidens* (Dobson, 1878) to one subgenus (*Vampyriscus*), and *V. nymphaea* and *V. brocki* to another (*Metavampyressa*) was not supported. Consequently, he treated *Metavampyressa* as a synonym of the subgenus *Vampyriscus* in the genus *Vampyressa*.

Pine, Miller, and Schamberger (1979) claimed that *Metavampyressa* R. L. Peterson, 1968 is a *nomen nudum* because Peterson did not comply with Article 13(a)(i) of the then current Code (ICZN 1964). Article 13(a)(i) required, that to be available, a name published after 1930 must be "accompanied by a statement that purports to give characters differentiating the taxon. . . ." Although only marginally a statement *per se*, R. L. Peterson (1968:[12] Fig. 6) did describe the third lower molars as "absent," and the dorsal stripe as "present," in his diagrammatic summary of major characters and, thereby, satisfied the requirements for availability.

Karyotypic analysis of the five then recognized species of the genus *Vampyressa* led Gardner (1977a) to state that *V. bidens*, *V. brocki*, and *V. nymphaea* are similar in having entirely biarmed autosomal complements in addition to having a comparatively low-crowned first lower premolar, a relatively weakly developed posterior cuspid on m2, dark body coloration, a pale mid-dorsal stripe, and conspicuous facial stripes. He followed W. B. Davis (1975) by relegating *Metavampyressa* to the synonymy of *Vampyriscus*, but retained the latter in *Vampyressa*. In his phylogenetic assessment of stenodermatines, R. D. Owen (1987:62) suggested establishing a new generic name for *Vampyressa melissa* and placing it in a separate subtribe, placing *V. nymphaea* in the genus *Mesophylla* under the subtribe Mesophyllatina (unavailable family-group name), and placing *V. bidens*, *V. brocki*, and *V. pusilla* in the subtribe Vampyrissatini (unavailable family-group name). Nevertheless, in his recommended generic classification, R. D. Owen (1987:65) retained *V. bidens*, *V. brocki*, and *V. pusilla*, along with *Mesophylla macconnelli*, in the genus *Vampyressa*. Simmons and Voss (1998), recognized *Vampyriscus* and *Metavampyressa* as subgenera of *Vampyressa*, but excluded *Mesophylla*. Wetterer, Rockman and Simmons (2000) deferred assessment of generic and subgeneric relationships within *Vampyressa*. When Lim, Pedro, and Passos (2003) recognized *Vampyressa thyone* and *V. pusilla* as full species, they retained *Vampyriscus brocki*, *V. nymphaea*, and *V bidens* in *Vampyressa*, although their cladogram (p. 19) showed clear separation of the latter three taxa from *Vampyressa* (*sensu stricto*). Based on mitochondrial DNA (mtDNA) and Recombination Activating Gene-2 (*RAG2*) analyses, R. J. Baker et al. (2003) essentially returned to the classification of

Cabrera (1958) by recognizing *Mesophylla*, *Vampyressa*, and *Vampyriscus* as separate genera, which they grouped with *Chiroderma*, *Platyrrhinus*, *Uroderma*, and *Vampyrodes* under the subtribe Vampyressina (unavailable family-group name). Porter and Baker (2004) also concluded that *Vampyriscus* warranted recognition at the generic level. The family-group names Vampyrissatini and Mesophyllatina listed by R. D. Owen (1987:62), Vampyressini used by Ferrarezzi and Gimenez (1996) and Wetterer, Rockman, and Simmons (2000), and Vampyressina as employed by R. J. Baker et al. (2003), are unavailable because they do not meet the requirements of Article 13 (ICZN 1985, 1999).

KEY TO SPECIES OF *VAMPYRISCUS*:

1. Dorsal stripe present (weakly defined in some *V. nymphaea* and *V. brocki*). 2
1′. Dorsal stripe appears to be absent . . . .
. . . . *Vampyriscus brocki*
2. One lower incisor on each side; m3 minute; greatest length of skull 19.5–20.6 mm. . . . *Vampyriscus bidens*
2′. Two lower incisors on each side; m3 absent; greatest length of skull either less than 19.5 mm or greater than 20.6 mm. 3
3. Forearm longer than 34.5 mm; greatest length of skull more than 20 mm. . . . . *Vampyriscus nymphaea*
3′. Forearm shorter than 35.5 mm; greatest length of skull less than 20 mm. . . . . *Vampyriscus brocki*

Vampyriscus bidens (Dobson, 1878)
Bidentate Yellow-eared Bat
SYNONYMS:

Arctibius Floresii Bonaparte, 1847:115; type locality "unexplored region of the Republic of Equatoria, which borders on the wilds of Brazil" (a *nomen oblitum*, see Remarks).
Chiroderma bidens Dobson, 1878:535; type locality "Huallaga," Peru; identified as the Río Huallaga by R. L. Peterson (1968:15), and restricted to Santa Cruz, Río Huallaga, Loreto, Peru, by D. C. Carter and Dolan (1978:56).
V[ampyrops]. bidens: O. Thomas, 1889:170; name combination.
V[ampyrops (Vampyriscus)]. bidens: O. Thomas, 1900b: 270; name combination.
Vampyriscus bidens: Miller, 1907b:157; first use of current name combination.
Vampyressa bidens: R. L. Peterson, 1968:1; name combination.

DISTRIBUTION: *Vampyriscus bidens* is found in Colombia, Venezuela, Guyana, Surinam, French Guiana, Ecuador, Peru, Brazil, and Bolivia.

MARGINAL LOCALITIES (Map 195): VENEZUELA: Barinas, Parque Nacional Tapo Caparo (Soriano, Ruiz,

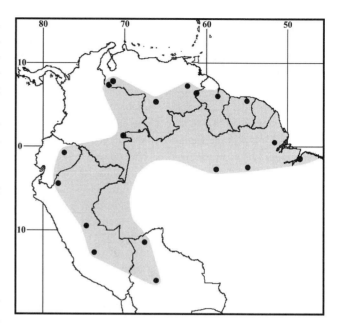

Map 195 Marginal localities for *Vampyriscus bidens* ●

and Zambrano 2005); Amazonas, San Juan de Manapiare (Lim and Engstrom 2001a); Bolívar, El Manaco, 59 km SE of El Dorado (Handley 1976); Bolívar, Los Patos (Handley 1976). GUYANA: Cuyuni-Mazaruni, forest reserve, 24 miles along the Potaro road from Bartica (Hill 1965). SURINAM: Para, 1 km S and 2 km E of Powaka (Genoways and Williams 1979b). FRENCH GUIANA: Les Nouragues? (not mapped, see Remarks). BRAZIL: Amapá, Santa Luzia do Pacuí (Peracchi, Raimundo, and Tannure 1984); Pará, Guama (Handley 1967); Pará, Alter do Chão (Bernard 2001a); Amazonas, Igarapé Aniba (C. O. C. Vieira 1942). BOLIVIA (S. Anderson 1997): Pando, right bank, Río Madre de Dios, opposite Independencia; Beni, Oromomo. PERU (Koopman 1978, except as noted): Ayacucho, Luisiana; Huánuco, Cerros del Sira; Amazonas, Huampami (J. L. Patton, Berlin, and Berlin 1982). ECUADOR: Napo, Concepción (Albuja 1983). COLOMBIA: Vaupés, Durania (Marinkelle and Cadena 1972). VENEZUELA: Apure, Nulita (Handley 1976).

SUBSPECIES: I consider *V. bidens* to be monotypic.

NATURAL HISTORY: Reporting on the nature and extent of individual and secondary-sexual variation in *V. bidens*, W. B. Davis (1975) stated that females average only slightly larger than males. He also noted that the number of lower incisors and lower molars is variable based on finding one specimen (out of 33 specimens) with two lower incisors and three lower molars on each side, and another with two lower incisors on one side, but only one on the other side. Although the normal number of lower molars is three, W. B. Davis (1975) found seven specimens with only two lower molars on one or both jaws. Although *V. bidens* is uncommon in collections, W. B. Davis (1975) and W. B. Davis and Dixon (1976) found the species composed a little over a third of the 447 stenodermatines Dixon netted at Mishana in northern Peru. Although their capture records showed a modal activity period, W. B. Davis and Dixon (1976) recorded individuals caught continually from dusk to near dawn. Handley (1976) reported Venezuelan *V. bidens* as mainly occurring in tropical humid forests; J. L. Patton, Berlin, and Berlin (1982) took this species from a variety of disturbed and undisturbed forest habitats in Peru. Koopman (1978) and Graham (1983) gave elevational limits in Peru as 200–1,000 m; 116 of 117 Venezuelan specimens were captured below 500 m (Handley 1976). These bats probably roost in trees and shrubs (Nowak and Paradiso 1983). Although poorly documented, the diet is primarily fruits (Gardner 1977c). Pregnant females have been taken in August in Surinam (Genoways and Williams 1979b) and in November in Peru (W. B. Davis 1975; W. B. Davis and Dixon 1976). R. Guerrero (1997) listed two species of streblid batflies recorded from *V. bidens*. T. E. Lee, Scott, and Marcum (2001) summarized most of the published information on the natural history of *V. bidens* in their *Mammalian Species* account. The karyotype is $2n = 26$, $FN = 48$, with a subtelocentric X chromosome (Gardner 1977a; Honeycutt, Baker, and Genoways 1980).

REMARKS: When Miller (1907b) elevated the subgenus *Vampyriscus* to generic rank, he stated that *V. bidens* had only two lower molars on each side. W. B. Davis (1975) noted variation in the number of incisors and molars in a sample of *V. bidens*, and concluded that characters other than dental formulae should be emphasized in differentiating species of *Vampyriscus* (which he treated as a subgenus of *Vampyressa*). Authors, subsequent to R. L. Peterson's (1968) inclusion of *V. bidens* in the genus *Vampyressa*, have almost universally referred to this species as *Vampyressa bidens*. W. B. Davis (1975) provided forearm lengths (34.0–37.2 mm) and cranial measurements for 13 males and 10 females from northern Peru. Swanepoel and Genoways (1979) listed forearm lengths (35.3–39.1 mm) and cranial measurements for four females and four males from Peru and Guyana. Albuja (1999) gave comparable measurements for Ecuadorian specimens.

Randolph L. Peterson (1968) erred in placing Hacienda Luisiana in Cuaco [*sic*; lapsus for Cusco], Peru, instead of in departamento Ayacucho. His Peruvian "Urubamba, Boca River" equals the north slope of the Cordillera Vilcabamba, departamento Cusco. Although Brosset and Charles-Dominique (1991) reported *V. bidens* from Les Nouragues, French Guiana, that record was based on a misidentification. However, the species recently has been confirmed for French Guiana (P. Charles-Dominique, pers. comm.).

Arctibeus floresii Bonaparte is a forgotten name (*nomen oblitum*) that could refer to any of several species of smaller stenodermatines that approximate the description (1847:115):

"Grey brown; beneath paler, with pale tips to the hair; two broad streaks on the face, and a narrow streak on the centre of the back, white. Arm-bone rather foliated, one inch four lines in length. Heel-bone very short. Second thumb-joint elongated, slender. Nose-leaf with a distinct central rib."

However, this description best fits *V. bidens*, which has the dorsal and facial stripes, hairy forearms, a short calcar, a forearm length averaging near 36 mm (1 inch, 4 lines), and occurs in eastern Ecuador.

Vampyriscus brocki (R. L. Peterson, 1968)
Brock's Yellow-eared Bat
SYNONYMS:

Vampyressa brocki R. L. Peterson, 1968:1; type locality "the upper headwaters of the Kuitaro River approximately 40 miles east of Dadanawa at Ow-wi-dy-wau (Oshi Wau head near Marurawaunowa), 2°50′N., 58°55′W., Rupununi [= Upper Takutu-Upper Essequibo] District, Guyana."

Vampyressa (Metavampyressa) brocki: R. L. Peterson, 1968:15; name combination in original description of subgenus and species.

Vampyressa [(Vampyriscus)] brocki: W. B. Davis, 1975: 264; name combination.

Vampyriscus brocki: R. J. Baker, Porter, Patton, and Van Den Bussche, 2000:25; first use of current name combination.

Distribution: *Vampyriscus brocki* is known from the northern and eastern Amazon basin in Colombia, Guyana, Surinam, French Guiana, Brazil, and Peru.

MARGINAL LOCALITIES (Map 196): GUYANA: Demerara-Mahaica, Ceiba Biological Center (Lim, Pedro, and Passos 2003); Upper Demerara-Berbice, Arampa, 3 miles E of Ituni (R. L. Peterson 1972). SURINAM: Sipaliwini, Grassalco (S. L. Williams and Genoways 1980a). FRENCH GUIANA: Paracou (Simmons and Voss 1998). BRAZIL: Pará, Rio Xingu, 52 km SSW of Altamira (USNM 549425); Amazonas, Município Rio Preto de Eva, Reserva Km 41 (INPA 2538). COLOMBIA: Amazonas, Leticia (R. J. Baker, Genoways, and Cadena 1972). PERU: Loreto, Reserva Nacional Allpahuayo-Mishana (Hice, Velazco, and Willig 2004); Loreto, Centro de Investigaciones "Jenaro Herrera" (Ascorra, Gorchov, and Cornejo 1994); Cusco, Ridge Camp (Emmons, Luna, and Romo 2001); Amazonas, Falso Paquisha (Vivar and Arana-Cardó 1994). GUYANA: Upper Takutu-Upper Essequibo, Ow-wi-dy-wau (type locality of *Vampyressa brocki* R. L.

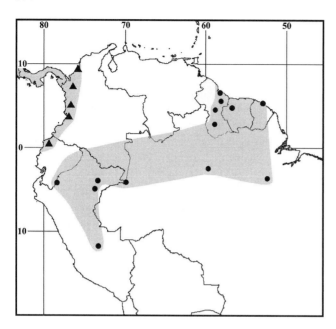

Map 196 Marginal localities for *Vampyriscus brocki* ● and *Vampyriscus nymphaea* ▲

Peterson); Potaro-Siparuni, Iwokrama Forest (Lim and Engstrom 2001b)

SUBSPECIES: I consider *V. brocki* to be monotypic.

NATURAL HISTORY: This species has been netted in mature tropical rainforest about 3.5 m above a forest trail in Colombia (R. J. Baker, Genoways, and Cadena 1972) and over a stream in secondary forest habitat in Surinam (S. L. Williams and Genoways 1980a). Simmons and Voss (1998) caught four in ground-level mist nets and three in elevated nets. In ground level nets they caught one in well-drained primary forest, two in creek-side primary forest, and one in a man-made clearing. The elevated nets were strung at heights of from 17 to 21 m above a dirt road in a gap in the forest canopy.

A female from Guyana was lactating when caught in March (R. L. Peterson 1968), and two females from Colombia each contained a minute embryo when taken, one on June 30, and the other on July 1 (R. J. Baker, Genoways, and Cadena 1972). Ascorra, Gorchov, and Cornejo (1994:544) recorded a pregnant female "early in the wet season" [October?] in Peru. The karyotype is $2n = 24$, FN = 44, the X chromosome is subtelocentric, and the Y is acrocentric (Honeycutt, Baker, and Genoways 1980).

REMARKS: Randolph L. Peterson (1968) placed *V. brocki* and *V. nymphaea* together in the subgenus *Metavampyressa*, a conclusion not supported by W. B. Davis (1975) or Gardner (1977a), who placed *V. brocki* in the expanded subgenus *Vampyriscus* along with *V. nymphaea* and *V. bidens*. Simmons and Voss (1998), however, listed their specimens as *Vampyressa (Metavampyressa) brocki*. Simmons and Voss

(1998) commented on the statement by S. L. Williams and Genoways (1980a) that *V. brocki* lacks a mid-dorsal stripe. The Paracou specimens have a faint dorsal stripe apparently similar to that described for the holotype by R. L. Peterson (1968). Nevertheless, the Rio Xingu specimen (USNM 549425) does not have a discernible dorsal stripe even when the fur is carefully brushed. Simmons and Voss (1998) pointed out that although *V. brocki* is similar in size to *V. pusilla*, the two are easily distinguished by the configuration of the upper inner incisors (simple cones in *V. brocki*, bilobed in *V. pusilla* and *V. thyone*), the crowns of first lower premolar (low blade in *brocki*, high narrow cone in *pusilla* and *V. thyone*), and the shape of the nasal aperture (wide and "straight" in *brocki*, narrow and V-shaped in *pusilla*). The facial stripes contrast strongly with the color of the head in *V. brocki*, but not in *V. pusilla* and *V. thyone*.

Muñoz (2001) lists records for *V. brocki* from the departments of El Cesar and Magdalena. Normally I would have mapped those records; however, because the work contains an unusually high number of errors and probable misidentifications, the identifications of these specimens need to be confirmed.

Vampyriscus nymphaea (O. Thomas, 1909)
Striped Yellow-eared Bat
SYNONYMS:

Vampyressa nymphaea O. Thomas, 1909b:230, type locality "Novita, Rio S. Juan, Chocó, W. Colombia."

Vampyressa (Metavampyressa) nymphaea: R. L. Peterson, 1968:15; name combination.

Vampyressa [(Vampyriscus)] nymphaea: W. B. Davis, 1975:264; name combination.

Mesophylla nymphaea: R. D. Owen, 1987:62; name combination.

Vampyressa nimphaea Albuja, 1989:109; incorrect subsequent spelling of *Vampyressa nymphaea* O. Thomas.

DISTRIBUTION: This species is found along the Pacific coastal plain of Colombia (Koopman 1976) and Ecuador (Albuja 1989, 1999). Elsewhere, the species has been recorded in Central America as far north as Nicaragua (J. K. Jones, Smith, and Turner 1971). There are two Amazon basin records (see Remarks), one in southeastern Peru (Terborgh, Fitzpatrick, and Emmons 1984; Hinchcliffe, Stachan, and Daniels 1989) and the other in Brazil (McNab 1969).

MARGINAL LOCALITIES (Map 196; localities listed from north to south): COLOMBIA: Córdoba, Sinú (Cuervo-Díaz, Hernández-Camacho, and Cadena 1986); Antioquia, Urabá (Cuervo-Díaz, Hernández-Camacho, and Cadena 1986); Chocó, Andagoya (Goodwin 1963); Valle del Cauca, Zabaletas (M. E. Thomas 1972). ECUADOR: Esmeraldas, Valle del Sade (Albuja 1989).

SUBSPECIES: *Vampyriscus nymphaea* is currently regarded as monotypic.

NATURAL HISTORY: The diet principally consists of fruit (Gardner 1977c; LaVal and Fitch 1977). T. H. Fleming, Hooper, and Wilson (1972) included insects as part of its diet. Forman, Phillips, and Rouk (1979) noted the similarity of the stomachs of *V. nymphaea* and *Vampyressa pusilla*. Swanepoel and Genoways (1979) summarized reports of measurements for the species and gave forearm lengths (34.9–39.0 mm) and cranial measurements of eight Colombian and Central American specimens. Rodríguez-Herrera and Tschapka (2005) described a leaf tent used as a day roost by seven bats they identified as *V. nymphaea*. The tent was located approximately 5 m above the ground in a *Cecropia insignis* in the eastern lowlands of Costa Rica.

This species exhibits bimodal polyestry (Wilson 1979) with birth peaks in February and August. R. Guerrero (1997) listed two species of streblid batflies known to parasitize *V. nymphaea*. The karyotype is $2n = 26$, FN = 48, with a medium-sized subtelocentric X and a small submetacentric or acrocentric Y chromosome (R. J. Baker 1973; R. J. Baker et al. 1973, 1982).

REMARKS: *Vampyriscus nymphaea* is the type species of *Metavampyressa* R. L. Peterson, which R. L. Peterson (1968) described as a subgenus of *Vampyressa*, and in which he also included *V. brocki*. W. B. Davis (1975) and Gardner (1977a) considered *Metavampyressa* to be a synonym of *Vampyriscus* and considered *V. nymphaea* to be allied with *V. brocki* and *V. bidens*. Although R. J. Baker et al. (2003) placed *Vampyriscus bidens* and *V. brocki* in a clade separate from *Mesophylla macconnelli* and *Vampyressa thyone*, they were unable to resolve the placement of *V. nymphaea*.

Reports of this species from the eastern lowlands of Peru (Terborgh, Fitzpatrick, and Emmons 1984; Hinchcliffe, Stachan, and Daniels 1989), and from the Amazon basin of Brazil (McNab 1969), probably were based on misidentified *V. bidens* or some other similar taxon. The Peruvian records are not vouchered by specimens.

Genus *Vampyrodes* O. Thomas, 1900

Alfred L. Gardner

Vampyrodes is a monotypic genus of large (forearms average 52.4 mm, range 46.8–56.7 mm, $n = 31$; mass averages 28.8 g, range 25.2–34.0 g, $n = 10$), pale to cinnamon-brown frugivorous bats having a prominent white dorsal stripe and conspicuous white facial stripes. The supraorbital stripes extend from the lateral margin of the noseleaf to the top of the head between the ears. Shorter white malar stripes extend from the corner of the mouth to the base of the ears. Although superficially similar, *Vampyrodes* easily can be distinguished from *Platyrrhinus* spp. by the conspicuous white facial stripes and two, instead of three, upper molars. The dental formula is 2/2, 1/1, 2/2, 2/3 × 2 = 30. There is no known fossil record.

SYNONYMS:

Vampyrops: O. Thomas, 1889:167; part; not *Vampyrops* W. Peters, 1865a.

Vampyrodes O. Thomas, 1900b:270; type species *Vampyrops caraccioli* O. Thomas, 1889, by original designation; described as a subgenus of *Vampyrops* W. Peters.

Vampyrodes caraccioli (O. Thomas, 1889)
Great Striped-faced Bat

SYNONYMS: See under Subspecies.

DISTRIBUTION: *Vampyrodes caraccioli* is known from Colombia, Venezuela, Trinidad and Tobago, Surinam, French Guiana, northern and western Brazil, eastern Ecuador, eastern Peru, and northern Bolivia. Elsewhere, the species is known from Central America north to southern Mexico.

MARGINAL LOCALITIES (Map 197; arranged by subspecies): *Vampyrodes c. caraccioli*. TRINIDAD AND TOBAGO: Tobago, Pigeon Peak (Husson 1954); Trinidad, Guayaguayare (C. H. Carter et al. 1981). VENEZUELA: Miranda, Birongo (Handley 1976). *Vampyrodes c. major*. COLOMBIA: Antioquia, La Tirana (USNM 499461). VENEZUELA: Bolívar, La Serranía de Los Pijiguaos (Ochoa et al. 1988); Bolívar, Km 125 (Handley 1976). GUYANA: Cowfly Camp (Lim and Engstrom 2001a). SURINAM: Sipaliwini, Raleigh Falls (Genoways, Williams, and Groen 1981). FRENCH GUIANA: l'Anse de Sinnamary (Brosset et al. 1996a). BRAZIL: Pará, Utinga (O. Thomas 1920c). FRENCH GUIANA: Trois Sauts (Brosset et al. 1996a:533). VENEZUELA: Amazonas, Tamatama

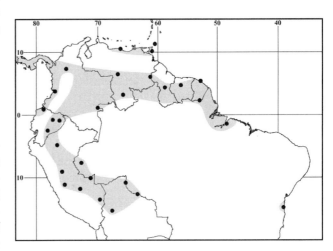

Map 197 Marginal localities for *Vampyrodes caraccioli* ●

(Handley 1976). COLOMBIA: Vaupés, Mitú (USNM 445471). ECUADOR: Orellana, Río Pinto (Albuja 1999). PERU: Loreto, San Lorenzo (O. Thomas 1924b). BRAZIL: Acre, Seringal Lagoinha (Taddei, Rezende, and Camora 1990). PERÙ: Ucayali, Balta (Voss and Emmons 1996). BOLIVIA: Beni, Guayaramerín (S. Anderson and Webster 1983); Beni, Mouth of Río Curiche (S. Anderson, Koopman, and Creighton 1982); La Paz, Sararia (S. Anderson, Koopman, and Creighton 1982). PERÙ: Puno, Fila Boca Guacamayo (USNM 579658); Cusco, Segakiato (S. Solari et al. 1998); Junín, San Ramón (Tuttle 1970); Ucayali, 3 km NE of Abra Divisoria (Graham and Barkley 1984). ECUADOR: Morona-Santiago, Sucúa (Webster and Jones 1984); Napo, Río Pucuno (Albuja 1999). COLOMBIA: Valle del Cauca, Río Zabaletas (M. E. Thomas 1972). ECUADOR: Esmeraldas, Luis Vargas Torres (Albuja 1999). *Isolated Brazilian record*: BRAZIL: Bahia, Una (Faria, Soares-Santos, and Sampaio 2006).

SUBSPECIES: I recognize two subspecies of *V. caraccioli*.

V. c. caraccioli (O. Thomas, 1889)

SYNONYMS:

Vampyrops Caracciolæ O. Thomas, 1889:167; type locality "Trinidad."

Vampyrops Caraccioli O. Thomas, 1893a:186; corrected original spelling, see Remarks.

V[ampyrops. (Vampyrodes)] Caraccioli: O. Thomas, 1900b:270; name combination.

Vampyrodes caracciolæ: Miller, 1907b:156; first use of current name combination and incorrect subsequent spelling of *Vampyrops caraccioli* O. Thomas.

V[ampyrodes]. caraccioloi Pittier and Tate, 1932:273; incorrect subsequent spelling of *Vampyrops caraccioli* O. Thomas.

Vampyrodes caraccioloi Goodwin and Greenhall, 1961:257; unjustified emendation of *Vampyrops caraccioli* O. Thomas.

The nominate subspecies is in Trinidad and Tobago, and adjacent Venezuela.

V. c. major G. M. Allen, 1908

SYNONYMS:

Vampyrodes major G. M. Allen, 1908:38; type locality "San Pablo, Isthmus of Panama," Canal Zone, Panama.

Vampyrodes ornatus O. Thomas, 1924b:532; type locality "San Lorenzo, Rio Marañon, nearly opposite mouth of Huallaga. Alt. 500'," Loreto, Peru.

This subspecies is found in Peru, Bolivia, Ecuador, Colombia, Venezuela, Surinam, French Guiana, and Brazil. Elsewhere, it occurs northward through Central America into southern Mexico.

NATURAL HISTORY: *Vampyrodes caraccioli* is a frugivore (Gardner 1977c) adapted to humid lowland forests (300 to 1,000 m) along the northern and western margins of the Amazon basin (e.g., see Koopman 1978), and the Caribbean lowlands of northern South America and Central America north into Mexico. The recent record from southern Bahia (Faria, Soares-Santos, and Sampaio 2006) is the first from eastern Brazil and extends the distribution over 3,000 km southeastward from the vicinity of Belém, Pará. Handley (1976) recorded the species in Venezuelan tropical and subtropical humid forest habitats, usually at elevations of less than 500 m, but occasionally as high as 1,000 m. The only specimens known from west of the Andes came from Pacific lowland rain forests of Colombia (M. E. Thomas 1972) and Ecuador (Albuja 1999). Muñoz (2001) cited J. A. Allen (1916b [= 1916c here]) as the authority for locality records from departamento El Chocó, Colombia; however, I have been unable to locate any reference to these records in any of J. A. Allen's reports. Reproductive data, summarized by Wilson (1979), suggest bimodal polyestry. Day roosts include foliage, branches, and palm fronds where groups of two to four have been found (Tuttle 1976; Morrison 1980). Willis, Willig, and Jones (1990) listed several species of mites known to parasitize *V. caraccioli* in their *Mammalian Species* account. R. Guerrero (1996) mentioned mites of the families Macronyssidae, Labidocarpidae, and Spinturnicidae recovered from two *V. caraccioli* netted in primary forest during the rainy season in the vicinity of Pakitza, Peru. He later (1997) listed two species of streblid flies found on this species. The karyotype is $2n = 30$, FN = 56 (R. J. Baker 1979).

REMARKS: This bat was originally described as *Vampyrops caracciolæ* by O. Thomas (1889), believing that Mr. Caracciolo's name was spelled Caracciola. O. Thomas (1893a:186) emended the spelling to *V. caraccioli*, stating "This should be *Vampyrops caraccioli*, instead of *V. caracciolae*, the mistake having been due to a misconception as to the name of the discoverer, Mr. Caracciolo, whose proper name is now well known to zoologists in connexion with the foundation of the Trinidad Field Naturalists' Club, of which he is president." Goodwin and Greenhall (1961) emended *caraccioli* to *caraccioloi* based upon Pittier and Tate's (1932) and Cabrera's (1958) use of the same spelling. Husson (1954), however, argued that *caracciolae* was the correct original spelling and that O. Thomas's (1893a) emendation to *caraccioli* was unjustified. In accordance with the explanation by D. C. Carter and Dolan (1978) and usage by others (e.g., R. Guerrero 1985a; Voss and Emmons 1996; Wetterer, Rockman, and Simmons 2000), I use the spelling *caraccioli* as correct for this taxon. Hall and Kelson (1959), Starrett and Casebeer (1968), and Hall (1981), treated *V. major* as a species distinct from *V. caraccioli*.

Starrett and Casebeer (1968:13), citing differences in size, color pattern, and dental and cranial morphology, argued for recognizing two species in *Vampyrodes*. Handley (1966c), Honacki, Kinman, and Koeppl (1982), and Koopman (1993, 1994) have treated the genus as monotypic. D. C. Carter and Dolan (1978) gave measurements for and other information on the holotype of *Vampyrops Caraccioli* O. Thomas.

Tribe Stenodermatini P. Gervais, 1856

Alfred L. Gardner

There are seven genera (*Ametrida, Ardops, Ariteus, Centurio, Phyllops, Sphaeronycteris,* and *Stenoderma*) in the Stenodermatini, all are characterized by white shoulder spots, a truncated rostrum, and globose braincase. *Ametrida, Centurio, Sphaeronycteris,* and *Stenoderma* are found in South America; the other genera are found in the Greater and Lesser Antilles. Several representatives have accessory dermal outgrowths on the face and head, and have numerous papillae if the buccal cavity. . Karyotypes among these genera are nearly identical (2*n* = 30–31, FN = 56), except for *Centurio* and *Sphaeronycteris,* which both lack the small pair of metacentric autosomes found in the other taxa (Myers 1981).

Genus *Ametrida* Gray, 1847

Ametrida is a monotypic genus of small, sexually dimorphic, brown to dark rusty-brown bats with the shortest forearm in the Phyllostomidae (forearm in males averages 25.4 mm; in females, 32.1 mm). The dorsal fur has four bands: a barely perceptible pale basal band, followed by brown and pale buff bands of approximately equal width, and a narrow terminal brown band that imparts the overall dorsal color. The ventral fur is unicolored and the same color as the dorsum over most of the chest and abdomen. The color of the throat, neck, and upper chest is paler and can be lightly frosted with white, especially under the forearms. The face, legs, and dorsal and ventral surfaces of the interfemoral membrane, which bears a thin fringe of hairs, are also furred. These bats have a conspicuous white spot on each shoulder and a smaller, less conspicuous, white spot on each side of the neck below the ear. Glandular masses are on the face on each side of noseleaf horseshoe and on the chest in males, where a pair protrudes through the fur (illustrated by R. L. Peterson 1965b). The chest glands of males are often surrounded by whitish to pale buff-colored fur. The skull has a globose braincase and a short rostrum. The upper outer incisors are smaller than the inner and all are evenly spaced between the canines. The lower in-

cisors are small, robust, and unequally trifid. The second upper premolar is about twice the size of the first, and its tip extends to or slightly beyond the height of the canine when viewed in the horizontal plane. The lower premolars are subequal, and the upper and lower third molars are minute. The dental formula is 2/2, 1/1, 2/2, 3/3 × 2 = 32.

The genus is known from Panama, South America, and the islands of Bonaire and Trinidad. There is no fossil record.

SYNONYMS:

Ametrida Gray, 1847a:15; type species *Ametrida centurio* Gray, 1847a, by monotypy.

Phyllostoma: J. A. Wagner, 1855:621; part; not *Phyllostoma* G. Cuvier, 1800.

Stenoderma: Winge, 1892:12; part; not *Stenoderma* É. Geoffroy St.-Hilaire, 1818b.

Ametrida centurio Gray, 1847

Little White-shouldered Bat

SYNONYMS:

A[metrida]. centurio Gray, 1847a:15; type locality "Brazils, Para"; restricted to Belém, Pará, Brazil by R. L. Peterson (1965b).

P[hyllostoma]. centurio: J. A. Wagner, 1855:629; name combination.

S[tenoderma]. ("Ametrida") centurio: Winge, 1892:12; name combination.

Ametrida minor H. Allen, 1894b:240; type locality unknown; identified by G. M. Allen (1902a) as Surinam, probably Paramaribo, to which place Husson (1962:182) restricted the type locality.

[Ametrida (Ametrida)] Centurio: Trouessart, 1897:163; name combination.

DISTRIBUTION: *Ametrida centurio* is in Venezuela, Guyana, Surinam, French Guiana, Brazil, Trinidad and Tobago (Trinidad), and the Netherlands Antilles (Bonaire). There is a single record from Barro Colorado Island, Panama (Handley, Wilson, and Gardner 1991).

MARGINAL LOCALITIES (Map 198; from R. L. Peterson 1965b, except as noted): NETHERLANDS ANTILLES: Bonaire, Kralendijk. VENEZUELA (Handley 1976): Distrito Federal, Pico Ávila, 5 km NNW Caracas; Sucre, Manacal. TRINIDAD AND TOBAGO: Trinidad, Las Cuevas. VENEZUELA: Delta Amacuro, Río Ibaruma (Linares and Rivas 2004). GUYANA: Cuyuni-Mazaruni, Kartabo. SURINAM: Marowijne, Moengo. FRENCH GUIANA: Cayenne (Brosset and Charles-Dominique 1991). BRAZIL: Amapá, Santa Luzia do Pacuí (Peracchi, Raimundo, and Tannure 1984); Pará, Belém; Mato Grosso, Serra do Roncador, 264 km N (by road) of Xavantina (Pine, Bishop, and Jackson 1970); Amazonas, Manaus. VENEZUELA: Amazonas, Cerro Neblina Base Camp (Gardner 1988);

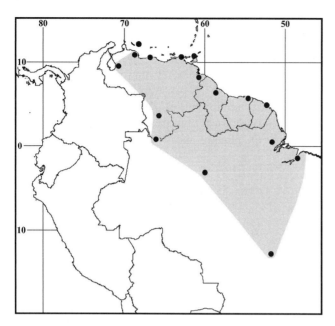

Map 198 Marginal localities for *Ametrida centurio* ●

Amazonas, Culebra (R. Guerrero, Hoogesteijin, and Soriano 1989); Trujillo, 25 km NW of Valera (Handley 1976); Falcón, Riecito (Handley 1976).

SUBSPECIES: I am treating *A. centurio* as monotypic.

NATURAL HISTORY: N. R. Reis and Peracchi (1987) found the seeds of *Piper aduncum* in the feces of two *A. centurio* captured in the vicinity of Manaus, Amazonas, Brazil. Handley (1976), N. R. Reis and Peracchi (1987), Gardner (1988), reported specimens caught in or near forest habitat, mainly adjacent to streams and river channels. Bernard and Fenton (2002) listed a specimen from a forest fragment and 45 captured in savanna habitat in the vicinity of Alter do Chão, Pará, Brazil. Brosset and Charles-Dominique (1991) and Simmons and Voss (1998) netted several specimens high (up to 37 m) above the ground in gaps in the canopy of tall primary forest in French Guiana. C. H. Carter et al. (1981) reported pregnant females from the island of Trinidad taken in July and August. Reproductive information, noted on specimen labels on *A. centurio* from Venezuela in the USNM, reveal that pregnant females were taken in April, May, June, and August, with a peak (8 of 11 pregnancies) in May. Lactating *A. centurio* were taken in April, May, July, and August, and their number also peaked in May (6 of 12). In addition, one juvenile (wing epiphyses unfused) was captured in August. V. R. McDaniel (1976) described the brain as similar to those of *Centurio senex* and *Stenoderma rufum*. Reed and Brennan (1975) reported a trombiculid mite, Wenzel (1976) reported a streblid batfly, R. Guerrero (1985a) added a spinturnicid mite, and R. Guerrero (1997) listed two streblid batflies as ectoparasites known from this species. T. E. Lee and Dominguez (2000)

provided a detailed description along with measurements and a summary of available natural history information in their *Mammalian Species* account. The karyotype is $2n = 30$–31, $FN = 56$, with a Y_1, Y_2 sex-chromosome system (X autosome fusion); the X is a subtelocentric, Y_1 a submetacentric, and Y_2 a metacentric (R. J. Baker and Hsu 1970; R. J. Baker 1979).

REMARKS: Sexual dimorphism is extreme, and males are significantly smaller than females (compare measurements in R. L. Peterson 1965b; Swanepoel and Genoways 1979). Each sex was regarded as a separate species until R. L. Peterson (1965b) synonymized *A. minor* with *A. centurio*. D. C. Carter and Dolan (1978) gave measurements of and commented on the holotype of *Ametrida centurio* Gray, which they and R. L. Peterson (1965b) identified as female.

Genus *Centurio* Gray, 1842

Centurio is a monotypic genus of medium-sized (forearm 42.1–47.0 mm, mass 17.0–25.5 g) stenodermatine bats that lack a typical noseleaf. Instead, the short, broad, and essentially naked face is covered by a regular series of wrinkled skin flaps and excrescent structures. These bats also have a loose ruff of skin under the chin that can be extended over the face to the level of the large inner lappet of each ear. The skin folds bear a small naked transverse pad that, when the ruff covers the face, lies over a medial vertical groove above the nostrils forming an unobstructed channel for breathing. The color pattern varies from pale brownish gray to yellow brown dorsally and there is a small white patch on each shoulder. Ventral coloration is similar to that of the dorsum, but paler. The interfemoral membrane is relatively large and especially hairy on the dorsal surface. The wing membrane is pale and nearly translucent between the metacarpals of digits III and IV, and has a conspicuous series of transverse, nearly transparent bars between the metacarpals and phalanges of digits IV and V and, to a lesser extent, between the metacarpal of digit V and the forearm. The skull has a high, rounded braincase, and an exceedingly short rostrum, on which the nares open at the level of the anterior root of the zygomatic arches (see Rehn 1901b; Miller 1907b; Paradiso 1968; and Snow, Jones, and Webster 1980, for more comprehensive descriptions). The dental formula is 2/2, 1/1, 2/2, 2/2 × 2 = 28. Koopman and Martin (1959) reported subfossil material from Tamaulipas, Mexico.

SYNONYMS:

Centurio Gray, 1842:259; type species *Centurio senex* Gray, 1842, by monotypy.

Trichocoryes H. Allen, 1861:360; type species *Centurio mcmurtrii* H. Allen, 1861, by monotypy; described as a subgenus of *Centurio* Gray.

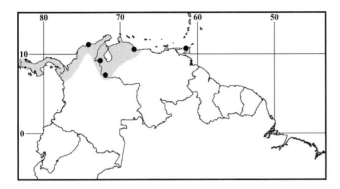

Map 199 Marginal localities for *Centurio senex* ●

Trichocorytes Gray, 1866b:118; incorrect subsequent spelling of *Trichocoryes* H. Allen.

Trichocoryctes Trouessart, 1897:163; incorrect subsequent spelling of *Trichocoryes* H. Allen.

Centurio senex Gray, 1842
Wrinkle-faced Bat

SYNONYMS: See under Subspecies.

DISTRIBUTION: In South America, *C. senex* occurs on the island of Trinidad and in northern Colombia and western Venezuela. Elsewhere, the species is found throughout Central America and northwestward into Jalisco, Mexico.

MARGINAL LOCALITIES (Map 199): VENEZUELA (Handley 1976): Falcón, Boca de Yaracuy; Apure, Nulita; Zulia, El Rosario, 42 km WNW of Encontrados. TRINIDAD AND TOBAGO: Trinidad; Port-of-Spain (Goodwin and Greenhall 1961). COLOMBIA: Magdalena, Finca El Recuerdo (Lemke et al. 1982).

SUBSPECIES: I recognize two subspecies of *C. senex*.

C. s. greenhalli Paradiso, 1968
SYNONYM:

Centurio senex greenhalli Paradiso, 1968:601; type locality "Port of Spain, St. George County [*sic*], Trinidad," Trinidad and Tobago.

This subspecies occurs only on the island of Trinidad.

C. s. senex Gray, 1842
SYNONYMS:

Centurio senex Gray, 1842:259; type locality "Amboyna," East Indies; subsequently suggested by Gray (1844) to be South America; restricted to Realejo, Nicaragua, by Goodwin (1946:327).

Centurio flavogularis Lichtenstein and Peters, 1854:335; type locality "Cuba."

Centurio mexicanus Saussure, 1860:381; type locality "les régions chaudes du Méxique."

Centurio mcmurtrii H. Allen, 1861:359; type locality "Mirador, [Veracruz,] Mexico."

Centurio minor Ward, 1891:750; type locality "Cerro de los Pájaros, Las Vigas," Veracruz, Mexico.

This subspecies is known from Venezuela and northern Colombia, and northwestward through Central America to western Mexico.

NATURAL HISTORY: *Centurio senex* occurs in lowland arid and evergreen subtropical and tropical humid forests. Gardner (1977c) wrote that the species may be an obligate frugivore. A captive in Panama readily ate several different fruit jams diluted with water. This bat dropped particulate items such as seeds and pieces of skin while feeding, ingesting only the soft, juicy, and semi-liquid matter. The naked chin appeared to function as a "drip tip." Fluids sometimes ran out of the mouth and formed a drop on the chin while the bat fed, and periodically, that drop was drawn back by the tongue into the mouth. Goodwin and Greenhall (1961) described the small opening to the esophagus and noted that a soft, mushy mixture of fruits was preferred by a captive. They found *C. senex*, on the island of Trinidad, hanging singly or in groups of two or, at most three, in tree foliage. J. K. Jones, Smith, and Turner (1971) reported collecting individuals hanging from small branches within "cubicles" formed by dense vine tangles, branches, and leaves. Paradiso (1968) reported a birth in January. Wilson (1979) reported pregnant females in every month between January and August, except May, and suggested asynchronous polyestry as the reproductive pattern. J. P. Webb and Loomis (1977) reported a nycteribiid batfly (*Basilia* sp.); and Wenzel, Tipton, and Kiewlicz (1966) mentioned a streblid batfly (*Strebla vespertilionis*), but suspected that *C. senex* was not its normal host. R. Guerrero (1997) did not list any batflies known from *C. senex*. Snow, Jones, and Webster (1980) provided illustrations, measurements, and a summary of the biology in their *Mammalian Species* account. The karyotype is $2n = 28$, $FN = 52$, with a subtelocentric X and a submetacentric Y chromosome (R. J. Baker and Hsu 1970; R. J. Baker 1973).

REMARKS: *Centurio senex* was unknown from South America when Cabrera's (1958) catalog was published. Paradiso (1968) discussed the problems concerning the provenance of the type specimen and fixation of a type locality. Although Lichtenstein and Peters (1854) gave Cuba as the source of the type of *C. flavogularis*, occurrence of the genus on the island has not been confirmed (Rehn 1901b; Silva-Taboada 1979). D. C. Carter and Dolan (1978) gave measurements for and other information on the holotype of *C. senex* Gray.

Genus *Pygoderma* W. Peters, 1863

Pygoderma is a monotypic genus of small, sexually dimorphic bats, endemic to South America. Females (forearm averaging 39.7 mm, $n = 22$) are larger than males (forearm averaging 36.9 mm, $n = 12$; measurements from Myers 1981). The dorsal pelage is tricolored with individual hairs brown at the base and tip, and buff medially. A small white patch of fur is present on each shoulder. The ventral pelage is gray brown, and sparse on the chest in males. The nose-leaf is prominent, the ears are broad and rounded, and the tragus is small. The interfemoral membrane is moderate in width and densely furred, as are the legs and feet. Specialized facial glands surround each eye, and are conspicuously swollen in males, but not as prominent in females. The deep, cuboid rostrum, is diagnostic of *Pygoderma*. Additional morphological details can be found in the descriptions by Miller (1907b), Myers (1981) and Webster and Owen (1984). The dental formula is 2/2, 1/1, 2/2, 2/2–3 × 2 = 28; a third lower molar is occasionally present (Webster and Owen 1984).

SYNONYMS:

Phyllostoma: J. A. Wagner, 1843a:366; part; not *Phyllostoma* G. Cuvier, 1800.

Stenoderma: W. Peters, 1863:83; part; not *Stenoderma* É. Geoffroy St.-Hilaire, 1818b.

Pygoderma W. Peters, 1863:83; type species *Stenoderma (Pygoderma) microdon* W. Peters, 1863, by monotypy; described as a subgenus of *Stenoderma* É. Geoffroy St.-Hilaire.

Sycophaga Winge, 1892:10; type species *Stenoderma humerale* Lund (=*Stenoderma humerale* Winge, 1892) by subsequent selection (Kretzoi and Kretzoi 2000:394); described as a subgenus of *Phyllostoma* G. Cuvier; preoccupied by *Sycophaga* Westwood, 1840 (Hymenoptera).

Pygoderma bilabiatum (J. A. Wagner, 1843)
Ipanema Bat

SYNONYMS: See under Subspecies.

DISTRIBUTION: *Pygoderma bilabiatum* is in Paraguay, Brazil, northern Argentina, and southern Bolivia. Records for Surinam are suspect (see Remarks).

MARGINAL LOCALITIES (Map 200; arranged by subspecies): *Pygoderma b. bilabiatum*. BRAZIL (C. O. C. Vieira 1942, except as noted): Pernambuco, Bonito (USNM 14816; not mapped, see Remarks); Bahia, Região de Conquista (Falcão, Soares-Santos, and Drummond 2005); Bahia, Itapebi (Faria, Soares-Santos, and Sampaio 2006); Espírito Santo, Município de Linhares (Peracchi and Albuquerque 1993); Rio de Janeiro, Macaé; São Paulo, Piquete; São Paulo, Primeiro Morro (Owen and Webster 1983); Paraná, Castro; Rio Grande do Sul, Parque Turvo (F. Silva 1975). ARGENTINA: Misiones, Posadas (Fornes and Delpietro 1969). PARAGUAY: Paraguarí, Sapucay (Podtiaguin 1944); Concepción, 8 km E of Con-

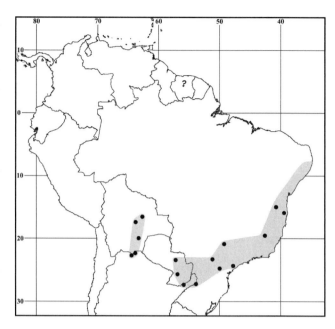

Map 200 Marginal localities for *Pygoderma bilabiatum* ●

cepción (Owen and Webster 1983). BRAZIL: Paraná, Parque Arthur Thomas (N. R. Reis, Peracchi, and Onuki 1993); São Paulo, Engenheiro Schmidt (Taddei 1979); Minas Gerais, Parque Estadual Coronel Fabricano (Owen and Webster 1983). *Pygoderma b. magna*. BOLIVIA: Santa Cruz, La Laguna (Ibáñez 1985a); Santa Cruz, 72 km ESE of Monteagudo (S. Anderson, Koopman, and Creighton 1982). ARGENTINA: Salta, Piquirenda Viejo (Owen and Webster 1983); Salta, Aguas Blancas (Olrog 1976). BOLIVIA: Santa Cruz, 25 km by road W of Buena Vista (S. Anderson, Koopman, and Creighton 1982).

SUBSPECIES: Owen and Webster (1983) recognized two subspecies of *P. bilabiatum*.

P. b. bilabiatum J. A. Wagner, 1843

SYNONYMS:

Phyllostoma humerale Lund, 1840c:313; *nomen nudum*.

Phyllostoma leucostigma Lund, 1840c:313; *nomen nudum*.

Phyllostoma bilabiatum J. A. Wagner, 1843a:366; type locality "Ypanema," São Paulo, Brazil.

Arctibeus leucomus Gray, 1849:57; type locality "Brazil."

Stenoderma (Pygoderma) microdon W. Peters, 1863:83, type locality "Surinam" (see Remarks).

P[ygoderma]. bilabiatum: W. Peters, 1865b:357; first use of current name combination.

Stenoderma humerale Winge, 1892:10; type locality "Lagoa Santa," Minas Gerais, Brazil.

Sycophaga humeralis: Winge, 1892:10; name combination.

Stenoderma bilabiatum: Ihering, 1894:23; name combination.

This subspecies is known from Brazil, northern Argentina (Misiones), and Paraguay. *Pygoderma bilabiatum* also has been reported from Surinam (see Remarks).

P. b. magna Owen and Webster, 1983

SYNONYM:

Pygoderma bilabiatum magna Owen and Webster, 1983: 146; type locality "Ichilo, 7 km N Santa Rosa, 800 m, Santa Cruz, Bolivia."

This subspecies occurs in southeastern Bolivia and adjacent Argentina.

NATURAL HISTORY: Relatively little is known about the natural history of *P. bilabiatum*. It has been captured in a variety of habitats in Argentina, including tropical and subtropical forests (Olrog 1967), and is known to roost in houses (Fornes and Delpietro 1969). Gardner (1977c) wrote that this bat may be an obligate frugivore. Barquez, Mares, and Braun (1999) reported liquids appearing to be fruit juices, but no solids, in the stomachs they examined. Pregnant females, each with a single embryo, have been captured in March, July, and August in Paraguay (Myers 1981), and in August in Brazil (Peracchi and Albuquerque 1971). Barquez, Mares, and Braun (1999) suggested the species may reproduce year-round. *Pygoderma bilabiatum* has a $2n = 30–31$, $FN = 56$ karyotype (Myers 1981). Webster and Owen (1984) provided a summary of the known natural history of *P. bilabiatum* in their *Mammalian Species* account.

REMARKS: Koopman (1958) pointed out that reports of this bat from Mexico (Dobson 1878:537; repeated by Miller 1907b:166; J. A. Allen 1910a:113; C. O. C. Vieira 1942; Podtiaguin 1944; and Cabrera 1958:91) were based upon a misinterpretation of W. Peters's footnote (1863:83), which referred to only the provenance of his specimen of "*Dermanura cinereum*." D. C. Carter and Dolan (1978:135, footnote) suggested that a specimen in the BM from Santa Catarina, Brazil, could be the holotype of *Arctibeus leucomus* Gray. They also measured and gave information for one of the two syntypes of *Stenoderma (Pygoderma) microdon* W. Peters; the other syntype, marked "*typus*" in the Berlin museum (ZMB) catalog, was presumed lost. Swanepoel and Genoways (1979:56) referred to the specimen D. C. Carter and Dolan (1978) reported on as the holotype; however, as the only remaining specimen of the two males W. Peters (1863) had at hand when describing the taxon, that specimen is a syntype and no lectotype has been selected.

Eisenberg's (1989) reference to specimens "recently" collected in Surinam is based on Webster and Owen's (1984) citation of a specimen in the National Museum of Natural History (USNM 14816) donated to the museum by Albert Koebele (misspelled Koeble on label) and cata-

logued 6 March 1885. Although this appears to be the only record for *P. bilabiatum* from Surinam subsequent to W. Peters's (1863) description of *Stenoderma microdon*, other evidence in the USNM suggests that the specimen did not come from Surinam. Koebele's specimen likely came from the vicinity of Bonito, Pernambuco, Brazil, where he collected several birds now in the National Museum of Natural History's Division of Birds. Husson (1962:187, 1978:155) cited an undated personal communication from Dr. G. H. H. Stein of the ZMB stating that the two syntypes of *Stenoderma microdon* W. Peters could not be found (also see Voss and Emmons 1996:12, footnote k). However, D. C. Carter and Dolan (1978:66) provided information on one of them, which Carter located in 1966, but he could not locate the other. Lim, Engstrom, and Ochoa (2005) cited Voss and Emmons (1996) when stating that the record for *P. bilabiatum* in Surinam was erroneous. The occurrence of *Pygoderma* in Surinam needs to be confirmed. Nevertheless, the report of *Platyrrhinus lineatus* in Surinam (see S. L. Williams, Genoways, and Groen 1983) suggests that a faunule of mammals with eastern Brazilian affinities may occur in Surinam.

Genus *Sphaeronycteris* W. Peters, 1882

This monotypic genus of small (forearm averages 36.5–40.5 mm, mass 11.0–20.8 g, includes pregnant females), pale brown to reddish-brown frugivorous bats is endemic to South America. The four-banded dorsal fur has a narrow whitish basal band (most conspicuous between shoulders and on nape) followed by a wide (1/3 to 1/2 of hair length) brown to reddish-brown band, an almost equally wide, white to pale buff band, and a terminal narrow, brown to reddish-brown band that imparts the overall color to the dorsum. The ventral fur lacks color bands and is the same color as the dorsum, except for the paler throat, upper chest, and undersurface of the forearms and adjacent flight membranes. Some individuals are lightly frosted with white; the color generally is darker in subadults. Most individuals have two pairs of white spots: a conspicuous spot on each shoulder at the origin of the antebrachial membrane and a larger, less distinct spot on each side of the neck below the ear. The face is well haired and the noseleaf is small, rounded, and lacks a well-defined median rib. The legs, feet, and both surfaces of the interfemoral membrane are lightly, but conspicuously, furred. The interfemoral membrane also has a sparse, ragged fringe of hair. Adult males have a conspicuous visor-shaped, fleshy protuberance on the forehead that extends forward above the eyes and noseleaf. A much smaller protuberance is present in females. Males can cover the lower face with a ruff of skin in a manner similar to that of *Centurio senex*. The cranium

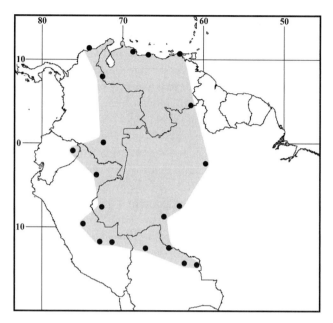

Map 201 Marginal localities for *Sphaeronycteris toxophyllum* ●

is relatively large and globose with a short rostrum. The anterior margin of the orbit consists of a conspicuously thin plate and the zygomatic arches are expanded laterally. The dental formula is 2/2, 1/1, 2/2, 3/3 × 2 = 32. Miller (1907b) and Husson (1959) provide more comprehensive descriptions of the skull and dentition. Linares (1968) reported sub-Recent fossil remains from Cueva de Quebrada, Aragua, Venezuela.

SYNONYMS:

Sphaeronycteris W. Peters, 1882:988; type species *Sphaeronycteris toxophyllum* W. Peters, 1882, by monotypy.
Stenoderma: Winge, 1892:11; not *Stenoderma* É. Geoffroy St.-Hilaire, 1818b.
Ametrida: Trouessart, 1897:163; not *Ametrida* Gray, 1847a.

Sphaeronycteris toxophyllum W. Peters, 1882
Visored Bat

SYNONYMS:

Sphaeronycteris toxophyllum W. Peters, 1882:989; type locality "*America tropicalis*"; restricted to Pebas, Loreto, Peru, by Cabrera (1958:92); also restricted by Husson (1959:119) to the "neighbourhood of Mérida, S.W. Venezuela, alt. 1600 to 3000 m" (see Remarks).
Stenoderma ("*Sphaeronycteris*") *toxophyllum*: Winge, 1892:11; name combination.
[*Ametrida (Sphaeronycteris)*] *toxophyllum*: Trouessart, 1897:163; name combination.

DISTRIBUTION: *Sphaeronycteris toxophyllum* is in Colombia, Venezuela, and the western Amazon basin of Ecuador, Peru, Brazil, and Bolivia.

MARGINAL LOCALITIES (Map 201): COLOMBIA: Magdalena, Parque Natural Nacional Tayrona (Cuervo Díaz, Hernández Camacho, and Cadena 1986). VENEZUELA (Handley 1976): Falcón, Riecito; Distrito Federal, Pico Ávila; Sucre, Manacal; Bolívar, 22.5 km NE of Icabarú. BRAZIL: Amazonas, Pedra do Gavião (Peracchi 1986); Amazonas, Humaitá (Peracchi 1986); Rondônia, Santo Antônio do Uayará (AMNH 92248). BOLIVIA: Beni, mouth of Río Baures (S. Anderson 1997); Santa Cruz, Parque Nacional Noël Kempff Mercado (L. H. Emmons, pers. comm.); Santa Cruz, "PRNB" (S. Anderson 1997); La Paz, Santa Ana de Madidi (S. Anderson 1997). PERU: Madre de Dios, Cocha Cashu (Terborgh, Fitzpatrick, and Emmons 1984); Cusco, Segakiato (S. Solari et al. 1998); Huánuco, Panguana (Hutterer et al. 1995). BRAZIL: Acre, Cruzeiro do Sul (Peracchi 1986). PERU: Loreto, Iquitos (Angulo and Díaz 2004). ECUADOR: Orellana, Campamento Petrolero Amo 2 (Albuja and Mena 1991). COLOMBIA: Caquetá, Estación Puerto Abeja (Montenegro and Romero-Ruiz 2000); Norte de Santander, Cúcuta (Sanborn 1941).

SUBSPECIES: I consider *S. toxophyllum* to be monotypic.

NATURAL HISTORY: Handley (1976) reported *S. toxophyllum* in a variety of forest habitats in Venezuela as well as in disturbed sites such as yards, pastures, orchards, and croplands. The species is frugivorous (Gardner 1977c). S. Anderson and Webster (1983) reported a Bolivian female pregnant in October. Ibáñez (1984a), reporting on a female from Venezuela that was both pregnant and lactating in October, suggested reproduction follows the pattern of bimodal polyestry and postpartum estrus, which is generally characteristic of stenodermatines. Venezuelan specimens of females in the USNM were pregnant when taken in February (1), April (5), July (3), August (3), and October (1). Lactating females were caught in April (2), May (4), July (11), August (2), and October (2). Juveniles (wing epiphyses unfused) were found in May (1), July (3), August (1), and October (1). A Colombian female in the USNM from Puerto Nariño, Amazonas, was pregnant when taken in April. According to Husson (1959), information on the label of a *S. toxophyllum* collected by Briceño Gabaldón at La Culata, near Mérida, Venezuela, gives eye color as blue, and the bat was found in a cavity in the ground. Information on the labels of two specimens in the USNM from Mérida, Venezuela, also collected by Gabaldón, has the eye color as green. These labels also contained the information "Casado en la Luz Electrica" and "Casado de mariposa," meaning that both bats were caught by butterfly net in the building housing the electrical generator. Specimens from southern Venezuela and a specimen from Cocha Cashu, Madre de Dios, Peru, (photographed by Louise H. Emmons) have a vermiculated gold-on-brown iris. The karyotype is $2n = 28$, FN = 52 (R. J. Baker 1973; R. J. Baker et al. 1982).

REMARKS: D. C. Carter and Dolan (1978) gave information on and measurements for the holotype of *Sphaeronycteris toxophyllum* W. Peters. Husson (1959) and Swanepoel and Genoways (1979) provided measurements of specimens from Venezuela, Bolivia, and Colombia. Cabrera's (1958:92) restriction of the type locality to Pebas, Loreto, Peru, is based on Rehn's (1901) report of a specimen from that locality. Peracchi (1986) argued that the type locality should be restricted to the vicinity of Mérida, Venezuela, as restricted by Husson (1959). Unless additional information becomes available that refutes the Peruvian locality, Cabrera's restriction to Pebas, Peru antedates Husson's (1959:119) restriction of the type locality to Mérida, Venezuela.

Tribe Sturnirini

Alfred L. Gardner

Taxa in Sturnirini are distinguished primarily on the basis of molar structure. The last upper premolar and the upper molars have the cusps shifted to the margins where they border a broad, longitudinal groove that is continuous from tooth to tooth. Lower molars are similar, except that the groove is interrupted by the paraconid of each molar. Other features of the Sturnirini include a narrow interfemoral membrane that often is hidden under the dense fur covering the membrane and legs, a sturnirine gland (de la Torre 1961) on each shoulder, and the lack of a tail and a true calcar. The Sturnirini occur from northern Mexico and the Lesser Antilles southward into Argentina.

Genus *Sturnira* Gray, 1842

Sturnira comprises at least 15 species in two subgenera (*Corvira* and *Sturnira*). Greatest diversity is on the forested slopes of the Andes where all 12 of the currently recognized species in South America occur. Members of the genus range from small to large (forearm 32–61 mm, mass 10–68 g) and can be identified externally by the greatly reduced and thickly haired interfemoral membrane; the usual presence of shoulder glands that stain adjacent hairs a conspicuous reddish, orange, or yellow color; and the lack of a tail. Shoulder gland activity may be correlated with breeding season and reproductive state. Body color varies from pale gray to dark gray or cinnamon brown, depending on the species. The pelage of adults of both sexes may be stained orange to yellow over the entire body by secretions from shoulder glands and, occasionally, by pollen. The noseleaf is normal for the subfamily and the ears are short with round tips. The braincase has a moderately developed sagittal crest and is conspicuously elevated above the rostrum, which is more than 1/2 as long as the braincase. The characteristic molars have the cusps and cuspids located on the extreme lingual and labial margins separated by a broad U-shaped median longitudinal groove, which is continuous from the last upper premolar through the upper molars, but is interrupted by the paraconid in lower molars. Sexual dimorphism is evident in the length of the canines, being considerably longer in males. The dental formula is 2/1–2, 1/1, 2/2, 3/3 × 2 = 30–32. The karyotype is $2n = 30$, FN = 56 in all species examined (summarized by R. J. Baker et al. 1982). The sex chromosomes are presumed to be neo-XY types derived from autosomal translocations involving both sex chromosomes (Tucker 1986).

The diet (summarized in Gardner 1977c) includes fruits and pollens from a variety of plants. Based on the frequency of seeds found in feces and digestive-tract contents, the fruits of *Solanum* spp. are preferred foods of those species whose diets are known (see Ibáñez 1984a).

SYNONYMS:

Phyllostoma: É. Geoffroy St.-Hilaire, 1810b:174; part; not *Phyllostoma* G. Cuvier, 1800.

Phyllostomus: Olfers, 1818:224; part; not *Phyllostomus* Lacépède, 1799b.

Sturnira Gray, 1842:257; type species *Sturnira spectrum* Gray, 1842, by monotypy.

Sturnia Gray, 1843:17; incorrect subsequent spelling of *Sturnira* Gray.

Stenoderma: Gay, 1847:30; not *Stenoderma* É. Geoffroy St.-Hilaire, 1818b.

Nyctiplanus Gray, 1849:58; type species *Nyctiplanus rotundatus* Gray, 1849, by monotypy.

Arctibeus: Tomes, 1860b:212; not *Arctibeus* Gray, 1838b; incorrect subsequent spelling of *Artibeus* Leach, 1821b.

Sycophaga Winge, 1892:10; part; preoccupied by *Sycophaga* Westwood, 1840 (Hymenoptera).

Sturmira O. Thomas, 1898b:3; incorrect subsequent spelling of *Sturnira* Gray.

Corvira O. Thomas, 1915b:310; type species *Corvira bidens* O. Thomas, 1915b, by original designation; recognized as a subgenus.

Sturnirops Goodwin, 1938:1; type species *Sturnirops mordax* Goodwin, 1938, by original designation.

Slumira Alberico and Orejuela, 1982:34; incorrect subsequent spelling of *Sturnira* Gray.

Sturnia Baud, 1986:46; incorrect subsequent spelling of *Sturnira* Gray.

REMARKS: Rodríguez-Mahecha et al. (1995:15, footnote) claimed the gender of *Sturnira* Gray, 1842, was neuter. The name *Sturnira* was based on the British ship HMS Starling, the companion ship to the HMS Sulphur.

The gender of *Sturnira* is feminine because all HMS ships are feminine.

KEY TO SOUTH AMERICAN SPECIES OF *STURNIRA* (MODIFIED FROM W. B. DAVIS 1980):

1. Lower incisors broad, normally one on each side; lower outer incisors, if present, reduced to slender spicules . . . . (subgenus *Corvira*) 2
1'. Lower incisors normally four, two on each side, and not reduced to spicules . . . . (subgenus *Sturnira*) 3
2. Forearm 39–44 mm; greatest length of skull averaging 21 mm . . . . *Sturnira bidens*
2'. Forearm 32–36 mm; greatest length of skull averaging 19 mm . . . . *Sturnira nana*
3. Molars and premolars usually in contact with adjacent teeth; upper inner incisors not conspicuously enlarged and usually not strongly procumbent 4
3'. Molars and premolars each separated by obvious gaps; upper inner incisors conspicuously enlarged and strongly procumbent . . . . *Sturnira koopmanhilli*
4. Lingual cuspids (metaconid and entoconid) of m1 and m2 poorly defined, usually forming a continuous sloping ridge 5
4'. Lingual cuspids (metaconid and entoconid) of m1 and m2 well defined and separated by a deep notch 9
5. Forearm longer than 51 mm; greatest length of skull more than 27 mm . . . . *Sturnira magna*
5'. Forearm shorter than 50 mm; greatest length of skull less than 27 mm 6
6. Forearm 41–47 mm; condylobasal length 20 mm or more; shoulder spots may or may not be conspicuous; lower incisors bilobed or trilobed 7
6'. Forearm 38–45 mm; condylobasal length usually less than 20 mm; shoulder spots inconspicuous or absent; lower incisors bilobed 8
7. Forearm 41–44 mm; upper incisors not obviously procumbent, anterior surface convex, and tips point downward; shoulder spots inconspicuous or absent; lower incisors trilobed . . . . *Sturnira sorianoi*
7'. Forearm 42–47 mm; upper incisors procumbent, anterior surface flat or slightly concave, and tips point forward; shoulder spots usually conspicuous; lower incisors bilobed . . . . *Sturnira oporaphilum*
8. Size larger; condylobasal length 19.5 mm or more; mastoidal breadth 11.6 mm or more; mandibular toothrow 7 mm or more . . . . *Sturnira bogotensis*
8'. Size smaller; condylobasal length 19.4 mm or less; mastoidal breadth less than11.6 mm; mandibular toothrow less than 7 mm . . . . *Sturnira erythromos*
9. Paraconulid present on m1 and m2 . . . . *Sturnira mistratensis*
9'. Paraconulid not present on m1 and m2 10

10. Forearm 54–62 mm; greatest length of skull approximately 29 mm . . . . *Sturnira aratathomasi*
10'. Forearm less than 55 mm; greatest length of skull less than 27 mm 11
11. Forearm 43–51 mm; greatest length of skull 24–26 mm; tips of inner upper incisors broad and weakly bilobed with lobes of equal size; zygomatic breadth about 14 mm . . . . *Sturnira tildae*
11'. Forearm 45.5 mm or less; greatest length of skull 20.0–24.5 mm; tips of inner upper incisors narrow, often pointed; zygomatic breadth less than 14 mm 12
12. Maxillary ramus of zygomatic arch not noticeably bowed outward; zygomatic arches strongly converging anteriorly; maxillary toothrows nearly parallel . . . .*Sturnira luisi*
12'. Maxillary ramus of zygomatic arch noticeably bowed outward; zygomatic arches not converging anteriorly; maxillary toothrows arched outward (not parallel) . . . . *Sturnira lilium*

Subgenus *Corvira* O. Thomas, 1915

Species of *Corvira* have a narrow skull with a long, narrow, and sloping rostrum, and weak (sometimes incomplete) zygomatic arches. The dentition includes a single lower incisor on each side, small molars and premolars separated by narrow gaps, and a triangular first upper molar with rounded angles. The subgenus contains two species, which differ mainly in size. There is no fossil record.

Sturnira bidens (O. Thomas, 1915)
Bidentate Yellow-shouldered Bat

SYNONYMS:

Corvira bidens O. Thomas, 1915b:311; type locality "Baeza, Upper Coca River, Oriente [Napo] of N. Ecuador. Alt. 6500'."
Sturnira bidens: de la Torre, 1961:iv; first use of current name combination.
Sturnira (Corvira) bidens: Gardner and O'Neill, 1969:2; name combination.

DISTRIBUTION: *Sturnira bidens* is known from Venezuela, Colombia, Ecuador, and Peru.

MARGINAL LOCALITIES (Map 202; localities listed from north to south): VENEZUELA: Mérida, Monte Zerpa, 4 km NW of Mérida (Soriano and Molinari 1984). COLOMBIA: Antioquia, La Candelaria (Muñoz 2001); Cauca, Munchique (Marinkelle and Cadena 1972); Huíla, El Parque Nacional Natural de la Cueva de los Guácharos (Lemke et al. 1982); Nariño, Reserva Natural La Planada (Ospina-Ante and Gomez 2000). ECUADOR: Napo, Baeza (type locality of *Corvira bidens* O. Thomas); Chimborazo, 4 km NE of Pallatanga (USNM 513447); Zamora-

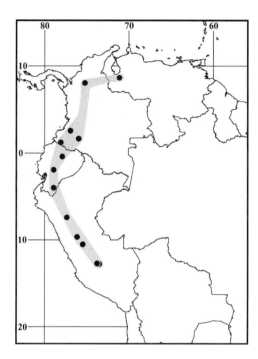

Map 202   Marginal localities for *Sturnira bidens* ●

Chinchipe, 4 km E of Sabanilla (USNM 513446). PERU: San Martín, Parque Nacional Río Abiseo (S. Solari et al. 2001b); Huánuco, eastern slope Cordillera Carpish (Gardner and O'Neill 1969); Pasco, Yanachaga-Chemillen (S. Solari et al. 2001b); Ayacucho, Huanhuachayo (Gardner 1976); Cusco, Wayrapata (S. Solari et al. 2001a).

SUBSPECIES: I am treating *S. bidens* as monotypic.

NATURAL HISTORY: This species inhabits Andean cloud forests at elevations from 1,700 to 3,000 m. Tamsitt, Cadena, and Villarraga (1986) reported netting a *S. bidens* in a cave in departamento Huila, Colombia. Fecal samples from 14 Venezuelan specimens contained remains of fruit representing at least six species (Molinari and Soriano 1987). Three females, each pregnant with a single embryo (CR 15–20 mm), were captured in August in Peru along with a male whose testes measured 3.5 × 4.5 mm (Gardner and O'Neill 1969). According to Molinari and Soriano (1987) the reproductive pattern is bimodal polyestry. They also reported a bat recovered 3 years after it had been banded.

REMARKS: *Sturnira bidens* can be distinguished from *S. nana* by larger size (forearm 39.3–43.3 mm, versus 32.6–35.7 mm; greatest length of skull 20.0–22.3 mm, versus 18.4–19.3 mm; Gardner and O'Neill 1971; Molinari and Soriano 1987; Swanepoel and Genoways 1979); well-haired legs, feet, and dorsal surface of uropatagium (sparsely haired in *S. nana*); and narrow tips of upper inner incisors (tips broad in *S. nana*). Molinari and Soriano (1987) illustrated the skull and dentition, summarized liter-

ature information, and added data from Venezuelan populations in their *Mammalian Species* account.

The report of *S. bidens* from Brazil (Marques and Oren 1987) was based on a misidentified *S. lilium* (MZUSP 13410) that lacked lower outer incisors. La Candelaria, the northernmost record in Colombia, cited by Muñoz (2001), is suspect because the elevation (120 m) is below the normal distributional range of the species. In the same account, Muñoz (2001) credited Marinkelle and Cadena (1972) for specimens from Mitú, departamento Vaupés; however, Marinkelle and Cadena (1972) cited ten specimens from Munchique, departamento Cauca, and gave the elevation as 3,000 m, which is appropriate for *S. bidens*; whereas Mitú, at 250 m is not.

Sturnira nana Gardner and O'Neill, 1971
Lesser Yellow-shouldered Bat

SYNONYM:

Sturnira nana Gardner and O'Neill, 1971:1; type locality "Huanhuachayo (12°44′S, 73°47′W), ca. 1,660 meters, Departamento de Ayacucho, Perú."

DISTRIBUTION: *Sturnira nana* is known only from the vicinity of the type locality.

MARGINAL LOCALITIES (Map 203): PERU: Ayacucho, San José, Río Santa Rosa (LSUMZ 16519); Ayacucho, Huanhuachayo (type locality of *Sturnira nana* Gardner and O'Neill).

SUBSPECIES: *Sturnira nana* is monotypic.

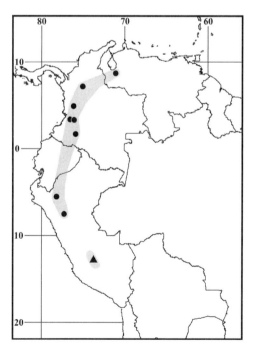

Map 203   Marginal localities for *Sturnira aratathomasi* ●
and *Sturnira nana* ▲

NATURAL HISTORY: Specimens were netted in lower montane forest and at the edges of clearings on the eastern slope of the Andes. The testes of one specimen measured 4.5 × 4.0 mm on 30 April. None of the females were pregnant or lactating when captured in late April and early May.

REMARKS: With an average forearm length of 34.3 mm and average greatest length of skull of 18.7 mm, *S. nana* is the smallest member of the genus and most closely resembles its sympatric congener *S. bidens* from which it can be identified by features summarized in the preceding account.

Subgenus *Sturnira* Gray, 1842

Sturnira (Sturnira) is characterized by a robust, narrow to broad skull having complete, well-ossified zygomatic arches. Dentition is normal for the genus, and two incisors are normally present in each lower jaw. The first upper molar is subquadrate in occlusal outline; molars and premolars are variable in size, but always in contact from canine to last molar. The subgenus contains at least 10 species. *Sturnira lilium* is known from late Pleistocene cave deposits in the vicinity of Lagoa Santa, Minas Gerais, Brazil (Winge 1892), and from subfossil material from Loltún, Yucatán, Mexico (T. Alvarez 1982).

Sturnira aratathomasi R. L. Peterson and Tamsitt, 1968
Giant Yellow-shouldered Bat
SYNONYM:

Sturnira aratathomasi R. L. Peterson and Tamsitt, 1968:1; type locality "2 km south of Pance (approximately 20 km southwest of Cali), Dept. of Valle, Colombia, at 1650 metres elevation."

DISTRIBUTION: *Sturnira aratathomasi* is known from Venezuela, Colombia, Ecuador, and Peru.

MARGINAL LOCALITIES (Map 203, localities listed from north to south): VENEZUELA: Mérida, Monte Zerpa, 4 km NW of Mérida (Soriano and Molinari 1984). COLOMBIA: Antioquia, Anorí (Muñoz 2001); Valle del Cauca, Paso de Galápagos (Alberico 1987); Valle del Cauca, Hacienda Los Alpes (Alberico 1987); Valle del Cauca, 2 km S of Pance (type locality of *Sturnira aratathomasi* R. L. Peterson and Tamsitt); Huila, El Parque Nacional Natural la Cueva de los Guacharos (Tamsitt, Cadena, and Villarraga 1986). ECUADOR: (locality unknown; R. L. Peterson and Tamsitt 1968). PERU: Amazonas, Cordillera Colán, east of La Peca (McCarthy, Barkley, and Albuja 1991); San Martín, Las Palmas (M. Romo pers. comm.).

SUBSPECIES: I treat *S. aratathomasi* as monotypic

NATURAL HISTORY: This rare species has been collected in tropical forests at elevations above 1,800 m (M. E. Thomas and McMurray 1974). Tamsitt, Cadena, and Villarraga (1986) reported specimens from montane and premontane wet forests. M. E. Thomas and McMurray (1974) reported finding unidentifiable fruit pulp in the stomachs of four *S. aratathomasi* from Colombia. They also reported a female pregnant when taken in February (fetus 34 mm CR) and another pregnant in August (fetus 43 mm CR). The karyotype is $2n = 30$, $FN = 56$ (Soriano and Molinari 1984). R. L. Peterson and Tamsitt (1968), M. E. Thomas and McMurray (1974), and Soriano and Molinari (1987) have provide illustrations, measurements, and additional information on reproduction, ecology, and behavior.

REMARKS: *Sturnira aratathomasi* is large for the genus (greatest length of skull 27.4–29.9 mm, forearm 54.8–62.0 mm, $n = 13$; M. E. Thomas and McMurray 1974; Soriano and Molinari 1987). Cranial characteristics include a long rostrum, procumbent and pointed inner upper incisors, and deep clefts separating the lingual cusps of m1 and m2 (R. L. Peterson and Tamsitt 1968; Tamsitt and Häuser 1985). The deep lingual clefts on m1 and m2 easily distinguish this species from the equally large *S. magna*.

Sturnira bogotensis Shamel, 1927
Bogota Yellow-shouldered Bat
SYNONYMS:

Sturnira lilium bogotensis Shamel, 1927:129; type locality "Bogota (Estacion 'La Uribe'), Colombia."
Sturnira ludovici: Hershkovitz, 1949c:441; part; not *S. ludovici* H. E. Anthony, 1924c.
Sturnira erythromos: de la Torre, 1961:124; part; not *S. erythromos* Tschudi, 1844b.
Sturnira bogotensis: Handley, 1976:25; first use of current name combination.

DISTRIBUTION: *Sturnira bogotensis* is known from Venezuela, Colombia, Ecuador, and Peru.

MARGINAL LOCALITIES (Map 204; localities listed from north to south): VENEZUELA: Mérida, Monte Zerpa, 4 km NW of Mérida (Soriano and Molinari 1984); Táchira, Buena Vista (Handley 1976). COLOMBIA: Cundinamarca, Sibaté (V. Pacheco and Patterson 1991). ECUADOR: Azuay, Yanasacha (V. Pacheco and Patterson 1992). PERU (V. Pacheco and Patterson 1992): Ancash, Huaylas; Lima, Bosque de Zárate; Lima, Asia.

SUBSPECIES: I consider *S. bogotensis* to be monotypic.

NATURAL HISTORY: This species is known from forested Andean slopes in Venezuela (Handley 1976), Colombia, and Ecuador. Peruvian specimens have been collected at lower elevations at sites in river valleys along the arid Pacific versant. Details of the diet, roosting preferences, and reproductive behavior remain unknown.

REMARKS: *Sturnira bogotensis* is not known from Bolivia and most reports of the species from Ecuador and Peru are based on misidentifications of *S. oporaphilum*.

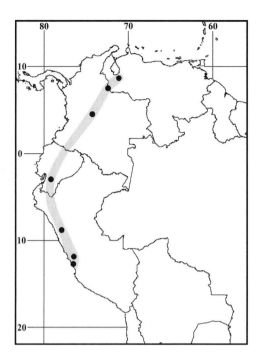

Map 204 Marginal localities for *Sturnira bogotensis* ●

Although originally described as a subspecies of *S. lilium*, several authors (e.g., Hall and Kelson 1959; Hall 1981) followed Hershkovitz (1949c) in regarding the name as a synonym of *S. ludovici* H. E. Anthony. Others (e.g., de la Torre 1961) treated *bogotensis* as a synonym of *S. erythromos* (Tschudi, 1844b). Handley (1976) first recognized *S. bogotensis* at the species level. S. Anderson, Koopman, and Creighton (1982) synonymized *S. bogotensis* under *S. oporaphilum* (Tschudi, 1844b). W. B. Davis (1980) recognized neither *S. bogotensis* nor *S. oporaphilum*, possibly because he considered them to be undeterminable. With the exception of the type locality, I have not included the records in Muñoz (2001), because he appears to have misidentified the specimens he cited as *S. bogotensis*. Muñoz (2001) said the lingual margins of the lower molars are notched as in *S. lilium*; whereas, actually the lingual margins are entire as in *S. erythromos* and *S. oporaphilum*. Apparently, Eisenberg and Redford (1999:167) also were confused about the identity of *S. bogotensis*, and included *S. oporaphilum* as a synonym even though *oporaphilum* has 83 years of priority.

This is a medium-sized member of the genus (greatest length of skull averages 22.6 mm, *n* = 5; Shamel 1927) similar to *S. erythromos* and *S. oporaphilum* in cranial and dental characteristics. Shamel's (1927) measurements of the forearm (averaging 45.2 mm) probably are too short, because in some of the specimens he examined, the proximal end of the ulna appears to have been cut off when the specimens were prepared. Specimens of *S. bogotensis* are larger

than specimens of *S. erythromos* and approach the size of *S. oporaphilum*. The upper inner incisors usually have convex anterior surfaces and downward-pointing tips; whereas the anterior surfaces are usually flat or slightly concave in *S. oporaphilum* and the tips point forward. The rostrum is short, comparable to that of *S. erythromos*, and conspicuously shorter than the rostrum of *S. oporaphilum*. The cranium is broader than that of *S. erythromos*, especially in mastoidal breadth (11.6–12.4 mm, *n* = 5; versus 10.8–11.5 mm, *n* = 17).

Sturnira erythromos (Tschudi, 1844)
Small Yellow-shouldered Bat

SYNONYMS:

Ph[yllostoma (Sturnira)]. erythromos Tschudi, 1844a:246; *nomen nudum.*

Ph[yllostoma] erythromos Tschudi, 1844b:64; type locality "der mittlern Waldregion zwischen 12 und 14° S.B.," Peru.

[*Sturnira*] *erythromos*: de la Torre, 1961:77; first use of current name combination.

Sturnira erithromos R. Guerrero, 1985:57; incorrect subsequent spelling of *Phyllostoma erythromos* Tschudi.

DISTRIBUTION: *Sturnira erythromos* has been recorded from Venezuela, Colombia, Ecuador, Peru, Bolivia, and Argentina.

MARGINAL LOCALITIES (Map 205): VENEZUELA (Handley 1976): Distrito Federal, Pico Ávila; Monagas, San Agustín, 3 km NW of Caripe. COLOMBIA: Meta, Cane Alto (V. Pacheco and Patterson 1992); Huila, El Parque Nacional Natural de la Cueva de los Guácharos (Lemke et al. 1982). ECUADOR: Chimborazo, Planchas (Albuja 1983); Morona-Santiago, Río San Vicente (Albuja 1999). PERU: Amazonas, 6 km by road SW of Laguna Pomacochas (V. Pacheco and Patterson 1992); Huánuco, Cerros del Sira (Koopman 1978); Cusco, Limacpunco (de la Torre 1961); Puno, Abra de Maruncunca (LSUMZ 26934). BOLIVIA (S. Anderson 1997): Beni, 5 km N of El Provenir; Santa Cruz, San José de Chiquitos. ARGENTINA (Barquez, Mares, and Braun 1999): Salta, Alto Macueta; Tucumán, Piedra Tendida; Catamarca, Cuesta del Clavillo, 3 km SW of Banderita; Jujuy, Abra de Cañas. BOLIVIA (S. Anderson, 1997): Tarija, 25 km NW of Entre Rios; Chuquisaca, 9 km N of Padilla; Cochabamba, 13 km N of Colomi; La Paz, Unduavi. PERU: Ayacucho, Huanhuachayo (Gardner 1976); Huánuco, eastern slope Cordillera Carpish (Gardner and O'Neill 1969); Piura, 15 km E of Canchaque (V. Pacheco and Patterson 1992). ECUADOR (Albuja 1999): El Oro, Trenchillas; Carchi, Pailón. COLOMBIA: Valle del Cauca, 2 km S of Pance (USNM 483451); Antioquia, Urrao (Muñoz 2001). VENEZUELA: Zulia, Novito (Handley 1976); Mérida, Monte Zerpa, 4 km NW of Mérida

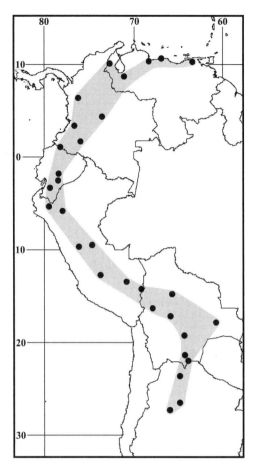

Map 205 Marginal localities for *Sturnira erythromus* ●

lingual margins of lower molars, lack of gaps between cheek teeth, and the usual lack of conspicuously pigmented shoulder spots. The species is most likely to be confused with *S. bogotensis*, which is a larger bat with a longer hind foot (exceeds 12 mm, dry measurement; see Shamel 1927), but with which it shares many cranial features. *Sturnira erythromos* usually is found at moderate to higher elevations above the normal range of *S. lilium* and *S. oporaphilum*, but within the habitat and range of *S. bogotensis*. Specimens from throughout the southern range of the species have been misidentified as *S. lilium*.

Sturnira koopmanhilli McCarthy, Albuja, and Alberico, 2006
Chocoan Yellow-shouldered Bat

SYNONYMS:

Sturnira mordax: Alberico, 1994:335; not *Sturnira mordax* Goodwin, 1938.

Sturnira koopmanhilli McCarthy, Albuja, and Alberico, 2006:97; type locality "Las Pambiles, Reserva Ecológica Cotacachi-Cayapas, Provincia de Esmeraldas, Ecuador, 00°32′N, 78°37′W, 1200 m."

DISTRIBUTION: *Sturnira koopmanhilli* is known only from western Colombia and Ecuador.

MARGINAL LOCALITIES (Map 210; from McCarthy, Albuja, and Alberico 2006): COLOMBIA: Chocó, 4 km N of La Italia; Valle del Cauca, Quebrada de la Delfina; Nariño, Reserva Natural La Planada. ECUADOR: Esmeraldas, Las Pambiles (type locality of *Sturnira koopmanhilli* McCarthy, Albuja, and Alberico); Chimborazo, Pallatanga.

SUBSPECIES: *Sturnira koopmanhilli* is monotypic.

NATURAL HISTORY: *Sturnira koopmanhilli* is found in the wet forests of western Colombia and Ecuador and the adjacent lower Pacific slope of the Andes. This region, usually referred to as the Chocó, is characterized by high rainfall and high biotic diversity and its flora and fauna have a high level of endemism. The greatly enlarged and procumbent upper incisors of *S. koopmanhilli* suggest specialized food habits. Details of its diet, behavior, reproduction, and other aspects of natural history are unknown.

REMARKS: Emphasis on the large, procumbent upper incisors misled Alberico (1994), Cadena, Anderson, and Rivas-Pava (1998), and Sánchez-Hernández, Romero-Almaraz, and Cuisin (2002) to identify specimens of *S. koopmanhilli* as *S. mordax*. Although V. Pacheco and Patterson (1991) recognized the form as undescribed (their *Sturnira* sp. A, p. 102), certain similarities with species in the subgenus *Corvira* (gaps between cheekteeth, elongated skull, and weak zygomatic arches) suggested that the species was intermediate between species of *Corvira* and *Sturnira*. Albuja (1999:127) also referred to this species as *Sturnira* sp. A, which he recognized on the basis of its bicolored

(Soriano and Molinari 1984); Carabobo, La Copa (Handley 1976).

SUBSPECIES: I am treating *S. erythromos* as monotypic.

NATURAL HISTORY: This species is known from forested habitats throughout is range (see Handley 1976, for habitats occupied in Venezuela). Barquez, Giannini, and Mares (1993) suggested foliage and hollow trees as roost sites. Details of its diet are unknown. M. E. Thomas (1972) reported two Colombian females that were pregnant when taken in December (Pance; reported as *S. lilium*). Gardner and O'Neill (1969) captured 10 pregnant females during August in Peru; each had a single embryo (2–18 mm CR). Additional information was provided by Giannini and Barquez (2003) in their *Mammalian Species* account.

REMARKS: A medium-sized *Sturnira* (forearm averaging 41.5 mm, greatest length of skull averaging 21.4 mm; see Gardner and O'Neill 1969, for additional measurements), *S. erythromos* has been confused with *S. lilium*, *S. ludovici*, *S. oporaphilum*, and *S. bogotensis*. Although not the smallest species of *Sturnira* (*contra* Eisenberg and Redford 1999:167) *S. erythromos* can be distinguished from other species of *Sturnira* on the basis of smaller size, smooth dorsal pelage, gaps between the cheek teeth, and procumbent upper incisors.

Sturnira lilium (É. Geoffroy St.-Hilaire, 1810)
Little Yellow-shouldered Bat

SYNONYMS: The following synonyms represent non-South American populations; see under Subspecies for additional synonyms.

Sturnira angeli de la Torre, 1961:109; unavailable name.

Sturnira angeli de la Torre, 1966:271; type locality "6 mi. NE Roseau (1000 ft), St. Paul Parish, Dominica, Windward Islands, West Indies."

Sturnira paulsoni de la Torre and Schwartz, 1966:301; type locality "Lowrt, 1000 ft, St. Andrew Parish, Saint Vincent, British Windward Islands, Lesser Antilles."

Sturnira lilium angeli: J. K. Jones and Phillips, 1976:10; name combination.

Sturnira lilium paulsoni: J. K. Jones and Phillips, 1976:11; name combination.

DISTRIBUTION: *Sturnira lilium* is in Colombia, Venezuela, the Guianas, Ecuador, Brazil, Peru, Bolivia, Paraguay, Uruguay, Argentina, and probably Chile (see Remarks). Elsewhere, the species is known from Mexico, Jamaica, Central America, and the Lesser Antilles.

MARGINAL LOCALITIES (Map 206): VENEZUELA (Handley 1976): Zulia, Cerro Azul, 33 km NW of La Paz; Falcón, Boca de Yaracuy; Distrito Federal, Pico

Ávila; Sucre, 21 km E of Cumaná. TRINIDAD AND TOBAGO: Trinidad, Blanchisseuse (C. H. Carter et al. 1981). VENEZUELA: Delta Amacuro, Río Ibaruma (Rivas 2000). GUYANA: Cuyuni-Mazaruni, Kartabu Point (USNM 545099). SURINAM: Commewijne, Nieuwe Grond Plantation (Williams and Genoways 1980a). FRENCH GUIANA: St. Elie (Brosset and Charles-Dominique 1991). BRAZIL: Amapá, Amapá (Peracchi, Raimundo, and Tannure 1984); Pará, Belém (Handley 1967); Pará, Quatipurú (O. Thomas 1920c); Ceará, Centro de Visitantes, Parque Nacional de Ubajara (S. S. P. Silva, Guedes, and Peracchi 2001); Ceará, Fortaleza (Piccinini 1974); Pernambuco, São Lourenço da Mata (Souza and Araújo 1990); Bahia, Bahia (P. Gervais 1856a); Espírito Santo, Mucurici (Ruschi 1965); Espírito Santo, Estação Experimental de Linhares (Peracchi and Albuquerque 1993); Rio de Janeiro, Rio de Janeiro (Pelzeln 1883); São Paulo, São Sebastião (McNab 1969); São Paulo, Iguapé (Pira 1904); Santa Catarina, Colonia Hansa (J. L. Lima 1926); Rio Grande do Sul, Taquara (Ihering 1892); Rio Grande do Sul, São Lourenço (Ihering 1892). URUGUAY: Tacuarembó, Arroyo Sauce de Tranquera (Ximénez, Langguth, and Praderi 1972); Paysandu, Paysandu (Podtiaguin 1944). ARGENTINA: Buenos Aires, Belgrano (Burmeister 1879); Santa Fe, Santa Fe (Barquez, Mares, and Braun 1999); Chaco, Resistencia (Podtiaguin 1944); Formosa, Clorinda (Podtiaguin 1944); Formosa, Río Porteño, 5 km S of Estancia Santa Catalina (Barquez and Ojeda 1992). PARAGUAY: Amamby, Parque Nacional Cerro Corá (USNM 554535). BOLIVIA: Santa Cruz, Santiago (S. Anderson, Koopman, and Creighton 1982); Santa Cruz, 15 km E of Ingeniero Mora (S. Anderson 1997); Santa Cruz, Parapita (S. Anderson 1997). ARGENTINA (Barquez, Mares, and Braun 1999, except as noted): Salta, Río Itiyuro, 1 km E of Tonono; Tucumán, Agua Colorada (Barquez and Ojeda 1992); Catamarca, La Merced; Catamarca, Cuesta del Clavillo, 5 km S of La Banderita; Salta, Seclantes (Yepes 1944); Jujuy, Abra de Cañas. BOLIVIA (S. Anderson 1997, except as noted): Tarija, San Lorenzo; Chuquisaca, Río Azuero, 70 km SE of Padilla (S. Anderson, Koopman, and Creighton 1982); Cochabamba, El Palmar; La Paz, 4 km by road NW of Alcoche. PERU: Puno, Río Tavara, SE of Fila Boca Guacamayo (USNM 579677); Ayacucho, Huanhuachayo (Gardner 1976); Pasco, San Juan (Tuttle 1970); Huánuco, Tingo María (LSUMZ 14169); Cajamarca, Río Zaña, 2 km N of Monteseco (V. Pacheco and Patterson 1991); Amazonas, Falso Paquisha (Vivar and Arana-Cardó 1994). ECUADOR: Morona-Santiago, Teniente Ortíz (Albuja 1983); Guayas, San Rafael (USNM 498904); Manabí, San Sebastián (Albuja 1999); Esmeraldas, La Tola (Dorst 1951). COLOMBIA: Valle del Cauca, Río Zabaletas (USNM 483496); Antioquia, Providencia (USNM 499424); El Cesar, Valledupar (USNM 281262).

Map 206 Marginal localities for *Sturnira lilium* ●

SUBSPECIES: Two of the four recognized subspecies of *S. lilium* occur in South America.

S. l. lilium É. Geoffroy St.-Hilaire, 1810
SYNONYMS:

Phyllostoma lilium É. Geoffroy St.-Hilaire, 1810b:181; type locality "Paraguay" (page 186); restricted to Asunción, Central, Paraguay, by Cabrera (1958:78); based on Azara's (1801b) "*chauve-souris quatrieme ou chauve-souris brun-rougeâtre.*"

[*Phyllostomus*] *spiculatus* Illiger, 1815:109; *nomen nudum.*

Ph[*yllostomus*]. *Lilium*: Olfers, 1818:224; name combination.

Phyllostoma spiculatum Lichtenstein, 1823:3; type locality not given; based on Azara's (1801b:277) "*chauvesouris brunrougeâtre*" from Paraguay, the same basis as *Phyllostoma lilium* É. Geoffroy St.-Hilaire, whose type locality was restricted to Asunción by Cabrera (1958:78).

Sturnira spectrum Gray, 1842:257; type locality "Brazils."

Phyllostoma excisum J. A. Wagner, 1842:358; type locality "Ypanema," São Paulo, Brazil.

Phyllostoma albescens J. A. Wagner, 1847:177; type locality "Ypanema," São Paulo, Brazil.

Phyllostoma fumarium J. A. Wagner, 1847:178; type locality "Brasilien."

Stenoderma chilense P. Gervais, 1847:30, in Gay 1847; type locality "Chile" (see Remarks).

Phyllostoma vampyrus Schinz, 1845:26; type locality "America tropicali."

Nyctiplanus rotundatus Gray, 1849:58; type locality "Brazils."

Stenoderma chiliensis P. Gervais, 1854:198; incorrect subsequent spelling of *Stenoderma chilense* P. Gervais.

Phyllostoma chrysocomos J. A. Wagner, 1855:635; a renaming of *Sturnira spectrum* Gray; therefore, the type locality is Brazil.

Ph[*yllostoma*] *canicula* P. Gervais, 1856a:37; unavailable name.

Sturnira lilium: P. Gervais, 1856a:39; first use of current name combination.

Sturnira chilensis: P. Gervais, 1856a:39; name combination.

Arctibeus lilium: Tomes, 1860b:212; name combination.

Stenoderma (Sturnira) excisum: Pelzeln, 1883:35; name combination.

Stenoderma (Sturnira) albescens: Pelzeln, 1883:35; name combination.

Phyllostoma chrysosema Natterer, 1883:35, in Pelzeln 1883; type locality "Rio Janeiro," Rio de Janeiro, Brazil.

Sturnira excisa: Ihering, 1894:23; name combination.

Sturnira lilium amazonica de la Torre, 1961:90; unavailable name.

Sturnira lilium andina de la Torre, 1961:95; unavailable name.

Sturnira lillium Shapley, Wilson, Warren, and Barnett, 2005:385; incorrect subsequent spelling of *Phyllostoma lilium* É. Geoffroy St.-Hilaire.

The nominate subspecies is in Venezuela, Trinidad and Tobago, Colombia, the Guianas, Ecuador, Brazil, Peru, Bolivia, Paraguay, Uruguay, northern Argentina, and is unconfirmed for Chile (see Remarks).

S. l. parvidens Goldman, 1917
SYNONYMS:

Sturnira lilium parvidens Goldman, 1917:116; type locality "Papayo (about 25 miles northwest of Acapulco), Guerrero, Mexico."

Sturnira lilium pallida de la Torre, 1961:106; unavailable name.

This subspecies is in northern and western Colombia and possibly northwestern Ecuador. Elsewhere, *S. l. parvidens* occurs northward through Central America to northern Mexico (Gannon, Willig, and Jones 1989).

NATURAL HISTORY: This species inhabits a variety of habitats from riparian gallery and humid tropical forests to seasonally arid tropical deciduous woodlands. The few documented instances of roosting sites include caves, buildings, and hollow trees (e.g., Goodwin and Greenhall 1961; Handley 1976). Natterer (in Pelzeln 1883) encountered nine in a hollow tree at Ypanema, São Paulo, Brazil. Peracchi and Albuquerque (1971) reported finding groups of from three to seven in hollow trees, also in Brazil. Evelyn and Stiles (2003) documented the preferred tree and cavity sizes for an eastern Mexican population. Among a wide variety of fruits and flower parts consumed, the fruits of *Cecropia* spp., *Piper* spp., and *Solanum* spp. appear to be more commonly eaten, and the latter may be preferred (Ibáñez 1984a). Wilson (1979) summarized reproductive data, which he said fit a pattern of bimodal polyestry. Ubelaker, Specian, and Duszynski (1977) listed three species of nematode endoparasites known to infect *S. lilium*. In their summary of ectoparasites, J. P. Webb and Loomis (1977) listed five species of streblid flies, three ticks (two families), and 12 mites (six families) known from *S. lilium*. R. Guerrero (1985a) added a flea, 11 species of mites, and 18 species of streblid flies. Rui and Graciolli (2005) recovered an additional two species of streblids from *S. lilium* they caught in the state of Rio Grande do Sul, Brazil.

REMARKS: *Sturnira lilium* has been confused with *S. tildae*, *S. luisi*, and *S. bogotensis*. It is most easily distinguishable from *S. tildae* by smaller size and more pointed incisors, from *S. luisi* by its anteriorly expanded zygomatic arches, and from *S. bogotensis* by its shorter forearm, smaller foot, and cusps separated by clefts on the lingual margin of the lower molars (entocristid is not separated by a cleft from the metaconid in *S. bogotensis*). *Sturnira lilium* consistently has trilobed lower incisors. Gannon, Willig, and Jones (1989) provide illustrations of the skull and karyotype, and other information in their *Mammalian Species* account.

Cabrera's (1958:78) assumption that É. Geoffroy St.-Hilaire (1810b) based his description of *Phyllostoma lilium* solely on Azara's (1801b:277) description of the "*chauvesouris quatrieme ou chauve-souris brun-rougeâtre*" is correct. The measurements and all other details came from Azara's account. By 1818, however, É. Geoffroy St.-Hilaire had acquired a specimen, which he discussed (1818:416) in terms of Azara's description. Therefore, because Geoffroy did not have a specimen at hand when he described *Phyllostoma lilium*, the specimen cited by Rode (1941) and D. C. Carter and Dolan (1978) is not the holotype. Also, as É. Geoffroy St.-Hilaire (1818a:416–17) later made clear, the bat he examined was not one of the specimens described by Azara. The specimen (MHMN 195), erroneously identified by Rode (1941:88) and D. C. Carter and Dolan (1978:51) as the holotype of *Sturnira lilium*, is most likely the specimen Geoffroy discussed and described in 1818. D. C. Carter and Dolan (1978) commented on and provided measurements for the types of *Phyllostoma albescens* J. A. Wagner, *P. excisum* J. A. Wagner, *P. fumarium* J. A. Wagner, and *Sturnira spectrum* Gray.

Gannon, Willig, and Jones (1989) gave the type locality of *Phyllostoma excisum* J. A. Wagner as Rio de Janeiro, Brazil. According to Pelzeln (1883:35), Natterer collected specimens of this taxon from Rio de Janeiro, Ypanema, and Ytararé. J. A. Wagner (1842:358) mentioned only Ypanema as the source of *P. excisum* and, thereby, restricted the type series (syntypes) to specimens from Ypanema, São Paulo, Brazil. D. C. Carter and Dolan (1978) provided information on the specimen from Rio de Janeiro they claimed was a syntype. Although the specimen likely was part of Natterer's collection examined and described by J. A. Wagner, that specimen is not a syntype for the reasons explained above.

Stenoderma chilensis P. Gervais usually is considered a synonym of *Sturnira lilium*; however, P. Gervais (1856a:39) and W. Peters (1865c:524) treated *Sturnira erythromos* (Tschudi) as a synonym of *Sturnira chilensis* (P. Gervais) although *S. erythromos* is the older name. Osgood (1943:237) doubted the occurrence of *S. chilensis* in Chile because the species has not been reported since its description. Mann (1978) did not mention any *Sturnira*; nor did Gantz and Martínez (2000). However, the Rijksmuseum van Natuurlijke Historie, Leiden, has a specimen of *S. lilium* (RMNH Reg. no. 24984) with "Chile" and "von d'Orbigny" on the label. This is specimen "i" on page 210 of Jentink's (1888) catalog. Several authors (e.g., Sherborn 1925) have cited Gay (1847) as the author of *S. chilensis*; however, P. Gervais (1856a:29, footnote) stated that he had provided Gay with descriptions and synonymies for most of the mammals and had overseen the preparation of the plates for the bats,

rodents, ruminants, and fossil vertebrates. Therefore, I have cited P. Gervais as the author of *Stenoderma chilensis*.

Cabrera (1958:78) disputed Ameghino's (1889:349) record of a specimen from Argentina as far south as the Río Negro. The record apparently is based on a fluid preserved *S. lilium* in the Natural History Museum, London (BM 14.4.4.3) labeled as from Patagonia. Barquez, Mares, and Braun (1999) also commented that the record is doubtful.

Sturnira luisi W. B. Davis, 1980
Luis's Yellow-shouldered Bat
SYNONYM:

Sturnira luisi W. B. Davis, 1980:1; type locality "Cariblanco, 3000 feet, 11 mi. NE Naranjo, Alajuela, Costa Rica."

DISTRIBUTION: *Sturnira luisi* occurs in Colombia, Ecuador, and Peru. Elsewhere, the species is known from Panama and Costa Rica (W. B. Davis 1980).

MARGINAL LOCALITIES (Map 207; from W. B. Davis 1980, except as noted; localities listed from north to south): COLOMBIA: Antioquia, near La Tirana, 24 km S and 22 km W of Zaragoza (USNM 499420); Valle del Cauca, Río Zabaletas (USNM 503418); Valle del Cauca, 2 km S of Pance (USNM 483497). ECUADOR: Imbabura, Chota; Esmeraldas, Estero Taquiama (Albuja 1999); Manabí, Estero Achiote (Albuja 1999); El Oro, Arenillas (Albuja 1999); El Oro, 9 miles S of Zaruma. PERU: Piura, 4 miles W of Suyo; Lambayeque, Las Juntas, Quebrada la Pachinga (LSUMZ 27256).

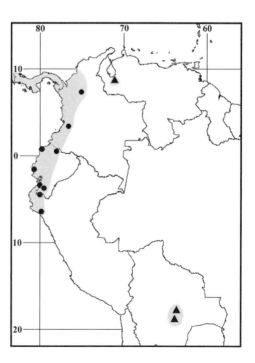

Map 207 Marginal localities for *Sturnira luisi* ● and *Sturnira sorianoi* ▲

SUBSPECIES: I consider *S. luisi* to be monotypic.

NATURAL HISTORY: This species occurs in forests from lower to moderate elevations (from near sea level to 1,800 m). Specimens of *S. luisi* have been misidentified as *S. lilium* and, except for information provided by W. B. Davis (1980), little is known about this species. Fruit pulp in stomachs is noted on the labels of several specimens of *S. luisi* in the USNM from Panama. Reproductive condition, also noted on specimen labels in the USNM, suggests bimodal polyestry. Pregnant females have been taken between 22 January and 30 March; lactating females between 27 February and 13 March, and again on 11 November. Colombian (Valle del Cauca) specimens were noted as pregnant on 14 March and on 14 and 15 November; lactating on 14 February; and juveniles (wing epiphyses open) noted on 12, 14, and 16 February.

REMARKS: This is a medium-sized *Sturnira* (forearm 41.0–45.5 mm, greatest length of skull 23.0–24.2 mm; W. B. Davis 1980). Tending to average larger than *S. lilium*, with which it is most commonly confused, *S. luisi* can be distinguished from *S. lilium* on the basis of its overall darker color and straight zygomatic arches that strongly converge anteriorly.

Brosset and Charles-Dominique (1991) compared lower dentitions of *Sturnira* from French Guiana with illustrations of *S. luisi* and *S. lilium* from W. B. Davis (1980) and suggested that *S. luisi* was closely related to *S. tildae*, if not a synonym of the latter. However, *S. luisi* is a Middle American Province taxon found from Costa Rica to northwestern Peru, whereas *S. tildae* occupies the greater Amazon basin and northeastern South America. Whatever their phylogenetic history and affinity, each taxon has a unique morphology warranting recognition as separate species. Eisenberg (1989) did not mention *S. luisi*.

Sturnira magna de la Torre, 1966
Greater Yellow-shouldered Bat

SYNONYMS:

Sturnira magna de la Torre, 1961:130; unavailable name.

Sturnira magna de la Torre, 1966:267, type locality "Santa Cecilia (100 m), Río Manití, Iquitos, Department of Loreto, Perú."

Sturnira magnum Rodríguez-Mahecha, Hernández-Camacho, Alberico, Mast, Mittermeier, and Cadena, 1995: 15, footnote; unjustified emendation of *Sturnira magna* de la Torre.

DISTRIBUTION: *Sturnira magna* is known from Colombia, Ecuador, Peru, Bolivia, and western Brazil.

MARGINAL LOCALITIES (Map 208): COLOMBIA: Guaviare, Salto Angostura II (Tamsitt, Cadena, and Villarraga 1986); Amazonas, Puerto Rastrojo (Tamsitt and Valdivieso 1986); Amazonas, Tarapacá (Marinkelle and Cadena 1972). PERU: Loreto, Centro de Investigaciones

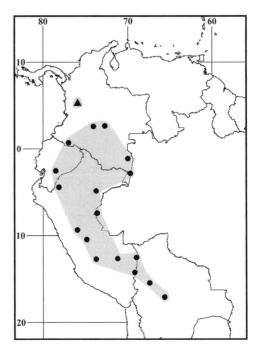

Map 208 Marginal localities for *Sturnira magna* ● and *Sturnira mistratensis* ▲

Jenaro Herrera (Ascorra, Gorchov, and Cornejo 1994). BRAZIL: Acre, Rio Moa, Parque Nacional da Serra do Divisor (M.R. Nogueira, Pol, and Peracchi 2000). PERU: Madre de Dios, Aguas Calientes (V. Pacheco et al. 1993); Madre de Dios, Reserva Cuzco Amazonico (Woodman et al. 1991). BOLIVIA: Cochabamba, El Sillar (Ibáñez 1985a); La Paz, Tomonoco (Webster and Jones 1980a). PERU: Puno, Abra de Maruncunca (LSUMZ 26912); Ayacucho, Huanhuachayo (Gardner 1976); Pasco, San Juan (Tuttle 1970); Huánuco, 9 km S and 2 km E of Tingo María (FMNH 98774); Amazonas, Huampami (J. L. Patton, Berlin, and Berlin 1982). ECUADOR: Azuay, San José Grande (Albuja and Mena 1991). COLOMBIA: Putumayo, Estación de Bombeo Guamués (Tamsitt, Cadena, and Villarraga 1986); Meta, Campamento Ardillas (Tamsitt and Valdivieso 1986).

SUBSPECIES: I treat *S. magna* as monotypic.

NATURAL HISTORY: This species inhabits mature forests on the eastern slopes of the northern Andes and adjacent Amazonian lowlands from Colombia to Bolivia (Koopman 1978; Tamsitt and Häuser 1985). *Sturnira magna* has an elevational range of from 200 m in Colombia to as high as 2,300 m in Peru (Tamsitt and Häuser 1985). Bolivian females were pregnant in November, December, February, and May (one noted for each month; CR lengths, 14, 45, 25, and 30 mm, respectively; Tamsitt and Häuser 1985). Tamsitt, Cadena, and Villarraga (1986) reported Colombian females pregnant in November and lactating in May, and juveniles in March and May. A lactating female and males with enlarged testes were taken in Peru during May (Gardner 1976). Bolivian males were reproductively active in July (Webster and Jones 1980a). Details of the diet are unknown. Tamsitt, Cadena, and Villarraga (1986) found males to average larger than females in cranial characters, but found no sexual dimorphism in external measurements.

REMARKS: This is one of the largest members of the genus and is exceeded in size only by *S. aratathomasi* whose measurements often overlap. The two species are easily distinguishable as follows: upper inner incisors are blunt and in contact in *S. magna*, but pointed and separate at the tips in *S. aratathomasi*; lingual cusps of lower molars form a low continuous ridge in *S. magna*, but lingual cusps are separated by a deep cleft in *S. aratathomasi*; and the posterior margin of the palate is broadly U-shaped in *S. magna*, but V-shaped in *S. aratathomasi*. Tamsitt and Häuser (1985) illustrated the cranium and mandibles, and provided measurements and additional information.

Sturnira mistratensis Contreras-Vega and Cadena, 2000
Mistrato Yellow-shouldered Bat

SYNONYM:

Sturnira mistratensis Contreras-Vega and Cadena, 2000: 286; type locality "COLOMBIA. Risaralda: Mistrató, corregiminto de Puerto de Oro, 980 m."

DISTRIBUTION: *Sturnira mistratensis* is known only from the type locality.

MARGINAL LOCALITIES (Map 208): COLOMBIA: Risaralda, Mistrató (type locality of *Sturnira mistratensis* Contreras-Vega and Cadena).

SUBSPECIES: *Sturnira mistratensis* is monotypic.

NATURAL HISTORY: Undoubtedly a frugivore, and probably with a preference for fruits of solanaceous plants in accord with its congeners, details of the natural history of *S. mistratensis* are unknown.

REMARKS: According to the original description (Contreras-Vega and Cadena 2000) the holotype is similar in color and size to *S. tildae*, but differs in having an entoconid on the lower molars (as in *S. aratathomasi*, *S. lilium*, *S. luisi*, and *S. tildae*) plus a paraconulid (not present in these four taxa); narrow, weakly bilobed upper inner incisors (variable and, except for *S. lilium*, not like other taxa); and a U-shaped posterior margin of the palate (V-shaped in these other four taxa). The forearm of the holotype measured 42.7 mm, greatest length of skull 22.4 mm, and zygomatic breadth 13.4 mm.

Sturnira oporaphilum (Tschudi, 1844)
Tschudi's Yellow-shouldered Bat

SYNONYMS: See under Subspecies.

DISTRIBUTION: *Sturnira oporaphilum* is in Venezuela, Colombia, Ecuador, Peru, Bolivia, and Argentina.

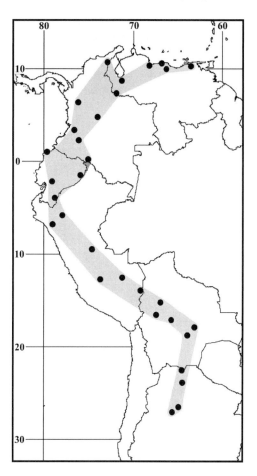

Map 209 Marginal localities for *Sturnira oporophilum* ●

MARGINAL LOCALITIES (Map 209; from V. Pacheco and Patterson 1991, except as noted): COLOMBIA: La Guajira, Villanueva (Hershkovitz 1949c). VENEZUELA (Handley 1976): Mérida, La Carbonera; Carabobo, 9 km NE of Montalbán; Distrito Federal, Los Venados; Monagas, San Agustín, 5 km NW of Caripe; Guárico, Hacienda La Elvira; Apure, Nulita. COLOMBIA: Cundinamarca, Suba (Valdivieso 1964); Huila, Río La Plata, 1 miles S of Moscopán; Putumayo, Parque Nacional Natural La Paya (Polanco-Ochoa, Jaimes, and Piragua 2000). ECUADOR: Pastaza, Lorocachi (USNM 548153); Zamora-Chinchipe, 3 km E de Cumbaratza. PERU: Amazonas, 6 road km SW of Lake Pomacochas; Huánuco, Cerros del Sira (Koopman 1978); Madre de Dios, Pantiacolla (V. Pacheco and Patterson 1992); Puno, Río Huari Huari, 5 km by road NE of San José (V. Pacheco and Patterson 1992). BOLIVIA (S. Anderson 1997): Beni, Serranía Pelón, 27 km by road N of Río Quiquibay; Santa Cruz, 15 km S of Santa Cruz. ARGENTINA (Barquez, Mares, and Braun 1999): Jujuy, Laguna La Brea; Tucumán, Piedra Tendida; Tucumán, Casa de Piedra; Salta, Serranía de Las Pavas. BOLIVIA (S. Anderson 1997): Santa Cruz, 13 km by road NW of Masicuri;

Cochabamba, El Sillar; La Paz, Pasto Grande. PERU: Ayacucho, Huanhuachayo (Gardner 1976); Cajamarca, Río Zaña, 2 km N of Monteseco. ECUADOR: Cañar, San Juan; Esmeraldas, Esmeraldas (de la Torre 1961). COLOMBIA: Valle del Cauca, 2 km S of Pance; Antioquia, Urrao (Muñoz 2001).

SUBSPECIES: I recognize two subspecies of *S. oporaphilum*.

S. o. ludovici (H. E. Anthony, 1924)
SYNONYMS:

Sturnira ludovici H. E. Anthony, 1924c:8, type locality "'near Gualea, elevation about 4000 feet,'" Pichincha, Ecuador.

Phyllostoma oporophilum Shamel, 1927:130; incorrect subsequent spelling of *Phyllostoma oporaphilum* Tschudi.

Sturnira ludivici Albuja, 1983:123; incorrect subsequent spelling of *Sturnira ludovici* H. E. Anthony.

This subspecies is in Colombia, Ecuador, and Venezuela.

S. o. oporaphilum (Tschudi, 1844)

Ph[yllostoma (Sturnira)]. oporaphilum Tschudi, 1844a:246; *nomen nudum*.

Ph[yllostoma]. oporaphilum Tschudi, 1844b:64; type locality "der mittlern Waldregion zwischen 12 und 14° S.B. [p. 68]," Peru.

Sturnira lilium: W. Peters, 1865c:524; part; not *S. lilium* É. Geoffroy St.-Hilaire, 1810b.

Sturnira oporaphilum: de la Torre, 1961:112; first modern use of current name combination.

The nominate subspecies is in Peru, Bolivia, and Argentina.

NATURAL HISTORY: Handley (1976) found this species in several forest habitats in Venezuela. Although *S. oporaphilum* is known to be a frugivore (M. E. Thomas 1972, reported under *S. ludovici*; Gardner 1977c), dietary details are unknown. The reproductive pattern appears to be bimodal polyestry (Wilson 1979). Wenzel (1976) reported six batflies (Streblidae) representing four genera found on Venezuelan bats identified as *S. ludovici*. J. P. Webb and Loomis's (1977) summary of phyllostomid ectoparasites known from *S. ludovici* added four species of streblid batflies as well as two ticks (Ixodidae and Argasidae) and seven mites (one macronyssid, two spinturnicids, and four trombiculids).

REMARKS: *Sturnira oporaphilum* can be confused with *S. erythromos* from which it is most easily distinguished on the basis of larger size, protruding (procumbent) upper inner incisors, and presence of gland-stained hair on the shoulders (shoulder spots absent in *S. erythromos*. Most of the literature on this species appears under the name *S. ludovici*. Hershkovitz (1949c) listed specimens of *S. lilium*

bogotensis as *S. ludovici*; while S. Anderson, Koopman, and Creighton (1982) included *bogotensis* and *ludovici* as synonyms of *S. oporaphilum*. Many authors (e.g., de la Torre 1952; Handley 1966c; Hall 1981) also have used the name *S. ludovici* for a species represented in the Mexican and Middle American fauna that I consider as the species *S. hondurensis* Goodwin, 1940. I follow Handley (1976) and V. Pacheco and Patterson (1991, 1992) in treating *S. bogotensis* as a separate species. Although V. Pacheco and Patterson (1991, 1992) treated *S. ludovici* and *S. oporaphilum* as separate species; I treat them here as northern and southern populations, respectively, of the same species. In his dissertation, de la Torre (1961) also distinguished between *S. ludovici* and *S. oporaphilum*, but he treated the latter as the name for the Central American and Mexican populations known as *S. hondurensis*. While V. Pacheco and Patterson (1991) confirmed a close morphological similarity between samples they identified as *S. ludovici* and *S. oporaphilum*, they noted what appeared to be a genetically closer relationship between *S. oporaphilum* and the *tildae-lilium-luisi* assemblage.

Ruschi's (1965) record for *S. ludovici* from Mucuricí, Espírito Santo, Brazil, obviously is a misidentification, possibly of *S. tildae*. Tucker and Bickham's (1986:36; number AK 8178) record for *S. ludovici* from Trinidad and Tobago (Trinidad) was based on a misidentified *S. tildae* (J. W. Bickham pers. comm.). Muñoz (2001) cited Shamel (1927) as the authority for his record from Villavicencio, Colombia. Shamel (1927) gave the record as representing *S. bogotensis*; however, the specimen (USNM 251990) is a skin lacking the skull and appears to be a *S. lilium*. Barquez, Mares, and Ojeda (1991) reported *S. oporaphilum* from provincia Tucumán, Argentina, but did not cite localities. Barquez, Giannini, and Mares (1993) cited the provinces of Salta, Jujuy, and Tucumán as the Argentine distribution of *S. oporaphilum*, but also did not mention localities (Map 209).

Sturnira sorianoi Sánchez-Hernández, Romero-Almaraz, and Schnell, 2005
Soriano's Yellow-shouldered Bat
SYNONYM:

Sturnira sorianoi Sánchez-Hernández, Romero-Almaraz, and Schnell, 2005:867; type locality "Venezuela, Asentamiento Monterrey, 8 km NNE from Mérida, ME. El Valle, 2,300 m, 8°37'N, 71°10'W."

DISTRIBUTION: *Sturnira sorianoi* is known from the state of Mérida, Venezuela, and departamento Santa Cruz, Bolivia.

MARGINAL LOCALITIES (Map 207; from Sánchez-Hernández, Romero-Almaraz, and Schnell 2005): VENEZUELA: Mérida, Asentamiento Monterrey (type locality of *Sturnira sorianoi* Sánchez-Hernández, Romero-Almaraz,

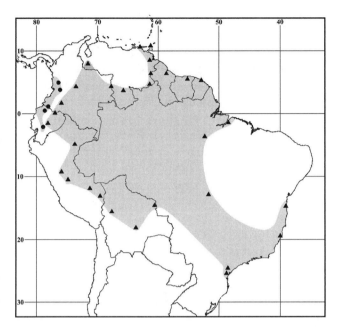

Map 210 Marginal localities for *Sturnira koopmanhilli* ● and *Sturnira tildae* ▲

and Schnell, 2005); Mérida, Monte Zerma [*sic*, = Monte Zerpa], 6 km N of Mérida. BOLIVIA: Santa Cruz, Río P. Tasama [*sic*, = Río Pitisama], 4.5 km N and 1.5 km E of Cerro Amborro [sic, = Amboró]; Santa Cruz, 13 km by road NW of Masicurs [*sic*, = Masicuri].

SUBSPECIES: I treat *S. sorianoi* as monotypic.

NATURAL HISTORY: The Venezuelan specimens were netted in montane cloud forest that supports a variety of epiphytes and arborescent ferns.

REMARKS: Four specimens of *S. sorianoi* are known, two from Venezuela and two from Bolivia. S. Anderson (1997) reported the Bolivian specimens under *S. oporaphilum*. According to the description (Sánchez-Hernández, Romero-Almaraz, and Schnell 2005), *S. sorianoi* is most easily distinguished from congeners by the smooth (non-cuspidate) lingual margins of the lower molars, the tricuspid lower incisors, and the apparent lack of stained fur on the shoulders.

Sturnira tildae de la Torre, 1959
Tilda's Yellow-shouldered Bat

SYNONYM:

Sturnira tildae de la Torre, 1959:1, type locality "Arima Valley, Trinidad," Trinidad and Tobago.

DISTRIBUTION: *Sturnira tildae* is on the island of Trinidad and in Colombia, Venezuela, Guyana, Surinam, French Guiana, Brazil, Ecuador, Peru, and northern Bolivia.

MARGINAL LOCALITIES (Map 210): TRINIDAD AND TOBAGO: Trinidad, Blanchisseuse (C. H. Carter et al. 1981). VENEZUELA: Delta Amacuro, Río Acoíma (Ochoa 1995). GUYANA: Cuyuni-Mazaruni, Kartabu Point (USNM 545103). SURINAM: Para, Zanderij (Genoways, Williams, and Groen 1981). FRENCH GUIANA: Paracou (Simmons and Voss 1998). BRAZIL: Pará, Belém (Handley 1967); Pará, Rio Xingu, 52 km SSW of Altamira (USNM 549498); Mato Grosso, Serra do Roncador, 264 km N (by road) of Xavantina (Pine, Bishop, and Jackson 1970); Bahia, Ilhéus (Faria, Soares-Santos, and Sampaio 2006); Espírito Santo, Município de Linhares (Peracchi and Albuquerque 1993); São Paulo, Iporanga (Trajano 1982); Paraná, Mae Catira (Miretzki, Peracchi, and Bianconi 2002). BOLIVIA (S. Anderson 1997): Santa Cruz, El Encanto; Santa Cruz, 1 km NE of Estancia Cuevas; La Paz, Alcoche. PERU: Madre de Dios, Ccolpa de Guacamayos (USNM 579679); Madre de Dios, Pakitza (Ascorra, Wilson, and Romo 1991); Pasco, San Juan (Tuttle 1970); Huánuco, 1 km S of Tingo María, (Tucker and Bickham 1986); Loreto, Centro de Investigaciones "Jenaro Herrera" (Ascorra, Gorchov, and Cornejo 1994). ECUADOR: Morona-Santiago, Río Llushín (Albuja 1999); Sucumbíos, Santa Cecilia (R. H. Baker 1974). COLOMBIA: Huila, El Parque Nacional Natural de la Cueva de los Guácharos (Lemke et al. 1982); Meta, Restrepo (Marinkelle and Cadena 1971). VENEZUELA: Táchira, Uribante-Caparo Dam (Ochoa et al. 1993). COLOMBIA: Vichada, Matavení (Marinkelle and Cadena 1971). VENEZUELA (Handley 1976): Amazonas, Belén; Bolívar, 45 km NE of Icabarú; Bolívar, El Manaco, 56 km SE of El Dorado; Sucre, Manacal.

SUBSPECIES: I am treating *S. tildae* as monotypic.

NATURAL HISTORY: This species occurs in a variety of humid forest habitats in the lower Orinoco and the greater Amazon basins. Preference appears to be for primary forest (Williams and Genoways 1980a; Brosset and Charles-Dominique 1991). Fruit pulp and juices have been found in the stomach contents (Goodwin and Greenhall 1961; Gardner 1977c). Uieda and Vasconcellos-Neto (1985) reported *Solanum asperum* and *S. grandiflorum* dispersed by *S. tildae* in the region around Manaus, Brazil. Goodwin and Greenhall (1961) reported a Trinidadian female pregnant in March. Wilson (1979) mentioned a Brazilian female pregnant in July. Label information for Brazilian specimens in the USNM note that some were pregnant when taken in August. A Peruvian female was pregnant when caught in August, another was lactating in November when a juvenile also was captured. Venezuelan females were pregnant in January, February, March, and May, and others were lactating in April and May. J. P. Webb and Loomis (1977) reported a tick from a Venezuelan specimen. R. Guerrero (1985a) added three mites and six streblid flies to the ectoparasites known from Venezuelan *S. tildae*.

REMARKS: With the exception of *S. magna*, this species is the largest *Sturnira* occurring in the Amazon basin.

Marinkelle and Cadena's (1971) illustration of the cranium and lower dentition shows the broad, weakly bilobed upper incisors characteristic of this species. Their measurements and description of color pattern and morphology were based on 154 specimens from Colombia. I do not agree with Brosset and Charles-Dominique's (1991) suggestion that *S. luisi* is a synonym of *S. tildae*. Simmons and Voss (1998) illustrated the skulls of *S. lilium* and *S. tildae* represented in their Paracou, French Guiana material. They also disagreed with Brosset and Charles-Dominique (1991) concerning the relationship of *S. luisi* to *S. tildae*. *Sturnira tildae* is larger than both *S. lilium* and *S. luisi*, has broad upper inner incisors, and lacks clefts between lingual cuspids of lower molars. The reports by Alberico (1994) and Cadena, Anderson, and Rivas-Pava (1998) of *S. tildae* from Nariño, Colombia, are of *S. koopmanhilli*.

Ruschi's (1965) report of *S. ludovici* from Mucuricí, Espírito Santo, Brazil, obviously is a misidentification and the specimen possibly represents *S. tildae*. The southernmost reports of *S. tildae* in eastern Brazil are of specimens from the states of Paraná (Miretzki, Peracchi, and Bianconi 2002) and São Paulo (Trajano 1982).

Sturnira species "A"

Albuja (1999:127) identified an undescribed species as "*Sturnira* Sp. B," from northwestern Ecuador. He said this species is medium-sized for the genus, grayish brown in color, and has a robust skull with widely curved zygomatic arches, high and near vertical coronoid processes, and large molars that show clear evidence of the cuspidate structure normal in other stenodermatines. I have seen a specimen of this species and Albuja's descriptive summary of this highly distinctive form, especially of the primitive morphology of the molars, is accurate.

Family Mormoopidae Saussure, 1860
James L. Patton and Alfred L. Gardner

The family Mormoopidae comprises the mustached, naked-backed, and ghost-faced bats in the genera *Pteronotus* and *Mormoops*. Restricted to the New World, mormoopids occur from the southwestern United States and the Greater Antilles to southern Brazil in humid tropical to semiarid and arid subtropical habitats below 3,000 m.

The Mormoopidae are small to medium-sized bats (forearm 35–65 mm) having either narrow (*Pteronotus*) or funnel-shaped (*Mormoops*) ears; wart-like bumps above the nostrils (sometimes interpreted as a rudimentary noseleaf); flap-like outgrowths fringing the lower lip; and a secondary fold on the tragus. The uropatagium is well developed and extends beyond the feet. The tail is about 1/2 as long as the uropatagium and the terminal 1/2 protrudes

above the membrane as a "free tail." The skull lacks postorbital processes, but has complete premaxillae, which are fused to each other and to the maxillae. The humerus has a well-developed trochiter, which does not articulate with the scapula. The capitulum is offset from the shaft axis and the well-developed distal spinous process projects beyond the medial epicondyle and distal articular surface of the humerus. The seventh cervical vertebra is free from the first thoracic. The pelvis is unmodified; but the sacral vertebrae are fused posteriorly to form a flattened, narrow urostyle. The baculum is either weakly developed or absent. The dental formula is 2/2, 1/1, 2/3, 3/3 × 2 = 34.

Following Miller's (1907b) review, the group was recognized until recently as the subfamily Chilonycterinae in the Phyllostomidae. Although several authors implied that the "chilonycterines" deserved higher taxonomic status (e.g., Machado-Allison 1967, on the basis of ectoparasites), Dalquest and Werner's (1954) elevation of the group to familial rank under the name Chilonycteridae was overlooked or ignored. Aellen (1970:9) used the name Mormoopinae (under the "Phyllostomatidae"), as also employed earlier by Rehn (1902a:162), but without comment. Most authors, since Miller (1907b), used Chilonycterinae as the name for this taxon. J. D. Smith (1972) also argued that the group warranted familial status, and applied to it the name Mormoopidae. This remains the modern usage, except by Hall (1981) who continued to treat the taxon under the name Chilonycterinae as a subfamily in the "Phyllostomatidae." Simmons and Conway (2001) reviewed the systematics and phylogeny of the family.

KEY TO GENERA AND SOUTH AMERICAN SPECIES OF MORMOOPIDAE:

1. Rostral portion of cranium strongly elevated, producing a pronounced angle between the rostrum and the abruptly rising forehead; p3 not noticeably reduced in size as compared to p2 and p4; tragus complex, with a pronounced secondary fold; ears round, funnel-shaped, and connected across rostrum by a prominent band of skin. . . . . *Mormoops megalophylla*
1′. Rostral portion of cranium not or only slightly elevated; p3 reduced to a small peg-like unicuspid sandwiched between p2 and p4; tragus with a small to moderate secondary fold; ears pointed and, although noticeably separated, may be connected by low, inconspicuous ridge of skin. 2
2. Wing membrane fused on middorsal line, giving a naked-backed appearance; rostral breadth always more than length of maxillary toothrow . . . . (subgenus *Pteronotus*) 3
2′. Wing membrane not fused on dorsal midline; rostral breadth equal to or less than length of maxillary toothrow. 4

3. Forearm longer than 49 mm; condylobasal length more than 15.5 mm; pubescence on dorsal surface of wing membrane short and dense . . . *Pteronotus gymnonotus*

3'. Forearm shorter than 49 mm; condylobasal length less than 15.4 mm; pubescence on dorsal surface of wing membrane a sparse mixture of long and short hairs . . . . *Pteronotus davyi*

4. Basioccipital narrowly constricted between auditory bullae; basisphenoid with two narrow and deep furrows; forearm usually longer than 50 mm . . . . (subgenus *Phyllodia*) *Pteronotus parnellii*

4'. Basioccipital not constricted between auditory bullae; furrows in basisphenoid wide and shallow; forearm usually shorter than 50 mm . . . . (subgenus *Chilonycteris*) *Pteronotus personatus*

Genus *Mormoops* Leach, 1821

The genus *Mormoops* contains three species, two restricted to the Greater Antilles and the third having disjunct distributions in Mexico and northern Central America, and in northern and western South America, including the island of Trinidad and the Netherlands Antilles. The fossil record consists of Pleistocene remains from cave deposits in Bermuda and the Greater Antilles (Koopman 1951; Koopman and Williams 1951; Silva-Taboada 1974, 1979).

Medium-size to large for the family (forearm 44–61 mm), these bats have long, lax pelage; short, rounded ears that are connected across the rostrum by two conspicuous bands of skin; and the posterior basal lobe of the pinna is continuous with the lateral skin flap of the lower lip. Ornamentation of the lower lip is complex and consists of a series of skin flaps. Each nostril opens in a separate pad on each side of a vertical, warty rhinolabial ridge. The tragus is complex and has a prominent secondary fold. The wing membrane is attached to the leg at the ankle. The skull is strongly flexed, with the slope of the frontals at almost at a 90° angle to the rostral plane. The mastoid is greatly reduced and the foramen magnum is above the level of the rostrum. The ramus curves upward following the flexion of the cranium. Outer upper incisors are reduced, delicate, and separated from the canines by a wide gap. The second lower premolar is long and narrow. See J. D. Smith (1972) for additional anatomical details.

SYNONYMS:

Aëllo Leach, 1821a:69; type species *Aëllo cuvieri* Leach, 1821a:71, by monotypy.

Mormoops Leach, 1821b:74; type species *Mormoops blainvillei* (corrected spelling of *Mormoops blainvillii* Leach, 1821b:77), by monotypy; placed on Official List of Generic Names in Zoology (name number 1077) as a name having precedence over *Aëllo* Leach, 1821a (ICZN 1957).

Lobostoma Gundlach, 1840:357; part.

Mormops F. Cuvier, 1829:422; unjustified emendation of *Mormoops* Leach, 1821a; placed on the Official Index of Rejected and Invalid Generic Names in Zoology (ICZN 1957).

Mormops Festa, 1906:7; incorrect subsequent spelling of *Mormoops* Leach.

REMARKS: Leach (1821a) named *Aello* (type species *A. cuvieri*) based on a mutilated specimen of unknown origin. Dobson (1878:458) equated *A. cuvieri* with *Mormoops blainvillei* Leach, from Cuba and Jamaica. Rehn (1902a:162–63) allied *A. cuvieri* with the nominate form of *M. blainvillei* from Jamaica, the place where J. D. Smith (1972:111) recommended restricting the type locality. Dobson (1878:454, footnote) was the first to realize that *Aello* and *Mormoops* were synonyms and, although published simultaneously, that the name *Aello* had page priority. Nevertheless, he retained *Mormoops* (using F. Cuvier's emendation *Mormops*) because Leach's description of *Aello* contained errors and because *Mormoops* "has the advantage of correct definition." Morrison-Scott (1955) petitioned the International Commission on Zoological Nomenclature to place *Mormoops* Leach, 1821, on the Official List of Generic Names in Zoology, citing Dobson's (1878:454) use of *Mormoops* over *Aello* as establishing priority for the former under the principle of the first reviser. As a result, *Mormoops* was placed on the Official List of Generic Names in Zoology as a name having precedence over *Aello* (ICZN 1957: Opinion 462). Opinion 462 also ruled that *Mormoops blainvillei* be placed on the Official List of Specific Names in Zoology as a name having precedence over *Aello cuvieri* Leach, 1821a; *Mormops* F. Cuvier, 1829, was placed on the Official Index of Rejected and Invalid Generic Names in Zoology; and *blainvillii* Leach was placed on the Official Index of Rejected and Invalid Specific Names in Zoology as an invalid original spelling for *blainvillei*. This last action seems to have been overlooked by J. D. Smith (1972). Hall (1981) ignored Opinion 462 (ICZN 1957) and used *Aello* as the generic name for the ghost-faced bats, and used the spelling "*blainvillii*" for the species on the Greater Antilles. *Mormoops* was not treated by Cabrera (1958). *Mormoops magna* Silva-Taboada, 1974, is known only from fossil remains from cave deposits in Cuba.

Mormoops megalophylla (W. Peters, 1864)
Ghost-faced Bat

SYNONYMS: The following synonyms are based on non-South American populations; see under Subspecies for additional synonyms.

Mormops megalophylla W. Peters, 1864b:381; type locality "Mexico"; restricted to southern Mexico by Rehn (1902a:167); later corrected to Parrás, Coahila, Mexico, by J. D. Smith (1972:116).

Mormoops megalophylla: Miller, 1900d:160; first use of current name combination.

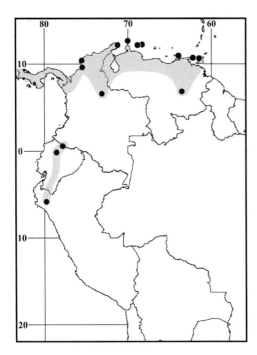

Map 211 Marginal localities for *Mormoops megalophylla* ●

Mormoops megalophylla senicula Rehn, 1902a:169; type locality "Fort Clark, Kinney County, Texas, USA."

Mormoops megalophylla rufescens W. B. Davis and Carter, 1962a:65; type locality "five miles west of Alamos, Sonora, México."

DISTRIBUTION: The South American distribution of *M. megalophylla* is discontinuous, with one segment of the population in northern Colombia, northern Venezuela, the island of Trinidad, and the Netherlands Antilles; the other segment is found at upper elevations in Ecuador and northern Peru. The species also occurs from the southern United States (Arizona and Texas), through Mexico and Central America into Honduras and El Salvador.

MARGINAL LOCALITIES (Map 211; from J. D. Smith 1972, except as noted): *Northern distribution*: NETHER-LANDS ANTILLES: Aruba, northwest end of island; Curaçao, Round Cliff; Bonaire, Grot Colombia. VENEZUELA: Nueva Esparta, Cueva Honda del Piache; Sucre, Enseñada Cauranta, 7 km N and 4 km E of Güiria. TRINIDAD AND TOBAGO: Trinidad, Port of Spain. VENEZUELA: Bolívar, Hato San José. COLOMBIA: Santander, San Gil; Sucre, Tulú Viejo; Bolívar, Cartagena; La Guajira, Nazaret. *Western distribution*: ECUADOR: Carchi, Gruta Rumichaca; Pichincha, Río Condor Huachana, 3.45 km NE of Lloa (USNM 513433). PERÚ: Lambayeque, 5.2 km N of Olmos (Graham and Barkley 1984).

SUBSPECIES: Two of the four subspecies of *M. mega-lophylla* that J. D. Smith (1972) recognized are found in mainland South America, and a third subspecies is on the Netherlands Antilles.

M. m. carteri J. D. Smith, 1972
SYNONYM:
Mormoops megalophylla carteri J. D. Smith, 1972:119; type locality "Gruta Rumichaca, 2 mi E La Paz, 8700 ft," Carchí, Ecuador.

This subspecies is known from the Colombian-Ecuadorian border (type locality) to northern Peru at elevations over 2,300 m in the drier western Andes of Ecuador (Albuja 1983, 1999) to as low as 150 m in the arid Pacific lowlands of northwestern Peru (Graham and Barkley 1984). Apparently, *M. m. carteri* has not been documented in adjacent Colombia (Muñoz 2001).

M. m. intermedia Miller, 1900
SYNONYMS:
Mormoops intermedia Miller, 1900d:160; type locality a "cave at Hatto, on the north coast of Curaçao," Netherlands Antilles.
Mormoops megalophylla intermedia: Rehn, 1902:170; first use of current name combination.

This subspecies inhabits the islands of Aruba, Curaçao, and Bonaire in the Netherlands Antilles.

M. m. tumidiceps Miller, 1902
SYNONYMS:
Mormoops tumidiceps Miller, 1902b:403; type locality "Point Gourde Caves, Trinidad."
Mormoops megalophylla tumidiceps: Goodwin and Greenhall, 1961:191; first use of current name combination.

This subspecies occurs on the island of Trinidad and in northern South America from Colombia to eastern Venezuela.

NATURAL HISTORY: *Mormoops megalophylla* roosts in a variety of natural and manmade structures, but apparently prefers warm, humid caves in which some aggregations have been estimated in the hundreds of thousands (Villa-R. 1967). Like many nonmolossid, neotropical insectivorous bats, *M. megalophylla* is more commonly caught in caves or in other day roosts than in mist nets. *Mormoops megalophylla* is most easily caught in mist nets at the entrances to caves or in arid regions where frequently they can be netted over water. The species is capable of rapid and enduring flight (Vaughan and Bateman 1970; Bateman and Vaughan 1974). Boada et al. (2003) identified the remains of Lepidoptera, Dermaptera, Coleoptera, and Diptera, in the feces of Ecuadorian *M. megalophylla*.

Albuja (1983:46) noted several females pregnant with 20–26 mm (CR) fetuses on 11 November in Ecuador. Boada et al. (2003) reported finding females lactating in January, March, and October, and one pregnant (24.5-mm CR) in May. Goodwin and Greenhall (1961) listed a tick and six trombiculid mites as parasitic on Trinidadian *M. megalophylla*. R. Guerrero (1985a) mentioned two ticks, two mites, and six streblid batflies as known ectoparasites.

He later (R. Guerrero 1997) added five streblid batflies to the list. This species, and other mormoopids (Goodwin and Greenhall 1961; Bateman and Vaughn 1974), form large aggregations that appear to shift seasonally from mixed to sexually segregated colonies, a shift that Silva-Taboada (1979) suggested is related to reproduction. Boada et al. (2003) caught 420 *M. megalophylla* in San Antonio de Pichincha Cave, of which 389 were males and 31 were females. *Mormoops megalophylla* has a $2n = 38$, FN $= 62$ karyotype (Sites, Bickham, and Haiduk 1981). Simmons and Conway (2001) recently reviewed the systematics of this species. Boada et al. (2003) provided information on behavior, ecology, and food habits in their study of the activity patterns of a population of *M. m. carteri* inhabiting Cueva de San Antonio de Pichincha, Ecuador.

REMARKS: South American *M. megalophylla* are characterized by a whitish to pink or red cape of long hairs over the shoulders that contrasts with the darker dorsal pelage. Measurements from J. D. Smith (1972) show that *Mormoops megalophylla carteri* is larger (forearm 56.6–60.6 mm, $n = 21$) than *M. m. intermedia* (forearm 49.8–54.2 mm, $n = 21$). This difference in size also is evident when comparing measurements in Eisenberg (1989:99) with corresponding measurements in Boada et al. (2003). Eisenberg's specimens (11 of each sex) represent Venezuelan *M. m. intermedia*, in which the forearm length in males averaged 55.5 mm, and that in females averaged 54.5 mm. The forearm lengths for 77 male and 9 female Ecuadorian *M. m. carteri* measured by Boada et al. (2003) had a mean and range of 58.4 mm (54.7–66.0 mm), and 57.7 mm (54.6–61.8 mm), respectively. D. C. Carter and Dolan (1978) gave information on and measurements of the holotype of *Mormops megalophylla* W. Peters.

Genus *Pteronotus* Gray, 1838

Bats of the genus *Pteronotus* range in size from the smallest (*P. quadridens* Gundlach; forearm 35.9–18.5 mm, $n = 15$; see Silva-Taboada 1976b, for use of this name) to the largest (*P. parnellii rubiginosus* J. A. Wagner; forearm 58.3–65.4 mm, $n = 126$) in the family (measurements from J. D. Smith 1972). These bats are characterized by short, stiff pelage; long, pointed ears that may be connected dorsally by a low ridge across the rostrum; entire (not lobed or incised) skin flaps on the lower lip; and a tragus that either is lanceolate and has a small secondary fold or is spatulate with a marked secondary fold. The wing membrane either attaches to the sides of the body (subgenera *Phyllodia* and *Chilonycteris*) or attaches over the back along the dorsal midline (subgenus *Pteronotus*). The cranium is "normal" in appearance (not strongly flexed as in *Mormoops*) and the rostrum, which is longer than wide, and braincase are on approximately the same plane. The zygomatic arches are well developed and lie nearly in same plane as the rostrum and braincase. The cranium has prominent lambdoidal crests that extend from the sagittal crest to the mastoid processes. The upper and lower incisors form a continuous series filling the space between canines. The second lower premolar is reduced to a small, peg-like unicuspid and displaced lingually in the toothrow. Color varies from gray to dark brownish gray in fresh pelage, to yellow, orange-yellow, and in extreme cases, to red in older fur. Color changes are attributed to bleaching from environmental sources (J. D. Smith 1972) and to pigment changes as the hair ages.

The fossil record consists of Pleistocene remains from Mexico (T. Alvarez 1963) and the Greater Antilles (Koopman and Williams 1951; Silva-Taboada 1979). *Pteronotus pristinus* Silva-Taboada, 1974, is known only from fossil remains from Cuba.

REMARKS: In his revision of the mormoopids, J. D. Smith (1972) combined the naked-backed bats, and the species formerly included in the genus *Chilonycteris*, together in the genus *Pteronotus*. J. D. Smith allocated these species among three subgenera: those with wings attached along the dorsal midline (the "naked-backed" species *Pteronotus davyi* and *P. suapurensis* [= *P. gymnonotus*]) under the subgenus *Pteronotus*; the large, normal-winged species (*Pteronotus parnellii*) under the subgenus *Phyllodia*; the medium-sized to small, normal-winged species (*Pteronotus macleayii*, *P. fuliginosus* [= *P. quadridens*], and *P. personatus*) in the subgenus *Chilonycteris*. Simmons and Conway (2001) validated this classification based on their cladistic analysis of 209 morphological characters and placed the extinct *P. pristinus* from Cuba in the subgenus *Phyllodia*. All but the West Indian endemics, *Pteronotus macleayii* and *P. quadridens*, are found in South America.

SYNONYMS:

Pteronotus Gray, 1838b:500; type species *Pteronotus davyi* Gray, 1838b, by monotypy.

Chilonycteris Gray, 1839; type species *Chilonycteris macleayii* Gray, 1839, by monotypy.

Lobostoma Gundlach, 1840:357; type species *Lobostoma quadridens* Gundlach, 1840, by subsequent designation (J. D. Smith 1972:55–56).

Phyllodia Gray, 1843a:50; type species *Phyllodia parnellii* Gray, 1843a, by monotypy.

Dermonotus Gill, 1872a:177; replacement name for *Pteronotus* Gray, 1838b, on the assumption that the name was preoccupied by *Pteronotus* Rafinesque, 1815 (a *nomen nudum*; Miller 1905).

Pteronotus davyi Gray, 1838
Davy's Naked-backed Bat
SYNONYMS: The following synonyms are based on non-South American populations; see under Subspecies for additional synonyms.

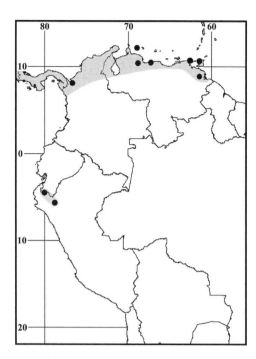

Map 212 Marginal localities for *Pteronotus davyi* ●

Chilonycteris Davyi fulvus O. Thomas, 1892c:410; type locality "Las Peñas, west coast of Jalisco," Mexico.

Pteronotus suapurensis calvus Goodwin, 1958a:1; type locality "Tehuantepec, Oaxaca, México."

DISTRIBUTION: The South American distribution of *P. davyi* is disjunct with one population found in northern Colombia and Venezuela and the adjacent islands of Trinidad and Curaçao, and the other in northwestern Perú. The species also occurs in Mexico, Central America, and the Lesser Antilles.

MARGINAL LOCALITIES (Map 212; from J. D. Smith 1972, except as noted): *Northern distribution*: NETHERLANDS ANTILLES: Curaçao, Kueba di Ratón (Petit, Rojer, and Pors 2006). TRINIDAD AND TOBAGO: Trinidad, Port of Spain. VENEZUELA: Delta Amacuro, Araguaimujo (Pirlot 1965b). COLOMBIA: Antioquia, Turbo (USNM 430075). VENEZUELA: Yaracuy, Minas de Aroa; Distrito Federal, Hacienda Carapiche (Handley 1976); Sucre, Las Melenas (Bisbal 1998). *Southern distribution*: PERU: Piura, 4 miles W of Suyo (type locality of *Pteronotus davyi incae* J. D. Smith); Cajamarca, Jaén.

SUBSPECIES: We recognize three subspecies of *P. davyi*, of which two are found in South America. *Pteronotus d. fulvus* (O. Thomas) occurs in Mexico and Central America.

P. d. davyi Gray, 1838

SYNONYMS:

Pteronotus davyi Gray, 1838b:500; type locality "Trinidad," Trinidad and Tobago.

Chilonycteris gymnonota: Tomes, 1863b:83; not *Chilonycteris gymnonotus* J. A. Wagner, 1843a.

Chilonycteris davyi: Dobson, 1878:453; name combination.

D[ermonotus]. davyi: Miller, 1902a:155; name combination.

The nominate subspecies is verified from only one Colombian locality, but is locally common in northern Venezuela and on the island of Trinidad. Elsewhere, it occurs northward through the Lesser Antilles to Martinique, and southward from Nicaragua through Panama in Central America. This subspecies was erroneously reported from the Brazilian states of Ceará, Piauí, and Pernambuco by Mares et al. (1981) and Willig (1983, 1985a); see Remarks.

P. d. incae J. D. Smith, 1972

SYNONYM:

Pteronotus davyi incae J. D. Smith, 1972:102; type locality "4 mi W Suyo, 1000 ft, Piura," Peru.

This subspecies is known from the departments of Piura, Cajamarca, and Lambayeque in northwestern Peru (J. D. Smith 1972; Graham and Barkley 1984).

NATURAL HISTORY: Bateman and Vaughan (1974) gave an interesting account of flyways, flight patterns, and roosting and nocturnal behavior of a mixed colony of four species of mormoopids in and around a cave in Mexico. They judged the size of the colony to be between 400,000 and 800,000 bats, of which an estimated 62% were *P. davyi*. Adams (1989) provided a detailed description in the *Mammalian Species* account that includes measurements and a summary of available information on natural history. R. Guerrero (1985a, 1997) listed a total of three species of mites, one tick, and eight species of streblid batflies known to parasitize this species. *Pteronotus davyi* has a $2n = 38$, FN = 60 karyotype (R. J. Baker 1967).

REMARKS: Aellen (1970) suggested that the *P. davyi* reported by Tesh, Arata, and Schneidau (1968) as negative for histoplasmosis came from the vicinity of Cartagena, Colombia. Muñoz (2001) also cited Tesh, Arata, and Schneidau (1968) as the authorities for his listing of *P davyi* in departamento Bolívar, Colombia; however, the report by Tesh, Arata, and Schneidau did not link the specimen they identified as *P. davyi* to any Colombian locality. The specimens reported from northeastern Brazil (Mares et al. 1981; Willig 1983, 1985a, 1985b) as *Pteronotus davyi* are *P. gymnonotus*. The two species are similar in appearance and are sometimes misidentified. A. Miranda-Ribeiro (1914) misidentified *P. gymnonotus* from Cáceres, Mato Grosso, Brazil, as *P. davyi*. Some reports of *P. davyi* from Brazil (e.g., C. O. C. Vieira 1955) were based on the opinion that *P. gymnonotus* was a junior synonym of *P. davyi* (see J. D. Smith 1977). *Pteronotus davyi* remains unknown south of the Rio Amazonas. D. C. Carter and Dolan (1978) gave information on and measurements of the type of *Pteronotus davyi* Gray.

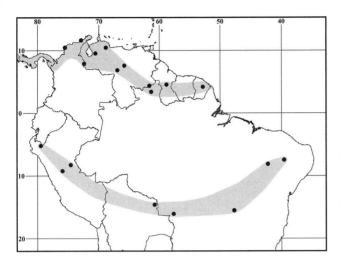

Map 213 Marginal localities for *Pteronotus gymnonotus* ●

Pteronotus gymnonotus (J. A. Wagner, 1843)
Big Naked-backed Bat
SYNONYMS:

Chilonycteris gymnonotus J. A. Wagner, 1843a:367; type locality "Cuyaba" (=Cuiabá), Mato Grosso, Brazil; see Remarks concerning authorship of name.

Dermonotus suapurensis J. A. Allen, 1904b:229; type locality "Suapure," Bolívar, Venezuela.

Pteronotus suapurensis: J. A. Allen, 1911:245; name combination.

Pteronotus davyi suapurensis: G. M. Allen, 1935:227; name combination.

Pteronotus suapurensis centralis Goodwin, 1942a:88; type locality "Matagalpa, Nicaragua, 3000 feet elevation."

Pteronotus gymnonotus: J. D. Smith, 1977:246; first use of current name combination.

Pteronotus gimnonotus Bredt and Uieda, 1996:56; incorrect subsequent spelling of *Chilonycteris gymnonotus* J. A. Wagner.

DISTRIBUTION: *Pteronotus gymnonotus* is known from northern Colombia, northwestern and central Venezuela, Guyana, French Guiana, Brazil, Peru, and eastern Bolivia. Elsewhere, the species occurs in Mexico and Central America.

MARGINAL LOCALITIES (Map 213; from J. D. Smith 1972 [as *P. saupurensis*], except as noted): *Northern distribution.* COLOMBIA: Bolívar, Cartagena; La Guajira, Ríohaca (Marinkelle and Cadena 1972); Bolívar, Summit of El Abismo (Ochoa, Castellanos, and Ibáñez 1988). VENEZUELA: Yaracuy, Minas de Aroa; Bolívar, Suapuré. GUYANA: Potaro-Siparuni, Iwokrama forest (Lim and Engstrom 2001a). FRENCH GUIANA: Régina and St. Georges road project (Brosset et al. 1996). BRAZIL: Roraima, Maracá Biological Station (F. Robinson 1998). VENEZUELA: Bolívar, Hato La Florida (T. E. Lee, Lim,

and Hanson 2000); Trujillo, 19 km N of Valera (Handley 1976); Táchira, El Palotal (Soriano, Ruiz, and Zambrano 2005). *Southern distribution* (localities listed from west to east). PERÚ: Piura, Huancabamba; Huánuco, 2 miles N of Tingo María; Ucayali, Yarinacocha. BOLIVIA: Santa Cruz, 38 km E of La Florida (Ibáñez and Ochoa 1989). BRAZIL: Mato Grosso, Cáceres (A. Miranda-Ribeiro 1914, as *P. davyi*); Distrito Federal, Gruta Dança dos Vampiros (Bredt, Uieda, and Magalhães 1999); Pernambuco, Fazenda Cantareno (Mares et al. 1981, as *P. davyi*); Ceará, Floresta Nacional do Araripe (MZUSP 16644, not mapped); Piauí, Fazenda Saquarema (Vizotto, Rodriques, and Dumbra 1980b).

SUBSPECIES: We treat *P. gymnonotus* as monotypic.

NATURAL HISTORY: Vizotto, Rodriques, and Dumbra (1980b) reported a colony of several thousand bats, of which the majority was *P. gymnonotus*. The colony, in a grotto in the Brazilian state of Piauí, contained lesser numbers of *P. parnellii*, *P. personatus*, and *Natalus stramineus*. *Pteronotus gymnonotus* shares large caves with *Phyllostomus hastatus* in Peru (Graham 1988). Willig (1983) found sexual dimorphism in size (males larger) in netted samples (misidentified as *P. davyi*) from the Northeast of Brazil. Lee, Lim, and Hanson (2000) reported a lactating female taken in either July or August in Venezuela. R. Guerrero (1985a, 1997) listed two species of mites, one of ticks, and eight species of streblid batflies that have been recovered from this species. *Pteronotus gymnonotus* has a $2n = 38$, $FN = 60$ karyotype (Sites, Bickham, and Haiduk 1981).

REMARKS: The specimens reported as *P. davyi* by Mares et al. (1981) and Willig (1983, 1985c), from the Brazilian states of Ceará, Pernambuco, and Piauí, are *P. gymnonotus*. When J. D. Smith (1977) clarified the identities of the two species of naked-backed bats, he said that J. A. Wagner (1843a) took the descriptions for *Chilonycteris gymnonotus* (a species J. A. Wagner attributed to Natterer) from Natterer's field diary. Therefore, J. D. Smith concluded that Natterer must be considered the author of those names, because Natterer was responsible both for the names and the conditions that made them available. D. C. Carter and Dolan (1978:15,27) argued that while J. A. Wagner (1843a) used Natterer's names, the descriptions do not match those in Natterer's field notes; therefore, J. D. Smith's (1977) claim was incorrect and Wagner alone must be considered the author of all of the names he had attributed to Natterer (including the names *Chilonycteris gymnonotus* and *C. rubiginosa*). Some authors continue to attribute the name *P. gymnonotus* to Natterer (e.g., T. E. Lee, Lim, and Hanson 2000). We agree with D. C. Carter and Dolan (1978) and attribute all of the names proposed in J. A. Wagner (1843a) to Wagner alone. D. C. Carter and Dolan (1978) gave information on and measurements of the holotype of *Chilonycteris gymnonotus* J. A. Wagner.

Pteronotus parnellii (Gray, 1843)

Parnell's Mustached Bat

SYNONYMS: The following synonyms are based on non-South American populations; see under Subspecies for additional synonyms.

Phyllodia parnellii Gray, 1843a:50; type locality "Jamaica."

Chilonycteris osburni Tomes, 1861:66; type locality "Sportsman's Cave," Jamaica.

Chilonycteris boothi Gundlach, 1861:154, in W. Peters 1861; type localities "Fundador, auch in Guines" (Matanzas and Habana provinces, respectively), Cuba.

Chilonycteris portoricensis Miller, 1902b:400; type locality "cave near Pueblo Viejo" (= Cueva di Fari), Puerto Rico.

Chilonycteris mexicana Miller, 1902b:401; type locality "San Blas," Nayarit, Mexico.

Chilonycteris parnellii pusillus G. M. Allen, 1917:168; type locality "Arroyo Salado, Santo Domingo," Dominican Republic.

Chilonycteris parnellii gonavensis Koopman, 1955:110; type locality "Cave near En Café, La Gonave island, Republic of Haiti."

Pteronotus rubiginosus: Burt and Stirton, 1961:25; name combination.

Pteronotus parnellii: Hall and Dalquest, 1963:218; first use of current name combination.

Pteronotus parnellii gonavensis: J. D. Smith, 1972:65; name combination.

Pteronotus parnellii pusillus: J. D. Smith, 1972:65; name combination.

Pteronotus parnellii portoricensis: J. D. Smith, 1972:68; name combination.

Pteronotus parnellii mexicanus: J. D. Smith, 1972:69; name combination.

Pteronotus parnellii mesoamericanus J. D. Smith, 1972:71; type locality "1 mi S and 0.75 mi E Yepocapa, 4280 ft, Chimaltenango, Guatemala."

DISTRIBUTION: *Pteronotus parnellii* occurs in Venezuela, the island of Trinidad, Guyana, Surinam, French Guiana, northeastern Colombia, Brazil east into Piauí and south into Mato Grosso do Sul, and in the eastern lowlands of Ecuador (unvouchered sight record by A. L. Gardner), Peru, and Bolivia. The species also occurs in Mexico, Central America, and the Greater Antilles.

MARGINAL LOCALITIES (Map 214; from J. D. Smith 1972, except as noted): *Northern distribution*. VENEZUELA: Falcón, La Cueva de Piedra Honda (Linares and Ojasti 1974); Aragua, Rancho Grande; Nueva Esparta, Cerro Mata Siete, 2 km N and 2 km E of La Asunción. TRINIDAD AND TOBAGO: Trinidad, Port of Spain. VENEZUELA: Delta Amacuro, Desembocadura del Río Acoíma (Belkis 2000). GUYANA: Barima-Waini,

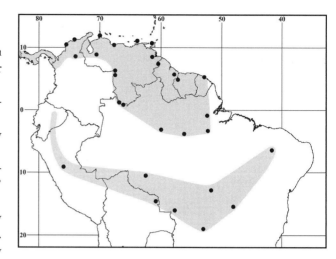

Map 214 Marginal localities for *Pteronotus parnellii* ●

Baramita (USNM 582260); Upper Demerara-Berbice, Dubulay Ranch (USNM 582261). SURINAM: Sipaliwini, Avanavero (Genoways, Williams, and Groen 1981). FRENCH GUIANA: Paracou (Simmons and Voss 1998). BRAZIL: Pará, Monte Dourado (USNM 544654); Pará, Caverna Pedra da Cachoeira (Trajano and Moreira 1991); Pará, Parque Nacional da Amazônia (Marques 1985); Amazonas, Manaus (N. R. Reis 1984). VENEZUELA: Amazonas, Cerro Neblina Base Camp (Gardner 1988). BRAZIL: Amazonas, Serra Cucuhy [= Cucuí]. VENEZUELA: Amazonas, 9 km SE of Puerto Ayacucho (USNM 407201); Apure, Cerro del Murciélago, 8 km NW of Puerto Páez; Barinas, 2 km SW of Altamira (USNM 418840). COLOMBIA: Bolívar, Norosí; Bolívar, Cartagena; Magdalena, Santa Marta. *Southern distribution*: BRAZIL: Piauí, Fazenda Olho da Agua (Mares et al. 1981); Goiás, Gruta Morro (Bredt, Uieda, and Magalhães 1999); Mato Grosso do Sul, Alto Sucuriú (Garutti, Cais, and Taddei 1984); Mato Grosso, Caiçara (type locality of *Chilonycteris rubiginosa* Wagner). BOLIVIA: Santa Cruz, 38 km E of La Florida (Ibáñez and Ochoa 1989). PERU: Huánuco, Cueva de Castillo (Bowles, Cope, and Cope 1979). BRAZIL: Rondônia, Ouro Preto D'Oeste (Marques 1989); Mato Grosso, Serra do Roncador, 264 km N (by road) of Xavantina.

SUBSPECIES: James D. Smith (1972) recognized eight subspecies of *P. parnellii*, of which two are found in South America. Linares and Ojasti (1974) increased the number in South America to three by describing *P. p. paraguanensis* from Venezuela.

P. p. fuscus (J. A. Allen, 1911)

SYNONYMS:

Chilonycteris rubiginosa fusca J. A. Allen, 1911:262; type locality "Las Quiguas, 5 miles south of Puerto Cabello, altitude 650 feet," Carabobo, Venezuela.

Pteronotus parnellii fuscus: J. D. Smith, 1972:77; first use of current name combination.

This subspecies occurs across northern Colombia and northern Venezuela (except the Península de Paraguaná) and on the island of Trinidad.

P. p. paraguanensis Linares and Ojasti, 1974

SYNONYM:

Pteronotus parnellii paraguanensis Linares and Ojasti, 1974:74; type locality "la cueva de Piedra Honda," 7 km SW of Pueblo Nuevo, 120 m, Península de Paraguaná, Falcón, Venezuela.

This subspecies is known from only the Península de Paraguaná, Venezuela.

P. p. rubiginosus (J. A. Wagner, 1843)

SYNONYMS:

Chilonycteris rubiginosa J. A. Wagner, 1843a:367; type locality "Caiçara," Mato Grosso, Brazil.

Ch[ilonycteris]. parnellii: J. A. Wagner, 1855:680; name combination.

Chilonycteris barbata Pelzeln, 1883:37; *nomen nudum*.

Chilonycteris personata: A. Miranda-Ribeiro, 1914:451; not *Chilonycteris personata* J. A. Wagner, 1843a.

Pteronotus parnellii rubiginosus: J. D. Smith, 1972:75; first use of current name combination.

This subspecies is known from eastern and southern Venezuela, Guyana, Surinam, French Guiana, Brazil, and eastern Peru and Bolivia.

NATURAL HISTORY: *Pteronotus parnellii* has been found in a variety of habitats ranging from lowland tropical wet forest to arid thorn forest at elevations from sea level to near 3,000 m. The species roosts in mine tunnels and caves in association with a large number of other species of bats (Graham 1988) and appears to prefer sites that have high humidity. Griffiths (1983b) described the laryngeal anatomy of *P. parnellii* and related its laryngeal structure to the production of constant frequency pulses. Herd's (1983) *Mammalian Species* account contains illustrations of this species and a summary of the published information on its natural history. R. Guerrero (1985a) listed two nematodes, two ticks, eight mites, a hemipteran, and eleven batflies known from this host. R. Guerrero (1997) increased the number of species of batfly ectoparasites known from *P. parnellii* to 27. The karyotype is $2n = 38$, $FN = 60$ (R. J. Baker 1967).

REMARKS: See Husson (1962:76) and J. D. Smith (1972:59) concerning the question of priority of *Phyllodia parnellii* over *Chilonycteris rubiginosa*; see Remarks in the account for *Pteronotus gymnonotus* concerning the authorship of names in J. A. Wagner (1843a). The species appears to be uncommon south and east of Venezuela. J.

D. Smith (1972:76–77) listed only three specimens from Guyana, for example, and only one from the Brazilian state of Mato Grosso (in addition to the type), although he was aware (J. D. Smith 1972:89) that the two specimens reported as *Chilonycteris personata* from Mato Grosso by C. O. C. Vieira (1942) represent *P. parnellii*. C. O. C. Vieira (1942) probably based his identification on the specimen A. Miranda-Ribeiro (1914) identified as *Chilonycteris personata* from Tapirapoã, Mato Grosso, the identity of which C. O. C. Vieira (1955) corrected to *Chilonycteris parnellii*. Vizotto, Rodriques, and Dumbra (1980b, 1980c) took 46 specimens from a mixed colony of *P. parnellii*, *P. gymnonotus*, *P. personatus*, and *Natalus stramineus* in a grotto in the state of Piauí, Brazil. Mares et al. (1981) also reported specimens from Piauí. Western Amazonian records vouchered by specimens are all from the environs of Tingo María, Peru (Bowles, Cope, and Cope 1979); however, Albuja (1983:42) mentioned a *P. parnellii* that escaped from one of us (ALG) while it was being removed from a net at Limoncocha, Sucumbios, Ecuador. Inexplicably, Albuja (1999:54) later identified the bat as a *P. personatus*. The *nomen nudum*, *Chilonycteris barbata*, listed by Pelzeln (1883:37), is one of Natterer's manuscript names for the species. The most recent taxonomic review was by Simmons and Conway (2001). D. C. Carter and Dolan (1978) gave information on and measurements of the holotype of *Chilonycteris rubiginosa* J. A. Wagner.

Pteronotus personatus (J. A. Wagner, 1843)
Wagner's Mustached Bat

SYNONYMS: The following synonyms are based on non-South American populations; see under Subspecies for additional synonyms.

Chilonycteris psilotis Dobson, 1878:451; type locality unknown; designated as "Tehuantepec, Oaxaca, Mexico," by de la Torre (1955:696).

Chilonycteris torrei continentis Sanborn, 1938:1; type locality "Laguna de Zotz, Petén, Guatemala."

DISTRIBUTION: *Pteronotus personatus* is known from northern Colombia, Venezuela, the island of Trinidad, Guyana, Surinam, and French Guiana; and a separate southern distribution from eastern Peru and eastern Bolivia, into Brazil as far as the states of Piauí and Pernambuco. The species also occurs in Mexico and Central America.

MARGINAL LOCALITIES (Map 215; from J. D. Smith 1972, except as noted): *Northern distribution.* COLOMBIA: La Guajira, Nazaret. VENEZUELA: Yaracuy, Minas de Aroa; Aragua, Rancho Grande (USNM 517304); Sucre, Finca Vuelta Larga (Bisbal 1998). TRINIDAD AND TOBAGO: Trinidad, Blanchisseuse (C. H. Carter et al. 1981). VENEZUELA: Anzoátegui, Mamo (Ochoa and Ibáñez 1985); Bolívar, Los Patos. GUYANA: Potaro-Siparuni,

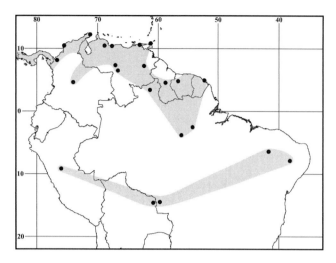

Map 215 Marginal localities for *Pteronotus personatus* ●

Iwokrama forest (Lim and Engstrom 2001a). SURINAM: Sipaliwini, Grassalco (S. L. Williams and Genoways 1980a). FRENCH GUIANA: 1 km N of Rémire (S. L. Williams, Phillips, and Pumo 1990). BRAZIL: Pará, Taperinha (Piccinini 1974); Pará, Parque Nacional da Amazônia (Marques 1985); Roraima, Maracá Biological Station (F. Robinson 1998). VENEZUELA: Bolívar, Mina de bauxita de la Serranía de los Pijiguaos (Ochoa et al. 1988); Apure, Caño Arauquita (Ochoa and Ibáñez 1985). COLOMBIA: Cundinamarca, Bogotá (Cuervo-Díaz, Hernández-Camacho, and Cadena 1986); Antioquia, Turbo; Bolívar, Cartagena (Grose and Marinkelle 1966). *Southern distribution.* BRAZIL: Piauí, Fazenda Olho da Agua (Mares et al. 1981); Pernambuco, Fazenda Saco (Mares et al. 1981); Mato Grosso, St. Vicente (type locality of *Chilonycteris personata* Wagner). BOLIVIA: Santa Cruz, 38 km E of La Florida (Ibáñez and Ochoa 1989). PERU: Huánuco, Cueva de las Lechuzas (Bowles, Cope, and Cope 1979).

SUBSPECIES: We recognize two subspecies of *P. personatus*, of which only *Pteronotus p. personatus* occurs in South America. *Pteronotus p. psilotis* (Dobson) is found in Mexico and Central America.

P. p. personatus (J. A. Wagner, 1843)
SYNONYMS:
Chilonycteris personata J. A. Wagner, 1843a:367; type locality "Mato grosso," Brazil; restricted to "St. Vicente," Mato Grosso, Brazil by J. A. Wagner (1847:186).
P[*teronotus*]. *personatus*: Vaughan and Bateman, 1970: 218; first use of current name combination.

The nominate subspecies is distributed across northern Colombia and Venezuela to the island of Trinidad, through Surinam and French Guiana into northeastern Brazil. In addition, this subspecies is known from the Brazilian states of Piauí, Pernambuco, and Mato Grosso, as well as from one

locality in eastern Bolivia (Ibáñez and Ochoa 1989) and one locality in east-central Perú (Bowles, Cope, and Cope 1979).

NATURAL HISTORY: Bateman and Vaughan (1974) described flyways, flight patterns, and roosting and nocturnal behavior of a mixed colony of four species of mormoopids in and around a cave in Mexico. They judged the size of the colony inhabiting the cave to exceed 400,000 bats, of which an estimated 4% were *P. personatus*. Graham's (1988) information on the species indicates similar habits in Peru. Grose, Marinkelle, and Striegel (1968) found the pathogenic yeast *Cryptococcus neoformans* in Colombian *P. psilotis* (= *P. personatus*). R. Guerrero (1985a) listed two species of ticks, two of mites, and one streblid batfly as known ectoparasites of *P. personatus*, and later (1997) increased the number of streblid batfly species to five. The karyotype is $2n = 38$, FN = 60 (R. J. Baker 1967; Sites, Bickham, and Haiduk 1981).

REMARKS: According to J. D. Smith (1972:89) the specimens from Cuiabá and Tapirapoã, Mato Grosso, Brazil, reported by C. O. C. Vieira (1942) as *Chilonycteris personata* were misidentified *Pteronotus parnellii rubiginosus*. Ochoa and Ibáñez (1985) reported *P. personatus* from the llanos of Venezuela. Albuja (1999) suggested that *P. personatus* was yet to be confirmed in Ecuador based on a sight record; however, the netted specimen he identified as having escaped from A. L. Gardner at Limoncocha was a *P. parnellii*, a much larger bat. We have no record of *P. personatus* from Ecuador. D. C. Carter and Dolan (1978) gave information on and measurements of one of the syntypes of *Chilonycteris personata* J. A. Wagner.

Family Noctilionidae Gray, 1821
Alfred L. Gardner

The Neotropical family Noctilionidae contains the genus *Noctilio* in which only two species, *N. albiventris* and *N. leporinus*, are recognized. The fossil record consists of Pleistocene (or sub-Recent) remains of *N. leporinus* from the West Indies (Martin 1972).

Noctilionids are medium-sized bats (forearm 54–92 mm) with narrow, pointed, and well-separated ears. The pointed muzzle lacks leaf-like structures, the nares open anteriorly from a well-developed and projecting rhinarium, and the chin has transverse ridges of skin. The mouth opening is transverse with the angle of the lips at about the level of the canines. The full lips and cheeks are elastic and the buccal cavity can be extended as cheek pouches. The upper lip is divided by two vertical grooves, one on each side of a prominent medial ridge that extends from the rhinarium to the mouth. The tail is well developed, about as long as the femur, and its terminal portion projects above the much longer uropatagium. The calcar is longer than the tibia.

The trochiter of humerus is smaller than the trochin and its articular contact with the scapula is not well defined. The trochin reaches the level of the head of the humerus, and the trochiter extends slightly beyond. Digits III and IV each have two phalanges, the second of which folds under the first when bat is at rest. The thoracic basket is relatively unspecialized and has a low keel on the mesosternum. The seventh cervical vertebra is free (not fused to the first thoracic). The ischium is fused to posterior end of a urostyle-like sacrum. The broad, massive skull has a high sagittal crest, but lacks postorbital processes. Premaxillae are fused together and with the maxillae, and palate is completely closed anteriorly. A baculum is lacking. The dental formula is 2/1, 1/1, 1/2, 3/3 × 2 = 28.

Genus *Noctilio* Linnaeus, 1766

SYNONYMS:

Vespertilio Linnaeus, 1758:32; part.

Noctilio Linnaeus, 1766:88; type species *Noctilio americanus* Linnaeus, 1766, by monotypy.

Pteropus Erxleben, 1777:130; part.

Noctileo Tiedemann, 1808:536; incorrect subsequent spelling of *Noctilio* Linnaeus.

Celaeno Leach, 1821a:69; type species *Celaeno brooksiana* Leach, 1821a, by monotypy.

Celano Gray, 1825a:243; incorrect subsequent spelling of *Celaeno* Leach.

Caelano Gray, 1825b:339; incorrect subsequent spelling of *Celaeno* Leach.

Noctillo Burmeister, 1879:3; incorrect subsequent spelling of *Noctilio* Linnaeus.

Noctitio Beckstein, 1801:213; incorrect subsequent spelling of *Noctilio*, but not *Noctilio* Linnaeus.

Dirias Miller, 1906b:84; type species *Noctilio albiventer* Spix, 1823, by original designation.

KEY TO SPECIES OF *NOCTILIO* (AFTER W. B. DAVIS 1976B):

1. Wing span about 400 mm; length of foot 20 mm or less; forearm less than 70 mm; length of maxillary toothrow 8.5 mm or less . . . . *Noctilio albiventris*

1'. Wing span about 500 mm; length of foot 25 mm or more; forearm more than 73 mm; length of maxillary toothrow more than 10 mm . . . . . . *Noctilio leporinus*

Noctilio albiventris Desmarest, 1818
Lesser Bulldog Bat

SYNONYMS: See under Subspecies.

DISTRIBUTION: *Noctilio albiventris* is in Colombia, Venezuela, the island of Trinidad, Guyana, Surinam, French Guiana, Brazil, Ecuador, Peru, Bolivia, and the upper Paraná drainage system of Paraguay, southwestern Brazil,

Map 216 Marginal localities for *Noctilio albiventris* ●

and northeastern Argentina. Elsewhere, the species is in Central America and southern Mexico (Polaco 1987).

MARGINAL LOCALITIES (Map 216): COLOMBIA: El Cesar, Río Guaimaral (Hershkovitz 1949c). VENEZUELA: Aragua, El Limón-Maracay (W. B. Davis 1976b); Monagas, Hato Mata de Bejuco (Handley 1976). GUYANA: Demerara-Mahaica, Georgetown (W. B. Davis 1976b). SURINAM: Nickerie, Wageningen (Husson 1978). FRENCH GUIANA: Saint-Laurent du Maroni (Brosset and Dubost 1968). BRAZIL: Amapá, Lago do Comprido (Piccinini 1974); Pará, Cabo do Maguari (Piccinini 1974); Ceará, Fortaleza (Piccinini 1974), Ceará, Russas (Taddei, Seixas, and Dias 1986); Piauí, Fronteiras (Taddei, Seixas, and Dias 1986); Bahia, Juàzeiro (C. O. C. Vieira 1942); Pará, Santa Júlia (type locality of *Dirias irex* O. Thomas); Mato Grosso, São Domingos (C. O. C. Vieira 1951); Minas Gerais, Uberlândia (Stutz et al. 2004); São Paulo, São José do Rio Prêto (McNab 1969); Paraná, Parque Estadual Mata dos Godoy (N. R. Reis et al. 2000). ARGENTINA (Barquez, Mares, and Braun 1999, except as noted): Misiones, Posadas; Santa Fe, Romang; Chaco, Resistencia (Fornes and Massoia 1978); Formosa, Parque Nacional Río Pilcomayo. PARAGUAY: Alto Paraguay, Fuerte Olimpo (type locality of *Noctilio albiventris cabrerai* W. B. Davis). BRAZIL: Mato Grosso do Sul, Corumbá (Taddei, Seixas, and Dias 1986); Mato Grosso, Cuyabá (Pelzeln 1883). BOLIVIA: Santa Cruz, Aserradero Moira (S. Anderson 1997); Santa Cruz, Buena Vista (S. Anderson, Koopman, and Creighton 1982); Beni, Río Curiraba (S. Anderson 1997). PERU: Madre de Dios, Pakitza (Ascorra, Wilson, and Romo 1991); Ayacucho, Luisiana (Tuttle 1970); Huánuco, 2 miles N of Tingo María (W. B. Davis 1976b); San Martín, Pachiza

(Koopman 1978); Amazonas, Huampami (J. L. Patton, Berlin, and Berlin 1982). ECUADOR: Pastaza, Río Acaro (Albuja 1999); Sucumbios, Limoncocha (R. H. Baker 1974). COLOMBIA: Cauca, Río Palo, 22 km S of Puerto Tejada (USNM 483297); Chocó, Noanamá (J. A. Allen 1916c); Chocó, Unguía (Wenzel, Tipton, and Kiewlicz 1966:578).

SUBSPECIES: We regognize three subspecies of *N. albiventris*.

N. a. albiventris Desmarest, 1818
SYNONYMS:

Noctilio albiventris Desmarest, 1818:15; type locality "l'Amérique méridionale"; restricted to Rio São Francisco, Bahia, Brazil, by Cabrera (1958), the locality stated by Spix (1823:58) as the origin of the material he described and figured as *Noctilio albiventer*.

Noctilio albiventer Desmarest, 1820:118; type locality "Inconnue, mais très-vraisemblablement l'Amérique méridionale" (see Remarks).

Noctilio albiventer Spix, 1823:58, pl. 36, Figs. 2 and 3; type locality "fluvem St. Francisci" (= Rio São Francisco), Bahia, Brazil.

Noctilio affinis d'Orbigny, 1837: pl. 10, Figs. 1–3; type locality not indicated on illustration; type locality "Concepcion," Beni, Bolivia, based on d'Orbigny and Gervais 1847:12.

Dirias albiventer: Miller, 1906b:84; name combination.

Noctilio zaparo Cabrera, 1907:57; type locality "Ahuano, on Río Napo," Napo, Ecuador.

Dirias zaparo: J. A. Allen, 1916c:225; name combination.

Dirias irex O. Thomas, 1920c:273; type locality "Santa Julia, Rio Iriri, Rio Xingú," Pará, Brazil.

Dirias albiventer albiventer: Cabrera, 1930:426; name combination.

Noctilio (Dirias) albiventer: Sanborn, 1949b:279; name combination.

Noctilio labialis: Hershkovitz, 1949c:433; not *Noctilio labialis* Kerr, 1792 (see Hershkovitz 1975).

Noctilio labialis labialis: Hershkovitz, 1949c:434; not *Noctilio labialis* Kerr, 1792.

Noctilio labialis albiventer: Hershkovitz, 1949c:434; name combination.

Noctilio labialis albiventris: Cabrera, 1958:55; name combination.

Noctilio labialis zaparo: Cabrera, 1958:56; name combination.

Noctilio labialis albiventris: Husson, 1962:63; name combination.

Noctilio albiventris albiventris: Hershkovitz, 1975:244; first modern use of current name combination.

Noctilio albiventris zaparo: Hershkovitz, 1975:244; name combination.

Noctilio albiventris affinis: W. B. Davis, 1976b:687; name combination.

The nominate subspecies occurs from northern Venezuela and the Guianas south through the greater Amazon basin of Brazil and western Colombia, Ecuador, Peru, and Bolivia; also southward along the Atlantic Coast of Brazil into the state of São Paulo.

N. a. cabrerai W. B. Davis, 1976
SYNONYMS:

Noctilio albiventris ruber: Hershkovitz, 1975:244; name combination; not *Vespertilio ruber* Rengger, 1830 (see Remarks).

Noctilio albiventris cabrerai W. B. Davis, 1976b:701 b:701; type locality "Fuerte Olimpo, Depto. de Olimpo [= Alto Paraguay], Paraguay."

This subspecies is distributed within the Río Paraná drainage basin of southwestern Brazil, Paraguay, and Argentina.

N. a. minor Osgood, 1910
SYNONYMS:

Noctilio minor Osgood, 1910:30; type locality "Encontrados, Zulia, Venezuela."

Noctilio albiventer minor: Osgood, 1912:62; name combination.

Dirias albiventer minor: Goldman, 1920:117; name combination.

Dirias minor: Goodwin, 1942c:121; name combination.

Noctilio labialis minor: Hershkovitz, 1949c:433; name combination.

Noctilio labialis labialis: Cabrera, 1958:56; name combination.

Noctilio albiventris minor: Hershkovitz, 1975:244; first use of current name combination.

This subspecies is found from western Colombia, north and west of the Andes, into western Venezuela. Elsewhere, it is known from Central America and Mexico as far north as the Pacific Coast of Chiapas (W. B. Davis 1976b; Hood and Pitocchelli 1983; Alberico 1987; Polaco 1987).

NATURAL HISTORY: *Noctilio albiventris* frequents humid tropical forest habitats where it forages in small groups over streams, rivers, marshes, and lake margins (W. B. Davis 1976b). Roost sites include hollow trees and buildings where the species has been found with other bats, usually species of *Molossus*. The species is considered to be insectivorous; however, Howell and Burch (1974) reported the remains of fish in the stomach contents of two specimens from Costa Rica. Reproduction appears to be highly synchronous with most births occurring in April and May in Panama (J. W. Anderson and Wimsatt 1963) or as late as July or August in Costa Rica (LaVal and Fitch 1977) and Peru (Tuttle 1970). Ectoparasites include mites

(B. McDaniel 1972; Yunker and Radovsky 1980), streblid batflies (Wenzel, Tipton, and Kiewlicz 1966; R. Guerrero 1997), and ticks (Fairchild, Kohls, and Tipton 1966). Hood and Pitocchelli (1983) illustrated the skull and summarized information on the natural history of *N. albiventris* in their *Mammalian Species* account. The karyotype is $2n = 34$, FN = 58 or 62 (R. J. Baker and Jordan 1970; J. L. Patton and Baker 1978).

REMARKS: *Noctilio albiventer* Desmarest, 1820, is based on É. Geoffroy St.-Hilaire's "*Noctilion à ventre blanc*" in the Paris Museum, and Pennant's (1771:365; 1781:554) "Peruvian bat, var. ß." Hershkovitz (1975) considered the name an emendation of *N. albiventris* Desmarest, 1818. Both names have the same basis; however, there is no way to determine whether *N. albiventer* is an emendation (unjustified) or an incorrect subsequent spelling of the earlier name. If an emendation, it invalidates *N. albiventer* Spix, 1823, which in any case is a junior synonym of *N. albiventris* Desmarest, 1818.

Cabrera (1958) included the population W. B. Davis (1976b) later named *N. a. cabrerai* under the name *N. labialis albiventris*, but Hershkovitz (1975) suggested that the population be called *N. albiventris ruber* Rengger, 1830. As pointed out by W. B. Davis (1976b) and discussed by Hershkovitz (1976a), Rengger (1830) coined the name *N. ruber* for certain Paraguayan bats that he believed were the same as Azara's (1801b) "*chauve-souris onzieme*," currently *Myotis ruber* (É. Geoffroy St.-Hilaire, 1806). Cabrera (1958) listed *N. ruber* in his synonymy of *N. labialis albiventris*, a contradiction of his earlier statement (1938:10) that "el *Noctilio ruber* de Rengger, que es el *Vespertilio ruber* de Geoffroy y por consiguiente un *Myotis*."

Noctilio leporinus (Linnaeus, 1758)
Greater Bulldog Bat
SYNONYMS: See under Subspecies.

DISTRIBUTION: *Noctilio leporinus* is found along the Pacific Coast of Colombia, Ecuador, and northern Peru, and from the Caribbean lowlands and eastern llanos of Colombia, through Venezuela, the island of Trinidad, Guyana, Surinam, French Guiana, and south through Brazil, eastern Peru, eastern Bolivia, Paraguay, into northern Argentina. Elsewhere, the species occurs northward through the Antilles to Cuba and through Central America to northwestern Mexico.

MARGINAL LOCALITIES (Map 217): VENEZUELA: Nueva Esparta, 1.5 km N of San Francisco de Macanao (J. D. Smith and Genoways 1974). TRINIDAD AND TOBAGO: Trinidad, Port-of-Spain (Goodwin and Greenhall 1961). VENEZUELA: Delta Amacuro, Guayo (Linares and Rivas 2004). GUYANA: Demerara-Mahaica, Demerara (Dobson 1878). SURINAM: Paramaribo, Paramaribo

Map 217 Marginal localities for *Noctilio leporhinus* ●

(Husson 1962). FRENCH GUIANA: Cayenne (W. B. Davis 1973). BRAZIL: Amapá, Lago Comprido (Piccinini 1974); Pará, Belém (USNM 460084); Ceará, Fortaleza (Piccinini 1974); Pernambuco, Açude (W. B. Davis 1973); Bahia, Baía (C. O. C. Vieira 1942); Espírito Santo, Gruta do Judeu (Ruschi 1951g); São Paulo, Campinas (Taddei, Seixas, and Dias 1986); Paraná, Guaratuba Bay (Bordignon and França 2002); Santa Catarina, São Francisco (C. O. C. Vieira 1942; not mapped); Rio Grande do Sul, Pôrto Alegre (C. O. C. Vieira 1955). ARGENTINA: Corrientes, Mercedes (W. B. Davis 1973); Santa Fe, Sauce Viejo (Barquez, Mares, and Braun 1999); Santiago del Estero, Bañado de Figueroa (Olrog 1976); Jujuy, Yuto (W. B. Davis 1973). BOLIVIA: Tarija, 1 km S of Camatindi (S. Anderson 1997); Santa Cruz, 10 km E of Ingeniero Mora (S. Anderson 1997); La Paz, Sararia (S. Anderson, Koopman, and Creighton 1982). PERU: Madre de Dios, Pakitza (V. Pacheco et al. 1993); Pasco, San Juan (Tuttle 1970). ECUADOR: Pastaza, Río Capahuari (Albuja 1983). COLOMBIA: Valle del Cauca, Punta Soldado, (Velasco-Abad and Alberico 1985); Boyacá, Orocué (Speiser 1900). ECUADOR: Esmeraldas, Esmeraldas (Tomes 1860c). PERU: Tumbes, Huásimo (Koopman 1978). COLOMBIA: El Cesar, El Orinoco (Hershkovitz 1949c). VENEZUELA: Miranda, Tacarigua de la Laguna (W. B. Davis 1973).

SUBSPECIES: W. B. Davis (1973) recognized three subspecies of *N. leporinus*.

N. l. leporinus (Linnaeus, 1758)

SYNONYMS:

[*Vespertilio*] *leporinus* Linnaeus, 1758:32; type locality "America"; restricted to Surinam by O. Thomas (1911b:131).

Vespertilio Minor Fermin, 1765:9; unavailable name (see Remarks).

Noctilio americanus Linnaeus, 1766:88; junior objective synonym of *Vespertilio leporinus* Linnaeus.

Pteropus leporinus: Erxleben, 1777:130; name combination.

Noctilio labialis Kerr, 1792:93; type localities "Peru, and the Musquito shore" (see Remarks); restricted to lower Ucayali region, Loreto, Peru, by Hershkovitz (1949c:434).

Noctilio novemboracensis Lacépède, 1799b:16; no type locality given.

Noctilio leporinus: Illiger, 1815:109; first use of current name combination.

Noctilio dorsatus Desmarest, 1818:15; type locality "l'Amérique méridionale."

Noctilio unicolor Desmarest, 1818:15; type locality "l'Amérique méridionale" ("probably Brazil," according to Hood and Jones 1984:1).

Celaeno brooksiana Leach, 1821a:70; type locality unknown.

Noctilio vittatus Schinz, 1821:870; type locality "Ostküste von Brasilien" (see Remarks).

Noctilio rufus Spix, 1823:57, pl. 35, Fig. 1; no type locality given in account (see Remarks), but stated as "apparently Brazil" by D. C. Carter and Dolan (1978:25), and "Amazonian Brazil" by Hood and Jones (1984:1).

Noctilio macropus Pelzeln, 1883:37; *nomen nudum*.

Noctilio longipes Pelzeln, 1883:37; *nomen nudum*.

Noctilio intermedius Pelzeln, 1883:38; *nomen nudum*.

The nominate subspecies is found in the Guianas and the Amazon lowlands of Peru, Ecuador, Colombia, and Brazil. It also occurs along the Atlantic drainage of coastal Brazil as far south as Espírito Santo.

N. l. mastivus (Vahl, 1797)

SYNONYMS:

[*Vespertilio*] *Mastivus* Vahl, 1797:132; type locality "Insula St. Crucis Americae" (= St. Croix), U.S. Virgin Islands.

Noctilio leporinus mastivus: True, 1884:603; first use of current name combination.

Noctilio leporinus mexicanus Goldman, 1915:136; type locality "Papayo, Guerrero, Mexico."

This subspecies occurs along the Pacific Coast of Colombia and northern Ecuador, and along the Caribbean lowlands of Colombia (including the Río Magdalena drainage system), eastward through Venezuela (including Isla Mar-

garita) and adjacent Trinidad into northwestern Guyana, and southward throughout the Río Orinoco and Río Negro drainage systems of southern Venezuela and eastern Colombia. Elsewhere, this subspecies occurs northward through the Antilles to Cuba in the east, and through Central America as far as Sinaloa, Mexico, in the west (W. B. Davis 1973; Hood and Jones 1984).

N. l. rufescens Olfers, 1818

SYNONYMS:

[*Noctilio*] ?*rufescens* Illiger, 1815:109; *nomen nudum*.

N[*octilio*]. *rufescens* Olfers, 1818:225; type locality "Paraguay"; name based exclusively on the "*chauve-souris rougeâtre*" of Azara (1801b).

Noctilio rufipes d'Orbigny, 1837: pl. 9, Figs. 1–4; type locality "les grandes forêts qui bordent le Rio de San-Miguel, au pays des sauvages Guarayos (Bolivia) [d'Orbigny and Gervais 1847:12]"; identified as Río San Miguel, Pampas de los Guarayos, Santa Cruz, Bolivia.

Noctilio leporinus rufipes: Cabrera, 1938:14; name combination.

Noctilio leporinus rufescens: Hershkovitz, 1959:340; first use of current name combination.

This subspecies is distributed throughout most of eastern Bolivia and the upper watershed of the Río Paraná drainage system in Argentina, Paraguay, and southern Brazil, mainly north of 30°S latitude (W. B. Davis 1973; Hood and Jones 1984).

NATURAL HISTORY: *Noctilio leporinus* is found in tropical lowland habitats where it usually forages over ponds, lakes, streams, rivers and their estuaries, and coastal bays and lagoons. Roosts have been located in caves, hollow trees, and rock crevices. Mares et al. (1981) reported a night roost located under a bridge in the Brazilian state of Pernambuco that was used by over 100 individuals, possibly for several years. *Noctilio leporinus* is one of the few piscivorous bats; although the species also consumes a variety of insects (see Silva-Taboada 1979:Appendix II). Bordignon and França (2002) found the remains of eight species of salt water fishes, representing six families, in the feces of *N. leporinus* netted in Guaratuba Bay, Paraná, Brazil.

Apparently, *N. leporinus* is reproductively active throughout the year, although reproduction within a particular colony may be synchronous (Silva-Taboada 1979). Ectoparasites include mites (Furman 1966; Goodwin and Greenhall 1961; Dusbabek 1969; B. McDaniel 1972; Herrin and Tipton 1975; Silva-Taboada 1979), ticks (Fairchild, Kohls, and Tipton 1966; E. K. Jones et al. 1972; Silva-Taboada 1979), bedbugs (Goodwin and Greenhall 1961), and batflies (Wenzel, Tipton, and Kiewlicz 1966; Wenzel 1976; R. Guerrero 1997). Hood and Jones (1984) illustrated the skull and summarized additional information on the natural history of *N. leporinus* in their *Mammalian Species* account. The karyotype is $2n = 34$, FN = 58 or 62 (Yonenaga, Frota-Pessoa, and Lewis 1969; R. J. Baker and Bickham 1980).

REMARKS: As commented on by W. B. Davis (1973), Cabrera's conclusions in his 1938 report on Argentine noctilionids formed the base for our understanding of the content and nomenclatural history of the genus. In his report, Cabrera pointed out that the names *N. macropus* and *N. longipes* were Natterer's manuscript names listed by Pelzeln (1883:37) as synonyms of *N. leporinus*. *Noctilio intermedius* is another of Natterer's names listed by Pelzeln, but not mentioned by Cabrera (1938, 1958). Natterer's names, as listed by Pelzeln (1883), are *nomina nuda*. Pelzeln (1883:38) attributed the name *N. rufescens* to Illiger and indicated that Natterer had used the name in an unpublished catalogue of his collections. However *N. rufescens* Illiger (1815) is a *nomen nudum* and, as correctly attributed by Hershkovitz (1959), the name dates from Olfers (1818). Hood and Jones (1984) listed "*Vespertilio Minor* Fermin, 1765:9, with type locality Suriname," in the synonymy of *N. leporinus*: apparently using the name uncritically from Husson (1962:65). *Vespertilio Minor* Fermin is not an available name because Fermin did not consistently apply the principles of binominal nomenclature (ICZN 1963). Therefore, *Vespertilio Minor* Fermin does not invalidate *N. albiventris minor* Osgood, 1910.

Apparently, Hood and Jones (1984) misinterpreted Cabrera's (1938) remarks regarding *N. vittatus*. When Cabrera (1938:12) said that *N. vittatus* Schinz was a new name for *N. dorsatus* Desmarest, he continued, "o sea para los incluidos de la forma tipica con la raya pálida en el lomo." Cabrera clearly meant that *N. vittatus* Schinz, 1821, was a name applied to the same phenotype described by Desmarest (1818) as *N. dorsatus*; not a replacement name, in the nomenclatural sense, for a preoccupied name.

Hood and Jones (1984) erred in giving "Vallee d'Ylo," Peru, as the type locality for *Vespertilio labialis* Kerr, 1792. As earlier stated by Hershkovitz (1975:243), Kerr assigned the bat from the Valley of Ylo (p. 93) to *Vespertilio leporinus*. Kerr said that *Vespertilio labialis* "Inhabits Peru, and the Musquito shore"; the former locality Hershkovitz (1949c) restricted to the lower Rio Ucayali, Loreto, Peru. Cabrera (1958:55) disagreed with Hershkovitz's restriction of the type locality of *V. labialis*, but Cabrera's argument suggests that he had confused Kerr's (1792) two accounts pertaining to *Noctilio*. Hershkovitz's (1949c) restriction of the type locality to the lower Río Ucayali is reasonable and, barring compelling evidence to the contrary, must stand.

As Husson (1962:115) noted in his account on *Trachops cirrhosus*, even though Spix (1823) did not always mention a precise locality for each of the bats he described, he said on page 53 that they came from Brazil. Therefore, D. C. Carter and Dolan (1978) were correct in giving Brazil as the origin of Spix's types of *N. albiventer* and *N. rufus*, but I do not know the basis for Hood and Jones's (1984) restriction of the type locality for the latter to "Amazonian Brazil."

Family Furipteridae Gray, 1866a
Alfred L. Gardner

The family Furipteridae comprises two Recent monotypic genera, *Amorphochilus* and *Furipterus*. These genera are exclusively Neotropical in distribution and have no fossil record. Furipterids are small, delicate bats with relatively long wings. The muzzle is blunt with anteriorly directed nostrils. The tail is encased in, but does not reach the posterior margin of the long uropatagium. The short, triangular tragus is characteristic of the family. Other familial characters include functional abdominal mammae and a reduced thumb, which is encased in the wing membrane to the base of the minute claw. The presternum is anteriorly broad, has a well-developed keel, and is slightly wider than long; the mesosternum has a reduced keel. The trochiter of humerus is only slightly longer than the trochin, and its articulation with the scapula is not well defined. The first of the two phalanges on digit III is conspicuously short. The braincase is large, globose, and elevated above the rostrum. Palatal branches of the premaxillaries are rudimentary. The dentary has a well-developed mental spur. Canines are short, their height about equal to the crown height of the tallest premolar. A baculum is present. The dental formula is 2/3, 1/1, 2/3, 3/3 × 2 = 36.

The family Furipteridae is closely related to the Natalidae with which it shares a number of structural features, including funnel-shaped ears and lack of an ossified third phalanx on digit III. The natalids, however, differ in a number of features, including a fully developed thumb and a tail that reaches the posterior margin of the uropatagium.

KEY TO THE GENERA AND SPECIES OF FURIPTERIDAE:
1. Muzzle complex and not thickly furred; nose bordered above and laterally by a conspicuous fleshy rim; a saclike structure under the chin is partly covered anteriorly by a split, fleshy protuberance; palate long, extending behind toothrow a distance exceeding length of last upper molar; mesopterygoid fossa wider than long . . . . *Amorphochilus schnablii*
1'. Muzzle relatively simple and well furred; no fleshy structures under chin; palate short, not extending much beyond last molar; mesopterygoid fossa longer than wide . . . . *Furipterus horrens*

Genus *Amorphochilus* W. Peters, 1877

A monotypic genus of small (forearm 36–37 mm, $n = 3$), delicate bats, *Amorphochilus* has long, dark smoky-gray fur that often has a brownish tinge. The snout is blunt and pig-like in appearance and the chin and lower lips bear fleshy wart-like structures. The tail extends nearly 4/5 the length of the uropatagium and is entirely enclosed by that membrane. Other characters are given in the key; see W. Peters (1877) and Osgood (1914b) for more comprehensive descriptions.

SYNONYM:

Amorphochilus W. Peters, 1877:185; type species *Amorphochilus schnablii* W. Peters, 1877, by monotypy.

Amorphochilus schnablii W. Peters, 1877
Schnabl's Smoky Bat

SYNONYMS:

Amorphochilus schnablii W. Peters, 1877:185; type locality "Tumbez im nördlichen Peru, an der Grenza von Ecuador"; restricted to Tumbes, departamento Tumbes, Peru, by Cabrera (1958:96).

Amorphochilus schnablii osgoodi J. A. Allen, 1914:381; type locality "Hacienda Limon, (altitude 3000 ft.), near Balsas, [Cajamarca,] Peru."

Amorphochilus chnablii Tovar, 1971:21; incorrect subsequent spelling of *Amorphochilus schnablii* W. Peters.

DISTRIBUTION: *Amorphochilus schnablii* occurs in western South America from coastal Ecuador (Guayas) through Peru into northern Chile.

MARGINAL LOCALITIES (Map 218): ECUADOR: Guayas, mouth of Río Javita, 2 km N of Monteverde (Ibáñez 1986); Guayas, Puná Island (J. A. Allen 1914). PERU: Tumbes, Tumbes (type locality of *Amorphochilus schnablii* W. Peters); Piura, 5 km W of San Isidro (Graham and Barkley 1984); Cajamarca, Hacienda Limón (type locality of *Amorphochilus schnablii osgoodi* J. A. Allen); Lima, Lima (Ortiz de la Puente 1951); Arequipa, Chucarapi (Sanborn 1941). CHILE: Tarapacá, Cuya (Mann 1950a).

SUBSPECIES: Although Koopman (1978) placed the Pacific coastal population in *A. s. schnablii* and the Cajamarca population in *A. s. osgoodi*, I am following Ortiz de la Puente (1951) and Cabrera (1958) in treating *A. schnablii* as monotypic.

NATURAL HISTORY: With the exception of two specimens from Hacienda Limón, Cajamarca, Perú, which is located in an arid hanging valley above the Río Marañon (Osgood 1914b), *A. schnablii* is known only from the xeric Pacific coastal lowlands of western Ecuador south into northern Chile. Specimens have been found in a road culvert (Ibáñez 1986), buildings, and an irrigation tunnel (Sanborn 1941). One was mistnetted at the entrance of a shallow, well-illuminated, fracture cave in desert habitat

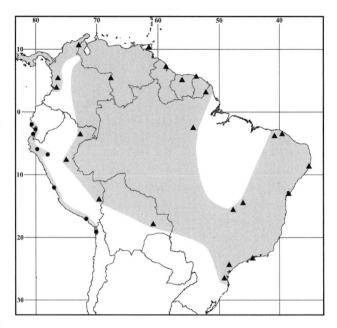

Map 218 Marginal localities for *Amorphochilus schnablii* ● and *Furipterus horrens* ▲

near San Isidro, Lambayeque, Perú (Graham and Barkley 1984). Limited evidence suggests that these bats feed exclusively on adult moths (Ortiz de la Puente 1951; Ibáñez 1986).

Ibáñez (1986) reported eight of ten females examined in Ecuador on 19 November were pregnant, each with a single fetus (2–6 mm CR). He suggested that births and lactation were synchronized with the brief rainy season (January to May), and confirmed the abdominal position of the mammae as reported for *Furipterus horrens* by Uieda, Sazima, and Storti-Filho (1980). Two females in the MVZ (125617 and 125619), collected in departamento Piura, Peru, on 18 August, 1967, each bear label notations "No Embs." Abdominal mammae and other similarities with *F. horrens* include sexual dimorphism in size (females have larger wings) and the suggestion of stable, sexually mixed colonies. R. Guerrero (1997) listed a streblid batfly, apparently unique to *A. schnablii*. Uchikawa (1988) described *Furipterobia chileensis* as a new species of myobiid mite from Chilean *A. schnablii*, and because the mite *Furipterobia amorphochilus* (Fain, 1976) was found on Peruvian specimens, but not on specimens from Chile, suggested that the Chilean and Peruvian populations represented different species. The karyotype is unknown.

Genus *Furipterus* Bonaparte, 1837

A monotypic genus of small, delicate bats (average mass and forearm length of 59 individuals: 3.3 mm and 34.3 mm, respectively; LaVal 1977), *Furipterus* has soft, thick, bluish-gray to slate-gray fur on the dorsum; the venter is slightly paler. The tail extends a little less than 2/3 the length of, and is completely encased in the uropatagium. The well-furred muzzle lacks the prominently rimmed, blunt snout, and the conspicuous fleshy lip and chin excrescences of *Amorphochilus schnablii*. The canines are longer and narrower, although their crown height is about equal in the two taxa. Other characters are given in the key; see Dobson (1878) and Miller (1907b) for more comprehensive descriptions.

SYNONYMS:

Furia F. Cuvier, 1828:150; type species *Furia horrens* F. Cuvier, 1828, by monotypy; preoccupied by *Furia* Linnaeus, 1758 (Annelida).

Furipterus Bonaparte, 1837b:fasc. 21 (under *Plecotus auritus*); type species *Furia horrens* F. Cuvier, 1828, by monotypy.

Furiella Gray, 1866a:91; based on "*Furia* Temm[inck], *Furipterus* Tomes, not Bonap[arte]"; no species mentioned; intended as a replacement name for *Furia* Temminck (= *Furia* F. Cuvier) on the assumption that *Furipterus* Bonaparte did not represent the same genus.

Furipterus horrens (F. Cuvier, 1828)
Smoky Bat

SYNONYMS:

Furia horrens F. Cuvier, 1828:155; type locality "la Mana," French Guiana.

Furipterus horrens: Tomes, 1856:175; first use of current name combination.

Furipterus caerulescens Tomes, 1856:176; type locality "St. Catharine, Brazil."

Furipterus torrens J. A. Allen, 1916c:226; incorrect subsequent spelling of *Furia horrens* F. Cuvier.

Furipterus harrens Uchikawa, 1988:163; incorrect subsequent spelling of *Furia horrens* F. Cuvier.

DISTRIBUTION: *Furipterus horrens* occurs in Colombia, the island of Trinidad, Venezuela south of the Río Orinoco, the Guianas, eastern Peru, and Brazil southward into the state of Santa Catarina. The species has not been recorded in Bolivia and Paraguay. Elsewhere, the species is known from Central America (Panama and Costa Rica; LaVal 1977).

MARGINAL LOCALITIES (Map 218): COLOMBIA: La Guajira, Villanueva (USNM 281947). VENEZUELA: Amazonas, 32 km S of Puerto Ayacucho (Handley 1976). TRINIDAD AND TOBAGO: Trinidad, San Fernando (W. Peters 1867a). GUYANA: Pomeroon-Supenaam, Maccasseema (Thomas 1887). SURINAM: Sipaliwini, Camp One, Coppename River (Sanborn 1941). FRENCH GUIANA: Mana (type locality of *Furia horrens* F. Cuvier); Camopi (Brosset and Dubost 1968). BRAZIL: Pará, Taperinha (Piccinini 1974); Ceará, Gruta de Ubajara (Uieda, Sazima, and Storti-Filho 1980); Ceará, Itapipoca (Piccinini 1974); Pernambuco, Rio Formoso (Bonato and Facure 2000); Bahia, Bahia (P. Gervais 1856a); Goiás, Município de Mambaí (Coimbra et al. 1982); Distrito Federal, Gruta Água Rasa (Bredt, Uieda, and Magalhães 1999); Rio de Janeiro, Praia da Sumaca (Pol, Nogueira, and Peracchi 2003); São Paulo, Fazenda Intervales (Fenton et al. 1999); Santa Catarina, Colonia Hansa (J. L. Lima 1926). BOLIVIA: Santa Cruz, San José de Chiquitos (MZUSP 7733). PERU: Puno, Bella Pampa (MVZ 116618); Ucayali, Río Disqui (Tuttle 1970); Loreto, Río Amazonas, ca. 10 km SSW of mouth of Río Napo on E bank of Quebrada Vainilla (LSUMZ 28445). COLOMBIA: Valle del Cauca, Quebrada La Cristalina (Camargo and Tamsitt 1990); Chocó, Andagada (J. A. Allen 1916c).

SUBSPECIES: I am treating *F. horrens* as monotypic.

NATURAL HISTORY: Long considered one of the rarer Neotropical bats, *F. horrens* has been discovered recently in relatively large aggregations. LaVal (1977) found 59 males roosting in small clusters inside a large hollow log (diameter of hollow approximately 1 m) in Costa Rica. The field catalog of Angelo P. Capparella (LSUMZ) contains the following notation from 5 August, 1983: "one of two flushed from inside large hollow log . . . shot in humid terra firme forest." Uieda, Sazima, and Storti-Filho (1980) studied colonies occupying two rock caves or grottos about 6 km apart in the vicinity of Ubajara, Ceará, Brazil. The larger grotto contained about 250 *F. horrens*; the smaller, about 150. A sample of 51 from the smaller grotto consisted of 29 males and 22 females. Adult females were significantly larger than adult males. Most of the adult females were lactating in late January and early February. Willig (1985a) reported a female from Exu, Pernambuco, Brazil, lactating in April. Uieda, Sazima, and Storti-Filho (1980) gave previously unknown details on the biology of *F. horrens*. They reported that the mammae are abdominal; females have a single young, which clings to the mother in a head-up position. The diet is insects; the stomach contents of three contained insect remains, principally the scales and hairs of lepidopterans. This was confirmed at Fazenda Intervales, São Paulo, Brazil by Fenton et al. (1999) who found that *F. horrens* fed extensively on moths, as they had predicted based on its echolocation calls, which they thought would be inaudible to moths that have hearing-based defenses. Flight activity began at twilight, but bats did not leave the grottos until after dark. Elsewhere, most specimens have been caught in buildings or netted in humid forest habitats. The two Exu, Pernambuco, Brazil specimens were caught at a mesic "serrote" in otherwise xeric caatinga habitat (Willig 1983). Also in Pernambuco, Bonato and Facure (2000) recovered the remains of a *F. horrens* from the stomach contents of the Fringe-lipped Bat, *Trachops cirrhosus*. R. Guerrero

(1997) listed a streblid batfly known from *F. horrens*. Uchikawa (1988) mentioned a myobiid mite found on a Guyanan *F. horrens*, but the host may have been a misidentified *Natalus tumidirostris*. The karyotype is $2n = 34$, FN $= 62$ (R. J. Baker, Genoways, and Seyfarth 1981).

REMARKS: Goodwin (1959b:12) pointed out that a *F. horrens* in the Leiden Museum (RMNH) from Guyana (British Guiana) was erroneously reported as *Natalus stramineus* by Beebe (1919:79), Jentink (1893:79), and Young (1896:44).

Family Thyropteridae Miller 1907

Don E. Wilson

The family Thyropteridae is monotypic and the genus *Thyroptera* is distributed from southern Mexico and Central America, throughout northern South America southward to southern Brazil. There are four species, all found in South America. The family is considered primitive, but has an unusual specialization in the form of a circular disk on the sole of each foot and an oval or circular disk attached by a short pedicle to the base of each thumb. The disks function as suction cups (Wimsatt and Villa-R. 1970; Riskin and Fenton 2002).

The Thyropteridae are small bats (forearm 31–38 mm) with an elongate, slender muzzle; circular and well-separated nares; and funnel-shaped ears. The wing membrane (plagiopatagium) attaches to along the leg and foot to the base of the claws, and the tail extends one to a few millimeters beyond the distal margin of the long interfemoral membrane. The toes have only two phalanges and the third and fourth toes (including claws) are fused. Wing digit II is reduced to an incomplete metacarpal; the third phalanx of digit III is ossified. The presternum is small, keeled, and its width is about 1/2 the combined length of the presternum and mesosternum. The mesosternum is broad, flat, and lacks a keel. The first and second thoracic vertebrae are fused. The trochiter of the humerus is larger than the trochin and articulates with the scapula. The skull has a rounded braincase elevated above the slender rostrum, and has complete premaxillae, but lacks postorbital processes. A baculum is present. The dental formula is 2/3, 1/1, 3/3, 3/3 $\times$ 2 = 38. *Thyroptera tricolor* and *T. lavali* (the latter also described as *T. robusta* Czaplewski, 1997) are known from the Middle Miocene La Venta fauna of northern Colombia (Czaplewski 1996b, 1997).

Genus *Thyroptera* Spix, 1823

SYNONYMS:

Thyroptera Spix, 1823:61; type species *Thyroptera tricolor* Spix, 1823, by monotypy.

Molossus: J. B. Fischer, 1829:90; part; not *Molossus* É. Geoffroy St.-Hilaire, 1805.

Thiroptera Lesson, 1842:19; incorrect subsequent spelling of *Thyroptera* Spix.

Dysopes: Schinz, 1844:138; part; not *Dysopes* Illiger, 1811.

Hyonycteris Lichtenstein and Peters, 1854:335; type species *Hyonycteris discifera* Lichtenstein and Peters, 1854, by monotypy.

REMARKS: The name *Thyroptera* is from the Greek, meaning "disk-winged." The only other bat with similar structures is *Myzopoda aurita* from Malagasy, but the suction disks of that species are not as specialized as those of *Thyroptera*.

KEY TO THE SPECIES OF *THYROPTERA*:

1. Disk at base of thumb obviously oval in outline; dorsal fur short and dense; small, linearly arranged wartlike structures on plagiopatagium and interfemoral membrane; ventral pelage bicolored 2
1'. Disk at base of thumb nearly circular in outline; dorsal fur long, not dense; wartlike structures on wing and tail membranes lacking; ventral pelage unicolored 3
2. Ventral pelage distinctly frosted, individual hairs dark brown to blackish basally with grayish brown to whitish tips; last lower incisor (i3) has well-defined accessory cuspulids . . . . *Thyroptera devivoi*
2'. Ventral pelage bicolored, but not appearing distinctly frosted; accessory cuspulids on third lower incisor (i3) either weakly developed or absent . . . *Thyroptera lavali*
3. Size larger, forearm usually more than 36 mm; ventral pelage unicolored white or pale gray; interfemoral membrane appearing nearly naked; inner upper incisor bifid . . . . *Thyroptera tricolor*
3'. Size smaller, forearm usually shorter than 36 mm; Ventral pelage unicolored yellowish brown (not white); interfemoral membrane usually conspicuously furred; both upper incisors bifid. . . . . *Thyroptera discifera*

Thyroptera devivoi Gregorin, Gonçalves, Lim, and Engstrom, 2006
De Vivo's Disk-winged Bat
SYNONYM:

Thyroptera devivoi Gregorin, Gonçalves, Lim, and Engstrom, 2006: 239; type locality "Estação Ecológica de Uruçuí-Una, municipality of Bom Jesus, state of Piauí, Brazil (08°52'S, 44°57'W."

DISTRIBUTION: *Thyroptera devivoi* is known from southwestern Guyana and northeastern Brazil.

MARGINAL LOCALITIES (Map 219; from Gregorin et al. 2006): GUYANA: Upper Takutu-Upper Essequibo, Tamton. BRAZIL: Piauí, Estação Ecológica de Uruçuí-Una (type locality of *Thyroptera devivoi*); Tocantins, Rio Novo, Estação Ecológica Serra Geral do Tocantins.

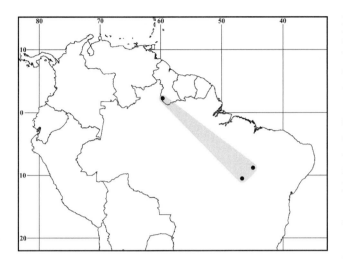

Map 219 Marginal localities for *Thyroptera devivoi*

SUBSPECIES: *Thyroptera devivoi* is considered to be monotypic.

NATURAL HISTORY: This species is known from four specimens, two from southwestern Guyana and two from northwestern Brazil. The Guyanan specimens were captured hanging under a palm leaf in savannah-gallery forest habitat. Both Brazilian specimens came from humid woodlands in Cerrado habitat.

Thyroptera discifera (Lichtenstein and Peters, 1854)
Peters's Disk-winged Bat

SYNONYMS:

Hyonycteris discifera Lichtenstein and Peters, 1854:336; type locality "Puerto Cabello (America Centralis)"; corrected to Puerto Cabello, Carabobo, Venezuela by Wilson (1976).

Th[*yroptera*]. *discifera*: J. A. Wagner, 1855:780; first use of current name combination.

Thyroptera discifera major Miller, 1931:411, type locality "San Julian, [Distrito Federal,] Venezuela."

DISTRIBUTION: *Thyroptera discifera* is in Colombia, Venezuela, Surinam, French Guiana, Brazil, Bolivia, and Peru. Elsewhere, an apparently isolated population occurs in eastern Nicaragua and Costa Rica (see Remarks).

MARGINAL LOCALITIES (Map 220): VENEZUELA: Carabobo, Puerto Cabello (type locality of *Hyonycteris discifera* Lichtenstein and Peters); Distrito Federal, San Julian (W. Robinson and Lyon 1901; type locality of *Thyroptera discifera major* Miller). SURINAM: Sipaliwini, Raleigh Falls (Genoways, Williams, and Groen 1981). FRENCH GUIANA: Cayenne (O. Thomas 1928c). BRAZIL: Pará, Belém (Pine 1993a); Bahia, Salvador (Gregorin et al. 2006); Pará, Prainha (B. D. Patterson 1992); Amazonas, Balbina (Pine 1993a); Mato Grosso, Aripuanã (Mok et al. 1982b); Mato Grosso, vicinity of Manso hydroelectric dam (Bezerra, Escarlate-Tavares, and Marinho-Filho 2005). BO-

LIVIA: Beni, La Cayoba (Torres, Rosas, and Tiranti 1988). PERU: Ucayali, Cumaría (O. Thomas 1928c); Loreto, Iquitos (O. Thomas 1928d). COLOMBIA: Nariño, La Guayacana (Pine 1993a); Cauca, Munchique (Pine 1993a); Cordoba, Socorré (Guimarães and d'Andretta 1956).

SUBSPECIES: Of the two subspecies recognized (Wilson 1976); only the nominate form, *T. d. discifera* Lichtenstein and Peters, 1854, occurs in South America. The other subspecies is *T. d. abdita* Wilson 1976:307, known from northeastern Costa Rica and from its type locality in Nicaragua.

NATURAL HISTORY: This species apparently roosts attached to the underside of leaves, in contrast to *T. tricolor*, which mainly roosts inside of fresh, rolled (unfurling) leaves of *Heliconia* spp. Torres, Rosas, and Tiranti (1988), however, reported on a group of 15 captured in a furled banana leaf 1.6 m above ground. Two species of nycteribiid bat flies are known from *T. discifera*, one from Colombian bats (Guimarães and d'Andretta 1956), and the other from Peruvian specimens (Theodor 1967). Wilson (1978) summarized information about the natural history of this species in his *Mammalian Species* account. *Thyroptera discifera* has a $2n = 32$, $FN = 38$ karyotype (R. J. Baker, Genoways, and Seyfarth 1981).

REMARKS: Early workers mistakenly assumed the type locality of *T. discifera* to be Puerto Caballos, Honduras. As a result, Miller named the South American populations *T. d. major* without realizing that the name was a junior synonym of the nominate form. Wilson (1976) reviewed the species, and described the Central American subspecies, *T. d. abdita*. Ochoa and Ibáñez (1985) questioned the validity of the Central American subspecies, but did not compare specimens. Rodríguez's (1993) report of *T. discifera* from Costa Rica was based on a specimen later reidentified by Timm and LaVal (1998) as a *T. tricolor*. Tschapka, Brooke,

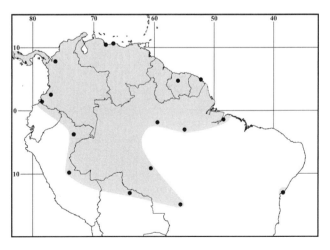

Map 220 Marginal localities for *Thyroptera discifera* ●

and Wasserthal (2000) have confirmed the presence of *T. discifera* in Costa Rica.

According to J. K. Jones, Swanepoel, and Carter (1977), Escondido River, 50 miles east of Bluefields, Nicaragua (type locality of *T. d. abdita*) is actually the I. P. Plantation, 3 km S and 13 km E of Rama. The epithet *discifera* refers to the suction disks, and *abdita* means hidden or secret, and refers to the confused nomenclatural history of the species and to the fact that the subspecies was not described until 84 years after the type series was collected in 1892. Some of the specimens in the Museu de Zoología, Universidade de São Paulo (MZUSP), labeled as collected in 1896 by Bricego in Bahia (= Salvador), Brazil and reported by Lima (1926:57) as *T. tricolor*, have been reidentified by Gregorin et al. (2006) as *T. discifera*. D. C. Carter and Dolan (1978) gave information on and measurements of the lectotype of *Hyonycteris discifera* Lichtenstein and Peters.

Thyroptera lavali Pine, 1993
LaVal's Disk-winged Bat

SYNONYMS:

Thyroptera lavali Pine, 1993a:213; type locality "'PERU: Loreto, Rio Yavari Mirim, Q[uebrada]. Esperanza 200 m. alt.'"

Thyroptera robusta Czaplewski, 1996b:153; unavailable name, published as a synonym (ICZN 1985: Art. 11[(e)(ii)]).

Thyroptera robusta Czaplewski, 1997:419; type locality "San Nicolás, Duke Locality 22," Honda Group, Villavieja Formation (Miocene), Monkey Beds, upper Magdalena River Valley, Huila, Colombia.

Thyroptera laveli Eisenberg and Redford, 1999:191; incorrect subsequent spelling of *Thyroptera lavali* Pine.

DISTRIBUTION: *Thyroptera lavali* is known from Peru, Brazil, Ecuador, and Venezuela. The species also is known from Miocene fossil remains in Colombia (Czaplewski 1996; Czaplewski et al. 2003).

MARGINAL LOCALITIES (Map 221): VENEZUELA: Anzoátegui, Río Morichal Largo (Linares 1998). BRAZIL: Pará, Alter do Chão (Graciolli and Bernard 2002). PERU: Loreto, Quebrada Esperanza (type locality of *Thyroptera lavali* Pine); Madre de Dios, Maskoitania (S. Solari et al. 2004); Cusco, Cashiriari-3 (S. Solari et al. 2004); Ucayali, Bosque Nacional Alexander von Humboldt (S. Solari, Pacheco, and Vivar 1999). ECUADOR: Orellana, 42 km S and 1 km E of Pompeya Sur (F. A. Reid, Engstrom, and Lim 2000).

SUBSPECIES: I am treating *T. lavali* as monotypic.

NATURAL HISTORY: Three females were caught in low-level mistnets in and around a *Mauritia* palm swamp in eastern Ecuador (F. A. Reid, Engstrom, and Lim 2000). One was a lactating female taken on 8 January; another

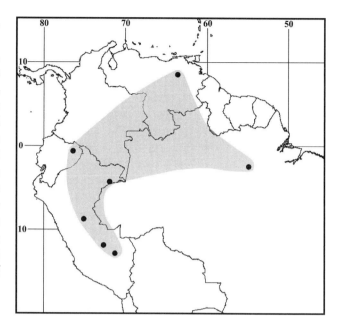

Map 221 Marginal localities for *Thyroptera lavali* ●

lactating female was netted on 19 February; and the third was pregnant with a single fetus (13 mm CR) when caught on 11 October. S. Solari, Pacheco, and Vivar (1999) captured a female carrying one nonflying young (forearm 18 mm) from a palm, in which the two were roosting more than 5 m above the ground. S. Solari et al. (2004) reported nycteribiid batflies of the genus *Hershkovitzia* along with macronyssid mites recovered from *T. lavali* in southeastern Peru.

REMARKS: Pine (1993a) reported four specimens, all taken on 16 September 1957 by Celestino Kalinowski at the type locality. Linares (1998) reported another specimen from northeastern Venezuela, and F. A. Reid, Engstrom, and Lim (2000) reported three from eastern Ecuador. S. Solari et al. (2004) reported on three collected in southeastern Peru and a fourth from Centro de Investigaciones Jennaro Herrera, Loreto, Peru, that had been misidentified as *T. discifera*. Pine (1993a) described *T. lavali* as larger and more robust than either previously known congener, and with shorter "more plush-like" fur. Among other distinctive features, *T. lavali* lacks the pale venter of *T. tricolor*, and the hairier uropatagium and forearms of *T. discifera*. The third lower incisor is unicuspidate in contrast to the obviously tricuspidate third lower incisors of *T. tricolor* and *T. discifera*. *Thyroptera lavali* also has a longer free portion of the tail (equaling as much as a fifth of total tail length) than do either *T. tricolor* or *T. discifera*. F. A. Reid, Engstrom, and Lim (2000) confirmed Pine's (1993a) observation that the suction disk on each wrist is oval, instead of nearly round as in the other two species.

Eisenberg and Redford (1999) claimed that *T. lavali* is known from several localities in Loreto, Peru. They may have been confused by Pine's (1993a:213–214) six iterations of the type locality, which is the only Peruvian locality that would have been known to them at that time. A second collection site (Bosque Nacional Alexander von Humboldt, departamento Ucayali) was reported by S. Solari, Pacheco, and Vivar (1999). Two additional localities from southeastern Peru, one based on a specimen previously reported as *T. discifera*, were discussed by S. Solari et al. (2004), who also provided a review of the species.

Apparently, when Czaplewski (1996) submitted his note placing *T. robusta* in the synonymy of *T. lavali*, he assumed that the original description of *T. robusta* would be published in 1996 before the appearance of the correction. The original description, however, was not published until 1997. While some readers might be inclined to date *T. robusta* from Czaplewski's note in *Mammalia* published June 28, 1996, because he provided the information required to make a new species-group available, Czaplewski did not state that he was proposing *T. robusta* as a new species. Instead, he clearly stated that *T. robusta* was a synonym of *T. lavali*, and Article 11(e)(ii) of the International Code of Zoological Nomenclature (ICZN 1985) in effect at that time, states "A name first published as a junior synonym after 1960 cannot be made available from that nomenclatural act."

Thyroptera tricolor Spix, 1823
Spix's Disk-winged Bat

SYNONYMS: See under Subspecies.

DISTRIBUTION: *Thyroptera tricolor* occurs from southern México through Central America and across South America, including the island of Trinidad, south into northern Bolivia and southeastern Brazil.

MARGINAL LOCALITIES (Map 222): COLOMBIA: Magdalena, Cacagualito (J. A. Allen 1900a). VENEZUELA: Mérida, Río Quebrada de Piedras, 4 km SSE of Nueva Bolivia (Soriano, Ruiz, and Zambrano 2005); Amazonas, Capibara (Handley 1976); Amazonas, Belén (Handley 1976); Bolívar, Parque Nacional Canaima (Ochoa, Molina, and Giner 1993); Bolívar, El Manaco, 59 km SE of El Dorado (Handley 1976). TRINIDAD AND TOBAGO: Trinidad, Las Cuevas (C. H. Carter et al. 1981). VENEZUELA: Delta Amacuro, Río Ibaruma, 38 km SE of Curiapo (Rivas 2000). GUYANA (Gregorin et al. 2006): Barimi-Waini, Baramita; Demerara-Mahaica, Loo Creek. SURINAM: Paramaribo, Paramaribo (Husson 1962). FRENCH GUIANA: Sinnamary (Brosset and Charles-Dominique 1991). BRAZIL: Amapá, Oiapoque, (Carvalho 1962); Pará, Abaetetuba (Piccinini 1974); Rio de Janeiro, Angra dos Reis (MZUSP 6585); São Paulo,

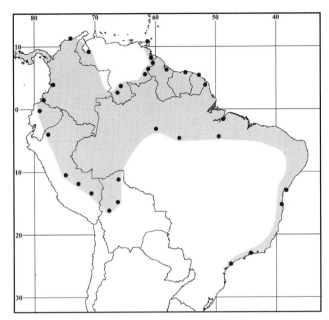

Map 222 Marginal localities for *Thyroptera tricolor* ●

Fazenda Poço Grande (type locality of *Thyroptera tricolor juquiaensis* C. O. C. Vieira); Pará, S end Ilha Tocantins (Pine 1993a); Pará, Uruá, Rio Tapajóz (B. V. Peterson and Lacey 1985); Amazonas, Dimona Reserve (Sampaio et al. 2003). BOLIVIA: Beni, Tumichucua (S. Anderson and Webster 1983); Beni, 500 m E of Campamento El Trapiche (S. Anderson 1997); La Paz, Chijchipa (S. Anderson 1997). PERU: Cusco, Hacienda Cadena (Sanborn 1951b); Cusco, Cashiriari-2 (S. Solari et al. 2001c); Pasco, San Juan (Tuttle 1970); Amazonas, La Poza (J. L. Patton, Berlin, and Berlin 1982). ECUADOR: Pichincha, Santo Domingo (Albuja 1983). COLOMBIA: Nariño, La Guayacana (Pine 1993a); Valle del Cauca, San José (J. A. Allen 1912).

SUBSPECIES: I recognize three subspecies of *T. tricolor*.

T. t. albiventer (Tomes, 1856)

SYNONYMS:

Hyonycteris albiventer Tomes, 1856:179; type locality "River Napo, near Quito," Napo, Ecuador.

Thy[roptera]. bicolor: Jiménez de la Espada, 1870:753; not *Thyroptera bicolor* Cantraine, 1845.

T[hyroptera]. albiventer: Miller, 1896:109; name combination.

Thyroptera tricolor albigula G. M. Allen, 1923b:1; type locality "Gutierrez in the mountains about twenty-five miles inland from Chiriquito," Chiriquí, Panama.

Thyroptera tricolor albiventris Dunn, 1931:430; *lapsus* for *Thyroptera tricolor albigula* G. M. Allen.

This subspecies is found in the lowlands of Colombia and Ecuador. Elsewhere, it occurs through Central America northward into southern Mexico.

T. t. juquiaensis C. O. C. Vieira, 1942

SYNONYM:

Thyroptera tricolor juquiaensis C. O. C. Vieira, 1942:391; type locality "Fazenda Poço Grande, rio Juquiá, município de Iguape, Estado de São Paulo," Brazil. This subspecies is known only by the holotype.

T. t. tricolor Spix, 1823

SYNONYMS:

Thyroptera tricolor Spix, 1823:61; type locality "ad littora fluminis Amazonum," Brazil; restricted to the "lower Amazon River below Santo Antônio Dolçá at the mouth of the Rio Icá," by Husson (1962:204).

M[olossus]? tricolor J. B. Fischer, 1829:97; name combination.

Dysopes thyropterus Schinz, 1844:148; type locality "ad flumen Amazonum."

Thyroptera bicolor Cantraine, 1845:489; type locality "Surinam."

The nominate subspecies inhabits the Amazon basin of Brazil and Peru, and the Guianan Shield of Venezuela, Guyana, Surinam, and French Guiana.

NATURAL HISTORY: The suction cups are used to hold the bats in place inside the rolled (furled) leaves of the *Heliconia* spp. and *Calathea* spp. they favor as roosting sites. Although Riskin and Fenton (2002) demonstrated how the thumb claws can be used to attach to rough surfaces, they concluded that the suction disks are used to adhere to smooth surfaces primarily by wet adhesion. Carvalho (1940) illustrated these bats roosting in the leaves of *Heliconia* sp. and *Ravenala* sp. near Piratúba, Pará, Brazil. Simmons and Voss (1998) reported finding three roosts in the scrolled dead leaves of *Phenakospermum guyannensis*. They also provided information on other roosting sites along with measurements of their 19 vouchers collected at Paracou, French Guiana. *Thyroptera tricolor* is the type host of the nycteribiid bat fly *Hershkovitzia cabala* B. V. Peterson and Lacey, 1985. Findley and Wilson (1974) studied the ecology of *T. tricolor* in Costa Rica, and Wilson and Findley (1977) summarized knowledge of its biology in their *Mammalian Species* account. *Thyroptera tricolor* has a 2*n* = 40, FN = 38 karyotype (R. J. Baker 1970).

REMARKS: The synonymies used here are adapted from Wilson and Findley (1977). The infraspecific classification is somewhat uncertain owing to some doubt concerning the validity of *T. t. juquiaensis*. Subspecific boundaries are not well established for the other two subspecies and Simmons and Voss (1998) suggested that subspecific names not be used because of the uncertainty of the identities of specimens assigned to these subspecies. The word *tricolor* refers to the pelage color, *albiventer* refers to the white venter of

that subspecies, and *juquiaensis* is named for the Rio Juquiá in Brazil. D. C. Carter and Dolan (1978) gave information on and measurements of the holotype of *Thyroptera bicolor* Cantraine. Some of the specimens from Baía (= Salvador, Bahia) reported by C. O. C. Vieira (1942) as *T. tricolor* were recently reidentified as *T. discifera* by Gregorin et al. (2006).

Family Natalidae Gray, 1866
Alfred L. Gardner

Natalidae is a Neotropical family of three genera and six currently recognized species known from northern Mexico and the West Indies south into Colombia, Venezuela, the Guianas, Brazil, and Paraguay (Simmons 2005; Tejedor 2005). Greatest diversity is in the West Indies where two genera are endemic and the family is represented in Pleistocene and Recent deposits (Koopman and Williams 1951; Koopman and Ruibal 1955; Koopman, Hecht, and Ledecky-Janecek 1957). Morgan and Czaplewski (2003) reported fossil remains from the Oligocene and Miocene of Florida, USA. Two species are found in South America, where the family is known from Pleistocene and Recent cave deposits in Brazil (Winge 1892). Natalids are small delicate bats structurally similar to some Old World vespertilionids (genus *Kerivoula*); however, Koopman and Cockrum (1967) considered the Natalidae most closely related to the New World families Furipteridae and Thyropteridae.

The Natalidae are characterized by a short thumb, the base of which is encased in the antebrachium; the lack of adhesive disks on wrist and foot; a cartilaginous third phalanx of digit III, except at its base; a long tail that is enclosed in the uropatagium; and a glandular structure (natalid organ) located on the forehead in adult males. The trochiter is nearly as large as the trochin and projects beyond the head of the humerus. The articulation facet of the trochiter with the scapula is approximately 1/2 the area of the glenoid fossa. The relatively large presternum is as wide as the combined lengths of the presternum and mesosternum. The last thoracic and all but last two lumbar vertebrae are fused. The skull lacks postorbital processes, and has complete and medially fused premaxillaries. Teeth are normal, and a baculum is present. The dental formula 2/3, 1/1, 3/3, 3/3 × 2 = 38.

More comprehensive descriptions can be found in Miller (1907b), Dalquest (1950b), Koopman and Cockrum (1967), Koopman (1984), and Morgan and Czaplewski (2003). Miller (1899c) reviewed the history and characteristics of the Natalidae, in which he included taxa now in the families Furipteridae and Thyropteridae. The most recent review (Tejedor 2005) includes the description of a new species from Mexico.

Genus *Natalus* Gray, 1838

Members of the genus *Natalus* are small (total length usually less than 110 mm; forearm, 27–45 mm), slender bats characterized by long wings, legs, and tail (legs and tail each approximately equal to or longer than head and body); large, funnel-shaped ears; small, comparatively inconspicuous eyes; and long, soft, and lax fur that varies from gray to yellowish and reddish buff or to chestnut brown. The skull is delicate with a globular braincase, a low sagittal crest, and a narrow, elongate, and somewhat tubular rostrum.

SYNONYMS:

Natalus Gray, 1838b:496; type species *Natalus stramineus* Gray, 1838b, by monotypy.

Vespertilio: P. Gervais, 1837:253; part; not *Vespertilio* Linnaeus, 1758.

Spectrellum P. Gervais, 1856a:51; type species *Spectrellum macrourum* P. Gervais, 1856a, by monotypy.

Nyctiellus P. Gervais, 1856a:84; type species *Vespertilio lepidus* P. Gervais, 1837, by monotypy.

Nycticellus Gray, 1866a:91; incorrect subsequent spelling of *Nyctiellus* P. Gervais.

Natalis Winge, 1892:36; unjustified emendation of *Natalus* Gray; junior homonym of *Natalis* Laporte, 1836 (Coleoptera).

Phodotes Miller, 1906b:85; type species *Natalus tumidirostris* Miller, 1900d, by original designation.

REMARKS: Miller (1907b) recognized four genera in the Natalidae (*Chilonatalus*, *Natalus*, *Nyctiellus*, and *Phodotes*). Until recently, arrangement of these taxa dates from Dalquest (1950b), who subsumed all four under *Natalus* and recognized three subgenera: *Natalus* (*Phodotes* a synonym), *Chilonatalus*, and *Nyctiellus*. Recently, Morgan and Czaplewski (2003), in their analysis of fossil and living forms, recommended treating these subgenera as full genera. South American species are representatives of *Natalus* in its restricted sense.

KEY TO THE SOUTH AMERICAN SPECIES OF *NATALUS*.

1. Rostrum swollen, overhanging molars, and usually hiding molars when viewed from above; palate deeply emarginate (to level of molars), its posterior border irregular and not well defined . . . . *Natalus tumidirostris*
1'. Rostrum not conspicuously swollen, its dorsal profile straight with upper surfaces of maxillae and premaxillae on same level and no hump evident on nasal behind external nares; palate not deeply emarginate, its poste-

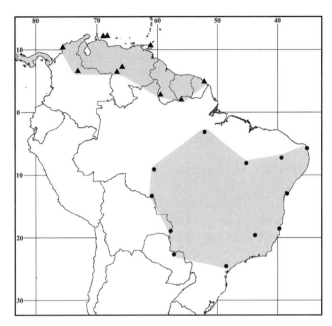

Map 223 Marginal localities for *Natalus stramineus* ● and *Natalus tumidirostris* ▲

rior border well defined and terminating well behind M3 . . . . *Natalus stramineus*

Natalus stramineus Gray, 1838
Gray's Funnel-eared Bat

SYNONYMS:

Natalus stramineus Gray, 1838b:496; type locality unknown; restricted to Lagoa Santa, Minas Gerais, Brazil by Cabrera (1958:95); however, not Brazil according to Goodwin (1959b) and Handley and Gardner (1990), but probably Antigua, British West Indies, as secondarily restricted by Goodwin (1959b:2; also see Remarks).

Vespertilio splendidus J. A. Wagner, 1845:148; type locality "Insel St. Thomas."

Natalis stramineus: Winge, 1892:13; name combination.

Natalus mexicanus Miller, 1902b:399; type locality "Santa Anita," Baja California Sur, Mexico.

Spectrellum stramineum: Elliot, 1905b:501; name combination.

Natalus dominicensis Shamel, 1928:67; type locality "Island of Dominica," Lesser Antilles.

Additional synonyms given under Subspecies.

DISTRIBUTION: The South American distribution is restricted to Brazil, Paraguay, and eastern Bolivia. Elsewhere, *N. stramineus* occurs in Mexico, Central America, Greater Antilles, and northern Lesser Antilles.

MARGINAL LOCALITIES (Map 223): BRAZIL: Pará, Caverna Planaltina (Trajano and Moreira 1991); Piauí, São João do Piauí (Taddei and Uieda 2001); Ceará,

Floresta Naçional Araripe-Apodí, 8 km S of Crato (Mares et al. 1981); Rio Grande do Norte, Natal (type locality of *N. s. natalensis* Goodwin); Bahia, Bahia (P. Gervais 1856a); Espírito Santo, Gruta do Rio Itaúnas (type locality of *Myotis espiritosantensis* Ruschi); Minas Gerais, Lagoa Santa (Winge 1892); São Paulo, Iporanga (Trajano 1982). PARAGUAY: Concepción, Parque Nacional Serranía de San Luis (López-González et al. 1998). BOLIVIA: Santa Cruz, Puerto Suarez (Taddei and Uieda 2001); Santa Cruz, Flor de Oro (Emmons 1998). BRAZIL: Mato Grosso, Aripuanã (Mok et al. 1982b).

SUBSPECIES: Only one subspecies of *N. stramineus* is currently recognized in South America; the species needs to be revised (Taddei and Uieda 2001).

N. s. macrourus P. Gervais, 1856

SYNONYMS:

Spectrellum macrourum P. Gervais, 1856a:51; no locality given; type locality later identified as Bahia, Brazil (P. Gervais 1856c:550).

Spectrellum macrurum P. Gervais, 1856c:550; incorrect subsequent spelling of *Spectrellum macrourum* P. Gervais.

Natalus stramineus natalensis Goodwin, 1959b:5; type locality "Natal, Rio Grande do Norte, Brazil."

Myotis espiritosantensis Ruschi, 1951d:7; type locality "Gruta do Rio Itaúnas, Municipio de Conceiçao da Barra," Espírito Santo, Brazil.

Natalus espiritosantensis: Ruschi, 1970:5; name combination.

This subspecies is known only from Brazil, Paraguay, and eastern Bolivia.

NATURAL HISTORY: *Natalus stramineus* is an agile, slow-flying insectivore that is rarely caught in nets, although it may be locally abundant based on numbers found in day roosts. The species roosts in wet caves, tunnels, and in abandoned buildings. Harrison and Pendleton (1973) described the baculum and palatal rugae of Central American *N. stramineus* and compared these structures with those of *N. (Chilonatalus) micropus* from Jamaica. R. Guerrero (1997 listed 11 species in 4 genera of streblid batflies that have been reported from *N. stramineus*. *Natalus stramineus* is the type host of *Natalimyobia handleyi* Uchikawa, 1988. The karyotype is $2n = 36$, FN = 56 (R. J. Baker 1970).

REMARKS: Dilford. C. Carter and Dolan (1978) stated that Lagoa Santa, Minas Gerais, Brazil, was the type locality of *N. stramineus* Gray, 1838b (as restricted by Cabrera 1958), not Antigua as restricted by Goodwin (1959b). Hall (1981) followed Goodwin's designation of Antigua; but Honacki, Kinman, and Koeppl (1982) followed D. C. Carter and Dolan (1978) and used Cabrera's (1958) restriction. According to Handley and Gardner (1990), how-

ever, data in the Natural History Museum, London associated with the type of *N. stramineus* is misleading because this specimen has been relabeled at least twice and information on its label has been altered in the process. Handley and Gardner (1990) reviewed the specimens in the British Museum along with related published and unpublished information in their investigations on the type of *N. stramineus* and its probable type locality. They concluded the following: Gray (1838b) had only one specimen of unknown origin when he named the taxon; the place-name Brazil on the current specimen label was the result of relabeling; and that the holotype most likely came from the Lesser Antilles (restricted type locality Antigua; Goodwin 1959b).

I am not aware of any South American mainland populations clearly assignable to *N. stramineus* except those in Brazil, Paraguay, and eastern Bolivia. Cuervo-Días, Hernández-Camacho, and Cadena (1986) and Alberico et al. (2000) listed *N. stramineus* for Colombia, but indicated neither specimens nor localities. Their listing of *Natalus micropus* Dobson, 1880a, with type locality "Environs of Kingston, Jamaica," is based on J. A. Allen's (1890) identification of *Natalus* from Isla de Providencia (which is extralimital). The Providencia population is known today as *Chilonatalus micropus brevimanus* Miller, 1898 (see Simmons 2005). Muñoz (2001) stated that *N. stramineus* was known from Cartagena, Bolívar, Colombia; that information needs to be confirmed.

The name *N. stramineus tronchonii* Linares, 1971, represents *N. tumidirostris*. Goodwin (1959b) reidentified Sanborn's (1941) two specimens of *N. stramineus* from the island of Trinidad as *N. tumidirostris*. Goodwin (1959b) also pointed out that the specimen from British Guiana (Guyana) identified as *N. stramineus* by Beebe (1919:219), Jentink (1893:79), and Young (1896:44) is a *Furipterus horrens*. Hall (1981) recognized six subspecies of *N. stramineus*, all extralimital to South America. Cabrera (1958) recognized the nominate subspecies (*N. s. stramineus*), but gave an incorrect distribution. Goodwin (1959b) provided the name *Natalus stramineus natalensis* for the South American population, apparently unaware that the name *Spectrellum macrourum* P. Gervais, 1856a, already was available. D. C. Carter and Dolan (1978) provided information from their examination of the holotype of *Natalus stramineus* Gray.

Natalus tumidirostris Miller, 1900
Miller's Funnel-eared Bat

SYNONYMS: See under Subspecies.

DISTRIBUTION: *Natalus tumidirostris* occurs in Colombia, Venezuela, Guyana, Surinam, French Guiana, the Netherlands Antilles, and the island of Trinidad.

MARGINAL LOCALITIES (Map 223): NETHER-LANDS ANTILLES: Curaçao, Hatto (type locality of *N. tumidirostris* Miller); Bonaire, 8.5 km N and 2 km W of Kralendijk (Genoways and Williams 1979a). TRINIDAD AND TOBAGO: Trinidad, Platanal (Goodwin 1959b). GUYANA: Upper Takutu-Upper Essequibo, vicinity of Dadanawa (Emmons 1993). FRENCH GUIANA: Montjoly (Brosset et al. 1996b). SURINAM: Sipaliwini, Sipaliwini airstrip (S. L. Williams, Genoways, and Groen 1983). VENEZUELA: Bolívar, Hato la Florida, 47 km SE of Caicara (Handley 1976); Bolívar, Mina de bauxita de la Serranía de los Pijiguaos (Ochoa et al. 1988). COLOMBIA: Santander, Macaregua Cave (Goodwin 1959b); Bolívar, Cartagena (USNM 433369).

SUBSPECIES: I recognize two subspecies of *N. tumidirostris*.

N. t. continentis (O. Thomas, 1911)

SYNONYMS:

Phodotes tumidirostris continentis O. Thomas, 1911c:513; type locality "San Esteban, Carabobo," Venezuela.

Natalus tumidirostris continentis: Goodwin, 1959b:11; name combination.

Natalus (Natalus) tumidirostris continentis: Aellen, 1970:23; name combination.

Natalus tumidirostris haymani Goodwin, 1959b:12; type locality "cave on the northwest side of Mt. Tamana, Trinidad," Trinidad and Tobago.

Natalus stramineus tronchonii Linares, 1971:81; type locality "Guavilanes cave..., Guasare river, Zulia, Venezuela."

This subspecies has been recorded from Colombia, Venezuela, the island of Trinidad, Guyana, Surinam, and French Guiana.

N. t. tumidirostris Miller, 1900

SYNONYMS:

Natalus tumidirostris Miller, 1900d:160; type locality "cave at Hatto, on north side of island of Curaçao," Netherlands Antilles.

Phodotes tumidirostris: Miller, 1906b:85; name combination.

Natalus tumidirostris tumidirostris: Goodwin, 1959b:11; name combination.

Natalus tumirostris Bisbal, 1998:179; incorrect subsequent spelling of *Natalus tumidirostris* Miller.

The nominate subspecies is known from only the Netherlands Antillean islands of Curaçao and Bonaire.

NATURAL HISTORY: Like other congeners, *N. tumidirostris* is most often encountered in caves or other day roosts. Eleven of the 160 Venezuelan specimens reported by Handley (1976) were netted over streams and ponds.

These were the only ones caught outside of caves. The only specimen known from Bonaire, Netherlands Antilles, was netted over water in otherwise xeric habitat (Genoways and Williams 1979a). Also, the single recorded specimen from Surinam was netted on a trail through tropical humid forest (S. L. Williams, Genoways, and Groen 1983). The record from French Guiana (Brosset et al. 1996b) also is of a single specimen collected in secondary forest. One reason these small insectivorous bats are rarely netted is that they are agile, relatively slow fliers.

Four females captured 19 April at Hato Florida, Bolívar, Venezuela, were pregnant, each with a single fetus (16–20 mm CR). Females from Venezuela in January and July showed no visible signs of reproductive activity. Testes of two males measured 3.0 × 2.5 mm on 9 January, whereas those of another on the same date measured 1.5 × 1.5 mm, which are the approximate testis measurements of males taken at other times of the year in Venezuela. Three lactating females were recorded on 6 August in the island of Trinidad (C. H. Carter et al. 1981); however, 3 females on 2 August and 39 on 6 August were neither lactating nor pregnant. Grose and Marinkelle (1968) found a yeast-like fungus (*Candida chiropterorum*) in the tissues of Colombian *N. tumidirostris*. *Natalus tumidirostris* has a $2n = 36$, $FN = 56$ karyotype (R. J. Baker and Jordan 1970; Linares and Löbig 1973).

REMARKS: Linares (1971) and Linares and Löbig. (1973) questioned the status of species included by Goodwin (1959b) in the subgenus *Natalus* and suggested that all named forms represent variation in a single species, *N. stramineus*. Although these suggestions have merit, the material at hand in the mammal collections of the USNM suggests that synonymizing *N. tumidirostris* with *N. stramineus* would be premature. Linares (1986 1998) recognized *N. stramineus* as the only *Natalus* in Venezuela. Although early reports by Beebe (1919:219), Jentink (1893:79), and Young (1896:44) of *Natalus* from Guyana proved to represent a *Furipterus horrens*, *N. tumidirostris* has been indicated in lists of the Guyanan fauna by Emmons (1993), Engstrom and Lim (2002), Lim, Engstrom, and Ochoa (2005) based on specimens in the ROM. D. C. Carter and Dolan (1978) provided information and measurements on the holotype of *Phodotes tumidirostris continentis* O. Thomas.

Family Molossidae P. Gervais, 1856
Judith L. Eger

As construed here, Western Hemisphere Molossidae P. Gervais (1862a) comprise two subfamilies: the monotypic South American endemic Tomopeatinae Miller, 1900f, and the highly diverse, nearly cosmopolitan Molossinae P. Gervais, 1856a. Unifying features include a humerus with trochiter much larger than the trochin; articular surface of trochin-scapula articulation nearly the same size as the glenoid fossa; the seventh cervical vertebra is fused with the first thoracic; the skull lacks postorbital processes, the antitragus is usually large, the tragus is reduced, and the ear lacks a basal lobe (Miller 1907b).

Subfamily Molossinae P. Gervais, 1856

The Molossinae, numbering at least 17 genera and 100 species, are found from southern Europe and Asia into Africa, Australia, and the Fiji Islands; and from the central United States south through Mexico, Central America, and the West Indies into the Patagonian region of Argentina and Chile. Ten genera occur in South America, where three of them (*Cabreramops*, *Molossops*, and *Neoplatymops*) and 13 species are endemic. The known fossil record dates from the late Eocene of Europe, the early Miocene of Africa, the late Oligocene, Miocene, and Pleistocene of South America, and from the Oligocene to the Pleistocene of North America (Legendre 1984c; Arroyo-Cabrales et al. 2002; Czaplewski 1996a, 1997; Czaplewski and Cartelle 1998; Czaplewski, Morgan, and Naeher 2003; Czaplewski et al. 2003). To date, four paleo-species of Molossinae have been recorded from the Oligocene and Miocene of South America: *Mormopterus faustoi*, *M. colombiensis*, *Eumops* sp., and *Potamops mascahehenes*. Four Recent species also have been recorded from the Quaternary of Brazil (see species accounts). The Molossinae probably originated in the Old World and may have invaded South America twice (Legendre 1984c).

Molossines have characteristically long and narrow wings, tough and leathery wing and tail membranes, and a thick tail that extends well beyond the posterior margin of the interfemoral membrane. The legs are short and robust, and the short, broad feet have long sensory hairs. The tragus is reduced, the antitragus usually large, and several species have a narrow, longitudinal, and downward-projecting "keel" on the inner fold of the ear. Many species have throat glands, which are more conspicuous in males than in females. These glands are assumed to be important in territoriality and other behaviors. Many species are sexually dimorphic, with males larger than females. Molossines are fast, strong flyers and insectivorous, catching insects on the wing. Miller (1907b) described many skeletal characters of the subfamily, which he treated as a family. For the most part, taxonomic listings follow Corbet and Hill (1991) and Koopman (1993). Based on a phenetic analysis of external and skull characters, Freeman

(1981) recognized two groups of molossines: the *Mormopterus* group with *Mormopterus* (including *Sauromys* and *Platymops*), *Molossops* (including *Cynomops* and *Neoplatymops*), *Myopterus*, and *Cheiromeles*; and the *Tadarida* group with *Tadarida*, *Chaerephon*, *Eumops*, *Molossus*, *Mops*, *Nyctinomops*, *Otomops*, and *Promops*. Legendre (1984a), in his study of the dental morphology of *Tadarida*, divided the restricted family Molossidae into three subfamilies: Molossinae, which included *Molossus*, *Eumops*, *Molossops*, *Cynomops*, *Neoplatymops*, *Myopterus*, and *Promops*; Cheiromelinae, which included *Cheiromeles*; and Tadaridinae, which included *Tadarida*, *Mormopterus*, *Nyctinomops*, *Otomops*, and *Rhizomops*. Legendre's subfamilies can be considered tribes in today's broader concept of the Molossidae. Sudman, Barkley, and Hafner (1994), in their study of the familial affinity of *Tomopeas ravus* (Vespertilionidae or Molossidae), found support for the part of Legendre's arrangement that grouped *Nyctinomops* and *Mormopterus* in a clade separate from *Eumops*, *Promops*, and *Molossus*.

Locality records plotted on the distribution maps in the following accounts are based primarily on specimens examined in the following institutions: American Museum of Natural History (AMNH), Institut Royal des Sciences Naturelles de Belgique (IRSNB), Michigan State University (MSU), Museum of Comparative Zoology, Harvard University (MCZ), Museum of Southwestern Biology, University of New Mexico (MSB), Natural History Museum, London (BM), Field Museum (FMNH), National Museum of Natural History (USNM), Riksmuseum van Naturlijke Historie, Leiden (RMNH), Texas A&M University (TCWC), University of Kansas (KU), Royal Ontario Museum (ROM), Zoologisches Museum der Humboldt-Universität zu Berlin (ZMB), and Zoologisches Staatssammlung, München (ZSM). I wish to express my appreciation to the many individuals representing these institutions, who have provided assistance over the years, and to M. Hafner, Louisiana State University, who kindly sent two loans. I am grateful to the late Professor R. L. Peterson whose many contributions, files, and notes about molossines greatly facilitated preparation of this account. Thanks also to B. Lim who provided hours of research, to A. L. Gardner for assistance with the taxonomy, to Barbarann Ruddell and Lorelie Mitchell for their support, and to the Library staff of the ROM for assistance with the literature. My research at the AMNH was partially funded by a Theodore Roosevelt Memorial Fund grant. Funding for research at other institutions was provided by the Royal Ontario Museum. This is contribution number 257 of the Centre for Biodiversity and Conservation Biology at the Royal Ontario Museum.

KEY TO THE SOUTH AMERICAN GENERA OF MOLOSSINAE:

1. Inner margins of ears separate, not arising from the same point on forehead, and not joined by a low band 2

1′. Inner margins of ears either arising from the same point on forehead, or joined by a low band 6

2. Anterior palate emarginate; antitragus low, wider (anteriorly-posteriorly) than high; length of third commissure of M3 equal to or longer than second . . . . *Mormopterus*

2′. Anterior palate not emarginate; antitragus distinct, usually higher than wide; third commissure of M3 shorter than second 3

3. Granulations on forearm; skull flattened; two upper premolars; upper incisors separated by a gap . . . . *Neoplatymops*

3′. No granulations on forearm; skull not flattened; one upper premolar; upper incisors in contact with each other 4

4. Lips wrinkled; upper incisors not projecting forward; lacrimal processes relatively undeveloped; lacrimal width of rostrum slightly greater than postorbital constriction . . . . *Cabreramops*

4′. Lips smooth; upper incisors projecting forward; lacrimal processes well developed; lacrimal width of rostrum considerably greater than postorbital constriction 5

5. Tips of ears elongated and pointed; ears joined to head by a flexible fold; one lower incisor in each ramus; third commissure of M3 well developed, about 1/2 the length of the second; palate distinctly domed; second phalanx of digits III and IV equal to or longer than first; basisphenoid pits clearly developed . . . . *Molossops*

5′. Tips of ears blunt and rounded; ears join head directly (connection not a flexible fold); normally two lower incisors in each ramus; third commissure of M3 greatly reduced; palate not distinctly domed; second phalanx of digits III and IV much shorter than first; basisphenoid pits absent or weakly developed and shallow . . . . *Cynomops*

6. Upper lip with deep vertical grooves or wrinkles; anterior border of hard palate emarginated; upper incisors separated by distinct gap 7

6′. Upper lip smooth, without distinct grooves or wrinkles; anterior border of palate without emargination; upper incisors in contact 8

7. Ears do not extend as far as tip of nose when laid forward; inner edges of ears not joined; three lower incisors in each ramus; short, thick, blunt-tipped bristles on face and chin . . . . *Tadarida*

7′. Ears extend beyond tip of nose when laid forward; inner edges of ears joined; two pairs of lower incisors; facial bristles few, long, and slender . . . . *Nyctinomops*

8. Ears large, extending almost to or beyond nose when laid forward; antitragus semicircular, wider than high; P3 usually present . . . . *Eumops*

8′. Ears small; antitragus pendant, higher than wide; P3 lacking 9

9. Two lower incisors in each ramus; upper incisors project anteriorly beyond canines (procumbent); palate distinctly dome shaped . . . . *Promops*

9′. One lower incisor in each ramus; upper incisors in line with canines; palate not distinctly domed . . . . *Molossus*

Genus *Cabreramops* Ibáñez, 1981

Cabreramops, a small (forearm 31–37 mm) molossid endemic to western South America, is distinguished by the following combination of characters. It has simple ears that arise from the same point on the head and do not reach the end of the muzzle when laid forward. The tragus is wide at its base and narrow at its tip, the antitragus is squarish in outline, and the upper lip is wrinkled. The lacrimal processes are relatively undeveloped; consequently the width of the rostrum across lacrimals is only slightly greater than the postorbital constriction. The hard palate is not emarginate between premaxillae. Upper incisors are widely separated at their base and converge toward their tips, and the M3 lacks the third commissure (foregoing description based in part on Ibáñez 1981). The dental formula is 1/2, 1/1, 1/2, 3/3 × 2 = 28.

SYNONYMS:

Molossops: Cabrera, 1917:20; not *Molossops* W. Peters, 1866a.

Cabreramops Ibáñez, 1981:105; type species *Molossops aequatorianus* Cabrera, 1917, by original designation.

[*Molossops (]Cabreramops[)]*: Koopman, 1993:234; treated as a subgenus of *Molossops* W. Peters.

Cabreramops aequatorianus (Cabrera, 1917)
Cabrera's Free-tailed Bat
SYNONYMS:

Molossus nasutus: Cabrera, 1901:370; not *Molossus nasutus* Spix, 1823.

Molossops aequatorianus Cabrera, 1917:20; type locality "Babahoyo sobre el río Guayas," Los Rios, Ecuador.

Molossops (Molossops) aequatorianus: Cabrera, 1958:117; name combination.

Cabreramops aequatorianus: Ibáñez, 1981:104; generic description and first use of current name combination.

Molossops [(Cabreramops)] aequatorianus: Koopman, 1993:234; name combination.

DISTRIBUTION: *Cabreramops aequatorianus* is known from only two localities in the Pacific lowlands of Ecuador.

Map 224 Marginal localities for *Cabreramops aequatorianus* ● and *Cynomops abrasus* ▲

MARGINAL LOCALITIES (Map 224): ECUADOR: Los Rios, Babahoyo (type locality of *Molossops aequatorianus* Cabrera); Guayas, 2.4 km SE of Chongon (USNM 513508).

SUBSPECIES: *Cabreramops aequatorianus* is monotypic.

NATURAL HISTORY: The species is known from five specimens. Four females were collected from a crevice in an old tree above the Río Guayas (Cabrera 1917). The fifth, a subadult male, was netted over a shallow stream in tropical dry forest.

REMARKS: The skull of the type (MNCNM 683) has been lost, and two of the paratypes are missing and presumed lost (Ibáñez 1981). Selected measurements (mm) of the holotype and a paratype (MNCNM 682) are as follows (Ibáñez 1981), respectively: forearm 35.9, 37.6; condylobasal length 13.4, 14.1. Other measurements are given by Ibáñez (1981, 1984b). Selected measurements (mm) of USNM 513508, an immature male with open epiphyseal sutures in the cranium and wing bones are: total length 70, tail 26, foot 8, ear 14, forearm 31.7, condylobasal length 13.0, and mass 5 g

Genus *Cynomops* O. Thomas, 1920

As interpreted herein, the genus *Cynomops* comprises five species traditionally included in the genus *Molossops* W. Peters, within which *Cynomops* has been treated as either a subgenus or a synonym. *Cynomops*, however, is clearly distinguishable from *Molossops* by the following charac-

teristics: the ears are bluntly rounded and lack a flexible fold where they are joined to the head; the antitragus is about as high as wide; and the tragus is relatively small, less than a third the size of the antitragus. The second phalanx is shorter than the first in digits III and IV. There are two pairs of lower incisors; the last upper molar lacks the third commissure; and the basisphenoid pits are absent, or at best, weakly developed. The dental formula is 1/2, 1/1, 2/2, 3/3 × 2 = 30. The karyotype is $2n = 34$, FN = 60 (Linares and Kiblisky 1969; J. W. Warner et al. 1974; Gardner 1977b). In a recent molecular analysis, S. L. Peters, Lim, and Engstrom (2002) confirmed *Molossops* and *Cynomops* as monophyletic sister lineages.

SYNONYMS:

Dysopes: Temminck, 1826:232; part; not *Dysopes* Illiger, 1811.

Molossus: W. Peters, 1866a:575; part; not *Molossus* É. Geoffroy St.-Hilaire, 1805a.

Molossops W. Peters, 1866a:575; part; described as a subgenus of *Molossus* É. Geoffroy St.-Hilaire.

Myopterus: W. Peters, 1869:402; not *Myopterus* É. Geoffroy St.-Hilaire, 1818b.

Molossus: O. Thomas, 1901h:190; not *Molossus* É. Geoffroy St.-Hilaire.

Molossus: O. Thomas, 1901j:440; not *Molossus* É. Geoffroy St.-Hilaire.

Molossops: O. Thomas, 1911a:113; not *Molossops* W. Peters.

Molossops: Osgood, 1914b:183; not *Molossops* W. Peters.

Cynomops O. Thomas, 1920a:189; type species *Molossus cerastes* O. Thomas, 1901j, by original designation.

Cynops Podtiaguin, 1944:28; incorrect subsequent spelling of *Cynomops* O. Thomas.

Cinomops Esbérard and Bergallo, 2005:514; incorrect subsequent spelling of *Cynomops* O. Thomas.

KEY TO THE SOUTH AMERICAN SPECIES OF *CYNOMOPS*:
1. Size larger: forearm longer than 40 mm; condylobasal length of males more than 19 mm, of females more than 18 mm . . . . *Cynomops abrasus*
1'. Size smaller: forearm shorter than 40 mm; condylobasal length of males less than 19 mm, of females less than 18 mm 2
2. Venter (gular and midventral region) much paler than dorsum, usually whitish or pale buff; basisphenoid pits weakly developed; rostrum relatively low and short, with anterior face of lacrimal ridges (including anterior rim of orbit and groove enclosing infraorbital canal opening) sloping posteriorly; length of forearm of males 30–36 mm, of females 29–35 mm; condylobasal length of males 14.3–17.5 mm, of females 13.6–16.4 mm . . . .
. . . . *Cynomops planirostris*

2'. Venter usually uniformly dark and lacking paler or whitish gular and midventral region; basisphenoid pits absent or obsolete; rostrum relatively deep with anterior face of lacrimal ridges (and groove enclosing infraorbital canal opening) arising more steeply, almost vertical 3

3. Size small: length of forearm of a male 33 mm, of females (*n*=2) 30.3–31.2 mm; braincase relatively short and broad; condylobasal length of a male 15.9 mm, of females 15.1–15.4 mm; zygomatic breadth of females 10.5–10.7 mm . . . . *Cynomops milleri*

3'. Size larger: length of forearm of males 34.0–36.6 mm, of females 31.7–36.0 mm; condylobasal length of males 16.6–18.3 mm, of females 15.3–16.9 mm; zygomatic breadth of males 11.5–12.5 mm, of females 10.7–11.7 mm 4

4. Size smaller: length of forearm of males 34.0–35.3 mm, of females 31.7–34.9 mm; cranium relatively long and narrow; condylobasal length of males 16.6–17.8 mm, of females 15.3–16.2 mm; zygomatic breadth of males 11.5–12.1 mm, of females 10.7–11.7 mm . . . .*Cynomops paranus*

4'. Size larger: length of forearm of males 34.0–36.6 mm, of females 33.0–36.0 mm; cranium relatively broad; condylobasal length of males 17.8–18.3 mm, of females 16.2–16.9 mm; zygomatic breadth of males 12.0–12.5 mm . . . . *Cynomops greenhalli*

Cynomops abrasus (Temminck, 1826)

Cinnamon Dog-faced Bat

SYNONYMS:

Dysopes abrasus Temminck, 1826:232; type locality "parties intérieures du Brésil"; herewith restricted to Votuporanga, São Paulo, Brazil.

M[*olossus* (*Promops*)]. *abrasus*: W. Peters, 1866a:574; name combination.

M[*olossus* (*Molossops*)]. *brachymeles* W. Peters, 1866a: 575, footnote; type locality "Peru"; herewith restricted to Marcapata, Cusco, Peru.

Molossus cerastes O. Thomas, 1901j:440; type locality "Villa Rica," Guairá, Paraguay.

[*Molossus* (*Myopterus*)] *cerastes*: Trouessart, 1904:101; name combination.

Molossops cerastes: Miller, 1907b:248; name combination.

Molossops mastivus O. Thomas, 1911a:113; type locality "Bartica Grove, Lower Essequibo," Cuyuni-Mazaruni, Guyana.

Cynomops cerastes: O. Thomas, 1920a:189; generic description and name combination.

Molossops [(*Cynomops*)] *brachymeles brachymeles*: Cabrera, 1958:118; name combination.

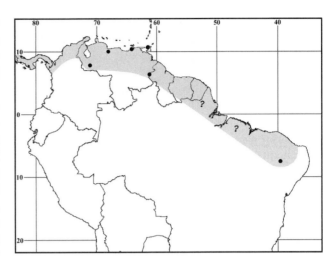

Map 225 Marginal localities for *Cynomops greenhalli* ●

Molossops [(*Cynomops*)] *brachymeles cerastes*: Cabrera, 1958:118; name combination.

Molossops [(*Cynomops*)] *brachymeles mastivus*: Cabrera, 1958:119; name combination.

Cynomops abrasus: Husson, 1962:246; first use of current name combination.

Molossops (*Cynomops*) *abrasus*: D. C. Carter and Dolan, 1978:84; name combination.

Cynomops abrasus cerastes: Barquez, Mares, and Braun, 1999:170; name combination.

DISTRIBUTION: *Cynomops abrasus* is known from northern and central South America on the eastern side of the Andes in Venezuela, Guyana, Surinam, French Guiana, Ecuador, Peru, and Brazil, south into Paraguay and northern Argentina. The only exception is Alberico and Naranjo's (1982) report of the species from southwestern Colombia, west of the Andes.

MARGINAL LOCALITIES (Map 224): GUYANA: Barima-Waini, 1.5 km E of Akwero (ROM 67507). SURINAM: Commewijne, Meerzorg (RMNH M119V232). FRENCH GUIANA: Paracou (Simmons and Voss 1998). BRAZIL: Piauí, Fazenda Olho do Aqua (Mares et al. 1981); Mato Grosso, Serra do Roncador, 280 km N by road from Xavantina (Pine, Bishop, and Jackson 1970); Rio de Janeiro, Rio de Janeiro (Esbérard and Bergallo 2005); São Paulo, São Sebastião (USNM 141441); Santa Catarina, Joinville (BM 14.1.27.1). ARGENTINA: Misiones, Tacuaruzú (Barquez, Mares, and Braun 1999). PARAGUAY: Central, San Lorenzo (USNM 461896). ARGENTINA: Santiago del Estero, Colonia Dora (Yepes 1944); Formosa, Puesto Divisadero (Barquez, Mares, and Braun 1999). BOLIVIA: Santa Cruz, 5 km S of Santa Rosa de Sara (Ibáñez 1985). PERU: Cusco, Marcapata (BM 4.12.4.3); Huánuco, 3 km N of Tingo María (TCWC 12433). ECUADOR: Morona Santiago, near Méndez (MCZ 27339); Pichincha,

Mt. Pichincha (IRSNB 9695). COLOMBIA: Valle del Cauca, Hacienda Jamaica (Alberico and Naranjo 1982). VENEZUELA: Bolívar, El Manaco, 59 km SE of El Dorado (USNM 387743).

SUBSPECIES: *Cynomops abrasus* is treated here as monotypic (see Remarks).

NATURAL HISTORY: The Venezuelan specimen was netted in tropical humid forest. One from Manaus, Brazil, was caught in the leaves of a buriti palm (*Maurita* sp.). Specimens from Paraguay were taken from the roof of a house, from a cave, but mostly from hollow trees. The most extensive report is that of Taddei, Vizotto, and Martins (1976), which includes measurements and information on taxonomy, distribution, and life history. Two males of *C. abrasus* from São Paulo, Brazil, weighed 36.4 g and 37.8 g, and two females, 24.4 g and 32.0 g (Taddei, Vizotto, and Martins 1976). Two females from Piauí, Brazil, weighed 30 g and 33 g. The mass of three males from Paraguay ranged from 35 to 44 g, while that of a female was 33 g. Other recorded weights of females are 31.9 g (Huánuco, Peru), and 26.7 g (Bolívar, Venezuela). Wingspans of 17 adults (male and females) from Paraguay ranged from 332 to 359 mm. Females are far more common in collections than males. Esbérard and Bergallo (2005) reported four females pregnant when taken in January in the state of Rio de Janeiro, and claimed that reproduction in *C. abrasus* is seasonal, at least in Brazil. J. W. Warner et al. (1974) reported a $2n = 34$, FN $= 60$ karyotype for a *C. abrasus* from Balta, Ucayali, Peru.

Remarks: Husson (1962:246) was the first to describe the skull of *Dysopes abrasus* Temminck. Although 80 specimens have been examined (45 from Paraguay, about 1/2 collected by W. T. Foster from 1900 to 1905), these are too few to adequately define the subspecific status of populations. The holotype of *Dysopes abrasus* Temminck is a young adult (see Husson 1962, pl. 26), but its pelage coloration, and external and cranial measurements, are a close match (except for a few age-related characters) for ROM 77315 from Votuporanga, São Paulo, Brazil, which place I propose as the restricted type locality. The series of specimens from São Paulo, reported by Taddei, Vizotto, and Martins (1976), also are similar to ROM 77315.

Two specimens in the BM from Marcapata, Cusco, Peru, closely match the published information for *C. brachymeles* W. Peters, the holotype of which appears to have been lost. One of these, a female (BM 4.12.4.3) represented by a skin and skull in good condition, conforms particularly closely to the original description.

A male (ROM 35637) from Machawira, Guyana, is a close match for the holotype of *C. mastivus* O. Thomas, also from Guyana. These and others from the Guianas and northern Brazil are consistently larger and darker than samples from other populations, and *mastivus* may represent a valid subspecies. Two males (one from Surinam and the other from the state of Amazonas, Brazil) each weighed 40 g. *Cynomops abrasus* is similar to both *C. brachymeles* and *C. cerastes* and I suggest that additional material will confirm their status as synonyms when additional specimens become available. D. C. Carter and Dolan (1978) provided information on and measurements of the holotypes of *Dysopes abrasus* Temminck, *Molossus cerastes* O. Thomas, and *Molossops mastivus* O. Thomas.

Cynomops greenhalli Goodwin, 1958
Greenhall's Dog-faced Bat
SYNONYMS:

Cynomops greenhalli Goodwin, 1958b:3; type locality "Royal Botanic Gardens, Port of Spain, Trinidad."
Molossops (Cynomops) greenhalli: Goodwin and Greenhall, 1961:282; name combination.

DISTRIBUTION: Although the species boundaries are not yet clearly defined, *C. greenhalli* has been found in Trinidad, Venezuela, and Brazil. Elsewhere, (depending on one's concept of the species limits in *C. greenhalli*) the species also occurs in Mexico and Central America.

MARGINAL LOCALITIES (Map 225): TRINIDAD AND TOBAGO: Trinidad, Port of Spain (type locality of *Cynomops greenhalli* Goodwin). VENEZUELA: Bolívar, El Manaco, 59 km SE of El Dorado (Handley 1976). BRAZIL: Pernambuco, Fazenda Pau Ferrado (S. L. Peters, Lim, and Engstrom 2002). VENEZUELA: Barinas, 8 km SW of Santa Bárbara (S. L. Peters, Lim, and Engstrom 2002); Carabobo, Sierra de Carabobo (BM 14.9.1.7); Sucre, Tacal (KU 119080).

SUBSPECIES: I regard *C. greenhalli* as monotypic.

NATURAL HISTORY: Goodwin (1958b) reported specimens taken from a hollow branch of a yellow Poui tree, *Tabebuia serratifolia*, in Trinidad. Venezuelan specimens came from a building (Linares and Kiblisky 1969), or were netted in tropical dry forest (Handley 1976). A female was pregnant female when taken on July 19 (Trinidad; C. H. Carter et al. 1981) and another lactating on July 28 (TCWC 22123). The karyotype is $2n = 34$, FN $= 60$ (Linares and Kiblisky 1969; J. W. Warner et al. 1974).

REMARKS: *Cynomops greenhalli* is difficult to differentiate from *C. paranus* (see S. L. Peters, Lim, and Engstrom 2002); both species occur sympatrically in Venezuela (Handley 1976; also see Remarks under *C. milleri*) and Brazil. The identifications of the *Cynomops* reported by Mares et al. (1981) should be verified. Simmons and Voss (1998) include *C. mexicanus* in this species. However, Gardner (1977b) reported differences in chromosomal morphology between *C. greenhalli* and *C. mexicanus*, and in their molecular and morphometric study of *Cynomops*, S. L. Peters,

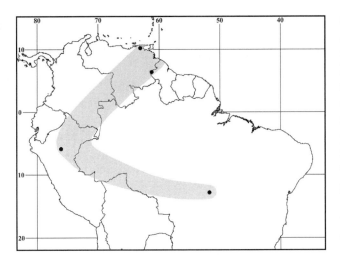

Map 226 Marginal localities for *Cynomops milleri* ●

Lim, and Engstrom (2002) found *C. mexicanus* to be the basal lineage for *Cynomops* and recognized *C. mexicanus* as a distinct species.

Cynomops milleri (Osgood, 1914)
Miller's Dog-faced Bat

SYNONYMS:

Molossops milleri Osgood, 1914b:183; type locality "Yurimaguas," Loreto, Peru.

Cynomops milleri: O. Thomas, 1920a:189; generic description and first use of current name combination.

Molossops (Cynomops) milleri: Cabrera, 1958:119; name combination.

DISTRIBUTION: *Cynomops milleri* occurs in Venezuela, Brazil, and Peru.

MARGINAL LOCALITIES (Map 226): VENEZUELA: Sucre, 10 km NW of Caripito [Monagas] (UK 119089); Bolívar, El Manaco, 59 km SE of El Dorado (USNM 387744). BRAZIL: Mato Grosso, Serra do Roncador, 264 km N of Xavantina (USNM 393769). PERU: Loreto, Yurimaguas (type locality of *Molossops milleri* Osgood).

SUBSPECIES: I consider *C. milleri* to be monotypic.

NATURAL HISTORY: The Venezuelan specimen from El Manaco was lactating when caught in humid tropical forest on 8 June; the Caripito specimen was pregnant with a 17-mm-CR fetus when taken on 8 May.

REMARKS: This species has not been reported since its original description in 1920. Koopman (1978) assumed that the holotype was a subadult and assigned it to *C. planirostris*, even though Osgood (1914b) and Sanborn (1947) had correctly aged it as adult. The only male known is USNM 393769 from Mato Grosso, Brazil, reported by Pine, Bishop, and Jackson (1970:669) as *Molossops planirostris*. Handley (1976:39) and S. L. Williams

and Genoways (1980b:491) identified a female (USNM 387744) as *Molossops paranus*, but this specimen is identified here as *C. milleri*. The record of *C. paranus* from Ecuador reported by F. A. Reid, Engstrom, and Lim (2000) and analyzed by S. L. Peters, Lim, and Engstrom (2002) is a small individual and also may represent *C. milleri*. This interpretation is consistent with the molecular results reported by S. L. Peters, Lim, and Engstrom (2002), indicating that the Ecuador specimen is basal to *C. paranus* from the Guianas. As S. L. Peters, Lim, and Engstrom (2002) suggested, morphological differences among the taxa of small *Cynomops* are subtle and larger sample sizes are needed to adequately diagnose species. *Cynomops milleri* is more closely related to *C. paranus* than to any other known taxon and appears to be broadly sympatric with it.

Cynomops paranus (O. Thomas, 1901)
Brown Dog-faced Bat

SYNONYMS:

Molossus planirostris paranus O. Thomas, 1901h:190; type locality "Para," Pará, Brazil.

[*Molossus (Myopterus) planirostris*] *paranus*: Trouessart, 1904:101; name combination.

Molossops paranus: Miller, 1907b:248; name combination.

Cynomops paranus: O. Thomas, 1920a:189; generic description and first use of current name combination.

Molossops planirostris paranus: C. O. C. Vieira, 1942:432; name combination.

Cynomops planirostris paranus: Goodwin, 1958b:5; name combination.

DISTRIBUTION: *Cynomops paranus* is recorded from Guyana, Surinam, French Guiana, Brazil, northern Argentina, and eastern Peru and Colombia. It also occurs in Central America as far north as Honduras (see Remarks).

MARGINAL LOCALITIES (Map 227): VENEZUELA: Yaracuy, El Central (USNM 374168). GUYANA: Upper Demerara-Berbice, Arampa (S. L. Peters, Lim, and Engstrom 2002). SURINAM: Paramaribo, Plantage Blauwgrond near Paramaribo (RMNH 25429). FRENCH GUIANA: Paracou (Simmons and Voss 1998). BRAZIL: Pará, Belém (type locality of *Molossus planirostris paranus* Thomas); Mato Grosso, Serra do Roncador, 280 km N of Xavantina (USNM 393769); Amazonas, Rio Negro, near Manaus (AMNH 79745). ARGENTINA: Corrientes, Laguna Paiva (Barquez, Mares, and Braun 1999). PERU: Ucayali, Balta (LSUMZ 12284); Ucayali, Bosque Nacional Alexander von Humboldt (S. Solari, Pacheco, and Vivar 1999). ECUADOR: Orellana, 38 km S of Pompeya Sur (F. A. Reid, Engstrom, and Lim 2000). COLOMBIA: Putumayo, Mocoa (ROM 41479); Norte de Santander, Cúcuta (Sanborn 1941).

SUBSPECIES: I am treating *C. paranus* as monotypic.

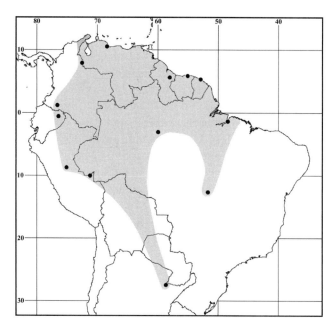

Map 227 Marginal localities for *Cynomops paranus* ●

NATURAL HISTORY: Handley (1976) reported specimens collected in dry and humid tropical forest.

REMARKS: Oldfield Thomas (1901h) originally described this taxon as a subspecies of *Molossus planirostris* W. Peters, an allocation that led to considerable subsequent confusion. Except for slightly larger size, the holotype of *Cynomops greenhalli* Goodwin, 1958b, is similar to the holotype of *Cynomops paranus* (forearm 36.0 versus 34.7 mm, greatest length of skull 19.0 versus 18.0 mm, condylobasal length 17.8 versus 16.5 mm, zygomatic breadth 12.4 versus 12.0 mm, breadth of braincase 8.7 versus 8.5 mm, and width across upper molars 7.9 versus 7.7 mm). Samples of *C. greenhalli* from northern Venezuela and Trinidad average slightly larger than *C. paranus* from the Guianas and Brazil. Larger samples might clarify the differences between these two taxa, especially where they are sympatric in Venezuela. In their molecular study, S. L. Peters, Lim, and Engstrom (2002) found *C. paranus* from Ecuador to be basal to *C. paranus* from the Guianas. Populations of *C. paranus* from western South America (Peru through Colombia and northward in Central America into Costa Rica) require further study, as does the status of *Molossops greenhalli mexicanus* J. K. Jones and Genoways, 1967. Although similar in size to the Trinidadian and Venezuelan populations, the Mexican populations share some characteristics with *C. planirostris* (see Gardner 1977b for a comparison of karyotypes). At present, it seems best to regard the Mexican population as a separate taxon, *Cynomops mexicanus*, particularly in light of the results of a molecular study in which *C. mexicanus* was determined as basal to other species of *Cynomops* (see S. L. Peters, Lim,

and Engstrom 2002). D. C. Carter and Dolan (1978) provided information on and measurements of the holotype of *Molossus planirostris paranus* O. Thomas.

Cynomops planirostris (W. Peters, 1866)
Southern Dog-faced Bat

SYNONYMS:

M[*olossus (Molossops)*]. *planirostris* W. Peters, 1866a:575, footnote; type localities "British Guiana, Barra do Rio Negro [Brazil], and Buenos Aires [Argentina]"; restricted by Cabrera (1958:119) to British Guiana, but later corrected by D. C. Carter and Dolan (1978:86) to Cayenne, French Guiana, based on information associated with the lectotype.

[*Molossus (Myopterus)*] *planirostris*: Trouessart, 1897: 142; name combination.

Molossops planirostris: Miller, 1907b:248; name combination.

Cynomops planirostris: O. Thomas, 1920a:189; generic description and first use of name combination.

Molossops [(Cynomops)] planirostris: Cabrera, 1958:119; name combination.

Molossops greenhalli: Mares, Willig, Streilein, and Lacher, 1981:112; not *Cynomops greenhalli* Goodwin, 1958.

DISTRIBUTION: *Cynomops planirostris* occurs east of the Andes from Colombia, Venezuela, and the Guianas, south through Peru and Bolivia into Paraguay and Argentina. Elsewhere, it has been recorded in Central America north into Honduras.

MARGINAL LOCALITIES (Map 228): VENEZUELA: Monagas, Hato Mata de Bejuco (USNM 496746). GUYANA:

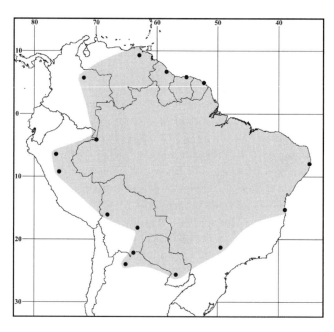

Map 228 Marginal localities for *Cynomops planirostris* ●

Essequibo Islands-West Demerara, 1 mile up Bonasika Creek (ROM 62768). SURINAM: Paramaribo, Paramaribo (RMNH 25435). FRENCH GUIANA: Cayenne (BM 5.1.8.4, topotype). BRAZIL: Pernambuco, Estacão Ecológica do Tapacurá (Mares et al. 1981, as *M. greenhalli*); Bahia, Una (McNab and Morrison 1963); São Paulo, Fazenda Esplanada (ROM 77312). PARAGUAY: Paraguarí, Sapucay (USNM 114908). ARGENTINA: Jujuy, Río Las Capillas, 15 km N Las Capillas (M. M. Díaz and Barquez 1999); Salta, Dique Itiyuro (Olrog and Barquez 1979). BOLIVIA: Santa Cruz, 10 km E of Ingeniero Mora (MSU 31015); La Paz, 0.5 km E of Saynani (S. Anderson 1997). PERU: Huánuco, Tingo María (MSB 49976); San Martín, Tarapoto (S. Solari, Pacheco, and Vivar 1999). COLOMBIA: Amazonas, Leticia (ROM 62577); Casanare, Pore (ROM 62520).

SUBSPECIES: *Cynomops planirostris* is treated here as monotypic.

NATURAL HISTORY: Occurring in a variety of habitats from open savannah and white-sand scrub, to tropical dry and tropical humid forests, *C. planirostris* has been found roosting in holes in trees, palms, and fence posts. See Vizotto and Taddei (1976) for additional natural history information. The specimen (USNM 555727) from Pernambuco, Brazil, was netted in a clearing at the edge of a remnant of Atlantic Rain Forest (Mares et al. 1981, as *Molossops greenhalli*). R. Guerrero (1985a) listed two ticks, two hemipterans, and two mites as known ectoparasites; to which he added (1997) a streblid batfly. Yunker, Lukoschus, and Giesen (1990) added three macronyssid mites. The karyotype is $2n = 34$, FN = 62; the X is a larger submetacentric and the Y is a medium-sized acrocentric chromosome (A. L. Gardner, pers. comm; specimen reported by Mares et al. [1981], from Estacão Ecológico do Tapacurá, Pernambuco, Brazil, as *Molossops greenhalli*).

REMARKS: As pointed out by Vizotto and Taddei (1976), and as also true for other species in the genus, there is significant sexual dimorphism in this species. A thorough analysis of geographic variation also is needed. Individuals of this species are much smaller in size (see measurements in key to species) than *C. abrasus*. Although similar in size to *C. milleri* and *C. paranus*, *C. planirostris* is readily distinguishable from them by its conspicuously paler venter.

Ceballos-Bendezu (1960), Koopman (1978), and others have used the name combination *Molossops planirostris paranus* for this species. Pine and Ruschi (1978) showed that *Molossops planirostris espiritosantensis* Ruschi, 1951f, is a *Molossus*, and not a *Molossops* or a *Cynomops*. D. C. Carter and Dolan (1978) provided information on and measurements of the specimen they designated as the lectotype of *Molossus (Molossops) planirostris* W. Peters.

Genus *Eumops* Miller, 1906

Eumops was revised by Eger (1977) and this account, including locality records, is based primarily on that revision. Bats of this genus differ greatly in size, with lengths of forearms ranging from 37 mm in *Eumops nanus* to 86 mm in *E. dabbenei*. Externally, *Eumops* can be distinguished from other molossid genera by its large, rounded pinnae that are joined medially on the forehead and have a greatly developed keel; the reduced and pointed or square-tipped tragus and the large, nearly oval antitragus; the smooth upper lips; and the cylindrical skull with well-developed basisphenoid pits and slightly arched palate. Dental features include the long, curved upper incisors; molars with the typical W-shaped ectoloph; and M3 with a variably developed third commissure. The dental formula is 1/2, 1/1, 2/2, 3/3 × 2 = 30; occasionally the first upper premolars is missing (premolars reduced to 1/2). The genus is known from the middle Miocene of Colombia (Czaplewski 1997).

SYNONYMS:

Vespertilio: G. Shaw, 1800:137; not *Vespertilio* Linnaeus, 1758.

Molossus É. Geoffroy St.-Hilaire, 1805a:279; part.

Molossus: Schinz, 1821:870; not *Molossus* É. Geoffroy St.-Hilaire.

Dysopes: Wied-Neuwied, 1826:226; not *Dysopes* Illiger, 1811.

Dysopes: J. A. Wagner, 1843a:367; not *Dysopes* Illiger.

Molossus: Gundlach in W. Peters, 1861:149; not *Molossus* É. Geoffroy St.-Hilaire.

Dysopes: W. Peters, 1864b:383; not *Dysopes* Illiger.

Promops: W. Peters, 1874:232; not *Promops* P. Gervais, 1856a.

Nyctinomus: H. Allen, 1889:561; not *Nyctinomus* É. Geoffroy St.-Hilaire, 1818b.

Molossus: Merriam, 1890:31; not *Molossus* É. Geoffroy St.-Hilaire.

Promops: J. A. Allen, 1900a:92; not *Promops* P. Gervais.

Promops: Miller, 1900e:471; not *Promops* P. Gervais.

Promops: O. Thomas, 1901b:190; not *Promops* P. Gervais.

Promops: J. A. Allen, 1904b:228; not *Promops* P. Gervais.

[*Molossus (| Myopterus[)]*: Trouessart, 1904:101; not *Myopterus* É. Geoffroy St.-Hilaire, 1818b.

Eumops Miller, 1906b:85; type species *Molossus californicus* Merriam, 1890, by original designation.

Molossides G. M. Allen, 1932:257; type species *Molossides floridanus* G. M. Allen, 1932, by original designation.

KEY TO THE SOUTH AMERICAN SPECIES OF *EUMOPS* (CAPITALIZED COLOR TERMS FROM RIDGWAY 1912):

1. Forearm shorter than 55 mm; greatest length of skull less than 22 mm 2

1′. Forearm longer than 55 mm, greatest length of skull more than 22 mm 7

2. A band of white hair (ca. 5 mm in width) on ventral surface of mesopatagium next to body between humerus and femur; remainder of body dark Chocolate; forearm 51–53 mm . . . . *Eumops maurus*

2′. No band of white hair on ventral surface of mesopatagium; forearm shorter than 50 mm 3

3. Greatest length of skull near 50% of forearm length; dorsal pelage short, rich Blackish Brown, ventral pelage white at base; basisphenoid pits deep and long; forearm 37–41 mm . . . . *Eumops hansae*

3′. Greatest length of skull less than 45 % of forearm length 4

4. Basisphenoid pits large, elongate, and shallow; dorsal pelage Snuff Brown to Blackish Brown, paler ventrally; forearm 44.9–47.1 mm; condylobasal length 17.3–17.8 mm; zygomatic breadth 10.6–11.0 mm . . . . *Eumops delticus*

4′. Basisphenoid pits small, oval, and shallow 5

5. Size larger: dorsal pelage Snuff Brown to Bister, ventral pelage Buffy Brown; forearm 43.5–49.0 mm; condylobasal length 18.7–19.7 mm; zygomatic breadth 11.8–12.3 mm . . . . *Eumops bonariensis*

5′. Size smaller: dorsal pelage Mummy Brown to Bister; ventral pelage paler; forearm 37.0–46.6 mm; condylobasal length 14.7–17.9 mm; zygomatic breadth 9.1–11.2 mm 6

6. Dorsal pelage Mummy Brown with gray hairs scattered throughout; forearm 40.6–46.6 mm; condylobasal length 14.9–17.9 mm; zygomatic breadth 9.7–11.2 mm; distributed south of the Amazon basin . . . . *Eumops patagonicus*

6′. Dorsal pelage Bister dorsally; forearm 37–42 mm; condylobasal length 14.7–16.5 mm; zygomatic breadth 9.1–10.0 mm; distributed north and west of the Amazon basin . . . . *Eumops nanus*

7. Ears long (35–44 mm), averaging 39.6 mm; tragus large, broad, and square; basisphenoid pits deep and elongate; mastoid breadth less than 52% of condyloincisive length 8

7′. Ears short (17–34 mm), averaging less than 34 mm; tragus small, pointed, or square; basisphenoid pits shallow; mastoid breadth greater than 52% of condyloincisive length 9

8. Third commissure of M3 1/4 the length of second; forearm 79.4 mm (73.9–83.3 mm); greatest length of skull 32.1 mm (31.5–34.4 mm); width across lacrimals 10.5 mm (9.2–11.5 mm) . . . . *Eumops perotis*

8′. Third commissure of M3 1/2 the length of second; forearm 70.9 mm (67.4–75.0 mm); greatest length of skull 28.7 mm (26.8–31.0 mm); width across lacrimals 8.1 mm (7.3–9.1 mm) . . . . *Eumops trumbulli*

9. Size larger: greatest length of skull more than 28 mm (males) or 27 mm (females); ear heavily keeled; dorsal pelage Cinnamon with buff basal band . . . . *Eumops dabbenei*

9′. Size smaller: greatest length of skull less than 28 mm (males) or 27 mm (females); ear not heavily keeled 10

10. Tragus small and pointed; dorsal pelage Blackish Brown; basisphenoid pits shallow; mastoid breadth less than 49% of greatest length of skull . . . . *Eumops auripendulus*

10′. Tragus broad and square; dorsal pelage Snuff Brown to Bister with white basal band; basisphenoid pits shallow, but well defined; mastoid breadth greater than 52% of greatest length of skull . . . . *Eumops glaucinus*

Eumops auripendulus (G. Shaw, 1800)
Black Bonneted Bat

SYNONYMS: See under subspecies.

DISTRIBUTION: *Eumops auripendulus* occurs on Trinidad and throughout northern South America southward into Bolivia, Paraguay, and northern Argentina. It is known elsewhere from Jamaica, Central America, and southern Mexico.

MARGINAL LOCALITIES (Map 229; from Eger 1977, except as noted): COLOMBIA: Magdalena, Don Diego. VENEZUELA: Carabobo, 10 km NW of Urama. TRINIDAD AND TOBAGO: Trinidad, Blanchisseuse.

Map 229 Marginal localities for *Eumops auripendulus* ●

GUYANA: Barima-Waini, Arakaka. SURINAM: Paramaribo, Paramaribo. FRENCH GUIANA: Cayenne. BRAZIL: Pará, Belém; Ceará, Baturité; Espírito Santo, Santa Teresa; São Paulo, Iporanga. ARGENTINA: Misiones, Bonpland; Santa Fe, Esperanza (Barquez, Mares, and Braun 1999); Chaco, Resistencia (Barquez, Mares, and Braun 1999). PARAGUAY: Central, Recoleta, near Asunción. BOLIVIA: Santa Cruz, 10 km E of Ingeniero Mora (S. Anderson, Koopman, and Creighton 1982); Beni, Camiaco (S. Anderson 1997); La Paz, Ixiamas. PERU: Junín, Perené; San Martín, Guayabamba (type locality of *Promops milleri* J.A. Allen); Piura, 6.4 km W of Suyo. ECUADOR: Guayas, Balzar. COLOMBIA: Nariño, Barbacoas; Antioquia, Medellín.

S U B S P E C I E S : I recognize two subspecies of *E. auripendulus* in South America.

E. a. auripendulus (G. Shaw, 1800)
S Y N O N Y M S :

Vespertilio Auripendulus G. Shaw, 1800:137; type locality "Guiana"; restricted to French Guiana by Husson (1962).

Molossus amplexicaudatus É. Geoffroy St.-Hilaire, 1805a: 279; type locality "Guiana."

Dysopes longimanus J. A. Wagner, 1843a:367; type localities "Villa Maria [Mato Grosso], Caiçara [Mato Grosso], Barra do Rio Negro [Amazonas], Brazil"; according to D. C. Carter and Dolan (1978:91), original labels of the two syntypes in the ZSM are both marked "Caicara" in Natterer's hand; therefore, I restrict the type locality to Caiçara, Mato Grosso, Brazil (see Remarks).

Dysopes leucopleura J. A. Wagner, 1843a:367; type locality "Caiçara," Mato Grosso, Brazil.

[*Molossus longimanus*] var. *leucopleura*: Pelzeln, 1883:42; name combination.

Promops milleri J. A. Allen, 1900a:92; type locality "Guayabamba," San Martín, Peru.

Promops barbatus J. A. Allen, 1904b:228; type locality "La Union," Bolívar, Venezuela.

Eumops abrasus: Miller, 1906b:85; not *Dysopes abrasus* Temminck, 1826.

Eumops milleri: Miller, 1906b:85; name combination.

Eumops abrasus milleri: Sanborn, 1932a:352; name combination.

Eumops abrasus oaxacensis Goodwin, 1956a:2; type locality "Mazatlán, about 3,000 feet elevation, District of Mixes, Oaxaca, Mexico."

Eumops maurus: Villa-R., 1956:543; not *Molossus maurus* O. Thomas, 1901g.

Eumops auripendulus: Goodwin, 1960:5; name combination.

Eumops auripendulus auripendulus: Husson, 1962:240; first use of current name combination.

Eumops auripendulus oaxacensis: Goodwin, 1969:112; name combination.

Eumops auropendulus Shapley, Wilson, Warren, and Barnett, 2005:382, 383; incorrect subsequent spelling of *Vespertilio auripendulus* G. Shaw.

This subspecies occurs in Trinidad and throughout northern South America southward into Bolivia and northwestern Argentina. It is known elsewhere from Jamaica, Central America, and southern Mexico.

E. a. major Eger, 1974
S Y N O N Y M S :

Eumops abrasus abrasus: Sanborn, 1932a, 351; not *Dysopes abrasus* Temminck, 1826.

Eumops auripendulus major Eger, 1974:2; type locality "Campo Viera, Misiones, Argentina."

This subspecies is found in eastern Brazil, southern Paraguay, and northeastern Argentina.

N A T U R A L   H I S T O R Y : *Eumops auripendulus* is found from sea level to as high as 2,000 m. Handley (1976) reported it from tropical dry forest in Venezuela. Elsewhere, the species has been collected in a variety of tropical forest habitats, including lowland rainforest in French Guiana (Brosset and Charles-Dominique 1991; Simmons and Voss 1998). These bats have been found roosting in hollow trees, walls and attics of houses, and under the corrugations of tin roofs. A female containing a 27-mm-CR fetus was taken on 26 February in the Panama Canal Zone. Of four females captured in the wall of a building in San Carlos de Río Negro, Amazonas, Venezuela, in March 1984, one was lactating and the other three (USNM 560642–560644) were pregnant, each with single fetus (CR lengths 15–37 mm). A female taken in Ecuador in February 1996 (ROM 105526) was not pregnant. In addition to the summary of parasites listed by Best et al. (2002) in their *Mammalian Species* account, R. Guerrero (1985a) listed a hemipteran and later (1997) a streblid batfly known from this species. *Eumops auripendulus* has a $2n = 42$, $FN = 62$ karyotype (J. W. Warner et al. 1974).

R E M A R K S : Étienne Geoffroy St.-Hilaire's (1805) *Molossus amplexicaudatus* was based on Buffon's (1789) description of the "Guiana" bat and, therefore, is an objective synonym of *Vespertilio auripendulus* G. Shaw, 1800. The specimen referred to as *Molossus rufus* by Temminck (1826) is a *Vespertilio auripendulus* according to Husson (1962). Four specimens identified as *Molossus nasutus* by J. A. Allen (1897b) from Guayabamba, Peru, are actually *E. auripendulus*. D. C. Carter and Dolan (1978) provided information on and measurements of two syntypes of *Dysopes longimanus* J. A. Wagner they located in the ZSM collections in Munich, Germany. I select ZSM 56, an adult female from Caiçara, Mato Grosso, Brazil, as the lectotype of *Dysopes longimanus* Wagner, 1843.

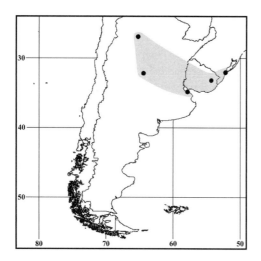

Map 230 Marginal localities for *Eumops bonariensis* ●

Eumops bonariensis (W. Peters, 1874)
Southern Bonneted Bat

SYNONYMS:

Promops bonariensis W. Peters, 1874:232; type locality "Buenos Aires," Argentina.

Molossus bonariensis: Dobson, 1876:715; name combination.

Promops bonaërensis Burmeister, 1879:89; unjustified emendation of *Promops bonariensis* W. Peters.

[*Molossus (Promops)*] *bonariensis*: Trouessart, 1897:144; name combination.

Eumops bonariensis: Miller, 1906b:85; first use of current name combination.

DISTRIBUTION: *Eumops bonariensis* is known from southeastern Brazil, Uruguay, and northern Argentina.

MARGINAL LOCALITIES (Map 230; from Eger 1977, except as noted): ARGENTINA: Tucumán, Las Talas (Barquez and Diaz 2001). BRAZIL: Rio Grande do Sul, Quinta. URUGUAY: Treinta y Tres, Treinta y Tres. ARGENTINA: Buenos Aires, La Plata; Córdoba, Segunda Usina (Barquez, Mares, and Braun 1999).

SUBSPECIES: I am treating *E. bonariensis* as monotypic.

NATURAL HISTORY: Barquez, Mares, and Braun (1999) reported *E. bonariensis* roosting in tree holes, under roofs, and in holes in bridges in transitional forest and in urban areas in provincia Tucumán, Argentina. Specimens have been netted over pools in forest habitat. Two females were lactating when taken in late December at Aguas Chiquitas, Tucumán. This species is preyed upon by barn owls (*Tyto alba*; Massoia, Chebez, and Heinonen-Fortabat 1989b).

REMARKS: Eger (1977) recognized three subspecies in addition to the nominate form of *E. bonariensis* (*E. b. beckeri* Sanborn, 1932a; *E. b. delticus* O. Thomas, 1923; and

E. b. nanus Miller, 1900e). These three taxa are now recognized as separate species (*beckeri* is a synonym of *E. patagonicus* O. Thomas, 1924). Uieda and Chaves (2005) reported *E. bonariensis* from the Botucatu region of the Brazilian state of São Paulo. That record is not mapped in this account because the specimen is more likely to represent *E. delticus* than *E. bonariensis* on geographic grounds. The identification needs to be confirmed. D. C. Carter and Dolan (1978) were unable to locate the type of *Promops bonariensis* W. Peters.

Eumops dabbenei O. Thomas, 1914
Dabbene's Bonneted Bat

SYNONYMS:

Eumops dabbenei O. Thomas, 1914a:481; type locality "Chaco," Argentina; identified by Barquez, Mares, and Braun (1999:197) as Tartagal, provincia Chaco.

Eumops perotis dabbenei: Sanborn, 1932a:350; name combination.

Eumops underwoodi mederai Massoia, 1976:264; type locality "San Javier," Santa Fe, Argentina.

DISTRIBUTION: *Eumops dabbenei* has been recorded from Venezuela, the Magdalena valley of Colombia (ANSP 5652), and from Paraguay and northern Argentina.

MARGINAL LOCALITIES (Map 231): *Northern distribution.* COLOMBIA: Rio Magdalena Valley (Eger 1977, specific locality unknown, not mapped). VENEZUELA: Aragua, Rancho Grande (Ochoa and Ibáñez 1985); Yaracuy, 10 km NW of Urama (Eger 1977); Apure, Hato El Frío (Ibáñez 1980). *Southern distribution.* PARAGUAY: Presidente Hayes, 2 km SE of Misión Inglesa

Map 231 Marginal localities for *Eumops dabbenei* ● and *Eumops delticus* ▲

(Harrison, Pendleton, and Harrison 1979); Presidente Hayes, 24 km WNW of Villa Hayes (Myers and Wetzel 1983). ARGENTINA (Barquez, Mares, and Braun 1999, except as noted): Santa Fe, San Javier (type locality of *Eumops underwoodi mederai* Massoia); Santiago del Estero, Sumampa; Tucumán, Tucumán; Salta, Güemes; Chaco, Tartagal (type locality of *Eumops dabbenei* O. Thomas).

SUBSPECIES: I am treating *E. dabbenei* as monotypic.

NATURAL HISTORY: Dabbene's Bonneted Bat has been found roosting in a house (Ibáñez 1980), netted in Venezuelan tropical dry forest (Handley 1976; Ibáñez 1980) and premontane humid forest (Ochoa and Ibáñez 1985), and in Argentine thorn scrub (Barquez, Mares, and Ojeda 1991), at elevations from sea level to 1,100 m. A female taken in April in Venezuela was neither pregnant nor lactating (August and Baker 1982). McWilliams et al. (2002) summarized the natural history information in their *Mammalian Species* account. *Eumops dabbenei* has a $2n = 48$, $FN = 56$ karyotype (August and Baker 1982).

REMARKS: This species is rare in collections. Although *E. dabbenei* is larger than *E. underwoodi*, which occurs from southern Arizona in the United States and into Costa Rica (Foster and Aguilar 1993), the two species share certain morphological characters such as heavily keeled, short ears; small pointed tragus; and shallow basisphenoid pits. Dolan and Carter (1979) suggested that there is a clinal increase in size from north to south in *E. underwoodi*, and *E. dabbenei* may represent a larger southern subspecies. However, I am treating *E. dabbenei* and *E. underwoodi* as separate species until there is more information on variation in both taxa. Harrison, Pendleton, and Harrison (1979) identified Misión Inglesa as in departamento Chaco (currently Alto Paraguay); however, in a letter (D. L. Harrison, *in litt.*) Harrison gave the coordinates as "58°25'W 23°50'S," which clearly place the locality in departamento Presidente Hayes. D. C. Carter and Dolan (1978) gave information on and measurements of the holotype of *Eumops dabbenei* O. Thomas.

Eumops delticus O. Thomas, 1923
Delta Bonneted Bat

SYNONYMS:

Eumops delticus O. Thomas, 1923d:341; type locality "Caldeirão," Ilha Marajó, Pará, Brazil.

Eumops bonariensis delticus: Sanborn, 1932a:355; name combination.

DISTRIBUTION: *Eumops delticus* is distributed in southeastern Colombia and along the Rio Amazonas of Brazil, south into the state of Bahia.

MARGINAL LOCALITIES (Map 231; from Eger 1977, except as noted): COLOMBIA: Caquetá, Tres Esquinas (ROM 83960). BRAZIL: Pará, Caldeirão; Bahia, Ilha

Madre de Dios (MZUSP 15375); Minas Gerais, Uberlândia (Stutz et al. 2004, as *E. bonariensis*); Bahia, São Marcelo; Pará, Boim.

SUBSPECIES: I am treating *E. delticus* as monotypic.

NATURAL HISTORY: Aside from general information based on other species of *Eumops*, nothing is known of the life history of *E. delticus*.

REMARKS: Eger (1977) treated *E. delticus* as a subspecies of *E. bonariensis*, but with evidence of sympatry between *E. bonarianesis* and *E. patagonicus* (originally considered a subspecies of *E. bonariensis*), it seems clear that these taxa represent different species. The two specimens from Uberlândia, Minas Gerais, reported by Stutz et al. (2004) as *E. bonariensis*, need to have their identity confirmed. D. C. Carter and Dolan (1978) provided information on and measurements of the holotype of *Eumops delticus* O. Thomas.

Eumops glaucinus (J. A. Wagner, 1843)
Wagner's Bonneted Bat

SYNONYMS:

Dysopes glaucinus J. A. Wagner, 1843a:368; type locality "Cuyaba," Mato Grosso, Brazil.

Molossus ferox Gundlach in W. Peters, 1861:149; type locality "Fundador," Camarioca, Cuba.

M[olossus (Promops)]. ferox: W. Peters, 1866a:574; name combination.

Molossus glaucinus: Dobson, 1876:714; name combination.

Nyctinomus orthotis H. Allen, 1889:561; type locality "Spanishtown," Jamaica.

[Molossus (Promops)] glaucinus: Trouessart, 1897:144; name combination.

Promops glaucinus: Miller, 1900e:471; name combination.

N[yctinomops]. orthotis: Miller, 1902b:393; name combination.

Promops orthotis: Miller, 1902c:250; name combination.

Eumops orthotis: Miller, 1906b:85; name combination.

Eumops glaucinus: Miller, 1906b:85; first use of current name combination.

Tadarida orthotis: Miller, 1924:87; name combination

DISTRIBUTION: *Eumops glaucinus* is from Colombia, Venezuela, Guyana, Brazil, Ecuador, Peru, Paraguay, and northern Argentina. Elsewhere, it occurs on Jamaica and Cuba, and in Central America, Mexico, and the United States (Florida).

MARGINAL LOCALITIES (Map 232; from Eger 1977, except as noted): COLOMBIA: Magdalena, Santa Marta. VENEZUELA: Falcón, Pedregal (USNM 522948); Carabobo, Montalbán (Handley 1976); Amazonas, San Juan (Handley 1976). GUYANA: Upper Takutu-Upper Essequibo, Aroquoi Tributary, Rupununi River. BRAZIL:

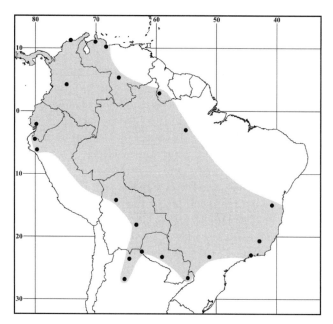

Map 232 Marginal localities for *Eumops glaucinus* ●

Pará, Tauari; Bahia, Região de Conquista (Falcão, Soares-Santos, and Drummond 2005); Minas Gerais, Viçosa (Mumford and Knudson 1978); Rio de Janeiro, Angra dos Reis; Paraná, Londrina (N. R. Reis et al. 1998). ARGENTINA: Misiones, Caraguatay (Vaccaro 1992). PARAGUAY: Presidente Hayes, Rincón Charrúa; Boquerón, Base Naval Pedro P. Peña (López-González et al. 1998). ARGENTINA: Tucumán, Tucumán (Barquez and Lougheed 1990); Jujuy, Yuto (Barquez and Ojeda 1992). BOLIVIA: Santa Cruz, Ingeniero Mora (S. Anderson, Koopman, and Creighton 1982); Beni, Espíritu (S. Anderson 1997). PERU: Lambayeque, Cerro la Vieja (LSUMZ 27225); Piura, 6 km W of Suyo. ECUADOR: Guayas, Guayaquil. COLOMBIA: Tolima, Espinal.

SUBSPECIES: I treat *E. glaucinus* as monotypic (see Remarks).

NATURAL HISTORY: *Eumops glaucinus* has been found roosting in hollow trees and in and under roofs of buildings. Handley (1976) reported the species in tropical humid forest, premontane dry forest, and premontane humid forest in Venezuela. Records indicate one young born in April or May and possibly a second parturition period in December. A pregnant female with a single embryo was captured in Bolivia in September (S. Anderson 1997). Aguirre (1994; *fide* S. Anderson 1997) identified insects of the orders Zygoptera, Orthoptera (Grillidae), and Coleoptera (Hydrophylidae) found in the stomachs of four individuals from Bolivia. In addition to the parasites Best, Kiser, and Rainey (1997) mention in their *Mammalian Species* account, R. Guerrero (1985a, 1997) added two species of hemipterans, two species of mites, and three

genera and species of streblid batflies. Two karyotypes are known for *E. glaucinus*: $2n = 40$, FN = 64, and $2n = 38$, FN = 64 (J. W. Warner et al. 1974), in addition to apparent geographic variation in the morphology of the X chromosome in the $2n = 38$ karyotype.

REMARKS: Timm and Genoways (2004) recognized *E. floridanus* (G. M. Allen, 1932) as a species of *Eumops* found only in the state of Florida, USA. Originally described from Pleistocene fossil material as *Molossides floridanus*, living specimens were first reported by Barbour (1936) as *Eumops glaucinus*. Koopman (1971) treated the Florida population as a subspecies of *E. glaucinus*. Consequently, most of the recent literature treated *E. glaucinus* as represented by two subspecies. D. C. Carter and Dolan (1978) provided information on and measurements of the holotype of *Dysopes glaucinus* J. A. Wagner and the lectotype of *Molossus ferox* Gundlach.

Eumops hansae Sanborn, 1932
Sanborn's Bonneted Bat

SYNONYMS:

Eumops hansae Sanborn, 1932a:356; type locality "Colonia Hansa, near Joinville," Santa Catarina, Brazil.

Eumops amazonicus Handley, 1955:177; type locality "Manaos, Amazonas," Brazil.

DISTRIBUTION: *Eumops hansae* occurs in Venezuela, Guyana, French Guiana, Brazil, Bolivia, Peru, and Ecuador. It also is known from Panama and Costa Rica.

MARGINAL LOCALITIES (Map 233): VENEZUELA: Aragua, Portachuelo (Ochoa, Castellanos, and Ibáñez 1988); Bolívar, El Manaco, 59 km SE of El Dorado (Eger

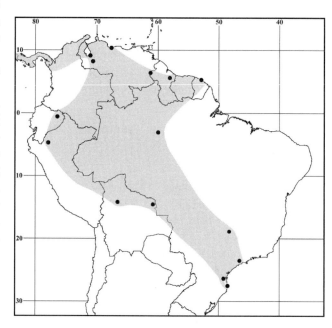

Map 233 Marginal localities for *Eumops hansae* ●

1977). GUYANA: Upper Demerara-Berbice, Arampa. FRENCH GUIANA: Paracou (Simmons and Voss 1998). BRAZIL: Amazonas, Manaus (type locality of *Eumops amazonicus* Handley); Minas Gerais, Uberlândia (Stutz et al. 2004); São Paulo, São Paulo (Gregorin 2001); Santa Catarina, Florinópolis (Cherem et al. 2005); Santa Catarina, Colonia Hansa (type locality of *Eumops hansae* Sanborn). BOLIVIA: Santa Cruz, 38 km E of La Florida (Ibáñez and Ochoa 1989); Beni, Espíritu (Aguirre and Urioste 1994). PERU: Amazonas, 43 km. (by road) NE of Chiriaco (Graham and Barkley 1984). ECUADOR: Orellana, Onkone Gare (F. A. Reid, Engstrom, and Lim 2000). VENEZUELA: Barinas, Unidad II, Reserva Forestal de Ticoporo (Ochoa et al. 1988); Mérida, Río Quebrada de Piedras, 4 km SSE of Nueva Bolivia (Soriano, Ruiz, and Zambrano 2005).

SUBSPECIES: *Eumops hansae* is treated here as monotypic.

NATURAL HISTORY: This species has been found roosting in hollow trees. Graham and Barkley (1984) reported it from tropical lowland forest in hilly terrain, and Handley (1976) also found it in tropical humid forest. Ibáñez and Ochoa (1989) collected specimens in savanna habitat near forest edge. F. A. Reid, Engstrom, and Lim (2000) recorded *E. hansae* from upland or terra firme primary forest. Brosset and Charles-Dominique (1991) and Simmons and Voss (1998) caught their French Guiana specimens in nets suspended in gaps high in the forest canopy. The stomach of a Bolivian specimen examined by Aguirre (1994; *fide* S. Anderson 1997) contained Orthoptera (Grillidae). *Eumops hansae* occurs over an elevational range of from 150 to 1,000 m. Best et al. (2001b) summarized available natural history information in their *Mammalian Species* account.

Eumops maurus (O. Thomas, 1901)
Guianan Bonneted Bat
SYNONYMS:

Molossus maurus O. Thomas, 1901g:141; type locality "Kanuku Mountains...savannahs near the base...240 feet," Upper Takutu-Upper Essequibo, Guyana.

[*Myopterus*] *maurus*: Trouessart, 1904:101; name combination.

Eumops maurus: Miller, 1906b:85; first use of current name combination.

Eumops geijskesi Husson, 1962:246; type locality "Suriname."

DISTRIBUTION: *Eumops maurus* occurs in Guyana, Surinam, Venezuela, and Ecuador.

MARGINAL LOCALITIES (Map 234): VENEZUELA: Monagas, near Uverito (J. Sánchez, Ochoa, and Ospino 1992). GUYANA: Upper Takutu-Upper Essequibo, savan-

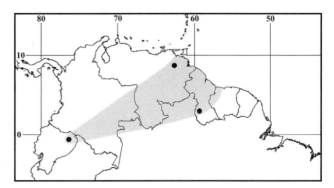

Map 234 Marginal localities for *Eumops maurus* ●

nas at base of Kanuku Mountains (type locality of *Molossus maurus* O. Thomas). SURINAM: (Husson 1962; no specific locality, not mapped). ECUADOR: Orellana, 42 km S and 12 km E of Pompeya Sur (F. A. Reid, Engstrom, and Lim 2000).

SUBSPECIES: I treat *E. maurus* as monotypic.

NATURAL HISTORY: The Ecuadorian specimen was pregnant when netted above a high bridge across the Río Tiputini in September (F. A. Reid, Engstrom, and Lim 2000). In Venezuela, a female was collected in a 15-year-old pine plantation (J. Sánchez, Ochoa, and Ospino 1992). Best et al. (2001a) presented measurements and additional information in their *Mammalian Species* account.

REMARKS: This rare species is presently known from only six specimens representing four localities: one is in Ecuador, another in Venezuela, plus the holotype from Guyana, and the three specimens from Surinam (exact locality unknown) that were the basis for the name *E. geijskesi*. D. C. Carter and Dolan (1978) provided information on and measurements of the holotypes of *Eumops geijskesi* Husson and *Molossus maurus* O. Thomas.

Eumops nanus (Miller, 1900)
Dwarf Bonneted Bat
SYNONYMS:

Promops nanus Miller, 1900e:471; type locality "Bogava, Chiriqui, Panama."

Eumops nanus: Miller, 1906b:85; first use of current name combination.

Eumops bonariensis nanus: Sanborn, 1932a:356; name combination.

DISTRIBUTION: *Eumops nanus* occurs in northern Colombia and Venezuela, Guyana, and Peru. Elsewhere, it has been recorded in Central America and southern Mexico.

MARGINAL LOCALITIES (Map 235; from Eger 1977, except as noted): COLOMBIA: La Guajira, La Isla. VENEZUELA: Falcón, Capatárida (Handley 1976).

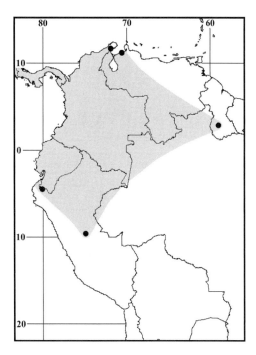

Map 235 Marginal localities for *Eumops nanus* ●

GUYANA: Upper Takutu-Upper Essequibo, Kuitaro River, 48 km E of Dadanawa. PERU: Huánuco, Panguana Biological Station (Hutterer et al. 1995, as *E. bonariensis*); Piura, 6.4 km W of Suyo.

SUBSPECIES: I regard *E. nanus* as monotypic.

NATURAL HISTORY: This species occurs in tropical thorn forest in Venezuela where it has been found in a tree hole (Handley 1976). Elsewhere, *E. nanus* has been netted in tropical humid forest habitat. Hutterer (1995) recorded *E. nanus* as a rare occupant of secondary forest and forest edges in eastern Peru. The karyotype is unknown.

REMARKS: Considered a subspecies of *E. bonariensis* by Eger (1977), *E. nanus* is herein accorded species ranking (see remarks under *E. delticus*). D. C. Carter and Dolan (1978) provided information on and measurements of the holotype of *Promops nanus* Miller.

Eumops patagonicus O. Thomas, 1924
Patagonian Bonneted Bat
SYNONYMS:

Eumops patagonicus O. Thomas, 1924a:234; type locality "Chubut," Argentina.

Eumops bonariensis beckeri Sanborn, 1932a:354; type locality "Trinidad, El Beni, Bolivia."

Eumops bonariensis patagonicus: Sanborn, 1932a:355; name combination.

Eumops bonariensis becheri Freeman, 1981:154; incorrect subsequent spelling of *Eumops bonariensis beckeri* Sanborn.

DISTRIBUTION: *Eumops patagonicus* occurs in Bolivia, Paraguay, Argentina, and southern Brazil.

MARGINAL LOCALITIES (Map 236): BOLIVIA: Beni, Acapulco (S. Anderson 1997). PARAGUAY (Eger 1977): Boquerón, Loma Plata; Concepción, 8 km E of Concepción. ARGENTINA: Misiones, Aristóbulo del Valle (Barquez, Mares, and Braun 1999). BRAZIL: Rio Grande do Sul, Município Garruchos (J. C. González 2004). URUGUAY: Artigas, Boca del Arroyo Mandiyú (Saralegui 1996). ARGENTINA: Corrientes, Goya (Barquez, Mares, and Braun 1999); Buenos Aires, Campo de Los Padres (Barquez, Mares, and Braun 1999); Chubut, Dolavón (Monjeau, Bonino, and Saba 1994); Santiago del Estero, 5 km E of Las Termas (Barquez and Ojeda 1992); Tucumán, Las Talas (Barquez and Ojeda 1992); Jujuy, Río Lavayén, 1 km N Santa Rita (M. M. Díaz and Barquez 1999); Salta, 15 km S and 15 km W of Orán (Barquez, Mares, and Braun 1999). BOLIVIA: Tarija, Estancia Bolívar (S. Anderson 1997); Santa Cruz, Ingeniero Mora (S. Anderson, Koopman, and Creighton 1982); Beni, Río Tijamuchi (S. Anderson 1997).

SUBSPECIES: I am treating *E. patagonicus* as monotypic.

NATURAL HISTORY: *Eumops patagonicus* is associated with chacoan thorn scrub (Mares, Ojeda, and Kosco 1981; Barquez and Díaz 2001) where it has been netted over pools and streams and found in hollow trees and in the roofs of houses. Its presence in the Yungas of Bolivia is associated with altered habitat (Barquez and Díaz 2001).

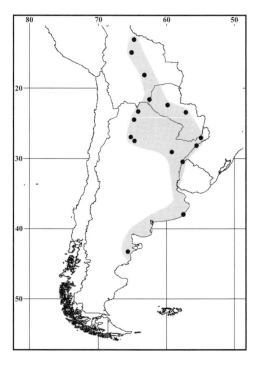

Map 236 Marginal localities for *Eumops patagonicus* ●

REMARKS: Sanborn (1932a) described *E. bonariensis beckeri* as a new subspecies and, while recognizing that *E. b. patagonicus* was about the same size, he questioned (p. 356) the validity of the type locality of *E. patagonicus*. Cabrera (1958) synonymized *E. patagonicus* with *E. bonariensis* and subsequent authors followed suit. Barquez (1987; as noted in Mares, Barquez, and Braun 1995) synonymized *E. patagonicus* with *E. bonariensis beckeri* because both taxa were sympatric in provincia Tucumán, Argentina. Barquez (1987; *fide* Mares, Barquez, and Braun 1995) and Barquez, Mares, and Braun (1999) recognized *E. patagonicus* and *E. bonariensis* as different species. D. C. Carter and Dolan (1978) provided information on and measurements of the holotype of *Eumops patagonicus* O. Thomas.

Eumops perotis (Schinz, 1821)
Greater Bonneted Bat
SYNONYMS:

Molossus perotis Schinz, 1821:870; type locality "von Brasilien"; identified by Wied-Neuwied (1826:231) as "Villa de S. Salvador dos Campos dos Goaytacases" [= Campos], Rio Paraiba, Rio de Janeiro, Brazil.

Dysopes perotis: Wied-Neuwied, 1826:227; name combination.

Dysopes (Molossus) gigas W. Peters, 1864b:383; type locality "Taburete, District Callajabas auf Cuba"; listed as Rio Negro, Amazonas, Brazil by Carter and Dolan (1978; see Remarks).

M[olossus (Promops)]. perotis: W. Peters, 1866a:574; name combination.

M[olossus (Promops)]. gigas: W. Peters, 1866a:574; name combination.

Molossus californicus Merriam, 1890:31; type locality "Alhambra, Los Angeles Co., California," U.S.A.

Promops perotis: O. Thomas, 1901b:191; name combination.

Eumops perotis: Miller, 1906b:85; first use of current name combination.

Eumops perotis renatae Pirlot, 1965a:5; type locality "Cumaná," Sucre, Venezuela.

E[umops]. renatae: Pirlot, 1968:90; name combination.

Eumops perotis perotis: Eger, 1977:48; name combination.

DISTRIBUTION: *Eumops perotis* is known from northern Venezuela, southern Brazil, Paraguay, Argentina, Bolivia, Peru, and Ecuador. A disjunct subspecies occurs in the southwestern United States and in northern Mexico.

MARGINAL LOCALITIES (Map 237; from Eger 1977, except as noted): VENEZUELA: Sucre, Cumaná (type locality of *Eumops perotis renatae* Pirlot). ECUADOR: Guayas, Guayaquil. BOLIVIA: Beni, San Joaquín (ROM 78139). BRAZIL: Maranhão, Barra do Corda; Rio de

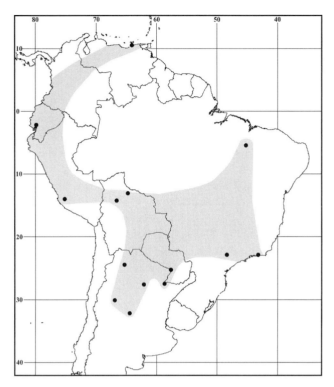

Map 237 Marginal localities for *Eumops perotis* ●

Janeiro, Rio de Janeiro; São Paulo, Botucatu region (Uieda and Chaves 2005). PARAGUAY: Central, Asunción. ARGENTINA (Barquez, Mares, and Braun 1999, except as noted): Corrientes, Laguna Paiva; Santiago del Estero, Girardet; Córdoba, Segunda Usina; La Rioja, Guayapa (Crespo 1958); Salta, 30 km NE of Salta. BOLIVIA: Beni, Espíritu (S. Anderson 1993; as *E. dabbenei*). PERU: Huancavelica, Córdova.

SUBSPECIES: I recognize two subspecies, of which only *E. p. perotis* occurs in South America.

NATURAL HISTORY: *Eumops perotis* is a common species in urban and suburban areas where it roosts in tree holes and in the roofs of houses (Barquez and Díaz 2001), and in rock crevices (Mares et al. 1996). Barquez and Díaz (2001) reported a young individual that had erupting teeth when captured in Tucumán in December, and a male from Jujuy with scrotal testes when captured in May. In addition to the parasites mentioned by Best, Kiser, and Freeman (1996) in their *Mammalian Species* account, R. Guerrero (1997) listed a streblid batfly known from this species. *Eumops perotis californicus* (from Alamos, Sonora, Mexico) has a $2n = 48$, $FN = 56$ karyotype (R. J. Baker 1970).

REMARKS: Pirlot (1965a) described *Eumops perotis renatae* and later (Pirlot 1968) elevated it to a species, based on a comparison with specimens of *E. perotis californicus* and *E. trumbulli*. However, until adequate numbers of specimens are available to determine the extent of geographic variation in *E. perotis*, I believe *E. renatae* should be treated as a synonym of *E. perotis*.

Dilford C. Carter and Dolan (1978) provided information on and measurements of the holotype of *Dysopes (Molossus) gigas* W. Peters. They also (1978) said that the holotype of *Dysopes (Molossus) gigas* W. Peters was not from Cuba, but from Brazil. When W. Peters (1864b) described *gigas*, he said it was the only specimen in the collection and that the epiphyses were incompletely ossified. Silva-Taboada (1979) also commented on the specimen, and suggested that an error in the catalog and subsequent labelling of the specimen had occurred, because a mixed lot of Brazilian and Cuban specimens had been cataloged at the same time. When Silva-Taboada (1979) examined the specimen, it was mounted with the skull still inside. *Eumops perotis* is recorded from the Quaternary of Brazil (Czaplewski and Cartelle 1998).

Eumops trumbulli (O. Thomas, 1901)
Trumbull's Bonneted Bat

SYNONYMS:

Promops trumbulli O. Thomas, 1901b:190; type locality "Para," Brazil.

Eumops trumbulli: Miller, 1906b:85; first use of current name combination.

Eumops perotis trumbulli: Sanborn, 1932a:350; name combination.

Eumops trumbuli N. R. Reis and Peracchi, 1987:180; incorrect subsequent spelling of *Promops trumbulli* O. Thomas.

DISTRIBUTION: *Eumops trumbulli* is known from the greater Amazon basin of Colombia, Venezuela, Guyana, Brazil, Bolivia, and Peru.

MARGINAL LOCALITIES (Map 238; from Eger 1977, except as noted): VENEZUELA: Bolívar, Maripa. GUYANA: Upper Takutu-Upper Essequibo, Rewa River. BRAZIL: Pará, Mocajuba. BOLIVIA: Beni, Providencia (S. Anderson 1997); Beni, Guayaramerín. PERU: Ucayali, Alto Río Tamaya. COLOMBIA: Amazonas, Leticia.

SUBSPECIES: *Eumops trumbulli* is treated here as monotypic.

NATURAL HISTORY: Individuals have been found in roofs of houses and hanging on dead leaves high in the forest canopy.

REMARKS: *Eumops trumbulli* is closely related to *E. perotis*, but the differences noted (see key to species of *Eumops*), in combination with distribution records in close proximity in northeastern Bolivia, justify separation as two species. The *Mammalian Species* account on *Eumops perotis* by Best, Kiser, and Freeman (1996) included *E. trumbulli* as a subspecies. D. C. Carter and Dolan (1978) provided

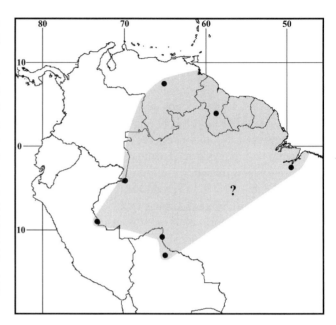

Map 238 Marginal localities for *Eumops trumbulli* ●

information on and measurements of the holotype of *Promops trumbulli* O. Thomas.

Genus *Molossops* W. Peters, 1866

The genus *Molossops* is characterized by a relatively large tragus (at least 1/2 the size of the antitragus), a short, wide antitragus (wider than high and notched posteriorly), and elongated, pointed ears that have a flexible fold where they join the head. The second phalanx of both digits III and IV is equal to or longer than the first phalanx. Dental and cranial features include a single pair of lower incisors, a well-developed third commissure on the last upper molar; and well-developed basisphenoid pits (usually elliptical in outline). The dental formula is 1/1, 1/1, 2/2, 3/3 × 2 = 28.

SYNONYMS:

Dysopes: Burmeister, 1854:72; part; not *Dysopes* Illiger, 1811.

Molossops W. Peters, 1866a:575; type species *Dysopes temminckii* Burmeister, 1854, by subsequent designation (Miller 1907b:247); described as a subgenus of *Molossus* É. Geoffroy St.-Hilaire, 1805a.

Myopterus: W. Peters, 1869:402; not *Myopterus* É. Geoffroy St.-Hilaire, 1818b.

Molossus: Winge, 1892:17; not *Molossus* É. Geoffroy St.-Hilaire, 1805a.

Mollossops Miretzki and Margarido, 1999:105; incorrect subsequent spelling of *Molossops* W. Peters.

REMARKS: Originally described as a subgenus of *Molossus*, this genus has been widely regarded as including *Cynomops* as either a subgenus or a junior synonym, despite O. Thomas's (1920a) diagnosis, which is adequate for distinguishing between the two genera. *Molossops* is readily separated from *Cynomops* by the features described above. Other taxa that have been incorrectly assigned to *Molossops* are *M. aequatorianus* Cabrera, 1917 (= *Cabreramops* Ibáñez, 1981) and *M. mattogrossensis* C. O. C. Vieira, 1942 (= *Neoplatymops* R. L. Peterson, 1965).

KEY TO THE SPECIES OF *MOLOSSOPS*:
1. Size larger; forearm of males 37.0–38.5 mm, of females 36–39 mm; condylobasal length of males 16.7–18.2 mm, of females 15.1–16.4 mm . . . . . . *Molossops neglectus*
1′. Size smaller; forearm less than 33 mm; condylobasal length less than 15 mm . . . . . . *Molossops temminckii*

Molossops neglectus S. L. Williams and Genoways, 1980
Rufous Dog-faced Bat
SYNONYMS:

Molossops temminckii: Tuttle, 1970:79; not *Dysopes temminckii* Burmeister, 1854.

Molossops (Molossops) neglectus S. L. Williams and Genoways, 1980b:489; type locality "1 km S and 2 km E of Powaka, Suriname, Suriname."

DISTRIBUTION: *Molossops neglectus* is known from Guyana, Surinam, Peru, northern and southeastern Brazil, southern Colombia, and northeastern Argentina (Misiones). The known distribution could be described as tripartite: Guyanan (including Belém, Brazil), southeastern (Brazil and northeastern Argentina), and western (Peru and southern Colombia).

MARGINAL LOCALITIES (Map 239): VENEZUELA: Bolívar, Unit V, Imataca Forest Reserve (Ochoa et al. 1993). GUYANA: Upper Demerara-Berbice, Arampa (Lim and Engstrom 2001a). SURINAM: Para, 1 km S and 2 km E of Powaka (type locality of *Molossops neglectus* S. L. Williams and Genoways). BRAZIL (Gregorin et al. 2004, except as noted): Pará, Belém (Ascorra, Wilson, and Handley 1991); São Paulo, Caetetus Ecological Station (Pedro, Passos, and Lim 2001); Rio de Janeiro, Paulo de Frontin; São Paulo, Salesópolis; São Paulo, Itú. ARGENTINA: Misiones, Parque Nacional Iguazú (Barquez, Giannini, and Mares 1993; Barquez, Mares, and Braun 1999). PERU: Pasco, San Juan (Tuttle 1970, as *M. temminckii*); Loreto, Jenaro Herrera (Ascorra, Gorchov, and Cornejo 1994). COLOMBIA: Putumayo, Puerto Leguizamo (Lim and Engstrom 2001a).

SUBSPECIES: I consider *M. neglectus* to be monotypic.

NATURAL HISTORY: M. Brock Fenton and associates collected four specimens in Guyana in July 1970 in nets set high in the forest canopy. Four Venezuelan specimens (one male and three females) also were caught in nets placed from 3 to 4 m above ground level in forest gaps left by logging. The holotype was taken at the boundary between

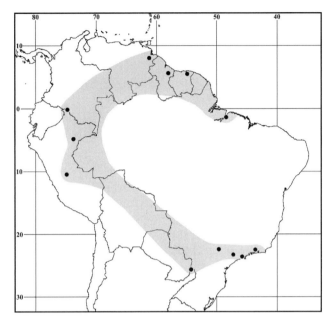

Map 239 Marginal localities for *Molossops neglectus* ●

savannah and forest in Surinam. The two females from Peru (taken 5 August 1964) were netted along the edge of newly cut mature forest; both were pregnant with fetuses measuring 20 mm CR (Tuttle 1970). Another pregnant female was captured at Centro de Investigaciones "Jenaro Herrera," Loreto, Peru, during the dry season (June-September; Ascorra, Gorchov, and Cornejo 1994). The Belém, Brazil, specimen is a subadult male taken on 11 March 1963. Gregorin et al. (2004) reported the species in the humid coastal forest and the drier semi-deciduous forest in the interior of southeastern Brazil. They also analyzed geographic variation within the species.

REMARKS: Ascorra, Wilson, and Handley (1991) reviewed *M. neglectus* and reported on specimens they found in museum collections and specimens misidentified in the literature. The sexes are strongly dimorphic cranially, but less so in wing dimensions. The wingspan of a male measured 272 mm, and that of a female, 270 mm. Unfortunately, S. L. Williams and Genoways (1980b) did not show the diagnostic basisphenoid pits in their illustration of the holotype. *Molossops neglectus* has eight ridges on the fleshy palate, whereas the palate of *M. temminckii* usually has six.

Molossops temminckii (Burmeister, 1854)
Dwarf Dog-faced Bat
SYNONYMS:

Dysopes temminckii Burmeister, 1854:72; type locality "Lagoa Santa," Minas Gerais, Brazil.

M[olossus (Molossops)]. Temminckii: W. Peters, 1866a: 575; subgeneric description and name combination.

Molossus hirtipes Winge, 1892:17; type locality "Lagoa Santa," Minas Gerais, Brazil.

[*Molossus (Myopterus)*] *Temminckii*: Trouessart, 1897: 142; name combination.

Molossops temminckii: Miller, 1907b:248; name combination.

Molossops temminckii sylvia O. Thomas, 1924a:234; type locality "Goya," Corrientes, Argentina; may be a valid subspecies.

Molossops temminckii griseiventer Sanborn, 1941:385; type locality "Espinal, west of Magdalena River on the plains of Tolima, Colombia"; may be a valid subspecies.

Molossops temminckii temminckii: Sanborn, 1941:386; name combination.

Molossops [(Molossops)] temminckii griseiventer: Cabrera, 1958:117; name combination.

Molossops [(Molossops)] temminckii sylvia: Cabrera, 1958:117; name combination.

Molossops [(Molossops)] temminckii temminckii: Cabrera, 1958:118; name combination.

Molossops teminckii Mares, Willig, Streilein, and Lacher, 1981:113; incorrect subsequent spelling of *Dysopes temminckii* Burmeister.

Molossops temninckii Coimbra, Borges, Guerra, and Mello, 1982:35; incorrect subsequent spelling of *Dysopes temminckii* Burmeister.

DISTRIBUTION: *Molossops temminckii* is found from Colombia, Venezuela, and Guyana, southwestward through Ecuador, Peru, Bolivia, Paraguay, and Brazil, into Uruguay and northern Argentina.

MARGINAL LOCALITIES (Map 240): VENEZUELA: Aragua, Las Delicias (USNM 522055). GUYANA: Upper Takutu-Upper Essequibo, near Kuitaro River, 64 km E of Dadanawa (ROM 57556). VENEZUELA: Guárico, Guyabal (USNM 522946); Apure, Hato "El Frio" (Ibáñez 1984a). PERU: Loreto, Curaray River mouth (AMNH 71634). BOLIVIA: Beni, San Joaquín (S. Anderson, Koopman, and Creighton 1982). BRAZIL: Mato Grosso, Serra do Roncador, approximately 264 km N of Xavantina (Pine, Bishop, and Jackson 1970); Tocantins, Município Babaçulândia (Nunes et al. 2005); Ceará, Floresta Nacional Araripe-Apodi, 8 km SSW of Crato (Mares et al. 1981); Bahia, Região de Conquista (Falcão, Soares-Santos, and Drummond 2005); Minas Gerais, Lagoa Santa (ZMB 5458, syntype of *Dysopes temminckii* Burmeister); São Paulo, Nova Aliancia (Vizotto and Taddei 1976); Paraná, Estação Ecológica do Caiuá (Miretzki and Margarido 1999). ARGENTINA: Misiones, Apósteles (Barquez, Mares, and Braun 1999). URUGUAY: Artigas, Río Cuareim, ca. 5 km above junction with Arroyo Yacaré (Ximénez 1969); Paysandú, near Arroyo Negro, 15 km S of Paysandu (S. L. Williams and Genoways 1980b). ARGENTINA (Barquez,

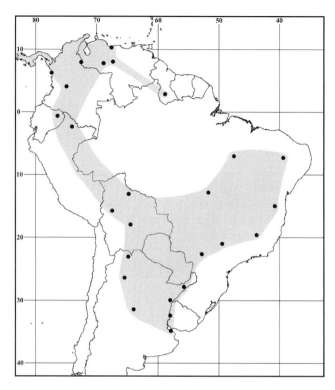

Map 240 Marginal localities for *Molossops temminckii* ●

Mares, and Braun 1999, except as noted): Buenos Aires, La Plata; Córdoba, Villa María; Tucumán, Las Mesadas; Salta, Río Zenta (ROM 53918). BOLIVIA (S. Anderson 1997): Santa Cruz, 5 km SW of Comarapa; La Paz, La Reserva. ECUADOR: Orellana, Onkone Gare, 38 km S of Pompeya Sur (F. A. Reid, Engstrom, and Lim 2000). COLOMBIA: Tolima, Guamo (ROM 44951); Chocó, Bahía Solano (ROM 69533); Norte de Santander, Cúcuta (ROM 84999).

SUBSPECIES: *Molossops temminckii* is treated here as monotypic pending revision (see Remarks).

NATURAL HISTORY: This species has been found roosting in a wide variety of situations, including buildings, rocky outcrops, hollow trees, and in holes in fence posts (Vizotto and Taddei 1976; Mares, Ojeda, and Kosco 1981; Mares et al. 1981; Myers and Wetzel 1983; Ibáñez 1984a). Aguirre (1994; *fide* S. Anderson 1997) reported on the contents of one stomach that contained the remains of Orthoptera (Grillidae), Hemiptera, Homoptera, Coleoptera, Lepidoptera, and Diptera (Nematocera); six measurable prey items were each less than 10 mm long. Two pregnant females, each with a single fetus, were taken in September (S. Anderson 1997). R. Guerrero (1997) listed a streblid batfly and Graciolli and Carvalho (2001) mentioned a nycteribiid batfly known from *M. temminckii*. Gardner (1977b) reported a $2n = 42$, FN $= 56$ karyotype for specimens from Colombia, east of the type locality of *Molossops temminckii griseiventer* Sanborn.

REMARKS: Based on discriminant function analysis, a sample representing the nominate subspecies is readily distinguished from samples of *M. t. sylvia* and *M. t. griseiventer*. However, the sample size for *M. t. griseiventer* (the most distinctive) was too small to be statistically reliable, and boundaries between these subspecies are not adequately established. Vizotto and Taddei (1976) found little sexual dimorphism in samples from southern Brazil, but Myers and Wetzel (1983) reported that sexual dimorphism in the Chaco Boreal population in Paraguay was striking (females average about 5% smaller), particularly in measurements of cranial breadth. I have not recognized subspecies in *M. temminckii* because of conflicting evidence on geographic variation; the species needs to be revised. D. C. Carter and Dolan (1978) gave information on and measurements of a syntype of *Dysopes temminckii* Burmeister and the holotype of *Molossops temminckii sylvia* O. Thomas.

Genus *Molossus* É. Geoffroy St.-Hilaire, 1805

The genus *Molossus* was reviewed by Miller (1913b) and more recently by Dolan (1982, 1989), who concentrated on the Middle American species. *Molossus* is found from Mexico south into northern Argentina and Uruguay; seven species are currently recognized. Similar to *Promops* externally, *Molossus* has smooth lips, a minute tragus, an antitragus that is constricted at its base, and short, rounded ears that arise from the same point on the forehead. The snout lacks a prominent longitudinal, medial ridge behind the nostrils (this ridge is conspicuous in *Promops*). The sagittal crest is well developed in adults, the palate is arched (but not as much as in *Promops*), and the basisphenoid pits are distinct. The upper incisors are short, relatively broad, and barely project forward of canines. Upper molars show little or no trace of a hypocone. The dental formula is 1/1, 1/1, 1/2, 3/3 × 2 = 26.

SYNONYMS:

Vespertilio: Kerr, 1792:97; not *Vespertilio* Linnaeus, 1758.

Molossus É. Geoffroy St.-Hilaire, 1805a:278; type species *Vespertilio molossus* Pallas, 1766, by absolute tautonomy (see Husson 1962:256–258).

Dysopes Illiger, 1811:76; type species *Vespertilio molossus*: Gmelin, 1788 (= *Vespertilio molossus* Pallas), by monotypy.

Dysopus Billberg, 1827:Table A; incorrect subsequent spelling of *Dysopes* Illiger.

Nolossus Miná-Palumbo, 1865:5; incorrect subsequent spelling of *Molossus* É. Geoffroy St.-Hilaire.

Disopes H. Gervais and Ameghino, 1880:8; incorrect subsequent spelling of *Dysopes* Illiger.

Mollossus Weithofer, 1887:285; incorrect subsequent spelling of *Molossus* É. Geoffroy St.-Hilaire.

Molossops: Ruschi, 1951f:2; not *Molossops* W. Peters, 1866a.

Mollossus Durrant, 1952:540; incorrect subsequent spelling of *Molossus* É. Geoffroy St.-Hilaire.

Molosus Polanco-Ochoa, Jaimes, and Piragua, 2000:675; incorrect subsequent spelling of *Molossus* É. Geoffroy St.-Hilaire.

KEY TO SOUTH AMERICAN SPECIES OF *MOLOSSUS* (MEASUREMENTS ARE MEANS AND EXTREMES).

1. Dorsal fur short, usually less than 3.5 mm (compare in shoulder region), and unicolored or indistinctly bicolored 2

1′. Dorsal fur long, usually more than 4 mm, and distinctly bicolored. 5

2. Forearm averaging less than 34.9 mm (33.2–36.0 mm); condylobasal length averaging less than 15.1 mm (13.9–15.5 mm). *Molossus coibensis*

2′. Forearm averaging more than 36 mm; condylobasal length more than 16 mm 3

3. Forearm averaging less than 40 mm in males (38.9–41.1 mm), less than 39 mm in females (38.4–40.5 mm); condylobasal length averaging less than 17.3 mm in males (16.9–17.6 mm), 16.2 mm in females (15.6–16.6 mm) . . . . *Molossus bondae*

3′. Forearm and condylobasal lengths longer than above 4

4. Forearm averaging 50.9 mm in males (48.5–54.0 mm), 50.3 mm in females (47–53 mm); condylobasal length averaging 20.8 mm in males (20.0–21.6 mm), 19.8 mm in females (19.1–20.6 mm) . . . . *Molossus rufus*

4′. Forearm averaging 45.6 mm in males (43–49 mm), 44.6 mm in females (41.0–48.5 mm); condylobasal length averaging 19 mm in males (17.5–20.1 mm), 17.8 mm in females (16.5–19.2 mm) . . . . *Molossus pretiosus*

5. Forearm 45–50 mm; condylobasal length 18.6–19.8 mm . . . . *Molossus sinaloae*

5′. Forearm less than 43 mm; condylobasal length less than 18 mm 6

6. Forearm averaging 41.7 mm in males (39.3–43.6 mm), 41 mm in females (39.0–42.9 mm); condylobasal length averaging 17.4 mm in males (16.7–18.0), 16.8 mm in females (15.9–17.6 mm) . . . . *Molossus currentium*

6′. Forearm 35–40 mm; and condylobasal length less than 16 mm . . . . *Molossus molossus*

Molossus bondae J. A. Allen, 1904
Bonda Mastiff Bat

SYNONYMS: See under Subspecies.

DISTRIBUTION: *Molossus bondae* is known from Colombia, Ecuador, and Venezuela. Elsewhere, the species is in Central America where Dolan (1989) and López-González

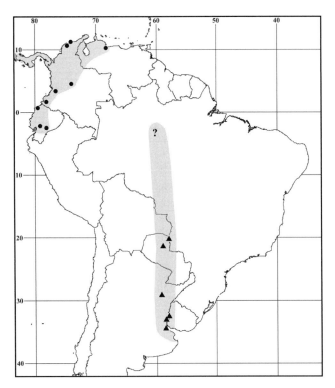

Map 241 Marginal localities for *Molossus bonda* ● *Molossus currentium* ▲

NATURAL HISTORY: These bats have been found roosting in and under different kinds of roofs, including thatch. The species has been netted over and adjacent to pools in streams and rivers. In South America, *M. bondae* usually is found at elevations below than 600 m. Handley (1976) recorded *M. bondae* from tropical and subtropical dry forest in Venezuela. In Central America, the species appears to be restricted to the humid lowlands of the Caribbean versant below 1,060 m (Dolan 1989). R. Guerrero (1985a) listed a nycteribiid bat fly, a hemipteran, two ticks, and four mites as known parasites; to which he later (1997) added two streblid batflies.

REMARKS: Pelage color varies from black to reddish brown. The reddish color is the result of wear (Gardner 1965) and bleaching, similar to that process described for the Mormoopidae by J. D. Smith (1972). *Molossus bondae* resembles *M. rufus* and *M. pretiosus*, but is smaller. Based on allozyme data, *M. bondae* also is close to these species phylogenetically (Dolan 1982, 1989). S. E. Burnett et al. 2001 provided measurements and biological information in their *Mammalian Species* account on *M. bondae*. In their recent revision, López-González and Presley (2001) described the Central American component as a new subspecies, *M. c. robustus*, and synonymized *M. bondae* under the name *M. currentium*.

and Presley (2001, as *Molossus currentium bondae*) record it from Panama northwest into Honduras.

MARGINAL LOCALITIES (Map 241): COLOMBIA: Magdalena, Bonda (type locality of *Molossus bondae* J. A. Allen). VENEZUELA: Carabobo, Montalbán (Handley 1976). COLOMBIA: Cundinamarca, Bogotá (H. E. Anthony 1923a). ECUADOR: Morona-Santiago, Sucua (López-González and Presley 2001); Guayas, Puente de Chimbo (AMNH 62104); Esmeraldas, 3 km W of Majua (USNM 513509). COLOMBIA: Nariño, Barbacoas (J. A. Allen 1916c); Valle del Cauca, 2 km S of Pance (M. E. Thomas 1972); Atlántico, Ponedera (ROM 68865).

SUBSPECIES: I recognize two subspecies, the nominate subspecies *M. b. bondae*, which occurs in South America, and *M. b. robustus* recently described by López-González and Presley (2001) and occurring in Central America.

M. b. bondae J. A. Allen, 1904

SYNONYMS:

Molossus bondae J. A. Allen, 1904b:228; type locality "Bonda," Río Manzanares, 11 km E of Santa Marta, Magdalena, Colombia.

M[olossus]. currentium bondae: López-González and Presley, 2001:769; name combination.

This subspecies occurs in Colombia, Venezuela, and Ecuador. Elsewhere, it is found in Panama.

Molossus coibensis J. A. Allen, 1904
Coiban Mastiff Bat

SYNONYMS:

Molossus coibensis J. A. Allen, 1904b:227; type locality "Coiba Island," Panama.

Molossus Burnesi O. Thomas, 1905a:584; type locality "Cayenne," French Guiana; incorrect original spelling of *Molossus barnesi* O. Thomas, 1905a.

Molossus barnesi O. Thomas, 1905a:585; selection of correct original spelling (Cabrera 1958:129).

Molossus cherriei J. A. Allen, 1916d:529; type locality "Tapirapoan," Mato Grosso, Brazil.

M[olossus]. burnsi: Hershkovitz, 1949c:454; incorrect subsequent spelling of *Molossus barnesi* O. Thomas.

Molossus burnesi: Husson, 1962:259; incorrect subsequent spelling of *Molossus barnesi* O. Thomas.

Molossus aztecus lambi Gardner, 1966:1; type locality "11 km. northwest from Esquintla, Chiapas, México."

DISTRIBUTION: *Molossus coibensis* is known from eastern Colombia, Peru, Ecuador, Venezuela, Guyana, French Guiana, and west-central Brazil. Elsewhere, it is known throughout Central America north into the state of Chiapas, Mexico.

MARGINAL LOCALITIES (Map 242): COLOMBIA: Atlántico, Ponedera (ROM 68864). VENEZUELA: Miranda, Guatopo National Park (Dolan 1982). GUYANA:

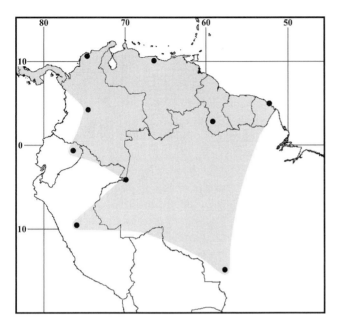

Map 242 Marginal localities for *Molossus coibensis* ●

Upper Takutu-Upper Essequibo, Kuitaro River, 30 km E of Dadanawa (Lim and Engstrom 2001a). FRENCH GUIANA: Cayenne (type locality of *Molossus barnesi* O. Thomas). BRAZIL: Mato Grosso, Tapirapoan (type locality of *Molossus cherriei* J. A. Allen). PERU: Huánuco, 19 miles S of Tingo María (Dolan 1982). COLOMBIA: Amazonas, Leticia (ROM 69583). ECUADOR: Orellana, Onkone Gare, 38 km S of Pompeya Sur (F. A. Reid, Engstrom, and Lim 2000). COLOMBIA: Tolima, Melgar (ROM 65473).

SUBSPECIES: I am treating *M. coibensis* as monotypic.

NATURAL HISTORY: In Mexico and Central America, *M. coibensis* occurs in seasonally dry, tropical deciduous forest habitats at low elevations (Dolan 1989). Specimens from Guyana came from "forest areas."

REMARKS: Although Freeman (1981) synonymized *M. barnesi* with *M. molossus*, the former represents a valid taxon based on records of sympatry of the two species in southern Guyana. Dolan (1982, 1989) found that *M. coibensis* carries a unique LDH allele, lacking in *M. molossus*, and that the two species are sympatric in Panama. I follow Dolan (1982, 1989) in treating *M. coibensis* and *M. barnesi* as conspecific. Should future work indicate otherwise, the name *M. barnesi* should be used for the Guyanan species. Specimens listed from Venezuela as *Molossus aztecus* by Handley (1976) are probably referable to *M. coibensis* (C. O. Handley Jr., pers. comm.). Simmons and Voss (1998) recognized *M. barnesi* as a species separate from *M. coibensis*, but I have included their record from French Guiana as representing *M. coibensis*. As indicated by Lim and Engstrom (2001a), a systematic study of the smaller-sized species of *Molossus* in northern South America is needed.

In the original description (O. Thomas 1905a), the spelling *M. Burnesi* was an obvious typesetter's error. Husson (1962) claimed that Miller (1913b), as first reviser, fixed the spelling as *M. burnesi*. Although Miller (1913b) revised the genus *Molossus*, he did not revise the spelling of the species name *M. barnesi*. Cabrera (1958:129), as the first reviser of the names *burnesi* and *barnesi*, selected the spelling *barnesi* and placed *M. burnesi* in the synonymy of *M. barnesi*. D. C. Carter and Dolan (1978) gave information on and measurements of the holotype of *Molossus barnesi* O. Thomas.

Molossus currentium O. Thomas, 1901
Corrientes Mastiff Bat

SYNONYMS:

Molossus obscurus currentium O. Thomas, 1901j:438; type locality "Goya, Corrientes," Argentina.

Molossus currentium: Miller, 1913b:89; first use of current name combination.

[*Molossus major*] *currentium*: Hershkovitz, 1949c:454; name combination.

Molossus major crassicaudatus: Cabrera, 1958:130; not *Molossus crassicaudatus* É. Geoffroy St.-Hilaire, 1805a.

[*Molossus molossus*] *currentium*: Koopman, 1993:235; name combination.

Molossus molossus crassicaudatus: Barquez, Mares, and Braun, 1999:221; part; not *Molossus crassicaudatus* É. Geoffroy St.-Hilaire.

Molossus currentium currentium: López-González and Presley, 2001:771; name combination.

DISTRIBUTION: *Molossus currentium* is known from Paraguay, northern Argentina, and Uruguay. The species probably occurs in southeastern Brazil and farther north in eastern Bolivia and Peru, as well as in the western Amazon basin of Brazil.

MARGINAL LOCALITIES (Map 241; from López-González and Presley 2001, except as noted): PARAGUAY: Alto Paraguay, Estancia Doña Julia; Alto Paraguay, Laguna General Díaz. URUGUAY: Paysandú, Arroyo Negro, 15 km S Paysandú (AMNH 205683). ARGENTINA: Corrientes, Goya (type locality of *Molossus obscurus currentium* O. Thomas); Entre Rios, Gualeguaychú; Buenos Aires, Tigre.

SUBSPECIES: I am treating *M. currentium* as monotypic.

NATURAL HISTORY: López-González and Presley (2001) pointed out that *M. currentium* occurs in grasslands (savannas) and palm forests. Willig et al. (2000) were the first to record *M. currentium* (as *M. bondae*) in the Matogrossense biome in eastern Paraguay.

REMARKS: Pelage color varies from black to reddish brown. The reddish color is the result of wear (Gardner 1965) and bleaching similar to the process described for the Mormoopidae by J. D. Smith (1972). In their recent revision, López-González and Presley (2001) synonymized *Molossus bondae* under the name *M. currentium*. However, I suggest that *M. bondae* and *M. currentium* represent different species. *Molossus bondae* has unicolored to indistinctly bicolored, short hair; whereas *M. currentium* has bicolored, long hair (D. C. Carter and Dolan, 1978). The holotype of *M. obscurus currentium* (BM 98.3.4.28) is a male; not a female as recorded by López-González and Presley (2001). They assigned a specimen from Manaus, Amazonas, Brazil (USNM 123828) to this species, but questioned its origin. The Manaus specimen is a juvenile and its age makes identification difficult. Although I indicate the Manaus record by a question mark, I have found specimens in the collection of the AMNH, collected on the Rio Madeira at Rosarinho, that are similar in size, pelage length, and color to *M. currentium* as described from Goya, Argentina. Therefore, the juvenile specimen from Manaus may prove to represent *M. currentium*. A detailed study of the small *Molossus* of South America is necessary to determine the species limits of *M. currentium* and *M. molossus*. There are two species of small *Molossus* in Paraguay (López-González and Presley, 2001; Myers and Wetzel, 1983), the larger *M. currentium* and the smaller *M. m. crassicaudatus*. Similarly, I have seen two sizes of small *Molossus* in collections from Bolivia and Peru. D. C. Carter and Dolan (1978) provided information on and measurements of the holotype of *Molossus obscurus currentium* O. Thomas.

Molossus molossus (Pallas, 1766)
Pallas's Mastiff Bat

SYNONYMS: See under Subspecies.

DISTRIBUTION: *Molossus molossus* occurs in the Netherlands Antilles and Trinidad and Tobago, and across northern South America, south throughout the greater Amazon basin into Ecuador, Peru, Bolivia, Paraguay, Uruguay, and northern Argentina. Elsewhere, it is known from the Lesser Antilles as far north as Guadeloupe and in Middle America as far north as southern Mexico.

MARGINAL LOCALITIES (Map 243): NETHERLANDS ANTILLES: Curaçao; Willemstad (type locality of *Molossus pygmaeus* Miller); Bonaire; 8.5 km N and 2 km W of Kralendijk (Genoways and Williams 1979a). VENEZUELA: Nueva Esparta, El Valle (J. D. Smith and Genoways 1974). TRINIDAD AND TOBAGO: Tobago, Charlotteville (Goodwin and Greenhall 1961). GUYANA: Essequibo Islands-West Demerara, Maria's Pleasure (ROM 64082). SURINAM: Paramaribo, Paramaribo (Husson

Map 243 Marginal localities for *Molossus molossus* ●

1962). FRENCH GUIANA: Cayenne (type locality of *Molossus longicaudatus* É. Geoffroy St.-Hilaire). BRAZIL: Amapá, Macapá (C. T. Carvalho 1962a); Ceará, Horto (S. S. P. Silva, Guedes, and Peracchi 2001); Pernambuco, Fazenda Saco (Mares et al. 1981); Bahia, Região de Conquista (Falcão, Soares-Santos, and Drummond 2005); Espírito Santo, Reserva Floresta de Nova Lombardia (type locality of *Molossops planirostris espiritosantensis* Ruschi); São Paulo, Campinas (ROM 75144); Rio Grande do Sul, Quinta (AMNH 235381). URUGUAY: Montevideo, Montevideo (FMNH 42425). ARGENTINA (Barquez, Mares, and Braun 1999): Buenos Aires, Energía; Córdoba, Río Cuarto; La Rioja, Villa Unión; Salta, Cerillos; Salta, Serranía de Las Pavas. BOLIVIA: Santa Cruz, Río Pitasama, 4.5 km N and 1.5 km E of Cerro Amboró (S. Anderson 1997); La Paz, Caranavi (S. Anderson, Koopman, and Creighton 1982). PERU: Cusco, Hacienda Cadena (Sanborn 1951b); Junín, Huacapistana (Koopman 1978); Lambayeque, Etén (Tuttle 1970). ECUADOR: Guayas, Daule (type locality of *Molossus daulensis* J. A. Allen). COLOMBIA: Cauca, El Bordo (ROM 69755); Valle del Cauca, Río Zabaletas (M. E. Thomas 1972); Atlántico, Barranquilla (ROM 45529); La Guajira, Nazareth (ROM 52707).

SUBSPECIES: I tentatively recognize four subspecies of *M. molossus* in South America.

M. m. crassicaudatus É. Geoffroy St.-Hilaire, 1805

SYNONYMS:

Molossus crassi-caudatus É. Geoffroy St.-Hilaire, 1805a: 279; type locality not given; based on Azara's (1801b) "*la dixième chauve-souris*" from Paraguay; restricted to Asunción, Central, Paraguay, by Cabrera (1958:130–131).

Molossus acuticaudatus Desmarest, 1820:116; type locality "Le Brésil."

Dysopes velox Temminck, 1826:234; type locality "Brésil."

Mol[ossus]. moxensis d'Orbigny, 1837:pl. 11, Figs. 1–4; type locality "Moxos," Beni, Bolivia.

Dysopes olivaceo-fuscus J. A. Wagner, 1847:202; type locality "Cuyaba," Mato Grosso, Brazil.

Dysopes amplexicaudatus J. A. Wagner, 1847:203; type locality "Caiçara," Mato Grosso, Brazil.

Molossus crassicaudatus tecticola Osgood, 1916:213; type locality "Juá, near Iguatú," Ceará, Brazil.

Molossus major crassicaudatus: Hershkovitz, 1949c:454; name combination.

Molossops planirostris espiritosantensis Ruschi, 1951f:2; type locality "Reserva Floresta de Nova Lombardia," Santa Teresa, Espírito Santo, Brazil.

Molossus molussus crassicaudatus: Koopman, 1978:21; first use of current name combination.

Molossus molussus Willig, 1985c:671; incorrect subsequent spelling of *Vespertilio molossus* Pallas.

Molosus molosus Polanco-Ochoa, Jaimes, and Piragua, 2000:675; incorrect subsequent spelling of *Vespertilio molossus* Pallas.

This subspecies is distributed east of the Andes from Brazil and southern Colombia into Peru and Argentina.

M. m. daulensis J. A. Allen, 1916

SYNONYM:

Molossus daulensis J. A. Allen, 1916d:530, type locality "Daule," Los Ríos, Ecuador.

This subspecies occurs west of the Andes in Ecuador and northern Peru.

M. m. molossus (Pallas, 1766)

SYNONYMS:

V[espertilio]. Molossus Pallas, 1766:49; part; type locality "America"; restricted to Martinique, West Indies, by Husson (1962:251).

V[espertilio]. Mol[ossus]. major Kerr, 1792:97; type locality "West India islands"; restricted to Martinique, West Indies, by Miller (1913b:90).

V[espertilio]. Mol[ossus]. minor Kerr, 1792:97; type locality "West India islands."

Molossus obscurus É. Geoffroy St.-Hilaire, 1805a:279; type locality not specified; restricted to Martinique, West Indies, by Husson (1962:258).

Molossus longicaudatus É. Geoffroy St.-Hilaire, 1805a: 279; type locality not specified; restricted to Cayenne, French Guiana, by Dolan (1982:131).

Molossus fusci-venter É. Geoffroy St.-Hilaire, 1805a:279; type locality not specified; probably Martinique according to Miller (1913b), and fixed as Martinique, West Indies, by Husson (1962:257).

Molossus fuliginosus Gray, 1838b:501; type locality not given, but identified as Bermuda by D. C. Carter and Dolan (1978:96); name preoccupied by *Molossus fuliginosus* Cooper, 1837 (a junior synonym of *Nycticea cynocephala* J. E. LeConte, 1831).

Molossus tropidorhynchus Gray, 1839:6; type locality "Cuba."

Molossus verrilli J. A. Allen, 1908a:581; type locality "Samana," Dominican Republic.

M[olossus (Molossus)]. obscurus: W. Peters, 1866a:575; name combination.

Molossus debilis Miller, 1913b:90; type locality "St. Kitts, Lesser Antilles."

Molossus fortis Miller, 1913b:89; type locality "Luquillo," Puerto Rico.

Molossus milleri D. H. Johnson, 1952:197, replacement name for *M. fuliginosus* Gray, preoccupied.

Molossus molossus: Husson, 1962:251; first use of current name combination.

This subspecies is common throughout northern South America, including the Guianas. Elsewhere, the nominate form is known from Bermuda and the Lesser Antilles as far north as Guadeloupe, and in Middle America north into Mexico.

M. m. pygmaeus Miller, 1900

SYNONYM:

Molossus pygmaeus Miller, 1900d:162; type locality "in an attic near Willemstad, Curaçao," Netherlands Antilles.

This subspecies is in the Netherlands Antillean islands of Curaçao and Bonaire.

NATURAL HISTORY: Throughout its range, *M. molossus* has been found roosting in houses, hollow trees, and palms, from sea level to as high as 1,250 m in habitats that vary from tropical arid forest to subtropical rain forest (e.g., see Handley 1976). Mares et al. (1981) reported the species from low and high caatinga habitats in the Northeast of Brazil. That is where Willig (1985a) recorded pregnant females from September to February and lactating females from January to May; with an "inactive" interval from June to August. Simmons and Voss (1998) reported two groups of *M. molossus*, each consisting of 11 females and 4 males, and a third group of 8 males and 1 female, roosting in a shed. The species is known to be insectivorous, and Barquez (1983) found that they feed heavily on beetles. Additionally, Aguirre 1994 (*fide* S. Anderson 1997) recorded Zygoptera, Orthoptera, Hemiptera, Homoptera, Diptera, and Hymenoptera from stomachs of Bolivian specimens; five measurable food items were from 5 to 25 mm in length. R. Guerrero (1985a) listed five mites, one hemipteran, and two fleas as known parasites of *M. molossus*; to which he later (1997) added 13 species of streblid batflies. Yunker, Lukoschus and Giesen (1990) added two macronyssid mites, and Graciolli and Carvalho (2001) mentioned a nycteribiid batfly, as known from this species. Three karyotypes have been reported for *M. molossus*, all have a diploid number or 48, but differ in the number of autosomal arms (FN). J. W. Warner et al. (1974) reported an FN = 58 karyotype for *M. molossus* from the West Indies and several localities in northern and western South America. R. J. Baker and Lopez (1968) gave the FN as 56 for Mexican and Central American samples; and T. R. O. Freitas, Bogo, and Christoff (1992) cited an FN count of 54 for samples from southern Brazil.

REMARKS: In an extensive discussion starting with *Vespertilio molossus* Pallas, 1766, Husson (1962) summarized the rather tortuous taxonomic history of *M. molossus*. Similarly, Dolan (1982, 1989) commented on the taxonomic history of small-sized species of *Molossus* and had some different interpretations. D. C. Carter and Dolan (1978) examined types of the following names I am treating here as synonyms of *M. molossus*: the paralectotype of *Molossus fuliginosus* Gray (= *M. milleri* D. H. Johnson), holotype of *Molossus longicaudatus* É. Geoffroy St.-Hilaire, syntypes of *Molossus obscurus* É. Geoffroy St.-Hilaire, holotype of *Molossus tropidorhynchus* Gray, and holotype of *Vespertilio molossus minor* Kerr.

The small members of the genus *Molossus* that have distinctly bicolored fur occur throughout the Antilles, and Middle and South America. Hall (1981) recognized six subspecies in the Antilles. I recognize three subspecies in mainland South America and another in the Netherlands Antilles. Dolan (1989) suggested that if there is a subspecific difference between *M. molossus* from the Lesser Antilles and populations from northern South America, the latter should be referred to as *M. m. minor*, with *M. longicaudatus*, *M. pygmaeus* and *M. daulensis* treated as junior synonyms. Lim and Engstrom (2001a) reported another small species with bicolored fur from Guyana that corresponds in size with *M. pygmaeus*. The collection at LSUMZ contains specimens of a similarly small species with bicolored fur that occurs sympatrically with a larger species in northeastern Peru. Clearly, a morphometric and genetic study of the species of small *Molossus*, including a complete review of the entire *M. molossus* group, is needed to clarify the status of the numerous available names. There are too many ambiguous records in the literature on small species

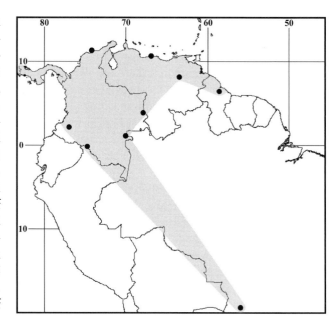

Map 244 Marginal localities for *Molossus pretiosus* ●

of *Molossus* to be able to confidently determine the distributional limits of species.

Molossus pretiosus Miller, 1902
Miller's Mastiff Bat

SYNONYM:

Molossus pretiosus Miller, 1902b:396; type locality "La Guaira," Distrito Federal, Venezuela.

DISTRIBUTION: *Molossus pretiosus* is found in Colombia, Venezuela, Guyana, and Brazil. Elsewhere, it occurs along the Pacific versant of Central America north into Nicaragua (Dolan 1982, 1989).

MARGINAL LOCALITIES (Map 244): COLOMBIA: Magdalena, Santa Marta (ROM 54650). VENEZUELA: Distrito Federal, La Guaira (type locality of *Molossus pretiosus* Miller); Bolívar, Bolívar (AMNH 16076). GUYANA: Cuyuni-Mazaruni, Kartabo (AMNH 14198). COLOMBIA: Guainia, Puerto Inírida (Marinkelle and Cadena 1972); Vaupés, Mitú (ROM 45144). BRAZIL: Mato Grosso do Sul, Fazenda Santa Terezinha (Gregorin and Taddei 2001). COLOMBIA: Putumayo, Puerto Leguizamo (Marinkelle and Cadena 1972); Cauca, El Bordo (ROM 69800).

SUBSPECIES: *Molossus pretiosus* is regarded here as monotypic.

NATURAL HISTORY: Similar to *M. rufus* but smaller, the two species are sympatric in Colombia, Venezuela, Guyana, Brazil, and northern Colombia. Marinkelle and Cadena (1972) reported *M. pretiosus* roosting in caves, hollow trees, and in roofs of buildings. Electrophoretic data confirm conspecificity between South and Central American populations of *M. pretiosus* (Dolan 1982, 1989). R. Guerrero (1985a) listed a mite, and later (1997), two streblid batflies as known ectoparasites.

REMARKS: Specimens of *M. pretiosus* from Colombia average smaller than those from Venezuela.

Molossus rufus É. Geoffroy St.-Hilaire, 1805
Black Mastiff Bat
SYNONYMS:

Molossus rufus É. Geoffroy St.-Hilaire, 1805a:279; type localities "de l'Amerique du nord, de Surinam, et principalement de Caïenne" (É. Geoffroy St.-Hilaire, 1805b: 154); restricted to Cayenne, French Guiana, by Miller (1913b:88).

Molossus castaneus É. Geoffroy St.-Hilaire, 1805a:279; no type locality given; based on Azara's (1801b) "*la chauve-souris châtaine ou la chauve-souris 9e*" from Paraguay; type locality restricted to Asunción, Central, Paraguay, by Cabrera (1958:132).

Molossus ursinus Spix, 1823:59; type locality "suburbiis Provinciae Para," Brazil.

Dysopes alecto Temminck, 1826:231; type locality "Les parties intérieures du Brésil."

Dysopes albus J. A. Wagner, 1843a:368; type locality "Capit. Mato grosso," Brazil; identified as "Engenho do Cap. Gama," Mato Grosso, by Pelzeln (1883:43).

Dysopes holosericeus J. A. Wagner, 1843a:368; type locality "Rio de Janeiro," Rio de Janeiro, Brazil.

Molossus myosurus Tschudi, 1845:83; type locality "Hacienda der Cejaregion 5000' ü. M. am Ostabhange der Binnencordillera [page 85]," Peru.

M[olossus (Molossus)]. rufus: W. Peters, 1866a:575; name combination.

Molossus albus: Pelzeln, 1883:43; name combination.

Molossus fluminensis Lataste, 1891:658; type locality "Rio Janeiro," Rio de Janeiro, Brazil.

[*Molossus (Molossus)*] *fluminensis*: Trouessart, 1897:143; name combination.

Molossus nigricans Miller, 1902b:395; type locality "Acaponeta, Tepic" (= Nayarit), Mexico.

Molossus pretiosus macdougalli Goodwin, 1956a:3; type locality "San Blas, 3 kilometers southeast of the city of Tehuantepec," Oaxaca, Mexico.

Molossus ater: Goodwin, 1960:4; not *Molossus ater* É. Geoffroy St.-Hilaire, 1805a.

DISTRIBUTION: *Molossus rufus* occurs throughout northern South America, including the island of Trinidad, south into Bolivia, Paraguay, southeastern Brazil, and northern Argentina. The species also is common in Central America and Mexico.

MARGINAL LOCALITIES (Map 245; most of the following records were reported as *M. ater*): TRINIDAD AND

Map 245 Marginal localities for *Molossus rufus* ●

TOBAGO: Trinidad, Port of Spain (Goodwin and Greenhall 1961). VENEZUELA: Delta Amacuro, Araguaimujo (Pirlot 1965b). GUYANA: Demerara-Mahaica, Buxton (FMNH 46333). SURINAM: Paramaribo, Paramaribo (Husson 1962). FRENCH GUIANA: Cayenne, (Brosset and Dubost 1968). BRAZIL: Amapá, Macapá (C. T. Carvalho 1962a); Pará, Belém (Piccinini 1974); Maranhão, Barra do Corda (C. O. C. Vieira 1957); Pernambuco, Dois Irmãos (Pohle 1927); Alagoas, Canoas (C. O. C. Vieira 1953); Espírito Santo, Campanhia Vale do Rio Dôce Forest Reserve (Pedro and Passos 1995); Rio de Janeiro, Rio de Janeiro (J. A. Wagner 1847); São Paulo, São Sebastião (USNM 141443); Paraná, Floresta Nacional do Irati (N. R. Reis et al. 2000); Rio Grande do Sul (C. O. C. Vieira 1955; no locality mentioned). ARGENTINA: Santa Fe, Santa Fe (Barquez, Mares, and Braun 1999); Córdoba, Alta Gracia (Barquez, Mares, and Braun 1999), Salta, Salta (Villa-R. and Villa-Cornejo 1969). BOLIVIA: Santa Cruz, 2 km S of Caranda (S. Anderson 1997); Beni, Rurrenabaque (S. Anderson, Koopman, and Creighton 1982). PERU: Cusco, Huajyumbe (Sanborn 1951b); Amazonas, Pomará (Koopman 1978). ECUADOR: Pastaza, Canelos (Albuja 1983). COLOMBIA: Norte de Santander, Cúcuta (AMNH 183317). VENEZUELA: Carabobo, Montalbán (Handley 1976).

SUBSPECIES: I am treating *M. rufus* as monotypic.

NATURAL HISTORY: Roosts have been found in houses, hollow trees, hollow logs, and in rock crevices at elevations from sea level to 1,500 m. Aguirre 1994 (*fide* S. Anderson 1997) recorded *M. rufus* roosting in palm fronds with *M. molossus*. *Molossus rufus* has been recorded from a variety of habitats and vegetative zones ranging from tropical dry forest to subtropical rain forest (Handley 1976). The species is sedentary and local demes vary morphologically and genetically (Dolan 1982). Dolan (1989) claimed that all species of *Molossus* appear to be polyestrous with two peaks of parturition per year, one in April-May and the second in July-August. Esbérard (2002) monitored reproduction in *M. rufus* throughout an entire year in Rio de Janeiro, Brazil, and found that females, and most males, left the roost he studied after the end of reproduction. He found pregnant females in September, October, November, and February; lactating females in August, October, November, December, and February. Only in July did he find a majority of males with abdominal testes; most if not all males examined in other months had descended testes, and presumably were reproductively active. In Bolivia, S. Anderson (1997) reported a pregnant female with a single embryo taken in November. Aguirre (1994; *fide* S. Anderson 1997) examined five stomachs of *M. rufus* taken in Espíritu, Bolivia, and found the remains of Zygoptera, Orthoptera, Hemiptera, and Coleoptera, of which six food items measured from 5 to 20 mm in length. New pelage is black, and the color of worn and ammonia-bleached pelage varies from red, through brown, to yellow. R. Guerrero (1985a, 1997) and Yunker, Lukoschus, and Giesen (1990) listed ectoparasites recovered from *M. rufus* (under *Molossus ater*).

REMARKS: Dolan (1989) recognized a single taxon of the Black Mastiff Bat because of the relative genic homogeneity and absence of geographic variation among Central American samples. However, she cautioned that there is an apparent hiatus between Middle and South American populations, indicating that further study is necessary to determine whether or not *M. rufus* is monotypic.

Étienne Geoffroy St.-Hilaire (1805a) named nine species of *Molossus* and, in his later paper published the same year (1805b), provided measurements and additional information. Miller (1913b) recognized *Molossus rufus* as a valid taxon, but Goodwin (1960) later incorrectly concluded that *Molossus rufus* was a synonym of *Eumops auripendulus* (G. Shaw, 1800). Goodwin believed the correct name for the common large *Molossus* was *M. ater* É. Geoffroy St.-Hilaire, 1805a, because he thought that Geoffroy St.-Hilaire's measurements of *M. rufus* indicated a bat too large to represent the large species of *Molossus*. Goodwin apparently did not consider Geoffroy St.-Hilaire's description of *Molossus ater* as having a more slender muzzle and larger, higher ears than *M. rufus* to be important. Geoffroy St.-Hilaire's description suggests that *M. ater* represents a

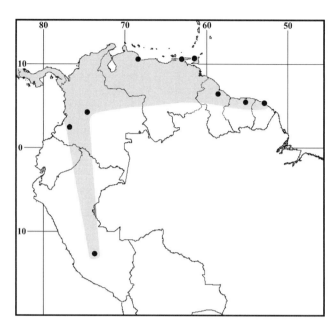

Map 246 Marginal localities for *Molossus sinaloae* ●

Eumops, but this cannot be verified because the holotype has been lost. D. C. Carter and Dolan (1978:99) verified the identity of *M. rufus* and designated a lectotype (MNHN A.428/224). They also examined and presented information on and measurements of the types of *Dysopes albus* J. A. Wagner, *Dysopes alecto* Temminck, and *Molossus fluminensis* Lataste.

Molossus sinaloae J. A. Allen, 1906
Sinaloan Mastiff Bat
SYNONYMS:

Molossus sinaloae J. A. Allen, 1906b:236; type locality "Escuinapa, Sinaloa," Mexico.

Molossus trinitatus Goodwin, 1959a:1; type locality "Belmont, Port of Spain, Trinidad."

M[*olossus*]. *s*[*inaloae*]. *trinitatus*: Handley, 1966c:773; name combination.

DISTRIBUTION: *Molossus sinaloae* occurs in western and northern Colombia, northern Venezuela, Surinam, French Guiana, and there is a single record from Peru. Elsewhere, it is known from Panama northwest into western-central Mexico.

MARGINAL LOCALITIES (Map 246; localities listed from west to east): COLOMBIA: Cauca, Mina California (Marinkelle and Cadena 1972); Tolima, Melgar (ROM 69582). PERU: Ayacucho, Hacienda Luisiana (LSUMZ 16649). VENEZUELA: Yaracuy, 10 km NW of Urama (Handley 1976); Sucre, Guaraúnos (Linares and Rivas 2004). TRINIDAD AND TOBAGO: Trinidad, Belmont (type locality of *Molossus trinitatus* Goodwin). GUYANA: Cuyuni-Mazaruni, Kartabo (Gardner 1965). SURINAM:

Para, Zanderijweg (Husson 1978). FRENCH GUIANA: Paracou (Simmons and Voss 1998).

SUBSPECIES: I am treating *M. sinaloae* as monotypic.

NATURAL HISTORY: Little is known about the ecology of *M. sinaloae*. The species has been found from sea level to 2,400 m, and is known to roost in houses (Goodwin 1959a). Most specimens have been caught in mist nets over ponds and pools in streams or over river beaches. This species occurs in tropical dry and subtropical humid forest (Handley 1976). R. Guerrero (1985a, under *M. trinitatus*) listed a mite, a tick and the protozoan, *Trypanosoma cruzi*, as known parasites. Jennings et al. (2002) summarized available information on this species in their *Mammalian Species* account. *Molossus sinaloae* has a $2n = 48$, $FN = 58$ karyotype (J. W. Warner et al. 1974).

REMARKS: Various authors have taken opposing views on the status of *Molossus trinitatus*. Handley (1966c) and Ojasti and Linares (1971) considered it a subspecies of *M. sinaloae*, while Husson (1978) and Freeman (1981) recognized *M. trinitatus* as a valid species. Both Koopman (1982) and Dolan (1989) include *M. trinitatus* in *M. sinaloae*. Simmons and Voss (1998) suggested that these taxa be recognized as subspecies. Jennings et al. (2002) recognized the populations from Mexico into Costa Rica as *M. s. sinaloae* J. A. Allen, and the Panamanian and South American population as *M. s. trinitatus* Goodwin. Genetically, Middle American samples of *M. sinaloae* are the most distinctive among species of *Molossus* (Dolan 1982, 1989). A subadult female (LSUMZ 16649) netted over a long, shallow pool in the gravel beach of the Río Apurímac at Hacienda Luisiana, Ayacucho, Peru, is tentatively identified as this species.

Genus *Mormopterus* W. Peters, 1865

Nearly pantropical, *Mormopterus* is found in the Neotropics, the Mascarene Islands, Madagascar, Moluccas, Sumatra, New Guinea, and Australia. Of the three Neotropical species, two are South American and one is Cuban. Although known from the late Oligocene of Brazil, the early and middle Miocene of Europe (Legendre 1984b), and the middle Miocene of Colombia (Czaplewski 1997), bats of this genus are uncommon and have been little studied. They are characterized by distinctly separated, simple ears having a reduced keel; a well-developed tragus; and a reduced antitragus that is wider than high. The lips are wrinkled. Basisphenoid pits are weakly developed or absent, the palate is emarginate anteriorly, and the third commissure of M3 is longer than the second. The dental formula is 1/2–3, 1/1, 1–2/2, 3/3 × 2 = 28 or 32.

SYNONYMS:

Vespertilio: Hermann, 1804:19; not *Vespertilio* Linnaeus, 1758.

Molossus: Gray, 1839:7; not *Molossus* É. Geoffroy St.-Hilaire, 1805a.

Dysopes: W. Peters, 1865a:258; part; not *Dysopes* Illiger, 1811.

Mormopterus W. Peters, 1865a:258; type species *Mormopterus jugularis* W. Peters, 1865a, by monotypy; described as a subgenus of *Dysopes* Illiger.

Nyctinomus: W. Peters in Sclater, 1865:468; not *Nyctinomus* É. Geoffroy St.-Hilaire, 1818b.

Nyctinomus: O. Thomas, 1893c:334; not *Nyctinomus* É. Geoffroy St.-Hilaire.

Micronomus Iredale and Troughton, 1934:100; *nomen nudum*.

Micronomus Troughton, 1943:361; type species *Micronomus norfolkensis* Gray, 1839, by original designation.

Tadarida: Koopman, 1978:21; not *Tadarida* Rafinesque, 1814.

REMARKS: de la Torre (1956) showed the holotype of *Mormopterus peruanus* J. A. Allen 1914, to be a specimen of *Tadarida brasiliensis* I. Geoffroy St.-Hilaire, 1824.

KEY TO THE SOUTH AMERICAN SPECIES OF *MORMOPTERUS*:

1. P3 present; pelage dark brown; brain case high (5.5–5.6 mm), dorsal profile of skull arched; anterior narial opening large; forearm 34.1–35.2 mm . . . . *Mormopterus phrudus*

1′. P3 absent; pelage pale gray; brain case low (5.0–5.2 mm), dorsal profile of skull flat; anterior narial opening smaller; forearm 34.3–39.0 mm . . . . *Mormopterus kalinowskii*

Mormopterus kalinowskii (O. Thomas, 1893)
Kalinowski's Mastiff Bat

SYNONYMS:

Nyctinomus kalinowskii O. Thomas, 1893c:334; type locality "central Peru."

[*Nyctinomus (Nyctinomus)*] *Kalinowskii*: Trouessart, 1897:148; name combination.

Mormopterus kalinowskii: Miller, 1907b:254; first use of current name combination.

Tadarida (Mormopterus) kalinowskii: Koopman, 1978:21; name combination.

Mormopterus (Mormopterus) kalinowskii: Freeman, 1981:161; name combination.

DISTRIBUTION: *Mormopterus kalinowskii* is known from the dry Pacific slope of the Andes and the arid inter-Andean river valleys of Peru and Chile.

MARGINAL LOCALITIES (Map 247): PERU: Lambayeque, 7 km S of Motupe (MVZ 154702); Cajamarca, Hacienda Limón (FMNH 19952); Cusco, Machu Picchu (AMNH 91553). CHILE: Tarapacá, Iquique (Mann

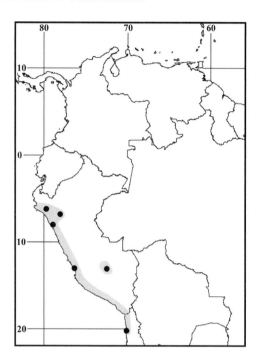

Map 247 Marginal localities for *Mormopterus kalinowskii* ●

1950b). PERU: Lima, Cañete (LSUMZ 16638); La Libertad, Trujillo (AMNH 165626).

SUBSPECIES: I consider *M. kalinowskii* to be monotypic.

NATURAL HISTORY: Essentially nothing is known of the life history of *M. kalinowskii*, except that it has been taken from near sea level to as high as 1,830 m. The karyotype for *M. kalinowskii* is $2n = 48$, FN $= 56$ (J. W. Warner et al. 1974).

REMARKS: Two specimens in the ROM collection have a small upper premolar supposedly characteristic of *M. phrudus*. As noted by Handley (1956), the association of the generic name *Mormopterus* with *M. kalinowskii* and *M. phrudus* is questionable. *Mormopterus* is characterized by distinctly separated simple ears as seen in *M. minutus* (Cuba) and *M. jugularis* (Madagascar). The ears of two specimens of *M. kalinowskii* preserved in alcohol (ROM 93774, 93775) resemble those of *Tadarida brasiliensis* more than they do those of South American *Mormopterus*. The ears are less than 1 mm apart at the base and are similar to *T. brasiliensis* in shape, although smaller. Characters shared by *Tadarida* and *Mormopterus* include a reduced ear keel, reduced antitragus, slightly wrinkled lips, and lack of basisphenoid pits. The specimen (AMNH 91553) listed by Freeman (1981:8) as representing "*Tadarida (Mormopterus) phrudus*" is a *Mormopterus kalinowskii*. D. C. Carter and Dolan (1978) gave information on and measurements of the holotype of *Nyctinomus kalinowskii* O. Thomas.

Mormopterus phrudus Handley, 1956
Incan Mastiff Bat

SYNONYMS:

Mormopterus phrudus Handley, 1956:197; type locality "San Miguel Bridge, Urubamba River, Machu Picchu, Cuzco, Peru."

Tadarida (Mormopterus) phrudus: Koopman, 1978:21; name combination.

Mormopterus (Mormopterus) phrudus: Freeman, 1981: 161; name combination.

DISTRIBUTION: *Mormopterus phrudus* is known from only the type locality in departamento Cusco, Peru.

MARGINAL LOCALITIES (Map 248): PERU: Cusco, Machu Picchu (type locality of *Mormopterus phrudus* Handley).

SUBSPECIES: *Mormopterus phrudus* is monotypic.

NATURAL HISTORY: The two known specimens of *M. phrudus* were captured from the San Miguel Bridge over the Río Urubamba, at an elevation of approximately 2,000 m, in lower subtropical forest habitat.

REMARKS: The species is known only from the holotype and one paratype (USNM 194449 and 194450, respectively). The specimen figured by Freeman (1981:77, Fig. 16) as a *M. phrudus* is a *M. kalinowskii*.

Genus *Neoplatymops* R. L. Peterson, 1965

Neoplatymops is a monotypic genus endemic to South America. Externally, *Neoplatymops* is a small molossid characterized by simple, widely separated ears, and by small

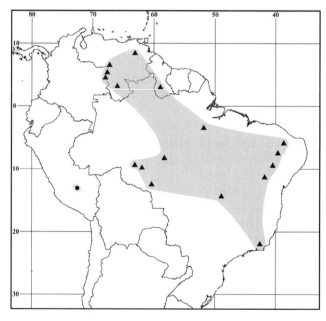

Map 248 Marginal localities for *Mormopterus phrudus* ● and *Neoplatymops mattogrossensis* ▲

wart-like granulations on dorsal surfaces of the forearms. The gular gland is present in both sexes. The skull is flattened and lacks a sagittal crest. Lacrimal ridges are well developed, and the infraorbital foramina are conspicuous and open dorso-anteriorly. The premaxillaries are joined anteriorly (no anterior palatine emargination). Upper incisors are hook shaped, separated from each other and from the canines by gaps; the lower incisors are deeply bifid and arranged in a semicircle; the first and second upper molars have reduced hypocones; and M3 has a well-developed third commissure. The dental formula is 1/2, 1/1, 2/2, 3/3 × 2 = 30.

Externally, *Neoplatymops* can be distinguished from *Molossops* and *Cynomops* by the presence of wart-like granulations on the forearm, two instead of one upper premolar in each maxilla, a gap between upper incisors, more widely separated lower canines, and larger infraorbital canals. The skull is flatter and has a relatively wider, shallower braincase. The body also is flattened. In addition, *Neoplatymops* differs from *Molossops* by having two pairs of lower incisors; a longer, narrower rostrum; and relatively narrower wing tips. *Neoplatymops* differs from *Cynomops* by having relatively narrower wing tips, M3 has a well-developed third commissure; and the lower incisors are deeply bifid and aligned in an arc. In contrast, *Cynomops* has comparatively broader wing tips, M3 lacks a third commissure, the lower incisors are not deeply bifid, and the outer lower incisors are crowded behind the inner incisors.

SYNONYMS:

Molossops: C. O. C. Vieira, 1942:430; not *Molossops* W. Peters, 1866a.

Neoplatymops R. L. Peterson, 1965a:3; type species *Molossops mattogrossensis* C. O. C. Vieira, 1942, by original designation.

REMARKS: This taxon was revised by R. L. Peterson (1965a) and part of this account is taken from that publication. Some authors (e.g., Freeman 1981; Koopman 1993) include *Neoplatymops* as a subgenus of *Molossops*. I suggest, however, that *Neoplatymops* is sufficiently distinct to warrant generic status.

Neoplatymops mattogrossensis (C. O. C. Vieira, 1942)
Mato Grosso Dog-faced Bat

SYNONYMS:

Molossops mattogrossensis C. O. C. Vieira, 1942:430; type locality "S. Simão, rio Juruena, norte de Mato Grosso," Brazil.

Molossops temminckii mattogrossensis: Cabrera, 1958: 117; name combination.

[*Cynomops*] *mattogrossensis*: Goodwin, 1958b:5; name combination.

Neoplatymops mattogrossensis: R. L. Peterson, 1965a:3; generic description and first use of current name combination.

Molossops (Neoplatymops) mattogrossensis: Freeman, 1981:156; name combination.

Neoplatymops mattogrossensis bolivarensis Linares and Escalante, 1992:422; type locality "Represa de Guri, Lower Caroni River, Bolivar State, Venezuela, 180 m."

DISTRIBUTION: *Neoplatymops mattogrossensis* is known from Venezuela, Guyana, Brazil, and eastern Colombia.

MARGINAL LOCALITIES (Map 248): VENEZUELA: Anzoátegui, Mamo (Ochoa and Ibáñez 1985). GUYANA: Upper Takutu-Upper Essequibo, Run-ny Decou (ROM 41567). BRAZIL (Gregorin 1998a, except as noted): Pará, Cachoeira Juruá; Ceará, Jaguaribe; Pernambuco, Fazenda Cantareno (Mares et al. 1981); Bahia, Joazeiro; Bahia, Irecê (I. Sazima and Taddei 1976); Rio de Janeiro, Fazenda Providência (Avilla, Rozenzstranch, and Abrantes 2001); Goiás, Cerro da Mesa (Avilla, Rozenzstranch, and Abrantes 2001); Rondônia, Km 575, BR 364; Mato Grosso, São Simão (type locality of *Molossops mattogrossensis* C. O. C. Vieira); Rondônia, Cachoeira Nazaré, W bank of Rio Ji-Paraná (MPEG 21262); Rondônia, BR 364, near Ariquemes. VENEZUELA: Amazonas, Tapara (R. L. Peterson 1965a). COLOMBIA: Vichada, 1 km N of the mouth of Caño Matavén (Ortiz-Von Halley and Alberico 1989). VENEZUELA: Amazonas, El Raudal (USNM 409576); Apure, Hato Cariben, 32 km NE of Puerto Paez (Handley 1976).

SUBSPECIES: *Neoplatymops mattogrossensis* is treated here as monotypic, but see Linares and Escalante (1992) who described a subspecies from Venezuela.

NATURAL HISTORY: The species has been recorded from tropical dry forest at elevations from 75 to 195 m, and in humid forest and savannas in Venezuela (Handley 1976; Linares and Escalante 1992). *Neoplatymops mattogrossensis* is common in rocky habitats, including the serrotes in the Caatinga of Brazil, where it roosts in narrow rock crevices (Mares et al. 1981; Willig 1985b), and underneath stones in savanna habitats (R. L. Peterson 1965a; Ortiz-Van Halley and Alberico 1989). The *N. mattogrossensis* that Avilla, Rozenzstranch, and Abrantes (2001) reported from the state of Rio de Janeiro was roosting with two others in a cavity, shared with a *Uroderma bilobatum,* in wooden fence post. In the state of Rondônia, Brazil, individuals were mistnetted as they flew through tree gaps and along the banks of the Rio Ji-Paraná in tropical forest habitat (A. L. Gardner, pers. comm.). In Venezuela, this species exhibits a harem social system with roosting groups consisting of a single male accompanied by two to four females (Linares and Escalante 1992). Linares and Escalente (1992) often found these bats associated with a small pallid scorpion, *Rhopalurus laticauda*, and a gekonid lizard, *Phyllodactylus* sp. Willig (1985b) reported that *N. mattogrossensis* in the Northeast of Brazil exhibited seasonal monoestry with parturition coinciding with the onset of the rainy season. The diet included a wide range of insects, with beetles and flies predominating (Willig 1985b). The species has a $2n = 48$, $FN = 60$ karyotype (or $FN = 62$, if the sex chromosomes are acrocentric; Willig and Jones 1985). R. Guerrero (1985a) listed a soft tick as an ectoparasite. Willig and Jones (1985) summarized available information on this species in their *Mammalian Species* account.

Genus *Nyctinomops* Miller, 1902

Species of the New World genus *Nyctinomops* can be distinguished from other molossids by their large ears, which are joined medially; an antitragus that is higher than wide; a small, rectangular tragus; and wrinkled lips. The second phalanx of digit IV is shorter than the first; the anterior palatal emargination is narrow; and the third commissure of M3 is well-developed and as long as the second. The dental formula is 1/2, 1/1, 2/2, 3/3 × 2 = 30.

SYNONYMS:

V[espertilio]: Pallas, 1766:49; not *Vespertilio* Linnaeus, 1758.

Molossus É. Geoffroy St.-Hilaire, 1805a:278; part.

Molossus: Rengger, 1830:88; not *Molossus* É. Geoffroy St.-Hilaire.

Nyctinomus: Gray, 1839:5; not *Nyctinomus* É. Geoffroy St.-Hilaire, 1818b.

Dysopes: J. A. Wagner, 1843a:368; not *Dysopes* Illiger, 1811.

Dysopes: T. R. Peale, 1848:21; not *Dysopes* Illiger.

Nyctinomus: Dobson, 1876:728; not *Nyctinomus* É. Geoffroy St.-Hilaire.

Nyctinomus: H. Allen, 1889:558; not *Nyctinomus* É. Geoffroy St.-Hilaire.

Nyctinomus: Ward, 1891:747; not *Nyctinomus* É. Geoffroy St.-Hilaire.

Nyctinomus: H. Allen, 1894a:171; not *Nyctinomus* É. Geoffroy St.-Hilaire.

Promops: J. A. Allen, 1900a:91; not *Promops* P. Gervais, 1856a.

Nyctinomops Miller, 1902b:393; type species *Nyctinomus femorosaccus* Merriam, 1889, by original designation.

Nyctinomus: J. A. Allen, 1914:386; not *Nyctinomus* É. Geoffroy St.-Hilaire.

Tadarida: Shamel, 1931:11; part (*macrotis* group); not *Tadarida* Rafinesque, 1814.

Tadarida: Ruschi, 1951f:19; not *Tadarida* Rafinesque.

Tadarida: Barriga-Bonilla, 1965:249; not *Tadarida* Rafinesque.

Nictynomops Esbérard and Bergallo, 2005:515; incorrect subsequent spelling of *Nyctinomops* Miller.

Nictynomops Esbérard and Bergallo, 2006:239; incorrect subsequent spelling of *Nyctinomops* Miller.

REMARKS: The genus *Nyctinomops* was resurrected by Freeman (1981) and, as currently understood, includes four New World species (*N. aurispinosus*, *N. femorosaccus*, *N. laticaudatus*, and *N. macrotis*) formerly treated as species of *Tadarida*. *Nyctinomops femorosaccus* is the only species not known from South America.

KEY TO THE SOUTH AMERICAN SPECIES OF *NYCTINOMOPS*:

1. Forearm longer than 55 mm; greatest length of skull more than 22 mm; basisphenoid pits large and relatively deep . . . . *Nyctinomops macrotis*
1'. Forearm shorter than 55 mm; greatest length of skull less than 22 mm; basisphenoid pits shallow 2
2. Forearm usually longer than 47 mm; greatest length of skull more than 19 mm . . . . *Nyctinomops aurispinosus*
2'. Forearm shorter (40–46 mm); greatest length of skull less than 19 mm . . . . *Nyctinomops laticaudatus*

Nyctinomops aurispinosus (T. R. Peale, 1848)
Peale's Free-tailed Bat

SYNONYMS:

Dysopes aurispinosus T. R. Peale, 1848:21; type locality "on board the U.S. Ship Peacock, off the coast of Brazil . . . about one hundred miles from land, south of Cape St. Roque."

Molossus aurispinosus: Cassin, 1858:5; name combination.

N[yctinomops]. aurispinosus: Miller, 1902b:393; first use of current name combination.

Tadarida aurispinosa: Shamel, 1931:11; name combination.

Tadarida similis Sanborn, 1941:386; type locality "Bogotá, Colombia."

DISTRIBUTION: *Nyctinomops aurispinosus* is known from Colombia, Venezuela, Peru, Bolivia, and Brazil. Elsewhere, it has been recorded in Central America and in Mexico as far north as Sinaloa.

MARGINAL LOCALITIES (Map 249): VENEZUELA: Aragua, Rancho Grande (Ochoa 1984). BRAZIL: off the coast of Brazil, 100 miles from land, S of Cape St. Roque (type locality of *Dysopes aurispinosus* T. R. Peale; not mapped); Piauí, Paulistana (Vizotto, Rodriques, and Dumbra 1980a); São Paulo, São José do Rio Prêto (Taddei and Garutti 1981). BOLIVIA: Santa Cruz, Hacienda Cerro Colorado (Ibáñez and Ochoa 1989); La Paz, La Reserva (S. Anderson 1997). PERU: Cusco, Huajyumbe (Sanborn 1951b); Lima, Lima (Ortiz de la Puente 1951); Lambayeque, 12 km N of Olmos (LSUMZ 25021). COLOMBIA:

Map 249 Marginal localities for *Nyctinomops aurispinosus* ●

Cundinamarca, Bogotá (type locality of *Tadarida similis* Sanborn).

SUBSPECIES: I am treating *N. aurispinosus* as monotypic.

NATURAL HISTORY: Most records are of specimens netted over pools in streams and rivers in thorn forest habitat (e.g., Ibáñez and Ochoa 1989). The species has been found in caves roosting with other species of *Nyctinomops*, such as *N. laticaudatus* (see D. C. Carter and Davis 1961; as *N. yucanicus*) and *N. macrotis* (see Sanborn 1951b). Females netted in late September in Bolivia were pregnant with fetuses measuring from 8 to 11 mm CR (Ibáñez and Ochoa 1989). J. K. Jones and Arroyo-Cabrales (1990) provided measurements and additional natural history information in their *Mammalian Species* account. *Nyctinomops aurispinosus* has a $2n = 48$, $FN = 58$ karyotype (J. W. Warner et al. 1974).

REMARKS: A small series of *N. aurispinosus* collected in Lambayeque, Peru (LSUMZ 25011–25014, 25021–25022, 25029) is slightly smaller (with no overlap in condylobasal length) than the series whose measurements were recorded by D. C. Carter and Davis (1961). Similarly, specimens identified as *N. laticaudatus europs* from Bolivia (S. Anderson, Koopman, and Creighton 1982) are slightly smaller than *N. aurispinosus* from Peru and specimens of *N. laticaudatus* from Guyana are slightly smaller than Bolivian and Paraguayan *N. laticaudatus*. Measurements of *N. aurispinosus* reported by Ochoa (1984) indicate that the species is largest in Venezuela. Clearly the two species can be confused when identification is based on measurements alone, and further study is needed to determine the extent

of variation within and between *N. aurispinosus* and *N. laticaudatus*.

Nyctinomops laticaudatus (É. Geoffroy St.-Hilaire, 1805)
Geoffroy's Free-tailed Bat

SYNONYMS: The following synonyms based on extralimital populations; additional synonyms are listed under Subspecies.

Nyctinomops yucatanicus Miller, 1902b:393; type locality "Chichen Itza, Yucatan," Mexico.

Tadarida yucatanica: Miller, 1924:87; name combination.

Tadarida laticaudata yucatanica: J. K. Jones and Alvarez, 1962:129; name combination.

DISTRIBUTION: *Nyctinomops laticaudatus* occurs in Colombia, Venezuela, Trinidad, the Guianas, and northern Brazil, southward into Paraguay, Bolivia, southeastern Brazil, and northern Argentina. There is a single record from western Peru, but the species has not been recorded from Ecuador and Uruguay. Elsewhere, *N. laticaudatus* occurs in Central America and Mexico.

MARGINAL LOCALITIES (Map 250): TRINIDAD AND TOBAGO: Trinidad, Port of Spain (Goodwin and Greenhall 1961). GUYANA: Upper Takutu-Upper Essequibo, Kuitaro River, 64 km E of Dadanawa (ROM 55028). SURINAM: Marowijne, near junction of Marowijne and Gonini Rivers (Husson 1962). FRENCH GUIANA: Camopi (Brosset and Dubost 1968). BRAZIL: Pernambuco, Exu (Mares et al. 1981); Bahia, Barra (FMNH 20988); Espírito Santo, Treis Barras (type locality of *Tadarida espiritosantensis* Ruschi); Rio de Janeiro (no specific locality; Esbérard and Bergallo 2006); São Paulo, Piracicaba (USNM

Map 250 Marginal localities for *Nyctinomops laticaudatus* ●

123830); Rio Grande do Sul, Tôrres (F. Silva and Souza 1980). ARGENTINA: Misiones, Arroyo Yabebyri (Massoia, Chebez, and Heinonen-Fortabat 1989b); Formosa, 13 km S of Clorinda (Barquez and Ojeda 1975); Tucumán, San Miguel de Tucumán (Barquez and Díaz 2001); Salta, 22 km SW of Orán (Mares, Ojeda, and Barquez 1989). BOLIVIA: Tarija, 8 km S and 10 km E of Villa Montes (S. Anderson, Koopman, and Creighton 1982); La Paz, Coroico (S. Anderson and Webster 1983). PERU: Lambayeque, 12 km N of Olmos (Sudman, Barkley, and Hafner 1994). COLOMBIA: Meta, La Angostura (type locality of *Tadarida laticaudata macarenensis* Barriga-Bonilla). VENEZUELA: Yaracuy, 10 km NW of Urama (Handley 1976).

SUBSPECIES: I recognize three subspecies of *N. laticaudatus* in South America.

N. l. europs (H. Allen, 1889)
SYNONYMS:

Nyctinomus europs H. Allen, 1889:558; type locality "Brazil"; restricted by Cabrera (1958:121) to Corumba, Mato Grosso [do Sul], Brazil.

[*Nyctinomus (Nyctinomus)*] *europs*: Trouessart, 1897:148; name combination.

Tadarida laticaudata macarenensis Barriga-Bonilla, 1965:249; type locality "margin derecha del Río Guayabero (= alto Río Guaviare), La Angostura [= Primera Angostura]," Meta, Colombia; may be valid as a subspecies.

This subspecies appears to be distributed in Venezuela, Colombia, Trinidad, the Guianas, and north central Brazil (see Remarks).

N. l. laticaudatus (É. Geoffroy St.-Hilaire, 1805)
SYNONYMS:

Molossus laticaudatus É. Geoffroy St.-Hilaire, 1805a:279; no locality given; based on Azara's (1801) "*chauve-souris obscure*"; identified as Paraguay by Oken (1816); restricted to Asunción, Central, Paraguay, by Cabrera (1958:121).

Molossus caecus Rengger, 1830:88; type locality "Asuncion," Central, Paraguay.

N[*yctinomops*]. *laticaudatus*: Miller, 1902b:393; first use of current name combination.

Tadarida laticaudata: Shamel, 1931:12; name combination.

This subspecies occurs in southern Brazil, eastern Paraguay, and eastern Bolivia (see Remarks).

N. l. gracilis (J. A. Wagner, 1843)
SYNONYMS:

Dysopes gracilis J. A. Wagner, 1843a:368; type locality "Cuyaba," Mato Grosso, Brazil.

N[*yctinomus (Nyctinomus)*]. *gracilis*: W. Peters, 1866a:573; name combination.

Tadarida gracilis: Miller, 1924:86; name combination.

Tadarida espiritosantensis Ruschi, 1951f:19; type locality "'Treis Barras,' no Município de Santa Teresa," Espírito Santo, Brazil.

This subspecies is found in central and eastern Brazil.

NATURAL HISTORY: Goodwin and Greenhall (1961) found colonies of more than 50 in rock crevices in Venezuela. Specimens in the MBUCV and the USNM taken in 1984 from the vicinity of San Carlos de Río Negro, Amazonas, Venezuela, came from a colony of several hundred occupying crevasses in a small rock island in the Río Negro (A. L. Gardner, pers. comm.). The species also has been found in cracks in the trunks of trees (Barriga-Bonilla 1965), in crevices between tightly packed fronds of the palm *Copernicia vespertilionum* in Cuba (Silva-Taboada and Koopman 1964), and in houses. Handley (1976) reported collections from dry to wet tropical forest habitats in Venezuela, mostly at lower elevations. R. Guerrero (1997) listed a streblid batfly that was not mentioned in the list of known parasites summarized by Avila-Flores, Flores-Martínez, and Ortega (2002) in their *Mammalian Species* account. *Nyctinomops laticaudatus* has a $2n = 48$, FN $= 58$ karyotype (J. W. Warner et al. 1974).

REMARKS: Barriga-Bonilla (1965:254) stated that the subspecies in northern Colombia is probably *N. l. yucatanicus*, but did not list either a specimen or any locality record. *Nyctinomops l. yucatanicus* (Miller, 1902b), Mexican and Central American in distribution, has been recorded in adjacent Panama (Silva-Taboada and Koopman 1964). Another subspecies, *N. l. ferrugineus* (Goodwin, 1954a), occurs in central Mexico.

The record of *N. laticaudatus* from Piura, Peru (TCWC 12443), listed by Koopman (1978) actually is based on a specimen of *N. aurispinosus*, as is the record in Graham and Barkley (1984). However, Sudman, Barkley, and Hafner (1994) listed both *N. laticaudatus* and *N. aurispinosus* from the same location in Peru (12 km N of Olmos, Lambayeque), and indicated a 5.88% sequence divergence of mitochondrial DNA between the two species. Their record of *N. laticaudatus* is the first from the Pacific slope of the Andes and could represent either the Central American subspecies *N. l. yucatanicus*, or an undescribed population.

The designation of subspecies in South America is problematic. Cabrera (1958) restricted the type locality of *T. europs* to Cuyabá, Mato Grosso, Brazil. Silva-Taboada and Koopman (1964) synonymized *Tadarida gracilis* with *T. l. laticaudata* and referred to *T. l. europs* from Venezuela and Brazil as the smallest subspecies. If the names *laticaudatus*, *europs*, and *gracilis* all refer to the same species, then Cabrera (1958) confused the picture by restricting the type locality of *Nyctinomus europs* H. Allen, to Cuyabá, when based on size, the type of *europs* is more similar to specimens from populations found in Venezuela, Guyana, and northern Brazil. Specimens from Guyana in the ROM have a forearm length of 39–42 mm. Although S. Anderson, Koopman, and Creighton (1982) reported specimens of *N. laticaudatus europs* from Bolivia, their specimens are closer in size to *N. l. laticaudatus*. Subsequently, S. Anderson (1997) recognized both subspecies in Bolivia. Handley (1976) recognized *N. gracilis* (J. A. Wagner) and *N. laticaudatus* as separate species. I am following Silva-Taboada and Koopman (1964) in recognizing a single species, *N. laticaudatus*. Silva-Taboada and Koopman (1964) claimed that size followed a bipolar cline within *N. laticaudatus*, and that specimens from the northern and southern limits of its distribution were the largest, whereas those from Venezuelan and Brazilian localities were the smallest.

Avila-Flores, Flores-Martínez, and Ortega (2002) recognized five subspecies. They listed *N. l. ferrugineus* Goodwin, and *N. l. yucatanicus* Miller, as representing populations from Mexico and Central America; and *N. l. europs* (H. Allen), *N. l. laticaudatus* (É. Geoffroy St.-Hilaire), and *N. l. macarenensis* Barriga-Bonilla, as representing the South American populations. They treated *Dysopes gracilis* J. A. Wagner, as a synonym of the nominate subspecies.

Nyctinomops macrotis (Gray, 1839)
Big Free-tailed Bat
SYNONYMS:

V[espertilio]. Molossus Pallas, 1766:49; part.

Nyctinomus macrotis Gray, 1839:5; type locality "Cuba."

Dysopes auritus J. A. Wagner, 1843a:368; type locality "Cuyaba," Mato Grosso, Brazil.

N[yctinomus (Nyctinomus)]. auritus: W. Peters, 1866a: 573; name combination.

Nyctinomus megalotis Dobson, 1876:728; type locality "Surinam."

Nyctinomus depressus Ward, 1891:747; type locality "Tacubaya," Distrito Federal, Mexico.

Nyctinomus macrotis nevadensis H. Allen, 1894a:171; type localities "Nevada and California," U.S.A.; restricted to California by J. A. Allen (1894:326, footnote).

Nyctinomus nevadensis: J. A. Allen, 1894:326; name combination.

[Nyctinomus (Nyctinomus)] megalotis: Trouessart, 1897: 147; name combination.

[Nyctinomus (Nyctinomus)] macrotis: Trouessart, 1897: 147; name combination.

Promops affinis J. A. Allen, 1900a:91; type locality "Taguaga," Magdalena, Colombia.

N[yctinomops]. depressus: Miller, 1902b: 393; name combination.

N[yctinomops]. affinis: Miller, 1902b: 393; name combination.

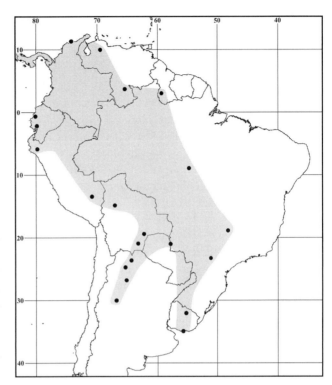

Map 251 Marginal localities for *Nyctinomops macrotis* ●

N[yctinomops]. macrotis: Miller, 1902b:393; first use of current name combination.

Nyctinomus molossus: Miller, 1913b:86; name combination; not *Vespertilio molossus* Pallas.

Nyctinomus æquatorialis J. A. Allen, 1914:386; type locality "Chone, Manavi [= Manabí], Ecuador."

Tadarida depressa: Miller, 1924:86; name combination.

Tadarida macrotis: Miller, 1924:86; name combination.

Tadarida nevadensis: Miller, 1924:87; name combination.

Tadarida molossa: Hershkovitz, 1949c:452; name combination; not *Vespertilio molossus* Pallas.

DISTRIBUTION: *Nyctinomops macrotis* occurs in South America from Colombia, Venezuela, and Guyana south into Ecuador, Peru, Brazil, Bolivia, Uruguay, and northern Argentina. Elsewhere, the species occurs in Mexico, Central America, the southern United States, and the Greater Antilles.

MARGINAL LOCALITIES (Map 251): COLOMBIA: Magdalena, Taguaga (type locality of *Promops affinis* J. A. Allen). VENEZUELA: Lara, Hacienda San Gerónimo (AMNH 132773); Amazonas, Caño Negro (Ochoa, Castellanos, and Ibáñez 1988). GUYANA: Upper Takutu-Upper Essequibo, Foot of Kanuku Mountains, 12 mi ENE of Dadanawa (ROM 61343). BRAZIL: Pará, Cachimbo (MZUSP 15380); Minas Gerais, Uberlândia (Stutz et al. 2004); Paraná, Londrina (N. R. Reis et al. 1998). URUGUAY: Tacuarembo, Pueblo del Barro (J. C. González

1977); Montevideo, Isla de Flores (Acosta y Lara 1950). PARAGUAY: Alto Paraguay, Fuerte Olimpo (Myers and Wetzel 1983). ARGENTINA (Barquez, Mares, and Braun 1999, except as noted): Tucumán, Tucumán; La Rioja, Cueva del Chacho; Salta, Parque San Martín, Salta; Jujuy, Yuto (Crespo 1958). BOLIVIA: Tarija, 1 km S of Canatindi (S. Anderson 1997); Santa Cruz, Hacienda Cerro Colorado (Ibáñez and Ochoa 1989); Beni, Km 35, NW of Yucumo (S. Anderson 1997). PERU: Cusco, near Marcapata (Sanborn 1951b); Lambayeque, 12 miles N of Olmos (Graham and Barkley 1984). ECUADOR: Guayas, Durán (Albuja 1999); Manabí, Chone (J. A. Allen 1914).

SUBSPECIES: I am treating *N. macrotis* as monotypic.

NATURAL HISTORY: Individuals have been found roosting in caves, hollow trees, and in rock fissures. Specimens have been taken in Guyana in dry tropical forest at elevations from 100 to 1,100 m, and in thorn forest in Bolivia (Ibáñez and Ochoa 1989). Two specimens were collected at Cerro de la Neblina, Amazonas, Venezuela (Gardner 1988). The first, a male was shot at dusk as it flew back and forth above a campsite located at 2,100 m on the top of the tepui; the second, a lactating female, was caught by a worker when it flew into his hammock on 19 February at the base camp at 140 m elevation on the Río Mawarinuma. Barquez and Díaz (2001) caught their specimens in buildings in the city of Tucumán, where *N. macrotis* is common. Milner, Jones, and Jones (1990) presented measurements and additional natural history information in their *Mammalian Species* account. *Nyctinomops macrotis* has a $2n = 48$, $FN = 58$ karyotype (J. W. Warner et al. 1974).

REMARKS: Husson (1962) reviewed the history of the names applied to this species when he reviewed the status of *Molossus molossus*. C. O. C. Vieira (1942, 1955) incorrectly listed *Tadarida macrotis* from Minas Gerais and Santa Catarina, Brazil. His measurements (C. O. C. Vieira 1955) indicated that his specimen represents a smaller species such as *Tadarida brasiliensis*. This was confirmed by A. L. Gardner (pers. comm.) who examined some of the material Vieira reported on in the collections in São Paulo. He also found that Vieira had misidentified one *Eumops hansae* and several *Tadarida brasiliensis* and *Nyctinomops laticaudatus* as *Tadarida macrotis*. Czaplewski and Cartelle (1998) reported a late Pleistocene record for *N. macrotis* from Bahia, Brazil and summarize other Quaternary records from South America.

Genus *Promops* P. Gervais, 1856

Promops, Neotropical in distribution, is represented by two species that are superficially similar to species of *Molossus*. The genus is distinguished by the following characters: the ears are short, rounded, have a narrow keel, and arise me-dially from the same point on the forehead; the antitragus is pendant and constricted at the base; and the tragus is minute. The lips are smooth, and the snout has a median ridge extending from behind the nares to the level of the ears. The upper incisors are curved, relatively slender, and less than half the height of the canines. Lower incisors are weakly bifid with the outer incisors crowded laterally behind the inner incisors. The anterior upper premolars are reduced to spicules and sometimes missing, and the third commissure of M3 is greatly reduced or absent. The skull has a highly domed palate and well-defined basisphenoid pits. The dental formula is 1/2, 1/1, 2/2, 3/3 × 2 = 30.

SYNONYMS:

Molossus: Spix, 1823:59; not *Molossus* É. Geoffroy St.-Hilaire, 1805a.

Promops P. Gervais, 1856a:58; type species *Promops ursinus* P. Gervais, 1856a, by monotypy.

Promops: W. Peters, 1866a: 574; used as a subgenus of *Molossus* É. Geoffroy St.-Hilaire.

Molossus: O. Thomas, 1901j:438; not *Molossus* É. Geoffroy St.-Hilaire.

Pomops Aguirre, 1999:108; incorrect subsequent spelling of *Promops* P. Gervais.

KEY TO THE SPECIES OF *PROMOPS*:

1. Forearm shorter than 50 mm; greatest length of skull shorter than 19.1 mm . . . . *Promops nasutus*
1'. Forearm longer than 51.5 mm; greatest length of skull 19.9 mm or longer . . . . *Promops centralis*

Promops centralis O. Thomas, 1915
Crested Mastiff Bat

SYNONYMS:

Promops centralis O. Thomas, 1915a:62; type locality "N. Yucatan," Mexico.

Promops occultus O. Thomas, 1915a:62; type locality "Sapucay," Paraguay; possibly valid as a subspecies.

Promops davisoni O. Thomas, 1921d:139; type locality "Chosica," Lima, Peru; possibly valid as a subspecies.

P[romops]. c[entralis]. occultus: Koopman, 1978:22; name combination.

P[romops]. c[entralis]. davisoni: Koopman, 1978:22; name combination.

DISTRIBUTION: *Promops centralis* is found in Venezuela, Guyana, Surinam, French Guiana, and Trinidad and Tobago, south through western Colombia into Ecuador and coastal Peru, and east of the Andes in Bolivia, western Brazil, Paraguay, and northern Argentina. Elsewhere, the species occurs northwestward through Central America into Mexico.

MARGINAL LOCALITIES (Map 252): TRINIDAD AND TOBAGO: Trinidad, Port of Spain (Goodwin and

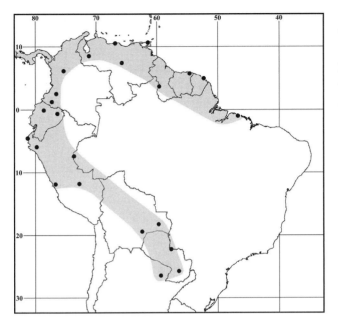

Map 252 Marginal localities for *Promops centralis* ●

Greenhall 1961). SURINAM: Marowijne, 10 km N and 24 km W of Moengo (Genoways and Williams 1979b). FRENCH GUIANA: Cayenne (Menegaux 1903, as *P. nasutus*). BRAZIL: Pará, Bragança (Gregorin and Taddei 2001). GUYANA: Upper Takutu-Upper Essequibo, Nappi Creek, Kanuku Mountains (Lim and Engstrom 2001a). VENEZUELA: Bolívar, Hato La Florida, 47 km ESE of Caicara (Handley 1976); Distrito Federal, Caracas (Ochoa and Ibáñez 1985); Mérida, 7 km SSW of Mérida (Soriano, Ruiz, and Zambrano 2005). COLOMBIA: Antioquia, Río Negro (AMNH 149260); Cauca, Popayán (Marinkelle and Cadena 1972); Nariño, Pasto (ROM 40361). ECUADOR: Orellana, 1 km. S of Estación Científica Yasuní (F.A. Reid, Engstrom, and Lim 2000). BRAZIL: Acre, Rio Moa (Gregorin and Taddei 2001). BOLIVIA: Santa Cruz, Guirapembi (Ibáñez and Ochoa 1989). PARAGUAY: Concepción, between Estrellas Ranch and Primavera Ranch (Baud 1981); Guaira, Villarica (Myers and Wetzel 1983). ARGENTINA: Formosa, El Colorado (Massoia 1976). BOLIVIA: Santa Cruz, Roboré (S. Anderson 1991). PERU: Cusco, Cashiriari-2 (S. Solari et al. 2001c); Lima, Chosica (type locality of *Promops davisoni* O. Thomas); Lambayeque, Olmos (Koopman 1978); Piura, Talara (Tuttle 1970). ECUADOR: Manabí, Mongoya (FMNH 53543).

SUBSPECIES: I am treating *P. centralis* as monotypic because of the unclear status of the population units that could represent subspecies

NATURAL HISTORY: *Promops centralis* has been found roosting on the underside of palm leaves in small colonies of up to six individuals (Goodwin and Greenhall

1961), in hollow trees, and under tree bark (Brosset 1966). They have been found lactating in April (Trinidad), August (Surinam), and November and December (southern Brazil). Handley (1976) reported *P. centralis* from Venezuelan dry tropical dry forest. Simmons and Voss (1998) caught two in nets stretched across a gap in the canopy between 17 and 21 m above a narrow dirt road in rainforest habitat in French Guiana. R. Guerrero (1985a) listed a hemipteran known to parasitize this species. *Promops centralis* has a $2n = 48$, FN = 58 karyotype (J. W. Warner et al. 1974).

REMARKS: Handley (1966c) considered *Promops occultus* and *P. davisoni* to be synonyms of *P. centralis*. Ojasti and Linares (1971) and Koopman (1978) regarded *P. occultus* and *P. davisoni* as subspecies of *P. centralis*. Koopman (1978, 1994) treated *P. c. occultus* as a subspecies of the Amazonian lowlands, and *P. c. davisoni* as a subspecies of the Pacific slope of the Andes; although he noted (1994) that *P. davisoni* might prove to be a synonym of *P. nasutus*. Gregorin and Taddei (2001), found *P. centralis* from Ecuador and Peru (sometimes referred to as *P. c. davisoni*) to be significantly smaller than *P. centralis* from Mexico and Guatemala. Cabrera (1958) did not include *P. centralis* in the South American mammal fauna. The species needs to be revised. Simmons and Voss (1998) provided measurements, a description, and drawings of the skull of a specimen from Paracou, French Guiana.

Promops nasutus (Spix, 1823)
Brown Mastiff Bat
SYNONYMS:

Molossus nasutus Spix, 1823:60; type locality "flumen St. Francisci," Bahia, Brazil.

Molossus fumarius Spix, 1823:60; type locality "ripam fluminis Itapicurú," Brazil.

Promops ursinus P. Gervais, 1856a:59; type locality "Miranda," Mato Grosso do Sul, Brazil.

M[olossus (Promops)]. nasutus: W. Peters, 1866a:574; name combination.

Molossus fosteri O. Thomas, 1901j:438; type locality "Villa Rica," Guaira, Paraguay.

Promops pamana Miller, 1913a:33; type locality "Hyutanaham, Upper Purus River," Amazonas, Brazil.

Promops ancilla O. Thomas, 1915a:63; type locality "Cachi, Salta," Argentina.

Promops nasutus ancilla: Cabrera, 1958:128; name combination.

Promops nasutus nasutus: Cabrera, 1958:128; name combination.

Promops nasutus downsi Goodwin and Greenhall, 1962:10; type locality "Memorial Park, Port-of-Spain, Trinidad."

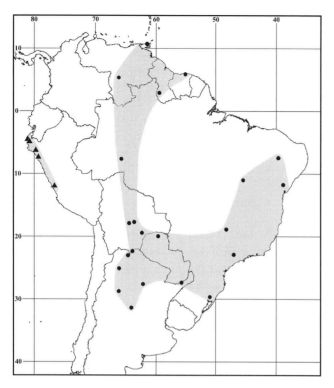

Map 253 Marginal localities for *Promops nasutus* ● and *Tomopeas ravus* ▲

DISTRIBUTION: *Promops nasutus* is on the island of Trinidad and in Venezuela, Guyana, Surinam, Brazil, Bolivia, Paraguay, and northern Argentina.

MARGINAL LOCALITIES (Map 253): TRINIDAD AND TOBAGO: Trinidad, Port of Spain (type locality of *Promops nasutus downsi* Goodwin and Greenhall). SURINAM: Paramaribo, Paramaribo (Genoways and Williams 1979b). GUYANA: Upper Takutu-Upper Essequibo, Dadanawa (ROM 48878). BOLIVIA: Santa Cruz, Río Pitasama, 4.5 km N and 1.5 km E of Cerro Amboró (S. Anderson 1997); Santa Cruz, Hacienda Cerro Colorado (Ibáñez and Ochoa 1989). PARAGUAY: Alto Paraguay, Agua Dulce (Myers and Wetzel 1983). BRAZIL: Minas Gerais, Uberlândia (Stutz et al. 2004); Bahia, São Marcelo (Goodwin and Greenhall 1962); Pernambuco, Exu (Mares et al. 1981, as *Promops* sp.); Bahia, Lamarão (Goodwin and Greenhall 1962); São Paulo, Campinas (I. Sazima and Uieda 1977); Rio Grande do Sul, Saparinga (F. Silva 1975). ARGENTINA (Barquez, Mares, and Braun 1999, except as noted): Misiones, Posadas (Crespo 1958); Santiago del Estero, Girardet; Córdoba, Córdoba; Catamarca, Balneario El Caolín, 6 km NW of Cumbicha; Salta, Cachi (type locality of *Promops ancilla* O. Thomas); Jujuy, Arroyo Sauzalito (Barquez and Díaz 2001). Salta, Quebrada de Acambuco. BOLIVIA: Santa Cruz, 3 km SE of Comarapa (S. Anderson 1997). BRAZIL: Amazonas, Hyutanaham (type locality of

Promops pamana Miller). VENEZUELA: Amazonas, San Juan (Handley 1976).

SUBSPECIES: I am treating *P. nasutus* as monotypic because of the unknown status of the several named populations that could represent subspecies (see Remarks).

NATURAL HISTORY: This species has been found roosting in palms, hollow trees, and the roofs of houses. It inhabits tropical and subtropical wet forest habitats in Venezuela (Handley 1976), thorn scrub habitat in Paraguay (Myers, White, and Stallings 1983), and the Yungas phytogeogrphic region of Argentina (Barquez and Díaz 2001). Barquez and Díaz (2001) reported *P. nasutus* in Tucumán, Argentina, roosting in crevices in a hillside above a river. A female taken in June in Salta was pregnant with a 3-g fetus. Wainberg (1966) reported a $2n = 40$ karyotype for *P. n. ancilla* from La Plata, Buenos Aires, Argentina.

REMARKS: There are five names available for subspecies of *P. nasutus*: *P. n. nasutus*, eastern Brazil; *P. n. fosteri*, Paraguay; *P. n. pamana*, western Brazil; *P. n. ancilla*, northwestern Argentina; and *P. n. downsi*, island of Trinidad. Genoways and Williams (1979b), on the basis of smaller size, suggested that *P. davisoni* (herein allied with *P. centralis*) from Ecuador and Peru may represent *P. nasutus*. The species needs to be revised. Czaplewski and Cartelle (1998) reported fossils identifies as *P. nasutus* from late Pleistocene deposits from the state of Bahia, Brazil.

Genus *Tadarida* Rafinesque, 1814

Bats of this genus are found in warm regions worldwide, from southern Oregon and Nebraska (U.S.A.) to south-central Argentina in the Western Hemisphere, and from southern Europe (Sicily), throughout Africa, and east to New Guinea and Australia in the Old World. The genus is characterized by the following features: the inner margins of ears arise from the same point, the antitragus is wider than high, and the tragus is square in outline and well developed. The lips are deeply wrinkled. Basisphenoid pits are shallow to moderate in depth; the palate is emarginate anteriorly; and the third commissure of M3 is variably developed, usually equal to or longer than the second. The dental formula is 1/2–3, 1/1, 2/2, 3/3 × 2 = 30 or 32.

SYNONYMS:

Cephalotes: Rafinesque, 1814:12; not *Cephalotes* É. Geoffroy, St.-Hilaire, 1810a.

Tadarida Rafinesque, 1814:55; type species *Cephalotes teniotis* Rafinesque, 1814, by monotypy.

Nyctinomus É. Geoffroy St.-Hilaire, 1818b:114; type species *Nyctinomus aegyptiacus* É. Geoffroy St.-Hilaire, 1818b, by monotypy.

Nyctinoma Bowdich, 1821:28; unjustified emendation of *Nyctinomus* É. Geoffroy St.-Hilaire.

Nyctinomes Gray, 1821:299; incorrect subsequent spelling of *Nyctinomus* É. Geoffroy St.-Hilaire.

Nyctinomia Fleming, 1822:178; incorrect subsequent spelling of *Nyctinomus* É. Geoffroy St.-Hilaire.

Dinops Savi, 1825:235; type species *D. cestoni* Savi, 1825, by monotypy.

Dysopes: Temminck, 1826:233; part; not *Dysopes* Illiger, 1811.

Nycticea: J. E. LeConte, 1831:432; not *Nycticea* (= *Nycticeius*) Rafinesque, 1819.

Molossus: Cooper, 1837:67; not *Molossus* É. Geoffroy St.-Hilaire, 1805a.

Molossus: d'Orbigny, 1837:pl. 10; not *Molossus* É. Geoffroy St.-Hilaire.

Rhinopoma: Gundlach, 1840:358; not *Rhinopoma* É. Geoffroy St.-Hilaire, 1818b.

Dysopes: J. A. Wagner, 1840:475; not *Dysopes* Illiger.

Molossus: Saussure, 1860:283–285; not *Molossus* É. Geoffroy St.-Hilaire.

Dysopes: Burmeister, 1861:391; not *Dysopes* Illiger.

Mormopterus: J. A. Allen, 1914:387; not *Mormopterus* W. Peters, 1865a.

Austronomus Iredale and Troughton, 1934:100; *nomen nudum*.

Austronomus Troughton, 1943:360; type species *Austronomus australis* (Gray, 1838b), by monotypy.

REMARKS: The genus formerly included *Chaerephon*, *Mops*, *Mormopterus*, and *Nyctinomops* as synonyms or subgenera (see Freeman 1981). Legendre (1984a), in a study of dental morphology, recognized four genera: *Nyctinomops*, *Tadarida*, *Mormopterus*, and *Rhizomops* nov. gen.; the latter a monotypic genus containing a single species, *R. brasiliensis*. The status of this new genus-level name awaits further study, and here I retain *T. brasiliensis* as a species of *Tadarida*.

Mahoney and Walton (1988:148–49) claimed that *Nyctinomus* É. Geoffroy St.-Hilaire has priority over *Tadarida* Rafinesque, 1814 and that Sherborn (1897a) and Rehn (1914) were mistaken in not accepting 1813 as the publication date for *Nyctinomus*. Part of their argument hinged on whether or not the copy É. Geoffroy St.-Hilaire sent to J. E. Gray, dated 1913, was a "proof" as stated by Sherborn (1897a:288) or was a preprint. The compilation in which the name *Nyctinomus* appears was not a journal or another serial publication; therefore, it probably does not qualify as a preprint. Sherborn (1897a) gave the publication date as 1818; whereas Rehn (1914) simply stated that *Nyctinomus* did not appear prior to 1816. The International Commission (ICZN 1987:Op. 1461) ruled that the publication dates for the zoological portions of

the *Histoire naturelle* sections of Savigny's *Description de l'Egypte* are to be taken from Sherborn (1897a). Therefore, the publication date had already been fixed as 1818 before the paper by Mahoney and Walton (1988) was published.

Tadarida brasiliensis (I. Geoffroy St.-Hilaire, 1824)
Brazilian Free-tailed Bat

SYNONYMS: The following synonyms represent populations outside of South America; additional synonyms are listed under Subspecies.

Nyctinomus murinus Gray, 1827:66; type locality "Jamaica"; valid as a subspecies.

Nyct[icea]. cynocephala J. E. LeConte, 1831:432; type locality not given; probably in the neighborhood of the LeConte plantation, Liberty County, Georgia, U.S.A., according to Shamel (1931:7); valid as a subspecies.

Molossus fuliginosus Cooper, 1837:67; type locality "Milledgeville, Georgia," U.S.A.

Rhinopoma carolinense: Gundlach, 1840:358; not *Rhinopoma carolinensis* É. Geoffroy St.-Hilaire, 1818b.

M[olossus]. mexicanus Saussure, 1860:283; type localities "Coffre de Perote, à 13,000 pieds…Ameca, au pied du Popocatepetl à…8,500 pieds"; restricted by Elliot (1904) to Cofre de Perote, Veracruz, Mexico (also see Benson 1944); valid as a subspecies.

Nyctinomus musculus Gundlach in W. Peters, 1861:149; type locality "Cuba"; restricted by Silva-Taboada (1976a) to San Antonio El Fundador, Canímar, provincia Matanzas; valid as a subspecies.

N[yctinomus (Nyctinomus)]. brasiliensis: W. Peters, 1866a: 573, footnote; name combination.

Nyctinomus mohavensis Merriam, 1889:25; type locality "Fort Mojave, Arizona," U.S.A.

Nyctinomus bahamensis Rehn, 1902b:641; type locality "Governor's Harbour, Eleuthera, Bahamas"; valid as a subspecies.

Nyctinomus mexicanus: Bailey, 1905:215; name combination.

Tadarida antillularum: Miller, 1924:85; name combination.

Tadarida cynocephala: Miller, 1924:85; name combination.

Tadarida muscula: Miller, 1924:86; name combination.

Tadarida mexicana: Miller, 1924:86; name combination.

Tadarida intermedia Shamel, 1931:7; type locality "Valley of Comitan, Chiapas, Mexico"; valid as a subspecies.

Tadarida constanzae Shamel, 1931:10; type locality "Constanza, Dominican Republic"; valid as a subspecies.

Tadarida texana Stager, 1942:49; type locality "Ney Cave, 20 miles north of Hondo, Medina County, Texas," U.S.A.

Map 254 Marginal localities for *Tadarida brasiliensis* ●

DISTRIBUTION: *Tadarida brasiliensis* has an unusual distribution in South America, where it is known from Trinidad and Tobago, western Venezuela, and Colombia, southward through the Andean countries of Ecuador, Peru, Bolivia, and Chile, and eastward across northern and central Argentina into Paraguay, Uruguay and southeastern Brazil. Elsewhere, it is known from Panama north through Mexico into the United States, and from the Lesser Antilles north into Cuba and the Bahamas (Hall 1981).

MARGINAL LOCALITIES (Map 254): TRINIDAD AND TOBAGO: Tobago, locality unknown (Goodwin and Greenhall 1961). VENEZUELA: Mérida, Tabay (Handley 1976). COLOMBIA: Vaupés, Mitú (ROM 69531). PERU: San Martín, San Martín (Tuttle 1970). BOLIVIA (S. Anderson 1997): La Paz, Coroico; Cochabamba, Yungas; Santa Cruz, 5.5 km by road NNE of Vallegrande; Santa Cruz, Boyuibe. PARAGUAY: Guairá, Villarica (Myers and Wetzel 1983). BRAZIL: Paraná, Parque Estadual Matas dos Godoy (N. R. Reis et al. 2000); São Paulo, Ipiranga (C. O. C. Vieira 1942); Minas Gerais, Uberlândia (Stutz et al 2004); Bahia, Barra (Shamel 1931); São Paulo, Paranapiacaba (MZUSP 15447); Santa Catarina, Hansa (MZUSP 15444); Rio Grande do Sul, Porto Alegre (S. M. Pacheco and Marques 1995). URUGUAY: Maldonado, Maldonado (Shamel 1931). ARGENTINA (Barquez, Mares, and Braun 1999, except as noted): Buenos Aires, La Plata; Buenos Aires, Mar del Plata; Buenos Aires, Bahía Blanca; Río Negro, Chimpay (FMNH 50929); Río Negro, San Carlos

de Bariloche. CHILE: Los Lagos, Rinihue (Osgood 1943); Biobío, Concepción (ROM 85240); Coquimbo, Paihuano (Osgood 1943); Tarapacá, Arica (Mann 1945). PERU: Arequipa, Arequipa (Tuttle 1970). ECUADOR: Carchi, Rumichaca (Albuja 1983).

SUBSPECIES: Two subspecies of *T. brasiliensis* are found in South America; others occur in the U.S.A., Mexico, Central America, the Greater Antilles, and the Bahamas (Hall 1981).

T. b. antillularum (Miller, 1902)

SYNONYMS:

Nyctinomus antillularum Miller, 1902b:398; type locality "Roseau, Dominica," Lesser Antilles.

Nyctinomus brasiliensis antillularum: G. M. Allen, 1911: 245; name combination.

Tadarida antillularum: Miller, 1924:85; name combination.

Tadarida brasiliensis antillularum: Schwartz, 1955:108; name combination.

This subspecies is known from the island of Tobago and northward through the Lesser Antilles into Puerto Rico.

T. b. brasiliensis (I. Geoffroy St.-Hilaire, 1824)

SYNONYMS:

Nyctinomus Brasiliensis I. Geoffroy St.-Hilaire, 1824a:343; type localities "province des Missions [estado do Rio Grande do Sul],...le district de Curityba [estado do Paraná]," Brazil; restricted to Curityba, Paraná, Brazil, by Shamel (1931:4).

Dysopes nasutus Temminck, 1826:233; type locality "Brésil."

Molossus rugosus d'Orbigny, 1837:pl. 10, Figs 3–5; type locality Corrientes, Argentina (d'Orbigny and Gervais 1847:13); but probably Tucumán according to D. C. Carter and Dolan 1978.

Dysopes naso J. A. Wagner, 1840:475; type locality "Brasílien."

Dysopes multispinosus Burmeister, 1861:391; type localities "von Mendoza bis Tecuman," Argentina.

Mormopterus peruanus J. A. Allen, 1914:387; type locality "Inca Mines," Puno, Peru.

Tadarida brasiliensis: O. Thomas, 1920d:222; first use of current name combination.

This subspecies is distributed from the island of Trinidad southward throughout mainland South America, west of the greater Orinoco and Amazon basins, into central Chile and Argentina.

NATURAL HISTORY: *Tadarida brasiliensis* ranges from sea level to 2,690 m (Albuja 1983). These bats have been found in caves, buildings, mine tunnels, under bridges, and between large stones on a riverine beach. *Tadarida brasiliensis* lives in large colonies, and maternity colonies can number in the multimillions (Villa-R. and Cockrum 1962). Barquez and Díaz (2001) estimated 12 million in one colony in Tucumán, Argentina. The species occurs in a wide variety of habitats ranging from deserts to lower montane humid forest. Two pregnant females, each with a single 12-mm–CR fetus, were caught on 23–24 August in Ecuador (Albuja 1983). Cartelle and Abuhid (1994) and Czaplewski and Cartelle (1998) reported fossil material representing this species from late Pleistocene cave deposits in the state of Bahia, Brazil.

Studies on *T. b. mexicana* indicate that a lactating female can discriminate her own pup from others on the basis of scent (Gustin and McCracken 1987), and that nursing is nonrandom and selective along kinship lines (McCracken 1984). Wilkins (1989) presented measurements and additional natural history information in his *Mammalian Species* account. *Tadarida brasiliensis* has a $2n = 48$, FN $= 56$ karyotype (J. W. Warner, et al. 1974).

REMARKS: The name *Rhinopoma carolinensis* (É. Geoffroy St.-Hilaire, 1818b) is likely an earlier name for *Tadarida brasiliensis cynocephala* J. E. LeConte, 1831. However, as there is some doubt about the origin of *Rhinopoma carolinensis* (according to Desmarest 1820), it seems appropriate to continue using *T. b. cynocephala* J. E. LeConte for the free-tailed bat from the southeastern United States.

Apparently, the records for *T. brasiliensis* from provincia Beni, Bolivia, listed by S. Anderson, Koopman, and Creighton (1982:15) were misidentifications. Later, S. Anderson (1997) listed these localities under *Nyctinomops laticaudatus* and did not record *T. brasiliensis* from Beni.

Subfamily Tomopeatinae Miller, 1907

Linda J. Barkley

The subfamily Tomopeatinae Miller, 1907b, contains the single monotypic genus *Tomopeas*. The subfamily is characterized by disc-shaped auditory bullae, reflecting the great expansion of the margin of its inner border. The ear lacks an anterior basal lobe, but has a distinct rudimentary keel. The upper lips are large and overhang the lower lips. The nostrils are tubular. The wings are broad, the flight membranes are thin, and the legs and feet are slender. The tail is enclosed in the uropatagium, except for the last two vertebrae, which extend free beyond the membrane margin (Miller 1907b).

Genus *Tomopeas* Miller, 1900

Tomopeas, the only member of the subfamily Tomopeatinae, is a monotypic genus whose familial relationships have been the subject of speculation since its description by Miller in 1900. The genus combines features of external morphology similar to both vespertilionid and molossid bats (Miller 1900f; Aellen 1966; W. B. Davis 1970b). The only known species, *Tomopeas ravus*, is endemic to the arid coastal region of northwestern Peru. There is no fossil record.

Tomopeas ravus is a small bat (mass 2–4 g, forearm 31–36 mm), with small feet (4–6 mm), long calcar (averages 7 mm), and dark wing and interfemoral membranes. Miller (1907b) and G. M. Allen (1939) described the tail as being entirely included within the uropatagium; however, the specimens examined by W. B. Davis (1970b) and Barkley (1984) have at least the distal two caudal vertebrae free of the tail membrane. The ear has a low, rounded tragus and a small, well-developed antitragus that lacks a basal lobe. The pinnae are separate and project anteriorly. The upper lip is broad, slightly wrinkled, conspicuously fringed with hair, and extends over the lower lip. The seventh cervical and first thoracic vertebrae are fused. The skull is dorsoventrally flattened and possesses a conspicuous shallow depression in the middle of the nasals that makes the anterior end of the rostrum appear to flare up. The disc-shaped tympanic is uniquely flattened. The dental formula is 1/2, 1/1, 1/2, 3/3 × 2 = 28

SYNONYM:

Tomopeas Miller, 1900f, 570; type species *Tomopeas ravus* Miller, 1900f, 570, by original designation.

REMARKS: Miller's (1900f) placement of *Tomopeas* in the Vespertilionidae became the traditional allocation and was followed, for example, by Cabrera (1958), Honacki et al. (1982), and Koopman (1993). However, the molossid-like characteristics of *T. ravus* have long been recognized (Miller 1907b; Aellen 1966; W. B. Davis 1970b) and the weight of evidence suggests that its affinities are closer to the Molossidae where it was placed by McKenna and Bell (1997). Benedict (1957) found the denticulate, coronal scales of the hair more characteristic of molossid hairs than of hairs from vespertilionid bats. Protein electrophoresis, albumin immunodiffusion comparisons, and cranial and postcranial morphological analyses (Barkley 1984) also support molossid affinity. The last cervical fused to the first thoracic vertebra is another feature shared with molossids that often is considered an important diagnostic feature for some groups of bats (Miller 1907b; W. B. Davis 1970b; Vaughan 1970).

Tomopeas ravus Miller, 1900
Peruvian Crevice-dwelling Bat

SYNONYM:

Tomopeas ravus Miller, 1900f:571; type locality "Yayan, Cajamarca, Peru (alt. 1000 metres)."

DISTRIBUTION: *Tomopeas ravus* is endemic to northwestern Peru.

MARGINAL LOCALITIES (Map 253): PERÚ: Piura, Monte Grande, 14 km N and 25 km E of Talara (W. B. Davis 1970b); Piura, Mallares (Aellen 1966); Lambayeque, Sierra Vieja, 7 km S Motupe (W. B. Davis 1970b); Cajamarca, Tolón (Miller 1900f); Lima, Chosica (Miller 1900f).

SUBSPECIES: *Tomopeas ravus* is monotypic.

NATURAL HISTORY (from Barkley 1984, except as noted): *Tomopeas ravus*, as its common name implies, is a crevice dweller. Most specimens (30 of 50 examined) were captured by hand from narrow rock crevices in granitic boulders; a few were mistnetted in scattered groves of acacia, mesquite, and palo verde trees (W. B. Davis 1970b). These bats usually roost alone and prefer crevices inclined approximately 50° from horizontal and near the apex of large boulders. No other species of bat has been found roosting with *T. ravus*, although *Mormopterus kalinowskii* has been captured from rock crevices in close proximity.

Analysis of stomach and fecal-pellet contents show a food preference for beetles (Coleoptera). Almost all *T. ravus* have been collected in July and August when the species apparently is reproductively active. Of 11 females examined from this time period, two contained large fetuses (16-mm CR) and four were lactating. A mite, *Loomisia peruviensis* (Acarinidae; Goff, Whitaker, and Barkley 1984), and a tick, *Ornithodoros* sp. (N. Wilson, *in litt.*) are known to infest *T. ravus*. A coccidian protozoan (*Eimeria tomopea*) has been recovered from the intestinal tract (Duszynski and Barkley 1985). External measurements of 22 *T. ravus*: total length 75–85 mm, tail 34–45 mm, ear 11–16 mm, and tragus 4–5 mm ($n = 13$).

Family Vespertilionidae Gray, 1821
Alfred L. Gardner

The family Vespertilionidae is nearly cosmopolitan in distribution and found on every continent except Antarctica and on all major islands except Greenland. Five genera (*Eptesicus, Histiotus, Lasiurus, Myotis,* and *Rhogeessa*) and at least 32 species are found in South America. These are small to large bats sharing the following characteristics: the trochiter is larger than trochin, projects beyond the head of the humerus, and broadly articulates with the scapula; digit II has a well-developed metacarpal and one small ossified phalanx; digit III has three phalanges and the third is ossified at its base; the seventh cervical vertebra is free; the lumbar vertebrae are free; the skull lacks postorbital processes; the premaxillae are separate and the palate is emarginate anteriorly; the tragus is simple and well developed; the muzzle lacks flaps and leaf-like appendages; and the tail is well developed and extends to the edge of, or

slightly beyond, the wide interfemoral membrane (Miller 1907b).

KEY TO SOUTH AMERICAN GENERA OF VESPERTILIONIDAE.

1. One upper and three lower incisors on each side 2
1'. Two upper and three lower incisors on each side 3
2. Size large to very large; emargination (gap) at anterior end of palate wide, approximately 1/3 the distance across canines; one or two upper premolars on each side; 1/2 or more of upper surface of uropatagium densely furred . . . . *Lasiurus*
2'. Size smaller; space between upper incisors narrow; one upper premolar on each side, upper surface of uropatagium not densely furred . . . . *Rhogeessa*
3. One upper premolar on each side 4
3'. Three upper premolars on each side . . . . *Myotis*
4. Pinnae normal, not extending well beyond muzzle when laid forward . . . . *Eptesicus*
4'. Pinna greatly enlarged, extending well beyond muzzle . . . . *Histiotus*

Genus *Eptesicus* Rafinesque, 1820
William B. Davis and Alfred L. Gardner

The genus *Eptesicus* contains two subgenera and approximately 18 species (Corbet and Hill 1980; Hill and Harrison 1987). *Eptesicus* occurs on both continents in the New World, but elsewhere the genus is primarily Palearctic with several species in Africa. The number of species recognized in the Western Hemisphere varies from five to nine, depending on the reviewer followed; here we recognize seven. All were assigned to the *serotinus* subgroup in the *serotinus* group of the nominate subgenus by Hill and Harrison (1987). *Eptesicus fuscus*, a medium-sized bat (forearm 48–54 mm) found mainly in North and Central America, and in the West Indies, is closely related to the common serotine *E. serotinus* of Europe. The other New World *Eptesicus* occur mainly in tropical and subtropical America. These are smaller bats (forearm 31–47 mm), of which seven species were recognized by W. B. Davis (1966). Koopman (1978, 1993, 1994) reduced the number of species to five. The systematic arrangement used here is similar to that proposed by W. B. Davis (1966), as amended by Simmons and Voss (1998). Characteristic features of the genus include well-developed upper incisors with the inner incisors larger than the outer; lower incisors are subequal, trifid, and the crown of the third is wider than the crowns of either of the first two; canines have a distinct cingulum, but lack secondary cuspules; and premolars and molars are normal for vespertilionids (Miller 1907b). The dental formula is 2/3, 1/1, 1/2, 3/3 × 2 = 32.

SYNONYMS:

Vespertilio: Schreber, 1774a:pl. 53; part; not *Vespertilio* Linnaeus, 1758.

Eptesicus Rafinesque, 1820:2; type species *Eptesicus melanops* Rafinesque, 1820 (= *Eptesicus fuscus* Palisot de Beauvois, 1796), by subsequent designation (Méhely 1900:206).

Cnephaeus Kaup, 1829:103; type species *Vespertilio serotinus* Schreber, 1774a, by monotypy.

Vesperugo Keyserling and Blasius, 1839:312; part.

Vesperus Keyserling and Blasius, 1839:313; part; proposed as a subgenus of *Vesperugo* Keyserling and Blasius, 1839; preoccupied by *Vesperus* Latreille, 1829 (Coleoptera, Cerambycidae).

Noctula Bonaparte, 1837b:fasc. xxi; type species *Noctula serotina* Bonaparte, 1837b, by monotypy.

Cateorus Kolenati, 1856a:131; type species *Vespertilio serotinus* Schreber, 1774a, by monotypy; proposed as a subgenus of *Vesperus* Keyserling and Blasius.

Meteorus Kolenati, 1856a:131; part; type species *Vespertilio discolor* Kuhl (*fide* Kretzoi and Kretzoi 2000) by subsequent designation; proposed as a subgenus of *Vesperus* Keyserling and Blasius; preoccupied by *Meteorus* Haliday, 1835 (Hymenoptera).

Amblyotus Kolenati, 1858:252; type species *Amblyotus latratus* Kolenati, 1858 (= *Vespertilio nilssoni* Keyserling and Blasius, 1839), by monotypy; proposed as a subgenus of *Exochura* Kolenati, 1858; preoccupied by *Amblyotus* Amyot and Serville, 1843 (Hemiptera).

Aristippe Kolenati, 1863:40; part.

Scotophilus: Tomes, 1857b:50; part; not *Scotophilus* Leach, 1821a.

Scotophilus: H. Allen, 1864:27; part; not *Scotophilus* Leach.

Pachyomus Gray, 1866a:90; type species *Scotophilus pachyomus* Tomes, 1857, by monotypy.

Nyctiptenus Fitzinger, 1870b:424; type species *Vespertilio smithi* Wagner, 1855, by monotypy.

Adelonycteris H. Allen, 1892a:466; part; replacement name for *Vesperus* Keyserling and Blasius, 1839, preoccupied.

Pareptesicus Bianchi, 1917:lxxvi 1820; type species *Vesperugo pachyotis* Dobson, 1871 by monotypy; proposed as a subgenus of *Eptesicus* Rafinesque.

Rhyneptesicus Bianchi, 1917: lxxvi; type species *Vesperugo nasutus* Dobson, 1877, by monotypy; proposed as a subgenus of *Eptesicus* Rafinesque.

Tuitatus Kishida and Mori, 1931:379; *nomen nudum*.

Vespadelus Iredale and Troughton, 1934:95; *nomen nudum*.

Rhineptesicus Horáek and Hanák, 1986:16; incorrect subsequent spelling of *Rhyneptesicus* Bianchi.

REMARKS: Cabrera (1958) listed the names *Neoromicia* Roberts, 1926; *Tuitatus* Kishida and Mori, 1931; and *Vespadellus* (incorrect subsequent spelling of *Vespadelus* Iredale and Troughton, 1934), in the synonymy of *Eptesicus*. Hill and Harrison (1987) treated *Neoromicia* as a subgenus of *Pipistrellus*. *Tuitatus* and *Vespadelus* are *nomina nuda* (ICZN 1999:Art. 13). Hoofer and Van Den Bussche (2003) suggested that *Histiotus* P. Gervais, 1856 should be treated as a subgenus of *Eptesicus*. We recognize *Histiotus* as a specialized eptesicoid, most closely related to New World *Eptesicus*, but treat the taxon as a separate genus.

KEY TO THE SOUTH AMERICAN SPECIES OF *EPTESICUS*.

1. Size large, forearm 48 mm or longer; greatest length of skull usually more than 19 mm; maxillary toothrow 7 mm or longer . . . . *Eptesicus fuscus*
1′. Size smaller, forearm usually less than 48 mm; greatest length of skull less than 19 mm; maxillary toothrow less than 7 mm. 2
2. Dorsal fur 8–10 mm or more in length. 3
2′. Dorsal fur less than 8 mm in length 4
3. Size larger, forearm 42 mm or longer; greatest length of skull 15.8 mm or more; maxillary toothrow 6.1 mm or more; sagittal and lambdoidal crests well developed and overhanging supraoccipital . . . . . *Eptesicus chiriquinus*
3′. Size smaller, forearm normally less than 43 mm; greatest length of skull usually 16 mm or less; maxillary toothrow 6.1 or less; sagittal and lambdoidal crests low, not well developed. . . . .*Eptesicus andinus*
4. Forearm 40 mm or longer; greatest length of skull 16–18 mm, usually more than 17 mm; maxillary toothrow 5.7–6.7 mm, usually more than 6 mm . . . .
. . . . *Eptesicus brasiliensis*
4′. Forearm 31–44 mm, usually less than 40 mm; greatest length of skull 13.9–17.1 mm, usually less than 16.8 mm; maxillary toothrow 6 mm or less. 5
5. Color pale grayish brown; distributed along arid Pacific versant of Ecuador and Peru . . . . . . *Eptesicus innoxius*
5′. Color dark brown to blackish brown; not distributed as above 6
6. Larger, forearm 37 mm or longer; greatest length of skull 15 mm or more; maxillary toothrow 5.4 mm or more . . . . *Eptesicus furinalis*
6′. Smaller, forearm 37 mm or less; greatest length of skull 15.2 mm or less; maxillary toothrow 5.4 mm or less . . . . *Eptesicus diminutus*

Eptesicus andinus J. A. Allen, 1914
Andean Brown Bat

SYNONYMS:

Eptesicus andinus J. A. Allen, 1914:332; type locality "Valle de las Papas," Huila, Colombia.

Eptesicus montosus O. Thomas, 1920b:363; type locality "Choro, north of Cochabamba, Highlands of Bolivia, on the upper waters of the R. Mamoré. Alt. 3600 m."

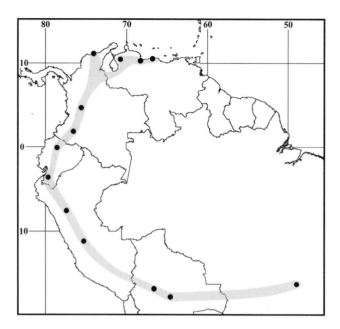

Map 255 Marginal localities for *Eptesicus andinus* ●

Eptesicus chiralensis H. E. Anthony, 1926:6; type locality "El Chiral," El Oro, Ecuador.

Eptesicus brasiliensis andinus: Hershkovitz, 1949c:451; name combination.

Eptesicus montosus chiralensis: W. B. Davis, 1966:255; name combination.

Eptesicus furinalis chiralensis: Koopman, 1978:19; name combination.

Eptesicus furinalis montosus: Koopman, 1978:19; name combination.

Eptesicus yinus Muñoz, 2001:209; incorrect subsequent spelling of *Eptesicus andinus* J. A. Allen.

DISTRIBUTION: *Eptesicus andinus* is distributed across the highlands of Bolivia northward at upper elevations along the Andes of Peru, Ecuador, and Colombia, into northwestern Venezuela.

MARGINAL LOCALITIES (Map 255; sequence of localities from east to west, north to south): VENEZUELA: Distrito Federal, Los Venados (USNM 370934); Carabobo, La Copa (USNM 441755); Falcón, Cerro Socopo (USNM 441764). COLOMBIA: Magdalena, Cincinati (AMNH 32671); Quindío, El Roble (AMNH 32802); Huila, Valle de las Papas (type locality of *Eptesicus andinus* J. A. Allen). ECUADOR: Pichincha, Yanacocha (Albuja 1999, as *E. brasiliensis*); El Oro, El Chiral (type locality of *Eptesicus chiralensis* H. E. Anthony). PERU: San Martín, Pampa del Cuy (MHNSM 7255); Junín, Chanchamayo, near Tarma (W. B. Davis 1966, as *E. montosus*). BOLIVIA: Cochabamba, Choro (type locality of *Eptesicus montosus* O. Thomas); Cochabamba, 25 km by road W of Comarapa [Santa Cruz] (S. Anderson, Koopman, and Creighton 1982,

as *E. furinalis montosus*). BRAZIL: Goiás, Anápolis (W. B. Davis 1966, as *E. montosus*).

SUBSPECIES: We are treating *E. andinus* as monotypic; the species needs revision.

NATURAL HISTORY: Handley (1976; as *E. montosus*) found *E. andinus* at high elevations in the Venezuelan state of Bolívar and in northern Venezuela where the 36 specimens he reported were netted over or near streams and in other areas near water. These sites were primarily openings in cloud and evergreen forest. Based on specimens in the USNM from Venezuela, one female was lactating in May, three in July, and four in early August; three were postlactating in July; and four in July and one in November showed no signs of reproductive activity. Three of the five females captured 13–14 August 1987 at Pampa del Cuy, ca. 3,300 m, San Martín, Peru, were pregnant, each with a single fetus. The fetuses of two females measured 11 mm CR; the fetus in the third measured 14 mm CR. The other two females showed no sign of reproductive activity.

REMARKS: The name *Eptesicus chiralensis* H. E. Anthony, apparently was overlooked by Cabrera (1958). W. B. Davis (1966) treated *E. chiralensis* as a subspecies of *E. montosus* and considered *E. montosus* to be distinct from *E. furinalis* and *E. andinus*. Koopman (1978, 1993, 1994) and Mies, Kurta, and King (1996) treated *E. chiralensis* and *E. montosus* as synonyms of *E. furinalis*. Simmons and Voss (1998) described the features useful for distinguishing *E. andinus* from *E. chiriquinus* (*E. andinus* has a relatively smooth braincase lacking well-developed sagittal and lambdoidal crests). Their interpretation of *E. furinalis* and the *Eptesicus andinus*-group, with a few exceptions, paralleled that of W. B. Davis (1966). Simmons and Voss (1998) also distinguished between the long-haired taxa *E. andinus* and *E. chiriquinus* and the widespread and similar, but short-haired, *E. brasiliensis*. Muñoz (2001) followed Koopman (1993) in treating *E. andinus* as a synonym of *E. brasiliensis*, and his locality records are not included in this account. D. C. Carter and Dolan (1978) provided information on and measurements of the holotype of *Eptesicus montosus* Thomas.

Eptesicus brasiliensis (Desmarest, 1819)
Brazilian Brown Bat

SYNONYMS: See under Subspecies.

DISTRIBUTION: *Eptesicus brasiliensis* is found on Trinidad and Tobago, and in Colombia, Venezuela, Guyana, and Surinam south through Ecuador, Peru, Brazil, and Paraguay into Uruguay and northern Argentina. Elsewhere, the species is found in Mexico and Central America.

MARGINAL LOCALITIES (Map 256; from W. B. Davis 1966, except as noted): TRINIDAD AND TOBAGO: Tobago (Goodwin and Greenhall 1961, no locality

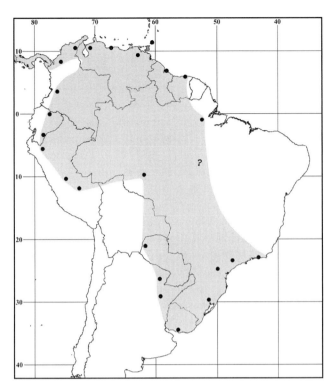

Map 256 Marginal localities for *Eptesicus brasiliensis* ●

mentioned). GUYANA: Demerara-Mahaica, Demerara. SURINAM: Paramaribo, Paramaribo. BRAZIL: Amapá, Recreio; Rio de Janeiro, Taquara; São Paulo, Ypanema (type locality of *Vespertilio nitens* J. A. Wagner); Paraná, Castro; Rio Grande do Sul, Montenegro (type locality of *Vesperus arge* Cope). URUGUAY: Montevideo, Santa Lucia. ARGENTINA: Corrientes, Goya (type locality of *E. argentinus* O. Thomas); Chaco, Puente sobre el Río Bermejo (Massoia 1976). PARAGUAY: Boquerón, Tajamar (USNM 555675). BRAZIL: Rondônia, Cachoeira Nazaré, W bank Rio Ji-Paraná (MPEG 21112). PERU: Cusco, Cashiriari-3 (S. Solari et al. 2001c); Pasco, Nevati (Tuttle 1970); Cajamarca, Bellavista. ECUADOR: Morona-Santiago, Gualaquiza; Napo, Cascada de San Rafael (Albuja 1999). COLOMBIA: Valle de Cauca, 6 miles N of Palmira; Córdoba, Catival; El Cesar, Valledupar, Sierra Negra (Hershkovitz 1949c). VENEZUELA (Handley 1976): Falcón, Cerro Socopo; Distrito Federal, Hacienda Carapiche; Monagas, Hato Mata de Bejuco.

SUBSPECIES: We recognize four subspecies of *E. brasiliensis*.

E. b. arge (Cope, 1889)

SYNONYMS:

Vesperus arge Cope, 1889a:131); type locality "Sao Joao" do Monte Negro (= Montenegro), Rio Grande do Sul, Brazil.

Eptesicus arge: Rehn, 1901a:755; name combination.

[*Vesperugo (Vesperus)*] *arge*: Trouessart, 1897:109; name combination.

[*Vespertilio (Vespertilio)*] *arge*: Trouessart, 1904:79; name combination.

Eptesicus argentinus O. Thomas, 1920b:365; type locality "Goya," Río Paraná, Corrientes, Argentina.

This subspecies is in the Río de la Plata basin of Uruguay, Paraguay, and northern Argentina, and is in southern Brazil (Paraná, Santa Catarina, Rio Grande do Sul, and southern Mato Grosso do Sul).

E. b. brasiliensis (Desmarest, 1819)

SYNONYMS:

Vespertilio brasiliensis Desmarest, 1819b:478; type locality "Le Brésil"; restricted to "Goiáz" by Cabrera (1958:105) on the basis of its mention as the first locality for the synonym *Vespertilio hilarii* I. Geoffroy St.-Hilaire.

Vespertilio Hilarii I. Geoffroy St.-Hilaire, 1824b:441; type localities "la capitainerie de Goyar [= estado do Goiás, Brazil] et la province des Missions [= Missões region, northwestern estado do Rio Grande do Sul, Brazil]."

Vespertilio derasus Burmeister, 1854:77; type locality "Brasilien."

Vespertilio arctoideus J. A. Wagner, 1855:758; type locality "Brasilien."

Vespertilio nitens J. A. Wagner, 1855:810; no type locality given; type locality identified as "Ypanema," São Paulo, Brazil, by Pelzeln (1883:45).

Vesperus Hilarii: Fitzinger, 1870a:139; name combination.

Vesperugo hilarii: Dobson, 1878:196; name combination.

[*Vesperugo (Vesperus)*] *Hilarii*: Trouessart, 1897:109; name combination.

[*Vespertilio (Vespertilio)*] *?nitens*: Trouessart, 1897:130; name combination.

[*Vespertilio (Vespertilio)*] *hilarii*: Trouessart, 1904:79; name combination.

Vesperugo (Eptesicus) fuscus: O. Thomas, 1898b:2; name combination; not *Vespertilio fuscus* Palisot de Beauvois, 1796.

Eptesicus hillarii Miller, 1907b:209; name combination and incorrect subsequent spelling of *Vespertilio hilarii* I. Geoffroy St.-Hilaire.

Eptesicus hilairei O. Thomas, 1920b:360; incorrect subsequent spelling of *Vespertilio hilarii* I. Geoffroy St.-Hilaire.

E[ptesicus]. fuscus hilarii: Pittier and Tate, 1932:275; name combination.

Eptesicus brasiliensis: O. Thomas, 1920b:367; name combination.

Eptesicus hilari Muñoz, 2001:209; incorrect subsequent spelling of *Vespertilio hilarii* I. Geoffroy St.-Hilaire.

E[ptesicus]. hillari Muñoz, 2001:209; incorrect subsequent spelling of *Vespertilio hilarii* I. Geoffroy St.-Hilaire.

The nominate subspecies is found in eastern Brazil from the state of Maranhão south through Goiás, Distrito Federal, and Minas Gerais, into São Paulo.

E. b. melanopterus (Jentink, 1904)

SYNONYMS:

Vesperus melanopterus Jentink, 1904:176); type locality "Paramaribo, [Paramaribo,] Suriname."

Eptesicus brasiliensis melanopterus: W. B. Davis, 1966: 261; first use of current name combination.

This subspecies is distributed throughout the lowlands of Colombia, Venezuela, Guyana, Surinam, and the Amazon basin of Brazil.

E. b. thomasi W. B. Davis, 1966

SYNONYM:

Eptesicus brasiliensis thomasi W. B. Davis, 1966:261; type locality "Canelos, ca. 500 m," Pastaza, Ecuador.

This subspecies is distributed within the western Amazon basin of eastern Ecuador, eastern Peru, and adjacent western Brazil; to be expected from northeastern Bolivia.

NATURAL HISTORY: *Eptesicus brasiliensis* has been found roosting in small groups in hollow trees and in houses. Of 64 specimens reported by Handley (1976) from Venezuela, 41 percent were captured by hand in holes in dead trees (snags) standing in lagoons, 6 percent by hand in houses, and almost all (94 percent) were taken near or over streams or other bodies of water. Of the eight females taken in 1973 by R. Mumford at Viçosa, Minas Gerais, Brazil, and examined by W. B. Davis (1966), one contained a 5-mm-CR embryo when taken on 18 September; two others were pregnant when captured on 5 November, one with a single fetus and the other with two. A juvenile, still unable to fly, was captured in a thatched hut at Nevati, Peru, in mid June (Tuttle 1970).

The dentition indicates a diet of insects, but feeding habits are unknown. *Eptesicus melanopterus* is the type host of the macronyssid mite *Steatonyssus surinamensis* Yunker, Lukoschus, and Giesen, 1990. *Eptesicus brasiliensis* has a $2n = 50$, $FN = 48$, karyotype (R. J. Baker and Patton 1967; R. J. Baker et al. 1982; T. R. O. Freitas, Bogo, and Christoff 1992).

REMARKS: Cabrera (1958:105) restricted the type locality of *E. brasiliensis* to "Goiaz [= Goiás], que es la primera localidad mencionada para *hilarii*, sinónimo indudable de *brasiliensis*." D. C. Carter and Dolan (1978:79) wrote the locality of a syntype of *Vespertilio hilarii* as "Goyar, Missiones, Brazil," which is a composite of the two locations "la capitainerie de Goyar et la province des Missions" given by I. Geoffroy St.-Hilaire (1824b). The material on which I. Geoffroy St.-Hilaire based his description of *Vespertilio hilarii* was collected in Brazil by Auguste de St.-Hilaire in the southern part of the state of Goiás ("Goyar") between 27 May and 4 September 1819, and in the Missoes region ("Missions") of northwestern Rio Grande do Sul between 9 February and 26 March 1821 (P. E. Vanzolini, *in litt.*). The state of Goiás is within the range of *E. b. brasiliensis* and "Missoes" is within the range of *E. b. arge*. Several authors (e.g., Cabrera 1958; W. B. Davis 1966; D. F. Williams 1978; Barquez, Mares, and Braun 1999) used the name combination *E. b. argentinus* for the subspecies we call *E. b. arge*.

According to W. B. Davis (1966:257), *Vespertilio hilarii* was based on a series of four specimens representing two or possibly three of the currently recognized South American species of *Eptesicus*. The lectotype, Paris Museum No. 834/639, identified as the holotype by Rode (1941:93; also see W. B. Davis 1966) has a forearm of 45 mm and a greatest length of skull of 20 mm. Measurements of one of the three paralectotypes are forearm 37 mm, greatest length of skull 15.5 mm. Measurements of the other two are forearm 34 and 33 mm, greatest length of skull 14.0 and 13.5 mm. W. B. Davis (1966) stated that the last two paralectotypes were referable to *E. diminutus*; the middle-sized one probably referable to *E. furinalis*; and the lectotype was an *E. brasiliensis*. D. C. Carter and Dolan (1978) could not locate these specimens. Instead they found only the specimen in the Zoologisches Museum der Humboldt-Universität zu Berlin (ZMB 3912) they listed as a syntype of *Vespertilio hilarii*. The external measurements of this specimen correspond with those of *E. brasiliensis*. D. C. Carter and Dolan (1978) gave information on and measurements for a syntype of *Vespertilio Hilarii* I. Geoffroy St.-Hilaire and for the holotypes of *Eptesicus argentinus* O. Thomas and *Vesperus melanopterus* Jentink.

The type locality for *E. brasiliensis*, as restricted by Cabrera (1958), depends on the identity and origin of the lectotype of *Vespertilio hilarii*. No type locality has been designated for *Vespertilio hilarii* and there seems to be uncertainty concerning identity of the lectotype (W. B. Davis 1966:257). Furthermore, W. B. Davis (1966) used a specimen in the British Museum (BM 7.1.1.365), originally in the Tomes collection, and on the label of which Tomes had noted as being identical with *Vespertilio hilarii*, to represent true *E. brasiliensis*. W. B. Davis (1966:256) interpreted "St. Cath. Brazil" on the label of BM 7.1.1.365 as meaning Santa Catharina, Minas Gerais; however, it more likely refers to the Brazilian state of Santa Catarina, which would place it within the range of the subspecies known as *E. b. arge*. We do not want to further complicate this situation by arbitrarily selecting a type locality for *Vespertilio hilarii* before examining the type material; therefore, we have followed the arrangement outlined by W. B. Davis (1966), except that we use the name *E. b. arge* for the southernmost population.

As is common in much of the literature, Muñoz's (2001) concept of *E. brasiliensis* is composite and includes *E. andinus* and *E. chiriquinus* (and the synonym *E. inca*). Therefore, we have not used his locality records for Colombia.

Barquez, Mares, and Braun (1999) considered *E. brasiliensis* to be uncommon in Paraguay and northern Argentina. They reidentified as *E. furinalis* several specimens previously reported as *E. brasiliensis* (e.g., Olrog 1960; Fornes and Massoia 1967; D. F. Williams 1978) in Argentina, and corrected the Argentine localities cited by W. B. Davis (1966).

Cabrera (1958) included *Vespertilio ferrugineus* Temminck, 1840:239; type locality "La Guyane Hollandaise," in his synonymy of *E. b. brasiliensis* spite of the association of this name with *Myotis* by Trouessart (1904:90). Husson (1962:218) also identified the specimens upon which Temminck based *Vespertilio ferrugineus* as belonging to the genus *Myotis* and proposed the replacement name *Myotis surinamensis*. One of us (ALG) has examined the type specimen of *Myotis surinamensis* and found it to be as described by D. C. Carter and Dolan (1978). We agree with them that the specimen does not represent a South American bat.

Eptesicus chiriquinus O. Thomas, 1920
Chiriqui Brown Bat

SYNONYMS:

Eptesicus chiriquinus O. Thomas, 1920b:362; type locality "Boquete," Chiriquí, Panama.

Eptesicus inca O. Thomas, 1920b:363; type locality "Chanchamayo, Cuzco, Peru"; corrected to Valle de Chanchamayo near Tarma, Junín, by Tuttle (1970:78).

E[ptesicus]. chiquinus: Muñoz, 2001:209; incorrect subsequent spelling of *Eptesicus chiriquinus* O. Thomas.

DISTRIBUTION: *Eptesicus chiriquinus* occurs at moderate to lower elevations in Colombia, Venezuela, the Guianas, Brazil, and eastern Ecuador, Peru, and Bolivia. Elsewhere, it occurs in Central America northwestward into the Chiapan highlands of Mexico.

MARGINAL LOCALITIES (Map 257): VENEZUELA: Zulia, La Rinconada (USNM 387728); Falcón, Cerro Socopo (Handley 1976, as *E. andinus*); Monagas, Hacienda San Fernando (USNM 409459); Bolívar, Imataca Forest Reserve (Ochoa et al. 1993, as *E. andinus*); Bolívar, 85 km SSE of El Dorado (USNM 387728). FRENCH GUIANA: Paracou (Simmons and Voss 1998). BRAZIL: Pará, Belém (USNM 460140); Amazonas, Santo Antônio do Guajará (Simmons and Voss 1998). BOLIVIA: Pando, 15 km NW of Puerto Camacho (S. Anderson 1993, as *E. andinus*); Santa Cruz, 5 km SE of Comarapa (S. Anderson 1993, as *E. and-*

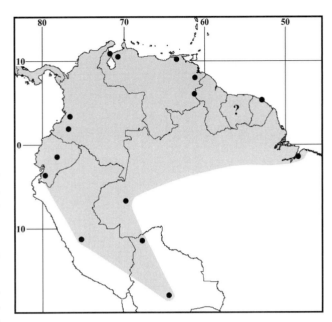

Map 257 Marginal localities for *Eptesicus chiriquinus* ●

inus). PERU: Junín, Valle de Chanchamayo, near Tarma (type locality of *Eptesicus inca* O. Thomas). ECUADOR: El Oro, El Chiral (AMNH 47217); Tungarahua, 1.5 km E of Mirador (USNM 513502). COLOMBIA: Cauca, Almaguer (W. B. Davis 1966); Valle del Cauca, 2 km S of Pance (USNM 483952).

SUBSPECIES: We treat *E. chiriquinus* as monotypic.

NATURAL HISTORY: Handley (1976) collected 13 specimens (reported under *E. andinus*) at elevations between 54 and 1,260 m in Venezuela. All were mistnetted; 75% near streams and other wetter areas; 25% in dry sites. Most (92%) were caught in evergreen forest; the remainder in open areas such as yards. In French Guiana, Simmons and Voss (1998) caught two in ground-level nets and four in elevated nets in modified forest habitats.

REMARKS: Oldfield Thomas (1920b:363) gave the type locality of *E. inca* (herein treated as a synonym of *E. chiriquinus*) as "Chanchamayo, Cuzco, Peru." Tuttle (1970:78) interpreted this locality as the "Valle de Chanchamayo, Departamento de Junín," and suggested that the holotype of *E. inca* was the bat earlier reported by O. Thomas as "*Vespertilio fuscus*" from "Chanchamayo, near Tarma, approximately in lat. 11°20'S, and long. 75°40'E [*sic*]" (O. Thomas 1893c:333). *Eptesicus chiriquinus* likely is more widely distributed in Colombia than our records indicate; Muñoz (2001) did not distinguish between *E. andinus*, *E. brasiliensis*, and *E. chiriquinus*. D. C. Carter and Dolan (1978) provided measurements on and additional information for the holotypes of *Eptesicus chiriquinus* O. Thomas and *Eptesicus inca* O. Thomas.

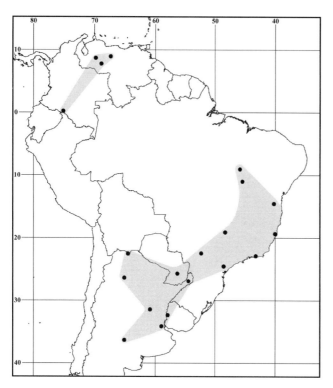

Map 258 Marginal localities for *Eptesicus diminutus* •

Eptesicus diminutus Osgood, 1915
Little Serotine

SYNONYMS: See under Subspecies.

DISTRIBUTION: *Eptesicus diminutus* has been recorded from Colombia, Venezuela, Brazil, Paraguay, Uruguay, and northern Argentina.

MARGINAL LOCALITIES (Map 258; from W. B. Davis's 1966 account for *E. dorianus*, except as noted): *Northern distribution.* VENEZUELA: Guárico, 9 km SE of Calabozo (Handley 1976); Apure, Hato El Frio (Ibáñez 1984a). COLOMBIA: Putumayo, Parque Nacional Natural La Paya (Polanco-Ochoa, Jaimes, and Piragua 2000). VENEZUELA: Barinas, Finca El Oasis (Ochoa, Castellanos, and Ibáñez 1988). *Southern distribution.* BRAZIL: Maranhão, Alto Parnaibo; Bahia, Região de Boa Nova (Falcão, Soares-Santos, and Drummond 2005); Espírito Santo, Município de Linares (Peracchi and Albuquerque 1993); Rio de Janeiro, Reserva Florestal do Grajaú (Esbérard 2003); São Paulo, Furnas do Yporanga. ARGENTINA: Misiones, Río Victoria (Massoia 1980a). URUGUAY: Paysandú, Quebracho Paysandú. ARGENTINA: Buenos Aires, Paraná de las Palmas y Canal 6 (Barquez, Mares, and Braun 1999); La Pampa, 25 km S of Luan Toro (DeSantis and Justo 1978); Santa Fe, Esperanza (type locality of *Eptesicus fidelis* O. Thomas); Tucumán, Aguas Chiquitas (D. F. Williams 1978); Salta, 24 km NW of Agua Blanca (Barquez, Mares, and Braun 1999).

PARAGUAY: Guairá, Villa Rica. BRAZIL: São Paulo, Parque Estadual Morro do Diabo (N. R. Reis et al. 1996); Minas Gerais, Panga Ecological Reserve (Pedro and Taddei 1997); Bahia, São Marcello.

SUBSPECIES: We recognize two subspecies of *E. diminutus*. The isolated population in Venezuela likely constitutes an as yet unnamed third subspecies.

E. d. diminutus Osgood, 1915
SYNONYMS:

Eptesicus diminutus Osgood, 1915:197; type locality "Saõ [*sic*] Marcello, Rio Preto, Bahia, Brazil."

Eptesicus dorianus: W. B. Davis, 1966:246; not *Vesperugo dorianus* Dobson, 1885, which is a *nomen dubium* according to D. F. Williams (1978).

Eptesicus dimidiatus Handley, 1976:37; incorrect subsequent spelling of *E. diminutus* Osgood, 1915.

This subspecies is known from five specimens from Venezuela (Handley, 1976) and eastern Brazil (W. B. Davis, 1966).

E. d. fidelis O. Thomas, 1920
SYNONYMS:

Eptesicus fidelis O. Thomas, 1920b:366; type locality "Esperanza," Santa Fé, Argentina.

E[*ptesicus*]. *fidens* Muñoz, 2001:210; incorrect subsequent spelling of *Eptesicus fidelis* O. Thomas.

This subspecies is known from southern Brazil, eastern Paraguay, western Uruguay and northern Argentina (W. B. Davis 1966; D. F. Williams 1978; Barquez, Mares, and Braun 1999).

NATURAL HISTORY: *Eptesicus diminutus* seems to be a bat of the drier tropical and subtropical forests of Brazil, Paraguay, and northern Argentina. The northern isolate occurs in the seasonally dry llanos of central Venezuela. Nothing is known of its biology, except that it is insectivorous. *Eptesicus diminutus* has a $2n = 50$, $FN = 48$ karyotype (D. F. Williams 1978).

REMARKS: This is the smallest *Eptesicus* in the Western Hemisphere. Measurements of the holotype of *Vespertilio dorianus* convinced D. F. Williams (1978) that the specimen, and hence the name *Eptesicus dorianus*, does not correspond to the smallest South American species of *Eptesicus*. Therefore, he applied the next available name, *E. diminutus*, to the species treated by W. B. Davis (1966) under the name *E. dorianus*. C. O. C. Vieira's (1942:408) forearm measurements of 34 and 35 mm for six *Eptesicus hilarii*, and 14.5 mm for the greatest length of skull for two of them (all from the Brazilian states of São Paulo, Santa Catarina, and Rio Grande do Sul), indicate that his specimens probably are representatives of *E. diminutus*. According to W. B. Davis (1966:257), I. Geoffroy St.-Hilaire's name *Vespertilio hilarii* was based on a series of four specimens representing two, or possibly three, of the currently recognized South American species of *Eptesicus* (see Remarks in the account for *E. brasiliensis*). Measurements of the smallest two (forearms 34 mm and 33 mm, length of skull 14.0 mm and 13.5 mm) also suggest that these specimens represent *E. diminutus*. According to D. F. Williams (1978), both the specimen (BM 1.8.1.1) identified as *E. dorianus* by W. B. Davis (1966), and the specimen (UCONN 15649) Wetzel and Lovett (1974) assigned to *E.* cf. *fidelis*, are *E. furinalis* (also see Remarks under *E. furinalis*). Villar-R. and Villa-Cornejo (1969) reported on this species in Argentina under the name *E. innoxius*. D. C. Carter and Dolan (1978) gave information on and measurements of the holotype of *Eptesicus fidelis* O. Thomas. The identity of the specimen from Parque Nacional Natural La Paya, Putumayo, Colombia (Polanco-Ochoa, Jaimes, and Piragua 2000) needs to be confirmed.

Eptesicus furinalis (d'Orbigny and Gervais, 1847)
Argentine Brown Bat

SYNONYMS: See under Subspecies.

DISTRIBUTION: *Eptesicus furinalis* is known from Colombia Venezuela, the Guianas, Brazil, Bolivia, Paraguay, Uruguay, and Argentina. Elsewhere, this species is found in Mexico and Central America.

MARGINAL LOCALITIES (Map 259): VENEZUELA (Handley 1976, except as noted): Falcón, 24 km NW

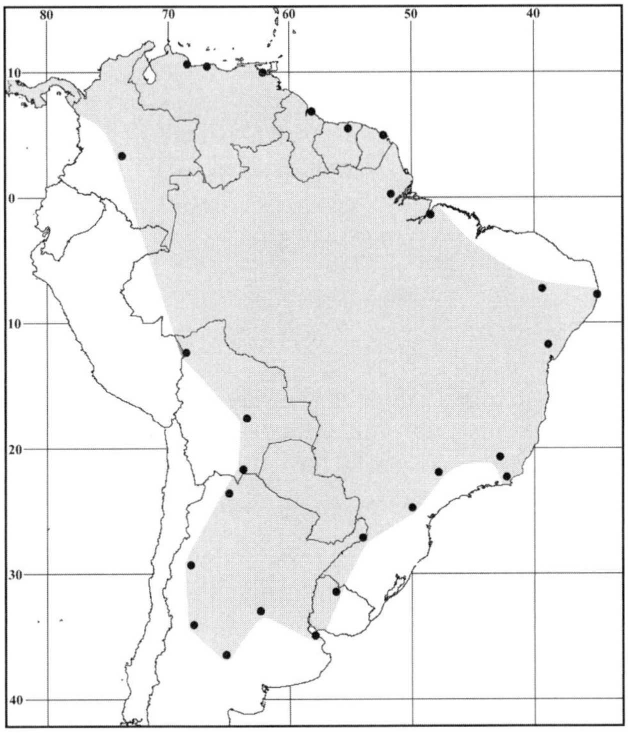

Map 259 Marginal localities for *Eptesicus furinalis* ●

of Urama; Miranda, San Andrés; Delta Amacuro, Caño Mánamo (Linares and Rivas 2004). GUYANA: Demerara-Mahaica, Georgetown (W. B. Davis 1966). SURINAM: Para, Zanderij (W. B. Davis 1966). FRENCH GUIANA: Cayenne (Brosset and Charles-Dominique 1991). BRAZIL: Amapá, Cachoeira Itaboca (Piccinini 1974, as *E. fuscus*); Pará, Belém (USNM 460141); Ceará, Floresta Nacional Araripe-Apodi, 9 km S of Crato (Mares et al. 1981); Pernambuco, Refúgio Ecológico Charles Darwin (L. A. M. Silva and Guerra 2000); Bahia, Lamarão (W. B. Davis 1966); Minas Gerais, Viçosa (USNM 541485); Rio de Janeiro, Serra Macaé (C. O. C. Vieira 1942); São Paulo, São Carlos (J. C. Motta and Taddei 1992); Paraná, Castro (W.B. Davis 1966). ARGENTINA: Misiones, 6 km NE by Highway 2 from Arroyo Paraíso (Barquez, Mares, and Braun 1999). URUGUAY: Tacuarembó, 40 km NW of Tacuarembó (Autino, Claps, and González 2004). ARGENTINA (Barquez, Mares, and Braun 1999): Buenos Aires, La Plata; Córdoba, La Maya; La Pampa, Carro Quemado; Mendoza, Ñacuñán; La Rioja, Villa Unión; Jujuy, Santa Bárbara. BOLIVIA: Tarija, Sierra Santa Rosa (O. Thomas, 1925); Santa Cruz, San Rafael de Amboró (S. Anderson 1997); Pando, Isla Gargantua (S. Anderson 1997). COLOMBIA: Meta, Los Micos (W. B. Davis 1966).

SUBSPECIES: We tentatively recognize two subspecies of *E. furinalis*.

E. f. furinalis (d'Orbigny and Gervais, 1847)
SYNONYMS:

Vespertilio furinalis d'Orbigny and Gervais, 1847:13; type locality "la province de Corrientes," Argentina.

Vesperugo (Vesperus) dorianus Dobson, 1885:17–18; type locality "Missiones Province," Argentina.

Eptesicus furinalis: O. Thomas, 1920b:365; name combination.

Eptesicus furinalis findleyi D. F. Williams, 1978:377; type locality "Aguas Chiquitas, about 800 m, Sierra de Medina, Provincia Tucumán, Argentina."

This subspecies is found in northern Argentina, Paraguay, Bolivia, Brazil, and Uruguay.

E. f. gaumeri (J. A. Allen, 1897)
SYNONYMS:

Adelonycteris gaumeri J. A. Allen, 1897c:231; type locality "Izamal," Yucatán, Mexico.

Eptesicus chapmani J. A. Allen, 1915:632; type locality "Lower Rio Solimoens"; identified as near Manaus, Amazonas, Brazil, by W. B. Davis (1966:267); later identified as "Manacaparú" (= Manacapurú) by Moojen (1948:345; also see Voss, Lunde, and Simmons 2001:118–19).

Eptesicus furinalis chapmani: W. B. Davis, 1966:267; name combination.

Eptesicus furinalis gaumeri: W. B. Davis, 1966:268; first use of current name combination.

E[*ptesicus*]. *f*[*urinalis*]. *guameri* Barquez, Mares, and Braun, 1999:124; incorrect subsequent spelling of *Adelonycteris gaumeri* J. A. Allen.

This subspecies is found in Colombia, Venezuela, Surinam, Guyana, French Guiana, and Venezuela, southward throughout the Amazon drainage of Brazil and Bolivia. Elsewhere, it occurs northwestward at low elevations through Central America into the Mexican states of Jalisco, Morelos, and Tamaulipas.

NATURAL HISTORY: This is a lowland species mainly found in forests, often in association with human habitations, at elevations below 1,000 m. *Eptesicus furinalis* usually roost in small groups, which have been located in hollow logs, tree holes, holes in standing snags, and in the attics of houses. Villa-R. (1967:405) reported a concentration of *E. f. gaumeri* estimated at 100,000 individuals in a cave in the Mexican state of Morelos. The only data we have on reproduction is from Central America where pregnant females have been taken in March and June, and lactating, nonpregnant females in July and August. All 35 female *E. f. gaumeri* captured on 21 March 1963 in a building on the outskirts of Turrialba, Costa Rica, were pregnant with fetus size ranging in CR length from 9 to 22 mm. Four were pregnant with a single fetus and the remainder was carrying twins. The dentition indicates a diet of insects, but food-habits analyses have not been done. Autino, Claps, and González (2004) recovered two species of nycteribiid bat flies from Uruguayan *E. furinalis*; one of which (*Basilia andersoni*) was described by B. V. Peterson and Maa (1970) from *E. furinalis* they had misidentified as *E. brasiliensis*. Mies, Kurta, and King's (1996) *Mammalian Species* account on *E. furinalis* includes undifferentiated information on *E. andinus*. *Eptesicus furinalis* has a $2n = 50$ FN $= 48$ karyotype (R. J. Baker and Patton 1967; D. F. Williams 1978).

REMARKS: *Adelonycteris gaumeri* was not mentioned by Cabrera (1958), probably because he considered it to be extralimital. Cabrera (1958:107) treated *Eptesicus chapmani* as a junior synonym of *Eptesicus melanopterus* (Jentink, 1904), which we treat as a subspecies of *E. brasiliensis*. W. B. Davis (1966) did not report any specimens of *E. furinalis* from Ecuador or Peru. Apparently, reports of this species from Ecuador and Peru (Koopman 1978; Albuja 1983, 1999; Mies, Kurta, and King 1996) are based on the inclusion of *E. chiralensis*, *E. inca*, and *E. montosus* as synonyms of *E. furinalis*. We treat these names as synonyms of *E. andinus* (see Simmons and Voss 1998). Cuervo, Hernández-Camacho, and Cadena (1986)

wrote that *E. furinalis* occurs throughout Colombia; however, the published record for Los Micos is the only one we have found.

In his review of South American small *Eptesicus*, D. F. Williams (1978) analyzed samples of *E. furinalis* from Brazil, Bolivia, Paraguay, and Argentina, and compared them with samples of *E. brasiliensis* and *E. diminutus*, and with measurements of the "holotype" of *E. dorianus*. On the basis of Dobson's (1885) measurements of *E. dorianus*, D. F. Williams (1978) concluded that *E. dorianus* was a junior synonym of *E. furinalis* (hence the inclusion in the synonymy of this species). However, the measurements provided later for the specimen presently identified as the holotype of *E. dorianus* suggested to Williams that the specimen either represented an *E. brasiliensis*, or a mixup of specimens has occurred since Dobson's description. Consequently, D. F. Williams (1978:381) considered the name *E. dorianus* to be a *nomen dubium*, because he was unable to assign the name to any known species with a high degree of certainty. D. C. Carter and Dolan (1978) gave information on and measurements of a specimen they identified as a syntype [the holotype?] of *Vesperugo (Vesperus) dorianus* Dobson, an adult female in alcohol that Carter examined and measure in the Natural History Museum, London (BM 86.11.3.13) that had been acquired from the Genoa Civic Museum. According to D. C. Carter and Dolan (1978:80), this specimen came from "San Ignazio, Missiones, Argentina," and was taken in November 1883 by Bore. Dobson (1885) mention only one specimen, an adult female, in the original description and all of Dobson's measurements of this bat were external, suggesting that it was preserved in fluid. This specimen has a different catalog number than the one W. B. Davis (1966) listed from the Natural History Museum, London (BM 1.8.1.1). The identity of the type specimen(s?) of *Vesperugo (Vesperus) dorianus* Dobson requires verification.

Eptesicus fuscus (Palisot de Beauvois, 1796)
Big Brown Bat
SYNONYMS:

Vespertilio fuscus Palisot de Beauvois, 1796:18; type locality "Philadelphia," Pennsylvania, U.S.A.

[*Eptesicus*] *fuscus*: Méhely, 1900:206, 338; first use of current name combination.

[*Vesperugo (Vesperus) serotinus*] *fuscus*: Trouessart, 1897:108; name combination.

[*Vespertilio (Eptesicus)*] *fuscus*: Trouessart, 1899:1280; name combination.

Eptesicus serotinus: Koopman, 1982:275; not *Vespertilio serotinus* Schreber, 1774a.

Additional synonyms for the South American distribution are listed under subspecies.

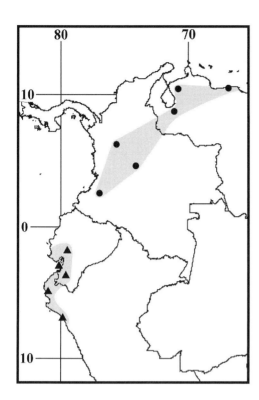

Map 260 Marginal localities for *Eptesicus fuscus* ● and *Eptesicus innoxius* ▲

DISTRIBUTION: *Eptesicus fuscus* occurs in the Andes of Colombia and Venezuela. Elsewhere, the species occurs in Canada, the United States, Mexico, Central America, and the West Indies.

MARGINAL LOCALITIES (Map 260): VENEZUELA: Distrito Federal, Los Venados (Handley 1976); Mérida, La Culata (type locality of *E. f. pelliceus* O. Thomas). COLOMBIA: Cundinamarca, Bogotá (W. B. Davis 1966); Cauca, Cerro Munchique (Alberico 1994); Antioquia, Medellín (W. B. Davis 1966). VENEZUELA: Falcón, Cerro Socopo (Handley 1976).

SUBSPECIES: We recognize 11 subspecies of *E. fuscus*, of which only *E. f. miradorensis* occurs in South America.

E. f. miradorensis (H. Allen, 1866)
SYNONYMS:

S[*cotophilus*]. *miradorensis* H. Allen, 1866:287; type locality "Mirador," Veracruz, Mexico.

[*Vespertilio (Eptesicus) fuscus*] *miradorensis*: Trouessart, 1899:1280; name combination.

Eptesicus fuscus miradorensis: Miller, 1912c:62; first use of current name combination.

Eptesicus fuscus pelliceus O. Thomas, 1920b:361; type locality "La Culata," Mérida, Venezuela.

E[*ptesicus*]. s[*erotinus*]. *miradorensis*: Koopman, 1994: 120; name combination.

This subspecies is found in the Andes of northern Colombia and Venezuela. Elsewhere, it occurs throughout most of Mexico and at higher elevations in Central America.

NATURAL HISTORY: *Eptesicus fuscus* is a relatively rare bat in South America. It occurs mainly in evergreen and cloud-forest habitats at elevations usually above 1,500 m. Individuals emerge well before dusk from their roosts in hollow trees, small caves, or man-made structures, to feed on a variety of night-flying insects, principally beetles. One or two young are produced per pregnancy with parturition taking place in April or May. The species is nonmigratory and, in North America, hibernates during inclement weather. Hibernation is unknown in South American representatives. R. W. Barbour and Davis (1969) provided details of ecology and behavior, based on studies of northern representatives of the species. T. A. Griffiths (1983b) described the laryngeal anatomy. Kurta and Baker (1990) summarized the known natural history information in their *Mammalian Species* account. *Eptesicus fuscus* has a $2n = 50$, FN $= 48$ karyotype (R. J. Baker and Patton, 1967; references in R. J. Baker et al., 1982).

REMARKS: We do not know the basis of Koopman's (1994) inclusion of "northeastern Brazil" in the distribution of *E. fuscus*. The listing of *E. fuscus* by G. A. B. Fonseca et al. (1996) for Brazil may have been based on Koopman (1994). The only large, long-haired *Eptesicus* known from lowland habitats is the smaller *E. chiriquinus*. According to Mok et al. (1982), the *Eptesicus* identified by Piccinini (1974) as *E. fuscus* from Amapá are *E. furinalis*. D. C. Carter and Dolan (1978) gave information on and measurements of the holotype of *Eptesicus fuscus pelliceus* O. Thomas. Czaplewski, Rincón, and Morgan (2005) recovered late Pleistocene remains of *Eptesicus fuscus* from a tar seep in the state of Zulia, Venezuela.

Eptesicus innoxius (P. Gervais, 1841)
Pacific Serotine
SYNONYMS:

Vespertilio innoxius P. Gervais, 1841:2; type locality "Omatope" (= Amotape), Piura, Peru.

Vespertilio espadae Cabrera, 1901:368; type locality "Babahoyo," Los Ríos, Ecuador.

Eptesicus punicus O. Thomas, 1920b:364; type locality "Puna," Isla Puna, Guayas, Ecuador.

Eptesicus innoxius: Cabrera, 1958:107; first use of current name combination.

DISTRIBUTION: *Eptesicus innoxius* is endemic to western South America where it is known only along the arid Pacific versant from Babahoyo, Ecuador, south to Puerto Eten on the coast of northwestern Peru.

MARGINAL LOCALITIES (Map 260): ECUADOR: Los Rios, Babahoyo (type locality of *Vespertilio espadae*

Cabrera); El Oro, Zaruma (W. B. Davis 1966). PERU: Lambayeque, Puerto Eten (W. B. Davis 1966); Piura, Amotape (type locality of *Vespertilio innoxius* P. Gervais). ECUADOR: Guayas, Puna (type locality of *Eptesicus punicus* O. Thomas).

SUBSPECIES: We treat *E. innoxius* as monotypic.

NATURAL HISTORY: No information seems to have been published on the behavior and ecology of *E. innoxius*. These bats probably roost in crevices in rocky outcrops, in cavities in man-made structures, and in holes in trees. The dentition indicates insectivory. Biologists from the Department of Wildlife and Fisheries Sciences, Texas A&M University, College Station, took three nonpregnant females on 29–30 July in mistnets set across a shallow stream. Field notes written by D. R. Patten (Texas A&M University) recorded that little green vegetation was present, except adjacent to the stream. Away from the stream the vegetation was either sparse grasses or leafless shrubs. Field notes written by R. W. Adams (Texas A&M University) described the vegetation as thorn forest.

REMARKS: D. C. Carter and Dolan (1978) gave information on and measurements of the holotype of *Eptesicus punicus* O. Thomas.

Genus *Histiotus* P. Gervais, 1856

Charles O. Handley, Jr., and Alfred L. Gardner

Histiotus comprises four species: *H. humboldti*, *H. macrotus*, *H. montanus*, and *H. velatus*. Although restricted in distribution to areas of cooler climate from the Andes of Colombia and Venezuela south to the Straits of Magellan, this South American endemic has been recorded from every country except the Guianas and Trinidad and Tobago. All four species are medium-sized (forearm 42–52 mm, mass 9–15 g) bats having unusually large ears (may exceed 35 mm, dry, in *H. macrotus*). The genus dates from the Late Pliocene or Early Pleistocene of North America (McKenna and Bell 1997). Pleistocene to Recent material representing *H. velatus* is known from Lagoa Santa, Minas Gerais, Brazil (Winge 1892).

SYNONYMS:

Plecotus: I. Geoffroy St.-Hilaire, 1824b:446; part; not *Plecotus* É. Geoffroy St.-Hilaire, 1818b.

Vespertilio: J. B. Fischer, 1829:118; part; not *Vespertilio* Linnaeus, 1758.

Nicticeius Pöppig, 1830:column 218; incorrect subsequent spelling of, but not *Nycticeius* Rafinesque, 1819.

Nycticeius: Poeppig, 1835:451; part; not *Nycticeius* Rafinesque.

Nycticeus: Lesson, 1836:120; part; incorrect subsequent spelling of, but not *Nycticeius* Rafinesque.

Nycticejus: Lesson, 1842:22; part; incorrect subsequent spelling of, but not *Nycticeius* Rafinesque.

Histiotus P. Gervais, 1856a:77, pl.13, Fig.6a & 6b (but not Fig. 6 [= *Plecotus auritus*]); type species *Plecotus velatus* I. Geoffroy St.-Hilaire, 1824b, by monotypy.

Vesperus: W. Peters, 1864b:383; part; not *Vesperus* Latreille, 1829 (Hymenoptera).

Vesperugo: Dobson, 1878:188; part; not *Vesperugo* Keyserling and Blasius, 1839.

Eptesicus: Olrog, 1951:508; part; not *Eptesicus* Rafinesque, 1820.

Histictus Ruschi and Bauer, 1957:40; incorrect subsequent spelling of *Histiotus* P. Gervais.

REMARKS: We have included names, in their respective synonymies in the following species and subspecies accounts, that either constitute usages before the valid name was proposed or represent usages that were based on partial or total misidentifications. This is contrary to general editorial policy for most accounts in this work; which has been either to give the author the benefit of the doubt on questions of identification or to ignore usages most likely or that are known to be based on errors (partial or total) of identification. However, Handley examined the *Histiotus* material in the Natural History Museum, London, and most of the material in the major museums in the United States and Canada, as well as some specimens in South American and continental European collections. These examinations, augmented by measurements and detailed descriptions made earlier in this century by Gerrit Miller of collections in France and Germany, have made it possible to correctly identify the specimens authors had on hand when they compiled their reports.

Hoofer and Van Den Bussch (2003) concluded that *Histiotus* is more closely related to New World species of *Eptesicus* than the latter are to Old World *Eptesicus*. They treated *Histiotus* as a subgenus of *Eptesicus* to avoid obvious paraphyly within *Eptesicus*, as currently understood. Although we accept Hoofer and Van Den Bussche's (2003) evidence as compelling, herein we arbitrarily treat *Histiotus* as a specialized estesicoid offshoot and separate from *Eptesicus*.

KEY TO THE SPECIES OF *HISTIOTUS*:

1. Width of anterior (medial) lobe of ear a third or more of total width of pinna and extending forward to near tip of muzzle . . . . *Histiotus velatus*

1'. Width of anterior medial lobe of ear a quarter or less of total width of pinna and not extending to level of muzzle
. . . . 2

2. Maxillary toothrow less than 5.7 mm; greatest length of skull less than 16.5 mm . . . . *Histiotus humboldti*

2'. Maxillary toothrow more than 5.7 mm; greatest length of skull more than 17 mm 3

3. Size large, greatest length of skull more than 18.4 mm, postpalatal length greater than 7 mm, length of ear from notch (dry) greater than 33 mm, and width of ear (dry) greater than 24 mm . . . . *Histiotus macrotus*

3'. Size smaller, greatest length of skull usually less (always less in zone of sympatry with *Histiotus macrotus* in Chile) than 18.4 mm, postpalatal length less than 7.2 mm, length of ear from notch (dry) less than 33 mm, and width of ear (dry) less than 24 mm . . . .
. . . . *Histiotus montanus*

Histiotus humboldti Handley, 1996
Humboldt's Leaf-eared Bat

SYNONYMS:

Histiotus montanus colombiae: Tamsitt and Valdivieso, 1966a:102; not *Histiotus colombiae* O. Thomas, 1916.

Histiotus humboldti Handley, 1996:2; type locality "Los Venados, 4 km NNW Caracas, 10°32'N, 66°54'W, 1498 m, Distrito Federal, Venezuela."

DISTRIBUTION: *Histiotus humboldti* has a fragmented range, suggesting a relictual distribution with isolated segments in southwestern Colombia (on the eastern flanks of the Western Andes at El Tambo and Quisquio, and near the head of the Cauca Valley at Popayán), north-central Colombia (northern extension of the Central Andes at La Ceja and Poblado), the Coast Range in northern Venezuela (Tierra Negra, Los Venados, and vicinity of Pico Ávila); and on Cerro Neblina in southernmost Venezuela. Elevational range is from 1,498 m at Los Venados, Venezuela, to 2,217 m at La Ceja, Colombia. The species undoubtedly occurs at higher elevations in northern Brazil and probably will be found elsewhere in the Guyanan Highlands.

MARGINAL LOCALITIES (Map 261; from Handley 1996): COLOMBIA: Antioquia, Poblado; Cauca, El Tambo. VENEZUELA: Mérida, Tierra Negra; Distrito Federal, 5 km NNE Caracas; Amazonas, Cerro de la Neblina, Camp II.

SUBSPECIES: We consider *H. humboldti* to be monotypic.

NATURAL HISTORY: *Histiotus humboldti* is a montane species that occurs at middle elevations, lower than *H. montanus*, which is usually found at this latitude. The species is sympatric with *H. montanus colombiae* in the Colombian Andes. Specimens from the Coast Range in northern Venezuela were taken in humid second-growth evergreen forest, which is fairly tall at Los Venados, but

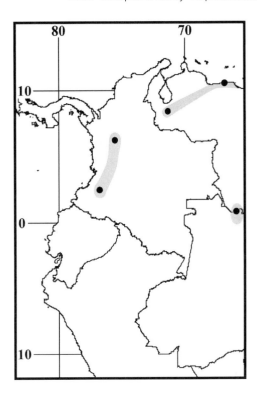

Map 261 Marginal localities for *Histiotus humboldti* •

low and dense on Pico Ávila. These localities are in Lower Montane humid forest in the Holdridge system (Ewel and Madriz 1968). The two *H. humboldti* from Cerro Neblina were netted in scrubby open tepuyan vegetation close to rocky sandstone bluffs. One was a male (mass 9.5 g) with testes and epididymides descended into the interfemoral membrane when caught on 18 March. The other was an extremely fat female (mass 11.5 g) that was neither pregnant nor lactating when caught on 20 March. Testes of a male from the Coast Range north of Caracas, Venezuela, measured 6 × 3 mm on 22 July, while those of another from the same region measured 7 × 4 mm on 23 August.

REMARKS: This is the species reported by Handley (1976) as "*Histiotus* sp. A" and by Gardner (1990) as "*Histiotus* sp." *Histiotus humboldti* can be distinguished from congeners by its smaller size, delicate rostrum, weaker zygomata, more inflated braincase, and smaller and weaker dentition. We treat the taxon as monotypic. The few available specimens do not show appreciable variation between isolated populations, which we consider relictual of a wider distribution.

Histiotus macrotus (Poeppig, 1835)
Greater Leaf-eared Bat

SYNONYMS:

N[*ycticeius*]. *macrotus* Poeppig, 1835:451, footnote; type locality "Antuco," Biobío, Chile.

Nycticeus chilensis Lesson, 1836:120; type locality "Chili méridional, dans les rochers subalpins d'Antaco [*sic*]"; based on "*N[ycticeius]. secunda* [Pöppig, 1830:column 218]"; therefore, an objective synonym of *Nycticeius macrotus* Poeppig, 1835.

Nycticejus chilensis: Lesson, 1842:22; name combination.

Nycticejus macrotis: Schinz, 1844:200; name combination.

Plecotus velatus: R. A. Philippi and Landbeck, 1861:289; not *Plecotus velatus* I. Geoffroy St.-Hilaire, 1824b.

Vespertilio velatus: R. A. Philippi and Landbeck, 1861:289; part; not *Plecotus velatus* I. Geoffroy St.-Hilaire.

Plecotus poeppigii Fitzinger, 1872:88; type locality "Antuco," Biobío, Chile; renaming of *Nycticeius macrotus* Poeppig.

Histiotus macrotus: W. Peters, 1876:788; part [Antuco]; first use of current name combination.

Vesperugo (Vesperus) macrotus: Dobson, 1878:189; part [Chile]; name combination.

[*Vesperugo (Vesperus)*] *velatus*: Trouessart, 1897:107; part: not *Plecotus velatus* I. Geoffroy St.-Hilaire.

[*Vesperugo (Vesperus)*] *macrotus*: Trouessart, 1897:107; name combination.

Vespertilio macrotus: Cabrera, 1903:284; name combination.

[*Vespertilio (Histiotus)*] *macrotus*: Trouessart, 1904:77; name combination.

Histiotus macrotus macrotus: Cabrera, 1958:108; part [Chile].

DISTRIBUTION: *Histiotus macrotus* occurs in western Argentina and central Chile at low elevations.

MARGINAL LOCALITIES (Map 262; localities listed from north to south): ARGENTINA: Jujuy, 8 km SE of Tres Cruces (Barquez and Lougheed 1990); Salta, 20 km NW of Cafayate (Barquez and Lougheed 1990); Catamarca, El Rodeo (Olrog 1960); Córdoba, Villa Cura Brochero (Barquez and Ojeda 1992); CHILE: Metropolitana de Santiago, Santiago (O. Thomas 1916); Biobío, Antuco (type locality of *Nycticeius macrotus* Poeppig). ARGENTINA: Río Negro, Estancia El Cóndor (Pearson and Pearson 1989). CHILE: Biobío, Concepción (Osgood 1943); Valparaiso, near Valparaiso (Mann 1978); Atacama, Chañaral (Mann 1978).

SUBSPECIES: We treat *H. macrotus* as monotypic.

NATURAL HISTORY: According to Mann (1978), *H. macrotus* is usually solitary and relatively uncommon in central Chile. In the vicinity of Santiago, the species is found in rock crevices and under roofs of houses. Farther north, however, especially near agricultural fields, Mann (1978) reported colonies numbering in the hundreds, if not thousands, roosting in abandoned mines in the Province of Chañaral. Based on the few records available (Barquez and Lougheed 1990; Pearson and Pearson 1990; Barquez,

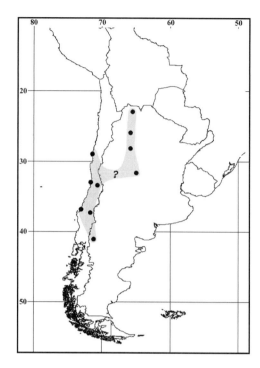

Map 262 Marginal localities for *Histiotus macrotus* ●

Mares, and Braun 1999), *H. macrotus* seems to be relatively uncommon in Argentina.

On 3 December 1976, Pearson and Pearson (1990) encountered two small, all-female clusters totaling 20 or fewer *H. macrotus* and one *H. montanus* in an attic at Estancia El Condor, Río Negro, Argentina. Of those captured, one showed no evidence of reproductive activity, whereas the others included one in late pregnancy and eight with nursing young. Among the three Pearson and Pearson (1990) encountered at this location 5 years later on 17 December, two had attached, naked young.

Brèthes (1913) reported a nycteribiid batfly from a specimen (misidentified as "*Vesperugo velatus*") from Santiago. The karyotype is unknown.

REMARKS: Although many authors have treated *H. laephotis* O. Thomas as conspecific with *H. macrotus*, we regard *H. laephotis* as a subspecies of *H. montanus*. Therefore, when we have not examined specimens reported as *laephotis*, we have assigned these records to *H. montanus*. *Histiotus macrotis* can be distinguished from congeners by its longer fur (11–12 mm mid-dorsally) and longer, darker ears. When laid forward, the tragus reaches beyond the muzzle (Mann 1978).

Histiotus montanus (R. A. Philippi and Landbeck, 1861)
Common Leaf-eared Bat

SYNONYMS: See under Subspecies.

DISTRIBUTION: *Histiotus montanus* is widespread in the Andes, where it can be found from Venezuela and

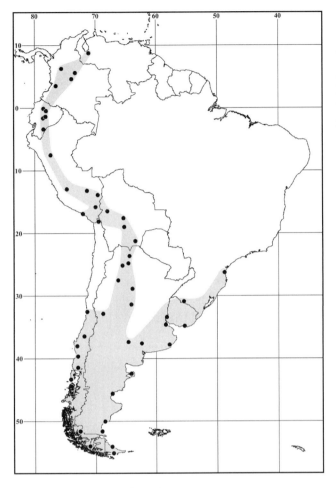

Map 263 Marginal localities for *Histiotus montanus* ●

Colombia through Ecuador, Peru, and Bolivia into Patagonia. The species also is found in the lowlands of Chile, Argentina, Uruguay, and southern Brazil, and occurs farther south (beyond the Beagle Channel) and higher (to 4,117 m) than any other South American bat. The southernmost vouchered record is from Puerto Toro, Isla de Navarino, Chile (Peña and Barria 1972); whereas the southernmost sight record is from Isla Grevy at 55°34′S (Olrog 1951).

MARGINAL LOCALITIES (Map 263): VENEZUELA: Mérida, Montes de la Hechicera (Linares 1973). COLOMBIA: Boyacá, Tunja (ROM 42784); Cundinamarca, Choachí (O. Thomas 1916). ECUADOR: Napo, Antisana Páramo, S slope Cerro Antisana, (R. H. Baker 1974); Tungurahua, 1.5 km E of Mirador (USNM 513495); Chimborazo, Riobamba (BM 99.9.9.141); Morona-Santiago, Gualaquiza (Tomes 1859, as *Vespertilio velatus*). PERU: San Martín, Pampa del Cuy (MHNSM 7265); Cusco, Huasampilla (Dobson 1878); Puno, Río Inambari (type locality of *Histiotus inambarus* H. E. Anthony). BOLIVIA: La Paz, Cota Cota in La Paz (S. Anderson 1997); Cochabamba, Pocona (S. Anderson 1997); Tarija, San Francisco Mission

(O. Thomas 1898b). ARGENTINA: Jujuy, Yuto (Linares 1973); Salta, Arroyo de los Puestos (CM 72360); Santiago del Estero, Río Saladillo (Barquez and Ojeda 1992); Córdoba, Córdoba (Guimarães and d'Andretta 1956); La Pampa, General Acha (ROM 75933). URUGUAY: Rivera, Rivera (Acosta y Lara 1955). BRAZIL: Santa Catarina, Joinville (type locality of *Histiotus alienus* O. Thomas). URUGUAY: Canelones, Jaureguiberry (AMNH 188780); Soriano, Soriano (BM 94.1.24.9). ARGENTINA: Buenos Aires, Buenos Aires (BM 4.8.8.1); Buenos Aires, near Balcarce (MSU 17883); Buenos Aires, Pigué (Acosta y Lara 1950); Chubut, Península Valdés (Daciuk 1977); Chubut, Pico Salamanca (O. Thomas 1929); Santa Cruz, Santa Cruz (Lataste 1892); Santa Cruz, Punta Loyola (BM 20.11.29.1); Tierra del Fuego, Viamonte (Linares 1973). CHILE: Magallanes, Puerto Toro (Peña and Barria 1972); Magallanes, Tierra del Fuego, Straits of Magellan (R. A. Philippi 1866); Magallanes, Puerto Prat (O. Thomas 1916); Los Lagos, Isla Chiloé, Río Inio (Osgood 1943); Los Lagos, Puerto Montt (W. Peters 1876); Los Lagos, Máfil (Osgood 1943); Araucania, 15 km E of Purén (Greer 1966); Biobío, San Carlos, Hacienda Zemita (MCZ 22164); Valparaiso, Zapallar (USNM 391789). ARGENTINA: Mendoza, Mendoza (USNM 236231); Catamarca, Potrero (D. F. Williams and Mares 1978); Salta, Toma de Los Laureles (Villa-R. and Villa-Cornejo 1969). BOLIVIA: Chuquisaca, Chuquisaca (d'Orbigny and Gervais 1847). CHILE: Tarapacá, Putre (Mann 1950b). PERU: Arequipa, Islay (Dobson 1878); Puno, Puno (FMNH 53004); Huancavelica, Huancavelica (BM 1938.9.26.3). ECUADOR: Pichincha, Pichincha (Linares 1973). COLOMBIA: Valle del Cauca, lime kiln, vicinity of Cali (Arata and Vaughn 1970); Antioquia, Poblado (AMNH 149245).

SUBSPECIES: We recognize six subspecies of *H. montanus*; also we include Natural History and Remarks sections following the distribution of each subspecies, instead of at the end of all subspecies accounts.

H. m. alienus O. Thomas, 1916

SYNONYMS:

Plecotus velatus: Burmeister, 1861:393; part (Mendoza); not *Plecotus velatus* I. Geoffroy St.-Hilaire.

Vesperugo montanus: Dobson, 1878:190; part (Mendoza).

Vesperugo velatus: Burmeister, 1879:101; part (Uruguayan specimens in the Berg Collection); not *Plecotus velatus* I. Geoffroy St.-Hilaire.

[*Vesperugo (Vesperus)*] *montanus*: Trouessart, 1897:107; part; name combination.

[*Vespertilio (Histiotus)*] *montanus*: Trouessart, 1904:77; part; name combination.

Histiotus velatus: J. A. Allen, 1905:187; not *Plecotus velatus* I. Geoffroy St.-Hilaire.

Histiotus montanus: O. Thomas, 1916:274; part (Uruguayan and Argentinean specimens).

Histiotus alienus O. Thomas, 1916:276; type locality "Joinville, Santa Catherina," Brazil.

Histiotus montanus montanus: Acosta y Lara, 1950:20; part.

This subspecies occurs in Uruguay, southern Brazil, and east-central Argentina, generally below 1,000 m.

REMARKS: Cabrera (1958) overlooked *H. alienus* O. Thomas, 1916, and did not include Brazil in the distribution of *H. montanus*. The references to Brazilian *H. montanus* by Trouessart (1897, 1904) likely refers to *H. velatus* (see Remarks under *H. velatus*). Corbet and Hill (1980, 1986, 1991), Honacki, Kinman, and Koeppl (1982), Linares (1973), and Koopman (1993) listed *H. alienus* as a species.

H. m. colombiae O. Thomas, 1916

SYNONYMS:

V[espertilio]. velatus: Tomes, 1859:546; not *Plecotus velatus* I. Geoffroy St.-Hilaire.

Vesperus montanus: W. Peters, 1876:790; part (Quito).

Histiotus colombiae O. Thomas, 1916:274; type locality "Choachi, near Bogotá," Cundinamarca, Colombia.

Histiotus montanus colombiae: Cabrera, 1958:109; first use of current name combination.

This subspecies is found in the Andes of Colombia, Venezuela, and Ecuador at elevations from 420 to 4,117 m.

NATURAL HISTORY: Two females were pregnant with a single fetus each when caught by one of us (ALG) on 18 August, 1.5 km E of Mirador, Tungurahua, Ecuador. The fetus of one measured 5 mm CR; the other, 11 mm CR. Albuja (1983:214) reported on a group of six *H. m. colombiae* captured during July in a hole in the wall of a canyon in the vicinity of Lincohuayco, Napo, Ecuador. The group consisted of one male and four females of which two were pregnant, and one was nursing young.

REMARKS: We suspect that the specimen reported by Tomes (1859) as acquired by Fraser in Gualaquiza (elevation 971 m), Ecuador, came from elsewhere at a higher elevation; perhaps from "Gualasio" (Fraser 1858:5942; = Gualaceo, Azuay), where Fraser visited along his journey from Cuenca to Gualaquiza.

H. m. inambarus H. E. Anthony, 1920

SYNONYMS:

Vespertilio peruvianus J. LeConte, 1858:74; type locality "Peru"; a forgotten name (*nomen oblitum*); see Remarks.

V[espertilio]. velatus: Tschudi, 1844b:74; not *Plecotus velatus* I. Geoffroy St.-Hilaire.

Plecotus peruvianus Fitzinger, 1872:84; type locality "Peru"; the name is unavailable because it is a junior secondary homonym of *Vespertilio peruvianus* J. LeConte, 1858; based on Tschudi's (1844b:74) Peruvian specimens for which Tschudi used the name *Vespertilio velatus*; not *Plecotus velatus* I. Geoffroy St.-Hilaire.

Vesperugo montanus: Dobson, 1878:190; part (Peru).

[*Vespertilio (Histiotus)*] *montanus*: Trouessart, 1904:77; part (Peru).

Histiotus inambarus H. E. Anthony, 1920:85; type locality "Río Inambari, (70°15′W, 13°55′S), altitude 2200 feet," Puno, Peru; restricted to "Segrario, Río Quiton" [*sic*] by Sanborn (1951b).

Histiotus macrotus: Sanborn, 1941:384; not *Nycticeius macrotus* Poeppig.

Histiotus montanus inambarus: Sanborn, 1951b:13; part (holotype only); first use of current name combination.

This subspecies is known from Peru and northern Chile, and probably occurs in western Bolivia. Its elevational distribution is from sea level, on the Pacific coast, up to about 4,000 m in the Andes, and down to about 700 m on the eastern slope.

NATURAL HISTORY: The majority of specimens have been taken in buildings (specimens identified as *H. macrotus* by Pearson [1951] and Sanborn [1941]).

REMARKS: Cabrera (1958:109) used the name *Histiotus montanus inambarus* H. E. Anthony, 1920, for this taxon; although he was aware of Fitzinger's publication (see Cabrera 1958:108, under *H. macrotus macrotus*), he overlooked both *Vespertilio peruvianus* J. LeConte, 1858, and *Plecotus peruvianus* Fitzinger, 1872. The latter was based on Tschudi's (1844b:75) description of a bat identified as *Vespertilio velatus*, which he stated was caught at 11,000 ft. Therefore, the bat clearly was a *H. montanus* and not a *H. velatus*, which is the only other species of *Histiotus* known from Peru, but occurring at lower elevations.

The name *Vespertilio peruvianus* J. LeConte, 1858, antedates *Vespertilio montanus* R. A. Philippi and Landbeck, 1861, by 3 years, and *Plecotus peruvianus* Fitzinger, 1872, by 14 years. However, because we have not found any reference to its use in the century and a half since its publication and, because the conditions of Arts. 23.9.1.1 and 23.9.2 have been met (ICZN 1999), we consider *Vespertilio peruvianus* J. LeConte to be a forgotten name (*nomen oblitum*). J. LeConte (1858:174) wrote "I received [it] from Mr. Cassin, who informed me that it was given to him as a native of Peru." We also reject the name *Plecotus peruvianus* Fitzinger, as unavailable because it is a junior subjective homonym of *Vespertilio peruvianus* J. LeConte.

All authors have mistakenly followed Sanborn's (1941) by referring specimens of *H. montanus inambarus* from southwestern Peru to *H. macrotus*. Excepting the holotype, the specimens Sanborn (1951b) identified as *H. montanus inambarus* represent *H. velatus*.

H. m. laephotis O. Thomas, 1916

SYNONYMS:

Plecotus velatus: d'Orbigny and Gervais, 1847:14; part (Chuquisaca, Bolivia); not *Plecotus velatus* I. Geoffroy St.-Hilaire.

Vesperugo montanus: Dobson, 1878:190; part (Bolivia).

[*Vesperugo (Vesperus)*] *velatus*: Trouessart, 1897:106; part (Bolivia); not *Plecotus velatus* I. Geoffroy St.-Hilaire.

Vesperugo (Histiotus) velatus: O. Thomas, 1898b:2; part (Bolivia and Argentina) not *Plecotus velatus* I. Geoffroy St.-Hilaire.

[*Vespertilio (Histiotus)*] *montanus*: Trouessart, 1904:77; part (Bolivia).

Histiotus laephotis O. Thomas, 1916:275; type locality "Caiza," Tarija, Bolivia.

[*Histiotus*] *macrotus*: H. E. Anthony, 1920:85; part (USNM 105105, Jujuy. Argentina); not *Nycticeius macrotus* Poeppig.

Histiotus macrotus laephotis: Cabrera, 1958:108; name combination.

Histiotus montanus montanus: Villa-R. and Villa-Cornejo, 1969:421; not *Vespertilio montanus* R. A. Philippi and Landbeck.

H[*istiotus*]. *montanus laephotis*: S. Anderson, 1997:12; first use of current name combination.

The subspecies *H. m. laephotis* occurs along the eastern flank of the Andes in northern Argentina and south-central Bolivia at elevations from 350 to 2,830 m.

REMARKS: Corbet and Hill treated *H. laephotis* as a species in their 1980 and 1986 editions of "A world list of mammalian species," but not in the 1991 third edition in which the name is not mentioned. Barquez, Mares, and Ojeda (1991) and Barquez, Mares, and Braun (1999) treated *laephotis* as a subspecies of *macrotus* but commented (1991:48), "The taxonomy of these two 'subspecies,' which have been captured within 10 km of one another in Tucumán, needs to be clarified." Several other authors (e.g., Honacki, Kinman, and Koeppl 1982; Koopman 1993) have followed Cabrera (1958) in treating *H. laephotis* O. Thomas, as conspecific with *H. macrotus*. We regard *H. laephotis* as a subspecies of *H. montanus*. If we have not examined and verified identities of specimens reported in the literature as *laephotis*, we have assigned these literature records to *H. montanus*.

H. m. magellanicus (R. A. Philippi, 1866)

SYNONYMS:

Vespertilio magellanicus R. A. Philippi, 1866:113; type locality "Magellanstrasse," Magallanes, Chile.

Vespertilio capucinus R. A. Philippi, 1866:114; type locality "Chile."

Vesperus magellanicus: W. Peters, 1876:790; name combination.

Vesperugo magellanicus: Dobson, 1878:190; name combination.

Vesperus montanus: W. Peters, 1876:790; part (Puerto Montt); not *Vespertilio montanus* R. A. Philippi and Landbeck.

[*Vesperugo (Vesperus)*] *magellanicus*: Trouessart, 1897:107; name combination.

[*Vespertilio (Histiotus)*] *magellanicus*: Trouessart, 1904:77; name combination.

Histiotus magellanicus: O. Thomas, 1916:273; name combination.

Histiotus montanus magellanicus: Osgood, 1943:61; first use of current name combination.

Histiotus montanus montanus: Osgood, 1943:59; part (Cautín, Malleco, and Temuco); not *Vespertilio montanus* R. A. Philippi and Landbeck.

Eptesicus montanus magellanicus: Olrog, 1951:508; name combination.

This southernmost subspecies of *H. montanus* occurs in southern Chile and insular portions of extreme southern Argentina, generally below 1,200 m.

NATURAL HISTORY: Peña and Barria (1972) found single individuals under loose bark on dead trees. Olrog (1951) reported seeing flying bats, which he identified as *Eptesicus montanus magellanicus*, on Isla Navarino (Wulaia at 55°03'S), Isla Bertrand (55°14'S), and Isla Grevy (55°34'S), all of which are farther south in Chile than "the southernmost record of any bat" cited by Koopman (1967). The specimen from Puerto Toro (Peña and Barria 1972), at 55°S, is the southernmost vouchered record.

REMARKS: Some authors (e.g., Barquez, Giannini, and Mares 1993; Barquez, Mares, and Braun 1999) have treated *magellanicus* as a species. Barquez, Mares, and Braun (1999:136) stated that the lack of any trace of a connecting band between the ears differentiates *magellanicus* from all other *Histiotus* in Argentina. Nevertheless, we found the band to be poorly developed or absent in other *H. montanus*. The darker color of *magellanicus* seems to be associated with the humid forest habitats that characterize its distribution in southern Chile and adjacent Argentina.

H. m. montanus (R. A. Philippi and Landbeck, 1861)

SYNONYMS:

Vespertilio velatus: Gay and Gervais, 1847:40, in Gay (1847); part (Santiago); not *Plecotus velatus* I. Geoffroy St.-Hilaire, 1824b.

Plecotus velatus: P. Gervais, 1854:215; part; not *Plecotus velatus* I. Geoffroy St.-Hilaire.

Histiotus velatus: P. Gervais, 1856a:77; part (Chile and Bolivia); not *Plecotus velatus* I. Geoffroy St.-Hilaire.

Vespertilio montanus R. A. Philippi and Landbeck, 1861: 289; type locality "Cordillera von Santiago," Chile.

Vesperus Segethii W. Peters, 1864b:383; type locality "Chili."

Vesperus montanus: W. Peters, 1876:789; part; name combination.

Vesperugo montanus: Dobson, 1878:189; part; name combination.

Vesperus velatus: Burmeister, 1879:101; part; not *Plecotus velatus* I. Geoffroy St.-Hilaire.

Vesperugo magellanicus: Lataste, 1892: 90; part (Santa Cruz); not *Vespertilio magellanicus* R. A. Philippi.

[*Vespertilio (Histiotus)*] *montanus*: Trouessart, 1904:77; part; name combination.

Vespertilio magellanicus: J. A. Allen, 1905:187; part (Santa Cruz); not *Vespertilio magellanicus* R. A. Philippi.

Histiotus montanus Miller, 1907b:214; name combination.

Histiotus montanus montanus: Osgood, 1943:59; first use of current name combination.

The nominate subspecies is found in central Chile and in north- and west-central Argentina south into Santa Cruz, and from sea level up to 2,100 m.

NATURAL HISTORY: The karyotype is $2n = 50$, FN $= 48$ (D. F. Williams and Mares 1978).

Histiotus velatus (I. Geoffroy St.-Hilaire, 1824)
Tropical Leaf-eared Bat

SYNONYMS:

Plecotus velatus I. Geoffroy St.-Hilaire, 1824b:446; type locality "le district de Curityba," Paraná, Brazil.

Vespertilio velatus: J. B. Fischer, 1829:118; name combination.

Vespertilio euryotis Temminck, 1840:241; *nomen nudum*.

Histiotus velatus: P. Gervais, 1856a:77; first use of current name combination.

Vesperugo velatus: Dobson, 1878:188; name combination.

[*Vespertilio (Vesperugo)*] *velatus*: Trouessart, 1897:106; part; name combination.

[*Vespertilio (Histiotus)*] *velatus*: Trouessart, 1904:77; part; name combination.

Histiotus velatus miotis O. Thomas, 1916:273; type locality "Chapada, Matto Grosso. Alt. 800 m," Brazil.

DISTRIBUTION: *Histiotus velatus* is found in northern Bolivia, southeastern Peru, Brazil (from Mato Grosso northeast into Maranhão and Ceará, and south into Rio Grande do Sul), Paraguay, and northern Argentina, and from near sea level up to 2,400 m.

MARGINAL LOCALITIES (Map 264): BRAZIL: Ceará, Faculdade de Veterinária do Ceará, Fortaleza (Piccinini 1974); Minas Gerais, Itinga (Guimarães and

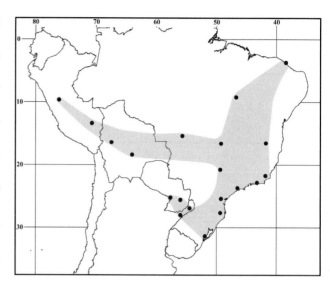

Map 264 Marginal localities for *Histiotus velatus* ●

d'Andretta 1956); Rio de Janeiro, Recreo do Mota (Peracchi 1968); Rio de Janeiro, Rio de Janeiro (C. O. C. Vieira 1942); São Paulo, São Bernardo (MPM 4901); Paraná, Curityba (type locality of *Plecotus velatus* I. Geoffroy St.-Hilaire); Santa Catarina, Catuíra (Amorim, Silva, and Silva 1970); Rio Grande do Sul, San Lorenzo (O. Thomas 1916). ARGENTINA: Corrientes, Virasoro (Vaccaro 1992). PARAGUAY: Central, Jardin Botánico (MCZ 42236); Guairá, Paso Yobay (MCZ 42235). ARGENTINA: Misiones, Río Victoria (Massoia 1980a). BRAZIL: São Paulo, São José do Rio Prêto (McNab 1969). BOLIVIA: Santa Cruz, 5.5 km by road NNE of Vallegrande (S. Anderson 1997); La Paz, Irupana (S. Anderson 1997). PERU: Huánuco, Cordillera Carpish, Carretera Central (LSUMZ 12587); Cusco, Hacienda Cadena (Sanborn 1951b, as *H. m. inambarus*). BRAZIL: Mato Grosso, Santa Anna de Chapada (type locality of *H. velatus miotis* O. Thomas); Goiás, Goiânia (MG 1332); Maranhão, Tranqueira (FMNH 26466).

SUBSPECIES: We treat *H. velatus* as monotypic.

NATURAL HISTORY: Peracchi (1968) reported finding females with young on 18 October in a building of the Universidade Rural of the state of Rio de Janeiro. Weights of four young ranged from 4 to 7 g. He also reported on a colony in a church in Recreio do Mota, Rio de Janeiro, estimated as containing from 30 to 35 bats; 12 females and 5 males were captured, none was reproductive. Peracchi and Albuquerque (1971) reported finding groups of 10 to 30 individuals in spaces in and under roofs at Universidade Rural, and stated that the species reproduces once a year in the middle of September. In the vicinity of Viçosa, Minas Gerais, Mumford and Knudson (1978) reported day roosts in attics and between rafters, and finding one bat hanging on the side of a building. Most were small colonies (6 to 12 individuals); females were not pregnant in July, but on 31 October, six had nursing young, two of which still had an attached umbilical cord and placenta. Males taken at the end of July had testes from 2 to 6 mm long; one taken 30 November had 2.5-mm testes.

Guimarães and d'Andretta (1956) reported nycteribiid batflies from Brazilian *H. velatus*. Amorim, Silva, and Silva (1970) isolated rabies virus from an individual from Catuíra, Santa Catarina, Brazil.

REMARKS: Cabrera (1958:110) repeated his earlier opinion (Cabrera, 1903:284) that the figures P. Gervais (1856, plate 13, Figs. 6–6b) presented as *H. velatus* did not show the characters of the species. Cabrera suggested that the figures may have been based on *H. macrotus*, but W. Peters (1876:786) had already pointed out that P. Gervais' figures are of *Plecotus auritus*. The photograph in Nowak (1999:445, and earlier editions) by O'Neill and identified as of a *H. montanus*, is of a *H. velatus* from the Cordillera Carpish, Huánuco, Peru. The photograph shows the greatly enlarged medial lobe of the ear that is diagnostic of this species. The misidentification of the bat in that photograph may be the basis for Eisenberg and Redford's (1999) mapping of the distribution of *H. macrotus* through Bolivia and into Peru.

Lataste (1892:88) is the source of the reference to Pará, Brazil, (cited under *Vesperugo (Vesperus) montanus* by Trouessart 1897:107; 1904:77; and others) in the distribution of *Histiotus*. However, as no *Histiotus* is known from the state of Pará, or from anywhere in the mid- or lower central Amazon basin, perhaps either Doria's "Para" in his letter to Lataste, or Lataste's spelling itself, was a lapsus for Paraná, where *H. velatus* is known to occur.

Despite several reports listing *H. velatus* from Uruguay (Burmeister 1879; Figueira 1894; Sanborn 1929a; Redford and Eisenberg 1992), the presence of the species has not been confirmed (Ximénez, Langguth, and Praderi 1972) and the records likely refer to *H. montanus*, which is relatively common there. Figueira (1894) and Sanborn (1929a) recorded *H. velatus* from Gruta de Arequita, Minas, Uruguay, but the specimens were later identified as *Myotis chiloensis* (see Devincenzi 1935).

Genus *Lasiurus* Gray, 1831

Alfred L. Gardner and Charles O. Handley, Jr.

Lasiurus includes at least 15 species, of which 5 of the 9 species recorded from South America are endemic. The species in South America are: *L. atratus, L. blossevillii, L.*

castaneus, L. cinereus, L. ebenus, L. ega, L. egregius, L. salinae, and *L. varius*. These are medium-sized to large vespertilionids (forearm 34.9–57.0 mm, mass 5.4–22.0 mm), and all have long, dense fur covering the proximal third or more of the upper surface of the uropatagium. The premaxillae and upper incisors are separated by a wide palatal emargination that is approximately a third the distance across the canines (widest of all American vespertilionids). The dental formula is 1/3, 1/1, 1–2/2, 3/3 = 30 or 32.

The genus is known from Canada and all of the conterminous United States, plus Hawaii, Mexico, Central America, and the West Indies, south through South America into Chile, and southern Argentina. There are extralimital records from Bermuda, Iceland, the Orkney Islands, and northern Canada (Great Slave Lake and Southampton Island, Northwest Territories). Fossil material is known from the Early Pliocene to Recent of North America and the Pleistocene to Recent of South America (Winge 1892; Czaplewski 1993; McKenna and Bell 1997).

SYNONYMS:

Vespertilio: Müller, 1776:20; part; not *Vespertilio* Linnaeus, 1758.

Nycteris Borkhausen, 1797:66; type species *Vespertilio noveboracensis* Erxleben, 1777; suppressed in favor of *Nycteris* É. Geoffroy St.-Hilaire and Cuvier, 1795 (ICZN, 1929:18).

Atalapha Rafinesque, 1814:12; part.

Nycticeius Rafinesque, 1819:417; part.

Taphozous: Lesson, 1827:84; part; not *Taphozous* É. Geoffroy St.-Hilaire, 1818b.

Nycticejus Temminck, 1824a:xviii; part; incorrect subsequent spelling of *Nycticeius* Rafinesque.

Lasiurus Gray, 1831:38; type species *Lasiurus borealis* (Müller, 1776), by subsequent designation (Miller and Rehn 1901:261); name originally applied to "hairy-tailed species of America," no species were mentioned.

Nyc[ticea]. J. E. LeConte, 1831:432; part; incorrect subsequent spelling of *Nycticeius* Rafinesque.

Nycticeius: Poeppig, 1835:451; part; not *Nycticeius* Rafinesque.

Scotophilus: Gray, 1838b:498; part; not *Scotophilus* Leach, 1821a.

Atalapha: P. Gervais, 1854:114; not *Atalapha* Rafinesque, 1814.

Aeorestes Fitzinger, 1870b:427; type species *Aeorestes villosissimus* (É. Geoffroy St.-Hilaire, 1806) by original designation.

Dasypterus W. Peters, 1870:912; type species *Atalapha intermedia* (H. Allen, 1862), by subsequent designation (Miller 1897a:115); described as a subgenus of *Atalapha* Rafinesque (*sensu* W. Peters 1870).

Lasyurus Cabrera, 1917:20; incorrect subsequent spelling of *Lasiurus* Gray.

Dasipterus Ruschi and Bauer, 1957:40; incorrect subsequent spelling of *Dasypterus* W. Peters.

Desípterus Ruschi and Bauer, 1957:41; incorrect subsequent spelling of *Dasypterus* W. Peters.

Lasirius Tamayo and Frassinetti, 1980:331; incorrect subsequent spelling of *Lasiurus* Gray.

Lasirurus Engstrom and Lim, 2002:375; incorrect subsequent spelling of *Lasiurus* Gray.

REMARKS: *Atalapha* Rafinesque, 1814, was used as the valid generic name during the latter part of the 19th century because it antedated *Lasiurus* Gray, 1831. However, Miller (1897a:13) claimed that because *Atalapha* "is clearly based on a Sicilian bat. . . . the use of the name for a genus confined to America is therefore impossible." In the appendix, Palmer (1904:126) listed *Atalapha sicula* from Sicily and *A. americana* (= *Vespertilio noveboracensis* Erxleben) from North America as the included species, but did not designate a type species. Palmer (1904:953) acknowledged Miller's (1897a) restriction of *Atalapha* Rafinesque, 1814, to a "Sicilian bat" as designating *Atalapha sicula* as the type species of *Atalapha*.

As pointed out by Miller (1909) and discussed by Hall and Kelson (1959:188), Hall in Hall and Jones (1961), and Hall (1981:219), *Nycteris* É. Geoffroy St.-Hilaire and Cuvier, 1795, is technically a *nomen nudum* and *Nycteris* Borkhausen, 1797, is the earliest available name for the American hairy-tailed bats, according to strict application of the Law of Priority. However, Opinion 111 (ICZN 1929) suspended the rules of nomenclature and adopted *Nycteris* É. Geoffroy St.-Hilaire and Cuvier, 1795, for the Old World slit-face bats (family Nycteridae), thereby making *Nycteris* Borkhausen a junior homonym and unavailable. The next available name is *Lasiurus* Gray, 1831. Hall (1981), believing the Law of Priority should not be suspended only to satisfy the desire by some zoologists to maintain familiar names even when applied erroneously, ignored Opinion 111 (ICZN 1929) and used *Nycteris* instead of *Lasiurus* for American bats.

Following H. Allen's (1894a) elevation of W. Peters's subgenus *Dasypterus* to generic level, most authors recognized two genera of lasiurine bats, *Lasiurus* and *Dasypterus*. The primary distinguishing character used by W. Peters (1870) and Miller (1907b) was the presence of two upper premolars on each side in *Lasiurus* and one in *Dasypterus*. Handley (1960) showed that presence or absence of the first upper premolar is variable in *Lasiurus* and, after emphasizing similarities between *Dasypterus* and *Lasiurus*, synonymized them. Although Handley (1960) did not recognize *Dasypterus*, even as a subgenus, several authors (e.g., Koopman 1993; Kurta and Lehr 1995) subse-

quently used *Dasypterus* as a subgenus to include only the "yellow bats," excluding *L. egregius*, which had been included in *Dasypterus* by W. Peters (1870) on the basis of a single upper premolar. Corbet and Hill (1991) included *L. egregius* in the subgenus *Dasypterus*. Some authors continued to treat *Dasypterus* as a full genus (Husson 1962, 1978; Peracchi and Albuquerque 1986; Barquez, Mares, and Braun 1999). In agreement with Handley (1960), R. J. Baker et al. (1988) did not find greater divergence between taxa formerly included in *Dasypterus* and those in *Lasiurus*, than they found between *L. borealis* and *L. cinereus*. As an alternative arrangement, R. J. Baker et al. (1988) suggested following Hall and Jones (1961), who had informally recognized three groups of Lasiurini: red bats, yellow bats, and hoary bats. Hoofer and Van Den Bussche (2003) considered the issue of relationships between Dasypterus and Lasiurus to be unresolved. Herein, we do not recognize *Dasypterus* as a valid genus-group taxon.

KEY TO THE SOUTH AMERICA SPECIES OF *LASIURUS*:

1. Wing membranes black 2
1'. Wing membranes usually mottled; conspicuously reddish along metacarpals and phalanges (appears paler than adjacent wing membranes in dried specimens) 7
2. Pelage either reddish (red, orange red, or chestnut), black, or a mixture of these colors; two upper premolars on each side (anterior premolar minute and displaced medially) 3
2'. Pelage uniformly either red or yellow; normally one upper premolar on each side 6
3. Dorsal fur black . . . . *Lasiurus ebenus*
3'. Dorsal fur reddish or chestnut 4
4. Dorsal fur reddish; contrasting black and white throat and upper chest; face and wings black; forearm 45.5–47.0 mm . . . . *Lasiurus atratus*
4'. Dorsal fur chestnut; color of head (including face) dark but not black; no contrasting black and white below; forearm 45 mm or shorter 5
5. Venter blackish; forearm 41.3–45.5 mm . . . . *Lasiurus castaneus*
5'. Venter orange-buff; forearm 38.7–43.4 mm . . . . *Lasiurus varius*
6. Dorsal and ventral fur yellowish . . . . *Lasiurus ega*
6'. Dorsal and ventral fur reddish . . . . . . *Lasiurus egregius*
7. Size large; forearm longer than 46 mm . . . . *Lasiurus cinereus*
7'. Size smaller; forearm shorter than 45 mm 8
8. Color pattern reddish above, paler below . . . . *Lasiurus blossevillii*
8'. Color pattern reddish to blackish brown above, dark below . . . . *Lasiurus salinae*

Lasiurus atratus Handley, 1996
Black-winged Red Bat

SYNONYMS:

Lasiurus spec.? Brosset and Charles-Dominique, 1991:543.

Lasiurus castaneus: Masson and Cosson, 1992:476; not *Lasiurus castaneus* Handley, 1960.

Lasiurus atratus Handley, 1996:5; type locality "Kaiserberg Airstrip, Zuid River, Suriname."

[*Lasiurus*] attratus Hoofer and Van Den Bussche, 2003:26; incorrect subsequent spelling of *Lasiurus atratus* Handley.

DISTRIBUTION: *Lasiurus atratus* is known from eastern Venezuela, central Guyana, southern Surinam, and central French Guiana.

MARGINAL LOCALITIES (Map 265; from Handley 1996, except as noted): VENEZUELA: Bolívar, Reserva Forestal Imataca (Unit 5); Bolívar, Km 55 on highway south of El Dorado. GUYANA: Potaro-Siparuni, Three Mile Camp (Lim et al. 1999). SURINAM: Sipaliwini, Kaiserberg Airstrip (type locality of *Lasiurus atratus* Handley). FRENCH GUIANA: 4 km N of Saül.

SUBSPECIES: We consider *L. atratus* to be monotypic.

NATURAL HISTORY: Sites in Surinam and Venezuela at which this bat was collected in 1961 and 1962 were in undisturbed lowland rainforest; Tropical Humid Forest (bh-T) in the Holdridge classification (Ewel and Madriz 1968). The specimen from Guyana was caught in a net placed in the forest canopy over a road in *terra firme* forest. The specimen from French Guiana was netted above a small stream between forest and a cultivated clearing (Masson and Cosson 1992). In the Imataca Forest Reserve, Venezuela, *L. atratus* were netted over pools along roadsides and in clearings in disturbed secondary forest (J. Ochoa, pers. comm.). The karyotype is unknown. Fossils are not known.

REMARKS: This is one of the few bats considered to be endemic to the Guyanan Subregion.

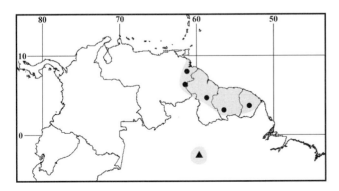

Map 265   Marginal localities for *Lasiurus atratus* ● and *Lasiurus castaneus* ▲

Map 266   Marginal localities for *Lasiurus blossevillii* ●

Lasiurus blossevillii [Lesson, 1826]
Southern Red Bat

SYNONYMS:

Atalapha teliotis H. Allen, 1891a:5; type locality unknown, but possibly California (Miller and Rehn, 1901); name is currently applied to a subspecies in the western United States, Canada, and Mexico.

Lasiurus borealis ornatus Hall, 1951:226; type locality "Penuela [Peñuela], Veracruz," Mexico; replacement name for *Lasiurus borealis mexicanus*: Miller, 1897a; not *Atalapha mexicana* Saussure, 1861 (= *Lasiurus cinereus* Palisot de Beauvois).

Additional synonyms under subspecies.

DISTRIBUTION: *Lasiurus blossevillii* has been recorded from the islands of Trinidad and Tobago and from every South American country except Chile (literature records for Chile pertain to *Lasiurus varius*). Elsewhere, the species is known from southern British Columbia, Canada, and the western United States (Genoways and Baker 1988) southward through Mexico and Central America.

MARGINAL LOCALITIES (Map 266): TRINIDAD AND TOBAGO: Tobago, Charlotteville (USNM 540697). GUYANA: Demerara-Mahaica, Demerara (BM 89.10. 26.4). SURINAM: Paramaribo, Coronie, near Totness (Husson 1962). FRENCH GUIANA: Paracou (Simmons and Voss 1998). BRAZIL: Amapá, Teresinha (USNM 391770); Pará, Pará (O. Thomas 1901h); Ceará, Aeroporto de Crato (Mares et al. 1981); Espírito Santo, Vitória (Ruschi 1954b); Minas Gerais, Viçosa (Mumford and Knudson 1978); Rio de Janeiro, Quinta da

Boa Vista (A. Miranda-Ribeiro 1907); São Paulo, São Sebastião (Handley 1960); Santa Catarina, Joinville (BM 9.11.19.2); Rio Grande do Sul, Quinta (AMNH 235380). URUGUAY: Montevideo, Montevideo (Ximénez, Langguth, and Praderi 1972). ARGENTINA: Buenos Aires, Rio La Plata, vicinity of Buenos Aires (Temminck 1840); Santa Fe, Santa Fe (Barquez, Mares, and Braun 1999); Corrientes, Laguna Paiva (Barquez and Ojeda 1992). PARAGUAY (Myers and Wetzel 1983): Cordillera, 12 km by road N of Tobatí; Presidente Hayes, Rincón Charrúa. BOLIVIA (S. Anderson 1997): Santa Cruz, 4 km E of Aserradero Moira; Santa Cruz, Santa Cruz. ARGENTINA: Salta, Quebrada de Acambuco (Mares, Ojeda, and Kosco 1981); Salta, Rosario de la Frontera (Barquez and Ojeda 1992); Córdoba, Río Ceballos (Barquez, Mares, and Braun 1999); Buenos Aires, Sierra de la Ventana (Vaccaro and Varela 2001); La Pampa, Laguna Chillhué (Tiranti and Torres 1988); Córdoba, La Paz (Barquez, Mares, and Braun 1999); Catamarca, Potrero, Río Potrero Dike (CM 42913). BOLIVIA: Cochabamba, Tinkusiri (S. Anderson 1997). PERU: Puno, "Juliaca" (Handley 1960); San Martín, Parque Nacional Río Abiseo (S. Solari et al. 2001b); Amazonas, Condechaca (O. Thomas 1926c). COLOMBIA: Cundinamarca, Sasaima (FMNH 72351); Cauca, Mazamorrero (ROM 69436). ECUADOR: Bolívar, Echeandía (Albuja 1983); Guayas, Río Chongón, 1.5 km SE of Chongón (USNM 513503). PERU: Lima, Callao (Ortiz de la Puente 1951); Lima, Cañete (LSUMZ 16635). COLOMBIA: Norte de Santander, Cúcuta (FMNH 72352). VENEZUELA: Zulia, El Rosario, 57 km WNW of Encontrados (Handley 1976); Carabobo, 10 km NW of Urama (Handley 1976); Distrito Federal, Macuto (W. Robinson and Lyon 1901). ECUADOR: Galápagos, Isla San Cristóbal (Chatham Island; J. A. Allen 1892: 47)

SUBSPECIES: We recognize three subspecies of *L. blossevillii* in South America.

L. b. blossevillii (Lesson, 1826)
SYNONYMS:

Vesp[ertilio]. *Lasiurus*: É. Geoffroy Saint-Hilaire, 1806: 200; not *Vespertilio lasiurus* Schreber, 1781.

Vespertilio blossevilii Anonymous [Lesson], 1826:95; type locality "Monte-Video," Uruguay; correct type locality is "la rivière de la Plata," Buenos Aires, Argentina (see Remarks).

Vespertilio bonariensis Lesson, 1827, in Lesson and Garnot, 1827:137; type locality "la rivière de la Plata," Buenos Aires, Argentina (see Remarks).

Scotophilus lasiurus: Gray, 1838b:498; not *Vespertilio lasiurus* Schreber.

Scotophilus Blossevilii: Gray, 1838b:498; name combination.

Nycticejus Lasiurus: Temminck, 1840:159; not *Vespertilio lasiurus* Schreber.

Vespertilio bursa Lund, 1842a:135; *nomen nudum*; part (see Winge 1892).

Nycticejus bonariensis: Temminck, 1840:159; name combination.

Lasiurus nattereri Fitzinger, 1870b:407; type locality "Cuyaba," Mato Grosso, Brazil.

[*Nycticejus (]Atalapha[)] bonaërensis* Burmeister, 1879:93; unjustified emendation of *V. bonariensis* Lesson.

Atalapha Frantzii: Hensel, 1872:25; not *Atalapha frantzii* W. Peters, 1870.

Atalapha borealis frantzii: O. Thomas, 1898b:2; name combination; not *Atalapha frantzii* W. Peters.

Lasiurus borealis bonariensis: O. Thomas, 1901j:435; name combination.

Vespertilio aurantius A. Miranda-Ribeiro, 1903:179; *nomen nudum*.

Lasiurus frantzii: Cabrera, 1903:293; name combination; not *Atalapha frantzii* W. Peters.

Lasiurus borealis mexicanus: J. L. Lima, 1926:108; part; not *Atalapha mexicana* Saussure, 1861.

Lasiurus borealis blossevillii: Cabrera, 1930:435; name combination.

Atalapha franzu Ximénez, Langguth, and Praderi, 1972:2; incorrect subsequent spelling of, but not *Atalapha frantzii* W. Peters.

L[asiurus]. borealis blosseivillii Barquez and Ojeda, 1992:248; incorrect subsequent spelling of *Vespertilio blossevillii* Lesson.

Lasiurus blossovillei Emmons, 1993:64; incorrect subsequent spelling of *Vespertilio blossevillii* Lesson.

This subspecies occurs in Peru, Guyana, Surinam, French Guiana, and the Amazon basin drainage of Colombia and Venezuela south through Brazil, Bolivia, Paraguay, Uruguay, into east-central Argentina.

L. b. brachyotis (J. A. Allen, 1892)
SYNONYMS:

Atalapha brachyotis J. A. Allen, 1892:47; type locality "Chatham Island" (Isla San Cristóbal), 1,700 ft., Galápagos, Ecuador.

Lasiurus bauri G. M. Allen, 1939:259; incorrect subsequent spelling (*lapsus*) for *A. brachyotis* J. A. Allen.

Lasiurus brachyotis: Heller, 1904:249; name combination.

Lasiurus borealis [brachyotis]: Niethammer, 1964:595; name combination.

Endemic to the Islas Galápagos, this subspecies has been recorded from Islas Floreana, Santa Cruz, and San Cristóbal (McCracken et al. 1997).

L. b. frantzii (W. Peters, 1870)

SYNONYMS:

Atalapha Frantzii W. Peters, 1870:908; type locality "Costa Rica."

Atalapha noveboracensis: Alston, 1879:22; not *Vespertilio noveboracensis* Erxleben, 1777.

Atalapha borealis frantzii: O. Thomas, 1898b:2; name combination.

Nycteris borealis mexicana: Goldman, 1920:216; not *Atalapha mexicana* Saussure.

Lasiurus borealis mexicana: J. L. Lima, 1926:108; part; not *Atalapha mexicana* Saussure.

Lasiurus borealis frantzii: Goldman, 1932:148; name combination.

Lasiurus blossevillii frantzii: Baker, Patton, Genoways, and Bickham, 1988; first use of current name combination.

This subspecies is in northern and western Colombia, northern Venezuela, and western Ecuador. Elsewhere, the subspecies occurs in Mexico and Central America.

NATURAL HISTORY: *Lasiurus blossevillii* roosts primarily in the foliage of trees and bushes (Shump and Shump 1982a). Tiranti and Torres (1998) found volant young and a lactating female in January and volant young in February in central Argentina. They also reported finding three clusters of *L. blossevillii* roosting in the upper branches of acacias. The clusters consisted of adult females with young; one cluster contained nine bats, and the other two contained four each. Albuja (1999) mentioned a pair of *L. b. brachyotis* copulating in February, and three females, each pregnant with twins in January and April, on Isla Santa Cruz. A female taken on 14 April 1971 at Cañete, Peru, was extremely fat and showed no evidence of reproductive activity.

Lasiurus blossevillii is migratory in southern Argentina (Pearson and Pearson 1990) and we suspect it is migratory elsewhere in South America. Graham (1983) reported an elevational range of from 200 to 2,400 m in Peru. The barn owl is one of the few known natural predators of *L. blossevillii*, especially on the Islas Galapagos (McCracken et al. 1997).

The karyotype is $2n = 28$, FN = 48 (R. J. Baker and Patton 1967; Bickham 1979). Wainberg (1966) reported a $2n = 22$ karyotype for Argentine *L. blossevillii*. The species is known from Quaternary cave deposits in Minas Gerais, Brazil (Czaplewski and Cartelle 1998) and from the Pliocene to Recent of North America (Czaplewski 1993). The Brazilian material was reported by Lund (1842a) from Lagoa Santa, under the name *Vespertilio bursa*; later identified as *Atalapha noveboracensis* by Winge (1892; also see Czaplewski and Cartelle 1998). The *Mammalian Species* account on *Lasiurus borealis* by Shump and Shump (1982a) also includes information on *L. blossevillii*.

REMARKS: With the exception of *L. brachyotis*, Cabrera (1958), as did many other authors (most recently, Koopman 1993; Koopman and McCracken 1998), included all red bats of South America under *L. borealis* Müller, 1776. However, we follow R. J. Baker et al. (1988) in treating the widely distributed *L. blossevillii* as a species distinct from *L. borealis* of southern Canada and the eastern United States. We also consider *L. salinae* and *L. varius* as separate from *L. blossevillii*, but treat *L. brachyotis* as a subspecies of *L. blossevillii*. Albuja (1999) retained *L. brachyotis* as a species. Barquez, Mares, and Braun (1999) treated *L. salinae* as a color variant of *L. blossevillii*. Eisenberg and Redford (1999) included both *L. borealis* and *L. blossevillii* as elements of the South American fauna.

Vespertilio blossevili is the original spelling in the unsigned (anonymous) abstracted description of *Vespertilio bonariensis* Lesson (in Lesson and Garnot 1827). Another description of *Vespertilio blossevillii* was provided by Lesson (in Lesson and Garnot 1827:137) under the name *Vespertilio bonariensis*. Lesson also wrote that this bat from "Buenos-Ayres" was captured by Mr. de Blosseville on a ship in the Río de la Plata. Therefore, the correct type locality is Buenos Aires, Río de la Plata, Buenos Aires, Argentina. With few exceptions, subsequent authors used *blossevillii* as the correct spelling and attributed the name to Lesson and Garnot. Although Lesson and Garnot are the authors of the volume "Zoologie," only Lesson is credited in the "Table des Matières," as the author of chapters I through V, volume 1, part 1, which includes the description of *Vespertilio bonariensis* (also see Woodward 1904:604). Lesson (1842:22) listed himself as the sole author of both *Vespertilio bonariensis* and "*Vespertilio Blossevillei*." Therefore, we attribute the anonymous authorship of *Vespertilio blossevillii* to Lesson alone. D. C. Carter and Dolan (1978) provided information on and measurements for two syntypes of *Atalapha frantzii* W. Peters.

Lasiurus castaneus Handley, 1960
Red-faced Red Bat

SYNONYM:

Lasiurus castaneus Handley, 1960:468; type locality "Tacarcuna Village, 3,200 ft., Río Pucro, Darién, Panama."

DISTRIBUTION: Although *L. castaneus* is listed as occurring in Colombia (Rodrígues-Mahecha et al. 1995), the only verified record for South America is from the vicinity of Manaus, Amazonas, Brazil. Muñoz (2001) indicated that *L. castaneus* was probable in Colombia. Elsewhere, it is known from Panama and Costa Rica.

MARGINAL LOCALITIES (Map 265): COLOMBIA (no specific locality; Rodríguez-Mahecha et al. 1995). BRAZIL: Amazonas, Km 41 Camp (Sampaio et al. 2003).

SUBSPECIES: *Lasiurus castaneus* is monotypic.

NATURAL HISTORY: Nothing is known about the natural history of *L. castaneus*. Seven Panamanian specimens are known. Four (including the holotype) from southern Panama taken at 1,050 m elevation, and three from the Caribbean coastal lowlands taken at Armila, provincia San Blas. Costa Rican specimens (a male and female; male banded and released) were netted in cloud forest near 1,500 m elevation in the Cordillera de Tilarán (Dinerstein 1985). The two specimens from the vicinity of Manaus, Amazonas, Brazil, were netted over a small pond in primary forest. Lengths of forearm range from 43.0 to 45.5 mm (Handley 1960; Dinerstein 1985) for Panamanian and Costa Rican specimens, with males smaller than females. The forearms of the male and female from the vicinity of Manaus, Amazonas, Brazil, measured 41.8 and 41.3 mm, respectively. The karyotype is unknown. Fossils are unknown.

REMARKS: Although the species is to be expected in Colombia (Eisenberg 1989; Muñoz 2001), and F. A. Reid (1997) implied a Colombian distribution in her range map, we are not aware of confirmed records for *L. castaneus* from that country. *Lasiurus castaneus* can be distinguished from other rufous-red lasiurines by its reddish face, nearly uniformly blackish underparts, and its forearm length of from 41.3 to 45.5 mm (longer than in *L. blossevillii* and *L. varius*; averaging shorter than in *L. atratus*, and much shorter than in *L. egregius*). It is likely that pregnant female "*Lasiurus borealis*" reported from Manaus by N. R. Reis and Peracchi (1987) represents the first record of *L. castaneus* for Brazil. *Lasiurus castaneus* and *L. atratus* may prove conspecific.

Lasiurus cinereus (Palisot de Beauvois, 1796)
Hoary Bat

SYNONYMS:

Vespertilio linereus Palisot de Beauvois, 1796:18 (spelling hand-corrected to *cinereus* before copies distributed); type locality Philadelphia, "Pennsylvania," U.S.A.

Vespertilio pruinosus Say in James, 1823:167; type locality "Engineer Cantonment," Washington Co., Nebraska, U.S.A.

A[*talapha*]. *mexicana* Saussure, 1861:97; type locality not given; presumed to be in southern Mexico (e.g., Miller 1897a).

Lasiurus cinereus: H. Allen, 1864:21; first use of current name combination.

Atalapha semota H. Allen, 1890a:173; type locality "Sandwich Islands" (= Hawaiian Islands), U.S.A.

Additional synonyms under subspecies.

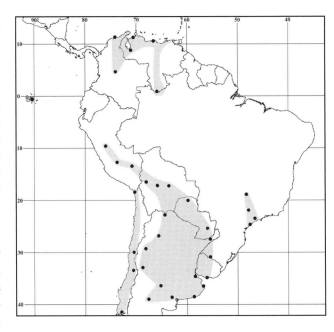

Map 267 Marginal localities for *Lasiurus cinereus* ●

DISTRIBUTION: In South America, *L. cinereus* has been recorded from Colombia, Venezuela, Ecuador (Islas Galápagos only), Peru, Chile, Bolivia, Paraguay, Argentina, and southeastern Brazil. Elsewhere, the species is known from Guatemala northward through the United States into northern Canada, and west into the Hawaiian Islands. Extralimital records also include Bermuda (G. M. Allen 1939), Iceland and the Orkney Islands (Koopman and Gudmundsson 1966), and the West Indies (Santo Domingo; Findley and Jones 1964).

MARGINAL LOCALITIES (Map 267): The distribution of *L. cinereus* appears to be regionalized as five discrete areas as follows: *Northern distribution.* COLOMBIA: Magdalena, Bonda (J. A. Allen 1900a). VENEZUELA: Falcón, Capatárida (Handley 1976); Miranda, Curvapao (Sanborn and Crespo 1957); Amazonas, Cerro de la Neblina, Camp II, ca. 2.5 km NE of Pico Phelps (Gardner 1988); Mérida, Paramo de la Culata (type locality of *Atalapha pallescens* W. Peters). COLOMBIA: Cundinamarca, Bogotá, Santa Isabel (Sanborn and Crespo 1957). *Galapagoan distribution.* ECUADOR: Galapagos, Isla Indefatigable (Niethammer 1964). *Southern distribution.* PERU: Huánuco, Zapatogocha (Gardner 1976); Ayacucho, San José (Gardner 1976); Cusco, Limacpunco (Sanborn 1953). BOLIVIA (S. Anderson 1997): La Paz, Calacoto; Cochabamba, 13 km N of Colomi; Santa Cruz, 4.5 km N of Buen Retiro. PARAGUAY: Alto Paraguay, 67 km N of Fortín Madrejón (Myers and Wetzel 1983); Caaguazú, Caaguazú (Sherman 1955). ARGENTINA: Misiones, Bonpland (Vaccaro and Massoia 1988a). BRAZIL:

Rio Grande do Sul, Livramento (Ruschi and Bauer 1957).
URUGUAY: Montevideo, Montevideo (Devincenzi 1935).
ARGENTINA: Buenos Aires, Buenos Aires (Vaccaro and
Varela 2001); Buenos Aires, Los Yngleses (O. Thomas
1910a); Buenos Aires, Quequén (Barquez, Mares, and
Braun 1999); Buenos Aires, Laguna Chasicó (Contreras
1973); Río Negro, Coronel J. F. Gómez (Vaccaro 1992);
La Pampa, Carro Quemado (Barquez, Mares, and Braun
1999); Mendoza, Godoy Cruz (Fornes and Massoia 1967);
La Rioja, Villa Unión (Sanborn and Crespo 1957); Tu-
cumán, Playa Larga (Mares et al. 1996); Salta, Río Pescado
(Barquez, Mares, and Braun 1999). *Brazilian distribution.*
BRAZIL: Minas Gerais, Uberlândia (Pedro and Taddei
1998); São Paulo, São Carlos (J. C. Motta and Taddei
1992); São Paulo, São Paulo (Sanborn and Crespo 1957);
São Paulo, Ignape [*sic*] (type locality of *Atalapha cinerea
brasiliensis* Pira). *Chilean distribution.* CHILE: Tarapacá,
Arica (Mann 1978); Coquimbo, Paiguano (Sanborn and
Crespo 1957); Metropolitana de Santiago, Santiago (Os-
good 1943); Los Lagos, Puerto Montt (Mann 1978).

SUBSPECIES: We recognize four subspecies of *L.
cinereus* and list the population on the Galapagos sepa-
rately, because we are unable to resolve its relationship to
recognized mainland subspecies at this time.

L. c. brasiliensis (Pira, 1904)

SYNONYMS:

Atalapha cinerea brasiliensis Pira, 1904:12; type locality
"Ignape," [= Iguape] São Paulo, Brazil.
Lasiurus cinereus: J. L. Lima, 1926:111; name combina-
tion.

This subspecies is known only from the Brazilian states
of Minas Gerais and São Paulo.

L. c. grayi Tomes, 1857

SYNONYMS:

Lasiurus Grayi Tomes, 1857a:40; type locality "Chili."
Atalapha Grayi: W. Peters, 1870:910; part; name combi-
nation.
[*Atalapha cinerea*] Var. (*Atalapha grayi*): Dobson, 1878:
273; part; name combination.
[*Lasiurus (Lasiurus) cinereus*] *grayi*: Trouessart, 1904:87;
name combination.
Dasypterus villosissimus: J. A. Allen, 1905:191, foot-
note; part; not *Vespertilio villosissimus* É. Geoffroy St.-
Hilaire.
Lasiurus cinereus villosissimus: Osgood, 1943:53; part; not
Vespertilio villosissimus É. Geoffroy St.-Hilaire.

The subspecies *L. c. grayi* appears to be found only in
Chile, from Arica south to Puerto Montt; but can be expected from adjacent southwestern Peru and western Bolivia and Argentina.

L. c. pallescens (W. Peters, 1870)

SYNONYMS:

Atalapha [*(Atalapha)*] *pallescens* W. Peters, 1870:914; type
locality "Paramo de la Culata, Andes de Merida (Region
frigida), Venezuela."
Lasiurus pallescens: J. A. Allen, 1900a:94; name combination.
Lasiurus cinereus pallescens: Hershkovitz, 1949c:452; first
use of current name combination.

This subspecies occurs in Colombia and Venezuela.

L. c. villosissimus (É. Geoffroy St.-Hilaire, 1806)

SYNONYMS:

Vesp[*ertilio*]. *villosissimus* É. Geoffroy St.-Hilaire, 1806:
204; type locality "Paraguay," and based on the *Chauve-
souris septième ou de chauve-souris brune-blanchâtre* of
Azara (1801b:284); restricted to Asunción by Cabrera
(1958:114).
Aeorestes villosissimus: Fitzinger, 1870b:427; name combination.
Atalapha cinerea: Hensel, 1872:25; name combination.
Atalapha villosissima: Burmeister, 1879:95; name combination.
[*Lasiurus cinereus*] *villosissimus*: O. Thomas, 1902c:238;
first use of current name combination.
Dasypterus villosissimus: J. A. Allen, 1905:191, footnote;
name combination.
Nycteris cinerea villosissimus: O. Thomas, 1910a:240;
name combination.

The most widely distributed of the South American subspecies, *L. c. villosissimus* is known from central Peru south
through Bolivia, Paraguay, Uruguay, and adjacent Brazil
into central Argentina.

L. cinereus (subspecies identity unknown)

DISTRIBUTION: This population of *L. cinereus* is
known from islas Floreana, Isabela, Santiago, Santa Cruz,
and San Cristobal in the Galapagos, Ecuador (Niethammer
1965; Orr 1966; McCracken et al. 1997).

NATURAL HISTORY: Like other *lasiurines*, *L. cinereus*
roosts primarily in the foliage of trees and bushes and occasionally has been found roosting on the bark of trees.
Sanborn and Crespo (1957) reported pregnant females
from northern Argentina collected in August, November,
and January; each carried two embryos. They suggested,
based on the approximate stage of development, the birth
dates would correspond to early September, December, and
March, respectively. Tiranti and Torres (1998) found a lactating female with two volant young roosting in an acacia
in early January in central Argentina. Graham (1983) reported an elevational range of from 1,000 to 3,300 m in
Peru. Although *L. cinereus* is migratory in North America,

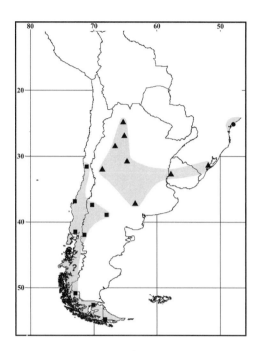

Map 268 Marginal localities for *Lasiurus ebenus* ●, *Lasiurus salinae* ▲, and *Lasiurus varius* ■

migratory behavior in South America has not been established.

The karyotype is $2n = 28$, FN $= 48$; the X chromosome is a medium-sized submetacentric, the Y a small acrocentric (R. J. Baker and Patton 1967; Bickham 1979). The species is known from the Pleistocene and Recent of North America (see references in Shump and Shump 1982b); fossils are unknown from South America.

REMARKS: While we recognize five subspecies of *L. cinereus* in South America, Shump and Shump (1982b) treated all South American hoary bats as *L. cinereus villosissimus* and did not mention the Islas Galapagos population. Redford and Eisenberg (1992:107, map 4.33) mapped two localities for *L. cinereus* in southern Chile; one appears to correspond to Chile Chico, XI Región Aisén, and the other to Puerto Natales, XII Región Magallanes. Our review of the literature, however, including references cited by Redford and Eisenberg (e.g., Sanborn and Crespo 1957; Mann 1978; Tamayo and Frassinetti 1980), has not revealed the source of those records, which we believe are errors. Eisenberg (1989) and Eisenberg and Redford (1999) showed a continuous distribution through Peru and Ecuador. However, as the localities they mapped indicate, and as is reflected in our distribution map for *L. cinereus* (Map 268), there are no records for southern Colombia, mainland Ecuador, and northern Peru. D. C. Carter and Dolan (1978) provided information and measurements for the holotype of *Atalapha pallescens* W. Peters.

Villa-R. and Villa-Cornejo (1969) mentioned catching a *L. c. villosissimus* at Finca Belgrano, Salta, Argentina, that had a well-worn band on its forearm. Unfortunately, the bat escaped before the band could be read. They raised the question: could that bat have been a migrant from North America (where numerous *L. cinereus* have been banded)? As Villa-R. and Villa-Cornejo commented, we have very little information on who may have been banding bats at that time (1965) in South America.

Lasiurus ebenus Fazzolari-Corrêa, 1994
Black Hairy-tailed Bat

SYNONYM:

Lasiurus ebenus Fazzolari-Corrêa, 1994:119; type locality "Brazil: São Paulo: Parque Estadual da Ilha do Cardoso (25°05′S, 47°59′W)."

DISTRIBUTION: *Lasiurus ebenus* is known only from the type locality in the Atlantic forest of southeastern Brazil.

MARGINAL LOCALITIES (Map 268): BRAZIL: São Paulo, Parque Estadual da Ilha do Cardoso (type locality of *Lasiurus ebenus* Fazzolari-Corrêa).

SUBSPECIES: *Lasiurus ebenus* is monotypic.

NATURAL HISTORY: The holotype was caught in a mistnet over a stream in premontane forest. It was the only *Lasiurus* captured in 13 months of field work in the park. Other bats with restricted distributions found at this locality were *Myotis ruber*, *Eptesicus diminutus*, and *Chiroderma doriae*. Among 27 species of bats known from Parque Estadual da Ilha do Cardoso, 6 were vespertilionids, of which *Myotis nigricans* was most common. The karyotype is unknown. There is no known fossil record.

Lasiurus ega (P. Gervais, 1856)
Southern Yellow Bat

SYNONYMS: See under subspecies.

DISTRIBUTION: *Lasiurus ega* has been recorded from the island of Trinidad and from every South American country except Chile and French Guiana. Elsewhere, the species is known from the southwestern United States into Panama.

MARGINAL LOCALITIES (Map 269; sequence of localities from Ecuador northward to the north coast of Colombia, thence clockwise): ECUADOR: Guayas, Puna Island (type locality of *Dasypterus ega punensis* J. A. Allen); Pichincha, Santo Domingo de los Colorados (Albuja 1983). COLOMBIA: Valle del Cauca, Cali (AMNH 14464); Antioquia, Turbo (FMNH 69900). VENEZUELA: Zulia, Lagunillas (FMNH 21980); Carabobo, El Central (Handley 1976); Miranda, 1 km S of Río Chico (Handley 1976). TRINIDAD AND TOBAGO: Trinidad, Sangre Grande (Goodwin and Greenhall 1961). GUYANA: Upper Demerara-Berbice, Berbice (Husson 1962). SURINAM:

Map 269 Marginal localities for *Lasiurus ega* ●

Sipaliwini, 1 km S and 3.5 km E of Sipaliwini Airstrip (S. L. Williams and Genoways 1980a). BRAZIL: Pará, Baião (Handley 1960); Ceará, Aeroporto de Crato (Mares et al. 1981); Pernambuco, Pernambuco (type locality of *Lasiurus caudatus* Tomes); Espírito Santo, Conceição da Barra (Ruschi 1954a); Minas Gerais, Viçosa (Mumford and Knudson 1978); Rio de Janeiro, Rio de Janeiro (C. O. C. Vieira 1942); São Paulo, Ypiranga (J. L. Lima 1926); Paraná, Londrina (N. R. Reis et al. 1998). ARGENTINA: Misiones, Los Helechos (Massoia, Chebez, and Heinonen-Fortabat 1989b). URUGUAY: Salto, Salto (Handley 1960); Montevideo, Pocitos (Acosta y Lara 1950). ARGENTINA: Buenos Aires, Wilde (Vaccaro and Varela 2001); Buenos Aires, Energía (Crespo 1974); La Pampa, Santa Rosa (Montalvo, Justo, and de Santis 1988); Córdoba, Río Cuarto (Tiranti and Torres 1988); Catamarca, Near Balneario El Caolín, 7 km N of Chumbicha (Barquez and Lougheed 1990); Jujuy, Arroyo La Urbana (Villa-R. and Villa-Cornejo 1969). BOLIVIA (S. Anderson 1997, except as noted): Tarija, Caiza (O. Thomas 1898b); Santa Cruz, Río Saguayo; La Paz, Sararia. PERU: Madre de Dios, Reserva Cuzco Amazónico (Woodman et al. 1991); Ucayali, Balta (Gardner 1976); Loreto, Sarayacu (AMNH 76257); Loreto, Puerto Indiana (Pirlot 1968). COLOMBIA: Meta, Caño Lozada (Barriga-Bonilla 1965); Cundinamarca, Pacho (Handley 1966).

SUBSPECIES: We recognize four subspecies of *L. ega* in South America.

L. e. argentinus (O. Thomas, 1901)

SYNONYMS:

Dasypterus ega argentinus O. Thomas, 1901i:247; type locality "Goya, Corrientes, Argentina."

[*Lasiurus (Dasypterus)*] *ega argentinus*: Trouessart, 1904:86; name combination.

Dasypterus intermedius: J. L. Lima, 1926:114; not *Lasiurus intermedius* H. Allen, 1862.

Dasypterus intermedius: C. O. C. Vieira, 1942:421; not *Lasiurus intermedius* H. Allen.

Lasiurus ega argentinus: Handley, 1960:473; first use of current name combination.

This subspecies occurs from southern Bolivia in the west, and the state of Bahía, Brazil in the east, southward through Paraguay, and western and southern Uruguay, to near 40°S in Argentina. The southernmost record is of a male caught on a ship off the coast of Argentina at 41°04.5'S, 56°21.5'W (Van Deusen 1961).

L. e. ega (P. Gervais, 1856)

SYNONYMS:

Vespertilio bursa Lund, 1842a:135; *nomen nudum*; part (see Winge 1892).

Nycticejus Ega P. Gervais, 1856a:73; type locality "Ega, ville du Brasil," Amazonas, Brazil.

Lasiurus caudatus Tomes, 1857a:42; type locality "Pernambuco," Brazil.

Lasiurus Aga Tomes, 1857a:43; incorrect subsequent spelling of *L. ega* P. Gervais.

Atalapha (Dasypterus) ega: W. Peters, 1870:914; name combination.

Atalapha (Dasypterus) caudata: W. Peters, 1870:914; name combination.

[*Lasiurus (Dasypterus)*] *ega*: Trouessart, 1904:86; name combination.

Lasiurus ega ega: Handley, 1960:474; name combination.

The nominate subspecies has been recorded from the island of Trinidad, southern Venezuela (Amazonas and Bolívar), Guyana, Surinam, eastern Brazil, and east of the Andes in Peru, and Bolivia.

L. e. fuscatus (O. Thomas, 1901)

SYNONYMS:

Dasypterus ega fuscatus O. Thomas, 1901i:246; type locality "Rio Cauquete, Cauca River, Colombia. Altitude 1000 m."

[*Lasiurus (Dasypterus)*] *ega fuscatus*: Trouessart, 1904:86; name combination.

Dasypterus ega punensis J. A. Allen, 1914:382; type locality "Puna Island," Guayas, Ecuador.

Dasypterus ega panamensis: Cabrera, 1958:115; name combination; part; not *Dasypterus ega panamensis* O. Thomas.

Lasiurus ega fuscatus Handley, 1960:474; first use of current name combination.

This subspecies occurs in western Colombia and western Ecuador.

L. e. panamensis (O. Thomas, 1901)

SYNONYMS:

Dasypterus ega panamensis O. Thomas, 1901i:246; type locality "Bogava, Chiriqui, Panama. Altitude 250 m."

[*Lasiurus (Dasypterus)*] *ega panamensis*: Trouessart, 1904:86; name combination.

Lasiurus ega panamensis: Handley, 1960:474; first use of current name combination.

This subspecies occurs in northern Colombia and northwestern Venezuela. Elsewhere, *L. e. panamensis* is known from Panama north through Central America, and southern and eastern Mexico, to the vicinity of Corpus Christi, Nueces Co., Texas, USA (Spencer, Choucair, and Chapman 1988).

NATURAL HISTORY: *Lasiurus ega* has a large geographic range encompassing a variety of habitats, from dry temperate and caatinga savannahs to tropical humid forests. The species shows a strong preference for palm trees (and secondarily for palm-frond-thatched roofs) as day roosts and palms may be a critical habitat element (see Kurta and Lehr 1995, and references therein).

Myers (1977) established that reproductive activity in eastern Paraguay begins in May, with ovulation approximately 3 months later (August). In the Paraguayan Chaco, early pregnancies were first found in September. Gestation ranged from 3 to 3.5 months; numbers of young ranged from two to four (average 2.9, $n = 17$). Males contained spermatozoa in the epididymides from August to at least October. Peracchi and Albuquerque (1986) reported a female with four young in November from the state of Rio de Janeiro, Brazil. Barquez and Lougheed (1990) reported three pregnant females taken in the province of Catamarca, Argentina, during mid-November; each carried two fetuses.

Vaucher (1981, 1985) reported on nematodes and cestodes found in Paraguayan *L. ega*. Although rabies has been reported in *L. xanthinus* (see Constantine, Humphrey, and Herbenick 1979), rabies in *L. ega* in South America remains unconfirmed. The report of rabies by Morales-Alarcón et al. (1968) is inconclusive because the positive sample comprised a mixed pool of 84 bats that included *Myotis nigricans* and *L. ega* taken from the thatched roofs of several houses.

The karyotype is $2n = 28$, FN $= 48$; the X chromosome is medium-sized acrocentric, the Y a minute acro-

centric (R. J. Baker and Patton 1967; R. J. Baker, Mollhagen, and Lopez 1971; Bickham 1987). Wainberg (1966) described the diploid number as 24 and the X chromosome as a large metacentric, based on specimens from Argentina. Quaternary fossil material from Minas Gerais, Brazil, was first reported by Lund (1842a) under the name *Vespertilio bursa* and later as *Atalapha ega* by Winge (1892; also see Czaplewski and Cartelle 1998).

REMARKS: We recognize four subspecies of *L. ega* (see subspecies synonymies above for content) and consider *L. xanthinus* (O. Thomas, 1897b) to be a separate species (R. J. Baker et al. 1988; Morales and Bickham 1995). Koopman (1993) and Kurta and Lehr (1995) treated *L. xanthinus* as a synonym of *L. ega*.

One of the two specimens Tomes (1857a:43) had at hand when he described *L. caudatus* was "a specimen in a bad state in spirit, from Chili." We have not included Chile in the distribution of *L. ega* because there have been no additional records confirming this species in the Chilean fauna (Osgood 1943; Cabrera 1958; Mann 1978; Tamayo and Frassinetti 1980). Van Deusen (1961) reported on a *L. ega* that flew aboard a ship that was about 335 km off the coast of Argentina. D. C. Carter and Dolan (1978) provided information on and measurements for the holotypes of *Dasypterus argentinus* O. Thomas, *Dasypterus ega fuscatus* O. Thomas, *Dasypterus ega panamensis* O. Thomas, and *Lasiurus caudatus* Tomes.

Lasiurus egregius (W. Peters, 1870)
Giant Red Bat

SYNONYMS:

Atalapha [*(Dasypterus)*] *egregia* W. Peters, 1870:912; type locality "Sta. Catharina," Brazil.

[*Lasiurus (Dasypterus)*] *egregius*: Trouessart, 1904:86; name combination.

Dasypterus egregius: Miller, 1907b:223; name combination.

Lasiurus egregius Handley, 1960:472; first use of current name combination.

DISTRIBUTION: South American records of *L. egregius* are from Brazil and French Guiana; elsewhere, the species is known from Panama (Handley 1966c).

MARGINAL LOCALITIES (Map 270): FRENCH GUIANA: 9.5 km S and 11.5 km W of Sinnamary (S. L. Williams, Phillips, and Pumo 1990); Pernambuco, Serra Negra de Floresta (Souza, Langguth, and Gimenez 2004); Santa Catarina (W. Peters 1870, specific locality unknown); Rio Grande do Sul, Município Livramento (Ruschi and Bauer 1957); Minas Gerais, Uberlândia (Stutz et al. 2004).

SUBSPECIES: We consider *L. egregius* to be monotypic.

NATURAL HISTORY: The species is known from single specimens from each of five localities as follows: one

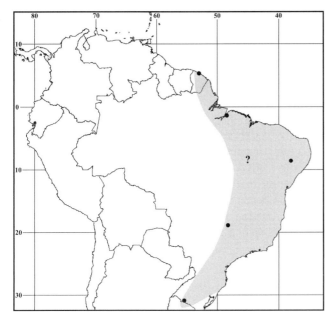

Map 270 Marginal localities for *Lasiurus egregius* ●

locality is in eastern Panama, one in French Guiana, one in northeastern Brazil, and two in southern Brazil. The specimen from French Guiana was netted over a stream through grassy flats backed by "mature secondary vegetation mixed with tropical forest" (S. L. Williams, Phillips, and Pumo 1990:206). Other species collected nearby include *Micronycteris megalotis*, *Phyllostomus elongatus*, *P. hastatus*, and *Myotis nigricans*. The bat from Belém, Brazil, a subadult female, was caught at 1950 hr. on 27 March 1968 in a mistnet stretched about 3 m above a narrow, but deep stream in tall, old, evergreen freshwater-swamp forest. The Panamanian specimen, an adult female, pregnant with two 5-mm-CR embryos, was mistnetted early in the evening of 23 February 1963, about 2 m above a shallow, 10-m-wide stream in tall, old, evergreen upland forest. The female from Belém, Brazil, was a subadult when caught on 27 March 1968. We lack ecological information for the holotype from Santa Catarina and the specimen from Rio Grande do Sul.

Streblid bat flies collected from the Panamanian *L. egregius*, represented *Strebla carolliae*, which commonly infests short-tailed fruit bats of the genus *Carollia*, and *S. vespertilionis*, which is commonly parasitic on the common vampire, *Desmodus rotundus* (see Wenzel, Tipton, and Kiewlicz 1966). Prior to these collections, Streblidae were not known from species of *Lasiurus*; it is possible that the flies were contaminants from other bats previously held in the collecting bag used to hold the *Lasiurus*. The karyotype is unknown. Fossils are not known.

REMARKS: Until 1960, authors followed W. Peters (1870) by including the Giant Red Bat with *Dasypterus*,

because it lacked the minute P1, even though in all other respects it resembled species of *Lasiurus*, as then defined. Handley (1960:473) showed that the presence or absence of P1 is variable in *Lasiurus* and synonymized *Dasypterus* in *Lasiurus*, declaring that he "did not believe that *Dasypterus* is useful even as a subgenus." D. C. Carter and Dolan (1978) provided information and measurements for the holotype of *Atalapha egregia* W. Peters.

Lasiurus salinae O. Thomas, 1902
Brown Hairy-tailed Bat

SYNONYMS:

Lasiurus borealis salinae O. Thomas, 1902c:238; type locality "Cruz del Eje, Central Cordova," Argentina.

[*Lasiurus (Lasiurus) borealis*] *salinae*: Trouessart, 1904:87; name combination.

Lasiurus enslenii J. L. Lima, 1926:113; type locality "Estado do rio Grande do Sul, São Lourenço," Brazil.

Lasiurus ensleveni G. M. Allen, 1939:155; incorrect subsequent spelling of *Lasiurus enslenii* J. L. Lima.

Lasiurus salinae: Mares, Barquez, and Braun 1995:229; first modern usage of current name combination.

DISTRIBUTION: *Lasiurus salinae* is in southeastern Brazil, Uruguay, Argentina, and probably Paraguay.

MARGINAL LOCALITIES (Map 268): ARGENTINA: Salta, Salta (TTU 32517); Tucumán, Los Basques (BM 3.6.6.3); Córdoba, Cruz del Eje (type locality of *Lasiurus borealis salinae* O. Thomas). URUGUAY: Artigas, 6 km NNW of Belén [Salto] (AMNH 205650). BRAZIL: Rio Grande do Sul, São Lourenço (type locality of *Lasiurus enslenii* J. L. Lima). ARGENTINA: La Pampa, N border of Salinas Grandes de Hidalgo (Tiranti and Torres 1998); San Juan, Pedernal (Mares, Barquez, and Braun 1995); La Rioja, Quebrada de Ascha (Yepes 1942).

SUBSPECIES: We treat *L. salinae* as monotypic.

NATURAL HISTORY: The type specimen is a female that was captured together with her with four young. Tiranti and Torres (1998) reported finding a specimen in a small hollow in an earthen bank bordering the salt flats of Salinas Grandes, La Pampa, Argentina. The karyotype is unknown. Fossils are not known.

REMARKS: When O. Thomas (1902c) described *L. borealis salinae*, he characterized the bat as dark, lacking the bright rufous coloration of the southern red bat, and having a larger foot. Mares, Barquez, and Braun (1995) tentatively recognized this species as distinct from "*borealis*" on the basis of darker and browner color. Subsequently, Morales and Bickham (1995) were unable to distinguish between the samples of red bats Mares, Barquez, and Braun (1995) had identified as *salinae* and those they also had identified as *borealis* (= *blossevillii*; also see Barquez, Mares, and Braun 1999). Nevertheless, we continue to recognize *L. salinae* on the basis of the widespread, apparently sympatric occurrence of dark individuals that are readily distinguishable from the brighter, more rufous *L. blossevillii*.

Lasiurus varius Poeppig, 1835
Chilean Red Bat

SYNONYMS:

Nycticeius, Raff. *Species prima* Poeppig [spelled Pöppig], 1830:column 217.

N[*ysticeius*]. *varius* Poeppig, 1835:451, footnote; type locality "Antuco," Biobío, Chile.

Nycticeus Pœpingii Lesson, 1836:324; type locality "Chili"; based on "*Nycticejus prima species*," Poeppig, 1830.

Nycticejus Pœpingii Lesson, 1842:22; name combination.

A[*talapha*]. *varia*: W. Peters, 1861:153; name combination.

Lasiurus varius: Fitzinger, 1870b:411; first use of current name combination.

[*Atalapha noveboracensis*] Var. ((*Atalapha varia*): Dobson, 1878; name combination.

[*Atalapha (Atalapha) borealis*] *varia*: Trouessart, 1897: 122; name combination.

[*Lasiurus (Lasiurus) borealis*] *varius*: Trouessart, 1904:87; name combination.

DISTRIBUTION: *Lasiurus varius* is known from west-central Argentina to Tierra del Fuego (Dabbene 1902), and Central and southern Chile.

MARGINAL LOCALITIES (Map 268): *Northern distribution.* CHILE: Coquimbo, Illapel (Cabrera 1903). ARGENTINA (Barquez, Mares, and Braun 1999): Neuquén, Chos Malal; Neuquén, Neuquén; Río Negro, El Bolsón. CHILE: Los Lagos, Puerto Montt (Mann 1978); Biobío, Concepción (Osgood 1943). *Southern distribution.* CHILE: Magallanes, Lago Paine, Parque Nacional Torres de Paine (Rau and Yáñez 1979); Magallanes, San Gregorio (Tamayo and Pérez 1979). ARGENTINA: Tierra del Fuego, Ushuaia (Dabbene 1902).

SUBSPECIES: We consider *L. varius* to be monotypic.

NATURAL HISTORY: Mann (1978) wrote that *L. varius* has been found in day roosts in trees and, less commonly, on rocks along the coast of Chile. He also found males with descended and enlarged testes in central Chile during August and September (austral spring). Although not citing duration or time of year when gestation is completed, Mann (1978) stated that normal births consisted of twins, which he claimed parallels numbers born to Chilean *L. cinereus*. Tamayo and Pérez (1979) claimed that at least the southern population is migratory.

REMARKS: *Lasiurus varius* is one of the southernmost occurring bats. Chilean and southwestern Argentine records mapped by Redford and Eisenberg (1992:104) for *L. borealis* actually represent *L. varius*.

Genus *Myotis* Kaup, 1829
Don E. Wilson

The genus *Myotis* is the most widely distributed genus of mammal other than man and his commensals, and contains about 100 species worldwide. Twelve species are found in South America of which seven are endemic. *Myotis* is considered a relatively primitive genus and can be distinguished from other members of the family by the retention of three premolars in each toothrow, by the great disparity in size between the first two and the third of these premolars, and by the tendency for the upper middle premolar to be displaced inward so that the first and third come together (Tate 1941). The dental formula is 2/3, 1/1, 3/3, 3/3 × 2 = 38.

SYNONYMS:

Myotis Kaup, 1829:106; type species *Vespertilio murinus* Schreber, 1774a, by monotypy (= *Vespertilio myotis* Borkhausen, 1797:80); see Miller 1912b; also see Remarks.

Nystactes Kaup, 1829:108; type species *Vespertilio bechsteinii* Kaup, 1829 (= *Vespertilio bechsteinii* Kuhl, 1818:30); preoccupied by *Nystactes* Gloger, 1827 (Aves).

Leuconöe Boie, 1830:256; type species *Vespertilio daubentonii* Kuhl, 1818, by subsequent designation (O. Thomas 1904d:382).

Vespertilio: Keyserling and Blasius, 1839:306; not *Vespertilio* Linnaeus, 1758.

Selysius Bonaparte, 1841a:3; type species *Vespertilio mystacinus* Kuhl, 1819, by monotypy.

Capaccinius Bonaparte, 1841b; type species *Vespertilio capaccinii* Bonaparte, 1837a, by tautonomy.

Trilatitus Gray, 1842:258; type species *Vespertilio hasseltii* Temminck, 1840, *fide* Kretzoi and Kretzoi (2000:415).

Tralatitus P. Gervais, 1849:213; incorrect subsequent spelling of *Trilatitus* Gray.

Brachyotus Kolenati, 1856a:131; type species *Vespertilio mystacinus* Kuhl, 1819, by subsequent designation (Ellerman and Morrison-Scott 1951:137); preoccupied by *Brachyotus* Gould, 1837 (Aves).

Isotus Kolenati, 1856a:131; type species *Vespertilio nattereri* Kuhl, 1818, by subsequent designation (Tate 1941:546).

Tralatitus Gray, 1866a:90; incorrect subsequent spelling of *Trilatitus* Gray.

Pternopterus W. Peters, 1867c:706; type species *Vespertilio lobipes* W. Peters, 1867c (= *Vespertilio muricola* Gray, 1846).

Exochurus Fitzinger, 1870a:75; type species *Vespertilio macrodactylus* Temminck, 1840, *fide* Kretzoi and Kretzoi (2000:136); preoccupied by *Exochura* Kolenati, 1858 (Vespertilionidae).

Aeorestes Fitzinger, 1870b:427; part.

Comastes Fitzinger, 1870b:565; type species *Vespertilio capaccinii* Bonaparte, 1837a, *fide* Kretzoi and Kretzoi (2000:85); preoccupied by *Comastes* Jan, 1863 (Reptilia).

Euvespertilio Acloque, 1899:38; contained five species, no type species selected.

Pizonyx Miller, 1906b:85; type species *Myotis vivesi* Menegaux, 1901, by original designation.

Chrysopteron Jentink, 1910a:74; type species *Kerivoula weberi* Jentink, 1890, by original designation.

Rickettia Bianchi, 1917:lxxviii; type species *Vespertilio rickettii* O. Thomas, 1894d, by monotypy.

Dichromyotis Bianchi, 1917:lxxviii; type species *Vespertilio formosus* Hodgson, 1835, by monotypy.

Paramyotis Bianchi, 1917:lxxix; replacement name for *Nystactes* Kaup.

Anamygdon Troughton, 1929:87; type species *Anamygdon solomonis* Troughton, 1929, by original designation.

Hesperomyotis Cabrera, 1958:103; type species *Myotis simus* O. Thomas, 1901f, by original designation.

M,rolis Alberico and Orejuela, 1982:34; incorrect subsequent spelling of *Myotis* Kaup.

Myottis J. C. González and Fabián, 1995:58; incorrect subsequent spelling of *Myotis* Kaup.

REMARKS: Linnaeus (1758) included all seven species of bats known to him in his genus *Vespertilio*. These included only two vespertilionids, the bats now known as *Vespertilio murinus* and *Plecotus auritus*. For most of the last half of the 18th century and the first quarter of the 19th, the name *Vespertilio murinus* Schreber (1774a) was applied to the bat we know today as *Myotis myotis*. By naming *Vespertilio myotis*, Borkhausen (1797) restricted the content of *Vespertilio murinus* Schreber to that one species. In subdividing the genus *Vespertilio* in 1829, Kaup based the genus *Myotis* on a single included species, "*Vespertilio murinus*." Nowhere in his curious little book does he mention the specific name *Myotis* Borkhausen. His intention that the name be applied to the bat called *Vespertilio murinus* by Schreber, and currently known as *Myotis myotis*, is clear from his description. Therefore, the type species of the genus *Myotis*, *Vespertilio murinus* Schreber, 1774a, is an unavailable junior homonym of *Vespertilio murinus* Linnaeus, 1758, and the first correctly applied available name is *Vespertilio myotis* Borkhausen, 1797. *Vespertilio murinus* Schreber, 1774, was placed on the Official Index of Rejected and Invalid Specific Names in Zoology (ICZN 1958a).

Neotropical species of *Myotis* were revised most recently by LaVal (1973a). Accounts are based, at least in part, on that report, except the accounts for *M. nesopolus* Miller (see Genoways and Williams 1979a) and *M. aelleni* Baud, 1979.

Ruschi (1951a) created the *nomen nudum*, *Myotis espiritosantensis*, based on specimens from the state of Espírito Santo, Brazil. Later (1951d) he made the name available, and later still (1970), correctly assigned it to the genus *Natalus*. *Natalus espiritosantensis* is a junior synonym of *Natalus stramineus macrourus* (P. Gervais, 1856a), and unrelated to the genus *Myotis* (see Pine and Ruschi 1976).

Husson (1962:218–21) used the name *Myotis surinamensis* as a replacement name for *Vespertilio ferrugineus* Temminck, 1840:239; type locality "La Guyane hollandaise." The holotype has long, bicolored fur unmatched by that of any known South American *Myotis*. The consensus is that the taxon probably does not represent a South American species (LaVal 1973a; D. C. Carter and Dolan 1978:72–73).

Vespertilio (*Leuconoe*) *pilosus* W. Peters, 1869:403, is another *Myotis* originally described as from Montevideo, Uruguay, that presently is not considered a member of the South American bat fauna (Acosta y Lara 1950; Cabrera 1958; D. C. Carter and Dolan 1978). Koopman (1993) and Simmons (2005) treated the name as a synonym of *Myotis rickettii* O. Thomas, 1894d, which occurs in China.

Woodman (1993), while commenting on the correct terminations required of adjectival forms of trivial names used in name combinations with genus-group names ending in *-otis*, stated that the gender of the generic name *Myotis* was feminine. Although his interpretation of gender is in doubt (see Pritchard 1994; ICZN 1999:Art. 30.1.4.2; David and Gosselin 2002:267; Gardner 2005), the gender of *Myotis* has been established as masculine (see ICZN 1958a, 1958b, 1987a).

Hoofer and Van Den Bussche (2003) considered *Myotis* to be the sole representative of the vespertilionid subfamily Myotinae Tate, 1943. They also suggested placing the New World species they sampled (*M. albescens*, *M. keaysi*, *M. nigricans*, *M. riparius*, and *M. ruber*) in the subgenus *Aeorestes* Fitzinger, 1870b, to distinguish them from Old World taxa. However, *Aeorestes* Fitzinger, with type species *Aeorestes villosissimus* (= *Lasiurus cinereus villosissimus*) by original designation, is not available for New World *Myotis*.

KEY TO SOUTH AMERICAN SPECIES OF *MYOTIS* (ADAPTED FROM LAVAL 1973A).

1. Sagittal crest usually present, often well developed; ratios of width across canines:post-orbital constriction usually greater than 1.0. 2

1'. Sagittal crest usually absent; if present, poorly developed; ratio of width across canines:post-orbital constriction usually less than 1.0. 6

2. Ratios of third metacarpal, tibia, and greatest length of skull high relative to forearm (ratios 0.96, 0.48, and 0.43, respectively); forearm 32 mm or less; dorsal fur conspicuously dark basally with paler tips . . . . *Myotis nesopolus*

2'. Ratios of third metacarpal, tibia, and greatest length of skull lower than those of *M. nesopolus* relative to forearm; forearm usually longer than 33 mm; dorsal fur almost unicolored 3

3. Fur on uropatagium extends along leg at least as far as, and usually beyond, knee; P3 may be crowded, but aligned in toothrow 4

3'. Fur does not reach knee; P3 usually crowded to lingual side of toothrow 5

4. Fur on uropatagium extends at least halfway from knee to foot along tibia, usually reaches foot; fur usually weakly bicolored; skull relatively small in all dimensions; greatest length of skull less than 15 mm, usually less than 14 mm . . . . *Myotis keaysi*

4'. Fur on uropatagium extends to or past knees, but no farther than halfway from knee to foot; fur usually unicolored; skull large in all dimensions; greatest length of skull 15.3 mm or more . . . . *Myotis ruber*

5. Fur extremely short, approximately 2 mm; post orbital constriction 3.8 mm or more . . . . *Myotis simus*

5'. Fur longer approximately 3 mm; post orbital constriction less than 3.8 mm . . . . *Myotis riparius*

6. Terminal half of dorsal hairs blond, with dark brown or black bases; forearm and greatest length of skull short (31.3 mm and 12.9 mm or less, respectively); rostrum short (maxillary toothrow 4.5 mm or less) . . . . *Myotis atacamensis*

6'. Tips of dorsal fur not as above; forearm and greatest length of skull longer (usually more than 33.0 mm and 13.1 mm, respectively); rostrum longer (maxillary toothrow 4.7 mm or more) 7

7. Uropatagium with a fringe, may not be easily visible without magnification; uropatagium also with pale border in many specimens; forearm usually 37–40 mm; greatest length of skull usually 14.5–16.0 mm . . . . *Myotis levis*

7'. Uropatagial fringe rarely present; no pale border on uropatagium 8

8. May have slight fringe on uropatagium, visible with magnification; dorsal fur dark (often blackish), many hairs with conspicuous white or yellowish tips, imparting a frosted appearance . . . . *Myotis albescens*

8'. No fringe on uropatagium; little or no frosting on dorsal hairs 9

9. Fur long (5–6 mm), bicolored; frontals comparatively steeply sloping; size large; forearm 40 mm or more;

greatest length of skull 14.5 mm or more . . . . *Myotis oxyotus*

9'. Fur variable in length; angle of slope of frontals variable; size moderate to small; forearm usually less than 40 mm; greatest length of skull usually less than 14.5 mm 10

10. Fur usually longer than 4 mm, often strongly bi- or tri-colored; color blond to dark brown, never blackish 11

10'. Fur rarely longer than 4 mm; fur weakly to moderately bicolored, often blackish . . . . *Myotis nigricans*

11. Fur bicolored; size smaller; forearm averaging 36 mm; greatest length of skull averaging 14.1 mm . . . . *Myotis chiloensis*

11'. Fur tricolored; size larger; forearm averaging 39 mm; greatest length of skull averaging 14.8 mm . . . . *Myotis aelleni*

Myotis aelleni Baud, 1979
Southern Myotis

SYNONYM:

Myotis aelleni Baud, 1979:268; type locality "El Hoyo de Epuyen, 230 m d'altitude, environ 42°10′S, 71°21′W, Province de Chubut, Argentine."

DISTRIBUTION: *Myotis aelleni* is known from the type locality in Chubut and in nearby Río Negro, Argentina.

MARGINAL LOCALITIES (Map 271): ARGENTINA: Río Negro, El Bolsón (Baud 1979); Chubut, El Hoyo de Epuyén (type locality of *Myotis aelleni* Baud).

Map 271 Marginal localities for *Myotis aelleni* ● and *Myotis atacamensis* ▲

SUBSPECIES: *Myotis aelleni* is monotypic.

NATURAL HISTORY: The type locality is in an area containing small, shallow lakes bordering the Río Epuyen. According to Topal (1963), the vegetation in the area is dominated by *Berberis buxifolia*. Much of the surrounding habitat has been converted to pastures and cultivated fields. The two known localities are 15 km apart and at similar elevations (230–350 m). The type series, collected 19 December 1975, included 107 females, 60% of which were lactating and the remainder was carrying mid- to late-term fetuses.

REMARKS: *Myotis aelleni* is the most recently described New World *Myotis*. It appears to be intermediate between *M. chiloensis* and *M. levis*. Pearson and Pearson (1990) examined a series of specimens taken from the same barn as the holotype. They were unable to separate *M. aelleni* from *M. chiloensis* with confidence, because the diagnostic characters did not appear to be consistent. Barquez, Mares, and Braun (1999:94) questioned the validity of this species, suggesting that it may prove to be a synonym of *M. chiloensis*.

Myotis albescens (É. Geoffroy St.-Hilaire, 1806)
Silver-tipped Myotis

SYNONYMS:

Vespertilio albescens É. Geoffroy St.-Hilaire, 1806:204–205; type locality "Yaguaron, Paraguari, Paraguay," by neotype designation (LaVal 1973a:26); original type locality "Paraguay," (É. Geoffroy St.-Hilaire, 1806:203, footnote) and based on Azara's (1801:294–295) *chauvesouris douzième*.

Vesp[*ertilio*]. *leucogaster* Schinz, 1821:180; type locality "Östküste von Brasilien," identified as "Flusse Mucuri," Bahia, Brazil by Wied-Neuwied (1826:279).

[*Vespertilio (Vespertilio)*] *albescens*: Trouessart, 1897:132; name combination.

DISTRIBUTION: *Myotis albescens* has been found in every South American country except Chile, and French Guiana. It also is unreported from the Netherlands Antilles and from Trinidad and Tobago. Elsewhere, *M. albescens* is found northward through Central America into Veracruz, Mexico.

MARGINAL LOCALITIES (Map 272; from LaVal 1973a, except as noted): VENEZUELA: Zulia, Río Aurare, 12 miles SE of Altagracia; Miranda, 5 km E of Río Chico; Monagas, 1 km N and 3 km W of Caripe. GUYANA: Demerara-Mahaica, Georgetown. SURINAM: Paramaribo, Paramaribo (Husson 1978). BRAZIL: Pará, Utinga; Bahia, Vila Nova (C. O. C. Vieira 1955); Bahia, Mucuri (type locality of *Vespertilio. leucogaster* Schinz). URUGUAY: San José, Rincón de Cufré. ARGENTINA (Barquez, Mares, and Braun 1999; except as noted): Buenos Aires, Bosch; Buenos Aires, Lobos; Entre Rios, Estación Paranacito; Corrientes,

Map 272 Marginal localities for *Myotis albescens* ●

Goya; Chaco, General Vedia; Chaco, Pozo del Gato; Santiago del Estero, La Banda; Tucumán, Dique San Ignacio; Salta, La Merced (LaVal 1973a); Salta, 27 km W of Aguas Blancas. BOLIVIA (S. Anderson 1997): Chuquisaca, Finca San Antonio; Santa Cruz, 1 km NE of Estancia Cuevas; La Paz, Cota Cota. PERU: Madre de Dios, Hacienda Erika (Ascorra, Wilson, and Romo 1991); Cusco, Segakiato (S. Solari et al. 1998); Pasco, San Juan; Huánuco, 19 miles S of Tingo María; Amazonas, La Poza (Patton, Berlin, and Berlin 1982); Piura, 4 miles W of Suyo. ECUADOR: Guayas, Isla Silva (Albuja 1983); Esmeraldas, San Miguel (Albuja 1999). COLOMBIA: Valle del Cauca, Río Zabaletas (M. Thomas 1972); Chocó, Condoto (Miller and Allen 1928).

SUBSPECIES: *Myotis albescens* is currently treated as monotypic.

NATURAL HISTORY: *Myotis albescens* is one of the more widely distributed species in the genus, and most specimens have been captured in lowland areas below 500 m elevation. It forages in forested and open grassland habitats, and is commonly netted along small streams. Roosts have been found in buildings, hollow trees, and rock crevices. Myers (1977) provided details of the reproductive cycle. Autino, Claps, and González (2004) reported a flea and a nycteribiid batfly as ectoparasites of Uruguayan *M. albescens*. Additional ecological and life history information can be found in Barlow (1965), Redford and Eisenberg (1992), S. Anderson (1997), and Linares (1998).

REMARKS: When É. Geoffroy St.-Hilaire (1806) described *Vespertilio albescens*, he did so without access to a specimen, basing his description on Azara's (1801) *chauve-souris douzième*, and mentioning only Paraguay as the source country. By 1900, when O. Thomas included the species in the genus *Myotis*, it was known from Costa Rica to Argentina, but lacked a precise type locality. Miller and Allen (1928:201) listed the type locality as "Paraguay," but in discussing the type specimen, stated, "None specified. Geoffroy's name is based on the '*chauve-souris douzième*' of Azara, hence the type locality is considered to be Paraguay, perhaps near Asunción where Azara resided." On 14 March 1790, the Real Gabinete de Historia Natural de Madrid received Azara's manuscript in which he described specimens that he later deposited there, as well as other mammals that had not been preserved (Morales-Agacino 1938). The specimens Azara had preserved in alcohol arrived at the Museum in Madrid on 8 February 1791, but only a few were accessioned because most were in poor condition. Azara described the species as "Morciélago 4" in his manuscript, but the Spanish publication of his *Quadrúpedos de Paraguay* (1802) described the species as *El Pardo Obscuro*, and the French edition (1801) as *chauve-souris douzième*. The manuscript was the guide to, and contained the descriptions of, the specimens sent by Azara, and in it he indicated that "Morciélago 4" was not represented by that number in the shipment, because after describing the bat, he had not saved it. Therefore, the only individual that could be the type was never preserved, and É. Geoffroy St.-Hilaire based his description on an incomplete French translation of Azara's description. Azara's original manuscript provides a clear type locality as part of the description: "Entre Una multitud de Morciélagos que Revoleteaban por me Quarto en la Estancia de Sn. Solano junto al Estero Yberá mate Dos identicos...," thus eliminating Asunción as the type locality (Morales-Agacino 1938). The Estero Yberá is a large marshy area in Corrientes Province, Argentina, separated from Paraguay by the Río Alto Paraná. The Estancia San Solano is located 18 km SW of the town of Yegros, about 100 km N of the Río Alto Paraná, but in Azara's time, it may have extended much farther south. LaVal (1973a) was unable to locate Estancia San Solano. Nevertheless, he fixed the type locality as Yaguaron, Paraguarí, Paraguay by designating a neotype from that locality.

Of the names listed by Cabrera (1958) as synonyms of *M. albescens*, *Vespertilio aenobarbus* has been identified as a *Nycticeius*; *Vespertilio isidori* identified as a *Pipistrellus* (see D. C. Carter and Dolan 1978); and *Vespertilio arsinoe*, *V. mundus*, and *Myotis punensis* are synonyms of *M. nigricans* (see LaVal 1973a). D. C. Carter and Dolan (l978)

provided measurements of and information on a paralectotype of *Vespertilio leucogaster*.

Myotis atacamensis (Lataste, 1892)
Atacaman Myotis
SYNONYMS:

Vespertilio atacamensis Lataste, 1892:79; type locality San Pedro de "Atacama," Antofagasta, Chile.

Myotis nigricans nicholsoni Sanborn, 1941:382; type locality "Hacienda Chucarapi, Tambo Valley, Department of Arequipa, Peru."

DISTRIBUTION: *Myotis atacamensis* is endemic to South America where it is known from western Peru and northern Chile.

MARGINAL LOCALITIES (Map 271; from LaVal 1973a, except as noted; localities listed from north to south): PERU: Lima, Bujama Baja; Arequipa, Chucarapi. CHILE: Tarapacá, Miñimiñi (Pine, Miller, and Schamberger 1979); Tarapacá, Canchones; Antofagasta, San Pedro de Atacama (type locality of *Vespertilio atacamensis* Lataste).

SUBSPECIES: *Myotis atacamensis* is currently treated as monotypic.

NATURAL HISTORY: Because the species is poorly understood systematically, the amount of natural history information clearly attributable to *Myotis atacamensis* is minuscule. Apparently, all of the specimens currently allocated to this species came from the coastal deserts of southern Peru and northern Chile at elevations from sea level to 2,436 m. The sites have forest vegetation at lower elevations and grade into scrub forest at higher elevations.

REMARKS: This poorly diagnosed species was based on three syntypes, which may no longer exist. Miller and Allen (1928), Osgood (1943), and LaVal (1973a) mentioned difficulty in distinguishing specimens of this taxon from *Myotis chiloensis*.

Myotis chiloensis (Waterhouse, 1840)
Chilean Myotis
SYNONYMS:

Vespertilio Chiloensis Waterhouse, 1838:5; type locality "Eastern side of Chiloe," Los Lagos, Chile.

Vespertilio gayi Lataste, 1892:79; type locality "Valdivia," Los Lagos, Chile.

[*Vespertilio (Vespertilio)*] *Gayi*: Trouessart, 1897:131; name combination.

[*Vespertilio (Vespertilio)*] *chiloensis*: Trouessart, 1897:131; name combination.

Myotis chiloensis arescens Osgood, 1943:55; type locality "Hacienda Limache, Province of Valparaiso, Chile."

DISTRIBUTION: *Myotis chiloensis* is known from Chile and western Argentina.

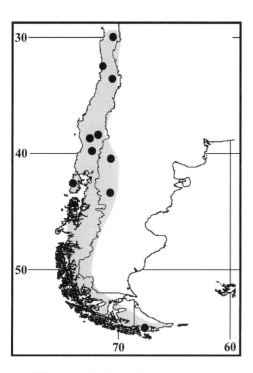

Map 273 Marginal localities for *Myotis chiloensis* ●

the former as a subspecies of *M. levis*, and the latter as a synonym of *M. oxyotus*.

Myotis keaysi J. A. Allen, 1914
Hairy-legged Myotis

SYNONYMS: See under Subspecies.

DISTRIBUTION: *Myotis keaysi* has been recorded from Colombia, Venezuela, Ecuador, Bolivia, Argentina, Peru, and the island of Trinidad. Elsewhere, it is known from southern Mexico and Central America.

MARGINAL LOCALITIES (Map 274; from LaVal 1973a except as noted; localities listed from north to south): TRINIDAD AND TOBAGO: Trinidad, Blanchisseuse. VENEZUELA: Distrito Federal, 9.4 km N of Caracas; Carabobo, La Copa; Barinas, Reserva Forestal de Ticoporo (Ochoa et al. 1988). COLOMBIA: Santander, Lebrija; Antioquia, El Porvenir (Muñoz 1990); Caldas, Vereda Quebrada Negra (Castaño et al. 2003); Cundinamarca, Hatogrande; Nariño, Reserva Natural La Planada (Ospina-Ante and Gómez 2000). ECUADOR: Tungurahua, Baños; Loja, Cajanuma. PERU: Piura, 25 km E de Olmos; San Martín, Parque Nacional Abiseo (S. Solari et al. 2001b));

MARGINAL LOCALITIES (Map 273; from LaVal 1973a, except as noted): CHILE: Coquimbo, Paihuano; Metropolitana de Santiago, Puente Alto; La Araucanía, Curacautín. ARGENTINA (Barquez, Mares, and Braun 1999): Neuquén, Estancia Alicura; Chubut, 3 km N of Tecka, along Highway 40. CHILE: Magallanes, Puerto Pescado; Los Lagos, Cucao; Los Lagos, Riñihue; La Araucanía, Temuco; Valparaiso, Zapallar.

SUBSPECIES: I am treating *M. chiloensis* as monotypic.

NATURAL HISTORY: *Myotis chiloensis* is one of the southernmost occurring species of bat (Koopman 1967), exceeded only by *Histiotus montanus magellanicus*, and is not known above 1,500 m elevation (LaVal 1973a; Pine, Miller, and Schamberger 1979).

REMARKS: The holotype was collected in January 1836 by Lieutenant Sullivan and given to Darwin during the voyage of the Beagle (Darwin in Waterhouse 1838:6). LaVal (1973a) concluded that the type was lost and designated a neotype, also from Chiloé Island. Cabrera (1958) recognized five subspecies: *M. c. alter* Miller and Allen, 1928; *M. c. arescens* Osgood, 1943; *M. c. atacamensis* (Lataste, 1892); *M. c. oxyotus* (W. Peters, 1866); and the nominate form. When LaVal (1973a) treated the species as monotypic, he synonymized *M. chiloensis alter* under *M. levis*, did not recognize *M. chiloensis arescens* as valid, and elevated *atacamensis* and *oxyotus* to specific status. Cabrera (1958) included *Myotis dinellii* and *Myotis thomasi* in the synonymy of *M. chiloensis*; however, LaVal (1973a) treated

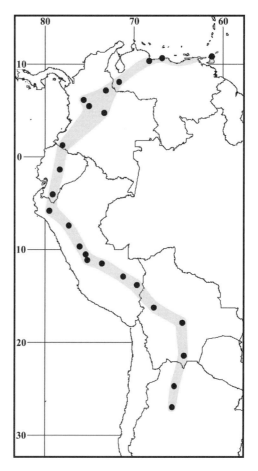

Map 274 Marginal localities for *Myotis keaysi* ●

Huánuco, E Slope Cordillera Carpish, Carretera Central; Pasco, Yanachaga-Chemillen (S. Solari et al. 2001b); Junín, Chanchamayo; Cusco, RAP Camp Two (Emmons, Luna, and Romo 2001; Madre de Dios, Hacienda Amazonía (V. Pacheco et al. 1993); Puno, Inca Mines. BOLIVIA (S. Anderson 1997): La Paz, Sacramento Alto; Santa Cruz, 5 km SW of Comarapa; Tarija, Rancho Tambo. ARGENTINA: Salta, Salta (Barquez and Ojeda 1992); Tucumán, El Nogalar, Ruta 307 (Barquez, Mares, and Braun 1999).

SUBSPECIES: I recognize two subspecies of *M. keaysi*.

M. k. keaysi J. A. Allen, 1914

SYNONYMS:

Myotis ruber keaysi J. A. Allen, 1914:383; type locality "Inca Mines (altitude 6000 feet), [Puno,] Peru (lat. 13°30'S., long. 70°W.)."

Myotis keaysi keaysi: LaVal, 1973a:22; first use of current name combination.

M,rolis [*sic*] *kearsi* Alberico and Orejuela, 1982:34; incorrect subsequent spelling of *Myotis keaysi* J. A. Allen.

Myotis keasy Ospina-Ante and Gómez, 2000:664; incorrect subsequent spelling of *Myotis keaysi* J. A. Allen.

This subspecies is known from the Andes of Colombia, Ecuador, Peru, Bolivia, and Argentina, above 1,100 m, with most specimens known from above 2,000 m.

M. k. pilosatibialis LaVal, 1973

SYNONYM:

Myotis keaysi pilosatibialis LaVal, 1973a:24; type locality "1 km W of Talanga, 750 m, Francisco Morazán, Honduras."

This subspecies occurs in northern Venezuela and on the island of Trinidad. Elsewhere, it is known from southern Mexico, southeastward through Central America into northwestern Panama.

NATURAL HISTORY: *Myotis keaysi* has been collected in both open areas and in closed-canopy forest at intermediate elevations, more commonly between 600 and 2,000 m. This species is known to form small colonies in hollow trees. Linares (1998) summarized the limited information available on Venezuelan populations.

REMARKS: Until LaVal's (1973a) revision, most specimens of *M. keaysi* had been identified as *M. nigricans*. Subspecific limits are still unclear; LaVal described four specimens from within the range of *M. k. pilosatibialis* that were indistinguishable from *M. k. keaysi*.

Myotis levis (I. Geoffroy St.-Hilaire, 1824)
Yellowish Myotis

SYNONYMS: See under Subspecies.

DISTRIBUTION: *Myotis levis* has been recorded from Uruguay, southern Brazil, southern Bolivia, Paraguay, and

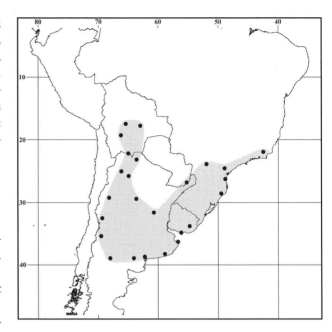

Map 275 Marginal localities for *Myotis levis* ●

northern Argentina. The species is not yet known from Chile.

MARGINAL LOCALITIES (Map 275): BOLIVIA (S. Anderson 1997): Cochabamba, Quebrada Mojon; Santa Cruz, Santa Cruz de la Sierra. ARGENTINA (Barquez, Mares, and Braun 1999): Salta, Hickman; Salta, Horcones; Santiago del Estero, Ojo de Agua; Santa Fe, Santa Fe. PARAGUAY: Itapúa, Río Pirayu-i (Baud and Menu 1993). BRAZIL: Paraná, Fazenda Cagibi (Bianconi, Mikich, and Pedro 2004); Rio de Janeiro (no specific locality; Esbérard and Bergallo 2006); Paraná, Palmeiras (type locality of *Myotis chiloensis alter* Miller and Allen); Santa Catarina, Joinville (Cherem et al. 2005); Santa Catarina, Nova Veneza (Cherem et al. 2005). URUGUAY (LaVal 1973a): Lavalleja, 9 miles S of Pirarajá; Montevideo, Colegio Pio-Colon, Montevideo. ARGENTINA (Barquez, Mares, and Braun 1999, except as noted): Buenos Aires, Punta Rosa; Buenos Aires, 15 km SW of Lobería; Buenos Aires, Bahia Blanca; La Pampa, Caleu Caleu, Pampa Central (LaVal 1973a); Neuquén, Neuquén (LaVal 1973a); Mendoza, Malargüe; Mendoza, Uspallata; La Rioja, Villa Unión; Salta, Cachi; Salta, Santa Victoria Oeste. BOLIVIA: Potosí, Finca Salo (S. Anderson, Koopman, and Creighton 1982, as *M. oxyotus*).

SUBSPECIES: I recognize two subspecies of *M. levis*.

M. l. dinellii O. Thomas, 1902

SYNONYMS:

Myotis Dinellii O. Thomas, 1902h:493; type locality "Tucuman," Tucumán, Argentina.

Myotis levis dinellii: LaVal, 1973a:39; first use of current name combination.

This subspecies is known from Argentina, southern Bolivia, and probably occurs in eastern Chile.

M. l. levis (I. Geoffroy St.-Hilaire, 1824

SYNONYMS:

Vespertilio polythrix I. Geoffroy St.-Hilaire, 1824b:443; type locality "la captainerie de Rio-Grande," Brazil; a *nomen oblitum*.

Vespertilio levis I. Geoffroy St.-Hilaire, 1824b:444; type locality "Brésil."

Scotophilus laevis: Gray, 1838b:498; name combination and incorrect subsequent spelling of *Vespertilio levis* I. Geoffroy St.-Hilaire.

Vespertilio laevis Temminck, 1840:249; incorrect subsequent spelling of *Vespertilio levis* I. Geoffroy St.-Hilaire.

V[espertilio]. laevis Schinz, 1844:187; incorrect subsequent spelling of *Vespertilio levis* I. Geoffroy St.-Hilaire.

Vespertilio nubilus J. A. Wagner, 1855:752; type locality "südlichen Brasilien"; identified as "Registo do Sai," Rio de Janeiro, and "Ypanema," São Paulo, Brazil by Pelzeln (1883:44); there has been no definitive selection of type locality.

Aeorestes levis: Fitzinger, 1870:435; name combination.

[Vespertilio (Vespertilio)] polythrix: Trouessart, 1897:131; name combination.

[Vespertilio (Vespertilio)] levis: Trouessart, 1897:131; name combination.

Myotis levis: J. L. Lima, 1926:103; first use of current name combination.

Myotis chiloensis alter Miller and Allen, 1928:194; type locality "Palmeiras, Paraná, Brazil."

The nominate subspecies occurs in southern Brazil and southeastern Paraguay, Uruguay, and in northeastern Argentina south into Provincia Buenos Aires.

NATURAL HISTORY: Autino, Claps, and González (2004) found a nycteribiid batfly as well as mentioning earlier reports on a flea as ectoparasites of Uruguayan *M. levis*. Tiranti (1996) reported a $2n = 44$, FN $= 50$ karyotype for *M. l. dinellii* from Argentina.

REMARKS: Dobson (1878) assigned a specimen from Uruguay to *M. levis*, but later authors either ignored the name *M. levis* (e.g., Miller and Allen 1928), or placed it in the synonymy of *M. ruber*, as did Cabrera (1958). Miller and Allen (1928) arranged *dinellii* as a subspecies of *M. chiloensis*; but LaVal (1973a), in recognizing *levis*, argued that the affinities of *dinellii* were with *levis*, although he also stated that it might prove to be a distinct species. Barquez, Mares, and Braun (1999) also treated *dinellii* as a subspecies of *M. levis*. D. C. Carter and Dolan (1978) provided information on and measurements of the types of *Vespertilio*

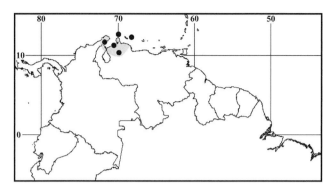

Map 276 Marginal localities for *Myotis nesopolus* ●

levis I. Geoffroy St.-Hilaire, *Vespertilio polythrix* I. Geoffroy St.-Hilaire, *Vespertilio nubilus* Wagner, *Myotis dinellii* O. Thomas, and *Myotis chiloensis alter* Miller and Allen.

Myotis nesopolus Miller, 1900
Curaçao Myotis

SYNONYMS: See under Subspecies.

DISTRIBUTION: *Myotis nesopolus* is found in northern Venezuela and on the Netherlands Antillean Islands of Curaçao and Bonaire.

MARGINAL LOCALITIES (Map 276): NETHERLANDS ANTILLES (Genoways and Williams 1979a): Curaçao, 2.8 km S and 4.5 km E of Westpunt; Bonaire, 8.5 km N and 2 km W of Kralendijk. VENEZUELA (LaVal 1973a): Lara, Río Tocuyo; Falcón, Capatárida; Zulia, 110 km N and 25 km W of Maracaibo.

SUBSPECIES: I recognize two subspecies of *M. nesopolus*.

M. n. nesopolus Miller, 1900

SYNONYMS:

Myotis nesopolus Miller, 1900a:123; type locality "near Willemstad, Curaçao, West Indies."

Myotis nesopolus nesopolus: Genoways and Williams, 1979a:316; first use of current name combination.

This subspecies is restricted to the islands of Curaçao and Bonaire, Netherlands Antilles.

M. n. larensis LaVal, 1973

SYNONYMS:

Myotis larensis LaVal, 1973a:44; type locality "Río Tocuyo, Lara, Venezuela."

Myotis nesopolus larensis: Genoways and Williams, 1979a: 319; first use of current name combination.

This is the mainland subspecies, which is known from only the Venezuelan states of Falcón, Lara, and Zulia (Linares 1998).

NATURAL HISTORY: According to Linares (1998) this species flies in open areas, primarily in arid, lowland

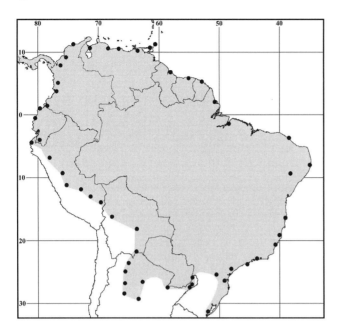

Map 277 Marginal localities for *Myotis nigricans* ●

habitats. They are known to roost in hollow trees, forage among columnar cacti, and feed on small insects such as flies and moths.

REMARKS: Miller and Allen (1928) relegated *M. nesopolus* to the synonymy of *M. nigricans*, a course followed by later workers. LaVal (1973a) did not treat *nesopolus*, but named *M. larensis* as a new species. Genoways and Williams (1979a) clarified the relationship between the island and mainland populations, treating *M. larensis* as a subspecies of *M. nesopolus*. Muñoz (2001:224) cited Hershkovitz (1949) as the source for Colombian records of *M. nesopolus*; however, there is no mention of the taxon in Hershkovitz's report.

Myotis nigricans (Schinz, 1821)
Black Myotis

SYNONYMS: See under Subspecies.

DISTRIBUTION: *Myotis nigricans* is known from Trinidad and Tobago, and from every South American country except Chile and Uruguay. Elsewhere, the species is in Mexico and Central America.

MARGINAL LOCALITIES (Map 277; from LaVal 1973a, except as noted). TRINIDAD AND TOBAGO: Tobago, no specific locality (Goodwin and Greenhall 1961); Trinidad, Botanic Gardens (Miller and Allen 1928). GUYANA: DemerarA&Minus;Mahaica, Georgetown. SURINAM: Paramaribo, Paramaribo (Husson 1978). FRENCH GUIANA: Paracou (Simmons and Voss 1998). BRAZIL: Amapá, Amapá (Peracchi, Raimundo, and Tannure 1984); Pará, Utinga; Ceará, Fortaleza (Piccinini 1974);

Pernambuco, Estação do Tapacurá (Mares et al. 1981); Bahia, Gruta dos Morcegos (McNab and Morrison 1963); Bahia, Pôrto Seguro (Faria, Soares-Santos, and Sampaio 2006); Espírito Santo, Companhia Vale do Rio Dôce Forest Reserve (Pedro and Passos 1995); Espírito Santo, Fazenda do Agá (original type locality of *M. nigricans* Schinz); Rio de Janeiro, km 42 antigua rodovia Rio-São Paulo; São Paulo, San Sebastian; São Paulo, Eldorado Paulista; Paraná, Floresta Nacional do Irati (N. R. Reis et al. 2000); Santa Catarina, Colonia Hansa (J. L. Lima 1926); Rio Grande do Sul, São Lourenço (C. O. C. Vieira 1942). ARGENTINA (Barquez, Mares, and Braun 1999, except as noted): Misiones, Libertad; Misiones, Dos de Mayo (Massoia and Chebez 1989); Misiones, Los Helechos; Corrientes, Laguna Paiva; Santiago del Estero, Campo Gallo; Santiago del Estero, Sumampa; Catamarca, Catamarca; Tucumán, Tafí del Valle; Salta, Rosario de Lerma; Jujuy, Santa Bárbara (LaVal 1973a). BOLIVIA: Tarija, Caraparí (O. Thomas 1925); Santa Cruz, 1 km NE of Estancia Cuevas (S. Anderson 1997); La Paz, Sacramento Alto (S. Anderson 1997). PERU: Puno, Bella Pampa; Cusco, Tono (V. Pacheco et al. 1993); Cusco, Las Malvinas (S. Solari et al. 1998); Junín, Chanchamayo; Huánuco, Tingo María; Cajamarca, Hacienda Limón; Piura, 7 km N and 15 km E of Talara. ECUADOR: Loja, Casanga; Manabí, Río Briseño; Esmeraldas, Esmeraldas. COLOMBIA: Nariño, La Guayacana; Valle del Cauca, 29 km E of Buenaventura; Chocó, Andagoya (Guimarães and d'Andretta 1956); Córdoba, Socorré (Guimarães and d'Andretta 1956); Bolívar, Sampués; Magdalena, Bonda. VENEZUELA: Zulia, Río Aurare; Falcón, 19 km NW of Urama; Miranda, 19 miles E of Caracas; Monagas, Hacienda San Fernando.

SUBSPECIES: Two of the three subspecies of *M. nigricans* currently recognized occur in South America.

M. n. nigricans (Schinz, 1821)
SYNONYMS:

Vespertilio nigricans Schinz, 1821:179; type locality "42 km S Rio de Janeiro, [Rio de Janeiro,] Brazil," by neotype designation (LaVal 1973:9); original type locality "Östküste von Brasilien," identified as "Fazenda de Agá, en der Gegend des Flusses Iritiba," Espírito Santo, by Wied-Neuwied (1826).

Vespertilio brasiliensis Spix, 1823:63; type locality not stated but known to be Brazil; preoccupied by *Vespertilio brasiliensis* Desmarest, 1822 (= *Eptesicus brasiliensis*).

Vespertilio spixii J. B. Fischer, 1829:111; replacement name for *Vespertilio brasiliensis* Spix.

Vespertilio parvulus Temminck, 1840:246; type locality "Brésil."

Vespertilio arsinoë Temminck, 1840:247; type locality "Surinam."

Vespertilio hypothrix d'Orbigny and Gervais, 1847:14 (footnote), 16; type locality "la province de Moxos [Beni,] (Bolivia)."

Vespertilio splendidus J. A. Wagner, 1855:148; type locality "St. Thomas," U.S. Virgin Islands.

Vespertilio concinnus H. Allen, 1866:281; type locality "San Salvador," El Salvador.

Vespertilio mundus H. Allen, 1866:280; type locality "Maracaibo," Zulia, Venezuela.

Vespertilio exiguus H. Allen, 1866:281; type locality "Aspinwall, N. G." (= Colón), Panamá, Panama.

Myotis nigricans: Miller, 1897a:74; first use of current name combination.

[*Vespertilio (Vespertilio)*] *nigricans*: Trouessart, 1897:130; name combination.

[*Vespertilio (Vespertilio)*] *arsinoë*: Trouessart, 1897:131; name combination.

Myotis chiriquensis J. A. Allen, 1904a:77; type locality "Boqueron, Chiriqui, Panama."

Myotis punensis J. A. Allen, 1914:383; type locality "Puna Island, [Guayas,] Ecuador."

Myotis bondae J. A. Allen, 1914:384; type locality "Bonda, Santa Marta, [Magdalena,] Colombia."

Myotis esmeraldae J. A. Allen, 1914:385; type locality "Esmeraldas, [Esmeraldas,] Ecuador."

Myotis maripensis J. A. Allen, 1914:385; type locality "Maripa, [Bolívar,] Venezuela."

Myotis nigracans R. J. Baker and Jordan, 1970:598, 600; incorrect subsequent spelling of *Vespertilio nigricans* Schinz

This subspecies occurs throughout South America north of about 26°S, except for the Andean highland area occupied by *M. n. osculatii*. Elsewhere, *M. n. nigricans* is found in Central America and in Mexico as far north as Tamaulipas.

M. n. osculatii (Cornalia, 1849)

SYNONYMS:

Vespertilio osculatii Cornalia, 1849:11; type locality "Regionibus equatorialibus, secus Fl. Napo decursum," (= Santa Rosa de Otas, Río Napo), Napo, Ecuador.

Phyllostomus quixensis Osculati, 1854:153; type locality "Rio Napo," Napo, Ecuador.

Myotis caucensis J. A. Allen, 1914:386; type locality "Rio Frio (altitude 3500 feet), Cauca River, [Valle del Cauca,] Colombia."

This subspecies in found throughout the Andes from about 05°N to about 15°S.

NATURAL HISTORY: Wilson (1971a) presented considerable ecological detail for *M. nigricans* on Barro Colorado Island, Panama. Myers (1977) outlined the reproductive cycle in Paraguay. A summary of the natural history of this species was provided by Wilson and LaVal (1974) in their *Mammalian Species* account.

REMARKS: The only other subspecies currently recognized (*M. n. extremus* Miller and Allen, 1928) is restricted to Mexico (Bogan 1978). *Myotis nigricans* is the most widely distributed and abundant species of *Myotis* in South America. Species limits remain poorly known, and there are many names currently in the synonymy of *Myotis nigricans nigricans* (see Cabrera 1958; LaVal 1973a). Changes from Cabrera (1958) include the recognition of *M. n. osculatii* as a subspecies distinct from the nominate form, and the allocation of *M. n. nicholsoni* Sanborn to *Myotis atacamensis* (Lataste) by LaVal (1973a). Specimens collected by the Italian explorer Gaetano Osculati near Santa Rosa d'Otas, a small village on the Río Napo located some 01°04′ south and 01°47′ east of Quito, were named *Vespertilio osculatii* by Cornalia (1849). Cabrera (1917, 1958) determined that *osculatii* was a synonym of *nigricans*, but believed that if the Andean animals were distinct, *osculatii* would have priority. LaVal (1973a) did not treat *osculatii*, and applied the name *M. n. caucensis* to these animals. D. C. Carter and Dolan (1978) were unable to locate the holotype of *Vespertilio osculatii*, perhaps because they attributed the name to Cabrera as *M. n. osculatii*. If the animals from this region are determined to represent a separate subspecies, the correct name has to be *Myotis nigricans osculatii* Cornalia (Ibáñez 1984b). LaVal (1973a) recognized *M. n. punensis* J. A. Allen, as a distinct subspecies, but Bogan (1978) returned it to the synonymy of *M. n. nigricans*. A future revision of this species will likely reveal *M. nigricans* to be composite, and additional names now in synonymy may be applicable to recognizable species and subspecies when the species limits are better understood and more narrowly defined. D. C. Carter and Dolan (1978) provided information on and measurements of the lectotype of *Vespertilio parvulus* Temminck, the holotype of *Vespertilio arsinoë* Temminck, and either the holotype or a syntype of *Vespertilio splendidus* Wagner.

Myotis oxyotus (W. Peters, 1866)
Montane Myotis

SYNONYMS: See under Subspecies.

DISTRIBUTION: *Myotis oxyotus* is known from the Andes of Venezuela, Colombia, Ecuador, Peru, and Bolivia. Elsewhere, the species is known from the mountains of Panama and Costa Rica.

MARGINAL LOCALITIES (Map 278; from LaVal 1973a, except as noted; localities listed clockwise in Venezuela and from north to south elsewhere): VENEZUELA: Distrito Federal, Alto Ño León (Handley 1976); Bolívar, 21 km NE of Icabarú (Handley 1976); Amazonas, Camp VII, Parque Nacional Serranía de la Neblina (Gardner

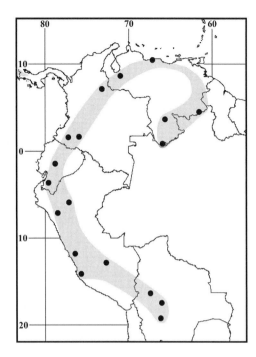

Map 278 Marginal localities for *Myotis oxyotus* ●

1988); Amazonas, Caño Culebra (Handley 1976); Mérida, La Mucuy. COLOMBIA: Santander, Lebrija; Huila, Parque Nacional Natural de la Cueva de los Guácharos (Lemke et al. 1982); Nariño, El Guabo. ECUADOR: Chimborazo, Mt. Chimborazo (original type locality of *Vespertilio oxyotus* W. Peters); El Oro, El Bosque. PERU: Cajamarca, Cajamarca (Koopman 1978); San Martín, Río Negro (Koopman 1978); Lima, Surco; Cusco, Santa Ana; Ica, Santiago. BOLIVIA: La Paz, Chulumani; Cochabamba, 1.3 km W of Jamachuma (S. Anderson 1997); Potosí, Finca Salo (S. Anderson 1997).

SUBSPECIES: I recognize two subspecies of *M. oxyotus*; only the nominate form occurs in South America.

M. o. oxyotus (W. Peters, 1866)
SYNONYMS:

Vespertilio oxyotus W. Peters, 1866b:19; type locality "Gruta Rumichaca, 2 mi E La Paz, Carchi, Ecuador, elevation 2600 m," by neotype designation (LaVal 1973:41); original type locality "Chimborazo, in einer Höhe von 9 bis 10,000 Fuss," Chimborazo, Ecuador.

[*Vespertilio (Vespertilio)*] *oxyotus*: Trouessart, 1897:130; name combination.

Myotis Thomasi Cabrera, 1901:370; type locality "muy probablemente procede del Brasil meridional"; corrected to "Archidona, sobre el citado río," Napo, Ecuador (Cabrera, 1902).

[*Myotis (Myotis)*] *oxyotus*: Trouessart, 1904:94; name combination.

Myotis chiloënsis oxyotus: Miller and Allen, 1928:193; name combination.

Myotis oxyotus oxyotus: LaVal, 1973a:42; first use of current name combination.

Myotis oxyotis: Lemke, Cadena, Pine, and Hernández-Camacho, 1982:230; incorrect subsequent spelling of *Vespertilio oxyotus* W. Peters.

M[*yotis*]. *o*[*xyotis*]. *oxyotis*: S. Anderson, 1993:25; incorrect subsequent spelling of *Vespertilio oxyotus* W. Peters.

[*Myotis*] *oxyota* Woodman, 1993:545; unjustified emendation of *Vespertilio oxyotus* W. Peters.

Myotis (Leuconoe) oxyotus oxyotus: S. Anderson, 1997: 275; name combination.

The nominate subspecies is in the Andes of Venezuela, Colombia, Ecuador, Peru, and Bolivia.

NATURAL HISTORY: *Myotis oxyotus* is primarily a highland species, found between 800 and 2,400 m. It apparently prefers forested areas in these cooler montane areas, where day roosts are known to include hollow trees. They have been mistnetted over roads and in clearings in the forest, as well as in more open areas adjacent to the forest (Linares 1998).

REMARKS: The holotype of *Myotis oxyotus* is presumed to have been destroyed during World War II, and LaVal (1973a) designated a neotype from Gruta Rumichaca, 2 miles E of La Paz, Carchi, Ecuador. Cabrera (1958) placed *M. oxyotus* as a subspecies of *M. chiloensis*, but LaVal elevated it to specific status, and named a new subspecies (*M. o. gardneri*) from Central America. LaVal (1973a) also suggested that specimens from the coast of Peru represented an undescribed subspecies.

Myotis riparius Handley 1960
Riparian Myotis
SYNONYMS:

Myotis simus riparius Handley, 1960:466; type locality "Tacarcuna Village, 3,200 ft., Río Pucro, Darién, Panama."

Myotis riparius: LaVal, 1973a:32; first use of current name combination.

[*Myotis*] *riparia* Woodman, 1993:545; unjustified emendation of *Myotis simus riparius* Handley.

Myotis (Leuconoe) riparius: S. Anderson, 1997:276; name combination.

DISTRIBUTION: *Myotis riparius* is known from Colombia, Venezuela, Trinidad and Tobago, Guyana, French Guiana, Brazil, Ecuador, Peru, Bolivia, Uruguay, Paraguay, and northern Argentina.

MARGINAL LOCALITIES (Map 279; from LaVal 1973a, except as noted): TRINIDAD AND TOBAGO: Trinidad, Blanchisseuse. VENEZUELA: Delta Amacuro,

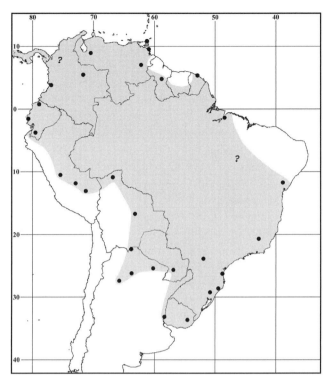

Map 279 Marginal localities for *Myotis riparius* ●

Caño Araguabisi (Linares and Rivas 2004); Bolívar, Río Supamo. GUYANA: Potaro-Siparuni, Clearwater Camp (Lim et al. 1999). FRENCH GUIANA: Paracou (Simmons and Voss 1998). BRAZIL: Pará, Utinga; Bahia, Lamarão; Minas Gerais, Viçosa; Paraná, Parque Estadual Vila Rica (Bianconi, Mikich, and Pedro 2004); Santa Catarina, Joinville (Cherem et al. 2005); Santa Catarina, Nova Veneza (Cherem et al. 2005); Rio Grande do Sul, Vila Oliva (J. C. González and Fabián 1995). URUGUAY (Ximénez, Langguth, and Praderi 1972): Rocha, Cerro de la Estancia San Miguel; Río Negro, boca del Arroyo Las Cañas. PARAGUAY: Paraguarí, Sapucay. ARGENTINA (Barquez, Mares, and Braun 1999): Formosa, 16 km S of Colonia Km 503; Santiago del Estero, Santo Domingo; Tucumán, Arroyo El Saltón; Salta, 6 km W of Piquirenda Viejo. BOLIVIA (S. Anderson 1997): Santa Cruz, Estancia Cachuela Esperanza; Pando, Ingavi. PERU: Cusco, Consuelo (V. Pacheco et al. 1993); Cusco, Las Malvinas (S. Solari et al. 1998); Pasco, Oxapampa. ECUADOR: El Oro, 9 miles S of Zaruma; Manabí, Estero Achiote (Albuja 1999); Esmeraldas, San Miguel (Albuja 1999). COLOMBIA: Valle del Cauca, Río Zabaletas (M. E. Thomas 1972, as *M. simus*); Boyacá, Trinidad. VENEZUELA: Barinas, Agua Fria.

SUBSPECIES: I consider *M. riparius* to be monotypic.

NATURAL HISTORY: This species occurs at low to intermediate elevations from sea level up to about 2,000 m. It is frequently found foraging along streams and rivers, as well as above trails in forested areas, and in clearings. Based on captures, Simmons and Voss (1998) reported that *M. riparius* preferred forested areas, as opposed to clearings, at Paracou, French Guiana. Although it inhabits both lowland rainforests and submontane forests, *M. riparius* is rarely found at higher elevations (Linares 1998). Autino, Claps, and González (2004) reported a flea and a nycteribiid batfly as ectoparasites of Uruguayan *M. riparius*.

REMARKS: Handley (1960) described *riparius* as a subspecies of *M. simus*. LaVal (1973a) elevated it to specific status. The bats identified by M. E. Thomas (1972:Table 1) as *M. simus* from Río Zabaletas, Colombia, represent *M. riparius*.

Myotis ruber (É. Geoffroy St.-Hilaire, 1806)
Red Myotis

SYNONYMS:

Vespertilio ruber É. Geoffroy St.-Hilaire, 1806:204; type locality "Sapucay, [Neembucú,] Paraguay," by neotype designation (LaVal 1973a:45); original type locality "Paraguay" (see É. Geoffroy St.-Hilaire, 1806:203), based on Azara's (1801:292–93) *chauve-souris onzième ou de chauve-souris cannelle*, and restricted to "near Asunción" by Miller and Allen (1928:198).

Vespertilio cinnamomeus J. A. Wagner, 1855:755; renaming of *V. ruber* É. Geoffroy St.-Hilaire.

Vespertilio kinnamon P. Gervais, 1856a:84; type locality "Capella Nova," Minas Gerais, Brazil.

Myotis ruber: O. Thomas, 1902h:493; first use of current name combination.

[*Myotis*] *rubra* Woodman, 1993:545; unjustified emendation of *Vespertilio ruber* É. Geoffroy St.-Hilaire.

DISTRIBUTION: *Myotis ruber* is endemic to South America where it is known from Brazil, Argentina, and Paraguay.

MARGINAL LOCALITIES (Map 280): BRAZIL: Alagoas, Quebrangulo (Baud and Menu 1993); Rio de Janeiro, Nova Friburgo (C. O. C. Vieira 1955); São Paulo, Piquete (C. O. C. Vieira 1955); Santa Catarina, Fazenda Palmital (Sipinski and Reis 1995); Santa Catarina, Santo Amaro da Imperatriz (Cherem et al. 2005); Rio Grande do Sul, São Laurenço (LaVal 1973a). ARGENTINA: Misiones, Jct. Hwy 21 and Arroyo Oveja Negra (Barquez, Mares, and Braun 1999). PARAGUAY: Itapúa, Salto Tembey (Baud and Menu 1993). ARGENTINA (Barquez, Mares, and Braun 1999): Corrientes, Corrientes; Formosa, Parque Nacional Río Pilcomayo. PARAGUAY: Presidente Hayes, 230 km NW of Villa Hayes (Baud and Menu 1993). BRAZIL: Paraná, Parque Estadual Mata dos Godoy (N. R. Reis, Peracchi, and Onuki 1993); São Paulo, Estação Ecológica dos Caetetus (Pedro, Passos, and Lim 2001); Minas Gerais, Viçosa (LaVal 1973a).

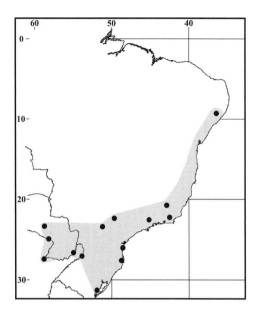

Map 280   Marginal localities for *Myotis ruber* ●

SUBSPECIES: I treat *M. ruber* as monotypic.

NATURAL HISTORY: *Myotis ruber* is rare throughout its range, and little is known of its natural history. According to López-González et al. (2001), in Paraguay, it is an inhabitant of lowland semideciduous forests.

REMARKS: J. A. Wagner (1855), erroneously believing that *Vespertilio ruber* É. Geoffroy St.-Hilaire, 1806, was based on the species known today as *Noctilio albiventris*, coined the name *cinnamomeus* for the red *Myotis* of Paraguay and northern Argentina. Cabrera (1958) included *Vespertilio polythrix* É. Geoffroy St.-Hilaire and *Vespertilio nubilis* J. A. Wagner in the synonymy of *M. ruber*, but LaVal (1973a) relegated both names to the synonymy of *Myotis levis*. No holotype was specified and, because Azara's specimens are unlikely to be found (see account of *Myotis albescens*), LaVal designated a neotype from Sapucay, Neembucú, Paraguay.

Myotis simus O. Thomas, 1901
Velvety Myotis
SYNONYMS:
Myotis simus O. Thomas, 1901f:541; type locality "Sarayacu, [Loreto,] Peru."
Myotis guaycuru Proença, 1943:313; type locality "Salobra," Rio Miranda, Mato Grosso do Sul, Brazil.
[*Myotis*] *sima* Woodman, 1993:545; unjustified emendation of *Myotis simus* O. Thomas.
Myotis (Leuconoe) simus: S. Anderson, 1997:276; name combination.

DISTRIBUTION: *Myotis simus* is endemic to South America where it is known from the Amazon basin of

Colombia, Bolivia, Brazil, Ecuador, and Peru, and farther south from Argentina and Paraguay.

MARGINAL LOCALITIES (Map 281; from LaVal 1973a, except as noted): *Amazon basin distribution.* COLOMBIA: Putumayo, Mocoa (Marinkelle and Cadena 1972). PERU: Loreto, Boca del Río Curaray. COLOMBIA: Amazonas, Leticia. BRAZIL: Amazonas, Manaus; Amazonas, Villa Bella Imperatriz; Pará, Igarapé Amorin; Amazonas, Auará Igarapé. BOLIVIA: Beni, 23 km W of San Javier (S. Anderson 1997); Beni, Río Curiraba (Wilson and Salazar 1990). PERU: Madre de Dios, Camp 5 (Baud 1986); Madre de Dios, Pakitza (V. Pacheco et al. 1993); Pasco, Oxapampa; Loreto, Orosa. ECUADOR: Napo, Río Capahuarí (Albuja 1999). *Southern distribution.* BRAZIL: Mato Grosso do Sul, Salobra (type locality of *Myotis guaycuru* Proença). PARAGUAY: Central, 17 km by road E of Luque (Myers and Wetzel 1979). ARGENTINA (Barquez, Mares, and Braun 1999): Corrientes, Isla Apipé; Formosa, Parque Nacional Pilcomayo.

SUBSPECIES: I am treating *M. simus* as monotypic, pending revision.

NATURAL HISTORY: Myers and Wetzel (1979) reported two pregnant *M. simus* caught in Paraguay in October as they emerged from a tree hole shared with *Noctilio albiventris*. Specimens from Balta, Ucayali, Peru, were netted over land at the margin of a small, isolated, shallow pool (a flood-filled, tree-less depression left by a meander of the Río Curanja) surrounded by primary tropical forest (A. L. Gardner, pers. comm.).

REMARKS: Cabrera (1958) erected the subgenus *Hesperomyotis* for *M. simus*, but Handley (1960) argued against subgeneric recognition. Findley (1972) also did not recognize *Hesperomyotis*, but LaVal (1973a) suggested

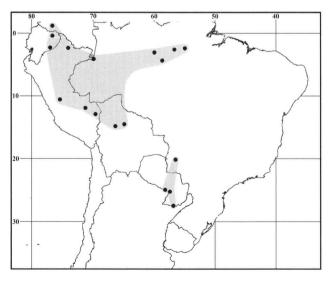

Map 281   Marginal localities for *Myotis simus* ●

that a thorough generic revision was needed to determine whether *M. simus* warranted recognition in a separate subgenus.

LaVal (1973) suggested that *Myotis guaycuru* Proença might be the senior synonym for *M. riparius*, but López-González et al. (2001) relegated *guaycuru* to the synonymy of *M. simus*. According to Baud and Menu (1993), at least some of the specimens reported as *M. simus* by Myers (1977) and Myers and Wetzel (1983) from departamento Presidente Hayes, Paraguay, represent *M. ruber*. Baud and Menu (1993) restricted the distribution of *M. simus* to the Amazon basin. Nevertheless, I have mapped Myers and Wetzel's (1979) Paraguayan records, and the Argentine records of Barquez, Mares, and Braun (1999), together with the type locality of *Myotis guaycuru* as a separate southern population. The bats M. E. Thomas (1972:Table 1) listed as *M. simus* from Río Zabaletas, Colombia, represent *M. riparius*. The Bahia, Brazil, record cited by C. O. C. Vieira (1955; not mapped) is considered suspect and may represent *M. ruber*. Clearly, the taxon needs revision. D. C. Carter and Dolan (1978) provided information on and measurements of the holotype of *Myotis simus* O. Thomas.

Genus *Rhogeessa* H. Allen, 1866

John W. Bickham and Luis A. Ruedas

The genus *Rhogeessa*, exclusively Neotropical in distribution, contains ten species (LaVal 1973b; Ramírez 1982; Genoways and Baker 1996), of which three, *R. hussoni*, *R. io*, and *R. minutilla* occur in South America. No fossils are known. This genus includes some the smallest of living mammals (LaVal 1973b); head and body length 36–53 mm, tail length 26–48 mm, forearm length 24.5–35.0 mm, length of skull including incisors 11.0–15.4 mm; adults weigh 3–10 g (Hall 1981; Nowak and Paradiso 1983). Otherwise externally similar to *Nycticeius*, species of *Rhogeessa* have well-developed perinasal glands that give the snout a puffy appearance in life. The lower third incisor lacks a lingual lobe; consequently, it varies from barely smaller than i2 to minute in size. Additional dental characters include tricuspid first and second lower incisors, with the outer cusp much lower than other two cusps; either a bicuspid or unicuspid i3; a well-developed lingual cingulum on the upper canine, usually with accessory cuspules; structures on the M3 are reduced to the parastyle, paracone, protocone, and first and second commissures; maxillary cheek teeth converge anteriorly; and the width of the posterior half of m3 is substantially less than the width of the anterior half. The cranium lacks basisphenoid pits, and the rostrum is narrower than the braincase. Wing and interfemoral membranes are relatively thin. The penis is short, 3–4 mm, and the baculum extends to within 0.5 mm of its distal end. The baculum varies in length from 0.5 to 0.8 mm; is saddle-shaped at its proximal end, and has well-developed proximal lateral knobs, poorly developed proximal median knob, and a long (usually narrow) shaft, which is circular or elliptical in cross section near its distal end (LaVal 1973b:11). The dental formula is 1/3, 1/1, 1/2, 3/3 × 2 = 30. Czaplewski, Rincón, and Morgan (2005) reported finding late Pleistocene remains of *Rhogeessa* from tar seep deposits in northern Venezuela, but did not allocate them to species.

SYNONYMS:

Rhogeëssa H. Allen, 1866:285; type species *Rhogeessa tumida* H. Allen, 1866, by subsequent designation (Miller 1897a:122).

Rhogöessa Marschall, 1873:11; incorrect subsequent spelling of *Rhogeessa* H. Allen.

Vesperugo (Rhogeëssa): Dobson, 1878:183; not *Vesperugo* Keyserling and Blasius, 1839.

Rhogoësa Alston, 1879:21; incorrect subsequent spelling of *Rhogeessa* H. Allen.

Rhogoessa Elliot, 1905b:489; unjustified emendation of *Rhogeessa* H. Allen.

Baeodon Miller, 1906b:85; type species *Rhogeessa alleni* O. Thomas, 1892d, by original designation (valid as a subgenus).

Rhogeesa S. Anderson, Koopman, and Creighton, 1982:1; incorrect subsequent spelling of *Rhogeessa* H. Allen.

Rhogeesa Koopman, 1982:275; incorrect subsequent spelling of *Rhogeessa* H. Allen.

Rogheessa N. R. Reis, Peracchi, Sekiama, and Lima, 2000:697,703; incorrect subsequent spelling of *Rhogeessa* H. Allen.

Rhogessa N. R. Reis, Peracchi, Sekiama, and Lima, 2000: 698; incorrect subsequent spelling of *Rhogeessa* H. Allen.

REMARKS: The most recent review of the genus was by LaVal (1973b) and much of the information provided herein is from his report.

KEY TO THE SOUTH AMERICAN SPECIES OF *RHOGEESSA*:

1. Color pale yellowish buff; forearm usually less than 27.6 mm; first phalanx of wing digit IV usually greater than 8.6 mm; cranium with a small, inflated area of the posterior medial parietals at the juncture of the sagittal crest with the lambdoidal crests (referred to as a helmet) . . . . *Rhogeessa minutilla*
1'. Color darker brownish; forearm usually greater than 27.1 mm; first phalanx of digit IV usually less than 8.8 mm; parietals not inflated at juncture of the sagittal crest with the lambdoidal crests (helmet lacking) 2

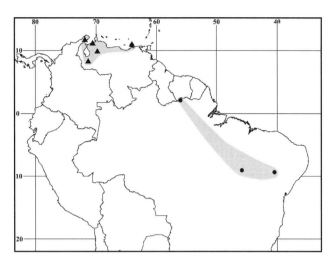

Map 282 Marginal localities for *Rhogeessa hussoni* ● and *Rhogeessa minutilla* ▲

2. Greatest length of skull more than 12.6 mm; length of mandibular toothrow less than 5.4 mm . . . . *Rhogeessa hussoni*

2'. Greatest length of skull less than 12.6 mm; length of mandibular toothrow 5.4 mm or more. . . *Rhogeessa io*

Rhogeessa hussoni Genoways and Baker, 1996
Eastern Little Yellow Bat

SYNONYMS:

Rhogeessa tumida: LaVal, 1973b:29; part; not *Rhogeessa tumida* H. Allen.

Rhogeessa tumida: S. L. Williams and Genoways, 1980a: 232; not *Rhogeessa tumida* H. Allen.

Rhogeessa hussoni Genoways and Baker, 1996:85; type locality "Suriname: Nickerie [*sic*, Sipaliwini] District, Sipaliwini Airstrip."

DISTRIBUTION: *Rhogeessa hussoni* is known from Surinam and Brazil.

MARGINAL LOCALITIES (Map 282; from Genoways and Baker 1996): SURINAM: Sipaliwini, Sipaliwini Airstrip (type locality of *Rhogeessa hussoni* Genoways and Baker). BRAZIL: Maranhão, Alto Parnaíba; Bahia, Fazenda São Raimundo.

SUBSPECIES: We treat *Rhogeessa hussoni* as monotypic.

NATURAL HISTORY: The holotype was mistnetted over the Sipaliwini River; vegetation along the bank was secondary tropical forest (S. L. Williams and Genoways 1980a). The karyotype is $2n = 52$, FN = 52 (Genoways and Baker 1996); originally reported by Honeycutt, Baker, and Genoways (1980) under the name *R. tumida*.

Rhogeessa io O. Thomas, 1903
Southern Little Yellow Bat

SYNONYMS:

Rhogeessa io O. Thomas, 1903c:382; type locality "Valencia, [Carabobo,] Venezuela."

Rhogeessa velilla O. Thomas, 1903c:383; type locality "Puná, Puná Island, Gulf of Guayaquil, [Guayas,] Ecuador."

Rhogeessa bombyx O. Thomas, 1913d:569; type locality "Condoto, Choco, Colombia. Alt. 300.'"

Rhogeëssa tumida riparia Goodwin, 1958c:5; type locality "Cuchivano, 3 miles west of Cumanocoa, Province of Sucre, northeastern Venezuela. Altitude 700 feet."

DISTRIBUTION: *Rhogeessa io* is in Colombia and Venezuela, with isolated records from Guyana, Surinam, Brazil, Bolivia, and Ecuador. Elsewhere, the species is known in Central America from Panama north into eastern Nicaragua (Genoways and Baker 1996).

MARGINAL LOCALITIES (Map 283; from LaVal 1973b [as *R. tumida*], except as noted): COLOMBIA: Atlántico, Barranquilla (Nicéforo-María 1947); El Cesar, Colonia Agrícola Caracolicito. VENEZUELA: Zulia, 30 mi E of Maracaibo; Miranda, near El Encantado (Handley 1976); Sucre, Manacal (Handley 1976). TRINIDAD AND TOBAGO: Trinidad, Maracas. GUYANA: Upper Takutu-Upper Essequibo, near Shea. BRAZIL: Mato Grosso, Serra do Roncador, 264 km (by road) N of Xavantina; Paraná, Londrina (N. R. Reis et al. 1998). BOLIVIA: Beni, Caravana (S. Anderson, Koopman, and Creighton 1982). BRAZIL: Amazonas, Manaus (Mok, Luizão, and Silva 1982). VENEZUELA: Apure, Hato Caribén (Handley 1976). COLOMBIA: Meta, Municipio de San Juan de Arama (Sánchez-Palomino and Rivas-Pava 1993); Huila, 16 km NE of Villavieja. ECUADOR: Guayas, San Ramón; Pichincha, Río Toachi (Albuja 1999). COLOMBIA: Valle de Cauca, Río Raposo; Chocó, Condoto (type locality of *Rhogeessa bombyx* O. Thomas).

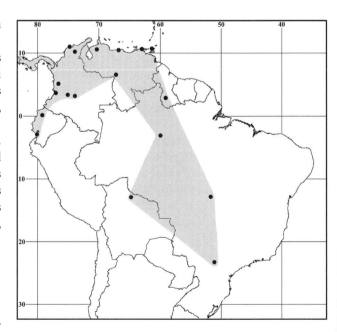

Map 283 Marginal localities for *Rhogeessa io* ●

SUBSPECIES: *Rhogeessa io* probably consists of two or more definable subspecies or species (see Remarks), but in this account we treat the taxon as monotypic.

NATURAL HISTORY: *Rhogeessa io* is the most widely distributed South American *Rhogeessa*. The species is known from every vegetation zone within an elevational range from sea level to approximately 1,500 m. These aerial insectivores have been netted along forest trails, over streams and pools of water, and in clearings, indicating that they forage for insects, often within a few meters of the ground and that they rely on streams and pools as sources of water. Their roosting habits are unknown, although they are presumed to roost in trees.

Pregnant females are known from February through April, and lactating females from March to July. Flying juveniles have been collected during May through August. Sperm production probably takes place during October to December (LaVal 1973, under *R. tumida*). Samples of populations from Trinidad and Venezuela have $2n = 30$ chromosomes (Bickham and Baker 1977).

REMARKS: Genoways and Baker (1996) applied the name *R. io* to the South American and southern Central American populations that LaVal (1973b), Bickham and Baker (1977), and Honeycutt, Baker, and Genoways (1980) had treated as southern populations of *R. tumida*. Genoways and Baker (1996) based their action on the high degree of morphological and chromosomal variability demonstrated by the population units they identified within the *tumida*-group. Muñoz (2001) recognized three species (*R. minutilla*, *R. parvula*, and *R. tumida*) as occurring in Colombia, and Castaño et al. (2003) listed two (*R. minutilla*, and *R. tumida*) in Departamento Caldas. Clearly, this species and other South American *Rhogeessa* are overdue for revision. D. C. Carter and Dolan (1978) provided notes on and measurements of the holotypes of *Rhogeessa io* O. Thomas, *Rhogeessa velilla* O. Thomas, and *Rhogeessa bombyx* O. Thomas.

Rhogeessa minutilla Miller, 1897
Tiny Yellow Bat

SYNONYMS: See under Subspecies.

distribution: *Rhogeessa minutilla* is on Isla Margarita, Venezuela, and in northeastern Colombia and northwestern Venezuela.

MARGINAL LOCALITIES (Map 282; from LaVal 1973b, except as noted): COLOMBIA: La Guajira, 119 km N and 32 km W of Maracaibo, Venezuela. VENEZUELA: Falcón, Capatárida; Nueva Esparta, 2 km N and 30 km W of Porlamar; Lara, Caserío Boro; Mérida, 3 km SE of San Juan de Lagunillas (Sosa, de Ascenção, and Soriano 1996).

SUBSPECIES: We recognize two subspecies of *R. minutilla*.

R. m. minutilla Miller, 1897
SYNONYMS:
Rhogeessa minutilla Miller, 1897b:139; type locality "Margarita Island, Venezuela."
Rhogeessa parvula minutilla: Goodwin, 1958c:7; name combination.

This subspecies is in the Guajira Peninsula of Colombia and Venezuela and elsewhere in Venezuela north of the state of Mérida.

R. m. cautiva Soriano, Naranjo, and Fariñas, 2004
SYNONYM:
R. m. cautiva Soriano, Naranjo, and Fariñas, 2004:442; type locality "Laguna de Caparú, 3 km SE San Juan de Lagunillas, Mérida state, Venezuela.

This subspecies is known only from the xeric intermontane basin of the rios Chama and Nuestra Señora, Mérida, Venezuela.

NATURAL HISTORY: *Rhogeessa minutilla* is known from sea level to over 900 m elevation in arid desert scrub and thorn forest, although one individual was captured in a mangrove swamp (Musso 1962). *Rhogeessa m. cautiva* occurs in xeric thorn-shrub forest in the Chama and Nuestra Señora river basin (Sosa, de Ascenção, and Soriano 1996; Soriano, Naranjo, and Fariñas 2004). Day roosts of this subspecies have been found in cavities in dead branches of columnar cacti; day roosts of *R. m. minutilla* are unknown. Sosa, de Ascenção, and Soriano (1996) found pregnant females in February and March (1 of 3) and in April and May (5 of 5), lactating females in June and July (1 of 1) and August and September (1 of 9). Elsewhere, juveniles have been observed in late June through the middle of July. Molting occurs in June and July. Sosa, de Ascenção, and Soriano (1996) recorded the remains of flying insects representing 29 families and 10 orders in their analysis of fecal samples from *R. m. cautiva* netted the vicinity of San Juan de Lagunillas. The four most common insect orders in their samples were Diptera, Hymenoptera, Lepidoptera, and Coleoptera. Frequency of occurrence at both the ordinal and family level showed strong seasonal variation, which could be interpreted as variation in either dietary preference or insect availability.

LaVal (1973b) illustrated the bacula, and stated that they resembled some bacula of *R. tumida*. However, in his comparison with bacula of South American "*R. tumida*" (= *R. io*), LaVal (1973b:37) described the bacula of *R. minutilla* as shorter, narrower, and with the lateral knobs extended at a higher angle from shaft, approaching a right angle in some specimens. *Rhogeessa minutilla* is most similar to, and presumably closely related to, *R. tumida*; even though Goodwin (1958c) considered *R. minutilla* to be a subspecies of *R. parvula*. The karyotype is not known.

REMARKS: Although the type locality is Margarita Island, off the coast of Venezuela, the greater part of the range of *R. minutilla* is along the mainland coast of northeastern Colombia and adjacent Venezuela. Ruedas and Bickham (1992), based on the examination of ten specimens each of *R. minutilla* and *R. tumida*, were able to readily distinguish the two species morphometrically. The question remains, however, as to whether *R. minutilla* is a distinct species or a subspecies of *R. tumida*, as suggested by J. D. Smith and Genoways (1974). An answer to this question will require larger samples of both putative species, and may have to rely heavily on chromosomal and molecular data, because *Rhogeessa* is such a morphologically complex group and includes cryptic species (R. J. Baker, 1984; also see Bickham and Baker 1977; R. J. Baker, Bickham, and Arnold 1985).

Literature Cited

Aagaard, E. M. J. 1982. Ecological distribution of mammals in the cloud forests and páramos of the Andes, Mérida, Venezuela. Ph.D. thesis, Colorado State University, Ft. Collins, 277 pp.

Abdala, F., D. A. Flores, and N. P. Giannini. 2001. Postweaning ontogeny of the skull of *Didelphis albiventris*. *J. Mammal.* 82:190–200.

Abravaya, J. P., and J. O. Matson. 1975. Notes on a Brazilian mouse, *Blarinomys breviceps* (Winge). *Contr. Sci. Nat. Hist. Mus. Los Angeles Co.* 270:1–8.

Ab'Sáber, A. N. 1971. A organização natural das paisagens inter e subtropicais brasileiras. In *III Simpósio sobre o cerrado*, coord. M. G. Ferri, 1–14. São Paulo: Edgard Blucher, EDUSP, xii + 239 pp.

Acha J., P., and J. Zapatel V. 1957. Estudio en quirópteros de la región de San Martín (Perú) como probables reservorios de rabia. *Bol. Oficina Sanitaria Panamer.* 47:211–22.

Acloque, A. 1900. *Mamifères*. Paris: J. B. Baillière et Fils, 84 pp.

Acosta, C. E., and R. D. Owen. 1993. Koopmania concolor. *Mammal. Species* 429:1–3.

Acosta y Lara, E. F. 1950. Quirópteros del Uruguay. *Communic. Zool., Mus. Hist. Nat. Montevideo* 3:1–73.

———. 1955. Algunos rasgos diferenciales entre *Histiotus montanus* e *Histiotus velatus*. *Communic. Zool., Mus. Hist. Nat. Montevideo* 4:1–8.

Adams, J. K. 1989. Pteronotus davyi. *Mammal. Species* 346:1–5.

Adler, G. H., J. J. Arboledo, and B. L. Travi. 1997. Diversity and abundance of small mammals in degraded tropical dry forest of northern Colombia. *Mammalia* 61:361–70.

Aellen, V. 1966. Sur une petite collection de chiroptères du nord-ouest du Perou. *Mammalia* 29:563–71, pl. 22. (Dated 1965; number 4 published in 1966.)

———. 1970. Catalogue raisonné des chiroptères de la Colombie. *Rev. Suisse Zool.* 77:1–37.

Agassiz, L. 1842. *Nomenclator zoologicus, continens nomina systematica generum animalium tam viventium quam fossilium, secundum ordinem alphabeticum deposita, adjectis auctoribus, libris in quibus reperiuntur, anno editionis, etymologia et familiis, ad quas pertinent, in variis classibus*. Fasciculus I. Continens Mammalia, Echinodermata et Acalephas. Soloduri: Jent et Gassmann, xii + 38 pp., iv + 14 pp., iv + 7 pp.

———. 1847. *Nomenclator zoologicus, continens nomina systematica generum animalium tam viventium quam fossilium, secundum ordinem alphabeticum deposita, adjectis auctoribus, libris in quibus reperiuntur, anno editionis, etymologia et familiis, ad quas pertinent, in variis classibus*. Fasciculus XII. Continens indicem universalem. Soloduri: Jent et Gassmann, viii + 393 pp. (Dated 1846 on inner title page, but 1847 on cover; title differs from that given above.)

———. 1848. *Nomenclatoris zoologici index universalis, continens nomina systematica classium, ordinum, familiarum et generum animalium omnium, tam viventium quam fossilium, secundum ordinem alphabeticum unicum deposita, adjectis homonymiis plantarum*. Soloduri: Jent et Gassmann, x + 1135 pp.

Aguiar, L. M. S. 2005. First record on the use of leaves of *Solanum lycocarpum* (Solanaceae) and fruits of *Emmotum nitens* (Icacinacea) by *Platyrrhinus lineatus*

(E. Geoffroy) (Chiroptera, Phyllostomidae) in the Brazilian cerrado. *Rev. Brasil. Zool.* 22:509–10.

Aguiar, L. M. S., W. R. de Camargo, and A. de S. Portella. 2006. Occurrence of white-winged vampire bat, *Diaemus youngi* (Mammalia, Chiroptera), in the Cerrado of Distrito Federal, Brazil. *Rev. Brasil. Zool.* 23:893–96.

Aguiar, L. M. S., M. Zortéa, and V. A. Taddei. 1996. New records of bats for the Brazilian Atlantic forest. *Mammalia* 59:667–71. [Dated 1995; number 4 marked "Achevé d'imprimer le 30 avril 1996."]

Aguirre, L. F. 1994. Estructura y ecología de las communidades de murciélagos del las sabana de Espíritu (Beni, Bolivia). Tesis de Licenciatura, Univ. Mayor de San Andres, La Paz, Bolivia, 166 pp.

———. 1999. Estado de conservación de los murciélagos de Bolivia. *Chiropt. Neotrop.* 5:108–12.

Aguirre, L. F., and R. J. de Urioste. 1994. Nuevos registros de murciélagos para Bolivia y los departamentos de Beni y Pando. *Ecol. Bolivia* 23:71–76.

Aiello, A. 1985. Sloth hair: Unanswered questions. In *The evolution and ecology of armadillos, sloths, and vermilinguas*, ed. G. G. Montgomery, 213–18. Washington, DC: The Smithsonian Institution Press, 10 (unnumbered) + 451 pp.

Alberico, M. S. 1987. Notes on distribution of some bats from southwestern Colombia. In *Studies in Neotropical mammalogy, essays in honor of Philip Hershkovitz*, ed. B. D. Patterson and R. M. Timm, 133–36. *Fieldiana Zool.*, 39:frontispiece, viii + 1–506.

———. 1990. Systematics and distribution of the genus *Vampyrops* (Chiroptera: Phyllostomidae) in northwestern South America. In *Vertebrates in the tropics*, ed. G. Peters and R. Hutterer, 345–54. Bonn: Museum Alexander Koenig.

———. 1994. First record of *Sturnira mordax* from Colombia with range extensions for other bat species. *Trianea* 5:335–41.

Alberico, M. S., and L. G. Naranjo H. 1982. Primer registro de *Molossops brachymeles* (Chiroptera: Molossidae) para Colombia. *Cespedesia* 11:141–43.

Alberico, M. S., and A. J. Negret. 1992. Primer aporte sobre los mamíferos del Valle del Patía (Cauca-Nariño). *Nov. Colombianas* 5:66–71.

Alberico, M. S., and J. E. Orejuela. 1982. Diversidad específica de dos comunidades de murciélagos en Nariño, Colombia. *Cespedesia*, supl., 3:31–40.

Alberico, M. S., and E. Velasco. 1991. Description of a new broad-nosed bat from Colombia. *Bonn. Zool. Beitr.* 42:237–39.

Alberico, M. S., A. Cadena, J. Hernández-Camacho, and Y. Muñoz-Saba. 2000. Mamíferos (Synapsida: Theria) de Colombia. *Biota Colomb.* 1:43–75.

Albuja V., L. 1983. *Murciélagos del Ecuador*. Quito, Ecuador: Escuela Politécnica Nac., Depto. Cien. Biol., xii + 285 pp. + map. [Publication dated 1982, but not published until 1983 according to L. Albuja V. (pers. comm.)].

———. 1989. Adiciones a la fauna de quirópteros del noroccidente del Ecuador. *Politécnica* 14:105–11.

———. 1991. Mamíferos. *In* Lista del Vertebrados del Ecuador. *Politécnica* 16:163–204.

———. 1999. *Murciélagos del Ecuador*. 2nd ed. Quito, Ecuador: Cicetrónica Cía. Ltda. Offset, 288 pp.

Albuja V., L., and A. L. Gardner. 2005. A new species of *Lonchophylla* Thomas (Chiroptera-Phyllostomidae) from Ecuador. *Proc. Biol. Soc. Washington* 118:442–49.

Albuja V., L., and P. Mena V. 1991. Adición de dos especies de quirópteros a la fauna del Ecuador. *Politécnica* 16:93–98.

———. 2004. Quirópteros de los bosques húmedos del occidente del Ecuador. *Politécnica* 25:19–96.

Albuja V., L., and B. D. Patterson. 1996. A new species of northern shrew-opossum (Paucituberculata: Caenolestidae) from the Cordillera del Cóndor, Ecuador. *J. Mammal.* 77:41–53.

Albuja V., L., and R. Rageot. 1986. Un mamífero nuevo para el Ecuador, Monodelphis adusta (Marsupialia: Didelphidae). *Politécnica* 11:97–103.

Albuja V., L., and P. Tapia. 2004. Hallazgo de una nueva especie Murciélago blanco (Emballonuridae: *Diclidurus scutatus*) en el Ecuador. *Politécnica* 25:152–55.

Albuja V., L., M. Ibarra, J. Urgilés, and R. Barriga. 1980. *Estudio preliminar de los vertebrados Ecuatorianos*. Escuela Politecnica Nacional, Departamento de Ciencias Biológicas, 143 pp.

Aléssio, F. M., A. R. M. Pontes, and V. L. da Silva. 2005. Feeding by *Didelphis albiventris* on tree gum in the northeastern Atlantic forest of Brazil. *Mastozool. Neotrop.* 12:53–56.

Alfaro, A. 1897. *Mamíferos de Costa Rica*. San José: Tipografía Nacional, vi +51 pp.

Alho, C. J. R., L. A. Pereira, and A. C. Paula. 1987. Patterns of habitat utilization by small mammal populations in cerrado biome of central Brazil. *Mammalia* 50:447–60. [Dated 1986; number 4 marked "Achevé d'imprimer le 15 avril 1987."]

Allen, G. M. 1902a. The type locality of Ametrida minor H. Allen. *Proc. Biol. Soc. Washington* 15:88–89.

———. 1902b. The mammals of Margarita Island, Venezuela. *Proc. Biol. Soc. Washington* 15:91–97.

———. 1908. Notes on Chiroptera. *Bull. Mus. Comp. Zoöl.* 52:25–62, 1 pl.

———. 1911. Mammals of the West Indies. *Bull. Mus. Comp. Zoöl.* 54:175–263.

————. 1914. A new bat from Mexico. *Proc. Biol. Soc. Washington* 27:109–11.

————. 1917. Two undescribed West Indian bats. *Proc. Biol. Soc. Washington* 30:165–70.

————. 1923a. A new shrew from Colombia. *Proc. New England Zool. Club* 8:37–38.

————. 1923b. A new disc-winged bat from Panama. *Proc. New England Zool. Club* 9:1–2.

————. 1932. A Pleistocene bat from Florida. *J. Mammal.* 13:256–59.

————. 1935. Bats from the Panama region. *J. Mammal.* 16:226–28.

————. 1939. *Bats.* Cambridge: Harvard Univ. Press, x + 368 pp. [Reprinted by Dover Publ., Inc., New York, 1967.]

Allen, H. 1861. Description of a new Mexican bat. *Proc. Acad. Nat. Sci. Philadelphia* 13:359–61.

————. 1862. Descriptions of two new species of Vespertilionidae, and some remarks on the genus Antrozous. *Proc. Acad. Nat. Sci. Philadelphia* 1862:246–48.

————. 1864. Monograph of the bats of North America. *Smithsonian Misc. Coll.* 165:xxiv + 85 pp.

————. 1866. Notes on the Vespertilionidae of tropical America. *Proc. Acad. Nat. Sci. Philadelphia* 1866:279–88.

————. 1889. On the genus *Nyctinomus* and description of two new species. *Proc. Amer. Philos. Soc.* 26:558–63.

————. 1890a. Description of a new species of bat, Atalapha semota. *Proc. U.S. Natl. Mus.* 13:173–75.

————. 1890b. Description of a new species of bat of the genus Carollia and remarks on Carollia brevicauda. *Proc. U.S. Natl. Mus.* 13:291–98.

————. 1891a. On a new species of Atalapha. *Proc. Amer. Phil. Soc.* 29:5–7.

————. 1891b. Description of a new species of *Vampyrops*. *Proc. Acad. Nat. Sci. Philadelphia* 1891:400–05.

————. 1892a. Change of name of a genus of bats. *Proc. Acad. Nat. Sci. Philadelphia* 1891:466. [Dated 1891; published 19 January 1892.]

————. 1892b. Description of a new genus of phyllostome bats. *Proc. U.S. Natl. Mus.* 15:141–42.

————. 1894a. A monograph of the bats of North America. *Bull. U.S. Natl. Mus.* 43:x + 198 pp., 38 pls. [Dated 1893; published 14 March 1894.]

————. 1894b. On a new species of Ametrida. *Proc. Boston Soc. Nat. Hist.* 26:240–46.

————. 1896a. Notes on the vampire bat (Diphylla ecaudata) with special reference to its relationship with Desmodus rufus. *Proc. U.S. Natl. Mus.* 18:769–77.

————. 1896b. Description of a new species of bat of the genus Glossophaga. *Proc. U.S. Natl. Mus.* 43:779–81.

————. 1897. Erratum. *Science* 5:153.

Allen, J. A. 1890. Notes on a small collection of West Indian bats, with descriptions of an apparently new species. *Bull. Amer. Mus. Nat. Hist.* 3:169–73.

————. 1891. Notes on a collection of mammals from Costa Rica. *Bull. Amer. Mus. Nat. Hist.* 3:203–18.

————. 1892. On a small collection of mammals from the Galapagos Islands, collected by Dr. G. Baur. *Bull. Amer. Mus. Nat. Hist.* 4:47–50.

————. 1893a. Description of a new species of opossum from the Isthmus of Tehuantepec, Mexico. *Bull. Amer. Mus. Nat. Hist.* 5:235–36.

————. 1893b. Further notes on Costa Rica mammals, with description of a new species of Oryzomys. *Bull. Amer. Mus. Nat. Hist.* 5:237–40.

————. 1894. Descriptions of ten new North American mammals, and remarks on others. *Bull. Amer. Mus. Nat. Hist.* 6:317–32.

————. 1895a. On the names of mammals given by Kerr in his 'Animal Kingdom,' published in 1792. *Bull. Amer. Mus. Nat. Hist.* 7:179–92.

————. 1895b. Descriptions of new American mammals. *Bull. Amer. Mus. Nat. Hist.* 7:327–40.

————. 1897a. Additional notes on Costa Rican mammals, with descriptions of new species. *Bull. Amer. Mus. Nat. Hist.* 9:31–44.

————. 1897b. On a small collection of mammals from Peru with descriptions of a new species. *Bull. Amer. Mus. Nat. Hist.* 9:115–19.

————. 1897c. Description of a new vespertilionine bat from Yucatan. *Bull. Amer. Mus. Nat. Hist.* 9:231–32.

————. 1900a. List of bats collected by Mr. H. H. Smith in the Santa Marta region of Colombia, with descriptions of new species. *Bull. Amer. Mus. Nat. Hist.* 13:87–94.

————. 1900b. Note on the generic names Didelphis and Philander. *Bull. Amer. Mus. Nat. Hist.* 13:185–90.

————. 1900c. Descriptions of new American marsupials. *Bull. Amer. Mus. Nat. Hist.* 13:191–99.

————. 1900d. On mammals collected in southeastern Peru by Mr. H. H. Keays, with descriptions of new species. *Bull. Amer. Mus. Nat. Hist.* 13:219–28.

————. 1901a. On a further collection of mammals from southeastern Peru, collected by Mr. H. H. Keays, with descriptions of new species. *Bull. Amer. Mus. Nat. Hist.* 14:41–46.

————. 1901b. Note on the names of a few South American mammals. *Proc. Biol. Soc. Washington* 14:183–85.

————. 1901c. A preliminary study of the North American opossums of the genus Didelphis. *Bull. Amer. Mus. Nat. Hist.* 14:149–188, pls. 22–25.

————. 1901d. Descriptions of two new opossums of the genus Metachirus. *Bull. Amer. Mus. Nat. Hist.* 14:213–18.

———. 1901e. New South American Muridae and a new Metachirus. *Bull. Amer. Mus. Nat. Hist.* 14:405–12.

———. 1902. A preliminary study of the South American opossums of the genus Didelphis. *Bull. Amer. Mus. Nat. Hist.* 16:249–79.

———. 1904a. Mammals from southern Mexico and Central and South America. *Bull. Amer. Mus. Nat. Hist.* 20:29–80.

———. 1904b. New bats from tropical America, with note on species of Otopterus. *Bull. Amer. Mus. Nat. Hist.* 20:227–37.

———. 1904c. New mammals from Venezuela and Colombia. *Bull. Amer. Mus. Nat. Hist.* 20:327–35.

———. 1904d. List of mammals from Venezuela, collected by Mr. Samuel M. Klages. *Bull. Amer. Mus. Nat. Hist.* 20:337–45.

———. 1904e. The tamandua anteaters. *Bull. Amer. Mus. Nat. Hist.* 20:385–98.

———. 1904f. Mammals from the District of Santa Marta, Colombia, collected by Mr. Herbert H. Smith, with field notes by Mr. Smith. *Bull. Amer. Mus. Nat. Hist.* 20:407–68.

———. 1905. Mammalia of southern Patagonia. In *Reports of the Princeton University Expeditions to Patagonia, 1896–1899*, ed. W. B. Scott, Part 1, 1–210, 29 pls. Stuttgart: E. Schweizwerbart'sche Verlagshandlung (E. Nägele), Vol. 3, Zoölogy.

———. 1906a. The proper name of the Mexican tamandua. *Proc. Biol. Soc. Washington* 19:200.

———. 1906b. Mammals from the states of Sinaloa and Jalisco, Mexico, collected by J. H. Batty during 1904 and 1905. *Bull. Amer. Mus. Nat. Hist.* 22:191–62.

———. 1908a. Mammalogical notes. *Bull. Amer. Mus. Nat. Hist.* 24:579–89.

———. 1908b. Mammals from Nicaragua. *Bull. Amer. Mus. Nat. Hist.* 24:647–70.

———. 1910a. Additional mammals from Nicaragua. *Bull. Amer. Mus. Nat. Hist.* 28:87–115.

———. 1910b. Mammals from the Caura District of Venezuela, with description of a new species of *Chrotopterus*. *Bull. Amer. Mus. Nat. Hist.* 28:145–49.

———. 1911. Mammals from Venezuela collected by Mr. M. A. Carriker, Jr., 1901–1911. *Bull. Amer. Mus. Nat. Hist.* 30:239–73.

———. 1912. Mammals from western Colombia. *Bull. Amer. Mus. Nat. Hist.* 31:71–95.

———. 1913. New mammals from Colombia and Ecuador. *Bull. Amer. Mus. Nat. Hist.*, 32:469–84.

———. 1914. New South American bats and a new octodont. *Bull. Amer. Mus. Nat. Hist.* 33:381–89.

———. 1915. New South American mammals. *Bull. Amer. Mus. Nat. Hist.* 34:625–34.

———. 1916a. New South American mammals. *Bull. Amer. Mus. Nat. Hist.* 35:83–87.

———. 1916b. List of mammals collected for the American Museum in Ecuador by William B. Richardson, 1912–1913. *Bull. Amer. Mus. Nat. Hist.* 35:113–25.

———. 1916c. List of mammals collected in Colombia by the American Museum of Natural History Expeditions, 1910–1915. *Bull. Amer. Mus. Nat. Hist.* 35:191–238.

———. 1916d. New mammals collected on the Roosevelt Brazilian Expedition. *Bull. Amer. Mus. Nat. Hist.* 35:523–30.

———. 1916e. Mammals collected on the Roosevelt Brazilian Expedition with field notes by Leo E. Miller. *Bull. Amer. Mus. Nat. Hist.* 35:559–610.

Allen, J. A., and F. M. Chapman. 1893. On a collection of mammals from the island of Trinidad, with descriptions of new species. *Bull. Amer. Mus. Nat. Hist.* 5:203–34.

———. 1897. On a second collection of mammals from the island of Trinidad, with descriptions of new species, and a note on some mammals from the island of Dominica, W.I. *Bull. Amer. Mus. Nat. Hist.* 9:13–30.

Alonso-Mejía, A., and R. A. Medellín. 1991. Micronycteris megalotis. *Mammal. Species* 376:1–6.

———. 1992. Marmosa mexicana. *Mammal. Species* 421:1–4.

Alston, E. R. 1879. Mammalia. In *Biologia Centrali-Americana*, ed. F. D. Godman and O. Salvin, part 1, pp.1–40; part 2, pp. 41–56. London: Taylor and Francis, 1:xx + 220 pp., 1879–1882. [Printed in nine parts between September 1879 and October 1881; prepage materials dated December 1882.]

———. 1880. Mammalia. In *Biologia Centrali-Americana*, ed. F. D. Godman and O. Salvin, part 8, pp. 177–200. London: Taylor and Francis, 1:xx + 220 pp., 1879–1882. [See above.]

Alvarez, J., M. R. Willig, J. K. Jones, Jr., and W. D. Webster. 1991. Glossophaga soricina. *Mammal. Species* 379:1–7.

Alvarez, T. 1963. Restos de mamíferos encontrados en una cueva de Valle Nacional, Oaxaca, México. *Rev. Biol. Trop.* 11:57–61.

———. 1968. Notas sobre una colección de mamíferos de la region costera del Río Balsas entre Michoacán y Guerrero. *Rev. Soc. Mex. Hist. Nat.* 29:21–35.

———. 1972. Nuevo registro para el vampiro del Pleistoceno, *Desmodus stocki* de Tlapacoya, México. *An. Esc. Nac. Cien. Biol., México* 19:163–65.

———. 1982. Restos de mamíferos recientes y pleistocénicos procedentes de las grutas de Loltún, Yucatán, México. *Inst. Nac. Antropol. Hist., Depto. Prehist., Cuaderno de Trabajo* 26:7–36.

Alvarez, T., and S. T. Alvarez-Castañeda. 1990. Cuatro nuevos registros de murciélagos (Chiroptera) del estado de Chiapas, México. *An. Esc. Nac. Cien. Biol., México* 33:157–61.

Alvarez del Toro, M. 1952. *Los animales silvestres de Chiapas.* Tuxtla Gutierrez, Chiapas: Edic. Gobierno del Estado, 247 pp.

Ameghino, F. 1887. Enumeración sistemática de las éspecies de mamíferos fósiles coleccionados por Carlos Ameghino en los terrenos eocenos de la Patagonia austral depositados en el Museo de la Plata. *Bol. Mus. La Plata* 1:1–26.

———. 1889. Contribución al conocimiento de los mamíferos fósiles de la República Argentina. *Acta Acad. Nac. Cien. Córdoba, Buenos Aires* 6:xxxii + 1–1027 pp.; atlas, 98 pls., each with text.

———. 1891. Mamíferos y aves fósiles Argentinas. Especies nuevas, adiciones y correcciones. *Rev. Argentina Hist. Nat., Buenos Aires* 1:240–59.

———. 1894. Enuméracion synoptique des espèces de mammifères fossiles des formations Eocenès di Patagonie. *Bol. Acad. Nac. Cien. Córdoba* 13:259–452.

———. 1898. Sinopis geológico-paleontológica de la Argentina. In *Segundo censo nacional de la República Argentina.* Buenos Aires. 1:111–255.

———. 1899. Sinopsis geológico-paleontológica [de la Argentina]. In *Segundo censo nacional de la República Argentina.* La Plata: Suplemento (Adiciones y Correcciones), 13 pp.

Amorim, A. F. de, R. A. da Silva, and N. M. da Silva. 1970. Isolamento do vírus rábico de morcêgo insectívoro, *Histiotus velatus,* capturado no estado de Santa Catarina. *Pesq. Agropec. Bras., Seç. Vet.* 5:433–35.

Andersen, K. 1906a. On the bats of the genera *Micronycteris* and *Glyphonycteris. Ann. Mag. Nat. Hist.,* ser. 7, 18:50–65.

———. 1906b. Brief diagnosis of a new genus and ten new forms of stenodermatous bats. *Ann. Mag. Nat. Hist.,* ser. 7, 18:419–23.

———. 1908a. On four little-known names of chiropteran genera. *Ann. Mag. Nat. Hist.,* ser. 8, 1:431–35.

———. 1908b. A monograph of the chiropteran genera *Uroderma, Enchisthenes* and *Artibeus. Proc. Zool. Soc. London* 1908:204–319.

Anderson, J. W., and W. A. Wimsatt. 1963. Placentation and fetal membranes of the Central American noctilionid bat, *Noctilio labialis minor. Amer. J. Anat.* 112:181–202.

Anderson, R. P., and C. O. Handley, Jr. 2001. A new species of three-toed sloth (Mammalia: Xenarthra) from Panamá, with a review of the genus *Bradypus. Proc. Biol. Soc. Washington* 114:1–33.

Anderson, S. 1960. Neotropical bats from western Mexico.

Univ. Kansas Publ., Mus. Nat. Hist. 14:1–8.

———. 1982. Monodelphis kunsi. *Mammal. Species* 190:1–3.

———. 1991. A brief history of Bolivian chiroptology and new records of bats. *Bull. Amer. Mus. Nat. Hist.* 206:138–44.

———. 1993. *Los mamíferos Bolivianos: Notas de distribución y claves de identificación.* La Paz: Publicación Especial del Instituto de Ecologí (Colección. Boliviana de Fauna), 159 pp.

———. 1997. Mammals of Bolivia, taxonomy and distribution. *Bull. Amer. Mus. Nat. Hist.* 231:1–652.

Anderson, S., and W. D. Webster. 1983. Notes on Bolivian mammals. 1. Additional records of bats. *Amer. Mus. Novit.,* no. 2766:1–3.

Anderson, S., K. F. Koopman, and G. K. Creighton. 1982. Bats of Bolivia: An annotated checklist. *Amer. Mus. Novit.,* no. 2750:1–24.

Anderson, S., B. R. Riddle, T. L. Yates, and J. A. Cook. 1993. *Los mamíferos del Parque Nacional Amboró y la Región de Santa Cruz de la Sierra, Bolivia.* Special Publication 2. Albuquerque, New Mexico: Museum of Southwestern Biology, 58 pp.

Anduze, P. J. 1956. Lista de los mamíferos señalados hasta el presente en Venezuela. *Mem. Soc. Cienc. Nat. La Salle* 16:5–18.

Angulo, S. R., and M. M. Díaz. 2004. Nuevos registros de *Sphaeronycteris toxophyllum* para la cuenca Amazónica de Perú. *Mastozool. Neotrop.* 11:233–36.

Aniskin, M., A. A. Varshavskii, S. I. Isaef, and V. M. Malyugin. 1991. Comparative analysis of differentially G- and C-stained chromosomes of five species of family Didelphidae (Marsupialia). Translated from *Genetika* 27:504–14.

Anonymous. 1994. Distribution and conservation status of the three-banded Armadillo. *Edentata* 1:17.

Anthony, H. E. 1919. Mammals collected in eastern Cuba in 1917. With descriptions of two new species. *Bull. Amer. Mus. Nat. Hist.* 41:625–43.

———. 1920. New rodents and new bats from Neotropical regions. *J. Mammal.* 1:81–86.

———. 1921a. Preliminary report on Ecuadorean mammals. No. 1. *Amer. Mus. Novit.,* no. 20:1–6.

———. 1921b. Mammals collected by William Beebe at the British Guiana Tropical Research Station. *Zoologica* 3:265–86.

———. 1922. Preliminary report on Ecuadorean mammals. No. 2. *Amer. Mus. Novit.,* no. 32:1–6.

———. 1923a. Mammals from Mexico and South America. *Amer. Mus. Novit.,* no. 54:1–10.

———. 1923b. Preliminary report on Ecuadorean mammals. No. 3. *Amer. Mus. Novit.,* no. 55:1–14.

———. 1924a. Preliminary report on Ecuadorean mammals. No. 4. *Amer. Mus. Novit.*, no. 114:1–6.

———. 1924b. Preliminary report on Ecuadorean mammals. No. 5. *Amer. Mus. Novit.*, no. 120:1–3.

———. 1924c. Preliminary report on Ecuadorean mammals. No. 6. *Amer. Mus. Novit.*, no. 139:1–9.

———. 1926. Preliminary report on Ecuadorean mammals. No. 7. *Amer. Mus. Novit.*, no. 240:1–6.

Anthony, R. 1906. Les coupures génériques de la famille des *Bradypodidae* (le genre *Hemibradypus* nov. g.). *Comptes Rendus, Acad. Sci., Paris* 142:292–94.

———. 1907. Les affinités des *Bradypodidae* (Paresseux) et, en particulier, de l'*Hemibradypus Mareyi* Anth. avec les Hapalopsidae du Santacruzien de l'Amérique de Sud. *Comptes Rendus, Acad. Sci., Paris* 144:219–21.

Aplin, K., and M. Archer. 1987. Recent advances in marsupial systematics, with a new, higher level classification of the Marsupialia. In *Possums and opossums: Studies in evolution*, ed. M. Archer, xv–lxxii. Sydney: Surrey Beatty and Sons Pty Limited and Royal Zool. Soc. New South Wales, 1:lxxii + 1–400, 4 pls.

Arata, A. A., and J. B. Vaughn. 1970. Analyses of the relative abundance and reproductive activity of bats in southwestern Colombia. *Caldasia* 10:517–28.

Arata, A. A., J. B. Vaughn, and M. E. Thomas. 1967. Food habits of certain Colombian bats. *J. Mammal.* 48:653–55.

Arata, A. A., J. B. Vaughn, K. W. Newell, R. A. J. Barth, and M. Gracian. 1968. *Salmonella* and *Shigella* infections in bats in selected areas of Colombia. *Amer. J. Trop. Med. Hyg.* 17:92–95.

Araujo, A., and J. Molinari. 2000. Presas de *Tyto alba* (Aves, Strigiformes) en una selva nublada Venezolana. In *Ecología Latinoamericana*, ed. J. E. Péfaur, 217–22. Mérida, Venezuela: Universidad de los Andes, 492 pp.

Archer, M. 1978. The nature of the molar-premolar boundary in marsupials and a reinterpretation of the homology of marsupial cheekteeth. *Mem. Queensland Mus.* 18:157–64.

Arends, A., F. J. Bonaccorso, and M. Genoud. 1995. Basal rates of metabolism of nectarivorous bats (Phyllostomidae) from a semiarid thorn forest in Venezuela. *J. Mammal.* 76:947–56.

Arias, J. R., and R. D. Naiff. 1981. The principal reservoir host of cutaneous leishmaniasis in the urban areas of Manaus, central Amazon of Brazil. *Mem. Inst. Oswaldo Cruz, Rio de Janeiro* 76:279–86.

Arita, H. T., and S. R. Humphrey. 1989. Revisión taxonómica de los murciélagos magueyeros del género *Leptonycteris* (Chiroptera: Phyllostomidae). *Acta Zool. Mex.*, 29:1–60. [Dated October 1988; published in 1989.]

Arnason, U., J. A. Adegoke, K. Bodin, E. W. Born, Y. B. Esa, A. Gullberg, M. Nilsson, R. V. Short, X. Xu, and A. Janke. 2002. Mammalian mitogenomic relationships and the root of the eutherian tree. *Proc. U.S. Natl. Acad. Sci.* 99:8151–6.

Arnold, M. L., R. J. Baker, and R. L. Honeycutt. 1983. Genic differentiation and phylogenetic relationships within two New World bat genera. *Biochem. Syst. Ecol.* 11:295–303.

Arroyo-Cabrales, J., and J. K. Jones, Jr. 1988. Balantiopteryx plicata. *Mammal. Species* 301:1–4.

———. 1988. Balantiopteryx io and Balantiopteryx infusca. *Mammal. Species* 313:1–3.

Arroyo-Cabrales, J., and A. L. Gardner. 2003. The type specimen of *Anoura geoffroyi lasiopyga* (Chiroptera: Phyllostomidae). *Proc. Biol. Soc. Washington* 116:737–41.

Arroyo-Cabrales, J., and R. D. Owen. 1996. Intraspecific variation and phenetic affinities of *Dermanura hartii*, with reapplication of the specific name *Enchisthenes hartii*. In *Contributions in Mammalogy: A Memorial Volume Honoring Dr. J. Knox Jones, Jr.*, ed. H. H. Genoways and R. J. Baker, 67–81. Lubbock: Museum of Texas Tech University, il + 315 pp.

———. 1997. Enchisthenes hartii. *Mammal. Species* 546:1–4.

Arroyo-Cabrales, J., R. Gregorin, D. A. Schlitter, and A. Walker. 2002. The oldest African molossid bat cranium (Chiroptera: Molossidae). *J. Vert. Paleo.* 22:380–87.

Ascorra, C. R., and D. E. Wilson. 1992. Bat frugivory and seed dispersal in the Amazon, Loreto, Peru. *Publ. Mus. Hist. Nat. UNMSM*, sér. A, 43:1–6.

Ascorra, C. F., D. L. Gorchov, and F. Cornejo. 1989. Observaciones en aves y murciélagos relacionadas con la dispersion de semillas en el valle Palcazu, Selva Central del Perú. *Bol. Lima* 62:91–95.

———. 1994. The bats from Jenaro Herrera, Loreto, Peru. *Mammalia* 57:533–52. [Dated 1993; number 4 marked "Achevé d'imprimer le 20 janvier 1994."]

Ascorra, C. F., S. Solari T., and D. Wilson. 1996. Diversidad y ecología de los quirópteros en Pakitza. In *Manu*, ed. D. E. Wilson and A. Sandoval, 593–612. Lima, Peru: Editorial Horizonte, 679 pp.

Ascorra, C. F., D. E. Wilson, and A. L. Gardner. 1991. Geographic distribution of *Micronycteris schmidtorum* Sanborn (Chiroptera: Phyllostomidae). *Proc. Biol. Soc. Washington* 104:351–55.

Ascorra, C. F., D. E. Wilson, and C. O. Handley, Jr. 1991. Geographic distribution of *Molossops neglectus* Williams and Genoways (Chiroptera: Molossidae). *J. Mammal.* 72:828–30.

Ascorra, C. F., D. E. Wilson, and M. Romo. 1991. Lista anotada de los quirópteros del Parque Nacional Manu, Perú. *Publ. Mus. Hist. Nat. UNMSM*, sér. A, 42:1–14.

Asdell, S. A. 1964. *Patterns of mammalian reproduction.* 2nd ed. Ithaca, New York: Cornell University Press, 670 pp.

Ashe, J. S., and R. M. Timm. 1988. *Chilamblyopinus piceus*, a new genus and species of amblyopinine (Coleoptera: Staphylinidae) from southern Chile, with a discussion of amblyopinine generic relationships. *J. Kansas Entomol. Soc.* 61:46–57.

Atramentowicz, M. 1986. Dynamique de population chez trois marsupiaux didelphidés de Guyane. *Biotropica* 18:136–49.

———. 1988. La frugivorie opportuniste de trois marsupiaux didelphidés de Guyane. *Rev. Ecol. (Terre et Vie)* 42:46–57.

August, P. V., and R. J. Baker. 1982. Observations on the reproductive ecology of some Neotropical bats. *Mammalia* 46:177–81.

Augustiny, G. 1943. Die Schwimmanpassung von Chironectes. *Zeitschr. Morph. Oekologie Tiere* 39:276–319.

Autino, A. G., G. L. Claps, and M. P. Bertolini. 1998. Primeros registros de insectos ectoparásitos (Diptera, Streblidae) de murciélagos de Parque Nacional Iguazú, Misiones, Argentina. *Rev. Bras. Ent.* 42:59–63.

Autino, A. G., G. L. Claps, and E. M. González. 2004. Nuevos registros de insectos (Diptera y Siphonaptera) ectoparásitos de murciélagos (Vespertilionidae) de norte de Uruguay. *Mastozool. Neotrop.*11:81–83.

Avila-Flores, R., J. J,. Flores-Martínez, and J. Ortega. 2002. Nyctinomops laticaudatus. *Mammal. Species* 697:1–6.

Ávila-Pires, F. D. de. 1958. Mamíferos colecionados nos arredores de Belém do Pará. *Bol. Mus. Paraense Emilio Goeldi*, Zool., 19:1–9.

———. 1964. Mamíferos colecionados na região do Rio Negro (Amazonas, Brasil). *Bol. Mus. Paraense Emílio Goeldi*, nov. sér., Zool., 42:1–23.

———. 1965. The type specimens of Brazilian mammals collected by Prince Maximilian zu Wied. *Amer. Mus. Novit.*, no. 2209:1–21.

———. 1968. Tipos de mamíferos recentes no Museu Nacional, Rio de Janeiro. *Arq. Mus. Nac.* 53:161–92.

———. 1994. Mamíferos descritos do estado do Rio Grande do Sul, Brasil. *Rev. Bras. Biol.* 54:367–84.

Ávila-Pires, F. D. de., and E. Gouvêa. 1977. Mamíferos do Parque Nacional do Itatiaia. *Bol. Mus. Nac., Rio de Janeiro*, n. sér., zool., 251:1–29.

Avilla, L, dos S, A. M. da S. Rozenzstranch, and É. A. L. Abrantes. 2001. First record of the South American flat-headed bat, *Neoplatymops mattogrossensis* (Vieira, 1942) in southeastern Brazil (Chiroptera, Molossidae). *Bol. Mus. Nac., Rio de Janeiro*, n. sér., zool., 463:1–6.

Ayala, S. C., A. d'Alessandro, R. Mackenzie, and D. Angel. 1973. Hemoparasite infections in 830 wild animals from the eastern llanos of Colombia. *J. Parasito.* 59:52–59.

Ayarzaguena, J. 1984. Sobre la distribución de *Dasypus sabanicola* y *D. novemcinctus* de Venezuela. *Mem. Soc. Cien. Nat. La Salle* 42:145–59. [Dated 1982; published in 1984.]

Azara, F. d.' 1801a. *Essais sur l'histoire naturelle des quadrupèdes de la province du Paraguay.* Traduits sur le manuscrit inédit de l'auteur, Pra. M. L. E. Moreau-Saint-Méry. Paris: Charles Pougens, 1:lxxx + 366 pp.

———. 1801b. *Essais sur l'histoire naturelle des quadrupèdes de la province du Paraguay.* Traduits sur le manuscrit inédit de l'auteur, Pra. M. L. E. Moreau-Saint-Méry. Paris: Charles Pougens, 2:1–499.

Azara, F. de. 1802a. *Apuntamientos para la historia natural de los quadrúpedos del Paraguay y Río de la Plata.* Madrid: La Imprinta de la Viuda de Ibarra, 1:xix + 1–318.

———. 1802b. *Apuntamientos para la historia natural de los quadrúpedos del Paraguay y Río de la Plata.* Madrid: La Imprinta de la Viuda de Ibarra, 2:2 (unnumbered), x + 1–328.

Azevedo, T. R. de, D. El Achkar, M. de Martins, and A. Ximenez. 1982. Lista sistemática dos mamíferos de Santa Catarina conservados nos principais museus do Estado. *Rev. Nordest. Biol.* 5:93–104.

Azócar, A., and M. Monasterio. 1980. Caracterización ecológicas del clima en el Páramo de Mucubají. In *Estudios ecológicos en los páramos andinos*, ed. M. Monasterio, 207–23. Mérida, Venezuela: Ed. Univ. Los Andes, 312 pp.

Bachman, J. 1837. Some remarks on the genus Sorex, with a monograph of the North American species. *J. Acad. Nat. Sci. Philadelphia* 7:362–403, 2 pls.

Bailey, V. 1905. Biological survey of Texas. *N. Amer. Fauna* 25:1–222, 16 pls.

Baird, S. F. 1857. Mammals. In *Reports of explorations and surveys to ascertain the most practicable and economical route for a railroad from the Mississippi River to the Pacific Ocean. Part I. General report upon the zoology of the several Pacific railroad routes.* Washington, D.C.: Beverly Tucker, 8(1):xix–xlviii, 1–757, 60 plates.

Baker, R. H. 1974. Records of mammals from Ecuador. *Publ. Mus., Michigan St. Univ.* 5:129–46.

Baker, R. J. 1967. Karyotypes of bats of the family Phyllostomidae and their taxonomic implications. *Southwest. Nat.* 12:407–28.

———. 1970. Karyotypic trends in bats. In *Biology of bats*, ed. W. A. Wimsatt, 65–96. New York: Academic Press, 1:xii + 406 pp.

————. 1973. Comparative cytogenetics of the New World leaf-nosed bats (Phyllostomatidae). *Period. Biologorum* 75:37–45.

————. 1979. Karyology. In *Biology of bats of the New World family Phyllostomatidae*. Part III, ed. R. J. Baker, J. K. Jones, Jr., and D. C. Carter, 107–55. Special Publications of the Museum 16. Lubbock: Texas Tech University Press, 441 pp.

————. 1984. A sympatric cryptic species of mammal: a new species of *Rhogeessa* (Chiroptera: Vespertilionidae). *Syst. Zool.* 33:178–83.

Baker, R. J., and J. W. Bickham. 1980. Karyotypic evolution in bats: Evidence of extensive and conservative chromosomal evolution in closely related taxa. *Syst. Zool.* 29:239–53.

Baker, R. J., and W. J. Bleier. 1971. Karyotypes of bats of the subfamily Carolliinae (Mammalia; Phyllostomatidae) and their evolutionary implications. *Experientia* 27:220.

Baker, R. J., and C. L. Clark. 1987. Uroderma bilobatum. *Mammal. Species* 279:1–4.

Baker, R. J., and T. C. Hsu. 1970. Further studies on the sex-chromosome systems of the American leaf-nosed bats (Chiroptera, Phyllostomatidae). *Cytogenetics* 9:131–38.

Baker, R. J., and J. K. Jones, Jr. 1975. Additional records of bats from Nicaragua, with a revised checklist of Chiroptera. *Occas. Papers Mus., Texas Tech Univ.*, no. 32:1–13.

Baker, R. J., and R. G. Jordan. 1970. Chromosomal studies of some Neotropical bats of the families Emballonuridae, Noctilionidae, Natalidae, and Vespertilionidae. *Caryologia* 23:595–604.

Baker, R. J., and G. Lopez. 1968. Notes on some bats from Tamaulipas. *Southwest. Nat.* 13:361–62.

————. 1970. Chromosomal variation in bats of the genus *Uroderma* (Phyllostomatidae). *J. Mammal.* 51:786–89.

Baker, R. J., and V. R. McDaniel. 1972. A new subspecies of Uroderma bilobatum (Chiroptera: Phyllostomatidae) from Middle America. *Occas. Papers Mus. Texas Tech Univ.*, no. 7:1–4.

Baker, R. J., and J. L. Patton. 1967. Karyotypes and karyotypic variation of North American vespertilionid bats. *J. Mammal.* 48:270–86.

Baker, R. J., W. R. Atchley, and V. R. McDaniel. 1972. Karyology and morphometrics of Peters' tent-making bat, *Uroderma bilobatum* Peters (Chiroptera, Phyllostomatidae). *Syst. Zool.* 21:414–29.

Baker, R. J., R. A. Bass, and M. A. Johnson. 1979. Evolutionary implications of chromosomal homology in four genera of stenodermine bats (Phyllostomatidae: Chiroptera). *Evolution* 33:220–26.

Baker, R. J., J. W. Bickham, and M. L. Arnold. 1985. Chromosomal evolution in *Rhogeessa*: possible speciation by centric fusions. *Evolution* 39:233–43.

Baker, R. J., C. G. Dunn, and K. Nelson. 1988. Allozymic study of the relationships of Phylloderma and four species of Phyllostomus. *Occas. Papers Mus. Texas Tech Univ.*, no. 125:1–14.

Baker, R. J., A. L. Gardner, and J. L. Patton. 1972. Chromosomal polymorphism of the phyllosomatid bat, *Mimon crenulatum. Experientia* 28:969–70.

Baker, R. J., H. H. Genoways, and A. Cadena. 1972. The phyllostomatid bat, *Vampyressa brocki*, in Colombia. *Bull. S. California Acad. Sci.* 71:54.

Baker, R. J., H. H. Genoways, and P. A. Seyfarth. 1981. Results of the Alcoa Foundation-Suriname Expeditions. VI. Additional chromosomal data for bats (Mammalia: Chiroptera) from Suriname. *Ann. Carnegie Mus.* 50:333–44.

Baker, R. J., R. L. Honeycutt, and R. A. Bass. 1988. Genetics. In *Natural history of vampire bats*, ed. A. M. Greenhall and U. Schmidt, 31–40. Boca Raton, Florida: CRC Press, frontispiece, 14 unnumbered + 246 pp.

Baker, R. J., C. S. Hood, and R. L. Honeycutt. 1989. Phylogenetic relationships and classification of the higher categories of the New World bat family Phyllostomidae. *Syst. Zool.* 38:228–38.

Baker, R. J., J. K. Jones, Jr., and D. C. Carter, eds. 1976. *Biology of bats of the New World family Phyllostomatidae.* Part I. *Special Publications of the Museum 10.* Lubbock: Texas Tech University Press, 218 pp.

Baker, R. J., J. K. Jones, Jr., and D. C. Carter, eds. 1977. *Biology of bats of the New World family Phyllostomatidae.* Part II. *Special Publications of the Museum 13.* Lubbock: Texas Tech University Press, 364 pp.

Baker, R. J., J. K. Jones, Jr., and D. C. Carter, eds. 1979. *Biology of bats of the New World family Phyllostomatidae.* Part III. *Special Publications of the Museum 16.* Lubbock: Texas Tech University Press, 441 pp.

Baker, R. J., T. Mollhagen, and G. Lopez. 1971. Notes on *Lasiurus ega. J. Mammal.* 52:849–52.

Baker, R. J., H. H. Genoways, W. J. Bleier, and J. W. Warner. 1973. Cytotypes and morphometrics of two phyllostomatid bats, Micronycteris hirsuta and Vampyressa pusilla. *Occas. Papers Mus., Texas Tech Univ.*, no. 17:1–10.

Baker, R. J., M. W. Haiduk, L. W. Robbins, A. Cadena, and B. F. Koop. 1982. Chromosomal studies of South American bats and their systematic implications. In *Mammalian biology in South America*, ed. M. A. Mares and H. H. Genoways, 303–327. The Pymatuning Symposia in Ecology 6. Special Publications Series. Pittsburgh:

Pymatuning Laboratory of Ecology, University of Pittsburgh, xii + 539 pp.

Baker, R. J., J. C. Patton, H. H. Genoways, J. W. Bickham. 1988. Genic studies of Lasiurus (Chiroptera: Vespertilionidae). *Occas. Papers Mus., Texas Tech Univ.*, no. 117:1–15.

Baker, R. J., C. A. Porter, J. C. Patton, and R. A. Van Den Bussche. 2000. Systematics of bats of the family Phyllostomidae based on RAG2 DNA sequences. *Occas. Papers Mus., Texas Tech Univ.*, no. 202:1–16.

Baker, R. J., S. R. Hoofer, C. A. Porter, and R. A. Van Den Bussche. 2003. Diversification among New World leaf-nosed bats: An evolutionary hypothesis and classification inferred from digenomic congruence of DNA sequence. *Occas. Papers Mus., Texas Tech Univ.*, no. 230:1–32.

Baker, R. J., R. M. Fonseca, D. A. Parish, C. J. Phillips, and F. G. Hoffmann. 2004. New bat of the genus *Lophostoma* (Phyllostomidae: Phyllostominae) from northwestern Ecuador. *Occas. Papers Mus., Texas Tech Univ.*, no. 232:i + 1–16.

Bangs, O. 1898a. A new murine opossum from Margarita Island. *Proc. Biol. Soc. Washington* 12:95–96.

———. 1898b. Descriptions of some new mammals from the Sierra Nevada de Sta. Marta, Colombia. *Proc. Biol. Soc. Washington* 12:161–65.

———. 1899. A new bat from Colombia. *Proc. New England Zoöl. Club* 1:73–74.

———. 1900. List of the mammals collected in the Santa Marta region of Colombia by W. W. Brown, Jr. *Proc. New England Zoöl. Club* 1:87–102.

———. 1901. The mammals collected in San Miguel Island, Panama, by W. W. Brown, Jr. *Amer. Nat.* 35:631–44.

———. 1902. Chiriqui Mammalia. *Bull. Mus. Comp. Zoöl.* 39:17–51.

———. 1906. Mammalia. Vertebrata from the savanna of Panama. *Bull. Mus. Comp. Zoöl.* 46:213–14.

Baptista, M., and M. A. R. Mello. 2001. Preliminary inventory of the bat species of the Poço das Antas Biological Reserve, RJ. *Chiropt. Neotrop.* 7:133–35.

Barbour, R. W., and W. H. Davis. 1969. *Bats of America.* Lexington: Univ. Press of Kentucky, 286 pp., 243 pls.

Barbour, T. 1932. A peculiar roosting habit of bats. *Quart. Rev. Biol.* 7:307–312.

———. 1936. *Eumops* in Florida. *J. Mammal.* 17:414.

Barclay, R. M. R., M. B. Fenton, M. D. Tuttle, and M. J. Ryan. 1981. Echolocation calls produced by *Trachops cirrhosus* (Chiroptera: Phyllostomatidae) while hunting for frogs. *Can. J. Zool.* 59:750–53.

Barkley, L. J. 1984. Evolutionary relationships and natural history of *Tomopeas ravus* (Mammalia: Chiroptera). M.S. thesis, Louisiana State Univ., Baton Rouge, 100 pp.

Barkley, L. J., and J. O. Whitaker, Jr. 1984. Confirmation of *Caenolestes* in Peru with information on diet. *J. Mammal.* 65:328–30.

Barlow, J. C. 1965. Land mammals from Uruguay: Ecology and zoogeography. Ph.D. thesis, Univ. of Kansas, Lawrence, 346 pp.

———. 1984. Xenarthrans and pholidotes. In *Orders and families of Recent mammals of the world*, ed. S. Anderson and J. K. Jones, Jr., 219–266. New York: John Wiley and Sons, xiv + 686 pp.

Barnett, A. A. 1991. Records of the grey-bellied shrew opossum, *Caenolestes caniventer* and Tate's shrew opossum, *Caenolestes tatei* (Caenolestidae, Marsupialia), from Ecuadorian montane forests. *Mammalia* 55:443–45.

———. 1993. Notes on the ecology of *Cryptotis montivaga* Anthony, 1921 (Insectivora, Soricidae), a high-altitude shrew from Ecuador. *Mammalia* 56:587–92. [Dated 1992; issue number 4 marked "Achevé d'imprimer 18 janvier 1993."]

———. 1999. Small mammals of the Cajas Plateau, southern Ecuador: Ecology and natural history. *Bull. Florida Mus. Nat. Hist.* 42:161–217.

Barquez, R. M. 1983. Breves comentarios sobre *Molossus molossus* (Chiroptera-Molossidae) de Bolivia. *Hist. Natural* 3:169–74.

———. 1984a. Morfométria y comentarios sobre la colección de murciélagos de la Fundación Miguel Lillo. Familias Emballonuridae, Noctilionidae, Mormoopidae, Phyllostomatidae, Furipteridae, Thyropteridae). (Mammalia, Chiroptera). *Hist. Natural* 3:213–23.

———. 1984b. Significativa extensión del rango de distribución de *Diaemus youngii* (Jentink, 1893) (Mammalia, Chiroptera, Phyllostomidae). *Hist. Natural* 4:67–68.

———. 1985. *Glossophaga soricina* (Pallas, 1766) en el noroeste Argentina (Chiroptera: Phyllostomidae). *Hist. Natural* 5:93–96.

———. 1987. Los Murciélagos de Argentina. Tesis doctoral. Facultad de Ciencias Naturales e Instituto Miguel Lillo, Universidad Nacional de Tucumán, Argentina, 525 pp.

———. 1988. Notes on identity, distribution, and ecology of some Argentine bats. *J. Mammal.* 69:873–76.

Barquez, R. M., and M. M. Díaz, 2001. Bats of the Argentine Yungas: A systematic and distributional analysis. *Acta Zool. Mex.*, 82:29–81.

Barquez, R. M., and S. C. Lougheed. 1990. New distributional records of some Argentine bat species. *J. Mammal.* 71:261–63.

Barquez, R. M., and R. A. Ojeda. 1975. *Tadarida laticaudata*, un nuevo molosido para la fauna Argentina (Chiroptera, Molossidae). *Neotrópica* 21:137–38.

————. 1979. Nueva subespecie de *Phylloderma* (Chiroptera, Phyllostomidae). *Neotrópica* 25:83–89.

————. 1992. The bats (Mammalia: Chiroptera) of the Argentine Chaco. *Ann. Carnegie Mus.* 61:239–261.

Barquez, R. M., and C. C. Olrog. 1980. Tres nuevas especies de *Vampyrops* para Bolivia (Chiroptera: Phyllostomatidae). *Neotrópica* 26:53–56.

————. 1985. *Anoura caudifer* nueva especie de murciélago para la Argentina. *Hist. Natural* 5:149–52.

Barquez, R. M., N. P. Giannini, and M. A. Mares. 1993. *Guide to the bats of Argentina*. Norman: Oklahoma Museum of Natural History, viii + 119 pp. 2 maps.

Barquez, R. M., M. A. Mares, and J. K. Braun. 1999. *The bats of Argentina. Special Publications of the Museum 42*. Lubbock: Texas Tech University Press, 10 (unnumbered) + 275 pp.

Barquez, R. M., M. A. Mares, and R. A. Ojeda. 1991. *Mammals of Tucuman*. Norman: Oklahoma Museum of Natural History, 282 pp., 2 maps.

Barquez, R. M., J. K. Braun, M. A. Mares, J. P. Jayat, and D. Flores. 2000. First record for Argentina for a bat in the genus *Micronycteris. Mammalia* 63:368–72. [Dated 1999; number 3 marked "Achevé d'imprimer le 3 mars 2000."]

Barreto, M., P. Barreto, and A. D'Alessandro. 1985. Colombian armadillos: Stomach contents and infection with *Trypanosoma cruzi. J. Mammal.* 66:188–93.

Barrett, T. V. 1979. The ecology of triatomine bugs (Hemiptera: Reduviidae) and their hosts in relation to the transmission of *Trypanosoma cruzi* Chagas, 1909, in the state of Bahia, Brazil. Ph.D. thesis, University of London, United Kingdom, 333 pp., 25 pls.

Barriga-Bonilla, E. 1965. Estudios mastozoológicos Colombianos, I *Chiroptera. Caldasia* 9:241–68.

Barroso, C. M. L., and H. Seuánez. 1991. Chromosome studies on *Dasypus, Euphractus* and *Cabassous* genera (Edentata: Dasypodidae). *Cytobios* 68:179–96.

Bateman, G. C., and T. A. Vaughn. 1974. Nightly activities of mormoopid bats. *J. Mammal.* 55:45–65.

Bates, M. 1944. Experiments with the virus of yellow fever in marsupials with special reference to brown and grey masked opossums. *Amer. J. Trop. Med. Hyg.* 24:91–103.

Baud, F. J. 1979. *Myotis aelleni*, nov. spec., chauve-souris nouvelle d'Argentine (Chiroptera: Vespertilionidae). *Revue Suisse Zool.* 86:267–78.

————. 1981. Expédition de Muséum de Genève au Paraguay: Chiroptères. *Revue Suisse Zool.* 88:567–81.

————. 1982. Présence de *Chrotopterus auritus* (Chiroptera, Phyllostomatinae) au Pérou. *Mammalia* 46:264–65.

————. 1982. Présence de *Rhinophylla alethina* (Mammalia, Chiroptera) en Equateur et répartition actuelle du genre en Amerique du Sud. *Revue Suisse Zool.* 89:815–21.

————. 1986. Chiroptères récoltés par l'expédition de l'IRSNB au sud-est du Pérou. *Bull. Inst. Roy. Sci. Nat Belg., Biol.* 56:45–49.

————. 1989. Présence de *Macrophyllum macrophyllum* Schinz (Chiroptera, Phyllostominae) au Paraguay. *Mammalia* 53:308–9.

Baud, F. J., and H. Menu. 1993. Paraguayan bats of the genus *Myotis*, with a redefinition of *M. simus* (Thomas, 1901). *Rev. Suisse Zool.* 100:595–607.

Baumgarten, J. E., and E. M. Vieira. 1994. Reproductive seasonality and development of *Anoura geoffroyi* (Chiroptera: Phyllostomidae) in central Brazil. *Mammalia* 58:415–22.

Beath, M. M., K. Benirschke, and L. E. Brownhill. 1962. The chromosomes of the nine-banded armadillo, *Dasypus novemcinctus. Chromosoma* 13:27.

Beaux, O. de. 1908. Über die schwarze Varietät der Tamandua longicaudata Gray. *Zool. Anz., Leipzig* 33:417–18.

Bechara, G. H., M. P. J. Szabo, W. V. Almeida Filho, J. N. Bechara, R. J. G. Pereira, J. E. Garcia, and M. C. Pereira. 2002. Ticks associated with armadillo (*Euphractus sexcinctus*) and anteater (*Myrmecophaga tridactyla*) of Emas National Park, state of Goiás. Brazil. *Ann. New York Acad. Sci.* 969:290–93.

Beckman J. 1772. *Caroli a Linné Systema naturae ex editione duodecima in epitomen redactum et praelectionibus academicis accommotatum a Iohanne Beckmanno. . . .* Gottingae: svmtv vidvae Vandenhoeck, 1:5 + 240 pp. + 10 unnumbered (Index).

Bechstein, J. M. 1800. *Thomas Pennant's Allgemeine uebersicht der vierfüssigen thiere. Aus dem Englischen übersetzt und mit anmerkungen und zusätzen versehen, von Johann Matthäus Bechstein*. Weimar: Im verlage des Industrie-Comptoir's, 2:x + 2 unnumbered, 323–666, pls. 35–54.

————. 1801. *Gemeinnützige naturgeschichte Deutschlands nach allen drei reichen*. Leipzig: Bey Siegfried Lebrecht Crusius, 1:frontispiece, xxx + 1370 pp., 24 plates.

Beebe, W. 1919. The higher vertebrates of British Guiana. *Zoologica* 2:205–27.

————. 1926. The three-toed sloth *Bradypus cuculliger cuculliger* Wagler. *Zoologica* 7:1–67.

Belkis A., R. 2000. Notas sobre los mamíferos de la planicie Amacuro (estado Delta Amacuro). *Mem. Soc. Cienc. Nat. La Salle* 58:43–59. [Dated "Enero/junio 1998;" published July 2000.]

Benedict, F. A. 1957. Hair structure as a generic character in bats. *Univ. California Publ. Zool.* 59:285–48.

Benirschke, K. 2006. Xenarthra. In *Atlas of mammalian chromosomes*, ed. S. R. O'Brian, J. C. Menniger, and W. G. Nash, 81–94. Hoboken, New Jersey: John Wiley & Sons, Inc., xlii + 1–714.

Benirschke, K., and D. H. Wurster. 1969. The chromosomes of the giant armadillo, Priodontes giganteus Geoffroy. *Acta Zool. Path., Antwerp* 49:125–30.

Benirschke, K., R. J. Low, and V. H. Ferm. 1969. Cytogenetic studies of some armadillos. In *Comparative mammalian cytogenetics*, ed. K. Benirschke, 330–45. New York: Springer-Verlag, xxi + 473 pp.

Benson, S. B. 1944. The type locality of *Tadarida mexicana* Saussure. *J. Washington Acad. Sci.* 34:159.

Bergallo, H. G. 1994. Ecology of a small mammal community in an Atlantic Forest area in southeastern Brazil. *Stud. Neotrop. Fauna Envir.* 29:197–217.

Bergallo, H. G., and R. Cerqueira. 1994. Reproduction and growth of the opossum *Monodelphis domestica* (Mammalia: Didelphidae) in northeastern Brazil. *J. Zool. London* 232:551–63.

Berkovitz, B. K. B. 1967. The dentition of a 25-day pouch-young specimen of *Didelphis virginiana* (Didelphidae: Marsupialia). *Arch. Oral Biol.* 12:1211–12.

———. 1978. Tooth ontogeny in *Didelphis virginiana* (*Marsupialia: Didelphidae*). *Australian J. Zool.* 26:61–68.

Bernard, E. 2001a. Species list of bats (Mammalia, Chiroptera) of Santarém area, Pará state, Brazil. *Rev. Bras. Zool.* 18:455–63.

———. 2001b. First capture of *Micronycteris homezi* Pirlot (Chiroptera, Phyllostomidae) in Brazil. *Rev. Bras. Zool.* 18:645–47.

———. 2003. Cormura brevirostris. *Mammal. Species* 737:1–3.

Bernard, E., and M. B. Fenton. 2002. Species diversity of bats (Mammalia: Chiroptera) in forest fragments, primary forests, and savannas in central Amazonia, Brazil. *Can. J. Zool.* 80:1124–40.

Bernardé, P. S., and V. A. Rocha. 2003. New record of *Glironia venusta* Thomas, 1912 (bushy-tailed opossum) (Mammalia: Glironiidae) for the state of Rondônia-Brazil. *Biociências* (Porto Alegre) 11:183–84.

Bertoni, A. de W. [1914]. Fauna Paraguaya. Catálogos sistemáticos de los vertebrados del Paraguay. Peces, batracios, reptiles, aves y mamíferos conocidos hasta 1914. In *Descripción física y económica del Paraguay*, ed. M. S. Bertoni, 1–86. Asunción: M. Brossa.

———. 1939. Catálogos sistemáticos de los vertebrados del Paraguay. *Rev. Soc. Cient. Paraguay* 4:3–60.

Best, R. C., and A. Y. Harada. 1985. Food habits of the silky anteater (*Cyclopes didactylus*) in the central Amazon. *J. Mammal.* 66:780–81.

Best, T. L., W. M. Kiser, and P. W. Freeman. 1996. Eumops perotis. *Mammal. Species* 534:1–8.

Best, T. L., W. M. Kiser, and J. C. Rainey. 1997. Eumops glaucinus. *Mammal. Species* 551:1–6.

Best, T. L., J. L. Hunt, L. A. McWilliams, and K. G. Smith. 2001a. Eumops maurus. *Mammal.* Species 667:1–3.

———. 2001b. Eumops hansae. *Mammal. Species* 687:1–3.

———. 2002. Eumops auripendulus. *Mammal. Species* 708:1–5.

Bezerra, A. M. R., F. Escarlate-Tavares, and J. Marinho-Filho. 2005. First record of *Thyroptera discifera* (Chiroptera: Thyropteridae) in the cerrado of central Brazil. *Acta Chiropterol.* 71:165–70.

Bezerra, A. M. R., F. H. G. Rodrigues, and A. P. Carmignotto. 2001. Predation of rodents by the yellow armadillo (*Euphractus sexcinctus*) in cerrado of the central Brazil. *Mammalia* 65:86–88.

Bianchi, V. 1917. Notes préliminaires sur les chauvesouries ou Chiroptères de la Russie [in Russian]. *Ann. Mus. Zool. l'Acad. Sci.*, Petrograd, 21:lxxiii–lxxxii.

Bianconi, G. V., S. B. Mikich, and W. A. Pedro. 2004. Diversidade de morcegos (Mammalia, Chiroptera) em remanescentes florestais do município de Fênix, noroeste do Paraná, Brasil. *Rev. Brasil. Zool.* 21:943–54.

Bickham, J. W. 1979. Chromosomal variation and evolutionary relationships of vespertilionid bats. *J. Mammal.* 60:350–63.

———. 1987. Chromosomal variation among seven species of lasiurine bats (Chiroptera: Vespertilionidae). *J. Mammal.* 68:837–842.

Bickham, J. W., and R. J. Baker. 1977. Implications of chromosomal variation in *Rhogeessa* (Chiroptera: Vespertilionidae). *J. Mammal.* 58:448–53.

Billberg, G. J. 1827. *Synopsis faunae scandinaviae. Mammalia.* Holmiae: Ordinum Equestrium., 1(1):viii + 55 pp., Tables A-E, I-XV + 3-page corrigenda et addenda.

Birney, E. C., J. A, Monjeau, C. J. Phillips, R. S. Sikes, and I. Kim. 1996a. *Lestodelphys halli*: new information on a poorly known Argentinian marsupial. *Mastozool. Neotrop.* 3:171–81.

Birney, E. C., R. S. Sikes, J. A. Monjeau, N. Guthmann, and C. J. Phillips. 1996b. Comments on Patagonian marsupials from Argentina. In *Contributions in Mammalogy: A Memorial Volume Honoring Dr. J. Knox Jones, Jr.*, ed. H. H. Genoways and R. J. Baker, 149–54. Lubbock: Museum of Texas Tech University, il + 315 pp.

Bisbal E., F. J. 1998. Mamíferos de la Península de Paria, estado Sucre, Venezuela y sus relaciones biogeográficas. *Interciencia* 23:176–81.

Blainville, M. H. M. D. de. 1840. *Ostéographie ou description iconographique comparée du squelette et du système dentaire des cinq classes d'animaux vertébrés récents et fossiles pour servir de base a la zoologie et a la géologie*. Mammifères.—Paresseaux.—G. *Bradypus*. Paris: Arthus Bertrand, Part 4 (fasc. 5, Atlas [fasc. 4]), 64 pp. + 6 pls. [See Sherborn 1898, for publication dates of parts.]

Blair, W. F., A. C. Hulse, and M. A. Mares. 1976. Origins and affinities of vertebrates of the North American Sonoran Desert and the Monte Desert of northwestern Argentina. *J. Biogeogr.* 3:1–18.

Blake, E. R. 1941. Two new birds from British Guiana. *Field Mus. Nat. Hist.*, zool. ser. 24:227–32.

Bloedel, P. 1955. Observations on the life histories of Panama bats. *J. Mammal.* 36:232–35.

Blumenbach, J. F. 1779. *Handbuch der Naturgeschichte*. Göttingen: Johann Christian Dieterich, 15 unnumbered + 448 pp., 2 pls.

———. 1797. *Handbuch der Naturgeschichte*. Fünfte Auflange. Göttingen: Johann Christian Dieterich, xviii + 714 pp., + 32 unnumbered, 2 pls.

Boada, C., S. Burneo, T. de Vries, and D. Tirira S. 2003. Notas ecológicas y reproductivas del murciélago rostro de fantasma *Mormoops megalophylla* (Chiroptera: Mormoopidae) en San Antonio de Pichincha, Pichincha, Ecuador. *Mastozool. Neotrop.* 10:21–26.

Boas, J. E. V. 1933. Der Hinterfuss von *Caenolestes*. *Det Kgl. Danske Videnskabernes Selskab.*, *Biologiske Meddelelser* 10(6):1–11.

Boddaert, P. 1784. *Elenchus animalium*. Roterdami: C. R. Hake, 1:xxxviii + 2 unnumbered + 174 pp.

Boddicker, M. 1998. Medium and large mammals: Biodiversity assessment in the lower Urubamba region. In *Biodiversity assessment and Monitoring of the lower Urubamba region, Peru. Cashiriari-3 Well Site and the Camisea and Urubamba Rivers*, ed. A. Alonso and F. Dalmeier, 219–44. Washington, DC: SI/MAB Series #2. Smithsonian Institution, xlii + 298 pp.

Boddicker, M., J. J. Rodríguez, and J. Amanzo. 1999. Medium and large mammals: Biodiversity assessment at the Pagoreni well site. In *Biodiversity assessment and monitoring of the lower Urubamba region, Peru*, ed. A. Alonso and F. Dallmeier, 151–66. Washington, DC: SI/MAB Series #3. Smithsonian Institution , xxxiv + 334 pp.

———. 2001. Assessment of the large mammals of the lower Urubamba region, Peru. In *Urubamba: The biodiversity of a Peruvian rainforest*, ed. A. Alonso, F. Dallmeier, and P. Campbell, 183–94. Washington, DC: SI/MAB Series #7. Smithsonian Institution, x + 204 pp.

Bogan, M. A. 1978. A new species of *Myotis* from the Islas Tres Marias, Nayarit, Mexico, with comments on variation in *Myotis nigricans*. *J. Mammal.* 59:519–30.

———. 1999. Family Vespertilionidae. In *Mamíferos del noroeste de México*, ed. S. T. Alvarez-Castañeda and J. L. Patton, 139–82. La Paz, México: Centro de Investigaciones Biológicas del Noroeste, S.C., 583 pp.

Boher-Bentti, S. 1988. Nuevos registros de distribución de *Lutreolina crassicaudata* (Desmarest, 1804) en Venezuela. *Trianea* 1:111–17.

Boie, F. 1830. Naturgeschichtliche Beyträge vermischten Inhalts. *Isis von Oken* 23:columns 256–58.

Boinski, S., and R. M. Timm. 1985. Predation by squirrel monkeys and double-toothed kites on tent-making bats. *Amer. J. Primat.* 9:121–27.

Boitard, P. 1842. *Le Jardin des Plantes, description et moeurs, des mammifères et la ménagerie et du Muséum d'Histoire Naturelle*. Paris: J.-J. Dubochet et Ce., lxiii + 472 pp., 56 pls.

Bokermann, W. 1957. Atualização do itinerário da viagem do Príncipe Wied ao Brasil (1815–1817). *Arq. Zool. São Paulo* 10:209–51.

Bonaccorso, F. J. 1979. Foraging and reproductive ecology in a Panamanian bat community. *Bull. Florida State Mus., Biol. Sci.* 24:359–408.

Bonaccorso, F. J., W. E. Glanz, and C. M. Sandford. 1980. Feeding assemblages of mammals at fruiting *Dipteryx panamensis* (Papilionaceae) trees in Panama: Seed predation, dispersal, and parasitism. *Rev. Biol. Trop.* 28:61–72.

Bonaparte, C. L. J. L. 1831. Saggio di una distribuzione metodica degli animali vertebrati. *Giorn. Arcadico Scien. Lett. Arti* 49:3–77.

———. 1837a. Fascicolo XX. In *Iconografia della fauna Italica per le quattro classi degli animali vertebrati*. Roma: Salvucci, vol. 1. [See Salvadori (1888) for date of publication.]

———. 1837b. Fascicolo XXI. In *Iconografia della fauna Italica per le quattro classi degli animali vertebrati*. Roma: Salvucci, vol. 1. [See Salvadori (1888) for date of publication.]

———. 1838. Synopsis Vertebratorum Systematis. *Nuovi Ann. Sci. Nat., Bologna* 2:105–33.

———. 1841a. Introduzione alla classe 1. Mammiferi [4 unnumbered pages]. *Iconografia della fauna Italica per le quattro classi degli animali vertebrati*. Roma: Salvucci, vol. 1. [See Salvadori (1888) for date of publication.]

———. 1841b. Indice distributivo del Tomo Primo = Mammiferi e Uccelli (che puó servire di avviso al legatore) [3 unnumbered pages]. *Iconografia della fauna italica per le quattro classi degli animali vertebrati*. 1. Roma: Salvucci, vol. 1. [See Salvadori (1888) for date of publication.]

———. 1845. *Catalogo metodico dei mammiferi Europei.* Milan: Coi Tipi di Luigi de Giacomo Pirola, 32 pp.

———. 1847. Description of a new species of bat. *Proc. Zool. Soc. London* 1847:115.

———. 1850. *Conspectus systematum. Mastozoölogiae. Ornithologiae. Herpetologiae. Ichthyologiae.* Lugduni-Batavorum: E. J. Brill, 4 pp.

Bonato, V., and K. G. Facure. 2000. Bat predation by the fringe-lipped bat *Trachops cirrhosus* (Phyllostomidae, Chiroptera). *Mammalia* 64:241–43.

Bonvicino, C. R., and A. M. R. Bezerra. 2003. Use of regurgitated pellets of Barn Owl (*Tyto alba*) for inventorying small mammals in the cerrado of central Brazil. *Studies Neotrop. Fauna Environ.* 38:1–5.

Bonvicino, C. R., A. Langguth, S. M. Lindbergh, and A. C. de Paula. 1998. An elevational gradient study of small mammals at Caparaó National Park, South eastern [*sic*] Brazil. *Mammalia* 61:547–60.[Dated 1997; number 4 marked "Achevé d'imprimer le 23 fevrier 1998."]

Bordignon, M. O., and A. de O. França. 2002. Fish consumption by Noctilio leporinus (Linnaeus, 1758) in Guaratuba Bay, southern Brazil. *Chirop. Neotrop.* 8:148–50.

Borkhausen, M. 1797. *Deutsche Fauna, oder Kurzgefasste Naturgeschichte der Thiere Deutschlands. Säugthiere und Vögel.* Frankfurt: Barrentrapp und Menner, 1:xxiv + 620 pp.

Bowdich, T. E. 1821. *An analysis of the natural classifications of Mammalia, for the use of students and travellers.* Paris: J. Smith, v + 6–115 pp., 15 pls.

Bowles, J. B., J. B. Cope, and E. A. Cope. 1979. Biological studies of selected Peruvian bats of Tingo Maria, Departmento de Huánuco. *Trans. Kansas Acad. Sci.* 82:1–10.

Brack E., A. J. 1974. Los vertebrados de las lomas costeras del Peru. *An. Cient., Lima* 12:85–92.

Bradbury, J. W., and L. H. Emmons. 1974. Social organization of some Trinidad bats. I. Emballonuridae. *Zeit. Tierpsychol.* 36:137–83.

Bradbury, J. W., and S. L. Vehrencamp. 1977a. Social organization and foraging in emballonurid bats. I. Field studies. *Behav. Ecol. Sociobiol.* 1:337–81.

———. 1977b. Social organization and foraging in emballonurid bats. IV. Mating systems. *Behav. Ecol. Sociobiol.* 2:1–17.

———. 1977c. Social organization and foraging in emballonurid bats. IV. Parental investment patterns. *Behav. Ecol. Sociobiol.* 2:19–29.

Braun, J. K., R. A. Van Den Bussche, P. K. Morton, and M. A. Mares. 2005. Phylogenetic and biogeographic relationships of mouse opossum *Thylamys* (Didelphimorphia, Didelphidae) in southern South America. *J. Mammal.* 86:147–59.

Braun, J. K., M. A. Mares, and J. S. Stafira. 2005. The chromosomes of some didelphid marsupials from Argentina. In *Contribuciones mastozoológicas en homenaje a Bernardo Villa,* ed. V. Sánchez-Cordero and R. A. Medellín, 59–66. Mexico: Instituto de Biología, UNAM; Instituto de Ecología, UNAM; CONABIO, xxiv + 680 pp.

Bredt, A., and W. Uieda. 1996. Bats from urban and rural environments of the Distrito Federal, mid-western Brazil. *Chiropt. Neotrop.* 2:54–57.

Bredt, A., W. Uieda, and E. D. Magalhães. 1999. Morcegos cavernícolas da região do Distrito Federal, centro-oeste do Brasil (Mammalia, Chiroptera). *Rev. Bras. Zool.* 16:731–70.

Brennan, J. M. 1970. Chiggers from the Bolivian-Brazilian border (Acarina: Trombiculidae). *J. Parasitol.* 56:807–12.

Brennan, J. M., and M. L. Goff. 1978. Two new species of *Crotiscus* (Acari: Trombiculidae) from northern South America. *J. Med. Entomol.* 14:569–72.

Brennan, J. M., and J. T. Reed. 1975. A list of Venezuela chiggers, particularly of small mammalian hosts (Acarina: Trombiculidae). *Brigham Young Univ. Sci. Bull.,* biol. ser., 20(1, part 2):45–75.

Brennan, J. M., and J. E. M. H. van Bronswijk. 1975. Parasitic mites of Surinam. XXI. New records of Surinam and certain French Guiana chiggers with the description of a new species of *Loomisia* Brennan and Reed, 1972 (Acarina: Trombiculidae). *J. Med. Entomol.* 12:243–49.

Bresslau, E. 1927. Ergebnisse einer zoologischen Forschungsreise in Brasilien 1913–1914 (Reisenbericht). *Abhandl. Senckenb. Naturf. Gesell.* 40:181–235, 2 pls.

Brèthes, J. 1913. Une nouvelle espèce de diptère pupipare du Chili. *Bol. Mus. Nacional, Buenos Aires* 5:297–99.

Brisson, M. J. 1762. *Regnum animale in classes IX.* Lugduni Batavorum: Theodorum Haak, 1–7 + 296 pp.

Brongniart, A. 1792. Catalogue de mammifères envoyés de Cayenne par M. le Blond. *Actes Soc. Nat. Hist. Paris* 1:115.

Brooks, D. M., T. Tarifa, J. M. Rojas, R. J. Vargas, and H. Aranibar. 2002. A preliminary assessment of the mammalian fauna of the eastern Bolivian panhandle. *Mammalia* 65:509–520. [Dated 2001; number 4 marked "Achevé d'imprimer le 11 01 2002."]

Broom, R. 1911. On the affinities of *Caenolestes. Proc. Linn. Soc. New South Wales* 36:315–320.

Brosset, A. 1963. Mammifères des îles Galapagos. Statut actuel des mammifères des îles Galapagos. *Mammalia* 27:323–38.

———. 1965. Contribution a l'etude des chiroptères de l'ouest de l'Ecuador. *Mammalia* 29:211–27.

————. 1966. *La biologie des chiroptères*. Paris: Masson et Cie., 240 pp.

Brosset, A., and P. Charles-Dominique. 1991. The bats from French Guiana: A taxonomic, faunistic and ecological approach. *Mammalia* 54:509–560. [Dated 1990; number 4 marked "Achevé d'imprimer le 27 fevrier 1991."]

Brosset, A., and G. Dubost. 1968. Chiroptères de la Guyane Française. *Mammalia* 31:583–94. [Dated 1967; published in 1968.]

Brosset, A., J. -F. Cosson, P. Gaucher, and D. Masson. 1996a. Les chiroptéres d'un marécage côtier de Guyane; composition du peuplement. *Mammalia* 59:527–37. [Dated 1995; number 4 marked "Achevé d'imprimer le 30 avril 1996."]

Brosset, A., P. Charles-Dominique, A. Cockle, J. F. Cosson, and D. Masson. 1996b. Bat communities and deforestation in French Guiana. *Can. J. Zool.* 74:1974–82.

Brown, B. E. 2004. Atlas of New World marsupials. *Fieldiana Zool.*, 102:viii + 1–308.

Brown, J. C., and D. W. Yalden. 1973. The description of mammals, 2. Limbs and locomotion of domestic mammals. *Mammal. Rev.* 3:107–34.

Bublitz, J. 1983. Beiträge zu Skelettbau, Verhalten und Lebensraum der Gattung *Caenolestes* Thomas 1895 (Marsupialia). *Zool. Anzeiger* 211:359–63.

————. 1987. Untersuchungen zur Systematik der Rezenten Caenolestidae Trouessart, 1898: Unter Verwendung craniometrischer Methoden. *Bonner Zool. Monogr.* 23:1–96.

Bucher, J. E., and R. S. Hoffmann. 1980. Caluromys derbianus. *Mammal. Species* 140:1–4.

Buck, C. D. 1933. *Comparative grammar of Greek and Latin*. Chicago: University of Chicago Press, xvi + 405 pp.

Buffon, G. L. le Clerc. 1763. *Histoire naturelle, générale et particulière, avec la description du cabinet du Roi*. Paris: L'Imprimerie Royale, 10:1–6 (unnumbered) + 368 pp., 57 pls.

————. 1765. *Histoire naturelle, générale et particulière, avec la description du cabinet du Roi*. Paris: L'Imprimerie Royale, 13:xx + 1–441, 58 pls.

————. 1767. *Histoire naturelle, générale et particulière, avec la description du cabinet du Roi*. Paris: L'Imprimerie Royale, 15:5 (unnumbered) + 1–207, 17 pls.

————. 1776. Histoire naturelle générale et particuliére, servant de suite à l'histoire des animaux quadrupèdes. In *Histoire naturelle, avec la description du cabinet du Roi*, ed. G. L. L. de Buffon. Paris: L'Imprimerie Royale, Supplement, 3:11 (unnumbered) + 330 + xxi pp., 65 pls.

————. 1782. Histoire naturelle générale et particuliére, servant de suite à l'histoire des animaux quadrupèdes. In *Histoire naturelle, avec la description du cabinet du Roi*, ed. G. L. L. de Buffon. Paris: L'Imprimerie Royale, Supplément, 6:viii + 405 + xxv pp., 49 pls.

————. 1789. Histoire naturelle générale et particuliére, servant de suite à l'histoire des animaux quadrupèdes. *In Histoire naturelle, avec la description du cabinet du Roi*, ed. G. L. L. de Buffon. Paris: L'Imprimerie Royale, Supplément, 7:1–8, ix–xx, + 364 pp., 81 pls.

Burmeister, H. 1848. Ueber Dasypus novemcinctus. *Zeit. Zool., Zoot. Palaeontol.* 1:199.

————. 1854. *Systematische Uebersicht der Thiere Brasiliens, welche während einer Reise durch die Provinzen von Rio de Janeiro und Minas Geraës gesammelt oder beobachtet wurden von Dr. Hermann Burmeister. Säugethiere (Mammalia)*. Berlin: Georg Reimer, 1:x + 342 pp.

————. 1856. *Erläuterungen zur Fauna Brasiliens, enthaltend Abbildungen und ausführliche Beschreibungen neuer oder ungenügend bekannter Thier-Arten*. Berlin: Georg Reimer, ix + 115 pp, 32 pls.

————. 1861. *Reise durch die La Plata-Staaten, mit besonderer Rücksicht auf die physische Beschaffenheit und den Culturzustand der argentinischen Republik ausgeführt in den Jahren 1857, 1858, 1859 und 1860*. Halle: H. W. Schmidt, 2:vi + 538 pp., 1 map.

————. 1862. Beschreibung eines behaarten Gurtelthieres Praopus hirsutus, aus dem National-Museum zu Lima. *Abhandl. Naturfors. Gesell. Halle* 6:145–48, 1 pl.

————. 1863. Ein neuer Chlamyphorus. *Abhandl. Naturfors. Gesell. Halle* 7:165–71, 1 pl.

————. 1869. [Untitled]. *Actas Soc. Paleontol, Buenos Aires* 1869:xxxii–xxxiv.

————. 1879. *Description physique de la République Argentine d'après des observations personnelles et étrangères*. Buenos Aires: Paul-Emile Coni, 3(1):vi + 556 pp.

Burnett, G. T., 1829. Illustrations of the Alipeda (alipeds), or bats and their allies; being the arrangement of the Cheiroptera, Volitantia, or wing-footed beasts, indicated in outline. *Quart. J. Sci.* 27:262–69.

————. 1830. Illustrations of the Quadrupeda, or quadrupeds, being the arrangement of the true four–footed beasts indicated in outline. *Quart. Jour. Sci. Lit. Art.* 1829:336–53.

Burnett, S. E., J. B. Jennings, J. C. Rainey, and T. L. Best. 2001. Molossus bondae. *Mammal. Species* 668:1–3.

Burt, W. H., and R. A. Stirton. 1961. The mammals of El Salvador. *Misc. Publ. Mus. Zool., Univ. Michigan* 117:1–69.

Busch, M., and F. O. Kravetz. 1992. Diet composition of *Monodelphis dimidiata* (Marsupialia, Didelphidae).

Mammalia 55:619–21. [Dated 1991; number 4 marked "Achevé d'imprimer le 20 mars 1992."]

Butterworth, B. B., and A. Starrett. 1964. Mammals collected by the Los Angeles County Museum Expedition to northeastern Venezuela. *Los Angeles Co. Mus., Contrib. Sci.* 85:1–8.

Cabanis, J. 1848. Saeugethiere. In *Versuch einer Fauna und Flora von Britisch-Guiana.* In *Reisen in Britisch-Guiana in den Jahren 1840–1844*, ed. R. Schomburgk, 766–86. Leipzig: J. J. Weber, 3:vii + 531–1260 pp.

Cabot N., J. 1989. Second record of *Chironectes minimus* (Marsupialia) in Bolivia. *Mammalia* 53:135–36.

Cabrera, A. 1901. Descripción de tres nuevos mamíferos americanos. *Bol. Soc. Española Hist. Nat.* 1:367–73.

———. 1902. Nota sobre el verdadero "habitat" del "Myotis Thomasi." *Bol. Soc. Española Hist. Nat.* 2:293.

———. 1903. Sinopsis de los quirópteros chilenos. *Rev. Chilena Hist. Nat.* 7:278–308.

———. 1907. A new South American bat. *Proc. Biol. Soc. Washington* 20:57–58.

———. 1913. Dos mamíferos nuevos de la fauna neotropical. *Trab. Mus. Nac. Cien. Nat.*, ser. zool., Madrid 9:1–15 + 1 pl.

———. 1916. El tipo de *Philander laniger* Desm. en el Museo de Ciencias Naturales de Madrid. *Bol. Real Soc. Española Hist. Nat.* 16:514–17.

———. 1917. Mamíferos del viaje al Pacífico verificado de 1862–1865 por una comisión de naturalistas enviada por el gobierno español. *Trab. Mus. Nac. Cien. Nat.*, ser. zool., Madrid 31:1–62.

———. 1919. *Genera Mammalium. Monotremata, Marsupialia.* Madrid: Museo Nacional de Ciencias Naturales, 177 pp., 19 pls.

———. 1925. *Genera Mammalium. Insectivora, Galeopithecia.* Madrid: Museo Nacional de Ciencias Naturales, 232 pp., 18 pls.

———. 1930. Breve sinopsis de los murciélagos argentinos. *Rev. Centro Estud. Agron. Veter., Buenos Aires* 23:418–42.

———. 1934. Dos nuevos micromamíferos del norte Argentino. *Notas Prelim. Mus. La Plata* 3:123–28.

———. 1938. Sobre dos murciélagos nuevos para la Argentina. *Notas Mus. La Plata, Zool.*, 3:5–14.

———. 1958. Catálogo de los mamíferos de América del Sur. *Rev. Mus. Argentino Cien. Nat. "Bernardino Rivadavia," Cien. Zool.* 4:xvi + iv + 308, 1957. [Dated 1957; published 27 March 1958, see notice on p. 308.]

Cabrera, A., and J. Yepes. 1940. *Mamíferos Sud Americanos (vida, costumbres y descripción).* Historia Natural Ediar. Buenos Aires: Compañia Argentina de Editores, 370 pp., 1 map, 78 pls.

Cáceres, N. C. 2004. Diet of three didelphid marsupials (Mammalia, Didelphimorphia) in southern Brazil. *Mamm. Biol. (Z. Säugetierk.)* 69:430–33.

Cáceres, N. C., and M. Pichorim. 2003. Use of an abandoned mottled piculet *Picumnus nebulosus* (Aves, Picidae) nest by the Brazilian Gracile Mouse Opossum *Gracilinanus microtarsus* (Mammalia, Didelphidae. *Biociências* (Porto Alegre) 11:97–99.

Cáceres, N. C., I. R. Ghizoni-jr, and M. E. Graipel. 2002. Diets of two marsupials, *Lutreolina crassicaudata* and *Micoureus demerarae*, in a coastal Atlantic Forest island of Brazil. *Mammalia* 66:331–40.

Cadena, A., R. P. Anderson, and P. Rivas-Pava. 1998. Colombian mammals from the Chocoan slopes of Nariño. *Occas. Papers Mus., Texas Tech Univ.*, no. 180:1–15.

Cagnolaro, L., and C. Violani. 1988. Introduction to the anastatic reprint of "Vertebratorum Synopsis . . ." by E. Cornalia (1849). *Atti Soc. Ital. Sci. Nat. Museo Civ. Stor. Nat. Milano* 129:433–34.

Câmara, E. M. V. C., L. C. Oliveira, and R. L. Meyer. 2004. Occurrence of the mouse opossum, *Marmosops incanus* in cerrado "stricto sensu" area, and new locality records for the cerrado and caatinga biomes in Minas Gerais state, Brazil. *Mammalia* 67:617–19. [Dated 2003; number 4 marked "Achevé d'imprimer le 31 mars 2004."]

Camardella, A. R., M. F. Abreu, and E. Wang. 2000. Marsupials found in felids scats in southeastern Brazil, and a range extension of *Monodelphis theresa*. *Mammalia* 64:379–82.

Camargo, L., and J. R. Tamsitt. 1990. Second occurrence of the smoky bat (*Furipterus horrens*) in Colombia. *Mammalia* 54:157–59.

Cantraine, F. 1845. Notice sur une nouvelle espece du genre Thyroptera Spix. *Bull. Acad. Roy. Sci. Bell.-lettr. Bruxelles* 12:489–95.

Carlini, A. A., and S. F. Vizcaíno. 1987. A new record of the armadillo *Chaetophractus vellerosus* (Gray, 1865) (Mammalia, Dasypodidae) in the Buenos Aires Province of Argentina: Possible causes for the disjunct distribution. *Studies Neotrop. Fauna Environ.* 22:53–56.

Carreira, J. C. A., A. M. Jansen, M. de N. Meirelles, F. Costa-e-Silva, and H. L. Lenzi. 2001. *Trypanosoma cruzi* in the scent glands of the opossum. *Exper. Parasitol.* 97:129–40.

Carrizo, L. V., M. S. Sánchez, M. I. Mollerach, and R. M. Barquez. 2006. Nuevo registro de *Chaeotophractus nationi* (Thomas, 1894) para Argentina; comentarios sobre su identidad sistemática y distribución. *Mastozool. Neotrop.* 12:233–236. [Dated 2005; statement on last page, "Se terminó de imprimir . . . en enero de 2006."]

Carroll, R. L. 1988. *Vertebrate paleontology and evolution.* New York: W. H. Freeman and Co., xiv + 698 pp.

Carstens, B. C., B. L. Lundrigan, and P. Myers. 2002. A phylogeny of the Neotropical nectar-feeding bats (Chirotera: Phyllostomidae) based on morphological and molecular data. *J. Mammal. Evol.* 9:23–53.

Cartelle, C., and V. S. Abuhid, 1994. Chiróptera do Pleistoceno final-Holoceno da Bahia. *Acta Geol. Leopoldensia* 39:429–40.

Carter, C. H., H. H. Genoways, R. S. Loregnard, and R. J. Baker. 1981. Observations on bats from Trinidad, with a checklist of species occurring on the island. *Occas. Papers Mus., Texas Tech Univ.*, no. 72:1–27.

Carter, D. C. 1966. A new species of *Rhinophylla* (Mammalia; Chiroptera; Phyllostomatidae) from South America. *Proc. Biol. Soc. Washington* 79:235–38.

———. 1968. A new species of *Anoura* (Mammalia: Chiroptera: Phyllostomidae) from South America. *Proc. Biol. Soc. Washington* 81:427–30.

Carter, D. C., and W. B. Davis. 1961. *Tadarida aurispinosa* (Peale) (Chiroptera: Molossidae) in North America. *Proc. Biol. Soc. Washington* 74:161–66.

Carter, D. C., and P. G. Dolan. 1978. *Catalog of type specimens of Neotropical bats in selected European museums.* Special Publications of the Museum 15. Lubbock: Texas Tech University Press, 136 pp.

Carter, D. C., and C. S. Rouk. 1973. Status of recently described species of *Vampyrops* (Chiroptera: Phyllostomatidae). *J. Mammal.* 54:975–77.

Carter, D. C., R. H. Pine, and W. B. Davis. 1966. Notes on Middle American bats. *Southwest. Nat.* 11:488–99.

Carter, T. S. 1983. The burrows of the giant armadillos, *Priodontes maximus* (Edentata: Dasypodidae). *Saugetierk. Mitteil.* 31:47–53.

Carter, T. S., and C. D. Encarnação. 1983. Characteristics and use of burrows by four species of armadillos in Brazil. *J. Mammal.* 64:103–8.

Carvalho, A. L. de. 1940. Zur Biologie einer Fledermaus (*Thyroptera tricolor* Spix) des Amazonas. *Sitzungsber. Gesells. Naturforsch. Freunde Berlin* 1939:249–53.

Carvalho, B. de A., L. F. B. Oliviera, A. P. Nunes, and M. S. Mattevi. 2002. Karyotypes of nineteen marsupial species from Brazil. *J. Mammal.* 83:58–70.

Carvalho, C. T. de. 1960a. Das visitas de morcegos ás flôres (Mammalia, Chiroptera). *Anal. Acad. Brasileira Ciênc.* 32:359–77.

———. 1960b. Sobre alguns mamíferos do sudeste do Pará. *Arq. Zool. Estado São Paulo* 11:121–32.

———. 1962. Lista preliminar dos mamíferos do Amapá. *Pap. Avulsos Depto. Zool., São Paulo* 15:283–97.

———. 1965. Comentários sobre os mamíferos descritos e figurados por Alexandre R. Ferreira em 1790. *Arq. Zool. São Paulo* 12:7–70.

Carvalho, C. T. de, and A. J. Toccheton. 1969. Mamíferos do nordeste do Pará, Brasil. *Rev. Biol. Trop.* 15:215–26.

Cassin, J. 1858. *Mammalogy and ornithology.* 2nd ed. Philadelphia: C. Sherman and Son, 466 pp.

Cassini, M. H. 1993. Searching strategies within food patches in the armadillo *Chaetophractus vellerosus.* *Anim. Behav.* 46:400–2.

Castaño, J. H., Y. Muñoz-Saba, J. E. Botero, and J. H. Vélez. 2003. Mamíferos del departamento de Caldas–Colombia. *Biota Colomb.* 4:247–59.

Castro-Arellano, I., H. Zarza, and R. A. Medellín. 2000. *Philander opossum. Mammal. Species* 638:1–8.

Catzeflis, F., C. Richard-Hansen, C. Fournier-Chambrillon, A. Lavergne, and J. Víe. 1997. Biométrie, reproduction et sympatrie chez *Didelphis marsupialis* et *D. albiventris* en Guyane française (Didelphidae: Marspialia). *Mammalia* 61:231–43.

Ceballos-Bendezu, I. 1959. Notas sobre los micromamíferos del Perú. *Rev. Universitaria (Univ. Nac. Cuzco)* 49:265–69.

———. 1960. Notas bioecologicas sobre algunos quirópteros del Brasil. *Bol. Fac. Cien., Univ. Nac. Cuzco* 1:1–9.

———. 1968. Quirópteros del departamento de Loreto (Perú). *Rev. Fac. Cien., Univ. Nac. San Antonio Abad Cuzco* 2:7–60.

———. 1981. Los mamíferos colectados en el Cuzco por Otto Garlepp. *Bol. Lima*, no. 16–18:1–12.

Cerqueira, R. 1980. A study of Neotropical *Didelphis* (Mammalia, Polyprotodontia, Didelphidae). Ph.D. thesis, University of London, United Kingdom, 414 pp.

———. 1984. Reproduction du *Didelphis albiventris* au nord-est du Brésil (Polyprotodontia, Didelphidae). *Mammalia* 48:95–104.

———. 1985. The distribution of *Didelphis* in South America (Polyprotodontia, Didelphidae). *J. Biogeogr.* 12:135–45.

Cerqueira, R., and B. Lemos. 2000. Morphometric differentiation between Neotropical black eared opossums *Didelphis marsupialis* and *Didelphis aurita* (Didelphimorphia, Didelphidae). *Mammalia* 64:319–27.

Cerqueira, R., R. Gentile, F. A. Fernandez, and P. S. D'Andrea. 1994. A five-year population study of an assemblage of small mammals in southeastern Brazil. *Mammalia* 57:507–17. [Dated 1993; number 4 marked "Achevé d'imprimer le 20 janvier 1994."]

Chapman, F. M. 1917. Distribution of bird life in Colombia. *Bull. Amer. Mus. Nat. Hist.* 36:1–656.

———. 1926. Distribution of bird life in Ecuador. *Bull. Amer. Mus. Nat. Hist.* 55:1–784.

Charles-Dominique, P. 1983. Ecology and social adaptations in didelphid marsupials: Comparison with eutherians of similar ecology. In *Recent advances in the study of mammalian behavior*, ed. J. F. Eisenberg and D. G. Kleiman, 395–422. Special Publication 7. American Society of Mammalogists, xvi + 753 pp.

————. 1993. Tent-use by the bat *Rhinophylla pumilio* (Phyllostomidae: Carolliinae) in French Guiana. *Biotropica* 25:111–16.

Charles-Dominique, P., M. Atramentowicz, M. Charles-Dominique, H. Gérard, A. Hladik, C. M. Hladik, and M. F. Prévost. 1981. Les mammifères frugivores arboricoles nocturnes d'une forêt guyanaise: Inter-relations plantes-animaux. *Rev. Ecol. (Terre et Vie)* 35:341–435.

Chavez, J. 1998. Ectoparasites of small mammals. In *Biodiversity Assessment in the Lower Urubamba Region*, ed. A. Alonso and F. Dalmeier, 245–56. Washington, DC: SI/MAB Series #2. Smithsonian Institution, xlii + 298 pp.

Chebez, J. C., and E. Massoia. 1996. Mamíferos de la provincia de Misiones. In *Fauna Misionera. Catálogo sistemático y zoogeográfico de los vertebrados de la provincia de Misiones (Argentina)*, ed. J. C. Chebez, 181–308. Buenos Aires: L. O. L. A. [Literature of Latin America], 1–320 + 1 p. errata.

Chenu, [J. C.]. 1853. *Encyclopédie d'histoire naturelle/ ou/ Traité complet de cette science/ d'après/ les travaux des naturalistes les plus éminents de tous les pays et de toutes les époques/Buffon, Daubenton, Lacépède, G. Cuvier, F. Cuvier, Geoffroy Saint-Hilaire, Latreille, De Jussieu, Brongniart, etc. etc./ Ouvrage résumant les observations des auteurs anciens et comprenant toutes les découvertes modernes jusqu'à nos jours. Carnassiers*. Paris: Chez Marescq et Compagnie, Chez Gustave Havard, 1:5 (unnumbered) + 312 pp., pls. 1–40.

Choate, J. R. 1970. Systematics and zoogeography of Middle American shrews of the genus Cryptotis. *Univ. Kansas Publ., Mus. Nat. Hist.* 19:195–317.

Choate, J. R., and E. D. Fleharty. 1974. Cryptotis goodwini. *Mammal. Species* 44:1–3.

Choe, J. C., and R. M. Timm. 1985. Roosting site selection by *Artibeus watsoni* (Chiroptera:Phyllostomidae) on *Anthurium ravenii* (Araceae) in Costa Rica. *J. Trop. Ecol.* 1:241–47.

Cherem, J. J., P. C. Simões-Lopes, S. Althoff, and M. E. Graipel. 2005. Lista dos mamíferos do estado de Santa Catarina, sul do Brasil. *Mastozool. Neotrop.* 11:151–184. [Dated 2004; published in February 2005, according to R. Barquez, pers. comm.]

Christison, D. 1880. A journey to central Uruguay. *Proc. Roy. Geograph. Soc.*, 2:663–89, 1 map.

Chubb, C. 1919. Notes on collections of birds in the British Museum from Ecuador, Peru, Bolivia, and Argentina. *Ibis* 1919:1–55, 256–90.

Cifelli, R. L. 1993. Theria of metatherian-eutherian grade and the origin of marsupials. In *Mammal phylogeny. Mesozoic differentiation, multituberculates, monotremes, early therians, and marsupials*, ed. F. S. Szalay, M. J. Novacek, and M. C. McKenna, 205–15. New York: Springer Verlag New York, Inc., x + 249 pp.

Clarke, F. M., and P. A. Racey. 2003. Discovery of the Bartica bat *Glyphonycteris daviesi* (Chiroptera: Phyllostomidae) in Trinidad, West Indies. *Acta Chiropterol.* 5:151–54.

Clemens, W. A., and L. G. Marshall. 1976. *American and European Marsupialia*. Fossilium Catalogus. 123:1–114. The Hague: W. Junk.

Cloutier, D., and D. W. Thomas. 1992. Carollia perspicillata. *Mammal. Species* 417:1–9.

Cockerell, T. D. A. 1930. An apparently extinct Euglandina from Texas. *Proc. Colorado Mus. Nat. Hist.* 9:52–53.

Cockrum, E. L. 1969. Migration in the guano bat, Tadarida brasiliensis. In *Contributions in mammalogy*, ed. J. K. Jones, Jr., 303–36. Museum of Natural History, Miscellaneus Publication No. 51. Lawrence: Univ. Kansas Printing Service, 428 pp..

Coelho, D. C., and J. Marinho-Filho. 2002. Diet and activity of *Lonchophylla dekeyseri* (Chiroptera, Phyllostomidae) in the Federal District, Brazil. *Mammalia* 66:319–30.

Coimbra, C. E. A., Jr., M. M. Borges, D. Q. Guerra, and D. A. Mello. 1982. Contribuição à zoogeografia e ecologia de morcegos em regiões de cerrado do Brasil central. *IBDF, Bol. Técnico* 7:33–38.

Coimbra-Filho, A. F. 1972. Mamíferos amaeçados de extinção no Brasil. *Anais Acad. Brasileira Cien.* 44(Suppl. 2):13–98.

Collar, N. J. 1996. Family Otididae (Bustards). In *Handbook of the birds of the world: Vol. III, Huatzin to auks*, ed. J. del Hoyo, A. Elliott, J. Sargatal, 240–73. Barcelona: Lynx Edicions, 821 pp.

Collins, L. R. 1973. *Monotremes and marsupials: A reference for zoological institutions*. Washington, DC: The Smithsonian Institution Press, vi + 323 pp.

Comiskey, J. A., J. P. Campbell, A. Alonso, S. Mistry, F. Dallmeier, P. Nuñez, H. Beltrán, S. Baldeón, W. Nauray, R. de la Colina, L. Acurio, and S. Udvardy. 2001. The vegetation communities of the Lower Urubamba region, Peru. In *Urubamba: The biodiversity of a Peruvian rainforest*, ed. A. Alonso, F. Dallmeier, and P. Campbell, 9–32. Washington, DC: Smithsonian Institution, x + 204 pp.

Constantine, D. G. 1970. Bats in relation to the health, welfare, and economy of man. In *Biology of bats*, Vol. II, ed.

W. A. Wimsatt, 319–449. New York: Academic Press, xv + 477 pp.

Constantine, D. G., G. L. Humphrey, and T. B. Herbenick. 1979. Rabies in *Myotis thysanodes, Lasiurus ega, Euderma maculatum*, and *Eumops perotis* in California. *J. Wildl. Diseases* 15:343–45.

Contreras, J. R. 1973. La mastofauna de la zona la laguna Chasico, provincia de Buenos Aires. *Physis, Buenos Aires* 32:215–19.

Contreras, J. R., and A. O. Contreras. 1992. Notas sobre mamíferos del Paraguay. I *Thylamys grisea* (Desmarest, 1827) (Marsupialia: Didelphidae). *Notul. Faunist.*, no. 27, 4 pp.

Contreras, J. R., and W. Silvera-Avalos. 1995. Incorporación del Pequeño marsupial *Monodelphis scalops* Thomas, 1988 [*sic*] a la mastofauna del Paraguay (Marsupialia: Didelphidae). *Notul. Faunist.*, no. 70, 2 pp.

Contreras-Vega, M., and A. Cadena. 2000. Una nueva especie del género *Sturnira* (Chiroptera: Phyllostomidae) de los Andes Colombianos. *Rev. Acad. Colomb. Cien.* 24:285–87.

Cooper, W. 1837. On two species of *Molossus* inhabiting the southern United States. *Ann. Lycaeum Nat. Hist. New York* 4:67–68.

Cope, E. D. 1889a. On the Mammalia obtained by the Naturalist Exploring Expedition to southern Brazil. *Amer. Nat.* 23:128–50.

———. 1889b. The Edentata of North America. *Amer. Nat.* 23:657–64.

Corbet, G. B., and J. E. Hill. 1980. *A world list of mammalian species*. London: British Museum (Natural History), viii + 226 pp.

———. 1986. *A world list of mammalian species*. 2nd ed. London: British Museum (Natural History), 7 (unnumbered) + 254 pp.

———. 1991. *A world list of mammalian species*. 3rd ed. London: Natural History Museum Publications & Oxford University Press, viii + 243 pp.

Cordero R., G. A. 2001. Ecological data on *Marmosops fuscatus* in the lowland tropical forest in northern Venezuela. *Mammalia* 65:228–31.

Cordero R., G. A., and R. A. Nicolas B. 1987. Feeding habits of the opossum (*Didelphis marsupialis*) in northern Venezuela. In *Studies in Neotropical mammalogy, essays in honor of Philip Hershkovitz*, ed. B. D. Patterson and R. M. Timm, 125–32, *Fieldiana Zool.*, 39:frontispiece, viii + 1–506.

Corin-Frederic, J. 1969. Les formules gonosomiques dites aberrantes chez les mammifères euthériens. Exemple particulier du paresseux *Choloepus hoffmanni* Peters (edente, xenarthre, famille des *Bradypodidae*). *Chromosoma (Berlin)* 27:268–87.

Cornalia, E. 1849. *Vertebratorum synopsis in Museo Mediolanensis extantium que per novam orbem Cajetanus Osculati collegit annis 1846–47–1848 speciebus novis vel minus cognitis adjectis nec non discriptionibus atque iconibus illustrata.* Modoetiae: Typogaphia Corbetta, 16 pp., 1 pl. [Reprinted in 1988 in *Atti Soc. Ital. Sci. Nat. Museo. Civ. Stor. Nat. Milano*, 129:435–52, but with original pagination, and with introduction by Cagnolaro and Violani.]

Costa, L. P., Y. L. R. Leite, and J. L. Patton. 2003. Phylogeography and systematic notes on two species of gracile mouse opossums, genus *Gracilinanus* (Marsupialia: Didelphidae) from Brazil. *Proc. Biol. Soc. Washington* 116:275–92.

Costa-Lima, A. da, and C. R. Hathaway. 1946. Pulgas: Bibliografia, catálogo e hospedadores. *Monogr. Inst. Oswaldo Cruz*, no. 4, 522 pp.

Cothran, E. G., M. J. Aivaliotis, and J. L. VandeBerg. 1985. The effect of diet on growth and reproduction in gray short-tailed opossums (*Monodelphis domestica*). *J. Exper. Zool.* 236:103–14.

Coues, E. 1877. Precursory notes on American insectivorous mammals, with descriptions of new species. *Bull. U.S. Geol. Geogr. Survey* 3:631–53.

Cramer, M. J., M. R. Willig, and C. Jones. 2001. Trachops cirrhosus. *Mammal. Species* 656:1–6.

Creighton, G. K. 1984. Systematic studies on opossums (Didelphidae) and rodents (Cricetidae). Ph.D. thesis, Univ. Michigan, Ann Arbor, 215 pp.

Crespo, J. A. 1950. Nota sobre mamíferos de Misiones. *Comun. Instit. Nac. Invest. Cien. Naturl., Mus. Argentino Cien. Natur. "Bernardino Rivadavia," Cien. Zool.* 1(4):1–14.

———. 1958. Nuevos especies y localidades de quirópteros para Argentina. *Neotropica* 4:27–32.

———. 1964. Dos mamíferos nuevos para la provincia de Córdoba. *Neotrópica*, 10:62.

———. 1974. Comentários sobre nuevas localidades para mamíferos de Argentina y de Bolivia. *Rev. Mus. Argentino Cien. Nat. "Bernardino Rivadavia,"* Zool. 11:1–31.

Crochet, J. Y. 1980. Les marsupiaux de Tertiaire d'Europe. Paris: Editions de la Foundation Singer-Polignac, 279 pp.

Croft, D. B. 2003. Behavior of carnivorous marsupials. In *Predators with pouches: The biology of carnivorous marsupials*, ed. M. E. Jones, C. R. Dickman, and M. Archer, 332–46. Collingwood, Australia: CSIRO Publishing, xix + 486 pp.

Cuartas, C. [A.], and J. Muñoz. 1999. Nemátodas en la cavidad abdominal y el tracto digestivo de algunos murciélagos Colombianos. *Caldasia* 21:10–25.

———. 2003a. *Marsupiales, cenoléstidos e insectivoros de Colombia.* Medellín, Colombia: Editorial Universidad de Antioquia, xiv + 227 pp.

———. 2003b. Lista de los mamíferos (Mammalia: Theria) del departamento de Antioquia, Colombia. *Biota Colomb.* 4:65–78.

Cuartas, C. A., J. Muñoz, and M. González. 2001. Una nueva especie de *Carollia* Gray, 1838 (Chiroptera: Phyllostomidae) from Colombia. *Actual Biol.* 23:63–73.

Cuéllar, E. 2001. The Tatujeikurajoyava (*Chlamyphorus retusus*) in the Izozog communities of the Bolivian Gran Chaco. *Edentata* 4:14–16.

Cuervo-Díaz, A., J. Hernández-Camacho, and A. Cadena G. 1986. Lista actualizada de los mamíferos de Colombia[:] anotaciones sobre su distribución. *Caldasia* 15:471–502.

Cunningham, R. O. 1871. *Notes on the natural history of the Straits of Magellan and west coast of Patagonia made during the voyage of H. M. S Nassau in the years 1866, 67, 68, & 69.* Edinburgh: Edmonston and Douglas, xvi +517pp.

Cushing, J. E., Jr. 1945. Pleistocene bats from San Josecito Cave, Nuevo Leon, Mexico. *J. Mammal.* 26:182–85.

Cuvier, F. 1825. *Des dents de mammifères, considérées comme caractères zoologiques.* Strasbourg and Paris: F. G. Levrault; Paris: Le Normant, lvi + 258 pp., 103 pls.

———. 1828. Description d'un nouveau genre de chauve-souris *sous le nom de Furie. Mém. Mus. Hist. Nat., Paris* 16:149–55, pl. 9.

———. 1829. Zoologie = Mammalogie. In *Dictionnaire des sciences naturelles, dans lequel on traite méthodiquement des différens êtres de la nature, considérés soit en eux-mêmes, d'après l'état actuel de nos connoissances, soit relativement à l'utilité qu'en peuvent retirer la médecine, l'agriculture, le commerce et les artes,* 357–519. Strasbourg and Paris: F. G. Levrault; Paris: Le Normant, 59:1–520.

Cuvier, G. 1798. *Tableau élémentaire de l'histoire naturelle des animaux.* Paris: Baudouin, xvi + 710 pp., 14 pls.

———. 1800. *Leçons d'anatomie comparée.* Paris: Baudouin, 1:xxxii + 521 pp., 9 tables.

———. 1817. *Le règne animal distribué d'après son organisation, pour servir de base à l'histoire naturelle des animaux et d'introduction à l'anatomie comparée.* Paris: Deterville, 1:xxxviii + 540 pp.

———. 1823. *Recherches sur les ossemens fossiles, ou l'on rétablit les charactères de plusieurs animaux dont les révolutions du globe one détruit les espèces.* Nouvelle edition. Paris et Amsterdam: G. Dufour et E. d'Ocagne, 5(1):1–405 + 27 pls.

Czaplewski, N. J. 1993. Late Tertiary bats (Mammalia, Chiroptera) from the southwestern United States. *Southwest. Nat.* 38:111–18.

———. 1996a. Opossums (Didelphidae) and bats (Noctilionidae and Molossidae) from the late Miocene of the Amazon basin. *J. Mammal.* 77:84–94.

———. 1996b. *Thyroptera robusta* Czaplewski, 1996, is a junior synonym of *Thyroptera lavali* Pine, 1993 (Mammalia: Chiroptera). *Mammalia* 60:153–55.

———. 1997. Chiroptera. In *Vertebrate paleontology in the Neotropics: The Miocene fauna of La Venta, Colombia,* ed. R. F. Kay, R. H. Madden, R. L. Cifelli, and J. J. Flynn, 408–29. Washington, DC: The Smithsonian Institution Press, xvi + 592 pp.

Czaplewski, N. J., and C. Cartelle. 1998. Pleistocene bats from cave deposits in Bahia, Brazil. *J. Mammal.* 79:784–803.

Czaplewski, N. J., G. S. Morgan, and T. Naeher. 2003. Molossid bats from the late Tertiary of Florida with a review of the Tertiary Molossidae of North America. *Acta. Chiropterol.* 5:61–74.

Czaplewski, N. J., A. D. Rincón, and G. S. Morgan. 2005. Fossil bat (Mammalia: Chiroptera) remains from Inciarte Tar Pit, Sierra de Perijá, Venezuela. *Carib. J. Sci.* 41:768–81.

Czaplewski, N. J., M. Takai, T. M. Naeher, N. Shigehara, and T. Setoguchi. 2003. Additional bats from the Middle Miocene La Venta Fauna of Colombia. *Rev. Acad. Colomb. Cienc.* 27: 263–82.

Dabbene, R. 1902. Fauna magellanica. Mamíferos y aves de la Tierra del Fuego é islas adyacentes. *An. Mus. Nac. Buenos Aires,* sér. 3, 1:341–409.

Daciuk, J. 1977. Notas faunísticas y bioecológicas de Peninsula Valdes y Patagonia XX. Presencia de *Histiotus montanus montanus* (Philippi y Landbeck), 1861 en la Peninsula Valdes (Chiroptera, Vespertilionidae). *Neotropica* 23:45–46.

Dalquest, W. W. 1950a. Records of mammals from the Mexican state of San Luis Potosi. *Occas. Papers Mus. Zool., Louisiana State Univ.,* no. 23:1–15.

———. 1950b. The genera of the chiropteran family Natalidae. *J. Mammal.* 31:436–43.

———. 1951. Bats from the island of Trinidad. *Proc. Louisiana Acad. Sci.* 14:26–33.

———. 1953a. Mexican bats of the genus *Artibeus. Proc. Biol. Soc. Washington* 66:61–66.

———. 1953b. Mammals of the Mexican state of San Luis Potosí. *Occas. Papers Mus. Zool., Louisiana State Univ.,* no. 23:1–15.

———. 1955. Natural history of the vampire bats of eastern Mexico. *Amer. Midl. Nat.* 53:79–87.

———. 1957a. American bats of the genus *Mimon. Proc. Biol. Soc. Washington* 70:45–47.

————. 1957b. Observations on the sharp-nosed bat, *Rhynchiscus naso* (Maximilian). *Texas J. Sci.* 9:219–26.

Dalquest, W. W., and E. R. Hall. 1947. Geographic range of the hairy-legged vampire in eastern Mexico. *Trans. Kansas Acad Sci.* 50:315–17.

Dalquest, W. W., and H. J. Werner. 1954. Histological aspects of the faces of North American bats. *J. Mammal.* 35:147–60.

Daubenton, L. J. M. 1763. Description des tatous. In *Histoire naturelle, générale et particuliére, avec la description du cabinet du Roi*, ed. G. L. le Clerc de Buffon and L. J. M. Daubenton, 232–78, pls. 37–42. Paris: L'Imprimerie Royale, 10:6 (unnumbered) + 368 pp., 57 pls.

Dávalos, L. M. 2004. A new Chocoan species of *Lonchophylla* (Chiroptera: Phyllostomidae). *Amer. Mus. Novit.*, no. 3426:1–14.

Dávalos, L. M., and J. A. Guerrero. 1999. The bat fauna of Tambito, Colombia. *Chirop. Neotrop.* 5:112–15.

David, N., and M. Gosselin. 2002. The grammatical gender of avian genera. *Bull. British Ornithol. Club* 122:257–82.

Davis, D. E. 1945a. The home range of some Brazilian mammals. *J. Mammal.* 26:119–27.

————. 1945b. The annual cycle of plants, mosquitoes, birds and mammals in two Brazilian forests. *Ecol. Monographs* 15:243–95.

————. 1947. Notes on the life histories of some Brazilian mammals. *Bol. Mus. Nac., Rio de Janeiro*, n. sér., zool. 76:1–8.

Davis, J. A., Jr. 1966. Maverick opossums. *Animal Kingd.* 69:112–17.

Davis, W. B. 1955. A new four-toed anteater from Mexico. *J. Mammal.* 36:557–59.

————. 1958. Review of Mexican bats of the Artibeus "cinereus" complex. *Proc. Biol. Soc. Washington* 71:163–66.

————. 1966. Review of South American bats of the genus Eptesicus. *Southwest. Nat.* 11:245–74.

————. 1968. A review of the genus *Uroderma* (Chiroptera). *J. Mammal.* 49:676–98.

————. 1969. A review of the small fruit bats (genus *Artibeus*) of Middle America. *Southwest. Nat.* 14:15–29.

————. 1970a. The large fruit bats (genus *Artibeus*) of Middle America with a review of the *Artibeus jamaicensis* complex. *J. Mammal.* 51:105–22.

————. 1970b. *Tomopeas ravus* Miller (Chiroptera). *J. Mammal.* 51:244–47.

————. 1970c. A review of the small fruit bats (genus *Artibeus*) of Middle America. Part II. *Southwest. Nat.* 14:389–402.

————. 1973. Geographic variation in the fishing bat, *Noctilio leporinus*. *J. Mammal.* 54:862–74.

————. 1975. Individual and sexual variation in *Vampyressa bidens*. *J. Mammal.* 56:262–65.

————. 1976a. Notes on the bats *Saccopteryx canescens* Thomas and *Micronycteris hirsuta* Peters. *J. Mammal.* 57:604–07.

————. 1976b. Geographic variation in the lesser noctilio, *Noctilio albiventris* (Chiroptera). *J. Mammal.* 57:687–707.

————. 1980. New Sturnira (Chiroptera: Phyllostomidae) from Central and South America, with key to currently recognized species. *Occas. Papers Mus., Texas Tech Univ.*, no. 70:1–5.

————. 1984. Review of the large fruit-eating bats of the *Artibeus "lituratus"* complex (Chiroptera: Phyllostomidae) in Middle America. *Occas. Papers Mus., Texas Tech Univ.*, no. 93:1–16.

Davis, W. B., and D. C. Carter. 1962a. Notes on Central American bats with description of a new subspecies of Mormoops. *Southwest. Nat.* 7:64–74.

————. 1962b. Review of the genus *Leptonycteris* (Mammalia: Chiroptera). *Proc. Biol. Soc. Washington* 75:193–98.

————. 1978. A review of the round-eared bats of the Tonatia silvicola complex, with descriptions of three new taxa. *Occas. Papers Mus., Texas Tech Univ.*, no. 53:1–12.

Davis, W. B., and J. R. Dixon. 1976. Activity of bats in a small village clearing near Iquitos, Peru. *J. Mammal.* 57:747–49.

Davis, W. B., D. C. Carter, and R. H. Pine. 1964. Noteworthy records of Mexican and Central American bats. *J. Mammal.* 45:375–87.

Deane, M. P., H. L. Lenzi, and A. M. Jansen. 1984. *Trypanosoma cruzi*: Vertebrate and invertebrate cycles in the same mammal host, the opossum *Didelphis marsupialis*. *Mem. Inst. Oswaldo Cruz, Rio de Janeiro* 79:513–15.

————. 1986. Double development cycle of *Trypanosoma cruzi* in the opossum. *Parasitol. Today* 2:146–47.

Dederer, P. H. 1909. Comparison of *Caenolestes* with Polyprotodonta and Diprotodonta. *Amer. Nat.* 43:614–18.

de la Torre, L. 1952. An additional record of the bat, *Sturnira ludovici*, in Mexico. *Nat. Hist. Misc.*, no. 105:1–2.

————. 1955. Bats from Guerrero, Jalisco and Oaxaca, Mexico. *Fieldiana Zool.* 37:695–701, 2 pls.

————. 1956. The status of *Mormopterus peruanus* J. A. Allen. *Proc. Biol. Soc. Washington* 69:187–88.

———. 1959. A new species of bat of the genus *Sturnira* (Phyllostomidae) from the Island of Trinidad, West Indies. *Nat. Hist. Misc.* 166:1–6.

———. 1961. The evolution, variation, and systematics of the Neotropical bats of the genus *Surnira*. Doctoral dissertation, University of Illinois, Urbana, v + 146 pp.

———. 1966. New bats of the genus *Sturnira* (Phyllostomidae) from the Amazonian lowlands of Perú and the Windward Islands, West Indies. *Proc. Biol. Soc. Washington* 79:267–72.

de la Torre, L., and A. Schwartz. 1966. New species of *Sturnira* (Chiroptera: Phyllostomidae) from the islands of Guadeloupe and Saint Vincent, Lesser Antilles. *Proc. Biol. Soc. Washington* 79:297–304.

de la Torre, L., and A. Starrett. 1959. Name changes and nomenclatural stability. *Nat. Hist.* 167:1–4.

Delgado, C. A. 2002. Food habits and habitat of the crab-eating fox, *Cerdocyon thous*, in the highlands of eastern Antioquia, Cordillera Central, Colombia. *Mammalia* 66:599–602.

Delpietro, H. A., J. R. Contreras, and J. F. Konolaisen. 1992. Algunas observaciones acerca del murciélago carnivoro *Chrotopterus auritus australis* (Thomas, 1905) en el noreste Argentino (Mammalia: Chiroptera, Phyllostominae. *Notul. Faunist.*, no. 26, 7 pp.

Delsuc, F., F. M. Catzeflis, M. J. Stanhope, and E. J. P. Douzery. 2001. The evolution of armadillos, anteaters and sloths depicted by nuclear and mitochondrial phylogenies: Implications for the status of the enigmatic fossil *Eurotamandua*. *Proc. Roy. Soc. London* 268:1605–15.

DeSantis, L. J. M., and E. R. Justo. 1978. Observaciones sobre algunos quirópteros de la provincia de La Pampa. *Neotrópica* 24:161–63.

Desmarest, A. G. 1803a. Sarigue ou Didelphe (*Didelphis*), genre de quadrupèdes de l'ordre des Carnassiers, sous-odre des Pédimanes. In *Nouveau dictionnaire d'histoire naturelle, appliquée aux arts, principalement à l'agriculture, à l'économie rurale et domestique: Par une société de naturalistes et d'agriculteurs: Avec des figures tirées des trois règnes de la nature*, 145–47. Paris: Deterville, 20:1–576.

———. 1803b. Tatou (*Dasypus*), famille et genre de quadrupèdes de l'ordre des Édentés, ayant pour caractères: Des dents molaires seulement; le corps couvert de tests ou de bandes écailleuses, formant une cuirasse. In *Nouveau dictionnaire d'histoire naturelle, appliquée aux arts, principalement à l'agriculture, à l'économie rurale et domestique: Par une société de naturalistes et d'agriculteurs: Avec des figures tirées des trois règnes de la nature*, 428–36. Paris: Deterville, 21:1–571.

———. 1804a. Yapock (*Lutra memmina* Boddaert; *Didelphis memmina* Cuv.), quadrupèdes de l'ordre des Carnassiers, sous orde des Pédimanes et du genre Sarigue. In *Nouveau dictionnaire d'histoire naturelle, appliquée aux arts, principalement à l'agriculture, à l'économie rurale et domestique:Ppar une société de naturalistes et d'agriculteurs: Avec des figures tirées des trois règnes de la nature*, 507–08. Paris: Deterville, 23:1–567.

———. 1804b. Tableau méthodique des mammifères. *In* Tableaux méthodiques d'histoire naturelle, 5–38. In *Nouveau dictionnaire d'histoire naturelle, appliquée aux arts, principalement à l'agriculture, à l'économie rurale et domestique: Par une société de naturalistes et d'agriculteurs: Avec des figures tirées des trois règnes de la nature.* Paris: Deterville, Vol. 24.

———. 1816. Bradype, *Bradypus*, Linn.; Erxleben; Cuv.; Illiger, etc.; *Tardigradus*, Brisson; *Choloepus* et *Prochilus*, Illiger. In *Nouveau dictionnaire d'histoire naturelle, appliquée aux arts, à l'agriculture, à l'économie rurale et domestique, à la médecine, etc. Par une société de naturalistes et d'agriculteurs*, 319–28. Nouv. éd. Paris: Deterville, 4:1–602.

———. 1817a. Didelphe, *Didelphis*, Linn., Cuv., Geoff., Lacép., Dumér., Illiger. In *Nouveau dictionnaire d'histoire naturelle, appliquée aux arts, à l'agriculture, à l'économie rurale et domestique, à la médecine, etc. Par une société de naturalistes et d'agriculteurs*, 417–33. Nouv. éd. Paris: Deterville, 9:1–6 (unnumbered) + 1–608, 6 pls.

———. 1817b. Fourmilier (*Myrmecophaga*, Linn., Briss., Schreb., Cuv., etc.). In *Nouveau dictionnaire d'histoire naturelle, appliquée aux arts, à l'agriculture, à l'économie rurale et domestique, à la médecine, etc. Par une société de naturalistes et d'agriculteurs*, 100–08. Nouv. éd. Paris: Deterville, 17:1–608.

———. 1818. Noctilion ou bec de lievre. In *Nouveau dictionnaire d'histoire naturelle, appliquée aux arts, à l'agriculture, à l'économie rurale et domestique, à la médecine, etc. Par une société de naturalistes et d'agriculteurs*, 14–16. Nouv. éd. Paris: Deterville, 23:1–612.

———. 1819a. Tatou. In *Nouveau dictionnaire d'histoire naturelle, appliquée aux arts, à l'agriculture, à l'économie rurale et domestique, à la médecine, etc. Par une société de naturalistes et d'agriculteurs*, 482–93. Nouv. éd. Paris: Deterville, 32:1–595.

———. 1819b. Vespertilion. In *Nouveau dictionnaire d'histoire naturelle, appliquée aux arts, à l'agriculture, à l'économie rurale et domestique, à la médecine, etc. Par une société de naturalistes et d'agriculteurs*, 461–81. Nouv. éd. Paris: Deterville, 35:1–572.

————. 1820. Mammalogie ou description des espèces de mammifères. Premier partie, contenant les ordres de Bimans, des Quadrumanes et des Carnassiers, viii + 1–276. In *Encylopédie Méthodique . . .* , Paris: Veuve Agasse, 196 vols.

————. 1822. *Mammalogie ou description des espèces de mammifères.* Seconde partie, contenant les ordres de Rongeurs, des Édentés, des Pachydermes, des Ruminans et des Cetacés, viii, 277–555. In *Encyclopédie Méthodique . . .* , Paris: Veuve Agasse, 196 vols.

————. 1827. Sarigue. In *Dictionnaire des sciences naturelles, dans lequel on traite méthodiquement des différens êtres de la nature, considérés soit en eux-mêmes, d'après l'état actuel de nos connoissances, soit relativement a l'utilité qu'en peuvent retirer la médecine, l'agriculture, le commerce et les artes*, ed. F. G. Cuvier, 377–400. Strasbourg and Paris: F. G. Levrault; Paris: Le Normant, 47:1–562.

Desmoulins, A. 1824. Didelphe. In *Dictionnaire classique d'histoire naturelle*, ed. J. B. G. M. Bory de Saint-Vincent, 485–94. Paris: Rey et Gravier, 5, 17 vols.

Devincenzi, G. J. 1935. Mamíferos del Uruguay. *Anal. Mus. Hist. Nat. Monetvideo*, sér. 2, 4:1–96, 12 pls.

Dias, D., A. L. Peracchi, and S. S. P. da Silva. 2002. Quirópteros do Parque Estadual da Pedra Branca, Rio de Janeiro, Brasil (Mammalia, Chiroptera). *Rev. Bras. Zool.* 19(Suppl. 2):113–40.

Dias, D., S. S. P. da Silva, and A. L. Peracchi. 2003. Ocurrência de *Glyphonycteris sylvestris* Thomas (Chiroptera, Phyllostomidae) no Estado do Rio de Janeiro, sudeste do Brasil. *Rev. Bras. Zool.* 20:365–66.

Díaz, A., J. E. Péfaur, and P. Durant. 1997. Ecology of South American Páramos, with emphasis on the fauna of the Venezuelan páramos. In *Polar and Alpine Tundra. Ecosystems of the world*, ed. F. E. Wielgolaski, 263–310. Amsterdam: Elsevier, 3:1–920.

Díaz, A., J. E. Péfaur, C. Bárnaba, and I. Correa. 1995. Éxito de captura en comunidades de micromamíferos epígeos en sectores boscosos de un páramo andino venezolano. *Resúmenes III Congreso Latinoamericano de Ecología*, Mérida, 24–27.

Díaz, M. M. 2000. Key to the mammals of Jujuy Province, Argentina. *Occ. Papers, Sam Noble Oklahoma Mus. Nat. Hist.* 7:1–29.

Díaz, M. M., and R. M. Barquez, 1999. Contributions to the knowledge of the mammals of Jujuy Province, Argentina. *Southwest. Nat.* 44:324–33.

Díaz, M. M., and M. R. Willig. 2004. Nuevos registros de *Glironia venusta* y *Didelphis albiventris* (Didelphimorphia) para Perú. *Mastozool. Neotrop.* 11:185–92.

Díaz, M. M., D. A. Flores, and R. M. Barquez. 2002. A new species of gracile mouse opossum, genus *Gracilinanus* (Didelphimorphia: Didelphidae), from Argentina. *J. Mammal.* 83:824–33.

Díaz, M. M., J. K. Braun, M. A. Mares, and R. M. Barquez. 2000. An update of the taxonomy, systematics, and distribution of the mammals of Salta Province, Argentina. *Occ. Papers, Sam Noble Oklahoma Mus. Nat. Hist.* 10:1–52.

Díaz de Pasqual, A. 1984. Componentes y variaciones numéricas en una comunidad submontana Andina de pequeños mamíferos Venezolanos. *Stud. Neotrop. Fauna Envir.* 19:89–98.

————. 1993. Caracterización del habitat de algunas especies de pequeños mamíferos de la selva nublada de Monte Zerpa, Mérida. *Ecotropicos* 6:1–9.

————. 1994. The rodent community of the Venezuelan Andes. *Polish Ecol. Stud.* 20:155–61.

Díaz de Pascual, A., and A. A. de Ascenção. 2000. Diet of the cloud forest shrew *Cryptotis meridensis* (Insectivora: Soricidae) in the Venezuelan Andes. *Acta Theriol.* 45:13–24.

Dickey, D. 1928. A new marsupial from El Salvador. *Proc. Biol. Soc. Washington* 41:15–16.

Dickson, J. M., and D. G. Green. 1970. The vampire bat (*Desmodus rotundus*): Improved methods of laboratory care and handling. *Lab. Animals* 4:37–44.

Dinerstein, E. 1985. First records of *Lasiurus castaneus* and *Antrozous dubiaquercus* from Costa Rica. *J. Mammal.* 66:411–12.

Ding, S. 1979. A new edentate from the Paleocene of Guang-dong. *Vert. Palasiatica* 12:62–64.

Dobson, G. E. 1871. Notes on nine new species of Indian and Indo-Chinese Vespertilionidae, with remarks on the synonymy and classification of some other species of the same family. *Proc. Asiatic Soc. Bengal* 1871:210–15.

————. 1876. A monograph of the group Molossi. *Proc. Zool. Soc. London* 1876:701–35.

————. 1877. Notes on a collection of Chiroptera from India and Burma, with descriptions of new species. *J. Asiatic Soc. Bengal* 46:310–13.

————. 1878. *Catalogue of the Chiroptera in the collection of the British Museum.* London: British Museum (Natural History), xlii + 567 pp., 29 pls.

————. 1879. Notes on recent additions to the collection of Chiroptera in the Muséum d'Histoire Naturelle at Paris, with descriptions of new and rare species. *Proc. Zool. Soc. London* 1878:873–80.

————. 1880a. Description of a new species of the genus *Natalus* (Vespertilionidae) from Jamaica. *Proc. Zool. Soc. London* 1880:443–44.

————. 1880b. On some new or rare species of Chiroptera in the collections of the Göttingen Museum. *Proc. Zool. Soc. London* 1880:461–65.

———. 1885. Notes on species of Chiroptera in the collection of the Genoa Civic Museum, with descriptions of new species. *Ann. Mus. Civico Stor. Nat. Genova*, ser. 2ᵃ, 2:16–19.

Dolan, P. G. 1982. Systematics of Middle American mastiff bats (*Molossus*). Ph.D. thesis, Texas Tech University, Lubbock, 151 pp.

———. 1989. *Systematics of Middle American mastiff bats of the genus Molossus*. Special Publications of the Museum 29. Lubbock: Texas Tech University Press., 71 pp.

Dolan, P. G., and D. C. Carter. 1979. Distribution notes and records for Middle American Chiroptera. *J. Mammal.* 60:644–49.

d'Orbigny, A., 1835. Mammifères. In *Voyage dans l'Amérique méridionale (le Brésil, la République orientale de l'Uruguay, la République Argentine, la Patagonie, la République du Chili, la République de Bolivia, la République du Pérou), exécuté pendant les années 1826, 1827, 1828, 1829, 1830, 1831, 1832 et 1833*, ed. A. d'Orbigny, Plate 8. Paris: P. Bertrand Paris; Strasbourg: V. Levrault, 4:1–32, 23 pls. [See Sherborn and Griffin 1934 for dates of publication.]

———. 1836. Mammifères. In *Voyage dans l'Amérique méridionale (le Brésil, la République orientale de l'Uruguay, la République Argentine, la Patagonie, la République du Chili, la République de Bolivia, la République du Pérou), exécuté pendant les années 1826, 1827, 1828, 1829, 1830, 1831, 1832 et 1833*, ed. A. d'Orbigny, Plate 6. Paris: P. Bertrand Paris; Strasbourg: V. Levrault, 4:1–32, 23 pls. [See Sherborn and Griffin 1934 for dates of publication.]

———. 1837. Mammifères. In *Voyage dans l'Amérique méridionale (le Brésil, la République orientale de l'Uruguay, la République Argentine, la Patagonie, la République du Chili, la République de Bolivia, la République du Pérou), exécuté pendant les années 1826, 1827, 1828, 1829, 1830, 1831, 1832 et 1833*, ed. A. d'Orbigny, Plates 9–11. Paris: P. Bertrand; Strasbourg: V. Levrault, 4:1–32, 23 pls. [See Sherborn and Griffin 1934 for dates of publication.]

d'Orbigny, A., and P. Gervais. 1847. Mammifères. In *Voyage dans l'Amérique méridionale (le Brésil, la République orientale de l'Uruguay, la République Argentine, la Patagonie, la République du Chili, la République de Bolivia, la République du Pérou), exécuté pendant les années 1826, 1827, 1828, 1829, 1830, 1831, 1832 et 1833*, ed. A. d'Orbigny. Paris: P. Bertrand; Strasbourg: V. Levrault, 4:1–32 + 23 pls. [See Sherborn and Griffin 1934 for dates of publication.]

Dorst, J. 1951. Étude d'une collection de chiroptères d'Ecuador. *Bull. Muséum, Paris*, 2ᵉ. ser., 23:602–06.

Doutt, J. K. 1938. Two new mammals from South America. *J. Mammal.* 19:100–01.

Duarte, M. A., and A. L. Viloria. 1992. Nuevo hallazgo de *Cryptotis thomasi* (Merriam, 1897) (Mammalia: Insectivora) en La Sierra de Perijá, noreste de Colombia. *Acta Cient. Venez.* 43:240–42.

Dunn, E. R. 1931. The disk-winged bat (Thyroptera) in Panama. *J. Mammal.* 12:429–30.

Durant, P., and A. Díaz. 1995. Aspectos de la ecología de roedores y musarañas de las cuencas hidrográficas Andino-Venezolanas. *Carib. J. Sci.* 31:83–94.

Durant, P., and J. E. Péfaur. 1984. Sistemática y ecología de la musaraña de Mérida. *Rev. Ecol. Conserv. Ornit. Latinoamer.* 1:3–14.

Durant, P., A. Díaz and A. Díaz de Pascual. 1994. Pequeños mamíferos alto-andinos, Mérida-Venezuela. *Rev. Forest. Latinoamer.* 14:103–31.

Durrant, S. D. 1952. Mammals of Utah. *Univ. Kansas Publ., Mus. Nat. Hist.* 6:1–549.

Dusbabek, F. 1969. Macronyssidae (Acarina: Mesostigmata) of Cuban bats. *Folia Parasitol.* 16:321–28.

Duszynski, D. W., and L. J. Barkley. 1985. *Eimeria* from bats of the world: A new species in *Tomopeas ravus* from Peru. *J. Parasitol.* 71:204–08.

Duvernoy, G. L. 1842. Notices pour servir a la monographie du genere musaraigne. *Sorex.* Cuv. *Mag. Zool. d'Anat. Comp. Palaeont.*, ser. 2, 4:1–48, pls. 38–54.

Eberhard, J. P. 1768. *Versuch eines neuen Entwurfs der Thiergeschichte. Nebst einem Anhang von einigen seltenen und noch wenig beschriebenen Thieren.* Halle: Rengerischen Buchhandlung, 14 (unnumbered) + 318 pp., 2 pls.

Eger, J. L. 1974. A new subspecies of the bat *Eumops auripendulus* (Chiroptera: Molossidae), from Argentina and eastern Brazil. *Life Sci. Occas. Papers, Roy. Ontario Mus.* 25:1–8.

———. 1977. Systematics of the genus *Eumops* (Chiroptera: Molossidae). *Life Sci. Contrib., Roy. Ontario Mus.* 110:1–69.

Eisenberg, J. F. 1989. *Mammals of the Neotropics. The northern Neotropics.* Chicago: The University of Chicago Press, 1:x + 449 pp., 21 pls.

Eisenberg, J. F., and E. Maliniak. 1985. Maintenance and reproduction of the two-toed sloth *Choloepus didactylus* in captivity. In *The evolution and ecology of armadillos, sloths, and vermilinguas*, ed. G. G. Montgomery, 327–342. Washington, DC: The Smithsonian Institution Press, 10 (unnumbered) + 451 pp.

Eisenberg, J. F., and K. F. Redford. 1999. *Mammals of the Neotropics. The central Neotropics.* Chicago: The University of Chicago Press, 1:x + 609 pp., 19 pls.

Eisenberg, J. F., and D. E. Wilson. 1981. Relative brain size and demographic strategies in didelphid marsupials. *Amer. Nat.* 118:1–15.

Eisenberg, J. F., M. A. O'Connell, and P. V. August. 1979. Density, productivity, and distribution of mammals in two Venezuelan habitats. In *Vertebrate ecology in the northern Neotropics*, ed. J. F. Eisenberg, 187–207. Washington, DC: The Smithsonian Institution Press, 271 pp.

Eisentraut, M. 1933. Biologische Studien im bolivianischen Chaco. III. Beitrag zur Biologie der Säugetierfauna. *Z. Säugetierk.* 8:47–69, pls. 9–11.

———. 1950. Die Ernährung der Fledermäuse (*Microchiroptera*). *Zool. Jarb. Jena* 79:114–77.

Ellerman, J. R., and T. C. S. Morrison-Scott. 1951. *Checklist of Palaearctic and Indian mammals 1758 to 1946*. London: British Museum (Natural History), map (foldout) + 810 pp.

Elliot, D. G. 1901. A synopsis of the mammals of North America and the adjacent seas. *Field Columbian Mus. Publ.*, zool. ser., 2:i–xiv, 1–471.

———. 1904. The land and sea mammals of Middle America and the West Indies. *Field Columbian Mus. Publ.*, zool. ser., 4:1–849, pls. 59–142.

———. 1905a. Descriptions of apparently new species and subspecies of mammals from Mexico and San Domingo. *Proc. Biol. Soc. Washington* 18:233–36.

———. 1905b. A check list of mammals of the North American continent the West Indies and the neighboring seas. *Field Columbian Mus. Publ.*, zool. ser., 6:iv + 761, frontispiece.

———. 1906. Descriptions of an apparently new species of monkey of the genus *Presbytis* from Sumatra, and of a bat of the genus *Dermanura* from Mexico. *Proc. Biol. Soc. Washington* 19:49–50.

———. 1907. A catalogue of the collection of mammals in the Field Columbian Museum. *Field Columbian Museum*, zool. ser., 8:viii + 1–694.

Emerson, K.C., and R. D. Price. 1975. Mallophaga of Venezuelan mammals. *Brigham Young Univ. Sci. Bull.*, biol. ser., 20(3):1–77.

Emmons, L. H. 1984. Geographic variation in densities and diversities of non-flying mammals in Amazonia. *Biotropica* 16:210–22.

———. 1991. Mammals of Alto Madidi *and* Mammal list. In *A biological assessment of the Alto Madidi Region and adjacent areas of northwest Bolivia, May 18–June 15, 1990*, ed. T. Parker and B. Bailey, 23–25, 72–73. *Conserv. Internat., RAP Working Papers* 1:1–108.

———. 1993. Mammal list: Kanuku mountain region. In *A biological assessment of the Kanuku Mountain region of southwestern Guyana*, ed. T. A. Parker, III, R. B. Foster,

L. H. Emmons, P. Freed, A. B. Forsyth, B. Hoffman, and B. D. Gill, Appendix 3, pp. 61–67. *Conserv. Internat., RAP Working Papers* 5:1–70.

———. 1998. Mammal fauna of Parque Nacional Noel Kempff Mercado. In *A biological assessment of Parque Nacional Noel Kempff Mercado, Bolivia*, ed. T. J. Killeen and T. S. Schulenberg, 129–35. *Conserv. Internat., RAP Working Papers* 10:1–372.

Emmons, L. H., and F. Feer. 1990. *Neotropical rainforest mammals: A field guide*. Chicago: The University of Chicago Press., xvi + 281 pp.

———. 1997. *Neotropical rainforest mammals: A field guide*. 2nd ed. Chicago: The University of Chicago Press, xvi + 307 pp., 29 + A–G pls.

Emmons, L. H., and M. Romo. 1994. Mammals of the upper Tambopata/Távara. In *The Tambopata-Candamo reserved zone of southeastern Perú: A biological assessment*, ed. R. B. Foster, J. L. Carr, and A. B. Forsyth, 140–43. *Conserv. Internat., RAP Working Papers* 6:1–184.

Emmons, L. H., L. Luna W., and M. Romo R. 2001. Preliminary list of mammals from three sites in the northern Cordillera de Vilcabamba, Peru. In *Biological and Social Assessments of the Cordillera de Vilcabamba, Peru*, ed. L. E. Alonso, A. Alonso, T. S. Schulenberg, and F. Dallmeier, Appendix 16, 255–57. RAP Working Papers 12 and SI/MAB Series #6. Washington, DC: Conservation International, 296 pp.

Enders, R. K. 1935. Mammalian life histories from Barro Colorado Island, Panama. *Bull. Mus. Comp. Zool.* 78:385–502.

———. 1937. Panniculus carnosus and formation of the pouch in didelphids. *J. Morphology* 61:1–26.

———. 1940. Observations on sloths in captivity in higher altitudes in the tropics and in Pennsylvania. *J. Mammal.* 21:5–7.

———. 1966. Attachment, nursing and survival of young in some didelphids. In *Comparative biology of reproduction in mammals*, ed. I. W. Rowlands, 195–203. New York: Academic Press, xxi + 559 pp.

Engelmann, G. F. 1985. The phylogeny of the Xenarthra. In *The evolution and ecology of armadillos, sloths, and vermilinguas*, ed. G. G. Montgomery, 51–64. Washington, DC: The Smithsonian Institution Press, 10 (unnumbered) + 451 pp.

Engstrom, M. D., and A. L. Gardner. 1988. Karyotype of *Marmosa canescens* (Marsupialia: Didelphidae): A mouse opossum with 22 chromosomes. *Southwest. Nat.* 33:231–33.

Engstrom, M. D., and B. K. Lim. 2002. Mamíferos de Guyana. In *Diversidad y conservación de los mamíferos neotropicales*, ed. G. Ceballos and J. A. Simonetti, 329–75. Mexico, DF: CONABIO-UNAM, 582 pp.

Engstrom, M. D., B. K. Lim, and F. A. Reid. 1999. *Guide to the mammals of the Iwokrama Forest*. Georgetown, Guyana: Iwokrama International Centre for Rain Forest Conservation and Development, 27 pp.

Erman, A. 1835. *Reise um die Erde durch Nord-Asien und die beiden Oceans in den Jahren 1828, 1829 und 1830*. Naturhistorischer Atlas. Berlin: G. Reimer, iv + 64 pp., 17 pls.

Erxleben, J. C. P. 1777. *Systema regni animalis per classes, ordines, genera, species, varietates cum synonymia et historia animalium. Classis I, Mammalia*. Lipsiae: Impensis Weygandianis, xlviii + 636 pp. + 64 (unnumbered).

Esbérard, C. [E. L.]. 2002. Composição de colônia e reprodução de *Molossus rufus* (E. Geoffroy) (Chiroptera, Molossidae) em um refúgio no sudeste do Brasil. *Rev. Bras. Zool.* 19:1153–60.

———. 2003. Diversidade de morcegos em área de Mata Atlântica regenerada no sudeste do Brasil. *Rev. Bras. Zoociênc.* 5:189–204.

———. 2004. Novo registro de *Micronycteris hirsuta* (Peters) (Mammalia, Chiroptera, Phyllostomidae) na Mata Atlântica, estado do Rio de Janeiro, Brasil. *Rev. Bras. Zool.* 21:403–04.

Esbérard, C. E. L., and H. G. Bergallo. 2004. Aspectos sobre a biologia de *Tonatia bidens* (Spix) no estado do Rio de Janeiro, sudeste do Brasil (Mammalia, Chiroptera, Phyllostomidae). *Mamm. Biol. (Z. Säugetierk.)* 69:253–59.

———. 2005. Nota sobre a biologia de *Cinomops*[sic] *abrasus* (Temminck) (Mammalia, Chiroptera, Molossidae) do Rio de Janeiro, Brasil. *Rev. Bras. Zool.* 22:514–16.

———. 2006. Research on bats in the state of Rio de Janeiro, southeastern Brazil. *Mastozool. Neotrop.* 12:237–43. [Dated 2005; statement on last page, "Se terminó de imprimir . . . en enero de 2006."]

Esbérard, C. E. L., and D. Faria. 2006. Novos registros de *Phylloderma stenops* Peters na Mata Atlântica, Brasil (Chiroptera, Phyllostomidae). *Biota Neotrop.* 6:1–5. http:///www.biotaneotropica.org.br/v6n2/pt/abstract? short-communication+bn02506022006.

Esbérard, C. E. L., A. de S. Chagas, M. Baptista, E. M. Luz, and C. S. Pereira. 1996. Observações sobre *Chiroderma doriae* Thomas, 1891 no Município do Rio de Janeiro, RJ (Mammalia, Chiroptera). *Rev. Bras. Biol.* 56:651–54.

Escarlate-Tavares, F., and L. M. Pessôa. 2005. Bats (Chiroptera, Mammalia) in barn owl (*Tyto alba*) pellets in northern Pantanal, Mato Grosso, Brazil. *Mastozool. Neotrop.* 12:61–67.

Esslinger, J. H. 1973. The genus *Litomosoides* Chandler, 1931 (Filarioidea: Onchocercidae) in Colombian bats and rats. *J. Parasitol.* 59:225–46.

Esslinger, J. H., J. B. Vaughn, and A. A. Arata. 1968. Filarial infections in Colombian bats. *Bull. Tulane Univ. Med. Fac.* 27:19–22.

Evelyn, M. J., and D. A. Stiles. 2003. Roosting requirements of two frugivorous bats (*Sturnira lilium* and *Arbiteus* [sic] *intermedius*) in fragmented Neotropical forest. *Biotropica* 35:405–18.

Ewel, J. J., and A. Madriz. 1968. *Zonas de vida de Venezuela*. Caracas: Ministerio de Agricultura y Cria. 265 pp., 119 figs., map.

Fabián, M. E., A. M. Rui, and K. P. de Oliveira. 1999. Distribução geográfica de morcegos Phyllostomidae (Mammalia: Chiroptera) no Rio Grande do Sul, Brasil. *Iheringia*, sér zool., 87:143–56.

Fabricius, J. C. 1801. *Systema eleutheratorum secundum ordines, genera, species, adjectis synonymis, locis, observationibus, descriptionibus*. Kiliae: Bibliopolii Academici Novi, 2:1–687.

Fadem, B. H., and E. A. Cole. 1985. Scent-marking in the grey short-tailed opossum (*Monodelphis domestica*). *Anim. behav.* 33:730–38.

Fadem, B. H., G. L. Trupin, E. Maliniak, J. L. VandeBerg, and V. Hayssen. 1982. Care and breeding of the gray short–tailed opossum (*Monodelphis domestica*). *Lab. Animal Sci.* 32:405–09.

Fain, A. 1976. Notes sur des Myobiidae parasites de rongeurs, d'insectivores et de chiropteres (Acarina: Prostigmata). *Acta Zool. Pathol. Antverp.* 64:3–32.

———. 1979. Les listrophorides d'Amerique neotropicale (Acarina: astigmates) II. Famille Atopomelidae. *Bull. Inst. Roy. Sci. Nat. Belgium* 51:1–158.

Fain, A., and F. S. Lukoschus. 1976a. Un nouveau genre et trois nouvelles espèces d'hypopes d'Echimyopinae (Acarina, Glycyphagidae). *Acarologia* 18:715–22.

———. 1976b. A new genus and species of Myobiidae from the marsupial *Lestoros inca* (Acarina: prostigmates). *Acarologia* 18:489–95.

———. 1976c. Two new species and two genera of Sarcoptidae from a marsupial and an insectivore. *Int. J. Acarol.* 2:1–8.

———. 1982. Diagnoses de nouveaux Listrophoridae néotropicaux. *Bull. Ann. Soc. Roy. Belge Ent.* 118:100–01.

———. 1984. New observations on the genus *Prolistrophorus* (Fain, 1970) (Acari, Astigmata, Listrophoridae). *Syst. Parasitol.* 6:161–89.

Fairchild, G. B., G. M. Kohls, and V. J. Tipton. 1966. The ticks of Panama. In *Ectoparasites of Panama*, ed. R. L. Wenzel and V. J. Tipton, 167–219. Chicago, Illinois: Field Museum of Natural History, 861 pp.

Falcão, F. C., B. Soares-Santos, and S. Drummond. 2005. Espécies de morcegos do Planalto da Conquista, Bahia, Brasil. *Chiropt. Neotrop.* 11:220–23.

Faria, D., B. Soares-Santos, and E. Sampaio. 2006. Bats from the Atlantic rainforest of southern Bahia, Brazil. *Biota. Neotrop.* 6:1–13—http://www.biotaneotropica .org.br/v6n2/pt/abstract?short-communication+ bn02506022006.

Fariña, R. A., and S. F. Vizcaíno. 2003. Slow moving or browsers? A note on nomenclature. In *Morphological studies on fossil and extant Xenarthra (Mammalia)*, ed. R. A. Fariña, S. F. Vizcaíno, and G. Storch, 3–4. *Senckenbergiana Biol.* 83:1–101.

Fazzolari-Corrêa, S. 1994. *Lasiurus ebenus*, a new vespertilionid bat from southeastern Brasil. *Mammalia* 58:119–23.

Felten, H. 1956a. Fledermäuse (Mammalia, Chiroptera) aus El Salvador. Teil IV. (Taf. 45–48). *Senckenbergiana Biol.* 37:341–67.

———. 1956b. Eine neue unterart von *Trachops cirrhosus* (Mammalia, Chiroptera) aus Brasilien. *Senckenbergiana Biol.* 37:369–70.

———. 1958. Weitere Säugetiere aus El Salvador (Mammalia: Marsupialia, Insectivora, Primates, Edentata, Lagomorpha, Carnivora und Artiodactyla). *Senckenbergiana Biol.* 39:213–26.

Fenton, M. B., J. O. Whitaker Jr., M. J. Vonhof, J. M. Waterman, W. A. Pedro, L. M. S. Aguiar, J. E. Baumgarten, S. Bouchard, D. M. Faria, C. V. Portfors, N. I. L. Rautenbach, W. Scully, and M. Zortéa. 1999. The diet of bats from southeastern Brazil: The relation to echolocation and foraging behavior. *Rev. Brasil. Zool.* 16:1081–85.

Fergusson-Laguna, A. 1981. Tasa metabólica y balance energético del armadillo *Dasypus sabanicola* (Mammalia-Edentata) en cautiverio. *Acta Cient. Venez.* 32:100–04.

———. 1984. *El Cachicamo Sabanero*. Caracas, Venezuela: Fondo Editorial, Acta Científica. Venezolana, 129 pp.

Fergusson-Laguna, A., and J. Pacheco. 1981. Calorimetria y composición corporal del armadillo *Dasypus sabanicola* (Mammalia-Edentata). *Acta Cient. Venez.* 32:239–43.

Fermin, P. 1765. *Histoire naturelle de la Hollande équinoxiale: Ou déscription des animaux, plantes, fruits, et autres curiosités naturelles, qui se trouvent dans la colonie de Surinam; avec leurs noms différents, tant François, que Latins, Hollandois, Indiens & Négre-Anglois*. Amsterdam: M. Macérus, frontispiece, xii + 240 pp., 2 unnumbered.

———. 1769. *Description générale, historique, géographique et physique de la colonie de Surinam, contenant ce qu'il y a de plus curieux & de plus remarquable, touchant sa situation, ses rivieres, ses forteresses; son gouvernement & sa police; avec les moeurs & les usages des habitants naturels du païs, & des Européens qui y sont établis; ainsi que des eclaircissements sur l'oeconomie générale des esclaves negres, sur les plantations & leurs produits, les arbres fruitiers, les plantes médécinales, & toutes les diverses especes d'animaux qu'on y trouve, &c*. Amsterdam: E. van Harrevelt, 2:1–352.

Ferrarezzi, H., and E. do A. Gimenez. 1996. Systematic patterns and the evolution of feeding habits in Chiroptera (Archonota: Mammalia). *J. Comp. Biol.* 1:75–94.

Ferrari-Pérez, F. 1886. Catalog of animals collected by the Geographical and Exploring Commission of the Republic of Mexico. *Proc. U.S. Natl. Mus.* 9:125–99.

Ferrell, C. S., and D. E. Wilson. 1991. Platyrrhinus helleri. *Mammal. Species* 373:1–5.

Festa, E. 1906. Viaggio del Dr. Enrico Festa nel Darien, nell'Ecuador e regioni vicine. *Boll. Musei Zool Anat., Torino* 21:1–8.

Figueira, J. H. 1894. Catálogo general de los animales y vegetales de la república oriental del Uruguay. Contribución a la fauna uruguaya. Enumeración de mamíferos. *Anal. Mus. Nac. Montevideo*, sér. 1, 1:187–217.

Findley, J. S. 1972. Phenetic relationships among bats of the genus *Myotis. Syst. Zool.* 21:31–52.

Findley, J. S., and C. Jones. 1964. Seasonal distribution of the hoary bat. *J. Mammal.* 45:461–70.

Findley, J. S., and D. E. Wilson 1974. Observations on the Neotropical disk-winged bat, *Thyroptera tricolor* Spix. *J. Mammal.* 55:562–71.

Fischer, E., W. Fischer, S. Borges, M. R. Pinheiro, and A. Vicentini. 1997. Predation of *Carollia perspicilata* [sic] by *Phyllostomus cf. elongatus* in central Amazonia. *Chirop. Neotrop.* 3:67–68.

Fischer, J. B. 1829. *Synopsis mammalium*. Stuttgardtiae: J. G. Cottae, xlii + 752 pp.

Fischer, G. 1803. *Das Nationalmuseum der Naturgeschichte zu Paris*. Frankfurt am Main: Friedrich Esslinger, 2:iv + 1–422, 3 (unnumbered), frontispiece, 4 pls, 1 map.

———. 1813. *Zoognosia tabulis synopticis illustrata*. Volumen primum. Tablas synopticas generales et comparativas, nec non characterum quorundam explicionem iconographicam continens. Mosquae: Nicolai Sergeidis Vsevolozsky, 1:xiv + 1–456, 8 pls.

———. 1814. *Zoognosia tabulis synopticis illustrata*. Volumen tertium. Quadrupedum reliquorum, cetorum et montrymatum descriptionem continens. Mosquae: Nicolai Sergeidis Vsevolozsky, 3:xxiv + 1–732.

———. 1817. Adversaria zoologica. Fasciculus primus. *Quaedam ad Mammalium systema et genera illustranda. Mem. Soc. Imp. Natur. Moscou* 5:357–446 [error for 428], 2 pls.

Fitzinger, L. J. 1856. [Untitled]. *Versamml. Deutscher Naturfors. Arzte, Wien, Tageblatt* 32:123.

———. 1860. Die Ausbeute der österreichischen Naturforscher an Säugethieren und Reptilien während der Weltumsegelung Sr. Majestät Fregatte Novara. *Sitzungsber. Kaiserl. Akad. Wiss., Wien* 42:383–416.

———. 1870a. Kritische Durchsict der Ordnung der Flatterthiere oder Handflügler (Chiroptera). Familie der Fledermäuse (Vespertiliones). III. Abtheilung. *Sitzungsber. Kaiserl. Akad. Wiss., Wien* 62:13–144.

———. 1870b. Kritische Durchsict der Ordnung der Flatterthiere oder Handflügler (Chiroptera). Familie der Fledermäuse (Vespertiliones). V. Abtheilung. *Sitzungsber. Kaiserl. Akad. Wiss., Wien* 62:253–438.

———. 1870c. Kritische Durchsict der Ordnung der Flatterthiere oder Handflügler (Chiroptera). Familie der Fledermäuse (Vespertiliones). VI. Abtheilung. *Sitzungsber. Kaiserl. Akad. Wiss., Wien* 62:527–79.

———. 1871a. Die Arten der natürlichen Familie der Faulthiere (Bradypodes), nach äusseren und osteologischen Merkmalen. *Sitzungsber. Kaiserl. Akad. Wiss., Wein* 63:331–405.

———. 1871b. Die natürliche Familie der Gürtelthiere (Dasypodes). I. Abtheilung. *Sitzungsber. Kaiserl. Akad. Wiss., Wien* 64:209–76.

———. 1871c. Die natürliche Familie der Gürtelthiere (Dasypodes). II. Abtheilung. *Sitzungsber. Kaiserl. Akad. Wiss., Wien* 64:329–90.

———. 1872. Kritische Durchsicht der Ordnung der Flatterthiere oder Handflügler (Chiroptera). Familie der Fledermäuse (Vespertiliones). VIII. Abtheilung. *Sitzungsber. Kaiserl. Akad. Wiss., Wien* 66:57–106.

Fleck, D. W., and J. D. Harder. 2000. Matse Indian rainforest habitat classification and mammalian diversity in Amazonian Peru. *J. Ethnobiol.* 20:1–36.

Fleming, J. 1822. *The philosophy of zoology; or a general view of the structure, functions, and classifications of animals.* Edinburgh: Archibald Constable and Co.; London: Hurst, Robinson and Co., 2:1–618.

Fleming, T. H. 1971. *Artibeus jamaicensis*: Delayed embryonic development in a Neotropical bat. *Science* 171:402–04.

———. 1972. Aspects of the population dynamics of three species of opossum in the Panama Canal Zone. *J. Mammal.* 53:619–23.

———. 1973. The reproductive cycles of three species of opossum and other mammals in the Panama Canal Zone. *J. Mammal.* 54:439–55.

———. 1988. *The short-tailed fruit bat.* Chicago: The University of Chicago Press, xvi + 365 pp.

———. 1995. Polinización y frugivoría en murciélagos filostomidos de regiones aridas. *Marmosiana* 1:87–93.

Fleming, T. H., and E. R. Heithaus. 1986. Seasonal foraging behavior of *Carollia perspicillata* (Chiroptera: Phyllostomidae). *J. Mammal.* 67:660–71.

Fleming, T. H., E. T. Hooper, and D. E. Wilson. 1972. Three Central American bat communities: structure, reproductive cycles, and movement patterns. *Ecology* 53:555–69.

Floch, H., and P. Fauran. 1958. Ixodides de la Guyane et das Antilles Françaises. *Arch. l'Instit. Pasteur Guyane Française l'Inini* 446:1–94.

Flores, D. A. 2004. [Review of] Atlas of New World marsupials. *Mastozool. Neotrop.* 11:133–36.

Flores, D. A., M. M. Díaz, and R. M. Barquez. 2000. Mouse opossums (Didelphimorphia, Didelphidae) of northwestern Argentina: Systematics and distribution. *Z. Säugetierk.* 65:321–39.

Flower, W. H. 1883. On the arrangement of the orders and families of existing Mammalia. *Proc. Zool. Soc. London* 1883:178–86.

Flower, W. H., and R. Lydekker. 1891. *An introduction to the study of mammals living and extinct.* London: Adam and Charles Black, xvi + 763 pp.

Fonseca, G. A. B. da, and M. C. M. Kierulff. 1988. Biology and natural history of Brazilian Atlantic Forest small mammals. *Bull. Florida State Mus. Biol. Sci.* 34:99–152.

Fonseca, G. A. B. da, G. Herrmann, Y. L. R. Leite, R. A. Mittermeier, A. B. Rylands, and J. L. Patton. 1996. Lista anotada dos mamíferos do Brasil. *Conserv. Internat., Occas. Papers,* no. 4:1–38.

Fonseca, G. A. B. da, A. P. Paglia, J. Sanderson, and R. A. Mittermeier. 2003. Marsupials of the New World: Status and conservation. In *Predators with pouches: The biology of carnivorous marsupials,* ed. M. E. Jones, C. R. Dickman, and M. Archer, 399–406. Collingwood, Australia: CSIRO Publishing, xix + 486 pp.

Fonseca, R. M., and C. M. Pinto. 2004. A new *Lophostoma* (Chiroptera: Phyllostomidae: Phyllostominae) from the Amazonia of Ecuador. *Occas. Papers Mus., Texas Tech Univ.,* no. 242:1–9.

Forman, G. L., R. J. Baker, and J. D. Gerber. 1968. Comments on the systematic status of vampire bats (Family Desmodontidae). *Syst. Zool.* 17:417–25.

Forman, G. L., C. J. Phillips, and C. S. Rouk. 1979. Alimentary tract. In *Biology of bats of the New World family Phyllostomatidae,* Part III, ed. R. J. Baker, J. K. Jones, Jr., and D. C. Carter, 205–28. Special Publications of the Museum 16. Lubbock: Texas Tech University Press, 441 pp.

Fornes, A. 1972. *Anoura geoffroyi geoffroyi* Gray, nuevo género para la Republica Argentina (*Chiroptera, Phyllostomidae, Glossophaginae*). *Physis, Buenos Aires* 31:51–53.

Fornes, A., and H. Delpietro. 1969. Sobre *Pygoderma bilabiatum* (Wagner) en la República Argentina (*Chiroptera, Phyllostomidae, Stenodermatinae*). *Physis, Buenos Aires* 29:141–44.

Fornes, A., and E. Massoia. 1965. Micromamíferos (*Marsupialia* y *Rodentia*) recolectados en la localidad bonaerense de Miramar. *Physis, Buenos Aires* 25:99–108.

———. 1966. Vampyrops lineatus (E. Geoffroy) nuevo género y espécie para la República Argentina (Chiroptera, Phyllostomidae). *Physis, Buenos Aires* 26:181–84.

———. 1967. Procedencias argentinas nuevas o poco conocidas para murciélagos (Noctilionidae, Phyllostomidae, Vespertilionidae y Molossidae). *Seg. Jorn. Entomoepid. Argentinas* (1965) 1:133–45.

———. 1968. Nuevos procedencias argentinas para *Noctilio labialis, Sturnira lilium, Molossops temmincki y Eumops abrasus (Mammalia: Chiroptera). Physis, Buenos Aires* 28:37–38.

———. 1969. *Macrophyllum macrophyllum* (Wied) nuevo género y especie para le República Argentina (Chiroptera, Phyllostominae). *Physis, Buenos Aires* 28:323–26.

Fornes, A., E. Massoia, and G. E. Forrest. 1967. *Tonatia sylvicola* (d'Orbigny) nuevo género y especie para la República Argentina (*Chiroptera Phyllostomidae*). *Physis, Buenos Aires* 27:149–52.

Foster, M. S., and R. Aguilar. 1993. Primer registro de *Eumops underwoodi* (Chiroptera: Molossidae) en Costa Rica. *Brenesia*, no. 39–40:179–80.

Foster, M. S., and R. M. Timm. 1976. Tent-making by *Artibeus jamaicensis* (Chiroptera: Phyllostomatidae) with comments on plants used by bats for tents. *Biotropica* 8:265–69.

Frantzius, A. 1869. Die Säugethiere Costaricas, ein Beitrag sur Kenntniss der geographischen Verbreitung der Säugethiere Amerikas. *Arch. Naturgesch*. 35:247–325.

Fraser, L. 1858. Mr. Louis Frazer [*sic*; extracts from his letters]. *Zoologist* 16:5939–42.

Frechkop, S., and J. Yepes. 1949. Étude systématique et zoogéographique des dasypodidés conservés a l'Institut. *Bull. Inst. Roy. Sci. Nat. Belgique* 25:1–56.

Freeman, P. W. 1981. A multivariate study of the family Molossidae (Mammalia, Chiroptera): morphology, ecology, evolution. *Fieldiana Zool.*, 7:vii + 173 pp.

Freitas, S. R., D. A. de Moraes, R. Santori, and R. Cerqueira. 1997. Habitat preference and food use by

Metachirus nudicaudatus and *Didelphis aurita* (Marsupialia, Didelphidae) in a restinga forest at Rio de Janeiro, Brazil. *Rev. Bras. Biol.* 57:93–98.

Freitas, T. R. O., M. R. Bogo, and A. U. Christoff. 1992. G-, C-bands and NOR studies in two species of bats from southern Brazil (Chirotera: Vespertilionidae, Molossidae). *Z. Säugetierk.* 57:330–34.

Frisch, J. L. 1775. *Das Natur-System der vierfüssigen Theire, in Tabellen, darinnen alle Ordnungen, Geschlechte und Arten, nicht nur mit bestimmenden Benennungen sondern beygesetzen unterscheidended kennseichen angezeigt werden, zum Nutzen der erwachsenen Schuljugend.* Glogau, iv + 30 + 4 unnumbered. [Not seen.]

Furman, D. P. 1966. The spinturnicid mites of Panama (Acarina: Spinturnicidae. In *Ectoparasites of Panama*, ed. R. L. Wenzel and V. J. Tipton, 125–66. Chicago, Illinois: Field Museum of Nat. Hist., xii + 861 pp.

———. 1972. Laelapid mites (Laelapidae: Laelapinae) of Venezuela. *Brigham Young Univ. Sci. Bull.*, biol. ser., 17(3):1–58 + 2 pp.

Galaz, J. L., J. C. Torres-Mura, and J. Yáñez. 1999. *Platalina genovensium* (Thomas, 1928), un quiróptero nuevo para la fauna de Chile (Phyllostomidae: Glossophaginae). *Notic. Mensual, Mus. Nac. Hist. Nat., Santiago* 337:6–12.

Gallardo, M. H., and B. D. Patterson. 1987. An additional 14–chromosome karyotype and sex-chromosome mosaicism in South American marsupials. In *Studies in Neotropical Mammalogy: Essays in honor of Philip Hershkovitz*, ed. B. D. Patterson and R. M. Timm, 111–16. *Fieldiana Zool.* 39:frontispiece, vii + 1–506.

Galliari, C. A., U. F. J. Pardiñas, and F. J. Goin. 1996. Lista comentada de los mamíferos argentinos. *Mastozool. Neotrop.* 3:39–62.

Gannon, M. R., M. R. Willig, and J. K. Jones, Jr. 1989. Sturnira lilium. *Mammal. Species* 333:1–5.

Gantz P., A, and D. R. Martínez. 2000. Orden Chiroptera. In *Mamíferos de Chile*, ed. A. Muñoz-Pedreros and J. Yañez-Valenzuela, 53–66. Valdivia, Chile: CEA Ediciones, viii + 463 pp.

Gardner, A. L. 1962. A new bat of the genus *Glossophaga* from Mexico. *Contr. Sci., Los Angeles Co. Mus.* 54:1–7.

———. 1965. The systematics of the bat genus *Molossus* (Chiroptera: Molossidae) in Mexico. M.Sc. thesis, University of Arizona, 89 pp.

———. 1966. A new subspecies of the Aztec mastiff bat, *Molossus aztecus* Saussure, from southern Mexico. *Contrib. Sci., Los Angeles Co. Mus.* 111:1–5.

———. 1973. *The systematics of the genus Didelphis (Marsupialia: Didelphidae) in North and Middle America.*

Special Publications of the Museum 4. Lubbock: Texas Tech University Press, 81 pp.

———. 1976. The distributional status of some Peruvian mammals. *Occas. Papers Mus. Zool., Louisiana State Univ.*, no. 48:1–18.

———. 1977a. Chromosomal variation in *Vampyressa* and a review of chromosomal evolution in the Phyllostomidae (Chiroptera). *Syst. Zool.* 26:300–18.

———. 1977b. Taxonomic implications of the karyotypes of *Molossops* and *Cynomops* (Mammalia: Chiroptera). *Proc. Biol. Soc. Washington* 89:545–50.

———. 1977c. Feeding habits. In *Biology of bats of the New World family Phyllostomatidae*, Part II, ed. R. J. Baker, J. K. Jones, Jr., and D. C. Carter, 293–350. Special Publications of the Museum 13. Lubbock: Texas Tech University Press, 364 pp.

———. 1981. [Review of] Husson, A. M. The Mammals of Suriname. *J. Mammal.* 62:445–48.

———. 1982. Virginia opossum. In *Wild mammals of North America*, ed. J. A. Chapman and G. A. Feldhamer, 3–36. Baltimore: Johns Hopkins Univ. Press, xiv + 1147 pp.

———. 1983. *Proechimys semispinosus* (Rodentia: Echimyidae): distribution, type locality, and taxonomic history. *Proc. Biol. Soc. Washington* 96:134–44.

———. 1988. The mammals of Parque Nacional Serranía de la Neblina, Territorio Federal Amazonas, Venezuela. In *Cerro de la Neblina. Resultados de la Expedición 1983–1987*, ed. C. Brewer-Carias, 695–765. Caracas: Editorial Sucre, viii + 922 pp.

———. 1990. Two new mammals from southern Venezuela and comments on the affinities of the highland fauna of Cerro de la Neblina. In *Advances in Neotropical mammalogy*, ed. K. H. Redford and J. F. Eisenberg, 411–24. Gainesville, Florida: The Sandhill Crane Press, Inc., x + 614 pp. [Dated 1989; published 6 February 1990.]

———. 1993a. Order Didelphimorphia. In *Mammal species of the world*, 2nd ed., ed. D. E. Wilson and D. M. Reeder, 15–24. Washington, DC: The Smithsonian Institution Press, xviii + 1206 pp.

———. 1993b. Order Paucituberculata. In *Mammal species of the world*, 2nd ed., ed. D. E. Wilson and D. M. Reeder, 25–26. Washington, DC: The Smithsonian Institution Press, xviii + 1206 pp.

———. 1993c. Order Microbiotheria. In *Mammal species of the world*, 2nd ed., ed. D. E. Wilson and D. M. Reeder, 27. Washington, DC: The Smithsonian Institution Press, xviii + 1206 pp.

———. 1993d. Order Xenarthra. In *Mammal species of the world*, 2nd ed., ed. D. E. Wilson and D. M. Reeder, 63–68. Washington, DC: The Smithsonian Institution Press, xviii + 1206 pp.

———. 2005a. Case 3328. *Didelphis* Linnaeus, 1758 (Mammalia, Didelphidae): Proposed correction of gender, and *Cryptotis* Pomel, 1848 (Mammalia, Soricidae): Proposed fixation of gender. *Bull. Zool. Nomencl.* 62:142–45.

———. 2005b. Order Didelphimorphia. In *Mammal species of the world*, 3rd ed., ed. D. E. Wilson and D. M. Reeder, 3–18. Baltimore: The Johns Hopkins Press, 1:xxxviii + 743 pp.

Gardner, A. L., and D. C. Carter. 1972a. A new stenodermine bat (Phyllostomatidae) from Peru. *Occas. Papers Mus., Texas Tech Univ.*, no. 2:1–4.

———. 1972b. A review of the Peruvian species of *Vampyrops* (Chiroptera: Phyllostomatidae). *J. Mammal.* 53:72–82.

Gardner, A. L., and G. K. Creighton. 1989. A new generic name for Tate's *microtarsus* group of South American mouse opossums (Marsupialia: Didelphidae). *Proc. Biol. Soc. Washington* 102:3–7.

Gardner, A. L., and C. S. Ferrell. 1990. Comments on the nomenclature of some Neotropical bats (Mammalia: Chiroptera). *Proc. Biol. Soc. Washington* 103:501–08.

Gardner, A. L., and J. P. O'Neill. 1969. The taxonomic status of *Sturnira bidens* (Chiroptera: Phyllostomidae) with notes on its karyotype and life history. *Occas. Papers Mus. Zool., Louisiana State Univ.*, no. 38:1–8.

———. 1971. A new species of *Sturnira* (Chiroptera: Phyllostomidae) from Peru. *Occas. Papers Mus. Zool., Louisiana State Univ.*, no. 42:1–7.

Gardner, A. L., and J. L. Patton. 1972. New species of *Philander* (Marsupialia: Didelphidae) and *Mimon* (Chiroptera: Phyllostomidae) from Peru. *Occas. Papers Mus. Zool., Louisiana State Univ.*, no. 43:1–12.

———. 1976. Karyotypic variation in oryzomyine rodents (Cricetinae) with comments on chromosomal evolution in the Neotropical cricetine complex. *Occas. Papers Mus. Zool., Louisiana State Univ.*, no. 49:1–48.

Gardner, A. L., and M. E. Sunquist. 2003. Opossum, *Didelphis virginiana*. In *Wild mammals of North America*, 2nd ed., ed. G. A. Feldhamer, B. C. Thompson, and J. A. Chapman, 3–29. Baltimore: The Johns Hopkins University Press, xiv + 1,216 pp.

Gardner, A. L., and D. E. Wilson. 1971. A melanized subcutaneous covering of the cranial musculature in the phyllostomid bat, *Ectophylla alba*. *J. Mammal.* 52:854–55.

Gardner, A. L., R. K. LaVal, and D. E. Wilson. 1970. The distributional status of some Costa Rican bats. *J. Mammal.* 51:712–29.

Garrod, A. H. 1878. Notes on the anatomy of Tolypeutes tricinctus, with remarks on other armadillos. *Proc. Zool. Soc. London* 1878:222–30.

Garutti, V., A. Cais, and V. A. Taddei. 1984. Notas sobre uma coleção de *Pteronotus parnellii rubiginosus* (Chiroptera, Mormoopidae) obtida no estado do Mato Grosso do Sul. *Ciênc. Cult.* 36:1589–92.

Gay, C. 1847. *Historia física y política de Chili según documentos adquiridos en este república durante doce años de residencia en ella y publicada bajo los auspicios del supremo gobierno. Zoología.* Vol. 1, Mammalia, 182 pp. Paris: En casa del autor; Chili: En el Museo de Historia Natural de Santiago, 496 pp.

Genoways, H. H., and R. J. Baker. 1988. *Lasiurus blossevillii* (Chiroptera: Vespertilionidae) in Texas. *Texas J. Sci.* 40:111–13.

———. 1996. A new species of the genus *Rhogeessa*, with comments on geographic distribution and speciation in the genus. In *Contributions in Mammalogy: A Memorial Volume Honoring Dr. J. Knox Jones, Jr.*, ed. H. H. Genoways and R. J. Baker, 83–87. Lubbock: Museum of Texas Tech University, il + 315 pp.

Genoways, H. H., and S. L. Williams. 1979a. Notes on bats (Mammalia: Chiroptera) from Bonaire and Curaçao, Dutch West Indies. *Ann. Carnegie Mus.* 48:311–21.

———. 1979b. Records of bats (Mammalia: Chiroptera) from Suriname. *Ann. Carnegie Mus.* 48:323–35.

———. 1980. Results of the Alcoa Foundation-Suriname Expeditions. I. A new species of bat of the genus *Tonatia* (Mammalia: Phyllostomatidae). *Ann. Carnegie Mus.* 49:203–11.

———. 1984. Results of the Alcoa Foundation-Suriname Expeditions. IX. Bats of the genus *Tonatia* (Mammalia: Chiroptera) in Suriname. *Ann. Carnegie Mus.* 53:327–46.

Genoways, H. H., S. L. Williams, and J. A. Groen. 1981. Results of the Alcoa Foundation-Suriname Expeditions. V. Noteworthy records of Surinamese mammals. *Ann. Carnegie Mus.* 50:319–32.

Gentile, R., and R. Cerqueira. 1995. Movement patterns of five species of small mammals in a Brazilian restinga. *J. Trop. Ecol.* 11:671–77.

Gentile, R., P. S. D'Andrea, and R. Cerqueira. 1995b. Age structure of two marsupial species in a Brazilian restinga. *J. Trop. Ecol.* 11:679–82.

Gentile, R., P. S. D'Andrea, R. Cerqueira, and L. S. Maroja. 2000. Population dynamics and reproduction of marsupials and rodents in a Brazilian rural area: A five-year study. *Stud. Neotrop. Fauna Envir.* 35: 1–9.

Geoffroy St.-Hilaire, É. 1803. *Catalogue des mammifères du Muséum National d'Histoire Naturelle*. Paris, 272 pp.

———. 1805a. Note sur une petite famille de chauve-souris d'Amérique, désignée sous le nom générique de Molossus. *Bull. Sci. Soc. Philom., Paris* 3:278–79.

———. 1805b. Mémoire sur quelques chauve-souris d'Amerique formant une petite famille sous le nom de Molossus. *Ann. Mus. Hist. Nat., Paris* 6:150–56.

———. 1806. Mémoire sur le genre et les espèces de Vespertilion, l'un des genres de la famille des chauve-souris. *Ann. Mus. Hist. Nat., Paris* 8:187–205, 3 pls.

———. 1810a. Description des rousettes et des céphalotes, deux nouveaux genres de la famille des chauve-souris. *Ann. Mus. Hist. Nat., Paris* 15:86–108, pls 4–7.

———. 1810b. Sur les phyllostomes et les mégadermes, deux genres de la famille des chauve-souris. *Ann. Mus. Hist. Nat., Paris* 15:157–98, pls 9–12.

———. 1818a. Sur de nouvelles chauve-souris, sous le nom de Glossophages. *Mem. Mus. Hist. Nat., Paris* 4:411–18, pls. 17–18.

———. 1818b. Description des mammifères qui se trouvent en Egypte. In *Description de l'Egypte, ou recueil des observations et des recherches qui ont été faites en Égypt pendant l'Expédition de l'Armée Française, publié par les ordres de sa Majeste l'Empereur Napoléon Le Grand*. Histoire Naturelle, 99–144. Paris: Commission D'Egypte, L'Imprimerie Impériale, 2:1–750. [Dated 1812; see Sherborn 1897 for dates of publication; also see ICZN 1987b.]

Geoffroy St.-Hilaire, I. 1824a. Mémoire sur une chauve-souris Américaine, formant une nouvelle espèce dans le genre Nyctinome. *Ann. Mus. Hist. Nat., Paris* 1:337–47, 1 pl.

———. 1824b. Sur les vespertilions du Brésil. *Ann. Sci. Nat., Paris* 3:440–47.

———. 1831. Plate CXLVIII, p. 139. In *Dictionaire classique d'histoire naturelle*, ed. J. B. G. M. Bory de Saint-Vincent. Paris: Rey et Gravier, Atlas, Vol. 17.

———. 1835. *Résumé des leçons de mammalogie, ou histoire naturelle des mammifères.* Paris, 68 pp.

———. 1839. *Tenrec.* Cuv. *Centetes.* Illig. et éricule. *Ericulus.* Is. Geoff. *Mag. Zool.*, ser. 2, mammifères, 1839(1):1–37, 4 pls.

———. 1847a. Note sur le genre Apar, sur ses espèces et sur ses caractères, établis jusqu'à présent d'après un animal factice. *Rev. Zool., Paris* 10:135–37.

———. 1847b. *Vie, travaux et doctrine scientifique d' Étienne Geoffroy Saint-Hilaire; par son fils M. Isidore Geoffroy Saint-Hilaire.* Paris: Chez P. Bertrand; Strasbourg: Chez Veuve Levrault, 4 (unnumbered) + 479 pp.

George, T. K., S. A. Marques, M. de Vivo, L. C. Branch, N. Gomes, and R. Rodrigues. 1988. Levantamento de mamíferos do Parna–Tapajós. *Brasil Florestal* 63:33–41.

Gerrard, E. 1862. *Catalogue of the bones of Mammalia in the collection of the British Museum.* London: British Museum (Natural History), iv + 296 pp.

Gervais, H., and F. Ameghino. 1880. *Les Mammifères fossiles de l'Amérique du Sud.* Paris: F. Savy, xi + 225 pp.

Gervais, P. 1837. Sur les animaux mammifères des Antilles. *L'Institut, Paris* 5(218):253–54.

———. 1841. Mammifères. In *Zoologie,* ed. Eydoux, J. F. T., and L. F. A. Souleyet, 1–68. In *Voyage autour du Monde exécuté pendant les années 1836 et 1837 sur la corvette La Bonite commandée par M. Vaillant.* Paris: Arthus Bertrand, 4:xl + 334 pp.; Atlas: plates 1–12 (Mammifères).

———. 1849. Vespertilion. In *Dictionnaire universel d'histoire naturelle,* ed. C. D. d'Orbigny, 211–15. Paris: Les Éditeurs, L. Houssiaux et Cia., 13:1–384.

———. 1854. *Histoire naturelle de mammifères, avec l'indication de leurs moeurs, et de leurs rapports avec les arts, le commerce et l'agriculture.* Paris: L. Curmer, 1:xxiv + 1–418, 19 pls.

———. 1855. *Histoire naturelle de mammifères, avec l'indication de leurs moeurs, et de leurs rapports avec les arts, le commerce et l'agriculture.* Paris: L. Curmer, 2:1–3 (unnumbered), 1–344, 69 pls.

———. 1856a. Deuxième mémoire. Documents zoologiques pour servir a la monographie des chéiroptères Sud-Américains. In *Mammifères,* ed. P. Gervais, 25–88. In *Animaux nouveaux ou rares recueillis pendant l'expédition dans les parties centrales de l'Amérique du Sud, de Rio de Janeiro a Lima, et de Lima au Para; exécutée par ordre du gouvernement français pendant les années 1843 a 1847, sous la direction du comte Francis de Castelnau,* ed. F. de Castelnau. Paris: P. Bertrand, 1:1–116, 20 pls., 1855. [Received by Académie Français on 30 June 1856; see Sherborn and Woodward 1901.]

———. 1856b. Quatrième mémoire. Sur quelques points de l'histoire zoologique des sarigues et plus particulièrement sur leur système dentaire. In *Mammifères,* ed P. Gervais, 95–103, 2 pls. In *Animaux nouveau ou rares recueillis pendant l'expédition dans les parties centrales de l'Amérique du Sud, de Río de Janeiro a Lima, et de Lima au Para; exécutée par ordre du gouvernement Français pendant les années 1843 a 1847, sous la direction du comte Francis de Castelnau,* ed. F. de Castelnau. Paris: P. Bertrand, 1:1–116, 20 pls., 1855. [Received by Académie Français on 30 June 1956; see Sherborn and Woodward 1901.]

———. 1856c. Documents pour servier à la monographie de chéiroptères Sud-Americaines. *Comptes Rendus, Acad. Sci., Paris* 42:547–50.

———. 1869. *Zoologie et paléontologie générales. Nouvelles recherches sur les animaux vertébrés vivants et fossiles.* Première Série. Paris: Arthus Bertrand., viii + 263 pp., 50 pls., 1867–1869.

Gettinger, D. 1987. Host associations of *Gigantolaelaps* (Acari: Laelapidae) in the Cerrado Province of central Brazil. *J. Med. Entomol.* 24:559–65.

———. 1992. Host specificity of *Laelaps* (Acari: Laelapidae) in central Brazil. *J. Med. Entomol.* 29:71–77.

———. 1997. *Androlaelaps cuicensis* (Acari: Laelapidae), A new species associated with *Monodelphis rubida* (Thomas, 1899) in the gallery forests of central Brazil. *Rev. Bras. Biol.* 57:345–48.

Gettinger, D., and R. Gribel. 1989. Spinturnicid mites (Gamasida: Spinturnicidae) associated with bats in central Brazil. *J. Med. Entomol.* 26:491–93.

Giannini, N. P., and R. M. Barquez. 2003. Sturnira erythromos. *Mammal. Species* 729:1–5.

Gill, T. 1872a. The bat genus Pteronotus renamed Dermonotus. *Proc. Biol. Soc. Washington* 14:177.

———. 1872b. Arrangement of the families of mammals with analytical tables. *Smithsonian Misc. Coll.* 11(1):vi + 98 pp.

———. 1884. Order IV.—Chiroptera. In *The standard natural history,* ed. J. S. Kingsley, 159–177. Boston: S. E. Cassino and Company, 5:frontispiece, viii + 535 pp., 41 unnumbered pls.

Gilmore, R. M. 1941. Zoology. In *The susceptibility to yellow fever of the vertebrates of eastern Colombia. I. Marsupialia,* ed. J. C. Bugher, J. Boshell-Manrique, M. Roca-Garcia, and R. M. Gilmore, 314–19. *Amer. J. Trop. Med.* 21:309–33.

Giral, G. E., M. S. Alberico, and L. M. Alvaré. 1991. Reproduction and social organization in *Peropteryx kappleri* (Chiroptera, Emballonuridae) in Colombia. *Bonn. Zool. Beitr.* 42:225–36.

Glass, B. P. 1985. History of classification and nomenclature in Xenarthra (Edentata). In *The evolution and ecology of armadillos, sloths, and vermilinguas,* ed. G. G. Montgomery, 1–4. Washington, DC: The Smithsonian Institution Press, 10 (unnumbered) + 451 pp.

Gloger, C. W. L. 1841–1842. *Gemeinnutziges Hand- und Hilfsbuch der Naturgeschichte. Für gebildete Leser aller Stände, besonders für die reifere Jugend und ihre Lehrer.* Breslau: U. Schulz und Co., 1:xxxxiv + 496. [Most of the section on mammals is included in the first 160 pages, which were published in 1841 according to O. Thomas 1895a.]

Gmelin, J. F. 1788. *Caroli a Linné . . . Systema naturae per regna tria naturae secundum classes ordines, genera, species cum characteribus, differentiis, synonymis, locis.* Editio decima tertia, aucta, reformata. Lipsiae: Georg. Emanuel. Beer, 1:1–2 + 10 (unnumbered), 3–1032 pp.

Gmelin, J. F. 1789. *Systema naturae per regna tria naturae, secundum classes, ordines, genera, species; cum characteribus, differentiis, synonymis, locis.* 13th ed. Volume I,

Regnume animale. Part 1, Mammalia. Lugdunum: J. B. Delamollierre, x + 500 pp.

Goeldi, E. A. 1893. *Os mammiferos do Brasil.* Rio de Janeiro: Livraría Classica de Alves & Co. iv + 182 pp.

Goeldi, E. A. 1894. Critical gleanings on the *Didelphyidae* of the Serra dos Orgãos, Brazil. *Proc. Zool. Soc. London* 1894:457–67.

Goeldi, E. A., and G. Hagmann. 1904. Prodromo de um catálogo crático, commentado da collecção de mammíferos no Museu do Pará. *Bol. Museu Goeldi, Pará* 4:38–122.

Goff, M. L. 1981. The genus *Teratothrix* (Acari: Trombculidae), with descriptions of three new species and a key to the species. *J. Med. Entomol.* 18:244–48.

———. 1982. A new species of *Sasacarus* (Acari: Trombiculidae) from an Incan 'rat' opossum (Marsupialia: Caenolestidae) in Peru. *J. Med. Entomol.* 19:60–62.

Goff, M. L., and R. M. Timm. 1985. A new species of *Peltoculus* (Acari: Trombiculidae) from Ecuador. *Internat. J. Acarol.* 11:233–35.

Goff, M. L., J. O. Whitaker, Jr., and L. J. Barkley. 1984. A new species of *Loomisia* (Acari: Trombiculidae) from a Peruvian bat. *J. Med. Entomol.* 21:80–81.

Goin, F. J., And A. A. Carlini. 1995. An early Tertiary microbiotheriid marsupial from Antarctica. *J. Vert. Paleo.* 15:205–07.

Goin, F. J., and P. Rey. 1997. Sobre las afinidades de *Monodelphis* Burnett, 1830 (Mammalia: Marsupialia: Didelphidae: Marmosinae). *Neotrópica* 43:93–98.

Goin, F. J., C. Velázquez, and O. Scaglia. 1992. Orientación de las crestas cortantes en el molar tribosfenico. Sus implicancias funcionales en didelfoideos (Marsupialia) fósiles y vivientes. *Rev. Mus. la Plata* (n. s.), 9, Paleontología, 57:183–98.

Goldfuss, G. A. 1812. *Vergleichende Naturbeschreibung der Säugethiere.* Erlangen: Verlage der Waltherschen Kunst- und Buchhandlung, 2:145–288.

Goldman, E. A. 1911. Three new mammals from Central and South America. *Proc. Biol. Soc. Washington* 24:237–40.

———. 1912a. Descriptions of twelve new species and subspecies of mammals from Panama. *Smithsonian Misc. Coll.* 56(36):1–11.

———. 1912b. New mammals from eastern Panama. *Smithsonian Misc. Coll.* 60(2):1–18.

———. 1913. Descriptions of new mammals from Panama and Mexico. *Smithsonian Misc. Coll.* 60(22):1–20.

———. 1914a. Descriptions of five new mammals from Panama. *Smithsonian Misc. Coll.* 63(5):1–7.

———. 1914b. A new bat of the genus Mimon from Mexico. *Proc. Biol. Soc. Washington* 27:75–76.

———. 1915. Five new mammals from Mexico and Arizona. *Proc. Biol. Soc. Washington* 28:133–38.

———. 1917. New mammals from North and Middle America. *Proc. Biol. Soc. Washington* 30:107–16.

———. 1920. Mammals of Panama. *Smithsonian Misc. Coll.* 69(5):1–309 + map + 39 pls.

———. 1932. The status of the Costa Rican red bat. *Proc. Biol. Soc. Washington* 45:148.

Gomes, N. F. 1991. Revisão sistemática do gênero *Monodelphis.* Master's Thesis, Universidade de São Paulo, 6 unnumbered pp. + 174 + 177–180 + 16 tables on 30 pp + 3 pls. + 34 figs + maps 1–3, 5–6.

González, E. M. 1996. *Comadreja colorada chica Monodelphis dimidiata (Wagner, 1847) Clase Mammalia orden Didelphimorphia Familia Didelphidae. Parques Municipales Protegidos, Serie Fauna Silvestre del Parque Lecocq.* Montevideo: Departamento de Cultura, IMM y Vida silvestre, Sociedad Uruguaya para la Conservación de la Naturaleza, 2:1–4.

———. 2001. *Guía de campo de los mamíferos de Uruguay. Introducción al estudio de los mamíferos.* Montivideo: Vida Silvestre, 339 pp.

González, E. M., and S. Claramunt. 2000. Behaviors of captive short-tailed opossums, *Monodelphis dimidiata* (Wagner, 1847) (Didelphimorphia, Didelphidae). *Mammalia* 64:271–85.

González, E. M., and G. Fregueiro. 1998. Primer registro de *Chironectes minimus* para Uruguay (Mammalia, Didelphidae). *Comunic. Zool. Mus. Hist. Nat. Montevideo* 12:1–6.

González, E. M., S. J. Claramunt, and A. M. Saralegui. 1999. Mamíferos hallados en egagrópilas de *Tyto alba* (Aves, Stigiformes, Tytonidae) en Bagé, Rio Grande do Sul, Brasil. *Iheringia*, sér. zool., 86:117–120.

González, E. M., A. M. Saralegui, and G. Fregueiro. 2000. The genus *Thylamys* Gray, 1843 in Uruguay (Didelphimorphia, Didelphidae). *Bol. Soc. Zool. Uruguay* 12:44–45.

González, J. C. 1973. Observaciones sobre algunos mamíferos de Bopicuá (Dpto. de Río Negro, Uruguay). *Comunic. Mus. Mun. Hist. Nat. Río Negro, Uruguay,* zool. 1:1–14.

———. 1977. Sobre la presencia de *Tadarida molossus* Pallas (Chiroptera, Molossidae) en el Uruguay. *Rev. Biol. Uruguay* 5:27–30.

———. 1985. Presencia de *Marmosa agilis chacoensis* el el Uruguay (Mammalia, Marsupialia, Didelphidae). *Comunic. Zool. Mus. Hist. Nat. Montevideo* 11:1–8 + 1 pl.

———. 2004. Primeiro registro de *Eumops patagonicus* Thomas, 1924 para o Brasil (Mammalia: Chiroptera: Molossidae). *Comun. Mus. Ciêc. Tecnol. PUCRS, Porto*

Alegre, sér. zool., 16:255–58. [Dated 2003; published 31 January 2004.]

González, J. C., and M. E. Fabián. 1995. Una nueva especie de murciélago para el estado de Rio Grande do Sul, Brasil: *Myotis riparius* Handley, 1960 (Chiroptera, Vespertilionidae). *Comun. Mus. Ciênc. Tecnol. PUCRS, Porto Alegre*, sér. zool., 8:55–59.

González, J. C., R. V. Marques, and S. M. Pacheco. 1997. Ocurrência de *Micoureus cinereus paraguayanus* (Tate), (Mammalia, Didelphidia, Marmosidae), no Rio Grande do Sul, Brasil. *Rev. Bras. Zool.* 14:195–200.

Goodwin, G. G. 1938. A new genus of bat from Costa Rica. *Amer. Mus. Novit.*, no. 976:1–2.

———. 1940. Three new bats from Honduras and the first record of *Enchisthenes harti* (Thomas) for North America. *Amer. Mus. Novit.*, no. 1075:1–3.

———. 1942a. New Pteronotus from Nicaragua. *J. Mammal.* 23:88.

———. 1942b. A summary of recognizable species of Tonatia, with descriptions of two new species. *J. Mammal.* 23:204–09.

———. 1942c. Mammals of Honduras. *Bull. Amer. Mus. Nat. Hist.* 79:107–95.

———. 1946. Mammals of Costa Rica. *Bull. Amer. Mus. Nat. Hist.* 87:271–473.

———. 1953. Catalogue of type specimens of Recent mammals in the American Museum of Natural History. *Bull. Amer. Mus. Nat. Hist.* 102:207–412.

———. 1954a. A new short-tailed shrew and a new free-tailed bat from Tamaulipas, Mexico. *Amer. Mus. Novit.*, no. 1670:1–3.

———. 1954b. A new short-tailed shrew from western Panama. *Amer. Mus. Novit.*, no. 1677:1–2.

———. 1956a. A preliminary report on the mammals collected by Thomas MacDougall in southeastern Oaxaca, Mexico. *Amer. Mus. Novit.*, no. 1757:1–15.

———. 1956b. Seven new mammals from Mexico. *Amer. Mus. Novit.*, no. 1791:1–10.

———. 1958a. Two new mammals from México. *Amer. Mus. Novit.*, no. 1871:1–3.

———. 1958b. Three new bats from Trinidad. *Amer. Mus. Novit.*, no. 1877:1–6.

———. 1958c. Bats of the genus *Rhogeessa*. *Amer. Mus. Novit.*, no. 1923:1–17.

———. 1959a. Descriptions of some new mammals. *Amer. Mus. Novit.*, no. 1967:1–8.

———. 1959b. Bats of the subgenus *Natalus*. *Amer. Mus. Novit.*, no. 1977:1–22.

———. 1960. The status of *Vespertilio auripendulus* Shaw, 1800, and *Molossus ater* Geoffroy, 1805. *Amer. Mus. Novit.*, no. 1994:1–6.

———. 1961. The murine opossums (genus *Marmosa*) of the West Indies, and the description of a new subspecies of *Rhipidomys* from Little Tobago. *Amer. Mus. Novit.*, no. 2070:1–20.

———. 1963. American bats of the genus *Vampyressa* with the description of a new species. *Amer. Mus. Novit.*, no. 2125:1–24.

———. 1969. Mammals from the state of Oaxaca, Mexico, in the American Museum of Natural History. *Bull. Amer. Mus. Nat. Hist.* 141:1–269, pls. 1–40.

Goodwin, G. G., and A. M. Greenhall. 1961. A review of the bats of Trinidad and Tobago. *Bull. Amer. Mus. Nat. Hist.* 122:187–302, pls. 7–46.

———. 1962. Two new bats from Trinidad, with comments on the status of the genus *Mesophylla*. *Amer. Mus. Novit.*, no. 2080:1–18.

———. 1964. New records of bats from Trinidad and comments on the status of *Molossus trinitatus* Goodwin. *Amer. Mus. Novit.*, no. 2195:1–23.

Gosse, P. H. 1851. *A naturalist's sojourn in Jamaica*. London: Longmans, Brown, Green, and Longmans, xxiv + 508 pp., 8 pls.

Graciolli, G. 2004. Nycteribiidae (Diptera, Hippoboscoidea) no sul do Brasil. *Rev. Bras. Zool.* 21:971–85.

Graciolli, G., and L. S. Aguiar. 2002. Ocurrência de moscas ectoparasitas (Diptera, Streblidae e Nycteribiidae) de morcegos (Mammalia, Chiroptera) no cerrado de Brasília, Distrito Federal, Brasil. *Rev. Bras. Zool.* 19(Suppl. 1):177–81.

Graciolli, G., and E. Bernard. 2002. Novos registros de moscas ectoparasitas (Diptera, Streblidae e Nycteribiidae) em morcegos (Mammalia, Chiroptera) do Amazonas e Pará, Brasil. *Rev. Bras. Zool.* 19(Suppl. 1):77–86.

Graciolli, G., and C. J. B. de Carvalho. 2001. Moscas estoparasitas (Diptera, Hippoboscoidea, Nycteribiidae) de morcegos (Mammalia, Chiroptera) do estado do Paraná, Brasil. I. *Basilia*, taxonomía e chave pictórica para as espécies. *Rev. Bras. Zool.* 18(Suppl. 1):33–49.

Graciolli, G., and D. C. Coelho. 2001. Streblidae (Diptera, Hippoboscoidea) sobre morcegos filostomídeos (Chiroptera, Phyllostomidae) em cavernas do Distrito Federal, Brasil. *Rev. Bras. Zool.* 18:965–70.

Graciolli, G., and A. M. Rui. 2001. Streblidae (Diptera, Hippoboscoidea) em morcegos (Chiroptera, Phyllostomidae) no nordeste do Rio Grande do Sul, Brasil. *Iheringia*, sér. zool., 90:85–92.

Graham, G. L. 1983. Changes in bat species diversity along an elevational gradient up the Peruvian Andes. *J. Mammal.* 64:559–71.

———. 1987. Seasonality of reproduction in Peruvian bats. In *Studies in Neotropical mammalogy, essays in honor of*

Philip Hershkovitz, ed. B. D. Patterson and R. M. Timm, 173–86. *Fieldiana Zool.* 39:frontispiece, viii + 1–506.

———. 1988. Interspecific associations among Peruvian bats at diurnal roosts and roost sites. *J. Mammal.* 69:711–20.

Graham, G. L., and L. J. Barkley. 1984. Noteworthy records of bats from Peru. *J. Mammal.* 65:709–11.

Graipel, M. E., J. J. Cherem, and A. Ximenez. 2001. Mamíferos terrestres náo voadores da Ilha de Santa Catarina, sul do Brasil. *Biotemas* 14:109–40.

Graipel, M. E., J J. Cherem, P. R. M. Miller, and L. Glock. 2004. Trapping small mammals in the forest understory: A comparison of three methods. *Mammalia* 67:551–58. [Dated 2003; number 4 marked "Achevé d'imprimer le 31 mars 2004."]

Grandidier, G. and M. Neveu-Lemaire. 1905. Description d'une nouvelle espèce de tatou, type d'un genre nouveau (Tolypoïdes bicinctus). *Bull. Mus. Hist. Nat., Paris* 7:370–72.

———. 1908. Observations relatives à quelques tatou rares ou inconnus habitant la "Puna" argentine et bolivienne. *Bull. Mus. Hist. Nat., Paris* 14:4–7, 2 pls.

Gray, J. E. 1821. On the natural arrangement of vertebrose animals. *London Med. Reposit.* 15:297–311.

———. 1825a. An attempt at a division of the family Vespertilionidae into groups. *Zool. J.* 2:242–43.

———. 1825b. An outline of an attempt at the disposition of Mammalia into tribes and families, with a list of the genera apparently appertaining to each tribe. *Ann. Philosophy,* 10:337–44.

———. 1827. A synopsis of the species of the class Mammalia. In *The animal kingdom arranged in conformity with its organization, by the Baron Cuvier, with additional descriptions of all the species hitherto named, and of many not before noticed, by Edward Griffith . . . and others,* ed. E. Griffith, 1–296. London: Geo. B. Whittaker, 5:1–392.

———. 1829. An attempt to improve the natural arrangement of the genera of bat, from actual examination; with some observations on the developement of their wings. *Phil. Mag.,* 6:28–36.

———. 1831. Descriptions of some new genera and species of bats. *Zoological Misc.* 1:37–38.

———. 1838a. [Revision of the genus *Sorex*, Linn.] *Proc. Zool. Soc. London* 1837:123–26.

———. 1838b. A revision of the genera of bats (Vespertilionidae), and the description of some new genera and species. *Mag. Zool. Bot.* 2:483–505.

———. 1839. Descriptions of some Mammalia discovered in Cuba by W. S. MacLeay, Esq. *Ann. Nat Hist.* 4:1–7.

———. 1841. Appendix C. Contributions towards the geographical distribution of the Mammalia of Australia, with notes on some recently discovered species. In *Journals of two expeditions of discovery in north-west and western Australia, during the years 1837, 38, and 39, under the authority of Her Majesty's Government,* ed. G. Grey, 397–414. London: T. and W. Boone, 2:frontispiece, xii + 482 pp.

———. 1842. Descriptions of some new genera and fifty unrecorded species of Mammalia. *Ann. Mag. Nat. Hist.,* ser. 1, 10:255–67.

———. 1843a. [Untitled letter addressed to the Curator]. *Proc. Zool. Soc. London* 1843:50.

———. 1843b. *List of the specimens of Mammalia in the collection of the British Museum.* London: British Museum (Natural History), xxviii + 216 pp.

———. 1843c. Specimens of Mammalia from Coban in Central America. *Proc. Zool. Soc. London* 1843:79

———. 1844. Mammalia. In *The zoology of the voyage of H.M.S. Sulfur, under the command of Captain Sir Edward Belcher, R.N., C.B., F.R.G.S., etc. during the years 1836–42,* ed. R. B. Hinds, 7–36, pls. 1–18. London: Smith, Elder and Co., 1:1–50, pls. 1–34.

———. 1846. *Catalog of the specimens and drawings of Mammalia and birds of Nepal and Thibet presented by B. H. Hodgson, Esq. to the British Museum.* London: British Museum (Natural History), xii + 156 pp.

———. 1847a. Characters of six new genera of bats not hitherto distinguished. *Proc. Zool. Soc. London* 1847:14–16.

———. 1847b. *List of the osteological specimens in the collection of the British Museum.* London: British Museum (Natural History), xxvi + 147 pp.

———. 1849. Observations on some Brazilian bats with the description of a new genus. *Proc. Zool. Soc. London* 1848:57–58. [Published 30 January 1849.]

———. 1850. On the genus *Bradypus* of Linnaeus. *Proc. Zool. Soc. London* 1849:65–73, pls. 10, 11.

———. 1865. Revision of the genera and species of entomophagous Edentata, founded on the examination of the specimens in the British Museum. *Proc. Zool. Soc. London* 1865:359–86.

———. 1866a. Synopsis of the genera of *Vespertilionidae* and *Noctilionidae. Ann. Mag. Nat. Hist.,* ser. 3, 17:89–93.

———. 1866b. Revision of the genera of Phyllostomidae, or leaf-nosed bats. *Proc. Zool. Soc. London* 1866:111–18.

———. 1869. *Catalogue of carnivorous, pachydermatous, and edentate Mammalia in the British Museum.* London: British Museum (Natural History), vii + 398 pp.

———. 1871a. On a new species of three-toed sloth from Costa Rica. *Ann. Mag. Nat. Hist.,* ser. 4, 7:302.

————. 1871b. Notes on the species of Bradypodidae in the British Museum. *Proc. Zool. Soc. London* 1871:428–49.

————. 1873a. *Hand-list of the edentate, thick-skinned and ruminant mammals in the British Museum.* London: British Museum (Natural History), viii + 176 pp., 42 pls.

————. 1873b. On Mammalia from the neighbourhood of Concordia, in New Granada. *Ann. Mag. Nat. Hist.*, ser. 4, 11:468–69.

————. 1874. On the short-tailed armadillo (*Muletia septemcincta*). *Proc. Zool. Soc. London* 1874:244–46, pl. 41.

Greegor, D. H., Jr. 1975. Renal capabilities of an Argentine desert armadillo. *J. Mammal.* 56:626–32.

————. 1985. Ecology of the little hairy armadillo *Chaetophractus vellerosus.* In *Evolution and ecology of armadillos, sloths, and vermilinguas,* ed. G. G. Montgomery, 397–406. Washington, DC: The Smithsonian Institution Press, 10 (unnumbered), 451 pp.

Greenbaum, I. F., and J. K. Jones, Jr. 1978. Noteworthy records of bats from El Salvador, Honduras, and Nicaragua. *Occas. Papers Mus., Texas Tech Univ.*, no. 55:1–7.

Greenbaum, I. F., R. J. Baker, and D. E. Wilson. 1975. Evolutionary implications of the karyotypes of the stenodermine genera *Ardops, Ariteus, Phyllops,* and *Ectophylla. Bull. S. California Acad. Sci.* 74:156–59.

Greene, H. W. 1989. Agonistic behavior by three-toed sloths, *Bradypus variegatus. Biotropica* 21:369–72.

Greenhall, A. M. 1956. The food of some Trinidad fruit bats (*Artibeus and Carollia*). *J. Agri. Soc. Trinidad Tobago*, no. 869:1–25.

————. 1957. Food preferences of Trinidad fruit bats. *J. Mammal.* 38:409–10.

————. 1970. The use of the precipitin test to determine host preferences of the vampire bats, *Desmodus rotundus* and *Diaemus youngi. Bijdr. Dierk.* 40:36–39.

————. 1972. The biting and feeding habits of the vampire bat, *Desmodus rotundus. J. Zool. Soc. London* 168:451–61.

————. 1988. Feeding behavior. In *Natural history of vampire bats,* ed. A. M. Greenhall and U. Schmidt, 111–32. Boca Raton, Florida: CRC Press, frontispiece, 14 unnumbered + 246 pp.

Greenhall, A. M., and J. L. Paradiso. 1968. Bats and bat banding. *Resource Publ., Bur. Sport Fish. Wildl.* 72:1–48.

Greenhall, A. M., and U. Schmidt, eds. 1988. *Natural history of vampire bats.* Boca Raton, Florida: CRC Press, frontispiece, 14 unnumbered + 246 pp.

Greenhall, A. M., and W. A. Schutt, Jr. 1996. Diamus youngi. *Mammal. Species* 533:1–7.

Greenhall, A. M., U. Schmidt, and G. Joermann. 1984. Diphylla ecaudata. *Mammal. Species* 227:1–3.

Greenhall, A. M., G. Joermann, U. Schmidt, and M. R. Seidel. 1983. Desmodus rotundus. *Mammal. Species* 202:1–6.

Greer, J. K. 1966. Mammals of Malleco Province, Chile. *Publ. Mus., Michigan State Univ.*, biol. ser., 3:49–152.

Gregorin, R. 1998a. Notes on the geographic distribution of Neoplatymops mattogrossensis (Vieira, 1942) (Chiroptera, Molossidae). *Chiropt. Neotrop.* 4:88–90.

————. 1998b. Extending geographic distribution of *Chiroderma doriae* Thomas, 1891 (Phyllostimidae, Stenodermatinae). *Chiropt. Neotrop.* 4:98–99.

————. 2001. Second record of *Eumops hansae* (Molossidae) in southeastern Brazil. *Bat Res. News* 42:50–51.

Gregorin, R, and A. D. Ditchfield. 2005. New genus and species of nectar-feeding bat in the tribe Lonchophyllini (Phyllostomidae: Glossophaginae) from northeastern Brazil. *J. Mammal.* 86:403–14.

Gregorin, R., and L. de F. Mendes. 1999. Sobre quirópteros (Emballonuridae, Phyllostomidae, Natalidae) de duas cavernas da Chapada Diamantina, Bahia, Brasil. *Iheringia*, sér. zool., 86:121–24.

Gregorin, R., and R. V. Rossi. 2005. *Glyphonycteris daviesi* (Hill, 1964), a rare Central American and Amazonian bat recorded for eastern Brazilian Atlantic forest (Chiroptera, Phyllostomidae). *Mammalia* 69:427–30.

Gregorin, R., and V. A. Taddei. 2001. New records of *Molossus* and *Promops* from Brazil (Chiroptera: Molossidae). *Mammalia* 64:471–76. [Dated 2000; number 4 marked "Achevé d'imprimer le 25 janvier 2001."]

Gregorin, R., B. K. Lim, W. A. Pedro, and F. C. Passos, V. A. Taddei. 2004. Distributional extension of *Molossops neglectus* (Chiroptera: Molossidae) into southeastern Brazil. *Mammalia* 68:233–37.

Gregorin, R., E. Gonçalves, B. K. Lim, and M. D. Engstrom. 2006. New species of disk-winged bat *Thyroptera* and range extension for *T. discifera. J. Mammal.* 87:238–46.

Gregory, W. K. 1910. The orders of mammals. *Bull. Amer. Mus. Nat. Hist.* 27:1–524.

————. 1922. On the "habitus" and "heritage" of Caenolestes. *J. Mammal.* 3:106–14.

Gribel, R. 1988. Visits of *Caluromys lanatus* (Didelphidae) to flowers of *Pseudobombax tomentosum* (Bombacaceae): A probable case of pollination by marsupials in central Brazil. *Biotropica* 20:344–47.

Griebel, R., and V. A. Taddei. 1989. Notes on the distribution of *Tonatia schulzi* and *Tonatia carrikeri* in the Brazilian Amazon. *J. Mammal.* 70:871–73.

Griffin, D. R. 1953. Acoustic orientation in tropical bats. *Science* 188:571.

Griffith, E., C. Hamilton-Smith, and E. Pidgeon. 1827. The class Mammalia arranged by the Baron Cuvier, with specific descriptions. In *The Animal Kingdom arranged in conformity with its organization, by the Baron Cuvier, . . . with additional descriptions of all the species hitherto named, and of many not before noticed, by Edward Griffith . . . and others*. London: Geo. B. Whittaker, 3:1–468.

Griffiths, T. A. 1978. Muscular and vascular adaptations for nectar-feeding in the glossophagine bats *Monophyllus* and *Glossophaga*. *J. Mammal.* 59:414–18.

———. 1982. Systematics of the New World nectar-feeding bats (Mammalia, Phyllostomidae), based on the morphology of the hyoid and lingual regions. *Amer. Mus. Novit.*, no. 2742:1–45.

———. 1983a. On the phylogeny of the Glossophaginae and the proper use of outgroup analysis. *Syst. Zool.* 32:283–85.

———. 1983b. Comparative laryngeal anatomy of the big brown bat, *Eptesicus fuscus*, and the mustached bat, *Pteronotus parnellii*. *Mammalia* 47:377–94.

Griffiths, T. A., and B. B. Criley. 1989. Comparative lingual anatomy of the bats *Desmodus rotundus* and *Lonchophylla robusta* (Chiroptera: Phyllostomidae). *J. Mammal.* 70:608–13.

Grimwood, I. R. 1969. Notes on the distribution and status of some Peruvian mammals 1968. *Amer. Comm. Internat. Wild Life Protection and New York Zool. Soc.*, Special Publ., no. 21:vi + 86 pp., 4 pls.

Grose, E. S., and C. J. Marinkelle. 1966. Species of *Sporotricum* [sic], *Trichophyton* and *Microsporum* from Columbian bats. *Trop. Geogr. Med.* 18:260–63.

———. 1968. A new species of *Candida* from Colombian bats. *Mycopathol. Mycol. Appl.* 36:225–27.

Grose, E. S., and J. R. Tamsitt. 1965. *Paracoccidioides brasiliensis* recovered from the intestinal tract of three bats (*Artibeus lituratus*) in Colombia, S.A. *Sabouraudia* 4:124–25.

Grose, E. S., C. J. Marinkelle, and C. Striegel. 1968. The use of tissue cultures in the identification of *Cryptococcus neoformans* isolated from Colombian bats. *Sabouraudia* 6:127–32.

Grubb, P. 2001. Case 3022, *Catalogue des mammifères du Muséum d'Histoire Naturelle* by Étienne Geoffroy Saint-Hilaire (1803): Proposed placement on the Official List of Works Approved as Available for Zoological Nomenclature. *Bull. Zool. Nomen.* 58:41–52.

Guedes, P. G., S. S. P. da Silva, A. R. Camardella, M. F. G. de Abreu, D. M. Borges-Nojosa, J. A. G. da Silva, and A. A. Silva. 2000. Diversidade de mamíferos do Parque Nacional de Ubajara (Ceará, Brasil). *Mastozool. Neotrop.* 1:95–100.

Guerra, D. Q. 1981a. *Peropteryx (Peronymus) leucopterus* Peters, 1867 no nordeste do Brasil (Chiroptera-Emballonuridae). *Rev. Nordestina Biol.* 3(especial):137–39. [Dated 1980; published February 1981.]

———. 1981b. Registro adicional de *Phylloderma stenops* Peters, 1865 (Chiroptera—Phyllostomatidae) para o Brasil. *Rev. Nordestina Biol.* 3(especial):141–43. [Dated 1980; published February 1981.]

———. 1981c. Presença de *Cabassous unicinctus* (Linné, 1758) (Edentata-Dasypodidae) no estado de Pernambuco, NE do Brasil. *Rev. Nordestina Biol.* 3(especial):181–83. [Dated 1980; published February 1981.]

Guerrero, J. A., E. de Luna, and D. González. 2004. Taxonomic status of *Artibeus jamaicensis triomylus* infered from molecular and morphometric data. *J. Mammal.* 85:866–74.

Guerrero, J. A., E. de Luna, and C. Sánchez-Hernández. 2003. Morphometrics in the quantification of character state identity for the assessment of primary homology: An analysis of character variation of the genus *Artibeus* (Chiroptera: Phyllostomidae). *Biol. J. Linnean Soc.* 80:45–5.

Guerrero, R. 1985a. Parasitología. In *El estudio de los mamíferos en Venezuela evaluación y perspectivos*, ed. M. Aguilera, 35–91. Caracas: Asociación Venezolana para el Estudio de los Mamíferos, 256 pp.

———. 1985b. Nematoda: Trichostrongyloidea parásitos de mamíferos silvestres de Venezuela. II Revisión del género *Viannaia* Travassos, 1914. *Mem. Soc. Cien. Nat. La Salle* 45:9–48.

———. 1996. Estudio preliminar de los ectoparásitos de los murciélagos de Pakitza, Parque Nacional Manu (Perú). In *Manu*, ed. D. E. Wilson and A. Sandoval, 643–58. Lima, Peru: Editorial Horizonte, 679 pp.

———. 1997. Catalogo de los Streblidae (Diptera: Pupipara) parasitos de murciélagos (Mammalia: Chiroptera) del Nuevo Mundo. VII. Lista de especies, hospedadores y paises. *Acta. Biol. Venez.* 17:9–24.

Guerrero, R., R. Hoogesteijn, and R. Soriano. 1989. Lista preliminary de los mamíferos del Cerro Marahuaca, T. F. Amazonas, Venezuela. *Acta Terramaris*, no. 1:71–77.

Guimarães, L. R. 1972. Nycteribiid batflies from Venezuela (Diptera: Nycteribiidae). *Brigham Young Univ. Sci. Bull.*, biol. ser., 17(1):1–11.

Guimarães, L. R., and M. A, V. d'Andretta. 1956. Sinopse dos *Nycteribiidae* (Diptera) do Novo Mundo. *Arq. Zool., Estado São Paulo* 10:1–184.

Gundlach, J. 1840. Beschreibung von vier auf Cuba gefangenen Fledermäusen. *Arch. Naturgesch.* 6(1):356–58.

Günther, A. 1879. Description of a new species of *Didelphys* from Demerara. *Ann. Mag. Nat. Hist.*, ser. 5, 4:108.

Gustin, M. K., and G. F. McCracken. 1987. Scent recognition between females and pups in the bat *Tadarida brasiliensis mexicana. Animal Behav.* 35:13–19.

Gut, H. J. 1959. A Pleistocene vampire bat from Florida. *J. Mammal.* 40:534–37.

Gut, H. J., and C. E. Ray. 1963. The Pleistocene vertebrate fauna of Reddick, Florida. *Quarterly J. Florida Acad. Sci.* 26:315–28.

Guthrie, D. A. 1980. Analysis of avifaunal and bat remains from midden sites on San Miguel Island. In *The California Islands: Proceedings of a multidisciplinary symposium*, ed. D. M. Power, 689–702. Santa Barbara: Santa Barbara Mus. Nat. Hist., 787 pp.

Hagmann, G. 1908. Die Landsäugetiere der Insel Mexiana. Als Beispiel der Einwirkung der Isolation auf die Umbildung der Arten. *Arch. Rass.- Gesell.-Biol., München* 5:1–31, 2 pls.

Hahn, W. L. 1905. A new bat from Mexico. *Proc. Biol. Soc. Washington* 18:247–48.

———. 1907. A review of the bats of the genus Hemiderma. *Proc. U.S. Natl. Mus.* 32:103–18.

Hall, E. R. 1951. A new name for the Mexican red bat. *Univ. Kansas Publ., Mus. Nat. Hist.* 5:223–26.

———. 1981. *The mammals of North America.* 2nd ed. New York: John Wiley and Sons, 1:xviii + 600 + 90 pp.

Hall, E. R., and W. W. Dalquest. 1963. The mammals of Veracruz. *Univ. Kansas Publ., Mus. Nat. Hist.* 14:165–362.

Hall, E. R., and W. B. Jackson. 1953. Seventeen species of bats recorded from Barro Colorado Island, Panama Canal Zone. *Univ. Kansas Publ., Mus. Nat. Hist.* 5:641–46.

Hall, E. R., and J. K. Jones, Jr. 1961. North American yellow bats, "Dasypterus," and a list of the named kinds of the genus Lasiurus Gray. *Univ. Kansas Publ., Mus. Nat. Hist.* 14:73–98.

Hall, E. R., and K. R. Kelson. 1952. The subspecific status of two Central American sloths. *Univ. Kansas Publ., Mus. Nat. Hist.* 5:313–17.

———. 1959. *The mammals of North America.* New York: Ronald Press Co., 1:xxx + 546 + 79 pp.

Hamlett, G. W. D. 1939. Identity of Dasypus septemcinctus Linnaeus with notes on some related species. *J. Mammal.* 20:328–36.

Handley, C. O., Jr. 1955. A new species of free-tailed bat (genus *Eumops*) from Brazil. *Proc. Biol. Soc. Washington* 68:177–78.

———. 1956. A new species of free-tailed bat (genus *Mormopterus*) from Peru. *Proc. Biol. Soc. Washington* 69:197–202.

———. 1957. A new species of murine opossum (genus *Marmosa*) from Peru. *J. Washington Acad. Sci.* 46:402–04. [Dated 1956; published 13 February 1957.]

———. 1960. Descriptions of new bats from Panama. *Proc. U.S. Natl. Mus.* 112:459–79.

———. 1966a. Descriptions of new bats (*Choeroniscus* and *Rhinophylla*) from Colombia. *Proc. Biol. Soc. Washington* 79:83–88. [Published 23 May 1966.]

———. 1966b. Descriptions of new bats (*Chiroderma* and *Artibeus*) from Mexico. *Anal. Inst. Biol. Univ. Nac. Autó. México* 36:297–301. [Dated 1965; published 20 June 1966.]

———. 1966c. Checklist of the mammals of Panama. In *Ectoparasites of Panama*, ed. R. L. Wenzel and V. J. Tipton, 753–795, map. Chicago: Field Museum of Natural History, xii + 861 pp. [Published 22 November 1966.]

———. 1967. Bats of the canopy of an Amazonian forest. *Atas do simpósio sôbre a biota Amazônica* 5:211–15.

———. 1976. Mammals of the Smithsonian Venezuelan Project. *Brigham Young Univ. Sci. Bull.*, biol. ser., 20(5):1–89, + 2, map.

———. 1980. Inconsistencies in formation of family-group and subfamily-group names in Chiroptera. In *Proceedings Fifth International Bat Research Conference*, ed. D. E. Wilson and A. L. Gardner, 9–13. Lubbock: Texas Tech University Press, 434 pp.

———. 1984. New species of mammals from northern South America: A long-tongued bat, genus *Anoura* Gray. *Proc. Biol. Soc. Washington* 97:513–21.

———. 1987. New species of mammals from northern South America: Fruit-eating bats, genus *Artibeus* Leach. In *Studies in Neotropical mammalogy, essays in honor of Philip Hershkovitz*, ed. B. D. Patterson and R. M. Timm, 163–72. *Fieldiana Zool.* 39:frontispiece, viii + 1–506.

———. 1990. The *Artibeus* of Gray 1838. In *Advances in Neotropical mammalogy* ed. J. F. Eisenberg, 443–68. Gainesville: Sandhill Crane Press. [Dated 1989; published 6 February 1990.]

———. 1991. The identity of *Phyllostoma planirostre* Spix, 1823 (Chiroptera: Stenodermatinae). In *Contributions to mammalogy in honor of Karl F. Koopman*, ed. T. A. Griffiths and D. Klingener, 12–17. *Bull. Amer. Mus. Nat. Hist.* 206:1–432.

———. 1996. New species of mammals from northern South America: Bats of the genera *Histiotus* Gervais and *Lasiurus* Gray (Mammalia: Chiroptera: Vespertilionidae). *Proc. Biol. Soc. Washington* 109:1–9.

Handley, C. O., Jr., and J. R. Choate. 1970. The correct name for the least short-tailed shrew (*Cryptotis parva*) of Guatemala (Mammalia: Insectivora). *Proc. Biol. Soc. Washington* 83:195–202.

Handley, C. O., Jr., and K. C. Ferris. 1972. Descriptions of new bats of the genus *Vampyrops*. *Proc. Biol. Soc. Washington* 84:519–24.

Handley, C. O., Jr., and A. L. Gardner. 1990. The holotype of *Natalus stramineus* Gray (Mammalia: Chiroptera: Natalidae). *Proc. Biol. Soc. Washington* 103:966–72.

Handley, C. O., Jr., and L. K. Gordon. 1979. New species of mammals from northern South America, genus *Marmosa* Gray. In *Vertebrate ecology in the northern Neotropics*, J. F. Eisenberg, 65–72. Washington, DC: The Smithsonian Institution Press, 271 pp.

Handley, C. O., Jr., and E. G. Leigh, Jr. 1991. Diet and food supply. In *Demography and natural history of the common fruit bat*, Artibeus jamaicensis, *on Barro Colorado Island, Panamá*, ed. C. O. Handley, Jr., D. E. Wilson, and A. L. Gardner, 147–49. *Smithsonian Contrib. Zool.* 511:iv + 1–173.

Handley, C. O., Jr., and D. W. Morrison. 1991. Foraging behavior. In *Demography and natural history of the common fruit bat*, Artibeus jamaicensis, *on Barro Colorado Island, Panamá*, ed. C. O. Handley, Jr., D. E. Wilson, and A. L. Gardner, 137–40. *Smithsonian Contrib. Zool.* 511:iv + 1–173.

Handley, C. O., Jr., and J. Ochoa G. 1997. New species of mammals from northern South America: A sword-nosed bat, genus *Lonchorhina* Tomes (Chiroptera: Phyllostomidae). *Mem. Soc. Cien. Nat. La Salle* 57:71–82.

Handley, C. O., Jr., A. L. Gardner, and D. E. Wilson. 1991. Food habits. In *Demography and natural history of the common fruit bat*, Artibeus jamaicensis, *on Barro Colorado Island, Panamá*, ed. C. O. Handley, Jr., D. E. Wilson, and A. L. Gardner, 141–46. *Smithsonian Contrib. Zool.* 511:iv + 1–173.

Handley, C. O., Jr., D. E. Wilson, and A. L. Gardner. 1991. Introduction. In *Demography and natural history of the common fruit bat*, Artibeus jamaicensis, *on Barro Colorado Island, Panamá*, ed. C. O. Handley, Jr., D. E. Wilson, and A. L. Gardner, 1–8. *Smithsonian Contrib. Zool.* 511:iv + 1–173.

Harlan, R. 1825. Description of a new genus of mammiferous quadrupeds, of the order Edentata. *Ann. Lyc. Nat. Hist. New York* 1:235–45.

Harrison, D. L. 1975. Macrophyllum macrophyllum. *Mammal. Species* 62:1–3.

Harrison, D. L., and J. H. M. Horne 1971. The palate of *Ectophylla* (*Mesophylla*) *macconnelli* Thomas, 1901 (*Chiroptera: Phyllostomatidae*), with comparative notes on the palates of some other phyllostomatid bats. *Mammalia* 35:245–53.

Harrison, D. L., and N. Pendleton. 1973. The palate and baculum of some funnel-eared bats (*Chiroptera: Natalidae*). *Mammalia* 37:427–32.

———. 1975. A second record of Wied's long-legged bat (*Macrophyllum macrophyllum* Schinz, 1821, Chiroptera: Phyllostomatidae) in El Salvador, with notes on the palate, reproduction and diet of the species. *Mammalia* 38:689–93. [Dated 1974; number 4 marked "Achevé d'imprimer le 30 juin 1975."]

Harrison, D. L., N. G. E. Pendleton, and G. C. D. Harrison. 1979. *Eumops dabbenei* Thomas, 1914 (Chiroptera: Molossidae), a free-tailed bat new to the fauna of Paraguay. *Mammalia* 43:251–52.

Hass, A., F. H. G. Rodrigues, and T. G. de Oliveira. 2003. The yellow armadillo, *Euphractus sexcinctus*, in the north/northeastern Brazilian coast. *Edentata* 5:46–47.

Hatt, R. T., H. I. Fisher, D. A. Langebartel, and G. W. Brainerd. 1953. Faunal and archaeological researches in Yucatan caves. *Bull. Cranbrook Inst. Sci.* 33:1–119.

Hayman, D. L. 1990. Marsupial cytogenetics. *Australian J. Zool.* 37:331–49.

Hayman, D. L., and P. G. Martin. 1974. Mammalia I: Monotremata and Marsupialia. *Animal Cytogenet.* 4:iv + 110 pp.

Hayman, D. L., J. A. W. Kirsch, P. G. Martin, and P. F. Waller. 1971. Chromosomal and serological studies of the Caenolestidae and their implications for marsupial evolution. *Nature* 231:194–95.

Haynes, M. A., and T. E. Lee, Jr. 2004. Artibeus obscurus. *Mammal. Species* 752:1–5.

Hayssen, V. 1980. Observations of the behaviour of the Brazilian short bare-tailed opossum (*Monodelphis domestica* Wagner) in captivity. *Bull. Australian Mammal Soc.* 6:37.

Heinonen-Fortabat, S., and J. C. Chebez. 1997. Los mamíferos de los parques nacionales de la Argentina. *Monografía Especial L. O. L. A.*no. 14:1–72.

Heithaus, E. R. 1982. Coevolution between bats and plants. In *Ecology of bats*, ed. T. H. Kunz, 327–67. New York: Plenum Press, xviii + 425 pp.

Heithaus, E. R., and T. H. Fleming. 1978. Foraging movements of a frugivorous bat *Carollia perspicillata* (Phyllostomatidae). *Ecol. Monogr.* 48:127–43.

Heithaus, E. R., T. H. Fleming, and P. A. Opler. 1975. Foraging patterns and resource utilization in seven species of bats in a seasonal tropical forest. *Ecology* 56:841–54.

Heller, E. 1904. Mammals of the Galapagos Archipelago, exclusive of the Cetacea. *Proc. California Acad. Sci.*, zool., 3:233–50, pl. 223.

Hensel, R. 1872. Beiträge zur Kenntniss der Säugethiere Süd-Brasiliens. *Abhandl. König. Akad. Wiss. Berlin* 1872:1–130, 3 pls.

Herd, R. M. 1983. Pteronotus parnellii. *Mammal. Species* 209:1–5.

Hermann, J. 1804. *Observationes zooligicae quibus novae complures, aliaeque animalium species describuntur et illustrantur. Opus posthumum edidit Fridericus Ludovicus Hammer. . . . Pars prior observationum quatuor centurias continens.* Argentorati: Amandum Koenig; Parisiis: Eundem, viii + 332 pp.

Hernández-Camacho, J. 1955. Una nueva especie Colombiana del género *Diclidurus* (Mammalia: Chiroptera): *Diclidurus ingens. Caldasia* 7:87–98.

Hernández-Camacho, J., and A. Cadena G. 1978. Notas para la revisión del género *Lonchorhina* (*Chiroptera, Phyllostomidae*). *Caldasia* 7:199–251.

Herrera, A. L. 1899. *Sinonimía vulgar y científica de los principales vertebrados Mexicanos.* Mexico: Oficina Tipográfica de la Secretaría de Fomento, 131 pp.

Herrera M., L. G., and C. Martínez del Río. 1998. Pollen digestion by New World bats: Effects of processing time and feeding habits. *Ecology* 79:2828–38.

Herrick, C. J. 1921. The brain of *Caenolestes obscurus. Field Mus. Nat. Hist.*, zool. ser., 14:157–62

Herrin, C. S., and V. J. Tipton. 1975. Spinturnicid mites of Venezuela (Acarina: Spinturnicidae). *Brigham Young Univ. Sci. Bull.*, biol. ser., 20(2):1–72.

Hershkovitz, P. 1949a. Generic names of the four-eyed pouch opossum and the woolly opossum (*Didelphidae*). *Proc. Biol. Soc. Washington* 62:11–12.

———. 1949b. Status of names credited to Oken, 1816. *J. Mammal.* 30:289–301.

———. 1949c. Mammals of northern Colombia, preliminary report no. 5: Bats (Chiroptera). *Proc. U.S. Natl. Mus.* 99:429–54.

———. 1951. Mammals from British Honduras, Mexico, Jamaica and Haiti. *Fieldiana Zool.* 31:547–69.

———. 1959. Nomenclature and taxonomy of the Neotropical mammals described by Olfers, 1818. *J. Mammal.* 40:337–53.

———. 1969. The evolution of mammals on southern continents. VI. The Recent mammals of the Neotropical region: A zoogeographical and ecological review. *Quart. Rev. Biol.* 44:1–70.

———. 1972. The Recent mammals of the Neotropical Region: A zoogeographic and ecological review. In *Evolution, mammals, and southern continents*, ed. A. Keast, F. C. Erk, and B. Glass, 311–431. Albany: State University of New York Press, 543 pp.

———. 1975. The scientific name of the lesser *Noctilio* (Chiroptera), with notes on the chauve-souris de la Vallee d'Ylo (Peru). *J. Mammal.* 56:242–47.

———. 1976a. The taxonomic status of "*Noctilio ruber* Rengger." *Mammalia* 40:164–66.

———. 1976b. Comments on generic names of four-eyed opossums (family Didelphidae). *Proc. Biol. Soc. Washington* 89:295–304.

———. 1981. *Philander* and four-eyed opossums once again. *Proc. Biol. Soc. Washington* 93:943–46.

———. 1987. A history of the Recent mammalogy of the Neotropical Region from 1492 to 1850. In *Studies in Neotropical mammalogy, essays in honor of Philip Hershkovitz*, ed. B. D. Patterson and R. M. Timm, 11–98. *Fieldiana Zool.* 39:frontispiece, viii + 1–506.

———. 1992a. Ankle bones: The Chilean opossum *Dromiciops gliroides* Thomas, and marsupial phylogeny. *Bonn. Zool. Beitr.* 43:181–213.

———. 1992b. The South American gracile mouse opossums, genus *Gracilinanus* Gardner and Creighton, 1989 (Marmosidae, Marsupialia): A taxonomic review with notes on general morphology and relationships. *Fieldiana Zool.*, 70:frontispiece, vi + 1–56.

———. 1997. Composition of the family Didelphidae Gray, 1821 (Didelphoidea: Marsupialia), with a review of the morphology and behavior of the included four-eyed opossums of the genus *Philander* Tiedemann, 1808. *Fieldiana Zool.*, 86:vi + 1–103.

———. 1999. *Dromiciops gliroides* Thomas, 1894, last of the Microbiotheria (Marsupialia), with a review of the family Microbiotheriidae. *Fieldiana Zool.*, 93:1–60.

Hibbard, C. W. 1953. The insectivores of the Rexroad fauna, upper Pliocene of Kansas. *J. Paleo.* 27:21–32.

Hice, C. L., and S. Solari. 2002. First record of *Centronycteris maximiliani* (Fisher, 1829) and two additional records of *C. centralis* Thomas, 1912 from Peru. *Acta Chiropterol.* 4:217–20.

Hice, C. L., P. M. Velazco, and M. R. Willig. 2004. Bats of the Reserva Nacional Allpahuayo-Mishana, northeastern Peru, with notes on community structure. *Acta Chiropterol.* 6:319–34.

Hildebrand, M. 1961. Body proportions of didelphid (and some others) marsupials, with emphasis on variability. *Am. J. Anat.* 109:239–49.

Hill, J. E. 1965. Notes on bats from British Guiana, with the description of a new genus and species of *Phyllostomidae. Mammalia* 28:553–72. [Dated 1964; number 4 published in 1965.]

———. 1966. The status of *Pipistrellus regulus* Thomas (Chiroptera, Vespertilionidae). *Mammalia* 30:302–07.

———. 1980. A note on *Lonchophylla* (Chiroptera: Phyllostomatidae) from Ecuador and Peru, with the description of a new species. *Bull. British Mus. Nat. Hist. (Zool.)* 38:233–36.

———. 1986. The status of *Lichonycteris degener* Miller, 1931 (Chiroptera: Phyllostomidae). *Mammalia*

49:579–82. [Dated 1985; number 4 marked "Achevé d'imprimer le 20 mars 1986."]

———. 1987. A note on *Balantiopteryx infusca* (Thomas, 1897) (Chiroptera: Emballonuridae). *Mammalia* 50:558–60. [Dated 1986; number 4 marked "Achevé d'imprimer le 15 avril 1987."

———. 1990. A memoir and bibliography of Michael Roger Oldfield Thomas, F.R.S. *Bull. Brit. Mus. (Nat. Hist.)*, hist. ser., 18:25–113.

Hill, J. E., and A. Bown. 1963. Occurrence of *Macrophyllum* in Ecuador. *J. Mammal.* 44:588.

Hill, J. E., and D. L. Harrison. 1987. The baculum in the Vespertilioninae (Chiroptera: Vespertilionidae) with a systematic review, a synopsis of *Pipistrellus* and *Eptesicus*, and the description of a new genus and subgenus. *Bull. Brit. Mus. (Nat. Hist.)*, zool. ser., 52:225–305.

Hill, J. E., and J. D. Smith. 1984. *Bats: A natural history.* Austin: Univ. Texas Press, 243 pp.

Hoare, C. A. 1972. *The trypanosomes of mammals.* Oxford: Blackwell Scientific Publications, xviii + 749 pp.

Hodgson, B. H. 1835. Classified catalogue of mammals delineated in the Nepalese Zoology now under publication. *J. Asiatic Soc. Bengal* 4:700.

Hoffmeister, D. F. 1957. Review of the long-nosed bats of the genus *Leptonycteris*. *J. Mammal.* 38:454–61.

Hoffstetter, R. 1961. Description d'une squelette de *Planops* (gravigrade du Miocène de Patagonie). *Mammalia* 25:57–96.

———. 1969. Remarques sur la phylogénie et la classification des Edentés Xenarthres (Mammifères) actuels et fossiles. *Bull. Mus. Natl. Hist. Nat., Paris* 41:92–103.

Holdridge, L. R. 1947. Determination of world plant formations from simple climatic data. *Science* 105:367–68.

Holdridge, L. R. 1967. *Life zone ecology: With photographic supplement prepared by Joseph A. Tosi, Jr.* rev. ed. San José, Costa Rica: Tropical Science Center, 206 pp.

Hollis, L. 2005. Artibeus planirostris. *Mammal. Species* 775:1–6.

Hollister, N. 1914. Four new mammals from tropical America. *Proc. Biol. Soc. Washington* 27:103–06.

———. 1925. The systematic name of the Texas armadillo. *J. Mammal.* 6:60.

Honacki, J. H., K. E. Kinman, and J. W. Koeppl, eds. 1982. *Mammal species of the world.* Lawrence, Kansas: Allen Press Inc. and Association of Systematics Collections, ix + 694 pp.

Honeycutt, R. L., R. J. Baker, and H. H. Genoways. 1980. Results of the Alcoa Foundation-Suriname Expeditions. III. Chromosomal data for bats (Mammalia: Chiroptera) from Suriname. *Ann. Carnegie Mus.* 49:237–50.

Hood, C. S., and R. J. Baker. 1986. G- and C-banding chromosomal studies of bats of the family Emballonuridae. *J. Mammal.* 67:705–11.

Hood, C. S., and J. K. Jones, Jr. 1984. Noctilio leporinus. *Mammal. Species* 216:1–7.

Hood, C. S., and J. Pitocchelli. 1983. Noctilio albiventris. *Mammal. Species* 197:1–5.

Hoofer, S.R., and R. A. Van Den Bussche. 2003. Molecular phylogentics of the chiropteran family Vespertilionidae. *Acta. Chiropterol.* 5(Suppl.):1–63.

Horácek, I., and V. Hanák. 1986. Generic status of *Pipistrellus savii* and comments on classification of the genus *Pipistrellus* (Chiroptera: Vespertilionidae). *Myotis.* nos. 23–24:9–16.

Housse, P. R. 1953. *Animales salvajes de Chile en su clasificación moderna. Su vida y costumbres.* Santiago: Universidad de Chile Publicaciones, 189 pp.

Howe, H. E. 1980. Monkey dispersal and waste of a Neotropical fruit. *Ecology* 61:944–59.

Howell, D. J., and D. Burch. 1974. Food habits of some Costa Rican bats. *Rev. Biol. Trop.* 21:281–94.

Hoyt, R. A., and R. J. Baker. 1980. Natalus major. *Mammal. Species* 130:1–3.

Hoyt, R. A., and J. S. Altenbach. 1981. Observations on *Diphylla ecaudata* in captivity. *J. Mammal.* 62:215–16.

Hsu, T. C. 1965. Chromosomes of two species of anteaters. *Mammal. Chromosomes Newsl.*, no. 14, 108, 1 fig.

Hsu, T. C., and K. Benirschke. 1967. *Dasypus novemcinctus* (nine-banded armadillo). In *An atlas of mammalian chromosomes.* New York: Springer-Verlag, Vol. 1, Folio 5.

———. 1969. *Cabassous centralis* (Central American soft-tailed armadillo). In *An atlas of mammalian chromosomes.* New York: Springer-Verlag, Vol. 3, Folio 107.

———. 1971a. *Euphractus sexcinctus* (six-banded armadillo). In *An atlas of mammalian chromosomes.* New York: Springer-Verlag, Vol. 6, Folio 258.

———. 1971b. Uroderma magnirostrum. In *An atlas of mammalian chromosomes.* New York: Springer-Verlag, Vol. 6, Folio 262.

Hsu, T. C., R. J. Baker, and T. Utakoji. 1968. The multiple sex chromosome system of American leaf-nosed bats (Chiroptera, Phyllostomidae). *Cytogenetics* 7:27–38.

Hummelinck, P. W. 1940. Studies on the fauna of Curaçao, Aruba, Bonaire and the Venezuelan Islands: No. 1–3. *Studies on the Fauna of Curaçao and the other Caribbean Islands* 1:1–130, 16 pls.

Humphrey, S. R., and L. N Brown. 1986. Report of a new bat (Chiroptera: *Artibeus jamaicensis*) in the United States is erroneous. *Florida Sci.* 49:262–63.

Hunsaker, D., II., ed. 1977a. *The biology of marsupials.* London: Academic Press, xv + 537 pp.

Hunsaker, D., II. 1977b. Ecology of New World marsupials. In *The Biology of Marsupials*, ed. D. Hunsaker II, 95–156. New York: Academic Press, xv + 537 pp.

Hunt, J. L., L. A. McWilliams, T. L. Best, and K. G. Smith. 2003. Eumops bonariensis. *Mammal. Species* 733:1–5.

Husband, T. P., G. D. Hobbs, C. N. Santos, and H. J. Stillwell. 1992. First record of *Metachirus nudicaudatus* for northeast Brazil. *Mammalia* 56:298–99.

Husson, A. M. 1954. On Vampyrodes caracciolae (Thomas) and some other bats from the Island of Tobago (British West Indies). *Zool. Med.* 33:63–67.

———. 1959. Notes on the Neotropical leaf-nosed bat *Sphaeronycteris toxophyllum* Peters. *Arch. Néerlandaises Zool.* 13(Suppl. 1):114–19. [Supplement 1 for Vol. 13, 1958 published in 1959.]

———. 1960. De zoogdieren van de Nederlandse Antillen. *Natuurwetens. Werkg. Ned. Ant., Curaçao* 12:viii + 1–170, 42 pls.

———. 1962. The bats of Suriname. *Zool. Verhand., Rijksmus. Nat. Hist., Leiden* 58:1–282, 30 pls.

———. 1963. On *Blarina pyrrhonota* and *Echimys macrourus*: Two mammals incorrectly assigned to the Suriname fauna. *Stud. Fauna Suriname Guyanas* 5:34–41, 2 pls.

———. 1973. Voorlopige lijst van zoogdieren van Suriname. *Zool. Bijdr., Rijksmus. Nat. Hist., Leiden* 14:1–15.

———. 1978. *The mammals of Suriname*. Zoölogische Monographieën van het Rijksmuseum van Natuurlijke Historie No. 2. Leiden: E. J. Brill, xxxiv + 569 pp. + 160 pls.

Hutchison, J. H. 1967. A Pleistocene vampire bat (*Desmodus stocki*) from Potter Creek Cave, Shasta County, California. *PaleoBios* 3:1–6.

Hutterer, R. 1993. Order Insectivora. In *Mammal species of the world*. 2nd ed, ed. D. E. Wilson and D. M. Reeder, 69–130. Washington, DC: The Smithsonian Institution Press.

Hutterer, R., N. Verhaagh, J. Diller, and R. Podloucky. 1995. An inventory of mammals observed at Panguana Biological Station, Amazonian Peru. *Ecotropica* 1:3–20.

Ibáñez, C. 1980. Nuevos datos sobre *Eumops dabbenei* Thomas, 1914 (Chiroptera, Molossidae). *Doñana, Acta Vert.* 6:248–52. [Dated June 1979; published in 1980.]

———. 1981. Descripción de un nuevo género de quiróptero neotropical de la familia Molossidae. *Doñana, Acta Vert.* 7:104–11. [Dated June 1980; published in 1981.]

———. 1984a. Biología y ecología de los murciélagos del Hato "El Frio" Apure, Venezuela. *Doñana, Acta Vert.* 8:i–xii, 1–271. [Dated 1981; published in 1984.]

———. 1984b. Quirópteros neotropicales en el Museo Nacional de Ciencias Naturales de Madrid. In *Actas II Reunion Iberoamericana de Conserv. Zool. Vert.*, ed. J. Castroviejo, 399–410, 1980. Cáceres, España, xiv + 631 pp.

———. 1985. Notas sobre distribución de quirópteros en Bolivia (Mammalia, Chiroptera). *Hist. Nat.* 5:329–36.

———. 1986. Notes on *Amorphochilus schnablii* Peters (Chiroptera, Furipteridae). *Mammalia* 49:584–87. [Dated 1985; number 4 marked "Achevé d'imprimer le 20 mars 1986."]

Ibáñez, C., and J. Ochoa G. 1989. New records of bats from Bolivia. *J. Mammal.* 70:216–19.

Ibáñez, C., J. Cabot, and S. Anderson. 1994. New records of Bolivian mammals in the collection of the Estación Biológica de Doñana. *Doñana Acta Vert.* 21:79–83.

ICZN. 1926. Opinion 91. Thirty-five generic names of mammals placed in the Official List of Generic Names. In Opinions rendered by the International Commission on Zoological Nomenclature. Opinions 91 to 97, pp. 1–2. *Smithsonian Misc. Coll.* 73(4):1–30.

———. 1929. Opinions rendered by the International Commission on Zoological Nomenclature. *Smithsonian Misc. Coll.* 73(6):18.

———. 1954. Opinion 258. Rejection for nomenclatural purposes of the work by Frisch (J.L.) published in 1775 un the title "Das natur-system de vierfüssign thiere." *Opinions and declarations rendered by the International Commission on Zoological Nomenclature, London* 5:245–52.

———. 1955. Direction 24. Completion of the entries relating to the names of certain genera in the class Mammalia made on the *Official List of Generic Names in Zoology* in the period up to the end of 1936. *Opinions and declarations rendered by the International Commission on Zoological Nomenclature, London* 1:219–46.

———. 1956. Opinion 417. Rejection for nomenclatural purposes of volume 3 (Zoologie) of the work by Lorenz Oken entitled Okens Lehrbuch der Naturgeschichte published in 1815–1816. *Opinions and declarations rendered by the International Commission on Zoological Nomenclature, London* 14:1–42.

———. 1957. Opinion 462. Addition to the "Official List of Generic Names in Zoology" of the generic name "Mormoops" Leach, 1821 (class Mammalia). *Opinions and declarations rendered by the International Commission on Zoological Nomenclature, London* 16:1–12.

———. 1958a. Direction 98. Interpretation under the plenary powers of the nominal species "Vespertilio murinus" Linnaeus, 1758, and insertion in the "Official List of Generic Names in Zoology' of a revised entry relating to the generic name "Vespertilio" Linnaeus, 1758 (Class

Mammalia) ("Direction" supplementary to "Opinion" 91). *Opinions and Declarations Rendered by the International Commission on Zoological Nomenclature, London* 1:127–60.

———. 1958b. *Official list of generic names in zoology.* London: The International Trust for Zoological Nomenclature, xxxviii + 220 pp.

———. 1958c. *Official index of rejected and invalid generic names in zoology.* London: The International Trust for Zoological Nomenclature, xiv + 132 pp.

———. 1961. *International Code of Zoological Nomenclature adopted by the XV International Congress of Zoology.* London: The International Trust for Zoological Nomenclature, xviii + 176 pp.

———. 1963. Opinion 660. Suppression under the plenary powers of seven specific names of turtles (Reptilia, testudines). *Bull. Zool. Nomencl.* 20:187–90.

———. 1964. *International Code of Zoological Nomenclature adopted by the XV International Congress of Zoology.* 2nd ed. London: The International Trust for Zoological Nomenclature, xviii + 176 pp.

———. 1985. *International Code of Zoological Nomenclature adopted by the XX General Assembly of the International Union of Biological Sciences.* London: The International Trust for Zoological Nomenclature, xx + 338 pp.

———. 1987a. *Official lists and indexes of names and works in zoology.* London: The International Trust for Zoological Nomenclature, 4 (unnumbered) + 365 pp.

———. 1987b. Opinion 1461. A ruling on the authorship and dates of the text volumes of the *Histoire naturelle* section of Savigny's *Description de l'Egypte. Bull. Zool. Nomencl.* 44:219–20.

———. 1998. Opinion 1894. *Regnum Animale . . . ,* ed. 2 (M. J. Brisson, 1762): Rejected for nomenclatural purposes, with the conservation of the mammalian generic names *Philander* (Marsupialia), *Pteropus* (Chiroptera), *Glis, Cuniculus,* and *Hydrochoerus* (Rodentia), *Meles, Lutra* and *Hyaena* (Carnivora), *Tapirus* (Perissodactyla), *Tragulus* and *Giraffa* (Artiodactyla). *Bull. Zool. Nomencl.* 55:64–71.

———. 1999. *International Code of Zoological Nomenclature.* 4th ed. London: The International Trust for Zoological Nomenclature, xxix + 306 pp.

———. 2002. Opinion 2005 (Case 3022). *Catalogue des mammifères du Muséum National d'Histoire Naturelle* by Étienne Geoffroy Saint-Hilaire (1803): Placed on the Official List of Works Approved as Available for Zoological Nomenclature. *Bull. Zool. Nomencl.* 59:153–54.

IGAC [Instituto Geográfico "Agustin Codazzi"]. 1982. Departamento de Cauca. Ministerio de Hacienda y Credito Publico, Colombia. (1:400,000 topographical map).

Ihering, H. von. 1892. Os mammiferos do Rio Grande do Sul. *Annuario do Estado do Rio Grande do Sul, para o anno 1893 de Graciano A. de Azambuja* 96–123.

———. 1894. *Os mammiferos de S. Paulo. Catálogo.* São Paulo: Typ. do "Diario Official," 30 pp.

———. 1914. Os gambas do Brazil. Marsupiaes do gen. Didelphis. *Rev. Mus. Paulista, São Paulo* 9:338–356.

Illiger, J. K. W. 1811. *Prodromus systematis mammalium et avium additis terminis zoographicis utriusque classis, eorumque versione germanica.* Berolini: C. Salfield, xviii + 302 pp.

———. 1815. Ueberblick der Säugthiere nach ihrer Vertheilung über die Welttheile. *Abhandl. König. Akad. Wiss. Berlin* 1804–1811:39–159.

Iredale, T., and E. Le G. Troughton. 1934. A check-list of the mammals recorded from Australia. *Mem. Australian Mus., Sydney* 6:xii + 122 pp.

Irwin, D. W., and R. J. Baker. 1967. Additional records of bats from Arizona and Sinaloa. *Southwest. Nat.* 12:195.

Izor, R. J. 1985. Sloths and other mammalian prey of the Harpy Eagle. In *The evolution and ecology of armadillos, sloths, and vermilinguas,* ed. G. G. Montgomery, 343–46. Washington, DC: The Smithsonian Institution Press, 10 (unnumbered) + 451 pp.

Izor, R. J., and R. H. Pine. 1987. Notes on the black-shouldered opossum, *Caluromysiops irrupta.* In *Studies in Neotropical mammalogy, essays in honor of Philip Hershkovitz,* ed. B. D. Patterson and R. M. Timm, 117–24. *Fieldiana Zool.* 39:frontispiece, viii + 1–506.

James, E. 1823. *Account of an expedition from Pittsburgh to the Rocky Mountains, performed in the years 1819 and '20, by order of the Hon. J. C. Calhoun, Sec'y of War: Under the command of Major Stephen H. Long.* Philadelphia: H. G. Carey and I. Lea, 1:1–8 + 1–503.

Jansa, S. A., and R. S. Voss. 2000. Phylogenetic studies on didelphid marsupials I. Introduction and preliminary results from nuclear IRBP gene sequences. *J. Mammal. Evol.* 7:43–77.

———. 2005. Phylogenetic relationships of the marsupial genus *Hyladelphys* based on nuclear gene sequences and morphology. *J. Mammal.* 86:853–65.

Jansen, A. M., F. Madeira, J. C. Carreira, E. Medina-Acosta, and M. P. Deane. 1997. *Trypanosoma cruzi* in the opossum *Didelphis marsupialis*: A study of the correlations and kinetics of the systemic and scent gland infections in naturally and experimentally infected animals. *Exp. Parasitol.* 86:37–44.

Jansen, A. M., A. P. de Pinho, C. V. Lisboa, E. Cupolillo, R. H. Mangia, and O. Fernandes. 1999 The sylvatic cycle of *Trypanosoma cruzi*: A still unsolved puzzle. *Mem. Inst. Oswaldo Cruz* 94(Suppl. 1):203–04.

Janson, C. H., J. Terborgh, and L. H. Emmons. 1981. Non-flying mammals as pollinating agents in the Amazonian forest. In *Reproductive Botany*, 1–6. Suppl. to *Biotropica* 13.

Jarrin, P. 2004. An unusual record of *Peropteryx macrotis* (Chiroptera: Emballonuridae) in the Andean highlands of Ecuador. *Mammalia* 67:613–15.[Dated 2003; number 4 marked "Achevé d'imprimer le 31 mars 2004."]

Jayat, J. P., and M. D. Miotti. 2006. Primer registro de *Monodelphis kunsi*(Didelphimorphia, Didelphidae) para Argentina. *Mastozool. Neotrop.* 12:253–56. [Dated 2005; statement on last page of number 2 says "Se terminó de imprimir . . . en enero de 2006."]

Jeanne, R. L. 1970. Note on a bat (*Phylloderma stenops*) preying upon the brood of a social wasp. *J. Mammal.* 51:624–25.

Jenkins, P. K., and L. Knutson. 1983. *A catalogue of the type specimens of Monotremata and Marsupialia in the British Museum (Natural History)*, London: British Museum (Natural History), 34 pp.

Jennings, J. B., T. L. Best, S. E. Burnett, and J. C. Rainey. 2002. Molossus sinaloae. *Mammal. Species* 691:1–5.

Jentink, F. A. 1888. Catalogue systématique des mammifères (Rongeurs, Insectivores, Cheiroptères, Edentés et Marsupiaux). *Mus. Hist. Nat. Pays Bas* 12:1–280.

———. 1890. Mammalia from the Malay Archipelago. In *Zoologishe Ergebnisse einer Reise in Niederländishce Ost-Indien*, Vol. 1, ed. M. W. C. Weber, Leiden: E. J. Brill, 460 pp, 25 pls.

———. 1893. On a collection of bats from West-Indies. *Notes Leyden Mus.* 15:278–83.

———. 1904. On *Kerivoula picta* and description of a new bat from Paramaribo. *Notes Leyden Mus.* 24:174–76.

———. 1910a. Chrysopteron bartelsii, novum genus et nova species from Java. *Notes Leyden Mus.* 32:73–77.

———. 1910b. Description of a shrew from Surinam. *Notes Leyden Mus.* 32:167–68.

Jiménez de la Espada, M. 1870. Algunos datos nuevos o curiosos acerca de la fauna de alto Amazonas (mamíferos). *Bol. Rev. Univ. Madrid* 1870:21–27.

Jiménez M., P., and J. Péfaur. 1982. Aspectos sistematicos y ecologicos de *Platalina genovensium* (Chiroptera: Mammalia). In *Zoología Neotropical*, ed. P. J. Salinas, 707–718. Caracas, Venezuela: *Actas del* VIII Congreso Latinoamericano de Zoología, 1:xx + 1–784.

Johnson, D. H. 1952. A new name for the Jamaican bat *Molossus fuliginosus* Gray. *Proc. Biol. Soc. Washington* 65:197–98.

Johnson, M. A. 1979. Evolutionary implications of G- and C- banded chromosomes of 13 species of stenodermine bats. M.Sc. thesis, Texas Tech University, Lubbock.

Jones, E. K., C. M. Clifford, J. E. Keirans, and G. M. Kohls. 1972. The ticks of Venezuela (Acarina: Ixodoidea) with a key to the species of *Amblyomma* in the Western Hemisphere. *Brigham Young Univ. Sci. Bull.*, biol. ser., 17(4):1–40.

Jones, J. K., Jr. 1958. Pleistocene bats from San Josecito Cave, Nuevo León, México. *Univ. Kansas Publ., Mus. Nat. Hist.* 9:389–96.

———. 1966. Bats from Guatemala. *Univ. Kansas Publ., Mus. Nat. Hist.* 16:439–72.

———. 1978. A new bat of the genus *Artibeus* from the Lesser Antillean Island of St. Vincent. *Occas. Papers Mus., Texas Tech Univ.*, no. 51:1–6.

Jones, J. K., Jr., and T. Alvarez. 1962. Taxonomic status of the free-tailed bat, Tadarida yucatanica Miller. *Univ. Kansas Publ., Mus. Nat. Hist.* 14:125–33.

Jones, J, K., Jr., and J. Arroyo-Cabrales. 1990. Nyctinomops aurispinosus. *Mammal. Species* 350:1–3.

Jones, J. K., Jr., and D. C. Carter. 1976. Annotated checklist, with keys to subfamilies and genera. In *Biology of bats of the New World family Phyllostomatidae*. Part I, ed. R. J. Baker, J. K. Jones, Jr., and D. C. Carter, 7–38. Special Publications of the Museum 10. Lubbock: Texas Tech University Press, 218 pp.

———. 1979. Systematic and distributional notes. In *Biology of bats of the New World family Phyllostomatidae*. Part III, ed. R. J. Baker, J. K. Jones, Jr., and D. C. Carter, 7–11. Special Publications of the Museum 16. Lubbock: Texas Tech University Press, 441 pp.

Jones, J. K., Jr., and H. H. Genoways. 1967. A new subspecies of the free-tailed bat, *Molossops greenhalli*, from western Mexico (Mammalia: Chiroptera). *Proc. Biol. Soc. Washington* 80:207–10.

Jones, J. K., Jr., and T. E. Lawlor. 1965. Mammals from Isla Cozumel, Mexico, with description of a new species of harvest mouse. *Univ. Kansas Publ., Mus. Nat. Hist.* 16:409–19.

Jones, J. K., Jr., and C. J. Phillips. 1970. Comments on systematics and zoogeography of bats in the Lesser Antilles. *Stud. Fauna Curaçao Carib. Isls.* 32:131–45.

———. 1976. Bats of the genus Sturnira in the Lesser Antilles. *Occas. Papers Mus., Texas Tech Univ.*, no. 40:1–16.

Jones, J. K., Jr., and C. S. Hood. 1993. Synopsis of South American bats of the family Emballonuridae. *Occas. Papers Mus., Texas Tech Univ.*, no. 155:1–32.

Jones, J. K., Jr., J. D. Smith, and H. H. Genoways. 1973. Annotated checklist of mammals of the Yucatan Peninsula, Mexico. I. Chiroptera. *Occas. Papers Mus., Texas Tech Univ.*, no. 13:1–31.

Jones, J. K., Jr., J. D. Smith, and R. W. Turner. 1971. Noteworthy records of bats from Nicaragua, with a checklist of the chiropteran fauna of the country. *Occas. Papers Mus. Nat. Hist., Univ. Kansas*, no. 2:1–35.

Jones, J. K., Jr., P. Swanepoel, and D. C. Carter. 1977. Annotated checklist of the bats of Mexico and Central America. *Occas. Papers Mus., Texas Tech Univ.*, no. 47:1–35.

Jones, K. E., A. Purvis, A. MacLarnon, O. R. P. Bininda-Emonds, and N. B. Simmons. 2002. A phylogenetic supertree of the bats (Mammalia: Chiroptera). *Biol. Rev.* 77:223–59.

Jones, M. E., C. R. Dickman, and M. Archer. 2003. Preface. In *Predators with pouches: the biology of carnivorous marsupials*, ed. M. E. Jones, C. R. Dickman, and M. Archer, xii–xvii. Collingwood, Australia: CSIRO Publishing, xix + 486 pp.

Jorge, W., R. C. Best, and R. M. Wetzel. 1986. Chromosome studies on the silky anteater *Cyclopes didactylus* L. (Myrmecophagidae: Xenarthra, Edentata). *Caryologia* 38:325–29.

Jorge, W., D. A. Meritt, and K. Benirschke. 1977. Chromosome studies in Edentata. *Cytobios* 18:157–72.

Jorge, W., A. T. Orsi-Souza, and R. Best. 1985. The somatic chromosomes of Xenarthra. In *The evolution and ecology of armadillos, sloths, and vermilinguas*, ed. G. G. Montgomery, 121–30. Washington, DC: The Smithsonian Institution Press, 10 (unnumbered) + 451 pp.

Julia, J. P., E. Richard, and J. Samaniego. 1994. Nota sobre la distribución geográfica del oso melero (*Tamandua tetradactyla*, Xenarthra: Myrmecophagidae) en el noroeste Argentino. *Notul. Faunist.* 66:1–4.

Julien-Laferrière, D. 1991. Organisation du peuplement de marsuiaux en Guyane française. *Rev. Ecol. (Terre Vie)* 46:125–44.

Julien-Laferrière, D., and M. Atramentowicz. 1990. Feeding and reproduction of three didelphid marsupials in two Neotropical forests (French Guiana). *Biotropica* 22:404–15.

Kalko, E. K. V., and M. A. Condon. 1998. Echolocation, olfaction and fruit display: How bats find fruit of flagellichorous cucurbits. *Funct. Ecol.* 12:364–72.

Kalko, E. K. V., D. Friemel, C. O. Handley, Jr., and H. Schnitzler. 1999. Roosting and foraging behavior of two Neotropical gleaning bats, *Tonatia silvicola* and *Trachops cirrhosus* (Phyllostomidae). *Biotropica* 31:344–53.

Kappler, A. 1881. *Holländisch-Guiana. Erlebnisse und Erfahrungen während eines 43 jährigen Aufenthalts in der Kolonie Surinam.* Stuttgart: W. Kohlhammer, x + 495 pp., 1 pl., 1 map.

Kaup, J. 1829. *Skizzirte Entwickelungs-Geschichte und natürliches System der europäischen Thierwelt.* Darmstadt und Leipzig: Carl Wilhelm Leste, xii + 204 pp.

Kelt, D. A., and D. R. Martínez. 1989. Notes on distribution and ecology of two marsupials endemic to the Valdivian forests of southern South America. *J. Mammal.* 70:220–24.

Kerr, R. 1792. *The animal kingdom or zoological system, of the celebrated Sir Charles Linnaeus. Class I. Mammalia: Containing a complete systematic description, arrangement, and nomenclature, of all the known species and varieties of the Mammalia, or animals which give suck to their young; being a translation of that part of the systema naturae, as lately published, with great improvements, by Professor Gmelin of Goettingen. Together with numerous additions from more recent zoological writers, and illustrated with copperplates.* Edinburgh: A. Strahan, T. Cadell, and W. Creech, xii + l–32 + 30 (unnumbered) + 33–400 pp., 7 pls.

Keyserling, A. G. von, and I. H. Blasius. 1839. Uebersicht der Gattungs- und Artcharaktere der europäischen Fledermäuse. *Arch. Naturgesch.* 5(1):293–331.

Kirsch, J. A. W. 1977a. The comparative serology of Marsupialia, and a classification of marsupials. *Aust. J. Zool.*, suppl. ser., 52:1–152.

———. 1977b. The classification of marsupials with special reference to karyotypes and serum proteins. In *The biology of marsupials*, ed. D. Hunsaker II, 1–50. New York: Academic Press, xviii + 537 pp.

Kirsch, J. A. W., and J. H. Calaby. 1977. The species of living marsupials: An annotated list. In *The biology of marsupials*, ed. B. Stonehouse and D. Gilmore, 9–26. Baltimore: University Park Press, viii + 486 pp.

Kirsch, J. A. W., and P. F. Waller. 1979. Notes on the trapping and behavior of the Caenolestidae (Marsupialia). *J. Mammal.* 60:390–95.

Kirsch, J. A. W., A. L. Dickerman, and O. A. Reig. 1995. Estudios de hibridización de marsupials carnivoros IV. Relaciones intergenericas en Didelphidae. *Marmosiana* 1:57–78.

Kishida, K., and T. Mori. 1931. On the distribution of terrestrial mammals in Korea [in Japanese]. *Dobuts. Zasshi, Tokyo* 43:372–91.

Koepcke, J. 1984. "Blattzelte" als Schlafplatze der Fledermaus *Ectophylla maconnelli* (Thomas, 1901) (Phyllostomidae) im tropischen Regenwald von Peru. *Saugertierk. Mitt.* 31:123–26.

Koepcke, J., and R. Kraft. 1984. Cranial and external characters of the larger fruit bats of the genus *Artibeus* from Amazonian Peru. *Spixiana* 7:75–84.

Kohls, G. M., D. E. Sonenshine, and C. M. Clifford. 1965. The systematics of the subfamily Ornithodorinae (Acarina: Argasidae). II. Identification of the larvae of the Western Hemisphere and descriptions of three new species. *Ann. Ent. Soc. Amer.* 58:331–64.

Kolenati, F. A. 1856a. Europa's Chiroptern. I. Synopsis der Europäischen Chiroptern. *Allgem. Deutsche Naturhist. Zeitung* 2:121–33.

———. 1856b. Europa's Chiroptern. II. Beschreibung der Europäischen lebenden Chiroptern. *Allgem. Deutsche Naturhist. Zeitung* 2:161–95.

———. 1858. Eine neue österreichische Fledermaus. *Akad. Wiss. Sitzungsber.* 29:250–56.

———. 1863. Beiträge sur Kenntniss der Phythirio-Myiarien. *Horae Soc. Entomolog. Rossicea* 2:9–110 + 15 pls.

Komeno, C. A., and A. X. Linhares. 1999. Batflies parasitic on some phyllostomid bats in southeastern Brazil: Parasitism rates and host-parasite relationships. *Mem. Inst. Oswaldo Cruz* 94:151–56.

Koop, B. F., and R. J. Baker. 1983. Electrophoretic studies of relationships of six species of *Artibeus* (Chiroptera: Phyllostomidae). *Occas. Papers Mus., Texas Tech Univ.*, no. 83:1–12.

Koopman, K. F. 1951. Fossil bats from the Bahamas. *J. Mammal.* 32:229.

———. 1955. A new subspecies of *Chilonycteris* from the West Indies and a discussion of the mammals of La Gonave. *J. Mammal.* 36:109–13.

———. 1956. Bats from San Luis Potosi with a new record for *Balantiopteryx plicata*. *J. Mammal.* 37:547–48.

———. 1958. Land bridges and ecology in bat distribution on islands off the northern coast of South America. *Evolution* 12:429–39.

———. 1967. The southernmost bats. *J. Mammal.* 48:487–88.

———. 1968. Taxonomic and distributional notes on Lesser Antillean bats. *Amer. Mus. Novit.*, no. 2333:1–13.

———. 1971. The systematic and historical status of the Florida *Eumops* (Chiroptera: Molossidae). *Amer. Mus. Novit.*, no. 2478:1–6.

———. 1976. Zoogeography. In *Biology of bats of the New World family Phyllostomatidae*. Part I, ed. R. J. Baker, J. K. Jones, Jr., and D. C. Carter, 39–47. Special Publications of the Museum 10. Lubbock: Texas Tech University Press, 218 pp.

———. 1978. Zoogeography of Peruvian bats with special emphasis on the role of the Andes. *Amer. Mus. Novit.*, no. 2651:1–33.

———. 1981. The distributional patterns of New World nectar-feeding bats. *Ann. Missouri Bot. Garden* 68:352–69.

———. 1982. Biogeography of the bats of South America. In *Mammalian biology in South America*, ed. M. A. Mares and H. H. Genoways, 273–302. The Pymatuning

Symposia in Ecology 6. Special Publications Series. Pittsburgh: Pymatuning Laboratory of Ecology, University of Pittsburgh, xii + 539 pp.

———. 1984. Bats. In *Orders and families of Recent mammals of the world*, ed. S. Anderson and J. K. Jones, Jr., 145–86. New York: John Wiley and Sons, xii + 686 pp.

———. 1988. Systematics and distribution. In *Natural history of vampire bats*, ed. A. M. Greenhall and U. Schmidt, 7–17. Boca Raton, Florida: CRC Press, frontispiece, 14 unnumbered + 246 pp.

———. 1989. A review and analysis of the bats of the West Indies. In *Biogeography of the West Indies: Past, present, and future*, ed. C. A. Woods, 635–44. Gainesville, Florida: Sandhill Crane Press, 878 pp.

———. 1993. Order Chiroptera. In *Mammal species of the world*, 2nd ed., ed. D. E. Wilson and D. M. Reeder, 137–242. Washington, DC: The Smithsonian Institution Press, xviii + 1207 pp.

———. 1994. Chiropteran systematics. Volume 8, Part 60, 6 (unnumbered) + 217 pp. in *Handbuch der Zoologie*, ed. J. Niethammer, H. Schliemann, and D. Starck. Berlin: Walter de Gruyter.

Koopman, K. F., and E. L. Cockrum. 1967. Bats. In *Recent mammals of the world*, ed. S. Anderson and J. K. Jones, Jr., 109–50. New York: Ronald Press Co., viii + 453 pp.

Koopman, K. F., and F. Gudmundsson. 1966. Bats in Iceland. *Amer. Mus. Novit.*, no. 2262:1–6.

Koopman, K. F., and J. K. Jones, Jr. 1970. Classification of bats. In *About bats*, ed. B. H. Slaughter and D. W. Walton, 22–28. Dallas: Southern Methodist University Press, vii + 339 pp.

Koopman, K. F., and P. S. Martin. 1959. Subfossil mammals from the Gómez Farías region and the tropical gradient of eastern Mexico. *J. Mammal.* 40:1–12.

Koopman, K. F., and G. F. McCracken. 1998. The taxonomic status of *Lasiurus* (Chiroptera: Vespertilionidae) in the Galapagos Islands. *Amer. Mus. Novit.*, no. 3243, 6 pp.

Koopman, K. F., and R. Ruibal. 1955. Cave-fossil vertebrates from Camaguey, Cuba. *Breviora, Mus. Comp. Zool.* 46:1–8.

Koopman, K. F., and E. E. Williams. 1951. Fossil chiroptera collected by H. E. Anthony in Jamaica, 1919–1920. *Amer. Mus. Novit.*, no. 1519:1–29.

Koopman, K. F., M. K. Hecht, and E. Ledecky-Janecek. 1957. Notes on the mammals of the Bahamas with special reference to bats. *J. Mammal.* 38:164–74.

Kraft, R. 1982. Notes on the type specimens of *Artibeus jamaicensis planirostris* (Spix, 1823). *Spixiana* 5:311–16.

Kraft R. 1995. Xenarthra. *Handb. Zoologie* 59:8 (unnumbered) + 1–80.

Krauss, F. 1862. Ueber ein neues Gürtelthier aus Surinam. *Arch. Naturgesch.* 28(1):19–34, pl. 3.

Kretzoi, M., and M. Kretzoi. 2000. *Index generum et subgenerum mammalium.* Fossilium Catalogus Animalia, Pars 137, Section 1:xvi + 1–433; Section 2:434–726. Leiden: Backhuys Publishers.

Krieg, H. 1929. Biologische Reisestudien in Südamerika. IX. Gürteltiere. *Zeits. Morphol. Okol. Tiere* 14:166–90.

Krumbiegel, I. 1940a. Die Säugetiere der Südamerika-Expeditionen Prof. Dr. Kriegs. I. Gürteltiere. *Zool. Anz.* 131:49–73.

———. 1940b. Die Säugetiere der Südamerika-Expeditionen Prof. Dr. Kriegs: 2. Ameisenbären. *Zool. Anz.* 131:161–88.

———. 1940c. Die Säugetiere der Südamerika-Expeditionen Prof. Dr. Kriegs. 5. Schwimmbeutler. *Zool. Anz.* 132:63–72, 11 figs.

———. 1941a. Die Säugethiere der Südamerika-Expeditionen Prof. Dr. Kriegs. 10. Opossums (*Didelphys*). *Zool. Anz.* 134:29–53.

———. 1941b. Die Säugetiere der Südamerika-Expeditionen Prof. Dr. Kriegs. 11. Mittelgrosse Didelphyiden (*Lutreolina* u. *Metachirus*). *Zool. Anz.* 134:189–211.

———. 1941c. Die Säugetiere der Südamerika-Expeditionen Prof. Dr. Kriegs. 14. Faultiere. *Zool. Anz.* 136:53–62.

Kuhl, H. 1818. Deutsche Fledermäuse. (*Neue*) *Annal. Wetterauer Geschichtsbl. Naturk.* 1:11–49, 185–225.

———. 1819. Die Deutschen Fledermäuse. *Wetterau. Gesell. N. Annal.* 1:11–49, 185–215.

———. 1820. Beiträge zur Zoologie. In *Beiträge zur Zoologie und vergleichenden Anatomie,* 1–152. Frankfurt am Main: Verlag der Hermannschen Buchhandlung, 6 unnumbered, 1:1–152.

Kühlhorn, F. 1953. Säugetierkündliche Studien aus Süd-Mattogrosso 1. Teil Marsupialia, Chiroptera. *Säugetierk. Mitt.* 1:115–22.

Kunz, T. H. 1982. Roosting ecology of bats. In *Ecology of bats,* ed. T. H. Kunz, 1–55. New York: Plenum Press, xviii + 425 pp.

Kunz, T. H., and I. M. Pena. 1992. Mesophylla macconnelli. *Mammal. Species* 405:1–5.

Kunz, T. H., P. V. August, and C. D. Burnett. 1983. Harem social organization in cave roosting *Artibeus jamaicensis* (Chiroptera: Phyllostomidae). *Biotropica* 15:133–38.

Kurta, A., and R. H. Baker. 1990. Eptesicus fuscus. *Mammal. Species* 356:1–10.

Kurta, A., and G. C. Lehr. 1995. Lasiurus ega. *Mammal. Species* 515:1–7.

Lacépède, B. G. E. de la V. 1799a. *Histoire naturelle par Buffon.* Paris: P. Didot et Firmin Didot 4:1–298, 21 pls.

[See Richmond 1899) for date of publication.]

———. 1799b. Tableau des divisions, sous-divisions, ordres et genres des Mammifères. Supplement to *Discours d'ouverture et de clôture du cours d'histoire naturelle donné dans le Muséum national d'Histoire naturelle, l'an VII de la République, et tableau méthodiques des mammifères et de oiseaux.* Paris: Plassan, 18 pp.

———. 1800. *Histoire naturelle par Buffon.* Paris: P. Didot et Firmin Didot, 7:1–347, 36 pls. [See Richmond 1899 for date of publication.]

———. 1801. *Histoire naturelle par Buffon.* Paris: P. Didot et Firmin Didot, 11:1–242, 29 pls. [See Richmond 1899 for date of publication.]

———. 1802. Tableau des divisions, sous-divisions, ordres et genres des mammifères, Par le C^en Lacépède; Avec l'indication de toutes les espèces décrites par Buffon, et leur distribution dans chacun des genres, par F. M. Daudin. In *Histoire naturelle par Buffon,* ed. B. G. E. de la V. Lacépède, 143–95. Paris: P. Didot et Firmin Didot, 14:1–346. [See Richmond 1899 for date of publication.]

Laemmert, H. W., Jr., L. da C. Ferreira, and R. M. Taylor. 1946. An epidemiological study of jungle yellow fever in Brazil. Part II. Investigations of vertebrate hosts and arthropod vectors. *Amer. J. Trop. Med.* 26(Suppl. 6):23–69.

Lahille, F. 1895. Contributions à l'étude des édentés à bandes mobiles de la République Argentine. *An. Mus. La Plata,* secc. zool., 2:3–32, 3 pls.

———. 1899. Ensayo sobre la distribución geográfica de los mamíferos en la República Argentina. In *III. Trabajos de la 2.a Sección (Ciencias Físico-Químicas y Naturales),* ed. C. Bery, M. B. Bahía, and F. B. Reyes, 165–206. *Primera reunión del Congreso Científico Latino Americano* [1898]. Buenos Aires: Sociedad Científica Argentina, 3:1–262..

Langguth, A., and F. Achaval. 1972. Notas ecologicas sobre el vampiro *Desmodus rotundus rotundus* (Geoffroy) en el Uruguay. *Neotropica* 18:45–53.

Langguth, A., and J. F. S. Lima. 1988. The karyotype of *Monodelphis americana* (Marsupialia-Didelphidae). *Rev. Nordestina Biol.* 6:1–5.

Langguth, A., V. L. A. G. Limeira, and S. Franco. 1997. Novo catálogo do material-tipo da coleção de mamífereos do Museu Nacional. *Publ. Avulsas Mus. Nac., Rio de Janeiro* 70:1–29.

Lapham, I. A. 1853. A systematic catalogue of the animals of Wisconsin. Mammalia. *Trans. Wisconsin State Agri. Soc.* 2:337–39.

Larrañaga, D. A. 1923. *Escritos.* Montevideo: Instituto Histórico y Geográfico del Uruguay, 2:1–512 + 2 tables.

Lassieur, S., and D. E. Wilson. 1989. Lonchorhina aurita. *Mammal. Species* 347:1–4.

Lasso, D. and P. Jarrín-V. 2005. Diet variability of *Micronycteris megalotis* in pristine and disturbed habitats of northwestern Ecuador. *Acta Chiropterol.* 7:121–30.

Lataste, F. 1891. Description d'une espece nouvelle ou mal connue de chauve-souris. *Ann. Mus. Civico Storia Nat. Genova*, ser. 2:658–64.

———. 1892. Etudes sur la faune chilienne. II—Note sur les chauve-souris. *Actes Soc. Sci. Chili, Santiago* 1:70–91.

Laurie, E. M. O. 1955. Notes on some mammals from Ecuador. *Ann. Mag. Nat. Hist.* ser. 12, 8:268–76.

LaVal, R. K. 1970. Banding returns and activity periods of some Costa Rican bats. *Southwest. Nat.* 15:1–10.

———. 1973a. A revision of the Neotropical bats of the genus *Myotis*. *Sci. Bull., Los Angeles Co. Mus.* 15:1–54.

———. 1973b. Systematics of the genus *Rhogeessa*. (Chiroptera: Vespertilionidae). *Occas. Papers, Mus. Nat. Hist., Univ. Kansas*. 19:1–47.

———. 1977. Notes on some Costa Rican bats. *Brenesia* nos. 10–11:77–83.

LaVal, R. K., and H. S. Fitch. 1977. Structure, movements, and reproduction in three Costa Rican bat communities. *Occas. Papers Mus. Nat. Hist., Univ. Kansas*, no. 69:1–28.

Lavergne, A., O. Verneau, J. L. Patton, and F. M. Catzeflis. 1997. Molecular discrimination of two sympatric species of opossum (genus *Didelphis*: Didelphidae) in French Guiana. *Molecular Ecol.* 6:889–91.

Lay, D. M. 1962. Seis mamíferos nuevos para la fauna de México. *An. Inst. Biol., México* 372–77.

Lazell, J. D., Jr., and K. F. Koopman. 1985. Notes on bats of Florida's lower Keys. *Florida Sci.* 48:37–42.

Leach, W. E. 1821a. The characters of three new genera of bats without foliaceous appendages to the nose. *Trans. Linnean Soc. London* 13:69–72.

———. 1821b. The characters of seven genera of bats with foliaceous appendages to the nose. *Trans. Linnean Soc. London* 13:73–82, pl. 7.

LeConte, J. 1858. On three new species of Vespertilionidae. *Proc. Acad. Nat. Sci. Philadelphia* 9:174–75, 1857.

LeConte, J. E. 1831. Appendix. In *The animal kingdom arranged in conformity with its organization by the Baron Cuvier*. Translated from the French with notes and additions by H. M'Murtrie, 431–39. New York: G. and C. and H. Carvill, 1:xxxii + 448 pp., 4 pls.

Ledru, A. -P. 1810. *Voyage aux iles de Ténériffe, la Trinité, Saint-Thomas, Sainte-Croix et Porta-Ricco, exécuté par ordre du gouvernement Français, depuis le 30 septembre 1796 jusqu'au 7 juin 1798, sous la direcion du Capitaine Baudin, pour faire de recherches et des collections relatives a l'histoire naturelle*. Paris: Arthus-Bertrand, 1:xlv + 315 pp.

Lee, A. K., and A. Cockburn. 1985. *Evolutionary ecology of marsupials*. Cambridge: Cambridge University Press, viii + 274 pp.

Lee, T. E., Jr., and D. J. Dominguez. 2000. Ametrida centurio. *Mammal. Species* 640:1–4.

Lee, T. E., Jr., S. R. Hoofer, and R. A. Van Den Bussche. 2002. Molecular phylogenetics and taxonomic revision of the genus *Tonatia* (Chirptera: Phyllostomidae). *J. Mammal.* 83:49–57.

Lee, T. E., Jr., B. K. Lim, and J. D. Hanson. 2000. Noteworthy records of mammals from the Orinoco River drainage of Venezuela. *Texas J. Sci.* 52:264–66.

Lee, T. E., Jr., J. B. Scott, and M. M. Marcum. 2001. Vampyressa bidens. *Mammal. Species* 684:1–3.

Legendre, S. 1984a. Étude odontologique des représentants actuels du groupe *Tadarida* (Chiroptera, Molossidae). Implications phylogéniques, systematiques et zoogéographiques. *Rev. Suisse Zool.* 91:399–442.

———. 1984b. Paleontologie–Identification de deux sousgenres fossiles et comprehension phylogenique du genre *Mormopterus* (Molossidae, Chiroptera). *C. R. Acad. Sc. Paris*, ser. 2, 298:715–20.

———. 1984c. Essai de biogeography phylogénique des molossides (Chiroptera). *Myotis* no. 21–22:30–36.

Lemke, T. O., and J. R. Tamsitt. 1980. Anoura cultrata (Chiroptera: Phyllostomatidae) from Colombia. *Mammalia* 43:579–80. [Dated 1979; number 4 marked "Achevé d'imprimer le 15 janvier 1980."

Lemke, T. O., A. Cadena, R. H. Pine, and J. Hernández-Camacho. 1982. Notes on opossums, bats, and rodents new to the fauna of Colombia. *Mammalia* 46:225–34.

Lemos, B., and R. Cerqueira. 2002. Morphological and morphometric differentiation in the white-eared opossum group (Didelphidae, *Didelphis*). *J. Mammal.* 83:354–69.

Lemos, B., G. Marroig, and R. Cerqueira. 2001. Evolutionary rate and morphological stasis in large-bodied opossum skulls (Didelphimorphia: Didelphidae). *J. Zool.* 255:181–89.

Lemos, B., M. Weksler, and C. R. Bonvicino. 2000. The taxonomic status of *Monodelphis umbristriata* (Didelphimorphia: Didelphidae). *Mammalia* 64:329–37.

Lenz, H. O. 1831. *Naturgeschichte der Säugethiere, nach Cuvier's Systeme bearbeitet*. Gotha: Becker'schen Buchhandlung, xiii + 524 pp.

Leo, M., and E. Ortiz. 1982. Un parque nacional "Gran Pajetén," justificación para su establecimiento. *Bol. Lima* 4:47–60.

Le Pont, F., and P. Desjeux. 1992. Présence de *Choloepus hoffmanni* dans les Yungas de La Paz (Bolivie). *Mammalia* 56:484–85.

Lesson, R.-P. 1826. Mammifères nouveaux ou peu connus, décrits et figures dans l'atlas zoologique du voyage autour du mone de la corvette la Coquille; par Mm. Lesson et Garnot. *Bull. Scienc. Nat. Géol.* 8:95–96.

———. 1827. *Manuel de mammalogie ou histoire naturelle des mammifères*. Paris: Roret, xv + 441 pp.

———. 1836. *Histoire naturelle générale et particulière des mammifères et des oiseaux découverts depuis la mort de Buffon*. Paris: Pourrat Frères, 5:1–512.

———. 1840. *Species des mammifères bimanes et quadrumanes; suivi d'un mémoire sur les oryctéropes*. Paris: J. B. Baillière, xiv + 292 pp

———. 1842. *Nouveau tableau do Règne Animal. Mammifères*. Paris: Arthus-Bertrand, 204 pp.

Lesson, R.-P., and P. Garnot. 1827. *Zoologie*, Vol. 1, part l, iv + 360 pp. In *Voyage autour du monde, exécuté par ordre du roi, sur la corvette de Sa Magesté, La Coquille, pendant les années 1822, 1823, 1824, et 1825, sous le Ministère et conformément aux instructions de A. E. M. le marquis Clermont-Tonnerre, ministre de la marine; et publié sous le auspices de son excellence Mgr le Cte de Chabrol, ministre de la marine et des colonies, par M. L.–I. Duperrey, capitaine de frégate, chevalier de Saint-Louis et membre de la Légion d'Honneur, commandant de l'Expédition*. Paris: A. Bertrand, 5 vols. [Volume 1 of the *Zoologie* dated 1826 with fascicles published from November 1826 to May 1830; publication date of fascicle consisting of pages 129–168 containing the description of *Vespertilio bonariensis* published 25 July 1827 according to Woodward 1904:604.]

Lew, D., and R. Pérez-Hernández. 2004. Una nueva especie del género *Monodelphis* (Didelphimorphia: Didelphidae) de la sierra de Lema, Venezuela. *Mem. Fund. La Salle Cienc. Nat.* 2004 ("2003"):7–25.

Lew, D., R. Pérez-Hernández, and J. Ventura. 2006. Two new species of *Philander* (Didelphimorphia, Didelphidae) from northern South America. *J. Mammal.* 87:224–37.

Lewis, R. E. 1974. Notes on the geographical distribution and host preferences in the order Siphonaptera. Part 4. Coptopsyllidae, Pygiopsyllidae, Stephanocircidae and Xiphiopsyllidae. *J. Med. Entomol.* 11:403–13.

Lewis, S. E., and D. E. Wilson. 1987. Vampyressa pusilla. *Mammal. Species* 292:1–5.

Liais, E. 1872. *Climats, géologie, faune et géographie botanique du Brésil*. Paris: Garnier Frères, viii + 640 pp., 1 map.

Lichtenstein, H. 1818a. *Das Zoologische Museum der Universität zu Berlin. Zweite Ausgabe*. Berlin: Ferdinand Dammler, 1–120, 1 pl.

———. 1818b. Die Werke von Marcgrave und Piso über die Naturgeschichte Brasiliens, erläutert aus den wieder aufgefundenen Originalseichnungen. *Abhandl. Akad. Wiss. Berlin* 1814–1815:201–22.

———. 1823. *Verzeichniss der Doubletten des Zoologischen Museums der Königl. Universität zu Berlin nebst Beschreibung vieler bisher unbekannter Arten von Säugethieren, Vögeln, Amphibien und Fischen*. Berlin: Commission bei T. Trautwein, x + 118 pp.

Lichtenstein, H., and W. Peters. 1854. Über neue merkwürdige Säugethiere des Königl. Zoologischen Museums. *Ber. Bekannt. Verhandl. Königl. Preuss. Akad. Wiss. Berlin* 1854:334–37, 3 pls.

Lim, B. K. 1993. Cladistic reappraisal of Neotropical stenodermatine bat phylogeny. *Cladistics* 9:147–65.

———. 1997. Morphometric differentiation and species status of the allopatric fruit-eating bats *Artibeus jamaicensis* and *A. planirostris* in Venezuela. *Studies Neotrop. Fauna Environ.* 32:65–71.

Lim, B. K., and M. D. Engstrom. 1998. Phylogeny of Neotropical short-tailed fruit bats, *Carollia* spp. Phylogenetic analysis of restriction site variation in mtDNA. In *Bat biology and conservation*, ed. T. H. Kunz and P. A. Racey, 43–58. Washington, DC: The Smithsonian Institution Press, xvi + 365 pp.

———. 2000. Preliminary survey of bats from the Upper Mazaruni of Guyana. *Chiropt. Neotrop.* 6:119–23.

———. 2001a. Species diversity of bats (Mammalia: Chiroptera) in Iwokrama Forest, Guyana, and the Guianan subregion: Implications for conservation. *Biodiv. Conserv.* 10:613–57.

———. 2001b. Bat community structure at Iwokrama Forest, Guyana. *J. Trop. Ecol.* 17:647–65.

Lim, B. K., and D. E. Wilson. 1993. Taxonomic status of *Artibeus amplus* (Chiroptera: Phyllostomidae) in northern South America. *J. Mammal.* 74:763–68.

Lim, B. K., M. D. Engstrom, and J. Ochoa G. 2005. Mammals. In *Checklist of the terrestrial vertebrates of the Guiana Shield*, ed. T. Hollowell and R. P. Reynolds, 77–92, pl. 6. *Bull. Biol. Soc. Washington* 13:x + 98 pp.

Lim, B. K., H. H. Genoways, and M. D. Engstrom. 2003. Results of the Alcoa Foundation-Suriname Expeditions. XII. First record of the giant fruit-eating bat, *Artibeus amplus*, (Mammalia·Chiroptera) from Suriname with a review of the species. *Ann. Carnegie Mus.* 72:99–107.

Lim, B. K., W. A. Pedro, and F. C. Passos. 2003. Differentiation and species status of the Neotropical yellow-eared bats *Vampyressa pusilla* and *V. thyone* (Phyllostomidae) with a molecular phylogeny and review of the genus. *Acta Chiropterol.* 5:15–29.

Lim, B. K., M. D. Engstrom, R. M. Timm, R. P. Anderson, and L. C. Watson. 1999. First records of 10 bat species in Guyana and comments on diversity of bats in Iwokrama Forest. *Acta Chiropterol.* 1:179–90.

Lim, B. K., M. D. Engstrom, T. E. Lee, Jr., J. C. Patton, and J. W. Bickham. 2004a. Molecular differentiation of large species of fruit-eating bats (*Artibeus*) and phylogenetic relationships based on the cytochrome *b* gene. *Acta Chiropterol.* 6:1–12.

Lim, B. K., M. D. Engstrom, N. B. Simmons, and J. M. Dunlop. 2004b. Phylogenetics and biogeography of least sac-winged bats (*Balantiopteryx*) based on morphological and molecular data. *Mamm. Biol. (Z. Säugetierk.)* 69:225–37.

Lima, J. L. 1926. Os morcegos da collecção do Museu Paulista. *Rev. Mus. Paulista, São Paulo* 14:43–128.

Lima, M., N. C. Stenseth, N. G. Yoccoz, and F. M. Jaksic. 2001. Demography and population dynamics of the mouse opossum (*Thylamys elegans*) in semi-arid Chile: Seasonality, feedback structure and climate. *Proc. Roy. Soc. London B* 268:2053–64.

Linares, O. J. 1966. Notas acerca de *Macrophyllum macrophyllum* (Wied). (Chiroptera). *Mem. Soc. Cien. Nat. La Salle* 26:53–61.

———. 1968. Quirópteres subfósiles encontrados en las cuevas venezolanas. *Bol. Soc. Venez. Espel.* 3:119–45.

———. 1969. Nuevos murciélagos para la fauna de Venezuela en el Museo de Historia Natural La Salle. *Mem. Soc. Cien. Nat. La Salle* 29:37–42.

———. 1971. A new subspecies of funnel-eared bat (*Natalus stramineus*) from western Venezuela. *Bull. S. California Acad. Sci.* 70:81–84.

———. 1973. Presence de l'oreillard d'Amerique du Sud dans les Andes venezueliennes (Chiroptères, *Vespertilionidae*). *Mammalia* 37:433–38.

———. 1986. *Murciélagos de Venezuela*. Caracas: Cuadernos Lagoven, 122 pp.

———. 1998. *Mamíferos de Venezuela*. Caracas: Sociedad Conservacionista Audubon de Venezuela, 691 pp.

Linares, O. J., and A. Escalante. 1992. A new subspecies of the South American flat-headed bat *(Neoplatymops mattogrossensis)* from southern Venezuela. *Mammalia* 56:417–24.

Linares, O. J., and P. Kiblisky. 1969. The karyotype and a new record of *Molossops greenhalli* from Venezuela. *J. Mammal.* 50:831–32.

Linares, O. J., and I. Löbig-A. 1973. El cariotipo del murciélago cavernicola *Natalus tumidirostris*, del norte de Venezuela, y observaciones sobre las afinidades de esta especie con *N. stramineus* (Chiroptera: Natalidae). *Bol. Soc. Venez. Espel.* 4:89–95.

Linares, O. J., and C. J. Naranjo. 1973. Notas acerca de una colección de murciélagos del género *Lonchorhina*, de la Cueva de Archidona, Ecuador (Chiroptera). *Bioespeleologia* 4:175–80.

Linares, O. J., and J. Ojasti. 1971. Una nueva especie de murciélago del género *Lonchorhina* (Chiroptera: Phyllostomatidae) del sur de Venezuela. *Noved. Cient.*, ser. zool. 36:1–8.

———. 1974. Una nueva subespecie del murciélago *Pteronotus parnellii*, en las cuevas de la Peninsula de Paraguana, Venezuela (Chiroptera: Mormoopidae). *Bol. Soc. Venez. Espel.* 5:73–78.

Linares, O. J., and B. Rivas A. 2004. Mamíferos del sistema deltaico (delta de Orinoco-golfo de Paria), Venezuela. *Mem. Fund. La Salle Cienc. Nat.*, nos. 159–160:27–104. [Dated 2003; published in 2004.]

Link, H. F. 1795. *Beyträge zur Naturgeschichte*. Rostock und Leipzig: Karl Christoph Stiller 1:1–8 (unnumbered) + 1–126.

Linnaeus, C. 1748. *Systema naturae; sistens regna tria naturae, in classes et ordines, genera et species redacta tabulisque aeneis illustrata*. Editio sexto, emendata et aucta. Stockholmiae: G. Kiesenetteri, 224 pp.

———. 1758. *Systema naturae per regna tria naturae, secundum classes, ordines, genera, species, cum characteribus, differentiis, synonymis, locis*. Editio decima, reformata. Holmiae: Laurentii Salvii, 1:1–824.

———. 1766. *Systema naturae per regna tria naturae, secundum classes, ordines, genera, species, cum characteribus, differentiis, synonymis, locis*. Editio duodecima, reformata. Holmiae: Laurentii Salvii, 1:1–532.

Lönnberg, E. 1913. Mammals from Ecuador and related forms. *Arkiv Zool., Stockholm* 8(16):1–36, 1 pl.

———. 1921. A second contribution to the mammalogy of Ecuador with some remarks on *Caenolestes*. *Arkiv Zool., Stockholm* 14(4):1–104, 1 pl.

———. 1922. A third contribution to the mammalogy of Ecuador. *Arkiv Zool., Stockholm* 14(20):1–23.

———. 1928. Notes on some South American edentates. *Arkiv Zool., Stockholm* 20(10):1–17, 1 pl.

———. 1937. Notes on some South-American mammals. *Arkiv Zool., Stockholm* 29A(19):1–29.

———. 1942. Notes on *Xenarthra* from Brazil and Bolivia. *Arkiv Zool., Stockholm* 34A(9):1–58, 1 pl.

López-Arevalo, H., O. Montenegro-Díaz, and A. Cadena. 1993. Ecología de los pequeños mamíferos de la Reserva Biológica Carpanta, en la Cordillera Oriental colombiana. *Studies Neotrop. Fauna Environ.* 28:193–210.

López-Fuster, M. J., R. Pérez-Hernández, J. Ventura, and M. Salazar. 2000. Effect of environment on skull-size variation in *Marmosa robinsoni* in Venezuela. *J. Mammal.* 81:829–37.

López-González, C. 1998. Micronycteris minuta. *Mammal. Species* 583:1–4.

López-González, C., and S. J. Presley. 2001. Taxonomic status of *Molossus bondae* J. A. Allen, 1904 (Chiroptera:

Molossidae), with description of a new subspecies. *J. Mammal.* 82:760–74.

López-González, C., S. J. Presley, R. D. Owen, M. R. Willig, and I. Gamarra de Fox. 1998. Noteworthy records of bats (Chiroptera) from Paraguay. *Mastozool. Neotrop.* 5:41–45.

López-González, C., S. J. Presley, R. D. Owen, and M. R. Willig. 2001. Taxonomic status of *Myotis* (Chiroptera: Vespertilionidae) in Paraguay. *J. Mammal.* 82:138–60.

Lord, R. D., H. Delpietro, and L. Lazaro. 1973. Vampiros que se alimentan de murciélagos. *Physis, Buenos Aires* 32:225.

Loretto, D., E. Ramalho, and M. V. Vieira. 2005. Defense behavior and nest architecture of *Metachirus nudicaudatus* Desmarest, 1817 (Marsupialia, Didelphidae). *Mammalia* 69:417–19.

Lubin, Y. D. 1983. *Tamandua mexicana.* In *Costa Rican natural history,* ed. D. H. Janzen, 494–496. Chicago: University of Chicago Press, xii + 816 pp.

Luckett, W. P. 1993. An ontogenetic assessment of dental homologies in therian mammals. In *Mammal phylogeny. Mesozoic differentiation, multituberculates, monotremes, early therians, and marsupials,* ed. F. S. Szalay, M. J. Novacek, and M. C. McKenna, 182–204. New York: Springer Verlag New York, Inc., x + 249 pp.

Lund, P. W. 1839a. Palaeontologie. *Écho du Monde Savant, Paris* 5:244–45. [Information extracted from a letter to Audouin from Lund, dated 5 Nov. 1838].

———. 1839b. Coup-d'oeil sur les espèces éteintes de mammifères du Brésil, extrait de quelques mémoires présentés à l'Académie royale des sciences de Copenhague. *Ann. Sci. Nat. (Zool.), Paris,* sér. 2, 11:214–34.

———. 1839c. Pattedyrene. *K. Danske Vidensk. Selskabs Naturv. Math. Afhandl.* 2:1–82, 13 pls. [Preprint of Lund, 1841a.]

———. 1840a. Fortsaettelse af Pattedyrene. *K. Danske Vidensk. Selskab. Naturv. Math. Afhandl.* 3:1–56, 11 pls. [Preprint of Lund, 1841b.]

———. 1840b. Tillaeg til de to sidste afhandlinger over Brasiliens Dyreverden för sidste Jordomvaeltning. *K. Danske Vidensk. Selskab. Naturv. Math. Afhandl.* 3:1–24, 3 pls. [Preprint of Lund, 1841c.]

———. 1840c. *Nouvelles recherches* sur la faune fossile du Brésil. *Ann. Sci. Nat.,* sér. 2 (zool.), 13:310–19.

———. 1841a. Blik paa Brasiliens Dyreverden för sidste Jordomvaeltning. Anden Afhandling: Pattedyrene. *K. Danske Vidensk. Selskabs Naturv. Math. Afhandl.* 8:61–144, pls. 1–13.

———. 1841b. Blik paa Brasiliens Dyreverden för sidste Jordomvaeltning. Tredie Afhandling: Fortsaettelse af Pattedyrene. *K. Danske Vidensk. Selskab. Naturv. Math. Afhandl.* 8:217–72, pls. 14–24.

———. 1841c. Blik paa Brasiliens Dyreverden för sidste Jordomvaeltning. Tillaeg til de to sidste afhandlinger over Brasiliens Dyreverden för sidste Jordomvaeltning. *K. Danske Vidensk. Selskab. Naturv. Math. Afhandl.* 8:273–96, pls. 25–27.

———. 1842a. Fortsatte Bemaerkninger over Brasiliens uddöde Dyrskagning. *K. Danske Vidensk. Selskabs Naturv. Math. Afhandl.* 9:123–36.

———. 1842b. Blik paa Brasiliens Dyreverden för sidste Jordomvaeltning. Fjerde Afhandling: Fortsaettelse af Pattedyrene. *K. Danske Vidensk. Selskab. Naturv. Math. Afhandl.* 9:137–208, pls. 28–38.

———. 1843. Conspectum dasypodum. *K. Danske Vidensk. Selskabs Naturv. Math. Afhandl.* 6 pp. [Preprint of Lund, 1845b.]

———. 1845a. Meddelelse af det Udbytte de I 1844 undersögte Knoglehuler have afgivet til Kundskaben om Brasiliens Dyreverden försidste Jordomvaeltning; I et brev. *K. Danske Vidensk. Selskabs Naturv. Math. Afhandl.* 5, 36 pp, 10 pls. [Preprint of Lund, 1846.]

———. 1845b. Conspectum dasypodum. *K. Danske Vidensk. Selskabs Naturv. Math. Afhandl.* 11:lxxxii–lxxxvi.

———. 1846. Meddelelse af det Udbytte de I 1844 undersögte Knoglehuler have afgivet til Kundskaben om Brasiliens Dyreverden försidste Jordomvaeltning; I et brev. *K. Danske Vidensk. Selskab. Naturv. Math. Afhandl.* 12:59–94, pls. 47–56.

Lunde, D. P., and W. A. Schutt, Jr. 2000. The peculiar carpal tubercles of male *Marmosops parvidens* and *Marmosa robinsoni* (Didelphidae: Didelphinae). *Mammalia* 63:495–504. [Dated 1999; number 4 marked "Achevé d'imprimer le 29 mai 2000."]

Lydekker, R. 1887. *Catalogue of the fossil Mammalia in the British Museum, (Natural History) Cromwell Road, S.W.* London: British Museum (Natural History), part V, vi + 345 pp.

———. 1890. Mammalia. *Zool. Record* 26:1–55.

———. 1894a. The La Plata Museum. *Nat. Sci.* 4:27–35, pl. 2.

———. 1894b. *A handbook of the Marsupialia and Monotremata.* Lloyd's Naturalist's Library. London: Edward Lloyd, Limited, xvi + 302 pp. 38 pls.

Lyne, A. G. 1959. The systematic and adaptive significance of the vibrissae in the Marsupialia. *Proc. Zool. Soc. London* 133:79–132.

Lyon, M. W., Jr. 1902. Description of a new phyllostome bat from the Isthmus of Panama. *Proc. Biol. Soc. Washington* 15:83–84.

———. 1906. Description of a new species of great anteater from Central America. *Proc. U.S. Natl. Mus.* 31:569–71, pl. 14.

Lyon, M. W., Jr., and W. H. Osgood. 1909. Catalogue of the type-specimens of the mammals in the United States National Museum, including the Biological Survey collection. *Bull. U.S. Natl. Mus.* 62:x + 1–325.

Machado-Allison, C. E. 1965. Las especies venezolanas del género *Periglischrus* Kolenati, 1857 (Acarina, Mesostigmata, Spinturnicidae). *Acta Biol. Venez.* 4:259–348.

———. 1967. The systematic position of the bats *Desmodus* and *Chilonycteris*, based on host-parasite relationships (Mammalia; Chiroptera). *Proc. Biol. Soc. Washington* 80:223–26.

MacPhee, R. D. E., and M. J. Novacek. 1993. Definition and relationships of Lipotyphla. In *Mammal phylogeny. Placentals*, ed. F. S. Szalay, M. J. Novacek, and M. C. McKenna, 81–102. New York: Springer Verlag New York, Inc., xii + 321 pp.

Macrini, T. E. 2004. Monodelphis domestica. *Mammal. Species* 760:1–8.

Mahoney, J. A., and D. W. Walton. Molossidae. In *Mammalia*, ed. D. W. Walton, 146–50. *Zoological catalogue of Australia*. Canberra: Australian Government Publishing Service, 5:x + 1–274.

Malcolm, J. R. 1990. Estimation of mammalian densities in continuous forest north of Manaus. In *Four Neotropical forests*, ed. A. H. Gentry, 339–57. New Haven: Yale University Press, xvi + 627 pp.

———. 1991. Comparative abundances of Neotropical small mammals by trap height. *J. Mammal.* 72:188–92.

Malia, M. J., R. M Adkins, and M. W. Allard. 2002. Molecular support for Afrotheria and the polyphyly of Lypotyphla based on analyses of the growth hormone receptor gene. *Mol. Phylogen. Evol.* 24:91–101.

Mann F., G. 1945. Mamíferos de Tarapacá. *Biologica, Santiago de Chile* 2: 23–98.

———. 1950a. Succión de sangre por *Desmodus*. *Investig. Zool. Chilenas* 1:7–8.

———. 1950b. Nuevos mamíferos de Tarapacá. *Investig. Zool. Chilenas* 2:4–6.

———. 1953. Filogenía y función de la musculatura de *Marmosa elegans* (Marsupialia, Didelphydae). *Investig. Zool. Chilenas* 1:3–15.

———. 1955. Monito del monte, *Dromiciops australis* Philippi. *Invest. Zool. Chilenas* 2:159–66.

———. 1956. Filogenía y función de la musculatura de *Marmosa elegans* (Marsupialia, Didelphydae). 2ª parte. *Investig. Zool. Chilenas* 3:3–28.

———. 1978. *Los pequeños mamíferos de Chile*. Con textos complementarios sistemáticos de Dr. *Roberto Donoso Barros*. Editor científico: Dr. Jorge N. Artigas. Gayana, Zool., no. 40, 342 pp. Santiago de Chile: Editorial de la Univ. de Concepción.

Mantilla-Meluk, H., and R. J. Baker. 2006. Systematics of small *Anoura* (Chrioptera: Phyllostomidae) from Colombia, with description of a new species. *Occas. Papers Mus., Texas Tech Univ. Press*, no. 261:1–18.

Marcgraf de Liebstad, G. 1648. Historiae rerum naturalium Brasiliae, libri octo. . . . Cum appendice de Tapuyis, et Chilensibus. Ioannes de Laet, Antvverpianus, In ordinem digessit & annotationes addidit, & varia ab auctore omissa supplevit & illustravit. In W. Piso. *Historia naturalis Brasiliae, in qua non tantum plantae et animalia, sed et indigenarum morbi, ingenia et mores discribuntur et iconibus supra quingentas illustrantur* Lugdun. Batavorum: apud Fransiscum Hackium, et Amstelodami: apud Lud. Elzevirium, 293 pp. + 13 (unnumbered).

Marelli, C. A. 1930. Importancia de la piel de comadreja negra en la industria peletera. *La Epoca* 9 de febrero de 1930:2.

———. 1932. Los vertebrados exhibidos en los zoológicos del Plata. *Mem. Jardín Zool. La Plata* 1930–1931, 4:1–275, 84 pls. + 3 unnumbered pp. + pls. A–F. ["Preparóse esta obra entre Septiembre de 1930 y Abril de 1931. Terminóse de imprimir en Julio de 1932."— printed on back of title page.]

Mares, M. A. 1986. [Letter to the editor]. *Science* 234:1311–12.

Mares, M. A., and J. K. Braun. 2000. Systematics and natural history of marsupials from Argentina. In *Reflections of a naturalist: Papers honoring Professor Eugene D. Fleharty*, ed. J. R. Choate, 23–46. Kansas: Fort Hays Studies, Special Issue no. 1, iv + 241.

Mares, M. A., and H. H. Genoways, eds. 1982. *Mammalian biology in South America*. The Pymatuning Symposia in Ecology 6. Special Publications Series. Pittsburgh: Pymatuning Laboratory of Ecology, University of Pittsburgh, xii + 539 pp.

Mares, M. A., R. M. Barquez, and J. K. Braun. 1995. Distribution and ecology of some Argentine bats (Mammalia). *Ann. Carnegie Mus.* 64:219–37.

Mares, M. A., J. K. Braun, and D. Gettinger. 1989. Observations on the distribution and ecology of the mammals of the cerrado grasslands of central Brazil. *Ann. Carnegie Mus.* 58:1–60.

Mares, M. A., J. Morello, and G. Goldstein. 1985. The Monte Desert and other subtropical semi-arid biomes of Argentina, with comments on their relation to North American arid areas. In *Hot deserts and arid shrublands*, ed. M. Evenari, I. Noy-Meir, and D. W. Goodall, 203–37. Amsterdam: Elsevier Scientific Publishing Company, 365 pp.

Mares, M. A., R. A. Ojeda, and R. M. Barquez. 1989. *Guide to the mammals of Salta Province, Argentina*. Norman: University of Oklahoma Press, xv + 303.

Mares, M. A., R. A. Ojeda, and M. P. Kosco. 1981. Observations on the distribution and ecology of the mammals of Salta Province, Argentina. *Ann. Carnegie Mus.* 50:151–206.

Mares, M. A., M. R. Willig, K. E. Streilein, and T. E. Lacher, Jr. 1981. The mammals of northeastern Brazil: A preliminary assessment. *Ann. Carnegie Mus.* 50:81–137.

Mares, M. A., R. M. Barquez, J. K. Braun, and R. A. Ojeda, 1996. Observations on the mammals of Tucumán Province, Argentina. I. Systematics, distribution, and ecology of the Didelphimorphia, Xenarthra, Chiroptera, Primates, Carnivora, Perissodactyla, Artiodactyla, and Lagomorpha. *Ann. Carnegie Mus.* 65:89–152.

Mares, M. A., R. A. Ojeda, J. K. Braun, and R. M. Barquez. 1997. Systematics, distribution, and ecology of the mammals of Catamarca Province, Argentina. In *Life among the muses: Papers in honor of James S. Findley*, ed. T. L. Yates, W. L. Gannon, and D. E. Wilson, 89–141. Special Publication, The Museum of Southwestern Biology 3:6(unnumbered) + 290 pp.

Marinho-Filho, J. 2003. Notes on the reproduction of six phyllostomid bat species in southeastern Brazil. *Chiropt. Neotrop.* 9:173–75.

Marinho-Filho, F., M. M. Guimarães, M. L. Reis, F. H. G. Rodrigues, O. Torres, and G. de Almeida. 1997. The discovery of the Brazilian three banded armadillo in the cerrado of central Brazil. *Edentata* 3:11–13.

Marinkelle, C. J. 1967. *Cimex hemipterus* (Fabr.) from bats in Colombia, South America (Hemiptera: Cimicidae). *Proc. Ent. Soc. Washington* 69:179–80.

———. 1970. *Vampyrops intermedius* sp. n. from Colombia (Chiroptera, Phyllostomatidae). *Rev. Bras. Biol.* 30:49–53.

Marinkelle, C. J., and A. Cadena. 1971. Remarks on *Sturnira tildae* in Colombia. *J. Mammal.* 52:235–37.

———. 1972. Notes on bats new to the fauna of Colombia. *Mammalia* 36:50–58.

Marinkelle, C. J., and E. Grose. 1966. Importancia de los murciélagos para la salud pública con especial referencia a las micosis zoonóticas. *Antioquia Med.* 16:179–94.

Marques, S. A. 1985. Novos registros de morcegos do Parque Nacional da Amazônia (Tapajós), com observações do perídodo de atividade noturna e reprodução. *Bol. Mus. Paraense Emílio Goeldi*, sér. zool., 2:71–83.

Marques, S. A. 1989. Ecologia animal. Levantamento faunístico da área sob influência da BR–364 (Cuiabá-Porto Velho). Programa Polonoroeste. *SCT/PR* CNPq, *Programa Polonoreste, Relatório de Pesquisa*, no. 4, 49 pp.

Marques, S. A., and D. C. Oren. 1987. First Brazilian record for *Tonatia schulzi* and *Sturnira bidens* (Chiroptera: Phyllostomidae). *Bol. Mus. Paraense Emílio Goeldi*, sér. zool., 3:159–60.

Marques-Aguiar, S. A. 1994. A systematic review of the large species of *Artibeus* Leach, 1821 (Mammalia: Chiroptera), with some phylogenetic inferences. *Bol. Mus. Paraense Emílio Goeldi*, sér. zool., 10:1–83.

Marques-Aguiar, S. A., C. C. S. Melo, G. F. S. Aguiar, and J. A. L. Queiróz. 2002. Levantamento preliminar da mastofauna da região de Anajás-Muaná, Ilha de Marajó, Pará, Brasil. *Rev. Bras. Zool.* 19:841–54.

Marschall, A. 1873. *Nomenclator zoologicus continens nomina systematica generum animalium tam viventium quam fossilium, secundum ordinem alphabeticum diposita*. Vindobonae: Caroli Ueberreuter (M. Salzer), iv + 482 pp.

Marshall, L. G. 1977a. Lestodelphys halli. *Mammal. Species* 81:1–3.

———. 1977b. First Pliocene record of the water opossum, *Chironectes minimus* (Didelphidae, Marsupialia). *J. Mammal.* 58:434–36.

———. 1978a. Lutreolina crassicaudata. *Mammal. Species* 91:1–4.

———. 1978b. Dromiciops australis. *Mammal. Species* 99:1–5.

———. 1978c. Glironia venusta. *Mammal. Species* 107:1–3.

———. 1978d. Chironectes minimus. *Mammal. Species* 109:1–6.

———. 1978e. Evolution of the Borhyaenidae, extinct South American predaceous marsupials. *Univ. Calif. Publ. Geol. Sci.* 117:1–89.

———. 1980. Systematics of the South American marsupial family Caenolestidae. *Fieldiana Geol.*, 5:1–145.

———. 1981. The families and genera of Marsupialia. *Fieldiana Geol.*, 8:vi + 1–65.

———. 1982a. Evolution of South American Marsupialia. In *Mammalian biology in South America*, ed. M. A. Mares and H. H. Genoways, 251–72. The Pymatuning Symposia in Ecology 6. Special Publications Series. Pittsburgh: Pymatuning Laboratory of Ecology, University of Pittsburgh, xii + 539 pp.

———. 1982b. Systematics of the South American family Microbiotheriidae. *Fieldiana Geol.*, 10:1–75.

———. 1982c. Systematics of the extinct South American marsupial family Polydolopidae. *Fieldiana Geol.*, 12:1–109.

Marshall, L. G., J. A. Case, and M. O. Woodburne. 1990. Phylogenetic relationships of the families of marsupials. In *Current Mammalogy*, ed. H. H. Genoways, 433–506. New York: Plenum Press, 2:xvi + 1–577.

Marshall, L. G., A. Berta, R. Hoffstetter, R. Pascual, O. A. Reig, M. Bombin, and A. Mones. 1984. Mammals and stratigraphy: Geochronology of the continental mammal-bearing Quaternary of South America. *Palaeovert. Mém. Extr.* 1984:1–76.

Martella, M. B., J. L. Navarro, and E. H. Bucher. 1985. Vertebrados asociados a los nidos de la catorra *Myiopsitta monachus* en Córdoba y La Rioja. *Physis, Buenos Aires* 43:49–51.

Martin, G. 2003. Nuevas localidades para marsupials Patagónicos (Didelphimorphia y Microbiotheria) en el noroeste de al Provincia del Chubut, Argentina. *Mastozool. Neotrop.* 10:148–53.

Martin, R. A. 1972. Synopsis of late Pliocene and Pleistocene bats of North America and the Antilles. *Amer. Midl. Nat.* 87:326–35.

Martinez, L., and B. Villa-R. 1940. Segunda contribución al conocimiento de los murciélagos mexicanos. II.–Estado de Guerrero. *An. Inst. Biol., México* 11:291–361.

Martino, A. M. G., J. O. Aranguren, and A. Arends. 2002. Feeding habits of *Leptonycteris curasoae* in northern Venezuela. *Southwest. Nat.* 47:78–85.

Martino, A., A. Arends, and J. Aranguren. 1998. Reproductive pattern of *Leptonycteris curasoae* Miller (Chiroptera: Phyllostomidae) in northern Venezuela. *Mammalia* 62:69–76.

Martins, E. G., and V. Bonato. 2004. On the diet of *Gracilinanus microtarsus* (Marsupialia, Didelphidae in an Atlantic rainforest fragment in southeastern Brazil. *Mamm. Biol. (Z. Säugetierk.)* 69:58–60.

Massoia, E. 1976. Cuatro notas sobre murciélagos de la República Argentina (Molossidae y Vespertilionidae). *Physis, Buenos Aires* 35:257–65.

———. 1980a. Mammalia de Argentina. 1. Los mamíferos silvestres de la provincia de Misiones. *Iguazú* 1:15–43.

———. 1980b. Un marsupial nuevo para la Argentina: *Monodelphis scalops* (Thomas) (Mammalia-Marsupialia). *Physis, Buenos Aires* 39:61–62.

———. 1982. Restos de mamíferos recolectados en el paraje Paso de los Molles, Pilcaniyeu, Río Negro. *Rev. Invest. Agropec. INTA, Buenos Aires* 17:39–53.

———. 1988. Presas de *Tyto alba* en Campo Ramón, departamento Obrera, provincia de Misiones—I. *Aprona,* bol. científ., no. 7:4–15.

Massoia, E., and J. C. Chebez. 1985. Hallazgo del "cabasu," *Cabassous tatouay* (Cingulata Dasypodidae) en Corrientes y nuevos datos sobre su distribución en Misiones. *Idia, Buenos Aires* 1985:56–58.

———. 1989. Notas zoogeográficas sobre algunos quirópteros misionereos. *Aprona,* bol. científ., no. 14:8.

Massoia, E., and A. Fornes. 1967. El estado sistemático, distribución geográfica y datos etoecológicos de algunos mamíferos neotropicales (Marsupialia y Rodentia) con la descripción de Cabreramys, género nuevo (Cricetidae). *Acta Zool. Lilloana* 23:407–30 + 14 pls.

———. 1972. Presencia y rasgos etoecológicos de *Marmosa agilis chacoensis* Tate en las provincias de Buenos Aires, Entre Ríos y Misiones (*Mammalia-Marsupialia-Didelphidae*). *Rev. Invest. Agropec. INTA, Buenos Aires* 9:71–82.

Massoia, E., J. C. Chebez, and S. Heinonen-Fortabat. 1989a. Segundo analisis comparado de egagropilas de *Tyto alba tuidara* en el departamento de Apóstoles, provincia de Misiones. *Aprona,* bol. científ., no.13:3–8.

———.1989b. Mamíferos y aves depredados for *Tyto alba tuidara* en Bonpland, departamento Candelaria, provincia de Misiones. *Aprona,* bol. científ., no. 15:19–24.

Massoia, E., A. Forasiepi, and P. Teta. 2000. *Los marsupiales de la Argentina.* Buenos Aires: Editorial L.O.L.A [Literture of Latin America], 71 pp.

Massoia, E., O. B. Vaccaro, C. Galliari, and S. Ambrosini. 1987. La mastofauna del Río Urugua-í, provincia de Misiones. *Rev. Mus. Argentino Cien. Nat. "Bernardino Rivadavia,"* zool. 14:111–24.

Masson, D., and J. F. Cosson. 1992. *Cyttarops alecto* (Emballonuridae) et *Lasiurus castaneus* (Vespertilionidae), Deux chiroptères nouveaux pour la Guyane française. *Mammalia* 56:475–78.

Matschie, P. 1894. Die von Herrn Paul Neumann in Argentinien gesammelten und beobachteten Säugethiere. *Sitzungsber. Gesells. Naturf. Freunde Berlin* 1894:57–64.

———. 1916. Bemerkungen über die Gattung *Didelphis* L. *Sitzungsber. Gesells. Naturf. Freunde Berlin* 1916:259–72, 3 pls.

———. 1917. Einige neue Formen der *Didelphis lanigera*-Gruppe. *Sitzungsber. Gesells. Naturf. Freunde Berlin* 1917:280–94.

Matson, J. O., and T. J. McCarthy. 2004. Sturnira mordax. *Mammal. Species* 755:1–3.

McBee, K., and R. J. Baker. 1982. Dasypus novemcinctus. *Mammal. Species* 162:1–9.

McCarthy, T. J. 1989. Human depredation by vampire bats (*Desmodus rotundus*) following a hog cholera campaign. *Amer. J. Trop. Med. Hyg.,* 40:320–22.

McCarthy, T. J., and C. O. Handley, Jr. 1988. Records of *Tonatia carrikeri* (Chiroptera: Phyllostomidae) from the Brazilian Amazon and *Tonatia schulzi* in Guyana. *Bat Res. News* 28:20–23. [Dated 1987; published in 1988.]

McCarthy, T. J., and J. Ochoa G. 1991. The presence of *Centronycteris maximiliani* and *Micronycteris daviesi* (Chiroptera) in Venezuela. *Texas J. Sci.* 43:332–34.

McCarthy, T. J., L. Albuja V., and M. S. Alberico. 2006. A new species of chocoan *Sturnira* (Chiroptera:

Phyllostomidae: Stenodermatinae) from western Ecuador and Colombia. *An. Carnegie Mus.* 75:97–110.

McCarthy, T. J., L. Albuja V., and I. Manzano. 2000. Rediscovery of the brown sac-winged bat, *Balantiopteryx infusca* (Thomas, 1897), in Ecuador. *J. Mammal.* 81:958–61.

McCarthy, T. J., L. J. Barkley, and L. Albuja V. 1991. Significant range extension of the giant Andean fruit bat, *Sturnira aratathomasi. Texas Jour. Sci.* 43:437–38.

McCarthy, T. J., A. Cadena G., and T. O. Lemke. 1983. Comments on the first *Tonatia carrikeri* (Chiroptera: Phyllostomatidae) from Colombia. *Lozania (Acta Zool. Colomb.)* 40:1–6.

McCarthy, T. J., A. L. Gardner, and C. O. Handley, Jr. 1992. Tonatia carrikeri. *Mammal. Species* 407:1–4.

McCarthy, T. J., P. Robertson, and J. Mitchell. 1989. The occurrence of *Tonatia schulzi* (Chiroptera: Phyllostomidae) in French Guiana with comments on the female genitalia. *Mammalia* 52:583–84. [Dated 1988; number 4 marked "Achevé d'imprimer le 7 avril 1989."

McCracken, G. F. 1984. Communal nursing in Mexican free-tailed bat maternity colonies. *Science* 223:1090–91.

McCracken, G. F., J. P. Hayes, J. Cevallos, S. Z. Guffey, and R. C. Romero. 1997. Observations on the distribution, ecology, and behaviour of bats in the Galapagos Islands. *J. Zool., London* 243:757–70.

McDaniel, B. 1972. Labidocarpid bat-mites of Venezuela (Listrophoroidea: Labidocarpidae). *Brigham Young Univ. Sci. Bull.*, biol. ser., 17:15–32.

McDaniel, V. R. 1976. Brain anatomy. In *Biology of bats of the New World family Phyllostomatidae. Part I*, ed. R. J. Baker, J. K. Jones, Jr., and D. C. Carter, 147–200. Special Publications of the Museum 10. Lubbock: Texas Tech University Press, 218 pp.

McKenna, M. C., and S. K. Bell. 1997. *Classification of mammals above the species level*. New York: Columbia University Press, xii + 631 pp.

McLellan, L. J. 1984. A morphometric analysis of *Carollia* (Chiroptera: Phyllostomidae). *Amer. Mus. Novit.*, no. 2651:1–33.

McMurtrie, H. 1831. *The animal kingdom arranged in conformity with its organization, by the Baron Cuvier, ... The Crustacea, Arachnides and Insecta, by P. A. Latreille. , ...* Translated from the French with notes and additions by H. M'Murtrie. New York: G. and C. and H. Carvill, 1:xxxii + 448 pp., 4 pls.

McNab, B. K. 1969. The economics of temperature regulation in Neotropical bats. *Comp. Biochem. Physiol.* 31:227–68.

———. 1978. Energetics of arboreal folivores: Physiological problems and ecological consequences of feeding on an ubiquitous food supply. In *The ecology of arboreal folivores*, ed. G. G. Montgomery, 153–62. Washington, DC: The Smithsonian Institution Press, 4 (unnumbered) + 574 pp.

———. 1982. The physiological ecology of South American mammals. In *Mammalian biology in South America*, ed. M. A. Mares and H. H. Genoways, 187–207. The Pymatuning Symposia in Ecology 6. Special Publications Series. Pittsburgh: Pymatuning Laboratory of Ecology, University of Pittsburgh, xii + 539 pp.

McNab, B. K., and P. Morrison. 1963. Observations on bats from Bahia, Brazil. *J. Mammal.* 44:21–23.

McWilliams, L. A., T. L. Best, J. L. Hunt, and K. G. Smith. 2002. Eumops dabbenei. *Mammal. Species* 707:1–3.

Medellín, R. A. 1988. Prey of *Chrotopterus auritus*, with notes on feeding behavior. *J. Mammal.* 69:841–44.

———. 1989. Chrotopterus auritus. *Mammal. Species* 343:1–5.

Medellín, R. A., and H. T. Arita. 1989. Tonatia evotis and Tonatia silvicola. *Mammal. Species* 334:1–5.

Medellín, R. A., D. E. Wilson, and D. Navarro L. 1985. Micronycteris brachyotis. *Mammal. Species* 251:1–4.

Méhely, K., L. 1900. *Magyarország denevéreinek monographiája*. Budapest: Monographia Chiropterorum Hungariae, xii + 372 pp., 22 pls.

Mello, D. A. 1977. *Trypanosoma (Megatrypanum) samueli* n. sp., a Trypanosomatidae isolated from *Monodelphis domesticus* (Wagner, 1842) (Marsupialia). *Ann. Parasitol. Hum. Comp.* 52:391–95.

Mello, D. A., and L. E. Moojen. 1979. Nota sobre uma coleção de roedores e marsupiais de algumas regiões do cerrado do Brasil central. *Rev. Brasil. Pesq. Méd. Biol.* 12:287–91.

Mello, M. A. R., and G. M. Schittini. 2005. Ecological analysis of three bat assemblages from conservation units in the lowland Atlantic forest of Rio de Janeiro, Brazil. *Chiropt. Neotrop.* 11:206–10.

Melville, R. V., and J. D. D. Smith, eds. 1987. *Official lists and indexes of names and works in Zoology*. London: The International Trust for Zoological Nomenclature, 366 pp.

Méndez, E. 1977. Mammalian siphonapteran associations, the environment, and biogeography of mammals of southwestern Colombia. *Quest. Entomol.* 13:81–182.

———. 1988. Parasites. In *Natural history of vampire bats*, ed. A. M. Greenhall and U. Schmidt, 191–206. Boca Raton, Florida: CRC Press, frontispiece, 14 unnumbered + 246 pp.

Menegaux, A. 1901. Description d'une variété et d'une espèce nouvelles de chiroptères rapportées du Mexique par M. Diquet. *Bull. Mus. Natl. Hist. Nat., Paris* 7:321–27.

——. 1902. Catalogue des mammifères rapportés par M. Geay de la Guyane Française en 1889 et 1900. *Bull. Mus. Natl. Hist. Nat., Paris* 8:490–96.

——. 1903. Catalogue des mammifères envoyés par M. Geay, de la Guyane Française, en 1902. *Bull. Mus. Natl. Hist. Nat., Paris* 9:114–16.

——. 1906. Description d'un bradypodidé nouveau (*Choloepus hoffmanni peruvianus* subsp. nov.) provenant du Pérou. *Bull. Mus. d'Hist. Nat.* 1906:460–64.

——. 1908. Les genres actuels de la famille des bradypodidés. *Comptes Rendus, Acad. Sci., Paris* 147:701–03.

——. 1909. A propos d'*Hemibradypus mareyi* Anth. = *Bradypus (Scaeopus) torquatus* (Ill.). *Bull. Soc. Zool. France* 34:27–32.

Meritt, D. A., Jr. 1973. In nature the behavior of armadillos. *Yearb. Amer. Philos. Soc.* 1972:383–84.

——. 1976. The La Plata three-banded armadillo *Tolypeutes matacus* in captivity. *Internat. Zoo Yearb.* 16:153–56.

——. 1985. Naked-tailed armadillos, *Cabassous* sp. In *Evolution and ecology of armadillos, sloths and vermilinguas*, ed. G. G. Montgomery, 389–92. Washington, DC: The Smithsonian Institution Press, 10 (unnumbered) + 451 pp.

Meritt, D. A., Jr., R. J. Low, and K. Benirschke. 1973. The chromosomes of *Zaedyus pichiy*. *Mammal. Chrom. Newsl.* 14:108–09.

Merriam, C. H. 1889. Descriptions of fourteen new species and one new genus of North American mammals. *N. Amer. Fauna* 2:5–37.

——. 1890. Description of a new species of *Molossus* from California (*Molossus californicus*). *N. Amer. Fauna* 4:31–32.

——. 1895. Revision of the shrews of the American genera *Blarina* and *Notiosorex*. *N. Amer. Fauna* 10:5–34.

——. 1897a. Descriptions of two new murine opossums from Mexico. *Proc. Biol. Soc. Washington* 11:43–44.

——. 1897b. Descriptions of five new shrews from Mexico, Guatemala, and Colombia. *Proc. Biol. Soc. Washington* 11:227–30.

——. 1898. Mammals of Tres Marias Islands, off western Mexico. *Proc. Biol. Soc. Washington* 12:13–19.

Merry, D. E., S. Pathak, and J. L. VandeBerg. 1983. Differential NOR activities in somatic and germ cells of *Monodelphis domestica* (Marsupialia, Mammalia). *Cytogent. Cell Genet.* 35:244–51.

Mertens, R. 1925. Verzeichniss der Säugetier-Typen des Senckenbergischen Museums. *Senckenbergiana* 7:18–37.

Meserve, P. L., D. A. Kelt, and D. R. Martínez. 1991. Geographical ecology of small mammals in continental Chile Chico, South America. *J. Biogeogr.* 18:179–87.

Meserve, P. L., B. K. Lang, and B. D. Patterson. 1988. Trophic relationships of small mammals in a Chilean temperate forest. *J. Mammal.* 69:721–30.

Meserve, P. L., R. Murúa, O. Lopetegui N., and J. R. Rau. 1982. Observations on the small mammal fauna of a primary temperate rain forest in southern chile. *J. Mammal.* 63:315–17.

Mies, R., A. Kurta, and D. G. King. 1996. Eptesicus furinalis. *Mammal. Species* 526:1–7.

Mikalauskas, J. S., R. Moratelli, and A. L. Peracchi. 2006. Ocorrência de *Chiroderma doriae* Thomas (Chiroptera, Phyllostomidae) no Estado de Sergipe, Brasil. *Rev. Brasil. Zool.* 23:877–78.

Miles, M. A., A. A. Almeida de Souza, and M. M. Póvoa. 1982. O ecótopo de *Panstrongylus megistus* (Hemiptera, Heteroptera, Reduviidae) na Floresta do Horto, Rio de Janeiro, Brasil. *Rev. Bras. Biol.* 42:31–36.

Miller, G. S., Jr. 1896. The Central American *Thyroptera*. *Proc. Biol. Soc. Washington* 10:109–12.

——. 1897a. Revision of the North American bats of the family Vespertilionidae. *N. Amer. Fauna* 13:1–140, 3 pls.

——. 1897b. Description of a new bat from Margarita Island, Venezuela. *Proc. Biol. Soc. Washington* 11:139.

——. 1898. Descriptions of five new phyllostome bats. *Proc. Acad. Nat. Sci. Philadelphia* 50:326–37.

——. 1899a. Notes on the naked-tailed armadillos. *Proc. Biol. Soc. Washington* 13:1–8.

——. 1899b. Descriptions of three new free-tailed bats. *Bull. Amer. Mus. Nat. Hist.* 12:173–81.

——. 1899c. History and characters of the family Natalidae. *Bull. Amer. Mus. Nat. Hist.* 12:245–53.

——. 1900a. Three new bats from the island of Curaçao. *Proc. Biol. Soc. Washington* 13:123–27.

——. 1900b. Note on Micronycteris brachyotis (Dobson) and M. microtis Miller. *Proc. Biol. Soc. Washington* 13:154–55.

——. 1900c. A bat of the genus Lichonycteris in South America. *Proc. Biol. Soc. Washington* 13:156.

——. 1900d. A second collection of bats from the island of Curaçao. *Proc. Biol. Soc. Washington* 13:159–62.

——. 1900e. A new free-tailed bat from Central America. *Ann. Mag. Nat. Hist.*, ser. 7, 6:471–72.

——. 1900f. A new bat from Peru. *Ann. Mag. Nat. Hist.*, ser. 7, 6:570–74.

——. 1902a. Note on the Chilonycteris davyi fulvus of Thomas. *Proc. Biol. Soc. Washington* 15:155.

——. 1902b. Twenty new American bats. *Proc. Acad. Nat. Sci. Philadelphia* 54:389–412.

——. 1902c. The generic position of Nyctinomus orthotis. *Proc. Biol. Soc. Washington* 15:250.

———. 1905. Note on the generic names *Pteronotus* and *Dermonotus*. *Proc. Biol. Soc. Washington* 18:223.

———. 1906a. A new genus of sac-winged bats. *Proc. Biol. Soc. Washington* 19:59–60.

———. 1906b. Twelve new genera of bats. *Proc. Biol. Soc. Washington* 19:83–87.

———. 1907a. A new name for the genus *Rhynchonycteris* Peters. *Proc. Biol. Soc. Washington* 20:65. [Published June 17, 1907.]

———. 1907b. The families and genera of bats. *Bull. U.S. Natl. Mus.* 57:xvii + 282 pp., 14 pls. [Published June 29, 1907.]

———. 1909. The generic name *Nycteris. Proc. Biol. Soc. Washington* 22:90.

———. 1911. Three new shrews of the genus *Cryptotis. Proc. Biol. Soc. Washington* 24:221–24.

———. 1912a. A small collection of bats from Panama. *Proc. U.S. Natl. Mus.* 42:21–26.

———. 1912b. *Catalogue of the mammals of western Europe (Europe exclusive of Russia) in the collection of the British Museum.* London: British Museum (Natural History), xvi + 1019 pp.

———. 1912c. List of North American land mammals in the United States National Museum, 1911. *Bull. U.S. Natl. Mus.* 79:xiv + 1–455.

———. 1913a. Five new mammals from tropical America. *Proc. Biol. Soc. Washington* 26:31–34.

———. 1913b. Notes on the bats of the genus Molossus. *Proc. U.S. Natl. Mus.* 46:85–92.

———. 1913c. Revision of the bats of the genus Glossophaga. *Proc. U.S. Natl. Mus.* 46:413–29.

———. 1924. List of North American Recent mammals 1923. *Bull. U.S. Natl. Mus.* 128:xvi + 1–673.

———. 1931. Two new South American bats. *J. Mammal.* 12:411–12.

Miller, G. S., Jr., and G. M. Allen. 1928. The American bats of the genera Myotis and Pizonyx. *Bull. U.S. Natl. Mus.* 144:1–218.

Miller, G. S., Jr., and R. Kellogg. 1955. List of North American Recent mammals. *Bull. U.S. Natl. Mus.* 205:xii + 1–954.

Miller, G. S., Jr., and J. A. G. Rehn. 1901. Systematic results of the study of North American land mammals to the close of the year 1900. *Proc. Boston Soc. Nat. Hist.* 30:1–352.

Milne-Edwards, A. 1871. Note sur une nouvelle espèce de tatou a cuirasse incomplète (Scleropleura bruneti). *Nouv. Arch. Mus. Hist. Nat. Paris* 7:177–179, pl. 12.

———. 1872a. Note sur une nouvelle espèce de tatou a cuirasse incomplète (*Scleropleura bruneti*). *Ann. Sci. Nat., Zool. Paléontol.*, ser. 5, 16:1.

———. 1872b. Mémoire sur la faune mammalogique du Tibet oriental et principalement de la principanté de Moupin. In *Recherches pour servir à l'histoire naturelle des mammifères comprenant des considérations sur la classification des ces animaux*, ed. H. Milne-Edwards and A. Milne-Edwards, 231–379. Paris: G. Masson, Éditeur, 1868–74, 1:ii + 394 pp.

Milner, J., C. Jones, and J. K. Jones, Jr. 1990. Nyctinomops macrotis. *Mammal. Species* 351:1–4.

Minoprio, J. D. L. 1945. Sobre el Chlamyphorus trunctus Harlan. *Acta Zool. Lilloana* 3:5–58, pls. 1–26.

Miranda-Ribeiro, A. de. 1903. *Basilia ferruginea* género novo da familia das nycteribias. *Achiv. Mus. Nac. Rio de Janeiro* 12:175–79.

———. 1907. Alguns dipteros interessantes. *Achiv. Mus. Nac., Rio de Janeiro* 14:229–39, 3 pls.

———. 1914. *Mammíferos. Cebidae, Hapalidae; Vespertilionidae, Emballonuridae, Phyllostomatidae; Felidae, Mustelidae, Canidae, Procyonidae Tapyridae; Suidae Cervidae Sciuridae, Muridae, Octodontidae, Coenduidae, Dasyproctidae, Caviidae e Leporidae; Platanistidae; Bradypodidae, Myrmecophagidae, Dasypodidae Didelphyidae.* Commissão de Linhas Telegraphicas Estrategicas de Matto-Grosso ao Amazonas, Annexo 5, 49 pp. + appendix (3 pp.), 25 pls.

———. 1935. Fauna de Terezopolis. *Bol. Mus. Nac., Rio de Janeiro* 11(3–4):1–40, 16 pls., 2 maps.

———. 1936. Didelphia ou Mammalia-ovovivipara. Marsupiaes, didelphos, pedimanos ou metatherios. *Rev. Mus. Paulista, São Paulo* 20:245–427, 8 pls.

Miranda-Ribeiro, P. de. 1955. Tipos das espécies y subespécies do Prof Alipio de Miranda Ribeiro depositados no Museu Nacional. *Arq. Mus. Nac., Rio de Janeiro* 42:389–417.

Miretzki, M., and T. C. C. Margarido. 1999. Morcegos da Estação Ecológica do Caiuá, Paraná (sul do Brasil). *Chiropt. Neotrop.* 5:105–08.

Miretzki, M., A. L. Perrachi, and G. V. Bianconi. 2002. Southernmost records of *Sturnira tildae* de la Torre, 1959 (Chiroptera: Phyllostomidae) in Brazil. *Mammalia* 66:306–09.

Moeller, W. 1968. Allometrische analyse der Gürteltierschädel ein Beiträg zur Phylogenie der Dasypodidae Bonaparte, 1838. *Zool. Jahrb. Anat.* 85:411–528.

———. 1975. Edentates. In *Grzimek's Animal Life Encyclopedia*, ed. B. Grzimek, 149–81 and 189–90. New York: Van Nostrand Reinhold Co., 11:1–635.

Mok, W. Y., and L. A. Lacey. 1980. Algumas considerações ecológicas sobre morcegos vampiros na epidemiologia da raiva humana na Bacia Amazônica. *Acta Amaz.* 10:335–42.

Mok, W. Y., R. C. C. Luizão, and M. do S. B. da Silva. 1982. Isolation of fungi from bats of the Amazon basin. *Appl. Environ. Microb.* 44:570–75.

Mok, W. Y., D. E. Wilson, L. A. Lacey, and R. C. C. Luizão. 1982. Lista atualizada de quirópteros da Amazônia brasileira. *Acta Amaz.* 12:814–23.

Molina, C., C. García, and J. Ochoa G. 1995. First record of *Mimon bennettii* (Chiroptera: Phyllostomidae) for Venezuela. *Mammalia* 59:263–65.

Molina, G. I. 1782. *Saggio sulla storia naturale del Chili.* Bologna: Stamperia di S. Tommaso d'Aquino, 367 pp., 1 map.

———. 1810. *Saggio sulla storia naturale del Chili.* Bologna: Fratelli Masi e Comp., frontispiece, vi + 306 pp., 1 map.

Molinari, J. 1994. A new species of *Anoura* (Mammalia Chiroptera Phyllostomidae) from the Andes of northern South America. *Trop. Zool.* 7:73–86.

Molinari, J., and P. J. Soriano. 1987. Sturnira bidens. *Mammal. Species* 276:1–4.

Molins de la Serna, M., and J. Lorenzo Prieto. 1982. Alimentación del rabipelado (*Didelphis marsupialis*) de la Sierra de Perijá. *Acta Cient. Venez.* 33:410.

Mondolfi, E. 1968. Descripción de un nuevo armadillo del género *Dasypus* de Venezuela (Mammalia-Edentata). *Mem. Soc. Cien. Nat. La Salle, Caracas* 27:149–67.

Mondolfi, E., and G. M. Padilla. 1957. Contribución al conocimiento del "perrito de agua" (*Chironectes minimus* Zimmermann). *Mem. Soc. Cien. Nat. La Salle, Caracas* 17:141–55.

Mondolfi, E., and R. Pérez-Hernández. 1984. Una nueva subespecie de zarigüeya del grupo *Didelphis albiventris* (Mammalia-Marsupialia). *Acta Cient. Venez.* 35:407–13.

Mones, A. 1980. Sobre una colección de vertebrados fósiles de Monte Hermoso (Plioceno superior), Argentina, con la descripión de una nueva especie de *Marmosa* (Marsupialia: Didelphidae). *Communic. Paleontol. Mus. Hist. Nat. Montevideo* 1:159–69.

Monjeau, A., N. Bonino, and S. Saba. 1994. Annotated checklist of the living land mammals in Patagonia, Argentina. *Mastozool. Neotrop.* 1:143–56.

Montalvo, C., E. Justo, and L. de Santis. 1988. Nota sobre algunos mamíferos de la provincia de La Pampa (Argentina). *Univ. Nac. La Pampa, Ser.* suppl., 4:177–80.

Montenegro, O. L., and M. Romero-Ruiz. 2000. Murciélagos del sector sur de la Serranía de Chiribiquete, Caquetá, Colombia. *Rev. Acad. Colomb. Cienc.* 23(Supl. Esp.):641–49. [Dated 1999; published February 2000.]

Montgomery, G. G. 1985a. Impact of vermilinguas (*Cyclopes, Tamandua*: Xenarthra=Edentata) on arboreal ant populations. In *The evolution and ecology of armadillos, sloths, and vermilinguas*, ed. G. G. Montgomery, 351–64. Washington, DC: The Smithsonian Institution Press, 10 (unnumbered) + 451 pp.

———. 1985b. Movements, foraging and food habits of the four extant species of Neotropical vermilinguas (Mammalia: Myrmecophagidae). In *The evolution and ecology of armadillos, sloths, and vermilinguas*, ed. G. G. Montgomery, 365–78. Washington, DC: The Smithsonian Institution Press, 10 (unnumbered) + 451 pp.

Montgomery, G. G., and M. E. Sunquist. 1975. Impact of sloths on Neotropical forest energy flow and nutrient cycling. In *Tropical ecological systems, trends in terrestrial and aquatic research*, ed. F. B. Golley and E. Medina, 69–98. New York: Springer-Verlag, New York, xvi + 398 pp.

———. 1978. Habitat selection and use by two-toed and three-toed sloths. In *The ecology of arboreal folivores*, ed. G. G. Montgomery, 329–59. Washington, DC: The Smithsonian Institution Press, Washington, DC, 4 (unnumbered) + 574 pp.

Moojen, J. 1943. Alguns mamíferos colecionados no nordeste do Brasil com a descrição de duas espécies novas e notas de campo. *Bol. Mus. Nac., Rio de Janeiro*, nova sér., zool., 5:1–14.

———. 1948. Speciation in the Brazilian spiny rats (genus *Proechimys*, family Echimyidae). *Univ. Kansas Publ. Mus. Nat. Hist.* 1:301–401.

Moraes, D. Astúa de. 2007. Range extension and first Brazilian record of the rare *Hyladelphys kalinowskii* (Hershkovitz, 1992) (Didelphimorphia, Didelphidae). *Mammalia* 69:174–76. [Dated 2006, published 2007.]

Morales, J. C., and J. W. Bickham. 1995. Molecular systematics of the genus *Lasiurus* (Chiroptera: Vespertilionidae) based on restriction site maps of the mitochondrial ribosomal genes. *J. Mammal.* 76:730–49.

Morales-Agacino, E., 1938. Sobre el tipo y la localidad tipica del Myotis albescens (E. Geoffroy). *Bol. Soc. Español Hist. Nat.* 37:17–19.

Morales-Alarcón, A., E. Osorno-M., C. Bernal C., and A. Llevas P. 1968. Aislamiento de virus rábico de murciélagos en Colombia, S. A. *Caldasia* 10:167–72.

Morando, M., and J. J. Polop. 1997. Annotated checklist of mammal species of Córdoba Province, Argentina. *Mastozool. Neotrop.* 4:129

Morgan, G. S., and N. J. Czaplewski. 1999. First fossil record of *Amorphochilus schnablii* (Chiroptera: Furipteridae), from the late Quaternary of Peru. *Acta Chiropterol.* 1:75–79.

———. 2003. A new bat (Chiroptera: Natalidae) from the early Miocene of Florida, with comments on natalid phylogeny. *J. Mammal.* 84:729–52.

Morgan, G. S., O. J. Linares, and C. E. Ray. 1988. New species of fossil vampire bats (Mammalia: Chiroptera: Desmodontidae) from Florida and Venezuela. *Proc. Biol. Soc. Washington* 101:912–28.

Morielle-Versute, E., V. A. Taddei, and M. Varella-Garcia. 1992. Chromosome banding studies of *Chrotopterus auritus* (Chiroptera: Phyllostomidae). *Rev. Bras. Genet.* 15:569–73.

Morrison, D. W. 1978. Foraging ecology of the frugivorous bat *Artibeus jamaicencis*. *Ecology* 59:716–23.

———. 1979. Apparent male defense of tree hollows in the fruit bat, *Artibeus jamaicensis*. *J. Mammal.* 60:11–15.

———. 1980. Foraging and day-roosting dynamics of canopy fruit-bats in Panama. *J. Mammal.* 61:20–29.

Morrison, D. W., and C. O Handley, Jr. 1991. Roosting behavior. In *Demography and natural history of the common fruit bat*, Artibeus jamaicensis, *on Barro Colorado Island, Panamá*, ed. C. O Handley, Jr., D. E. Wilson, and A. L. Gardner, 131–35. *Smithsonian Contrib. Zool.* 511:iv + 1–173.

Morrison-Scott, T. C. S. 1955. Proposed addition to the "Official List of Generic Names in Zoology" of the generic name "Mormoops" Leach, 1821 (Class Mammalia). *Bull. Zool. Nomen.* 11:183–85.

Motta, J. C., Jr., and V. A. Taddei. 1992. Bats as prey of Stygian owls in southeastern Brazil. *J. Raptor Res.* 26:259–60.

Motta, M. F. D. 1988. Estudo do desenvolvimento extrauterino de *Didelphis aurita* Wied, 1826, em cativeiro—Investigação de critérios para estimativa de idade. M.S. thesis, Universidade Federal do Rio de Janeiro/Museu Nacional, Rio de Janeiro, Brasil, 105 pp.

Motta, M. F. D., J. C. A. Carreira, and A. M. R. Franco. 1983. A note on reproduction of *Didelphis marsupialis* in captivity. *Mem. Inst. Oswaldo Cruz* 78:507–09.

Mouchaty, S. K., A. Gullberg, A. Janke, and U. Arnason. 2000. The phylogenetic position of the Talpidae within Eutheria based on analysis of complete mitochondrial sequences. *Mol. Biol. Evol.* 17:60–67.

Muchhala, N., P. Mena V., and L. Albuja V. 2005. A new species of *Anoura* (Chiroptera: Phyllostomidae) from the Ecuadorian Andes. *J. Mammal.* 86:457–61.

Muirhead, L. 1819. Mazology. In *The Edinburgh encyclopaedia*, 4th ed., ed. D. Brewster, 393–486, pls. 353–58. Edinburgh: William Blackwood, 13:1–744, pls. 347–71, 1830. [Dated 1830, but apparently printed in 1819 see p. 744; also see Stone 1900.]

Muizon, C. de, and C. Argot. 2003. comparative anatomy of the Tiupampa didelphimorphs; an approach to locomotory habits of early marsupials. In *Predators with pouches: the biology of carnivorous marsupials*, ed.

M. E. Jones, C. R. Dickman, and M. Archer, 43–62. Collingwood, Australia: CSIRO Publishing, xix + 486 pp.

Müller, P. L. S. 1776. *Des Ritters Carl von Linné Königlich Schwedischen Leibarztes u. u. vollständiges Natursystem nach der zwölften lateinischen Ausgabe und nach Anleitung des holländischen Houttuynischen Werks mit einer ausführlichen Erklärung*. Supl. Erste Classe Säugende Thiere. Nürenberg: Gabriel Nicolaus Raspe, 62 pp. + 2 unnumbered, 3 pls.

Mumford, R. E., and D. M. Knudson. 1978. Ecology of bats at Vicosa, Brazil. In *Proceedings of the Fourth International Bat Research Conference, Nairobi*, ed. R. J. Olembo, J. B. Castelino, and F. A. Mutere, 287–95. Nairobi: Kenya Literature Bureau, 328 pp.

Muñoz, J. 1986. Murciélagos del parque natural "El Refugio" (Antioquia, Colombia). *Actual. Biol.* 15:66–76.

———. 1990. Diversidad y hábitos alimenticios de murciélagos en transectos altitudinales a través de la Cordillera Central de los Andes en Colombia. *Studies Neotrop. Fauna Environ.* 25:1–17.

———. 1993. Murciélagos del norte de Antioquia (Colombia). *Studies Neotrop. Fauna Environ.* 28:83–93.

———. 2001. *Los murciélagos de Colombia. Systemática, distribucion, descripción, historia natural y ecología*. Medellín, Colombia: Editorial Universidad de Antioquia, xvii + 391 pp.

Muñoz, J., and C. A. Cuartas. 2001. *Saccopteryx antioquensis* n. sp. (Chiroptera: Emballonuridae) from northwestern Colombia. *Actual Biol.* 23:53–61.

Muñoz-Pedreros, A., and R. E. Palma. 2000. Marsupiales. In *Mamíferos de Chile*, ed. A. Muñoz-Pedreros and J. Yañez-Valenzuela, 43–52. Valdivia, Chile: CEA Ediciones, viii + 463 pp.

Musso, A. 1962. Lista de los mamíferos conocidos de la Isla de Margarita. *Mem. Soc. Cien. Nat. La Salle* 22:163–80.

Mustrangi, M. A., and J. L. Patton. 1997. Phylogeography and systematics of the slender mouse opossum *Marmosops* (Marsupialia, Didelphidae). *Univ. California Publ. Zool.* 130:x + 86 pp.

Myers, P. 1977. Patterns of reproduction of four species of vespertilionid bats in Paraguay. *Univ. California Publ. Zool.* 107:iv + 41 pp., 11 pls.

———. 1981. Observations on *Pygoderma bilabiatum* (Wagner). *Z. Säugetierk.* 46:146–51.

———. 1982. Origin and affinities of the mammal fauna of Paraguay. In *Mammalian biology in South America*, ed. M. A. Mares and H. H. Genoways, 85–93. The Pymatuning Symposia in Ecology 6. Special Publications Series. Pittsburgh: Pymatuning Laboratory of Ecology, University of Pittsburgh, xii + 539 pp.

Myers, P., and J. L. Patton. (this volume). Genus *Lestoros* Oehser, 1934. In The mammals of South America,

Vol. 1. ed. A. L. Gardner, 124–26. Chicago: University of Chicago Press.

Myers, P., and R. M. Wetzel. 1979. New records of mammals from Paraguay. *J. Mammal.* 60:638–41.

———. 1983. Systematics and zoogeography of the bats of the Chaco Boreal. *Misc. Publ. Mus. Zool., Univ. Michigan* 165:iv + 1–59.

Myers, P., R. White, and J. Stallings. 1983. Additional records of bats from Paraguay. *J. Mammal.* 64:143–45.

Nagorsen, D., and J. R. Tamsitt. 1981. Systematics of *Anoura cultrata, A. brevirostrum,* and *A. werckleae. J. Mammal.* 62:82–100.

Naples, V. L. 1982. Cranial osteology and function in the tree sloths, *Choloepus* and *Bradypus. Amer. Mus. Novit.,* no. 2739:1–41.

———. 1985. Form and function of the masticatory musculature in the tree sloths, *Bradypus* and *Choloepus. J. Morphol.* 183:25–50.

Navarro L., D., and D. E. Wilson. 1982. Vampyrum spectrum. *Mammal. Species* 184:1–4.

Nelson, E. W. 1912. Two genera of bats new to Middle America. *Proc. Biol. Soc. Washington* 25:93.

Neveu-Lemaire, M., and G. Grandidier. 1911. *Notes sur les mammifères des hauts plateaux de l'Amerique du Sud.* Imprimerie Nationale, Paris, viii + 127 pp.

Nicéforo-María, H. 1947. Quirópteros de Colombia. *Bol. Inst. La Salle, Bogotá* 34:34–47.

Niethammer, J. 1965. Contribution à la connaissance des mammifères terrestres de L'ile Indefatigable (= Santa Cruz), Galapagos. Résyktats de l'Expédition allenmande aux Galapagos 1962/63. No. VIII. *Mammalia* 28:593–606. [Dated 1964; number 4 published in 1965.]

Nitikman, L. Z., and M. A. Mares. 1987. Ecology of small mammals in a gallery forest of central Brazil. *Ann. Carnegie Mus.* 56:75–95.

Nogueira, J. C., and A. C. S. Castro. 2003. Male genital system of South American didelphids. In *Predators with pouches: The biology of carnivorous marsupials,* ed. M. E. Jones, C. R. Dickman, and M. Archer, 183–204. Collingwood, Australia: CSIRO Publishing, xix + 486 pp.

Nogueira, M. R., and A. L. Peracchi. 2002. The feeding specialization in *Chiroderma doriae* (Phyllostomidae, Stenodermatinae) with comments on its conservation implications. *Chirop. Neotrop.* 8:143–48.

Nogueira, M. R., A. L. Peracchi, and A. Pol. 2002. Notes on the lesser white-lined bat, *Saccopteryx leptura* (Schreber) (Chiroptera, Emballonuridae), from southeastern Brazil. *Rev. Bras. Zool.* 19:1123–30.

Nogueira, M. R., A. Pol, and A. L. Peracchi. 2000. New records of bats from Brazil with a list of additional species for the chiropteran fauna of the state of Acre,

western Amazon basin. *Mammalia* 63:363–68. [Dated 1999; number 3 marked "Achevé d'imprimer le 3 mars 2000."]

Nogueira, M. R., M. N. F. da Silva, G. G. O. Câmara. 1999. Morphology of the male genital system of the bushy-tailed opossum *Glironia venusta* Thomas, 1912 (Didelphimorphia, Didelphidae). *Mammalia* 63:231–36.

Nogueira, M. R., V. C. Tavares, and A. L. Peracchi. 2003. New records of *Uroderma magnirostrum* Davis (Mammalia, Chiroptera) from southeastern Brazil, with comments on its natural history. *Rev. Bras. Zool.* 20:691–97.

Noronha, R. C. R., C. Y. Nagamachi, J. C. Pieczarka, S. Marques-Aguiar, and R. M. S. Barros. 2001. Sex-autosome translocations: Meiotic behaviour suggests an inactivation block with permanence of autosomal gene activity in phyllostomid bats. *Caryologia* 54:267–77.

Noronha, R. C. R., C. Y. Nagamachi, J. C. Pieczarka, S. Marques-Aguiar, M. F. L. de Assis, and R. M. S. Barros. 2004. Meiotic analyses of the sex chromosomes in Carolliinae-Phyllostomidae (Chiroptera): NOR separates the XY_1Y_2 into two independent parts. *Caryologia* 57:1–9.

Notman, H. 1923. A new genus and species of Staphylinidae parasitic on a South American opossum. *Amer. Mus. Novit.,* no. 68:1–3.

Novacek, M. J. 1990. Morphology, paleontology, and the higher clades of mammals. In *Current mammalogy,* ed. H. H. Genoways, 507–44. New York: Plenum Press, 2:xvi + 1–577.

Nowak, R. M. 1991. *Walker's mammals of the world.* 5th ed. Baltimore: The Johns Hopkins Univ. Press, 1:lxiii + 642 pp.

———. 1999. *Walker's mammals of the world.* 6th ed. Baltimore: The Johns Hopkins Univ. Press, 1:lxiv + 836 pp. + lv–lxx.

Nowak, R. M., and J. L. Paradiso. 1983. *Walker's mammals of the world.* 4th ed. Baltimore: The Johns Hopkins Univ. Press, 1:xlvi + 1–568 + xlvii–lxi.

Nunes, A., S. Marques-Aguiar, N. Saldanha, R. S. Silva, and A. Bezerra. 2005. New records on the geographic distribution of bat species in the Brazilian Amazonia. *Mammalia* 69:109–15.

Obenchain, J. B. 1925. The brains of the South American marsupials Caenolestes and Orolestes. *Field Mus. Nat. Hist.,* zool. ser., 14:175–232 + 13 pl.

Ochoa G., J. 1984. Presencia de *Nyctinomops aurispinosa* en Venezuela (Chiroptera: Molossidae). *Acta Cient. Venez.* 35:147–50.

———. 1985. Nueva localidad para *Marmosa tyleriana* (Marsupialia: Didelphidae) en Venezuela. *Doñana, Acta Vert.* 12:183–85.

————. 1995. Los mamíferos de la Región de Imataca, Venezuela. *Acta Cient. Venez.* 46:274–87.

Ochoa G., J., and C. Ibáñez. 1984. Nuevo murciélago del género *Lonchorhina* (Chiroptera: Phyllostomidae). *Mem. Soc. Cien. Nat. La Salle* 42:145–59. [Dated 1982; published 1984.]

————. 1985. Distributional status of some bats from Venezuela. *Mammalia* 49:65–73.

Ochoa G., J., and J. Sánchez H. 1988. Nuevos registros de *Lonchorhina fernandezi* (Chiroptera: Phyllostomidae) para Venezuela, con algunas anotaciones sobre su biología. *Mem. Soc. Cien. Nat. La Salle* 48:133–54.

Ochoa, J. G. [*sic*], and J. H. [*sic*] Sánchez. 2005. Taxonomic status of *Micronycteris homezi* (Chiroptera, Phyllostomidae). *Mammalia* 69:323–36.

Ochoa G., J., H. Castellanos, and C. Ibáñez. 1988. Records of bats and rodents from Venezuela. *Mammalia* 52:175–80.

Ochoa G., J., C. Molina, and S. Giner. 1993. Inventário y estudio comunitario de los mamíferos del Parque Nacional Canaima, con una lista de las especies registradas para la Guayana Venezolana. *Acta Cient. Venez.* 44:245–62.

Ochoa G., J., and J. Sanchez H., and C. Ibáñez. 1988. Records of bats and rodents from Venezuela. *Mammalia* 52:175–80.

Ochoa G., J., P. J. Soriano, and J. Hernández-Camacho. 1994. Sobre la presencia de *Cyttarops alecto* (Chiroptera: Emballonuridae) en Colombia. *Trianea* 5:411–14.

Ochoa G., J., J. Sánchez H., M. Bevilacqua, and R. Rivero. 1988. Inventário de los mamíferos de la Reserva Forestal de Ticoporo y la Serranía de Los Pijiguaos, Venezuela. *Acta Cient. Venez.* 39:269–80.

Ochoa G., J., P. J. Soriano, D. Lew, and M. Ojeda C. 1993. Taxonomic and distrubutional notes on some bats and rodents from Venezuela. *Mammalia* 57:393–400.

O'Connell, M. A. 1979. Ecology of didelphid marsupials from northern Venezuela. In *Vertebrate ecology in the northern Neotropics*, ed. J. F. Eisenberg, 73–88. Washington, DC: The Smithsonian Institution Press, 271 pp.

————. 1983. Marmosa robinsoni. *Mammal. Species* 203:1–6.

————. 1989. Population dynamics of Neotropical small mammals in seasonal habitats. *J. Mammal.* 70:532–48.

Oehser, P. H. 1934. Another new generic name for a South American marsupial. *J. Mammal.* 15:240.

Ojasti, J. 1966. Cuatro nuevos murciélagos para la fauna venezolana. *Acta Biol. Venez.* 5:91–97.

Ojasti, J., and O. J. Linares. 1971. Adiciones a la fauna de murciélagos de Venezuela con notas sobre las especies del género *Diclidurus* (Chiroptera). *Acta Biol. Venez.* 7:421–41.

Ojasti, J., and E. Mondolfi. 1968. Esbozo de la fauna de mamíferos de Caracas. In *Estudio de Caracas*, Vol. 3. Ecología vegetal y fauna, ed. A. Tovar, 441–61. Caracas: Ediciones de la Biblioteca de la Universidad, Central de Venezuela, 467 pp.

Ojasti, J., and C. J. Naranjo. 1974. First record of *Tonatia nicaraguae* in Venezuela. *J. Mammal.* 55:249.

Ojasti, J., R. Guerrero, and O. E. Hernández P. 1992. Mamíferos de la Expedición de Tapirapecó, estado Amazonas, Venezuela. *Acta Biol. Venez.* 14:27–40.

Ojeda, R. A., and R. M. Barquez. 1978. Contribución al conocimiento de los quirópteros de Bolivia. *Neotrópica* 24:33–38.

Ojeda, R. A., and M. A. Mares. 1989. *A biogeographic analysis of the mammals of Salta Province, Argentina. Patterns of species assemblage in the Neotropics.* Special Publications of the Museum 27. Lubbock: Texas Tech University Press, 66 pp.

Oken, L. 1816. *Lehrbuch der Naturgeschichte.* Dritter Theil. Zoologie. Jena: August Schmid und Comp., 3:xvi + 1270 pp., 1 table.

Olfers, I. von. 1818. Bemerkungen zu Illiger's Ueberblick der Säugthiere, nach ihrer Vertheilung über die Welttheile, rücksichtlich der Südamerikanischen Arten (Species). In *Journal von Brasilien, oder vermischte Nachrichten aus Brasilien, auf wissenschaftlichen Reisen gesammelt*, W. L. Eschwege, 192–237. In *Neue Bibliothek der wichtigsten Reisebeschreibungen zur Erweiterung der Erd– und Völkerkunde; in Verbindung mit einigen anderen Gelehrten gesammelt und herausgegeben*, ed. F. I. Bertuch. Weimar: Verlage des Landes–Industrie–Comptoirs, 15(2):xii + 304 pp., 6 pls.

Oliveira, J. A. de, M. L. Lorini, and V. G. Persson. 1992. Pelage variation in *Marmosa incana* (Didelphidae, Marsupialia) with notes on taxonomy. *Z. Saugetierk.* 57:129–36.

Oliveira, T. G. de. 1995. The Brazilian tree-banded armadillo *Tolypeutes tricinctus* in Maranhão. *Edentata* 2:18–19.

Oliver, W. L. R. 1976. The management of yapoks (*Chironectes minimus*) at Jersey Zoo, with observations on their behavior. *Ann. Report, Jersey Wildl. Preserv. Trust* 13:32–36.

Oliver-Schneider, C. 1919. Sobre la distribución geográfica del *Dromiciops australis. Fed. Phil. Act. Soc. Scient. Chili* 27:53.

————. 1946. Catálago de los mamíferos de la provincia de Concepción. *Bol. Soc. Biol. Concepción, Chile* 21:67–83.

Olmos, F. 1995. Edentates in the caatinga of Serra da Capivara National Park. *Edentata* 2:16–16

Olrog, C. C. 1951. Notas sobre mamíferos y aves del Archipielago de Cabo de Hornos. *Acta Zool. Lilloana* 9:505–32, 1950. [Dated 1950; print date "el día 25 de Octubre de 1951."]

———. 1960. Notas mastozoológicas. II. Sobre la colección del Instituto Miguel Lillo. *Acta Zool. Lilloana*, 17:403–20, 1959. [Dated 1959; print date "el día 10 de Junio de 1960."]

———. 1967. *Pygoderma bilabiatum*, un murciélago nuevo para la fauna Argentina (Mammalia, Chiroptera, Phyllostomidae). *Neotrópica* 13:104.

———. 1976. Sobre mamíferos del noroeste Argentino. *Acta Zool. Lilloana* 32:5–14.

Olrog, C. C., and R. M. Barquez. 1979. Dos quirópteros nuevos para la fauna Argentina. *Neotrópica* 25:185–86.

Olrog, C. C., and M. M. Lucero. 1981. *Guía de los mamíferos Argentinos*. San Miguel de Tucumán, Argentina: Ministerio de Cultura y Educación, Fundación Miguel Lillo, 151 pp.

Orr, R. T. 1966. Evolutionary aspects of the mammalian fauna of the Galapagos. In *The Galapagos. Proceedings of the symposium of the Galapagos International Scientific Project*, ed. R. I. Bowman, 276–81. Berkeley: The University of California Press, xvii + 318 pp.

Ortega, J., and H. T. Arita. 1997. Mimon bennettii. *Mammal. Species* 549:1–4.

Ortega, J., and I. Castro-Arellano. 2001. Artibeus jamaicensis. *Mammal. Species* 662:1–9.

Ortiz de la Puente D., J. 1951. Estudio monografico de los quirópteros de Lima y alrededores. *Publ. Mus. Hist. Nat. "Javier Prado,"* ser. A, zool. ser., 7:1–48, 9 pls.

Ortiz-Von Halley, B, and M. Alberico. 1989. Primer reqistro de *Neoplatymops mattogrossensis* (Cunha Vieira 1942) (Chiroptera: Molossidae) en territorio Colombiano. *Trianea* 3:263.

Osculati, G. 1854. *Esplorazione delle regione equatoriali*. Milano: I. Fratelli Centenari e Comp., 344 pp.

Osgood, W. H. 1910. Mammals from the coast and islands of northern South America. *Field Mus. Nat. Hist.*, zool. ser., 10:23–32.

———. 1912. Mammals from western Venezuela and eastern Colombia. *Field Mus. Nat. Hist.*, zool. ser., 10:33–66, 2 pls.

———. 1913. New Peruvian mammals. *Field Mus. Nat. Hist.*, zool. ser., 10:93–100.

———. 1914a. Four new mammals from Venezuela. *Field Mus. Nat. Hist.*, zool. ser., 10:135–41.

———. 1914b. Mammals of an expedition across northern Peru. *Field Mus. Nat. Hist.*, zool. ser., 10:143–85.

———. 1915. New mammals from Brazil and Peru. *Field Mus. Nat. Hist.*, zool. ser., 10:187–98.

———. 1916. Mammals of the Collins-Day South American Expedition. *Field Mus. Nat. Hist.*, zool. ser., 10:199–216.

———. 1919. Names of some South American mammals. *J. Mammal.* 1:33–36.

———. 1921. A monographic study of the American marsupial, *Caenolestes. Field Mus. Nat. Hist.*, zool. ser., 14:1–156, 22 pls.

———. 1924. Review of living caenolestids with description of a new genus from Chile. *Field Mus. Nat. Hist.*, zool. ser., 14:165–73.

———. 1943. The mammals of Chile. *Field Mus. Nat. Hist.*, zool. ser., 30:1–268.

Ospina-Ante, O., and L. G. Gómez. 2000. Riqueza, abundancia relativa y patrones de actividad temporal de la comunidad de los murciélagos quirópteros de la Reserva Natural La Planada, Nariño, Colombia. *Rev. Acad Colomb. Cien.* 23(Supl. Esp.):659–69. [Dated 1999; published February 2000.]

Owen, J. G., D. J. Schmidly, and W. B. Davis. 1984. A morphometic analysis of three species of *Carollia* (Chiroptera, Glossophaginae) from Middle America. *Mammalia* 48:85–93.

Owen, R. 1845. *Odontography or, a treatise on the comparative anatomy of the teeth; their physiological relations, mode of development, and microscopic structure, in the vertebrate animals*. Atlas. London: Hippolyte Bailliere, 2:1–37 + 168 pls., 1840–45.

Owen, R. 1852. Teeth. In *The cyclopedia of anatomy and physiology*, ed. R. B. Todd, 864–935. London: Longman, Brown, Green, and longmans, 4:5 (unnumbered) + 1,543 pp.

Owen, R. D. 1987. *Phylogenetic analyses of the bat subfamily Stenodermatinae (Mammalia: Chiroptera)*. Special Publications of the Museum 26. Lubbock: Texas Tech University Press, 65 pp.

———. 1988. Phenetic analyses of the bat subfamily Stenodermatinae (Chiroptera: Phyllostomidae). *J. Mammal.* 69:795–810.

———. 1991. The systematic status of *Dermanura concolor* (Peters, 1865) (Chiroptera: Phyllostomidae), with description of a new genus. In *Contributions to mammalogy in honor of Karl F. Koopman*, ed. T. A. Griffiths and D. Klingener, 18–25. *Bull. Amer. Mus. Nat. Hist.* 206:1–432.

Owen, R. D., and W. D. Webster. 1983. Morphological variation in the Ipanema bat, *Pygoderma bilabiatum*, with a description of a new subspecies. *J. Mammal.* 64:146–49.

Owen, R. D., R. K. Chesser, and D. C. Carter. 1990. The systematic status of Tadarida brasiliensis cynocephala and Antillean members of the Tadarida brasiliensis group, with comments on the generic name Rhizomops Legendre. *Occas. Papers Mus., Texas Tech Univ.*, no. 133:1–7.

Pacheco, S. M., and R. V. Marques. 1995. Observações sobre o parto en *Tadarida brasiliensis* (I. Geoffroy, 1824) (Mammalia, Chiroptera, Molossidae) em Porto Alegre, Rio Grande do Sul, Brasil. *Comun. Mus. Ciênc. Tecnol. PUCRS, Porto Alegre*, sér. zool., 8:3–11.

Pacheco, V., and B. D. Patterson. 1991. Phylogenetic relationships of the New World bat genus *Sturnira* (Chiroptera: Phyllostomidae). In *Contributions to mammalogy in honor of Karl F. Koopman*, ed. T. A. Griffiths and D. Klingener, 101–21. *Bull. Amer. Mus. Nat. Hist.* 206:1–432.

———. 1992. Systematics and biogeographic analyses of four species of *Sturnira* (Chiroptera: Phyllostomidae), with emphasis on Peruvian forms. *Mem. Mus. Hist. Nat., Lima* 21:57–81.

Pacheco, V., B. D. Patterson, J. L. Patton, L. H. Emmons, S. Solari, and C. F. Ascorra. 1993. List of mammal species known to occur in Manu Biosphere Reserve, Peru. *Publ. Mus. Hist. Nat. UNMSM*, ser. A, zool., 44:1–12.

Pacheco, V., H. Macedo, E. Vivar, C. Ascorra, R. Arana-Cardó, and S. Solari. 1995. Lista anotada de los mamíferos peruanos. *Conserv. Internat., Occas. Papers Conserv. Biol.*, no. 2:1–35.

Palisot de Beauvois, A. M. F. J. 1796. *Catalogue raisonné du museum, de Mr. C. W. Peale, membre de la société philosophique de Pensylvanie*. Philadelphie: De l'Imprimerie de Parent, xiv + 42 pp. [Incomplete; only the first fascicle was published.]

Pallas, P. S. 1766. *Miscellanea zoologica quibus novae imprimis atque obscurae animalium species describuntur et observationibus iconibusque illustrantur*. Hague Comitum: P. van Cleef, xii + 224 pp., 14 pls.

———. 1767. Vespertiliones in genre. In *Spicilegia Zoologica quibus novae et obscurae animalium species iconibus, descriptionibus atque commentariis illustrantur*. Berolini: G. A. Lange, Fasc. 3, 35 pp., 4 pls.

Palma, R. E. 1994. Historical relationships of South American mouse opossums (*Thylamys*, Didelphidae): Evidence from molecular systematics and historical biogeography. Ph.D. thesis, University of New Mexico, Albuquerque, xiii + 112 pp.

———. 1995a. The karyotypes of two South American mouse opossums of the genus *Thylamys* (Marsupialia: Didelphidae), from the Andes, and eastern Paraguay. *Proc. Biol. Soc. Washington* 108:1–5.

———. 1995b. Range expansion of two South American mouse opossums (*Thylamys*, Didelphidae) and their biogeographic implications. *Rev. Chilena Hist. Nat.* 68:515–22.

———. 1997. Thylamys elegans. *Mammal. Species* 572:1–4.

Palma, R. E., and A. E. Spotorno. 1999. Molecular systematics of marsupials based on the rRNA 12S mitochondrial gene: The phylogeny of Didelphimorphia and the living fossil microbiotheriid *Dromiciops gliroides* Thomas. *Mol. Phylogenet. Evol.* 13:525–35.

Palma, R. E., and T. L. Yates. 1996. The chromosomes of Bolivian didelphid marsupials. *Occas. Papers Mus., Texas Tech Univ.*, no. 162:1–20.

Palmer, T. S. 1898. Random notes on the nomenclature of the Chiroptera. *Proc. Biol. Soc. Washington* 12:109–14.

———. 1899. Notes on *Tatoua* and other genera of edentates. *Proc. Biol. Soc. Washington* 13:71–73.

———. 1904. Index generum mammalium: A list of the genera and families of mammals. *N. Amer. Fauna* 23:1–984.

Papavero, N. 1971. *Essays on the history of Neotropical dipterology, with special reference to collectors (1750–1905)*. São Paulo: Museu de Zoología, Universidade de São Paulo, I:viii + 1–216, 10 figs., 12 maps.

Paradiso, J. L. 1968. A review of the wrinkle-faced bats (*Centurio senex* Gray) with description of a new subspecies. *Mammalia* 31:595–604. [Dated 1967; number 4 printed in 1968.]

Pardiñas, U. F. J., A. M. Abba, and M. L. Merino. 2004. Micromamíferos (Didelphimorphia y Rodentia) del sudoeste de la provincia de Buenos Aires (Argentina): Taxonomía y distribución. *Mastozool. Neotrop.* 11:211–32.

Passos, F. C., W. R. Silva, W. A. Pedro, and M. R. Bonin. 2003. Frugivoria em morcegos (Mammalia, Chiroptera) no Parque Estadual Intervales, sudeste do Brasil. *Rev. Brasil. Zool.* 20:511–17.

Patten, D. R. 1971. A review of the large species of *Artibeus* (Chiroptera: Phyllostomatidae) from western South America. Ph.D. thesis, Texas A&M University, College Station, 175 pp. + xvii.

Patterson, B., and R. Pascual. 1968. The fossil mammal fauna of South America. V. Evolution of mammals on southern continents. *Quart. Rev. Biol.* 43:409–51.

Patterson, B., and R. Pascual. 1972. The fossil mammal fauna of South America. In *Evolution, mammals, and southern continents*, ed. A. Keast, F. C. Erk, and B. Glass, 247–310. Albany: State University of New York Press, 543 pp.

Patterson, B. D. 1992. Mammals in the Royal Natural History Museum, Stockholm, collected in Brazil and Bolivia by A. M. Olalla during 1934–1938. *Fieldiana Zool.*, 66:1–42.

Patterson, B. D., and M. H. Gallardo. 1987. Rhyncholestes raphanurus. *Mammal. Species* 286:1–5.

Patterson, B. D., P. L. Meserve, and B. K. Lang. 1989. Distribution and abundance of small mammals along an elevational transect in temperate rainforests of Chile. *J. Mammal.* 70:67–78.

———. 1990. Quantitative habitat associations of small mammals along an elevational transect in temperate rainforests of Chile. *J. Mammal.* 71:620–33.

Patton, J. C., and R. J. Baker. 1978. Chromosomal homology and evolution of phyllostomatoid bats. *Syst. Zool.* 27:449–62.

Patton, J. L., and A. L. Gardner. 1971. Parallel evolution of multiple sex–chromosome systems in the phyllostomatid bats, *Carollia* and *Choeroniscus. Experientia* 27:105–06.

Patton, J. L., and M. N. F. da Silva, 1997. Definition of species of pouched four-eyed opossums (Didelphidae, *Philander). J. Mammal.* 78:90–102.

Patton, J. L., and L. P. Costa. 2003. Molecular phylogeography and species limits in rainforest didelphid marsupials of South America. In *Predators with pouches: The biology of carnivorous marsupials,* ed. M. E. Jones, C. R. Dickman, and M. Archer, 63–81. Collingwood, Australia: CSIRO Publishing, xix + 486 pp.

Patton, J. L., B. Berlin, and E. A. Berlin. 1982. Aboriginal perspectives of a mammal community in Amazonian Perú: Knowledge and utilization patterns among the Aguaruna Jívaro. In *Mammalian biology in South America,* ed. M. A. Mares and H. H. Genoways, 111–28. The Pymatuning Symposia in Ecology 6. Special Publications Series. Pittsburgh: Pymatuning Laboratory of Ecology, University of Pittsburgh, xii + 539 pp.

Patton, J. L., S. F. dos Reis, and M. N. F. da Silva. 1995. Relationships among didelphid marsupials based on sequence variation in the mitochondrial cytochrome b gene. *J. Mammal. Evol.* 3:3–29. [Vol. 3, no. 1 is dated March 1996; however this number was issued in 1995; Copy in Smithsonian Libraries is stamped as received "Nov 28, 1995."]

Patton, J. L., M. N. F. da Silva, and J. R. Malcolm. 2000. Mammals of the Rio Juruá and the evolutionary and ecological diversification of Amazonia. *Bull. Amer. Mus. Nat. Hist.* 244:1–306.

Paula-Couto, C. de. 1950. Revistas e comentadas. In *Peter Wilhelm Lund, Memórias sôbre a Paleontologia Brasileira.* Rio de Janeiro, Brasil: Instituto Nacional do Livro, 592 pp., 56 pls.

———. 1973. Edentados fósseis de São Paulo. *Anal. Acad. Brasileira Ciênc.* 45:261–75.

———. 1979. *Tratado de Paleomastozoología.* Rio de Janeiro: Academia Brasileira de Ciências, 590 pp.

Paynter, R. A., Jr. 1982. *Ornithological gazetteer of Venezuela.* Cambridge: Harvard College Press, iv + 245 pp.

———. 1985. *Ornithological gazetteer of Argentina.* Cambridge: Harvard College Press, vi + 507 pp., 1 map.

———. 1988. *Ornithological gazetteer of Chile.* Cambridge: Harvard College Press, vi + 329 pp., 1 map.

———. 1993. *Ornithological gazetteer of Ecuador.* 2nd ed. Harvard University, Cambridge, xii + 247 pp., 1 map.

———. 1997. *Ornithological gazetteer of Colombia.* 2nd ed. Harvard University, Cambridge, x + 537 pp., 1 map.

Paynter, R. A., Jr., and A. M. G. Caperton. 1977. *Ornithological gazetteer of Paraguay.* Cambridge: Harvard College Press, iv + 43 pp., 1 map.

Paynter, R. A., Jr., and M. A. Traylor, Jr. 1977. *Ornithological gazetteer of Ecuador.* Cambridge: Harvard College Press, viii + 155 pp., 1 map.

———. 1981. *Ornithological gazetteer of Colombia.* Cambridge: Harvard College Press, vi + 311 pp.

———. 1991. *Ornithological gazetteer of Brazil.* Cambridge: Harvard College Press, 1:viii + 1–352; 2:353–788, 1 map.

Paynter, R. A., Jr., M. A. Traylor, Jr., and B. Winter. 1981. *Ornithological gazetteer of Bolivia.* Cambridge: Harvard College Press, vi + 80 pp., 1 map.

Peale, C. W., and A. M. F. J. [Palisot de] Beauvois. 1796. *Scientific and descriptive catalogue of Peale's Museum.* Philadelphia: Samuel H. Smith, xii + 44 pp.

Peale, T. R. 1848. *Mammalia and ornithology.* Philadelphia: C. Sherman, xvi + 338 pp.

Pearson, O. P. 1951. Mammals in the highlands of southern Peru. *Bull. Mus. Comp. Zool.* 106:117–74.

———. 1957. Addition to the mammalian fauna of Peru and notes on some other Peruvian mammals. *Breviora, Mus. Comp. Zool.* 73:1–7.

———. 1983. Characteristics of a mammalian fauna from forests in Patagonia, South America. *J. Mammal.* 64:476–92.

Pearson, O. P., and A. K. Pearson. 1990. Reproduction of bats in southern Argentina. In *Advances in Neotropical mammalogy,* ed. K. H. Redford and J. F. Eisenberg, 549–66. Gainesville, Florida: The Sandhill Crane Press, Inc., x + 614 pp. [Dated 1989; published February 6, 1990.]

Pearson, O. P., and C. P. Ralph. 1978. The diversity and abundance of vertebrates along an altitudinal gradient in Peru. *Memorias, Mus. Hist. Nat. "Javier Prado," Lima* 18:1–97.

Pedro, W. A., and F. C. Passos. 1995. Occurrence and food habits of some bat species from the Linhares Forest Reserve, Espirito Santo, Brazil. *Bat Res. News* 36:1–2.

Pedro, W. A., and V. A. Taddei. 1997. Taxonomic assemblage of bats from Panga Reserve, southeastern Brazil:

Abundance patterns and trophic relations in the Phyllostomidae (Chiroptera). *Bol. Mus. Biol. Mello Leitão*, nov. sér., 6:3–21.

———. 1998. Bats from southwestern Minas Gerais, Brazil (Mammalia: Chiroptera). *Chiropt. Neotrop.* 4:85–88.

Pedro, W. A., C. A. K. Komeno, and V. A. Taddei. 1994. Morphometrics and biological notes on *Mimon crenulatum* (Chiroptera: Phyllostomidae). *Bol. Mus. Para. Emilio Goeldi*, sér. zool., 10:107–12.

Pedro, W. A., F. C. Passos, and B. K. Lim. 2001. Morcegos (Chiroptera; Mammalia) da Estação Ecológica dos Caetetus, estado de São Paulo. *Chiropt. Neotrop.* 7:136–40.

Pelzeln, A. von. 1883. *Brasilische Säugethiere. Resultate von Johann Natterer's Reisen in den Jahren 1817 bis 1835. Verhandl. Kaiserl.-Königl. Zool.-bot. Gesellsch., Wien* 33(Suppl.):1–140.

Peña G., L. E., and G. Barria P. 1972. Presencia de *Histiotus montanus magellanicus* Philp. y de *Myotis chiloensis chiloensis* Waterh. (Chiroptera), al sur del estrecho de Magallanes. *An. Mus. Hist. Nat. Valparaiso* 5:201–02.

Pennant, T. 1771. *Synopsis of quadrupeds*. Chester: J. Monk, xxv + 382 pp., 31 pls.

———. 1781. *History of quadrupeds*. London: B. White, 2:285–566 + 14 (unnumbered), pls. 32–52.

Peracchi, A. L. 1968. Sôbre os hábitos de "Histiotus velatus" (Geoffroy, 1824) (Chiroptera, Vespertilionidae). *Rev. Bras. Biol.* 28:469–73.

———. 1986. Considerações sobre a distribuição e a localidade-tipo de *Sphaeronycteris toxophyllum* Peters, 1882 (Chiroptera, Phyllostomidae). *Publ. Avulsas Mus. Nac., Rio de Janeiro* 65:97–100.

Peracchi, A. L., and S. T. de Albuquerque. 1971. Lista provisória dos quirópteros dos estados do Rio de Janeiro e Guanabara, Brasil. *Rev. Bras. Biol.* 31:405–13.

———. 1985. Consideraçoes sobre a distribuição geográfica de algumas espécies do género *Micronycteris* Gray, 1866 (Mammalia, Chiroptera, Phyllostomidae). *Arq. Univ. Fed. Rur. Rio de Janeiro, Itaguai* 8:23–26.

———. 1986. Quirópteros do estado do Rio de Janeiro, Brasil. *Publ. Avulsas Mus. Nac., Rio de Janeiro* 66:63–69.

———. 1993. Quirópteros do município de Linhares, estado do Espírito Santo, Brasil (Mammalia, Chiroptera). *Rev. Bras. Biol.* 53:575–81.

Peracchi, A. L., S. T. de Albuquerque, and S. D. L. Raimundo. 1982. Contribuição ao conhecimento dos hábitos alimentares de *Trachops cirrhosus* (Spix, 1823) (Mammalia, Chiroptera, Phyllostomidae). *Arq. Univ. Fed. Rur. Rio de Janeiro, Itaguai* 5:1–5.

Peracchi, A. L., S. D. L. Raimundo, and A. M. Tannure. 1984. Quirópteros do Território Federal do Amapá, Brasil (Mammalia, Chiroptera). *Arq. Univ. Fed. Rural Rio de Janeiro, Itaguaí* 7:89–100.

Pereira, L. G., S. E. M. Torres, H. S. da Silva, and L. Giese. 2001. Non-volant mammals of Ilha Grande and adjacent areas in southern Rio de Janeiro state, Brazil. *Bol. Mus. Nac., Rio de Janeiro*, zool., no. 459:1–15.

Pereira Júnior, H. R. J., W. Jorge, and M. E. L. T. da Costa. 2004. Chromosome study of anteaters (Myrmecophagideae [*sic*], Xenarthra)–a preliminary report. *Gen. Mol. Biol.* 27:391–94.

Pérez-Hernández, R. 1985. Notas preliminares acerca de la taxonomía de la familia Didelphidae (Mammalia-Marsupialia) en Venezuela. *Mem. Soc. Cien. Nat. La Salle* 45:47–76.

———. 1990. Distribution of the family Didelphidae (Mammalia-Marsupialia) in Venezuela. In *Advances in Neotropical mammalogy*, ed. K. H. Redford and J. F. Eisenberg, 363–410. Gainesville, Florida: The Sandhill Crane Press, Inc., x + 614 pp. [Dated 1989; published February 6, 1990.]

Pérez-Hernández, R., P. Soriano, and D. Lew. 1994. *Marsupiales de Venezuela*. Caracas: Cuadernos Lagoven, 76 pp.

Peters, J. L. 1934. *Check-list of birds of the world*. Vol. 2. Cambridge: Harvard University Press, xii + 401 pp.

Peters, S. L., B. K. Lim, and M. D. Engstrom. 2002. Systematics of dog-faced bats (*Cynomops*) based on molecular and morphometric data. *J. Mammal.* 83:1097–110.

Peters, W. 1856a. Über die Chiropterengattungen *Mormops* und *Phyllostoma*. *Abhandl. König. Preuss. Akad. Wiss. Berlin* 1857:287–310, 2 pls.

———. 1856b. Ueber die systematische Stellung der Gattung *Mormops* Leach und über die Classification de *Phyllostomata* so wie über eine neue Art der Gattung *Vampyrus*, von welcher hier ein kurzer Bericht gegeben wird. *Monatsber. König. Preuss. Akad. Wiss. Berlin* 1857:409–15.

———. 1858. [Charakterist eines neuen zweizehigen Faulthiers.] *Monatsber. König. Preuss. Akad. Wiss. Berlin* 1859:128.

———. 1859. Neue Beiträge zur Kenntniss der Chiropteren. *Monatsber. König. Preuss. Akad. Wiss. Berlin* 1860:222–25.

———. 1860. Eine neue Gattung von Flederthieren, *Chiroderma villosum*, aus Brasilien. *Monatsber. König. Preuss. Akad. Wiss. Berlin* 1861:747–55.

———. 1861. Eine Ubersicht der von Hrn. Dr. Gundlach beobachteten Fledertheire auf Cuba mit. Die nach dem Leben gemachten Bescreibungen de Hrn. Dr. Gundlach wurden vorgelegt, welche nebst den Originalexemplaren durch Hrn. Geheimerath Sezekorn in Kassel zur näheren

Vergleichung übersandt worden waren. *Monatsber. König. Preuss. Akad. Wiss. Berlin* 1862:149–56.

———. 1863. Nachricht von einem neuen frugivoren Flederthiere, *Stenoderma (Pygoderma) microdon* aus Surinam. *Monatsber. König. Preuss. Akad. Wiss. Berlin* 1864:83–85.

———. 1864a. Über neue Arten der Säugethiergattungen *Geomys, Haplodon* und *Dasypus. Monatsber. König. Preuss. Akad. Wiss. Berlin* 1865:177–81.

———. 1864b. Über einege neue Säugethiere (*Mormops, Macrotus, Vesperus, Molossus, Capromys*), Amphibien (*Platydactylus, Otoctyptis, Euprepes, Ungalia, Dromicus, Tropidonotus, Xenodon, Hylodes*) und Fische (*Sillago, Sebastes, Channa, Myctophum, Carassius, Barbus, Capoeta, Poecilia, Saurenchelys, Leptocephalus*). *Monatsber. König. Preuss. Akad. Wiss. Berlin* 1865:381–99.

———. 1864c. Über das normale Vorkommen von nur sechs Halswirbeln bei *Choloepus Hoffmanni* Ptrs. *Monatsber. König. Preuss. Akad. Wiss. Berlin* 1865:678–80.

———. 1865a. Abbildungen zu einer Monographie der Chiropteren vor und gab eine Übersicht der von ihm befolgten systematischen Ordnung der hieher gehörigen Gattungen. *Monatsber. König. Preuss. Akad. Wiss. Berlin* 1866:256–58.

———. 1865b. Über Flederthiere (*Vespertilio soricinus* Pallas, *Choeronycteris* Lichtenst., *Rhinophylla pumilio* nov. gen., *Artibeus fallax* nov. sp., *A. concolor* nov. sp., *Dermanura quadrivittatum* nov. sp., *Nycteris grandis* n. sp.). *Monatsber. König. Preuss. Akad. Wiss. Berlin* 1866:351–59.

———. 1865c. Über die zu den *Vampyri* gehörigen Flederthiere und über die natürliche Stellung der Gattung *Antrozous. Monatsber. König. Preuss. Akad. Wiss. Berlin* 1866:503–25.

———. 1866a. Über die brasilianischen, von Spix beschriebenen Flederthiere. *Monatsber. König. Preuss. Akad. Wiss. Berlin* 1866:568–88, 1 pl. [Pages 563 to end of volume were published in 1866.]

———. 1866b. Über einige neue oder weniger bekannte Flederthiere. *Monatsber. König. Preuss. Akad. Wiss. Berlin* 1867:16–25.

———. 1866c. Über neue oder ungenügend bekannte Flederthiere (*Vampyrops, Uroderma, Chiroderma, Ametrida, Tylostoma, Vespertilio, Vesperugo*) und Nager (*Tylomys, Lasiomys*). *Monatsber. König. Preuss. Akad. Wiss. Berlin* 1867:392–411, 2 pls.

———. 1867a. Fernere Mittheilungen zur Kenntniss der Flederthiere, namentlich über Arten des Leidener und Britischen Museums. *Monatsber. König. Preuss. Akad. Wiss. Berlin* 1867:672–81.

———. 1867b. Note on a collection of bats from Trinidad. *Proc. Zool. Soc. London* 1866:430–31. [Published April 1867.]

———. 1867c. Über die zu den Gattungen *Mimon* und *Saccopteryx* gehörigen Flederthiere. *Monatsber. König. Preuss. Akad. Wiss. Berlin* 1867:469–81.

———. 1867d. Über Flederthiere (*Pteropus Gouldii, Rhinolophus Deckenii, Vespertilio lobipes, Vesperugo Temminckii*) und Amphibien (*Hypsilurus Godeffroyi, Lygosoma scutatum, Stenostoma narirostre, Onychocephalus unguirostris, Ahaetulla polylepis, Pseudechis scutellatus, Hoplobatrachus Reinhardtii, Hyla coriacea*). *Monatsber. König. Preuss. Akad. Wiss. Berlin* 1868:703–12, 1 pl.

———. 1868a. Über eine neue Untergattung der Flederthiere, so wie über neue Gattungen und Arten von Fischen. *Monatsber. König. Preuss. Akad. Wiss. Berlin* 1869:145–48.

———. 1868b. Über die zu den *Glossophagæ* gehörigen Flederthiere und über eine neue Art der Gattung *Colëura. Monatsber. König. Preuss. Akad. Wiss. Berlin* 1869:361–68.

———. 1869. Bemerkungen über neue oder weniger bekannte Flederthiere, besonder des Pariser Museums. *Monatsber. König. Preuss. Akad. Wiss. Berlin* 1870:391–406.

———. 1870. Eine Monographiscen Übersicht der Chiropterengattungen *Nycteris* und *Atalapha* vor. *Monatsber. König. Preuss. Akad. Wiss. Berlin* 1871:900–14.

———. 1872. Über den *Vespertilio calcaratus* Prinz zu Wied und eine neue Gattung der Flederthiere, *Tylonycteris. Monatsber. König. Preuss. Akad. Wiss. Berlin* 1873:699–706.

———. 1874. Über eine neue Art von Flederthieren, *Promops bonariensis* und über *Lophuromys*, eine Nagergattung von Westafrika. *Monatsber. König. Preuss. Akad. Wiss. Berlin* 1875:232–34, 1 pl.

———. 1876. Über die mit *Histiotus velatus* verwandten Flederthiere aus Chile. *Monatsber. König. Preuss. Akad. Wiss. Berlin* 1876:785–92, 1 pl.

———. 1877. Über eine neue Gattung von Flederthieren, *Amorphochilus*, aus Peru und über eine neue *Crocidura* aus Liberia. *Monatsber. König. Preuss. Akad. Wiss. Berlin* 1878:184–88, 1 pl.

———. 1880. Über neue Flederthiere (*Vesperus, Vampyrops*). *Monatsber. König. Preuss. Akad. Wiss. Berlin* 1881:257–59.

———. 1882. Über Sphaeronycteris toxophyllum, eine neue Gattung und Art der frugivoren blattnasigen Flederthiere, aus dem tropischen Amerika. *Sitzunsber. König. Preuss. Akad. Wiss. Berlin* 1882:987–90, 1 pl.

Peterson, B. V., and L. A. Lacey. 1985. A new species of *Hershkovitzia* (Diptera: Nycteribiidae) from Brazil, with a key to the described species of the genus. *Proc. Ent. Soc. Washington* 87:578–82.

Peterson, B. V., and T. C. Maa. 1970. One new and one previously unrecorded species of *Basilia* (Diptera: Nycteribiidae) from Uruguay. *Can. Entomol.* 102:1480–87.

Peterson, N. E., and R. H. Pine. 1982. Chave para identificação de mamíferos da região amazônica brasileira com exceção dos quirópteros e primatas. *Acta Amaz.* 12:465–82.

Peterson, R. L. 1965a. A review of the flat–headed bats of the family Molossidae from South America and Africa. *Life Sci. Contrib., Roy. Ontario Mus.* 64:1–32.

———. 1965b. A review of the bats of the genus Ametrida, family Phyllostomidae. *Life Sci. Contrib., Roy. Ontario Mus.* 65:1–13.

———. 1968. A new bat of the genus Vampyressa from Guyana, South America, with a brief systematic review of the genus. *Life Sci. Contrib., Roy. Ontario Mus.* 73:1–17.

———. 1972. A second specimen of *Vampyressa brocki* (Stenoderminae: Phyllostomatidae) from Guyana, South America, with further notes on the systematic affinities of the genus. *Can. J. Zool.* 50:457–69.

Peterson, R. L., and P. Kirmse. 1969. Notes on *Vampyrum spectrum*, the false vampire bat, in Panama. *Can. J. Zool.* 47:140–42.

Peterson, R. L., and J. R. Tamsitt. 1968. A new species of bat of the genus *Sturnira* (family Phyllostomatidae) from northwestern South America. *Life Sci. Occas. Papers, Roy. Ontario Mus.* 12:1–8.

Petit, S. 1997. The diet and reproductive schedules of *Leptonycteris curasoae curasoae* and *Glossophaga longirostris elongata* (Chiroptera: Glossophaginae) on Curaçao. *Biotropica* 29:214–23.

Petit, S., A. Rojer, and L. Pors. 2006. Surveying bats for conservation: The status of cave-dwelling bats on Curaçao from 1993 to 2003. *Animal Conserv.* 9:207–17.

Petter, F. 1968. Une sarigue nouvelle du nord–est du Bresil, *Marmosa karimii* sp. nov. (Marsupiaux, Didelphides). *Mammalia* 32:313–16, pl. 22.

Philippi, F. 1893. Un nuevo marsupial chileno. *Anal. Univ. Chile* 86:31–34, 1 pl.

———. 1894. Ein neues Beutelthier Chile's. *Arch. Naturgesch.* 60(1):33–35, pl. 4, Figs. 2, 2a.

Philippi, R. A. 1866. Ueber ein paar neue Chilenische Säugthiere. *Arch. Naturgesch.* 32(1):113–117.

———. 1870. Ueber ein neues Faulthier. *Arch. Naturgesch.* 1(1):263–67, 1 fig.

———. 1894. Beschreibung einer dritten Beutelmaus aus

Chile. *Arch. Naturgesch.* 60(1):36; pl. 4, Figs. 1, 1a.

Philippi, R. A., and L. Landbeck. 1861. Neue Wirbelthiere von Chile. *Arch. Naturgesch.* 27(1):289–301.

Phillips, C. J. 1971. The dentition of glossophagine bats: Development, morphological characteristics, variation, pathology, and evolution. *Misc. Publ. Mus. Nat. Hist., Univ. Kansas* 54:1–138.

Phillips, C. J., T. Nagato, and B. Tandler. 1987. Comparative ultrastructure and evolutionary patterns of acinar secretory glands in Neotropical bats. In *Studies in Neotropical mammalogy, essays in honor of Philip Hershkovitz,* ed. B. D. Patterson and R. M. Timm, 213–30. *Fieldiana Zool.,* 39:frontispiece, viii + 1–506.

Piccinini, R. S. 1971. Estudo sistemático e bionômico dos quirópteros (Chiroptera) do estado do Ceará: I-Quirópteros coletados na área onde está localizada a Faculdade de Veterinária do Ceará. *Rev. Med. Vet.* 7:39–52.

———. 1974. Lista provisória dos quirópteros da coleção do Museu Paraense Emílio Goeldi (*Chiroptera*). *Bol. Mus. Paraense Emílio Goeldi,* zool. 77:1–32.

Pinder, L. 1993. Body measurements, karyotype, and birth frequencies of Maned Sloth (*Bradypus torquatus*). *Mammalia* 57:43–48.

Pine, R. H. 1972a. A new subgenus and species of murine opossum (genus *Marmosa*) from Peru. *J. Mammal.* 53:279–82.

———. 1972b. *The bats of the genus* Carollia. Texas A&M Univ., Technical Monograph, 8:1–125.

———. 1973a. Anatomical and nomenclatural notes on opossums. *Proc. Biol. Soc. Washington* 86:391–402.

———. 1973b. Mammals (exclusive of bats) of Belém, Pará, Brazil. *Acta Amaz.* 3:47–79.

———. 1975. A new species of *Monodelphis* (Mammalia: Marsupialia: Didelphidae) from Bolivia. *Mammalia* 39:320–22.

———. 1976. *Monodelphis umbristriata* (A. de Miranda-Ribeiro) is a distinct species of opossum. *J. Mammal.* 57:785–87.

———. 1977. *Monodelphis iheringi* (Thomas) is a recognizable species of Brazilian opossum (Mammalia: Marsupialia: Didelphidae). *Mammalia* 41:235–37.

———. 1980. Taxonomic notes on "*Monodelphis dimidiata itatiayae* (Miranda-Ribeiro)," *Monodelphis domestica* (Wagner) and *Monodelphis maraxina* Thomas (Mammalia: Marsupialia: Didelphidae). *Mammalia* 43:495–99. [Dated 1979; number 4 marked "Achevé d'imprimer le 15 janvier 1980."]

———. 1981. Review of the mouse opossums *Marmosa parvidens* Tate and *Marmosa invicta* Goldman (Mammalia: Marsupialia: Didelphidae) with description of a new species. *Mammalia* 45:55–70.

———. 1993a. A new species of *Thyroptera* Spix (Mammalia: Chiroptera: Thyropteridae) from the Amazon Basin of northeastern Perú. *Mammalia* 57:213–25.

———. 1993b. Review of: K. H. Redford and J. F. Eisenberg. 1992. Mammals of the Neotropics, Volume 2. *J. Mammal.* 74:1079–83.

Pine, R. H., and J. P. Abravaya. 1978. Notes on the Brazilian opossum *Monodelphis scalops* (Thomas) (Mammalia: Didelphidae). *Mammalia* 42:379–82.

Pine, R. H., and J. E. Anderson. 1980. Notes on stomach contents in *Trachops cirrhosus* (Chiroptera: Phyllostomatidae). *Mammalia* 43:568–70. [Dated 1979; number 4 marked "Achevé d'imprimer le 15 janvier 1980."]

Pine, R. H., and C. O. Handley, Jr. 1984. A review of the Amazonian short-tailed opossum *Monodelphis emiliae* (Thomas). *Mammalia* 48:239–45.

Pine, R. H., and A. Ruschi. 1978. Concerning certain bats described and recorded from Espírito Santo, Brazil. *An. Inst. Biol., Univ. Natl. Autón. México*, sér. zool., 47:183–96. [Dated 1976; published in 1978.]

Pine, R. H., I. R. Bishop, and R. L. Jackson. 1970. Preliminary list of mammals of the Xavantina/Cachimbo Expedition (central Brazil). *Trans. Roy. Soc. Trop. Med. Hyg.* 64:668–70.

Pine, R. H., P. L. Dalby, and J. O. Matson. 1985. Ecology, postnatal development, morphometrics, and taxonomic status of the short-tailed opossum, *Monodelphis dimidiata*, an apparently semelparous annual marsupial. *Ann. Carnegie Mus.* 54:195–231.

Pine, R. H., S. D. Miller, and M. L. Schamberger. 1979. Contributions to the mammalogy of Chile. *Mammalia* 43:339–76.

Pine, R. H., N. E. Pine, and S. D. Bruner. 1981. Mammalia. In: *Aquatic biota of tropical South America. Part 2: Anarthropoda*, ed. S. H. Hurlbert, G. Rodríguez, and N. D. dos Santos, 267–98. San Diego: San Diego State University Press, xi + 298 pp.

Pine, R. H., J. E. Rice, J. E. Bucher, D. H. Tank, Jr., and A. M. Greenhall. 1985. Labile pigments and flourescent pelage in didelphid marsupials. *Mammalia* 49:249–56.

Pine, R. H., R. K. LaVal, D. C. Carter, and W. Y. Mok. 1996. Notes on the graybeard bat, *Micronycteris daviesi* (Hill) (Mammalia: Chiroptera: Phyllostomidae), with the first records from Ecuador and Brazil. In *Contributions in Mammalogy: A Memorial Volume Honoring Dr. J. Knox Jones, Jr.*, ed. H. H. Genoways and R. J. Baker, 183–90. Lubbock: Museum of Texas Tech University, il + 315 pp.

Pinho, A.P. de, E. Cupolillo, R. H. Mangia, O. Fernandes, and A. M. Jansen. 2000. *Trypanosoma cruzi* in the sylvatic environment: distinct transmission cycles involving two sympatric marsupials. *Trans. Roy. Soc. Trop. Med. Hyg.* 94:1–6.

Pira, A. 1904. Über Fledermäuse von São Paulo. *Zool. Anz.* 28:12–19, 1905. [Published 23 August 1904.]

Pirlot, P. 1965a. Deux formes nouvelles de chiroptères des genres Eumops et Leptonycteris. *Le Natur. Can.* 112:5–7.

———. 1965b. Chiroptères de l'est du Venezuela. *Mammalia* 29:375–89.

———. 1967. Nouvelle recolte dechiroptères dans l'ouest du Venezuela. *Mammalia* 31:260–74.

———. 1968. Chiroptères du Perou, specialement de haute-Amazonie. *Mammalia* 32:86–96.

———. 1972. Chiroptères de moyenne Amazonie. *Mammalia* 36:71–85.

Pirlot, P., and J. R. León. 1965. Chiroptères de l'est du Venezuela. I. Région de Cumaná et ile de Margarita. *Mammalia* 29:367–74.

Pittier, H., and H. H. Tate. 1932. Sobre fauna Venezolana. *Bol. Soc. Venez. Cien. Nat.* 1:249–78.

Plumpton, D. L., and J. K. Jones, Jr. 1992. Rhynchonycteris naso. *Mammal. Species* 413:1–5.

Poche, F. 1908. Über die Anatomie und die systematische Stellung von Bradypus torquatus (Ill.). *Zool. Anz.* 33:567–80.

———. 1912. Über den Inhalt und die Erscheinungszeiten der einzelnen Teile, Hefte etc. und die verschiedenen Ausgaben des Schreber'schen Säugetierwerkes (1774–1855). *Arch. Naturgesch.* 77(suppl. 4):124–83.

Pocock, R. I. 1924. The external characters of the South American edentates. *Proc. Zool. Soc. London* 1924:983–1031.

Podtiaguin, B. 1944. Contribuciones al conocimiento de los murciélagos del Paraguay. *Rev. Soc. Cient. Paraguay* 6:25–62.

Poeppig, E. L. 1830. Doctor Pöppig's naturhistorische Berichte aus Chile. *Floriep's Not. Geb. Natur- und Heilk.*, no. 586:, columns 215–18.

———. 1835. *Reise in Chile, Peru, und auf dem Amazonenströme während de Jahre 1827–1832*. Leipzig: F. Fleischer, 1:xii + 466 pp.

Pohle, H. 1927. Über die von Prof. Bresslau in Brasilien gesammelten Säugetiere (ausser den Nagethieren). *Abhandl. Senckenberg. Naturf. Gesell., Frankfurt* 40:239–47.

Pol, A., M. R. Nogueira, and A. L. Peracchi. 2003. Primeiro registro da família Furipteridae (Mammalia, Chiroptera) para o estado do Rio de Janeiro, Brasil. *Rev. Bras. Zool.* 20:561–563.

Polaco, O. J. 1987. First record of *Noctilio albiventris* (Chiroptera: Noctilionidae) in Mexico. *Southwest. Nat.* 32:508–09.

Polanco-Ochoa, R. V. Jaimes, and W. Piragua. 2000. Los mamíferos del Parque Nacional Natural La Paya, Amazonia Colombiana. *Rev. Acad. Colomb. Cienc.* 23 (Supl. Esp.):671–82. [Dated December 1999; published February 2000.]

Pomel, A. 1848. Études sur les carnassiers insectivores (extrait). Seconde partie, Classification des insectivores. *Arch. Sci. Phys. Nat., Genève* 9:244–51.

Porini, G. 2001. Tatú carreta *(Priodontes maximus)* en Argentina. *Edentata* 4:9–14.

Porter, C. A., and R. J. Baker. 2004. Systematics of *Vampyressa* and related genera of phyllostomid bats as determined by cytochrome-*b* sequences. *J. Mammal.* 85:126–32.

Portfors, C. V., M. B. Fenton, L. M. de S. Aguiar, J. E. Baumgarten, M. J. Vonhof, S. Bouchard, D. M. de Faria, W. A. Pedro, N. I. L. Rautenbach, and M. Zortéa. 2000. Bats from Fazenda Intervales, southeastern Brazil—Species account and comparison between different sampling methods. *Rev. Brasil. Zool.* 17:533–38.

Power, D. M., and J. R. Tamsitt. 1973. Variation in *Phyllostomus discolor* (Chiroptera: Phyllostomatidae). *Can. J. Zool.* 51:461–68.

Price, R. D., and K. C. Emerson. 1986. New species of *Cummingsia* Ferris (Mallophaga: Trinenoponidae) from Peru and Venezuela. *Proc. Biol. Soc. Washington* 94:748–52.

Pridmore, P. A. 1994. Locomotion in *Dromiciops australis* (Marsupialia: Microbiotheriidae). *Austr. J. Zool.* 42:679–99.

Pritchard, P. C. H. 1994. Comment on gender and declension of generic names. *J. Mammal.* 75:549–50.

Proença, M. C. 1943. "*Myotis guaycurú*" n. sp., morcego proveniente de Salobra, estado de Mato Grosso (Microchiroptera, Vespertilionidae). *Rev. Bras. Biol.* 3:313–15.

Pumo, D. E., I. Kim, J. Remsen, C. J. Phillips, and H. H. Genoways. 1996. Molecular systematics of the fruit bat, *Artibeus jamaicensis*: Origin of an unusual island population. *J. Mammal.* 77:491–503.

Rafinesque, C. S. 1814. *Précis des découvertes et travaux somiologiques de Mr. C. S. Rafinesque–Schmaltz, entre 1800 et 1814 ou choix raisonné de ses principales découvertes en zoologie et en botanique, pour servir d'introduction à ses ouvrages futurs.* Palerme: Royale Typographie Militaire, aux dépen de l'Auteur, 56 pp.

———. 1815. *Analyse de la nature, ou tableau de l'univers et des corps organisés.* Palerme: l'Imprimerie de Jean Barravecchia, 224 pp.

———. 1818. Further discoveries in natural history, made during a journey through the western region of the United States. *Amer. Monthly Mag.* 3:445–47.

———. 1819. Prodrome de soixante-dix nouveaux genres d'animaux découverts dans l'intérieur des États-Unis d'Amérique durant l'année 1818. *J. Phys. Chim. d'Hist. Nat. Arts* 88:417–29.

———. 1820. *Annals of nature; or, Annual synopsis of new genera and species of animals, plants, &c. discovered in North America.* Lexington, Kentucky: T. Smith, 16 pp.

Rageot, R., and L. Albuja. 1994. Mamíferos de un sector de la Alta Amazonia Ecuatoriana: Mera, provincia de Pastaza. *Politecnica* 19:165–208.

Ramírez P., J. 1982. *Rhogeessa.* In *Mammal Species of the world*, ed. J. H. Honacki, K. E. Kinman, and J. W. Koeppl, 201–02. Lawrence, Kansas: Allen Press and The Association of Systematics Collections, ix+694 pp.

Ramoni-Perazzi, P., G. Bianchi, and M. Molina. 1994. Hallazgo de la comadreja colicorta *Monodelphis adusta* (Thomas, 1897) en la cuenca del Lago de Maracaibo, Venezuela. *Acta Cient. Venez.* 45:325–26.

Rand, D. M., and R. A. Paynter, Jr. 1981. *Ornithological gazetteer of Uruguay.* Cambridge: Harvard College Press, iv + 75 pp., 1 map.

Ranzani, C. 1820. *Elementi di zoologia.* Bologna: A. Nobili, 2:167–405, 1 unnumbered errata sheet, pls. 5–9.

Rapp, W. 1852. *Anatomische Untersuchungen über die Edentaten. Zweite verbesserte und vermehrte Auflage.* Tübingen: Ludwig Friedrich Fues, 4 (unnumbered) + 108 pp., 9 pls.

Rasweiler, J. J., IV. 1982. The contribution of observations on early pregnancy in the little sac-winged bat, *Peropteryx kappleri*, to an understanding of the evolution of reproductive mechanisms in monovular bats. *Biol. Repro.* 27:681–702.

Rau, J. R., and J. Yáñez. 1979. Nuevos registros de *Lasiurus borealis* en Magallanes. *Not. Mensual, Mus. Nac. Hist. Nat., Santiago* 23:12–14.

Ray, C. E., O. J. Linares, and G. S. Morgan. 1988. Paleontology. In *Natural history of vampire bats*, ed. A. M. Greenhall and U. Schmidt, 19–30. Boca Raton, Florida: CRC Press, frontispiece, 14 unnumbered + 246 pp.

Reddell, J. R. 1968. The hairy-legged vampire, *Diphylla ecaudata*, in Texas. *J. Mammal.* 49:769.

Redford, K. H. 1985a. Feeding and food preference in captive and wild giant anteaters (*Myrmecophaga tridactyla*). *J. Zool., London* 205:559–72.

———. 1985b. Food habits of armadillos (Xenarthra: Dasypodidae). In *Evolution and ecology of armadillos, sloths and vermilinguas*, ed. G. G. Montgomery, 429–38. Washington, DC: The Smithsonian Institution Press, 10 (unnumbered) + 451 pp.

———. 1986. Dietary specialization and variation in two mammalian myrmecopages (variation in mammalian myrmecophagy). *Rev. Chilena Hist. Nat.* 59:201–08.

Redford, K. H., and J. F. Eisenberg. 1992. *Mammals of the Neotropics. The Southern Cone.* Chicago: The University of Chicago Press, 2:x + 430 pp., 18 pls.

Redford, K. H., and G. A. B. da Fonseca. 1986. The role of gallery forests in the zoogeography of the cerrado's non-volant mammalian fauna. *Biotropica* 18:126–35.

Redford, K. H., and R. M. Wetzel. 1985. Euphractus sexcinctus. *Mammal. Species* 252:1–4.

Reed, J. T., and J. M. Brennan. 1975. The subfamily Leeuwenhoekinae in the Neotropics (Acarina: Trombiculidae). *Brigham Young Univ. Sci. Bull.*, biol. ser., 20(1, pt. 1):1–42.

Reeve, E. C. R. 1942. A statistical analysis of taxonomic differences within the genus *Tamandua* Gray (Xenarthra). *Proc. Zool. Soc. London* 111A:279–302.

Regidor, H. A., and M. Gorostiague. 1996. Reproduction in the white eared opossum (*Didelphis albiventris*) under temperate conditions in Argentina. *Stud. Neotrop. Fauna Envir.* 31:133–36.

Rehn, J. A. G. 1900. On the Linnaean genera Myrmecophaga and Didelphis. *Amer. Nat.* 34:575–78.

———. 1901a. Notes on Chiroptera. *Proc. Acad. Nat. Sci. Philadelphia* 52:755–59. [Dated 1900; published 9 February 1901.]

———. 1901b. A study of the genus Centurio. *Proc. Acad. Nat. Sci. Philadelphia* 53:295–302.

———. 1902a. A revision of the genus Mormoops. *Proc. Acad. Nat. Sci. Philadelphia* 54:160–72.

———. 1902b. Three new American bats. *Proc. Acad. Nat. Sci. Philadelphia* 54:638–41.

———. 1904. A revision of the mammalian genus Macrotus. *Proc. Acad. Nat. Sci. Philadelphia* 56:427–46.

———. 1914. Tadarida Rafinesque versus Nyctinomus Geoffroy. *Proc. Biol. Soc. Washington* 27:217–18.

Reichenbach, H. G. L. 1836. *Das Königlich sächsische naturhistorische Museum in Dresden:. Ein Leitfaden bie Beschauung der Schätze desselben.* Leipzig: das Universum der Natur zur Unterhaltung un Belehrung über Vor- und mit Welt, Lief. 5, viii + 64 pp.

Reid, F. A. 1997. *A field guide to the mammals of Central America & southeast Mexico.* New York: Oxford University Press, xiv + 4 unnumbered + 334 pp, 52 pls.

Reid, F. A., and C. A. Langtimm. 1993. Distributional and natural history notes for selected mammals from Costa Rica. *Southwest. Nat.* 38:299–302.

Reid, F. A., M. D. Engstrom, and B. K. Lim. 2000. Noteworthy records of bats from Ecuador. *Acta Chiropterol.* 2:37–51.

Reid, J. 1837. [untitled]. *Proc. Zool. Soc. London* 1837:4.

Reig, O. A. 1955. Noticia preliminar sobre la presencia de microbiotherinos vivientes en la fauna Sudamericana. *Invest. Zool. Chilenas* 2:121–30.

———. 1957. Sobre la posición sistematica de *Zygolestes paranensis* Amegh. y de *Zygolestes entrerrianus* Amegh., con una reconsideración sobre la edad y la correlación del Mesopotamiense. *Holmbergia* 5:209–26.

———. 1958. Notas para un actualización del conocimiento de la fauna de la formación Chapadmala. II. Amphibia, Reptilia, Aves, Mammalia (Marsupialia, Didelphidae, Borhyaenidae). *Acta Geol. Lilloana* 2:255–83.

———. 1959. El segundo ejemplar conocido de *Lestodelphys halli* (Thomas). *Neotropica* 5:57–58.

———. 1964. Roedores y marsupiales del Partido de General Pueyrredón y regiones adyacentes (provincia de Buenos Aires, Argentina). *Publ. Mus. Munic. Cien. Nat. Mar del Plata* 1:203–23.

———. 1981. Teoría del origen y desarollo de la fauna de mamíferos de America del Sur. *Monogr. Nat., Mus. Munic. Cien. Natur. "Lorenzo Scaglia"* 1:1–162.

Reig, O. A., and N. O. Bianchi. 1969. The occurrence of an intermediate didelphid karyotype in the short-tailed opossum (genus *Monodelphis*). *Experientia* 25:1210–11.

Reig, O. A., and I. Löbig. 1970. Estudios citogenéticos en marsupiales didélfidos de la fauna de Venezuela. *Acta Cient. Venez.* 21(Supl. 1):35.

Reig, O. A., and C. Sonnenschein. 1970. The chromosomes of *Marmosa fuscata* Thomas, from northern Venezuela (Marsupialia, Didelphidae). *Experientia* 26:199–201.

Reig, O. A., J. A. W. Kirsch, and L. G. Marshall. 1985. New conclusions on the relationships of the opossum-like marsupials with an annotated classification of the Didelphimorphia. *Ameghiniana* 21:335–43.

———. 1987. Systematic relationships of the living and Neocenozoic American opossum-like marsupials (Suborder Didelphimorphia) with comments on the classification of these and of the Cretaceous and Paleogene New World and European metatherians. In *Possums and opossums: Studies in evolution*, ed. M. Archer, 1–92. Sydney: Surrey Beatty and Sons Pty Limited and Royal Zool. Soc. New South Wales, 1:lxxii + 1–400, 4 pls.

Reig, O. A., A. L. Gardner, N. O. Bianchi, and J. L. Patton. 1977. The chromosomes of the Didelphidae (Marsupialia) and their evolutionary significance. *Biol. J. Linn. Soc. London* 9:191–216.

Reinhardt, J. 1851. "d. 12te Decbr." [summary of meeting on 12 December 1849]. *Videnskab. Meddel., Kjöbenhavn* 1:v.

———. 1872. [untitled; summary of meeting on 10 May 1872] *Videnskab. Meddel., Kjöbenhavn* 3(4):III.

Reis, S. F. dos. 1989. Biología reproductiva de *Artibeus lituratus* (Olfers, 1818) (Chiroptera: Phyllostomidae). *Rev. Brasil. Biol.* 49:369–72.

Reis, N. R. dos. 1984. Estrutura de comunidade de morcegos na região de Manaus, Amazonas. *Rev. Bras. Biol.* 44:247–54.

Reis, N. R. dos, and J.-L. Guillaumet. 1983. Les chauves-souris frugivores de la région de Manaus et leur rôle dans la dissémination des espèces végétales. *Rev. Ecol. (Terre Vie)* 38:147–69.

Reis, N. R. dos, and M. F. Muller. 1995. Bat diversity of forests and open areas in a subtropical region of south Brazil. *Ecol. Austral* 5:31–36.

Reis, N. R. dos, and A. L. Peracchi. 1987. Quirópteros da região de Manaus, Amazonas, Brasil (Mammalia, Chiroptera). *Bol. Mus. Paraense Emílio Goeldi, sér zool.* 3:161–82.

Reis, N. R. dos, and H. O. R. Schubart. 1979. Notas preliminares sobre os morcegos do Parque Nacional da Amazônia (Médio Tapajós). *Acta Amaz.* 9:507–15.

Reis, N. R. dos, A. L. Peracchi, and M. K. Onuki. 1993. Quiróperos de Londrina, Paraná, Brasil (Mammalia, Chiroptera). *Rev. Bras. Zool.* 10:371–81.

Reis, N. R. dos, A. L. Peracchi, and M. L. Sekiama. 1999. Morcegos da Fazenda Monte Alegre, Telêmaco Borba, Paraná (Mammalia, Chiroptera). *Rev. Bras. Zool.* 16:501–05.

Reis, N. R. dos, A. L. Peracchi, M. F. Muller, E. A. Bastos, and E. S. Soares. 1996. Quirópteros do Parque Estadual Morro do Diabo, São Paulo, Brasil (Mammalia, Chiroptera). *Rev. Bras. Biol.* 56:87–92.

Reis, N. R. dos, A. L. Peracchi, I. P. de Lima, M. L. Sekiama, and V. J. Rocha. 1998. Updated list of the Chiroptera of the city of Londrina, Paraná, Brasil. *Chiropt. Neotrop.* 4:96–98.

Reis, N. R. dos, A. L. Peracchi, M. L. Sekiama, and I. P. de Lima. 2000. Diversidade de morcegos (Chiroptera, Mammalia) em fragmentos florestais no estado do Paraná, Brasil. *Rev. Bras. Zool.* 17:697–704.

Rengger, J. R. 1830. *Naturgeschichte der Saeugethiere von Paraguay.* Basel: Schweghauserschen, xvi + 394 pp.

Renner, S. S. 1989. Floral biological observations on *Heliamphora tatei* (Sarraceniaceae) and other plants from Cerro de la Neblina in Venezuela. *Plant Syst. Evol.* 163:21–29.

Repenning, C. A. 1967. Subfamilies and genera of the Soricidae. *Geol. Surv. Prof. Paper* 565:iv + 74 pp.

Rezsutek, M., and G. N. Cameron. 1993. Mormoops megalophylla. *Mammal. Species* 448:1–5.

Rhoads, S. N. 1894. Description of a new armadillo, with remarks on the genus Muletia Gray. *Proc. Acad. Nat. Sci. Philadelphia* 1894:111–14.

Ribeiro, R. D., M. P. Barreto, C. A. de Camargo, and G. K. F. Takeda. 1985. Estudo comparativo entre a eficiência de hemoculturas e xenodiagnósticos seriados efetuados em gambás do género *Didelphis*, naturalmente infectados pelo *Trypanosoma cruzi. Rev. Bras. Biol.* 44:389–94. [Dated November 1984; published in 1985.]

Richardson, K. C., T. A. J. Bowden, and P. Myers. 1987. The cardiogastric gland and alimentary tract of caenolestid marsupials. *Acta Zool.* 68:65–70.

Ridgway, R. 1912. *Color standards and color nomenclature.* Washington, DC: Author, 43 pp.

Rincón R., D. A. 2001. Quirópteros subfosiles presentes en los depósitos de guano de la Cueva de Los Murciélagos, Isla de Toas, Estado Zulia, Venezuela. *Anartia* 13:1–13.

Ringuelet, A. B. de. 1953. Revisión de los didelfidos fosiles Argentina. *Rev. Mus. La Plata*, 3:265–308.

Riskin, D. K., and B. B. Fenton. 2002. Sticking ability in Spix's disk-winged bat, *Thyroptera tricolor* (Microchiroptera: Thyropteridae). *Can. J. Zool.* 79:2261–67. [Dated 2001; published 30 January 2002.]

Rivas A., B. 2000. Notas sobre los mamíferos de la planicie Amacuro (estado Delta Amacuro). *Mem. Soc. Cien. Nat. La Salle* 58:43–59. [Dated 1998; published in July 2000.]

Rivas-Pava, P., P. Sánchez-Palomino, and A. Cadena. 1996. Estructura trófica de la comunidad de quirópteros en bosques de galería de la Serranía de la Macarena (Meta–Colombia). In *Contributions in Mammalogy: A Memorial Volume Honoring Dr. J. Knox Jones, Jr.*, ed. H. H. Genoways and R. J. Baker, 237–48. Lubbock: Museum of Texas Tech University, il + 315 pp.

Rivillas, C., F. Caro, H. Carvajal, and I Vélez. 2004. Algunos tremátodos digéneos (Rhopaliasidae, Opistorchiidae) de *Phillander* [sic] *opossum* (Marsupialia: Mammalia) de la costa Pacífica colombiana, incuyendo *Rhopalias caucensis* n. sp. *Rev. Acad. Colomb. Cienc.* 28:591–600.

Roberts, A. 1924. Some additions to the list of South African mammals. *Ann. Transvaal Mus.* 10:59–76.

———. 1926. Some new S. African mammals and some changes in nomenclature. *Ann. Transvaal Mus.* 11:245–63.

Roberts, M., L. Newman, and G. Peterson. 1982. The management and reproduction of the large hairy armadillo, *Chaetophractus villosus*, at the National Zoological Park, Washington, DC *Internat. Zoo Yearb.* 22:185–94.

Robinson, E. S., P. B. Samollow, and J. L. VandeBerg. 1993. Developmental profiles of X-chromosome replication patterns in opossums (*Monodelphis domestica* and *Didelphis virginiana*). *Abstr. 6th Int. Theriol. Congr., Sydney*, ed. M. L. Augee, 263. Sydney, Australia: University of New South Wales.

Robinson, F. 1998. The bats of the Ilha de Maracá. In *Maracá: The biodiversity and environment of an Amazonian rainforest*, ed. W. Milliken and J. A.

Ratter, 165–87. Chichester: John Wiley & Sons, xxii + 508 pp.

Robinson, W., and M. W. Lyon, Jr. 1901. An annotated list of mammals collected in the vicinity of La Guaira, Venezuela. *Proc. U.S. Natl. Mus.* 24:135–62.

Rode, P. 1937. Étude d'une collection de mammifères de l'Ecuador oriental (Mission Flornoy). *Bull. Mus. Natl. Hist. Nat., Paris*, ser. 2, 9:342–46.

———. 1941. Catalogue des types de mammifères du Musum National d'Histoire Naturelle. II. Ordre des chiroptères. *Bull. Mus. Natl. Hist. Nat., Paris*, ser. 2, 13:227–52.

Rodger, J. C. 1982. The testis and its excurrent ducts in American caenolestid and didelphid marsupials. *Amer. J. Anat.* 163:269–82.

Rodrigues, L. R. R., R. M. S. Barros, S. Marques-Aguiar, M. F. L. Assis, J. C. Pieczarka, and C. Y. Nagamachi. 2003. Comparative cytogenetics of two phyllostomids bats. A new hypothesis to the origin of the rearrangeed X chromosome from *Artibeus lituratus*. *Caryologia* 56:413–19.

Rodrigues, M. G. R. , A. Bredt, and W. Uieda. 1994. Arborizacão de Brasília, Distrito Federal, e possíveis fontes de alimento para morcegos fitófagos. *An. II Congr. Bras. Arborização Urbana* 2:311–26.

Rodríguez F., J. 1993. *Thyroptera discifera* (Chiroptera: Thyropteridae) en Costa Rica. *Rev. Biol. Trop.* 41:929.

Rodríguez-Herrera, B., and M. Tschapka. 2005. Tent use by *Vampyressa nymphaea* (Chiroptera: Phyllostomidae) in *Cecropia insignis* (Moraceae) in Costa Rica. *Acta Chiropterol.* 7:171–74.

Rodríguez-Mahecha, J. V., J. I. Hernández-Camacho, T. R. Defler, M. Alberico, R. B. Mast, R. A. Mittermeier, and A. Cadena. 1995. Mamíferos Colombianos: Sus nombres comunes e indígenas. *Conserv. Internat., Occas. Papers Conserv. Biol.*, 3:1–56.

Roguin, L. de. 1986. Les mammifères du Paraguay dans les collections du Muséum de Genève. *Rev. Suisse Zool.* 93:1009–22.

Roig, V. G. 1971. La presencia de estados de hibernación en *Marmosa elegans* (Marsupialia–Didelphidae). *Acta Zool. Lilloana* 28:5–12.

Romaña, C., and J. W. Abalos. 1950. Lista de los quirópteros de la colección del Instituto de Medicina Regional, y sus parasitos. *Anal. Inst. Med. Reg.* 3:111–17.

Rood, J. P. 1970. Notes on the behavior of the pygmy armadillo. *J. Mammal.* 51:179.

Rose, K. D., and R. J. Emry. 1993. Relationships of Xenarthra, Pholidota, and fossil "edentates": The morphological evidence. In *Mammal phylogeny. Placentals*, ed.

F. S. Szalay, M. J. Novacek, and M. C. McKenna, 81–102. New York: Springer Verlag New York, Inc., xii + 321 pp.

Rosenthal, M. A. 1957. Observations on the water opossum or yapok *Chironectes minimus* in captivity. *Internat. Zoo. Yearb.* 15:4–6.

Rouk, C. S., and D. C. Carter. 1972. A new species of *Vampyrops* (Chiroptera: Phyllostomatidae) from South America. *Occas. Papers Mus., Texas Tech Univ.*, no. 1:1–7.

Ruedas, L. A., and J. W. Bickham. 1992. Morphological differentiation between *Rhogeessa minutilla* and *R. tumida* (Mammalia: Chiroptera: Vespertilionidae). *Proc. Biol. Soc. Washington* 105:403–09.

Rui, A. M., and G. Graciolli. 2005. Moscas ectoparasitas (Diptera, Streblidae) de morcegos (Chiroptera, Phyllostomidae) no sul do Brasil: Associações hosedeiros-parasitos e taxas de infestação. *Rev. Brasil. Zool.* 22:438–45.

Rüppell, E. 1842a. Beschreibung mehrerer neuer Säugethiere, in der zoologischen Sammlung der Senckenbergischen naturforschenden Gesellschaft befindlich. *Mus. Senckenb.* 3:129–44, 2 pls. [Although volume 3 bears date of 1845, it was issued in three "hefts" between 1839 and 1845; heft 2, pp. 91–196 was issued in 1842.]

———. 1842b. Verzeichniss der in dem Museum der Senckenbergischen naturforschenden Gesellschaft aufgestellten Sammlungen. *Mus. Senckenb.* 3:145–96. [Although volume 3 bears date of 1845, it was issued in three "hefts" between 1839 and 1845; heft 2 pp. 91–196 was issued in 1842.]

Ruschi, A. 1951a. Morcegos do estado do Espírito Santo. Introdução e consideracões gerais. *Bol. Mus. Biol. Prof. Mello Leitão, Santa Teresa, E. E. Santo*, zool., no. 1:1–16.

———. 1951b. Morcegos do estado do Espírito Santo. Família *Desmodontidae*, chave analítica para os gêneros e espécies representadas no E. E. Santo. Descrição de *Desmodus rotundus rotundus* e algumas observações a seu respeito. *Bol. Mus. Biol. Prof. Mello Leitão, Santa Teresa, E. E. Santo*, zool., no. 2:1–7 + 3 unnumbered pp., 4 figs.

———. 1951c. Morcegos do estado do Espírito Santo. Descrição de *Diphylla ecaudata* Spix e algumas observações a seu respeito. *Bol. Mus. Biol. Prof. Mello Leitão, Santa Teresa, E. E. Santo*, zool., no.3:1–3 + 3 unnumbered pp., 3 figs.

———. 1951d. Morcegos do estado do Espírito Santo. Família *Vespertilionidae*, chave analítica para os gêneros e espécies representadas no E. Santo. Descrição de *Myotis nigricans nigricans* e *Myotis espiritosantensis* n. sp. e algumas observações a seu respeito. *Bol. Mus. Biol.*

Prof. Mello Leitão, Santa Teresa, E. E. Santo, zool., no. 4:1–11 + 4 unnumbered pp., 4 figs.

———. 1951e. Morcegos do estado do Espírito Santo. Família *Vespertilionidae*. Descrição das espécies: *Lasiurus borealis mexicanus* e *Dasypterus intermedius*, com algumas observações biológicas a respeito. *Bol. Mus. Biol. Prof. Mello Leitão, Santa Teresa, E. E. Santo*, zool., no. 5:1–3 + 3 unnumbered pp., 4 figs.

———. 1951f. Morcegos do estado do Espírito Santo. Família Molossidae. Chave analítica para os gêneros e espécies representadas no E. E. Santo. Descriação de *Molossus rufus rufus, Molossops planirostris espiritosantensis* n. sub sp, e *Tadarida espiritosantensis* n. sp. e dados biológicos a respeito. *Bol. Mus. Biol. Prof. Mello Leitão, Santa Teresa, E. E. Santo*, zool., no. 6:1–7.

———. 1951g. Morcegos do estado do Espírito Santo. Família Noctilionidae. Chave analítica para os gêneros representadas no Brasil, com a descriação da única espécie representada no Espírito Santo: *Noctilio leporinus leporinus* (Linnaeus). *Bol. Mus. Biol. Prof. Mello Leitão, Santa Teresa, E. E. Santo*, zool., no. 7:1–7.

———. 1951h. Morcegos do estado do Espírito Santo. Família Emballonuridae, chave analítica para os gêneros, espécies, e subspecies representados no estado do Espírito Santo. Descrição de *Peropteryx macrotis macrotis* e *Peropteryx kappleri*. *Bol. Mus. Biol. Prof. Mello Leitão, Santa Teresa, E. E. Santo*, zool., no. 8:1–13.

———. 1952. Morcegos do estado do Espírito Santo. IXa. Família Emballonuridae. Chave analítica para os gêneros espécies e subspecies representados no E. E. Santo. *Bol. Mus. Biol. Prof. Mello Leitão, Santa Teresa, E. E. Santo*, zool., no. 10, 19 pp.

———. 1953a. Dois casos de sanguivorismo de Desmodus rotundus rotundus (E. Geoffroy) e Diphylla ecaudata Spix, no homem, e outras observações sobre os quirópteros hematófagos e acidentalmente hematófagos. *Bol. Mus. Biol. Prof. Mello Leitão, Santa Teresa, E. E. Santo*, biol., no. 13:1–8.

———. 1953b. Morcegos do estado do Espírito Santo. XVII. Família Phyllostomidae. Descrição das espécies: *Lonchophylla mordax* e *Hemiderma perspicillatum*, com algumas abservações. *Bol. Mus. Biol. Prof. Mello Leitão, Santa Teresa, E. E. Santo*, zool., no. 19:1–7.

———. 1953c. Morcegos do estado do Espírito Santo. XVIII. Família Phyllostomidae. Descrição das espécies *Artibeus jamaicensis lituratus* e *Vampyrops lineatus*, com algumas abservaçõoes. *Bol. Mus. Biol. Prof. Mello Leitão, Santa Teresa, E. E. Santo*, zool., no. 20:1–11 + 5 unnumbered pages, 4 figs.

———. 1954a. Morcegos do estado do Espírito Santo. XX.

Chaves analíticas e artificiais para a determinação das famílias, gêneros, espécies e sub-espécies dos morcegos representados no estado do Espírito Santo, e a lista atualizada das mesmas. *Bol. Mus. Biol. Prof. Mello Leitão, Santa Teresa, E. E. Santo*, zool., no. 22A:1–17 + 4 unnumbered.

———. 1954b. Algumas espécies zoológicas e botanicas em vias de extinção no estado do E. E. Santo. *Bol. Mus. Biol. Prof. Mello Leitão, Santa Teresa, E. E. Santo*, prot. natureza, no. 16A:45.

———. 1965. Lista dos mamíferos do estado do Espírito Santo. *Bol. Mus. Biol. Prof. Mello Leitão, Santa Teresa, E. E. Santo*, zool., no. 24A:1–40.

———. 1970. Morcegos do estado do Espirito Santo. Chaves analíticas e artificiais para a determinação das famílias, géneros, espécies, e subspecies dos morcegos representados no E. E. Santo, com a descrição de uma nova espécie da família Natalidae Miller, 1899 da região do Rio Itaunas, em conceição de Barra e Rio Mucuré ao sul da Bahia: *Natalus espiritosantensis* n. sp.—2 pranchas em nanfein. *Bol. Mus. Biol. Prof. Mello Leitão, Santa Teresa, E. E. Santo*, zool., no. 43:1–11, 2 figs.

Ruschi, A., and A. G. Bauer. 1957. Classificação de quirópteros do Rio Grande do Sul. *Inst. Perq. Veterin. "Desiderio Finamor," Porto Alegre*, 2:38–41.

Russell, R. J. 1953. Description of a new armadillo (Dasypus novemcinctus) from Mexico with remarks on geographic variation of the species. *Proc. Biol. Soc. Washington* 66:21–26.

Saavedra, B. and J. A. Simonetti. 2001. New records of *Dromiciops gliroides* (Microbiotheria: Microbiotheriidae) and *Geoxus valdivianus* (Rodentia: Muridae) in central Chile: Their implications for biogeography and conservation. *Mammalia* 65:96–100.

Saban, R. 1958. Insectivora. In *Traité de Paléontologie*, ed. J. Piveteau, 822–901. Paris: Masson et Cie, Éditeurs, 6(2):1–962.

Saez, F. A., M. E. Drets, and N. Brum. 1964. The chromosomes of the mulita (*Dasypus hybridus*): A mammalian edentate of South America. In *Mammalian cytogenetics and related problems in radiobiology*, ed. C. Pavan, C. Chagas, O. Frota-Pessoa, and L. R. Caldas, 163–70. New York: Pergamon Press, xviii + 427 pp.

Sahley, C., and L. Baraybar. 1996. Natural history of the long-snouted bat, *Platalina genovensium* (Phyllostomatidae: Glossophaginae) in southwestern Peru. *Vida Silv. Neotrop.* 5:1–9.

Salazar[-]B[ravo]., J., K. H. Redford, and A. M. Stearman. El perezoso de dos dedos de Hoffmann (*Choloepus hoffmanni* Peters, 1859; Megalonychidae) en Bolivia. *Mus. Nac. Hist. Nat. (Bolivia) Comunic.* 9:18–21.

Salazar[-Bravo], J. A. [*sic*], M. L. Campbell, S. Anderson, S. L. Gardner, and J. L. Dunnum. 1994. New records of Bolivian mammals. *Mammalia* 58:125–30.

Salazar-Bravo, J., E. Yensen, T. Tarifa, and T. L. Yates. 2002. Distributional records of Bolivian mammals. *Mastozool. Neotrop.* 9:70–78.

Salazar-Bravo, J., T. Tarifa, L. F. Aguirre, E. Yensen, and T. L. Yates. 2003. Revised checklist of Bolivian mammals. *Occas. Papers Mus., Texas Tech Univ.*, no. 220:1–27.

Salvadori, T. 1888. Le date della pubblicazione della "Iconografia della fauna Italica" del Bonaparte ed indice delle specie illustrate in detta opera. *Boll. Mus. Zool. Anat. Comp.* 39:1–25.

Sampaio, E. M., E. K. V. Kalko, E. Bernard, B. Rodríguez-Herrera, and C. O. Handley, Jr. 2003. A biodiversity assessment of bats (Chiroptera) in a tropical lowland rainforest of central Amazonia, including methodological and conservation considerations. *Stud. Neotrop. Fauna Envir.* 38:17–31.

Sanborn, C. C. 1929a. The land mammals of Uruguay. *Field Mus. Nat. Hist.*, zool. ser., 17:147–65.

———. 1929b. Records of Dasypus kappleri and Dasypus pastasae. *J. Mammal.* 10:258.

———. 1930. Distribution and habits of the three-banded armadillo (Tolypeutes). *J. Mammal.* 11:61–68, pl. 4.

———. 1932a. The bats of the genus Eumops. *J. Mammal.* 13:347–57.

———. 1932b. Neotropical bats in the Carnegie Museum. *Ann. Carnegie Mus.* 21:171–83.

———. 1933. Bats of the genera Anoura and Lonchoglossa. *Field Mus. Nat. Hist.*, zool. ser., 20:23–28.

———. 1935. New mammals from Guatemala and Honduras. *Field Mus. Nat. Hist.*, zool. ser., 20:81–85.

———. 1936. Records and measurements of Neotropical bats. *Field Mus. Nat. Hist.*, zool. ser., 20:93–106.

———. 1937. American bats of the subfamily Emballonurinae. *Field Mus. Nat. Hist.*, zool. ser., 20:321–354.

———. 1938. Notes on Neotropical bats. *Occas. Papers Mus. Zool., Univ. Michigan*, no. 373:1–5.

———. 1941. Descriptions and records of Neotropical bats. *Field Mus. Nat. Hist.*, zool. ser., 27:371–87.

———. 1943. External characters of the bats of the subfamily Glossophaginae. *Field Mus. Nat. Hist.*, zool. ser., 24:271–77.

———. 1947. Catalogue of type specimens of mammals in Chicago Natural History Museum. *Fieldiana Zool.* 32:209–93.

———. 1949a. Bats of the genus Micronycteris and its subgenera. *Fieldiana Zool.* 31:215–33.

———. 1949b. Mammals from the Rio Ucayali, Peru. *J. Mammal.* 30:277–88.

———. 1951a. Two new mammals from southern Peru. *Fieldiana Zool.* 31:473–77.

———. 1951b. Mammals from Marcapata, southeastern Perú. *Publ. Mus. Hist. Nat. "Javier Prado,"* ser. A, zool., 6:1–26.

———. 1953. Mammals from the departments of Cuzco and Puno, Perú. *Publ. Mus. Hist. Nat. "Javier Prado,"* ser. A, zool. 12:1–8.

———. 1954. Bats from Chimantá-Tepuí, Venezuela with remarks on *Choeroniscus*. *Fieldiana Zool.* 34:289–93.

———. 1955. Remarks on the bats of the genus *Vampyrops*. *Fieldiana Zool.* 37:403–13.

Sanborn, C. C., and J. A. Crespo. 1957. El murciélago blanquizco (*Lasiurus cinereus*) y sus subespecies. *Bol. Mus. Argentino Cien. Nat. "Bernardino Rivadavia," e Inst. Nac. Investig. Cienc. Nat.*, no. 4:1–13.

Sánchez, F., and M. Alvear. 2003. Comentarios sobre el uso de hábitat, dieta y conocimiento popular de los mamíferos en un bosque andino de Caldas, Colombia. *Bol. Cient., Mus. Hist. Nat., Univ. Caldas* 7:121–44.

Sánchez, F., and A. Cadena. 2000. Migración de *Leptonycteris curasoae* (Chiroptera: Phyllostomidae) en las zonas áridas del norte de Colombia. *Rev. Acad. Colomb. Cienc.*, 23(Supl. Esp.):683–86. [Dated 1999; published in February 2000.]

Sánchez, F., P. Sánchez-Palomino, and A. Cadena. 2004. Inventario de mamíferos en un bosque de los Andes centrales de Colombia. *Caldasia* 26:291–309. .

Sánchez H., J., J. Ochoa G., and A. Ospino. 1992. First record of *Eumops maurus* (Chiroptera: Molossidae) for Venezuela. *Mammalia* 56:151–52.

Sánchez-Hernández, C., and C. B. Chávez T. 1984. Observaciones sobre la biología del murciélago de cápula *Diclidurus virgo* Thomas. *Actas II Reunión Iberoamer. Conserv. Zool. Vert.* 1:411–16.

Sánchez-Hernández, C., M. de L. Romero-Almarez, and M. Aguilar-Morales. 1990. Anatomia e histólogia de la cápsula del murciélago blanca *Diclidurus albus virgo* Thomas. *Southwest. Nat.* 35:241–44.

Sánchez-Hernández, C., M. de L. Romero-Almaraz, and J. Cuisin. 2002. *Sturnira mordax* (Chiroptera, Phyllostomidae), in Ecuador. *Mammalia* 66:439–41.

Sánchez-Palomino, P., P. Rivas-Pava, and Cadena. 1993. Composición, abundancia y riqueza de especies de la comunidad de murciélagos en bosques de galería en la Serranía de la Macarena (Meta-Colombia). *Caldasia* 17:301–12.

Sánchez-Palomino, P., M. del P. Rivas-Pava, and A. Cadena. 1996. Diversidad biologica de una comunidad de quirópteros y su relación con la estructura del habitat de bosque de galería, Serranía de la Macarena, Colombia. *Caldasia* 18:343–53.

Sanderson, I. T. 1949. A brief review of the mammals of Suriname (Dutch Guiana), based upon a collection made in 1938. *Proc. Zool. Soc. London* 119:755–89, 5 pls.

Santoa, I. B., G. A. B. da Fonseca, S. Rigueira, and R. B. Machado. 1994. The rediscovery of the Brazilian three banded armadillo and notes on its conservation status. *Edentata* 1:11–15.

Santori, R. T., D. A. de Moraes, and R. Cerqueira. 1996. Diet composition of *Metachirus nudicaudatus* E. Geoffroy, 1803 and *Didelphis aurita* Wied, 1826 (Marsupialia, Didelphoidea) *Mammalia* 59:511–15. [Dated 1995; number 4 marked "Achevé d'imprimer le 30 avril 1996."]

Santos, I. B., G. A. B. da Fonseca, S. E. Rigueira, and R. B. Machado. 1994. The rediscovery of the Brazilian three banded armadillo and notes on its conservation status. *Edentata* 1:11–15.

Santos, M., L. F. Aguirre, L. B. Vázquez, and J. Ortega. 2003. Phyllostomus hastatus. *Mammal. Species* 722:1–6.

Saralegui, A. M. 1996. *Eumops patagonicus*, Thomas, 1924, en el Uruguay (Mammalia: Chiroptera: Molossidae). *Com. Zool. Mus. Hist. Nat., Montevideo* 12:1–4.

Saunders, R. 1975. Venezuelan Macronyssidae (Acarina: Mesostigmata). *Brigham Young Univ. Sci. Bull.*, biol. ser., 20(2):75–90.

Saussure, H. de. 1860. Note sur quelques mammifères du Mexique. *Rev. Mag. Zool.* ser. 2, 12:1–11, 53–57, 97–110, 241–54, 281–93, 377–83, 425–31, 479–94, 4 pls. [Also printed as an independently-paginated, 82–page separate.]

———. 1861. Diagnosis Cheiropteræ Mexicanæ e familia *Vespertilionidarum*. *Rev. Mag. Zool., Paris*, ser. 2, 13:97.

Sauthier, D. E. U., A. M. Abba, L. G. Pagano, and U. F. J. Pardiñas. 2005. Ingreso de micromamíferos Brasílicos en la provincia de Buenos Aires, Argentina. *Mastozool. Neotrop.* 12:91–95.

Savi, P. 1825. Descrizione del *Dinops Cestoni* nuovo animale della famiglia dei Pipistrelli. *Nuovo Gior. Letter., Pisa* 10:229–36.

Sawada, I., and M. Harada. 1986. Bat cestodes from Bolivia, South America with descriptions of six new species. *Zool. Sci., Tokyo* 3:367–77.

Sazima, I. 1976. Observations on the feeding habits of phyllostomatid bats (*Carollia, Anoura*, and *Vampyrops*) in southeastern Brazil. *J. Mammal.* 57:381–82.

———. 1978. Vertebrates as food items of the woolly false vampire, *Chrotopterus auritus*. *J. Mammal.* 59:617–18.

———. 1980. Observations on the feeding habits of phyllostomatid bats (*Carollia, Anoura*, and *Vampyrops*) in southeastern Brazil. *J. Mammal.* 57:381–82.

Sazima, I., and V. A. Taddei. 1976. A second Brazilian record of the South American flat-headed bat, *Neoplatymops mattogrossensis*. *J. Mammal.* 57:757–58.

Sazima, I., and W. Uieda. 1977. O morcego *Promops nasutus* no sudeste brasileiro (Chiroptera, Molossidae). *Ciênc. Cult.* 29:312–14.

———. 1978. Présence d'*Artibeus concolor* dans le Nord-Est du Brésil (Chiroptères, Phyllostomidae). *Mammalia* 42:255–56.

———. 1980. Feeding behavior of the white-winged vampire bat, *Diaemus youngii* on poultry. *J. Mammal.* 61:102–04.

Sazima, I., L. D. Vizotto, and V. A. Taddei. 1978. Uma nova espécie de *Lonchophylla* da Serra do Cipé, Minas Gerais, Brasil (Mammalia, Chiroptera, Phyllostomidae). *Rev. Bras. Biol.* 38:81–89.

Sazima, I., S. Vogel, and M. Sazima. 1989. Bat pollination of *Encholirium glaziovii*, a terrestrial bromeliad. *Plant Syst. Evol.* 168:167–79.

Sazima, M., and I. Sazima. 1975. Quiropterofilia en *Lafoensia pacari* St. Hil. (Lythraceae), na Serra do Cipó, Minas Gerais. *Ciênc. Cult.* 27:405–16.

Sazima, M., and I. Sazima. 1987. Additional observations on *Passiflora mucronata*, the bat-pollinated passion flower. *Ciênc. Cult.* 39:310–12.

Sazima, M., S. Buzato, and I. Sazima. 1999. Bat-pollinated flower assemblages and bat visitors at two Atlantic forest sites in Brazil. *An. Botany* 83:705–12.

Schaldach, W. J., Jr. 1965. Notas breves sobre algunos mamíferos del sur de México. *An. Inst. Biol., Univ. Nac. Autón. México* 35:129–37.

———. 1966. New forms of mammals from southern Oaxaca, Mexico, with notes on some mammals of the coastal range. *Säugtierk. Mitt.* 14:286–97.

Schaller, G. B. 1983. Mammals and their biomass on a Brazilian ranch. *Arq. Zool., São Paulo* 31(2):1–36.

Schellenberg, J. R. 1798. *Helvetische Entomologie; oder Verzeichniss der schweizerischen insekten nach einer methode geordnet*. Zürich: Orell, Füssli und Compagnie, 1:1–149 + 4 unnumbered, 16 pls. [Translated by J. P. de Clairville, 1798. *Entomologie Helvetique ou catalogue des insects de la Suisse rangées d'apres une nouvelle methode*, with translation on facing pages.]

Schinz, H. R. 1821. *Das Thierreich eingetheilt nach dem Bau der Thiere als Grundlage ihrer Naturgeschichte und der vergleichenden Anatomie von dem Herrn Ritter von Cuvier*. Erster band. Säugethiere und Vögel. Stuttgart und Tübingen: J. G. Cotta'schen Buchhandlung, 1:xxxviii + 894 pp.

———. 1824. *Naturgeschichte und Abbildungen der Säugethiere*. Zürich: Brodtmann's Lithographischer Kunstanstalt, vi + 417 pp., 8 (unnumbered), 178 pls.

————. 1825. *Das Thierreich eingetheilt nach dem Bau der Thiere als Grundlage ihrer Naturgeschichte und der vergleichenden Anatomie von dem Herrn Ritter von Cuvier.* Vierter band. Zoophyten. Stuttgart und Tübingen: J. G. Cotta'schen Buchhandlung, 4:xix + 793 pp.

————. 1844. *Systematisches Verzeichniss aller bis jetzt bekannten Säugethiere oder Synopsis Mammalium nach dem Cuvier'schen System.* Solothurn: Jent und Gassmann, 1:xvi + 1–587.

————. 1845. *Systematisches Verzeichniss aller bis jetzt bekannten Säugethiere oder Synopsis Mammalium nach dem Cuvier'schen System.* Solothurn: Jent und Gassmann, 2:iv + 1–574.

Schmid, K. 1818. *Naturhistorische Beschreibung der Säugthiere.* München: Verlage der Litographischen Kunst-Anstalt, viii + 9–192, 130 pls.

Schmidt, C. 1988. Reproduction. In *Natural history of vampire bats*, ed. A. M. Greenhall and U. Schmidt, 99–110. Boca Raton, Florida: CRC Press, frontispiece, 14 unnumbered + 246 pp.

Schomburgk, R. H. 1840. Mr. Schomburgk's recent expedition in Guiana. *An. Nat. Hist.* 5:29–35.

Schreber, J. C. D. von. 1774a. *Die Säugthiere in Abbildungen nach der Natur mit Beschreibungen.* Erlangen: Wolfgang Walther, 1(1–9):1–190, pls. 1–62. [See Poche 1912 and Sherborn 1891 for date of publication.]

————. 1774b. *Die Säugthiere in Abbildungen nach der Natur mit Beschreibungen.* Erlangen: Wolfgang Walther, 2:191–280, pls. 63–93, 95–107. [Poche 1912 gave 1774 for pages 191–254 and associated plates, and 1775 for pages 255–80 and plates 101–07A, B; Sherborn 1891 gave 1775, but 1774 for plates 63–80, as dates of publication.]

————. 1777. *Die Säugthiere in Abbildungen nach der Natur mit Beschreibungen.* Erlangen: Wolfgang Walther, 3:377–440, pls. 104B, 107Aa, 109B, 110B, 115B, 125B, 127B, 136, 146A–65; 1776–78. [This part of Volume 3 was published in 1777; see Poche 1912 and Sherborn 1891 for dates of publication of parts; Volume 3 is dated 1778.]

————. 1778. *Die Säugthiere in Abbildungen nach der Natur mit Beschreibungen.* Erlangen: Wolfgang Walther, 3:441–590. [This part of Volume 3 was published in 1778; see Poche 1912 and Sherborn 1891 for dates of publication of parts; Volume 3 is dated 1778.]

————. 1781. *Die Säugthiere in Abbildungen nach der Natur mit Beschreibungen.* Erlangen: Wolfgang Walther, 4:683–90, pls 58B, 62B, 159B, C, D, 212–22. [This part of Volume 4 was published in 1781; see Sherborn 1891 for dates of publication of parts.]

Schulze, F. E., W. Kükenthal, and K. Heider. 1926. *Nomenclator animalium generum et subgenerum.*

Berlin: Verlage der Preussischen Akademie der Wissenschaften, 1:1–476.

Schutt, W. A., Jr., and J. S. Altenbach. 1997. A sixth digit in *Diphylla ecaudata*, the hairy legged vampire bat (Chiroptera, Phyllostomidae). *Mammalia* 61:280–85.

Schwartz, A. 1955. The status of the species of the *brasiliensis* group of the genus *Tadarida. J. Mammal.* 36:106–09.

Sclater, P. L. 1856. List of mammals and birds collected by Mr. Bridges in the vicinity of the town of David in the province of Chiriqui in the state of Panama. *Proc. Zool. Soc. London* 1856:138–43.

————. 1865. Report on a collection of animals from Madagascar, transmitted to the Society by Mr. J. Caldwell. *Proc. Zool. Soc. London* 1865:467–70, pl. 27.

————. 1871. [Report on additions to the Society's menagerie in May 1871]. *Proc. Zool. Soc. London* 1871:543–46, pl. 43.

————. 1873. [Report on the additions to the Society's menagerie in October and November 1872.] *Proc. Zool. Soc. London* 1872:860–62, pl. 72.

Seba, A. 1734. *Locupletissimi rerum naturalium thesauri accurata descriptio, et iconibus artificiosissimis expressio, per universam physices historiam. Opus, cui, in hoc rerum genere, nullum par exstitit. Ex toto terrarum orbe collegit, digessit, descripsit, et depingendum curavit.* Amstelaedami: Apud J. Wetstenium, & Gul. Smith, & Janssonio-Waesbergios, 1:1–178 + 38 (unnumbered), 111 pls.

Segall, W. 1969. The middle ear region of *Dromiciops. Acta Anat.* 72:489–501.

Seluja, G. A., M. V. Di Tomaso, N. Brum-Zorilla, and H. Cardoso. 1984. Low karyotypic variation in two didelphids (Marsupialia): Karyogram and chromosome banding analysis. *J. Mammal.* 65:702–07.

Shamel, H. H. 1927. A new bat from Colombia. *Proc. Biol. Soc. Washington* 40:129–30.

————. 1928. A new bat from Dominica. *Proc. Biol. Soc. Washington* 41:67–68.

————. 1930a. A new murine opossum from Argentina. *J. Washington Acad. Sci.* 20:83.

————. 1930b. A new name for Marmosa muscula Shamel. *J. Mammal.* 11:311.

————. 1931. Notes on the American bats of the genus Tadarida. *Proc. U.S. Natl. Mus.* 2862:1–27.

Shapley, R. L., D. E. Wilson, A. N. Warren, and A. A. Barnett. 2005. Bats of the Potaro Plateau region, western Guyana. *Mammalia* 69:375–94.

Shaw, G. 1800. *General zoology or systematic natural history.* London: G. Kearsley, 1:xv +1–552, 121 pls.

Shaw, J. H., T. S. Carter, and J. C. Machado-Neto. 1985. Ecology of the giant anteater *Myrmecophaga tridactyla* in Serra da Canastra, Minas Gerais, Brazil: A pilot study.

In *The evolution and ecology of armadillos, sloths, and vermilinguas*, ed. G. G. Montgomery, 379–84. Washington, DC: The Smithsonian Institution Press, 10 (unnumbered) + 451 pp.

Shaw, J. J. 1985. The hemoflagellates of sloths, vermilinguas (anteaters), and armadillos. In *The evolution and ecology of armadillos, sloths, and vermilinguas*, ed. G. G. Montgomery, 279–92. Washington, DC: The Smithsonian Institution Press, 10 (unnumbered) + 451 pp.

Sherborn, C. D. 1891. On the dates of the parts, plates, and text of Schreber's 'Säugthiere.' *Proc. Zool. Soc. London* 1891:587–92.

———. 1897a. On the dates of the natural history portion of Savigny's 'Description de l'Égypte.' *Proc. Zool. Soc. London* 1897:285–88.

———. 1897b. Note on the dates of "The Zoology of the 'Beagle.'" *Ann. Mag. Nat. Hist.*, ser. 6, 20:483.

———. 1898. Dates of Blainville's 'Ostéographie.' *Ann. Mag. Nat. Hist.*, ser. 7, 2:76.

———. 1902. *Index animalium sive index nominum quae ab A.D. MDCCLVIII generibus et speciebus animalium imposita sunt.* London: J. and C. F. Clay, lx + 1195 pp.

———. 1915. [Untitled note]. In *Catalogue of the books, manuscripts, maps and drawings in the British Museum (Natural History)*, p. 2082. London: British Museum (Natural History), 5: 4 pp. (unnumbered) + 1957–2403.

———. 1922. *Index animalium sive index nominum quae ab A.D. MDCCLVIII generibus et speciebus animalium imposita sunt.* Part I. Introduction, bibliography and index A-Aff.. London: British Museum (Natural History), cxxxii + 128 pp.

———. 1925. *Index animalium sive index nominum quae ab A.D. MDCCLVIII generibus et speciebus animalium imposita sunt.* Part VI. Index *Ceyl.*–Concolor. London: British Museum (Natural History), pp. 1197–452.

———. 1932. *Index animalium sive index nominum quae ab A.D. MDCCLVIII generibus et speciebus animalium imposita sunt.* Epilogue, additions to bibliography, additions and corrections, and index to trivialia. London: British Museum (Natural History), i–vii + cxxxiii–cxlviii, 1–208 pp.

Sherborn, C. D., and F. J. Griffin. 1934. On the dates of publication of the natural history portions of Alcide d'Orbigny's 'Voyage Amérique méridionale.' *Ann. Mag. Nat. Hist.*, ser. 10, 13:130–34.

Sherborn, C. D., and B. B. Woodward. 1893. On the dates of the 'Encyclopédie Méthodique' (Zoology). *Proc. Zool. Soc. London* 1893:582–84.

———. 1901. Dates of publication of the zoological and botanical portions of some French voyages.—*Part II.* Ferret and Galinier's 'Voyage en Abyssinie'; Lefebvre's 'Voyage en Abyssinie'; 'Exploration scientifique de l'Algérie'; Castelnau's 'Amérique du Sud'; Dumont d'Urville's 'Voyage de l'Astrolabe'; Laplace's 'Voyage sur la Favorite'; Jacquemont's 'Voyage dans l'Inde'; Tréhouart's 'Commission scientifique d'Islande'; Cailliaud, 'Voyage à Méroé'; 'Expédition scientifique de Morée'; Fabre, 'Commission scientifique du Nord'; Du Petit-Thouars, 'Voyage de la Vénus'; and on the dates of the 'Faune Française.' *Ann. Mag. Nat. Hist.*, ser. 7, 8:161–64, 333–36, and 491–94.

Sherman, H. B. 1955. A record of *Lasiurus* and of *Vampyrops* from Paraguay. *J. Mammal.* 36:130.

Shufeldt, R. W. 1926a. Observacões sobre certos peixes e mammíferos do Brasil e mais particularmente sobre su osteologia. *Rev. Mus. Paulista* 14:502–61, 23 pls.

———. 1926b. Observations upon certain fishes and mammals of Brazil, more particularly their osteology. *Rev. Mus. Paulista* 14:563–614, 23 pls.

Shump, K. A., Jr., and A. U. Shump. 1982a. Lasiurus borealis. *Mammal. Species*, 183:1–6.

———. 1982b. Lasiurus cinereus. *Mammal. Species*, 185:1–5.

Sigé, B. 1972. La faunule du mamifères du Cretace superieur de Laguna Umayo (Andes peruviennes). *Bull. Mus. Natl. Hist. Nat., Paris*, sci. terre, 19:375–405.

Silva, F. 1975. Três novas ocorrências de quirópteros para o Rio Grande do Sul, Brasil (*Mammalia, Chiroptera*). *Iheringia*, sér. zool., 46:51–53.

Silva, F., and M. F. B. Souza. 1980. *Tadarida laticaudata* Geoffroy, 1805, nova ocorrência para o estado de Rio Grande do Sul, Brasil (Chiroptera, Mammalia). *Iheringia*, sér. zool., 56:3–5.

Silva, J. M. C. da, and D. C. Oren. 1993. Observations on the habitat and distribution of the Brazilian three-banded armadillo *Tolypeutes tricinctus*, a threatened caatinga endemic. *Mammalia* 57:149–52.

Silva, L. A. M. da, and D. de Q. Guerra. 2000. Bats from a remnant of Atlantic forest in Northeast Brazil. *Chiropt. Neotrop.* 6:125–26.

Silva, M. M. S., N. M. S. Harmani, and E. F. B. Gonçalves. 1996. Bats from the metropolitan region of São Paulo, southeastern Brazil. *Chiropt. Neotrop.* 2:39–41.

Silva, M. N. F. da, and A. Langguth. 1989. A new record of *Glironia venusta* from the lower Amazon, Brazil. *J. Mammal.* 70:873–75.

Silva, S. S. P. da, P. G. Guedes, and A. L. Peracchi. 2001. Levantamento preliminar dos morcegos do Parque Nacional de Ubajara (Mammalia, Chiroptera), Ceará, Brasil. *Rev. Bras. Zool.* 18:139–44.

Silva, S. S. P. da, P. G. Guedes, A. R. Camardella, and A. L. Peracchi. 2004. Survey of bats (Mammalia, Chiroptera), with comments on reproduction status, in Serra Das Almas Private Heritage Reserve, in the state of Ceará,

Northwestern [*sic*] of Brazil. *Chirop. Neotrop.* 10):191–95.

Silva-Júnior, J. de S., and A. P. Nunes. 2001. The disjunct geographical distribution of the yellow armadillo, *Euphractus sexcinctus* (Xenarthra, Dasypodidae). *Edentata* 4:16–18.

Silva-Júnior, J. de S., M. E. B. Fernandes, and R. Cerqueira. 2001. New records of the yellow armadillo *(Euphractus sexcinctus)* in the state of Maranhão, Brazil (Xenarthra, Dasypodidae). *Edentata* 18–23.

Silva-Taboada, G. 1974. Fossil Chiroptera from cave deposits in central Cuba, with description of two new species (genera *Pteronotus* and *Mormoops*) and the first West Indian record of *Mormoops megalophylla*. *Acta Zool. Cracov.* 19:33–73.

———. 1976a. La localidad tipo de algunos murciélagos cubanos descritos in el siglo XIX. *Misc. Zool.* 5:2–3.

———. 1976b. Historia y actualización taxonómica de algunas especies antillanas de murciélagos de los géneros *Pteronotus*, *Brachyphylla*, *Lasiurus*, y *Antrozous* (Mammalia: Chiroptera). *Poeyana* 153:1–24.

———. 1979. *Los murciélagos de Cuba*. La Habana, Cuba: Editorial Academia, La Habana, xiv + 425 pp., 15 pls.

Silva-Taboada, G., and K. F. Koopman. 1964. Notes on the occurrence and ecology of *Tadarida laticaudata yucatanica* in Eastern Cuba. *Amer. Mus. Novit.*, no. 2174:1–6.

Silveira, E. K. P. da. 1968. Notas sôbre a história natural do tamanduá-mirim *(Tamandua tetradactyla chiriquensis* J. A. Allen, 1904, Myrmecophagidae), com referências a fauna do istmo de Panamá. *Vellozia* 6:9–31.

Simmons, N. B. 1996. A new species of *Micronycteris* (Chiroptera: Phyllostomidae) from northeastern Brazil, with comments on phylogenetic relationships. *Amer. Mus. Novit.*, no. 3158:1–34.

Simmons, N. B. 2005. Order Chiroptera. In *Mammal species of the world*, 3rd ed., ed. D. E. Wilson and D. M. Reeder, 312–529. Baltimore: The Johns Hopkins Press, 1:xxxviii + 743 pp.

Simmons, N. B., and T. M. Conway. 2001. Phylogentic relationships of mormoopid bats (Chiroptera: Mormoopidae) based on morphological data. *Bull. Amer. Mus. Nat. Hist.* 258:1–97.

Simmons, N. B., and R. S. Voss. 1998. The mammals of Paracou, French Guiana: A Neotropical lowland rainforest fauna, Part 1. Bats. *Bull. Amer. Mus. Nat. Hist.* 237:1–219.

Simmons, N. B., R. S. Voss, and D. W. Fleck. 2002. A new Amazonian species of *Micronycteris* (Chiroptera: Phyllostomidae) with notes on the roosting behavior of sympatric congeners. *Amer. Mus. Novit.*, no. 3358:1–14.

Simmons, N. B., R. S. Voss, and H. C. Peckham. 2000. The bat fauna of the Saül region, French Guiana. *Acta Chiropterol.* 2:23–36.

Simonetta, A. M. 1979. First record of *Caluromysiops* from Colombia. *Mammalia* 43:247–248.

Simpson, G. G. 1945. The principles of classification and a classification of mammals. *Bull. Amer. Mus. Nat. Hist.* 85:xvi + 350 pp.

———. 1970. The Argyrolagidae, extinct South American marsupials. *Bull. Mus. Comp. Zool.* 139:1–86.

———. 1980. *Splendid isolation: The curious history of South American mammals*. New Haven: Yale Univ. Press, x + 266 pp.

Sipinski, E. A. B., and N. R. do Reis. 1995. Dados ecológicos dos quirópteros da Reserva Volta Velha, Itapoá, Santa Catarina, Brasil. *Rev. Bras. Zool.* 12:519–28.

Sites, J. W., Jr., J. W. Bickham, and M. W. Haiduk. 1981. Conservative chromosomal change in the bat family Mormoopidae. *Can. J. Genet. Cytol.* 23:459–67.

Slaughter, B. H. 1970. Evolutionary trends of chiropteran dentitions. In *About bats*, ed. B. H. Slaughter and D. W. Walton, 51–83. Dallas, Texas: Southern Methodist Univ. Press, vii + 339 pp.

Smit, F. G. A. M. 1953. Descriptions of new and little-known Siphonaptera. *British Mus. (Nat. Hist.), Entomol. Bull.* 3:187–219.

Smith, J. D. 1972. Systematics of the chiropteran family Mormoopidae. *Misc. Publ. Mus. Nat. Hist., Univ. Kansas* 56:1–132.

———. 1976. Chiropteran evolution. In *Biology of bats of the New World family Phyllostomatidae*. Part I, ed. R. J. Baker, J. K. Jones, Jr., and D. C. Carter, 49–69. Special Publications of the Museum 10. Lubbock: Texas Tech University Press, 218 pp.

———. 1977. On the nomenclatorial status of *Chilonycteris gymnonotus* Natterer, 1843. *J. Mammal.* 58:245–46.

Smith, J. D., and H. H. Genoways. 1974. Bats of Margarita Island, Venezuela, with zoogeographic comments. *Bull. S. California Acad. Sci.* 73:64–79.

Smith, J. D., and C. Hood. 1984. Genealogy of the New World nectar-feeding bats reexamined: A reply to Griffiths. *Syst. Zool.* 33:435–60.

Smith, P. G., and S. M. Kerry. 1996. The Iwokrama rain forest programme for sustainable development: How much of Guyana's bat (Chiroptera) diversity does it encompass? *Biodiv. Conserv.* 5:921–42.

Snow, J. L., J. K. Jones, Jr., and W. D. Webster. 1980. Centurio senex. *Mammal. Species* 138:1–3.

Soares, C. A., and R. S. Carneiro. 2002. Social behavior between mothers x young of sloths *Bradypus variegatus*

Schinz, 1825 (Xenarthra: Bradypodidae). *Brazil. J. Biol.* 62:249–52.

Sodré, M. M., and W. Uieda. 2006. First record of the ghost bat *Diclidurus scutatus* Peters (Mammalia, Chiroptera, Emballonuridae) in São Paulo city, Brazil. *Rev Brasil. Zool.* 23:897–98.

Solari, A. J., and N. O. Bianchi. 1975. The synaptic behavior of the X and Y chromosomes in the marsupial *Monodelphis dimidiata*. *Chromosoma* 52:11–25.

Solari, S. 2003. Diversity and distribution of *Thylamys* (Didelphidae) in South America, with emphasis on species from the western side of the Andes. In *Predators with pouches: The biology of carnivorous marsupials*, ed. M. E. Jones, C. R. Dickman, and M. Archer, 82–101. Collingwood, Australia: CSIRO Publishing, xix + 486 pp.

———. 2004. A new species of *Monodelphis* (Didelphimorphia: Didelphidae) from southeastern Peru. *Mamm. Biol. (Z. Säugetierk.)* 69:145–52.

Solari, S., and R. J. Baker. 2006. Mitochondrial DNA sequence, karyotypic, and morphological variation in the *Carollia castanea* species complex (Chiroptera: Phyllostomidae) with description of a new species. *Occas. Papers Mus., Texas Univ.*, no. 254:1 +1–16.

Solari, S., V. Pacheco, and E. Vivar. 1999. New distribution records of Peruvian bats. *Rev. Peru. Biol.* 6:152–59.

Solari, S., E. Vivar, J. J. Rodrígues, and J. L. Mena. 1998. Small mammals: Biodiversity assessment in the Lower Urubamba Region. In *Biodiversity assessment and monitoring of the Lower Urubamba Region, Peru. Cashiriari-3 Well Site and the Camisea and Urubamba Rivers*, ed. A. Alonso and F. Dalmeier, 209–18. Washington, DC: SI/MAB Series #2. Smithsonian Institution, xlii + 298 pp.

Solari, S., E. Vivar, J. J. Rodríguez, P. Velazco, and E. Montesinos. 1999. Small mammals: Biodiversity assessment at the Pagoreni well site. In *Biodiversity assessment and long-term monitoring, Lower Urubamba Region, Perú*, ed. A. Alonso and F. Dalmeier, 137–50. Washington, DC: SI/MAB Series #3. Smithsonian Institution, xxxiv + 334 pp.

Solari, S., E. Vivar, P. Velazco, and J. J. Rodríguez. 2001a. Small mammals of the southern Vilcabamba Region, Peru. In *Biological and social assessments of the Cordillera de Vilcabamba, Peru*, ed. L. A. Alonso, A. Alonso, T. S. Schulenberg, and F. Dallmeier, 110–16. RAP Working Papers 12 and SI/MAB Series #6. Washington, DC: Conservation International, 296 pp.

———. 2001b. Appendix 19. Small mammal diversity from several montane forest localities (1300–2800 m) on the eastern slope of the Peruvian Andes. In *Biological and social assessments of the Cordillera de Vilcabamba,*

Peru, ed. L. A. Alonso, A. Alonso, T. S. Schulenberg, and F. Dallmeier, 262–64. RAP Working Papers 12 and SI/MAB Series #6. Washington, DC: Conservation International, 296 pp.

Solari, S., E. Vivar, P. M. Velazco, J. J. Rodríguez, D. E. Wilson, R. J. Baker, and J. L. Mena. 2001c. The small mammal community of the Lower Urubamba Region, Peru. In *Urubamba: The biodiversity of a Peruvian rainforest*, ed. A. Alonso, F. Dallmeier, and P. Campbell, 171–181. Washington, DC: SI/MAB Series #7. The Smithsonian Institution, x + 204 pp.

Solari, S., J. J. Rodriguez, E. Vivar, and P. M. Velazco. 2002. A framework for assessment and monitoring of small mammals in a lowland tropical forest. *Envir. Monitor. Assessm.* 76:89–104.

Solari, S., R. A. Van Den Bussche, S. R. Hoofer, and B. D. Patterson. 2004. Geographic distribution, ecology, and phylogenetic affinities of *Thyroptera lavali* Pine 1993. *Acta Chiropterol.* 6:293–302.

Solmsen, E.-H. 1985. *Lonchorhina aurita* Tomes, 1863 (Phyllostominae, Phyllostomidae, Chiroptera) in westlichen Ecuador. *Z. Säugetierk.* 50:329–37.

Soriano, P. J., 1987. On the presence of the short-tailed opossum *Monodelphis adusta* (Thomas) in Venezuela. *Mammalia* 51:321–24.

Soriano, P. J., and J. Molinari. 1984. Hallazgo de *Sturnira aratathomasi* (Mammalia: Chiroptera) en Venezuela y descripción de su cariotipo. *Acta Cient. Venez.* 35:310–11.

———. 1987. Sturnira aratathomasi. *Mammal. Species* 284:1–4.

Soriano, P. J., M. R. Fariñas, and M. E. Naranjo. 2000. A new subspecies of Miller's long-tongued bat (*Glossophaga longirostris*) from a semiarid enclave of the Venezuelan Andes. *Z. Säugetierk.* 65:369–74.

Soriano, P. J., M. E. Naranjo, and M. R. Fariñas. 2004. A new subspecies of the little desert bat (*Rhogeessa minutilla*) from a Venezuelan semiarid enclave. *Mamm. Biol. (Z. Säugetierk.)* 69:439–43.

Soriano, P. J., A. Ruiz, and Z. Zambrano. 2005. New noteworthy records of bats for the Andean region of Venezuela and Colombia. *Mammalia* 69:251–55.

Soriano, P. J., M. Sosa, and O. Rossell. 1991. Hábitos alimentarios de *Glossophaga longirostris* Miller (Chiroptera: Phyllostomidae) en una zona árida de los Andes venezolanos. *Rev. Biol. Trop.* 39:263–68.

Soriano, P. J., A. Utrera, and M. Sosa. 1990. Inventario preliminar de los mamíferos del Parque Nacional General Cruz Carrillo (Guaramacal), estado Trujillo, Venezuela. *Biollania* 7:83–99.

Sosa, M., A. de Ascenção, and P. J. Soriano. 1996. Dieta y patrón reproductivo de *Rhogeessa minutilla*

(Chiroptera: Vespertilionidae) en una zona árida de los Andes de Venezuela. *Rev. Biol. Trop.* 44:867–75.

Sousa, M.A. N. de, A. Langguth, and E. do A. Gimenez. 2004. Mamíferos dos brejos de altitude Paraíba e Pernambuco. In *Brejos de altitude em Pernambuco e Paraíba, historia natural, ecologia e conservação*, ed. K. C. Porto, J. J. P. Cabral, and M. Tabarelli, 229–54. Brasilia: Ministério do Meio Ambiente, Séria Biodiversidade 9:1–324.

Souza, M. J. de, and M. C. P. de Araújo. 1990. Conservative pattern of the G-bands and diversity of C-banding patterns and NORs in Stenodermatinae (Chiroptera-Phyllostomatidae). *Rev. Bras. Genet.* 13:255–68.

Souza-Lopes, M. J. de. 1978. Cariótipo de duas espécies de morcegos de Pernambuco (Chiroptera-Phyllostomatidae). *Rev. Nordest. Biol.* 1:113–17.

Speiser, P. 1900. Ueber die Strebliden, Fledermausparasiten aus der Gruppe der pupiparen Dipteren. *Arch. Naturgesch.* 66(1):31–70.

Spencer, S. G., P. C. Choucair, and B. R. Chapman. 1988. Northward expansion of the southern yellow bat, *Lasiurus ega*, in Texas. *Southwest. Nat.* 33:493.

Spillmann, F. 1927. Sobre dos nuevas especies de "Bradypus" de la region costeña de la Republica del Ecuador. *An. Univ. Central, Quito* 39:317–22, 1 pl.

Spix, J. B. von. 1823. *Simiarum et vespertilionum Brasiliensium species novae, ou, Histoire naturelle des espèces nouvelles de singes et de chauves-souris observées et recueilles pendant le voyage dans l'intérieur du Brésil exécuté par ordre de S. M. le Roi de Bavière dans les années 1817, 1818, 1819, 1820.* Monachii: Francisci Seraphici Hübschmanni, vii + 72 pp., 38 pls.

Springer, M. S., M. Westerman, and J. A. W. Kirsch. 1994. Relationships among orders and families of marsupials based on 12S Ribosomal DNA sequences and the timing of the marsupial radiation. *J. Mammal. Evol.*, 2:85–115.

Squarcia, S. M., and E. B. Casanave. 1999. Discriminación entre las subspecies de *Zaedyus pichiy* (Desmarest, 1804) (Mammalia, Dasypodidae), utilizando caracteres morfométricos craneanos. *Physis, Buenos Aires* 57:19–24.

Srbek-Araujo, A. C., and A. G. Chiarello. 2005. Is camera-trapping an efficient method for surveying mammals in Neotropical forests? A case study in south-eastern Brazil. *J. Trop. Ecol.* 21:121–25.

Stager, K. E. 1942. A new free-tailed bat from Texas. *Bull. S. California Acad. Sci.* 61:49–50.

Stallings, J. R. 1989. Small mammal inventories in an eastern Brazilian park. *Bull. Florida State Mus.*, biol. sci., 34:153–200. [Dated 1988; published 10 May 1989.]

Stanhope, M. J., V. G. Waddell, O. Madsen, W. de Jong, S. B. Hedges, G. C. Cleven, D. Kao, and M. S. Springer. 1998. Molecular evidence for multiple origins of Insectivora and for a new order of endemic African insectivore mammals. *Proc. U.S. Natl. Acad. Sci.* 95:9967–72.

Starrett, A. 1969. A new species of *Anoura* (Chiroptera: Phyllostomatidae) from Costa Rica. *Contr. Sci., Los Angeles Co. Mus.* 157:1–9

———. 1972. Cyttarops alecto. *Mammal. Species* 13:1–2.

———. 1976. Comments on bats newly recorded from Costa Rica. *Contr. Sci., Los Angeles Co. Mus.* 277:1–5.

Starrett, A., and R. S. Casebeer. 1968. Records of bats from Costa Rica. *Contr. Sci., Los Angeles Co. Mus.*148:1–21.

Stein, B. R. 1981. Comparative limb myology of two opossums, *Didelphis* and *Chironectes*. *J. Morph.* 169:113–40.

Stephens, L., and M. A. Traylor, Jr. 1983. *Ornithological gazetteer of Peru.* Cambridge: Harvard College Press, vi + 270 pp., 1 map.

Stiglich, G. 1918. *Diccionario geográfico peruano y almanaque de "La Cronica" para 1918.* Lima: Casa Editora, N. Moral, 38 unnumbered + 483 pp.

———. 1922. *Diccionario geográfico del Perú.* Lima: Imp. Torres Aguirre, vi + 1193 pp.

Stiles, C. W., and M. O. Nolan. 1931. Key catalogue of parasites reported for Chiroptera (bats) with their possible public health importance. *Bull. Natl. Inst. Health* 155:iii + 603–742.

Stock, A. D. 1975. Chromosome banding pattern homology and its phylogenetic implications in the bat genera *Carollia* and *Choeroniscus*. *Cytogenet. Cell Genet.* 14:34–41.

Stock, C. 1925. *Cenozoic gravigrade edentates of western North America, with special reference to the Pleistocene Megalonychinae and Mylodontidae of Rancho La Brea.* Washington, DC: Carnegie Inst. of Washington, Publ. 331, frontispiece, xiv + 206, 47 pls.

Stone, W. 1900. The date of publication of Brewster's American edition of the Edinburgh Encyclopaedia. *Science* 10:685–86.

———. 1914. On a collection of mammals from Ecuador. *Proc. Acad. Nat. Sci. Philadelphia* 1914:9–19.

Storch, G. 1981. *Eurotamandua joresi*, ein Myrmecophagide aus dem Eozän der "Grube Messel" bei Darmstadt (Mammalia, Xenarthra). *Senckenberg. Lethaea* 61:247–89.

Storr, G. C. C. 1780. *Prodromus methodi Mammalium.* Tübingen: Litteris Reissianis, 43 pp., 4 tables.

Straney, D. O. 1981. The stream of heredity: Genetics in the study of phylogeny. In *Mammalian population genetics*, ed. M. H. Smith, and J. Joule, 100–38. Athens: The University of Georgia Press, i–xi + 380 pp.

Straney, D. O., M. H. Smith, I. F. Greenbaum, and R. J. Baker. 1979. Biochemical genetics. In *Biology of bats of the New World family Phyllostomatidae.* Part III, ed. R. J. Baker, J. K. Jones, Jr., and D. C. Carter, 157–176. Special Publications of the Museum 16. Lubbock: Texas Tech University Press, 441 pp.

Streilein, K. E. 1982a. Behavior, ecology, and distribution of South American marsupials. In *Mammalian biology in South America*, ed. M. A. Mares and H. H. Genoways, 231–50. The Pymatuning Symposia in Ecology 6. Special Publications Series. Pittsburgh: Pymatuning Laboratory of Ecology, University of Pittsburgh, xii + 539 pp.

———. 1982b. Ecology of small mammals in the semiarid Brazilian caatinga. I. Climate and faunal composition. *Ann. Carnegie Mus.* 51:79–107.

———. 1982c. The ecology of small mammals in the semi-arid Brazilian caatinga. III. Reproductive biology and population ecology. *Ann. Carnegie Mus.* 51:251–69.

———. 1982d. The ecology of small mammals in the semi-arid Brazilian caatinga. IV. Habitat selection. *Ann. Carnegie Mus.* 51:331–43.

———. 1982e. The ecology of small mammals in the semi-arid Brazilian caatinga. V. Agonistic behaviour and overview. *Ann. Carnegie Mus.* 51:345–69.

Stutz, W. H., M. C. de Albuquerque, W. Uieda, E. M. de Macedo, and C. B. Fraça. 2004. Updated list of bats from Uberlandia, state of Minas Gerais, southeastern Brazil. *Chiropt. Neotrop.*10:188–90.

Sudman, P. D., L. J. Barkley, and M. S. Hafner. 1994. Familial affinity of *Tomopeas ravus* (Chiroptera) based on protein electrophoretic and cytochrome *b* sequence data. *J. Mammal.* 75:365–77.

Sunquist, M. E., and G. G. Montgomery. 1973a. Activity pattern of a translocated silky anteater (*Cyclopes didactylus*). *J. Mammal.* 54:782

———. 1973b. Activity patterns and rates of movement of two-toed and three-toed sloths (*Choloepus hoffmanni* and *Bradypus infuscatus*). *J. Mammal.* 54:946–54.

Swainson, W. 1835. *On the natural history and classification of quadrupeds.* In *The cabinet cyclopaedia. Natural history*, ed. D. Lardner. London: Longman, Rees, Orme, Brown, Green, and Longman; and John Taylor, 121: viii + 397 pp. ["A new edition" of volume 121 with the same pagination was published in 1845.]

Swanepoel, P., and H. H. Genoways. 1979. Morphometrics. In *Biology of bats of the New World family Phyllostomatidae.* Part III, ed. R. J. Baker, J. K. Jones, Jr., and D. C. Carter, 13–106. Special Publications of the Museum 16. Lubbock: Texas Tech University Press, 441 pp.

Szalay, F. 1982. A new appraisal of marsupial phylogeny and classification. In *Carnivorous marsupials*, ed. M. Archer, 621–40. Mosman: Royal Zoological Society of New South Wales, 2:iv + 397–804.

Szeplaki O., E., J. Ochoa G., and J. Clavijo A. 1988. Stomach contents of the greater long-nosed armadillo (*Dasypus kappleri*) in Venezuela. *Mammalia* 52:422–25.

Szyszlo, V. de. 1955. *La naturaleza en la América ecuatorial. Descripción de la naturaleza de la región Amazónica del Perú, Brasil, Bolivia, Ecuador, Colombia, Venezuela y de la Guayana. Observaciones hechas durante doce viajes en los años 1904 a 1953.* Lima: Sanmarti y Cia., vii + 528 pp., 16 pls.

Taddei, V. A. 1969. Aspectos da biologia de *Artibeus lituratus lituratus* (Lichtenstein, 1823) (Chiroptera, Phyllostomidae). *Ciênc. Cult.* 21:451–52.

———. 1975. Phyllostomidae (Chiroptera) do norte-ocidental do estado de São Paulo. I–Phyllostominae. *Ciênc. Cult.* 27:621–32.

———. 1976. The reproduction of some Phyllostomidae (Chiroptera) from the northwestern region of the state of São Paulo. *Bol. Zool., Univ. São Paulo* 1:313–30.

———. 1979. Phyllostomidae (Chiroptera) do norte-ocidental do estado de São Paulo. III–Stenodermatinae. *Ciênc. Cult.* 31:900–14.

———. 1980. Aspectos da biologia de *Chiroderma doriae*, Thomas, 1891 (Chiroptera, Phyllostomidae). *An. Acad. Brasil. Ciênc.* 52:643–44.

Taddei, V. A., and V. Garutti. 1981. The southernmost record of the free-tailed bat, *Tadarida aurispinosa.* *J. Mammal.* 62:851–52.

Taddei, V. A., and W. A. Pedro. 1993. A record of *Lichonycteris* (Chiroptera: Phyllostomidae) from Northeast Brazil. *Mammalia* 57:454–56.

———. 1996. *Micronycteris brachyotis* (Chiroptera, Phyllostomidae) from the state of São Paulo, Brazil. *Rev. Bras. Biol.* 56:217–22.

Taddei, V. A., and N. R. dos Reis. 1980. Notas sobre alguns morcegos da Ilha de Maracá, Território Federal de Roraima (Mammalia, Chiroptera). *Acta Amaz.* 10:363–68.

———. 1983. Uma nova espécie de *Lonchophylla* do Brasil e chave para identificação das especies do género (Chiroptera, Phyllostomidae). *Ciênc. Cult.* 35:625–29.

Taddei, V. A., and W. Uieda. 2001. Distribution and morphometrics of *Natalus stramineus* from South America (Chiroptera, Natalidae). *Iheringia*, sér. zool., 91:123–32.

Taddei, V. A., and E. C. Vicente-Tranjan. 1998. Biological and distributional notes on *Platyrrhinus helleri* (Chiroptera: Phyllostomidae) in Brazil. *Mammalia* 62:112–17.

Taddei, V. A., I. M. de Rezende, and D. Camora. 1990. Notas sobre uma coleção de morcegos de Cruzeiro do Sul, Rio Juruá, estado do Acre (Mammalia, Chiroptera). *Bol. Mus. Paraense Emilio Goeldi*, zool., 6:75–88.

Taddei, V. A., R. B. de Seixas, and A. L. Dias. 1986. Noctilionidae (Mammalia, Chiroptera) do sudeste Brasileiro. *Ciênc. Cult.* 38:904–16.

Taddei, V. A., S. A. de Souza, and J. L. Manuzzi. 1988. Notas sôbre uma coleção de *Lonchophylla bokermanni* de Ilha Grande, sudeste do Brasil (Mammalia, Chiroptera). *Rev. Bras. Biol.* 48:851–55.

Taddei, V. A., L. D. Vizotto, and S. M. Martins. 1976. Notas taxonomicas e biologicas sobre *Molossops brachymeles cerastes* (Thomas, 1901) (Chiroptera—Molossidae). *Naturalia* 2:61–69.

Taddei, V. A., L. D. Vizotto, and I. Sazima. 1978. Notas sobre *Lionycteris* e *Lonchophylla* nas coleções do Museu Paraense Emílio Goeldi (Mammalia, Chiroptera, Phyllostomidae). *Bol. Mus. Paraense Emílio Goeldi*, zool., 92:1–14.

———. 1983. Uma nova espécie de *Lonchophylla* do Brasil e chave para identificação das especies do género (Chiroptera, Phyllostomidae). *Ciênc. Cult.* 35:625–29.

Taggart, D. A., G. A. Shimmin, C. R. Dickman, and W. G. Breed. 2003. Reproductive biology of carnivorous marsupials: Clues to the likelihood of sperm competition. In *Predators with pouches:The biology of carnivorous marsupials*, ed. M. E. Jones, C. R. Dickman, and M. Archer, 358–75. Collingwood, Australia: CSIRO Publishing, xix–486 pp.

Tálice, R. V., S. Laffite de Mosera, and T. Machado. 1960. Observaciones sobre *Monodelphis dimidiata*. *Actas y Trabajos del Primer Congreso Sudamericano de Zoología*, La Plata, 4:149–56.

Talmage, R. V., and G. D. Buchanan. 1954. The armadillo (Dasypus novemcinctus). A review of its natural history, ecology, anatomy and reproductive physiology. *Rice Inst. Pamphlet, Monogr. Biol.* 41(2):1–135.

Tamayo H., M. 1968. Los armadillos descritos por J. I. Molina. *Notic. Mensual, Mus. Nac. Hist. Nat., Santiago* 12:3–11.

———. 1973. Los armadillos en Chile. Situación de *Euphractus sexcinctus* (Linnaeus, 1758) (Mammalia, Edentata, Dasypodidae. *Notic. Mensual, Mus. Nac. Hist. Nat., Santiago* 17:3–6.

———. 2000. Orden Xenarthra. In *Mamíferos de Chile*, ed. A. Muñoz-Pedreros and J. Yañez-Valenzuela, 67–72. Valdivia, Chile: CEA Ediciones, viii + 463 pp.

Tamayo H., M., and D. Frassinetti C. 1980. Catálogo de los mamíferos fosiles y vivientes de Chile. *Bol. Mus. Nac. Hist. Nat. Chile* 37:323–99.

Tamayo H., M., and V. Pérez d'A. 1979. Hallazgo del murciélago colorado, *Lasiurus borealis varius* (Poeppig, 1835) en Magallanes y consideraciones acerca de la distribución de los *Lasiurus* (Chiroptera, Vespertilionidae). *Notic. Mensual, Mus. Nac. Hist. Nat., Santiago* 23:3–10.

Tamsitt, J. R. 1966. Altitudinal distribution, ecology, and general life history of bats in the Andes of Colombia. *Amer. Phil. Soc. Yearb.*1966:372–73.

Tamsitt, J. R., and C. Häuser. 1985. Sturnira magna. *Mammal. Species* 240:1–4.

Tamsitt, J. R., and D. Nagorsen. 1982. Anoura cultrata. *Mammal. Species* 179:1–5.

Tamsitt, J. R., and D. Valdivieso. 1962. *Desmodus rotundus rotundus* from a high altitude in southern Colombia. *J. Mammal.* 43:106–07.

———. 1963a. Records and observations on Colombian bats. *J. Mammal.* 44:168–80.

———. 1963b. Notes on bats from Leticia, Amazonas, Colombia. *J. Mammal.* 44:263.

———. 1963c. Reproductive cycle of the big fruit-eating bat, *Artibeus lituratus* Olfers. *Nature* 198:104.

———. 1965. Reproduction of the female big fruit-eating bat, *Artibeus lituratus palmarum*, in Colombia. *Carib. J. Sci.* 5:157–66.

———. 1966a. Bats from Colombia in the Swedish Museum of Natural History, Stockholm. *Mammalia* 30:97–104.

———. 1966b. Taxonomic comments on *Anoura caudifer*, *Artibeus lituratus*, and *Molossus molossus*. *J. Mammal.* 47:230–38.

———. 1986. Variación morfométrica en el murciélago *Sturnira magna* (Chiroptera: Phyllostomidae). *Caldasia* 15:743–60.

Tamsitt, J. R., A. Cadena, and E. Villarraga. 1986. Records of bats (*Sturnira magna* and *Sturnira aratathomasi*) from Colombia. *J. Mammal.* 67:754–57.

Tamsitt, J. R., D. Valdivieso, and J. Hernández Camacho. 1964. Bats of the Bogota savanna, Colombia, with notes on altitudinal distribution of Neotropical bats. *Rev. Biol. Trop.*, 12:107–15.

Tandler, B., T. Nagato, and C. J. Phillips. 1986. Systematic implications of comparative ultrastructure of secretory acini in the submandibular salivary gland in *Artibeus* (Chiroptera: Phyllostomidae). *J. Mammal.* 67:81–90.

Tarifa, T., and S. Anderson. 1997. Two additional records of *Glironia venusta* Thomas, 1912 (Marsupialia, Didelphidae) for Bolivia. *Mammalia* 61:111–13.

Tate, G. H. H. 1931. Brief diagnoses of twenty-six apparently new forms of *Marmosa* (Marsupialia) from South America. *Amer. Mus. Novit.*, no. 493:1–14.

———. 1932. Distribution of the South American shrews. *J. Mammal.* 13:223–28.

———. 1933. A systematic revision of the marsupial genus *Marmosa*. *Bull. Amer. Mus. Nat. Hist.* 66:1–250, 26 pls., 1 table (9 sections, pocketed).

————. 1934. New generic names for two South American marsupials. *J. Mammal.* 15:154.

————. 1939. The mammals of the Guiana region. *Bull. Amer. Mus. Nat. Hist.* 76:151–229.

————. 1941. Results of the Archbold Expeditions. No. 39. Review of *Myotis* of Eurasia. *Bull. Amer. Mus. Nat. Hist.* 78:537–65.

Tavares, V. C., and V. A. Taddei. 2003. Range extension of *Micronycteris schmidtorum* Sanborn, 1935 (Chiroptera: Phyllostomidae) to the Brazilian Atlantic forest, with comments on taxonomy. *Mammalia* 67:463–67.

Tejedor, A. 2003. First record of *Saccopteryx canescens* (Chiroptera: Emballonuridae) for southeastern Peru. *Chirop. Neotrop.* 9:162–64.

————. 2005. A new species of funnel-ear bat (Natalidae: *Natalus*) from Mexico. *J. Mammal.* 86:1109–20.

Telford, S. R., Jr., R. J. Tonn, J. J. Gonzalez, and P. Betancourt. 1981. Dinámica de las infecciones tripanosómicas entre la comunidad de los bosques tropicales secos de los llanos altos de Venezuela. *Bol. Direcc. Malariol. San. Amb.* 21:196–209.

Temminck, C. J. 1824a. Tableau méthodique de mammifères, répartis en ordres, genres et sections, avec une énumération approximative des espèces comprises dans les groupes, suivant le relevé le plus récent dans cette classe du régne animal. In *Monographies de mammalogie ou description de quelques genres de mammifères dont les espèces ont été observées dans les différens musées de l'Europe*, xiii–xxxii. Paris: G. Dufour et E. d'Ocagne, 1:xxxii + 268 pp., 25 pls., 1824–27.

————. 1824b. Deuxième monographie sur le genre sarigue.—*Didelphis* (Linn.). In *Monographies de mammalogie ou description de quelques genres de mammifères dont les espèces ont été observées dans les différens musées de l'Europe*, 21–54, pls. 5 and 6. Paris: G. Dufour et E. d'Ocagne, 1:xxxii + 268 pp., 25 pls., 1824–27.

————. 1826. Sixième monographie. Sur le genre molosse.—*Dysopes* (Illig.). In *Monographies de mammalogie ou description de quelques genres de mammifères dont les espèces ont été observées dans les différens musées de l'Europe*, 205–44, pls. 17–24. Paris: G. Dufour et E. d'Ocogne, 1:xxxii + 268 pp., 25 pls., 1824–27. [See Sherborn 1915 for dates of publication.]

————. 1838. Over de geslachten Taphozous, Emballonura, Urocryptus en Diclidurus. *Tijdschr. Nat. Gesch. Physiol.* 5:1–34, 1 pl.

————. 1840. Treizième monographies sur les cheiroptères vespertilionides formant les genres Nyctice, Vespertilion et Furie. In *Monographies de mammalogie ou description de quelques genres de mammifères dont les espèces ont été observées dans les différens musées de l'Europe*, 141–272. Leiden: C. C. van der Hock; Paris: E. d'Ocagne et A. Bertrand, 2:1–392 pp., pls. 28–70, 1835–41. [See Sherborn 1915 for date of publication.]

————. 1841. Quatorzième monographie. Sur les genres taphien-queue-en-fourreau-queue-cache-et queue-bivalve. In *Monographies de mammalogie ou description de quelques genres de mammifères dont les espèces ont été observées dans les différens musées de l'Europe*, 273–304, pls. 60–61. Leiden: C. C. van der Hock; Paris: E. d'Ocagne et A. Bertrand, 2:1–392, pls 28–70, 1835–41. [See Sherborn 1915 for date of publication.]

Temple-Smith, P. 1987. Sperm structure and marsupial phylogeny. In *Possums and Opossums: Studies in evolution*, ed. M. Archer, 171–93. Sydney: Surrey Beatty and Sons Pty Limited and Royal Zool. Soc. New South Wales, 1:lxxii + 1–400, 4 pls.

Terborgh, J. W., J. W. Fitzpatrick, and L. Emmons. 1984. Annotated checklist of bird and mammal species of Cocha Cashu Biological Station, Manu National Park, Peru. *Fieldiana Zool.*, 21:1–29.

Tesh, R. B., A. A. Arata, and J. D. Schneidau, Jr. 1968. Histoplasmosis in Colombian bats with a consideration of some of the factors influencing the prevalence of natural infection in Chiroptera. *Amer. J. Trop. Med. Hyg.* 17:102–06.

Teta, P., U. F. J. Pardiñas, and G. D'Elía. 2005. 80 años después, redescubrimiento de *Chacodelphys formosa* (Marsupialia, Didelphidae) en el Chaco Húmedo de Argentina. XX Jornada de Mastozoología, Buenos Aires, Argentina [1 page Abstract].

————. 2006. Rediscovery of *Chacodelphys*: A South American marsupial genus previously known from a single specimen. *Mammal. Biol.* 71:309–14.

Texera, W. A. 1973. *Zaedyus pichiy* (Edentata, Dasypodidae) nueva especie en la provincia de Magallanes, Chile. *An. Inst. Patagonia, Punta Arenas (Chile)* 4:335–37.

Thenius, E. 1969. Stammesgeschichte der Säugetiere (einschliesslich der Hominiden). *Handb. Zoologie* 8:1–368.

————. 1989. Zähne und Gebiss de Säugetiere. *Handb. Zoologie* 8:xi + 1–513.

Theodor, O. 1967. An illustrated catalogue of the Rothschild collection of Nycteribiidae (Diptera) in the British Museum (Natural History) with keys and short descriptions for the identification of subfamilies, genera, species and subspecies. *Brit. Mus. (Nat. Hist.) Publ.* 655:1–506, 5 pls.

Thomas, M. E. 1972. Preliminary study of the annual breeding patterns and population fluctuations of bats in three ecologically distinct habitats in southwestern Colombia. Ph.D. thesis, Tulane University, New Orleans, Louisiana, 161 pp.

Thomas, M. E., and D. N. McMurray. 1974. Observations on *Sturnira aratathomasi* from Colombia. *J. Mammal.* 55:834–36.

Thomas, O. 1880. On mammals from Ecuador. *Proc. Zool. Soc. London* 1880:393–403, pl. 38.

———. 1882. On a collection of rodents from north Peru. *Proc. Zool. Soc. London* 1882:98–111, 1 pl.

———. 1887. On the small Mammalia collected in Demerara by Mr. W. L. Sclater. *Proc. Zool. Soc. London* 1887:150–53.

———. 1888a. Diagnoses of four new species of *Didelphys*. *Ann. Mag. Nat. Hist.*, ser. 6, 1:158–59.

———. 1888b. *Catalogue of the Marsupialia and Monotremata in the collection of the British Museum (Natural History)*. London: British Museum (Natural History), xiv + 401 pp., 38 pls.

———. 1889. Description of a new stenodermatous bat from Trinidad. *Ann. Mag. Nat. Hist.*, ser. 6, 4:167–70.

———. 1891. Note on *Chiroderma villosum*, Peters, with the description of a new species of the genus. *Ann. Mus. Civ. Stor. Nat. Genova*, ser. 2, 10:881–83.

———. 1892a. On the probable identity of certain specimens formerly in the Lidth de Jeude collection and now in the British Museum, with those figured by Albert Seba in his 'Thesaurus' of 1734. *Proc. Zool. Soc. London* 1892:309–18.

———. 1892b. Description of a new bat of the genus *Artibeus* from Trinidad. *Ann. Mag. Nat. Hist.*, ser. 6, 10:408–10.

———. 1892c. Note on Mexican examples of *Chilonycteris davyi*, Gray. *Ann. Mag. Nat. Hist.*, ser. 6, 10:410.

———. 1892d. Description of a new Mexican bat. *Ann. Mag. Nat. Hist.*, ser. 6, 10:477–78.

———. 1893a. Further notes on the genus *Chiroderma*. *Ann. Mag. Nat. Hist.*, ser. 6, 9:186–87.

———. 1893b. A preliminary list of the mammals of Trinidad. *J. Trinidad Field Nat. Club* 1:158–68.

———. 1893c. On some mammals from central Peru. *Proc. Zool. Soc. London* 1893:333–41, 2 pls.

———. 1894a. On a new species of armadillo from Bolivia. *Ann. Mag. Nat. Hist.*, ser. 6, 13:70–72.

———. 1894b. On two new Neotropical mammals. *Ann. Mag. Nat. Hist.*, ser. 6, 13:436–39.

———. 1894c. On *Micoureus griseus*, Desm., with description of a new genus and species of Didelphydae. *Ann. Mag. Nat. Hist.*, ser. 6, 14:184–88.

———. 1894d. Description of a new species of *Vespertilio* from China. *Ann. Mag. Nat. Hist.*, ser. 6, 14:300–01.

———. 1895a. An analysis of the mammalian generic names given in Dr. C. W. Gloger's 'Naturgeschichte' (1841). *Ann. Mag. Nat. Hist.*, ser. 6, 15:189–93.

———. 1895b. On small mammals from Nicaragua and Bogota. *Ann. Mag. Nat. Hist.*, ser. 6, 16:55–60.

———. 1895c. Descriptions of four small mammals from South America, including one belonging to the peculiar marsupial genus "*Hyracodon*," Tomes. *Ann. Mag. Nat. Hist.*, ser. 6, 16:367–70.

———. 1896a. On *Caenolestes*, a still existing survivor of the Epanorthidae of Ameghino, and the representative of a new family of Recent marsupials. *Proc. Zool. Soc. London* 1895:870–78.

———. 1896b. On new small mammals from the Neotropical Region. *Ann. Mag. Nat. Hist.*, ser. 6, 18:301–14.

———. 1897a. Descriptions of four new South-American mammals. *Ann. Mag. Nat. Hist.*, ser. 6, 20:218–21.

———. 1897b. Descriptions of new bats and rodents from America. *Ann. Mag. Nat. Hist.*, ser. 6, 20:544–53.

———. 1898a. Description of a new *Echimys* from the neighbourhood of Bogota. *Ann. Mag. Nat. Hist.*, ser. 7, 1:243–45.

———. 1898b. On the small mammals collected by Dr. Borelli in Bolivia and northern Argentina. *Boll. Mus. Zool. ed Anat. Comp. Univ. Torino* 13:1–4.

———. 1898c. On seven new small mammals from Ecuador and Venezuela. *Ann. Mag. Nat. Hist.*, ser. 7, 1:451–57.

———. 1898d. Descriptions of new mammals from South America. *Ann. Mag. Nat. Hist.*, ser. 7, 2:265–75.

———. 1898e. Notes on various American mammals. *Ann. Mag. Nat. Hist.*, ser. 7, 2:318–20.

———. 1899a. On some small mammals from District Cuzco, Peru. *Ann. Mag. Nat. Hist.*, ser. 7, 3:40–44.

———. 1899b. On a new species of *Marmosa*. *Ann. Mag. Nat. Hist.*, ser. 7, 3:44–45.

———. 1899c. On new small mammals from South America. *Ann. Mag. Nat. Hist.*, ser. 7, 3:152–55.

———. 1899d. Descriptions of new Neotropical mammals. *Ann. Mag. Nat. Hist.*, ser. 7, 4:278–88.

———. 1900a. Descriptions of new Neotropical mammals. *Ann. Mag. Nat. Hist.*, ser. 7, 5:217–22.

———. 1900b. Descriptions of new Neotropical mammals. *Ann. Mag. Nat. Hist.*, ser. 7, 5:269–74.

———. 1900c. List of the mammals obtained by Dr. G. Franco Grillo in the province of Parana, Brazil. *Ann. Mus. Civico Storia Nat. Genova*, ser. 2, 20:546–49.

———. 1900d. Descriptions of new rodents from western South America. *Ann. Mag. Nat. Hist.*, ser. 7, 6:294–302.

———. 1901a. New mammals from Peru and Bolivia, with a list of those recorded from the Inambari River, upper Madre de Dios. *Ann. Mag. Nat. Hist.*, ser. 7, 7:178–90.

———. 1901b. A new free-tailed bat from the lower Amazons. *Ann. Mag. Nat. Hist.*, ser. 7, 7:190–91.

———. 1901c. New South-American *Sciuri*, *Heteromys*, *Cavia*, and *Caluromys*. *Ann. Mag. Nat. Hist.*, ser. 7, 7:192–96.

———. 1901d. The generic names Myrmecophaga and Didelphis. *Amer. Nat.* 35:143–45.

———. 1901e. New species of *Saccopteryx*, *Sciurus*, *Rhipidomys*, and *Tatu* from South America. *Ann. Mag. Nat. Hist.*, ser. 7, 7:366–71.

———. 1901f. New *Myotis*, *Artibeus*, *Sylvilagus*, and *Metachirus* from Central and South America. *Ann. Mag. Nat. Hist.*, ser. 7, 7:541–45.

———. 1901g. On a collection of mammals from the Kanuku Mountains, British Guiana. *Ann. Mag. Nat. Hist.*, ser. 7, 8:139–54.

———. 1901h. On a collection of bats from Para. *Ann. Mag. Nat. Hist.*, ser. 7, 8:189–93.

———. 1901i. New Neotropical mammals, with a note on the species of *Reithrodon*. *Ann. Mag. Nat. Hist.*, ser. 7, 8:246–55.

———. 1901j. On a collection of bats from Paraguay. *Ann. Mag. Nat. Hist.*, ser. 7, 8:435–43.

———. 1901k. On mammals obtained by Mr. Alphonse Robert on the Rio Jordão, S. W. Minas Geraes. *Ann. Mag. Nat. Hist.*, ser. 7, 8:526–36.

———. 1902a. On mammals from the Serra do Mar of Parana, collected by Mr. Alphonse Robert. *Ann. Mag. Nat. Hist.*, ser. 7, 9:59–64.

———. 1902b. On mammals from Cochabamba, Bolivia, and the region north of that place. *Ann. Mag. Nat. Hist.*, ser. 7, 9:125–43.

———. 1902c. On mammals collected at Cruz del Eje, central Cordova, by Mr. P. O. Simons. *Ann. Mag. Nat. Hist.*, ser. 7, 9:237–45.

———. 1902d. On some mammals from Coiba Island, off the west coast of Panama. *Novit. Zoologicae.*, no. 9:135–37.

———. 1902e. Notes on the phyllostomatous genera *Mimon* and *Tonatia*. *Ann. Mag. Nat. Hist.*, ser. 7, 10:53–54.

———. 1902f. On *Marmosa marmota* and *elegans* with descriptions of new subspecies of the latter. *Ann. Mag. Nat. Hist.*, ser. 7, 10:158–62.

———. 1902g. New forms of *Saimiri*, *Oryzomys*, *Phyllotis*, *Coendou* and *Cyclopes*. *Ann. Mag. Nat. Hist.*, ser. 7, 10:246–50.

———. 1902h. On *Azara's "chauve-souris onzieme"* (*Myotis ruber*, Geoff.) and a new species allied to it. *Ann. Mag. Nat. Hist.*, ser. 7, 10:493–94.

———. 1903a. Two new glossophagine bats from Central America. *Ann. Mag. Nat. Hist.*, ser. 7, 11:286–89.

———. 1903b. New mammals from Chiriqui. *Ann. Mag. Nat. Hist.*, ser. 7, 11:376–82.

———. 1903c. Two South American forms of *Rhogeessa*. *Ann. Mag. Nat. Hist.*, ser. 7, 11:382–83.

———. 1903d. On a collection of mammals from the small islands off the coast of western Panama. *Novit. Zool.*, no. 10:39–42.

———. 1903e. New forms of *Sciurus*, *Oxymycterus*, *Kannabateomys*, *Proechimys*, *Dasyprocta*, and *Caluromys* from South America. *Ann. Mag. Nat. Hist.*, ser. 7, 11:487–93.

———. 1903f. Notes on South-American monkeys, bats, carnivores, and rodents, with descriptions of new species. *Ann. Mag. Nat. Hist.*, ser. 7, 12:455–64.

———. 1904a. Two new mammals from South America. *Ann. Mag. Nat. Hist.*, ser. 7, 13:142–44.

———. 1904b. On the mammals collected by Mr. A. Robert at Chapada, Matto Grosso (Percy Sladen Expedition to Central Brazil). *Proc. Zool. Soc. London* 1903(2):232–44, 1 pl.

———. 1904c. New forms of *Saimiri*, *Saccopteryx*, *Balantiopteryx* and *Thrichomys* from the Neotropical region. *Ann. Mag. Nat. Hist.*, ser. 7, 13:250–55.

———. 1904d. A new bat from the United States, representing the European *Myotis (Leuconoe) Daubentoni*. *Ann. Mag. Nat. Hist.*, ser. 7, 13:382–84.

———. 1904e. New *Sciurus*, *Rhipidomys*, *Sylvilagus*, and *Caluromys* from Venezuela. *Ann. Mag. Nat. Hist.*, ser. 7, 14:33–37.

———. 1905a. New Neotropical *Molossus*, *Conepatus*, *Nectomys*, *Proechimys*, and *Agouti*, with a note on the genus *Mesomys*. *Ann. Mag. Nat. Hist.*, ser. 7, 15:584–91.

———. 1905b. New Neotropical *Chrotopterus*, *Sciurus*, *Neacomys*, *Coendou*, *Proechimys*, and *Marmosa*. *Ann. Mag. Nat. Hist.*, ser. 7, 16:308–14.

———. 1905c. The generic names given by Frisch in 1775. *Ann. Mag. Nat. Hist.*, ser. 7, 16:461–64.

———. 1906. On mammals collected in south-west Australia for Mr. W. E. Balston. *Proc. Zool. Soc. London* 1906:468–78.

———. 1907. On Neotropical mammals of the genera *Callicebus*, *Reithrodontomys*, *Ctenomys*, *Dasypus*, and *Marmosa*. *Ann. Mag. Nat. Hist.*, ser. 7, 20:161–68.

———. 1908. The missing premolar of the Chiroptera. *Ann. Mag. Nat. Hist.*, ser. 8, 1:346–48.

———. 1909a. New species of *Oecomys* and *Marmosa* from Amazonia. *Ann. Mag. Nat. Hist.*, ser. 8, 3:378–80.

———. 1909b. Notes on some South-American mammals with descriptions of new species. *Ann. Mag. Nat. Hist.*, ser. 8, 4:230–42.

———. 1910a. A collection of mammals from eastern Buenos Ayres, with descriptions of related new mammals from other localities. *Ann. Mag. Nat. Hist.*, ser. 8, 5:239–47.

———. 1910b. Mammals from the River Supinaam, Demerara, presented by Mr. F. V. McConnell to the British Museum. *Ann. Mag. Nat. Hist.*, ser. 8, 6:184–89.

———. 1910c. On mammals collected in Ceará, N.E. Brazil by Fräulein Dr. Snethlage. *Ann. Mag. Nat. Hist.*, ser. 8, 6:500–03.

———. 1911a. Three new South-American mammals. *Ann. Mag. Nat. Hist.*, ser. 8, 7:113–15.

———. 1911b. The mammals of the tenth edition of Linnaeus; an attempt to fix the types of the genera and the exact bases and localities of the species. *Proc. Zool. Soc. London* 1911:120–58.

———. 1911c. New mammals from tropical South America. *Ann. Mag. Nat. Hist.*, ser. 8, 7:513–17.

———. 1912a. On small mammals from the lower Amazon. *Ann. Mag. Nat. Hist.*, ser. 8, 9:84–90.

———. 1912b. A new genus of opossums and a new tucotuco. *Ann. Mag. Nat. Hist.*, ser. 8, 9:239–41.

———. 1912c. Three small mammals from South America. *Ann. Mag. Nat. Hist.*, ser. 8, 9:408–10.

———. 1912d. Small mammals from South America. *Ann. Mag. Nat. Hist.*, ser. 8, 10:44–48.

———. 1912e. A new vespertilionine bat from Angola. *Ann. Mag. Nat. Hist.*, ser. 8, 10:204–06.

———. 1912f. New bats and rodents from S. America. *Ann. Mag. Nat. Hist.*, ser. 8, 10:403–11.

———. 1912g. New *Centronycteris* and *Ctenomys* from S. America. *Ann. Mag. Nat. Hist.*, ser. 8, 10:638–40.

———. 1913a. On some rare Amazonian mammals from the collection of the Para Museum. *Ann. Mag. Nat. Hist.*, ser. 8, 11:130–36.

———. 1913b. A new genus of glossophagine bat from Colombia. *Ann. Mag. Nat. Hist.*, ser. 8, 12:270–71.

———. 1913c. The geographical races of the woolly opossum (*Philander laniger*). *Ann. Mag. Nat. Hist.*, ser. 8, 12:358–61.

———. 1913d. New mammals from South America. *Ann. Mag. Nat. Hist.*, ser. 8, 12:567–74.

———. 1914a. New *Callicebus* and *Eumops* from S. America. *Ann. Mag. Nat. Hist.*, ser. 8, 13:480–81.

———. 1914b. Four new small mammals from Venezuela. *Ann. Mag. Nat. Hist.*, ser. 8, 14:410–14.

———. 1915a. On bats of the genus *Promops*. *Ann. Mag. Nat. Hist.*, ser. 8, 16:61–64.

———. 1915b. A new genus of phyllostome bats and a new *Rhipidomys* from Ecuador. *Ann. Mag. Nat. Hist.*, ser. 8, 16:310–12.

———. 1916. Notes on bats of the genus *Histiotus*. *Ann. Mag. Nat. Hist.*, ser. 8, 17:272–76.

———. 1917a. Preliminary diagnoses of new mammals obtained by the Yale-National Geographic Society Peruvian Expedition. *Smithsonian Misc. Coll.* 68(4):1–3.

———. 1917b. Some notes on three-toed sloths. *Ann. Mag. Nat. Hist.*, ser. 8, 19:352–57.

———. 1918. On small mammals from Salta and Jujuy collected by Mr. E. Budin. *Ann. Mag. Nat. Hist.*, ser. 9, 1:186–93.

———. 1919a. On some small mammals from Catamarca. *Ann. Mag. Nat. Hist.*, ser. 9, 3:115–18.

———. 1919b. On small mammals collected by Sr. E. Budin in northwestern Patagonia. Ann. Mag. Nat. Hist., ser. 9, 3:199–212.

———. 1919c. List of mammals from the highlands of Jujuy, north Argentina, collected by Sr. E. Budin. *Ann. Mag. Nat. Hist.*, ser. 9, 4:128–35.

———. 1920a. A further collection of mammals from Jujuy. *Ann. Mag. Nat. Hist.*, ser. 9, 5:188–96.

———. 1920b. On Neotropical bats of the genus *Eptesicus*. *Ann. Mag. Nat. Hist.*, ser. 9, 5:360–67.

———. 1920c. On mammals from the lower Amazons in the Goeldi Museum, Para. *Ann. Mag. Nat. Hist.*, ser. 9, 6:266–83.

———. 1920d. Report on the Mammalia collected by Mr. Edmund Heller during the Peruvian Expedition of 1915 under the auspices of Yale University and the National Geographic Society. *Proc. U.S. Natl. Mus.* 58:217–49, 2 pls.

———. 1921a. New *Rhipidomys*, *Akodon*, *Ctenomys*, and *Marmosa* from the Sierra Santa Barbara, S. E. Jujuy. *Ann. Mag. Nat. Hist.*, ser. 9, 7:183–87.

———. 1921b. Three new species of *Marmosa* with note on *Didelphys waterhousei* Tomes. *Ann. Mag. Nat. Hist.*, ser. 9, 7:519–23.

———. 1921c. A new genus of opossum from southern Patagonia. *Ann. Mag. Nat. Hist.*, ser. 9, 8:136–39.

———. 1921d. A new bat of the genus *Promops* from Peru. *Ann. Mag. Nat. Hist.*, ser. 9, 8:139–43.

———. 1921e. On mammals from the Province of San Juan, western Argentina. *Ann. Mag. Nat. Hist.*, ser. 9, 8:214–21.

———. 1921f. New *Cryptotis*, *Thomasomys*, and *Oryzomys* from Colombia. *Ann. Mag. Nat. Hist.*, ser. 9, 8:354–57.

———. 1921g. A new short-tailed opossum from Brazil. *Ann. Mag. Nat. Hist.*, ser. 9, 8:441–42.

———. 1923a. The geographical races of *Lutreolina crassicaudata*. *Ann. Mag. Nat. Hist.*, ser. 9, 11:583–85.

———. 1923b. New subspecies of *Metachirus*. *Ann. Mag. Nat. Hist.*, ser. 9, 11:602–07.

———. 1923c. A new short-tailed opossum from Marajó, Amazonia. *Ann. Mag. Nat. Hist.*, ser. 9, 12:157.

———. 1923d. Two new mammals from Marajó Island. *Ann. Mag. Nat. Hist.*, ser. 9, 12:341–42.

———. 1923e. Three new mammals from Peru. *Ann. Mag. Nat. Hist.*, ser. 9, 12:692–94.

———. 1924a. New South American small mammals. *Ann. Mag. Nat. Hist.*, ser. 9, 13:234–37.

———. 1924b. On a collection of mammals made by Mr. Latham Rutter in the Peruvian Amazons. *Ann. Mag. Nat. Hist.*, ser. 9, 13:530–38.

———. 1924c. A new short-tailed opossum from Argentina. *Ann. Mag. Nat. Hist.*, ser. 9, 13:586.

———. 1924d. Nomina conservanda in Mammalia. *Proc. Zool. Soc. London* 1924:345–48.

———. 1924e. A new subspecies of *Nyctinomus australis*. *Ann. Mag. Nat. Hist.*, ser. 9, 14:455–56.

———. 1925. The Spedan Lewis South American Exploration. I. On mammals from southern Bolivia. *Ann. Mag. Nat. Hist.*, ser. 9, 15:575–82.

———. 1926a. On mammals from Gorgona Island with the description of a new sloth. *Ann. Mag. Nat. Hist.*, ser. 9, 17:309–11.

———. 1926b. The Spedan Lewis South American Exploration. II. On mammals collected in the Tarija Dept., southern Bolivia. *Ann. Mag. Nat. Hist.*, ser. 9, 17:318–28.

———. 1926c. The Godman-Thomas Expedition to Peru—III. On mammals collected by Mr. R. W. Hendee in the Chachapoyas region of north Peru. *Ann. Mag. Nat. Hist.*, ser. 9, 18:156–67.

———. 1927a. The Godman-Thomas Expedition to Peru. V. Mammals collected by Mr. R. W. Hendee in the province of San Martin, N. Peru, mostly at Yurac Yacu. *Ann. Mag. Nat. Hist.*, ser. 9, 19:361–375.

———. 1927b. The Godman-Thomas Expedition to Peru. VI. On mammals from the upper Huallaga and neighboring highlands. *Ann. Mag. Nat. Hist.*, ser. 9, 20:594–608.

———. 1928a. A new genus and species of glossophagine bat, with a subdivision of the genus *Choeronycteris*. *Ann. Mag. Nat. Hist.*, ser. 10, 1:120–23.

———. 1928b. Size differences in the little "pichi" armadillos. *Ann. Mag. Nat. Hist.*, ser. 10, 1:526–27.

———. 1928c. The Godman-Thomas Expedition to Peru. VII. The mammals of the Rio Ucayali. *Ann. Mag. Nat. Hist.*, ser. 10, 2:249–65.

———. 1928d. The Godman-Thomas Expedition to Peru. VIII. On mammals obtained by Mr. Hendee at Pebas and Iquitos, upper Amazons. *Ann. Mag. Nat. Hist.*, ser. 10, 2:285–94.

———. 1929. The mammals of Señor Budin's Patagonian Expedition, 1927–28. *Ann. Mag. Nat. Hist.*, ser. 10, 4:35–45.

Thomas, O., and J. St. Leger. 1926. The Godman-Thomas Expedition to Peru—IV. On mammals collected by Mr. R. W. Hendee north of Chachapoyas, province of Amazonas, north Peru. *Ann. Mag. Nat. Hist.*, ser. 9, 18:345–49.

Thomas, R., and K. R. Thomas. 1977. A small-vertebrate thanatocenosis from northern Peru. *Biotropica* 9:131–32.

Thompson, S. D. 1989. Thermoregulation in the water opossum (*Chironectes minimus*): An exception that "proves" the rule. *Physiol. Zoology* 61:450–60.

Thrasher, J. D., M. Barenfus, S. T. Rich, and D. V. Shupe. 1971. The colony management of *Marmosa mitis*, the pouchless opossum. *Lab. Animal Sci.* 21:526–36.

Thunberg, C. P. 1818. Tekning *ef en ny art* Myr-Ätare *eller* Bälta *ifrån Brasilien. Kongl. Vetensk. Acad. Handl., Stockholm*, 1818:65–68, 1 pl.

Tiedemann, F. 1808. *Zoologie. Zu seinen Vorlesungen entworfen. Allgemeine Zoologie, Mensch und Säugthiere*. Landshut: Webershen Buchhandlung, 1:xvi + 610 pp., + 2 (unnumbered).

Timm, R. M. 1982. Ectophylla alba. *Mammal. Species* 166:1–4.

———. 1985. Artibeus phaeotis. *Mammal. Species* 235:1–6.

———. 1987. Tent construction by bats of the genera *Artibeus* and *Uroderma*. In *Studies in Neotropical mammalogy, essays in honor of Philip Hershkovitz*, ed. B. D. Patterson and R. M. Timm, 187–212. *Fieldiana Zool.*, 39:frontispiece, viii + 1–506.

Timm, R. M., and H. H. Genoways. 2004. The Florida bonneted bat, *Eumops floridanus* (Chiroptera: Molossidae): Distribution, morphometrics, systematics, and ecology. *J. Mammal.* 85:852–65.

Timm, R. M., and R. K. LaVal. 1998. A field key to the bats of Costa Rica. *Occas. Publ. Ser., Center Latin Amer. Stud., Univ. Kansas*, 22:1–30.

Timm, R. M., and R. K. LaVal. 2000. Mammals. In *Monteverde: Ecology and conservation a a tropical cloud forest*, ed. N. M. Nadkarni and N. T. Wheelwright, 223–44, 408–09, 553–60. New York: Oxford University Press, xxiii + 573 pp.

Timm, R. M., and R. D. Price. 1985. A review of *Cummingsia* Ferris (Mallophaga: Trimenoponidae), with a description of two new species. *Proc. Biol. Soc. Washington* 98:391–402.

———. 1988. A new *Cummingsia* (Mallophaga: Trimenoponidae) from a Peruvian mouse-opossum (Marsupialia). *J. Kansas Entomol. Soc.* 61:76–79.

———. 1989. *Cummingsia micheneri*, a new species of Mallophaga (Trimenoponidae) from a Venezuelan mouse-opossum (Marsupialia). *J. Kansas Entomol. Soc.* 62:575–80.

Timm, R. M., R. K. LaVal, and B. Rodrígues-H. 1999. Clave de campo para los murciélagos de Costa Rica. *Brenesia* 52:1–32.

Tipton, V. J., and C. E. Machado-Allison. 1972. Fleas of Venezuela. *Brigham Young Univ. Sci. Bull.*, biol., ser., 17(6):1–115.

Tiranti, S. I. 1996. The karyotype of *Myotis levis dinellii* (Chiroptera: Vespertilionidae) from South America. *Texas J. Sci.* 48:143–46.

Tiranti, S. I., and M. P. Torres. 1988. Observaciones sobre murciélagos de la provincia de Cordoba. *IV Jornadas Argentinas de Mastozoología, 6–9 de Noviembre de 1988, Tucumán, Libro de Resumes* [abstracts], 61.

———. 1998. Observations on bats of Córdoba and La Pampa Provinces, Argentina. *Occas. Papers Mus., Texas Tech Univ.*, no. 175:1–13.

Toldt, K. 1908. Die Chiropterenausbeute. *Kais. Acad. Wiss., Wein* 76:43–53

Tomes, R. F. 1856. On three genera of Vespertilionidae, Furipterus, Natalus and Hyonycteris, with the descriptions of two new species. *Proc. Zool. Soc. London* 1856:172–81, 2 pls.

———. 1857a. A monograph of the genus Lasiurus. *Proc. Zool. Soc. London* 1857:34–45.

———. 1857b. Descriptions of four undescribed species of bats. *Proc. Zool. Soc. London* 1857:50–54.

———. 1859. Notes on a collection of Mammalia made by Mr. Fraser at Gualaquiza. *Proc. Zool. Soc. London* 1858:546–49.

———. 1860a. Description of a new species of opossum obtained by Mr. Fraser in Ecuador. *Proc. Zool. Soc. London* 1860:58–60.

———. 1860b. Notes on a second collection of Mammalia made by Mr. Fraser in the Republic of Ecuador. *Proc. Zool. Soc. London* 1860:211–21.

———. 1860c. Notes on a third collection of Mammalia made by Mr. Fraser in the Republic of Ecuador. *Proc. Zool. Soc. London* 1860:260–68.

———. 1861. Notes on a collection of mammals made by the late Mr. Osburn in Jamaica. *Proc. Zool. Soc. London* 1861:63–69.

———. 1863a. Notice of a new American form of marsupial. *Proc. Zool. Soc. London* 1863:50–51.

———. 1863b. On a new genus and species of leaf-nosed bats in the Museum at Fort Pitt. *Proc. Zool. Soc. London* 1863:81–84.

Topal, G. 1963. The zoological results of Gy. Topal's collectings in South America. 1. Preliminary report. *Ann. Hist., Nat. Mus. Nat. Hungary,* 55:233–41.

Torres, M. P., T. Rosas, and S. I. Tiranti. 1988. *Thyroptera discifera* (Chiroptera, Thyropteridae) in Bolivia. *J. Mammal.* 69:434–35.

Tosi, J. A., Jr. 1971. *Inventaración y demonstraciones forestales, Panama. Zonas de vida.* United Nations development plan, Rome: UN, OAN, x + 123 pp.

Tovar S., A. 1971. Catalogo de mamíferos Peruanos. *An. Cientif.* 9:18–37.

Trajano, E. 1982. New records of bats from southeastern Brazil. *J. Mammal.* 63:529.

Trajano, E., and E. A. Gimenez. 1998. Bat community in a cave from eastern Brazil, including a new record of *Lionycteris* (Phyllostomidae, Glossophaginae). *Stud. Neotrop. Fauna Envir.* 33:69–75.

Trajano, E., and J. R. de A. Moreira. 1991. Estudo da fauna de cavernas da Província Espeleológica Arenítica Altamira-Itaituba, Pará. *Rev. Brasil. Biol.* 51:13–29.

Travassos, L. 1955. Sôbre dois novos Dicrocoeliidae de Chiroptera. *An. Acad. Bras. Cien.* 27:561–65.

Tribe, C. J. 1990. Dental age classes in *Marmosa incana* and other didelphoids. *J. Mammal.* 71:566–69.

Trouessart, E.-L. 1878. Catalogue de mammifères vivants et fossiles. Ordo III. Chiroptera. *Rev. Mag. Zool.*, 3ᵉ sér., 6:201–54.

———. 1897. *Catalogus mammalium tam viventium quam fossilium.* Fasciculus I. Primates, Prosimiae, Chiroptera, Insectivora. Berolini: R. Friedländer & Sohn, l:vi + 218 pp.

———. 1897. *Catalogus mammalium tam viventium quam fossilium.* Fasciculus III. Rodentia II (Myomorpha, Histricomorpha, Lagomorpha). Berolini: R. Friedländer & Sohn, l:vi + 218 pp.

———. 1898. *Catalogus mammalium tam viventium quam fossilium.* Fasciculus V. Sirenia, Cetacea, Edentata, Marsupialia, Allotheria, Monotremata. Berolini: R. Friedländer & Sohn, 2:999–1264.

———. 1904. *Catalogus mammalium tam viventium quam fossilium. Quinquennale supplementium, anno 1904,* Fasc. 1, 1–288. Berolini: R. Friedländer & Sohn, vii + 929 pp.

———. 1905. *Catalogus mammalium tam viventium quam fossilium. Quinquennale supplementium (1899–1904).* Cetacea, Edentata, Marsupialia, Allotheria, Monotremata.—Index alphabeticus, Fasc. 4, 753–929. Berolini: R. Friedländer & Sohn, vii + 929 pp.

———. 1910. Mammifères de la Mission de l'Équateur, d'après les collections formées par le Dr. Rivet. In *Mission du Service Géographique de l'Armée pour la mesure d'un arc de méridien équatoriale en Amérique du Sud sous le contróle scientifique de l'Académie de Sciences, 1899–1906.* Paris: Gauthier-Villars, et Cie., Vol. 9, Zoologie, Fasc. 1.—Mammifères, Oiseaux, Trochilidae,

A.1–A.31, 8 pls. [Dated 1911; but issued on 24 November 1910.]

Troughton, E. 1929. A new genus and species of bat (Kerivoulinæ) from the Solomons, with a review of the genera of the subfamily. *Records Australia Mus.* 17:85–99.

———. 1943. *Furred animals of Australia.* 2nd ed. Sydney: Angus and Robertson, Ltd, xxviii + 374 pp., 75 pls.

True, F. W. 1884. A provisional list of the mammals of North and Central America, and the West Indian Islands. *Proc. U.S. Natl. Mus.* 7:587–611. [Released as an 18-page pamphlet in 1885.]

———. 1896. Note on the occurrence of an armadillo of the genus Xenurus in Honduras. *Proc. U.S. Natl. Mus.* 18:345–47. [Dated 1895; published 8 July 1896.]

Trupin, G. L., and B. H. Fadem. 1982. Sexual behavior of the gray short-tailed opossum (*Monodelphis domestica*). *J. Mammal.* 63:409–14.

Tschapka, M., A. P. Brooke, and L. H. Wasserthal. 2000. *Thyroptera discifera* (Chiroptera: Thyropteridae): A new record for Costa Rica and observations on echolocation. *Z. Säugetierk.* 65:193–98.

Tschudi, J. J. von. 1844a. Mammalium conspectus. *Archiv Naturgesch.* 10(1):244–55.

———. 1844b. *Untersuchungen über die Fauna peruana.* Therologie, [part 2;, 21–76]. St. Gallen: Scheitlin und Zollikofer. [See Sherborn 1922:cxxiv for dates of publication of parts.]

———. 1845. *Untersuchungen über die Fauna peruana.* Therologie, [parts 3, 4, and 5;77–244]. St. Gallen: Scheitlin und Zollikofer. [See Sherborn 1922:cxxiv for dates of publication of parts.]

Tucker, P. K. 1986. Sex chromosome-autosome translocations in the leaf-nosed bats, family Phyllostomidae. *Cytogenet. Cell Genet.* 43:19–27.

Tucker, P. K., and J. W. Bickham. 1986. Sex chromosome-autosome translocations in the leaf-nosed bats, family Phyllostomidae II. Meiotic analyses of the subfamilies Stenodermatinae and Phyllostominae. *Cytogenet. Cell Genet.* 43:28–37.

Turner, H. N., Jr. 1853. On the arrangement of the edentate Mammalia. *Proc. Zool. Soc. London* 1851:205–21.

Turner-Erfort, V. L. 1994. Gastrointestinal morphology of the South American caenolestid marsupials (Mammalia: Paucituberculata). M.Sc. Thesis, Department of Biological Sciences, University of Illinois at Chicago, Chicago, Illinois, 74 pp.

Turton, W. 1800. *A general system of nature, through the three grand kingdoms of animals, vegetables and minerals: Systematically divided into their several classes, orders, genera, species and varieties, with their habitations, manners, economy, structure and peculiarities. Translated from Gmelin's last edition of the celebrated Systema Naturae by Sir Charles Linné, amended and enlarged by the improvements and discoveries of later naturalists and societies, with appropriate copper plates.* Swansea: Lackington, Allen and Co., 1:vii + 1–943.

Tuttle, M. D. 1967. Predation by *Chrotopterus auritus* on geckos. *J. Mammal.* 48:319.

———. 1968. Feeding habits of *Artibeus jamaicensis*. *J. Mammal.* 49:787.

———. 1970. Distribution and zoogeography of Peruvian bats, with comments on natural history. *Univ. Kansas Sci. Bull.* 49:45–86.

———. 1976. Collecting techniques. In *Biology of bats of the New World family Phyllostomatidae.* Part I, ed. R. J. Baker, J. K. Jones, Jr., and D. C. Carter, 71–88. Special Publications of the Museum 10. Lubbock: Texas Tech University Press, 218 pp.

Tuttle, M. D., L. K. Taft, and M. J. Ryan. 1981. Acoustical location of calling frogs by philander opossums. *Biotropica* 13:233–34.

Tyndale-Biscoe, H. 2003. Dedication to Pat Woolley. In *Predators with pouches: The biology of carnivorous marsupials,* ed. M. E. Jones, C. R. Dickman, and M. Archer, v. Collingwood, Australia: CSIRO Publishing, xix + 486 pp.

Tyndale-Biscoe, H., and R. B. Mackenzie. 1976. Reproduction in *Didelphis marsupialis* and *D. albiventris* in Colombia. *J. Mammal.* 57:249–65.

Tyson, E. 1698. Carigueya, seu marsupiale americanum, or, the anatomy of an opossum, dissected at Gresham-College. *Phil. Trans. Roy. Soc. London* 20:105–64, 2 pls.

Ubelaker, J. E., R. D. Specian, and D. W. Duszynski. 1977. Endoparasites. In *Biology of bats of the New World family Phyllostomatidae.* Part II, ed. R. J. Baker, J. K. Jones, Jr., and D. C. Carter, 7–56. Special Publications of the Museum 13. Lubbock: Texas Tech University Press, 364 pp.

Uchikawa, K. 1988. Mybiidae (Acarina, Trombidiformes) associated with minor families of Chiroptera (Mammalia) and a discussion of phylogeny of chiropteran myobiid genera. *J. Parasit.* 74:159–76.

Ueshima, N. 1972. New World Polyctenidae (Hemiptera), with special reference to Venezuelan species. *Brigham Young Univ. Sci. Bull.,* biol. ser., 17(1):13–21.

Uieda, W. 1980. Ocorrência de *Carollia castanea* na Amazônia Brasileira (Chiroptera, Phyllostomidae). *Acta Amaz.* 10:936–38.

———. 1986. Aspectos da morfologia lingual das três espécies de morcegos hematófagos (Chiroptera, Phyllostomidae). *Rev. Bras. Biol.* 46:581–87.

———. 1987. Morcegos hematófagos e a raiva do herbívoros no Brasil. *An. Semin. Cien. FIUBE, Uberaba* 1:13–29.

Uieda, W., and V. F. de Araújo. 1987. Manutenção dos morcegos hematófagos *Diaemus youngii* e *Diphylla ecaudata* (Chiroptera, Phyllostomidae), em cautiveiro. *An. Semin. Cien. FIUBE, Uberaba* 1:30–42.

Uieda, W., and M. E. Chaves. 2005. Bats from Botucatu region, state of São Paulo, southeastern Brazil. *Chiropt. Neotrop.* 11:224–26.

Uieda, W., and J. Vasconcellos-Neto. 1985. Dispersão de *Solanum* spp. (Solanaceae) por morcégos na região de Manaus, AM, Brasil. *Rev. Bras. Zool.* 2:449–58.

Uieda, W., I. Sazima, and A. Storti-Filho. 1980. Aspectos da biologia do morcego *Furipterus horrens* (Mammalia, Chiroptera, Furipteridae). *Rev. Bras. Biol.* 40:59–66.

Unger, K. L. 1982. Nest-building behavior of the Brazilian bare-tailed opossum, *Monodelphis domestica*. *J. Mammal.* 63:160–62.

USBGN. 1987. *Gazetteer of Ecuador*, 2nd ed. Defense Mapping Agency, Washington, DC, xiv + 735 pp.

Utrera, A., and C. Ramo. 1989. Ordenamiento de la fauna silvestre de Apuroquia. *Biollania* 6:51–76.

Vaccaro, O. B. 1992. Comentarios sobre nuevas localidades para quirópteros de Argentina. *Rev. Mus. Argentino Cien. Nat. "Bernardino Rivadavia,"* zool. 16:27–36.

Vaccaro, O. B., and E. Massoia. 1988a. Nueva especie para la provincia de Misiones, Argentina: Lasiurus cinereus villosissimus (Geoffroy, 1806) (Chiroptera, Vespertilionidae). *Rev. Mus. Argentino Cien. Nat. "Bernardino Rivadavia,"* zool. 15:41–45.

———. 1988b. La presencia de Glossophaga soricina soricina (Pallas, 1766) en la provincia de Misiones, Argentina (Chiroptera, Phyllostomidae). *Rev. Mus. Argentino Cien. Nat. "Bernardino Rivadavia,"* zool. 15:49–53.

Vaccaro, O. B., and M. J. Piantanida. 1998. Type specimens of Recent mammals housed in national collections of Argentina. *Iheringia*, sér. zool., 85:67–73.

Vaccaro, O. B., and E. A. Varela. 2001. Quirópteros de la ciudad de Buenos Aires y de la provincia de Buenos Aires, Argentina. *Rev. Mus. Argentino Cien. Nat.*, 3:181–93.

Vahl, M. 1797. Beskrivelse paa tre nye arter flagermuse. *Skrivt. Naturhist.–Selskabet, Kiobenhavn* 4:121–38.

Valdez, R. 1970. Taxonomy and geographic variation of the bats of the genus *Phyllostomus*. Ph.D. thesis, Texas A&M University, College Station, xi + 131 pp.

Valdivieso, D. 1964. La fauna quiróptera del departamento de Cundinamarca, Colombia. *Rev. Biol. Trop.* 12:19–45.

Valdivieso, D., and J. R. Tamsitt. 1962. First records of the pale spear-nosed bat in Colombia. *J. Mammal.* 43:422–23.

Valenciennes, A. 1838. Observations sur les mâchoires fossiles des couches oolithiques de Stonefield, nommées Didelphis Prevostii et D. Bucklandii. *Comptes Rendus, Acad. Sci., Paris* 7:572–80.

VandeBerg, J. L. 1983. The gray short-tailed opossum: A new laboratory animal. *ILAR News* 26:9–12.

———. 1990. The gray short-tailed opossum (*Monodelphis domestica*) as a model didelphid species for genetic research. *Australian J. Zool.* 37:235–47.

VandeBerg, J. L., and E. S. Robinson. 1997. The laboratory opossum (*Monodelphis domestica*) in biomedical research. In *Marsupial Biology. Recent research, new perspectives*, ed. N. Saunders and L. Hinds, 238–53. Sydney: University of New South Wales Press Ltd, xii + 413 pp.

Van Den Bussche, R. A. 1992. Restriction-site variation and molecular systematics of New World leaf-nosed bats. *J. Mammal.* 73:29–42.

Van Den Bussche, R. A., J. L. Hudgeons, and R. J. Baker. 1998. Phylogenetic accuracy, stability and congruence. Relationships within and among the New World bat genera *Artibeus*, *Dermanura*, and *Koopmania*. In *Bat biology and conservation*, ed. T. H. Kunz and P. A. Racey, 59–71. Washington, DC: The Smithsonian Institution Press, xvi + 365 pp.

Van Den Bussche, R. A., R. J. Baker, H. A. Wichman, and M. J. Hamilton. 1993. Molecular phylogenetics of Stenodermatini bat genera: Congruence of data from nuclear and mitochondrial DNA. *Mol. Biol. Evol.* 10:944–59.

Van Deusen, H. M. 1961. Yellow bat collected over South Atlantic. *J. Mammal.* 42:530–31.

Varejão, J. B. M., and C. M. C. Valle. 1982. Contribuição ao estudo da distribuição geográfica das espécies do género *Didelphis* (Mammalia, Marsupialia) no estado de Minas Gerais, Brasil. *Lundiana* 2:5–55.

Vargas, J., and J. A. Simonetti. 2001. New distributional records of small mammals at Beni Biosphere Reserve, Bolivia. *Mamm. Biol. (Z. Säugetierk.)* 66:379–82.

———. 2004. Small mammals in a tropical fragmented landscape in Beni, Bolivia. *Mamm. Biol. (Z. Säugetierk.)* 69:65–69.

Varona, L. S. 1974. *Catálogo de los mamíferos vivientes y extinguidos de las Antillas*. Habana: Academia de Ciencias de Cuba, viii + 139 pp.

Vaucher, C. 1981. Helminthes parasites du Paraguay. II. *Postorchigenes mbopi* n. sp. (Trematoda: Lecithodendriidae) chez *Lasiurus ega argentinus* (Thomas). *Bull. Soc. Neuchateloise Sci. Nat.* 104:47–51.

———. 1985. Helminthes parasites du Paraguay. X. *Hymenolepis dasipteris* n. sp. (Cestoda: Hymenolepididae) chez *Dasipterus* [*sic*] *ega argentinus* Thomas

(Chiroptera, Vespertilionidae). *Bull. Soc. Neuchateloise Sci. Nat.* 108:23–27.

Vaughan, T. A. 1970. The skeletal system. In *Biology of bats*, ed. W. A. Wimsatt, 97–138. New York: Academic Press, 1:1–406.

Vaughan, T. A., and G. C. Bateman. 1970. Functional morphology of the forelimb of mormoopid bats. *J. Mammal.* 51:217–35.

Vaz, S. M. 2003. A localidade tipo da preguiça-de coleira, *Bradypus torquatus* Illiger, 1811 (Xenarthra, Bradypodidae). *Edentata* 5:1–4.

Velasco Abad, E., and M. Alberico. 1985. Notas sobre algunos mamíferos nuevos de la fauna vallecaucana. *Cespedesia* 13:291–95. [Dated 1984; print date on back cover "26 de Septiembre de 1985."]

Velazco, P. M. 2005. Morphological phylogeny of the bat genus *Platyrrhinus* Saussure, 1860 (Chiroptera: Phyllostomidae) with the description of four new species. *Fieldiana Zool.* 105:1–53.

Velazco, P. M., and S. Solari. 2003. Taxonomía de *Platyrrhinus dorsalis* y *Platyrrhinus lineatus* (Chiroptera: Phyllostomidae) en Perú. *Masto. Neotrop.* 10:303–19.

Ventura, J., R. Pérez-Hernández, and M. J. Lopez-Fuster. 1998. Morphometric assessment of the *Monodelphis brevicaudata* group (Didelphimorphia: Didelphidae) in Venezuela. *J. Mammal.* 79:104–17.

Ventura, J., D. Lew, R. Pérez-Hernández, and M. J. López-Fuster. 2005. Skull size and shape relationships between Venezuelan *Monodelphis* taxa (Didelphimorphia Didelphidae), including the recently described species M. *reigi* Lew & Pérez-Hernández 2004. *Trop. Zool.* 18:227–35.

Vesey-FitzGerald, D. 1936. Trinidad mammals. *Tropical Agri., Trinidad* 8:161–65.

Vieira, C. O. da C. 1942. Ensaio monográfico sobre os quirópteros do Brasil. *Arq. Zool. Estado São Paulo* 3:219–471.

———. 1945. Sôbre uma coleção de mamíferos de Mato Grosso. *Arq. Zool. Estado São Paulo* 4:395–429.

———. 1949. Xenarthros e marsupiais do estado de São Paulo. *Arq. Zool. Estado São Paulo* 7:325–62.

———. 1950. Xenartros e marsupiais do estado de São Paulo. *Arq. Zool. Estado São Paulo* 7:325–62. [Dated 1949; published in 1950.]

———. 1951. Notas sobre os mamíferos obtidos pela expedição do Insituto Butantá ao Rio das Mortes e Serra do Roncador. *Papéis Avulsos Dept. Zool., São Paulo* 10:105–25.

———. 1953. Sobre una colecção de mamíferos do estado de Alagoas. *Arq. Zool. Estado São Paulo* 8:209–24.

———. 1955. Lista remissiva dos mamíferos do Brasil. *Arq. Zool. Estado São Paulo* 8:341–474.

———. 1957. Sobre mamíferos do estado do Maranhão. *Papéis Avulsos Dept. Zool., São Paulo* 13:125–32.

Vieira, E. M., and A. R. T. Palma. 1996. Natural history of Thylamys velutinus (Marsupialia, Didelphidae) in central Brazil. *Mammalia* 60:481–84.

Villa-R., B. 1956. Otros murciélagos nuevos para la fauna de México. *An. Inst. Biol., México* 26:543–45.

———. 1967. *Los murciélagos de México*. México: Inst. Biol., Univ. Nac. Autón México, xvi + 491 pp. [Dated 1966; published in 1967.]

Villa-R, B., and E. L. Cockrum. 1962. Migration in the guano bat *Tadarida brasiliensis mexicana* (Saussure). *J. Mammal.* 43:43–64.

Villa-R., B., and M. Villa-Cornejo. 1969. Algunos murciélagos del norte de Argentina. In *Contributions in mammalogy*, ed. J. K. Jones, Jr., 407–28. Museum of Natural History, Miscellaneus Publication No. 51, 428 pp. Lawrence: Univ. Kansas Printing Service.

———. 1973. Observaciones acerca de algunos murciélagos del norte de Argentina, especialmente de la biología del vampiro *Desmodus r. rotundus*. *An. Inst. Biol., México*, sér. zool., 42:107–48. [Dated 1971; published on 31 January 1973.]

Villa-R., B., N. M. da Silva, and B. Villa-Cornejo. 1969. Estudio del contenido estomacal de los murciélagos hematófagos *Desmodus rotundus rotundus* (Geoffroy) y *Diphylla ecaudata ecaudata* Spix (Phyllostomatidae, Desmodinae). *An. Inst. Biol., México* 40:291–97.

Virey, J. 1819. Singes. In *Nouveau dictionnaire d'histoire naturelle, appliquée aux arts, à l'agriculture, à l'économie rurale et domestique, à la médecine, etc. par une société de naturalistes et d'agriculteurs*, 257–99. Nouv. éd. Paris: Deterville, 31:1–577.

Vivar, E., and R. Arana-Cardó. 1994. Lista preliminar de los mamíferos de la Cordillera del Condor, Amazonas, Perú. *Publ. Mus. Hist Nat. UNMSM*, ser. A, zool., 46:1–6.

Vivar, E., V. Pacheco, and M. Valqui. 1997. A new species of *Cryptotis* (Insectivora: Soricidae) from northern Peru. *Amer. Mus. Novit.*, no. 3202:1–15.

Vivo, M. de, and N. F. Gomes. 1989. First record of *Caluromysiops irrupta* Sanborn, 1951 (Didelphidae) from Brasil. *Mammalia* 53:310–11.

Vizcaíno, S. F. 1995. Identificación específica de las "mulitas," género *Dasypus* L. (Mammalia, Dasypodidae), del noroeste Argentino. Descripción de una nueva especie. *Mastozool. Neotrop.* 2:5–13.

Vizcaíno, S. F. 1997. Armadillos del noroeste argentino (provincias de Jujuy y Salta). *Edentata* 3:7–10.

Vizcaíno, S. F., and A. Giallombardo. 2001. Armadillos del noroeste Argentino (provincias de Jujuy y Salta). *Edentata* 4:5–9.

Vizotto, L. D., and V. A. Taddei. 1976. Notas sobre *Molossops temminckii temminckii* e *Molossops planirostris* (Chiroptera–Molossidae). *Naturalia* 2:47–59.

Vizotto, L. D., A. J. Dumbra, and V. Rodrigues. 1980. Primeira occurência no Brasil de *Tonatia carrikeri* (Allen, 1910) (Chiroptera–Phyllostominae). VII Congresso Brasileiro de Zoología, Mossoró, 1:98–99.

Vizotto, L. D., V. Rodriques, and A. J. Dumbra. 1980a. Terceiro registro brasileiro de *Neoplatymops mattogrossensis* (Vieira, 1942) (Chiroptera-Molossidae). *Rev. Nordestina Biol.* 3(especial):244–46.

———. 1980b. Sobre ocorréncia e dados biométricos de *Pteronotus (Pteronotus) gymnonotus* (Natterer, in Wagner, 1843), no estado do Piauí (Chiroptera-Mormoopidae). *Rev. Nordestina Biol.* 3(especial):246–47.

———. 1980c. Notas sobre *Pteronotus (Phyllodia) parnellii rubiginosus* (Wagner, 1843) (Chiroptera-Mormoopidae) e sua ocurréncia no estado do Piauí. *Rev. Nordestina Biol.* 3(especial):248–49.

Vogel, S. 1969. Chirpterophilie in der neotropischen Flora. *Neue Mitteil. 2 u. 3. Flora* 158:289–323.

Vogel, S., A. V. Lopes, and I. C. Machado. 2005. Bat pollination in the NE Brazilian endemic *Mimosa lewisii*: An unusual case and first report for the genus. *Taxon* 54:693–700.

Voigt, F. S. 1831. *Das Thierreich, geordnet nach seiner Organisation, als Grundlage der Naturgeschichte der Thier und Einleitung in die vergleichende Anatomie von Baron von Cuvier*. Erster Band, die Säugthiere und Vögel enthaltend. Leipzig: F. A. Brockhaus, 1:xlviii + 975 pp.

Voss, R. S., and L. H. Emmons. 1996. Mammalian diversity in Neotropical lowland rainforests: A preliminary assessment. *Bull. Amer. Mus. Nat. Hist.* 230:1–115.

Voss, R. S., and S. A. Jansa. . 2003. Phylogenetic studies on didelphid marsupials II. Nonmolecular data and new IRBP sequences: Separate and combined analyses of didelphine relationships with denser taxon sampling. *Bull. Amer. Mus. Nat. Hist.* 276:1–82.

Voss, R. S., A. L. Gardner, and S. A. Jansa. 2004. On the relationships of "*Marmosa*" *formosa* Shamel, 1930 (Marsupialia: Didelphidae), a phylogenetic puzzle from the Chaco of northern Argentina. *Amer. Mus. Novit.*, no. 3442:1–18.

Voss, R. A., D. P. Lunde, and S. A. Jansa. 2005. On the contents of *Gracilinanus* Gardner and Creighton, 1989, with the description of a previously unrecognized clade of small didelphid marsupials. *Amer. Mus. Novit.*, no. 3482:1–34.

Voss, R. S., D. P. Lunde, and N. B. Simmons. 2001. Mammals of Paracou, French Guiana: A Neotropical lowland rainforest fauna. Part 2: nonvolant species. *Bull. Amer. Mus. Nat. Hist.* 263:1–236.

Voss, R. S, T. Tarifa, and E. Yensen. 2004. An introduction to *Marmosops* (Marsupialia: Didelphidae), with the description of a new species from Bolivia and notes on the taxonomy and distribution of other Bolivian forms. *Amer. Mus. Novit.*, no. 3466:1–40.

Waage, J. K., and R. Best. 1985. Arthropod associates of sloths. In *The evolution and ecology of armadillos, sloths, and vermilinguas*, ed. G. G. Montgomery, 297–312. Washington, DC: The Smithsonian Institution Press, 10 (unnumbered) + 451 pp.

Wagler, J. G. 1830. *Natürliches System der Amphibien, mit vorangehender Classification der Säugthiere und Vögel*. München, Stuttgart und Tübingen: J. G. Cotta'schen Buchhandlung, vi + 354 pp., 2 foldouts.

Wagler, J. G. 1831. Mittheilungen über die Gattungen der Sippe *Bradypus*. *Isis von Oken* 24:columns 604–12.

Wagner, J. A. 1840. *Die Säugthiere in Abbildungen nach der Natur mit Beschreibungen von Dr. Johann Christian Daniel von Schreber. Supplementband. Erste Abtheilung: Die Affen und Flederthiere*. Erlangen: Expedition das Schreber'schen Säugthier- und des Esper'sschen Schmetterlingswerkes, und in Commission der Voss'schen Buchhandlung in Leipzig, 1:i–xiv, i–vi + 551 pp., pls.1–62. [All parts completed in 1839, complete volume issued in 1840 (Poche 1912; Sherborn 1891)].

———. 1841. Bericht über die Leistungen in der Naturgeschichte der Säugthiere während der beiden Jahre 1839 und 1840. *Arch. Naturgesch.* 7(2):1–58.

———. 1842. Diagnosen neuer Arten brasilischer Säugthiere. *Arch. Naturgesch.* 8(1):356–62.

———. 1843a. Diagnosen neuer Arten brasilischer Handflügler. *Arch. Naturgesch.* 9(1):365–68.

———. 1843b. *Die Säugthiere in Abbildungen nach der Natur mit Beschreibungen von Dr. Johann Christian Daniel von Schreber. Supplementband. Dritter Abtheilung: Die Beutelthiere und Nager (erster Abschnitt)*. Erlangen: Expedition das Schreber'schen Säugthier- und des Esper'sschen Schmetterlingswerkes, und in Commission der Voss'schen Buchhandlung in Leipzig, 3:xiv + 614 pp., pls. 85–165. [See Poche 1912 and Sherborn 1891 for dates of publication.]

———. 1844. *Die Säugthiere in Abbildungen nach der Natur mit Beschreibungen von Dr. Johann Christian Daniel von Schreber. Supplementband. Vierte Abtheilung: Die Nager (zweiter Abschnitt), Zahnlücker, Einhufer, Dickhäuter und Wiederkäuer*. Erlangen: Expedition das Schreber'schen Säugthier- und des Esper'sschen Schmetterlingswerkes, und in Commission der Voss'schen Buchhandlung in Leipzig, 4:xii + 523 pp., pls. 168–327. [See Poche 1912 and Sherborn 1891 for date of publication.]

———. 1845. Diagnosen einiger neuen Arten von Nagern und Handflüglern. *Arch. Naturgesch.* 11(1):145–49.

———. 1847. Beiträge zur Kenntniss der Säugthiere Amerika's. *Abhandl. Math.–Physik. König. Bayer. Akad. Wiss., München* 5:121–208, 3 pls.

———. 1855. *Die Säugthiere in Abbildungen nach der Natur mit Beschreibungen von Dr. Johann Christian Daniel von Schreber. Supplementband. Fünfte Abtheilung: Die Affen, Zahnlücker, Beutelthiere, Hufthiere, Insektenfresser und Handflügler.* Leipzig: T. O. Weigel, 5:xxvi + 810 pp., pls. 1–51. [See Poche 1912 and Sherborn 1891 for date of publication.]

Wagner, P. C. 1763. *Abbildungen der seltensten un schönsten Stücke des Hochfüstlichen Naturalienkabinets in Bayreuth.* Bayreuth. [Dated 20 April 1762; no publisher or printer indicated; copy of pages 4 and 5 (*Armodillus*) examined, remainder of volume not seen.]

Wainberg, R. L. 1966. Cytotaxonomy of South-American Chiroptera. *Arch. Biol. (Liège)* 77:411–23.

———. 1972. Cariología y craniometría de *Monodelphis dimidiata* Wagner (Marsupialia, Didelphidae). *Physis, Buenos Aires* 31:327–36.

Walker, E. P., F. Warnick, S. E. Hamlet, K. I. Lange, M. A. Davis, H. E. Uible, and P. F. Wright. l964. *Mammals of the world.* Baltimore: The Johns Hopkins University Press, 1:xlx + 644 pp.

Walsh, J., and R. Gannon. 1967. *Time is short and the water rises.* New York: E. P. Dutton and Co., Inc., 224 pp., 40 pls., 2 maps.

Ward, H. L. 1891. Descriptions of three new species of Mexican bats. *Amer. Nat.* 24:743–53.

Warner, J. W., J. L. Patton, A. L. Gardner, and R. J. Baker. 1974. Karyotpic analysis of twenty–one species of molossid bats (Molossidae: Chiroptera). *Can. J. Genet. Cytol.* 16:167–76.

Warner, R. M. 1983. Karyotypic megaevolution and phylogenetic analysis: New World nectar–feeding bats revisited. *Syst. Zool.* 32:279–82.

Waterhouse, G. R. 1838. Mammalia. In *The zoology of the voyage of the H.M.S. Beagle under the command of Captain Fitzroy, R. N., during the years 1832–1836,* ed. C. Darwin, Fascicles 2, 4, and 5 (pages i–vi + 1–48, pls. 1–24, 33). London: Smith, Elder and Co, 2:xii + 97 pp., 35 pls, 1838–1839. [See Sherborn 1897b for dates of publication of parts.]

———. 1839. Mammalia. In *The zoology of the voyage of the H.M.S. Beagle under the command of Captain Fitzroy, R. N., during the years 1832–1836,* ed. C. Darwin, Fascicle 10 (pages vii–ix + 49–97, pls. 25–32, 34).

London: Smith, Elder and Co, 2:xii + 97 pp., 35 pls, 1838–39. [See Sherborn 1897b for dates of publication of parts.]

———. 1841. *Marsupialia or pouched animals.* In *The Naturalist's Library,* ed. W. Jardine. Mammalia. Edinburgh: W. H. Lizars, 11:frontispiece, i–xvi, 17–323 pp. + 35 pls.

———. 1846. *A natural history of the Mammalia.* London: Hippolyte Bailliere, Publisher, 1:2 (unnumbered) + 1–553, 22 pls.

Webb, J. P., Jr., and R. B. Loomis. 1977. Ectoparasites. In *Biology of bats of the New World family Phyllostomatidae.* Part II, ed. R. J. Baker, J. K. Jones, Jr., and D. C. Carter, 57–119. Special Publications of the Museum 13. Lubbock: Texas Tech University Press, 364 pp.

Webb, S. D. 1985. The interrelationships of tree sloths and ground sloths. In *The evolution and ecology of armadillos, sloths, and vermilinguas,* ed. G. G. Montgomery, 105–12. Washington, DC: The Smithsonian Institution Press, 10 (unnumbered) + 451 pp.

Weber, M. 1928. *Die Säugethiere einführung in die Anatomie und Systematik der recenten und fossilen Mammalia.* Systematischer Teil. Jena: Gustav Fischer, 2:xiv + 1–898 pp.

Webster, W. D. 1983. Systematics and evolution of bats of the genus Glossophaga. Ph.D. thesis, Texas Tech University, Lubbock, 332 pp.

———. 1993. *Systematics and evolution of bats of the genus Glossophaga.* Special Publications of the Museum 36. Lubbock: Texas Tech University Press, 184 pp.

Webster, W. D., and C. M. Fugler. 1984. Lista de quirópteros de las regiones norteñas de Bolivia. *Comun. Mus. Nac. Hist. Nat. (La Paz)* 3:13–19.

Webster, W. D., and C. O. Handley, Jr. 1986. Systematics of Miller's long-tongued bat *Glossophaga longirostris,* with description of two new subspecies. *Occas. Papers Mus., Texas Tech Univ.,* no. 100:1–22.

Webster, W. D., and J. K. Jones, Jr. 1980a. Noteworthy records of bats from Bolivia. *Occas. Papers Mus., Texas Tech Univ.,* no. 68:1–6.

———. 1980b. Taxonomic and nomenclatorial notes on bats of the genus *Glossophaga* in North America, with description of a new species. *Occas. Papers Mus., Texas Tech Univ.,* no. 71:1–12.

———. 1983. The first record of *Glossophaga commissarisi* (Chiroptera: Phyllostomidae) from South America. *J. Mammal.* 64:150.

———. 1984. Notes on a collection of bats from Amazonian Ecuador. *Mammalia* 48:247–52.

———. 1993. Glossophaga commissarisi. *Mammal. Species* 446:1–4.

Webster, W. D., and W. B. McGillivray. 1984. Additional records of bats from French Guiana. *Mammalia* 48:463–65.

Webster, W. D., and R. D. Owen. 1984. Pygoderma bilabiatum. *Mammal. Species* 220:1–3.

Webster, W. D., C. O. Handley, Jr., and P. J. Soriano. 1998. Glossophaga longirostris. *Mammal. Species* 576:1–5.

Weinbeer, M., and E. K. V. Kalko. 2004. Morphological characteristics predict alternate foraging strategy and microhabitat selection in the orange-bellied bat, *Lampronycteris brachyotis*. *J. Mammal.* 86:1116–23.

Weithofer, A. 1887. [Untitled]. " . . . ausgeführte Arbeit über fossile Cheiropteren der französischen Phosphorite." *Anz. Kaiserl. Akad. Wiss. Wien*, 24:285–86.

Wenzel, R. L. 1976. The streblid batflies of Venezuela (Diptera: Streblidae). *Brigham Young Univ. Sci. Bull.*, biol. ser., 20(4):1–177.

Wenzel, R. L., V. J. Tipton, and A. Kiewlicz. 1966. The streblid batflies of Panama (Diptera: Calypterae: Streblidae). In *Ectoparasites of Panama*, ed. R. L. Wenzel and V. J. Tipton, 405–676. Chicago: Field Museum of Natural History, xii + 861 pp.

Wetterer, A. L., M. V. Rockman, and N. B. Simmons. 2000. Phylogeny of phyllostomid bats (Mammalia: Chiroptera): Data from diverse morphological systems, sex chromosomes, and restriction sites. *Bull. Amer. Mus. Nat. Hist.* 248:1–200.

Wetzel, R. M. 1974. A review of the Recent species of *Bradypus* L. and *Choloepus* Illiger (Edentata). In *First International Theriological Congress, Transactions*, ed. A. V. Yablokov, 294–95. Moscow: Publishing House "Nauka," 2:1–339.

———. 1975. The species of *Tamandua* Gray (Edentata, Myrmecophagidae). *Proc. Biol. Soc. Washington* 88:95–112.

———. 1980. Revision of the naked–tailed armadillos, genus *Cabassous* McMurtrie. *Ann. Carnegie Mus.* 49:323–57.

———. 1982. Systematics, distribution, ecology, and conservation of South American edentates. In *Mammalian biology in South America*, ed. M. A. Mares and H. H. Genoways, 345–75. The Pymatuning Symposia in Ecology 6. Special Publications Series. Pittsburgh: Pymatuning Laboratory of Ecology, University of Pittsburgh, xii + 539 pp.

———. 1985a. The identification and distribution of Recent Xenarthra (= Edentata). In *The evolution and ecology of armadillos, sloths, and vermilinguas*, ed. G. G. Montgomery, 5–22. Washington, DC: The Smithsonian Institution Press, 10 (unnumbered) + 451 pp.

———. 1985b. Taxonomy and distribution of armadillos, Dasypodidae. In *The evolution and ecology of armadillos, sloths, and vermilinguas*, ed. G. G. Montgomery, 23–48. Washington, DC: The Smithsonian Institution Press, 10 (unnumbered) + 451 pp.

Wetzel, R. M., and F. D. de Ávila Pires. 1980. Identification and distribution of the Recent sloths of Brazil (Edentata). *Rev. Bras. Biol.* 40:831–36.

Wetzel, R. M., and D. Kock. 1973. The identity of *Bradypus variegatus* Schinz (Mammalia, Edentata). *Proc. Biol. Soc. Washington* 86:25–34.

Wetzel, R. M., and J. W. Lovett. 1974. A collection of mammals from the Chaco of Paraguay. *Univ. Connecticut Occas. Papers*, biol. sci. ser., 2:203–16.

Wetzel, R. M., and E. Mondolfi. 1979. The subgenera and species of long–nosed armadillos, genus *Dasypus* L. *In Vertebrate ecology in the northern Neotropics*, ed. J. F. Eisenberg, 43–63. Washington, DC: The Smithsonian Institution Press, 271 pp.

Whitaker, J. O., Jr., and R. E. Mumford. 1977. Records of ectoparasites from Brazilian mammals. *Ent. News* 88:255–58.

White, E. W. 1880. Notes on *Chlamydophorus truncatus*. *Proc. Zool. Soc. London* 1880:8–11.

Wied-Neuwied, M. P. zu. 1820a. *Diclidurus* Klappenschwanz. Ein neues genus der Chiropteren aus Brasilien. *Isis von Oken* 4:columns 1629–1630.

———. 1820b. *Reise nach Brasilien in den Jahren 1815 bis 1817*. Frankfurt a. M.: Heinrich Ludwig Brönner, 1:xxxvi + 380 pp. + 5 (unnumbered)., 25 pls., 2 maps. [Quarto edition.]

———. 1821. *Reise nach Brasilien in de Jahren 1815 bis 1817*. Frankfurt a. M.: Heinrich Ludwig Brönner, 2:xviii + 345 pp. + 1 (unnumbered), 16 pls., 1 map. [Quarto edition.]

———. 1826. *Beiträge zur Naturgeschichte von Brasilien. Verzeichniss der* Amphibien, Säugthiere *und* Vögel, *welche auf einer Reise zwischen dem 13ten und dem 23sten Grade südlicher Breite im östlichen Brasilien beobachtet wurden.* II. Abtheilung. *Mammalia.* Säugthiere. Weimar: Gr. H. S. priv. Landes–Industrie–Comptoirs, 2:1–622, 5 pls.

Wilkins, K. T. 1989. Tadarida brasiliensis. *Mammal. Species* 331:1–10.

Williams, D. F. 1978. Taxonomic and karyologic comments on small brown bats, genus *Eptesicus*, from South America. *Ann. Carnegie Mus.* 47:361–83.

Williams, D. F., and M. A. Mares. 1978. Karyological affinities of the South American big-eared bat, *Histiotus montanus* (Chiroptera, Vespertilionidae). *J. Mammal.* 59:844–46.

Williams, S. L., and H. H. Genoways. 1980a. Results of the Alcoa Foundation-Suriname Expeditions. II. Additional records of bats (Mammalia: Chiroptera) from Suriname. *Ann. Carnegie Mus.* 49:213–36.

———. 1980b. Results of the Alcoa Foundation-Suriname Expeditions. IV. A new species of bat of the genus *Molossops* (Mammalia: Molossidae). *Ann. Carnegie Mus.* 49:487–98.

Williams, S. L., H. H. Genoways, and J. A. Groen. 1983. Results of the Alcoa Foundation-Suriname Expeditions. VII. Records of mammals from central and southern Suriname. *Ann. Carnegie Mus.* 52:329–36.

Williams, S. L., C. J. Phillips, and D. E. Pumo. 1990. New records of bats from French Guiana. *Texas J. Sci.* 42:204–06.

Williams, S. L., M. R. Willig, and F. A. Reid. 1995. Review of the *Tonatia bidens* complex (Mammalia: Chiroptera), with descriptions of two new subspecies. *J. Mammal.* 76:612–26.

Willig, M. R. 1983. Composition, microgeographic variation, and sexual dimorphism in caatingas and cerrado bat communities from northeast Brazil. *Bull. Carnegie Mus. Nat. Hist.* 23:1–131.

———. 1985a. Reproductive activity of female bats from northeast Brazil. *Bat Res. News* 26:17–20.

———. 1985b. Ecology, reproductive biology, and systematics of *Neoplatymops mattogrossensis* (Chiroptera: Molossidae). *J. Mammal.* 66:618–28.

———. 1985c. Reproductive patterns in bats from caatingas and cerrado biomes of northeast Brasil. *J. Mammal.* 66:668–81.

Willig, M. R., and R. R. Hollander. 1986. Vampyrops lineatus. *Mammal. Species* 275:1–5.

Willig, M. R., and J. K. Jones, Jr. 1985. Neoplatymops mattogrossensis. *Mammal. Species* 244:1–3.

Willig, M. R., S. J. Presley, R. D. Owen, and C. Lopez-Gonzálea. 2000. Composition and structure of bat assemblages in Paraguay: A subtropical-temperate interface. *J. Mammal.* 81:386–401.

Willis, J. K. B., M. R. Willig, and J. K. Jones, Jr. 1990. Vampyrodes caraccioli. *Mammal. Species* 359:1–4.

Wilson, D. E. 1971a. Ecology of *Myotis nigricans* (Mammalia: Chiroptera) on Barro Colorado Island, Panama Canal Zone. *J. Zool.* 163:1–13.

———. 1971b. Food habits of *Micronycteris hirsuta* (Chiroptera: Phyllostomidae). *Mammalia* 35:107–10.

———. 1976. The subspecies of *Thyroptera discifera* (Lichtenstein and Peters). *Proc. Biol. Soc. Washington* 89:305–12.

———. 1978. Thyroptera discifera. *Mammal. Species* 104:1–3.

———. 1979. Reproductive patterns. In *Biology of bats of the New World family Phyllostomatidae. Part III*, ed. R. J. Baker, J. K. Jones, Jr., and D. C. Carter, 317–78.

Special Publications of the Museum 16. Lubbock: Texas Tech University Press, 441 pp.

Wilson, D. E., and J. S. Findley. 1977. Thyroptera tricolor. *Mammal. Species* 71:1–3.

Wilson, D. E., and I. Gamarra de Fox. 1991. El murciélago Macrophyllum macrophyllum (Chiroptera: Phyllostomidae) en Paraguay. *Bol. Mus. Nac. Hist. Nat. Paraguay* 10:33–35.

Wilson, D. E., and R. K. LaVal. 1974. Myotis nigricans. *Mammal. Species* 39:1–3.

Wilson, D. E., and D. M. Reeder, eds. 1993. *Mammal species of the world*, 2nd ed. Washington, DC: The Smithsonian Institution Press, xviii + 1207 pp.

Wilson, D. E., and J. A. Salazar. 1990. Los murciélagos de la Reserva de la Biosfera "Estación Biológica Beni," Bolivia. *Ecol. Bolivia* 13:47–56. [Dated May 1989; published in 1990; see S. Anderson (1997:566).]

Wilson, D. E., C. O. Handley, Jr., and A. L. Gardner. 1991. Reproduction on Barro Colorado Island. In *Demography and natural history of the common fruit bat, Artibeus jamaicensis, on Barro Colorado Island, Panamá*, ed. C. O. Handley, Jr., D. E. Wilson, and A. L. Gardner, 43–52. *Smithsonian Contrib. Zool.* 511:iv + 1–173.

Wilson, D. E., R. Baker, S. Solari, and J. J. Rodríguez. 1997. Bats: Biodiversity assessment in the Lower Urubamba Region. In *Biodiversity assessment and monitoring of the Lower Urubamba Region, Peru. San Martin-3 and Cashiriari-2 well sites*, ed. F. Dalmeier and A. Alonso, 293–302. Washington, DC: SI/MAB Series #1. Smithsonian Institution, lxxii + 1–368.

Wimsatt, W. A. 1959. Portrait of a vampire. *Ward's Nat. Sci. Bull.* 32:35–39, 62–63.

Wimsatt, W. A., and H. Trapido. 1952. Reproduction and the female reproductive cycle in the tropical American vampire bat, *Desmodus rotundus murinus*. *Amer. J. Anat.* 91:415–46.

Wimsatt, W. A., and B. Villa-R. 1970. Locomotor adaptations in the disc-winged bat *Thyroptera tricolor*. I. Functional organization of the adhesive discs. *Amer. J. Anat.* 129:89–119.

Winge, H. 1892. Jordfundne og nulevende Flagermus (Chiroptera) fra Lagoa Santa, Minas Geraes, Brasilien. *E Museo Lundii, Kjöbenhavn* 2:1–65, 2 pls. [Often cited as from 1893, but copy in the Zoologisk Museum, Copenhagen, bears date 3/12 1892, = 3 December 1892 in Winge's handwriting.]

———. 1893. Jordfundne og nulevende Pungdyr (Marsupialia) fra Lagoa Santa, Minas Geraes, Brasiliens. *E Museo Lundii, Kjöbenhavn* 2(2):1–133, pls. 35–37.

———. 1915. Jordfundne og nulevende gumlere (*Edentata*) fra Lagoa Santa, Minas Geraies, Brasilien. Med udsigt over gumlernes inbyrdes slaegtskab. *E Museu Lundii, Kjöbenhavn* 3(2):1–321, 42 pls.

———. 1923. *Pattedyr-Slægter. I. Monotremata, Marsupialia, Insectivora, Chiroptera, Edentata.* Copenhagen: H. Hagerups Forlag, viii + 1–361, 1 pl.

Wolffsohn, J. A. 1921. Catálogo de craneos de mamíferos de Chile colectados entre los años 1896–1918. *Rev. Chilena Hist. Nat.* 25:511–29.

Wolfgang, R. W. 1954. Studies on the endoparasitic fauna of Trinidad mammals. X. Parasites of Cheiroptera. *Can. J. Zool.* 32:20–24.

Woloszyn, B. W., and N. A. Mayo. 1974. Postglacial remains of a vampire bat (Chiroptera: *Desmodus*) from Cuba. *Acta Zool. Cracov.* 19:253–66, 1 pl.

Woodman, N. 1993. The correct gender of mammalian generic names ending in–*otis. J. Mammal.* 74:544–46.

———. 1996. Taxonomic status of the enigmatic *Cryptotis avia* (Mammalia: Insectivora: Soricidae), with comments on the distribution of the Colombian small-eared shrew, *Cryptotis colombiana. Proc. Biol. Soc. Washington* 109:409–18.

———. 2002. A new species of small-eared shrew from Colombia and Venezuela (Mammalia: Soricomorpha: Soricidae: Genus *Cryptotis*). *Proc. Biol. Soc. Washington* 115:249–72.

———. 2003a. New record of the rare emballonurid bat *Centronycteris centralis* Thomas, 1912 in Costa Rica, with notes on feeding habits. *Carib. J. Sci.* 39:399–402.

———. 2003b. A new small-eared shrew of the *Cryptotis nigrescens*-group from Colombia (Mammalia: Soricomorpha: Soricidae). *Proc. Biol. Soc. Washington* 116:853–72.

———. 2004. Designation of the type species of *Musaraneus* Pomel, 1848 (Mammalia: Soricomorpha: Soricidae). *Proc. Biol. Soc. Washington* 117:266–70.

Woodman, N., and A. Díaz de Pascual. 2004. Cryptotis meridensis. *Mammal. Species* 761:1–5.

Woodman, N., and J. P. J. Morgan. 2005. Skeletal morphology of the forefoot in shrews (Mammalia: Soricidae) of the genus *Cryptotis*, as revealed by digital x-rays. *J Morph.* 266:60–73.

Woodman, N., and R. M. Timm. 1992. A new species of small–eared shrew, genus *Cryptotis* (Insectivora: Soricidae), from Honduras. *Proc. Biol. Soc. Washington* 105:1–12.

———. 1993. Intraspecific and interspecific variation in the *Cryptotis nigrescens* species complex of small-eared shrews (Insectivora: Soricidae), with the description of a new species from Colombia. *Fieldiana Zool.* 74:1–30.

———. 1999. Geographic variation and evolutionary relationships among broad-clawed shrews of the *Cryptotis*

goldmani-group (Mammalia: Insectivora: Soricidae). *Fieldiana Zool.* 91:1–35.

———. 2000. Taxonomy and evolutionary relationships of Phillips' small-eared shrew, *Cryptotis phillipsii* (Schaldach, 1966), from Oaxaca, Mexico (Mammalia: Insectivora: Soricidae). *Proc. Biol. Soc. Washington* 113:339–55

Woodman, N., C. A. Cuartas-Calle, and C. A. Delgado 2003. The humerus of *Cryptotis colombiana* and its bearing on the phylogenetic relationships of the species (Soricomorpha: Soricidae). *J. Mammal.* 84:832–39.

Woodman, N., R. M. Timm, R. Arana C., V. Pacheco, C. A. Schmidt, E. D. Hooper, and C. Pacheco A. 1991. Annotated checklist of the mammals of Cuzco Amazonico, Peru. *Occas. Papers, Univ. Kansas Mus. Nat. Hist.* 145:1–12.

Woodman, N., N. A. Slade, R. M. Timm, and C. A. Schmidt. 1995. Mammalian community structure in lowland tropical Peru, as determined by removal trapping. *Zool. J. Linnean Soc.* 113:1–20.

Woodward, B. B., ed. 1904. France. [*Voyages &c.- Coquille.*], Voyage autour du monde . . . sur . . . la Coquille pendant . . . 1822–25 . . . P. M. L. I. Duperry, &c. [page 604]. *Catalogue of the books, manuscripts, maps and drawings in the British Museum (Natural History).* London: British Museum (Natural History), Vol. 2, E– K:501–1038.

Wujek, D. E., and J. M. Cocuzza. 1986. Morphology of hair of two- and three-toed sloths (Edentata: Bradypodidae). *Rev. Biol. Trop.* 34:243–46.

Ximénez, A. 1967. Contribución al conocimiento de *Lutreolina crassicaudata* (Desmarest, 1804) y sus formas geográficas (Mammalia-Didelphidae). *Comunic. Zool. Mus. Hist. Nat. Montevideo* 9:1–7.

———. 1969. Dos nuevos géneros de quirópteros para el Uruguay (Phyllostomidae – Molossidae). *Comunic. Zool. Mus. Hist. Nat. Montevideo* 10:1–8.

———. 1972. Hallazgo de *Tamandua tetradactyla* (Linné, 1758) en el Uruguay (Edentata-Myrmecophagidae). *Neotropica* 18:134–38.

Ximénez, A., A. Langguth, and R. Praderi. 1972. Lista sistemática de los mamíferos del Uruguay. *An. Mus. Nac. Hist. Nat., Montevideo*, sér. 2a, 7(5):1–49.

Yancey, F. D., II, J, R. Goetze, and C. Jones. 1998a. Saccopteryx bilineata. *Mammal. Species* 581:1–5.

———. 1998b. Saccopteryx leptura. *Mammal. Species* 582:1–3.

Yanis, E., J. Cayon, and E. Ramíres. 1973. The chromosomes of *Metachirus nudicaudatus* (Marsupialia: Didelphidae). *Austr. J. Zool.* 21:369–73.

Yates, T. L. 1984. Insectivores, elephant shrews, tree shrews, and dermopterans. In *Orders and families of*

Recent mammals of the world, ed. S. Anderson and J. K. Jones, Jr., 117–44. New York: John Wiley and Sons, xix + 686 pp.

Yee, D. A. 2000. Peropteryx macrotis. *Mammal. Species* 643:1–4.

Yensen, E., T. Tarifa, and S. Anderson. 1994. New distributional records of some Bolivian mammals. *Mammalia* 58:405–13.

Yepes, J. 1928. Los "Edentata" argentinos. Sistemática y distribución. *Rev. Univ. Buenos Aires*, sér. 2a, 1:461–515, 6 figs. [Reprint separate is paginated independently as pp. 1–55 + 6 pls.]

———. 1929. Notas sobre la distribución geográfica del pichi ciego menor ("Chlamyphorus trunctus") y pichi llorón ("Chaetophractus vellerosus"). *Physis, Buenos Aires* 9:439–46.

———. 1931. El escudete cefalico del pichi ciego menor (*Chlamyphorus truncatus* Harl.). *Rev. Chilena Hist. Nat.* 35:107–12.

———. 1932. Las formas geográficas del "pichi ciego" menor ("Chlamyphorus truncatus" Harl.). *Physis, Buenos Aires* 11:9–18, 2 pls.

———. 1933. Una especie nueva de "mulita" (Dasipodinae[*sic*]) para el norte argentino. *Physis, Buenos Aires* 11:225–32.

———. 1935. Las especies argentinas del género "Cabassous" (Dasypodidae). *Physis, Buenos Aires* 11:438–44, 3 pls.

———. 1936. Mamíferos coleccionados en la parte central y occidental de la provincia de La Rioja. *Physis, Buenos Aires* 12:31–42, 3 pls.

———. 1937. Los mamíferos de Mendoza y sus relaciones con la faunas limítrofes. *Novena Reunión Soc. Argentina Patol. Reg. Mendoza, Buenos Aires* 2:689–725, 1 map.

———. 1939. Una nueva subespecie de "pichi ciego" mayor (Chlamyphorinae) y su probable distribucion geografica. *Physis, Buenos Aires* 16:35–39.

———. 1942. Sobre la distribución occidental del murciélago ceniciento (*Lasiurus cinereus villossimus*). *Rev. Argentina Zoogeogr.* 2:159.

———. 1944. Comentarios sobre cien localidades nuevas para mamíferos sudamericanos. *Rev. Argentina Zoogeogr.* 4:59–71.

Yonenaga [-Yassuda], Y., O. Frota–Pessoa, and K. R. Lewis. 1969. Karyotypes of seven species of Brazilian bats. *Caryologia* 22:63–80.

Yonenaga-Yassuda, Y., S. Kasahara, M. J. Souza, and M. L'Abbate. 1982. Constituent heterochromatin, G bands and nucleolus—organizer regions in four species of Didelphidae (Marsupialia). *Genetica* 58:71–77.

Young, C. G. 1896. Notes on Berbice bats. *Timehri*, 10:44–46.

Yunker, C. E., and F. J. Radovsky. 1980. Parasitic mites of Suriname, XXXVI. A new genus and two new species of the Neotropical Macronyssidae (Acari: Mesostigamata). *J. Med. Entomol.* 17:545–54.

Yunker, C. E., F. S. Lukoschus, and K. M. T. Giesen. 1990. Parasitic mites of Surinam, XXIV. The subfamily Ornithonyssinae, with descriptions of a new genus and treee new species (Acari: Mesostigmata: Macronyssidae). *Zool. Med.* 63:169–86.

Zetek, J. 1930. The water opossum—Chironectes panamensis Goldman. *J. Mammal.* 11:470–71.

Ziegler, B. 1989. Staatliches Museum für Naturkunde in Stuttgart 1989. *Jh. Gess. Naturkde. Württenberg* 145:315–39.

Zimmermann, E. A. W. 1780. *Geographische Geschichte der Menschen, und der algemein verbreiteten vierfüssigen Thiere*. Zweiter Band. Leipzig: Wenganschen Buchhandlung, 2:6(unnumbered) + 1–432.

Zortéa, M. 1995. Observations on tent-using in the carolline bat *Rhinopylla* [*sic*] *pumilo* in southeastern Brazil. *Chirop. Neotrop.* 1:2–4.

———. 1996. Folivory in *Platyrrhinus* (*Vampyrops*) *lineatus*. *Bat Res. News* 34:59–60.

———. 2003. Reproductive patterns and feeding habits of three nectarivourous bats (Phyllostomidae: Glossophagnae) from the Brazilian Cerrado. *Brazil. J. Biol.* 63:159–68.

Zortéa, M., and B. F. A. de Brito. 2000. Tents used by *Vampyressa pusilla* (Chiroptera: Phyllostomidae) in southeastern Brazil. *J. Trop. Ecol.* 16:475–80.

Zortéa, M., and S. L. Mendes. 1993. Folivory in the big fruit-eating bat, *Artibeus lituratus* (Chiroptera: Phyllostomidae) in eastern Brazil. *J. Trop. Ecol.* 9:117–20.

Zortéa, M. and V. A. Taddei. 1995. Taxonomic status of *Tadarida espiritosantensis* Ruschi, 1951 (Chiroptera: Molossidae). *Bol. Mus. Biol. Mello Leitão*, 2:15–21.

Zortéa, M., R. Gregorin, and A. D. Ditchfield. 1998. *Lichonycteris obscura* from Espírito Santo state, southeastern Brazil. *Chirop. Neotrop.* 4:95–96.

Gazetteer of Marginal Localities

THE FOLLOWING LOCALITIES are arranged alphabetically by country and by major political unit (state, district, province, region, or department) within each country. Localities are not segregated by political units for French Guiana or Trinidad and Tobago. Place names are arranged alphabetically according to the name of the reported site of capture or to the nearest place name. Different sites of capture referenced to the same place name are arranged from north to south and from west to east. Variants of place names, including errors and alternate spellings used by the authors cited, are placed in their appropriate alphabetical sequence.

The majority of the coordinates come from the latest country gazetteers produced by the United States Board on Geographic Names. Some coordinates are based on information contained in the Ornithological Gazetteers of the Neotropics series available through the Bird Department, Museum of Camparative Zoology, Harvard University, Cambridge, Massachusetts 02138. This series of gazetteers is particularly useful for older place names that may no longer be in use. A few coordinates are based on information on Geonet. Other coordinates are those provided by the authors of the reports citing the locality. A few entries are annotated.

Argentina

Buenos Aires (Provincia de)

Arroyo El Pescado, 38°27'S, 58°15'W
Bahia Blanca, 20 m, 38°43'S, 62°17'W
Balcarce, 15 km NW, 37°46'S, 57°55'W
Balcarce, near, 37°48'S, 57°51'W
Belgrano, 34°34'S, 58°28'W
Bosch, 15 km NW Balcarce, 37°38'S, 58°14'W
Buenos Aires, 34°36'S, 58°27'W
Campana, 34°10'S, 58°57'W
Campo de los Padres, Mar del Plata, sea level, 38°00'S, 57°34'W
Carmen de Patagones, 40 m, 40°48'S, 62°59'W
Energía, 38°44'S, 58°44'W
Espigas, 36°25'S, 60°40'W
Estancia Luis Chico, 35°20'S, 57°11'W
General Lavalle, sea level, 36°24'S, 56°58'W
General Lavalle, 10 miles S, ca. 100 m, 36°32'S, 56°58'W
Juancho, 6 m, 37°09'S, 57°05'W
Laguna Chasicó, 25 m, 38°38'S, 63°06'W
La Plata, 19 m, 34°55'S, 57°57'W
Lobería, 15 km SE, 38°18'S, 58°56'W
Lobos, 35°12'S, 59°06'W
Los Ingleses de Ajo (= Estancia los Ingleses), 36°31'S, 56°53'W
Los Yngleses (=Estancia Los Ingleses), 36°31'S, 56°53'W
Mar del Plata, sea level, 38°00'S, 57°33'W
Mar del Sur, 38°20'S, 57°59'W
Necochea, 10 m, 38°33'S, 58°45'W
Paraná de las Palmas, sea level, 34°18'S, 58°33'W
Paraná de las Palmas y Canal 6, 34°07'S, 59°02'W
Pergamino, 33°53'S, 60°36'W
Pigué, 280 m, 37°37'S, 62°25'W
Puerto Indio, 35°15'S, 57°20'W

Punta Rosa, Cabo San Antonio, 36°23′S, 56°43′W
Quequén, 38°32′S, 58°42′W
Río La Plata, vicinity of Buenos Aires, 34°36′S, 58°27′W
Sierra de la Ventana, 38°09′S, 61°48′W
Tigre, 34°25′S, 58°34′W
Urdampilleta, 36°27′S, 61°26′W
Wilde, sea level, 34°42′S, 58°19′W
25 de Mayo, 52 m, 35°26′S, 60°10′W

Catamarca (Provincia de)

Ancasti, 1,000 m, 28°49′S, 65°30′W
Andalgalá, 1,060 m, 27°36′S, 66°19′W
Balneario El Caolín, near, 7 km N of Chumbicha, 28°48′S, 66°14′W
Balneario El Caolín, 6 km NW of Cumbicha, 28°48′S, 66°14′W
Belén, 27°39′S, 67°02′W
Catamarca, 28°28′S, 65°47′W
Choya, 13 km NNW of Andalgalá, 4,000 m, 27°32′S, 66°24′W
Cuesta del Clavillo, 3 km SW of La Banderita, 27°20′S, 65°59′W
Cuesta del Clavillo, 5 km S of La Banderita, 27°22′S, 66°00′W
El Rodeo, 28°13′S, 65°52′W
La Guardia, 29°33′S, 65°27′W
La Merced, 28°10′S, 65°41′W
Pastos Largos, 27°39′S, 68°09′W
Potrero, Río Potrero Dike, 27°32′S, 66°20′W
Potrero Dike (= Potrero River Dike), El Potrero, 27°32′S, 66°20′W
Potrero River Dike, El Potrero, 27°32′S, 66°20′W
Quirós, 28°45′S, 65°07′W

Chaco (Provincia de)

Avia Terai, 100 m, 26°42′S, 60°44′W
Campo del Cielo, 27°53′S, 61°49′W
Colonia Benitez, 27°20′S, 58°57′W
El Mangrullo, 2 km NNW and 11 km NE (by road), 26°09′S, 61°15′W
General Vedia, 5 km N, 26°54′S, 58°37′W
General Vedia, Río de Oro, 26°56′S, 58°41′W
Isla del Cerritos, confluence of Ríos Paraguay and Paraná, 27°19′S, 58°40′W
Las Palmas, 27°04′S, 58°42′W
Pampa del Indio, 26°02′S, 59°55′W
Pozo del Gato, 24°40′S, 61°58′W
Puente sobre el Río Bermejo, 26°20′S, 59°22′W
Resistencia, 50 m, 27°27′S, 58°59′W
San Carlos, 27°05′S, 59°29′W

Selvas del Río de Oro, 26°48′S, 58°57′W
Tartagal, 24°11′S, 62°07′W

Chubut (Provincia de)

Dolavón, approximately 280 km W, ca. 43°45′S, 68°57′W
Dolavón, 43°18′S, 65°42′W
El Hoyo de Epuyén, 42°04′S, 71°30′W
La Concepción, 200 m, 44°14′S, 66°09′W
Nahuel Pan, 842 m, 42°59′S, 71°11′W
Península Valdez, 42°33′S, 64°00′W
Península Valdés, Golfo San José, 42°25′S, 64°07′W
Pico Salamanca, 45°35′S, 67°20′W
Punta Norte, 3 km S, Peninsula Valdés, 42°06′S, 63°45′W
Río Azul, 275 m, 42°01′S, 71°06′W
Tecka, 3 km N, along Highway 40, 43°27′S, 70°48′W

Córdoba (Provincia de)

Alta Gracia, 31°40′S, 64°26′W
Corazón de María, 31°30′S, 64°00′W
Córdoba, 500 m, 31°25′S, 64°10′W
Cruz del Eje, 30°44′S, 64°48′W
Espinillo, 33°01′S, 64°21′W
La Maya, 45 m SE of Bell Ville, 33°00′S, 62°28′W
La Paz, 32°13′S, 65°03′W
Noetinger, 32°22′S, 62°19′W
Río Ceballos, 500 m, 31°10′S, 64°20′W
Río Cuarto, 33°08′S, 64°21′W
Segunda Usina, 32°13′S, 64°31′W
Villa Cura Brochero, 31°41′S, 65°01′W
Villa Dolores, 529 m, 31°56′S, 65°12′W
Villa María, 31°30′S, 64°00′W
Villa Nueva, 200 m, 32°26′S, 63°15′W
Villa Valeria, 34°20′S, 64°55′W

Corrientes (Provincia de)

Corrientes, 27°28′S, 58°50′W
Goya, 600 m, 29°08′S, 59°16′W
Isla Apipé, 27°30′S, 56°53′W
Itá Ibaté, 27°26′S, 57°20′W
Ituzaingó, 27°35′S, 56°41′W
Ituzaingó, 15 km W, 27°36′S, 56°49′W
Laguna Paiva, 27°30′S, 58°45′W
Mercedes, 29°12′S, 58°05′W
Virasoro, 28°03′S, 56°02′W

Entre Rios (Provincia de)

Brazo Largo, 33°47′S, 58°36′W
Estación Paranacito, Islas de Ibicuy, 33°42′S, 59°01′W

Gualeguaychú, Islas del Ibicui, Río Gualeguaychú, ca. 33°01′S, 58°31′W

La Paz, Río Paraná, 50 m, 30°45′S, 59°39′W

Parque Nacional El Palmar, 31°59′S, 58°18′W

Sauce de Luna, 31°14′S, 59°13′W

Formosa (Provincia de)

Bouvier, 25°27′S, 57°35′W

Clorinda, 75 m, 25°17′S, 57°43′W

Clorinda, 13 km S, 25°23′S, 57°44′W

Colonia Km 503, 16 km S, 25°26′S, 60°15′W

El Colorado, 26°28′S, 59°22′W

Estancia Linda Vista, 100 m, ca. 25°13′S, 59°47′W

Herradura, 26°29′S, 59°15′W

Ingeniero Guillermo N. Juárez, 100 m, 23°54′S, 61°51′W

La Urbana, 24°54′S, 59°13′W

La Victoria, 24°29′S, 59°34′W

Parque Nacional Río Pilcomayo, 25°00′S, 58°15′W

Pozo del Tigre, 124 m, 24°54′S, 60°19′W

Puesto Divisadero, 35 km S, 5 km E of Ing. Guillermo N. Juárez, 24°14′S, 61°56′W

Reserva Ecológica El Bagual, 26°10′S, 58°56′W

Reserva Natural Formosa, 24°19′S, 61°43′W

Riacho Pilaga (= Estancia Linda Vista), 10 miles NW of Km 182 (= Fontana), 100 m, 25°13′S, 59°47′W

Río Porteño, 5 km S of Estancia Santa Catalina, 24°56′S, 59°12′W

Tapia, 24°16′S, 60°09′W

Jujuy (Provincia de)

Abra de Cañas, 1,724 m, 23°40′S, 64°54′W

Agua Salada, 23°49′S, 64°36′W

Altura de Yuto, Río San Francisco, 23°38′S, 64°28′W

Arroyo La Urbana, 24°20′S, 64°50′W

Arroyo Sauzalito, Parque Nacional Calilegua, 23°04′S, 64°42′W

Bárcena, 9 km NW, ca. 23°54′S, 65°33′W

Campground on Highway 9 at provincial border with Salta, 24°28′S, 65°21′W

Cerro Santa Bárbara, 1,800 m, 24°07′S, 64°29′W

El Caulario, Río Caulario, 23°55′S, 65°05′W

El Palmar, 500 m, 24°12′S, 64°44′W

Finca La Carolina, 1,310 m, 24°11′S, 65°19′W

Ingenio La Esperanza, 24°14′S, 64°52′W

Laguna La Brea, 23°56′S, 64°28′W

La Quiaca, 22°07′S, 65°36′W

León, 24°03′S, 65°26′W

Maimará, 23°37′S, 65°28′W

Palma Sola, 550 m, 24°00′S, 64°19′W

Parque Nacional Calilegua, ca. 23°45′S, 64°56′W

Reyes, 24°10′S, 65°22′W

Río Coyaguima, 4,000 m, 22°44′S, 66°29′W

Rio Las Capillas, 15 km N of Las Capillas on provincial Route 20, 24°02′S, 65°07′W

Río Lavayén, 1 km N of Santa Rita, 24°28′S, 64°48′W

Salar Cauchari, 31 km N of Cauchari on provincial road 70, 3,840 m, 23°50′S, 66°47′W

Salinas Grandes, 2,400 m, 23°34′S, 65°41′W

San Antonio, 1,500 m, 24°22′S, 65°20′W

San Salvador de Jujuy, 20 km W, 24°10′S, 65°28′W

Santa Bárbara, W of San Pedro, 1,000 m, 23°36′S, 65°04′W

Sunchal, 23°34′S, 65°00′W

Tres Cruces, 8 km SE, 22°58′S, 65°32′W

Yuto, 350 m, 23°38′S, 64°28′W

Zapla, 950 m, 24°15′S, 65°08′W

La Pampa (Provincia de)

Caleu Caleu, Pampa Central, 38°59′S, 64°04′W

Caleufú, 206 m, 35°35′S, 64°33′W

Carro Quemado, 36°28′S, 65°20′W

General Acha, 37°23′S, 64°36′W

Laguna Chillhué, 37°15′S, 64°09′W

Lihue Calel, 38°02′S, 65°33′W

Luan Toro, 25 km S, 36°20′S, 65°07′W

Salinas Grandes de Hidalgo, N border, ca. 37°10′S, 63°32′W

Santa Rosa, 36°37′S, 64°17′W

La Rioja (Provincia de)

Ávila (= Pozo de Ávila), 400 m, 29°22′S, 66°47′W

Cueva del Chacho, Patquía, 30°03′S, 66°53′W

Guayapa, 30°07′S, 66°57′W

El Barreal, 29°38′S, 66°02′W

La Rioja, 490 m, 29°26′S, 66°51′W

Potrerillo, 28°51′S, 67°55′W

Quebrada de Ascha, 15 km NE of Aimogasta, ca. 28°25′S, 66°42′W

Villa Castelli, 29°00′S, 68°11′W

Villa Unión, 29°18′S, 68°12′W

Mendoza (Provincia de)

Chacras de Coria, 33°00′S, 68°52′W

El Sosneado, 8 km NW, 1,635 m, 35°00′S, 69°39′W

Godoy Cruz, 32°55′S, 68°50′W

Guayquerías, 36°52′S, 69°15′W

Horno del Gringo (= Hornito del Gringo), 1,100 m, 32°48′S, 68°49′W

Lavalle, 600 m, 32°43′S, 68°35′W

Malargüe, 1,450 m, 35°28′S, 69°35′W

Mendoza, ca. 900 m, 32°53′S, 68°49′W
Ñacuñán, 34°03′S, 67°58′W
Rivadavia, 654 m, 33°11′S, 68°28′W
San Rafael, 680 m, 34°36′S, 68°20′W
Tupungato, 1,067 m, 33°16′S, 69°05′W
Uspallata, 32°35′S, 69°22′W

Misiones (Provincia de)

Aristóbulo del Valle, 27°06′S, 54°57′W
Arroyo Paraíso, 6 km NE by Highway 2, 27°08′S, 54°00′W
Arroyo Yabebyri, between Loreto and El Arroyo Yabebyri on National Highway 12, 27°15′S, 55°32′W
Bonpland, ca. 100 m, 27°29′S, 55°29′W
Campamento Yacú-poi, 25°57′S, 54°25′W
Caraguatay, Río Paraná, 100 m, 26°37′S, 54°46′W
Cataratas del Iguazú, 125 m, 25°41′S, 54°26′W
Colonia Mártires, 27°26′S, 55°23′W
Dos de Mayo, 27°02′S, 54°39′W
Fracrán, 26°46′S, 54°16′W
Gobernador Lanusse, 25°58′S, 54°17′W
Hwy 21 and Arroyo Oveja Negra, Jct., approx 2 km W of Parque Provincial Moconá, 27°08′S, 53°54′W
Leandro N. Alem, 27°36′S, 55°19′W
Libertad, 25°55′S, 54°36′W
Los Helechos, 27°33′S, 55°03′W
Montecarlo, 26°34′S, 54°47′W
Parque Nacional Iguazú, 25°41′S, 54°27′W
Posadas, 81 m, 27°23′S, 55°53′W
Puerto Gisela, 50 m, 27°01′S, 55°27′W
Puerto Libertad, Río Paraná, 52 km S of Iguaçú Falls, 25°55′S, 54°36′W
Reserva Natural Estricta San Antonio, 26°05′S, 53°46′W
Río Urugua-í, 30 km from Puerto Libertad, ca. 25°58′S, 54°30′W
Río Victoria, 26°58′S, 54°30′W
San Ignacio, 27°16′S, 55°32′W
San Javier, Río Paraguay, 27°53′S, 55°08′W
San Pedro, 26°38′S, 54°08′W
Tacuaruzú, 27°37′S, 55°39′W
Teyú Cuaré, 27°05′S, 55°23′W

Neuquén (Provincia de)

Chos-Malal, 2,500 m, 37°23′S, 70°16′W
Collon-Curá, 800 m, 40°07′S, 70°44′W
Confluencia, 40°04′S, 70°04′W
Estancia Alicura, 40°30′S, 70°45′W
Estancia Tehuel Malal, 41°02′S, 71°11′W
Nahuel Huapi National Park, vicinity, 41°00′S, 71°30′W
Neuquén, ca. 400 m, 38°57′S, 68°04′W
Río Collón Curá, 40°28′S, 70°39′W
Zapala, 7 km NW, 38°51′S, 70°07′W

Río Negro (Provincia de)

Arroyo Pilcaniyeu, 41°07′S, 70°51′W
Cerro Leones, 40°30′S, 71°04′W
Chimpay, 175 m, 39°10′S, 66°09′W
Comallo, 10 km WSW, 41°02′S, 70°16′W
Coronel J. F. Gómez, 39°02′S, 67°39′W
El Bolson, 41°58′S, 71°31′W
Estación Perito Moreno, 2 km E, 41°04′S, 71°01′W
Estancia El Cóndor, 41°07′S, 71°13′W
Hotel Tunquelen, 25 km E of San Carlos de Bariloche, ca. 41°10′S, 71°04′W
Los Menucos, 15 km SE, 40°55′S, 68°01′W
Paso de Los Molles, 40°56′S, 70°43′W
Puerto Blest, Parque Nacional Nahuel Huapi, 41°02′S, 71°49′W
San Carlos de Bariloche, 4 km SW, 41°08′S, 71°17′W

Salta (Provincia de)

Aguaray, 22°16′S, 63°44′W
Aguas Blancas, 24 km NW (= Agua Blanca), 22°33′S, 64°32′W
Aguas Blancas, 27 km W, 22°45′S, 64°45′W
Aguas Blancas (= Agua Blanca), 22°44′S, 64°28′W
Alto Macueta, 2 km N del Cruce de Macueta y Campo Lago, 22°02′S, 63°59′W
Antilla, 26°07′S, 64°36′W
Arroyo de los Puestos, Parque Nacional "El Rey," 24°49′S, 64°37′W
Arroyo Santelmita, naciente de, Parque Nacional Baritú, 900 m, 22°31′S, 64°34′W
Cachi, 25°06′S, 66°11′W
Cachi, 30 km E, 3,000 m, 25°09′S, 66°10′W
Cafayate, 20 km NW, 25°58′S, 65°51′W
Cafayate, 26°05′S, 65°58′W
Cerrillos, 24°55′S, 65°30′W
Cueva del Indio, Cafayate, 26°02′S, 66°18′W
Cueva del Murcielagallo, 15 km SSW of Santa Victoria, 2,000 m, 22°15′S, 65°02′W
Dique Itiyuro, 22°10′S, 63°50′W
Finca Falcón, 700 m, 22°19′S, 63°58′W
Güemes, 734 m, 24°31′S, 65°02′W
Hickman, 23°13′S, 63°34′W
Horcones, 25°48′S, 64°55′W
Ingenio San Martín del Tabacal, 23°16′S, 64°15′W
La Angostura, 22°12′S, 63°38′W
La Merced, ca. 1,200 m, 24°58′S, 65°29′W
Molinos, 25°25′S, 66°19′W
Orán (= San Ramon de la Nueva Orán), 23°08′S, 64°20′W
Orán, 15 km S and 15 km W, Río Santa María, 23°19′S, 64°14′W
Orán, 22 km SW, Río Santa María, 23°12′S, 64°24′W

Parque San Martín, Salta, 1,190 m, 24°47′S, 65°25′W
Piquirenda Viejo, 22°21′S, 63°50′W
Piquirenda Viejo, 6 km W, 22°21′S, 63°53′W
Puesto Campo Grande, 17 km E of Santo Domingo, 24°37′S, 63°20′W
Quebrachal, 27°17′S, 64°04′W
Quebrada de Acambuco, 5 km W of Dique Itiyuro, 22°21′S, 63°59′W
Quebrada de San Lorenzo, 12 km NW of Salta, 24°43′S, 65°28′W
Quebrada Tartagal, Finca Abra Grande, 23°15′S, 64°26′W
Río de las Conchas, 2 km N and 6 km W of Metán, 25°27′S, 65°01′W
Río Pescado, 22°53′S, 64°27′W
Río Zenta, 23°05′S, 64°53′W
Rosario de la Frontera, 25°48′S, 64°58′W
Rosario de Lerma, 24°59′S, 65°35′W
Salta, 30 km NE, 24°30′S, 65°22′W
Salta, 1,190 m, 24°47′S, 65°25′W
San Andrés, 1,800 m, 23°04′S, 64°36′W
San Antonia de los Cobres, 3,650 m, 24°11′S, 66°21′W
Santa Bárbara, 25°56′S, 65°45′W
Santa Victoria Oeste, 22°15′S, 64°58′W
Serranía de Las Pavas, 22°33′S, 64°32′W
Tabacal, 23°16′S, 64°15′W
Tartagal, 30 km N, 22°16′S, 63°44′W
Tartagal, 22°32′S, 63°49′W
Tartagal, 27 km E, along Tonono Rd., 22°25′S, 63°38′W
Toma de Los Laureles, 6 km SSW of Chicoana, 1,400 m, 25°09′S, 65°35′W
Río Itiyuro, 1 km E of Tonono, 22°23′S, 63°29′W
Vado de Arrazayal, 20 km NW of Aguas Blancas, 22°33′S, 64°32′W

San Juan (Provincia de)

Cañada Honda, 500 m, 31°59′S, 68°33′W
Castaño Nuevo, 9 km NW of Villa Nueva, 1,535 m, 31°02′S, 69°33′W
La Iglesia (= Iglesia), 1,990 m, 30°24′S, 69°13′W

San Luis (Provincia de)

Potrero de los Funes, 15 km by road NE of San Luis, 800 m, 33°08′S, 66°17′W
San Francisco del Monte de Oro, 7 km E, 32°36′S, 66°07′W
San Luis, 708 m, 33°18′S, 66°21′W
San Martín, 32°31′S, 66°03′W

Santa Cruz (Provincia de)

Comodoro Rivadavia, 45°52′S, 67°30′W
Estancia La Madrugada, 47°08′S, 66°27′W

Estancia Madujada (= Estancia La Madrugada), 47°08′S, 66°27′W
Meseta El Pedrero, 46°46′S, 69°38′W
Port Désiré (Puerto Deseado), sea level, 47°45′S, 65°54′W
Punta Loyola, 51°37′S, 69°01′W
San Julián, sea level, 49°18′S, 67°43′W
Santa Cruz, sea level, 50°02′S, 68°30′W

Santa Fe (Provincia de)

Calchaquí, 56 m, 29°54′S, 60°18′W
Esperanza, 31°27′S, 60°56′W
Malabrigo, 29°21′S, 59°59′W
Romang, Isla El Laurel, 29°30′S, 59°46′W
San Javier, 24 m, 30°35′S, 59°57′W
Santa Fe, 31°39′S, 60°43′W
Sauce Viejo, 31°50′S, 60°52′W

Santiago del Estero (Provincia de)

Ahí Veremos, 200 m, 25°52′S, 63°58′W
Bañado de Figueroa, Río Salado, 100 m, 27°27′S, 63°36′W
Campo Alegre, 28°19′S, 61°54′W
Campo Gallo, 26°36′S, 62°50′W
Colonia Dora, 28°37′S, 62°57′W
Girardet, 27°38′S, 62°11′W
La Banda, 27°42′S, 64°14′W
Las Termas, 5 km E, Río Dulce, 27°30′S, 64°46′W
Lavalle, 28°12′S, 68°08′W
Ojo de Agua, 29°28′S, 63°39′W
Quimilí, 27°38′S, 62°25′W
Río Saladillo, 28°52′S, 63°59′W
Robles, 28°03′S, 63°59′W
Santa Isabel, 26°20′S, 64°20′W
Santiago de Estero, 27°47′S, 64°16′W
Santo Domingo, 26°12′S, 63°46′W
Sumampa, 29°22′S, 63°27′W
Villa La Punta, 28°23′S, 64°45′W

Tierra del Fuego (Territorio Nacional de la)

Straits of Magellan, 54°00′S, 71°00′W
Ushuaia, 54°48′S, 68°18′W
Viamonte, sea level, 54°02′S, 67°22′W

Tucumán (Provincia de)

Agua Colorada, 26°26′S, 64°53′W
Aguas Chiquitas, Sierra de Medina, 800 m, 26°24′S, 65°07′W
Aguas Chiquitas, 4 km NE of Fishery Station El Cadillal, 26°37′S, 65°12′W

Arroyo El Saltón, Reserva Provincial Santa Ana, 27°26′S, 65°46′W

Casa de Piedra, Río Los Sosa, ruta 307, Km 24.9, 830 m, 27°03′S, 65°37′W

Concepción, ca 300 m, 27°20′S, 65°35′W

Dique San Ignacio, La Cocha, 27°44′S, 65°40′W

El Nogalar, Ruta 307, 1,700 m, 27°01′S, 65°40′W

Finca El Jagüel, 26°28′S, 64°48′W

Km 76, Ruta 307, 26°48′S, 65°43′W

La Cienega, 3,000 m, 26°46′S, 65°39′W

Las Mesadas, 26°27′S, 65°30′W

Las Talas, 4 km N of Bella Vista, 27°00′S, 65°17′W

Los Basques (= Los Vásquez), 445 m, 26°50′S, 65°13′W

Parque Provincial La Florida, 27°18′S, 65°54′W

Piedra Tendida, 12 km WNW of Barruyacú, Río Cajon, 760 m, 26°30′S, 64°52′W

Piedra Tendida, 8 km W of Dique El Cajón, 26°30′S, 64°52′W

Playa Larga, Río Los Sosa, Ruta 307, Km 19.7 (Monteros-Tucumán), 27°03′S, 65°40′W

Raco, 1,000 m, 26°38′S, 65°26′W

Río Los Sosa, Rt. 307, Km 23, ca. 27°08′S, 65°33′W

Río Pueblo Viejo, Reserva Provincial La Florida, 27°13′S, 65°37′W

San Javier, 26°48′S, 65°23′W

San Miguel de Tucumán, 26°49′S, 65°13′W

San Pedro de Colalao, 26°14′S, 65°29′W

Tafí del Valle, 2,000 m, 26°52′S, 65°41′W

Tapia, 689 m, 26°36′S, 65°18′W

Tucumán, 26°49′S, 65°13′W

Bolivia
Beni (Departamento de El)

Acapulco, 13°04′S, 64°49′W

Aguadulce, 13°00′S, 64°49′W

Arruda, 4 km NE of San Joaquín, 13°04′S, 64°49′W

Bresta, 196 m, 14°33′S, 67°20′W

Camiaco, 16 km N of Limoquije, 16°19′S, 64°44′W

Campamento El Trapiche, 500 m E, 14°48′S, 66°19′W

Campamento 6 de Agosto, 15°17′S, 67°04′W

Campo Monos, 130 m, 14°40′S, 66°05′W

Caravana, 12°55′S, 64°49′W

Cascajal, across river from, 12°13′S, 65°13′W

Charuplaya, on Río Sécure, ca. 15°55′S, 65°55′W

Costa Marques [Brazil], 4 km above, Río Iténez, 12°29′S, 64°15′W

Costa Marques [Brazil], 1.5 km below, Río Iténez, 12°29′S, 64°18′W

El Consuelo, 14°20′S, 67°15′W

El Porvenir, 5 km N, 14°50′S, 65°44′W

Espíritu, 14°13′S, 66°40′W

Estación Biológica del Beni, 14°51′S, 66°21′W

Estancia Yutiole, 13°15′S, 64°49′W

Guayaramerín (= Puerto Sucre), 10°49′S, 65°25′W

Guayaramerín, 5 km S, 10°52′S, 65°25′W

Iténez (= Magdalena), 13°20′S, 64°08′W

Km 35, NW of Yucumo, 14°52′S, 67°07′W

La Cayoba, Río Itonamas, ca. 13°04′S, 64°10′W ("Ca. 30 km N Magdalena on E bank of the Itonamas River" [Torres, Rosas, and Tiranti 1988].)

La Embocada, 1 km E, 15°03′S, 66°58′W

La Granja, 4 km N of Magdalena, west bank of Río Itonamas, 200 m, 13°18′S, 64°09′W

Espíritu, near, 14°11′S, 66°38′W

Oromomo, 16°02′S, 66°10′W

Pampa de Meio, 12°30′S, 64°15′W

Piedras Blancas, 13°15′S, 64°70′W

Providencia, 13°01′S, 65°03′W

Puerto Caballo, 13°43′S, 65°21′W

Puerto Salinas, 14°20′S, 67°33′W

Puerto San Lorenzo, Río Sécure, 15°45′S, 65°23′W

Remanso, 13°34′S, 61°54′W

Río Baures, mouth, 12°30′S, 64°18′W

Río Curiche, mouth, 12°40′S, 63°30′W

Río Curiraba, 14°50′S, 66°23′W

Río Grande, 5 km NW of mouth, on Río Mamoré, 15°50′S, 64°41′W

Río Grande, 5 km below mouth, on Río Mamoré, 15°50′S, 64°41′W

Río Iténez, 4 km above Costa Marquez (= Costa Marques), Brazil, 12°29′S, 64°15′W

Río Iténez, bank opposite Costa Marquez (= Costa Marques), Brazil, 12°29′S, 64°17′W

Río Iténez, 1.5 km below Costa Marquez (= Costa Marques), Brazil, 12°29′S, 64°18′W

Río Mamoré, 12°26′S, 65°11′W

Río Matos, 14°51′S, 66°17′W

Río Tijamuchi, 14°56′S, 65°09′W

Río Yacuma, 2 km from mouth, 13°38′S, 65°25′W

Rurrenabaque, 14°28′S, 67°34′W

San Borja, 75 km S, Río Maniqui, 15°24′S, 66°28′W

San Joaquín, near Lago Saramuchyqui, 13°04′S, 64°49′W

San Juan, 4 km S of San Joaquín, 13°06′S, 64°49′W

San Pedro, 10 km W, 14°20′S, 64°55′W

San Xavier, 23 km W, 14°34′S, 64°55′W

Serranía Eva Eva, 15°36′S, 66°38′W

Serranía Pilón, 27 km by road N of Río Quiquibay, 15°15′S, 67°02′W

Totaisal, 1 km SW of Porvenir, 14°51′S, 66°21′W

Tumichucua, 11°13′S, 66°14′

Yacuma, 45 km (by road) N, 400 m, 14°42′S, 67°04′W

Versalles, 12°44'S, 63°18'W
Villa Montes, 13°06'S, 65°25'W

Chuquisaca (Departamento de)

Bolivian border near Sargento Rodriguez, Nueva Asunción, Paraguay, 20°33'S, 62°17'W
Carandaity (= Carandaiti), 10 km SE, 20°49'S, 63°08'W
Carandaity (= Carandaiti), 30 km SE, 20°56'S, 62°53'W
Chuquisaca (= Sucre), 19°02'S, 65°17'W
Finca San Antonio, 21°00'S, 65°23'W
Padilla, 9 km N, 19°18'S, 64°22'W
Padilla, 34 km SE, 2,380 m, 19°32'S, 64°07'W
Padilla, 70 km SE, along Río Azuero, 1,100 m, 19°46'S, 63°53'W
Río Azuero, 70 km SE of Padilla, 1,100 m, 19°46'S, 63°53'W
Río Limón, 19°33'S, 64°08'W
Tarabuco, 12 km N and 11 km E, 19°04'S, 64°49'W
Tarabuco, 4 km N, 19°08'S, 64°56'W
Tihumayu, 19°11'S, 64°30'W

Cochabamba (Departamento de)

Campamento Yuqui, 16°47'S, 64°57'W
Chaparé, 17°11'S, 65°49'W
Choro (= El Choro), 3,500 m, 16°56'S, 66°42'W
Cochabamba, 17°24'S, 66°09'W
Colomi, 13 km N, 17°13'S, 65°54'W
Comarapa (Depto. Santa Cruz), 25 km by road W, 2,800 m, 17°51'S, 64°40'W
El Palmar, Río Cochi Mayu, 17°06'S, 65°33'W
El Sillar, 45 km SW of Tunaré, 1,850 m, 17°07'S, 65°44'W
Incachaca, 17°14'S, 65°41'W
Jamachuma, 1.3 km W, 17°32'S, 66°07'W
Laguna Alalay, 17°25'S, 66°09'W
Misión San Antonio (= San Antonio del Chimoré), Río Chimoré, 400 m, 16°43'S, 65°07'W
Parque Nacional Carrasco, 12.5 km SW of Villa Tunari, 17°04'S, 65°29'W
Pocona, 17°39'S, 65°24'W
Puerto Patiño, 50 km NW of Villa Tunari, 16°37'S, 65°47'W
Quebrada Mojon, 6.6 km by road NW of López Mendoza, at Km 98 from Cochabamba, 17°28'S, 65°27'W
Río Yanimayo, 80 km N of Monte Punco, 17°25'S, 64°59'W
Rodeo, 7.5 km SE, 17°41'S, 65°36'W
Sajta, 17°06'S, 64°47'W
San Antonio (= Villa Tunari), 16°57'S, 65°23'W
("San Antonio, later called Villa Tunari, or the confluence of the San Antonio and Espiritu Santo rivers, or 0.5 km NE of Villa Tunari" [Anderson, Koopman, and Creighton 1982:18].)
Seque Rancho, 16°35'S, 66°45'W
Serranía Mosetenes, 16°14'S, 66°25'W
Siberia Cloud Forest, 2,800 m, 17°51'S, 64°40'W
Tablas Monte, 4.4 km by road N, 17°04'S, 66°01'W
Tinkusiri, 17 km E of Totora, 17°45'S, 65°02'W
Todos Santos, 16°48'S, 65°08'W
Villa Tunari, 0.5 km NE, 16°57'S, 65°24'W
Villa Tunari, 12.5 km SW, Parque Nacional Carrasco, 17°04'S, 65°29'W
Villa Tunari, 13 km SW, 17°02'S, 65°29'W
Vinto, 2,621 m, 17°26'S, 66°19'W
Yungas, 17°00'S, 65°50'W
Yungas (region), 16°20'S, 66°45'W
Yungas de Totora, Río Yanimayo, 80 km N of Monte Punco, 17°25'S, 64°59'W

La Paz (Departamento de)

Alcoche, 4 km by road NW, 425 m, 15°41'S, 67°42'W
Alcoche, 4 km (by road) N, 15°40'S, 67°40'W
Alcoche, 15°40'S, 67°40'W
Alto Madidi, Río Madidi, 270 m, 13°34'S, 68°44'W
(Site is a lumber camp or saw mill (abandoned) on the upper Río Madidi. Emmons [1991] called the saw mill "Aserradero Moira," while S. Anderson [1997] used the name "Moire" for this site.)
Alto Río Madidi, 13°34'S, 68°44'W
Aserradero Moira, Alto Madidi, 270 m, 13°34'S, 68°44'W
Bellavista (= Bella Vista), 1,400 m, 15°20'S, 68°13'W
Berlin, 16°30'S, 68°09'W
Calacoto, 16°32'S, 68°06'W
Caracato, 16°59'S, 67°00'W
Caranavi, 47 km by road N, at Serranía Bella Vista, 1,350 m, 15°38'S, 67°32'W
Caranavi, 35 km by road N, at Serranía Bella Vista, 1,650 m, 15°40'S, 67°35'W
Caranavi, 20 km by road NNE, 15°42'S, 67°35'W
Caranavi, 6.6 km by road downstream, 653 m, 15°38'S, 67°39'W
Caranavi, 920 m, 15°46'S, 67°36'W
Chijchipa, 1,114 m, 16°09'S, 67°45'W
Chijchipani, 16°08'S, 67°44'W
Chulumani, 1,810 m, 16°24'S, 67°31'W
Chuspipata, 1 km S, 3,050 m, 16°21'S, 67°47'W
Coroico, 16°10'S, 67°44'W
Cota Cota in La Paz, 16°33'S, 68°05'W
Cuticucho (= Cuti Khuchu), 2,970 m, ca. 16°08'S, 68°07'W
El Vertigo, Zongo Valley, 16°05'S, 68°02'W

Guanay, 5 km by road SE, 15°30'S, 67°50'W

Huajchilla, 16°37'S, 68°03'W

Irupana, 16°29'S, 67°28'W

Irupana, 3 km S, 16°30'S, 67°28'W

Ixiamas, 25 km W, 13°53'S, 68°21'W

Ixiamas, 13°45'S, 68°09'W

La Paz Valley, 3,600 m, 16°30'S, 68°00'W

La Reserva, 15°45'S, 67°31'W

Lavi Grande, 6 km S of Irupana, 16°32'S, 67°28'W

Llamachaque, 5 km NE of Pelechuco, 3,150 m, 14°48'S, 69°02'W

Mapiri, 491 m, 15°15'S, 68°10'W

Mururata, 1,400 m, 16°08'S, 67°45'W

Ñequejahuira, Río Unduavi, 2,450 m, 16°20'S, 67°50'W

Pasto Grande, 16°36'S, 67°29'W

Pitiguaya, Río Unduavi, 16°21'S, 67°46'W

Puerto Linares, 1 mile W, 350 m, 15°29'S, 67°31'W

Río Beni, 13°15'S, 67°18'W

Río Challana, 5 km (by road) SE of Guamay 15°28'S, 67°52'W

Río Machariapo, 14°36'S, 68°35'W

Rios Aceromarca and Unduavi, near confluence, 16°18'S, 67°54'W

Río Solocama, 14 km on road Chulmani to Irupana, 16°18'S, 67°32'W

Río Solocame (= Río Solocama), 16°18'S, 67°32'W

Sacramento Alto, 8 km N of Chuspipata, 2,575 m, 16°16'S, 67°47'W

Santa Ana de Madidi, 12°34'S, 67°10'W

Santa Rosa, 8 km SW, 12°16'S, 68°27'W

Sararia, ca. 20 km by river N of Puerto Linares, 277–600 m, 15°17'S, 67°37'W

Saynani, 0.5 km E, 16°07'S, 68°05'W

Saynani hydroelectric generating station, near, 2,500 m, ca. 16°07'S, 68°05'W

Serranía Bella Vista, 35 to 38 km by road N of Caranavi, 15°40'S, 67°35'W

Suapi and Mururata, between, 16°08'S, 67°46'W

Ticunhuaya, Río Tipuani, 1,500 m, 15°28'S, 68°18'W

Tomonoco, 2 km W of Puerto Linares, 350 m, 15°29'S, 67°31'W

("A small military outpost, enveloped by virgin tropical forest, adjacent to the Río Beni" [Webster and Jones 1980a:2].)

Tomonoco, 1 mile W of Puerto Linares, 350 m, 15°29'S, 67°31'W

Tumupasa, 436 m, 14°09'S, 67°55'W

Unduavi, 59 km NE of La Paz, 16°19'S, 67°54'W

Viacha, 3,851 m, 16°39'S, 68°18'W

Yanacachi, 16°23'S, 67°43'W

Yungas, centered at 16°10'S, 67°30'W

Zongo, 30 km by road N, 2,000 m, 15°53'S, 67°52'W

Oruro (Departamento de)

Finca Santa Helena, ca. 10 km by road SW of Pazña, 3,750 m, 18°37'S, 66°59'W

Mount Sajama, 18°07'S, 69°00'W

Oruro, 17°59'S, 67°09'W

Oruro, 64 km S, 18°34'S, 67°09'W

Pando (Departamento de)

Centro Dieciocho, 10°36'S, 66°47'W

Chivé, 12°23'S, 68°35'W

Cobija, 11°02'S, 68°44'W

Independencia, 11°26'S, 67°34'W

Ingavi, 10°57'S, 66°50'W

Isla Gargantua, 12°23'S, 68°35'W

La Cruz, 11°24'S, 67°13'W

Palmira, 180 m, 11°42'S, 67°56'W

Puerto Camacho, 15 km NW, 11°28'S, 67°50'W

Río Madre de Dios, Right bank, opposite Independencia, 11°26'S, 67°34'W

Río Nareuda, 11°17'S, 68°55'W

San Miguel, 11°40'S, 67°43'W

Santa Rosa, Río Madre de Dios, 12°13'S, 68°24'W

Victoria, 170 m, 10°59'S, 66°10'W

Potosí (Departamento de)

Finca Salo, Río Cachimayo, 19°18'S, 66°12'W

Uyuni, 3,669 m, 20°28'S, 66°50'W

Santa Cruz (Departamento de)

Arroyo San Francisco, 13°33'S, 68°00'W

Ascención, 6 km by road W, 15°43'S, 63°09'W

Aserradero Moira, 14°33'S, 61°11'W

Aserradero Moira, 4 km E, 14°33'S, 61°10'W

Aserradero Moira, 45 km E, 14°37'S, 60°48'W

Aserradero Moira, 55 km E, 14°38'S, 60°44'W

Aserradero Pontons, 17°05'S, 59°34'W

Basilio, 18°08'S, 63°19'W

Boyuibe, 20°25'S, 63°17'W

Buena Vista, 25 km by road W, 17°24'S, 63°46'W

Buenavista (= Buena Vista), 400 m, 17°27'S, 63°40'W

Buena Vista, 400 m, 17°27'S, 63°40'W

Buen Retiro, 4.5 km N, 17°14'S, 63°38'W

Campamento Los Fierros, 3 km S, 14°34'S, 60°53'W

Campamento Los Fierros, 17 km S, 14°33'S, 60°49'W

Campamento Los Fierros, 23 km S, 14°38'S, 60°45'W

Campamento Los Fierros, 27.5 km S, 14°40, 60°44'W

Candelaria, 16°46'S, 58°26'W

Caranda, 2 km S, 17°33'S, 63°32'W

Castedo (= Estancia Castedo), 14 km by road NE of San Ramón, 16°35'S, 62°25'W

Caucaya, 875 m, 19°25'S, 63°29'W

Cerro Amboró, 4.5 km N and 1.5 km E, Río Pitasama, 17°45'S, 63°40'W

Cerro Colorado, 19°27'S, 62°21'W

Comarapa, 17°54'S, 64°29'W

Comarapa, 3 km SE, 17°57'S, 64°30'W

Comarapa, 5 km SE, 17°58'S, 64°29'W

Comarapa, 5 km SW, 17°57'S, 64°32'W

El Encanto, 14°38'S, 60°42'W

El Refugio, 14°45'S, 61°01'W

El Tunal, 17°53'S, 64°33'W

Estancia Cachuela Esperanza, 16°47'S, 63°14'W

Estancia Cuevas, 1 km NE, 18°11'S, 63°44'W

Estancia Patuju, 17°37'S, 59°32'W

Flor de Oro, Río Iténez, Parque Nacional Noël Kempff Mercado, 13°35'S, 61°34'W

Guirapembi, 19°26'S, 62°31'W

Hacienda Cerro Colorado, 19°27'S, 62°21'W

Hacienda El Pelicano, 3 km N of Zanja Honda, 500 m, 18°16'S, 63°11'W

Huanchaca I, Parque Nacional Noël Kempff Mercado, 13°55'S, 60°49'W

Independencia, 11°26'S, 67°34'W

Ingeniero Mora, 18°10'S, 63°16'W

Ingeniero Mora, 10 km E, 18°10'S, 63°11'W

Ingeniero Mora, 15 km E, 18°10'S, 63°08'W

La Florida, 38 km E, 14°36'S, 60°53'W

Lago Caimán, 200 m, 13°36'S, 60°55'W

La Laguna, 10 km N of San Ramón, 16°33'S, 62°41'W

La Laguna, 4 km N of San Ramón, 16°36'S, 62°41'W

Las Cruces, 2 km SW, 17°47'S, 63°22'W

Los Fierros, Parque Nacional Noël Kempff Mercado, 14°32'S, 60°53'W

Los Palmares, 15°25'S, 61°00'W

Mangabalito, 200 m, 13°47'S, 60°33'W

Masicuri, 14.5 km by road NW, 18°45'S, 63°53'W

Masicuri, 13 km by road NW, 18°46'S, 63°52'W

Meseta de Huanchaca, Parque Nacional Noël Kempff Mercado, 14°25'S, 60°50'W

Monteagudo, 72 km ESE, 1,220 m, 19°59'S, 63°18'W

Palmarito, Río San Julian, 16°49'S, 62°37'W

Parapita, 20°00'S, 63°00'W

Parque Nacional Noël Kempff Mercado, 14°32'S, 60°53'W

Pampa de la Isla, 480 m, 17°48'S, 63°10'W

Perseverencia, 14°38'S, 62°37'W

PRNB, 14°20'S, 62°25'W

Puerto Suárez, 18°57'S, 57°51'W

Punta Rieles, 16°35'S, 64°12'W

Quiñe, 5 km NE, 18°03'S, 67°28'W

Río Ichilo, 52 km S of mouth of Río Chapare, 16°28'S, 64°44'W

Río Negrillo, 15°03'S, 62°45'W

Río Palometillas (in Provincia de Sara), 16°40'S, 63°40'W

Río Pitasama, 4.5 km N and 1.5 km E of Cerro Amboró, Parque Nacional de Amboró, 17°45'S, 63°40'W

Río Saguayo, 17°34'S, 63°48'W

Río Surutú, 17°24'S, 63°51'W

Río Tucavaca, 24 km by road N of Santiago de Chiquitos, 18°07'S, 59°37'W

Roboré, 29.5 km W, 18°19'S, 60°02'W

Roboré, Río Roboré, 300 m, 18°20'S, 59°45'W

San Carlos, 80 km N, 16°40'S, 63°45'W

San Ignacio de Velasco, 300 m, 16°23'S, 60°59'W

San Javier, 12 km S, 16°27'S, 62°39'W

San José (= San José de Chiquitos), 17°51'S, 60°47'W

San José de Chiquitos, 17°51'S, 60°47'W

San Matías, 16°22'S, 58°24'W

San Miguel Rincón, 17°23'S, 63°32'W

San Rafael de Amboró, 17°36'S, 63°36'W

San Rafael de Amboró, 4 km S and 13 km W, 17°39'S, 63°48'W

San Ramón, 10 km N, 16°36'S, 62°42'W

San Ramón, Castedo, 16°35'S, 62°25'W

San Ramón, 16°44'S, 62°41'W

Santa Cruz, 17°48'S, 63°10'W

Santa Cruz de la Sierra, 480 m, 17°48'S, 63°10'W

Santa Cruz, 15 km S, 17°55'S, 63°08'W

Santa Cruz, 27 km SE, 17°58'S, 63°03'W

Santa Rosa, 7 km N, 17°03'S, 63°35'W

Santa Rosa de la Roca, 15°50'S, 61°27'W

Santa Rosa de Sara, 25 km NW, 16°58'S, 63°49'W

Santa Rosa de Sara, 5 km S, 17°07'S, 63°35'W

Santiago, Chiquitos, 700 m, 18°19'S, 59°34'W

Serranía Negra, meseta above Lago Caimán, ca. 13°37'S, 60°57'W

Surutu, 17°24'S, 63°51'W

Tita, 18°25'S, 62°10'W

Tita, 8 km SE, 18°28'S, 62°07'W

Vallegrande (Río Cienega), 5.5 km by road NNE, 18°27'S, 64°04'W

Warnes, 19°30'S, 63°10'W

Tarija (Departamento de)

Caiza (= Villa Ingavi), 600 m, 21°49'S, 63°34'W

Camataqui (= Villa Abecia), 25 km SSE, 3,500 m, 21°14'S, 65°15'W

Camatindi, 1 km S, 21°00'S, 63°23'W

Camiri, 10 km S, 20°10'S, 63°34'W

Capirenda, 21°07'S, 63°00'W

Caraparí, 21°49'S, 63°46'W

Carlazo, 21°28'S, 64°32'W

Cuyambuyo, 4 km by road N, 980 m, 22°13'S, 64°36'W

Entre Rios, 25 km NW, 21°23'S, 64°21'W

Entre Rios, 5 km NNW, 21°29'S, 64°12'W

Entre Rios, 21°32'S, 64°12'W

Estancia Bolívar, 21°38'S, 62°34'W

Rancho Tambo, 61 km by road E of Tarija, 21°27'S, 64°19'W

Río Lipeo, 640 m, 22°41'S, 64°26'W

Samuhuate (= Samayhuate), 21°48'S, 62°55'W

San Francisco Mission (= Villa Montes), 500 m, Río Pilcomayo, 21°15'S, 63°30'W

San Lorenzo, 21°26'S, 64°47'W

Serranía Sama, 21°27'S, 64°52'W

Sierra Santa Rosa, 21°42'S, 63°54'W

Tablada, 21°33'S, 64°47'W

Tapecua, 21°26'S, 63°55'W

Tarija, 21°31'S, 64°45'W

Villa Montes, 21°15'S, 63°30'W

Villa Montes, 8 km S and 10 km E, east bank Río Pilcomayo, 21°19'S, 63°25'W

Villa Montes, 90 km SE, 21°50'S, 62°55'W

Yacuiba, 30 km NW, 830 m, 21°50'S, 63°33'W

Brazil

Acre (Estado de)

Alto Acre (= Alto Rio Acre near Rio Branco, 09°58'S, 67°48'W).

Cruzeiro do Sol, Rio Juruá, 188 m, 07°38'S, 72°36'W

Igarapé Porongaba, right bank Rio Juruá, 08°40'S, 72°47'W

Nova Vida, Rio Juruá, 08°22'S, 72°49'W

Rio Branco, 09°58'S, 67°48'W

Rio Moa, Parque Nacional da Serra do Divisor, 07°26'S, 73°39'W

Sena Madureira, 135 m, 09°04'S, 68°40'W

Seringal Lagoinha, Rio Juruá, approx. 30 km E of Cruzeiro do Sul, ca. 07°40'S, 72°40'W

Serra da Jaquirana, 07°27S, 73°41'W

Alagoas (Estado de)

Canoas, Rio Largo, ca. 09°30'S, 35°45'W

Limoeiro da Anadia, 09°44'S, 36°31'W

Mangabeira, sea level, 09°55'S, 36°08'W

Mangabeiras (= Mangabeira), sea level, 09°55'S, 36°08'W

Manimbu, 10°10'S, 36°22'W

Pôrto Calvo, 09°04'S, 35°24'W

Quebrangulo, 411 m, 09°20'S, 36°29'W

Rio Largo, 09°29'S, 35°51'W

São Miguel dos Campos, 09°47'S, 36°05'W

Sinimbú, 09°55'S, 36°08'W

Usina Sinimbú, 09°55'S, 36°08'W

Amapá (Território do)

Amapá, 02°03'N, 50°48'W

Cachoeira Itaboca, ca. 00°13'N, 51°44'W

Colônia Agrícola de Matapi, 00°41'N, 51°26'W

Colônia Torrão, Calçoene, 02°30'N, 50°57'W

Fazenda Itapuã, 02°05'N, 50°55'W

Ferreira Gomes, 00°48'N, 51°08'W

Horto Florestal de Macapá, 00°02'N, 51°03'W

Igarapé Novo, afluente esquerdo Igarapé Amazonas, afluente esquerdo Rio Iratapuru, 00°20'S, 52°23'W

Igarapé Rio Branco, alto Rio Maracá, 00°32'N, 52°12'W

Km 160, Rodovia Perimetral Norte, ca. 00°29N, 50°36'W

Lago do Comprido, 01°42'N, 50°50'W

Macapá, 00°02'N, 51°03'W

Macapá, Florestal, 00°02'N, 51°03'W

Mazagão, Rio Vila Nova, 00°07'S, 51°17'W

Oiapoque, 03°50'N, 51°50'W

Piratuba (= Lago Piratuba), 01°37'N, 50°10'W

Recreio, Rio Jari, 01°00'S, 52°30'W

Rio Amapari, Serra do Navio, 00°59'N, 52°03'W

Rio Araguarí, Município de Marcapá (not located; mouth of river at 01°15'N, 49°55'W)

Rio Maruanum, 00°11'N, 51°16'W

Rio Tracajatuba, 00°56'N, 51°00'W

Santa Luzia do Pacuí, 00°30'N, 51°40'W

São Joaquim do Pacuí, 00°26'N, 50°54'W

Serra do Navio, Rio Amapari, 00°59'N, 52°03'W

Taperebá, Rio Cassiporé, 03°39'N, 51°12'W

Teresinha, Serra do Navio, 00°58'N, 52°02'W

Tracajatuba, 01°00'N, 51°06'W

Vila Velha do Cacìporé, 03°13'N, 51°13'W

Vila Velha do Cassiporé, 03°13'N, 51°13'W

Amazonas (Estado do)

Altamira, right bank Rio Juruá, 06°35'S, 68°54'W

Alto Rio Urucú, 04°51'S, 65°16'W

Aurá Igarapé, 04°22'S, 58°43'W

Ayapuá (= Aiapuá), Rio Purus, 04°29'S, 62°04'W

Balbina, ca. 01°50'S, 59°30'W

Barro Vermelho, Left bank Rio Juruá, 06°28'S, 68°46'W

Borba, 25 m, 04°24'S, 59°35'W

Castanhal, Rio Jamundá, near Faro, 01°50'S, 56°58'W

Comunidade Colina, Rio Tiquié, Município São Gabriel da Cachoeira, ca. 00°05'N, 68°25'W

Codajás, 03°50'S, 62°05'W

Codajaz (= Codajás), 03°50'S, 62°05'W

Colocação Vira-Volta, Rio Juruá, 03°17'S, 66°14'W

Serra Cucuhy (= Cucuí), 76 m, 01°12'N, 66°50'W

Dimona Reserve, ca. 80 km N of Manaus, 02°19'S, 60°06'W

Ega (= Tefé), 03°22'S, 64°42'W

Fazenda Esteio, Cabo Frio Reserve, ca 72 km N of Manaus, 02°23'S, 59°56'W

Florestal Reserve, ca 74 km N of Manaus, ca. 02°24'S, 59°51'W

Gavião, ca. 02°27'S, 59°47'W

Humaitá, 07°31'S, 63°02'W

Hyutanaham (= Huitanaã), upper Purus River, 07°40'S, 65°46'W

Iaunari, 00°31'S, 64°50'W

Igarapé Acará, ca. 03°05'S, 60°01'W

Igarapé Aniba, 02°43'S, 58°49'W

Igarapé Auará, Rio Madeira, 04°22'S, 59°43'W

Igarapé Cacão Pereira, 03°09'S, 60°07'W

Igarapé Grande, Rio Juruá, ca. 06°35'S, 69°50'W

Igarapé Nova Empresa, left bank Rio Juruá, ca. 06°48'S, 70°44'W

Ilha Paxiuba, right bank Rio Juruá, 03°19'S, 66°00'W

Itacoatiara, 03°08'S, 58°25'W

Iucali, Rio Negro, 00°13'S, 66°49'W

Jauareté, Rio Uaupés (= Iauaretê) 00°36'N, 69°12'W

Jainu, right bank Rio Juruá, 06°28'S, 68°46'W

Juruá, 03°27'S, 66°03'W

Km 41 Camp (=Reserva Km 41), Município de Rio Preto da Eva, ca. 02°28'S, 59°43'W,

Lago Baptista, 03°18'S, 58°15'W

Lago de Arara, 03°26'S, 61°22'W

Lago de Capiranga, 03°14'S, 67°49'W

Lago de Tefé, 03°27'S, 64°47'W

Lago Janauacá, 03°08'S, 60°01'W

Lago Vai-Quem-Quer, right bank Rio Juruá, 03°19'S, 66°01'W

Macaco, left bank Rio Jaú, 02°05'S, 62°07'W

Manacapurú, 03°20'S, 60°40'W

Manaos (= Manaus), Rio Negro, 50 m, 03°08'S, 60°01'W

Manaus, ca. 80 km N, 02°24'S, 59°43'W

Manaus, 03°08'S, 60°01'W

Manaus, Rio Negro, INPA, 50 m, 03°08'S, 60°01'W

Manaus, Zoológico, 03°08'S, 60°01'W

Pedra do Gavião, 01°27'S, 61°38'W
("Próximo de Moura, margem direita do Rio Negro...." [Peracchi 1986:97].)

Presidente Figueiredo (Km 110 on highway BR 174), 02°65'S, 59°41'W

Purus river (Hyutanaham = Huitanaã), 07°40'S, 65°46'W

Reserva Ducke, 80 m, 03°08'S, 60°02'W

Reserva Florestal, PDBFF, ca. 70 km N of Manaus, ca. 02°30'S, 65°00'W

Reserva Km 41, Município de Rio Preto da Eva, ca. 02°28'S, 59°43'W

Rio Andira, mouth, 02°45'S, 56°49'W

Rio Aripuana, tributary of lower Rio Madeira, 05°07'S, 60°24'W

Rio Jaú, above mouth, ca. 01°53S, 61°27W

Rio Juruá, ca. 06°45'S, 69°00'W
(A Garbe collecting site on the Rio Juruá [Lima 1926].)

Rio Negro, near Manaus, 03°06'S, 60°00'W

Rio Negro near junction with Rio Branco, 01°24'S, 61°58'W

Rio Piratucu, Boca, 01°59'S, 56°58'W

Rio Xiriviny, 80 m, 00°59'S, 61°53'W

Rosarinho, L. Miguel, Rio Madeira, 03°44'S, 59°06'W

Rosariulla (= Rosarinho), Rio Madeira, 03°44'S, 59°06'W
(An Olalla locality, based on original field sketch map, on left [east] bank of the Rio Madeira near the mouth of stream draining Lago Sampaio.)

Santa Clara, Vila Bela Imperatriz (near Parintins), 02°50'S, 56°55'W

Santa Cruz, Rio Eiru, 07°23'S, 70°47'W

Santarém, 02°26'S, 54°42'W

Santarém, Mojui dos Campos, 02°36'S, 54°43'W

Santo Antônio (= Belo Horizonte), 07°18'S, 69°35'W

Santo Antônio de Amatary, ca. 03°00'S, 58°00'W

Santo Antônio do Guajará, 06°42'S, 69°52'W

Santo Isadoro, Tefé, 03°27'S, 64°47'W

Seringal Condor, Rio Juruá, 06°45'S, 70°51'W

Sitio Isidoro, near Tefé, 04°00'S, 64°30'W

Tahuapunta, Rio Vaupés, at the Colombian border, 03°36'N, 69°12'W
(See Aellen [1970:6] and Paynter and Traylor (1981: 252) for discussions on this locality; other spellings include Tahuapunto and Tauapunto [e.g., Hershkovitz 1987:48], and Tauá.)

Tapaiúna, 03°23'S, 58°16'W

Tapatinga, 04°16'N, 69°56'W

Taracuá (= Taraquá), Rio Negro (= Rio Vaupés), 00°06'N, 68°28'W

Tauá, Rio Uaupes, 03°36'N, 69°12'W

Tefé, ca. 180 km S, 4 km inland from the upper Urucú river, 04°53'S, 65°16'W

Villa Bella Imperatrix (= Parintins; Lago Açu Andirá), 02°36'S, 56°44'W

Bahia (Estado da)

Andaraí, 12°48'S, 41°20'W

Aritaguá, 14°43'S, 39°06'W

Bahia (= Salvador), 12°59'S, 38°31'W

Baía (= Salvador), 12°59'S, 38°31'W

Barra, 12°42'S, 41°33'W

Barra, Rio São Francisco, 12°42'S, 41°33'W

Belmonte, municipality of, 15°52'S, 38°53'W

Boca d'Abu (= Boca do Ubu), near, ca. 15°51'S, 38°56'W

Bom Jesus da Lapa, 13°15'S, 43°25'W

BR–367 (Brazil Highway 367), between Porto Seguro and BR–101, ca. 16°27'S, 39°05'W

Canavieiras, Rio Pardo, 15°39'S, 38°57'W

Canudos, 11°32'S, 45°05'W

Caverna Poço Encantado, 12°57'S, 41°06'W

Cocorobó, 09°53'S, 39°02'W

Comechatiba (= Cumuruxatiba), 17°06'S, 39°11'W

Fazenda Barrinha (= Barrinha), Juazeiro da Bahia, 371 m, 09°25'S, 40°30'W

Fazenda Beijo Grande, 12 km S, 1.1 km E of Itabuna, 14°48'S, 39°16'W

Fazenda Boa Vista, Palmas de Monte Alto, 14°17'S, 43°20'W

Fazenda Formoso, Coribe, 13°45'S, 44°28'W

Fazenda Lajeido, km 147 on Route BA 130, 10°30'S, 40°20'W

Fazenda Rio Pratudão, 14°14'S, 45°56'W

Fazenda São Raimundo, Juàzeiro da Bahia, ca. 09°25'S, 40°30'W

Fazenda Serra do Teimoso, 15°09'S, 39°32'W

Fazenda Sertão do Formoso, 14°38'S, 45°48'W

Feira, 12°15'S, 38°57'W

Feira de Santana, 12°15'S, 38°57'W

Gruta dos Morcegos, near Paulo Afonso, canyon of Rio São Francisco (Paulo Afonso = 09°21'S, 38°14'W)

Helvécia, Nova Viçosa, 17°53'S, 39°22'W

Ilha Madre de Deus, 12°45'S, 38°38'W

Ilhéus, municipality of, 14°47'S, 39°03'W

Ipitinga, 12°54'S, 38°20'W

Irecê, 11°18'S, 41°52'W

Itabuna, 14°48'S, 39°16'W

Itagibá, 14°17'S, 39°51'W

Itaparica, 12°54'S, 38°42'W

Itapebi, municipality of, 15°57'S, 39°32'W

Jaguaquara, 600 m, 13°32'S, 39°58'W

Jequié, 13°51'S, 40°05'W

Juàzeiro, Rio San Francisco, 371 m, 09°25'S, 40°30'W

Lamarão, 11°45'S, 38°55'W

Machado Portela, 13°09'S, 40°45'W

Morro d'Arara, 18°06'S, 39°35'W

Mucurí, 18°05'S, 39°34'W

Nova Viçosa, 17°53'S, 39°22'W

Poliraguá, municipality of, 15°36'S, 39°53'W

Pôrto Seguro, 3 miles W, ca. 16°26'S, 39°08'W

Pôrto Seguro, municipality of, 16°27'S, 39°03'W

Pôrto Seguro, sea level, 16°26'S, 39°05'W

Queimadas, 10°58'S, 39°38'W

Região de Boa Nova, ca. 14°35'S, 40°16'W

Região de Conquista, ca. 15°00'S, 40°50'W

Reserva Biológica de Una, ca. 15°10'S, 39°03'W

Rio Mucurí, 18°05'S, 39°34'W

Rio Prêto, Formosa (= Formosa do Rio Prêto), 500 m, 11°03'S, 45°12'W

Salvador (= Bahia of Gervais [1856a]), 12°59'S, 38°31'W

Samarão (= Lamarão), 300 m, 11°45'S, 38°55'W

São Gonçalo dos Campos, 12°25'S, 38°58'W

São Marcella, junction with Rio Prêto, 11°12'S, 43°48'W.

São Marcelo, Rio Prêto, 11°02'S, 45°32'W

São Marcello, 11°02'S, 45°32'W

Suburbiis Bahiae (= Salvador), 12°59'S, 38°31'W (Identified as "Salvador, Estado da Bahia" by Carvalho [1965:61].)

Una, 15°18'S, 39°04'W

Una, municipality of, 15°18'S, 39°05'W

Urucutuca, Ilheus–Urucutuca–Capueira, 14°39'S, 39°03'W

Vila Nova, 10°27'S, 40°11'W

Villa Viçosa (= Nova Viçosa), Rio Peruhype (= Rio Peruípe), sea level, 17°53'S, 39°22'W (Villa Viçosa = Marobá according to Bokermann [1957: 223].)

Vitória da Conquista, 14°51'S, 40°51'W

Ceará (Estado do)

Aeroporto de Crato, 10 km SW of Crato, 07°17'S, 39°27'W

Baturité, 04°20'S, 38°53'W

Centro de Visitantes, Parque Nacional de Ubajara, ca. 03°48S, 40°52'W

Colegio Agricola de Crato, 5 km W of Crato, 07°14'S, 39°26'W

Crato, 07°14'S, 39°23'W

Faculdade de Veterinária do Ceará, Sítio do Itaperí, Fortaleza, 03°46'S, 38°30'W

Floresta Nacional Araripe-Apodí, 8 km SSW of Crato, 07°18'S, 39°25'W

Floresta Nacional Araripe-Apodi, 8 km S of Crato, 07°19'S, 39°23'W

Floresta Nacional Araripe-Apodi, 9 km S of Crato, 07°20'S, 39°23'W

Floresta Nacional Araripe-Apodí, 21 kms SSW of Crato, 07°24'S, 39°27'W

Fortaleza, sea level, 03°43'S, 38°30'W

Gruta de Ubajara, Parque Nacional de Ubajara, 03°48'S, 40°52'W

Gruta do Tião, Parque Nacional de Ubajara, ca. 03°48'S, 40°52'W

Horto, Parque Nacional de Ubajara, ca. 03°48'S, 40°52'W

Ipu, 04°20'S, 40°42'W

Itapagé, 03°41'S, 39°34'W

Itapipoca, 03°30'S, 39°35'W

Jaguaribe, 05°52'S, 38°39'W

Juá (near Iguatú, 06°22'S, 39°18'W)

Km 19 on Route CE 96, 4 km SE of Nova Olinda, 07°09'S, 40°34'W

Parque Nacional de Ubajara, 03°49'S, 40°53'W

Quixada, 04°58'S, 39°01'W

Reserva Particular do Patrimônio Natural Serra das Almas, ca. 05°10'S, 40°55'W

Russas, 15 m, 04°56'S, 37°58'W

San Antônio (= Santo Antônio), 05°20'S, 40°40'W

Santana do Cariri, 07°11'S, 39°44'W

São Benedito, 895 m, 04°03'S, 40°53'W

S. Paulo (= São Paulo), on the top of Serra da Ibiapaba, 900 m, 04°00'S, 41°00'W

São Paulo, 900 m, 04°00'S, 41°00'W

Sitio Luanda, 4 km S of Crato, 07°17'S, 38°20'W

Distrito Federal

Distrito Federal, 15°45'S, 47°45'W

Estação Ecológica do Jardim Botânico de Brasília, 15°56'S, 47°51'W

Parque Nacional de Brasília, ca. 8 km N of Brasília, 1,100 m, 15°43'S, 47°54'W

Brasília, 40 km N, ca. 15°29'S, 47°53'W

Brasília, 15°47'S, 47°55'W

Brasília, 20 km S, 15°58'S, 47°55'W

Fazenda Agua Limpa, 18 km S of Brasília, 1,000 m, 15°57'S, 47°56'W

Gruta Água Rasa, 860 m, 15°32'S, 47°44'W

Gruta Dança dos Vampiros, 940 m, 15°33'S, 47°45'W

Gruta do Sal, Fazenda Palestina, 805 m, 15°30'S, 48°10'W

Gruta Toca do Falcão, Fazenda Três Lagoas, 800 m, 15°52'S, 48°11'W

Espírito Santo (Estado do)

Castelo, 110 m, 20°36'S, 41°12'W

Colatina, 19°32'S, 40°37'W

Companhia Vale do Rio Dôce Forest Reserve, ca. 19°12'S, 40°06'W

Conceição da Barra, 18°35'S, 39°45'W

Duas Bocas Biological Reserve, 20°16'S, 40°29'W

Engenheiro Reeve (= Rive), 20°46'S, 41°28'W

Estação Experimental de Linhares, 19°25'S, 40°03'W

Fazenda Coroaba, Rio Jucú, Rio Espírito Santo, 20°28'S, 40°30'W

Fazenda do Agá, Rio Iritiba, 20°42'S, 40°45'W

Fazenda do Caboclo, Rio Itaúnas, ca. 18°20'S, 39°50'W

Fazenda Santa Terezinha, 33 km NE of Linhares, 19°08'S, 39°57'W

Gruta do Judeu, Ilha do Francés, 20°54'S, 40°45'W

Gruta do Limoeiro, Castelo, 20°33'S, 41°14'W

Gruta do Rio Itaúnas, Município de Conceiçao da Barra, 18°34'S, 39°46'W

Japaraná (Sant'Anna) (= Juparaná), 19°35'S, 40°18'W

Juparanã, 19°35'S, 40°18'W

Lagoa de Juparanin da Praya (= Lagoa de Juparanã), 19°35'S, 40°18'W

Lagoon of Juaparanã (= Lagoa de Juparanã), 19°35'S, 40°18'W

Linhares, 19°25'S, 40°04'W

Linhares Forest Reserve, ca. 19°12'S, 40°02'W

Mato do Lava d'Agua, 1.7 km by road from Santa Tereza, ca. 19°55'S, 40°36'W

Mucuricí, 90 m, 18°06'S, 40°31'W

Município de Linhares, ca. 19°25'S, 40°03'W

M7, Aracruz Florestal, 16 km ENE of Aracruz, 19°47'S, 40°08'W

Reserva Floresta de Nova Lombardia, ca. 19°50'S, 40°32'W

Rio Dôce, 19°37'S, 39°49'W

Rio Espírito Santo, 20°20'S, 40°20'W

Rio São José, 20°00'S, 40°40'W

Santa Leopoldina, 20°06'S, 40°32'W

Santa Lúcia Biological Station, 19°58'S, 40°32'W

Santa Teresa, 659 m, 19°55'S, 40°36'W

São Pedro de Itabapoana, 21°05'S, 41°29'W

Tres Barras, 20°58'S, 41°26'W

Vitória, 20°19'S, 40°21'W

Goiás (Estado de)

Annapolis (= Anápolis), 1,000 m, 16°20'S, 48°58'W

Anápolis, 1,000 m, 16°20'S, 48°58'W

Aragarças, 15°55'S, 52°15'W

Colinas do Sul, 20 km NW, 14°04'S, 48°12'W

Cristalina, 7 km SE, 16°47'S, 47°33'W

Fazenda Moinho, Município de Niquelândia, ca. 14°29'S, 48°37'W

Formosa, 15°32'S, 47°20'W

Goiânia, 730 m, 16°40'S, 49°16'W

Gruta Morro, Fazenda Cristal, 840 m, 15°27'S, 48°09'W

Inhumas, 16°22'S, 49°30'W

Minaçu, 40 km SW, ca. 13°43'S, 48°13'W

Município de Mambaí, 14°28'S, 46°07'W

Niquelândia, 55 km N, ca. 14°08'S, 48°28'W

São Bento, 18°27'S, 51°47'W

Serra da Mesa, Rio Peixe/Rio Maranhão, ca. 14°20'S, 48°59'W

Serra do Jaraguá, 15°48'S, 49°21'W

Veadeiros, 14°07'S, 47°31'W

Maranhão (Estado do)

Alto Parnahyba (= region along the Alto Rio Parnaíba, centered at 09°06'S, 45°56'W)

Alto Parnaíba (= region along the Alto Rio Parnaíba, centered at 09°06'S, 45°56'W)

Alto Parnaibo (= Alto Parnahyba, region along the Alto Rio Parnaíba, centered at 09°06'S, 45°56'W)

Amarante do Maranhão, Rio Pindaré, 05°36'S, 46°45'W

Anil, sea level, 02°32'S, 44°14'W

Arari, near, 03°28'S, 44°47'W

Barra do Corda, Rio Mearim, 05°30'S, 45°15'W

Caju Island, 02°47'S, 42°05'W

Cajual Island, 02°26'S, 44°03'W

Carolina, 170 m, 07°20'S, 47°28'W

Côcos, 04°35'S, 43°40'W

Humberto de Campos, 02°37'S, 43°27'W

Imperatriz, 100 m, 05°32'S, 47°29'W

Juryassú (= Turiaçu), 01°41'S, 45°21'W

Maranhão City (= São Luís), 02°31S, 44°16'W

Mirador State Park, 06°26'S, 44°48'W

Miritiba (= Humberto de Campos), 02°37'S, 43°27'W

Posto Indígena Awá, Reserva Indígena Caru, upper Rio Turiaçu, ca. 03°054S, 46°35'W

Rosário, 200 m, 01°50'S, 44°50'W

São João dos Patos, 06°30'S, 43°42'W

São Luis, sea level, 02°31'S, 44°16'W

São Miguel, left bank Rio Parnaíba (opposite União, Piauí), 04°39'S, 43°36'W

Tranqueira, ca. 100 m, ca. 09°17'S, 46°48'W

Tury-assú (= Turiaçu), 01°41'S, 45°21'W

Mato Grosso (Estado do)

Araguaiana, 15°43'S, 51°51'W

Aripuanã, 09°10'S, 60°38'W

Aripuanan (= Aripuanã), 09°10'S, 60°38'W

Acorizal, 15°12'S, 56°22'W

Barra do Arica, 15°59'S, 55°54'W

Caicara (= Caiçara), 16°04'S, 57°43'W

Cáceres, 16°04'S, 57°41'W

Chapada (= Chapada dos Guimarães), 800 m, 15°26'S, 55°45'W

Chavantina das Mortes (= Chavantina), 14°40'S, 52°21'W

Chavantina, 14°40'S, 52°21'W

Cuiabá, 15°35'S, 56°05'W

Cuyabá (= Cuiabá), 15°35'S, 56°05'W

Fazenda Acorizal, 15°12'S, 55°22'W

Fazenda São José, ca. 15°50'S, 55°31'W

Fazenda São Luis, 30 km N of Barra do Garças, 15°38'S, 52°21'W

Jaurú (= Rio Jaurú), 16°22'S, 57°46'W

Lava Pés, between 3 and 15 km SE of Cáceres, ca. 13°07'S, 57°37'W

Manso Hydroelectric Dam, 14°52'S, 55°48'W

Mato Grosso, 250 m, 15°00'S, 59°57'W

Reserva Ecológica Cristalino, 40 km N of Alta Floresta, 09°36'S, 55°56'W

Reserva Particular do Patrimônio Natural, ca. 16°30'S, 55°40'W

Salto da Felicidade, 14°41'S, 57°45'W

Salto do Sepotuba, Rio Sepotuba, 15°55'S, 57°37'W

Salto Utiariry, 360 m, 13°02'S, 58°17'W

Santa Anna de Chapada (= Chapada dos Guimarãos), 800 m, 15°26'S, 55°45'W

São Domingos, Rio das Mortes, 13°30'S, 51°23'W

São Marcos Rd., 10 km NE of Urucún, 200 m, 19°05'S, 57°34'W

São Simão, Rio Juruena, 08°13'S, 58°16'W

Serra da Chapada, 870 m, 15°25'S, 55°45'W

Serra do Roncador, 12°00'S, 52°00'W

Serra do Roncador, 264 km (by road) N of Xavantina, 500 m, 12°51'S, 51°46'W

Serra do Roncador, 280 km (by road) N of Xavantina, ca. 12°42'S, 51°44'W

St. Vicente (= São Vicente), 14°30'S, 59°45'W

Suia Missu Road, 284 km N of Xavantina, 12°40'S, 51°46'W

Tapirapoan (= Tapirapuã), 280 m, 14°51'S, 57°45'W

Utiariti, 13°02'S, 58°17'W

Utiarity (= Utiariti), Rio Papagaio, 13°02'S, 58°17'W

Villa María (= Cáceres), 16°04'S, 57°41'W

Mato Grosso do Sul (Estado do)

Alto Sucuriú, afluente da margem direita do Rio Paraná, 19°03'S, 53°04'W

Aporé, ca. 18°57'S, 52°01'W

Campo Grande, 20°27'S, 54°37'W

Corumbá, 19°01'S, 57°39'W

Corumbá, São Marcus Rd., 19°01'S, 57°39'W

Corumbá, 22 km S, ca. 19°00'S, 57°51'W

Coxim, 25 km N, ca. 18°17'S, 54°43'W

Dourados, 12 km N, 22°07'S, 54°47'W

Fazenda Acurizal (= Acurizal), Rio Paraguai, 17°51'S, 57°35'W

Fazenda Cedro, 14 km SW of Dourados, 22°17'S, 54°54'W

Fazenda Santa Terezinha, Pantanal da Nhecolândia, Aquidauana, 19°22'S, 56°03'W

Maracajú, 21°38'S, 55°09'W

Miranda, 20°14'S, 56°22'W

Pantanal, 50 m, 18°00'S, 56°00'W

Rio Ivinheima, 23°14'S, 53°42'W

Rio Pardo, 21°46'S, 52°09'W

Salobra, 125 m, 20°10'S, 56°31'W

São João do Monte Negro, 29°42'S, 51°28'W

São Marcos Rd., 10 km NE of Urucúm, ca. 19°06'S, 57°35'W

Urucúm, 19°09'S, 57°38'W

Xavantina, ca. 21°15'S, 52°47'W

Minas Gerais (Estado de)

Alfenas, 843 m, 21°26'S, 45°57'W

Alpinópolis, 20°52'S, 46°23'W

Barra do Paraopeba, 510 m, 18°50'S, 45°11'W

Brumado, 21°07'S, 44°18'W

Conceição do Mato Dentro, 19°01'S, 43°25'W

Capella-Nova (= Capela Nova), 20°55'S, 43°37'W

Caratinga, 19°47'S, 42°08'W

Caratinga Biological Station, 480 m, 19°50'S, 41°50'W

Fazenda Esmeralda, 30 km E and 4 km N (by road) from Rio Casca, 19°20'S, 42°50'W

Fazenda Monte Claros, ca. 19°00'S, 41°00'W

Frutal, 20°02'S, 48°55'W

Governador Valadares, 18°51'S, 41°56'W

Gruta de Palhares, 19°52'S, 47°22'W

Itinga, 16°36'S, 41°47'W

Januária, 15°29'S, 44°22'W

Jequitínhonha, 16°26'S, 41°00'W

Lagoa Santa, Rio das Velhas, 19°38'S, 43°49'W

Lagoa Santo, 19°38'S, 43°49'W

Marliéria, 13 km E, Parque Estadual do Rio Dôce, ca. 19°43'S, 42°37'W

Mata da Prefeitura, 6 km SW of Viçosa, 20°47'S, 42°55'W

Matias Barbosa, 21°53'S, 43°20'W

Mocambinho, 16°10'S, 43°21'W

Monte Bello, 26°20'S, 46°20'W

Morro Solto, Município de Rio Dôce, 510 m, ca 15°15'S, 43°51'W

Panga Ecological Reserve, 800 m, 19°11'S, 48°25'W

Parque Estadual da Floresta do Rio Dôce, 19°46'S, 42°38'W

Parque Estadual do Rio Dôce ("Campolina"), 19°48'S, 42°28'W

Parque Nacional do Caparaó, Serra do Caparaó, 1,000 m, 20°30'S, 41°40'W

Parque Estadual Rio Dôce, Coronel Fabriciano, 19°31'S, 42°38'W

Passa Quatro, 1,500 m, 22°28'S, 45°05'W

Passos, 728 m, 20°43'S, 46°37'W

Pirapora, Rio São Francisco, 17°21'S, 44°56'W

Pouso Alegre, 22°13'S, 45°56'W

Ribeirão Pedra, 18°05'S, 43°40'W

Rio Dôce State Park, 19°46'S, 42°38'W

Rio Dôce, Parque Estadual Coronel Fabricano, 19°31'S, 42°38'W

Rio Jordão, 18°26'S, 48°06'W

Salto da Água Vermelha, 19°52'S, 50°21'W

São Gonçalo do Rio Preto, 15 km S, Parque Estadual do Rio Preto, ca. 22°02'S, 45°35'W

Serra Caparão, Fazenda Cardoso, 2 leagues NE of Caparão, 1,020 m, 20°31'S, 41°54'W

Serra da Canastra, 20°08'S, 46°39'W

Serra do Cipó, Jaboticatubas, 900 m, 19°16'S, 43°36'W

Sete Lagoas, 19°27'S, 44°14'W

Sete Lagoas, 3 miles ESE, 19°28'S, 44°12'W

Três Marías, 18°10'S, 45°15'W

Uberaba, 19°45'S, 47°55'W

Uberlândia, 18°56'S, 48°18'W

Viçosa, ca. 700 m, 20°45'S, 42°53'W

Pará (Estado do)

Abaetetuba, 01°40'S, 48°48'W

Altamira, 9 km S, ca. 03°18'S, 52°11'W

Altamira, 52 km SSW, east bank Rio Xingu, 03°39'S, 52°22'W

Alter do Chão, right bank Rio Tapajoz, 02°30'S, 54°57'W

Anajá and Muaná, between, Ilha de Marajó, ca. 01°13'S, 49°30'W

Aramanaí, 02°49'S, 54°59'

Aramanay (= Aramanaí), Rio Tapajós, 02°49'S, 54°59'W

Area do Caraipe, ca. 17 km SSW of Tucuruí, west side Rio Tocantins, ca. 03°50'S, 49°30'W

Aruá, 02°39'S, 55°38'W

Baião, 02°41'S, 49°41'W

Baião, Rio Tocantins, 02°44'S, 49°35'W

Barcarena, 01°30'S, 48°40'W

Belém, 14 miles N, 01°16'S, 48°29'W

Belém, 01°27'S, 48°29'W

Belém, Utinga, sea level, 01°27'S, 48°29'W

(Wooded area around waterworks on eastern edge of Belém.)

Belém, Varzea, 01°27'S, 48°29'W

Belterra, 30 km SW of Santarém, 02°38S, 54°57'W

Benevides, 01°22'S, 48°15'W

Boca de Igarapé Piaba, Ilha Marajó, 01°00'S, 50°00'W

Boim, Rio Tapajóz, 50 m, 02°49'S, 55°10'W

Bosque Rodrigues Alves, Belém, 01°27'S, 48°29'W

Bragança, 01°03'S, 46°46'W

Cabo do Maguari, 00°18'S, 48°22'W

Cachimbo, 08°57'S, 54°54'W

Cachoeira (= Cachoeira do Arari; also called Arariuna), Ilha de Marajó, 01°01'S, 48°58'W

Cachoeira da Porteira, Rio Trombetas, 01°05'S, 57°02'W

Cachoeira do Espelho, east bank Rio Xingu, 52 km SSW of Altamira, 03°39'S, 52°22'W

Cachoeira Juruá, Rio Xingu, 03°25'S, 51°53'W

Cachoeira Porteira, mouth of Rio Mapuera, 00°56'S, 57°02'W

Cachoeira Porteira (near Oriximiná, 01°45'S, 55°52'W)

Caldeirão, Ilha Marajó, 00°37'S, 51°03'W

Cametá, Rio Tocantins, 02°15'S, 49°30'W

Canudos, 07°16'S, 58°07'W

Capim, 01°40'S, 47°47'W

Capipi, 07°33'S, 57°42'W

Carajás, Floresta Nacional Tapirapé-Aquiri, 05°48'S, 50°31'W

Caverna do Tatajuba, 17 km S of Altamira, 03°20'S, 52°12'W

Caverna do Valdeci, 9 km by road SE of Altamira, ca. 03°15'S, 52°09'W

Caverna Pedra da Cachoeira, 03°19'S, 52°20'W

Caverna Planaltina, Município Medicilândia, 03°12'S, 52°12'W

Caxiricatuba, Rio Tapajóz, 200 m, 02°50'S, 55°08'W

Faro, 02°11'S, 56°44'W

Fazenda Velho, Belém, 01°27'S, 48°29'W

Faz Tesa, Soure, Ilha de Marajó, 00°44'S, 48°31'W

Flexal, 05°34'S, 57°13'W

Floresta Nacional Tapirapé-Aquiri, 05°48'S, 50°31'W

Fordlândia, Rio Tapajós, 03°40'S, 55°30'W

Gradaús, 07°43'S, 51°11'W

Guama Ecological Research Area, near Belem, 01°27'S, 48°29'W

Igarapé Amorim, Rio Tapajos, ca. 100 m, 02°26'S, 55°00'W

Igarapé Amorin (= Igarapé Amorim), Rio Tapajos, ca. 100 m, 02°26'S, 55°00'W

Igarapé Assu (= Igarapé Açu), 50 m, 01°07'S, 47°37'W

Igarapé Brabo (= Igarapé Amarim), Rio Tapajóz, 02°26'S, 55°00'W

Igarapé Maroi, 02°51'S, 55°03'W

Ilha Caratateua (= Ilha das Barreiras), Belém, 01°15'S, 48°27'W

Ilha das Onças, 10 m, 01°27'S, 48°33'W

Ilha da Urucurituba, Rio Amazonas, mouth of Rio Tapajós, ca. 02°24'S, 54°41'W

Ilha de Marajó, 01°00'S, 49°30'W

Ilha do Taiuna (= Ilha do Tayaúna), Rio Tocantins, 02°15'S, 49°29'W

Ilha Marajó, 01°00'S, 49°30'W

Ilha Mexiana, 00°02'S, 49°35'W

Ilha Tocantins, S end, 04°25'S, 49°32'W

Ipitinga, Rio Acará, 03°42'S, 48°21'W

Jandiaí-Caratateua, 00°59'S, 46°42'W

Jatobal, 73 km N, 45 km W of Marabá, 04°32'S, 49°32'W

Km 217 (5 km by road SW of a point at 04°05'S, 55°06'W), 04°07'S, 55°08'W

Km 446, BR 165 (= Brazil Highway 165), Santarém-Cuiabá, near Itaituba, 05°10'S, 56°00'W

Lago Arary (= Lago Arari), Ilha Marajó, 00°37'S, 49°07'W

Mangabeira, Rio Tocantins, 02°33'S, 49°35'W

Marajó Island, 01°00'S, 49°30'W

Mocajuba, 02°35'S, 49°30'W

Mojuí dos Campos, 22 km S of Santarém, 02°36'S, 54°43'W

Monte Dourado, Rio Jari, 00°53'S, 52°28'W

Mucajatuba, 01°35'S, 48°12'W

Óbidos, 01°55'S, 55°31'W

Pará (= Belém), 01°27'S, 48°29'W

Parque Nacional da Amazônia, 54 km by road S of Itaituba, 03°50'S, 56°15'W

Parque Nacional da Amazônia, km 65 da Rodôvia Itaituba a Jacareacanga, Medio Tapajós, 04°37'S, 56°15'W

Parú Savanna, ca. 01°50'N, 55°30'W

Peixe bois (= Peixe-boi), 01°12'S, 47°18'W

Piquiatuba, 02°34'S, 54°42'W

Pôrto de Moz, Rio Xingu, 01°45'S, 52°14'W

Posto Indígena Kô Kraimôrô, Rio Xingu, ca 07°45'S, 51°12'W

Prainha, Rio Tapajóz, ca. 03°00'S, 55°07'W

Quatipurú, 00°52'S, 46°59'W

Rio Cupari (junction with Rio Tapajós), 03°41'S, 55°25'W

Rio Iriri, E bank, 85 km SW of Altamira, 03°50'S, 52°40'W

Rio Tapajós, 02°24'S, 54°43'W

Rio Xingu, east bank, 52 km SSW of Altamira (= Cachoeira do Espelho), 03°39'S, 52°22'W

Santa Júlia, 03°54'S, 53°21'W

Santa María, 01°03'S, 46°46'W

Santarém, 02°26'S, 54°42'W

S. Antonio do Prata (= Santo Antônio do Prata), 45 m, 01°20'S, 47°37'W

Serra do Tumucumaque, near Triós, 02°18'N, 55°16'W

Serra Norte, 145 km SW of Marabá, 06°00'S, 50°18'W

Soure, Ilha Marajó, 00°44'S, 48°31'W

Tamaru (lagoa), 02°21'S, 53°58'W

Taperinha, 02°32'S, 54°17'W

Tauari, 03°05'S, 55°06'W

Tauary (= Tauari), Rio Tapajós, 03°05'S, 55°06'W

Uruá, Rio Tapajóz, Parque Nacional da Amazônia, 04°37'S, 56°15'W

(Also called Acampamento Uruá, the headquarters of Parque Nacional da Amazônia; = km 65 da Rodôvia Itaituba a Jacareacanga.)

Utinga, 01°27'S, 48°29'W

Vila Braga, 04°25'S, 56°17'W

Villa Braga (= Vila Braga), 04°25'S, 56°17'W

Paraíba (Estado da)

Alagôa do Monteiro, 07°53'S, 37°07'W
Camaratuba, 06°39'S, 35°08'W
Fazenda Espírito Santo, 07°05'S, 36°21'W
João Pessoa, 07°07'S, 34°52'W
Paraíba, 07°07'S, 34°52'W
Penha, 07°10'S, 34°48'W

Paraná (Estado do)

Alto Tibagi, Floresta Nacional do Irati, 890 m, 25°27'S, 50°35'W
Arthur Thomas Municipal Park, Londrina, 23°18'S, 51°09'W
Castro, 980 m, 24°47'S, 50°00'W
Curityba (=Curitiba), 25°25'S, 49°15'W
Estação Ecológica do Caiuá, Rio Paranapanema, 22°39'S, 52°51'W
Fazenda Cagibi, 23°55'S, 51°57'W
Fazenda Guajuvira, 23°53'S, 51°57'W
Fazenda Monte Alegre, 880 m, 24°12'S, 50°33'W
Floresta Nacional do Irati, 890 m, 25°27'S, 50°35'W
Gruta da Lancinha II, Rio Branco do Sul, 25°10'S, 49°18'W
Guaratuba, 25°54'S, 48°34'W
Guaratuba Bay, 25°52'S, 48°39'W
Londrina, 23°18'S, 51°09'W
Mae Catira, 35 km NE of Curitiba, 400 m, 25°25'S, 48°52'W
Morretes, coastal foot of Serra do Mar, 8 m, 25°28'S, 48°49'W
Palmeira, 25°25'S, 50°00'W
Palmeiras, ca 750 m, 24°35'S, 48°57'W
Parque Estadual Mata dos Godoy, 700 m, 23°27'S, 51°15'W
Parque Estadual Vila Rica, 23°55'S, 51°57'W
Piraquara, 25°26'S, 49°04'W
Rio Paracaí, 23°41'S, 53°57'W
Roça Nova, 25°28'S, 48°58'W
Salto Grande, 25°59'S, 52°43'W
Umbará, 25°35'S, 49°17'W

Pernambuco (Estado de)

Açude, 08°02'S, 37°40'W
Bonito, 08°29'S, 35°44'W
Buique, 08°37'S, 37°09'W
Cariris Velhos (= Serra dos Cariris Velhos), 07°30'S, 37°00'W
Caruaru, 08°14'S, 35°55'W
Dois Irmãos, 08°50'S, 35°57'W

Estação Ecológica do Tapacurá, Município de São Lourenço da Mata, 170 m, 08°00'S, 35°03'W
Estação Florestal de Experimentação de Saltinho, 08°45'S, 35°06'W
Exu, 07°31'S, 39°43'W
Exu, 17 km S, 07°41'S, 39°42'W
Escola Agricola de Exu, 0.7 km S of Exu, 07°31'S, 39°43'W
Fazenda Batente, 6 km SE of Exu, 07°33'S, 39°41'W
Fazenda Cantareno, 4.5 km NNE of Exu, 07°29'S, 39°41'W
Fazenda Alto do Ferreira, 5 km SW of Exu, 07°33'S, 39°45'W
Fazenda Maniçoba, 13.7 km SSW of Exu, 07°37'S, 39°47'W
Fazenda Pau Ferrado, 2.6 km E of Exu, 07°31'S, 39°40'W
Fazenda Paus Grandes, 14.2 km E of Exu, 07°31'S, 39°36'W
Fazenda Pomonha, 21 km SSW of Exu, ca. 07°41'S, 39°50'W
Fazenda Saco, 6.6 km NNE of Serra Talhada, 07°56'S, 38°16'W
Garanhuns, 850 m, 08°54'S, 36°29'W
Mato do Camocim, Município de São Lourenço da Mata, 170 m, 07°58'S, 35°02'W
Pedrados dos Pontais, Fazenda Matumbo, Toritama (Toritama = 08°01'S, 36°04'W)
Pernambuco (= Recife), 08°03'S, 34°54'W
Pesqueira, 08°22'S, 36°42'W
Poção, 08°11'S, 36°42'W
Recife, 08°03'S, 34°54'W
Refúgio Ecológico Charles Darwin, 07°49'S, 34°55'W
Rio Formoso, 08°39'S, 35°09'W
Saltinho, Rio Formoso, 08°45'S, 35°06'W
São Lorenço, 08°00'S, 35°03'W
São Lorenço da Mata, 31 m, 08°00'S, 35°03'W
São Lourenço da Mata, 08°00'S, 35°03'W
Serra da Jabitacá, 07°58'S, 37°20'W
Serra de Gritadeira, 18 km SSW of Exu, 07°40'S, 39°47'W
Serra Negra de Floresta, 08°39'S, 38°03'W
Serrote das Lajes, 17 km S of Exu, ca. 07°39'S, 39°42'W
Sete Barras, 24°23'S, 47°55'W

Piauí (Estado do)

Deserto, 03°39'S, 41°36'W
Estação Ecológica de Uruçuí-Una, 08°52'S, 44°57'W
Fazenda Olho da Agua, 2 km N of Valença do Piauí, 06°26'S, 41°45'W
Fazenda Saguarema, Município de São Jorão do Piauí, 08°10'S, 42°15'W
Fronteiras, 530 m, 07°05'S, 40°37'W

Km 18 on Route BR 316 (= 18 km S of Teresina on Brazil Highway 316), 05°14'S, 42°47'W

Município de Teresina, 05°05'S, 42°49'W

Parahyba (= Parnaíba), 02°54'S, 41°46'W

Paranaguá (= Parnaguá), 410 m, 10°13'S, 44°38'W

Parque National de Sete Cidades, Piracuruca, 03°56'S, 41°44'W

Paulistana, 08°09'S, 41°09'W

Piracuruca, Parque National de Sete Cidades, 03°56'S, 41°44'W

São João do Piauí, 08°10'S, 45°15'W

São Paulo, Serra de Ibiapaba, ca. 04°00'S, 41°00'W

Serra da Capivara National Park, ca. 08°30'S, 42°20'W

Serra Ibiapaba, 04°00'S, 41°00'W

Sete Cidades, Piracuruca, 03°56'S, 41°42'W

Rio de Janeiro (Estado do)

Alto Itatiaya (Alto Itatiaia, = Pico das Agulhas Negras), 2,180 m, 22°23'S, 44°38'W

Angra dos Reis, 23°00'S, 44°18'W

Bara Mansa (= Barra Mansa), 22°32'S, 44°11'W

Bonsucesso, Serra dos Orgãos, 1,000 m, 22°12'S, 42°44'W

Campo Bello (= Itatiaia), 22°30'S, 44°34'W

Estação Ecológica Estadual de Paraíso, 80 m, 22°18'S, 42°07'W

Fazenda da Lapa, São João Marcos, 22°54'S, 43°58'W

Fazenda Providência, Rio Paraíba do Sul, 250 m, 21°56'S, 42°37'W

Fazena Reunidas, Morro de São João, Casimiro de Abreu, 22°30'S, 41°59'W

Fazenda von Tapebuçú, 22°26'S, 41°52'W
("Antiga fazenda proximo à lagoa do Iriri no município de Casimiro de Abreu" [Bokermann 1957:214]; "an old farm situated between the towns of the Barra de São João and Macaé, near the coast" [Ávila Pires 1965:7].)

Floresta da Tijuca, 22°56'S, 43°17'W

Ilha Grande, 23°07'S, 44°10'W

Itatiaya (= Pico de Agulhas Negras), 22°23'S, 44°38'W

Km 42 antigua rodovia Rio-São Paulo, Município de Itaguaí, 22°52'S, 43°47'W

Lagoa de Araruama, 22°53'S, 42°12'W

Macaé, 22°23'S, 41°42'W

Mangaratiba, Ilha da Marambaia, 23°04'S, 43°58'W

Maricá, 22°55'S, 42°49'W

Nova Friburgo, 22°16'S, 42°32'W

Paraíso do Tobias, 630 m, 21°15'S, 42°04'W

Parati, 23°13'S, 44°43'W

Parque Arruda Câmara, 12 m, 23°00'S, 43°23'W

Parque Estadual da Pedra Branca, 100 m, ca. 23°58'S, 43°28'W

Paulo de Frontin, 22°32'S, 43°41'W

Petrópolis, 22°31'S, 43°10'W

Pico de Agulhas Negras, 22°23'S, 44°38'W

Piraí, 22°38'S, 43°54'W

Poço das Antas Biological Reserve, 22°35'S, 42°17'W

Praia da Sumaca, 80 m, 23°17'S, 44°32'W

Quinta da Boa Vista, 22°54'S, 43°14'W

Recreio do Mota, 21°47'S, 41°51'W

Reserva Florestal do Grajaú, 22°56'S, 43°16'W

Rio de Janeiro, 22°54'S, 43°14'W

Rio de Janeiro, Estado do, 22°00'S, 42°30'W

Sacra Família do Tinguá, 22°29'S, 43°36'W

Santo Antônio de Pádua, 21°31'S, 42°12'W

Sapitiba (= Sepetiba), 22°58'S, 43°42'W

Sapitiva (= Sepetiba), 22°58'S, 43°42'W

Sepetiba, 22°58'S, 43°42'W

Serra de Macaé, 22°19'S, 42°20'W

Taquara (= Floresta da Tijuca), 22°56'S, 43°17'W

Teresópolis (= Teresópolis), 22°26'S, 42°59'W

Terezópolis (= Teresópolis), 22°26'S, 42°59'W

Theresópolis, 5 miles N, 1,300 m, 22°22'S, 42°58'W

Theresópolis, Organ Mts., 22°26'S, 42°59'W

Therezopolis (= Teresópolis), 22°26'S, 42°59'W

Tinguá, 22°36'S, 43°26'W

Vila do Abraão, Ilha Grande, sea level, 23°07'S, 44°10'W

Rio Grande do Norte (Estado do)

Natal, 05°47'S, 35°13'W

Rio Grande do Sul (Estado do)

Barma Farm, Brasilândia, 21°35'S, 52°07'W

Caçapava do Sul, 30°30'S, 53°30'W

Dom Pedro de Alcântara, 29°18'S, 49°44'W

Guahyba Island (= Ilha Guaíba), near Pôrto Alegre, ca. 30°06'S, 51°16'W

Guarita, 27°22'S, 53°45'W

Livramento (= Santana do Livramento), 30°53'S, 55°31'W

Maquiné, 29°41'S, 50°11'W

Montenegro, 29°42'S, 51°28'W

Mundo Novo (= Três Coroas), 29°32'S, 50°48'W

Município de Garruchos, 28°11'S, 55°38'W

Município do Livramento (Santana do Livramento = 30°53'S, 55°31'W)

Parque Turvo, Município de Tenente Portela, 27°16'S, 53°51'W

Paso del Duraznero, Município de Bagé, 200 m, 31°54'S, 54°06'W

Passo Fundo, 28°15'S, 52°24'W

Pôrto Alegre, 30°04'S, 51°11'W

Quinta, 32°05'S, 52°17'W

Restinga Seca, 36 m, 29°49'S, 53°23'W

Rio Cahy (= Rio Caí), 29°56'S, 51°16'W

Rio Grande, 32°02'S, 52°05'W

Rolante, left bank Rio Riozinho, ca. 6 km from the sede de Município de Rolante, 29°45'S, 50°30'W

San Lorenzo, 31°22'S, 51°58'W

Santa Maria, 29°41'S, 53°48'W

Santo Cristo, 27°50'S, 54°40'W

São João do Monte Negro (= Montenegro), 29°42'S, 51°28'W

São Laurenço do Sul, 31°22'S, 51°58'W

São Lourenço, sea level, 31°22'S, 51°58'W

São Lourenco do Sul (= São Lourenço), 31°22'S, 51°58'W

Saparinga, 29°40'S, 51°00'W

Taquara, 29°39'S, 50°47'W

Taquara do Mundo Novo, 29°39'S, 50°47'W

Tôrres, 29°21'S, 49°44'W

Unidade Maquiné, 29°40'S, 50°13'W

Viamão, 125 m, 30°05'S, 51°02'W

Vila Oliva, 29°14'S, 50°53'W

Rondônia (Estado de)

Bananeira, Rio Mamoré, 10°39'S, 65°23'W

BR 364 (Brazil Highway 364), near Ariquemes, 09°19'S, 63°08'W

Cachoeira do Pau grande am Mamoré, 10°28'S, 65°24'W

Cachoeira Nazaré, Rio Ji-Parana, 100 m, 09°45'S, 61°55'W

Calama, Rio Madeira (opposite the mouth of Rio Gy-Paraná (= Rio Ji-Paraná), 50 m, 08°03'S, 62°53'W

Jaburi farm, 280 m, ca. 11°36'S, 60°43'W

Ji-Paraná (= Rondônia), 200 m, 10°52'S, 61°57'W ("Current name, contra USBGN, is doubtless 'Ji-Paraná'" [Paynter and Traylor 1991:528].)

Km 575 on BR-364 (Brazil Highway 364), 12°31'S, 60°26'W

Ouro Preto D'Oeste, 10°30'S, 62°28'W

Pôrto Velho, 08°46'S, 63°54'W

Pôrto Velho, 20 km SW, on road to Cachoeira Teotônio, ca. 08°52'S, 64°03'W

Pyrineus, 10°30'S, 61°46'W

Rio Jaru, upper, ca. 10°56'S, 63°04'W

São Antônio de Guaporé, 12°32'S, 63°34'W

Santo Antônio do Uayará (= Aroaya), Rio Madeira (= Santo Antônio do Rio Madeira), 08°48'S, 64°56'W

Roraima (Território de)

Boa Vista, Rio Branco, 02°49'N, 60°40'W

Ilha de Maracá, 03°25'N, 61°40'W

Lucetania Ranch (Webster and Handley 1986), not located

Maracá Biological Station, ca. 03°21'N, 61°24'W

Monte Roraima, 05°12'N, 60°44'W

Paulo, Mount Roraima, 2,772 m (at border of Brazil, Venezuela, and Guyana), 05°12'N, 60°44'W

Serra da Lua, near Boa Vista, Rio Branco, 02°15'N, 60°45'W

Santa Catarina (Estado de)

Araranguá, 28°56'S, 49°29'W

Blumenau, 26°56'S, 49°03'W

Catuíra, 400 m, 27°40'S, 49°24'W

Colonia Hansa, (= Corupá), near Joinville, 26°26'S, 49°14'W

Concordia, 500 m, 27°14'S, 52°01'W

Fazenda Palmital, Reserva Natural Volta Velha, 26°04'S, 48°37'W

Florianópolis, sea level, 27°35'S, 48°34'W

Hansa (= Corupá), 62 m, 26°26'S, 49°14'W

Humboldt (= Colonia Hansa), 62 m, 26°26'S, 49°14'W

Ilha Santa Catarina, 27°36'S, 48°30'W

Joinvile, 4 m, 26°18'S, 48°50'W

Joinville (= Joinvile), 4 m, 26°18'S, 48°50'W

Nova Veneza, 28°39'S, 49°30'W

Parque Municipal da Lagoa do Peri, 27°43'S, 48°32'W

Pôrto Belo, sea level, 27°10'S, 48°33'W

Rio Novo do Itapocú, 26°34'S, 48°40'W

Santo Amaro da Imperatriz, 27°41'S, 48°46'W

São Francisco, 26°14'S, 48°39'W

São Paulo (Estado de)

Adamantina, ca. 21°42'S, 51°05'W

Araraquara, 850 m, 21°05'S, 47°19'W

Avanhandava, Rio Tieté, 21°28'S, 49°57'W

Barra, Rio Juquiá, 23°45'S, 47°48'W

Baurú, 22°19'S, 49°04'W

Biritiba Mirim, Rio Paraíba valley, 23°35'S, 46°02'W

Boracéia, 22°39'S, 45°54'W

Botucatu region, 22°52'S, 48°26'W

Caetetus Ecological Station (formerly Fazenda Paraíso), ca. 22°25'S, 49°42'W

Campinas, 22°54'S, 47°05'W

Cananéia, 25°01'S, 47°57'W

Cantareira, 21°51'S, 47°29'W

Caraguatatuba, sea level, 22°37'S, 45°25'W

Casa Grande, 23°58'S, 46°16'W

Cotia, 23°37'S, 46°56'W

Cruzeiro, 20°42'S, 47°20'W

Eldorado Paulista, sea level, 24°32'S, 48°06'W

Emas, 21°56'S, 47°22'W

Engenheiro Schmidt, 20°52'S, 49°18'W

Estação Biológica de Boracéia, 23°39'S, 45°54'W

Estação Ecológica dos Caetetus (formerly Fazenda Paraíso), ca. 22°25′S, 49°42′W

Fazenda da Toca, Ilha de São Sebastião, 150 m, 23°50′S, 45°18′W

Fazenda Esplanada, Sales, 21°19′S, 49°30′W

Fazenda Intervales, Serra de Paranapiacaba, 24°20′S, 48°25′W

Fazenda Paraguassu, 22°27′S, 47°32′W

Fazenda Poço Grande, 24°42′S, 47°33′W

Furnas do Yporanga (= Iporanga), 24°35′S, 48°35′W

Galia, Caetutus Ecological Station, 22°23′S, 49°40′W

Grota de Mirassol, 20°46′S, 49°28′W

Gruta de Mirossol, 20°46′S, 49°28′W

Gruta São José do Ribeira, Pedro Cubas, ca. 24°32′S, 48°18′W

Guaratuba (= Varjão do Guaratuba), ca. 23°45′S, 45°55′W

Guaratuba (Varjão), ca. 23°45′S, 45°55′W

Guariba, 21°21′S, 48°14′W

Ibití (= Monte Alegre do Sul), 22°40′S, 46°41′W

Ignape (= Iguapé), 24°43′S, 47°33′W

Iguapé, sea level, 24°43′S, 47°33′W

Ilha de São Sebastião, 23°50′S, 45°18′W

Ilha Victoria, 23°45′S, 45°01′W

Ipanema (= Bacaetava), 23°26′S, 47°36′W

Ipiranga, 21°48′S, 47°42′W

Iporanga, 24°35′S, 48°35′W

Irapuã, 21°18′S, 49°24′W

Itapetininga, 23°36′S, 48°03′W

Itararé, 24°07′S, 49°20′W

Itú, 23°17′S, 47°18′W

Ituverava, 635 m, 20°20′S, 47°47′W

Juquiá, 24°18′S, 47°33′W

Lins, 21°40′S, 49°45′W

Mirassol, Sitio Progresso, 20°47′S, 49°32′W

Mogi das Cruzes, 23°31′S, 46°11′W

Nova Aliancia, 21°02′S, 49°30′W

Olimpia, 20°44′S, 48°54′W

Panga Ecological Reserve, 800 m, 19°10′S, 48°23′W

Paranapiacaba, 23°47′S, 46°19′W

Parque Estadual do Ilha do Cardoso, 25°08′S, 47°58′W

Parque Estadual do Morro do Diabo, 22°30′S, 52°20′W

Parque Florestal do Itapetinga, 800–1,350 m, 23°10′S, 46°25′W

Peruíbe, 24°19′S, 47°00′W

Piquete, 900 m, 22°36′S, 45°11′W

Piracicaba, 22°43′S, 47°38′W

Pirituba, 23°38′S, 47°21′W

Pontal do Paranapanema, Parque Estadual Morro do Diabo, Município de Teodoro Sampaio, 22°30′S, 52°20′W

Primeiro Morro, 24°22′S, 47°50′W

Quadro Pentrado (= Quadro Penteado), 50 m, ca. 24°23′S, 48°12′W

(Quadro Penteado is "31 km W of Sete Barras," Paynter and Traylor 1991:502.)

Registro, Gruta do Diablo, 24°30′S, 47°50′W

Ribeirão da Lagoa, 23°41′S, 45°56′W

Ribeirão Prêto, 21°10′S, 47°48′W

Rio Juquiá, 24°22S, 47°49′W

Rocha, ca. 40 km NNW of Juquiá, ca. 50 m, 24°00′S, 47°46′W

Salesópolis, 23°38′S, 45°52′W

Salto Grande, 368 m, 22°54′S, 49°59′W

San Sebastian (= São Sebastião), sea level, 23°48′S, 45°25′W

San Sebastião (= São Sebastião), sea level, 23°48′S, 45°25′W

Santa Gertrudes, 22°28′S, 47°31′W

Santa Virgínia Nucleous, Parque Estadual da Serra do Mar, 900 m, ca. 23°20′S, 45°23′W

São Bernardo, 23°42′S, 46°33′W

São Carlos, 840 m, 21°58′S, 47°52′W

São José do Rio Prêto, 20°48′S, 49°23′W

São Paulo, 23°32′S, 46°37′W

São Sebastião, 23°48′S, 45°25′W

Serrinha, Serra do Bananal, 29°09′S, 50°47′W

Sorocaba, 23°29′S, 47°27′W

Tanabí, 20°37′S, 49°37′W

Varjão, 5 km S of Casa Grande, ca. 23°45′S, 45°55′W

Victoria, 570 m, 22°47′S, 48°24′W

Ypanema (= Bacaetava), 950 m, 23°26′S, 47°36′W

Ypiranga (= Ipiranga), 23°36′S, 46°37′W

Sergipe (Estado do)

Brejo Grande, 10°26′S, 36°28′W

Crasto, 10 m, 11°15′S, 37°25′W

Estação Ecológica Serra de Itabaiana, 10°40′S, 37°25′W

Tocantins (Estado do)

Araguatins, 05°38′S, 48°08′W

Cana Brava, 12°54′S, 46°46′W

Município Aliança do Tocantins, 11°08′S, 48°48′W

Município Babaçulândia, 07°03′S, 47°36′W

Município Palmeirante, 07°52′S, 47°56′W

Município Sucupira, 11°58′S, 48°50′W

Palma, (= Paraná), 12°33′S, 47°52′W

Paraíso do Tocantins, 10°05′S, 48°25′W

Rio Novo, Estação Ecológica Serra Geral do Tocantins, 10°33′S, 46°45′W

Chile

Aisén (Región de)

Chile Chico, 46°33′S, 71°44′W

Antofagasta (Región de)

San Pedro de Atacama, 22°55′S, 68°13′W

Araucania (Region de la)

Curacautín, 300 m, 38°26′S, 71°53′W
Paso Pino Hachado, 15.2 km W, 1,460 m, 38°39′S, 70°54′W
Purén, 15 km E, 61 m, 38°01′S, 73°05′W
Temuco, ca. 200 m, 38°44′S, 72°36′W

Atacama (Región de)

Chañaral, 29°00′S, 71°25′W
Domeyko, 28°57′S, 70°54′W
Mina Altamira, 26°27′S, 70°19′W

Biobío (Region de)

Antuco, 650 m, 37°20′S, 71°41′W
Chaitén, 42°55′S, 72°43′W
Concepción, 36°50′S, 73°03′W
Huépil, 37°14′S, 71°56′W
Lota, 37°05′S, 73°10′W
Río Itata, 36°23′S, 72°52′W
San Fabián de Alico, 36°33′S, 71°33′W
San Carlos, Hacienda Zemita, 400 m, 36°25′S, 71°58′W

Coquimbo (Región de)

Aucó, Las Chinchillas National Reserve, 31°30′S, 71°06′W
Coquimbo, 29°58′S, 71°21′W
Illapel, 310 m, 31°38′S, 71°10′W
Illapel, 7.5 km E, 31°39′S, 71°08′W
Los Vilos, 4 km S, 31°57′S, 71°31′W
Paiguano, 1,000 m, 30°01′S, 70°32′W
Paihuano (= Paiguano), 30°01′S, 70°32′W

Los Lagos (Región de)

Cucao, Isla de Chiloe, sea level, 42°38′S, 74°07′W
Huite (= Puerto Huite), Isla Chiloé, 42°07′S, 73°27′W
Máfil, 39°39′S, 72°57′W
Maicolpue, 67 km W of Osorno, 50–110 m, 40°33′S, 73°46′W
Río Inio, mouth, Isla Chiloé, 43°21′S, 74°07′W
Parque Nacional Puyehue, 40°39′S, 72°19′W
Peulla, 41°06′S, 72°02′W
Puerto Carmen, Isla Chiloé, 43°08′S, 73°43′W

Puerto Montt, 41°28′S, 72°57′W
Quellón, 15 km N, near mouth of Río Yaldad, 20 m, 43°08′S, 73°44′W
Rinihue, ca. 100 m, 39°49′S, 72°27′W
Río Inio, Isla Chiloé, 43°21′S, 74°07′W
Seno Reloncaví, 40 km S, ca. 42°00′S, 72°45′W
Valdivia, 39°48′S, 73°14′W
Valle de La Picada, 425–1,135 m, 41°02′S, 72°30′W

Magallanes (Región de)

Dazy Harbour, sea level, 52°41′S, 70°34′W
Lago Paine, Parque Nacional Torres de Paine, 50°51′S, 72°57′W
Mina Pecket, 52°53′S, 71°06′W
Puerto Pescado, Isla de Navarino, 55°05′S, 67°40′W
Puerto Prat, Last Hope Inlet (= Bahía Hope), 5 m, 51°38′S, 72°38′W
Puerto Toro, Isla de Navarino, 55°05′S, 67°06′W
San Gregorio, 52°37′S, 70°08′W
Straits of Magellan, Tierra del Fuego, 54°00′S, 71°00′W

Maule (Región de)

Río Maule, junction of Río Claro, ca. 1,000 m, 35°43′S, 71°05′W
Los Queules National Reserve, 540–750 m, 35°59′S, 72°41′W

Santiago (Región Metropolitana de)

Las Condes, 700 m, 33°22′S, 70°31′W
Puente Alto, 700 m, 33°37′S, 70°35′W
Río Colorado, 30 km E of Guayacán, ca. 33°29′S, 70°03′W
Santiago, 33°27′S, 70°40′W

Tarapacá (Región de)

Arica, 18°29′S, 70°20′W
Belén, 1 km W, 18°29′S, 69°33′W
Caleta Cuya (= Caleta Camarones), 19°12′S, 70°17′W
Canchones, ca. 1,000 m, 20°27′S, 69°37′W
Cuya, 19°07′S, 70°08′W
Esquiña, Camarones Valley, 18°58′S, 69°32′W
Guanillos (Punta de), sea level, 21°12′S, 70°06′W
Iquique, 20°13′S, 70°10′W
Miñimiñi, 1,800 m, 19°11′S, 69°41′W
Parinacota, 4,390 m, 18°12′S, 69°16′W
Putre, 3,500 m, 18°12′S, 69°34′W
Valle de Azapa, 100 m, 18°30′S, 70°15′W
Zapahuira, 7 km SE of Socorama, 3,100 m, 18°16′S, 69°35′W

Valparaíso (Región de)

Curaumilla, 33°06′S, 71°45′W

Papudo, 32°31′S, 71°27′W

Río Colorado, 1,100 m, 32°52′S, 70°25′W

Valparaíso, sea level, 33°02′S, 71°38′W

Zapallar, sea level, 32°33′S, 71°28′W

Colombia

Amazonas (Comisaría de)

Caserío Araracuara, 00°24′S, 72°17′W

Isla Santa Sofia, 3 miles W, ca. 04°00′S, 70°17′W

Isla Santa Sofia, 30 km NW of Leticia, 03°57′S, 70°13′W

Leticia, 7 km N, 04°06′S, 69°57′W

Leticia, ca. 100 m, 04°09′S, 69°57′W

Leticia, 35-40 km SW, on road to Calderón, ca. 04°10′S, 69°59′W

Puerto Patiño, 03°49′S, 70°26′W

Puerto Rastrojo, Río Mirití Paraná, 150 m, 01°11′S, 70°02′W

Río Amacayacu (= Río Amaca-Yacú), 100 m, 03°47′S, 70°21′W

Tarapacá, 02°52′S, 69°44′W

Antioquia (Departamento de)

Aljibes, 26 km S and 22 km W of Zaragoza, 630 m, 07°16′N, 75°04′W

Alto Bonito, upper Río Sucio, 400 m, 07°01′N, 76°17′W

Amalfi, La Vetilla, 1,500 m, 06°55′N, 75°04′W

Anorí, 1,535 m, 07°05′N, 75°08′W

Antioquia, 700 m, 06°33′N, 75°50′W

Barbosa, 06°26′N, 75°20′W

Betania, 06°58′N, 75°24′W

Buenos Aires, 24 km S, and 22 km W of Zaragoza, 450 m, ca. 07°18′N, 75°05′W

Cáceres, Río Cauca, 75 m, 07°35′N, 75°20′W

Caicedo, Páramo Frontino, 1,800, m, 06°25′N, 76°00′W

Cavernas del Nus, 06°25′N, 74°03′W

Chigorodó, ca. 100 m, 07°41′N, 76°42′W

Concordia, 2,020 m, 06°03′N, 75°55′W

Dabeiba, Río Sucio, 600 m, 07°00′N, 76°16′W

El Doce, 200 m, 07°29′N, 75°15′W

El Porvenir, 06°03′N, 75°42′W

Finca Los Sauces, 05°40′N, 75°07′W

Guapá, 07°34′N, 76°38′W

Jericó, 1,967 m, 05°47′N, 75°47′W

La Bodega, 1,750 m, 05°52′N, 75°07′W

La Candelaria, 120 m, 08°01′N, 75°13′W

La Ceja, 06°02′N, 75°26′W

La Popala, 450 m, 05°58′N, 75°44′W

La Tebaida, 1,000 m, 06°03′N, 75°00′W

La Tirana, 25 km S, 27 km W of Zaragoza (33 km SW of Zaragoza), 520 m, 07°18′N, 75°05′W

Los Ríos, 06°04′N, 75°11′W

Medellín, 1,500 m, 06°15′N, 75°35′W

Paramillo, 06°59′N, 75°51′W

Páramo de Sonsón, 7 km E of Sonsón, 3,100 m, 05°42′N, 75°14′W

Páramo Frontino, 06°28′N, 76°04′W

Parque Natural El Refugio, Vereda Río Claro, 05°54′N, 74°40′W

Poblado, 5 km S of Medellín, 1,600 m, 06°13′N, 75°34′W

Providencia, 23 km S and 22 km W of Zaragoza, 07°16′N, 75°04′W

Puerto Nare, La Sierra–Cavernas del Nus, 400 m, 06°25′N, 74°33′W

Puerto Triunfo, Río Claro, 250 m, 05°54′N, 74°39′W

Puerto Valdivia, ca. 180 m, 07°18′N, 75°23′W

Puri, ca. 170 m, 07°25′N, 75°20′W

Río Negro (= Rionegro), 2,120 m, 06°09′N, 75°22′W

San Francisco, ca. 2,500 m, 06°28′N, 75°25′W

San Pedro, 30 km N of Medellín, 2,450 m, 06°28′N, 75°33′W

Santa Barbara, Río Urrao, 2,700 m, 06°25′N, 76°15′W

Santa Elena, 2,620–2,750 m, 06°13′N, 75°30′W

Santa Teresa, near Mutata, 07°15′N, 76°20′W

Sonsón, 05°42′N, 75°18′W

Turbo, 08°06′N, 76°43′W

Urabá, 130 m, 07°20′N, 76°26W

Urrao, Río Ana, 06°20′N, 76°11′W

Valdivia, 07°11′N, 75°27′W

Valdivia, 9 km S, 1,200 m, 07°07′N, 75°27′W

Ventanas, ca. 07°05′N, 75°27′W

Vereda Las Confusas, right margin of quebrada La Cristalina, 25 km NW of municipio San Luis, 650 m, 06°03′N, 75°45′W

Vereda La Soledad, Río Verde de los Montes, 15 km E of municipio Sonsón, 1,200 m, 05°40′N, 75°05′W

Vereda Pajarito, 06°45′N, 75°55′W

Vereda Río Claro, ca. 05°54′N, 74°39′W

Vereda San Antonio del Prado, 06°11′N, 75°40′W

Villa Arteaga, 07°20′N, 76°26′W

Villa Marina, 1,750 m, 06°10′N, 75°39′W

Zaragoza, 100 m, 07°30′N, 74°52′W

Zaragoza, 24 km S and 22 km W, near La Tirana, 520 m, ca. 07°18′N, 75°05′W

Arauca (Intendencia de)

Arauca, Intendencia de, 06°40′N, 71°00′W

Puerto Gaitán, 06°15′N, 71°27′W

Río Arauca, ca. 07°02′N 72°00′W

Atlántico (Departamento de)

Barranquilla, 100 m, 11°00′N, 74°48′W
La Peña, 10°35′N, 75°02′W
Ponedera, 8 m, 10°38′N, 74°45′W

Bolívar (Departamento de)

Cartagena, 10°25′N, 75°32′W
Norosí, 120 m, 08°32′N, 74°02′W
Old fortifications at Cartagena, 10°25′N, 75°32′W
Río San Pedro, 08°38′N, 74°04′W
Sampués, 25 km S of Sincelejo, 170 m, 09°11′N, 75°23′W
San Martín de Loba, 20 m, 09°03′N, 74°08′W

Boyacá (Departamento de)

Guaicaramo, 04°43′N, 73°02′W
Hacienda La Primavera; 07°00′N, 72°20′W
Miraflores, 5°12′N, 73°12′W
Muzo, 1,300 m, 05°30′N, 74°10′W
Orocué, 04°48′N, 71°20′W
Río Covaría, near mouth, 07°03′N, 72°04′W
Trinidad, 200 m, 05°25′N, 71°40′W
Tunja, 2826 m, 05°31′N, 73°22′W

Caldas (Departamento de)

Cenicafé Reserva Natural Planalto, 1,400 m, 05°00′N, 75°36′W
Charca de Guarinocito, 200 m, 05°30′N, 75°05′W
Finca Los Naranjos, Vereda Kilómetro 41, 930 m, 05°07′N, 75°40′W
Hacienda Riomanso, 240 m, near 05°34′N, 74°53′W (precise locality not found)
La Dorada, 05°27′N, 74°40′W
Pennsylvania (= Pensilvania), 2,100 m, 05°23′N, 75°10′W
Quebrada Guayabal, Recinto del Pensamiento Barrio Maltería, 3,020 m, 05°02′N, 75°26′W
Samaná, Río Honda, 2,500 m, 05°15′N, 75°15′W
Vereda Quebrada Negra, ca. 05°23′N, 75°06′W

Caquetá (Intendencia del)

Estación Puerto Abeja, Río Mesay, 240 m, 00°04′N, 72°26′W
Florencia, Río Bodoquera, 01°36′N, 75°36′W
Muralla, 01°31′N, 75°41′W
Raudal el Tubo, Río Cuñaré, 00°28′N, 72°33′W
Tres Esquinas, 190 m, 00°43′N, 75°16′W
Tres Troncos, 00°08′N, 74°41′W
Villaraz, 1,200 m, 01°37′N, 75°40′W

Casanare (Intendencia del)

Pore, 05°43′N, 72°00′W

Cauca (Departamento del)

Almaguer, 01°55′N, 76°50′W
Bellavista, 02°56′N, 76°39′W
Betania, 02°41′N, 76°44′W
Boca del Río Saijá, 02°52′N, 77°41′W
Cerro Munchique, 02°32′N, 76°57′W
Cocal, 150 m, 02°31′N, 77°00′W
El Bordo, ca. 1,000 m, 02°07′N, 76°59′W
El Papayo, Río Saija, ca. 02°48′N, 77°26′W
El Tambo, 1,700 m, 02°27′N, 76°49′W
Gorgona Island, sea level, 03°00′N, 78°12′W
Isla Gorgona, sea level, 03°00′N, 78°12′W
La Costa, 900 m, (not located; near El Tambo, 02°25′N, 76°50′W)
Mazamorrero, 03°00′N, 76°35′W
Mina California, 2,400 m, 02°25′N, 76°49′W
    (Same coordinates as those for El Tambo; Marinkelle and Cadena [1972].)
Moscopas (= Moscopán?), 1 mile N, ca. 02°16′S, 76°10′W
Munchique (= Cerro Munchique), near El Tambo, 02°32′N, 76°59′W
Páramo de Puracé, 1 km E of Laguna San Raphael, 3,000 m, 02°25′N, 76°25′W
Páramo de Santo Domingo, Belalcázar (Páez), km 44, 02°30′N, 76°35′W
Popayán, 65 km W, 3,230 m, 02°50′N, 77°00′W
Popayán, 1,760 m, 02°27′N, 76°36′W
Río Cauca, 08°54′N, 74°28′W
Río Guachicono, ca. 02°12′N, 76°50′W
Río Palo, 22 km S of Puerto Tejada, 1,000 m, 03°02′N, 76°22′W
Quisquio (= Chisquío), 1,700 m, 02°29′N, 76°52′W
    ("A small town 6 km northwest of El Tambo" [Tamsitt and Valdivieso 1966b:97].)
Tambito Nature Preserve, 02°30′N, 77°00′W
Watering Bay, Gorgona Island, sea level, 02°59′N, 78°12′W

Chocó (Departamento de El)

Andagada (= Río Andagueda), 05°25′N, 76°24′W
    (Coordinates for Bagado, one of the localities on the Río Andagueda where Kerr collected.)
Andagoya, 65 m, 05°06′N, 76°41′W
Andaguya (=Andagoya), Río Condoto, 65 m, 05°06′N, 76°41′W
Arusi, 30 m, 05°32′N, 77°31′W
Bagadó (= Andagada of J. A. Allen [1916c]), Río Andagueda, 05°25′N, 76°24′W

Bahía Solano, 06°14′N, 77°25′W

Boquerón, 06°43′N, 77°30′W

Condoto, 90 m, 05°06′N, 76°37′W

Ensenada de Utría, 06°02′N, 77°21′W

Istmina, 65 m, 05°09′N, 76°40′W

La Italia, 4 km N, 35 m, 04°57′N, 76°25′W

Noanamá, 53 m, 04°42′N, 76°56′W

Nóvita, 70 m, 04°57′N, 76°34′W

Parque Nacional Natural Los Katíos, sea level, 07°50′N, 77°10′W

Playa de Oro, Río San Juan, 210 m, 05°19′N, 76°24′W

Playa de Oro, 2 km above, Río San Juan, 210 m, 05°19′N, 76°22′W

Quibdó, 54 m, 05°42′N, 76°40′W

Quebrada El Platinero, 12 km W of Istmina (by road), 100 m, 05°00′N, 76°45′W

Río Atrato, 08°17′N, 76°58′W

Río Bando, Río Sandó, 04°47′N, 77°22′W

Río Curiche, 07°02′N, 77°40′W

Rio Patio, near, approximately 15 km S and 30 km W of Quibdó, ca 200 m, 05°35′N, 76°55′W

Río Sandó, 160 m, 05°03′N, 76°57′W

Sautata, 10 m, 07°50′N, 77°04′W

Sipí, 45 m, 04°39′N, 76°36′W

Unguía, sea level, 08°01′N, 77°04′W

Córdoba (Departamento de)

Catival, upper Río San Jorge, 120 m, 08°17′N, 75°41′W

El Contento, 09°08′N, 75°32′W

Jaraquiel, 20 m, 08°42′N, 75°57′W

Río Ure, 07°50′N, 75°30′W

Sinú (= Río Sinú?), ca. 09°24′N, 75°49′W

Socorré, Río Sinú, 110 m, 07°51′N, 76°17′W

Cundinamarca (Departamento de)

Bogotá, 2,590 m, 04°36′N, 74°05′W

Bogotá, El Verjón (= Páramo del Verjón), 2,600 m, 04°32′N, 74°04′W

Bogotá, plains of, 04°36′N, 74°05′W

Bogotá, Santa Isabel, 04°36′N, 74°05′W

Boquerón, San Francisco, 3,000 m, 04°16′N, 74°33′W

Choachí, 1,966 m, 04°32′N, 73°56′W

Fusagusugá, near Bogotá (40 km SW of), 1,750 m, 04°21′N, 74°22′W

Girardot, ca. 300 m, 04°18′N, 74°49′W

Guasca, 04°40′N, 73°33′W

Hatogrande, 13 km E of Gachalá, 2,735 m, 04°41′N, 73°24′W

La Mesa, 1,320 m, 04°38′N, 74°28′W

La Reserva Biológica Carpanta, 04°34′N, 73°41′W

La Selva, near Bogotá (= Estancia La Selva), 04°36′N, 74°05′W

Medina, 576 m, 04°30′N, 73°21′W

Mesitas del Colegio, 04°35′N, 74°27′W

Pacho, 1,860 m, 05°08′N, 74°10′W

Paime, 05°22′N, 74°10′W

Páramo de Chisacá, 3,100 m, 04°30′N, 74°20′W

Represa de Neusa, 05°10′N, 73°57′W

San Antonio de Tena, 04°37′N, 74°21′W

San Juan de Río Seco, 04°51′N, 74°38W

San Juan de Ríoseco, 04°51′N, 74°38W

Santandercito, 04°36′N, 74°21′N

Sasaima, 1,221 m, 04°58′N, 74°26′W

Sibaté, 2,613 m, 04°30′N, 74°16′W

Suba, 2,583 m, 04°44′N, 74°05′W

Susumuco, 04°12′N, 73°44′W

Villeta, 842 m, 05°01′N, 74°28′W

El Cesar (Departamento de)

Caracolicito, 100 m, 10°12′N, 73°58′W

Colonia Agrícola de Caracolicito, 100 m, 10°12′N, 73°58′W

El Orinoco, 10°10′N, 73°24′W

El Paso, 698 m, 09°40′N, 73°45′W

La Palma, 07°45′N, 73°30′W

Pueblo Bello, 10°25′N, 73°35′W

Pueblo Viejo, 2,440 m, 10°32′N, 73°25′W

Río Guaimaral, (Municipio de) Valledupar, 128 m, 10°05′N, 73°32′W

San Sebastián (= San Sebastián de Rábago), Sierra Nevada de Santa Marta, 2,040 m, 10°34′N, 73°36′W

Valledupar, Sierra Negra, 10°29′N, 73°15′W

Guainía (Comisaría del)

Puerto Inírida, 100 m, 03°51′N, 67°55′W

Raudal Mabajate, Río Negro, 01°59′N, 67°07′W

Guaviare (Comisaría del)

Salto Angostura II, Río Guayabero, 200 m, 02°35′N, 72°45′W

San José del Guaviare, 02°35′N, 72°38′W

Huila (Departamento del)

Acevedo, 1,700 m, 01°49′N, 75°52′W

Aguas Claras, Río Suazo, ca. 01°37′N, 75°59′W

Cueva del Indio, 01°36′N, 75°56′W

El Parque Nacional Natural de la Cueva de los Guácharos, 1,900 m, 01°38′N, 75°58′W

Isnos, Parque Nacional Natural de Puracé, 01°55'N, 76°15'W

La Candela, 01°50'N, 76°20'W

La Plata, 1,054 m, 02°23'N, 75°54'W

Pitalito, 1,250 m, 01°51'N, 76°02'W

Río La Plata, 1 mile S of Moscopán, 02°14'N, 76°09'W

San Adolfo, 1,500 m, 01°37'N, 75°59'W

San Augustin (= San Agustín), 1,690 m, 01°53'N, 76°16'W

Valle de las Papas, 01°50'N, 76°35'W

Villavieja, 16 km NE, ca. 03°19'N, 75°08'W

La Guajira (Departamento de)

Bahía Honda, 12°21'N, 71°47'W

Dibulla, sea level, 11°17'N, 73°19'W

Fonseca, 192 m, 10°54'N, 72°51'W

La Concepción, 800 m, ca. 11°03'N, 73°27'W

La Isla, near Cojoro, 37 km NNE of Paraguaipoa, 15 m, 11°41'N, 71°55'W

Las Marimondas, Sierra de Perijá, 1,000 m, 10°35'N, 72°45'W
    (Labels include the name "Fonseca," but see Hershkovitz [1949:383].)

Maracaibo, Venezuela, 119 km N and 32 km W, ca., 11°43'N, 71°54'W

Nasaret (= Nazareth), 200 m, 12°11'N, 71°17'W

Nazaret (= Nazareth), 200 m, 12°11'N, 71°17'W

Nazareth, 5 km W, 12°11'N, 71°20'W

Nazareth, 200 m, 12°11'N, 71°17'W

Puerto Estrella, 12°21'N, 71°18'W

Ríohaca, sea level, 11°33'N, 72°55'W

San Miguel, 10°58'N, 73°29'W

Sierra de Perijá, 10°35'N, 72°45'W

Sierra Negra (Serranía de Perijá), 10°37'N, 72°54'W

Villanueva, Sierra Negra, 280 m, 10°37'N, 72°59'W

Magdalena (Departamento del)

Bonda, 50 m, 11°14'N, 74°07'W

Cacagualito, a plantation 20 miles E of Santa Marta, 450 m, 11°17'N, 74°00'W

Cincinati, 1,480 m, 11°06'N, 74°06'W

Donamo (= Don Amo), 300 m, 11°15'N, 73°58'W

Don Diego, sea level, 11°14'N, 73°41'W

Finca El Recuerdo, Minca, 11°09'N, 74°07'W

Hacienda El Recuerdo (= El Recuerdo), Minca, 850 m, 11°07'N, 74°06'W

Isla Salamanca, sea level, 10°59'N, 74°27'W

Mamatoco, 35 m, 11°14'N, 74°10'W

Minca, 11°09'N, 74°07'W

Palamina (= Palomino), 11°02'N, 73°39'W

Palomino, 11°02'N, 73°39'W

Parque Natural Nacional Tayrona, 11°18'N, 74°10'W

Río Frio, 30 m, 10°55'N, 74°10'W

Santa Marta, sea level, 11°15'N, 74°12'W

Tagauga (= Taganga), 11°16'N, 74°12'W

Tusma, Río Magdalena, 10°00'N, 75°00'W

Valparaiso (= Cincinati), 11°06'N, 74°06'W

Meta (Departamento del)

Barrigón (= Puerto Barrigón), 240 m, 04°10'N, 73°01'W

Cabaña Duda, junction Río Duda and Río Guayabero, 02°33'N, 74°03'W

Campamento Ardillas, 5 km E of Cabaña Duda, 250 m, 02°33'N, 74°03'W

Campamento Chamusa (= Campamento Izawa), 250 m, 02°45'N, 74°l0'W

Cane Alto, NW of Restrepo, 04°17'N, 73°36'W

Caño Entrada, Sierranía de La Macarena, 550 m, 03°08'N, 72°52'W

Carimagua, 04°34'N, 71°20'W

Caño Lozada, Río Guayabero, 02°12'N, 73°53'W

El Porvenir, 04°46'N, 71°23'W

Finca Cháviva, 04°21'N, 72°19'W

Finca "El Buque," Villavicencio, 04°09'N, 73°39'W

Finca El Capricho, 38 km E of Villavicencio, 04°09'N, 73°16'W

Hacienda Los Guaduales, near San Martín, approximately 38 km NNE of San Juan de Arama, ca. 04°22'N, 73°35'W

La Angostura, Río Guayabero, 250 m, 02°20'N, 73°55'W
    (= Primera Angostura on Río Guayabero [= alto Río Guaviare]; Barriga-Bonilla 1965.)

Los Micos, 450 m, 03°17'N, 73°53'W

Mina de Upin, near Restrepo, 550 m, 04°18'N, 73°34'W

Municipio de San Juan de Arama, La Serranía de La Macarena, Reserva Nacional Natural La Macarena, 450–500 m, 03°10'N, 73°57'W

Puerto López, 200 m, 04°05'N, 72°58'W

Reserva Nacional de La Macarena, cerca de Caño Cristales, 02°45'N, 73°55'W

Reserva National Natural La Macarena (sector norte), Município de San Juan de Arama, 450–500 m, 03°10'N, 73°57'W

Restrepo, ca. 500 m, 04°15'N, 73°34'W

Río Guapaya, Serranía de La Macarena, 02°54'N, 73°39'W

Río Ocoa, 04°08'N, 73°15'W

Salinas de Upin, 04°15'N, 73°36'W

San Juan de Arama, 400 m, 03°24'N, 73°49'W

Serranía de La Macarena, 02°45'N, 76°08'W

Upin Salt Mine, 04°15'N, 73°36'W

Villavicencio, 500 m, 04°09'N, 73°37'W

Nariño (Departamento de)

Altaquer, 1,075 m, 01°15′N, 78°07′W
Barbacoas, 35 m, 01°41′N, 78°09′W
Buenavista, 350 m, 01°29′N, 78°05′W
El Carmen, 1,500 m, ca. 00°40′N, 77°10′W
El Guabo, 2,150 m, 01°35′N, 77°14′W
Finca San Marino, 640 m, ca. 01°55′N, 77°14′W
Hacienda Buenos Aires, 01°07′N, 77°53′W
Junín, 01°20′N, 78°09′W
Junín, 5 km E, 850 m, 01°20′N, 78°08′W
La Guarapería, 900 m, 01°21′N, 78°08′W
La Guayacana, 250 m, 01°26′N, 78°27′W
La Victoria, 30 m, 01°32′N, 78°45′W
Pasto, 2,600 m, 01°13′N, 77°17′W
Quebrada La Ensillada, ca. 1 km SE of Altaquer, 1,400 m, ca. 01°15′N, 78°07′W
Recaurte (= Ricaurte), 1,250 m, 01°13′N, 77°59′W
Reserva Natural La Planada, 1,300–2,100 m, 01°10′N, 78°05′W
Ricuarte (= Ricaurte), 1,250 m, 01°13′N, 77°59′W
San Filipe, 01°08′N, 77°25′W
San Pablo, 1,400 m, 01°06′N, 78°01′W
Tumaco, 30 km E, on Tumaco-Tuquerres Road, 01°50′N, 78°38′W

Norte de Santander (Departamento de)

Cúcuta, 215 m, 07°54′N, 72°31′W
El Guayabal (= Guayabal), 200 m, 08°01′N, 72°30′W
Gramales (= Gramalote), 07°53′N, 72°48′W
(SW of Bucaramanga according to D. C. Carter [1968]; also see Nagorsen and Tamsitt [1981:96].)
Hacienda La Primavera, 2,150 m, 07°00′N, 72°20′W
Pamplona, 07°23′N, 72°39′W
Río Zulia, 08°12′N, 72°30′W

Putumayo (Intendencia del)

Estación de Bombeo Guamués, 915 m, 00°40′N, 77°00′W
Mocoa, 580 m, 01°09′N, 76°37′W
Parque Nacional Natural La Paya, 00°10′N, 75°05′W
Puerto Leguizamo, 100 m, 00°12′S, 74°46′W
Quebrada El Hacha, 00°01′S, 75°31′W
Río Mecaya, 00°28′N, 75°20′W
(Hershkovitz locality; see Paynter and Traylor [1981: 157].)

Quindío (Departamento del)

Armenia, 1,500 m, 04°31′N, 75°41′W
Calarca, ca. 1,500 m, 04°31′N, 75°38′W

El Roble, 2,200 m, 04°41′N, 75°36′W
Finca Rincón Santo, 04°38′N, 75°34′W

Risaralda (Departamento de)

Apía, Río Apía, 05°05′N, 75°56′W
La Pastora, 04°42′N, 75°30′W
Mistrató, 05°18′N, 75°53′W
Pueblorrico, 1,560 m, 05°12′N, 76°08′W

Santander (Departamento de)

Barichara, 1340 m, 06°38′N, 73°14′W
Caño Muerto, 07°10′N, 73°55′W
Charalá, 06°17′N, 73°09′W
El Hoyo, 06°28′N, 73°00′W
(West of Mogotes according to C. J. Marinkelle, *in lit.*)
Finca El Rasgón, 06°55′N, 73°03′W
Lebrija, ca. 1,100 m, 07°07′N, 73°13′W
Macaregua Cave, between Barichara and San Gil, 06°36′N, 73°06′W
San Gil, 1,095 m, 06°33′N, 73°08′W
Suratá, above, 07°22′N, 73°00′W
Tierrabuena, 06°55′N, 73°40′W

Sucre (Departamento de)

Tulú Viejo (= Tuluviejo), 09°37′N, 75°27′W

Tolima (Departamento del)

Chicoral, Río Coello, 500 m, 04°13′N, 74°59′W
El Boquerón, 5.5 km SE of Ibagué, 1,430 m, 04°25′N, 75°08′W
Espinal, ca. 200 m, 04°09′N, 74°53′W
Guamo, 402 m, 04°01′N, 74°58′W
Mariquita, 535 m, 05°12′N, 74°54′W
Melgar, 430 m, 04°12′N, 74°39′W
Natagaima, 03°37′N, 75°06′W
Purificación, 300 m, 03°51′N, 74°57′W

Valle del Cauca (Departamento del)

Alto de Galápagos, 04°53′N, 76°13′W
Alto del Oso, 1,400 m, 03°45′N, 76°32′W
Anchicayá (= Ladrilleros?), 03°39′N, 76°56′W
Bahía Málaga, north side, 40 m, 04°08′N, 78°12′W
Bahía Málaga Naval Base, 40 m, 03°59′N, 77°20′W
Bajo Calima, E of Buenaventura, 03°56′N, 76°50′W
Bellavista, 03°37′N, 76°39′W
Buenaventura, 6 km N, 03°56′N, 77°04′W
Buenaventura, 03°53′N, 77°04′W

Buenaventura, 29 Km E (= Río Zabaletas, 29 km SE of Buenaventura), 75 m, 03°44′N, 76°57′W

Cali, 1,060 m, 03°27′N, 76°31′W

Calima, 03°56′N, 76°31′W

Cauca, 04°43′N, 76°01′W

Coal mines, west of Cali, 03°27′N, 76°32W

Cuartel B-V-83, Concesión de Bajo Calima, 40 m, 03°57′N, 77°08′W

Dagua, 816 m, 03°40′N, 76°42′W

El Silencio, ca. 04°15′N, 76°15′W

El Tambor, ca. 10 mi NE of Vijes, 03°45′N, 76°22′W

Finca Hollanda, 03°37′N, 76°14′W

Finca "Zingara," 03°30′N, 76°38′W

Hacienda Jamaica, 11 km S, and 2 km W of Cartago, 04°39′N, 75°56′W

Hacienda Los Alpes, 6 km S, 11 km E of Florida, 2,400 m, 03°16′N, 76°09′W

Hormiguero, approx. 20 km SE of Cali, ca 1,000 m, 03°19′N, 76°29′W

La Bocana, 7 km SW of Buenaventura, 03°50′N, 77°06′W

La Guarapería, 01°21′N, 78°08′W

Las Lomitas (= Lomitas), 1,400 m, 03°38′N, 76°38′W

Lime kiln, vicinity of Cali, 03°27′N, 76°31′W

Morga, 7 km SE of Cali, 1,000 m, 03°23′N, 76°27′W

Palmira, 6 miles N, 1,067 m, 03°35′N, 76°15′W.

Palmira, 1,066 m, 03°32′N, 76°16′W

Pance, 2 km S, 1,560 m, 03°21′N, 76°38′W
   (= Finca El Topacio; [M. Thomas 1972:13].)

Páramo de Barragán, 00°40′N, 75°40′W

Paso de Galápagos, 8 km N and 4 km E of El Cairo, 1,800 m, 04°50′N, 76°12′W

Pichindé, 10 km SW of Cali, 03°26′N, 76°37′W

Punta Soldado, sea level, 03°48′N, 77°11′W

Quebrada de la Delfina, Vereda Aguas Lindas, 03°50′N, 76°14′W

Quebrada La Cristalina, near left bank of Río Calima, Vereda Río Bravo, Municipalidad de Calima, 800 m, 03°54′09″N, 76°39′24″W

Río Achicaya, 8 km W of Danubio, 300 m, 03°37′N, 76°56′W

Río Cauquitá, near Cali, 1,000 m, 03°15′N, 76°40′W

Río Cayambre, approx. 60 km S of Buenaventura, 480 m, 03°20′N, 77°00′W

Río Chanco (= Río El Chanco), 03°57′N, 76°43′W

Río Frio, 04°09′N, 76°15′W

Río Raposo, ca. 50 m, 03°38′N, 77°05′W

Río Raposo, 27 km S of Buenaventura, near sea level, 03°38′N, 77°05′W

Río Zabaletas, 29 km SE of Buenaventura, 75 m, 03°44′N, 76°57′W

San Antonio, 4 km NW, 03°30′N, 76°38′W

San José, 115 m, 03°51′N, 76°52′W

Tenerife, 03°44′N, 76°07′W

Zabaletas, Río Zabaletas, 75 m, 03°44′N, 76°57′W

Vaupés (Comisaría del)

Caño Cubiyu (= Cubigu), near mouth, Río Vaupés, 50 km W of Mitú, ca. 01°02′N, 70°12′W

El Internado de María Reina, Mitú, 01°08′N, 70°03′W

Mitú, Río Vaupés, 01°08′N, 70°03′W

Cerro de las Pinturas, 02°10′N, 71°30′W

Durania, 01°16′N, 70°11′W
   (Correctly known as Urania according to Hernández-Camacho and Cadena [1978:200].)

Río Vaupés, Right bank, Tahuapunta, 03°36′N, 69°12′W
   (See Tahuapunta, Amazonas, Brazil.)

Vichada (Comisaría del)

Caño Avispas, 100 m, 05°15′N, 67°52′W

Caño Matavén, 1 km N, 90 m, 04°33′N, 67°54′W

Finca La Arepa (not located)

Maipures, Río Orinoco, 05°11′N, 67°51′W

Matavení, 04°17′N, 67°49′W

Santa Teresita, 100 m, 04°21′N, 69°50′W

Ecuador
Azuay (Provincia de)

Baños, 8 km SW of Cuenca, 2,540 m, 02°57′S, 79°04′W

Bestión, 3,125 m, 03°10′S, 79°13′W

Cuenca, 02°53′S, 79°00′W

Manta Real, 650 m, 02°34′S, 79°21′W

Mazán, 2,700 m, 02°49′S, 79°13′W

Molleturo, 2,315 m, 02°48′S, 79°26′W

San José Grande, Río Paute, 1,300 m, 02°35′S, 78°30′W

Valle Yunguilla, 03°18′S, 79°18′W

Yanasacha, 03°02′S, 79°10′W

Bolívar (Provincia de)

Barraganete, 430 m, 01°26′S, 79°17′W

Echeandía, 01°28′S, 79°12′W

Echeandía, 3 km SW, ca. 01°30′S, 79°13′W

Sinche (= Hacienda Sinche), 3,200 m, 01°33′S, 78°57′W

Cañar (Provincia de)

Chical, 02°24′S, 78°58′W

San Juan, 15 miles W of Huigra, 02°14′S, 79°08′W

Carchi (Provincia de)

Gruta Rumichaca, 00°30′N, 77°50′W
(= Gruta de la Paz, according to Albuja [1983:268], 2 miles E of La Paz.)
Maldonado, 1,500 m, 00°54′N, 78°06′W
Puente Piedra, 900 m, 01°02′N, 78°17′W
Rumichaca, 2,690 m, 00°49′N, 77°40′W
(= Gruta de Rumichaca [Albuja 1983:275].)
San Gabriel, 00°36′N, 77°49′W

Chimborazo (Provincia de)

Carmen, near Sinche, 01°38′S, 78°40′W
La Isla, 02°15′S, 72°14′W
Mt. Chimborazo, 2,743–3,048 m, 01°28′S, 78°48′W
Pallatanga, 4 km NE, 2,225 m, 01°57′S, 78°55′W
Pallatanga, 1,650 m, 02°00′S, 78°57′W
Paujchi, 02°25′S, 78°58′W
Planchas, 2,900 m, 01°49′S, 78°24′W
Riobamba, 2,750 m, 01°40′S, 78°38′W
San Antonio, 02°17′S, 78°59′W
Urbina, 01°30′S, 78°44′W
Volcán Chimborazo, 01°28′S, 78°48′W

El Oro (Provincia de)

Arenillas, 03°27′S, 80°04′W
Cayancas, 35 m, 03°29′S, 80°06′W
El Bosque, 1,600 m, 03°39′S, 79°38′W
El Chiral, 1,630 m, 03°38′S, 79°41′W
Piñas, 03°42′S, 79°42′W
Portovelo, 03°43′S, 79°39′W
Puente de Moromoro, 1 km SW, 920 m, 03°44′S, 79°44′W
Zaruma, 1,200 m, 03°41′S, 79°37′W
Zaruma, 9 miles S, 600 m, 03°48′S, 79°34′W

Esmeraldas (Provincia de)

Alto Tambo, Hacienda La Granada (3 km via Lita-San Lorenzo), 600 m, 00°52′N, 78°32′W
Alto Tambo, 2 km S, 700 m, 00°54′N, 78°33′W
Bilsa, 800 m, 00°37′N, 79°51′W
Borbón, 01°05′N, 78°59′W
Cachavi (= Cachabí), 00°58′N, 78°48′W
Cachabí (= Urbina), 00°58′N, 78°48′W
Chontaduro, 00°56′N, 79°27′W
Concepción, sea level, 01°02′N, 78°49′W
El Porvenir, 18 m, 00°56′N, 79°27′W
Esmeraldas, sea level, 00°59′N, 79°42′W
Estación Experimental La Chiquita, 01°17′N, 78°50′W
Estero Taquiama, 40 m, 00°43′N, 79°48′W
Hacienda La Granada, 560 m, 00°52′N, 78°31′W
La Chiquita, 12 km S of San Lorenzo, 60 m, 01°14′N, 78°46′W
La Tola, sea level, 01°12′N, 79°06′W
Los Pambiles, Cordillera de Taisán, 00°32′N, 78°38′W
Luis Vargas Torres, Playa del Oro, Río Santiago, 50 m, 00°53′N, 78°48′W
Majua, 3 km W, 00°42′N, 79°34′W
Malimpia, 00°24′N, 79°27′W
Mataje, 50 m, 01°22′N, 78°43′W
Montañas de Chancameta, 00°37′N, 79°22′W
Playa de Oro, 00°53N, 78°48′W
Quinindé (= Rosa Zárate), 00°20′N, 79°28′W
Río Bravo, 00°41′S, 78°57′W
Río Quinindé, 00°20′N, 79°28′W
Río Tiaone, 40 m, 00°56′N, 79°42′W
Rosa Zárate, 00°20′N, 79°28′W
Sade, 00°30′N, 79°23′W
San Lorenzo, 01°17N, 78°50′W
San Miguel, 125 m, 00°44′N, 78°55′W
San Miguel de Cayapas, 3 km S, 00°38′N, 78°57′W
St. Javier (= San Javier), 100 m, 01°04′N, 78°48′W
Tabiazo, 40 m, 00°58′N, 79°42′W
Urbina (= Cachabí), 150 m, 00°58′N, 78°48′W
Valle del Sade, 200 m, 00°31′N, 79°20′W
Viche, 42 m, 00°38′N, 79°33′W

Galápagos (Territorio de)

Isla Indefatigable (= Isla Santa Cruz), 00°38′S, 90°23′W
Isla San Cristóbal, 00°50′S, 89°26′W

Guayas (Provincia de)

Ancón, Santa Elena (= Península de Santa Elena), sea level, 02°20′S, 80°52′W
Balzar, Río Daule, 100 m, 01°22′S, 79°54′W
Balzar Mts. (= Balzar), 01°22′S, 79°54′W
Bucay, Río Chimbo, 312 m, 02°10′S, 79°06′W
Cerro Baja Verde (= Cerro de Bajo Verde), 32 km NW of Guayaquil, 150 m, 01°59′S, 80°02′W
Cerro Manglar Alto (= Cerro Manglaralto, headwaters of Río Manglaralto), 460 m, 01°47′S, 80°38′W
Cerro Manglaralto, 460 m, 01°47′S, 80°38′W
Chongón, 40 m, 02°14′S, 80°04′W
Chongón, 2.4 km SE (= Chongoncito), 20 m, 02°15′S, 80°05′W
Daule, 01°50′S, 79°56′W
Durán, sea level, 02°12′S, 79°50′W
Guayaquil, 02°10′S, 79°50′W
Huerta Negra, near Tenque, 10 km ESE of Baláo, 03°00′S, 79°47′W

Isla Puná, 02°50′S, 80°08′W

Isla Silva (= Isla de Silva), Río Babahoyo, 01°55′S, 79°42′W

La Unión, 01°58′S, 80°01′W

Manglar Alto, near, 01°50′S, 80°44′W

Pacaritambo (= Hacienda Pacaritambo), 01°22′S, 79°53′W

Puente de Chimbo, 370 m, 02°10′S, 79°10′W

Puná, Isla Puná, sea level, 02°58′S, 80°11′W

Puna Island, 02°50′S, 80°08′W

Reserva Ecológica Manglares Churute, 10 m, 02°30′S, 79°45′W

Río Chongón, 1.5 km SE of Chongón, 02°15′S, 80°04′W

Río Javita, mouth, 2 km N of Monteverde, 02°02′S, 80°43′W

Sanbrondon (= Samborondón), sea level, 01°57′S, 79°44′W

San Rafael, 7 km S of Balao, 02°58′S, 79°47′W

San Ramón, Isla Puná, 02°56′S, 80°08′W

Imbabura (Provincia de)

Chota, 14 km NNE of Ibarra, 1,768 m, 00°28′N, 78°04′W

Hacienda La Vega, 5 km ESE of San Pablo del Lago, 2,600 m, 00°12′N, 78°10′W

Ibarra, 00°21′N, 78°07′W

Lita, 510 m, 00°52′N, 78°28′W

Paramba, 700 m, 00°49′N, 78°21′W

Loja (Provincia de)

Cajanuma, 2,200 m, 04°05′S, 79°12′W

Casanga, 870 m, 04°01′S, 79°45′W

Cebollal, 968 m, 04°02′S, 80°02′W

Celica, 2,100 m, 04°07′S, 79°59′W

Loja, 04°00′S, 79°12′W

Malacatos, 1,600 m, 04°14′S, 79°15′W

Masanamaca, 12 km S of Vilcabamba, ca. 04°21′S, 79°14′W

Podocarpus National Park, 04°55′S, 79°12′W

Punta Santa Ana, 1,100 m, 03°50′S, 79°33′W

San Pedro de Vilcabamba, 3 km N of Vilcabamba, ca. 04°13′S, 79°15′W

Valle, 03°58′S, 79°14′W

Valley of Casanga, 04°05′S, 79°45′W

Los Ríos (Provincia de)

Abras de Mantequilla, 12 km NE of Vinces, 01°28′S, 79°39′W

Babahoyo, 5 m, 01°49′S, 79°31′W

El Papayo, 01°36′S, 79°35′W

Vinces, 01°32′S, 79°45′W

Manabí (Provincia de)

Bahía de Caráquez, 00°36′S, 80°26′W

Chone, 45 km NE, 00°23′S, 79°55′W

Chone, 00°41′S, 80°06′W

Chontillal, 00°04′S, 79°06′W

Cuaque (= Coaque), sea level, 00°00,′ 80°06′W

Espuela Perdida, 500 m, 01°36′S, 80°41′W

Estero Achiote, 550 m, 01°36′S, 80°42′W

Mongoya (Río), 00°10′S, 79°38′W

Puerto López, sea level, 01°34′S, 80°49′W

Río Blanco, Puerto Lopez, 200 m, 01°35′S, 80°45′W

Río Briseño, Bahia de Caráquez, sea level, 00°31′S, 80°27′W

Río de Oro, 02°10′S, 79°22′W

San José, 110 m, 01°29′S, 80°20′W

San Sebastián, 650 m, 01°35′S, 80°42′W

San Vicente, Punto López, 00°35′S, 80°24′W

Vueltas Largas, 01°33′S, 80°42′W

Morona-Santiago (Provincia de)

Achupallas, 2,080 m, 03°27′S, 78°21′W

Gualaquiza, 750 m, 03°24′S, 78°33′W

Los Tayos, 03°05′S, 78°13′W

Macas, 1,015 m, 02°19′S, 78°07′W

Méndez, Río Paute, a tributary of the Río Upano, 750 m, 02°43′S, 78°18′W

Méndez, near, 02°44′S, 78°19′W

Río Llushín, 01°35′S, 78°09′W

San Carlos de Limón, 800 m, 03°13′S, 78°26′W

Sucúa, 02°28′S, 78°10′W

Teniente Ortíz, 03°03′S, 78°06′W

Uuntsuants, 1,300 m, 02°33′S, 77°54′W

Yaupi, 03°01′S, 77°51′W

Napo (Provincia de)

Alto Coca, 00°05′S, 77°15′W

Antisana Páramo, S slope Cerro Antisana, Parque Nacional de Cotopaxi, 4,117 m, 00°30′S, 78°08′W

Avila, 600 m, 00°38′S, 77°25′W

Baeza, 00°27′S, 77°53′W

Cascada de San Rafael, 400 m, 00°06′S, 77°34′W

Cavernas de Jumandi, 431 m, 00°56′S, 77°50′W

Cerro Huataracu, 00°41′S, 77°33′W

Concepción, 00°48′S, 77°25′W

Cosanga, Río Aliso; 00°34′S, 77°52′W

El Salado, Alto Coca, 1,800 m, 00°15′S, 77°41′W

Huamaní, 1,250 m, 00°43′S, 77°37′W

Huataracu, 500 m, 00°42′S, 77°20′W

Loreto, 00°38′S, 77°20′W

Papallacta, 00°22′S, 78°08′W

Páramo de Papallacta, 3,300–3,500 m, ca. 00°21′S, 78°11′W

Río Pucuno, 00°46′S, 77°18′W

San José (= San José Nuevo), 00°26′S, 77°20′W

San José (abajo) (= San José Nuevo), Mt. Sumaco, 00°26′S, 77°20′W

San José de Sumaco (= San José Nuevo), 00°26′S, 77°20′W

Santa Rosa de Otas, 00°58′S, 77°28′W

Orellana (Provincia de)

Campamento Petrolero Amo 2, Parque Nacional Yasuní, 200 m, 00°55′S, 76°13′W

Coca, 260 m, 00°28′S, 76°58′W

Confluence of rivers Coca and Napo (= Coca), 260 m, 00°28′S, 76°58′W

Estació Científica Yasuní, 1 km S, 00°41′S, 76°24′W

Ishpingo I, 179 m, 01°06′S, 75°39′W

La Coca, 260 m, 00°28′S, 76°58′W

Laguna Garzacocha, 180 m, 00°29′S, 76°21′W

Onkone Gare, 38 km S of Pompeya Sur, 00°39′S, 76°27′W

Onkone Gare, 30 km SW, 00°45′S, 76°46′W

Pompeya Sur, 38 km S, 00°39′S, 76°27′W

Pompeya Sur, 42 km S and 1 km E, 00°41′S, 76°26′W

Pompeya Sur, 42 km S and 12 km E, 00°40′S, 76°22′W

Río Pinto, 00°54′S, 76°13′W

San José de Payamino, 00°30′S, 77°17′W

Santa Rosa de Arapino, 570 m, 00°50′S, 77°28′W

Pastaza (Provincia de)

Alto Pastaza (= upper Río Pastaza), 01°24′S, 78°25′W (The Río Pastaza arises near Baños [01°24′S, 78°25′W] and flows southeastward into Peru. "Alto Pastaza" is interpreted as "near Baños.")

Andoas, 200 m, 02°34′S, 76°48′W

Canelos, 01°35′S, 77°45′W

Cavernas de Mera, 4 km NE of Mera, 01°27′S, 78°07′W

Lorocachi, 01°32′S, 75°57′W

Mera, Río Alpuyuca (= Río Alpayacu), 1,160 m, 01°28′S, 78°08′W

Montalvo (= Andoas), Río Bobonazo, 250 m, 02°04′S, 76°58′W

Palmera, ca. 25 km E of Baños (or 16 km E of Mirador), Río Pastaza, 1,220 m, 01°25′S, 78°12′W

Putsu, between Canelos and Puyo, 01°50′S, 77°50′W

Puyo, 8 miles WNW, 1,200 m, 01°37′S, 78°07′W

Puyo, 975 m, 01°28′S, 77°59′W

Río Acaro, 500 m, 01°23′S, 77°25′W

Río Capahuari, 02°03′S, 77°10′W

Río Copotaza, 02°07′S, 77°27′W

Río Pindo Yacu, 02°08′S, 76°03′W

Sarayacu, Río Bobonaza, 700 m, 01°44′S, 77°29′W

Tiguino, 130 km S of Coca, 300 m, 01°07′S, 76°57′W

Pichincha (Provincia de)

Aloag, 11 km W, ca. 00°29′S, 78°41′W

Bellavista, 2,200 m, 00°00′S, 78°41′W

Centro Científico Río Palenque, 00°33′S, 79°22′W

Cerro Iliniza, 3,750 m, 00°40′S, 78°42′W

El Paraíso, 00°11′N, 78°46′W

Estación Científica Río Palenque, 00°35′S, 79°30′W

Estación Forestal La Favorita, 00°16′S, 78°46′W

Finca La Esperanza, 500 m, 00°15′S, 79°09′W

Gualea, 1,500 m, 00°07′N, 78°44′W

Gualea, below, 1,000 m, west slope of Andes, 0°07′N, 78°50′W

Hacienda Garzón, 3,100 m, 00°13′S, 78°34′W

Las Máquinas, on trail from Aloag to Santo Domingo de los Colorados, 2,180 m, 00°26′S, 78°44′W

Mindo, 1,260 m, 00°03′S, 78°48′W

Mojanda, western side, 00°08′N, 78°17′W

Mt. Pichincha, 00°10′S, 78°35′W

Nanegal, 00°07′N, 78°46′W

Niebli, 2,300 m, 00°06′N, 78°31′W

Pacta (= Pacto), 1,200 m, 00°12′N, 78°52′W

Patricia Pilar, 220 m, 00°30′N, 79°22′W

Pichincha, 3,353 m, 00°10′S, 78°33′W

Plan Piloto, 00°04′S, 79°19′W

Playa Rica, 00°50′S, 78°47′W

Región de Gualea (Gualea = 00°07′S, 78°44′W)

Río Blanco, 00°24′S, 78°48′W

Río Condor Huachana, 3.5 km NE of Lloa, 3,250 m, 00°13′S, 78°33′W

Río Palenque Field Station, 00°35′S, 79°22′W

Río Saloya, near Mt. Cayambe, 1,750 m, 00°01′S, 78°53′W

Río Toachi, 00°06′N, 79°17′W

Río Tulipe, near Gualea, ca. 00°07′S, 78°44′W

Saloya West, 1,750 m, 00°01′S, 78°53′W

San Antonio de Pichincha, 2,397 m, 00°01′S, 78°27′W

Santo Domingo, 00°15′S, 79°09′W

Santo Domingo de los Colorados, 00°15′S, 79°09′W

Yanacocha, 3,250 m, 00°07S, 78°33′W

Yanayacu, 2,075 m, 00°35′S, 78°52′W

Sucumbíos (Provincia de)

Duvuno, 340 m, 00°02′S, 77°07W

Lago Agrio, 300 m, 00°06′N, 76°53′W

Largato Cocha (= Río Largatocacha), 300 m, 00°39′S, 75°16′W

Limón Cocha, 300 m, 00°24′S, 76°38′W

("Missionary station established 1955 on unspoiled lemon-colored lake in tropical forest, 2 km from mouth of Río Jivino, tributary of the Río Napo" [Paynter and Traylor 1977:66].)

Limoncocha, 300 m, 00°24′S, 76°38′W
Santa Cecilia, 340 m, 00°03′N, 76°58′W
Zafiro, ca. 00°08′S, 76°50′W

Tungurahua (Provincia de)

Baños, 1,800 m, 01°24′S, 78°25′W
Mirador, 01°26′S, 78°15′W
Mirador, 1.5 km E, 1,430 m, 01°26′S, 78°14′W
San Antonio, 01°09′S, 78°28′W
San Francisco, E of Ambato, ca. 3,000 m, 01°18′S, 78°30′W

Zamora-Chinchipe (Provincia de)

Cuevas de Numbala, 1,890 m, 04°33′S, 79°04′W
Cumbaratza, 3 km E, 1,000 m, 03°58′S, 78°50′W
Destacamento Militar Condor Mirador, 1,750 m, 03°38′S, 78°23′W
Los Encuentros, 4 km ENE, 850 m, 03°44′S, 78°37′W
Miazi, 04°19′S, 78°39′W
Sabanilla, 04°02′S, 79°01′W
Sabanilla, 4 km E, 1,585 m, 04°01′S, 78°57′W
Zamora, 1,000 m, 04°04′S, 78°58′W

French Guiana

Arataye, 03°59′N, 52°34′W
Armontabo, 03°49′N, 52°03′W
(Along the Armontado river, a tributary of Oyapock (Brosset and Charles-Dominique 1990:511).)
Cabassou, 04°53′N, 52°18′W
Camopi, 03°10′N, 52°11′W
Cayenne, 04°56′N, 52°20′W
Grotte du Bassin du Tapir, Les Nouragues, 04°05′N, 52°40′W
I'ilot le Père, 04°56′N, 52°12′W
Inini Territory, Arataye River, 04°00′N, 52°40′W
Ipousin, 04°09′N, 52°24′W
Isle le Père, 04°56′N, 52°12′W
Kaw, 100 m, 04°29′N, 52°02′W
Kourou, sea level, 05°09′N, 52°39′W
L'Anse de Sinnamary, 05°22′N, 52°55′W
Les Eaux Claires, ca. 7 km N of Saül, 200 m, 03°37′N, 53°12′W
Les Nouragues, 04°05′N, 52°40′W
Mana, 05°39′N, 53°47′W
Montjoly, sea level, 04°55′N, 52°16′W

Nouragues, 04°05′N, 52°40′W
Ouanary, 04°13′N, 51°40′W
Paracou, near Sinnamary, 05°17′N, 52°55′W
Petit-Saut, 05°04′N, 53°03′W
Piste de Saint Élie, 05°18′N, 53°04′W
Pointe Combi, 05°19′N, 52°57′W
Régina and St. Georges road project, ca. 04°07′N, 52°54′W
Rémire, 1 km N, 04°54′N, 52°17′W
Rémire, 1.5 km NE, 04°54′N, 52°16′W
River Arataye, right bank, ca. 5 km downstream from Saut Pararé, 30 m, 04°02′N, 52°40′W
Route de Kaw, 04°31′N, 52°05′W
Saint-Laurent du Maroni, 05°30′N, 54°02′W
Saül, 4 km N, ca. 03°39′N, 53°12′W
Saül, 03°37′N, 53°12′W
Saut Pararé, 04°02′N, 52°42′W
Sinnamary, 05°23′N, 52°57′W
Sinnamary, 3.5 km S and 10 km W, 05°21′N, 53°02′W
Sinnamary, 9.5 km S and 11.5 km W, 05°18′N, 53°04′W
St. Elie (= Piste de Saint Élie), 05°18′N, 53°04′W
Table du Mahury, 171 m, 04°52′N, 52°16′W
Tamanoir, Mana River, 05°09′N, 53°45′W
Trois Sauts, Flewe Oyapock, 02°15′N, 53°06′W

Guyana
Barima-Waini (District)

Akwero, 1.5 km E, 07°39′N, 58°55′W
Arakaka, 122 m, 07°35′N, 60°01′W
Baramita, north of airstrip, 07°22′N, 60°29′W
Cart Market, 1 mile E of Aquero (= Akwero), 07°39′N, 58°55′W
Warapoco (= Waropoko) Mission, Waropoko Creek, 07°48′N, 59°15′W

Cuyuni-Mazaruni (District)

Bartica Grove, 24 miles along the Potaro road from Bartica, 3 m, 06°24′N, 58°37′W
Bartica, 24 miles from (= 38.4 km S of Bartica), 06°00′N, 58°40′W
Bartica, 24 miles along Potaro Road, 06°00′N, 58°40′W
Issano Road, 12 miles W of Bartica-Potaro Road, 05°35′N, 59°01′W
Kalacoon (= Kalakun), Essequibo River, 06°24′N, 58°39′W
Kamakusa, 05°57′N, 59°54′W
Karowrieng River, within 12 miles of Maipuri Falls, ca. 05°41′N, 60°17′W
Kartabo (= Kartabu), 06°23′N, 58°41′W
Kartabu Point, 06°23′N, 58°41′W
Kuwaima Falls, 400 m, 06°03′N, 60°39′W

Mazaruni (= Penal Settlement), 100 m, 06°24′N, 58°39′W

Mazaruni River, 06°25′N, 58°38′W

Namai Creek, 5 km W of Paruima, 05°48′N, 61°06′W

27–Mile Camp, 27 miles along the Potaro Road from Bartica, 05°58′N, 58°41′W

Demerara-Mahaica (District)

Buxton, 06°47′N, 58°02′W

Ceiba Biological Center, 06°30′N, 58°13′W

Ceiba Biological Station, 06°30′N, 58°13′W

Demerara (= Georgetown), 06°48′N, 58°10′W

Georgetown, sea level, 06°48′N, 58°10′W

Hyde Park, Demerara River, 100 m, 06°30′N, 58°16′W

Loo Creek, 68 km by road S of Georgetown, 06°14′N, 58°15′W

East Berbice-Corentyne (District)

Boundary Camp, Itabu Creek Head, Corantijn River, 610 m, ca. 01°33′N, 58°10′W

(Acary Mts., 2,000 ft. (Blake 1941:227); "ca. 183 m from Brasilian border at head of Itabu Creek in Acarai Mountains. . . ." [Stephens and Traylor 1985:25].)

Itabu Creek Head, Corantijn River, ca. 01°33′N, 58°10′W

Mango Landing, 30 m, 05°10′N, 57°18′W

Upper Canje Creek, 05°07′N, 57°37′W

Essequibo Islands-West Demerara (District)

Bonasika Creek, 1.4 km up, 06°42′N, 58°29′W

Bonasika Creek, 1 mile up, 06°42′N, 58°29′W

Bonasica (= Bonasika River), Essequibo River, 06°45′N, 58°30′W

Buck Hall, 06°56′N, 58°33′W

Marias Pleasure, Wakenaam Island, 06°58′N, 58°26′W

New River Falls (= Drios Falls), 03°21′N, 57°35′W

Wai Patush Island (= Waipotosi Island), 11 km S of Fort Island, 06°45′N, 58°33′W

Pomeroon-Supenaam (District)

Better Hope, 07°23′N, 58°31′W

Supinaam (= Supenaam) River, 06°59′N, 58°31′W

Maccasseema (= Makasima), 100 m, 07°15′N, 58°43′W

Potaro-Siparuni (District)

Clearwater Camp, 90 m, 04°44′N, 58°51′W

Cowfly Camp, Iwokrama Forest, 80 m, 04°20′N, 58°49′W

Echerak River, 460 m, 05°10′N, 59°30′W

Iwokrama Forest, centered at 04°30′N, 58°50′W

Iwokrama Reserve, S. Falls, Siparuni River, 04°32′N, 59°05′W

Kaieteur Falls, Potaro River, 366 m, 05°10′N, 59°28′W

Kato, Kawa Valley, 04°40′N, 59°49′W

Kurupukari, 04°40′N, 58°40′W

Minnehaha Creek, 05°08′N, 59°07′W

Mount Ayanganna, 05°23′N, 59°56′W

Mount Kowa, 1,300 m, 04°51′N, 59°41′W

Pakatau Falls, 04°45′N, 59°01′W

Paramakatoi, 210 m, 04°42′N, 59°43′W

Three Mile Camp, 75 m, 04°38′N, 58°43′W

Upper Demerara-Berbice (District)

Arampa, 3 miles E of Ituni, 05°33′N, 58°06′W

Arampa, 5 km S of Ituni, 05°28′N, 58°15′W

Bada Creek, 50 m, 05°35′N, 58°08′W

Berbice, 05°27′N, 57°57′W

Comaccka (= Takama), 100 m, 05°34′N, 57°55′W

Dubulay Ranch (= Du Bu Lay Ranch), 05°41′N, 57°52′W

Ituni, 10 miles NE, 05°38′N, 58°09′W

Ituni, 3 km SSW, ca. 05°31′N, 58°15′W

Linden, 6 miles NE, 06°04′N, 58°14′W

Linden, 3 miles S, 05°57′N, 58°18′W

Lucky Spot, 3 miles S of Linden, 05°59′N, 58°18′W

Kaow Island, Essequibo River, 06°25′N, 58°37′W

Wikki River, 05°33′N, 57°55′W

Upper Takutu-Upper Essequibo (District)

Arakwai River, 5 km upstream from Dadanawa, 02°48′N, 59°31′W

Aroquoi Tributary, Rupununi River 02°45′N, 59°30′W (= Arawai River Area, 3.2 km S of Dadanawa)

Dadanawa, 150 m, 02°50′N, 59°30′W

Dadanawa, Bush Island, 02°50′N, 59°29′W

Dadanawa, 12 miles E, ca. 02°50′N, 59°13′W

Dadanawa, 15 miles E, ca. 02°50′N, 59°16′W

Foot of Kanuku Mountains, 12 miles ENE of Dadanawa, 02°57′N, 59°25′W

Kamoa River, 50 km WSW of Gunn'S Strip, 01°35′N, 58°38′W

Kanuku Mountains, 03°12′N, 59°30′W

Karanambo, 03°45′N, 59°18′W

Komawariwau river, 20 miles E of Dadanawa, ca. 02°54′N, 59°19′W

Kuitaro (= Kwitaro) River, 30 km E of Dadanawa, 02°50′N, 59°13′W

Kuitaro (= Kwitaro) River, 48 km E of Dadanawa, 02°50′N, 59°02′W

Kuitaro (= Kwitaro) River, 64 km E of Dadanawa, 02°50′N, 58°54′W

Maruranowa (= Marurawaunwa Village), 02°45'N, 59°10'W

Marurawaunwa Village, 02°45'N, 59°10'W

Nappi Creek, Kanuku Mountains, 03°36'N, 59°37'W

Onera Falls, Onora River, 01°37'N, 58°38'W

Rewa River, 03°53'N, 58°48'W

Run-ny Decou, on Kuitaro River, above Peter'S Camp (= vicinity of Peter'S Field), 03°02'N, 58°57'W

Shea, 02°49'N, 59°08'W

Tamton, 02°21'N, 59°42'W

Ow-wi-dy-wau, upper headwaters of the Kwitaro River, ca. 40 miles E of Dadanawa, 02°50'N, 58°55'W

Netherlands Antilles

Aruba

Ex Cave (cave not located; coordinates for Aruba, 12°30'N, 69°58'W)

Northwest end of island, 12°38'N, 70°03'W

Quaridikiri Cave (cave not located; coordinates for Aruba, 12°30'N, 69°58'W)

Bonaire

Kralendijk, 8.5 km N and 2 km W, 12°14'N, 68°17'W

Kralendijk, 12°10'N, 68°16'W

Oeroesjam Blanco, Grot Colombia, near Barcadera, 12°13'N, 68°19'W

Curaçao

Fort Beckenburg, 12°04'N, 68°52'W

Hatto (= Hato), 12°11'N, 68°58'W

Klein St. Martha, 12°10'N, 69°00'W

Kueba di Ratón, ca. 12°13'N, 68°58'W

Round Cliff (= Roude Klip), 12°09'N, 68°52'W

Westpunt, 2.8 km S and 4.5 km E, ca. 12°35'N, 70°00'W

Willemstad, 12°07'N, 68°57'W

Paraguay

Alto Paraná (Departamento de)

Puerto Bertoni, 91 m, 25°38'S, 54°40'W

Alto Paraguay (Departamento de)

Agua Dulce, 170 km W of Bahia Negra, 20°00'S, 59°40'W

Bahía Negra, 20°11'S, 58°09'W

Cerro León, 50 km (by road) WNW of Fortín Madrejón, 20°23'S, 60°19'W

Estancia Doña Julia, 20°11'S, 58°04'W

Fortín Madrejón, 67 km N, 20°05'S, 59°51'W

Fuerte Olimpo, Río Paraguay, ca. 200 m, 21°02'S, 57°54'W

Laguna General Díaz, 21°18'S, 59°01'W

Madrejón (= Fortín Madrejón), 60 km (by road) N, 20°30'S, 59°52'W

Mayor Pablo Lagarenza, 28 km N and 50 km W, 19°42'S, 61°13'W

Palmar de las Islas, 19°38'S, 60°37'W

Puerto Guaraní, ca. 100 m, 21°28'S, 57°55'W

Puerto Sastre, ca. 150 m, 22°06'S, 57°59'W

Amambay (Departamento de)

Arroyo Tacuara, 2 km N of Potrerito, ca. 22°30'S, 56°00'W

Cerro Corá, 4 km (by road) SW, 22°40'S, 56°00'W

Parque Nacional Cerro Corá, 22°37'S, 55°59'W

Pedro Juan Caballero, 28 km SW, 22°40'S, 56°00'W

Boquerón (Departamento de)

Base Naval Pedro P. Peña, 240 m, 22°27'S, 62°21'W

Copagro, Km 589, Transchaco Highway, 21°25'S, 61°25'W

Filadelfia, 19 km N, 22°09'S, 60°02'W

Filadelfia (= Colonia Filadelfia), 175 m, 22°21'S, 60°02'W

Filadelfia, 15 km S, 22°29'S, 60°02'W

Fortín Nueve, 23°52'S, 60°53'W

Km 588, 9 km by rd. WSW, Trans Chaco Highway, 21°29'S, 61°30'W

Loma Plata, 22°23'S, 59°51'W

Orloff, 22°18'S, 59°56'W

Parque Nacional Teniente (Agripino) Enciso, 21°05'S, 61°37'W

Tajamar, 200 m E of Fortín Teniente Agripino Enciso, 21°05'S, 61°46'W

Teniente Enciso, 21°05'S, 61°46'W

Teniente Ochoa, 24 km NW, 21°46'S, 61°09'W

Caaguazú (Departamento de)

Caaguazú, 25°26'S, 56°02'W

Caazapá (Departamento de)

Estancia Dos Marías, Río Tebicuray, 26°46'S, 56°32'W

Canendiyu (Departamento de)

Curuguaty, 13.3 km by road N, ca. 24°25'S, 55°41'W

Curuguaty, 24°31'S, 55°42'W

Igatimí, 215 m, 24°05'S, 55°30'W

Central (Departamento de)

Asunción, 25°16′S, 57°40′W
Asunción, vicinity, 25°16′S, 57°40′W
Jardín Botanico, Río Miranda, Asunción, 25°16′S, 57°40′W
Luque, 17 km by road E, ca. 25°16′S, 57°26′W
San Lorenzo, 25°20′S, 57°32′W
Tapuá, 25°11′S, 57°49′W
(See Tate 1933:218, footnote.)

Concepción (Departamento de)

Belén, Río Ypané, 23°30′S, 57°06′W
Concepción, 175 m, 23°25′S, 57°17′W
Concepción, 8 km E, 23°25′S, 57°13′W
Estancia Estrellas et Estancia Primavera, entre, 22°17′S, 57°40′W
Estrellas Ranch and Primavera Ranch, between, 22°17′S, 57°40′W
Parque Nacional Serranía de San Luis, 170 m, 22°40′S, 57°21′W
San Lázaro, 1 km NE, 22°10′S, 57°54′W

Cordillera (Departamento de La)

Estancia Sombrero, 110 m, 25°05′S, 56°36′W
Salto de Pirateta, 10 km S of Piribebuy, 25°26′S, 57°02′W
Tobatí, 12 km (by road) N, 25°10′S, 57°04′W

Guairá (Departamento de)

Colonia Independencia, 200 m, 25°43′S, 56°15′W
Itapé, 25°51′S, 56°38′W
Paso Yobay, ca. 20 km S of Caáguazu (= Paso Yuvay or Yjhovy), 25°37′S, 56°02′W
Villa Rica (= Villarrica), 25°45′S, 56°26′W
Villarica (= Villarrica), 25°45′S, 56°26′W

Itapúa (Departamento de)

Arroyo San Rafael, 27°08′S, 56°23′W
Río Pirapó, near mouth, 26°44′S, 56°36′W
Río Pirayu-i, 26°51′S, 55°14′W
Salto Tembey, Río Pirayu, 26°42′S, 55°02′W
San Benito, 26°38′S, 55°43′W
Santa María, 27°18′S, 55°54′W
San Rafael, 8 km N, ca. 26°36′S, 54°53′W

Misiones (Departamento de)

Curupayty, 26°50′S, 56°50′W

San Ignacio, 75 m, 26°52′S, 57°03′W
(See Tate 1933:22, footnote.)

Ñeembucú (Departamento de)

Tacuaraí, 27°01′S, 57°44′W

Paraguarí (Departamento de)

Cordillera de los Altos, ca. 7 km NW of Paraguarí, ca. 26°36′S, 57°10′W
Parque Nacional Ybycuí, 25°46′S, 57°03′W
Sapucaí, ca. 300 m, 25°40′S, 56°55′W
Sapucay (= Sapucaí), ca. 300 m, 25°40′S, 56°55′W

Presidente Hayes (Departamento de)

Colonia Benjamín Aceval, 24°58′S, 57°34′W
Estancia La Victoria, 120 m, 23°29′S, 58°35′W
Juan de Zalazar, 5–7 km W, ca. 23°06′S, 60°01′W
Misión Inglesa, 2 km SE, 23°28′S, 58°15′W
Puerto Cooper, 23°03′S, 57°43′W
Puerto Pinasco, 150 m, 22°43′S, 57°50′W
Rincón Charrúa, 275 km by road NW of Villa Hayes, 23°18′S, 59°02′W
Villa Hayes, 230 km by road NW, ca. 23°23′S, 58°46′W
Villa Hayes, 24 km WNW, 24°59′S, 57°43′W

San Pedro (Departamento de)

Nueva Germania, 23°54′S, 56°34′W
Tacuatí, Acá-poí, 23°27′S, 56°35′W

Peru
Amazonas (Departamento de)

Bagua Grande, 15 km WNW, 05°33′S, 78°33′W
Balsas, 3 km E, 06°50′S, 77°59′W
Bellavista, right bank Río Marañon, 05°30′S, 78°40′W
Chiriaco, 43 km NE, 04°47′S, 77°56′W
Chiriaco, 43 km (by road) NE, 04°47′S, 77°56′W
Chiriaco, 20 km SW, 04°57′S, 78°15′W
Chirimoto, 2,100 m, 06°31′S, 77°24′W
Condechaca, Río Utcubamba, 06°20′S, 77°52′W
Cordillera Colán, east of La Peca, 3,165 m, ca. 05°34′S, 78°17′W
Cordillera del Cóndor, ca. 04°50′S, 78°35′W
Corosha, 1,500 m, 05°50′S, 77°45′W
Falso Paquisha, Río Comaina, Cordillera del Cóndor, ca. 04°10′S, 78°23′W

Río Kagka, headwaters, a small tributary of the Río Cenepa, 790 m, 04°16′S, 78°10′W

Huampami (= Huampam), Río Cenepa, 210 m, 04°28′S, 78°10′W

Laguna Pomacocha, 6 km by road SW, 1,800 m, ca. 05°52′S, 77°57′W

La Peca, 12 km by trail E (= La Peca Nueva), ca. 2,000 m, ca. 05°34′S, 78°17′W

La Peca, 10 km by trail SE, 1,325 m, 05°36′S, 78°17′W

La Peca, 10 trail km SW, 05°39′S, 78°20′W

La Poza, 170 m, 04°02′S, 77°44′W

Pomacochas, 5 km N and 5 km E, 1,800 m, 05°47′S, 77°53′W

Pomará, 05°16′S, 78°26′W

Puesto Vigilancia 3, Alfonso Ugarte, Cordillera del Condor, 03°54′S, 78°25′W

Quebrada Kagka, 790 m, 04°16′S, 78°10′W

Río Cenepa, mouth of, 04°40′S, 78°12′W

Río Huampami, headwaters of, east of Huampami, Río Cenepa, 04°47′S, 78°17′W

Río Kagka, 790 m, 04°15′S, 78°10′W

Santa Rosa, 04°01′S, 77°47′W

Soledad, 03°30, 77°44W

Tambo Carrizal, 2,000 m, 06°47′S, 77°52′W

Ancash (Departamento de)

Chasquitambo, 10°18′S, 77°36′W

Huaylas, 2,720 m, 08°52′S, 77°54′W

Pariacoto, 1 km N and 12 km E, ca. 09°30′S, 77°46′W

Río Pira Valley, 1 km N and 12 km E of Pariacoto, 2,500 m, ca. 09°30′S, 77°46′W

Apurímac (Departamento de)

Quebrada Matará, 13°45′S, 72°54′W

Arequipa (Departamento de)

Arequipa, 16°24′S, 71°33′W

Arequipa, 12 miles E, 16°23′S, 71°21′W

Arequipa, 18 km E, 16°23′S, 71°20′W

Cariveli (= Caraveli), 15°46′S, 73°22′W

Chucarapi, 120 m, 17°04′S, 71°44′W

Islay, sea level, 16°59′S, 72°07′W

Mollendo, 3 miles N, 17°00′S, 72°01′W

Valle del Tambo, 17°10′S, 71°48′W

Ayacucho (Departamento de)

Apurimac (= Río Apurimac at Hacienda Luisiana), 12°40′S, 73°44′W

Ayacucho, 13°09′S, 74°13′W

Estera Ruana, 1,900 m, ca. 12°43′S, 73°49′W

Hacienda Luisiana, 12°40′S, 73°44′W

Huahuanchayo (= Huanhuachayo), ca. 1,660 m, 12°44′S, 73°47′W

(Huanhuachayo is "a clearing along the Andean mule trail connecting Hacienda Luisiana and nearby communities along the Río Apurimac and Río Santa Rosa with the mountain town of Tambo" [Gardner and Patton 1976:42].)

Huanhuachayo, 1,660 m, 12°44′S, 73°47′W

Luisiana (= Hacienda Luisiana), 12°40′S, 73°44′W

Puquio, 15 miles WNW, 3,660 m, 14°37′S, 74°21′W

San José, Río Santa Rosa above Hacienda Luisiana, 1,000 m, 12°44′S, 73°46′W

Yuraccyacu, 2,600 m, 12°45′S, 73°48′W

Cajamarca (Departamento de)

Bellavista, 05°37′S, 78°42′W

Cajamarca, 35 mi WNW, 1,830 m, ca. 06°48′S, 78°44′W

Cajamarca, 07°10′S, 78°31′W

Grutas de Ninabamba, Hualgayoc, 06°40′S, 78°47′W

Gruta de Niñabamba, Hacienda Niñabamba (Niñabamba = Minabamba), 06°40′S, 78°47′W

Hacienda Limón (= Limón), Río Marañón, ca. 2,000 m, 06°52′S, 78°05′W

Jaén, 2,000 m, 05°42′S, 78°50′W

Jaén, 9 km S, 05°44′S, 78°45′W

Las Ashitas, ca. 42 km W of Jaén, 3,150 m, 05°42′S, 79°08′W

Quebrada Huarandosa, Río Chinchipe, 05°12′S, 78°48′W

Río Zaña, 2 km N of Monteseco, 06°51′S, 79°05′W

San Ignacio, 05°08′S, 78°59′W

Tambillo, ca. 06°10′S, 78°45′W

Taulís, 2670 m, 06°54′S, 79°03′W

Tolón, 07°18′S, 79°19′W

Yayen (= Yagen), 06°31′S, 78°16′W

Cusco (Departamento de)

Bosque Aputinye, ca. 13°00′S, 72°32′W

Bosque de Nubes, km 135 on carretera Paucartambo-Pillcopata, ca. 13°16′S, 71°35′W

Calca, 55.4 km by road N, 3,560 m, 13°10′S, 72°00′W

Cashiriari-2, 469 m, 11°52′S, 72°46′W

Cashiriari-3, 579 m, 11°53′S, 72°39′W

Consuelo, 17 km by road W of Pilcopata at Km 165, ca 1,200 m, ca. 13°08′S, 71°15′W

Cordillera Vilcabamba, 13°00′S, 73°00′W

Cordillera Vilcabamba, west side, 12°39′S, 73°30′W

Cordillera Vilcabamba, W side, 12°39′S, 73°30′W

Cordillera Vilcabamba (summit), 3,540 m, 13°00′S, 73°00′W

Cosñipata, 13°04′S, 71°11′W

Hacienda Cadena, 13°24′S, 70°43′W

Hacienda Erika, 13°55′S, 71°12′W

Hacienda Huyro, between Huayopata and Quillabamba, ca 1,830 m, ca. 12°58′S, 72°36′W

Hacienda Villa Carmen, Cosñipata (= Río Cosñipata), 600 m, 12°50′S, 71°15′W

Huadquiña, 1,500 m, 13°07′S, 72°39′W

Huajyumbe (= Huayjllumbe), 13°15′S, 70°30′W

Huasampilla (= Huaisampillo), ca. 13°14′S, 71°26′W-S&T

Itahuanía, 450 m, 12°47′S, 71°13′W

Km 134, 54 km by road NE of Paucartambo, 2,190 m, 13°08′S, 71°34′W

Km marker 112, 32 km by road NE of Pucartambo, below Acjanaco, 3,200 m, ca. 13°25′S, 71°30′W

Konkariari, 300 m, 11°48′S, 72°52′W

La Convención, 695 m, 11°53′S, 72°39′W

Las Malvinas, 480 m,11°53′S, 72°56′W

Limacpunco, 2,400 m, 13°28′S, 70°55′W

Limbani, vicinity, 14°08′S, 69°42′W

Llactahuamán, ca. 12°52′S, 73°31′W

Machu Picchu, 1,830 m, 13°07′S, 72°34′W

Marcapata, 13°30′S, 70°55′W

Nusiniscato River, 13°15′S, 70°34′W

Ocobamba, 1,900 m, 12°52′S, 72°22′W

Ollantaytambo (= Ollantaitambo), 13°16′S, 72°16′W

Pacuaratambo, 72 km by road NE, 1,460 m, ca. 13°11′S, 71°33′W

Paucartambo, 10 km by road N, 3,400 m, 13°14′S, 71°38′W

Pagoreni, 465 m, 11°42′S, 72°54′W

Peruanita, 350 m, 11°41′S, 73°00′W

Pillahuata, 13°08′S, 71°25′W

Pozo Cashiriari C. (= Cashiriari-3), Camisea, 579 m, 11°53′S, 72°39′W

Puente Unión, Km marker 150, Albergue Bosque de las Nubes, 1,700 m, ca. 13°09′S, 71°20′W

Puquiura (= Pucyura), 3,351 m, 13°29′S, 72°07′W

Quillabamba, 90 km by road SE (below Abra Málaga), 3,540 m, ca. 13°10′S, 72°32′W

Quince Mil (= Quincemil), 13°13′S, 70°42′W

Quincemil, 13°13′S, 70°42′W

RAP Camp 1, north slope of Cordillera Vilcabamba, 3,350 m, 11°40′S, 73°38′W

RAP Camp Two, north slope of Cordillera Vilcabamba, 2,050 m, 11°33′S, 73°38′W

Ridge Camp (RAP Camp 3), 1,000 m, 11°46′S, 73°20′W

Río Cosñipata, 13°02′S, 71°10′W

Río Mapitunare (= Río Mapitonoare), 12°24′S, 73°45′W

Río Mapitunari (= Río Mapitonoare), 12°24′S, 73°45′W

Río Pagoreni, 2 km SW of Tangoshiari, 1,000 m, 11°46′S, 73°27′W

San Jerónimo, 13°34′S, 71°54′W

San Martín-3, La Convención, 487 m,11°47′S, 72°42′W

Santa Ana, 1,050 m, 12°52′S, 72°43′W

Segakiato, 330 m, 11°48′S, 72°53′W

Tangoshiari, 2 km SW, Río Pagoreni, 11°46′S, 73°26′W

Tocopoqueu (= Tocopogueyu), Ocobamba Valley, 2,780 m, 12°53′S, 72°21′W

Tono, 5 km S of Río Tono, 18 road km W of Patria, ca. 875 m, ca. 13°01′S, 71°19′W

Torontoy, 2,400 m, 13°10′S, 72°30′W

Wayrapata, La Convención, 2,445 m, 12°50′S, 73°30′W

Huancavelica (Departamento de)

Córdova, 14°01′S, 75°10′W

Huancavelica, 3,660 m, 12°46′S, 75°03′W

Lircay, ca. 3,500 m, 13°00′S, 74°43′W

Ticrapo, 2 km E, 13°21′S, 75°24′W

Huánuco (Departamento de)

Carpish Pass, 09°42′S, 76°09′W

Cerros del Sira, 09°30′S, 74°40′W

Chinchavita (= Chinchavito), 1,000 m, 09°29′S, 75°55′W

Cordillera Carpish, Carretera Central, 2,400 m, 09°42′S, 76°09′W

Cordillera Carpish, eastern slope, 2,400 m, 09°42′S, 76°09′W

Cueva de Castillo, 5 km NW of Tingo María, 09°07′S, 75°58′W

Cueva de las Lechuzas, 5 km SW of Tingo María, 09°20′S, 76°01′W

Cueva de Lechuzas, 5 km SW of Tingo María, 09°20′S, 76°01′W

Divisoria (= La Divisoria), Cordillera Azul, 1,600 m, ca. 09°05′S, 75°46′W

Hacienda Buena Vista, Río Chinchao, ca. 1,070 m, 09°31′S, 75°52′W

Hacienda Éxito, Río Cayumba, 915 m, 09°26′S, 76°00′W

Hacienda Paty, trail to, NE of Carpish Pass, ca. 09°38′S, 76°08′W

Huánuco, 09°56′S, 76°15′W

Leguía, Pachitea drainage, 09°20′S, 75°00′W

Panguana, Río Llullapichis, near Río Pachitea, 260 m, 09°37′S, 74°56′W

Panguana Biological Station, Río Llullapichis, 09°37′S, 74°56′W

Paso Carpish, 09°42′S, 76°09′W

Puerto Márquez (= San Márquez), Río Pachitea, ca. 100 m, 09°32′S, 74°56′W

Río Pachitea (Junction with Río Ucayali, 08°45′S, 74°32′W).

(Type locality of *Anthorina peruana* O. Thomas, 1923. The holotype was acquired by L. Rutter on Río Pachitea at "1500 ft" in the general vicinity of its confluence with Río Palcazu. The mapped place name closest to this site within the Departamento de Huánuco on the upper Río Pachitea is Puerto Victoria, 09°53′S, 74°47′W.)

San Antonio, Río Chinchao, ca. 09°33′S, 75°52′W

Santa Elena, 09°07′S, 75°51′W

Tingo María, 3 km N, 09°16′S, 75°58′W

Tingo María, 3 miles N, 800 m, 09°17′S, 75°58′W

Tingo María, 2 miles N, 500 m, 09°17′S, 75°59′W

Tingo María, 600 m, 09°18′S, 75°59′W

Tingo María, Hotel Turistas, 09°19′S, 75°59′W

("1 km S of the center of Tingo María and adjacent to the east bank of the Río Huallaga" [Bowles, Cope, and Cope 1979:3].)

Tingo María, 1 km S, 09°19′S, 75°59′W

Tingo María, 3 miles S, 730 m, 09°21′S, 75°59′W

Tingo María, 2 km E and 9 km S, 09°23′S, 75°58′W

Tingo María, 19 km S, 09°28′S, 75°53′W

Tingo María, 19 miles S, 840 m, 09°32′S, 76°00′W

Tingo María, 31 km S, 850 m, 09°32′S, 76°00′W

Zapatogocha (= Bosque Zapatogocha), Cordillera Carpish, ca. 09°40′S, 76°03′W

Ica (Departamento de)

Santiago, 400 m, 14°11′S, 75°44′W

Vieja Island (= Isla Independencia), sea level, 14°15′S, 76°12′W

Junín (Departamento de)

Acobamba, 3,500 m, 11 20′S, 75°41′W

Chanchamayo, near Tarma, 900 m, 11°15′S, 75°19′W

("Chanchamayo, near Tarma, approximately in lat. 11° 20′ S., and long. 75° 40′ E. about 3000 feet" [O. Thomas 1893b].)

Chanchamayo (= Pueblo Nuevo), 600 m, 11°10′S, 75°16′W (see Gardner and Patton, 1976:42)

Chanchamayo Valley, 3 miles SW of San Ramón, 11°10′S, 75°22′W

Huacapistana, 11°14′S, 75°29′W

Maraynioc, 11°22′S, 75°24′W

Perené, 10°58′S, 75°18′W

Río Perené, ca. 32 km N of Satipo, 10°54′S, 74°51′W

San Ramón, 11°08′S, 75°20′W

(San Ramón of Tuttle [1970:82], "3 km. NW San Ramón....," 11°07′S, 75°23′W)

San Ramón, 3 km NW, 11°07′S, 75°23′W (= San Ramón of Tuttle (1970:82).)

San Ramón, 2 km NW, 884 m, 11°07′S, 75°22′W

Tarma, (of Tuttle 1970, = 11°08′S, 75°22′W)

(= "32–35 km. NE Tarma.... Netting was done near and over a stream 12–15° feet wide at places 600–800 feet from where it joined the Río Palca. Nearly vertical canyon walls extend upwards on both sides of the stream valley for more than 1000 feet. Brushy forest [10–35 feet tall] bordered the stream, giving way to scattered shrubs, grass and rocky cliffs within 150 feet" [Tuttle 1970:82].)

Tarma, near (= San Bartolomé; Vaurie [1972:32]), ca. 11°15′S, 75°19′W

Utcuyacu, Perené drainage, 1,465 m, 11°12′S, 75°28′W

Vitoc, 3.2 km N, Río Tulumayo, 850 m, 11°10′S, 75°15′W

Vitoc, 3 km N, Río Tulumayo, 850 m, 11°10′S, 75°15′W

Vitoc Valley, ca. 11°18′S, 75°20′W

La Libertad (Departamento de)

Machi (= Mache), 3,000 m, 08°02′S, 78°32′W

Menochucho, 20 miles E of Trujillo, 08°01′S, 78°50′W

Otuzco, 2,641 m, 07°54′S, 78°35′W

Pacasmayo, 8 m, 07°24′S, 79°34′W

Santiago, 3,100 m, 08°09′S, 78°11′W

Trujillo, 08°05′S, 78°55′W

Lambayeque (Departamento de)

Cabache, 06°33′S, 79°15′W

Cerro la Vieja (= Sierra Vieja), 7 km S of Motupe, 06°13′S, 79°44′W

Eten, 06°54′S, 79°52′W

Lambayeque, 06°42′S, 79°55′W

Las Juntas, Quebrada Pachinga, ca. 06°27′S, 79°52′W

Motupe, 7 km S, 06°13′S, 79°44′W

Olmos, 12 km N, 150 m, 05°55′S, 79°47′W

Olmos, 5.2 km N, 05°52′S, 79°47′W

Olmos, 05°59′S, 79°46′W

Puerto Eten, 06°56′S, 79°52′W

Sierra Vieja (= Cerro la Vieja), 7 miles S of Motupe, 06°13′S, 79°44′W

Lima (Departamento de)

Asia, 46 m, 12°47′S, 76°30′W

Bosque de Zárate, ca. 11°53′S, 76°27′W

Buhama Baja (= Bujama Baja), 95 km by road S of Lima, sea level, 12°43′S, 76°39′W

Callao, sea level, 12°04′S, 77°09′W

Cañete, 13°05′S, 76°24′W

Canta, 1 mile W, ca. 11°28′S, 76°39′W

Cerro Agustino, Lima, 12°04′S, 77°00′W
Chosica, 11°54′S, 76°42′W
Lima, 12°03′S, 77°03′W
Lomas de Lachay, 11°21′S, 77°23′W
Matucana, 3 miles E, 11°51′S, 76°21′W
Surco, 1,900 m, 11°52′S, 76°28′W
Yangas, 11°43′S, 76°51′W

Loreto (Departamento de)

Apayacu, 03°19′S, 72°07′W
Boca del Río Curaray, 200 m, 02°22′S, 74°05′W
Centro de Investigaciones "Jenaro Herrera," 04°55′S, 73°45′W
Contamana, 134 m, 07°20′S, 75°01′W
Curaray River mouth (= Boca del Río Curaray), 200 m, 02°22′S, 74°05′W
Estación Biológica Allpahuayo, 25 km SW of Iquitos, 171 m, 03°58′S, 73°25′W
Estación Biológica Pithecia, Río Samiri, 220 km SW of Iquitos, ca. 04°42′S, 74°13′W
Iquitos, 03°46′S, 73°15′W
Jenaro Herrera, 04°55′S, 73°45′W
Jenaro Herrera, Requena, 05°03′S, 73°50′W
Mishana, ca. 03°48′S, 73°32′W
Nuevo San Juan, Río Gálvez, 150 m, 05°15′S, 73°10′W
Orosa, ca. 200 m, 03°26′S, 72°08′W
Parinari, 04°34′S, 74°26′W
Pebas, 100 m, 03°20′S, 71°49′W
Puerto Almendra, km 6 de la carretera Iquitos-Nauta, 120 m, 03°30′S, 73°14′W
Puerto Arturo, 100 m, 05°50′S, 76°03′W
Puerto Indiana, 03°29′S, 73°03′W
Puerto Meléndez, 500 m, 04°27′S, 77°31′W
Quebrada Aucayo, 03°50′S, 73°05′W
Quebrada Esperanza, Río Yavarí Mirim, 200 m, 04°21′S, 71°58′W
(First major left-bank tributary of the Río Yarvarí Mirim; Quebrada Esperanza is probably near mouth. This is a Kalinowski locality [R. S. Voss, pers. comm.].)
Quebrada Pushaga (= Río Puchuya), Río Morona, 220 m, 03°56′S, 77°14′W
Quebrada Sucusari, 03°15′S, 72°54′W
Reserva Nacional Allpahuayo-Mishana, 03°58′S, 73°25′W
Río Amazonas, ca. 10 km SSW of mouth Río Napo on E bank of Quebrada Vainilla (= Río Marupa Caño), 100 m, 03°35′S, 72°45′W
Río Curaray, junction with Río Napo, 200 m, 02°22′S, 74°05′W
Río Curaray, mouth, 200 m, 02°22′S, 74°05′W
Río Gálvez, 73°05′S, 05°15′W

Río Nanay, lower, a few kilometers W of Iquitos, ca. 03°45′S, 73°20′W
Río Santa María, left bank of Río Napo, 01°20′S, 74°40′W
San Lorenzo, Río Marañón, 04°49′S, 76°36′W
Santa Cruz, near Yurimaguas, Río Huallaga, 150 m, 05°33′S, 75°48′W
Sarayacu, Río Ucayali, 06°44′S, 75°06′W
Yurimaguas, 05°54′S, 76°05′W

Madre de Dios (Departamento de)

Aguas Calientes, ca. 1 km below Shintuya on left bank Río Madre de Dios, 450 m, ca. 12°41S, 71°18′W
Alberque, Río Madre de Dios, ca. 12 km E of Puerto Maldonado, 200 m, 12°33′S, 69°05′W
Camp 5, left bank Río Molinowski, 12°54′S, 69°39′W
Camp 8, left bank Río Molinowski, 13°01′S, 69°55′W
Ccolpa de Guacamayos, west bank Río Tambopata, 188 m, 13°09′S, 69°36′W
Cerro de Pantiacolla, 600 m, 12°35′S, 71°18′W
("Cerro de Pantiacolla lies above the Río Palota, 10–15 km NNW of Shintuya" [Timm 1987:190].)
Cerros de Távara, 500 m, 13°02′S, 69°41′W
Cocha Cashu, 45 km NW (= 80 river km upstream) of the mouth of the Río Manu and about 8 km inside the border of Manu National Park, 380 m, 11°51′S, 71°19′W
Cocha Salvador, ca. 12°00′S, 71°28′W
Hacienda Amazonía, 500–600 m, ca. 12°56′S, 71°11′W
Hacienda Erika, Río Alto Madre de Dios, 450 m, 12°54′S, 71°12′W
Itahuana, 12°47′S, 71°13′W
Manú, 12°15′S, 70°54′W
Maskoitania, ca. 12°52′S, 71°17′W
Pakitza, 350 m, 11°55′S, 71°15′W
Pampas de Heath (= Pampa del Heath), ca. 50 km (by river) S of Puerto Pardo, 160 m, ca. 13°00′S, 69°00′W
Pantiacolla (= Cerro de Pantiacolla), 12°35′S, 71°18′W
Reserva Cuzco Amazónico, north bank of Río Madre de Dios, 14 km E of Puerto Maldonado, ca. 200 m, 12°33′S, 69°03′W
Río La Torre, 12°51′S, 69°17′W
Río Palotoa (= Río Pantiacolla), left bank, 12 km from mouth, 490 m, ca. 12°35′S, 71°26′W

Pasco (Departamento de)

Cerro Jonatán, 750 m, ca. 10°19′S, 75°15′W
Huancabamba, 10°21′S, 75°32′W
Iscozacín, 10°11′S, 75°10′W
Nevati, 300 m, 10°21′S, 74°51′W
("Nevati is a mission station surrounded by a village of about 250 Campa Indian inhabitants. It is located

on the north bank of the Río Pichis, about 10 km. SE Puerto Bermudez. The surrounding land is undulating and the virgin evergreen forest is 60–120 feet tall" [Tuttle 1970:81].)

Oxapampa, 300 m, 10°34'S, 75°24'W

Palmira, 10°27'S, 75°27'W

Posuzo (= Pozuzo), upper Río Pachitea (Río Pozuzo), 1,000 m, 10°04'S, 75°32'W

Pozuzo (formerly in Depto. Huánuco), 1,000 m, 10°04'S, 75°32'W

Puerto Victoria, Río Pachitea, 09°54'S, 74°58'W

San Alberto, ca. 10°34'S, 75°30'W

San Juan, 274 m, 10°30'S, 74°53'W

("San Juan is a Campa Indian village, population about 75, located on the east bank of the Río Azupizu about 5 km. from San Pablo. Immediately behind the village to the southeast is a sharply rising hill that reaches an elevation of about 2000 feet. Beyond, the country becomes rugged, with large hills and many cliffs. San Juan is surrounded by virgin evergreen forest 60–120 feet tall" [Tuttle 1970:82].)

San Pablo, 10°27'S, 74°52'W

("San Pablo is a Campa Indian village of about 175 inhabitants located in undulating country on the east bank of the Río Azupizu" (Tuttle 1970:81).)

Yanachaga-Chemillen, 2,600 m, ca. 10°34'S, 75°30'W

(Collecting site at San Alberto, located on eastern limit of Parque Nacional Yanachaga-Chemillen, a few kilometers west of Oxapampa (S. Solari, pers. comm.).)

Piura (Departamento de)

Amotape, 12 m, 04°52'S, 81°01'W

Angola (= Angolo), 04°50'S, 80°45'W

Canchaque, 05°24'S, 79°36'W

Canchaque, 15 km by road E, 1,740 m, ca. 05°24'S, 79°30'W

Canchaque, 15 km E, 1,740 m, ca. 05°24'S, 79°28'W

Cerro Chinguela, 2,900 m, 05°07'S, 79°23'W

Fondo (= Fondos), 14 km N and 13 km E of Talara, 04°27'S, 81°10'W

Hacienda Bigotes (= Bigote), 05°19'S, 79°48'W

Huancabamba, 05°14'S, 79°28'W

Km 30, on road from Huancabamba to San Ignacio, 3,000 m, ca. 05°13'S, 79°17'W

Las Juntas, Río La Pachinga, ca. 14 km N and 25 km E of Olmos (Depto. Lambayeque), 300 m, 05°45'S, 79°42'W

Machete, on Zapalache-Carmen trail, ca. 05°04'S, 79°24'W

Mallares, 04°51'S, 80°46'W

Monte Grande, 14 km N and 25 km E of Talara, 04°28'S, 81°03'W

Olmos (Depto. Lambayeque), 25 km E, 1,290 m, 05°52'S, 79°34'W

Palambla, 05°23'S, 79°37'W

Papayal, 05°14'S, 79°39'W

Piura, 05°12'S, 80°38'W

Salitral, Río Chira, 04°51'S, 80°42'W

San Isidro, 5 km W, 05°59'S, 79°47'W

Sullana, 04°53'S, 80°41'W

Suyo, 6.4 km W, 04°32'S, 80°05'W

Suyo, 6 km W, 04°33'S, 80°05'W

Suyo, 4 miles W, 305 m, 04°31'S, 80°04'W

Talara, 7 km N and 15 km E, 04°25'S, 81°03'W

Talara, 04°34'S, 81°17'W

Puno (Departamento de)

Abra de Maruncunca, 10 km SW of San Juan del Oro, 3,000 m, ca. 14°16'S, 69°14'W

Abra Marracunco, 14 km W of Yanahuaya, 2,210 m, 16°24'S, 71°41'W

Agualani, 9 km N of Limbani, 2,840 m, 14°06'S, 69°42'W

Bella Pampa, 16 miles N and 2 miles E of Limbani, 13°55'S, 69°38'W

Coasa, 14°00'S, 69°88'W

Inca Mines (= Santo Domingo), 1,800 m, 13°51'S, 69°41'W

Fila Boca Guacamayo, Río Tavara, 13°30'S, 69°41'W

"Juliaca" (=Inca Mines), 13°51'S, 69°41'W

("Juliaca," as used by J. A. Allen [1900:219–28], and Handley [1960], actually is Inca Mines on the Río Inambari according to J. A. Allen [1901:41–46].)

La Pampa, 573 m, 13°39'S, 69°36'W

Limbani, vicinity, 2,600 to 3,150 m, 14°08'S, 69°42'W

Oconeque, Río Quitún, 1,956 m, 14°03'S, 69°42'W

Ollachea, 11 km NNE, 1,875 m, 13°44'S, 70°26'W

Puno, 15°50'S, 70°02'W

Río Huari Huari, 5 km by road NE of San José, ca. 14°00'S, 69°15'W

Río Inambari (= Sagrario, Río Quitún according to Sanborn [1951]), 670 m, 13°55'S, 69°41'W

Río Tavara, SE of Fila Boca Guacamayo, 13°30'S, 69°41'W

Río Yahuaramayo, 13°18'S, 70°18'W

Sagraria (= Sagrario), 13°55'S, 69°41'W

Sandia, 14°17'S, 69°26'W

Santo Domingo, 1,800 m, 13°51'S, 69°41'W

Valle Grande, 14°15'S, 69°12'W

Zona Reservada Tambopata-Candamo, Sandia, 500 m, 13°21'S, 69°59'W

San Martín (Departamento de)

Calaveras, 06°19'S, 76°49'W

Gran Pajatén, 07°29'S, 77°22'W

Guayabamba (= Santa Rosa de Huayabamba), 1,500 m, 06°22'S, 77°25'W

La Playa, ca. 28 km NE of Pataz (Depto. La Libertad), ca. 07°33'S, 77°26'W

Las Palmas, Parque Nacional del Río Abiseo, 2,100 m, ca. 07°33'S, 77°23'W

Moyabamba (= Moyobamba), 860 m, 06°03'S, 76°58'W

Moyobamba, 860 m, 06°03'S, 76°58'W

Pachiza, 07°16'S, 76°46'W

Pampa del Cuy, Parque Nacional Río Abiseo, 07°34'S, 77°27'W

Parque Nacional Abiseo, ca. 07°29'S, 77°22'W

Puca Tambo, 1,480 m, 06°10'S, 77°16'W

Rioja, 06°10'S, 77°10'W

Río Negro, ca. 800 m, 05°56'S, 77°09'W

Roque, 1,097 m, 06°24'S, 76°48'W

San Martín, 06°30'S, 76°45'W

Shapaja, 06°36'S, 76°16'W

Tarapoto, 360 m, 06°30'S, 76°25'W

Yurac Yacu, 05°57'S, 77°11'W

("Yurac Yacu is a rather large village at the junction of the Yurac Yacu stream with the Río Mayo. It is 35 miles east of Jumbilla on the opposite side of the mountains, and a little over 20 miles WNW of Moyobamba" [O. Thomas 1927a:361].)

Tacna (Departamento de)

Tarata, 10 miles S, 17°36'S, 70°01'W

Tumbes (Departamento de)

Campo Verde, 03°51'S, 80°10'W

Corrales (= San Pedro de los Incas), 03°36'S, 80°29'W

Huásimo (= El Huásimo), 04°01'S, 80°31'W

Matapalo, 03°41'S, 80°12'W

Quebrada Faical, 2 km E of El Caucho, 03°50'S, 80°15'W,

Tumbes, 03°34'S, 80°28'W

Zorritos, 03°40'S, 80°40'W

Ucayali (Departamento de)

Alto Río Tamaya, 08°59'S, 73°19'W

Abra Divisoria, 3 km NE, 09°03'S, 75°43'W

Balta, Río Curanja, 300 m, 10°06'S, 71°14'W

Bosque Nacional Alexander von Humboldt, 250 m, 08°47'S, 75°08'W

Cerro Tihuayo, Pucallpa, not located (Pucallpa = 08°23'S, 74°32'W)

Chicosa, 10°28'S, 74°03'W

Cumaría, 100 m, 09°51'S, 74°01'W

Cumeria (= Cumaría), 09°51'S, 74°01'W

IVITA Biological Station, 59 km SW of Pucallpa on highway to Lima, ca. 08°35'S, 74°52'W

Lake Yarinacocha, 100 m, 08°15'S, 74°43'W

Largato, Upper Río Ucayali, 10°40'S, 73°54'W

Pucallpa, 08°23'S, 74°32'W

Pucallpa, 11 km SE, 08°27'S, 74°28'W

Pucallpa, 59 km W, 08°35'S, 74°52'W

Pucallpa, 59 km SW, ca. 08°35'S, 74°52'W

Pucallpa, 61 miles SE, 180 m, ca. 08°50'S, 73°55'W

Río Disqui (= Río Pisqui), confluence with Río Ucayali, 07°39'S, 75°03'W

Santa Rosa, Alto Ucayali, 300 m, 10°42'S, 73°50'W

Tushemo (= Tushma), near Masisea, Río Ucayali, 225 m, 08°37'S, 74°21'W

(See O. Thomas [1928c:250] for location; specimens from this place often reported as from Masisea.)

Yarinacocha, Río Ucayali, 08°17'S, 74°39'W

Surinam

Brokopondo (District)

Brokopondo, 70 m, 05°04'N, 54°58'W

Brownsberg Nature Park, 3 km S and 20 km W of Afobakka, 04°55'N, 55°12'W

Brownsberg (= Browns Berg), 514 m, 04°53'N, 55°13'W

Brownsweg, 8 km S and 2 km W, 04°55'N, 55°11'W

Finisanti (= Finisanti Presie), Saramacca River, 05°08'N, 55°29'W

Gros, about 100 km S of Paramaribo, 05°06'N, 55°15'W

Ligolio, east bank of Gran Rio, 03°54'N, 55°33'W

Rudi Kappelvliegveld, 1.5 km W, 330 m, 03°47'N, 56°10'W

Rudi Kappelvliegveld, 03°47'N, 56°09'W

Rudi Kappelvliegveld, 3 km SW, 320 m, 03°46'N, 56°10'W

Commewijne (District)

Bergendal (= Berg en Dal), 50 m, 05°09'N, 55°04'W

Coropina Kreek, near Republiek, 05°32'N, 55°10'W

Meerzorg, 05°49'N, 55°09'W

Nieuwe Grond Plantation, 05°51'N, 54°57'W

Plantation Clevia, west bank Suriname River, 05°52'N, 55°06'W

Coronie (District)

Totness, 05°53'N, 56°19'W

Marowijne (District)

Albina, sea level, 05°30'N, 54°03'W

Albina, 3 km SW, 05°29'N, 54°04'W

Galibi, near mouth of Marowijne River, sea level, 05°43′N, 54°01′W

Langamankondre, 05°43′N, 54°01′W

Marowijne and Gonini river junction, 04°10′N, 54°26′W

Moengo, 10 km N and 24 km W, 05°43′N, 54°38′W

Moengo, 05°37′N, 54°24′W

Perica, 05°44′N, 54°43′W

Nickerie (District)

Wageningen, 05°46′N, 56°41′W

Para (District)

Carolina Kreek, 05°24′N, 55°11′W

Jodensavanne, Surinam River, 05°25′N, 54°59′W

Powaka (= Powakka), 05°27′N, 55°05′W

Powakka, 05°27′N, 55°05′W

Powaka, 1 km S and 2 km E, 05°26′N, 55°03′W

Zanderij, 05°27′N, 55°12′W

Zanderijweg, near Zanderij, 05°26′N, 55°14′W

Paramaribo (District)

Coronie, near Totness, 05°53′N, 56°19′W

Leonsberg, sea level, 05°53′N, 55°07′W

Paramaribo, 05°50′N, 55°10′W

Plantage Blauwgrond, near Paramaribo, 05°50′N, 55°08′W

Saramacca (District)

Geyskes Creek, Tafelberg, 700 m, 03°55′N, 56°10′W

Groningen, 05°48′N, 55°28′W

La Poule, 05°47′N, 55°25′W

Tafelberg, 700 m, 03°55′N, 56°10′W

Tibiti River, 05°26′N, 55°55′W

Sipaliwini (District)

Anton van Aerde Cave, Tafelberg Mtn., 03°56′N, 56°09′W

Avanavero, 04°50′N, 57°14′W

Bitagron (= Witagron), 05°10′N, 56°06′W

Camp One, Coppename River, above Kaaimanston, 05°04′N, 56°06′W

Coppename River, above Kaaimanston, 05°07′N, 56°09′W

Grassalco, 04°46′N, 56°46′W

Kaaimanston, 05°06′N, 56°05′W

Kaiserberg Airstrip, Zuid River, 280 m, 03°06′N, 56°28′W

Kayserberg Airstrip, 03°06′N, 56°28′W

King Frederick Wilhelm Falls, Corentyne River, 112 m, 03°28′N, 57°37′W

Marowijne and Gonini rivers, near junction, 04°15′N, 54°22′W

Oelemarie, 03°06′N, 54°31′W

Paloemeu Airstrip (= Vincent Fajks), 210 m, 03°21′N, 55°27′W

Raleigh Falls, Coppename River, 04°43′N, 56°12′W

Sipaliwini, 02°02N, 56°08W

Sipaliwini Airstrip, 02°02′N, 56°08′W

Sipaliwini Airstrip, 1 km S and 3.5 km E, 02°01′N, 56°05′W

Sipaliwini Savanna, 02°06′N, 56°02′W

Stondansi Falls (= Stonedansi Falls), Nickerie River, 05°09′N, 56°29′W

Voltzberg (= Voltz Berg), 04°41′N, 56°10′W

Wanica (District)

Kwatta, sea level, 05°51′N, 55°18′W

Santo Boma Locks, about 12 km SW of Paramaribo, 05°47′N, 55°17′W

Trinidad and Tobago

Tobago

Charlotteville, sea level, 11°19′N, 60°33′W

Pigeon Peak, 525 m, 11°18′N, 60°33′W

Pirate'S Cay, 1 km N of Charlotteville, 11°19′N, 60°33′W

Richmond, 11°13′N, 60°37′W

Robinson Crusoe'S Cave, 11°13′N, 60°40′W

Runnemede, 11°15′N, 60°42′W

Speyside, 11°17′N, 60°32′W

Tobago (island), 11°15′N, 60°40′W

Trinidad

Arima, 7 miles N, 10°44′N, 61°17′W

Barataria, 10°39′N, 61°28′W

Belmont, 10°39′N, 61°31′W

Blanchisseuse, 10°47′N, 61°18′W

Botanic Gardens, 10°39′N, 61°31′W

Caparo, 10°27′N, 61°20′W

Carapichaima Village, 10°25′N, 61°21′W

Caura, 10°14′N, 61°23′W

Cumaca, 10°42′N, 61°09′W

Cunapo, Sangre Grande, 10°55′N, 61°07′W

Fillette Pt., 10°48′N, 61°21′W

Guayaguayare, 10°08′N, 61°02′W

Las Cuevas, 10°46′N, 61°23′W

Maracas, 10°41′N, 61°24′W

Platanal, 10°44′N, 61°09′W

Port of Spain (var. Port-of-Spain), 10°39′N, 61°31′W

Princes Town, 10°16′N, 61°23′W

Princestown (= Princes Town), 10°16′N, 61°23′W

Saint Anns, 10°40′N, 61°31′W

Salazar Trace, Point Fortin, 10°11′N, 61°41′W
San Fernando, 10°17′N, 61°28′W
Sangre Grande, 10°35′N, 61°07′W
Santa Cruz, 10°43′N, 61°28′W
Santa Maria (= Saint Mary's), 10°13′N, 61°32′W
Saut d'Eau Cave, 10°45′N, 61°31′W
Talparo, 10°29′N, 61°18′W
Victoria-Mayaro Forest Reserve, 10°08′N, 61°08′W
Waterloo, 10°28′N, 61°28′W

Uruguay
Artigas (Departamento de)

Arroyo Mandiyú, boca del, 30°30′S, 57°42′W
Belén (Salto), 6 km NNW, ca. 32°44′S, 57°48′W
Cañada de La Laguna y Cañada del Potrero, Estancia El Retiro, 30°23′S, 57°19′W
Isla Rica, Río Uruguay, 30°32′S, 57°53′W.
Río Cuareim, 5 km above junction with Arroyo Yacaré, ca. 30°05′S, 57°00′W

Canelones (Departamento de)

Jaureguiberry, 34°48′S, 55°24′W
Los Titanes Balneario, 34°48′S, 55°33′W

Cerro Largo (Departamento de)

Estancia La Formosa, 31°48′S, 54°27′W

Durazno (Departamento de)

Estancia San Jorge, 32°55′S, 55°55′W

Lavalleja (Departamento de)

Gruta de Arequita, 34°17′S, 55°14′W
Pirarajá, 9 miles S, 100 m, 33°52′S, 54°46′W
Río Olimar Chico, 33°14′S, 54°31′W

Maldonado (Departamento de)

Maldonado, sea level, 34°54′S, 54°57′W

Montevideo (Departamento de)

Colegio Pio-Colon, Montevideo, sea level, 34°53′S, 56°11′W
Colón, 10 m, 34°48′S, 56°14′W
Isla de Flores, 34°56′S, 55°53′W
Montevideo, sea level, 34°53′S, 56°11′W

Pocitos, 34°55′S, 56°09′W
Santa Lucia, 34°27′S, 56°24′W.

Paysandú (Departamento de)

Arroyo Negro, 15 km S of Paysandú, 32°27′S, 58°05′W
Quebracho Paysandú, 32°22′S, 58°01′W

Río Negro (Departamento de)

Boca del Arroyo Las Cañas, 7 km SW of Fray Bentos, 33°10′S, 58°21′W
Bopicuá, 10 m, 33°06′S, 58°01′W

Rivera (Departamento de)

Rivera, 200 m, 30°54′S, 55°31′W

Rocha (Departamento de)

Camino del Indio, ca. 34°04′S, 53°42′W
Cerro de la Estancia San Miguel, 33°39′N, 54°35′W
San Vicente, 15 miles N, 34°00′S, 53°49′W

Salto (Departamento de)

Salto, 31°23′S, 57°58′W

San José (Departamento de)

Rincón de Cufré (= Cufré), 34°12′S, 57°06′W

Soriano (Departamento de)

Soriano, 33°24′S, 58°19′W

Tacuarembó (Departamento de)

Arroyo Sauce de Tranquera, 31°47′S, 55°53′W
Pueblo del Barro, Río Tacuarembó, 32°03′S, 55°23′W
Tacuarembó, 40 km NW, ca. 31°29′S, 56°14′W

Trenta y Tres (Departamento de)

Trenta y Tres, 33°14′S, 54°23′W

Venezuela
Amazonas (Estado de)

Belén (= Culebra), Río Cunucunuma, 56 km NNW of Esmeralda, 150 m, 03°39′N, 65°46′W

Boca Mavaca, 84 km SSE of Esmeralda, 138 m, 02°30'N, 65°13'W

Buena Vista, Río Casiquiare, 01°57'N, 66°42'W

Cabecera del Caño Culebra, Cerro Duida, 40 km NNW of Esmeralda, 1,200–1,400 m, 03°30'N, 65°43'W

Cabecera del Caño Negro, Cerro Duida, 1,400 m, 03°30'N, 65°43'W

Cacurí, Río Ventuari, 04°49'N, 65°21'W

Campamento 2, Alto Río Atacavi, 140 m, 03°05'N, 66°54'W

Caño Culebra, Cerro Duida, 50 km NNW of Esmeralda, 800 m, 03°37'N, 65°41'W

Caño Majagua, Río Ventuari, 05°20'N, 65°40'W

Caño Negro, Cerro Marahuaca, 1,200 m, 03°38'N, 65°28'W

Capibara, Brazo Casiquiare, 130 m, 02°38'N, 66°19'W

Cerro de Tamacuare, 1,550 m, Serranía de Tapirapeco, 01°15'N, 64°47'W

Cerro Duida, 1,370-2,100 m, 03°25'N, 65°40'W

Cerro Duida, 8 miles N of Esmeralda, 03°15'N, 65°50'W

Cerro Neblina Base Camp, left (west) bank Río Baria (= Río Mawarinuma), 140 m, 00°49'50"N, 66°09'40"W

Cerro de la Neblina, Camp II, 2,100 m, 2.8 km NE Pico Phelps, 00°50'N, 65°59'W

Cerro de la Neblina, Camp VI, 1,800 m, 00°52'N, 65°56'W

Cerro de la Neblina, Camp VII, 1,800 m, 00°50'40"N, 65°58'10"W

Clearwater Stream, 1.5 km S of Cerro de la Neblina Base Camp, 140 m, 03°09'N, 65°33'W

Coyowateri, Sierra Parima, 02°26'N, 64°16'W

Culebra, Río Cunucunuma, 56 km NNW of Esmeralda, 150 m, 03°39'N, 65°46'W

El Raudal, 33 km S of Puerto Ayacucho, ca. 05°24'N, 67°38'W

Esmeralda, near foot of Mt. Duida, 03°10'N, 65°33'W

Ilha Paxiuba, 03°19'S, 66°00'W

Merey, Brazo Casiquiare, 02°17'N, 67°11'W

Paría, 25 km S of Puerto Ayacucho, 114 m, 05°27'N, 67°35'W

Puerto Ayacucho, 40–50 km NE, between Puerto Ayacucho and El Burro, 100 m, ca. 05°58'N, 67°18'W

Puerto Ayacucho, 119–161 m, 05°40'N, 67°39'W

Puerto Ayacucho, 9 km SE, 05°35'N, 67°30'W

Puerto Ayacucho, 14 km SSE, 135 m, 05°33'N, 67°36'W

Puerto Ayacucho, 15 km SSE, 135 m, 05°33'N, 67°36'W

Puerto Ayachucho, 25 km S, 05°27'N, 67°38'W

Puerto Ayacucho, 28 km SSE, 05°27'N, 67°37'W

Puerto Ayacucho, 30 km S, 05°26'N, 67°39'W

Puerto Ayacucho, 32 km S, 05°25'N, 67°40'W

Puerto Ayacucho, 32 km SSE, 05°31'N, 67°38'W

Raudal de Atures, Río Orinoco, 05°35'N, 67°36'W

Río Manapiare, 163 km ESE of Puerto Ayacucho, 05°18'N, 66°13'W

Río Mavaca, western border, 160 m, 02°21'N, 65°29'W

Río Mavaca, 108 km SSE of Esmeralda, 140 m, 02°15'N, 65°17'W

Río Mawarinuma, Serranía de la Neblina, (= Cerro Neblina Base Camp), 140 m, 00°49'50"N, 66°09'40"W

Río Ocamo, mouth, 02°48'N, 64°15'W

Salto Yureba, Cano Yureba, 04°11'N, 66°42'W

San Carlos de Río Negro, 120 m, 01°55'N, 67°04'W

San Juan (= Manapiare), Río Manapiare, 163 km ESE of Puerto Ayacucho, 155 m, 05°18'N, 66°13'W

Serra de Neblina, 00°50'N, 66°10'W

Serranía de la Neblina, Río Mawarinuma (= Cerro Neblina Base Camp), 140 m, 00°49'50"N, 66°09'40"W

Sierra Parima, 200 m, 02°40'N, 64°20'W

Tamatama, Río Orinoco, 135 m, 03°10'N, 65°49'W

Tapara, NW of Cerro Divida (= Cerro Duida), Upper Río Cunucunuma, 03°13'N, 65°58'W

Anzoátegui (Estado de)

Campo Petrolero Zumo de MENEVEN, ca. 09°07'N, 63°52'W

Fundo la Unión, Paso Bajito, Río Moquete, 08°30'N, 64°05'W

La Viuda, 08°41'N, 63°52'W

Mamo, 08°28'N, 63°06'W

Paso Bajito, Río Moquete, 08°30'N, 64°05'W

Río Morichal Largo, 08°29'N, 63°41'W

Apure (Estado de)

Caño Arauquita, 07°20'N, 67°07'W

Cerro del Murciélago, 8 km NW of Puerto Páez, 76 m, 06°16'N, 67°31'W

Hato Acapulco, 07°00'N, 69°05'W

Hato Caribén, Río Cinaruco, 32 km NE of Puerto Páez, 06°33'N, 67°13'W

Hato El Cedral, 07°26'N, 69°20'W

Hato El Frío, 30 km W of El Samán, 07°43'N, 68°54'W

Hato Matapalos, 07°41'N, 68°30'W

La Blanquita, 4 km N of Nulo, 07°12'N, 71°45'W

La Villa, 60 km NE of Puerto Páez, ca. 06°26'N, 67°11'W

Nula (= Nulita), 3 km N, 24 m, 07°20'N, 71°54'W

Nulita, Selvas de San Camilo, 29 km SSW of Santo Domingo, 24 m, 07°19'N, 71°57'W

Puerto Páez, 8 km NW, ca. 06°16'N, 67°31'W

Puerto Páez, 1 km W, 100 m, 06°13'N, 67°29W

Puerto Páez, 75 m, 06°13'N, 67°28'W

Río Cinaruco, 65 km NW of Puerto Páez, 76 m, 06°33'N, 67°55'W

San Camilo, 100 m, 07°21'N, 71°0l'W

Aragua (Estado de)

El Limón-Maracay, 10°17'N, 67°37'W

Estación Biológica Rancho Grande, 13 km NW of Maracay, 1,050 m, 10°21'N, 67°40'W

Hacienda La Esperanza, 10°06'N, 67°37'W

Las Delicias, Maracay, 445 m, 10°15'N, 67°36'W

Maracay, 445 m, 10°15'N, 67°36'W

Portachuelo, Rancho Grande, 1,100 m, 10°21'N, 67°40'W

Rancho Grande Biological Station, 13 km NW of Maracay, 10°21'N, 67°40'W

Barinas (Estado de)

Agua Fria, 7 km NNE of Altamira, 1,070 m, 08°53'N, 70°29'W

Altamira, 700 m, 08°50'N, 70°30'W

Altamira, 2 km SW, 611 m, 08°49'N, 70°31'W

Barinas, 180 m, 08°38'N, 70°12'W

Barinitas, 500 m, 08°47'N, 70°25'W

Finca El Oasis, 42 km E of Barinas, 100 m, 08°38'N, 69°48'W

La Lengüeta, 08°30'N, 70°23'W

La Vega del Río Santo Antonio, 2 km SW of Altamira, 611 m, 08°50'N, 70°30'W

Parque Nacional Tapo Caparo, 4 km N of Santa María de Caparo, 340 m, 07°45'N, 71°28'W

Reserva Forestal de Ticoporo, 07°48'N, 69°55'W

Reserva Forestal de Ticoporo, Unidad II, 08°02'N, 71°41'W

Río Santo Domingo, 2 km SW of Altamira, 08°50'N, 70°30'W

Santa Bárbara, 8 km SW, 07°48'N, 71°10'W

Unidad II, Reserva Forestal de Ticoporo, 100 m, 08°10'N, 70°40'W

Bolívar (Estado de)

Arabupu (= Arabopó), Mt. Roraima, 05°06'N, 60°44'W

Auyán-tepuí, 1,850 m, 05°55'N, 62°32'W

Boca de Villacoa, Río Orinoco, 65 m, 06°23'N, 67°14'W

Bolívar, 08°08'N, 63°33'W

Caicara, 50 m, 07°38'N, 66°10'W

Caicara del Orinoco, 50 m, 07°38'N, 66°10'W

Chimantá Tepuy, Parque Nacional Canaima, 05°18'N, 62°10'W

Ciudad Bolívar, 08°08'N, 63°33'W

Churi-tepuí, 05°13'N, 61°54'W

Cumbre Cerro Auyantepuy (= Auyán-tepuí), 1,850 m, 05°55'N, 62°32'W

El Abismo, summit, near Icabarú, 960 m, 04°20'N, 61°44'W

El Dorado, 85 km SSE (= Km 125), 06°02'N, 61°22'W

El Dorado–Santa Elena Highway, Km 134, 05°54'N, 62°26'W

El Manteco, 28 km NE of Los Patos, 07°11'N, 62°22'W

El Manteco, 07°21'N, 62°32'W

El Manteco, 12 km S, ca. 07°27'N, 62°32'W

El Manaco, 56 km SE of El Dorado, 150 m, 06°22'N, 61°15'W

El Manaco, 59 km SE of El Dorado, 150 m, 06°21'N, 61°17'W

El Manaco, 63 km SE of El Dorado, 150 m, 06°13'N, 61°07'W

(Km 63 en la carretera El Dorado—Santa Elena de Uairen according to R. Guerrero (pers. comm.).)

El Manaco, 68 km SE of El Dorado, 06°11'N, 61°05'W

El Palmar, 500 m, 07°58'N, 61°53'W

El Paují, carretera Santa Elena de Uairen-Icabarú, 900 m, 04°23'N, 61°29'W

El Raudal, Caño El Cambur, 15 km S of Maripa, 07°22'N, 65°18'W

Florida (= Hato La Florida), 35 km ESE of Caicara, 07°34'N, 65°52'W

Guaniamo, 20 km S, 06°26'N, 66°07'W

Guasipati, 5 km NNW, 190 m, 07°31'N, 61°55'W

Guri, 07°39'N, 62°50'W

Hacienda Sagitario, Km 90 on Puerto Ayacucho-Caicara highway, 90 m, ca. 06°09'N, 67°13'W

Hato El Torete, 170 m, ca. 06°54'N, 63°14'W

Hato La Florida, 35 km ESE of Caicara, 07°34'N, 65°52'W

Hato La Florida, 47 km ESE of Caicara, 50 m, 07°26'N, 65°48'W

Hato La Florida, 47 km SE of Caicara, 07°19'N, 65°52'W

Hato San José, 20 km W of La Paragua, 300 m, 06°49'N, 63°31'W

Icabarú, 45 km NE, 04°37'N, 61°29'W

Icabarú, 40 km NE, 964 m, 04°35'N, 61°21'W

Icaburú, 22.5 km NE (= Independencia), 04°29'N, 61°37'W

Icabarú, 21 km NE, 851 m, 04°28'N, 61°38'W

Icabarú, 13 km NE, 817 m, 04°25'N, 61°40'W

Icabarú, 04°20'N, 61°46'W

Independencia, 22.5 km NE of Icabarú, 824 m, 04°29'N, 61°37'W

Km 38, SE of El Dorado, 100 m, 06°29'N, 61°23'W

Km 55 on highway south of El Dorado, ca. 06°23'N, 61°30

Km 85, about 65 km SSE of El Dorado, ca. 05°59'N, 61°26'W

Km 125, 85 km SSE of El Dorado, 1,030 m, 06°02′N, 61°22′W

(Km 85 en la carretera El Dorado-Santa Elena de Uairén; near Salto de El Danto according to R. Guerrero [pers. comm.].)

Km 134, carretera El Dorado-Santa Elena de Uairén, Paso del Alto La Escalera, 1,300 m, 05°54′N, 61°26′W

La Tigra, Parcela 18 El Palmar, 250 m, 07°31′N, 61°31′W

La Paragua (Lote 6), 06°50′N, 63°20′W

Los Patos, 07°11′N, 62°22′W

La Unión, 06°55′N, 64°55′W

Maripa, 07°26′N, 65°09′W

Meseta de Jaua, Río Marajano, 05°15′N, 64°28′W

Mina de bauxita de la Serranía de los Pijiguaos, 06°28′N, 66°43′W

Parque Nacional Canaima, centered at ca. 05°30′N, 61°51′W

Piedra Virgin, 70 km SSE of El Dorado, 375 m, ca. 06°28′N, 61°29′W

Raudal Chalimana, Río Paramichí, 04°10′N, 62°59′W

Río Marajano, Meseta de Jaua, 05°15′N, 64°28′W

Río Mocho, 06°57′N, 65°20′W

(Tributary of Río Mato, which is "the largest western tributary of the Rio Caura" [J. A. Allen 1910:145].)

Río Suapuré, 06°48′N, 67°01′W

Río Supamo, 50 km SE of El Manteco, 350 m, 07°00′N, 62°15′W

Río Yuruan, 19 km from mouth, 06°48′N, 61°50′W

Salto Ichun, Río Paragua, 04°44′N, 63°20′W

San Ignacio de Yuruaní, 930 m, 05°00′N, 61°00′W

Santa Lucía de Surukún, 45 km NE of Icabarú, 851 m, ca. 05°17′N, 61°47′W

Santa Mario de Erebato, 04°58′N, 64°50′W

San Martín de Turumbán, Isla de Anacoco, Río Cuyuní, 06°42′N, 61°05′W

Serranía de los Pijiguaos, 06°38′N, 66°44′W

Suapuré, 07°14′N, 65°10′W

Carabobo (Estado de)

El Central, 10 km NW of Urama, 25 m, 10°33′N, 68°25′W

Embalse Río Morón, 10°27′N, 68°11′W

La Copa, 4 km NW of Montalbán, 1,537 m, 10°15′N, 68°22′W

Montalbán, 9 km NE, 773 m, 10°16′N, 68°16′W

Montalbán, 598–900 m, 10°13′N, 68°20′W

Puerto Cabello, 10°28′N, 68°01′W

San Esteban, 10°26′N, 68°01′W

Sierra de Carabobo, ca. 10°02′N, 68°08′W

Urama, 10 km NW, 25 m, 10°33′N, 68°25′W

Cojedes (Estado de)

El Candelo, SW of La Sierra, 640 m, 09°53′N, 68°39′W

Hato Nuevo, 09°13′N, 68°05′W

Delta Amacuro (Estado de)

Araguaimujo, 08°54′N, 61°32′W

Caño Araguabisi, W end Isla Tobejuba, 09°29′N, 60°57′W

Caño Araguaito, 08°58′N, 61°25′W

Caño Mánamo, 09°55′N, 62°16′W

Caño Mariusa, ca. 09°22′N, 61°15′W

Curiapo, 08°33′N, 61°00′W

Desembocadura del Río Acoíma, 5 km S of El Toro, 5 m, 08°29′N, 61°29′W

El Toro, 08°31′N, 61°30′W

Guayo, 09°00′N, 60°50′W

Güiniquina, 09°10′N, 61°03′W

Imataca Forest Reserve, Unit V, 08°00′N, 61°18′W

Isla Cocuina, 09°47′N, 61°49′W

Los Güires, 09°15′N, 61°54′W

Punta Culebra, Caño Mánamo, SE Isla Mánamo, 09°55′N, 62°16′W

Raudales del Río Acoíma, ca. 08°22′N, 61°29′W

Reserva Forestal Imataca (Unit V), ca. 28 km E Tumeremo, 08°00′N, 61°18′W

Reserva Forestal de Imataca, 180 m, 07°25′N, 61°10′W

Río Acoíma, 15 km S of El Toro, ca. 08°22′N, 61°29′W

Río Ibaruma, 38 km SE of Curiapo, 08°10′N, 60°47′W

Unidad V, Reserva Forestal de Imataca, 08°00′N, 61°18′W

Winikina, ca. 09°12′N, 61°08′W

Distrito Federal

Alto Ño León, 33 km WSW of Caracas, 1,665–1,950 m, 10°26′N, 67°10′W

Boca del Tigre Valley, 5 km NW of Caracas, 1,394 m, 10°32′N, 66°57′W

Caracas, 9.4 km N, 2,092 m, 10°35′N, 66°54′W

Caracas, 5 km N, 10°33′N, 66°54′W

Caracas, 5 km NNE, 10°33′S, 66°52′W

Caracas, 900 m, 10°30′N, 66°53′W

Caracas (Santa Monica), 900 m, 10°30′N, 66°53′W

El Junquito, 2,000 m, 10°28′N, 67°05′W

Hacienda Carapiche, 390 m, 10°29′N, 67°19′W

La Guaira, 10°36′N, 66°56′W

Los Venados, 4 km NNW of Caracas, 1,465–1,524 m, 10°32′N, 66°54′W

Macuto, 10°37′N, 66°52′W

Pico Ávila, 6 km NNW of Caracas, 2,150 m, 10°33′N, 66°52′W

Pico Ávila, 5 km NNW of Caracas, 2,150 m, 10°33'N, 66°52'W

San Julián, sea level, 10°37'N, 66°50'W

Falcón (Estado de)

Boca de Yaracuy, 28 km WNW of Puerto Cabello, 2 m, 10°35'N, 68°15'W

Capatárida, 40 m, 11°10'N, 70°37'W

Capatárida, 6 km SE, 11°07'N, 70°35'W

Cerro Santa Ana, Peninsula de Paraguaná, 15 km SSW of Pueblo Nuevo, 500 m, 11°50'N, 69°59'W

Cerro Socopo, 84 km NW Carora, 1,250 m, 10°28'N, 70°48'W.

La Cueva de Piedra Honda, 7 km SW of Pueblo Nuevo, 120 m, Península de Paraguaná, 11°52'N, 69°58'W

La Pastora, 16 km ENE of Mirimire, 70 m, 11°13'N, 68°37'W

Mirimire, 20 km NNE, 11°20'N, 68°41'W

Mirimire, 13 km N and 13 km E, 75 m, 11°16'N, 68°38'W

Mirimire, 5 km N and 13 km E (near La Pastora), 11°13'N, 68°36'W

Mirimire, 16 km ENE, near La Pastora, 70 m, 11°12'N, 68°37'W

Mirimire, 14 km ENE, 60 m, 11°12'N, 68°38'W

Mirimire, 13 km ENE, 270 m, 11°12'N, 68°38'W

Mirimire, 12 km ENE, 11°11'N, 68°39'W

Mirimire, 250 m, 11°10'N, 68°43'W

Pastora, 14 km ENE of Mirimire, 150 m, 11°12'N, 68°37'W

Pedregal, 20 km S of Caldera, 11°01'N, 70°08'W

Pueblo Nuevo, 7 km SW, Península de Paraguaná, 11°52'N, 69°58'W

Pueblo Nuevo, 15 km SW, Península de Paraguaná, 13 m, 11°50'N, 69°59'W

Riecito, 30 km S of Mirimire, 300 m, 10°54'N, 68°46'W

Río Socopito, 80 km NW of Carora, 480 m, 10°30'N, 70°44'W

Urama, 24 km NW, 10°35'N, 68°27'W

Urama, 19 km NW, 10°33'N, 68°24'W

Guárico (Estado de)

Calabozo, 9 km SE, 100 m, 08°51'N, 67°23'W

Dos Caminos, 09°35'N, 67°18'W

Embalse de Guárico, 10 km N of Calabozo, 09°01'N, 67°26'W

Estación Biológica de los Llanos, 9 km SE of Calabozo, 08°51'N, 67°23'W

Guyabal, 08°00'N, 67°24'W

Hacienda La Elvira, 10 km NE of Altagracia, 630 m, 09°55'N, 66°19'W

Hato La Palmita, 09°37'N, 67°29'W

Hato Las Palmitas, 35 km SSW of San Juan de los Morros, 180 m, 09°36'N, 67°27'W

Km 47, NNW of San Fernando (Estado Apure), on carretera Calabozo, ca. 08°04'N, 67°32'W

Masaguaral, 08°34'N, 67°35'W

Santa Rita, cerca del Río Manapire, 08°08'N, 66°15'W

Lara (Estado de)

Caserío Boro, 10 km N of El Tocuyo, 520 m, 09°53'N, 69°47'W

Hacienda San Geronimo, San Miguel, 960 m, 09°53'N, 69°31'W

Río Tocuyo, 500 m, 10°16'N, 69°56'W

Mérida (Estado de)

Asentamiento Monterrey, 8 km NNE of Mérida, 2,300 m, 08°40'N, 71°07'W

Cafetal de Milla, 1,630 m, 08°30'N, 71°22'W

Cueva de la Azulita, La Azulita, 08°43'N, 71°27'W

Cueva del Salado, 4 km E of Bailadores, 2,000 m, ca. 08°15'N, 71°47'W

Hacienda San Pedro, 08°24'N, 71°38'W

La Azulita, 1,135 m, 08°43'N, 71°27'W

La Carbonera, 12 km SE of La Azulita, 2,150–2,190 m, 08°38'N, 71°21'W

La Culata, 4,487 m, 08°45'N, 71°05'W

La Mucuy, 4 km E of Tabay, 2,107 m, 08°36'N, 71°01'W

Mérida, 7 km SSW, along road from Mérida to El Morro, 2,000 m, 08°32'N, 71°11'W

Montes de la Hechicera, near Mérida, 2,100 m, ca. 08°40'N, 71°10'W

Monte Zerpa, 6 km N of Mérida, 2,000 m, 08°39'N, 71°08'W

Monte Zerpa, 4 km NW of Mérida, 2,000 m, 08°37'N, 71°09'W

Paramito, 3 km W of Timotes, 08°59'N, 70°46'W

Páramo de Mariño, 08°16'N, 71°56'W

Río Albarregas, 1,630 m, 08°35'N, 71°09'W

Río Quebrada de Piedras, 4 km SSE of Nueva Bolivia, 240 m, 09°07'N, 71°04'W

San Juan de Lagunillas, 3 km SE, 820 m, 08°19'N, 71°20'W

Santa Rosa, 1 km N of Mérida, 8°37'N, 71°09'W

Tabay, 08°36'N, 71°01'W

Tabay, 4 km E (= La Mucuy), 2,107 m, 08°36'N, 71°01'W

Tabay, 6 km ESE, 2,630 m, 08°35'N, 71°03'W

Tierra Negra, 1,550 m, ca. 12 km S San Juan de Lagunillas, 08°25'N, 71°22'W

Zea, near Quebrada Los Churruscos, 1,050 m, 08°22'N, 71°47'W

Miranda (Estado de)

Birongo, 60 m, 10°29′N, 66°16′W

Birongo, 4 km SW, 10°28′N, 66°17′W

Caracas, 19 miles E, 10°31′N, 66°38′W

Cueva Ricardo Zuloaga, near El Encantado, 15 km SE of Caracas, 548 m, 10°27′N, 66°47′W

Cueva Walter Dupouy, 4 km SW of Birongo, 10°28′N, 66°16′W

Curupao, 5 km NNW of Guarenas, 1,150 m, 10°31′N, 66°38′W

Curvapao (= Curupao), 10°31′N, 66°38′W

El Encantado, near, 13 km SE of Caracas, 570 m, 10°27′N, 66°48′W

Guatopo National Park, 10°03′N, 66°27′W

Parque Nacional Guatopo, 21 km NW of Altagracia, 630 m, 10°03′N, 66°27′W

Río Chico, sea level, 10°19′N, 65°58′W

Río Chico, 5 km E, sea level, 10°19′N, 65°55′W

Río Chico, 7 km E, 10°19′N, 65°54′W

Río Chico, 1 km S, 10°18′N, 65°58′W

Río Chico, 6 km SSE, 1 m, 10°17′N, 65°58′W

San Andrés, 16 km SSE of Caracas, 10°22′N, 66°50′W

Tacarigua de la Laguna, 10°19′N, 65°52′W

Turgua, near, 10°22′N, 66°50′W

Monagas (Estado de)

Cachipo, 09°54′N, 63°08′W

Caicara (= Caicara de Maturín), 175 m, 09°49′N, 63°36′W

Caño Colorado, 100 m, 09°55′N, 62°55′W

Caripe, 1 km N and 3 km W, 1,190 m, 10°12′N, 63°33′W

Caripe, 3 km SW, 10°10′N, 63°31′W

Caripito, 100 m, 10°08′N, 63°06′W

Cueva del Guacharo, 09°09′N, 63°32′W

Cueva de Saffont, ca 10°10′N, 63°32′W

Hacienda San Fernando, 1 km N and 3 km W of Caripe, 1,190 m, 10°12′N, 63°33′W

Hato Mata de Bejuco, 55 km SSE of Maturín, 18 m, 09°19′N, 62°56′W

Isla Tigre, 09°36′N, 62°25′W

Laguna Guasacónica, 09°12′N, 62°40′W

Los Araguaneyes, ca. 09°47′N, 62°51′W

Maturín, 74 m, 09°45′N, 63°11′W

Maturín, 47 km SE, ca. 09°30′N, 62°55′W

Paso Nuevo, 08°27′N, 62°44′W

Quebrada San Pablo, 10–15 km SE of Caripito, ca. 10°03N, 63°01′W

Río Guanipa, 2 km W of mouth, ca. 09°56′N, 62°26′W

Río Morichal Largo, 09°27′N, 62°25′W

San Agustín, 5 km N of Caripe, 1,180 m, 10°15′N, 63°29′W

San Agustín, 5 km NW of Caripe, 1,150 m, 10°13′N, 63°32′W

San Agustín, 3 km NW of Caripe, 10°13′N, 63°30′W

San Agustín, 3 km SW of Caripe, 854 m, 10°10′N, 63°31′W

Uverito, 35 km S of Temblador, 08°40′N, 62°37′W

Nueva Esparta (Estado de)
Isla Margarita

Cerro Mata Siete (= Cerro Matasiete), 2 km N and 2 km E of La Asunción, 350 m., 11°03′N, 63°53′W

Cueva Honda del Piache, ca. 10°57′N, 63°50′W

El Valle, 50 m, 10°59′N, 63°52′W

La Asunción, 2 km N and 2 km E, 350 m, 11°03′N, 63°53′W

La Asunción, 3 km NNE, 11°05′N, 63°52′W

La Asunción, 3 km NE, 11°04′N, 63°52′W

La Asunción, 10 km WSW, 11°04′N, 63°58′W

Las Piedras, 11°04′N, 63°59′W

Porlamar, 2 km N and 30 km W, 10°59′N, 64°05′W

Porlamar, 31 km W, 10°58′N, 64°06′W

San Francisco de Macanao, 1.5 km N, 75 m, 11°03′N, 64°18′W

Sucre (Estado de)

Caño Brea, 10°06′N, 62°38′W

Caripito (Monagas), 40 km NW, ca. 10°20′N, 63°18′W

Caripito (Monagas), 10 km NW, 10°11′N, 63°06′W

Carúpano, 10°40′N, 63°14′W

Cerro Papelón, 900 m, 10°11′N, 63°49′W

Cristóbal Colón (= Macuro), 150 m, 10°39′N, 61°56′W

Cumaná, sea level, 10°28′N, 64°10′W

Cumaná, 2 km E, 10°28′N, 64°09′W

Cumaná, 16 km E, 10°27′N, 64°02′W

Cumaná, 21 km E, 10°27′N, 63°58′W

Ensenada Cauranta, 12 km NE of Güiria, 90 m, 10°40′N, 62°15′W

Ensenada Cauranta, 11 km NE Güiria, 40 m, 10°40′N, 62°15′W

Ensenada Cauranta, 9 km NE of Güiria, 4–7 m, 10°38′N, 62°15′W

Finca Vuelta Larga, 9.7 km SE of Guaraúnos, 20 m, 10°33′N, 63°07′W

Guaraúnos, 10°33′N, 63°07′W

Las Melenas, 800 m, 10°41′N, 62°37′W

Manacal, 26 km ESE of Carúpano, 200–366 m, 10°17′N, 63°03′W

Marigüitar, 10°27′N, 63°55′W

Mt. Turumiquire, 10°07′N, 63°52′W

Santa Rosa, 20 km (by road) SE of Casanay, ca. 10°25′N, 63°20′W

Tacal, 11 km SSW of Cumaná, 10°23′N, 64°13′W

Táchira (Estado de)

Betania, 26.5 km SSW of Rubio, 2,250 m, ca. 07°31′N, 72°33′W

Buena Vista, near Páramo de Tamá, 41 km SW of San Cristobal, 2,395 m, 07°27′N, 72°26′W

El Hatico, 3 km E of Seboruco, 1,220 m, 08°09′N, 72°02′W

El Palotal, 6 km SSE of Ureña, 320 m, 07°52′N, 72°26′W

La Fría, 08°13′N, 72°15′W

Las Mesas, 17 km NE of San Juan de Colón, 460 m, 08°09′N, 72°10′W

Las Mesas, 45 km N and 6 km E of San Cristóbal, 460 m, 08°09′N, 72°10′W

Páramo de Tamá, 3,329 m, 07°25′N, 72°26′W

Páramo Zumbador, 07°58′N, 72°04′W

Potosí, 14 km SE of Pregonero, 1,020 m, 07°56′N, 71°41′W

San Cristóbal, 41 km SW, 07°27′N, 72°26′W

San Juan de Colón, 797 m, 08°02′N, 72°16′W

Uribante-Caparo Dam, 10 km SE of Preonero, 1,100 m, 07°57′N, 71°39′W

Trujillo (Estado de)

Agua Viva, 164 m, 09°34′N, 70°36′W

El Dividive, near, 30 km NW of Valera, 09°41′N, 70°29′W

Hacienda Misisí, 15 km E of Trujillo, 2,360 m, 09°21′N, 70°18′W

La Ceiba, 52 km WNW of Valera, 29 m, 09°28′N, 71°04′W

Macizo de Guaramacal, 6.5 km ENE of Boconó, Parque Nacional General Cruz Carrillo, 2,220 m, ca. 09°18′N, 70°13′W

Río Motatán, 9.8 km NNE of Motatán, 230 m, 09°24′N, 70°36′W

Río Motatán, site E-II, 09°05′N, 70°39′W

Trujillo, 09°22′N, 70°26′W

Trujillo, 13 km E, 09°21′N, 70°18′W

Trujillo, 15 km E, ca. 09°21′N, 70°16′W

Valera, 25 km NW, 90 m, 09°32′N, 70°43′W

Valera, 23 km NW, 90 m, 09°31′N, 70°42′W

Valera, 19 km N, 164 m, ca. 09°30′N, 70°36′W

Yaracuy (Estado de)

El Central, 10 km NW of Urama, 25 m, 10°31′N, 68°23′W

Minas de Aroa, Bolívar R.R., 10°25′N, 68°54′W

("Aroa is the name now used to designate the district of which Pueblo Nuevo is the center, but it was formerly applied to the copper mine situated about three miles up the gorge of the Aroa River. It is the headquarters of the Bolivar Railway" [Carriker in J. A. Allen 1911:241].)

Minas de Aroa, 20 km NW of San Felipe, 395–400 m, 10°25′N, 68°54′W

Urama, 10 km NW, 25 m, 10°31′N, 68°23′W

Zulia (Estado de)

Altagracia, 12 miles SE, Río Aurare (= Río Anaure), sea level, 10°37′N, 71°25′W

Boca del Río de Oro, 60 km WNW of Encontrados, 73 m, 09°09′N, 72°36′W

Cerro Azul, 10°51′N, 72°16′W

Cerro Azul, 40 km NW of La Paz, 80 m, 10°56′N, 72°16′W

Cerro Azul, 33 km NW of La Paz, 75 m, 10°53′N, 72°14′W

Cojoro, 11°40′N, 71°48′W

El Laberinto, 10°33′N, 72°13′W

El Rosario, 65 km WNW of Los Encontrados, ca. 09°18′N, 72°46′W

El Rosario, 57 km WNW of Los Encontrados, 61 m, ca. 09°15′N, 72°42′W

El Rosario, 48 km WNW of Los Encontrados, 54 m, ca. 09°11′N, 72°38′W

El Rosario, 42 km WNW of Los Encontrados, ca. 09°10′N, 72°36′W

El Rosario, 39 km WNW of Los Encontrados, 24 m, ca. 09°09′N, 72°35′W

El Tukuco, 46 km SSW of Machiques, 400 m, 09°43′N, 72°47′W

Empalado Savannas (= Empalado Sabana), 10°40′N, 71°30′W

Hacienda Platanal, 33 km NW of La Paz, 10°51′N, 72°15′W

Kasmera, 21 km SW of Machiques, 270 m, 09°59′N, 72°43′W

Kunana, Río Negro, 1,140 m, 09°25′N, 72°25′W

Lagunillas, 10°08′N, 71°16′W

La Paz, 35 km NW, near Cerro Azul, 80 m, 10°51′N, 72°16′W

La Rinconada, ca. 09°13′N, 72°41′W

Maracaibo, 110 km N and 25 km W, 5 m, 11°38′N, 71°50′W

Maracaibo, sea level, 10°40′N, 71°37′W

Maracaibo, 30 miles E, ca. 10°36′N, 70°20′W

Machiques, 15 km W, 10°04′N, 72°42′W

Novito, 19 km WSW of Machiques, 1,135 m, 10°02′N, 72°43′W

Río Aurare (= Río Anaure), 10°37′N, 71°25′W

List of Taxa

Class Mammalia
Cohort Marsupialia
Order Didelphimorphia
Family Didelphidae
Subfamily Caluromyinae
Genus *Caluromys*
C. derbianus . . . . Derby's Woolly Opossum
C. lanatus . . . . . . Brown-eared Woolly Opossum
C. philander . . . . Bare-tailed Woolly Opossum
Genus *Caluromysiops*
C. irrupta . . . . Black-shouldered Opossum
Genus *Glironia*
G. venusta . . . . Bushy-tailed Opossum
Subfamily Didelphinae
Tribe Didelphini
Genus *Chironectes*
C. minimus . . . . Water Opossum
Genus *Didelphis*
D. albiventris . . . . White-eared Opossum
D. aurita . . . . Big-eared Opossum
D. imperfecta . . . . Guianan White-eared Opossum
D. marsupialis . . . . Common Opossum
D. pernigra . . . . . . Andean White-eared Opossum
Genus *Lutreolina*
L. crassicaudata . . . . Lutrine Opossum
Genus *Philander*
P. andersoni . . . . . Anderson's Four-eyed Opossum
P. deltae . . . . Orinoco Four-eyed Opossum
P. frenatus . . . . . Southeastern Four-eyed Opossum
P. mcilhennyi . . . McIlhenny's Four-eyed Opossum
P. mondolfii . . . . . Mondolfi's Four-eyed Opossum

P. opossum . . . . Gray Four-eyed Opossum
Tribe Metachirini
Genus *Metachirus*
M. nudicaudatus . . . . . Brown Four-eyed Opossum
Tribe Monodelphini
Genus *Chacodelphys*
C. formosa . . . . Pygmy Opossum
Genus *Cryptonanus*
C. agricolai . . . . Agricola's Mouse Opossum
C. chacoensis . . . . Chacoan Mouse Opossum
C. guahybae . . . . Guahyba Mouse Opossum
C. ignitus . . . . Red-bellied Mouse Opossum
C. unduaviensis . . . . . . Unduave Mouse Opossum
Genus *Gracilinanus*
G. aceramarcae . . . . Aceramarca Opossum
G. agilis . . . . Agile Opossum
G. dryas . . . . Wood sprite Opossum
G. emiliae . . . . Emilia's Opossum
G. marica . . . . Northern Gracile Opossum
G. microtarsus . . . . Brazilian Gracile Opossum
Genus *Hyladelphys*
H. kalinowskii . . . . Kalinowski's Opossum
Genus *Lestodelphys*
L. halli . . . . Patagonian Opossum
Genus *Marmosa*
M. andersoni . . . . . . Anderson's Mouse Opossum
M. lepida . . . . Rufus Mouse Opossum
M. murina . . . . Murine Opossum
M. quichua . . . . Quechuan Mouse Opossum
M. robinsoni . . . . Robinson's Mouse Opossum
M. rubra . . . . Red Mouse Opossum

M. tyleriana . . . . Tyleria Mouse Opossum
M. xerophila . . . . Guajira Mouse Opossum
Genus *Marmosops*
 M. bishopi . . . . Bishop's Slender Opossum
 M. cracens . . . . Narrow-headed Slender Opossum
 M. creightoni . . . . . . Creighton's Slender Opossum
 M. fuscatus . . . . Dusky Slender Opossum
 M. handleyi . . . . Handley's Slender Opossum
 M. impavidus . . . . . . Tschudi's Slender Opossum
 M. incanus . . . . Gray Slender Opossum
 M. juninensis . . . . Junin Slender Opossum
 M. neblina . . . . Neblina Slender Opossum
 M. noctivagus . . . . White-bellied Slender Opossum
 M. ocellatus . . . . Spectacled Slender Opossum
 M. parvidens . . . . Delicate Slender Opossum
 M. paulensis . . . . Brazilian Slender Opossum
 M. pinheiroi . . . . Pinheiro's Slender Opossum
Genus *Micoureus*
 M. alstoni . . . . Alston's Mouse Opossum
 M. constantiae . . . . White-bellied Woolly Mouse
 Opossum
 M. demerarae . . . . Woolly Mouse Opossum
 M. paraguayensis . Tate's Woolly Mouse Opossum
 M. phaeus . . . . Little Woolly Mouse Opossum
 M. regina . . . Bare-tailed Woolly Mouse Opossum
Genus *Monodelphis*
 M. adusta . . . . Sepia Short-tailed Opossum
 M. americana . . Northern Three-striped Opossum
 M. brevicaudata . . . Northern Red-sided Opossum
 M. dimidiata . . . . Yellow-sided Opossum
 M. domestica . . . . Gray Short-tailed Opossum
 M. emiliae . . . . Emilia's Short-tailed Opossum
 M. glirina . . . . . Amazonian Red-sided Opossum
 M. iheringi . . . . . Ihering's Three-striped Opossum
 M. kunsi . . . . Pygmy Short-tailed Opossum
 M. maraxina . . . . . Marajo Short-tailed Opossum
 M. osgoodi . . . . . . Osgood's Short-tailed Opossum
 M. palliolata . . . . . . Hooded Red-sided Opossum
 M. reigi . . . . Reig's Opossum
 M. ronaldi . . . . Ronald's Opossum
 M. rubida . . . . Red Short-tailed Opossum
 M. scalops . . . . Tawny-headed Opossum
 M. sorex . . . . Southern Red-sided Opossum
 M. theresa . . . . Southern Three-striped Opossum
 M. umbristriata . . . . Faint-striped Opossum
 M. unistriata . . One-striped Short-tailed Opossum
Genus *Thylamys*
 T. cinderella . . . . Cinderella Thylamys
 T. elegans . . . . Elegant Thylamys
 T. karimii . . . . Karimi's Thylamys
 T. macrurus . . . . Paraguayan Thylamys
 T. pallidior . . . . White-bellied Thylamys

T. pusillus . . . . Chacoan Thylamys
T. sponsorius . . . . Argentine Thylamys
T. tatei . . . . Tate's Thylamys
T. velutinus . . . . Dwarf Thylamys
T. venustus . . . . Buff-bellied Thylamys
Order Microbiotheria
Family Microbiotheriidae
 Genus *Dromiciops*
 D. gliroides . . . . Monito del Monte
Order Paucituberculata
Family Caenolestidae
 Genus *Caenolestes*
 C. caniventer . . . . Gray-bellied Caenolestid
 C. condorensis . . . . Andean Caenolestid
 C. convelatus . . . . Northern Caenolestid
 C. fuliginosus . . . . Dusky Caenolestid
 Genus *Lestoros*
 L. inca . . . . Incan Caenolestid
 Genus *Rhyncholestes*
 R. raphanurus . . . . Long-nosed Caenolestid
Cohort Placentalia
Magnorder Xenarthra
Order Cingulata
Family Dasypodidae
 Subfamily Dasypodinae
 Genus *Dasypus*
 D. hybridus . . . . . Southern Long-nosed Armadillo
 D. kappleri . . . . . . Greater Long-nosed Armadillo
 D. novemcinctus . . . . Nine-banded Armadillo
 D. pilosus . . . . Woolly Armadillo
 D. sabanicola . . . . . Llanos Long-nosed Armadillo
 D. septemcinctus . . . . Seven-banded Armadillo
 D. yepesi . . . . Yepe's Mulita
 Subfamily Euphractinae
 Tribe Chlamyphorini
 Genus *Calyptophractus*
 C. retusus . . . . Greater Fairy Armadillo
 Genus *Chlamyphorus*
 C. truncatus . . . . Pink Fairy Armadillo
 Tribe Euphractini
 Genus *Chaetophractus*
 C. nationi . . . . Andean Hairy Armadillo
 C. vellerosus . . . . Screaming Hairy Armadillo
 C. villosus . . . . Big Hairy Armadillo
 Genus *Euphractus*
 E. sexcinctus . . . . Yellow Armadillo
 Genus *Zaedyus*
 Z. pichiy . . . . Pichi
 Subfamily Tolypeutinae
 Tribe Priodontini
 Genus *Cabassous*
 C. centralis . . . . Northern Naked-tailed Armadillo

A. luismanueli . . . . Molinari's Tailless Bat
Genus *Choeroniscus*
   C. godmani . . . . Godman's Long-nosed Bat
   C. minor . . . . Little Long-nosed Bat
   C. periosus . . . . Handley's Long-nosed Bat
Genus *Glossophaga*
   G. commissarisi . Commissaris's Long-tongued Bat
   G. longirostris . . . . Greater Long-tongued Bat
   G. soricina . . . . Pallas's Long-tongued Bat
Genus *Leptonycteris*
   L. curasoae . . . . Curaçaoan Long-nosed Bat
Genus *Lichonycteris*
   L. degener . . . . Pale Brown Long-nosed Bat
   L. obscura . . . . Dark Brown Long-nosed Bat
Genus *Scleronycteris*
   S. ega . . . . Ega Long-tongued Bat
Subfamily Lonchophyllinae
Genus *Lionycteris*
   L. spurrelli . . . . Spurrell's Long-nosed Bat
Genus *Lonchophylla*
   L. bokermanni . . . . Bokermann's Nectar Bat
   L. chocoana . . . . Chocoan Nectar Bat
   L. concava . . . . Goldman's Nectar Bat
   L. dekeyseri . . . . Dekeyser's Nectar Bat
   L. handleyi . . . . Handley's Nectar Bat
   L. hesperia . . . . Western Nectar Bat
   L. mordax . . . . Brazilian Nectar Bat
   L. orcesi . . . . Orces's Nectar Bat
   L. robusta . . . . Big Nectar Bat
   L. thomasi . . . . Thomas's Nectar Bat
Genus *Platalina*
   P. genovensium . . . . . Peruvian Long-tongued Bat
Genus *Xeronycteris*
   X. vieirai . . . . Vieira's Flower Bat
Subfamily Phyllostominae
Genus *Chrotopterus*
   C. auritus . . . . Great Woolly Bat
Genus *Glyphonycteris*
   G. behnii . . . . Behn's Graybeard Bat
   G. daviesi . . . . Davies's Graybeard Bat
   G. sylvestris . . . . Little Graybeard Bat
Genus *Lampronycteris*
   L. brachyotis . . . . Yellow-throated Bat
Genus *Lonchorhina*
   L. aurita . . . . Tomes's Sword-nosed Bat
   L. fernandezi . . . . . Fernandez's Sword-nosed Bat
   L. inusitata . . . . Hairy-faced Sword-nosed Bat
   L. marinkellei . . . . Marinkelle's Sword-nosed Bat
   L. orinocensis . . . . Orinoco Sword-nosed Bat
Genus *Lophostoma*
   L. aequatorialis . . . . Ecuadorian Round-eared Bat
   L. brasiliense . . . . Pygmy Round-eared Bat

L. carrikeri . . . . Carriker's Round-eared Bat
L. schulzi . . . . Schulz's Round-eared Bat
L. silvicolum . . . . Orbigny's Round-eared Bat
L. yasuní . . . . Yasuni Round-eared Bat
Genus *Macrophyllum*
   M. macrophyllum . . . . Long-legged Bat
Genus *Micronycteris*
   M. brosseti . . . . Brosset's Big-eared Bat
   M. hirsuta . . . . Crested Big-eared Bat
   M. homezi . . . . Pirlot's Big-eared Bat
   M. matses . . . . Matses Big-eared Bat
   M. megalotis . . . . Brazilian Big-eared Bat
   M. microtis . . . . Little Big-eared Bat
   M. minuta . . . . White-bellied Big-eared Bat
   M. sanborni . . . . Sanborn's Big-eared Bat
   M. schmidtorum . . . . Schmidts' Big-eared Bat
Genus *Mimon*
   M. bennettii . . . . Bennett's Spear-nosed Bat
   M. cozumelae . . . . Cozumel Spear-nosed Bat
   M. crenulatum . . . . Striped Spear-nosed Bat
   M. koepckeae . . . . . Koepcke's Spear-nosed Bat
Genus *Neonycteris*
   N. pusilla . . . . Least Big-eared Bat
Genus *Phylloderma*
   P. stenops . . . . Spear-nosed Bat
Genus *Phyllostomus*
   P. discolor . . . . Pale Spear-nosed Bat
   P. elongatus . . . . Lesser Spear-nosed Bat
   P. hastatus . . . . Greater Spear-nosed Bat
   P. latifolius . . . . Guianan Spear-nosed Bat
Genus *Tonatia*
   T. bidens . . . . Spix's Round-eared Bat
   T. saurophila . . . . . Lizard-eating Round-eared Bat
Genus *Trachops*
   T. cirrhosus . . . . Fringe-lipped Bat
Genus *Trinycteris*
   T. nicefori . . . . Nicefori's Bat
Genus *Vampyrum*
   V. spectrum . . . . Great Spectral Bat
Subfamily Stenodermatinae
Tribe Ectophyllini
Genus *Artibeus (Artibeus)*
   A. amplus . . . . Giant Artibeus
   A. fimbriatus . . . . Fringed-lipped Artibeus
   A. fraterculus . . . . Fraternal Artibeus
   A. jamaicensis . . . . Jamaican Artibeus
   A. lituratus . . . . Great Artibeus
   A. obscurus . . . . Black Artibeus
   A. planirostris . . . . Spix's Artibeus
Genus *Artibeus (Dermanura)*
   A. anderseni . . . . Andersen's Fruit-eating Bat
   A. cinereus . . . . Gervais's Fruit-eating Bat

A. glaucus . . . . Silver Fruit-eating Bat
A. gnomus . . . . Dwarf Fruit-eating Bat
A. phaeotis . . . . Pygmy Fruit-eating Bat
A. watsoni . . . . Watson's Fruit-eating Bat
Genus *Artibeus (Koopmania)*
    A. concolor . . . . Plain Fruit-eating Bat
Genus *Chiroderma*
    C. doriae . . . . Brazilian Big-eyed Bat
    C. salvini . . . . Salvin's Big-eyed Bat
    C. trinitatum . . . . Little Big-eyed Bat
    C. villosum . . . . Hairy Big-eyed Bat
Genus *Enchisthenes*
    E. hartii . . . . Hart's Fruit-eating Bat
Genus *Mesophylla*
    M. macconnellii . . . . MacConnell's Bat
Genus *Platyrrhinus*
    P. albericoi . . . . Alberico's Broad-nosed Bat
    P. aurarius . . . . Eldorado Broad-nosed Bat
    P. brachycephalus . Short-headed Broad-nosed Bat
    P. chocoensis . . . . Chocoan Broad-nosed Bat
    P. dorsalis . . . . Thomas's Broad-nosed Bat
    P. helleri . . . . Heller's Broad-nosed Bat
    P. infuscus . . . . Buffy Broad-nosed Bat
    P. ismaeli . . . . Ismael's Broad-nosed Bat
    P. lineatus . . . . White-lined Broad-nosed Bat
    P. masu . . . . Quechuan Broad-nosed Bat
    P. matapalensis . . . . . Matapalo Broad-nosed Bat
    P. nigellus . . . . Little Black Broad-nosed Bat
    P. recifinus . . . . Recife Broad-nosed Bat
    P. vittatus . . . . Greater Broad-nosed Bat
Genus *Uroderma*
    U. bilobatum . . . . Common Tent-making Bat
    U. magnirostrum . . . . Brown Tent-making Bat
Genus *Vampyressa*
    V. melissa . . . . Melissa's Yellow-eared Bat
    V. pusilla . . . . Southern Yellow-eared Bat
    V. thyone . . . . Little Yellow-eared Bat
Genus *Vampyriscus*
    V. bidens . . . . Bidentate Yellow-eared Bat
    V. brocki . . . . Brock's Yellow-eared Bat
    V. nymphaea . . . . Striped Yellow-eared Bat
Genus *Vampyrodes*
    V. caraccioli . . . . Great Striped-faced Bat
Tribe Stenodermatini
Genus *Ametrida*
    A. centurio . . . . Little White-shouldered Bat
Genus *Centurio*
    C. senex . . . . Wrinkle-faced Bat
Genus *Pygoderma*
    P. bilabiatum . . . . Ipanema Bat
Genus *Sphaeronycteris*
    S. toxophyllum . . . . Visored Bat

Tribe Sturnirini
Genus *Sturnira (Corvira)*
    S. bidens . . . . Bidentate Yellow-shouldered Bat
    S. nana . . . . Lesser Yellow-shouldered Bat
Genus *Sturnira (Sturnira)*
    S. aratathomasi . . . . Giant Yellow-shouldered Bat
    S. bogotensis . . . . . Bogota Yellow-shouldered Bat
    S. erythromus . . . . . . Small Yellow-shouldered Bat
    S. koopmanhilli . . Chocoan Yellow-shouldered Bat
    S. lilium . . . . Little Yellow-shouldered Bat
    S. luisi . . . . Luis's Yellow-shouldered Bat
    S. magna . . . . Greater Yellow-shouldered Bat
    S. mistratensis . . . Mistrato Yellow-shouldered Bat
    S. oporaphilum . . Tschudi's Yellow-shouldered Bat
    S. sorianoi . . . . . . Soriano's Yellow-shouldered Bat
    S. tildae . . . . Tilda's Yellow-shouldered Bat
Family Mormoopidae
Genus *Mormoops*
    M. megalophylla . . . . Ghost-faced Bat
Genus *Pteronotus*
    P. davyi . . . . Davy's Naked-backed Bat
    P. gymnonotus . . . . Big Naked-backed Bat
    P. parnellii . . . . Parnell's Mustached Bat
    P. personatus . . . . Wagner's Mustached Bat
Family Noctilionidae
Genus *Noctilio*
    N. albiventris . . . . Lesser Bulldog Bat
    N. leporhinus . . . . Greater Bulldog Bat
Family Furipteridae
Genus *Amorphochilus*
    A. schnablii . . . . Smoky Bat
Genus *Furipterus*
    F. horrens . . . . Thumbless Bat
Family Thyropteridae
Genus *Thyroptera*
    T. devivoi . . . . De Vivo's Disk-winged Bat
    T. discifera . . . . Peters's Disk-winged Bat
    T. lavali . . . . LaVal's Disk-winged Bat
    T. tricolor . . . . Spix's Disk-winged Bat
Family Natalidae
Genus *Natalus*
    N. stramineus . . . . Gray's Funnel-eared Bat
    N. tumidirostris . . . . Miller's Funnel-eared Bat
Family Molossidae
Subfamily Tomopeatinae
Genus *Tomopeas*
    T. ravus . . . . Peruvian Crevice-dwelling Bat
Subfamily Molossinae
Genus *Cabreramops*
    C. aequatorianus . . . . Cabrera's Free-tailed Bat
Genus *Cynomops*
    C. abrasus . . . . Cinnamon Dog-faced Bat

C. *greenhalli* . . . . Greenhall's Dog-faced Bat
C. *milleri* . . . . Miller's Dog-faced Bat
C. *paranus* . . . . Brown Dog-faced Bat
C. *planirostris* . . . . Southern Dog-faced Bat
Genus *Eumops*
   E. *auripendulus* . . . . Black Bonneted Bat
   E. *bonariensis* . . . . Southern Bonneted Bat
   E. *dabbenei* . . . . Dabbene's Bonneted Bat
   E. *delticus* . . . . Delta Bonneted Bat
   E. *glaucinus* . . . . Wagner's Bonneted Bat
   E. *hansae* . . . . Sanborn's Bonneted Bat
   E. *maurus* . . . . Guianan Bonneted Bat
   E. *nanus* . . . . Dwarf Bonneted Bat
   E. *patagonicus* . . . . Patagonian Bonneted Bat
   E. *perotis* . . . . Greater Bonneted Bat
   E. *trumbulli* . . . . Trumbull's Bonneted Bat
Genus *Molossops*
   M. *neglectus* . . . . Rufous Dog-faced Bat
   M. *temminckii* . . . . Dwarf Dog-faced Bat
Genus *Molossus*
   M. *bondae* . . . . Bonda Mastiff Bat
   M. *coibensis* . . . . Coiban Mastiff Bat
   M. *currentium* . . . . Corrientes Mastiff Bat
   M. *molossus* . . . . Pallas's Mastiff Bat
   M. *pretiosus* . . . . Miller's Mastiff Bat
   M. *rufus* . . . . Black Mastiff Bat
   M. *sinaloae* . . . . Sinaloan Mastiff Bat
Genus *Mormopterus*
   M. *kalinowskii* . . . . Kalinowski's Little Mastiff Bat
   M. *phrudus* . . . . Incan Little Mastiff Bat
Genus *Neoplatymops*
   N. *mattogrossensis* . . . Mato Grosso Dog-faced Bat
Genus *Nyctinomops*
   N. *aurispinosus* . . . . Peale's Free-tailed Bat
   N. *laticaudatus* . . . . Geoffroy's Free-tailed Bat
   N. *macrotis* . . . . Big Free-tailed Bat
Genus *Promops*
   P. *centralis* . . . . Crested Mastiff Bat
   P. *nasutus* . . . . Brown Mastiff Bat
Genus *Tadarida*
   T. *brasiliensis* . . . . Brazilian Free-tailed Bat

Family Vespertilionidae
Genus *Eptesicus*
   E. *andinus* . . . . Andean Brown Bat
   E. *brasiliensis* . . . . Brazilian Brown Bat
   E. *chiriquinus* . . . . Chiriqui Brown Bat
   E. *diminutus* . . . . Little Serotine
   E. *furinalis* . . . . Argentine Brown Bat
   E. *fuscus* . . . . Big Brown Bat
   E. *innoxius* . . . . Pacific Serotine
Genus *Histiotus*
   H. *humboldti* . . . . Humboldt's Leaf-eared Bat
   H. *macrotus* . . . . Greater Leaf-eared Bat
   H. *montanus* . . . . Common Leaf-eared Bat
   H. *velatus* . . . . Tropical Leaf-eared Bat
Genus *Lasiurus*
   L. *atratus* . . . . Black-winged Red Bat
   L. *blossevillii* . . . . Southern Red Bat
   L. *castaneus* . . . . Red-faced Red Bat
   L. *cinereus* . . . . Hoary Bat
   L. *ebenus* . . . . Black Hairy-tailed Bat
   L. *ega* . . . . Southern Yellow Bat
   L. *egregius* . . . . Giant Red Bat
   L. *salinae* . . . . Brown hairy-tailed Bat
   L. *varius* . . . . Chilean Red Bat
Genus *Myotis*
   M. *aelleni* . . . . Southern Myotis
   M. *albescens* . . . . Silver-tipped Myotis
   M. *atacamensis* . . . . Atacama Myotis
   M. *chiloensis* . . . . Chilean Myotis
   M. *keaysi* . . . . Hairy-legged Myotis
   M. *levis* . . . . Yellowish Myotis
   M. *nesopolus* . . . . Curaçao Myotis
   M. *nigricans* . . . . Black Myotis
   M. *oxyotus* . . . . Montane Myotis
   M. *riparius* . . . . Riparian Myotis
   M. *ruber* . . . . Red Myotis
   M. *simus* . . . . Velvety Myotis
Genus *Rhogeessa*
   R. *hussoni* . . . . Eastern Little Yellow Bat
   R. *io* . . . . Southern Little Yellow Bat
   R. *minutilla* . . . . Tiny Yellow Bat

Contributors

Joaquin Arroyo-Cabrales
Laboratorio de Arqueozoología, INAH
Moneda #16, Col. Centro
06060 México, D.F., Mexico

Linda J. Barkley
4485 SW 160th Avenue
Aloha, OR 97007

John W. Bickham
Department of Wildlife and Fisheries Science
210 Nagle Hall
Texas A & M University
College Station, TX 77843-2258
Current Address
Center for the Environment
Purdue University
503 Nortwestern Avenue
West Lafayette, IN 47907-2966

Rui Cerqueira
Departamento de Ecologia, Laboratorio de
Vertebrados
Universidade Federal Rio de Janeiro
Caixa Postal 68020
CEP 21941-590, Rio de Janeiro, RJ, Brazil

G. Ken Creighton
World Wildlife Fund
1250 24th Street, NW
Washington, DC 20037

Marian Dagosto
Northwestern University Medical School
303 E. Chicago Ave.
Chicago, IL 60611-3008

William B. Davis (deceased)
Department of Wildlife and Fisheries Science
Texas A&M University
College Station, TX 77843

Judith L. Eger
Department of Natural History
Royal Ontario Museum
100 Queen's Park
Toronto, Ontario
Canada M5S 2C6

John F. Eisenberg (deceased)
1401 Edwards Street
Bellingham, WA 98226

Louise H. Emmons
Division of Mammals, MRC 108
National Museum of Natural History
Smithsonian Institution
Washington, DC 20013-7012

Alfred L. Gardner
Biological Survey Unit
USGS Patuxent Wildlife Research Center
National Museum of Natural History
PO Box 37012
NHB MRC-111
Washington, DC 20013-7012

Hugh H. Genoways
University of Nebraska State Museum
W436 Nebraska Hall
Lincoln, NE 68588-0514

Thomas A. Griffiths
Department of Biology
Illinois Wesleyan University
Bloomington, IL 61702

Charles O Handley, Jr. (deceased)
Division of Mammals
National Museum of Natural History
Washington, DC 20560-0108

Craig S. Hood
Department of Biological Sciences
Loyola University
New Orleans, LA 70118

Karl F. Koopman (deceased)
Department of Mammalogy
American Museum of Natural History
Central Park West at 79th Street
New York, NY 10024

Miriam Kwon
15 Eagle Heights Drive
Orchard Park, NY 14127

Suely A. Marques-Aguiar
Museu Paraense E. Goeldi/MCT
Caixa Postal 399
CEP-66.040-170 Belém, Pará, Brazil

Laura J. McLellan
Ocean County College
College Drive

PO BOX 2001
Toms River, NJ 08753

Philip Myers
Museum of Zoology
University of Michigan
Ann Arbor, MI 48109-1079

Virginia L. Naples
Department of Biological Sciences
Northern Illinois University
DeKalb, IL 60115-2861

Bruce D. Patterson
Division of Mammals
Field Museum of Natural History
Chicago, IL 60605-2496

James L. Patton
Museum of Vertebrate Zoology
3101 Valley Life Sciences Building
University of California
Berkeley, CA 94720-3160

Oliver P. Pearson (deceased)
Museum of Vertebrate Zoology
3101 Valley Life Sciences Building
University of California
Berkeley, CA 94720-3160

Jaime E. Péfaur
Grupo de Ecología Animal
Departamento de Biología
Facultad de Ciencias
Universidad de Los Andes
Mérida, 5101 Venezuela

Ronald H Pine
928 Hilltop Drive
Lawrence, KS 66044

Kent H. Redford
Director, WCS Institute
Wildlife Conservation Society
2300 Southern Blvd.
Bronx, N.Y. 10460

Mary Anne Rogers
Division of Fishes
Field Museum of Natural History
1400 S. Lake Shore Dr.
Chicago, IL 60605-2496

Luis A. Ruedas
Department of Biology
Museum of Vertebrate Biology
Portland State University
Portland, OR 97207-0751

Maria Nazareth F. da Silva
Instituto Nacional de Pesquisas da Amazônia
Caixa Postal 478
CEP 69083-000 Manaus, AM, Brazil

Barbara R. Stein
Museum of Vertebrate Zoology
3101 Valley Life Sciences Building
University of California
Berkeley, CA 94720-3160

Robert M. Timm
Natural History Museum & Biodiversity Research Center
Dyche Hall, 1345 Jayhawk Blvd.
University of Kansas
Lawrence, KS 66045-7561

Christopher J. Tribe
15 Little Lane
Melbourn
Royston, Herts
SG8 6BU, UK

Ralph M. Wetzel (deceased)
Florida State Museum
University of Florida
Gainesville, FL 32611

Stephen L. Williams
Mayborn Museum Complex
One Bear Place #97154
Baylor University
Waco, TX 76798-7154

Don E. Wilson
Division of Mammals, MRC 108
National Museum of Natural History
Smithsonian Institution
Washington, DC 20013-7012

Neal Woodman
Biological Survey Unit
USGS Patuxent Wildlife Research Center
Smithsonian Institution
PO Box 37012
NHB MRC 111
Washington, DC 20013-7012

Index of Scientific Names